**Summaries** Where to Look for Help

PREREQUISITES

Your algebra, trigonometry, or precalculus textbooks, if you kept them, are good places to look because you are used to the notation and presentation. If you do not have these books for reference, we have provided the necessary review in *Student Mathematics Handbook*. Also see

TECHNOLOGY

See *Technology Manuals for Maple, Mathematica,* or *Mathlab.* We have also found the TI-92 to be of great value in working many calculus problems. Inexpensive software includes *Derive* (which is included with the TI-92), and *Converge* for graphical representation of many calculus concepts.

DIFFERENTIATION AND GRAPHING

INTEGRATION

INFINITE SERIES

ANALYTIC GEOMETRY, VECTORS, AND THREE DIMENSIONS

FUNCTIONS OF TWO VARIABLES

DIFFERENTIAL EQUATIONS

CALCULUS

SECOND EDITION

INSTRUCTOR'S EDITION

Ceiling interior of the Saminid Mausoleum, built around 900 A.D. in Bukhara, Uzbekistan. This small cubical tomb (about 9 1/2 meters for the length of each side), built entirely of small yellow bricks, is the oldest surviving building, and probably the first, to demonstrate how to put a perfect hemisphere dome on a square base. The Iranian architects—Bukhara back then was part of a geographically larger Iranian culture—who designed this building were well versed in math, as were their later western counterparts who put up the medieval cathedrals of Europe, but these Islamic architects worked from a strong knowledge of geometry and algebra. The "squinch" above the corner of the square supports the weight of the dome—this was an innovation. Note the geometry in action: Take the midpoint of each side at the square base, and create eight line segments. Then take the midpoint of each of these eight line segments and connect adjacent midpoints at a higher level (you can see this in the photo above the squinch). So a four-sided object moves upward to one with eight sides. Repeat this process again and you have a 16-sided shape. Notice the circle of the hemisphere sits easily on the 16 sides. In summary, the square approaches the circle as a limit. In writing of another, far larger building with such a dome, built in 1088 in Isfahan, Iran, architecture critic Eric Schroeder has written: "European dome-builders never approached their skill. How ingeniously the Western builder compensated his ignorance of the mechanics of dome construction is attested by the ten chains round the base of St. Peter's, and the concealed cone which fastens the haunch of St. Paul's. But engineers could not hope to prescribe an ideally light dome of plain masonry before Newton's work on the calculus (late in the seventeenth century)." Read on, and upon completion of this text you will be qualified, at least mathematically, to do many marvelous things with the practical tool called calculus.

CALCULUS

SECOND EDITION
INSTRUCTOR'S EDITION

GERALD L. BRADLEY

Claremont McKenna College

KARL J. SMITH

Santa Rosa Junior College

PRENTICE HALL

Upper Saddle River, New Jersey 07458

Acquisitions Editor: George Lobell
Editor-in-Chief: Jerome Grant
Editorial Director: Tim Bozik
Associate Editor-in-Chief, Development: Carol Trueheart
Production Editor: Barbara Mack
Senior Managing Editor: Linda Mihatov Behrens
Executive Managing Editor: Kathleen Schiaparelli
Assistant Vice President of Production and Manufacturing: David W. Riccardi
Marketing Manager: Melody Marcus
Manufacturing Buyer: Alan Fischer
Manufacturing Manager: Trudy Pisciotti
Supplements Editor/Editorial Assistant: Gale Epps
Art Director: Maureen Eide
Associate Creative Director: Amy Rosen
Director of Creative Services: Paula Maylahn
Assistant to Art Director: John Christiana
Art Manager: Gus Vibal
Art Editor: Rhoda Sidney
Interior Designer: Geri Davis, The Davis Group, Inc.
Cover Designer: Joseph Sengotta
Cover Photo: Louvre Museum and Pyramid. Tom Craig/FPG International
Photo Researcher: Beth Boyd
Photo Research Administrator: Melinda Reo
Art Studio: Monotype Composition
Historical Quest Portraits: Steven S. Nau
Copy Editor: Susan Reiland

© 1999, 1995 by Prentice-Hall, Inc.
Simon & Schuster/A Viacom Company
Upper Saddle River, New Jersey 07458

Printed in the United States of America

10 9 8 7 6 5 4 3 2 1

ISBN 0-13-081055-X

Prentice-Hall International (UK) Limited, *London*
Prentice-Hall of Australia Pty. Limited, *Sydney*
Prentice-Hall Canada Inc., *Toronto*
Prentice-Hall Hispanoamericana, S.A., *Mexico*
Prentice-Hall of India Private Limited, *New Delhi*
Prentice-Hall of Japan, Inc., *Tokyo*
Simon & Schuster Asia Pte. Ltd., *Singapore*
Editora Prentice-Hall do Brasil, Ltda., *Rio de Janeiro*

Contents

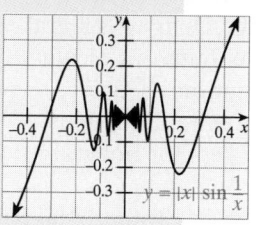

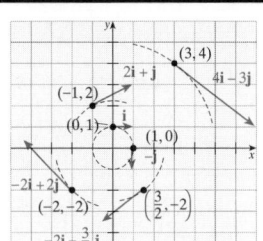
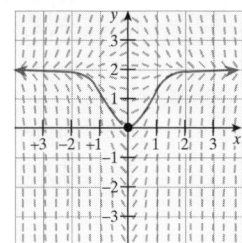
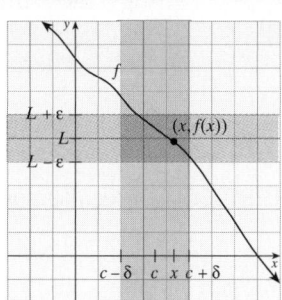

About the Authors

Gerald L. Bradley

Karl J. Smith

Gerald L. Bradley received his B.S. degree in mathematics from Harvey Mudd College in 1962, as a member of the second full graduating class, and his Ph.D. from the California Institute of Technology in 1966. A former NSF and Woodrow Wilson Fellow, Professor Bradley has taught at Claremont McKenna College since 1966, and twice has served as chairman of the mathematics department. His primary field of interest and research is matrix theory, and he is the author of *A Primer of Linear Algebra*, also with Prentice Hall. He has a strong interest in the development of undergraduate mathematics, and has co-authored a top-selling business calculus text as well as a basic text in finite mathematics. His personal interests include history, archaeology, and bridge, but he spends most of his spare time with his passion: writing science fiction novels.

Karl J. Smith received his B.A. and M.A. (in 1967) degrees in mathematics from UCLA. Then he moved in 1968 to northern California to teach at Santa Rosa Junior College, where he has been ever since. Along the way, he served as department chair, and he received a Ph.D. in 1979 in mathematics education at Southeastern University. A past president of the American Mathematical Association of Two-Year Colleges, Professor Smith is very active nationally in mathematics education. He was founding editor of *Western AMATYC News*, a chairperson of the committee on Mathematics Excellence, and a NSF grant reviewer. He was a recipient in 1979 of an Outstanding Young Men of America Award, in 1980 of an Outstanding Educator Award, and in 1989 of an Outstanding Teacher Award. Professor Smith is the author of several successful textbooks. In fact, over one million students have learned mathematics from his textbooks.

Foreword

Over the years I have seen and used a number of calculus texts. I have not always viewed "improvements" in these as real improvements, at least for my students. However, when I was asked to take a look at the manuscript for the 2nd edition of the Bradley/Smith text, I was struck by how readable it is and how basically friendly. It is by no means a "once-over-lightly" treatment of calculus, but at the same time it does not wear out the reader with excessive detail and mind-numbing rigor. This is, after all, a text intended for a course at a level that does not require all that much rigor. The explanations are careful without being pedantic and fussy, well illustrated with examples without exhausting the reader with an overabundance, and uncluttered with alternate ways of doing things.

In particular, I like the historical-biographical essays with associated projects and the calculator/computer projects, all placed where the instructor can use them or not, depending on taste or what the length of the term will allow. Some are provocative and open-ended, allowing for additional time to be spent by students who are ready for more of a challenge. In fact, this expandability would even allow the use of the text in an honors section while the text is also used for regular sections, thus allowing students to transfer in and out of honors sections during the calculus sequence.

The applications are interesting and wide-ranging. A number of them were new to me. Yet I was convinced that they would be understandable to students. In my experience with a number of recent calculus texts, some applications are so far from the experience of most of my students that they end up being mainly frustrating, rather than beneficial. Such texts also use language students don't know and ideas from other fields in which students have no background. This is not the case with Bradley/Smith.

The historical and biographical sketches, along with associated projects involving further reading and exploration, lend a humane touch to a subject this is often viewed by the unconverted as cold, mechanical, and not the work of real human beings. The authors here have done a nice job of making it clear that the calculus as we know it has been developed over many years by a variety of investigators. It is the work of some of the best scientific minds of history. And if the subject does not come easily and without help, it is not surprising. The subject is clear to us now but it is the result of centuries of human effort and ingenuity. Calculus remains one of the greatest of scientific achievements, not only a useful subject but one that forms a profound and cohesive body of knowledge. It's worth the effort it takes to master it.

Gerald L. Alexanderson
Santa Clara University

Preface

Calculus teaching is undergoing great changes, most of which will be of lasting benefit to students. We applaud teaching by using numerical, graphical, and analytical approaches to important concepts. The goal is clear student understanding of concepts, not simply the ability to "manipulate symbols." You will find this book rich with tables, graphs, and algebraic characterizations of each main concept. Our goal in writing this text was to blend the best aspects of calculus reform with the reasonable goals and methodology of traditional calculus. In incorporating so much of calculus reform, we made a deliberate effort not "to throw the baby out with the bath water." Calculus should not be a terminal course, but rather one that prepares students in engineering, science, math, and other related areas to move on to more advanced and necessary career or professional courses. Unlike some reform books, this text addresses topics such as continuity, the mean value theorem, l'Hopital's rule, parametric equations, polar coordinates, sequences, and series. In short, this text is an attempt at Reform with Reason.

The acceptance and response from our first edition has been most gratifying. In spite of the fact that many professors are reluctant to adopt a first edition textbook, we found widespread acceptance for our book and appreciate the many suggestions we received. With this second edition, we checked and rechecked the accuracy of the text material, and have taken extraordinary effort to ensure the accuracy of the answers. We incorporated many of your suggested improvements:

- This edition correctly reflects the precalculus mathematics being taught at most colleges and universities. We assume knowledge of the trigonometric functions, as well as e^x and $\ln x$. These functions are reviewed in Chapter 1. We also assume a knowledge of the conic sections and their graphs.

- It is possible to begin the course with Chapter 1 or Chapter 2 (where the calculus topics begin).

- Modeling was added as a major theme in this edition. Modeling is introduced in Section 2.1, and then appears in almost every section of the book. These applications are designed MODELING PROBLEMS.

- We have taken the introduction of differential equations seriously. Students in many allied disciplines need to use differential equations early in their studies, and consequently cannot wait for a post calculus course. In this edition, we introduce differential equations in a natural and reasonable way. Slope fields are introduced as a geometric view of anti-differentiation in Section 5.1, and then are used to introduce a graphical solution to differential equations in Section 5.6. We consider separable differential equations in Chapter 5 and first-order linear equations in Chapter 7, and demonstrate the use of both modeling a variety of applied situations. Exact and homogeneous differential equations appear in Chapter 15, along with an introduction to second-order linear equations. The "early and often" approach to differential equations is intended to illustrate their value in continuous modeling and to provide a solid foundation for further study.

- This edition offers an early presentation of transcendental functions: logarithmic, exponential, and trigonometric functions are heavily integrated into all chapters of this book (especially Chapters 1–5).

- We continue to exploit the *humanness* of mathematics, but instead of simply including *Historical Notes*, as we did in the first edition, we have transformed these notes into *problems* that lead the reader from the development of a concept to actually participating in the discovery process. The problems are designated as Historical Quest problems. The problems are not designed to be "add-on or challenge problems," but rather are intended to become an integral part of the usual assignment. The level of difficulty of Quest problems ranges from easy to difficult. An extensive selection of biographies of noted mathematicians can be found on the Internet site accompanying this text.

- We are aware that most calculus books grow "bigger and bigger" as they progress from one edition to the next. As they grow, they add "more features" and become more and more expensive for the student. We are determined that this does not happen with this book. In an effort to keep down costs, we made the decision that a full color edition, while visually appealing, does not add to *mathematical* understanding enough to justify the added cost. Consequently, this edition uses color functionally as a pointer and not as a decoration. We ask for your feedback regarding our decision.

The major issue driving calculus reform is the poor performance of students trying to master the concepts of calculus. Much of this failure can be attributed to the way most students learn mathematics in high school, which often involves stressing rote memorization over insight and understanding. On the other hand, some reform texts are perceived as spending so much time with the development of insight and understanding that students are not given enough exposure to important computational and problem-solving skills to perform well in more advanced courses. In our view, it is equally wrong to foster a situation in which the student understands too little about a lot or a lot about too little. This text aims at a middle ground by providing sound development, stimulating problems, and well-developed pedagogy within a framework of a traditional topic structure. "Think then do" is a fair summary of our approach.

Conceptual Understanding Through Verbalization

Besides developing some minimal skills in algebraic manipulation and problem solving, today's calculus text should require students to cultivate verbal skills in a mathematical setting. This is not just because real mathematics wields its words precisely and compactly, but because verbalization should help students think conceptually.

COOPERATIVE LEARNING (GROUP RESEARCH PROJECTS)   In July 1991, the National Science Foundation funded a group of instructors who met on the campus of New Mexico State University to discuss the topic, "Discovering Calculus through Student Projects." The participants concluded that encouraging students to work on significant projects in small groups acts as a counterbalance to traditional lecture methods and can serve to foster both conceptual understanding and the development of technical skills. However, many instructors still believe that mathematics can be learned only through independent work. We feel that independent work is of primary importance, but that students must also learn to work with others in group projects. After all, an individual's work in the "real world" is often done as part of a group and almost always involves solving problems for which there is no answer "in the back of the book." In response to this need, we have included challenging exercises (the Journal and Putnam problems) and a number of group research projects, each of which appears at the end of a chapter and involves intriguing questions whose mathematical content is tied loosely to the chapter just concluded. These projects have been developed and class-tested by their individual authors, to whom we are greatly indebted. Note that the complexity of these projects increases as we progress through the book, and the mathematical maturity of the student is developed.

MATHEMATICAL COMMUNICATION   We have included several opportunities for mathematical communication in terms that can be understood by nonprofessionals. The guest essays provide alternate viewpoints. The questions that follow are called MATHEMATICAL ESSAYS and are included to encourage individual writing assignments and mathematical exposition. We believe that students will benefit from individual writing and research in mathematics. Shorter problems encouraging written communication are included in the problem sets and are designed by the logo WHAT DOES THIS SAY? Another pedagogical feature is the **"What this says:"** box in which we rephrase mathematical ideas in everyday language. In the problem sets we encourage students to summarize procedures, processes, or to describe a mathematical result in everyday terms. Concept problems are found throughout the book as well as at the end of each chapter, and these problems, as well as the **mathematical essay** problems, are included to prove that mathematics is more than "working problems and getting answers." Mathematics education *must* include the communication of mathematical ideas.

Integration of Technology

COMPUTATIONAL WINDOWS   Reform is driven partly by the need to embrace the benefits technology brings to the learning of mathematics. Simply adding a lab course to the traditional calculus is possible, but this may lead to unacceptable work loads for all involved. We choose to include technology as an aid to the understanding of calculus, rather than to write a calculus course developed around the technology. While we have included over 130 windows devoted to the use of technology, we strive to keep such references "platform neutral" because specific calculators and computer programs frequently change and are better considered in separate technology manuals. The technology in the book is organized under the title "TECHNOLOGY WINDOWS" to give insight into how technological advances can be used to help understand calculus. TECHNOLOGY WINDOWS also appear in the exercises and involve problems requiring a graphing calculator or software and computer. On the other hand, problems that are not specially designated may still use technology (for example to solve a higher-degree equation).

TECHNOLOGY LABORATORY MANUALS   For those with personal access to a computer, companion Technological Manuals (available wrapped with the text at a small charge) discuss TI graphing calculators, HP graphing calculators, *Mathematica*, *Maple*, and *MATLAB*.

SIGNIFICANT DIGITS   We have included a brief treatment in Appendix C. On occasion, we show the entire calculator or computer output of 12 digits for clarity even though such a display may exceed the requisite number of significant digits.

GREATER TEXT VISUALIZATION   Related to, but not exclusively driven by, the use of technology, is the greater use of graphs and other mathematical pictures throughout this text. Over 1,900 graphs appear—more than nearly any other calculus text. This increased visualization is intended to help develop greater student intuition. Much of this visualization appears in the wide margins to accompany the text. Its purpose is to provide explanation to supplement and/or replace that of the text prose. Also, since many tough calculus problems are often tough geometry (and algebra) problems, this increased emphasis on graphs will help students' problem-solving skills. Additional graphs are related to the student problems, including answer art.

Problem Solving

PROBLEMS   We believe that students *learn* mathematics by *doing* mathematics. Therefore, the problems and applications are perhaps the most important feature of any calculus book, and you find the problems in this book extend from routine practice to challenging. The problem sets are divided into *A* Problems (routine), *B* Problems (requiring independent thought), and *C* Problems (theory problems). In this book we also include past Putnam examination problems as well as problems found in current mathematical journals. You will find the scope and depth of the problems in this book to be extraordinary. Even though engineering and physics examples and problems play a prominent role, we include applications from a wide variety of fields such as biology, economics, ecology, psychology, and sociology. The problems have been in the development stages for over ten years and most of them have been class tested. In addition, the chapter summaries provide not only topical review, but also many miscellaneous exercises. Although the chapter reviews are typical of examinations, the miscellaneous problems are not presented as graded problems, but rather as a random list of problems loosely tied to the ideas of that chapter. In addition, there are cumulative reviews located at natural subdivision points in the text: Chapters 1-6, Chapters 7-11, and Chapters 12-15.

JOURNAL PROBLEMS   In an effort to show that "mathematicians work problems too," we have reprinted problems from leading mathematics journals. We have chosen problems which are within reason of the intended audience of this book. If students need help or hints for these problems, they can search out the original presentation and solution in the cited journal. In addition, we have included problems from the **Putnam Examinations**. These problems, which are more challenging, are offered in the miscellaneous problems at the end of various chapters and are provided to give insight into the type of problems that are asked in mathematical competitions. The annual national competition is given under the auspices of the Mathematical Association of America. The problems are designed to recognize mathematically talented college and university students.

THINK TANK PROBLEMS   It has been said that mathematical discovery is directed toward two major goals—the formulation of proofs and the construction of counterexamples. Most calculus books focus only on the first goal (the body of proofs and true statements), but we feel that some attention should be paid to the formulation of counterexamples for false statements. Throughout this book we ask the student to formulate an example satisfying certain conditions. We have designated this type of problem as a *think tank* problem.

Topics

STUDENT MATHEMATICS HANDBOOK   The content of this text adapts itself to either semester or quarter systems, and both differentiation and integration can be introduced in the first course. We begin calculus with a minimum of review. The prerequisite material which is often included in a calculus textbook has been bound separately in a companion book, *Student Mathematics Handbook*. Our handbook is offered *free of charge* with every *new* copy of the textbook. The Handbook not only includes the necessary review material and formulas, but also contains a catalog of curves and a complete integral table. We feel this is an important supplement to the textbook because we have found that the majority of errors our students make in a calculus class are not errors in calculus, but errors in basic algebra and trigonometry. A unified and complete treatment of this prerequisite material, easily referenced and keyed to the textbook, has been a valuable tool for our students taking calculus. Those portions of the text that benefit from an appropriate precalculus review are marked by the symbol ⬭SMH

SEQUENCE OF TOPICS   We resisted the temptation to label certain sections as optional, because that is a prerogative of individual instructors and schools. However, the following sections could be skipped without any difficulty: 4.7, 5.9, 6.5 (delay until Sec. 13.6), 7.8, 12.8, and 13.8. To assist instructors with the pacing of the course, we have written the material so each section reasonably can be covered in one classroom day, but to do so requires that the students read the text in order to tie together the ideas which might be discussed in a classroom setting.

PROOFS   One of the trends in the move to reform calculus is to minimize the role of a mathematical theorem. We do not agree with this aspect of reform. Precise reasoning has been, and we believe will continue to be, the backbone of good mathematics. While never sacrificing good pedagogy and student understanding, we present important results as *theorems*. We do not pretend to prove every theorem in this book; in fact we often only outline the steps one would take in proving a theorem. We refer the reader to Appendix B for certain longer proofs, or sometimes to an advanced calculus text. Why then do we include the heading "PROOF" after each theorem? It is because we want the student to know for a result to be a theorem there must be a proof. We use the heading not necessarily to give a complete proof in the text, but to give some direction to where a proof can be found, or an indication of how it can be constructed.

Supplementary Materials

 Student Mathematics Handbook and Integration Table for CALCULUS by Karl J. Smith offers a review of prerequisite material, a catalog of curves, and a complete integral table. This handbook is presented free of charge along with the purchase of a new book.

 A *Student Survival and Solutions Manual* by Karl J. Smith offers a running commentary of hints and suggestions to help ensure the students' success in calculus. Since this manual is written by one of the authors of this text, the solutions given in the manual complement all of the procedures and development of the textbook. The problem numbers in the book shown in color indicate the solutions included in the *Survival Manual.*

Technology Manuals by John Gresser offers computer applications keyed to the sections in the book. These manuals are identical except for the specific keystrokes on TI calculators, *Mathematica*, and *Maple*. There are similar manuals for the HP and MATLAB by Frank Hagin and Jack Cohen.

A Complete Solutions Manual by Karl J. Smith contains a brief solution for every problem in the book.

An Answer Book by Karl J. Smith contains only the answers to most problems in the book.

An Instructor's Guide offers sample tests and reviews for each chapter in the book. This guide also includes sample transparencies.

Computerized Computer Testing Program is available in both IBM and Macintosh formats.

Resources for Calculus, Volumes 1-5, A. Wayne Roberts (Project Director) available from the Mathematical Association of America, 1993:

> Dudley, Underwood (ed), *Readings for Calculus*, MAA Notes, No. 28
>
> Fraga, Robert (ed), *Calculus Problems for a New Century*, MAA Notes, No. 28
>
> Jackson, Michael B., and Ramsay, John (eds), *Problems for Student Investigation*, MAA Notes, No. 30
>
> Snow, Anita E. (ed), *Learning for Discovery: A Lab Manual for Calculus*, MAA Notes, No. 27
>
> Straffin, Philip (ed), *Applications of Calculus*, MAA Notes, No. 29

This is a valuable collection of resource materials for calculus instructors. All material may be reproduced for classroom use.

Acknowledgments

The writing and publishing of a calculus book is a tremendous undertaking. We take this responsibility very seriously because a calculus book is instrumental in transmitting knowledge from one generation to the next. We would like to thank the many people who helped us in the preparation of this book. First, we thank our editor,

George Lobell, who led us masterfully through the development and publication of this book. We sincerely appreciate Donald Gecewicz, who read and critiqued each word of the manuscript. We also appreciate the work of Barbara Mack in college production, who kept us all on track, and we especially thank Susan Reiland for her meticulous attention to detail.

Of primary concern is the accuracy of the book. We had the assistance of many: Jerry Alexanderson and Mike Ecker, who read the entire manuscript and offered us many valuable suggestions; the accuracy checkers of the first edition, Ken Sydel, Diana Gerardi, Kurt Norlin, Terri Bittner, and Mary Toscano; and of the second edition, Nancy and Mary Toscano. We also would like to thank the following readers of the text for their many suggestions for improvement:

REVIEWERS OF THE SECOND EDITION:

Gerald Alexanderson, Santa Clara University

David Arterburn, New Mexico Tech

Linda A. Bolte, Eastern Washington University

Brian Borchers, New Mexico Tech

Mark Farris, Midwestern State University

Sally Fieschbeck, Rochester Institute of Technology

Mike Ecker, Pennsylvania State University, Wilkes-Barre Campus

Stuart Goldenberg, California Polytechnic Institute

Roger Jay, Tomball College

John H. Jenkins, Embry-Riddle Aeronautical University

Kathy Kepner, Paducah Community College

Daniel King, Sarah Lawrence College

Don Leftwich, Oklahoma Christian University

Ching Lu, Southern Illinois University at Edwardsville

Ann Morlet, Cleveland State University

Dena Jo Perkins, Oklahoma Christian University

Judith Reeves, California Polytechnic Institute

Jim Roznowski, Delta College

Lowell Stultz, Kalamazoo Valley Community College

Tingxiu Wang, Oakton Community College

REVIEWERS OF THE FIRST EDITION:

Neil Berger, University of Illinois at Chicago

Michael L. Berry, West Virginia Wesleyan College

Barbara H. Briggs, Tennessee Technical University

Robert Broschat, South Dakota State University

Robert D. Brown, University of Kansas

Dan Chiddix, Ricks College

Philip Crooke, Vanderbilt University

Ken Dunn, Dalhousie University

John H. Ellison, Grove City College

William P. Francis, Michigan Technological University

Harvey Greenwald, California Polytechnic San Luis Obispo

Richard Hitt, University of South Alabama

Joel W. Irish, University of Southern Maine

Clement T. Jeske, University of Wisconsin-Platteville

Lawrence Kratz, Idaho State University

Sam Lessing, Northeast Missouri University

Estela S. Llinas, University of Pittsburgh at Greensburg

Pauline Lawman, Western Kentucky University

William E. Mastrocola, Colgate University

Philip W. McCartney, Northern Kentucky University

E. D. McCune, Stephen F., Austin State University

John C. Michels, Chemeketa Community College

Pamela B. Pierce, College of Wooster

Connie Schrock, Emporia State University

Tatiana Shubin, San Jose State University

Tingxiu Wang, Oakton Community College

Gerald L. Bradley
Karl J. Smith

The rotunda of City Hall, in Manhattan, New York City.
Designed by John McComb, Jr., reputedly the first full-time, American-born architect, and built in the early nineteenth century in the neoclassical style. (photo credit: Behrenholtz)

1 Functions and Graphs

PREVIEW

This chapter introduces several topics from algebra and trigonometry that are essential for the study of calculus. The necessary prerequisites for this course are reviewed in a companion book, *Student Mathematics Handbook,* and are briefly discussed in this chapter. This handbook is provided free with new copies of this book.

The concept of function is an especially important prerequisite for calculus. Even if you wish to skip most of this initial chapter, you should review the basic notions of functions and inverse functions in Sections 1.3 and 1.5, and the notation and terminology for exponential and logarithmic functions in Section 1.6.

PERSPECTIVE

Calculus is often the gateway course for many careers, and this chapter reviews some of the prerequisite ideas needed to study calculus. Although modern science requires the use of many different skills and procedures, calculus is the primary mathematical tool for dealing with change. Sir Isaac Newton, one of the discoverers of calculus, once remarked that to accomplish his results, he "stood on the shoulders of giants." Indeed, calculus was not born in a moment of divine inspiration but developed gradually, as a variety of apparently different ideas and methods merged into a coherent pattern. The purpose of this initial chapter is to lay the foundation for the development of calculus.

1.1 Preliminaries

IN THIS SECTION **distance on a number line, absolute value, distance in the plane, trigonometry**

Every mathematical book that is worth reading must be read "backwards and forwards," if I may use the expression. I would modify Lagrange's advice a little and say, "Go on, but often return to strengthen your faith." When you come on a hard and dreary passage, pass it over; come back to it after you have seen its importance or found its importance or found the need for it further on.

GEORGE CHRYSTAL, *ALGEBRA, PART 2* (EDINBURGH, 1889)

This section provides a quick review of some fundamental concepts and techniques from precalculus mathematics. If you have recently had a precalculus course, you may skip over this material.

Algebra, geometry, and trigonometry are important ingredients of calculus. Even though we will review many ideas from algebra, geometry, and trigonometry, we will not be able to develop every idea from these courses before we use it in calculus. You may need, from time to time, to refer to a precalculus book. For example, the law of cosines from trigonometry may be needed to solve a problem in a section that never mentions trigonometry in the exposition. For this reason, we have made available a separate reference manual, *Student Mathematics Handbook,* which includes the background material you will need for this course. We suggest that you keep it close at hand. References to this handbook are indicated by the logo SMH

DISTANCE ON A NUMBER LINE

You are probably familiar with the set of **real numbers** and several of its subsets, including the counting or natural numbers, the integers, the rational numbers, and the irrational numbers.

■ FIGURE 1.1 Real number line

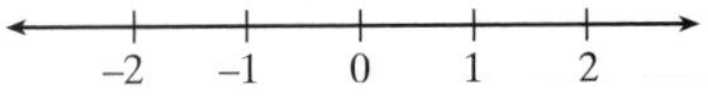

The real numbers can most easily be visualized by using a **one-dimensional coordinate system** called a **real number line,** as shown in Figure 1.1.

Notice that a number a is less than a number b if it is to the left of b on a real number line, as shown in Figure 1.2. Similar definitions can be given for $a > b$, $a \le b$, and $a \ge b$.

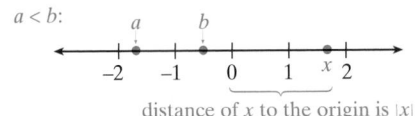

■ FIGURE 1.2 Geometric definition of *less than*

The location of the number 0 is chosen arbitrarily, and a unit distance is picked (meters, feet, inches, ...). Numbers are ordered on the real number line according to the following order properties.

Order Properties

For all real numbers a, b, and c:

Trichotomy law: Exactly one of the following is true:

$$a < b, \quad a > b, \quad \text{or} \quad a = b$$

Transitive law of inequality: If $a < b$ and $b < c$, then $a < c$.

Additive law of inequality: If $a < c$ and $b < d$, then $a + b < c + d$.

Multiplicative law of inequality: If $a < b$, then

$$ac < bc \text{ if } c > 0 \quad \text{and} \quad ac > bc \text{ if } c < 0$$

ABSOLUTE VALUE

> ### Absolute Value
>
> The **absolute value** of a real number a, denoted by $|a|$, is
>
> $$|a| = \begin{cases} a & \text{if } a \geq 0 \\ -a & \text{if } a < 0 \end{cases}$$

WARNING $|a|$ is NOT the number a without its sign.

The number x is located $|x|$ units away from 0—to the right if $x > 0$ and to the left if $x < 0$.

Absolute value is used to describe the distance between points on a number line.

> ### Distance between Two Points on a Number Line
>
> The **distance** between the numbers x_1 and x_2 on a number line is
>
> $$|x_2 - x_1|$$

WARNING Note that $|x_2 - x_1| = |x_1 - x_2|$

For example, the distance between 2 and -3 is $|2 - (-3)| = 5$ units.

Several properties of absolute value that you will need in this course are summarized in Table 1.1.

■ TABLE 1.1

Properties of Absolute Value

WARNING "p if and only if q" is used to mean that both a statement and its converse are true. That is:

If p, then q, *and*

if q, then p.

For example, property 8 has two parts:

(i) If $|a| < b$, then $-b < a < b$.

(ii) If $-b < a < b$, then $|a| < b$.

Let a and b be any real numbers.	
Property	Comment
1. $|a| \geq 0$	**1.** Absolute value is nonnegative.
2. $|-a| = |a|$	**2.** The absolute value of a number and the absolute value of its opposite are equal.
3. $|a|^2 = a^2$	**3.** If an absolute value is squared, the absolute value can be dropped because both squares are nonnegative.
4. $|ab| = |a||b|$	**4.** The absolute value of a product is the product of the absolute values.
5. $\left|\dfrac{a}{b}\right| = \dfrac{|a|}{|b|}, \quad b \neq 0$	**5.** The absolute value of a quotient is the quotient of the absolute values.
6. $-|a| \leq a \leq |a|$	**6.** Any number a is between the absolute value of that number and its opposite, inclusive.
7. Let $b \geq 0$; $|a| = b$ if and only if $a = \pm b$	**7.** This property is useful in solving absolute value equations. See Example 1.
8. Let $b > 0$; $|a| < b$ if and only if $-b < a < b$	**8. and 9.** These are the main properties used in solving absolute value inequalities. See Example 2.
9. Let $b > 0$; $|a| > b$ if and only if $a > b$ or $a < -b$	
10. $|a + b| \leq |a| + |b|$	**10.** This property is called the **triangle inequality.** It is used in both theory and numerical computations involving inequalities.

Property 7 is sometimes stated as $|a| = |b|$ if and only if $a = \pm b$. Since $|b| = \pm b$, it follows that this property is equivalent to property 7. Also, properties 8 and 9 are true for $\leq$ and $\geq$ inequalities. Specifically, if $b > 0$, then

$$|a| \leq b \quad \text{if and only if} \quad -b \leq a \leq b$$

and

$$|a| \geq b \quad \text{if and only if} \quad a \geq b \text{ or } a \leq -b$$

A convenient notation for representing intervals on a number line is called **interval notation** and is summarized in the accompanying table. Note that a solid dot (•) at an endpoint of an interval indicates that it is included in the interval, whereas an open dot (○) indicates that it is excluded. An interval is **bounded** if both of its endpoints are real numbers. A bounded interval is **open** if it includes neither of its endpoints, **half-open** if it includes only one endpoint, and **closed** if it includes both endpoints. The symbol "∞" (pronounced *infinity*) is used for intervals that are not limited in one direction or another. In particular, $(-\infty, \infty)$ denotes the entire real number line.

If this notation is new to you, please check the *Handbook* (Section 2.6) for further examples. We will use this interval notation to write the solutions of absolute value equations and absolute value inequality problems.

Name of Interval	Inequality Notation	Interval Notation	Graph
Closed interval	$a \leq x \leq b$	$[a, b]$	
	$a \leq x$	$[a, \infty)$	
	$x \leq b$	$(-\infty, b]$	
Open interval	$a < x < b$	(a, b)	
	$a < x$	(a, ∞)	
	$x < b$	$(-\infty, b)$	
Half-open interval	$a < x \leq b$	$(a, b]$	
	$a \leq x < b$	$[a, b)$	
Real number line	All real numbers	$(-\infty, \infty)$	

Absolute Value Equations   Absolute value property 7 allows us to solve absolute value equations easily. For that reason, this property is called the **absolute value equation property.**

EXAMPLE 1   *Solving an equation with an absolute value on one side*

Solve $|2x - 6| = x$.

Solution   If $2x - 6 \geq 0$, then $2x - 6 = x$ or $x = 6$ *Property 7*

If $2x - 6 < 0$, then $-(2x - 6) = x$

$$-3x = -6$$

$$x = 2$$

The solutions are $x = 6$ and $x = 2$.   ■

Technology Window—Graphing Calculators

Many calculators have equation-solving capabilities. For Example 1 (a sample calculator output is shown at the right), the equation is input into the *solver utility,* the variable is identified, and then the calculator gives a solution.

If you do not have a solver utility, but you do have a graphing calculator, you can solve equations to any reasonable degree of accuracy by graphing on the calculator. For example, look at Example 1. You can graph two functions, $y_1 = |2x - 6|$ and $y_2 = x$, and then estimate their intersection.

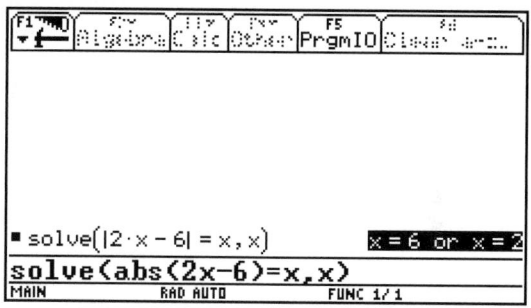

When we show calculator graphs in this book, we will show them just as you will see them on a calculator, without scale and without labeling. There are many considerations of scale and limitations on the x- and y-axes, which we will discuss from time to time in these special **Technology Window** boxes.

The TRACE key allows us to find approximate coordinates of points on the curves. For Example 1, we find X=2, Y=2 and X=6, Y=6 as possible points of intersection. Checking $x = 2$ and $x = 6$ in the original equation, we can verify the solution. There is also a ZOOM utility that allows you to enlarge (zoom in) or shrink (zoom out). Finally, you can press RANGE to reset the scale. Example 1, with a scale with $0 \leq x \leq 10$ and $0 \leq y \leq 10$, is shown below.

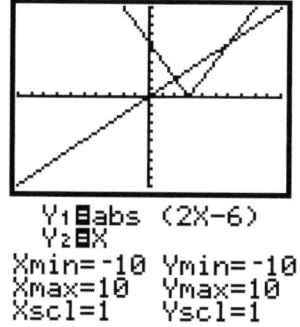

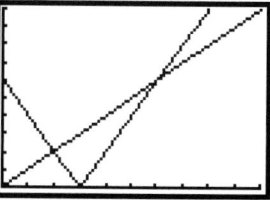

Some graphing calculators have an *intersect utility*. Check your owner's manual, but most work the same way. You place the cursor on the first curve and press ENTER, then on the second curve and press ENTER. Then enter a guess and press INTERSECT. The result of this operation for the first intersection in Example 1 is shown at the right.

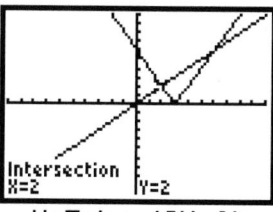

The absolute value expression $|x - a|$ can be interpreted as the distance between x and a on a number line. An equation of the form

$$|x - a| = b$$

is satisfied by two values of x that are a given distance b from a when represented on a number line. For example, $|x - 5| = 3$ states that x is 3 units from 5 on a number line. Thus, x is either 2 or 8.

Geometric representation	Algebraic representation

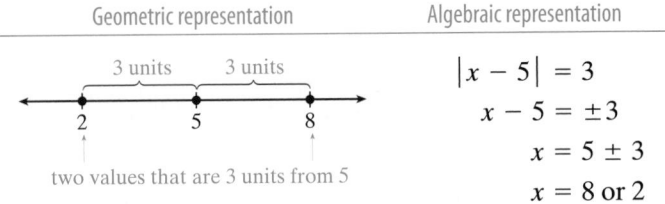

Absolute Value Inequalities   Because $|x - 5| = 3$ states that the distance from x to 5 is 3 units, the inequality $|x - 5| < 3$ states that the distance from x to 5 is less than 3 units, whereas $|x - 5| > 3$ states that the distance from x to 5 is greater than 3 units.

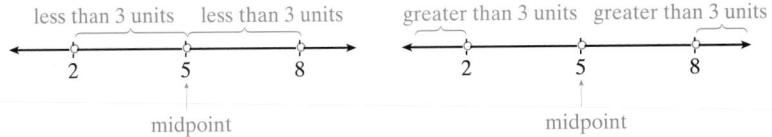

EXAMPLE 2	*Solving an absolute value inequality*

Solve $|2x - 3| \le 4$.

Solution   **Algebraic solution:** $-4 \le 2x - 3 \le 4$ *Property 8*

$$-4 + 3 \le 2x - 3 + 3 \le 4 + 3$$

$$-1 \le 2x \le 7$$

$$-\frac{1}{2} \le \frac{2x}{2} \le \frac{7}{2}$$

$$-\frac{1}{2} \le x \le \frac{7}{2}$$

The solution is the interval $[-\frac{1}{2}, \frac{7}{2}]$.

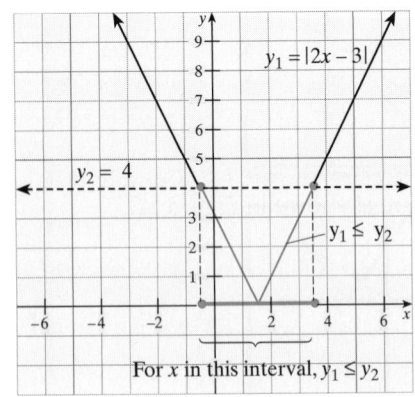

Geometric solution: Graph $y_1 = |2x - 3|$ and $y_2 = 4$. Because we are looking for $|2x - 3| \le 4$, we note those x-values on the real number line for which the graph of y_1 is below the graph of y_2. We see that the interval is $[-0.5, 3.5]$ or $[-\frac{1}{2}, \frac{7}{2}]$. ∎

When absolute value is applied to measurement, it is called **tolerance.** Tolerance is an allowable deviation from a standard. For example, a cement bag whose weight w lb is "90 lb plus or minus 2 lb" might be described as having a weight given by $|w - 90| \le 2$. When considered as a tolerance, the expression $|x - a| \le b$ may be interpreted as x being compared to a with an **absolute error** of measurement of b units. Consider the following example.

EXAMPLE 3   *Absolute value as a tolerance*

Suppose you purchase a 90-lb bag of cement. It will not weigh exactly 90 lb. The material must be measured, and the measurement is approximate. Some bags will weigh as much as 2 lb over 90 lb, and some will weigh as much as 2 lb under 90 lb. If so, the bag could weigh as much as 92 lb and as little as 88 lb. State this as an absolute value inequality.

Solution   Let w = weight of the bag of cement in pounds. Then

$$90 - 2 \le \quad w \quad \le 90 + 2$$
$$-2 \le w - 90 \le 2$$

Equivalently, $\left| w - 90 \right| \le 2$.   ∎

DISTANCE IN THE PLANE

Absolute value is used to find the distance between two points on a number line. To find the distance between two points in a coordinate plane, we use the *distance formula*, which is derived by using the Pythagorean theorem.

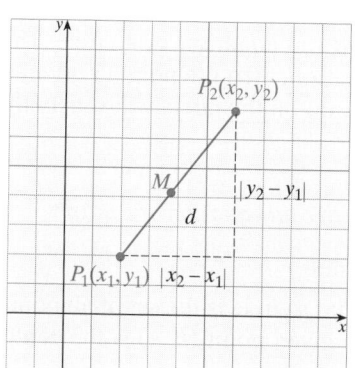

■ **FIGURE 1.3**   Distance formula

THEOREM 1.1   *Distance between two points in the plane*

The distance d between the points $P_1(x_1, y_1)$ and $P_2(x_2, y_2)$ in the plane is given by

$$d = \sqrt{(\Delta x)^2 + (\Delta y)^2} = \sqrt{(x_2 - x_1)^2 + (y_2 - y_1)^2}$$

where Δx (read "delta x") is the **horizontal change** $x_2 - x_1$ (sometimes called *run*) and Δy (read "delta y") is the **vertical change** $y_2 - y_1$ (sometimes called *rise*).

Proof   Using the two points, form a right triangle by drawing lines through the given points parallel to the coordinate axes, as shown in Figure 1.3. The length of the horizontal side of the triangle is $\left| x_2 - x_1 \right| = \left| \Delta x \right|$, and the length of the vertical side is $\left| y_2 - y_1 \right| = \left| \Delta y \right|$. Then

$$d^2 = \left| \Delta x \right|^2 + \left| \Delta y \right|^2 \qquad \text{Pythagorean theorem}$$
$$d^2 = (\Delta x)^2 + (\Delta y)^2 \qquad \text{Absolute value Property 3}$$
$$d = \sqrt{(\Delta x)^2 + (\Delta y)^2} \qquad \text{Solve for } d$$

Midpoint Formula   Related to the formula for the distance between two points is the formula for finding the midpoint of a line segment, as shown in Figure 1.3.

> **Midpoint Formula**
>
> The **midpoint,** M, of the segment with endpoints $P_1(x_1, y_1)$ and $P_2(x_2, y_2)$ has coordinates
>
> $$M\left(\frac{x_1 + x_2}{2}, \frac{y_1 + y_2}{2} \right)$$

Notice that the coordinates of the midpoint of a segment are found by averaging the first and second components of the coordinates of the endpoints, respectively. You are asked to derive this formula in Problem 74.

Relationship between an Equation and a Graph   **Analytic geometry** is that branch of geometry that ties together the geometric concept of position with an algebraic representation—namely, coordinates. For example, you remember from algebra that a line can be represented by an equation. Precisely what does this mean? Can we make a statement that is true for any curve, not just for lines? We answer in the affirmative with the following definition.

> ### Graph of an Equation
>
> The **graph of an equation** in two variables x and y is the collection of all points $P(x, y)$ whose coordinates (x, y) satisfy the equation.

There are two frequently asked questions in analytic geometry:

1. Given a graph (a geometrical representation), find the corresponding equation.
2. Given an equation (an algebraic representation), find the corresponding graph.

In Example 4, we use the distance formula to derive the equation of a circle. This means that if x and y are numbers that satisfy the equation, then the point (x, y) must lie on the circle. Conversely, the coordinates of any point on the circle will satisfy the equation.

| **EXAMPLE 4** | *Using the distance formula to derive an equation of a graph* |

Find the equation of a circle with center (h, k) and radius r.

Solution   Let (x, y) be any point on a circle. Recall that a circle is the set of all points in the plane a given distance from a given point. The given point (h, k) is the center and the given distance is the radius r.

$$r = \text{DISTANCE FROM } (h, k) \text{ TO } (x, y)$$
$$r = \sqrt{(x - h)^2 + (y - k)^2}$$
$$r^2 = (x - h)^2 + (y - k)^2, \quad \text{or} \quad (x - h)^2 + (y - k)^2 = r^2 \qquad \blacksquare$$

A **unit circle** is a circle of radius 1 and center at the origin, so its equation is $x^2 + y^2 = 1$.

| **EXAMPLE 5** | *Finding the equation of a circle* |

Find the equation of the circle with center $(3, -5)$ that passes through the point $(1, 8)$.

Solution   See Figure 1.4. The radius is the distance from the center to the given point:

$$r = \sqrt{(1 - 3)^2 + [8 - (-5)]^2}$$
$$= \sqrt{4 + 169}$$
$$= \sqrt{173}$$

Thus, the equation of the circle is

$$(x - 3)^2 + [y - (-5)]^2 = (\sqrt{173})^2$$
$$(x - 3)^2 + (y + 5)^2 = 173 \qquad \blacksquare$$

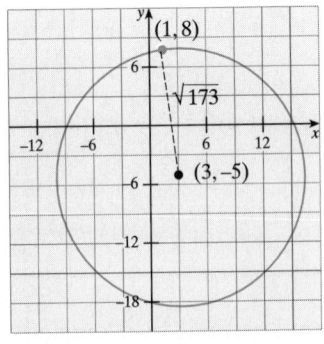

■ **FIGURE 1.4**   Circle with center $(3, -5)$ passing through $(1, 8)$

| EXAMPLE 6 | *Graphing a circle given its equation* |

Sketch the graph of the circle whose equation is

$$4x^2 + 4y^2 - 4x + 8y - 5 = 0$$

Solution   We need to convert this equation into standard form. To do this we use a process called **completing the square** (see Section 2.4 of the *Student Mathematics Handbook*).

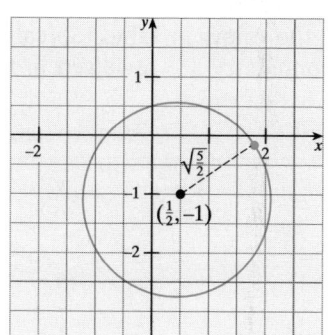

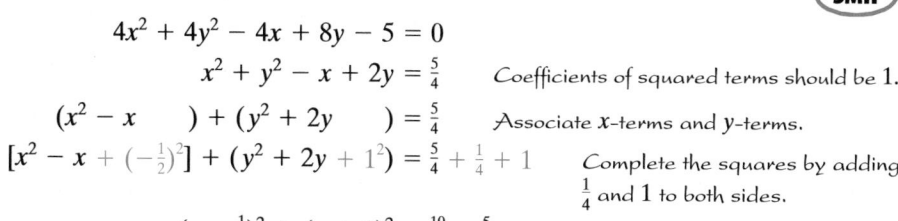

$$4x^2 + 4y^2 - 4x + 8y - 5 = 0$$

$$x^2 + y^2 - x + 2y = \tfrac{5}{4} \qquad \text{\textit{Coefficients of squared terms should be 1.}}$$

$$(x^2 - x \qquad) + (y^2 + 2y \qquad) = \tfrac{5}{4} \qquad \text{\textit{Associate x-terms and y-terms.}}$$

$$[x^2 - x + (-\tfrac{1}{2})^2] + (y^2 + 2y + 1^2) = \tfrac{5}{4} + \tfrac{1}{4} + 1 \qquad \text{\textit{Complete the squares by adding}} \atop \text{\textit{\tfrac{1}{4} and 1 to both sides.}}$$

$$(x - \tfrac{1}{2})^2 + (y + 1)^2 = \tfrac{10}{4} = \tfrac{5}{2}$$

This is a circle with center at $(\tfrac{1}{2}, -1)$ and radius $\sqrt{\tfrac{5}{2}}$. The graph is shown in Figure 1.5.

■ FIGURE 1.5   Sketch of $4x^2 + 4y^2 - 4x + 8y - 5 = 0$

TRIGONOMETRY

SMH   One of the prerequisites for this book is trigonometry. If you need some review, consult a text on trigonometry or Chapter 3 of the *Student Mathematics Handbook* that accompanies this book.

Angles are commonly measured in degrees and radians. A **degree** is defined to be $\tfrac{1}{360}$ revolution and a **radian** $\tfrac{1}{2\pi}$ revolution. Thus, to convert between degree and radian measure, use the following formula:

$$\frac{\theta \text{ measured in degrees}}{360} = \frac{\theta \text{ measured in radians}}{2\pi}$$

WARNING   When angles are measured in calculus, we shall see that radian measure is generally preferable to degree measure. For example, in Chapter 2, we will use the formula for the area of a sector, which requires that the angle be measured in radians. You may assume that when we write expressions such as $\sin x$, $\cos x$, and $\tan x$, the angle x is in radians unless otherwise specified by a degree symbol.

| EXAMPLE 7 | *Converting degree measure to radian measure* |

Convert $255°$ to radian measure.

Solution $\dfrac{255}{360} = \dfrac{\theta}{2\pi}$

$$\theta = \left(\frac{\pi}{180}\right)(255) \approx 4.450589593$$

| EXAMPLE 8 | *Converting radian measure to degree measure* |

Express 1 radian in terms of degrees.

Solution $\dfrac{\theta}{360} = \dfrac{1}{2\pi}$

$$\theta = \left(\frac{180}{\pi}\right)(1)$$

$$\approx 57.29577951°$$

Solving Trigonometric Equations   There will be many times in calculus when you will need to solve a trigonometric equation. As you may remember from trigonometry, solving a trigonometric equation is equivalent to evaluating an inverse trigonometric relation. Inverse functions will be introduced in Section 1.5; for now, we will solve trigonometric equations whose solutions involve the values in Table 1.2 (called a **table of exact values**). These exact values from trigonometry are reviewed in the *Student Mathematics Handbook*.

■ TABLE 1.2
Exact Trigonometric Values

Angle θ Function	0	$\frac{\pi}{6}$	$\frac{\pi}{4}$	$\frac{\pi}{3}$	$\frac{\pi}{2}$
$\cos\theta$	1	$\frac{\sqrt{3}}{2}$	$\frac{\sqrt{2}}{2}$	$\frac{1}{2}$	0
$\sin\theta$	0	$\frac{1}{2}$	$\frac{\sqrt{2}}{2}$	$\frac{\sqrt{3}}{2}$	1
$\tan\theta$	0	$\frac{\sqrt{3}}{3}$	1	$\sqrt{3}$	undefined

It is customary to use the values from Table 1.2 whenever possible. The approximate calculator values will be given only when necessary.

EXAMPLE 9   *Evaluating sine, cosine, tangent, secant, cosecant, and cotangent*

Evaluate $\cos\frac{\pi}{3}$; $\sin\frac{5\pi}{6}$; $\tan(-\frac{5\pi}{4})$; $\sec 1.2$; $\csc(-4.5)$; and $\cot 180°$.

Solution $\cos\dfrac{\pi}{3} = \dfrac{1}{2}$ Exact value; Quadrant I

$\sin\dfrac{5\pi}{6} = \dfrac{1}{2}$ Exact value; Quadrant II

$\tan\left(-\dfrac{5\pi}{4}\right) = -1$ Exact value; Quadrant II

$\sec 1.2 \approx 2.759703601$ Approximate calculator value

$\csc(-4.5) \approx 1.022986384$ Approximate calculator value

$\cot 180°$ is not defined.   ■

EXAMPLE 10   *Solving a trigonometric equation by factoring*

Solve $2\cos\theta\sin\theta = \sin\theta$ on $[0, 2\pi)$.

Solution $2\cos\theta\sin\theta - \sin\theta = 0$
$\sin\theta(2\cos\theta - 1) = 0$
$\sin\theta = 0 \qquad 2\cos\theta - 1 = 0$
$\theta = 0, \pi \qquad\qquad \cos\theta = \dfrac{1}{2}$
$\theta = \dfrac{\pi}{3}, \dfrac{5\pi}{3}$

WARNING   Do not divide both sides by $\sin\theta$, because you might lose a solution. Notice that if $\theta = 0$ or π, then $\sin\theta = 0$. You cannot divide by 0.   ←   ■

Technology Window—Graphing Calculators

The procedure for solving trigonometric equations using calculators is quite similar to solving equations with absolute values. For Example 10, graph Y1=2 COS(X) SIN(X) and Y2 = SIN(X). We see that there are four points of intersection of the curves on the interval $[0, 2\pi)$: By inspection we can see that the curves intersect at 0 and π. (Note that the curves also intersect at $x = 2\pi$, but that value is not in the domain.) By using the ZOOM or INTERSECT, we can find answers with better accuracy. We see that the approximate solution is X $\approx$ 0, 1.0, 3.1, 5.2.

If your calculator has a solver, you might see an expression such as $\theta = \dfrac{(6 \cdot @n19 + 1) \cdot \pi}{3}$; the symbol @n19 represents any integer, which means this expression should be interpreted as $\theta = \dfrac{(6n + 1)\pi}{3}$. Suppose your solver utility gives a solution

$$\theta = (6n + 1)\tfrac{\pi}{3}, (6n - 1)\tfrac{\pi}{3}, n\pi$$

If $n = 0$: $\theta = \tfrac{\pi}{3}, -\tfrac{\pi}{3}, 0$

If $n = 1$: $\theta = \tfrac{7\pi}{3}, \tfrac{5\pi}{3}, \pi$

If $n = 2$: $\theta = \tfrac{13\pi}{3}, \tfrac{11\pi}{3}, 2\pi$

$\vdots$

On $[0, 2\pi)$, the solution is

$$\theta = 0, \tfrac{\pi}{3}, \pi, \tfrac{5\pi}{3}$$

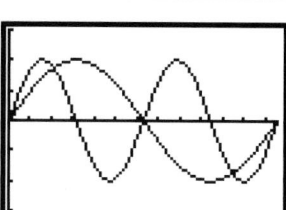

```
Y₁◻2cos Xsin X
Y₂◻sin X
Xmin=0
Xmax=6.2831853...
Xscl=.5
Ymin=-1.5
Ymax=1.5
Yscl=.5
```
TRACE: X=1.0476976 Y=.8660254
X=5.2359878 Y=-.8660254

```
F1▼◻) F2▼   F3▼  F4▼   F5     F6
▼ ◻ Algebra Calc Other PrgmIO Clear a-z...



■ solve(2·cos(θ)·sin(θ) = sin(θ), θ)
    θ = (6·@n3 + 1)·π    or  θ = (6·@n3 − 1)·π   or ▶
          3                        3
solve(2*cos(θ)*sin(θ)=sin(θ),...
MAIN        RAD AUTO        FUNC 1/30
```

EXAMPLE 11 *Solving a trigonometric equation using identities*

Solve $\sin x + \sqrt{3} \cos x = 1$ on $[0, 2\pi)$

Solution

$$\sqrt{3} \cos x = 1 - \sin x$$
$$3 \cos^2 x = 1 - 2 \sin x + \sin^2 x \qquad \textit{Square both sides.}$$
$$3(1 - \sin^2 x) = 1 - 2 \sin x + \sin^2 x$$
$$2 \sin^2 x - \sin x - 1 = 0$$
$$(2 \sin x + 1)(\sin x - 1) = 0$$
$$\sin x = -\tfrac{1}{2} \qquad \sin x = 1$$
$$x = \tfrac{7\pi}{6}, \tfrac{11\pi}{6} \qquad x = \tfrac{\pi}{2}$$

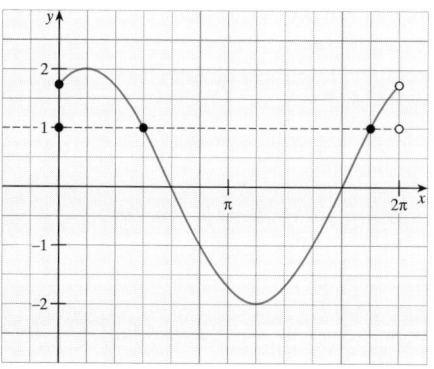

There are two intersection points, at $x = \pi/2$ and $x = 11\pi/6$

However, since we squared both sides, we need to check for extraneous roots by substituting into the original equation. Checking, we see that $x = \tfrac{7\pi}{6}$ is extraneous, and the solution is $x = \tfrac{\pi}{2}, \tfrac{11\pi}{6}$. ∎

1.1 Problem Set

A 1. Fill in the missing parts in the table.

Inequality Notation	Interval Notation
$-3 < x < 4$	**a.**
b.	$[3, 5]$
c.	$[-2, 1)$
$2 < x \leq 7$	**d.**

2. Fill in the missing parts in the table.

Inequality Notation	Interval Notation
a.	$(-\infty, -2)$
b.	$[\frac{\pi}{4}, \sqrt{2}]$
$x > -3$	**c.**
$-1 \leq x \leq 5$	**d.**

3. Represent each of the following on a number line.
   a. $-1 \leq x < 4$ b. $-1 \leq x \leq 3$
   c. $(0, 2)$ d. $[-2, 1)$

4. Represent each of the following on a number line.
   a. $x > 2$ b. $x \leq 4$
   c. $(-\infty, 2) \cup (2, \infty)$ d. $-2 < x \leq 3$ or $x \geq 5$

In Problems 5–6, plot the given points P and Q on a Cartesian plane, find the distance between them, and find the coordinates of the midpoint of the line segment $\overline{PQ}$.

5.* a. $P(2, 3), Q(-2, 5)$ b. $P(-2, 3), Q(4, 1)$
6. a. $P(-5, 3), Q(-5, -7)$ b. $P(-4, 3), Q(3, -4)$

SMH *Solve each equation in Problems 7–24. Assume that a, b, and c are known constants.*

7. $x^2 - x = 0$ 8. $2y^2 + y - 3 = 0$
9. $y^2 - 5y + 3 = 17$ 10. $x^2 + 5x + a = 0$
11. $3x^2 - bx = c$ 12. $4x^2 + 20x + 25 = 0$
13. $|2x + 4| = 16$ 14. $|5y + 2| = 12$
15. $|3 - 2w| = 7$ 16. $|5 - 3t| = 14$
17. $|3x + 1| = -4$ 18. $|1 - 5x| = -2$
19. $\sin x = -\frac{1}{2}$ on $[0, 2\pi)$
20. $(\sin x)(\cos x) = 0$ on $[0, 2\pi)$

21. $(2\cos x + \sqrt{2})(2\cos x - 1) = 0$ on $[0, 2\pi)$
22. $(3\tan x + \sqrt{3})(3\tan x - \sqrt{3}) = 0$ on $[0, 2\pi)$
23. $\cot x + \sqrt{3} = \csc x$ on $[0, 2\pi)$
24. $\sec^2 x - 1 = \sqrt{3}\tan x$ on $[0, 2\pi)$

Solve each inequality in Problems 25–34, and give your answer using interval notation.

25. $3x + 7 < 2$ 26. $5(3 - x) > 3x - 1$
27. $-5 < 3x < 0$ 28. $-3 < y - 5 \leq 2$
29. $3 \leq -y < 8$ 30. $-5 \leq 3 - 2x < 18$
31. $t^2 - 2t \leq 3$ 32. $s^2 + 3s - 4 > 0$
33. $|x - 8| \leq 0.001$ 34. $|x - 5| < 0.01$

In Problems 35–38, find an equation of a circle with given center C and radius r.

35. $C(-1, 2); r = 3$ 36. $C(3, 0); r = 2$
37. $C(0, 1.5); r = 0.25$ 38. $C(-1, -5); r = 4.1$

Graph the circles in Problems 39–42.

39. $x^2 - 2x + y^2 + 2y + 1 = 0$
40. $4x^2 + 4y^2 + 4y - 15 = 0$
41. $x^2 + y^2 + 2x - 10y + 25 = 0$
42. $2x^2 + 2y^2 + 2x - 6y - 9 = 0$

SMH *Use the sum and difference formulas from trigonometry to find the exact values of the expressions in Problems 43–46. Check by finding a calculator approximation.*

43. $\sin(-\frac{\pi}{12})$ 44. $\cos\frac{7\pi}{12}$
45. $\tan\frac{\pi}{12}$ 46. $\sin 165°$

B 47. ■ **What Does This Say?**† Describe a process for solving a quadratic equation.
48. ■ **What Does This Say?** Describe a process for solving absolute value equations.
49. ■ **What Does This Say?** Describe a process for solving absolute value inequalities.
50. ■ **What Does This Say?** Describe a process for solving trigonometric equations.

Specify the period for each graph in Problems 51–56. Also graph each curve.

51. a. $y = \sin x$ b. $y = \cos x$ c. $y = \tan x$

52. a. $y = 2\sin 2\pi x$ b. $y = 3\cos 3\pi x$ c. $y = 4\tan\left(\frac{\pi x}{5}\right)$

---

†Many problems in this book are labeled **What Does This Say?** Following the question will be a question for you to answer in your own words, or a statement for you to rephrase in your own words. These problems are intended to be similar to the "What This Says" boxes in the text.

SSM   *Problem numbers marked in color denote problems with solutions in the Student Survival Manual.* Answers to odd-numbered problems are given in Appendix E.

53. $y = \tan(2x - \frac{\pi}{2})$

54. $y = 2\cos(3x + 2\pi) - 2$

55. $y = 4\sin(\frac{1}{2}x + 2) - 1$

56. $y = \tan\left(\dfrac{x}{2} + \dfrac{\pi}{3}\right)$

57. The current I (in amperes) in a certain circuit (for some convenient unit of time) which generates the following set of data points:

Time	Current	Time	Current
0	−60.000000	10	30.00000
1	−58.68886	11	40.14784
2	−54.81273	12	48.54102
3	−48.54102	13	54.81273
4	−40.14784	14	58.68886
5	−30.00000	15	60.00000
6	−18.54102	16	58.68886
7	−6.27171	17	54.81273
8	6.27171	18	48.54102
9	18.54102	19	40.14784
		20	30.00000

Plot the data points and draw a smooth curve passing through these points. Determine possible values of A, B, C, and D so that these data are modeled by the equation

$$y - A = B\sin C(x - D).$$

58. Suppose that a point P on a waterwheel with a 30-ft radius is d units from the water. If the waterwheel turns at 6 revolutions per minute, the height of the point P above the water is given by the following set of data points:

Time	Height	Time	Height
0	−1.000	10	−1.000
1	4.729	11	4.729
2	19.729	12	19.729
3	38.271	13	38.270
4	53.271	14	53.270
5	59.000	15	59.000
6	53.271	16	53.271
7	38.271	17	38.271
8	19.729	18	19.730
9	4.729	19	4.730
		20	−1.000

Plot the data points and draw a smooth curve passing through these points. Determine possible values of A, B, C, and D so that these data are modeled by the equation

$$y - A = B\cos C(x - D)$$

Determine a possible equation of a curve that is generated by this waterwheel.

59. The sun and moon tide curves are shown here.* During a new moon, the sun and moon tidal bulges are centered at the same longitude, so their effects are added to produce maximum high tides and minimum low tides. This produces maximum tidal range (the distance between the low and high tides).

NEW MOON (Spring tide)

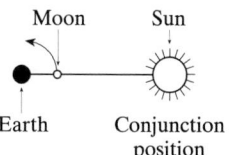

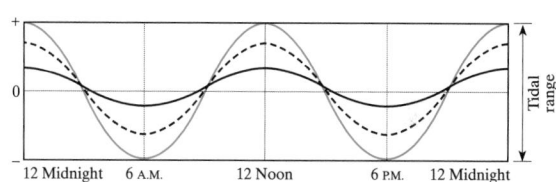

Write possible equations for the sun curve, the moon curve, and the combined curve. Assume the tidal range is 10 ft and the period is 2.

60. The sun and moon tide curves are shown in the figure. During the third quarter (neap tide), the sun and moon tidal bulges are at right angles to each other. Thus, the sun tide reduces the effect of the moon tide.

THIRD QUARTER (Neap tide)

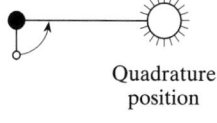

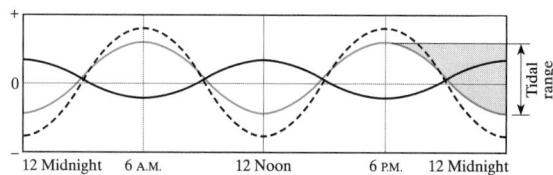

*From *Introductory Oceanography*, 5th ed., by H. V. Thurman, p. 253. Reprinted with permission of Merrill, an imprint of Macmillan Publishing Company. Copyright © 1988 Merrill Publishing Company, Columbus, Ohio.

Write possible equations for the sun curve, the moon curve, and the combined curve. Assume the tidal range is 4 ft and the period is 12.

61. A person is swimming at a depth of d units beneath the surface of the water.

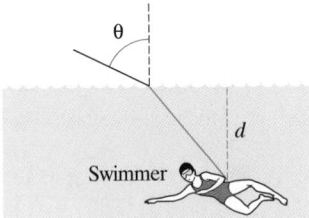

Because light is refracted by the water, a viewer standing above the waterline sees the swimmer at an apparent depth of s units. In physics,* it is shown that if the person is viewed from an angle of incidence θ, then

$$s = \frac{3d \cos \theta}{\sqrt{7 + 9 \cos^2 \theta}}$$

a. If $d = 5.0$ meters and $\theta = 37°$, what is the apparent depth of the swimmer?
b. If the actual depth is $d = 5.0$ meters, what angle of incidence yields an apparent depth of $s = 2.5$ meters?

62. JOURNAL PROBLEM *The Mathematics Student Journal,* by Murray Klamkin.[†] Determine all the roots of the quartic equation $x^4 - 4x = 1$.

63. Hɪꜱᴛᴏʀɪᴄᴀʟ Qᴜᴇꜱᴛ   The division of one revolution into 360 equal parts (called degrees) is no doubt due to the sexagesimal (base 60) numeration system used by the Babylonians. Several explanations have been put

---

*R. A. Serway, *Physics,* 3rd ed. Philadelphia: Saunders, 1992, p. 1007.

†Most mathematics journals have problem sections that solicit interesting problems and solutions for publication. From time to time, we will reprint a problem from a mathematics journal. If you have difficulty solving a journal problem, you may wish to use a library to find the problem and solution as printed in the journal. The title of the journal is included as part of the problem, and we will generally give you a reference as a footnote. This problem is found in Volume 28 (1980), issue 3, p. 2.

forward to account for the choice of this number. (For example, see Howard Eve's *In Mathematical Circles.*) One possible explanation is put forth by Otto Neugebauer, scholar and authority on early Babylonian mathematics and astronomy. In early Sumerian times, there existed a *Babylonian mile,* equal to about seven of our miles. Sometime in the first millennium B.C., when Babylonian astronomy reached the stage in which systematic records of the stars were kept, the Babylonian mile was adapted for measuring spans of time. Since a complete day was found to be equal to 12 time-miles, and one complete day is equivalent to one revolution of the sky, a complete circuit was divided into 12 equal parts. Then, for convenience, the Babylonian mile was subdivided into 30 equal parts. One complete circuit therefore has $(12)(30) = 360$ equal parts.

Show that the radius of a circle can be applied exactly six times to its circumference as a chord.

64. Hɪꜱᴛᴏʀɪᴄᴀʟ Qᴜᴇꜱᴛ   The numeration system we use (base 10) evolved over a long period of time. It is often called the *Hindu-Arabic* system because its origins can be traced back to the Hindus in Bactria (now Afghanistan). Later, in A.D. 700, India was invaded by the Arabs, who used and modified the Hindu numeration system, and, in turn, introduced it to Western civilization. The Hindu Brahmagupta stated the rules for operations with positive and negative numbers in the seventh century A.D. There are some indications that the Chinese had some knowledge of negative numbers as early as 200 B.C. On the other hand, the Western mathematician Girolamo Cardan (1501–1576) was calling numbers such as (-1) absurd as late as 1545.

Write a paper on the history of the real number system.

65. ■ **What Does This Say?** If $ax + b = 0$, what effect does changing b have on the solution $(a \neq 0)$?

66. ■ **What Does This Say?** If $ax^2 + bx + c = 0$, what effect does changing c have on the solution $(a \neq 0)$?

67. ■ **What Does This Say?** If $\sin ax = b$, what effect does changing a have on the solution $(a \neq 0)$?

68. If $c \geq 0$, show that $|x| \leq c$ if and only if $-c \leq x \leq c$.

69. Show that $-|x| \leq x \leq |x|$ for any number x.

70. Prove that $|a| = |b|$ if and only if $a = b$ or $a = -b$.

71. Prove that if $|a| < b$ and $b > 0$, then $-b < a < b$.

72. Prove the triangle inequality:

$$|x + y| \leq |x| + |y|$$

73. Show that $|x| - |y| \leq |x - y|$ for all x and y.

74. Derive the **midpoint formula**

$$M\left(\frac{x_1 + x_2}{2}, \frac{y_1 + y_2}{2}\right)$$

for the midpoint of a segment with endpoints $P(x_1, y_1)$ and $Q(x_2, y_2)$.

1.2 *Lines in the Plane*

IN THIS SECTION **slope of a line, forms for the equation of a line, parallel and perpendicular lines, best-fitting line**

SLOPE OF A LINE

A distinguishing feature of a line is the fact that its *inclination* with respect to the horizontal is constant. It is common practice to specify inclination by means of a concept called *slope*. A carpenter might describe a roof line that rises 1 ft for every 3 ft of horizontal "run" as having a slope or pitch of 1 to 3.

Let Δx and Δy represent, as before, the amount of change in the variables x and y, respectively. Then a nonvertical line L that rises (or falls) Δy units (measured from bottom to top) for every Δx units of run (measured from left to right) is said to have a *slope* of $m = \Delta y / \Delta x$. (If Δy is negative, then the "rise" is actually a fall; and if Δx is negative, then the run is actually right to left.) In particular, if $P(x_1, y_1)$ and $Q(x_2, y_2)$ are two distinct points on L, then the changes in the variables x and y are given by $\Delta x = x_2 - x_1$ and $\Delta y = y_2 - y_1$, and the slope of L is

$$m = \frac{\Delta y}{\Delta x} = \frac{y_2 - y_1}{x_2 - x_1} \quad \text{for } \Delta x \neq 0 \qquad See\ Figure\ 1.6.$$

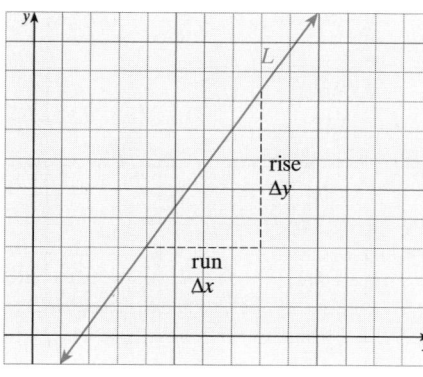

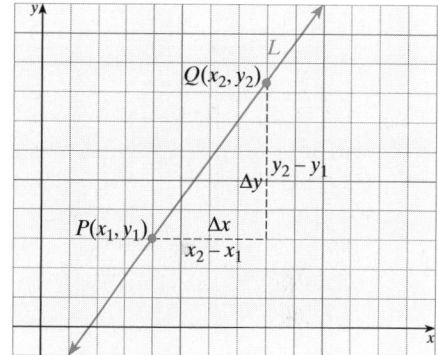

a. The slope of L is $m = \dfrac{\Delta y}{\Delta x}$.   **b.** The slope is given by $m = \dfrac{y_2 - y_1}{x_2 - x_1}$.

■ **FIGURE 1.6**   The slope of a line

Slope of a Line

A nonvertical line that contains the points $P(x_1, y_1)$ and $Q(x_2, y_2)$ has **slope**

$$m = \frac{\Delta y}{\Delta x} = \frac{y_2 - y_1}{x_2 - x_1}$$

We say that a line with slope m is *rising* (when viewed from left to right) if $m > 0$, *falling* if $m < 0$, and *horizontal* if $m = 0$.

There is a useful trigonometric formulation of slope. The **angle of inclination** of a line L is defined to be the nonnegative angle ϕ ($0 \leq \phi < \pi$) formed between L and the positively directed x-axis.

Angle of Inclination

The **angle of inclination** of a line L is the angle ϕ ($0 \le \phi < \pi$) between L and the positive x-axis. Then the **slope** of L with inclination ϕ is

$$m = \tan \phi$$

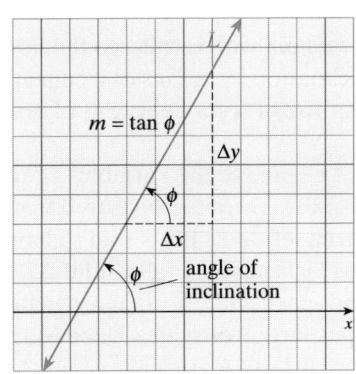

■ **FIGURE 1.7** Trigonometric form for the slope of a line

We see that the line L is *rising* if $0 < \phi < \frac{\pi}{2}$ and is *falling* if $\frac{\pi}{2} < \phi < \pi$. The line is *horizontal* if $\phi = 0$ and is *vertical* if $\phi = \frac{\pi}{2}$. Notice that if $\phi = \frac{\pi}{2}$, $\tan \phi$ is not defined; therefore m is not defined for a vertical line. A **vertical line** is said to have **no slope.**

WARNING ➤ Note that saying a line has *no slope* (vertical line) is not the same as saying it has *zero slope* (horizontal line). Sometimes we say a vertical line has **infinite slope.** ◄

To derive the trigonometric representation for slope, we need to find the slope of the line through $P(x_1, y_1)$ and $Q(x_2, y_2)$, where ϕ is the angle of inclination. From the definition of the tangent, we have

$$\tan \phi = \frac{y_2 - y_1}{x_2 - x_1} = \frac{\Delta y}{\Delta x} = m \qquad \textit{See Figure 1.7.}$$

Lines with various slopes are shown in Figure 1.8.

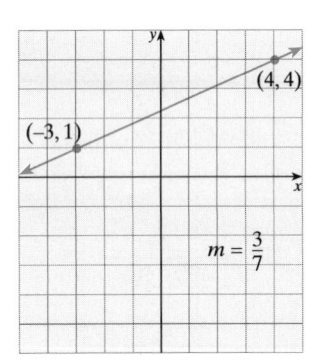

Positive slope; line rises.

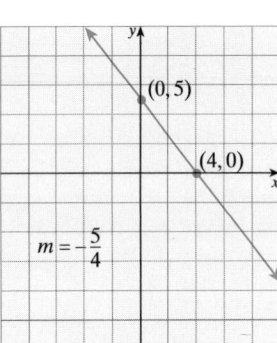

Negative slope; line falls.

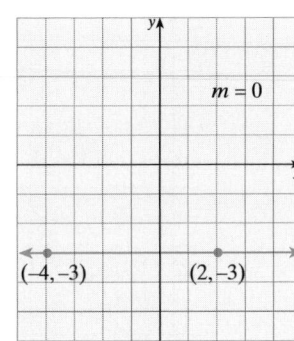

Zero slope ($\Delta y = 0$); line is horizontal.

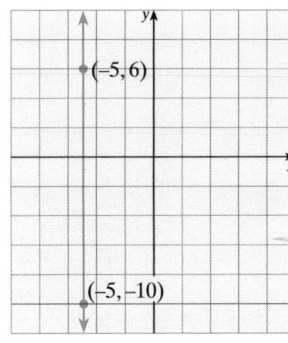

Slope is undefined; ($\Delta x = 0$) line is vertical.

■ **FIGURE 1.8**   Examples of slope

FORMS FOR THE EQUATION OF A LINE

In algebra you studied several forms of the equation of a line. The derivations of some of these are reviewed in the problems. Here is a summary of the forms most frequently used in calculus.

Forms of a Linear Equation

STANDARD FORM:	$Ax + By + C = 0$	A, B, C constants (A and B not both 0)
SLOPE–INTERCEPT FORM:	$y = mx + b$	Slope m, y-intercept $(0, b)$
POINT–SLOPE FORM:	$y - k = m(x - h)$	Slope m, through point (h, k)
HORIZONTAL LINE:	$y = k$	Slope 0
VERTICAL LINE:	$x = h$	Slope undefined

> **EXAMPLE 1**   *Deriving the two-intercept form of the equation of a line*

Derive the equation of the line with intercepts $(a, 0)$ and $(0, b)$, $a \neq 0$, $b \neq 0$.

Solution   The slope of the equation passing through the given points is

$$m = \frac{b - 0}{0 - a} = -\frac{b}{a}$$

Use the point–slope form with $h = 0$, $k = b$. (You can use either of the given points.)

$$y - b = -\frac{b}{a}(x - 0)$$

$$ay - ab = -bx$$

$$bx + ay = ab$$

$$\frac{x}{a} + \frac{y}{b} = 1 \qquad \text{Divide both sides by } ab. \quad \blacksquare$$

Two quantities x and y that satisfy a linear equation $Ax + By + C = 0$ (A and B not both 0) are said to be *linearly related*. This terminology is illustrated in Example 2.

> **EXAMPLE 2**   *Linearly related variables*

When a weight is attached to a helical spring, it causes the spring to lengthen. According to Hooke's law, the length d of the spring is linearly related to the weight w.* If $d = 4$ cm when $w = 3$ g and $d = 6$ cm when $w = 6$ g, what is the original length of the spring, and what weight will cause the spring to lengthen to 5 cm?

Solution   Because d is linearly related to w, we know that points (w, d) lie on a line, and the given information tells us that two such points are $(3, 4)$ and $(6, 6)$ as shown in Figure 1.9.

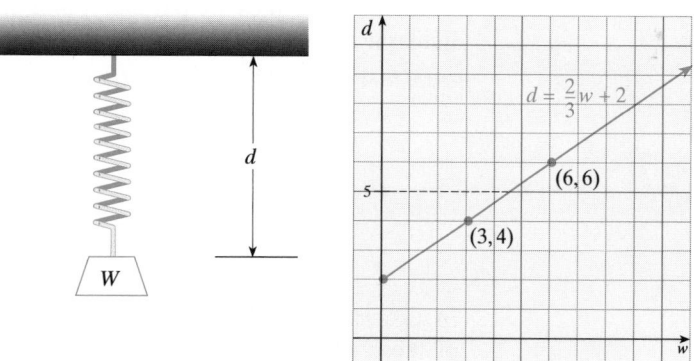

■ **FIGURE 1.9**   The length d of the spring is linearly related to the weight w of the attached object.

We first find the slope of the line and then use the point–slope form to derive its equation.

$$m = \frac{6 - 4}{6 - 3} = \frac{2}{3}$$

---

*Hooke's law is useful for small displacements, but for larger displacements, it may not be a good model.

Next, substitute into the point–slope form with $h = 3$ and $k = 4$:

$$d - 4 = \tfrac{2}{3}(w - 3)$$
$$d = \tfrac{2}{3}w + 2$$

The original length of the spring is found for $w = 0$:

$$d = \tfrac{2}{3}(0) + 2 = 2$$

The original length was 2 cm. To find the weight that corresponds to $d = 5$, we solve the equation

$$5 = \tfrac{2}{3}w + 2$$
$$\tfrac{9}{2} = w$$

Therefore, the weight that corresponds to a length of 5 cm is 4.5 g.   ■

PARALLEL AND PERPENDICULAR LINES

It is often useful to know whether two given lines are either parallel or perpendicular. A vertical line can be parallel only to other vertical lines and perpendicular only to horizontal lines. Cases involving nonvertical lines may be handled by the criteria given in the following theorem.

THEOREM 1.2   *Slope criteria for parallel and perpendicular lines*

If L_1 and L_2 are nonvertical lines with slopes m_1 and m_2, then

L_1 and L_2 are **parallel** if and only if $m_1 = m_2$;

L_1 and L_2 are **perpendicular** if and only if $m_1 m_2 = -1$, or $m_1 = -\dfrac{1}{m_2}$.

In other words, lines are parallel if and only if their slopes are equal and are perpendicular if and only if their slopes are negative reciprocals of each other.

Proof   The key ideas behind these two slope criteria are displayed in Figure 1.10.

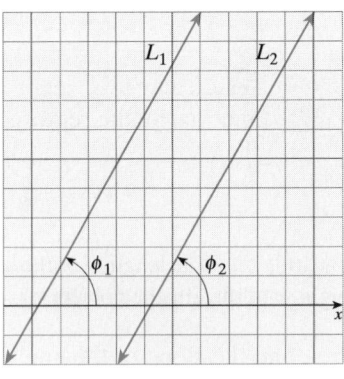

a. Parallel lines have equal slope: $m_1 = m_2$.

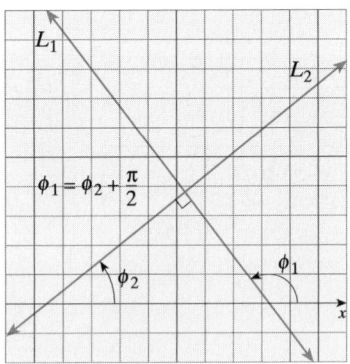

b. Perpendicular lines have negative reciprocal slopes: $m_1 m_2 = -1$.

■ FIGURE 1.10   Parallel and perpendicular lines

The criterion for parallel lines follows from the fact that lines are parallel whenever their angles of inclination are equal (see Figure 1.10a). On the other hand, if L_1

and L_2 are perpendicular (see Figure 1.10b), one angle of inclination, say ϕ_2, must be an acute angle and must satisfy

$$\phi_1 = \phi_2 + \tfrac{\pi}{2}$$

Then

$$
\begin{aligned}
m_1 &= \tan \phi_1 \\
&= \tan(\phi_2 + \tfrac{\pi}{2}) \\
&= \tan[\tfrac{\pi}{2} - (-\phi_2)] \\
&= \cot(-\phi_2) \\
&= -\cot \phi_2 \\
&= \frac{-1}{\tan \phi_2} \\
&= \frac{-1}{m_2}
\end{aligned}
$$

Thus, $m_1 m_2 = -1$. We leave it to you to show that L_1 and L_2 are perpendicular if $m_1 m_2 = -1$.

EXAMPLE 3   *Finding equations for parallel and perpendicular lines*

Let L be the line $3x + 2y = 5$.

a. Find an equation of the line that is parallel to L and passes through $P(4, 7)$.

b. Find an equation of the line that is perpendicular to L and passes through $P(4, 7)$.

Solution   By rewriting the equation of L as $y = -\tfrac{3}{2}x + \tfrac{5}{2}$, we see that the slope of L is $m = -\tfrac{3}{2}$.

a. Any line that is parallel to L must also have slope $m_1 = -\tfrac{3}{2}$. The required line contains the point $P(4, 7)$. Use the point–slope form to find the equation and write your answer in standard form:

$$
\begin{aligned}
y - 7 &= -\tfrac{3}{2}(x - 4) \\
2y - 14 &= -3x + 12 \\
3x + 2y - 26 &= 0
\end{aligned}
$$

b. Any line perpendicular to L must have slope $m_2 = \tfrac{2}{3}$ (negative reciprocal of the slope of L). Once again, the required line contains the point $P(4, 7)$, and we find

$$
\begin{aligned}
y - 7 &= \tfrac{2}{3}(x - 4) \\
3y - 21 &= 2x - 8 \\
2x - 3y + 13 &= 0
\end{aligned}
$$

These lines are shown in Figure 1.11.

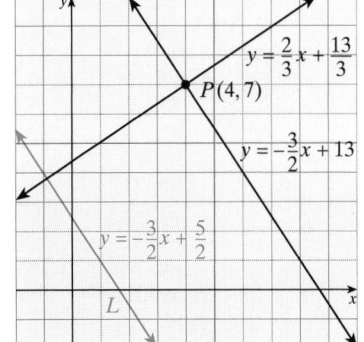

■ **FIGURE 1.11**   Graph of lines parallel and perpendicular to the given line $3x + 2y = 5$

BEST-FITTING LINE

In mathematical modeling, it is often necessary to deal with numerical data and make assumptions regarding the relationship between two variables. For example,

> IQ and salary
>
> Study time and grades
>
> Age and heart disease

Runner's speed and runner's brand of shoe

Math grades in the 8th grade and amount of TV viewing

Teachers' salaries and beer consumption

All are attempts to relate two variables in some way or another. If it is established that there is a **correlation,** then the next step in the modeling process is to identify the nature of the relationship. This is called **regression analysis.** In this section, we consider only linear relationships; that is, we are looking for the *best-fitting line* for a given set of data.

The first consideration is one of correlation. We want to know whether two variables are related—that is, dependent on one another. Let us call one variable x and the other y. These variables can be represented as ordered pairs (x, y) in a graph called a **scatter diagram.**

> **EXAMPLE 4**   *Best-fitting line*

A survey of 20 students compared the grade received on an examination with the length of time the student studied. Draw a scatter diagram to represent the data in the table, where x is the study time (in minutes) and y is the grade. Later, in Chapter 12, we shall show that the best-fitting line is

$$y = 1.27x + 34.81 \quad \text{for } 0 \le x \le 50$$

Draw this line and then use this equation to predict the grade for a student who studies for 30 minutes.

Student number	Length of study time (nearest 5 min.)	Grade (100 possible)
1	30	72
2	40	85
3	30	75
4	35	78
5	45	89
6	15	58
7	15	71
8	50	94
9	30	78
10	0	10
11	20	75
12	10	43
13	25	68
14	25	60
15	25	70
16	30	68
17	40	82
18	35	75
19	20	65
20	15	62

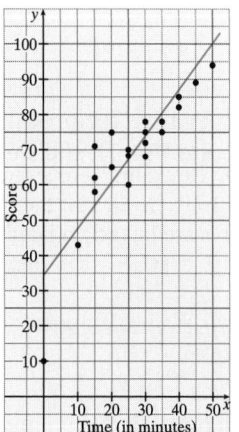

■ **FIGURE 1.12**   Best-fitting line

Solution   First, you need to select a scale for the x- and y-axes, and then plot the points corresponding to the data points shown in the table. In drawing the best-fitting line, first plot the y-intercept: $(0, 34.81)$. Next, to use the slope, $m = 1.27$, write

$$m = \frac{1.27}{1} = \frac{12.7}{10} \approx \frac{4}{3}$$

Using the scale shown in Figure 1.12, we count up 4 squares and over 3 squares to plot a second point so that we can draw the line.

Finally, we use $y = f(x) = 1.27x + 34.81$ to find

$$f(30) = 1.27(30) + 34.81 = 72.91$$

Using this equation, we find that for $x = 30$ minutes of study time, the expected grade is 73.   ■

1.2 Problem Set

 1. ■ **What Does This Say?** Outline a procedure for graphing a linear equation.

In Problems 2–15, find the equation in standard form for the line that satisfies the given requirements.

2. passing through $(1, 4)$ and $(3, 6)$
3. passing through $(-1, 7)$ and $(-2, 9)$
4. horizontal line through $(-2, -5)$
5. passing through the point $(1, \frac{1}{2})$ with slope 0
6. slope 2 and y-intercept $(0, 5)$
7. vertical line through $(-2, -5)$
8. slope -3, x-intercept $(5, 0)$
9. x-intercept $(7, 0)$ and y-intercept $(0, -8)$
10. x-intercept $(4.5, 0)$ and y-intercept $(0, -5.4)$
11. passing through $(-1, 8)$ parallel to $3x + y = 7$
12. passing through $(4, 5)$ parallel to the line passing through $(2, 1)$ and $(5, 9)$
13. passing through $(3, -2)$ perpendicular to $4x - 3y + 2 = 0$
14. passing through $(-1, 6)$ perpendicular to the line through the origin with slope 0.5
15. perpendicular to the line whose equation is $x - 4y + 5 = 0$ where it intersects the line whose equation is $2x + 3y - 1 = 0$

In Problems 16–29, find, if possible, the slope, the y-intercept, and the x-intercept of the line whose equation is given. Sketch the graph of each equation.

16. $y = \frac{2}{3}x - 8$
17. $y = -\frac{5}{7}x + 3$
18. $y - 4 = 4.001(x - 2)$
19. $y - 9 = 6.001(x - 3)$
20. $5x + 3y - 15 = 0$
21. $3x + 5y + 15 = 0$
22. $2x - 3y - 2,550 = 0$
23. $6x - 10y - 3 = 0$
24. $\frac{x}{2} + \frac{y}{3} = 2$
25. $\frac{x}{2} - \frac{y}{3} = 1$
26. $y = 2x$
27. $x = 5y$
28. $y - 5 = 0$
29. $x + 3 = 0$

30. Find an equation for a vertical line L such that a region bounded by L, the x-axis, and the line $2y - 3x = 6$ has area 3.

31. Find an equation for a horizontal line M such that the region bounded by M, the y-axis, and the line $2y - 3x = 6$ has an area of 3.

32. Find an equation of the perpendicular bisector of the line segment connecting $(-3, 7)$ and $(4, -1)$.

33. Three vertices of a parallelogram are $(1, 3), (4, 11)$, and $(3, -2)$. If $(1, 3)$ and $(3, -2)$ lie on the same side, what is the fourth vertex?

Technology Window

Use a graphing calculator in Problems 34–45 to solve the systems of equations graphically, and approximate the coordinates of the intersection point to the nearest tenth. Some calculators have an INTERSECT *feature. Check with your owner's manual. Then solve the system algebraically and compare the results.*

34. $\begin{cases} 2x - 3y = -8 \\ x + y = 6 \end{cases}$

35. $\begin{cases} 2x - 3y = -8 \\ 4x - 6y = 0 \end{cases}$

36. $\begin{cases} 2x - 3y = -8 \\ y = \frac{2}{3}x + \frac{8}{3} \end{cases}$

37. $\begin{cases} 3x - 4y = 16 \\ -x + 2y = -6 \end{cases}$

38. $\begin{cases} y = 3x + 1 \\ x - 2y = 8 \end{cases}$

39. $\begin{cases} 2x + 3y = 12 \\ -4x + 6y = 18 \end{cases}$

40. $\begin{cases} 2x + 3y = 15 \\ y = \frac{2}{3}x - 20 \end{cases}$

41. $\begin{cases} x + y = 12 \\ 0.6y = 0.5(40) \end{cases}$

42. $\begin{cases} x + 3y = \cos^2 \frac{\pi}{3} \\ x + y = -\sin^2 \frac{\pi}{3} \end{cases}$

43. $\begin{cases} x^2 + y^2 = 4 \\ x - y = 0 \end{cases}$

44. $\begin{cases} x^2 + y^2 = 10 \\ (x - 3)^2 + y^2 = 9 \end{cases}$

45. $\begin{cases} x^2 + y^2 = 15 \\ (x + 4)^2 + y^2 = 16 \end{cases}$

---

46. A life insurance table indicates that a woman who is now A years old can expect to live E years longer. Suppose that A and E are linearly related and that $E = 50$ when $A = 24$ and $E = 20$ when $A = 60$.
   a. At what age may a woman expect to live 30 years longer?
   b. What is the life expectancy of a newborn female child?
   c. At what age is the life expectancy zero?

47. On the Fahrenheit temperature scale, water freezes at $32°$ and boils at $212°$; the corresponding temperatures on the Celsius scale are $0°$ and $100°$. Given that the Fahrenheit and Celsius temperatures are linearly related, first find numbers r and s so that $F = rC + s$, and then answer these questions.
   a. Mercury freezes at -39 °C. What is the corresponding Fahrenheit temperature?
   b. For what value of C is $F = 0$?
   c. What temperature is the same in both scales?

48. The average SAT mathematics scores of incoming students at an eastern liberal arts college have been declining in recent years. In 1992, the average SAT score was 575; in 1997, it was 545. Assuming the average SAT score varies linearly with time, answer these questions.
   a. Express the average SAT score in terms of time measured from 1992.

b. If the trend continues, what will the average SAT score of incoming students be at the turn of the century?

c. When will the average SAT score be 497?

49. A manufacturer's total cost consists of a fixed overhead of $5,000 plus production costs of $60 per unit. Assuming the cost varies linearly with the level of production, express the total cost in terms of the number of units produced and draw the graph.

50. A certain car rental agency charges $40 per day with 100 free miles plus 34¢ per mile after the first 100 miles. First express the cost of renting a car from this agency for one day in terms of the number of miles driven. Then draw the graph and use it to check your answers to these questions.

a. How much does it cost to rent a car for a 1-day trip of 50 mi?

b. How many miles were driven if the daily rental cost was $92.36?

51. A manufacturer buys $200,000 worth of machinery that depreciates linearly so that its trade-in value after 10 years will be $10,000. Express the value of the machinery as a function of its age and draw the graph. What is the value of the machinery after 4 years?

52. ■ **What Does This Say?** Discuss the correlation shown by the following chart taken from the November 1987 issue of *Scientific American*.

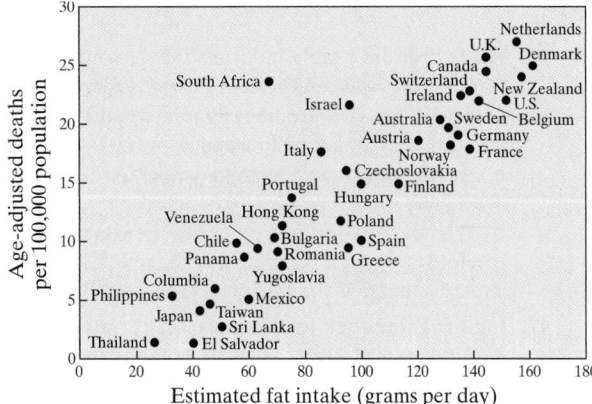

It can be shown that the best-fitting line is one of the following:

A. $y = 0.139x$ **B.** $y = -3 + 0.231x$
C. $y = 1 + 0.981x$

Which do you think is the correct one? Use your choice to estimate the number of deaths per 100,000 population to be expected from an average fat intake of 150 g/day (roughly the average fat intake in the United States).

53. ■ **What Does This Say?** Discuss the correlation shown by the following chart taken from the April 1991 issue of *Scientific American*.

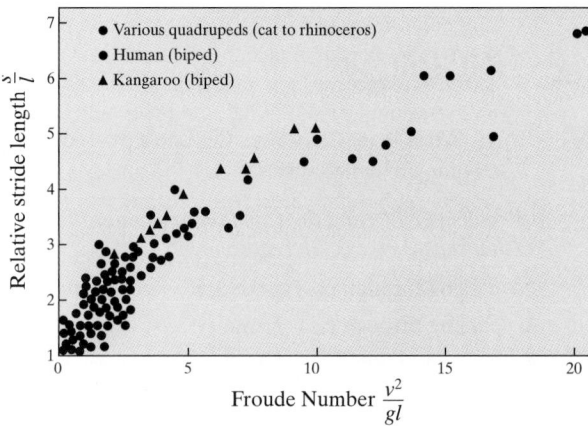

It can be shown that the best-fitting line is one of the following:

A. $y = 0.31x$
B. $y = 0.22x + 2$
C. $y = 1 + 0.29x$

Which do you think is correct? Use your choice to estimate the relative stride length that corresponds to a Froude number $x = 4$.

54. Show that if the point $P(x, y)$ is equidistant from $A(1, 3)$ and $B(-1, 2)$, its coordinates must satisfy the equation $4x + 2y - 5 = 0$. Sketch the graph of this equation.

55. Let $P_1(2, 6)$, $P_2(-1, 3)$, $P_3(0, -2)$, and $P_4(a, b)$ be points in the plane that are located so that $P_1P_2P_3P_4$ is a parallelogram.

a. There are three possible choices for P_4. One is $A(3, 1)$. What are the others, which we call B and C?

b. The center of a triangle is the point where its three medians intersect. Find the center of $\triangle ABC$ and of $\triangle P_1P_2P_3$. Do you notice anything interesting? We will revisit this problem using vectors in Chapter 10.

56. ■ **What Does This Say?** Ethyl alcohol is metabolized by the human body at a constant rate (independent of concentration). Suppose the rate is 10 mL per hour.

a. Express the time t (in hours) required to metabolize the effects of drinking ethyl alcohol in terms of the amount A of ethyl alcohol consumed.

b. How much time is required to eliminate the effects of a liter of beer containing 3% ethyl alcohol?

c. Discuss how the function in part **a** can be used to determine a reasonable "cutoff" value for the amount of ethyl alcohol A that each individual may be served at a party.

57. Since the beginning of the month, a local reservoir has been losing water at a constant rate (that is, the amount of water in the reservoir is a linear function of time). On the 12th of the month, the reservoir held 200 million gallons of water; on the 21st, it held only 164 million gallons. How much water was in the reservoir on the 8th of the month?

58. To encourage motorists to form car pools, the transit authority in a certain metropolitan area has been offering a special reduced rate at toll bridges for vehicles containing four or more persons. When the program began 30 days ago, 157 vehicles qualified for the reduced rate during the morning rush hour. Since then, the number of vehicles qualifying has increased at a constant rate (that is, the number is a linear function of time), and 247 vehicles qualified today. If the trend continues, how many vehicles will qualify during the morning rush hour 14 days from now?

59. ■ **What Does This Say?** The value of a certain rare book doubles every 10 years. The book was originally worth $3.
   a. How much is the book worth when it is 30 years old? When it is 40 years old?
   b. Is the relationship between the value of the book and its age linear? Explain.

60. HISTORICAL QUEST The region between the Tigris and Euphrates Rivers (present-day Iraq) is rightly known as the Cradle of Civilization. During the so-called Babylonian period (roughly 2000–600 B.C.),* important mathematical ideas began to germinate in the region, including positional notation for numeration. Unlike their Egyptian contemporaries who usually wrote on fragile papyrus, Babylonian mathematicians recorded their ideas on clay tablets. One of these tablets, in the Yale Collection, shows a system equivalent to

$$\begin{cases} xy = 600 \\ (x + y)^2 - 150(x - y)^2 = 100 \end{cases}$$

Find a positive solution $(x > 0, y > 0)$ for this system correct to the nearest tenth.

61. HISTORICAL QUEST The Louvre Tablet from the Babylonian civilization is dated about 1500 B.C. It shows a system equivalent to

$$\begin{cases} xy = 1 \\ x + y = a \end{cases}$$

Solve this system for x and y in terms of a.

62. Show that, in general, a line passing through $P(h, k)$ with slope m has the equation
$$y - k = m(x - h)$$

63. If $A(x_1, y_1)$ and $B(x_2, y_2)$ with $x_1 \neq x_2$ are two points on the graph of the line $y = mx + b$, show that
$$m = \frac{y_2 - y_1}{x_2 - x_1}$$

*For an interesting discussion of Mesopotamian mathematics, see *A History of Mathematics,* 2nd ed., by Carl B. Boyer, revised by Uta C. Merzbach, John Wiley and Sons, Inc., New York, 1968, pp. 23–42.

Use this fact to show that the graph of $y = mx + b$ is a line with slope m and then show that the line has y-intercept $(0, b)$.

64. Show that the distance s from the point (x_0, y_0) to the line $Ax + By + C = 0$ is given by the formula
$$s = \left| \frac{Ax_0 + By_0 + C}{\sqrt{A^2 + B^2}} \right|$$

65. Let L_1 and L_2 have slopes m_1 and m_2, respectively, and let ϕ be the angle between L_1 and L_2, as shown in Figure 1.13. Show that
$$\tan \phi = \frac{m_2 - m_1}{1 + m_1 m_2}$$

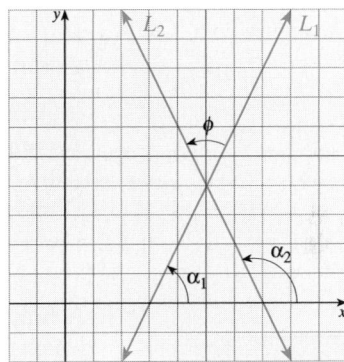

■ **FIGURE 1.13** Angle ϕ between given lines

66. **JOURNAL PROBLEM** *Ontario Secondary School Mathematics Bulletin.*[†] Show that there is just one line in the family $y - 8 = m(x + 1)$ that is five units from the point $(2, 4)$.

67. The *center* of a triangle with vertices $A(x_1, y_1)$, $B(x_2, y_2)$, $C(x_3, y_3)$ is the point where its three medians intersect. It can be shown that the center $P(x_0, y_0)$ is located 2/3 the distance from each vertex to the midpoint of the opposite side. Use this fact to show that
$$x_0 = \frac{x_1 + x_2 + x_3}{3}$$
and
$$y_0 = \frac{y_1 + y_2 + y_3}{3}$$

[†]Volume 18 (1982), issue 2, p. 7.

1.3 Functions

IN THIS SECTION   definition of a function, functional notation, domain and range of a function, composition of functions, classification of functions ■

Scientists, economists, and other researchers study relationships between quantities. For example, an engineer may need to know how the illumination from a light source on an object is related to the distance between the object and the source; a biologist may wish to investigate how the population of a bacterial colony varies with time in the presence of a toxin; an economist may wish to determine the relationship between consumer demand for a certain commodity and its market price. The mathematical study of such relationships involves the concept of a *function.*

DEFINITION OF A FUNCTION

> **Function**
>
> A **function** f is a rule that assigns to each element x of a set X a unique element y of a set Y. The element y is called the **image** of x under f and is denoted by $f(x)$ (read as "f of x"). The set X is called the **domain** of f, and the set of all images of elements of X is called the **range** of the function.

Input value x

Square

Multiply by 5

Add 2

Output value
$5x^2 + 2$

A function whose *name* is f can be thought of as the set of ordered pairs (x, y) for which each member x of the domain is associated with exactly one member $y = f(x)$. The function can also be regarded as a rule that assigns a unique "output" in the set Y to each "input" from the set X.

> ■ *What This Says:* To be called a function, the rule must have the property that it assigns one to each input and only one output.

A visual representation of a function is shown in Figure 1.14. Note that it is quite possible for two different elements in the domain X to map into the same element in the range, and that it is possible for Y to include elements not in the range of f. If the range of f does, however, consist of all of Y, then f is said to map **X onto Y.** Furthermore, if each element in the range is the image of one and only one element in the domain, then f is said to be a **one-to-one function.** We shall have more to say about these terms in Section 1.5. A function f is said to be **bounded** on $[a, b]$ if there exists a number B so that $\left| f(x) \right| \le B$ for all x in $[a, b]$.

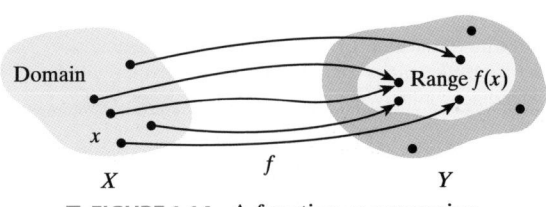

Domain

Range $f(x)$

x

X f Y

■ FIGURE 1.14 A function as a mapping

Most of our work will be with real-valued functions of a real variable, that is, functions whose domain and range are both sets of real numbers.*

FUNCTIONAL NOTATION

Functions can be represented in various ways, but usually they are specified by using a mathematical formula. It is traditional to let x denote the input and y the corresponding output, and to write an equation relating x and y. The letters x and y that appear in such an equation are called **variables.** Because the value of the variable y is determined by that of the variable x, we call y the **dependent variable** and x the **independent variable.**

In this book, when we define functions by expressions such as

$$f(x) = 2x + 3 \qquad \text{or} \qquad g(x) = x^2 + 4x + 5$$

we mean the functions f and g are the sets of all ordered pairs (x, y), satisfying the equations $y = 2x + 3$ and $y = x^2 + 4x + 5$, respectively. To **evaluate** a function f means to find the value of f for some particular value in the domain. For example, to evaluate f at $x = 2$ is to find $f(2)$.

EXAMPLE 1   *Using functional notation*

Suppose $f(x) = 2x^2 - x$. Find $f(-1), f(0), f(2), f(\pi), f(x + h)$, and $\dfrac{f(x + h) - f(x)}{h}$, where x and h are real numbers and $h \neq 0$.

Solution   In this case, the defined function f tells us to subtract the independent variable x from twice its square. Thus, we have

$$f(-1) = 2(-1)^2 - (-1) = 3$$
$$f(0) = 2(0)^2 - (0) = 0$$
$$f(2) = 2(2)^2 - 2 = 6$$
$$f(\pi) = 2\pi^2 - \pi$$

To find $f(x + h)$, we begin by writing the formula for f in more neutral terms, say as

$$f(\ \ ) = 2(\ \ )^2 - (\ \ )$$

Then we insert the expression $x + h$ inside each box, obtaining

$$f(\boxed{x + h}) = 2(\boxed{x + h})^2 - (\boxed{x + h})$$
$$= 2(x^2 + 2xh + h^2) - (x + h)$$
$$= 2x^2 + 4xh + 2h^2 - x - h$$

Finally, if $h \neq 0$,

$$\frac{f(x + h) - f(x)}{h} = \frac{[2x^2 + 4xh + 2h^2 - x - h] - [2x^2 - x]}{h}$$
$$= \frac{4xh + 2h^2 - h}{h} = 4x + 2h - 1 \qquad \blacksquare$$

---

*The functions that appear in this book belong to a very special class of **elementary functions,** defined by Joseph Liouville (1809–1882). You will study some nonelementary functions in complex analysis and in advanced calculus. Certain functions that appear in physics and higher mathematics are not elementary functions.

The expression $\dfrac{f(x + h) - f(x)}{h}$ is called a **difference quotient** and is used in Chapter 2 to compute the *derivative*.

Technology Window

Graphing calculators are very good at evaluating functions. Most enter functions via a key labeled $\boxed{Y =}$. Consider Example 1 and note that we are concerned with three functions:

$$f(x) = 2x^2 - x$$
$$f(x + h) = 2(x + h)^2 - (x + h)$$
$$\frac{f(x + h) - f(x)}{h}$$

F1	F2▼	F3	F4	F5▼	F6▼		
▼ ┏	Zoom	Edit	✓	All	Style	▷ ◁ . . .	

▲PLOTS

✓y1 = 2·x² − x

✓y2 = 2·(x + h)² − (x + h)

✓y3 = $\dfrac{y2(x) - y1(x)}{h}$

y4 = ■
y5 =
y6 =
y7 =

y4(x)=

MAIN   RAD AUTO   FUNC

$h = 0.1$

At the right we have shown these functions as Y1, Y2, and Y3. Next, most calculators have TABLE capabilities that evaluate functions for given values. At the right, for example, we have shown these functions evaluated for integers from -1 to 6 and $h = 0.1$.

x	y1	y2	y3		
-1.	3.	2.52	-4.8		
0.	0.	-.08	-.8		
1.	1.	1.32	3.2		
2.	**6.**	6.72	7.2		
3.	15.	16.12	11.2		
4.	28.	29.52	15.2		
5.	45.	46.92	19.2		
6.	66.	68.32	23.2		

x=2.

MAIN   RAD AUTO   FUNC

$h = 0.01$

These steps may seem puzzling when you see them here, but experiment to see what you can do.

Here is an example for you to try. Store the value $h = 0.01$ and look at the resulting table outputs. Can you find any patterns, or make any generalizations? Check your table values with the ones shown here.

x	y1	y2	y3		
-1.	3.	2.9502	-4.98		
0.	0.	-.0098	-.98		
1.	1.	1.0302	3.02		
2.	**6.**	6.0702	7.02		
3.	15.	15.11	11.02		
4.	28.	28.15	15.02		
5.	45.	45.19	19.02		
6.	66.	66.23	23.02		

x=2.

MAIN   RAD AUTO   FUNC

In calculus, we sometimes find the slope of a line passing through points that are *very* close together. This idea is illustrated with Example 2.

    EXAMPLE 2   *Slope of a line through points on a given curve*

Let $f(x) = x^2 - 3x - 4$. Find the slope of a line passing through the given points on the graph of f.

a. $x_1 = 2$ and $x_2 = 3$ b. $x_1 = 2$ and $x_3 = 2.001$ c. $x_1 = 2$ and $x_4 = 2 + \delta$

Solution   Recall the formula for the slope of the line passing through $(x, f(x))$ and $(x + \delta, f(x + \delta))$ is

$$m = \frac{f(x + \delta) - f(x)}{(x + \delta) - x} = \frac{f(x + \delta) - f(x)}{\delta}$$

$f(2) = 2^2 - 3(2) - 4 = -6$; the point on f is $(2, -6)$, as shown in Figure 1.15.

a. $f(3) = 3^2 - 3(3) - 4 = -4$; point is $(3, -4)$.

$$m = \frac{-4 - (-6)}{3 - 2} = 2$$

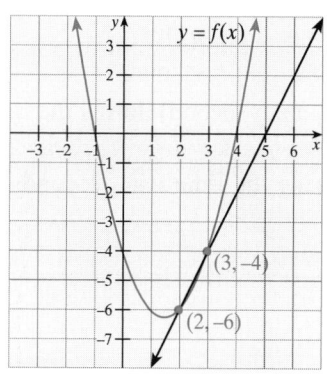

■ **FIGURE 1.15**   Graph of f and one secant line

b. $f(2.001) = 2.001^2 - 3(2.001) - 4 = -5.998999$; point is $(2.001, -5.998999)$.

$$m = \frac{-5.998999 - (-6)}{2.001 - 2} = 1.001$$

c. $f(2 + \delta) = (2 + \delta)^2 - 3(2 + \delta) - 4 = \delta^2 + \delta - 6$; point is $(2 + \delta, \delta^2 + \delta - 6)$.

$$m = \frac{(\delta^2 + \delta - 6) - (-6)}{(2 + \delta) - 2} = \frac{\delta^2 + \delta}{\delta} = \delta + 1$$

Reconcile parts a and b: If $\delta = 0.001, m = 0.001 + 1 = 1.001$.   ■

Sometimes functions must be defined in pieces because they have a split domain. These functions require more than one formula to define the function and therefore are called **piecewise-defined functions.**

EXAMPLE 3   *Evaluating a piecewise-defined function*

If $f(x) = \begin{cases} x \sin x & \text{if } x < 2 \\ 3x^2 + 1 & \text{if } x \geq 2 \end{cases}$ find $f(-0.5), f(\frac{\pi}{2})$, and $f(2)$.

Solution   To find $f(-0.5)$, we use the first line of the formula because $-0.5 < 2$:

$$f(-0.5) = -0.5 \sin(-0.5) = 0.5 \sin 0.5 \qquad \textit{This is the exact value.}$$
$$\approx 0.2397 \qquad \textit{This is the approximate value.}$$

To find $f(\frac{\pi}{2})$, we use the first line of the formula because $\frac{\pi}{2} \approx 1.57 < 2$:

$$f(\tfrac{\pi}{2}) = \tfrac{\pi}{2} \sin \tfrac{\pi}{2} = \tfrac{\pi}{2}$$

Finally, because $2 \geq 2$, we use the second line of the formula to find $f(2)$:

$$f(2) = 3(2)^2 + 1 = 13 \qquad ■$$

Functional notation can be used in a wide variety of applied problems, as shown by Example 4 and again in the problem set.

EXAMPLE 4   *Applying functional notation*

It is known that an object dropped from a height in a vacuum will fall a distance of s ft in t seconds according to the formula

$$s(t) = 16t^2, \quad t \geq 0$$

a. How far will the object fall in the first second? In the *next* 2 seconds?

b. How far will it fall during the time interval $t = 1$ to $t = 1 + h$ seconds?

c. What is the average rate of change of distance (in feet per second) during the time $t = 1$ sec to $t = 3$ sec?

d. What is the average rate of change of distance during the time $t = x$ seconds to $t = x + h$ seconds?

Solution

a. $s(1) = 16(1)^2 = 16$

In the first second, the object will fall 16 ft.
In the next two seconds, the object will fall

$$s(1 + 2) - s(1) = s(3) - s(1) = 16(3)^2 - 16(1)^2 = 128$$

The object will fall 128 ft in the next 2 sec.

b. $s(1 + h) - s(1) = 16(1 + h)^2 - 16(1)^2$
$$= 16 + 32h + 16h^2 - 16 = 32h + 16h^2$$

c. $\text{AVERAGE RATE} = \dfrac{\text{CHANGE IN DISTANCE}}{\text{CHANGE IN TIME}} = \dfrac{s(3) - s(1)}{3 - 1} = \dfrac{128}{2} = 64$

The average rate of change is 64 ft/sec.

d. $\dfrac{s(x + h) - s(x)}{(x + h) - x} = \dfrac{s(x + h) - s(x)}{h}$ *Does this look familiar? See Example 1.*

$$= \dfrac{16(x + h)^2 - 16x^2}{h} = \dfrac{16x^2 + 32xh + 16h^2 - 16x^2}{h} = 32x + 16h \quad\blacksquare$$

DOMAIN AND RANGE OF A FUNCTION

WARNING This domain convention will be used throughout this text.

In this book, unless otherwise specified, the domain of a function is the set of real numbers for which the function is defined. We call this the **domain convention.** If a function f is **undefined** at x, it means that x is not in the domain of f. The most frequent exclusions from the domain are those values that cause division by 0 or negative values under a square root. In applications, the domain is often specified by the context. For example, if x is the number of people on an elevator, the context requires that negative numbers and nonintegers be excluded from the domain; therefore, x must be an integer such that $0 \le x \le c$ where c is the maximum capacity of the elevator.

EXAMPLE 5   *Domain of a function*

Find the domain for the given functions.

a. $f(x) = 2x - 1$ b. $g(x) = 2x - 1, x \ne -3$

c. $h(x) = \dfrac{(2x - 1)(x + 3)}{x + 3}$ d. $F(x) = \sqrt{x + 2}$ e. $G(x) = \dfrac{4}{5 - \cos x}$

Solution

a. All real numbers; $(-\infty, \infty)$

b. All real numbers except -3

c. Because the expression is meaningful for all $x \ne -3$, the domain is all real numbers except -3.

d. F has meaning if and only if $x + 2$ is nonnegative; therefore, the domain is $x \ge -2$, or $D = [-2, \infty)$.

e. G is defined whenever $5 - \cos x \neq 0$. This imposes no restriction on x since $|\cos x| \leq 1$. Thus, the domain of G is the set of all real numbers; $D = (-\infty, \infty)$. ∎

Equality of Functions

Two functions f and g are **equal** if and only if

1. f and g have the same domain.
2. $f(x) = g(x)$ for all x in the domain.

WARNING ➤
$$f(x) = 2x - 1$$
$$g(x) = 2x - 1, x \neq -3$$
$$h(x) = \frac{(2x - 1)(x + 3)}{x + 3}$$

WARNING ➤ In Example 5 (repeated in the margin here), the functions g and h are equal. A common mistake is to "reduce" the function h to the function f:

WRONG: $h(x) = \dfrac{(2x - 1)(x + 3)}{x + 3} = 2x - 1 = f(x)$

RIGHT: $h(x) = \dfrac{(2x - 1)(x + 3)}{x + 3} = 2x - 1, x \neq -3$; therefore, $h(x) = g(x)$.

COMPOSITION OF FUNCTIONS

There are many situations in which a quantity is given as a function of one variable that, in turn, can be written as a function of a second variable. Suppose, for example, that your job is to ship x packages of a product via Federal Express to a variety of addresses. Let x be the number of packages to ship, and let f be the weight of the x objects and g be the cost of shipping. Then

Weight is a function $f(x)$ of the number of objects x.

Cost is a function $g[f(x)]$ of the weight.

So we have expressed cost as a function of the number of packages. This process of evaluating a function of a function is known as *functional composition*.

Composition of Functions

The **composite function** $f \circ g$ is defined by

$$(f \circ g)(x) = f[g(x)]$$

for each x in the domain of g for which $g(x)$ is in the domain of f.

■ *What This Says:* To visualize how functional composition works, think of $f \circ g$ in terms of an "assembly line" in which g and f are arranged in series, with output $g(x)$ becoming the input of f, as illustrated in Figure 1.16.

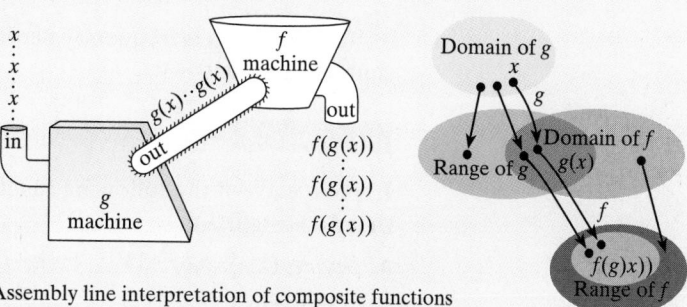

Assembly line interpretation of composite functions

■ **FIGURE 1.16** Composition of functions

EXAMPLE 6   *Finding the composition of functions*

If $f(x) = 3x + 5$ and $g(x) = \sqrt{x}$, find the composite functions $f \circ g$ and $g \circ f$.

Solution   The function $f \circ g$ is defined by $f[g(x)]$:

$$(f \circ g)(x) = f[g(x)] = f(\sqrt{x}) = 3\sqrt{x} + 5$$

The function $g \circ f$ is defined by $g[f(x)]$:

$$(g \circ f)(x) = g[f(x)] = g(3x + 5) = \sqrt{3x + 5}$$ ■

WARNING ▶ Example 6 illustrates that *functional composition is not commutative.* ◀

That is, $f \circ g$ is not, in general, the same as $g \circ f$.

EXAMPLE 7   *An application of composite functions*

Air pollution is a problem for many metropolitan areas. Suppose that carbon monoxide is measured as a function of the number of people according to the following information:

Number of People	Daily Carbon Monoxide Level (in parts per million)
100,000	1.41
200,000	1.83
300,000	2.43
400,000	3.05
500,000	3.72

Studies show that a refined formula for the average daily level of carbon monoxide in the air is

$$L(p) = 0.70\sqrt{p^2 + 3}$$

Further assume that the population of a given metropolitan area is growing according to the formula $p(t) = 1 + 0.02t^3$, where t is the time from now (in years) and p is the population (in hundred thousands). Based on these assumptions, what level of air pollution should be expected in 4 years?

Solution   The level of pollution is $L(p) = 0.70\sqrt{p^2 + 3}$, where $p(t) = 1 + 0.02t^3$. Thus, the pollution level at time t is given by the composite function

$$(L \circ p)(t) = L[p(t)] = L(1 + 0.02t^3) = 0.70\sqrt{(1 + 0.02t^3)^2 + 3}$$

In particular, when $t = 4$, we have

$$(L \circ p)(4) = 0.70\sqrt{[1 + 0.02(4)^3]^2 + 3} \approx 2.00 \text{ ppm}$$ ■

In calculus, it is frequently necessary to express a function as the composite of two simpler functions.

EXAMPLE 8   *Expressing a given function as a composite of two functions*

Express each of the following functions as the composite of two functions u and g so that $f(x) = g[u(x)]$.

a. $f(x) = (x^2 + 5x + 1)^5$ b. $f(x) = \cos^3 x$

c. $f(x) = \sin x^3$ d. $f(x) = \sqrt{5x^2 - x}$

Solution   There are often many ways to express $f(x)$ as a composite $g[u(x)]$. Perhaps the most natural is to choose u to represent the "inner" portion of f and g as the "outer" portion. Such choices are indicated in the table below.

Given Function $f(x) = g[u(x)]$	Inner Function $u(x)$	Outer Function $g[u(x)]$
a. $f(x) = (x^2 + 5x + 1)^5$	$u(x) = x^2 + 5x + 1$	$g[u(x)] = [u(x)]^5$
b. $f(x) = \cos^3 x$	$u(x) = \cos x$	$g[u(x)] = [u(x)]^3$
c. $f(x) = \sin x^3$	$u(x) = x^3$	$g[u(x)] = \sin [u(x)]$
d. $f(x) = \sqrt{5x^2 - x}$	$u(x) = 5x^2 - x$	$g[u(x)] = \sqrt{u(x)}$

WARNING   Note how composite functions are formed from an "inner" function and an "outer" function.

CLASSIFICATION OF FUNCTIONS

We will now describe some of the common types of functions used in this text.

Polynomial Function

A **polynomial function** is a function of the form

$$f(x) = a_n x^n + a_{n-1} x^{n-1} + \cdots + a_2 x^2 + a_1 x + a_0$$

where n is a nonnegative integer and $a_n, \ldots, a_2, a_1, a_0$ are constants. If $a_n \neq 0$, the integer n is called the **degree** of the polynomial. The constant a_n is called the **leading coefficient** and the constant a_0 is called the **constant term** of the polynomial function. In particular,

A **constant function** is zero degree: $f(x) = a$
A **linear function** is first degree: $f(x) = ax + b$
A **quadratic function** is second degree: $f(x) = ax^2 + bx + c$
A **cubic function** is third degree: $f(x) = ax^3 + bx^2 + cx + d$
A **quartic function** is fourth degree:

$$f(x) = ax^4 + bx^3 + cx^2 + dx + e$$

Examples of polynomial functions:

$f(x) = 5$
$f(x) = 2x - \sqrt{2}$
$f(x) = 3x^2 + 5x - \frac{1}{2}$
$f(x) = \sqrt{2}x^3 - \pi x$

Rational Function

A **rational function** is the quotient of two polynomial functions, $p(x)$ and $d(x)$:

$$f(x) = \frac{p(x)}{d(x)}, \quad d(x) \neq 0$$

Examples:

$f(x) = x^{-1}$
$f(x) = \dfrac{x - 5}{x^2 + 2x - 3}$
$f(x) = x^{-3} + \sqrt{2}x$

When we write $d(x) \neq 0$, we mean that all values c for which $d(c) = 0$ are excluded from the domain of d.

If r is any nonzero real number, the function $f(x) = x^r$ is called a **power function** with exponent r. You should be familiar with the following cases:

Integral powers ($r = n$, a positive integer): $f(x) = x^n = \underbrace{x \cdot x \cdot \cdots \cdot x}_{n \text{ factors}}$

Reciprocal powers (r is a negative integer): $f(x) = x^{-n} = 1/x^n$ for $x \neq 0$
Roots ($r = m/n$ is a rational number): $f(x) = x^{m/n} = \sqrt[n]{x^m} = (\sqrt[n]{x})^m$ for

$$x \geq 0 \text{ if } n \text{ even}, n \neq 0 \quad (m/n \text{ is reduced})$$

Examples:

$f(x) = x^6$
$f(x) = x^{-4}$
$f(x) = x^{3/4}$
$f(x) = \sqrt[3]{x^2}$

Power functions can also have irrational exponents (such as $\sqrt{2}$ or π), but such functions must be defined in a special way (see Section 1.6).

A function is called **algebraic** if it can be constructed using algebraic operations (such as adding, subtracting, multiplying, dividing, or taking roots) starting with polynomials. Functions that are not algebraic are called **transcendental.** The following are transcendental functions:

(SMH)

Trigonometric functions are the functions sine, cosine, tangent, secant, cosecant, and cotangent. The basic forms of these functions are reviewed in Chapter 3 of the *Student Mathematics Handbook.* You can also review these functions by consulting a trigonometry or precalculus textbook.

Exponential functions are functions of the form $f(x) = b^x$, where b is a positive constant. We will introduce these functions in Section 1.6.

Logarithmic functions are functions of the form $f(x) = \log_b x$, where b is a positive constant. We will also study these functions in Section 1.6.

1.3 Problem Set

In Problems 1–12, find the domain of f and compute the indicated values or state that the corresponding x-value is not in the domain. Tell whether any of the indicated values are zeros of the function, that is, values of x that cause the functional value to be 0.

A 1. $f(x) = 2x + 3; f(-2), f(1), f(0)$

2. $f(x) = -x^2 + 2x + 3; f(0), f(1), f(-2)$

3. $f(x) = 3x^2 + 5x - 2; f(1), f(0), f(-2)$

4. $f(x) = x + \dfrac{1}{x}; f(-1), f(1), f(2)$

5. $f(x) = \dfrac{(x + 3)(x - 2)}{x + 3}; f(2), f(0), f(-3)$

6. $f(x) = (2x - 1)^{-3/2}; f(1), f(\frac{1}{2}), f(13)$

7. $f(x) = \sqrt{x^2 + 2x}; f(-1), f(\frac{1}{2}), f(1)$

8. $f(x) = \sqrt{x^2 + 5x + 6}; f(0), f(1), f(-2)$

9. $f(x) = \sin(1 - 2x); f(-1), f(\frac{1}{2}), f(1)$

10. $f(x) = \sin x - \cos x; f(0), f\left(-\frac{\pi}{2}\right), f(\pi)$

11. $f(x) = \begin{cases} -2x + 4 & \text{if } x \le 1 \\ x + 1 & \text{if } x > 1 \end{cases}$

$f(3), f(1), f(0)$

12. $f(x) = \begin{cases} 3 & \text{if } x < -5 \\ x + 1 & \text{if } -5 \le x \le 5 \\ \sqrt{x} & \text{if } x > 5 \end{cases}$

$f(-6), f(-5), f(16)$

In Problems 13–20, evaluate the difference quotient
$$\frac{f(x + h) - f(x)}{h} \text{ for the given function.}$$

13. $f(x) = 9x + 3$

14. $f(x) = 5 - 2x$

15. $f(x) = 5x^2$

16. $f(x) = 3x^2 + 2x$

17. $f(x) = |x|$ if $x < -1$ and $0 < h < 1$

18. $f(x) = |x|$ if $x > 1$ and $0 < h < 1$

19. $f(x) = \dfrac{1}{x}$

20. $f(x) = \dfrac{x + 1}{x - 1}$

Find the slope of the lines passing through the given points on the given curves in Problems 21–26.

21. $f(x) = 2x^2 - 5x$, for $x_1 = 2$ and $x_2 = 3$

22. $f(x) = 3x^2 + 2x$, for $x_1 = -2$ and $x_2 = -1.5$

23. $f(x) = -2x^2 + 3x$, for $x_1 = 1$ and $x_2 = 1.001$

24. $f(x) = -2x^2 - 3x$, for $x_1 = 1$ and $x_2 = 1.01$

25. $f(x) = 2 - x^3$, for $x_1 = 1$ and $x_2 = 1 + \delta$

26. $f(x) = 1 + x^3$, for $x_1 = -1$ and $x_2 = -1 + \delta$

State whether the functions f and g in Problems 27–32 are equal.

27. $f(x) = \dfrac{2x^2 + x}{x}; g(x) = 2x + 1$

28. $f(x) = \dfrac{2x^2 + x}{x}; g(x) = 2x + 1, x \ne 0$

29. $f(x) = \dfrac{2x^2 - x - 6}{x - 2}$; $g(x) = 2x + 3, x \neq 2$

30. $f(x) = \dfrac{3x^2 - 7x - 6}{x - 3}$; $g(x) = 3x + 2, x \neq 3$

31. $f(x) = \dfrac{3x^2 - 5x - 2}{x - 2}$; $g(x) = 3x + 1$

32. $f(x) = \dfrac{(3x + 1)(x - 2)}{x - 2}, x \neq 6$;

   $g(x) = \dfrac{(3x + 1)(x - 6)}{x - 6}, x \neq 2$

In Problems 33–38, find the composite functions $f \circ g$ and $g \circ f$.

33. $f(x) = x^2 + 1$ and $g(x) = 2x$

34. $f(x) = \sin x$ and $g(x) = 1 - x^2$

35. $f(t) = \sqrt{t}$ and $g(t) = t^2$

36. $f(u) = \dfrac{u - 1}{u + 1}$ and $g(u) = \dfrac{u + 1}{1 - u}$

37. $f(x) = \sin x$ and $g(x) = 2x + 3$

38. $f(x) = \dfrac{1}{x}$ and $g(x) = \tan x$

In Problems 39–48, express f as the composition of two functions u and g such that $f(x) = g[u(x)]$.

39. $f(x) = (2x^2 - 1)^4$

40. $f(x) = (x^2 + 1)^3$

41. $f(x) = |2x + 3|$

42. $f(x) = \sqrt{5x - 1}$

43. $f(x) = \tan^2 x$

44. $f(x) = \tan x^2$

45. $f(x) = \sin \sqrt{x}$

46. $f(x) = \sqrt{\sin x}$

47. $f(x) = \sin\left(\dfrac{x + 1}{2 - x}\right)$

48. $f(x) = \tan\left(\dfrac{2x}{1 - x}\right)$

Ⓑ 49. Suppose the total cost (in dollars) of manufacturing q units of a certain commodity is given by

$$C(q) = q^3 - 30q^2 + 400q + 500$$

   a. Compute the cost of manufacturing 20 units.
   b. Compute the cost of manufacturing the 20th unit.

50. An efficiency study of the morning shift at a certain factory indicates that an average worker who arrives on the job at 8:00 A.M. will have assembled

$$f(x) = -x^3 + 6x + 15x^2$$

   CD players x hours later ($0 \leq x \leq 8$).
   a. How many players will such a worker have assembled by 10:00 A.M.?
   b. How many players will such a worker assemble between 9:00 A.M. and 10:00 A.M.?

51. In physics, a light source of luminous intensity K candles is said to have *illuminance* $I = K/s^2$ on a flat surface s ft away. Suppose a small, unshaded lamp of luminous intensity 30 candles is connected to a rope that allows it to

be raised and lowered between the floor and the top of a 10-ft-high ceiling. Assume that the lamp is being raised and lowered in such a way that at time t (in min) it is $s = 6t - t^2$ ft above the floor.
   a. Express the illuminance on the floor as a composite function of t for $0 < t < 6$.
   b. What is the illuminance when $t = 1$? When $t = 4$?

52. Biologists have found that the speed of blood in an artery is a function of the distance of the blood from the artery's central axis. According to *Poiseuille's law,* the speed (cm/sec) of blood that is r cm from the central axis of an artery is given by the function

$$S(r) = C(R^2 - r^2)$$

   where C is a constant and R is the radius of the artery.* Suppose that for a certain artery, $C = 1.76 \times 10^5$ cm/sec² and $R = 1.2 \times 10^{-2}$ cm.

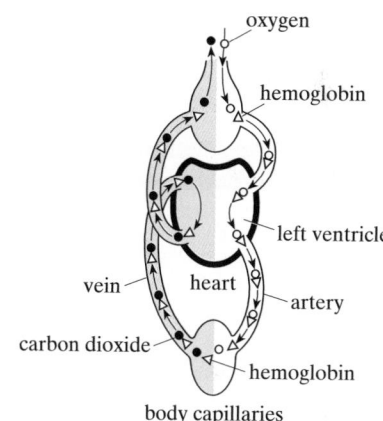

   a. Compute the speed of the blood at the central axis of this artery.
   b. Compute the speed of the blood midway between the artery's wall and central axis.

53. At a certain factory, the total cost of manufacturing q units during the daily production run is $C(q) = q^2 + q + 900$ dollars. On a typical workday, the numbers of units manufactured during the first t hours of a production run can be modeled by the function $q(t) = 25t$.
   a. Express the total manufacturing cost as a function of t.
   b. How much will have been spent on production by the end of the third hour?
   c. When will the total manufacturing cost reach $11,000?

---

*The law and the unit *poise,* a unit of viscosity, are both named for the French physician Jean Louis Poiseuille (1799–1869).

60. JOURNAL PROBLEM *The Mathematics Student Journal.** Given that $f(11) = 11$ and

$$f(x + 3) = \frac{f(x) - 1}{f(x) + 1}$$

for all x, find $f(2000)$.

*Volume 28 (1980), issue 3, p. 2. Note that the journal problem requests $f(1979)$, which, no doubt, was related to the publication date. We have taken the liberty of updating the requested value.

1.4   Graphs of Functions

IN THIS SECTION   **graph of a function (vertical line test, intercepts, symmetry), new functions from old—transformation of functions**

GRAPH OF A FUNCTION

Graphs have visual impact. They also reveal information that may not be evident from verbal or algebraic descriptions. Two graphs depicting practical relationships are shown in Figure 1.17.

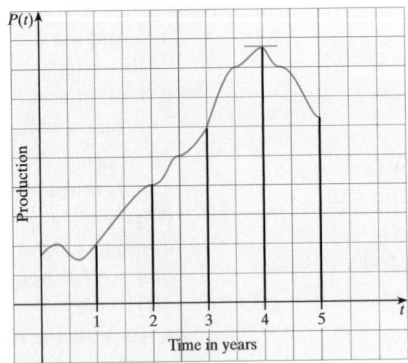

a. A production function

This graph describes the variation in total industrial production in a certain country over a five year time span. The fact that the graph has a peak suggests that production is greatest at the corresponding time.

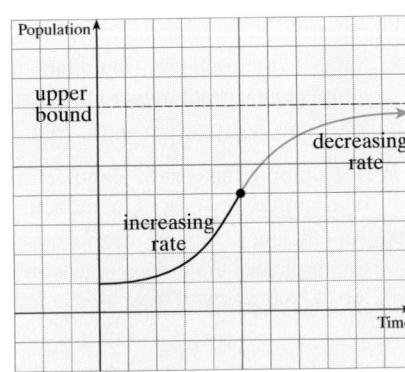

b. Bounded population growth

This graph represents the growth of a population when environmental factors impose an upper bound on the possible size of the population. It indicates that the rate of population growth increases at first and then decreases as the size of the population gets closer and closer to the upper bound.

■ **FIGURE 1.17**   Two graphs with practical interpretations

To represent a function $y = f(x)$ geometrically as a graph, it is traditional to use a Cartesian coordinate system on which units for the independent variable x are marked on the horizontal axis and units for the dependent variable y on the vertical axis.

Graph of a Function

The **graph** of a function f consists of points whose coordinates (x, y) satisfy $y = f(x)$, for all x in the domain of f.

Technology Window – Computer or Calculator Graphing

If you use a computer or a calculator to help you draw graphs, you will need to pay particular attention to Xmin, Xmax, and Xscl, as well as Ymin, Ymax, and Yscl. These selections will determine the region of the plane you will see on your screen. Computers and graphing calculators rely heavily on the technique of graphing by plotting points. When you use a computer or a calculator, the graph is quite often the easiest part of the task; the difficult part is choosing a scale on the axes so that the shape of the graph can be ascertained. In this text, we will frequently show you how the graph looks when using a graphing calculator.

a. $f(x) = 2x - 1$

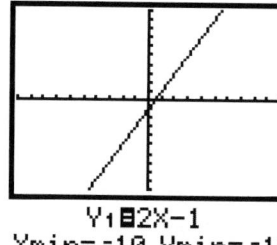

Y₁■2X-1
Xmin=-10 Ymin=-10
Xmax=10   Ymax=10
Xscl=1   Yscl=1

Domain: $(-\infty, \infty)$
Range: $(-\infty, \infty)$

b. $g(x) = 2x - 1, x \neq -3$

Calculators do not always show deleted points. The graph looks the same as the one in part **a**. *You* must add the circle at the point $(-3, -7)$.

Domain: $x \neq -3$
Range: $y \neq -7$

c. $h(x) = \dfrac{(2x - 1)(x + 3)}{x + 3}$

The graph is the same as the one in part **b**.

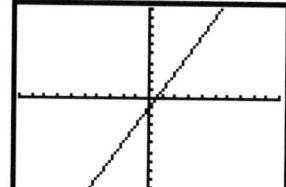

d. $F(x) = \sqrt{x + 2}$

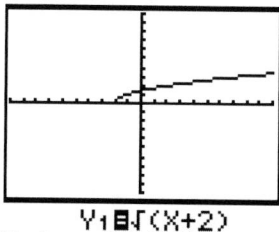

Y₁■√(X+2)
Xmin=-10 Ymin=-10
Xmax=10   Ymax=10
Xscl=1   Yscl=1

Domain: $[-2, \infty)$
Range: $[0, \infty)$

e. $G(x) = \dfrac{4}{5 - \cos x}$

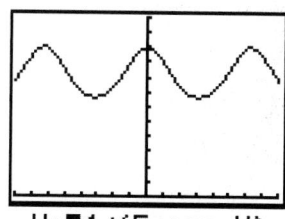

Y₁■4/(5-cos X)
Xmin=-8 Ymin=0
Xmax=8   Ymax=1.2
Xscl=1   Yscl=.1

Domain: $(-\infty, \infty)$
Range: $[0.667, 1]$

In part **e**, it is easy to see that the maximum value of y is 1, but the minimum value cannot be easily seen from the graph. However, by using the TRACE function on a calculator, we can find an approximate minimum value. Later in the course, we will discuss better ways for finding the maximum and minimum values.

In Chapter 4, we will discuss efficient techniques involving calculus that you can use to draw accurate graphs of functions. In algebra you began sketching lines by plotting points, but you quickly discovered that this is not a very efficient way to draw more complicated graphs, especially without the aid of a graphing calculator or computer. Table 1.3 includes a few common graphs you have probably encountered in previous courses. We will assume that you are familiar with their general shape and know how to sketch each of them either by hand or with the assistance of your graphing calculator.

Vertical Line Test   By definition of a function, for a given x in the domain there is only one number y in the range. Geometrically, this means that any vertical line $x = a$ crosses the graph of a function at most once. This observation leads to the following useful criterion.

> **The Vertical Line Test**
>
> A curve in the plane is the graph of a function if and only if it intersects no vertical line more than once.

Look at Figure 1.18 for examples of the vertical line test.

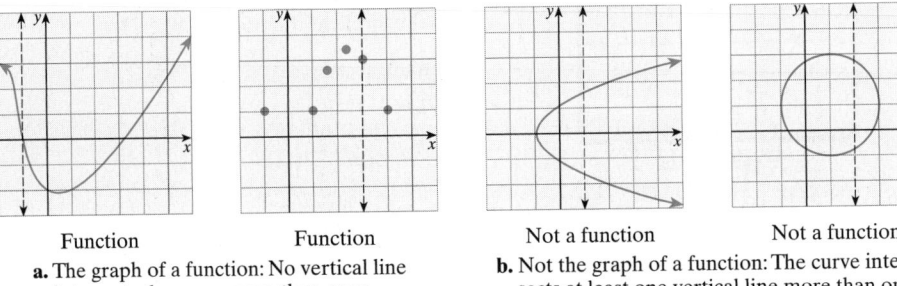

Function | Function | Not a function | Not a function

a. The graph of a function: No vertical line intersects the curve more than once.

b. Not the graph of a function: The curve intersects at least one vertical line more than once.

■ **FIGURE 1.18**   The vertical line test

Intercepts   The points where a graph intersects the coordinate axes are called *intercepts*. Here is a definition.

> **Intercepts**
>
> If the number zero is in the domain of f and $f(0) = b$, then the point $(0, b)$ is called the **y-intercept** of the graph of f.
>
> If a is a real number in the domain of f such that $f(a) = 0$, then $(a, 0)$ is an **x-intercept** of f.
>
> > ■ *What This Says:*   To find the x-intercepts, set y equal to 0 and solve for x. To find the y-intercept, set x equal to 0 and solve for y.

■ **TABLE 1.3** **Directory of Curves**

Identity Function $y = x$	Standard Quadratic Function $y = x^2$	Standard Cubic Function $y = x^3$
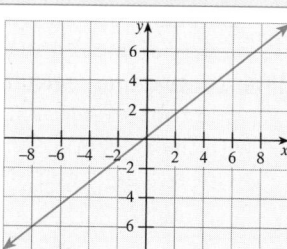	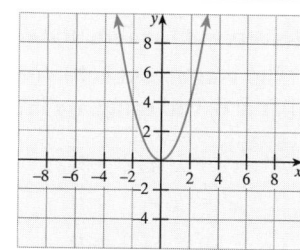	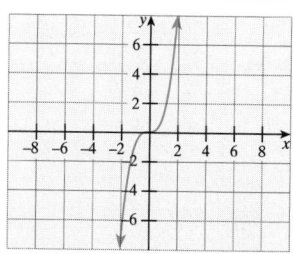

Absolute Value Function $y = |x| = \sqrt{x^2}$	Square Root Function $y = \sqrt{x}$	Cube Root Function $y = \sqrt[3]{x}$
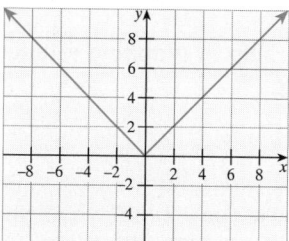	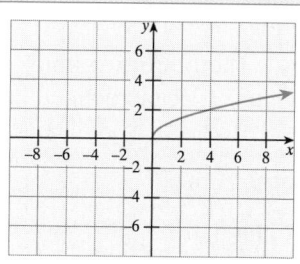	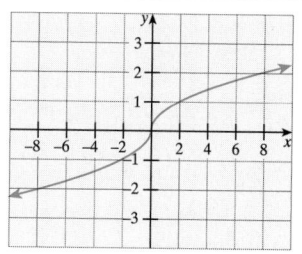

Standard Reciprocal $y = \dfrac{1}{x}$	Standard Reciprocal Squared $y = \dfrac{1}{x^2}$	Standard Square Root Reciprocal $y = \dfrac{1}{\sqrt{x}}$
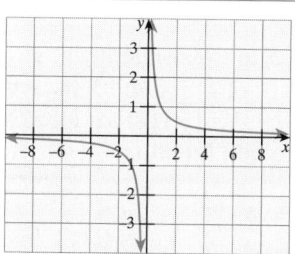	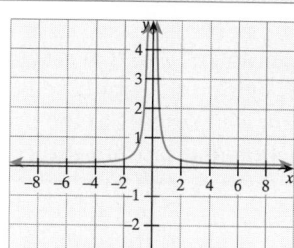	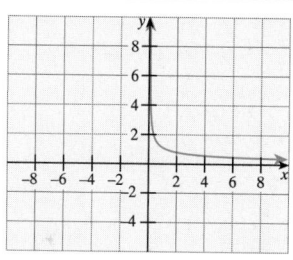

Cosine Function $y = \cos x$	Sine Function $y = \sin x$	Tangent Function $y = \tan x$
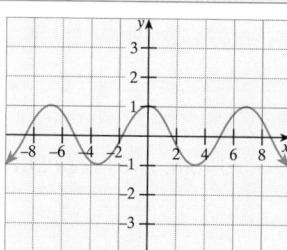	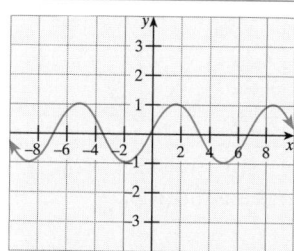	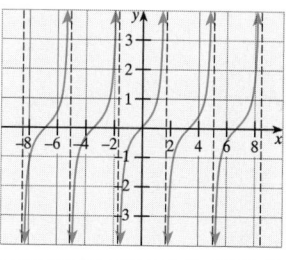

Secant Function $y = \sec x$	Cosecant Function $y = \csc x$	Cotangent Function $y = \cot x$
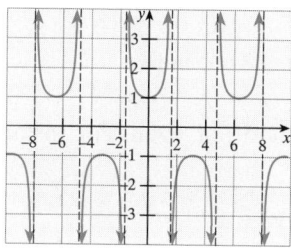	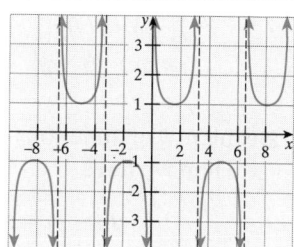	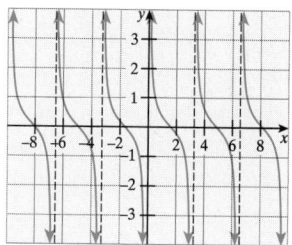

37

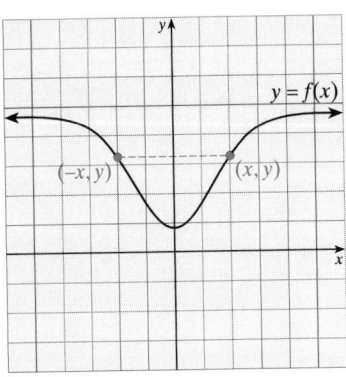

| EXAMPLE 1 | *Finding the intercepts* |

Find all intercepts of the function $f(x) = -x^2 + x + 2$.

Solution   The y-intercept is $(0, f(0)) = (0, 2)$. To find the x-intercepts, solve the equation $f(x) = 0$. Factoring, we find that

$$-x^2 + x + 2 = 0$$
$$x^2 - x - 2 = 0$$
$$(x + 1)(x - 2) = 0$$
$$x = -1 \text{ or } x = 2$$

Thus, the intercepts are $(0, 2), (-1, 0)$ and $(2, 0)$.   ∎

Symmetry   There are two kinds of symmetry that help in graphing a function, as shown in Figure 1.19 and defined in the following box.

Symmetry

The graph of $y = f(x)$ is **symmetric with respect to the y-axis** if whenever $P(x, y)$ is a point on the graph, so is the point $(-x, y)$ that is the mirror image of P in the y-axis. Thus, y-axis symmetry occurs if and only if $f(-x) = f(x)$ for all x in the domain of f. A function with this property is called an **even function.**

The graph of $y = f(x)$ is **symmetric with respect to the origin** if whenever $P(x, y)$ is on the graph, so is $(-x, -y)$, the mirror of P in the origin. Symmetry with respect to the origin occurs when $f(-x) = -f(x)$ for all x. A function that satisfies this condition is called an **odd function.**

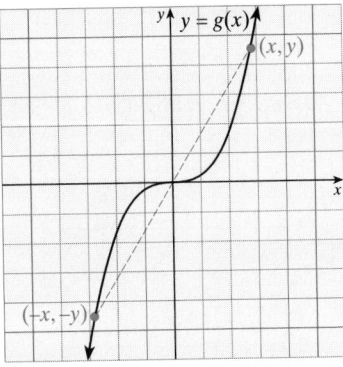

a. Graph of an even function f Symmetry with respect to the y-axis:
$$f(-x) = f(x)$$

b. Graph of an odd function g Symmetry with respect to the origin:
$$f(-x) = -f(x)$$

■ **FIGURE 1.19**   Graphs of even and odd functions

There are plenty of functions that are neither odd nor even. For instance, let $h(x) = x^2 + x$. Then we have

$$h(-x) = (-x)^2 + (-x) = x^2 - x$$

which equals neither $h(x)$ nor $-h(x)$.

WARNING The graph of a nonzero function cannot be symmetric with respect to the x-axis.

You may wonder why we have said nothing about symmetry with respect to the x-axis, but such symmetry would require $f(x) = -f(x)$, which is precluded by the vertical line test (do you see why?).

EXAMPLE 2   *Even and odd functions*

Classify the given functions as even, odd, or neither.
a. $f(x) = x^2$ b. $g(x) = x^3$ c. $h(x) = x^2 + 5x$

Solution

a. $f(x) = x^2$ is *even*, because
$$f(-x) = (-x)^2 = x^2 = f(x)$$

The graph below shows that the graph of the even function $f(x) = x^2$ is symmetric with respect to the y-axis.

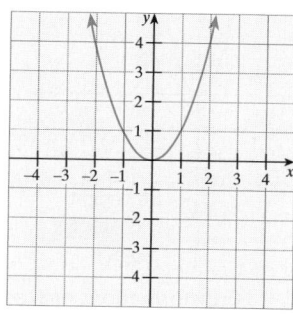

b. $g(x) = x^3$ is *odd*, because
$$g(-x) = (-x)^3 = -x^3 = -g(x)$$

The graph below shows that the graph of the odd function $g(x) = x^3$ is symmetric with respect to the origin.

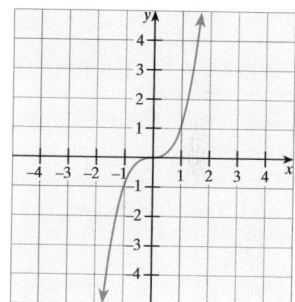

c. $h(x) = x^2 + 5x$ is *neither* even nor odd because
$$h(-x) = (-x)^2 + 5(-x) = x^2 - 5x$$

Note that $h(-x) \neq h(x)$ and $h(-x) \neq -h(x)$.

The graph below is not symmetric with respect to the x-axis, the y-axis, or the origin.

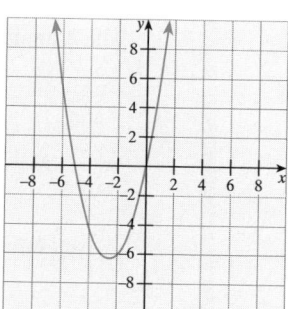

NEW FUNCTIONS FROM OLD—TRANSFORMATION OF FUNCTIONS

Sometimes the graph of a function can be obtained by altering the graph of another function by either translation or reflection. We call these translations and reflections *transformations* of a function. This procedure is illustrated in Figure 1.20, in which we have sketched the graph of $y = x^2$ and then translated and reflected that graph.

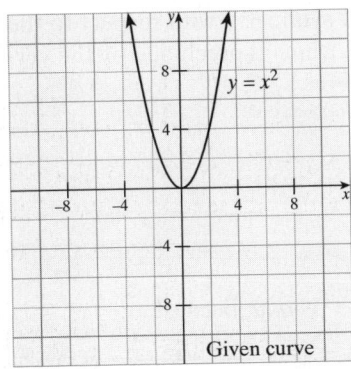

Given curve

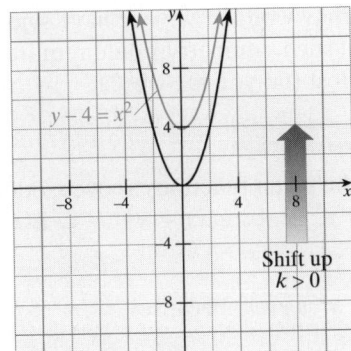

$y - 4 = x^2$

Shift up
$k > 0$

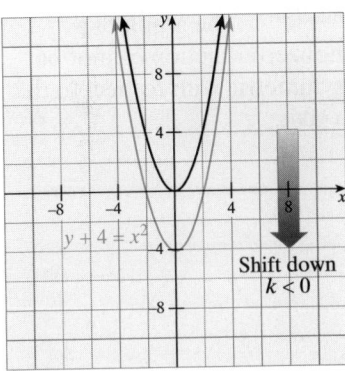

$y + 4 = x^2$

Shift down
$k < 0$

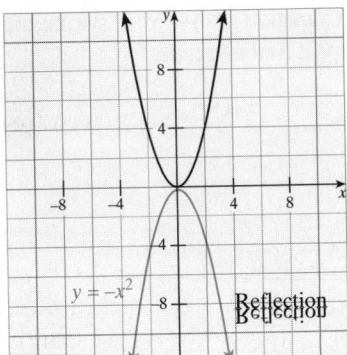

$y = -x^2$

Reflection

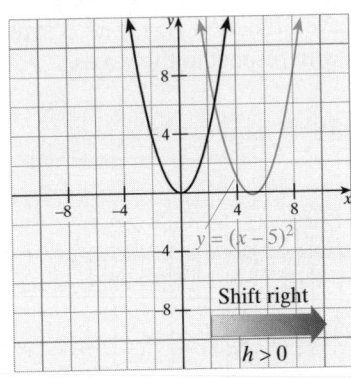

$y = (x - 5)^2$

Shift right

$h > 0$

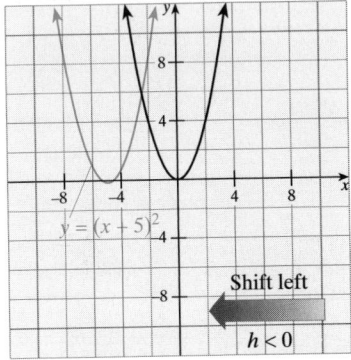

$y = (x + 5)^2$

Shift left

$h < 0$

■ **FIGURE 1.20**   Transformations of $y = x^2$

Functional Transformations

The graph defined by the equation

$$y - k = f(x - h)$$

is said to be a **translation** of the graph defined by $y = f(x)$.

The translation (shift, as shown in Figure 1.20) is

to the right if $h > 0$

to the left if $h < 0$

up if $k > 0$

down if $k < 0$

A **reflection in the x-axis** of the graph of $y = f(x)$ is the graph of

$$y = -f(x)$$

A **reflection in the y-axis** of the graph of $y = f(x)$ is the graph of

$$y = f(-x)$$

■ *What This Says:*   If we replace x by $x - h$ and y by $y - k$, the graph is translated so the origin $(0, 0)$ is moved to the point (h, k). Also, the graph is reflected in the x-axis if we replace y by $-y$ in its equation; it is reflected in the y-axis if we replace x by $-x$.

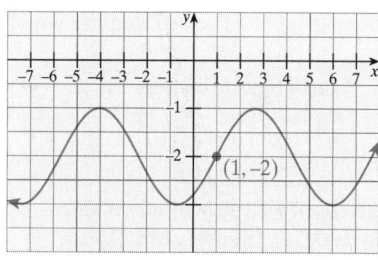

■ **FIGURE 1.21** Translation of $y = \sin x$ from $(0,0)$ to $(h,k) = (1,-2)$

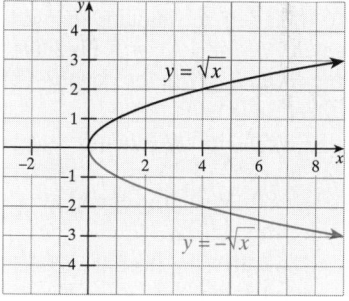

■ **FIGURE 1.22** Reflection in the x-axis of $y = \sqrt{x}$

| EXAMPLE 3 | *Graphing with a translation* |

Graph $y + 2 = \sin(x - 1)$.

Solution   By inspection, the desired graph is the graph of the standard sine function $y = \sin x$ that has been translated to the point $(h, k) = (1, -2)$. The graph is shown in Figure 1.21.

■

| EXAMPLE 4 | *Graphing with a reflection* |

Graph $y = -\sqrt{x}$.

Solution   The graph of this function is a reflection in the x-axis of the graph of $y = \sqrt{x}$, as shown in Figure 1.22.

■

1.4 Problem Set

Classify the functions defined in Problems 1–8 as even, odd, or neither.

 1. $f_1(x) = x^2 + 1$ **2.** $f_2(x) = \sqrt{x^2}$

3. $f_3(x) = \dfrac{1}{3x^3 - 4}$ **4.** $f_4(x) = x^3 + x$

5. $f_5(x) = \dfrac{1}{(x^3 + 3)^2}$ **6.** $f_6(x) = \dfrac{1}{(x^3 + x)^2}$

7. $f_7(x) = |x|$ **8.** $f_8(x) = |x + x^3|$

Use the directory of curves (Table 1.3) and the ideas of translation and reflection to sketch the graphs of the functions given in Problems 9–24.

9. $f(x) = x^2 + 4$ **10.** $f(x) = (x + 4)^2$

11. $y = -x^3$ **12.** $y = -|x|$

13. $y = \dfrac{1}{x - 3}$ **14.** $y + 1 = \dfrac{1}{x}$

15. $y = \sqrt{x - 1} + 2$ **16.** $y = 2 - \sqrt{x + 1}$

17. $y = \sqrt[3]{x + 2}$ **18.** $y = \sqrt[3]{x} + 2$

SMH **19.** $y = \cos(x - 1)$ **20.** $y = \cos x - 1$

21. $y = \sin(x + 2)$ **22.** $y = \sin x + 2$

23. $y = \tan(x + 1)$ **24.** $y = \tan x + 1$

25. If point A in Figure 1.23 has coordinates $(2, f(2))$, what are the coordinates of P and Q?

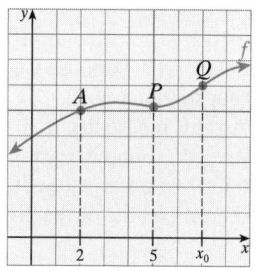   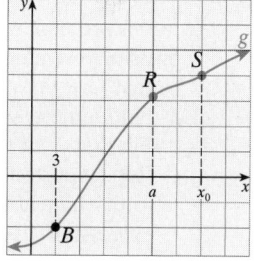

Problem 25   **Problem 26**

■ **FIGURE 1.23**

26. If point B in Figure 1.23 has coordinates $(3, g(3))$, what are the coordinates of R and S?

Find the x-intercepts, if any, for the functions given in Problems 27–36.

27. $f(x) = 3x^2 - 5x - 2$

28. $f(x) = 6x^2 + 5x - 6$

29. $f(x) = (x - 15)(2x + 25)(3x - 65)(4x + 1)$

30. $f(x) = (x^2 - 10)(x^2 - 12)(x^2 - 20)$
31. $f(x) = 5x^2 - 2x - 1$
32. $f(x) = 4x^2 + 3x - 5$
33. $f(x) = 5x^3 - 3x^2 + 2x$
34. $f(x) = x^4 - 41x^2 + 400$
35. $f(x) = \dfrac{x^2 - 1}{x^2 + 2}$
36. $f(x) = \dfrac{x(x^2 - 3)}{x^2 + 5}$

Technology Window

Use numerical or graphical methods to estimate (correct to two decimal places) the x-intercepts, if any, for the functions given in Problems 37–44. Some calculators have a root finder feature (check your owner's manual).

37. $f(x) = \cos x + x^2 - 30$
38. $f(x) = \sin x + x^2 - 20$
39. $f(x) = \dfrac{x^2 - 5}{x^2 + 1}$
40. $f(x) = x^3 - 3\sqrt{x} - 5$
41. $f(x) = x^3 - 13x^2 - 20x + 84$
42. $f(x) = x^3 + 8x^2 - 53x - 60$
43. $f(x) = x^4 - 12x^3 - 396x^2 + 2{,}592x + 46{,}656$
44. $f(x) = 6x^3 - 65x^2 - 2{,}512x + 27{,}511$

B **45.** The function $f(x) = \dfrac{x^2 - 9}{x - 3}$ is not defined when $x = 3$. In this problem we investigate the behavior of this expression for values of x close to 3. Suppose we choose x so that $|x - 3| < 0.1$ (but so that $x \neq 3$).
  a. Find some x_0 satisfying this inequality.
  b. Find $f(x_0)$ for the value you found in part **a**.
  c. Repeat parts **a** and **b** for two other values. Can you form a conclusion about the value of the given expression?
  d. How could factoring have been used to simplify your work in part **b**?

46. The function $g(x) = \dfrac{x^3 - 27}{x - 3}$ is not defined when $x = 3$. In this problem we investigate the behavior of this expression for values of x close to 3. Suppose we choose x so that $|x - 3| < 0.01$ (but so that $x \neq 3$).
  a. Find some x_0 satisfying this inequality.
  b. Find $g(x_0)$ for the value you found in part **a**.
  c. Repeat parts **a** and **b** for two other values. Can you form a conclusion about the value of the given expression?

  d. How could factoring have been used to simplify your work in part **b**?

47. Charles's law for gases states that if the pressure remains constant, then
$$V(T) = V_0\left(1 + \frac{T}{273}\right)$$
where V is the volume (in.³), V_0 is the initial volume (in.³), and T is the temperature (in degrees Celsius).
  a. Sketch the graph of $V(T)$ for $V_0 = 100$ and $T \geq -273$.
  b. What is the temperature needed for the volume to double?

48. A particle starts at $P(0, 0)$ and its coordinates change every second by increments $\Delta x = 3$, $\Delta y = 5$. Find its new position after three seconds. Write the equation of the line described by this problem.

49. A particle starts at $P(-3, 5)$ and its coordinates change every second by increments $\Delta x = 5$, $\Delta y = -2$. Find its new position after two seconds. Write the equation of the line described by this problem.

50. A ball is thrown directly upward from the edge of a cliff in such a way that t seconds later, it is
$$s = -16t^2 + 96t + 144$$
feet above the ground at the base of the cliff. Sketch the graph of this equation (making the t-axis the horizontal axis) and then answer these questions.
  a. How high is the cliff?
  b. When (to the nearest tenth of a second) does the ball hit the ground at the base of the cliff?
  c. Estimate the time it takes for the ball to reach its maximum height. What is the maximum height?

The trajectory of a cannonball shot from the origin with initial velocity v and initial angle of inclination α (measured from level ground), is given by the equation
$$y = mx - 16v^{-2}(1 + m^2)x^2$$
where $m = \tan\alpha$. Let $v = 200$ ft/s to answer the questions in Problems 51–54.

51. If the angle of inclination is 42°, estimate the point where the cannonball will hit the ground.

52. Using a graphing utility of a calculator, determine the maximum height reached by the cannonball.

53. If $\alpha = 47°$, draw the path of the cannonball. Which graph in the directory of curves (Table 1.3) shows the basic shape of this graph?

54. Using a graphing utility of a calculator, determine the angle α that will maximize the distance the cannonball will travel.

55. It is estimated that t years from now, the population of a certain suburban community will be
$$P(t) = 20 - \frac{6}{t + 1}$$
thousand people.

a. What will the population of the community be nine years from now?

b. By how much will the population increase during the ninth year?

c. What will happen to the size of the population in the "long run"?

56. To study the rate at which animals learn, a psychology student performed an experiment in which a rat was sent repeatedly through a laboratory maze. Suppose that the time (in minutes) required for the rat to traverse the maze on the nth trial was approximately

$$f(n) = 3 + \frac{12}{n}$$

a. What is the domain of the function

$$f(x) = 3 + \frac{12}{x}?$$

b. For what values of n does $f(n)$ have meaning in the context of the psychology experiment?

c. How long did it take the rat to traverse the maze on the third trial?

d. On which trial did the rat first traverse the maze in 4 minutes or less?

e. According to the function f, what will happen to the time required for the rat to traverse the maze as the number of trials increases? Will the rat ever be able to traverse the maze in less than 3 minutes?

Technology Window

In Problems 57–58, let $f(x) = x^2 + 1$ and $g(x) = x^3 - x^2 - 9x + 9$.

57. Plot f and g on the coordinate system. From your graph, estimate the three values of x where the plots intersect.

58. Form the rational function $r(x) = \dfrac{g(x)}{f(x)}$ and plot it on $[-20, 20]$. Make a hand sketch of what you see. The graph of $r(x)$ looks linear for "large" x. Can you figure out *which* linear function approximates r for large x?

59. Define the function $G(x) = x^5 + 2x^4 - 9x - 18$ and plot G for several x-values to get a good idea of its behavior.

a. Zoom in on the largest x-value for which $G = 0$. Does the function become "almost linear" as you zoom in?

b. Factor G to find the exact x-values at which $G = 0$. [*Hint*: $G(-2) = 0$.]

c. Define a new rational function, $R = G/F$, where $F(x) = x^3 + 3$. Decide whether R is "almost linear" for x large (as was the case with the r discussed in Problem 58). Discuss briefly.

Technology Window

60. Imagine a sphere of unknown radius r with a meter stick sitting upright at the "north pole." A wire is strung from the top of the stick to a point of tangency of the sphere; when paced off along the surface back to the pole, the distance on the sphere's surface is 25 meters.

a. Show that one gets the following equation for r:

$$\cos \frac{25}{r} - \frac{r}{r+1} = 0$$

b. Graphically estimate, to one decimal place accuracy, the meaningful value of r. (*Note*: There are many solutions, but the one that you want is greater than 25. Why?)

61. **H**ISTORICAL **Q**UEST   One of the best known mathematical theorems is the *Pythagorean theorem*, named after the Greek philosopher Pythagoras. Very little is known about the life of Pythagoras, but we do know he was born on the island of Samos. He founded a secret brotherhood called the Pythagoreans that continued for at least 100 years after Pythagoras was murdered for political reasons. Even though the cult was

PYTHAGORAS
ca. 500 B.C.

called a "brotherhood," it did admit women. According to Lynn Osen in *Women in Mathematics*, the order was carried on by his wife and daughters after his death. In fact, women were probably more welcome in the centers of learning in ancient Greece than in any other age from that time until now.

State and prove the Pythagorean theorem.

62. The graph of an equation is *symmetric with respect to the y-axis* if whenever (x, y) satisfies the equation, so does $(-x, y)$.

a. Show that the graph of an even function must be symmetric with respect to the y-axis.

b. State an analogous criterion for symmetry with respect to the x-axis.

c. What kind of symmetry does the graph of an odd function have? Use your graphing calculator to explore with functions $y = x$, $y = x^3$, $y = \sin x$.

1.5 Inverse Functions; Inverse Trigonometric Functions

INVERSE FUNCTIONS

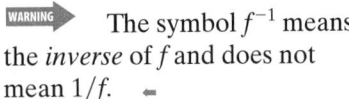

WARNING   The symbol f^{-1} means the *inverse* of f and does not mean $1/f$.   ◄

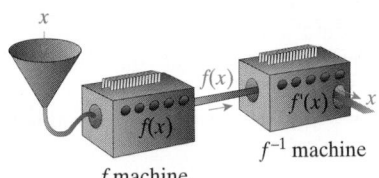

f machine f^{-1} machine

The inverse function f^{-1} reverses the effect of f. This relationship can be illustrated by function "machines."

For a given function f, we write $y_0 = f(x_0)$ to indicate that f maps the number x_0 in its domain into the corresponding number y_0 in the range. If f has an inverse f^{-1}, it is the function that reverses the effect of f in the sense that

$$f^{-1}(y_0) = x_0$$

For example, if

$$f(x) = 2x - 3, \text{ then } f(0) = -3, \ f(1) = -1, f(2) = 1,$$

and the inverse f^{-1} reverses f so that

$f(0) = -3$	$\boldsymbol{f^{-1}(-3) = 0}$	that is, $f^{-1}[f(0)] = 0$
$f(1) = -1$	$\boldsymbol{f^{-1}(-1) = 1}$	that is, $f^{-1}[f(1)] = 1$
$f(2) = 1$	$\boldsymbol{f^{-1}(1) = 2}$	that is, $f^{-1}[f(2)] = 2$

In the case where the inverse of a function is itself a function, we have the following definition.

> ### Inverse Function
>
> Let f be a function with domain D and range R. Then the function f^{-1} with domain R and range D is the **inverse of** f if
>
> $$f^{-1}[f(x)] = x \quad \text{for all } x \text{ in } D$$
>
> $$f[f^{-1}(y)] = y \quad \text{for all } y \text{ in } R$$
>
> ■ *What This Says:*   Suppose we consider a function defined by a set of ordered pairs $y = f(x)$. The image of x is y, as shown in Figure 1.24. If y is a member of the domain of the function $g = f^{-1}$, then $g(y) = x$. This means that f matches each element of x to exactly one y, and g matches those same elements of y back to the original values of x. If you think of a function as a set of ordered pairs (x, y), the inverse of f is the set of ordered pairs with the components (y, x).

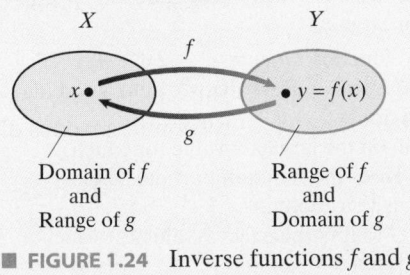

■ **FIGURE 1.24**   Inverse functions f and g

The language of this definition suggests that there is only one inverse function of f. Indeed, it can be shown (see Problem 60) that if f has an inverse, then the inverse is unique.

EXAMPLE 1   *Inverse of a given function defined as a set of ordered pairs*

Let $f = \{(0,3), (1,5), (3,9), (5,13)\}$; find f^{-1}, if it exists.

Solution   The inverse simply reverses the ordered pairs:

$$f^{-1} = \{(3,0), (5,1), (9,3), (13,5)\}$$

EXAMPLE 2   *Inverse of a given function defined by an equation*

Let $f(x) = 2x - 3$; find f^{-1}, if it exists.

Solution   To find f^{-1}, let $y = f(x)$ and interchange the x and y variables, and *then* solve for y.

Given function: $y = 2x - 3$ Inverse: $x = 2y - 3$

$$2y = x + 3$$
$$y = \tfrac{1}{2}(x + 3)$$

Thus, we represent the inverse function as $f^{-1}(x) = \tfrac{1}{2}(x + 3)$. To verify that these functions are inverses of each other, we note that

$$f[f^{-1}(x)] = f[\tfrac{1}{2}(x + 3)] = 2[\tfrac{1}{2}(x + 3)] - 3 = x + 3 - 3 = x$$

and

$$f^{-1}[f(x)] = f^{-1}(2x - 3) = \tfrac{1}{2}[(2x - 3) + 3] = \tfrac{1}{2}(2x) = x$$

for all x.

CRITERIA FOR EXISTENCE OF AN INVERSE f^{-1}

The inverse of a function may not exist. For example,

$$f = \{(0,0), (1,1), (-1,1), (2,4), (-2,4)\} \quad \text{and} \quad g(x) = x^2$$

do not have inverses because if we attempt to find the inverses, we obtain relations that are not functions. In the first case, we find

Possible inverse of f: $\{(0,0), (1,1), (1,-1), (4,2), (4,-2)\}$

This is not a function because not every member of the domain is associated with a single member in the range: $(1,1)$ and $(1,-2)$, for example.

In the second case, if we interchange the x and y in the equation for the function g where $y = x^2$ and then solve for y, we find:

$$x = y^2 \quad \text{or} \quad y = \pm\sqrt{x} \quad \text{for } x \geq 0$$

But this is not a function of x, because for any positive value of x, there are two corresponding values of y, namely, $\sqrt{x}$ and $-\sqrt{x}$.

A function f will have an inverse f^{-1} on the interval I when there is exactly one number in the domain associated with each number in the range. That is, f^{-1} exists if $f(x_1)$ and $f(x_2)$ are equal only when $x_1 = x_2$. A function with this property is said to be **one-to-one.** This is equivalent to the graphical criterion shown in Figure 1.25.

Horizontal Line Test

A function f has an inverse if and only if no horizontal line intersects the graph of $y = f(x)$ at more than one point.

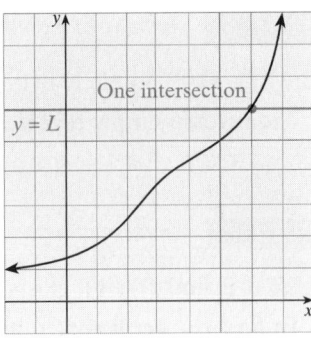

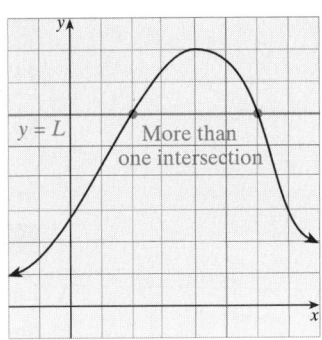

a. A function that has an inverse

b. A function that does not have an inverse

■ FIGURE 1.25   Horizontal line test

A function is said to be **strictly increasing** on an interval I if its graph is always rising on I, and **strictly decreasing** on I if the graph always falls on I. It is called **strictly monotonic** on I if it is either strictly increasing or strictly decreasing throughout that interval. A strictly monotonic function must be one-to-one and hence must have an inverse. For example, if f is strictly increasing on the interval I, we know that

$$x_1 > x_2 \text{ implies } f(x_1) > f(x_2)$$

so there is no way to have $f(x_1) = f(x_2)$ unless $x_1 = x_2$. This observation is formalized in the following theorem.

THEOREM 1.3   *A monotonic function has an inverse*

Let f be a function that is continuous and strictly monotonic on an interval I. Then f^{-1} exists and is monotonic on I (increasing if f is increasing and decreasing if f is decreasing).

Proof   We have already commented on why f^{-1} exists. To show that f^{-1} is strictly increasing whenever f is increasing, let y_1 and y_2 be numbers in the range of f, with $y_2 > y_1$. We shall show that $f^{-1}(y_2) > f^{-1}(y_1)$. Because y_1, y_2 are in the range of f, there exist numbers x_1, x_2 in the domain I such that $y_1 = f(x_1)$ and $y_2 = f(x_2)$. Because $y_2 > y_1$, it follows that $f(x_2) > f(x_1)$, and because f is strictly increasing, we must have $x_2 > x_1$. Thus, $f^{-1}(y_2) > f^{-1}(y_1)$, and f^{-1} is strictly increasing. Similarly, if f is strictly decreasing, then so is f^{-1}. (The details are left for the reader.)   ▄

GRAPH OF f^{-1}

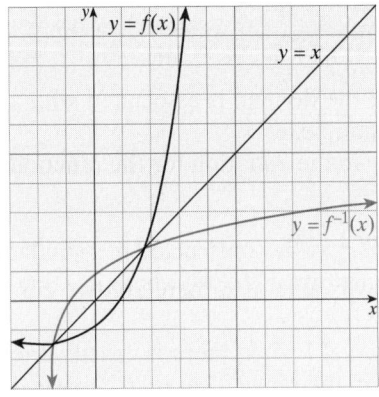

■ FIGURE 1.26   The graphs of f and f^{-1} are reflections in the line $y = x$.

The graphs of f and its inverse f^{-1} are closely related. In particular, if (a, b) is a point on the graph of f, then $b = f(a)$ and $a = f^{-1}(b)$, so (b, a) is on the graph of f^{-1}. It can be shown that (a, b) and (b, a) are reflections of one another in the line $y = x$. (See Figure 1.26.) These observations yield the following procedure for sketching the graph of an inverse function.

> **Procedure for Obtaining the Graph of f^{-1}**
>
> If f^{-1} exists, its graph may be obtained by reflecting the graph of f in the line $y = x$.

INVERSE TRIGONOMETRIC FUNCTIONS

The trigonometric functions are not one-to-one, so their inverses do not exist. However, if we restrict the domains of the trigonometric functions, then the inverses exist. In trigonometry, you probably distinguished between the sine curve with unrestricted domain and the sine curve with restricted domain by writing $y = \sin x$ and $y = \text{Sin } x$, respectively. In calculus, it is not customary to make such a distinction by using a capital letter.

Let us consider the sine function first. We know that the sine function is strictly increasing on the closed interval $[-\frac{\pi}{2}, \frac{\pi}{2}]$, and if we restrict $\sin x$ to this interval, it does have an inverse, as shown in Figure 1.27.

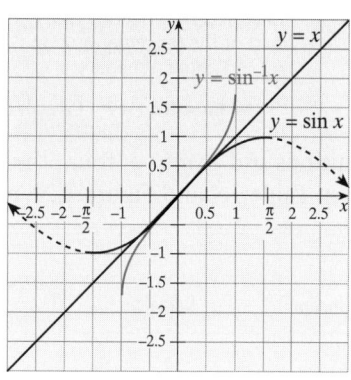

a. The graph of $\sin^{-1}x$ is obtained by reflecting the part of the sine on $[-\frac{\pi}{2}, \frac{\pi}{2}]$ about $y = x$.

b. The graph of the inverse sine function, $y = \sin^{-1}x$.

■ **FIGURE 1.27**   Inverse sine function

Inverse Sine Function

$$y = \sin^{-1}x \quad \text{if and only if} \quad x = \sin y \quad \text{and} \quad -\frac{\pi}{2} \le y \le \frac{\pi}{2}$$

The function $\sin^{-1} x$ is sometimes written arcsin x.

WARNING ➤   The function $\sin^{-1} x$ is *NOT* the reciprocal of $\sin x$. To denote the reciprocal, write $(\sin x)^{-1}$. ←

Inverses of the other five trigonometric functions may be constructed in a similar manner. For example, by restricting $\tan x$ to the open interval $(-\frac{\pi}{2}, \frac{\pi}{2})$ where it is one-to-one, we can define the inverse tangent function as follows.

Inverse Tangent Function

$$y = \tan^{-1}x \quad \text{if and only if} \quad x = \tan y \quad \text{and} \quad -\frac{\pi}{2} < y < \frac{\pi}{2}$$

The function $\tan^{-1} x$ is sometimes written arctan x.

The graph of $y = \tan^{-1} x$ is shown in Figure 1.28.

WARNING It is easier to remember the restrictions on the domain and the range if you do so in terms of quadrants, as shown in Table 1.4. Note the last column, which gives the range (or value of the angle y). For example, if $y = \sin^{-1} x$ and if x is positive, then y terminates in Quadrant I; in other words, $0 \leq y \leq \frac{\pi}{2} \approx 1.57$; on the other hand, if x is negative, then the terminal side of angle y is in Quadrant IV, with $-1.57 \approx -\frac{\pi}{2} \leq y \leq 0$. Finally, if x is 0, then $y = 0$.

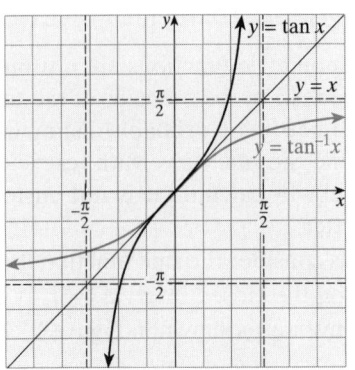
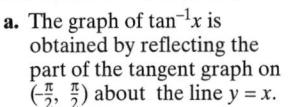

a. The graph of $\tan^{-1}x$ is obtained by reflecting the part of the tangent graph on $(-\frac{\pi}{2}, \frac{\pi}{2})$ about the line $y = x$.

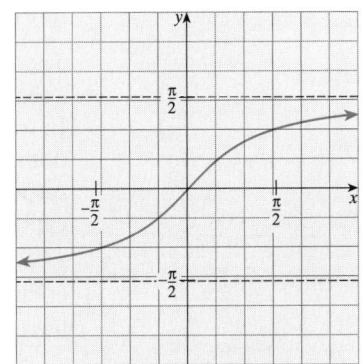

b. The graph of $\tan^{-1}x$

■ **FIGURE 1.28**   Graph of the inverse tangent function

Definitions and graphs of four other fundamental inverse trigonometric functions are given in Table 1.4 and Figure 1.29, respectively.

■ **TABLE 1.4**

Definition of Inverse Trigonometric Functions

Inverse Function	Domain	Range	Value of x can be Pos.	Value of x can be Neg.	Value of x can be Zero
			Quadrant		
$y = \sin^{-1} x$	$-1 \leq x \leq 1$	$-\frac{\pi}{2} \leq y \leq \frac{\pi}{2}$	I	IV	0
		Quadrants I and IV			
$y = \cos^{-1} x$	$-1 \leq x \leq 1$	$0 \leq y \leq \pi$	I	II	$\frac{\pi}{2}$
		Quadrants I and II			
$y = \tan^{-1} x$	$-\infty < x < +\infty$	$-\frac{\pi}{2} < y < \frac{\pi}{2}$	I	IV	0
		Quadrants I and IV			
$y = \sec^{-1} x$	$x \geq 1$ or $x \leq -1$	$0 \leq y \leq \pi, y \neq \frac{\pi}{2}$	I	II	undefined
		Quadrants I and II			
$y = \csc^{-1} x$	$x \geq 1$ or $x \leq -1$	$-\frac{\pi}{2} \leq y \leq \frac{\pi}{2}, y \neq 0$	I	IV	undefined
		Quadrants I and IV			
$y = \cot^{-1} x$	$-\infty < x < +\infty$	$0 < y < \pi$	I	II	$\frac{\pi}{2}$
		Quadrants I and II			

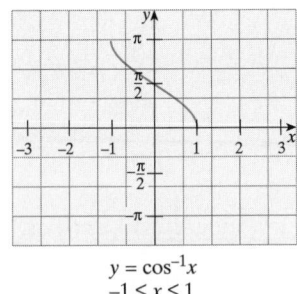

$y = \cos^{-1}x$
$-1 \leq x \leq 1$
$0 \leq y \leq \pi$

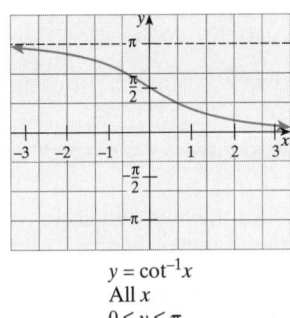

$y = \cot^{-1}x$
All x
$0 \leq y \leq \pi$

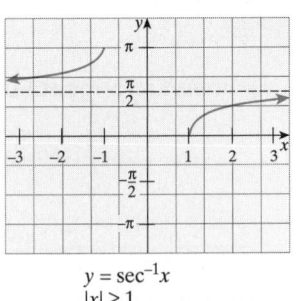

$y = \sec^{-1}x$
$|x| \geq 1$
$0 \leq y \leq \pi, y \neq \frac{\pi}{2}$

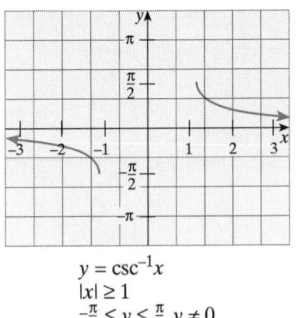

$y = \csc^{-1}x$
$|x| \geq 1$
$-\frac{\pi}{2} \leq y \leq \frac{\pi}{2}, y \neq 0$

■ **FIGURE 1.29**   Graphs of four inverse trigonometric functions

> **EXAMPLE 3**   *Evaluating inverse trigonometric functions*

Evaluate the given functions.

a. $\sin^{-1}\left(\frac{-\sqrt{2}}{2}\right)$ b. $\sin^{-1}0.21$ c. $\cos^{-1}0$ d. $\tan^{-1}\left(\frac{1}{\sqrt{3}}\right)$

Solution

a. $\sin^{-1}\left(\frac{-\sqrt{2}}{2}\right) = -\frac{\pi}{4}$;   Think: $x = \frac{-\sqrt{2}}{2}$ is negative, so y is in Quadrant **IV**; the reference angle is the angle whose sine is $\frac{\sqrt{2}}{2}$, it is $\frac{\pi}{4}$, so in Quadrant **IV** the angle is $-\frac{\pi}{4}$.

b. $\sin^{-1}0.21 \approx 0.2115750$ By calculator; be sure to use radian mode and inverse sine (not reciprocal).

c. $\cos^{-1}0 = \frac{\pi}{2}$ Memorized exact value.

d. $\tan^{-1}\left(\frac{1}{\sqrt{3}}\right) = \frac{\pi}{6}$ Think: $x = \left(\frac{1}{\sqrt{3}}\right)$ is positive, so y is in Quadrant **I**; the reference angle is the same as the value of the inverse tangent in Quadrant **I**. ∎

INVERSE TRIGONOMETRIC IDENTITIES

The definition of inverse functions yields 2 formulas, which we call the inversion formulas for sine and tangent.

WARNING ▶   The inversion formulas for $\sin^{-1}$ and $\tan^{-1}$ are valid only on the specified domains. ◀

> **Inversion Formulas**
>
> $\sin(\sin^{-1}x) = x$ for $-1 \le x \le 1$
> $\sin^{-1}(\sin y) = y$ for $-\frac{\pi}{2} \le y \le \frac{\pi}{2}$
>
> $\tan(\tan^{-1}x) = x$ for all x
> $\tan^{-1}(\tan y) = y$ for $-\frac{\pi}{2} < y < \frac{\pi}{2}$

> **EXAMPLE 4**   *Inversion formula for x inside and outside domain*

Evaluate the given functions.

a. $\sin(\sin^{-1}0.5)$ b. $\sin(\sin^{-1}2)$ c. $\sin^{-1}(\sin 0.5)$ d. $\sin^{-1}(\sin 2)$

Solution

a. $\sin(\sin^{-1}0.5) = 0.5$, because $-1 \le 0.5 \le 1$.

b. $\sin(\sin^{-1}2)$ does not exist, because 2 is not between -1 and 1.

c. $\sin^{-1}(\sin 0.5) = 0.5$, because $-\frac{\pi}{2} \le 0.5 \le \frac{\pi}{2}$.

d. $\sin^{-1}(\sin 2) = 1.1415927$, by calculator

For exact values, notice that (reduction principle)

$$\sin 2 = \sin(\pi - 2) \quad \text{(so that } -\frac{\pi}{2} \le \pi - 2 \le \frac{\pi}{2})$$

and we have $\sin^{-1}(\sin 2) = \sin^{-1}[\sin(\pi - 2)] = \pi - 2$. ∎

Some trigonometric identities correspond to inverse trigonometric identities, but others do not. For example,

$$\sin(-x) = -\sin x \quad \text{and} \quad \cos(-x) = \cos x$$

It is true that

$$\sin^{-1}(-x) = -\sin^{-1}(x)$$

but in general,

$$\cos^{-1}(-x) \neq \cos^{-1}x$$

(For a counterexample, try $x = 1$: $\cos^{-1}(-1) = \pi$ and $\cos^{-1}1 = 0$.)

EXAMPLE 5   *Proving inverse trigonometric identities*

For $-1 \leq x \leq 1$, show that:

a. $\sin^{-1}(-x) = -\sin^{-1}x$ b. $\cos(\sin^{-1}x) = \sqrt{1 - x^2}$

Solution

a. Let $y = \sin^{-1}(-x)$

$$\begin{aligned}
\sin y &= -x &&\text{\textit{Definition of inverse sine}}\\
-\sin y &= x \\
\sin(-y) &= x &&\text{\textit{Opposite angle identity}}\\
-y &= \sin^{-1}x &&\text{\textit{Definition of inverse sine}}\\
y &= -\sin^{-1}x
\end{aligned}$$

Thus, $\sin^{-1}(-x) = -\sin^{-1}x$.

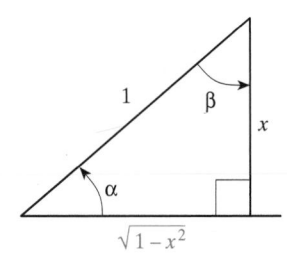

■ **FIGURE 1.30**   A reference triangle

b. Let $\alpha = \sin^{-1}x$ so $\sin \alpha = x$, where $0 \leq \alpha \leq \frac{\pi}{2}$. Construct a right triangle with an acute angle α and hypotenuse 1, as shown in Figure 1.30. We call this triangle a **reference triangle.** The side opposite α is x (since $\sin \alpha = x$), and by the Pythagorean theorem, the adjacent side is $\sqrt{1 - x^2}$. Thus, we have

$$\begin{aligned}
\cos(\sin^{-1}x) &= \cos \alpha \\
&= \frac{\pm \sqrt{1 - x^2}}{1} \\
&= \sqrt{1 - x^2} &&\text{\textit{Substitute} } \cos \alpha = x \text{ \textit{and choose the positive value}}\\
& &&\text{\textit{for the radical because} } \alpha \text{ \textit{is in Quadrant} } \text{I.} ■
\end{aligned}$$

Reference triangles, such as the one shown in Figure 1.30, are extremely useful devices for obtaining inverse trigonometric identities. For instance, let α and β be angles of a right triangle with hypotenuse 1. If the side opposite α is x (so that $\sin \alpha = x$), then

$$\cos^{-1}x + \sin^{-1}x = \alpha + \beta = \tfrac{\pi}{2}   \text{ for } 0 \leq x \leq 1$$

since the acute angles of a right triangle must sum to $\frac{\pi}{2}$. The same reasoning can also be used to show that

$$\tan^{-1}x + \cot^{-1}x = \tfrac{\pi}{2}$$

and

$$\sec^{-1}x + \csc^{-1}x = \tfrac{\pi}{2}$$

Identities involving inverse trigonometric functions have a variety of uses. For instance, most calculators have keys for evaluating $\sin^{-1}x$, $\cos^{-1}x$, and $\tan^{-1}x$, but what about the other three inverse trigonometric functions? The answer is given by the following theorem, which allows us to compute $\sec^{-1}x$, $\csc^{-1}x$, and $\cot^{-1}x$ using reciprocal identities involving the three inverse trigonometric functions the calculator does have.

THEOREM 1.4   *Reciprocal identities for inverse trigonometric functions*

$$\sec^{-1}x = \cos^{-1}\frac{1}{x} \quad \text{if } |x| \geq 1$$

$$\csc^{-1}x = \sin^{-1}\frac{1}{x} \quad \text{if } |x| \geq 1$$

$$\cot^{-1}x = \begin{cases} \tan^{-1}\dfrac{1}{x} & \text{if } x \text{ is positive} \\ \tan^{-1}\dfrac{1}{x} + \pi & \text{if } x \text{ is negative} \\ \dfrac{\pi}{2} & \text{if } x = 0 \end{cases}$$

Proof   We shall prove that $\cot^{-1}x = \tan^{-1}\frac{1}{x}$ if $x > 0$, and leave the rest of the proof as an exercise. Let $\beta = \cot^{-1}x$, as indicated by the reference triangle shown in Figure 1.30. Then,

$$\cot \beta = x \quad \text{for } 0 \leq \beta \leq \frac{\pi}{2} \qquad \textit{Definition of } \cot^{-1}\beta$$

$$\frac{1}{\tan \beta} = x \qquad\qquad \textit{Reciprocal identity}$$

$$\tan \beta = \frac{1}{x}$$

$$\beta = \tan^{-1}\frac{1}{x}$$

Thus, $\cot^{-1}x = \beta = \tan^{-1}\frac{1}{x}$, as claimed.

EXAMPLE 6   *Evaluating inverse reciprocal functions*

Evaluate the given inverse functions using the inverse identities and a calculator.
a. $\sec^{-1}(-3)$ b. $\csc^{-1}7.5$ c. $\cot^{-1}2.4747$ d. $\cot^{-1}(-4.852)$

Solution

a. $\sec^{-1}(-3) = \cos^{-1}\left(-\frac{1}{3}\right) \approx 1.910633236$

b. $\csc^{-1}7.5 = \sin^{-1}\left(\frac{1}{7.5}\right) \approx 0.1337315894$

c. Since 2.4747 is positive, we have $\cot^{-1}2.4747 = \tan^{-1}\left(\frac{1}{2.4747}\right) \approx 0.3840267299$.

d. Since -4.852 is negative, $\cot^{-1}(-4.852) = \tan^{-1}\left(\frac{1}{-4.852}\right) + \pi \approx 2.938338095$. ∎

Note that $\cot^{-1}x$ can also be computed using the identity

$$\tan^{-1}x + \cot^{-1}x = \frac{\pi}{2}$$

For instance, the inverse cotangent computed in Example **6d** can be found as

$$\cot^{-1}(-4.852) = \frac{\pi}{2} - \tan^{-1}(-4.852) \approx 2.938338095$$

1.5 Problem Set

Ⓐ 1. ■ **What Does This Say?** Discuss the restrictions on the domain and range in the definition of the inverse trigonometric functions.

2. ■ **What Does This Say?** Discuss the use of reference triangles with respect to the inverse trigonometric functions.

Determine which pairs of functions defined by the equations in Problems 3–8 are inverses of each other.

3. $f(x) = 5x + 3$; $g(x) = \dfrac{x - 3}{5}$

4. $f(x) = \frac{2}{3}x + 2$; $g(x) = \frac{3}{2}x + 3$

5. $f(x) = \frac{4}{5}x + 4$; $g(x) = \frac{5}{4}x + 3$

6. $f(x) = \dfrac{1}{x}, x \neq 0$; $g(x) = \dfrac{1}{x}, x \neq 0$

7. $f(x) = x^2, x < 0$; $g(x) = \sqrt{x}, x > 0$

8. $f(x) = x^2, x \geq 0$; $g(x) = \sqrt{x}, x \geq 0$

Find the inverse (if it exists) of each function given in Problems 9–18.

9. $f = \{(4, 5), (6, 3), (7, 1), (2, 4)\}$

10. $g = \{(3, 9), (-3, 9), (4, 16), (-4, 16)\}$

11. $f(x) = 2x + 3$ **12.** $g(x) = -3x + 2$

13. $f(x) = x^2 - 5, x \geq 0$ **14.** $g(x) = x^2 - 5, x < 0$

15. $F(x) = \sqrt{x} + 5$ **16.** $G(x) = 10 - \sqrt{x}$

17. $h(x) = \dfrac{2x - 6}{3x + 3}$ **18.** $h(x) = \dfrac{2x + 1}{x}$

Give the exact values for functions in Problems 19–30.

19. a. $\cos^{-1}\frac{1}{2}$ **b.** $\sin^{-1}\left(-\frac{\sqrt{3}}{2}\right)$

20. a. $\sin^{-1}\left(-\frac{1}{2}\right)$ **b.** $\cos^{-1}\left(-\frac{1}{2}\right)$

21. a. $\tan^{-1}(-1)$ **b.** $\cot^{-1}(-\sqrt{3})$

22. a. $\sec^{-1}(-\sqrt{2})$ **b.** $\csc^{-1}(-\sqrt{2})$

23. a. $\sin^{-1}\left(-\frac{\sqrt{3}}{2}\right)$ **b.** $\sec^{-1}(-1)$

24. a. $\sec^{-1}\left(\frac{2}{\sqrt{3}}\right)$ **b.** $\cot^{-1}(-1)$

25. $\cos\left(\sin^{-1}\frac{1}{2}\right)$ **26.** $\sin\left(\cos^{-1}\frac{1}{\sqrt{2}}\right)$

27. $\cot\left(\tan^{-1}\frac{1}{3}\right)$ **28.** $\tan\left(\sin^{-1}\frac{1}{3}\right)$

29. $\cos\left(\sin^{-1}\frac{1}{5} + 2\cos^{-1}\frac{1}{5}\right)$

SMH   [*Hint:* Use the addition law for $\cos(\alpha + \beta)$.]

30. $\sin\left(\sin^{-1}\frac{1}{5} + \cos^{-1}\frac{1}{4}\right)$

SMH   [*Hint:* Use the addition law for $\sin(\alpha + \beta)$.]

31. Suppose that α is an acute angle of a right triangle where
$$\sin \alpha = \frac{s^2 - t^2}{s^2 + t^2} \quad (s > t > 0)$$
Show $\alpha = \tan^{-1}\left(\dfrac{s^2 - t^2}{2st}\right)$

32. If $\sin \alpha + \cos \alpha = s$ and $\sin \alpha - \cos \alpha = t$, show that
$$\alpha = \tan^{-1}\left(\frac{s + t}{s - t}\right)$$

Ⓑ *Sketch the graph of f in Problems 33–40 and then use the horizontal line test to determine whether f has an inverse. If f^{-1} exists, sketch its graph.*

33. $f(x) = x^2$, for all x **34.** $f(x) = x^2, x \leq 0$

35. $f(x) = 10^x$, for all x **36.** $f(x) = e^x, x \geq 0$

37. $f(x) = \sqrt{1 - x^2}$, on $(-1, 1)$

38. $f(x) = x(x - 1)(x - 2)$, on $[1, 2]$

39. $f(x) = \cos x$, on $[0, \pi]$ **40.** $f(x) = \tan x$, on $\left(-\frac{\pi}{2}, \frac{\pi}{2}\right)$

Simplify each expression in Problems 41–46.

41. $\sin(2 \tan^{-1}x)$ **42.** $\tan(2 \tan^{-1}x)$

43. $\tan(\cos^{-1}x)$ **44.** $\cos(2 \sin^{-1}x)$

45. $\sin(\sin^{-1}x + \cos^{-1}x)$ **46.** $\cos 2(\sin^{-1}x + \cos^{-1}x)$

47. Use reference triangles, if necessary, to justify each of the following identities.

   a. $\cot^{-1}x = \dfrac{\pi}{2} - \tan^{-1}x$ for all x

   b. $\sec^{-1}x = \cos^{-1}\left(\dfrac{1}{x}\right)$ for all $|x| \geq 1$

   c. $\csc^{-1}x = \sin^{-1}\left(\dfrac{1}{x}\right)$ for all $|x| \geq 1$

48. Use the identities in Problem 47 to evaluate each function rounded to four decimal places.

   a. $\cot^{-1}0.67$ **b.** $\sec^{-1}1.34$

   c. $\csc^{-1}2.59$ **d.** $\cot^{-1}(-1.54)$

49. Use the identities in Problem 47 to evaluate each function rounded to four decimal places.

   a. $\cot^{-1}1.5$ **b.** $\cot^{-1}(-1.5)$

   c. $\sec^{-1}(-1.7)$ **d.** $\csc^{-1}(-1.84)$

50. A painting 3 ft high is hung on a wall in such a way that its lower edge is 7 ft above the floor. An observer whose eyes are 5 ft above the floor stands x ft away from the wall. Express the angle θ subtended by the painting as a function of x.

51. To determine the height of a building (see Figure 1.31), select a point P and find the angle of elevation to be α. Then move out a distance of x units (on a level plane) to point Q and find that the angle of elevation is now β. Find the height h of the building, as a function of x.

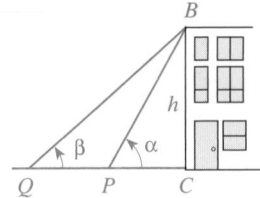

■ **FIGURE 1.31**
Determining height

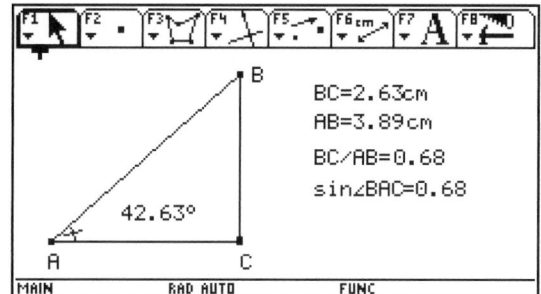

Technology Window

52. A TI-92 or an interactive geometric program (such as *The Geometer's Sketchpad* 3) is necessary for this problem.* Consider a triangle labeled as shown in Figure 1.32. Construct a table showing the lengths of the sides, namely $|\overline{BC}|$ and $|\overline{AB}|$, relative to $\angle BAC$. Also compare $\sin A$ and the ratio $|\overline{BC}|/|\overline{AB}|$. One possible display is shown in Figure 1.32.

BC=2.63cm
AB=3.89cm
BC/AB=0.68
sin∠BAC=0.68

42.63°

DATA	∠BAC	BC	AB	BC/AB	sin∠A
	c1	c2	c3	c4	c5
1	18.123	.93673	3.0115	.31105	.31105
2	34.422	1.9613	3.4696	.56529	.56529
3	42.714	2.6424	3.8953	.67834	.67834
4	55.744	4.2826	5.8846	.82654	.82654
5	79.852	15.989	16.243	.98435	.98435
6	85.451	35.972	36.086	.99685	.99685
7	89.729	606.11	606.12	.99999	.99999

c1=

■ **FIGURE 1.32**   Sample TI-92 output

  a. What will happen if $m\angle BAC$ reaches 90°?
  b. What will happen if $m\angle BAC$ is between 90° and 180°?
  c. Reformulate the questions in parts **a** and **b** for the ratio $|\overline{AC}|/|\overline{AB}|$.
  d. Reformulate the question in parts **a** and **b** for the ratio $|\overline{BC}|/|\overline{AC}|$.

———

*See "Using Interactive-Geometry Software for Right-Angle Trigonometry," by Charles Embse and Arne Engebretsen, *The Mathematics Teacher*, October 1996, pp. 602–605.

C 53. a. Find $\tan^{-1}1 + \tan^{-1}2 + \tan^{-1}3$ using a calculator. Make a conjecture about the exact value.
  b. Prove the conjecture from part **a** using trigonometric identities. [*Hint:* Find $\tan(\tan^{-1}1 + \tan^{-1}2 + \tan^{-1}3)$.]

54. Prove the conjecture of Problem 53a using reference triangles.

55. Prove the conjecture from Problem 53a using right triangles as follows. You may use the figure shown in Figure 1.33; assume that $\triangle ABC, \triangle ABD$, and $\triangle DEF$ are all right triangles with lengths of sides as shown in the figure.

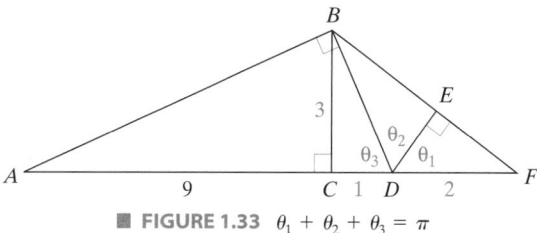

■ **FIGURE 1.33** $\theta_1 + \theta_2 + \theta_3 = \pi$

56. In Problem 47 we proved the identity.
$$\cot^{-1}x = \frac{\pi}{2} - \tan^{-1}x$$
by using a reference triangle. In the text, we stated the identity
$$\cot^{-1}x = \begin{cases} \tan^{-1}\dfrac{1}{x} & \text{if } x \text{ is positive} \\ \tan^{-1}\dfrac{1}{x} + \pi & \text{if } x \text{ is negative} \\ \dfrac{\pi}{2} & \text{if } x = 0 \end{cases}$$
Show that these identities are equivalent.

57. In Problem 47 we proved the identity
$$\sec^{-1}x = \cos^{-1}\left(\frac{1}{x}\right) \text{ for all } |x| \geq 1$$
The graph of $y = \sec^{-1}x$ suggests that
$$\sec^{-1}(-x) = \pi - \sec^{-1}x \text{ for } x > 1.$$
Either prove this identity or find a counterexample.

58. In Problem 47 we proved the identity
$$\csc^{-1}x = \sin^{-1}\left(\frac{1}{x}\right) \text{ for all } |x| \geq 1$$
By examining the graph of $y = \csc^{-1}x$, conjecture an identity of the form
$$\csc^{-1}(-x) = A + B\csc^{-1}x \text{ for } x > 1$$
Prove this identity, and then use it to evaluate $\csc^{-1}(-9.38)$.

59. Suppose that $\triangle ABC$ is *not* a right triangle, but has an obtuse angle β. Draw $\overline{BD}$ perpendicular to $\overline{AC}$, forming right triangles $\triangle ABD$ and $\triangle BDC$ (with right angles at D). Show that
$$\frac{\sin \alpha}{a} = \frac{\sin \gamma}{c}$$

60. Show that if f^{-1} exists, it is unique.
  [*Hint:* Show that if g_1 and g_2 both satisfy
$$g_1[f(x)] = x = f[g_1(x)]$$
$$g_2[f(x)] = x = f[g_2(x)]$$
then $g_1(x) = g_2(x)$.]

1.6 *Exponential and Logarithmic Functions*

exponential functions, logarithmic functions, the natural base *e*, the natural logarithms, continuous compounding of interest

EXPONENTIAL FUNCTIONS

> **Exponential Function**
>
> The function *f* is an **exponential function** if
> $$f(x) = b^x$$
> where *b* is a positive constant other than 1 and *x* is any real number.

Recall that if *n* is a natural number, then

$$b^n = \underbrace{b \cdot b \cdot b \cdot \cdots \cdot b}_{n \text{ factors}}$$

Furthermore, if $b \neq 0$, then

$$b^0 = 1, \quad b^{-n} = \frac{1}{b^n}, \quad \text{and} \quad b^{1/n} = \sqrt[n]{b}$$

Also, if *m* and *n* are any integers, and *m/n* is a reduced fraction, then

$$b^{m/n} = (b^{1/n})^m$$

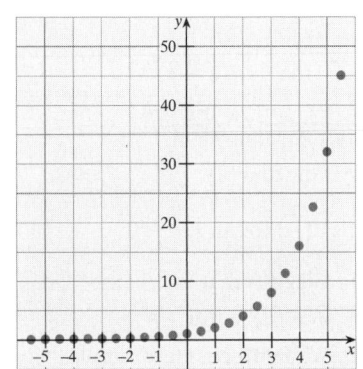

■ FIGURE 1.34 Graph of $y = 2^x$ for rational exponents

This definition tells us what b^x means for rational values of *x*. However, we now wish to enlarge the domain of *x* to include all real numbers. To get a feeling for what is involved in this problem, let us examine the special case where $b = 2$. In Figure 1.34, we have plotted several points with coordinates $(r, 2^r)$, where *r* is a rational number. We must now attach meaning to b^x if *x* is not a rational number. To enlarge the domain of *x* to include all real numbers, we need the help of the following theorem.

THEOREM 1.5   *Bracketing theorem for exponents*

Suppose *b* is a real number greater than 1. Then for any real number *x*, there is a unique real number b^x. Moreover, if *p* and *q* are any two rational numbers such that $p < x < q$, then

$$b^p < b^x < b^q$$

Proof   A formal proof of this theorem is beyond the scope of this course, but we can outline the needed steps. The proof depends on a property of real numbers called the *completeness property of the real numbers*. What the completeness property says is that each real number can be approximated to any prescribed degree of accuracy by a rational number. This is equivalent to saying that if *x* is a real number, then for every positive integer *n* there exists a rational number r_n such that $|x - r_n| < \frac{1}{n}$.

The bracketing theorem gives meaning to expressions such as $2^{\sqrt{3}}$, since

$$1.732 < \sqrt{3} < 1.733$$

implies

$$2^{1.732} < 2^{\sqrt{3}} < 2^{1.733}$$

> **EXAMPLE 1**   *Graph of an exponential function*

Graph $f(x) = 2^x$.

Solution   Begin by plotting points using rational x-values. We see that the graph has a fairly well defined shape (as shown in Figure 1.35a), but it is riddled with "holes" that correspond to those points whose x-coordinates are irrational numbers.

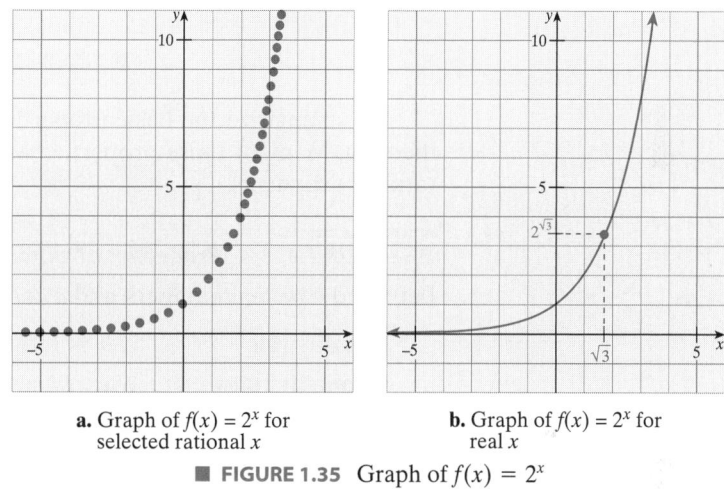

a. Graph of $f(x) = 2^x$ for selected rational x

b. Graph of $f(x) = 2^x$ for real x

■ **FIGURE 1.35**   Graph of $f(x) = 2^x$

We complete the graph by connecting the points with a smooth curve as shown in Figure 1.35b. However, let us take a moment to use the bracketing theorem for exponents to relate the completed graph to the definition of 2^x for an irrational x-value. Consider the point $(x, 2^x)$ for $x = \sqrt{3}$. Because $\sqrt{3}$ is irrational, the decimal representation is nonterminating and nonrepeating. In other words, $\sqrt{3}$ is the limit of a sequence of rational numbers; specifically,

$$1, 1.7, 1.73, 1.732, 1.7320, 1.73205, 1.732050, 1.7320508, \ldots$$

Then, $2^{\sqrt{3}}$ is the limit of the sequence of numbers

$$2^1, 2^{1.7}, 2^{1.73}, 2^{1.732}, 2^{1.7320}, 2^{1.73205}, 2^{1.732050}, 2^{1.7320508}, \ldots$$

Graphically, this means that as the rational numbers $1, 1.7, 1.73, \ldots$ tend toward $\sqrt{3}$, the points $(1, 2)$, $(1.7, 2^{1.7})$, $(1.73, 2^{1.73})$, ... tend toward the "hole" in the graph of $y = 2^x$ that corresponds to $x = \sqrt{3}$.   ■

The shape of the graph of $y = b^x$ for any $b > 1$ is essentially the same as that of $y = 2^x$. The graph of $y = b^x$ for a typical base $b > 1$ is shown in Figure 1.36a, and by reflecting this graph about the y-axis, we obtain the graph of $y = b^{-x}$. The case where

$0 < b < 1$ is shown in Figure 1.36b. In particular, notice that in this case, the reciprocal $c = \dfrac{1}{b}$ satisfies $c > 1$, and $y = b^x$ and $y = b^{-x}$ are thus the same as $y = c^{-x}$ and $y = c^x$, respectively.

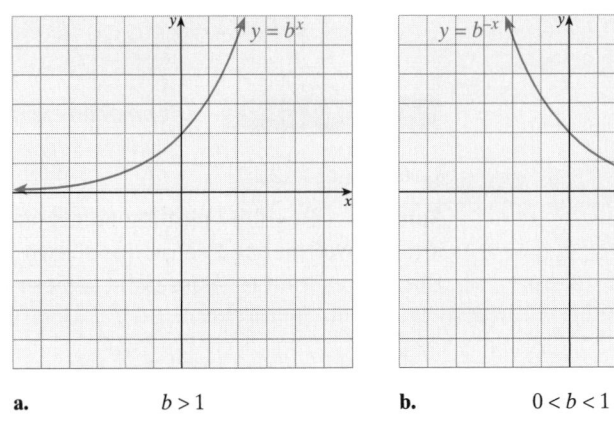

a. $b > 1$ **b.** $0 < b < 1$

■ **FIGURE 1.36** Graph of $y = b^x$

We summarize the basic properties of exponential functions with the following theorem. Many of these properties were proved in previous courses for bases with rational exponents.

THEOREM 1.6   *Properties of exponential functions*

Let x and y be real numbers, and a and b positive real numbers.

Equality rule	If $b \neq 1$, then $b^x = b^y$ if and only if $x = y$.
Inequality rules	If $x > y$ and $b > 1$, then $b^x > b^y$.
	If $x > y$ and $0 < b < 1$, then $b^x < b^y$.
Product rule	$b^x b^y = b^{x+y}$
Quotient rule	$\dfrac{b^x}{b^y} = b^{x-y}$
Power rules	$(b^x)^y = b^{xy}; \quad (ab)^x = a^x b^x; \quad \left(\dfrac{a}{b}\right)^x = \dfrac{a^x}{b^x}$

Graphical Properties $y = b^x$ is always above the x-axis ($b^x > 0$), is rising for $b > 1$ (function increasing), and falling for $0 < b < 1$ (function decreasing). That is, the function $f(x) = b^x$ is monotonic. (See Figure 1.36.)

Several parts of this theorem are used in the following example.

EXAMPLE 2   *Exponential equations*

Solve each of the following exponential equations.

a. $2^{x^2+3} = 16$ b. $2^x 3^{x+1} = 108$ c. $(\sqrt{2})^{x^2} = \dfrac{8^x}{4}$

Solution

a. $2^{x^2+3} = 16$

         $2^{x^2+3} = 2^4$ Write 16 as 2^4 so that the equality rule can be used.

     $x^2 + 3 = 4$ or $x = \pm 1$

b.
$$2^x 3^{x+1} = 108$$
$$2^x 3^x 3 = 3 \cdot 36 \qquad \text{Product rule}$$
$$(2 \cdot 3)^x = 36 \qquad \text{Divide both sides by 3 and then use power rule.}$$
$$6^x = 6^2$$
$$x = 2 \qquad \text{Equality rule}$$

c.
$$(\sqrt{2})^{x^2} = \frac{8^x}{4}$$
$$(2^{1/2})^{x^2} = \frac{(2^3)^x}{2^2}$$
$$2^{x^2/2} = 2^{3x-2}$$
$$\frac{x^2}{2} = 3x - 2 \qquad \text{Equality rule}$$
$$x^2 - 6x + 4 = 0$$
$$x = \frac{6 \pm \sqrt{36 - 4(1)(4)}}{2} = 3 \pm \sqrt{5}$$

∎

Technology Window

Many calculators and computer programs have solver utilities that will solve equations such as those shown in Example 2. One such example is shown here.

It is also instructive to consider the relationship between the solution to an equation and its graph. You can use a graphing calculator to estimate solutions to equations such as those in Example 2. For example, in part **a**, you can graph both the left and right sides of the equation and then look for the intersection point(s), if any. The TRACE with the cursor on the curve Y1 = 2^(X^2+3) gives the following values:

$X = -1.021277 \quad Y = 16.48413 \quad \leftarrow$ *A root is between*

$X = -.9787234 \quad Y = 15.53983 \quad \leftarrow$ *these x-values.*

and

$X = .97872348 \quad Y = 15.539837 \quad \leftarrow$ *A root is between*

$X = 1.0212766 \quad Y = 16.48431 \quad \leftarrow$ *these x-values.*

It looks as though the roots are $x = -1$ and $x = 1$. This conclusion can be confirmed by the intermediate value theorem, which we discuss in Section 2.4, but for now these values can be checked by substitution.

Finally, many calculators have an INTERSECT key. If you use this feature, you can obtain the intersection $(-1, 16)$ and $(1, 16)$ directly. Check your owner's manual to see if your calculator has this feature.

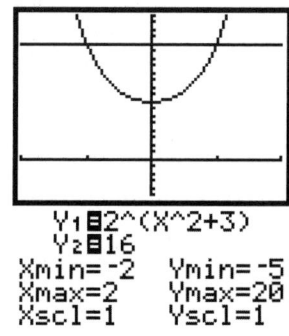

LOGARITHMIC FUNCTIONS

It can be shown that if $b > 0$, $b \neq 1$, the exponential function $f(x) = b^x$ is monotonic and has an inverse which itself is monotonic. We call this inverse function the **logarithm of x to the base b.** Here is a definition, with some notation.

Logarithmic Function

If $b > 0$ and $b \neq 1$, the **logarithm of x to the base b** is the function $y = \log_b x$ that satisfies $b^y = x$; that is,

$$y = \log_b x \quad \text{means} \quad b^y = x$$

■ *What This Says:*   It is useful to think of a logarithm as an exponent. That is, consider the following sequence of interpretations:

$$y = \log_b x$$

y is the logarithm to the base of b of x.

y is the **exponent on a base b** that gives x.

$$b^y = x$$

Notice that $y = \log_b x$ is defined only for $x > 0$ because $b^y > 0$ for all y. We have sketched the graph of $y = \log_b x$ for $b > 1$ in Figure 1.37 by reflecting the graph of $y = b^x$ in the line $y = x$.

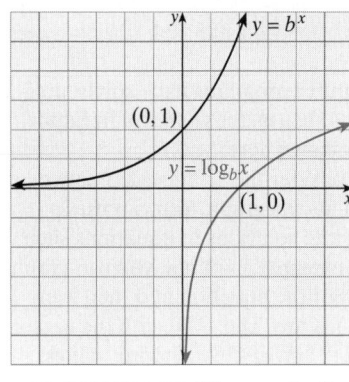

■ **FIGURE 1.37** The graph of
$y = \log_b x$ for $b > 1$

Because $y = b^x$ is a continuous, increasing function that satisfies $b^x > 0$ for all x, $\log_b x$ must also be continuous and increasing, and its graph must lie entirely to the right of the y-axis.

Also, because $b^0 = 1$ and $b^1 = b$, we have

$$\log_b 1 = 0 \quad \text{and} \quad \log_b b = 1$$

so $(1, 0)$ and $(b, 1)$ lie on the logarithmic curve.

Theorem 1.7 lists several general properties of logarithms.

THEOREM 1.7   *Basic properties of logarithmic functions*

Assume $b > 0$ and $b \neq 1$.

WARNING Remember that $\log_b x$ is defined only for $x > 0$. ⬅

Equality rule	$\log_b x = \log_b y$ if and only if $x = y$
Inequality rules	If $x > y$ and $b > 1$, then $\log_b x > \log_b y$
	If $x > y$ and $0 < b < 1$, then $\log_b x < \log_b y$
Product rule	$\log_b (xy) = \log_b x + \log_b y$

Quotient rule	$\log_b \dfrac{x}{y} = \log_b x - \log_b y$
Power rule	$\log_b x^p = p \log_b x$ for any real number p
Inversion rules	$b^{\log_b x} = x$ and $\log_b b^x = x$
Special values	$\log_b b = 1$ and $\log_b 1 = 0$

Proof   Each part of this theorem can be derived by using the definition of logarithm in conjunction with a suitable property of exponentials. (See Problems 74 and 75.) ═══

EXAMPLE 3   *Evaluate a logarithmic expression*

Evaluate $\log_2\left(\frac{1}{8}\right) + \log_2 128$.

Solution   You can evaluate this on your calculator, or you can use Theorem 1.7:

$$\log_2\left(\frac{1}{8}\right) + \log_2 128 = \log_2(2)^{-3} + \log_2 2^7$$
$$= \log_2[2^{-3}(2^7)] = \log_2 2^4 = 4\log_2 2 = 4(1) = 4$$

What we have shown here is longer than what you would do in your work, which would probably be shortened as follows:

$$\log_2(\tfrac{1}{8}) + \log_2 128 = \log_2 16 \qquad \textit{Because } (\tfrac{1}{8})(128) = 16$$
$$= 4 \qquad \textit{Because 4 is the \underline{exponent} on 2}$$
$$\textit{that yields 16} \qquad \blacksquare$$

EXAMPLE 4   *Logarithmic equation*

Solve the equation $\log_3(2x + 1) - 2\log_3(x - 3) = 2$.

Solution   Use Theorem 1.7, and remember that both $2x + 1 > 0$ and $x - 3 > 0$.

$$\log_3(2x + 1) - 2\log_3(x - 3) = 2$$
$$\log_3(2x + 1) - \log_3(x - 3)^2 = 2$$
$$\log_3\frac{2x + 1}{(x - 3)^2} = 2$$
$$3^2 = \frac{2x + 1}{(x - 3)^2}$$
$$9(x - 3)^2 = 2x + 1$$
$$9x^2 - 56x + 80 = 0$$
$$(x - 4)(9x - 20) = 0$$
$$x = 4, \tfrac{20}{9}$$

Notice that $x - 3 < 0$ if $x = \frac{20}{9}$. The given logarithmic equation has only $x = 4$ as a solution because we cannot take the logarithm of a negative number.   ■

EXAMPLE 5   *Exponential equation with base 10*

Solve $10^{5x+3} = 195$.

Technology Window

For a graphical solution, you must use the change of base theorem from algebra to input a base other than base e or base 10:

$$\log_3 N = \log N / \log 3$$

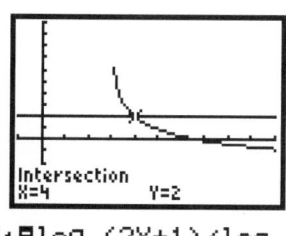

```
Intersection
X=4      Y=2
Y₁☐log (2X+1)/log 3
-2*log (X-3)/log 3
Y₂☐2
 Xmin=-1   Ymin=-5
 Xmax=10   Ymax=10
 Xscl=1    Yscl=1
```

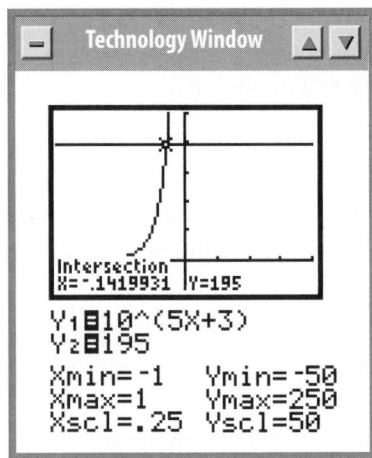

Y₁■10^(5X+3)
Y₂■195

Xmin=-1   Ymin=-50
Xmax=1   Ymax=250
Xscl=.25   Yscl=50

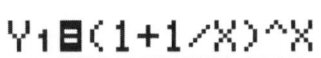

X	Y₁
1	2
2	2.25
3	2.3704
4	2.4414
5	2.4883
10	2.5937
50	2.6916
100	2.7048
1000	2.7169
5000	2.718
10000	2.7181
1E6	2.7183
1E12	2.7183

Solution

$$10^{5x+3} = 195$$
$$5x + 3 = \log_{10} 195$$

Remember that $5x + 3$ is the exponent on a base 10 that gives 195.

$$5x = \log_{10} 195 - 3$$
$$x = \frac{\log_{10} 195 - 3}{5}$$
$$\approx -0.1419930777 \quad \text{By calculator} \qquad ■$$

NATURAL BASE e

In elementary algebra you may have used exponential bases of 2 or 10, but in calculus we use as a base a certain irrational number e whose decimal expansion is

$$e \approx 2.71828182845 \ldots$$

This number, called the **natural exponential base,** can be defined by the limit:

e is the limiting value of $\left(1 + \frac{1}{x}\right)^x$ as x becomes large

We shall study limits in the next chapter, and in Chapter 4 (after we discuss limits involving infinity), we will write

$$e = \lim_{x \to +\infty} \left(1 + \frac{1}{x}\right)^x$$

At first glance you may think this limit has the value 1. After all, $1 + 1/x$ clearly approaches 1 as x increases without bound, and $1^x = 1$ for all x. But the limit process does not work that way, as indicated by the tabular computations in the margin.

The letter e was chosen in honor of the great Swiss mathematician Leonhard Euler (1707–1783), who investigated this limit and explored a number of applications in which this limit plays a useful role.

The function $f(x) = e^x$ is called the **natural exponential function.** This function obeys all the basic rules of Theorem 1.6 for exponential functions with base $b > 1$. Because many of the applications we consider in calculus will have complicated exponents, such as

$$e^{3x^2 - 2\sin x + 8}$$

we sometimes streamline this form by writing

$$\exp(3x^2 - 2\sin x + 8)$$

Exp Notation

If $f(x)$ is an expression in x, the notation

$$\exp(f(x))$$

means $e^{f(x)}$.

NATURAL LOGARITHMS

There are two frequently used bases for logarithms. If the base 10 is used, then the logarithm is called a **common logarithm,** and if the base e is used, the logarithm is called a **natural logarithm.** We use a special notation for each of these logarithms.

Common Logarithm

The **common logarithm**, $\log_{10} x$, is denoted by **log x**.

Natural Logarithm

The **natural logarithm**, $\log_e x$, is denoted by **ln x**.

(ln x is pronounced "ell n x" or "lawn x")

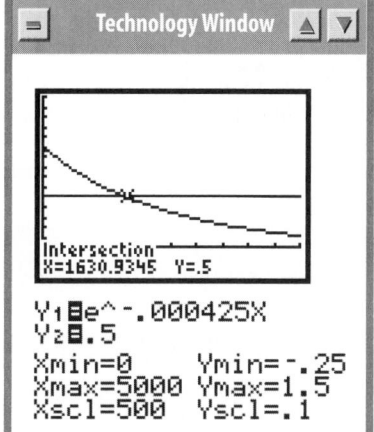

Technology Window

Intersection
X=1630.9345 Y=.5

Y₁■e^-.000425X
Y₂■.5

Xmin=0   Ymin=-.25
Xmax=5000 Ymax=1.5
Xscl=500   Yscl=.1

EXAMPLE 6 *Solving an exponential equation with base e*

Solve $\frac{1}{2} = e^{-0.000425t}$.

Solution We have

$$\frac{1}{2} = e^{-0.000425t}$$

$$-0.000425t = \ln 0.5 \qquad \text{Remember that } -0.000425t \text{ is the } \underline{\text{exponent on}} \underline{\text{base } e} \text{ that yields } \frac{1}{2} = 0.5.$$

$$t = \frac{\ln 0.5}{-0.000425} \approx 1,630.934542 \qquad \blacksquare$$

If a logarithmic equation has a base other than base 10 or e, you can solve it using the equality rule of Theorem 1.7. The following theorem provides a useful formula for converting from one base to another.

THEOREM 1.8 *Change-of-base theorem*

$$\log_b x = \frac{\log_a x}{\log_a b} \qquad \text{Remember, the definition of logarithms} \atop \text{requires that } b > 0, b \neq 1.$$

In particular, when $a = e$,

$$\log_b x = \frac{\ln x}{\ln b} \quad \text{for any } b > 0 \; (b \neq 1)$$

Proof Let $y = \log_b x$. Then

$$b^y = x \qquad \text{Definition of logarithm}$$

$$\log_a b^y = \log_a x \qquad \text{Equality rule of logarithms}$$

$$y \log_a b = \log_a x \qquad \text{Power rule of logarithms}$$

$$y = \frac{\log_a x}{\log_a b} \qquad \text{Divide both sides by } \log_a b.$$

EXAMPLE 7 *Solving an exponential equation using the change-of-base theorem*

Solve $6^{3x+2} = 200$.

Solution $6^{3x+2} = 200$

$$3x + 2 = \log_6 200$$

$$3x = \log_6 200 - 2$$

$$x = \frac{\log_6 200 - 2}{3} \qquad \text{To evaluate } \log_6 200, \text{ use } \log_6 200 = \frac{\ln 200}{\ln 6}.$$

$$\approx 0.3190157417 \qquad \blacksquare$$

Because we most often use base e, it is worthwhile to state some useful properties of natural logarithms.

THEOREM 1.9   *Basic properties of the natural logarithm*

a. $\ln 1 = 0$

b. $\ln e = 1$

c. $e^{\ln x} = x$ for all $x > 0$

d. $\ln e^y = y$ for all y

e. $b^x = e^{x \ln b}$ for any $b > 0$ $(b \neq 1)$

Proof   Parts **a** and **b** follow immediately from the definitions of $\ln x$ and e^x. Parts **c** and **d** are just the inversion rules for base e.

We show the proof of part **e**:

$$
\begin{aligned}
b^x &= y && \text{Let } y = b^x. \\
&= e^{\ln y} && \text{Property } \boldsymbol{c} \text{ of this theorem} \\
&= e^{x \ln b} && \text{If } y = b^x, \text{ then } x = \log_b y = \frac{\ln y}{\ln b}, \text{ so that} \\
&&& \ln y = x \ln b.
\end{aligned}
$$

Technology Window

If you are using a software package such as *Mathematica*, *Derive*, or *Maple*, sometimes referred to as CAS programs, be careful about the notation. Some versions do not distinguish between $\log x$ (common logarithm) and $\ln x$ (natural logarithm). All logarithms on these versions are assumed to be natural logarithms, so to evaluate a common logarithm requires using the change-of-base theorem.

EXAMPLE 8   *Exponential growth*

A biological colony grows in such a way that at time t (in minutes), the population is

$$P(t) = P_0 e^{kt}$$

where P_0 is the initial population and k is a positive constant. Suppose the colony begins with 5,000 individuals and contains a population of 7,000 after 20 min. Find k and determine the population (rounded to the nearest hundred individuals) after 30 min.

Solution   Because $P_0 = 5,000$, the population after t minutes will be

$$P(t) = 5,000 e^{kt}$$

In particular, because the population is 7,000 after 20 min,

$$
\begin{aligned}
P(20) &= 5,000 e^{k(20)} \\
7,000 &= 5,000 e^{20k} \\
\tfrac{7}{5} &= e^{20k} \\
20k &= \ln\left(\tfrac{7}{5}\right) \\
k &= \tfrac{1}{20} \ln\left(\tfrac{7}{5}\right) \\
&\approx 0.0168236
\end{aligned}
$$

Finally, to determine the population after 30 min, substitute this value for k to find:

$$
\begin{aligned}
P(30) &= 5,000 e^{30k} \\
&\approx 8,282.5117
\end{aligned}
$$

The expected population is approximately 8,300.   ∎

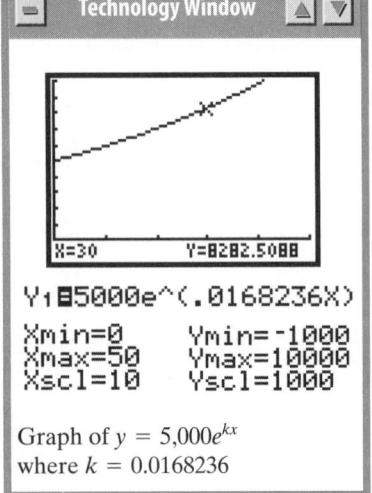

Technology Window

X=30   Y=8282.5088

Y₁■5000e^(.0168236X)

Xmin=0   Ymin=-1000
Xmax=50   Ymax=10000
Xscl=10   Yscl=1000

Graph of $y = 5{,}000 e^{kx}$
where $k = 0.0168236$

CONTINUOUS COMPOUNDING OF INTEREST

One reason e is called the "natural" exponential base is that many natural growth phenomena can be described in terms of e^x. As an illustration of this fact, we close this section by showing how e^x can be used to describe the accounting procedure called *continuous compounding of interest.*

If a sum of money, called the **present value** or **principal,** is denoted by P and invested at an annual rate of r for t years, then the **future value** is denoted by A and is found by

$$A = P + I$$

where I denotes the amount of interest. **Interest** is an amount of money paid for the use of another's money. **Simple interest** is found by multiplication: $I = Prt$. For example, \$1,000 invested for 3 years at a 15% simple annual interest rate generates $I = \$1,000(0.15)(3) = \$450$, so the future value in 3 years is $A = \$1,000 + \$450 = \$1,450$.

Most businesses, however, pay interest on the interest as well as on the principal, and when this is done, it is called **compound interest.** For example, the future value of \$1,000 invested at 15% annual interest compounded annually for 3 years can be found as follows:

First Year:

$$A = P + I$$
$$= P \cdot 1 + Pr \qquad I = Prt \text{ and } t = 1$$
$$= \boldsymbol{P(1 + r)} \qquad \text{For this example, } A = \$1,000(1 + 0.15)$$
$$= \$1,150$$

Second Year:

$$A = \boldsymbol{P(1 + r)} + I \qquad \text{The total amount from the first year}$$
$$\text{becomes the principal for the second year.}$$

$$A = \boldsymbol{P(1 + r)} \cdot 1 + \boldsymbol{P(1 + r)} \cdot r$$
$$= P(1 + r)(1 + r)$$
$$= P(1 + r)^2 \qquad \text{For this example, } A = \$1,000(1 + 0.15)^2$$
$$= \$1,322.50$$

Third Year:

$$A = P(1 + r)^2 + P(1 + r)^2 \cdot r$$
$$= P(1 + r)^2(1 + r)$$
$$= P(1 + r)^3 \qquad \text{For this example, } A = \$1,000(1 + 0.15)^3$$
$$\approx \$1,520.88$$

Notice that with simple interest the amount in 3 years is \$1,450, as compared with \$1,520.88 when interest is compounded annually.

This discussion leads to the following **compound interest future value formula.** If a principal of P dollars is invested at an interest rate of i per period for a total of N periods, then the future amount A is given by the formula

$$A = P(1 + i)^N$$

Compound interest is usually stated in terms of an annual interest rate r and a given number of years t. The frequency of compounding (that is, the number of

compoundings per year) is denoted by n. Therefore, $i = \dfrac{r}{n}$ and $N = nt$ in the formula for A.

The first two graphs in Figure 1.38 show how an amount of money in an account over a one-year period of time grows, first with quarterly compounding and then with monthly compounding. Notice that these are "step" graphs, with jumps occurring at the end of each compounding period.

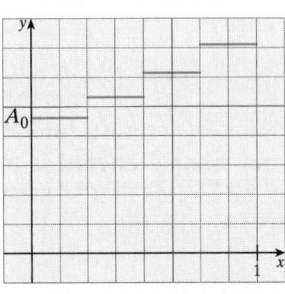   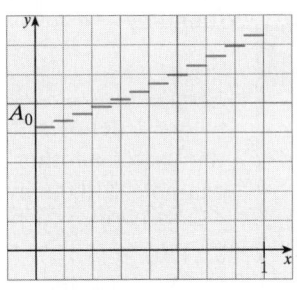   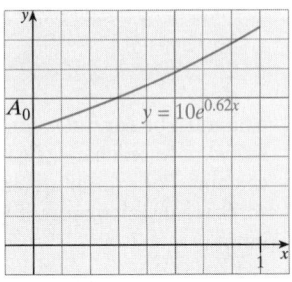

a. Quarterly compounding   **b.** Monthly compounding   **c.** Continuous compounding

■ **FIGURE 1.38**   The growth of an account over a one-year period with different compounding frequencies

With continuous compounding, we compound interest not quarterly, or monthly, or daily, or even every second, but *instantaneously,* so that the future amount of money A in the account grows continuously, as shown in Figure 1.38c. In other words, we compute A as the limiting value of

$$P\left(1 + \frac{r}{x}\right)^x$$

as the number of compounding periods x grows without bound. We will use properties of limits in Chapter 4 to derive the formula in the following box.

Future Value with Continuous Compounding

If P dollars are invested at an annual interest rate r **compounded continuously,** then the future value after t years is

$$A = Pe^{rt}$$

EXAMPLE 9   *Compounding of interest*

If \$12,000 is invested for 5 years at 18%, find the future value at the end of 5 years if interest is compounded

a. monthly

b. continuously

Solution $P = \$12{,}000; t = 5;$ and $r = 18\% = 0.18$ are given.

a. $n = 12; A = P(1 + i)^N = \$12{,}000\left(1 + \dfrac{0.18}{12}\right)^{12 \cdot 5} \approx \$29{,}318.64$

b. $A = \$12{,}000e^{0.18(5)} \approx \$29{,}515.24$ ■

1.6 Problem Set

Ⓐ *Sketch the graph of the functions in Problems* 1–4.

1. $y = 3^x$

2. $y = 4^{-x}$

3. $y = -e^{-x}$

4. $y = -e^x$

Evaluate the expressions (with calculator accuracy) in Problems 5–12.

5. $32^{2/5} + 9^{3/2}$

6. $(1 + 4^{3/2})^{-1/2}$

7. $e^3 e^{2.3}$

8. $\dfrac{(e^{1.3})^2}{e^{1.3} + \sqrt{e^{-1.4}}}$

9. $5{,}000\left(1 + \dfrac{0.135}{12}\right)^{12(5)}$

10. $145{,}000\left(1 + \dfrac{0.073}{365}\right)^{-365(5)}$

11. $2{,}589e^{0.45(6)}$

12. $850{,}000e^{-0.04(10)}$

Evaluate the expressions given in Problems 13–22.

13. $\log_2 4 + \log_3 \dfrac{1}{9}$

14. $2^{\log_2 3 - \log_2 5}$

15. $5 \log_3 9 - 2 \log_2 16$

16. $(\log_2 \tfrac{1}{8})(\log_3 27)$

17. $(3^{\log_3 7})(\log_5 0.04)$

18. $e^{5 \ln 2}$

19. $\log_3 3^4 - \ln e^{0.5}$

20. $\ln(\log 10^e)$

21. $\exp(\ln 3 - \ln 10)$

22. $\exp(\log_{e^2} 25)$

Solve the logarithmic and exponential equations to calculator accuracy in Problems 23–41.

23. $\log_x 16 = 2$

24. $\log x = 5.1$

25. $e^{-3x} = 0.5$

26. $\ln x^2 = 9$

27. $7^{-x} = 15$

28. $e^{2x} = \ln(4 + e)$

29. $\dfrac{1}{2} \log_3 x = \log_2 8$

30. $\log_2 (x^{\log_2 x}) = 4$

31. $3^{x^2 - x} = 9$

32. $4^{x^2 + x} = 16$

33. $2^x 5^{x+2} = 25{,}000$

34. $3^x 4^{x+1/2} = 3{,}456$

35. $(\sqrt[3]{2})^{x+10} = 2^{x^2}$

36. $(\sqrt[3]{5})^{x+2} = 5^{x^2}$

37. $e^{2x+3} = 1$

38. $\dfrac{e^{x^2}}{e^{x+6}} = 1$

39. $\log_3 x + \log_3 (2x+1) = 1$

40. $\ln\left(\dfrac{x^2}{1 - x}\right) = \ln x + \ln\left(\dfrac{2x}{1 + x}\right)$

41. $2^{3 \log_2 x} = 4 \log_3 9$

Ⓑ **42.** If $\log_b 1{,}296 = 4$, what is $\left(\dfrac{3}{2}b\right)^{3/2}$?

43. If $\log_{\sqrt{b}} 106 = 2$, what is $\sqrt{b - 25}$?

44. Solve $\log_2 x + \log_5 (2x + 1) = \ln x$ correct to the nearest tenth.

Decide whether each of the curves given in Problems 45–48 *is exponential or logarithmic.*

45.
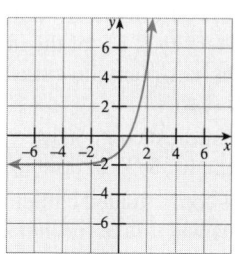

46.
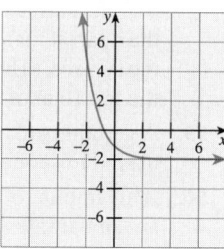

47.
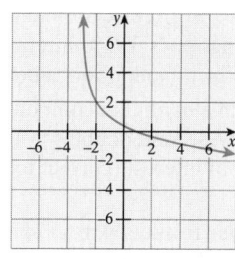

48.
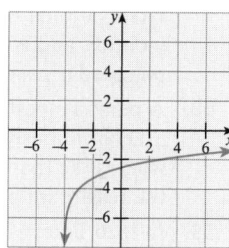

Data points with a curve fit to those points are shown in Problems 49–52. *Decide whether the data are better modeled by an exponential or a logarithmic function.*

49.

50.
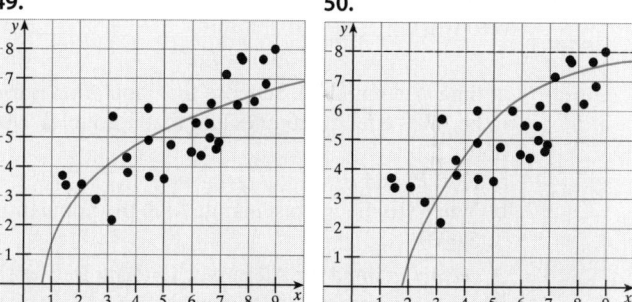

51.

52.
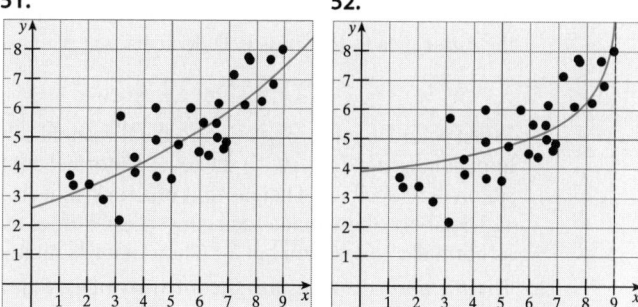

53. Solve $\log_x 2 = \log_3 x$ correct to the nearest tenth.

54. According to the *Bouguer–Lambert law*, a beam of light that strikes the surface of a body of water with intensity

I_0 will have intensity I at a depth of x meters, where

$$I = I_0 e^{kx}, \quad k > 0$$

The constant k, called the *absorption coefficient*, depends on such things as the wavelength of the beam of light and the purity of the water. Suppose a given beam of light is only 5% as intense at a depth of 2 meters as it is at the surface. Find k and determine at what depth (to the nearest meter) the intensity is 1% of the intensity at the surface. (This explains why plant life exists only in the top 10 m of a lake or sea.)

55. A certain bank pays 6% interest compounded continuously. How long will it take for $835 to double?

56. If an amount of money is compounded continuously at an annual rate r, how long will it take for that amount of money to double?

57. First National Bank pays 7% interest compounded monthly, and World Savings pays 6.95% interest compounded continuously. Which bank offers a better deal?

58. A person invests $8,500 in a bank that compounds interest continuously. If the investment doubles in 10 years, what is the (annual) rate of interest (correct to the nearest hundredth percent)?

59. In 1626, Peter Minuit traded trinkets worth $24 for land on Manhattan Island. Assume that in 1990 the same land was worth $25.2 billion. Find the annual rate of interest compounded continuously at which the $24 would have had to be invested during this time to yield the same amount.

60. Biologists estimate that the population of a bacterial colony is

$$P(t) = P_0 2^{kt}$$

at time t (in minutes). Suppose the population is found to be 1,000 after 20 minutes and that it doubles every hour.
   a. Find P_0 and k.
   b. When (to the nearest minute) will the population be 5,000?

61. A *decibel* (named for Alexander Graham Bell) is the smallest increase of the loudness of a sound that is detectable by the human ear. In physics, it is shown that when two sounds of intensity I_1 and I_2 (watts/cm³) occur, the difference in loudness is D decibels, where

$$D = 10 \log\left(\frac{I_1}{I_2}\right)$$

When sound is rated in relation to the threshold of human hearing ($I_0 = 10^{-16}$) the level of normal conversation is 50 decibels, whereas that of a rock concert is 110 decibels. Show that a rock concert is 60 times as loud as normal conversation but a million times as intense.

62. The *Richter scale* measures the intensity of earthquakes. Specifically, if E is the energy (watt/cm³) released by a quake, then it is said to have *magnitude M*, where

$$M = \frac{\log E - 11.4}{1.5}$$

   a. Express E in terms of M.

   b. How much more energy is released in an $M = 8.5$ earthquake (such as the devastating Alaska quake of 1964) than in an average quake of magnitude $M = 6.5$ (the Los Angeles quake of 1994)?

63. **THINK TANK PROBLEM** In the definition of the exponential function $f(x) = b^x$, we require that b be a positive constant. What happens if $b < 0$, for example, if $b = -2$? For what values of x is f defined? Describe the graph of f in this case.

64. Let $E(x) = 2^{x^2 - 2x}$.
   a. Use a calculator to sketch $E(x)$.
   b. Where does the graph cross the y-axis? What happens to the graph as $x \to +\infty$ and as $x \to -\infty$?
   c. Use a calculator utility to find the value of x that minimizes $E(x)$. What is the minimum value?

65. A manufacturer of car batteries estimates that p percent of the batteries will work for at least t months, where

$$p(t) = 100 e^{-0.03t}$$

   a. What percent of the batteries can be expected to last at least 40 months?
   b. What percent can be expected to *fail* before 50 months?
   c. What percent can be expected to fail between the 40th and 50th months?

66. If $3,600 is invested at 15% compounded daily, how much money will there be in 7 years?
   a. Use a 365-day year; this is known as *exact interest*.
   b. Use a 360-day year; this is known as *ordinary interest*.

67. If P dollars are borrowed for N months compounded monthly at an annual interest rate of r, then the monthly payment is found by the formula

$$m = \frac{P\left(\dfrac{r}{12}\right)}{1 - \left(1 + \dfrac{r}{12}\right)^{-N}}$$

   a. Use this formula to determine the monthly car payment for a new car costing $17,487 with a down payment of $7,487. The car is financed for 4 years at 12%.
   b. A home loan is made for $210,000 at 8% interest for 30 years. What is the monthly payment with a 20% down payment?

68. A cool drink is removed from a refrigerator and is placed in a room where the temperature is 70 °F. According to a result in physics known as *Newton's law of cooling*, the temperature of the drink in t minutes will be

$$F(t) = 70 - A e^{-kt}$$

where A and k are positive constants. Suppose the temperature of the drink was 35 °F when it left the refrigerator, and 30 minutes later, it was 50 °F (that is, $F(0) = 35$ and $F(30) = 50$).
   a. Find A and e^{-30k}.
   b. What will the temperature (to the nearest degree) of the drink be after one hour?
   c. What would you expect to happen to the temperature as $t \to +\infty$?

69. **SPY PROBLEM** An internationally famous spy receives an e-mail message that his best friend, Sigmund ("Siggy") Leiter, has been found murdered, the body stuffed unceremoniously in a freezer. Fighting back tears of grief, he remembers that t hours after death, a body has temperature

$$T = A + (B - A)e^{-0.03t}$$

where A is the air temperature and B is the temperature of the body at the time of death. The police inform the Spy that at the time of discovery, 1:00 P.M. on Thursday, the corpse had temperature 40 °F, and the freezer 10 °F. He knows the deed was done by either Coldfinger or André Scélérat. If Coldfinger was in jail from Monday until Wednesday noon and Scélérat was at a villains' convention in Las Vegas from noon on Wednesday until Friday, who "iced" Siggy and when? (By the way, one of the few normal things about Siggy was his body temperature, 98.6 °F.)

70. **HISTORICAL QUEST**   John Napier was a Scottish landowner who was the Isaac Asimov of his day, having envisioned the tank, the machine gun, and the submarine. He also predicted that the end of the world would occur between 1688 and 1700. He is best known today as the inventor of logarithms, which until the advent of the calculator were used extensively with complicated calculations. Napier's logarithms are not identical to the logarithms we use today. Napier chose to use $1 - 10^{-7}$ as his given number, and then he multiplied by 10^7. That is, if

JOHN NAPIER
1550–1617

$$N = 10^7\left(1 - \frac{1}{10^7}\right)^L$$

then L is Napier's logarithm of the number N; that is

$$L = \text{nog } N \text{ means } N = 10^7(1 - 10^{-7})^L.$$

One difference between Napier's logarithms (which for clarity we will call a nog) and modern logarithms is apparent when stating the product, quotient, and power rules for logarithms.
a. Show that if $L_1 = \text{nog } N_1$ and $L_2 = \text{nog } N_2$, then

$$\text{nog } N_1 + \text{nog } N_2 = \text{nog}\left(\frac{N_1 N_2}{10^7}\right)$$

b. State and prove similar results for a quotient rule and a power rule for Napier logarithms.
c. What is nog 10^7?
d. If nog $N = 10^7$, then $N \approx e^r$ for some rational number r. What is r?
e. Fortunately, Napier's 1614 paper on logarithms was read by a true mathematician, Henry Briggs (1561–1630), and together they decided that base 10 made a lot more sense. In the year Napier died, Briggs published a table of common logarithms (base 10) that at the time was a major accomplishment. In this paper he used the words "mantissa" and "characteristic." What is the meaning that Briggs gave to these words?

f. Who was the first person to publish a table of natural logarithms (base e)?

71. **HISTORICAL QUEST**   It is commonly acknowledged that the time for logarithms to be invented was right around the beginning of the 17th century. The Swiss instrument-maker Jobst Bürgi chose as a base a number a little greater than the number 1 (that is, $1 + 10^{-4} > 1$) rather than a number a little smaller than 1 (namely, $1 - 10^{-7} < 1$ chosen by Napier). Bürgi multiplied all of his powers by 10^8 and came up with a system that is very similar to the natural logarithms we use today. The following table compares Napier's and Bürgi's tables.

JOBST BÜRGI
1552–1632

Napier's Table		Bürgi's Table	
1. $10^7(1 - 10^{-7})$		**10(1)**	$10^8(1 + 10^{-4})$
2. $10^7(1 - 10^{-7})^2$		**10(2)**	$10^8(1 + 10^{-4})^2$
3. $10^7(1 - 10^{-7})^3$		**10(3)**	$10^8(1 + 10^{-4})^3$
$\vdots$			$\vdots$
n $10^7(1 - 10^{-7})^n$		**10(n)**	$10^8(1 + 10^{-4})^n$

Bürgi continued his table to 23,027 entries.
a. Why do you think Bürgi stopped at 23,027?
b. Let $m = n \times 10^{-4}$ so that

$$x = (1 + 10^{-4})^n = [(1 + 10^{-4})^{10^4}]^m$$

If we let bog x be Bürgi's logarithms, what is bog x? How is this related to e?
c. Prove bog $x^n = n$ bog x.

ⓒ 72. For $b > 0$ and all positive integers m and n, show that:

a. $b^m b^n = b^{m+n}$ b. $\dfrac{b^m}{b^n} = b^{m-n}$

73. For $b > 0$ and all positve integers m and n, show that
a. $(b^m)^n = b^{mn}$
b. $(\sqrt[n]{b})^m = \sqrt[n]{b^m}$

74. Prove:
a. $\log_b x + \log_b y = \log_b(xy)$
b. $\log_b x - \log_b y = \log_b\left(\dfrac{x}{y}\right)$

75. Let b be any positive number other than 1. Show that

$$x^x = b^{x \log_b x}$$

76. Let a and b be any positive numbers other than 1. Show that

a. $\log_a x = \dfrac{\log_b x}{\log_b a}$

b. $(\log_a b)(\log_b a) = 1$

Chapter 1   Review

Proficiency Examination

Concept Problems

1. Characterize the following sets of numbers: natural numbers ($\mathbb{N}$), whole numbers ($\mathbb{W}$), integers ($\mathbb{J}$), rational numbers ($\mathbb{Q}$), irrational numbers ($\mathbb{Q}'$), and real numbers ($\mathbb{R}$).

2. Define absolute value.

3. State the triangle inequality.

4. State the distance formula for points $P(x_1, y_1)$ and $Q(x_2, y_2)$.

5. Define slope in terms of angle of inclination.

6. List the following forms of the equation of a line:
   a. standard form   b. slope–intercept form
   c. point–slope form   d. horizontal line
   e. vertical line

7. State the slope criteria for parallel and for perpendicular lines.

8. Define function.

9. Define the composition of functions.

10. What is meant by the graph of a function?

11. Draw a quick sketch of an example of each function.
    a. identity function
    b. standard quadratic function
    c. standard cubic function
    d. absolute value function
    e. cube root function
    f. standard reciprocal
    g. standard reciprocal squared
    h. cosine function
    i. sine function
    j. tangent function
    k. secant function
    l. cosecant function
    m. cotangent function
    n. exponential function ($b > 1$)
    o. exponential function ($0 < b < 1$)
    p. logarithmic function ($b > 1$)
    q. logarithmic function ($0 < b < 1$)
    r. inverse cosine function
    s. inverse sine function
    t. inverse tangent function
    u. inverse secant function
    v. inverse cosecant function
    w. inverse cotangent function

12. What is a polynomial function?

13. What is a rational function?

14. a. What is an exponential function?
    b. How is such a function related to a logarithmic function?

15. a. What is a logarithmic function?
    b. What is a common logarithm?
    c. What is a natural logarithm?

16. a. Define an inverse function.
    b. What is the procedure for graphing the inverse of a given function?

17. What is the horizontal line test?

18. State the change-of-base theorem for logarithms.

19. State the inversion formulas for sine and tangent.

20. State the reciprocal inverse trigonometric identities.

Practice Problems

21. Find an equation for the lines satisfying the given conditions:
    a. through $\left(-\frac{1}{2}, 5\right)$ with slope $m = -\frac{3}{4}$
    b. through $(-3, 5)$ and $(7, 2)$
    c. with x-intercept $(4, 0)$ and y-intercept $\left(0, -\frac{3}{7}\right)$
    d. through $\left(-\frac{1}{2}, 5\right)$ and parallel to the line $2x + 5y - 11 = 0$
    e. the perpendicular bisector of the line segment joining $P(-3, 7)$ and $Q(5, 1)$

Sketch the graph of each of the equations in Problems 22–33.

22. $3x + 2y - 12 = 0$ 23. $y - 3 = |x + 1|$

24. $y - 3 = -2(x - 1)^2$ 25. $y = x^2 - 4x - 10$

26. $y = 2\cos(x - 1)$ 27. $y + 1 = \tan(2x + 3)$

28. $y = \sin^{-1}(2x)$ 29. $y = \tan^{-1}x^2$

30. $y = e^{-x} + e^x$ 31. $y = \ln(1 - x)$

32. $y = e^{2x} + \ln x$ 33. $y = e^x - \ln x + 15$

34. If $f(x) = \dfrac{1}{x + 1}$, what value(s) of x satisfy
$$f\left(\frac{1}{x + 1}\right) = f\left(\frac{2x + 1}{2x + 4}\right)?$$

35. $f(x) = \sqrt{\dfrac{x}{x - 1}}$ and $g(x) = \dfrac{\sqrt{x}}{\sqrt{x - 1}}$, does $f = g$? Why or why not?

36. If $f(x) = \sin x$ and $g(x) = \sqrt{1 - x^2}$, find the composite functions $f \circ g$ and $g \circ f$.

37. Solve: $\log_2 x + \log_3 x^2 = 5$

38. Solve: $e^{2x} = 2e^x + 3$

39. An open box with a square base is to be built for $96. The sides of the box will cost $3/ft^2 and the base will cost $8/ft^2. Express the volume of the box as a function of the length of its base.

40. How quickly will $2,000 grow to $5,000 when invested at an annual rate of 8% if interest is compounded:
    a. quarterly?   b. monthly?   c. continuously?

Supplementary Problems*

In Problems 1–3, find the perimeter and area of the given figure.

1. the right triangle with vertices $(-1,3),(-1,8)$, and $(11,8)$
2. the triangle bounded by the lines $y = 5, 3y - 4x = 11$, and $12x + 5y = 25$
3. the trapezoid with vertices $A(-3,0), B(5,0), C(2,8)$, and $D(0,8)$

In Problems 4–7, find an equation for the indicated line or circle.

4. the vertical line through the point where the line $y = 2x - 7$ is tangent to the parabola $y = x^2 + 6x - 3$
5. the circle that is tangent to the x-axis and is centered at $(5,4)$
6. the circle with center on the y-axis that passes through the origin and is tangent to the line $3x + 4y - 40 = 0$
7. the line through the two points where the circles $x^2 + y^2 - 5x + 7y = 3$ and $x^2 + y^2 + 4y = 0$ intersect
8. Find constants A and B so that $\tan(x + \frac{\pi}{3}) = \dfrac{A + \tan x}{1 + B \tan x}$.
9. Find constants A and B so that $\sin^3 x = A \sin 3x + B \sin x$.
10. In a triangle, the perpendicular segment drawn from a given vertex to the opposite side is called the *altitude* on that side. Consider the triangle with vertices $(-2, 1)$, $(5, 6)$, and $(3, -2)$. Find an equation for each line containing an altitude of this triangle.
11. Show that the three lines found in Problem 10 intersect at the same point. Find the coordinates of this point. Is this the same point where the three medians meet?
12. Let $f(x) = x^2 + 5x - 9$. For what values of x is it true that $f(2x) = f(3x)$?
13. If an object is shot upward from the ground with an initial velocity of 256 ft/sec, its distance in feet above the ground at the end of t seconds is given by $s(t) = 256t - 16t^2$ (neglecting air resistance). What is the highest point for this projectile?
14. It is estimated that t years from now, the population of a certain suburban community will be

$$P(t) = \frac{11t + 12}{2t + 3}$$

thousand people. What is the current population of the community? What will the population be in 6 years? When will there be 5,000 people in the community?

15. Evaluate each of the given numbers (calculator approximations).
    a. $\ln 4.5$ b. $e^{2.8}$ c. $\tan^{-1} 2$ d. $\sec^{-1}(-3.1)$

_____
*The supplementary problems are presented in a somewhat random order, not necessarily in order of difficulty.

16. Evaluate each of the given numbers (exact values).
    a. $e^{\ln \pi}$ b. $\ln e^{\sqrt{2}}$ c. $\sin(\cos^{-1} \frac{\sqrt{5}}{4})$
17. Find the exact value of $\sin(2 \tan^{-1} 3)$.
18. Find the exact value of $\sin(\cos^{-1} \frac{3}{5} + \sin^{-1} \frac{5}{13})$.

Solve each equation in Problems 19–26 for x.

19. $4^{x-1} = 8$
20. $2^{x^2+4x} = \frac{1}{16}$
21. $\log_2 2^{x^2} = 4$
22. $\log_4 \sqrt{x(x - 15)} = 1$
23. $\log_2 x + \log_2(x - 15) = 4$
24. $3^{2x-1} = 6^x 3^{1-x}$
25. $\ln(x - 1) + \ln(x + 1) = 2 \ln \sqrt{12}$
26. $\sqrt{x} = \cos^{-1} 0.317 + \sin^{-1} 0.317$

In Problems 27–30, find f^{-1}, if it exists.

27. $f(x) = 2x^3 - 7$
28. $f(x) = \sqrt[7]{2x + 1}, x \geq -\frac{1}{2}$
29. $f(x) = \sqrt{e^x - 1}, x \geq 0$
30. $f(x) = \dfrac{x + 5}{x - 7}, x \neq 7$
31. Show that for any $a \neq 1$, the function $f(x) = \dfrac{x + a}{x - 1}$ is its own inverse.
32. Let $f(x) = \dfrac{ax + b}{cx + d}$. Find $f^{-1}(x)$ in terms of a, b, c, and d. Under what conditions does f^{-1} exist?
33. Find $f^{-1}(x)$ if $f(x) = \dfrac{x + 1}{x - 1}$. What is the domain of f^{-1}?
34. First show that $\tan^{-1} x + \tan^{-1} y = \tan^{-1}\left(\dfrac{x + y}{1 - xy}\right)$ for $xy \neq 1$ whenever $-\dfrac{\pi}{2} < \tan^{-1}\left(\dfrac{x + y}{1 - xy}\right) < \dfrac{\pi}{2}$. Then establish the following equations.
    a. $\tan^{-1}(\frac{1}{2}) + \tan^{-1}(\frac{1}{3}) = \frac{\pi}{4}$
    b. $2 \tan^{-1}(\frac{1}{3}) + \tan^{-1}(\frac{1}{7}) = \frac{\pi}{4}$
    c. $4 \tan^{-1}(\frac{1}{5}) - \tan^{-1}(\frac{1}{239}) = \frac{\pi}{4}$
    Note: The identity in part *c* will be used in Chapter 8 to estimate the value of π.
35. **THINK TANK PROBLEM:** Each of the following equations may be either true or false. In each case, either show that the equation is generally true or find a counterexample.
    a. $\tan^{-1} x = \dfrac{\sin^{-1} x}{\cos^{-1} x}$
    b. $e^{1/x} = \dfrac{1}{e^x}$ for $x > 1$
    c. $\tan^{-1} x = \dfrac{1}{\tan x}$
    d. $\cot^{-1} x = \dfrac{\pi}{2} - \tan^{-1} x$
    e. $\cos(\sin^{-1} x) = \sqrt{1 - x^2}$
    f. $\sec^{-1}\left(\dfrac{1}{x}\right) = \cos^{-1} x$

36. First sketch the graph of $y = -x^2 + 5x - 6$. Next, use your graph to obtain the graphs of the related functions:
a. $y = -x^2 + 5x$
b. $y = x^2 - 5x + 6$
c. $y = -(x + 1)^2 + 5(x + 1) - 6$

37. A manufacturer of lightbulbs estimates that the fraction $F(t)$ of bulbs that remain burning after t weeks is given by

$$F(t) = e^{-kt}$$

where k is a positive constant. Suppose twice as many bulbs are burning after 5 wk as after 9 wk.
a. Find k and determine the fraction of bulbs still burning after 7 wk.
b. What fraction of the bulbs burn out before 10 wk?
c. What fraction of the bulbs can be expected to burn out between the 4th and 5th weeks?

38. A bus charter company offers a travel club the following arrangements: If no more than 100 people go on a certain tour, the cost will be $500 per person, but the cost per person will be reduced by $4 for each person in excess of 100 who takes the tour.
a. Express the total revenue R obtained by the charter company as a function of the number of people who go on the tour.
b. Sketch the graph of R. Estimate the number of people that results in the greatest total revenue for the charter company.

39. How much should you invest now at an annual interest rate of 6.25% so that your balance 10 years from now will be $2,000 if interest is compounded
a. quarterly?   b. continuously?

40. First National Bank pays 7% interest compounded monthly and Fells Cargo Bank pays 6.95% interest compounded continuously. Which bank offers the better deal?

41. Many materials, such as brick, steel, aluminum, and concrete, expand with increases in temperature. This is why spaces are placed between the cement slabs in sidewalks. Suppose you have a 100-ft length of material securely fastened at both ends, and assume that the buckle is linear. (It is not, but this assumption will serve as a worthwhile approximation.) If the height of the buckle is x ft and the percentage of swelling is y, then x and y are related as shown in Figure 1.39.

Find the amount of buckling (to the nearest inch) for the following materials:
a. brick; $y = 0.03$ [This means $(0.03\%)(100 \text{ ft}) = 0.03$ ft, which is y in Figure 1.39.]
b. steel; $y = 0.06$
c. aluminum; $y = 0.12$ d. concrete; $y = 0.05$

42. A ball has been dropped from the top of a building. Its height (in feet) after t seconds is given by the function

$$h(t) = -16t^2 + 256$$

a. How high will the ball be after 2 sec?
b. How far will the ball travel during the third second?
c. How tall is the building?
d. When will the ball hit the ground?

43. Suppose the number of worker-hours required to distribute new telephone books to x percent of the households in a certain rural community is given by the function

$$f(x) = \frac{600x}{300 - x}$$

a. What is the domain of the function f?
b. For what values of x does $f(x)$ have a practical interpretation in this context?
c. How many worker-hours were required to distribute new telephone books to the first 50% of the households?
d. How many worker-hours were required to distribute new telephone books to the entire community?
e. What percentage of the households in the community had received new telephone books by the time 150 worker-hours had been expended?

44. Find the area of each of the following plane figures:
a. the circle with $P(0,0)$ and $Q(2,3)$ endpoints of a diameter
b. the trapezoid with vertices $A(0,0)$, $B(4,0)$, $C(1,3)$, and $D(2,3)$

45. Find the volume and the surface area of each of the following solid figures:
a. a sphere with radius 4
b. a rectangular parallelepiped (box) with sides of length 2, 3, and 5
c. the right circular cylinder (including top and bottom) with height 4 and radius 2
d. the inverted cone with height 5 and top radius 3 (lateral surface only)

46. Consider the triangle with vertices $A(-1, 4)$, $B(3, 2)$, and $C(3, -6)$. Determine the midpoints M_1 and M_2 of sides $\overline{AB}$ and $\overline{AC}$, respectively, and show that the line segment $\overline{M_1 M_2}$ is parallel to side $\overline{BC}$ with half its length.

47. Generalize the procedure of Problem 46 to show that the line segment joining the midpoints of any two sides of a given triangle is parallel to the third side and has half its length.

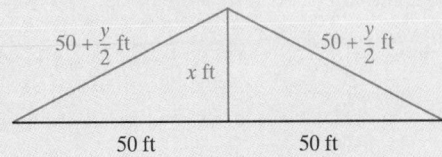

■ **FIGURE 1.39** Buckling of a given material

48. Let $ABCD$ be a quadrilateral in the plane, and let $P, Q, R,$ and S be the midpoints of sides $\overline{AB}, \overline{BC}, \overline{CD},$ and $\overline{DA}$, as shown in Figure 1.40. Show that $PQRS$ is a parallelogram.

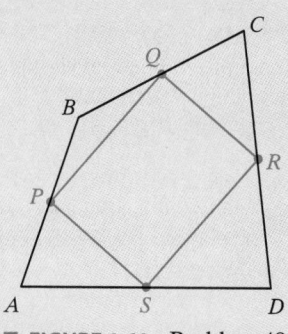

■ **FIGURE 1.40**   Problem 48

49. Find a constant c that guarantees that the graph of the equation

$$x^2 + xy + cy = 4$$

will have a y-intercept of $(0, -5)$. What are the x-intercepts of the graph?

50. A manufacturer estimates that when the price for each unit is p dollars, the profit will be $N = -p^2 + 14p - 48$ thousand dollars. Sketch the graph of the profit formula and answer these questions.
  a. For what values of p is this a profitable operation? (That is, when is $N > 0$?)
  b. What price results in maximum profit? What is the maximum profit?

51. A mural 7 feet high is hung on a wall in such a way that its lower edge is 5 feet higher than the eye of an observer standing 12 feet from the wall. (See Figure 1.41.) Find the angle θ subtended by the mural at the observer's eye.

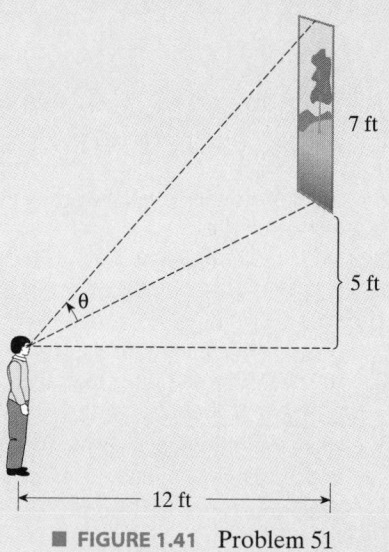

■ **FIGURE 1.41**   Problem 51

52. In Figure 1.42, ship A is at point P at noon and sails due east at 9 km/hr. Ship B arrives at point P at 1:00 P.M. and sails at 7 km/hr along a course that forms an angle of 60° with the course of ship A. Find a formula for the distance $s(t)$ separating the ships t hours after noon. Approximately how far apart (to the nearest kilometer) are the ships at 4:00 P.M.?

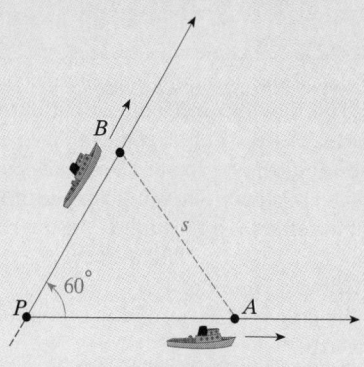

■ **FIGURE 1.42**   Problem 52

53. Under the provisions of a proposed property tax bill, a homeowner will pay $100 plus 8% of the assessed value of the house. Under the provisions of a competing bill, the homeowner will pay $1,900 plus 2% of the assessed value. If only financial considerations are taken into account, how should a homeowner decide which bill to support?

54. Two jets bound for Los Angeles leave New York 30 minutes apart. The first travels 550 mph, and the second goes 650 mph. How long will it take the second plane to pass the first?

55. To raise money, a service club has been collecting used bottles that it plans to deliver to a local glass company for recycling. Since the project began 8 days ago, the club has collected 2,400 pounds of glass, for which the glass company currently offers 15¢ per pound. However, since bottles are accumulating faster than they can be recycled, the company plans to reduce the price it pays by 1¢ per pound each day until the price reaches 0¢ fifteen days from now. Assuming that the club can continue to collect bottles at the same rate and that transportation costs make more than one trip to the glass company unfeasible, express the club's revenue from its recycling project as a function of the number of additional days the project runs. Draw the graph and estimate when the club should conclude the project and deliver the bottles to maximize its revenue.

56. An open box with a square base is to be built for $48. The sides of the box will cost $3/ft^2 and the base will cost $4/ft^2. Express the volume of the box as a function of the length of its base.

57. A closed box with a square base is to have a volume of 250 cubic feet. The material for the top and bottom of the box costs $2/ft², and the material for the sides costs $1/ft². Express the construction cost of the box as a function of the length of its base.

58. The famous author John Uptight must decide between two publishers who are vying for the rights to his new book, *Zen and the Art of Taxidermy*. Publisher A offers royalties of 1% of net proceeds on the first 30,000 copies sold and 3.5% on all copies in excess of that figure, and expects to net $2 on each copy sold. Publisher B will pay no royalties on the first 4,000 copies sold but will pay 2% on the net proceeds of all copies sold in excess of 4,000 copies, and expects to net $3 on each copy sold.
   a. Express the revenue John should expect if he signs with Publisher A (as a function P_A of the number of books sold, x). Likewise, find the revenue function P_B associated with Publisher B.
   b. Sketch the graphs of $P_A(x)$ and $P_B(x)$ on the same coordinate axes.
   c. For what values of x are the two offers equivalent?

   d. With whom should he sign if he expects to sell 4,000 copies? 5,000 copies?
   e. State a simple criterion for determining which publisher he should choose if he expects to sell N copies.

*59. PUTNAM EXAMINATION PROBLEM: If f and g are real-valued functions of one real variable, show that there exist numbers x, y such that

$$0 \le x \le 1, 0 \le y \le 1, \text{ and } |xy - f(x) - g(y)| \ge \tfrac{1}{4}$$

*60. PUTNAM EXAMINATION PROBLEM: Consider a polynomial $f(x)$ with real coefficients having the property $f(g(x)) = g(f(x))$ for every polynomial $g(x)$ with real coefficients. What is $f(x)$?

———

*The Putnam Examination is a national annual examination given under the auspices of the Mathematical Association of America and is designed to recognize mathematically talented college and university students. We include Putnam examination problems, which should be considered optional, to encourage students to consider taking the examination. Putnam problems used by permission of the American Mathematical Association.

2 Limits and Continuity

PREVIEW

We begin this chapter by asking the question, "What Is Calculus?" We shall see that calculus is used to model many aspects of the world about us, so we spend a little time discussing what is meant by mathematical modeling. The necessary prerequisites for this course are reviewed in Chapter 1 and in a companion book, *Student Mathematics Handbook*.

An essential feature of calculus involves making "infinitesimally small" changes in a quantity. We give precise meaning to this notion by introducing and exploring the *limit of a function* and a related concept known as *continuity*.

PERSPECTIVE

Change is a fact of our daily lives. Physical scientists use mathematical models to investigate phenomena such as the motion of planets, the decay of radioactive substances, chemical reaction rates, ocean currents, and weather patterns. Economists and business managers examine consumer trends; psychologists study learning tendencies; and ecologists explore patterns of pollution and population changes involving complex relationships among species. Even areas such as political science and medicine use mathematical models in which change is the key ingredient.

Calculus involves three great ideas: limit, derivative, and integral. The limit concept is discussed in this chapter. This concept is of fundamental importance because both the derivative and the integral use the limit as an essential part of their definitions.

2.1 What Is Calculus?

the limit: Zeno's paradox; the derivative: the tangent problem; the integral: the area problem; introduction to mathematical modeling ■

If there is an event that marked the coming of age of mathematics in Western culture, it must surely be the essentially simultaneous development of the calculus by Newton and Leibniz in the seventeenth century.* Before this remarkable synthesis, mathematics had often been viewed as merely a strange but harmless pursuit, indulged in by those with an excess of leisure time. After calculus, mathematics became virtually the only acceptable language for describing the physical universe. This view of mathematics and its association with the scientific method has come to dominate the Western view of how the world ought to be explained. This domination is so complete that it is virtually impossible for us to understand how earlier cultures explained what happened around them.

ELEMENTARY MATHEMATICS

1. Slope of a line

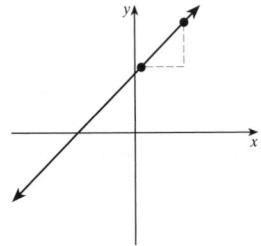

2. Tangent line to a circle

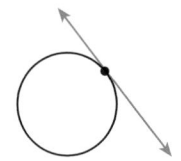

3. Area of a region bounded by line segments

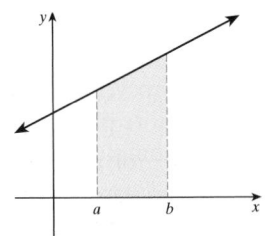

4. Average changes in position and velocity

5. Average of a finite collection of numbers

■ **FIGURE 2.1** Topics from elementary mathematics

What distinguishes calculus from your previous mathematics courses of algebra, geometry, and trigonometry is the transition from static or discrete applications (see Figure 2.1) to those that are dynamic or continuous (see Figure 2.2). For example, in elementary mathematics you considered the slope of a line, but in calculus we define the (nonconstant) slope of a nonlinear curve. In elementary mathematics you found average changes in quantities such as the position and velocity of a moving object, but in calculus we can find instantaneous changes in the same quantities. In elementary mathematics you found the average of a finite collection of numbers, but in calculus we can find the average value of a function with infinitely many values over an interval.

You might think of calculus as the culmination of all of your mathematical studies. To a certain extent that is true, but it is also the beginning of your study of mathematics as it applies to the real world around us. Calculus is a three-semester or four-quarter course that *begins* your college work in mathematics. All your prior work in mathematics is considered elementary mathematics, with calculus the dividing line between elementary mathematics and mathematics as it is used in a variety of theoretical and applied topics.

Calculus is the mathematics of motion and change. Its development in the seventeenth century by Newton and Leibniz was the result of their attempts to answer some fundamental questions about the world and the way things work. These investigations led to two fundamental concepts—namely, the *derivative* and the *integral*. The breakthrough in the development of these concepts was the formulation of a mathematical tool called a *limit*.

1. **Limit:** The limit is a mathematical tool for studying the *tendency* of a function as its variable *approaches* some value. Calculus is based on the concept of limit. We introduce the limit of a function informally in Section 2.2 and then examine the concept more formally in the proofs of Appendix B.

2. **Derivative:** The derivative is defined as a limit, and it is used initially to compute rates of change and slopes of tangent lines to curves. The study of derivatives is called *differential calculus*. Derivatives can be used in sketching graphs and in finding the extreme (largest and smallest) values of functions. The derivative is introduced and developed in Chapter 3, and its applications are examined in Chapter 4.

3. **Integral:** The integral is found by taking a special limit of a sum of terms, and the study of this process is called *integral calculus*. Area, volume, arc length,

*See the guest essay at the end of this chapter.

work, and hydrostatic force are a few of the many quantities that can be expressed as integrals. Integrals and their applications are studied in Chapters 5 and 6.

We will briefly describe each of these concepts later in this section.

One might naturally suppose that an event so momentous must involve ideas so profound that average mortals can hardly hope to comprehend them. In fact, nothing could be further from the truth. The essential ideas of calculus—the derivative and the integral—are quite straightforward and had been known prior to either Newton or Leibniz. The contribution of Newton and Leibniz was to recognize that the idea of finding tangents (the derivative) and the idea of finding areas (the integral) are related and that this relation can be used to give a simple and unified description of both processes.

Let us begin by taking an intuitive look at each of these three essential ideas of calculus.

THE LIMIT: ZENO'S PARADOX

In the guest essay at the end of this chapter, John Troutman mentions Zeno's paradoxes, which are concerned with infinite processes. Zeno (ca. 500 B.C.) was a Greek philosopher who is known primarily for his famous paradoxes. One of those concerns a race between Achilles, a legendary Greek hero, and a tortoise. When the race begins, the (slower) tortoise is given a head start, as shown in Figure 2.3.

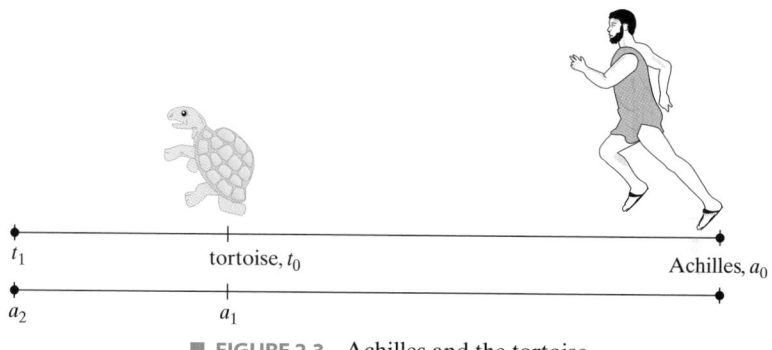

■ **FIGURE 2.3**   Achilles and the tortoise

Is it possible for Achilles to overtake the tortoise? Zeno pointed out that by the time Achilles reaches the tortoise's starting point, $a_1 = t_0$, the tortoise will have moved ahead to a new point t_1. When Achilles gets to this next point, a_2, the tortoise will be at a new point t_2. The tortoise, even though much slower than Achilles, keeps moving forward. Although the distance between Achilles and the tortoise is getting smaller and smaller, the tortoise will apparently always be ahead.

Of course, common sense tells us that Achilles will overtake the slow tortoise, but where is the error in reasoning? The error is in the assumption that the sum of an infinite number of finite time intervals must itself be infinite. This discussion is getting at an essential idea in calculus, the notion of a limit.

Consider the successive positions for both Achilles and the tortoise:

Starting position

Achilles:	$a_0,$	$a_1,$	$a_2,$	$a_3,$	$a_4, \ldots$	
Tortoise:	$t_0,$	$t_1,$	$t_2,$	$t_3,$	$t_4, \ldots$	

CALCULUS

1. Slope of a curve

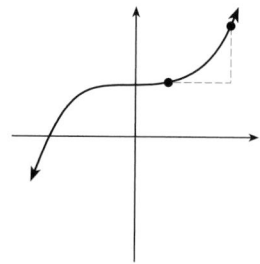

2. Tangent line to a general curve

3. Area of a region bounded by curves

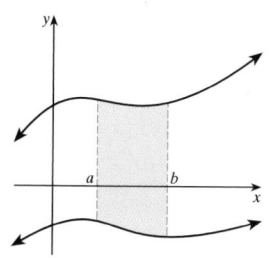

4. Instantaneous changes in position and velocity
5. Average of an infinite collection of numbers

■ **FIGURE 2.2**   Topics from calculus

Zeno was concerned with three problems. . . . These are the problem of the infinitesimal, the infinite, and continuity. . . . From him to our own day, the finest intellects of each generation in turn attacked these problems but achieved broadly speaking nothing. . . . The problem of the infinitesimal was solved by Weierstrass, the solution of the other two was begun by Dedekind and definitely accomplished by Cantor.

BERTRAND RUSSELL
INTERNATIONAL MONTHLY, 1901

The positions for Achilles, as well as those for the tortoise, are ordered with positive integers. Such ordered listings are called *sequences.*

For Achilles and the tortoise, we have two sequences, $\{a_1, a_2, a_3, \ldots, a_n, \ldots\}$ and $\{t_1, t_2, t_3, \ldots, t_n, \ldots\}$, where $a_n < t_n$ for all values of n. We will see in Chapter 8 that both the sequence for Achilles' position and the sequence for the tortoise's position have limits, and it is precisely at that limit point that Achilles overtakes the tortoise.

The idea of limit is introduced in this chapter and is used to define the other two basic concepts of calculus: the derivative and the integral. Even if the solution to Zeno's paradox using limits seems unnatural at first, do not be discouraged. It took over 2,000 years to refine the ideas of Zeno and provide conclusive answers to those questions about limits that will be introduced in this chapter. We revisit Zeno's paradoxes in Problem Set 8.2.

EXAMPLE 1   *An intuitive preview of a limit*

The sequence $\frac{1}{2}, \frac{2}{3}, \frac{3}{4}, \frac{4}{5}, \ldots$ can be described by writing a *general term*: $\frac{n}{n+1}$ where $n = 1, 2, 3, 4, \ldots$. Can you guess the limit, L, of this sequence?

Solution   PREVIEW   The *limit* is an important idea in calculus, and we discuss this concept extensively later in this chapter. We will say that L is the number that the sequence with general term $\frac{n}{n+1}$ tends toward as n becomes large without bound. We will define a notation to summarize this idea:

$$L = \lim_{n \to \infty} \frac{n}{n+1}$$

As you consider larger and larger values for n, you find a sequence of fractions:

$$\frac{1}{2}, \frac{2}{3}, \frac{3}{4}, \ldots, \frac{1{,}000}{1{,}001}, \frac{1{,}001}{1{,}002}, \ldots, \frac{9{,}999{,}999}{10{,}000{,}000}, \cdots$$

It is reasonable to guess that the sequence of fractions is approaching the number 1.   ∎

THE DERIVATIVE: THE TANGENT PROBLEM

A **tangent line** (or, if the context is clear, simply *tangent*) to a circle at a given point P is a line that intersects the circle at P and only at P. This characterization does not apply for curves in general, as you can see by looking at Figure 2.4.

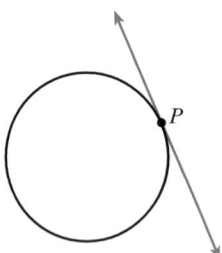

At each point P on a circle, there is only one line that intersects the circle exactly once.

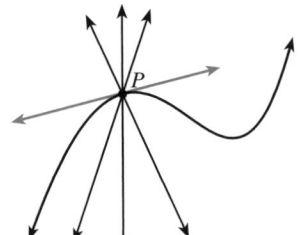

At a point P on a curve, there may be several lines that intersect the curve only once.

■ **FIGURE 2.4**   Tangent line

To find a tangent line, begin by considering a line that passes through two points on the curve, as shown in Figure 2.5a. This line is called a **secant line.**

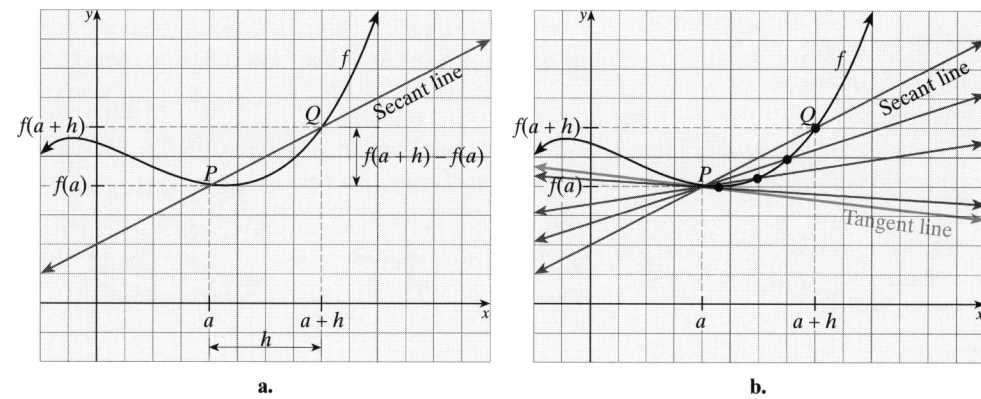

a.   **b.**

■ **FIGURE 2.5**   Secant line

The coordinates of the two points P and Q are $P(a, f(a))$ and $Q(a + h, f(a + h))$. The slope of the secant line is

$$m = \frac{f(a + h) - f(a)}{h}$$

Recall from Chapter 1 that the slope of any line is defined to be rise/run. Can you see how this formula for the slope of the secant line fits the model of "rise/run"? Now imagine that Q moves along the curve toward P, as shown in Figure 2.5b. You can see that the secant line approaches a limiting position as h approaches zero. In Section 3.1 we define this limiting position to be the tangent line. The slope of the tangent line is defined as a limit of the sequence of slopes of a set of secant lines.

> **PREVIEW**   Once again, we can use limit notation to summarize this idea:
> We say that the slopes of the secant lines, as h becomes small, tend toward a number that we call the slope of the tangent line. We will define the following notation to summarize this idea:
>
> $$\begin{bmatrix} \text{slope of} \\ \text{tangent at } P \end{bmatrix} = \lim_{h \to 0} \frac{f(a + h) - f(a)}{h}$$

THE INTEGRAL: THE AREA PROBLEM

You probably know the formula for the area of a circle with radius r:

$$A = \pi r^2$$

The Egyptians were the first to find areas of circles over 5,000 years ago, but the Greek Archimedes (ca. 200 B.C.; see Historical Quest Problem 54) showed how to derive the formula for the area of a circle by using a limiting process. Consider the area of inscribed polygons, as shown in Figure 2.6.

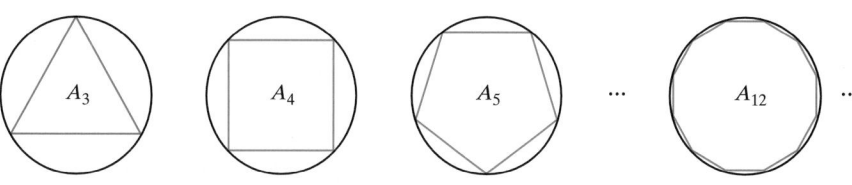

■ **FIGURE 2.6**   Approximating the area of a circle

Even though Archimedes did not use the following notation, here is the essence of what he did, using a method called "exhaustion":

Let A_3 be the area of the inscribed equilateral triangle;

A_4 be the area of the inscribed square; and

A_5 be the area of the inscribed regular pentagon.

How can we find the area of this circle? As you can see from Figure 2.6, if we consider the area of A_3, then A_4, then $A_5, \ldots$, we should have a sequence of areas such that each successive area more closely approximates that of the circle. Later in this book, we will write this idea as a limit statement:

$$A = \lim_{n \to \infty} A_n$$

In this course we will use limits in yet a different way to find the area of regions enclosed by curves. For example, consider the area shown in color in Figure 2.7. We can approximate the area by using rectangles (see Figure 2.8). If R_n is the area of the nth rectangle, then the total area can be approximated by finding the sum

$$R_1 + R_2 + R_3 + \cdots + R_{n-1} + R_n$$

This process is shown in Figure 2.8.

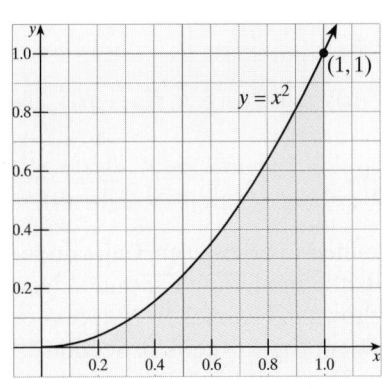

■ **FIGURE 2.7**   Area under a curve

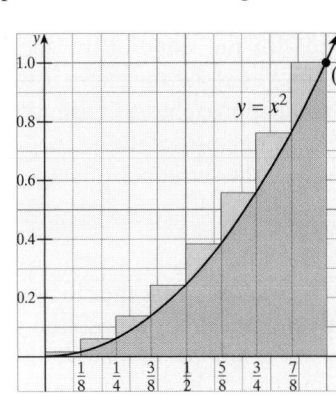

a. 8 approximating rectangles

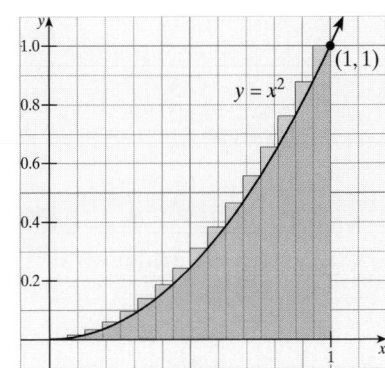

b. 16 approximating rectangles

■ **FIGURE 2.8**   Approximating the area using circumscribed rectangles

The area problem leads to a process called *integration,* and the study of integration forms what is called **integral calculus.** Similar reasoning allows us to calculate such things as volumes, the length of a curve, the average value of a function, or the amount of work required for a particular task.

INTRODUCTION TO MATHEMATICAL MODELING

A real-life situation is usually far too complicated to be precisely and mathematically defined. When confronted with a problem in the real world, therefore, it is usually necessary to develop a mathematical framework based on certain assumptions about the real world. This framework can then be used to find a solution to the real-world problem. The process of developing this body of mathematics is referred to as **mathematical modeling.**

Some mathematical models are quite accurate, particularly those used in the physical sciences. For example, one of the first models we will consider in calculus is a model for the path of a projectile. Other rather precise models predict such things as the time of sunrise and sunset, or the speed at which an object falls in a vacuum. Other mathematical models, however, are less accurate, especially those that involve

HOW GLOBAL CLIMATE IS MODELED

We find a good example of mathematical modeling by looking at the work being done with weather prediction. In theory, if the correct assumptions could be programmed into a computer, along with appropriate mathematical statements of the ways global climate conditions operate, we would have a model to predict the weather throughout the world. In the global climate model, a system of equations calculates time-dependent changes in wind as well as temperature and moisture changes in the atmosphere and on the land. The model may also predict alterations in the temperature of the ocean's surface. At the National Center for Atmospheric Research, a CRAY supercomputer is used to do this modeling.

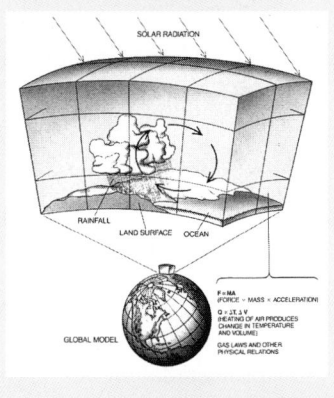

FROM *SCIENTIFIC AMERICAN*, MARCH 1991.

examples from the life sciences and social sciences. Only recently has modeling in these disciplines become precise enough to be expressed in terms of calculus.

What, precisely, is a mathematical model? Sometimes, mathematical modeling can mean nothing more than a textbook word problem. But mathematical modeling can also mean choosing appropriate mathematics to solve a problem that has previously been unsolved. In this book, we use the term mathematical modeling to mean something between these two extremes. That is, it is a process we will apply to some real-life problem that does not have an obvious solution. It usually cannot be solved by applying a single formula.

The first step of what we call mathematical modeling involves *abstraction*.

With the method of abstraction, certain assumptions about the real world are made, variables are defined, and appropriate mathematics is developed. The next step is to simplify the mathematics or derive related mathematical facts from the mathematical model.

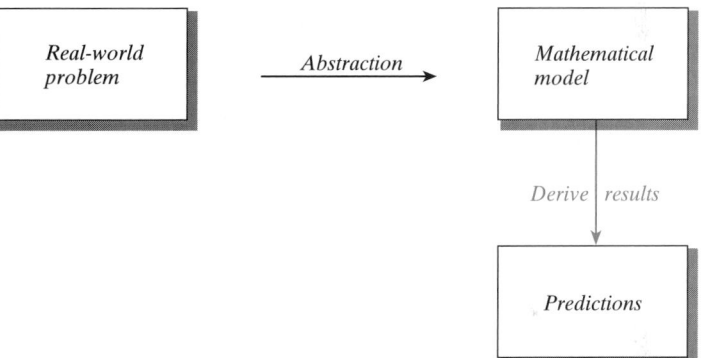

The results derived from the mathematical model should lead us to some predictions about the real world. The next step is to gather data from the situation being modeled and then to compare those data with the predictions. If the two do not agree, then the gathered data are used to modify the assumptions used in the model.

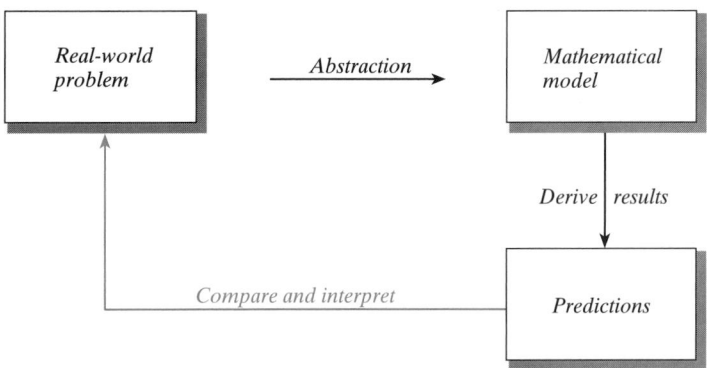

Mathematical modeling is an ongoing process. As long as the predictions match the real world, the assumptions made about the real world are regarded as correct, as are the defined variables. On the other hand, as discrepancies are noticed, it is necessary to construct a closer and more dependable mathematical model. You might wish to read the article from *Scientific American* quoted in the margin on page 79.

EXAMPLE 2 *Identifying models*

It is common to gather data when studying most phenomena. Classifying the data by comparing it with known models is an important mathematical skill. Identify each of the data sets as an example of a linear model, a quadratic model, a cubic model, a logarithmic model, an exponential model, or a sinusoidal model.

a. x	y	b. x	y	c. x	y	d. x	y	e. x	y
-4	2.00	-4	-20.7	0.0	0.0	-4	0.27	3	0.0
-3	1.28	-3	-16.5	1.0	1.7	-3	0.30	5	1.10
-2	0.72	-2	-12.3	1.5	2.0	-2	0.33	6	1.39
-1	0.32	-1	-8.2	2.5	1.2	-1	0.36	7	1.61
0	0.08	0	-3.8	3.5	-0.7	0	0.40	8	1.79
1	0.00	1	0.4	4.5	-1.9	1	0.44	9	1.95
2	0.08	2	4.7	5.5	-1.4	2	0.49	10	2.08
3	0.32	3	8.9	6.0	-0.6	3	0.52	11	2.20
4	0.72	4	13.12	6.5	0.4	4	0.60	12	2.30

Solution Sketch each graph and observe the pattern (if any) of the graphs.

a.

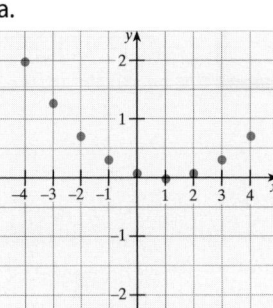

Quadratic model

b.

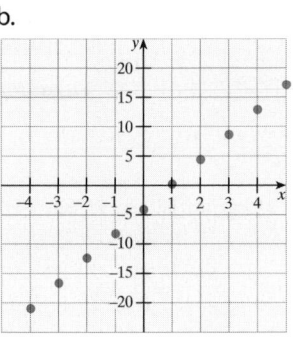

Linear model

c.

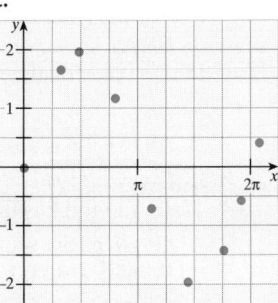

The data points seem to form a pattern, and we hypothesize that the data points are periodic; this is a sinusoidal model.

d.

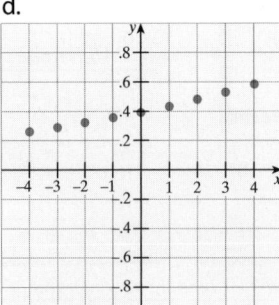

The data points look almost linear (but not quite), so we hypothesize that the data points may indicate an exponential model.

e.

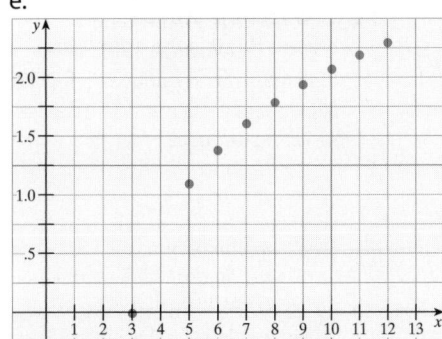

The data points seem to indicate a logarithmic model.

2.1 Problem Set

(A) 1.* ■ **What Does This Say?**[†] What is a mathematical model?

2. ■ **What Does This Say?** Why are mathematical models necessary or useful?

3. ■ **What Does This Say?** An analogy to Zeno's tortoise paradox can be made as follows. A woman standing in a room cannot walk to a wall. To do so, she would have to go half the distance, then half the remaining distance, and then again half of what still remains. This process can always be continued and thus never ends. Draw an appropriate figure for this problem and then present an argument using sequences to show that the woman will, indeed, reach the wall.

4. ■ **What Does This Say?** Zeno's paradoxes remind us of an argument that might lead to an absurd conclusion:

   Suppose I am playing baseball and decide to steal second base. To run from first to second base I must first go half the distance, then half the remaining distance, and then again half of what remains. This process is continued so that I never reach second base. Therefore it is pointless to steal base.

   Draw an appropriate figure for this problem and then present an argument using sequences to show that the conclusion is absurd.

5. Consider the sequence $0.3, 0.33, 0.333, 0.3333, \ldots$. What do you think is the appropriate limit of this sequence?

6. Consider the sequence $6, 6.6, 6.66, 6.666, 6.6666, \ldots$. What do you think is the appropriate limit of this sequence?

7. Consider the sequence $0.9, 0.99, 0.999, 0.9999, \ldots$. What do you think is the appropriate limit of this sequence?

8. Consider the sequence $9.9, 9.99, 9.999, 9.9999, \ldots$. What do you think is the appropriate limit of this sequence?

9. Consider the sequence $0.2, 0.27, 0.272, 0.2727, \ldots$. What do you think is the appropriate limit of this sequence?

10. Consider the sequence $0.4, 0.45, 0.454, 0.4545, \ldots$. What do you think is the appropriate limit of this sequence?

11. Consider the sequence $3, 3.1, 3.14, 3.141, 3.1415, 3.14159, 3.141592, \ldots$. What do you think is the appropriate limit of this sequence?

12. Consider the sequence $1, 1.4, 1.41, 1.414, 1.4142, 1.41421, 1.414213, 1.4142135, \ldots$. What do you think is the appropriate limit of this sequence?

13. Copy the following figures on your paper. Draw what you think is an appropriate tangent line for each curve at the point P.

    a.
    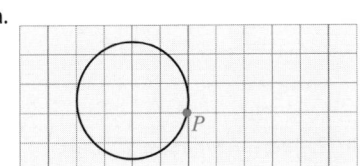

    b.
    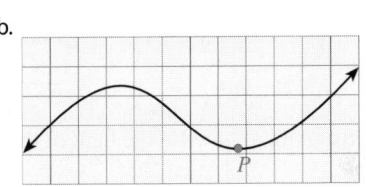

14. Copy the following figures on your paper. Draw what you think is an appropriate tangent line for each curve at the point P.

    a.   b.
    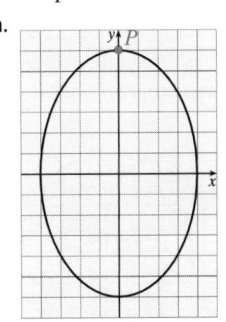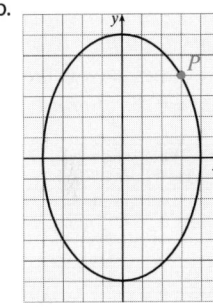

15. Copy the following figures on your paper. Draw what you think is an appropriate tangent line for each curve at the point P.

    a.
    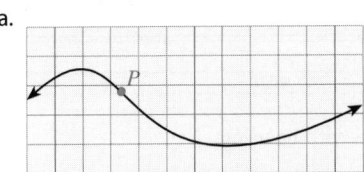

    b.
    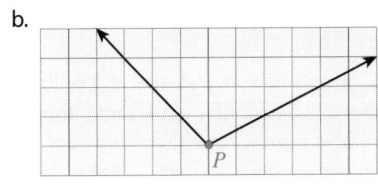

*Problem numbers marked in color denote problems with solutions in the *Student Survival Manual.* Answers to odd-numbered problems are given in Appendix E.

[†]Many problems in this book are labeled **What Does This Say?** Following the question will be a question for you to answer in your own words, or a statement for you to rephrase in your own words. These problems are intended to be similar to the "What This Says" boxes that appear throughout the book.

16. Copy the following figures on your paper. Draw what you think is an appropriate tangent line for each curve at the point P.

a.   b.

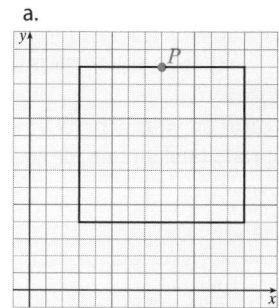

*In Problems 17–22, **guess** the requested limits.*

17. $\lim_{n \to \infty} \dfrac{2n}{n+4}$ **18.** $\lim_{n \to \infty} \dfrac{2n}{3n+1}$ **19.** $\lim_{n \to \infty} \dfrac{n+1}{n+2}$

20. $\lim_{n \to \infty} \dfrac{n+1}{2n}$ **21.** $\lim_{n \to \infty} \dfrac{3n}{n^2+2}$ **22.** $\lim_{n \to \infty} \dfrac{3n^2+1}{2n^2-1}$

23. Use a limit statement to write the limit of the sequence
$$2, \tfrac{3}{2}, \tfrac{4}{3}, \tfrac{5}{4}, \ldots$$

24. Use a limit statement to write the limit of the sequence
$$2, \tfrac{3}{4}, \tfrac{4}{9}, \tfrac{5}{16}, \tfrac{6}{25}, \ldots$$

In Problems 25–28, use the following calculator table display. Notice that the entries in the column headed x seem to be getting closer to 1. Using the patterns displayed by the table, give the anticipated values for the missing entry at the bottom of the requested column.

x	y1	y2	y3	y4
.91	1.0989	7.4945	1.8281	2.73
.92	1.087	7.4348	1.8464	2.76
.93	1.0753	7.3763	1.8649	2.79
.94	1.0638	7.3191	1.8836	2.82
.95	1.0526	7.2632	1.9025	2.85
.96	1.0417	7.2083	1.9216	2.88
.97	1.0309	7.1546	1.9409	2.91
.98	1.0204	7.102	1.9604	2.94
.99	1.0101	7.0505	1.9801	2.97
1.				

x=1.

MAIN   RAD AUTO   FUNC

25. the column with heading y_1

26. the column with heading y_2

27. the column with heading y_3

28. the column with heading y_4

29. a. Draw the graph of $y = x^2$. Label the points on this curve where $x_0 = 1$ and $x_1 = 3$. Draw the secant line through these points. Find the slope of this secant line.

b. Find the slope of the secant line passing through points on the curve $y = x^2$ where $x_0 = 1$ and $x_2 = 2$.

c. Complete the following table for finding the slope of various secant lines for $x_0 = 1$ and x_n as given. The answers for parts **a** and **b** have been filled in for you.

n	x_n	point	slope
1	3	$(3, 9)$	$m = 4$
2	2	$(2, 4)$	$m = 3$
3	1.5	$(1.5, 2.25)$	$m = ?$
4	1.1	$(1.1, 1.21)$	$m = ?$

d. On your graph from part **a**, draw a tangent line at $x_0 = 1$. Guess the slope of the tangent line. Compare this answer with the sequence of answers for slope in part **c**.

30. a. Draw the graph of $y = x^2$. Label the points on this curve where $x_0 = -2$ and $x_1 = 1$. Draw the secant line through these points. Find the slope of this secant line.

b. Find the slope of the secant line passing through points on the curve $y = x^2$ where $x_0 = -2$ and $x_2 = -1$.

c. Complete the following table for finding the slope of various secant lines for $x_0 = -2$ and x_n as given. The answers for parts **a** and **b** have been filled in for you.

n	x_n	point	slope
1	1	$(1, 1)$	$m = -1$
2	-1	$(-1, 1)$	$m = -3$
3	-1.5	$(-1.5, 2.25)$	$m = ?$
4	-1.9	$(-1.9, 3.61)$	$m = ?$

d. On your graph from part **a**, draw a tangent line at $x_0 = -2$. Guess the slope of the tangent line. Compare this answer with the sequence of answers for slope in part **c**.

In Problems 31–36, estimate the enclosed area in each figure.

31.

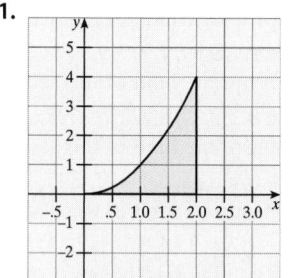

32.

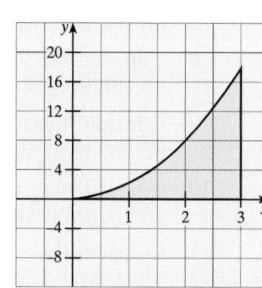

33.

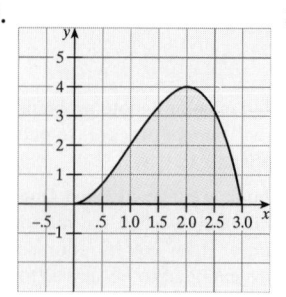

34.

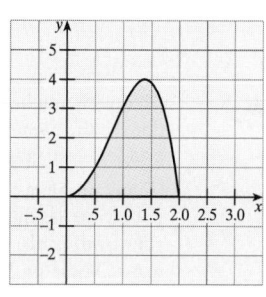

35. 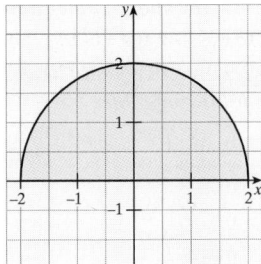   **36.**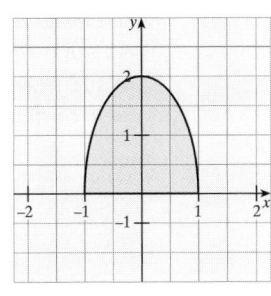

MODELING PROBLEMS   *Identify each of the data sets in Problems 37–46 as an example of a linear model, a quadratic model, a cubic model, a logarithmic model, an exponential model, or a sinusoidal model.*

37.

x	y
−4	−140
−3	−96
−2	−60
−1	−32
0	−12
1	0
2	4
3	0
4	−12

38.

x	y
−4	−10.2
−3	−8.0
−2	−5.8
−1	−3.6
0	−1.4
1	0.8
2	3.0
3	5.2
4	7.4

39.

x	y
−4	−0.19531
−3	−0.078125
−2	−0.3125
−1	−1.25
0	−5
1	−20
2	−80
3	−320
4	−1,280

40.

x	y
−4	−0.0003
−3	−0.0009
−2	−0.0027
−1	−0.0082
0	−0.0247
1	−0.0741
2	−0.2222
3	−0.6667
4	−2.0000

41.

x	y
−4	238
−3	173
−2	118
−1	73
0	38
1	13
2	−2
3	−7
4	−2

42.

x	y
−4	4.84
−3	2.46
−2	−1.41
−1	−2.42
0	−4.76
1	−2.20
2	1.68
3	4.55
4	4.66

43.

x	y
−2.0	−28.0
−1.5	−8.5
−1.0	1.0
−0.5	3.5
0.0	2.0
0.5	−0.5
1.0	−1.0
1.5	3.5
2.0	16.0

44.

x	y
−4	−3.5
−3	−2.4
−2	−1.3
−1	−0.2
0	0.9
1	2.0
2	3.1
3	4.2
4	5.3

45.

x	y
2	0.0
3	4.0
4	6.4
5	8.0
6	9.3
7	10.3
8	11.2
9	12.0
10	12.7

46.

x	y
1	4.2
2	5.6
3	6.4
4	7.0
5	7.4
6	7.8
7	8.1
8	8.4
9	8.6

B **47.** **H**ISTORICAL **Q**UEST   There are a number of Egyptian papyri that date back over five millennia. One of the most famous is called the Ahmes Papyrus after the Egyptian scribe who copied it about 1650 B.C. According to the historian Carl B. Boyer, the scribe tells us that the material is from the Middle Kingdom of about 2000 to 1800 B.C. This scroll is usually called the Rhind Papyrus because it was discovered when it was purchased in a Nile resort town by Henry Rhind.

A method for finding the area of a circle is described on this papyrus roll, a method considered to be one of the outstanding achievements of the time. This method has been designated Problem 50 of the papyrus. In this problem, Ahmes assumed that the area of a circular field with diameter of nine units is the same as the area of a square with a side of eight units.

Using Ahmes' assumptions, known today as the Egyptian rule for finding the area of a circle, comment on what must have been the Egyptians' approximation for π.

48. **H**ISTORICAL **Q**UEST   The Egyptian method for calculating the area of a circle (see Problem 47) may have been derived from a method illustrated in Problem 48 of the Rhind Papyrus. In this problem a scribe formed an octagon from a square of side nine units by trisecting the sides and cutting off the four corners of isosceles triangles, each having an area of $4\frac{1}{2}$ units.*

Compare the area of this octagon with the area of a circle inscribed within the square and with the area of a square with eight units on a side.

---

*Carl B. Boyer, *A History of Mathematics.* New York: John Wiley & Sons, 1968, p. 18.

49. Suppose the circle in Figure 2.6 has radius 1. We know that the area of the circle is $A = \pi(1)^2 = \pi$. Find the sequence of areas for the inscribed polygons $A_3, A_4, A_5, \ldots$, and show that these areas form a sequence of numbers that seems to have a limit π.

50. Repeat Problem 49 for a circle with radius 2. What is the apparent value of the new limit?

51. Calculate the sum of the areas of the rectangles shown in Figure 2.8a.

52. Calculate the sum of the areas of the rectangles shown in Figure 2.8b.

53. Use the results of Problems 51 and 52 to make a guess about the shaded area under the curve.

54. **HISTORICAL QUEST**

ARCHIMEDES OF SYRACUSE
287–212 B.C.

Archimedes was one of the greatest mathematicians of all time, and is known for his work with levers, floating bodies, spirals, as well as with all sorts of two- and three-dimensional geometrical figures. For this Historical Quest, we focus on the problem of approximating the area of a circle. Archimedes' approach uses what we have described as "exhaustion" or "method of compression." Instead of using only inscribed polygons (as shown in Figure 2.6), he uses both inscribed and circumscribed polygons. The area of the circle is then "compressed" between the areas of the inscribed and circumscribed polygons.

Archimedes showed that the area of a circle is equal to that of a triangle with base equal to its circumference and height equal to its radius. In modern symbols,

$$A = \tfrac{1}{2} rC$$

At the time Archimedes did his work, it was known that the area of a circle is proportional to the square of its radius; that is, $A = c_1 r^2$, for some constant c_1. Similarly, it was known that the circumference C and diameter d are related by $C = c_2 d$. Use Archimedes' formula to show that $c_1 = c_2$.

Archimedes then showed that

$$3\tfrac{10}{71} < \pi < 3\tfrac{1}{7}$$

by explicitly obtaining the ratio of the circumference of a circle to its diameter.*

55. **MODELING PROBLEM**   Suppose you roll a pair of dice 100 times. The possible outcomes are $2, 3, \ldots, 11, 12$. Several possible models are proposed:

*See C. H. Edwards, Jr., *The Historical Development of Calculus*, Springer-Verlag, 1979, pp. 31–35, for an outline of this method. Edwards cites an article by W. R. Knorr which states that Archimedes actually found a more accurate approximation to π by starting with inscribed and circumscribed polygons with 640 sides.

Model 1:   *The outcomes are equally likely. That is, if $P(x)$ is the probability that the sum of the tossed dice is x, then $P(2) = P(3) = P(4) = \cdots = P(12)$.*

Model 2:   *The outcomes are not equally likely, and occur according to the following table:*

Outcome	Probability
2	0.0278
3	0.0556
4	0.0833
5	0.1111
6	0.1389
7	0.1667
8	0.1389
9	0.1111
10	0.0833
11	0.0556
12	0.0278

a. Carry out the experiment described by this modeling problem. Which of the two models seems to be more descriptive?

b. Repeat the experiment described five more times. Submit a record of the results.

c. Do you think you can refine your answer to part **b** to build a better model? Describe this model.

MODELING PROBLEMS   *Suppose we want to build a closed rectangular box with a square base that will hold 1 gallon (231 $in.^3$). In Problems 56–57, we wish to find the size of the box that minimizes the cost.*

56. We find that the cost of each side of the box is \$1.00/in.^{2} for materials, and the cost for assembling the box is \$0.10/in. for gluing an edge. Construct a table showing the cost for the box for integral values of the base in the domain [1, 10]. Using this table of values, what is the size of the box that minimizes the cost?

57. a. If the cost of each side of the box is \$$a$/in.^{2} for materials, and the cost for assembling the box is \$$b$/in. for gluing an edge, write a formula showing the cost of the box.

b. Apply the formula to find the cost if each side of the box is \$1.00/in.^{2} for materials, and the cost for assembling the box is \$1.00/in. for gluing an edge. What is the size of the box that minimizes the cost?

c. Hypothesize the cost of materials for the side of a box that you would want to build in your town (other than \$1.00/in.^{2} used in part **b**). Also hypothesize the cost of assembling the box in your town (other than the \$1.00/in. used in part **b**).

d. Based on your results from parts **a** and **b,** make a conjecture about the effect of the cost of materials and assembly on the minimum cost for the construction of the box.

2.2 The Limit of a Function

intuitive notion of limits, limits by graphing, limits by table, limits that do not exist, limit of a function (formal definition) ∎

Our goal in this section is to give an introduction to the concept of the *limit of a function.*

The development of the limit concept was a major breakthrough in the history of mathematics, and it is unrealistic for you to expect to understand everything about this concept immediately. Have patience, read the examples carefully, and work as many problems as possible. Eventually, the limit concept will become a useful part of your mathematical toolkit.

INTUITIVE NOTION OF A LIMIT

The limit of a function f is a tool for investigating the behavior of $f(x)$ as x gets closer and closer to a particular number c. To visualize this concept, we begin with an example.

EXAMPLE 1   *Modeling velocity*

A freely falling body experiencing no air resistance falls $s(t) = 16t^2$ feet in t seconds. Express the body's velocity at time $t = 2$ as a limit.

Solution   We need to define some sort of "mathematical speedometer" for measuring the *instantaneous velocity* of the body at time $t = 2$. Toward this end, we first compute the *average velocity* $\bar{v}(t)$ of the body between time $t = 2$ and any other time t by the formula

$$\bar{v}(t) = \frac{\text{DISTANCE TRAVELED}}{\text{ELAPSED TIME}} = \frac{s(t) - s(2)}{t - 2}$$

$$= \frac{16t^2 - 16(2)^2}{t - 2} = \frac{16t^2 - 64}{t - 2}$$

As t gets closer and closer to 2, it is reasonable to expect the average velocity $\bar{v}(t)$ to approach the value of the required instantaneous velocity at time $t = 2$. We write

$$\lim_{t \to 2} \bar{v}(t) = \underbrace{\lim_{t \to 2} \frac{16t^2 - 64}{t - 2}}_{\text{This is the instantaneous velocity at } t = 2.}$$

Notice that we cannot find the instantaneous velocity at time $t = 2$ by simply substituting $t = 2$ into the average velocity formula because this would yield the meaningless form 0/0. ∎

Next, we introduce an intuitive definition of limit.

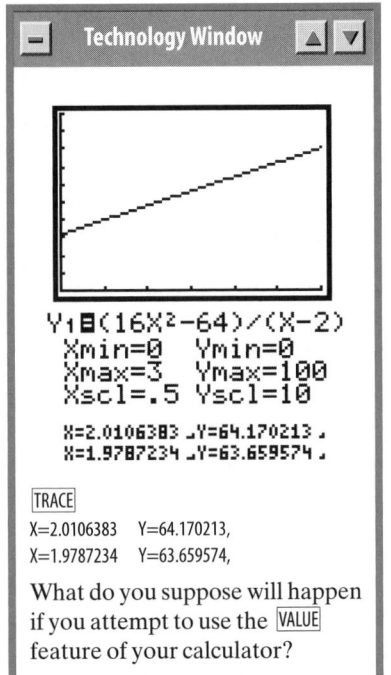

Technology Window ▲ ▼

Y₁⊟(16X²-64)/(X-2)
Xmin=0   Ymin=0
Xmax=3   Ymax=100
Xscl=.5   Yscl=10

X=2.0106383 .Y=64.170213 .
X=1.9787234 .Y=63.659574 .

[TRACE]
X=2.0106383   Y=64.170213,
X=1.9787234   Y=63.659574,

What do you suppose will happen if you attempt to use the [VALUE] feature of your calculator?

Limit of a Function (Informal Definition)

The notation

$$\lim_{x \to c} f(x) = L$$

is read "the limit of $f(x)$ as x approaches c is L" and means that the functional values $f(x)$ can be made arbitrarily close to L by choosing x sufficiently close to c (but not equal to c).

∎ *What This Says:* If $f(x)$ becomes arbitrarily close to a single number L as x approaches c from either side, then we say that L is the limit of $f(x)$ as x approaches c.

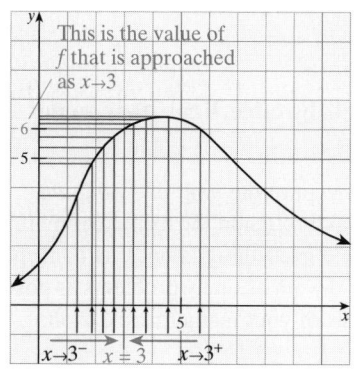

■ **FIGURE 2.9**   Limit as x approaches c

This informal definition of limit cannot be used in proofs until we give precise meaning to terms such as "arbitrarily close to L" and "sufficiently close to c." This will be done at the end of this section. For now, we shall use this informal definition to gain a working knowledge of limits.

LIMITS BY GRAPHING

Figure 2.9 shows the graph of a function f and the number $c = 3$. The arrowheads are used to illustrate possible sequences of numbers along the x-axis, approaching from both the left and the right. As x approaches $c = 3$, $f(x)$ gets closer and closer to 5. We write this as

$$\lim_{x \to 3} f(x) = 5$$

As x approaches 3 from the left, we write $x \to 3^-$, and as x approaches 3 from the right we write $x \to 3^+$. We say that the limit at $x = 3$ exists only if the value approached from the left is the same as the value approached from the right.

One-Sided Limits

Right-hand limit: We write $\lim_{x \to c^+} f(x) = L$ if we can make the number $f(x)$ as close to L as we please by choosing x sufficiently close to c on an interval (c, b) immediately to the right of c.

Left-hand limit: We write $\lim_{x \to c^-} f(x) = L$ if we can make the number $f(x)$ as close to L as we please by choosing x sufficiently close to c on the interval (a, c) immediately to the left of c.

Two-Sided Limits

One-sided and two-sided limits: We write $\lim_{x \to c} f(x) = L$ if and only if the one-sided limits

$$\lim_{x \to c^-} f(x) \quad \text{and} \quad \lim_{x \to c^+} f(x)$$

both exist and are equal to L. The limit $\lim_{x \to c} f(x) = L$ is called a **two-sided limit.**

These notions of one- and two-sided limits are demonstrated in Figure 2.10.

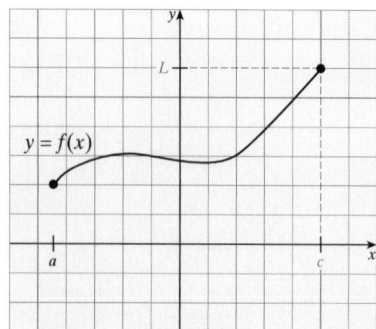

a. Left-handed limit
$$\lim_{x \to c^-} f(x) = L$$

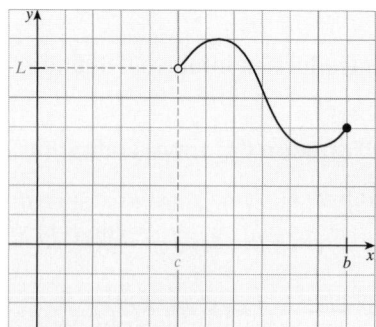

b. Right-handed limit
$$\lim_{x \to c^+} f(x) = L$$

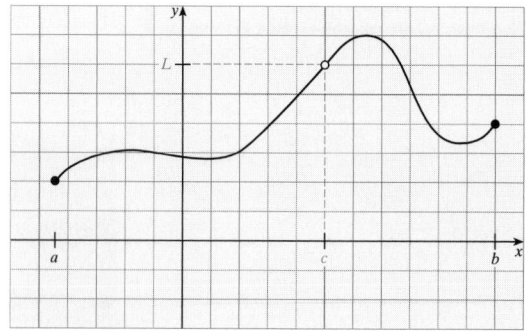

c. Two-sided limit
$$\lim_{x \to c} f(x) = L$$

■ **FIGURE 2.10**   We say that $\lim_{x \to c} f(x) = L$ if and only if $\lim_{x \to c^-} f(x) = \lim_{x \to c^+} f(x)$.

EXAMPLE 2   *Estimating limits by graphing*

Given the functions defined by the graphs in Figure 2.11, find the requested limits by inspection, if they exist.

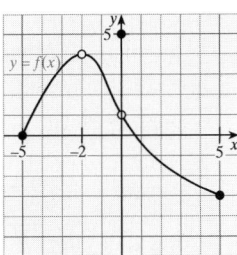
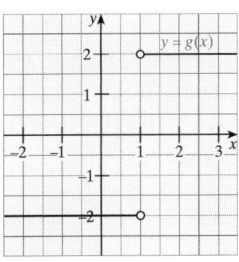
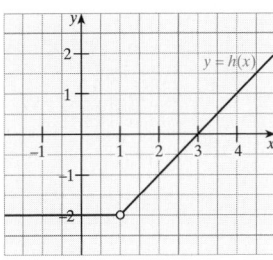

a. $\lim\limits_{x \to 0} f(x)$ b. $\lim\limits_{x \to 1} g(x)$ c. $\lim\limits_{x \to 1} h(x)$

■ **FIGURE 2.11**   Limits from a graph

Solution

a. Take a good look at the given graph; notice the open circles on the graph at $x = 0$ and $x = -2$ and also notice that $f(0) = 5$. To find $\lim\limits_{x \to 0} f(x)$ we need to look at both the left and right limits. Look at Figure 2.11a to find

$$\lim_{x \to 0^-} f(x) = 1 \quad \text{and} \quad \lim_{x \to 0^+} f(x) = 1$$

so $\lim\limits_{x \to 0} f(x)$ exists and $\lim\limits_{x \to 0} f(x) = 1$.

   Notice here that *the value of the limit as $x \to 0$ is not the same as the value of the function at $x = 0$.*

b. Look at Figure 2.11b to find

$$\lim_{x \to 1^-} g(x) = -2 \quad \text{and} \quad \lim_{x \to 1^+} g(x) = 2$$

so the limit of $g(x)$ as $x \to 1$ does not exist.

c. Look at Figure 2.11c to find

$$\lim_{x \to 1^-} h(x) = -2 \quad \text{and} \quad \lim_{x \to 1^+} h(x) = -2$$

so $\lim\limits_{x \to 1} h(x) = -2$.   ■

EXAMPLE 3   *Finding the limit from Example 1 by graphing*

Find $\lim\limits_{t \to 2} \dfrac{16t^2 - 64}{t - 2}$ by graphing.

Solution

$$\bar{v}(t) = \frac{16t^2 - 64}{t - 2} = \frac{16(t^2 - 4)}{t - 2} = \frac{16(t - 2)(t + 2)}{t - 2} = 16(t + 2), t \neq 2$$

The graph of $\bar{v}(t)$ is a line with a deleted point, as shown in Figure 2.12. If you have a graphing calculator, compare this with the graph shown on your calculator (see the technology window for Example 1).

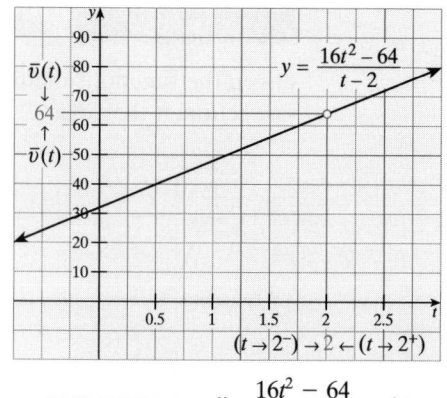

■ **FIGURE 2.12** $\displaystyle\lim_{t \to 2} \frac{16t^2 - 64}{t - 2} = 64$

The limit can now be seen:

$$\lim_{t \to 2} \bar{v}(t) = 64$$

That is, the instantaneous velocity of the falling body in Example 1 is 64 ft/sec.   ■

It is important to remember that when we write

$$\lim_{x \to c} f(x) = L$$

WARNING   The limit of a function as the independent variable approaches a point does not depend on the value of the function at that point.   ◂

we do not require c itself to be in the domain of f, nor do we require $f(c)$, if it is defined, to be equal to the limit. Functions with the special property that

$$\lim_{x \to c} f(x) = f(c)$$

are said to be **continuous at $x = c$.** Continuity is considered in Section 2.4.

LIMITS BY TABLE

It is not always convenient (or even possible) to first draw a graph to find limits. You can also use a calculator or a computer to construct a table of values for $f(x)$ as $x \to c$.

EXAMPLE 4   *Finding a limit from a table*

Find $\displaystyle\lim_{t \to 2} \frac{16t^2 - 64}{t - 2}$ by using a table.

Solution   You will recognize this limit from Examples 1 and 3. We need to begin by selecting sequences of numbers for $t \to 2^-$ and $t \to 2^+$:

		t approaches from the left; $t \to 2^-$.			t approaches from the right; $t \to 2^+$.		
t	1.950	1.995	1.999	2	2.001	2.015	2.100
$\bar{v}(t)$	63.200	63.920	63.984	Undefined	64.016	64.240	65.600
		$\bar{v}(t)$ approaches 64 from the left.			$\bar{v}(t)$ approaches 64 from the right.		

That is, the pattern of numbers suggests

$$\lim_{t \to 2} \frac{16t^2 - 64}{t - 2} = 64$$

as we found using a graphical approach in Example 3.   ■

> **EXAMPLE 5**   *Finding limits of trigonometric functions*

Evaluate $\lim\limits_{x\to 0} \sin x$ and $\lim\limits_{x\to 0} \cos x$.

Solution   We can evaluate these limits by table or by graph.

By table:

x	1	0.5	0.1	0.01	-0.5	-0.1	-0.01
$\sin x$	0.84	0.48	0.0998	0.0099998	-0.48	-0.0998	-0.0099998
$\cos x$	0.54	0.88	0.9950	0.9999500	0.88	0.9950	0.9999500

The pattern of numbers in the table suggests that

$$\lim_{x\to 0} \sin x = 0 \quad \text{and} \quad \lim_{x\to 0} \cos x = 1 \qquad \blacksquare$$

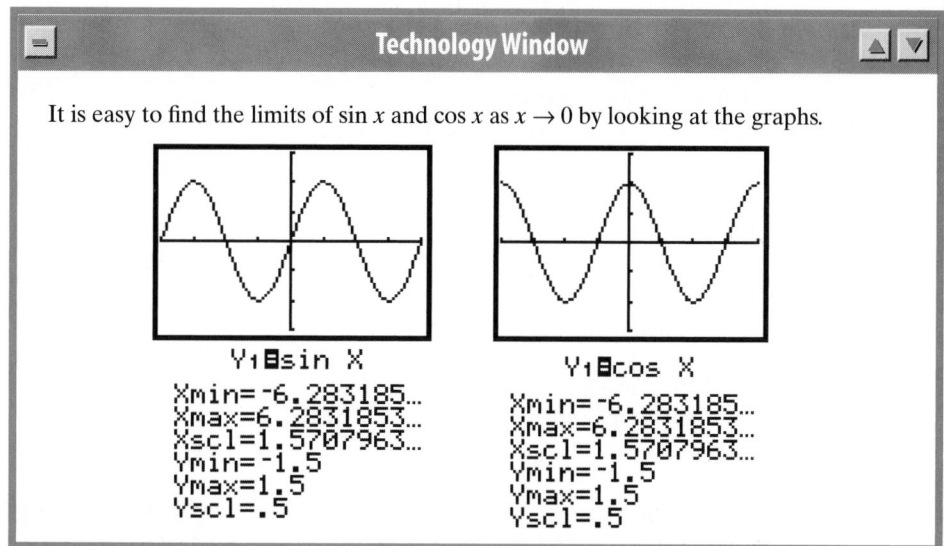

Technology Window

It is easy to find the limits of $\sin x$ and $\cos x$ as $x \to 0$ by looking at the graphs.

> **EXAMPLE 6**   *Evaluating a trigonometric limit using a table*

Evaluate $\lim\limits_{x\to 0} \dfrac{\sin x}{x}$.

Solution $f(x) = \dfrac{\sin x}{x}$ is an even function because

$$f(-x) = \frac{\sin(-x)}{-x} = \frac{-\sin x}{-x} = \frac{\sin x}{x} = f(x)$$

This means that we need to find only the right-hand limit at 0 because the limiting behavior from the left will be the same as that from the right. Consider the following table.

x	0.1	0.05	0.01	0.001	0
$f(x)$	0.998334	0.999583	0.9999833	0.999999833	Undefined

The table suggests that $\lim\limits_{x\to 0^+} \dfrac{\sin x}{x}$ and hence that $\lim\limits_{x\to 0} \dfrac{\sin x}{x} = 1$. We shall revisit this limit in Section 2.3. $\qquad \blacksquare$

Technology Window

There are several ways you can evaluate a limit, depending on the model of the calculator you have available. Here we look at the limit by constructing a table, by drawing a graph, and by using a utility to evaluate the limit directly. We verify the result of Example 6.

Numerical Approach

t	$x = \dfrac{1}{t}$	$\dfrac{\sin x}{x}$
10	0.100	0.998334
20	0.050	0.999583
30	0.033	0.999815
40	0.025	0.999896
50	0.020	0.999933
60	0.017	0.999954
70	0.014	0.999966
80	0.013	0.999974
90	0.011	0.999979
100	0.010	0.999983
120	0.008	0.999988

From the table, we conclude the limit is 1.

We should note that when we use a table (using either calculator or computer values), we may be misled. All we can really say is that the table indicates a certain limiting behavior. The limit may or may not exist (see Problem 64).

Graphical Approach

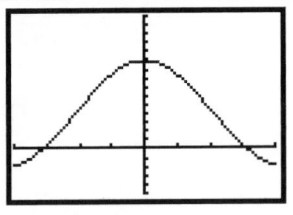

Y₁=sin X/X

Xmin=-4   Ymin=-.5
Xmax=4   Ymax=1.5
Xscl=1   Yscl=.1

A graphing calculator does not show that there is a "hole" at $x = 0$.

Technological Approach

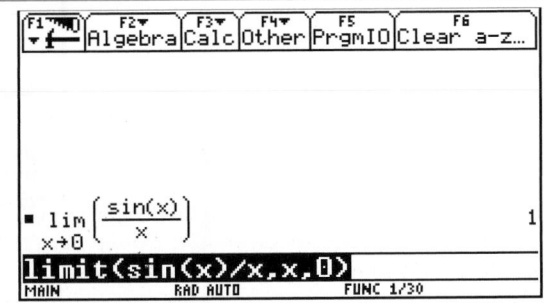

LIMITS THAT DO NOT EXIST

It may happen that a function f does not have a (finite) limit as $x \to c$. When $\lim\limits_{x \to c} f(x)$ fails to exist, the functional values $f(x)$ are said to **diverge** as $x \to c$.

EXAMPLE 7   *Functional values that diverge*

Evaluate $\lim\limits_{x \to 0} \dfrac{1}{x^2}$.

Solution   As $x \to 0$, the corresponding functional values of $f(x) = \dfrac{1}{x^2}$ grow arbitrarily large, as indicated in the table.

	x approaches 0 from the left; $x \to 0^-$.			x approaches 0 from the right; $x \to 0^+$.		
x	−0.1	−0.05	−0.001	0	0.001	0.005	0.01
$f(x) = \dfrac{1}{x^2}$	100	400	1×10^6	undefined	1×10^6	4×10^4	1×10^4

The graph of f is shown in Figure 2.13.

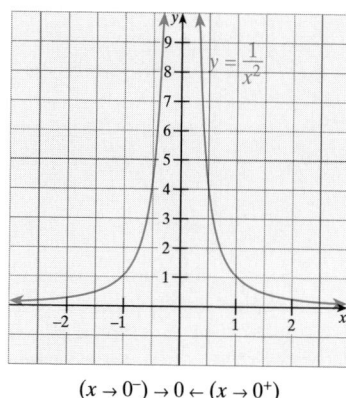

$(x \to 0^-) \to 0 \leftarrow (x \to 0^+)$

■ **FIGURE 2.13** $\displaystyle\lim_{x \to 0} \frac{1}{x^2}$ does not exist, and the graph illustrates that f rises without bound.

Geometrically, the graph of $y = f(x)$ rises without bound as $x \to 0$. Thus, $\displaystyle\lim_{x \to 0} \frac{1}{x^2}$ does not exist.

A Limit That Tends to Infinity

A function f that increases or decreases without bound as x approaches c is said to **tend to infinity** (∞) at c. We indicate this behavior by writing

$$\lim_{x \to 0} f(x) = +\infty \quad \text{if } f \text{ increases without bound}$$

and by

$$\lim_{x \to c} f(x) = -\infty \quad \text{if } f \text{ decreases without bound}$$

WARNING   It is important to remember that ∞ is **not** a number, but is merely a symbol denoting unrestricted growth in the magnitude of the function.

Geometrically, the graph of $y = f(x)$ rises without bound as $x \to 0$. Thus, $\displaystyle\lim_{x \to 0} \frac{1}{x^2}$ does not exist.

Using this notation, we can rewrite the answer to Example 7 as

$$\lim_{x \to 0} \frac{1}{x^2} = +\infty$$

EXAMPLE 8   *A limit that diverges by oscillation*

Evaluate $\displaystyle\lim_{x \to 0} \sin \frac{1}{x}$.

Solution   Note this is not the same as $\displaystyle\lim_{x \to 0} \frac{\sin x}{x}$. The values of $f(x) = \sin \dfrac{1}{x}$ oscillate infinitely often between 1 and -1 as x approaches 0. For example, $f(x) = 1$ for $x = \dfrac{2}{\pi}, \dfrac{2}{5\pi}, \dfrac{2}{9\pi}, \dots$ and $f(x) = -1$ for $x = \dfrac{2}{3\pi}, \dfrac{2}{7\pi}, \dfrac{2}{11\pi}, \dots$. The graph of $f(x)$ is shown in Figure 2.14.

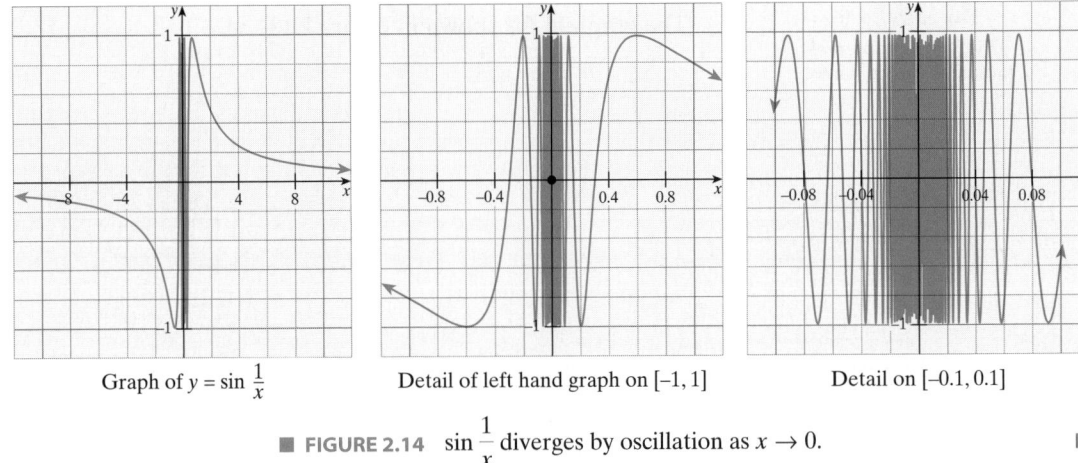

Graph of $y = \sin \frac{1}{x}$ Detail of left hand graph on $[-1, 1]$ Detail on $[-0.1, 0.1]$

■ **FIGURE 2.14** $\sin \dfrac{1}{x}$ diverges by oscillation as $x \to 0$.   ■

Because the values of $f(x)$ do not approach a unique number L as $x \to 0$, the limit does not exist. This kind of limiting behavior is called **divergence by oscillation.**

In the next section, we will introduce some properties of limits that will help us evaluate limits efficiently. In the following problem set, remember that the emphasis is on an intuitive understanding of limits, including their evaluation by graphing and by table.

LIMIT OF A FUNCTION (FORMAL DEFINITION)*

Our informal definition of the limit provides valuable intuition and allows you to develop a working knowledge of this fundamental concept. For theoretical work, however, the intuitive definition will not suffice, because it gives no precise, quantifiable meaning to the terms "arbitrarily close to L" and "sufficiently close to c." In the nineteenth century, leading mathematicians, including Augustin-Louis Cauchy (1789–1857) and Karl Weierstrass (1815–1897) sought to put calculus on a sound logical foundation by giving precise definitions for the foundational ideas of calculus. The following definition, derived from the work of Cauchy and Weierstrass, gives precision to the limit notion.

> **Limit of a Function (Formal definition)**
>
> The limit statement
> $$\lim_{x \to c} f(x) = L$$
> means that for each $\epsilon > 0$, there corresponds a number $\delta > 0$ with the property that
> $$|f(x) - L| < \epsilon \quad \text{whenever} \quad 0 < |x - c| < \delta$$

We show this definition graphically in Figure 2.15.

Because the Greek letters ϵ (epsilon) and δ (delta) are traditionally used in this context, the formal definition of limit is sometimes called the **delta–epsilon** definition of the limit. The goal of this section is to show how this formal definition embodies our intuitive understanding of the limit process and how it can be used rigorously to establish a variety of results.

---

*Optional.

For each ε > 0	**there is a δ > 0**	**such that**	**if 0 < |x − c| < δ,**	**then |f(x) − L| < ε.**
This forms an interval around L on the y-axis.	This forms an interval around c on the x-axis.		This says that if x is in the δ-interval on the x-axis...	...then $f(x)$ is in the ε-interval on the y-axis.

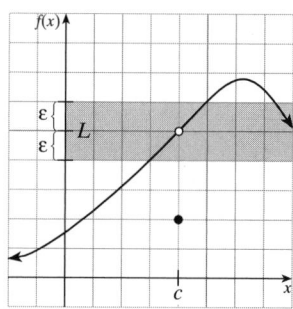 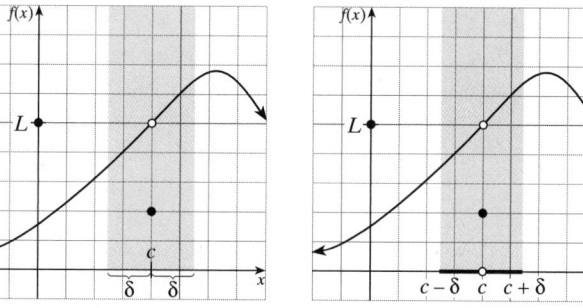 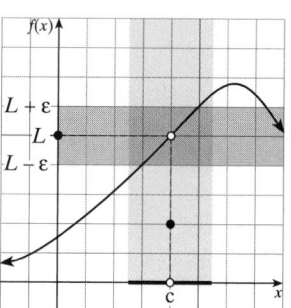

■ **FIGURE 2.15**   Formal definition of limit: $\lim\limits_{x \to c} f(x) = L$

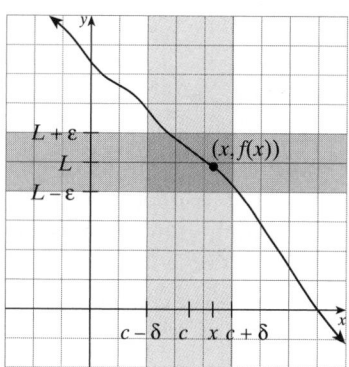

■ **FIGURE 2.16**   The delta–epsilon definition of limit

Do not be discouraged if this material seems difficult—it is. Probably your best course of action is to read this section carefully and examine the details of a few examples closely. Then, using the examples as models, try some of the exercises. This material often takes several attempts, but if you persevere, you should come away with an appreciation of the delta–epsilon process—and with it, a better understanding of calculus.

Behind the formal language is a fairly straightforward idea. In particular, to establish a specific limit, say $\lim\limits_{x \to c} f(x) = L$, a number $\epsilon > 0$ is chosen first to establish a desired degree of proximity to L, and then a number $\delta > 0$ is found that determines how close x must be to c to ensure that $f(x)$ is within ϵ units of L.

The situation is summarized in Figure 2.16, which shows a function that satisfies the conditions of the definition. Notice that whenever x is within δ units of c (but not equal to c), the point $(x, f(x))$ on the graph of f must lie in the rectangle formed by the intersection of the horizontal band of width 2ϵ (gray screen) centered at L and the vertical band of width 2δ (blue region) centered at c. The smaller the ϵ-interval around the proposed limit L, generally the smaller the δ-interval will need to be for L to lie in the ϵ-interval. If such a δ can be found no matter how small ϵ is, then L must be the limit.

The following examples illustrate delta–epsilon proofs, one in which the limit exists and one in which it does not.

EXAMPLE 9   *A delta–epsilon proof of a limit statement*

Show that $\lim\limits_{x \to 2} (4x - 3) = 5$.

Solution   From the graph of $f(x) = 4x - 3$ (see Figure 2.17), we guess that the limit as $x \to 2$ is 5. The object of this example is to *prove* that the limit is 5.

We have

$$
\begin{aligned}
\left| f(x) - L \right| &= \left| (4x - 3) - 5 \right| \\
&= \left| 4x - 8 \right| \\
&= \underline{4 \left| x - 2 \right|}
\end{aligned}
$$

This must be less than ϵ whenever $|x - 2| < \delta$.

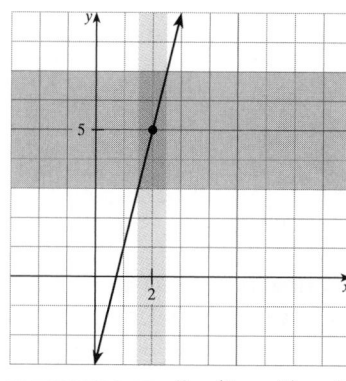

■ **FIGURE 2.17** $\lim\limits_{x \to 2} (4x - 3) = 5$

Choose $\delta = \dfrac{\epsilon}{4}$; then

$$
\left| f(x) - L \right| = 4 \left| x - 2 \right| < 4\delta = 4 \left(\frac{\epsilon}{4} \right) = \epsilon
$$

■

EXAMPLE 10   *A delta–epsilon proof that a limit does not exist*

Show that $\displaystyle\lim_{x\to 0}\frac{1}{x}$ does not exist.

Solution   Let $f(x) = \dfrac{1}{x}$ and L be any number. Suppose that $\displaystyle\lim_{x\to 0} f(x) = L$. Look at the graph of f, as shown in Figure 2.18.

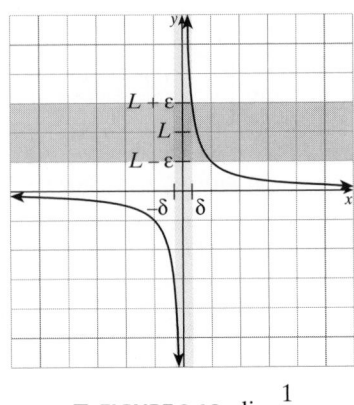

■ **FIGURE 2.18** $\displaystyle\lim_{x\to 0}\frac{1}{x}$

It would seem that no matter what value of ϵ is chosen, it would be impossible to find a corresponding δ. Consider the absolute value expression required by the definition of limit: If

$$\left| f(x) - L \right| < \epsilon, \quad \text{or for this example,} \quad \left| \frac{1}{x} - L \right| < \epsilon$$

then $-\epsilon < \dfrac{1}{x} - L < \epsilon$ *Property of absolute value (Table 1.1, p. 3)*

and $L - \epsilon < \dfrac{1}{x} < L + \epsilon$

If $\epsilon = 1$ (not a particularly small ϵ), then

$$\left| \frac{1}{x} \right| < |L| + 1$$

$$|x| > \frac{1}{|L| + 1}$$

Since L was chosen arbitrarily, this proves that $\displaystyle\lim_{x\to 0}\frac{1}{x}$ does not exist. In other words, since ϵ can be chosen very small, $|x|$ will be very large, and it will be impossible to squeeze $\dfrac{1}{x}$ between $L - \epsilon$ and $L + \epsilon$.   ■

Appendix B gives many of the important proofs in calculus, and if you look there, you will see many of them are given in δ–ϵ form. In the following problem set, remember that the emphasis is on an intuitive understanding of limits, including their evaluation by graphing and by table.

2.2 Problem Set

A *Given the functions defined by the graphs in Figure 2.19, find the limits in Problems 1–6.*

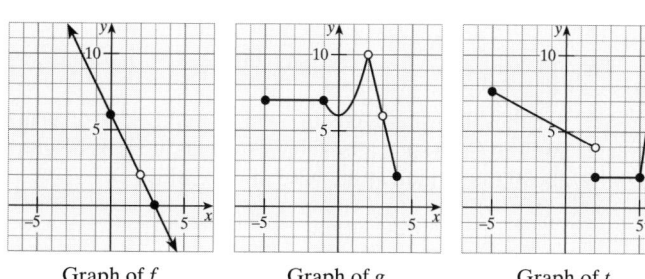

Graph of f   Graph of g   Graph of t

■ **FIGURE 2.19**   Graphs of the functions f, g, and t

1. a. $\lim\limits_{x \to 3} f(x)$ b. $\lim\limits_{x \to 2} f(x)$ c. $\lim\limits_{x \to 0} f(x)$

2. a. $\lim\limits_{x \to -3} g(x)$ b. $\lim\limits_{x \to -1} g(x)$ c. $\lim\limits_{x \to 4^+} g(x)$

3. a. $\lim\limits_{x \to 4} t(x)$ b. $\lim\limits_{x \to -4} t(x)$ c. $\lim\limits_{x \to -5^+} t(x)$

4. a. $\lim\limits_{x \to 2^-} f(x)$ b. $\lim\limits_{x \to 2^+} f(x)$ c. $\lim\limits_{x \to 2} f(x)$

5. a. $\lim\limits_{x \to 3^-} g(x)$ b. $\lim\limits_{x \to 3^+} g(x)$ c. $\lim\limits_{x \to 3} g(x)$

6. a. $\lim\limits_{x \to 2^-} t(x)$ b. $\lim\limits_{x \to 2^+} t(x)$ c. $\lim\limits_{x \to 2} t(x)$

Find the limits by filling in the appropriate values in the tables in Problems 7–9.

7. $\lim\limits_{x \to 5} f(x)$, where $f(x) = 4x - 5$

$x \to 5^-$

x	2	3	4	4.5	4.9	4.99
$f(x)$	3					

$f(x) \to\ ?$

8. $\lim\limits_{x \to 2^-} g(x)$, where $g(x) = \dfrac{x^3 - 8}{x^2 + 2x + 4}$

$x \to 2^-$

x	1	1.5	1.9	1.99	1.999	1.9999
$g(x)$	−1					

$g(x) \to\ ?$

9. $\lim\limits_{x \to 2} h(x)$, where $h(x) = \dfrac{3x^2 - 2x - 8}{x - 2}$

	$x \to 2^-$				$2^+ \leftarrow x$			
x	1	1.9	1.99	1.999	2.001	2.1	2.5	3
$h(x)$	7							

$h(x) \to\ ?\ \leftarrow h(x)$

10. Find $\lim\limits_{x \to 0} \dfrac{\tan 2x}{\tan 3x}$ using the following procedure based on the fact that $f(x) = \tan x$ is an odd function.

If $f(x) = \dfrac{\tan 2x}{\tan 3x}$, then

$$f(-x) = \frac{\tan(-2x)}{\tan(-3x)} = \frac{-\tan 2x}{-\tan 3x} = f(x).$$

Thus, we simply need to check for $x \to 0^+$. Find the limit by completing the following table.

$x \to 0^+$

x	1	0.5	0.1	0.01	0.001	0.0001
$f(x)$	15.33					

$f(x) \to\ ?$

Describe each illustration in Problems 11–16 using a limit statement.

11.   12.

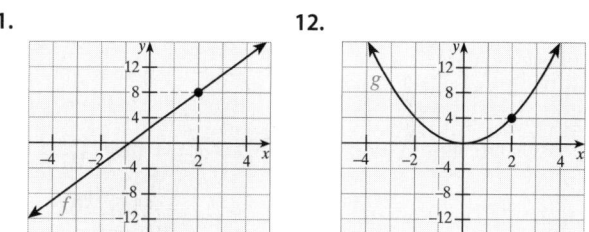

13.   14.

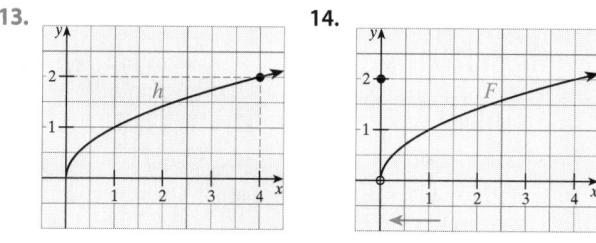

15.   16.

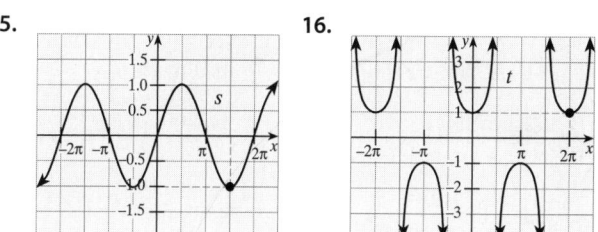

B 17. ■ **What Does This Say?** Explain a process for finding a limit.

Evaluate the limits in Problems 18–50 to two decimal places by graphing or by using a table of values, and check your answer using a calculator. If the limit does not exist, explain why.

18. $\lim\limits_{x \to 0^+} x^4$ 19. $\lim\limits_{x \to 0^+} \cos x$

20. $\lim\limits_{x \to 2^-} (x^2 - 4)$ 21. $\lim\limits_{x \to 3^-} (x^2 - 4)$

22. $\lim\limits_{x \to 1^+} \dfrac{1}{x - 3}$

23. $\lim\limits_{x \to -3^+} \dfrac{1}{x - 3}$

24. $\lim\limits_{x \to 3} \dfrac{1}{x - 3}$

25. $\lim\limits_{x \to \pi/2} \tan x$

26. a. $\lim\limits_{x \to 0} \dfrac{\cos x}{x}$

  b. $\lim\limits_{x \to \pi} \dfrac{\cos x}{x}$

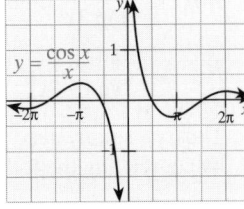

27. a. $\lim\limits_{x \to 1^+} \ln(x - 1)$

  b. $\lim\limits_{x \to 2} \ln(x - 1)$

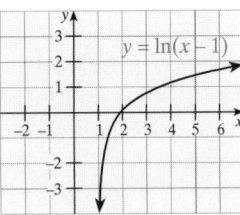

28. a. $\lim\limits_{x \to 0.4} |x| \sin \dfrac{1}{x}$

  b. $\lim\limits_{x \to 0} |x| \sin \dfrac{1}{x}$

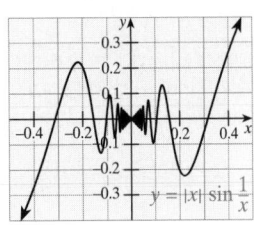

29. a. $\lim\limits_{x \to 0} \dfrac{1 - \cos x}{x}$

  b. $\lim\limits_{x \to \pi} \dfrac{1 - \cos x}{x}$

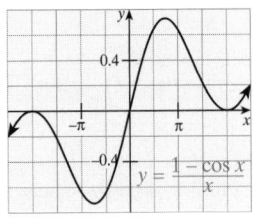

30. a. $\lim\limits_{x \to 3} \dfrac{x^2 + 3x - 10}{x - 2}$ b. $\lim\limits_{x \to 3} \dfrac{x^2 + 3x - 10}{x - 3}$

31. a. $\lim\limits_{x \to 0} x^2 e^{-x}$ b. $\lim\limits_{x \to 1} x^2 e^{-x}$

32. a. $\lim\limits_{x \to 0} e^{-x^3}$ b. $\lim\limits_{x \to 1} e^{-x^3}$

33. $\lim\limits_{x \to \pi/2} \dfrac{2x - \pi}{\cos x}$ **34.** $\lim\limits_{x \to 1} \dfrac{\sin \dfrac{\pi}{x}}{x - 1}$

35. $\lim\limits_{x \to 9} \dfrac{\sqrt{x} - 3}{x - 9}$ **36.** $\lim\limits_{x \to 9} \dfrac{\sqrt{x} - 3}{x - 3}$

37. $\lim\limits_{x \to 2} \dfrac{\sqrt{x + 2} - 2}{x - 2}$ **38.** $\lim\limits_{x \to 1} \dfrac{\sqrt[3]{x} - 1}{\sqrt{x} - 1}$

39. $\lim\limits_{x \to 3^+} \dfrac{\sqrt{x - 3} + x}{3 - x}$ **40.** $\lim\limits_{x \to 4^+} \dfrac{\dfrac{1}{\sqrt{x}} - \dfrac{1}{2}}{x - 4}$

41. $\lim\limits_{x \to 0} \dfrac{\sin 2x}{x}$ **42.** $\lim\limits_{x \to 0} \dfrac{\sin 3x}{x}$

43. $\lim\limits_{x \to 0} \dfrac{1 - \dfrac{1}{x + 1}}{x}$ **44.** $\lim\limits_{x \to 1} \dfrac{1 - \dfrac{1}{x}}{x - 1}$

45. $\lim\limits_{x \to 0} (1 + x)^{1/x}$ **46.** $\lim\limits_{x \to 1} (1 + x)^{1/x}$

47. $\lim\limits_{x \to 0} \left(x^2 - \dfrac{2^x}{2,000} \right)$ **48.** $\lim\limits_{x \to 0} \dfrac{\tan x - x}{x^2}$

49. $\lim\limits_{x \to 0} \cos \dfrac{1}{x}$ **50.** $\lim\limits_{x \to 0} \sin \dfrac{1}{x}$

51. A ball is thrown directly upward from the edge of a cliff and travels in such a way that t seconds later, its height above the ground at the base of the cliff is

$$s(t) = -16t^2 + 40t + 24 \text{ ft}$$

  a. Compute the limit

$$v(t) = \lim\limits_{x \to t} \dfrac{s(x) - s(t)}{x - t}$$

  to find the instantaneous velocity of the ball at time t.

  b. What is the ball's initial velocity?

  c. When does the ball hit the ground, and what is its impact velocity?

  d. When does the ball have velocity 0? What physical interpretation should be given to this time?

52. **MODELING PROBLEM**   Tom and Sue are driving along a straight, level road in a car whose speedometer needle is broken but that has a trip odometer that can measure the distance traveled from an arbitrary starting point in tenths of a mile. At 2:50 P.M., Tom says he would like to know how fast they are traveling at 3:00 P.M., so Sue takes down the odometer readings listed in the table, makes a few calculations, and announces the desired velocity. What is her result?

time t		2:50	2:55	2:59
odometer reading		33.9	38.2	41.5

time t	3:00	3:01	3:03	3:06
odometer	42.4	43.2	44.9	47.4

Technology Window

In Problems 53–56, estimate the limits by plotting points or by using tables.

53. $\lim\limits_{x \to 13} \dfrac{x^3 - 9x^2 - 45x - 91}{x - 13}$

54. $\lim\limits_{x \to 13} \dfrac{x^3 - 9x^2 - 39x - 86}{x - 13}$

55. $\lim\limits_{x \to 13} \dfrac{x^4 - 26x^3 + 178x^2 - 234x + 1,521}{x - 13}$

56. $\lim\limits_{x \to 0} (\sin x)^x$

57. ⱧISTORICAL ⱭUEST   By the second half of the 1700s, it was generally accepted that without logical underpinnings, calculus would be limited. Augustin-Louis Cauchy developed an acceptable theory of limits, and in doing so removed much doubt about the logical validity of calculus. Cauchy is described by the historian Howard Eves not only as a first-rate mathematician with tremendous mathematical productivity but also as

AUGUSTIN-LOUIS CAUCHY
(1789–1857)

a lawyer (he practiced law for 14 years), a mountain climber, and a painter (he worked in watercolors). Among other characteristics that distinguished him from his contemporaries, he advocated respect for the environment.

Cauchy wrote a treatise on integrals in 1814 that was considered a classic, and in 1816 his paper on wave propagation in liquids won a prize from the French Academy. It has been said that with his work the modern era of analysis began. In all, he wrote over 700 papers, which are, today, considered no less than brilliant.

Cauchy did not formulate the δ–ϵ definition of limit that we use today, but formulated instead a purely arithmetical definition. Consult some history of mathematics books to find a translation of Cauchy's definition, which appeared in his monumental treatise, *Cours d'Analyse de l'Ecole Royale Polytechnique* (1821). As part of your research, find when and where the δ–ϵ definition of limit was first used.

Ⓒ In Problems 58–63, *use the formal definition of the limit to prove or disprove the given limit statement.*

58. $\lim_{x \to 2} (x + 3) = 5$

59. $\lim_{t \to 0} (3t - 1) = 0$

60. $\lim_{x \to -2} (3x + 7) = 1$

61. $\lim_{x \to 1} (2x - 5) = -3$

62. $\lim_{x \to 2} (x^2 + 2) = 6$

63. $\lim_{x \to 2} \dfrac{1}{x} = \dfrac{1}{2}$

64. The tabular approach is a convenient device for discussing limits informally, but if it is not used very carefully, it can be misleading. For example, for $x \neq 0$, let

$$f(x) = \sin \frac{1}{x}$$

a. Construct a table showing the values of $f(x)$ for $x = \dfrac{-2}{\pi}, \dfrac{-2}{9\pi}, \dfrac{-2}{13\pi}, \dfrac{2}{19\pi}, \dfrac{2}{7\pi}, \dfrac{2}{3\pi}$. Based on this table, what would you say about $\lim_{x \to 0} f(x)$?

b. Construct a second table, this time showing the values of $f(x)$ for $x = \dfrac{1}{2\pi}, \dfrac{-1}{11\pi}, \dfrac{-1}{20\pi}, \dfrac{1}{50\pi}, \dfrac{1}{30\pi}, \dfrac{1}{5\pi}$. Now, what would you say about $\lim_{x \to 0} f(x)$?

c. Based on the results in parts **a** and **b,** what do you conclude about $\lim_{x \to 0} \sin \dfrac{1}{x}$?

2.3   Properties of Limits

IN THIS SECTION   computations with limits, using algebra to find limits, two special trigonometric limits, limits of piecewise-defined functions ∎

COMPUTATIONS WITH LIMITS

Here is a list of properties that can be used to evaluate a variety of limits.

Basic Properties and Rules for Limits

For any real number c, suppose the functions f and g both have limits at $x = c$.

Constant rule	$\lim\limits_{x \to c} k = k$ for any constant k
Limit of x rule	$\lim\limits_{x \to c} x = c$
Multiple rule	$\lim\limits_{x \to c} [sf(x)] = s \lim\limits_{x \to c} f(x)$ for any constant s
	The limit of a constant times a function is the constant times the limit of the function.
Sum rule	$\lim\limits_{x \to c} [f(x) + g(x)] = \lim\limits_{x \to c} f(x) + \lim\limits_{x \to c} g(x)$
	The limit of a sum is the sum of the limits.
Difference rule	$\lim\limits_{x \to c} [f(x) - g(x)] = \lim\limits_{x \to c} f(x) - \lim\limits_{x \to c} g(x)$
	The limit of a difference is the difference of the limits.

Product rule	$\displaystyle\lim_{x \to c} [f(x)g(x)] = [\lim_{x \to c} f(x)][\lim_{x \to c} g(x)]$
	The limit of a product is the product of the limits.
Quotient rule	$\displaystyle\lim_{x \to c} \frac{f(x)}{g(x)} = \frac{\displaystyle\lim_{x \to c} f(x)}{\displaystyle\lim_{x \to c} g(x)}$ if $\displaystyle\lim_{x \to c} g(x) \neq 0$
	The limit of a quotient is the quotient of the limits, as long as the limit of the denominator is not zero.
Power rule	$\displaystyle\lim_{x \to c} [f(x)]^n = [\lim_{x \to c} f(x)]^n$ n is a rational number and the limit on the right exists.
	The limit of a power is the power of the limit.

It is fairly easy graphically to justify the rules for the limit of a constant and the limit of x, as shown in Figure 2.20.

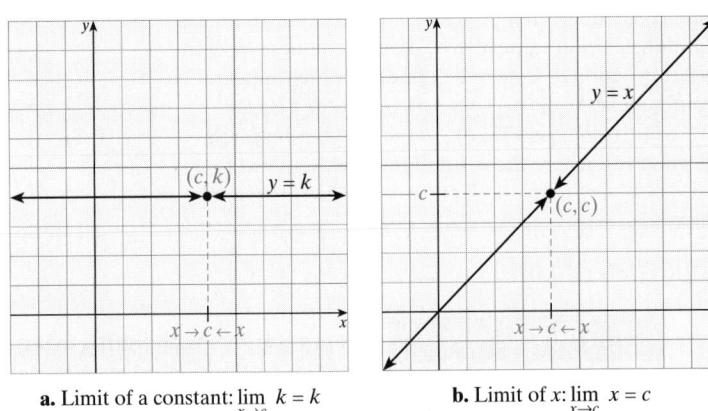

a. Limit of a constant: $\displaystyle\lim_{x \to c} k = k$ **b.** Limit of x: $\displaystyle\lim_{x \to c} x = c$

■ **FIGURE 2.20**   Two basic limits

All of these properties of limits (including the two basic limits shown in Figure 2.20) can be proved using the definition of limit.

> **EXAMPLE 1**   *Finding the limit of a polynomial function*

Evaluate $\displaystyle\lim_{x \to 2} (2x^5 - 9x^3 + 3x^2 - 11)$.

Solution

$$\lim_{x \to 2} (2x^5 - 9x^3 + 3x^2 - 11) = \lim_{x \to 2} (2x^5) - \lim_{x \to 2} (9x^3) + \lim_{x \to 2} (3x^2) - \lim_{x \to 2} (11)$$

Sum and difference rules

$$= 2[\lim_{x \to 2} x^5] - 9[\lim_{x \to 2} x^3] + 3[\lim_{x \to 2} x^2] - 11$$

Multiple and constant rules

$$= 2[\lim_{x \to 2} x]^5 - 9[\lim_{x \to 2} x]^3 + 3[\lim_{x \to 2} x]^2 - 11$$

Power rule

$$= 2(2)^5 - 9(2)^3 + 3(2)^2 - 11 = -7$$

Limit of x rule   ■

COMMENT   If you consider Example 1 carefully, it is easy to see that if f is any polynomial, then the limit at $x = c$ can be found by substituting $x = c$ into the formula for $f(x)$.

Limit of a Polynomial Function

If P is a polynomial function, then

$$\lim_{x \to c} P(x) = P(c)$$

EXAMPLE 2   *Finding the limit of a rational function*

Evaluate $\displaystyle\lim_{z \to -1} \frac{z^3 - 3z + 7}{5z^2 + 9z + 6}$.

Solution $\displaystyle\lim_{z \to -1} \frac{z^3 - 3z + 7}{5z^2 + 9z + 6} = \frac{\displaystyle\lim_{z \to -1}(z^3 - 3z + 7)}{\displaystyle\lim_{z \to -1}(5z^2 + 9z + 6)}$ *Quotient rule*

$$= \frac{(-1)^3 - 3(-1) + 7}{5(-1)^2 + 9(-1) + 6}$$

$$= \frac{9}{2} \quad \text{\small Both numerator and denominator are polynomial functions.}$$

> WARNING   You must be careful about when you write the word "limit" and when you do not; pay particular attention to this when looking at the examples in this section.

Notice that if the denominator of the rational function is not zero, the limit can be found by substitution.

Limit of a Rational Function

If Q is a rational function defined by $Q(x) = \dfrac{P(x)}{D(x)}$, then

$$\lim_{x \to c} Q(x) = \frac{P(c)}{D(c)}$$

provided $\displaystyle\lim_{x \to c} D(x) \neq 0$.

EXAMPLE 3   *Finding the limit of a power (or root) function*

Evaluate $\displaystyle\lim_{x \to -2} \sqrt[3]{x^2 - 3x - 2}$.

Solution $\displaystyle\lim_{x \to -2} \sqrt[3]{x^2 - 3x - 2} = \lim_{x \to -2}(x^2 - 3x - 2)^{1/3}$

$$= \left[\lim_{x \to -2}(x^2 - 3x - 2)\right]^{1/3} \quad \text{\small Power rule}$$

$$= [(-2)^2 - 3(-2) - 2]^{1/3} = 8^{1/3} = 2$$

Once again, for values of the function for which $f(c)$ is defined, the limit can be found by substitution.

In the previous section we used a table to find that $\displaystyle\lim_{x \to 0} \sin x = 0$ and $\displaystyle\lim_{x \to 0} \cos x = 1$. In the following example we use this information, along with the properties of limits, to find other trigonometric limits.

| EXAMPLE 4 | *Finding trigonometric limits algebraically* |

Given that $\lim\limits_{x\to 0} \sin x = 0$ and $\lim\limits_{x\to 0} \cos x = 1$, evaluate:

a. $\lim\limits_{x\to 0} \sin^2 x$ b. $\lim\limits_{x\to 0} (1 - \cos x)$

Solution

a.
$$
\begin{aligned}
\lim_{x\to 0} \sin^2 x &= [\lim_{x\to 0} \sin x]^2 &&\text{Power rule}\\
&= 0^2 &&\lim_{x\to 0} \sin x = 0\\
&= 0
\end{aligned}
$$

b.
$$
\begin{aligned}
\lim_{x\to 0} (1 - \cos x) &= \lim_{x\to 0} 1 - \lim_{x\to 0} \cos x &&\text{Difference rule}\\
&= 1 - 1 &&\text{Constant rule and } \lim_{x\to 0} \cos x = 1\\
&= 0
\end{aligned}
$$
∎

The following theorem states that we can find limits of trigonometric, exponential, and logarithmic functions by direct substitution, as long as the number that x is approaching is in the domain of the given function. Proofs of several parts are outlined in the problem set.

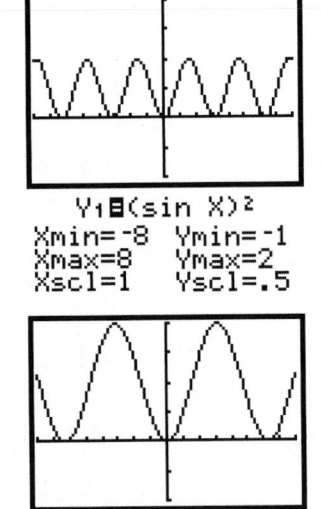
THEOREM 2.1 **Limits of transcendental functions**

If c is any number in the domain of the given function, then

$$\lim_{x\to c} \cos x = \cos c \qquad \lim_{x\to c} \sin x = \sin c \qquad \lim_{x\to c} \tan x = \tan c \qquad \lim_{x\to c} b^x = c^x$$

$$\lim_{x\to c} \sec x = \sec c \qquad \lim_{x\to c} \csc x = \csc c \qquad \lim_{x\to c} \cot x = \cot c \qquad \lim_{x\to c} \ln x = \ln c$$

Proof We shall show that $\lim\limits_{x\to c} \sin x = \sin c$. The other limit formulas may be proved in a similar fashion (see Problems 66–67). Let $h = x - c$. Then $x = h + c$, so $x \to c$ as $h \to 0$. Thus,

$$\lim_{x\to c} \sin x = \lim_{h\to 0} \sin(h + c)$$

Using the trigonometric identity $\sin(A + B) = \sin A \cos B + \cos A \sin B$ and the limit formulas for sums and products, we find that

$$
\begin{aligned}
\lim_{x\to c} \sin x &= \lim_{h\to 0} \sin(h + c)\\
&= \lim_{h\to 0} [\sin h \cos c + \cos h \sin c]\\
&= \lim_{h\to 0} \sin h \cdot \lim_{h\to 0} \cos c + \lim_{h\to 0} \cos h \cdot \lim_{h\to 0} \sin c\\
&= 0 \cdot \cos c + 1 \cdot \sin c &&\lim_{h\to 0} \sin h = 0 \text{ and } \lim_{h\to 0} \cos h = 1\\
&= \sin c
\end{aligned}
$$

Note that $\sin c$ and $\cos c$ do not change as $h \to 0$ because these are constants with respect to h.
═══

| EXAMPLE 5 | *Limits of transcendental functions* |

Evaluate the following limits:

a. $\lim\limits_{x\to 1} (x^2 \cos \pi x)$ b. $\lim\limits_{x\to 2} (x + \ln \sqrt{x})$

c. $\lim\limits_{x\to 0} \dfrac{\sec x}{e^{-x} + 1}$ d. $\lim\limits_{x\to \pi/3} (e^{3x} \tan 2x)$

Solution

a. $\lim\limits_{x\to 1} (x^2 \cos \pi x) = [\lim\limits_{x\to 1} x]^2 [\lim\limits_{x\to 1} \cos \pi x] = 1^2 \cos \pi = -1$

b. $\lim\limits_{x\to 2} (x + \ln\sqrt{x}) = \lim\limits_{x\to 2} x + \lim\limits_{x\to 2} \ln\sqrt{x} = 2 + \ln\sqrt{2}$

c. $\lim\limits_{x\to 0} \dfrac{\sec x}{e^{-x} + 1} = \dfrac{\lim\limits_{x\to 0} \sec x}{\lim\limits_{x\to 0} (e^{-x} + 1)} = \dfrac{1}{e^0 + 1} = \dfrac{1}{2}$

d. $\lim\limits_{x\to \pi/3} (e^{3x} \tan 2x) = [\lim\limits_{x\to \pi/3} e^{3x}][\lim\limits_{x\to \pi/3} \tan 2x] = e^{\pi}(-\sqrt{3}) = -\sqrt{3}\, e^{\pi}$ ∎

USING ALGEBRA TO FIND LIMITS

Sometimes the limit of $f(x)$ as $x \to c$ *cannot* be evaluated by direct substitution. In such a case, we look for another function that agrees with f for all values of x *except at the troublesome value $x = c$.* We illustrate with some examples.

EXAMPLE 6 *Evaluating a limit using fraction reduction*

Evaluate $\lim\limits_{x\to 2} \dfrac{x^2 + x - 6}{x - 2}$.

Solution If you try substitution on this limit, you will obtain:

$$\text{If } x = 2, \text{ then } x^2 + x - 6 = 0.$$
$$\downarrow$$
$$\lim\limits_{x\to 2} \dfrac{x^2 + x - 6}{x - 2} = \dfrac{0}{0}$$
$$\uparrow$$
$$\text{If } x = 2, \text{ then } x - 2 = 0.$$

The form $\frac{0}{0}$ is called an **indeterminate form** because the value of the limit cannot be determined without further analysis.

If the expression is a rational expression, the next step is to simplify the function by factoring and simplifying to see if the reduced form is a polynomial.

$$\lim\limits_{x\to 2} \dfrac{x^2 + x - 6}{x - 2} = \lim\limits_{x\to 2} \dfrac{(x + 3)(x - 2)}{x - 2} = \lim\limits_{x\to 2} (x + 3)$$

This simplification is valid only if $x \neq 2$. Now complete the evaluation of the reduced function by direct substitution. This is not a problem, because $\lim\limits_{x\to 2}$ is concerned with values *as x approaches* 2, not the value where $x = 2$.

$$\lim\limits_{x\to 2} \dfrac{x^2 + x - 6}{x - 2} = \lim\limits_{x\to 2} (x + 3) = 5$$ ∎

Another algebraic technique for finding limits is to rationalize either the numerator or the denominator to obtain an algebraic form that is not indeterminate.

EXAMPLE 7 *Evaluating a limit by rationalizing*

Evaluate $\lim\limits_{x\to 4} \dfrac{\sqrt{x} - 2}{x - 4}$.

Solution Once again, notice that both the numerator and denominator of this rational expression are 0 when $x = 4$, so we cannot evaluate the limit by direct

Technology Window

Y₁ ▊ (X²+X−6)/(X−2)

Xmin=-10 Ymin=-10
Xmax=10   Ymax=10
Xscl=1   Yscl=1

The blank next to the Y= when the value X=2 is input means that the function is not defined when $x = 2$.

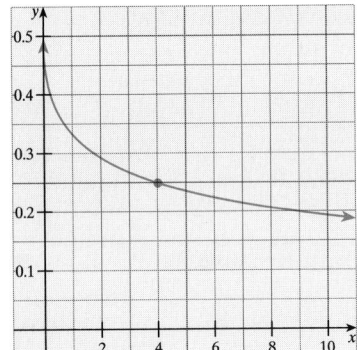

substitution. Instead, rationalize the numerator:

$$\lim_{x \to 4} \frac{\sqrt{x} - 2}{x - 4} = \lim_{x \to 4} \frac{\sqrt{x} - 2}{x - 4} \cdot \frac{\sqrt{x} + 2}{\sqrt{x} + 2} \qquad \text{Multiply by 1.}$$

$$= \lim_{x \to 4} \frac{x - 4}{(x - 4)(\sqrt{x} + 2)}$$

WARNING ▷ This method will work only if the resulting numerator allows the fraction to be simplified. ◄

$$= \lim_{x \to 4} \frac{1}{\sqrt{x} + 2}$$

$$= \frac{1}{\sqrt{4} + 2} = \frac{1}{4}$$ ∎

EXAMPLE 8   *Evaluating a limit involving an exponential function*

Evaluate $\displaystyle\lim_{x \to 0} \frac{e^{2x} + e^x - 2}{e^x - 1}$.

Solution   Factor the numerator:

$$\lim_{x \to 0} \frac{e^{2x} + e^x - 2}{e^x - 1} = \lim_{x \to 0} \frac{(e^x + 2)(e^x - 1)}{e^x - 1} = e^0 + 2 = 3$$ ∎

TWO SPECIAL TRIGONOMETRIC LIMITS

Another property that will be especially important in our future work is given in the following box.

> **Squeeze Rule**
>
> If $g(x) \leq f(x) \leq h(x)$ on an open interval containing c, and if
>
> $$\lim_{x \to c} g(x) = \lim_{x \to c} h(x) = L,$$
>
> then $\displaystyle\lim_{x \to c} f(x) = L$.

> ■ *What This Says:* If a function can be squeezed between two functions with equal limits, then that function must also have that same limit.

THEOREM 2.2   *Special limits involving sine and cosine*

$$\lim_{h \to 0} \frac{\sin h}{h} = 1 \qquad \lim_{h \to 0} \frac{\cos h - 1}{h} = 0$$

Proof   To prove the sine limit theorem requires some principles that are not entirely obvious. However, we can demonstrate its plausibility by considering Figure 2.21, in which $\overparen{AOC}$ is a sector of a circle of radius 1 measured in radians. The line segments $\overline{AD}$ and $\overline{BC}$ are drawn perpendicular to segment $\overline{OC}$.

Assume $0 < h < \frac{\pi}{2}$; that is, h is in Quadrant I.

$$\left| \overparen{AC} \right| = h$$

$$\left| \overline{AD} \right| = \sin h$$

$$\left| \overline{BC} \right| = \tan h = \frac{\sin h}{\cos h}$$

$$\left| \overline{OD} \right| = \cos h$$

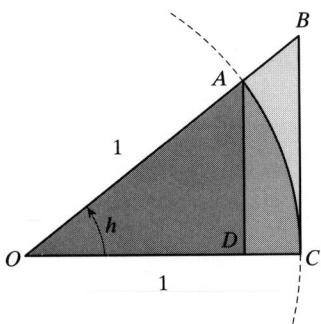

■ **FIGURE 2.21** Trigonometric relationships in the proof that

$$\lim_{h \to 0} \frac{\sin h}{h} = 1$$

Now, compare the area of the sector AOC with those of $\triangle AOD$ and $\triangle BOC$. In particular, since the area of the circular sector of radius r and central angle θ is $\frac{1}{2}r^2\theta$ (see the *Student Mathematics Handbook*), sector AOC must have area

$$\frac{1}{2}(1)^2 h = \frac{1}{2}h$$

We also find that $\triangle AOD$ has area

$$\frac{1}{2}\left|\overline{OD}\right|\left|\overline{AD}\right| = \frac{1}{2}\cos h \sin h,$$

and $\triangle BOD$ has area

$$\frac{1}{2}\left|\overline{BC}\right|\left|\overline{OC}\right| = \frac{1}{2}\left|\overline{BC}\right|(1) = \frac{1}{2}\frac{\sin h}{\cos h}$$

By comparing areas (see Figure 2.21), we have:

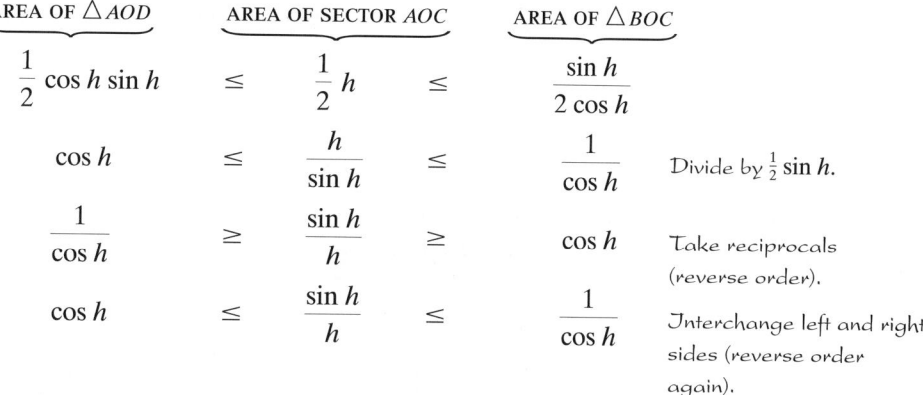

AREA OF $\triangle AOD$		AREA OF SECTOR AOC		AREA OF $\triangle BOC$	
$\dfrac{1}{2}\cos h \sin h$	$\leq$	$\dfrac{1}{2}h$	$\leq$	$\dfrac{\sin h}{2\cos h}$	
$\cos h$	$\leq$	$\dfrac{h}{\sin h}$	$\leq$	$\dfrac{1}{\cos h}$	Divide by $\frac{1}{2}\sin h$.
$\dfrac{1}{\cos h}$	$\geq$	$\dfrac{\sin h}{h}$	$\geq$	$\cos h$	Take reciprocals (reverse order).
$\cos h$	$\leq$	$\dfrac{\sin h}{h}$	$\leq$	$\dfrac{1}{\cos h}$	Interchange left and right sides (reverse order again).

This same inequality holds in the interval $(-\frac{\pi}{2}, 0)$, which is Quadrant IV for the angle h. This can be shown by using the trigonometric identities $\cos(-h) = \cos h$ and $\sin(-h) = -\sin h$. Finally, we take the limit of all parts as $h \to 0$ to find

$$\lim_{h\to 0}\cos h \leq \lim_{h\to 0}\frac{\sin h}{h} \leq \lim_{h\to 0}\frac{1}{\cos h}$$

By Theorem 2.1, $\lim_{h\to 0}\cos h = \cos 0 = 1$. Thus,

$$1 \leq \lim_{h\to 0}\frac{\sin h}{h} \leq \frac{1}{1}$$

From the squeeze rule, we conclude that $\lim_{h\to 0}\dfrac{\sin h}{h} = 1$.

You are asked to prove the second part of this theorem in Problem 62.   ━

EXAMPLE 9   *Evaluation of trigonometric and inverse trigonometric limits*

Find the given limits.

a. $\displaystyle\lim_{x\to 0}\frac{\sin 3x}{5x}$ b. $\displaystyle\lim_{x\to 0}\frac{\sin^{-1}x}{x}$

Technology Window ▲ ▼ ▬

You might want to reinforce your intuition for $\lim_{x\to 0}\dfrac{\sin x}{x}$ by looking at the graph.

Y₁=sin X/X
Xmin=-5   Ymin=-1
Xmax=5   Ymax=1
Xscl=1   Yscl=.1

SMH

Solution

a. We prepare the limit for evaluation by Theorem 2.2 by writing

$$\frac{\sin 3x}{5x} = \frac{3}{5}\left(\frac{\sin 3x}{3x}\right)$$

Since $3x \to 0$ as $x \to 0$, we can set $h = 3x$ in Theorem 2.2 to obtain

$$\lim_{x \to 0} \frac{\sin 3x}{5x} = \lim_{x \to 0} \frac{3}{5}\left(\frac{\sin 3x}{3x}\right) = \frac{3}{5}\lim_{h \to 0} \frac{\sin h}{h} = \frac{3}{5}(1) = \frac{3}{5}$$

b. Let $u = \sin^{-1} x$, so $\sin u = x$. Thus $u \to 0$ as $x \to 0$, and

$$\lim_{x \to 0} \frac{\sin^{-1} x}{x} = \lim_{u \to 0} \frac{u}{\sin u} = 1$$ ∎

LIMITS OF PIECEWISE-DEFINED FUNCTIONS

In Section 1.3 we defined a *piecewise-defined function*. To evaluate

$$\lim_{x \to c} f(x)$$

where the domain of f is divided into pieces, we first look to see whether c is a critical value separating two of the pieces. If so, we need to consider one-sided limits, as illustrated by the following examples.

It is easy to see that the left- and right-hand limits are not the same.

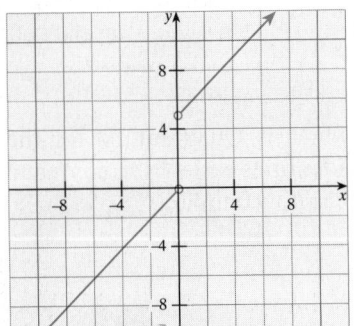

EXAMPLE 10   *Limit of a piecewise-defined function*

Find $\lim_{x \to 0} f(x)$ where $f(x) = \begin{cases} x + 5 & \text{if } x > 0 \\ x & \text{if } x < 0 \end{cases}$.

Solution   Notice that $f(0)$ is not defined, and that it is necessary to consider left- and right-hand limits.

$$\lim_{x \to 0^-} f(x) = \lim_{x \to 0^-} x \qquad\qquad f(x) = x \text{ to the left of } 0.$$
$$= 0$$
$$\lim_{x \to 0^+} f(x) = \lim_{x \to 0^+} (x + 5) \qquad f(x) = x + 5 \text{ to the right of } 0.$$
$$= 5$$

Because the left- and right-hand limits are not the same, we conclude that $\lim_{x \to 0} f(x)$ does not exist.   ∎

EXAMPLE 11   *Limit of a piecewise-defined function*

Find $\lim_{x \to 0} g(x)$ where $g(x) = \begin{cases} x + 1 & \text{if } x > 0 \\ x^2 + 1 & \text{if } x < 0 \end{cases}$

Solution
$$\lim_{x \to 0^-} g(x) = \lim_{x \to 0^-} (x^2 + 1) = 1$$
$$\lim_{x \to 0^+} g(x) = \lim_{x \to 0^+} (x + 1) = 1$$

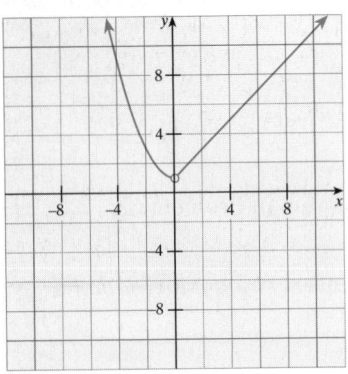

Compare this graph with the graph in Example 10.

Because the left- and right-hand limits are equal, we conclude that $\lim_{x \to 0} g(x) = 1$.   ∎

2.3 Problem Set

A *In Problems 1–32, evaluate each limit.*

1. $\lim\limits_{x \to -2} (x^2 + 3x - 7)$

2. $\lim\limits_{t \to 0} (t^3 - 5t^2 + 4)$

3. $\lim\limits_{x \to 3} (x + 5)(2x - 7)$

4. $\lim\limits_{x \to 0} \dfrac{e^{-2x} + \sin^{-1} x}{x - 1}$

5. $\lim\limits_{z \to 1} \dfrac{z^2 + z - 3}{z + 1}$

6. $\lim\limits_{x \to 3} \dfrac{x^2 + 3x - 10}{3x^2 + 5x - 7}$

7. $\lim\limits_{x \to \pi/3} \sec x$

8. $\lim\limits_{x \to \pi/4} \dfrac{1 + \tan x}{\csc x + 2}$

9. $\lim\limits_{x \to 1/3} \dfrac{x \sin \pi x}{1 + \cos \pi x}$

10. $\lim\limits_{x \to 6} \dfrac{\tan(\pi/x)}{x - 1}$

11. $\lim\limits_{u \to -2} \dfrac{4 - u^2}{2 + u}$

12. $\lim\limits_{x \to 2} \dfrac{x^2 - 4x + 4}{x^2 - x - 2}$

13. $\lim\limits_{x \to 1} \dfrac{\frac{1}{x} - 1}{x - 1}$

14. $\lim\limits_{x \to 0} \dfrac{(x + 1)^2 - 1}{x}$

15. $\lim\limits_{x \to 1} \left(\dfrac{x^2 - 3x + 2}{x^2 + x - 2} \right)^2$

16. $\lim\limits_{x \to 3} \sqrt{\dfrac{x^2 - 2x - 3}{x - 3}}$

17. $\lim\limits_{x \to 1} \dfrac{\sqrt{x} - 1}{x - 1}$

18. $\lim\limits_{y \to 2} \dfrac{\sqrt{y + 2} - 2}{y - 2}$

19. $\lim\limits_{x \to 0} \dfrac{\sin 2x}{x}$

20. $\lim\limits_{x \to 0} \dfrac{\sin 4x}{9x}$

21. $\lim\limits_{t \to 0} \dfrac{\tan 5t}{\tan 2t}$

22. $\lim\limits_{x \to 0} \dfrac{\cot 3x}{\cot x}$

23. $\lim\limits_{x \to 0} \dfrac{1 - \cos x}{\sin x}$

24. $\lim\limits_{x \to 0} \dfrac{\sin^2 x}{2x}$

25. $\lim\limits_{x \to 0} \dfrac{\sin^2 x}{x^2}$

26. $\lim\limits_{x \to 0} \dfrac{x^2 \cos 2x}{1 - \cos x}$

27. $\lim\limits_{x \to 0} e^{-x} \sin x$

28. $\lim\limits_{x \to 1} (\ln x) \cos x$

29. $\lim\limits_{x \to 0} \dfrac{\sec x - 1}{x \sec x}$

30. $\lim\limits_{x \to \pi/4} \dfrac{1 - \tan x}{\sin x - \cos x}$

31. $\lim\limits_{x \to 0} \dfrac{e^{-4x} - 1}{e^{-2x} + e^{-x} - 2}$

32. $\lim\limits_{x \to 0} \dfrac{\sin^{-1} x + \tan^{-1}(2x)}{x}$

33. ■ **What Does This Say?** How do you find the limit of a polynomial function?

34. ■ **What Does This Say?** How do you find the limit of a rational function?

35. ■ **What Does This Say?** How do you find
$$\lim\limits_{x \to 0} \dfrac{\sin ax}{x}$$
for $a \neq 0$?

B *In Problems 36–43, compute the one-sided limit or use one-sided limits to find the given limit, if it exists.*

36. $\lim\limits_{x \to 2^-} (x^2 - 2x)$

37. $\lim\limits_{x \to 1^+} \dfrac{\sqrt{x - 1} + x}{1 - 2x}$

38. $\lim\limits_{x \to 2} |x - 2|$

39. $\lim\limits_{x \to 3} |3 - x|$

40. $\lim\limits_{x \to 0} \dfrac{|x|}{x}$

41. $\lim\limits_{x \to -2} \dfrac{|x + 2|}{x + 2}$

42. $\lim\limits_{x \to 2} f(x)$ where $f(x) = \begin{cases} 3 - 2x & \text{if } x \leq 2 \\ x^2 - 5 & \text{if } x > 2 \end{cases}$

43. $\lim\limits_{s \to 1} g(s)$ where $g(s) = \begin{cases} \dfrac{s^2 - s}{s - 1} & \text{if } s < 1 \\ \sqrt{1 - s} & \text{if } s \geq 1 \end{cases}$

■ **What Does This Say?** *In Problems 44–53, explain why the given limit does not exist.*

44. $\lim\limits_{x \to 1} \dfrac{1}{x - 1}$

45. $\lim\limits_{x \to 2^+} \dfrac{1}{\sqrt{x - 2}}$

46. $\lim\limits_{t \to 2} \dfrac{t^2 - 4}{t^2 - 4t + 4}$

47. $\lim\limits_{x \to 3} \dfrac{x^2 + 4x + 3}{x - 3}$

48. $\lim\limits_{x \to 1} f(x)$ where $f(x) = \begin{cases} 2 & \text{if } x \geq 1 \\ -5 & \text{if } x < 1 \end{cases}$

49. $\lim\limits_{t \to -1} g(t)$ where $g(t) = \begin{cases} 2t + 1 & \text{if } t \geq -1 \\ 5t^2 & \text{if } t < -1 \end{cases}$

50. $\lim\limits_{x \to \pi/2} \tan x$

51. $\lim\limits_{x \to 1} \csc \pi x$

52. $\lim\limits_{x \to 0} e^{1/x}$

53. $\lim\limits_{x \to 0} \ln x$

In Problems 54–59, either evaluate the limit or explain why it does not exist.

54. $\lim\limits_{x \to 1} \dfrac{\frac{1}{x} - 1}{\sqrt{x} - 1}$

55. $\lim\limits_{x \to 0} \left(\dfrac{1}{x} - \dfrac{1}{x^2} \right)$

56. $\lim\limits_{x \to 5} f(x)$ where $f(x) = \begin{cases} x + 3 & \text{if } x \neq 5 \\ 4 & \text{if } x = 5 \end{cases}$

57. $\lim\limits_{t \to 2} g(t)$ where $g(t) = \begin{cases} t^2 & \text{if } -1 \leq t < 2 \\ 3t - 2 & \text{if } t \geq 2 \end{cases}$

58. $\lim\limits_{x \to 2} f(x)$ where $f(x) = \begin{cases} 2(x + 1) & \text{if } x < 3 \\ 4 & \text{if } x = 3 \\ x^2 - 1 & \text{if } x > 3 \end{cases}$

59. $\lim\limits_{x \to 3} f(x)$ where $f(x) = \begin{cases} 2(x+1) & \text{if } x < 3 \\ 4 & \text{if } x = 3 \\ x^2 - 1 & \text{if } x > 3 \end{cases}$

60. **THINK TANK PROBLEM**   Evaluate

$$\lim_{x \to 0} \left[x^2 - \frac{\cos x}{1,000,000,000} \right]$$

Explain why a calculator solution may lead to an incorrect conclusion about the limit.

61. **HISTORICAL QUEST**   Karl Gauss is considered to be one of the four greatest mathematicians of all time, along with Archimedes (Historical Quest Problem 54, Section 2.1), Newton (Historical Quest Problem 1, Mathematical Essays, p. 124), and Euler (Historical Quest Chapter 4 Supplementary Problem 84). Gauss graduated from college at the age of 15 and proved what was to become the fundamental theorem of algebra for his doctoral thesis at the age of 22. He published only

KARL GAUSS
1777–1855

a small portion of the ideas that seemed to storm his mind, because he believed that each published result had to be complete, concise, polished, and convincing. His motto was "Few, but ripe." Carl B. Boyer, in *A History of Mathematics*, describes Gauss as the last mathematician to know everything in his subject, and we could relate Gauss to nearly every topic of this book. Such a generalization is bound to be inexact, but it does emphasize the breadth of interest Gauss displayed.

In 1793, Gauss suggested that for large n, the density of primes behaves like the function $\dfrac{x}{(\ln x)}$. A **prime number** is a counting number that has exactly two divisors. Euclid (ca. 365–300 B.C.) investigated primes and proved that there are infinitely many prime numbers.

In this Quest we investigate Gauss' suggestion. Let $\pi(n)$ be the number of primes less than or equal to x.

$$\pi(10) = 4$$

Primes less than or equal to 10: 2, 3, 5, and 7

$$\pi(25) = 9$$

Primes less than or equal to 25: 2, 3, 5, 7, 11, 13, 17, 19, and 23

History records many who constructed tables for $\pi(n)$. Figure 2.22 shows values for $\pi(x)$. Draw the graph of $f(x) = \dfrac{x}{\ln x}$ to test Gauss' suggestion, and formulate a conclusion.

In 1949, elementary proofs of the *prime number theorem* were published separately by Alte Selberg and Paul Erdös. Look up the statement of this theorem. How is it related to Gauss' conjecture?

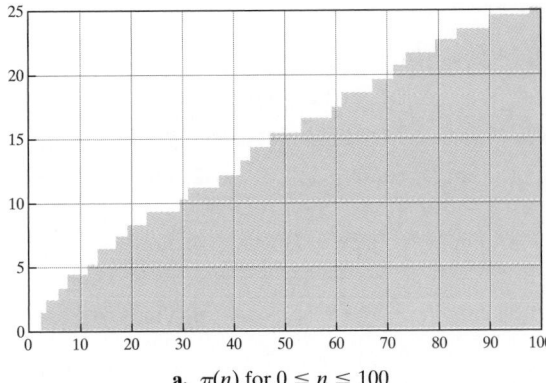

a. $\pi(n)$ for $0 \le n \le 100$

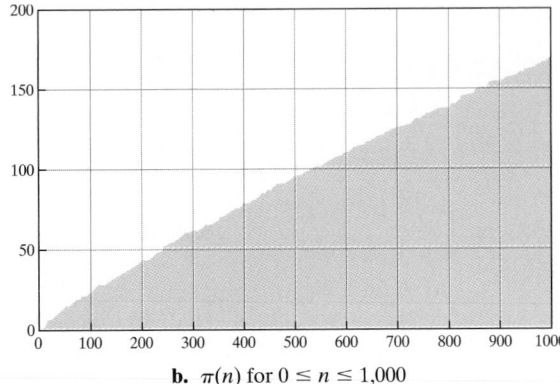

b. $\pi(n)$ for $0 \le n \le 1,000$

■ **FIGURE 2.22**   Graphs of $\pi(n)$

62. Prove the second part of Theorem 2.2:

$$\lim_{h \to 0} \frac{\cos h - 1}{h} = 0$$

63. Let $f(x) = \dfrac{1}{x^2}$ with $x \neq 0$, and let L be any fixed positive integer. Show that

$$f(x) > 100L \text{ if } |x| < \frac{1}{10\sqrt{L}}$$

What does this imply about $\lim\limits_{x \to 0} f(x)$?

64. **THINK TANK PROBLEM**   Give an example for which neither $\lim\limits_{x \to c} f(x)$ nor $\lim\limits_{x \to c} g(x)$ exists, but $\lim\limits_{x \to c} [f(x) + g(x)]$ does exist.

65. Use the sum rule to show that if $\lim\limits_{x \to c} [f(x) + g(x)]$ and $\lim\limits_{x \to c} f(x)$ both exist, then so does $\lim\limits_{x \to c} g(x)$.

66. Show that $\lim\limits_{x \to x_0} \cos x = \cos x_0$.
[*Hint:* You will need to use the trigonometric identity $\cos(A + B) = \cos A \cos B - \sin A \sin B$.]

67. Show that $\lim\limits_{x \to x_0} \tan x = \tan x_0$ whenever $\cos x_0 \neq 0$.

2.4 Continuity

INTUITIVE NOTION OF CONTINUITY

Continuous at $x = c$:
$\lim_{x \to c} f(x)$ exists and is equal to $f(c)$.

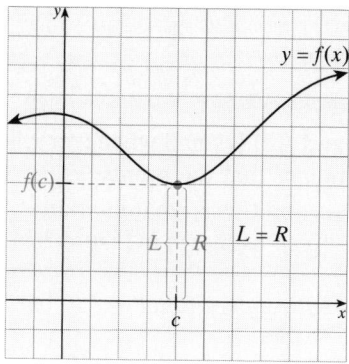

The idea of *continuity* may be thought of informally as the quality of having parts that are in immediate connection with one another. The idea evolved from the vague or intuitive notion of a curve "without breaks or jumps" to a rigorous definition first given toward the end of the nineteenth century (see Historical Quest Problem 54).

We begin with a discussion of *continuity at a point*. It may seem strange to talk about continuity *at a point*, but it should seem natural to talk about a curve being "discontinuous at a point." A few such discontinuities are illustrated by Figure 2.23.

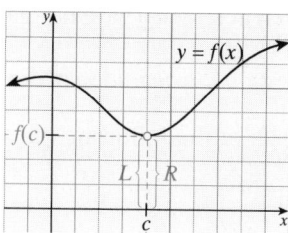

a. HOLE: $f(c)$ is not defined
$\lim_{x \to c} f(x)$ exists

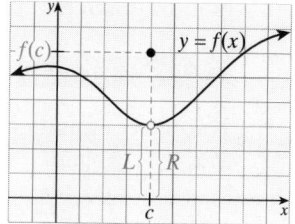

b. HOLE: $f(c)$ is defined
$\lim_{x \to c} f(x)$ exists and is not equal to $f(c)$

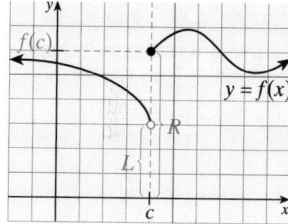

c. JUMP: $\lim_{x \to c^-} f(x)$ is not the same as $\lim_{x \to c^+} f(x)$

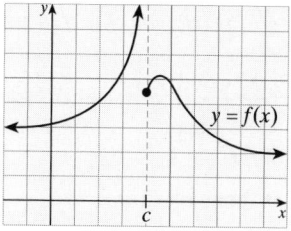

d. POLE: $f(x)$ is defined at $x = c$; $\lim_{x \to c^-} f(x) = +\infty$

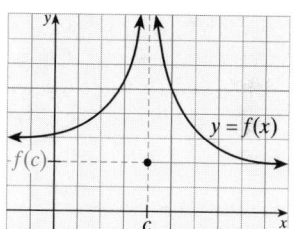

e. POLE: $f(x)$ is defined at $x = c$; $\lim_{x \to c^-} f(x) = +\infty$ and $\lim_{x \to c^+} f(x) = +\infty$

■ **FIGURE 2.23** Types of discontinuity: Holes, poles, and jumps

DEFINITION OF CONTINUITY

Let us consider the conditions that must be satisfied for a function f to be continuous at a point c. First, $f(c)$ must be defined or we have a "hole" in the graph, as shown in Figure 2.23a. (An open dot indicates an excluded point.) If $\lim_{x \to c} f(x)$ has one value as $x \to c^-$ and another as $x \to c^+$, then $\lim_{x \to c} f(x)$ does not exist and there will be a "jump" in the graph of f, as shown in Figure 2.23c. Finally, if one or both of the one-sided limits at c are infinite ($+\infty$ or $-\infty$), there will be a "pole" at $x = c$, as shown in Figure 2.23d and e.

If f is continuous at c, the points $(x, f(x))$ converge to $(c, f(c))$ as $x \to c$.

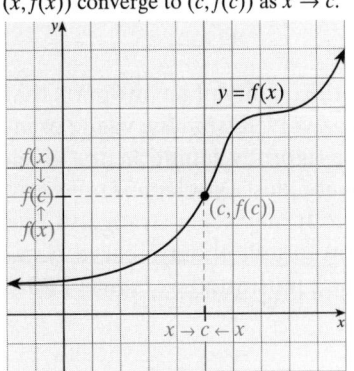

■ **FIGURE 2.24** The geometric interpretation of continuity

Continuity of a Function at a Point

A function f is **continuous at a point** $x = c$ if

1. $f(c)$ is defined;
2. $\lim\limits_{x \to c} f(x)$ exists;
3. $\lim\limits_{x \to c} f(x) = f(c)$.

A function that is not continuous at c is said to have a **discontinuity** at that point.

■ *What This Says:*   The third condition, $\lim\limits_{x \to c} f(x) = f(c)$, summarizes the idea behind continuity. It says that if x is close to c, then $f(x)$ must be close to $f(c)$.

If f is continuous at $x = c$, the difference between $f(x)$ and $f(c)$ is small whenever x is close to c because $\lim\limits_{x \to c} f(x) = f(c)$. Geometrically, this means that the points $(x, f(x))$ on the graph of f converge to the point $(c, f(c))$ as $x \to c$, and this is what guarantees that the graph is unbroken at $(c, f(c))$ with no "gap" or "hole," as shown in Figure 2.24.

EXAMPLE 1   *Testing the definition of continuity with a given function*

Test the continuity of each of the following functions at $x = 1$. If it is not continuous at $x = 1$, explain.

a. $f(x) = \dfrac{x^2 + 2x - 3}{x - 1}$

b. $g(x) = \dfrac{x^2 + 2x - 3}{x - 1}$ if $x \neq 1$ and $g(x) = 6$ if $x = 1$

c. $h(x) = \dfrac{x^2 + 2x - 3}{x - 1}$ if $x \neq 1$ and $h(x) = 4$ if $x = 1$

d. $F(x) = \dfrac{x + 3}{x - 1}$ if $x \neq 1$ and $F(x) = 4$ if $x = 1$

e. $G(x) = 7x^3 + 3x^2 - 2$

f. $H(x) = 2 \sin x - \tan x$

Solution

a. The function f is not continuous at $x = 1$ (hole; $f(1)$ not defined) because it is not defined at this point.

b. 1. $g(1)$ is defined; $g(1) = 6$.

   2. $\lim\limits_{x \to 1} g(x) = \lim\limits_{x \to 1} \dfrac{x^2 + 2x - 3}{x - 1}$

$$= \lim\limits_{x \to 1} \frac{(x - 1)(x + 3)}{x - 1}$$

$$= \lim\limits_{x \to 1} (x + 3) = 4$$

   3. $\lim\limits_{x \to 1} g(x) \neq g(1)$, so g is not continuous at $x = 1$ (hole; $g(c)$ defined).

c. Compare h with g of part **b.** We see that all three conditions of continuity are satisfied, so h is continuous at $x = 1$.

d. 1. $F(1)$ is defined; $F(1) = 4$.

    2. $\lim\limits_{x \to 1} F(x) = \lim\limits_{x \to 1} \dfrac{x+3}{x-1}$; the limit does not exist.

    The function F is not continuous at $x = 1$ (pole).

e. 1. $G(1)$ is defined; $G(1) = 8$.

    2. $\lim\limits_{x \to 1} G(x) = 7(\lim\limits_{x \to 1} x)^3 + 3(\lim\limits_{x \to 1} x)^2 - \lim\limits_{x \to 1} 2 = 8$.

    3. $\lim\limits_{x \to 1} G(x) = G(1)$.

    Because the three conditions of continuity are satisfied, G is continuous at $x = 1$.

f. 1. $H(1)$ is defined; $H(1) = 2 \sin 1 - \tan 1$

    2. $\lim\limits_{x \to 1} H(x) = 2 \lim\limits_{x \to 1} \sin x - \lim\limits_{x \to 1} \tan x = 2 \sin 1 - \tan 1$

    3. $\lim\limits_{x \to 1} H(x) = H(1)$.

    Because the three conditions of continuity are satisfied, H is continuous at $x = 1$.

■

CONTINUITY THEOREMS

It is often difficult to determine whether a given function is continuous at a specified number. However, many common functions are continuous wherever they are defined.

THEOREM 2.3   *Continuity theorem*

If f is a polynomial or a rational function, a power function, a trigonometric function, an inverse trigonometric function, an exponential function, or a logarithmic function, then f is continuous at any number $x = c$ for which $f(c)$ is defined.

Proof   The proof of the continuity theorem is based on the limit properties stated in the previous section. For instance, a polynomial is a function of the form

$$P(x) = a_n x^n + a_{n-1} x^{n-1} + \cdots + a_1 x + a_0$$

where $a_0, a_1, \ldots, a_n$ are constants. We know that $\lim\limits_{x \to c} a_0 = a_0$ and that $\lim\limits_{x \to c} x^m = c^m$ for $m = 1, 2, \ldots, n$. This is precisely the statement that the function $g(x) = ax^m$ is continuous at any number $x = c$, or simply continuous. Because P is a sum of functions of this form and a constant function, it follows from the limit properties that P is continuous.

The proofs of the other parts follow similarly.   ═

The limit properties of the previous section can also be used to prove a second continuity theorem. This theorem tells us that continuous functions may be combined in various ways *without creating a discontinuity*.

THEOREM 2.4   *Properties of continuous functions*

If f and g are functions that are continuous at $x = c$, then the following functions are also continuous at $x = c$.

Scalar multiple	sf	for any numbers s (called a *scalar*)
Sum and difference	$f + g$ and $f - g$	
Product	fg	

| **Quotient** | $\dfrac{f}{g}$ | provided $g(c) \neq 0$ |
| **Composition** | $f \circ g$ | provided g is continuous at c and f is continuous at $g(c)$ |

Proof   The first four properties in this theorem follow directly from the basic limit rules given in Section 2.3. For instance, to prove the product property, note that since f and g are given to be continuous at $x = c$, we have

$$\lim_{x \to c} f(x) = f(c) \text{ and } \lim_{x \to c} g(x) = g(c)$$

If $P(x) = f(x)g(x)$, then

$$\lim_{x \to c} P(x) = \lim_{x \to c} f(x)g(x)$$
$$= [\lim_{x \to c} f(x)][\lim_{x \to c} g(x)]$$
$$= f(c)g(c) = P(c)$$

so $P(x)$ is continuous at $x = c$, as required.

The continuous composition property is proved in a similar fashion, but requires the following limit rule.

Composition Limit Rule

If $\lim_{x \to c} g(x) = L$ and f is a function that is continuous at L, then $\lim_{x \to c} f[g(x)] = f(L)$. That is,

$$\lim_{x \to c} f[g(x)] = f(L) = f(\lim_{x \to c} g(x))$$

This property applies in the same way to other kinds of limits, in particular to one-sided limits.

Now we can prove the continuous composition property of Theorem 2.4. Let $h(x) = (f \circ g)(x)$. Then we have

$$\lim_{x \to c} (f \circ g)(x) = \lim_{x \to c} f[g(x)] \qquad \textit{Definition of composition}$$
$$= f[\lim_{x \to c} g(x)] \qquad \textit{Composition limit rule}$$
$$= f[g(c)] \qquad \textit{g is continuous at } x = c$$
$$= (f \circ g)(c) \qquad \textit{Definition of composition}$$

■ *What This Says:*   The limit of a continuous function is the function of the limiting value. The continuous composition property says that a continuous function of a continuous function is continuous.

We need to talk about a function being continuous on an interval. To do so, we must first know how to handle continuity at the endpoints of the interval, which leads to the following definition.

One-Sided Continuity

The function f is **continuous from the right at a** if and only if

$$\lim_{x \to a^+} f(x) = f(a)$$

and it is **continuous from the left at b** if and only if

$$\lim_{x \to b^-} f(x) = f(b)$$

CONTINUITY ON AN INTERVAL

The function f is said to be **continuous on the open interval (a, b)** if it is continuous at each number in this interval. If f is also continuous from the right at a, we say it is **continuous on the half-open interval $[a, b)$.** Similarly, f is **continuous on the half-open interval $(a, b]$** if it is continuous at each number between a and b and is continuous from the left at the endpoint b. Finally, f is **continuous on the closed interval $[a, b]$** if it is continuous at each number between a and b and is both continuous from the right at a and continuous from the left at b.

EXAMPLE 2   *Testing for continuity on an interval*

Find the intervals on which each of the given functions is continuous.

a. $f_1(x) = \dfrac{x^2 - 1}{x^2 - 4}$ b. $f_2(x) = |x^2 - 4|$ c. $f_3(x) = \csc x$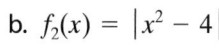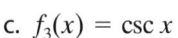

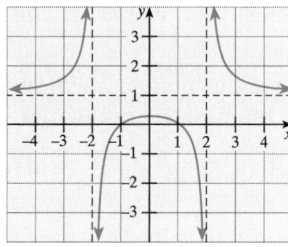 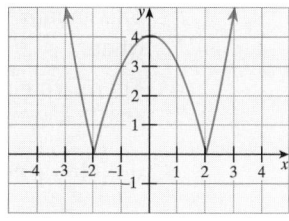 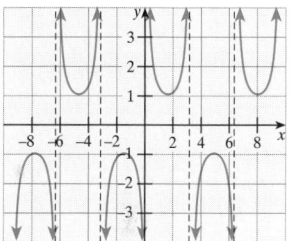

d. $f_4(x) = \sin \dfrac{1}{x}$ 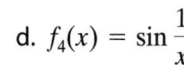   e. $f_5(x) = \begin{cases} x \sin \frac{1}{x} & \text{if } x \neq 0 \\ 0 & \text{if } x = 0 \end{cases}$

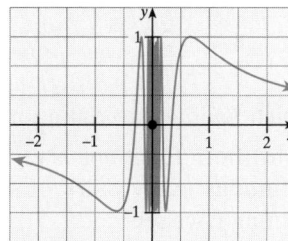 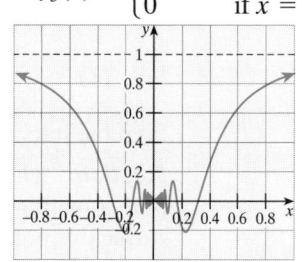

Solution

a. Function f_1 is not defined when $x^2 - 4 = 0$ or when $x = 2$ or $x = -2$. The curve is continuous on $(-\infty, -2) \cup (-2, 2) \cup (2, \infty)$.

b. Function f_2 is continuous on $(-\infty, \infty)$.

c. The cosecant function is not defined at $x = n\pi$, n an integer. At all other points it is continuous.

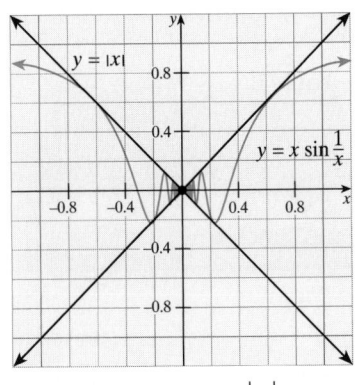

■ **FIGURE 2.25** $y = -|x|$

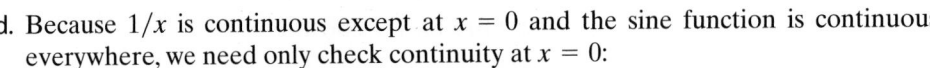

d. Because $1/x$ is continuous except at $x = 0$ and the sine function is continuous everywhere, we need only check continuity at $x = 0$:

$$\lim_{x \to 0} \sin \frac{1}{x} \text{ does not exist.}$$

Therefore, $f(x) = \sin \frac{1}{x}$ is continuous on $(-\infty, 0) \cup (0, \infty)$.

e. It can be shown that

$$-|x| \le x \sin \frac{1}{x} \le |x|, \quad x \ne 0$$

(see Figure 2.25 to see the plausibility of this inequality). We can now use the squeeze rule. Because $\lim_{x \to 0} |x| = 0$ and $\lim_{x \to 0} (-|x|) = 0$, it follows that $\lim_{x \to 0} x \sin \frac{1}{x} = 0$. Because $f_5(0) = 0$ we see that f is continuous at $x = 0$ and therefore is continuous on $(-\infty, \infty)$. ■

Usually there are only a few points in the domain of a given function f where a discontinuity can occur. We use the term **suspicious point** for a number c where either:

1. The domain of f splits; or

2. Substitution of $x = c$ causes division by 0 in the function.

For Example 2, the suspicious points can be listed:

a. $\dfrac{x^2 - 1}{x^2 - 4}$ has suspicious points for division by zero when $x = 2$ and $x = -2$.

b. $|x^2 - 4| = x^2 - 4$ when $x^2 - 4 \ge 0$ and $|x^2 - 4| = 4 - x^2$ when $x^2 - 4 < 0$. This means the definition of the function changes when $x^2 - 4 = 0$, namely, when $x = 2$ and $x = -2$.

c. There are no suspicious points; we know the function cannot be continuous at places where the function is not defined.

d. $\sin \dfrac{1}{x}$ has a suspicious point when $x = 0$ (division by 0).

e. $x \sin \dfrac{1}{x}$ has a suspicious point when $x = 0$ (division by 0).

EXAMPLE 3   *Checking continuity at suspicious points*

Let $f(x) = \begin{cases} 3 - x & \text{if } -5 \le x < 2 \\ x - 2 & \text{if } 2 \le x < 5 \end{cases}$ and $g(x) = \begin{cases} 2 - x & \text{if } -5 \le x < 2 \\ x - 2 & \text{if } 2 \le x < 5 \end{cases}$.

Find the intervals on which f and g are continuous.

Solution   The domain for both functions is $[-5, 5)$; the continuity theorem tells us both functions are continuous everywhere on that interval except possibly at the suspicious points. Examining f, we see

$$\lim_{x \to 2^-} f(x) = \lim_{x \to 2^-} (3 - x) = 1 \text{ and } \lim_{x \to 2^+} f(x) = \lim_{x \to 2^+} (x - 2) = 0$$

so $\lim_{x \to 2} f(x)$ does not exist and f is discontinuous at $x = 2$. Thus, f is continuous for $-5 \le x < 2$ and for $2 < x < 5$.

For g, the domain is also $-5 \le x < 5$, and again, the only suspicious point is $x = 2$. We have $g(2) = 0$ and

$$\lim_{x \to 2^-} g(x) = \lim_{x \to 2^-} (2 - x) = 0 \text{ and } \lim_{x \to 2^+} g(x) = \lim_{x \to 2^+} (x - 2) = 0$$

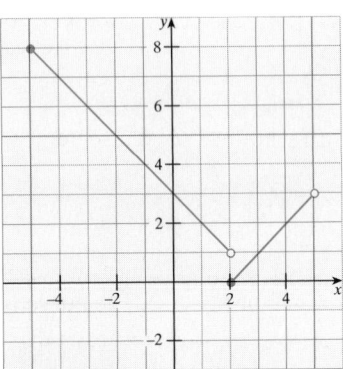

Although a graph is not part of the derivation, it can often be helpful in finding suspicious points. This is the graph of function f.

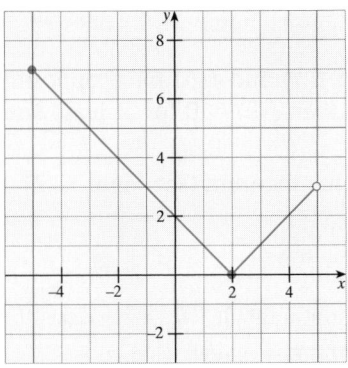

This is the graph of g. Does the graph reinforce our conclusions?

Therefore, $\lim_{x \to 2} g(x) = 0 = g(2)$, and g is continuous at $x = 2$. Hence, g is continuous throughout the interval $[-5, 5)$. ∎

THE INTERMEDIATE VALUE THEOREM

Intuitively, if f is continuous throughout an entire interval, its graph on that interval may be drawn "without the pencil leaving the paper." That is, if $f(x)$ varies continuously from $f(a)$ to $f(b)$ as x increases from a to b, then it must hit every number L between $f(a)$ and $f(b)$, as shown in Figure 2.26.

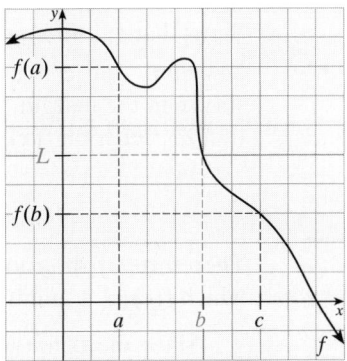

■ FIGURE 2.26 If L lies between $f(a)$ and $f(b)$, then $f(c) = L$ for some c between a and b.

To illustrate the property shown in Figure 2.26, suppose f is a function defined as the weight of a person at age x. If we assume that weight varies continuously with time, a person who weighs 50 pounds at age 6 and 120 pounds at age 15 must weigh 100 pounds at some time between ages 6 and 15.

This feature of continuous functions is known as the *intermediate value property*. A formal statement of this property is contained in the following theorem.

THEOREM 2.5 *The intermediate value theorem*

If f is a continuous function on the closed interval $[a, b]$ and L is some number strictly between $f(a)$ and $f(b)$, then there exists at least one number c on the open interval (a, b) such that $f(c) = L$.

Proof This theorem is intuitively obvious, but it is not at all easy to prove. A proof may be found in most advanced calculus textbooks. ═

> ■ *What This Says:* If f is a continuous function (with emphasis on the word *continuous*) on some *closed* interval, then $f(x)$ must take on all values between $f(a)$ and $f(b)$.

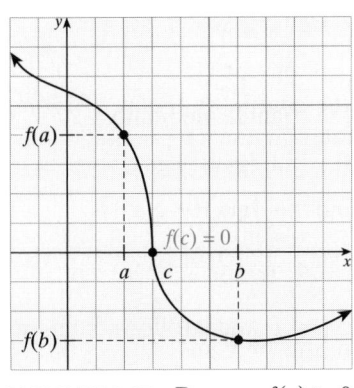

■ FIGURE 2.27 Because $f(a) > 0$ and $f(b) < 0$, then $f(c) = 0$ for some c between a and b

The intermediate value theorem can be used to estimate roots of the equation $f(x) = 0$. Suppose $f(a) > 0$ and $f(b) < 0$, so the graph of f is above the x-axis at $x = a$ and below for $x = b$. Then, if f is continuous on the closed interval $[a, b]$, there must be a point $x = c$ between a and b where the graph crosses the x-axis, that is, where $f(c) = 0$. The same conclusion would be drawn if $f(a) < 0$ and $f(b) > 0$. The key is that $f(x)$ changes sign between $x = a$ and $x = b$. This is shown in Figure 2.27 and is summarized in the following theorem.

===

THEOREM 2.6   *Root location theorem*

If f is continuous on the closed interval $[a, b]$ and if $f(a)$ and $f(b)$ have opposite algebraic signs (one positive and the other negative), then $f(c) = 0$ for at least one number c on the open interval (a, b).

Proof   This follows directly from the intermediate value theorem (see Figure 2.27). The details of this proof are left as a problem.

===

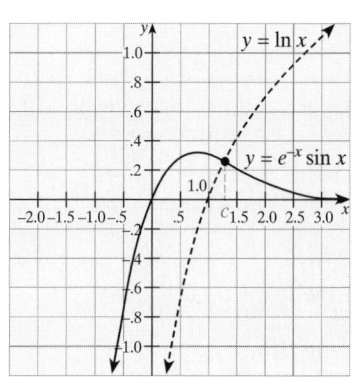

By calculator, we estimate $c \approx 1.3$.

EXAMPLE 4   *Using the root location theorem*

Show that $e^{-x} \sin x = \ln x$ has at least one solution on the interval $[1, 2]$.

Solution   We can use a graphing utility to estimate the point of intersection, as shown in the margin. The root location theorem will confirm this observation. Notice that the function $f(x) = e^{-x} \sin x - \ln x$ is continuous on $[1, 2]$. We find that

$$f(1) = e^{-1}\sin 1 - \ln 1 \approx 0.31 > 0 \quad \text{and}$$
$$f(2) = e^{-2}\sin 2 - \ln 2 \approx -0.57 < 0$$

Therefore, by the root location theorem, there is at least one number c on $(1, 2)$ for which $f(c) = 0$, and it follows that $e^{-c}\sin c = \ln c$. ∎

Since the definition of continuity involves the limit, formal proofs of continuity theorems can be given in terms of the δ–ϵ definition of limit. Here is an example.

EXAMPLE 5   *Formal proof of the sum rule for continuity*

Show that if f and g are continuous at $x = x_0$, then $f + g$ is also continuous at $x = x_0$.

Solution   The continuity of f and g at $x = x_0$ says that

$$\lim_{x \to x_0} f(x) = f(x_0) \text{ and } \lim_{x \to x_0} g(x) = g(x_0)$$

This means that if $\epsilon > 0$ is given, there exist numbers δ_1 and δ_2 such that

$$\left| f(x) - f(x_0) \right| < \frac{\epsilon}{2} \quad \text{whenever} \quad \left| x - x_0 \right| < \delta_1$$

and

$$\left| g(x) - g(x_0) \right| < \frac{\epsilon}{2} \quad \text{whenever} \quad \left| x - x_0 \right| < \delta_2$$

Let δ be the smaller of δ_1 and δ_2; that is, $\delta = \min(\delta_1, \delta_2)$. Then

$$\left| x - x_0 \right| < \delta_1 \quad \text{and} \quad \left| x - x_0 \right| < \delta_2$$

must both be true whenever $\left| x - x_0 \right| < \delta$ so that (by the triangle inequality)

$$\left| [f(x) + g(x)] - [f(x_0) + g(x_0)] \right| = \left| [f(x) - f(x_0)] + [g(x) - g(x_0)] \right|$$
$$\leq \left| f(x) - f(x_0) \right| + \left| g(x) - g(x_0) \right|$$
$$< \frac{\epsilon}{2} + \frac{\epsilon}{2} = \epsilon$$

Thus, $\left| (f + g)(x) - (f + g)(x_0) \right| < \epsilon$ whenever $\left| x - x_0 \right| < \delta$, and it follows from the definition of limit that

$$\lim_{x \to x_0} [f(x) + g(x)] = f(x_0) + g(x_0)$$

In other words, $f + g$ is continuous at $x = x_0$, as claimed. ∎

2.4 Problem Set

Which of the functions described in Problems 1–6 represent continuous functions? State the domain, if possible, for each example.

1. the temperature on a specific day at a given location considered as a function of time

2. the humidity on a specific day at a given location considered as a function of time

3. the selling price of ATT stock on a specific day considered as a function of time

4. the number of unemployed people in the United States during January 1997 considered as a function of time

5. the charges for a taxi ride across town considered as a function of mileage

6. the charges to mail a package as a function of its weight

Identify all suspicious points and determine all points of discontinuity in Problems 7–18.

7. $f(x) = x^3 - 7x + 3$

8. $f(x) = \dfrac{3x + 5}{2x - 1}$

9. $f(x) = \dfrac{3x}{x^2 - x}$

10. $f(t) = 3 - (5 + 2t)^3$

11. $h(x) = \sqrt{x} + \dfrac{3}{x}$

12. $f(u) = \sqrt[3]{u^2 - 1}$

13. $f(x) = \begin{cases} x^2 - 2 & \text{if } x > 1 \\ 2x - 3 & \text{if } x \le 1 \end{cases}$

14. $g(t) = \begin{cases} 3t + 2 & \text{if } t \le 1 \\ 5 & \text{if } 1 < t \le 3 \\ 3t^2 - 1 & \text{if } t > 3 \end{cases}$

15. $f(x) = 3 \tan x - 5 \sin x \cos x$

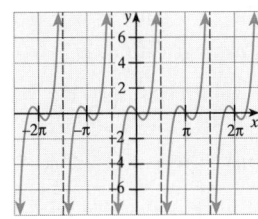

16. $g(x) = \dfrac{\cot x}{\sin x - \cos x}$

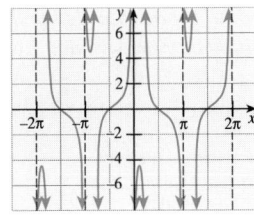

17. $f(x) = \dfrac{e^x}{x}$

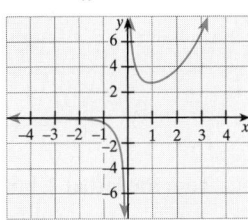

18. $g(x) = \dfrac{\ln x}{x - 1}$

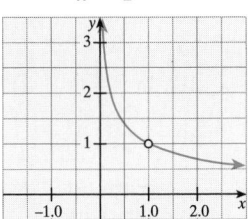

In Problems 19–24, the given function is defined everywhere except at $x = 2$. In each case, find the value that should be assigned to $f(2)$, if any, to guarantee that f will be continuous at 2.

19. $f(x) = \dfrac{x^2 - x - 2}{x - 2}$

20. $f(x) = \sqrt{\dfrac{x^2 - 4}{x - 2}}$

21. $f(x) = \dfrac{\sin (\pi x)}{x - 2}$

22. $f(x) = \dfrac{\cos \dfrac{\pi}{x}}{x - 2}$

23. $f(x) = \begin{cases} 2x + 5 & \text{if } x > 2 \\ 15 - x^2 & \text{if } x < 2 \end{cases}$

24. $f(x) = \dfrac{\dfrac{1}{x} - 1}{x - 2}$

In Problems 25–29, determine whether the given function is continuous on the prescribed interval.

25. a. $f(x) = \dfrac{1}{x}$ on $[1, 2]$ b. $f(x) = \dfrac{1}{x}$ on $[0, 1]$

26. $f(x) = \begin{cases} x^2 & \text{if } 0 \le x < 2 \\ 3x + 1 & \text{if } 2 \le x < 5 \end{cases}$

27. $g(t) = \begin{cases} 15 - t^2 & \text{if } -3 < t \le 0 \\ 2t & \text{if } 0 < t \le 1 \end{cases}$

28. $f(x) = x \sin x$ on $(0, \pi)$

29. $f(x) = \dfrac{\sin x}{e^x}$ on $[0, \pi]$

Ⓑ 30. Let $f(x) = \begin{cases} x & \text{if } x \neq 0 \\ 2 & \text{if } x = 0 \end{cases}$ and

$g(x) = \begin{cases} 3x & \text{if } x \neq 0 \\ -2 & \text{if } x = 0 \end{cases}$

Show that $f + g$ is continuous at $x = 0$ even though f and g are both discontinuous there.

In Problems 31–38, show that the given equation has at least one solution on the indicated interval.

31. $\sqrt[3]{x} = x^2 + 2x - 1$ on $[0, 1]$

32. $\dfrac{1}{x + 1} = x^2 - x - 1$ on $[1, 2]$

33. $\sqrt[3]{x-8}+9x^{2/3}=29$ on $[0,8]$

34. $\tan x = 2x^2 - 1$ on $\left[-\dfrac{\pi}{4}, 0\right]$

35. $\cos x - \sin x = x$ on $\left[0, \dfrac{\pi}{2}\right]$

36. $\cos x = x^2 - 1$ on $[0, \pi]$

37. $e^{-x} = x^3$ on $[0, 1]$

38. $\ln x = (x-2)^2$ on $[1, 2]$

39. Let $f(x) = \begin{cases} x^2 & \text{if } x > 2 \\ x+1 & \text{if } x \leq 2 \end{cases}$

    Show that f is continuous from the left at 2, but not from the right.

40. Find constants a and b such that $f(2) = f(0)$ and f is continuous at $x = 1$.

    $$f(x) = \begin{cases} ax + b & \text{if } x > 1 \\ 3 & \text{if } x = 1 \\ x^2 + b & \text{if } x < 1 \end{cases}$$

41. ■ **What Does This Say?** Use the intermediate value theorem to explain why the hands of a clock coincide at least once every hour.

42. ■ **What Does This Say?** The graph shown in Figure 2.28 models how the growth rate of a bacterial colony changes with temperature.*

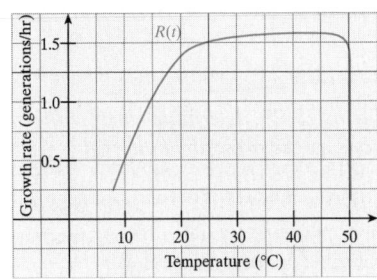

■ **FIGURE 2.28**   Growth rate of a bacterial colony

What happens when the temperature reaches 45 °C? Does it make sense to compute
$$\lim_{t \to 50} R(t)?$$
Write a paragraph describing how temperature affects the growth rate of a species.

43. **MODELING PROBLEM**   A fish swims upstream at a constant speed v relative to the water, which in turn flows at a constant speed v_w ($v_w < v$) relative to the ground. The energy expended by the fish in traveling to a point upstream is given by
$$E(v) = \frac{Cv^k}{v - v_w}$$
where $C > 0$ is a physical constant and $k > 0$ is a number that depends on the type of fish.†

*Michael D. La Grega, Philip L. Buckingham, and Jeffery C. Evans, *Hazardous Waste Management*. New York: McGraw-Hill, 1994, pp. 565–566.

a. Compute $\lim\limits_{v \to v_w} E(v)$. Interpret your result in words.

b. What happens to $E(v)$ as $v \to +\infty$? Interpret your result in words.

44. The population (in thousands) of a colony of bacteria t minutes after the introduction of a toxin is given by the function
$$P(t) = \begin{cases} t^2 + 1 & \text{if } 0 \leq t < 5 \\ -8t + 66 & \text{if } t \geq 5 \end{cases}$$

a. When does the colony die out?

b. Show that at some time between $t = 2$ and $t = 7$, the population is 9,000.

In Problems 45–51, find constants a and b so that the given function will be continuous for all x.

45. $f(x) = \begin{cases} ax + 3 & \text{if } x > 5 \\ 8 & \text{if } x = 5 \\ x^2 + bx + 1 & \text{if } x < 5 \end{cases}$

46. $f(t) = \begin{cases} \dfrac{at - 4}{t - 2} & \text{if } t \neq 2 \\ b & \text{if } t = 2 \end{cases}$

47. $f(t) = \begin{cases} \dfrac{\sqrt{x} - a}{x - 1} & \text{if } x > 0, x \neq 1 \\ b & \text{if } x = 1 \end{cases}$

48. $f(x) = \begin{cases} e^x - a & \text{if } x = 0 \\ b & \text{if } x > 0 \\ \dfrac{x^3 - 4x}{x^2 - 2x} & \text{if } x < 0 \end{cases}$

49. $f(x) = \begin{cases} \dfrac{\sin ax}{x} & \text{if } x < 0 \\ 5 & \text{if } x = 0 \\ x + b & \text{if } x > 0 \end{cases}$

50. $f(x) = \begin{cases} \dfrac{\tan ax}{\tan bx} & \text{if } x < 0 \\ 4 & \text{if } x = 0 \\ ax + b & \text{if } x > 0 \end{cases}$

51. $f(x) = \begin{cases} 2\sin(a\cos^{-1}x) & \text{if } x < 0 \\ \sqrt{3} & \text{if } x = 0 \\ ax + b & \text{if } x > 0 \end{cases}$

▬	**Technology Window**	▲ ▼

The functions in Problems 52–53 are undefined at an obvious point. If possible, define the function there so it is continuous for all x. If not possible, explain why.

52. $\dfrac{\sin x - x}{x^3}$

53. $\dfrac{x^4 - 2x^3 + 3x^2 - 5x - 2}{|x - 2|}$

†E. Batschelet, *Introduction to Mathematics for Life Scientists*, 2nd ed. New York: Springer-Verlag, 1976, p. 280.

54. HISTORICAL QUEST   The first modern formulation of the notion of continuity appeared in a pamphlet published by Bernard Bolzano, a Czechoslovakian priest whose mathematical work was, for the most part, overlooked by his contemporaries. In explaining the concept of continuity, Bolzano said that one must understand the phrase, "A function $f(x)$ (that) varies according to the law of continuity for all values of x which lie inside or outside certain limits, is nothing other than this: If x is any such value, the difference

BERNARD BOLZANO
1781–1848

$$f(x + \omega) - f(x)$$

can be made smaller than any given quantity, if one makes ω as small as one wishes."

In his book, *Cours d'Analyse,* Cauchy (see HISTORICAL QUEST Problem 57, Section 2.2) introduces the concept of continuity for a function defined on an interval in essentially the same way as Bolzano. In this book, Cauchy points out that the continuity of many functions is easily verified. As an example, he argues that sin x is continuous on every interval because "the numerical value of sin $\frac{1}{2}\alpha$, and consequently that of the difference

$$\sin(x + \alpha) - \sin x = 2 \sin\tfrac{1}{2}\alpha \cos(x + \tfrac{1}{2}\alpha)$$

decreases indefinitely with that of α."*

Show that sin x is continuous, and also show that the given expression decreases as claimed by Cauchy.

We conclude this HISTORICAL QUEST by noting that the formal $\delta-\epsilon$ definition of limits and continuity that we use today was first done by Karl Weierstrass (1815–1897), a German secondary school teacher who did his research at night. The historian David Burton describes Weierstrass as "the world's greatest analyst during the last third of the nineteenth century—the father of modern analysis."

55. Let f be a continuous function and suppose that $f(a)$ and $f(b)$ have opposite signs. The root location theorem tells

_____

*A. L. Cauchy, *Cours d'Analyse de l'Ecole Royale Polytechnique, Oeuvres,* Ser. 2, Vol. 3. Paris: Gauthier-Villars, 1897, p. 44.

us that at least one root of $f(x) = 0$ lies between $x = a$ and $x = b$.

a. Let $c = (a + b)/2$ be the midpoint of the interval $[a, b]$. Explain how the root location theorem can be used to determine whether the root lies in $[a, c]$ or in $[c, b]$. Does anything special have to be said about the case where the interval $[a, b]$ contains more than one root?

b. Based on your observation in part **a,** describe a procedure for approximating a root of $f(x) = 0$ more and more accurately. This is called the *bisection method* for root location.

c. Apply the bisection method to locate at least one root of $x^3 + x - 1 = 0$ on $[0, 1]$. Check your answer using your calculator.

56. Apply the bisection method (see Problem 55) to locate at least one root of each of the given equations, and then check your answer using your calculator.
a. $\cos x = \ln(2x + 1)$ on $[0, 1]$
b. $e^x + e^{-x} = 4x^2$ on $[0, 1]$

57. THINK TANK PROBLEM   Find functions f and g such that f is discontinuous at $x = 1$ but fg is continuous there.

58. THINK TANK PROBLEM   Give an example of a function defined for all real numbers that is continuous at only one point.

59. Prove the root location theorem, assuming the intermediate value theorem.

60. Show that f is continuous at c if and only if it is both continuous from the right and continuous from the left at c.

61. Show that if f and g are both continuous at $x = c$ and $g(c) \neq 0$, then f/g must also be continuous there.

62. Let $u(x) = x$ and $f(x) = \begin{cases} 0 & \text{if } x \neq 0 \\ 1 & \text{if } x = 0 \end{cases}$.

Show that $\lim\limits_{x \to 0} f[u(x)] \neq f[\lim\limits_{x \to 0} u(x)]$.

Chapter 2   Review

Proficiency Examination

Concept Problems

1. What are the three main topics in calculus? Briefly describe each.

2. What is a mathematical model?

3. State the informal definition of a limit of a function. Discuss this informal definition.

4. State the formal definition of a limit.

5. State the following basic rules for limits:
a. limit of a constant   b. multiple rule
c. sum rule   d. difference rule
e. product rule   f. quotient rule
g. power rule

h. limit of a polynomial function
i. limit of a rational function
j. limit of transcental functions

6. State the squeeze rule.

7. What are the values of the given limits?
a. $\lim\limits_{x \to 0} \dfrac{\sin x}{x}$ b. $\lim\limits_{x \to 0} \dfrac{\cos x - 1}{x}$

8. Define the continuity of a function at a point and discuss.

9. State the continuity theorem as well as the properties of continuous functions.

10. State the intermediate value theorem.

Practice Problems

Evaluate the limits in Problems 11–16.

11. $\lim\limits_{x \to 3} \dfrac{x^2 - 4x + 9}{x^2 + x - 8}$ **12.** $\lim\limits_{x \to 4} \dfrac{\sqrt{x} - 2}{x - 4}$

13. $\lim\limits_{x \to 2} \dfrac{x^2 - 5x + 6}{x^2 - 4}$ **14.** $\lim\limits_{x \to 0} \dfrac{1 - \cos x}{2 \tan x}$

15. $\lim\limits_{x \to 0} \dfrac{\sin 9x}{\sin 5x}$ **16.** $\lim\limits_{x \to (1/2)^-} \dfrac{|2x - 1|}{2x - 1}$

Determine whether the functions given in Problems 17–18 are continuous on the interval $[-5, 5]$.

17. $f(t) = \dfrac{1}{t} - \dfrac{3}{t + 1}$ **18.** $g(x) = \dfrac{x^2 - 1}{x^2 + x - 2}$

19. Find constants A and B such that f is continuous for all x:

$$f(x) = \begin{cases} Ax + 3 & \text{if } x < 1 \\ 2 & \text{if } x = 1 \\ x^2 + B & \text{if } x > 1 \end{cases}$$

20. Show that the equation

$$x + \sin x = \dfrac{1}{\sqrt{x} + 3}$$

has at least one solution on the interval $[0, \pi]$.

Supplementary Problems*

In Problems 1–20, evaluate the given limit.

1. $\lim\limits_{x \to 2} \dfrac{3x^2 - 7x + 2}{x - 2}$ **2.** $\lim\limits_{x \to -3} \dfrac{4x^2 + 11x - 3}{x^2 - x - 12}$

3. $\lim\limits_{x \to 1^+} \sqrt{\dfrac{x^2 - x}{x - 1}}$ **4.** $\lim\limits_{x \to 1} \dfrac{x^3 - 1}{x^2 - 1}$

5. $\lim\limits_{x \to 4} |4 - x|$ **6.** $\lim\limits_{x \to 0^-} \dfrac{|x|}{x}$

7. $\lim\limits_{x \to 1} \dfrac{x^2 - 3x + 2}{x^2 - 1}$ **8.** $\lim\limits_{x \to 1/\pi} \dfrac{1 + \cos\frac{1}{x}}{\pi x - 1}$

9. $\lim\limits_{x \to 0^+} (1 + x)^{4/x}$ **10.** $\lim\limits_{x \to 0} (1 + 2x)^{1/x}$

11. $\lim\limits_{x \to 1} \dfrac{x^5 - 1}{x - 1}$ **12.** $\lim\limits_{x \to e^2} \dfrac{(\ln x)^3 - 8}{\ln x - 2}$

13. $\lim\limits_{x \to 0} \dfrac{\sin 3x}{\sin 2x}$ **14.** $\lim\limits_{t \to 0} \dfrac{\tan^{-1} t}{\sin^{-1} t}$

15. $\lim\limits_{x \to 0} \dfrac{\sin(\cos x)}{\sec x}$ **16.** $\lim\limits_{x \to 0} \tan 2x \cot x$

17. $\lim\limits_{x \to 0} \dfrac{1 - \sin x}{\cos^2 x}$ **18.** $\lim\limits_{x \to 0} \dfrac{1 - 2 \cos x}{\sqrt{3} - 2 \sin x}$

19. $\lim\limits_{x \to 0} \dfrac{e^{3x} - 1}{e^x - 1}$ **20.** $\lim\limits_{x \to 3} \dfrac{\frac{1}{x} - \frac{1}{3}}{x - 3}$

Evaluate $\lim\limits_{\Delta x \to 0} \dfrac{f(x + \Delta x) - f(x)}{\Delta x}$ *for the function f in Problems 21–28.*

21. $f(x) = 7$ **22.** $f(x) = 3x + 5$ **23.** $f(x) = \sqrt{2x}$

24. $f(x) = x(x + 1)$ **25.** $f(x) = \dfrac{4}{x}$ **26.** $f(x) = \sin x$

27. $f(x) = e^x$ **28.** $f(x) = \ln x$

Decide whether the functions in Problems 29–32 are continuous on the given intervals. If not, redefine the function to make it continuous everywhere on the given interval, or else explain why that cannot be done.

29. $f(x) = \dfrac{x + 4}{x - 8}$ on $[-5, 5]$ **30.** $f(x) = \dfrac{x + 4}{x - 8}$ on $[0, 10]$

31. $f(x) = \dfrac{\sqrt{x} - 8}{x - 64}$ on $\mathbb{R}$ **32.** $f(x) = \dfrac{\sqrt{x} - 6}{x - 36}$ on $[-5, 5]$

Decide whether the functions in Problems 33–36 are continuous on the given intervals. Check all of the suspicious points.

33. $r(x) = \begin{cases} 1 & \text{if } x \text{ is rational} \\ -1 & \text{if } x \text{ is irrational} \end{cases}$

34. $f(x) = |x - 2|$ on $[-5, 5]$

35. a. $f(x) = \dfrac{x^2 - x - 6}{x + 2}$ on $[0, 5]$

b. $f(x) = \dfrac{x^2 - x - 6}{x + 2}$ on $[-5, 5]$

c. $f(x) = \begin{cases} \dfrac{x^2 - x - 6}{x + 2} & \text{on } [-5, 5], x \neq -2 \\ -4 & \text{for } x = -2 \end{cases}$

d. $f(x) = \begin{cases} \dfrac{x^2 - x - 6}{x + 2} & \text{on } [-5, 5], x \neq -2 \\ -5 & \text{for } x = -2 \end{cases}$

36. a. $g(x) = \dfrac{x^2 - 3x - 10}{x + 2}$ on $[0, 5]$

b. $g(x) = \dfrac{x^2 - 3x - 10}{x + 2}$ on $[-5, 5]$

*The supplementary problems are presented in a somewhat random order, not necessarily in order of difficulty.

c. $g(x) = \begin{cases} \dfrac{x^2 - 3x - 10}{x + 2} & \text{on } -5 < -2 \\ -7 & \text{for } x = -2 \\ x - 5 & \text{for } -2 < x \le 5 \end{cases}$

The greatest integer function $f(x) = [\![x]\!]$ is the largest integer that is less than or equal to x. This definition is used in Problems 37–39.

37. a. Graph $f(x) = [\![x]\!]$ on $[-3, 6]$.

 b. Find $\lim_{x \to 3} f(x)$.

 c. For what values of a does $\lim_{x \to a} f(x)$ exist?

38. Repeat Problem 37 for $f(x) = \left[\!\!\left[\dfrac{x}{2}\right]\!\!\right]$.

39. Let $f(x) = [\![x^2 + 1]\!]^{[\![x + 1]\!]}$; find $\lim_{x \to 1} f(x)$.

40. Find constants A and B so that the following function $f(x)$ will be continuous for all x:

$$f(x) = \begin{cases} \dfrac{x^2 - Ax - 6}{x - 2} & \text{if } x > 2 \\ x^2 + B & \text{if } x \le 2 \end{cases}$$

41. Find numbers a and b so that

$$\lim_{x \to 0} \frac{\sqrt{ax + b} - 1}{x} = 1$$

42. Find a number c so that

$$\lim_{x \to 3} \frac{x^3 + cx^2 + 5x + 12}{x^2 - 7x + 12}$$

exists. Then find the corresponding limit.

43. If a function f is not continuous at $x = c$, but can be made continuous at c by being given a new value at the point, it is said to have a *removable discontinuity* at $x = c$. Which of the following functions has a removable discontinuity at $x = c$?

 a. $f(x) = \dfrac{2x^2 + x - 15}{x + 3}$ at $c = -3$

 b. $f(x) = \dfrac{x - 2}{|x - 2|}$ at $c = 2$

 c. $f(x) = \dfrac{2 - \sqrt{x}}{4 - x}$ at $c = 4$

44. If $f(x) = x^3 - x^2 + x$, show that there is a number c such that $f(c) = 0$.

45. Prove that $\sqrt{x + 3} = e^x$ has at least one real root, and then use a graphing utility to find the root correct to the nearest tenth.

46. Show that the tangent line to the circle $x^2 + y^2 = r^2$ at the point (x_0, y_0) has the equation $y_0 y + x_0 x = r^2$. *Hint:* Recall that the tangent line at a point P on a circle with center C is the line that passes through P and is perpendicular to the line segment $\overline{CP}$.

47. **THINK TANK PROBLEM**   It is not necessarily true that $\lim_{x \to c} f(x)$ and $\lim_{x \to c} g(x)$ exist whenever $\lim_{x \to c} [f(x) \cdot g(x)]$

exists. Find functions f and g such that $\lim_{x \to c} [f(x) \cdot g(x)]$ exists and $\lim_{x \to c} g(x) = 0$, but $\lim_{x \to c} f(x)$ does not exist.

48. **THINK TANK PROBLEM**   It is not necessarily true that $\lim_{x \to c} f(x)$ and $\lim_{x \to c} g(x)$ exist whenever $\lim_{x \to c} \dfrac{f(x)}{g(x)}$ exists.

 Find functions f and g such that $\lim_{x \to 0} \dfrac{f(x)}{g(x)}$ exists, but neither $\lim_{x \to 0} f(x)$ nor $\lim_{x \to 0} g(x)$ exists.

49. **MODELING PROBLEM**   When analyzing experimental data involving two variables, a useful procedure is to pass a smooth curve through a number of plotted data points and then perform computations as if the curve were the graph of an equation relating the variables. Suppose the following data are gathered as a result of a physiological experiment in which skin tissue is subjected to external heat for t seconds, and a measurement is made of the change in temperature $\triangle T$ required to cause a change of 2.5 °C at a depth of 0.5 mm in the skin:

t (sec)	1	2	3	4	10	20	25	30
$\triangle T$ (°C)	12.5	7.5	5.8	5	4.5	3.5	2.9	2.8

Plot the data points in a coordinate plane and then answer these questions.

 a. What temperature difference $\triangle T$ would be expected if the exposure time is 2.2 sec?

 b. Approximately what exposure time t corresponds to a temperature difference $\triangle T = 8$ °C?

50. **SPY PROBLEM**   With grief and the desire for revenge burning in his soul, the Spy confronts Siggy's slayer (recall Problem 69, Section 1.6) and is immediately captured. He escapes in a stolen milk truck, driving at 72 km/h, and has a 40-minute head start on his pursuers, who are chasing him in a Ferrari going 168 km/h. The distance from the smugglers' headquarters to the border (and freedom) is 83.8 km. Does he make it?

51. **JOURNAL PROBLEM**   The theorem

$$\lim_{h \to 0} \frac{\sin h}{h} = 1$$

stated in this chapter has been the subject of much discussion in mathematical journals:

W. B. Gearhart and H. S. Shultz, "The Function sin x/x," *College Mathematics Journal* (1990): 90–99.

L. Gillman, "π and the Limit of (sin $\alpha)/\alpha$," *American Mathematical Monthly* (1991): 345–348.

F. Richman, "A Circular Argument," *College Mathematics Journal* (1993): 160–162.

D. A. Rose, "The Differentiability of sin x," *College Mathematics Journal* (1991): 139–142.

P. Ungar, "Reviews," *American Mathematical Monthly* (1986): 221–230.

Some of these articles argue that the demonstration shown in the text is circular, since we use the fact that the area of a circle is πr^2. How do we know the area of circle? The answer, of course, is that we learned it in elementary school, but that does not constitute a proof. On the other hand, some of these articles argue that the reasoning is not necessarily circular. Read one or more of these journal articles and write a report.

52. a. Show that $\sin x < x$ if $0 < x \leq \frac{\pi}{2}$. Refer to Figure 2.29. *Hint:* Compare the area of an appropriate triangle and sector.

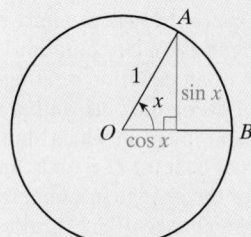

■ **FIGURE 2.29**   Figure for Problems 52 and 53

   b. Show that $|\sin x| < |x|$ if $0 < |x| < \frac{\pi}{2}$.
   c. Use the definition of continuity to show that $\sin x$ is continuous at $x = 0$.
   d. Use the formula
   $$\sin(\alpha + \beta) = \sin \alpha \cos \beta + \sin \beta \cos \alpha$$
   to show that $\sin x$ is continuous for all real x.

53. Follow the procedure outlined in Problem 52 to show that $\cos x$ is continuous for all x. *Hint:* After showing that
$$\lim_{x \to 0} \cos x = 1,$$
you may need the identity
$$\cos 2x = 1 - 2 \sin^2 x$$
to show that $\cos x$ is continuous for $x \neq 0$.

54. A regular polygon of n sides is inscribed in a circle of radius R, as shown in Figure 2.30.

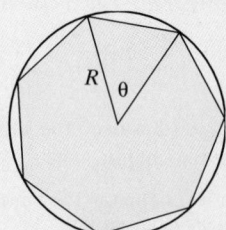

■ **FIGURE 2.30**   A regular polygon of seven sides with central angle θ (Problem 54)

   a. Show that the perimeter of the polygon is given by
   $$P(\theta) = \frac{4\pi R}{\theta} \sin \frac{\theta}{2} \text{ where } \theta = \frac{2\pi}{n} \text{ is the central angle}$$
   subtended by one side of the polygon.

   b. Use the formula in part **a** together with the fact that
   $$\lim_{x \to 0} \frac{\sin x}{x} = 1 \text{ to show that a circle of radius } R \text{ has}$$
   circumference $2\pi R$.
   c. Modify the approach suggested in parts **a** and **b** to show that a circle of radius R has area πR^2. *Hint:* First express the area of the shaded polygon in Figure 2.30 as a function of θ.

55. ■ **What Does This Say?**   A cylindrical tank containing 50L of water is drained into an empty rectangular trough that can hold 75L. Explain why there must be a time when the height of water in the tank is the same as that in the trough.

56. The radius of the earth is roughly 4,000 mi, and an object located x miles from the center of the earth weighs w lb, where
$$w(x) = \begin{cases} Ax & \text{if } x \leq 4{,}000 \\ \dfrac{B}{x^2} & \text{if } x > 4{,}000 \end{cases}$$
where A and B are positive constants. Assuming that w is continuous for all x, what must be true about A and B?

57. **MODELING PROBLEM**   The windchill temperature in degrees is a function of the air temperature T (in degrees Fahrenheit) and the wind speed v (in mi/h).* If we hold T constant and consider the windchill as a function of v, we have
$$w(v) = \begin{cases} T & \text{if } 0 \leq v \leq 4 \\ 91.4 + (91.4 - T)(0.0203v - 0.304\sqrt{v} - 0.474) \\ & \text{if } 4 < v < 45 \\ 1.6T - 55 & \text{if } v \geq 45 \end{cases}$$
   a. If $T = 30$, what is the windchill for $v = 20$? What is it for $v = 50$?
   b. For $T = 30$, what wind speed corresponds to a windchill temperature of $0°$ Fahrenheit?
   c. For what value of T is the windchill function continuous at $v = 4$? At $v = 45$?

58. Based on the estimate that there are 10 billion acres of arable land on the earth and that each acre can produce enough food to feed 4 people, some demographers believe that the earth can support a population of no more than 40 billion people. The population of the earth was approximately 5 billion in 1986 and 6 billion in 1997. If the population of the earth were growing according to the formula
$$P(t) = P_0 e^{rt}$$
where t is the time after the population is P_0 and r is the growth rate, when would the population reach the theoretical limit of 40 billion?

---

*From William Bosch and L. G. Cobb, "Windchill," *UMAP Module No. 658* (1984), pp. 244–247.

59. A function f is said to satisfy a *Lipschitz condition* (named for the 19th century mathematician, Rudolf Lipschitz, 1832–1903) on a given interval if there is a positive constant M such that

$$\left| f(x) - f(y) \right| < M \left| x - y \right|$$

for all x and y in the interval (with $x \neq y$). Suppose f satisfies a Lipschitz condition on an interval and let c be a fixed number chosen arbitrarily from the interval. Use the formal definition of limit to prove that f is continuous at c.

60. MODELING PROBLEM A population model employed at one time by the U.S. Census Bureau uses the formula

$$P(t) = \frac{202.31}{1 + e^{3.98 - 0.314t}}$$

to estimate the population of the United States (in millions) for every tenth year from the base year of 1790. For example, if $t = 0$, then the year is 1790, and if $t = 20$, the year is 1990.

a. Draw the graph using this population model and predict the population in the year 2000.

b. Consult an almanac or some other source to find the actual population figures for the years from 1790 to present.

c. What happens to the population P "in the long run," that is, as t increases without bound?

Technology Window

61. In this problem, you are to find the tangent line to a curve using the idea that the tangent line is the limit of secant lines. As a nontrivial example, consider the function defined by

$$f(x) = \frac{(x^3 - 5)(x^2 - 1)}{x^2 + 1}$$

and graphically find the tangent line at the specific point $(2, f(2)) = (2, 1.8)$. To this end, consider two points on the graph, namely $(2, f(2))$ and $(b, f(b))$, where b is close to, but not equal to 2. The slope of the secant line (the line that connects the two points) is

$$\frac{f(b) - f(2)}{b - 2.0}$$

Technology Window

To find the secant line for a given b, the slope is

$$m = \frac{f(b) - f(2)}{b - 2}$$

and hence the secant is expressed

$$y = f(2) + m(x - 2.0)$$

Experimentally find b so that the secant is effectively tangent to the curve at $(2, f(2))$. Find the equation of the tangent line.

62. Here you are to investigate a function whose tangent at a particular point is problematic. Consider

$$f(x) = (x - 2)^{2/3} + 2x^3$$

near $x = 2$. As in Problem 61, choose b close to, but not equal to 2 (try both $b < 2$ and $b > 2$). Report on your findings. Next try to explain what is happening by zooming the graph of $f(x)$ near $(2, f(2))$.

*63. PUTNAM EXAMINATION PROBLEM Evaluate

$$\lim_{x \to +\infty} \left[\frac{1}{x} \cdot \frac{a^x - 1}{a - 1} \right]^{1/x} \quad \text{where } a > 0, a \neq 1$$

*64. PUTNAM EXAMINATION PROBLEM Figure 2.31 shows a rectangle of base b and height h inscribed in a circle of radius 1, surmounted by an isosceles triangle. Find

$$\lim_{h \to 2/5} \frac{A(h)}{bh}$$

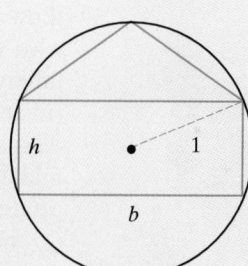

■ **FIGURE 2.31** Putnam problem

---

*Putnam examination problems should be considered optional.

Calculus Was Inevitable

By John L. Troutman, Professor of Mathematics at Syracuse University

The invention of calculus is now credited jointly to Isaac Newton (1642–1727) and Gottfried W. Leibniz (1646–1716); when they produced their separate publications, however (around 1685), there was a bitter controversy throughout Europe as to whose work had been done first. Part of the explanation for this is the fact that each had done his actual work earlier (Newton in 1669; Leibniz, a little later); some can be attributed to the rivalry between scientists in England who championed Newton and those in Europe who supported Leibniz; but much simply reflects how ripe the intellectual climate was for the blooming of calculus. Indeed, even if neither of these great mathematicians had existed, it seems almost certain that the principles of calculus—including the fundamental theorem (introduced in Chapter 5)—would have been announced by the end of the seventeenth century.

Emergence of calculus was effectively demanded by the philosophical spirit of the times. Natural philosophers had long believed that the universe was constructed according to understandable mathematical principles, although they disagreed about just what these principles were and how they might be formulated. For example, the Pythagorean school (ca. 600 B.C.) maintained that everything consisted of (whole) numbers and their ratios; hence their consternation upon discovering that different entities such as $\sqrt{2}$ could be constructed. Next, the early astronomers announced that heavenly bodies move in circular orbits around the Earth as center, and later that the Earth itself must be a perfect sphere to reflect the divine hand of its Creator. Both of these assertions are now known to be false, and by 1612, Kepler had already explained why. Galileo (1564–1642) announced that the distance traveled by a heavy body falling from rest is proportional to the square of the elapsed time, and Fermat asserted (in 1657) that light moves along those paths that minimize the time of travel (we will discuss this principle in Chapter 4). The question was whether such laws could be formulated and justified mathematically, and what kind of mathematics would be appropriate to describe these phenomena.

For already from antiquity there was the warning of Zeno (ca. 450 B.C.), in the form of paradoxes, against unwise speculation about phenomena whose analysis involved infinite processes. In particular, he "proved" that motion was impossible if time consisted of individual instants, and conversely, that covering a given distance was impossible if length was capable of infinite subdivision (see Chapter 8). Thus, although seventeenth-century scientists and philosophers might propose such principles, it was evident that there must be underlying subtleties in the mathematics required to support them (that, in fact, required another two centuries for satisfactory clarification).

To understand better how the mathematics developed, we must examine some of the previous attempts to solve the twin problems of classical origin that motivated the emergence of the calculus—those of finding the tangent line to a given planar geometrical curve and finding the area under the curve. Newton himself acknowledged that his greater vision resulted from his "standing on the shoulders of giants." Who were these giants, and what had they contributed? First, there were the efforts of the Greek mathematicians, principally Eudoxus, Euclid, and Archimedes, who had originated the geometric concepts of tangency and area between 400 B.C. and 200 B.C., together with examples of each, such as the construction of tangent lines to a circle and the area under a parabolic curve. The Hindu and Arabic mathematicians had extended the number system and the formal language of algebra with certain of its laws by A.D. 1300, but it was not until Newton's own era that the methods of algebra and geometry were combined satisfactorily in the analytic geometry of René Descartes (1596–1650) to produce the recognition that a geometrical curve could be regarded as the locus of points whose coordinates satisfied an algebraic equation. This provided potential numerical exactitude to geometric constructions as well as the possibility of giving geometric proofs for limiting algebraic arguments. Since Euclidian geometry was then generally regarded as the only reliable mathematics, it was the latter direction that was most frequently taken. And it was this direction that was taken in the seventeenth century by de Roberval, Fermat, Cavalieri, Huygens, Wallis, and others, not the least of whom was Newton's own teacher at Cambridge, Isaac Barrow (1630–1677). These giants, as Newton called them, obtained equations for tangent lines to, and the (correct) areas under, polynomial curves with equations $y = x^n$ for $n = 1, 2, 3, \ldots, 9$, and certain other geometrically defined curves such as the spiral and the cycloid.

In his analysis of tangency, Barrow incorporated the approximating infinitesimal triangle with sides $\Delta x, \Delta y, \Delta s$ that is now standard in expositions of calculus, and Cavalieri attempted to "count" an indefinite number of parallel equidistant lines to obtain areas. It was known that in specific cases these problems were related, and that they were equivalent, respectively, to the kinematic problems of characterizing velocity and distance traveled during a motion, problems which directly confronted the paradoxes of Zeno.

All that remained was for some mathematicians to sense the generality underlying these specific constructions and to devise a usable notation for presenting the results. This was accomplished essentially independently by Newton (who, justly mistrusting the required limiting arguments, suppressed his own contributions until he could validate them geometrically) and by the only slightly less cautious Leibniz. However, as we have argued, by this time (about 1670), it was almost inevitable that someone should do so.

What calculus has provided is a mathematical language that, by means of the derivative, can describe the rates of change used to characterize various physical processes (such as velocity) and, by means of the integral, can show how macroscopic entities (such as area or distance) can emerge from properly assembled microscopic elements. Moreover, the fundamental theorem, which states that these are inverse operations, supplies an exact method for passing between these types of description. Finally, the ability to relate the results of limiting arguments by simple algebraic formulas permits the correct use of calculus while retaining skepticism regarding its foundations. This has enabled applications to go forward while mathematicians have sought an appropriate axiomatic basis.

Our present technological age attests to the success of this endeavor, and to the value of calculus.

Mathematical Essays

Use a library or references other than this textbook to research the information necessary to answer the questions in Problems 1–11.

1. **HISTORICAL QUEST**   Sir Isaac Newton was one of the greatest mathematicians of all time. He was a genius of the highest order but was often absent-minded. One story about Newton is that, when he was a boy, he was sent to cut a hole in the bottom of the barn door for the cats to go in and out. He cut two holes—a large one for the cat and a small one for the kittens.

ISAAC NEWTON
1642–1727

    Newton considered himself a theologian rather than a mathematician or a physicist. He spent years searching for clues about the end of the world and the geography of hell. One of Newton's quotations about himself is, "I seem to have been only like a boy playing on the seashore and diverting myself in now and then finding a smoother pebble or prettier shell than ordinary, whilst the great ocean of truth lay all undiscovered before me."

    Write an essay about Isaac Newton and his discovery of calculus. This essay should be at least 500 words.

2. **HISTORICAL QUEST**   At the age of 14, Gottfried Leibniz attempted to reform Aristotelian logic. He wanted to create a general method of reasoning by calculation. At the age of 20, he applied for his doctorate at the university in Leipzig and was refused (because, the officials said, he was too young). He received his doctorate the next year at the University of Altdorf, where he made such a favorable impression that he was offered a professorship, which he declined, saying he

GOTTFRIED LEIBNIZ
1646–1716

had very different things in view. Leibniz went on to invent calculus, but not without a bitter controversy developing between Leibniz and Newton. Most historians agree that the bitterness over who invented calculus materially affected the history of mathematics. J. S. Mill characterized Leibniz by saying, "It would be difficult to name a man more remarkable for the greatness and universality of his intellectual powers than Leibniz."

    Write a 500-word essay about Gottfried Leibniz and his discovery of calculus.

3. **HISTORICAL QUEST**   Write a 500-word essay about the controversy surrounding the discovery of calculus by Newton and Leibniz.

4. The Greek mathematicians mentioned in this guest essay include Eudoxus, Euclid, and Archimedes. Write a short paper about contributions they made that might have been used by Newton.

5. What is the definition of elementary functions as given by Joseph Liouville?

6. The guest essay mentions the contributions of de Roberval, Fermat, Cavalieri, Huygens, Wallis, and Barrow toward the invention of calculus. Write a short paper about these contributions.

7. In this guest essay, Troutman argues that the invention of calculus was inevitable, and even if Newton and Leibniz had not invented it, someone else would have. Write a 500-word essay either defending or refuting this thesis.

8. **HISTORICAL QUEST**   Sophie Germain is one of the first women to publish original mathematical research in number theory. In her time, women were not admitted to first-rate universities and were not, for the most part, taken seriously, so she wrote at first under the pseudonym LeBlanc. The situation is not too different from that portrayed by Barbra Streisand in the movie *Yentl*. Even though Germain's most important research was in number theory, she was awarded the prize of the French Academy for a paper entitled "Memoir on the Vibrations of Elastic Plates."

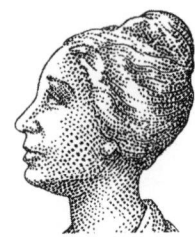

SOPHIE GERMAIN
1776–1831

    As we progress through this book, we will profile many mathematicians in the history of mathematics, and you will notice that most of them are white males. Why? Write a paper on the history of women mathematicians and their achievements. Your paper should include a list of many prominent women mathematicians and their primary contributions. It should also include a lengthy profile of at least one woman mathematician.

9. **HISTORICAL QUEST**   *The Navajo are a Native American people who, despite considerable interchange and assimilation with the surrounding dominant culture, maintain a world view that remains vital and distinctive. The Navajo believe in a dynamic universe. Rather than consisting of objects and situations, the universe is made up of processes. Central to our Western mode of thought is the idea that things are separable entities that can be subdivided into smaller discrete units. For us, things that change through time do so by going from one specific state to another specific state. While we believe time to be continuous, we often even break it into discrete units or freeze it and talk about an instant or point in time. Among the Navajo, where the focus is on process, change is ever present; interrelationship and motion are of primary significance. These incorporate and subsume space and time.*

    There are, in every culture, groups or individuals who think more about some ideas than do others. For other cultures, we know about the ideas of some professional groups or some ideas of the culture at large. We know little, however, about the mathematical thoughts of individuals in those cultures who are specially inclined toward mathematical ideas. In Western culture, on the other hand, we focus on, and record much about, those special individuals while including little about everyone else. Realization of this difference should make us particularly wary of any comparisons across cultures. Even more important, it should encourage finding out more about the ideas of mathematically oriented innovators in other cultures and, simultaneously, encourage expanding the scope of Western history to recognize and include mathematical ideas held by different groups within our culture or by our culture as a whole.

From *Ethnomathematics* by
Marcia Ascher, pp. 128–129 and 188–189.

Write a paper discussing this quotation.

10. **Book Report** "Ethnomathematics, as it is being addressed here, has the goal of broadening the history of mathematics to one that has a multicultural, global perspective." Read the book *Ethnomathematics* by Marcia Ascher (Pacific Grove: Brooks/Cole, 1991), and prepare a book report.

11. Make up a word problem involving a limit or continuity. Send your problems to:

Bradley and Smith
Prentice Hall Publishing Company
1 Lake Street
Upper Saddle River, NJ 07458

The best ones submitted will appear in the next edition (along with credit to the problem poser).

CONTENTS

3

Differentiation

PREVIEW

In this chapter, we develop the main ideas of differential calculus. We begin by defining the *derivative,* which is the central concept of differential calculus. Then we develop a list of rules and formulas for finding the derivative of a variety of expressions involving polynomials, rational and root functions, and trigonometric, logarithmic, and inverse trigonometric functions. Along the way, we shall see how the derivative can be used to find slopes of tangent lines and rates of change.

PERSPECTIVE

A *calculus* is a body of calculation or reasoning associated with a certain concept. For *differential calculus,* that concept is the *derivative,* one of the fundamental ideas in all mathematics and, arguably, a cornerstone of modern scientific thought. The basic ideas of what we now call calculus had been fermenting in intellectual circles throughout much of the seventeenth century. The genius of Newton and Leibniz (see the guest essay at the end of Chapter 2) centered not so much on the discovery of those ideas as on their systematization.

In this chapter, we shall consider various ways of efficiently computing derivatives. We shall also see how the derivative can be used to find rates of change and to measure the direction (slope) at each point on a graph. Falling body problems in physics and marginal analysis from economics are examples of applied topics to be discussed, and we shall also explore several topics involving basic concepts. It is fair to say that your success with differential calculus hinges on understanding this material.

3.1   *An Introduction to the Derivative: Tangents*

tangent lines, slope of a tangent, the derivative, existence of derivatives, continuity and differentiability, and derivative notation

TANGENT LINES

In elementary mathematics a **tangent to a circle** is defined as a line in the plane of the circle that intersects the circle in exactly one point. However, this is much too narrow a view for our purposes in calculus. We will now consider the concept of a line tangent to a given curve (not necessarily a circle) at a given point. In general, it is not a simple matter to find the slope of a tangent line at a given point $P_0(x_0, y_0)$. This is because the formula

$$\text{slope} = m = \frac{\Delta y}{\Delta x} = \frac{y_1 - y_0}{x_1 - x_0}$$

requires knowledge not only of the point of tangency (x_0, y_0), but of at least one other point (x_1, y_1) on the line as well.

SLOPE OF A TANGENT

The limit procedure for finding the slope of a tangent was originally developed by Pierre de Fermat and was later used by Isaac Newton. The novelty of the Fermat–Newton approach was the use of the "dynamic" limit process to attack the "static" problem of finding tangents.

Suppose we wish to find the slope of the tangent to $y = f(x)$ at the point $P(x_0, f(x_0))$. The strategy is to approximate the tangent by other lines whose slopes can be computed directly. In particular, consider the line joining the given point P to the neighboring point Q on the graph of f, as shown in Figure 3.1. This line is called a **secant** (a line that intersects, but is not tangent to, a curve). Compare the secants shown in Figure 3.1.

WARNING   Do not confuse *secant line* (a line that intersects a curve in two or more points) with the *secant function* of trigonometry. ←

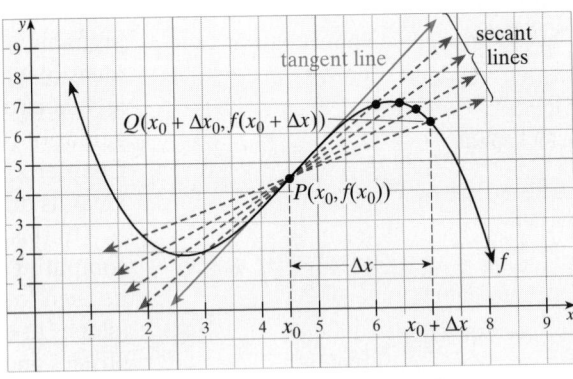

■ **FIGURE 3.1**   The secant $\overline{PQ}$

Notice that a secant is a good approximation to the tangent at point P as long as Q is close to P.

To compute the slope of a secant, first label the coordinates of the neighboring point Q, as indicated in Figure 3.1. In particular, let Δx denote the change in the x-coordinate between the given point $P(x_0, f(x_0))$ and the neighboring point

WARNING Δx is a single symbol and does not mean delta times x. Do not forget that as $\Delta x \to 0$, Δx is getting close to 0, but is not equal to 0.

$Q(x_0 + \Delta x, f(x_0 + \Delta x))$. The slope of this secant, m_{sec}, is easy to calculate:

$$m_{\text{sec}} = \frac{\Delta y}{\Delta x} = \frac{f(x_0 + \Delta x) - f(x_0)}{\Delta x}$$

To bring the secant closer to the tangent, let Q approach P *on the graph of f* by letting Δx approach 0. As this happens, the slope of the secant should approach the slope of the tangent at P. We denote the slope of the tangent by m_{tan} to distinguish it from the slope of a secant. These observations suggest the following definition.

Slope of a Line Tangent to a Graph at a Point

At the point $P(x_0, f(x_0))$, the tangent line to the graph of f has **slope** given by the formula

$$m_{\text{tan}} = \lim_{\Delta x \to 0} \frac{f(x_0 + \Delta x) - f(x_0)}{\Delta x}$$

provided this limit exists.

EXAMPLE 1   *Slope of a tangent at a particular point*

Find the slope of the tangent to the graph of $f(x) = x^2$ at the point $P(-1, 1)$.
Solution   Figure 3.2a shows the tangent to f at $x = -1$.

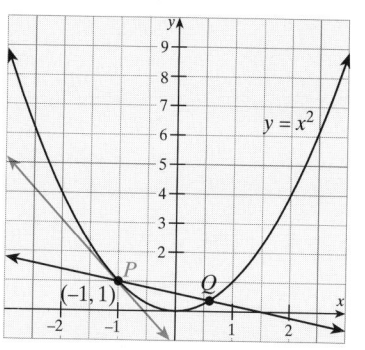

a. Tangent line at $(-1, 1)$

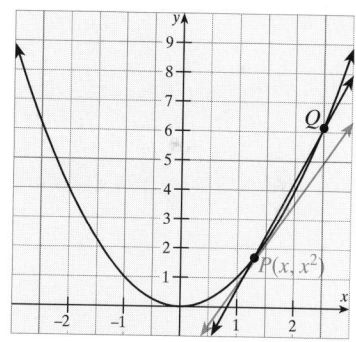

b. Tangent line at (x, x^2)

■ **FIGURE 3.2**   Tangent lines to the graph of $y = x^2$

The slope of the tangent is given by

$$m_{\text{tan}} = \lim_{\Delta x \to 0} \frac{f(-1 + \Delta x) - f(-1)}{\Delta x} \qquad \textit{Because } f(x) = x^2,$$
$$\textit{f}(-1 + \Delta x) = (-1 + \Delta x)^2.$$

$$= \lim_{\Delta x \to 0} \frac{(-1 + \Delta x)^2 - (-1)^2}{\Delta x}$$

$$= \lim_{\Delta x \to 0} \frac{1 - 2\Delta x + (\Delta x)^2 - 1}{\Delta x}$$

$$= \lim_{\Delta x \to 0} \frac{-2\Delta x + (\Delta x)^2}{\Delta x}$$

$$= \lim_{\Delta x \to 0} \frac{(-2 + \Delta x)\Delta x}{\Delta x} \qquad \text{Factor out } \Delta x \text{ and reduce.}$$

$$= \lim_{\Delta x \to 0} (-2 + \Delta x) = -2 \qquad \blacksquare$$

In Example 1 we found the slope of the tangent to the graph of $y = x^2$ at the point $(-1, 1)$. In Example 2, we perform the same calculation again, this time representing the given point algebraically as (x, x^2). This is the situation shown in Figure 3.2b for the slope of the tangent to $y = x^2$ at *any* point (x, x^2).

EXAMPLE 2   *Slope of a tangent at an arbitrary point*

Derive a formula for the slope of the tangent to the graph of $f(x) = x^2$, and then use the formula to compute the slope at $(4, 16)$.

Solution   Figure 3.2b shows a tangent at an arbitrary point $P(x, x^2)$ on the curve. From the definition of slope of the tangent,

$$m_{\text{tan}} = \lim_{\Delta x \to 0} \frac{f(x + \Delta x) - f(x)}{\Delta x} \qquad \begin{array}{l} \text{Because } f(x) = x^2, \\ f(x + \Delta x) = (x + \Delta x)^2. \end{array}$$

$$= \lim_{\Delta x \to 0} \frac{(x + \Delta x)^2 - x^2}{\Delta x}$$

$$= \lim_{\Delta x \to 0} \frac{x^2 + 2x\Delta x + (\Delta x)^2 - x^2}{\Delta x}$$

$$= \lim_{\Delta x \to 0} \frac{2x\Delta x + (\Delta x)^2}{\Delta x}$$

$$= \lim_{\Delta x \to 0} \frac{(2x + \Delta x)\Delta x}{\Delta x} \qquad \text{Factor and reduce.}$$

$$= \lim_{\Delta x \to 0} (2x + \Delta x) = 2x$$

At the point $(4, 16)$, $x = 4$, so $m_{\text{tan}} = 2(4) = 8$. $\qquad \blacksquare$

The result of Example 2 gives a general formula for the slope of a line tangent to the graph of $f(x) = x^2$, namely, $m_{\text{tan}} = 2x$. The answer from Example 1 can now be verified using this formula; if $x = -1$, then $m_{\text{tan}} = 2(-1) = -2$.

Technology Window—Tangent Lines Using a Graphing Calculator

Many calculators will graph tangent lines at particular points for a given function. For example, if we enter the function $f(x) = x^2$ from Example 2, we can graph the function f and then request the utility to draw the tangent line at a requested point. The calculator output shown at the top of p. 131 graphs not only the function $y = x^2$, but also the tangent line at $x = -1$.

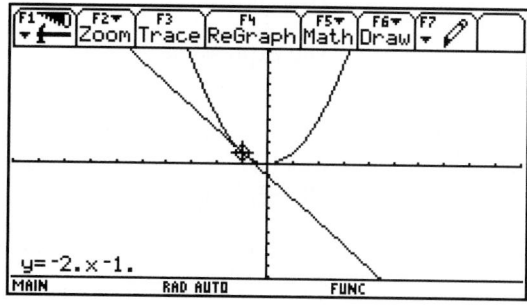

Not all graphing calculators will draw tangent lines. If you have a calculator that does not do so, you can still use your graphing calculator to graph

$$\frac{f(x_0 + \Delta x) - f(x_0)}{\Delta x}$$

for a small value of Δx. The resulting graph can be used to approximate the slope of the tangent line at a point x_0. Using $f(x) = x^2$ from Example 2, we created the graph below by graphing

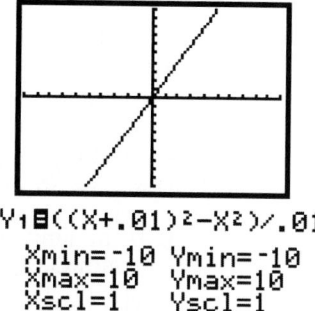

$$Y = ((X + 0.01)^2 - X^2)/.01$$

Note this is

$$\frac{f(x_0 + 0.01) - f(x_0)}{0.01} = \frac{(x_0 + 0.01)^2 - x_0^2}{0.01}$$

Use the TRACE to find the approximate slope of f at various values of x. Compare the results with those obtained in Example 2. *Hint:* Try resetting the RANGE value Xmin $= -9$ and note the trace values. On some models, you can enter this information as Y1 $= $ X^2 and Y2 $= $ (Y1(X + 0.01) $-$ Y1(X))/0.01. In this way, if you want to work another similar problem you need only change Y1.

PREVIEW

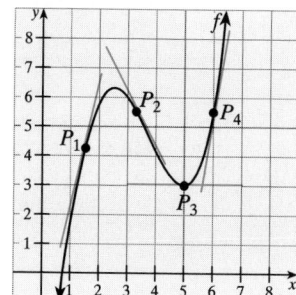

Knowledge of tangent lines at various points of a function f will give us some idea about the shape of the curve, as shown in Figure 3.3. We will consider this concept later in Chapter 4.

■ **FIGURE 3.3**   Tangent lines to a curve indicate the shape of the curve.

THE DERIVATIVE

The expression

$$\frac{f(x + \Delta x) - f(x)}{\Delta x}$$

which gives a formula for the slope of a secant to the graph of a function f, is called the **difference quotient** of f. The limit of the difference quotient

$$\lim_{\Delta x \to 0} \frac{f(x + \Delta x) - f(x)}{\Delta x}$$

which gives a formula for the slope of the tangent to the graph of f at the point $(x, f(x))$ is called the **derivative** of f and is frequently denoted by the symbol $f'(x)$ (read "eff prime of x"). To **differentiate** a function f at x means to find its derivative at the point $(x, f(x))$.

Derivative

The **derivative** of f at x is given by

$$f'(x) = \lim_{\Delta x \to 0} \frac{f(x + \Delta x) - f(x)}{\Delta x}$$

provided this limit exists.

The derivative is one of the fundamental concepts in calculus, and it is important to make some observations regarding this definition.

1. If the limit for the difference quotient exists, then we say that the function f is **differentiable at x.**

2. The value of a derivative depends only on the limit process and not on the symbols used in that process. In Example 2, we found that if $f(x) = x^2$, then $f'(x) = 2x$. This means that we also know

$$\text{If } g(t) = t^2, \text{ then } g'(t) = 2t.$$
$$\text{If } h(u) = u^2, \text{ then } h'(u) = 2u.$$
$$\vdots$$

3. Notice that the derivative of a function is itself a function.

Finding the slope of a tangent line is just one of several applications of the derivative that we shall discuss in this chapter. In Section 3.4, we shall examine rectilinear motion and other rates of change, and in Section 3.8, marginal analysis from economics. In Chapter 4 we shall examine more complex applications such as curve sketching and optimization.

EXAMPLE 3   *Derivative using the definition*

Differentiate $f(t) = \sqrt{t}$.

Solution
$$f'(t) = \lim_{\Delta t \to 0} \frac{f(t + \Delta t) - f(t)}{\Delta t}$$
$$= \lim_{\Delta t \to 0} \frac{\sqrt{t + \Delta t} - \sqrt{t}}{\Delta t}$$

$$= \lim_{\Delta t \to 0} \frac{\sqrt{t + \Delta t} - \sqrt{t}}{\Delta t} \left(\frac{\sqrt{t + \Delta t} + \sqrt{t}}{\sqrt{t + \Delta t} + \sqrt{t}} \right) \qquad \text{Rationalize numerator.}$$

$$= \lim_{\Delta t \to 0} \frac{(t + \Delta t) - t}{\Delta t (\sqrt{t + \Delta t} + \sqrt{t})}$$

$$= \lim_{\Delta t \to 0} \frac{\Delta t}{\Delta t (\sqrt{t + \Delta t} + \sqrt{t})}$$

$$= \lim_{\Delta t \to 0} \frac{1}{\sqrt{t + \Delta t} + \sqrt{t}} \qquad \text{Reduce fraction.}$$

$$= \frac{1}{2\sqrt{t}} \qquad \text{For } t > 0 \qquad \blacksquare$$

WARNING Notice that $f(t) = \sqrt{t}$ is defined for all $t \geq 0$, whereas its derivative $f'(t) = \dfrac{1}{2\sqrt{t}}$ is defined for all $t > 0$. This shows that a function need not be differentiable throughout its entire domain. ◄

PREVIEW An encouraging word is necessary after reading this first example of finding a derivative. By now you are aware of the fact that the derivative concept is one of the main ideas of calculus, and the previous example indicates that finding a derivative can be long and tedious. In the next section, we will begin to simplify the *process* of finding derivatives, so that you can quickly and efficiently find the derivative of a given function (without using this definition directly). For now, however, we focus on the derivative *concept* and *definition*.

THEOREM 3.1   *Equation of a line tangent to a curve at a point*

If f is a differentiable function at x_0, the graph of $y = f(x)$ has a tangent line at the point $P(x_0, f(x_0))$ with slope $f'(x_0)$ and equation

$$y = f'(x_0)(x - x_0) + f(x_0)$$

Proof   To find the equation of the tangent to the curve $y = f(x)$ at the point $P(x_0, y_0)$, we use the fact that the slope of the tangent is the derivative $f'(x_0)$ and apply the point–slope formula for the equation of a line:

$$y - k = m(x - h) \qquad \text{Point–slope formula}$$
$$y - y_0 = m(x - x_0) \qquad \text{Given point } (x_0, y_0)$$
$$y - f(x_0) = f'(x_0)(x - x_0) \qquad \begin{array}{l} y_0 = f(x_0) \text{ and} \\ m_{\tan} = f'(x_0) \end{array}$$
$$y = f'(x_0)(x - x_0) + f(x_0) \qquad \begin{array}{l} \text{Add } f(x_0) \text{ to both} \\ \text{sides.} \end{array}$$

EXAMPLE 4   *Equation of a tangent*

Find an equation for the tangent to the graph of $f(x) = \dfrac{1}{x}$ at the point where $x = 2$.

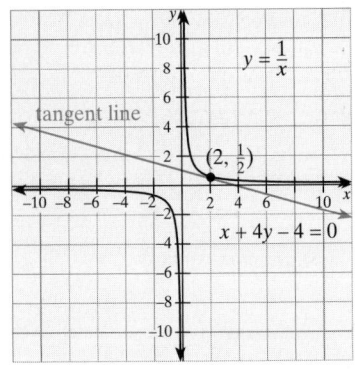

■ **FIGURE 3.4** Tangent to $y = \dfrac{1}{x}$
at $(2, \tfrac{1}{2})$

Solution   The graph of the function $y = \dfrac{1}{x}$, the point where $x = 2$, and the tangent at the point are shown in Figure 3.4. First, find $f'(x)$:

$$f'(x) = \lim_{\Delta x \to 0} \frac{f(x + \Delta x) - f(x)}{\Delta x} \qquad \textit{Definition of derivative}$$

$$= \lim_{\Delta x \to 0} \frac{\dfrac{1}{x + \Delta x} - \dfrac{1}{x}}{\Delta x} \qquad f(x) = \frac{1}{x}; f(x + \Delta x) = \frac{1}{x + \Delta x}$$

$$= \lim_{\Delta x \to 0} \frac{x - (x + \Delta x)}{x \Delta x (x + \Delta x)} \qquad \textit{Simplify the fraction.}$$

$$= \lim_{\Delta x \to 0} \frac{-1}{x(x + \Delta x)}$$

$$= \frac{-1}{x^2}$$

Next, find the slope of the tangent at $x = 2$: $m_{\tan} = f'(2) = -\tfrac{1}{4}$. Since $f(2) = \tfrac{1}{2}$, the equation of the tangent can now be found by using Theorem 3.1:

$$y = -\tfrac{1}{4}(x - 2) + \tfrac{1}{2} \quad \text{or, in standard form,} \quad x + 4y - 4 = 0 \qquad ■$$

WARNING   The slope of f at x_0 is not the derivative f' but the *value* of the derivative at x_0. In Example 4 the function is defined by $f(x) = 1/x$, the derivative is $f'(x) = -1/x^2$, and the slope at $x = 2$ is the *number* $f'(2) = -\tfrac{1}{4}$.

EXAMPLE 5   *A line that is perpendicular to a tangent*

Find the equation of the line that is perpendicular to the tangent of $f(x) = \dfrac{1}{x}$ at $x = 2$ and intersects it at the point of tangency.

Solution   From Example 4, we found that the slope of the tangent is $f'(2) = -\tfrac{1}{4}$ and that the point of tangency is $(2, \tfrac{1}{2})$. In Section 1.2, we saw that two lines are perpendicular if and only if their slopes are negative reciprocals of each other. Thus, the perpendicular line we seek has slope 4 (the negative reciprocal of $m = -\tfrac{1}{4}$), as shown in Figure 3.5. The desired equation is

$$y - \tfrac{1}{2} = 4(x - 2) \qquad \textit{Point–slope formula}$$

In standard form, the equation is $4x - y - \tfrac{15}{2} = 0$; compare the coefficients of the variables in the tangent and perpendicular lines.   ■

The perpendicular line we found in Example 5 has a name:

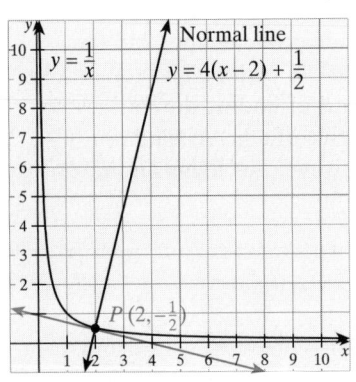

■ **FIGURE 3.5** Graph for Example 5

Normal Line to a Graph

The **normal line** to the graph of f at the point P is the line that is perpendicular to the tangent to the graph at P.

It is important to take some time to study the relationship between the graph of a function and its derivative. Figure 3.6 shows a function f and its derivative f'. Notice that when the graph of f is rising, the derivative f' is positive, and when the graph of f is falling, the derivative f' is negative. The graph of f' crosses the x-axis when the derivative of f is 0.

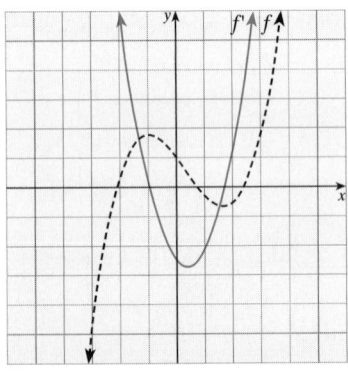

■ **FIGURE 3.6** Graphical relation of an equation and its graph

| EXAMPLE 6 | *Drawing the graph of a function, given the graph of its derivative* |

Consider the graph of f', as shown in Figure 3.7. Sketch a possible graph of f.

Solution   When the derivative is positive, the graph of f is rising, and when the derivative is negative, the graph of f is falling. For this example, we also see that at places where the derivative is zero, the graph has a relative maximum to the left of the y-axis and a relative minimum to the right of the y-axis, as shown in Figure 3.8.

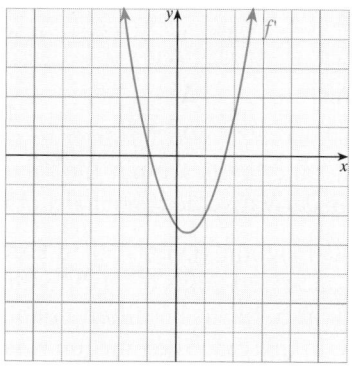

■ FIGURE 3.7   Graph of derivative of a function f

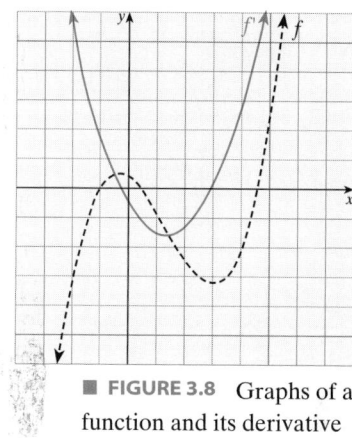

■ FIGURE 3.8   Graphs of a function and its derivative   ■

EXISTENCE OF DERIVATIVES

We observed that a function is differentiable only if the limit in the definition of derivative exists. At points where a function f is not differentiable, we say that *the derivative of f does not exist.* The three common ways for a derivative to fail to exist at a point $(c, f(c))$ in the domain of f are shown in Figure 3.9.

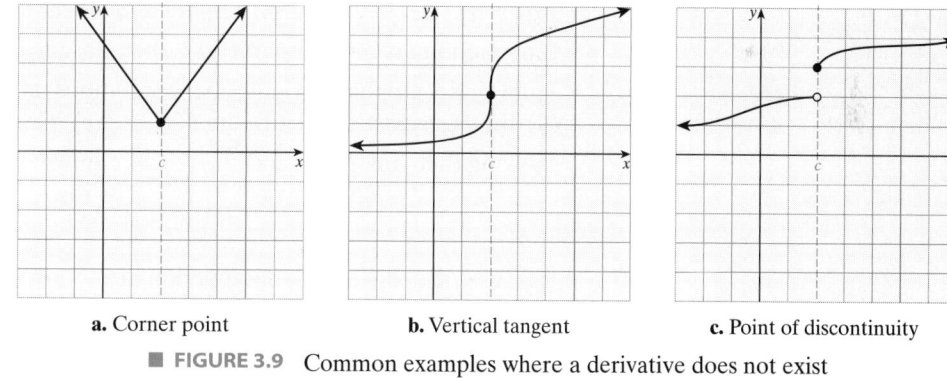

a. Corner point   **b.** Vertical tangent   **c.** Point of discontinuity

■ FIGURE 3.9   Common examples where a derivative does not exist

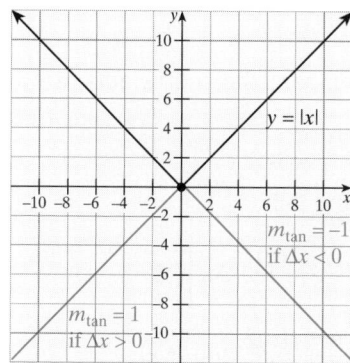

■ FIGURE 3.10 $f(x) = |x|$ is not differentiable at $x = 0$ because the slope from the left does not equal the slope from the right.

| EXAMPLE 7 | *A function that does not have a derivative because of a corner* |

Show that the absolute value function $f(x) = |x|$ is not differentiable at $x = 0$.

Solution   The graph of $f(x) = |x|$ is shown in Figure 3.10. Note that because the slope "from the left" at $x = 0$ is -1 while the slope "from the right" is $+1$, the graph has a corner at the origin, which prevents a unique tangent from being drawn there.

We can show this algebraically by using the definition of derivative:

$$f'(0) = \lim_{\Delta x \to 0} \frac{f(0 + \Delta x) - f(0)}{\Delta x}$$

$$= \lim_{\Delta x \to 0} \frac{f(\Delta x) - f(0)}{\Delta x}$$

$$= \lim_{\Delta x \to 0} \frac{|\Delta x|}{\Delta x}$$

We must now consider one-sided limits, because

$$|\Delta x| = \Delta x \text{ when } \Delta x > 0, \text{ and } |\Delta x| = -\Delta x \text{ when } \Delta x < 0.$$

$$\lim_{\Delta x \to 0^-} \frac{|\Delta x|}{\Delta x} = \lim_{\Delta x \to 0^-} \frac{-\Delta x}{\Delta x} = -1 \qquad \textit{Derivative from the left}$$

$$\lim_{\Delta x \to 0^+} \frac{|\Delta x|}{\Delta x} = \lim_{\Delta x \to 0^+} \frac{\Delta x}{\Delta x} = 1 \qquad \textit{Derivative from the right}$$

The left- and right-hand limits are not the same; therefore, the limit does not exist. This means that the derivative does not exist at $x = 0$. ■

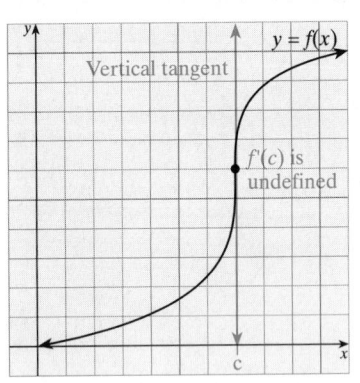

■ **FIGURE 3.11** Vertical tangent line at $x = c$

The continuous function $f(x) = |x|$ in Example 7 failed to be differentiable at $x = 0$ because the one-sided limits of its difference quotients were unequal. A continuous function may also fail to be differentiable at $x = c$ if its difference quotient diverges to infinity. In this case, the function is said to have a *vertical tangent* at $x = c$, as illustrated in Figure 3.11. We will have more to say about such functions when we discuss curve sketching with derivatives in Chapter 4.

CONTINUITY AND DIFFERENTIABILITY

If the graph of a function has a tangent at a point, we would expect to be able to draw the graph continuously (without the pencil leaving the paper). In other words, we expect the following theorem to be true.

THEOREM 3.2   *Differentiability implies continuity*

If a function f is differentiable at c, then it is also continuous at c.

Proof   Recall that for f to be continuous at $x = c$: (1) $f(c)$ must be defined; (2) $\lim_{x \to c} f(x)$ exists; and (3) $\lim_{x \to c} f(x) = f(c)$ Thus, continuity can be established by showing that $\lim_{\Delta x \to 0} f(c + \Delta x) = f(c)$ or, equivalently,

$$\lim_{\Delta x \to 0} [f(c + \Delta x) - f(c)] = 0$$

Because f is a differentiable function at $x = c$, $f'(c)$ exists and

$$\lim_{\Delta x \to 0} \frac{f(c + \Delta x) - f(c)}{\Delta x} = f'(c)$$

Therefore, by applying the product rule for limits, we find that

$$\lim_{\Delta x \to 0} [f(c + \Delta x) - f(c)] = \lim_{\Delta x \to 0} \left[\frac{f(c + \Delta x) - f(c)}{\Delta x} \cdot \Delta x \right]$$

$$= \left[\lim_{\Delta x \to 0} \frac{f(c + \Delta x) - f(c)}{\Delta x} \right] [\lim_{\Delta x \to 0} \Delta x]$$

$$= f'(c) \cdot 0 = 0$$

Thus, $\lim_{x \to c} f(x) = f(c)$, and we see that the conditions for continuity are satisfied.

WARNING   Be sure you understand what we have just shown with Example 7 and Theorem 3.2: If a function is differentiable at $x = c$, then it must be continuous at that point. The converse is not true: If a function is continuous at $x = c$, then it may or may not be differentiable at that point. Finally, if a function is discontinuous at $x = c$, then it cannot possibly have a derivative at that point. (See Figure 3.12c.)

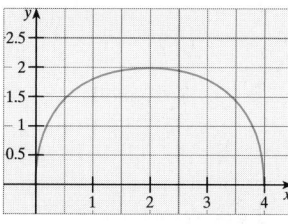

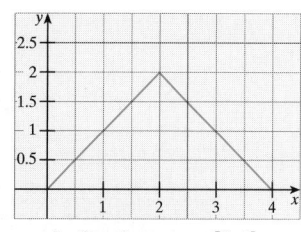

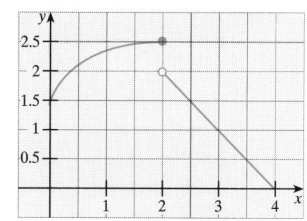

a. Continuous on $[0, 4]$; differentiable on $(0, 4)$

b. Continuous on $[0, 4]$; not differentiable at $x = 2$

c. Discontinuous at $x = 2$; cannot be differentiable at $x = 2$

■ **FIGURE 3.12**   A function continuous at $x = 2$ may or may not be differentiable at $x = 2$. A function discontinuous at $x = 2$ cannot be differentiable at $x = 2$.

DERIVATIVE NOTATION

In certain situations, it is convenient or suggestive to denote the derivative of $y = f(x)$ by $\dfrac{dy}{dx}$ instead of $f'(x)$. This notation is called the *Leibniz notation* because Leibniz was the first to use it.

For example, if $y = x^2$, the derivative is $y' = 2x$ and the Leibniz notation is $\dfrac{dy}{dx} = 2x$. The symbol $\dfrac{dy}{dx}$ is read "the derivative of y with respect to x." When we wish to denote the value of the derivative at c in the Leibniz notation, we shall write

$$\left. \frac{dy}{dx} \right|_{x=c}$$

For instance, we would evaluate $\dfrac{dy}{dx} = 4x^2$ at $x = 3$ by writing

$$\left. \frac{dy}{dx} \right|_{x=3} = 4x^2 \big|_{x=3} = 4(3)^2 = 36$$

Another notation omits reference to y and f altogether, and we can write

$$\frac{d}{dx}(x^2) = 2x$$

which is read "the derivative of x^2 with respect to x is $2x$."

WARNING   Despite its appearance, $\dfrac{dy}{dx}$ is a single symbol and is *not* a fraction. In Section 3.8, we introduce a concept called a *differential* that will provide independent meaning to symbols like dy and dx, but for now, these symbols have meaning only in connection with the Leibniz derivative symbol $\dfrac{dy}{dx}$.

EXAMPLE 8   *Derivative at a point with Leibniz notation*

Find $\left. \dfrac{dy}{dx} \right|_{x=-1}$ if $y = x^3$.

Solution
$$\frac{dy}{dx} = \frac{d}{dx}(x^3)$$

$$= \lim_{\Delta x \to 0} \frac{(x + \Delta x)^3 - x^3}{\Delta x}$$

$$= \lim_{\Delta x \to 0} \frac{[x^3 + 3x^2\Delta x + 3x(\Delta x)^2 + (\Delta x)^3] - x^3}{\Delta x}$$

$$= \lim_{\Delta x \to 0} [3x^2 + 3x(\Delta x) + (\Delta x)^2]$$

$$= 3x^2$$

At $x = -1$, $\quad \dfrac{dy}{dx}\bigg|_{x=-1} = 3x^2\big|_{x=-1} = 3.$ ∎

Technology Window

Derivative notation with technology requires that you input not only the function, but the variable in the expression and the value at which you wish to find the derivative. The usual format is

nDeriv(*expression, variable, value*)

The symbol "nDeriv" depends on the calculator or software, and sometimes is "nDer" (TI-85/86), "d" (TI-92), "diff" (Maple V), or "Dif" (Derive). This notation is used in the illustration at the right. Some software and calculators will find the exact value of a derivative, whereas some will do a numerical evaluation (as shown here).

```
nDeriv(X^3,X,-1)
                3.000001
```

EXAMPLE 9   *Estimating a derivative using a table*

Estimate the derivative of $f(x) = \cos x$ at $x = \frac{\pi}{6}$ by evaluating the difference quotient

$$\frac{\Delta y}{\Delta x} = \frac{f(x + \Delta x) - f(x)}{\Delta x}$$

near the point $x = \frac{\pi}{6}$.

Solution $\quad \dfrac{\Delta y}{\Delta x} = \dfrac{\cos(\frac{\pi}{6} + \Delta x) - \cos\frac{\pi}{6}}{\Delta x}$

Choose a sequence of values for $\Delta x \to 0$: say $1, \frac{1}{2}, \frac{1}{4}, \frac{1}{8}, \ldots$, and use a calculator (or a computer) to estimate the difference quotient by table. We show selected elements from this sequence of calculations:

Δx	$\dfrac{\pi}{6} + \Delta x$	Difference quotient $\dfrac{\Delta y}{\Delta x}$
1	1.523598776	−0.81885
0.5	1.023598776	−0.69146
0.125	0.648598776	−0.55276
0.0625	0.586098776	−0.52673
0.015625	0.539223776	−0.50675
0.00195313	0.525551906	−0.50085
0.00012207	0.523720846	−0.50005
0.000007629	0.523606405	−0.50000
↓	↓	↓
0	$\frac{\pi}{6} \approx 0.5235987756$	$-\frac{1}{2}$

A similar table for negative values of Δx should also be considered. From the table, we would guess that $f'(\frac{\pi}{6}) = -0.5$.   ■

Technology Window

Some calculators will evaluate limits directly, whereas others may require an investigation of the possible answer by a table of values or graph with trace.

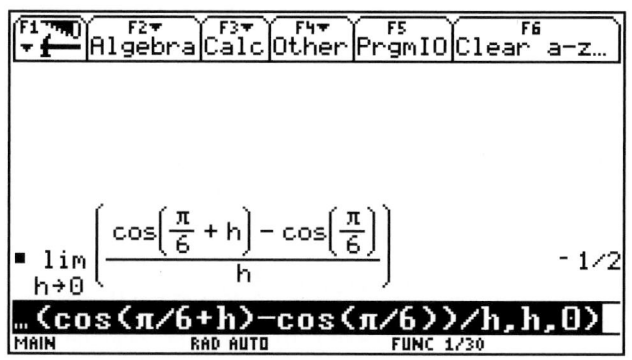

3.1   Problem Set

A
1. ■ **What Does This Say?** Describe the process of finding a derivative using the definition.

2. ■ **What Does This Say?** What is the definition of a derivative?

3. ■ **What Does This Say?** Discuss the truth or falsity of the following statements:

If a function f is continuous on (a, b), then it is differentiable on (a, b).

If a function f is differentiable on (a, b), then it is continuous on (a, b).

4. ■ **What Does This Say?** Discuss the relationship between the derivative of a function f at a point $x = x_0$ and the tangent line at that same point.

In each of Problems 5–10, the graph of a function f' is given. Draw a possible graph of f.

5.
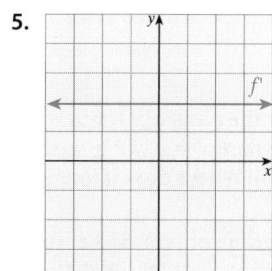

6.
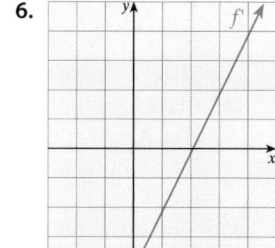

7.
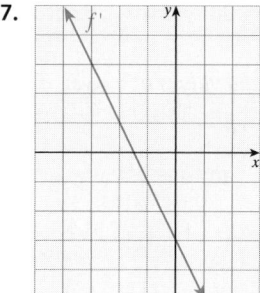

8.
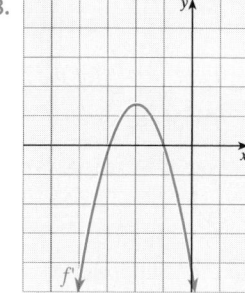

9.
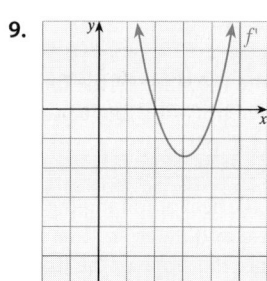

10.
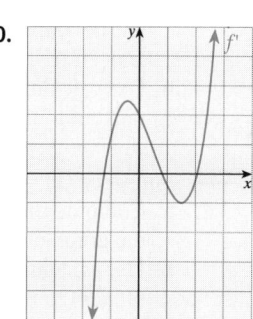

In each of Problems 11–16, a function f is given along with a number c in its domain.
a. *Find the difference quotient of f.*
b. *Find $f'(x)$ by computing the limit of the difference quotient.*

11. $f(x) = 3$ at $c = -5$
12. $f(x) = x$ at $c = 2$
13. $f(x) = 2x$ at $c = 1$
14. $f(x) = 2x^2$ at $c = 1$
15. $f(x) = 2 - x^2$ at $c = 0$
16. $f(x) = -x^2$ at $c = 2$

Use the definition to differentiate the functions given in Problems 17–28, and then describe the set of all numbers for which the function is differentiable.

17. $f(x) = 5$
18. $g(x) = 3x$
19. $f(x) = 3x - 7$
20. $g(x) = 4 - 5x$
21. $g(x) = 3x^2$
22. $h(x) = 2x^2 + 3$
23. $f(x) = x^2 - x$
24. $g(t) = 4 - t^2$
25. $f(s) = (s - 1)^2$
26. $f(x) = \dfrac{1}{2x}$
27. $f(x) = \sqrt{5x}$
28. $f(x) = \sqrt{x + 1}$

Find an equation for the tangent to the graph of the function at the specified point in Problems 29–34.

29. $f(x) = 3x - 7$ at $(3, 2)$
30. $g(x) = 3x^2$ at $(-2, 12)$
31. $f(s) = s^3$ at $s = -\frac{1}{2}$
32. $g(t) = 4 - t^2$ at $t = 0$
33. $f(x) = \dfrac{1}{x + 3}$ at $x = 2$
34. $g(x) = \sqrt{x - 5}$ at $x = 9$

Find an equation of the normal line to the graph of the function at the specified point in Problems 35–38.

35. $f(x) = 3x - 7$ at $(3, 2)$
36. $g(x) = 4 - 5x$ at $(0, 4)$
37. $f(x) = \dfrac{1}{x + 3}$ at $x = 3$
38. $f(x) = \sqrt{5x}$ at $x = 5$

Find $\left.\dfrac{dy}{dx}\right|_{x=c}$ for the functions and values of c given in Problems 39–42.

39. $y = 2x, c = -1$
40. $y = 4 - x, c = 2$
41. $y = 1 - x^2, c = 0$
42. $y = \dfrac{4}{x}, c = 1$

Ⓑ 43. Suppose $f(x) = x^2$.
   a. Compute the slope of the secant joining the points on the graph of f whose x-coordinates are -2 and -1.9.
   b. Use calculus to compute the slope of the line that is tangent to the graph when $x = -2$ and compare this slope with your answer in part **a**.

44. Suppose $f(x) = x^3$.
   a. Compute the slope of the secant joining the points on the graph of f whose x-coordinates are 1 and 1.1.
   b. Use calculus to compute the slope of the line that is tangent to the graph when $x = 1$ and compare this slope to your answer from part **a**.

45. Sketch the graph of the function $y = x^2 - x$. Determine the value of x for which the derivative is 0. What happens to the graph at the corresponding point(s)?

46. a. Find the derivative of $f(x) = x^2 - 3x$.

b. Show that the parabola whose equation is $y = x^2 - 3x$ has one horizontal tangent. Find the equation of this line.
   c. Find a point on the graph of f where the tangent is parallel to the line $3x + y = 11$.
   d. Sketch the graph of the parabola whose equation is $y = x^2 - 3x$. Display the horizontal tangent and the tangent found in part **c**.

47. a. Find the derivative of $f(x) = 4 - 2x^2$.
   b. The graph of f has one horizontal tangent. What is its equation?
   c. At what point on the graph of f is the tangent parallel to the line $8x + 3y = 4$?

48. Show that the function $f(x) = |x - 2|$ is not differentiable at $x = 2$.

49. Is the function $f(x) = 2|x + 1|$ differentiable at $x = 1$?

50. Let $f(x) = \begin{cases} -x^2 & \text{if } x < 0 \\ x^2 & \text{if } x \geq 0 \end{cases}$

Does $f'(0)$ exist? *Hint:* Find the difference quotient and take the limit as $\Delta x \to 0$ from the left and from the right.

51. Let $f(x) = \begin{cases} -2x & \text{if } x < 1 \\ \sqrt{x} - 3 & \text{if } x \geq 1 \end{cases}$

a. Sketch the graph of f.
   b. Show that f is continuous but not differentiable at $x = 1$.

52. **THINK TANK PROBLEM** Give an example of a function that is continuous on $(-\infty, \infty)$ but is not differentiable at $x = 5$.

Estimate the derivative $f'(c)$ in Problems 53–58 by evaluating the difference quotient

$$\frac{\Delta y}{\Delta x} = \frac{f(c + \Delta x) - f(c)}{\Delta x}$$

at a succession of numbers near c.

53. $f(x) = (2x - 1)^2$ for $c = 1$
54. $f(x) = \dfrac{1}{x + 1}$ for $c = 2$
55. $f(x) = \sin x$ for $c = \frac{\pi}{3}$
56. $f(x) = \cos x$ for $c = \frac{\pi}{3}$
57. $f(x) = \sqrt{x}$ for $c = 4$
58. $f(x) = \sqrt[3]{x}$ for $c = 8$

59. Show that the tangent to the parabola $y = Ax^2$ (for $A \neq 0$) at the point where $x = c$ will intersect the x-axis at the point $(c/2, 0)$. Where does it intersect the y-axis?

60. Find the point(s) on the graph of $f(x) = -x^2$ such that the tangent at that point passes through the point $(0, 9)$.

61. ■ **What Does This Say?**
  a. Find the derivatives of the functions $y = x^2$ and $y = x^2 - 3$ and account geometrically for their similarity.
  b. Without further computation, find the derivative of $y = x^2 + 5$.

62. ■ **What Does This Say?**
  a. Find the derivative of $f(x) = x^2 + 3x$.
  b. Find the derivatives of the functions $g(x) = x^2$ and $h(x) = 3x$ separately. How are these derivatives related to the derivative in part **a**?
  c. In general, if $f(x) = g(x) + h(x)$, what would you guess is the relationship between the derivative of f and the derivatives of g and h?

Technology Window

63. Consider a graph of the function defined by $f(x) = x^{2/3}$.

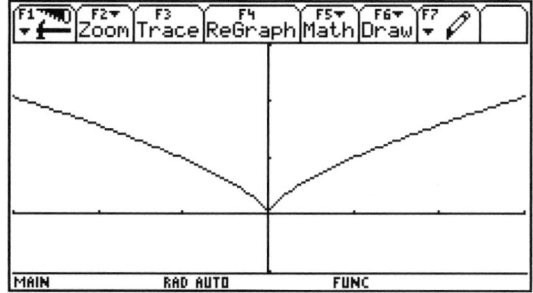

There is a tangent utility on many calculators that will draw tangent lines at a given point. Use this utility, or use the above simulated graph to draw tangent lines as $x \to 0^-$. Next, draw tangent lines as $x \to 0^+$. Describe what happens, and use this description to support the conclusion that there is no tangent line at $x = 0$.

64. Return to another function seen earlier (Problem 62, Chapter 2 Supplementary Problems)

$$f(x) = (x - 2)^{2/3} + 2x^3$$

which gives trouble in seeking the tangent at $(2, f(2))$. Attempt to compute $f'(2)$, either "by hand" or using a computer. Describe what happens; in particular, do you see why the tangent there is meaningless?

65. Compute the difference quotient for the function defined by

$$f(x) = \begin{cases} \dfrac{\sin x}{x} & \text{if } x \neq 0 \\ 1 & \text{if } x = 0 \end{cases}$$

Do you think $f(x)$ is differentiable at $x = 0$? If so, what is the equation of the tangent line at $x = 0$?

---

66. HISTORICAL QUEST In the quest essay at the end of Chapter 2, it was noted that Isaac Newton, who invented calculus, considered Fermat as "one of the giants" on whose shoulders he stood. Fermat was a lawyer by profession, but he liked to do mathematics in his spare time. He wrote well over 3,000 mathematical papers and notes. Fermat developed a general procedure for finding tangent lines that is a precursor to the methods of Newton and Leibniz.

PIERRE DE FERMAT
(1601–1665)

We shall explore this procedure by finding a tangent line to the curve

$$x^3 + y^3 - 2xy = 0$$

A graph is shown in Figure 3.13.

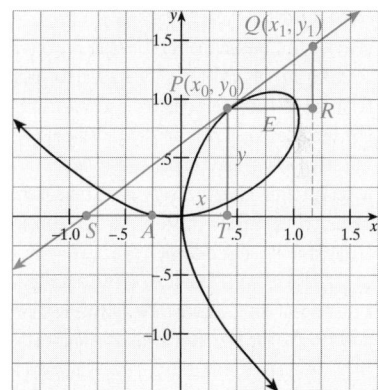

■ **FIGURE 3.13** Fermat's method of subtangents

  a. Let $P(x_0, y_0)$ be a given point on the curve, and let S be the x-axis intercept of the tangent line at P. Let $Q(x_1, y_1)$ be another point on the tangent line and let T and R be located so $\triangle STP$ and $\triangle PRQ$ are right triangles. Finally, let $A = |ST|$ and $E = |PR|$. Express x_1 and y_1 in terms of x_0, y_0, A, and E.
  b. Fermat reasoned that if E were very small, then Q would "almost" be on the curve. Substitute the values for x_1 and y_1 you found in part **a** into the expression

$$x^3 + y^3 - 2xy$$

That is, substitute the (x, y) values you found in part **a** into this expression.
  c. With the answer from part **b,** you can follow the steps of Fermat by dividing both sides by E. Fermat reasoned that since E was close to 0, this step should be permitted. Now, after doing this, Fermat further reasoned that since E was close to 0, he could now set $E = 0$ and solve for A. Carry out these steps to write A in terms of x_0 and y_0.

d. Fermat then constructed the tangent at P by joining P to S. Draw the tangent line for the given curve at the point $(0.5, 0.93)$ by plotting the point corresponding to the calculated value of A. Find an equation for the tangent line to the curve

$$x^3 + y^3 - 2xy = 0$$

at the point $P(0.5, 0.93)$.

67. **HISTORICAL QUEST**   The groundwork for much of the mathematics we do today, and certainly a necessity for calculus, is the development of *analytic geometry* by Descartes and Pierre de Fermat (see Problem 66). Descartes' ideas for analytic geometry were published in 1637 as one of three appendices to his *Discourse on the Method* (of *Reasoning Well and Seeking Truth in the Sciences*). In that same year, Fermat sent an essay entitled "Introduction to Plane and Solid Loci" to Paris, and in this essay he laid the foundation for analytic geometry. Fermat's paper was more complete and systematic, but Descartes' was published first. Descartes is generally credited with the discovery of analytic geometry, and we speak today of the Cartesian coordinate system and Cartesian geometry to honor Descartes' discovery. Today we describe analytic geometry from two viewpoints: (1) given a curve, describe it by an equation (Descartes' viewpoint); and (2) given an equation, describe it by a curve (Fermat's viewpoint).

RENÉ DESCARTES
(1596–1650)

In this Historical Quest we will describe Descartes' circle method for finding a tangent line to a given curve. This method uses algebra and geometry rather than limits. Descartes' method for finding the tangent line to the curve $y = f(x)$ at the point $P(x_0, f(x_0))$ involved first finding the point $Q(x_1, 0)$, which is the point of intersection of the normal line to the curve at P with the x-axis, as shown in Figure 3.14.

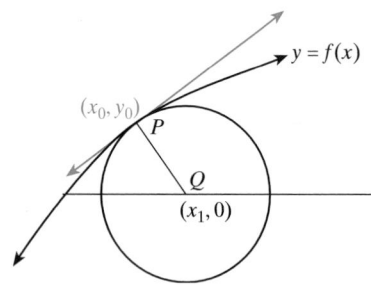

■ **FIGURE 3.14** Descartes' circle method for finding tangents

Descartes then wrote the equation of the circle with center Q passing through P:

$$(x - x_1)^2 + y^2 = (x_0 - x_1)^2 + y_0^2$$

Descartes' next step was to use the equation of the given curve $y = f(x)$ (actually any equation involving two variables) to eliminate one of the variables (usually y) from the equation of the circle. Descartes reasoned that the circle will cut the given curve in two places *except* when $\overline{PQ}$ is normal, in which case the two intersection points will coincide and the circle will be tangent to the curve at P. That is, Descartes imposed the condition that the resulting equation (after substitution) has only one root to solve for x_1. The point Q is thus found and the normal line was then known to Descartes. The tangent line can be taken as the perpendicular through P to the normal line.

Carry out Descartes' circle method for the parabola $y^2 = 4x$ at the point $(4, 4)$.

68. **Book Report**   **HISTORICAL QUEST**   Pierre de Fermat (see Problem 66) also obtained the first method for differentiating polynomials, but his real love was number theory. His most famous problem has come to be known as Fermat's last theorem. He wrote in the margin of a text:

"To divide a cube into two cubes, a fourth power, or in general, any power whatever above the second, into powers of the same denominations, is impossible, and I have assuredly found an admirable proof of this, but the margin is too narrow to contain it."

In 1993 an imaginative, but lengthy, proof was constructed by Andrew Wiles, and in 1997 Wiles was awarded the Wolfskehl Prize for proving this theorem. An engaging new book, *Fermat's Enigma*, shows how the greatest mathematical puzzle of our age was solved: Lock self in a room and emerge seven years later. Write a paper on the history of Fermat's last theorem, including some discussion of the current status of its proof. [You might begin with "Fermat's Last Theorem, The Four Color Conjecture, and Bill Clinton for April Fools' Day," by Edward B. Burger and Frank Morgan, *American Mathematical Monthly,* March 1997, pp. 246–250. Next, see the NOVA film, *The Proof,* which was shown in 1997 on many PBS stations (check with your library or PBS Online). Finally, see "Fermat's Last Stand" by Simon Singh and Kenneth A. Riber, *Scientific American,* November 1997, pp. 68–73.]

69. If $f'(c) \neq 0$, what is the equation of the normal line to $y = f(x)$ at the point $P(c, f(c))$? What is the equation if $f'(c) = 0$?

70. Suppose a parabola is given in the plane along with its axis of symmetry. Explain how you could construct the tangent at a given point P on the parabola using only compass and straightedge methods. *Hint:* You may assume that the parabola has an equation of the form $y = Ax^2$ in which the y-axis is the axis of symmetry and the vertex of the parabola is at the origin. Then use the result of Problem 66.

3.2 *Techniques of Differentiation*

derivative of a constant function, derivative of a power function, procedural rules for finding derivatives, higher derivatives   ∎

DERIVATIVE OF A CONSTANT FUNCTION

We begin by proving that the derivative of any constant function is zero. Notice that this is plausible because the graph of the constant function $f(x) = k$ is a horizontal line, and its slope is zero. Thus, for example, if $f(x) = 5$, then $f'(x) = 0$.

THEOREM 3.3 *Constant rule*

A constant function $f(x) = k$ has derivative $f'(x) = 0$; in Leibniz notation,

$$\frac{d}{dx}(k) = 0$$

Proof Note that if $f(x) = k$, then $f(x + \Delta x) = k$ for all Δx. Therefore, the difference quotient is

$$\frac{f(x + \Delta x) - f(x)}{\Delta x} = \frac{k - k}{\Delta x} = 0$$

and

WARNING ➤ Remember $\Delta x \neq 0$. even though $\Delta x \to 0$. ←

$$f'(x) = \lim_{\Delta x \to 0} \frac{f(x + \Delta x) - f(x)}{\Delta x} = \lim_{\Delta x \to 0} 0 = 0$$

as claimed.

DERIVATIVE OF A POWER FUNCTION

Recall that a **power function** is a function of the form $f(x) = x^n$ where n is a real number. For example, $f(x) = x^2$, $g(x) = x^{-3}$, $h(x) = x^{1/2}$ are all power functions. So are

$$F(x) = \frac{1}{x^2} = x^{-2} \quad \text{and} \quad G(x) = \sqrt[3]{x^2} = x^{2/3}$$

Here is a simple rule for finding the derivative of any power function.

THEOREM 3.4 *Power rule*

For any real number n, the power function $f(x) = x^n$ has the derivative $f'(x) = nx^{n-1}$; in Leibniz notation,

$$\frac{d}{dx}(x^n) = nx^{n-1}$$

Proof If the exponent n is a positive integer, we can prove the power rule by using the binomial theorem with the definition of derivative. Begin with the difference quotient:

$$\frac{f(x + \Delta x) - f(x)}{\Delta x} = \frac{(x + \Delta x)^n - x^n}{\Delta x}$$

$$= \frac{\left[x^n + nx^{n-1}\Delta x + \dfrac{n(n-1)}{2}x^{n-2}(\Delta x)^2 + \cdots + (\Delta x)^n\right] - x^n}{\Delta x}$$

$$= \frac{nx^{n-1}\Delta x + \frac{n(n-1)}{2}x^{n-2}(\Delta x)^2 + \cdots + (\Delta x)^n}{\Delta x}$$

$$= nx^{n-1} + \frac{n(n-1)}{2}x^{n-2}\Delta x + \cdots + (\Delta x)^{n-1}$$

Note that Δx is a factor of every term in this expression except the first. Hence, as $\Delta x \to 0$, we have

$$f'(x) = \lim_{\Delta x \to 0} \frac{f(x + \Delta x) - f(x)}{\Delta x}$$

$$\lim_{\Delta x \to 0} = \left[nx^{n-1} + \frac{n(n-1)}{2}x^{n-2}\Delta x + \cdots + (\Delta x)^{n-1}\right]$$

$$= nx^{n-1}$$

If $n = 0$, then $f(x) = x^0 = 1$, so $f'(x) = 0$. We shall prove the power rule for negative integer exponents later in this section, and we shall deal with the case in which the exponent is any real number in Section 3.6. Note, however, that we have already verified the power rule for the rational exponent $\frac{1}{2}$ in Example 3 of Section 3.1, when we showed that the derivative of $f(t) = \sqrt{t} = t^{1/2}$ is

$$f'(t) = \frac{1}{2}t^{-1/2} = \frac{1}{2\sqrt{t}} \quad \text{for } t > 0$$

For the following examples, and the problems at the end of this section, you may assume that the power rule is valid when the exponent n is any real number. ══

EXAMPLE 1   *Using the power rule to find a derivative*

Differentiate each of the following functions.

a. $f(x) = x^8$ b. $g(x) = x^{3/2}$ c. $h(x) = \dfrac{\sqrt[3]{x}}{x^2}$

Solution

a. Applying the power rule with $n = 8$, we find that

$$\frac{d}{dx}(x^8) = 8x^{8-1} = 8x^7$$

b. Applying the power rule with $n = \frac{3}{2}$, we get

$$\frac{d}{dx}(x^{3/2}) = \frac{3}{2}x^{3/2-1} = \frac{3}{2}x^{1/2} = \frac{3}{2}\sqrt{x}$$

c. For this part you need to recognize that $h(x) = \dfrac{x^{1/3}}{x^2} = x^{-5/3}$. Applying the power rule with $n = -\frac{5}{3}$, we find that

$$\frac{d}{dx}(x^{-5/3}) = -\frac{5}{3}x^{-5/3-1} = -\frac{5}{3}x^{-8/3}$$ ■

PROCEDURAL RULES FOR FINDING DERIVATIVES

The next theorem expands the class of functions that we can differentiate easily by giving rules for differentiating certain combinations of functions, such as sums, differences, products, and quotients. We shall see that the derivative of a sum (difference) is the sum (difference) of derivatives, but the derivative of a product (or quotient) does not have such a simple form. For example, to convince yourself that the derivative of a product is not the product of the separate derivatives, consider the power functions

$$f(x) = x \quad \text{and} \quad g(x) = x^2$$

> **WARNING** Note the product and quotient rules do not "behave" as you might expect.

and their product

$$p(x) = f(x)g(x) = x^3$$

Because $f'(x) = 1$ and $g'(x) = 2x$, the product of the derivatives is

$$f'(x)g'(x) = (1)(2x) = 2x$$

whereas the actual derivative of $p(x) = x^3$ is $p'(x) = 3x^2$. The product rule tells us how to find the derivative of a product.

THEOREM 3.5   *Basic rules for combining derivatives—Procedural forms*

If f and g are differentiable functions at all x, and a, b, and c are any real numbers, then the functions cf, $f + g$, fg, and f/g (for $g(x) \neq 0$) are also differentiable, and their derivatives satisfy the following formulas:

Name of Rule	Function Notation	Leibniz Notation
Constant multiple	$[cf(x)]' = cf'(x)$	$\dfrac{d}{dx}(cf) = c\dfrac{df}{dx}$
Sum rule	$[f(x) + g(x)]' = f'(x) + g'(x)$	$\dfrac{d}{dx}(f + g) = \dfrac{df}{dx} + \dfrac{dg}{dx}$
Difference rule	$[f(x) - g(x)]' = f'(x) - g'(x)$	$\dfrac{d}{dx}(f - g) = \dfrac{df}{dx} - \dfrac{dg}{dx}$

The constant multiple, sum, and difference rules can be combined into a single rule, which is called the *linearity rule*.

Linearity rule	$[af(x) + bg(x)]' = af'(x) + bg'(x)$	$\dfrac{d}{dx}(af + bg) = a\dfrac{df}{dx} + b\dfrac{dg}{dx}$
Product rule	$[f(x)g(x)]' = f(x)g'(x) + f'(x)g(x)$	$\dfrac{d}{dx}(fg) = f\dfrac{dg}{dx} + g\dfrac{df}{dx}$
Quotient rule	$\left[\dfrac{f(x)}{g(x)}\right]' = \dfrac{g(x)f'(x) - f(x)g'(x)}{[g(x)]^2}$	$\dfrac{d}{dx}\left(\dfrac{f}{g}\right) = \dfrac{g\dfrac{df}{dx} - f\dfrac{dg}{dx}}{g^2}$

Proof of the Product Rule   We shall prove the product rule in detail, leaving the other rules as problems.

Let $f(x)$ and $g(x)$ be differentiable functions of x and let $p(x) = f(x)g(x)$. We shall add and subtract the term $f(x + \Delta x)g(x)$ to the numerator of the difference quotient for $p(x)$ to create difference quotients for $f(x)$ and $g(x)$. Thus,

$$p'(x) = \frac{dp}{dx} = \lim_{\Delta x \to 0} \frac{p(x + \Delta x) - p(x)}{\Delta x}$$

$$= \lim_{\Delta x \to 0} \frac{f(x + \Delta x)g(x + \Delta x) - f(x)g(x)}{\Delta x}$$

$$= \lim_{\Delta x \to 0} \frac{f(x + \Delta x)g(x + \Delta x) - f(x + \Delta x)g(x) + f(x + \Delta x)g(x) - f(x)g(x)}{\Delta x}$$

$$= \lim_{\Delta x \to 0} \left[f(x + \Delta x)\left[\frac{g(x + \Delta x) - g(x)}{\Delta x} \right] + g(x)\left[\frac{f(x + \Delta x) - f(x)}{\Delta x} \right] \right]$$

$$= \lim_{\Delta x \to 0} f(x + \Delta x) \underbrace{\lim_{\Delta x \to 0} \left[\frac{g(x + \Delta x) - g(x)}{\Delta x} \right]}_{\text{This is the derivative of } g.} + \lim_{\Delta x \to 0} g(x) \underbrace{\lim_{\Delta x \to 0} \left[\frac{f(x + \Delta x) - f(x)}{\Delta x} \right]}_{\text{This is the derivative of } f.}$$

$$= f(x)g'(x) + g(x)f'(x) \qquad \lim_{\Delta x \to 0} f(x + \Delta x) = f(x) \text{ because } f \text{ is continuous.} \qquad \blacksquare$$

Note that in each part of Theorem 3.5, we prove the differentiability of the appropriate functional combination at the same time we are establishing the differentiation formula.

EXAMPLE 2   *Using the basic rules to find a derivative*

Differentiate each of the following functions.

a. $f(x) = 2x^2 - 5\sqrt{x}$

b. $p(x) = (3x^2 - 1)(7 + 2x^3)$

c. $q(x) = \dfrac{4x - 7}{3 - x^2}$

d. $g(x) = (4x + 3)^2$

e. $F(x) = \dfrac{2}{3x^2} - \dfrac{x}{3} + \dfrac{4}{5} + \dfrac{x + 1}{x}$

Solution

a. Apply the linearity rule (constant multiple, sum, and difference), and power rules:
$$f'(x) = 2(x^2)' - 5(x^{1/2})' = 2(2x) - 5(\tfrac{1}{2})(x^{-1/2}) = 4x - \tfrac{5}{2}x^{-1/2}$$

b. Apply the product rule; then apply the linearity and power rules:
$$\begin{aligned} p'(x) &= (3x^2 - 1)(7 + 2x^3)' + (3x^2 - 1)'(7 + 2x^3) \\ &= (3x^2 - 1)[0 + 2(3x^2)] + [3(2x) - 0](7 + 2x^3) \\ &= (3x^2 - 1)(6x^2) + (6x)(7 + 2x^3) \\ &= 6x(5x^3 - x + 7) \end{aligned}$$

c. Apply the quotient rule, then the linearity and power rules:
$$\begin{aligned} q'(x) &= \frac{(3 - x^2)(4x - 7)' - (4x - 7)(3 - x^2)'}{(3 - x^2)^2} \\ &= \frac{(3 - x^2)(4 - 0) - (4x - 7)(0 - 2x)}{(3 - x^2)^2} \\ &= \frac{12 - 4x^2 + 8x^2 - 14x}{(3 - x^2)^2} = \frac{4x^2 - 14x + 12}{(3 - x^2)^2} \end{aligned}$$

d. Apply the product rule:

$$g'(x) = (4x + 3)(4x + 3)' + (4x + 3)'(4x + 3)$$
$$= (4x + 3)(4) + (4)(4x + 3) = 8(4x + 3)$$

Sometimes when the exponent is 2, it is easier to expand before differentiating:

$$g(x) = (4x + 3)^2 = 16x^2 + 24x + 9$$
$$g'(x) = 32x + 24$$

e. Write the function using negative exponents for rational expressions:

$$F(x) = \tfrac{2}{3}x^{-2} - \tfrac{1}{3}x + \tfrac{4}{5} + 1 + x^{-1}$$

Then apply the power rule term by term to obtain

$$F'(x) = \tfrac{2}{3}(-2x^{-3}) - \tfrac{1}{3} + 0 + 0 + (-1)x^{-2}$$
$$= -\tfrac{4}{3}x^{-3} - \tfrac{1}{3} - x^{-2} \qquad\blacksquare$$

In applying the power rule term by term in Example 2e, we really used the following generalization of the linearity rule.

COROLLARY TO THEOREM 3.5   *The extended linearity rule*

If $f_1, f_2, \ldots, f_n$ are differentiable functions and $a_1, a_2, \ldots, a_n$ are constants, then

$$\frac{d}{dx}[a_1 f_1 + a_2 f_2 + \cdots + a_n f_n] = a_1 \frac{df_1}{dx} + a_2 \frac{df_2}{dx} + \cdots + a_n \frac{df_n}{dx}$$

Proof   The proof is a straightforward extension (using mathematical induction) of the proof of the linearity rule of Theorem 3.5.   ═

Example 3 illustrates how the extended linearity rule can be used to differentiate a polynomial.

EXAMPLE 3   *Derivative of a polynomial function*

Differentiate the polynomial function $p(x) = 2x^5 - 3x^2 + 8x - 5$.

Solution $p'(x) = \dfrac{d}{dx}[2x^5 - 3x^2 + 8x - 5]$

$$= 2\frac{d}{dx}(x^5) - 3\frac{d}{dx}(x^2) + 8\frac{d}{dx}(x) - \frac{d}{dx}(5) \qquad \textit{Extended linearity rule}$$
$$= 2(5x^4) - 3(2x) + 8(1) - 0 \qquad\qquad \textit{Power rule; constant rule}$$
$$= 10x^4 - 6x + 8 \qquad\qquad\qquad\qquad\qquad \blacksquare$$

EXAMPLE 4   *Derivative of a product of polynomials*

Differentiate $p(x) = (x^3 - 4x + 7)(3x^5 - x^2 + 6x)$.

Solution   We could expand the product function $p(x)$ as a polynomial and proceed as in Example 3, but it is easier to use the product rule:

$$p'(x) = (x^3 - 4x + 7)(3x^5 - x^2 + 6x)' + (x^3 - 4x + 7)'(3x^5 - x^2 + 6x)$$
$$= (x^3 - 4x + 7)(15x^4 - 2x + 6) + (3x^2 - 4)(3x^5 - x^2 + 6x)$$

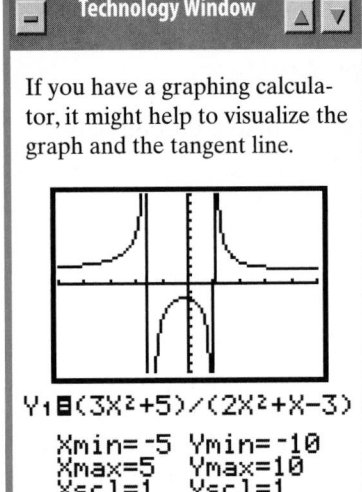

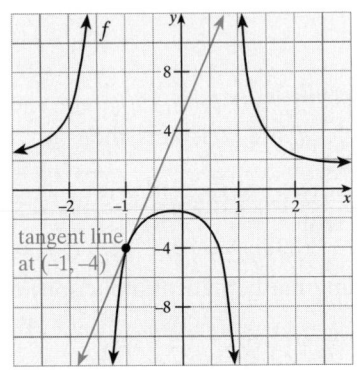

■ **FIGURE 3.15** Graph of f and the tangent line at the point $(-1, -4)$

This form is an acceptable answer, but if you are using software or an algebraic calculator, more than likely you will obtain

$$p'(x) = 24x^7 - 72x^5 + 100x^4 + 24x^3 + 12x^2 - 62x + 42$$ ■

EXAMPLE 5   *Equation of a tangent line*

Find the standard form equation for the line tangent to the graph of

$$f(x) = \frac{3x^2 + 5}{2x^2 + x - 3}$$

at the point where $x = -1$.

Solution   Evaluating $f(x)$ at $x = -1$, we find that $f(-1) = -4$ (verify); therefore, the point of tangency is $(-1, -4)$. The slope of the tangent line at $(-1, -4)$ is $f'(-1)$. Find $f'(x)$ by applying the quotient rule:

$$f'(x) = \frac{(2x^2 + x - 3)(3x^2 + 5)' - (3x^2 + 5)(2x^2 + x - 3)'}{(2x^2 + x - 3)^2}$$

$$= \frac{(2x^2 + x - 3)(6x) - (3x^2 + 5)(4x + 1)}{(2x^2 + x - 3)^2}$$

The slope of the tangent line is

$$f'(-1) = \frac{(2 - 1 - 3)(-6) - (3 + 5)(-4 + 1)}{(2 - 1 - 3)^2} = \frac{(-2)(-6) - (8)(-3)}{(-2)^2} = 9$$

From the formula (Theorem 3.1) $y = f'(x_0)(x - x_0) + f(x_0)$, we find that an equation for the tangent line is

$$y = 9(x + 1) + (-4)$$

or, in standard form, $9x - y + 5 = 0$. The graphs of both f and its tangent line at $(-1, -4)$ are shown in Figure 3.15.   ■

EXAMPLE 6   *Finding horizontal tangent lines*

Let $y = (x - 2)(x^2 + 4x - 7)$. Find all points on this curve where the tangent line is horizontal.

Solution   The tangent line will be horizontal when $dy/dx = 0$, because the derivative dy/dx measures the slope and a horizontal line has slope 0 (see Figure 3.16). Applying the product rule, we find

$$\frac{dy}{dx} = (x - 2)(x^2 + 4x - 7)' + (x - 2)'(x^2 + 4x - 7)$$

$$= (x - 2)(2x + 4) + (1)(x^2 + 4x - 7) = 2x^2 - 8 + x^2 + 4x - 7$$

$$= 3x^2 + 4x - 15 = (3x - 5)(x + 3)$$

Thus, $\frac{dy}{dx} = 0$ when $x = \frac{5}{3}$ or $x = -3$. The corresponding points $\left(\frac{5}{3}, \frac{-22}{27}\right)$ and $(-3, 50)$ are the points on the curve at which the tangent is horizontal.   ■

In the following example, we use the quotient rule to extend the proof of the power rule to the case in which the exponent n is a negative integer.

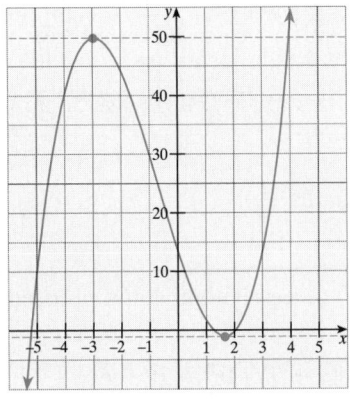

■ **FIGURE 3.16** Graph of curve f and the horizontal tangent lines

EXAMPLE 7 *Proof of the power rule for negative exponents*

Show that $\dfrac{d}{dx}(x^n) = nx^{n-1}$ if $n = -m$, where m is a positive integer.

Solution   We have $f(x) = x^n = x^{-m} = 1/x^m$, so apply the quotient rule:

$$\frac{d}{dx}(x^n) = \frac{d}{dx}\left(\frac{1}{x^m}\right) = \frac{x^m(1)' - (1)(x^m)'}{(x^m)^2} = \frac{x^m(0) - mx^{m-1}}{x^{2m}}$$

$$= -mx^{(m-1)-2m} = -mx^{-m-1} = nx^{n-1} \qquad \text{Substitute } -m = n. \qquad \blacksquare$$

HIGHER DERIVATIVES

Occasionally, it is useful to differentiate the derivative of a function. In this context, we shall refer to f' as the **first derivative** of f and to the derivative of f' as the **second derivative** of f. We could denote the second derivative by $(f')'$, but for simplicity we write f''. Other higher-order derivatives are defined and denoted similarly. Thus, the **third derivative** of f is the derivative of f'' and is denoted by f'''. In general, for $n > 3$, the nth derivative of f is denoted by $f^{(n)}$, for example, $f^{(4)}$ or $f^{(5)}$. In Leibniz notation, higher derivatives for $y = f(x)$ are denoted as follows:

Leibniz notation					
First derivative:	y'	$f'(x)$		$\dfrac{dy}{dx}$	or $\dfrac{d}{dx}f(x)$
Second derivative:	y''	$f''(x)$	$\dfrac{d}{dx}\left(\dfrac{dy}{dx}\right) = \dfrac{d^2y}{dx^2}$		or $\dfrac{d^2}{dx^2}f(x)$
Third derivative:	y'''	$f'''(x)$	$\dfrac{d}{dx}\left(\dfrac{d^2y}{dx^2}\right) = \dfrac{d^3y}{dx^3}$		or $\dfrac{d^3}{dx^3}f(x)$
Fourth derivative:	$y^{(4)}$	$f^{(4)}(x)$		$\dfrac{d^4y}{dx^4}$	or $\dfrac{d^4}{dx^4}f(x)$
$\vdots$	$\vdots$	$\vdots$		$\vdots$	$\vdots$
nth derivative	$y^{(n)}$	$f^{(n)}(x)$		$\dfrac{d^ny}{dx^n}$	or $\dfrac{d^n}{dx^n}f(x)$

> ■ *What This Says:*   Because the derivative of a function is a function, differentiation can be applied over and over, as long as the derivative itself is a differentiable function.
>
> Notice also that for derivatives higher than the third, the parentheses distinguish a derivative from a power. For example, $f^4 \neq f^{(4)}$.

You should also note that all higher derivatives of a polynomial $p(x)$ will also be polynomials, and if p has degree n, then $p^{(k)}(x) = 0$ for $k \geq n + 1$.

EXAMPLE 8 *Higher derivatives for a polynomial function*

Find the first and all higher-order derivatives of

$$p(x) = -2x^4 + 9x^3 - 5x^2 + 7$$

Solution $p'(x) = -8x^3 + 27x^2 - 10x;$ $p''(x) = -24x^2 + 54x - 10;$
$p'''(x) = -48x + 54;$ $p^{(4)}(x) = -48;$ $p^{(5)}(x) = 0;$ $\ldots$
$p^{(n)}(x) = 0$ $(n \geq 5)$ $\blacksquare$

Technology Window △ ▽

Many graphing calculators can find numerical values for higher derivatives. Check your owner's manual.

3.2 Problem Set

Ⓐ *To demonstrate the power of the theorems of this section, Problems 1–4 ask you to go back and rework some problems in Section 3.1, using the material of this section instead of the definition of derivative.*

1. Find the derivatives of the functions given in Problems 11–16 of Problem Set 3.1.

2. Find the derivatives of the functions given in Problems 17–22 of Problem Set 3.1.

3. Find the derivatives of the functions given in Problems 23–27 of Problem Set 3.1.

4. Find the derivatives of the functions given in Problems 39–42 of Problem Set 3.1.

Differentiate the functions given in Problems 5–20. Assume that C is a constant.

5. a. $f(x) = 3x^4 - 9$ b. $g(x) = 3(9)^4 - x$
6. a. $f(x) = 5x^2 + x$ b. $g(x) = \pi^3$
7. a. $f(x) = x^3 + C$ b. $g(x) = C^2 + x$

8. a. $f(t) = 10t^{-1}$ b. $g(t) = \dfrac{7}{t}$

9. $r(t) = t^2 - \dfrac{1}{t^2} + \dfrac{5}{t^4}$

10. $f(x) = \pi^3 - 3\pi^2$

11. $f(x) = \dfrac{7}{x^2} + x^{2/3} + C$

12. $g(x) = \dfrac{1}{2\sqrt{x}} + \dfrac{x^2}{4} + C$

13. $f(x) = \dfrac{x^3 + x^2 + x - 7}{x^2}$

14. $g(x) = \dfrac{2x^5 - 3x^2 + 11}{x^3}$

15. $f(x) = (2x + 1)(1 - 4x^3)$
16. $g(x) = (x + 2)(2\sqrt{x} + x^2)$

17. $f(x) = \dfrac{3x + 5}{x + 9}$

18. $f(x) = \dfrac{x^2 + 3}{x^2 + 5}$

19. $g(x) = x^2(x + 2)^2$
20. $f(x) = x^2(2x + 1)^2$

In Problems 21–24, find f', f'', f''', and $f^{(4)}$.

21. $f(x) = x^5 - 5x^3 + x + 12$

22. $f(x) = \dfrac{1}{4}x^8 - \dfrac{1}{2}x^6 - x^2 + 2$

23. $f(x) = \dfrac{-2}{x^2}$

24. $f(x) = \dfrac{4}{\sqrt{x}}$

25. Find $\dfrac{d^2y}{dx^2}$, where $y = 3x^3 - 7x^2 + 2x - 3$.

26. Find $\dfrac{d^2y}{dx^2}$, where $y = (x^2 + 4)(1 - 3x^3)$.

In Problems 27–32, find the standard form equation for the tangent line to $y = f(x)$ at the specified point.

27. $f(x) = x^2 - 3x - 5$, where $x = -2$
28. $f(x) = x^5 - 3x^3 - 5x + 2$, where $x = 1$
29. $f(x) = (x^2 + 1)(1 - x^3)$, where $x = 1$

30. $f(x) = \dfrac{x + 1}{x - 1}$, where $x = 0$

31. $f(x) = \dfrac{x^2 + 5}{x + 5}$, where $x = 1$

32. $f(x) = 1 - \dfrac{1}{x} + \dfrac{2}{\sqrt{x}}$, where $x = 4$

Find the coordinates of each point of the graph of the given function in Problems 33–39 where the tangent line is horizontal

33. $f(x) = 2x^3 - 7x^2 + 8x - 3$
34. $f(t) = t^4 + 4t^3 - 8t^2 + 3$
35. $g(x) = (3x - 5)(x - 8)$

36. $f(t) = \dfrac{1}{t^2} - \dfrac{1}{t^3}$

37. $f(x) = \sqrt{x}(x - 3)$

38. $h(u) = \dfrac{1}{\sqrt{u}}(u + 9)$

39. $h(x) = \dfrac{4x^2 + 12x + 9}{2x + 3}$

Ⓑ 40. a. Differentiate the function

$$f(x) = 2x^2 - 5x - 3$$

   b. Factor the function in part **a** and differentiate by using the product rule. Show that the two answers are the same.

41. a. Use the quotient rule to differentiate

$$f(x) = \dfrac{2x - 3}{x^3}$$

   b. Rewrite the function in part **a** as

$$f(x) = x^{-3}(2x - 3)$$

   and differentiate by using the product rule.

c. Rewrite the function in part **a** as

$$f(x) = 2x^{-2} - 3x^{-3}$$

and differentiate.
d. Show that the answers to parts **a, b,** and **c** are all the same.

42. Find numbers $a, b,$ and c that guarantee that the graph of the function $f(x) = ax^2 + bx + c$ will have x-intercepts at $(0,0)$ and $(5,0)$ and a tangent with slope 1 where $x = 2$.

43. Find the equation for the tangent to the curve with equation $y = x^4 - 2x + 1$ that is parallel to the line $2x - y - 3 = 0$.

44. Find equations for two tangent lines to the graph of

$$f(x) = \frac{3x + 5}{1 + x}$$

that are perpendicular to the line $2x - y = 1$.

45. Let $f(x) = (x^3 - 2x^2)(x + 2)$.
a. Find an equation for the tangent to the graph of f at the point where $x = 1$.
b. Find an equation for the normal line to the graph of f at the point where $x = 0$.

46. Find an equation for a normal line to the graph of $f(x) = (x^3 - 2x^2)(x + 2)$ that is parallel to the line $x - 16y + 17 = 0$.

47. Find all points (x, y) on the graph of $y = 4x^2$ with the property that the tangent at (x, y) passes through the point $(2, 0)$.

48. Find the equations of all the tangents to the graph of the function

$$f(x) = x^2 - 4x + 25$$

that pass through the origin.

Determine which (if any) of the functions $y = f(x)$ given in Problems 49–52 satisfy the equation

$$y''' + y'' + y' = x + 1$$

49. $f(x) = x^2 + 2x - 3$
50. $f(x) = x^3 + x^2 + x$
51. $f(x) = \frac{1}{2}x^2 + 3$
52. $f(x) = 2x^2 + x$

53. **HISTORICAL QUEST** When working with rational forms, we need to be careful about division by zero. One of the earliest recorded treatments of division by zero is attributed to the Hindu mathematician Āryabhata (476–499). He also gave rules for approximations of square roots and sums of arithmetic progressions as well as rules for basic algebraic manipulations. One example of his work is the following calculation for π: "Add four to one hundred, multiply by eight and add again sixty-two thousand; the result is the approximate value of the circumference of a circle whose diameter is twenty-thousand."

Follow the steps of Āryabhata's approximation for π. After you have completed this demonstration, discuss the procedure and technology you used and contrast it with the tools that Ārryabhata must have had available. *The first Indian satellite was named ARYABHAT in his honor.*

54. What is the relationship between the degree of a polynomial function P and the value of k for which $P^{(k)}(x)$ is first equal to 0?

55. Prove the constant multiple rule $(cf)' = cf'$.

56. Prove the sum rule $(f + g)' = f' + g'$.

57. Use the definition of derivative to find the derivative of f^2, given that f is a differentiable function.

58. Prove the product rule by using the result of Problem 57 and the identity

$$fg = \tfrac{1}{2}[(f + g)^2 - f^2 - g^2]$$

59. Prove the quotient rule

$$\left(\frac{f}{g}\right)' = \frac{gf' - fg'}{g^2}$$

where $g(x) \neq 0$. *Hint:* First show that the difference quotient for f/g can be expressed as

$$\frac{\frac{f}{g}(x + \Delta x) - \frac{f}{g}(x)}{\Delta x}$$

$$= \frac{f(x + \Delta x)g(x) - f(x)g(x + \Delta x)}{(\Delta x)g(x + \Delta x)g(x)}$$

and then subtract and add the term $g(x)f(x)$ in the numerator.

60. Show that the reciprocal function $r(x) = 1/f(x)$ has the derivative $r'(x) = -f'(x)/[f(x)]^2$ at each point x where f is differentiable and $f(x) \neq 0$.

61. If $f, g,$ and h are differentiable functions, show that the product fgh is also differentiable and

$$(fgh)' = fgh' + fg'h + f'gh$$

62. Let f be a function that is differentiable at x.
a. If $g(x) = [f(x)]^3$, show that

$$g'(x) = 3[f(x)]^2 f'(x)$$

Hint: Write $g(x) = [f(x)]^2 f(x)$ and use the product rule.
b. Show that $p(x) = [f(x)]^4$ has the derivative

$$p'(x) = 4[f(x)]^3 f'(x)$$

63. Find constants $A, B,$ and C so that

$$y = Ax^3 + Bx + C$$

satisfies the equation

$$y''' + 2y'' - 3y' + y = x$$

3.3 *Derivatives of Trigonometric, Exponential, and Logarithmic Functions*

derivatives of the sine and the cosine, differentiation of the other trigonometric functions, derivatives of exponential and logarithmic functions ■

DERIVATIVES OF THE SINE AND THE COSINE

In calculus we assume that the trigonometric functions are functions of real numbers or of angles measured in radians. We make this assumption because the trigonometric differentiation formulas rely on limit formulas that become complicated if degree measurement is used instead of radian measure.

WARNING ▶ Trigonometric formulas use radian measure unless otherwise stated. ◂

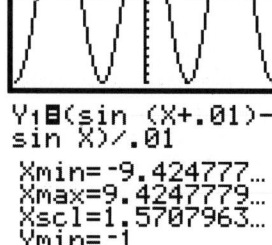

```
Y₁ᴮ(sin (X+.01)-
sin X)/.01
Xmin=-9.424777…
Xmax=9.4247779…
Xscl=1.5707963…
Ymin=-1
Ymax=1
Yscl=.1
```

Before stating the theorem that derives the derivative of sine and cosine, suppose we look at the graph of the difference quotient. Consider $f(x) = \sin x$. Then

$$\frac{\sin(x + \Delta x) - \sin x}{\Delta x} = \frac{\sin(x + 0.01) - \sin x}{0.01}$$

is the difference quotient for $\Delta x = 0.01$. The graph of this difference quotient, shown in the margin, is input as

$$Y1=(\text{SIN}(X + .01) - \text{SIN } X)/.01$$

From the graph of this difference quotient, it appears that the derivative of $f(x) = \sin x$ is $f'(x) = \cos x$. We now verify this with the following theorem. Before studying the derivation of this theorem, you might need to recall the following limits, which were established in Theorem 2.2:

$$\lim_{h \to 0} \frac{\sin h}{h} = 1 \qquad \lim_{h \to 0} \frac{\cos h - 1}{h} = 0$$

THEOREM 3.6   *Derivatives of the sine and cosine functions*

The functions $\sin x$ and $\cos x$ are differentiable for all x and

$$\frac{d}{dx} \sin x = \cos x \qquad \frac{d}{dx} \cos x = -\sin x$$

Proof   The proofs of these two formulas are similar. We shall prove the first using the trigonometric identity

$$\sin(\alpha + \beta) = \sin \alpha \cos \beta + \cos \alpha \sin \beta$$

and leave the proof of the second formula as a problem. From the definition of the derivative,

$$\frac{d}{dx} \sin x = \lim_{\Delta x \to 0} \frac{\sin(x + \Delta x) - \sin x}{\Delta x}$$

$$= \lim_{\Delta x \to 0} \frac{\sin x \cos \Delta x + \cos x \sin \Delta x - \sin x}{\Delta x}$$

$$= \lim_{\Delta x \to 0} \left[\sin x \left(\frac{\cos \Delta x}{\Delta x} \right) + \cos x \left(\frac{\sin \Delta x}{\Delta x} \right) - \frac{\sin x}{\Delta x} \right]$$

$$= (\sin x) \lim_{\Delta x \to 0} \left(\frac{\cos \Delta x - 1}{\Delta x} \right) + (\cos x) \lim_{\Delta x \to 0} \frac{\sin \Delta x}{\Delta x}$$

$$= \sin x(0) + \cos x(1)$$

$$= \cos x$$

EXAMPLE 1   *Derivative involving a trigonometric function*

Differentiate $f(x) = 2x^4 + 3\cos x + \sin a$, for constant a.

Solution $f'(x) = \dfrac{d}{dx}(2x^4 + 3\cos x + \sin a)$

$$= 2\frac{d}{dx}(x^4) + 3\frac{d}{dx}(\cos x) + \frac{d}{dx}(\sin a) \qquad \text{Extended linearity rule}$$

$$= 2(4x^3) + 3(-\sin x) + 0 \qquad \text{Power rule, derivative of cosine, and}$$
$$\text{derivative of a constant}$$

$$= 8x^3 - 3\sin x$$

EXAMPLE 2   *Derivative of a trigonometric function with product rule*

Differentiate $f(x) = x^2 \sin x$.

Solution $f'(x) = \dfrac{d}{dx}(x^2 \sin x)$

$$= x^2\frac{d}{dx}(\sin x) + \sin x\frac{d}{dx}(x^2) \qquad \text{Product rule}$$

$$= x^2 \cos x + 2x \sin x \qquad \text{Power rule and derivative of sine}$$

EXAMPLE 3   *Derivative of a trigonometric function with quotient rule*

Differentiate $h(t) = \dfrac{\sqrt{t}}{\cos t}$.

Solution   Write $\sqrt{t}$ as $t^{1/2}$. Then

$$h'(t) = \frac{d}{dt}\left[\frac{t^{1/2}}{\cos t}\right]$$

$$= \frac{\cos t\frac{d}{dt}(t^{1/2}) - t^{1/2}\frac{d}{dt}\cos t}{\cos^2 t} \qquad \text{Quotient rule}$$

$$= \frac{\frac{1}{2}t^{-1/2}\cos t - t^{1/2}(-\sin t)}{\cos^2 t} \qquad \text{Power rule and derivative of sine}$$

$$= \frac{\frac{1}{2}t^{-1/2}(\cos t + 2t \sin t)}{\cos^2 t} \qquad \text{Common factor } \frac{1}{2}t^{-1/2}$$

$$= \frac{\cos t + 2t \sin t}{2\sqrt{t}\cos^2 t}$$

DIFFERENTIATION OF THE OTHER TRIGONOMETRIC FUNCTIONS

You will need to be able to differentiate not only the sine and cosine functions, but also the other trigonometric functions. To find the derivatives of these functions

(SMH) you will need the following identities, which are given in the *Student Mathematics Handbook*.

$$\tan x = \frac{\sin x}{\cos x} \qquad \cot x = \frac{\cos x}{\sin x}$$

$$\sec x = \frac{1}{\cos x} \qquad \csc x = \frac{1}{\sin x}$$

You will also need the following identities:

$$\cos^2 x + \sin^2 x = 1 \qquad 1 + \tan^2 x = \sec^2 x \qquad \cot^2 x + 1 = \csc^2 x$$

THEOREM 3.7 *Derivatives of the trigonometric functions*

The six basic trigonometric functions $\sin x$, $\cos x$, $\tan x$, $\csc x$, $\sec x$, and $\cot x$ are all differentiable wherever they are defined, and

$$\frac{d}{dx}\sin x = \cos x \qquad \frac{d}{dx}\cos x = -\sin x$$

$$\frac{d}{dx}\tan x = \sec^2 x \qquad \frac{d}{dx}\cot x = -\csc^2 x$$

$$\frac{d}{dx}\sec x = \sec x \tan x \qquad \frac{d}{dx}\csc x = -\csc x \cot x$$

Proof The derivatives for sine and cosine were given in Theorem 3.6. All the other derivatives in this theorem are proved by using the appropriate quotient rules along with formulas for the derivatives of the sine and cosine. We will obtain the derivative of the tangent function and leave the rest as problems.

$$\frac{d}{dx}\tan x = \frac{d}{dx}\frac{\sin x}{\cos x} \qquad \text{Trigonometric identity}$$

$$= \frac{\cos x \dfrac{d}{dx}\sin x - \sin x \dfrac{d}{dx}\cos x}{\cos^2 x} \qquad \text{Quotient rule}$$

$$= \frac{\cos x(\cos x) - \sin x(-\sin x)}{\cos^2 x} \qquad \text{Derivative of } \sin x \text{ and } \cos x$$

$$= \frac{\cos^2 x + \sin^2 x}{\cos^2 x}$$

$$= \frac{1}{\cos^2 x} \qquad \cos^2 x + \sin^2 x = 1$$

$$= \sec^2 x \qquad 1/\cos^2 x = \sec^2 x$$

EXAMPLE 4 *Derivative of a trigonometric function with the product rule*

Differentiate $f(\theta) = 3\theta \sec \theta$.

Solution

$$f'(\theta) = \frac{d}{d\theta}(3\theta \sec \theta)$$

$$= 3\theta \frac{d}{d\theta}\sec \theta + \sec \theta \frac{d}{d\theta}(3\theta) \qquad \text{Product rule}$$

$$= 3\theta \sec \theta \tan \theta + 3 \sec \theta$$

EXAMPLE 5   *Derivative of a product of trigonometric functions*

Differentiate $f(x) = \sec x \tan x$.

Solution
$$f'(x) = \frac{d}{dx}(\sec x \tan x)$$

$$= \sec x \frac{d}{dx}\tan x + \tan x \frac{d}{dx}\sec x \qquad \text{Product rule}$$

$$= \sec x (\sec^2 x) + \tan x (\sec x \tan x)$$

$$= \sec^3 x + \sec x \tan^2 x \qquad\qquad\qquad \blacksquare$$

EXAMPLE 6   *Equation of a tangent line involving a trigonometric function*

Find the equation of the tangent line to the curve $y = \cot x - 2 \csc x$ at the point where $x = \frac{2\pi}{3}$.

Solution   The slope of the tangent line is the derivative of y with respect to x at the point $x = \frac{2\pi}{3}$.

$$\frac{dy}{dx} = \frac{d}{dx}(\cot x - 2 \csc x)$$

$$= \frac{d}{dx}\cot x - 2\frac{d}{dx}\csc x \qquad \text{Linearity rule}$$

$$= -\csc^2 x - 2(-\csc x \cot x)$$

$$= 2 \csc x \cot x - \csc^2 x$$

Writing this expression in terms of sine and cosine, we find

$$2 \csc x \cot x - \csc^2 x = 2\left(\frac{1}{\sin x}\right)\left(\frac{\cos x}{\sin x}\right) - \frac{1}{\sin^2 x}$$

$$= \frac{2 \cos x - 1}{\sin^2 x}$$

$$\left.\frac{dy}{dx}\right|_{x=2\pi/3} = \frac{2 \cos \frac{2\pi}{3} - 1}{\sin^2 \frac{2\pi}{3}} = \frac{2\left(-\frac{1}{2}\right) - 1}{\left(\frac{\sqrt{3}}{2}\right)^2} = -\frac{8}{3}$$

When $x = \frac{2\pi}{3}$, we have

$$y = \cot \frac{2\pi}{3} - 2 \csc \frac{2\pi}{3} = -\frac{\sqrt{3}}{3} - 2\left(\frac{2\sqrt{3}}{3}\right) = \frac{-5\sqrt{3}}{3}$$

so the point of tangency is $\left(\frac{2\pi}{3}, \frac{-5\sqrt{3}}{3}\right)$. The desired equation is

$$y + \frac{5\sqrt{3}}{3} = -\frac{8}{3}\left(x - \frac{2\pi}{3}\right)$$

$$24x + 9y + 15\sqrt{3} - 16\pi = 0 \qquad\qquad\qquad \blacksquare$$

DERIVATIVES OF EXPONENTIAL AND LOGARITHMIC FUNCTIONS

The next theorem, which is easy to prove and remember, is one of the most important results in all of differential calculus.

THEOREM 3.8   *Derivative rule for the natural exponential*

The natural exponential function e^x is differentiable for all x, with derivative

$$\frac{d}{dx}(e^x) = e^x$$

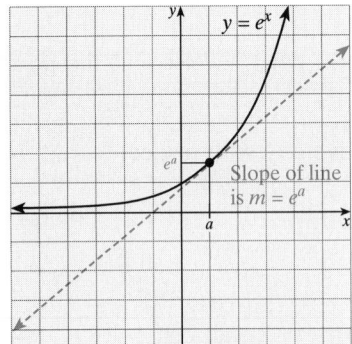

■ **FIGURE 3.17** The slope of $y = e^x$ at each point (a, e^a) is $m = e^a$.

Proof   We shall proceed informally. Recall the definition of e:

$$\lim_{n \to \infty}\left(1 + \frac{1}{n}\right)^n = e$$

Let $n = \dfrac{1}{\Delta x}$, so that $\lim\limits_{\Delta x \to 0}(1 + \Delta x)^{1/\Delta x} = e$. This means that for Δx very small, $e \approx (1 + \Delta x)^{1/\Delta x}$ or $e^{\Delta x} \approx 1 + \Delta x$ so that $e^{\Delta x} - 1 \approx \Delta x$. Thus, $\lim\limits_{\Delta x \to 0}\dfrac{e^{\Delta x} - 1}{\Delta x} = 1$. Finally, using the limit in the definition of derivative for e^x, we obtain

$$\begin{aligned}
\frac{d}{dx}(e^x) &= \lim_{\Delta x \to 0}\frac{e^{(x + \Delta x)} - e^x}{\Delta x}\\[2mm]
&= \lim_{\Delta x \to 0}\frac{e^x(e^{\Delta x} - 1)}{\Delta x}\\[2mm]
&= e^x \lim_{\Delta x \to 0}\frac{e^{\Delta x} - 1}{\Delta x}\\[2mm]
&= e^x(1)\\[2mm]
&= e^x
\end{aligned}$$

The fact that $\dfrac{d}{dx}(e^x) = e^x$ means that the slope of the graph of $y = e^x$ at any point $x = a$ is $m = e^a$, the y-coordinate of the point, as shown in Figure 3.17. This is one of the features of the exponential function $y = e^x$ that makes it "natural."

EXAMPLE 7   *A second derivative involving e^x*

For $f(x) = e^x \sin x$, find $f'(x)$ and $f''(x)$.

Solution   Apply the product rule twice:

$$\begin{aligned}
f'(x) &= e^x(\sin x)' + (e^x)' \sin x\\
&= e^x(\cos x) + e^x(\sin x)\\
&= e^x(\cos x + \sin x)\\[2mm]
f''(x) &= e^x(\cos x + \sin x)' + (e^x)'(\cos x + \sin x)\\
&= e^x(-\sin x + \cos x) + e^x(\cos x + \sin x)\\
&= 2e^x \cos x
\end{aligned}$$

THEOREM 3.9   *Derivative rule for the natural logarithmic function*

The natural logarithmic function $\ln x$ is differentiable for all $x > 0$, with derivative

$$\frac{d}{dx}(\ln x) = \frac{1}{x}$$

Proof   According to the definition of the derivative, we have

$$\frac{d}{dx}(\ln x) = \lim_{\Delta x \to 0} \frac{\ln(x + \Delta x) - \ln x}{\Delta x}$$

$$= \lim_{\Delta x \to 0} \frac{1}{\Delta x} \ln\left(\frac{x + \Delta x}{x}\right)$$

$$= \lim_{\Delta x \to 0} \frac{1}{\Delta x} \ln\left(1 + \frac{\Delta x}{x}\right) \qquad \text{Let } h = \frac{x}{\Delta x}, \text{ so } \Delta x = \frac{x}{h}.$$
$$\text{As } \Delta x \to 0, h \to +\infty.$$

$$= \lim_{h \to +\infty} \frac{h}{x} \ln\left(1 + \frac{1}{h}\right)$$

$$= \frac{1}{x}\left[\lim_{h \to +\infty} h \ln\left(1 + \frac{1}{h}\right)\right]$$

$$= \frac{1}{x}\left[\lim_{h \to +\infty} \ln\left(1 + \frac{1}{h}\right)^h\right] \qquad \text{Power rule for logarithms}$$

$$= \frac{1}{x}\left[\ln \lim_{h \to +\infty} \left(1 + \frac{1}{h}\right)^h\right] \qquad \text{Since } \ln x \text{ is continuous, we}$$
$$\text{use the composition limit rule.}$$

$$= \frac{1}{x}\ln e \qquad \text{Definition of } e$$

$$= \frac{1}{x} \qquad \ln e = 1$$

EXAMPLE 8   *Derivative of a quotient involving a natural logarithm*

Differentiate $f(x) = \dfrac{\ln x}{\sin x}$.

Solution   We use the quotient rule:

$$f'(x) = \frac{(\sin x)\dfrac{d}{dx}(\ln x) - (\ln x)\dfrac{d}{dx}(\sin x)}{\sin^2 x}$$

$$= \frac{(\sin x)\left(\dfrac{1}{x}\right) - (\ln x)(\cos x)}{\sin^2 x}$$

$$= \frac{\sin x - x \ln x \cos x}{x \sin^2 x}$$

Technology Window

When you are using calculators or computer programs such as *Mathematica, Derive,* or *Maple,* the form of the derivative may vary. For example, you might obtain

$$\frac{d}{dx}(\tan x) = \tan^2 x + 1 \quad \text{instead of } \sec^2 x \text{ as found in this section;}$$

$$\frac{d}{dx}(\cot x) = -\cot^2 x - 1 \quad \text{instead of } -\csc^2 x;$$

$$\frac{d}{dx}(\sec x) = \frac{\sin x}{\cos^2 x} \quad \text{instead of } \sec x \tan x; \quad \text{and}$$

$$\frac{d}{dx}(\csc x) = -\frac{\cos x}{\sin^2 x} \quad \text{instead of } -\csc x \cot x.$$

Example 8 using a TI-92 is shown at the right (notice that the answer differs from that shown in the text).

Although these forms are different from those shown in this section, you should notice that they are equivalent by recalling some of the fundamental identities from trigonometry.

A good test for your calculator is to try to find the derivative of $|x|$ at $x = 0$, which we know does not exist. Many calculators will give the incorrect answer of 0.

3.3 Problem Set

Ⓐ *Differentiate the functions given in Problems 1–32.*

1. $f(x) = \sin x + \cos x$
2. $f(x) = 2 \sin x + \tan x$
3. $g(t) = t^2 + \cos t + \cos \frac{\pi}{4}$
4. $g(t) = 2 \sec t + 3 \tan t - \tan \frac{\pi}{3}$
5. $f(t) = \sin^2 t$ (*Hint:* Use the product rule.)
6. $g(x) = \cos^2 x$ (*Hint:* Use the product rule.)
7. $f(x) = \sqrt{x} \cos x + x \cot x$
8. $f(x) = 2x^3 \sin x - 3x \cos x$
9. $p(x) = x^2 \cos x$
10. $p(t) = (t^2 + 2) \sin t$
11. $q(x) = \dfrac{\sin x}{x}$
12. $r(x) = \dfrac{e^x}{\sin x}$
13. $h(t) = e^t \csc t$
14. $f(\theta) = \dfrac{\sec \theta}{2 - \cos \theta}$
15. $f(x) = x^2 \ln x$
16. $g(x) = \dfrac{\ln x}{x^2}$
17. $h(x) = e^x(\cos x + \sin x)$
18. $f(x) = x^{-1} \ln x$
19. $f(x) = e^{-x} \sin x$
20. $g(x) = xe^{-x} \cos x$
21. $f(x) = \dfrac{\tan x}{1 - 2x}$
22. $g(t) = \dfrac{1 + \sin t}{\sqrt{t}}$

23. $f(x) = \dfrac{2 + \sin t}{t + 2}$
24. $f(\theta) = \dfrac{\theta - 1}{2 + \cos \theta}$
25. $f(x) = \dfrac{\sin x}{1 - \cos x}$
26. $f(x) = \dfrac{x}{1 - \sin x}$

Ⓢ🅜🅗 27. $f(x) = \dfrac{1 + \sin x}{2 - \cos x}$
28. $g(x) = \dfrac{\cos x}{1 + \cos x}$
29. $f(x) = \dfrac{\sin x + \cos x}{\sin x - \cos x}$
30. $f(x) = \dfrac{x^2 + \tan x}{3x + 2 \tan x}$
31. $g(x) = \sec^2 x - \tan^2 x + \cos x$
32. $g(x) = \cos^2 x + \sin^2 x + \sin x$

Find the second derivative of each function given in Problems 33–44.

33. $f(\theta) = \sin \theta$
34. $f(\theta) = \cos \theta$
35. $f(\theta) = \tan \theta$
36. $f(\theta) = \cot \theta$
37. $f(\theta) = \sec \theta$
38. $f(\theta) = \csc \theta$
39. $f(x) = \sin x + \cos x$
40. $f(x) = x \sin x$
41. $f(x) = e^x \cos x$
42. $g(t) = t^3 e^t$
43. $h(t) = \sqrt{t} \ln t$
44. $f(t) = \dfrac{\ln t}{t}$

B *Find an equation for the tangent line at the prescribed point for each function in Problems 47–54.*

47. $f(\theta) = \tan \theta$ at $(\frac{\pi}{4}, 1)$ **48.** $f(\theta) = \sec \theta$ at $(\frac{\pi}{3}, 2)$

49. $f(x) = \sin x$, where $x = \frac{\pi}{6}$

50. $f(x) = \cos x$, where $x = \frac{\pi}{3}$

51. $y = x + \sin x$, where $x = 0$

52. $y = x \sec x$, where $x = 0$

53. $y = e^x \cos x$, where $x = 0$

54. $y = x \ln x$, where $x = 1$

55. Which of the following functions satisfy $y'' + y = 0$?
  a. $y_1 = 2 \sin x + 3 \cos x$ b. $y_2 = 4 \sin x - \pi \cos x$
  c. $y_3 = x \sin x$ d. $y_4 = e^x \cos x$

56. For what values of A and B does

$$y = A \cos x + B \sin x$$

satisfy $y'' + 2y' + 3y = 2 \sin x$?

57. For what values of A and B does

$$y = Ax \cos x + Bx \sin x$$

satisfy $y'' + y = -3 \cos x$?

C 58. THINK TANK PROBLEM   Give an example of a differentiable function with a discontinuous derivative.

59. Complete the proof of Theorem 3.6 by showing that

$$\frac{d}{dx} \cos x = -\sin x.$$

Hint: You will need to use the identity

$$\cos(\alpha + \beta) = \cos \alpha \cos \beta - \sin \alpha \sin \beta$$

Prove the requested parts of Theorem 3.7 in Problems 60–62.

60. $\dfrac{d}{dx} \cot x = -\csc^2 x$

61. $\dfrac{d}{dx} \sec x = \sec x \tan x$

62. $\dfrac{d}{dx} \csc x = -\csc x \cot x$

63. Use the limit of a difference quotient to prove that

$$\frac{d}{dx} \tan x = \sec^2 x$$

64. JOURNAL PROBLEM   Write a short paper on the difficulties of differentiating trigonometric functions measured in degrees.*

———

*See, for example, "Fallacies, Flaws, and Flimflam," *The College Mathematics Journal,* Vol. 23, No. 3, May 1992, and Vol. 24, No. 4, September 1993.

3.4 Rates of Change: Rectilinear Motion

IN THIS SECTION   **rate of change (geometric preview), average and instantaneous rate of change, rectilinear motion (modeling in physics), falling body problem, relative rates of change**

RATE OF CHANGE—GEOMETRIC PREVIEW

The graph of a linear function $f(x) = ax + b$ is the line $y = ax + b$, whose slope $m = a$ can be thought of as the rate at which y is changing with respect to x (see Figure 3.18a). However, for another function g that is *not* linear, the rate of change of $y = g(x)$ with respect to x varies from point to point, as shown in Figure 3.18b.

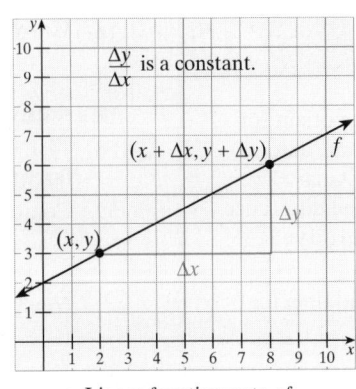
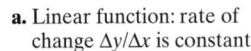
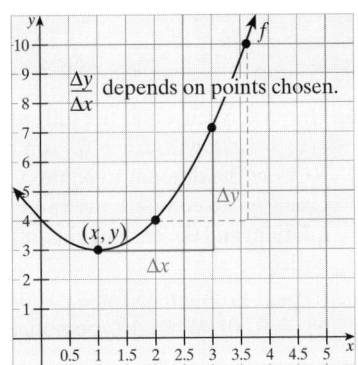

a. Linear function: rate of change $\Delta y/\Delta x$ is constant.

b. Nonlinear function: rate of change $\Delta y/\Delta x$ depends on chosen points.

■ **FIGURE 3.18** Rate of change is measured by the slope of a tangent line.

Because the slope of the tangent is given by the derivative of the function, the preceding geometric observations suggest that the rate of change of a function is measured by its derivative. This connection will be made more precise in the following discussion.

AVERAGE AND INSTANTANEOUS RATE OF CHANGE

Suppose y is a function of x, say $y = f(x)$. Corresponding to a change from x to $x + \Delta x$, the variable y changes from $f(x)$ to $f(x + \Delta x)$. The change in y is $\Delta y = f(x + \Delta x) - f(x)$, and the **average rate of change of y with respect to x is**

WARNING This formula for the change in y is important. Find Δy in Figure 3.19.

$$\text{AVERAGE RATE OF CHANGE} = \frac{\text{change in } y}{\text{change in } x} = \frac{\Delta y}{\Delta x} = \frac{f(x + \Delta x) - f(x)}{\Delta x}$$

As the interval over which we are averaging becomes shorter (that is, as $\Delta x \to 0$), the average rate of change approaches what we would intuitively call the **instantaneous rate of change of y with respect to x,** and the difference quotient approaches the derivative $\dfrac{dy}{dx}$. Thus, we have

$$\text{INSTANTANEOUS RATE OF CHANGE} = \lim_{\Delta x \to 0} \frac{\Delta y}{\Delta x} = \lim_{\Delta x \to 0} \frac{f(x + \Delta x) - f(x)}{\Delta x} = f'(x)$$

To summarize:

Instantaneous Rate of Change

Suppose $f(x)$ is differentiable at $x = x_0$. Then the **instantaneous rate of change** of $y = f(x)$ with respect to x at x_0 is the value of the derivative of f at x_0. That is,

$$\text{INSTANTANEOUS RATE OF CHANGE} = f'(x_0) = \frac{dy}{dx}\bigg|_{x = x_0}$$

EXAMPLE 1 *Instantaneous rate of change*

Find the rate at which the function $y = x^2 \sin x$ is changing with respect to x when $x = \pi$.

■ **FIGURE 3.19** A change in Δy corresponding to a change Δx

Solution   For any x, the instantaneous rate of change is the derivative,

$$\frac{dy}{dx} = 2x \sin x + x^2 \cos x$$

Thus, the rate when $x = \pi$ is

$$\left.\frac{dy}{dx}\right|_{x=\pi} = 2\pi \sin \pi + \pi^2 \cos \pi = 2\pi(0) + \pi^2(-1) = -\pi^2$$

The negative sign indicates that when $x = \pi$, the function is *decreasing* at the rate of $\pi^2 \approx 9.9$ units of y for each one-unit increase in x. ∎

    Let us consider an example comparing the average rate of change and the instantaneous rate of change.

EXAMPLE 2   *Comparison between average rate and instantaneous rate of change*

Let $f(x) = x^2 - 4x + 7$.

a. Find the instantaneous rate of change of f at $x = 3$.

b. Find the average rate of change of f with respect to x between $x = 3$ and 5.

Solution

a. The derivative of the function is

$$f'(x) = 2x - 4$$

Thus, the instantaneous rate of change of f at $x = 3$ is

$$f'(3) = 2(3) - 4 = 2$$

The tangent line at $x = 3$ has slope 2, as shown in Figure 3.20.

b. The (average) rate of change from $x = 3$ to $x = 5$ is found by dividing the change in f by the change in x. The change in f from $x = 3$ to $x = 5$ is

$$f(5) - f(3) = [5^2 - 4(5) + 7] - [3^2 - 4(3) + 7] = 8$$

Thus, the average rate of change is

$$\frac{f(5) - f(3)}{5 - 3} = \frac{8}{2} = 4$$

The slope of the secant line is 4, as shown in Figure 3.20. ∎

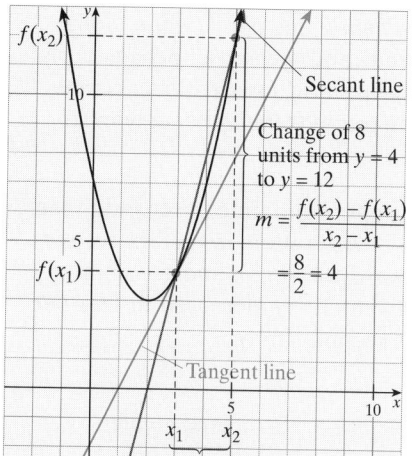

■ **FIGURE 3.20** Comparison of instantaneous rate of change and average rate of change from x_1 to x_2

RECTILINEAR MOTION (Modeling in Physics)

Rectilinear motion is motion along a straight line. For example, the up and down motion of a yo-yo may be regarded as rectilinear, as may the motion of a rocket early in its flight.

    When studying rectilinear motion, we may assume that the object is moving along a coordinate line. The *position* of the object on the line is a function of time t and is often expressed as $s(t)$. The rate of change of $s(t)$ with respect to time is the object's **velocity** $v(t)$, and the rate of change of the velocity with respect to t is its **acceleration** $a(t)$. The absolute value of the velocity is called **speed.** Thus,

$$\text{Velocity is } v(t) = \frac{ds}{dt} \qquad \text{Speed is } |v(t)| = \left|\frac{ds}{dt}\right|$$

$$\text{Acceleration is } a(t) = \frac{dv}{dt} = \frac{d^2s}{dt^2}$$

If $v(t) > 0$, we say that the object is *advancing* and if $v(t) < 0$, it is *retreating*. If $v(t) = 0$, the object is neither advancing nor retreating, and we say it is *stationary*. The object is *accelerating* when $a(t) > 0$ and is *decelerating* when $a(t) < 0$. The significance of the acceleration is that it gives the rate at which the velocity is changing. These ideas are summarized in the following box.

Rectilinear Motion

An object that moves along a straight line with *position s(t)* has *velocity*
$$v(t) = \frac{ds}{dt} \text{ and } acceleration \ a(t) = \frac{dv}{dt} = \frac{d^2s}{dt^2} \text{ when these derivatives exist. The}$$
speed of the object is $\left| v(t) \right|$.

WARNING ▶   Rectilinear motion involves *position, velocity,* and *acceleration.* Sometimes there is confusion between the words "speed" and "velocity." Because speed is the absolute value of the velocity, it indicates how fast an object is moving, whereas velocity indicates both speed and direction (relative to a given coordinate system). ◂

NOTATIONAL COMMENT   If distance is measured in meters and time in seconds, then velocity is measured in meters per second (m/s) and acceleration, in meters per second per second (m/s/s). The notation m/s/s is awkward, so m/s^2 is more commonly used. Similarly, if distance is measured in feet, then velocity is measured in feet per second (ft/s) and acceleration, in feet per second per second (ft/s^2).

When you are riding in a car, you do not feel the velocity, but you do feel the acceleration. That is, you feel *changes* in the velocity.

EXAMPLE 3   *The position, velocity, and acceleration of a moving object*

Assume that the position at time t of an object moving along a line is given by
$$s(t) = 3t^3 - 40.5t^2 + 162t$$
for t on $[0, 8]$. Find the initial position, velocity, and acceleration for the object and discuss the motion.

Solution   The position at time t is given by the function s. The initial position occurs at time $t = 0$, so

$$s(0) = 0 \qquad \text{The object starts at the origin.}$$

The velocity is determined by finding the derivative of the position function:

$$
\begin{aligned}
v(t) = s'(t) &= 9t^2 - 81t + 162 \\
&= 9(t^2 - 9t + 18) \\
&= 9(t-3)(t-6) \qquad \text{The initial velocity is } v(0) = 162.
\end{aligned}
$$

When $t = 3$ and when $t = 6$, the velocity v is 0, which means the *object is stationary* at those times. Furthermore,

$$v(t) > 0 \text{ on } [0, 3) \qquad \text{Object is advancing.}$$
$$v(t) < 0 \text{ on } (3, 6) \qquad \text{Object is retreating.}$$
$$v(t) > 0 \text{ on } (6, 8] \qquad \text{Object is advancing.}$$

For the acceleration,

$$a(t) = s''(t) = v'(t) = 18t - 81$$
$$= 18(t - 4.5) \qquad \text{The initial acceleration is } a(0) = -81.$$

We see that $a(t) < 0$ on $[0, 4.5)$ *Velocity is decreasing; that is, the object is decelerating.*

$a(t) = 0$ at $t = 4.5$ *Velocity is not changing.*

$a(t) > 0$ on $(4.5, 8]$ *Velocity is increasing; so the object is accelerating.*

The table (technology window shown in the margin) gives us values for s, v, and a. We use these values to plot a few points, as shown in Figure 3.21. The actual path of the object is back and forth on the axis and the figure is for clarification only.

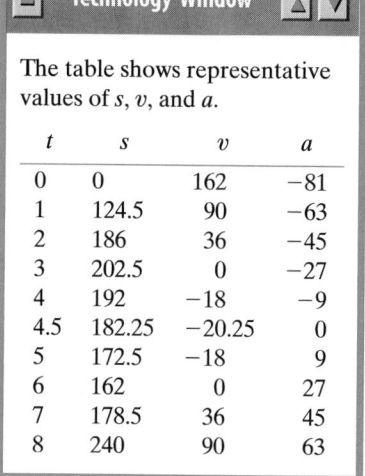

Technology Window

The table shows representative values of s, v, and a.

t	s	v	a
0	0	162	−81
1	124.5	90	−63
2	186	36	−45
3	202.5	0	−27
4	192	−18	−9
4.5	182.25	−20.25	0
5	172.5	−18	9
6	162	0	27
7	178.5	36	45
8	240	90	63

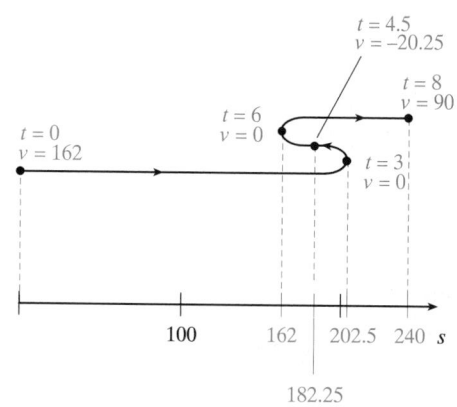

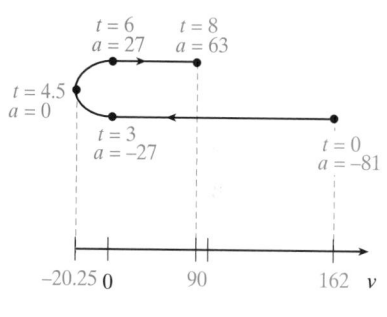

a. Position of the object: when $v > 0$, the object advances, and when $v < 0$, the object retreats.

b. Velocity of the object: when $a > 0$, the velocity increases, and when $a < 0$, the velocity decreases.

■ **FIGURE 3.21**   Analysis of rectilinear motion

Recall that the *speed* of the particle is the absolute value of its velocity. The speed decreases from 162 to 0 between $t = 0$ and $t = 3$, and increases from 0 to 20.25 as the velocity becomes negative between $t = 3$ and $t = 4.5$. Then for $4.5 < t < 6$, the particle slows down again, from 20.25 to 0, after which it speeds up.   ■

A common mistake is to think that a particle moving on a straight line speeds up when its acceleration is positive and slows down when the acceleration is negative, but this is not quite correct. Instead, the following is generally true:

The speed *increases* (particle speeds up) when the velocity and acceleration have the same signs.

The speed *decreases* (particle slows down) when the velocity and acceleration have opposite signs.

Technology Window

You can use a graphing calculator or computer to look at the graphs of the position, velocity, and acceleration functions.

Y1=3X^3−40.5X^2+162X

Y2=9X^2−81X+162

Y3=18x−81

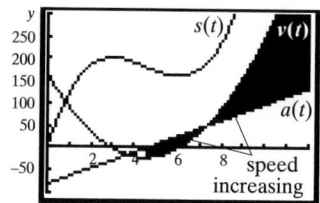

Notice that the speed is increasing when both the velocity and acleration functions have the same sign.

FALLING BODY PROBLEM

As a second example of rectilinear motion, we shall consider a *falling body problem*. In such a problem, it is assumed that an object is projected (that is, thrown, fired, dropped, etc.) vertically in such a way that the only acceleration acting on the object is the constant downward acceleration g due to gravity, which on the earth near sea level is approximately 32 ft/s^2 or 9.8 m/s^2. At time t, the height of the object is given by the following formula:

Formula for the height of an object

v_0 is the initial velocity.
↓
$$h(t) = -\tfrac{1}{2}gt^2 + v_0 t + s_0 \qquad \{s_0 \text{ is the initial height.}$$
↑
g is the acceleration due to gravity.

where s_0 and v_0 are the object's initial height and velocity, respectively. We shall derive this formula in Chapter 4.

EXAMPLE 4   *Position, velocity, and acceleration of a falling object*

Suppose a person standing at the top of the Tower of Pisa (176 ft high) throws a ball directly upward with an initial speed of 96 ft/s.

a. Find the ball's height, its velocity, and acceleration at time t.

b. When does the ball hit the ground, and what is its impact velocity?

c. How far does the ball travel during its flight?

Solution   First, draw a picture such as the one shown in Figure 3.22 to help you understand the problem.

a. Substitute the known values into the formula for the height of an object:

$v_0 = 96$ ft/s: Initial velocity
↓
$$h(t) = -\tfrac{1}{2}(32)t^2 + 96t + 176 \leftarrow h_0 = 176 \text{ ft is height of tower}$$
↑
$g = 32$ ft/s^2: Constant downward gravitational acceleration

$$h(t) = -16t^2 + 96t + 176 \qquad \text{This is the position, or position, function.}$$
$$\text{It gives the height of the ball.}$$

The velocity at time t is the derivative:

$$v(t) = \frac{dh}{dt} = -32t + 96$$

The acceleration is the derivative of the velocity function:

$$a(t) = \frac{dv}{dt} = \frac{d^2h}{dt^2} = -32$$

This means that the velocity of the ball is always decreasing at the rate of 32 ft/s^2.

b. The ball hits the ground when $h(t) = 0$. Solve the equation

$$-16t^2 + 96t + 176 = 0$$

to find that this occurs when $t \approx -1.47$ and $t \approx 7.47$. Disregarding the negative value, we see that impact occurs at $t \approx 7.47$ s. The impact velocity is

$$v(7.47) \approx -143 \text{ ft/s}$$

The negative sign here means the ball is coming down at the moment of impact.

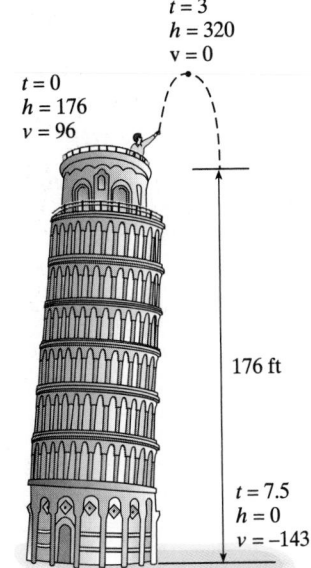

$t = 3$
$h = 320$
$v = 0$

$t = 0$
$h = 176$
$v = 96$

176 ft

$t = 7.5$
$h = 0$
$v = -143$

FIGURE 3.22   The motion of a ball thrown upward from the Tower of Pisa

c. The ball travels upward for some time and then falls downward to the ground, as shown in Figure 3.22. We need to find the distance it travels upward plus the distance it falls to the ground. The turning point at the top (the highest point) occurs when the velocity is zero. Solve the equation

$$-32t + 96 = 0$$

to find that this occurs when $t = 3$. The ball starts at $h(0) = 176$ ft and rises to a maximum height when $t = 3$. Thus, the maximum height is

$$h(3) = -16(3)^2 + 96(3) + 176 = 320$$

and the total distance traveled is

$$\underbrace{(320 - 176)}_{\text{Upward distance}} + \underbrace{320}_{\text{Downward distance}} = 464$$

Initial height

The total distance traveled is 464 ft.   ∎

RELATIVE RATES OF CHANGE

Often we are not as interested in the instantaneous rate of change of a quantity as its relative rate of change, defined as follows:

$$\text{RELATIVE RATE} = \frac{\text{INSTANTANEOUS RATE OF CHANGE}}{\text{SIZE OF QUANTITY}}$$

For instance, if you are earning \$5,000/yr and receive a \$1,000 raise, you would probably be very pleased. However, that \$1,000 raise might be something of an insult if you were earning \$100,000/yr. In both cases, the change is the quantity \$1,000, but in the first case the relative rate of change is $1,000/5,000 = 20\%$, whereas in the second, it is a paltry $1,000/100,000 = 1\%$.

In terms of the derivative, the relative rate of change may be defined as follows:

Relative Rate of Change

The **relative rate of change** of $y = f(x)$ at $x = x_0$ is given by the ratio

$$\text{RELATIVE RATE OF CHANGE} = \frac{f'(x_0)}{f(x_0)}$$

EXAMPLE 5   *Relative rate of change*

The aerobic rating of a person x years old is

$$A(x) = 110\left[\frac{\ln x - 2}{x}\right]$$

What is the relative rate of change of the aerobic rating of a 20-year-old person? A 50-year-old person?

Solution   Applying the quotient rule, we obtain the derivative of $A(x)$:

$$A'(x) = 110\left[\frac{x(1/x) - \ln x - 2)(1)}{x^2}\right] = 110\left[\frac{3 - \ln x}{x^2}\right]$$

Thus, the relative rate of change of A at age x is

$$\text{RELATIVE RATE} = \frac{A'(x)}{A(x)} = \frac{110\left[\dfrac{3 - \ln x}{x^2}\right]}{110\left[\dfrac{\ln x - 2}{x}\right]} = \frac{3 - \ln x}{(\ln x - 2)x}$$

In particular, when $x = 20$, the relative rate is

$$\frac{A'(20)}{A(20)} = \frac{3 - \ln 20}{(\ln 20 - 2)20} \approx 0.0002143$$

In other words, at age 20, the person's aerobic rating is increasing at the rate of 0.02%/yr. However, at age 50, the relative rate is

$$\frac{A'(50)}{A(50)} = \frac{3 - \ln 50}{(\ln 50 - 2)50} \approx -0.00954$$

and the person's aerobic rating is *decreasing* (because of the negative sign) at the relative rate of 0.95%/yr.   ∎

3.4 Problem Set

Ⓐ *For each function f given in Problems 1–14, find the rate of change with respect to x at x = x_0.*

1. $f(x) = x^2 - 3x + 5$ when $x_0 = 2$

2. $f(x) = 14 + x - x^2$ when $x_0 = 1$

3. $f(x) = -2x^2 + x + 4$ for $x_0 = 1$

4. $f(x) = \dfrac{-2}{x + 1}$ for $x_0 = 1$

5. $f(x) = \dfrac{2x - 1}{3x + 5}$ when $x_0 = -1$

6. $f(x) = (x^2 + 2)(x + \sqrt{x})$ when $x_0 = 4$

7. $f(x) = x \cos x$ when $x_0 = \pi$

8. $f(x) = (x + 1) \sin x$ when $x_0 = \frac{\pi}{2}$

9. $f(x) = x \ln \sqrt{x}$ when $x_0 = 1$

10. $f(x) = x^2 e^{-x}$ when $x_0 = 0$

11. $f(x) = \sin x \cos x$ when $x_0 = \frac{\pi}{2}$

12. $f(x) = \dfrac{x^2}{x^2 + 1}$ when $x_0 = 1$

13. $f(x) = \left(x - \dfrac{2}{x}\right)^2$ when $x_0 = 1$

14. $f(x) = \sin^2 x$ when $x_0 = \frac{\pi}{4}$

The function s(t) in Problems 15–22 gives the position of an object moving along a line. In each case:

a. *Find the velocity at time t.*

b. *Find the acceleration at time t.*

c. *Describe the motion of the object; that is, tell when it is advancing and when it is retreating. Compute the total distance traveled by the object during the indicated time interval.*

d. *Tell when the object is accelerating and when it is decelerating.*

SMH *For a review of solving equations, see Sections 2.5, 2.6, and 3.7 of the* Student Mathematics Handbook.

15. $s(t) = t^2 - 2t + 6$ on $[0, 2]$

16. $s(t) = 3t^2 + 2t - 5$ on $[0, 1]$

17. $s(t) = t^3 - 9t^2 + 15t + 25$ on $[0, 6]$

18. $s(t) = t^4 - 4t^3 + 8t$ on $[0, 4]$

19. $s(t) = \dfrac{2t + 1}{t^2}$ for $1 \le t \le 3$

20. $s(t) = t^2 + t \ln t$ for $1 \le t \le e$

21. $s(t) = 3 \cos t$ for $0 \le t \le 2\pi$

22. $s(t) = 1 + \sec t$ for $0 \le t \le \frac{\pi}{4}$

Ⓑ 23. It is estimated that x years from now, $0 \le x \le 10$, the average SAT mathematics score of the incoming students at a certain eastern liberal arts college will be

$$f(x) = -6x + 582$$

a. Derive an expression for the rate at which the average SAT score will be changing with respect to time x years from now.

b. What is the significance of the fact that the expression in part **a** is a negative constant?

24. A particle moving on the x-axis has position

$$x(t) = 2t^3 + 3t^2 - 36t + 40$$

after an elapsed time of t seconds.
a. Find the velocity of the particle at time t.
b. Find the acceleration at time t.
c. What is the total distance traveled by the particle during the first 3 seconds?

25. An object moving on the x-axis has position

$$x(t) = t^3 - 9t^2 + 24t + 20$$

after t seconds. What is the total distance traveled by the object during the first 8 seconds?

26. A car has position

$$s(t) = 50t \ln t + 50$$

ft/s after t seconds of motion. What is its acceleration (to the nearest hundredth ft/s^2) after 10 seconds?

27. An object moves along a straight line so that after t minutes, its position relative to its starting point (in meters) is

$$s(t) = 10t + te^{-t}$$

a. At what speed (to the nearest thousandth m/min) is the object moving at the end of 4 min?
b. How far (to the nearest thousandth m) does the object actually travel during the fifth minute?

28. A bucket containing 5 gal of water has a leak. After t seconds, there are

$$Q(t) = 5\left(1 - \frac{t}{25}\right)^2$$

gallons of water in the bucket.
a. At what rate (to the nearest hundredth gal) is water leaking from the bucket after 2 seconds?
b. How long does it take for all the water to leak out of the bucket?
c. At what rate is the water leaking when the last drop leaks out?

MODELING PROBLEMS   *In Problems 29–50, set up an appropriate model to answer the given question. Be sure to state your assumptions.*

29. A person standing at the edge of a cliff throws a rock directly upward. It is observed that 2 seconds later the rock is at its maximum height (in ft) and that 5 seconds after that, it hits the ground at the base of the cliff.
a. What is the initial velocity of the rock?
b. How high is the cliff?
c. What is the velocity of the rock at time t?
d. With what velocity does the rock hit the ground?

30. A projectile is shot upward from the earth with an initial velocity of 320 ft/s.
a. What is its velocity after 5 seconds?
b. What is its acceleration after 3 seconds?

31. A rock is dropped from a height of 90 ft. One second later another rock is dropped from height H. What is H

(to the nearest foot) if the two rocks hit the ground at the same time?

32. A ball is thrown vertically upward from the ground with an initial velocity of 160 ft/s.
a. When will the ball hit the ground?
b. With what speed will the ball hit the ground?
c. When will the ball reach its maximum height?

33. An object is dropped (initial velocity $v_0 = 0$) from the top of a building and falls 3 seconds before hitting the pavement below. Determine the height of the building in ft.

34. An astronaut standing at the edge of a cliff on the moon throws a rock directly upward and observes that it passes her on the way down exactly 4 seconds later. Three seconds after that, the rock hits the ground at the base of the cliff. Use this information to determine the initial velocity v_0 and the height of the cliff. *Note:* $g = 5.5$ ft/s^2 on the moon.

35. Answer the question in Problem 34 assuming the astronaut is on Mars, where $g = 12$ ft/s^2.

36. A car is traveling at 88 ft/s (60 mi/h) when the driver applies the brakes to avoid hitting a child. After t seconds, the car is $s(t) = 88t - 8t^2$ feet from the point where the brakes were first applied. How long does it take for the car to come to a stop, and how far does it travel before stopping?

37. It is estimated that t years from now, the circulation of a local newspaper can be modeled by the formula

$$C(t) = 100t^2 + 400t + 50t \ln t$$

a. Find an expression for the rate at which the circulation will be changing with respect to time t years from now.
b. At what rate will the circulation be changing with respect to time 5 years from now?
c. By how much will the circulation actually change during the sixth year?

38. An efficiency study of the morning shift at a certain factory indicates that an average worker who arrives on the job at 8:00 A.M. will have assembled

$$f(x) = -\tfrac{1}{3}x^3 + \tfrac{1}{2}x^2 + 50x$$

units x hours later.
a. Find a formula for the rate at which the worker will be assembling the units after x hours.
b. At what rate will the worker be assembling units at 9:00 A.M.?
c. How many units will the worker actually assemble between 9:00 A.M. and 10:00 A.M.?

39. An environmental study of a certain suburban community suggests that t years from now, the average level of carbon monoxide in the air can be modeled by the formula

$$q(t) = 0.05t^2 + 0.1t + 3.4$$

parts per million.

a. At what rate will the carbon monoxide level be changing with respect to time one year from now?

b. By how much will the carbon monoxide level change in the first year?

c. By how much will the carbon monoxide level change over the next (second) year?

40. According to *Newton's law of universal gravitation,* if an object of mass M is separated by a distance r from a second object of mass m, then the two objects are attracted to one another by a force that acts along the line joining them and has magnitude

$$F = \frac{GmM}{r^2}$$

where G is a positive constant. Show that the rate of change of F with respect to r is inversely proportional to r^3.

41. The population of a bacterial colony is approximately

$$P(t) = P_0 + 61t + 3t^2$$

thousand t hours after observation begins, where P_0 is the initial population. Find the rate at which the colony is growing after 5 hours.

42. The gross domestic product (GDP) of a certain country is

$$g(t) = t^2 + 5t + 106$$

billion dollars t years after 1995.

a. At what rate was the GDP changing in 1997?

b. At what percentage rate was the GDP changing in 1997?

43. It is projected that x months from now, the population of a certain town will be

$$P(x) = 2x + 4x^{3/2} + 5,000$$

a. At what rate will the population be changing with respect to time 9 months from now?

b. At what percentage rate will the population be changing with respect to time 9 months from now?

44. Assume that your starting salary is $30,000 and you get a raise of $3,000 each year.

a. Express the percentage rate of change of your salary as a function of time.

b. At what percentage rate will your salary be increasing after one year?

c. What will happen to the percentage rate of change of your salary in the long run?

45. The gross domestic product (GDP) of a certain country is growing at a constant rate. In 1995 the GDP was 125 billion dollars, and in 1997 it was 155 billion dollars. At what percentage rate will the GDP be growing in 2000?

46. If y is a linear function of x, what will happen to the percentage rate of change of y with respect to x as x increases without bound?

47. According to Debye's formula in physical chemistry, the orientation polarization P of a gas satisfies

$$P = \frac{4}{3}\pi\left(\frac{\mu^2}{3kT}\right)N$$

where μ, k, and N are constants and T is the temperature of the gas. Find the rate of change of P with respect to T.

Technology Window ▲ ▼

48. A disease is spreading in such a way that after t weeks, for $0 \le t \le 6$, it has affected

$$N(t) = 5 - t^2(t - 6)$$

hundred people. Health officials declare that this disease will reach epidemic proportions when the percentage rate of increase of $N(t)$ at the start of a particular week is at least 30% per week. The epidemic designation level is dropped when the percentage rate falls below this level.

a. Find the percentage rate of change of $N(t)$ at time t.

b. Between what weeks is the disease at the epidemic level?

---

49. An object attached to a helical spring is pulled down from its equilibrium position and then released, as shown in Figure 3.23.

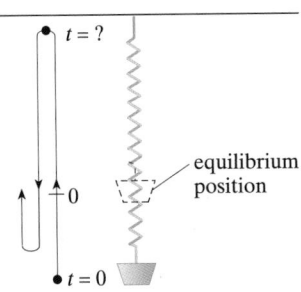

■ **FIGURE 3.23**   Helical spring

Suppose that t seconds later, its position (in centimeters, measured relative to the equilibrium position) is modeled by

$$s(t) = 7 \cos t$$

a. Find the velocity and acceleration of the object at time t.

b. Find the length of time required for one complete oscillation. This is called the *period* of the motion.

c. What is the distance between the highest point reached by the object and the lowest point? Half of this distance is called the *amplitude* of the motion.

50. Two cars leave a town at the same time and travel at constant speeds along straight roads that meet at an angle of 60° in the town. If one car travels twice as fast as the other and the distance between them increases at the rate of 45 mi/h, how fast is the slower car traveling?

51. SPY PROBLEM   The Spy safely crosses the border to the friendly nation of Azusa (see Problem 50 of Chapter 2 Supplementary Problems) and is immediately contacted by his superior, Lord Notso Sweething (code name "N"). He is sent on a special mission in space. An encounter with an enemy agent leaves him with a mild concussion that causes him to forget where he is. Fortunately, he remembers the formula for the height of a projectile,

$$h(t) = -\tfrac{1}{2}gt^2 + v_0 t + s_0$$

and the values of g for various heavenly bodies. To deduce his whereabouts, he throws a rock directly upward (from ground level) and notes that it reaches a maxi-

mum height of 37.5 ft and hits the ground 5 s after leaving his hand. Where is he? *Note:* You will need to know that g is 32 ft/s^2 on earth, 5.5 ft/s^2 on the moon, 12 ft/s^2 on Mars, and 28 ft/s^2 on Venus.

C 52. Find the rate of change of the volume of a cube with respect to the length of one of its edges. How is this rate related to the surface area of the cube?

53. Show that the rate of change of the volume of a sphere with respect to its radius is equal to its surface area.

54. Van der Waal's equation states that a gas that occupies a volume V at temperature T (Kelvin) exerts pressure P, where

$$\left(P + \frac{A}{V^2}\right)(V - B) = kT$$

and A, B, k are physical constants. Find the rate of change of pressure with respect to volume, assuming fixed temperature.

3.5 The Chain Rule

IN THIS SECTION   **introduction to the chain rule, extended derivative formulas, justification of the chain rule** ■

INTRODUCTION TO THE CHAIN RULE

Suppose it is known that the carbon monoxide pollution in the air is changing at the rate of 0.02 ppm (parts per million) for each person in a town whose population is growing at the rate of 1,000 people per year. To find the rate at which the level of pollution is increasing with respect to time, we form the product

$$(0.02 \text{ ppm/person})(1,000 \text{ people/year}) = 20 \text{ ppm/year}$$

In this example, the level of pollution L is a function of the population P, which is itself a function of time t. Thus, L is a composite function of t, and

$$\begin{bmatrix}\text{RATE OF CHANGE OF } L \\ \text{WITH RESPECT TO } t\end{bmatrix} = \begin{bmatrix}\text{RATE OF CHANGE OF } L \\ \text{WITH RESPECT TO } P\end{bmatrix}\begin{bmatrix}\text{RATE OF CHANGE OF } P \\ \text{WITH RESPECT TO } t\end{bmatrix}$$

Expressing each of these rates in terms of an appropriate derivative in Leibniz form, we obtain the following equation:

$$\frac{dL}{dt} = \frac{dL}{dP}\frac{dP}{dt}$$

These observations anticipate the following important theorem.

THEOREM 3.10   *Chain rule*

If $y = f(u)$ is a differentiable function of u and u in turn is a differentiable function of x, then $y = f(u(x))$ is a differentiable function of x and its derivative is given by the product

$$\frac{dy}{dx} = \frac{dy}{du}\frac{du}{dx}$$

Proof A rigorous proof involves a few details that make it inappropriate to include at this point in the text. A justification (partial proof) is given at the end of this section, and a full proof of the chain rule is included in Appendix B.

WARNING ▶ Recall our earlier warning in Section 3.1 against thinking of dy/dx as a fraction. That said, there are certain times when this incorrect reasoning can be used as a mnemonic device. For instance, "canceling du," as indicated below, makes it easy to remember the chain rule:

$$\frac{dy}{dx} = \frac{dy}{d\cancel{u}}\frac{d\cancel{u}}{dx}$$

EXAMPLE 1 *The chain rule*

Find dy/dx if $y = u^3 - 3u^2 + 1$ and $u = x^2 + 2$.

Solution Because $dy/du = 3u^2 - 6u$ and $du/dx = 2x$, it follows from the chain rule that

$$\frac{dy}{dx} = \frac{dy}{du}\frac{du}{dx} = (3u^2 - 6u)(2x)$$

Notice that this derivative is expressed in terms of the variables x and u. To express dy/dx in terms of x alone, we substitute $u = x^2 + 2$ as follows:

$$\frac{dy}{dx} = [3(x^2 + 2)^2 - 6(x^2 + 2)](2x) = 6x^3(x^2 + 2)$$ ■

The chain rule is actually a rule for differentiating composite functions. In particular, if $y = f(u)$ and $u = u(x)$, then y is the composite function $y = (f \circ u)(x) = f[u(x)]$, and the chain rule can be rewritten as follows:

═══

THEOREM 3.10a *The chain rule (alternate form)*

If u is differentiable at x and f is differentiable at $u(x)$, then the composite function $f \circ u$ is differentiable at x and

$$\frac{d}{dx}f[u(x)] = \frac{d}{du}f(u)\frac{du}{dx}$$

or

$$(f \circ u)'(x) = f'[u(x)]u'(x)$$

> ■ *What This Says:* In Section 1.3 when we introduced composite functions, we talked about the "inner" and "outer" functions. With this terminology, the chain rule says that the derivative of the composite function $f[u(x)]$ is equal to the derivative of the inner function u times the derivative of the outer function f evaluated at the inner function.

EXAMPLE 2 *The chain rule applied to a power*

Differentiate $y = (3x^4 - 7x + 5)^3$.

Solution Here, the "inner" function is $u(x) = 3x^4 - 7x + 5$ and the "outer" function is u^3, so we have

$$\begin{aligned} y' &= (u^3)'[u(x)]' \\ &= (3u^2)(12x^3 - 7) \\ &= 3(3x^4 - 7x + 5)^2(12x^3 - 7) \end{aligned}$$ ■

You could, with a lot of work, have found the derivative in Example 2 without using the chain rule either by expanding the polynomial or by using the product rule. The answer would be the same but would involve much more algebra. The chain rule allows us to find derivatives that would otherwise be very difficult to handle.

EXAMPLE 3   *Differentiation with quotient rule inside the chain rule*

Differentiate $g(x) = \sqrt[4]{\dfrac{x}{1 - 3x}}$.

Solution   Write $g(x) = \left(\dfrac{x}{1 - 3x}\right)^{1/4} = u^{1/4}$ where $u = \dfrac{x}{1 - 3x}$ is the inner function and $u^{1/4}$ is the outer function. Then

$$g'(x) = (u^{1/4})'u'(x) = \tfrac{1}{4}u^{-3/4}u'(x)$$

and we have

$$g'(x) = \frac{1}{4}\left(\frac{x}{1 - 3x}\right)^{-3/4}\left(\frac{x}{1 - 3x}\right)'$$

$$= \frac{1}{4}\left(\frac{x}{1 - 3x}\right)^{-3/4}\left[\frac{(1 - 3x)(1) - x(-3)}{(1 - 3x)^2}\right] \qquad \text{Quotient rule}$$

$$= \frac{1}{4}\left(\frac{x}{1 - 3x}\right)^{-3/4}\left[\frac{1}{(1 - 3x)^2}\right] = \frac{1}{4x^{3/4}(1 - 3x)^{5/4}}$$

■

EXTENDED DERIVATIVE FORMULAS

The chain rule can be used to obtain generalized differentiation formulas for the standard functions, as displayed in the following box.

If u is a differentiable function of x, then

Extended Power Rule

$$\frac{d}{dx}u^n = nu^{n-1}\frac{du}{dx}$$

Extended Trigonometric Rules

$$\frac{d}{dx}\sin u = \cos u\,\frac{du}{dx} \qquad\qquad \frac{d}{dx}\cos u = -\sin u\,\frac{du}{dx}$$

$$\frac{d}{dx}\tan u = \sec^2 u\,\frac{du}{dx} \qquad\qquad \frac{d}{dx}\cot u = -\csc^2 u\,\frac{du}{dx}$$

$$\frac{d}{dx}\sec u = \sec u \tan u\,\frac{du}{dx} \qquad\qquad \frac{d}{dx}\csc u = -\csc u \cot u\,\frac{du}{dx}$$

Extended Exponential and Logarithmic Rules

$$\frac{d}{dx}e^u = e^u\frac{du}{dx} \qquad\qquad \frac{d}{dx}\ln u = \frac{1}{u}\frac{du}{dx}$$

Chain rule with a trigonometric function

Differentiate $f(x) = \sin(3x^2 + 5x - 7)$.

Solution   Think of this as $f(u) = \sin u$, where $u = 3x^2 + 5x - 7$, and apply the chain rule:

$$f'(x) = \cos(3x^2 + 5x - 7) \cdot (3x^2 + 5x - 7)'$$
$$= (6x + 5)\cos(3x^2 + 5x - 7)$$

■

Chain rule with other rules

Differentiate $g(x) = \cos x^2 + 5\left(\dfrac{3}{x} + 4\right)^6$.

Solution $\dfrac{dg}{dx} = \dfrac{d}{dx}\cos x^2 + 5\dfrac{d}{dx}(3x^{-1} + 4)^6$

$$= -\sin x^2 \dfrac{d}{dx}(x^2) + 5\left[6(3x^{-1} + 4)^5 \dfrac{d}{dx}(3x^{-1} + 4)\right]$$

$$= (-\sin x^2)(2x) + 30(3x^{-1} + 4)^5(-3x^{-2})$$

$$= -2x \sin x^2 - 90x^{-2}(3x^{-1} + 4)^5$$

■

Extended power and cosine rules

Differentiate $y = \cos^4(3x + 1)^2$.

Solution $\dfrac{dy}{dx} = 4\cos^3(3x + 1)^2 \dfrac{d}{dx}\cos(3x + 1)^2$ *Extended power rule*

$$= 4\cos^3(3x + 1)^2 \cdot [-\sin(3x + 1)^2] \cdot \dfrac{d}{dx}(3x + 1)^2 \quad \text{\textit{Extended cosine rule}}$$

$$= -4\cos^3(3x + 1)^2 \sin(3x + 1)^2 \cdot 2(3x + 1)(3) \quad \text{\textit{Extended power rule}}$$

$$= -24(3x + 1)\cos^3(3x + 1)^2 \sin(3x + 1)^2$$

■

Extended power rule inside quotient rule

Differentiate $p(x) = \dfrac{\tan 7x}{(1 - 4x)^5}$.

Solution $\dfrac{dp}{dx} = \dfrac{(1 - 4x)^5\left[\dfrac{d}{dx}\tan 7x\right] - \tan 7x\left[\dfrac{d}{dx}(1 - 4x)^5\right]}{[(1 - 4x)^5]^2}$

$$= \dfrac{(1 - 4x)^5(\sec^2 7x)\dfrac{d}{dx}(7x) - (\tan 7x)\left[5(1 - 4x)^4\dfrac{d}{dx}(1 - 4x)\right]}{(1 - 4x)^{10}}$$

$$= \dfrac{(1 - 4x)^5(\sec^2 7x)(7) - (\tan 7x)(5)(1 - 4x)^4(-4)}{(1 - 4x)^{10}}$$

$$= \dfrac{(1 - 4x)^4[7(1 - 4x)\sec^2 7x + 20 \tan 7x]}{(1 - 4x)^{10}}$$

$$= \dfrac{7(1 - 4x)\sec^2 7x + 20 \tan 7x}{(1 - 4x)^6}$$

■

Differentiate $e^{-3x} \sin x$.

Solution   Use the product rule:

$$\frac{d}{dx}[e^{-3x} \sin x] = e^{-3x}\left[\frac{d}{dx}(\sin x)\right] + \left[\frac{d}{dx}(e^{-3x})\right] \sin x \qquad \text{Product rule}$$

$$= e^{-3x}(\cos x) + [e^{-3x}(-3)] \sin x$$

$$= e^{-3x}(\cos x - 3 \sin x) \qquad\qquad\blacksquare$$

Find the x-coordinate of each point on the curve

$$y = (x + 1)^3(2x + 3)^2$$

where the tangent is horizontal.

Solution   Because horizontal tangents have zero slope, we need to solve the equation $f'(x) = 0$. First use the product rule and the extended power rule to find the derivative:

$$f'(x) = (x + 1)^3\left[\frac{d}{dx}(2x + 3)^2\right] + (2x + 3)^2\left[\frac{d}{dx}(x + 1)^3\right]$$

$$= (x + 1)^3\left[(2)(2x + 3)^1 \frac{d}{dx}(2x + 3)\right] + (2x + 3)^2\left[(3)(x + 1)^2 \frac{d}{dx}(x + 1)\right]$$

$$= (x + 1)^3(2)(2x + 3)(2) + (2x + 3)^2(3)(x + 1)^2(1)$$

$$= (x + 1)^2(2x + 3)[4(x + 1) + 3(2x + 3)]$$

$$= (x + 1)^2(2x + 3)(10x + 13)$$

$\boxed{\text{SMH}}$ *See Problem 23, p. 192.*

From the final factored form, we see that $f'(x) = 0$ when $x = -1, -\frac{3}{2}$, or $-\frac{13}{10}$, and these are the x-coordinates of all the points on the graph where horizontal tangents occur. $\blacksquare$

An environmental study of a certain suburban community suggests that the average daily level of carbon monoxide in the air may be modeled by the formula

$$C(p) = \sqrt{0.5p^2 + 17}$$

parts per million when the population is p thousand. It is estimated that t years from now, the population of the community will be

$$p(t) = 3.1 + 0.1t^2$$

thousand. At what rate will the carbon monoxide level be changing with respect to time 3 years from now?

Solution $\dfrac{dC}{dt} = \dfrac{dC}{dp}\dfrac{dp}{dt} = \left[\dfrac{1}{2}(0.5p^2 + 17)^{-1/2}(0.5)(2p)\right][0.2t]$

When $t = 3, p(3) = 3.1 + 0.1(3)^2 = 4$, so

$$\left.\frac{dC}{dt}\right|_{t=3} = \left[\frac{1}{2}(0.5 \cdot 4^2 + 17)^{-1/2}(4)\right][0.2(3)] = 0.24$$

Technology Window

It is worth taking a moment to discuss graphing the function from Example 9 to verify the answer graphically. If you input Y=(X+1)^3(2X+3)^2 and graph with a standard scale, the graph is not satisfactory.

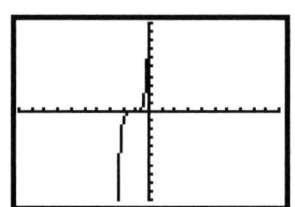

Y₁＝(X+1)^3(2X+3)^2
Xmin=-10 Ymin=-10
Xmax=10   Ymax=10
Xscl=1   Yscl=1

This might be a good place to experiment with the $\boxed{\text{ZOOM}}$ and $\boxed{\text{WINDOW}}$ features of your calculator. What scale is necessary to obtain a graph similar to the following?

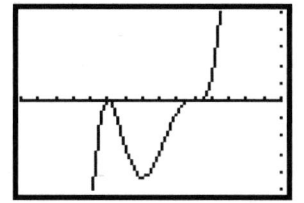

The carbon monoxide level will be changing at the rate of 0.24 part per million. It will be increasing because the sign of dC/dt is positive. ∎

JUSTIFICATION OF THE CHAIN RULE

To get a better feel for why the chain rule is true, suppose x is changed by a small amount Δx. This will cause u to change by an amount Δu, which, in turn, will cause y to change by an amount Δy. *If Δu is not zero,* we can write

$$\frac{\Delta y}{\Delta x} = \frac{\Delta y}{\Delta u}\frac{\Delta u}{\Delta x}$$

By letting $\Delta x \to 0$, we force Δu to approach zero as well, since

$$\Delta u = \left(\frac{\Delta u}{\Delta x}\right)\Delta x \quad \text{so} \quad \lim_{\Delta x \to 0} \Delta u = \left(\frac{du}{dx}\right)(0)$$

It follows that

$$\lim_{\Delta x \to 0} \frac{\Delta y}{\Delta x} = \left(\lim_{\Delta u \to 0} \frac{\Delta y}{\Delta u}\right)\left(\lim_{\Delta x \to 0} \frac{\Delta u}{\Delta x}\right)$$

or, equivalently,

$$\frac{dy}{dx} = \frac{dy}{du}\frac{du}{dx}$$

Unfortunately, there is a flaw in this "proof" of the chain rule. At the beginning we assumed that $\Delta u \neq 0$. However, it is theoretically possible for a small change in x to produce no change in u so that $\Delta u = 0$. This is the case that we consider in the proof in Appendix B.

> **HISTORICAL NOTE**
>
> A calculus book written by the famous mathematician G. H. Hardy (1877–1947) contained essentially this "incorrect" proof rather than the one given in Appendix B. It is even more remarkable that the error was not noticed until the fourth edition.

3.5 Problem Set

 1. ■ **What Does This Say?** What is the chain rule?

2. ■ **What Does This Say?** When do you need to use the chain rule?

In Problems 3–9, use the chain rule to compute the derivative dy/dx and write your answer in terms of x only.

3. $y = u^2 + 1; u = 3x - 2$

4. $y = 2u^2 - u + 5; u = 1 - x^2$

5. $y = \dfrac{2}{u^2}; u = x^2 - 9$

6. $y = \cos u; u = x^2 + 7$

7. $y = u \tan u; u = 3x + \dfrac{6}{x}$ **8.** $y = u^2; u = \ln x$

9. $y = e^u; u = \sec x$

Differentiate each function in Problems 10–32 with respect to the given variable of the function.

10. a. $g(u) = u^5$ **b.** $u(x) = 3x - 1$
   c. $f(x) = (3x - 1)^5$

11. a. $g(u) = u^3$ **b.** $u(x) = x^2 + 1$
   c. $f(x) = (x^2 + 1)^3$

12. a. $g(u) = u^{15}$ **b.** $u(x) = 3x^2 + 5x - 7$
   c. $f(x) = (3x^2 + 5x - 7)^{15}$

13. a. $g(u) = u^7$ **b.** $u(x) = 5 - 8x - 12x^2$
   c. $f(x) = (5 - 8x - 12x^2)^7$

14. a. $f(x) = \sin^2 x(\cos x)$
   b. $g(x) = \sin^2 \theta(\cos x)$, θ a constant

15. a. $f(\theta) = \sin(\sin \theta)$ **b.** $f(\theta) = \sin(\cos \theta)$

16. a. $f(x) = \sqrt{\sin x^2}$ **b.** $g(x) = \sin^2(\sqrt{x})$

17. $s(\theta) = \sin(4\theta + 2)$ **18.** $c(\theta) = \cos(5 - 3\theta)$

19. $f(x) = xe^{1-2x}$ **20.** $g(x) = \ln(3x^4 + 5x)$

21. $p(x) = \sin x^2 \cos x^2$ **22.** $f(x) = \csc^2(\sqrt{x})$

23. $f(x) = (2x^2 + 1)^4(x^2 - 2)^5$

24. $f(x) = (x^3 + 1)^5(2x^3 - 1)^6$

25. $f(x) = \sqrt{\dfrac{x^2 + 3}{x^2 - 5}}$ **26.** $f(x) = \sqrt{\dfrac{2x^2 - 1}{3x^2 + 2}}$

27. $f(x) = \sqrt[3]{x + \sqrt{2x}}$ **28.** $g(x) = \ln(\ln x)$

29. $f(t) = \exp(t^2 + t + 5)$ **30.** $g(t) = t^2e^{-t} + (\ln t)^2$

31. $f(x) = \ln(\sin x + \cos x)$ **32.** $T(x) = \ln(\sec x + \tan x)$

For each function in Problems 33–38, find an equation for the tangent line to the graph at the prescribed point.

33. $f(x) = \sqrt{x^2 + 5}$ at $(2, 3)$

34. $f(x) = (5x + 4)^3$ at $(-1, -1)$

35. $f(x) = x^2(x - 1)^2$ where $x = \frac{1}{2}$

36. $f(x) = \sin(3x - \pi)$ where $x = \frac{\pi}{2}$

37. $f(x) = xe^{2-3x}$ where $x = 0$

38. $f(x) = \dfrac{\ln \sqrt[3]{x}}{x}$ where $x = 1$

Find the x-coordinate of each point in Problems 39–44 where the graph of the given function has a horizontal tangent line.

39. $f(x) = x\sqrt{1 - 3x}$ **40.** $g(x) = x^2(2x + 3)^2$

41. $g(x) = \dfrac{(x - 1)^2}{(x + 2)^3}$ **42.** $f(x) = (2x^2 - 7)^3$

43. $T(x) = x^2e^{1-3x}$ **44.** $V(x) = \dfrac{\ln\sqrt{x}}{x^2}$

B 45. The graphs of $u = g(x)$ and $y = f(u)$ are shown in Figure 3.24.

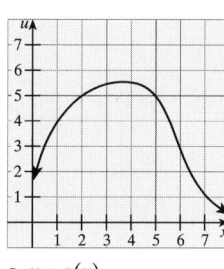   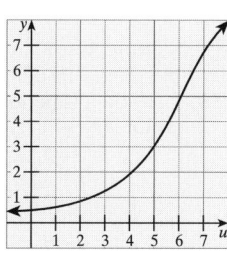

  a. $u = g(x)$ **b.** $y = f(u)$

  ■ **FIGURE 3.24** Find slope of $y = f[g(x)]$.

  a. Find the approximate value of u at $x = 2$. What is the slope of the tangent line at that point?

  b. Find the approximate value of y at $x = 5$. What is the slope of the tangent line at that point?

  c. Find the slope of $y = f[g(x)]$ at $x = 2$.

46. MODELING PROBLEM   Assume that a spherical snowball melts in such a way that its radius decreases at a constant rate (that is, the radius is a linear function of time). Suppose it begins as a sphere with radius 10 cm and takes 2 hours to disappear.

  a. What is the rate of change of its volume after 1 hour? (Recall $V = \frac{4}{3}\pi r^3$.)

  b. At what rate is its surface area changing after 1 hour? (Recall $S = 4\pi r^2$.)

47. MODELING PROBLEM   It is estimated that t years from now, the population of a certain suburban commu-

nity is modeled by the formula

$$p(t) = 20 - \frac{6}{t + 1}$$

where $p(t)$ is in thousands of people. A separate environmental study indicates that the average daily level of carbon monoxide in the air will be

$$L(p) = 0.5\sqrt{p^2 + p + 58}$$

ppm when the population is p thousand. Find the rate at which the level of carbon monoxide will be changing with respect to time two years from now.

48. At a certain factory, the total cost of manufacturing q units during the daily production run is

$$C(q) = 0.2q^2 + q + 900$$

dollars. From experience, it has been determined that approximately

$$q(t) = t^2 + 100t$$

units are manufactured during the first t hours of a production run. Compute the rate at which the total manufacturing cost is changing with respect to time one hour after production begins.

49. An importer of Brazilian coffee estimates that local consumers will buy approximately

$$D(p) = \frac{4{,}374}{p^2}$$

pounds of the coffee per week when the price is p dollars per pound. It is estimated that t weeks from now, the price of Brazilian coffee will be

$$p(t) = 0.02t^2 + 0.1t + 6$$

dollars per pound. At what rate will the weekly demand for the coffee be changing with respect to time 10 weeks from now? Will the demand be increasing or decreasing?

50. When electric blenders are sold for p dollars apiece, local consumers will buy

$$D(p) = \frac{8{,}000}{p}$$

blenders per month. It is estimated that t months from now, the price of the blenders will be

$$p(t) = 0.04t^{3/2} + 15$$

dollars. Compute the rate at which the monthly demand for the blenders will be changing with respect to time 25 months from now. Will the demand be increasing or decreasing?

MODELING PROBLEMS   *When a point source of light of luminous intensity K (candles) shines directly on a point on a surface s meters away, the illuminance on the surface is given by the formula $I = Ks^{-2}$. Use this formula to answer the questions in Problems 51–52. (Note that 1 lux = 1 candle/m².)*

51. Suppose a person carrying a 20-candlepower light walks toward a wall in such a way that at time t (seconds) the distance to the wall is $s(t) = 28 - t^2$ meters.
    a. How fast is the illuminance on the wall increasing when the person is 19 m from the wall?
    b. How far is the person from the wall when the illuminance is changing at the rate of 1 lux/s?

52. A lamp of luminous intensity 40 candles is 20 m above the floor of a room and is being lowered at the constant rate of 2 m/s. At what rate will the illuminance at the point on the floor directly under the lamp be increasing when the lamp is 15 m above the floor? Your answer should be in lux/s rounded to the nearest hundredth.

53. To form a *simple pendulum*, a weight is attached to a rod that is then suspended by one end in such a way that it can swing freely in a vertical plane. Let θ be the angular displacement of the rod from the vertical, as shown in Figure 3.25.

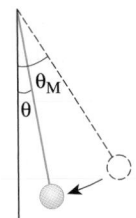

■ **FIGURE 3.25**   Problem 53

It can be shown that as long as the maximum displacement θ_M is small, it is reasonable to assume that $\theta = \theta_M \sin kt$ at time t, where k is a constant that depends on the length of the rod. Show that

$$\frac{d^2\theta}{dt^2} + k^2\theta = 0$$

54. A lighthouse is located 2 km directly across the sea from a point O on the shoreline, as shown in Figure 3.26.

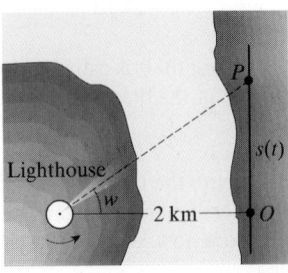

■ **FIGURE 3.26**   Problem 54

A beacon in the lighthouse makes 3 complete revolutions (that is, 6π radians) each minute, and during part of each revolution, the light sweeps across the face of a row of cliffs lining the shore.
    a. Show that t minutes after it passes point O, the beam of light is at a point P located $s(t) = 2\tan(6\pi t)$ km from O.
    b. How fast is the beam of light moving at the time it passes a point on the cliff that is 4 km from the lighthouse?

55. Using only the formula

$$\frac{d}{dx}\sin u = \cos u \,\frac{du}{dx}$$

and the identities $\cos x = \sin(\frac{\pi}{2} - x)$ and $\sin x = \cos(\frac{\pi}{2} - x)$, show that

$$\frac{d}{dx}\cos u = -\sin u \,\frac{du}{dx}$$

56. **THINK TANK PROBLEM**   Let $g(x) = f[u(x)]$, where $u(-3) = 5$, $u'(-3) = 2$, $f(5) = 3$, and $f'(5) = -3$. Find an equation for the tangent to the graph of g at the point where $x = -3$.

57. Let f be a function for which

$$f'(x) = \frac{1}{x^2 + 1}$$

    a. If $g(x) = f(3x - 1)$, what is $g'(x)$?
    b. If $h(x) = f\left(\frac{1}{x}\right)$, what is $h'(x)$?

58. Let f be a function for which $f(2) = -3$ and $f'(x) = \sqrt{x^2 + 5}$. If

$$g(x) = x^2 f\left(\frac{x}{x - 1}\right),$$

    what is $g'(2)$?

59. a. If $F(x) = \ln|\cos x|$, show that $F'(x) = -\tan x$.
    b. If $F(x) = \ln|\sec x + \tan x|$, show that $F'(x) = \sec x$.

60. If $\dfrac{df}{dx} = \dfrac{\sin x}{x}$ and $u(x) = \cot x$, what is $\dfrac{df}{du}$?

61. Use the chain rule to find

$$\frac{d}{dx}f'[f(x)] \quad \text{and} \quad \frac{d}{dx}f[f'(x)]$$

    assuming these derivatives exist.

62. Show that if a particle moves along a straight line with position $s(t)$ and velocity $v(t)$, then its acceleration satisfies

$$a(t) = v(t)\frac{dv}{ds}$$

    Use this formula to find $\dfrac{dv}{ds}$ in the case where $s(t) = -2t^3 + 4t^2 + t - 3$.

3.6 *Implicit Differentiation*

IN THIS SECTION **general procedure for implicit differentiation, derivative formulas for the inverse trigonometric functions, logarithmic differentiation** ■

GENERAL PROCEDURE FOR IMPLICIT DIFFERENTIATION

The equation $y = \sqrt{1 - x^2}$ **explicitly** defines $f(x) = \sqrt{1 - x^2}$ as a function of x for $-1 \le x \le 1$, but the same function can also be defined **implicitly** by the equation $x^2 + y^2 = 1$, as long as we restrict y by $0 \le y \le 1$ so the vertical line test is satisfied. To find the derivative of the explicit form, we use the chain rule:

$$\frac{d}{dx}\sqrt{1 - x^2} = \frac{d}{dx}(1 - x^2)^{1/2} = \tfrac{1}{2}(1 - x^2)^{-1/2}(-2x) = \frac{-x}{\sqrt{1 - x^2}}$$

To obtain the derivative of the same function in its implicit form, we simply differentiate across the equation $x^2 + y^2 = 1$, remembering that y is a function of x:

$$\frac{d}{dx}(x^2 + y^2) = \frac{d}{dx}(1) \qquad \text{Derivative of both sides}$$

$$2x + 2y\frac{dy}{dx} = 0 \qquad \text{Chain rule for the derivative of } y$$

$$\frac{dy}{dx} = -\frac{x}{y} \qquad \text{Solve for } \frac{dy}{dx}.$$

$$= -\frac{x}{\sqrt{1 - x^2}} \qquad \text{Write as a function of } x, \text{ if desired.}$$

The procedure we have just illustrated is called **implicit differentiation.** Our illustrative example was simple, but consider a differentiable function $y = f(x)$ defined by the equation

$$x^2y^3 - 6 = 5y^3 + x$$

Implicit differentiation tells us that

$$x^2\frac{d}{dx}(y^3) + y^3\frac{d}{dx}(x^2) - \frac{d}{dx}(6) = 5\frac{d}{dx}(y^3) + \frac{d}{dx}(x)$$

$$x^2\left(3y^2\frac{dy}{dx}\right) + y^3(2x) - 0 = 5\left(3y^2\frac{dy}{dx}\right) + 1$$

$$(3x^2y^2 - 15y^2)\frac{dy}{dx} = 1 - 2xy^3$$

$$\frac{dy}{dx} = \frac{1 - 2xy^3}{3x^2y^2 - 15y^2}$$

You may think that we are not finished since the derivative involves both x and y, but for many applications, that is enough. In this example, we could have found y as an explicit function of x, namely,

$$y = \left(\frac{x + 6}{x^2 - 5}\right)^{1/3}$$

and then found dy/dx by the chain rule, but consider a differentiable function $y = f(x)$ defined by the equation

$$x^2y + 2y^3 = 3x + 2y$$

Finding y as an explicit function of x is very difficult in this case, and it is not at all hard to imagine similar situations where solving for y in terms of x would be impossible or at least not worth the effort. We begin our work with implicit differentiation by using it to find the slope of a tangent line at a point on a circle.

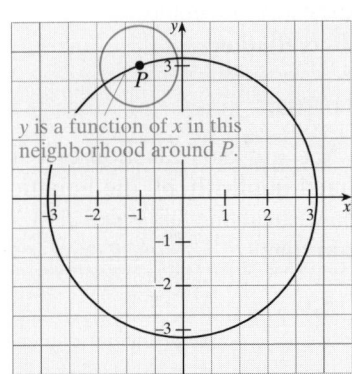

y is a function of x in this neighborhood around P.

■ **FIGURE 3.27** Graph of $x^2 + y^2 = 10$ showing a neighborhood about $P(-1, 3)$

EXAMPLE 1   *Slope of a tangent using implicit differentiation*

Find the slope of the tangent to the circle $x^2 + y^2 = 10$ at the point $P(-1, 3)$.

Solution   We recognize the graph of $x^2 + y^2 = 10$ as not being the graph of a function. However, if we look at a small neighborhood around the point $(-1, 3)$, as shown in Figure 3.27, we see that this part of the graph *does pass the vertical line test* for functions. Thus, the required slope can be found by evaluating the derivative dy/dx at $(-1, 3)$. Instead of solving for y and finding the derivative, we *take the derivative of both sides of the equation:*

$$x^2 + y^2 = 10 \qquad \text{Given equation}$$

$$\frac{d}{dx}[x^2 + y^2] = \frac{d}{dx}(10) \qquad \text{Derivative of both sides}$$

$$2x + 2y\frac{dy}{dx} = 0 \qquad \text{Do not forget that y is a function of x.}$$

$$\frac{dy}{dx} = -\frac{x}{y} \qquad \text{Solve the equation for } \frac{dy}{dx}.$$

The slope of the tangent at $P(-1, 3)$ is

$$\frac{dy}{dx}\bigg|_{(x, y) = (-1, 3)} = -\frac{x}{y}\bigg|_{(-1, 3)} = -\frac{-1}{3} = \frac{1}{3}$$

Here is a general description of the procedure for implicit differentiation.

Procedure for Implicit Differentiation

Suppose an equation defines y implicitly as a differentiable function of x. To find $\dfrac{dy}{dx}$:

Step 1.   Differentiate both sides of the equation with respect to x. Remember that y is really a function of x for part of the curve and use the chain rule when differentiating terms containing y.

Step 2.   Solve the differentiated equation algebraically for $\dfrac{dy}{dx}$.

EXAMPLE 2   *Implicit differentiation*

If $y = f(x)$ is a differentiable function of x such that

$$x^2y + 2y^3 = 3x + 2y$$

find $\dfrac{dy}{dx}$.

Solution   The process is to differentiate both sides of the given equation with respect to x. To help remember that y is a function of x, replace y by the symbol $y(x)$:

$$x^2 y(x) + 2[y(x)]^3 = 3x + 2y(x)$$

Differentiate both sides of this equation term by term with respect to x:

$$\frac{d}{dx}\{x^2 y(x) + 2[y(x)]^3\} = \frac{d}{dx}\{3x + 2y(x)\}$$

$$\frac{d}{dx}[x^2 y(x)] + 2\frac{d}{dx}[y(x)]^3 = 3\frac{d}{dx}x + 2\frac{d}{dx}y(x)$$

$$\underbrace{x^2 \frac{d}{dx}y(x) + y(x)\frac{d}{dx}x^2}_{\text{Product rule}} + \underbrace{2\left\{3[y(x)]^2 \frac{d}{dx}y(x)\right\}}_{\text{Extended power rule}} = 3 + 2\frac{d}{dx}y(x)$$

$$x^2 \frac{d}{dx}y(x) + 2xy(x) + 6[y(x)]^2 \frac{d}{dx}y(x) = 3 + 2\frac{d}{dx}y(x)$$

Now replace $y(x)$ by y and $\frac{d}{dx}y(x)$ by $\frac{dy}{dx}$ and rewrite the equation:

$$x^2 \frac{dy}{dx} + 2xy + 6y^2 \frac{dy}{dx} = 3 + 2\frac{dy}{dx}$$

Finally, solve this equation for $\frac{dy}{dx}$:

$$x^2 \frac{dy}{dx} + 6y^2 \frac{dy}{dx} - 2\frac{dy}{dx} = 3 - 2xy$$

$$(x^2 + 6y^2 - 2)\frac{dy}{dx} = 3 - 2xy$$

$$\frac{dy}{dx} = \frac{3 - 2xy}{x^2 + 6y^2 - 2}$$

Notice that the formula for dy/dx contains both the independent variable x and the dependent variable y. This is usual when derivatives are computed implicitly. ∎

> **WARNING**   It is important to realize that implicit differentiation is a technique for finding dy/dx that is valid only if y is a differentiable function of x, and careless application of the technique can lead to errors. For example, there is clearly no real-valued function $y = f(x)$ that satisfies the equation $x^2 + y^2 = -1$, yet formal application of implicit differentiation yields the "derivative" $dy/dx = -x/y$. To be able to evaluate this "derivative," we must find some values for which $x^2 + y^2 = -1$. Because no such values exist, the derivative does *not* exist.

In Example 2 it was suggested that you temporarily replace y by $y(x)$, so you would not forget to use the chain rule when first learning implicit differentiation. In the following example, we eliminate this unnecessary step and differentiate the given equation directly. Just keep in mind that y is really a function of x and remember to use the chain rule (or extended power rule) when it is appropriate.

| EXAMPLE 3 | *Implicit differentiation; simplified notation* |

Find $\dfrac{dy}{dx}$ if y is a differentiable function of x that satisfies

$$\sin(x^2 + y) = y^2(3x + 1)$$

Solution   There is no obvious way to solve the given equation explicitly for y. Differentiate implicitly to obtain

$$\frac{d}{dx}[\sin(x^2 + y)] = \frac{d}{dx}[y^2(3x + 1)]$$

$$\underbrace{\cos(x^2 + y)\frac{d}{dx}(x^2 + y)}_{\text{Chain rule}} = \underbrace{y^2\frac{d}{dx}(3x + 1) + (3x + 1)\frac{d}{dx}y^2}_{\text{Product rule}}$$

$$\cos(x^2 + y)\left(2x + \frac{dy}{dx}\right) = y^2(3) + (3x + 1)\left(2y\,\frac{dy}{dx}\right)$$

Finally, solve for $\dfrac{dy}{dx}$:

$$2x\cos(x^2 + y) + \cos(x^2 + y)\frac{dy}{dx} = 3y^2 + 2y(3x + 1)\frac{dy}{dx}$$

$$[\cos(x^2 + y) - 2y(3x + 1)]\frac{dy}{dx} = 3y^2 - 2x\cos(x^2 + y)$$

$$\frac{dy}{dx} = \frac{3y^2 - 2x\cos(x^2 + y)}{\cos(x^2 + y) - 2y(3x + 1)}$$

∎

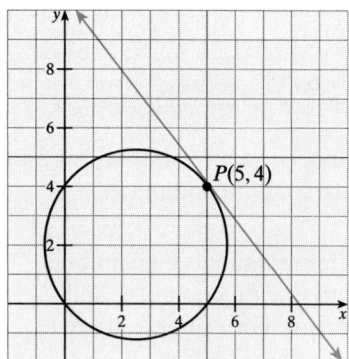

| EXAMPLE 4 | *Slope of a tangent line using implicit differentiation* |

Find the slope of a tangent line to the circle $x^2 + y^2 = 5x + 4y$ at the point $P(5, 4)$.

Solution   The slope of a curve $y = f(x)$ is $\dfrac{dy}{dx}$, which we find implicitly:

$$x^2 + y^2 = 5x + 4y$$

$$\frac{d}{dx}(x^2 + y^2) = \frac{d}{dx}(5x + 4y)$$

$$2x + 2y\frac{dy}{dx} = 5 + 4\frac{dy}{dx}$$

$$2y\frac{dy}{dx} - 4\frac{dy}{dx} = 5 - 2x$$

$$(2y - 4)\frac{dy}{dx} = 5 - 2x$$

$$\frac{dy}{dx} = \frac{5 - 2x}{2y - 4}$$

Note that the expression is undefined at $y = 2$. This makes sense when you see that the tangent is vertical there. Look at the graph and see whether you should exclude any other values.

At $(5, 4)$, the slope of the tangent line is

$$\left.\frac{dy}{dx}\right|_{(5,\,4)} = \left.\frac{5 - 2x}{2y - 4}\right|_{(5,\,4)} = \frac{5 - 2(5)}{2(4) - 4} = \frac{-5}{4}$$

∎

EXAMPLE 5   *Second derivative by implicit differentiation*

Find $\dfrac{d^2y}{dx^2}$ if $x^2 + y^2 = 10$.

Solution   In Example 1 we found (implicitly) that $\dfrac{dy}{dx} = -\dfrac{x}{y}$. Thus,

$$\frac{d^2y}{dx^2} = \frac{d}{dx}\left(\frac{-x}{y}\right) = \underbrace{\frac{y\dfrac{d}{dx}(-x) - (-x)\dfrac{d}{dx}y}{y^2}}_{\text{Quotient rule}} = \frac{-y + x\dfrac{dy}{dx}}{y^2}$$

Note that the expression for the second derivative contains the first derivative dy/dx. To simplify the answer, substitute the algebraic expression previously found for dy/dx:

$$\frac{-y + x\dfrac{dy}{dx}}{y^2} = \frac{-y + x\dfrac{-x}{y}}{y^2} \qquad \text{Substitute } \frac{dy}{dx} = -\frac{x}{y}.$$

$$= \frac{-y^2 - x^2}{y^3}$$

$$= \frac{-(x^2 + y^2)}{y^3}$$

$$= \frac{-10}{y^3} \qquad \text{Substitute } x^2 + y^2 = 10.$$

Thus, $\dfrac{d^2y}{dx^2} = \dfrac{-10}{y^3}$ ■

Implicit differentiation is a valuable theoretical tool. For example, in Section 3.2 we proved the power rule for the case where the exponent is an integer; implicit differentiation now allows us to extend the proof for all real exponents.

EXAMPLE 6   *Proof of power rule for real (rational and irrational) exponents*

Prove $\dfrac{d}{dx}(x^r) = rx^{r-1}$ for all real numbers r.

Solution   If $y = x^r$, then $y = e^{r\ln x}$, so that

$$\ln y = r\ln x \qquad \text{Definition of exponent}$$

$$\frac{1}{y}\frac{dy}{dx} = r\left(\frac{1}{x}\right) \qquad \text{Implicit differentiation}$$

$$\frac{dy}{dx} = y\left(\frac{r}{x}\right) \qquad \text{Solve for } \frac{dy}{dx}.$$

$$= x^r\left(\frac{r}{x}\right) \qquad \text{Substitute } y = x^r.$$

$$= rx^{r-1} \qquad \text{Property of exponents}$$ ■

Notice in Example 6 that $y = x^r = e^{r\ln x}$ is differentiable because it is defined as the composition of the differentiable functions $y = e^u$ and $u = r\ln x$.

In Section 3.3, we found the derivative of $f(x) = e^x$ (Theorem 3.8). This derivative is easily found using implicit differentiation, as shown in the following example.

EXAMPLE 7 *Derivative rule for the natural exponential using implicit differentiation*

$$\frac{d}{dx}(e^x) = e^x$$

Solution Let $v = e^x$, so that $x = \ln v$. Then

$$\frac{d}{dx}(x) = \frac{d}{dx}(\ln v) \qquad \text{Derivative of both sides of } x = \ln v.$$

$$1 = \frac{1}{v}\frac{dv}{dx} \qquad \text{Implicit differentiation}$$

$$v = \frac{dv}{dx}$$

$$e^x = \frac{dv}{dx} = \frac{d}{dx}(e^x) \qquad \text{Because } v = e^x \qquad \blacksquare$$

DERIVATIVE FORMULAS FOR THE INVERSE TRIGONOMETRIC FUNCTIONS

Next, we shall use implicit differentiation to obtain differentiation formulas for the six inverse trigonometric functions. Note that these derivatives are not inverse trigonometric functions or even trigonometric functions, but are instead rational functions or roots of rational functions.

THEOREM 3.11 *Differentiation formulas for six inverse trigonometric functions*

If u is a differentiable function of x, then

$$\frac{d}{dx}(\sin^{-1}u) = \frac{1}{\sqrt{1-u^2}}\frac{du}{dx} \qquad \frac{d}{dx}(\cos^{-1}u) = \frac{-1}{\sqrt{1-u^2}}\frac{du}{dx}$$

$$\frac{d}{dx}(\tan^{-1}u) = \frac{1}{1+u^2}\frac{du}{dx} \qquad \frac{d}{dx}(\cot^{-1}u) = \frac{-1}{1+u^2}\frac{du}{dx}$$

$$\frac{d}{dx}(\sec^{-1}u) = \frac{1}{|u|\sqrt{u^2-1}}\frac{du}{dx} \qquad \frac{d}{dx}(\csc^{-1}u) = \frac{-1}{|u|\sqrt{u^2-1}}\frac{du}{dx}$$

WARNING Note that the derivative of each inverse trigonometric function $y = \cos^{-1}x$, $y = \cot^{-1}x$, and $y = \csc^{-1}x$, is the *opposite* of the derivative of the corresponding inverse cofunction $y = \sin^{-1}x$, $y = \tan^{-1}x$, and $y = \sec^{-1}x$. ←

Proof We shall prove the first formula and leave the others as problems. Let $\alpha = \sin^{-1}x$, so $x = \sin\alpha$. Because the sine function is one-to-one and differentiable on $[-\pi/2, \pi/2]$, the inverse sine is also differentiable. To find its derivative, we proceed implicitly:

$$\sin\alpha = x$$

$$\frac{d}{dx}(\sin\alpha) = \frac{d}{dx}(x)$$

$$\cos\alpha\frac{d\alpha}{dx} = 1$$

Since $-\dfrac{\pi}{2} \le \alpha \le \dfrac{\pi}{2}$, $\cos\alpha \ge 0$, so

$$\frac{d\alpha}{dx} = \frac{1}{\cos\alpha} = \frac{1}{\sqrt{1-\sin^2\alpha}} = \frac{1}{\sqrt{1-x^2}} \qquad \text{Note reference triangle.}$$

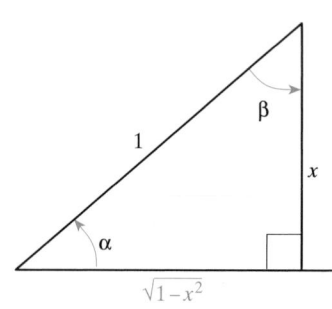

$\sin\alpha = x$ and
$\cos\alpha = \sqrt{1-x^2}$

If u is a differentiable function of x, then the chain rule gives

$$\frac{d}{dx}(\sin^{-1}u) = \frac{d}{du}(\sin^{-1}u)\frac{du}{dx} = \frac{1}{\sqrt{1-u^2}}\frac{du}{dx}$$

EXAMPLE 8   *Derivatives involving inverse trigonometric functions*

Differentiate each of the following functions.

a. $f(x) = \tan^{-1}\sqrt{x}$ b. $g(t) = \sin^{-1}(1-t)$ c. $h(x) = \sec^{-1}e^{2x}$

Solution

a. Let $u = \sqrt{x}$ in the formula for $\frac{d}{dx}(\tan^{-1}u)$:

$$f'(x) = \frac{d}{dx}(\tan^{-1}\sqrt{x}) = \frac{1}{1+(\sqrt{x})^2}\frac{d}{dx}(\sqrt{x}) = \frac{1}{1+x}\left(\frac{1}{2}\frac{1}{\sqrt{x}}\right) = \frac{1}{2\sqrt{x}(1+x)}$$

b. Let $u = (1-t)$;

$$g'(t) = \frac{d}{dt}[\sin^{-1}(1-t)] = \frac{1}{\sqrt{1-(1-t)^2}}\frac{d}{dt}(1-t) = \frac{-1}{\sqrt{1-(1-t)^2}} = \frac{-1}{\sqrt{2t-t^2}}$$

c. Let $u = e^{2x}$;

$$h'(x) = \frac{d}{dx}[\sec^{-1}e^{2x}] = \frac{1}{|e^{2x}|\sqrt{(e^{2x})^2-1}}\frac{d}{dx}(e^{2x})$$

$$= \frac{1}{e^{2x}\sqrt{e^{4x}-1}}(2e^{2x}) = \frac{2}{\sqrt{e^{4x}-1}}$$

The derivatives of b^u and $\log_b u$ for a differentiable function $u = u(x)$ and a base b other than e can be obtained using the chain rule and the change-of-base formulas from Chapter 1. We summarize the results in the following theorem.

THEOREM 3.12   *Derivatives of exponential and logarithmic functions with base b*

Let u be a differentiable function of x and b be a positive number (other than 1). Then

$$\frac{d}{dx}b^u = (\ln b)b^u\frac{du}{dx} \qquad \frac{d}{dx}(\log_b u) = \frac{1}{\ln b}\cdot\frac{1}{u}\frac{du}{dx}$$

Proof   Because $b^u = e^{u\ln b}$, we can apply the chain rule as follows:

$$\frac{d}{dx}b^u = \frac{d}{dx}(e^{u\ln b}) = e^{u\ln b}\frac{d}{dx}(u\ln b) = e^{u\ln b}\left(\ln b\frac{du}{dx}\right) = (\ln b)b^u\frac{du}{dx}$$

To differentiate the logarithm, recall the change-of-base formula $\log_b u = \frac{\ln u}{\ln b}$ so that

$$\frac{d}{dx}\log_b u = \frac{d}{dx}\left(\frac{\ln u}{\ln b}\right) = \frac{1}{\ln b}\cdot\frac{1}{u}\frac{du}{dx}$$

> ■ *What This Says:* The derivatives of b^x and $\log_b x$ are the same as the derivatives of e^x and $\ln x$, respectively, except for a factor of $\ln b$ that appears as a multiplier in the formula
>
> $$\frac{d}{dx}(b^x) = (\ln b)b^x$$
>
> and as a divisor in the formula
>
> $$\frac{d}{dx}(\log_b x) = \frac{1}{(\ln b)x}$$

EXAMPLE 9   *Derivative of an exponential function with base $b \neq e$*

Differentiate $f(x) = x(2^{1-x})$.

Solution   Apply the product rule:

$$f'(x) = \frac{d}{dx}(x\,2^{1-x}) = x\frac{d}{dx}(2^{1-x}) + 2^{1-x}\frac{d}{dx}(x)$$
$$= x(\ln 2)(2^{1-x})(-1) + 2^{1-x}(1) = 2^{1-x}(1 - x\ln 2)$$ ■

The following theorem will prove useful in Chapter 5.

THEOREM 3.13   *Derivative of $\ln |u|$*

If $f(x) = \ln|x|$, $x \neq 0$, then $f'(x) = \frac{1}{x}$.

Also, if u is a differentiable function of x, then $\frac{d}{dx}\ln|u| = \frac{1}{u}\frac{du}{dx}$.

Proof   Using the definition of absolute value,

$$f(x) = \begin{cases} \ln x & \text{if } x > 0 \\ \ln(-x) & \text{if } x < 0 \end{cases}$$

so $$f'(x) = \begin{cases} \dfrac{1}{x} & \text{if } x > 0 \\ \dfrac{1}{-x}(-1) = \dfrac{1}{x} & \text{if } x < 0 \end{cases}$$

Thus, $f'(x) = \frac{1}{x}$ for all $x \neq 0$.

The second part of the theorem (for u, a differentiable function of x) follows from the chain rule.   ◻

LOGARITHMIC DIFFERENTIATION

Logarithmic differentiation is a procedure in which logarithms are used to trade the task of differentiating products and quotients for that of differentiating sums and differences. It is especially valuable as a means for handling complicated product or quotient functions and power functions where variables appear in both the base and the exponent.

EXAMPLE 10   *Logarithmic differentiation*

Find the derivative of $y = \dfrac{e^{2x}(2x - 1)^6}{(x^3 + 5)^2(4 - 7x)}$.

Solution   The procedure called logarithmic differentiation requires that we first take the logarithm of both sides and then apply properties of logarithms before attempting to take the derivative.

$$y = \frac{e^{2x}(2x - 1)^6}{(x^3 + 5)^2(4 - 7x)}$$

$$\ln y = \ln\left[\frac{e^{2x}(2x - 1)^6}{(x^3 + 5)^2(4 - 7x)}\right]$$

$$= \ln e^{2x} + \ln(2x - 1)^6 - \ln(x^3 + 5)^2 - \ln(4 - 7x)$$

$$= 2x + 6\ln(2x - 1) - 2\ln(x^3 + 5) - \ln(4 - 7x)$$

Next, differentiate both sides with respect to x and then solve for $\dfrac{dy}{dx}$:

$$\frac{1}{y}\frac{dy}{dx} = 2 + 6\left[\frac{1}{2x - 1}(2)\right] - 2\left[\frac{1}{x^3 + 5}(3x^2)\right] - \left[\frac{1}{4 - 7x}(-7)\right]$$

$$\frac{dy}{dx} = y\left[2 + \frac{12}{2x - 1} - \frac{6x^2}{x^3 + 5} + \frac{7}{4 - 7x}\right]$$

This is the derivative in terms of x and y. If we want the derivative in terms of x alone, we can substitute the value of y:

$$\frac{dy}{dx} = \frac{e^{2x}(2x - 1)^6}{(x^3 + 5)^2(4 - 7x)}\left[2 + \frac{12}{2x - 1} - \frac{6x^2}{x^3 + 5} + \frac{7}{4 - 7x}\right] \qquad \blacksquare$$

EXAMPLE 11   *Derivative of a function with a variable in both the base term and the exponent*

Find $\dfrac{dy}{dx}$, where $y = (x + 1)^{2x}$.

Solution
$$y = (x + 1)^{2x}$$
$$\ln y = \ln[(x + 1)^{2x}] = 2x\ln(x + 1)$$

Differentiate both sides of this equation:

$$\frac{1}{y}\frac{dy}{dx} = 2x\left\{\frac{d}{dx}[\ln(x + 1)]\right\} + \left[\frac{d}{dx}(2x)\right]\ln(x + 1) \qquad \text{Product rule}$$

$$= 2x\left[\frac{1}{x + 1}(1)\right] + 2\ln(x + 1) = \frac{2x}{x + 1} + 2\ln(x + 1)$$

Finally, multiply both sides by $y = (x + 1)^{2x}$:

$$\frac{dy}{dx} = \left[\frac{2x}{x + 1} + 2\ln(x + 1)\right](x + 1)^{2x} \qquad \blacksquare$$

3.6 Problem Set

A *Find $\dfrac{dy}{dx}$ by implicit differentiation in Problems 1–14.*

1. $x^2 + y^2 = 25$
2. $x^2 + y = x^3 + y^3$
3. $xy = 25$
4. $xy(2x + 3y) = 2$
5. $x^2 + 3xy + y^2 = 15$
6. $x^3 + y^3 = x + y$
7. $\dfrac{1}{y} + \dfrac{1}{x} = 1$
8. $(2x + 3y)^2 = 10$
9. $\sin(x + y) = x - y$
10. $\tan \dfrac{x}{y} = y$
11. $\cos xy = 1 - x^2$
12. $e^{xy} + 1 = x^2$
13. $\ln(xy) = e^{2x}$
14. $e^{xy} + \ln y^2 = x$

In Problems 15–18, find $\dfrac{dy}{dx}$ two ways:

a. *By implicit differentiation of the equation*
b. *By differentiating an explicit formula for y*

15. $x^2 + y^3 = 12$
16. $xy + 2y = x^2$
17. $x + \dfrac{1}{y} = 5$
18. $xy - x = y + 2$

Find the derivative $\dfrac{dy}{dx}$ in Problems 19–32.

19. $y = \sin^{-1}(2x + 1)$
20. $y = \cos^{-1}(4x + 3)$
21. $y = \tan^{-1}\sqrt{x^2 + 1}$
22. $y = \cot^{-1}x^2$
23. $y = (\sin^{-1}2x)^3$
24. $y = (\tan^{-1}x^2)^4$
25. $y = \sec^{-1}(e^{-x})$
26. $y = \ln|\sin^{-1}x|$
27. $y = \tan^{-1}\left(\dfrac{1}{x}\right)$
28. $y = \sec^{-1}(3x)$
29. $y = \csc^{-1}(2x^2)$
30. $y = \ln[\sin^{-1}(e^x)]$
31. $x \sin^{-1}y + y \tan^{-1}x = x$
32. $\sin^{-1}y + y = 2xy$

In Problems 33–38, find an equation of the tangent line to the graph of each equation at the prescribed point.

33. $x^2 + y^2 = 13$ at $(-2, 3)$
34. $x^3 + y^3 = y + 21$ at $(3, -2)$
35. $\sin(x - y) = xy$ at $(0, \pi)$
36. $3^x + \log_3(xy) = 10$ at $(2, 1)$
37. $x \tan^{-1}y = x^2 + y$ at $(0, 0)$
38. $\sin^{-1}(xy) + \dfrac{\pi}{2} = \cos^{-1}y$ at $(1, 0)$

Find the slope of the tangent to the graph at the points indicated in Problems 39–42.

39. bifolium: $(x^2 + y^2)^2 = 4x^2y$ at $(1, 1)$

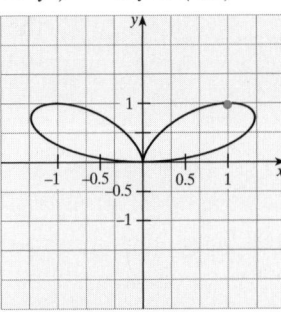

40. lemniscate of Bernoulli:
$(x^2 + y^2)^2 = \dfrac{25}{3}(x^2 - y^2)$ at $(2, 1)$

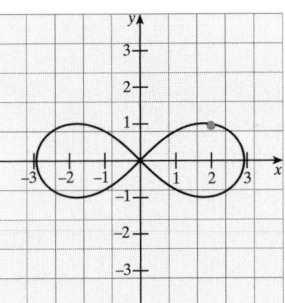

41. folium of Descartes:
$x^3 + y^3 - \dfrac{9}{2}xy = 0$ at $(2, 1)$

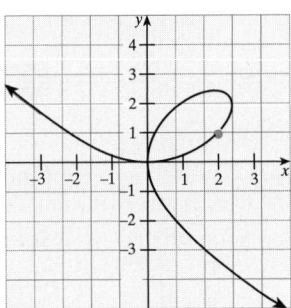

42. cissoid of Diocles:
$y^2(6 - x) = x^3$ at $(3, 3)$

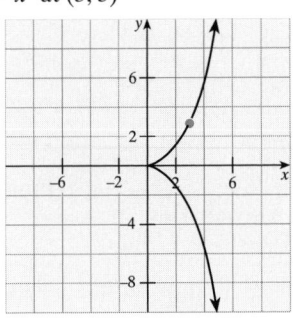

43. Find an equation of the normal line to the curve $x^2 + 2xy = y^3$ at $(1, -1)$.

44. Find an equation of the normal line to the curve $x^2\sqrt{y - 2} = y^2 - 3x - 5$ at $(1, 3)$.

Use implicit differentiation to find the second derivative y'' of the functions given in Problems 45–46.

45. $7x + 5y^2 = 1$ **46.** $x^2 + 2y^3 = 4$

Ⓑ 47. ■ **What Does This Say?** Compare and contrast the derivatives of the following functions.
   a. $y = x^2$ **b.** $y = 2^x$ **c.** $y = e^x$ **d.** $y = x^e$

48. ■ **What Does This Say?** Compare and contrast the derivatives of the following functions.
   a. $y = \log x$ **b.** $y = \ln x$

49. ■ **What Does This Say?** Discuss logarithmic differentiation.

Use logarithmic differentiation in Problems 50–55 to find dy/dx. You may express your answer in terms of both x and y, and you do not need to simplify the resulting rational expressions.

50. $y = \sqrt[18]{(x^{10} + 1)^3(x^7 - 3)^8}$

51. $y = \dfrac{(2x - 1)^5}{\sqrt{x - 9}(x + 3)^2}$

52. $y = \dfrac{e^{2x}}{(x^2 - 3)^2 \ln \sqrt{x}}$

53. $y = \dfrac{e^{3x^2}}{(x^3 + 1)^2(4x - 7)^{-2}}$

54. $y = x^x$ **55.** $y = x^{\ln\sqrt{x}}$

56. Let $\dfrac{u^2}{a^2} + \dfrac{v^2}{b^2} = 1$, where a and b are nonzero constants.

   Find   **a.** $\dfrac{du}{dv}$ **b.** $\dfrac{dv}{du}$

57. Show that the tangent at the point (a, b) on the curve whose equation is $2x^2 + 3xy + y^2 = -2$ is horizontal if $4a + 3b = 0$. Find two such points on the curve.

58. Find two points on the curve whose equation is $x^2 - 3xy + 2y^2 = -2$ where the tangent is vertical.

59. Let g be a differentiable function of x that satisfies $g(x) < 0$ and $x^2 + g^2(x) = 10$ for all x.

   a. Use implicit differentiation to show that
$$\frac{dg}{dx} = \frac{-x}{g(x)}$$

   b. Show that
$$g(x) = -\sqrt{10 - x^2}$$
   satisfies the given requirements. Then use the chain rule to verify that
$$\frac{dg}{dx} = \frac{-x}{g(x)}$$

60. Find the equations of the tangent and the normal line to the curve $x^3 + y^3 = 2Axy$ at the point (A, A), where A is a constant.

61. THINK TANK PROBLEM

   a. If $x^2 + y^2 = 6y - 10$ and $\dfrac{dy}{dx}$ exists, show that
$$\frac{dy}{dx} = \frac{x}{3 - y}.$$

   b. Show that there are no real numbers x, y that satisfy the equation
$$x^2 + y^2 = 6y - 10$$

   c. What can you conclude from the result found in part **a** in light of this observation in part **b**?

62. Find all points on the lemniscate
$$(x^2 + y^2)^2 = 4(x^2 - y^2)$$
where the tangent is horizontal. (See Figure 3.28.)

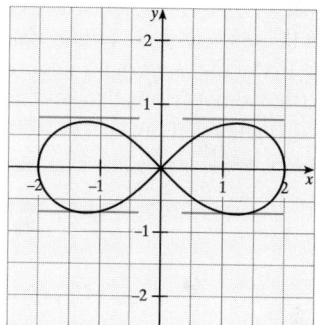

■ **FIGURE 3.28** Lemniscate $(x^2 + y^2)^2 = 4(x^2 - y^2)$

63. Find all points on the cardioid
$$(x^2 + y^2)^{3/2} = \sqrt{x^2 + y^2} + x$$
where the tangent is vertical. (See Figure 3.29.)

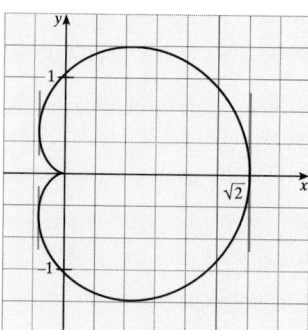

■ **FIGURE 3.29** Cardioid $(x^2 + y^2)^{3/2} = \sqrt{x^2 + y^2} + x$

64. The tangent to the curve

$$x^{2/3} + y^{2/3} = 8$$

at the point $(8, 8)$ and the coordinate axes form a triangle, as shown in Figure 3.30. What is the area of this triangle?

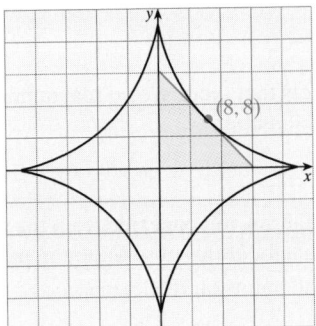

■ **FIGURE 3.30**   Problem 64

65. **MODELING PROBLEM**   A worker stands 4 m from a hoist being raised at the rate of 2 m/s, as shown in Figure 3.31. Model the worker's angle of sight θ using an inverse trigonometric function, and then determine how fast θ is changing at the instant when the hoist is 1.5 m above eye level.

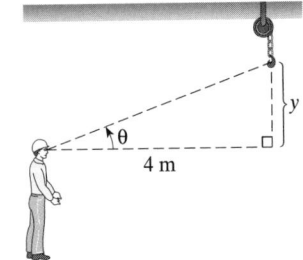

■ **FIGURE 3.31**   Modeling an angle of elevation

66. **THINK TANK PROBLEM**   Find two differentiable functions f that satisfy the equation

$$x - [f(x)]^2 = 9$$

Give the explicit form of each function, and sketch its graph.

67. Show that the tangent to the ellipse

$$\frac{x^2}{a^2} + \frac{y^2}{b^2} = 1$$

at the point (x_0, y_0) is

$$\frac{x_0 x}{a^2} + \frac{y_0 y}{b^2} = 1$$

(See Figure 3.32.)

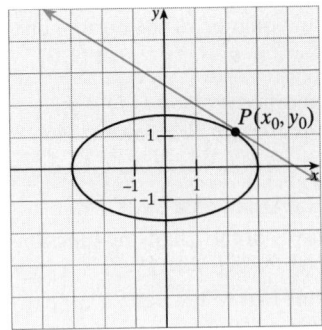

■ **FIGURE 3.32**   Ellipse $\dfrac{x^2}{a^2} + \dfrac{y^2}{b^2} = 1$

68. Find an equation of the tangent to the hyperbola

$$\frac{x^2}{a^2} - \frac{y^2}{b^2} = 1$$

at the point (x_0, y_0). (See Figure 3.33.)

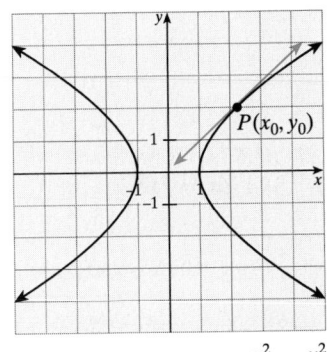

■ **FIGURE 3.33**   Hyperbola $\dfrac{x^2}{a^2} - \dfrac{y^2}{b^2} = 1$

69. Use implicit differentiation to find the second derivative y'', where y is a differentiable function of x that satisfies

$$ax^2 + by^2 = c$$

($a, b,$ and c are constants).

70. Show that the sum of the x-intercept and the y-intercept of any tangent to the curve

$$\sqrt{x} + \sqrt{y} = C$$

is equal to C^2.

*The **angle between curves C_1 and C_2** at the point of intersection P is defined as the angle $0 \le \theta \le \frac{\pi}{2}$ between the tangent lines at P. Specifically, the angle from C_1 to C_2 is the angle from the tangent line to C_1 at P, to the tangent line to C_2 at P as shown in Figure 3.34.*

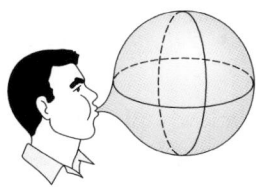

■ FIGURE 3.34   The angle θ from C_1 to C_1 at P

Use this information for Problems 71–72.

71. If θ is the acute angle from curve C_1 to curve C_2 at P and the tangent lines to C_1 and C_2 at P have slopes m_1 and m_2, respectively, show that

$$\tan \theta = \frac{|m_2 - m_1|}{1 + m_1 m_2}$$

72. Find the angle from the circle

$$x^2 + y^2 = 1$$

to the circle

$$x^2 + (y - 1)^2 = 1$$

at each of the two points of intersection.

73. JOURNAL PROBLEM (The Pi Mu Epsilon Journal), by Bruce W. King*   When a professor asked the calculus class to find the derivative of y^2 with respect to x^2 for the function $y = x^2 - x$, one student found

$$\frac{dy}{dx} \cdot \frac{y}{x}$$

Was this answer correct? Suppose

$$y = x^3 - 3x^2 + \frac{7}{x}$$

What is the derivative of y^2 with respect to x^2 for this function?

_____

*Volume 7 (1981), p. 346.

3.7 Related Rates and Applications

IN THIS SECTION   **related rates and applications**   ■

When working a related-rate problem, you must distinguish between the general situation and the specific situation. The *general situation* comprises properties that are true at *every* instant of time, whereas the *specific situation* refers to those properties that are guaranteed to be true only at the *particular* instant of time that the problem investigates. Here is an example.

EXAMPLE 1   *An application involving related rates*

A spherical balloon is being filled with a gas in such a way that when the radius is 2 ft, the radius is increasing at the rate of 1/6 ft/min. How fast is the volume changing at this time?

Solution   THE GENERAL SITUATION   Let V denote the volume and r the radius, both of which are functions of time t (minutes). Since the container is a sphere, its volume is given by

$$V = \tfrac{4}{3}\pi r^3$$

Differentiating both sides implicitly with respect to time t yields

$$\frac{dV}{dt} = \frac{d}{dt}\left(\frac{4}{3}\pi r^3\right)$$

$$= 4\pi r^2 \frac{dr}{dt} \qquad \text{Do not forget to use the chain rule because } r \text{ is also a function of time.}$$

THE SPECIFIC SITUATION   Our goal is to find $\dfrac{dV}{dt}$ at the time when $r = 2$ and $\dfrac{dr}{dt} = \dfrac{1}{6}$.

$$\left.\frac{dV}{dt}\right|_{r=2} = 4\pi(2)^2\left(\frac{1}{6}\right) = \frac{8\pi}{3} \approx 8.37758041$$

This means that the rate of change of volume of the container with respect to time is increasing at about 8.38 ft³/min. ∎

Although each related-rate problem has its own "personality," many can be handled by the following summary:

Procedure for Solving Related-Rate Problems

The General Situation

Step 1. *Draw a figure, if appropriate, and assign variables to the quantities that vary.* Be careful not to label a quantity with a number unless it *never* changes in the problem.

Step 2. *Find a formula or equation that relates the variables.* Eliminate unnecessary variables; some of these "extra" variables may be constants, but others may be eliminated because of given relationships among the variables.

Step 3. *Differentiate the equations.* You will usually differentiate implicitly with respect to time.

The Specific Situation

Step 4. *Substitute specific numerical values and solve algebraically for any required rate.* List the known quantities; list as unknown the quantity you wish to find. Substitute all values into the formula. The only remaining variable should be the unknown, which may be a variable or a rate. Solve for the unknown.

EXAMPLE 2   *Moving shadow problem*

A person 6 ft tall is walking away from a streetlight 20 ft high at the rate of 7 ft/s. At what rate is the length of the person's shadow increasing?

(SMH)   **Solution**   THE GENERAL SITUATION (See *Student Mathematics Handbook*, Problem 37, p. 14.)

Step 1. Let x denote the length (in feet) of the person's shadow, and y, the distance between the person and the streetlight, as shown in Figure 3.35. Let t denote the time (in seconds).

Step 2. Since $\triangle ABC$ and $\triangle DEC$ are similar, we have

$$\frac{x + y}{20} = \frac{x}{6}$$

Step 3. Write this equation as $x + y = \frac{20}{6}x$, or $y = \frac{7}{3}x$, and differentiate both sides with respect to t:

$$\frac{dy}{dt} = \frac{7}{3}\frac{dx}{dt}$$

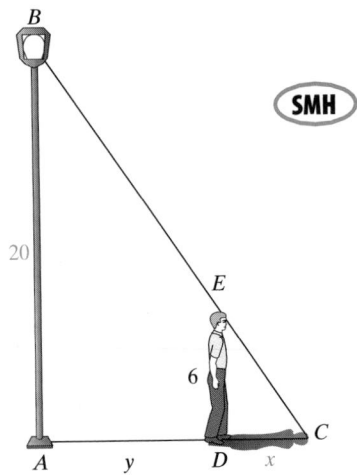

■ **FIGURE 3.35**   A person walking away from a streetlamp.

THE SPECIFIC SITUATION

Step 4. List the known quantities. We know that $dy/dt = 7$. Our goal is to find dx/dt. Substitute and then solve for the unknown value:

$$\frac{dy}{dt} = \frac{7}{3}\frac{dx}{dt}$$

$$7 = \frac{7}{3}\frac{dx}{dt} \qquad \text{Substitute.}$$

$$3 = \frac{dx}{dt} \qquad \text{Multiply both sides by } \tfrac{3}{7}.$$

The length of the person's shadow is increasing at the rate of 3 ft/s.   ∎

EXAMPLE 3   *Leaning ladder problem*

A bag is tied to the top of a 5-m ladder resting against a vertical wall. Suppose the ladder begins sliding down the wall in such a way that the foot of the ladder is moving away from the wall. How fast is the bag descending at the instant the foot of the ladder is 4 m from the wall and the foot is moving away at the rate of 2 m/s?

Solution   THE GENERAL SITUATION   (See *Student Mathematics Handbook*, Problem 38, p. 14.) Let x and y be the distances from the base of the wall to the foot and top of the ladder, respectively, as shown in Figure 3.36.

Notice that $\triangle TOB$ is a right triangle, so a relevant formula is the Pythagorean theorem:

$$x^2 + y^2 = 25$$

Differentiate both sides of this equation with respect to t:

$$2x\frac{dx}{dt} + 2y\frac{dy}{dt} = 0$$

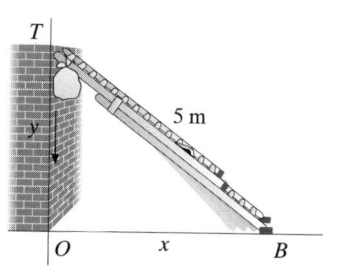

■ **FIGURE 3.36**   A ladder sliding down a wall.

THE SPECIFIC SITUATION   At the particular instant in question, $x = 4$ and $y = \sqrt{25 - 4^2} = 3$. We also know that $\frac{dx}{dt} = 2$, and the goal is to find $\frac{dy}{dt}$ at this instant. We have

$$2(4)(2) + 2(3)\frac{dy}{dt} = 0$$

$$\frac{dy}{dt} = -\frac{8}{3}$$

This tells us that at the instant in question, the bag is descending (since dy/dt is negative) at the rate of $8/3 \approx 2.7$ m/sec.   ∎

EXAMPLE 4   *Modeling a physical application involving related rates*

When air expands *adiabatically* (that is, with no change in heat), the pressure P and the volume V satisfy the relationship

$$PV^{1.4} = C$$

where C is a constant. At a certain instant, the pressure is 20 lb/in.^{2} and the volume is 280 in.^{3}. If the volume is decreasing at the rate of 5 in.^{3}/s at this instant, what is the rate of change of the pressure?

Solution   THE GENERAL SITUATION   The required equation was given, so we begin by differentiating both sides with respect to t. Remember, because C is a constant, its derivative with respect to t is zero.

$$1.4PV^{0.4}\frac{dV}{dt} + V^{1.4}\frac{dP}{dt} = 0 \qquad \text{Product rule}$$

THE SPECIFIC SITUATION   At the instant in question, $P = 20, V = 280,$ and $dV/dt = -5$ (negative because the volume is decreasing). The goal is to find dP/dt. First substitute:

$$(20)(1.4)(280)^{0.4}(-5) + (280)^{1.4}\frac{dP}{dt} = 0$$

Now, solve for $\dfrac{dP}{dt}$:

$$\frac{dP}{dt} = \frac{5(20)(1.4)(280)^{0.4}}{(280)^{1.4}} = 0.5$$

Thus, at the instant in question, the pressure in increasing (because its derivative is positive) at the rate of 0.5 lb/in.^{2} per second.   ∎

EXAMPLE 5   *The water level in a cone-shaped tank*

A tank filled with water is in the shape of an inverted cone 20 ft high with a circular base (on top) whose radius is 5 ft. Water is running out of the bottom of the tank at the constant rate of 2 ft^3/min. How fast is the water level falling when the water is 8 ft deep?

Solution   THE GENERAL SITUATION   Consider a conical tank with height 20 ft and circular base of radius 5 ft, as shown in Figure 3.37. Suppose that the water level is h ft and that the radius of the surface of the water is r.

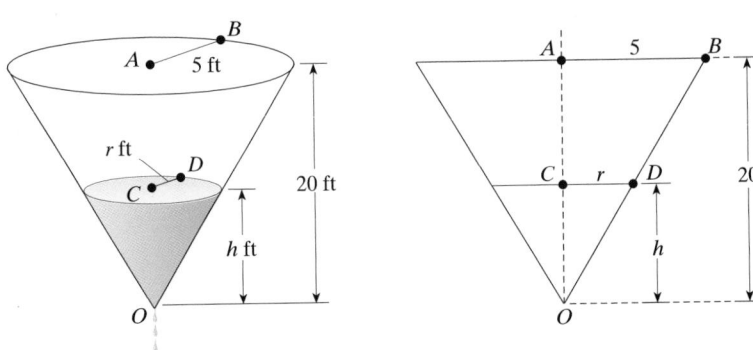

■ FIGURE 3.37   A conical water tank

Let V denote the volume of water in the tank after t minutes. We know that

$$V = \tfrac{1}{3}\pi r^2 h$$

(SMH) (See *Student Mathematics Handbook*, Problem 39, p. 14.) Once again, we use similar triangles (see Figure 3.37) to write $\frac{5}{20} = \frac{r}{h}$, or $r = \frac{h}{4}$. We substitute this into the formula to obtain

$$V = \tfrac{1}{3}\pi\left(\frac{h}{4}\right)^2 h = \tfrac{1}{48}\pi h^3$$

and then differentiate both sides of this equation with respect to t:

$$\frac{dV}{dt} = \frac{\pi}{16}h^2\frac{dh}{dt}$$

THE SPECIFIC SITUATION   Begin with the known quantities: We know that $dV/dt = -2$ (negative, because the volume is decreasing). The goal is to find dh/dt. At the particular instant in question, $h = 8$; we substitute to find:

$$-2 = \frac{\pi}{16}(8)^2 \frac{dh}{dt}$$

$$\frac{-1}{2\pi} = \frac{dh}{dt}$$

At the instant when the water is 8 ft deep, the water level is falling (since dh/dt is negative) at a rate of $\frac{1}{2\pi} \approx 0.16$ ft/min ≈ 2 in./min.   ∎

WARNING   A common error in solving related-rate problems is to substitute numerical values too soon or, equivalently, to use relationships that apply only at a particular moment in time. This is the reason we have separated related-rate problems into two distinct parts. Be careful to work with general relationships among the variables and substitute specific numerical values only after you have found general rate relationships by differentiation.   ▬

EXAMPLE 6   *Modeling with an angle of elevation*

Every day, a flight from Los Angeles to New York flies directly over my home at a constant altitude of 4 mi. If I assume that the plane is flying at a constant speed of 400 mi/h, at what rate is the angle of elevation of my line of sight changing with respect to time when the horizontal distance between the approaching plane and my location is exactly 3 mi?

Solution   THE GENERAL SITUATION   Let x denote the horizontal distance between the plane and the observer, as shown in Figure 3.38a. The height of the observer is insignificant when compared to the height of the plane.

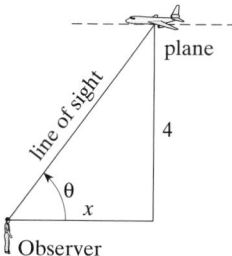
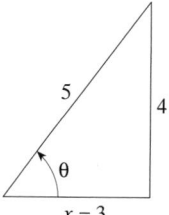

a. Observer of an approaching plane   **b.** Triangle for computing $\sin\theta$ when $x = 3$

■ **FIGURE 3.38**   Angle of elevation problem

Then the angle of observation θ can be modeled by

$$\cot \theta = \frac{x}{4} \quad \text{or} \quad \theta = \cot^{-1}\left(\frac{x}{4}\right)$$

Differentiate both sides of this equation with respect to t to obtain

$$\frac{d\theta}{dt} = \frac{-1}{1 + (x/4)^2}\left(\frac{1}{4}\right)\frac{dx}{dt} = \frac{-4}{16 + x^2}\frac{dx}{dt}$$

THE SPECIFIC SITUATION   At the instant when $x = 3$, we are given that $(dx/dt) = -400$ (negative because the distance is decreasing). Thus, the angle of elevation is changing at the rate of

$$\frac{d\theta}{dt} = \frac{-4}{16 + 3^2}(-400) = 64 \text{ rad/h}$$

The angle of elevation is changing at the rate of 64 radians per hour or, equivalently,

$$(64 \text{ rad/h})\left(\frac{360 \text{ deg}}{2\pi \text{ rad}}\right)\left(\frac{1 \text{ h}}{3,600 \text{ s}}\right) \approx 1.02 \text{ deg/s} \qquad \blacksquare$$

3.7   Problem Set

A *Find the indicated rate in Problems 1–9, given the other information. Assume $x > 0$ and $y > 0$.*

1. Find $\dfrac{dy}{dt}$ where $x^2 + y^2 = 25$, and $\dfrac{dx}{dt} = 4$ when $x = 3$.

2. Find $\dfrac{dx}{dt}$ where $x^2 + y^2 = 25$, and $\dfrac{dy}{dt} = 2$ when $x = 4$.

3. Find $\dfrac{dy}{dt}$ where $5x^2 - y = 100$, and $\dfrac{dx}{dt} = 10$ when $x = 10$.

4. Find $\dfrac{dx}{dt}$ where $4x^2 - y = 100$, and $\dfrac{dy}{dt} = -6$ when $x = 1$.

5. Find $\dfrac{dx}{dt}$ where $y = 2\sqrt{x} - 9$, and $\dfrac{dy}{dt} = 5$ when $x = 9$.

6. Find $\dfrac{dy}{dt}$ where $y = 5\sqrt{x + 9}$, and $\dfrac{dx}{dt} = 2$ when $x = 7$.

7. Find $\dfrac{dy}{dt}$ where $xy = 10$, and $\dfrac{dx}{dt} = -2$ when $x = 5$.

8. Find $\dfrac{dy}{dt}$ where $5xy = 10$, and $\dfrac{dx}{dt} = -2$ when $x = 1$.

9. Find $\dfrac{dx}{dt}$ where $x^2 + xy - y^2 = 11$, and $\dfrac{dy}{dt} = 5$ when $x = 4$ and $y > 0$.

MODELING PROBLEM   *In physics, Hooke's law says that when a spring is stretched x units beyond its natural length, the elastic force $F(x)$ exerted by the spring is $F(x) = -kx$, where k is a constant that depends on the spring. Assume $k = 12$ in Problems 10 and 11.*

10. If a spring is stretched at the constant rate of $\frac{1}{4}$ in./s, how fast is the force $F(x)$ changing when $x = 2$ in.?

11. If a spring is stretched at the constant rate of $\frac{1}{4}$ in./s, how fast is the force $F(x)$ changing when $x = 3$ in.?

12. A particle moves along the parabolic path given by $y^2 = 4x$ in such a way that when it is at the point $(1, -2)$, its horizontal velocity (in the direction of the x-axis) is 3 ft/s. What is its vertical velocity (in the direction of the y-axis) at this instant?

13. A particle moves along the elliptical path given by $4x^2 + y^2 = 4$ in such a way that when it is at the point $(\sqrt{3}/2, 1)$, its x-coordinate is increasing at the rate of 5 units per second. How fast is the y-coordinate changing at that instant?

14. A rock is dropped into a lake and an expanding circular ripple results. When the radius of the ripple is 8 in., the radius is increasing at a rate of 3 in./s. At what rate is the area enclosed by the ripple changing at this time?

15. A pebble dropped into a pond causes a circular ripple. Find the rate at which the radius of the ripple is changing at a time when the radius is one foot and the area enclosed by the ripple is increasing at the rate of 4 ft²/s.

16. An environmental study of a certain community indicates that there will be $Q(p) = p^2 + 3p + 1,200$ units of a harmful pollutant in the air when the population is p thousand. The population is currently 30,000 and is increasing at a rate of 2,000 per year. At what rate is the level of air pollution increasing?

17. It is estimated that the annual advertising revenue received by a certain news paper will be $R(x) = 0.5x^2 + 3x + 160$ thousand dollars when its circulation is x thousand. The circulation of the paper is currently 10,000 and is increasing at a rate of 2,000 per year. At what rate will the annual advertising revenue be increasing with respect to time 2 years from now?

18. Hospital officials estimate that approximately $N(p) = p^2 + 5p + 900$ people will seek treatment in the emergency room each year if the population of the community is p thousand. The population is currently 20,000 and is growing at the rate of 1,200 per year. At what rate is the number of people seeking emergency room treatment increasing?

19. Boyle's law states that when gas is compressed at constant temperature, the pressure P of a given sample satisfies the equation $PV = C$, where V is the volume of the sample and C is a constant. Suppose that at a certain

time the volume is 30 in.^{3}, the pressure is 90 lb/in.^{2}, and the volume is increasing at the rate of 10 in.^{3}/s. How fast is the pressure changing at this instant? Is it increasing or decreasing?

B 20. ■ **What Does This Say?** What do we mean by a related-rate problem?

21. ■ **What Does This Say?** Outline a procedure for solving related-rate problems.

SMH

22. The volume of a spherical balloon is increasing at a constant rate of 3 in.^{3}/s. At what rate is the radius of the balloon increasing when the radius is 2 in.?

SMH

23. The surface area of a sphere is decreasing at the constant rate of 3π cm^2/s. At what rate is the volume of the sphere decreasing at the instant its radius is 2 cm?

24. A person 6 ft tall walks away from a streetlight at the rate of 5 ft/s. If the light is 18 ft above ground level, how fast is the person's shadow lengthening?

25. A ladder 13 ft long rests against a vertical wall and is sliding down the wall at the rate of 3 ft/s at the instant the foot of the ladder is 5 ft from the base of the wall. At this instant, how fast is the foot of the ladder moving away from the wall?

26. A car traveling north at 40 mi/h and a truck traveling east at 30 mi/h leave an intersection at the same time. At what rate will the distance between them be changing 3 hours later?

27. A person is standing at the end of a pier 12 ft above the water and is pulling in a rope attached to a rowboat at the waterline at the rate of 6 ft of rope per minute, as shown in Figure 3.39. How fast is the boat moving in the water when it is 16 ft from the pier?

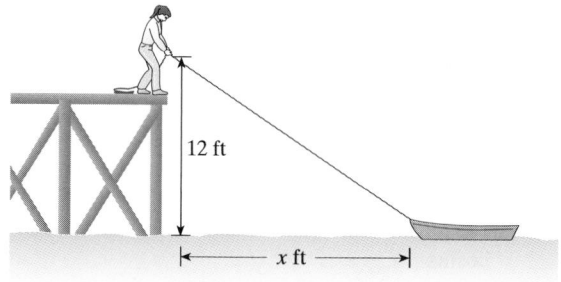

■ **FIGURE 3.39** Problem 27

28. One end of a rope is fastened to a boat and the other end is wound around a windlass located on a dock at a point 4 m above the level of the boat. If the boat is drifting away from the dock at the rate of 2 m/min, how fast is the rope unwinding at the instant when the length of the rope is 5 m?

29. A ball is dropped from a height of 160 ft. A light is located at the same level, 10 ft away from the initial position of the ball. How fast is the ball's shadow moving along the ground one second after the ball is dropped?

30. A person 6 ft tall stands 10 ft from point P directly beneath a lantern hanging 30 ft above the ground, as shown in Figure 3.40.

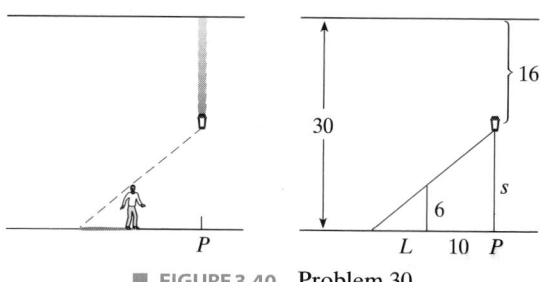

■ **FIGURE 3.40**   Problem 30

The lantern starts to fall, thus causing the person's shadow to lengthen. Given that the lantern falls $16t^2$ ft in t seconds, how fast will the shadow be lengthening when $t = 1$?

31. A race official is watching a race car approach the finish line at the rate of 200 km/h. Suppose the official is sitting at the finish line, 20 m from the point where the car will cross, and let θ be the angle between the finish line and the official's line of sight to the car, as shown as Figure 3.41. At what rate is θ changing when the car crosses the finish line? Give your answer in terms of rad/s.

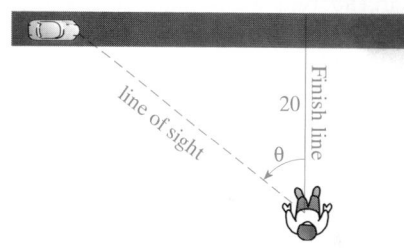

■ **FIGURE 3.41**   Problem 31

MODELING PROBLEMS   *In Problems 32–36, set up an appropriate model to answer the given question. Be sure to state your assumptions.*

32. Consider a piece of ice in the shape of a sphere that is melting at the rate of 5 in.^{3}/min. Model the volume of ice by a function of the radius r. How fast is the radius changing at the instant when the radius is 4 in.? How fast is the surface area of the sphere changing at the same instant? What assumption are you making in this model about the shape of the ice?

33. A certain medical procedure requires that a balloon be inserted into the stomach and then inflated. Model the shape of the balloon by a sphere of radius r. If r is increasing at the rate of 0.3 cm/min, how fast is the volume changing when the radius is 4 cm?

34. Model a water tank by a cone 40 ft high with a circular base of radius 20 ft. at the top. Water is flowing into the tank at a constant rate of 80 ft^3/min. How fast is the water level rising when the water is 12 ft deep? Give your answer to the nearest foot per minute.

35. In Problem 34, suppose that whater is also flowing out the bottom of the tank. At what rate should the water be allowed to flow out so that the water level will be rising at a rate of only 0.05 ft/min when the water is 12 ft deep? Give your answer to the nearest hundredth of a cubic foot per minute.

36. The air pressure $p(s)$ at a height of s meters above sea level is modeled by the formula

$$p(s) = e^{-0.000125s} \text{ atmospheres}$$

An instrument box carrying a device for measuring pressure is dropped into the ocean from a plane and falls in such a way that after t seconds it is

$$s(t) = 3{,}000 - 49t - 245(e^{-t/5} - 1)$$

meters above the ocean's surface.
   a. Find ds/dt and then use the chain rule to find dp/dt. How fast is the air pressure changing 2 seconds after the box begins to fall?
   b. When (to the nearest second) does the box hit the water? How fast is the air pressure changing at the time of impact?
   What assumptions are you making in this model?

37. At noon on a certain day, a truck is 250 mi due east of a car. The truck is traveling west at a constant speed of 25 mi/h, while the car is traveling north at 50 mi/h.
   a. At what rate is the distance between them changing at time t?
   b. At what time is the distance between the car and the truck neither increasing nor decreasing?
   c. What is the minimal distance between the car and the truck? *Hint:* This distance must occur at the time found in part **b**. Do you see why?

38. A weather balloon is rising vertically at the rate of 10 ft/s. An observer is standing on the ground 300 ft horizontally from the point where the balloon was released. At what rate is the distance between the observer and the balloon changing when the balloon is 400 ft high?

39. An observer watches a plane approach at a speed of 500 mi/h and an altitude of 3 mi. At what rate is the angle of elevation of the observer's line of sight changing with respect to time when the horizontal distance between the plane and the observer is 4 mi? Give your answer in radians per minute.

40. A person 6 ft tall is watching a streetlight 18 ft high while walking toward it at a speed of 5 ft/s, as shown in Figure 3.42. At what rate is the angle of elevation of the person's line of sight changing with respect to time when the person is 9 ft from the base of the light?

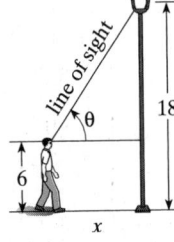

■ FIGURE 3.42

41. A revolving searchlight in a lighthouse 2 mi offshore is following a beachcomber along the shore, as shown in Figure 3.43.

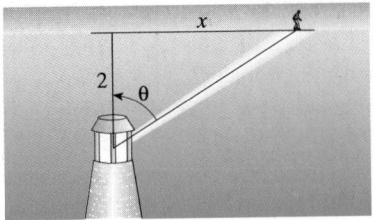

■ FIGURE 3.43 Problem 41

When the beachcomber is 1 mi from the point on the shore that is closest to the ligthhouse, the searchlight is turning at the rate of 0.25 rev/h. How fast is the beachcomber walking at that moment? *Hint:* Note that 0.25 rev/h is the same as $\frac{\pi}{2}$ rad/h.

42. A water trough is 2 ft deep and 10 ft long, and it has a trapezoidal cross section with base lengths 2 ft and 5 ft, as shown in Figure 3.44.

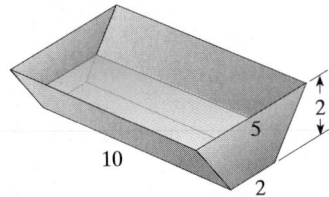

■ FIGURE 3.44 Problem 42

   a. Find a relationship between the volume of water in the trough at any given time and the depth of the water at that time.
   b. If the water enters the trough at the rate of 10 ft³/min, how fast is the water level rising (to the nearest $\frac{1}{2}$ in./min) when the water is 1 ft deep?

43. At noon, a ship sails due north from a point P at 8 knots (nautical miles per hour). Another ship, sailing at 12 knots, leaves the same point 1 h later on a course 60° east of north. How fast is the distance between the ships increasing at 2 P.M.? At 5 P.M.? *Hint:* Use the law of cosines.

44. A swimming pool is 60 ft long and 25 ft wide. Its depth varies uniformly from 3 ft at the shallow end to 15 ft at the deep end, as shown in Figure 3.45.

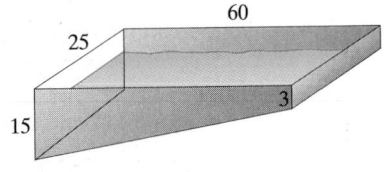

■ FIGURE 3.45 Problem 44

Suppose the pool is being filled with water at the rate of 800 ft³/min. At what rate is the depth of water increasing at the deep end when it is 5 ft deep at that end?

45. MODELING PROBLEM   Suppose a water bucket is modeled by the frustum of a cone with height 1 ft and upper and lower radii of 1 ft and 9 in., respectively, as shown in Figure 3.46.

■ **FIGURE 3.46**   Problem 45

If water is leaking from the bottom of the bucket at the rate of 8 in.³/min, at what rate is the water level falling when the depth of water in the bucket is 6 in.? *Hint:* The volume of the frustum of a cone with height h and base radii r and R is

$$V = \frac{\pi h}{3}(R^2 + rR + r^2)$$

46. A lighthouse is located 2 km directly across the sea from a point S on the Shoreline. A beacon in the lighthouse makes 3 complete revolutions (6π radians) each minute, and during part of each revolution, the light sweeps across the face of a row of cliffs lining the shore.

a. Show that t minutes after it passes point S, the beam of light is at a point P located $s(t) = 2\tan 6\pi t$ km from S.

b. How fast is the beam of light moving at the time it passes a point on the cliff located 4 km from the lighthouse?

47. MODELING PROBLEM   A car is traveling at the rate of 40 ft/s along a straight, level road that parallels the seashore. A rock with a family of seals is located 50 yd offshore.

a. Model the angle θ between the road and the driver's line of sight as a function of the distance x to the point P directly opposite the rocks.

b. As the distance x in Figure 3.47 approaches 0, what happens to $d\theta/dt$?

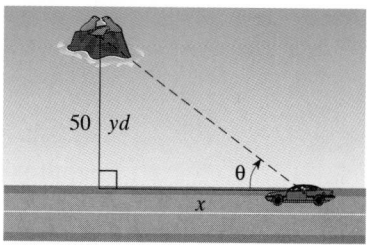

■ **FIGURE 3.47**   Problem 47

c. Suppose the car is traveling at v ft/s. Now what happens to $d\theta/dt$ as $x \to 0$? What effect does this have on a passenger looking at the seals if the car is traveling at a high rate of speed?

3.8 Linear Approximation and Differentials

IN THIS SECTION   tangent line approximation, the differential, error propagation, marginal analysis in economics, the Newton–Raphson method for approximating roots

TANGENT LINE APPROXIMATION

If $f(x)$ is differentiable at $x = a$, the tangent line at a point $P(a, f(a))$ on the graph of $y = f(x)$ has slope $m = f'(a)$ and equation

$$\frac{y - f(a)}{x - a} = f'(a) \quad \text{or} \quad y = f(a) + f'(a)(x - a)$$

In the immediate vicinity of P, the tangent line closely approximates the shape of the curve $y = f(x)$. For instance, if $f(x) = x^3 - 2x + 5$, the tangent line at $P(1, 4)$ has slope $f'(1) = 3(1)^2 - 2 = 1$ and equation

$$y = 4 + (1)(x - 1) = x + 3$$

The graph of $y = f(x)$, the tangent line at $P(1, 4)$, and two enlargements showing how the tangent line approximates the graph of f near P are shown in Figure 3.48.

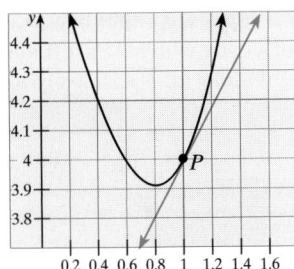 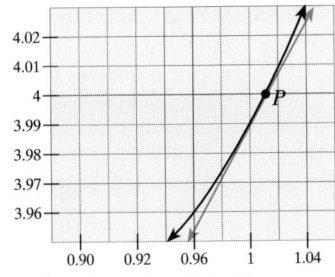

■ **FIGURE 3.48**   Tangent line approximation of $f(x) = x^3 - 2x + 5$ at $P(1, 4)$

Our observation about tangent lines suggests that if x_1 is near a, then $f(x_1)$ must be close to the point on the tangent line to $y = f(x)$ at $x = x_1$. That is,

$$f(x_1) \approx f(a) + f'(a)(x_1 - a)$$

We refer to this as a *linear approximation* of $f(x)$ at $x = a$ and the function

$$L(x) = f(a) + f'(a)(x - a)$$

is called a **linearization** of the function at a point $x = a$. We can use this line as an approximation of f as long as the line remains close to the graph of f, as shown in Figure 3.49.

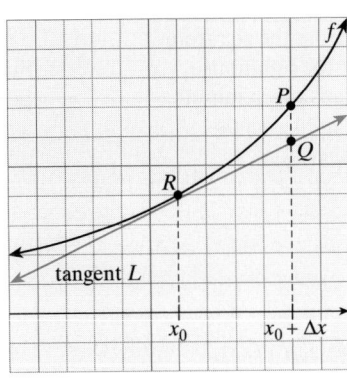 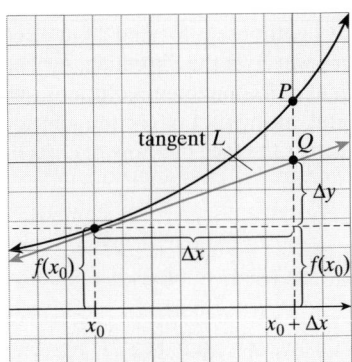

a. Tangent line to f at R **b.** Tangent line approximation

■ **FIGURE 3.49**   Tangent line approximation

Recall, from Section 1.1, the notation Δx for the difference $x_1 - a$ and the corresponding notation Δy for $f(x_1) - f(a)$. Then the linear approximation formula can be written in the form

$$f(x_1) - f(a) \approx f'(a)(x - a) \qquad \textit{Equation of tangent line}$$

$$\Delta y \approx f'(a)\Delta x \qquad \textit{Also, } \frac{\Delta y}{\Delta x} \approx f'(a)$$

In words, the change in y near a is approximately equal to $f'(a)$ times the corresponding change in x. This version of linear approximation is sometimes called the *incremental approximation formula.*

   EXAMPLE 1   *Incremental approximation*

Show that if $f(x) = \sin x$, the function $\dfrac{\Delta f}{\Delta x}$ approximates the function $f'(x) = \cos x$ for small values of Δx.

Solution   The approximation formula $\Delta f = f(x_0 + \Delta x) - f(x_0) \approx f'(x_0)\Delta x$ implies that

$$\frac{\Delta f}{\Delta x} = \frac{f(x_0 + \Delta x) - f(x_0)}{\Delta x} \approx f'(x_0)$$

Because $f(x) = \sin x$, $f'(x) = \cos x$, and

$$\frac{\sin(x + \Delta x) - \sin x}{\Delta x} \approx \cos x$$

Figure 3.50 shows the graphs for three different choices of Δx. Notice as Δx becomes smaller, it is more difficult to see the difference between f and g; in fact, for very small Δx, the graphs are indistinguishable.

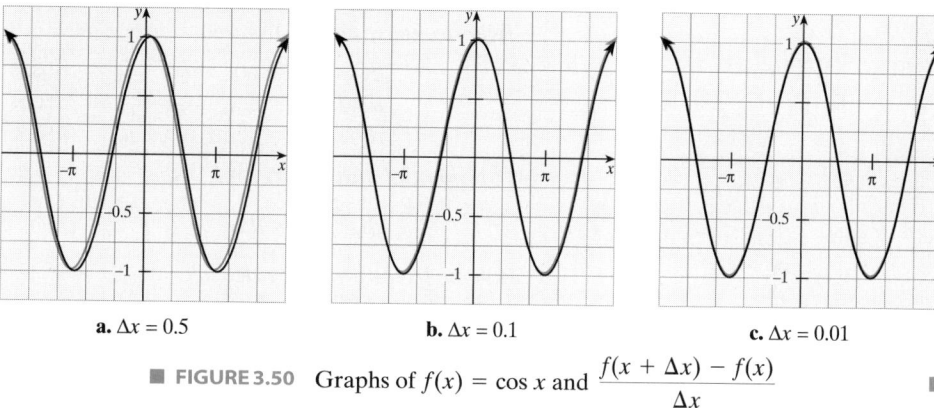

a. $\Delta x = 0.5$ **b.** $\Delta x = 0.1$ **c.** $\Delta x = 0.01$

■ **FIGURE 3.50**   Graphs of $f(x) = \cos x$ and $\dfrac{f(x + \Delta x) - f(x)}{\Delta x}$ ■

THE DIFFERENTIAL

We have already observed that writing the derivative of $f(x)$ in the Leibniz notation df/dx suggests that the derivative may be incorrectly regarded as a quotient of "df" by "dx." It is a tribute to the genius of Leibniz that this erroneous interpretation of his notation often turns out to make good sense.

To give dx and dy meaning as separate quantities, let x be fixed and define dx to be an independent variable equal to Δx, the change in x. That is, define dx, called the **differential of x,** to be an independent variable equal to the change in x. That is, we define dx to be Δx. Then, if f is differentiable at x, we define dy, called the **differential of y,** by the formula

$$dy = f'(x)dx \quad \text{or, equivalently,} \quad df = f'(x)\, dx$$

If we relate differentials to Figure 3.51, we see that $dx = \Delta x$ and that Δy is the rise of f that occurs for a change of Δx, whereas dy is the rise of a tangent relative to the same change in x (Δy and dy are not the same thing). This is shown in Figure 3.51.

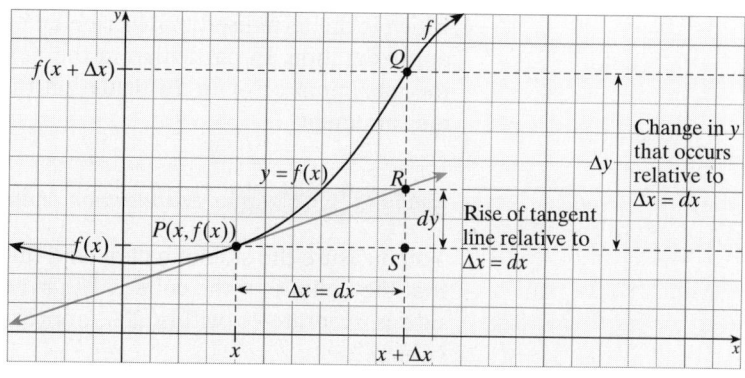

■ **FIGURE 3.51**   Geometrical definition of dx and dy

We now restate the standard rules and formulas for differentiation in terms of differentials. Remember that a and b are constants, while f and g are functions.

Differential Rules

Linearity rule $\quad d(af + bg) = a\,df + b\,dg$

Product rule $\quad d(fg) = f\,dg + g\,df$

Quotient rule $\quad d\left(\dfrac{f}{g}\right) = \dfrac{g\,df - f\,dg}{g^2} \quad (g \neq 0)$

Power rule $\quad d(x^n) = nx^{n-1}\,dx$

Trigonometric rules

$$d(\sin x) = \cos x\,dx \qquad d(\cos x) = -\sin x\,dx$$
$$d(\tan x) = \sec^2 x\,dx \qquad d(\cot x) = -\csc^2 x\,dx$$
$$d(\sec x) = \sec x \tan x\,dx \qquad d(\csc x) = -\csc x \cot x\,dx$$

Exponential and logarithmic rules

$$d(e^x) = e^x\,dx \qquad d(\ln x) = \frac{1}{x}\,dx$$

Inverse trigonometric rules

$$d(\sin^{-1}x) = \frac{dx}{\sqrt{1 - x^2}} \qquad d(\cos^{-1}x) = \frac{-dx}{\sqrt{1 - x^2}}$$
$$d(\tan^{-1}x) = \frac{dx}{\sqrt{1 + x^2}} \qquad d(\cot^{-1}x) = \frac{-dx}{\sqrt{1 + x^2}}$$
$$d(\sec^{-1}x) = \frac{dx}{|x|\sqrt{x^2 - 1}} \qquad d(\csc^{-1}x) = \frac{-dx}{|x|\sqrt{x^2 - 1}}$$

EXAMPLE 2 *Differential involving a product and a trigonometric function*

Find $d(x^2 \sin x)$.

Solution
$$d(x^2 \sin x) = x^2\,d(\sin x) + \sin x\,d(x^2)$$
$$= x^2(\cos x\,dx) + \sin x\,(2x\,dx)$$
$$= (x^2 \cos x + 2x \sin x)dx \qquad \blacksquare$$

ERROR PROPAGATION

In the next example, the approximation formula is used to study **propagation of error,** which is the term used to describe an error that accumulates from other errors in an approximation. In particular, in the next example, the derivative is used to estimate the maximum error in a calculation that is based on figures obtained by imperfect measurement.

EXAMPLE 3 *Propagation of error in a volume measurement*

You measure the side of a cube and find it to be 10 cm long. From this you conclude that the volume of the cube is $10^3 = 1,000$ cm^3. If your original measurement of the side is accurate to within 2%, approximately how accurate is your calculation of volume?

Solution   The volume of the cube is $V(x) = x^3$, where x is the length of a side. If you take the length of a side to be 10 when it is really $10 + \Delta x$, your error is Δx; and your corresponding error when computing the volume will be ΔV, given by

$$\Delta V = V(10 + \Delta x) - V(10) \approx V'(10)\Delta x$$

Now, $V'(x) = 3x^2$, so $V'(10) = 300$. Also, your measurement of the side can be off by as much as 2%—that is, by as much as $0.02(10) = 0.2$ cm in either direction. Substituting $\Delta x = \pm 0.2$ in the incremental approximation formula for ΔV, we get

$$\Delta V = 3(10)^2(\pm 0.2) \approx \pm 60$$

Thus, the propagated error in computing the volume is approximately ± 60 cm^3. Hence the *maximum* error in your measurement of the side is $|\Delta x| = 0.2$ and the corresponding maximum error in your calculation for the volume is

$$|\Delta V| \approx V'(10)|\Delta x| = 300(0.2) = 60$$

This says that, at worst, your calculation of the volume as 1,000 cm^3 is off by 60 cm^3, or 6% of the calculated volume.   ■

Error Propagation

If x_0 represents the measured value of a variable and $x_0 + \Delta x$ represents the exact value, then Δx is the **error in measurement.** The difference between $f(x + \Delta x)$ and $f(x)$ is called the **propagated error** and is defined by

$$\Delta f = f(x + \Delta x) - f(x)$$

Relative Error

The **relative error** is $\dfrac{\Delta f}{f} \approx \dfrac{df}{f}$.

Percentage Error

The **percentage error** is $100\left(\dfrac{\Delta f}{f}\right)\%$.

In Example 3, the approximate propagated error in measuring volume is ± 60, and the approximate relative error is $\Delta V/V = \pm 60/10^3 = \pm 0.06$.

>   **EXAMPLE 4**   *Modeling relative error and percentage error*

A certain container is modeled by a right circular cylinder whose height is twice the radius of the base. The radius is measured to be 17.3 cm, with a maximum measurement error of 0.02 cm. Estimate the corresponding propagated error, the relative error, and the percentage error when calculating the surface area S.

Solution   We have (Figure 3.52)

$$S = \underbrace{2\pi r}_{\text{Circumference}}\overbrace{2r}^{\text{Height}} + \underbrace{\pi r^2}_{\text{Top}} + \underbrace{\pi r^2}_{\text{Bottom}} = 6\pi r^2$$
$$\underbrace{}_{\text{Lateral side}}$$

where r is the radius of the cylinder's base. Then the approximate *propagated error* is

$$\Delta S \approx S'(r)\Delta r = 12\pi r\Delta r = 12\pi(17.3)(\pm 0.02) \approx \pm 13.0438927$$

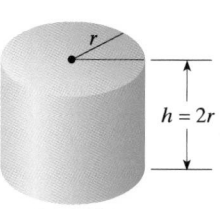

$$S = 6\pi r^2$$

■ **FIGURE 3.52** Surface area of a right circular cylinder

Thus, the maximum error in the measurement of the surface area is about 13.04 cm^2. Is this a large or a small error? The *relative error* is found by computing the ratio

$$\frac{\Delta S}{S} = \frac{12\pi r \Delta r}{6\pi r^2} = 2r^{-1}\Delta r = 2(17.3)^{-1}(\pm 0.02) \approx \pm 0.0023121$$

This tells us that the maximum error of approximately 13.04 is fairly small relative to the surface area S. The corresponding *percentage error* is found by

$$100\left(\frac{\Delta S}{S}\right)\% = 100(\pm 0.0023121387)\% = \pm 0.23121387\%$$

This means that the percentage error is about $\pm 0.23\%$. ∎

MARGINAL ANALYSIS IN ECONOMICS

In economics, the use of the derivative to approximate the change in a function produced by a one-unit change in its independent variable is called **marginal analysis.** In particular, if $C(x)$ is the total cost and $R(x)$ is the total revenue when x units are produced, then $C'(x)$ is called the **marginal cost** and $R'(x)$, the **marginal revenue.**

If production (sales) is increased by 1 unit, then $\Delta x = 1$, and the approximation formula

$$\Delta C = C(x + \Delta x) - C(x) \approx C'(x)\Delta x$$

becomes

$$\Delta C = C(x + 1) - C(x) \approx C'(x)$$

Similarly,

$$\Delta R = R(x + \Delta x) - R(x) \approx R'(x)\Delta x$$

becomes

$$\Delta R = R(x + 1) - R(x) \approx R'(x)$$

That is, the marginal cost $C'(x)$ is an approximation to the cost $C(x + 1) - C(x)$ of producing the $(x + 1)$st unit, and similarly, the marginal revenue $R'(x)$ is an approximation to the revenue derived from the sale of the $(x + 1)$st unit. To summarize:

> ■ *What This Says:*   If $C(x)$ is the total cost of manufacturing x units and $R(x)$ is the total revenue from the sale of x units, then:
>
> The **marginal cost,** $C'(x)$, approximates the cost of producing the $(x + 1)$st unit. The **marginal revenue,** $R'(x)$, approximates the revenue derived from the sale of the $(x + 1)$st unit.

EXAMPLE 5   *Modeling change in cost and revenue*

A manufacturer models the total cost (in dollars) of a particular commodity by the function

$$C(x) = \tfrac{1}{8}x^2 + 3x + 98$$

and the price per item (in dollars) by

$$p(x) = \tfrac{1}{3}(75 - x)$$

where x is the number of items produced ($0 \leq x \leq 50$).

a. Find the marginal cost and the marginal revenue.

b. Use marginal cost to estimate the cost of producing the 9th unit. What is the actual cost of producing the 9th unit?

c. Use marginal revenue to estimate the revenue derived from producing the 9th unit. What is the actual revenue derived from producing the 9th unit?

Solution

a. The marginal cost is $C'(x) = \frac{1}{4}x + 3$.

To find the marginal revenue, we must first find the revenue function:

$$R(x) = xp(x) = x(\tfrac{1}{3})(75 - x) = -\tfrac{1}{3}x^2 + 25x$$

Thus, the marginal revenue is $R'(x) = -\frac{2}{3}x + 25$.

b. The cost of producing the 9th unit is the change in cost as x increases from 8 to 9 and is estimated by

$$C'(8) = \tfrac{1}{4}(8) + 3 = 5$$

We estimate the cost of producing the 9th unit to be $5. The actual cost is

$$\Delta C = C(9) - C(8) = [\tfrac{1}{8}(9)^2 + 3(9) + 98] - [\tfrac{1}{8}(8)^2 + 3(8) + 98]$$
$$= 5\tfrac{1}{8} = 5.125 \quad \text{(that is, \$5.13)}$$

c. The revenue (to the nearest cent) obtained from the sale of the 9th unit is approximated by the marginal revenue:

$$R'(8) = -\tfrac{2}{3}(8) + 25 = \tfrac{59}{3} \approx 19.67 \quad \text{(that is, \$19.67)}$$

The actual revenue (nearest cent) obtained from the sale of the 9th unit is

$$\Delta R = R(9) - R(8) = \tfrac{58}{3} \approx 19.33 \quad \text{(that is, \$19.33)} \qquad \blacksquare$$

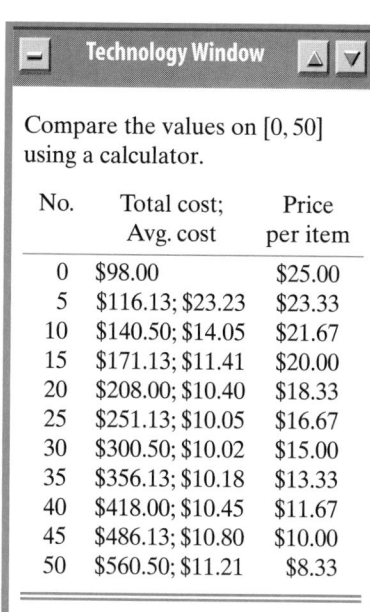

Technology Window

Compare the values on $[0, 50]$ using a calculator.

No.	Total cost; Avg. cost	Price per item
0	$98.00	$25.00
5	$116.13; $23.23	$23.33
10	$140.50; $14.05	$21.67
15	$171.13; $11.41	$20.00
20	$208.00; $10.40	$18.33
25	$251.13; $10.05	$16.67
30	$300.50; $10.02	$15.00
35	$356.13; $10.18	$13.33
40	$418.00; $10.45	$11.67
45	$486.13; $10.80	$10.00
50	$560.50; $11.21	$8.33

WARNING   Remember that the revenue from the sale of the 9th item is *not* the price of one item if 9 are sold. Rather, it is the additional revenue the company has earned by selling the 9th item—that is, the total revenue of 9 items minus the total revenue of 8 items.

THE NEWTON–RAPHSON METHOD FOR APPROXIMATING ROOTS

The Newton–Raphson method is a different kind of tangent line approximation, one that uses tangents as a means for estimating roots of equations. The basic idea behind the procedure is illustrated in Figure 3.53. In this figure, r is a root of the equation $f(x) = 0$, x_0 is an approximation to r, and x_1 is a better approximation obtained by taking the x-intercept of the line that is tangent to the graph of f at $(x_0, f(x_0))$.

THEOREM 3.14   *The Newton–Raphson method*

To approximate a root of the equation $f(x) = 0$, start with a preliminary estimate x_0 and generate a sequence $x_1, x_2, x_3, \ldots$ using the formula

$$x_{n+1} = x_n - \frac{f(x_n)}{f'(x_n)} \qquad f'(x_n) \neq 0$$

Either this sequence of approximations will approach a limit that is a root of the equation or else the sequence does not have a limit.

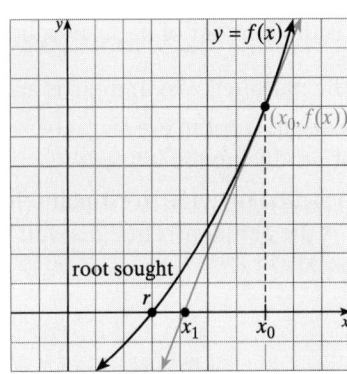
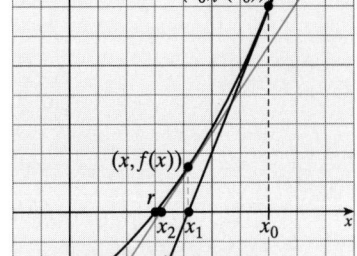

a. Estimating a root, r, of $y = f(x)$

b. First, second, and third estimates ($x_0, x_1,$ and x_2, respectively)

■ **FIGURE 3.53** The Newton–Raphson method

Proof Rather than present a formal proof, we will present a geometric description of the procedure to help you understand what is happening. Let x_0 be an initial approximation such that $f'(x_0) \neq 0$. To find a formual for the improved approximation x_1, recall that the slope of the tangent line through $(x_0, f(x_0))$ is the derivative $f'(x_0)$. Therefore (see Figure 3.53b),

$$\underbrace{f'(x_0)}_{\text{Slope of the tangent through }(x_0, f(x_0))} = \frac{\Delta y}{\Delta x} = \frac{f(x_0) - 0}{x_0 - x_1}$$

or, equivalently (by solving the equation for x_1),

$$x_1 = x_0 - \frac{f(x_0)}{f'(x_0)} \qquad f'(x_0) \neq 0$$

If this procedure is repeated using x_1 as the initial approximation, an even better approximation may often be obtained (see Figure 3.53b). This approximation, x_2, is related to x_1 as x_1 was related to x_0. That is,

$$x_2 = x_1 - \frac{f(x_1)}{f'(x_1)}$$

If this process produces a limit, it can be continued until the desired degree of accuracy is obtained. In general, the nth approximation x_n is related to the $(n - 1)$st by the formula

$$x_n = x_{n-1} - \frac{f(x_{n-1})}{f'(x_{n-1})}$$

Here is a step-by-step procedure for applying the Newton–Raphson method. A flowchart for the method appears in Figure 3.54.

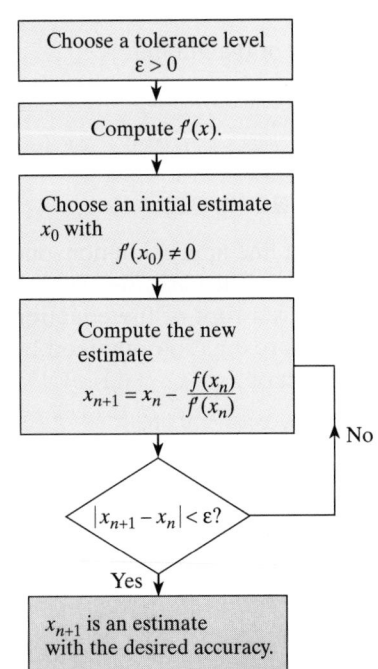

■ **FIGURE 3.54** Flowchart for the Newton–Raphson method

Procedure for Applying the Newton–Raphson Method to Solve the Equation $f(x) = 0$

1. Choose a number $\epsilon > 0$ that determines the allowable tolerance for estimated solutions.
2. Compute $f'(x)$ and choose a number x_0 (with $f'(x_0) \neq 0$) "close" to a solution of $f(x) = 0$ as an initial estimate.
3. Compute a new approximation with the formula

$$x_{n+1} = x_n - \frac{f(x_n)}{f'(x_n)}$$

4. Repeat step 3 unitl $\left| x_{n+1} - x_n \right| < \epsilon$. The estimate $\bar{x} = x_n$ then has the required accuracy.

EXAMPLE 6   *Estimating a root with the Newton–Raphson method*

Approximate a real root of the equation $x^3 + x + 1 = 0$ on $[-2, 2]$.

Solution   Let $f(x) = x^3 + x + 1$. Our goal is to find the root of the equation $f(x) = 0$. The derivative of f is $f'(x) = 3x^2 + 1$, and so

$$x - \frac{f(x)}{f'(x)} = x - \frac{x^3 + x + 1}{3x^2 + 1} = \frac{2x^3 - 1}{3x^2 + 1}$$

Thus, for $n = 1, 2, 3, \ldots,$

$$x_{n+1} = x_n - \frac{f(x_n)}{f'(x_n)} = \frac{2x_n^3 - 1}{3x_n^2 + 1}$$

A convenient choice for the preliminary estimate is $x_0 = -1$. Then

$$x_1 = \frac{2x_0^3 - 1}{3x_0^2 + 1} = -0.75 \qquad \textit{You will need a calculator or a spreadsheet to help you with these calculations.}$$

$$x_2 = \frac{2x_1^3 - 1}{3x_1^2 + 1} \approx -0.6860465$$

$$x_3 = \frac{2x_2^3 - 1}{3x_2^2 + 1} \approx -0.6823396$$

So to two decimal places, the root seems to be approximately $x_n \approx -0.68$. ∎

In general, we shall stop finding new estimates when successive approximations x_n and x_{n+1} are within a desired tolerance of each other. Specifically, if we wish to have the solutions to be within ϵ ($\epsilon > 0$) of each other, we compute approximations until $\left| x_{n+1} - x_n \right| < \epsilon$ is satisfied. Using the Newton–Raphson method (Theorem 3.14), we see that this condition is equivalent to

$$\left| x_{n+1} - x_n \right| = \left| \frac{-f(x_n)}{f'(x_n)} \right| < \epsilon$$

Technology Window

If you have a graphing calculator, you can use a graph and the TRACE instead of the Newton-Raphson method. For Example 6 define f: Y1=X^3+X+1

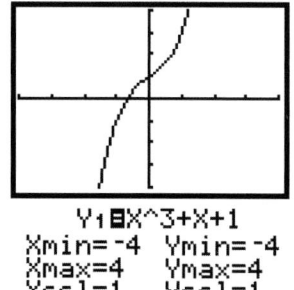

```
Y1=X^3+X+1
Xmin=-4    Ymin=-4
Xmax=4     Ymax=4
Xscl=1     Yscl=1

X=-.7659574  Y=-.2153376
X=-.6808511  Y=.00353486
```

We see the zero is between -0.52 and -0.74.

Check using ROOT to find $x = -.6823278$. If your calculator does not have a root utility you can use the ZOOM to find the root to any desired degree of accuracy.

3.8 Problem Set

Ⓐ *Find the differentials indicated in Problems* 1–16.

1. $d(2x^3)$
2. $d(3 - 5x^2)$
3. $d(2\sqrt{x})$
4. $d(x^5 + \sqrt{x^2 + 5})$
5. $d(x \cos x)$
6. $d(x \sin 2x)$
7. $d\left(\dfrac{\tan 3x}{2x}\right)$
8. $d(xe^{-2x})$
9. $d(\ln|\sin x|)$
10. $d(x \tan^{-1} x)$
11. $d(e^x \ln x)$
12. $d\left(\dfrac{\sin^{-1} x}{e}\right)$
13. $d\left(\dfrac{x^2 \sec x}{x - 3}\right)$
14. $d(x\sqrt{x^2 - 1})$
15. $d\left(\dfrac{x - 5}{\sqrt{x + 4}}\right)$
16. $d\left(\dfrac{\ln\sqrt{x}}{x}\right)$

17. ■ **What Does This Say?** What is a differential?
18. ■ **What Does This Say?** Discuss error propagation, including relative error and percentage error.

Ⓑ *Use differentials to approximate the requested values in Problems* 19–22, *and then determine the error as compared to the calculator value.*

19. $\sqrt{0.99}$
20. $\cos(\frac{\pi}{2} + 0.01)$
21. $(3.01)^5 - 2(3.01)^3 + 3(3.01)^2 - 2$
22. $\sqrt[4]{4{,}100} + \sqrt[3]{4{,}100} + 3\sqrt{4{,}100}$

23. You measure the radius of a circle to be 12 cm and use the formula $A = \pi r^2$ to calculate the area. If your measurement of the radius is accurate to within 3%, approximately how accurate (to the nearest percent) is your calculation of the area?

24. **EXPERIMENT** Suppose a 12-oz can of Coke has a height of 4.5 in. Determine the radius with an accuracy to within 1%. Check your answer by examining a Coke can.

25. **MODELING PROBLEM** You measure the radius of a sphere to be 6 in. and use the formula $V = \frac{4}{3}\pi r^3$ to calculate the volume. If your measurement of the radius is accurate to within 1%, approximately how accurate (to the nearest percent) is your calculation of the volume?

26. It is projected that t years from now the circulation of a local newspaper will be

$$C(t) = 100t^2 + 400t + 5{,}000$$

Estimate the amount by which the circulation will increase during the next 6 months.

27. An environmental study suggests that t years from now, the average level of carbon monoxide in the air will be

$$Q(t) = 0.05t^2 + 0.1t + 3.4$$

parts per million (ppm). By aproximately how much will the carbon monoxide level change during the next 6 months?

28. A manufacturer's total cost (in dollars) is

$$C(q) = 0.1q^3 - 0.5q^2 + 500q + 200$$

when the level of production is q units. The current level of production is 4 units, and the manufacturer is planning to decrease this to 3.9 units. Estimate how the total cost will change as a result.

29. At a certain factory, the daily output is

$$Q(L) = 60{,}000L^{1/3}$$

units, where L denotes the size of the labor force measured in worker-hours. Currently 1,000 worker-hours of labor are used each day. Estimate the effect on output that will be produced if the labor force is cut to 940 worker-hours.

30. **MODELING PROBLEM** In a model developed by John Helms, the water evaporation $E(T)$ for a ponderosa pine is given by

$$E(T) = 4.6e^{17.3T/(T + 237)}$$

where T (degrees Celsius) is the surrounding air temperature.* If the temperature is increased by 5% from 30 °C, use differentials to estimate the corresponding percentage change in $E(T)$.

31. **MODELING PROBLEM** A soccer ball made of leather $\frac{1}{8}$ in. thick has an inner diameter of $8\frac{1}{2}$ in. Model the ball as a hollow sphere and estimate the volume of its leather shell.

32. A cubical box is to be constructed from three kinds of building materials. The material used in the four sides of the box costs 2¢/in.^{2}, the material in the bottom costs 3¢/in.^{2}, and the material used for the lid costs 4¢/in.^{2}. Estimate the additional total cost of all the building materials if the length of a side is increased from 20 in. to 21 in.

33. **MODELING PROBLEM** In a healthy person of height x in., the average pulse rate in beats per minute is modeled by the formula

$$P(x) = \frac{596}{\sqrt{x}} \qquad 30 \le x \le 100$$

Estimate the change in pulse rate that corresponds to a height change from 59 to 60 in.

---

*John A. Helms, "Environmental Control of Net Photosynthesis in Naturally Grown Pinus Ponderosa Nets," *Ecology* (Winter 1972) p. 92.

34. **MODELING PROBLEM**   A drug is injected into a patient's bloodstream. The concentration of the drug in the bloodstream t hours after the drug is injected is modeled by the formula

$$C(t) = \frac{0.12t}{t^2 + t + 1}$$

milligrams per cubic centimeter. Estimate the change in concentration over the time period from 30 to 35 minutes after injection.

35. **MODELING PROBLEM**   According to Poiseuille's law, the speed of blood flowing along the central axis of an artery of radius R is modeled by the formula $S(R) = cR^2$, where c is a constant.* What percentage error (rounded to the nearest percent) will you make in the calculation of $S(R)$ from this formula if you make a 1% error in the measurement of R?

36. **MODELING PROBLEM**   A second law attributed to Poiseuille models the volume of a fluid flowing through a small tube in unit time under fixed pressure by the formula $V = kR^4$, where k is a positive constant and R is the radius of the tube. This formula is used in medicine to determine how wide a clogged artery must be opened to restore a healthy flow of blood. Suppose the radius of a certain artery is increased by 5%. Approximately what effect does this have on the volume of the blood flowing through the artery?[†]

37. **MODELING PROBLEM**   A certain cell is modeled as a sphere. If the formulas $S = 4\pi r^2$ and $V = \frac{4}{3}\pi r^3$ are used to compute the surface area and volume of the sphere, respectively, estimate the effect on S and V produced by a 1% increase in the radius r.

38. **MODELING PROBLEM**   The period of a pendulum is given by the formula

$$T = 2\pi\sqrt{\frac{L}{g}}$$

where L is the length of the pendulum in feet, $g = 32$ ft/s^2 is the acceleration due to gravity, and T is time in seconds. If the pendulum has been heated enough to increase its length by 0.4%, what is the approximate percentage change in its period?

39. The *thermal expansion coefficient* of an object is defined to be

$$\sigma = \frac{L'(T)}{L(T)}$$

where $L(T)$ is the length of the object when the temperature is T. Suppose a 75-ft span of a bridge is built with steel with $\sigma = 1.4 \times 10^{-5}$ per degree Celsius. Approximately how much will the length change during a year when the temperature varies from -10 °C in winter to 40 °C in summer?

40. The radius R of a spherical ball is measured as 14 in.
  a. Use differentials to estimate the maximum propagated error in computing volume V if R is measured with a maximum error of $\frac{1}{8}$ inch.
  b. With what accuracy must the radius R be measured to guarantee an error of at most 2 in.^{3} in the calculated volume?

41. **MODELING PROBLEM**   A thin horizontal beam of alpha particles strikes a thin vertical foil, and the scattered alpha particles will travel along a cone of vertex angle θ, as shown in Figure 3.55.

Time 1: Stream is focused.

Time 2: Stream hits foil.

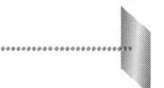

Time 3: Stream is scattered after colliding with foil and disperses in the shape of a cone.

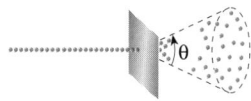

■ **FIGURE 3.55**   Paths of alpha particles

A vertical screen is placed at a fixed distance from the point of scattering. Physical theory predicts that the number N of alpha particles falling on a unit area of the screen is inversely proportional to $\sin^4(\theta/2)$. Suppose N is modeled by the formula

$$N = \frac{1}{\sin^4(\theta/2)}$$

Estimate the change in the number of alpha particles per unit area of the screen if θ changes from 1 to 1.1.

42. Suppose the total cost of manufacturing q units is

$$C(q) = 3q^2 + q + 500$$

dollars.
  a. Use marginal analysis to estimate the cost of manufacturing the 41st unit.
  b. Compute the actual cost of manufacturing the 41st unit.

43. A manufacturer's total cost is

$$C(q) = 0.1q^3 - 5q^2 + 500q + 200$$

dollars, where q is the number of units produced.
  a. Use marginal analysis to estimate the cost of manufacturing the 4th unit.
  b. Compute the actual cost of manufacturing the 4th unit.

44. **MODELING PROBLEM**   Suppose the total cost of producing x units of a particular commodity is modeled by

$$C(x) = \tfrac{1}{7}x^2 + 4x + 100$$

---

*See "Introduction to Mathematics for Life Scientists," 2nd edition. New York: Springer-Verlag (1976): pp. 102–103.
†Ibid.

and that each unit of the commodity can be sold for

$$p(x) = \tfrac{1}{4}(80 - x)$$

dollars.

a. What is the marginal cost?
b. What is the price when the marginal cost is 10?
c. Estimate the cost of producing the 11th unit.
d. Find the actual cost of producing the 11th unit.

45. **MODELING PROBLEM** At a certain factory, the daily output is modeled by the formula

$$Q(L) = 360 \, L^{1/3}$$

units, where L is the size of the labor force measured in worker-hours. Currently, 1,000 worker-hours of labor are used each day. Use differentials to estimate the effect that one additional worker-hour will have on the daily output.

46. Use the Newton–Raphson method to estimate a root of the equation

$$x^6 - x^5 + x^3 = 3$$

47. Let $f(x) = -2x^4 + 3x^2 + \tfrac{11}{8}$.
a. Show that the equation $f(x) = 0$ has at least two solutions. *Hint:* Use the intermediate value theorem.
b. Use $x_0 = 2$ in the Newton–Raphson method to find a root of the equation $f(x) = 0$.
c. Show that the Newton–Raphson method fails if you choose $x_0 = \tfrac{1}{2}$ as the initial estimate. *Hint:* You should obtain $x_1 = -x_0, x_2 = x_0, \ldots$.

48. In Chapter 13, we will show that the volume of a spherical segment is given by

$$V = \frac{\pi}{3} H^2 (3R - H)$$

where R is the radius of the sphere and H is the height of the segment, as shown in Figure 3.56.

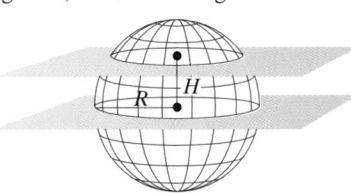

■ **FIGURE 3.56** Problem 48

If $V = 8$ and $R = 2$, use the Newton–Raphson method to estimate the corresponding H.

49. **H**ISTORICAL **Q**UEST The Greek geometer Archimedes (see Historical Quest Problem 54 in Section 2.1) is acknowledged to be one of the greatest mathematicians of all time. Ten treatises of Archimedes have survived the rigors of time (as well as traces of some lost works), and are masterpieces of mathematical exposition. In one of these works, *On the Sphere and Cylinder*, Archimedes asks where a sphere should be cut in order to divide it into two pieces whose volumes have a given ratio.

ARCHIMEDES
287–212 B.C.

Show that if a plane at distance x_c from the center of a sphere with $R = 1$ divides the sphere into two parts, one with volume *twice* that of the other, then

$$3x_c^3 - 9x_c^2 + 2 = 0$$

Use the Newton–Raphson method to estimate x_c.

50. Suppose the plane described in Problem 49 is located so that it divides the sphere in the ratio 1:3. Find an equation of x_c, and estimate the value of x_c by the Newton–Raphson method. (*Hint:* You may need the result of Problem 48.)

Technology Window

In Problems 51–52, we look at the speed at which the Newton–Raphson method converges. Let $e_n = \left| x_{\text{exact}} - x_n \right|$ denote the error after n applications of the Newton–Raphson method. Typically, the error in each iteration tends to be the square of the previous error; that is, $e_{n+1} = K e_n^2$, where K is some constant. You are to investigate this relationship.

51. Do three or four iterations of the Newton–Raphson method toward finding the positive root of $x^2 - 2 = 0$, starting with $x_0 = 2$. You should see the above convergence pattern. What is K?

52. Take several iterations of the Newton–Raphson method in looking for the obvious solution of $(x - 5)^2 = 0$, starting with $x_0 = 4$. Comment on the rate of convergence. Why is this problem not "typical" in its rate of convergence?

C 53. Show that if h is sufficiently small, then
a. $\sqrt{1 + h}$ is approximately equal to $1 + \dfrac{h}{2}$.
b. $\dfrac{1}{1 + h}$ is approximately equal to $1 - h$.

54. If h is sufficiently small, find an approximate value for $\sqrt[n]{A^n + h}$ for constant A.

55. Tangent line approximations are useful only if Δx is small. Illustrate this fact by trying to approximate $\sqrt{97}$ by regarding 97 as being near 81 (instead of 100).

56. **THINK TANK PROBLEM** Let

$$f(x) = -x^4 + x^2 + A$$

for constant A. What value of A should be chosen that guarantees that if $x_0 = \tfrac{1}{3}$ is chosen as the initial estimate, the Newton–Raphson method produces

$$x_1 = -x_0, x_2 = x_0, x_3 = -x_0, \ldots?$$

57. Suppose that when x units of a certain commodity are produced, the total cost is $C(x)$ and the total revenue is $R(x)$. Let $P(x)$ denote the total profit, and let

$$A(x) = \frac{C(x)}{x}$$

be the average cost.

a. Show that $P'(x) = 0$ when marginal revenue equals marginal cost.

b. Show that $A'(x) = 0$ when average cost equals marginal cost.

58. Can you solve the equation $\sqrt{x} = 0$ using the Newton–Raphson method with the initial estimate $x_0 = 0.05$? Does it make any difference if we choose another initial estimate (other than $x_0 = 0$)?

59. ■ **What Does This Say?** Suppose that when we try to use the Newton–Raphson method to approximate a solution of $f(x) = 0$, we find that $f(x_n) = 0$ but $f'(x_n) \neq 0$ for some x_n. What does this imply about $x_{n+1}, x_{n+2}, \ldots$? Explain.

60. ⓗISTORICAL ⓠUEST   Among the peoples of the region between the Tigris and Euphrates Rivers (at different times) during the period

2000–600 B.C. were Sumerians, Akkadians, Chaldeans, and Assyrians. Since the middle half of the nineteenth century, archeologists have found well over 50,000 clay tablets describing these great civilizations. Records show that they had highly developed religion, history, and science (including alchemy, astronomy, botany, chemistry, math, and zoology).

Mesopotamian culture had iterative formulas for computing algebraic quantities such as roots. In particular, they approximated $\sqrt{N}$ by repeatedly applying the formula

$$x_{n+1} = \frac{1}{2}\left(x_n + \frac{N}{x_n}\right) \quad \text{for} \quad n = 1, 2, 3, \ldots$$

a. Apply the Newton–Raphson method to $f(x) = x^2 - N$ to justify this formula.

b. Apply the formula to estimate $\sqrt{1{,}265}$ correct to five decimal places.

Chapter 3   Review

Proficiency Examination

Concept Problems

1. What is the slope of a tangent? How does this compare to the slope of a secant?

2. Define the derivative of a function.

3. What is a normal line to a graph?

4. What is the relationship between continuity and differentiability?

5. List and explain some of the notations for derivative.

6. State the following procedural rules for finding derivatives:
   a. constant multiple   b. sum rule   c. difference rule
   d. linearity rule   e. product rule   f. quotient rule
   g. State each of these rules again, this time in differential form.

7. State the following derivative rules:
   a. constant rule   b. power rule
   c. trigonometric rules   d. exponential rule
   e. logarithmic rule   f. inverse trigonometric rules

8. What is a higher derivative? List some of the different notations for higher derivatives.

9. What is meant by rate of change? Distinguish between average and instantaneous rate of change.

10. What is relative rate of change?

11. How do you find the velocity and the acceleration for an object with position $s(t)$? What is speed?

12. State the chain rule.

13. Outline a procedure for logarithmic differentiation.

14. Outline a procedure for implicit differentiation.

15. Outline a procedure for solving related-rate problems.

16. What is meant by tangent line approximation?

17. Define the differential of x and the differential of y for a function $y = f(x)$. Draw a sketch showing Δx, Δy, dx, and dy.

18. Define the terms propagated error, relative error, and percentage error.

19. What is meant by marginal analysis?

20. What is the Newton–Raphson method?

Practice Problems

Find $\dfrac{dy}{dx}$ *in Problems* 21–30.

21. $y = x^3 + x\sqrt{x} + \cos 2x$ **22.** $y = \sqrt{3x} + \dfrac{3}{x^2}$

23. $y = \sqrt{\sin(3 - x^2)}$ **24.** $xy + y^3 = 10$

25. $y = x^2 e^{-\sqrt{x}}$ **26.** $y = \dfrac{\ln 2x}{\ln 3x}$

27. $y = \sin^{-1}(3x + 2)$ **28.** $y = \tan^{-1} 2x$

29. $y = \sin^2(x^{10} + \sqrt{x}) + \cos^2(x^{10} + \sqrt{x})$

30. $y = \dfrac{\ln(x^2 - 1)}{\sqrt[3]{x}(1 - 3x)^3}$

31. Find $\dfrac{d^2y}{dx^2}$, the second derivative where $y = x^2(2x - 3)^3$.

32. Use the definition of derivative to find $\dfrac{d}{dx}(x - 3x^2)$.

33. Find the equation of the tangent line to the graph of
$$y = (x^2 + 3x - 2)(7 - 3x)$$
at the point where $x = 1$.

34. Let $f(x) = \sin^2\left(\dfrac{\pi x}{4}\right)$. Find equations of the tangent line

and the normal line to the graph of f at $x = 1$.

35. A rock tossed into a stream causes a circular ripple of water whose radius increases at a constant rate of 0.5 ft/s. How fast is the area contained inside of the ripple changing when the radius is 2 ft?

Supplementary Problems

Find dy/dx in Problems 1–36.

1. $y = x^4 + 3x^2 - 7x + 5$ **2.** $y = x^5 + 3x^3 - 11$

3. $y = \sqrt{\dfrac{x^2 - 1}{x^2 - 5}}$ **4.** $y = \dfrac{\cos x}{x + \sin x}$

5. $2x^2 - xy + 2y = 5$ **6.** $y = (x^2 + 3x - 5)^7$

7. $y = (x^3 + x)^{10}$ **8.** $y = \sqrt{x}(x^2 + 5)^{10}$

9. $y = \sqrt[3]{x}(x^3 + 1)^5$ **10.** $y = (x^2 + 3)^5(x^3 - 5)^8$

11. $y = (x^4 - 1)^{10}(2x^4 + 3)^7$ **12.** $y = \sqrt{\sin 5x}$

13. $y = \sqrt{\cos \sqrt{x}}$ **14.** $y = (\sin x + \cos x)^3$

15. $y = (\sqrt{x} + \sqrt[3]{x})^5$ **16.** $y = \sqrt{\dfrac{x^3 - x}{4 - x^2}}$

17. $y = \exp(2x^2 + 5x - 3)$ **18.** $y = \ln(x^2 - 1)$

19. $y = x3^{2-x}$ **20.** $y = \log_3(x^2 - 1)$

21. $e^{xy} + 2 = \ln\dfrac{y}{x}$ **22.** $y = \sqrt{x}\sin^{-1}(3x + 2)$

23. $y = e^{\sin x}$ **24.** $y = 2^x\log_2 x$

25. $y = e^{-x}\log_5 3x$ **26.** $x2^y + y2^x = 3$

27. $\ln(x + y^2) = x^2 + 2y$ **28.** $y = e^{-x}\sqrt{\ln 2x}$

29. $y = \sin(\sin x)$ **30.** $y = \cos(\sin x)$

31. $x^{1/2} + y^{1/2} = x$ **32.** $4x^2 - 16y^2 = 64$

33. $\sin xy = y + x$

34. $\sin(x + y) + \cos(x - y) = xy$

35. $y = \dfrac{x}{\sin^{-1} x} + \dfrac{\tan^{-1} x}{x}$

36. $y = (\sin x)(\sin^{-1} x) + x\cot^{-1} x$

Find d^2y/dx^2 in Problems 37–41.

37. $y = x^5 - 5x^4 + 7x^3 - 3x^2 + 17$

38. $y = \dfrac{x - 5}{2x + 3} + (3x - 1)^2$ **39.** $x^2 + y^3 = 10$

40. $x^2 + \sin y = 2$ **41.** $x^2 + \tan^{-1} y = 2$

In Problems 42–54, find an equation of the tangent to the curve at the indicated point.

42. $y = x^4 - 7x^3 + x^2 - 3$ at $(0, -3)$

43. $y = (3x^2 + 5x - 7)^3$ where $x = 1$

44. $y = (x^3 - 3x^2 + 3)^2$ where $x = -1$

45. $y = x\cos x$ where $x = \frac{\pi}{2}$

46. $y = \dfrac{\sin x}{\sec x\tan x}$ where $x = \pi$

47. $xy^2 + x^2y = 2$ at $(1, 1)$ **48.** $y = x\ln ex$ where $x = 1$

49. $y = xe^{2x-1}$ where $x = \frac{1}{2}$ **50.** $e^{xy} = x - y$ at $(1, 0)$

51. $y = (1 - x)^x$ where $x = 0$

52. $y = 2^x - \log_2 x$ where $x = 1$

53. $y = \dfrac{3x - 4}{3x^2 + x - 5}$ where $x = 1$

54. $x^{2/3} + y^{2/3} = 2$ at $(1, 1)$

55. Let $f(x) = (x^3 - x^2 + 2x - 1)^4$. Find equations of the tangent and normal lines to the graph of f at $x = 1$.

56. Find equations for the tangent and normal lines to the graph of $y = \left(2x + \dfrac{1}{x}\right)^3$ at the point $(1, 27)$.

57. Use the chain rule to find $\dfrac{dy}{dt}$ when $y = x^3 - 7x$ and $x = t\sin t$.

58. Find $f''(x)$ if $f(x) = x^2\sin x^2$.

59. Find f', f'', f''', and $f^{(4)}$ if $f(x) = x^4 - \dfrac{1}{x^4}$.

60. Find f', f'', and f''' if $f(x) = x(x^2 + 1)^{7/2}$.

61. Let $f(x) = \sqrt[3]{\dfrac{x^4 + 1}{x^4 - 2}}$. Find $f'(x)$ by using implicit differentiation to differentiate $[f(x)]^3$.

62. Find y' if $x^3y^3 + x - y = 1$. Leave your answer in terms of x and y.

63. Find y' and y'' if $x^2 + 4xy - y^2 = 8$. Your answer may involve x and y, but not y'.

64. Find the derivative of
$$f(x) = \begin{cases} x^2 + 5x + 4 & \text{for } x \le 0 \\ 5x + 4 & \text{for } 0 < x < 6 \\ x^2 - 2 & \text{for } x \ge 6 \end{cases}$$

65. Find equations of the tangent and normal lines to the curve given by $x^3 - y^3 = 2xy$ at the point $(-1, 1)$.

66. Use differentials to approximate $(16.01)^{3/2} + 2\sqrt{16.01}$.

67. Use differentials to approximate $\cos\dfrac{101\pi}{600}$.

68. Use differentials to estimate the change in the volume of a cone if the height of the cone is increased from 10 cm to 10.01 cm while the radius of the base stays fixed at 2 cm.

69. On New Year's Eve, a network TV camera is focusing on the descent of a lighted ball from the top of a building that is 600 ft away. The ball is falling at the rate of 20 ft/min. At what rate is the angle of elevation of the camera's line of sight changing with respect to time when the ball is 800 ft from the ground?

70. Suppose f is a differentiable function whose derivative satisfies $f'(x) = 2x^2 + 3$. Find $\dfrac{d}{dx}f(x^3 - 1)$.

71. Suppose f is a differentiable function such that $f'(x) = x^2 + x$. Find $\dfrac{d}{dx}f(x^2 + x)$.

72. Let $f(x) = 3x^2 + 1$ for all x. Use the chain rule to find $\dfrac{d}{dx}(f \circ f)(x)$.

73. Let $f(x) = \sin 2x + \cos 3x$ and $g(x) = x^2$. Use the chain rule to find $\dfrac{d}{dx}(f \circ g)(x)$.

74. Let $f(x) = \begin{cases} x \sin \dfrac{1}{x} & \text{if } x \neq 0 \\ 0 & \text{if } x = 0 \end{cases}$

    Use the definition of the derivative to find $f'(x)$ if it exists.

75. A car and a truck leave an intersection at the same time. The car travels north at 60 mi/h and the truck travels east at 45 mi/h. How fast is the distance between them changing after 45 minutes?

76. A spherical balloon is being filled with air in such a way that its radius is increasing at a constant rate of 2 cm/s. At what rate is the volume of the balloon increasing at the instant when its surface has area 4π cm²?

77. Suppose the total cost of producing x units of a particular commodity is

    $$C(x) = \tfrac{2}{5}x^2 + 3x + 10$$

    and that each unit of the commodity can be sold for

    $$p(x) = \tfrac{1}{5}(45 - x)$$

    dollars.
    a. What is the marginal cost?
    b. What is the price when the marginal cost is 23?
    c. Estimate the cost of producing the 11th unit.
    d. Find the actual cost of producing the 11th unit.

78. A block of ice in the shape of a cube originally having volume 1,000 cm³ is melting in such a way that the length of each of its edges is decreasing at the rate of 1 cm/hr. At what rate is its surface area decreasing at the time its volume is 27 cm³? Assume that the block of ice maintains its cubical shape.

79. Show that the rate of change of the area of a circle with respect to its radius is equal to the circumference.

80. A charged particle is projected into a linear accelerator. The particle undergoes a constant acceleration that changes its velocity from 1,200 m/s to 6,000 m/s in 2×10^{-3} seconds. Find the acceleration of the particle.

81. A rocket is launched vertically from a point on the ground that is 3,000 horizontal feet from an observer with binoculars. If the rocket is rising vertically at the rate of 750 ft/s at the instant it is 4,000 ft above the ground, how fast must the observer change the angle of elevation of her line of sight to keep the rocket in sight at that instant?

82. **MODELING PROBLEM** Assume that a certain artery in the body is modeled by a circular tube whose cross section has radius 1.2 mm. Fat deposits are observed to build up uniformly on the inside wall of the artery. Find the rate at which the cross-sectional area of the artery is decreasing relative to the thickness of the fat deposit at the instant when the deposit is 0.3 mm thick.

83. **MODELING PROBLEM** A processor who sells a certain raw material has analyzed the market and determined that the unit price should be modeled by the formula

    $$p(x) = 60 - x^2$$

    (thousand dollars) for x tons ($0 \leq x \leq 7$) produced. Estimate the change in the unit price that accompanies each change in sales:
    a. from 2 tons to 2.05 tons   b. from 1 ton to 1.1 tons
    c. from 3 tons to 2.95 tons

84. **MODELING PROBLEM** A company sends out a truck to deliver its products. To estimate costs, the manager models gas consumption by the formula

    $$G(x) = \frac{1}{300}\left(\frac{1{,}500}{x} + x\right)$$

    gal/mi, under the assumption that the truck travels at a constant rate of x mi/h ($x \geq 5$). The driver is paid \$16 per hour to drive the truck 300 mi. Gasoline costs \$2 per gallon.
    a. Find an expression for the total cost $C(x)$ of the trip.
    b. Use differentials to estimate the additional cost if the truck is driven at 57 mi/h instead of 55 mi/h.

85. A viewer standing at ground level and 30 ft from a platform watches an object rise from that platform at the constant rate of 3 ft/s. (See Figure 3.57.) How fast is the angle of sight between the viewer and the object changing at the instant when $\theta = \frac{\pi}{4}$?

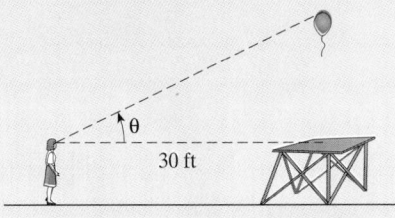

■ **FIGURE 3.57** Problem 85

86. A lighthouse is 4,000 ft from a straight shore. Watching the beam on the shore from the point P on the shore that is closest to the lighthouse, an observer notes that the light is moving at the rate of 3 ft/s when it is 1,000 ft from P. How fast is the light revolving at this instant?

87. A light is 4 miles from a straight shoreline. The light revolves at the rate of 2 rev/min. Find the speed of the spot of light along the shore when the light spot is 2 miles past the point on the shore closest to the source of light.

88. A particle of mass m moves along the x-axis. The velocity $v = \dfrac{dx}{dt}$ and position $x = x(t)$ satisfy the equation
$$m(v^2 - v_0^2) = k(x_0^2 - x^2)$$
where k, x_0, and v_0 are positive constants. The force F acting on the object is defined by $F = ma$, where a is the object's acceleration. Show that $F = -kx$.

89. The equation $\dfrac{d^2s}{dt^2} + ks = 0$ is called a **differential equation of simple harmonic motion.** Let A be any number. Show that the function $s(t) = A \sin 2t$ satisfies the equation
$$\frac{d^2s}{dt^2} + 4s = 0$$

90. Suppose $L(x)$ is a function with the property that $L'(x) = x^{-1}$. Use the chain rule to find the derivatives of the following functions.

a. $f(x) = L(x^2)$ b. $f(x) = L\left(\dfrac{1}{x}\right)$

c. $f(x) = L\left(\dfrac{2}{3\sqrt{x}}\right)$ d. $f(x) = L\left(\dfrac{2x+1}{1-x}\right)$

91. Let f and g be differentiable functions such that $f[g(x)] = x$ and $g[f(x)] = x$ for all x (that is, f and g are inverses). Show that
$$\frac{dg}{dx} = \frac{1}{\dfrac{df}{dx}}$$

92. A baseball player is stealing second base. He runs at 30 ft/s and when he is 25 ft from second base, the catcher, while standing at home plate, throws the ball toward the base at a speed of 120 ft/s. At what rate is the distance between the ball and the player changing at the time the ball is thrown?

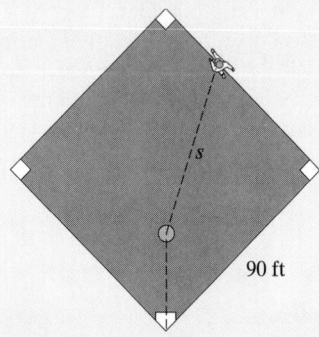

90 ft

93. A rod $\overline{OA}$ 2 m long is rotating counterclockwise in a plane about O at the rate of 3 rev/s, as shown in Figure 3.58.

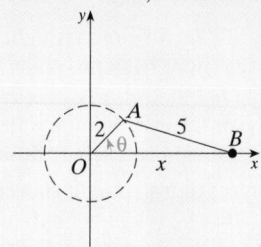

■ **FIGURE 3.58** Problem 93

The rod $\overline{AB}$ is attached to $\overline{OA}$ at A, and the end B slides along the x-axis. Suppose $\overline{AB}$ is 5 meters long. What are the velocity and the acceleration of the motion of the point B along the x-axis?

94. MODELING PROBLEM   The ideal gas law states that for an ideal gas, pressure P is modeled by the formula $P = kT/V$, where V is the volume, T is the temperature, and k is a constant. Suppose the temperature is kept fixed at 100 °C, and the pressure decreases at the rate of 7 lb/in.^{2} per min. At what rate is the volume changing at the instant the pressure is 25 lb/in.^{2} and the volume is 30 in.^{3}?

95. This problem deals with the function
$$f(x) = \frac{x^3 - 2.1x^2 + x - 2}{x^6 + 1}$$

a. Prove that f has at least one zero on the interval $[0, 5]$.

b. Graphically show that f has only one zero in this interval. *Note:* This will require some careful specification of domains and ranges.

c. Pick a reasonable starting x_0-value (for example, the closest integer), and do enough iterations of the Newton–Raphson method to get at least six significant figure precision. Hand in computer outputs showing all the x_n values.

d. Now pick x_0 larger (for example, $x_0 = 3$, or larger). Do several iterations and explain the results.

e. Try x_0 around 0.5; explain what happens and why it happens.

f. Try x_0 around -1.0; tell what happens, and why.

96. PUTNAM EXAMINATION PROBLEM   Suppose $f(x) = ax^2 + bx + c$, where a, b, and c are real numbers and $|f(x)| \le 1$ for $x \le 1$. Show that $|f'(x)| \le 4$ for $|x| \le 1$.

97. PUTNAM EXAMINATION PROBLEM   A point P is taken on the curve $y = x^3$. The tangent at P meets the curve again at Q. Prove that the slope of the curve at Q is four times the slope at P.

98. PUTNAM EXAMINATION PROBLEM   A particle of unit mass moves on a straight line under the action of a force that is a function $f(v)$ of the velocity v of the particle, but the form of this function is not known. A motion is observed, and the distance x covered in time t is found to be related to t by the formula $x = at + bt^2 + ct^3$, where a, b, and c have numerical values determined by observation of the motion. Find the function $f(v)$ for the range of v covered by this experiment.

Chaos

This project is to be done in groups of three or four students. Each group will submit a single written report.

An exciting new topic in mathematics attempts to bring order to the universe by considering disorder. This topic is called **chaos theory,** and it shows how structures of incredible complexity and disorder really exhibit beauty and order.

> Water flowing through a pipe offers one of the simplest physical models of chaos. Pressure is applied to the end of the pipe and the water flows in straight lines. More pressure increases the speed of the laminar flow until the pressure reaches a critical value, and a radically new situation evolves—turbulence. A simple laminar flow suddenly changes to a flow of beautiful complexity consisting of swirls within swirls. Before turbulence, the path of any particle was quite predictable. After a minute change in the pressure, turbulence occurs and predictability is lost. **Chaos** is concerned with systems in which minute changes suddenly transform predictability into unpredictability.*

For this research project begin with

$$f(x) = x^3 - x = x(x - 1)(x + 1)$$

Use the Newton–Raphson method to investigate what happens as you change the starting value, x_0, to solve $f(x) = 0$. What would you select as a starting value? Two important x-values in our study are

$$s_3 = \frac{1}{\sqrt{3}} \quad \text{and} \quad s_5 = \frac{1}{\sqrt{5}}$$

One would *not* want to pick x_0 as either $\pm s_3$. Why not? Also, what happens if one picks $x_0 = s_5$? That is, what are $x_1, x_2, \ldots$?

Generate a good plot of $f(x)$ on $[-2, 2]$. Explain from the plots why you would *expect* that an initial value $x_0 > s_3$ would lead to $\lim_{x \to +\infty} x_n = 1$. (Also, by symmetry, $x_0 < -s_3$ leads to $\lim_{x \to +\infty} x_n = -1$.)

Explain why if $|x_0| < s_5$, you would expect $x_n \to 0$. Numerically verify your assertions. Now to see the "chaos," use the following x_0-values and take 6 to 10 iterations, until convergence occurs, and report what happens: 0.448955; 0.447503; 0.447262; 0.447222; 0.4472215; 0.4472213.

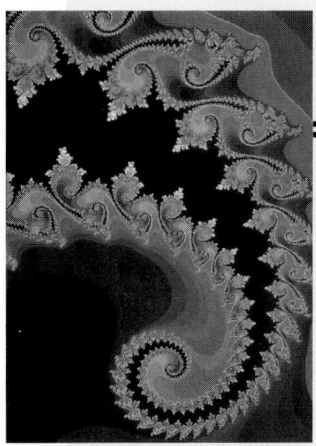

This fractal image is an example of what is called mathematical chaos.

Mathematics is one of the oldest of the sciences; it is also one of the most active, for its strength is the vigor of perpetual youth.

A. R. FORSYTHE
NATURE 84 (1910): 285.

---

*Thanks to Jack Wadhams of Golden West College for this paragraph.

The whole of mathematics consists in the organization of a series of aids to the imagination in the process of reasoning.

A. N. WHITEHEAD
UNIVERSAL ALGEBRA
(CAMBRIDGE, 1898), P. 12.

Write a paper about chaos. Your paper should include, but not be limited to, answers to the questions on this page. Some references (to get you started) are listed:

Chaos: Making a New Science, James Gleich, Penguin Books, 1987.

Chaos and Fractals: New Frontiers of Science, H. O. Peitgen, H. Jurgens, and D. Saupe, Springer Verlag, 1992.

Newton's Method and Fractal Patterns, Philip D. Straffin, Jr., UMAP Module 716, COMAP, 1991.

4

Additional Applications of the Derivative

PREVIEW

In Chapter 3, we used the derivative to find tangent lines and to compute rates of change. The primary goal of this chapter is to examine the use of calculus in curve sketching, optimization, and other applications.

PERSPECTIVE

Homing pigeons and certain other birds are known to avoid flying over large bodies of water whenever possible. The reason for this behavior is not entirely known. However, it is reasonable to speculate that it may have something to do with minimizing the energy expended in flight, because the air over a lake is often "heavier" than that over land. Suppose a pigeon is released from a boat at point B on the lake shown in the accompanying figure.* It will fly to its loft at point L on the lakeshore by heading across water to a point P on the shore and then flying directly from P to L along the shore. If the pigeon expends e_w units of energy per mile over water and e_L units over land, where should P be located to minimize the total energy expended in flight?†

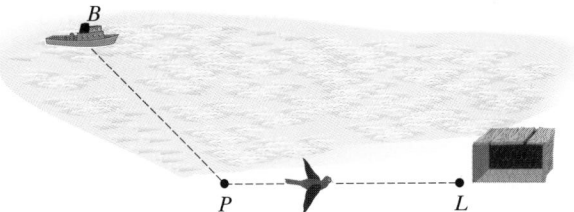

Optimization (finding the maximum or minimum values) is one of the most important applications we will study in calculus, from maximizing profit and minimizing cost, to maximizing the strength of a structure or minimizing the distance traveled. Optimization problems are considered in Sections 4.6 and 4.7.

---

*Edward Batschelet, *Introduction to Mathematics for Life Scientists,* 2nd ed. (New York: Springer-Verlag, 1979), pp. 276–277.

†See Problem 32, Section 4.7.

215

4.1   Extreme Values of a Continuous Function

EXTREME VALUE THEOREM

One of the principal goals of calculus is to investigate the behavior of various functions. As part of this investigation, we will be laying the groundwork for solving a large class of problems that involve finding the maximum or minimum value of a function. Such problems are called **optimization problems.** Let us begin by introducing some useful terminology.

Absolute Maximum and Minimum

Let f be a function defined on a set D that contains the number c. Then

$f(c)$ is the **absolute maximum** of f on D if

$$f(c) \geq f(x) \text{ for all } x \text{ in } D;$$

$f(c)$ is the **absolute minimum** of f on D if

$$f(c) \leq f(x) \text{ for all } x \text{ in } D.$$

Together, the absolute maximum and minimum of f on the interval I are called the **extreme values,** or the **absolute extrema,** of f on I. A function does not necessarily have extreme values on a given interval. For instance, the continuous function $g(x) = x$ has neither a maximum nor a minimum on the open interval $(0, 1)$, as shown in Figure 4.1a.

The discontinuous function defined by

$$h(x) = \begin{cases} x^2 & \text{for } x \neq 0 \\ 1 & \text{for } x = 0 \end{cases}$$

has a maximum on the closed interval $[-1, 1]$ but no minimum, as shown in Figure 4.1b. Incidentally, this graph also illustrates the fact that a function may assume an absolute extremum at more than one point. In this case, the maximum is at $(-1, 1), (0, 1),$ and $(1, 1)$.

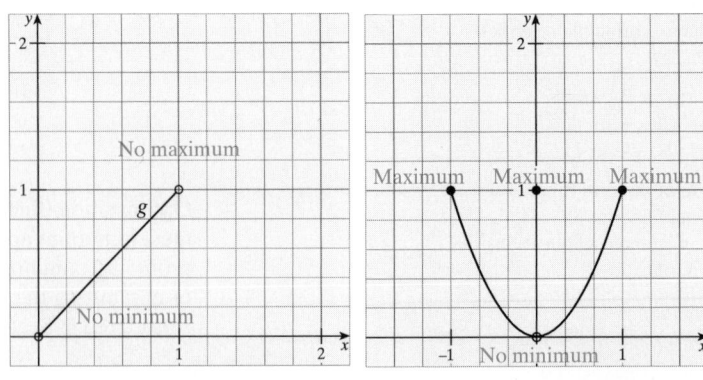

a. The continuous function $g(x) = x$ has no extrema on the open interval $(0, 1)$.

b. The discontinuous function h has a maximum, but not a minimum on the closed interval $[-1, 1]$.

■ **FIGURE 4.1**   Functions that lack one or both extreme values

If a function f is continuous and the interval I is closed and bounded, it can be shown that both an absolute maximum and an absolute minimum *must* occur. This result, called the **extreme value theorem,** plays an important role in our work.

THEOREM 4.1   *The extreme value theorem*

A continuous function f on a closed, bounded interval $[a, b]$ has an absolute maximum and an absolute minimum.

Proof   Even though this result may seem quite reasonable (see Figure 4.2), its proof requires concepts beyond the scope of this text and will be omitted.

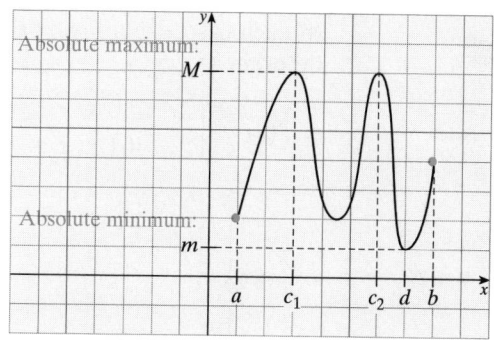

Absolute maximum at $x = c_1$ and $x = c_2$
Absolute minimum at $x = d$

■ **FIGURE 4.2**   Extreme value theorem

This theorem does *not* apply if the function is not continuous; neither does the theorem apply if the interval is not closed or not bounded. You will be asked for appropriate counterexamples in the problem set.

Note that the maximum of a function occurs at the highest point on its graph and the minimum occurs at the lowest point. These properties are illustrated in Example 1.

EXAMPLE 1   *Extreme values of a continuous function*

The graph of a function f is shown in Figure 4.3. Locate the extreme values of f defined on the closed interval $[a, b]$.

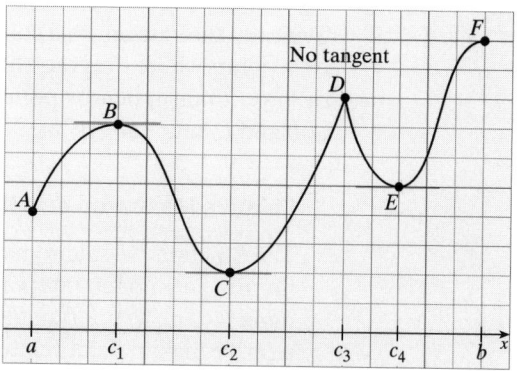

■ **FIGURE 4.3**   A continuous function on a closed interval $[a, b]$

Solution   The highest point on the graph occurs at the right endpoint F, and the lowest point occurs at C. Thus, the absolute maximum is $f(b)$, and the absolute minimum is $f(c_2)$.   ∎

In Example 1, the existence of maxima and minima as required by the extreme value theorem may seem obvious, but there are times when it seems as if the extreme value theorem fails. When this occurs, you need to be sure the conditions of the extreme value theorem are satisfied—namely, that the function f is continuous on a closed, bounded interval. Consider Example 2.

EXAMPLE 2   *Conditions of the extreme value theorem*

In each case, explain why the given function does not contradict the extreme value theorem.

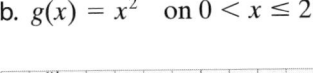

a. $f(x) = \begin{cases} 2x & \text{if } 0 \le x < 1 \\ 1 & \text{if } 1 \le x \le 2 \end{cases}$ b. $g(x) = x^2$ on $0 < x \le 2$

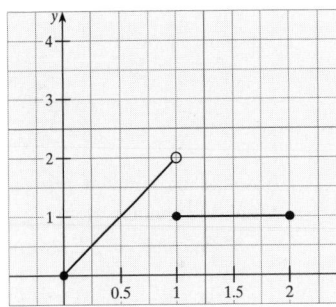

Does not have a maximum value.

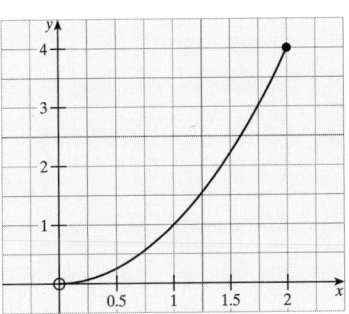

g does not have a minimum value
(but it does have a maximum value).

Solution
a. The function f has no maximum. It takes on all values arbitrarily close to 2, but it never reaches the value 2. The extreme value theorem is not violated because f is not continuous on $[0, 2]$.
b. Although the functional values of $g(x)$ become arbitrarily small as x approaches 0, it never reaches the value 0, so g has no minimum. The function g is continuous on the interval $(0, 2]$, but the extreme value theorem is not violated because the interval is not closed.   ∎

RELATIVE EXTREMA

Typically, the extrema of a continuous function occur either at endpoints of the interval or at points where the graph has a "peak" or a "valley" (points where the graph is higher or lower than all nearby points). For example, the function f in Figure 4.3 has "peaks" at B and D and "valleys" at C and E. Peaks and valleys are what we call *relative extrema*.

Relative Maximum and Relative Minimum

The function f is said to have a **relative maximum** at the point c if $f(c) \ge f(x)$ for all x in an open interval containing c. Also, f is said to have a **relative minimum** at d if $f(d) \le f(x)$ for all x in an open interval containing d. Collectively, relative maxima and relative minima are called **relative extrema.**

Next we shall formulate a procedure for finding relative extrema. By looking at Figure 4.3, we see that there are horizontal tangents at B, C, and E, whereas a tangent does not exist at D. This suggests that the relative extrema of f occur either where the derivative is zero (horizontal tangent) or where the derivative does not exist (no vertical tangent). This notion leads us to the following definition.

Critical Numbers and Critical Points

Suppose f is defined at c and either $f'(c) = 0$ or $f'(c)$ does not exist. Then the number c is called a **critical number** of f, and the point $P(c, f(c))$ on the graph of f is called a **critical point.**

If there is a relative maximum at c, then the functional value $f(c)$ is that maximum value. Similarly, if there is a relative minimum at c, then the minimum value is $f(c)$.

WARNING Note that if $f(c)$ is not defined, then c *cannot* be a critical number.

EXAMPLE 3   *Finding critical numbers*

Find the critical numbers for the given functions.

a. $f(x) = 4x^3 - 5x^2 - 8x + 20$ b. $f(x) = \dfrac{e^x}{x - 2}$ c. $f(x) = 2\sqrt{x}(6 - x)$

Solution
a. $f'(x) = 12x^2 - 10x - 8$ is defined for all values of x. Solve

$$12x^2 - 10x - 8 = 0$$
$$2(3x - 4)(2x + 1) = 0$$
$$x = \tfrac{4}{3}, -\tfrac{1}{2} \qquad \text{These are the critical numbers.}$$

b. $f'(x) = \dfrac{(x - 2)e^x - e^x(1)}{(x - 2)^2} = \dfrac{e^x(x - 3)}{(x - 2)^2}$

The derivative is not defined at $x = 2$, but f is not defined at 2 either, so $x = 2$ is not a critical number. The actual critical numbers are found by solving $f'(x) = 0$:

$$\frac{e^x(x - 3)}{(x - 2)^2} = 0$$
$$x = 3 \qquad \text{This is the only critical number since } e^x > 0.$$

c. Write $f(x) = 12x^{1/2} - 2x^{3/2}$ so $f'(x) = 6x^{-1/2} - 3x^{1/2}$.

The derivative is not defined at $x = 0$. We have $f(0) = 12(0)^{1/2} - 2(0)^{3/2} = 0$, so we see that f *is* defined at $x = 0$, which means that $x = 0$ is a critical number. For other critical numbers, solve $f'(x) = 0$:

$$6x^{-1/2} - 3x^{1/2} = 0$$
$$3x^{-1/2}(2 - x) = 0$$
$$x = 2 \qquad \text{The critical numbers are } x = 0, 2. \qquad \blacksquare$$

EXAMPLE 4   *Critical numbers and critical points*

Find the critical numbers and the critical points for the function

$$f(x) = (x - 1)^2(x + 2)$$

Solution   Because the function f is a polynomial, we know that it is continuous and that the derivative exists for all x. Thus, we find the critical numbers by using the product rule and extended power rule to solve the equation $f'(x) = 0$.

$$f'(x) = (x - 1)^2(1) + 2(x - 1)(1)(x + 2)$$
$$= (x - 1)[(x - 1) + 2(x + 2)]$$
$$= (x - 1)(3x + 3)$$
$$= 3(x - 1)(x + 1)$$

The critical numbers are $x = \pm 1$. To find the critical points, we need to find the y-component for each critical number:

$$f(1) = (1 - 1)^2(1 + 2) = 0$$
$$f(-1) = (-1 - 1)^2(-1 + 2) = 4$$

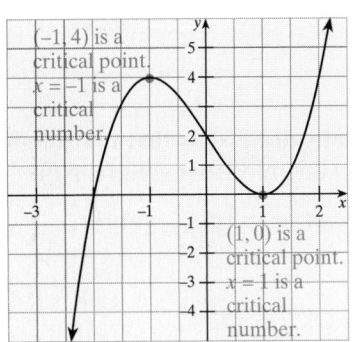

■ **FIGURE 4.4**   The graph of $f(x) = (x - 1)^2(x + 2)$

Thus, the critical points are $(1, 0)$ and $(-1, 4)$. The graph of $f(x) = (x - 1)^2(x + 2)$ is shown in Figure 4.4.   ■

Note how the relative extrema occur at the critical points. Our observation that the relative extrema occur only at points on a graph where there is either a horizontal tangent line or no tangent at all is equivalent to the following result.

THEOREM 4.2   *Critical number theorem*

If a continuous function f has a relative extremum at c, then c must be a critical number of f.

> ■ *What This Says:*   If a point is a relative maximum or a relative minimum value for a function, then either the derivative is 0 or it does not exist at that point.

Proof   Since f is continuous, $f(c)$ is defined. If $f'(x)$ does not exist, then c is a critical number by definition. We shall show that if $f'(c)$ exists and a relative maximum occurs at c, then $f'(c) = 0$. Our approach will be to examine the difference quotient. (The case where $f'(c)$ exists and a relative minimum occurs at c is handled similarly in Problem 65.)

Because a relative maximum occurs at c, we have $f(c) \geq f(x)$ for every number x in an open interval (a, b) containing c. Therefore, if Δx is small enough so $c + \Delta x$ is in (a, b), then

$$f(c) \geq f(c + \Delta x) \qquad \textit{Because a relative maximum occurs at c}$$

$$f(c) - f(c + \Delta x) \geq 0$$
$$f(c + \Delta x) - f(c) \leq 0 \qquad \textit{Multiply both sides by } -1, \textit{reversing the inequality.}$$

For the next step we want to divide both sides by Δx (to write the left side as a difference quotient). However, as this is an inequality, we need to consider two possibilities:

1. Suppose $\Delta x > 0$ (the inequality does not reverse):

$$\frac{f(c + \Delta x) - f(c)}{\Delta x} \le 0 \qquad \text{Divide both sides by } \Delta x.$$

Now we take the limit of both sides as Δx approaches from the right (because Δx is positive).

$$\underbrace{\lim_{\Delta x \to 0^+} \frac{f(c + \Delta x) - f(c)}{\Delta x}}_{f'(c)} \le \underbrace{\lim_{\Delta x \to 0^+} 0}_{0}$$
$$f'(c) \le 0$$

Thus, $f'(x) \le 0$.

2. Next, suppose $\Delta x < 0$ (the inequality reverses). Then

$$\frac{f(c + \Delta x) - f(c)}{\Delta x} \ge 0$$

This time we take the limit of both sides as Δx approaches 0 from the left (because Δx is negative):

$$\lim_{\Delta x \to 0^-} \frac{f(c + \Delta x) - f(c)}{\Delta x} \ge \lim_{\Delta x \to 0^-} 0$$
$$f'(c) \ge 0$$

Because we have shown that $f'(c) \le 0$ and $f'(c) \ge 0$, it follows that $f'(c) = 0$.

WARNING   Theorem 4.2 tells us that a relative extremum of a continuous function f can occur *only* at a critical number, but *it does not say that a relative extremum must occur at each critical number.*

For example, if $f(x) = x^3$, then $f'(x) = 3x^2$ and $f'(0) = 0$, but there is no relative extremum at $c = 0$ on the graph of f because the graph is rising for $x < 0$ and also for $x > 0$, as shown in Figure 4.5a. It is also quite possible for a continuous function g to have no relative extremum at a point c where $g'(x)$ does not exist (see Figure 4.5b).

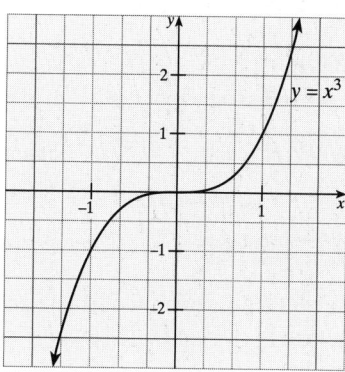

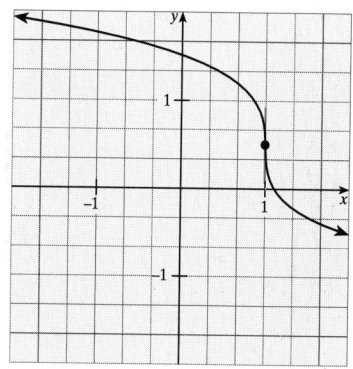

a. The graph of $f(x) = x^3$
No relative extremum occurs
at $c = 0$ even though $f'(0) = 0$

b. Although $g'(1)$ does not exist,
no relative extremum occurs
at $c = 1$.

■ **FIGURE 4.5**   A relative extremum may not occur at each critical number.

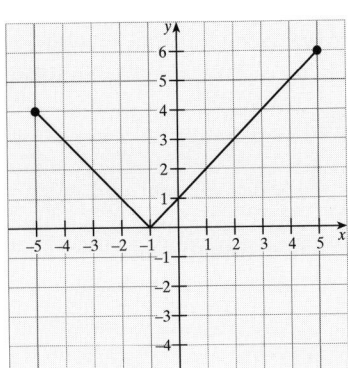

EXAMPLE 5   *Critical numbers where the derivative does not exist*

Find the critical numbers for $f(x) = |x + 1|$ on $[-5, 5]$.

Solution   If $x > -1$, then $f(x) = x + 1$ and $f'(x) = 1$. However, if $x < -1$, then $f(x) = -x - 1$ and $f'(x) = -1$. Because these are not the same,

$$\lim_{x \to -1^-} f'(x) \neq \lim_{x \to -1^+} f'(x)$$

we see that the derivative does not exist at $x = c = -1$. Because $f(-1)$ is defined, it follows that -1 is a critical number.   ∎

ABSOLUTE EXTREMA

Suppose we are looking for the absolute extrema of a continuous function f on the closed, bounded interval $[a, b]$. We know that these extrema exist by Theorem 4.1. Also, Theorem 4.2 enables us to narrow the list of "candidates" for points where extrema can occur from the entire interval $[a, b]$ to just the endpoints $x = a, x = b$, and the critical numbers c between a and b. This suggests the following procedure.

Procedure for Finding Absolute Extrema

To find the absolute extrema of a continuous function f on $[a, b]$:

Step 1.   Compute $f'(x)$ and find all critical numbers of f on $[a, b]$.

Step 2.   Evaluate f at the endpoints a and b and at each critical number c.

Step 3.   Compare the values in step 2.
The largest value is the absolute maximum of f on $[a, b]$.
The smallest value is the absolute minimum of f on $[a, b]$.

Figure 4.6 shows some of the possibilities in the application of this procedure.

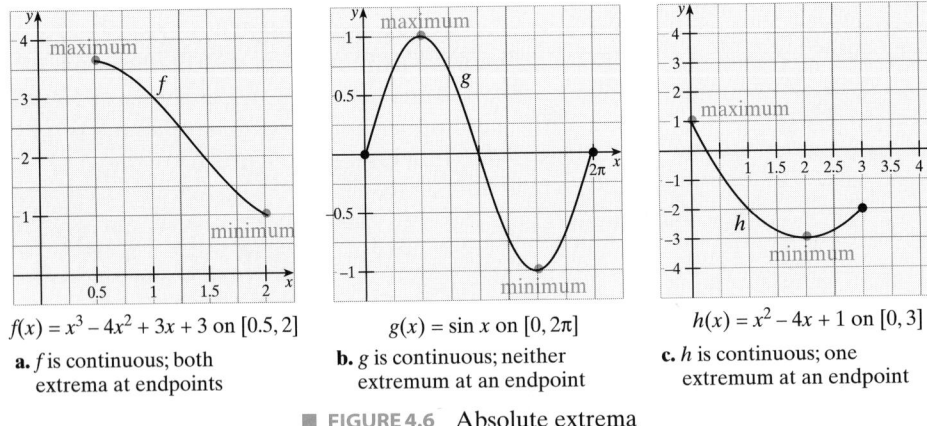

$f(x) = x^3 - 4x^2 + 3x + 3$ on $[0.5, 2]$

a. f is continuous; both extrema at endpoints

$g(x) = \sin x$ on $[0, 2\pi]$

b. g is continuous; neither extremum at an endpoint

$h(x) = x^2 - 4x + 1$ on $[0, 3]$

c. h is continuous; one extremum at an endpoint

■ **FIGURE 4.6**   Absolute extrema

EXAMPLE 6   *Absolute extrema of a polynomial function*

Find the absolute extrema of the function defined by the equation $f(x) = x^4 - 2x^2 + 3$ on the closed interval $[-1, 2]$.

Solution   Because f is a polynomial function, it is continuous on the closed interval $[-1, 2]$. Theorem 4.1 tells us that there must be an absolute maximum and an absolute minimum on the interval.

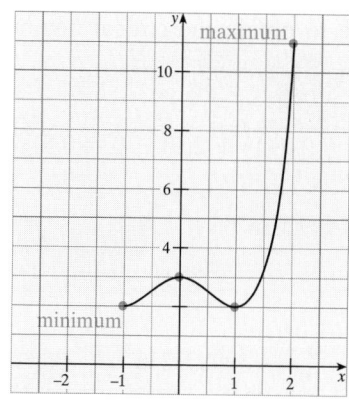

■ **FIGURE 4.7**   The graph of $f(x) = x^4 - 2x^2 + 3$ on $[-1, 2]$

Step 1. $f'(x) = 4x^3 - 2(2x)$
$$= 4x(x^2 - 1)$$
$$= 4x(x - 1)(x + 1).$$

The critical numbers are $x = 0, 1,$ and -1.

Step 2. Values at endpoints: $f(-1) = 2$
$$f(2) = 11$$

Critical numbers: $f(0) = 3$
$$f(1) = 2$$

Step 3. The absolute maximum of f occurs at $x = 2$ and is $f(2) = 11$; the absolute minima of f occur at $x = 1$ and $x = -1$ and are $f(1) = f(-1) = 2$. The graph of f is shown in Figure 4.7.   ■

EXAMPLE 7   *Absolute extrema when the derivative does not exist*

Find the absolute extrema of $f(x) = x^{2/3}(5 - 2x)$ on the interval $[-1, 2]$.

Solution

Step 1. To find the derivative, rewrite the given function as
$f(x) = 5x^{2/3} - 2x^{5/3}$. Then
$$f'(x) = \tfrac{10}{3}x^{-1/3} - \tfrac{10}{3}x^{2/3} = \tfrac{10}{3}x^{-1/3}(1 - x)$$

Critical numbers are found by solving $f'(x) = 0$ and by locating the places where the derivative does not exist. First,
$$f'(x) = 0 \text{ when } x = 1$$

Even though $f(0)$ exists, we note that $f'(x)$ does not exist at $x = 0$ (notice the division by zero when $x = 0$). Thus, the critical numbers are $x = 0$ and $x = 1$.

Step 2. Values at endpoints: $f(-1) = 7$
$$f(2) = 2^{2/3} \approx 1.587401052$$

Critical numbers: $f(0) = 0$
$$f(1) = 3$$

Step 3. The absolute maximum of f occurs at $x = -1$ and is $f(-1) = 7$; the absolute minimum of f occurs at $x = 0$ and is $f(0) = 0$. The graph of f is shown in Figure 4.8.   ■

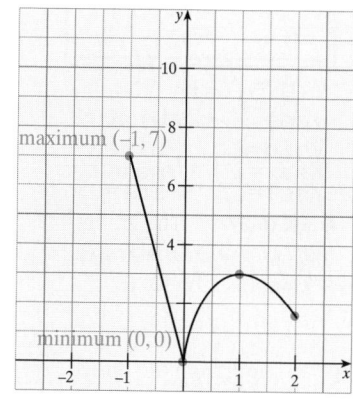

■ **FIGURE 4.8**   Graph of $f(x) = 5x^{2/3} - 2x^{5/3}$ on $[-1, 2]$

EXAMPLE 8   *Absolute extrema for a trigonometric function*

Find the absolute extrema of the following continuous function on $[0, \frac{\pi}{2}]$:
$T(x) = \tfrac{1}{2}(\sin^2 x + \cos x) + 2\sin x - x$.

Solution

Step 1. We have

$$T'(x) = \tfrac{1}{2}(2\sin x \cos x - \sin x) + 2(\cos x) - 1$$
$$= \tfrac{1}{2}(2\sin x \cos x - \sin x + 4\cos x - 2)$$
$$= \tfrac{1}{2}[2\cos x(\sin x + 2) - (\sin x + 2)]$$
$$= \tfrac{1}{2}[(\sin x + 2)(2\cos x - 1)]$$

Set each factor equal to zero and solve to obtain the critical number $x = \pi/3$.

Step 2. Evaluate the function at the endpoints:

$$T(0) = \tfrac{1}{2}(\sin^2 0 + \cos 0) + 2 \sin 0 - 0 = \tfrac{1}{2}(0 + 1) + 2(0) - 0 = 0.5$$

$$T(\tfrac{\pi}{2}) = \tfrac{1}{2}(\sin^2 \tfrac{\pi}{2} + \cos \tfrac{\pi}{2}) + 2 \sin \tfrac{\pi}{2} - \tfrac{\pi}{2}$$

$$= \tfrac{1}{2}(1 + 0) + 2(1) - \tfrac{\pi}{2} = \tfrac{5}{2} - \tfrac{\pi}{2} \approx 0.9292036732$$

Maximum value:

$$T(\tfrac{\pi}{3}) = \tfrac{1}{2}(\sin^2 \tfrac{\pi}{3} + \cos \tfrac{\pi}{3}) + 2 \sin \tfrac{\pi}{3} - \tfrac{\pi}{3}$$

$$= \tfrac{1}{2}(\tfrac{3}{4} + \tfrac{1}{2}) + 2(\tfrac{\sqrt{3}}{2}) - \tfrac{\pi}{3} = \tfrac{5}{8} + \sqrt{3} - \tfrac{\pi}{3} \approx 1.309853256$$

Step 3. The absolute maximum of T is approximately 1.31 at $x = \tfrac{\pi}{3}$ and the absolute minimum of T is 0.5 at $x = 0$. The graph is shown in Figure 4.9 as part of the technology window.   ■

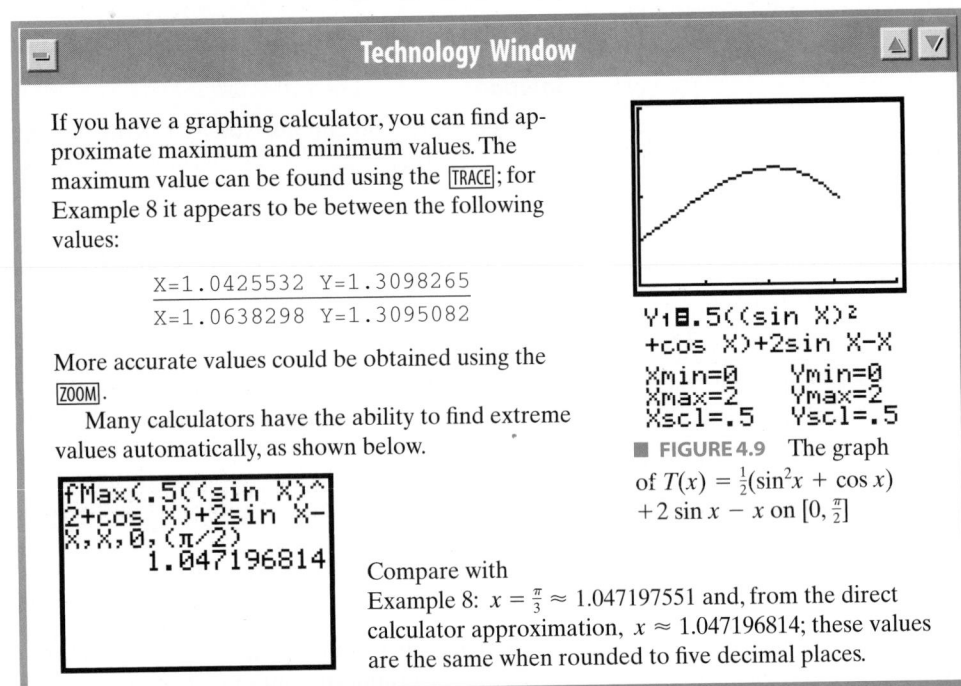

Technology Window

If you have a graphing calculator, you can find approximate maximum and minimum values. The maximum value can be found using the [TRACE]; for Example 8 it appears to be between the following values:

```
X=1.0425532  Y=1.3098265
X=1.0638298  Y=1.3095082
```

More accurate values could be obtained using the [ZOOM].

Many calculators have the ability to find extreme values automatically, as shown below.

```
fMax(.5((sin X)^
2+cos X)+2sin X-
X,X,0,(π/2)
            1.047196814
```

```
Y₁■.5((sin X)²
+cos X)+2sin X-X
Xmin=0      Ymin=0
Xmax=2      Ymax=2
Xscl=.5     Yscl=.5
```

■ **FIGURE 4.9** The graph of $T(x) = \tfrac{1}{2}(\sin^2 x + \cos x) + 2 \sin x - x$ on $[0, \tfrac{\pi}{2}]$

Compare with Example 8: $x = \tfrac{\pi}{3} \approx 1.047197551$ and, from the direct calculator approximation, $x \approx 1.047196814$; these values are the same when rounded to five decimal places.

OPTIMIZATION

In our next two examples, we examine applications involving optimization. Such problems are investigated in more depth in Sections 4.6 and 4.7.

EXAMPLE 9   *Maximum and minimum velocity of a moving particle*

A particle moves along the t-axis with position

$$s(t) = t^4 - 8t^3 + 18t^2 + 60t - 8$$

Find the largest and smallest values of its velocity for $1 \leq t \leq 5$.

Solution   The velocity is

$$v(t) = s'(t) = 4t^3 - 24t^2 + 36t + 60$$

To find the largest value of $v(t)$, we compute the derivative of v:

$$v'(t) = 12t^2 - 48t + 36$$
$$= 12(t-3)(t-1)$$

Setting $v'(t) = 0$, we find that the critical numbers of $v(t)$ are $t = 3, 1$. (*Note:* v is a polynomial function, so it has a derivative for all t.) Now we evaluate v at the critical numbers and endpoints.

$t = 1$ is both a critical number and an endpoint:

$$v(1) = 4(1)^3 - 24(1)^2 + 36(1) + 60 = 76$$

$t = 5$ is an endpoint:

$$v(5) = 4(5)^3 - 24(5)^2 + 36(5) + 60 = 140$$

$t = 3$ is a critical number:

$$v(3) = 4(3)^3 - 24(3)^2 + 36(3) + 60 = 60$$

The largest value of the velocity is 140 at the endpoint where $t = 5$, and the smallest value is 60 when $t = 3$.

EXAMPLE 10   *An applied maximum value problem*

A box with a square base is constructed so that the length of one side of the base plus the height is 10 in. What is the largest possible volume of such a box?

Solution   We let b be the length of one side of the base and h be the height of the box, as shown in Figure 4.10. The volume, V, is

$$V = b^2 h$$

Because our methods apply only to functions of one variable, it may seem that we cannot deal with V as a function of two variables. However, we know that $b + h = 10$; therefore, $h = 10 - b$, and we can now write V as a function of b alone:

$$V(b) = b^2(10 - b)$$

The domain is not stated, but we must have $b \geq 0$ and $10 - b = h \geq 0$, so that $0 \leq b \leq 10$. First, we find the critical numbers. Note that V is a polynomial function, so the derivative exists everywhere in the domain. Write $V(b) = 10b^2 - b^3$ and find $V'(b) = 20b - 3b^2$. Then

$$V'(b) = 0$$
$$20b - 3b^2 = 0$$
$$b(20 - 3b) = 0$$
$$b = 0, \tfrac{20}{3}$$

Checking the endpoints and the critical numbers, we have

$$V(0) = 0$$
$$V(10) = 10^2(10 - 10) = 0$$
$$V(\tfrac{20}{3}) = (\tfrac{20}{3})^2(10 - \tfrac{20}{3}) = \tfrac{4,000}{27}$$

Thus, the largest value for the volume V is $\frac{4,000}{27} \approx 148.1$ in.^{3}. It occurs when the square base has a side of length $\frac{20}{3}$ in. and the height is $h = 10 - \frac{20}{3} = \frac{10}{3}$ in.

Technology Window

$Y_1 = 4X^3 - 24X^2 + 36X + 60$

Xmin=0 Ymin=0
Xmax=6 Ymax=200
Xscl=1 Yscl=25

Using VALUE or TRACE, find that the maximum is 140 when $t = 5$ and the minimum is 60 when $t = 3$.

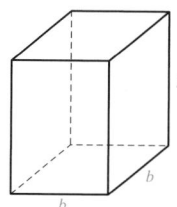

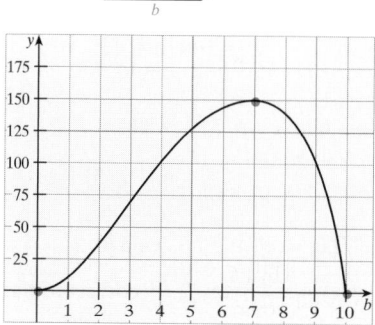

■ **FIGURE 4.10** Volume of a box

4.1 Problem Set

Ⓐ *In Problems 1–14, find the largest and smallest values of each continuous function on the given closed, bounded interval.*

1. $f(x) = 5 + 10x - x^2$ on $[-3, 3]$
2. $f(x) = 10 + 6x - x^2$ on $[-4, 4]$
3. $f(x) = x^3 - 3x$ on $[-1, 3]$
4. $f(t) = t^4 - 8t^2$ on $[-3, 3]$
5. $f(x) = x^3$ on $[-\frac{1}{2}, 1]$
6. $g(x) = x^3 - 3x$ on $[-2, 2]$
7. $f(x) = x^5 - x^4$ on $[-1, 1]$
8. $g(t) = 3t^5 - 20t^3$ on $[-1, 2]$
9. $h(t) = te^{-t}$ on $[0, 2]$ 10. $s(x) = \dfrac{\ln\sqrt{x}}{x}$ on $[1, 3]$
11. $f(x) = |x|$ on $[-1, 1]$
12. $f(x) = |x - 3|$ on $[-4, 4]$
13. $f(u) = \sin^2 u + \cos u$ on $[0, 2]$
14. $g(u) = \sin u - \cos u$ on $[0, \pi]$
15. ■ **What Does This Say?** Outline a procedure for finding the absolute extrema of a continuous function on a closed, bounded interval. Include in your outline a discussion of what is meant by critical numbers.

Technology Window

16. In Example 7, we found that the maximum value of $f(x) = x^{2/3}(5 - 2x)$ on $[-1, 2]$ is 7 and occurs at $x = -1$. The graph of this function on a leading brand of graphing calculator is shown:

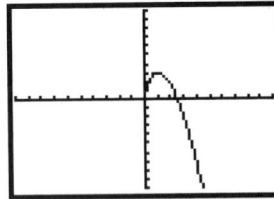

```
Y1☐X^(2/3)(5-2X)
Xmin=-10  Ymin=-10
Xmax=10   Ymax=10
Xscl=1    Yscl=1
```

This graph is not correct. Can you explain the discrepancy? It is not our intent to "make up" problems to use with a calculator, but *whenever* you use a calculator or computer to assist you with calculus, you must understand the nature of the functions with which you are working, and not rely only on the calculator or computer output.

In Problems 17–28, find the largest and smallest values of each continuous function on the closed, bounded interval. If the function is not continuous on the interval, so state.

17. $f(u) = 1 - u^{2/3}$ on $[-1, 1]$
18. $g(t) = (50 + t)^{2/3}$ on $[-50, 14]$
19. $g(x) = 2x^3 - 3x^2 - 36x + 4$ on $[-4, 4]$
20. $g(x) = x^3 + 3x^2 - 24x - 4$ on $[-4, 4]$
21. $f(x) = \frac{8}{3}x^3 - 5x^2 + 8x - 5$ on $[-4, 4]$
22. $f(x) = \frac{1}{6}(x^3 - 6x^2 + 9x + 1)$ on $[0, 2]$
23. $h(x) = \tan x + \sec x$ on $[0, 2\pi]$
24. $s(t) = t \cos t - \sin t$ on $[0, 2\pi]$
25. $f(x) = e^{-x} \sin x$ on $[0, 2\pi]$
26. $g(x) = \cot^{-1}\left(\dfrac{x}{9}\right) - \cot^{-1}\left(\dfrac{x}{5}\right)$ on $[0, 10]$
27. $f(x) = \begin{cases} 9 - 4x & \text{if } x < 1 \\ -x^2 + 6x & \text{if } x \geq 1 \end{cases}$ on $[1, 4]$
28. $f(x) = \begin{cases} 8 - 3x & \text{if } x < 2 \\ -x^2 + 3x & \text{if } x \geq 2 \end{cases}$ on $[-1, 4]$

Find the required extremum in Problems 29–35, or explain why it does not exist.

29. The smallest value of $f(x) = x^2$ on $[-1, 1]$
30. The largest value of $f(x) = \dfrac{1}{x(x + 1)}$ on $[-0.5, 0]$
31. The smallest value of $g(x) = \dfrac{9}{x} + x - 3$ on $[1, 9]$
32. The smallest value of $g(x) = \dfrac{x^2 - 1}{x^2 + 1}$ on $[-1, 1]$
33. The largest value of
$$f(t) = \begin{cases} -t^2 - t + 2 & \text{if } t < 1 \\ 3 - t & \text{if } t \geq 1 \end{cases} \text{ on } [-2, 3]$$
34. The smallest value of $f(x) = e^x + e^{-x} - x$ on $[0, 2]$
35. The largest value of $g(x) = \dfrac{\ln x}{\cos x}$ on $[2, 3]$, correct to the nearest tenth

Ⓑ *In Problems 36–43, find the extrema.*

36. $f(\theta) = \cos^3 \theta - 4\cos^2 \theta$ on $[-0.1, \pi + 0.1]$
37. $g(\theta) = \theta \sin \theta$ on $[-2, 2]$
38. $f(x) = 20 \sin(378\pi x)$ on $[-1, 1]$
39. $g(u) = 98u^3 - 4u^2 + 72u$ on $[0, 4]$
40. $f(w) = \sqrt{w}(w - 5)^{1/3}$ on $[0, 4]$
41. $h(x) = \sqrt[3]{x} \sqrt[3]{(x - 3)^2}$ on $[-1, 4]$

42. $h(x) = \cos^{-1} x \tan^{-1} x$ on $[0, 1]$

43. $f(x) = e^{-x}(\cos x + \sin x)$ on $[0, 2\pi]$

Find a function that not only meets the conditions specified in Problems 44–47, but also meets each of the following conditions:

    a. *a minimum but no maximum*
    b. *a maximum but no minimum*
    c. *both a maximum and a minimum*
    d. *neither a maximum nor a minimum*

*That is, each of these parts **a–d** may need a different example to satisfy the conditions specified in Problems 44–47.*

44. THINK TANK PROBLEM   For each of the four given conditions, find a function that is discontinuous and defined on an open interval.

45. THINK TANK PROBLEM   For each of the four given conditions, find a function that is discontinuous and defined on a closed interval.

46. THINK TANK PROBLEM   For each of the four given conditions, find a function that is continuous and defined on an open interval.

47. THINK TANK PROBLEM   For each of the four given conditions, find a function that is continuous and defined on a closed interval.

48. THINK TANK PROBLEM   Give a counterexample to show that the extreme value theorem does not necessarily apply if one disregards the condition that f be continuous (that is, f need not be continuous).

49. THINK TANK PROBLEM   Give a counterexample to show that the extreme value theorem does not necessarily apply if one disregards the condition that f be defined on a closed interval (that is, f may be defined on an open interval).

50. An object moves along the t-axis with position

$$s(t) = t^3 - 6t^2 - 15t + 11$$

Find the largest value of its velocity on $[0, 4]$.

51. An object moves along the t-axis with position

$$s(t) = t^4 - 2t^3 - 12t^2 + 60t - 10$$

Find the largest value of its velocity on $[0, 3]$.

52. Find two nonnegative numbers whose sum is 8 and the product of whose squares is as large as possible.

53. Find two nonnegative numbers such that the sum of one and twice the other is 12 if it is required that their product be as large as possible.

54. Under the condition that $3x + y = 80$, maximize the product xy^3 when $x \geq 0$, $y \geq 0$.

55. Under the condition that $3x + y = 126$, maximize xy when $x \geq 0$ and $y \geq 0$.

56. Under the condition that $2x - 5y = 18$, maximize x^2y when $x \geq 0$ and $y \leq 0$.

57. Show that if a rectangle with fixed perimeter P is to enclose the greatest area, it must be a square.

58. Find all points on the circle $x^2 + y^2 = a^2$ $(a \geq 0)$ such that the product of the x-coordinate and the y-coordinate is as large as possible.

59. MODELING PROBLEM   A business manager models the cost of a certain commodity by the formula

$$C(x) = 0.125x^2 + 20,000$$

where x is the number of units produced. Show that the average cost,

$$A(x) = \frac{C(x)}{x}$$

is minimized when the average cost is equal to $C'(x)$. (Recall from Chapter 3 that $C'(x)$ is called the marginal cost.)

60. Generalize the result of Problem 59 by showing that if $C(x)$ is the cost of a commodity and $A(x) = C(x)/x$ is the average cost, then $A(x)$ is minimized when average cost equals marginal cost.

61. **a.** Show that $\frac{1}{2}$ is the number that exceeds its own square by the greatest amount.
    b. Which nonnegative number exceeds its own cube by the greatest amount?
    c. Which nonnegative number exceeds its nth power $(n > 0)$ by the greatest amount?

62. Given the constants $a_1, a_2, \ldots, a_n$, find the value of x that guarantees that the sum

$$S(x) = (a_1 - x)^2 + (a_2 - x)^2 + \cdots + (a_n - x)^2$$

will be as small as possible.

63. Find the smallest value of m so that the line $y = mx$ will lie above the curve

$$y = 1 - \frac{1}{x}$$

for all $x > 0$. *Hint:* Begin by locating the value of x that minimizes

$$f(x) = mx - 1 + \frac{1}{x}$$

64. Explain why the function

$$f(x) = \frac{8}{\sin x} + \frac{27}{\cos x}$$

must attain a minimum in the open interval $(0, \frac{\pi}{2})$. Show that if the minimum is attained at $x = \theta$, then $\tan \theta = \frac{2}{3}$.

65. Show that if $f'(c)$ exists and a relative minimum occurs at c, then c must be a critical number of f.

4.2   *The Mean Value Theorem*

mean value theorem, Rolle's theorem; proof of the MVT, some uses of the mean value theorem

MEAN VALUE THEOREM

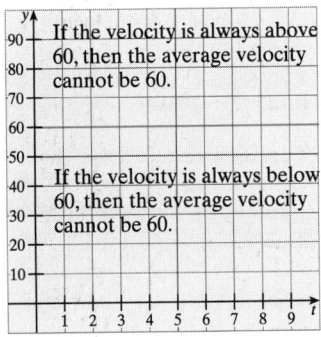

If the velocity is always above 60, then the average velocity cannot be 60.

If the velocity is always below 60, then the average velocity cannot be 60.

If an automobile averages 60 mi/h on a trip, it is reasonable to expect that the speedometer must read exactly 60 *at least once* during the trip. More generally, suppose an object moves along a straight line so that its position is $s(t)$ from its starting point at time t. The average velocity of the object from time $t = a$ to time $t = b$ is given by the ratio

$$\frac{s(b) - s(a)}{b - a}$$

Just as in the case of the automobile, we expect that there should be at least one time t_0 between a and b when the instantaneous velocity equals this average velocity. That is, there exists some t_0 such that

$$\underbrace{s'(t_0)}_{\text{Instantaneous velocity}} = \underbrace{\frac{s(b) - s(a)}{b - a}}_{\text{Average velocity}}$$

This is an example of the *mean value theorem for derivatives,* which we shall refer to simply as the **mean value theorem** (abbreviated MVT).

THEOREM 4.3   *The mean value theorem for derivatives (MVT)*

If f is continuous on the closed interval $[a, b]$ and differentiable on the open interval (a, b), then there exists at least one number c on (a, b) such that

$$\frac{f(b) - f(a)}{b - a} = f'(c)$$

Proof   The proof of the MVT follows later in this section.

> ■   *What This Says:*   Under reasonable conditions, there must be at least one number c between a and b such that the derivative $f'(c)$ actually equals the difference quotient
>
> $$\frac{f(b) - f(a)}{b - a}$$

This equation may or may not be easy to solve. Example 1 illustrates one method of finding such a number c.

EXAMPLE 1   *Finding the number c specified by the MVT*

Show that the function $f(x) = x^3 + x^2$ satisfies the hypotheses of the MVT on the closed interval $[1, 2]$, and find a number c between 1 and 2 so that

$$f'(c) = \frac{f(2) - f(1)}{2 - 1}$$

Solution   Because f is a polynomial function, it is differentiable and hence also continuous on the entire interval $[1, 2]$. Thus, the hypotheses of the MVT are satisfied.

By differentiating f, we find that

$$f'(x) = 3x^2 + 2x$$

for all x. Therefore, we have $f'(c) = 3c^2 + 2c$, and the MVT equation

$$f'(c) = \frac{f(2) - f(1)}{2 - 1}$$

is satisfied when

$$3c^2 + 2c = \frac{f(2) - f(1)}{2 - 1} = \frac{12 - 2}{1} = 10$$

Solving the resulting equation $3c^2 + 2c - 10 = 0$ by the quadratic formula, we obtain

$$c = \frac{-1 \pm \sqrt{31}}{3}$$

The negative value is not in the open interval $(1, 2)$, but the positive value is:

$$c = \frac{-1 + \sqrt{31}}{3} \approx 1.522588121$$

satisfies the requirements of the MVT.   ∎

ROLLE'S THEOREM

The key to the proof of the mean value theorem is the following result, which is really just the MVT in the special case where $f(b) = f(a)$.

THEOREM 4.4   Rolle's theorem*

Suppose f is continuous on the closed interval $[a, b]$ and differentiable on the open interval (a, b). Then if $f(a) = f(b)$, there exists at least one number c between a and b such that $f'(c) = 0$.

Proof   If f is constant on the closed interval $[a, b]$, then $f'(x) = 0$ for *all* x between a and b. If f is not constant throughout the interval $[a, b]$, the largest and smallest values of f on $[a, b]$ cannot be the same. Because $f(a) = f(b)$, at least one extreme value of f does not occur at an endpoint. According to the extreme value theorem and the critical number theorem (Theorems 4.1 and 4.2 of Section 4.1), such an extreme value must occur at a critical number between a and b. Recall that a critical number c has the property that $f'(c) = 0$ or else $f'(c)$ does not exist. Because $f'(x)$ is assumed to exist throughout the interval (a, b), it follows that $f'(c) = 0$ for some number between a and b, as claimed.

The proof of Rolle's theorem makes much more sense if it is considered geometrically, as shown in Figure 4.11. Figure 4.11a shows that if f is constant on the closed interval $[a, b]$, then $f'(x) = 0$ for all x in the interval.

---

*Rolle's theorem is named for Michel Rolle (1652–1719), a French mathematician who investigated a special case of the mean value theorem in a book on the algebra of equations published in 1690.

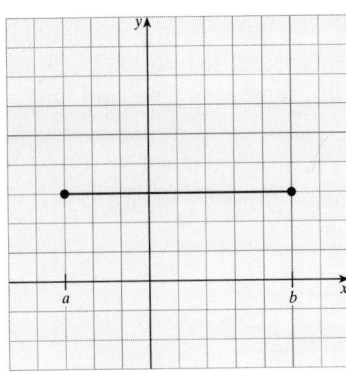
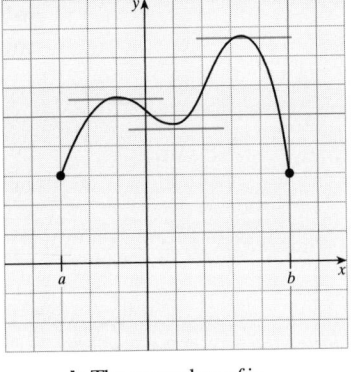

a. The special case where
$f(a) = f(b)$; f is constant
for this example.

b. The case where f is
not constant on $[a, b]$

■ **FIGURE 4.11**   A geometric interpretation of Rolle's theorem

However, if f is not constant, as shown in Figure 4.11b, its graph must change direction at least once in order to begin and end on the same level. [Remember that f is continuous and $f(a) = f(b)$.] The graph must have a horizontal tangent (with a derivative of zero) wherever the direction changes. Figure 4.11b shows that it is quite possible for the graph to have several such transition points, even though the theorem asserts only that at least one exists.

The mean value theorem has a similar geometric interpretation, as shown in Figure 4.12. Rolle's theorem can be viewed as saying that if $f(a) = f(b) = d$, then for at least one number c between a and b, the tangent line at $(c, f(c))$ is parallel to the line $y = d$ through the endpoints (a, d) and (b, d). It is reasonable to expect a similar result to hold if the endpoints of the graph are not necessarily at the same height. Note that in Figure 4.12b, the graph of Figure 4.12a has been "lifted and tilted," but the tangent line at $(c, f(c))$ is still parallel to the line L through the endpoints $P(a, f(a))$ and $Q(b, f(b))$. Because the tangent line at $(c, f(c))$ has slope $f'(c)$, it follows that

$$\underbrace{f'(c)}_{\text{Slope at } (c, f(c))} = \underbrace{\frac{f(b) - f(a)}{b - a}}_{\text{Slope of } L}$$

Now we are ready to prove the mean value theorem.

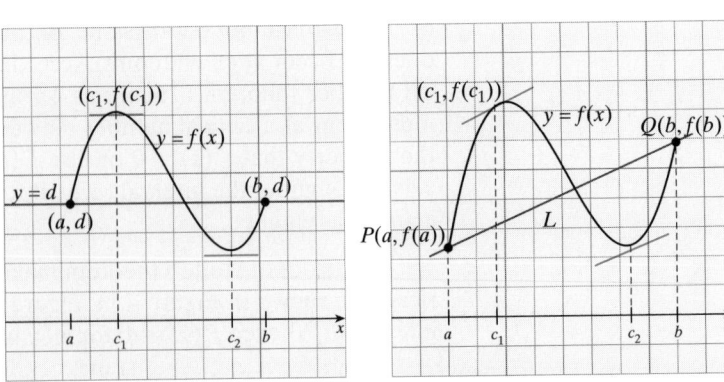

a. Rolle's theorem

b. Mean value theorem

■ **FIGURE 4.12**   A geometric comparison of Rolle's theorem and the mean value theorem

PROOF OF THE MVT

We now restate the mean value theorem, and complete its proof.

If f is continuous on the closed interval $[a, b]$ and differentiable on the open interval (a, b), then there exists at least one number c such that

$$\frac{f(b) - f(a)}{b - a} = f'(c) \quad \text{for } a < c < b$$

Proof   The proof of the MVT uses Rolle's theorem. Suppose f is continuous on $[a, b]$ and differentiable on (a, b). Define a new function g as follows:

$$g(x) = \left[\frac{f(b) - f(a)}{b - a}\right](x - a) + f(a) - f(x)$$

for $a \leq x \leq b$. Because f satisfies the hypotheses of the MVT, the function g is also continuous on the closed interval $[a, b]$ and differentiable on the open interval (a, b). In addition, we find that

$$g(a) = \left[\frac{f(b) - f(a)}{b - a}\right](a - a) + f(a) - f(a) = 0$$

$$g(b) = \left[\frac{f(b) - f(a)}{b - a}\right](b - a) + f(a) - f(b)$$

$$= [f(b) - f(a)] + f(a) - f(b) = 0$$

Thus, g satisfies the hypotheses of Rolle's theorem, so there exists at least one number c between a and b for which $g'(c) = 0$. Differentiating the function g, we find that

$$g'(x) = \frac{f(b) - f(a)}{b - a} - f'(x)$$

Because $g'(c) = 0$, we have

$$0 = g'(c) = \frac{f(b) - f(a)}{b - a} - f'(c)$$

This means that

$$f'(c) = \frac{f(b) - f(a)}{b - a}$$

as required. $=$

SOME USES OF THE MEAN VALUE THEOREM

The MVT has many different uses. In Example 2, we use it to establish a trigonometric inequality.

EXAMPLE 2   *Using the MVT to prove a trigonometric inequality*

Show that $\left|\sin x - \sin y\right| \leq \left|x - y\right|$ for numbers x and y by applying the mean value theorem.

Solution   The inequality is true if $x = y$. Suppose $x \neq y$; then $f(\theta) = \sin \theta$ is differentiable and hence continuous for all θ, with $f'(\theta) = \cos \theta$. By applying the MVT to f on the closed interval with endpoints x and y, we see that

$$\frac{f(x) - f(y)}{x - y} = f'(c)$$

for some c between x and y. Because $f'(c) = \cos c$ and

$$f(x) - f(y) = \sin x - \sin y$$

it follows that

$$\frac{\sin x - \sin y}{x - y} = \cos c$$

Finally, we take the absolute value of the expression on each side, remembering that $\left| \cos c \right| \leq 1$ for any number c:

$$\left| \frac{\sin x - \sin y}{x - y} \right| = \left| \cos c \right| \leq 1$$

Thus, $\left| \sin x - \sin y \right| \leq \left| x - y \right|$, as claimed.   ■

The primary use of the MVT is as a tool for proving certain key theoretical results of calculus. For example, in Theorem 4.5, we use the MVT to prove that a function whose derivative is always zero on an interval must be constant on that interval. This apparently simple result and its corollary (Theorem 4.6) turn out to be crucial to our development of the integral in Chapter 5.

THEOREM 4.5   *Zero derivative theorem*

Suppose f is a continuous function on the closed interval $[a, b]$ and is differentiable on the open interval (a, b), with $f'(x) = 0$ for all x on (a, b). Then the function f is constant on $[a, b]$.

Proof   Let x_1 and x_2 be two distinct numbers $(x_1 \neq x_2)$ chosen arbitrarily from the closed interval $[a, b]$. The function f satisfies the requirements of the MVT on the interval with endpoints x_1 and x_2, which means that there exists a number c between x_1 and x_2 such that

$$\frac{f(x_2) - f(x_1)}{x_2 - x_1} = f'(c)$$

By hypothesis, $f'(x) = 0$ throughout the open interval (a, b), and because c lies within this interval, we have $f'(c) = 0$. Thus, by substitution we have

$$\frac{f(x_2) - f(x_1)}{x_2 - x_1} = 0$$

$$f(x_2) = f(x_1)$$

Because x_1 and x_2 were chosen arbitrarily from $[a, b]$, we conclude that $f(x) = k$, a constant, for all x, as required.

THEOREM 4.6   *Constant difference theorem*

Suppose the functions f and g are continuous on the closed interval $[a, b]$ and differentiable on the open interval (a, b). Then if $f'(x) = g'(x)$ for all x in (a, b), there exists a constant C such that

$$f(x) = g(x) + C$$

for all x on $[a, b]$.

Proof   Let $h(x) = f(x) - g(x)$; then

$$h'(x) = f'(x) - g'(x)$$
$$= 0 \qquad \text{\textit{Because} } f'(x) = g'(x)$$

Thus, by Theorem 4.5, $h(x) = C$ for some constant C and all x on $[a, b]$, and because $h(x) = f(x) - g(x)$, it follows that

$$f(x) - g(x) = C$$
$$f(x) = g(x) + C \qquad \text{\textit{Add} } g(x) \text{ \textit{to both sides.}}$$

■ *What This Says:*   Two functions with equal derivatives on an open interval differ by a constant on that interval.

4.2   *Problem Set*

Ⓐ **1.** ■ **What Does This Say?** What does Rolle's theorem say, and why is it important?

2. ■ **What Does This Say?** Without looking, state the hypotheses used in the proof of the MVT. How are the hypotheses used in the proof? Can the conclusion of the MVT be true if any or all of the hypotheses are not satisfied?

In Problems 3–20, verify that the given function f satisfies the hypotheses of the MVT on the given interval $[a, b]$. Then find all numbers c between a and b for which

$$\frac{f(b) - f(a)}{b - a} = f'(c)$$

3. $f(x) = 2x^2 + 1$ on $[0, 2]$

4. $f(x) = -x^2 + 4$ on $[-1, 0]$

5. $f(x) = x^3 + x$ on $[1, 2]$

6. $f(x) = 2x^3 - x^2$ on $[0, 2]$

7. $f(x) = x^4 + 2$ on $[-1, 2]$

8. $f(x) = x^5 + 3$ on $[2, 4]$

9. $f(x) = \sqrt{x}$ on $[1, 4]$

10. $f(x) = \dfrac{1}{\sqrt{x}}$ on $[1, 4]$

11. $f(x) = \dfrac{1}{x + 1}$ on $[0, 2]$

12. $f(x) = 1 + \dfrac{1}{x}$ on $[1, 4]$

13. $f(x) = \cos x$ on $[0, \frac{\pi}{2}]$

14. $f(x) = \sin x + \cos x$ on $[0, 2\pi]$

15. $f(x) = e^x$ on $[0, 1]$

16. $f(x) = \frac{1}{2}(e^x + e^{-x})$ on $[0, 1]$

17. $f(x) = \ln x$ on $[\frac{1}{2}, 2]$

18. $f(x) = \dfrac{\ln\sqrt{x}}{x}$ on $[1, 3]$

19. $f(x) = \tan^{-1}x$ on $[0, 1]$

20. $f(x) = x\sin^{-1}x$ on $[0, 1]$

Decide whether Rolle's theorem can be applied to f on the interval indicated in Problems 21–30.

21. $f(x) = |x - 2|$ on $[0, 4]$ **22.** $f(x) = \tan x$ on $[0, 2\pi]$

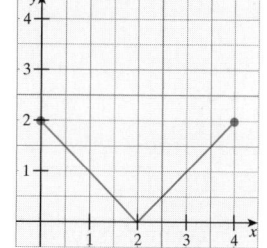 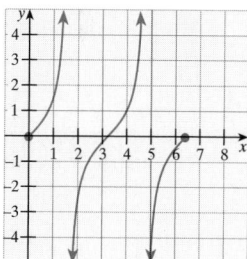

23. $f(x) = \sin x$ on $[0, 2\pi]$

24. $f(x) = |x| - 2$ on $[0, 4]$

25. $f(x) = \sqrt[3]{x} - 1$ on $[-8, 8]$

26. $f(x) = \dfrac{1}{x - 2}$ on $[-1, 1]$

27. $f(x) = \dfrac{1}{x - 2}$ on $[1, 2]$

28. $f(x) = 3x + \sec x$ on $[-\pi, \pi]$

29. $f(x) = \sin^2 x$ on $[-\frac{\pi}{2}, \frac{\pi}{2}]$

30. $f(x) = \sqrt{\ln x}$ on $[1, 2]$

31. Let $g(x) = 8x^3 - 6x + 8$. Find a function f so that $f'(x) = g'(x)$ and $f(1) = 12$.

32. Let $g(x) = \sqrt{x^2 + 5}$. Find a function f so that $f'(x) = g'(x)$ and $f(2) = 1$.

33. Show that $f(x) = \dfrac{x+4}{5-x}$ and $g(x) = \dfrac{-9}{x-5}$ differ by a constant. Are the conditions of the constant difference theorem satisfied? Does $f'(x) = g'(x)$?

34. Show that $f(x) = (x-2)^3$ and $g(x) = (x^2 + 12)(x-6)$ differ by a constant. Use f and g to demonstrate the constant difference theorem.

35. Let $f(x) = (x-1)^3$ and $g(x) = (x^2 + 3)(x-3)$. Use f and g to demonstrate the constant difference theorem.

36. Let f be defined as shown in Figure 4.13.

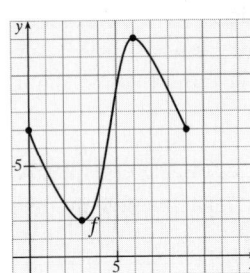

■ **FIGURE 4.13**   Function f on $[0, 9]$

Use f to estimate the values of c that satisfy the conclusion of Rolle's theorem on $[0, 9]$. What theorem would apply for the interval $[0, 5]$?

37. Let g be defined as shown in Figure 4.14.

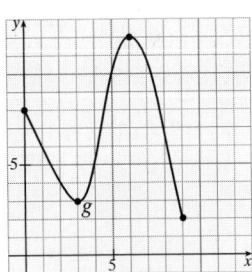

■ **FIGURE 4.14**   Function g on $[0, 9]$

Use g to estimate the values of c that satisfy the conclusion of the mean value theorem on $[0, 9]$. What theorem would apply for the interval $[4, 8]$?

Ⓑ 38. **Alternative form of the mean value theorem**   If f is continuous on $[a, b]$ and differentiable on (a, b), then there exists a number c in (a, b) such that

$$f(b) = f(a) + (b-a)f'(c)$$

Derive this alternative form of the MVT.

39. Let u and v be any two numbers between $-\frac{\pi}{2}$ and $\frac{\pi}{2}$. Use the MVT to show that

$$|\tan u - \tan v| \geq |u - v|$$

40. If $f(x) = \dfrac{1}{x}$ on $[-1, 1]$, does the mean value theorem apply? Why or why not?

41. If $g(x) = |x|$ on $[-2, 2]$, does the mean value theorem apply? Why or why not?

42. **THINK TANK PROBLEM**   Is it true that

$$|\cos x - \cos y| \leq |x - y|$$

for all x and y? Either prove that the inequality is always valid or find a counterexample.

43. **THINK TANK PROBLEM**   Consider

$$f(x) = \begin{cases} 1 & \text{if } x \geq 0 \\ -1 & \text{if } x < 0 \end{cases}$$

$f'(x) = 0$ for all x in the domain, but f is not a constant. Does this example contradict the zero derivative theorem? Why or why not?

44. a. Let n be a positive integer. Show that there is a number c between 0 and x for which

$$\frac{(1+x)^n - 1}{x} = n(1+c)^{n-1}$$

b. Use part **a** to evaluate

$$\lim_{x \to 0} \frac{(1+x)^n - 1}{x}$$

45. a. Show that there is a number w between 0 and x for which

$$\frac{\cos x - 1}{x} = -\sin w$$

b. Use part **a** to evaluate

$$\lim_{x \to 0} \frac{\cos x - 1}{x}$$

46. Use the MVT to evaluate

$$\lim_{x \to \pi^+} \frac{\cos x + 1}{x - \pi}$$

47. Let $f(x) = 1 + \dfrac{1}{x}$. If a and b are constants such that $a < 0$ and $b > 0$, show that there is no number w between a and b for which

$$f(b) - f(a) = f'(w)(b - a)$$

48. Show that for any $x > 4$, there is a number w between 4 and x such that

$$\frac{\sqrt{x} - 2}{x - 4} = \frac{1}{2\sqrt{w}}$$

Use this fact to show that if $x > 4$, then

$$\sqrt{x} < 1 + \frac{x}{4}$$

49. Show that if an object moves along a straight line in such a way that its velocity is the same at two different

times (that is, for a differentiable function v, we are given $v(t_1) = v(t_2)$ for $t_1 \neq t_2$), then there is some intermediate time when the acceleration is zero.

50. **MODELING PROBLEM**   Two radar patrol cars are located at fixed positions 6 mi apart on a long, straight road where the speed limit is 55 mi/h. A sports car passes the first patrol car traveling at 53 mi/h, and then 5 min later, it passes the second patrol car going 48 mi/h. Analyze a model of this situation to show that at some time between the two clockings, the sports car exceeded the speed limit. (*Hint:* Use the MVT.)

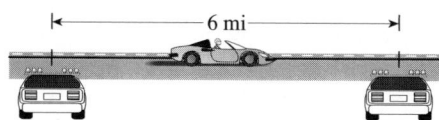

51. **MODELING PROBLEM**   Suppose two race cars begin at the same time and finish at the same time. Analyze a model to show that at some point in the race they had the same speed.

52. Use Rolle's theorem with

$$f(x) = (x - 1) \sin x$$

to show that the equation $\tan x = 1 - x$ has at least one solution for $0 < x < 1$.

53. Use the MVT to show that

$$\sqrt{1 + x} - 4 < \tfrac{1}{8}(x - 15)$$

if $x > 15$.   *Hint:* Let $f(x) = \sqrt{1 + x}$.

54. Use the MVT to show that

$$\frac{1}{2x + 1} > \frac{1}{5} + \frac{2}{25}(2 - x)$$

if $0 < x < 2$.

C 55. Let $f(x) = \tan x$. Note that

$$f(\pi) = f(0) = 0$$

Show that there is no number w between 0 and π for which $f'(w) = 0$. Why does this fact not contradict the MVT?

56. Use Rolle's theorem or the MVT to show that there is no number a for which the equation

$$x^3 - 3x + a = 0$$

has *two* distinct solutions in the interval $[-1, 1]$.

57. If $a > 0$ is a constant, show that the equation

$$x^3 + ax - 1 = 0$$

has exactly one real solution.   *Hint:* Let $f(x) = x^3 + ax - 1$ and use the intermediate value theorem to show that there is at least one root. Then assume there are two roots, and use Rolle's theorem to obtain a contradiction.

58. If $a > 0$ and n is a positive integer, use Rolle's theorem or the MVT to show that the polynomial

$$p(x) = x^{2n+1} + ax + b$$

can have at most one real root for constants a and b.

59. Show that if $f''(x) = 0$ for all x, then f is a linear function. (That is, $f(x) = Ax + B$ for constants $A \neq 0$ and B.)

60. Show that if $f'(x) = Ax + B$ for constants $A \neq 0$ and B, then $f(x)$ is a quadratic function. (That is, $f(x) = ax^2 + bx + c$ for constants $a, b,$ and c, where $a \neq 0$.)

61. The total profit from sales of a commodity is given by

$$P(x) = R(x) - C(x)$$

where $R(x)$ is the revenue and $C(x)$ is the cost associated with a production level of x units. Suppose that a and b are equally profitable levels of production—that is, $P(a) = P(b)$. If $R(x)$ and $C(x)$ are both differentiable for $a < x < b$, show that there is at least one level of production between a and b where marginal cost equals marginal revenue—that is, $C'(x) = R'(x)$.

4.3   First-Derivative Test

IN THIS SECTION   **increasing and decreasing functions, the first-derivative test, curve sketching with the first derivative**

INCREASING AND DECREASING FUNCTIONS

Suppose an ecologist has modeled the size of a population of a certain species as a function f of time t (months). If it turns out that the population is increasing until the end of the first year and is decreasing thereafter, it is reasonable to expect the population to be maximized at time $t = 12$ and for the population curve to have a high

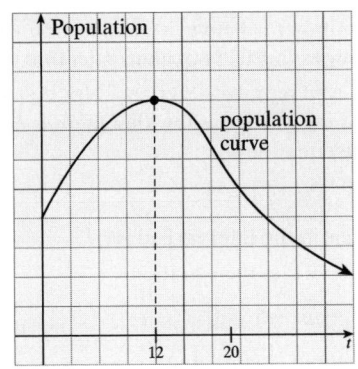

■ **FIGURE 4.15** A population curve

point at $t = 12$, as shown in Figure 4.15. If the graph of a function f, such as this population curve, is rising throughout the interval $0 < x < 12$ (and never flattens out on that interval), we say that f is *strictly increasing* on that interval. Similarly, the graph of the function in Figure 4.15 is *strictly decreasing* on the interval $12 < t < 20$. These terms may be defined more formally as follows:

Strictly Increasing and Strictly Decreasing

The function f is **strictly increasing** on an interval I if

$$f(x_1) < f(x_2) \quad \text{whenever} \quad x_1 < x_2$$

for x_1 and x_2 on I. Likewise, f is **strictly decreasing** on I if

$$f(x_1) > f(x_2) \quad \text{whenever} \quad x_1 < x_2$$

for x_1 and x_2 on I. (See Figure 4.16.)

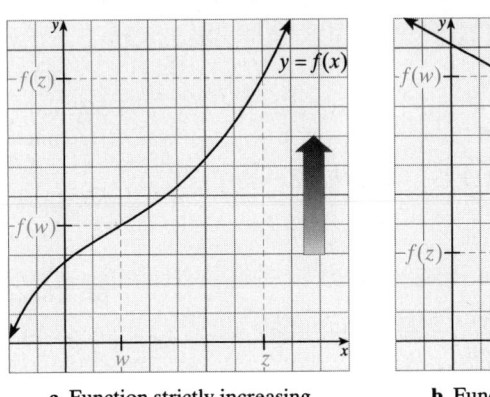
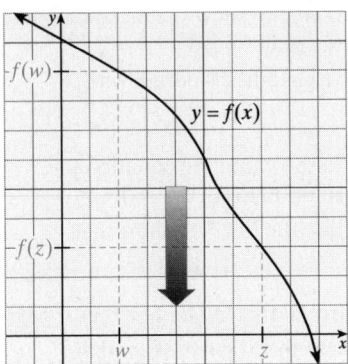

a. Function strictly increasing   **b.** Function strictly decreasing

■ **FIGURE 4.16** Increasing and decreasing functions

A function f is said to be (strictly) **monotonic** on an interval I if it is either strictly increasing on all of I or strictly decreasing on I. Monotonic behavior is closely related to the sign of the derivative $f'(x)$. In particular, if the graph of a function has tangent lines with positive slope on I, the graph will be inclined upward and f will be increasing on I (see Figure 4.17). Since the slope of the tangent at each

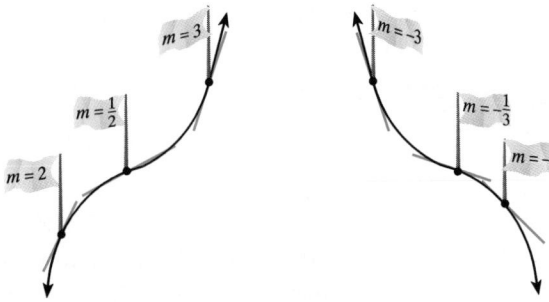

■ **FIGURE 4.17** The graph is rising where $f' > 0$ and falling where $f' < 0$. Notice that the small flags indicate the slope at various points on the graph.

point on the graph is measured by the derivative f', it is reasonable to expect f to be increasing on intervals where $f' > 0$. Similarly, it is reasonable to expect f to be decreasing on an interval when $f' < 0$. These observations are established formally in Theorem 4.7.

THEOREM 4.7   *Monotone function theorem*

Let f be differentiable on the open interval (a, b).

If $f'(x) > 0$ on (a, b), then f is strictly increasing on (a, b)

and

If $f'(x) < 0$ on (a, b), then f is strictly decreasing on (a, b).

Proof   We shall prove that f is strictly increasing on (a, b) if $f'(x) > 0$ throughout the interval. The strictly decreasing case is similar and is left as an exercise for the reader.

Suppose $f'(x) > 0$ throughout the interval (a, b), and let x_1 and x_2 be two numbers chosen arbitrarily from this interval, with $x_1 < x_2$. The MVT tells us that

$$\frac{f(x_2) - f(x_1)}{x_2 - x_1} = f'(c) \quad \text{or} \quad f(x_2) - f(x_1) = f'(c)(x_2 - x_1)$$

for some number c between x_1 and x_2. Because both $f'(c) > 0$ and $x_2 - x_1 > 0$, it follows that $f'(c)(x_2 - x_1) > 0$, and therefore

$$f(x_2) - f(x_1) > 0 \quad \text{or} \quad f(x_2) > f(x_1)$$

That is, if x_1 and x_2 are any two numbers in (a, b) such that $x_1 < x_2$, then $f(x_1) < f(x_2)$, which means that f is strictly increasing on (a, b).

To determine where a function f is increasing or decreasing, we begin by finding the critical numbers (where the derivative is zero or does not exist). These numbers divide the x-axis into intervals, and we test the sign of $f'(x)$ in each of these intervals. Finally, if $f'(x) > 0$ in an interval, then f is increasing in that same interval; and if $f'(x) < 0$ in an interval, then f is decreasing in that same interval. This procedure is illustrated in the following example.

EXAMPLE 1   *Finding intervals of increase and decrease*

Determine where the function defined by $f(x) = x^3 - 3x^2 - 9x + 1$ is strictly increasing and where it is strictly decreasing.

Solution   First, we find the derivative:

$$f'(x) = 3x^2 - 6x - 9 = 3(x + 1)(x - 3)$$

Next, we determine the critical numbers: $f'(x)$ exists for all x and $f'(x) = 0$ at $x = -1$ and $x = 3$ (by inspection). These critical numbers divide the x-axis into three parts, as shown in Figure 4.18a, and we select a typical number from each of these intervals. For example, we select -2, 0, and 4, evaluate the derivative at these numbers, and mark each interval as increasing ($\uparrow$) or decreasing ($\downarrow$), according to whether the derivative is positive or negative, respectively. This is shown in Figure 4.18b.

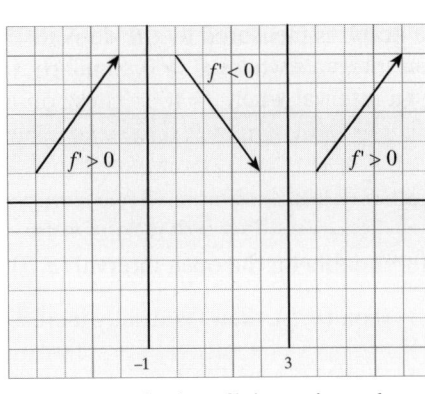

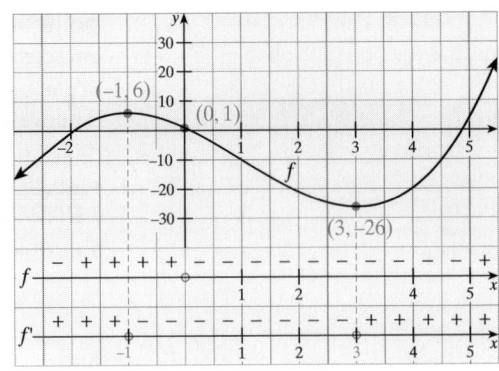

a. Intervals where f is increasing and where it is decreasing

b. The graph of f along with the sign graphs for f and f'

■ **FIGURE 4.18** $f(x) = x^3 - 3x^2 - 9x + 1$

The function f is increasing for $x < -1$ and for $x > 3$; f is decreasing for $-1 < x < 3$.   ■

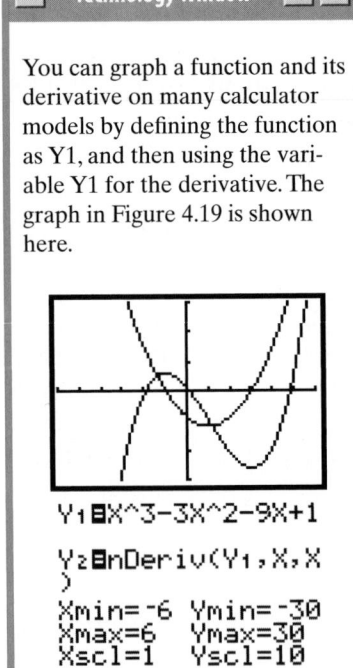
EXAMPLE 2   *Comparison of the graphs of a function and its derivative*

Graph $f(x) = x^3 - 3x^2 - 9x + 1$ and $f'(x) = 3x^2 - 6x - 9$ and compare.

a. When $f'(x) > 0$, what can be said about the graph of f?
b. When the graph of f is falling, what can be said about the graph of f'?
c. Where do the critical numbers of f appear on the graph of f?

Solution   The graphs of f and f' are shown in Figure 4.19.

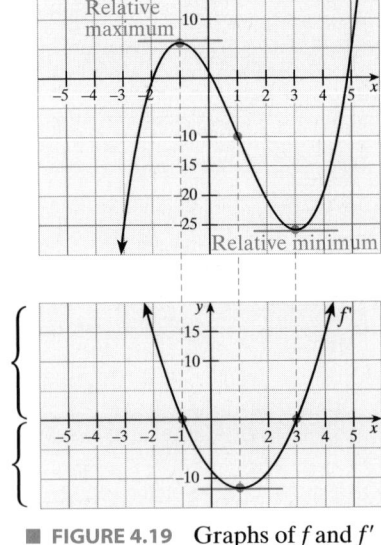

a. When $f'(x) > 0$, the graph of f is rising.

b. When the graph of f is falling, we have $f'(x) < 0$, so the graph of f' is below the x-axis.

■ **FIGURE 4.19**   Graphs of f and f'

c. The critical numbers of f are where $f'(x) = 0$, so they are the x-intercepts of the graph of f'.   ■

THE FIRST-DERIVATIVE TEST

Every relative extremum is a critical point. However, as we saw in Section 4.1, not every critical point is necessarily a relative extremum. If the derivative is positive to

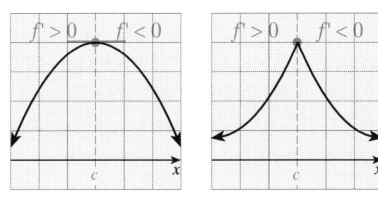

a. A relative maximum

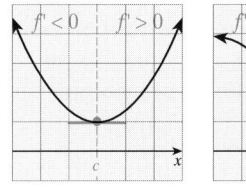
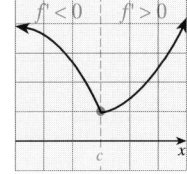

b. A relative minimum

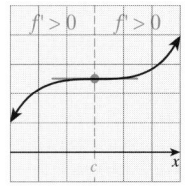
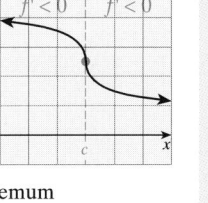

c. No extremum

■ **FIGURE 4.20** Three patterns of behavior near a critical number

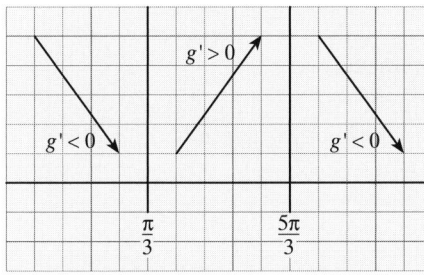

a. Intervals where $g(t) = t - 2 \sin t$ is increasing or decreasing

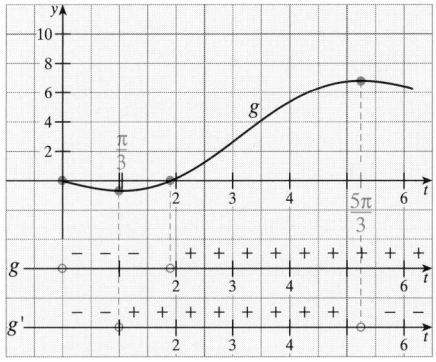

b. The graph of $g(t) = t - 2 \sin t$ for $0 \le t \le 2\pi$

■ **FIGURE 4.21** The first-derivative test for $g(t) = t - 2 \sin t$

the left of a critical number and negative to its right, the graph changes from increasing to decreasing and the critical point must be a relative maximum, as shown in Figure 4.20a. If the derivative is negative to the left of a critical number and positive to its right, the graph changes from decreasing to increasing and the critical point is a relative minimum (Figure 4.20b). However, if the sign of the derivative is the same on both sides of the critical number, then it is neither a relative maximum nor a relative minimum (Figure 4.20c). These observations are summarized in a procedure called the *first-derivative test for relative extrema.*

The First-Derivative Test for Relative Extrema

Step 1. Find all critical numbers of f. That is, find all numbers c such that $f(c)$ is defined and either $f'(c) = 0$ or $f'(c)$ does not exist.

Step 2. Classify each critical point $(c, f(c))$ as follows:

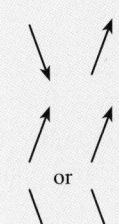

a. The point $(c, f(c))$ is a **relative maximum** if $f'(x) > 0$ (rising) for all x in an open interval (a, c) to the left of c, and $f'(x) < 0$ (falling) for all x in an open interval (c, b) to the right of c.

b. The point $(c, f(c))$ is a **relative minimum** if $f'(x) < 0$ (falling) for all x in an open interval (a, c) to the left of c, and $f'(x) > 0$ (rising) for all x in an open interval (c, b) to the right of c.

or

c. The point $(c, f(c))$ is **not an extremum** if the derivative $f'(x)$ has the same sign in open intervals (a, c) and (c, b) on both sides of c.

Suppose we apply this first-derivative test to the polynomial

$$f(x) = x^3 - 3x^2 - 9x + 1$$

In Example 2 we found that this function has the critical numbers -1 and 3, and that f is increasing when $x < -1$ and $x > 3$ and decreasing when $-1 < x < 3$ (see the arrow pattern above). The first-derivative test tells us there is a relative maximum at -1 ($\uparrow \downarrow$ pattern) and a relative minimum at 3 ($\downarrow \uparrow$ pattern).

EXAMPLE 3   *Relative extrema using the first-derivative test*

Find all critical numbers of $g(t) = t - 2 \sin t$ for $0 \le t \le 2\pi$, and determine whether each corresponds to a relative maximum, a relative minimum, or neither.

Solution   Because $g'(t) = 1 - 2 \cos t$ exists for all t, the only critical numbers occur when $g'(t) = 0$, that is, when $\cos t = \frac{1}{2}$. Solving, we find that the critical numbers for $g(t)$ on the interval $[0, 2\pi]$ are $\frac{\pi}{3}$ and $\frac{5\pi}{3}$.

Next, we examine the sign of $g'(t)$. Because $g'(t)$ is continuous, it is enough to check the sign of $g'(t)$ at convenient numbers on each side of the critical numbers, as shown in Figure 4.21a. Notice that the arrows show the increasing and decreasing pattern for g. According to the first-derivative test, there is a relative minimum at $\frac{\pi}{3}$ and a relative maximum at $\frac{5\pi}{3}$. The graph of g is shown in Figure 4.21b.   ■

CURVE SKETCHING WITH THE FIRST DERIVATIVE

The ability to draw a quick and accurate sketch is one of the most important skills you can learn in mathematics. It has been said that René Descartes revolutionized *thinking* when he introduced the rectangular coordinate system. Today, nearly every aspect of mathematics, engineering, physics, industry, education, and the social sciences is enhanced by the use of graphs.

In calculus, we use graphs to analyze problems and we use analytic methods to help us understand graphs. Plotting points, although useful in introducing a curve, is a very poor method for drawing a graph. As we have seen, there are many properties of curves that require calculus, and those properties could easily be overlooked if we relied only on graphing technology. Software and graphing calculators are helpful but incomplete ways of visualizing mathematics. For example, the relatively simple function $y = \sin\left(\dfrac{x}{50\pi}\right)$ is difficult to sketch using most graphing calculators. We now consider some graphing techniques that should help you to sketch curves quickly and accurately.

The First-Derivative Procedure for Sketching the Graph of a Continuous Function

Step 1. Compute the derivative $f'(x)$ and determine the critical numbers of f—that is, where $f'(x) = 0$ or $f'(x)$ does not exist.

Step 2. Substitute each critical number into $f(x)$ to find the y-coordinate of the corresponding critical point. Plot these critical points on a coordinate plane.

Step 3. Determine where the function is increasing or decreasing by checking the sign of the derivative on the intervals whose endpoints are the critical numbers found in step 2.

Step 4. Sketch the graph so that it rises on the intervals where $f'(x) > 0$, falls on the intervals where $f'(x) < 0$, passes through the critical points, and has a horizontal tangent where $f'(x) = 0$.

EXAMPLE 4   *Sketching a polynomial using the first-derivative procedure*

Sketch the graph of $f(x) = 2x^3 + 3x^2 - 12x - 5$.

Solution   We begin by computing and factoring the derivative:

$$f'(x) = 6x^2 + 6x - 12 = 6(x + 2)(x - 1)$$

We see that $f'(x)$ exists for all x, and from the factored form of the derivative we see that $f'(x) = 0$ when $x = -2$ and when $x = 1$. The corresponding critical points are found as follows:

$$f(-2) = 2(-2)^3 + 3(-2)^2 - 12(-2) - 5 = 15; \quad \text{critical point } (-2, 15)$$

$$f(1) = 2(1)^3 + 3(1)^2 - 12(1) - 5 = -12; \quad \text{critical point } (1, -12)$$

Next, to find the intervals of increase and decrease of the function, we plot the critical numbers on a number line and check the sign of the derivative at values to the left and right of -2 and 1. We find that f is increasing and decreasing, as indicated in Figure 4.22.

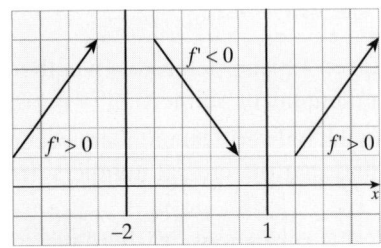

■ **FIGURE 4.22** Intervals of increase and decrease for $f(x) = 2x^3 + 3x^2 - 12x - 5$

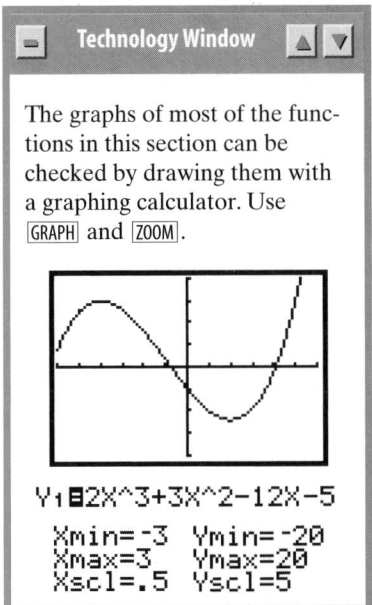
The arrow pattern in Figure 4.22 suggests that the graph of f has a relative maximum at $(-2, 15)$ and a relative minimum at $(1, -12)$. We begin the sketch by plotting these points on a coordinate plane. We put a "cap" at the relative maximum $(-2, 15)$ and a "cup" at the relative minimum $(1, -12)$, as shown in Figure 4.23a. Finally, we note that because $f(0) = -5$, the graph has its y-intercept at $(0, -5)$. We complete the sketch by drawing the curve so that it increases for $x < -2$ to the relative maximum at $(-2, 15)$, decreases for $-2 < x < 1$, passing through the y-intercept $(0, -5)$ on its way to the relative minimum at $(1, -12)$, and then increases again for $x > 1$. The completed graph is shown in Figure 4.23b. ∎

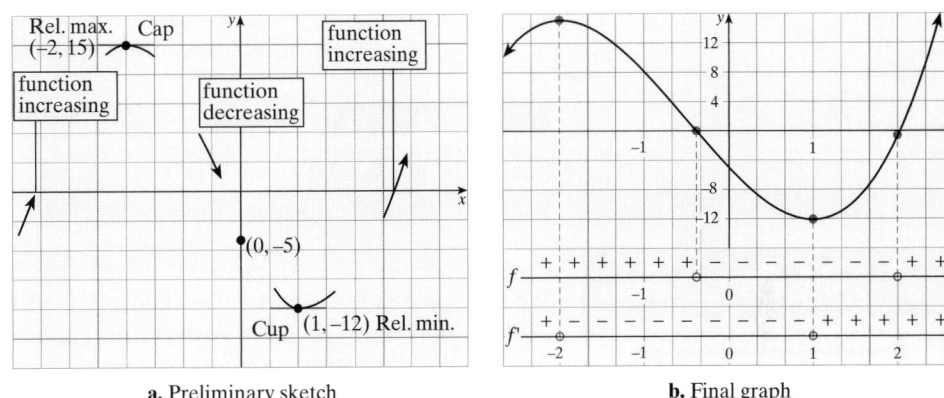

a. Preliminary sketch   **b.** Final graph

■ **FIGURE 4.23** The graph of $f(x) = 2x^3 + 3x^2 - 12x - 5$

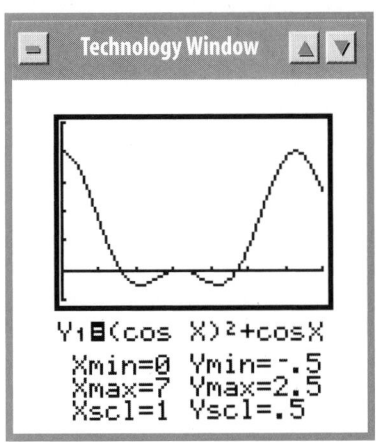
EXAMPLE 5   *Sketching a trigonometric function using the first-derivative procedure*

Sketch the graph of $f(x) = \cos^2 x + \cos x$ on $[0, 2\pi]$.

Solution The derivative of f is

$$f'(x) = -2 \cos x \sin x - \sin x$$

Because $f'(x)$ exists for all x, we solve $f'(x) = 0$ to find the critical numbers:

$$-2 \cos x \sin x - \sin x = 0$$
$$\sin x(2 \cos x + 1) = 0$$
$$\sin x = 0 \quad \text{or} \quad \cos x = -\tfrac{1}{2}$$

The critical numbers on the interior of the interval $[0, 2\pi]$ are $x = \frac{2\pi}{3}, \pi,$ and $\frac{4\pi}{3}$. Because $f(\frac{2\pi}{3}) = -\frac{1}{4}, f(\pi) = 0$, and $f(\frac{4\pi}{3}) = -\frac{1}{4}$, the corresponding critical points are $(\frac{2\pi}{3}, -\frac{1}{4}), (\pi, 0),$ and $(\frac{4\pi}{3}, -\frac{1}{4})$.

Next, we plot the critical numbers on a number line and determine the sign of the derivative $f'(x)$ in each interval, as indicated in Figure 4.24.

The arrow pattern in the diagram indicates that we have relative minima at $(\frac{2\pi}{3}, -\frac{1}{4})$ and $(\frac{4\pi}{3}, -\frac{1}{4})$, and a relative maximum at $(\pi, 0)$. We plot these points on a coordinate plane and place "cups" at the relative minima and a "cap" at the relative maximum. Finally, we sketch a smooth curve through these points, obtaining the graph shown in Figure 4.25. If the curve is defined over a particular interval, then you should also plot the endpoints.

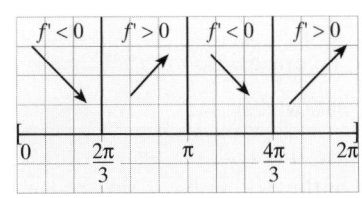

■ **FIGURE 4.24** Intervals of increase and decrease for $f(x) = \cos^2 x + \cos x$

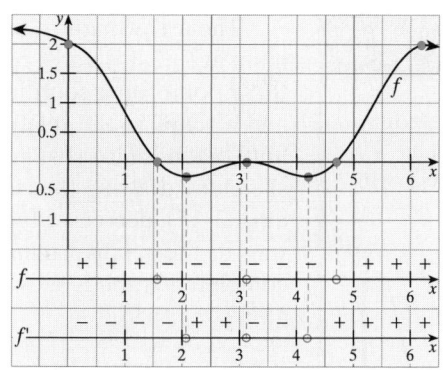

■ **FIGURE 4.25**   The graph of $f(x) = \cos^2 x + \cos x$

> **EXAMPLE 6**   *Sketching a graph using the first-derivative procedure*

Sketch the graph of $f(x) = x^{1/3}(x - 4)$.

Solution   Rewrite the given function as $f(x) = x^{4/3} - 4x^{1/3}$ and differentiate:

$$f'(x) = \tfrac{4}{3}x^{1/3} - \tfrac{4}{3}x^{-2/3} = \tfrac{4}{3}x^{-2/3}(x - 1)$$

We see that $f'(x)$ does not exist when $x = 0$ and that $f'(x) = 0$ only when $x = 1$. Because $f(0) = 0$ and $f(1) = -3$, the critical points are $(0, 0)$ and $(1, -3)$.

     We plot the critical numbers on a number line and determine the sign of $f'(x)$ for each interval determined by the critical numbers, as shown in Figure 4.26.

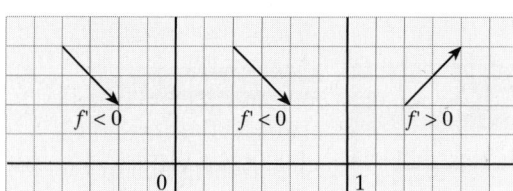

■ **FIGURE 4.26**   Intervals of increase and decrease for $f(x) = x^{1/3}(x - 4)$

     We see that there is a relative minimum at $(1, -3)$ and neither kind of extremum at $(0, 0)$. We plot the critical points on a coordinate plane with a "cup" at $(1, -3)$, as shown in Figure 4.27a. Finally, we pass a smooth curve through these points, obtaining the graph shown in Figure 4.27b. Note that at $(0, 0)$ the curve looks as if it has a vertical tangent.

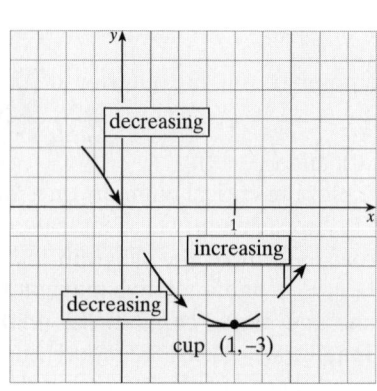

a. Preliminary sketch

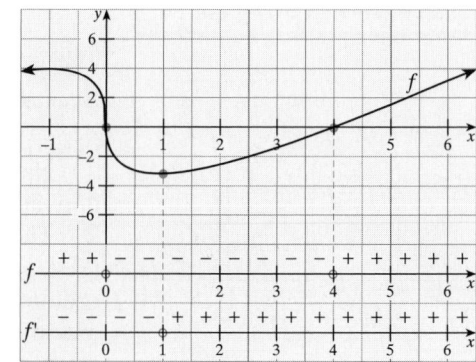

b. Final graph

■ **FIGURE 4.27**   The graph of $f(x) = x^{1/3}(x - 4)$

Technology Window

```
Y₁▤X^(1/3)(X-4)

Xmin=-10  Ymin=-10
Xmax=10   Ymax=10
Xscl=1    Yscl=1
```

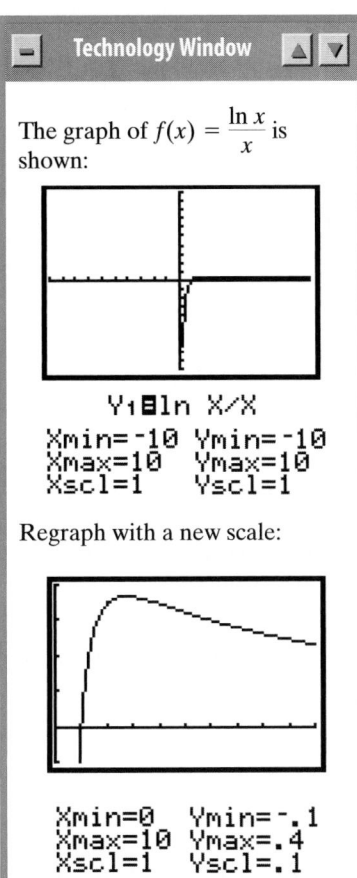

The graph of $f(x) = \dfrac{\ln x}{x}$ is shown:

Y₁▤ln X/X
Xmin=-10 Ymin=-10
Xmax=10 Ymax=10
Xscl=1 Yscl=1

Regraph with a new scale:

Xmin=0 Ymin=-.1
Xmax=10 Ymax=.4
Xscl=1 Yscl=.1

WARNING ▶ In Section 4.5, we shall develop a general procedure for finding limits such as

$$\lim_{x \to +\infty} \frac{\ln x}{x}$$

For now, we rely on calculator accuracy. ◀

EXAMPLE 7   *Graphing a function involving a logarithmic function*

Determine where the function $f(x) = (\ln x)/x$ is increasing and where it is decreasing. Sketch the graph of f.

Solution   First of all, note that f is defined only for $x > 0$ and that

$$f'(x) = \frac{x(1/x) - (\ln x) \cdot 1}{x^2} = \frac{1 - \ln x}{x^2}$$

Because $x^2 > 0$ for all $x > 0$, it follows that $f'(x) = 0$ only when $\ln x = 1$; thus, e is the only critical number of f. We use the first-derivative test and look at the sign of $f'(x)$ to the left of e (at $x = 1$, for instance) and to the right (at $x = 3$, for example). We see that

$$f'(x) > 0 \quad \text{for } x < e \text{ and}$$
$$f'(x) < 0 \quad \text{for } x > e.$$

Thus, f is increasing to the left of e and decreasing to the right, as shown in Figure 4.28a.

We now have most of the information we need to sketch the graph of f. In particular, we know that the graph lies entirely to the right of the y-axis (because f is defined only for $x > 0$) and that it rises to a relative maximum at $x = e$, after which it falls indefinitely. Moreover, because we obtain

$$\frac{\ln x}{x} = 0 \quad \text{only for } x = 1,$$

it follows that the graph crosses the x-axis on the way up but not on the way down, and this causes us to suspect that the graph must "flatten out" in some way as $x \to +\infty$ (see Figure 4.28b).

But what happens at the two "ends" of the graph, as $x \to 0^+$ and as $x \to +\infty$? As x approaches 0 from the right, $\ln x$ decreases without bound and $f(x) = (\ln x)/x$ does the same, so the graph approaches the negative y-axis asymptotically. As x increases without bound, we have seen that $f(x)$ decreases, but by how much? The graph cannot cross the x-axis a second time because $\ln x = 0$ only for $x = 1$, so the graph must "flatten out" in some way (see Figure 4.28b). Using a calculator, we find that

$$\lim_{x \to +\infty} \frac{\ln x}{x} = 0$$

so the graph must approach the x-axis asymptotically as $x \to +\infty$. The completed graph is shown in Figure 4.28c.

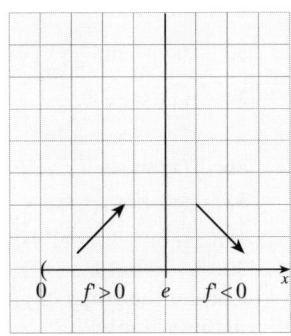

a. Tendency diagram for $f(x) = \dfrac{\ln x}{x}$

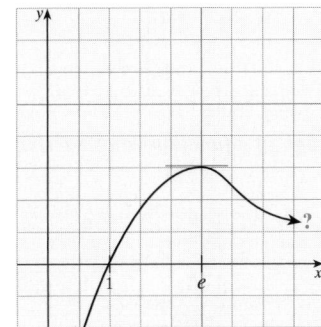

b. The graph rises to a maximum at e and then "flattens out" as it falls.

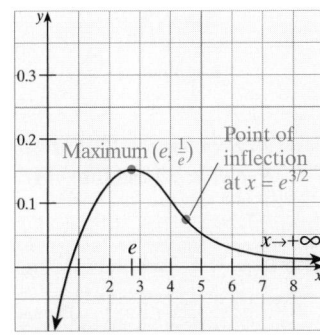

c. Further analysis shows that the graph approaches the x-axis as $x \to +\infty$.

■ **FIGURE 4.28**   The graph of $f(x) = \dfrac{\ln x}{x}$ ■

4.3 Problem Set

1. ■ **What Does This Say?** What is the first-derivative test?

2. ■ **What Does This Say?** What is the relationship between the graph of a function and the graph of its derivative?

In Problems 3–4, identify which curve represents a function $y = f(x)$ and which curve represents its derivative $y = f'(x)$.

3. 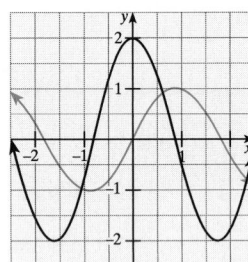 **4.**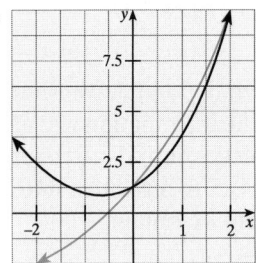

Draw a curve that represents the derivative of the function defined by the curves shown in Problems 5–8.

5. 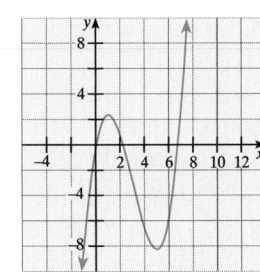 **6.**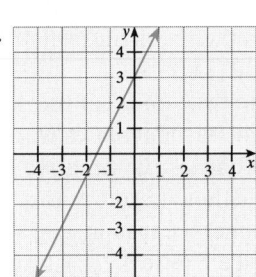

7. 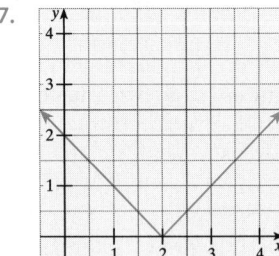 **8.**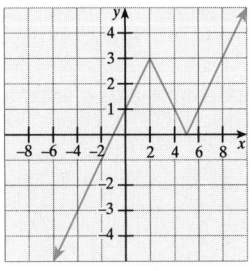

In Problems 9–12, draw the graph of a function whose derivative matches the shown graph.

9. 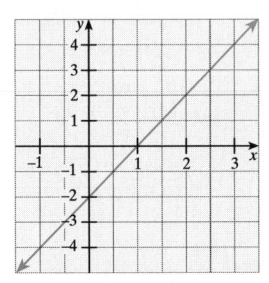 **10.**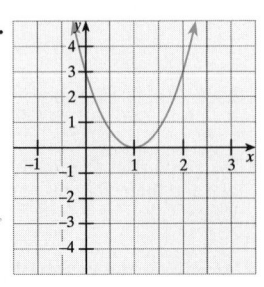

11. 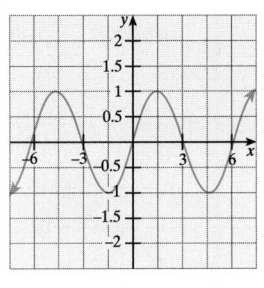 **12.**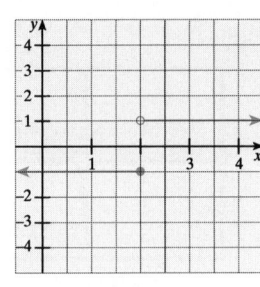

In Problems 13–38,
  a. *find the critical numbers;*
  b. *show where the function is increasing and where it is decreasing;*
  c. *plot each critical point and label it as a relative maximum, a relative minimum, or neither;*
  d. *sketch the graph.*

13. $f(x) = x^3 + 3x^2 + 1$ **14.** $f(x) = \frac{1}{3}x^3 - 9x + 2$

15. $f(x) = x^3 + 35x^2 - 125x - 9,375$

16. $f(x) = x^5 + 5x^4 - 550x^3 - 2,000x^2 + 60,000x$

17. $f(x) = x^5 - 5x^4 + 100$

18. $f(x) = x + \frac{1}{x}$ **19.** $f(x) = \frac{x - 1}{x^2 + 3}$

20. $f(x) = x \ln x$ **21.** $f(t) = (t + 1)^2(t - 5)$

22. $f(t) = (2t - 1)^2(t^2 - 9)$

23. $f(x) = (x - 3)(x - 7)(2x + 1)$

24. $f(x) = 16(x - 3)^2(x - 7)^2$

25. $f(x) = \sqrt{x^2 + 1}$ **26.** $f(x) = \sqrt{x^2 - 2x + 2}$

27. $g(x) = x^{2/3}(2x - 5)$ **28.** $g(x) = x^{1/3}\sqrt{x + 15}$

29. $f(\theta) = 2\cos\theta - \theta$ on $[0, 2\pi]$

30. $c(\theta) = \theta + \cos 2\theta$ for $0 \le \theta \le \pi$

31. $t(x) = \tan^2 x$ for $-\frac{\pi}{4} \le x \le \frac{\pi}{4}$

32. $f(x) = 9\cos x - 4\cos^2 x$ on $[0, \pi]$

33. $f(x) = \sin\left(\frac{x}{50\pi}\right)$ on $[0, 100\pi^2]$

34. $f(x) = xe^{-x}$

35. $f(x) = x + 2\tan^{-1} x$

36. $f(x) = \sin^{-1} x - \ln x$ on $(0, 1]$

37. $f(x) = x^3 - \cos x$

38. $f(x) = x^5 + \sin x^2$

39. ■ **What Does This Say?** For n a nonnegative integer, consider the graph of $f(x) = (x - 2)^n$. What is the effect of n on these graphs?

In Problems 40–43, determine whether the given function has a relative maximum, a relative minimum, or neither at the given critical numbers(s).

40. $f(x) = (x^3 - 3x + 1)^7$ at $x = 1, x = -1$

41. $f(x) = \dfrac{e^{-x^2}}{3 - 2x}$ at $x = 1, x = \dfrac{1}{2}$

42. $f(x) = (x^2 - 4)^4(x^2 - 1)^3$ at $x = 1, x = 2$

43. $f(x) = \sqrt[3]{x^3 - 48x}$ at $x = 4$

44. Suppose f is a differentiable function with derivative

$$f'(x) = (x - 1)^2(x - 2)(x - 4)(x + 5)^4$$

Find all critical numbers of f and determine whether each corresponds to a relative maximum, a relative minimum, or neither.

45. Suppose f is a differentiable function with derivative

$$f'(x) = \dfrac{(2x - 1)\ln(2x^2 - 3x + 2)}{(x - 2)^2}$$

Find all critical numbers of f and determine whether each corresponds to a relative maximum, a relative minimum, or neither.

46. Sketch a graph of a function f that is differentiable on the interval $[-1, 4]$ and satisfies the following conditions:
(i) The function f is decreasing on $(1, 3)$ and increasing elsewhere on $[-1, 4]$.
(ii) The largest value of f is 5 and the smallest is 0.
(iii) The graph of f has relative extrema at $(1, 5)$ and $(3, 4)$.

47. Sketch a graph of a function f that satisfies the following conditions:
(i) $f'(x) > 0$ when $x < -5$ and when $x > 1$.
(ii) $f'(x) < 0$ when $-5 < x < 1$.
(iii) $f(-5) = 4$ and $f(1) = -1$.

48. Sketch a graph of a function f that satisfies the following conditions:
(i) $f'(x) < 0$ when $x < -1$.
(ii) $f'(x) > 0$ when $-1 < x < 3$ and when $x > 3$.
(iii) $f'(-1) = 0$ and $f'(3) = 0$.

49. Find constants $a, b,$ and c such that the graph of $f(x) = ax^2 + bx + c$ has a relative maximum at $(5, 12)$ and crosses the y-axis at $(0, 3)$.

50. Use calculus to prove that the relative extremum of the quadratic function

$$y = ax^2 + bx + c \quad (a \neq 0)$$

occurs at $x = -b/(2a)$.

51. Use calculus to prove that the relative extremum of the quadratic function

$$f(x) = (x - p)(x - q)$$

occurs midway between its x-intercepts.

52. MODELING PROBLEM   In physics, the formula

$$I = I_0\left(\dfrac{\sin\theta}{\theta}\right)^2$$

is used to model light intensity in the study of Fraunhofer diffraction.
a. Show that $I(0) = I_0$.

b. Sketch the graph for $[-3\pi, 3\pi]$. What are the critical numbers on this interval?

53. MODELING PROBLEM   At a temperature of T (in degrees Celsius), the speed of sound in air is modeled by the formula

$$v = v_0\sqrt{1 + \tfrac{1}{273}T}$$

where v_0 is the speed at $0\,°C$. Sketch the graph of v for $T > 0$.

54. Find constants $a, b,$ and c that guarantee that the graph of

$$f(x) = x^3 + ax^2 + bx + c$$

will have a relative maximum at $(-3, 18)$ and a relative minimum at $(1, -14)$.

55. Find constants $A, B, C,$ and D that guarantee that the graph of

$$f(x) = 3x^4 + Ax^3 + Bx^2 + Cx + D$$

will have horizontal tangents at $(2, -3)$ and $(0, 7)$. There is a third point that has a horizontal tangent. Find this point. Then, for all three points, determine whether each corresponds to a relative maximum, a relative minimum, or neither.

56. Let

$$f(x) = (x - A)^m(x - B)^n$$

where A and B are real numbers, and m and n are positive integers with $m > 1$ and $n > 1$. Find the critical numbers of f.

57. Let

$$f(x) = x^{1/3}\sqrt{Ax + B}$$

where A and B are positive constants. Find all critical numbers of f and classify the corresponding value of f as a relative maximum, a relative minimum, or neither.

58. Suppose $f(x)$ and $g(x)$ are both continuous for all x and differentiable at $x = c$. Show that if c is a critical number for both f and g, then it is also a critical number for the product function fg. If a relative maximum occurs at c for both f and g, is it true that a relative maximum of fg occurs at c?

59. Show that if $f(x)$ is strictly increasing on an interval I where it is differentiable, then $f'(x) \geq 0$ for all x in I. *Hint:* You may use the fact that if $f(x) > 0$ for every x in an interval except perhaps for c, and if $\lim_{x \to c} f(x)$ exists, then $\lim_{x \to c} f(x) \geq 0$.

60. ■ What Does This Say?   Let f and g be strictly increasing functions on the interval $[a, b]$.
a. If $f(x) > 0$ and $g(x) > 0$ on $[a, b]$, show that the product fg is also strictly increasing on $[a, b]$.
b. If $f(x) < 0$ and $g(x) < 0$ on $[a, b]$, is fg strictly increasing, strictly decreasing, or neither? Explain.

61. ■ What Does This Say?
a. Suppose f is strictly increasing for $x > 0$. Let n be a positive integer and define $g(x) = x^n f(x)$ for $x > 0$. If

$f(x) > 0$ on an interval $[a, b]$ with $a > 0$, show that g is strictly increasing on the same interval.

b. Let $g(x) = x^n \sin x$ for $0 < x < \frac{\pi}{2}$, where n is a positive integer. Explain why g is strictly increasing on this interval.

62. Complete the proof of the monotone function theorem (Theorem 4.7). Specifically, show that if f is differentiable on (a, b) and $f'(x) < 0$ throughout the interval $[a, b]$, then f is strictly decreasing on $[a, b]$.

THINK TANK PROBLEMS   *Problems 63–69 are statements about continuous functions. In each case, either show that the statement is generally true or find a counterexample.*

63. If f and g both have a relative maximum at c, then so does $f + g$.

64. If f and g are both differentiable and f and g both have a relative extremum at c, then so does $f + g$.

65. If f is strictly increasing for $x > c$ and strictly decreasing for $x < c$, then $f'(c) = 0$.

66. If $f'(0) = 0$, $f'(-\frac{1}{2}) < 0$, and $f'(\frac{1}{2}) > 0$, then f has a relative minimum at $(0, f(0))$.

67. If f' is nonnegative on an interval I, then f is strictly increasing on I.

68. If f is strictly decreasing on an interval I, then $-f$ is strictly increasing on I.

69. If f and g are each strictly increasing on an interval I, then $f - g$ is also strictly increasing on I.

70. **JOURNAL PROBLEM**   *Mathematics Magazine.** Give an elementary proof that

$$f(x) = \frac{1}{\sin x} - \frac{1}{x}, \quad 0 < x \le \frac{\pi}{2}$$

is positive and increasing.

———

*Volume 55 (1982), p. 300. "Elementary proof" in the question means that you should use only techniques from beginning calculus. For our purposes, you simply need to give a reasonable argument to justify the conclusion.

4.4 Concavity and the Second-Derivative Test

IN THIS SECTION   concavity, inflection points, curve sketching with the second derivative, the second-derivative test for relative extrema

Knowing where a given graph is rising and falling gives only a partial picture of the graph. For example, suppose we wish to sketch the graph of $f(x) = x^3 + 3x + 1$. The derivative $f'(x) = 3x^2 + 3$ is positive for all x so the graph is always rising. But in what *way* is it rising? Each of the graphs in Figure 4.29 is a possible graph of f, but they are quite different from one another. Later in this section we shall show that the correct graph is the one in Figure 4.29c. The focus of our work is a characteristic of graphs called *concavity* that will allow us to distinguish among graphs such as these. In addition, we shall develop a *second-derivative test* that enables us to classify a critical point P of a function f as a relative maximum or a relative minimum by examining the sign of the second derivative f'' at P.

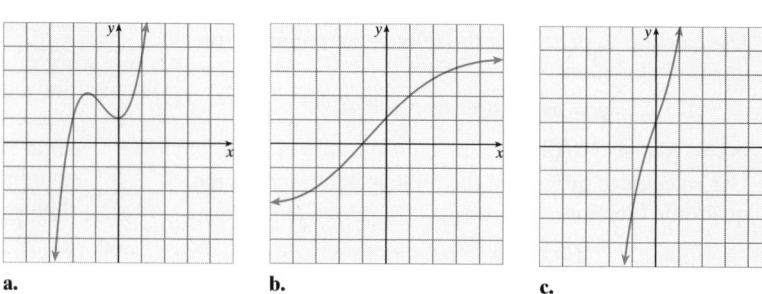

a.   b.   c.

■ **FIGURE 4.29**   Which curve is the graph of $f(x) = x^3 + 3x + 1$?

CONCAVITY

A portion of graph that is cupped upward is called *concave up,* and a portion that is cupped downward is *concave down.* Figure 4.30 shows a graph that is concave up between A and C and concave down between C and E. At various points on the graph,

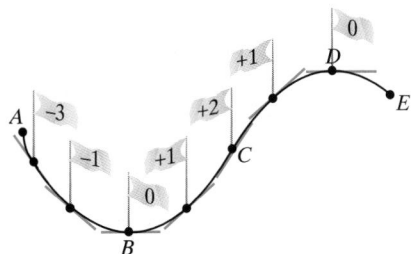

■ **FIGURE 4.30**   The slope of a graph increases or decreases, depending on its concavity

the slope is indicated by "flags," and we observe that the slope increases from A to C and decreases from C to E. This is no accident! *The slope of a graph increases on an interval where the graph is concave up and decreases where the graph is concave down.*

Conversely, a graph will be concave up on any interval where the slope is increasing and concave down where the slope is decreasing. Because the slope is found by computing the derivative, it is reasonable to expect the graph of a given function f to be concave up where the derivative f' is strictly increasing. According to the monotone function theorem (Theorem 4.7), this occurs when $(f')' > 0$, which means that the graph of f is concave up where the *second derivative f''* satisfies $f'' > 0$. Similarly, the graph is concave down where $f'' < 0$. We use this observation to *define* concavity.

Concavity

The graph of a function f is **concave upward** on any open interval I where $f''(x) > 0$, and it is **concave downward** where $f''(x) < 0$.

EXAMPLE 1   *Concavity for a polynomial function*

Find where the graph of $f(x) = x^3 + 3x + 1$ is concave up and where it is concave down.

Solution   We find that $f'(x) = 3x^2 + 3$ and $f''(x) = 6x$. Therefore, $f''(x) < 0$ if $x < 0$ and $f''(x) > 0$ if $x > 0$, so the graph of f is concave down for $x < 0$ and concave up for $x > 0$. Using this information, we can now answer the question asked in Figure 4.29—the correct graph is c.   ■

INFLECTION POINTS

In Figure 4.30, notice that the graph changes from concave up to concave down at the point C. It will be convenient to give a name to such a transition point.

Inflection Point

Suppose the graph of a function f has a tangent line (possibly vertical) at the point $P(c, f(c))$ and that the graph is concave up on one side of P and concave down on the other side. Then P is called an **inflection point** of the graph.

Returning to Example 1, notice that the graph of $f(x) = x^3 + 3x + 1$ has exactly one inflection point, at $(0, 1)$, where the concavity changes from down to up.

Various kinds of graphical behavior are illustrated in Figure 4.31. Note that the graph is rising on the interval $[a, c_1]$, falling on $[c_1, c_2]$, rising on $[c_2, c_3]$, falling on $[c_3, c_4]$, rising on $[c_4, c_5]$, and falling on $[c_5, b]$. The concavity is up on $[p_1, c_1]$, $[c_1, c_3]$, and $[c_3, p_2]$ and down otherwise.

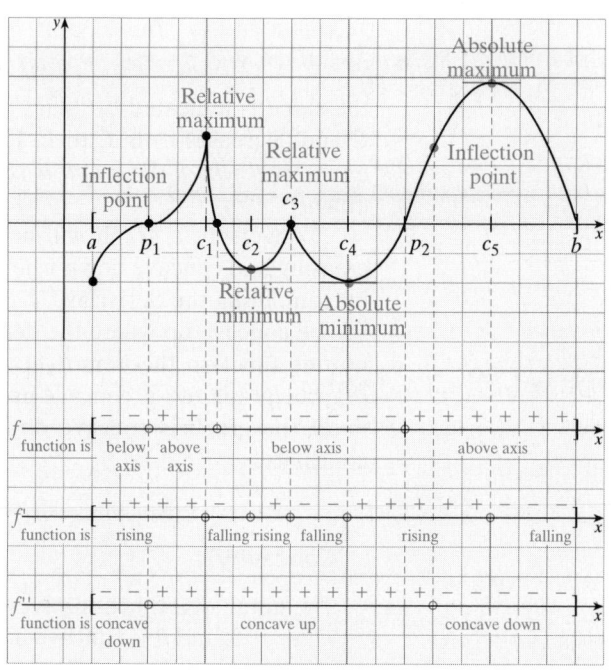

■ **FIGURE 4.31**   A graph of a function showing critical points and inflection points

The graph has relative maxima at c_1, c_3, and c_5, and relative minima at c_2 and c_4. There are horizontal tangents $(f'(x) = 0)$ at all of these points except c_1 and c_3, where there are sharp points ($f'(c_1)$ and $f'(c_3)$ do not exist). There is only a horizontal tangent at p_1, that is, $f'(p_1) = 0$, but no relative extremum appears there. Instead, we have points of inflection at p_1 and p_2, because the concavity changes direction at each of these points.

In general, the concavity of the graph of f will change only at points where $f''(x) = 0$ or $f''(x)$ does not exist—that is, at critical numbers of the derivative f'. We shall call the number c a **second-order critical number** if $f''(c) = 0$ or $f''(c)$ does not exist, and in this context an "ordinary" critical number (where $f'(x) = 0$ or $f'(x)$ does not exist) will be referred to as a **first-order critical number.** Inflection points correspond to second-order critical numbers and must actually be on the graph of f. Specifically, a number c such that $f''(c)$ is not defined and the concavity of f changes at c will correspond to an inflection point if and only if $f(c)$ is defined.

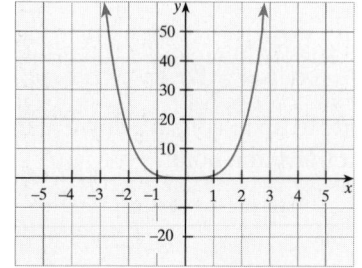

■ **FIGURE 4.32**   The graph of $f(x) = x^4$ has no inflection point at $(0, 0)$ even though $f''(0) = 0$.

WARNING   A continuous function f need not have an inflection point where $f'' = 0$. For instance, if $f(x) = x^4$, we have $f''(0) = 0$, but the graph of f is always concave up (see Figure 4.32). ←

EXAMPLE 2   *Concavity and inflection points*

Discuss the concavity of $f(x) = \dfrac{x^2}{2\sqrt{2}} + \sin x$ on the interval $\left[0, \dfrac{\pi}{2}\right]$ and find all the inflection points of f.

Solution   Find f' and f'':

$$f'(x) = \frac{x}{\sqrt{2}} + \cos x \qquad f''(x) = \frac{1}{\sqrt{2}} - \sin x$$

Both $f'(x)$ and $f''(x)$ are defined for all x on the interval $[0, \frac{\pi}{2}]$ and $f''(x) = 0$ when $\sin x = \frac{1}{\sqrt{2}}$. Thus, the only possible location for an inflection point is where $x = \frac{\pi}{4}$. Testing to the left and right of $\frac{\pi}{4}$, we see that the concavity changes (from up to down) at $x = \frac{\pi}{4}$, so a point of inflection occurs there, and this is the only such point on the interval $[0, \frac{\pi}{2}]$. The graph of f is shown in Figure 4.33.

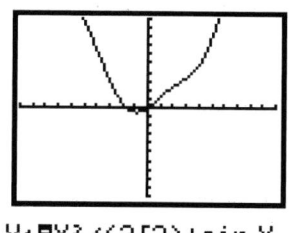
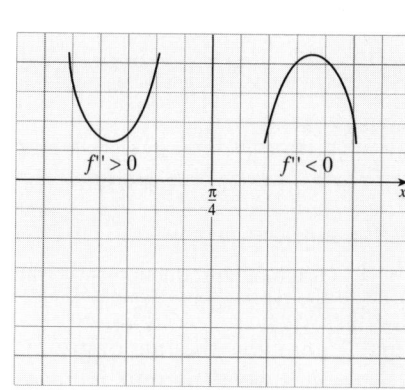

a. Concavity

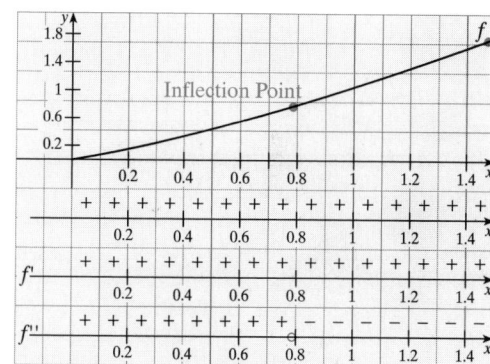

b. Graph

■ **FIGURE 4.33**   Concavity and graph of $f(x) = \dfrac{x^2}{2\sqrt{2}} + \sin x$ ■

WARNING   TECHNOLOGY TIP   In spite of the warning shown in the technology window in the margin, on some graphing calculator models you can find where $f''(x) = 0$ by graphing nDeriv(nDeriv(Y1,X,X),X,X) and then finding the zeros using the [ROOT] feature of your calculator.   ←

Here is an example of one way inflection points may appear in practical applications.

EXAMPLE 3   *Modeling worker efficiency*

An efficiency study of the morning shift at a factory indicates that the number of units produced by an average worker t hours after 8:00 A.M. may be modeled by the formula $Q(t) = -t^3 + 9t^2 + 12t$. At what time in the morning is the worker performing most efficiently?

Solution   We assume that the morning shift runs from 8:00 A.M. until noon and that worker efficiency is maximized when the rate of production

$$R(t) = Q'(t) = -3t^2 + 18t + 12$$

is as large as possible for $0 \le t \le 4$. The derivative of R is

$$R'(t) = Q''(t) = -6t + 18$$

which is zero when $t = 3$; this is the critical number. Using the optimization criterion of Section 4.1, we know that the extrema of $R(t)$ on the closed interval $[0, 4]$ must occur at either the interior critical number 3 or at one (or both) of the endpoints (which are 0 and 4). We find that

$$R(0) = 12 \qquad R(3) = 39 \qquad R(4) = 36$$

so the rate of production $R(t)$ is greatest and the worker is performing most efficiently when $t = 3$, that is, at 11:00 A.M. The graphs of the production function Q and its derivative, the rate-of-production function R, are shown in Figure 4.34. Notice that the production curve is steepest and the rate of production is greatest when $t = 3$.

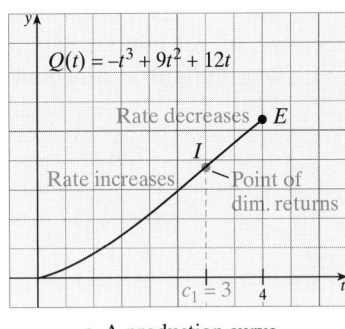
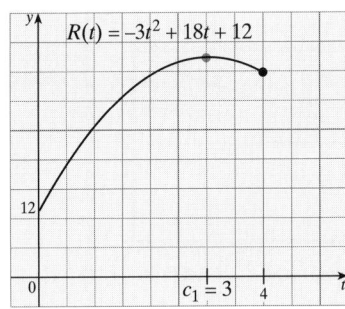

a. A production curve   **b.** Rate of production

■ **FIGURE 4.34**   Graph of a production curve showing the point of diminishing returns ■

In Example 3, note how the rate of production, as measured by the slope of the graph of the average worker's output, increases from 0 to the inflection point I and then decreases from I to E, as shown in Figure 4.34a. Because the point I marks the point where the rate of production "peaks out," it is natural to refer to I as a point of **diminishing returns.** It is also an inflection point on the graph of Q. Knowing that this point occurs at 11.00 A.M., the manager of the factory might be able to increase the overall output of the labor force by scheduling a break near this time.

CURVE SKETCHING WITH THE SECOND DERIVATIVE

In Section 4.3, we learned how to find where a graph is rising and where it is falling by examining the sign of the first derivative. Now, by examining the sign of the second derivative, we can determine the concavity of the graph, which gives us a more refined picture of the graph's appearance.

> **EXAMPLE 4**   *Sketching the graph of a polynomial function*

Determine where the function $f(x) = x^4 - 4x^3 + 10$ is increasing and decreasing and where its graph is concave up and concave down. Find the relative extrema and inflection points and sketch the graph of f.

Solution   The first derivative,

$$f'(x) = 4x^3 - 12x^2 = 4x^2(x - 3)$$

is zero when $x = 0$ and when $x = 3$. Because $4x^2 > 0$ for $x \neq 0$, we have $f'(x) < 0$ for $x < 3$ (except for $x = 0$) and $f'(x) > 0$ for $x > 3$. The pattern showing where f is increasing and where it is decreasing is displayed in Figure 4.35a.

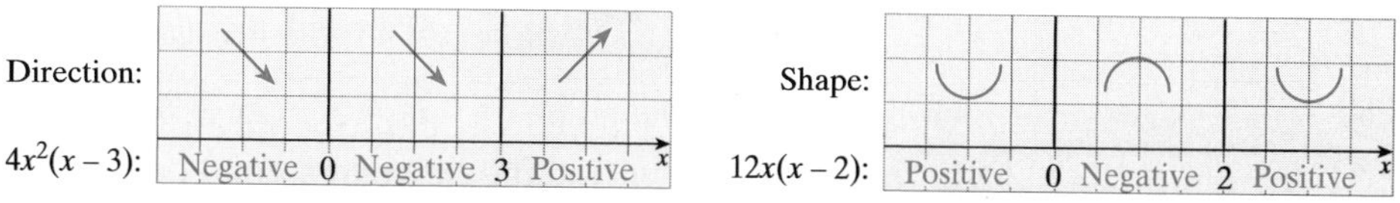

Direction:

$4x^2(x-3)$: Negative 0 Negative 3 Positive

a. First derivative signs.

Shape:

$12x(x-2)$: Positive 0 Negative 2 Positive

b. Second derivative signs.

■ **FIGURE 4.35**   First and second derivatives of $f(x) = x^4 - 4x^3 + 10$

Next, to determine the concavity of the graph we compute

$$f''(x) = 12x^2 - 24x = 12x(x - 2)$$

If $x < 0$ or $x > 2$, then $f''(x) > 0$ and the graph is concave up. It is concave down when $0 < x < 2$, because $f''(x) < 0$ on this interval. The concavity of the graph of f is shown in Figure 4.35b.

The two diagrams in Figure 4.35 tell us that there is a relative minimum at $x = 3$ and inflection points at $x = 0$ and $x = 2$ (because the second derivative changes sign at these points).

To find the y-values of the critical points and the inflection points, evaluate f at $x = 0, 2,$ and 3:

$$f(0) = (0)^4 - 4(0)^3 + 10 = 10$$
$$f(2) = (2)^4 - 4(2)^3 + 10 = -6$$
$$f(3) = (3)^4 - 4(3)^3 + 10 = -17$$

Finally, to sketch the graph of f, we first place a "cup" ($\cup$) at the minimum point $(3, -17)$ and note that $(0, 10)$ and $(2, -6)$ are inflection points; remember there is also a horizontal tangent at $(0, 10)$. The preliminary graph is shown in Figure 4.36a. Complete the sketch by passing a smooth curve through these points, using the two diagrams in Figure 4.35 as a guide for determining where the graph is rising and falling and where it is concave up and down. The completed graph is shown in Figure 4.36b.

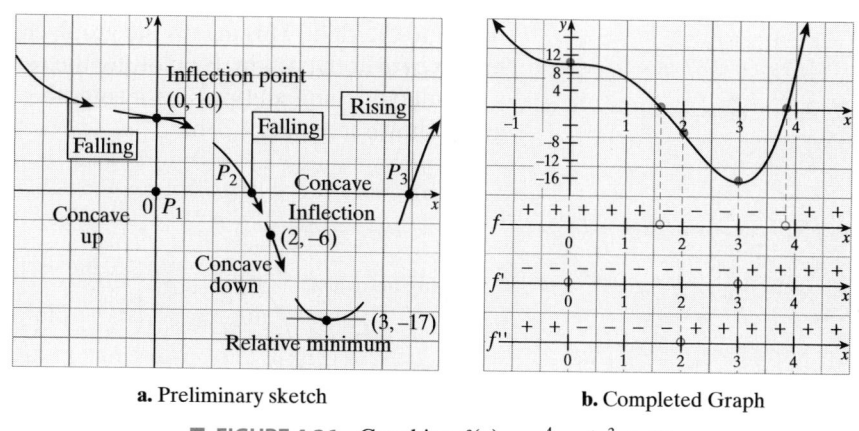

a. Preliminary sketch

b. Completed Graph

■ **FIGURE 4.36**   Graphing $f(x) = x^4 - 4x^3 + 10$ ■

A point on a graph where the curve makes an abrupt change in direction is called a **cusp.** Our next example features a graph with such a point.

| EXAMPLE 5 | *Sketching a graph with a cusp* |

Sketch the graph of $f(x) = x^{2/3}(2x + 5)$.

Solution   Find the first and second derivatives and write them in factored form. To do this, write $f(x) = 2x^{5/3} + 5x^{2/3}$.

$$f'(x) = 2(\tfrac{5}{3})x^{2/3} + 5(\tfrac{2}{3})x^{-1/3} = \tfrac{10}{3}x^{-1/3}(x + 1)$$
$$f''(x) = \tfrac{10}{3}(\tfrac{2}{3})x^{-1/3} + \tfrac{10}{3}(-\tfrac{1}{3})x^{-4/3} = \tfrac{10}{9}x^{-4/3}(2x - 1)$$

We see

$f'(x) = 0$ when $x = -1$; $f'(x)$ does not exist at $x = 0$. The first-order critical numbers are -1 and 0.

$f''(x) = 0$ when $x = \tfrac{1}{2}$; $f''(x)$ does not exist at $x = 0$. The second-order critical numbers are $\tfrac{1}{2}$ and 0.

Check the intervals of increase and decrease (Figure 4.37a) and of concavity (Figure 4.37b).

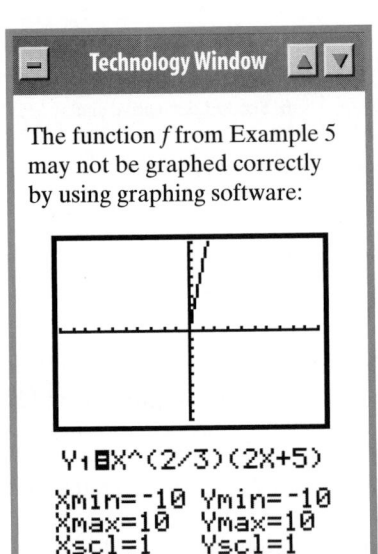
Direction:

$\tfrac{10}{3}x^{-1/3}(x + 1)$: Positive -1 Negative 0 Positive

a. First derivative shows intervals of increase and decrease.

Shape:

$\tfrac{10}{9}x^{-4/3}(2x - 1)$: Negative 0 Negative $\tfrac{1}{2}$ Positive

b. Second derivative shows concavity pattern for the graph.

■ **FIGURE 4.37**   Preliminary work for graphing $f(x) = x^{2/3}(2x + 5)$

The diagrams in Figure 4.37 suggest that the graph of f has a relative maximum at $x = -1$, a relative minimum at $x = 0$, and an inflection point at $x = \tfrac{1}{2}$. We must find the y-coordinates of these points by evaluating f:

$$f(-1) = 3 \qquad f(0) = 0 \qquad f(\tfrac{1}{2}) = \frac{6}{\sqrt[3]{4}} \approx 3.78$$

Plot and label the points $(-1, 3)$, $(0, 0)$, and $(0.5, 3.8)$ on the preliminary graph as shown in Figure 4.38a. Note that the graph is concave down on both sides of $x = 0$ and that the slope $f'(x)$ *decreases* without bound to the left of $x = 0$ and *increases* without bound to the right. This means the graph changes direction abruptly at $x = 0$, and we have a *cusp* at the origin. By plotting the relative extrema and the inflection point on a coordinate plane and passing a smooth curve through these points, we obtain the graph shown in Figure 4.38b.

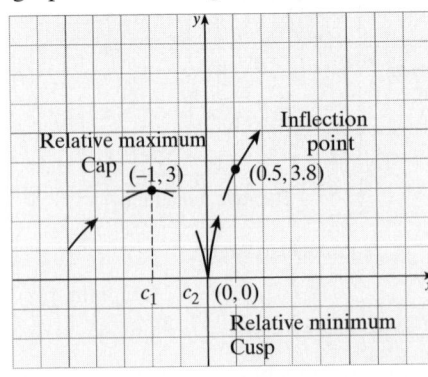

a. Preliminary sketch

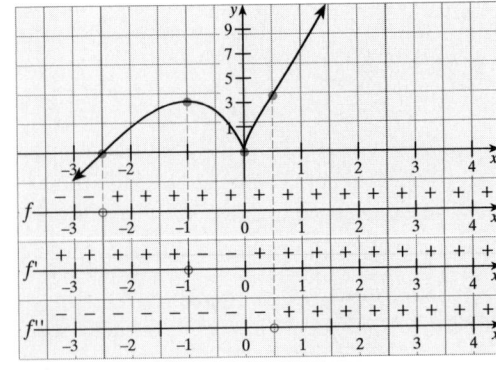

b. Completed graph

■ **FIGURE 4.38**   Graph of $f(x) = x^{2/3}(2x + 5)$

EXAMPLE 6   *Sketching the graph of a trigonometric function*

Sketch the graph of $T(x) = \sin x + \cos x$ on $[0, 2\pi]$.

Solution   You probably graphed this function in trigonometry by adding ordinates. However, with this example we wish to illustrate the power of calculus to complete the graph. Thus, we begin by finding the first and second derivatives:

$$T'(x) = \cos x - \sin x \qquad T''(x) = -\sin x - \cos x$$

We find the critical numbers (both T' and T'' are defined for all values of x):

$$T'(x) = 0 \text{ when } \cos x = \sin x; \quad \text{thus, } x = \tfrac{\pi}{4} \text{ and } x = \tfrac{5\pi}{4}.$$
$$T''(x) = 0 \text{ when } \cos x = -\sin x; \quad \text{thus, } x = \tfrac{3\pi}{4} \text{ and } x = \tfrac{7\pi}{4}.$$

We check the intervals of increase and decrease as well as the concavity pattern, as shown in Figure 4.39.

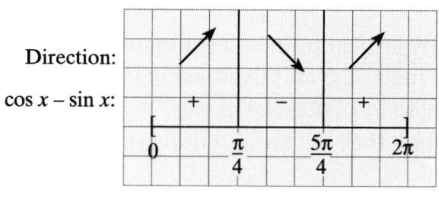 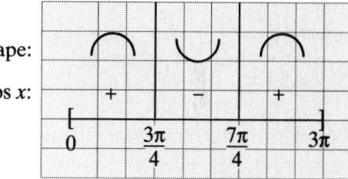

a. Intervals of increase and decrease

b. Concavity pattern

■ **FIGURE 4.39**   Preliminary work for sketching $T(x) = \sin x + \cos x$

Find the critical points, as well as the points of inflection.

Relative maximum: $T(\tfrac{\pi}{4}) = \sqrt{2}$; the critical point is $(\tfrac{\pi}{4}, \sqrt{2})$.

Relative minimum: $T(\tfrac{5\pi}{4}) = -\sqrt{2}$; the critical point is $(\tfrac{5\pi}{4}, -\sqrt{2})$.

Inflection: $T(\tfrac{3\pi}{4}) = 0$; the inflection point is $(\tfrac{3\pi}{4}, 0)$.

$T(\tfrac{7\pi}{4}) = 0$; the inflection point is $(\tfrac{7\pi}{4}, 0)$.

Finally, find the value of T at the endpoints: $T(0) = 1$ and $T(2\pi) = 1$. This information is shown in Figure 4.40.

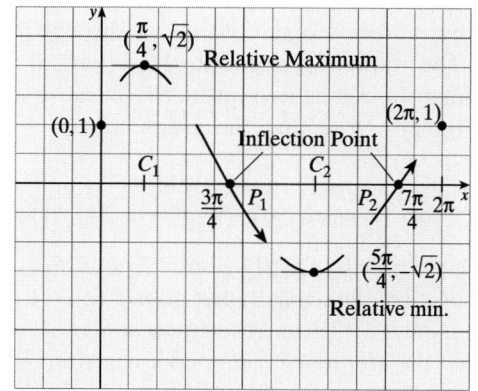 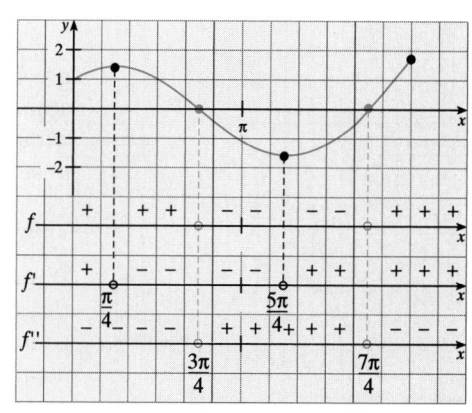

a. Preliminary sketch

b. Completed sketch

■ **FIGURE 4.40**   Graph of $T(x) = \sin x + \cos x$ on $[0, 2\pi]$

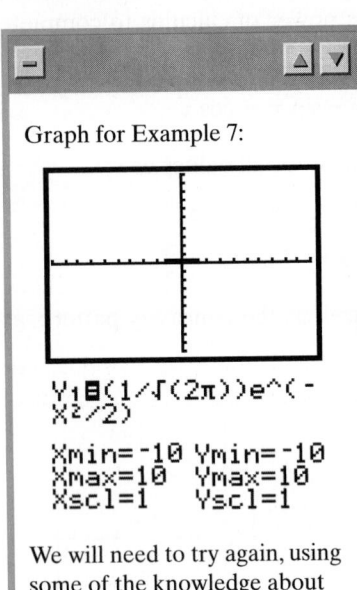

Graph for Example 7:

Y₁⊟(1/√(2π))e^(-X²/2)

Xmin=-10 Ymin=-10
Xmax=10   Ymax=10
Xscl=1   Yscl=1

We will need to try again, using some of the knowledge about the curve shown in the solution of this example.

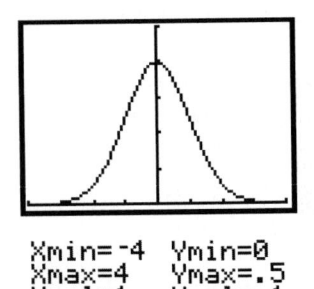

Xmin=-4 Ymin=0
Xmax=4   Ymax=.5
Xscl=1   Yscl=.1

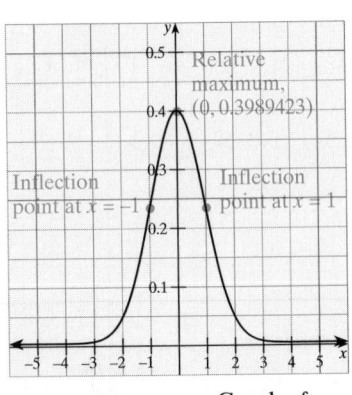

■ **FIGURE 4.42** Graph of
$$f(x) = \frac{1}{\sqrt{2\pi}} e^{-x^2/2}$$

■ **EXAMPLE 7**   *Sketching the graph of an exponential function*

Determine where the function

$$f(x) = \frac{1}{\sqrt{2\pi}} e^{-x^2/2}$$

is increasing, decreasing, concave up, and concave down. Find the relative extrema and inflection points and sketch the graph. This function plays an important role in statistics, where it is called the *standard normal density function.*

Solution   The first derivative is

$$f'(x) = \frac{-x}{\sqrt{2\pi}} e^{-x^2/2}$$

Because $e^{-x^2/2}$ is always positive, $f'(x) = 0$ if and only if $x = 0$. Hence, the corresponding point

$$\left(0, \frac{1}{\sqrt{2\pi}}\right) \approx (0, 0.4)$$

is the only critical point. Checking the sign of f' on each side of 0, we find that f is increasing for $x < 0$ and decreasing for $x > 0$, so there is a relative maximum at $x = 0$, as indicated in Figure 4.41a.

We find that the second derivative of f is

$$f''(x) = \frac{x^2}{\sqrt{2\pi}} e^{-x^2/2} - \frac{1}{\sqrt{2\pi}} e^{-x^2/2} = \frac{1}{\sqrt{2\pi}}(x^2 - 1)e^{-x^2/2}$$

which is zero when $x = \pm 1$. We find that $f(1) = f(-1) \approx 0.24$, and that the concavity of the graph of f is as indicated in Figure 4.41b.

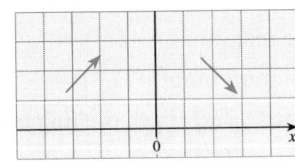

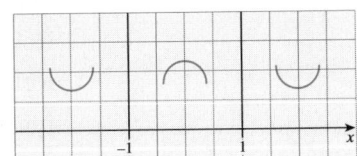

a. Intervals of increase and decrease for $f(x)$

b. Concavity for $f(x)$

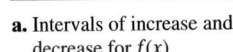

■ **FIGURE 4.41**   Preliminary sketches for the graph of $f(x) = \frac{1}{\sqrt{2\pi}} e^{-x^2/2}$

Finally, we draw a smooth curve through the known points, as shown in Figure 4.42. The graph of f rises to the high point at approximately $(0, 0.4)$ and then falls, approaching the x-axis asymptotically because $e^{-x^2/2}$ approaches 0 as $|x|$ increases without bound. Note that the graph has no x-intercepts, because $e^{-x^2/2}$ is always positive.   ■

THE SECOND-DERIVATIVE TEST FOR RELATIVE EXTREMA

It is often possible to classify a critical point $P(c, f(c))$ on the graph of f by examining the sign of $f''(c)$. Specifically, suppose $f'(c) = 0$ and $f''(c) > 0$. Then there is a horizontal tangent line at P and the graph of f is concave up in the neighborhood of P. This means that the graph of f is cupped upward from the horizontal tangent at P, and it is reasonable to expect P to be a relative minimum, as shown in Figure 4.43a. Similarly, we expect P to be a relative maximum if $f'(c) = 0$ and $f''(c) < 0$, because the graph is cupped down beneath the critical point P, as shown in Figure 4.43b.

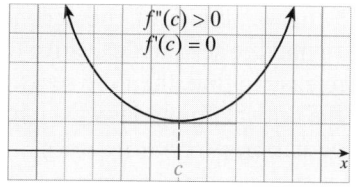

a. Relative minimum
$f'(c) = 0$ and $f''(c) > 0$

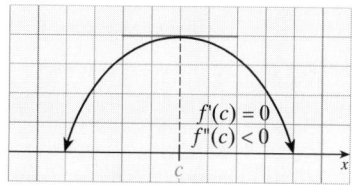

b. Relative maximum
$f'(c) = 0$ and $f''(c) < 0$

■ **FIGURE 4.43**   Second-derivative test for relative extrema

These observations lead to the *second-derivative test* for relative extrema.

The Second-Derivative Test for Relative Extrema

Let f be a function such that $f'(c) = 0$ and the second derivative exists on an open interval containing c.

If $f''(c) > 0$, there is a **relative minimum** at $x = c$.

If $f''(c) < 0$, there is a **relative maximum** at $x = c$.

If $f''(c) = 0$, then the second-derivative test fails and gives no information, so the first-derivative test (or some other test) must be used.

EXAMPLE 8   *Using the second-derivative test*

Use the second-derivative test to determine whether each critical number of the function $f(x) = 3x^5 - 5x^3 + 2$ corresponds to a relative maximum, a relative minimum, or neither.

Solution   Once again, we begin by finding the first and second derivatives:

$$f'(x) = 15x^4 - 15x^2 = 15x^2(x - 1)(x + 1)$$
$$f''(x) = 60x^3 - 30x = 30x(2x^2 - 1)$$

To apply the second-derivative test, evaluate the second derivative for the critical numbers $x = 0, 1,$ and -1:

$f''(0) = 0$;   test fails at $x = 0$.

$f''(1) = 30$;   positive, so test tells us that there is a relative minimum at $x = 1$.

$f''(-1) = -30$;   negative, so test tells us that there is a relative maximum at $x = -1$.

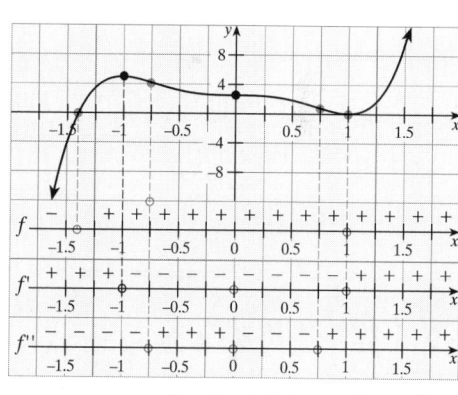

■ **FIGURE 4.44**   Graph of $f(x) = 3x^5 - 5x^3 + 2$

When the second-derivative test fails (in this case at $x = 0$) you should revert to the first-derivative test:

The derivative is negative to the right of 0 and negative to the left of 0. Neither kind of extremum occurs at $x = 0$ ($\downarrow \downarrow$ pattern). Actually, there is an inflection point at $x = 0$ because the second derivative is negative to the right of 0 and positive to the left of 0. The graph is shown in Figure 4.44. ∎

Example 8 demonstrates both the strength and the weakness of the second-derivative test. In particular, when it is relatively easy to find the second derivative (as with a polynomial) and if the zeros of this function are easy to find, then the second-derivative test provides a quick means for classifying the critical points. However, if it is difficult to compute $f''(c)$ or if $f''(c) = 0$, it may be easier, or even necessary, to apply the first-derivative test.

We conclude this section with a summary of the first-derivative and second-derivative tests (Table 4.1).

■ TABLE 4.1

Summary of the First-Derivative and Second-Derivative Tests

Find the critical numbers c such that $f(c)$ is defined and either $f'(c) = 0$ or $f'(c)$ does not exist.

First-Derivative Test
If $f'(x)$ changes from positive to negative (left to right) at c ($\uparrow\downarrow$ pattern), then f has a relative maximum at $x = c$.

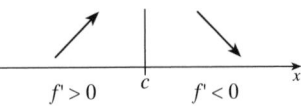

$f' > 0$ c $f' < 0$

If $f'(x)$ changes from negative to positive (left to right) at c ($\downarrow\uparrow$ pattern), then f has a relative minimum at $x = c$.

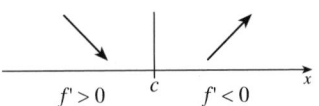

$f' > 0$ c $f' < 0$

WARNING ➡ If it is easy to find the second derivative, then it is better to bypass the first-derivative test and move directly to the second-derivative test. ←

Second-Derivative Test
If $f''(c) < 0$, relative maximum at $x = c$.

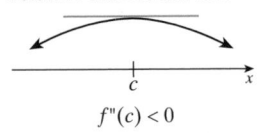

$f''(c) < 0$

If $f''(c) > 0$, relative minimum at $x = c$.

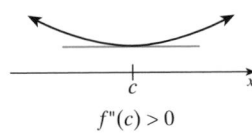

$f''(c) > 0$

If $f''(c) = 0$, second-derivative test fails.

Inflection Points
There is an inflection point at c if f'' (the concavity) changes sign there. This may occur if $f''(c) = 0$ or if $f''(c)$ is not defined.

4.4 *Problem Set*

 1. ■ **What Does This Say?** What is the second-derivative test?

2. ■ **What Does This Say?** What is the relationship among concavity, points of inflection, and the second derivative?

3. ■ **What Does This Say?** The cartoon on page 250 exclaims, "Our prices are rising slower than any place in town." Restate using calculus.

Find all critical points of the given function in Problems 4–38, and determine where the graph of the function is rising, falling, concave up, or concave down. Sketch the graph.

4. $f(x) = x^2 + 5x - 3$

5. $f(x) = 2(x + 20)^2 - 8(x + 20) + 7$

6. $f(x) = x^3 - 3x - 4$

7. $f(x) = \frac{1}{3}x^3 - 9x + 2$

8. $f(x) = (x - 12)^4 - 2(x - 12)^3$

9. $f(x) = 1 + 2x + 18/x$

10. $f(u) = 3u^4 - 2u^3 - 12u^2 + 18u - 5$

11. $g(u) = u^4 + 6u^3 - 24u^2 + 26$

12. $f(x) = \sqrt{x^2 + 1}$

13. $g(t) = (t^3 + t)^2$

14. $f(t) = (t^3 + 3t^2)^3$

15. $f(t) = t^3 - 3t^4$

16. $f(t) = 4t^5 - 5t^4$

17. $g(t) = 5t^6 - 6t^3$

18. $f(x) = \dfrac{1}{x^2 + 3}$

19. $f(x) = \dfrac{x}{x^2 + 1}$

20. $f(t) = t - \ln t$

21. $f(t) = t^2 e^{-3t}$

22. $f(x) = e^x + e^{-x}$

23. $f(x) = \dfrac{e^x - e^{-x}}{e^x + e^{-x}}$

24. $f(x) = (\ln x)^2$

25. $f(x) = \ln x^2$

26. $f(t) = 2t^{4/3} + 9t + 1$

27. $f(x) = x^{4/3}(x - 27)$

28. $f(x) = x^4 - 3x^3 + x - 5$

29. $f(x) = x^5 + 2x^3 - x^2 + 11$

30. $t(\theta) = \sin\theta - 2\cos\theta$ for $0 \le \theta \le 2\pi$

31. $t(\theta) = \theta + \cos 2\theta$ for $0 \le \theta \le \pi$

32. $h(u) = \dfrac{\sin u}{2 + \cos u}$ for $0 \le u \le 2\pi$

33. $f(t) = -\frac{1}{4}\sin 2t + \cos t$ for $-\pi \le t \le \pi$

34. $f(x) = (x + 1)\tan^{-1} x$ for $-\frac{\pi}{2} < x < \frac{\pi}{2}$

35. $g(x) = 2x - \sin^{-1} x$ for $-1 \le x \le 1$

36. $f(x) = x^3 + \sin x$ on $[-\pi, \pi]$

37. $f(x) = \tan x - x^2 + 3$ on $\left(\frac{\pi}{2}, \frac{3\pi}{2}\right)$

38. $f(x) = \cos^{-1} x + \sin^{-1} x$

Ⓑ **39. THINK TANK PROBLEM** Sketch the graph of a function with the following properties:

$f'(x) > 0$ when $x < -1$
$f'(x) > 0$ when $x > 3$
$f'(x) < 0$ when $-1 < x < 3$
$f''(x) < 0$ when $x < 2$
$f''(x) > 0$ when $x > 2$

40. THINK TANK PROBLEM Sketch the graph of a function with the following properties:

$f'(x) > 0$ when $x < 2$ and when $2 < x < 5$
$f'(x) < 0$ when $x > 5$
$f'(2) = 0$
$f''(x) < 0$ when $x < 2$ and when $4 < x < 7$
$f''(x) > 0$ when $2 < x < 4$ and when $x > 7$

41. THINK TANK PROBLEM Sketch the graph of a function with the following properties:

$f'(x) > 0$ when $x < 1$
$f'(x) < 0$ when $x > 1$
$f''(x) > 0$ when $x < 1$
$f''(x) > 0$ when $x > 1$
What can you say about the derivative of f when $x = 1$?

42. THINK TANK PROBLEM Sketch the graph of a function with the following properties: There are relative extrema at $(-1, 7)$ and $(3, 2)$. There is an inflection point at $(1, 4)$. The graph is concave down only when $x < 1$. The x-intercept is $(-4, 0)$ and the y-intercept is $(0, 5)$.

43. ■ **What Does This Say?** Examine the graphs:
a. $f(x) = (x + 1)^{1/3}$
b. $f(x) = (x + 1)^{2/3}$
c. $f(x) = (x + 1)^{4/3}$
d. $f(x) = (x + 1)^{5/3}$

Generalize to make a statement about the effect of a positive integer n on the graph of $f(x) = (x + 1)^{n/3}$.

44. Use calculus to show that the graph of the quadratic function $y = Ax^2 + Bx + C$ is concave up if $A > 0$ and concave down if $A < 0$.

45. Find constants A, B, and C that guarantee that the function $f(x) = Ax^3 + Bx^2 + C$ will have a relative extremum at $(2, 11)$ and an inflection point at $(1, 5)$. Sketch the graph of f.

46. ■ **What Does This Say?** Consider the graph of $y = x^3 + bx^2 + cx + d$ for constants $b, c,$ and d. What happens to the graph as b changes?

47. Let $S(x) = \frac{1}{2}(e^x - e^{-x})$ and $C(x) = \frac{1}{2}(e^x + e^{-x})$. These functions are known as the *hyperbolic sine* and *hyperbolic cosine*, respectively. These functions are examined in Section 7.8.
a. Show that $S'(x) = C(x)$ and $C'(x) = S(x)$.
b. Sketch the graphs of S and C.

48. Find all points of intersection of the curves

$$y = 2^{x^2 - 2x} \quad \text{and} \quad y = 4^x$$

Sketch the two curves on the same set of coordinate axes.

49. A function that plays a prominent role in statistics is

$$D(t) = \frac{1}{\sqrt{2\pi}\,\sigma} \exp\left[\frac{-1}{2}\left(\frac{t - m}{\sigma}\right)^2\right]$$

where m is a real number and σ is a positive constant.
a. Find $D'(t)$ and determine all relative extrema of $D(t)$.
b. What happens to $D(t)$ as $t \to +\infty$ and as $t \to -\infty$?
c. Sketch the graph of $D(t)$.

50. A *Gompertz curve* is the graph of a function of the form

$$G(x) = A \exp(-Be^{-kx}) \quad \text{for } x > 0$$

where A, B, and k are positive constants. Such graphs appear in certain population studies. Find $G'(x)$ and $G''(x)$, and sketch the graph of G.

MODELING PROBLEMS *In Problems 51–58, set up an appropriate model to answer the given question. Be sure to state your assumptions.*

51. At noon on a certain day, Frank sets out to assemble five stereo sets. His rate of assembly increases steadily throughout the afternoon until 4:00 P.M., at which time he has completed three sets. After that, he assembles sets at a slower and slower rate until he finally completes the fifth set at 8:00 P.M. Sketch a rough graph of a function that represents the number of sets Frank has completed after t hours of work.

52. The deflection of a hardwood beam of length ℓ is given by

$$D(x) = \frac{9}{4}x^4 - 7\ell x^3 + 5\ell^2 x^2$$

where x is the distance from one end of the beam. What value of x yields the maximum deflection?

53. A manufacturer estimates that if x thousand units of a particular commodity are produced, the total cost (in dollars) will be

$$C(x) = x^3 - 24x^2 + 350x + 400$$

At what level of production will the marginal cost $C'(x)$ be minimized?

54. The total cost of producing x thousand units of a certain commodity is

$$C(x) = 2x^4 - 6x^3 - 12x^2 - 2x + 1$$

Determine the largest and smallest values of the marginal cost $C'(x)$ for $0 \le x \le 3$.

55. An industrial psychologist conducts two efficiency studies at the Chilco appliance factory. The first study indicates that the average worker who arrives on the job at 8:00 A.M. will have assembled

$$-t^3 + 6t^2 + 13t$$

blenders in t hours (without a break), for $0 \le t \le 4$. The second study suggests that after a 15-minute coffee break, the average worker can assemble

$$-\tfrac{1}{3}t^3 + \tfrac{1}{2}t^2 + 25t$$

blenders in t hours after the break for $0 < t \le 4$.
Note: The 15-minute break is not part of the work time.

   a. Verify that if the coffee break occurs at 10:00 A.M., the average worker will assemble 42 blenders before the break and $49\tfrac{1}{3}$ blenders for the two hours after the break.
   b. Suppose the coffee break is scheduled to begin x hours after 8:00 A.M. Find an expression for the total number of blenders $N(x)$ assembled by the average worker during the morning shift (8 A.M. to 12:15 P.M.).
   c. At what time should the coffee break be scheduled so that the average worker will produce the maximum number of blenders during the morning shift? How is this optimum time related to the point of diminishing returns?

56. An important formula in physical chemistry is *van der Waals' equation*, which says that

$$\left(P + \frac{a}{V^2}\right)(V - b) = nRT$$

where P, V, and T are the pressure, volume, and temperature, respectively, of a gas, and a, b, n, and R are positive constants. The *critical temperature* T_C of the gas is the highest temperature at which the gaseous and liquid phases can exist as separate states.

   a. When $T = T_C$, the pressure P can be expressed as a function $P(V)$ of V alone. Show how this can be done, and then find $P'(V)$ and $P''(V)$.
   b. The *critical volume* V_C is the volume that satisfies $P'(V_C) = 0$ and $P''(V_C) = 0$. Find V_C.
   c. Find T_C, the point where $P''(V) = 0$, using the V_C from part **b** to write it in terms of a, b, n, and R. Finally, find the *critical pressure* $P_C = P(V_C)$ in terms of a, b, n, and R.
   d. Draw a sketch of P as a function of V.

57. In an experiment, a biologist introduces a toxin into a bacterial colony and then measures the effect on the population of the colony. Suppose that at time t (in minutes) the population is

$$P(t) = 5 + e^{-0.04t}(t + 1)$$

thousand. At what time will the population be the largest? What happens to the population in the long run (as $t \to +\infty$)? Find where the graph of P has an inflection point, and interpret this point in terms of the population. Sketch the graph of P.

58. Research indicates that the power P required by a bird to maintain flight is given by the formula

$$P = \frac{w^2}{2\rho S v} + \frac{1}{2}\rho A v^3$$

where v is the relative speed of the bird, w is its weight, ρ is the density of air, and S and A are constants associated with the bird's size and shape.* What speed will minimize the power? You may assume that w, ρ, S, and A are all positive.

59. **HISTORICAL QUEST** One of the most famous women in the history of mathematics is Maria Gaëtana Agnesi (pronounced än yā′ zē). She was born in Milan, the first of 21 children. Her first publication was at age 9, when she wrote a Latin discourse defending higher education for women. Her most important work was a now-classic calculus textbook published in 1748. Maria Agnesi is also remembered for a curve called the witch of Agnesi. This curve is defined ($a > 0$) by the equation

**MARIA AGNESI
(1718–1799)**

$$y = \frac{a^3}{x^2 + a^2}$$

The curve was named *versiera* (from the Latin verb *to turn*) by Agnesi, but John Colson, an Englishman who translated her work, confused the word *versiera* with the word *avversiera*, which means "wife of the devil" in Latin; the curve has ever since been called the "witch of Agnesi." This was particularly unfortunate because Colson wanted Agnesi's work to serve as a model for budding young mathematicians, especially young women.

   Graph this curve and find the critical numbers, extrema, and points of inflection.

60. Let f be a function that is differentiable at 0 and satisfies $f(1) = -1$. If $y = f(x)$, suppose

$$\frac{dy}{dx} = \frac{3y^2 + x}{y^2 + 2}$$

   a. Find an equation for the tangent line to the graph of f at the point where $x = 1$.

---

*C. J. Pennycuick, "The Mechanics of Bird Migration," *Ibis* III, pp. 525–556.

b. Note that the origin is a critical point. What kind of relative extremum (if any) occurs at this point?

61. Show that if f is concave up on an interval I, then the graph of f lies above all its tangents on I. *Hint:* Use the mean value theorem to show that if c is a point in I, then

$$f(x) > f(c) + f'(c)(x - c)$$

for all x in I. This property is sometimes used as the definition of concave up.

62. ■ **What Does This Say?** Rephrase in your own words what Problem 61 is all about.

4.5 Curve Sketching: Limits Involving Infinity and Asymptotes

IN THIS SECTION limits to infinity, infinite limits, graphs with asymptotes, graphing strategy ■

LIMITS TO INFINITY

In applications, we are often concerned with what happens "in the long run"—that is, as time t grows large without bound. To examine such limits precisely, we use the following definition.

WARNING Even though the symbols ∞ and $+\infty$ mean the same thing, for the time being we use $+\infty$ to help distinguish between $+\infty$ and $-\infty$. ←

Limits to Infinity

The limit statement $\lim_{x \to +\infty} f(x) = L$ means that for any number $\epsilon > 0$, there exists a number N_1 such that

$$\left| f(x) - L \right| < \epsilon \text{ whenever } x > N_1$$

for x in the domain of f. Similarly, $\lim_{x \to -\infty} f(x) = M$ means that for any $\epsilon > 0$, there exists a number N_2 such that

$$\left| f(x) - M \right| < \epsilon \text{ whenever } x < N_2$$

This definition can be illustrated graphically, as shown in Figure 4.45.

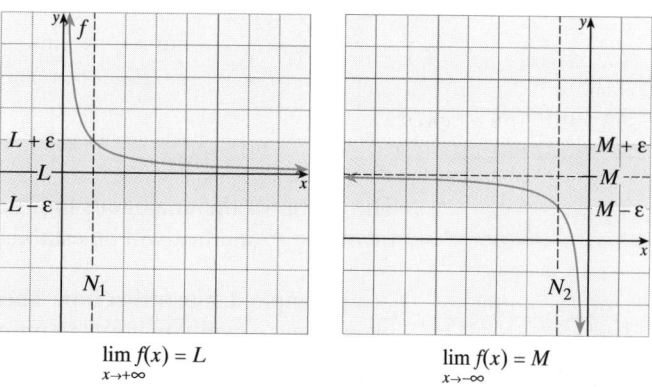

$$\lim_{x \to +\infty} f(x) = L \qquad \lim_{x \to -\infty} f(x) = M$$

■ **FIGURE 4.45** Graphical representation of limits to infinity

With this formal definition, we can show that all the rules for limits established in Chapter 1 also apply to $\lim_{x \to +\infty} f(x)$ and $\lim_{x \to -\infty} f(x)$.

Rules Involving Limits to Infinity

If $\lim\limits_{x \to +\infty} f(x)$ and $\lim\limits_{x \to +\infty} g(x)$ exist, then for constants a and b:

Power rule: $\quad \lim\limits_{x \to +\infty} [f(x)]^n = \left[\lim\limits_{x \to +\infty} f(x)\right]^n$

Linearity rule: $\quad \lim\limits_{x \to +\infty} [af(x) + bg(x)] = a \lim\limits_{x \to +\infty} f(x) + b \lim\limits_{x \to +\infty} g(x)$

Product rule: $\quad \lim\limits_{x \to +\infty} [f(x)g(x)] = \left[\lim\limits_{x \to +\infty} f(x)\right]\left[\lim\limits_{x \to +\infty} g(x)\right]$

Quotient rule: $\quad \lim\limits_{x \to +\infty} \dfrac{f(x)}{g(x)} = \dfrac{\lim\limits_{x \to +\infty} f(x)}{\lim\limits_{x \to +\infty} g(x)}$ if $\lim\limits_{x \to +\infty} g(x) \neq 0$

The same results hold for $\lim\limits_{x \to -\infty} f(x)$, if it exists.

The following theorem will allow us to evaluate certain limits to infinity with ease.

THEOREM 4.8   *Special limits to infinity*

If n is a positive rational number, A is any real number, and x^n is defined, then

$$\lim_{x \to +\infty} \frac{A}{x^n} = \lim_{x \to -\infty} \frac{A}{x^n} = 0$$

Furthermore, for $k > 0$,

$$\lim_{x \to +\infty} \frac{\ln x^k}{x^n} = 0, \qquad \lim_{x \to +\infty} \frac{x^n}{e^{kx}} = 0$$

Proof   We begin by proving that $\lim\limits_{x \to +\infty} \dfrac{1}{x} = 0$. For $\epsilon > 0$, let $N = \dfrac{1}{\epsilon}$. Then for $x > N$, we have

$$x > N = \frac{1}{\epsilon} \quad \text{so that} \quad \frac{1}{x} < \epsilon$$

This means that $\left|\dfrac{1}{x} - 0\right| < \epsilon$ so that from the definition of limit we have $\lim\limits_{x \to +\infty} \dfrac{1}{x} = 0$. Now let n be a rational number, say $n = p/q$. Then

$$\lim_{x \to +\infty} \frac{A}{x^n} = \lim_{x \to +\infty} \frac{A}{x^{p/q}} = A \lim_{x \to +\infty} \left[\frac{1}{\sqrt[q]{x}}\right]^p$$

$$= A\left[\sqrt[q]{\lim_{x \to +\infty} \frac{1}{x}}\right]^p = A[\sqrt[q]{0}]^p = A \cdot 0 = 0$$

The proof for the analogous limits as $x \to -\infty$ follows similarly. The limit statements involving e^x and $\ln x$ will be established in Section 4.8.

Example 1 illustrates how Theorem 4.8 can be used along with the other limit properties to evaluate limits to infinity.

EXAMPLE 1   *Evaluating limits to infinity*

Evaluate

$$\lim_{x \to +\infty} \sqrt{\frac{3x - 5}{x - 2}} \qquad \text{and} \qquad \lim_{x \to -\infty} \left(\frac{3x - 5}{x - 2}\right)^3$$

Solution   Notice that for $x \neq 0$,

$$\frac{3x - 5}{x - 2} = \frac{x(3 - \frac{5}{x})}{x(1 - \frac{2}{x})} = \frac{3 - \frac{5}{x}}{1 - \frac{2}{x}}$$

Also, according to Theorem 4.8, we know that

$$\lim_{x \to +\infty} \frac{5}{x} = \lim_{x \to +\infty} \frac{2}{x} = 0$$

We now find the limits using the quotient rule, the power rule, and Theorem 4.8:

$$\lim_{x \to +\infty} \sqrt{\frac{3x - 5}{x - 2}} = \lim_{x \to +\infty} \left(\frac{3x - 5}{x - 2} \right)^{1/2} = \left(\lim_{x \to +\infty} \frac{3x - 5}{x - 2} \right)^{1/2}$$

$$= \left(\frac{\lim_{x \to +\infty} (3 - \frac{5}{x})}{\lim_{x \to +\infty} (1 - \frac{2}{x})} \right)^{1/2} = \left(\frac{3 - 0}{1 - 0} \right)^{1/2} = \sqrt{3}$$

Similarly,

$$\lim_{x \to \infty} \left(\frac{3x - 5}{x - 2} \right)^3 = 3^3 = 27$$

Technology Window

You can use certain computer programs or calculators to help you with limits to infinity.

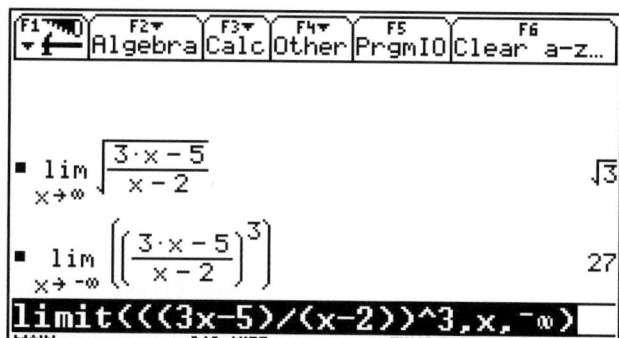

You can also use a graphing calculator to help with limits to infinity.

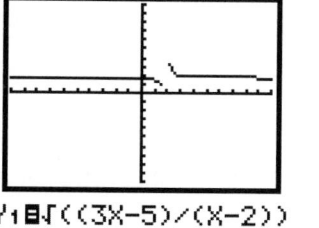

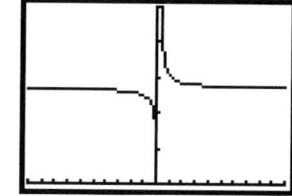

If you move the cursor along the curve (toward $-\infty$ or $+\infty$), the TRACE values may approach the desired limit.

| EXAMPLE 2 | *Finding a formula for future value with continuous compounding* |

In Chapter 1, we showed that the future value A of P dollars invested at an annual rate r compounded continuously after t years is

$$A = \lim_{n \to +\infty} P\left(1 + \frac{r}{n}\right)^n$$

Show that $A = Pe^{rt}$.

Solution   To evaluate the given limit, let $m = n/r$ so that $r/n = 1/m$ and $nt = mrt$. Then,

$$A = \lim_{n \to +\infty} P\left(1 + \frac{r}{n}\right)^{nt} \qquad m \to +\infty \text{ as } n \to +\infty.$$

$$= \lim_{m \to +\infty} P\left(1 + \frac{1}{m}\right)^{mrt}$$

$$= \lim_{m \to +\infty} P\left[\left(1 + \frac{1}{m}\right)^m\right]^{rt} \qquad \textit{Properties of exponents}$$

$$= P\left[\lim_{m \to +\infty}\left(1 + \frac{1}{m}\right)^m\right]^{rt} \qquad \textit{Linearity rule}$$

$$= Pe^{rt} \qquad\qquad\qquad\qquad\qquad\blacksquare$$

When evaluating a limit of the form

$$\lim_{x \to +\infty} \frac{p(x)}{d(x)} \quad \text{or} \quad \lim_{x \to -\infty} \frac{p(x)}{d(x)},$$

where $p(x)$ and $d(x)$ are polynomials, it is often useful to divide both $p(x)$ and $d(x)$ by the highest power of x that occurs in either. The limit can then be found by applying Theorem 4.8. This process is illustrated by the following examples.

| EXAMPLE 3 | *Evaluating a limit to infinity* |

Evaluate $\lim_{x \to +\infty} \dfrac{3x^3 - 5x + 9}{5x^3 + 2x^2 - 7}$.

Solution   We may assume that $x \neq 0$, because we are interested only in very large values of x. Dividing both the numerator and denominator of the given expressions by x^3, we find

$$\frac{3x^3 - 5x + 9}{5x^3 + 2x^2 - 7} = \frac{3x^3 - 5x + 9}{5x^3 + 2x^2 - 7} \cdot \frac{\frac{1}{x^3}}{\frac{1}{x^3}} = \frac{3 - \frac{5}{x^2} + \frac{9}{x^3}}{5 + \frac{2}{x} - \frac{7}{x^3}}$$

Thus,

$$\lim_{x \to +\infty} \frac{3x^3 - 5x + 9}{5x^3 + 2x^2 - 7} = \lim_{x \to +\infty} \frac{3 - \dfrac{5}{x^2} + \dfrac{9}{x^3}}{5 + \dfrac{2}{x} - \dfrac{7}{x^3}}$$

$$= \frac{\lim\limits_{x \to +\infty} \left(3 - \dfrac{5}{x^2} + \dfrac{9}{x^3}\right)}{\lim\limits_{x \to +\infty} \left(5 - \dfrac{2}{x} - \dfrac{7}{x^3}\right)}$$

$$= \frac{3 - 0 + 0}{5 + 0 - 0} = \frac{3}{5} \qquad \blacksquare$$

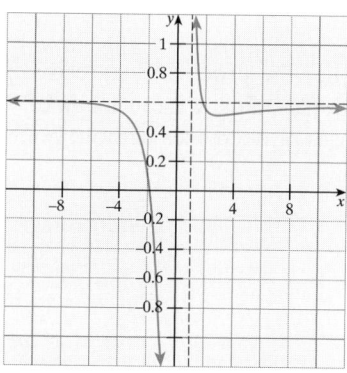

Graph of $y = \dfrac{3x^3 - 5x + 9}{5x^3 + 2x^2 - 7}$

EXAMPLE 4   *Evaluating a limit to negative infinity*

Evaluate $\lim\limits_{x \to -\infty} \dfrac{95x^3 + 57x + 30}{x^5 - 1,000}$.

Solution   Dividing the numerator and the denominator by the highest power, x^5, we find that

$$\lim_{x \to -\infty} \frac{95x^3 + 57x + 30}{x^5 - 1,000} = \lim_{x \to -\infty} \frac{95x^3 + 57x + 30}{x^5 - 1,000} \cdot \frac{\dfrac{1}{x^5}}{\dfrac{1}{x^5}}$$

$$= \lim_{x \to -\infty} \frac{\dfrac{95}{x^2} + \dfrac{57}{x^4} + \dfrac{30}{x^5}}{1 - \dfrac{1,000}{x^5}}$$

$$= \frac{0 + 0 + 0}{1 - 0}$$

$$= 0 \qquad \blacksquare$$

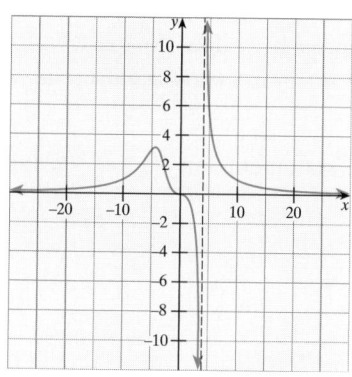

Graph of $y = \dfrac{95x^3 + 57x + 30}{x^5 - 1,000}$

EXAMPLE 5   *Evaluating a limit to infinity involving e^x*

Find $\lim\limits_{x \to +\infty} e^{-x} \cos x$.

Solution   We cannot use the product rule since $\lim\limits_{x \to +\infty} \cos x$ does not exist (it diverges by oscillation—do you see why?). Note, however, that

$$e^{-x} \cos x = \frac{\cos x}{e^x}$$

must become smaller and smaller as $x \to +\infty$ since the numerator $\cos x$ is trapped between -1 and 1, while the denominator e^x grows relentlessly larger with x. Thus,

$$\lim_{x \to +\infty} e^{-x} \cos x = 0$$

The graph shown in Figure 4.46 confirms this conclusion. We shall establish this limit more informally in Section 4.8. $\blacksquare$

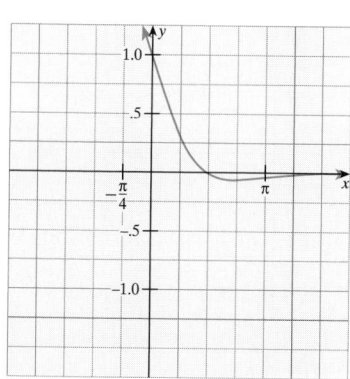

■ **FIGURE 4.46**   Graph of $y = e^{-x} \cos x$

INFINITE LIMITS

In mathematics, we use the symbol ∞ to describe either the process of unrestricted growth or the end result of such growth. With this understanding, we can speak of *infinite limits*—that is, the result when $f(x)$ either increases or decreases without bound as $x \to c$. The limit statement

$$\lim_{x \to c} f(x) = +\infty$$

may be defined formally as follows.

> **Infinite Limit**
>
> We write $\lim_{x \to c} f(x) = +\infty$ if for any number $N > 0$ (no matter how large), it is possible to find a number $\delta > 0$ such that $f(x) > N$ whenever $0 < |x - c| < \delta$.

> **WARNING** ■ *What This Does Not Say:* An infinite limit does not exist in the sense that limits were defined in Chapter 1. The symbolism $\lim_{x \to c} f(x) = +\infty$ *does not mean* that $f(x)$ approaches a *number* $+\infty$ as x approaches c. That is, ∞ is **not** a number. Nevertheless, writing $\lim_{x \to c} f(x) = +\infty$ or $\lim_{x \to c} f(x) = -\infty$ conveys more specific information than simply saying "the limit of $f(x)$ as x approaches c does not exist." ←

A similar definition holds for the infinite limit statement $\lim_{x \to c} f(x) = -\infty$ (see Problem 61). For example,

$$\lim_{x \to 5^-} \frac{4}{x - 5} = -\infty \quad \text{and} \quad \lim_{x \to 5^+} \frac{4}{x - 5} = +\infty$$

or

$$\lim_{x \to -4^-} \frac{-7}{x + 4} = +\infty \quad \text{and} \quad \lim_{x \to -4^+} \frac{-7}{x + 4} = -\infty$$

are infinite limits. The graphs of these functions are shown in Figure 4.47.

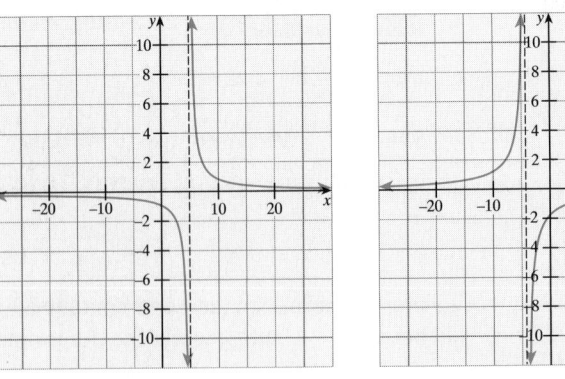

a. Graph of $y = \dfrac{4}{x - 5}$ 　　**b.** Graph of $y = \dfrac{-7}{x + 4}$

■ **FIGURE 4.47** Graphs showing infinite limits

EXAMPLE 6   *Infinite limit*

Find $\lim\limits_{x\to2^-} \dfrac{3x-5}{x-2}$ and $\lim\limits_{x\to2^+} \dfrac{3x-5}{x-2}$.

Solution   Notice that $\dfrac{1}{x-2}$ increases without bound as x approaches 2 from the right and $\dfrac{1}{x-2}$ decreases without bound as x approaches 2 from the left. That is,

$$\lim_{x\to2^+} \frac{1}{x-2} = +\infty \quad \text{and} \quad \lim_{x\to2^-} \frac{1}{x-2} = -\infty$$

We also have $\lim\limits_{x\to2}(3x-5) = 1$, and it follows that

$$\lim_{x\to2^+} \frac{3x-5}{x-2} = +\infty \quad \text{and} \quad \lim_{x\to2^-} \frac{3x-5}{x-2} = -\infty$$ ∎

GRAPHS WITH ASYMPTOTES

Figure 4.48 shows a graph that approaches the horizontal line $y = 2$ as $x \to -\infty$, the vertical line $x = 3$ as x approaches 3 from either side, and the slanting line $x - 2y = 2$ as $x \to +\infty$. This is referred to as *asymptotic behavior*, and the lines $y = 2$, $x = 3$, and $y = \frac{1}{2}x - 1$ are called **asymptotes** of the graph. Here is a formal definition of the three kinds of asymptotes we shall consider.

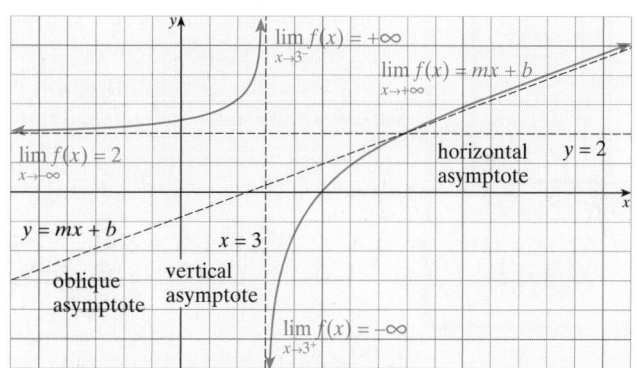

■ **FIGURE 4.48**   A typical graph with asymptotes

Vertical Asymptote

The line $x = c$ is a **vertical asymptote** of the graph of f if either of the one-sided limits

$$\lim_{x\to c^-} f(x) \quad \text{or} \quad \lim_{x\to c^+} f(x)$$

is infinite.

Horizontal Asymptote

The line $y = L$ is a **horizontal asymptote** of the graph of f if

$$\lim_{x\to+\infty} f(x) = L \quad \text{or} \quad \lim_{x\to-\infty} f(x) = L$$

Oblique Asymptote

The line $y = mx + b$ is an **oblique asymptote** of the graph of f if f is a reduced rational function and

$$f(x) = \frac{p(x)}{d(x)} = mx + b + \frac{r}{d(x)}$$

where $\lim\limits_{x \to +\infty} \dfrac{r}{d(x)} = 0$.

EXAMPLE 7   *Graphing a rational function with asymptotes*

Sketch the graph of $f(x) = \dfrac{3x - 5}{x - 2}$.

Solution   VERTICAL ASYMPTOTES   First, make sure the given function is written in simplified form. Because vertical asymptotes for $f(x) = \dfrac{3x - 5}{x - 2}$ occur at values of x for which f is not defined, we look for values that cause the denominator to be zero; that is, we solve $d(c) = 0$ and then evaluate $\lim\limits_{x \to c^-} f(x)$ and $\lim\limits_{x \to c^+} f(x)$ to ascertain the behavior of the function at $x = c$. For this example, $x = 2$ is a value that causes division by zero, so we find

$$\lim_{x \to 2^+} \frac{3x - 5}{x - 2} = +\infty \quad \text{and} \quad \lim_{x \to 2^-} \frac{3x - 5}{x - 2} = -\infty$$

(We found these limits in Example 5.) This means that $x = 2$ is a vertical asymptote and that the graph is moving downward as $x \to 2$ from the left and upward as $x \to 2$ from the right. This information is recorded on the preliminary graph shown in Figure 4.49a by a dashed vertical line with upward ($\uparrow$) and downward ($\downarrow$) arrows.

HORIZONTAL ASYMPTOTES   To find the horizontal asymptotes, we compute

$$\lim_{x \to +\infty} \frac{3x - 5}{x - 2} = \lim_{x \to +\infty} \frac{3x - 5}{x - 2} \cdot \frac{\frac{1}{x}}{\frac{1}{x}} = \lim_{x \to +\infty} \frac{3 - \frac{5}{x}}{1 - \frac{2}{x}} = \frac{3 - 0}{1 - 0} = 3$$

and

$$\lim_{x \to -\infty} \frac{3x - 5}{x - 2} = 3 \quad \text{(The steps here are the same as for } x \to +\infty.\text{)}$$

This means that $y = 3$ is a horizontal asymptote. This information is recorded on the preliminary graph shown in Figure 4.49a by a dashed horizontal line with outbound arrows ($\leftarrow, \rightarrow$).

OBLIQUE ASYMPTOTES   This function does not have oblique asymptotes because the degree of the numerator is not 1 more than the degree of the denominator.

The preliminary sketch gives us some valuable information about the graph, but it does not present the entire picture. Next, we use calculus to find where the function is increasing and decreasing (first derivative) and where it is concave up and concave down (second derivative).

$$f'(x) = \frac{-1}{(x - 2)^2} \quad \text{and} \quad f''(x) = \frac{2}{(x - 2)^3}$$

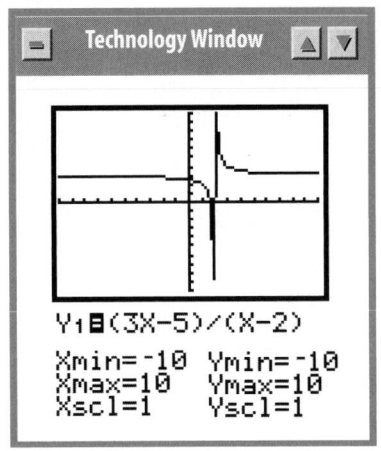

Technology Window

Y₁ ▣ (3X-5)/(X-2)
Xmin=⁻10 Ymin=⁻10
Xmax=10   Ymax=10
Xscl=1   Yscl=1

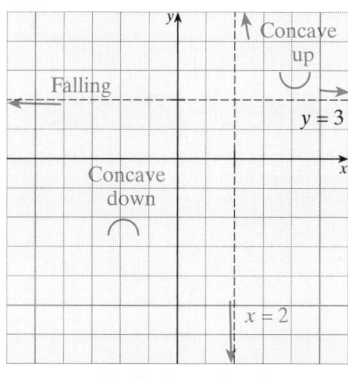
a. Preliminary sketch

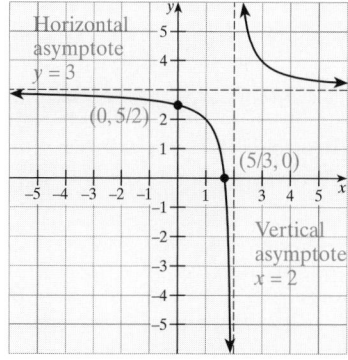
b. Completed sketch

■ **FIGURE 4.49** Graph of $f(x) = \dfrac{3x - 5}{x - 2}$

Neither derivative is ever zero, and both are undefined at $x = 2$. Checking the signs of the first and second derivatives, we find that this curve has no points of inflection (function is not defined at $x = 2$). This information is added to the preliminary sketch shown in Figure 4.49a. The completed graph is shown in Figure 4.49b. ■

EXAMPLE 8   *Sketching a curve with an oblique asymptote*

Discuss and sketch the graph of $y = \dfrac{x^2 - x - 2}{x - 3}$.

Solution   Performing the division, we write

$$y = \frac{x^2 - x - 2}{x - 3} = x + 2 + \frac{4}{x - 3}$$

Find the first and second derivatives, along with the critical numbers for y' and y'':

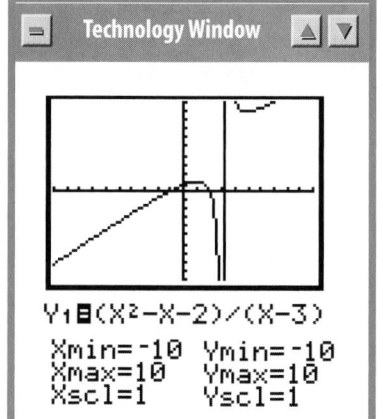

Y₁ = (X²-X-2)/(X-3)
Xmin=-10 Ymin=-10
Xmax=10   Ymax=10
Xscl=1   Yscl=1

Derivatives	Critical Numbers
$y' = 1 - 4(x - 3)^{-2}$	$1 - 4(x - 3)^{-2} = 0$
	$1 = 4(x - 3)^{-2}$
	$(x - 3)^2 = 4$
	$x - 3 = \pm 2$
	$x = 5, 1$
$y'' = 8(x - 3)^{-3}$	$8(x - 3)^{-3} = 0$ *No second-order critical numbers*

The second-derivative test: $f''(5) > 0$; relative minimum at $x = 5$.
$f''(1) < 0$; relative maximum at $x = 1$.

The intervals of increase and decrease, relative extrema, and concavity are indicated in the preliminary sketch in Figure 4.50a. Note that there is a vertical asymptote at $x = 3$, and an oblique asymptote at $y = x + 2$, because $y = x + 2 + \dfrac{4}{x - 3}$ and $\dfrac{4}{x - 3} \to 0$ as $x \to +\infty$. There are no horizontal asymptotes. Finally, we plot a few additional points and then draw a smooth curve, as shown in Figure 4.50b.

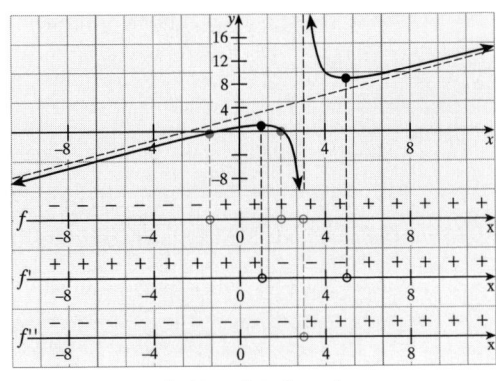

a. Preliminary sketch   **b.** Completed graph

■ FIGURE 4.50   Graph of $y = \dfrac{x^2 - x - 2}{x - 3}$ ■

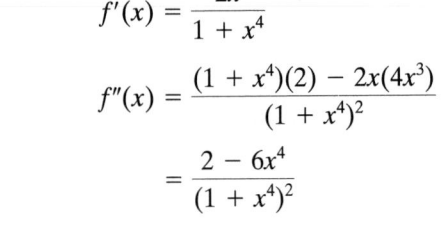

EXAMPLE 9   *Graph of an inverse trigonometric function*

Sketch the graph of $f(x) = \tan^{-1} x^2$.

Solution   We find that

$$f'(x) = \frac{2x}{1 + x^4}$$

$$f''(x) = \frac{(1 + x^4)(2) - 2x(4x^3)}{(1 + x^4)^2}$$

$$= \frac{2 - 6x^4}{(1 + x^4)^2}$$

Thus, $f'(x) = 0$ only when $x = 0$, and because $f''(0) > 0$, the graph is concave up, so there is a relative minimum at $x = 0$. Because $f(0) = \tan^{-1}(0) = 0$, the minimum occurs at $(0, 0)$.

Next, note that $f''(x) = 0$ when $6x^4 = 2$—that is, when $x = \pm \sqrt[4]{\tfrac{1}{3}} \approx \pm 0.7598357$. Because $f''(x) < 0$ for $|x| > \sqrt[4]{\tfrac{1}{3}}$ and $f''(x) > 0$ for $-\sqrt[4]{\tfrac{1}{3}} < x < \sqrt[4]{\tfrac{1}{3}}$, the concavity of the graph changes at $x = \pm \sqrt[4]{\tfrac{1}{3}}$, and points of inflection must occur there. Finally, since $\lim\limits_{x \to +\infty} \tan^{-1} x^2 = \tfrac{\pi}{2} = \lim\limits_{x \to -\infty} \tan^{-1} x^2$, the graph has $y = \tfrac{\pi}{2}$ as a horizontal asymptote. The graph of f is shown in Figure 4.51.   ■

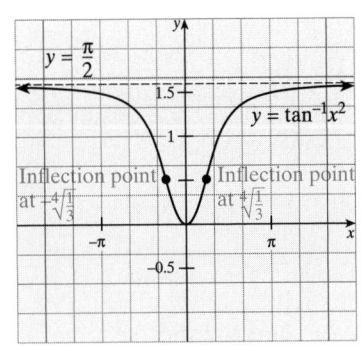

■ FIGURE 4.51   Graph of $f(x) = \tan^{-1} x^2$

GRAPHING STRATEGY

It is worthwhile to combine the techniques of curve sketching from calculus with those techniques studied in precalculus courses. You may be familiar with **extent** (finding the domain and the range of the function) and **symmetry** (with respect to the *x*-axis, *y*-axis, or origin). These features are reviewed in Chapter 4 of the *Student Mathematics Handbook*. We now have all the tools we need to describe a general procedure for curve sketching, and this procedure is summarized in Table 4.2.

(SMH)

■ **TABLE 4.2   Graphing Strategy for a Function Defined by** $y = f(x)$

Step	Procedure
Simplify.	If possible, simplify algebraically the function you wish to graph.
Find derivatives and critical numbers.	Compute the first and second derivatives; factor, if possible, and find the first- and second-order critical numbers.
Apply the second-derivative test.	Use the second-derivative test to find the relative maxima or minima: Substitute the first-order critical numbers, c_1, with the following test: $f''(c_1) > 0$, relative minimum $f''(c_1) < 0$, relative maximum $f''(c_1) = 0$, test fails; use first-derivative test
Determine concavity and points of inflection.	Use the second-order critical numbers, c_2: Substitute $f''(c_2) > 0$, concave up; $f''(c_2) < 0$, concave down. Find the points of inflection. These are located where the concavity changes from up to down or down to up, and are found at the x-intercepts of the second derivative.

Apply the first-derivative test.
↗ ↘
Maximum
↘ ↗
Minimum
	Use the first-derivative test if the second-derivative test fails or is too complicated.
	1. Let c be a first-order critical number of f.
	2. a. $(c, f(c))$ is a relative minimum if
	  $f'(x) < 0$ for all x in an open interval (a, c) to the left of c and
	  $f'(x) > 0$ for all x in an open interval (c, b) to the right of c.
	  b. $(c, f(c))$ is a relative maximum if
	  $f'(x) > 0$ for all x in an open interval (a, c) to the left of c and
	  $f'(x) < 0$ for all x in an open interval (c, b) to the right of c.
	  c. $(c, f(c))$ is not an extremum if $f'(x)$ has the same sign in open intervals (a, c) and (c, b) on both sides of c.

Determine intervals of increase and decrease.	Use the first-order critical numbers of f:
	  $f'(x) > 0$, curve rising (indicate these regions by ↑)
	  $f'(x) < 0$, curve falling (indicate these regions by ↓)

Find asymptotes.	**1.** *Vertical asymptote*—The vertical asymptote, if one exists, will occur at a value $x = c$ for which f is not defined. Use the limits $\lim_{x \to c^-} f(x)$ and $\lim_{x \to c^+} f(x)$ to determine the behavior of the graph near $x = c$. Show this behavior with arrows (↑ ↓). **2.** *Horizontal asymptote*—Compute $\lim_{x \to +\infty} f(x)$ and $\lim_{x \to +\infty} f(x)$. If either is finite, plot the associated horizontal asymptote. **3.** *Oblique asymptote*—If f is a rational function with no common factors and with the degree of the numerator 1 more than the degree of the denominator, then write $f(x) = mx + b + r(x)/d(x)$ and plot the asymptote $y = mx + b$.
Plot points.	**1.** If c is a relative maximum, relative minimum, or inflection point, plot $(c, f(c))$. Show the relative maximum points by using a "cap" (⌢), and relative minimum points by using a "cup" (⌣). **2.** x-intercepts: Set $y = 0$ and if $x = a$ is a solution, plot $(a, 0)$. y-intercepts: Set $x = 0$ and if $y = b$ is a solution, plot $(0, b)$; a function will have, at most, one y-intercept. **3.** Asymptote intercepts: Plot those points where f intersects horizontal or oblique asymptotes.
Sketch the curve.	Draw a smooth curve through the plotted points.

4.5 Problem Set

Ⓐ 1. ■ **What Does This Say?** Outline a method for curve sketching.

2. ■ **What Does This Say?** What are critical numbers? Discuss the importance of critical numbers in curve sketching.

3. ■ **What Does This Say?** Discuss the importance of concavity and points of inflection in curve sketching.

4. ■ **What Does This Say?** Discuss the importance of asymptotes in curve sketching.

Evaluate the limits in Problems 5–24.

5. $\lim\limits_{x \to +\infty} \dfrac{2{,}000}{x + 1}$

6. $\lim\limits_{x \to +\infty} \dfrac{7{,}000}{\sqrt{x} + 1}$

7. $\lim\limits_{x \to +\infty} \dfrac{3x + 5}{x - 2}$

8. $\lim\limits_{x \to +\infty} \dfrac{x + 2}{3x - 5}$

9. $\lim\limits_{t \to +\infty} \dfrac{9t^5 + 50t^2 + 800}{t^5 - 1{,}000}$

10. $\lim\limits_{x \to -\infty} \dfrac{(2x + 5)(x - 3)}{(7x - 2)(4x + 1)}$

11. $\lim\limits_{x \to +\infty} \dfrac{x}{\sqrt{x^2 + 1{,}000}}$

12. $\lim\limits_{x \to -\infty} \dfrac{3x}{\sqrt{4x^2 + 10}}$

13. $\lim\limits_{x \to +\infty} \dfrac{x^{5.916} + 1}{x^{\sqrt{35}}}$

14. $\lim\limits_{x \to +\infty} \dfrac{x^{6.083} + 1}{x^{\sqrt{37}}}$

15. $\lim\limits_{x \to 1^-} \dfrac{x - 1}{|x^2 - 1|}$

16. $\lim\limits_{x \to 3^+} \dfrac{x^2 - 4x + 3}{x^2 - 6x + 9}$

17. $\lim\limits_{x \to 0^+} \dfrac{x^2 - x + 1}{x - \sin x}$

18. $\lim\limits_{x \to \pi/4^+} \dfrac{\sec x}{\tan x - 1}$

19. $\lim\limits_{x \to +\infty} \left(x \sin \dfrac{1}{x} \right)$

20. $\lim\limits_{x \to 0^+} \dfrac{x^2}{1 - \cos x}$

21. $\lim\limits_{x \to 0^+} \dfrac{\ln \sqrt[3]{x}}{\sin x}$

22. $\lim\limits_{x \to +\infty} \dfrac{\ln \sqrt[3]{x}}{\sin x}$

23. $\lim\limits_{x \to -\infty} e^x \sin x$

24. $\lim\limits_{x \to +\infty} \dfrac{\tan^{-1} x}{e^{0.1x}}$

Ⓑ *Find all vertical and horizontal asymptotes of the graph of the functions given in Problems 25–44. Find where each graph is rising and where it is falling, determine concavity, and locate all critical points and points of inflection. Finally, sketch the graph.*

25. $f(x) = \dfrac{3x + 5}{7 - x}$

26. $g(x) = \dfrac{15}{x + 4}$

27. $f(x) = 4 + \dfrac{2x}{x - 3}$

28. $g(x) = x - \dfrac{x}{4 - x}$

29. $f(x) = \dfrac{x^3 + 1}{x^3 - 8}$

30. $f(x) = \dfrac{2x^2 - 5x + 7}{x^2 - 9}$

31. $g(x) = \dfrac{8}{x - 1} + \dfrac{27}{x + 4}$

32. $f(x) = \dfrac{1}{x + 1} + \dfrac{1}{x - 1}$

33. $g(t) = (t^3 + t)^2$

34. $g(t) = t^{-1/2} + \frac{1}{3}t^{3/2}$

35. $f(x) = (x^2 - 9)^2$

36. $g(x) = x(x^2 - 12)$

37. $f(x) = x^{1/3}(x - 4)$

38. $f(u) = u^{2/3}(u - 7)$

39. $f(x) = (2x^2 + 3x)e^{-x}$

40. $f(x) = \ln(4 - x^2)$

41. $T(\theta) = \sin \theta - \cos \theta$ for $0 \le \theta \le 2\pi$

42. $f(x) = x - \sin 2x$ for $0 \le x \le \pi$

43. $f(x) = \sin^2 x - 2 \sin x + 1$ for $0 \le x \le \pi$

44. $T(\theta) = \tan^{-1} \theta - \tan^{-1} \dfrac{\theta}{3}$

45. **MODELING PROBLEM** The *ideal speed v* for a banked curve on a highway is modeled by the equation

$$v^2 = gr \tan \theta$$

where g is the constant acceleration due to gravity, r is the radius of the curve, and θ is the angle of the bank. Assuming that r is constant, sketch the graph of v as a function of θ for $0 \le \theta \le \frac{\pi}{2}$.*

46. **MODELING PROBLEM** According to Einstein's special theory of relativity, the mass of a body is modeled by the expression

$$m = \dfrac{m_0}{\sqrt{1 - \dfrac{v^2}{c^2}}}$$

where m_0 is the mass of the body at rest in relation to the observer, m is the mass of the body when it moves with speed v in relation to the observer, and c is the speed of light. Sketch the graph of m as a function of v. What happens as $v \to c$?

47. **THINK TANK PROBLEM** Sketch a graph of a function f with all the following properties:

The graph has $y = 1$ and $x = 3$ as asymptotes;

f is increasing for $x < 3$ and $3 < x < 5$ and is decreasing elsewhere;

The graph is concave up for $x < 3$ and for $x > 7$ and concave down for $3 < x < 7$;

$$f(0) = 4 = f(5) \text{ and } f(7) = 2.$$

48. **THINK TANK PROBLEM** Sketch a graph of a function g with all the following properties:

(1) g is increasing for $x < -1$ and decreasing only for $x > 3$;

(2) The graph has only one critical point $(1, -1)$, and no inflection points;

(3) $\lim\limits_{x \to -\infty} g(x) = -1$; $\lim\limits_{x \to +\infty} g(x) = 2$;
   $\lim\limits_{x \to -1^+} g(x) = \lim\limits_{x \to 3} g(x) = -\infty$

---

*In physics it is shown that if one travels around the curve at the ideal speed, no frictional force is required to prevent slipping. This greatly reduces wear on tires and contributes to safet

49. ■ **What Does This Say?** Frank Kornerkutter has put off doing his math homework until the last minute, and he is now trying to evaluate

$$\lim_{x\to 0^+}\left(\frac{1}{x^2}-\frac{1}{x}\right)$$

At first he is stumped, but suddenly he has an idea: Because

$$\lim_{x\to 0^+}\frac{1}{x^2}=+\infty \quad\text{and}\quad \lim_{x\to 0^+}\frac{1}{x}=+\infty$$

it must surely be true that the limit in question has the value $+\infty-(+\infty)=0$. Having thus "solved" his problem, he celebrates by taking a nap. Is he right, and if not, what is wrong with his argument?

50. Find constants a and b that guarantee that the graph of the function defined by

$$f(x)=\frac{ax+5}{3-bx}$$

will have a vertical asymptote at $x=5$ and a horizontal asymptote at $y=-3$.

51. JOURNAL PROBLEM   *College Mathematics Journal** by Michael G. Murphy.

Find the oblique asymptote of the curve with equation

$$y=\frac{x^2+3x+7}{x+2}$$

Solution 1. By division,

$$\frac{x^2+3x+7}{x+2}=x+1+\frac{5}{x+2}$$

Because the final term tends to zero as x grows, the asymptote is the line of the equation $y=x+1$.

Solution 2. Follow a procedure frequently used in calculating limits at infinity:

$$\frac{x^2+3x+7}{x+2}=\frac{x+3+\frac{7}{x}}{1+\frac{2}{x}}$$

For large x, the value is approximately $x+3$, so the asymptote should be the line of the equation $y=x+3$.

Reconcile these solutions. What is wrong?

52. JOURNAL PROBLEM   *Parabola.*[†]

Draw a careful sketch of the curve $y=\dfrac{x^2}{x^2-1}$, indicating clearly any vertical or horizontal asymptotes, turning points, or points of inflection.

53. Let

$$P(x)=a_nx^n+a_{n-1}x^{n-1}+\cdots+a_1x+a_0$$

be a polynomial with $a_n\neq 0$ and let

$$L=\lim_{x\to-\infty}P(x) \text{ and } M=\lim_{x\to+\infty}P(x).$$

Fill in the missing entries in the following table:

Sign of a_n	n	L	M
+	even	**a.**	$+\infty$
+	odd	$-\infty$	**b.**
−	even	**c.**	**d.**
−	odd	**e.**	$-\infty$

54. a. Show that, in general, the graph of the function

$$f(x)=\frac{ax^2+bx+c}{rx^2+sx+t}$$

will have $y=\dfrac{a}{r}$ as a horizontal asymptote and that when $br\neq as$, the graph will cross this asymptote at the point where

$$x=\frac{at-cr}{br-as}$$

b. Sketch the graph of each of the following functions:

$$g(x)=\frac{x^2-4x-5}{2x^2+x-10}$$

$$h(x)=\frac{3x^2-x-7}{-12x^2+4x+8}$$

55. HISTORICAL QUEST   The possibility of division by zero is a fact that causes special concern to mathematicians. One of the first recorded observations of division by zero comes from the twelfth-century Hindu mathematician Bhaskaracharya (also known as Bhaskara), who made the following observation: "The fraction, whose denominator is zero, is termed an infinite quantity." Bhaskaracharya then went on to give a very beautiful conception of infinity that involved his view of God and creation. Bhaskaracharya gave a solution for the so-called Pell equation

$$x^2=1+py^2$$

which is related to the problem of cutting a given sphere so the volumes of the two parts have a specified ratio.

Solve this equation for $p=2/3$ and write this Pell equation as $y=f(x)$.[‡] Find

$$\lim_{x\to+\infty}f(x)$$

C THINK TANK PROBLEMS   *In Problems* 56-59, *either show that the statement is generally true or find a counter example.*

56. If f and g are concave up on the interval I, then so is $f + g$.

57. If f is concave up and g is concave down on an interval I then fg is neither concave up nor concave down on I.

58. If $f(x) < 0$ and $f''(x) > 0$ for all x on I, then the function $g = f^2$ is concave up on I.

59. If $f(x) > 0$ and $g(x) > 0$ for all x on I and if f and g are concave up on I, then fg is also concave up on I.

60. Consider the rational function

$$f(x) = \frac{a_n x^n + a_{n-1}x^{n-1} + \cdots + a_1 x + a_0}{b_m x^m + b_{m-1}x^{m-1} + \cdots + b_1 x + b_0}$$

a. If $m > n$ and $b_m \neq 0$, show that the x-axis is the only horizontal asymptote of the graph of f.

b. If $m = n$, show that the line $y = a_n/b_m$ is the only horizontal asymptote of the graph of f.

c. If $m < n$, is it possible for the graph to have a horizontal asymptote? Explain.

61. ■ **What Does This Say?** State what you think should be the formal definition of each of the following limit statements:

a. $\lim\limits_{x \to c^+} f(x) = -\infty$ b. $\lim\limits_{x \to c^-} f(x) = +\infty$

62. Prove that if

$$\lim_{x \to +\infty} f(x) \quad \text{and} \quad \lim_{x \to +\infty} g(x)$$

both exist, so does

$$\lim_{x \to +\infty} [f(x) + g(x)]$$

and

$$\lim_{x \to +\infty} [f(x) + g(x)] = \lim_{x \to +\infty} f(x) + \lim_{x \to +\infty} g(x)$$

Hint: The key is to show that if

$$\left| f(x) - L \right| < \frac{\epsilon}{2}$$

for $x > N_1$ and

$$\left| g(x) - M \right| < \frac{\epsilon}{2}$$

for $x > N_2$, then

$$\left| [f(x) + g(x)] - (L + M) \right| < \epsilon$$

whenever $x > N$ for some number N. You should also show that N relates to N_1 and N_2.

63. Prove the following limit rule. If

$$\lim_{x \to c} f(x) = +\infty \quad \text{and} \quad \lim_{x \to c} g(x) = A \ (A > 0)$$

then

$$\lim_{x \to c} [f(x)g(x)] = +\infty$$

Hint: Notice that because $\lim\limits_{x \to c} g(x) = A$, the function $g(x)$ is near A when x is near c. Therefore, because $\lim\limits_{x \to +\infty} f(x) = +\infty$, the product $f(x)g(x)$ is large if x is near c. Formalize these observations for the proof.

4.6   *Optimization in the Physical Sciences and Engineering*

IN THIS SECTION   **optimization procedure, Fermat's principle of optics and Snell's law**   ■

In Chapter 2, we introduced the process involved with model building. Now that we have developed the necessary calculus skills, we shall use those skills to solve practical optimization problems.

OPTIMIZATION PROCEDURE

Nothing takes place in the world whose meaning is not that of some maximum or minimum.

LEONHARD EULER

It is common to ask for the best procedure, the greatest value, the least cost, or the shortest path. The process of developing something to the utmost extent is called **optimization.** Entire courses in mathematics are devoted to this topic, and in this section and the next we will develop procedures involving calculus to solve real-life problems that seek the maximum or minimum value.

Optimization Procedure (Modeled After Pólya's Procedure)

Step 1. Understand the problem. Ask yourself if you can separate the given quantities and those you must find. What is unknown? Draw a picture to help you understand the problem.

Step 2. Choose the variables. Decide which quantity is to be optimized (that is, maximized or minimized) and call it Q. Choose other variables for unknown quantities and label your diagram using these symbols.

Step 3. Express Q in terms of the variables defined in step 2. Use the information given in the problem and the principle of substitution to rewrite Q in terms of a single variable, say x. **In other words, Q may begin as a formula involving several variables, but by using given information or known formulas, the goal is to write Q as a function of *one* variable so that $Q = f(x)$.**

Step 4. Find the domain for the function $Q = f(x)$.

Step 5. Use calculus to find the *absolute* maximum or minimum value of f. In particular, if the domain of f is a closed interval $[a, b]$, then the procedure outlined in Section 4.1 can be used:

   a. Compute $f'(x)$ and find all critical numbers of f on $[a, b]$;
   b. Evaluate f at the endpoints a, b, and at each critical number, c;
   c. Compare the values in step **b** to find either the largest or smallest value.

Step 6. Convert the result obtained in step 5 back into the context of the original problem, making all appropriate interpretations. Be sure to answer the question asked.

EXAMPLE 1   *Maximizing a constrained area*

You need to fence in a rectangular play zone for children, to fit into a right-triangular plot with sides measuring 4 m and 12 m. What is the maximum area for this play zone?

Solution   A picture of the play zone is shown in Figure 4.52. Let x and y denote the length and width of the inscribed rectangular. The appropriate formula for the area is

$$A = \ell w = xy$$

We wish to maximize Q where $Q = A = xy$. Now, write this as a function of a single variable. To do this, note that

$$\triangle ABC \sim \triangle ADF$$

This means that corresponding sides of these triangles are proportional; therefore

$$\frac{4 - y}{4} = \frac{x}{12}$$
$$4 - y = \tfrac{1}{3}x$$
$$y = 4 - \tfrac{1}{3}x$$

We can now write Q as a function of x alone:

$$Q(x) = x(4 - \tfrac{1}{3}x) = 4x - \tfrac{1}{3}x^2$$

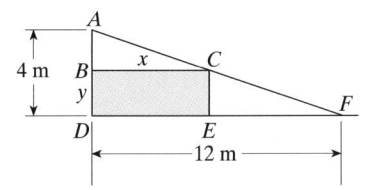

■ **FIGURE 4.52**   Children's play zone

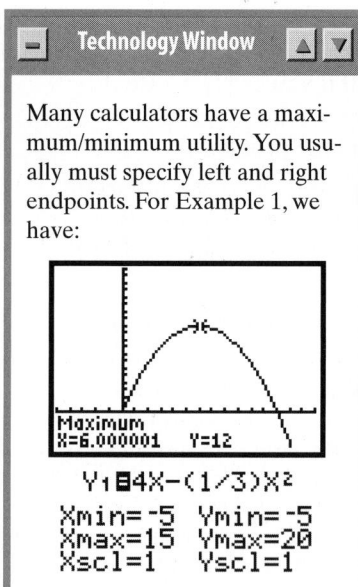
The domain for x is $0 \le x \le 12$. The critical numbers for Q are values such that $Q'(x) = 0$ (because there are no places where the derivative does not exist). Since

$$Q'(x) = 4 - \tfrac{2}{3}x$$

the critical number is $x = 6$. Evaluate $Q(x)$ at the endpoints and the critical number:

$$Q(6) = 4(6) - \tfrac{1}{3}(6)^2 = 12; \quad Q(0) = 0; \quad Q(12) = 0$$

The maximum area occurs when $x = 6$. This means that

$$y = 4 - \tfrac{1}{3}(6) = 2$$

The largest rectangular play zone that can be built in the triangular plot is a rectangle 6 m long and 2 m wide. ∎

EXAMPLE 2   *Maximizing a volume*

A carpenter wants to make an open-topped box out of a rectangular sheet of tin 24 in. wide and 45 in. long. The carpenter plans to cut congruent squares out of each corner of the sheet and then bend the edges of the sheet upward to form the sides of the box, as shown in Figure 4.53. If the box is to have the greatest possible volume, what should its dimensions be?

Solution   If each square corner cut out has side x, the box will be x inches deep, $45 - 2x$ inches long, and $24 - 2x$ inches wide.

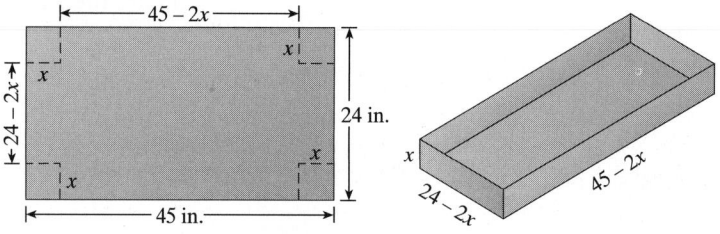

■ **FIGURE 4.53**   A box cut from a 24 in. by 45 in. piece of tin

The volume of the box shown in Figure 4.53 is

$$V(x) = x(45 - 2x)(24 - 2x) = 4x^3 - 138x^2 + 1{,}080x$$

To find the domain, we note that the dimensions must all be nonnegative; therefore, $x \ge 0$, $45 - 2x \ge 0$ (or $x \le 22.5$), and $24 - 2x \ge 0$ (or $x \le 12$). This implies that the domain is $[0, 12]$.

To find the critical numbers (the derivative is defined everywhere in the domain), we find values for which the derivative is 0:

$$V'(x) = 12x^2 - 276x + 1{,}080 = 12(x - 18)(x - 5)$$

The critical numbers are $x = 5$ and $x = 18$, but $x = 18$ is not in the domain, so the only relevant critical number is $x = 5$. Evaluating $V(x)$ at the critical number $x = 5$ and the endpoints $x = 0, x = 12$, we find

$$V(5) = 5(45 - 10)(24 - 10) = 2{,}450; \quad V(0) = 0; \quad V(12) = 0$$

Thus, the box with the largest volume is found when $x = 5$. Such a box has dimensions 5 in. $\times$ 14 in. $\times$ 35 in. ∎

WARNING   If the hypotheses of the extreme value theorem apply to a particular problem, we can often avoid testing a candidate to see if it is a maximum or a minimum. Suppose we have three candidates and two are obviously minima—the third *must be a maximum.* ◂

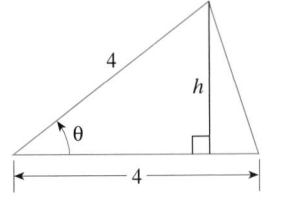

■ **FIGURE 4.54**   Triangle with 4-in. sides

| **EXAMPLE 3** | *Maximizing an area using trigonometry* |

Two sides of a triangle are 4 inches long. What should the angle between these sides be to make the area of the triangle as large as possible?

Solution   The triangle is shown in Figure 4.54. In general, the area of a triangle is given by

$$A = \tfrac{1}{2}bh$$

In this case, the base b is 4 and, because $\sin \theta = h/4$, the height h is $4 \sin \theta$. Thus,

$$A = \tfrac{1}{2}(4)(4 \sin \theta) = 8 \sin \theta$$

To find the domain, we note that the largest angle in a triangle is π; therefore $0 \le \theta \le \pi$. Also, because $A'(\theta)$ is defined throughout this interval, the critical numbers are found when the derivative is 0. We have $A'(\theta) = 8 \cos \theta$, which is zero only when $\theta = \frac{\pi}{2}$. Because

$$A(\tfrac{\pi}{2}) = 8 \sin \tfrac{\pi}{2} = 8; \quad A(0) = 8 \sin 0 = 0; \quad A(\pi) = 8 \sin \pi = 0$$

we see that the area is maximized when the angle θ is $\frac{\pi}{2}$, that is, when the triangle is a right triangle. ■

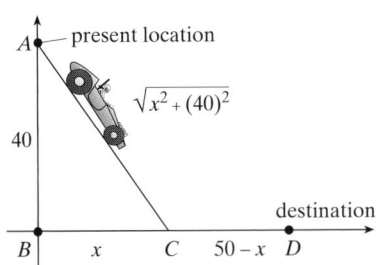

■ **FIGURE 4.55**   A path traveled by a dune buggy

| **EXAMPLE 4** | *Modeling Problem: Minimize time of travel* |

A dune buggy is on the desert at a point A located 40 km from a point B, which lies on a long, straight road, as shown in Figure 4.55. The driver can travel at 45 km/h on the desert and 75 km/h on the road. The driver will win a prize if she arrives at the finish line at point D, 50 km from B, in 85 min or less. Set up and analyze a model to help her decide on a route to minimize the time of travel. Does she win the prize?

Solution   Suppose the driver heads for a point C located x km down the road from B toward her destination, as shown in Figure 4.55. We want to minimize the time. We will need to remember the formula $d = rt$, or in terms of time, $t = d/r$.

$$
\begin{aligned}
\text{TIME} &= \text{TIME FROM } A \text{ TO } C + \text{TIME FROM } C \text{ TO } D \\[4pt]
&= \frac{\text{DISTANCE FROM } A \text{ TO } C}{\text{RATE FROM } A \text{ TO } C} + \frac{\text{DISTANCE FROM } C \text{ TO } D}{\text{RATE FROM } C \text{ TO } D} \\[4pt]
T(x) &= \frac{\sqrt{x^2 + 1,600}}{45} + \frac{50 - x}{75}
\end{aligned}
$$

The domain of T is $[0, 50]$. Next, find the derivative of time T with respect to x:

$$
\begin{aligned}
T'(x) &= \tfrac{1}{45}[\tfrac{1}{2}(x^2 + 1,600)^{-1/2}(2x)] + \tfrac{1}{75}(-1) \\[4pt]
&= \frac{x}{45\sqrt{x^2 + 1,600}} - \frac{1}{75} \\[4pt]
&= \frac{5x - 3\sqrt{x^2 + 1,600}}{225\sqrt{x^2 + 1,600}}
\end{aligned}
$$

The derivative exists for all x and is zero when

$$5x - 3\sqrt{x^2 + 1,600} = 0$$

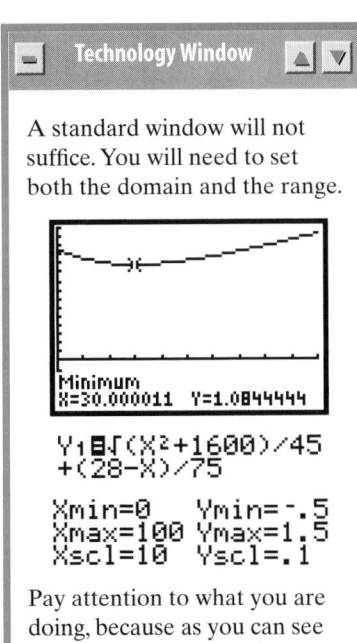

Technology Window

A standard window will not suffice. You will need to set both the domain and the range.

Minimum
X=30.000011   Y=1.0844444

Y1=√(X²+1600)/45
+(28-X)/75

Xmin=0   Ymin=-.5
Xmax=100 Ymax=1.5
Xscl=10   Yscl=.1

Pay attention to what you are doing, because as you can see here, the output may give only an approximate value (30.000011). Note from Example 4 that the minimum value is $x = 30$.

Solving this equation, we find $x = 30$ (-30 is extraneous). Evaluating $T(x)$ here and at the endpoints, we find that

$$T(30) = \frac{\sqrt{30^2 + 1{,}600}}{45} + \frac{50 - 30}{75} \approx 1.3778 \approx 1 \text{ hr } 23 \text{ min}$$

$$T(0) = \frac{\sqrt{0^2 + 1{,}600}}{45} + \frac{50 - 0}{75} \approx 1.5556 \approx 1 \text{ hr } 33 \text{ min}$$

$$T(50) = \frac{\sqrt{50^2 + 1{,}600}}{45} + \frac{50 - 50}{75} \approx 1.4229 \approx 1 \text{ hr } 25 \text{ min}$$

The driver can minimize the total driving time by heading for a point that is 30 miles from the point B and then traveling on the road to point D to win the prize. ∎

EXAMPLE 5   *Modeling Problem: Optimizing a constrained area*

A wire of length L is to be cut into two pieces, one of which will be bent to form a circle and the other to form a square. Study a model of the total area to determine how the wire should be cut to:
a. maximize the sum of the areas enclosed by the two pieces.
b. minimize the sum of the areas enclosed by the two pieces.

Solution   To understand the problem, we draw a sketch as shown in Figure 4.56, and label the radius of the circle r and the side of the square s.

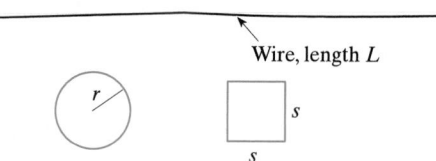

■ **FIGURE 4.56**   Forming a circle and a square from a wire with length L

We find that the combined area is

$$\text{AREA} = \text{AREA OF CIRCLE} + \text{AREA OF SQUARE} = \pi r^2 + s^2$$

We need to write the radius r and the side s in terms of the length of wire, L:

$$L = \text{CIRCUMFERENCE OF CIRCLE} + \text{PERIMETER OF SQUARE}$$
$$= 2\pi r + 4s$$

Thus, $s = \frac{1}{4}(L - 2\pi r)$. Remember, L is a given constant, so the variable for the area function is r (although the problem could just as easily have been done in terms of s). By substitution,

$$A(r) = \pi r^2 + [\tfrac{1}{4}(L - 2\pi r)]^2 = \pi r^2 + \tfrac{1}{16}(L - 2\pi r)^2$$

To find the domain, we note that $r \geq 0$ and that $L - 2\pi r \geq 0$, so the domain is $0 \leq r \leq \dfrac{L}{2\pi}$. Note when $r = 0$, there is no circle, and when $r = \dfrac{L}{2\pi}$ there is no square. The derivative of $A(r)$ is

$$A'(r) = 2\pi r + \tfrac{1}{8}(L - 2\pi r)(-2\pi)$$
$$= 2\pi r - \tfrac{\pi}{4}(L - 2\pi r)$$
$$= \tfrac{\pi}{4}(8r - L + 2\pi r)$$

Solve $A'(r) = 0$ to find $r = \dfrac{L}{2(\pi + 4)}$.

Thus, the extreme values of the area function on $\left[0, \dfrac{L}{2\pi}\right]$ must occur either at the endpoints or at the critical number $\dfrac{L}{2\pi + 8}$. Evaluating $A(r)$ at each of these numbers, we find

$$A(0) = \pi(0)^2 + \frac{1}{16}[L - 2\pi(0)]^2 = \frac{L^2}{16} = 0.0625L^2$$

$$A\left(\frac{L}{2\pi}\right) = \pi\left(\frac{L}{2\pi}\right)^2 + \frac{1}{16}\left[L - 2\pi\left(\frac{L}{2\pi}\right)\right]^2 = \frac{L^2}{4\pi} \approx 0.079\,577\,471L^2$$

$$A\left(\frac{L}{2\pi + 8}\right) = \pi\left(\frac{L}{2\pi + 8}\right)^2 + \frac{1}{16}\left[L - 2\pi\left(\frac{L}{2\pi + 8}\right)\right]^2 = \frac{L^2}{4(\pi + 4)}$$
$$\approx 0.035\,006\,197L^2$$

Comparing these values, we see that the smallest area occurs at $r = \dfrac{L}{2\pi + 8}$ and the largest area occurs at $r = \dfrac{L}{2\pi}$. To summarize:

1. To maximize the sum of the areas, do not cut the wire at all. Bend the wire to form a circle of radius $r = \dfrac{L}{2\pi}$.

2. To minimize the sum of the areas, cut the wire at the point located $\dfrac{2\pi L}{2\pi + 8}$
$= \dfrac{\pi L}{\pi + 4}$ units from one end, and form the circular part with radius $r = \dfrac{L}{2\pi + 8}$.

If you use a graphing calculator or a computer, you can verify this result (for $L = 1$) by looking at the graph of the area function, as shown in Figure 4.57. ∎

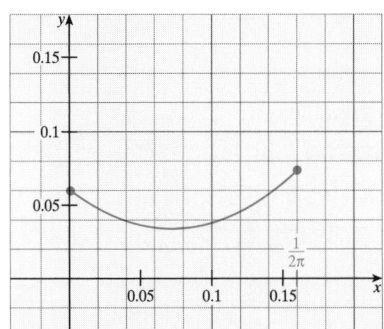

■ **FIGURE 4.57** Graph of $A(r) = \pi r^2 + \frac{1}{16}(1 - 2\pi r)^2$

EXAMPLE 6   *Modeling Problem: Optimizing an angle of observation*

A painting is hung on a wall in such a way that its upper and lower edges are 10 ft and 7 ft above the floor, respectively. An observer whose eyes are 5 ft above the floor stands x feet from the wall, as shown in Figure 4.58. How far away from the wall should the observer stand to maximize the angle subtended by the painting?

Solution   In Figure 4.58, θ is the angle whose vertex occurs at the observer's eyes at a point O located x feet from the wall. Note that α is the angle between the line $\overline{OA}$ drawn directly from the observer's eyes to the wall and the line $\overline{OB}$ from the eyes to the bottom edge of the painting. In $\triangle OAB$ the angle at O is α, with $\cot \alpha = \dfrac{x}{2}$. In $\triangle OAC$, the angle at O is $(\alpha + \theta)$, and $\cot(\alpha + \theta) = \dfrac{x}{5}$. By using the definition of inverse cotangent, we have

$$\theta = (\alpha + \theta) - \alpha = \cot^{-1}\left(\frac{x}{5}\right) - \cot^{-1}\left(\frac{x}{2}\right)$$

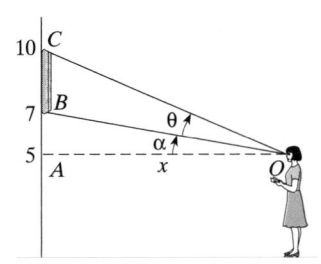

■ **FIGURE 4.58** The angle θ is subtended by observer's eye.

To maximize θ, we first compute the derivative

$$\frac{d\theta}{dx} = \frac{-1}{\left[1 + \left(\frac{x}{5}\right)^2\right]}\left(\frac{1}{5}\right) - \frac{-1}{\left[1 + \left(\frac{x}{2}\right)^2\right]}\left(\frac{1}{2}\right) = \frac{-5}{25 + x^2} + \frac{2}{4 + x^2}$$

Solving the equation $\dfrac{d\theta}{dx} = 0$ yields

$$-5(4 + x^2) + 2(25 + x^2) = 0$$
$$-3x^2 + 30 = 0$$
$$x = \pm\sqrt{10}$$

Because distance must be nonnegative, we reject the negative value. We apply the first-derivative test to show that the positive critical number $\sqrt{10}$ corresponds to a relative maximum (verify). Thus, the angle θ is maximized when the observer stands $\sqrt{10}$ ft away from the wall. ∎

PHYSICS: FERMAT'S PRINCIPLE OF OPTICS AND SNELL'S LAW

Light travels at different rates in different media; the more optically dense the medium, the slower the speed of transit. Consider the situation shown in Figure 4.59, in which a beam of light originates at a point A in one medium, then strikes the upper surface of a second, denser medium at a point P, and is refracted to a point B in the second medium.

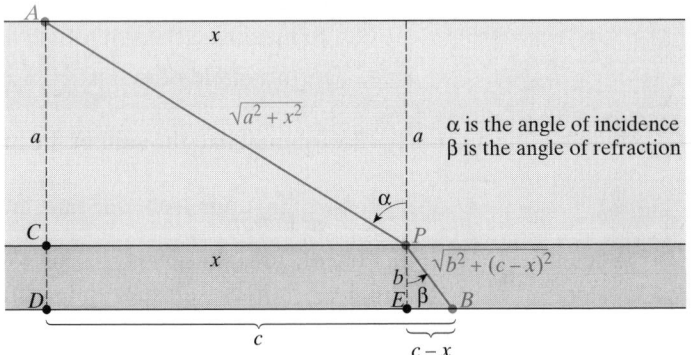

■ FIGURE 4.59   The path of a light beam through two media of different density

Suppose light travels with speed v_1 in the first medium and speed v_2 in the second. What can be said about the path followed by the beam of light?

Our method of investigating this question is based on the following optical property.

Fermat's Principle of Optics

Light travels between two points in such a way as to minimize the time of transit.

This problem is very similar to the dune buggy problem of Example 4: Light minimizes the time, and minimum time was the goal of Example 4. In Figure 4.59, let $a = |\overline{AC}|, b = |\overline{CD}|$, and $c = |\overline{DB}|$, and let x denote the distance from C to P. Because A and B are fixed points, the path APB is determined by the location of P, which, in turn, is determined by x. Because v_1 is a constant, the time required for the light to travel from A to P is given by

$$T_1 = \frac{|\overline{AP}|}{v_1} = \frac{\sqrt{a^2 + x^2}}{v_1}$$

and the time required for the light to go from P to B is

$$T_2 = \frac{|\overline{PB}|}{v_2} = \frac{\sqrt{b^2 + (c-x)^2}}{v^2}$$

Therefore, the total time of transit is

$$T = T_1 + T_2 = \frac{\sqrt{a^2 + x^2}}{v_1} + \frac{\sqrt{b^2 + (c-x)^2}}{v_2}$$

where it is clear that $0 \le x \le c$. According to Fermat's principle, the path followed by the beam of light is the one that corresponds to the smallest possible value of T; that is, we want to minimize T as a function of x.

Toward this end, we begin by finding $\dfrac{dT}{dx}$:

$$\frac{dT}{dx} = \frac{x}{v_1 \sqrt{a^2 + x^2}} - \frac{c-x}{v_2 \sqrt{b^2 + (c-x)^2}}$$

Note from Figure 4.59 that if α is the angle of incidence of the beam of light and β is the angle of refraction, then (from the definition of sine)

$$\sin \alpha = \frac{x}{\sqrt{a^2 + x^2}} \quad \text{and} \quad \sin \beta = \frac{c-x}{\sqrt{b^2 + (c-x)^2}}$$

Therefore (by substitution), we see the derivative of T can be expressed as

$$\frac{dT}{dx} = \frac{\sin \alpha}{v_1} - \frac{\sin \beta}{v_2}$$

and it follows that the only critical number occurs when

$$\frac{\sin \alpha}{\sin \beta} = \frac{v_1}{v_2}$$

By using the second-derivative test in the interval $[0, c]$, it can be shown that this critical number corresponds to an absolute minimum. The corresponding value of x enables us to locate P and hence to determine the path followed by the beam of light. We have established the following law of optics.

Snell's Law of Refraction

If a beam of light strikes the boundary between two media with angle of incidence α and is refracted through an angle β, then

$$\frac{\sin \alpha}{\sin \beta} = \frac{v_1}{v_2}$$

where v_1 and v_2 are the rates at which light travels through the first and second medium, respectively. The constant ratio

$$n = \frac{\sin \alpha}{\sin \beta}$$

is called the **relative index of refraction** of the two media.

4.6 Problem Set

Ⓐ 1. ■ **What Does This Say?** Describe an optimization procedure.

2. ■ **What Does This Say?** Why is it important to check endpoints when finding an optimum value?

Ⓑ 3. A woman plans to fence off a rectangular garden whose area is 64 ft². What should be the dimensions of the garden if she wants to minimize the amount of fencing used?

4. The highway department is planning to build a rectangular picnic area for motorists along a major highway. It is to have an area of 5,000 yd² and is to be fenced off on the three sides not adjacent to the highway. What is the least amount of fencing that will be needed to complete the job?

5. **EXPERIMENT** Pull out a sheet of $8\frac{1}{2}$-in. by 11-in. binder paper. Cut squares from the corners and fold the sides up to form a container. Show that the maximum volume of such a container is about 1 liter.

6. **JOURNAL PROBLEM** *Parabola.** Farmer Jones has to build a fence to enclose a 1,200 m² rectangular area *ABCD*. Fencing costs \$3 per meter, but Farmer Smith has agreed to pay half the cost of fencing $\overline{CD}$, which borders the property. Given *x* is the length of side $\overline{CD}$, what is the minimum amount (to the nearest cent) Jones has to pay?

7. Find the rectangle of largest area that can be inscribed in a semicircle of radius *R*, assuming that one side of the rectangle lies on the diameter of the semicircle.

8. A tinsmith wants to make an open-topped box out of a rectangular sheet of tin 24 in. wide and 45 in. long. The tinsmith plans to cut congruent squares out of each corner of the sheet and then bend the edges of the sheet upward to form the sides of the box. What are the dimensions of the largest box that can be made in this fashion?

9. Find the dimensions of the right circular cylinder of largest volume that can be inscribed in a sphere of radius *R*.

10. Given a sphere of radius *R*, find the radius *r* and altitude 2*h* of the right circular cylinder with largest lateral surface area that can be inscribed in the sphere. *Hint:* The lateral surface area is

$$S = 2\pi r 2h$$

11. Each edge of a square has length *L*. Determine the edge of the square of largest area that can be circumscribed about the given square.

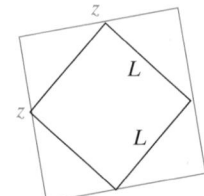

12. Find the dimensions of the right circular cylinder of largest volume that can be inscribed in a right circular cone of radius *R* and altitude *H*.

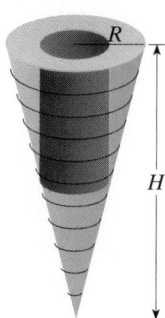

13. A truck is 250 mi due east of a sports car and is traveling west at a constant speed of 60 mi/h. Meanwhile, the sports car is going north at 80 mi/h. When will the truck and the car be closest to each other? What is the minimum distance between them? *Hint:* Minimize the square of the distance.

14. Show that of all rectangles with a given perimeter, the square has the largest area.

15. Show that of all rectangles with a given area, the square has the smallest perimeter.

16. A closed box with square base is to be built to house an ant colony. The bottom of the box and all four sides are to be made of material costing \$1/ft², and the top is to be constructed of glass costing \$5/ft². What are the dimensions of the box of greatest volume that can be constructed for \$72?

17. According to postal regulations, the girth plus the length of a parcel sent by fourth-class mail may not exceed 108 in. What is the largest possible volume of a rectangular parcel with two square sides that can be sent by fourth-class mail?

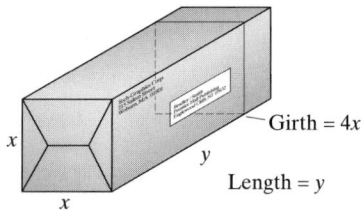

18. The bottom of an 8-ft-high mural painted on a vertical wall is 13 ft above the ground. The lens of a camera fixed to a tripod is 4 ft above the ground. How far from the wall should the camera be placed to photograph the mural with the largest possible angle?

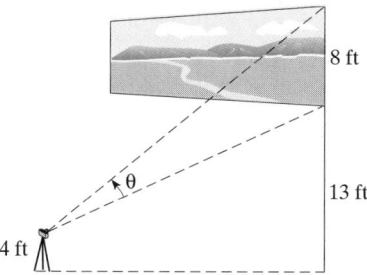

19. Missy Smith is at a point A on the north bank of a long, straight river 6 mi wide. Directly across from her on the south bank is a point B, and she wishes to reach a cabin C located s mi down the river from B. Given that Missy can row at 6 mi/h (including the effect of the current) and run at 10 mi/h, what is the minimum time (to the nearest minute) required for her to travel from A to C in each case?

a. $s = 4$ b. $s = 6$

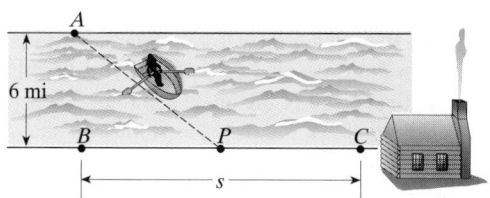

20. MODELING PROBLEM Two towns A and B are 12 mi apart and are located 5 and 3 mi, respectively, from a long, straight highway. A construction company has a contract to build a road from A to the highway and then to B. Analyze a model to determine the length (to the nearest mile) of the *shortest* road that meets these requirements.

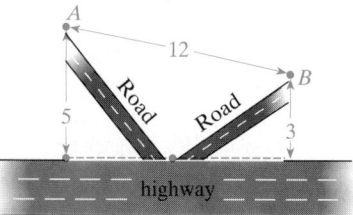

21. A poster is to contain 108 cm² of printed matter, with margins of 6 cm each at top and bottom and 2 cm on the sides. What is the minimum cost of the poster if it is to be made of material costing 20¢/cm²?

22. An isosceles trapezoid has a base of 14 cm and slant sides of 6 cm, as shown in Figure 4.60. What is the largest area of such a trapezoid?

■ **FIGURE 4.60** Area of a trapezoid

23. SPY PROBLEM It is noon. The spy has returned from space (Problem 51 of Section 3.4) and is driving a jeep through the sandy desert in the tiny principality of Alta Loma. He is 32 km from the nearest point on a straight, paved road. Down the road 16 km is a power plant in which a band of international terrorists has placed a time bomb set to explode at 12:50 P.M. The jeep can travel at 48 km/h in the sand and at 80 km/h on the paved road. If he arrives at the power plant in the shortest possible time, how long will our hero have to defuse the bomb?

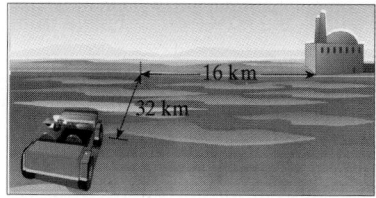

24. A cylindrical container with no top is to be constructed to hold a fixed volume of liquid. The cost of the material used for the bottom is 50¢/in.², and the cost of the material used for the curved face is 30¢/in.². Use calculus to find the radius of the least expensive container.

25. EXPERIMENT Use the fact that 12 oz $\approx$ 355 mℓ = 355 cm³ to find the dimensions of the 12-oz Coke® can that can be constructed using the least amount of metal. Compare these dimensions with a Coke from your refrigerator. What do you think accounts for the difference? An interesting article that discusses a similar question regarding tuna fish cans and the resulting responses is "What Manufacturers Say about a Max/Min Application," by Robert F. Cunningham, Trenton State College, *The Mathematics Teacher,* March 1994, pp. 172–175.

26. A stained glass window in the form of an equilateral triangle is built on top of a rectangular window, as shown in Figure 4.61. The rectangular part of the window is of clear glass and transmits twice as much light per square foot as the triangular part, which is made of stained glass. If the entire window has a perimeter of 20 ft, find the dimensions (to the nearest ft) of the window that will admit the most light.

■ **FIGURE 4.61** Maxmizing the amount of light

27. Figure 4.62 shows a thin lens located p cm from an object AB and q cm from the image RS.

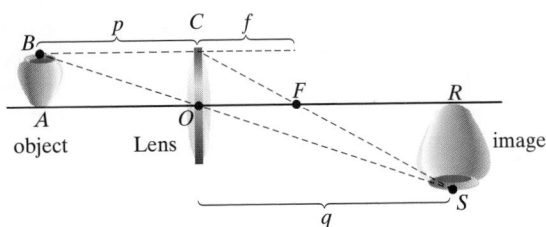

■ **FIGURE 4.62**   Image from a lens

The distance f from the center O of the lens to the point labeled F is called the **focal length** of the lens.

a. Using similar triangles, show that

$$\frac{1}{p} + \frac{1}{q} = \frac{1}{f}$$

b. Suppose a lens maker wished to have $p + q = 24$. What is the largest value of f for which this condition can be satisfied?

28. **MODELING PROBLEM**   One end of a cantilever beam of length L is built into a wall, and the other end is supported by a single post. The deflection, or "sag," of the beam at a point located x units from the built-in end is modeled by the formula

$$D = k(2x^4 - 5Lx^3 + 3L^2x^2)$$

where k is a positve constant. Where does the maximum deflection occur on the beam?

29. **MODELING PROBLEM**   When a mechanical system is at rest in an equilibrium position, its potential energy is minimized with respect to any small change in its position. Figure 4.63 shows a system involving a pulley, two small weights of mass m, and a larger weight of mass M.

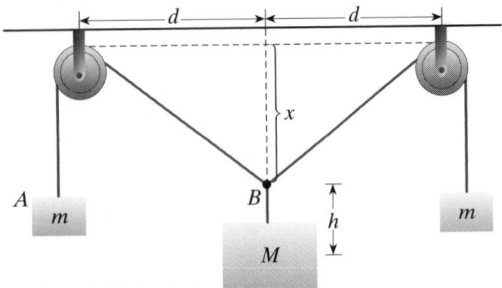

■ **FIGURE 4.63**   A pulley system

In physics, the total potential energy of the system is modeled by

$$E = -Mg(x + h) - 2mg(L - \sqrt{x^2 + d^2})$$

where x, h, and d are the distances shown in Figure 4.63, L is the length of the cord from A around one pulley to B, and g is the constant acceleration due to gravity. All the other symbols represent constants. Use this information to find the value of x for which E is minimized.

30. **MODELING PROBLEM**   A resistor of R ohms is connected across a battery of E volts whose internal resistance is r ohms. According to the principles of electricity, the formula

$$I = \frac{E}{R + r}$$

models the current (amperes) in the circuit, while

$$P = I^2R$$

models the power (watts) in the external resistor. Assuming that E and r are constant, what value of R will result in maximum power in the external resistor?

31. **MODELING PROBLEM**   A lamp with adjustable height hangs directly above the center of a circular kitchen table that is 8 ft in diameter. Model the illumination I at the edge of the table to be directly proportional to the cosine of the angle θ and inversely proportional to the square of the distance d, where θ and d are as shown in Figure 4.64.

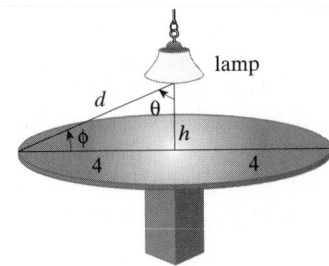

■ **FIGURE 4.64**   Illumination on a kitchen table

a. Show that

$$I(\theta) = \frac{k}{16} \cos \theta \sin^2 \theta$$

and find $I'(\theta)$.

b. Show that $I'(\theta) = 0$ when $\tan \theta_0 = \sqrt{2}$. What are $\sin \theta_0$ and $\cos \theta_0$?

c. How close to the table (to the nearest ft) should the lamp be pulled to maximize the illumination at the edge of the table?

32. In Problem 31, suppose the illumination is directly proportional to the sine of the angle ϕ and inversely proportional to the square of the distance d, where ϕ is the angle at which the ray of light meets the table. How close to the table (to the nearest ft) should the lamp be pulled to maximize the illumination at the edge of the table? Is your answer the same as in Problem 31? Explain why or why not.

33. **MODELING PROBLEM**   One model of a computer disk storage system uses the function

$$T(x) = N\left(k + \frac{c}{x}\right)p^{-x}$$

for the average time needed to send a file correctly by modem (including all retransmission of messages in which errors are detected), where x is the number of information bits, p is the (fixed) probability that any particular file will be received correctly, and N, k, and c are positive constants.*
a. Find $T'(x)$.
b. For what value of x is $T(x)$ minimized?
c. Sketch the graph of T.

34. **MODELING PROBLEM**   In crystallography, a fundamental problem is the determination of the *packing fraction* of a crystal lattice, which is defined as the fraction of space occupied by the atoms in the lattice, assuming the atoms are hard spheres. When the lattice contains exactly two different kinds of atoms, the packing fraction is modeled by the formula

$$F(x) = \frac{K(1 + c^2 x^3)}{(1 + x)^3}$$

where $x = \dfrac{r}{R}$ is the ratio of the radii of the two kinds of atoms in the lattice and c and K are positive constants.[†]

a. The function F has exactly one critical number. Find it and use the second-derivative test to determine whether it is a relative maximum or a relative minimum.

b. The numbers c and K and the domain of F depend on the cell structure in the lattice. For ordinary rock salt, it turns out that $c = 1$, $K = \dfrac{2\pi}{3}$, and the domain is $[\sqrt{2} - 1, 1]$. Find the largest and smallest values of F on this domain.

c. Repeat part **b** for β-cristobalite, for which $c = \sqrt{2}$, $K = \dfrac{\sqrt{3}\pi}{6}$, and the domain is $[0, 1]$.

35. If air resistance is neglected, it can be shown that the stream of water emitted by a fire hose will have height

$$y = -16(1 + m^2)(x/v)^2 + mx$$

above a point located x ft from the nozzle, where m is the slope of the nozzle, and v is the velocity of the stream of

water as it leaves the nozzle. (See Figure 4.65.) Assume v is constant.

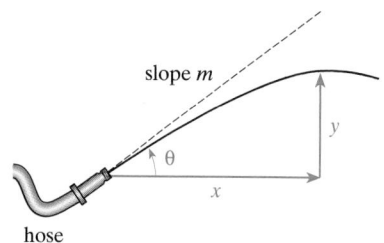

slope m

θ

y

x

hose

■ **FIGURE 4.65**   Stream of water from a fire hose

a. For fixed m, determine the distance x that results in maximum height.
b. If m is allowed to vary but the firefighter must stand $x = x_0$ ft from the base of a burning building, what is the highest point on the building that the firefighter can reach with the water from her hose?

36. A telecommunications satellite is located

$$r(t) = \frac{4{,}831}{1 + 0.15 \cos 0.06t}$$

miles above the center of the earth t minutes after achieving orbit.
a. Sketch the graph of $r(t)$.
b. What are the lowest and highest points on the satellite's orbit? These are called the *perigee* and *apogee* positions, respectively.

37. **MODELING PROBLEM**   In relativity theory, when a particle's mass is relatively large and its velocity is near the speed of light c, the relationship between its wavelength λ and total energy E is modeled by the formula

$$E = \sqrt{\left(\frac{hc}{\lambda}\right)^2 + m_0^2 c^4}$$

where m_0 is the rest mass of the particle and h is a constant (Planck's constant). Sketch the graph of $E(\lambda)$. What happens to E as $\lambda \to +\infty$?

38. **MODELING PROBLEM**   When the velocity of an object becomes comparable to the speed of light, the theory of relativity models the Doppler shift s by the formula

$$s = \sqrt{\frac{c + v}{c - v}} - 1$$

where c is the speed of light and $s \geq 0$.
a. Express v as a function of s and sketch the graph of $v(s)$.
b. Certain stellar objects called quasars appear to be moving away at velocities approaching c. The fastest one has a redshift of $s = 3.78$. What is the velocity in relation to c?

*Paul J. Campbell, "Calculus Optimization in Information Technology," UMAP Module 1991: *Tool for Teaching*. Lexington, MA: CUPM Inc., 1992; pp. 175–199.
[†]John C. Lewis and Peter P. Gillis, "Packing Factors in Diatomic Crystals," *American Journal of Physics*, Vol. 61, No. 5 (1993), pp. 434–438.

c. How fast is the velocity changing with respect to s when $s = 3.78$?

39. The lower right-hand corner of a piece of paper is folded over to reach the leftmost edge, as shown in Figure 4.66.

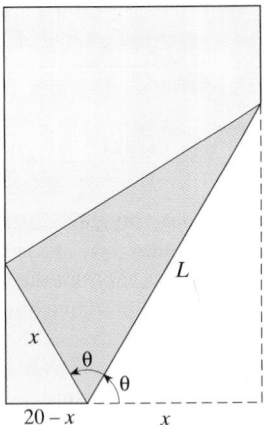

■ **FIGURE 4.66** Paper folding problem

If the page is 20 cm wide and 30 cm long, what is the length L of the shortest possible crease? *Hint:* First express $\cos \theta$ and $\cos 2\theta$ in terms of x and then use the trigonometric identity

$$\cos 2\theta = 2 \cos^2 \theta - 1$$

to eliminate θ and express L in terms of x alone.

40. Find the length of the longest pipe that can be carried horizontally around a corner joining two corridors that are $2\sqrt{2}$ ft wide. *Hint:* Show that the length L can be written as

$$L(\theta) = \frac{2\sqrt{2}}{\sin \theta} + \frac{2\sqrt{2}}{\cos \theta}$$

and find the absolute minimum of $L(\theta)$ on an appropriate interval, as shown in Figure 4.67.

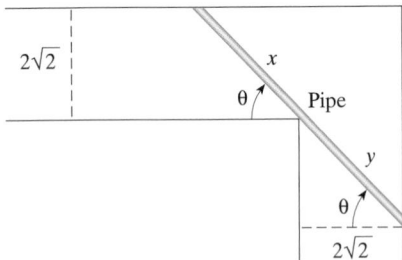

■ **FIGURE 4.67** Corner problem

41. **MODELING PROBLEM** It is known that water expands and contracts according to its temperature. Physical experiments suggest that an amount of water

that occupies 1 liter at 0 °C will occupy

$$V(T) = 1 - 6.42 \times 10^{-5}T$$
$$+ 8.51 \times 10^{-6}T^2 - 6.79 \times 10^{-8}T^3$$

liters when the temperature is T °C. At what temperature is $V(T)$ minimized? How is this result related to the fact that ice forms only at the upper levels of a lake during winter?

42. **MODELING PROBLEM** Light emanating from a source A is reflected by a mirror to a point B, as shown in Figure 4.68. Use Fermat's principle of optics to show that the angle of incidence α equals the angle of reflection β.

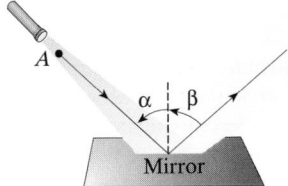

■ **FIGURE 4.68** Angle of incidence is equal to the angle of reflection

43. Congruent triangles are cut out of a square piece of paper 20 in. on a side, leaving a star-like figure that can be folded to form a pyramid, as indicated in Figure 4.69.

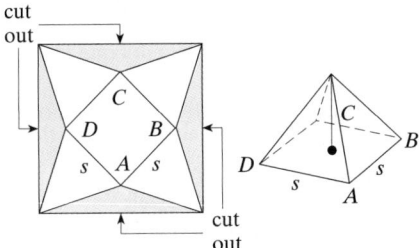

■ **FIGURE 4.69** Constructing a pyramid

What is the largest volume of the pyramid that can be formed in this manner?

44. **MODELING PROBLEM** In the study of Fraunhofer diffraction in optics, a light beam of intensity I_0 from a source L passes through a narrow slit and is diffracted onto a screen, as shown in Figure 4.70.

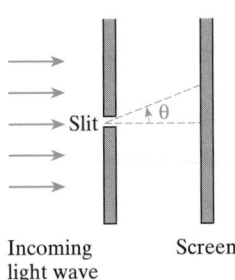

■ **FIGURE 4.70** Fraunhofer diffraction

Experiments suggest that the intensity $I(\theta)$ of light on the screen may be modeled by the formula

$$I(\theta) = I_0\left(\frac{\sin(\beta/2)}{\beta/2}\right)^2$$

where $\beta = 2\pi a \sin\left(\dfrac{\theta}{\lambda}\right)$, λ is the wavelength of the light, and a is related to the width of the slit.

a. Use the chain rule to find $\dfrac{dI}{d\theta}$.

b. Sketch the graph of $I(\beta)$. For what values of β is the intensity $I(\beta) = 0$? The corresponding angles result in black bands in the diffraction pattern.

45. MODELING PROBLEM A universal joint is a coupling used in cars and other mechanical systems to join rotating shafts together at an angle. In a model designed to explore the mechanics of the universal joint mechanism, the angular velocity $\beta(t)$ of the output (driven) shaft is modeled by the formula

$$\beta(t) = \frac{\alpha \cos \gamma}{1 - \sin^2 \gamma \, \sin^2(\alpha t)}$$

where α (in rad/s) is the angular velocity of the input (driving) shaft, and γ is the angle between the two shafts, as shown in Figure 4.71.*

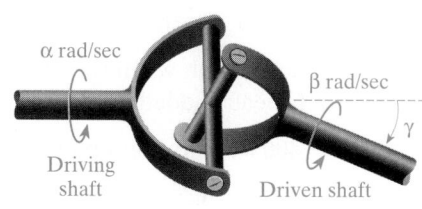

■ FIGURE 4.71 Universal joint

a. Sketch the graph of $\beta(t)$ for the case where α and γ are both positive constants.

b. What are the largest and smallest values of $\beta(t)$?

46. MODELING PROBLEM Rainbows are formed when sunlight traveling through the air is both reflected and refracted by raindrops.[†] Figure 4.72 shows a raindrop,

*Thomas O'Neil, "A Mathematical Model of a Universal Joint," *UMAP Modules 1982: Tools for Teaching.* Lexington, MA: Consortium for Mathematics and Its Applications, Inc., 1983, pp. 393–405.

[†]This problem is based on an article by Steve Janke, "Somewhere Within the Rainbow," *UMAP Modules 1992: Tools for Teaching.* Lexington, MA: Consortium for Mathematics and Its Applications, Inc., 1993.

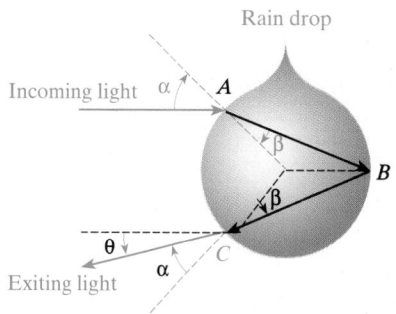

■ FIGURE 4.72 Light passing through a raindrop

which we assume to be a sphere for simplicity. An incoming beam of sunlight strikes the raindrop at point A with angle of incidence α, and some of the light is refracted through the angle β to point B. The process is then reversed as indicated in the figure, and the light finally exits the drop at point C.

a. At the interface points, A, B, and C, the light beam is deflected (from a straight-line path) and loses intensity. For instance, at A, the incoming beam is deflected through a clockwise rotation of $(\alpha - \beta)$ radians. Show that the total deflection (at all three interfaces) is

$$D = \pi + 2\alpha - 4\beta$$

b. For raindrops falling through air, Snell's law of refraction has the form $\sin \alpha = 1.33 \sin \beta$. Use this fact to express the derivative $D'(\alpha)$ in terms of α and β.

c. The intensity of rainbow light reaching the eye of an observer will be greatest when the total deflection $D(\alpha)$ is minimized. If the minimum deflection occurs when $\alpha = \alpha_0$, the corresponding angle of observation $\theta = \pi - D(\alpha_0)$ is called the *rainbow angle* because the rainbow is brightest when viewed in that direction (see Figure 4.73).

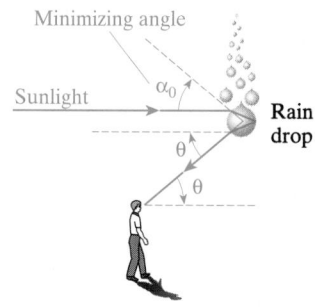

■ FIGURE 4.73 A rainbow is brightest when viewed at the rainbow angle

Solve $D'(\alpha) = 0$ and show that the critical number you find minimizes $D(\alpha)$. Then find the rainbow angle θ. (*Hint:* You may assume $0 \le \alpha - \beta \le \dfrac{\pi}{2}$.)

4.7 Optimization in Business, Economics, and the Life Sciences

IN THIS SECTION   optimization of discrete functions; *economics:* **two general principles of marginal analysis;** *business management:* **an inventory model and optimal holding time;** *physiology:* **concentration of drug in the bloodstream and optimal angle for vascular branching** ■

We begin with an example that illustrates how calculus can be used to model the price of a commodity that maximizes profit.

EXAMPLE 1   *Maximizing profits*

A manufacturer can produce a pair of earrings at a cost of $3. The earrings have been selling for $5 per pair and, at this price, consumers have been buying 4,000 pairs per month. The manufacturer is planning to raise the price of the earrings and estimates that for each $1 increase in the price, 400 fewer pairs of earrings will be sold each month. At what price should the manufacturer sell the earrings to maximize profit?

Solution   Let x denote the number of $1 price increases, and let $P(x)$ represent the corresponding profit.

$$\text{PROFIT} = \text{REVENUE} - \text{COST}$$
$$= (\text{NUMBER SOLD})(\text{PRICE PER PAIR}) - (\text{NUMBER SOLD})(\text{COST PER PAIR})$$
$$= (\text{NUMBER SOLD})(\text{PRICE PER PAIR} - \text{COST PER PAIR})$$

Recall that 4,000 pairs of earrings are sold each month when the price is $5 per pair and 400 fewer pairs will be sold each month for each added dollar in the price. Thus,

$$\text{NUMBER OF PAIRS SOLD} = 4,000 - 400 (\text{NUMBER OF \$ INCREASES})$$
$$= 4,000 - 400x$$

Knowing that the price per pair is $5 + x$, we can now write the profit as a function of x:

$$P(x) = (\text{NUMBER SOLD})(\text{PRICE PER PAIR} - \text{COST PER PAIR})$$
$$= (4,000 - 400x)[(5 + x) - 3] = 400(10 - x)(2 + x)$$

To find the domain, we note that $x \geq 0$. And $400(10 - x)$, the number of pairs sold, should be nonnegative, so $x \leq 10$. Thus, the domain is $[0, 10]$.

The critical numbers are found when the derivative is 0 (P is a polynomial function, so there are no values for which the derivative is not defined):

$$P'(x) = 400(10 - x)(1) + 400(-1)(2 + x)$$
$$= 400(8 - 2x) = 800(4 - x)$$

The critical number is $x = 4$ and the endpoints are $x = 0$ and $x = 10$. Checking for the maximum profit:

$$P(4) = 400(10 - 4)(2 + 4) = 14,400; \quad P(0) = 8,000; \quad P(10) = 0$$

The maximum possible profit is $14,400, which will be generated if the earrings are sold for $9.00 per pair. The graph of the profit function is shown in Figure 4.74. ■

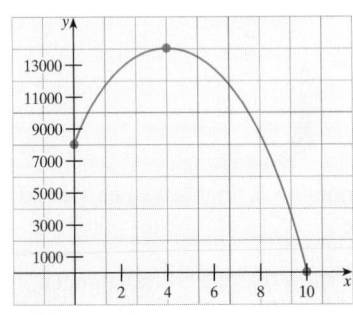

■ **FIGURE 4.74**   The profit function $P(x)$

OPTIMIZATION OF DISCRETE FUNCTIONS

Sometimes the function to be optimized has practical meaning only when its independent variable is a positive integer. This can lead to certain problems when applying the processes of calculus, because the theorems we have developed require continuous functions. If we model a function whose variable is defined as a positive integer, but as part of the modeling process we assume the function is defined for all real values (so that it is continuous), it may happen that the optimization procedure leads to a nonintegral value of the independent variable, and additional analysis is needed to obtain a meaningful solution.

EXAMPLE 2   *Maximizing a discrete revenue function*

A bus company will charter a bus that holds 50 people to groups of 35 or more. If a group contains exactly 35 people, each person pays $60. In larger groups, everybody's fare is reduced by $1 for each person in excess of 35. Determine the size of the group for which the bus company's revenue will be the greatest.

Solution   We wish to maximize the revenue:

REVENUE $=$ (NUMBER OF PEOPLE IN THE GROUP)(FARE PER PERSON)

Let x be the number of people in excess of 35 who take the trip. Then,

NUMBER OF PEOPLE IN THE GROUP $= 35 + x$

FARE PER PERSON $= 60 - x$

Let $R(x)$ be the revenue for the bus company:

$$R(x) = (35 + x)(60 - x) = 2{,}100 + 25x - x^2$$

Next, find the domain: We note that there must be at least 35 people ($x = 0$) and at most 50 people ($x = 15$); thus $0 \le x \le 15$, *but because x represents a number of people, it must also be an integer.*

The critical numbers are found by solving

$$R'(x) = 25 - 2x = 0$$

Because the derivative exists throughout the interval, the only critical number is $x = 12.5$. But x must be an integer, so $x = 12.5$ is not in the domain. To find the optimal *integer* solution, observe that R is increasing on $(0, 12.5)$ and decreasing on $(12.5, 15)$, as shown in Figure 4.75.

■ TABLE 4.3   **Finding the number of persons to maximize the revenue**

x	No.	Fare	Revenue
0	35	60	2,100
1	36	59	2,124
2	37	58	2,146
3	38	57	2,166
4	39	56	2,184
5	40	55	2,200
6	41	54	2,214
7	42	53	2,226
8	43	52	2,236
9	44	51	2,244
10	45	50	2,250
11	46	49	2,254
12	47	48	2,256
13	48	47	2,256
14	49	46	2,254
15	50	45	2,250

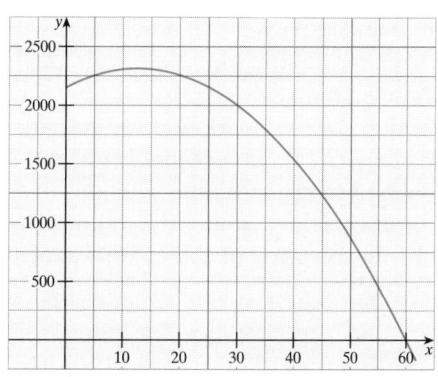

a. The continuous revenue function

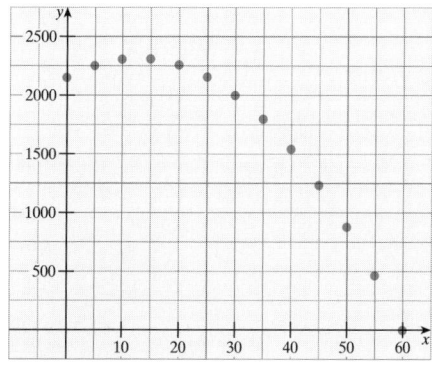

b. The discrete revenue function

■ FIGURE 4.75   Graphs of $R(x) = -x^2 + 25x + 2{,}100$

It follows that the optimal integer value of x is either $x = 12$ or $x = 13$. Because

$$R(12) = 2{,}256 \quad \text{and} \quad R(13) = 2{,}256$$

we conclude that the bus company's revenue will be greatest when the group contains either 12 or 13 people in excess of 35—that is, for groups of 47 or 48. In either case, the revenue will be $2,256.

The graph of revenue as a function of x is a collection of discrete points corresponding to the integer values of x, as indicated in Figure 4.75b. Technically, calculus cannot be used to study such a function, so instead, we worked with the differentiable function $R(x) = -x^2 + 25x + 2{,}100$, which is defined for all values of x and whose graph "connects" the points in the discrete graph. After applying calculus to this continuous model, we obtained a mathematical solution that was not the solution of the discrete practical problem, but that did suggest where to look for the required solution. To verify that this integer solution is correct, we can look at Table 4.3 in the margin, which shows all possibilities from $x = 0$ (35 people) to $x = 15$ (50 people). For each number of travelers, the total revenue is calculated, and we see that the maximum value 2,256 is obtained when $x = 12$ and again when $x = 13$.   ∎

ECONOMICS APPLICATIONS

In Section 3.8, we described *marginal analysis* as that branch of economics that is concerned with the way quantities such as price, cost, revenue, and profit vary with small changes in the level of production. Specifically, recall that if $C(x)$ is the total cost and $p(x)$ is the market price per unit when x units of a particular commodity are produced, then $R(x) = xp(x)$ is the **total revenue** and $P(x) = R(x) - C(x)$ is the profit function. These functions are shown in Figure 4.76.

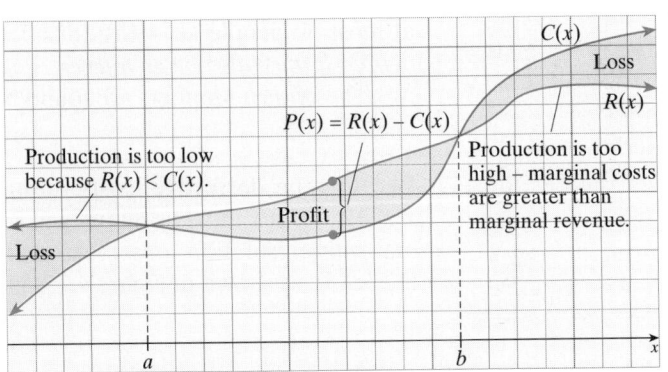

■ **FIGURE 4.76** Cost, revenue, and profit functions

In Section 3.8, we worked with marginal quantities and their role as rates of change, namely, $C'(x)$, the *marginal cost*, and $R'(x)$, the *marginal revenue*.*

We will now consider these functions in optimization problems. For example, the manufacturer certainly would like to know what level of production results in maximum profit. To solve this problem, we want to maximize the profit function

$$P(x) = R(x) - C(x)$$

---

*Recall (from Section 3.8) that marginal cost is the instantaneous rate of change of production cost C with respect to the number of units x_0 produced and the marginal revenue is the instantaneous rate of change of revenue with respect to the number of units x_0 produced.

We differentiate with respect to x and find that

$$P'(x) = R'(x) - C'(x)$$

Thus, $P'(x) = 0$ when $R'(x) = C'(x)$, and by using economic arguments we can show that a maximum occurs at the corresponding critical point.

Maximum Profit

Profit is maximized when marginal revenue equals marginal cost.

> **EXAMPLE 3**   *Maximizing profit*

A manufacturer estimates that when x units of a particular commodity are produced each month, the total cost (in dollars) will be

$$C(x) = \tfrac{1}{8}x^2 + 4x + 200$$

and all units can be sold at a price of $p(x) = 49 - x$ dollars per unit. Determine the price that corresponds to the maximum profit.

Solution   The marginal cost is $C'(x) = \tfrac{1}{4}x + 4$. The revenue is

$$\begin{aligned} R(x) &= xp(x) \\ &= x(49 - x) \\ &= 49x - x^2 \end{aligned}$$

The marginal revenue is $R'(x) = 49 - 2x$. The profit is maximized when $R'(x) = C'(x)$:

$$\begin{aligned} R'(x) &= C'(x) \\ 49 - 2x &= \tfrac{1}{4}x + 4 \\ x &= 20 \end{aligned}$$

Thus, the price that corresponds to the maximum profit is

$$p(20) = 49 - 20 = 29 \text{ dollars}$$ ∎

A second general principle of economics involves the following relationship between marginal cost and the *average cost* $A(x) = \dfrac{C(x)}{x}$.

Minimum Average Cost

Average cost is minimized at the level of production where the marginal cost equals the average cost.

To justify this second business principle, find the derivative of the average cost function:

$$A'(x) = \frac{xC'(x) - C(x)}{x^2} = \frac{C'(x) - \dfrac{C(x)}{x}}{x} = \frac{C'(x) - A(x)}{x}$$

Thus, $A'(x) = 0$ when $C'(x) = A(x)$, and once again, economic theory justifies this result as a minimum (rather than a maximum, which would seem possible using only calculus).

EXAMPLE 4   *Minimizing average cost*

A manufacturer estimates that when x units of a particular commodity are produced each month, the total cost (in dollars) will be

$$C(x) = \tfrac{1}{8}x^2 + 4x + 200$$

and they can all be sold at a price of $p(x) = 49 - x$ dollars per unit. Determine the level of production where the average cost is minimized.

Solution   The average cost is $A(x) = \dfrac{C(x)}{x} = \dfrac{1}{8}x + 4 + 200x^{-1}$ and $A(x)$ is minimized when $C'(x) = A(x)$. Thus,

$$\tfrac{1}{4}x + 4 = \tfrac{1}{8}x + 4 + 200x^{-1}$$
$$x^2 = 1{,}600 \qquad \text{Multiply both sides by } 8x \text{ and simplify.}$$

If we disregard the negative solution, it follows that the minimal average cost occurs when $x = 40$ units. You might have noticed that we did not use the information $p(x) = 49 - x$ in arriving at the solution. It is included because when doing real-world modeling, you must often make a choice about which parts of the available information are necessary to solve the problem. ∎

The average cost, marginal cost, marginal revenue, and total profit graphs for Examples 2 and 3 are shown in Figure 4.77.

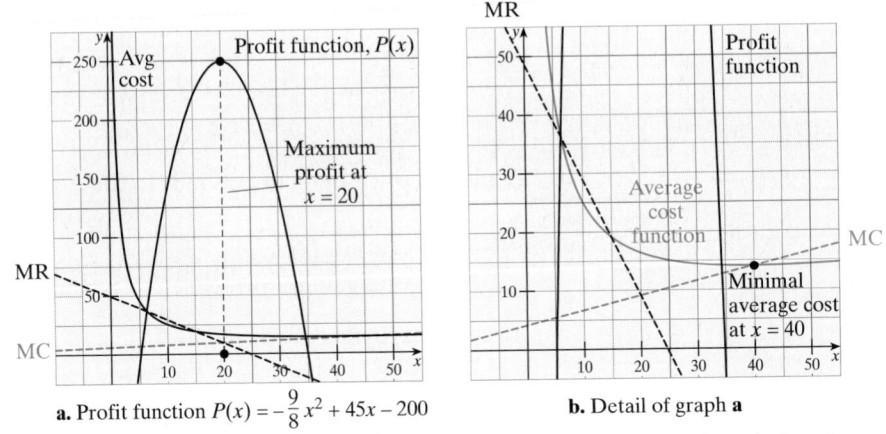

a. Profit function $P(x) = -\dfrac{9}{8}x^2 + 45x - 200$

b. Detail of graph **a**

■ **FIGURE 4.77**   Average cost, marginal cost, marginal revenue, and profit functions

Notice that the marginal revenue and marginal cost graphs intersect below the peak of the profit graph and that the marginal cost graph intersects the average cost graph at the lowest point of the average cost graph, as predicted by the theory.

BUSINESS MANAGEMENT APPLICATIONS

Sometimes mathematical methods can be used to assist managers in making certain business decisions. As an illustration, we shall show how calculus can be applied to a problem involving inventory control.

An Inventory Model   For each shipment of raw materials, a manufacturer must pay an ordering fee to cover handling and transportation. When the raw materials arrive, they must be stored until needed and storage costs result. If each shipment of raw materials is large, few shipments will be needed and ordering costs will be low, but storage costs will be high. If each shipment is small, ordering costs will be high because many shipments will be needed, but storage cost will be low. Managers want to determine the shipment size that will minimize the total cost. Here is an example of how the problem may be solved using calculus.

EXAMPLE 5   *Modeling Problem: Managing inventory to minimize cost*

A retailer orders 6,000 calculator batteries a year from a distributor and is trying to decide how often to order the batteries. The ordering fee is $20 per shipment, the storage cost is $0.96 per battery per year, and each battery costs the retailer $0.25. Suppose that the batteries are sold at a constant rate throughout the year and that each shipment arrives just as the preceding shipment has been used up. How many batteries should the retailer order each time to minimize the total cost?

Solution   We begin by writing the cost function:

$$\text{TOTAL COST} = \text{STORAGE COST} + \text{ORDERING COST} + \text{COST OF BATTERIES}$$

We need to find an expression for each of these unknowns. Assume that the same number of batteries must be ordered each time an order is placed; denote this number by x so that $C(x)$ is the corresponding total cost.

$$\text{STORAGE COST} = \left(\begin{array}{c}\text{AVERAGE NUMBER}\\ \text{IN STORAGE PER YR}\end{array}\right)\left(\begin{array}{c}\text{COST OF STORING 1}\\ \text{BATTERY FOR 1 YR}\end{array}\right) = \left(\frac{x}{2}\right)(0.96) = 0.48x$$

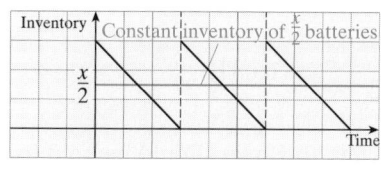

■ **FIGURE 4.78** Inventory graph

The average number of batteries in storage during the year is half of a given order (that is, $x/2$), and we assume that the total yearly storage cost is the same as if the $x/2$ batteries were kept in storage for the entire year. This situation is shown in Figure 4.78. To find the total ordering cost, we can multiply the ordering cost per shipment by the number of shipments. We also note that because 6,000 batteries are ordered during the year and because each shipment contains x batteries, the number of shipments is $6,000/x$.

$$\text{ORDERING COST} = \left(\begin{array}{c}\text{ORDERING COST}\\ \text{PER SHIPMENT}\end{array}\right)\left(\begin{array}{c}\text{NUMBER OF}\\ \text{SHIPMENTS}\end{array}\right) = (20)\left(\frac{6,000}{x}\right) = \frac{120,000}{x}$$

$$\text{COST OF BATTERIES} = \left(\begin{array}{c}\text{TOTAL NUMBER}\\ \text{OF BATTERIES}\end{array}\right)\left(\begin{array}{c}\text{COST PER}\\ \text{BATTERY}\end{array}\right) = 6,000(0.25) = 1,500$$

Thus, we model the total cost by the function

$$C(x) = 0.48x + \frac{120,000}{x} + 1,500$$

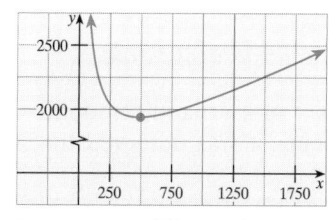

■ **FIGURE 4.79** The total cost function
$C(x) = 0.48x + 120,000x^{-1} + 1,500$

The goal is to minimize $C(x)$ on $(0, 6000]$. To obtain the critical numbers, we find the derivative

$$C'(x) = 0.48 - 120,000x^{-2}$$

and then solve $C'(x) = 0$:

$$0.48 - 120,000x^{-2} = 0$$
$$0.48x^2 = 120,000$$
$$x = \pm 500$$

The root $x = -500$ does not lie in the interval $(0,6000]$. It is easy to check that C is decreasing on $(0, 500)$ and increasing on $(500, 6000]$, as shown in Figure 4.79. Thus, the absolute minimum of C on the interval $(0, 6000]$ occurs when $x = 500$, and we conclude that to minimize cost, the manufacturer should order the batteries in lots of 500.

Optimal Holding Time   Even with an asset that increases in value, there often comes a time when continuing to hold the asset is less advantageous than selling it and investing the proceeds of the sale. How should the investor decide when to sell? One way is to hold the asset until the time its present value at the current prevailing rate of interest is maximized. In other words, hold until today's dollar equivalent of the selling price is as large as possible, then sell. Here is an example that illustrates this strategy.

EXAMPLE 6   *Modeling Problem: Optimal holding time*

Suppose you own an asset whose market price t years from now is estimated to be $V(t) = 10,000e^{\sqrt{t}}$ dollars. If the prevailing rate of interest is 8% compounded continuously, when should the asset be sold?

Solution   The present value of the asset in t years is modeled by the function $P(t) = V(t)e^{-rt}$ where r is the annual interest rate and t is the time in years. Thus,

$$P(t) = 10,000e^{\sqrt{t}}e^{-0.08t}$$
$$= 10,000 \exp(\sqrt{t} - 0.08t)$$

To maximize P, find $P'(t)$ and solve $P'(t) = 0$:

$$P'(t) = 10,000 \exp(\sqrt{t} - 0.08t)\left(\frac{1}{2} \cdot \frac{1}{\sqrt{t}} - 0.08\right)$$

$P'(t) = 0$ when $\dfrac{1}{2\sqrt{t}} - 0.08 = 0$ or $t \approx 39.06$ years. Thus, the asset should be held for 39 years and then sold.

PHYSIOLOGY APPLICATIONS

Calculus can be used to model a variety of situations in the biological and life sciences. We shall consider two examples from physiology. In the first, we use a model for the concentration of drug in the bloodstream of a patient to determine when the maximum concentration occurs.

Concentration of a Drug in the Bloodstream

EXAMPLE 7   *Modeling Problem: Maximum concentration of a drug*

Let $C(t)$ denote the concentration in the blood at time t of a drug injected into the body intramuscularly. In a now-classic paper by E. Heinz, it was observed that the concentration is given by

$$C(t) = \frac{k}{b-a}(e^{-at} - e^{-bt}) \qquad t \geq 0$$

where a, b (with $b > a$), and k are positive constants that depend on the drug.* At what time does the largest concentration occur? What happens to the concentration at $t \to +\infty$?

Solution   To locate the extrema, we solve $C'(t) = 0$:

$$C'(t) = \frac{d}{dt}\left[\frac{k}{b-a}(e^{-at} - e^{-bt})\right]$$

$$= \frac{k}{b-a}[(-a)e^{-at} - (-b)e^{-bt}] = \frac{k}{b-a}(be^{-bt} - ae^{-at})$$

We see that $C'(t) = 0$ when

$$be^{-bt} = ae^{-at}$$

$$e^{at-bt} = \frac{a}{b}$$

$$at - bt = \ln\frac{a}{b} \qquad \text{Definition of logarithm}$$

$$t = \frac{1}{a-b}\ln\frac{a}{b}$$

The second-derivative test can be used to show that the largest value of $C(t)$ occurs at t.

To see what happens to the concentration as $t \to +\infty$, we compute the limit

$$\lim_{t\to+\infty} C(t) = \lim_{t\to+\infty} \frac{k}{b-a}[e^{-at} - e^{-bt}]$$

$$= \frac{k}{b-a}\left[\lim_{t\to+\infty}\frac{1}{e^{at}} - \lim_{t\to+\infty}\frac{1}{e^{bt}}\right] = \frac{k}{b-a}[0-0]$$

This tells us that the longer the drug is in the blood, the closer the concentration is to 0. The graph of C is shown in Figure 4.80.

Intuitively, we would expect the Heinz concentration function to begin at 0, increase to a maximum, and then gradually drop off to 0. Figure 4.80 indicates that $C(t)$ does not have these characteristics, because it does not quite get back to 0 in finite time. This suggests that the Heinz model may apply most reliably to the period of time right after the drug has been injected.   ∎

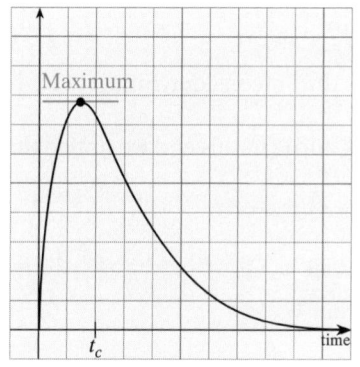

■ **FIGURE 4.80**   Graph of $c(t) = \frac{k}{b-a}(e^{-at} - c^{-bt})$

Modeling the Optimal Angle for Vascular Branching   The blood vascular system operates in such a way that the circulation of blood—from the heart, through the organs of the body, and back to the heart—is accomplished with as little expenditure of energy as possible. Thus, it is reasonable to expect that when an artery branches, the angle between the "parent" artery and its "daughter" should minimize the total resistance to the flow of blood. Figure 4.81 shows a small artery of radius r branching from a larger artery of radius R. Blood flows in the direction of the arrows from point A to the branch at B and then to points C and D. For simplicity, we assume that C and D are located in such a way that $\overline{CB}$ is perpendicular to the main line through A, B,

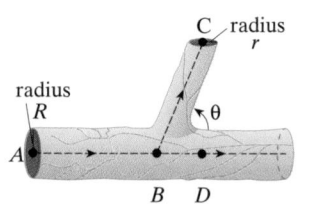

■ **FIGURE 4.81**   Vascular branching

*E. Heinz, "Problems bei der Diffusion kleiner Substanzmengen innerhelf des menschlichen Körpers," *Biochem.*, Volume 319 (1949), pp. 482–492.

and D. We wish to find the value of the branching angle θ that minimizes the total resistance to the flow of blood as it moves from A to B and then to point C, which is located a fixed perpendicular distance h from the line through A and B.*

Poiseuille's Law

The resistance to the flow of blood in an artery is directly proportional to the artery's length and inversely proportional to the fourth power of its radius.

According to Poiseuille's law, the resistance to flow from A to B is

$$f_1 = \frac{ks_1}{R^4}$$

and the resistance from B to C is

$$f_2 = \frac{ks_2}{r^4}$$

where k is a viscosity constant, $s_1 = \left|\overline{AB}\right|$, and $s_2 = \left|\overline{BC}\right|$. Thus, the total resistance to flow may be modeled by the sum

$$f = f_1 + f_2 + \frac{ks_1}{R^4} + \frac{ks_2}{r^4} = k\left(\frac{s_1}{R^4} + \frac{s_2}{r^4}\right)$$

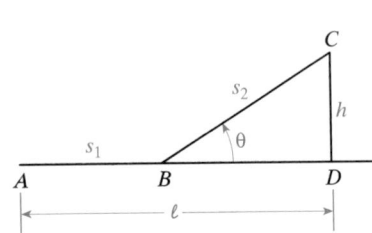

■ **FIGURE 4.82** Minimizing the resistance to blood flow

The next task is to write f as a function of θ. To do this, reconsider Figure 4.81 by labeling s_1, s_2, h, and ℓ as shown in Figure 4.82.

We want to find equations for s_1 and s_2 in terms of h and θ; to this end we notice that

$$\sin \theta = \frac{h}{s_2} \qquad\qquad \tan \theta = \frac{h}{\ell - s_1}$$

$$s_2 = \frac{h}{\sin \theta} \qquad\qquad \ell - s_1 = \frac{h}{\tan \theta}$$

$$= h \csc \theta \qquad\qquad s_1 = \ell - \frac{h}{\tan \theta}$$

$$= \ell - h \cot \theta$$

We can now write f as a function of θ:

$$f(\theta) = k\left(\frac{s_1}{R^4} + \frac{s_2}{r^4}\right) = k\left(\frac{\ell - h \cot \theta}{R^4} + \frac{h \csc \theta}{r^4}\right)$$

To minimize f we need to find the critical numbers (remember that h and k are constants):

$$\frac{df}{d\theta} = k\left[\frac{-h(-\csc^2 \theta)}{R^4} + \frac{h(-\csc \theta \cot \theta)}{r^4}\right] = kh \csc \theta \left(\frac{\csc \theta}{R^4} - \frac{\cot \theta}{r^4}\right)$$

---

*The key to solving this problem is a result due to work of the 19th-century French physiologist, Jean Louis Poiseuille (1799–1869). Our discussion of vascular branching is adapted from *Introduction to Mathematics for Life Scientists*, 2nd edition, by Edward Batschelet (New York: Springer-Verlag, 1976, pp. 278–280). In this excellent little book, Batschelet develops a number of interesting applications of calculus, several of which appear in the problem set.

Because $\csc\theta, k$, and h are never zero, we see that the derivative is zero when

$$\left(\frac{\csc\theta}{R^4} - \frac{\cot\theta}{r^4}\right) = 0$$

$$\frac{\csc\theta}{R^4} = \frac{\cot\theta}{r^4}$$

$$\frac{1}{R^4 \sin\theta} = \frac{\cos\theta}{r^4 \sin\theta}$$

$$\frac{r^4}{R^4} = \cos\theta$$

By finding the second derivative and noting that both r and R are positive, we can show that any value θ_m that satisfies this equation yields a value $f(\theta_m)$ that is a minimum. Thus, the required optimal angle is $\theta_m = \cos^{-1}(r^4/R^4)$.

4.7 Problem Set

Ⓐ *In Problems 1–3, we give the cost C of producing x units of a particular commodity and the selling price p when x units are produced. In each case, determine the level of production that maximizes profit.*

1. $C(x) = \frac{1}{8}x^2 + 5x + 98$ and $p(x) = \frac{1}{2}(75 - x)$

2. $C(x) = \frac{2}{5}x^2 + 3x + 10$ and $p(x) = \frac{1}{5}(45 - x)$

3. $C(x) = \frac{1}{5}(x + 30)$ and $p(x) = \dfrac{70 - x}{x + 30}$

4. Suppose the total cost of producing x units of a certain commodity is

$$C(x) = 2x^4 - 10x^3 - 18x^2 + x + 5$$

Determine the largest and smallest values of the marginal cost for $0 < x < 5$.

5. Suppose the total cost of manufacturing x units of a certain commodity is

$$C(x) = 3x^2 + x + 48$$

dollars. Determine the minimum average cost.

6. MODELING PROBLEM   A toy manufacturer produces an inexpensive doll (Flopsy) and an expensive doll (Mopsy) in units of x hundreds and y hundreds, respectively. Suppose it is possible to produce the dolls in such a way that

$$y = \frac{82 - 10x}{10 - x} \qquad 0 \le x \le 8$$

and that the company receives *twice* as much for selling a Mopsy doll as for selling a Flopsy doll. Find the level of production for both x and y for which the total revenue derived from selling these dolls is maximized. What vital assumption must be made about sales in this model?

7. MODELING PROBLEM   A business manager estimates that when p dollars are charged for every unit of a product, the sales will be $x = 380 - 20p$ units. At this level of production, the average cost is modeled by

$$A(x) = 5 + \frac{x}{50}$$

a. Find the total revenue and total cost functions, and express the profit as a function of x.

b. What price should the manufacturer charge to maximize profit? What is the maximum profit?

8. Suppose a manufacturer estimates that, when the market price of a certain product is p, the number of units sold will be

$$x = -6\ln\left(\frac{p}{40}\right)$$

It is also estimated that the cost of producing these x units will be

$$C(x) = 4xe^{-x/6} + 30$$

a. Find the average cost, the marginal cost, and the marginal revenue for this production process.

b. What level of production x corresponds to maximum profit?

9. A manufacturer can produce shoes at a cost of \$50 a pair and estimates that if they are sold for x dollars a pair, consumers will buy approximately

$$s(x) = 1,000e^{-0.1x}$$

pairs of shoes per week. At what price should the manufacturer sell the shoes to maximize profit?

10. A certain industrial machine depreciates in such a way that its value (in dollars) after t years is

$$Q(t) = 20{,}000e^{-0.4t}$$

   a. At what rate is the value of the machine changing with respect to time after 5 years?
   b. At what percentage rate is the value of the machine changing with respect to time after t years?

11. It is projected that t years from now, the population of a certain country will be $P(t) = 50e^{0.02t}$ million.
   a. At what rate will the population be changing with respect to time 10 years from now?
   b. At what percentage rate will the population be changing with respect to time t years from now?

12. **MODELING PROBLEM**   Some psychologists model a child's ability to memorize by a function of the form

$$g(t) = \begin{cases} t \ln t + 1 & \text{if } 0 < t \le 4 \\ 1 & \text{if } t = 0 \end{cases}$$

   where t is time, measured in years. Determine when the largest and smallest values of g occur.

13. It is estimated that t years from now the population of a certain country will be

$$p(t) = \frac{160}{1 + 8e^{-0.01t}}$$

   million. When will the population be growing most rapidly?

B 14. **MODELING PROBLEM**   The owner of the Pill Boxx drugstore expects to sell 600 bottles of hair spray each year. Each bottle costs $4, and the ordering fee is $30 per shipment. In addition, it costs 90¢ per year to store each bottle. Assuming that the hair spray sells at a uniform rate throughout the year and that each shipment arrives just as the last bottle from the previous shipment is sold, how frequently should shipments of hair spray be ordered to minimize the total cost?

15. **MODELING PROBLEM**   An electronics firm uses 18,000 cases of connectors each year. The cost of storing one case for a year is $4.50, and the ordering fee is $20 per shipment. Assume that the connectors are used at a constant rate throughout the year and that each shipment arrives just as the preceding shipment has been used up. How many cases should the firm order each time to keep total cost to a minimum?

16. Suppose the total cost (in dollars) of manufacturing x units of a certain commodity is

$$C(x) = 3x^2 + 5x + 75$$

   a. At what level of production is the average cost per unit the smallest?
   b. At what level of production is the average cost per unit equal to the marginal cost?
   c. Graph the average cost and the marginal cost functions on the same set of axes, for $x > 0$.

17. Suppose the total revenue (in dollars) from the sale of x units of a certain commodity is

$$R(x) = -2x^2 + 68x - 128$$

   a. At what level of sales is the average revenue per unit equal to the marginal revenue?
   b. Verify that the average revenue is increasing if the level of sales is less than the level in part **a** and decreasing if the level of sales is greater than the level in part **a.**
   c. On the same set of axes, graph the relevant portions of the average and marginal revenue functions.

18. A manufacturer finds that the demand function for a certain product is

$$x(p) = \frac{73}{\sqrt{p}}$$

   Should the price p be raised or lowered to increase consumer expenditure? Explain your answer.

19. **MODELING PROBLEM**   Suppose you own a rare book whose value t years from now is modeled as $300e^{\sqrt{3t}}$. If the prevailing rate of interest remains constant at 8% compounded continuously, when will be the most advantageous time to sell?

20. **MODELING PROBLEM**   Suppose you own a parcel of land whose value t years from now is modeled as

$$V(t) = 200 \ln \sqrt{2t}$$

   thousand dollars. If the prevailing rate of interest is 10% compounded continuously, when will be the most advantageous time to sell?

21. A store has been selling skateboards at the price of $40 per board, and at this price skaters have been buying 45 boards a month. The owner of the store wishes to raise the price and estimates that for each $1 increase in price, 3 fewer boards will be sold each month. If each board costs the store $29, at which price should the store sell the boards to maximize profit?

22. **MODELING PROBLEM**   As more and more industrial areas are constructed, there is a growing need for standards ensuring control of the pollutants released into the air. Suppose that the pollution at a particular location is based on the distance from the source of the pollution according to the principle that for distances greater than or equal to 1 mi, the concentration of particulate matter (in parts per million, ppm) decreases as the reciprocal of the distance from the source. This means that if you live 3 mi from a plant emitting 60 ppm, the pollution at your home is $\frac{60}{3} = 20$ ppm. On the other hand, if you live 10 miles from the plant, the pollution at your home is $\frac{60}{10} = 6$ ppm. Suppose that two plants 10 mi apart are releasing 60 and 240 ppm, respectively. At what point between the plants is the pollution a minimum? Where is it a maximum?

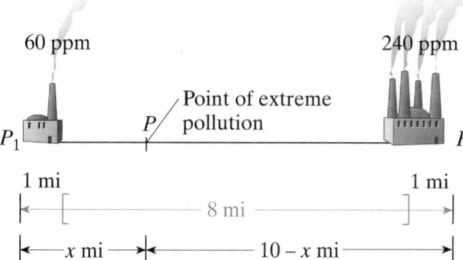

23. A tour agency is booking a tour and has 100 people signed up. The price of a ticket is $2,000 per person. The agency has booked a plane seating 150 people at a cost of $125,000. Additional costs to the agency are incidental fees of $500 per person. For each $10 that the price is lowered, a new person will sign up. How much should the price be lowered for all participants to maximize the profit to the tour agency?

24. A bookstore can obtain the best-seller *20,000 Leagues Under the Majors* from the publisher at a cost of $6 per book. The store has been offering the book at a price of $30 per copy and has been selling 200 copies a month at this price. The bookstore is planning to lower its price to stimulate sales and estimates that for each $2 reduction in the price, 20 more books will be sold per month. At what price should the bookstore sell the book to generate the greatest possible profit?

25. A Florida citrus grower estimates that if 60 orange trees are planted, the average yield per tree will be 400 oranges. The average yield will decrease by 4 oranges for each additional tree planted on the same acreage. How many trees should the grower plant to maximize the total yield?

26. Farmers can get $2 per bushel for their potatoes on July 1, and after that the price drops 2¢ per bushel per day. On July 1, a farmer has 80 bu of potatoes in the field and estimates that the crop is increasing at the rate of 1 bu per day. When should the farmer harvest the potatoes to maximize revenue?

27. A viticulturist estimates that if 50 grapevines are planted per acre, each grapevine will produce 150 lb of grapes. Each additional grapevine planted per acre (up to 20) reduces the average yield per vine by 2 lb. How many grapevines should be planted to maximize the yield per acre?

28. A commuter train carries 600 passengers each day from a suburb to a city. It now costs $5 per person to ride the train. A study shows that 50 additional people will ride the train for each 25¢ reduction in fare. What fare should be charged to maximize total revenue?

29. **MODELING PROBLEM** To raise money, a service club has been collecting used bottles, which it plans to deliver to a local glass company for recycling. Since the project began 80 days ago, the club has collected 24,000 pounds of glass, for which the glass company offers 1¢ per pound. However, because bottles are accumulating faster than they can be recycled, the company plans to reduce the price it will pay by 1¢ per 100 pounds of used glass.
   a. What is the most advantageous time for the club to conclude its project and deliver all the bottles?
   b. What assumptions must be made in part **a** to solve the problem with the given information?

30. Suppose that the demand function for a certain commodity is expressed as

$$p(x) = \sqrt{\frac{120 - x}{0.1}} \quad \text{for} \quad 0 \le x \le 120$$

where x is the number of items sold.
   a. Find the total revenue function explicitly and use its first derivative to determine the price at which revenue is maximized.
   b. Graph the relevant portions of the demand and revenue functions.

31. Suppose the demand equation for a certain commodity is linear, that is,

$$p(x) = \frac{b - x}{a} \quad \text{for} \quad 0 \le x \le b$$

where a and b are positive constants.
   a. Find the total revenue function explicitly and use its first derivative to determine its intervals of increase and decrease.
   b. Graph the relevant portions of the demand and revenue functions.

32. **MODELING PROBLEM** Homing pigeons rarely fly over large bodies of water unless forced to do so, presumably because it requires more energy to maintain altitude in flight over the cool water. Suppose a pigeon is released from a boat floating on a lake 3 mi from a point A on the shore and 10 mi away from the pigeon's loft, as shown in Figure 4.83.

■ **FIGURE 4.83** Flight path for a pigeon

Assuming the pigeon requires twice as much energy to fly over water as over land and it follows a path that minimizes total energy, find the angle θ of its heading as it leaves the boat. (This is the situation in the perspective section of the chapter introduction.)

33. **MODELING PROBLEM** In an experiment, a fish swims s meters upstream at a constant velocity v m/sec relative to the water, which itself has velocity v_1 relative

to the ground. The results of the experiment suggest that if the fish takes t seconds to reach its goal (that is, to swim s meters), the energy it expends is given by

$$E = cv^k t$$

where $c > 0$ and $k > 2$ are physical constants. Assuming v_1 is known, what velocity v minimizes the energy?

34. In a learning model, two responses (A and B) are possible for each of a series of observations. If there is a probability p of getting response A in any observation, the probability of getting a response A exactly n times in a series of M observations is

$$F(p) = p^n(1 - p)^{M-n}$$

The *maximum likelihood estimate* is the value of p that maximizes $F(p)$ on $[0, 1]$. For what value of p does this occur?

35. **MODELING PROBLEM**   The production of blood cells plays an important role in medical research involving leukemia and other so-called *dynamical diseases*. In 1977, a mathematical model was developed by A. Lasota that involved the cell production function

$$P(x) = Ax^s e^{-sx/r}$$

where A, s, and r are positive constants and x is the number of granulocytes (a type of white blood cell) present.*

a. Find the granulocyte level x that maximizes the production function P. How do you know it is a maximum?

b. If $s > 1$, show that the graph of P has two inflection points. Sketch the graph of P and give a brief interpretation of the inflection points.

c. Sketch the graph for the case where $0 < s < 1$. What is different about this case?

36. **MODELING PROBLEM**   During a cough, the diameter of the trachea decreases. The velocity v of air in the trachea during a cough may be modeled by the formula

$$v = Ar^2(r_0 - r)$$

where A is a constant, r is the radius of the trachea during the cough, and r_0 is the radius of the trachea in a relaxed state. Find the radius of the trachea when the velocity is greatest, and find the maximum velocity of air. (Notice that $0 \le r \le r_0$.)

37. **MODELING PROBLEM**   The work of V. A. Tucker and K. Schmidt-Koenig[†] models the energy expended in

flight by a certain kind of bird by the function

$$E = \frac{1}{v}[a(v - b)^2 + c]$$

where a, b, and c are positive constants, v is the velocity of the bird, and the domain of v is $[16, 60]$. What value of v will minimize the energy expenditure in the case where $a = 0.04$, $b = 36$, and $c = 9$?

38. **MODELING PROBLEM**   A plastics firm has received an order from the city recreation department to manufacture 8,000 special Styrofoam kickboards for its summer swimming program. The firm owns 10 machines, each of which can produce 50 kickboards per hour. The cost of setting up the machines to produce the kickboards is $800 per machine. Once the machines have been set up, the operation is fully automated and can be overseen by a single production supervisor earning $35 per hour.

a. How many machines should be used to minimize the cost of production?

b. How much will the supervisor earn during the production run if the optimal number of machines is used?

39. **MODELING PROBLEM**   After hatching, the larva of the codling moth goes looking for food. The period between hatching and finding food is called the *searching period*. According to a model developed by P. L. Shaffer and H. J. Gold,[‡] the length (in days) of the searching period is given by

$$S(T) = (-0.03T^2 + 1.67T - 13.67)^{-1}$$

where T (°C) is the air temperature ($20 \le T \le 30$), and the percentage of larvae that survive the searching period is

$$N(T) = -0.85T^2 + 45.45T - 547$$

a. Sketch the graph of N and find the largest and smallest survival percentages for the allowable range of temperatures, $20 \le T \le 30$.

b. Find $S'(T)$ and solve $S'(T) = 0$. Sketch the graph of S for $20 \le T \le 30$.

c. The percentage of codling moth eggs that hatch at a given temperature T is given by

$$H(T) = -0.53T^2 + 25T - 209$$

for $20 \le T \le 30$. Sketch the graph of H and determine the temperatures at which the largest and smallest hatching occur.

40. **MODELING PROBLEM**   According to a certain logistic model, the world's population (in billions) t years after 1960 is modeled by the function

$$P(t) = \frac{40}{1 + 12e^{-0.08t}}$$

*See "A Blood Cell Population Model, Dynamical Diseases, and Chaos," by W. B. Gearhart and M. Martelli, *UMAP Modules 1990: Tools for Teaching*. Arlington, MA: Consortium for Mathematics and Its Applications (CUPM) Inc., 1991.

†V. A. Tucker and K. Schmidt-Koenig, "Flight of Birds in Relation to Energetics and Wind Directions," *The Auk* 88 (1971), pp. 97–107.

‡P. L. Shaffer and H.J. Gold, "A Simulation Model of Population Dynamics of the Codling Moth *Cydia Pomonella*," *Ecological Modeling*, Vol. 30, 1985, pp. 247–274.

a. If this model is correct, at what rate will the world's population be increasing with respect to time in the year 2000? At what percentage rate will it be increasing at this time?

b. Sketch the graph of P. What feature on the graph corresponds to the time when the population is growing most rapidly? What happens to $P(t)$ as $t \to +\infty$ (that is, "in the long run")?

41. MODELING PROBLEM   Generalize Problem 38. Specifically, a manufacturing firm received an order for Q units of a certain commodity. Each of the firm's machines can produce n units per hour. The setup cost is S dollars per machine, and the operating cost is p dollars per hour.

a. Derive a formula for the number of machines that should be used to minimize the total cost of filling the order.

b. Show that when the total cost is minimal, the setup cost is equal to the cost of operating the machines.

42. MODELING PROBLEM   (Continuation of Problem 39) Suppose you have 10,000 codling moth eggs. Find a function F for the number of moths that hatch and survive to have their first meal when the temperature is T. For what temperature T on $[20, 30]$ is the number of dining moths the greatest? How large is the optimum dining party and how long did it take to find its meal?

43. MODELING PROBLEM   An epidemic spreads through a community in such a way that t weeks after its outbreak, the number of residents who have been infected is given by a function of the form

$$f(t) = \frac{A}{1 + Ce^{-kt}}$$

where A is the total number of susceptible residents. Show that the epidemic is spreading most rapidly when half the susceptible residents have been infected.

44. MODELING PROBLEM   In certain tissues, cells exist in the shape of circular cylinders. Suppose such a cylinder has radius r and height h. If the volume is fixed (say, at v_0), find the value of r that minimizes the total surface area ($S = 2\pi rh + 2\pi r^2$) of the cell.

45. MODELING PROBLEM   A store owner expects to sell Q units of a certain commodity each year. It costs S dollars to order each new shipment of x units; and it costs t dollars to store each unit for a year. Assuming that the commodity is used at a constant rate throughout the year and that each shipment arrives just as the preceding shipment has been used up, show that the total cost of maintaining inventory is minimized when the ordering cost equals the storage cost.

46. MODELING PROBLEM   Sometimes investment managers determine the optimal holding time of an asset worth $V(t)$ dollars by finding when the relative rate of change $V'(t)/V(t)$ equals the prevailing rate of interest r.

Show that the optimal time determined by this criterion is the same as that found by maximizing the present value of $V(t)$.

47. An important quantity in economic analysis is *elasticity of demand*, defined by

$$E(x) = \frac{p}{x}\frac{dx}{dp},$$

where x is the number of units of a commodity demanded when the price is p dollars per unit.* Show that

$$\frac{dR}{dx} = \frac{R}{x}\left[1 + \frac{1}{E(x)}\right]$$

That is, marginal revenue is $[1 + 1/E(x)]$ times average revenue.

48. MODELING PROBLEM   There are alternatives to the inventory model analyzed in Example 5. Suppose a company must supply N units/month at a uniform rate. Assume the storage cost/unit is S_1 dollars/month and that the setup cost is S_2 dollars. Further assume that production is at a uniform rate of m units/month (with no units left over in inventory at the end of the month). Let x be the number of items produced in each run.

a. Explain why the total average cost per month may be modeled by

$$C(x) = \frac{S_1 x}{2}\left(1 - \frac{N}{m}\right) + \frac{S_2 N}{x}$$

b. Find an expression for the number of items that should be produced in each run to minimize the total average cost C.

Economists refer to the optimum found in this inventory model as the *economic production quantity* (EPQ), whereas the optimum found in the model analyzed in Example 5 is called the *economic order quantity* (EOQ).

49. MODELING PROBLEM   In this section we showed that the Heinz concentration function,

$$C(t) = \frac{k}{b - a}(e^{-at} - e^{-bt})$$

for $b > a$ has exactly one critical number, namely,

$$t_c = \frac{1}{b - a}\ln\left(\frac{b}{a}\right)$$

Find $C''(t)$ and use the second-derivative test to show that a relative maximum for the concentration $C(t)$ occurs at time $t = t_c$.

---

*See the module with the intriguing title, "Price Elasticity of Demand: Gambling, Heroin, Marijuana, Prostitution, and Fish," by Yves Nievergelt, *UMAP Modules 1987: Tools for Teaching.* Arlington, MA: CUPM Inc., 1988, pp. 153–181.

50. **MODELING PROBLEM**   Let $f(\theta)$ be the resistance to blood flow obtained on page 296, namely

$$f(\theta) = k\left(\frac{\ell - h\cot\theta}{R^4} + \frac{h\csc\theta}{r^4}\right)$$

We found that $\theta_n = \cos^{-1}(r^4/R^4)$ is a critical number of f. Find $f''(\theta)$ and use the second-derivative test to verify that a relative minimum occurs at $\theta = \theta_m$. Explain why this must be an *absolute* minimum.

51. **MODELING PROBLEM**   Beehives are formed by packing together cells that may be modeled as regular hexagonal prisms open at one end, as shown in Figure 4.84.

It can be shown that a cell with hexagonal side of length s and prism height h has surface area

$$S(\theta) = 6sh + 1.5s^2(-\cot\theta + \sqrt{3}\csc\theta)$$

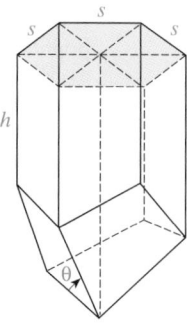

■ **FIGURE 4.84**   Beehive and beehive cell

for $0 < \theta < \frac{\pi}{2}$. What is the angle θ that minimizes the surface area of the cell (assuming that s and h are fixed)?

4.8   l'Hôpital's Rule

IN THIS SECTION
l'Hôpital's rule, indeterminate forms 0/0 and ∞/∞, other indeterminate forms, special limits involving e^x and $\ln x$

l'HÔPITAL'S RULE

In curve sketching, optimization, and other applications, it is often necessary to evaluate a limit of the form $\lim f(x)/g(x)$, where $\lim f(x)$ and $\lim g(x)$ are either both 0 or both ∞. Such limits are called 0/0 *indeterminate form* and ∞/∞ *indeterminate form*, respectively, because their values cannot be determined without further analysis.

In Chapter 2, certain limits were considered (such as Example 6, Section 2.3):

$$\text{Evaluate } \lim_{x\to 2}\frac{x^2 + x - 6}{x - 2}.$$

We noted that this is the indeterminate form 0/0, and we found the limit by factoring and simplifying. Let $f(x) = x^2 + x - 6$, and $g(x) = x - 2$, and consider the following table:

$x \to 2^-$:	1.9	1.99	1.9999	1.999	
$f(x)$:	-4.9	-0.499	-0.005	-0.0005	$\ldots$
$g(x)$:	-0.1	-0.01	-0.001	$-.0004$	$\ldots$
$\dfrac{f(x)}{g(x)}$:	4.9	4.99	4.999	4.9999	$\ldots$

This table suggests that the ratio $f(x)/g(x)$ tends to 5 as $x \to 2^-$. Take another look at the table. It appears that the values of $f(x)$ tend toward zero *five* times as fast as those of $g(x)$. Since $f'(x)$ and $g'(x)$ measure the rates of change of f and g, respectively, this observation suggests that

$$\lim_{x\to 2}\frac{f(x)}{g(x)} = 5 = \lim_{x\to 2}\frac{f'(x)}{g'(x)}$$

By formalizing such considerations, we are led to the following theorem, which is known as *l'Hôpital's rule* after Guillaume François Antoine de l'Hôpital (1661–1704), although it was actually discovered by Johann Bernoulli (see Historical Quest in Problem 63).

THEOREM 4.9   *l'Hôpital's Rule*

Let f and g be differentiable functions on an open interval containing c (except possibly at c itself). Suppose $\lim\limits_{x \to c} \dfrac{f(x)}{g(x)}$ produces an indeterminate form $\dfrac{0}{0}$ or $\dfrac{\infty}{\infty}$ and that

$$\lim_{x \to c} \frac{f'(x)}{g'(x)} = L$$

where L is either a finite number, $+\infty$, or $-\infty$. Then

$$\lim_{x \to c} \frac{f(x)}{g(x)} = L$$

The theorem also applies to one-sided limits and to limits at infinity (where $x \to +\infty$ and $x \to -\infty$).

Proof   The general proof is given in Appendix B. However, we can obtain a sense of why it is true. Suppose $f(x)$ and $g(x)$ are differentiable functions such that $f(a) = g(a) = 0$. Then, using the linearization formula

$$F(x) \approx F(a) + F'(a)(x - a) \qquad \textit{See Section 3.8}$$

We can write

$$\frac{f(x)}{g(x)} \approx \frac{f(a) + f'(a)(x - a)}{g(a) + g'(a)(x - a)} = \frac{0 + f'(a)(x - a)}{0 + g'(a)(x - a)} = \frac{f'(a)}{g'(a)}$$

so

$$\lim_{x \to a} \frac{f(x)}{g(x)} = \frac{f'(a)}{g'(a)}$$

INDETERMINATE FORMS $0/0$ AND ∞/∞

We now consider a variety of problems involving indeterminate forms. We again consider a limit first computed in Chapter 2, but instead of using a geometric argument together with the squeeze property of limits, we use l'Hôpital's rule.

EXAMPLE 1   *Using l'Hôpital's rule to compute a familiar trigonometric limit*

Evaluate $\lim\limits_{x \to 0} \dfrac{\sin x}{x}$.

Solution   Note that this is of indeterminate form because $\sin x$ and x both approach 0 as $x \to 0$. This means that l'Hôpital's rule applies:

$$\lim_{x \to 0} \frac{\sin x}{x} = \lim_{x \to 0} \frac{\cos x}{1} = 1$$

EXAMPLE 2   *l'Hôpital's rule with a $0/0$ form*

Evaluate $\lim\limits_{x \to 2} \dfrac{x^7 - 128}{x^3 - 8}$.

Solution   For this example, $f(x) = x^7 - 128$ and $g(x) = x^3 - 8$, and the form is 0/0.

$$\lim_{x \to 2} \frac{x^7 - 128}{x^3 - 8} = \lim_{x \to 2} \frac{7x^6}{3x^2} \qquad \textit{l'Hôpital's rule}$$

$$= \lim_{x \to 2} \frac{7x^4}{3} \qquad \textit{Simplify.}$$

$$= \frac{7(2)^4}{3} = \frac{112}{3} \qquad \textit{Limit of a quotient} \qquad ■$$

EXAMPLE 3   *Limit is not an indeterminate form*

Evaluate $\lim\limits_{x \to 0} \dfrac{1 - \cos x}{\sec x}$.

Solution   You must always remember to check that you have an indeterminate form before applying l'Hôpital's rule. The limit is

$$\lim_{x \to 0} \frac{1 - \cos x}{\sec x} = \frac{\displaystyle\lim_{x \to 0} (1 - \cos x)}{\displaystyle\lim_{x \to 0} \sec x} = \frac{0}{1} = 0 \qquad ■$$

WARNING ➤   If you blindly apply l'Hôpital's rule in Example 3, you obtain the WRONG answer:

$$\lim_{x \to 0} \frac{1 - \cos x}{\sec x} = \lim_{x \to 0} \frac{\sin x}{\sec x \tan x} \qquad \textit{This is NOT correct.}$$

$$= \lim_{x \to 0} \frac{\cos x}{\sec x} = \frac{1}{1} = 1$$

This answer is blatantly WRONG, as you can see by looking at the following technology window. ◄

Technology Window

You can use a graphing calculator to help find many indeterminate-form limits. For instance, the limit in Example 3 can easily be checked by looking at the graph. If you have a graphing calculator, you can see that as $x \to 0$ from either the left or the right, the limit looks the same—namely, 0. The graphs of the functions for the remainder of the examples of this section are shown so that you can note the relationship between the requested limit, the function, and l'Hôpital's rule.

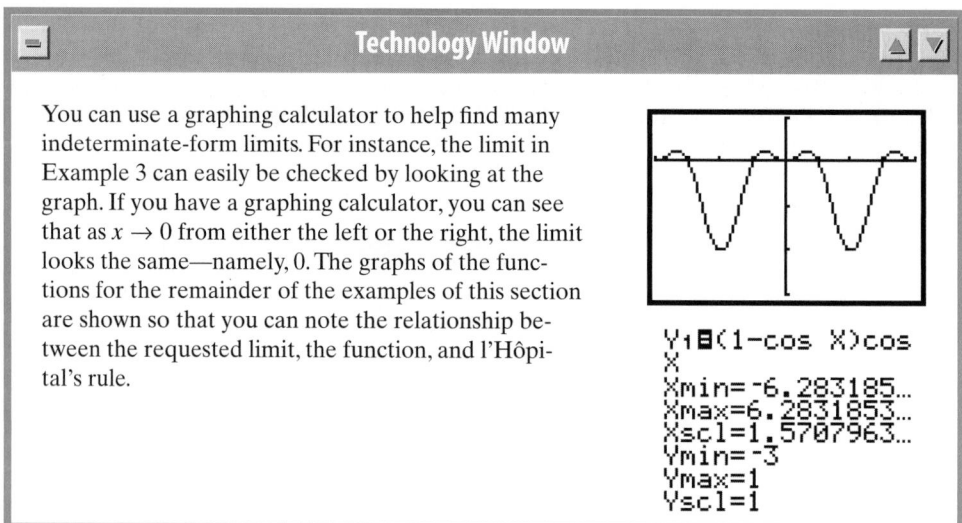

```
Y₁◼(1-cos X)cos
X
Xmin=-6.283185…
Xmax=6.2831853…
Xscl=1.5707963…
Ymin=-3
Ymax=1
Yscl=1
```

EXAMPLE 4   *l'Hôpital's rule applied more than once*

Evaluate $\lim\limits_{x \to 0} \dfrac{x - \sin x}{x^3}$.

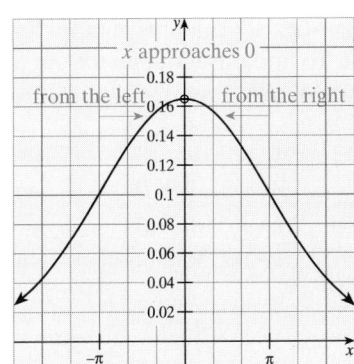

By looking at the graph, you can reinforce the result obtained using l'Hôpital's rule.

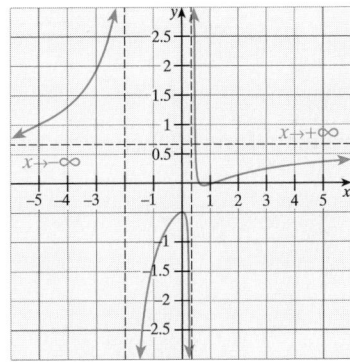

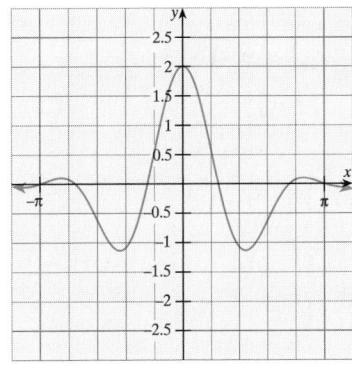

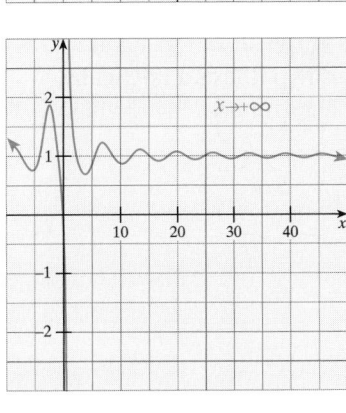

Solution   This is a 0/0 indeterminate form, and we find that

$$\lim_{x \to 0} \frac{x - \sin x}{x^3} = \lim_{x \to 0} \frac{1 - \cos x}{3x^2}$$

This is still the indeterminate form 0/0, so l'Hôpital's rule can be applied once again:

$$\lim_{x \to 0} \frac{1 - \cos x}{3x^2} = \lim_{x \to 0} \frac{-(-\sin x)}{6x} = \frac{1}{6} \lim_{x \to 0} \frac{\sin x}{x} = \frac{1}{6}(1) = \frac{1}{6}$$ ■

EXAMPLE 5   *l'Hôpital's rule with an ∞/∞ form*

Evaluate $\lim\limits_{x \to +\infty} \dfrac{2x^2 - 3x + 1}{3x^2 + 5x - 2}$.

Solution   Using the methods of Section 2.3, we could compute this limit by multiplying by $(1/x^2)/(1/x^2)$. Instead, we note that this is of the form ∞/∞ and apply l'Hôpital's rule:

$$\lim_{x \to +\infty} \frac{2x^2 - 3x + 1}{3x^2 + 5x - 2} = \lim_{x \to +\infty} \frac{4x - 3}{6x + 5} \qquad \text{Apply l'Hôpital's rule again.}$$

$$= \lim_{x \to +\infty} \frac{4}{6} = \frac{2}{3}$$ ■

It may happen that even when l'Hôpital's rule applies to a limit, it is not the best way to proceed, as illustrated by the following example.

EXAMPLE 6   *Using l'Hôpital's rule with other limit properties*

Evaluate $\lim\limits_{x \to 0} \dfrac{(1 - \cos x) \sin 4x}{x^3 \cos x}$.

Solution   This limit has the form 0/0, but direct application of l'Hôpital's rule leads to a real mess (try it!). Instead, we compute the given limit by using the product rule for limits along with two simple applications of l'Hôpital's rule. Specifically, using the product rule for limits (assuming the limits exist), we have

$$\lim_{x \to 0} \frac{(1 - \cos x) \sin 4x}{x^3 \cos x} = \left[\lim_{x \to 0} \frac{1 - \cos x}{x^2} \right] \left[\lim_{x \to 0} \frac{\sin 4x}{x} \right] \left[\lim_{x \to 0} \frac{1}{\cos x} \right]$$

$$= \left[\lim_{x \to 0} \frac{\sin x}{2x} \right] \left[\lim_{x \to 0} \frac{4 \cos 4x}{1} \right] \left[\lim_{x \to 0} \frac{1}{\cos x} \right] = (\tfrac{1}{2})(4)(1) = 2$$ ■

EXAMPLE 7   *Hypotheses of l'Hôpital's rule are not satisfied*

Evaluate $\lim\limits_{x \to +\infty} \dfrac{x + \sin x}{x - \cos x}$.

Solution   This limit has the indeterminate form ∞/∞. If you try to apply l'Hôpital's rule, you find

$$\lim_{x \to +\infty} \frac{x + \sin x}{x - \cos x} = \lim_{x \to +\infty} \frac{1 + \cos x}{1 + \sin x}$$

The limit on the right does not exist, because both $\sin x$ and $\cos x$ oscillate between -1 and 1 as $x \to +\infty$. Recall that l'Hôpital's rule applies only if this limit exists. This does not mean that the limit of the original expression does not exist or that we cannot find it; it simply means that we cannot apply l'Hôpital's rule. To find this limit, factor out an x from the numerator and denominator and proceed as follows:

$$\lim_{x \to +\infty} \frac{x + \sin x}{x - \cos x} = \lim_{x \to +\infty} \frac{x\left(1 + \dfrac{\sin x}{x}\right)}{x\left(1 - \dfrac{\cos x}{x}\right)} = \lim_{x \to +\infty} \frac{1 + \dfrac{\sin x}{x}}{1 - \dfrac{\cos x}{x}} = \frac{1 + 0}{1 - 0} = 1 \qquad \blacksquare$$

OTHER INDETERMINATE FORMS

l'Hôpital's rule itself applies only to the indeterminate forms $0/0$ and ∞/∞. However, other indeterminate forms, such as $1^{\infty}, 0^{\infty}, \infty^{0}, \infty - \infty$, and $0 \cdot \infty$, can often be manipulated algebraically into one of the standard forms $0/0$ or ∞/∞, and then evaluated using l'Hôpital's rule. The following examples illustrate this procedure.

EXAMPLE 8   *Limit of the form 1^{∞}*

Find $\lim\limits_{x \to +0} (e^x + 2x)^{3/x}$.

Solution   Note that this limit is indeed of the indeterminate form 1^{∞}. Let

$$L = \lim_{x \to 0} (e^x + 2x)^{3/x}$$

Take the logarithm of both sides:

$$\begin{aligned}
\ln L &= \ln[\lim_{x \to 0} (e^x + 2x)^{3/x}] \\
&= \lim_{x \to 0} \ln(e^x + 2x)^{3/x} & &\text{\small ln } x \text{ \small is continuous.} \\
&= \lim_{x \to 0} \frac{3\ln(e^x + 2x)}{x} & &\frac{0}{0} \text{ \small form} \\
&= \lim_{x \to 0} \frac{3\left(\dfrac{e^x + 2}{e^x + 2x}\right)}{1} & &\text{\small l'Hôpital's rule} \\
&= 9
\end{aligned}$$

Thus, $\ln L = 9$ and $L = e^9$. $\qquad \blacksquare$

EXAMPLE 9   *Limit of the form 0^0*

Find $\lim\limits_{x \to 0^+} x^{\sin x}$.

Solution   This is a 0^0 indeterminate form. From the graph shown in Figure 4.85, it looks as though the desired limit is 1. We can verify this hypothesis analytically.
As in Example 8, we begin by using properties of logarithms:

$$\begin{aligned}
L &= \lim_{x \to 0^+} x^{\sin x} \\
\ln L &= \ln \lim_{x \to 0^+} x^{\sin x} \\
&= \lim_{x \to 0^+} \ln x^{\sin x} \\
&= \lim_{x \to 0^+} (\sin x)\ln x \\
&= \lim_{x \to 0^+} \frac{\ln x}{\csc x} & &\text{\small This is } \frac{\infty}{\infty} \text{ \small form.}
\end{aligned}$$

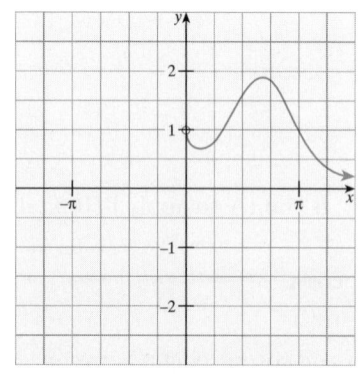

■ **FIGURE 4.85**   Graph of $x^{\sin x}$

$$= \lim_{x\to 0^+} \frac{1/x}{-\csc x \cot x} \quad \text{l'Hôpital's rule}$$

$$= \lim_{x\to 0^+} \frac{-\sin^2 x}{x \cos x} = \lim_{x\to 0^+} \left(\frac{\sin x}{x}\right)\left(\frac{-\sin x}{\cos x}\right)$$

$$= (1)(0) = 0$$

Thus, $L = e^0 = 1$.

EXAMPLE 10   *Limit of the form ∞^0*

Find $\lim_{x\to +\infty} x^{1/x}$.

Solution   This is a limit of the indeterminate form ∞^0. If $L = \lim_{x\to +\infty} x^{1/x}$, then

$$\ln L = \ln \lim_{x\to +\infty} x^{1/x}$$

$$= \lim_{x\to +\infty} \ln x^{1/x} \quad \text{The limit of a log is the log of a limit.}$$

$$= \lim_{x\to +\infty} \frac{1}{x} \ln x$$

$$= \lim_{x\to +\infty} \frac{\ln x}{x}$$

$$= \lim_{x\to \infty} \frac{1/x}{1} \quad \text{l'Hôpital's rule}$$

$$= 0$$

Thus, we have $\ln L = 0$; therefore, $L = e^0 = 1$.

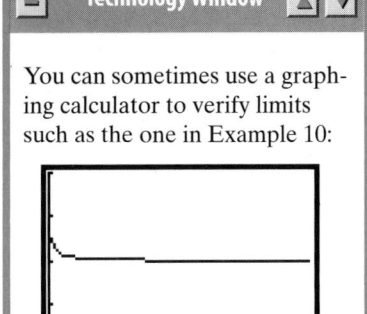

You can sometimes use a graphing calculator to verify limits such as the one in Example 10:

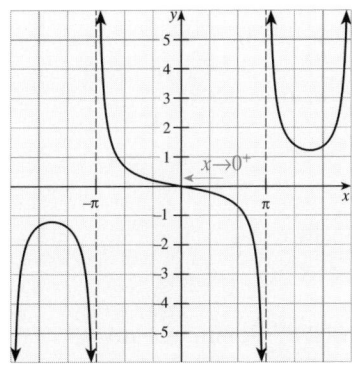

Notice that the graph looks like the line $y = 1$, which confirms that $L = 1$.

EXAMPLE 11   *Substitution with l'Hôpital's rule*

Evaluate $\lim_{x\to +\infty} x \tan \frac{1}{x}$.

Solution   This is an $\infty \cdot 0$ indeterminate form, but we can convert it into a $0/0$ form by using the substitution $u = 1/x$ and noting that $u \to 0$ as $x \to +\infty$.

$$\lim_{x\to +\infty} x \tan \frac{1}{x} = \lim_{u\to 0} \frac{1}{u} \tan u = \lim_{u\to 0} \frac{\tan u}{u} \quad \text{This is a } \frac{0}{0} \text{ form.}$$

$$= \lim_{u\to 0} \frac{\sec^2 u}{1} = 1$$

EXAMPLE 12   *l'Hôpital's rule with the form $\infty - \infty$*

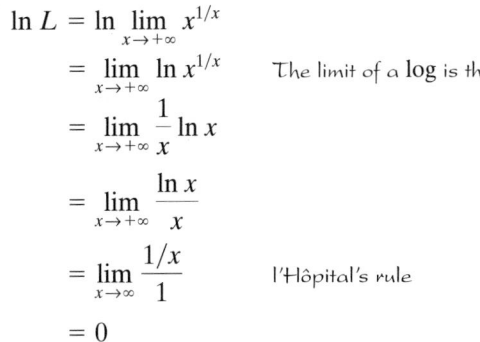

Evaluate $\lim_{x\to 0^+} \left(\frac{1}{x} - \frac{1}{\sin x}\right)$.

Solution   As it stands, this has the form $\infty - \infty$, because

$$\frac{1}{x} \to +\infty \text{ and } \frac{1}{\sin x} \to +\infty \text{ as } x \to 0 \text{ from the right.}$$

However, using a little algebra, we find

$$\lim_{x\to 0^+} \left(\frac{1}{x} - \frac{1}{\sin x}\right) = \lim_{x\to 0^+} \frac{\sin x - x}{x \sin x}$$

This limit is now of the form 0/0, so the hypotheses of l'Hôpital's rule are satisfied. Thus,

$$\lim_{x \to 0^+} \frac{\sin x - x}{x \sin x} = \lim_{x \to 0^+} \frac{\cos x - 1}{\sin x + x \cos x} \qquad \textit{Again, the form } \frac{0}{0}$$

$$= \lim_{x \to 0^+} \frac{-\sin x}{\cos x + x(-\sin x) + \cos x} = \frac{0}{2} = 0 \qquad \blacksquare$$

EXAMPLE 13   *l'Hôpital's rule with the form 0 · ∞*

Evaluate $\displaystyle\lim_{x \to \pi/2^-} \left(x - \frac{\pi}{2}\right) \tan x$.

Solution   This limit has the form $0 \cdot \infty$, because

$$\lim_{x \to \pi/2^-} \left(x - \frac{\pi}{2}\right) = 0 \quad \text{and} \quad \lim_{x \to \pi/2^-} \tan x = +\infty$$

Write $\tan x = \dfrac{1}{\cot x}$ to obtain

$$\lim_{x \to \pi/2^-} \left(x - \frac{\pi}{2}\right) \tan x = \lim_{x \to \pi/2^-} \frac{x - \frac{\pi}{2}}{\cot x} \qquad \textit{Form 0/0}$$

$$= \lim_{x \to \pi/2^-} \frac{1}{-\csc^2 x} \qquad \textit{l'Hôpital's rule}$$

$$= \lim_{x \to \pi/2^-} (-\sin^2 x) = -1 \qquad \blacksquare$$

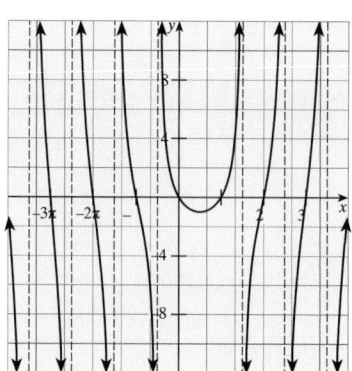

Look at the graph as $x \to \frac{\pi}{2}$ from both the left and the right.

SPECIAL LIMITS INVOLVING e^x AND $\ln x$

We close this section with a theorem that summarizes the behavior of certain important special functions involving e^x and $\ln x$ near 0 and at $\pm\infty$.

THEOREM 4.10   *Limits involving natural logarithms and exponentials*

If k and n are positive integers, then

$$\lim_{x \to 0^+} \frac{\ln x}{x^n} = -\infty \qquad\qquad \lim_{x \to +\infty} \frac{\ln x}{x^n} = 0$$

$$\lim_{x \to +\infty} \frac{e^{kx}}{x^n} = +\infty \qquad\qquad \lim_{x \to +\infty} x^n e^{-kx} = 0$$

Proof   These can all be verified directly or by applying l'Hôpital's rule. For example,

$$\lim_{x \to +\infty} \frac{\ln x}{x^n} = \lim_{x \to +\infty} \frac{1/x}{nx^{n-1}} = \lim_{x \to +\infty} \frac{1}{nx^n} = 0$$

The other parts are left for you to verify. $=$

■   *What This Says:*   The limit statements

$$\lim_{x \to +\infty} \frac{e^{kx}}{x^n} = +\infty \quad \text{and} \quad \lim_{x \to +\infty} \frac{\ln x}{x^n} = 0$$

are especially important. They tell us that, in the long run, any exponential e^{kx} dominates any

4.8   Problem Set

A 1. ■ **What Does This Say?** An incorrect use of l'Hôpital's rule is illustrated in the following limit computations. In each case, explain what is wrong and find the correct value of the limit.

a. $\lim_{x \to \pi} \dfrac{1 - \cos x}{x} = \lim_{x \to \pi} \dfrac{\sin x}{1} = 0$

b. $\lim_{x \to \pi/2} \dfrac{\sin x}{x} = \lim_{x \to \pi/2} \dfrac{\cos x}{1} = 0$

2. ■ **What Does This Say?** Sometimes l'Hôpital's rule leads to inconclusive computation. For example, observe what happens when the rule is applied to

$$\lim_{x \to +\infty} \frac{x}{\sqrt{x^2 - 1}}$$

Use any method you wish to evaluate this limit.

Find each of the limits in Problem 3–53.

3. $\lim_{x \to 1} \dfrac{x^3 - 1}{x^2 - 1}$

4. $\lim_{x \to 2} \dfrac{x^3 - 27}{x^2 - 9}$

5. $\lim_{x \to 1} \dfrac{x^{10} - 1}{x - 1}$

6. $\lim_{x \to -1} \dfrac{x^{10} - 1}{x + 1}$

7. $\lim_{x \to 0} \dfrac{1 - \cos^2 x}{\sin^3 x}$

8. $\lim_{x \to 0} \dfrac{1 - \cos^2 x}{3 \sin x}$

9. $\lim_{x \to \pi} \dfrac{\cos \frac{x}{2}}{\pi - x}$

10. $\lim_{x \to 0} \dfrac{1 - \cos x}{x^2}$

11. $\lim_{x \to 0} \dfrac{\sin ax}{\cos bx}, \; ab \neq 0$

12. $\lim_{x \to 0} \dfrac{\tan 3x}{\sin 5x}$

13. $\lim_{x \to 0} \dfrac{x - \sin x}{\tan x - x}$

14. $\lim_{x \to 0} \dfrac{1 - \cos^2 x}{x \tan x}$

15. $\lim_{x \to \pi/2} \dfrac{3 \sec x}{2 + \tan x}$

16. $\lim_{x \to 0} \dfrac{x + \sin^3 x}{x^2 + 2x}$

17. $\lim_{x \to 0} \dfrac{\sin 3x \sin 2x}{x \sin 4x}$

18. $\lim_{x \to \pi/2} \dfrac{\sin 2x \cos x}{x \sin 4x}$

19. $\lim_{x \to 0} \dfrac{x^2 + \sin x^2}{x^2 + x^3}$

20. $\lim_{x \to 0} x^2 \sin \frac{1}{x}$

21. $\lim_{x \to +\infty} x^{3/2} \sin \frac{1}{x}$

22. $\lim_{x \to 1} (1 - \cos x) \cot x$

23. $\lim_{x \to 1} \dfrac{(x - 1)\sin(x - 1)}{1 - \cos(x - 1)}$

24. $\lim_{\theta \to 0} \dfrac{\theta - 1 + \cos^2 \theta}{\theta^2 + 5\theta}$

25. $\lim_{x \to 0} \dfrac{x + \sin(x^2 + x)}{3x + \sin x}$

26. $\lim_{x \to \pi/2^-} \sec 3x \cos 9x$

27. $\lim_{x \to 0} \left(\dfrac{1}{\sin 2x} - \dfrac{1}{2x} \right)$

28. $\lim_{x \to +\infty} x^{-5} \ln x$

29. $\lim_{x \to 0^+} x^{-5} \ln x$

30. $\lim_{x \to 0^+} (\sin x) \ln x$

31. $\lim_{x \to +\infty} \dfrac{\ln(\ln x)}{x}$

32. $\lim_{x \to +\infty} \left(1 - \dfrac{3}{x} \right)^{2x}$

33. $\lim_{x \to +\infty} \left(1 + \dfrac{1}{2x} \right)^{3x}$

34. $\lim_{x \to +\infty} (\ln x)^{1/x}$

35. $\lim_{x \to 0} (e^x + x)^{1/x}$

36. $\lim_{x \to 0} (\sin x)^{1/\ln \sqrt{x}}$

37. $\lim_{x \to 0} \left(\cot x - \dfrac{1}{x} \right)$

38. $\lim_{x \to 0^+} \left(\dfrac{1}{x^2} + \ln x \right)$

B 39. $\lim_{x \to (\pi/2)^-} \left(\dfrac{1}{\pi - 2x} + \tan x \right)$

40. $\lim_{x \to +\infty} \left(\sqrt{x^2 - x} - x \right)$

41. $\lim_{x \to +\infty} [x - \ln(x^3 - 1)]$ *Hint:* $\ln e^x = x$

42. $\lim_{x \to +\infty} [x - \ln(e^x + e^{-x})]$ *Hint:* $\ln e^x = x$

43. $\lim_{x \to 0^+} \left(\dfrac{1}{x^2} - \ln \sqrt{x} \right)$

44. $\lim_{x \to 0} (e^x - 1 - x)^x$

45. $\lim_{x \to 0^+} (\ln x)(\cot x)$

46. $\lim_{x \to 0^+} (e^x - 1)^{1/\ln x}$

47. $\lim_{x \to +\infty} \dfrac{x + \sin 3x}{x}$

48. $\lim_{x \to +\infty} \dfrac{x(\pi + \sin x)}{x^2 + 1}$

49. $\lim_{x \to 0^+} \left(\dfrac{2 \cos x}{\sin 2x} - \dfrac{1}{x} \right)$

50. $\lim_{x \to +\infty} \left(\dfrac{x^3}{x^2 - x + 1} - \dfrac{x^3}{x^2 + x - 1} \right)$

51. $\lim_{x \to 0} \dfrac{(2 - x)(e^x - x - 2)}{x^3}$

52. $\lim_{x \to 0} \dfrac{\tan^{-1}(3x) - 3 \tan^{-1} x}{x^3}$

53. $\lim_{x \to +\infty} x^5 \left[\sin\left(\dfrac{1}{x} \right) - \dfrac{1}{x} + \dfrac{1}{6x^3} \right]$

54. Find A so $\lim_{x \to +\infty} \left(\dfrac{x + A}{x - 2A} \right)^x = 5$

55. For a certain value of B, the limit

$$\lim_{x \to +\infty} (x^4 + 5x^3 + 3)^B - x$$

is finite and nonzero. Find B and then compute the limit.

56. For what values of constants C and D is it true that

$$\lim_{x \to 0} (x^{-3} \sin 7x + Cx^{-2} + D) = -2$$

Technology Window

Find the limits in Problems 57–59 using the following methods.
a. *graphically*   **b.** *analytically*   **c.** *numerically*
Compare, contrast, and reconcile the three methods.

57. $\lim_{x \to 0^+} x^x$

58. $\lim_{x \to 0^+} x^{x^x}$

59. $\lim_{x \to +\infty} \left[x \sin^{-1}\left(\dfrac{1}{x} \right) \right]^{x^2}$

60. ■ **What Does This Say?** Write a paper on using technology to evaluate limits.

61. **H**ISTORICAL **Q**UEST   The French mathematician Guillaume François Antoine Marquis de l'Hôpital wrote the world's first elementary calculus text, *Analyse des Infiniment Petits,* in which he developed Leibniz's "analysis of the infinitely small" with elegance and clarity.* We consider a problem from this book.

GUILLAUME l'HÔPITAL
1661–1704

At a point C on the horizontal ceiling of a room, attach a cord of length r that has a pulley affixed at its other end, the point F. At a second point B on the ceiling, a distance d from point C, attach a cord of length l, pass it through the pulley at F, and connect a weight at the other end, at D. Release the weight and allow the system to achieve its equilibrium position as shown in Figure 4.86. l'Hôpital supposes that the attached weight is heavy enough that the weights of the cords and the pulley can be ignored, and he asks: What is the geometry of the equilibrium configuration?

a. Assume $r \geq d$. Draw an illustration to illustrate the equilibrium for this case.

b. Assume $r < d$. This is the situation shown in Figure 4.86. Let E be the intersection of the extension of the segment $\overline{DE}$ with $\overline{BC}$, and let x denote the distance $\overline{EC}$. Find a function f so that $f(x)$ is the length of the distance $\overline{ED}$.

c. Find a maximum value of f on $0 \leq x \leq r$.

d. Use the result of part **c** to show, as l'Hôpital did, that the equilibrium configuration is the same regardless of the exact amount of the weight W.

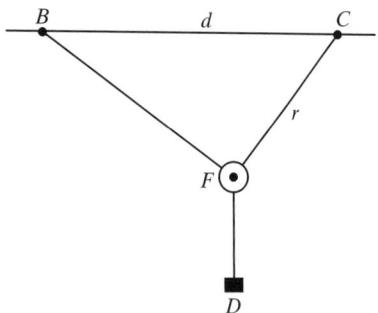

■ **FIGURE 4.86**   l'Hôpital's Pulley Problem

62. **H**ISTORICAL **Q**UEST   The French mathematician Guillaume François Antoine Marquis de l'Hôpital (see Problem 61) is best known today for the rule that bears his name.

Actually, the rule was discovered by l'Hôpital's teacher, Johann Bernoulli. Not only did l'Hôpital neglect to cite his sources in his book, but there is also evidence that he paid Bernoulli for his results and for keeping their arrangements for payment confidential. In a letter dated March 17, 1694, he asked Bernoulli "to communicate to me your discoveries…"—with the request not to mention them to others— "…it would not please me if they were made public." l'Hôpital's argument, which was originally given without using functional notation, can easily be reproduced:*

$$\frac{f(a + dx)}{g(a + dx)} = \frac{f(a) + f'(a)\,dx}{g(a) + g'(a)\,dx}$$

$$= \frac{f'(a)\,dx}{g'(a)\,dx}$$

$$= \frac{f'(a)}{g'(a)}$$

Supply reasons for this argument, and give necessary conditions for the functions f and g.

63. **H**ISTORICAL **Q**UEST   The remarkable Bernoulli family of Switzerland produced at least eight noted mathematicians over three generations. Two brothers, Jacob (1654–1705) and Johann (1667–1748), were bitter rivals. These brothers were extremely influential advocates of the newly born calculus. Johann was the most prolific of the clan and was responsible for the discovery of l'Hôpital's rule (see Problem 61), Bernoulli numbers, Bernoulli polynomials, the lemniscate of Bernoulli, the Bernoulli equation, the Bernoulli theorem, and the Bernoulli distribution. He did a great deal of work with differential equations. Johann was jealous and cantankerous; he tossed a son (Daniel) out of the house for winning an award he had expected to win himself.

Write a report on the Bernoulli family.

64. Find constants a and b so that

$$\lim_{x \to 0}\left(\frac{\sin 2x}{x^3} + \frac{a}{x^2} + b\right) = 1$$

65. A weight hanging by a spring is made to vibrate by applying a sinusoidal force, and the displacement at time t is given by

$$f(t) = \frac{C}{\beta^2 - \alpha^2}(\sin \alpha t - \sin \beta t)$$

where C, α, and β are constants such that $\alpha \neq \beta$. What happens to the displacement as $\beta \to \alpha$? You may assume that α is fixed.

66. For A and B positive constants, define

$$f(x) = (e^x + Ax)^{B/x}$$

a. Compute

$$L_1 = \lim_{x \to 0} f(x) \quad \text{and} \quad L_2 = \lim_{x \to +\infty} f(x)$$

b. What is the largest value of A for which the equation $L_1 = BL_2$ has a solution? What are L_1 and L_2 in this case?

---

*From "Two Historical Applications of Calculus" by Alexander J. Hahn, *The College Mathematics Journal*, Vol. 29, No. 2, March 1998, pp. 93–97.

---

†D. J. Struik, *A Source Book in Mathematics, 1200–1800.* Cambridge, MA: Harvard University Press, 1969, pp. 313–316.

Chapter 4　Review

Proficiency Examination

Concept Problems

1. What is the difference between absolute and relative extrema of a function?

2. State the extreme value theorem.

3. What are the critical numbers of a function? What is the difference between critical numbers and critical points?

4. Outline a procedure for finding the absolute extrema of a continuous function on a closed interval $[a, b]$.

5. State both Rolle's theorem and the mean value theorem, and discuss the relationship between the two.

6. State the first-derivative test.

7. State the second-derivative test.

8. What is an asymptote?

9. Define $\lim\limits_{x \to +\infty} f(x) = L$ and $\lim\limits_{x \to c} f(x) = +\infty$.

10. Outline a graphing strategy for a function defined by $y = f(x)$.

11. What do we mean by optimization? Outline an optimization procedure.

12. State l'Hôpital's rule.

Practice Problems

Evaluate the limits in Problems 13–16.

13. $\lim\limits_{x \to \pi/2} \dfrac{\sin 2x}{\cos x}$

14. $\lim\limits_{x \to 1} \dfrac{1 - \sqrt{x}}{x - 1}$

15. $\lim\limits_{x \to +\infty} \left(\dfrac{1}{x} - \dfrac{1}{\sqrt{x}} \right)$

16. $\lim\limits_{x \to +\infty} \left(1 - \dfrac{2}{x} \right)^{3x}$

Sketch the graph of each function in Problems 17–22. Include all key features, such as relative extrema, inflection points, asymptotes, and intercepts.

17. $f(x) = x^3 + 3x^2 - 9x + 2$

18. $f(x) = x^{1/3}(27 - x)$

19. $f(x) = \dfrac{x^2 - 1}{x^2 - 4}$

20. $f(x) = (x^2 - 3)e^{-x}$

21. $f(x) = x + \tan^{-1} x$

22. $f(x) = \sin^2 x - 2 \cos x$ on $[0, 2\pi]$

23. Determine the largest and smallest values of
$$f(x) = x^4 - 2x^5 + 5$$
on the closed interval $[0, 1]$.

24. A box is to have a square base, an open top, and volume of 2 ft^3. Find the dimensions of the box (to the nearest inch) that uses the least amount of material.

25. Find the area of the rectangle of maximum area that can be inscribed in the semicircle defined by the equation
$$y = \sqrt{a^2 - x^2}$$
for fixed $a > 0$.

Supplementary Problems

Sketch the graph of each function in Problems 1–20. Use as many as possible of the key features such as relative extrema, inflection points, concavity, asymptotes, and intercepts.

1. $f(x) = x^3 + 6x^2 + 9x - 1$　　2. $f(x) = x^4 + 4x^3 + 4x^2 + 1$

3. $f(x) = 3x^4 - 4x^3 + 1$　　4. $f(x) = 3x^4 - 4x^2 + 1$

5. $f(x) = 6x^5 - 15x^4 + 10x^3$

6. $f(x) = 3x^5 - 10x^3 + 15x + 1$

7. $f(x) = \dfrac{9 - x^2}{3 + x^2}$　　8. $f(x) = \dfrac{x}{1 - x}$

9. $f(x) = \sin 2x - \sin x, [-\pi, \pi]$

10. $f(x) = \sin x \sin 2x, [-\pi, \pi]$

11. $f(x) = \dfrac{x^2 - 4}{x^2}$　　12. $f(x) = \dfrac{x^2 + 2x - 3}{x^2 - 3x + 2}$

13. $f(x) = \dfrac{3x - 2}{(x + 1)^2(x - 2)}$　　14. $f(x) = \dfrac{x^3 + 3}{x(x + 1)(x + 2)}$

15. $f(x) = x^2 \ln \sqrt{x}$

16. $f(x) = \sin^{-1} x + \cos^{-1} x$

17. $f(x) = x(e^{-2x} + e^{-x})$

18. $f(x) = \ln\left(\dfrac{x - 1}{x + 1} \right)$

19. $f(x) = \dfrac{5}{1 + e^{-x}}$

20. $f(x) = xe^{1/x}$

In Problems 21–24, the graph of the given function $f(x)$ for $x > 0$ is one of the six curves shown in Figure 4.87. In each case, match the function to a graph.

21. $f(x) = x2^{-x}$

22. $f(x) = \dfrac{\ln \sqrt{x}}{x}$

23. $f(x) = \dfrac{e^x}{x}$

24. $f(x) = e^{-x} \sin x$

a.
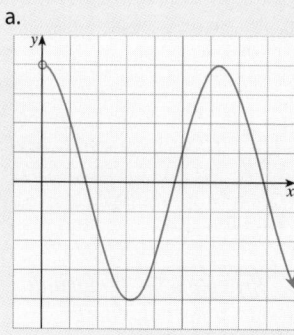

b.
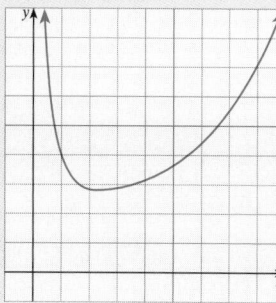

c.
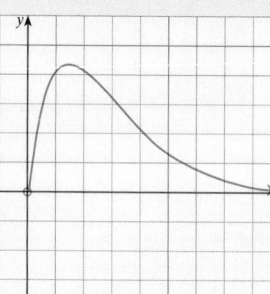

d.
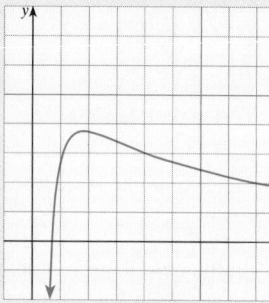

e.
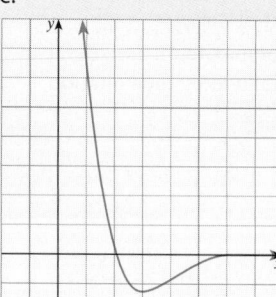

f.
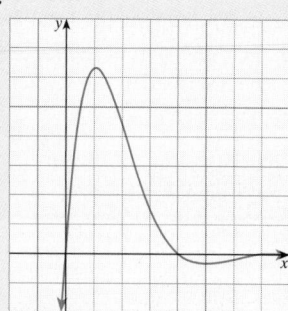

■ **FIGURE 4.87**   Problems 21–24

Determine the maximum and minimum value of each function on the interval given in Problems 25–28.

25. $f(x) = x^4 - 8x^2 + 12$ on $[-1, 2]$

26. $f(x) = \sqrt{x}(x - 5)^{1/3}$ on $[0, 6]$

27. $f(x) = 2x - \sin^{-1} x$ on $[0, 1]$

28. $f(x) = e^{-x} \ln x$ on $[\frac{1}{2}, 2]$

Evaluate the limits in Problems 29–50.

29. $\lim\limits_{x \to +\infty} \dfrac{x \sin^2 x}{x^2 + 1}$

30. $\lim\limits_{x \to +\infty} \dfrac{2x^4 - 7}{6x^4 + 7}$

31. $\lim\limits_{x \to +\infty} (\sqrt{x^2 - x} - x)$

32. $\lim\limits_{x \to +\infty} [\sqrt{x(x + b)} - x]$

33. $\lim\limits_{x \to 0} \dfrac{x \sin x}{x + \sin^3 x}$

34. $\lim\limits_{x \to 0} \dfrac{x \sin x}{x^2 - \sin^3 x}$

35. $\lim\limits_{x \to 0} \dfrac{x \sin^2 x}{x^2 - \sin^2 x}$

36. $\lim\limits_{x \to 0} \dfrac{x - \sin x}{\tan^3 x}$

37. $\lim\limits_{x \to 0} \dfrac{\sin^2 x}{\sin x^2}$

38. $\lim\limits_{x \to 0} \left(\dfrac{1}{x^2} - \dfrac{1}{x^2 \sec x} \right)$

39. $\lim\limits_{x \to \pi/2^-} \dfrac{\sec^2 x}{\sec^2 3x}$

40. $\lim\limits_{x \to \pi/2^-} (1 - \sin x) \tan x$

41. $\lim\limits_{x \to \pi/2^-} (\sec x - \tan x)$

42. $\lim\limits_{x \to +\infty} (\sqrt{x^2 + 4} - \sqrt{x^2 - 4})$

43. $\lim\limits_{x \to 0^+} (1 + x)^{4/x}$

44. $\lim\limits_{x \to 1^-} \left(\dfrac{1}{1 - x} \right)^x$

45. $\lim\limits_{x \to 0^+} x^{\tan x}$

46. $\lim\limits_{x \to 0} \dfrac{5^x - 1}{x}$

47. $\lim\limits_{x \to 0} \dfrac{\ln(x^2 + 1)}{x}$

48. $\lim\limits_{x \to +\infty} \left(\dfrac{1}{x} \right)^x$

49. $\lim\limits_{x \to +\infty} \left(4 - \dfrac{1}{x} \right)^x$

50. $\lim\limits_{x \to +\infty} \dfrac{e^x \cos x - 1}{x}$

51. **JOURNAL PROBLEM**   *Mathematics Teacher.** Which of the graphs in Figure 4.88 is the derivative and which is the function?

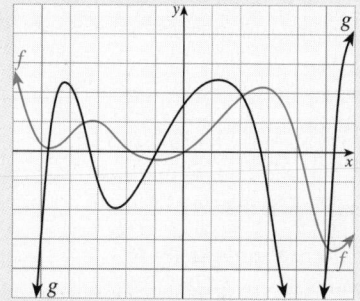

■ **FIGURE 4.88**   A function and its derivative

52. Determine $a, b,$ and c such that the graph of $f(x) = ax^3 + bx^2 + c$ has an inflection point and slope 1 at $(-1, 2)$.

53. ■   **What Does This Say?**   Explain why the graph of a quadratic polynomial cannot have a point of inflection. How many inflection points does the graph of a cubic polynomial have?

54. Find the points on the hyperbola $x^2 - y^2 = 4$ that are closest to the point $(0, 1)$.

55. Find numbers $A, B, C,$ and D that guarantee that the function

$$f(x) = Ax^3 + Bx^2 + Cx + D$$

will have a relative maximum at $(-1, 1)$ and a relative minimum at $(1, -1)$.

56. A Norman window consists of a rectangle with a semi-circle surmounted on the top. What are the dimensions of the Norman window of largest area with a fixed perimeter of P_0 meters?

---

*December 1990, p. 718.

57. An apartment complex has 200 units. When the monthly rent for each unit is $600, all units are occupied. Experience indicates that for each $20-per-month increase in rent, 5 units will become vacant. Each rented apartment costs the owners of the complex $80 per month to maintain. What monthly rent should be charged to maximize the owner's profit? What is the maximum profit? How many units are rented when profit is a maximum?

58. A farmer wishes to enclose a rectangular pasture with 320 ft of fence. Find the dimensions that give the maximum area in these situations:
a. The fence is on all four sides of the pasture.
b. The fence is on three sides of the pasture and the fourth side is bounded by a wall.

59. A peach grower has determined that if 30 trees are planted per acre, each tree will average 200 lb of peaches per season. However, for each tree grown in addition to the 30 trees, the average yield for each of the trees in the grove drops by 5 lb per tree. How many peach trees should be planted on each acre to maximize the yield of peaches per acre? What is the maximum yield?

60. MODELING PROBLEM   To obtain the maximum price, a shipment of fruit should reach the market as early as possible after the fruit has been picked. If a grower picks the fruit immediately for shipment, 100 cases can be shipped at a profit of $10 per case. By waiting, the grower estimates that the crop will yield an additional 25 cases per week, but because the competitor's yield will also increase, the grower's profit will decrease by $1 per case per week. Use calculus to determine when the grower should ship the fruit to maximize profit. What will be the maximum profit?

61. MODELING PROBLEM   Oil from an offshore rig located 3 mi from the shore is to be pumped to a location on the edge of the shore that is 8 mi east of the rig. The cost of constructing a pipe in the ocean from the rig to the shore is 1.5 times as expensive as the cost of construction on land. Set up and analyze a model to determine how the pipe should be laid to minimize cost.

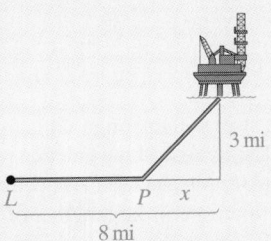

62. The owner of a novelty store can obtain joy buzzers from the manufacturer for 40¢ each. It is estimated that 60 buzzers will be sold when the price is $1.20 per buzzer and that 10 more buzzers will be sold for every 10¢ decrease in price. What price should be charged to maximize profit?

63. Westel Corporation manufactures telephones and has developed a new cellular phone. Production analysis shows that its price must not be less than $50; if x units are sold, then the price is given by the formula $p(x) = 150 - x$. The total cost of producing x units is given by the formula $C(x) = 2,500 + 30x$. Find the maximum profit, and determine the price that should be charged to maximize the profits.

64. MODELING PROBLEM   A manufacturer receives an order for 5,000 items. There are 12 machines available, each of which can produce 25 items per hour. The cost of setting up a machine for a production run is $50. Once the machines are in operation, the procedure is fully automated and can be supervised by a single worker earning $20 per hour. Set up and analyze a model to determine the number of machines that should be used to minimize the total cost of filling the order. State any assumptions that must be made.

65. MODELING PROBLEM   The personnel manager of a department store estimates that if N temporary salespersons are hired for the holiday season, the total net revenue derived (in hundreds of dollars) from their efforts may be modeled by the function

$$R(N) = -3N^4 + 50N^3 - 261N^2 + 540N$$

for $0 \leq N \leq 9$. How many salespersons should be hired to maximize total net revenue?

66. Show that the graph of a polynomial of degree n, with $n > 2$, has at most $n - 2$ inflection points.

67. Show that the graph of the function $f(x) = x^n$, with $n > 1$, has either one or no inflection points, depending on whether n is odd or even.

68. Each tangent line to the circle $x^2 + y^2 = 1$ at a point in the first quadrant will intersect the coordinate axes at points $(x_1, 0)$ and $(0, y_1)$. Determine the line for which $x_1 + y_1$ is a minimum.

69. MODELING PROBLEM   An accelerated particle moving at speed close to the speed of light emits power P in the direction θ given by

$$P(\theta) = \frac{a \sin \theta}{(1 - b \cos \theta)^5} \quad \text{for } 0 < b < 1 \text{ and } a > 0$$

Find the value of θ ($0 \leq \theta \leq \pi$) for which P has the greatest value.

70. Suppose that f is a continuous function defined on the closed interval $[a, b]$ and that $f'(x) = c$ on the open interval (a, b) for some constant c. Use MVT to show that

$$f(x) = c(x - a) + f(a)$$

for all x in the interval $[a, b]$.

71. Find the point of inflection of the curve

$$y = (x + 1) \tan^{-1} x$$

72. Find the critical numbers for the function

$$f(x) = \tan^{-1}\left(\frac{x}{a}\right) - \tan^{-1}\left(\frac{x}{b}\right), \quad a > b$$

Classify each as a relative maximum, a relative minimum, or neither.

73. Suppose that $f''(x)$ exists for all x and that $f''(x) + c^2 f(x) = 0$ for some number c with $f'(0) = 1$. Show that for any $x = x_0$, there is a number w between 0 and x_0 for which

$$f'(x_0) + c^2 f(w)x_0 = 1$$

74. MODELING PROBLEM According to the *Mortality and Morbidity Report* of the U.S. Centers for Disease Control, the following table gives the number of annual deaths from acquired immune deficiency syndrome (AIDS) in the United States for the years 1982–1991:*

Year	Deaths	Year	Deaths
1982	400	1987	13,900
1983	1,400	1988	17,300
1984	3,200	1989	32,000
1985	6,200	1990	30,000
1986	10,660	1991	45,000

a. To model the number of annual AIDS deaths, a data analysis program produced the cubic polynomial

$$N(x) = -8.58197x^3 + 732.727x^2 - 3,189.9x + 4,375.09$$

Sketch the graph of N and determine the time when its highest point occurs for $0 \le x \le 20$ where $x = 0$ represents 1982.

b. The same report gives the number of cases of reported AIDS for the period 1984–1991:

Year	Deaths	Year	Deaths
1984	4,445	1988	31,001
1985	8,249	1989	33,722
1986	12,932	1990	41,595
1987	21,070	1991	43,672

The same data analysis program used in part **a** yields

$$C(x) = -171.247x^3 + 3,770.90x^2 - 19,965.1x + 34,893.9$$

as a modeling polynomial for the number of cases as a function of time x. When does this model predict the number of reported cases will be the largest? When does it predict the number of reported cases will drop to 0?

*For an interesting discussion of how mathematical modeling can be applied to a problem of great public and personal interest, see "Modeling the AIDS Epidemic," by Allyn Jackson, *Notices of the American Mathematical Society*, Vol. 36, No. 8 (October 1989), pp. 981–983.

c. When the data in part **b** are modeled exponentially, the data analysis program produces the function

$$y = 1{,}676e^{0.3256x}$$

Sketch the cubic modeling formula $y = C(x)$ from part **b** and this exponential formula on the same set of coordinate axes along with the data points from the table in part **b**. Which formula do you think does a better job of fitting the data?

d. Explore other modeling formulas for the data given in part **a**.

e. Call the Centers for Disease Control or check the World Wide Web (www.cdc.gov) for the most recent updates for the data in parts **a** and **b**. Do the formulas in this problem correctly model the current information? Explain the discrepancies.

f. Based on the information you obtained in part **e**, find new models for the information in parts **a** and **b** using all the data you have available.

Technology Window

Using the graphing and differentiation programs of your computer or calculator, graph, $f(x)$, $f'(x)$, $f''(x)$ for each function in Problems 75–78. Print out a copy, if possible. On each graph, indicate the intervals where

a. $f'(x) > 0$ **b.** $f'(x) < 0$ **c.** $f''(x) > 0$
d. $f''(x) < 0$ **e.** $f'(x) = 0$ **f.** $f'(x)$ *does not exist*
g. $f''(x) = 0$

Describe how these inequalities qualitatively determine the shape of the graph of the functions over the given interval.

75. $f(x) = \sin 2x$ on $[-\pi, \pi]$

76. $f(x) = x^3 - x^2 - x + 1$ on $[-\frac{3}{2}, 2]$

77. $f(x) = x^4 - 2x^2$ on $[-2, 2]$

78. $f(x) = x^3 - x + \dfrac{1}{x} + 1$ on $[-2, 2]$

79. Consider a string 60 in. long that is formed into a rectangle. Using a graphing calculator or a graphing program, graph the area $A(x)$ enclosed by the string as a function of the length x of a given side $(0 < x < 30)$. From the graph, deduce that the maximum area is enclosed when the rectangle is a square. What is the minimum area enclosed? Compare this problem to the exact solution. What conclusions do you draw about the desirability of analytical solutions and the role of the computer?

Technology Window

80. MODELING PROBLEM: The beach of a lake follows contours that are approximated by the curve $4x^2 + y^2 = 1$, and a nearby road lies along the curve $y = 1/x$ for $x > 0$. Using your graphing calculator or computer software, determine the closest approach of the road to the lake in the north–south direction. Take the positive y-axis as pointing north.

81. Using your graphing and differentiation programs, locate and identify all the relative extrema for the function defined by

$$f(x) = \tfrac{1}{5}x^5 - \tfrac{5}{4}x^4 + 2x^2$$

This may require judicious choices of the window settings.

82. MODELING PROBLEM Suppose you are a manager of a fleet of delivery trucks. Each truck is driven at a constant speed of x mi/h ($15 \le x \le 55$), and gas consumption (gal/mi) is modeled by the function

$$\frac{1}{250}\left(\frac{750}{x} + x\right)$$

Using a graphing and differentiation program (and/or root-solving program on your computer or calculator), answer the following questions:
a. If gas costs $1.70/gal, estimate the steady speed that will minimize the cost of fuel for a 500-mi trip.
b. Estimate the steady speed that minimizes the cost if the driver is paid $28 per hour and the price of gasoline remains constant at $1.70/gal.

83. MODELING PROBLEM The sketch shows a circular wheel of radius 1 ft moving at an angular speed of 2 radians/s in the counterclockwise direction. A rod of length L is attached to the circumference and the other end of the rod can only move vertically, as shown in Figure 4.89.

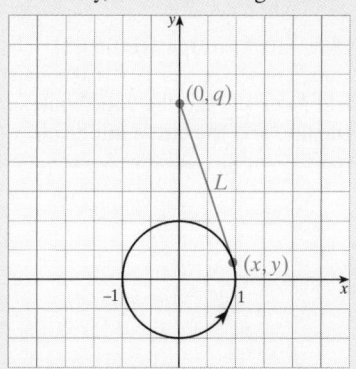

■ **FIGURE 4.89** A rod attached to a circular wheel

a. Argue that a point on the circumference can be expressed as

$$(x, y) = (\cos 2t, \sin 2t)$$

b. Derive the equation for q:

$$q(t) = \sin 2t + \sqrt{L^2 - \cos^2 2t}$$

c. Compute $q'(t)$.
d. From the physics of the problem, argue that velocity $q'(t) = 0$ when the (x, y) point is at the top or the bottom of the wheel. Then *conjecture* at what point(s) q' will achieve its maximum value.

From here on, use some computing help.

e. Compute $q'(t)$ and compare it with your answer in part **c.**
f. For $L = 2$, plot $q(t)$ and $q'(t)$ and comment on what you see. Are there any surprises?
g. Now, for L slightly larger than 1 (that is, 1.1 or 1.01) repeat part **f.**
h. Let $L = 2.02$; for what values of t is the velocity $q'(t)$ the greatest? Specifically, find at least one t value to two-decimal-place accuracy. Compare with your answer in part **d.**
i. Compare the maximum speed of point $(0, q)$ with that of the other end of the rod, (x, y).
j. **Exploratory** What do you think happens to q'_{max} as $L \to 1.0$?

84. HISTORICAL QUEST Leonhard Euler is one of the giants in the history of mathematics. His name is attached to something in almost every branch of mathematics. He was the most prolific writer on the subject of mathematics, and his mathematical textbooks were masterfully written. His writing was not at all slowed down by his total blindness for the last 17 years of his life. He possessed a phenomenal memory, had almost total recall, and could mentally calculate solutions to long and complicated problems. The basis for the historical development of calculus, as well as modern-day analysis, is the notion of a *function*. Euler's book *Introductio in analysin infinitorum* (1784) first used the function concept as the basic idea. It was the identification of functions, rather than curves, as the principal object of study, that permitted the advancement of mathematics in general, and calculus in particular. In Chapter 1, we noted that the number e is named in honor of Euler. Euler did not use the definition we use today, namely

$$e = \lim_{n \to +\infty} \left(1 + \frac{1}{n}\right)^n$$

LEONHARD EULER
1707–1783

Instead, Euler used series in his work (which we will study in Chapter 8).

In this Historical Quest problem we will lay the groundwork for a Historical Quest in Chapter 8. Euler introduced $\log_a x$ (which he wrote as ℓx) as that exponent y such that $a^y = x$. This was done in 1748, which makes it the first appearance of a logarithm interpreted explicitly as an exponent. He does not define $a^0 = 1$, but instead writes

$$a^\epsilon = 1 + k\epsilon$$

for an infinitely small number ϵ. In other words,

$$k = \lim_{\epsilon \to 0} \frac{a^\epsilon - 1}{\epsilon}$$

Explain why $k = \ln a$.

When Euler was 13, he registered at the University of Basel and was introduced to another famous mathematician, Johann Bernoulli, who was an instructor there at the time (see Historical Quest, Section 4.8, Problem 62).

If Bernoulli thought that a student was promising, he would provide, sometimes gratis, private instruction. Here is Euler's own account of his first encounter with Bernoulli:

"I soon found an opportunity to gain introduction to the famous professor Johann Bernoulli, whose good pleasure it was to advance me further in the mathematical sciences. True, because of his business he flatly refused me private lessons, but he gave me much wiser advice, namely, to get some more difficult mathematical books and work through them with all industry, and wherever I should find some check or difficulties, he gave me free access to him every Saturday afternoon and was so kind as to elucidate all difficulties, which happened with such greatly desired advantage that whenever he had obviated one check for me, because of that ten others disappeared right away, which is certainly the way to make a happy advance in the mathematical sciences." *

85. **PUTNAM EXAMINATION PROBLEM**   Given the parabola $y = 2mx$, what is the length of the shortest chord that is normal to the curve at one end? *Hint*: If $\overline{AB}$ is normal to the parabola, where A and B are the points $(2mt^2, 2mt)$ and $(2ms^2, 2ms)$, show that the slope of the tangent at A is $1/(2t)$.

_____

*From *Elements of Algebra* by Leonhard Euler.

86. **PUTNAM EXAMINATION PROBLEM**   Prove that the polynomial

$$(a - x)^6 - 3a(a - x)^5 + \tfrac{5}{2}a^2(a - x)^4 - \tfrac{1}{2}a^4(a - x)^2$$

has only negative values for $0 < x < a$. *Hint:* Show that if $x = g(1 - y)$, the polynomial becomes $a^6 y^2 g(y)$ where

$$g(y) = y^4 - 3y^3 + \tfrac{5}{2}y^2 - \tfrac{1}{2}$$

and then prove that $g(y) < 0$ for $0 < y < 1$.

87. **PUTNAM EXAMINATION PROBLEM**   Find the maximum value of $f(x) = x^3 - 3x$ on the set of all real numbers x satisfying $x^4 + 36 \le 13x^2$.

88. **PUTNAM EXAMINATION PROBLEM**   Let T be an acute triangle. Inscribe a pair R and S of rectangles in T, as shown in Figure 4.90.

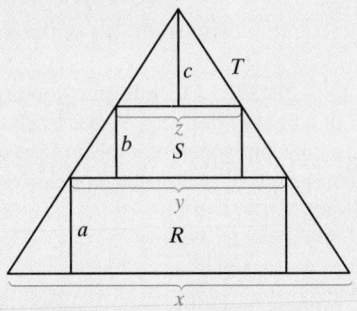

■ **FIGURE 4.90**   Problem 88

Let $A(x)$ denote the area of polygon X. Find the maximum value, or show that no maximum exists of the ratio

$$\frac{A(R) + A(S)}{A(T)}$$

where T ranges over all triangles R, and S ranges over all rectangles as shown above.

89. **PUTNAM EXAMINATION PROBLEM**   Which is greater,

$$(\sqrt{n})^{\sqrt{n+1}} \quad \text{or} \quad (\sqrt{n+1})^{\sqrt{n}}$$

where $n > 8$? *Hint:* Use the function $f(x) = \dfrac{\ln x}{x}$, and show that $x^y > y^x$ when $e \le x < y$.

90. **PUTNAM EXAMINATION PROBLEM**   The graph of the equation $x^y = y^x$ for $x > 0, y > 0$ consists of a straight line and a curve. Find the coordinates of the point where the line and the curve intersect.

Wine Barrel Capacity

This project is to be done in groups of three or four students. Each group will submit a single written report.

A wine barrel has a hole in the middle of its side called a **bung hole.** To determine the volume of wine in the barrel, a **bung rod** is inserted in the hole until it hits the lower seam. Determine how to calibrate such a rod so that it will measure the volume of the wine in the barrel.

You should make the following assumptions:

1. The barrel is cylindrical.
2. The distance from the bung hole to the corner is λ.
3. The ratio of the height to the diameter of the barrel is t. This ratio should be chosen so that for a given λ value, the volume of the barrel is maximal.

Your paper is not limited to the following questions, but it should include these concerns: You should show that the volume of the cylindrical barrel is $V = 2\pi\lambda^3 t(4 + t^2)^{-3/2}$, and you should find the approximate ideal value for t. Johannes Kepler was the first person to show mathematically why coopers were guided in their construction of wine barrels by one rule: *make the staves* (the boards that make up the sides of the barrel) *one and one-half times as long as the diameter.* (This is the approximate *t*-value.) You should provide dimensions for the barrel as well as for the bung rod.

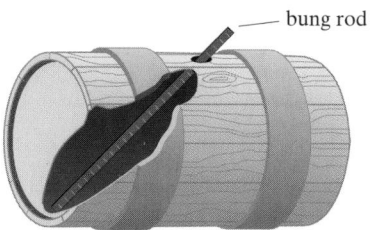

bung rod

Johannes Kepler (1571–1630) is usually remembered for his work in astronomy, in particular for his three laws of planetary motion. Tycho Brahe (1546–1601) was working for Rudolf II, Holy Roman Emperor in Prague in 1599, and he asked Kepler to work with him. The historian Burton describes this as a fortunate alliance. "Tycho was a splendid observer but a poor mathematician, while Kepler was a splendid mathematician but a poor observer." Because Kepler was a Protestant during a time when most intellectuals were required to be Catholic, Kepler had trouble supporting himself, and consequently worked for many benefactors. While serving the Austrian emperor Matthew I, Kepler observed with admiration the ability of a young vintner to declare quickly and easily the capacities of a number of different wine casks. He describes how this can be done in his book The New Stereometry of Wine Barrels, Mostly Austrian.

---

*The idea for this group research project comes from research done at Iowa State University as part of a National Science Foundation grant. Our thanks to Elgin Johnston of Iowa State University.

CONTENTS

5

Integration

PREVIEW

If an object moves along a straight line with velocity $v(t)$, what is its position $s(t)$ at time t? On Christmas Day, it starts snowing at a steady rate. A snowplow starts out at noon, going 2 miles during the first hour and 1 mile during the second. What time did it start snowing? At first glance, these problems (see Problem 70 of the supplementary problems, for example) may appear to have little in common, but they are all typical of what we shall encounter in our study of *integral calculus*, the second of the basic areas discussed in Section 2.1.

PERSPECTIVE

The key concept in integral calculus is *integration*, a procedure that involves computing a special kind of limit of sums called the *definite integral*. We shall find that such limits can often be computed by reversing the process of differentiation; that is, given a function f, we find a function F such that $F' = f$. This is called *indefinite integration*, and the equation $F' = f$ is an example of a *differential equation*.

Finding integrals and solving differential equations are extremely important processes in calculus. We begin our study of these topics by defining definite and indefinite integration and showing how they are connected by a remarkable result called the *fundamental theorem of calculus*. Then we examine several techniques of integration and show how area, average value, and other quantities can be set up and analyzed by integration. Our study of differential equations begins in this chapter and will continue in appropriate sections throughout this text. We also establish a mean value theorem for integrals and develop numerical procedures for estimating the value of a definite integral.

5.1 Antidifferentiation

reversing differentiation, antiderivative notation, antidifferentiation formulas, applications, area as an antiderivative

REVERSING DIFFERENTIATION

A physicist who knows the acceleration of a particle may want to determine its velocity or its position at a particular time. An ecologist who knows the rate at which a certain pollutant is being absorbed by a particular species of fish might want to know the actual amount of pollutant in the fish's system at a given time. In each of these cases, a derivative f' is given and the problem is that of finding the corresponding function f. Toward this end, we make the following definition.

Antiderivative

An **antiderivative** of a function f is a function F that satisfies

$$F' = f$$

Suppose we know $f(x) = 3x^2$. We want to find a function $F(x)$ so that $F'(x) = 3x^2$. It is not difficult to use the power rule in reverse to discover that $F(x) = x^3$ is such a function. However, that is not the only possibility:

Given:	$F(x) = x^3$	$G(x) = x^3 - 5$	$H(x) = x^3 + \pi^2$
Find:	$F'(x) = 3x^2$	$G'(x) = 3x^2$	$H'(x) = 3x^2$

Are there other functions that have a derivative equal to f? Clearly, there are infinitely many, and they all seem to differ by a constant. If F is an antiderivative of f, then so is $F + C$ for any constant C, because

$$[F(x) + C]' = F'(x) + 0 = f(x)$$

Conversely, the following theorem shows that any antiderivative of f can be expressed in this form. This is the constant difference theorem we proved in Section 4.2.

THEOREM 5.1   *Antiderivatives differ by a constant*

If F is an antiderivative of the continuous function f, then any other antiderivative of f must have the form

$$G(x) = F(x) + C$$

■ *What This Says:*   Two antiderivatives of the same function differ by a constant.

Proof   If F and G are both antiderivatives of f, then $F' = f$ and $G' = f$ and Theorem 4.6 (the constant difference theorem) tells us that

$$G(x) - F(x) = C$$

so $G(x) = F(x) + C$.

EXAMPLE 1	*Finding antiderivatives*

Find general antiderivatives for the given functions.

a. $f(x) = x^5$ b. $s(x) = \sin x$

Solution

a. If $F(x) = x^6$, then $F'(x) = 6x^5$, so we see that a particular antiderivative of f is $F(x) = \dfrac{x^6}{6}$ to obtain $F'(x) = \dfrac{6x^5}{6} = x^5$. By Theorem 5.1, the most general antiderivative is $G(x) = \dfrac{x^6}{6} + C$.

b. If $S(x) = -\cos x$, then $S'(x) = \sin x$, so $G(x) = -\cos x + C$. ∎

Recall that the slope of a function $y = f(x)$ at each point (x, y) on its graph is given by the derivative $f'(x)$. We can exploit this fact to obtain a "picture" of the graph of f. Reconsider Example 1c where $y' = 1/x$. There is an antiderivative $F(x)$ of $1/x$ such that the slope of F at each point x is $1/x$ for each nonzero value of x. Let us draw a graph of these slopes:

> If $x = 1$, then the slope is $\frac{1}{1} = 1$. Draw short line segments at $x = 1$, each with slope 1, for different y-values as shown in Figure 5.1**a**.
>
> If $x = -3$, then the slope is $-1/3$, so draw short line segments at $x = -3$, each with slope $-1/3$, also shown in Figure 5.1**a**.

If we continue to plot these slope points for different values of x, we obtain many little slope lines. The resulting graph shown in Figure 5.1**c** is known as a **slope field** for the equation $y' = 1/x$.

Finally, notice the relationship between the slope field for $y' = 1/x$ and its antiderivative $y = \ln|x| + C$ (found in Example 1c). If we choose particular values for C, say $C = 0$, $C = -\ln 2$, or $C = 2$ and draw these particular antiderivatives in Figure 5.1**c**, we notice that these particular solutions are anticipated by the slope field drawn in part **b**. That is, the slope field shows the entire family of antiderivatives of the original equation.

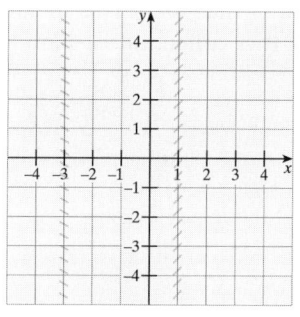

a. Slopes for $x = 1$ and $x = -3$

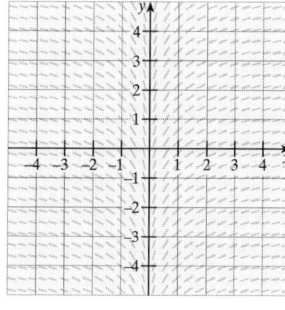

b. Slope field

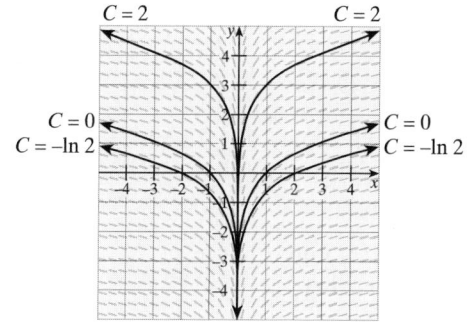

c. Sample antiderivatives

■ **FIGURE 5.1** Slope field for the equation $y' = \dfrac{1}{x}$

Slope fields will be discussed in more detail in Section 5.6. In general, slope fields and antiderivatives obtained as flow curves are usually generated using technology in

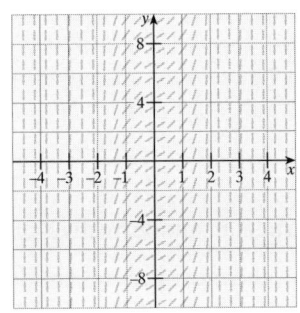

■ **FIGURE 5.2**   Slope field for $y' = e^{x^2}$

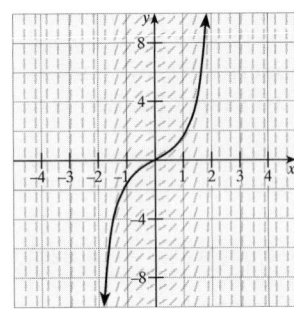

■ **FIGURE 5.3**   Antiderivative of $y' = e^{x^2}$ passing through $(0,0)$

computers and calculators. Here is an example in which the antiderivatives cannot be obtained as elementary functions.

> **EXAMPLE 2**   *Finding an antiderivative using a slope field*

Consider the slope field for $y' = e^{x^2}$, which is shown in Figure 5.2. Draw a possible graph of the antiderivative of e^{x^2} which passes through the point $(0,0)$.

Solution   Each little segment represents the slope of e^{x^2} for a particular x value. For example, if $x = 0$, the slope is 1, if $x = 1$ the slope is e, and if $x = 2$, the slope is e^4 (quite steep). However, to draw the antiderivative, step back and take a large view of the graph, and "go with the flow." We sketch the apparent graph that passes through $(0,0)$, as shown in Figure 5.3.   ■

ANTIDERIVATIVE NOTATION

It is worthwhile to define a notation to indicate the operation of antidifferentiation.

> **Indefinite Integral**
>
> The notation
>
> $$\int f(x)\, dx = F(x) + C$$
>
> where C is an arbitrary constant means that F is an antiderivative of f. It is called the **indefinite integral of f** and satisfies the condition that $F'(x) = f(x)$ for all x in the domain of f.

> ■ *What This Says:*   This is nothing more than the definition of antiderivative, along with a convenient notation. We also agree that in the context of antidifferentiation, C is an arbitrary constant.

The graph of $F(x) + C$ for different values of C is called **a family of functions** (see Figure 5.4). Because each member of the family $y = F(x) + C$ has the same derivative at x, the slope of each graph at x is the same. This means that the graph of all functions of the form $y = F(x) + C$ is a collection of parallel curves, as shown in Figure 5.4.

WARNING   It is important to remember that

$$\int f(x)\, dx$$

represents a family of functions. ◄

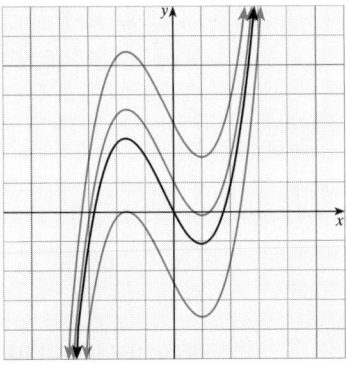

■ **FIGURE 5.4**   Several members of the family of curves $y = F(x) + C$

The process of finding indefinite integrals is called **indefinite integration.** Notice that this process amounts to finding an antiderivative of f and adding an arbitrary constant C, which is called the **constant of integration.**

| **EXAMPLE 3** | *Antidifferentiation* |

Find each of the following indefinite integrals.

a. $\displaystyle\int 5x^3\, dx$ b. $\displaystyle\int \sec^2 x\, dx$ c. $\displaystyle\int e^x\, dx$

Solution

a. Since $\dfrac{d}{dx}(x^4) = 4x^3$, it follows that $\dfrac{d}{dx}\left(\dfrac{5x^4}{4}\right) = 5x^3$. Thus,

$$\int 5x^3\, dx = \frac{5x^4}{4} + C$$

b. Because $\dfrac{d}{dx}(\tan x) = \sec^2 x$, we have $\displaystyle\int \sec^2 x\, dx = \tan x + C$

c. Since $\dfrac{d}{dx}(e^x) = e^x$, we have $\displaystyle\int e^x\, dx = e^x + C$ ∎

ANTIDIFFERENTIATION FORMULAS

Example 3 leads us to state formulas for antidifferentiation. Theorem 5.2 summarizes several fundamental properties of indefinite integrals, each of which can be derived by reversing an appropriate differentiation formula.

THEOREM 5.2   *Basic integration rules*

	Differentiation Formulas	Integration Formulas
PROCEDURAL RULES		
Constant multiple:	$\dfrac{d}{du}(cf) = c\dfrac{df}{du}$	$\displaystyle\int cf(u)\, du = c\int f(u)\, du$
Sum rule:	$\dfrac{d}{du}(f + g) = \dfrac{df}{du} + \dfrac{dg}{du}$	$\displaystyle\int [f(u) + g(u)]\, du = \int f(u)\, du + \int g(u)\, du$
Difference rule:	$\dfrac{d}{du}(f - g) = \dfrac{df}{du} - \dfrac{dg}{du}$	$\displaystyle\int [f(u) - g(u)]\, du = \int f(u)\, du - \int g(u)\, du$
Linearity rule:	$\dfrac{d}{du}(af + bg) = a\dfrac{df}{du} + b\dfrac{dg}{du}$	$\displaystyle\int [af(u) + bg(u)]\, du = a\int f(u)\,du + b\int g(u)\, du$
BASIC FORMULAS		
Constant rule:	$\dfrac{d}{du}(c) = 0$	$\displaystyle\int 0\, du = c$
Exponential rule:	$\dfrac{d}{du}(e^u) = e^u$	$\displaystyle\int e^u\, du = e^u + C$
Power rule:	$\dfrac{d}{du}(u^n) = nu^{n-1}$	$\displaystyle\int u^n\, du = \begin{cases} \dfrac{u^{n+1}}{n+1} + C; & n \neq -1 \\ \ln|u| + C; & n = -1 \end{cases}$
Logarithmic rule:	$\dfrac{d}{du}(\ln|u|) = \dfrac{1}{u}$	

Trigonometric rules:

$$\frac{d}{du}(\cos u) = -\sin u \qquad \int \sin u \, du = -\cos u + C$$

$$\frac{d}{du}(\sin u) = \cos u \qquad \int \cos u \, du = \sin u + C$$

$$\frac{d}{du}(\tan u) = \sec^2 u \qquad \int \sec^2 u \, du = \tan u + C$$

$$\frac{d}{du}(\sec u) = \sec u \tan u \qquad \int \sec u \tan u \, du = \sec u + C$$

$$\frac{d}{du}(\csc u) = -\csc u \cot u \qquad \int \csc u \cot u \, du = -\csc u + C$$

$$\frac{d}{du}(\cot u) = -\csc^2 u \qquad \int \csc^2 u \, du = -\cot u + C$$

Inverse trigonometric rules:

$$\frac{d}{du}(\sin^{-1} u) = \frac{1}{\sqrt{1-u^2}} \qquad \int \frac{du}{\sqrt{1-u^2}} = \sin^{-1} u + C$$

$$\frac{d}{du}(\tan^{-1} u) = \frac{1}{1+u^2} \qquad \int \frac{du}{1+u^2} = \tan^{-1} u + C$$

$$\frac{d}{du}(\sec^{-1} u) = \frac{1}{|u|\sqrt{u^2-1}} \qquad \int \frac{du}{|u|\sqrt{u^2-1}} = \sec^{-1} u + C$$

Proof   Each of these parts can be derived by reversing the accompanying derivative formula. For example, to obtain the power rule, note that if n is any number other than -1, then

$$\frac{d}{du}\left[\frac{1}{n+1}u^{n+1}\right] = \frac{1}{n+1}[(n+1)u^n] = u^n$$

so that $\dfrac{1}{n+1}u^{n+1}$ is an antiderivative of u^n and

$$\int u^n \, du = \frac{1}{n+1}u^{n+1} + C \quad \text{for } n \neq -1$$

Now we shall use these rules to compute a number of indefinite integrals.

EXAMPLE 4   *Indefinite integral of a polynomial function*

Evaluate $\displaystyle\int (x^5 - 3x^2 - 7)\, dx$.

Solution   The first two steps are usually done mentally:

$$\int (x^5 - 3x^2 - 7)\, dx = \int x^5 \, dx - 3\int x^2 \, dx - 7\int dx$$

$$= \frac{x^{5+1}}{5+1} - 3\frac{x^{2+1}}{2+1} - 7x + C$$

$$= \tfrac{1}{6}x^6 - x^3 - 7x + C$$

EXAMPLE 5   *Indefinite integral with a mixture of forms*

Evaluate $\displaystyle\int (5\sqrt{x} + 4\sin x)\, dx$.

Solution

$$\int (5\sqrt{x} + 4 \sin x)\, dx = 5 \int x^{1/2}\, dx + 4 \int \sin x\, dx$$

$$= 5 \frac{x^{3/2}}{\frac{3}{2}} + 4(-\cos x) + C$$

$$= \tfrac{10}{3} x^{3/2} - 4 \cos x + C \quad\blacksquare$$

Antiderivatives will be used extensively in integration in connection with a marvelous result called the fundamental theorem of calculus (Section 5.4).

APPLICATIONS

In Chapter 3, we used differentiation to compute the slope at each point on the graph of a function. Example 6 shows how this procedure can be reversed.

> **EXAMPLE 6**   *Given the slope, find the function*

The graph of a certain function F has slope $4x^3 - 5$ at each point (x, y) and contains the point $(1, 2)$. Find the function F.

Analytic Solution   Because the slope of the tangent at each point (x, y) is given by $F'(x)$, we have

$$F'(x) = 4x^3 - 5$$

and it follows that

$$\int F'(x)\, dx = \int (4x^3 - 5)\, dx$$

$$F(x) = 4\left(\frac{x^4}{4}\right) - 5x + C$$

$$= x^4 - 5x + C$$

The family of curves is $y = x^4 - 5x + C$. To find the one that passes through $(1, 2)$, substitute:

$$2 = 1^4 - 5(1) + C$$

$$6 = C$$

The curve is $y = x^4 - 5x + 6$. $\blacksquare$

In Section 3.4, we observed that an object moving along a straight line with position $s(t)$ has velocity $v(t) = \dfrac{ds}{dt}$ and acceleration $a(t) = \dfrac{ds}{dt}$. Thus, we have

$$v(t) = \int a(t)\, dt \quad\text{and}\quad s(t) = \int v(t)\, dt$$

These formulas are used in Example 7.

> **EXAMPLE 7**   *Modeling Problem: Stopping distance for an automobile*

The brakes of a certain automobile produce a constant deceleration of 22 ft/s². If the car is traveling at 60 mi/h (88 ft/s) when the brakes are applied, how far will it travel before coming to a complete stop?

Solution

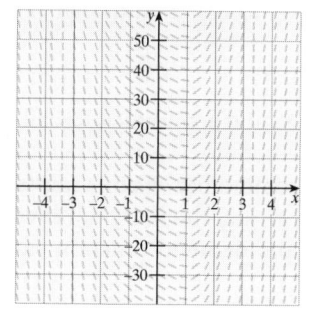

Technology Window

The slope field for $y' = 4x^3 - 5$ is drawn using technology.

We are interested in drawing the solution passing through $(1, 2)$. Remember to "go with the flow."

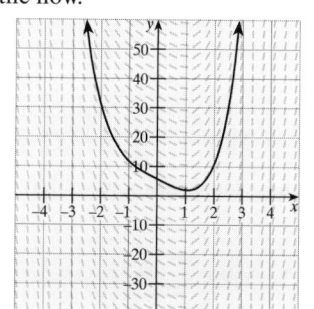

Solution   Let $a(t)$, $v(t)$, and $s(t)$ denote the acceleration, velocity, and position of the car t seconds after the brakes are applied. We shall assume that s is measured from the point where the brakes are applied, so that $s(0) = 0$.

$$v(t) = \int a(t)\, dt$$

$$= \int (-22)\, dt \qquad \textit{Negative because the car is decelerating}$$

$$= -22t + C_1 \qquad \textit{v(0) = -22(0) + C}_1 \textit{ = 88, so that C}_1 \textit{ = 88. Starting velocity is 88.}$$

$$= -22t + 88$$

Similarly,

$$s(t) = \int v(t)\, dt$$

$$= \int (-22t + 88)\, dt$$

$$= -11t^2 + 88t + C_2 \qquad \textit{s(0) = -11(0)}^2 \textit{ + 88(0) + C}_2 \textit{ = 0 so that C}_2 \textit{ = 0. Starting position is 0.}$$

$$= -11t^2 + 88t$$

Finally, the car comes to rest when its velocity is 0, so we need to solve $v(t) = 0$ for t:

$$-22t + 88 = 0$$

$$t = 4$$

This means that the car decelerates for 4 s before coming to rest, and in that time it travels

$$s(4) = -11(4)^2 + 88(4) = 176 \text{ ft} \qquad \blacksquare$$

Indefinite integration also has applications in business and economics. Recall that the *demand function* for a particular commodity is the function $p(x)$, which gives the price p that consumers will pay for each unit of the commodity when x units are brought to market. Then the total revenue is $R(x) = xp(x)$, and the marginal revenue is $R'(x)$. Our final example of this section shows how the demand function can be determined from the marginal revenue.

> **EXAMPLE 8**   *Finding the demand function given the marginal revenue*

A manufacturer estimates that the marginal revenue of a certain commodity is $R'(x) = 240 + 0.1x$ when x units are produced. Find the demand function $p(x)$.

Solution

$$R(x) = \int R'(x)\, dx$$

$$= \int (240 + 0.1x)\, dx = 240x + 0.1(\tfrac{1}{2}x^2) + C = 240x + 0.05x^2 + C$$

Because $R(x) = xp(x)$ where $p(x)$ is the demand function, we must have $R(0) = 0$ so that

$$240(0) + 0.05(0)^2 + C = 0 \quad \text{or} \quad C = 0$$

Thus, $R(x) = 240x + 0.05x^2$. It follows that the demand function is

$$p(x) = \frac{R(x)}{x} = \frac{240x + 0.05x^2}{x} = 240 + 0.05x \qquad \blacksquare$$

EXAMPLE 9   *Modeling Problem: The motion of a particle*

A particle moves along a coordinate axis in such a way that its acceleration is modeled by $a(t) = 2t^{-2}$ for time $t > 0$. If the particle is at $s = 5$ when $t = 1$ and has velocity $v = -3$ at that time, where is it when $t = 4$?

Solution   Because $a(t) = v'(t)$, it follows that

$$v(t) = \int a(t)\,dt = \int 2t^{-2}\,dt = -2t^{-1} + C_1$$

and since $v(1) = -3$, we have

$$-3 = v(1) = \frac{-2}{1} + C_1 \quad \text{so} \quad C_1 = -3 + 2 = -1$$

We also know $v(t) = s'(t)$, so

$$s(t) = \int v(t)\,dt = \int (-2t^{-1} - 1)\,dt = -2\ln|t| - t + C_2$$

Since $s(1) = 5$ we have

$$5 = s(1) = -2\ln|1| - 1 + C_2 \quad \text{or} \quad C_2 = 6$$

Thus, $s(t) = -2\ln|t| - t + 6$ so that $s(4) \approx -0.7726$.
   The particle is at -0.7726 when $t = 4$.   ■

AREA AS AN ANTIDERIVATIVE

In the next section, we will consider area as the limit of a sum, and we conclude this section by showing how area can be computed by antidifferentiation. The connection between area as a limit and area as an antiderivative is then made by a result called the *fundamental theorem of calculus* (see Section 5.4).

THEOREM 5.3   *Area as an antiderivative*

If f is a continuous function such that $f(x) \geq 0$ for all x on the closed interval $[a, b]$, then the area bounded by the curve $y = f(x)$, the x-axis, and the vertical lines $x = a$, $x = t$ is an antiderivative of $f(t)$ on $[a, b]$.

Proof   Define an **area function,** $A(t)$, as the area of the region bounded by the curve $y = f(x)$, the x-axis, and the vertical lines $x = a$, $x = t$ for $a \leq t \leq b$, as shown in Figure 5.5.

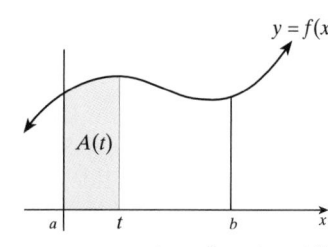

■ **FIGURE 5.5**   Area function $A(t)$

We need to show that $A(t)$ is an antiderivative of f on the interval $[a, b]$; that is, we need to show that $A'(t) = f(t)$.
   Let $h > 0$ be small enough so that $t + h < b$ and consider the numerator of the difference quotient for $A(t)$, namely, the difference $A(t + h) - A(t)$. Geometrically

this difference is the area under the curve $y = f(x)$ between $x = t$ and $x = t + h$, as shown in Figure 5.6a.

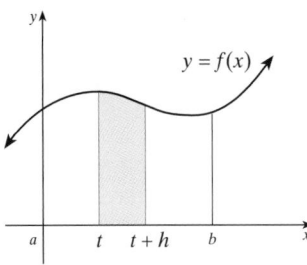
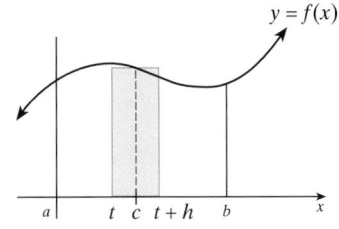

a. $A(t + h) - A(t)$ is the area over $[t, t + h]$

b. $A(t + h) - A(t)$ is approximated by the area of rectangle

■ **FIGURE 5.6**   The area under the curve $y = f(x)$

If h is small enough, this area is approximately the same as the area of a rectangle with base h and height $f(c)$, where c is the midpoint of the interval $[t, t + h]$, as shown in Figure 5.6b. Thus, we have

$$\underbrace{A(t + h) - A(t)}_{\text{Area under the curve on } [t, t + h]} \approx \underbrace{hf(c)}_{\text{Area of rectangle}}$$

The difference quotient for $A(t)$ satisfies

$$\frac{A(t + h) - A(t)}{h} \approx f(c)$$

Finally, by taking the limit as $h \to 0$, we find the derivative of the area function $A(t)$ satisfies

$$\lim_{h \to 0} \frac{A(t + h) - A(t)}{h} = \lim_{h \to 0} \frac{hf(c)}{h}$$

$$A'(t) = f(t)$$

The limit on the left is the definition of derivative, and on the right we see that since f is continuous and c is the midpoint of the interval $[t, t + h]$, c must approach t as $h \to 0$. Thus, $A(t)$ is an antiderivative of $f(t)$. ══

EXAMPLE 10   *Area as an antiderivative*

Find the area under the parabola $y = x^2$ over the interval $[0, 1]$. This area is shown in Figure 5.7.

Solution   Let $A(t)$ be the area function for this example—namely, the area under $y = x^2$ on $[0, t]$. Since f is continuous and $f(x) \geq 0$ on $[0, 1]$, Theorem 5.3 tells us that $A(t)$ is an antiderivative of $f(t) = t^2$ on $[0, 1]$. That is,

$$A(t) = \int t^2 \, dt = \tfrac{1}{3}t^3 + C$$

for all t in the interval $[0, 1]$. Clearly, $A(0) = 0$, so

$$A(0) = \tfrac{1}{3}(0)^3 + C \quad \text{or} \quad C = 0$$

and the area under the curve is

$$A(1) = \tfrac{1}{3}(1)^3 + 0 = \tfrac{1}{3} \qquad ■$$

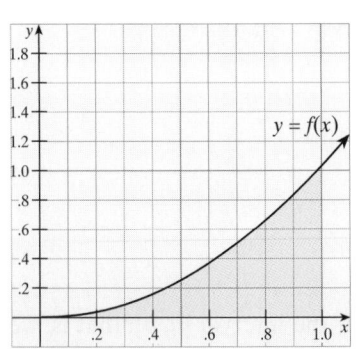

■ **FIGURE 5.7**   Area of $y = f(x)$ over $[0, 1]$

5.1 *Problem Set*

Ⓐ *Find the indefinite integral in Problems 1–30.*

1. $\int 2\, dx$

2. $\int -4\, dx$

3. $\int (2x + 3)\, dx$

4. $\int (4 - 5x)\, dx$

5. $\int (4t^3 + 3t^2)\, dt$

6. $\int (-8t^3 + 15t^5)\, dt$

7. $\int \dfrac{dx}{2x}$

8. $\int 14e^x\, dx$

9. $\int (6u^2 - 3\cos u)\, du$

10. $\int (5t^3 - \sqrt{t})\, dt$

11. $\int \sec^2 \theta\, d\theta$

12. $\int \sec \theta \tan \theta\, d\theta$

13. $\int 2\sin \theta\, d\theta$

14. $\int \dfrac{\cos \theta}{3}\, d\theta$

15. $\int \dfrac{5}{\sqrt{1 - y^2}}\, dy$

16. $\int \dfrac{dx}{10(1 + x^2)}$

17. $\int (u^{3/2} - u^{1/2} + u^{-10})\, du$

18. $\int (x^3 - 3x + \sqrt[4]{x} - 5)\, dx$

19. $\int x(x + \sqrt{x})\, dx$

20. $\int y(y^2 - 3y)\, dy$

21. $\int \left(\dfrac{1}{t^2} - \dfrac{1}{t^3} + \dfrac{1}{t^4} \right) dt$

22. $\int \dfrac{1}{t}\left(\dfrac{2}{t^2} - \dfrac{3}{t^3} \right) dt$

23. $\int (2x^2 + 5)^2\, dx$

24. $\int (3 - 4x^3)^2\, dx$

25. $\int \left(\dfrac{x^2 + 3x - 1}{x^4} \right) dx$

26. $\int \dfrac{x^2 + \sqrt{x} + 1}{x^2}\, dx$

27. $\int \dfrac{x^2 + x - 2}{x^2}\, dx$

28. $\int \left(1 + \dfrac{1}{x} \right)\left(1 - \dfrac{4}{x^2} \right) dx$

29. $\int \dfrac{\sqrt{1 - x^2} - 1}{\sqrt{1 - x^2}}\, dx$

30. $\int \dfrac{x^2}{x^2 + 1}\, dx$

The slope $F'(x)$ at each point on a graph is given in Problems 31–38 along with one point (x_0, y_0) on the graph. Use this information to graph F.

31. slope $x^2 + 3x$ with point $(0, 0)$

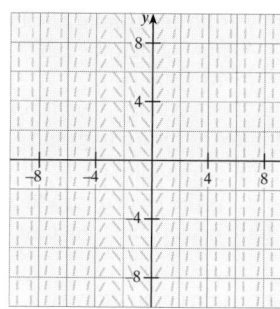

32. slope $(2x - 1)^2$ with point $(1, 3)$

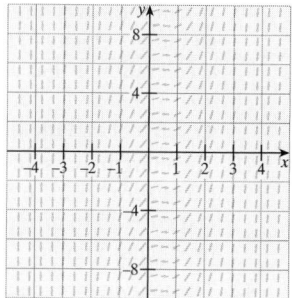

33. slope $(\sqrt{x} + 3)^2$ with point $(4, 36)$

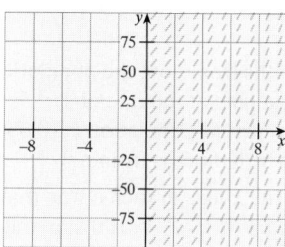

34. slope $3 - 2\sin x$ with point $(0, 0)$

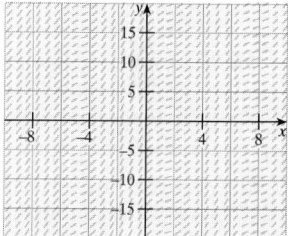

35. slope $\dfrac{x + 1}{x^2}$ with point $(1, -2)$

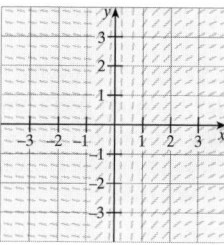

36. slope $\dfrac{2}{x\sqrt{x^2 - 1}}$ with point $(4, 1)$

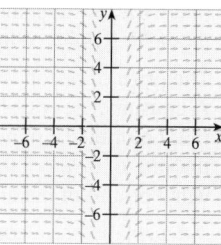

37. slope $x + e^x$ with point $(0, 2)$

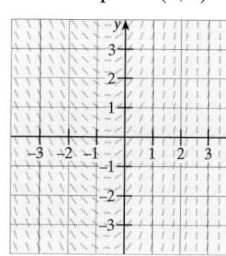

38. slope $\dfrac{x^2 - 1}{x^2 + 1}$ with point $(0, 0)$

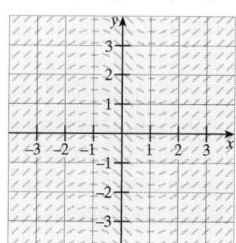

B 39. a. If $F(x) = \int \left(\frac{1}{\sqrt{x}} - 4 \right) dx$, find F so that $F(1) = 0$.

   b. Sketch the graphs of $y = F(x)$, $y = F(x) + 3$, and $y = F(x) - 1$.

   c. Find a constant C_0 so that the largest value of $G(x) = F(x) + C_0$ is 0.

40. A ball is thrown directly upward from ground level with an initial velocity of 96 ft/s. Assuming that the ball's only acceleration is that due to gravity (that is, $a(t) = -32$ ft/s^2), determine the maximum height reached by the ball and the time it takes to return to ground level.

41. The marginal cost of a certain commodity is $C'(x) = 6x^2 - 2x + 5$, where x is the level of production. If it costs \$5 to produce 1 unit, what is the total cost of producing 5 units?

42. The marginal revenue of a certain commodity is $R'(x) = -3x^2 + 4x + 32$, where x is the level of production (in thousands). Assume $R(x) = 0$ when $x = 0$.
   a. Find the demand function $p(x)$.
   b. Find the level of production that results in maximum revenue. What is the market price per unit at this level of production?

43. It is estimated that t months from now, the population of a certain town will be changing at the rate of $4 + 5t^{2/3}$ people per month. If the current population is 10,000, what will the population be 8 months from now?

44. A particle travels along the x-axis in such a way that its acceleration at time t is $a(t) = \sqrt{t} + t^2$. If it starts at the origin with an initial velocity of 2 (that is, $s(0) = 0$ and $v(0) = 2$), determine its position and velocity when $t = 4$.

45. An automobile starts from rest (that is, $v(0) = 0$) and travels with constant acceleration $a(t) = k$ in such a way that 6 s after it begins to move, it has traveled 360 ft from its starting point. What is k?

46. The price of bacon is currently \$1.80/lb in Styxville. A consumer service has conducted a study predicting that t months from now, the price will be changing at the rate of $0.984 + 0.012\sqrt{t}$ cents per month. How much will a pound of bacon cost 4 months from now?

47. An airplane has a constant acceleration while moving down the runway from rest. What is the acceleration of the plane at liftoff if the plane requires 900 ft of runway before lifting off at 88 ft/s (60 mi/h)?

48. The brakes of a certain automobile produce a constant deceleration of k ft/s^2. The car is traveling at 60 mi/h (88 ft/s) when the driver is forced to hit the brakes, and it comes to rest at a point 121 ft from the point where the brakes were applied. What is k?

49. After its brakes are applied, a certain sports car decelerates at a constant rate of 28 ft/s^2. Compute the stopping distance if the car is going 60 mi/h (88 ft/s) when the brakes are applied.

50. **SPY PROBLEM**   The Spy, having defused the bomb in Problem 23, Section 4.6, is driving the sports car in Problem 49 at a speed of 60 mi/h on Highway 1 in the remote republic of San Dimas. Suddenly he sees a camel in the road 199 ft in front of the car. After a reaction time of 00.7 seconds, he steps on the brakes. Will he stop before hitting the camel?

51. A particle moves along the x-axis in such a way that at time $t > 0$, its velocity (in ft/s) is
$$v(t) = t^{-1} + t$$
   How far does it move between times $t = 1$ and $t = e^2$?

52. A manufacturer estimates that the marginal cost in a certain production process is
$$C'(x) = 0.1e^x + 21\sqrt{x}$$
   when x units are produced. If the cost of producing 1 unit is \$100, what does it cost (to the nearest dollar) to produce 4 units?

In Problems 53–58, find the area under the curve defined by the given equation, above the x-axis, and over the given interval.

53. $y = x^2$ over $[1, 4]$ 54. $y = \sqrt{x}$ over $[1, 4]$

55. $y = e^x - x$ over $[0, 2]$ 56. $y = \dfrac{x + 1}{x}$ over $[1, 2]$

57. $y = \cos x$ over $[0, \frac{\pi}{2}]$ 58. $y = (1 - x^2)^{-1/2}$ over $[0, \frac{1}{2}]$

Technology Window

59. Evaluate $\displaystyle\int \frac{dy}{2y\sqrt{y^2 - 1}}$

   using the indicated methods.
   a. Use an inverse trigonometric differentiation rule.
   b. Use technology (TI-92, *Maple*, *Mathematica*, or *Derive*) to evaluate this integral.
   c. Reconcile your answers for parts **a** and **b**.

C 60. If $a, b,$ and c are constants, use the linearity rule twice to show that

$$\int [af(x) + bg(x) + ch(x)]\, dx$$
$$= a \int f(x)\, dx + b \int g(x)\, dx + c \int h(x)\, dx$$

61. Use the area as an antiderivative theorem to find the area under the line $y = mx + b$ over the interval $[c, d]$ where $m > 0$ and $mc + b > 0$. Check your result by using geometry to find the area of a trapezoid.

5.2   Area as the Limit of a Sum

area as the limit of a sum, the general approximation scheme, summation notation, area using summation formulas

AREA AS THE LIMIT OF A SUM

Computing area has been a problem of both theoretical and practical interest since ancient times; except for a few special cases, the problem is not easy. For example, you may know the formulas for computing the area of a rectangle, square, triangle, circle, and even a trapezoid. You have probably found the areas of regions that were more complicated but could be broken up into parts using these formulas. However, how would you find the area between the parabola $y = x^2$ and the x-axis on the interval $[0, 1]$? (See Figure 5.8a.) By solving this problem, we shall demonstrate a general procedure for computing area.

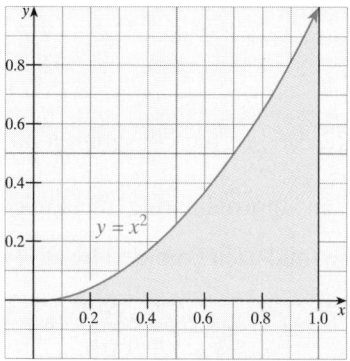

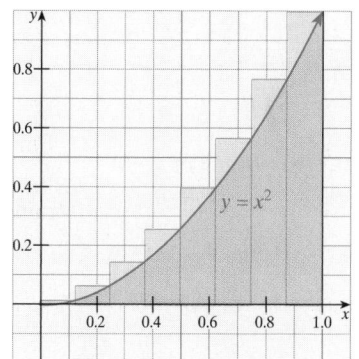

a. Problem: Compute the area under the curve $y = x^2$, above the x-axis and between the lines $x = 0$ and $x = 1$.

b. The required area is approximately the same as the total area bounded by the shaded rectangles.

■ **FIGURE 5.8**   Example of the area problem

EXAMPLE 1   *Estimating the area under a parabola using rectangles and right endpoints*

Estimate the area under the parabola $y = x^2$ on the interval $[0, 1]$.

Solution   Observe that although we cannot compute the area by applying a simple formula, we can certainly *estimate* the area by adding the areas of approximating rectangles constructed on subintervals of $[0, 1]$, as shown in Figure 5.8b. To simplify computations, we shall require all approximating rectangles to have the same width and shall take the height of each rectangle to be the y-coordinate of the parabola above the *right endpoint* of the subinterval on which it is based.*

---

*Actually, there is nothing special about right endpoints, and we could just as easily have used any other well-defined point in the base subinterval—say, the left endpoint or the midpoint

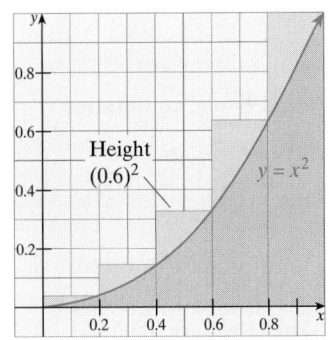

■ **FIGURE 5.9a**   Partitioning of Figure 5.8a into 5 subdivisions

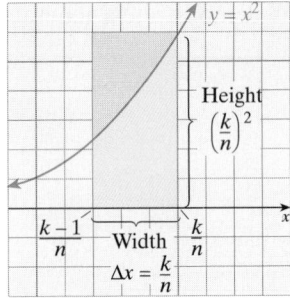

■ **FIGURE 5.9b**   Detail of Figure 5.9a

For the first estimate, we divide the interval $[0, 1]$ into 5 subintervals, as shown in Figure 5.9a. Because the approximating rectangles all have the same width, the right endpoints are $x_1 = 0.2, x_2 = 0.4, x_3 = 0.6, x_4 = 0.8$, and $x_5 = 1$. This subdivision is called a *partition of the interval*. The width of each subdivision is denoted by Δx and is found by dividing the length of the interval by the number of subintervals:

$$\Delta x = \frac{1 - 0}{5} = \frac{1}{5} = 0.2$$

If we let S_n be the total area of n rectangles, we find for the case where $n = 5$ that

$$\begin{aligned} S_5 &= f(x_1)\Delta x + f(x_2)\Delta x + f(x_3)\Delta x + f(x_4)\Delta x + f(x_5)\Delta x \\ &= [f(x_1) + f(x_2) + f(x_3) + f(x_4) + f(x_5)]\Delta x \\ &= [f(0.2) + f(0.4) + f(0.6) + f(0.8) + f(1)](0.2) \\ &= [0.2^2 + 0.4^2 + 0.6^2 + 0.8^2 + 1^2](0.2) = 0.44 \end{aligned}$$

Even though $S_5 = 0.44$ serves as a reasonable approximation of the area, we see from Figure 5.9b in the margin that this approximation is too large. Let us rework Example 1 using a general scheme rather than a specified number of rectangles. Partition the interval $[0, 1]$ into n equal parts, each with width

$$\Delta x = \frac{1 - 0}{n} = \frac{1}{n}$$

For $k = 1, 2, 3, \ldots, n$, the kth subinterval is $\left[\frac{k - 1}{n}, \frac{k}{n}\right]$, and on this subinterval, we then construct an approximating rectangle with width $\Delta x = \frac{1}{n}$ and height $\left(\frac{k}{n}\right)^2$, since $y = x^2$. The total area bounded by all n rectangles is

$$S_n = \left[\left(\frac{1}{n}\right)^2 + \left(\frac{2}{n}\right)^2 + \cdots + \left(\frac{n}{n}\right)^2\right]\left(\frac{1}{n}\right)$$

Consider different choices for n, as shown in Figure 5.10.

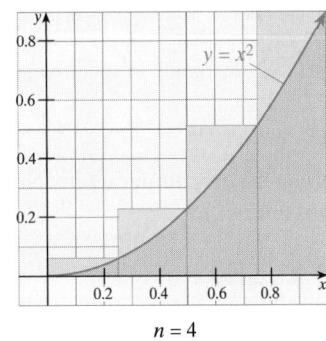
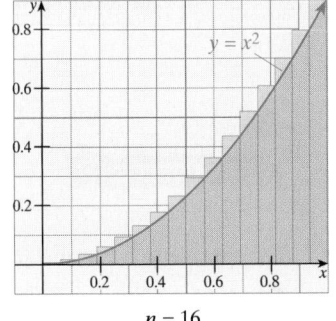
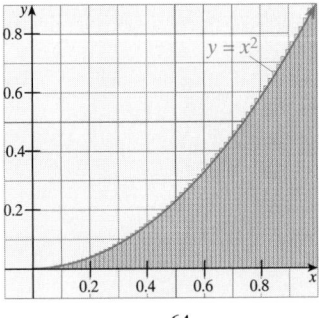

$n = 4$ $n = 16$ $n = 64$

■ **FIGURE 5.10**   The area estimate is improved by taking more rectangles.

n	S_n
5	0.440
10	0.385
50	0.343
100	0.338
1,000	0.334
5,000	0.333

If we increase the number of subdivisions n, the width $\Delta x = \frac{1}{n}$ of each approximating rectangle will decrease, and we would expect the area estimates S_n to improve. Thus, it is reasonable to *define* the area A under the parabola to be the *limit of* S_n as $\Delta x \to 0$ or, equivalently, as $n \to +\infty$. We can attempt to predict its value by seeing what happens to the sum as n grows large without bound. It is both tedious and difficult to evaluate such sums by hand, but fortunately we can use a calculator or computer to obtain some (rounded) values for S_n as shown in the margin. Notice that for $n = 5$, the value 0.44 corresponds to the calculation in Example 1.

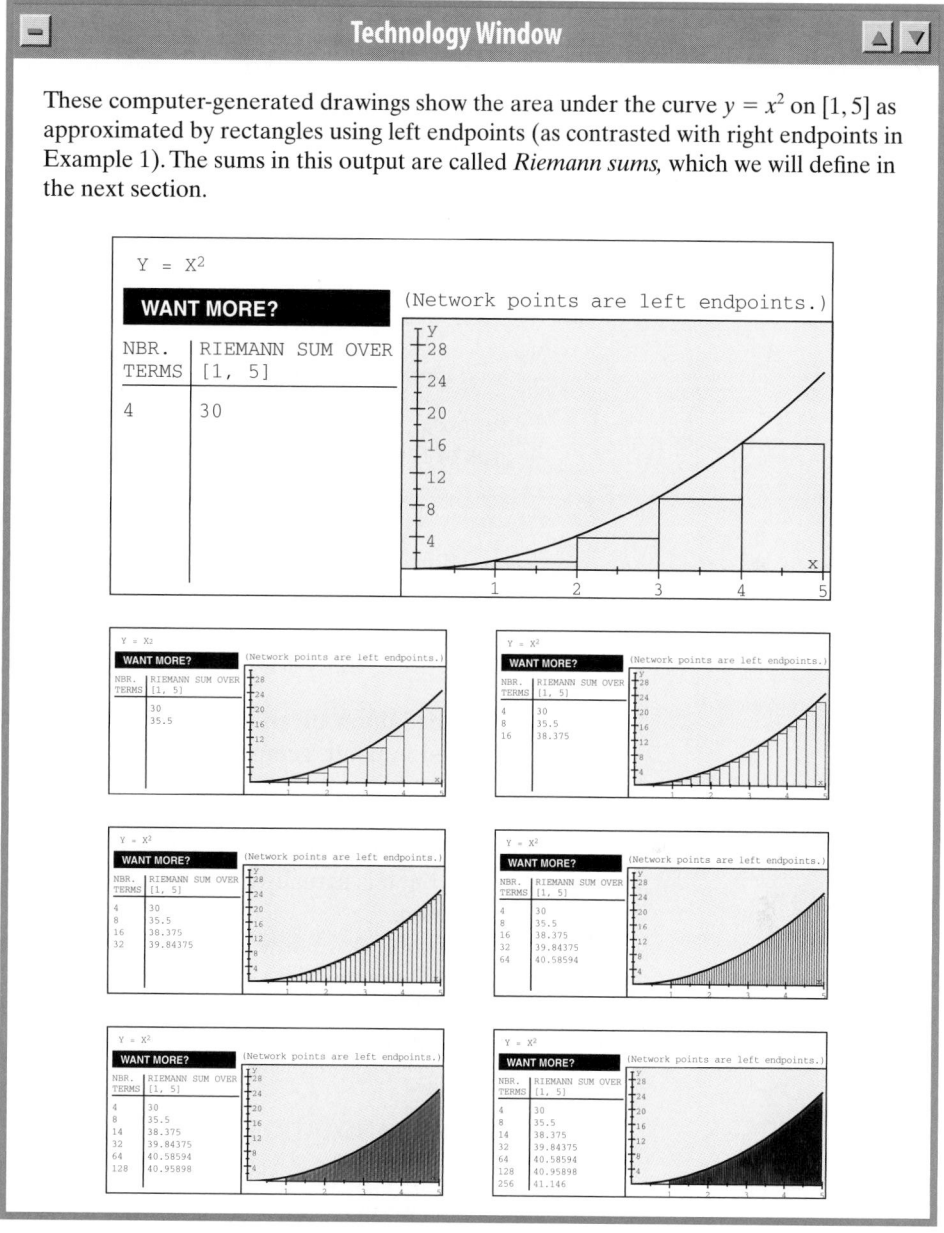

THE GENERAL APPROXIMATION SCHEME

We now compute the area under any curve $y = f(x)$ on an interval $[a, b]$ where f is a nonnegative continuous function. We first partition the interval $[a, b]$ into n equal subintervals, each of width

$$\Delta x = \frac{b - a}{n}$$

For $k = 1, 2, 3, \ldots, n$ the kth subinterval is $[a + (k-1)\Delta x, a + k\Delta x]$, and the kth approximating rectangle is constructed with width Δx and height $f(a + k\Delta x)$ equal to the altitude of the curve $y = f(x)$ above the right endpoint of the subinterval. Adding the area of these n rectangles, we obtain

$$S_n = \overbrace{f(a + \Delta x)\Delta x}^{\substack{\text{Area of}\\\text{first rectangle}}} + \overbrace{f(a + 2\Delta x)\Delta x}^{\substack{\text{Area of}\\\text{second rectangle}}} + \cdots + \overbrace{f(a + n\Delta x)\Delta x}^{\substack{\text{Area of}\\n\text{th rectangle}}}$$

as an estimate of the area under the curve. In advanced calculus, it is shown that the continuity of f guarantees the existence of $\lim_{\Delta x \to 0} S_n$, and we use this limit to define the required area, as shown in the following box.

Area as the Limit of a Sum

Suppose f is continuous and $f(x) \geq 0$ throughout the interval $[a, b]$. Then the **area** of the region under the curve $y = f(x)$ over this interval is

$$A = \lim_{\Delta x \to 0} [f(a + \Delta x) + f(a + 2\Delta x) + \cdots + f(a + n\Delta x)]\Delta x$$

where $\Delta x = \dfrac{b - a}{n}$.

The definition of area as a limit of a sum is consistent with the area concept introduced in plane geometry. For example, it would not be difficult to use this formula to show that a rectangle has area $A = \ell w$ or that a triangle has area $A = \frac{1}{2}bh$. You will also note that we maintained everyday usage in saying that the *formula* for the area of a rectangle is $A = \ell w$ or the *area* of the region under the curve $y = f(x)$ is a limit, but it is actually more proper to say that the limit *is the definition* of the area of the region.

The problem we now face is how to implement this definition of area as the limit of a sum. The immediate answer (discussed in this section) is to use summation formulas and technology. The long-range goal is to develop integral calculus, which is discussed in the next section.

SUMMATION NOTATION

The expanded form of the sum for the definition of area makes it awkward to use. Therefore, we shall digress to introduce a more compact notation for sums. Using this **summation notation,** we express the sum $a_1 + a_2 + \cdots + a_n$ as follows:

$$a_1 + a_2 + \cdots + a_n = \sum_{k=1}^{n} a_k$$

The summation notation is sometimes called the **sigma notation** because the uppercase Greek letter sigma (Σ) is used to denote the summation process. The **index** k is called the **index of summation** (or *running index*). The terminology used in connection with the summation notation is shown:

$$\text{Upper limit of summation} \quad \text{General term}$$

$$\sum_{k=1}^{n} a_k$$

Lower limit of summation   Index of summation

Note that in the summation process, the choice of summation index is immaterial. For example, the following sums are all exactly the same:

$$\sum_{k=3}^{7} k^2 = \sum_{j=3}^{7} j^2 = \sum_{i=3}^{7} i^2 = \sum_{\lambda=3}^{7} \lambda^2$$

In general, an index ($k, j, i,$ or λ) that represents a process in which it has no direct effect on the result is called a **dummy variable.**

Several useful properties of sums and sum formulas are listed in Theorem 5.4. We shall use the summation notation throughout the rest of this text, especially in this chapter, Chapter 6, and Chapter 8.

THEOREM 5.4   *Basic rules and formulas for sums*

For any numbers c and d, and positive integers $k, m,$ and n:

1. **Constant term rule** $\displaystyle\sum_{k=1}^{n} c = \underbrace{c + c + \cdots + c}_{n \text{ terms}} = nc$

2. **Sum rule** $\displaystyle\sum_{k=1}^{n} (a_k + b_k) = \sum_{k=1}^{n} a_k + \sum_{k=1}^{n} b_k$

3. **Scalar multiple rule** $\displaystyle\sum_{k=1}^{n} ca_k = c\sum_{k=1}^{n} a_k = \left(\sum_{k=1}^{n} a_k\right)c$

4. **Linearity rule** $\displaystyle\sum_{k=1}^{n} (ca_k + db_k) = c\sum_{k=1}^{n} a_k + d\sum_{k=1}^{n} b_k$

5. **Subtotal rule**   If $1 < m < n$, then
$$\sum_{k=1}^{n} a_k = \sum_{k=1}^{m} a_k + \sum_{k=m+1}^{n} a_k$$

6. **Dominance rule**   If $a_k \le b_k$ for $k = 1, 2, \ldots, n$, then
$$\sum_{k=1}^{n} a_k \le \sum_{k=1}^{n} b_k$$

Proof   These properties can all be established by applying well-known algebraic rules (see Problem 50).

AREA USING SUMMATION FORMULAS

Using summation notation, we can streamline the symbolism in the formula for the area under the curve $y = f(x), f(x) \ge 0$ on the interval $[a, b]$. In particular, note that the approximating sum S_n is

$$S_n = [f(a + \Delta x) + f(a + 2\Delta x) + \cdots + f(a + n\Delta x)]\Delta x$$

$$= \sum_{k=1}^{n} f(a + k\Delta x)\Delta x$$

where $\Delta x = \dfrac{b - a}{n}$. Thus, the formula for the definition of area is shown:

Area under $y = f(x)$ above the x-axis
between $x = a$ and $x = b$

n is the number of approximating rectangles

Width of each rectangle

$$A = \lim_{n \to +\infty} S_n = \lim_{\Delta x \to 0} \sum_{k=1}^{n} f(a + k\Delta x)\,\Delta x$$

Height of the kth rectangle

From algebra we recall certain summation formulas (proved using mathematical induction) that we will need in order to find areas using the limit definition.

Summation Formulas

$$\sum_{k=1}^{n} 1 = n$$

$$\sum_{k=1}^{n} k = 1 + 2 + 3 + \cdots + n = \frac{n(n + 1)}{2}$$

$$\sum_{k=1}^{n} k^2 = 1^2 + 2^2 + 3^2 + \cdots + n^2 = \frac{n(n + 1)(2n + 1)}{6}$$

$$\sum_{k=1}^{n} k^3 = 1^3 + 2^3 + 3^3 + \cdots + n^3 = \frac{n^2(n + 1)^2}{4}$$

EXAMPLE 2   *Area using the definition and summation formulas*

Use the summation definition of area to find the area under the parabola $y = x^2$ on the interval $[0, 1]$. You estimated this area in Example 1.

Solution   Partition the interval $[0, 1]$ into n subintervals with width $\Delta x = \dfrac{1 - 0}{n}$.

The right endpoint of the kth subinterval is $a + k\Delta x = \dfrac{k}{n}$, and $f\left(\dfrac{k}{n}\right) = \dfrac{k^2}{n^2}$. Thus, from the definition of area we have

$$A = \lim_{\Delta x \to 0} \sum_{k=1}^{n} f(a + k\Delta x)\Delta x = \lim_{n \to +\infty} \sum_{k=1}^{n} \left(\frac{k^2}{n^2}\right)\left(\frac{1}{n}\right)$$

$$= \lim_{n \to +\infty} \sum_{k=1}^{n} \frac{k^2}{n^3}$$

$$= \lim_{n \to +\infty} \frac{1}{n^3} \sum_{k=1}^{n} k^2 \qquad\qquad \textit{Scalar multiple rule}$$

$$= \lim_{n \to +\infty} \frac{1}{n^3}\left[\frac{n(n + 1)(2n + 1)}{6}\right] \qquad \textit{Summation formula for squares}$$

$$= \lim_{n \to +\infty} \frac{1}{6}\left[2 + \frac{3}{n} + \frac{1}{n^2}\right] = \frac{1}{3}$$

This is the same as the answer found by antidifferentiation in Example 10 of Section 5.1.

Technology Window

Sometimes using summation as in Example 2 is difficult, or even impossible. However, with a computer or programmable calculator, we can often find a pattern leading to the appropriate limit, just as we used tables to find limits in Chapter 2. Consider the following example.

EXAMPLE 3 *Tabular approach for finding area*

Use a computer to estimate the area under the curve $y = \sin x$ on the interval $[0, \frac{\pi}{2}]$.

Solution We see $a = 0$ and $b = \frac{\pi}{2}$, so $\Delta x = \frac{\frac{\pi}{2} - 0}{n} = \frac{\pi}{2n}$. The right endpoints are

$$a + \Delta x = 0 + \left(\frac{\pi}{2n}\right) = \frac{\pi}{2n}; \quad a + 2\Delta x = \frac{2\pi}{2n} = \frac{\pi}{n}; \quad a + 3\Delta x = \frac{3\pi}{2n}; \quad \cdots$$

$$a + n\Delta x = \frac{n\pi}{2n} = \frac{\pi}{2} = b$$

Thus, $S_n = \sum_{k=1}^{n} f\left(0 + \frac{k\pi}{2n}\right)\left(\frac{\pi}{2n}\right) = \sum_{k=1}^{n} \left[\sin\left(\frac{k\pi}{2n}\right)\right]\left(\frac{\pi}{2n}\right)$

Now we know from the definition that the actual area is

$$S = \lim_{n\to+\infty} \sum_{k=1}^{n} \left[\sin\left(\frac{k\pi}{2n}\right)\right]\left(\frac{\pi}{2n}\right)$$

which we can estimate by computing S_n for successively large values of n, as summarized in the table in the margin. Note that the table suggests that

$$\lim_{\Delta x \to 0} S_n = \lim_{\Delta x \to +0} S_n = 1$$

Thus, we expect the actual area under the curve to be 1 square unit.

n	S_n
10	1.07648
20	1.03876
50	1.01563
100	1.00783
500	1.00157

5.2 Problem Set

A *Evaluate the sums in Problems 1–8 by using the summation formulas.*

1. $\sum_{k=1}^{6} 1$ **2.** $\sum_{k=1}^{250} 2$ **3.** $\sum_{k=1}^{15} k$

4. $\sum_{k=1}^{10} (k+1)$ **5.** $\sum_{k=1}^{5} k^3$ **6.** $\sum_{k=1}^{7} k^2$

7. $\sum_{k=1}^{100} (2k-3)$ **8.** $\sum_{k=1}^{100} (k-1)^2$

Use the properties of summation notation in Problems 9–12 to evaluate the given limits.

9. $\lim_{n\to+\infty} \sum_{k=1}^{n} \frac{k}{n^2}$ **10.** $\lim_{n\to+\infty} \sum_{k=1}^{n} \frac{k^2}{n^3}$

11. $\lim_{n\to+\infty} \sum_{k=1}^{n} \left(1 + \frac{k}{n}\right)\left(\frac{2}{n}\right)$ **12.** $\lim_{n\to+\infty} \sum_{k=1}^{n} \left(1 + \frac{2k}{n}\right)^2\left(\frac{2}{n}\right)$

First sketch the region under the graph of $y = f(x)$ on the interval $[a, b]$ in Problems 13–21. Then approximate the area of each region by using right endpoints and the formula

$$S_n = \sum_{k=1}^{n} f(a + k\Delta x)\Delta x$$

for $\Delta x = \frac{b-a}{n}$ and the indicated values of n.

13. $f(x) = 4x + 1$ on $[0, 1]$ for
    a. $n = 4$ b. $n = 8$

14. $f(x) = 3 - 2x$ on $[0, 1]$ for
    a. $n = 3$ **b.** $n = 6$

15. $f(x) = x^2$ on $[1, 2]$ for
    a. $n = 4$ **b.** $n = 6$

16. $f(x) = \cos x$ on $[-\frac{\pi}{2}, 0]$ for $n = 4$

17. $f(x) = x + \sin x$ on $[0, \frac{\pi}{4}]$ for $n = 3$

18. $f(x) = \dfrac{1}{x^2}$ on $[1, 2]$ for $n = 4$

19. $f(x) = \dfrac{2}{x}$ on $[1, 2]$ for $n = 4$

20. $f(x) = \sqrt{x}$ on $[1, 4]$ for $n = 4$

21. $f(x) = \sqrt{1 + x^2}$ on $[0, 1]$ for $n = 4$

Ⓑ *Find the exact area under the given curve on the interval pre-scribed in Problems 22–27 by using the summation formulas.*

22. $y = 4x^3 + 2x$ on $[0, 2]$

23. $y = 4x^3 + 2x$ on $[1, 2]$

24. $y = 6x^2 + 2x + 4$ on $[0, 3]$

25. $y = 6x^2 + 2x + 4$ on $[1, 3]$

26. $y = 3x^2 + 2x + 1$ on $[0, 1]$

27. $y = 4x^3 + 3x^2$ on $[0, 1]$

THINK TANK PROBLEMS *Show that each statement about area in Problems 28–33 is generally true or provide a counter example. It will probably help to sketch the indicated region for each problem.*

28. If $C > 0$ is a constant, the region under the line $y = C$ on the interval $[a, b]$ has area $A = C(b - a)$.

29. If $C > 0$ is a constant and $b > a \geq 0$, the region under the line $y = Cx$ on the interval $[a, b]$ has area $A = \frac{1}{2}C(b - a)$.

30. The region under the parabola $y = x^2$ on the interval $[a, b]$ has area less than

$$\tfrac{1}{2}(b^2 + a^2)(b - a)$$

31. The region under the curve $y = \sqrt{1 - x^2}$ on the interval $[-1, 1]$ has area $A = \frac{\pi}{2}$.

32. Let f be a function that satisfies $f(x) \geq 0$ for x in the interval $[a, b]$. Then the area under the curve $y = f^2(x)$ on the interval $[a, b]$ must always be greater than the area under $y = f(x)$ on the same interval.

33. A function f is said to be *even* if $f(-x) = f(x)$. If f is even and $f(x) \geq 0$ throughout the interval $[-a, a]$, then the area under the curve $y = f(x)$ on this interval is *twice* the area under $y = f(x)$ on $[0, a]$.

34. Show that the region under the curve $y = x^3$ on the interval $[0, 1]$ has area $\frac{1}{4}$ square units.

35. Use the definition of area to show that the area of a rectangle equals the product of its length ℓ and its width w.

36. Show that the triangle with vertices $(0, 0), (0, h)$, and $(b, 0)$ has area $A = \frac{1}{2}bh$.

37. a. Compute the area under the parabola $y = 2x^2$ on the interval $[1, 2]$ as the limit of a sum.
   b. Let $f(x) = 2x^2$ and note that $g(x) = \frac{2}{3}x^3$ defines a function that satisfies $g'(x) = f(x)$ on the interval $[1, 2]$. Verify that the area computed in part **a** satisfies $A = g(2) - g(1)$.
   c. The function defined by

$$h(x) = \tfrac{2}{3}x^3 + C$$

   for any constant C also satisfies $h'(x) = f(x)$. Is it true that the area in part **a** satisfies $A = h(2) - h(1)$?

| **Technology Window** |

Use a tabular approach to compute the area under the curve $y = f(x)$ on each interval given in Problems 38–44 as the limit of a sum of terms.

38. $f(x) = 4x$ on $[0, 1]$

39. $f(x) = x^2$ on $[0, 4]$

40. $f(x) = \cos x$ on $[-\frac{\pi}{2}, 0]$
   (Compare with Problem 16.)

41. $f(x) = x + \sin x$ on $[0, \frac{\pi}{4}]$
   (Compare with Problem 17.)

42. $f(x) = \ln(x^2 + 1)$ on $[0, 3]$

43. $f(x) = e^{-3x^2}$ on $[0, 1]$

44. $f(x) = \cos^{-1}(x + 1)$ on $[-1, 0]$

45. a. Use the tabular approach to compute the area under the curve $y = \sin x + \cos x$ on the interval $[0, \frac{\pi}{2}]$ as the limit of a sum.
   b. Let $f(x) = \sin x + \cos x$ and note that $g(x) = -\cos x + \sin x$ satisfies $g'(x) = f(x)$ on the interval $[0, \frac{\pi}{2}]$. Verify that the area computed in part **a** satisfies $A = g(\frac{\pi}{2}) - g(0)$.
   c. The function

$$h(x) = -\cos x + \sin x + C$$

   for constant C also satisfies $h'(x) = f(x)$. Is it true that the area in part **a** satisfies $A = h(\frac{\pi}{2}) - h(0)$?

Ⓒ 46. Derive the formula

$$\sum_{k=1}^{n} k = 1 + 2 + 3 + \cdots + n = \frac{n(n + 1)}{2}$$

by completing these steps:

a. Use the properties of summation formulas to show that

$$\sum_{k=1}^{n} k = \frac{1}{2}\sum_{k=1}^{n}[k^2 - (k - 1)^2] + \frac{1}{2}\sum_{k=1}^{n} 1$$

$$= \frac{1}{2}\sum_{k=1}^{n}[k^2 - (k - 1)^2] + \frac{1}{2}n$$

b. Show that $\sum_{k=1}^{n} [k^2 - (k-1)^2] = n^2$

Hint: Expand the sum by writing out a few terms. Note the internal cancellation.

c. Combine parts **a** and **b** to show that

$$\sum_{k=1}^{n} k = \frac{n(n+1)}{2}.$$

47. First find constants $a, b, c,$ and d such that

$$k^3 = a[k^4 - (k-1)^4] + bk^2 + ck + d$$

and then modify the approach outlined in Problem 46 to establish the formula

$$\sum_{k=1}^{n} k^2 = \frac{n(n+1)(2n+1)}{6}$$

48. The purpose of this problem is to verify the results of the technology window on page 331. Specifically, we shall find the area A under the parabola $y = x^2$ on the interval $[0, 1]$ using approximating rectangles with heights taken at the *left* endpoints.

Verify that

$$\lim_{n \to +\infty} \sum_{k=1}^{n} \left(\frac{k-1}{n}\right)^2 \left(\frac{1}{n}\right) = \frac{1}{3}$$

Compare this with the procedure outlined in Example 1. Note that when the interval $[0, 1]$ is subdivided into n equal parts, the kth subinterval is

$$\left(\frac{k-1}{n}, \frac{k}{n}\right)$$

49. Develop a formula for area based on approximating rectangles with heights taken at the *midpoints* of subintervals.

50. Use the properties of real numbers to establish the summation formulas in Theorem 5.4. For example, to prove the linearity rule, use the associative, commutative, and distributive properties of real numbers to note that

$$\sum_{k=1}^{n} (ca_k + db_k)$$

$$= (ca_1 + db_1) + (ca_2 + db_2) + \cdots + (ca_n + db_n)$$

$$= (ca_1 + ca_2 + \cdots + ca_n) + (db_1 + db_2 + \cdots + db_n)$$

$$= c(a_1 + a_2 + \cdots + a_n) + d(b_1 + b_2 + \cdots + b_n)$$

$$= c \sum_{k=1}^{n} a_k + d \sum_{k=1}^{n} b_k$$

5.3   Riemann Sums and the Definite Integral

IN THIS SECTION   Riemann sums, the definite integral, area as an integral, properties of the definite integral, distance as an integral

We shall soon discover that not just area, but also such useful quantities as distance, volume, mass, and work, can be first approximated by sums and then obtained exactly by taking a limit involving the approximating sums. The special kind of limit of a sum that appears in this context is called the *definite integral,* and the process of finding integrals is called *definite integration* or *Riemann integration* in honor of the German mathematician Bernhard Riemann (1826–1866), who pioneered the modern approach to integration theory. We begin by introducing some special notation and terminology.

RIEMANN SUMS

Recall from Section 5.2 that to find the area under the graph of the function $y = f(x)$ on the closed interval $[a, b]$ where f is continuous and $f(x) \geq 0$, we proceed as follows:

1. Partition the interval into n subintervals of equal width $\Delta x = \dfrac{b-a}{n}$.

2. Evaluate f at the right endpoint $a + k\Delta x$ of the kth subinterval for $k = 1, 2, \ldots, n$.

3. Form the approximating sum of the areas of the n rectangles, which we denote by $S_n = \lim_{n \to \infty} \sum_{k=1}^{n} f(a + k\Delta x)\Delta x$.

4. Because we expect the estimates S_n to improve as Δx decreases, we *define* the area A under the curve, above the x-axis, and bounded by the lines $x = a$ and $x = b$, to be the limit of S_n as $\Delta x \to 0$. Thus, we write

$$A = \lim_{n \to +\infty} \sum_{k=1}^{n} f(a + k\Delta x)\Delta x$$

if this limit exists. This means that A can be estimated to any desired degree of accuracy by approximating the sum S_n with Δx sufficiently small (or, equivalently, n sufficiently large).

This approach to the area problem contains the essentials of integration, but there is no compelling reason for the partition points to be evenly spaced or to insist on evaluating f at right endpoints. These conventions are for convenience of computation, but to accommodate applications other than area, it is necessary to consider a more general type of approximating sum and to specify what is meant by the limit of such sums. The approximating sums that occur in integration problems are called **Riemann sums,** and the following definition contains a step-by-step description of how such sums are formed.

Riemann Sum

Suppose a function f is given, along with a closed interval $[a, b]$ on which f is defined. Then:

Step 1. Partition the interval $[a, b]$ into n subintervals by choosing points $\{x_0, x_1, \ldots, x_n\}$ arranged in such a way that

$$a = x_0 < x_1 < x_2 < \cdots < x_{n-1} < x_n = b$$

Call this partition P. For $k = 1, 2, \ldots, n$, the kth subinterval width is $\Delta x_k = x_k - x_{k-1}$. The largest of these widths is called the **norm** of the partition P and is denoted by $\|P\|$; that is,

$$\|P\| = \max_{k=1,2,\ldots,n} \{\Delta x_k\}$$

Step 2. Choose a number arbitrarily from each subinterval. For $k = 1, 2, \ldots, n$, the number x_k^* chosen from the kth subinterval is called the *kth subinterval representative* of the partition P.

Step 3. Form the sum

$$R_n = f(x_1^*)\Delta x_1 + f(x_2^*)\Delta x_2 + \cdots + f(x_n^*)\Delta x_n = \sum_{k=1}^{n} f(x_k^*)\Delta x_k$$

This is the **Riemann sum** associated with f, the given partition P, and the chosen subinterval representatives $x_1^*, x_2^*, \ldots, x_n^*$.

■ *What This Says:* We will express quantities from geometry, physics, economics, and other applications in terms of the Riemann sum

$$\sum_{k=1}^{n} f(x_k^*)\Delta x_k$$

Riemann sums are generally used to produce the correct integral for a particular application. Note that the Riemann sum *does not* require that the function *f* be nonnegative. In addition, x_k^* is any point in the *k*th subinterval; it does not need to be something "nice" like the left or right endpoint, or the midpoint.

EXAMPLE 1   *Formation of the Riemann sum for a given function*

Suppose the interval $[-2, 1]$ is partitioned into 6 subintervals with subdivision points $a = x_0 = -2, x_1 = -1.6, x_2 = -0.93, x_3 = -0.21, x_4 = 0.35, x_5 = 0.82,$ and $x_6 = 1 = b$. Find the norm of this partition P and the Riemann sum associated with the function $f(x) = 2x$, the given partition, and the subinterval representatives $x_1^* = -1.81$, $x_2^* = -1.12, x_3^* = -0.55, x_4^* = -0.17, x_5^* = 0.43,$ and $x_6^* = 0.94$.

Solution   Before we can find the norm of the partition or the required Riemann sum, we must compute the subinterval width Δx_k and evaluate f at each subinterval representative x_k^*. These values are shown in Figure 5.11 and the computations are shown as follows.

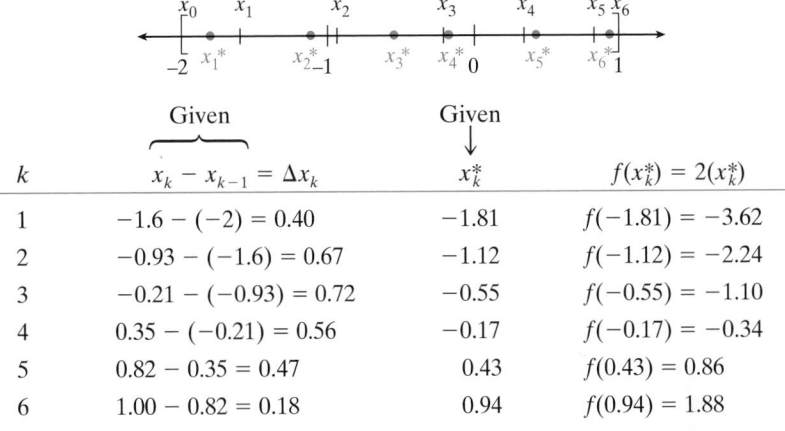

k	$x_k - x_{k-1} = \Delta x_k$	x_k^*	$f(x_k^*) = 2(x_k^*)$
1	$-1.6 - (-2) = 0.40$	-1.81	$f(-1.81) = -3.62$
2	$-0.93 - (-1.6) = 0.67$	-1.12	$f(-1.12) = -2.24$
3	$-0.21 - (-0.93) = 0.72$	-0.55	$f(-0.55) = -1.10$
4	$0.35 - (-0.21) = 0.56$	-0.17	$f(-0.17) = -0.34$
5	$0.82 - 0.35 = 0.47$	0.43	$f(0.43) = 0.86$
6	$1.00 - 0.82 = 0.18$	0.94	$f(0.94) = 1.88$

■ **FIGURE 5.11**   Riemann sum

WARNING ➤   Notice from Example 1 that the Riemann sum does not necessarily represent an area. The sum found is negative (and area must be nonnegative) ✏

From the table, we see that the largest subinterval width is $\Delta x_3 = 0.72$, so the partition has norm $\|P\| = 0.72$. Finally, by using the definition, we compute the Riemann sum:

$$R_6 = (-3.62)(0.40) + (-2.24)(0.67) + (-1.10)(0.72) + (-0.34)(0.56)$$
$$+ (0.86)(0.47) + (1.88)(0.18) \approx -3.1886 \approx -3.19$$ ■

THE DEFINITE INTEGRAL

By comparing the formula for Riemann sum with that of area in the previous section, we recognize that the sum S_n used to approximate area is actually a special kind of Riemann sum which has

$$\Delta x_k = \Delta x = \frac{b - a}{n} \quad \text{and} \quad x_k^* = a + k\Delta x$$

for $k = 1, 2, \ldots, n$. Because the subintervals in the partition P associated with S_n are equally spaced, it is called a **regular partition.** When we express the area under the curve $y = f(x)$ as $A = \lim_{\Delta x \to 0} S_n$, we are actually saying that A can be estimated to any desired accuracy by finding a Riemann sum of the form S_n with norm

$$\|P\| = \frac{b - a}{n}$$

sufficiently small. We use this interpretation as a model for the following definition.

Definite Integral

If f is defined on the closed interval $[a, b]$, we say f is **integrable on $[a, b]$** if

$$I = \lim_{\|P\| \to 0} \sum_{k=1}^{n} f(x_k^*)\Delta x_k$$

exists. This limit is called the **definite integral** of f from a to b. The definite integral is denoted by

$$I = \int_a^b f(x)\, dx \quad \text{or} \quad I = \int_{x=a}^{x=b} f(x)\, dx$$

The latter is used if we wish to emphasize the variable of integration.

■ *What This Says:* To say that f is *integrable* with definite integral I means the number I can be approximated to any prescribed degree of accuracy by any Riemann sum of f with norm sufficiently small. As long as the conditions of this definition are satisfied (that is, f is defined on $[a, b]$ and the Riemann sum exists), we can write

$$\int_a^b f(x)\, dx = \lim_{\|P\| \to 0} \sum_{k=1}^{n} f(x_k^*)\Delta x_k$$

Formally, I is the definite integral of f on $[a, b]$ if for each number $\epsilon > 0$, there exists a number $\delta > 0$ such that if

$$\sum_{k=1}^{n} f(x_k^*)\Delta x_k$$

is any Riemann sum of f whose norm satisfies $\|P\| < \delta$, then

$$\left| I - \sum_{k=1}^{n} f(x_k^*)\Delta_k \right| < \epsilon$$

WARNING   ↓ Upper limit of integration
Integrand

$$\int_a^b f(x)\, \overbrace{dx}$$

↑ Lower limit of integration ←

In advanced calculus, it is shown that when this limit exists, it is unique. Moreover, its value is independent of the particular way in which the partitions of $[a, b]$ and the subinterval representatives x_k^* are chosen.

Consider the notation used in the margin. The function f that is being integrated is called the **integrand;** the interval $[a, b]$ is the **interval of integration;** and the endpoints a and b are called, respectively, the **lower and upper limits of integration.**

In the special case where $a = b$, the interval of integration $[a, b]$ is really just a point, and the integral of any function on this "interval" is defined to be 0; that is,

$$\int_a^a f(x)\, dx = 0$$

Also, at times, we shall consider integrals in which the lower limit of integration is a larger number than the upper limit. To handle this case, we specify that the integral from b to a is the *opposite* of the integral from a to b:

$$\int_b^a f(x)\, dx = -\int_a^b f(x)\, dx$$

To summarize:

Definite Integral at a Point

$$\int_a^a f(x)\, dx = 0$$

Opposite of a Definite Integral

$$\int_b^a f(x)\, dx = -\int_a^b f(x)\, dx$$

At first, the definition of the definite integral may seem rather imposing. How are we to tell whether a given function f is integrable on an interval $[a, b]$? If f is integrable, how are we supposed to compute the definite integral? Answering these questions is not easy, but in advanced calculus, it is shown that f is integrable on a closed interval $[a, b]$ if it is continuous on the interval except at a finite number of points and if it is bounded on the interval (that is, there is a number $A > 0$ such that $|f(x)| < A$ for all x in the interval). We will state a special case of this result as a theorem.

THEOREM 5.5   *Integrability of a continuous function*

If f is continuous on an interval $[a, b]$, then f is integrable on $[a, b]$.

Proof   The proof requires the methods of advanced calculus and is omitted here.

Our next example illustrates how to use the definition to find a definite integral.

EXAMPLE 2   *Evaluating a definite integral using the definition*

Evaluate $\displaystyle\int_{-2}^1 4x\, dx$.

Solution   We note the variable of integration is x, so we interpret this as

$$\int_{x=-2}^{x=1} 4x\, dx$$

The integral exists because $f(x) = 4x$ is continuous on $[-2, 1]$. Because the integral can be computed by any partition whose norm approaches 0 (that is, the integral is independent of the sequence of partitions *and* the subinterval representatives), we shall simplify matters by choosing a partition in which the points are evenly spaced. Specifically, we divide the interval $[-2, 1]$ into n subintervals, each of width

$$\Delta x = \frac{1 - (-2)}{n} = \frac{3}{n}$$

For each k, we choose the kth subinterval representative to be the right endpoint of the kth subinterval; that is,

$$x_k^* = -2 + k\Delta x = -2 + k\left(\frac{3}{n}\right)$$

Finally, we form the Riemann sum

$$\int_{-2}^{1} 4x \, dx = \lim_{\|P\| \to 0} \sum_{k=1}^{n} f(x_k^*)\Delta x$$

$$= \lim_{n \to +\infty} \sum_{k=1}^{n} 4\left(-2 + \frac{3k}{n}\right)\left(\frac{3}{n}\right) \qquad n \to +\infty \text{ as } \|P\| \to 0$$

$$= \lim_{n \to +\infty} \frac{12}{n^2} \sum_{k=1}^{n} (-2n + 3k)$$

$$= \lim_{n \to +\infty} \frac{12}{n^2} \left(\sum_{k=1}^{n} (-2n) + \sum_{k=1}^{n} 3k \right)$$

$$= \lim_{n \to +\infty} \frac{12}{n^2} \left((-2n)n + 3\left[\frac{n(n+1)}{2}\right] \right) \qquad \text{Form for the sum of integers}$$

$$= \lim_{n \to +\infty} \frac{12}{n^2} \left(\frac{-4n^2 + 3n^2 + 3n}{2} \right)$$

$$= \lim_{n \to +\infty} \frac{-6n^2 + 18n}{n^2} = -6 \qquad \blacksquare$$

AREA AS AN INTEGRAL

Because we have used the development of area in Section 5.2 as the model for our definition of the definite integral, it is no surprise to discover that the area under a curve can be expressed as a definite integral. However, integrals can be positive, zero, or negative (as in Example 2), and we certainly would not expect the area under a curve to be a negative number! The actual relationship between integrals and area under a curve is contained in the following observation, which follows from the definition of area as the limit of a sum along with Theorem 5.5.

WARNING   We will find areas using a definite integral, but not every definite integral can be interpreted as an area.

Area as an Integral

Suppose f is continuous and $f(x) \geq 0$ on the closed interval $[a, b]$. Then the area under the curve $y = f(x)$ on $[a, b]$ is given by the definite integral of f on $[a, b]$. That is,

$$\text{Area} = \int_{a}^{b} f(x) \, dx$$

Usually we find area by evaluating a definite integral, but sometimes area can be used to help us evaluate the integral. At this stage of our study, it is not easy to evaluate Riemann sums, so if you happen to recognize that the integral represents the area of some common geometric figure, you can use the known formula instead of the definite integral, as shown in Example 3.

EXAMPLE 3   *Evaluating an integral using an area formula*

Evaluate $\displaystyle\int_{-3}^{3} \sqrt{9 - x^2}\, dx$.

Solution   Let $f(x) = \sqrt{9 - x^2}$. The curve $y = \sqrt{9 - x^2}$ is a semicircle centered at the origin of radius 3, as shown in Figure 5.12. From geometry, we know the area of the circle is $A = \pi r^2 = \pi(3)^2 = 9\pi$. Thus, the area of the semicircle is $\frac{9\pi}{2}$; therefore,

$$\int_{-3}^{3} \sqrt{9 - x^2}\, dx = \frac{9\pi}{2}$$

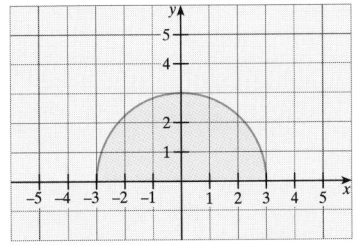

■ **FIGURE 5.12**   The curve $y = \sqrt{9 - x^2}$ is a semicircle of radius 3.

PROPERTIES OF THE DEFINITE INTEGRAL

In computations involving integrals, it is often helpful to use the three general properties listed in the following theorem.

THEOREM 5.6   *General Properties of the Definite Integral*

Linearity rule   If f and g are integrable on $[a, b]$, then so is $rf + sg$ for constants r, s.

$$\int_{a}^{b} [rf(x) + sg(x)]\, dx = r\int_{a}^{b} f(x)\, dx + s\int_{a}^{b} g(x)\, dx$$

Dominance rule   If f and g are integrable on $[a, b]$ and $f(x) \le g(x)$ throughout this interval, then

$$\int_{a}^{b} f(x)\, dx \le \int_{a}^{b} g(x)\, dx$$

Subdivision rule   For any number c such that $a < c < b$,

$$\int_{a}^{b} f(x)\, dx = \int_{a}^{c} f(x)\, dx + \int_{c}^{b} f(x)\, dx$$

assuming all three integrals exist.

Proof   Each of these rules can be established by using a familiar property of sums with the definition of the definite integral. For example, to derive the linearity rule, we note that any Riemann sum of the function $rf + sg$ can be expressed as

$$\sum_{k=1}^{n} [rf(x_k^*) + sg(x_k^*)]\Delta x_k = r\left[\sum_{k=1}^{n} f(x_k^*)\Delta x_k\right] + s\left[\sum_{k=1}^{n} g(x_k^*)\Delta x_k\right]$$

and the linearity rule then follows by taking the limit on each side of this equation as the norm of the partition tends to 0.

The dominance rule and the subdivision rule are interpreted geometrically for nonnegative functions in Figure 5.13.

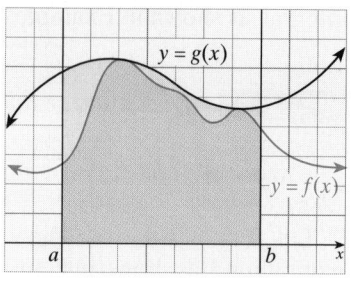
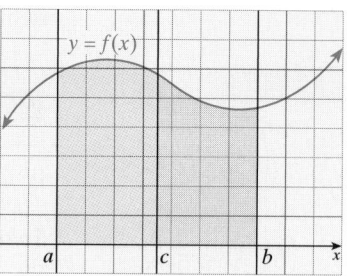

a. The dominance rule
If $g(x) \geq f(x)$, the area under $y = g(x)$ can be no less than the area under $y = f(x)$.

b. The subdivision rule
The area under $y = f(x)$ on $[a, b]$ equals the sum of the areas on $[a, c]$ and $[c, b]$.

■ **FIGURE 5.13**   Comparison of dominance and subdivision rules

Notice that if $g(x) \geq f(x) \geq 0$, the curve $y = g(x)$ is always above (or touching) the curve $y = f(x)$, and the dominance rule expresses the fact that the area under the upper curve $y = g(x)$ cannot be less than the area under $y = f(x)$. The subdivision rule says that the area under $y = f(x)$ above $[a, b]$ is the sum of the area on $[a, c]$ and the area on $[c, b]$. The following example illustrates one way of using the subdivision rule.

> **EXAMPLE 4**   *Subdivision rule*

If $\int_{-2}^{1} f(x)\, dx = 3$ and $\int_{-2}^{7} f(x)\, dx = -5$, what is $\int_{1}^{7} f(x)\, dx$?

Solution   According to the subdivision rule, we have

$$\int_{-2}^{7} f(x)\, dx = \int_{-2}^{1} f(x)\, dx + \int_{1}^{7} f(x)\, dx$$

Therefore,

$$\int_{1}^{7} f(x)\, dx = \int_{-2}^{7} f(x)\, dx - \int_{-2}^{1} f(x)\, dx = (-5) - 3 = -8$$ ■

DISTANCE AS AN INTEGRAL

Many quantities other than area can be computed as the limit of a sum. For example, suppose an object moving along a line is known to have continuous velocity $v(t)$ for each time t between $t = a$ and $t = b$, and we wish to compute the total distance traveled by the object during this time period.

Let the interval $[a, b]$ be partitioned into n equal subintervals, each of length $\Delta t = \dfrac{b - a}{n}$ as shown in Figure 5.14.

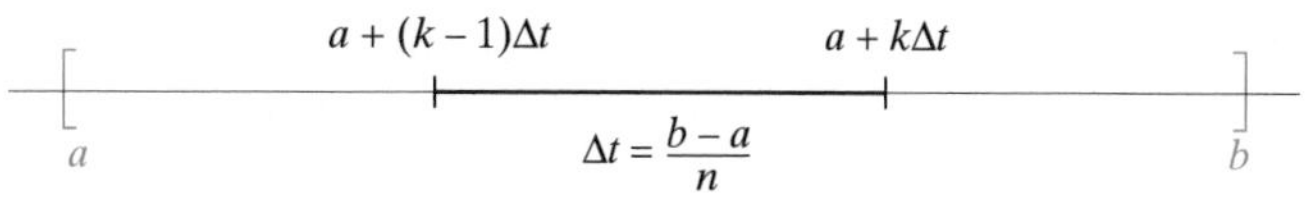

■ **FIGURE 5.14**   The distance traveled during the kth time subinterval

The kth subinterval is $[a + (k - 1)\Delta t, a + k\Delta t]$ and, if Δt is small enough, the velocity $v(t)$ will not change much over the subinterval so it is reasonable to approximate $v(t)$ by the constant velocity of $v[a + (k - 1)\Delta t]$ throughout the entire subinterval.

The corresponding change in the object's position will be approximated by the product

$$v[a + (k - 1)\Delta t]\Delta t$$

and will be positive if $v[a + (k - 1)\Delta t]$ is positive and negative otherwise. Both cases may be summarized by the formula

$$\big|v[a + (k - 1)\Delta t]\big|\Delta t$$

and the total distance traveled by the object as t varies from $t = a$ to $t = b$ is given by the sum

$$S_n = \sum_{k=1}^{n} \big|v[a + (k - 1)\Delta t]\big|\Delta t$$

which we recognize as a Riemann sum. We can make the approximation more precise by taking more refined partitions (that is, shorter and shorter time intervals Δt). Therefore, it is reasonable to *define* the exact distance S traveled as the *limit* of the sum S_n as $\Delta t \to 0$ or, equivalently, as $n \to +\infty$, so that

$$S = \lim_{n \to +\infty} \sum_{k=1}^{n} \big|v[a + (k - 1)\Delta t]\big|\Delta t = \int_a^b \big|v(t)\big|\, dt$$

Distance

The **distance traveled** by an object with continuous velocity $v(t)$ along a straight line from time $t = a$ to $t = b$ is

$$S = \int_a^b \big|v(t)\big|\, dt$$

Note: There is a difference between the *total distance* traveled by an object over a given time interval $[a, b]$, and the **displacement** of the object over the same interval, which is defined as the difference between the object's final and initial positions. It is easy to see that displacement is given by

$$D = \int_a^b v(t)\, dt$$

Thus, an object that moves forward 2 units and back 3 on a given time interval has moved a total distance of 5 units, but its displacement is -1 because it ends up 1 unit to the left of its initial position.

EXAMPLE 5   *Distance moved by an object whose velocity is known*

An object moves along a straight line with velocity $v(t) = t^2$ for $t > 0$. How far does the object travel between times $t = 1$ and $t = 2$?

Solution   We have $a = 1$, $b = 2$, and $\Delta t = \dfrac{2-1}{n} = \dfrac{1}{n}$; therefore, the required distance is

$$
\begin{aligned}
S &= \int_{t=1}^{t=2} |v(t)|\, dt = \lim_{n \to +\infty} \sum_{k=1}^{n} \left| v\left[1 + (k-1)\left(\frac{1}{n}\right)\right] \right| \left(\frac{1}{n}\right) \\
&= \lim_{n \to \infty} \sum_{k=1}^{n} \left| v\left(\frac{n+k-1}{n}\right) \right| \left(\frac{1}{n}\right) = \lim_{n \to \infty} \sum_{k=1}^{n} \frac{(n+k-1)^2}{n^2}\left(\frac{1}{n}\right) \\
&= \lim_{n \to \infty} \frac{1}{n^3} \sum_{k=1}^{n} [(n^2 - 2n + 1) + k^2 + 2(n-1)k] \\
&= \lim_{n \to \infty} \frac{1}{n^3}\left[(n^2 - 2n + 1)\sum_{k=1}^{n} 1 + \sum_{k=1}^{n} k^2 + 2(n-1)\sum_{k=1}^{n} k \right] \\
&= \lim_{n \to \infty} \frac{1}{n^3}\left[(n^2 - 2n + 1)n + \frac{n(n+1)(2n+1)}{6} + 2(n-1)\frac{n(n+1)}{2} \right] \\
&= \lim_{n \to \infty} \frac{14n^3 - 9n^2 + n}{6n^3} = \frac{14}{6} = \frac{7}{3}
\end{aligned}
$$

Thus, we expect the object to travel $\frac{7}{3}$ units during the time interval $[1, 2]$ ∎

By considering the distance as an integral, we see that zero or negative values of a definite integral can also be interpreted geometrically. When $v(t) > 0$ on a time interval $[a, b]$, then the total distance S traveled by the object between times $t = a$ and $t = b$ is the same as the area under the graph of $v(t)$ on $[a, b]$.

When $v(t) > 0$, the object moves forward, but when $v(t) < 0$, it reverses direction and moves backward (to the left). In the general case where $v(t)$ changes sign on the time interval $[a, b]$, the integral

$$
\int_a^b v(t)\, dt
$$

measures the net displacement of the object, taking into account both forward and backward motion.

For example, for the velocity function $v(t)$ graphed in Figure 5.15a, the net displacement is 0 because the area above the t-axis is the same as the area below, but in Figure 5.15b, there is more area below the t-axis, which means the net displacement is negative, and the object ends up "behind" (to the left of) its starting position.

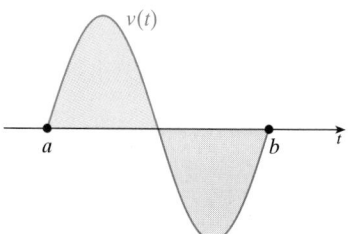
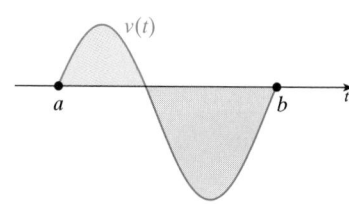

a. Net displacement of 0

There is as much area under the positive part of the velocity curve as there is above the negative part.

b. Negative net displacement

There is more area above the negative part of the velocity curve than there is under the positive part.

■ **FIGURE 5.15**   Definite integral in terms of displacement

---

PREVIEW   Students often become discouraged at this point, thinking they will have to compute all definite integrals using the limit of a Riemann sum definition, so we offer an encouraging word. Recall the definition of derivative in Chapter 3. It was difficult to find derivatives using the definition, but we soon proved some theorems to make it easier to find and evaluate derivatives. The same is true of integration. It is difficult to apply the definition of a definite integral, but you will soon discover that computing most definite integrals is no harder than finding an antiderivative.

As a preview of this result, consider an object moving along a straight line. We assume its position is given by $s(t)$ and that its velocity $v(t) = s'(t)$ is positive ($v(t) > 0$) so that it is always moving forward. The total distance traveled by such an object between times $t = a$ and $t = b$ is clearly $S = s(b) - s(a)$, but earlier in this section, we showed that this distance is also given by the definite integral of $v(t)$ over the interval $[a, b]$. Thus we have

$$\text{TOTAL DISTANCE} = \int_a^b v(t)\, dt = s(b) - s(a)$$

where $s(t)$ is an antiderivative of $v(t)$.

This observation anticipates the *fundamental theorem of calculus,* which provides a vital link between differential and integral calculus. We shall formally introduce the fundamental theorem in the next section. To illustrate, notice how our observation applies to Example 5, where $v(t) = t^2$ and the interval is $[1, 2]$. Since an antiderivative of $v(t)$ is $s(t) = \frac{1}{3}t^3$, we have

$$S = \int_1^2 t^2\, dt = s(2) - s(1) = \left[\frac{1}{3}(2)^3 - \frac{1}{3}(1)^3\right] = \frac{7}{3}$$

which coincides with the result found numerically in Example 5.

5.3   Problem Set

A *In Problems 1–10, estimate (using right endpoints) the given*

integral $\int_a^b f(x)\,dx$ *by using a Riemann sum*

$$S_n = \sum_{k=1}^{n} f(a + k\Delta x)\Delta x \ for\ n = 4.$$

1. $\int_0^1 (2x + 1)\,dx$ **2.** $\int_0^1 (4x^2 + 2)\,dx$

3. $\int_1^3 x^2\,dx$ **4.** $\int_0^2 x^3\,dx$

5. $\int_0^1 (1 - 3x)\,dx$ **6.** $\int_1^3 (x^2 - x^3)\,dx$

7. $\int_{-\pi/2}^0 \cos x\,dx$ **8.** $\int_0^{\pi/4} (x + \sin x)\,dx$

9. $\int_0^1 e^x\,dx$ **10.** $\int_1^2 \dfrac{dx}{x}$

In Problems 11–16, $v(t)$ is the velocity of an object moving along a straight line. Use the formula

$$S_n = \sum_{k=1}^{n} \left|v[a + (k - 1)\Delta t]\right|\Delta t$$

where $\Delta t = \dfrac{b - a}{n}$ to estimate (using right endpoints) the total distance traveled by the object during the time interval $[a, b]$. Let $n = 4$ for Problems 11–16.

11. $v(t) = 3t + 1$ on $[1, 4]$ **12.** $v(t) = 1 + 2t$ on $[1, 2]$

13. $v(t) = \sin t$ on $[0, \pi]$ **14.** $v(t) = \cos t$ on $[0, \frac{\pi}{2}]$

15. $v(t) = e^{-t}$ on $[0, 1]$ **16.** $v(t) = \dfrac{1}{t + 1}$ on $[0, 1]$

Evaluate each of the integrals in Problems 17–22 by using the following information together with the linearity and subdivision properties:

$$\int_{-1}^2 x^2\,dx = 3; \quad \int_{-1}^0 x^2\,dx = \frac{1}{3}; \quad \int_{-1}^2 x\,dx = \frac{3}{2}; \quad \int_0^2 x\,dx = 2$$

17. $\int_0^{-1} x^2\,dx$ **18.** $\int_{-1}^2 (x^2 + x)\,dx$

19. $\int_{-1}^2 (2x^2 - 3x)\,dx$ **20.** $\int_0^2 x^2\,dx$

21. $\int_{-1}^0 x\,dx$ **22.** $\int_{-1}^0 (3x^2 - 5x)\,dx$

Use the dominance property of integrals to establish the given inequality in Problems 23–24.

23. $\int_0^1 x^3\,dx \le \dfrac{1}{2}$ *Hint: Note that $x^3 \le x$ on $[0, 1]$.*

24. $\int_0^\pi \sin x\,dx \le \pi$ *Hint: $\sin x \le 1$ for all x.*

B **25.** Given $\int_{-2}^4 [5f(x) + 2g(x)]\,dx = 7$ and

$\int_{-2}^4 [3f(x) + g(x)]\,dx = 4$, find

$$\int_{-2}^4 f(x)\,dx \quad \text{and} \quad \int_{-2}^4 g(x)\,dx$$

26. Suppose $\int_0^2 f(x)\,dx = 3$, $\int_0^2 g(x)\,dx = -1$,

and $\int_0^2 h(x)\,dx = 3$.

a. Evaluate $\int_0^2 [2f(x) + 5g(x) - 7h(x)]\,dx$.

b. Find s such that

$$\int_0^2 [5f(x) + sg(x) - 6h(x)]\,dx = 0$$

27. Evaluate $\int_{-1}^2 f(x)\,dx$ given that

$$\int_{-1}^1 f(x)\,dx = 3; \quad \int_2^3 f(x)\,dx = -2; \quad \int_1^3 f(x)\,dx = 5$$

28. Let $f(x) = \begin{cases} 2 & \text{for } -1 \le x \le 1 \\ 3 - x & \text{for } 1 < x < 4 \\ 2x - 9 & \text{for } 4 \le x \le 5 \end{cases}$

Sketch the graph of f on the interval $[-1, 5]$ and show that f is continuous on this interval. Then use Theorem 5.6 to evaluate

$$\int_{-1}^5 f(x)\,dx$$

29. Let $f(x) = \begin{cases} 5 & \text{for } -3 \le x \le -1 \\ 4 - x & \text{for } -1 < x < 2 \\ 2x - 2 & \text{for } 2 \le x \le 5 \end{cases}$

Sketch the graph of f on the interval $[-3, 5]$ and show that f is continuous on this interval. Then use Theorem 5.6 to evaluate

$$\int_{-3}^5 f(x)\,dx$$

30. ⱧISTORICAL ℚUEST
During the eighteenth century, integration was considered simply as an antiderivative. That is, there were no underpinnings for the concept of an integral until Cauchy formulated the definition of the integral in 1823 (see Ⱨistorical ℚuest, Problem Set 2.2). This formulation was later completed by Georg Friedrich Rie-

GEORG FRIEDRICH RIEMANN
1826–1866

mann. In this section, we see that history honored Riemann by naming the process after him. In his personal life he was frail, bashful, and timid, but in his professional life he was one of the giants in mathematical history. Riemann used what are called topological methods in his theory of functions and in his work with surfaces and spaces. Riemann is remembered for his work in geometry (Riemann surfaces) and analysis. In his book *Space Through the Ages*, Cornelius Lanczos wrote, "Although Riemann's collected papers fill only one single volume of 538 pages, this volume weighs tons if measured intellectually. Every one of his many discoveries was destined to change the course of mathematical science."

For this $\mathbb{H}$istorical $\mathbb{Q}$uest, investigate the Königsberg bridge problem, and its solution.

This famous problem was formulated by Leonhard Euler (1707–1783). The branch of mathematics known today as *topology* began with Euler's work on the bridge problem and other related questions, and was extended in the nineteenth century by Riemann and others.

C **31.** Generalize the subdivision property by showing that for $a \leq c \leq d \leq b$

$$\int_a^b f(x)\, dx = \int_a^c f(x)\, dx + \int_c^d f(x)\, dx + \int_d^b f(x)\, dx$$

whenever all these integrals exist.

32. If $Cx + D \geq 0$ for $a \leq x \leq b$, show that

$$\int_a^b (Cx + D)\, dx = (b - a)\left[\frac{C}{2}(b + a) + D\right]$$

Hint: Sketch the region under the line $y = Cx + D$, and express the integral as an area.

33. For $b > a > 0$, show that

$$\int_a^b x^2\, dx = \frac{1}{3}(b^3 - a^3)$$

In Problems 34–36, use the partition
$P = \{-1, -0.2, 0.9, 1.3, 1.7, 2\}$ *on the interval* $[-1, 2]$.

34. Find the subinterval widths

$$\Delta x_k = x_k - x_{k-1}$$

for $k = 1, 2, \ldots, 5$. What is the norm of P?

35. Compute the Riemann sum associated with $f(x) = 4 - 5x$, the partition P, and the subinterval representatives

$$x_1^* = -0.5, x_2^* = 0.8, x_3^* = 1, x_4^* = 1.3, x_5^* = 1.8.$$

36. Compute the Riemann sum associated with $f(x) = x^3$, the partition P, and subinterval representatives

$$x_1^* = -1, x_2^* = 0, x_3^* = 1, x_4^* = \frac{128}{81}, \text{ and } x_5^* = \frac{125}{64}.$$

37. If the numbers a_k and b_k satisfy $a_k \leq b_k$ for $k = 1, 2, \ldots, n$, then $\sum_{k=1}^{n} a_k \leq \sum_{k=1}^{n} b_k$.
Use this dominance property of sums to establish the dominance property of integrals.

38. THINK TANK PROBLEM   Either prove that the following result is generally true or find a counterexample: If $f(x)$ is not identically zero on the interval $[a, b]$, then

$$\int_a^b f(x)\, dx \neq 0$$

for some x-value on $[a, b]$.

39. THINK TANK PROBLEM   Either prove that the following result is generally true or find a counterexample: If f and g are continuous on $[a, b]$, then

$$\int_a^b f(x)g(x)\, dx = \left[\int_a^b f(x)\, dx\right]\left[\int_a^b g(x)\, dx\right]$$

40. a. If $f(x) < 0$ on the interval $[a, b]$, show that

$$\int_a^b f(x)\, dx = -A$$

where A is the area under the graph of $y = -f(x)$ on $[a, b]$.

b. Combine the formula in part **a** with the formula of Theorem 5.6 to show that

$$\int_a^b f(x)\, dx = P - N$$

where P is the absolute value of the sum of the area of all positive regions ($f(x) > 0$) and N is the sum of the area of all negative regions ($f(x) < 0$).

c. Use the formula in part **b** to evaluate

$$\int_{-2}^{3} (2x + 1)\, dx$$

41. Use the definition of the definite integral to prove that

$$\int_a^b C\, dx = C(b - a)$$

42. Prove the *bounding rule* for definite integrals:
If f is integrable on the closed interval $[a, b]$ and $m \leq f(x) \leq M$ for constants m, M, and all x in the closed interval, then

$$m(b - a) \leq \int_a^b f(x)\, dx \leq M(b - a)$$

5.4 The Fundamental Theorems of Calculus

the first fundamental theorem of calculus, the second fundamental theorem of calculus

THE FIRST FUNDAMENTAL THEOREM OF CALCULUS

In the previous section we observed that if $v(t)$ is the velocity of an object at time t as it moves along a straight line, then

$$\int_a^b v(t)\, dt = s(b) - s(a)$$

where $s(t)$ is the displacement of the object and satisfies $s'(t) = v(t)$. This result is an application of a general theorem discovered by the English mathematician Isaac Barrow (1630–1677), who was Newton's mentor at Cambridge.

THEOREM 5.7   *The first fundamental theorem of calculus*

If f is continuous on the interval $[a, b]$ and F is any function that satisfies $F'(x) = f(x)$ throughout this interval, then

$$\int_a^b f(x)\, dx = F(b) - F(a)$$

Proof   Let $P = \{x_0, x_1, x_2, \ldots, x_n\}$ be a regular partition of the interval, with subinterval widths $\Delta x = \dfrac{b - a}{n}$. Note that F satisfies the hypotheses of the mean value theorem (Theorem 4.3, Section 4.2) on each of the closed subintervals $[x_{k-1}, x_k]$. Thus, the MVT tells us that there is a point x_k^* in each open subinterval (x_{k-1}, x_k) for which

$$\frac{F(x_k) - F(x_{k-1})}{x_k - x_{k-1}} = F'(x_k^*) \quad \text{or} \quad F(x_k) - F(x_{k-1}) = F'(x_k^*)(x_k - x_{k-1})$$

Because $F'(x_k^*) = f(x_k^*)$ and $x_k - x_{k-1} = \Delta x = \dfrac{b - a}{n}$, we can write
$F(x_k) - F(x_{k-1}) = f(x_k^*)\Delta x$, so that

$$F(x_1) - F(x_0) = f(x_1^*)\Delta x$$
$$F(x_2) - F(x_1) = f(x_2^*)\Delta x$$
$$\vdots$$
$$F(x_n) - F(x_{n-1}) = f(x_n^*)\Delta x$$

Thus, by adding both sides of all the equations, we obtain

$$\sum_{k=1}^n f(x_k^*)\Delta x = f(x_1^*)\Delta x + f(x_2^*)\Delta x + \cdots + f(x_n^*)\Delta x$$
$$= [F(x_1) - F(x_0)] + [F(x_2) - F(x_1)] + \cdots + [F(x_n) - F(x_{n-1})]$$
$$= F(x_n) - F(x_0)$$

Because $x_0 = a$ and $x_n = b$, we have

$$\sum_{k=1}^n f(x_k^*)\Delta x = F(b) - F(a)$$

Finally, we take the limit as $\|P\| \to 0$, and because $F(b) - F(a)$ is a constant, we have

$$\int_a^b f(x)\, dx = F(b) - F(a)$$

Recall (from Section 5.1) that an *antiderivative of a function f* is a function F that satisfies $F' = f$.

■ *What This Says:*   The definite integral $\int_a^b f(x)\, dx$ can be computed by finding an antiderivative F on the interval $[a, b]$ and evaluating it at the limits of integration a and b. Also notice that this theorem does not say *how* to find the antiderivative, nor does it say that an antiderivative F *exists*. The *second* fundamental theorem of calculus asserts the existence of F (page 353).

To give you some insight into why this theorem is important enough to be named *the fundamental theorem of calculus,* we repeat Example 2 from Section 5.3 and then work the same example using the fundamental theorem.

EXAMPLE 1   *Evaluating a definite integral*

Evaluate $\displaystyle\int_{-2}^1 4x\, dx$ using the definition of the definite integral and also using the fundamental theorem.

Solution

Solution from Section 5.3 using Riemann sums:

$$\int_{-2}^1 4x\, dx = \lim_{\|P\| \to 0} \sum_{k=1}^n f(x_k^*)\Delta x$$

$$= \lim_{n \to +\infty} \sum_{k=1}^n 4\left(-2 + \frac{3k}{n}\right)\left(\frac{3}{n}\right)$$

$$= \lim_{n \to +\infty} \frac{12}{n^2} \sum_{k=1}^n (-2n + 3k)$$

$$= \lim_{n \to +\infty} \frac{12}{n^2} \left(\sum_{k=1}^n (-2n) + \sum_{k=1}^n 3k\right)$$

$$= \lim_{n \to +\infty} \frac{12}{n^2}\left((-2n)n + 3\left[\frac{n(n+1)}{2}\right]\right)$$

$$= \lim_{n \to +\infty} \frac{12}{n^2}\left(\frac{-4n^2 + 3n^2 + 3n}{2}\right)$$

$$= \lim_{n \to +\infty} \frac{-6n^2 + 18n}{n^2}$$

$$= -6$$

Solution using the fundamental theorem of calculus:

If $F(x) = 2x^2$, then $F'(x) = 4x$, so F is an antiderivative of f. Thus,

$$\int_{-2}^1 4x\, dx = F(1) - F(-2)$$

$$= 2(1)^2 - 2(-2)^2 = -6$$

■

The variable used in a definite integral is a **dummy variable** in the sense that it can be replaced by any other variable with no effect on the value of the integral. For instance, we have just found that

$$\int_{-2}^{1} 4x \, dx = -6$$

and, without further computation, it follows that

$$\int_{-2}^{1} 4t \, dt = -6 \qquad \int_{-2}^{1} 4u \, du = -6 \qquad \int_{-2}^{1} 4N \, dN = -6$$

Henceforth, when evaluating an integral by the fundamental theorem, we shall denote the difference

$$F(b) - F(a) \text{ by } F(x)\big|_{a}^{b}, \text{ which means } F(x)\big|_{x=a}^{x=b}$$

Sometimes we also write $[F(x)]_{a}^{b}$, where $F'(x) = f(x)$ on $[a, b]$. This notation is illustrated in Example 2.

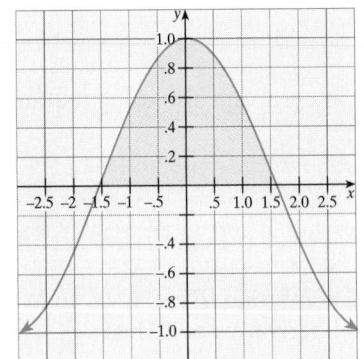

If you look at the graph of $f(x) = \cos x$ on $\left[-\dfrac{\pi}{2}, \dfrac{\pi}{2}\right]$, you can see that an area of 2 seems reasonable.

EXAMPLE 2   *Finding the area under a curve using the fundamental theorem of calculus*

Find the area under the curve $y = \cos x$ on $\left[-\frac{\pi}{2}, \frac{\pi}{2}\right]$.

Solution   Because $f(x) = \cos x$ is continuous on $\left[-\frac{\pi}{2}, \frac{\pi}{2}\right]$, and because the derivative of $\sin x$ is $\cos x$, it follows (from the definition of antiderivative) that $\sin x$ is an antiderivative of $\cos x$. Since $f(x) \geq 0$, the required area is given by the integral

$$A = \int_{-\pi/2}^{\pi/2} \cos x \, dx = \sin x \big|_{-\pi/2}^{\pi/2} = \sin \frac{\pi}{2} - \sin\left(-\frac{\pi}{2}\right) = 1 - (-1) = 2$$

The region has area $A = 2$ square units.   ■

WARNING   It is important to remember that the *definite* integral $\int_{a}^{b} f(x) \, dx$ for fixed numerical values of a and b is a number, whereas the *indefinite* integral $\int f(x) \, dx$ is a family of functions.   ◄

The relationship between the indefinite and definite integral is given by

$$\int_{a}^{b} f(x) \, dx = \left[\int f(x) \, dx \right]\Bigg|_{a}^{b}$$

EXAMPLE 3   *Evaluating an integral using the fundamental theorem*

Evaluate   a. $\displaystyle\int_{-3}^{5} (-10) \, dx$ b. $\displaystyle\int_{4}^{9} \left[\frac{1}{\sqrt{x}} + x\right] dx$ c. $\displaystyle\int_{-2}^{2} |x| \, dx$

Solution

a. $\displaystyle\int_{-3}^{5} (-10) \, dx = -10x \big|_{-3}^{5} = -10(5 + 3) = -80$

b. $\displaystyle\int_4^9 \left[\frac{1}{\sqrt{x}} + x\right] dx = \int_4^9 [x^{-1/2} + x]\, dx = \left[\left(\frac{x^{1/2}}{\frac{1}{2}}\right) + \frac{x^2}{2}\right]\Bigg|_4^9$

$$= (6 + \tfrac{81}{2}) - (4 + \tfrac{16}{2}) = 34\tfrac{1}{2}$$

c. $\displaystyle\int_{-2}^2 |x|\, dx = \int_{-2}^0 |x|\, dx + \int_0^2 |x|\, dx$ *Subdivision rule*

$$= \int_{-2}^0 (-x)\, dx + \int_0^2 x\, dx \qquad \text{Recall, } |x| = x \quad \text{if } x \ge 0$$
$$|x| = -x \quad \text{if } x < 0$$

$$= -\frac{x^2}{2}\bigg|_{-2}^0 + \frac{x^2}{2}\bigg|_0^2 = -\left(0 - \frac{4}{2}\right) + \left(\frac{4}{2} + 0\right) = 4 \qquad \blacksquare$$

THE SECOND FUNDAMENTAL THEOREM OF CALCULUS

In certain circumstances, it is useful to consider an integral of the form

$$\int_a^x f(t)\, dt$$

where the upper limit of integration is a variable instead of a constant. As x varies, so does the value of the integral, and

$$F(x) = \int_a^x f(t)\, dt$$

is a function of the variable x. For instance,

$$F(x) = \int_2^x t^2\, dt = \tfrac{1}{3}x^3 - \tfrac{1}{3}(2)^3 = \tfrac{1}{3}x^3 - \tfrac{8}{3}$$

The fundamental theorem of calculus tells us that if f is continuous on $[a, b]$, then

$$\int_a^b f(x)\, dx = F(b) - F(a)$$

where F is an antiderivative of f on $[a, b]$. But in general, what guarantee do we have that such an antiderivative even exists? The answer is provided by the following theorem, which is often referred to as the second fundamental theorem of calculus.

THEOREM 5.8   *Second fundamental theorem of calculus*

Let $f(t)$ be continuous on the interval $[a, b]$ and define the function G by the integral equation

$$G(x) = \int_a^x f(t)\, dt$$

for $a \le x \le b$. Then G is an antiderivative of f on $[a, b]$; that is,

$$G'(x) = \frac{d}{dx}\left[\int_a^x f(t)\, dt\right] = f(x)$$

on $[a, b]$.

Proof Apply the definition of derivative to G:

$$G'(x) = \lim_{\Delta x \to 0} \frac{G(x + \Delta x) - G(x)}{\Delta x}$$

$$= \lim_{\Delta x \to 0} \frac{1}{\Delta x}[G(x + \Delta x) - G(x)]$$

$$= \lim_{\Delta x \to 0} \frac{1}{\Delta x}\left[\int_a^{x+\Delta x} f(t)\, dt - \int_a^x f(t)\, dt\right]$$

$$= \lim_{\Delta x \to 0} \frac{1}{\Delta x}\left[\int_a^{x+\Delta x} f(t)\, dt + \int_x^a f(t)\, dt\right]$$

$$= \lim_{\Delta x \to 0} \frac{1}{\Delta x}\left[\int_x^{x+\Delta x} f(t)\, dt\right] \qquad \text{Subdivision rule}$$

Since f is continuous on $[a, b]$, it is continuous on $[x, x + \Delta x]$ for any x. Let $m(x)$ and $M(x)$ be the smallest and largest values, respectively, for $f(t)$ on $[x, x + \Delta x]$. (*Note:* In general, m and M depend on x, but are constants as far as t-integration is concerned.) Since $m(x) \le f(t) \le M(x)$, we have

$$\int_x^{x+\Delta x} m(x)\, dt \le \int_x^{x+\Delta x} f(t)\, dt \le \int_x^{x+\Delta x} M(x)\, dt$$

so (by the dominance rule because $m \le f(x) \le M$ on $[a, b]$)

$$m(x)[(x + \Delta x) - x] \le \int_x^{x+\Delta x} f(t)\, dx \le M(x)[(x + \Delta x) - x]$$

or

$$m(x) \le \frac{1}{\Delta x}\int_x^{x+\Delta x} f(t)\, dt \le M(x)$$

The continuity of f guarantees that $m(x) \to f(x)$ as $\Delta x \to 0$ and $M(x) \to f(x)$ as $\Delta x \to 0$, so the integral in the above inequality is "squeezed" toward $f(x)$ as $\Delta x \to 0$ and we have

$$G'(x) = \lim_{\Delta x \to 0} \frac{1}{\Delta x}\int_x^{x+\Delta x} f(t)\, dt = f(x)$$

Notice that if F is *any* antiderivative of f on the interval $[a, b]$, then the antiderivative

$$G(x) = \int_a^x f(t)\, dt$$

found in Theorem 5.8 satisfies $G(x) = F(x) + C$ for some constant C and all x on the interval $[a, b]$. In particular, when $x = a$, we have

$$0 = \int_a^a f(t)\, dt = G(a) = F(a) + C$$

so that $C = -F(a)$. Finally, by letting $x = b$, we find that

$$\int_a^b f(t)\, dt = G(b) = F(b) + C = F(b) + [-F(a)] = F(b) - F(a)$$

This provides an alternative proof of the first fundamental theorem.

EXAMPLE 4   *Using the second fundamental theorem*

Differentiate $F(x) = \int_7^x (2t - 3)\, dt$.

Solution   From the second fundamental theorem, we can obtain $F'(x)$ by simply replacing t with x in the integrand $f(t) = 2t - 3$. Thus,

$$F'(x) = \frac{d}{dx}\left[\int_7^x (2t - 3)\, dt\right] = 2x - 3$$

The second fundamental theorem of calculus can also be applied to an integral function with a variable *lower* limit of integration. For example, to differentiate

$$G(z) = \int_z^5 \frac{\sin u}{u}\, du$$

reverse the order of integration and apply the second fundamental theorem of calculus as before:

$$G'(z) = \frac{d}{dz}\left[\int_z^5 \frac{\sin u}{u}\, du\right] = \frac{d}{dz}\left[-\int_5^z \frac{\sin u}{u}\, du\right] = -\frac{d}{du}\left[\int_5^z \frac{\sin u}{u}\, du\right] = -\frac{\sin z}{z}$$

5.4 Problem Set

A *In Problems 1–30, evaluate the definite integral.*

1. $\int_{-10}^{10} 7\, dx$

2. $\int_{-5}^{7} (-3)\, dx$

3. $\int_{-3}^{5} (2x + a)\, dx$

4. $\int_{-2}^{2} (b - x)\, dx$

5. $\int_{-1}^{2} ax^3\, dx$

6. $\int_{-1}^{1} (x^3 + bx^2)\, dx$

7. $\int_1^2 \frac{c}{x^3}\, dx$

8. $\int_{-2}^{-1} \frac{p}{x^2}\, dx$

9. $\int_0^9 \sqrt{x}\, dx$

10. $\int_0^{27} \sqrt[3]{x}\, dx$

11. $\int_0^1 (5u^7 + \pi^2)\, du$

12. $\int_0^1 (7x^8 + \sqrt{\pi})\, dx$

13. $\int_1^2 x^{2a}\, dx, a \neq -\frac{1}{2}$

14. $\int_1^2 (2x)^\pi\, dx$

15. $\int_{\ln 2}^{\ln 5} 5e^x\, dx$

16. $\int_{e^{-2}}^{e} \frac{dx}{x}$

17. $\int_0^4 \sqrt{x}\,(x + 1)\, dx$

18. $\int_0^1 \sqrt{t}\,(t - \sqrt{t})\, dt$

19. $\int_1^2 \frac{x^3 + 1}{x^2}\, dx$

20. $\int_1^4 \frac{x^2 + x - 1}{\sqrt{x}}\, dx$

21. $\int_1^{\sqrt{3}} \frac{6a}{1 + x^2}\, dx$

22. $\int_0^{0.5} \frac{b\, dx}{\sqrt{1 - x^2}}$

23. $\int_{-2}^3 (\sin^2 x + \cos^2 x)\, dx$

24. $\int_0^{\pi/4} (\sec^2 x - \tan^2 x)\, dx$

25. $\int_0^1 (1 - e^t)\, dt$

26. $\int_1^2 \frac{x^3 + 1}{x}\, dx$

27. $\int_0^1 \frac{x^2 - 4}{x - 2}\, dx$

28. $\int_0^1 \frac{x^2 - 1}{x^2 + 1}\, dx$

29. $\int_{-1}^2 (x + |x|)\, dx$

30. $\int_0^2 (x - |x - 1|)\, dx$

In Problems 31–38, find the area of the region under the given curve over the prescribed interval.

31. $y = x^2 + 1$ on $[-1, 1]$

32. $y = \sqrt{t}$ on $[0, 1]$

33. $y = \sec^2 x$ on $[0, \frac{\pi}{4}]$

34. $y = \sin x + \cos x$ on $[0, \frac{\pi}{2}]$

35. $y = e^t - t$ on $[0, 1]$

36. $y = (x^2 + x + 1)\sqrt{x}$ on $[1, 4]$

37. $y = \frac{x^2 - 2x + 3}{x}$ on $[1, 2]$

38. $y = \frac{2}{1 + t^2}$ on $[0, 1]$

In Problems 39–44, find the derivative of the given function.

39. $F(x) = \int_0^x \dfrac{t^2 - 1}{\sqrt{t + 1}}\, dt$

40. $F(x) = \int_{-2}^x (t + 1)\,\sqrt[3]{t}\, dt$

41. $F(t) = \int_1^t \dfrac{\sin x}{x}\, dx$

42. $F(t) = \int_t^2 \dfrac{e^x}{x}\, dx$

43. $F(x) = \int_x^1 \dfrac{dt}{\sqrt{1 + 3t^2}}$

44. $F(x) = \int_{\pi/3}^x \sec^2 t \tan t\, dt$

Ⓑ *The formulas in Problems 45–50 are taken from a table of integrals. In each case, use differentiation to verify that the formula is correct.*

45. $\displaystyle\int \cos^2 au\, du = \dfrac{u}{2} + \dfrac{\sin 2au}{4a} + C$

46. $\displaystyle\int u \cos^2 au\, du = \dfrac{u^2}{4} + \dfrac{u \sin 2au}{4a} + \dfrac{\cos 2au}{8a^2} + C$

47. $\displaystyle\int \dfrac{u\, du}{(a^2 - u^2)^{3/2}} = \dfrac{1}{\sqrt{a^2 - u^2}} + C$

48. $\displaystyle\int \dfrac{du}{u^2 - a^2} = \dfrac{1}{2a} \ln\left|\dfrac{u - a}{u + a}\right| + C$

49. $\displaystyle\int \dfrac{u\, du}{\sqrt{a^2 - u^2}} = -\sqrt{a^2 - u^2} + C$

50. $\displaystyle\int (\ln|u|)^2\, du = u(\ln|u|)^2 - 2u \ln|u| + 2u + C$

51. ■ **What Does This Say?** What is the relationship between finding an area and evaluating an integral?

52. ■ **What Does This Say?** Discuss $\displaystyle\int_{-4}^4 x^{1/2}\, dx$.

53. THINK TANK PROBLEM If you use the first fundamental theorem for the following integral, you find

$$\int_{-1}^1 \dfrac{dx}{x^2} = \left[-\dfrac{1}{x}\right]\Bigg|_{-1}^1 = -1 - 1 = -2$$

But the function $y = \dfrac{1}{x^2}$ is never negative, so the above "evaluation" cannot be correct. Describe the error.

54. Evaluate

$$\int_0^2 f(x)\, dx \quad \text{where} \quad f(x) = \begin{cases} x^3 & \text{if } 0 \le x < 1 \\ x^4 & \text{if } 1 \le x \le 2 \end{cases}$$

55. Evaluate

$$\int_0^\pi f(x)\, dx \quad \text{where} \quad f(x) = \begin{cases} \cos x & \text{if } 0 \le x < \pi/2 \\ x & \text{if } \pi/2 \le x \le \pi \end{cases}$$

56. a. If $F(x) = \int \left(\dfrac{1}{\sqrt{x}} - 4\right) dx$, find F so that $F(1) = 0$.

b. Sketch the graphs of $y = F(x), y = F(x) + 3,$ and $y = F(x) - 1$.

c. Find a constant C such that the largest value of $G(x) = F(x) + C$ is 0.

57. Let $g(x) = \int_0^x f(t)\, dt$, where f is the function defined by the following graph. Note that f crosses the x-axis at three points on $[0, 2]$; label these (from left to right), $x = a, x = b,$ and $x = c$

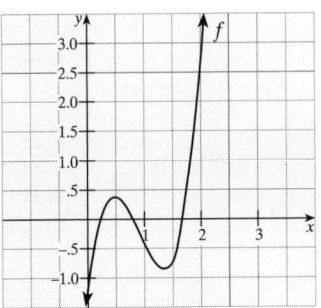

a. What can you say about $g(a)$?

b. Estimate $g(1)$.

c. Where does g have a maximum value on $[0, 2]$?

d. Sketch a rough graph of g.

58. Let $g(x) = \int_0^x f(t)\, dt$, where f is the function defined by the following graph.

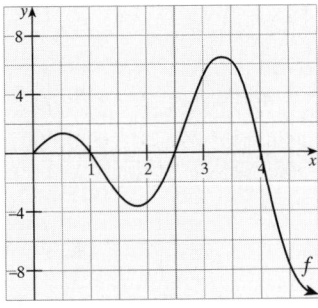

a. Where does g have a relative minimum on $[0, 5]$?

b. Where does g have a relative maximum value on $[0, 5]$?

c. If $g(1) = 1, g(2.5) = -2.5,$ and $g(4) = 4,$ sketch a rough graph of g.

59. THINK TANK PROBLEM The purpose of this problem is to provide a counterexample showing that the integral of a product (or quotient) is not necessarily equal to the product (quotient) of the respective integrals.

a. Show that $\int x\sqrt{x}\, dx \neq \left(\int x\, dx\right)\left(\int \sqrt{x}\, dx\right)$.

b. Show that $\displaystyle\int \dfrac{\sqrt{x}}{x}\, dx \neq \dfrac{\int \sqrt{x}\, dx}{\int x\, dx}$.

60. Suppose f is a function with the property that $f'(x) = f(x)$ for all x.

a. Show that $\displaystyle\int_a^b f(x)\,dx = f(b) - f(a)$.

b. Show that

$$\int_a^b [f(x)]^2\,dx = \frac{1}{2}\{[f(b)]^2 - [f(a)]^2\}$$

61. Let $\displaystyle F(x) = \int_1^x \frac{e^t}{t+1}\,dt$.

Note that $\displaystyle F(2x) = \int_1^{2x} \frac{e^t}{t+1}\,dt$.

What is $F'(4)$? *Hint:* Use the chain rule with $u = 2x$.

62. Find an equation for the tangent line to the curve $y = F(x)$ at the point P where $x = 1$ if

$$F(x) = \int_1^{\sqrt{x}} \frac{2t+1}{t+2}\,dt$$

63. Suppose $F(x)$ is the integral function

$$F(x) = \int_{u(x)}^{v(x)} f(t)\,dt$$

What is $F'(x)$? This result is called *Leibniz' rule*.

5.5 Integration by Substitution

IN THIS SECTION **substitution with indefinite integration, substitution with definite integration**

SUBSTITUTION WITH INDEFINITE INTEGRATION

The method of substitution is the integration version of the chain rule. Recall that according to the chain rule, the derivative of $(x^2 + 3x + 5)^9$ is

$$\frac{d}{dx}(x^2 + 3x + 5)^9 = 9(x^2 + 3x + 5)^8(2x + 3)$$

Thus, $\displaystyle\int 9(x^2 + 3x + 5)^8(2x + 3)\,dx = (x^2 + 3x + 5)^9 + C$

Note that the product is of the form $g(u)\dfrac{du}{dx}$ where, in this case, $g(u) = 9u^8$ and $u = x^2 + 3x + 5$.

You can integrate many products of the form $g(u)\dfrac{du}{dx}$ by applying the chain rule in reverse, as indicated by the following theorem.

THEOREM 5.9 *Integration by substitution*

Let f, g, and u be differentiable functions of x such that

$$f(x) = g(u)\frac{du}{dx}$$

Then

$$\int f(x)\,dx = \int g(u)\frac{du}{dx}\,dx = \int g(u)\,du = G(u) + C$$

where G is an antiderivative of g.

Proof   If G is an antiderivative of g, then $G'(u) = g(u)$ and, by the chain rule,

$$f(x) = \frac{d}{dx}[G(u)] = G'(u)\frac{du}{dx} = g(u)\frac{du}{dx}$$

Integrating both sides of this equation, we obtain

$$\int f(x)\ dx = \int \left[g(u)\frac{du}{dx} \right] dx = \int \left[\frac{d}{dx}G(u) \right] dx = G(u) + C$$

as required.

=====

EXAMPLE 1   *Integration of an indefinite integral by substitution*

Find $\displaystyle\int 9(x^2 + 3x + 5)^8(2x + 3)\ dx$.

Solution   Look at the problem and make the observations as shown in the boxes:

Let $u = x^2 + 3x + 5$. $\leftarrow$ If $u = x^2 + 3x + 5$,
                              then $du = (2x + 3)\ dx$.

$$\int \overbrace{9(x^2 + 3x + 5)^8}\ \underbrace{(2x + 3)\ dx} = \int 9u^8\ du$$

This is $g(u)$.   This is du.

Now complete the integration of $g(u)$ and, when you are finished, back-substitute to express u in terms of x:

$$\int 9u^8\ du = 9\left(\frac{u^9}{9}\right) + C = (x^2 + 3x + 5)^9 + C$$ ∎

Sometimes the expression for du is not quite as obvious as that shown in Example 1, so a more general procedure is shown next:

$$\int 9(x^2 + 3x + 5)^8(2x + 3)\ dx$$

Let $u = x^2 + 3x + 5$, so $\dfrac{du}{dx} = 2x + 3$, which implies $dx = \dfrac{du}{2x + 3}$.

Substitute these values:

$$\int 9\overbrace{(x^2 + 3x + 5)^8}^{u}(2x + 3)\overbrace{dx}^{\frac{du}{2x+3}} = \int 9u^8(2x + 3)\frac{du}{2x + 3}$$

$$= \underbrace{\int 9u^8\ du}$$

The goal here is to have an expression of the form $\int g(u)\ du$—that is, all terms involving x and dx should be eliminated.

EXAMPLE 2   *Substitution with a radical function*

Find $\displaystyle\int \sqrt{3x + 7}\ dx$.

Solution   Let $u = 3x + 7$, so $du = 3\ dx$ or $dx = \dfrac{du}{3}$; substitute:

$$\int \sqrt{3x + 7}\ dx = \int \sqrt{u}\ \frac{du}{3} = \frac{1}{3}\int u^{1/2}\ du$$

$$= \frac{1}{3}\left(\frac{u^{3/2}}{\frac{3}{2}}\right) + C = \frac{2}{9}u^{3/2} + C = \frac{2}{9}(3x + 7)^{3/2} + C$$ ∎

WARNING   After you have made your substitution and simplified, there should be no leftover *x*-values in the integrand. ◄

EXAMPLE 3   *Substitution with leftover x-values*

Find $\int x(4x - 5)^3 \, dx$.

Solution   Let $u = 4x - 5$, so $du = 4 \, dx$ implies $dx = \dfrac{du}{4}$; substitute:

$$\int x(4x - 5)^3 \, dx = \int xu^3\left(\frac{du}{4}\right) = \frac{1}{4}\int xu^3 \, du$$

There is a leftover x-value.

We are not ready to integrate until the leftover *x*-term has been eliminated. Because $u = 4x - 5$, we can solve for *x*:

$$x = \frac{u + 5}{4}$$

so

$$\frac{1}{4}\int xu^3 \, du = \frac{1}{4}\int \left(\frac{u + 5}{4}\right)u^3 \, du = \frac{1}{16}\int (u^4 + 5u^3) \, du$$

$$= \frac{1}{16}\left(\frac{u^5}{5} + 5\frac{u^4}{4}\right) + C = \frac{1}{80}(4x - 5)^5 + \frac{5}{64}(4x - 5)^4 + C \quad \blacksquare$$

EXAMPLE 4   *Change of variable by substitution*

Find $\int \dfrac{x \, dx}{\sqrt{x^2 + 1}}$.

Solution   Let $u = x^2 + 1$, so $du = 2x \, dx$ and $dx = \dfrac{du}{2x}$; substitute:

$$\int \frac{x \, dx}{\sqrt{x^2 + 1}} = \int \frac{x\left(\frac{du}{2x}\right)}{\sqrt{u}} = \frac{1}{2}\int u^{-1/2} \, du = \frac{1}{2}\left(\frac{u^{1/2}}{\frac{1}{2}}\right) + C = (x^2 + 1)^{1/2} + C \quad \blacksquare$$

EXAMPLE 5   *Substitution with a trigonometric function*

Find $\int (4 - 2\cos\theta)^3 \sin\theta \, d\theta$.

Solution   Let $u = 4 - 2\cos\theta$, so $\dfrac{du}{d\theta} = -2(-\sin\theta)$ and $d\theta = \dfrac{du}{2\sin\theta}$; substitute:

$$\int (4 - 2\cos\theta)^3 \sin\theta \, d\theta = \int u^3 \sin\theta \frac{du}{2\sin\theta} = \frac{1}{2}\int u^3 \, du$$

$$= \frac{1}{2}\left(\frac{u^4}{4}\right) + C = \frac{1}{8}(4 - 2\cos\theta)^4 + C \quad \blacksquare$$

Technology Window

As we have previously noted, the forms obtained by using software will sometimes vary considerably. For example, using software, we obtain the following form for Example 5:

$$2 \cos^4 \theta - 16 \cos^3 \theta - 64 \cos \theta - 48 \sin^2 \theta$$

To show the forms are equivalent, we can change this form to cosines:

$$2 \cos^4 \theta - 16 \cos^3 \theta + 48 \cos^2 \theta - 64 \cos \theta - 48$$

Now, we expand the answer shown in Example 5 to obtain

$$\tfrac{1}{8}(4 - 2 \cos \theta)^4 + C = 2 \cos^4 \theta - 16 \cos^3 \theta + 48 \cos^2 \theta - 64 \cos \theta + 32 + C$$

You can now see that for an appropriate choice of C, these forms are the same. The process of reconciling a software answer and an answer from human calculation is not very different from the process of proving that equations in trigonometry are identities.

EXAMPLE 6   *Using a trigonometric identity with substitution*

Find $\displaystyle\int \sin 2x \, dx$.

Solution   Let $u = 2x$, so $du = 2 \, dx$ so that $dx = \dfrac{du}{2}$; substitute:

$$\int \sin 2x \, dx = \int \sin u \frac{du}{2} = \frac{1}{2} \int \sin u \, du = -\frac{1}{2} \cos u + C = -\frac{1}{2} \cos 2x + C$$

If you wish to check your work with integration problems, you can check by using differentiation:

$$\frac{d}{dx}\left(-\frac{1}{2} \cos 2x + C\right) = -\frac{1}{2}(-2 \sin 2x) + 0 = \sin 2x \qquad \blacksquare$$

Here is a problem in which the rate of change of a quantity is known and we use the method of substitution to find an expression for the quantity itself.

EXAMPLE 7   *Find the volume when the rate of flow is known*

Water is flowing into a tank at the rate of $\sqrt{3t + 1}$ ft³/min. If the tank is empty when $t = 0$, how much water does it contain 5 min later?

Solution   Because the rate at which the volume V is changing is dV/dt,

$$\frac{dV}{dt} = \sqrt{3t + 1}$$

$$V = \int \sqrt{3t + 1} \, dt = \int \sqrt{u} \, \frac{du}{3}$$

$$\boxed{\begin{aligned} u &= 3t + 1 \\ du &= 3 \, dt \end{aligned}}$$

$$V = \frac{1}{3} \int u^{1/2} \, du$$

$$= \frac{1}{3} \cdot \frac{2}{3} u^{3/2} + C$$

$$V = \tfrac{2}{9}(3t + 1)^{3/2} + C$$

$$V(0) = \tfrac{2}{9}(3 \cdot 0 + 1)^{3/2} + C = 0 \text{ so that } C = -\tfrac{2}{9} \qquad \text{The initial volume is } 0.$$

$$V(t) = \tfrac{2}{9}(3t + 1)^{3/2} - \tfrac{2}{9} \text{ and } V(5) = \tfrac{2}{9}(16)^{3/2} - \tfrac{2}{9} = 14$$

The tank contains 14 ft³ of water.   ∎

SUBSTITUTION WITH DEFINITE INTEGRATION

Example 7 could be considered as a definite integral:

$$\int_0^5 \sqrt{3t + 1}\, dt = \tfrac{2}{9}(3t + 1)^{3/2}\big|_0^5 = \tfrac{2}{9}(16)^{3/2} - \tfrac{2}{9}(1)^{3/2} = 14$$

Notice that the definite integral eliminates the need for finding C. Furthermore, the following theorem eliminates the need for returning to the original variable.

THEOREM 5.10   *Substitution with the definite integral*

If $f(u)$ is a continuous function of u and $u(x)$ is a differentiable function of x, then

$$\int_a^b f[u(x)]u'(x)\, dx = \int_{u(a)}^{u(b)} f(u)\, du$$

Proof   Let $F(u) = \int f(u)\, du$ be an antiderivative of $f(u)$. Then $F'(u) = f(u)$, and by the chain rule

$$\frac{dF(u)}{dx} = \frac{dF}{du}\frac{du}{dx} = f(u)\frac{du}{dx}$$

Thus, we have

$$\int f[u(x)]u'(x)\, dx = \int \frac{dF(u)}{dx}\, dx = F(u) = \int f(u)\, du$$

as claimed. The statement about the definite integrals follows from the fundamental theorem of calculus.   ═

EXAMPLE 8   *Substitution with the definite integral*

Evaluate $\int_1^2 (4x - 5)^3\, dx$.

Solution   Let $u = 4x - 5$, $du = 4\, dx$

If $x = 2$, then $u = 4(2) - 5 = 3$.

$$\int_1^2 (4x - 5)^3\, dx = \int_{-1}^3 u^3 \frac{du}{4}$$

WARNING ▶ You cannot change variables and keep the original limits of integration.

If $x = 1$, then $u = 4(1) - 5 = -1$.

$$= \frac{1}{4} \cdot \frac{u^4}{4}\bigg|_{-1}^3$$

$$= \tfrac{1}{16}(81 - 1) = 5 \qquad ∎$$

Notice that substitution with the definite integral does not require that you return to the original variable.

5.5   Problem Set

A   *Problems 1–8 present pairs of integration problems, one of which will use substitution and one of which will not. As you are working these problems, think about when substitution may be appropriate.*

1. a. $\int_0^4 (2t + 4)\, dt$ **b.** $\int_0^4 (2t + 4)^{-1/2}\, dt$

2. a. $\int_0^{\pi/2} \sin\theta\, d\theta$ **b.** $\int_0^{\pi/2} \sin 2\theta\, d\theta$

3. a. $\int_0^{\pi} \cos t\, dt$ **b.** $\int_0^{\sqrt{\pi}} t\cos t^2\, dt$

4. a. $\int_0^4 \sqrt{x}\, dx$ **b.** $\int_{-4}^0 \sqrt{-x}\, dx$

5. a. $\int_0^{16} \sqrt[4]{x}\, dx$ **b.** $\int_{-16}^0 \sqrt[4]{-x}\, dx$

6. a. $\int x(3x^2 - 5)\, dx$ **b.** $\int x(3x^2 - 5)^5\, dx$

7. a. $\int x^2\sqrt{2x^3}\, dx$ **b.** $\int x^2\sqrt{2x^3 - 5}\, dx$

8. a. $\int \dfrac{dx}{\sqrt{1 - x^2}}$ **b.** $\int \dfrac{x\, dx}{\sqrt{1 - x^2}}$

Use substitution to evaluate the indefinite integrals in Problems 9–34.

9. $\int (2x + 3)^4\, dx$ **10.** $\int \sqrt{3t - 5}\, dt$

11. $\int (x - 27)^{2/3}\, dx$ **12.** $\int (11 - 2x)^{-4/5}\, dx$

13. $\int (x^2 - \cos 3x)\, dx$ **14.** $\int \csc^2 5t\, dt$

15. $\int \sin(4 - x)\, dx$ **16.** $\int s\sqrt{s^2 + 4}\, ds$

17. $\int \sqrt{t}(t^{3/2} + 5)^3\, dt$ **18.** $\int \dfrac{(6x - 9)\, dx}{(x^2 - 3x + 5)^3}$

19. $\int x\sin(3 + x^2)\, dx$ **20.** $\int \sin^3 t\cos t\, dt$

21. $\int \dfrac{x\, dx}{2x^2 + 3}$ **22.** $\int \dfrac{x^2\, dx}{x^3 + 1}$

23. $\int x\sqrt{2x^2 + 1}\, dx$ **24.** $\int \dfrac{4x\, dx}{2x + 1}$

25. $\int \sqrt{x}\, e^{x\sqrt{x}}\, dx$ **26.** $\int \dfrac{e^{\sqrt[3]{x}}\, dx}{x^{2/3}}$

27. $\int x(x^2 + 4)^{1/2}\, dx$ **28.** $\int x^3(x^2 + 4)^{1/2}\, dx$

29. $\int \dfrac{\ln x}{x}\, dx$ **30.** $\int \dfrac{\ln(x + 1)}{x + 1}\, dx$

31. $\int \dfrac{dx}{\sqrt{x}(\sqrt{x} + 7)}$ **32.** $\int \dfrac{dx}{x^{2/3}(\sqrt[3]{x} + 1)}$

33. $\int \dfrac{e^t\, dt}{e^t + 1}$ **34.** $\int \dfrac{e^{\sqrt{t}}\, dt}{\sqrt{t}(e^{\sqrt{t}} + 1)}$

B   *Evaluate the definite integrals given in Problems 35–44. Approximate the answers to Problems 43 and 44 to two significant digits.*

35. $\int_0^1 \dfrac{5x^2\, dx}{2x^3 + 1}$ **36.** $\int_1^4 \dfrac{e^{-\sqrt{x}}\, dx}{\sqrt{x}}$

37. $\int_{-\ln 2}^{\ln 2} \dfrac{1}{2}(e^x - e^{-x})\, dx$ **38.** $\int_0^2 (e^x - e^{-x})^2\, dx$

39. $\int_1^2 \dfrac{e^{1/x}\, dx}{x^2}$ **40.** $\int_0^2 x\sqrt{2x + 1}\, dx$

41. $\int_0^{\pi/6} \tan 2x\, dx$ **42.** $\int_0^1 x^2(x^3 + 9)^{1/2}\, dx$

43. $\int_0^5 \dfrac{0.58}{1 + e^{-0.2x}}\, dx$ **44.** $\int_0^{12} \dfrac{5{,}000}{1 + 10e^{-t/5}}\, dt$

45. **HISTORICAL QUEST**   Johann Peter Gustav Lejeune Dirichlet was a professor of mathematics at the University of Berlin and is known for his role in formulating a rigorous foundation for calculus. He was not known as a good teacher. His nephew wrote that the mathematics instruction he received from Dirichlet was the most dreadful experience of his life. Howard Eves tells of the time Dirichlet was to deliver a lecture on definite integrals, but because of illness he posted the following note:

LEJEUNE DIRICHLET
1805–1859

> *Because of illness 9*
> *cannot lecture today*
>
> *Dirichlet*

The students then doctored the note to read:

> *Michaelmas*
> $\displaystyle\int$ *Because of illness 9*
> *cannot lecture today*   *d (1 Frdor)*
> *Easter*
>
> *Dirichlet*

Michaelmas and Easter were school holidays, and 1 Frdor (Fredrichsd'or) was the customary honorarium for a semester's worth of lectures.

a. What is the answer when you integrate the student-doctored note?

b. The so-called *Dirichlet function* is often used for counterexamples in calculus. (We use it several times in this text.) Look up the definition of this function. What special property does it have?

B *Find the area of the region under the curves given in Problems 46–49.*

46. $y = t\sqrt{t^2 + 9}$ on $[0, 4]$

47. $y = \dfrac{1}{t^2}\sqrt{5 - \dfrac{1}{t}}$ on $[\frac{1}{5}, 1]$

48. $y = x(x - 1)^{1/3}$ on $[2, 9]$

49. $y = |x|$ on $[2, 3]$

50. a. Show that if f is continuous and *odd* [that is, $f(-x) = -f(x)$] on the interval $[-a, a]$, then

$$\int_{-a}^{a} f(x)\,dx = 0$$

b. Show that if f is continuous and *even* $[f(-x) = f(x)]$ on the interval $[-a, a]$, then

$$\int_{-a}^{a} f(x)\,dx = 2\int_{0}^{a} f(x)\,dx = 2\int_{-a}^{0} f(x)\,dx$$

Use the results of Problem 50 to evaluate the integrals given in Problems 51–54.

51. $\displaystyle\int_{-\pi}^{\pi} \sin x\,dx$

52. $\displaystyle\int_{-\pi/2}^{\pi/2} \cos x\,dx$

53. $\displaystyle\int_{-3}^{3} x\sqrt{x^4 + 1}\,dx$

54. $\displaystyle\int_{-1}^{1} \frac{\sin x\,dx}{x^2 + 1}$

55. In each of the following cases, determine whether the given relationship is true or false.

a. $\displaystyle\int_{-175}^{175} (7x^{1001} + 14x^{99})\,dx = 0$

b. $\displaystyle\int_{0}^{\pi} \sin^2 x\,dx = \int_{0}^{\pi} \cos^2 x\,dx$

c. $\displaystyle\int_{-\pi/2}^{\pi/2} \cos x\,dx = \int_{-\pi}^{0} \sin x\,dx$

56. The slope at each point (x, y) on the graph of $y = F(x)$ is given by $x(x^2 - 1)^{1/3}$, and the graph passes through the point $(3, 1)$. Use this information to find F. Sketch the graph of F.

57. The slope at each point (x, y) on the graph of $y = F(x)$ is given by

$$\frac{2x}{1 - 3x^2}$$

What is $F(x)$ if the graph passes through $(0, 5)$?

58. A particle moves along the x-axis in such a way that at time t, its velocity is $v(t) = t^2(t^3 - 8)^{1/3}$.
a. At what time does the particle turn around?
b. If the particle starts at $x = 1$, where does it turn around?

59. MODELING PROBLEM   A rectangular storage tank has a square base 10 ft on a side. Water is flowing into the tank at the rate modeled by the function

$$R(t) = t(3t^2 + 1)^{-1/2} \text{ ft}^3/\text{s}$$

at time t seconds. If the tank is empty at time $t = 0$, how much water does it contain 4 s later? What is the depth of the water (to the nearest quarter inch) at that time?

60. JOURNAL PROBLEM   *College Mathematics Journal.** Evaluate

$$\int [(x^2 - 1)(x + 1)]^{-2/3}\,dx$$

61. MODELING PROBLEM   Environmentalists model the rate at which the ozone level is changing in a suburb of Los Angeles by the function

$$L'(t) = \frac{0.24 - 0.03t}{\sqrt{36 + 16t - t^2}}$$

parts per million per hour (ppm/h) t hours after 7:00 A.M.
a. Express the ozone level $L(t)$ as a function of t if L is 4 ppm at 7:00 A.M.
b. Use the graphing utility of your calculator to find the time between 7:00 A.M. and 7:00 P.M. when the highest level of ozone occurs. What is the highest level?
c. Use your graphing utility or another utility of your calculator to determine a second time during the day when the ozone level is the same as it is at 11:00 A.M.

C **62.** A *logistic function* is one of the form

$$Q(t) = \frac{B}{1 + Ae^{-rt}}$$

Evaluate

$$\int Q(t)\,dt$$

___
*From the *College Mathematics Journal*, Sept. 1989, p. 343. Problem by Murray Klamkin.

5.6 *Introduction to Differential Equations*

IN THIS SECTION introduction and terminology, direction fields, separable differential equations, modeling exponential growth and decay, orthogonal trajectories, modeling fluid flow through an orifice, modeling the motion of a projectile: escape velocity

The study of differential equations is as old as calculus itself. Today, it would be virtually impossible to make a serious study of physics, astronomy, chemistry, or engineering without encountering physical models based on differential equations. In addition, differential equations are beginning to appear more frequently in the biological and social sciences, especially in economics. We begin by introducing some basic terminology and examining a few modeling procedures.

INTRODUCTION AND TERMINOLOGY

Any equation that contains a derivative or differential is called a **differential equation.** For example, the equations

$$\frac{dy}{dx} = 3x^3 + 5 \qquad \frac{dP}{dt} = kP^2 \qquad \left(\frac{dy}{dx}\right)^2 + 3\frac{dy}{dx} + 2y = xy \qquad \frac{d^2x}{dt^2} + 2\frac{dx}{dt} + 5t = \sin t$$

are all differential equations.

Many practical situations, especially those involving rates, can be described mathematically by differential equations. For example, the assumption that population P grows at a rate proportional to its size can be expressed by the differential equation

$$\frac{dP}{dt} = kP$$

where t is time and k is the constant of proportionality.

A **solution** of a given differential equation is a function that satisfies the equation. A **general solution** is a characterization of all possible solutions of the equation. We say that the equation is **solved** when we find a general solution.

For example, $y = x^2$ is a solution of the differential equation

$$\frac{dy}{dx} = 2x$$

because

$$\frac{dy}{dx} = \frac{d}{dx}(y) = \frac{d}{dx}(x^2) = 2x$$

Moreover, because any solution of this equation must be an indefinite integral of $2x$, it follows that

$$y = \int 2x \, dx = x^2 + C$$

is the general solution of the differential equation.

EXAMPLE 1   *Finding future revenue*

An oil well that yields 300 barrels of crude oil a day will run dry in 3 years. It is estimated that t days from now, the price of the crude oil will be $p(t) = 30 + 0.3\sqrt{t}$ dollars per barrel. If the oil is sold as soon as it is extracted from the ground, what will be the total future revenue from the well?

Solution   Let $R(t)$ denote the total revenue up to time t. Then the rate of change of revenue is $\frac{dR}{dt}$, the number of dollars received per barrel is $p(t) = 30 + 0.3\sqrt{t}$, and the number of barrels sold per day is 300. Thus, we have

$$\begin{bmatrix} \text{RATE OF CHANGE} \\ \text{OF TOTAL REVENUE} \end{bmatrix} = \begin{bmatrix} \text{NUMBER OF DOLLARS} \\ \text{PER BARREL} \end{bmatrix}\begin{bmatrix} \text{NUMBER OF BARRELS} \\ \text{SOLD PER DAY} \end{bmatrix}$$

$$\frac{dR}{dt} = (30 + 0.3\sqrt{t})(300)$$

$$= 9{,}000 + 90\sqrt{t}$$

This is actually a statement of the chain rule: $\frac{dR}{dt} = \frac{dR}{dB} \cdot \frac{dB}{dt}$, where B denotes the number of barrels extracted and R denotes the revenue. It is often helpful to use the chain rule in this way when setting up differential equations.

We solve this differential equation by integration:

$$R = \int \frac{dR}{dt}\, dt = \int (9{,}000 + 90\sqrt{t})\, dt$$

$$R(t) = 9{,}000t + 60t^{3/2} + C$$

Because $R(0) = 0$, it follows that $C = 0$. We also are given that the well will run dry in 3 years or 1,095 days, so that the total revenue obtained during the life of the well is

$$R(1{,}095) = 9{,}000(1{,}095) + 60(1{,}095)^{3/2}$$

$$\approx 12{,}029{,}064.52$$

The total future revenue is approximately \$12 million.   ∎

DIRECTION FIELDS

We can use slope fields to help with a visualization of a differential equation. In Section 5.1 we looked at small segments with slope defined by the derivative of some function. The collection of all such line segments is called the *slope field* or **direction field** of the differential equation. To construct a direction field, it is often useful to examine those points (x, y) where the slope y' has a constant value c. For each fixed c these points lie on a curve called an **isocline**. The following example illustrates the procedure for finding isoclines and determining the direction field for a particular equation.

EXAMPLE 2   *The direction field of a given differential equation*

Describe the isoclines of the differential equation

$$\frac{dy}{dx} = x^2 + y^2$$

Sketch the direction field along with a few solution curves.

Solution   The isoclines will be circles of the form $x^2 + y^2 = C$ for C constant and $C \geq 0$. When $C = 1$, the corresponding isocline is a circle centered at the origin with radius 1. For $C = 1/4$, the circle will have radius 1/2, and for $C = 4$, the radius is 2. Several isoclines are drawn in Figure 5.16a, with enough points to indicate the nature of the direction field. To draw the solution curve $y = y(x)$ that passes through a given point (x_0, y_0), start at (x_0, y_0) and move in the direction indicated by the direction field, as illustrated in Figure 5.16b for the solution containing the point $(1, 2)$.

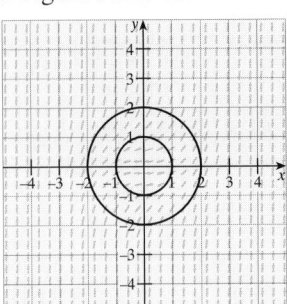
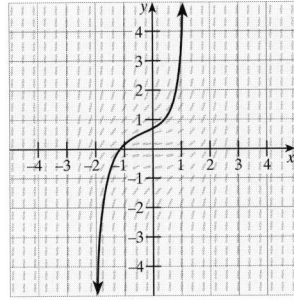

a. The isoclines for $\dfrac{dy}{dx} = x^2 + y^2$.   **b.** Solution curves are drawn using the indicated direction at each point. The particular solution through $(1, 2)$ is shown.

■ **FIGURE 5.16**   Direction field for a differential equation

Today, with the assistance of computers, we can sometimes draw the direction field in order to draw a particular solution (or a family of solutions) as shown in the following example.

EXAMPLE 3   *Finding a solution, given the direction field*

The direction field for the differential equation

$$\frac{dy}{dx} = y - x^2$$

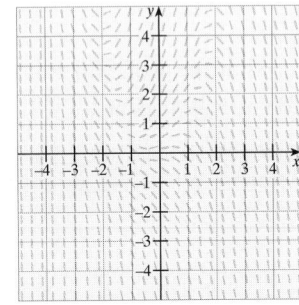

■ **FIGURE 5.17**   Direction field for $y' = y - x^2$

is shown in Figure 5.17.
a. Sketch a solution to the initial value problem passing through $(1, 2)$.
b. Sketch a solution to the initial value problem passing through $(1, 0)$.

Solution
a. The initial value $y(2) = 1$ means the solution passes through $(2, 1)$, and is shown in Figure 5.18a.
b. Since $y(0) = 1$, the solution passes through $(0, 1)$, as shown in Figure 5.18b.

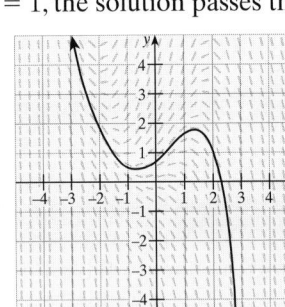
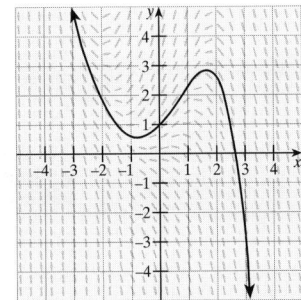

a. Solution passes through $(2, 1)$ **b.** Solution passes through $(0, 1)$

■ **FIGURE 5.18**   Particular solutions from a given direction field

SEPARABLE DIFFERENTIAL EQUATIONS

Solving a differential equation is often a complicated process. However, many important equations can be expressed in the form

$$\frac{dy}{dx} = \frac{g(x)}{f(y)}$$

To solve such an equation, first separate the variables into the differential form

$$f(y)\, dy = g(x)\, dx$$

and then integrate both sides separately to obtain

$$\int f(y)\, dy = \int g(x)\, dx$$

A differentiable equation of the form $dy/dx = g(x)/f(y)$ is said to be **separable.** This procedure is illustrated in Example 4.

EXAMPLE 4   *Separable differential equation*

Solve $\dfrac{dy}{dx} = \dfrac{x}{y}$.

Solution
$$\frac{dy}{dx} = \frac{x}{y}$$
$$y\, dy = x\, dx$$
$$\int y\, dy = \int x\, dx$$
$$\tfrac{1}{2}y^2 + C_1 = \tfrac{1}{2}x^2 + C_2$$
$$x^2 - y^2 = C \quad \text{where } C = 2(C_1 - C_2)$$ ■

Notice the treatment of constants in Example 2. Because all constants can be combined into a single constant, it is customary not to write $C = 2(C_1 - C_2)$, but rather to simply replace all the arbitrary constants in the problem by a single arbitrary constant after the last integral is found.

The remainder of this section is devoted to selected applications involving separable differential equations.

MODELING EXPONENTIAL GROWTH AND DECAY

A process is said to undergo **exponential change** if the relative rate of change of the process is modeled by a constant; in other words,

$$\frac{Q'(t)}{Q(t)} = k$$

or

$$\frac{dQ}{dt} = kQ(t)$$

If the constant k is positive, the exponential change is called **growth,** and if k is negative, it is called **decay.** Exponential growth occurs in certain populations, and exponential decay, in the disintegration of radioactive substances.

To solve the growth/decay equation, separate the variables and integrate both sides:

$$\frac{dQ}{dt} = kQ$$
$$\int \frac{dQ}{Q} = \int k\, dt$$
$$\ln|Q| = kt + C_1$$
$$e^{kt+C_1} = Q \qquad \textit{Definition of natural logarithm}$$
$$e^{kt}e^{C_1} = Q$$

Thus, $Q = Ce^{kt}$, where $C = e^{C_1}$. Finally, if we let Q_0 be the initial amount, we see that

$$Q_0 = Ce^0 \quad \text{or} \quad C = Q_0$$

Growth/Decay Equation

The **growth/decay equation** of a substance is

$$Q(t) = Q_0 e^{kt}$$

where $Q(t)$ is the amount of the substance present at time t, Q_0 is the initial amount of the substance, and k is a constant that depends on the substance. If $k > 0$, it is a growth equation; if $k < 0$, it is a decay equation.

The graph of $Q = Q_0 e^{kt}$ for $k < 0$ is shown in Figure 5.19.

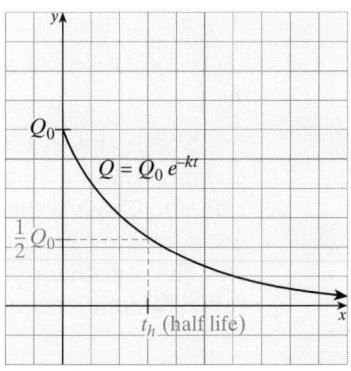

■ **FIGURE 5.19**   The decay curve for a radioactive substance

In Figure 5.19 we have also indicated the time t_h required for half of a given substance to disintegrate. The time t_h is called the **half-life** of the substance, and it provides a measure of the substance's rate of disintegration.

EXAMPLE 5   *Amount of a radioactive substance present*

A particular radioactive substance has a half-life of 600 yr. Find k for this substance and determine how much of a 50-g sample will remain after 125 yr.

Solution   From the decay equation,

$$Q(t) = Q_0 e^{kt}$$

$$\frac{Q(t)}{Q_0} = e^{kt}$$

$$\frac{1}{2} = e^{k(600)} \qquad \text{Half-life is 600 years.}$$

$$k(600) = \ln \tfrac{1}{2} \qquad \text{Definition of logarithm}$$

$$k = \frac{\ln \tfrac{1}{2}}{600} \approx -0.0011552453$$

Next, to see how much of a 50-g sample will remain after 125 yr, substitute $Q_0 = 50$, $k = -0.0011552453$, and $t = 125$ into the decay equation:

$$Q(125) \approx 50e^{-0.0011552453(125)} \approx 43.27682805$$

There will be about 43 g present. ∎

One of the more interesting applications of radioactive decay is a technique known as **carbon dating,** which is used by geologists, anthropologists, and archaeologists to estimate the age of fossils and other objects.* The technique is based on the fact that all animal and vegetable systems (whether living or dead) contain both stable carbon ^{12}C and a radioactive isotope ^{14}C. Scientists assume that the ratio of ^{14}C to ^{12}C in the air has remained approximately constant throughout history. Living systems absorb carbon dioxide from the air, so the ratio of ^{14}C to ^{12}C in a living system is the same as that in the air itself. When a living system dies, the absorption of carbon dioxide ceases. The ^{12}C already in the system remains while the ^{14}C decays, and the ratio of ^{14}C to ^{12}C decreases exponentially. The half-life of ^{14}C is approximately 5,730 years. The ratio of ^{14}C to ^{12}C in a fossil t years after it was alive is approximately

$$R = R_0 e^{kt}$$

where $k = \dfrac{\ln(1/2)}{5{,}730}$ and R_0 is the ratio of ^{14}C to ^{12}C in the atmosphere. By comparing $R(t)$ with R_0, scientists can estimate the age of the object. Here is an example.

EXAMPLE 6   *Modeling Problem: Carbon dating*

An archaeologist has found a fossil in which the ratio of ^{14}C to ^{12}C is 20% of the ratio found in the atmosphere. Approximately how old is the fossil?

Solution   The age of the fossil is the value of t for which $R(t) = 0.20R_0$:

$$0.20R_0 = R_0 e^{kt}$$
$$0.20 = e^{kt}$$
$$kt = \ln 0.20$$
$$t = \frac{1}{k}\ln 0.20 \qquad k = \frac{\ln(1/2)}{5{,}730}$$
$$\approx 13{,}304.64798$$

The fossil is approximately 13,000 yr old. ∎

ORTHOGONAL TRAJECTORIES

A curve that intersects each member of a given family of curves at right angles is called an **orthogonal trajectory** of that family. Orthogonal families arise in many applications. For example, in thermodynamics, the best flow across a planar surface is orthogonal to the curves of constant temperature, called *isotherms*. In the theory of fluid flow, the flow lines are orthogonal trajectories of *velocity potential curves*. The basic

*Carbon dating is used primarily for estimating the age of relatively "recent" specimens. For example, it was used (along with other methods) to determine that the Dead Sea Scrolls were written and deposited in the Caves of Qumran approximately 2,000 years ago. For dating older specimens, it is better to use techniques based on radioactive substances with longer half-lives. In particular, potassium–40 is often used as a "clock" for events that occurred between 5 and 15 million years ago. Paleoanthropologists find this substance especially valuable because it often occurs in volcanic deposits and can be used to date fossils trapped in such deposits. See Mathematical Essay 5 on page 402.

procedure for finding orthogonal trajectories involves differential equations and is demonstrated in Example 7.

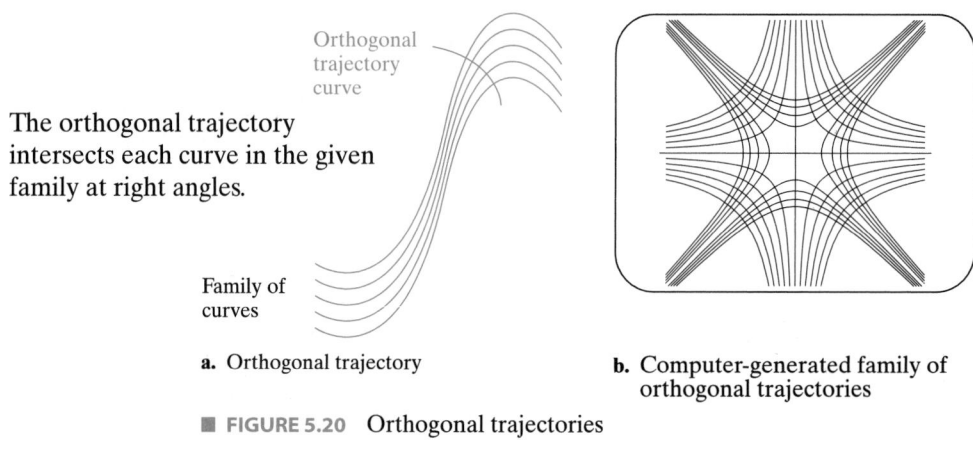

The orthogonal trajectory intersects each curve in the given family at right angles.

Orthogonal trajectory curve

Family of curves

a. Orthogonal trajectory

b. Computer-generated family of orthogonal trajectories

■ **FIGURE 5.20**   Orthogonal trajectories

EXAMPLE 7   *Finding orthogonal trajectories*

Find the orthogonal trajectories of the family of curves of the form

$$xy = C$$

Solution   We are seeking a family of curves. Each curve in that family intersects each curve in the family $xy = C$ at right angles, as shown in Figure 5.21. Assume that a typical point on a given curve in the family $xy = C$ has coordinates (x, y) and that a typical point on the orthogonal trajectory curve has coordinates (X, Y).*

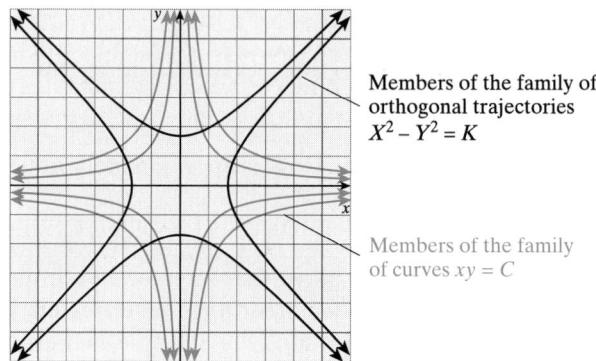

Members of the family of orthogonal trajectories $X^2 - Y^2 = K$

Members of the family of curves $xy = C$

■ **FIGURE 5.21**   The family of curves $xy = C$ and their orthogonal trajectories

Let P be a point where a particular curve of the form $xy = C$ intersects the orthogonal trajectory curve. At P, we have $x = X$ and $y = Y$, and the slope dY/dX of the orthogonal trajectory is the same as the negative reciprocal of the slope dy/dx of the

———

*We use uppercase letters for one kind of curve and lowercase for the other to make it easier to tell which curve is being mentioned at each stage of the following discussion.

curve $xy = C$. Using implicit differentiation, we find

$$xy = C$$

$$x\frac{dy}{dx} + y = 0 \qquad \text{Product rule}$$

$$\frac{dy}{dx} = \frac{-y}{x}$$

Thus, at the point of intersection P, the slope $\frac{dY}{dX}$ of the orthogonal trajectory is

$$\frac{dY}{dX} = -\frac{1}{\dfrac{dy}{dx}} = -\frac{1}{\dfrac{-y}{x}} = \frac{x}{y} = \frac{X}{Y}$$

According to this equation, the coordinates (X, Y) of the orthogonal trajectory curve satisfy the separable differential equation

$$\frac{dY}{dX} = \frac{X}{Y}$$

discussed in Example 4. Using the result of that example, we see that the orthogonal trajectories of the family $xy = C$ are the curves in the family

$$X^2 - Y^2 = K$$

where K is a constant. The given family of curves $xy = C$ and the family of orthogonal trajectory curves $X^2 - Y^2 = K$ are shown in Figure 5.21. ∎

MODELING FLUID FLOW THROUGH AN ORIFICE

Consider a tank that is filled with a fluid being slowly drained through a small, sharp-edged hole in its base, as shown in Figure 5.22.

By using a principle of physics known as Torricelli's law,* we can show that the rate of discharge dV/dt at time t is proportional to the square root of the depth h at that time. Specifically, if all dimensions are given in terms of feet, the drain hole has area A_0, and the height above the hole is h at time t (seconds), then

$$\frac{dV}{dt} = -4.8\,A_0\sqrt{h}$$

is the rate of flow of water in cubic feet per second. This formula is used in Example 8.

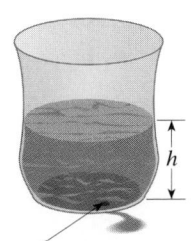

Hole has area A_0 and the rate of discharge is $\frac{dV}{dt}$.

■ **FIGURE 5.22** The flow of a fluid through an orifce

> **EXAMPLE 8**   *Fluid flow through an orifice*

A cylindrical tank is filled with a liquid that is draining through a small circular hole in its base. If the tank is 9 ft high with radius 4 ft, and the drain hole has radius 1 in., how long does it take for the tank to empty?

_____

*Torricelli's law says that the stream of liquid through the orifice has velocity $v = \sqrt{2gh}$, where $g = 32$ ft/s^2 is the acceleration due to gravity and h is the height of the liquid above the orifice. The factor 4.8 that appears in the rate of flow equation is required to compensate for the effect of friction.

Solution   Because the drain hole is a circle of radius $\frac{1}{12}$ ft ($= 1$ in.), its area is $\pi r^2 = \pi\left(\frac{1}{12}\right)^2$, and the rate of flow according to Torricelli's law is

$$\frac{dV}{dt} = -4.8\left(\frac{\pi}{144}\right)\sqrt{h}$$

Because the tank is cylindrical, the amount of fluid in the tank at any particular time will form a cylinder of radius 4 ft and height h. Also, because we are using $g = 32$ ft/s^2, it is implied that the time, t, is measured in seconds. The volume of such a liquid cylinder is

$$V = \pi r^2 h = \pi(4)^2 h = 16\pi h$$

and by differentiating both sides of this equation with respect to t, we obtain

$$\frac{dV}{dt} = 16\pi \frac{dh}{dt}$$

$$-4.8\left(\frac{\pi}{144}\right)\sqrt{h} = 16\pi \frac{dh}{dt} \qquad \text{Tarricelli's law}$$

$$\frac{dh}{dt} = \frac{-4.8\sqrt{h}}{144(16)} \approx -0.0021\sqrt{h}$$

$$\frac{1}{\sqrt{h}}\, dh = -0.0021\, dt \qquad \text{Separate the variables.}$$

$$\int h^{-1/2}\, dh = \int (-0.0021)\, dt \qquad \text{Integrate both sides.}$$

$$2h^{1/2} + C_1 = -0.0021t + C_2$$

$$2\sqrt{h} = -0.0021t + C$$

To evaluate C, recall that the tank is full at time $t = 0$. Thus, $h = 9$ when $t = 0$, so that

$$2\sqrt{9} = -0.0021(0) + C \quad \text{so that} \quad C = 6$$

Thus, the general formula is

$$2\sqrt{h} = -0.0021t + 6$$

where time t is in seconds (because g is 32 ft/sec^2).

Now we can find the depth of the fluid at any given time or the time at which a prescribed depth occurs. In particular, the tank is empty at the time t_e when $h = 0$. By substituting $h = 0$ into this formula, we find

$$2\sqrt{0} = -0.0021t_e + 6 \quad \text{so that} \quad t_e \approx 2,880 \text{ seconds}$$

Thus, roughly 48 min are required to drain the tank.   ■

MODELING THE MOTION OF A PROJECTILE: ESCAPE VELOCITY

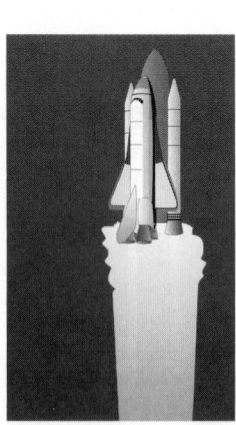

On April 12, 1981, the United States launched the world's first space shuttle, *Columbia*.

Consider a projectile that is launched with initial velocity v_0 from a planet's surface along a direct line through the center of the planet, as shown in Figure 5.23.

We shall find a general formula for the velocity of the projectile and the minimal value of v_0 required to guarantee that the projectile will escape the planet's gravitational attraction.

We assume that the only force acting on the projectile is that due to gravity, although in practice, factors such as air resistance must also be considered. With this assumption, Newton's law of gravitation can be used to show that when the projectile

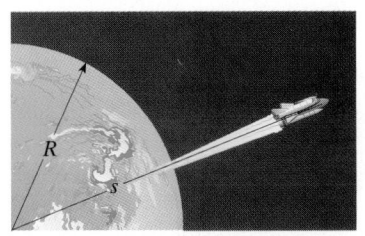

FIGURE 5.23 A projectile launched from the surface of a planet

is at a distance s from the center of the planet, its acceleration is given by the formula

$$a = \frac{-gR^2}{s^2}$$

where R is the radius of the planet and g is the acceleration due to gravity at the planet's surface (see Problem 68).*

Our first goal is to express the velocity v of the projectile in terms of the distance s. Because the projectile travels along a straight line, we know that

$$a = \frac{dv}{dt} \quad \text{and} \quad v = \frac{ds}{dt}$$

and by applying the chain rule, we see that

$$a = \frac{dv}{dt} = \frac{dv}{ds} \cdot \frac{ds}{dt} = \frac{dv}{ds} v$$

Therefore, by substitution for a we have

$$\frac{dv}{ds} v = \frac{-gR^2}{s^2}$$
$$v \, dv = -gR^2 s^{-2} \, ds$$
$$\int v \, dv = \int -gR^2 s^{-2} \, ds$$
$$\tfrac{1}{2}v^2 + C_1 = gR^2 s^{-1} + C_2$$
$$v^2 = 2gR^2 s^{-1} + C$$

To evaluate the constant C, recall that the projectile was fired from the planet's surface with initial velocity v_0. Thus, $v = v_0$ when $s = R$, and by substitution:

$$v_0^2 = 2gR^2 R^{-1} + C \quad \text{which implies} \quad C = v_0^2 - 2gR$$

so

$$v^2 = 2gR^2 s^{-1} + v_0^2 - 2gR$$

Because the projectile is launched in a direction away from the center of the planet, we would expect it to keep moving in that direction until it stops. In other words, *the projectile will keep moving away from the planet until it reaches a point where $v = 0$*. Because $2gR^2 s^{-1} > 0$ for all $s > 0$, v^2 will always be positive if $v_0^2 - 2gR \geq 0$. On the other hand, if $v_0^2 - 2gR < 0$, then sooner or later v will become 0 and the projectile will eventually fall back to the surface of the planet.

Therefore, we conclude that the projectile will escape from the planet's gravitational attraction if $v_0^2 \geq 2gR$, that is, if $v_0 \geq \sqrt{2gR}$. For this reason, the minimum speed for which this can occur, namely,

$$v_0 = \sqrt{2gR}$$

is called the **escape velocity** of the planet. In particular, for the earth, $R = 3{,}956$ mi and $g = 32$ ft/s$^2 = 0.00606$ mi/s^2, and the escape velocity is

$$v_0 = \sqrt{2gR} \approx \sqrt{2(0.00606)(3{,}956)} \approx 6.924357$$

The escape velocity for the earth is 6.92 mi/s.

---

*According to Newton's law of gravitation, the force of gravity acting on a projectile of mass m has magnitude $F = mk/s^2$, where k is constant. If this is the only force acting on the projectile, then $F = ma$, and we have $ma = F = mk/s^2$. By canceling the m's on each side of the equation, we obtain $a = k/s^2$.

5.6   Problem Set

Ⓐ *Verify in Problems 1–8 that if y satisfies the prescribed relationship with x, then it will be a solution of the given differential equation.*

1. If $x^2 + y^2 = 7$, then $\dfrac{dy}{dx} = -\dfrac{x}{y}$.

2. If $5x^2 - 2y^2 = 3$, then $\dfrac{dy}{dx} = \dfrac{5x}{2y}$.

3. If $xy = C$, then $\dfrac{dy}{dx} = \dfrac{-y}{x}$.

4. If $x^2 - 3xy + y^2 = 5$, then
$$(2x - 3y)dx + (2y - 3x)dy = 0$$

5. If $y = \sin(Ax + B)$, then
$$\dfrac{d^2y}{dx^2} + A^2y = 0$$

6. If $y = \dfrac{x^4}{20} - \dfrac{A}{x} + B$, then
$$x\dfrac{d^2y}{dx^2} + 2\dfrac{dy}{dx} = x^3$$

7. If $y = 2e^{-x} + 3e^{2x}$, then
$$y'' - y' - 2y = 0$$

8. If $y = Ae^x + Be^x \ln x$, then
$$xy'' + (1 - 2x)y' - (1 - x)y = 0$$

Find the particular solution of the first-order linear differential equations in Problems 9–14. A graphical solution within its direction field is shown.

9. $\dfrac{dy}{dx} = -\dfrac{x}{y}$
passing through $(2, 2)$

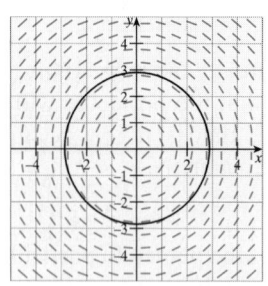

10. $\dfrac{dy}{dx} - y = 10$
passing through $(-2, -9)$

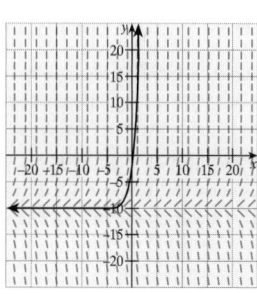

11. $\dfrac{dy}{dx} - y^2 = 1$
passing through $(\pi, 1)$

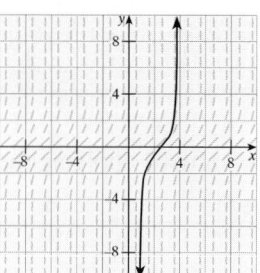

12. $\dfrac{dy}{dx} = e^{x+y}$
passing through $(0, 0)$

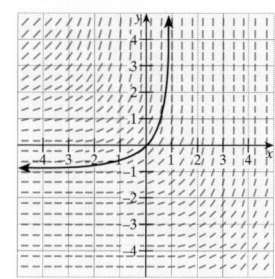

13. $\dfrac{dy}{dx} = \sqrt{\dfrac{x}{y}}$
passing through $(4, 1)$

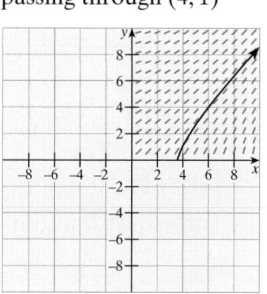

14. $\dfrac{dy}{dx} = y^2\sqrt{x}$
passing through $(9, -\tfrac{1}{18})$

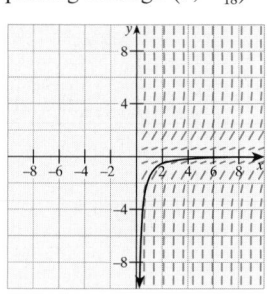

In Problems 15–20, draw the indicated particular solution for the differential equations whose direction fields are given.

15. $(0, 1)$

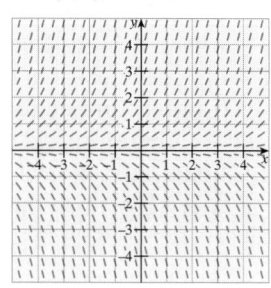

16. $(0, 1)$

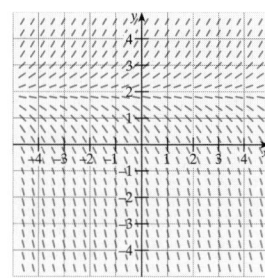

17. $(0, 0)$

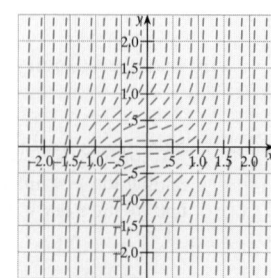

18. $(3, 3)$

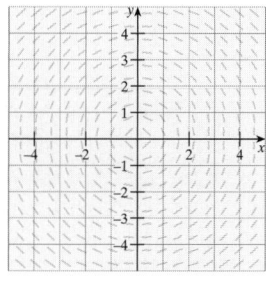

19. $(1, 0)$ **20.** $(0, 0)$

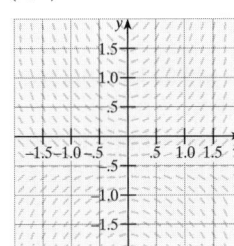   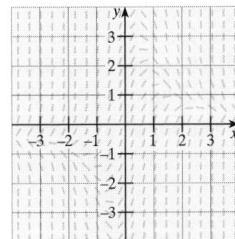

Find the general solution of the separable differential equations given in Problems 21–28.

21. $\dfrac{dy}{dx} = 3xy$

22. $\dfrac{dy}{dx} = \sqrt{\dfrac{y}{x}}$

23. $\dfrac{dy}{dx} = \dfrac{x}{y}\sqrt{1 - x^2}$

24. $\dfrac{dy}{dx} = (y - 4)^2$

25. $xy \, dx + \sqrt{xy} \, dy = 0$

26. $\dfrac{dy}{dx} = \dfrac{y}{\sqrt{1 - 2x^2}}$

27. $\dfrac{dy}{dx} = \dfrac{\sin x}{\cos y}$

28. $x^2 \, dy + \sec y \, dx = 0$

In Problems 29–32, find the general solution of the given differential equation by using either the product or the quotient rule.

29. $x \, dy + y \, dx = 0$

30. $\dfrac{x \, dy - y \, dx}{x^2} = 0$

31. $y \, dx = x \, dy,$ $x > 0, y > 0$

32. $x^2 y \, dy + xy^2 \, dx = 0$

B *Find the orthogonal trajectories of the family of curves given in Problems 33–41. In each case, sketch several members of the given family of curves and several members of the family of orthogonal trajectories on the same coordinate axes.*

33. the lines $2x - 3y = C$

34. the lines $y = x + C$

35. the curves $y = x^3 + C$

36. the curves $y = x^4 + C$

37. the curves $xy^2 = C$

38. the parabolas $y^2 = 4kx$

39. the curves $y^2 = Cx^3$

40. the circles $x^2 + y^2 = r^2$

41. the exponential curves $y = Ce^{-x}$

MODELING PROBLEMS *Write a differential equation to model the situation given in each of Problems 42–47. Do not solve.*

42. The number of bacteria in a culture grows at a rate that is proportional to the number present.

43. A sample of radium decays at a rate that is proportional to the amount of radium present in the sample.

44. The rate at which the temperature of an object changes is proportional to the difference between its own temperature and the temperature of the surrounding medium.

45. When a person is asked to recall a set of facts, the rate at which the facts are recalled is proportional to the number of relevant facts in the person's memory that have not yet been recalled.

46. The rate at which an epidemic spreads through a community of P susceptible people is proportional to the product of the number of people who have caught the disease and the number who have not.

47. The rate at which people are implicated in a government scandal is proportional to the product of the number of people already implicated and the number of people involved who have not yet been implicated.

48. What do you think the orthogonal trajectories of the family of curves

$$x^2 - y^2 = C$$

will be? Verify your conjecture. *Hint:* Take another look at Example 7.

49. The following is a list of six families of curves. Sketch several members of each family and then determine which pairs are orthogonal trajectories of one other.
 a. the circles $x^2 + y^2 = A$
 b. the ellipses $2x^2 + y^2 = B^2$
 c. the ellipses $x^2 + 2y^2 = C$
 d. the lines $y = Cx$
 e. the parabolas $y^2 = Cx$
 f. the parabolas $y = Cx^2$

50. The Dead Sea Scrolls were written on parchment at about 100 B.C. What percentage of ^{14}C originally contained in the parchment remained when the scrolls were discovered in 1947?

51. Tests of an artifact discovered at the Debert site in Nova Scotia show that 28% of the original ^{14}C is still present. What is the probable age of the artifact?

52. **HISTORICAL QUEST**

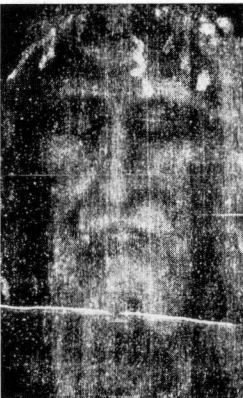

The Shroud of Turin is a rectangular linen cloth kept in the Chapel of the Holy Shroud in the cathedral of St. John the Baptist in Turin, Italy. It shows the image of a man whose wounds correspond with the biblical accounts of the crucifixion.

   In 1389, Pierre d'Arcis, the Bishop of Troyes, wrote a memo to the Pope, accusing a colleague of passing off "a certain cloth, cunningly painted" as the burial shroud of Jesus Christ. Despite this early testimony of forgery, this so-called Shroud of Turin has survived as a famous relic. In 1988, a small sample of the Shroud of Turin was taken and scientists from Oxford University, the University of Arizona, and the Swiss Federal Institute of Technology were permitted to test it. Suppose the cloth contained 90.7% of the original amount of carbon.

   According to this information, how old is the Shroud?

53. A cylindrical tank of radius 3 ft is filled with water to a depth of 5 ft. Determine how long (to the nearest minute) it takes to drain the tank through a sharp-edged circular hole in the bottom with radius 2 in.

54. Rework Problem 53 for a tank with a sharp-edged drain hole that is square with side of length 1.5 in.

55. EXPERIMENT   A rectangular tank has a square base 2 ft on a side that is filled with water to a depth of 4 ft. It is being drained from the bottom of the tank through a sharp-edged square hole that is 2 in. on a side.
    a. Show that at time t, the depth h satisfies the differential equation
    $$\frac{dh}{dt} = -\frac{1}{30}\sqrt{h}$$
    b. How long will it take to empty the tank?
    c. Construct this tank and then drain it out of the 4 in.^{2} hole. Is the time that it takes consistent with your answer to part **b**?

56. A toy rocket is launched from the surface of the earth with initial velocity $v_0 = 150$ ft/s. (The radius of the earth is roughly 3,956 mi, and $g = 32$ ft/s^2.)
    a. Determine the velocity (to the nearest ft/s) of the rocket when it is first 200 feet above the ground. (Remember, this is not the same as 200 ft from the center of the earth.)
    b. What is s when $v = 0$? Determine the maximum height above the ground that is attained by the rocket.

57. Determine the escape velocity of each of the following heavenly bodies:
    a. moon ($R = 1,080$ mi; $g = 5.5$ ft/s^2)
    b. Mars ($R = 2,050$ mi; $g = 12$ ft/s^2)
    c. Venus ($R = 3,800$ mi; $g = 28$ ft/s^2)

58. Population statistics indicate that t years after 1990, a certain city was growing at a rate of approximately $1,500t^{-1/2}$ people per year. In 1994, the population of the city was 39,000.
    a. What was the population in 1990?
    b. If this pattern continues, how many people will be living in the city in the year 1999?

59. The radius of planet X is one-fourth that of planet Y, and the acceleration due to gravity at the surface of X is eight-ninths that at the surface of Y. If the escape velocity of planet X is 6 ft/s, what is the escape velocity of planet Y?

60. A survey indicates that the population of a certain town is growing in such a way that the rate of growth at time t is proportional to the square root of the population P at the time. If the population was 4,000 ten years ago and is observed to be 9,000 now, how long will it take before 16,000 people live in the town?

61. A scientist has discovered a radioactive substance that disintegrates in such a way that at time t, the rate of disintegration is proportional to the *square* of the amount present.
    a. If a 100-g sample of the substance dwindles to only 80 g in 1 day, how much will be left after 6 days?
    b. When will only 10 g be left?

62. The shape of a tank is such that when it is filled to a depth of h feet, it contains
    $$V = 9\pi h^3$$
    ft^3 of water. The tank is being drained through a sharp-edged circular hole of radius 1 in. If the tank is originally filled to a depth of 4 ft, how long does it take for the tank to empty?

63. A rectangular tank has a square base 4 ft on a side and is 10 ft high. Originally, it was filled with water to a depth of 6 feet, but now is being drained from the bottom of the tank through a sharp-edged square hole 1 in. on a side.
    a. Find an equation involving the rate dh/dt.
    b. How long will it take to drain the tank?

64. The radioactive substance neptunium-139 decays to 73.36% of its original amount after 24 hours. How long would it take for 43% of the original neptunium to be present? What is the half-life of neptunium-139?

65. A certain artifact is tested by carbon dating and found to contain 73% of its original carbon-14. As a cross-check, it is also dated using radium, and was found to contain 32% of the original amount. Assuming the dating procedures were accurate, what is the half-life of radium?

66. The radioactive isotope gallium-67 (symbol ^{67}Ga) used in the diagnosis of malignant tumors has a half-life of 46.5 hours. If we start with 100 mg of ^{67}Ga, what percent is lost between the 30th and 35th hours? Is this the same as the percent lost over any other 5-hour period?

67. MODELING PROBLEM   A projectile is launched from the surface of a planet whose radius is R and where the acceleration due to gravity at the surface is g.
    a. If the initial velocity v_0 of the projectile satisfies $v_0 < \sqrt{2gR}$, show that the maximum height above the surface of the planet reached by the projectile is
    $$h = \frac{v_0^2 R}{2gR - v_0^2}$$
    b. On a certain planet, it is known that $g = 25$ ft/s^2. A projectile is fired with an initial velocity of $v_0 = 2$ mi/sec and attains a maximum height of 450 mi. What is the radius of the planet?

68. A projectile is launched from the surface of a planet whose radius is R and where the constant acceleration due to gravity is g. According to Newton's law of gravitation, the force of gravity acting on a projectile of mass m has magnitude
    $$F = \frac{mk}{s^2}$$

where k is a constant. If this is the only force acting on the projectile, then $F = ma$ where a is the acceleration of the projectile. Show that

$$a = \frac{-gR^2}{s^2}$$

69. MODELING PROBLEM   In physics, it is shown that the amount of heat Q (calories) that flows through an object by conduction will satisfy the differential equation

$$\frac{dQ}{dt} = -kA\,\frac{dT}{ds}$$

where t (seconds) is the time of flow, k is a physical constant (the *thermal conductivity* of the object), A is the surface area of the object measured at right angles to the direction of flow, and T is the temperature at a point s centimeters within the object, measured in the direction of the flow, as shown in Figure 5.24.

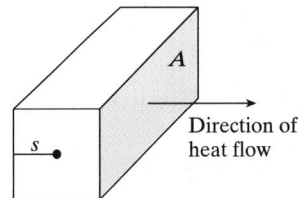

■ **FIGURE 5.24**   Heat conduction through an object

Under certain conditions (equilibrium), the rate of heat flow dQ/dt will be constant. Assuming these conditions to exist, find the number of calories that will flow each second across the face of a square pane of glass 2 cm thick and 50 cm on a side if the temperature on one side of the pane is 5° and on the other side is 60°. The thermal conductivity of glass is approximately $k = 0.0025$.

5.7 The Mean Value Theorem for Integrals; Average Value

IN THIS SECTION   mean value theorem for integrals, modeling average value of a function   ■

MEAN VALUE THEOREM FOR INTEGRALS

In Section 4.2 we established a very useful theoretical tool called the mean value theorem, which said that under reasonable conditions, there is at least one number c on the interval $[a, b]$ such that the value of the derivative at $x = c$ is exactly the same as

$$\frac{f(b) - f(a)}{b - a}$$

The mean value theorem for integrals is similar, and in the special case where $f(x) \geq 0$ it has a geometric interpretation that makes the theorem easy to understand. In particular, the theorem says that it is possible to find at least one number c on the interval $[a, b]$ such that the area of the rectangle with height $f(c)$ and width $(b - a)$ has exactly the same area as the region under the curve $y = f(x)$ on $[a, b]$. This is illustrated in Figure 5.25.

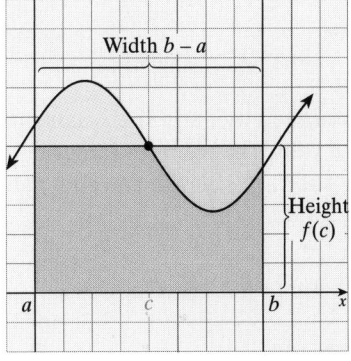

The shaded rectangle has the same area as the region under the curve $y = f(x)$ on $[a, b]$.

■ **FIGURE 5.25**   Geometric interpretation of the mean value theorem for integrals

---

THEOREM 5.11   *Mean value theorem for integrals*

If f is continuous on the interval $[a, b]$, there is at least one number c between a and b such that

$$\int_a^b f(x)\, dx = f(c)(b - a)$$

Proof   Suppose M and m are the largest and smallest values of f, respectively, on $[a, b]$. This means that

$$m \le \quad f(x) \quad \le M \quad \text{when } a \le x \le b$$

$$\int_a^b m\, dx \le \quad \int_a^b f(x)\, dx \quad \le \int_a^b M\, dx \qquad \text{Dominance rule}$$

$$m(b - a) \le \quad \int_a^b f(x)\, dx \quad \le M(b - a)$$

$$m \le \frac{1}{b - a}\int_a^b f(x)\, dx \le M$$

Because f is continuous on the closed interval $[a, b]$ and because the number

$$I = \frac{1}{b - a}\int_a^b f(x)\, dx$$

lies between m and M, the intermediate value theorem says there exists a number c between a and b for which $f(c) = I$; that is,

$$\frac{1}{b - a}\int_a^b f(x)\, dx = f(c)$$

$$\int_a^b f(x)\, dx = f(c)(b - a)$$

The mean value theorem for integrals does not specify how to determine c. It simply guarantees the existence of at least one number c in the interval. However, Example 1 shows how to find a value of c guaranteed by this theorem.

EXAMPLE 1   *Finding c in the mean value theorem for integrals*

Find a value of c guaranteed by the mean value theorem for integrals for $f(x) = \sin x$ on $[0, \pi]$.

Solution

$$\int_0^\pi \sin x\, dx = -\cos x \Big|_0^\pi = -\cos \pi + \cos 0 = -(-1) + 1 = 2$$

The region bounded by f and the x-axis on $[0, \pi]$ is shaded in Figure 5.26.

The mean value theorem for integrals asserts the existence of a number c on $[0, \pi]$ such that $f(c)(b - a) = 2$. We can solve an equation to find this value:

$$f(c)(b - a) = 2$$

$$(\sin c)(\pi - 0) = 2$$

$$\sin c = \frac{2}{\pi}$$

$$c \approx 0.690107 \quad \text{or} \quad 2.451486$$

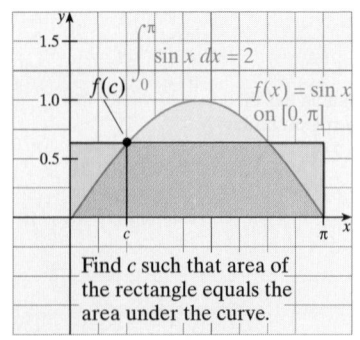

■ FIGURE 5.26   Graph of $f(x) = \sin x$ illustrating the mean value theorem on $[0, \pi]$

Because each choice of c is between 0 and π, we have found the value c guaranteed by the mean value theorem for integrals. ∎

MODELING AVERAGE VALUE OF A FUNCTION

There are many practical situations in which one is interested in the *average value* of a continuous function on an interval, such as the average level of air pollution over a 24-hour period, the average speed of a truck during a 3-hour trip, or the average productivity of a worker during a production run.

You probably know that the average value of n numbers $x_1, x_2, \ldots, x_n$ is

$$\frac{x_1 + x_2 + \cdots + x_n}{n}$$

but what if there are infinitely many numbers? Specifically, what is the average value of $f(x)$ on the interval $a \leq x \leq b$? To see how the definition of finite average value can be used, imagine that the interval $[a, b]$ is divided into n equal subintervals, each of width

$$\Delta x = \frac{b - a}{n}$$

Then for $k = 1, 2, \ldots, n$, let x_k^* be a number chosen arbitrarily from the kth subinterval. Then the average value V of f on $[a, b]$ is estimated by the weighted sum

$$S_n = \frac{f(x_1^*) + f(x_2^*) + \cdots + f(x_n^*)}{n} = \frac{1}{n}\sum_{k=1}^{n} f(x_k^*)$$

Because $\Delta x = \dfrac{b - a}{n}$, we know that $\dfrac{1}{n} = \dfrac{1}{b - a}\Delta x$ and

$$S_n = \frac{1}{n}\sum_{k=1}^{n} f(x_k^*) = \left[\frac{1}{b-a}\Delta x\right]\sum_{k=1}^{n} f(x_k^*) = \frac{1}{b-a}\sum_{k=1}^{n} f(x_k^*)\Delta x$$

The sum on the right is a Riemann sum with norm $\|P\| = \dfrac{b - a}{n}$. It is reasonable to expect the estimating average S_n to approach the "true" average value A of $f(x)$ on $[a, b]$ as $n \to +\infty$. Thus, we model average value by

$$A = \lim_{n\to\infty} \frac{1}{b - a}\sum_{k=1}^{n} f(x_k^*)\,\Delta x = \frac{1}{b - a}\int_a^b f(x)\,dx$$

We use this integral as a definition of average value.

Average Value

If f is continuous on the interval $[a, b]$, the **average value** of f on this interval is given by the integral

$$\frac{1}{b - a}\int_a^b f(x)\,dx$$

EXAMPLE 2   *Modeling Problem: Average speed of traffic*

Suppose a study suggests that between the hours of 1:00 P.M. and 4:00 P.M. on a normal weekday, the speed of the traffic at a certain expressway exit is modeled by the formula

$$S(t) = 2t^3 - 21t^2 + 60t + 20$$

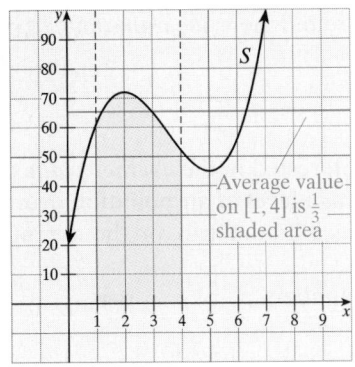

Average value on $[1, 4]$ is $\frac{1}{3}$ shaded area

kilometers per hour, where t is the number of hours past noon. Compute the average speed of the traffic between the hours of 1:00 P.M. and 4:00 P.M.

Solution   Our goal is to find the average value of $S(t)$ on the interval $[1, 4]$. The average speed is

$$\frac{1}{4-1}\int_1^4 (2t^3 - 21t^2 + 60t + 20)\,dt = \tfrac{1}{3}[\tfrac{1}{2}t^4 - 7t^3 + 30t^2 + 20t]_1^4$$

$$= \tfrac{1}{3}(240 - 43.5) = 65.5 \qquad \blacksquare$$

Note that the mean value theorem for integrals can be interpreted as saying that the average value of $f(x)$ on $[a, b]$ equals the value of f for some number c between a and b. That is, *a function must assume its average value at least once on any closed, bounded interval $[a, b]$ where it is continuous.*

Actually, this is quite reasonable, because the intermediate value theorem of continuous functions assures us that a continuous function f assumes every value between its maximum M and minimum m, and we would expect the average value V to be between these two extremes. Example 3 illustrates one way of using these ideas.

EXAMPLE 3   *Modeling Problem: Average temperature*

Suppose that x hours after midnight, the temperature (in degrees Celsius) in Minneapolis one night can be modeled by the formula

$$T(x) = 2 - \tfrac{1}{7}(x - 13)^2$$

Find the average temperature between 2:00 A.M. and 2:00 P.M. and a time when the average temperature actually occurs.

Solution   We wish to find the average temperature T on the interval $[2, 14]$ (because 2 P.M. is 14 h after midnight). The average value is

$$T = \frac{1}{14-2}\int_2^{14}[2 - \tfrac{1}{7}(x - 13)^2]\,dx = \tfrac{1}{12}[2x - \tfrac{1}{7}\cdot\tfrac{1}{3}(x-13)^3]\Big|_2^{14}$$

$$= \tfrac{1}{12}\Big[\tfrac{587}{21} - \tfrac{1,415}{21}\Big] \approx -3.2857143$$

Thus, the average temperature on the given time period is approximately 3.286 °C below zero. To determine when this temperature actually occurs, solve the equation

AVERAGE TEMPERATURE = TEMPERATURE AT TIME x

$$-3.2857143 = 2 - \tfrac{1}{7}(x - 13)^2$$

$$37 = (x - 13)^2$$

$$x = 13 \pm \sqrt{37} \approx 19.082763 \quad \text{or} \quad 6.9172375$$

The first value is to be rejected because it is not in the interval $[2, 14]$, so we find that the average temperature occurs 6.917 h after midnight, or at approximately 6:55 A.M. $\qquad \blacksquare$

EXAMPLE 4   *Modeling Problem: Average population*

The logistic formula

$$P(t) = \frac{202.31}{1 + e^{3.938 - 0.314t}}$$

was developed by the United States Bureau of the Census to represent the population of the United States (in millions) during the period 1790–1990. Time t in the formula is the number of decades after 1790. Thus, $t = 0$ for 1790, $t = 20$ for 1990. Use this formula to compute the average population of the United States between 1790 and 1990. When did the average population actually occur?

Solution The average population is given by the integral

$$A = \frac{1}{20 - 0} \int_0^{20} \frac{202.31 \, dt}{1 + e^{3.938 - 0.314t}} \qquad \text{Multiply by } \frac{e^{0.314t}}{e^{0.314t}}.$$

$$= \frac{1}{20} \int_0^{20} \frac{202.31 \, e^{0.314t}}{e^{0.314t} + e^{3.938}} \, dt \qquad \begin{array}{l} \text{Let } u = e^{0.314t} + e^{3.938}; \\ du = 0.314 e^{0.314t} \, dt \end{array}$$

$$= \frac{1}{20} \left(\frac{202.31}{0.314} \right) \int_{t=0}^{t=20} \frac{du}{u}$$

$$= \frac{1}{20} \left(\frac{202.31}{0.314} \right) \ln \left| e^{0.314t} + e^{3.938} \right| \Big|_0^{20} \approx 77.7827445296$$

To find when the average population of 77.7827 million actually occurred, solve $P(t) = 77.7827$:

$$\frac{202.31}{1 + e^{3.938 - 0.314t}} = 77.7827$$

$$e^{3.938 - 0.314t} = \frac{202.31}{77.7827} - 1$$

$$3.938 - 0.314t = \ln\left(\frac{202.31}{77.7827} - 1 \right)$$

$$t \approx 11.043$$

The average population occurred 11 decades after 1790, roughly in the year 1900. ∎

5.7 Problem Set

Ⓐ *In Problems 1–10, find c such that*

$$\int_a^b f(x) \, dx = f(c)(b - a)$$

as guaranteed by the mean value theorem for integrals. If you cannot find such a value, explain why the theorem does not apply.

1. $f(x) = 4x^3$ on $[1, 2]$
2. $f(x) = x^2 + 4x + 1$ on $[0, 2]$
3. $f(x) = 15x^{-2}$ on $[1, 5]$
4. $f(x) = 12x^{-3}$ on $(0, 3)$
5. $f(x) = 2 \csc^2 x$ on $[-\frac{\pi}{3}, \frac{\pi}{3}]$

6. $f(x) = \cos x$ on $[-\frac{\pi}{2}, \frac{\pi}{2}]$
7. $f(x) = e^{2x}$ on $[-\frac{1}{2}, \frac{1}{2}]$
8. $f(x) = \dfrac{x}{1 + x}$ on $[0, 1]$
9. $f(x) = \dfrac{x + 1}{1 + x^2}$ on $[-1, 1]$
10. $f(x) = \tan x$ on $[0, 2]$

Determine the area of the indicated region in Problems 11–16 and then draw a rectangle with base $(b - a)$ and height $f(c)$ for some c on $[a, b]$ so that the area of the rectangle is equal to the area of the given region.

11. $y = \frac{1}{2}x$ on $[0, 10]$

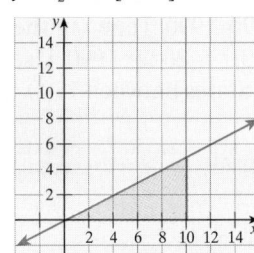

12. $y = x^2$ on $[0, 3]$

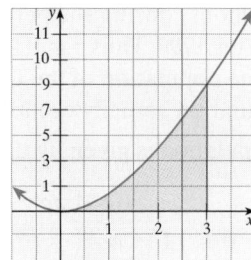

13. $y = x^2 + 2x + 3$ on $[0, 2]$

14. $y = \frac{1}{x^2}$ on $[0.5, 2]$

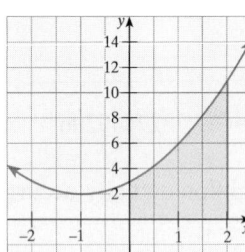

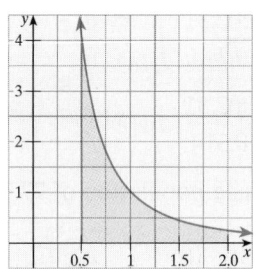

15. $y = \cos x$ on $[-1, 1.5]$

16. $y = x + \sin x$ on $[0.5, 2]$

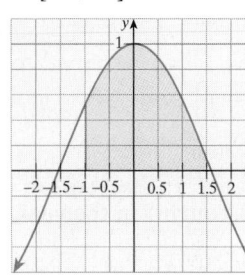

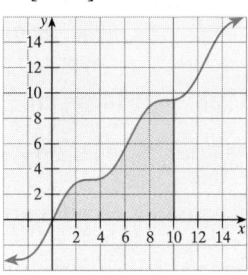

Find the average value of the function given in Problems 17–30 on the prescribed interval.

17. $f(x) = x^2 - x + 1$ on $[-1, 2]$

18. $f(x) = x^3 - 3x^2$ on $[-2, 1]$

19. $f(x) = e^x - e^{-x}$ on $[-1, 1]$

20. $f(x) = \dfrac{x}{2x + 3}$ on $[0, 1]$

21. $f(x) = \sin x$ on $[0, \frac{\pi}{4}]$

22. $f(x) = 2\sin x - \cos x$ on $[0, \frac{\pi}{2}]$

23. $f(x) = \sqrt{4 - x}$ on $[0, 4]$

24. $f(x) = \sqrt[3]{1 - x}$ on $[-7, 0]$

25. $f(x) = (2x - 3)^3$ on $[0, 1]$

26. $f(x) = x\sqrt{2x^2 + 7}$ on $[0, 1]$

27. $f(x) = x(x^2 + 1)^3$ on $[-2, 1]$

28. $f(x) = \dfrac{x}{\sqrt{x^2 + 1}}$ on $[0, 3]$

29. $f(x) = \sqrt{9 - x^2}$ on $[-3, 3]$

Hint: The integral can be evaluated as part of a circle.

30. $f(x) = \sqrt{2x - x^2}$ on $[0, 2]$

Hint: The integral can be evaluated as part of a circle.

Ⓑ 31. If an object is propelled upward from ground level with an initial velocity v_0, then its height at time t is given by

$$s = -\frac{1}{2}gt^2 + v_0 t$$

where g is the constant acceleration due to gravity. Show that between times t_0 and t_1, the average height of the object is

$$s = -\frac{1}{6}g[t_1^2 + t_1 t_0 + t_0^2] + \frac{1}{2}v_0(t_1 + t_0)$$

32. What is the average velocity for the object described in Problem 31 during the same time period?

33. Records indicate that t hours past midnight, the temperature at the local airport was

$$f(t) = -0.3t^2 + 4t + 10$$

degrees Celsius. What was the average temperature at the airport between 9:00 A.M. and noon?

34. Suppose a study indicates that t years from now, the level of carbon dioxide in the air of a certain city will be

$$L(t) = te^{-0.01t^2}$$

parts per million (ppm) for $0 \le t \le 20$.

a. What is the average level of carbon dioxide in the first 3 years?

b. At what time (or times) does the average level of carbon dioxide actually occur? Answer to the nearest month.

35. MODELING PROBLEM The number of bacteria (in thousands) present in a certain culture after t minutes is modeled by

$$Q(t) = \frac{2{,}000}{1 + 0.3e^{-0.276t}}$$

a. What was the average population during the *second* ten minutes ($10 \le t \le 20$)?

b. At what time during the period $10 \le t \le 20$ is the average population actually attained?

Ⓒ 36. Let $f(t)$ be a function that is continuous and satisfies $f(t) \ge 0$ on the interval $[0, \frac{\pi}{2}]$. Suppose it is known that for any number x between 0 and $\frac{\pi}{2}$, the region under the graph of f on $[0, x]$ has area $A(x) = \tan x$.

a. Explain why $\displaystyle\int_0^x f(t)\,dt = \tan x$ for $0 \le x \le \frac{\pi}{2}$.

b. Differentiate both sides of the equation in part **a** and deduce the identity of f.

37. THINK TANK PROBLEM Suppose that $f(t)$ is continuous for all t and that for any number x it is known that the average value of f on $[-1, x]$ is

$$A(x) = \sin x$$

Use this information to deduce the identity of f.

38. MODELING PROBLEM   In Example 7 of Section 4.7, we gave the Heinz function

$$f(t) = \frac{k}{b-a}(e^{-at} - e^{-bt}) \qquad t \ge 0$$

where $f(t)$ is the concentration of a drug in a person's bloodstream t hours after an intramuscular injection. The coefficients a and b ($b > a$) are characteristics of the drug and the patient's metabolism, and are called the *absorption* and *diffusion* rates, respectively.

a. Show that for each fixed t, $f(t)$ can be thought of as the average value of a function of the form

$$g(\lambda) = (At^2 + Bt + C)e^{-\lambda t}$$

over the interval $a \le \lambda \le b$. Find A, B, and C.
b. ■ **What Does This Say?** So what? In particular, can you see any value in the interpretation of the Heinz concentration function as an average value? Explain.

5.8 Numerical Integration: The Trapezoidal Rule and Simpson's Rule

IN THIS SECTION   approximation by rectangles, trapezoidal rule, Simpson's rule, error estimation   ■

The fundamental theorem of calculus can be used to evaluate an integral whenever an appropriate antiderivative is known. However, certain functions have no simple antiderivative. To find a definite integral of such a function, it is often necessary to use numerical approximation.

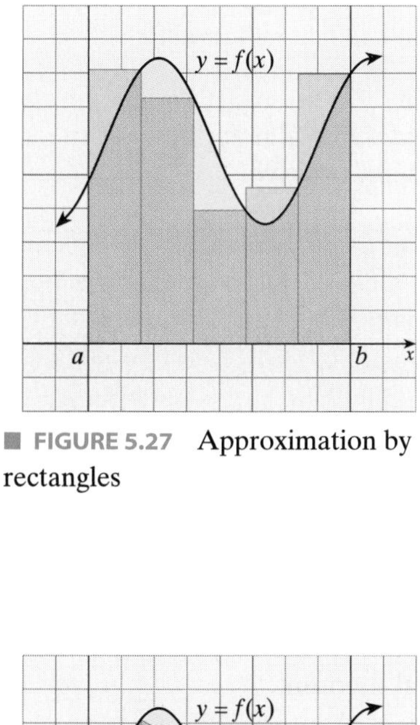

APPROXIMATION BY RECTANGLES

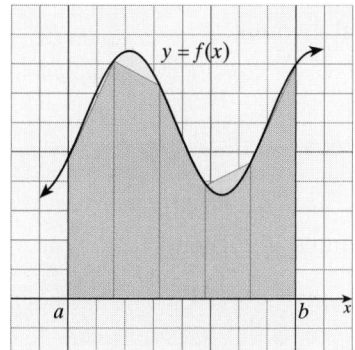

■ **FIGURE 5.27** Approximation by rectangles

If $f(x) \ge 0$ on the interval $[a, b]$, the definite integral $\int_a^b f(x)\,dx$ is equal to the area under the graph of f on $[a, b]$. As we saw in Section 5.2, one way to approximate this area is to use n rectangles, as shown in Figure 5.27. In particular, divide the interval $[a, b]$ into n subintervals, each of width $\Delta x = \dfrac{b-a}{n}$, and let x_k^* denote the right endpoint of the kth subinterval. The base of the kth rectangle is the kth subinterval, and its height is $f(x_k^*)$. Hence, the area of the kth rectangle is $f(x_k^*)\Delta x$. The sum of the areas of all n rectangles is an approximation for the area under the curve and hence an approximation for the corresponding definite integral. Thus,

$$\int_a^b f(x)\,dx \approx f(x_1^*)\Delta x + f(x_2^*)\Delta x + \cdots + f(x_n^*)\Delta x$$

This approximation improves as the number of rectangles increases, and we can estimate the integral to any desired degree of accuracy by taking n large enough. However, because fairly large values of n are usually required to achieve reasonable accuracy, approximation by rectangles is rarely used in practice.

TRAPEZOIDAL RULE

■ **FIGURE 5.28** Approximation by trapezoids

The accuracy of the approximation generally improves significantly if trapezoids are used instead of rectangles. Figure 5.28 shows the area from Figure 5.27 approximated by n trapezoids instead of rectangles. Even from these rough illustrations you can see how much better the approximation is in this case.

Suppose the interval $[a, b]$ is partitioned into n equal parts by the subdivision points $x_0, x_1, \ldots, x_n$, where $x_0 = a$ and $x_n = b$. The kth trapezoid is shown in greater detail in Figure 5.29.

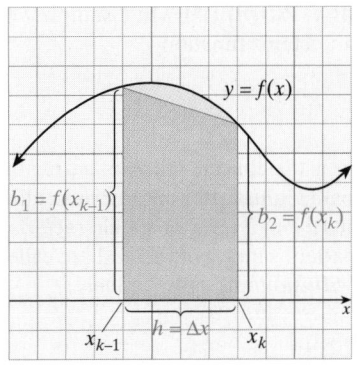

Recall the formula for the area of a trapezoid:

$$A = \frac{1}{2}(b_1 + b_2)h$$

■ **FIGURE 5.29**   The kth trapezoid has area $\frac{1}{2}[f(x_{k-1}) + f(x_k)]\Delta x$.

If we let T_n denote the sum of the areas of n trapezoids, we see that

$$T_n = \frac{1}{2}[f(x_0) + f(x_1)]\Delta x + \frac{1}{2}[f(x_1) + f(x_2)]\Delta x + \cdots + \frac{1}{2}[f(x_{n-1}) + f(x_n)]\Delta x$$
$$= \frac{1}{2}[f(x_0) + 2f(x_1) + 2f(x_2) + \cdots + 2f(x_{n-1}) + f(x_n)]\Delta x$$

The sum T_n estimates the total area under the curve $y = f(x)$ on the interval $[a, b]$ and hence also estimates the integral

$$\int_a^b f(x)\, dx$$

This approximation formula is known as the *trapezoidal rule* and applies as a means of approximating the integral, even if the function f is not positive.

Trapezoidal Rule

Let f be continuous on $[a, b]$. The **trapezoidal rule** is

$$\int_a^b f(x)\, dx \approx \frac{1}{2}[f(x_0) + 2f(x_1) + 2f(x_2) + \cdots + 2f(x_{n-1}) + f(x_n)]\Delta x$$

where $\Delta x = \dfrac{b-a}{n}$ and, for the kth subinterval, $x_k = a + k\Delta x$. Moreover, the larger the value for n, the better the approximation.

Our first example uses the trapezoidal rule to estimate the value of an integral that we can compute exactly by using the fundamental theorem.

EXAMPLE 1   *Trapezoidal rule approximation*

Use the trapezoidal rule with $n = 4$ to estimate $\displaystyle\int_{-1}^{2} x^2\, dx$.

Solution   The interval is $[a, b] = [-1, 2]$, so $a = -1$ and $b = 2$. Then, $\Delta x = \dfrac{2 - (-1)}{4} = \dfrac{3}{4} = 0.75$. Thus,

$$x_0 = a = -1 \qquad\qquad f(x_0) = f(-1) = (-1)^2 = 1$$
$$x_1 = a + 1 \cdot \Delta x = -1 + \tfrac{3}{4} = -\tfrac{1}{4} \qquad\qquad 2f(x_1) = 2(-\tfrac{1}{4})^2 = \tfrac{1}{8} = 0.125$$
$$x_2 = a + 2 \cdot \Delta x = -1 + \tfrac{6}{4} = \tfrac{1}{2} \qquad\qquad 2f(x_2) = 2(\tfrac{1}{2})^2 = \tfrac{1}{2} = 0.5$$

$$x_3 = a + 3 \cdot \Delta x = -1 + \tfrac{9}{4} = \tfrac{5}{4} \qquad\qquad 2f(x_3) = 2(\tfrac{5}{4})^2 = \tfrac{25}{8} = 3.125$$

$$x_4 = a + 4 \cdot \Delta x = b = 2 \qquad\qquad\qquad f(x_4) = 2^2 = 4$$

$$T_4 = \tfrac{1}{2}[1 + 0.125 + 0.5 + 3.125 + 4](0.75) = 3.28125 \qquad\qquad \blacksquare$$

The exact value of the integral in Example 1 is

$$\int_{-1}^{2} x^2 \, dx = \frac{x^3}{3}\bigg|_{-1}^{2} = \frac{8}{3} - \frac{-1}{3} = 3$$

Therefore, the trapezoidal estimate T_4 involves an error, which we denote by E_4. We find that

$$E_4 = \int_{-1}^{2} x^2 \, dx - T_4 = 3 - 3.28125 = -0.28125$$

The negative sign indicates that the trapezoidal formula *overestimated* the true value of the integral in Example 1.

SIMPSON'S RULE

Roughly speaking, the accuracy of a procedure for estimating the area under a curve depends on how well the upper boundary of each approximating strip fits the shape of the given curve. Trapezoidal strips often result in a better approximation than rectangles, and it is reasonable to expect even greater accuracy to occur if the approximating strips have curved upper boundaries, as shown in Figure 5.30.

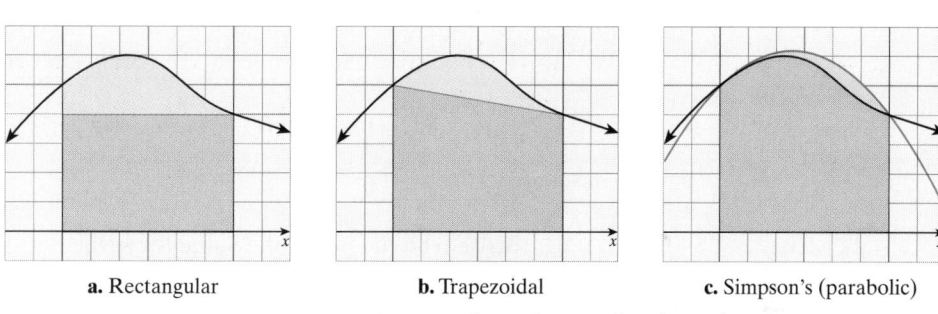

a. Rectangular   **b.** Trapezoidal   **c.** Simpson's (parabolic)

■ **FIGURE 5.30**   A comparison of approximating strips

The name given to the procedure in which the approximating strip has a parabolic arc for its upper boundary is called **Simpson's rule.***

We shall derive Simpson's rule as a means for approximating, as before, the area under the curve $y = f(x)$ on the interval $[a, b]$, where f is continuous and satisfies $f(x) \geq 0$. First, we partition the given interval into a number of equal subintervals, but this time, we require the number of subdivisions to be an *even* number (because this requirement will simplify the formula associated with the final result). If $x_0, x_1, \ldots, x_n$ are the subdivision points in our partition (with $x_0 = a$ and $x_n = b$), we pass a parabolic arc through the points, three at a time (the points with x-coordinates x_0, x_1, x_2, then those with x_2, x_3, x_4, and so on). It can be shown (see Problem 42) that the region under the parabolic curve $y = f(x)$ on the interval $[x_{2k-2}, x_{2k}]$ has area

$$\tfrac{1}{3}[f(x_{2k-2}) + 4f(x_{2k-1}) + f(x_{2k})]\Delta x$$

where $\Delta x = \dfrac{b - a}{n}$. This procedure is illustrated in Figure 5.31.

---

*This rule is named for Thomas Simpson (1710–1761), an English mathematician who, curiously, neither discovered nor made any special use of the formula that bears his name.

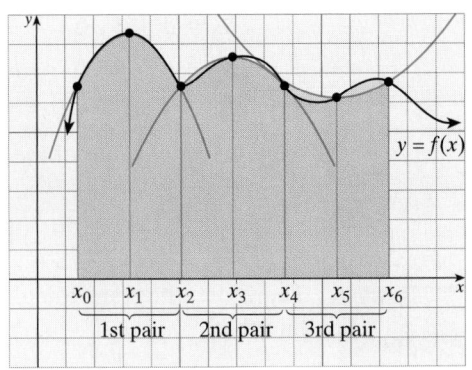

■ **FIGURE 5.31** Approximation using parabolas

By adding the area of the approximating parabolic strips and combining terms, we obtain the sum S_n of n parabolic regions:

$$S_n = \tfrac{1}{3}[f(x_0) + 4f(x_1) + f(x_2)]\Delta x + \tfrac{1}{3}[f(x_2) + 4f(x_3) + f(x_4)]\Delta x$$
$$+ \cdots + \tfrac{1}{3}[f(x_{n-2}) + 4f(x_{n-1}) + f(x_n)]\Delta x$$
$$= \tfrac{1}{3}[f(x_0) + 4f(x_1) + 2f(x_2) + 4f(x_3) + \cdots + 4f(x_{n-1}) + f(x_n)]\Delta x$$

These observations are summarized in the following box:

Simpson's Rule

Let f be continuous on $[a, b]$. The **Simpson's rule** is

$$\int_a^b f(x)\, dx \approx \tfrac{1}{3}[f(x_0) + 4f(x_1) + 2f(x_2) + \cdots + 4f(x_{n-1}) + f(x_n)]\Delta x$$

where $\Delta x = \dfrac{b-a}{n}, x_k = a + k\Delta x, k$ an integer, and n an even integer.
Moreover, the larger the value for n, the better the approximation.

Technology Window ▲ ▼

Computer output for Example 2 is shown:

Simpson's rule to calculate estimate.

Type of estimate	# of sub-intervals	Estimate over [1, 2]
Simpson	10	0.6931502
Trapezoid	10	0.6937714
Trapezoid	100	0.6931534
Simpson	100	0.6931472
Simpson	1000	0.6931472

EXAMPLE 2 *Approximation by Simpson's rule*

Use Simpson's rule with $n = 10$ to approximate $\displaystyle\int_1^2 \frac{dx}{x}$.

Solution We have $\Delta x = \dfrac{2-1}{10} = 0.1$, and $x_0 = a = 1, x_1 = 1.1, x_2 = 1.2, \ldots,$ $x_9 = 1.9, x_{10} = 2$. Then

$$\int_1^2 \frac{1}{x}\, dx \approx S_{10}$$

$$= \frac{1}{3}\left(\frac{1}{1} + \frac{4}{1.1} + \frac{2}{1.2} + \frac{4}{1.3} + \frac{2}{1.4} + \frac{4}{1.5} + \frac{2}{1.6} + \frac{4}{1.7} + \frac{2}{1.8} + \frac{4}{1.9} + \frac{1}{2}\right)(0.1)$$

$$\approx 0.6931502 \qquad\blacksquare$$

ERROR ESTIMATION

The difference between the value of an integral and its estimated value is called its **error.** Since this error is a function of n we denote it by E_n.

THEOREM 5.12   *Error in the trapezoidal rule and Simpson's rule*

If f has a continuous second derivative on $[a, b]$, then the error E_n in approximating $\int_a^b f(x)\, dx$ by the trapezoidal rule satisfies

Trapezoidal error: $|E_n| \leq \dfrac{(b-a)^3}{12n^2} M$ where M is the maximum value of $|f''(x)|$ on $[a, b]$.

Moreover, if f has a continuous fourth derivative on $[a, b]$, then the error E_n (n even) in approximating $\int_a^b f(x)\, dx$ by Simpson's rule satisfies

Simpson's error: $|E_n| \leq \dfrac{(b-a)^5}{180n^4} K$ where K is the maximum value of $|f^{(4)}(x)|$ on $[a, b]$.

Proof   The proofs of these error estimates are beyond the scope of this book and can be found in many advanced calculus textbooks and in most numerical analysis books.

EXAMPLE 3   *Estimate of error when using Simpson's rule*

Estimate the accuracy of the approximation of $\displaystyle\int_1^2 \frac{dx}{x}$ by Simpson's rule with $n = 10$ in Example 2.

Solution   If $f(x) = \dfrac{1}{x}$, we find that $f^{(4)}(x) = 24x^{-5}$. The maximum value of this function will occur at a critical number (there are none on $[1, 2]$) or at an endpoint. Thus, the largest value of $|f^{(4)}(x)|$ on $[1, 2]$ is $|f^{(4)}(1)| = 24$. Now, apply the error formula with $K = 24$, $a = 1$, $b = 2$, and $n = 10$ to obtain

$$|E_n| \leq \frac{K(b-a)^5}{180n^4} = \frac{24(2-1)^5}{180(10)^4} \approx 0.0000133$$

That is, the error in the approximation in Example 2 is guaranteed to be no greater than 0.0000133.   ■

With the aid of the error estimates, we can decide in advance how many subintervals to use to achieve a desired degree of accuracy.

EXAMPLE 4   *Choosing the number of subintervals to guarantee given accuracy*

How many subintervals are required to guarantee that the error will be less than 0.00005 in the approximation of

$$\int_1^2 \frac{dx}{x}$$

on $[1, 2]$ using the trapezoidal rule?

Solution   Because $f(x) = x^{-1}$, we have $f''(x) = 2x^{-3}$. On $[1, 2]$ the largest value of $|f''(x)|$ is $|f''(1)| = 2$, so $M = 2$, $a = 1$, $b = 2$, and

$$|E_n| \leq \frac{2(2-1)^3}{12n^2} = \frac{1}{6n^2}$$

The goal is to find the smallest positive integer n for which

$$\frac{1}{6n^2} < 0.00005$$

$$10,000 < 3n^2 \quad \text{Multiply by the positive number } 60,000n^2.$$

$$10,000 - 3n^2 < 0$$

$$(100 - \sqrt{3}n)(100 + \sqrt{3}n) < 0$$

$$n < -\frac{100}{\sqrt{3}} \quad \text{or} \quad n > \frac{100}{\sqrt{3}} \approx 57.735$$

The smallest positive integer that satisfies this condition is $n = 58$; therefore, 58 subintervals are required to ensure the desired accuracy. ∎

If f is the linear function $f(x) = Ax + B$, then $f''(x) = 0$ and we can take $M = 0$ as the error estimate. In this case, the error in applying the trapezoidal rule satisfies $|E_n| \le 0$. That is, the trapezoidal rule is *exact* for a linear function, which is what we would expect, because the region under a line on an interval is a trapezoid.

In discussing the accuracy of the trapezoidal rule as a means of estimating the value of the definite integral

$$I = \int_a^b f(x)\, dx$$

we have focused attention on the "error term," but this only measures the error that comes from estimating I by the trapezoidal or Simpson approximation sum. There are other kinds of error that must be considered in this or any other method of approximation. In particular, each time we cut off digits from a decimal, we incur what is known as a round-off error. For example, a round-off error occurs when we use 0.66666667 in place of $\frac{2}{3}$ or 3.1415927 for the number π. Round-off errors occur even in large computers and can accumulate to cause real problems. Specialized methods for dealing with these and other errors are studied in numerical analysis.

The examples of numerical integration examined in this section are intended as illustrations and thus involve relatively simple computations, whereas in practice, such computations often involve hundreds of terms and can be quite tedious. Fortunately, these computations are extremely well suited to automatic computing, and the reader who is interested in pursuing computer methods in numerical integration will find an introduction to this topic in the computer supplement.

5.8 Problem Set

A *Approximate the integrals in Problems 1–2 using the trapezoidal rule and Simpson's rule with the specified number of subintervals and then compare your answers with the exact value of the definite integral.*

1. $\int_1^2 x^2\, dx$ with $n = 4$ 2. $\int_0^4 \sqrt{x}\, dx$ with $n = 6$

Approximate the integrals given in Problems 3–8 with the specified number of subintervals using:
a. *the trapezoidal rule;* **b.** *Simpson's rule.*

3. $\int_0^1 \frac{dx}{1 + x^2}$ with $n = 4$ 4. $\int_{-1}^0 \sqrt{1 + x^2}\, dx$ with $n = 4$

5. $\int_2^4 \sqrt{1 + \sin x}\, dx$ with $n = 4$

6. $\int_0^2 x \cos x\, dx$ with $n = 6$

7. $\int_0^2 x\, e^{-x}\, dx$ with $n = 6$ 8. $\int_1^2 \ln x\, dx$ with $n = 6$

Ⓑ **9.** ■ **What Does This Say?** Describe the trapezoidal rule.

10. ■ **What Does This Say?** Describe Simpson's rule.

Estimate the value of the integrals in Problems 11–18 to within the prescribed accuracy.

11. $\displaystyle\int_0^1 \frac{dx}{x^2 + 1}$ with error less than 0.05

12. $\displaystyle\int_{-1}^2 \sqrt{1 + x^2}\, dx$ with error less than 0.05

13. $\displaystyle\int_0^1 \cos 2x\, dx$ accurate to three decimal places

14. $\displaystyle\int_1^2 x^{-1}\, dx$ accurate to three decimal places

15. $\displaystyle\int_0^2 x\sqrt{4 - x}\, dx$ with error less than 0.01

16. $\displaystyle\int_0^\pi \theta \cos^2 \theta\, d\theta$ with error less than 0.01

17. $\displaystyle\int_0^1 \tan^{-1} x\, dx$ with error less than 0.01

18. $\displaystyle\int_0^\pi e^{-x} \sin x\, dx$ to three decimal places

In Problems 19–24, determine how many subintervals are required to guarantee accuracy to within 0.00005 using:
a. *the trapezoidal rule;* **b.** *Simpson's rule*

19. $\displaystyle\int_1^3 x^{-1}\, dx$

20. $\displaystyle\int_{-1}^4 (x^3 + 2x^2 + 1)\, dx$

21. $\displaystyle\int_1^4 \frac{dx}{\sqrt{x}}$

22. $\displaystyle\int_0^2 \cos x\, dx$

23. $\displaystyle\int_0^1 e^{-2x}\, dx$

24. $\displaystyle\int_1^2 \ln \sqrt{x}\, dx$

25. A quarter-circle of radius 1 has the equation $y = \sqrt{1 - x^2}$ for $0 \le x \le 1$, which implies that

$$\int_0^1 \sqrt{1 - x^2}\, dx = \tfrac{\pi}{4}$$

  a. Estimate π correct to one decimal place by applying the trapezoidal rule to this integral.
  b. Estimate π correct to one decimal place by applying Simpson's rule to this integral.

26. Find the smallest value of n for which the trapezoidal rule estimates the value of the integral

$$\int_1^2 x^{-1}\, dx$$

with six-decimal-place accuracy.

27. The width of an irregularly shaped dam is measured at 5-m intervals, with the results indicated in Figure 5.32. Use the trapezoidal rule to estimate the area of the face of the dam.

■ **FIGURE 5.32**   Area of the face of a dam

28. Jack and Jill are traveling in a car with a broken odometer. To determine the distance they traveled between noon and 1:00 P.M., Jack (the passenger) takes a speedometer reading every 5 minutes.

Minutes (after noon)	0	5	10	15	20
Speedometer reading	54	57	50	51	55

Minutes	25	30	35	40	45	50	55	60
Speedometer	60	49	53	47	39	42	48	53

Use the trapezoidal rule to estimate the total distance traveled by the couple from noon to 1:00 P.M.

29. An industrial plant spills pollutant into a lake. The pollutant spread out to form the pattern shown in Figure 5.33. All distances are in feet.

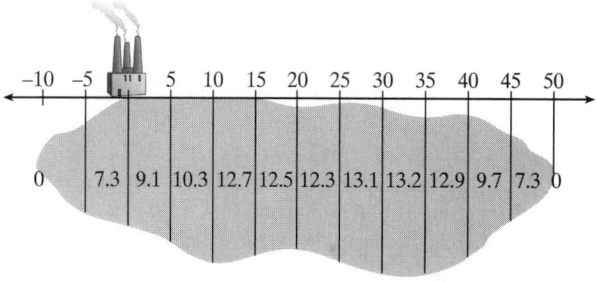

■ **FIGURE 5.33**   Pollutant spill

Use Simpson's rule to estimate the area of the spill.

Technology Window ▲ ▼

30. Apply the trapezoidal rule to the following data found on a spreadsheet (table):

	A	B
	x	$f(x)$
1	0	3.7
2	0.3	3.9
3	0.6	4.1
4	0.9	4.1
5	1.2	4.2
6	1.5	4.4
7	1.8	4.6
8	2.1	4.9
9	2.4	5.2
10	2.7	5.5
11	3	6

31. Apply Simpson's rule to the following data found on a spreadsheet (table):

	A	B
1	x	$f(x)$
2	0	10
3	0.5	9.75
4	1	10
5	1.5	10.75
6	2	12
7	2.5	13.75
8	3	16
9	3.5	18.75
10	4	22
11	4.5	25.75
12	5	30

32. In this problem, we explore the "order of convergence" of three numerical integration methods. A method is said to have "order of convergence n^k" if $E_n \cdot n^k = C$, where n is the number of intervals in the approximation, and k is a constant power. In other words, the error $E_n \to 0$ as $C/n^k \to 0$. In each of the following cases, use the fact that

$$I = \int_0^\pi \sin x \, dx = -\cos x \Big|_0^\pi = 2$$

  a. Use the trapezoidal rule to estimate I for $n = 10, 20, 40,$ and 80. Compute the error E_n in each case and compute $E_n \cdot n^k$ for $k = 1, 2, 3, 4$. Based on your results, you should be able to conclude that the order of convergence of the trapezoidal approximation is n^2.

  b. Repeat part **a** using Simpson's rule. Based on your results, what is the order of convergence for Simpson's rule?

  c. Repeat part **b** using a rectangular approximation with right endpoints. What is the order of convergence for this method?

33. Let $I = \int_0^\pi (9x - x^3) \, dx$

  a. Estimate I using rectangles, the trapezoidal rule, and Simpson's rule, for $n = 10, 20, 40, 80$. Something interesting happens with Simpson's rule. Explain.

  b. **Simpson struggles!**   In contrast to part **a,** here is an example where Simpson's rule does not live up to expectations. For $n = 10, 20, 40,$ and 80, use Simpson's rule and the rectangular rule (select midpoints) for this integral and make a table of errors. Then try to explain Simpson's poor performance. *Hint:* Look at the formula for the error.

$$\int_0^2 \sqrt{4 - x^2} \, dx = \pi$$

34. **HISTORICAL QUEST**   The mathematician Seki Kōwa was born in Fujioka, Japan, the son of a samurai, but was adopted by a patriarch of the Seki family. Seki invented and used an early form of determinants for solving systems of equations, and he also invented a method for approximating areas that is very similar to the rectangular method introduced in this section. This method, known as the *yenri* (circle principle), found the area of a circle by dividing the circle into small rectangles, as shown in Figure 5.34.

TAKAKAZU SEKI KŌWA
1642–1708

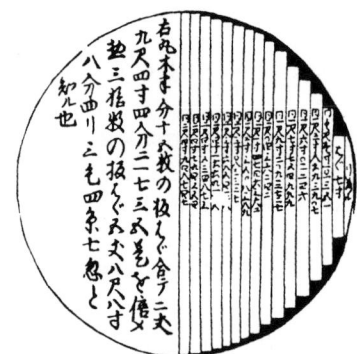

■ **FIGURE 5.34**   Early Asian Calculus

The sample shown in Figure 5.34 was drawn by a student of Seki Kōwa.

For this quest, draw a circle with radius 10 cm. Draw vertical chords through each centimeter on a diameter (you should have 18 rectangles). Measure the heights of the rectangles and approximate the area of the circle by adding the areas of the rectangles. Compare this with the formula for the area of this circle.

35. **HISTORICAL QUEST**   In 1670, a predecessor of Seki Kōwa (see Problem 34), Kazuyuki Sawaguchi, wrote seven volumes that concluded with fifteen problems that he believed were unsolvable. In 1674, Seki Kōwa published solutions to all fifteen of Kazuyuki's unsolvable problems.

One of the "unsolvable" problems was the following: Three circles are inscribed in a circle, each tangent to the other two and to the original circle. All three cover all but 120 square units of the circumscribing circle. The diameters of the two smaller inscribed circles are equal, and each is five units less than the diameter of the larger inscribed circle. Find the diameters of the three inscribed circles. Explain how Seki Kōwa might have gone about solving this problem.

C **36.** **HISTORICAL QUEST**   Isaac Newton (see Historical Quest on page 124) invented a preliminary version of Simpson's rule. In 1779, Newton wrote an article as an addendum to *Methodus Differentials* (1711) in which he gave the following example: If there are four ordinates at equal intervals, let A be the sum of the first and fourth, B the sum of the second and third, and R the interval between the first and fourth; then . . .

ROGER COTES
1682–1716

the area between the first and fourth ordinates is approximated by $\frac{1}{8}(A + 3B)R$. This is known today as the "Newton–Cotes three-eighths rule," which we can state in modern terms.

$$\int_{x_0}^{x_3} f(x)\, dx \approx \frac{3}{8}(y_0 + 3y_1 + 3y_2 + y_3)\Delta x$$

Roger Cotes and James Stirling (1692–1770) both knew this formula, as well as what we called in this section Simpson's rule. In 1743, this rule was rediscovered by Thomas Simpson (1710–1761).

Estimate the integral

$$\int_0^3 \tan^{-1} x\, dx$$

using the Newton–Cotes three-eighths rule, then compare with approximation using left endpoints (rectangles) and trapezoids with $n = 4$. Which of the three rules gives the most accurate estimate?

37. Show that if $p(x)$ is any polynomial of degree less than or equal to 3, then

$$\int_a^b p(x)\, dx = \frac{b - a}{6}\left[p(a) + 4p\left(\frac{a + b}{2}\right) + p(b)\right]$$

This result is often called the *prismoidal rule*.

38. Use the prismoidal rule (Problem 37) to evaluate

$$\int_{-1}^2 (x^3 - 3x + 4)\, dx$$

39. Use the prismoidal rule (Problem 37) to evaluate

$$\int_{-1}^3 (x^3 + 2x^2 - 7)\, dx$$

40. Let $p(x)$ be a polynomial of degree at most 3.

a. Show that there is a number c between 0 and 1 such that

$$\int_{-1}^1 p(x)\, dx = p(c) + p(-c)$$

b. Show that there is a number c such that

$$\int_{-1/2}^{1/2} p(x)\, dx = \frac{1}{3}[p(-c) + p(0) + p(c)]$$

41. Let $p(x) = Ax^3 + Bx^2 + Cx + D$ be a cubic polynomial. Show that Simpson's rule gives the exact value for

$$\int_a^b p(x)\, dx$$

42. The object of this exercise is to prove Simpson's rule for the special case involving three points.

a. Let $P_1(-h, f(-h)), P_2(0, f(0)), P_3(h, f(h))$. Find the equation of the form $y = Ax^2 + Bx + C$ for the parabola through the points P_1, P_2, and P_3.

b. If $y = p(x)$ is the quadratic function found in part **a,** show that

$$\int_{-h}^h p(x)\, dx = \frac{h}{3}[p(-h) + 4p(0) + p(h)]$$

c. Let $Q_1(x_1, f(x_1)), Q_2(x_2, f(x_2)), Q_3(x_3, f(x_3))$ be points with $x_2 = x_1 + h$, and $x_3 = x_1 + 2h$. Explain why

$$\int_{x_1}^{x_3} p(x)\, dx = \frac{h}{3}[p(x_1) + 4p(x_2) + p(x_3)]$$

5.9   An Alternative Approach: The Logarithm as an Integral

| IN THIS SECTION | natural logarithm as an integral, geometric interpretation, the natural exponential function |

NATURAL LOGARITHM AS AN INTEGRAL

You may have noticed in Section 1.6 that we did not prove the properties of exponential functions for all real number exponents. To treat exponentials and logarithms *rigorously*, we use the alternative approach provided in this section. Specifically, we use a definite integral to introduce the *natural logarithmic function* and then use this function to *define* the *natural exponential function*. In this way we can develop precise meanings for b^x and $\log_b x$.

Natural Logarithm

The **natural logarithm** is the function defined by

$$\ln x = \int_1^x \frac{dt}{t} \qquad x > 0$$

At first glance, it appears there is nothing "natural" about this definition, but if this integral function has the properties of a logarithm, why should we not call it a logarithm? We begin with a theorem that shows that $\ln x$ does indeed have the properties we would expect of a logarithm.

THEOREM 5.13   *Properties of a logarithm as defined by* $\ln x = \int_1^x \dfrac{dt}{t}$

Let $x > 0$ and $y > 0$ be positive numbers. Then

a. $\ln 1 = 0$

b. $\ln xy = \ln x + \ln y$

c. $\ln \dfrac{x}{y} = \ln x - \ln y$

d. $\ln x^p = p \ln x$ for all rational numbers p

Proof

a. Let $x = 1$. Then $\displaystyle\int_1^1 \frac{dt}{t} = 0$.

b. For fixed positive numbers x and y, we use the additive property of integrals as follows:

$$\ln(xy) = \int_1^{xy} \frac{dt}{t} = \int_1^x \frac{dt}{t} + \int_x^{xy} \frac{dt}{t}$$

$$= \ln x + \int_1^y \frac{x \, du}{ux}$$

$$= \ln x + \ln y$$

> Let $u = \dfrac{t}{x}$, so $t = ux$;
> $dt = x \, du$.
> If $t = x$, then $u = 1$;
> If $t = xy$, then $u = y$.

c. and d. The proofs are outlined in the problem set.   ═

GEOMETRIC INTERPRETATION

An advantage of defining the logarithm by the integral formula is that calculus can be used to study the properties of $\ln x$ from the beginning. For example, note that if $x > 1$, the integral

$$\ln x = \int_1^x \frac{dt}{t}$$

may be interpreted geometrically as the area under the graph of $y = \dfrac{1}{t}$ from $t = 1$ to $t = x$, as shown in Figure 5.35. Because the area shown in Figure 5.35 is positive, we have $\ln x > 0$ if $x > 1$. If $0 < x < 1$, then

$$\ln x = \int_1^x \frac{dt}{t} = -\int_x^1 \frac{dt}{t} < 0$$

so that

$$\ln x > 0 \text{ if } x > 1$$
$$\ln 1 = 0$$
$$\ln x < 0 \text{ if } 0 < x < 1$$

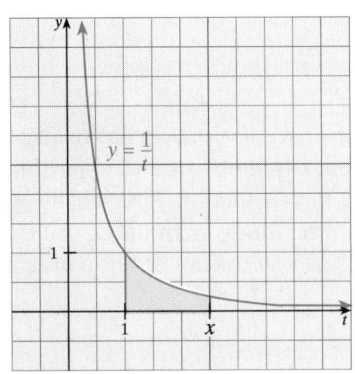

■ **FIGURE 5.35** If $x > 1$, $\ln x = \displaystyle\int_1^x \frac{dt}{t}$ is the area under $y = \dfrac{1}{t}$ on $[1, x]$

The definition $\ln x = \int_1^x \dfrac{dt}{t}$ makes it easy to differentiate $\ln x$. Recall from Section 5.4 that according to the second fundamental theorem of calculus, if f is continuous on $[a, b]$, then

$$F(x) = \int_a^x f(t)\, dt$$

is a differentiable function of x with derivative $\dfrac{dF}{dx} = f(x)$ on any interval $[a, x]$. Therefore, because $\dfrac{1}{t}$ is continuous for all $t > 0$, it follows that $\ln x = \int_1^x \dfrac{dt}{t}$ is differentiable for all $x > 0$ with derivative $\dfrac{d}{dx}(\ln x) = \dfrac{1}{x}$. By applying the chain rule, we also find that

$$\frac{d}{dx}(\ln u) = \frac{1}{u}\frac{du}{dx}$$

for any differentiable function u of x with $u > 0$.

To analyze the graph of $f(x) = \ln x$, we use the curve-sketching methods of Chapter 4:

a. $\ln x$ is continuous for all $x > 0$ (because it is differentiable), so its graph is "unbroken."

b. The graph of $\ln x$ is always *rising*, because the derivative

$$\frac{d}{dx}(\ln x) = \frac{1}{x}$$

is positive for $x > 0$. (Recall that the natural logarithm is defined only for $x > 0$.)

c. The graph of $\ln x$ is *concave down*, because the second derivative

$$\frac{d^2}{dx^2}(\ln x) = \frac{d}{dx}\left(\frac{1}{x}\right) = \frac{-1}{x^2}$$

is negative for all $x > 0$.

d. Note that $\ln 2 > 0$ $\left(\text{because } \int_1^2 \dfrac{dt}{t} > 0\right)$ and because $\ln 2^p = p \ln 2$, it follows that $\lim\limits_{p \to +\infty} \ln 2^p = +\infty$. But the graph of $f(x) = \ln x$ is always rising, and thus $\lim\limits_{x \to +\infty} \ln x = +\infty$. Similarly, it can be shown that $\lim\limits_{x \to 0^+} \ln x = -\infty$.

e. If b is any positive number, there is exactly one number a such that $\ln a = b$ (because the graph of $\ln x$ is always rising for $x > 0$). In particular, we define $x = e$ as the unique number that satisfies $\ln x = 1$.

These features are shown in Figure 5.36.

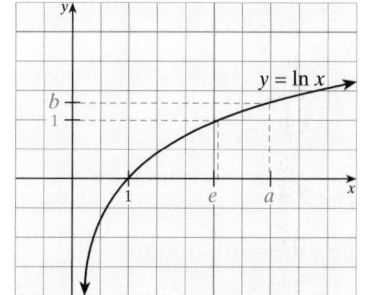

■ **FIGURE 5.36**   The graph of the natural logarithm function, $\ln x$

THE NATURAL EXPONENTIAL FUNCTION

Originally, we introduced the natural exponential function e^x and then defined the natural logarithm $\ln x$ as the inverse of e^x. In this alternative approach, we note that because the natural logarithm is an increasing function, it must be one-to-one. Therefore it has an inverse function, which we denote by $E(x)$.

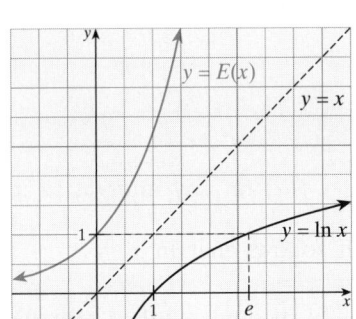

■ **FIGURE 5.37** The graph of $E(x)$ is the reflection of the graph of $\ln x$ in line $y = x$.

Because $\ln x$ and $E(x)$ are inverses, we have

$$E(x) = y \quad \text{if and only if} \quad \ln y = x$$

From the definition of inverse formulas we have

$$E(\ln x) = x \quad \text{and} \quad \ln[E(x)] = x$$

We call these formulas the **inversion formulas.** Therefore,

$$E(0) = E(\ln 1) \qquad \textit{Because } \ln 1 = 0$$
$$= 1 \qquad\qquad \textit{Because } E(\ln x) = x$$

and

$$E(1) = E(\ln e) \qquad \textit{Because } \ln e = 1$$
$$= e \qquad\qquad \textit{Because } E(\ln e) = e$$

To obtain the graph of $E(x)$, we reflect the graph of $y = \ln x$ in the line $y = x$. The graph is shown in Figure 5.37. Notice that $\ln x = 1$ has the unique solution $x = e$.

The following algebraic properties of the natural exponential function can be obtained by using the properties of the natural logarithm given in Theorem 5.13 along with the inversion formulas.

THEOREM 5.14   *Properties of the exponential as defined by $E(x)$*

For any numbers x and y,

a. $E(x + y) = E(x)E(y)$

b. $E(x - y) = \dfrac{E(x)}{E(y)}$

c. $[E(x)]^p = E(px)$

Proof

a. We use the fact that $\ln AB = \ln A + \ln B$ to show that

$$\ln[E(x)E(y)] = \ln E(x) + \ln E(y) \qquad \textit{Property of logarithms}$$
$$= x + y \qquad\qquad\qquad \textit{Inversion formula}$$
$$= \ln E(x + y) \qquad\quad \textit{Inversion formula}$$

Because $\ln x$ is a one-to-one function, we conclude that

$$E(x)E(y) = E(x + y)$$

b. and c. The proofs are similar and are left as exercises.

Next, we use implicit differentiation along with the differentiation formula

$$\frac{d}{dx}(\ln x) = \frac{1}{x}$$

to obtain a differentiation formula for $E(x)$.

THEOREM 5.15   *Derivative of $E(x)$*

The function defined by $E(x)$ is differentiable and

$$\frac{dE}{dx} = E(x)$$

Proof   It can be shown that the inverse of any differentiable function is also differentiable (see Appendix B). Since ln x is differentiable, it follows that its inverse $E(x)$ is also differentiable, and using implicit differentiation, we find that

$$\ln y = \ln[E(x)] \qquad y = E(x)$$
$$\ln y = x \qquad \text{Inversion formula}$$
$$\frac{1}{y}\frac{dy}{dx} = 1 \qquad \text{Differentiate implicitly.}$$
$$\frac{dy}{dx} = y$$
$$\frac{dE}{dx} = E(x) \qquad \text{Because } \frac{dy}{dx} = \frac{dE}{dx} \text{ and } E(x) = y$$

Finally, we observe that if r is a rational number, then

$$\ln(e^r) = r \ln e = r$$

so that

$$E(r) = e^r$$

This means that $E(x) = e^x$ when x is a rational number, and we **define** $E(x)$ to be e^x for irrational x as well. In particular, we now have an alternative definition for e: $E(1) = e^1 = e$.

5.9 Problem Set

1. Use the integral definition

$$\ln x = \int_1^x \frac{dt}{t}$$

to show that $\ln x \to -\infty$ as $x \to 0^+$. *Hint:* What happens to $\ln(2^{-N})$ as N grows large without bound?

2. Use Simpson's rule with $n = 8$ subintervals to estimate

$$\ln 3 = \int_1^3 \frac{dt}{t}$$

Compare your estimate with the value of ln 3 obtained from your calculator.

3. Use the error estimate for Simpson's rule to determine the accuracy of the estimate in Problem 2. How many subintervals should be used in Simpson's rule to estimate ln 3 with an error not greater than 0.00005?

4. Prove the quotient rule for logarithms,

$$\ln\left(\frac{x}{y}\right) = \ln x - \ln y$$

for all $x > 0$, $y > 0$. *Hint:* Use the product rule for logarithms.

5. Show that $\ln(x^p) = p \ln x$ for $x > 0$ and all rational exponents p by completing the following steps.
  a. Let $F(x) = \ln(x^p)$ and $G(x) = p \ln x$. Show that $F'(x) = G'(x)$ for $x > 0$. Conclude that $F(x) = G(x) + C$.

  b. Let $x = 1$ and conclude that
  $$F(x) = G(x)$$
  that is, that $\ln(x^p) = p \ln x$.

6. Use Rolle's theorem to show that
$$\ln M = \ln N$$
if and only if $M = N$. *Hint:* Show that if $M \neq N$, Rolle's theorem implies that
$$\frac{d}{dx}(\ln x) = 0$$
for some number c between M and N. Why is this impossible?

7. The product rule for logarithms states
$$\log_b(MN) = \log_b M + \log_b N$$
For this reason, we want the natural logarithm function ln x to satisfy the functional equation
$$f(xy) = f(x) + f(y)$$
Suppose $f(x)$ is a function that satisfies this equation throughout its domain D.
  a. Show that if $f(1)$ is defined, then $f(1) = 0$.
  b. Show that if $f(-1)$ is defined, then $f(-1) = 0$.
  c. Show that $f(-x) = f(x)$ for all x in D.

d. If $f'(x)$ is defined for each $x \neq 0$, show that

$$f'(x) = \frac{f'(1)}{x}$$

Then show that

$$f(x) = f'(1) \int_1^x \frac{dt}{t} \quad \text{for } x > 0$$

e. Conclude that any solution of

$$f(xy) = f(x) + f(y)$$

that is not identically 0 and has a derivative for all $x \neq 0$ must be a multiple of

$$L(x) = \int_1^{|x|} \frac{dt}{t} \quad \text{for } x \neq 0$$

8. Show that for each number A, there is only one number x for which $\ln x = A$. *Hint:* If not, then $\ln x = \ln y = A$ for $x \neq y$. Why is this impossible?

9. a. Use an area argument to show that $\ln 2 < 1$.
   b. Show that $\ln 3 > 1$, then explain why
   $$2 < e < 3$$

Chapter 5 Review

Proficiency Examination

Concept Problems

1. What is an antiderivative?
2. State the integration rule for powers.
3. State the exponential rule for integration.
4. State the integration rules for the trigonometric functions.
5. State the inverse trigonometric rules for integration.
6. What is an area function? What do we mean by area as an antiderivative? What conditions are necessary for an integral to represent an area?
7. What is the formula for area as the limit of a sum?
8. What is a Riemann sum?
9. Define a definite integral.
10. Complete these statements summarizing the general properties of the definite integral.

    a. Definite integral at a point: $\int_a^a f(x)\, dx = $ _____

    b. Opposite of a definite integral: $\int_b^a f(x)\, dx = $ _____

11. How can distance be expressed as an integral?
12. State the first fundamental theorem of calculus.
13. State the second fundamental theorem of calculus.
14. Describe in your own words the process of integration by substitution.
15. What is a differential equation?
16. What is a separable differential equation? How do you solve such an equation?
17. What is the growth/decay formula? Describe carbon dating.
18. What is an orthogonal trajectory?
19. State the mean value theorem for integrals.
20. What is the formula for the average value of a continuous function?

21. State the following approximation rules:
    a. rectangular approximation
    b. trapezoidal rule
    c. Simpson's rule

Practice Problems

22. Given that $\int_0^1 x^4\, dx = \frac{1}{5}$, $\int_0^1 x^2\, dx = \frac{1}{3}$, and $\int_0^1 dx = 1$, find

    $$\int_0^1 [x^2(2x^2 - 3)]\, dx$$

23. Find $F'(x)$ if $F(x) = \int_3^x t^5 \sqrt{\cos(2t + 1)}\, dt$.

Evaluate the integrals in Problems 24–29.

24. $\displaystyle\int \frac{dx}{1 + 4x^2}$

25. $\displaystyle\int xe^{-x^2}\, dx$

26. $\displaystyle\int_1^4 (\sqrt{x} + x^{-3/2})\, dx$

27. $\displaystyle\int_0^{\pi/2} \frac{\sin x\, dx}{(1 + \cos x)^2}$

28. $\displaystyle\int_0^1 (2x - 6)(x^2 - 6x + 2)^2\, dx$

29. $\displaystyle\int_{-2}^1 (2x + 1)\sqrt{2x^2 + 2x + 5}\, dx$

30. Find the area under the curve $f(x) = 3x^2 + 2$ over $[-1, 3]$.

31. Find the average value of $y = \cos 2x$ on the interval $[0, \frac{\pi}{2}]$.

32. An object experiences linear acceleration given by

    $$a(t) = 2t + 1 \quad \text{ft/s}^2$$

    Find the velocity and position of the object, given that it starts its motion (at $t = 0$) at $s = 4$ with initial velocity 2 ft/s.

33. In 1995, a team of archaeologists led by Michel Brunet of the University of Poitiers announced the identification

in Chad of *Australopithecus* fossils believed to be about 3.5 Myr (million years ago) old.* Using the decay formula, explain why the archaeologists were reluctant to use carbon-14 to date their find.

34. When it is x years old, a certain industrial machine generates revenue at the rate of $R'(x) = 1{,}575 - 5x^2$ thousand dollars per year. Find a function that measures the amount of revenue, and find the revenue for the first five years.

35. A slope field for the differential equation
$$y' = x + y$$
in given in Figure 5.38. Draw the particular solutions passing through the points requested in parts a–d.
   a. $(0, -4)$ b. $(0, 2)$ c. $(4, 0)$ d. $(-4, 0)$
   e. All of the curves you have drawn in parts a–d seem to have a common asymptote. This asymptote is the solution of the given differential equation. From the slope field, write the equation for the solution.

_____

*"Early Hominid Fossils from Africa," by Meave Leakey and Alan Walker. *Scientific American*, June 1997, p. 79.

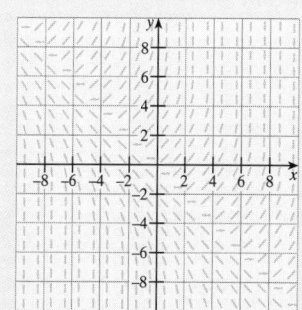

■ **FIGURE 5.38**   Direction field

36. Solve the differential equation $\dfrac{dy}{dx} = y^2 \sin 3x$.

37. a. Find the necessary n to estimate the value of
   $$\int_0^{\pi/2} \cos x \, dx$$
   to within 0.0005 of its correct value using the trapezoidal rule.
   b. What is n if Simpson's rule is used?

Supplementary Problems

Find the indefinite integrals of the functions in Problems 1–3.

1. Given that $\displaystyle\int_{-1}^{0} f(x)\, dx = 3$, $\displaystyle\int_{0}^{1} f(x)\, dx = -1$, and $\displaystyle\int_{-1}^{1} g(x)\, dx = 7$ find $\displaystyle\int_{-1}^{1} [3g(x) + 2f(x)]\, dx$.

2. Use the definition of definite integral to find
   $$\int_0^1 (3x^2 + 2x - 1)\, dx.$$

3. Use the definition of definite integral to find
   $$\int_0^1 (4x^3 + 6x^2 + 3)\, dx.$$

Find the definite and indefinite integrals in Problems 4–28. If you do not have a technique for finding a closed (exact) answer, approximate the integral using numerical integration.

4. $\displaystyle\int_0^1 (5x^4 - 8x^3 + 1)\, dx$ 5. $\displaystyle\int_{-1}^2 30(5x - 2)^2\, dx$

6. $\displaystyle\int_0^1 (x\sqrt{x} + 2)^2\, dx$ 7. $\displaystyle\int_1^2 \frac{x^2\, dx}{\sqrt{x^3 + 1}}$

8. $\displaystyle\int_2^2 (x + \sin x)^3\, dx$ 9. $\displaystyle\int_{-1}^0 \frac{dx}{\sqrt{1 - 2x}}$

10. $\displaystyle\int_1^2 \frac{dx}{\sqrt{3x - 1}}$ 11. $\displaystyle\int_{-1}^0 \frac{dx}{\sqrt[3]{1 - 2x}}$

12. $\displaystyle\int \frac{1}{x^3}\, dx$ 13. $\displaystyle\int \frac{5x^2 - 2x + 1}{\sqrt{x}}\, dx$

14. $\displaystyle\int \frac{x + 1}{2x}\, dx$ 15. $\displaystyle\int \frac{3 - x}{\sqrt{1 - x^2}}\, dx$

16. $\displaystyle\int (e^{-x} + 1)e^x\, dx$ 17. $\displaystyle\int \frac{\sin x - \cos x}{\sin x + \cos x}\, dx$

18. $\displaystyle\int \sqrt{x}(x^2 + \sqrt{x} + 1)\, dx$ 19. $\displaystyle\int (x - 1)^2\, dx$

20. $\displaystyle\int \frac{x^2 + 1}{x^2}\, dx$ 21. $\displaystyle\int (\sin^2 x + \cos^2 x)\, dx$

22. $\displaystyle\int x(x + 4)\sqrt{x^3 + 6x^2 + 2}\, dx$

23. $\displaystyle\int x(2x^2 + 1)\sqrt{x^4 + x^2}\, dx$

24. $\displaystyle\int \frac{dx}{\sqrt{x}(\sqrt{x} + 1)^2}$ 25. $\displaystyle\int x\sqrt{1 - 5x^2}\, dx$

26. $\displaystyle\int \sqrt{\sin x - \cos x}(\sin x + \cos x)\, dx$

27. $\displaystyle\int_{-10}^{10} [3 + 7x^{73} - 100x^{101}]\, dx$

28. $\displaystyle\int_{-\pi/4}^{\pi/4} [\sin(4x) + 2\cos(4x)]\, dx$

In Problems 29–32, draw the indicated particular solution for the differential equations whose direction fields are given.

29. $y(4) = 0$

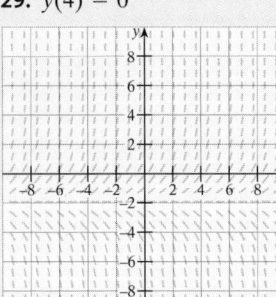

30. $y(10) = 1$

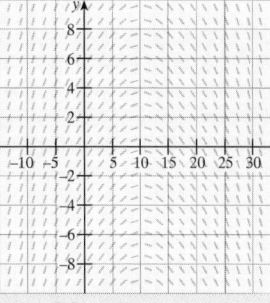

31. $y(0) = 2$

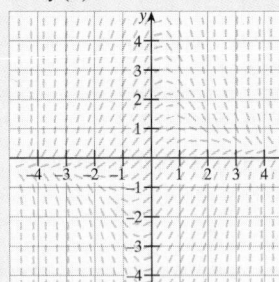

32. $y(0) = 0$

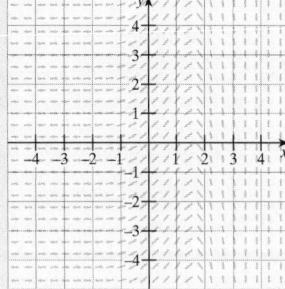

33. Find $F'(x)$ where $F(x) = \displaystyle\int_5^x t^2 \cos^4 t\, dt$

34. Find the area under $f(x) = x^{-1}$ on $[1, 4]$.

35. Find the area under $f(x) = 2 + x - x^2$ on $[-1, 1]$.

36. Find the area under $f(x) = e^{4x}$ on $[0, 2]$.

37. Find the area bounded by the curve $y = x\sqrt{x^2 + 5}$, the x-axis, and the vertical lines $x = -1$ and $x = 2$.

38. Find $f(t)$ if $f''(t) = \sin 4t - \cos 2t$ and $f(\frac{\pi}{2}) = f'(\frac{\pi}{2}) = 1$.

39. Find $f(x)$ if $f'''(x) = 2x^3 + x^2$, given that $f''(1) = 2$, $f'(1) = 1$, and $f(1) = 0$.

Solve the differential equations in Problems 40–47.

40. $\dfrac{dy}{dx} = (1 - y)^2$

41. $\dfrac{dy}{dx} = \dfrac{\cos 4x}{y}$

42. $\dfrac{dy}{dx} = \left(\dfrac{\cos y}{\sin x}\right)^2$

43. $\dfrac{dy}{dx} = \dfrac{x}{y}$

44. $\dfrac{dy}{dx} = y(x^2 + 1)$

45. $\dfrac{dy}{dx} = \dfrac{x}{y}\sqrt{\dfrac{y^2 + 2}{x^2 + 1}}$

46. $\dfrac{dy}{dx} = \dfrac{\cos^2 y}{\cot x}$

47. $\dfrac{dy}{dx} = \sqrt{\dfrac{x}{y}}$

48. Find the average value of $f(x) = \dfrac{\sin x}{\cos^2 x}$ on the interval $[0, \frac{\pi}{4}]$.

49. Find the average value of $f(x) = \sin x$
   a. on $[0, \pi]$ b. on $[0, 2\pi]$

50. Use the trapezoidal rule with $n = 6$ to approximate $\displaystyle\int_0^\pi \sin x\, dx$. Compare your result with the exact value of this integral.

51. Estimate $\displaystyle\int_0^1 \sqrt{1 + x^3}\, dx$ using the trapezoidal rule with $n = 6$.

52. Estimate $\displaystyle\int_0^1 \dfrac{dx}{\sqrt{1 + x^3}}$ using the trapezoidal rule with $n = 8$.

53. Estimate $\displaystyle\int_0^1 \sqrt{1 + x^3}\, dx$ using Simpson's rule with $n = 6$.

54. Estimate $\displaystyle\int_0^1 \dfrac{dx}{\sqrt{1 + x^3}}$ using Simpson's rule with $n = 8$.

55. Use the trapezoidal rule to estimate to within 0.00005 the value of the integral
$$\int_1^2 \sqrt{x + \frac{1}{x}}\, dx$$

56. Use the trapezoidal rule to approximate $\displaystyle\int_0^1 \dfrac{x^2\, dx}{1 + x^2}$ with an error no greater than 0.005.

57. Use Simpson's rule to estimate to within 0.00005 the value of the integral
$$\int_1^2 \sqrt{x + \frac{1}{x}}\, dx$$

58. **H**ISTORICAL **Q**UEST   In Section 2.3 (Problem 61), we met Karl Gauss and the prime number theorem. Recall that $\pi(x)$ is the number of primes less than or equal to x. Adrien-Marie Legendre is best known for his work with elliptic integrals and mathematical physics. In 1794, he proved that π was an irrational number and formulated a conjecture that is equivalent to the prime number theorem. His conjecture was

$$\pi(x) \approx \frac{x}{\ln x - 1.08366}$$

ADRIEN-MARIE LEGENDRE 1752–1833

Gauss came up with a different estimate:

$$\pi(x) \approx \int_0^x \frac{1}{\ln u}\, du$$

For this Quest, use Table 5.1 to compare $\pi(x)$, Gauss' approximation, Legendre's approximation, and the approximation $x/(\ln x - 1)$. Which approximation do you think is the "best"?

■ **TABLE 5.1   Actual Number of Primes Less Than x**

x	$\pi(x)$
10^3	168
10^4	1,229
10^5	9,592
10^6	78,498
10^7	664,579
10^8	5,761,455
10^9	50,847,478
10^{10}	455,052,511

59. Find the average value of the function defined by

$$f(x) = \frac{\cos x}{1 - \dfrac{x^2}{2}}$$

on $[0, 1]$. *Hint:* Use the trapezoidal rule with $n = 6$.

60. The brakes of a certain automobile produce a constant deceleration of k m/s^2. The car is traveling at 25 m/s when the driver is forced to hit the brakes. If it comes to rest at a point 50 m from the point where the brakes are applied, what is k?

61. A particle moves along the x-axis in such a way that $a(t) = -4s(t)$, where $s(t)$ and $a(t)$ are its position and acceleration, respectively, at time t. The particle starts from rest at $s = 5$.
   a. Show that $v^2 + 4t^2 = 100$. *Hint:* First use the chain rule to show that $a(t) = v(dv/ds)$.
   b. What is the velocity when the particle first reaches $t = 3$? *Note:* The sign of v is determined by the direction the particle is moving at the time in question.

62. A manufacturer estimates marginal revenue to be $100x^{-1/3}$ dollars per unit when the level of production is x units. The corresponding marginal cost is found to be $0.4x$ dollars per unit. Suppose the manufacturer's profit is \$520 when the level of production is 16 units. What is the manufacturer's profit when the level of production is 25 units?

63. A tree has been transplanted and after t years is growing at a rate of

$$1 + \frac{1}{(t + 1)^2}$$

feet per year. After 2 years it has reached a height of 5 ft. How tall was the tree when it was transplanted?

64. A manufacturer estimates that the marginal revenue of a certain commodity is

$$R'(x) = \sqrt{x}(x^{3/2} + 1)^{-1/2}$$

dollars per unit when x units are produced. Assuming no revenue is obtained when $x = 0$, how much revenue is obtained from producing $x = 4$ units?

65. Find a function whose tangent has slope $x\sqrt{x^2 + 5}$ for each value of x and whose graph passes through the point $(2, 10)$.

66. A particle moves along the x-axis in such a way that after t seconds its acceleration is $a(t) = 12(2t + 1)^{-3/2}$. If it starts at rest at $x = 3$, where will it be 4 seconds later?

67. An environmental study of a certain community suggests that t years from now, the level of carbon monoxide in the air will be changing at the rate of $0.1t + 0.2$ parts per million per year. If the current level of carbon monoxide in the air is 3.4 parts per million, what will the level be 3 years from now?

68. A woman, driving on a straight, level road at the constant speed v_0 is forced to apply her brakes to avoid hitting a cow, and the car comes to a stop 3 s later and s_0 ft from

the point where the brakes were applied. Continuing on her way, she increases her speed by 20 ft/s to make up time but is again forced to hit the brakes, and this time it takes her 5 s and s_1 feet to come to a full stop. Assuming that her brakes supplied a constant deceleration d each time they were used, find d and determine v_0, s_0, and s_1.

69. A study indicates that x months from now the population of a certain town will be increasing at the rate of $10 + 2\sqrt{x}$ people per month. By how much will the population of the town increase over the next 9 mo?

70. It is estimated that t days from now a farmer's crop will be increasing at the rate of $0.3t^2 + 0.6t + 1$ bushels per day. By how much will the value of the crop increase during the next 6 days if the market price remains fixed at \$2 per bushel?

71. Records indicate that t months after the beginning of the year, the price of turkey in local supermarkets was

$$P(t) = 0.06t^2 - 0.2t + 1.2$$

dollars per pound. What was the average price of turkey during the first 6 mo of the year?

72. MODELING PROBLEM   V. A. Tucker and K. Schmidt-Koenig have investigated the relationship between the velocity v (km/h) of a bird in flight and the energy $E(v)$ expended by the bird.[*] Their study showed that for a certain kind of parakeet, the rate of change of the energy expended with respect to velocity is modeled (for $v > 0$) by

$$\frac{dE}{dv} = \frac{0.074v^2 - 112.65}{v^2}$$

   a. What is the most economical velocity for the parakeet? That is, find the velocity v_0 that minimizes the energy.
   b. Suppose it is known that $E = E_0$ when $v = v_0$. Express E in terms of v_0 and E_0.
   c. Express the average energy expended as the parakeet's velocity ranges from $v = \frac{1}{2}v_0$ to $v = v_0$ in terms of E_0.

73. MODELING PROBLEM   A toxin is introduced into a bacterial culture, and t hours later, the population $P(t)$ of the culture is found to be changing at the rate

$$\frac{dP}{dt} = -(\ln 2)2^{5-t}$$

If there were 1 million bacteria in the culture when the toxin was introduced, when will the culture die out?

74. MODELING PROBLEM   **The Snowplow Problem of R. P. Agnew** (This problem was mentioned in the preview for this chapter.) One day it starts snowing at a steady rate sometime before noon. At noon, a snowplow starts to clear a straight, level section of road. If the plow clears 1 mile of road during the first hour but requires 2 hours to clear the second mile, at what time did it start snowing? Answer this question by completing the following steps:

---

[*]Adapted from "Flight Speeds of Birds in Relation to Energies and Wind Directions," by V. A. Tucker and K. Schmidt-Koenig. *The Auk,* Vol. 88 (1971), pp. 97–107.

a. Let t be the time (in hours) from noon. Let h be the depth of the snow at time t, and let s be the distance moved by the plow. If the plow has width w and clears snow at the constant rate p, explain why $wh\dfrac{ds}{dt} = p$.

b. Suppose it started snowing t_0 hours before noon. Let r denote the (constant) rate of snowfall. Explain why

$$h(t) = r(t + t_0)$$

By combining this equation with the differential equation in part **a,** note that

$$wr(t + t_0)\frac{ds}{dt} = p$$

Solve this differential equation (with appropriate conditions) to obtain t_0 and answer the question posed in the problem.

75. a. If f is continuous on $[a, b]$, show that

$$\left|\int_a^b f(x)\,dx\right| \le \int_a^b |f(x)|\,dx$$

b. Show that $\left|\displaystyle\int_1^4 \frac{\sin x}{x}\,dx\right| \le \dfrac{3}{2}$.

76. A company plans to hire additional advertising personnel. Suppose it is estimated that if x new people are hired, they will bring in additional revenue of $R(x) = \sqrt{2x}$ thousand dollars and that the cost of adding these x people will be $C(x) = \frac{1}{3}x$ thousand dollars. How many new people should be hired? How much total net revenue (that is, revenue minus cost) is gained by hiring these people?

77. The half-life of the radioactive isotope cobalt-60 is 5.25 years.
   a. What percentage of a given sample of cobalt-60 remains after 5 years?
   b. How long will it take for 90% of a given sample to disintegrate?

78. The rate at which salt dissolves in water is directly proportional to the amount that remains undissolved. If 8 lb of salt is placed in a container of water and 2 lb dissolves in 30 min, how long will it take for 1 lb to dissolve?

79. Scientists are observing a species of insect in a certain swamp region. The insect population is estimated to be 10 million and is expected to grow at the rate of 2% per year. Assuming that the growth is exponential and stays that way for a period of years, what will the insect population be in 10 yr? How long will it take to double?

80. Solve the system of differential equations

$$\begin{cases} 2\dfrac{dx}{dt} + 5\dfrac{dy}{dt} = t \\[2mm] \dfrac{dx}{dt} + 3\dfrac{dy}{dt} = 7\cos t \end{cases}$$

Hint: Solve for dx/dt and dy/dt algebraically, and then integrate.

81. **SPY PROBLEM**   While lunching on cassoulet de chameau at his favorite restaurant in San Dimas (recall Problem 50, Section 5.1), the Spy finds a message from "N" spelled out in his alphabet soup:
"The average of the temperature $F(t) = at^3 + bt^2 + ct + d$ between 9 and 3 is that between two fixed times t_1 and t_2. Mother waits at the well."

The Spy knows "Mother" is "N" himself, and "the well" is a sleazy bar at the edge of town. He decides that time t is measured from noon because the soup was served at that time. The cryptic nature of the message suggests that t_1 and t_2 are independent of $a, b, c,$ and d and that one of them is the rendezvous time. When should he arrive for the meeting? (Remember, it's dangerous to go to the well too often.)

82. **PUTNAM EXAMINATION PROBLEM**   Where on the parabola $4ay = x^2$ $(a > 0)$ should a chord be drawn so that it will be normal to the parabola and cut off a parabolic sector of minimum area? That is, find P so that the shaded area in Figure 5.39 is as small as possible.

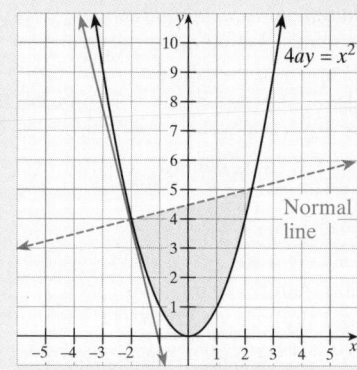

■ **FIGURE 5.39**   Minimum area problem

83. **PUTNAM EXAMINATION PROBLEM**   If $a_0, a_1, \ldots, a_n$ are real numbers that satisfy

$$\frac{a_0}{1} + \frac{a_1}{2} + \cdots + \frac{a_n}{n + 1} = 0$$

show that the equation $a_0 + a_1x + a_2x^2 + \cdots + a_nx^n = 0$ has at least one real root. *Hint:* Use the mean value theorem for integrals.

84. **PUTNAM EXAMINATION PROBLEM**   A heavy object is attached to end A of a light rod AB of length a. The rod is hinged at B so that it can turn freely in a vertical plane. If the rod is balanced in the vertical position above the hinge and then slightly disturbed, show that the time required for the rod to pass from the horizontal position to the lowest position is

$$T = \sqrt{\frac{a}{g}}\ln(1 + \sqrt{2})$$

Kinematics of Jogging

Ralph Boas (1912–1992), who was a professor of mathematics at Northwestern University, wrote this guest essay. Professor Boas is well known for his many papers and professional activities. In addition to his work in real and complex analysis, he wrote many expository articles, such as this guest essay, about teaching or using mathematics.

Nature herself exhibits to us measurable and observable quantities in definite mathematical dependence; the conception of a function is suggested by all the processes of nature where we observe natural phenomena varying according to distance or to time. Nearly all the "known" functions have presented themselves in the attempt to solve geometrical, mechanical, or physical problems.

J. T. MERTZ,
A HISTORY OF EUROPEAN THOUGHT IN THE NINETEENTH CENTURY
(EDINBURGH AND LONDON, 1903),
p. 696.

Some people think that calculus is dull, but it did not seem so three centuries ago, when it was invented. Then, it produced unexpected results; and, now and then, it still does. This essay is about such a result.

You have learned about the intermediate value theorem (see Section 2.4), which tells you, for instance, that if you jog at 8 min per mile, there must be some instant when your speed is exactly $\frac{1}{8}$ mi per minute—assuming, as is only natural in a course in calculus, that your elapsed time is a continuous function of the distance covered. This principle is very intuitive and was recognized before calculus was invented: Galileo was aware of it in 1638, and thought that it had been known to Plato. On the other hand, there is a question with a much less intuitive answer that was noticed only recently (and, as happens more often than mathematicians like to admit, by a physicist). Suppose that you average 8 min/mi, must you cover some one continuous mile (such as a "measured mile" on a highway) in exactly 8 min? The answer is not intuitive at all: It depends on whether or not your total distance was an integral number of miles. More precisely, if you cover an integral number of miles, then you cover exactly one mile in some 8 min. However, if you cover a nonintegral number of miles, there is not necessarily any one continuous mile that you cover in 8 min.

To prove this, let x be the distance (in miles) covered at any point during your trip, and suppose that when you stop you have covered an integral number of n miles. Let $f(x)$ be the time (in minutes) that it took to cover the first x miles; we will suppose that f is a continuous function. If you averaged 8 min/mi, then $f(x) - 8x = 0$ when $x = 0$ and when $x = n$. Now suppose that you never did cover any consecutive mile in 8 min; in mathematical terms,

$$f(x + 1) - f(x) \neq 8$$

Because

$$f(x + 1) - f(x) - 8$$

is continuous and never 0, it must either always be positive, or else always negative; let us suppose the former. Write the corresponding facts for $x = 0, 1, \ldots, n$:

$$f(1) - f(0) > 8$$
$$f(2) - f(1) > 8$$
$$\vdots$$
$$f(n) - f(n - 1) > 8$$

If we add these inequalities we obtain

$$f(n) - f(0) > 8n$$

But we started with the assumption that $f(n) = 8n$ and $f(0) = 0$, so assuming that $f(x + 1) - f(x)$ is never 8 leads to a contradiction.

It is somewhat harder to show that only integral values of n will work. Suppose you jog so that your time to cover x miles is

$$J(x) = k \sin^2 \frac{\pi x}{n} + 8x$$

where n is *not* an integer and k is a small number. This is a legitimate assumption, because J is an increasing function (as a time has to be), if k is small enough. To be sure of this, we calculate $J'(x)$—and here we actually have to use some calculus (or have a calculator that will do it for us). We find

$$J'(x) = \frac{k\pi}{n} \sin \frac{2\pi x}{n} + 8$$

If k is small enough $\left(k < \frac{8n}{\pi}\right)$, then $J'(x) > 0$. This shows not only that J increases but also that

$$J(x + 1) - J(x)$$

cannot be eight. Because $J(x + 1) - J(x)$ is never negative, if you jog so that your time is $J(x)$ you will never even cover a whole mile in less than 8 min.

Mathematical Essays

Use a library or references other than this textbook to research the information necessary to answer questions in Problems 1–11.

1. **HISTORICAL QUEST** The derivative is one of the great ideas of calculus. Write an essay of at least 500 words about some application of the derivative that is not discussed in this text.

2. **HISTORICAL QUEST** The concept of the integral is one of the great ideas of calculus. Write an essay of at least 500 words about the relationship of integration and differentiation as it relates to the history of calculus.

3. **HISTORICAL QUEST** Write a report on Georg Riemann. (See Historical Quest in Section 5.3.) Include the 1984 development in the solution of the Riemann hypotheses.

4. **HISTORICAL QUEST** As we saw in an Historical Quest in Section 5.8, the mathematician Seki Kōwa was doing a form of integration at about the same time that Newton and Leibniz were inventing the calculus. Write a paper on the history of calculus from the Eastern viewpoint.

5. "Lucy," the famous prehuman whose skeleton was discovered in East Africa, has been found to be approximately 3.8 million years old. About what percentage of ^{14}C would you expect to find if you tried to "date" Lucy by the usual carbon dating procedure? The answer you get to this question illustrates why it is reasonable to use carbon dating only on more recent artifacts, usually less than 50,000 years old (roughly the time since the last major ice age). Read an article on alternative dating procedures such as potassium–argon and rubidium–strontium dating. Write a paper comparing and contrasting such methods.

6. In the guest essay it was assumed that the time to cover x miles is

$$J(x) = k \sin^2 \frac{\pi x}{n} + 8x$$

Suppose that $n = 5$. What choices for k seem reasonable?

7. Suppose that the time to cover x miles is given by

$$J(x) = \sin^2 \frac{\pi x}{5} + 8x$$

Graph this function on $[0, 8]$.

8. Use calculus to find how small k from the guest essay needs to be in the expression

$$J(x) = k \sin^2 \frac{\pi x}{n} + 8x$$

so that $J'(x) > 0$.

9. a. Find a number $x, x \neq 0$, that, when divided by 2, gives a display of 0 on your calculator. Write a paper describing your work as well as the processes of your calculator.

   b. Calculate $\sqrt{2}$ using a calculator. Next, repeatedly subtract the integer part of the displayed number and multiply the result by 10. Describe the outcome, and then devise a method for finding $\sqrt{2}$ using calculus. Write a paper comparing these answers.

10. **Book Report** Eli Maor, a native of Israel, has a long-standing interest in the relations between mathematics and the arts. Read the fascinating book *To Infinity and Beyond, A Cultural History of the Infinite* (Boston: Birkhäuser, 1987), and prepare a book report.

11. **Book Report** In Section 3.1, we told the story of Fermat's last theorem. The recent book *Fermat's Enigma* by Simon Singh (New York: Walker & Company, 1997) tells the story of how Wiles solved the greatest mathematical puzzle of our age: he locked himself in a room and emerged seven years later. Read this book and prepare a book report.

12. Make up a word problem involving an application of the integral. Send your problem to:

    Bradley and Smith
    Prentice Hall Publishing Company
    1 Lake Street
    Upper Saddle River, NJ 07458

    The best one submitted will appear in the next edition (along with credit to the problem-poser).

6 Additional Applications of the Integral

PREVIEW

In Chapter 5, we found that area and average value can be expressed in terms of the definite integral. The goal of this chapter is to consider various other applications of integration such as computing volume, arc length, surface area, work, hydrostatic force, and centroids of planar regions.

PERSPECTIVE

What is the volume of the material that remains when a hole of radius of $0.5R$ is drilled into a sphere of radius R? The "center" of a rectangle is the point where its diagonals intersect, but what is the center of a quarter circle? If a reservoir is filled to the top of a dam shaped like a parabola, what is the total force of the water on the face of the dam? If a worker is carrying a bag of sand up a ladder and the bag is leaking sand through a hole in such a way that all the sand is gone when the worker reaches the top, how much work is done by the worker? These questions are typical of the kind we shall answer in this chapter.

6.1 Area Between Two Curves

area between curves, area by vertical strips, area by horizontal strips, two applications to economics ∎

AREA BETWEEN CURVES

In some practical problems, you may have to compute the area between two curves. Suppose f and g are functions such that $f(x) \geq g(x)$ on the interval $[a, b]$, as shown in Figure 6.1. We do not insist that both f and g be nonnegative functions, but we begin by showing that case in Figure 6.1.

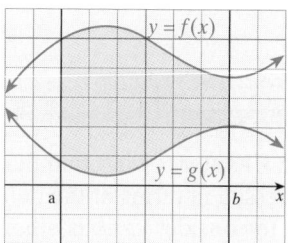 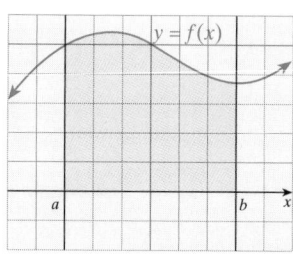 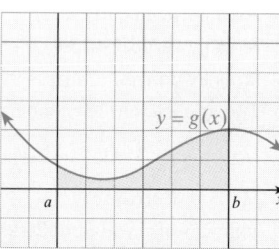

Area between $y = f(x)$ and $y = g(x)$ is equal to the area under $y = f(x)$ minus the area under $y = g(x)$

∎ **FIGURE 6.1**   Area between two curves

To find the area of the region R between the curves from $x = a$ to $x = b$, we subtract the area between the lower curve $y = g(x)$ and the x-axis from the area between the upper curve $y = f(x)$ and the x-axis; that is,

$$\text{AREA OF } R = \int_a^b f(x)\,dx - \int_a^b g(x)\,dx = \int_a^b [f(x) - g(x)]\,dx$$

This formula seems obvious in the situation where both $f(x) \geq 0$ and $g(x) \geq 0$, as shown in Figure 6.1. However, the following derivation requires only that f and g be continuous and satisfy $f(x) \geq g(x)$ on the interval $[a, b]$. We wish to find the area between the curves $y = f(x)$ and $y = g(x)$ on this interval. Choose a partition $\{x_0, x_1, x_2, \ldots, x_n\}$ of the interval $[a, b]$ and a representative number x_k^* from each subinterval $[x_{k-1}, x_k]$. Next, for each index k, with $k = 1, 2, \ldots, n$, construct a rectangle of width

$$\Delta x_k = x_k - x_{k-1}$$

and height

$$f(x_k^*) - g(x_k^*)$$

equal to the vertical distance between the two curves at $x = x_k^*$. A typical approximating rectangle is shown in Figure 6.2b. We refer to this approximating rectangle as a **vertical strip.**

The representative rectangle has area

$$[f(x_k^*) - g(x_k^*)]\Delta x_k$$

and the total area between the curves $y = f(x)$ and $y = g(x)$ can be estimated by the Riemann sum

$$\sum_{k=1}^{n} [f(x_k^*) - g(x_k^*)]\Delta x_k$$

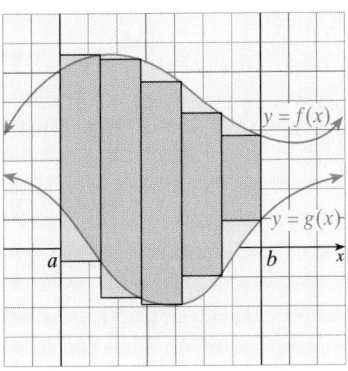
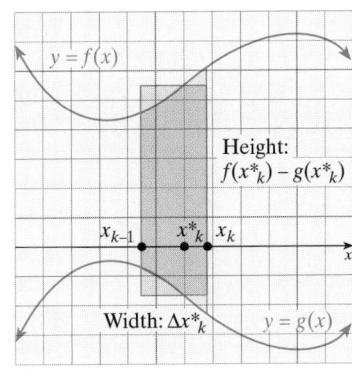

a. Approximating rectangles   **b.** A typical vertical strip

■ **FIGURE 6.2**   Using Riemann sums to find the area between two curves

It is reasonable to expect this estimate to improve if we increase the number of subdivision points in the partition P in such a way that the norm $\|P\|$ approaches 0. Thus, the region between the two curves has area

$$A = \lim_{\|P\| \to 0} \sum_{k=1}^{n} [f(x_k^*) - g(x_k^*)] \Delta x_k$$

which we recognize as the integral of the function $f(x) - g(x)$ on the interval $[a, b]$. These observations may be used to define the area between the curves.

Area Between Two Curves

If f and g are continuous and satisfy $f(x) \geq g(x)$ on the closed interval $[a, b]$, then the **area between the two curves** $y = f(x)$ and $y = g(x)$ is given by

$$A = \int_a^b [f(x) - g(x)] \, dx$$

■ *What This Says:*   To find the area between two curves on a given closed interval $[a, b]$ use the formula

$$A = \int_a^b [\text{TOP CURVE} - \text{BOTTOM CURVE}] \, dx$$

It is no longer necessary to require either curve to be above the x-axis. In fact, we will see later in this section that the curves might cross somewhere in the domain so that one curve is on top for part of the interval and the other curve is on top for the rest.

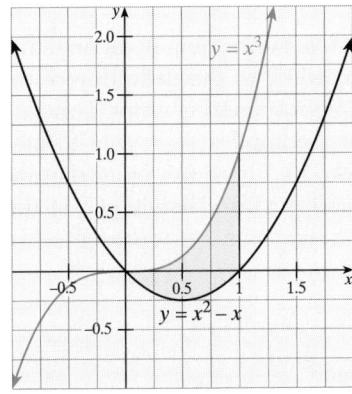

■ **FIGURE 6.3**   The area between the curves $y = x^3$ and $y = x^2 - x$ on $[0, 1]$

EXAMPLE 1   *Area between two curves*

Find the area of the region between the curves $y = x^3$ and $y = x^2 - x$ on the interval $[0, 1]$.

Solution   The region is shown in Figure 6.3. We need to know which curve is the *top curve* on $[0, 1]$. Solve

$$x^3 = x^2 - x \quad \text{or} \quad x(x^2 - x + 1) = 0$$

The only real root is $x = 0$ ($x^2 - x + 1 = 0$ has no real roots). Thus, the same curve is on top throughout the interval $[0, 1]$. To see which curve is on top, take some representative value, such as $x = 0.5$, and note that because $(0.5)^3 > 0.5^2 - 0.5$, the curve $y = x^3$ must be above $y = x^2 - x$. Thus, the required area is given by

$$A = \int_0^1 [\underbrace{x^3}_{\substack{\text{Top} \\ \text{curve}}} - \underbrace{(x^2 - x)}_{\substack{\text{Bottom} \\ \text{curve}}}]\, dx = (\tfrac{1}{4}x^4 - \tfrac{1}{3}x^3 + \tfrac{1}{2}x^2)\big|_0^1 = \tfrac{5}{12} \qquad \blacksquare$$

In Section 5.1, we defined area for a continuous function f with the restriction that $f(x) \geq 0$. Example 1 shows us that when considering the area between two curves, we need to be concerned no longer with the nonnegative restriction but only with whether $f(x) \geq g(x)$. We now use this idea to find the area for a function that is negative.

EXAMPLE 2	*Area with a negative function*

Find the area of the region bounded by the curve $y = e^{2x} - 3e^x + 2$ and the x-axis.

Solution   We find the points of intersection of the curve and the x-axis:

$$e^{2x} - 3e^x + 2 = 0 \qquad \text{The curve intersects the}$$
$$\text{x-axis where $y = 0$.}$$

$$(e^x - 1)(e^x - 2) = 0 \qquad \text{so that} \quad x = 0, \ln 2$$

The graph of $f(x) = e^{2x} - 3e^x + 2$ is shown in Figure 6.4. We see that on the interval $[0, \ln 2]$, $f(x) \leq 0$, but we can find the area of the given region by considering the area between the curves defined by equations $y = 0$ and $y = e^{2x} - 3e^x + 2$:

$$A = \int_0^{\ln 2} [\underbrace{0}_{\substack{\text{Top} \\ \text{curve}}} - \underbrace{(e^{2x} - 3e^x + 2)}_{\substack{\text{Bottom} \\ \text{curve}}}]\, dx = \left(-\frac{e^{2x}}{2} + 3e^x - 2x\right)\Big|_0^{\ln 2}$$

$$= \tfrac{3}{2} - 2\ln 2 \approx 0.114 \qquad \blacksquare$$

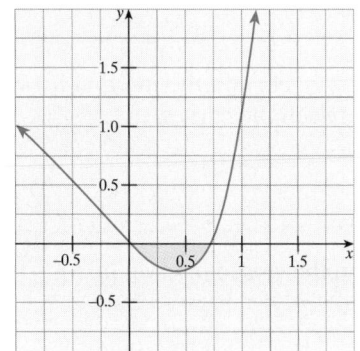

■ **FIGURE 6.4**   Area of region

AREA BY VERTICAL STRIPS

Although the only mathematically correct way to establish a formula involving integrals is to form Riemann sums and take the limit according to the definition of the definite integral, we can simulate this procedure with approximating strips. This simplification is especially useful for finding the area of a complicated region formed when two curves intersect one or more times, as shown in Figure 6.5. Note that the vertical strip has height $f(x) - g(x)$ if $y = f(x)$ is above $y = g(x)$, and height $g(x) - f(x)$ if $y = g(x)$ is above $y = f(x)$. In either case, the height can be represented by $|f(x) - g(x)|$, and the area of the vertical strip is

$$\Delta A = |f(x) - g(x)|\,\Delta x = |f(x) - g(x)|\,dx$$

Thus, we have a new integration formula for area; namely,

$$A = \int_a^b |f(x) - g(x)|\, dx$$

WARNING ➤ You cannot use the formula $A = \int [f - g]\, dx$ directly here because the hypothesis $f \geq g$ is not satisfied. In order to use $A = \int |f - g|\, dx$ over the entire interval, you must remember that $|f - g|$ might be $f - g$ over part of the interval and $g - f$ over another part (see Figure 6.5). Make sure you check to see which curve is on top. Because the curves cross in Example 3, the interval must be subdivided accordingly. ←

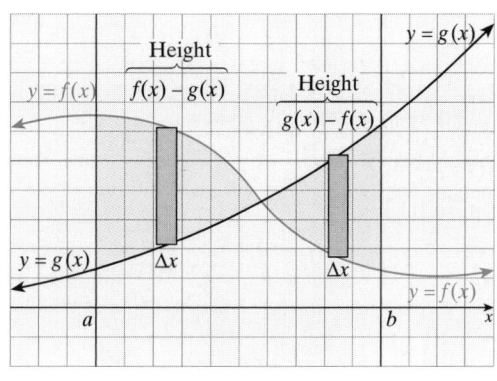
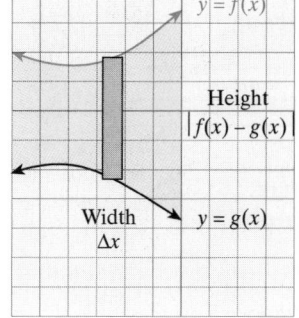

a. Approximation by strips   **b.** A typical vertical strip

■ **FIGURE 6.5**   Area by vertical strips

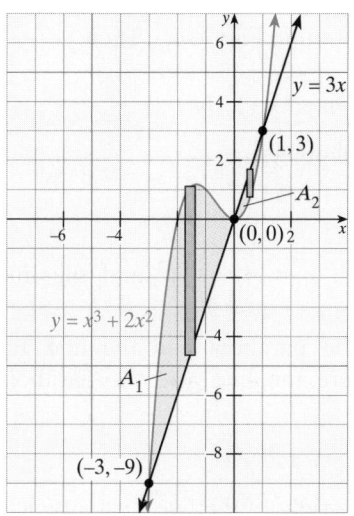

■ **FIGURE 6.6**   The area between the curve $y = x^3 + 2x^2$ and the line $y = 3x$

EXAMPLE 3   *Area between intersecting curves*

Find the area of the region bounded by the line $y = 3x$ and the curve $y = x^3 + 2x^2$.

Solution   The region between the curve and the line is the shaded portion of Figure 6.6. Part of the process of graphing these curves is to find which is the top curve and which is the bottom. To do this we need to find where the curves intersect:

$$x^3 + 2x^2 = 3x \quad \text{or} \quad x(x + 3)(x - 1) = 0$$

The points of intersection occur at $x = -3, 0,$ and 1. In the subinterval $[-3, 0]$, labeled A_1 in Figure 6.6, the curve $y = x^3 + 2x^2$ is on top (test a typical point in the subinterval, such as $x = -1$), and on $[0, 1]$, the region labeled A_2, curve $y = 3x$ is on top. The representative vertical strips are shown in Figure 6.6, and the area between the curve and the line is given by the sum

$$A = \int_{-3}^{0} [(x^3 + 2x^2) - (3x)]\, dx + \int_{0}^{1} [(3x) - (x^3 + 2x^2)]\, dx$$

$$= \left(\tfrac{1}{4}x^4 + \tfrac{2}{3}x^3 - \tfrac{3}{2}x^2\right)\Big|_{-3}^{0} + \left(\tfrac{3}{2}x^2 - \tfrac{1}{4}x^4 - \tfrac{2}{3}x^3\right)\Big|_{0}^{1}$$

$$= 0 - \left(\tfrac{81}{4} - \tfrac{54}{3} - \tfrac{27}{2}\right) + \left(\tfrac{3}{2} - \tfrac{1}{4} - \tfrac{2}{3}\right) - 0$$

$$= \tfrac{71}{6} \text{ (or 11.83333333 by calculator)}$$ ■

AREA BY HORIZONTAL STRIPS

For many regions, it is easier to form horizontal strips rather than vertical strips. The procedure for horizontal strips duplicates the procedure for vertical strips. If we want to find the area between two curves of the form $x = F(y)$ and $x = G(y)$ on the interval $[c, d]$, we form horizontal strips. Such a region is shown in Figure 6.7, together with a typical horizontal approximating rectangle of width Δy, which we refer to as a **horizontal strip.**

Note that regardless of which curve is "ahead" or "behind," the horizontal strip has length $|F(y) - G(y)|$ and area

$$\Delta A = |F(y) - G(y)|\,\Delta y$$

However, in practice, you must make sure to find the points of intersection of the curves and divide the integrals so that in each region one curve is the *leading curve*

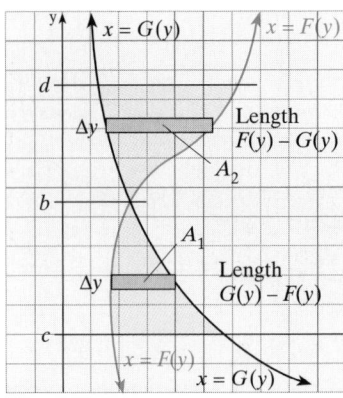
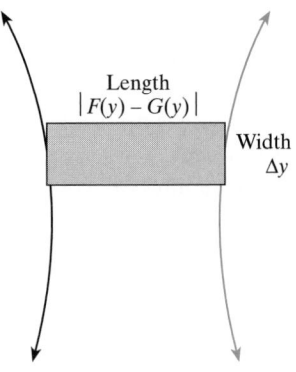

a. Approximation by horizontal strips of width Δy **b.** A typical horizontal strip

■ **FIGURE 6.7**   Area by horizontal strips

("right curve") and the other is the *trailing curve* ("left curve"). Suppose the curves intersect where $y = b$ for b on the interval $[c, d]$, as shown in Figure 6.7. Then

$$A = \int_c^b \underbrace{[G(y) - F(y)]}_{G \text{ ahead of } F}\, dy + \int_b^d \underbrace{[F(y) - G(y)]}_{F \text{ ahead of } G}\, dy$$

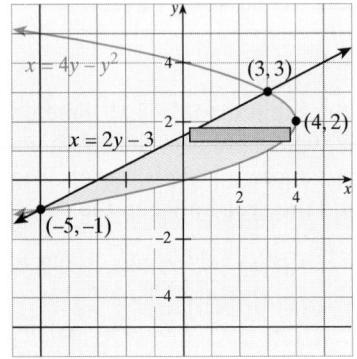

■ **FIGURE 6.8**   The area between the curves $x = 4y - y^2$ and $x = 2y - 3$ using horizontal strips

EXAMPLE 4   *Area by horizontal strips*

Find the area of the region between the parabola $x = 4y - y^2$ and the line $x = 2y - 3$.

Solution   Figure 6.8 shows the region between the parabola and the line, together with a typical horizontal strip. To find where the line and the parabola intersect, solve

$$4y - y^2 = 2y - 3 \quad \text{to obtain} \quad y = -1 \text{ and } y = 3$$

Throughout the interval $[-1, 3]$, the parabola is to the right of the line (test a typical point between -1 and 3, such as $y = 0$). Thus, the horizontal strip has area

$$\Delta A = [\underbrace{(4y - y^2)}_{\text{Right curve}} - \underbrace{(2y - 3)}_{\text{Left curve}}]\Delta y$$

and the area between the parabola and the line is given by

$$A = \int_{-1}^{3} [(4y - y^2) - (2y - 3)]\, dy = \int_{-1}^{3} (3 + 2y - y^2)\, dy$$
$$= (3y + y^2 - \tfrac{1}{3}y^3)\big|_{-1}^{3} = (9 + 9 - 9) - (-3 + 1 + \tfrac{1}{3}) = 10\tfrac{2}{3} \qquad ■$$

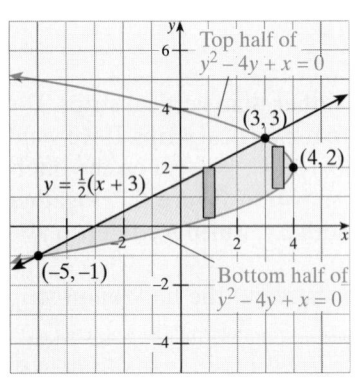

■ **FIGURE 6.9**   The area between the curves $x = 4y - y^2$ and $x = 2y - 3$ using vertical strips

In Example 4, the area can also be found by using vertical strips, but the procedure is more complicated. Note in Figure 6.9 that on the interval $[-5, 3]$, a representative vertical strip would extend from the bottom half of the parabola $y^2 - 4y + x = 0$ to the line $y = \frac{1}{2}(x + 3)$, whereas on the interval $[3, 4]$, a typical vertical strip would extend from the bottom half of the parabola $y^2 - 4y + x = 0$ to the top half. Thus, the area is given by the sum of two integrals. It can be shown that the computation of area by vertical strips gives the same result as that found by horizontal strips in Example 4.

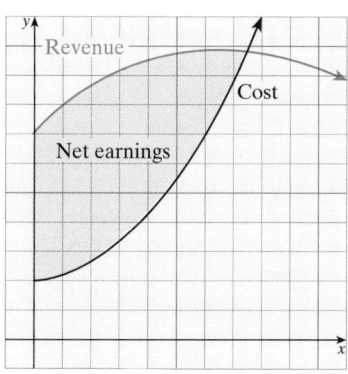

FIGURE 6.10 Net earnings is the difference between total revenue and total cost.

TWO APPLICATIONS TO ECONOMICS

Next, we shall see how definite integration can be used in economics to compute *net earnings* generated on time and a special quantity called *consumer's surplus.*

Net Earnings   The net earnings generated by an industrial machine for a period of time is the difference between the total revenue generated by the machine and the total cost of operating and servicing the machine, as shown in Figure 6.10.

In the following example, the net earnings of a machine are calculated as a definite integral and interpreted as the area between two curves.

> **EXAMPLE 5**   *Modeling Problem: Profitability of a piece of equipment*

Suppose that a piece of equipment is purchased in 1990 and will generate revenue of $R(x) = 5,000 - 20x^2$ dollars per year for x years after 1990. At the same time, the cost of maintaining and operating the equipment is $C(x) = 2,000 + 10x^2$ dollars.
a. For how many years will the use of the equipment remain profitable?
b. What are the net earnings generated by the machine during its period of profitability?

Solution   The cost and revenue curves are sketched in Figure 6.11.
a. Note that the revenue curve is above the cost curve until they cross, so the use of the equipment will be profitable as long as the revenue exceeds the cost—that is, until

$$5,000 - 20x^2 = 2,000 + 10x^2$$
$$30x^2 = 3,000$$
$$x = \pm 10 \quad (-10 \text{ is not in the domain})$$

The equipment will be profitable for 10 yr.

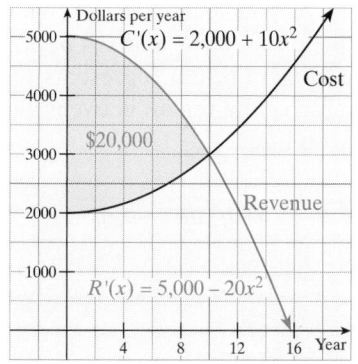

FIGURE 6.11 The net earnings from a piece of equipment

b. The difference $P(x) = R(x) - C(x)$ represents the net earnings of the machine after x years after 1990. To obtain the *total* net earnings, we integrate over the period of profitability—that is, from $x = 0$ to $x = 10$:

$$\text{NET EARNINGS} = \int_0^{10} [R(x) - C(x)] \, dx$$
$$= \int_0^{10} [(5,000 - 20x^2) - (2,000 + 10x^2)] \, dx$$
$$= \int_0^{10} (3,000 - 30x^2) \, dx$$
$$= (3,000x - 10x^3) \big|_0^{10} = 20,000$$

This piece of equipment should generate $20,000 of net earnings for the years 1990–2000 (its 10-year period of profitability).   ∎

Consumer's Surplus   In a competitive economy, the total amount that consumers actually spend on a commodity is usually less than the total amount they would have been willing to spend. The difference between the two amounts can be thought of as a savings realized by consumers and is known in economics as the **consumer's surplus.**

To get a better feel for the concept of consumer's surplus, consider an example of a couple who are willing to spend $500 for their first TV set, $300 for a second set, and only $50 for a third set. If the market price is $300, then the couple would buy only two sets and would spend a total of 2 × $300 = $600. This is less than the $500 + $300 = $800 that the couple would have been willing to spend to get the two sets. The savings of $800 − $600 = $200 is the couple's consumer's surplus. Consumer's surplus has a simple geometric interpretation, which is illustrated in Figure 6.12.

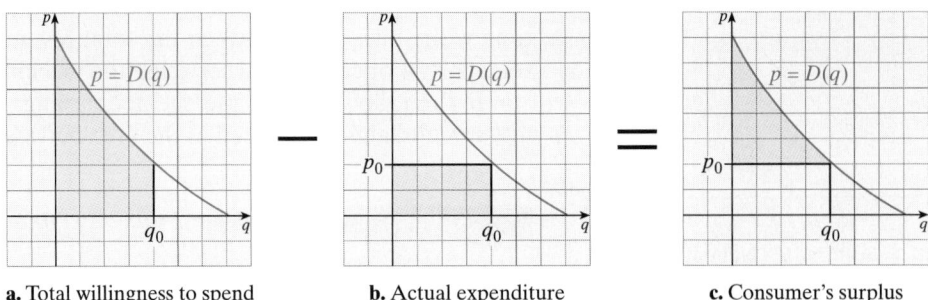

a. Total willingness to spend   **b.** Actual expenditure   **c.** Consumer's surplus

■ **FIGURE 6.12**   Geometric interpretation of consumer's surplus

Note that p and q denote the market price and the corresponding demand, respectively. Figure 6.12a shows the region under the demand curve $p = D(q)$ from $q = 0$ to $q = q_0$, and the area of this region represents the total amount that consumers are willing to spend to get q_0 units of the commodity. The rectangle in Figure 6.12b has an area of $p_0 q_0$ and represents the actual consumer expenditure for q_0 units at p_0 dollars per unit. The difference between these two areas (Figure 6.12c) represents the consumer's surplus. That is, consumer's surplus is the area of the region between the demand curve $p = D(q)$ and the horizontal line $p = p_0$, so that

$$\int_0^{q_0} [D(q) - p_0]\, dq = \int_0^{q_0} D(q)\, dq - \int_0^{q_0} p_0\, dq = \int_0^{q_0} D(q)\, dq - p_0 q_0$$

Consumer's Surplus

If q_0 units of a commodity are sold at a price of p_0 dollars per unit, and if $p = D(q)$ is the consumer's demand function for the commodity, then

$$\begin{bmatrix} \text{CONSUMER'S} \\ \text{SURPLUS} \end{bmatrix} = \begin{bmatrix} \text{TOTAL AMOUNT CONSUMERS} \\ \text{ARE WILLING TO} \\ \text{SPEND FOR } q_0 \text{ UNITS} \end{bmatrix} - \begin{bmatrix} \text{ACTUAL CONSUMER} \\ \text{EXPENDITURE FOR} \\ q_0 \text{ UNITS} \end{bmatrix}$$

$$= \int_0^{q_0} D(q)\, dq - p_0 q_0$$

EXAMPLE 6   *Modeling consumer's surplus*

Suppose the consumer's demand function for a certain commodity is $D(q) = 4(25 - q^2)$ dollars per unit. Find the consumer's surplus if the commodity is sold for $64 per unit.

Solution   First find the number of units that will be bought by solving the demand equation $p = D(q)$ for q when $p = \$64$:

$$64 = 4(25 - q^2) \quad \text{so that} \quad q = 3 \qquad \textit{Disregard} -3.$$

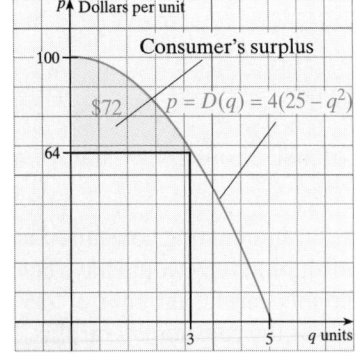

■ **FIGURE 6.13**   Consumer's demand curve $p = 4(25 - q^2)$ showing consumer's surplus

This says that $q_0 = 3$ units will be bought when the price is $p_0 = \$64$ per unit. The corresponding consumer's surplus is

$$\int_0^3 D(q)\, dq - 64(3) = \int_0^3 4(25 - q^2)\, dq - 192$$
$$= 4(25q - \tfrac{1}{3}q^3)\big|_0^3 - 192 = 72$$

The consumer's surplus is the shaded area in Figure 6.13. ∎

6.1 Problem Set

A *Sketch a representative vertical or horizontal strip and find the area of the given region in Problems 1–6.*

1. $y = -x^2 + 6x - 5$
$y = \tfrac{3}{2}x - \tfrac{3}{2}$

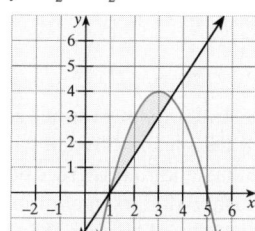

2. $x = y^2 - 6y$
$x = -y$

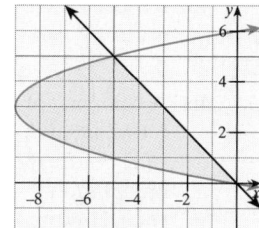

3. $x = y^2 - 5y$
$x = 0$

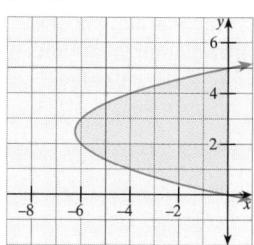

4. $y = x^2 - 8x$
$y = 0$

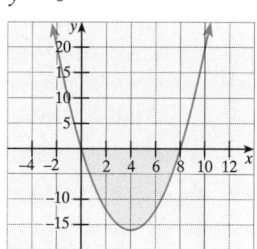

5. $y = \sin 2x$ on $[0, \pi]$
$y = 0$

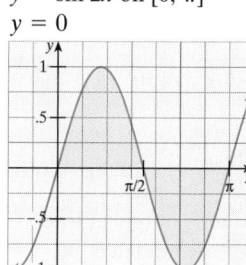

6. $y = (x - 1)^3$
$y = x - 1$

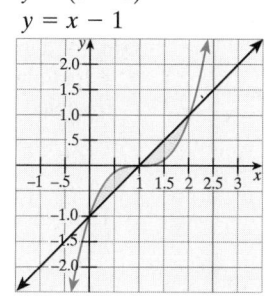

Sketch the area of the region between the given curves; then find the area of each region in Problems 7–25.

7. $y = x^2, y = x, x = -1, x = 1$

8. $y = x^3, y = x, x = -1, x = 1$

9. $y = x^2, y = x^3$ **10.** $y = x^2, y = \sqrt[3]{x}$

11. $y = x^2 - 1, x = -1, x = 2, y = 0$

12. $y = 4x^2 - 9, x = 3, x = 0, y = 0$

13. $y = x^4 - 3x^2, y = 6x^2$ **14.** $x = 8 - y^2, x = y^2$

15. $x = 2 - y^2, x = y$

16. $y = x^2 + 3x - 5, y = -x^2 + x + 7$

17. $y = 2x^3 + x^2 - x - 1, y = x^3 + 2x^2 + 5x - 1$

18. $y = \sin x, y = \cos x, x = 0, x = \frac{\pi}{4}$

19. $y = \sin x, y = \sin 2x, x = 0, x = \pi$

20. $y = |x|, y = x^2 - 6$

21. $y = |4x - 1|, y = x^2 - 5, x = 0, x = 4$

22. x-axis, $y = x^3 - 2x^2 - x + 2$

23. y-axis, $x = y^3 - 3y^2 - 4y + 12$

24. $y = e^x, y = \tfrac{1}{2}e^x + \tfrac{1}{2}, x = -2, x = 2$

25. $y = \dfrac{1}{\sqrt{1 - x^2}}, y = \dfrac{2}{x + 1}, y$-axis

Find the consumer's surplus for the given demand function defined by $D(q)$ at the point that corresponds to the sales level q as given in Problems 26–29.

26. $D(q) = 3.5 - 0.5q$
 a. $q_0 = 1$
 b. $q_0 = 1.5$

27. $D(q) = 2.5 - 1.5q$
 a. $q_0 = 1$
 b. $q_0 = 0$

28. $D(q) = 100 - 8q$
 a. $q_0 = 4$
 b. $q_0 = 10$

29. $D(q) = 150 - 6q$
 a. $q_0 = 5$
 b. $q_0 = 12$

B **30.** ■ **What Does This Say?** When finding the area between two curves, discuss reasons for deciding between vertical and horizontal strips.

31. ■ **What Does This Say?** What is consumer's surplus?

In Problems 32–35, find the consumer's surplus at the point of market equilibrium [the level of production where supply $S(q)$ equals demand $D(q)$].

Demand function	Supply function
32. $D(q) = 14 - q^2$	$S(q) = 2q^2 + 2$
33. $D(q) = 25 - q^2$	$S(q) = 5q^2 + 1$
34. $D(q) = 32 - 2q^2$	$S(q) = \frac{1}{3}q^2 + 2q + 5$
35. $D(q) = 27 - q^2$	$S(q) = \frac{1}{4}q^2 + \frac{1}{2}q + 5$

36. Find the area of the region that contains the origin and is bounded by the lines $2y = 11 - x$ and $y = 7x + 13$ and the curve $y = x^2 - 5$.

37. Show that the region defined by the inequalities $x^2 + y^2 \leq 8$, $x \geq y$, and $y \geq 0$ has area π.

38. Find the area of the region bounded by the curve $\sqrt{x} + \sqrt{y} = 1$ and the coordinate axes.

39. **MODELING PROBLEM**   Suppose an industrial machine that is x years old generates revenue
$$R(x) = 6{,}025 - 10x^2$$
dollars and costs
$$C(x) = 4{,}000 + 15x^2$$
dollars to operate and maintain.
   a. For how many years is it profitable to use this machine? [Recall that profit is $P(x) = R(x) - C(x)$.]
   b. What are the net earnings generated by the machine during its period of profitability? Interpret this amount as the area between two curves.

40. After t hours on the job, one factory worker is producing
$$Q_1(t) = 60 - 2(t - 1)^2$$
units per hour, and a second is producing
$$Q_2(t) = 50 - 5t$$
units per hour. If both arrive on the job at 8:00 A.M., how many more units (to the nearest unit) will the first worker have produced by noon than the second worker? Interpret your answer as the area between two curves.

41. Suppose that x years from now, one investment plan will be generating profit at the rate of
$$P_1'(x) = 100 + x^2$$
dollars per year, whereas a second plan will be generating profit at the rate of
$$P_2'(x) = 190 + 2x$$
dollars per year. Neither plan generates a profit in the beginning (when $x = 0$).
   a. For how many years will the second plan be more profitable?
   b. How much excess profit will you earn if you invest in the second plan instead of the first for the period of time in part **a**? Interpret the excess profit as the area between two curves.

42. Parts for a piece of heavy machinery are sold in units of 1,000. The demand for the parts (in dollars) is given by $p(x) = 110 - x$. The total cost is given by $C(x) = x^3 - 25x^2 + 2x + 30$ (dollars).

   a. For what value of x is the profit
$$P(x) = xp(x) - C(x)$$
   maximized?
   b. Find the consumer's surplus with respect to the price that corresponds to maximum profit.

43. Repeat Problem 42 with $p(x) = 124 - 2x$ and $C(x) = 2x^3 - 59x^2 + 4x + 76$.

44. Suppose when q units of a commodity are produced, the demand is $p(q) = 45 - q^2$ dollars per unit, and the marginal cost is
$$\frac{dC}{dq} = 6 + \frac{1}{4}q^2$$
Assume there is no overhead (so $C(0) = 0$).
   a. Find the total revenue and the marginal revenue.
   b. Find the value of q (to the nearest unit) that maximizes profit.
   c. Find the consumer's surplus at the value of q where profit is maximized. (Use the exact value of q.)

45. Repeat Problem 44 with
$$p(q) = \frac{1}{4}(10 - q)^2 \quad \text{and} \quad \frac{dC}{dq} = \frac{3}{4}q^2 + 5$$

46. A company plans to hire additional personnel. Suppose it is estimated that as x new people are hired, it will cost $C(x) = 0.2x$ thousand dollars and that these x people will bring in $R(x) = \sqrt{3x}$ thousand dollars in additional revenue. How many new people would be hired to maximize the profit? How much net revenue would the company gain by hiring these people?

47. Suppose the demand function for a certain commodity is $D(q) = 20 - 4q^2$, and that the marginal cost is $C'(q) = 2q + 6$, where q is the number of units produced. Find the consumer's surplus at the sales level q_0 where profit is maximized.

Technology Window

48. **MODELING PROBLEM**   Imagine a cylindrical fuel tank lying on its side (of length $L = 20$ ft); the ends are circular with radius b. You will soon be asked to compute the amount of fuel in the tank for a given level.
   a. Explain why the volume of the tank may be modeled by
$$V = 2L \int_{-b}^{b} \sqrt{b^2 - y^2}\, dy$$
   b. Explain why the volume of fuel at level h ($-b \leq h \leq b$) may be modeled by
$$V(h) = 2L \int_{-b}^{h} \sqrt{b^2 - y^2}\, dy$$
   c. Finally, for $b = 4$, numerically compute $V(h)$ for $h = -3, -2, \ldots, 4$. *Note:* $V(0)$ and $V(4)$ will serve as a check on your work.

C **49. MODELING PROBLEM**   The *supply function* represents the amount of a commodity that would be supplied to the market at a given price. If the market price is s_0, then those producers who would be willing to supply the commodity for a price less than s_0 realize a gain. For instance, if the market price is \$12 and the corresponding supply is only \$10, then those producers who are willing to supply the commodity at \$10 gain from the fact that the price is actually \$12. The **producer's surplus** is defined to be the total gain realized by all producers who are willing to supply the commodity for a price that is less than the market price.

Suppose the supply function for a certain commodity is $s(q)$, where q is the number of units of the commodity that will be supplied to the market when the price is s dollars per unit. Show that the producer's surplus with respect to the fixed price s_0 is modeled by

$$\int_0^{q_0} [s_0 - s(q)]\, dq$$

where q_0 is the sales level that corresponds to the price s_0; that is, $s_0 = s(q_0)$.

6.2   Volume by Disks and Washers

IN THIS SECTION   method of cross sections, volumes of revolution: disks and washers

b. Top of a single coin (a cross-section) has area B. The volume of the stack of coins is Bh.

a. Stack of coins with height h

■ **FIGURE 6.14**   The volume of a cylinder

METHOD OF CROSS SECTIONS

A number describing the three-dimensional extent of a set is called *volume* and is measured in cubic units. We say that a cube with side 1 has *unit* volume. Table 6.1 reviews some of the common solids whose volume formulas you may remember from precalculus.

A right cylinder can be regarded as a number of congruent disks stacked in a vertical pile, as shown in Figure 6.14a, and its volume can be computed by taking the product of the common cross-sectional area B and the height h of the stack. For example, a right circular cylinder of height h and radius r has circular cross sections of area $B = \pi r^2$, and its volume is $V = Bh = \pi r^2 h$.

A similar approach can be used to find the volume of other solids whose cross sections are known. However, when the cross-sectional areas are not constant, it may be necessary to use calculus.

■ **TABLE 6.1**
Volume Formulas

Name of Solid	Characteristics	Volume Formula	Picture
Cube	Side s	$V = s^3$	
Sphere	Radius r	$V = \dfrac{4}{3}\pi r^3$	
Right circular cone	Height h Circular base of radius r	$V = \dfrac{1}{3}\pi r^2 h$	

**■ TABLE 6.1
(continued)**

Name of Solid	Characteristics	Volume Formula	Picture
Regular tetrahedron	A pyramid formed by 4 equilateral triangles, each of side a	$V = \dfrac{1}{12}\sqrt{2}a^3$	
Cylinder	Height h Area of base B	$V = Bh$	
Circular cylinder	Height h Circular base of radius r	$V = \pi r^2 h$	

Let S be a solid and suppose that for $a \leq x \leq b$, the cross section of S that is perpendicular to the x-axis at x has area $A(x)$. Think of cutting the solid with a knife and removing a very thin slab whose face has area $A(x)$ and whose thickness is Δx, as shown in Figure 6.15.

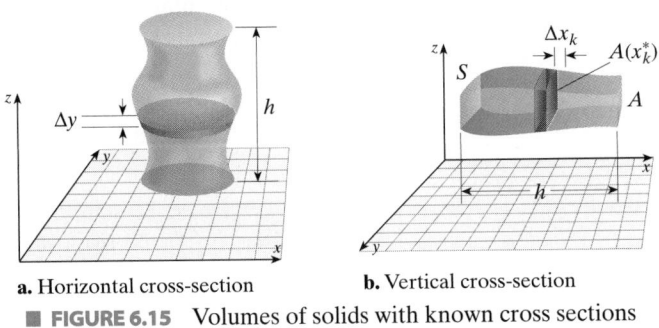

a. Horizontal cross-section   **b.** Vertical cross-section

■ FIGURE 6.15   Volumes of solids with known cross sections

To find the volume of S, we first take a partition

$$P = \{x_0, x_1, \ldots, x_n\} \quad \text{of the interval } [a, b]$$

and choose a representative number x_k^* in each subinterval $[x_{k-1}, x_k]$.
Next, we consider a cylindrical slab with width

$$\Delta x_k = x_k - x_{k-1}$$

and constant cross-sectional area $A(x_k^*)$. This slab has volume

$$\Delta V_k = A(x_k^*)\Delta x_k$$

and by adding up the volumes of all such slabs, we obtain an approximation to the volume of the solid S:

$$\Delta V = \sum_{k=1}^{n} A(x_k^*)\Delta x_k$$

The approximation improves as the number of partition points increases, and it is reasonable to *define* the volume V of the solid S as the limit of ΔV as the norm of the partition $\|P\|$ tends to 0. That is,

$$V = \lim_{\|P\|\to 0} \sum_{k=1}^{n} A(x_k^*)\Delta x_k$$

which we recognize as the definite integral $\int_a^b A(x)\,dx$. To summarize:

Volume of a Solid with Known Cross-Sectional Area

A solid S with cross-sectional area $A(x)$ at each point perpendicular to the x-axis on the interval $[a, b]$ has **volume**

$$V = \int_a^b A(x)\,dx$$

EXAMPLE 1   *Volume of a solid using square cross sections*

The base of a solid is the region in the xy-plane bounded by the y-axis and the lines $y = 1 - x$, $y = 2x + 5$, and $x = 3$. Each cross section perpendicular to the x-axis is a square. Find the volume of the solid.

Solution   The solid resembles a tapered brick, and it may be constructed by "gluing" together a number of thin slabs with square cross sections, like the one shown in Figure 6.16. We begin by drawing the base in two dimensions and then find the volume of the kth slice.

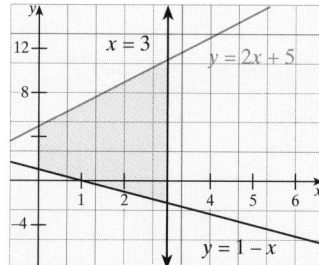 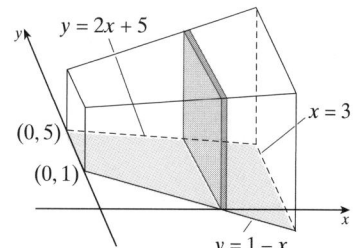 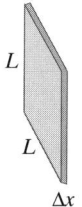

a. Two-dimensional graph of the given base

b. Three-dimensional solid

c. Cross-sectional representative element: $\Delta V = L^2\,\Delta x$

■ **FIGURE 6.16**   A solid with a square cross section

To model this construction mathematically, we subdivide the interval $[0, 3]$, form a vertical approximating rectangle on each resulting subinterval, and then construct a slab with square cross section on each approximating rectangle. If we choose the width of a typical slab to be Δx and the height to be L, the slab will have volume

ΔV, where

$$\begin{aligned}
\Delta V &= L^2 \Delta x \\
&= L^2(x) \Delta x \\
&= [(2x + 5) - (1 - x)]^2 \Delta x \\
&= (3x + 4)^2 \Delta x
\end{aligned}$$

The volume of the entire solid is obtained by integrating to "add up" all the volumes ΔV, and we find that the volume of the solid is

$$\begin{aligned}
V &= \int_0^3 (3x + 4)^2 \, dx \\
&= \int_0^3 (9x^2 + 24x + 16) \, dx \\
&= (3x^3 + 12x^2 + 16x)\Big|_0^3 = 237
\end{aligned}$$

DRAWING LESSON 1: SKETCHING A PRISM

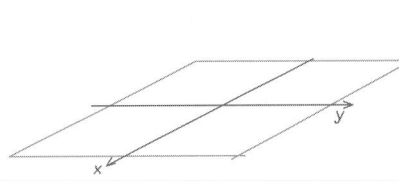

a. Draw the y-axis as a horizontal, and draw the x-axis pointing down and to the left. Lightly outline the xy-plane.

b. Sketch an axis perpendicular to the xy-plane. Use dashed segments for hidden parts.

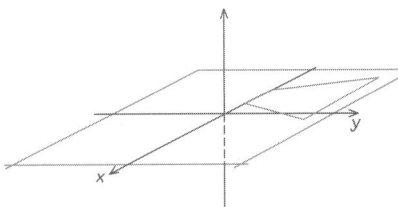

c. Sketch an outline of the base on the xy-plane.

d. Draw short vertical segments from the vertices of the base. These will be the edges of the prism.

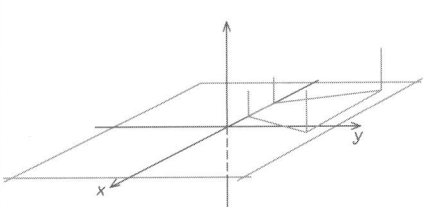

e. Connect the endpoints of these vertical segments to form the top base of the prism. Use your eraser to dash any segments that are now hidden.

f. Use colored pencils or a highlighter to shade the prism and the xy-plane.

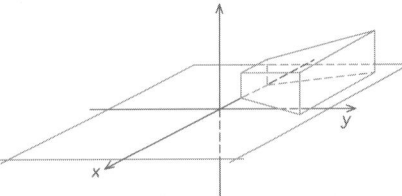

EXAMPLE 2   *Volume of a regular pyramid with a square base*

A regular pyramid has a square base of side L and has its apex located H units above the center of its base. Derive the formula $V = \frac{1}{3}HL^2$.

Solution   The pyramid is shown in Figure 6.17.

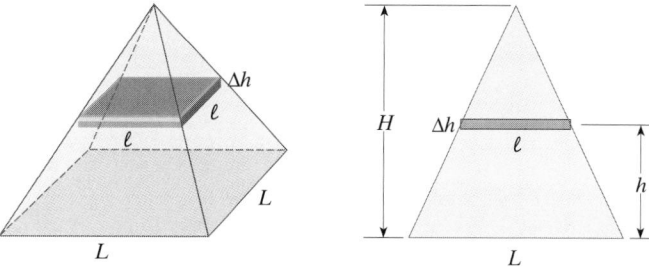

■ **FIGURE 6.17**   The volume of a pyramid

This pyramid can be constructed by stacking a number of thin square slabs. Suppose that a representative slab has side ℓ and thickness Δh, and that it is located h units above the base of the pyramid as shown in Figure 6.17. By creating a proportion from corresponding parts of similar triangles, we see that

$$\frac{\ell}{L} = \frac{H-h}{H} \quad \text{so that} \quad \ell = L\left(1 - \frac{h}{H}\right)$$

Therefore, the volume of the representative slab is

$$\Delta V = \ell^2 \Delta h$$
$$= L^2\left(1 - \frac{h}{H}\right)^2 \Delta h$$

To compute the volume V of the entire pyramid, we integrate with respect to h from the base of the pyramid ($h = 0$) to the apex ($h = H$). Thus,

$$V = \int_0^H L^2\left(1 - \frac{h}{H}\right)^2 dh = L^2 \int_0^H \left[1 - \frac{2}{H}h + \frac{1}{H^2}h^2\right] dh$$
$$= L^2\left(h - \frac{h^2}{H} + \frac{h^3}{3H^2}\right)\Bigg|_0^H = L^2\left(H - \frac{H^2}{H} + \frac{H^3}{3H^2}\right) = \frac{1}{3}HL^2 \qquad ■$$

Other volume formulas (including those in Table 6.1) may be found in a similar fashion (see Problems 60–61).

VOLUMES OF REVOLUTION: DISKS AND WASHERS

A **solid of revolution** is a solid figure S obtained by revolving a region R in the xy-plane about a line L (called the **axis of revolution**) that lies outside R. Note that such a solid S may be thought of as having circular cross sections in the direction perpendicular to L.

Suppose the function f is continuous and satisfies $f(x) \geq 0$ on the interval $[a, b]$, and suppose we wish to find the volume of the solid S generated when the region R under the curve $y = f(x)$ on $[a, b]$ is revolved about the x-axis. That is, *the axis of revolution is horizontal and it is a boundary of the region R.* Our strategy will be to form vertical strips and revolve them about the x-axis, generating what are called **disks** (that is, thin right circular cylinders) that approximate a portion of the solid of revolution S, as shown in Figure 6.18.

Now we can compute the total volume of S by using integration to sum the volumes of all the approximating disks. Recall that the formula for the volume of a

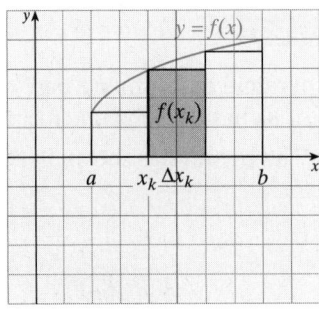 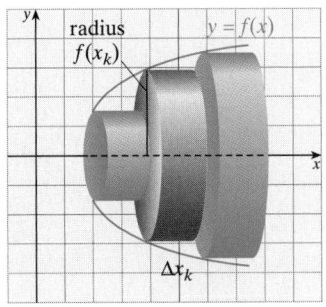

a. A representative vertical strip has height $f(x_k)$ and width Δx_k.

b. The representative disk is formed by revolving the representative strip about the x-axis. This disk has radius $f(x_k)$ and thickness Δx_k.

■ **FIGURE 6.18**   The disk method

cylinder of height h and cross-sectional area A is Ah. Figure 6.18a shows a typical vertical strip with height $f(x_k)$ and width Δx_k. You might notice that the width of the strip is the same as the thickness of the disk. The solid of revolution can be thought of as having cross sections perpendicular to the x-axis that are circular disks of volume

$$\Delta V(x) = \underbrace{\pi[f(x_k)]^2}\Delta x_k$$

Area of circular cross section

The total volume may be found by integration:

$$V = \int_a^b \overbrace{A(x)}^{\text{Area of base}} \; \overbrace{dx}^{\text{Thickness}} = \int_a^b \pi[f(x)]^2 \, dx$$

This procedure may be summarized as follows:

The Disk Method

The **disk method** is used to find a volume generated when a region R is revolved about an axis L that is *perpendicular* to a typical approximating strip in R. Suppose R is the region bounded by the curve $y = f(x)$, the x-axis, and the vertical lines $x = a$ and $x = b$. Then if R is revolved about the x-axis, it generates a solid with volume

$$V = \int_a^b \pi y^2 \, dx = \int_a^b \pi[f(x)]^2 \, dx$$

■   *What This Says:*   The following diagram may help you remember the key ideas behind the disk method.

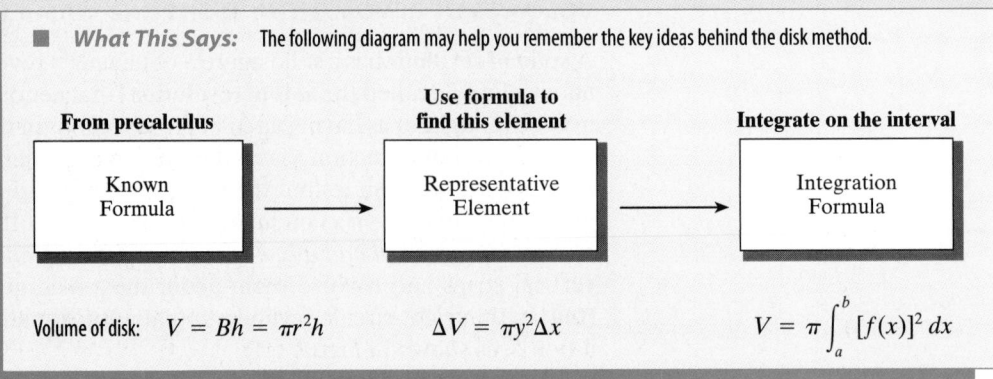

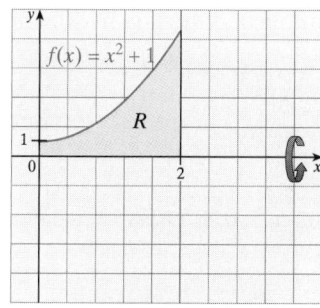

a. The region under $y = x^2 + 1$ on $[0, 2]$

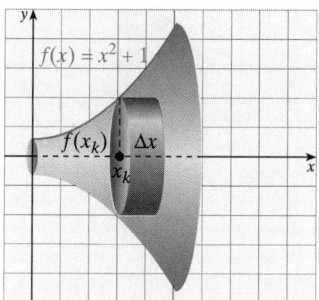

b. The solid of revolution

■ **FIGURE 6.19**   Volume of a solid of revolution: Disk method

EXAMPLE 3   *Volume of a solid of revolution: Disk method*

Find the volume of the solid S formed by revolving the region under the curve $y = x^2 + 1$ on the interval $[0, 2]$ about the x-axis.

Solution   The region is shown in Figure 6.19.

$$V = \pi \int_0^2 (x^2 + 1)^2 \, dx = \pi \int_0^2 (x^4 + 2x^2 + 1) \, dx$$

$$= \pi \left(\frac{1}{5}x^5 + \frac{2}{3}x^3 + x \right)\Big|_0^2 = \frac{206}{15}\pi \approx 43.14453911 \quad ■$$

With a small modification of the disk method, we can find the volume of a solid figure generated by revolving about the x-axis the region between two curves $y = f(x)$ and $y = g(x)$ where $f(x) \geq g(x)$ for $a \leq x \leq b$. When a typical vertical strip is revolved about the x-axis, a "washer" with cross-sectional area $\pi([f(x)]^2 - [g(x)]^2)$ is formed, as shown in Figure 6.20.

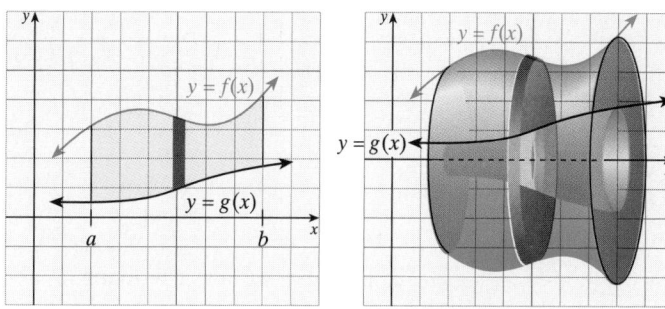

■ **FIGURE 6.20**   The washer method

The volume of the solid of revolution is found by the formula given in the following box.

The Washer Method

The **washer method** is used to find volume when a region between two curves is revolved about an external axis perpendicular to the approximating strip. In particular, suppose f and g are continuous functions on $[a, b]$ with $f(x) \geq g(x) \geq 0$. Then if R is the region bounded above by $y = f(x)$, below by $y = g(x)$, and on the sides by $x = a$ and $x = b$, the solid formed by revolving R about the x-axis has volume

$$V = \int_a^b \pi (\underbrace{[f(x)]^2}_{\text{Outer radius}} - \underbrace{[g(x)]^2}_{\text{Inner radius}}) \, dx$$

WARNING $f^2 - g^2 \neq (f - g)^2$

The disk method and the washer method also apply when the axis of revolution is a line other than the x-axis. In Example 4, we consider what happens when a particular region R is revolved not only about the x-axis, but also about other axes.

EXAMPLE 4   *Volume of a solid of revolution: Washer method*

Let R be the solid region bounded by the parabola $y = x^2$ and the line $y = x$. Find the volume of the solid generated when R is revolved about the
a. x-axis   b. y-axis   c. line $y = 2$

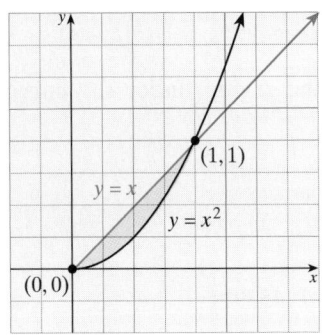

■ **FIGURE 6.21** The region R bounded by $y = x^2$ and $y = x$

Solution   First, find the points of intersection of the parabola and the line by solving the following system of equations:

$$\begin{cases} y = x^2 \\ y = x \end{cases}$$

This is equivalent to solving $x = x^2$, which has solution $x = 0, 1$. Draw the region R in the xy-plane as shown in Figure 6.21.

a. We form the solid by rotating the region R about the x-axis. Note that the line $y = x$ is always above the parabola on the interval $[0, 1]$, so when we form a washer to approximate the volume of revolution, the outer radius is $y = R(x) = x$ and the inner radius is $y = r(x) = x^2$. Thus, the required volume is

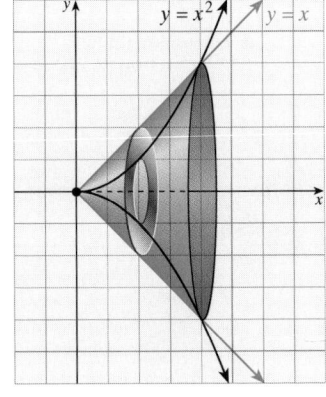

$$V = \pi \int_0^1 [x^2 - (x^2)^2]\, dx$$

$$= \pi \int_0^1 (x^2 - x^4)\, dx$$

$$= \pi\left(\frac{1}{3}x^3 - \frac{1}{5}x^5\right)\Big|_0^1 = \frac{2\pi}{15}$$

WARNING   For the disk method or the washer method, make sure that the approximating strips are perpendicular to the axis of revolution. ◄

b. Because we are revolving R about the y-axis, we use *horizontal* strips to approximate the solid of revolution, as shown at the right. Note that the parabola $x = \sqrt{y}$ is to the right of the line $x = y$ on the interval $[0, 1]$, so the approximating washer has outer radius $R = \sqrt{y}$ and inner radius $r = y$; thus,

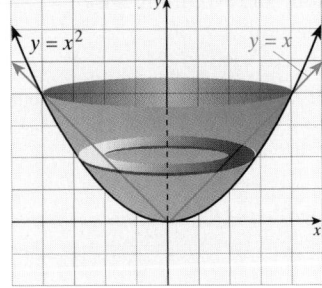

$$A = \pi \int_0^1 [(\sqrt{y})^2 - (y)^2]\, dy$$

$$= \pi \int_0^1 (y - y^2)\, dy$$

$$= \pi\left[\frac{y^2}{2} - \frac{y^3}{3}\right]_0^1 = \frac{\pi}{6}$$

c. The outer radius is $R = 2 - x^2$ and the inner radius is $r = 2 - x$, as shown at the right. The volume is

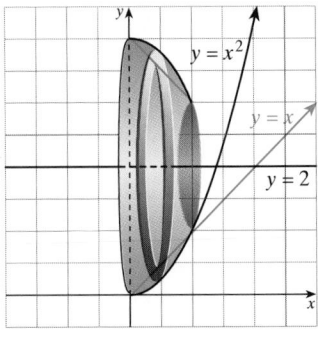

$$A = \pi \int_0^1 [(2 - x^2)^2 - (2 - x)^2]\, dx$$

$$= \pi \int_0^1 (x^4 - 5x^2 + 4x)\, dx$$

$$= \pi\left[\frac{x^5}{5} - \frac{5x^3}{3} + \frac{4x^2}{2}\right]_0^1 = \frac{8\pi}{15}$$

■

6.2   Problem Set

A *In Problems 1–6, sketch the given region and then find the volume of the solid whose base is the given region and which has the property that each cross section perpendicular to the x-axis is a* **square.**

1. the triangular region bounded by the coordinates axes and the line $y = 3 - x$

2. the region bounded by the x-axis and the semicircle $y = \sqrt{16 - x^2}$

3. the region bounded by the line $y = x + 1$ and the curve $y = x^2 - 2x + 3$

4. the region bounded above by $y = \sqrt{\sin x}$ and below by the x-axis on the interval $[0, \pi]$

5. the region bounded above by $y = \sqrt{\cos x}$ and below by the x-axis on the interval $\left[-\dfrac{\pi}{2}, \dfrac{\pi}{2} \right]$

6. the triangular region with vertices $(1, 1), (3, 5)$, and $(3, -2)$

In Problems 7–12, sketch the region and then find the volume of the solid whose base is the given region and which has the property that each cross section perpendicular to the x-axis is an **equilateral triangle.**

7. the region bounded by the circle $x^2 + y^2 = 9$

8. the region bounded by the curves $y = x^3$ and $y = x^2$

9. the region bounded by the y-axis, the parabola $y = x^2$, and the line $2x + y - 3 = 0$

10. the region bounded above by $y = \cos x$, below by $y = \sin x$, and on the left by the y-axis

11. the region bounded above by the curve $y = \tan x$ and below by the x-axis, on the interval $\left[0, \dfrac{\pi}{4} \right]$

12. the region bounded by the x-axis and the curve $y = e^x$ between $x = 1$ and $x = 3$

In Problems 13–20, draw a representative strip and set up an integral for the volume of the solid formed by revolving the given region:

   a. *about the x-axis*
   b. *about the y-axis*

Set up the integral only; DO NOT EVALUATE.

13. the region bounded by $x = y^2$ and $y = x^2$

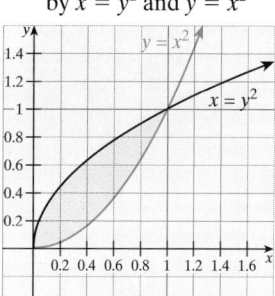

14. the region bounded by $y = \frac{1}{3}x$ and $x = y^2$

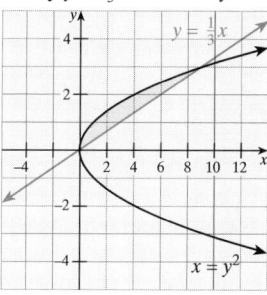

15. the region bounded by the lines $y = 1, x = 2$, and the curve $y = x^2 + 1$

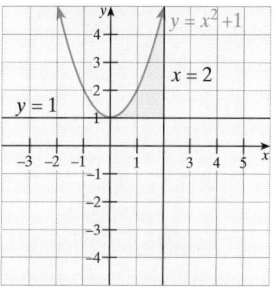

16. the region bounded by the curves $y = x^2$ and $y = -x^2 - 4x$

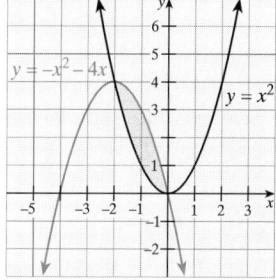

17. the region bounded by $y = 0.1x^2$ and $y = \ln x$

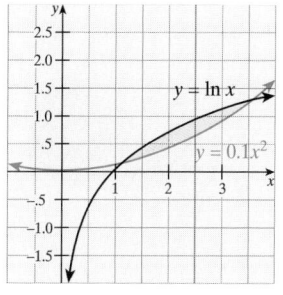

18. the region bounded by $y = \pi x/2$ and $y = \sin^{-1} x$

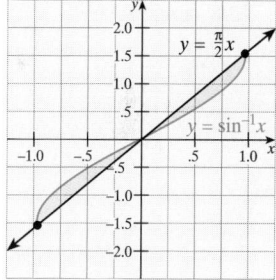

19. the region bounded by $y = e^x - 1$, $y = 2e^{-x}$, and $x = 0$

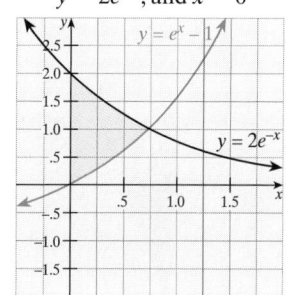

20. the region bounded by $y = \tan^{-1} x$ and $y = 0.1(e^x - 1)$

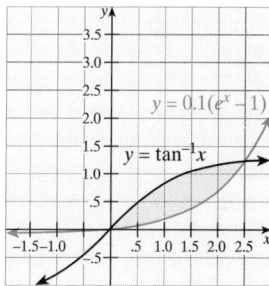

B *In Problems 21–24, find the volume of the solid whose base is bounded by the circle $x^2 + y^2 = 9$ with the indicated cross sections taken perpendicular to the x-axis.*

21. squares

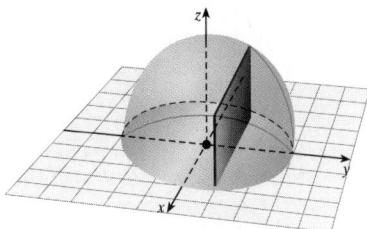

22. equilateral triangles

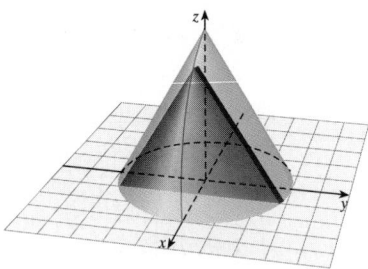

23. isosceles right triangles

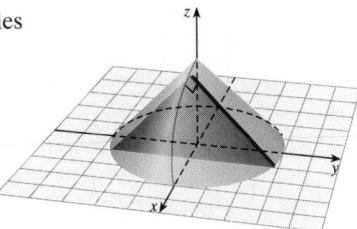

24. semicircles

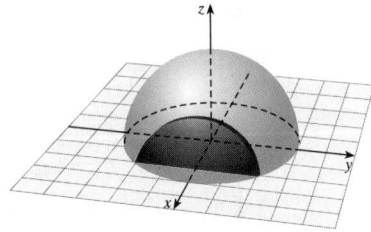

In Problems 25–28, find the volume of the solid whose base is bounded by the graphs of $y = x + 1$ and $y = x^2 - 1$ with the indicated cross sections taken perpendicular to the x-axis.

25. squares

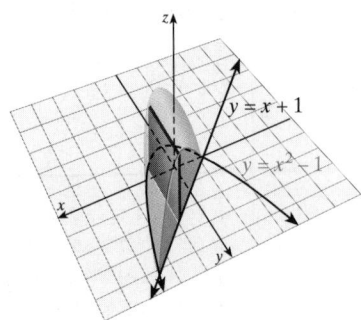

26. equilateral triangles

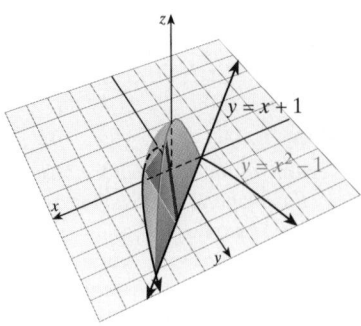

27. rectangles of height 1

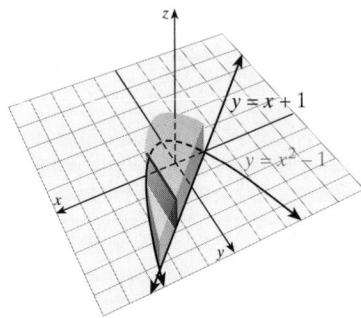

28. semicircles

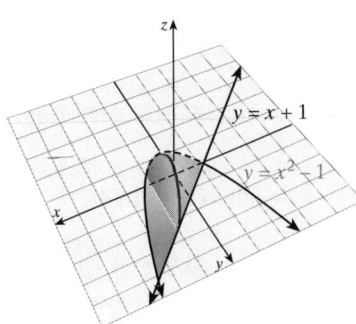

In Problems 29–38, find the volume of the solid formed when the region described is revolved about the x-axis.

29. the region under the curve $y = \sqrt{x}$ on the interval $0 \le x \le 1$

30. the region under the curve $y = \sqrt[3]{x}$ on the interval $0 \le x \le 8$

31. the region bounded by the lines $y = x, y = 2x$, and $x = 1$

32. the region bounded by the lines $x = 0, x = 1, y = x + 1$, and $y = x + 2$

33. the region under the curve $y = x^2 + x^3$ on the interval $0 \le x \le \pi$; approximate your answer to the nearest unit

34. the region bounded by the curves $y = 2 - x^2$ and $y = x^2$

35. the region bounded by the curves $y = x^2$ and $y = x^3$

36. the region under the curve $y = \sqrt{2} \sin x$ on the interval $0 \le x \le \pi$

37. the region between $y = \sin x$ and $y = \cos x$ on $0 \le x \le \pi/4$

38. the region bounded by the curves $y = e^x$ and $y = e^{-x}$ on $[0, 2]$.

In Problems 39–46, find the volume of the solid formed by revolving the given region about the y-axis.

39. the region bounded by the lines $y = 2x$, the y-axis, and $y = 1$

40. the region bounded by the curve $y = \sqrt{x}$, the y-axis, and the line $y = 1$

41. the region bounded by the curves $y = x^2$ and $y = x^3$

42. the region bounded by the lines $y = x$, $y = 2x$, and $y = 1$

43. the region bounded by the parabola $y = 1 - x^2$, the y-axis, and the positive x-axis

44. the region bounded by the parabolas $y = x^2$, $y = 1 - x^2$, and the y-axis

45. the region bounded by $y = \tan^{-1} x$, the y-axis, and $y = \frac{\pi}{4}$

46. the region bounded in the first quadrant by $y = \sin^{-1} x$, $y = \cos^{-1} x$, and the y-axis

In Problems 47–50, find the volume of the solid generated by revolving each region about the prescribed axis.

47. the region bounded by $y = x^2$ and $y = x^3$, about the line $y = -1$

48. the region bounded by $x = \sqrt{4 - y}$, $x = 0$, and $y = -1$, about the y-axis

49. the region bounded by $y = x^3$, $y = 12 - x^2$, and $y = 0$, about the line $x = -1$

50. the region bounded by $y = \sin^{-1} x$, $y = \cos^{-1} x$, and the x-axis, about the line $x = -1$

In Problems 51–53, find the volume V of the solid with the given information regarding its cross section.

51. The base of the solid is the hyperbola
$$\frac{x^2}{4} - \frac{y^2}{9} = 1$$
for $2 \le x \le 5$, and the cross sections perpendicular to the x-axis are squares.

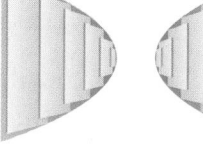

52. The base of the solid is an equilateral triangle, each side of which has length 4. The cross sections perpendicular to a given altitude of the triangle are squares.

53. The base of the solid is an isosceles right triangle whose legs are each 4 units long. Each cross section perpendicular to a side is a semicircle.

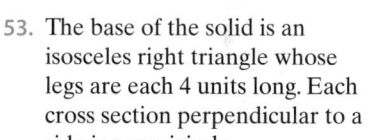

54. The great pyramid of Cheops is approximately 480 ft tall and 750 ft square at the base. Find the volume of this pyramid by using the cross section method.

55. When viewed from above, a swimming pool has the shape of the ellipse
$$\frac{x^2}{900} + \frac{y^2}{400} = 1$$
The cross sections of the pool perpendicular to the ground and parallel to the y-axis are squares. If the units are in feet, what is the volume of the pool?

Technology Window

56. Repeat Example 1 using a spreadsheet to find the volume.

57. Repeat Example 3 using a spreadsheet to find the volume.

58. Cross-sectional areas are measured at 1-foot intervals along the length of an irregularly shaped object, with the results listed in the following table (x in ft and A in ft^2):

x	0	1	2	3	4	5
A	1.12	1.09	1.05	1.03	0.99	1.01

x	6	7	8	9	10
A	0.98	0.99	0.96	0.93	0.91

Estimate the volume (correct to the nearest hundredth) by using the trapezoidal rule.

59. Cross-sectional areas are measured at 2-meter intervals along the length of an irregularly shaped object, with the results listed in the following table (x in meters and A in m^2):

x	0	2	4	6	8	10
A	1.12	1.09	1.05	1.03	0.99	1.01

x	12	14	16	18	20
A	0.98	0.99	0.96	0.93	0.91

Estimate the volume (correct to the nearest hundredth) by using Simpson's rule.

C 60. A hemisphere of radius r may be regarded as a solid whose base is the region bounded by the circle $x^2 + y^2 = r^2$ and with the property that each cross section perpendicular to the x-axis is a semicircle with a diameter in the base. Use this characterization and the method of cross sections to show that a sphere of radius r has volume $V = \frac{4}{3}\pi r^3$. *Note:* This is a formula given in Table 6.1.

61. Use the method of cross sections to show that the volume of a regular tetrahedron of side a is $\frac{1}{12}\sqrt{2}a^3$. *Note*: This is a formula given in Table 6.1.

62. Our old friend, Frank Kornercutter, conjectures that a tetrahedron (Problem 61) is nothing more than a solid figure with an equilateral triangular base and cross sections perpendicular to that base that are also equilateral triangles. Is Frank correct this time? Either prove that he is or show that his conjecture must be false.

6.3 Volume by Shells

the method of cylindrical shells, summary of methods for computing volume ■

THE METHOD OF CYLINDRICAL SHELLS

Sometimes it is easier (or even necessary) to compute a volume by taking the approximating strip parallel to the axis of rotation instead of perpendicular to the axis as in the disk and washer methods. Figure 6.22a shows a region R under the curve $y = f(x)$ on the interval $[a, b]$, together with a representative vertical strip. When this strip is revolved about the y-axis, it forms a solid called a **cylindrical shell.**

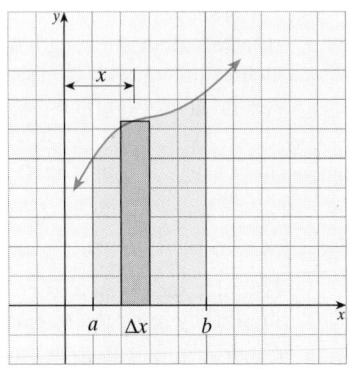

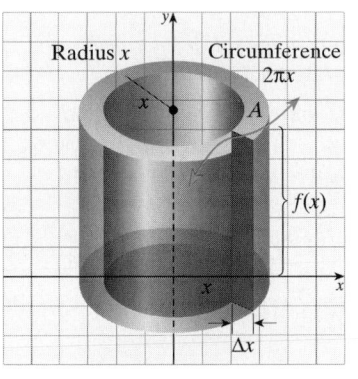

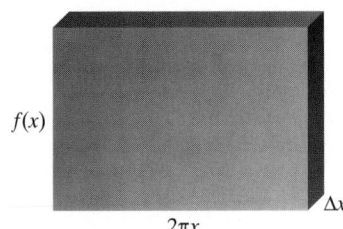

a. A vertical strip in the region R under the curve $y = f(x)$ on the interval $[a, b]$

b. When the strip is revolved about the y-axis, a shell is generated.

c. The unwrapped "flattened" shell has volume $\Delta V = 2\pi x f(x)\Delta x$.

■ **FIGURE 6.22**   Method of cylindrical shells

When the approximating strip is rotated about the y-axis, it generates a cylindrical shell of height $f(x)$ and thickness Δx, as shown in Figure 6.22b. Because the strip is x units from the axis of rotation and is assumed to be very thin, the cross section of the shell (perpendicular to the y-axis) will be a circle of radius x and circumference $2\pi x$. If we imagine the shell to be cut and flattened out, it is seen to be a rectangular slab of volume

$$\Delta V = \underbrace{2\pi x f(x)}_{\text{Area of the rectangular slab}} \cdot \overbrace{\Delta x}^{\text{Thickness}}$$

(See Figure 6.22c.) Thus, the total volume of the solid is given by the integral

$$V = \int_a^b 2\pi x f(x)\, dx$$

The approximating shells are shown in Figure 6.23 and the formula is repeated in the following box.

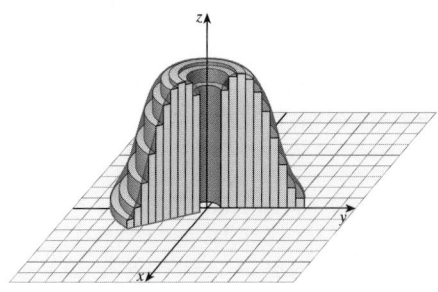

■ **FIGURE 6.23** Approximating a solid of revolution by cylindrical shells

Method of Cylindrical Shells

The **shell method** is used to find a volume generated when a region R is revolved about an axis L *parallel* to a typical approximating strip in R. In particular, if R is the region bounded by the curve $y = f(x)$, the x-axis, and the vertical lines $x = a$ and $x = b$ where $0 \le a \le b$, then the solid generated by revolving R about the y-axis has volume

$$V = \int_a^b 2\pi x f(x)\, dx$$

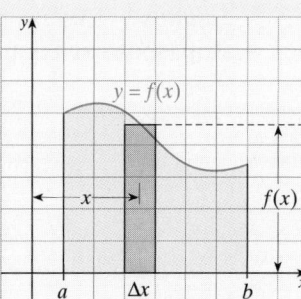

EXAMPLE 1   *Volume using cylindrical shells*

Find the volume of the solid formed by revolving the region bounded by the curve $y = x^{-2}$ and the x-axis for $1 \le x \le 2$ about:
a. the y-axis
b. the line $x = -1$

Solution

a. A typical vertical strip has height $f(x) = x^{-2}$ (Figure 6.24a). The volume is given by

$$V = 2\pi \int_1^2 x\left(\frac{1}{x^2}\right) dx$$

$$= 2\pi \int_1^2 x^{-1}\, dx$$

$$= 2\pi(\ln 2 - \ln 1) \approx 4.3552$$

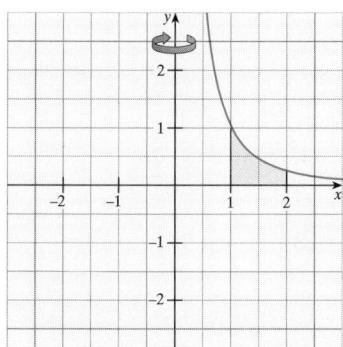

a. Rotation about the y-axis

■ **FIGURE 6.24** Volume by shells

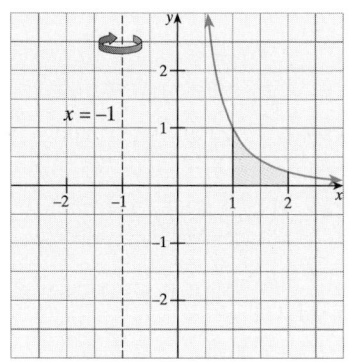

b. Rotation about $x = -1$

■ **FIGURE 6.24** Volume by shells

b. When the region is revolved about the line $x = -1$ (Figure 6.24b), the distance from the axis to the typical vertical strip is $(x + 1)$, so the volume is

$$V = 2\pi \int_1^2 (x + 1)\left(\frac{1}{x^2}\right) dx$$

$$= 2\pi \int_1^2 (x^{-1} + x^{-2}) \, dx$$

$$= 2\pi \left[\ln x - x^{-1}\right]\Big|_1^2$$

$$= 2\pi \left[\ln 2 + \frac{1}{2}\right]$$

$$\approx 7.4968$$

Example 1 could conceivably be done using the washer method, but sometimes it is not even possible to solve an equation of the form $y = f(x)$ for x. The next example illustrates such a situation.

EXAMPLE 2 *Volume of a solid of revolution using shell method*

Find the volume of the solid formed by revolving the region bounded by the graph of $y = x^3 + x^2 + 1$, the x-axis, $x = 1$, and $x = 3$ about the y-axis.

Solution This example illustrates why we need the shell method. A typical vertical strip (shown in Figure 6.25b) has height $f(x) = x^3 + x^2 + 1$ and width dx.

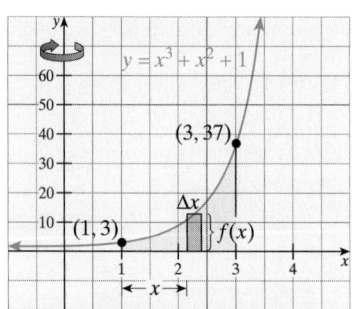

■ **FIGURE 6.25** A volume of revolution by the shell method

The volume, by the method of shells, is

$$V = 2\pi \int_1^3 x(x^3 + x^2 + 1) \, dx = 2\pi \int_1^3 (x^4 + x^3 + x) \, dx$$

$$= 2\pi \left(\frac{1}{5}x^5 + \frac{1}{4}x^4 + \frac{1}{2}x^2\right)\Big|_1^3 = 144.8\pi \approx 454.9$$

SUMMARY OF METHODS FOR COMPUTING VOLUME

Table 6.2 compares and contrasts the disk, washer, and shell methods for computing volume.

■ **TABLE 6.2**   **Volumes of Revolution for which the axis of revolution is either the *x*-axis or the *y*-axis**

Summary

a. Disk method: A representative rectangle is **perpendicular** to the axis of revolution. The axis of revolution is a boundary of the region.

b. Washer method: A representative rectangle is **perpendicular** to the axis of revolution. The axis of revolution is *not* part of the boundary.

| **Horizontal axis of revolution** | **Vertical axis of revolution** | **Horizontal axis of revolution** | **Vertical axis of revolution** |

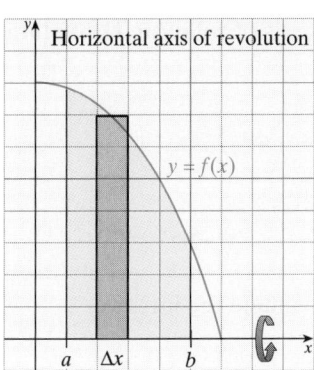 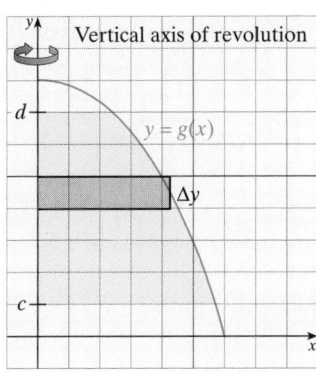 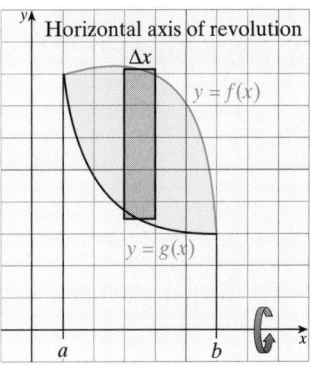 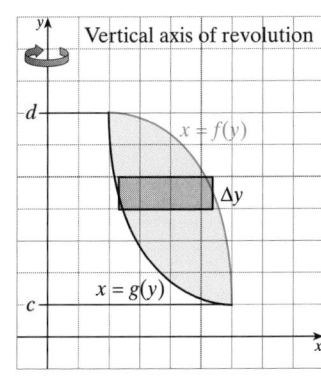

Width of rectangle is Δx.

$$V = \pi \int_a^b [f(x)]^2 \, \overbrace{dx}^{}$$

Length of rectangle

Width of rectangle is Δy.

$$V = \pi \int_c^d [g(y)]^2 \, \overbrace{dy}^{}$$

Length of rectangle

Top curve   Bottom curve
Width

$$V = \pi \int_a^b ([f(x)]^2 - [g(x)]^2) \, \overbrace{dx}^{}$$

Right curve   Left curve
Width

$$V = \pi \int_c^d ([f(y)]^2 - [g(y)]^2) \, \overbrace{dy}^{}$$

c. Shell method: A representative rectangle is **parallel** to the axis of revolution.

| **Horizontal axis of revolution** | **Vertical axis of revolution** |

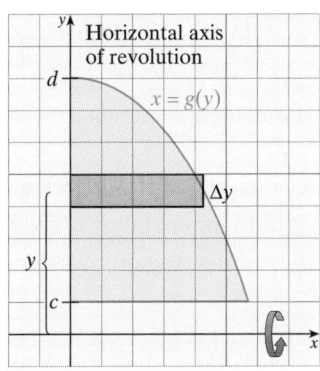 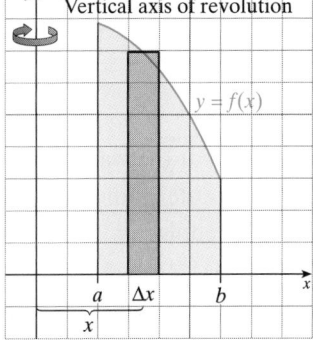

Width

$$V = 2\pi \int_c^d y \, [g(y)] \, \overbrace{dy}^{}$$

↑ Length
Distance to axis

Width

$$V = 2\pi \int_a^b x \, [f(x)] \, \overbrace{dx}^{}$$

↑ Length
Distance to axis

We conclude this section with an example.

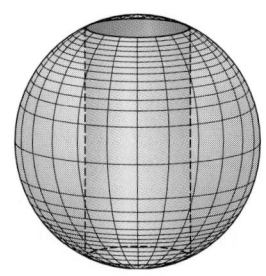

FIGURE 6.26 Drill a hole in sphere

| **EXAMPLE 3** | *Comparing the method of shells and washers* |

A hole of radius r is bored through the center of a solid sphere of radius R ($r < R$), as shown in Figure 6.26. Find the volume of the solid that remains by using:

a. cylindrical shells   b. washers

Solution

a. The required volume can be thought of as twice the volume when the shaded region of Figure 6.27 is <u>revolved</u> about the y-axis. If we use vertical strips, a typical strip has height $y = \sqrt{R^2 - x^2}$ and is x units from the axis of rotation. The volume is given by

$$V = 2\int_r^R 2\pi xy \, dx = 4\pi \int_r^R x\sqrt{R^2 - x^2} \, dx$$

$$= 4\pi \left[-\frac{1}{3}(R^2 - x^2)^{3/2} \right]_r^R = \frac{4\pi}{3}(R^2 - r^2)^{3/2}$$

where we have used the substitution $u = R^2 - x^2$ to evaluate the integral.

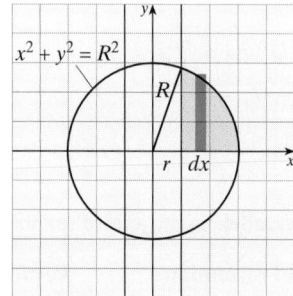

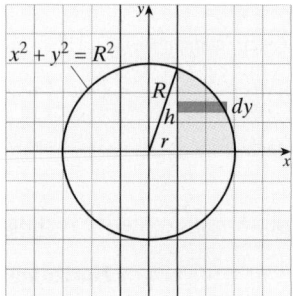

a. Cylindrical shells; strip
   parallel to axis of rotation

b. Washer method; strip
   perpendicular to axis of
   rotation

FIGURE 6.27 Volume of solid

b. For the washer method, we use horizontal strips and note that a typical washer has outer radius $x = \sqrt{R^2 - y^2}$ and inner radius r (see Figure 6.27b). The height of the cut is $h = \sqrt{R^2 - r^2}$, so the volume (again by symmetry) is given by

$$V = 2\int_0^h \pi[(\sqrt{R^2 - y^2})^2 - r^2] \, dy$$

$$= 2\pi \int_0^{\sqrt{R^2 - r^2}} [(R^2 - y^2) - r^2] \, dy$$

$$= 2\pi \left[(R^2 - r^2)y - \frac{y^3}{3} \right]_0^{\sqrt{R^2 - r^2}}$$

$$= 2\pi \left[(R^2 - r^2)^{3/2} - \frac{1}{3}(R^2 - r^2)^{3/2} \right]$$

$$= \frac{4\pi}{3}(R^2 - r^2)^{3/2}$$

6.3   Problem Set

A *In Problems 1–8, set up but do not evaluate an integral using the shaded strips for the volume generated when the given region is revolved about:*

a. *the x-axis*   **b.** *the y-axis*

c. *the line* $y = -1$ **d.** *the line* $x = -2$

1. $y = 4 - x, 0 \le x \le 4$ **2.** $y = 4 - x^2, 0 \le x \le 2$

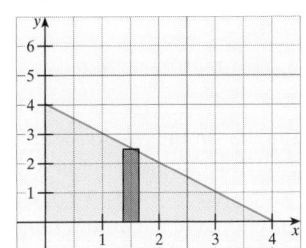 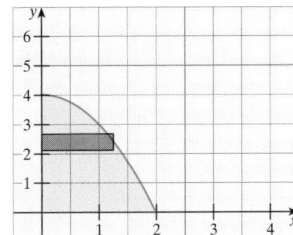

3. $y = \sqrt{4 - x^2}, 0 \le x \le 2$ **4.** $y = \sqrt{4 - x^2}, -2 \le x \le 2$

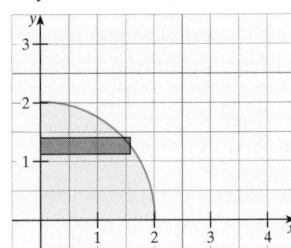 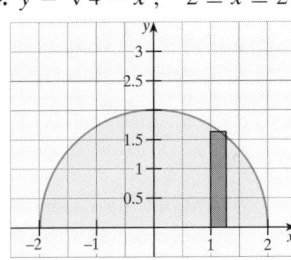

5. $y = e^{-x}, 0 \le x \le 1$ **6.** $y = \dfrac{x}{x + 1}, 0 \le x \le 1$

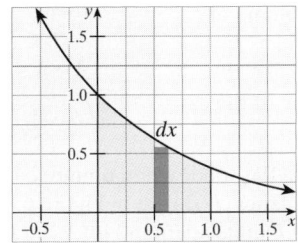 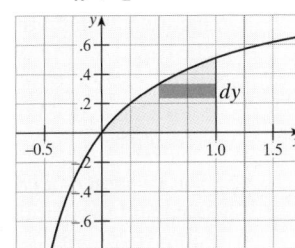

7. $y = \sin^{-1} x, 0 \le x \le 1$ **8.** $y = \ln x, 1 \le x \le 2$

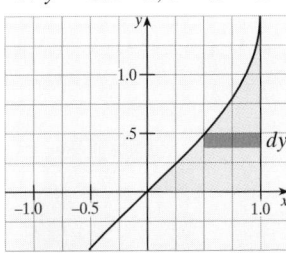 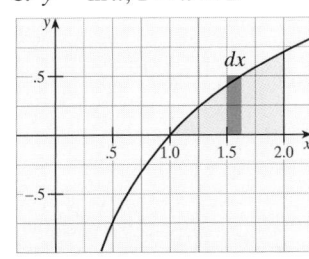

In Problems 9–16, use the shell method to find the volume of the solid formed by revolving the given region about the y-axis.

9. the region bounded by the curve $y = 2x$, the y-axis, and the line $y = 1$

10. the region bounded by the curve $y = \sqrt{x}$, the y-axis, and the line $y = 1$

11. the region bounded by the curves $y = x^2$ and $y = x^3$

12. the region bounded by the lines $y = x, y = 2x$, and $y = 1$

13. the region bounded by the parabola $y = 1 - x^2$, the y-axis, and the positive x-axis

14. the region bounded by the parabolas $y = x^2, y = 1 - x^2$, and the y-axis

15. the region bounded by the curve $y = e^{-x^2}$, the y-axis, and the line $y = \frac{1}{2}$

16. the region bounded by the curve $y = \sqrt{\tan^{-1}x}$, the y-axis, and the line $y = \sqrt{\pi}/2$

In Problems 17–24, find the volume (to four decimal places) of the solid formed by revolving the given region about the y-axis.

17. the region under the curve $y = x^{-2}$ on the interval $[1, 2]$

18. the region under the curve $y = \dfrac{1}{x(1 + x^2)}$ on the interval $[1, 2]$

19. the region under the curve $y = \dfrac{1}{x\sqrt{1 - x^2}}$ on the interval $[\frac{1}{4}, \frac{1}{2}]$

20. the region under the curve $y = \dfrac{1}{\sqrt{1 - x^4}}$ on $[\frac{1}{4}, \frac{1}{2}]$

21. the region under the curve $y = \dfrac{1}{1 + x^4}$ on $[0, 1]$

22. the region under the curve $y = \dfrac{1}{x^2\sqrt{x^2 - 1}}$ on $[2, 3]$

23. the region under the curve $y = \dfrac{e^x}{x}$ on $[1, 2]$

24. the region under the curve $y = \dfrac{1}{x^4}\sqrt{x^{-2} - 1}$

In Problems 25–34, find the volume (to four decimal places) of the solid generated by revolving each region about the prescribed axis.

25. the region bounded by the curves $x = 0, y = \sqrt{1 - x^2}$, and $y = x^2$; about the y-axis

26. the region bounded by the lines $y = 1, x = 1$, and the curve $y = x^3 + 2x + 1$; about the line $x = 1$

27. the region bounded by the curve $x = y(1 - y)$ and the y-axis; about the y-axis

28. the region between the curves $x = y(4 - y)$ and $x - 2y + 3 = 0$; about the line $y = -1$

29. the region inside the ellipse

$$2(x - 3)^2 + 3(y - 2)^2 = 6$$

about the y-axis

30. the region between the curves $y = e^x$ and $y = -x^2 + 4$; about the line $x = 2$

31. the region bounded by the curves $y = \dfrac{2}{1 + x^2}$ and $y = x^2$; about the line $x = -2$

32. the region bounded by the curves $y = \ln x$ and $y = x^2 - 4$; about the y-axis

33. the triangular region with vertices $(1, 1), (2, 5),$ and $(4, 1)$; about the y-axis

34. the triangular region with vertices $(1, 1), (3, 4),$ and $(5, 1)$, about the x-axis

Ⓑ 35. Find the volume of the solid generated when the region $y = x^{-1/2}$ on the interval $[1, 4]$ is revolved about
   a. the x-axis
   b. the y-axis
   c. the line $y = -2$

36. Let R be the region bounded by the curve $y = kx$ and the line $x = k$. Find the volume when R is revolved about:
   a. the y-axis
   b. the line $x = 2k$

37. The portion on the ellipse

$$\frac{x^2}{9} + \frac{y^2}{4} = 1$$

with $x \geq 0$ is rotated about the y-axis to form a solid S. A hole of radius 1 is drilled through the center of S, along the y-axis. Find the volume of the part of S that remains.

38. Suppose a hole of radius 1 is drilled through the center of the solid S of Problem 37 along the x-axis. How much volume remains after this drilling?

39. **HISTORICAL QUEST** Johannes Kepler is usually remembered for his work in astronomy (see Section 11.3, and Historical Quest on p. 785). Kepler made other interesting mathematical discoveries. In fact, he has been described as "number-intoxicated." He was looking for mathematical harmonies in the physical universe. He is quoted in *The World of Mathematics*: "Nothing holds me; I will indulge my sacred fury; I will triumph over mankind by the honest confession that I have stolen the golden vases of the Egyptians to build up a tabernacle for my God far away from the confines of Egypt."

JOHANNES KEPLER
1571–1630

In this Historical Quest we look at Kepler's derivation for the volume of a torus, generated by revolving a circle of radius a around a vertical axis at a distance b from its center. He found

$$V = (\pi a^2)(2\pi b) = 2\pi^2 a^2 b$$

He derived this formula by dissecting a torus into infinitely many thin vertical circular slices by considering planes perpendicular to the axis of revolution, as shown in Figure 6.28.

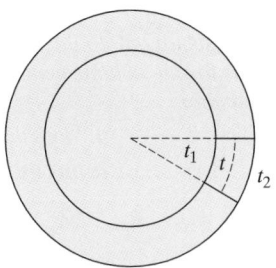

■ **FIGURE 6.28**   Torus cross section

Note that each slice is thinner on the inside (nearer the axis) and thicker on the outside. Kepler assumed that the volume of each slice is $\pi a^2 t$ where $t = \frac{1}{2}(t_1 + t_2)$. Because t is the average of its minimum and maximum thickness, it must be the thickness of the slice at its center. Use washers or shells to verify Kepler's formula for the volume of a torus.

Ⓒ 40. Find the volume of the football-shaped solid (called an *ellipsoid*) formed by revolving the ellipse

$$\frac{x^2}{a^2} + \frac{y^2}{b^2} = 1$$

about the x-axis. What is the volume if the ellipse is revolved about the y-axis?

41. A "cap" is formed by truncating a sphere of radius R at a point h units from the center, as shown in Figure 6.29. Find the volume of the cap.

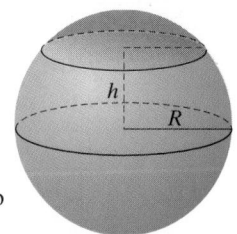

■ **FIGURE 6.29**   Volume of the cap of a sphere

42. The *frustum of a cone* is the solid region bounded by a cone and two parallel planes, as shown in Figure 6.30. Suppose the planes are h units apart and intersect the cone in plane regions of area A_1 and A_2. Use integration to show that the frustum has volume

$$V = \frac{h}{3}(A_1 + \sqrt{A_1 A_2} + A_2)$$

■ **FIGURE 6.30**   Volume of the frustum of a cone

43. A spherical sector (or a "gem") is formed from a section of a sphere as indicated in Figure 6.31. Assume that the spherical base has radius R and that the parallel truncating planes are h_1 and h_2 units from the center of the sphere, where $0 < h_1 < h_2 < R$. Find the volume of the gem.

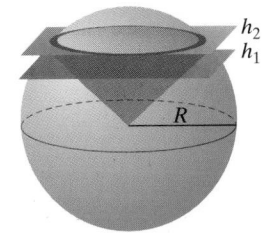

■ **FIGURE 6.31**   Volume of a section of a sphere

6.4   *Arc Length and Surface Area*

IN THIS SECTION   **the arc length of a graph, the area of a surface of revolution**   ■

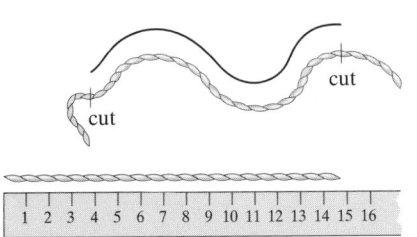

Physical method for measuring the length of a curve

THE ARC LENGTH OF A GRAPH

If a function f has a derivative that is continuous on some interval, then f is said to be **continuously differentiable** on the inteval. The portion of the graph of a continuously differentiable function f that lies between $x = a$ and $x = b$ is called the **arc** of the graph on the interval $[a, b]$. To find the length of this arc, let P be a partition of the interval $[a, b]$, with subdivision points $x_0, x_1, \ldots, x_n$, where $x_0 = a$ and $x_n = b$. Let P_k denote the point (x_k, y_k) on the graph, where $y_k = f(x_k)$. By joining the points $P_0, P_1, \ldots, P_n$, we obtain a polygonal path whose length approximates that of the arc. Figure 6.32 demonstrates the labeling for the case where $n = 6$.

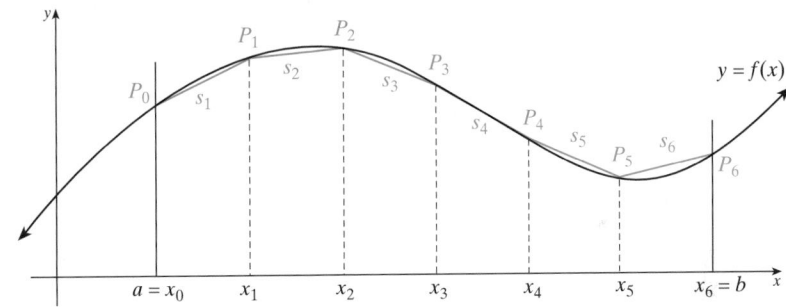

■ **FIGURE 6.32**   A polygonal path approximating an arc of a curve

The length of the polygonal path connecting the points $P_0, P_1, \ldots, P_n$ on the graph of f is the sum

$$\sum_{k=1}^{n} \Delta s_k$$

where Δs_k is the length of the segment joining P_{k-1} to P_k. By applying the distance formula and rearranging terms with $\Delta x_k = x_k - x_{k-1}$ and $\Delta y_k = y_k - y_{k-1}$, we find that

$$\Delta s_k = \sqrt{(x_k - x_{k-1})^2 + (y_k - y_{k-1})^2}$$
$$= \sqrt{(\Delta x_k)^2 + (\Delta y_k)^2}$$
$$= \sqrt{\frac{(\Delta x_k)^2 + (\Delta y_k)^2}{(\Delta x_k)^2}} \, \Delta x_k$$
$$= \sqrt{1 + \left(\frac{\Delta y_k}{\Delta x_k}\right)^2} \, \Delta x_k$$

It is reasonable to expect a connection between the ratio $\Delta y_k / \Delta x_k$ and the derivative $dy/dx = f'(x)$. Indeed, it can be shown that

$$\Delta s_k = \sqrt{1 + [f'(x_k^*)]^2}\, \Delta x_k$$

for some number x_k^* between x_{k-1} and x_k. Therefore, the arc length of the graph of f on the interval $[a, b]$ may be estimated by the Riemann sum

$$\sum_{k=1}^{n} \sqrt{1 + [f'(x_k^*)]^2}\, \Delta x_k$$

This estimate may be improved by increasing the number of subdivision points in the partition P of the interval $[a, b]$ in such a way that the subinterval lengths tend to zero. Thus, it is reasonable to define the actual arc length to be the limit

$$\lim_{\|P\| \to 0} \sum_{k=1}^{n} \sqrt{1 + [f'(x_k^*)]^2}\, \Delta x_k$$

Notice that if f' is continuous on the interval $[a, b]$, then so is $\sqrt{1 + [f'(x)]^2}$. Thus, this limit exists and is the integral of $\sqrt{1 + [f'(x)]^2}$ with respect to x on the interval $[a, b]$. These observations lead us to the following definition.*

Arc Length

Let f be a function whose derivative f' is continuous on the interval $[a, b]$. Then the **arc length,** s, of the graph of $y = f(x)$ between $x = a$ and $x = b$ is given by the integral

$$s = \int_a^b \sqrt{1 + [f'(x)]^2}\, dx$$

Similarly, for the graph of $x = g(y)$, where g' is continuous on the interval $[c, d]$, the arc length from $y = c$ to $y = d$ is

$$s = \int_c^d \sqrt{1 + [g'(y)]^2}\, dy$$

EXAMPLE 1   *Arc length of a curve*

Find the arc length of the curve $y = x^{3/2}$ on the interval $[0, 4]$.

Solution   Let $f(x) = x^{3/2}$; therefore, $f'(x) = \frac{3}{2}x^{1/2}$.

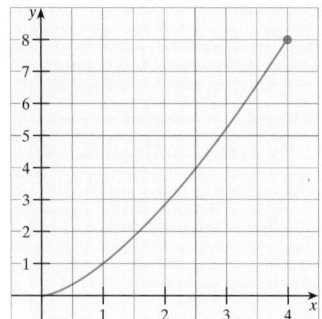

$$s = \int_0^4 \sqrt{1 + [\tfrac{3}{2}x^{1/2}]^2}\, dx$$

$$= \int_0^4 \sqrt{1 + \tfrac{9}{4}x}\, dx = [\tfrac{4}{9} \cdot \tfrac{2}{3}(1 + \tfrac{9}{4}x)^{3/2}]_0^4$$

$$= \tfrac{8}{27}[(10)^{3/2} - (1)^{3/2}] \approx 9.073415289 \qquad \blacksquare$$

---

*We have used integration to obtain a meaningful definition of arc length for a function $f(x)$ with a continuous derivative $f'(x)$. It is possible to define arc length for certain curves $y = f(x)$ when $f(x)$ does not have a continuous derivative, but we will not pursue this more general topic.

EXAMPLE 2   *Arc length of a curve x = g(y)*

Find the arc length of the curve $x = \frac{1}{3}y^3 + \frac{1}{4}y^{-1}$ from $y = 1$ to $y = 3$.

Solution   Because $g(y) = \frac{1}{3}y^3 + \frac{1}{4}y^{-1}$, we have $g'(y) = y^2 - \frac{1}{4}y^{-2} = \dfrac{4y^4 - 1}{4y^2}$,

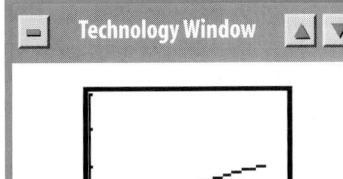
which is continuous throughout the interval from $y = 1$ to $y = 3$. Therefore, the arc length is

$$
\begin{aligned}
\int_1^3 \sqrt{1 + [g'(y)]^2}\, dy &= \int_1^3 \sqrt{1 + \left(\frac{4y^4 - 1}{4y^2}\right)^2}\, dy \\
&= \int_1^3 \sqrt{1 + \frac{16y^8 - 8y^4 + 1}{16y^4}}\, dy \\
&= \int_1^3 \sqrt{\frac{16y^4 + 16y^8 - 8y^4 + 1}{16y^4}}\, dy \\
&= \int_1^3 \sqrt{\frac{(4y^4 + 1)^2}{(4y^2)^2}}\, dy = \int_1^3 \frac{4y^4 + 1}{4y^2}\, dy = \int_1^3 \left(y^2 + \frac{1}{4}y^{-2}\right) dy \\
&= \left(\frac{1}{3}y^3 + \frac{1}{4} \cdot \frac{y^{-1}}{-1}\right)\Big|_1^3 = \frac{53}{6} \approx 8.833333333 \quad \blacksquare
\end{aligned}
$$

EXAMPLE 3   *Estimating arc length using numerical integration*

Find the length of the curve defined by $y = \sin x$ on $[0, 2\pi]$.

Solution   Because $y' = \cos x$, we have, from the arc length formula,

$$s = \int_0^{2\pi} \sqrt{1 + \cos^2 x}\, dx$$

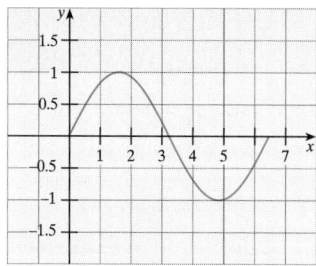

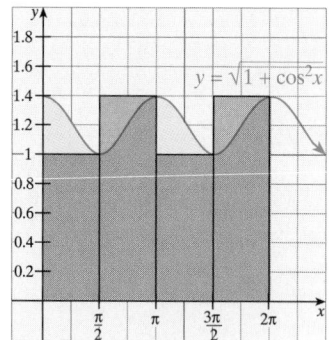

■ **FIGURE 6.33**   Right-endpoint approximation for $n = 4$

We do not have techniques to allow us to evaluate this integral, so we turn to numerical integration. The region is shown in Figure 6.33. If we consider four rectangles, we find, by various methods:

Method ($n = 4$)	Approximation
Rectangles	
Left endpoints	7.584476
Right endpoints	7.584476
Midpoints	7.695299
Trapezoids	7.584476
Simpson's rule	7.150712
Calculator (n not specified)	7.640396

In practice, you would choose just *one* of the methods whose approximate values are given. You might also note that, for this example, Simpson's rule performs worse than the trapezoidal, midpoint, or even the left- and right-endpoint methods. For more accurate results, you might wish to use a computer program. A simple output is shown in the following technology window.   ■

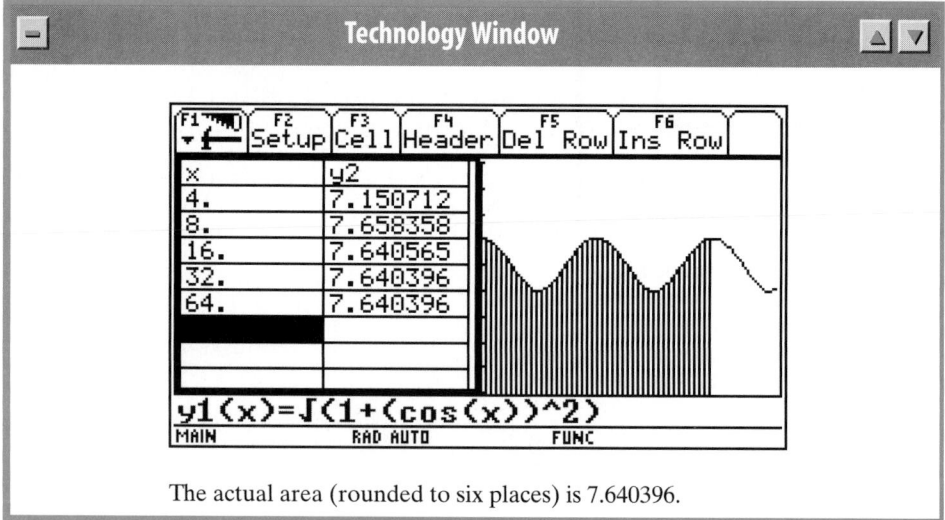

The actual area (rounded to six places) is 7.640396.

THE AREA OF A SURFACE OF REVOLUTION

When the arc of a graph is revolved about a line L, it generates a surface that we shall refer to as a **surface of revolution.** Figure 6.34 shows an arc, together with a typical approximating line segment. Notice that when the segment is revolved about L, it generates the surface of the frustum of a right circular cone.

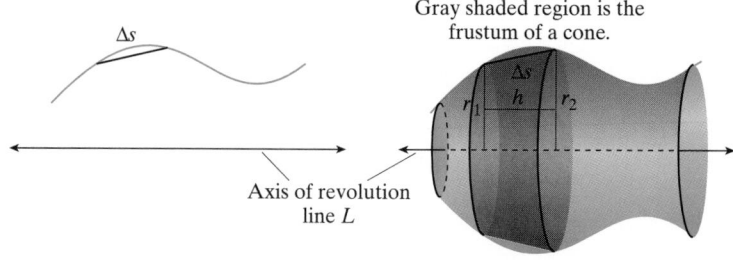

■ **FIGURE 6.34**   A surface of revolution

Rather than argue formally with Riemann sums, we will proceed intuitively. It can be shown that a frustum with slant height ℓ and base radii r_1 and r_2 has lateral surface area $\pi(r_1 + r_2)\ell$ (see Problem 33). Figure 6.34 shows an approximating line segment we shall assume has length Δs and is located h units above the axis of revolution L. Then, by revolving this segment about L, we generate a frustum whose slant height is Δs and which is so thin (because Δs is very small) that the radii of both its circular bases are essentially the same as h.

To find the area of the surface of revolution, let the radii of the bases be r_1 and r_2. Then the approximating frustum has surface area $\pi(r_1 + r_2)\Delta s$. Because both r_1 and r_2 are essentially the same as h, we can estimate the surface area of the frustum by

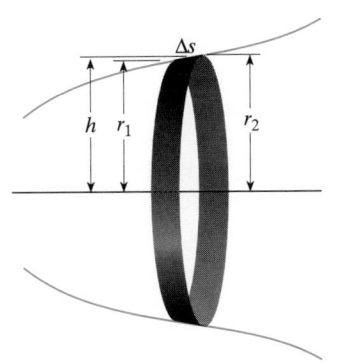

$$\underset{\text{Lateral length}}{\Delta S = \pi(h + h)\overline{\Delta s} = 2\pi h\Delta s = 2\pi f(x)\sqrt{1 + [f'(x)]^2}\,\Delta x}$$

Distance to axis of revolution is $h = f(x)$.

An increment of surface area ΔS

The surface area S can be obtained by integrating this expression on the interval $[a, b]$. These considerations lead us to the following definition.

Surface Area

Suppose f' is continuous on the interval $[a, b]$. Then the surface generated by revolving about the x-axis the arc of the curve $y = f(x)$ on $[a, b]$ has **surface area**

$$S = 2\pi \int_a^b f(x)\sqrt{1 + [f'(x)]^2}\,dx$$

An easy-to-remember form for surface area is $S = 2\pi\int y\,ds$ since $ds = \sqrt{1 + [f'(x)]^2}\,dx$.

You may find it instructive to derive this formula by the more rigorous approach outlined in Problem 35. We close with two examples that illustrate the use of the surface area formula.

> **EXAMPLE 4**   *Area of a surface of revolution*

Find the area of the surface generated by revolving about the x-axis the arc of the curve $y = x^3$ on $[0, 1]$.

Solution   Because $f(x) = x^3$, we have $f'(x) = 3x^2$, which is certainly continuous on the interval $[0, 1]$.

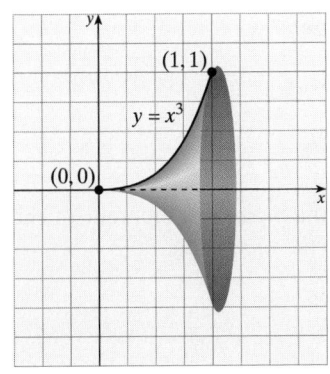

$$S = 2\pi \int_0^1 x^3\sqrt{1 + (3x^2)^2}\,dx = 2\pi \int_0^1 x^3\sqrt{1 + 9x^4}\,dx = 2\pi \int_1^{10} u^{1/2}\left(\frac{du}{36}\right)$$

> Let $u = 1 + 9x^4$; $du = 36\,x^3\,dx$.
> If $x = 1$, then $u = 10$.
> If $x = 0$, then $u = 1$.

$$= \frac{\pi}{18}\left(\frac{2}{3}u^{3/2}\right)\Big|_1^{10} = \frac{\pi}{27}[10\sqrt{10} - 1] \approx 3.5631 \qquad \blacksquare$$

| | EXAMPLE 5 | *Derive the formula for the surface area of a sphere* |

Find a formula for the surface area of a sphere of radius r.

Solution   We can generate the surface of the sphere by revolving the semicircle $y = \sqrt{r^2 - x^2}$ about the x-axis. We find that

$$y' = -x(r^2 - x^2)^{-1/2} = \frac{-x}{\sqrt{r^2 - x^2}}$$

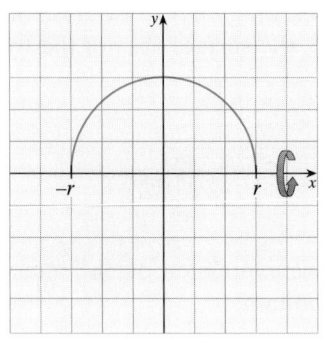

Because the semicircle intersects the x-axis at $x = r$ and $x = -r$, the interval of integration will be $[-r, r]$. Finally, by applying the formula for the surface area, we find

$$S = 2\pi \int_{-r}^{r} \sqrt{r^2 - x^2} \sqrt{1 + \left(\frac{-x}{\sqrt{r^2 - x^2}}\right)^2} \, dx$$

$$= 2\pi \int_{-r}^{r} \sqrt{(r^2 - x^2)\left(\frac{r^2 - x^2 + x^2}{r^2 - x^2}\right)} \, dx$$

$$= 2\pi \int_{-r}^{r} r \, dx = 2\pi r(2r) = 4\pi r^2 \qquad\blacksquare$$

To generalize the formula for surface area to apply to any vertical or horizontal axis of revolution, suppose that an axis of revolution is $R(x)$ units from a typical element of arc on the graph of $y = f(x)$, as shown in Figure 6.35.

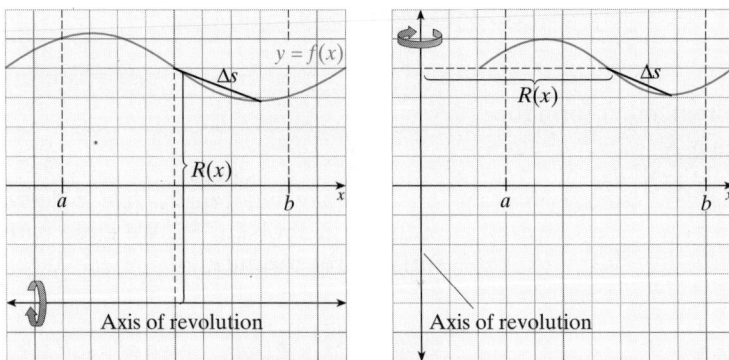

■ **FIGURE 6.35**   Surface of revolution about a general vertical or horizontal axis

Then $2\pi R(x)$ is the circumference of a circle of radius $R(x)$, and it can be shown that an element of surface area is

$$\Delta S = 2\pi R(x)\Delta s = 2\pi R(x)\sqrt{1 + [f'(x)]^2} \, \Delta x$$

Thus, the surface of revolution on the interval $[a, b]$ has area

$$S = 2\pi \int_{a}^{b} R(x)\sqrt{1 + [f'(x)]^2} \, dx$$

In particular, if the graph of $y = f(x)$ is revolved about the y-axis, an element of the arc is $R(x) = x$ units from the y-axis, and the resulting surface has area

$$S = 2\pi \int_{a}^{b} x\sqrt{1 + [f'(x)]^2} \, dx$$

6.4   Problem Set

A *Find the length of the arc of the curve $y = f(x)$ on the intervals given in Problems 1–12.*

1. $f(x) = 3x + 2$ on $[-1, 2]$
2. $f(x) = 5 - 4x$ on $[-2, 0]$
3. $f(x) = 1 - 2x$ on $[1, 3]$
4. $f(x) = x^{3/2}$ on $[0, 4]$
5. $f(x) = \frac{2}{3}x^{3/2} + 1$ on $[0, 4]$
6. $f(x) = \frac{1}{3}(2 + x^2)^{3/2}$ on $[0, 3]$
7. $f(x) = \frac{1}{12}x^5 + \frac{1}{5}x^{-3}$ on $[1, 2]$
8. $f(x) = \frac{1}{3}x^3 + \frac{1}{4}x^{-1}$ on $[1, 4]$
9. $f(x) = \frac{1}{4}x^4 + \frac{1}{8}x^{-2}$ on $[1, 2]$
10. $f(x) = \sqrt{e^{2x} - 1} - \sec^{-1}(e^x)$ on $[0, \ln 2]$
11. Find the length of the curve defined by $9x^2 = 4y^3$ between the points $(0, 0)$ and $(2\sqrt{3}, 3)$.
12. Find the length of the curve defined by $(y + 1)^2 = 4x^3$ between the points $(0, -1)$ and $(1, 1)$.

B

15. Find the area of the surface generated when the arc of the curve
$$y = \frac{1}{3}x^3 + (4x)^{-1}$$
between $x = 1$ and $x = 3$ is revolved about
   a. the x-axis   b. the y-axis

16. Find the area of the surface generated when the arc of the curve
$$x = \frac{3}{5}y^{5/3} - \frac{3}{4}y^{1/3}$$
between $y = 0$ and $y = 1$ is revolved about
   a. the y-axis   b. the x-axis
   c. the line $y = -1$

Find the surface area generated when the graph of each function given in Problems 17–20 on the prescribed interval is revolved about the x-axis.

17. $f(x) = 2x + 1$ on $[0, 2]$
18. $f(x) = \sqrt{x}$ on $[2, 6]$

19. $f(x) = \frac{1}{3}x^3 + \frac{1}{4}x^{-1}$ on $[1, 2]$

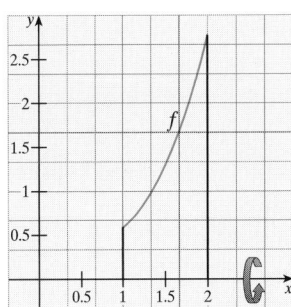

Give your answer correct to the nearest hundredth.

20. $f(x) = \frac{1}{4}x^4 + \frac{1}{8}x^{-2}$ on $[1, 2]$

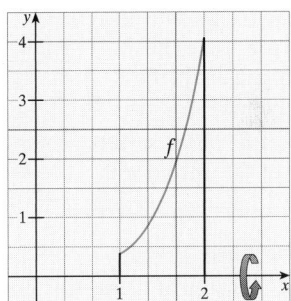

Give your answer correct to the nearest hundredth.

Find the surface area generated when the graph of each function given in Problems 21–24 on the prescribed interval is revolved about the y-axis.

21. $f(x) = \frac{1}{3}(12 - x)$ on $[0, 3]$
22. $f(x) = \frac{2}{3}x^{3/2}$ on $[0, 3]$
23. $f(x) = \frac{1}{3}\sqrt{x}\,(3 - x)$ on $[1, 3]$

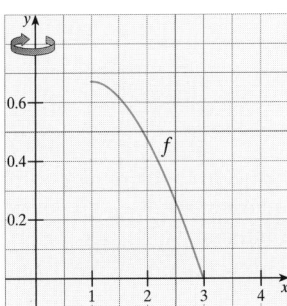

Give your answer correct to the nearest hundredth.

24. $f(x) = 2\sqrt{4 - x}$
on $[1, 3]$

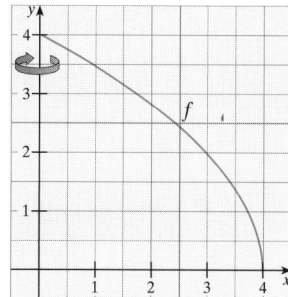

Give your answer correct to the nearest hundredth.

25. The graph of the equation
$$x^{2/3} + y^{2/3} = 1$$
is an **astroid.**

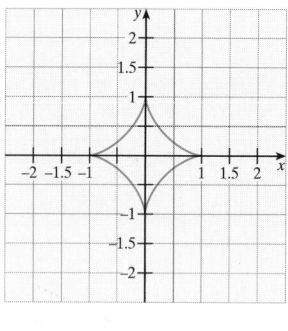

   Find the length of this particular astroid by finding the length of the first quadrant portion, $y = (1 - x^{2/3})^{3/2}$ on $[0, 1]$. By symmetry, the entire curve is four times the length of the arc in the first quadrant.

26. Find the area of the surface generated by revolving the portion of the astroid with $y \geq 0$ (see Problem 25)
$$x^{2/3} + y^{2/3} = 1 \text{ on } [-1, 1]$$
about the x-axis.

27. Estimate the area of the surface obtained by revolving $y = \tan x$ about the x-axis on the interval $[0, 1]$.

28. The following table gives the slope $f'(x_k)$ for points x_k located at 0.3-unit intervals on a certain curve defined by $y = f(x)$. Use the trapezoidal rule to estimate the arc length of $y = f(x)$ over the interval $[0, 3]$.

x_k	$f'(x_k)$
0.0	3.7
0.3	3.9
0.6	4.1
0.9	4.1
1.2	4.2
1.5	4.4
1.8	4.6
2.1	4.9
2.4	5.2
2.7	5.5
3.0	6.0

29. Use the table in Problem 28 to estimate the surface area generated when $y = f(x)$ is revolved about the y-axis over the interval $[0, 3]$.

Ⓒ 30. Show that the formula for the surface area of a sphere of radius r is
$$A = 4\pi r^2$$

31. Show that a cone of radius r and height h has surface area
$$S = \pi r \sqrt{r^2 + h^2}$$
Hint: Revolve part of the line $hy = rx$ about the x-axis.

32. Show that when the arc of the graph of $f(x)$ between $x = a$ and $x = b$ is revolved about the y-axis, the surface generated has the area
$$S = 2\pi \int_a^b x\sqrt{1 + [f'(x)]^2}\, dx$$

33. Figure 6.36 shows the frustum of a cone with base radii r_1 and r_2 and slant height ℓ.

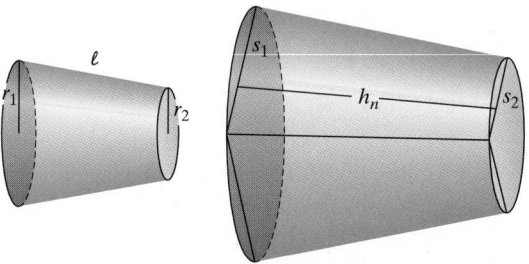

■ **FIGURE 6.36**   Frustum of a cone

Suppose that a regular polygon of n sides is inscribed in each circular base. Each side of the polygon in the upper base has length s_1 and each side of the polygon in the lower base has length s_2. Notice that a trapezoid is formed when the ends of corresponding sides in these polygons are joined by line segments.

a. The surface area of the frustum can be estimated by adding the areas of the approximating trapezoids. Show that this estimate is
$$\tfrac{1}{2}n(s_1 + s_2)h_n$$
where h_n is the distance between corresponding sides in the polygons inscribed in the circular bases.

b. Show that the polygon inscribed in the larger base has perimeter
$$\lim_{n \to +\infty} ns_1 = 2\pi r$$
Use similar reasoning to establish the following limit statements:
$$\lim_{n \to +\infty} ns_2 = 2\pi r_2 \text{ and } \lim_{n \to +\infty} h_n = \ell$$

c. Compare the results of parts **a** and **b** to show that the frustum has surface area
$$A = \pi(r_1 + r_2)\ell$$

34. If $f'(x)$ exists throughout the interval $[x_{k-1}, x_k]$, show that there exists a number x_k^* in this interval for which
$$\sqrt{(\Delta x_k)^2 + (\Delta y_k)^2} = \sqrt{1 + [f'(x_k^*)]^2}\,\Delta x_k$$
where $\Delta x_k = x_k - x_{k-1}$ and $\Delta y_k = f(x_k) - f(x_{k-1})$.
Hint: Use the mean value theorem.

35. Verify the surface area formula by completing the following steps (see Figure 6.37).

a. Let $P = \{x_0, x_1, \dots, x_n\}$ be a partition of the interval $[a, b]$, and for $k = 0, 1, \dots, n$, let P_k denote the point (x_k, y_k) where $y_k = f(x_k)$. Show that when the line segment between P_{k-1} and P_k is revolved about the x-axis, it generates a frustum of a cone with surface area

$$\pi(y_{k-1} + y_k)\ell_k$$

where ℓ_k is the distance between P_{k-1} and P_k.

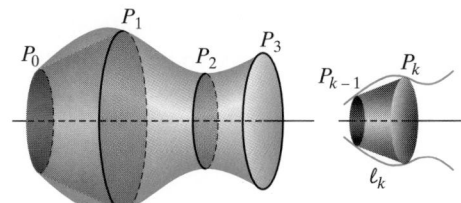

■ **FIGURE 6.37**   Surface area formula

b. Notice that $\frac{1}{2}(y_{k-1} + y_k)$ is a number between y_{k-1} and y_k. Use the intermediate value theorem (Section 2.4) to show that

$$y_{k-1} + y_k = 2f(c_k)$$

for some number c_k between x_{k-1} and x_k.

c. Explain why the surface generated when the arc of the graph of f between $x = a$ and $x = b$ is revolved about the x-axis has area

$$\lim_{\|P\|\to 0} \sum_{k=1}^{n} 2\pi f(c_k)\sqrt{1 + [f'(x_k^*)]^2}\, \Delta x_k$$

d. Notice that as $\|P\| \to 0$, the number c_k and x_k^* are "squeezed" together. Use this observation to show that the surface area is given by the integral

$$S = 2\pi \int_a^b f(x)\sqrt{1 + [f'(x)]^2}\, dx$$

36. The center of a circle of radius r is located at $(R, 0)$, where $R > r$. When the circle is revolved about the y-axis, it generates a *torus* (a doughnut-shaped solid). What is the surface area of the torus?

37. Let n be a positive integer and C, D constants such that

$$n(n-1) = \frac{1}{16CD}$$

Find the length of the curve

$$y = Cx^{2n} + Dx^{2(1-n)}$$

from $x = a$ to $x = b$.

38. Find the length of

$$y = \frac{x^3}{24} + \frac{2}{x}$$

from $x = 2$ to $x = 3$. Note that this curve does not fit the pattern in Problem 37. Find a general formula for the length of the curve

$$y = Cx^n + Dx^{2-n}$$

from $x = a$ to $x = b$ for

$$n(n-2) = \frac{1}{4CD}$$

39. Suppose it is known that the arc length of the curve $y = f(t)$ on the interval $0 \le t \le x$ is

$$L(x) = \ln(\sec x + \tan x)$$

for every x on $0 \le x \le 1$. If the curve $y = f(x)$ passes through the origin, what is $f(x)$?

6.5   Physical Applications: Work, Liquid Force, and Centroids

IN THIS SECTION   **modeling work, modeling fluid pressure and force, modeling the centroid of a plane region, the volume theorem of Pappus**   ■

MODELING WORK

In physics, "force" is an influence that tends to cause motion in a body. When a constant force of magnitude F is applied to an object through a distance d, it performs **work** equal to the product $W = Fd$.

Work Done by a Constant Force

If a body moves a distance d in the direction of an applied constant force F, the **work** W done is

$$W = Fd$$

Common Units of Work and Force

Mass	Dist.	Force	Work
kg	m	newton (N)	joule
g	cm	dyne (dyn)	erg
slug	ft	pound	ft-lb

If there is no movement, there is no work!

Reprinted with special permission
of King Features Syndicate.

In the U.S. system of measurements, work is typically expressed in *newton-meters* (called *joules*); in the *centimeter-gram-second* (CGS) system, the basic unit of work is the *dyne-centimeter*, called an erg, or the *ft-lb.*

For example, the work done in lifting a 90-lb bag of concrete 3 ft is $W = Fd = (90 \text{ lb})(3 \text{ ft}) = 270$ ft-lb. Notice that this definition does not conform to everyday use of the word *work*. If you work at lifting the concrete all day, but you are not able to move the sack of concrete, then no work has been done.

If F is a variable force given by the continuous function $F(x)$, then calculus is needed to find the work done in moving a distance d along the x-axis against F. Suppose F is a continuous function and $F(x)$ is a variable force that acts on an object moving along the x-axis from $x = a$ to $x = b$. We shall define the work done by this force. Partition the interval $[a, b]$ into subintervals, and let Δx_k be the length of the kth subinterval I_k. If Δx_k is sufficiently small, we can expect the force to be essentially constant on I_k, equal, for instance, to $F(x_k^*)$, where x_k^* is a point chosen arbitrarily from the interval I_k. It is reasonable to expect that the work Δw_k required to move the object on the interval I_k is approximately $\Delta w_k = F(x_k^*)\Delta x_k$, and the total work is estimated by the sum

$$\sum_{k=1}^{n} F(x_k^*)\Delta x_k$$

as indicated in Figure 6.38

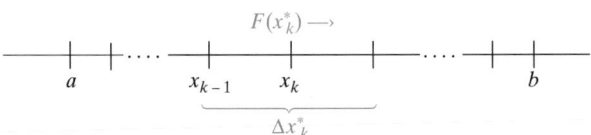

■ **FIGURE 6.38**   Work performed by a variable force

By taking the limit as the norm of the partition approaches 0, we find the total work may be modeled by

$$W = \lim_{\|P\| \to 0} \sum_{k=1}^{n} F(x_k^*)\Delta x_k = \int_{a}^{b} F(x)\, dx$$

Work Done by a Variable Force

The work done by the variable force $F(x)$ in moving an object along the x-axis from $x = a$ to $x = b$ is given by

$$W = \int_{a}^{b} F(x)\, dx$$

EXAMPLE 1   *Work with a variable force*

An object located x ft from a fixed starting position is moved along a straight road by a force of $F(x) = 3x^2 + 5$ lb. What work is done by the force to move the object
a. through the first 4 ft?   b. from 1 ft to 4 ft?

Solution

a. $W = \displaystyle\int_{0}^{4} (3x^2 + 5)dx = (x^3 + 5x)\Big|_{0}^{4} = 84$ ft-lb

b. $W = \displaystyle\int_{1}^{4} (3x^2 + 5)dx = (x^3 + 5x)\Big|_{1}^{4} = 78$ ft-lb   ■

Hooke's law, named for the English physicist Robert Hooke (1635–1703), states that *when a spring is pulled x units past its equilibrium* (*rest*) *position, there is a restoring force F(x) = kx that pulls the spring back toward equilibrium.** (See Figure 6.39.) The constant k in this formula is called the **spring constant.**

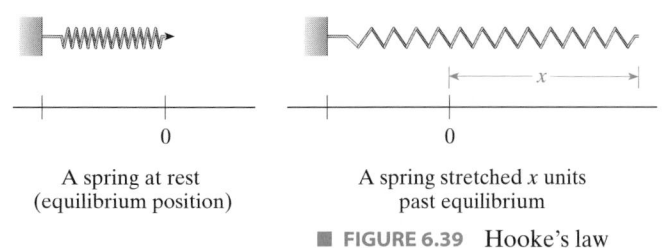

Hooke's law: A force $F(x) = kx$ acts to restore the spring to its equilibrium position.

A spring at rest
(equilibrium position)

A spring stretched x units
past equilibrium

■ **FIGURE 6.39**   Hooke's law

| EXAMPLE 2 | *Modeling work using Hooke's law* |

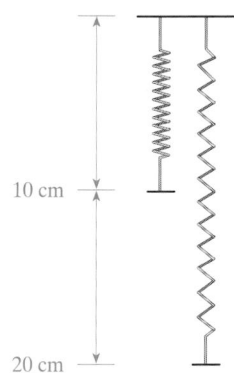

The natural length of a certain spring is 10 cm. If it requires 2 dyn · cm of work to stretch the spring to a total length of 18 cm, how much work will be performed in stretching the spring to a total length of 20 cm?

Solution   Assume that the point of equilibrium is at 0 on a number line, and let x be the length the free end of the spring is extended past equilibrium. Because the stretching force of the spring is $F(x) = kx$, the work done in stretching the spring b cm beyond equilibrium is

$$W = \int_0^b F(x)dx = \int_0^b kx\, dx = \tfrac{1}{2}kb^2$$

We are given that $W = 2$ when $b = 8$, so

$$2 = \tfrac{1}{2}k(8)^2 \quad \text{implies} \quad k = \tfrac{1}{16}$$

Thus, the work done in stretching the spring b cm beyond equilibrium is

$$W = \tfrac{1}{2}(\tfrac{1}{16})b^2 = \tfrac{1}{32}b^2 \text{ dyn·cm}$$

In particular, when the total length of the spring is 20 cm, it is extended $b = 10$ cm, and the required work is

$$W = \tfrac{1}{32}(10)^2 = \tfrac{25}{8} = 3.125 \text{ dyn·cm}$$ ■

| EXAMPLE 3 | *Modeling the work performed in pumping out a tank of water* |

A tank in the shape of a right circular cone of height 12 ft and radius 3 ft is inserted into the ground with its vertex pointing down and its top at ground level, as shown in Figure 6.40a. If the tank is half-filled with water (density $\rho = 62.4$ lb/ft^3), how much work is performed in pumping all the water in the tank to ground level? What changes if the water is pumped to a height of 3 ft above ground level?

Solution   Set up a coordinate system with the origin at the vertex of the cone and the y-axis as the axis of symmetry (Figure 6.40b). Partition the interval $0 \leq y \leq 6$ (remember, the tank is only half full). Choose a representative point y_k^* in the kth subinterval, and construct a thin disk-like slab of water y_k^* units above the vertex of the cone. Our plan is to think of the water in the tank as a collection of these slabs

---

*Actually, Hooke's law applies only in ideal circumstances, and a spring for which the law applies is sometimes called an ideal (or linear) spring.

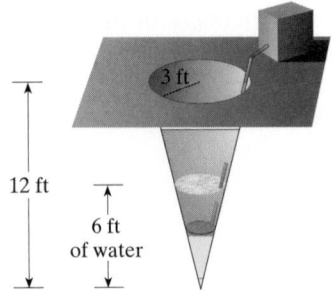
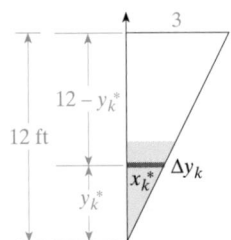

a. A conical water tank

b. A disk of water is $12 - y_k^*$ units from the top.

■ **FIGURE 6.40**   Work in pumping water

of water piled on top of each other. We shall find the work done to raise a typical water disk and then compute the total work by using integration to add the contributions of all such slabs.

Note that the force required to lift the slab is equal to the weight of the slab, which equals its volume multiplied by the weight per cubic foot of water. Let x_k^* be the radius of the kth slab. By similar triangles (Figure 6.40b), we have

$$\frac{x_k^*}{y_k^*} = \frac{3}{12} \Rightarrow x_k^* = \frac{y_k^*}{4}$$

Hence, the volume of the cylindrical slab is

$$\Delta V_k = \pi x_k^{*2} \Delta y_k = \pi \underbrace{\left(\frac{y_k^*}{4}\right)^2}_{\text{Radius}} \overbrace{\Delta y_k}^{\text{Thickness}}$$

so that the weight of the slab is modeled by

$$62.4 \pi \left(\frac{y_k^*}{4}\right)^2 \Delta y_k \, \text{lb}$$

From Figure 6.40b, we see that the slab of water must be raised $12 - y_k^*$ feet, which means that the work ΔW_k required to raise this slab is

$$\Delta W_k = 62.4 \pi \left(\frac{y_k^*}{4}\right)^2 (12 - y_k^*) \Delta y_k \quad \text{Weight-density of water is 62.4 lb/ft}^3.$$

Finally, to compute the total work, we add the work required to lift each thin slab and take the limit of the sum as the norm of the partition approaches 0:

$$W = \lim_{n \to \infty} 62.4 \pi \sum_{k=1}^{n} \left(\frac{y_k^*}{4}\right)^2 (12 - y_k^*) \Delta y_k$$

$$= 62.4 \pi \int_0^6 \left(\frac{y}{4}\right)^2 (12 - y) dy = \frac{62.4}{16} \pi \int_0^6 (12y^2 - y^3) dy$$

$$= 3.9 \pi (4y^3 - \tfrac{1}{4}y^4) \big|_0^6 = 2{,}106\pi \approx 6{,}616 \text{ ft-lb}$$

If the water is pumped to a height of 3 ft above ground level, all that changes is the distance moved by the slab of water. It becomes $12 + 3 - y_k^* = 15 - y_k^*$, and the work is given by

$$W = 62.4 \pi \int_0^6 \left(\frac{y}{4}\right)^2 (15 - y) dy \approx 9{,}263 \text{ ft-lb}$$

■

MODELING FLUID PRESSURE AND FORCE

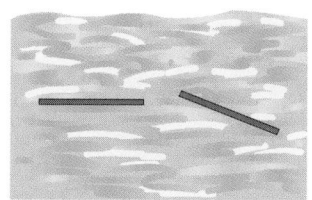

Pascal's principle: In an equilibrium state, the fluid pressure is the same in all directions.

Weight-density, ρ (lb/ft^3)	
Water	62.4
Seawater	64.0
Gasoline	42.0
Kerosene	51.2
Milk	64.5
Mercury	849.0

Anyone who has dived into water has probably noticed that the pressure (that is, the force per unit area) due to the water's weight increases with depth. Careful observations show that the water pressure at any given point is directly proportional to the depth at that point. The same principle applies to other fluids.

In physics, **Pascal's principle** (named for Blaise Pascal, 1623–1662, a French mathematician and scientist) states that fluid pressure is the same in all directions. This means that the pressure must be the same at all points on a surface submerged horizontally, and in this case, the fluid force on the surface is given by the formula in the following box.

Fluid Force

If a surface of area A is submerged horizontally at a depth h in a fluid, the weight of the fluid exerts a force of

$$F = (\text{pressure})(\text{area}) = \rho h A$$

on the surface, where ρ is the fluid density (weight per unit volume). This is called **fluid force** or **hydrostatic force.**

It turns out that this fluid force does not depend on the shape of the container or its size. For instance, each of the containers in Figure 6.41 has the same pressure on its base because the fluid depth h and the base area A are the same in each case.

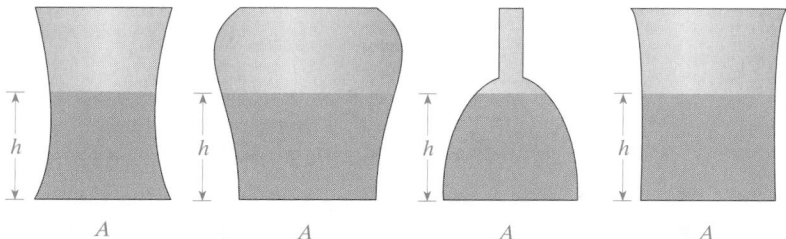

■ FIGURE 6.41   The fluid force $F = \rho h A$ does not depend on the shape or size of the container.

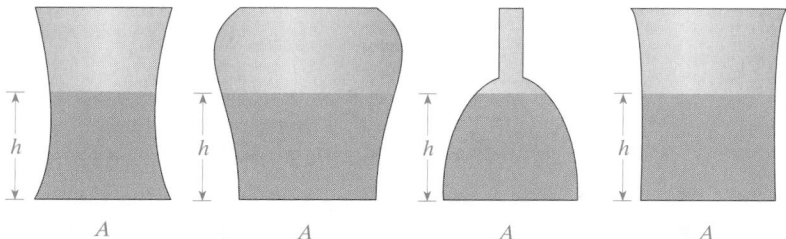

■ FIGURE 6.42   Pressure on a submerged object

When the surface is submerged vertically or at an angle, however, this simple formula does not apply because different parts of the surface are at different depths, as shown in Figure 6.42. We shall use integration to compute the fluid force in such cases. Consider a plate that is submerged vertically in a fluid of density ρ, as shown in Figure 6.43a. We set up a coordinate system with the horizontal axis on the surface of the fluid and the positive h-axis (depth) pointing down. Thus, greater depths correspond to larger values of h (see Figure 6.43b). For simplicity, we assume the plate is oriented so its top and bottom are located a units and b units below the surface, respectively.

Our strategy will be to think of the plate as a pile of subplates or slabs, each so thin that we may regard it as being at a constant depth below the surface. A typical slab (a horizontal strip) is shown in Figure 6.43b. Note that it has thickness Δh and length $L = L(h)$, so that its area is $\Delta A = L\Delta h$. Because the slab is at a constant depth h below the surface, the fluid force on its surface is

$$\Delta F = (\text{pressure})(\text{area}) = \rho h \Delta A = \rho h L(h)\Delta h$$

and by integrating as h varies from a to b, we can model the total force on the plate.

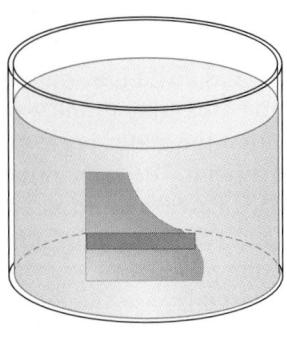 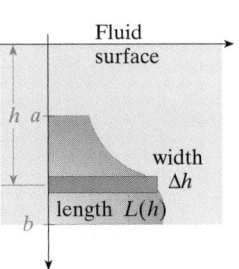

a. A plate submerged vertically in a fluid of density ρ

b. A typical slab has area $A = L(h)\Delta h$.

■ **FIGURE 6.43**   Hydrostatic force

Fluid (Hydrostatic) Force

Suppose a flat surface (a plate) is submerged vertically in a fluid of weight-density ρ (lb/ft³) and that the submerged portion of the plate extends from $x = a$ to $x = b$ on a vertical axis. Then the total force F exerted by the fluid is given by

$$F = \int_a^b \rho h(x)L(x)\,dx$$

where $h(x)$ is the depth at x and $L(x)$ is the corresponding length of a typical horizontal approximating strip.

| EXAMPLE 4 | *Fluid force on a vertical surface* |

The cross sections of a certain trough are inverted isosceles triangles with height 6 ft and base 4 ft, as shown in Figure 6.44. Suppose the trough contains water to a depth of 3 ft. Find the total fluid force on one end.

Solution   *First,* we set up a coordinate system in which the horizontal axis lies at the fluid surface and the positive vertical axis (the h-axis) points down (see Figure 6.44.b). *Next,* we find expressions for the length and depth of a typical thin slab (horizontal strip) in terms of the variables. We assume the slab has thickness Δh and length L. Then by similar triangles (see Figure 6.44c) we have

$$\frac{L}{4} = \frac{3-h}{6} \quad \text{so that} \quad L = \frac{2}{3}(3-h)$$

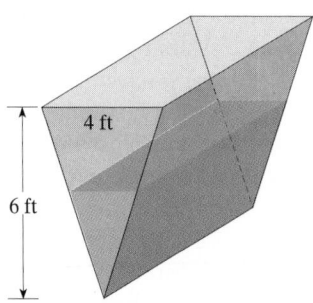 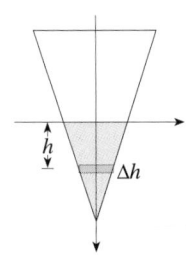 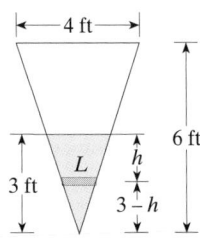

a. A trough with half-filled triangular cross sections

b. Side view of trough

c. Use similar triangles to find $\dfrac{L}{4} = \dfrac{3-h}{6}$.

■ **FIGURE 6.44**   Fluid force

We see that the approximating slab has area $\Delta A = L\Delta h = \frac{2}{3}(3 - h)\Delta h$. *Finally,* multiply the product of the fluid's weight-density at a given depth (ρh) and the area of the approximating slab (ΔA) and integrate on the interval of depths occupied by the vertical plate:

$$F = \int_0^3 \underbrace{\tfrac{2}{3}\rho h(3 - h)dh}_{\Delta F \,=\, \text{force on a typical slab}}$$

$$= \tfrac{2}{3}(62.4)\int_0^3 (3h - h^2)dh \qquad \rho = 62.4 \text{ lb/ft}^3 \text{ for water}$$

$$= 41.6\left(\tfrac{3}{2}h^2 - \tfrac{1}{3}h^3\right)\Big|_0^3 = 187.2 \text{ lb}$$ ■

EXAMPLE 5   *Modeling the force on one face of a dam*

A reservoir is filled with water to the top of a dam. If the dam is in the shape of a parabola 40 ft high and 20 ft wide at the top, as shown in Figure 6.45, what is the total fluid force on the face of the dam?

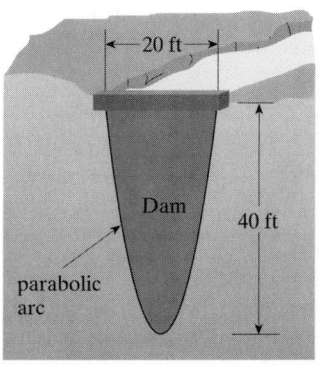

a. Cross section of a dam

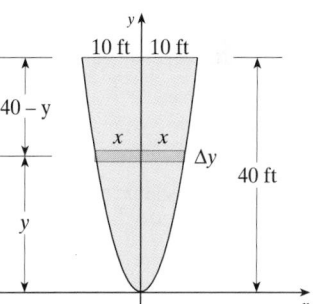

b. A typical horizontal strip is $40 - y$ ft below the water surface.

■ **FIGURE 6.45**   Force on one face of a dam

Solution   Instead of putting the x-axis on the surface of the water (at the top of the dam), we place the origin at the vertex of the parabola and the positive y-axis along the parabola's axis of symmetry (Figure 6.45b). The advantage of this choice of axes is that the parabola can be represented by an equation of the form $y = cx^2$. Because we know that $y = 40$ when $x = 10$, it follows that

$$40 = c(10)^2 \quad \text{so that} \quad c = \tfrac{2}{5} \quad \text{and thus} \quad y = \tfrac{2}{5}x^2$$

The typical horizontal strip on the face of the dam is located y feet above the x-axis, which means it is $h = 40 - y$ feet below the surface of the water. The strip is Δy feet wide and $L = 2x$ feet long. Write L as a function of y:

$$y = \tfrac{2}{5}x^2 \quad \text{so that} \quad x = \sqrt{\tfrac{5}{2}y}$$

and

$$L = 2x = 2\left(\sqrt{\tfrac{5}{2}y}\right) = \sqrt{10y}$$

Therefore

$$\Delta A = L\Delta y = 2\sqrt{\tfrac{5}{2}y}\,\Delta y = \sqrt{10y}\,\Delta y$$

and the total force may be modeled by the integral

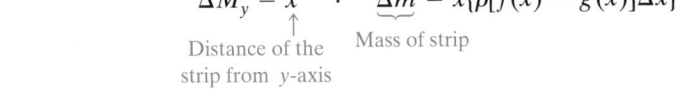

Depth below water: $h = 40 - y$

$$F = \int_0^{40} \rho h \underbrace{L(h)\, dh}_{\text{Area of the strip}} = \int_0^{40} (62.4)(\sqrt{10})(40 - y)\sqrt{y}\, dy$$

Density of water = 62.4

$$= 62.4\sqrt{10} \int_0^{40} (40y^{1/2} - y^{3/2})dy$$

$$= 62.4\sqrt{10} \left(\tfrac{80}{3}y^{3/2} - \tfrac{2}{5}y^{5/2}\right)\Big|_0^{40}$$

$$= 62.4\sqrt{10}\, (y^{3/2})(\tfrac{80}{3} - \tfrac{2}{5}y)\Big|_0^{40}$$

$$= 532{,}480 \text{ lb}$$

∎

MODELING THE CENTROID OF A PLANE REGION

In mechanics, it is often important to determine the point where an irregularly shaped plate will balance. The **moment of a force** measures its tendency to produce rotation in an object and depends on the magnitude of the force and the point on the object where it is applied. Since the time of Archimedes (287–212 B.C.), it has been known that the balance point of an object occurs where all its moments cancel out (so there is no rotation).

The **mass** of an object is a measure of its **inertia**—that is, its propensity to maintain a state of rest or uniform motion. A thin plate whose material is distributed uniformly, so that its density ρ (mass per unit area) is constant, is called a **homogeneous lamina.** The balance point of such a lamina is called its **centroid** and may be thought of as its geometrical center. We shall see how centroids can be computed by integration in this section, and shall examine the topic even further in Section 13.6.

Consider a homogeneous lamina that covers a region R bounded by the curves $y = f(x)$ and $y = g(x)$ on the interval $[a, b]$, and consider a thin, vertical approximating strip within R, as shown in Figure 6.46.

The mass of the strip is given by

$$\Delta m = \rho \cdot \underbrace{[f(x) - g(x)]\Delta x}_{\text{Area of strip}}$$

Density

and the total mass of the lamina may be found by integration:

$$m = \rho \int_a^b [f(x) - g(x)]dx$$

The geometrical center of the approximating strip is $(\tilde{x}, \tilde{y})$, where $\tilde{x} = x$ and $\tilde{y} = \frac{1}{2}[f(x) + g(x)]$. The **moment of the approximating strip about the y-axis** is defined to be the product

$$\Delta M_y = \tilde{x} \cdot \underbrace{\Delta m}_{\text{Mass of strip}} = \tilde{x}\{\rho[f(x) - g(x)]\Delta x\}$$

Distance of the
strip from y-axis

This product provides a measure of the tendency of the strip to rotate about the y-axis. Similarly, the **moment of the strip about the x-axis** is defined to be the product

$$\Delta M_x = \tilde{y}\Delta m = \tfrac{1}{2}\underbrace{[f(x) + g(x)]}\{\rho[f(x) - g(x)]\Delta x\}$$

Average distance of the strip from x-axis

$$= \tfrac{1}{2}\rho\{[f(x)]^2 - [g(x)]^2\}\Delta x$$

Centroid

Theoretically, the lamina should balance on a point placed at its centroid.

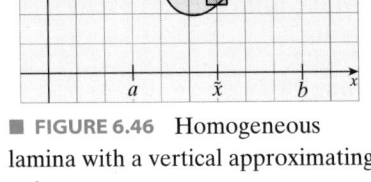

■ **FIGURE 6.46** Homogeneous lamina with a vertical approximating strip

This product provides a measure of the tendency of the strip to rotate about the x-axis.

Integrating on $[a, b]$, we find the moments of the entire lamina R about the y-axis and x-axis may be modeled by

$$M_y = \rho \int_a^b x[f(x) - g(x)]dx \quad \text{and} \quad M_x = \frac{1}{2}\rho \int_a^b \{[f(x)]^2 - [g(x)]^2\}dx$$

If the entire mass m of the lamina R were located at the point $(\bar{x}, \bar{y})$, then its moments about the x-axis and y-axis would be $m\bar{y}$ and $m\bar{x}$, respectively. Thus, we have $m\bar{x} = M_y$ and $m\bar{y} = M_x$, so that

$$\bar{x} = \frac{M_y}{m} = \frac{\rho \int_a^b x[f(x) - g(x)]\, dx}{\rho \int_a^b [f(x) - g(x)]\, dx} \quad \text{and} \quad \bar{y} = \frac{M_x}{m} = \frac{\frac{1}{2}\rho \int_a^b \{[f(x)]^2 - [g(x)]^2\}\, dx}{\rho \int_a^b [f(x) - g(x)]\, dx}$$

We can now summarize these results.

Mass and Centroid

Let f and g be continuous and satisfy $f(x) \geq g(x)$ on the interval $[a, b]$, and consider a thin plate (lamina) of uniform density ρ that covers the region R between the graphs of $y = f(x)$ and $y = g(x)$ on the interval $[a, b]$. Then

The **mass** of R is $m = \rho \int_a^b [f(x) - g(x)]\, dx$

The **centroid** of R is the point $(\bar{x}, \bar{y})$ such that

$$\bar{x} = \frac{M_y}{m} = \frac{\rho \int_a^b x[f(x) - g(x)]\, dx}{\rho \int_a^b [f(x) - g(x)]\, dx} \quad \text{and} \quad \bar{y} = \frac{M_x}{m} = \frac{\frac{1}{2}\rho \int_a^b \{[f(x)]^2 - [g(x)]^2\}dx}{\rho \int_a^b [f(x) - g(x)]\, dx}$$

EXAMPLE 6   *Centroid of a thin plate*

A homogeneous lamina R has constant density $\rho = 1$ and is bounded by the parabola $y = x^2$ and the line $y = x$. Find the mass and the centroid of R.

Solution   We see that the line and the parabola intersect at the origin and at the point $(1, 1)$, as shown in Figure 6.47.

Because $\rho = 1$, the mass is the same as the area of the region R. That is,

$$m = A = \int_0^1 (x - x^2)dx = \left(\frac{x^2}{2} - \frac{x^3}{3}\right)\Big|_0^1 = \frac{1}{6}$$

and we find that the region has moments M_y and M_x about the y-axis and x-axis, respectively, where

$$M_y = \int_0^1 x(x - x^2)dx = \left(\frac{x^3}{3} - \frac{x^4}{4}\right)\Big|_0^1 = \frac{1}{12}$$

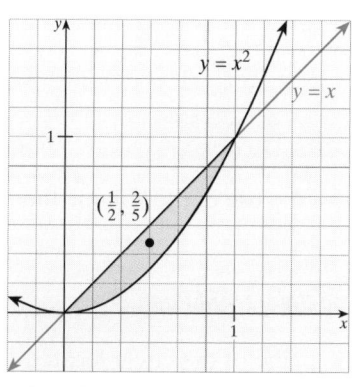

■ FIGURE 6.47 The centroid of a planar region

and

$$M_x = \frac{1}{2}\int_0^1 (x + x^2)(x - x^2)dx = \frac{1}{2}\int_0^1 (x^2 - x^4)dx$$

$$= \frac{1}{2}\left(\frac{x^3}{3} - \frac{x^5}{5}\right)\bigg|_0^1 = \frac{1}{2}\left(\frac{1}{3} - \frac{1}{5}\right) = \frac{1}{15}$$

Thus, the centroid of the region R has coordinates

$$\bar{x} = \frac{M_y}{m} = \frac{\frac{1}{12}}{\frac{1}{6}} = \frac{1}{2} \quad \text{and} \quad \bar{y} = \frac{M_x}{m} = \frac{\frac{1}{15}}{\frac{1}{6}} = \frac{2}{5}$$ ∎

VOLUME THEOREM OF PAPPUS

In certain geometric applications, it is convenient to speak of the center of mass (x, y) without referring to a plate covering R. In this context, the center of mass is the centroid of the region R. We shall use this interpretation in the following theorem, which is usually attributed to Pappus of Alexandria (ca. 300 A.D.), whom many regard as the last great Greek geometer.

THEOREM 6.1   *Volume theorem of Pappus*

The solid generated by revolving a region R about a line outside its boundary (but in the same plane) has volume $V = As$, where A is the area of R and s is the distance traveled by the centroid of R.

Proof   First, choose a coordinate system in which the y-axis coincides with the axis of revolution. Figure 6.48 shows a typical region R together with a vertical approximating rectangle. We shall assume that this rectangle has area ΔA and that it is located x units from the y-axis.

Notice that when the rectangle is revolved about the y-axis, it generates a shell of volume $\Delta V = 2\pi x \Delta A$. Thus, by partitioning the region R into a number of rectangles and taking the limit of the sum of the volumes of all the related approximating shells as the norm of the partition approaches 0, we find that

$$V = \lim_{\|P\| \to 0} \sum_{k=1}^n 2\pi x_k \Delta A_k = \int 2\pi x \, dA = 2\pi \int x \, dA$$

$$= 2\pi \bar{x} \int dA \qquad \text{Because } \bar{x} = \frac{\int x \, dA}{\int dA}$$

$$= 2\pi \bar{x} A \qquad \text{Because } \int dA = A$$

$$= As \qquad \text{Where } s = 2\pi\bar{x} \text{ is the distance traveled by the}$$
$$\text{centroid (the circumference of a circle of radius } \bar{x}\text{)}$$ ▭

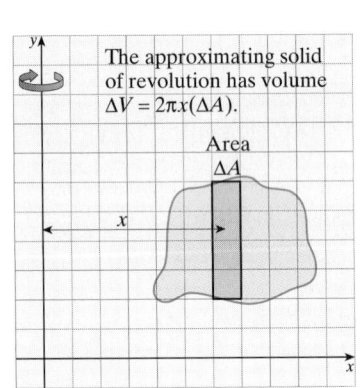

The approximating solid of revolution has volume $\Delta V = 2\pi x(\Delta A)$.

Area ΔA

■ **FIGURE 6.48**   The volume theorem of Pappus

We close this section with an example that illustrates the use of the volume theorem of Pappus.

EXAMPLE 7   *Volume of a torus using the volume theorem of Pappus*

When a circle of radius r is revolved about a line in the plane of the circle located R units from its center ($R > r$), the solid figure so generated is called a **torus,** as shown in Figure 6.49. Show that the torus has volume $V = 2\pi^2 r^2 R$.

Axis of revolution

The centroid travels $2\pi R$ units.
■ **FIGURE 6.49**   The volume of a torus

Solution   The circle has area $A = \pi r^2$ and its center (which is its centroid) travels $2\pi R$ units. Comparing this example to the volume theorem of Pappus, we see $s = R$. Therefore, the torus has volume

$$V = (2\pi R)A = 2\pi R(\pi r^2) = 2\pi^2 r^2 R \qquad ■$$

Incidentally, the volume found in Example 7 using the theorem of Pappus was worked in Problem 39 of Section 6.3 by using shells.

6.5 Problem Set

Ⓐ 1. ■ **What Does This Say?** Discuss work and how to find it.

2. ■ **What Does This Say?** Discuss fluid force and how to find it.

3. ■ **What Does This Say?** What is meant by a centroid? Outline a procedure for finding both mass and centroid.

4. What is the work done in lifting a 50-lb bag of salt 5 ft?

5. What is the work done in lifting an 850-lb billiard table 15 ft?

6. A spring whose natural length is 10 cm exerts a force of 30 N when stretched to a length of 15 cm. Find the spring constant, and then determine the work done in stretching the spring 7 cm beyond its natural length.

7. A 5-lb force will stretch a spring 9 in. beyond its natural length. How much work is required to stretch it 1 ft beyond its natural length?

8. Suppose it takes 4 dyn-cm of work to stretch a spring 10 cm beyond its natural length. How much work is needed to stretch it 4 cm further?

9. A bucket weighing 75 lb when filled and 10 lb when empty is pulled up the side of a 100-ft building. How much more work is done in pulling up the full bucket than the empty bucket?

10. A 30-ft rope weighing 0.4 lb/ft hangs over the edge of a building 100 ft high. How much work is done in pulling the rope to the top of the building?

11. A 30-lb ball hangs at the bottom of a cable that is 50 ft long and weighs 20 lb. The entire length of cable hangs

over a cliff. Find the work done to raise the cable and get the ball to the top of the cliff.

12. An object moving along the x-axis is acted upon by a force $F(x) = x^4 + 2x^2$. Find the total work done by the force in moving the object from $x = 1$ cm to $x = 2$ cm. *Note:* Distance is in cm, so force is in dynes.

13. An object moving along the x-axis is acted upon by a force $F(x) = |\sin x|$. Find the total work done by the force in moving the object from $x = 0$ cm to $x = 2\pi$ cm. *Note:* Distance is in cm, so force is in dynes.

In Problems 14–17, *the given figure is the vertical cross section of a tank containing the indicated substance. Find the fluid force against the end of the tank. The weight densities are given on page* 445.

14. water
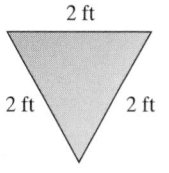
2 ft
2 ft   2 ft

15. seawater
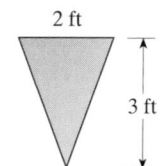
2 ft
3 ft

16. gasoline
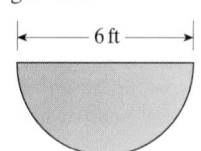
|— 6 ft —|

17. kerosene
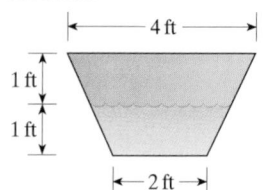
|— 4 ft —|
1 ft
1 ft
|— 2 ft —|

In Problems 18–21, set up, but do not evaluate an integral for the fluid force against the indicated vertical plate.

18. water

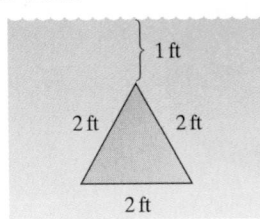

19. milk (half cookie)

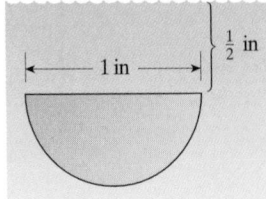

20. mercury

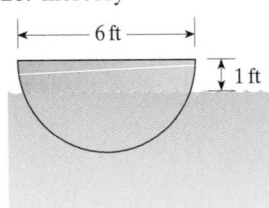

21. kerosene

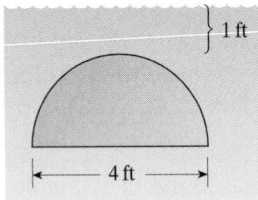

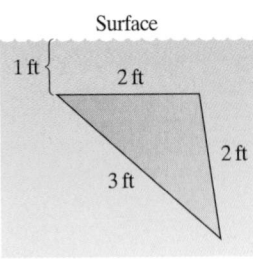

■ **FIGURE 6.50**   Fluid force on a submerged plate

34. A tank in the shape of an inverted right circular cone of height 6 ft and radius 3 ft is half full of water. How much work is performed in pumping all the water in the tank over the top edge?

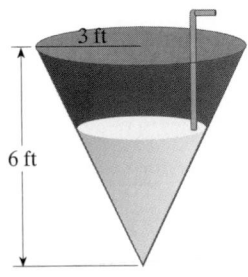

Find the centroid of each planar region in Problems 22–27.

22. the region in the first quadrant bounded by the curves $y = x^3$ and $y = \sqrt[3]{x}$

23. the region bounded by the parabola $y = x^2 - 9$ and the x-axis

24. the region bounded by the parabola $y = 4 - x^2$ and the line $y = x + 2$

25. the region bounded by the curve $y = x^{-1}$, the x-axis, and the lines $x = 1$ and $x = 2$

26. the region bounded by the curve $y = x^{-1}$ and the line $2x + 2y = 5$

27. the region bounded by $y = \dfrac{1}{\sqrt{x}}$ and the x-axis on $[1, 4]$

Use the theorem of Pappus to compute the volume of the solids generated by revolving the regions given in Problems 28–31 about the prescribed axis.

28. the region bounded by the parabola $y = \sqrt{x}$, the x-axis, and the line $x = 4$, about the line $x = -1$

29. the triangular region with vertices $(-3, 0)$, $(0, 5)$, and $(2, 0)$, about the line $y = -1$ and above the line $x = 3$

30. the semicircular region $y = \sqrt{1 - x^2}$, about the line $y = -1$

31. the semicircular region $x = \sqrt{4 - y^2}$, about the line $x = -2$

Ⓑ **32.** The centroid of any triangle lies at the intersection of the medians, two-thirds the distance from each vertex to the midpoint of the opposite side. Locate the centroid of the triangle with vertices $(0, 0)$, $(7, 3)$, and $(7, -2)$ using this method, and then check your result by calculus.

33. A triangular plate is submerged in water as indicated in Figure 6.50. Find the fluid force on the plate.

35. If water leaks from the bottom of the tank in Problem 34, how much work is done?

36. A cylindrical tank with base radius 2 ft and height 12 ft is sunk into the ground so that the top of the tank is at surface level. Find the work done to pump the liquid to the top of the tank if the liquid density is 80 lb/ft³ and the tank is filled to the top.

37. A hemispherical bowl with radius 10 ft is filled with water to a level of 6 ft. Find the work done to the nearest ft-lb to pump all the water to the top of the bowl.

38. A holding tank has the shape of a rectangular parallelepiped 20 ft by 30 ft by 10 ft.
   a. How much work is done in pumping all the water to the top of the tank?
   b. How much work is done in pumping all the water out of the tank to a height of 2 ft above the top of the tank?

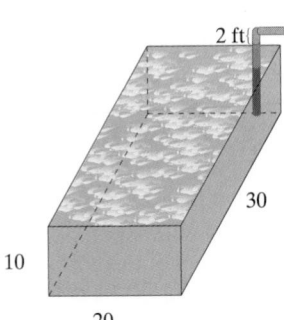

39. A cylindrical tank of radius 3 ft and height 10 ft is filled to a depth of 2 ft with a liquid of density $\rho = 40 \text{ lb/ft}^3$. Find the work done in pumping all the liquid to a height of 2 ft above the top of the tank.

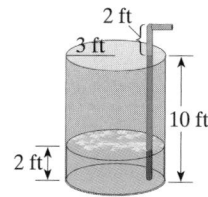

40. An oil can in the shape of a rectangular parallelepiped is filled with oil of density 0.87 g/cm^3. What is the total force on a side of the can if it is 20 cm high and has a square base 15 cm on a side?

41. A swimming pool 20 ft by 15 ft by 10 ft deep is filled with water. A cube 1 ft on a side lies on the bottom of the pool. Find the total fluid force on the five exposed faces of the cube.

42. Suppose a log of radius 1 ft lies at the bottom of the pool in Problem 41. What is the total fluid force on one end of the log?

43. MODELING PROBLEM An object weighing 800 lb on the surface of the earth is propelled to a height of 200 mi above the earth. How much work is done against gravity? *Hint:* Assume the radius of the earth is 4,000 mi, and use Newton's law $F = -k/x^2$, for the force on an object x miles from the center of the earth.

44. A dam is in the shape of an isosceles trapezoid 200 ft at the top, 100 ft at the bottom, and 75 ft high. When water is up to the top of the dam, what is the force on its face?

45. MODELING PROBLEM According to Coulomb's law in physics, two similarly charged particles repel each other with a force inversely proportional to the square of the distance between them. Suppose the force is 12 dynes when they are 5 cm apart.
  a. How much work is done in moving one particle from a distance of 10 cm to a distance of 8 cm from the other?
  b. Set up and analyze a model to determine the amount of work performed in moving one particle from an "infinite" distance to a distance of 8 cm from the other? State any assumptions that you must make.

46. MODELING PROBLEM Figure 6.51 shows a dam whose face against the water is an inclined rectangle.
  a. Set up and analyze a model for determining the fluid force against the face of the dam when the water is level with the top.

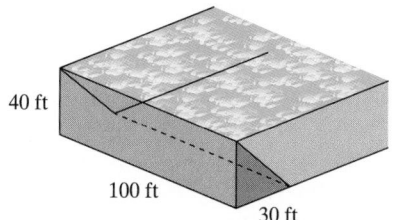

■ **FIGURE 6.51**   Dam with face at an inclined rectangle

  b. What would the fluid force be if the face of the dam were a vertical rectangle (not inclined)?

47. MODELING PROBLEM Figure 6.52 shows a swimming pool, part of whose bottom is an inclined plane. Set up and analyze a model for determining the fluid force on the bottom when the pool is filled to the top with water.

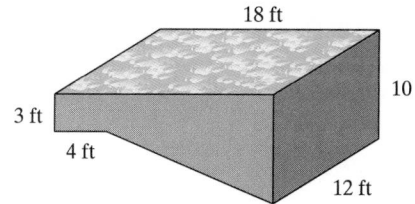

■ **FIGURE 6.52**   Swimming pool dimensions

48. An irregular plate is measured at 0.2-ft intervals from top to bottom, and the corresponding widths noted. The plate is then submerged vertically in water with its top 1 ft below the surface. Use Simpson's rule to estimate the total force on the face of the plate.

depth, x_k	width, L_k
1.0	0.0
1.2	3.3
1.4	3.6
1.6	3.7
1.8	3.7
2.0	3.6
2.2	3.4
2.4	2.9
2.6	0.0

*A region R with uniform density is said to have **moments of inertia** I_x about the x-axis and I_y about the y-axis, where*

$$I_x = \int y^2 \, dA \quad \text{and} \quad I_y = \int x^2 \, dA$$

The dA represents the area of an approximating strip. In Problems 49–50, find I_x and I_y for the given region.

49. the region bounded by the parabola $y = \sqrt{x}$, the x-axis, and the line $x = 4$

50. the region bounded by the curve $y = \sqrt[3]{1 - x^3}$ and the coordinate axes

51. MODELING PROBLEM If a region has area A and moment of inertia I_y about the y-axis, then it is said to have a **radius of gyration**

$$\rho_y = \sqrt{\frac{I_y}{A}}$$

with respect to the y-axis. Set up and analyze a model to determine the radius of gyration for the triangular region bounded by the line $y = 4 - x$ and the coordinate axes.

52. Show that the centroid of a rectangle is the point of intersection of its diagonals.

53. A right circular cone is generated when the region bounded by the line $y = x$ and the vertical lines $x = 0$ and $x = r$ is revolved about the x-axis. Use Pappus' theorem to show that the volume of this cone is

$$V = \tfrac{1}{3}\pi r^3$$

54. Let R be the triangular region bounded by the line $y = x$, the x-axis, and the vertical line $x = r$. When R is rotated about the x-axis, it generates a cone of volume $V = \frac{1}{3}\pi r^3$. Use Pappus' theorem to determine $\bar{y}$, the y-coordinate of the centroid of R. Then use similar reasoning to find $\bar{x}$.

55. Find the volume of the solid figure generated when a square of side L is revolved about a line that is outside the square, parallel to two of its sides, and located s units from the closer side.

56. Find the volume of the solid figure generated by revolving an equilateral triangle of side L about one of its sides.

57. Let R be a region in the plane, and suppose that R can be partitioned into two subregions R_1 and R_2. Let $(\bar{x}_1, \bar{y}_1)$ and $(\bar{x}_2, \bar{y}_2)$ be the centroids of R_1 and R_2, respectively, and let A_1 and A_2 be the areas of the subregions. Show that $(\bar{x}, \bar{y})$ is the centroid of R, where

$$\bar{x} = \frac{A_1 x_1 + A_2 x_2}{A_1 + A_2} \quad \text{and} \quad \bar{y} = \frac{A_1 y_1 + A_2 y_2}{A_1 + A_2}$$

Chapter 6   Review

Proficiency Examination

Concept Problems

1. Describe the process for finding the area between two curves.

2. Describe the process for finding volumes of solids with a known cross-sectional area.

3. Compare and contrast the methods of finding volumes by disks, washers, and shells.

4. What is the formula for arc length of a graph?

5. What is the formula for area of a surface of revolution?

6. How do you find the work done by a variable force?

7. How do you find the force exerted by a liquid on an object submerged vertically?

8. What are the formulas for the mass and centroid of a homogeneous lamina?

9. State the volume theorem of Pappus.

10. State Hooke's law.

11. What is Pascal's principle?

Practice Problems

JOURNAL PROBLEM   *Mathematics Teacher,* December 1990, p. 695. *Use the integrals A–G to answer the questions in Problems 12–17.*

A. $\pi \displaystyle\int_a^b [f(x)]^2 \, dx$

B. $\pi \displaystyle\int_c^d [f(y)]^2 \, dy$

C. $\displaystyle\int_a^b A(x) \, dx$

D. $\displaystyle\int_c^d A(y) \, dy$

E. $\pi \displaystyle\int_a^b [f(x) - g(x)] \, dx$

F. $\pi \displaystyle\int_a^b \{[f(x)]^2 - [g(x)]^2\} \, dx$

G. $\pi \displaystyle\int_c^d \{[f(y)]^2 - [g(y)]^2\} \, dy$

12. Which formula does not seem to belong to this list?

13. Which expressions represent volumes of solids of revolution?

14. Which expressions represent volumes of solids containing holes?

15. Which expressions represent volumes of a solid with cross sections that are perpendicular to an axis?

16. Which expressions represent volumes of solids of revolution revolved about the x-axis?

17. Which expressions represent volumes of solids of revolution revolved about the y-axis?

18. Find the area under the curve $f(x) = 3x^2 + 2$ on $[-1, 3]$.

19. Find the area between the curves $y = x^2$ and $y^3 = x$.

20. The base of a solid in the xy-plane is the circular region given by $x^2 + y^2 = 4$. Every cross section perpendicular to the x-axis is a rectangle whose height is twice the length of the side that lies in the xy-plane. Find the volume of this solid.

21. Let R be a region bounded by the graphs of the equations $y = (x - 2)^2$ and $y = 4$.

  a. Sketch the graph of R and find its area.

  b. Find the volume of a solid of revolution of R about the x-axis.

  c. Find the volume of a solid of revolution of R about the y-axis.

22. Find the arc length of the curve $y = 2 - x^{3/2}$ from $(0, 2)$ to $(1, 1)$. Give the exact value.

23. Find the surface area obtained by revolving the parabolic arc $y^2 = x$ from $(0, 0)$ to $(1, 1)$ about the x-axis. Give the answer to the nearest hundredth unit.

24. Find the centroid of the region between $y = x^2 - x$ and $y = x - x^3$ on $[0, 1]$.

25. An observation porthole on a ship is circular with diameter 2 feet and is located so its center is 3 feet below the waterline. Find the fluid force on the porthole, assuming that the boat is in seawater (density is 64 lb/ft^3). Set up the integral only.

Supplementary Problems

Find the area of the regions bounded by the curves and lines in Problems 1–4.

1. $4y^2 = x$ and $2y = x - 2$
2. $y = x^3, y = x^4$
3. $x = y^{2/3}, x = y^2$
4. $y = \sqrt{3} \sin x, y = \cos x$, and the vertical lines $x = 0$ and $x = \pi/2$

Find the volume of a solid of revolution obtained by revolving the region R about the indicated axis in Problems 5–8.

5. R is bounded by $y = \sqrt{x}, x = 9, y = 0$; about the x-axis
6. R is bounded by $y = x^2, y = 9 - x^2$, and $x = 0$; about the y-axis
7. R is bounded by $y = x^2$ and $y = x^4$; about the x-axis
8. R is bounded by $y = \sqrt{\cos x}, x = \frac{\pi}{4}, x = \frac{\pi}{3}$, and $y = 0$; about the x-axis

9. The base of a solid is the region in the xy-plane that is bounded by the curve $y = 4 - x^2$ and the x-axis. Every cross section of the solid perpendicular to the y-axis is a rectangle whose height is twice the length of the side that lies in the xy-plane. Find the volume of the solid.

10. The base of a solid is the region R in the xy-plane bounded by the curve $y^2 = x$ and the line $y = x$. Find the volume of the solid if each cross section perpendicular to the x-axis is a semicircle with a diameter in the xy plane.

11. Find the centroid of a homogeneous lamina covering the region R that is bounded by the parabola $y = 2 - x^2$ and the line $y = x$.

12. A force of 100 lb is required to compress a spring from its natural length of 10 in. to a length of 8 in. Find the work required to compress it to a length of 7 in.

13. A force of 1,000 lb is required to compress a spring from its natural length of 10 in. to a length of 8 in. Find the work required to stretch it to a length of 11 in.

14. A bowl is formed by revolving the region bounded by the curve $y = 2x^2$ and the lines $x = 0$ and $y = 1$ about the y-axis (units are in ft). The bowl is filled with water to a height of 9 in.
    a. How much work is done in pumping the water to the top of the bowl?
    b. How much work is done in pumping the water to a point 0.5 ft above the top of the bowl?

15. Find the volume of a sphere of radius a by revolving the region bounded by the semicircle $y = \sqrt{a^2 - x^2}$ about the x-axis.

16. Two solids S_1 and S_2 have as their base the region in the xy-plane bounded by the parabola $y = x^2$ and the line $y = 1$. For solid S_1, each cross section perpendicular to the x-axis is a square, and for S_2, each cross section perpendicular to the y-axis is a square. Find the volumes of S_1 and S_2.

17. Use calculus to find the lateral surface area of a right circular cylinder with radius r and height h.

18. Use calculus to find the surface area of a right circular cone with (top) radius r and height h.

19. A 20-lb bucket in a well is attached to the end of a 60-ft chain that weighs 16 lb. If the bucket is filled with 50 lb of water, how much work is done in lifting the bucket and chain to the top of the well?

20. A conical tank whose vertex points down is 10 ft high with (top) diameter 4 ft. The tank is filled with a fluid of density $\rho = 22$ lb/ft³. Find the work required to pump all the fluid to the top of the tank.

21. **MODELING PROBLEM**   A swimming pool is 3 ft deep at one end and 8 ft deep at the other, as shown in Figure 6.53. The pool is 30 ft long and 25 ft wide with vertical sides. Set up and analyze a model to find the fluid force against one of the 30-ft sides.

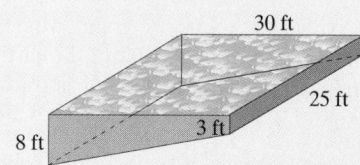

■ **FIGURE 6.53**   Cross section of a swimming pool

22. The ends of a container filled with water have the shape of the region R bounded by the curve $y = x^4$ and the line $y = 16$. Find the fluid force (to two decimal places) exerted by the water on one end of the container.

23. A conical tank whose vertex points down is 12 ft high with (top) diameter 6 ft. The tank is filled with a fluid of density $\rho = 22$ lb/ft³. If the tank is only half full, find the work required to pump all the fluid to a height of 2 ft above the top of the tank.

24. Each end of a container filled with water has the shape of an isosceles triangle, as shown in Figure 6.54. Find the force exerted by the water on an end of the container.

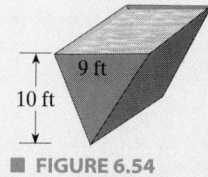

■ **FIGURE 6.54**

Water pressure

25. A container has the shape of the solid formed by rotating the region bounded by the curve $y = x^3$ ft and the lines $x = 0$ ft and $y = 1$ ft about the y-axis. If the container is filled with water, how much work is required to pump all the water out of the container?

26. A gate in a water main has the shape of a circle with diameter 10 ft. If water stands 2 ft high in the main, what is the force exerted by the water on the gate?

Suppose a particle of mass m moves along a number line under the influence of a variable force F which always acts in a direction parallel to the line of motion. At a given time, the particle is said to have **kinetic energy**

$$K = \tfrac{1}{2}mv^2$$

where v is the particle's velocity, and it has **potential energy**

$$P = -\int_{s_0}^{s} F\, dx$$

relative to the point s on the line of motion. Use these definitions in Problems 27–28.

27. A spring at rest has potential energy $P = 0$. If the spring constant is 30 lb/ft, how far must the spring be stretched so that its potential energy is 20 ft-lb?

28. A particle with mass m falls freely near the earth's surface. At time $t = 0$, the particle is s_0 units above the ground and is falling with velocity v_0.
  a. Find the potential energy P of the particle in terms of s if the potential energy is 0 when $s = s_0$.
  b. Find the potential energy P and the kinetic energy K of the particle in terms of time t.

29. An object weighs $w = 1{,}000r^{-2}$ grams, where r is the object's distance from the center of the earth. What work is required to lift the object from the surface of the earth ($r = 6{,}400$ km) to a point 3,600 km above the surface of the earth?

30. MODELING PROBLEM   A certain species has population $P(t)$ at time t whose growth rate is affected by seasonal variations in the food supply. Suppose the population may be modeled by the differential equation

$$\frac{dP}{dt} = k(P - A)\cos t$$

where k and A are physical constants. Solve the differential equation and express $P(t)$ in terms of k, A, and P_0, the initial population.

31. Find the center of mass of a thin plate of constant density if the plate occupies the region bounded by the lines $2y = x, x + 3y = 5$, and the x-axis.

32. The portion of the ellipse
$$\frac{x^2}{a^2} + \frac{y^2}{b^2} = 1$$
that lies in the first quadrant between $x = r$ and $x = a$, for $0 < r < a$, is revolved about the x-axis to form a solid S_e. Find the volume of S_e.

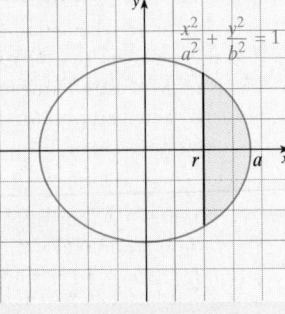

33. The portion of the hyperbola
$$\frac{x^2}{a^2} - \frac{y^2}{b^2} = 1$$
that lies in the first quadrant between $x = a$ and $x = r$, for $0 < a < r$, is revolved about the x-axis to form a solid S_h. Find the volume of S_h.

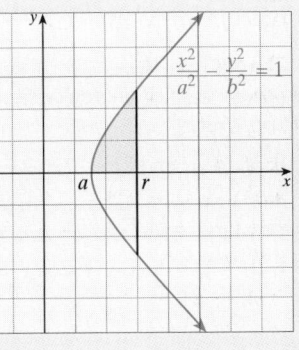

34. Let R be the part of the ellipse
$$\frac{x^2}{a^2} + \frac{y^2}{b^2} = 1$$
that lies in the first quadrant. Use the theorem of Pappus to find the coordinates of the centroid of R. You may use the facts that R has area $A = \tfrac{1}{4}\pi ab$, and the semiellipsoid it generates when it is revolved about the x-axis has volume $V = \tfrac{2}{3}\pi ab^2$.

35. MODELING PROBLEM   A container of a certain fluid is raised from the ground on a pulley. While resting on the ground, the container and fluid together weigh 200 lb. However, as the container is raised, fluid leaks out in such a way that at height x feet above the ground, the total weight of the container and fluid is $200 - 0.5x$ lb. Set up and analyze a model to find the work done in raising the container 100 ft.

36. A bag of sand is being lifted vertically upward at the rate of 31 ft/min. If the bag originally weighed 40 lb and leaks at the rate of 0.2 lb/s, how much work is done in raising it until it is empty?

37. The vertical end of a trough is an equilateral triangle, 3 ft on a side. Assuming the trough is filled with a fluid of density $\rho = 40$ lb/ft³, find the force on the end of the trough.

38. The radius of a circular water main is 2 ft. Assuming the main is half full of water, find the total force exerted by the water on a gate that crosses the main at one end.

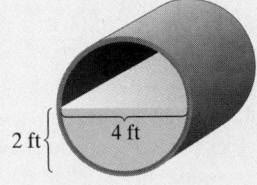

39. The force acting on an object moving along a straight line is known to be a linear function of its position. Find the force function F if it is known that 13 ft-lb of work is required to move the object over 6 ft and 44 ft-lb of work is required to move it 12 ft.

40. A tank has the shape of a rectangular parallelepiped with length 10 ft, width 6 ft, and height 8 ft. The tank contains a liquid of density $\rho = 20$ lb/ft³. Find the level of the liquid in the tank if it will require 2,400 ft-lb of work to move all the liquid in the tank to the top.

41. A tank is constructed by placing a right circular cylinder of height 10 ft and radius 2 ft on top of a hemisphere with the same radius. If the tank is filled with a fluid of density $\rho = 24$ lb/ft^3, how much work is done in bringing all the fluid to the top of the tank?

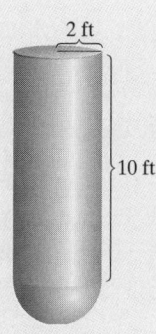

42. MODELING PROBLEM   A piston in a cylinder causes a gas either to expand or to contract. Assume the pressure P and the volume of gas V in the cylinder satisfy the adiabatic law (no exchange of heat)

$$PV^{1.4} = C$$

where C is a constant.
  a. If the gas goes from volume V_1 to volume V_2, show that the work done may be modeled by
  $$W = \int_{V_1}^{V_2} CV^{-1.4}\, dV$$
  b. How much work is done if 0.6 m^3 of steam at a pressure of 2,500 Newtons/m^2 expands 50%?

43. A thin sheet of tin is in the shape of a square, 16 cm on a side. A rectangular corner of area 156 cm^2 is cut from the square, and the centroid of the portion that remains is found to be 4.88 cm from one side of the original square, as shown in Figure 6.55. How far is it from the other three sides?

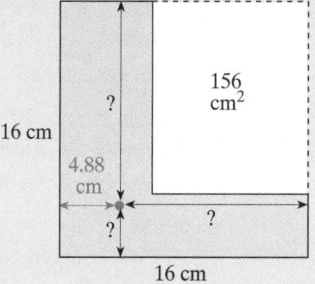

■ **FIGURE 6.55**
Centroid of a square

44. The front and back faces of a container are parabolic surfaces with vertical axes and the vertex of the parabola at the bottom (assume the coordinate axes are located so the parabola has the form $y = kx^2$). Each parabolic face is 10 ft across the top and 15 ft deep. Suppose the container is filled with a liquid of density $\rho = 40$ lb/ft^3. Find the force on either parabolic face.

45. Let R be the region bounded by the semicircle $y = \sqrt{4 - x^2}$ and the x-axis, and let S be the solid with base R and trapezoidal cross sections perpendicular to the x-axis. Assume that the two slant sides and the shorter parallel side are all half the length that lies in the base, as shown in Figure 6.56. Set up and evaluate an integral for the volume of S.

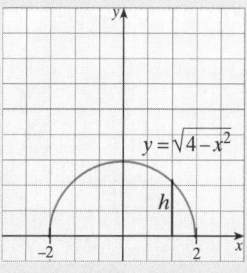
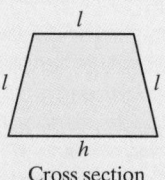

■ **FIGURE 6.56**   Volume of a region S

46. A child is building a sand structure at the beach.
  a. Find the work done if the density of the sand is $\rho = 140$ lb/ft^3 and the structure is a cone of height 1 ft and diameter 2 ft.
  b. What if the structure is a tower (cylinder) of height 1 ft and radius 4 in.?

47. Let S be the solid generated by revolving about the x-axis the region bounded by the parabola $y = x^2$ and the line $y = x$. Find the number A so that the plane perpendicular to the x-axis at $x = A$ divides S in half. That is, the volume on one side of A equals the volume on the other side.

48. A plate in the shape of an isosceles triangle is submerged vertically in water, as shown in Figure 6.57. Find the force on one side of the plate if the two equal sides have length 3, the third side has length 5, and the top vertex of the triangle is 4 units below the surface.

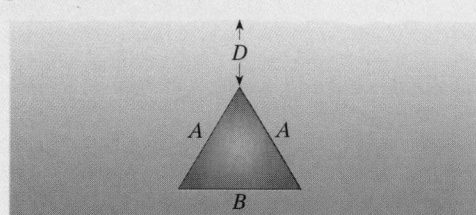

■ **FIGURE 6.57**   Hydrostatic force

49. Repeat Problem 48 for equal sides of length A, the third side with length B, and the top vertex D units below the surface, as shown in Figure 6.57.

50. Find the arc length of the curve
$$y = Ax^3 + \frac{B}{x}$$
on the interval $[a, b]$, where A and B are positive constants with $AB = 1/12$.

51. THINK TANK PROBLEM   When the line segment $y = x - 2$ between $x = 1$ and $x = 3$ is revolved about the x-axis, it generates part of a cone. Explain why the surface area of this cone is *not* given by the integral

$$\int_1^3 2\pi(x - 2)\sqrt{2}\, dx$$

What is the actual surface area?

52. HISTORICAL QUEST

AMALIE ("EMMY") NOETHER
1882–1935

Emmy Noether has been called the greatest woman mathematician in the history of mathematics. Once again, we see a woman overcoming great obstacles in achieving her education. At the time she received her Ph.D. at the University of Erlangen (in 1907), the Academic Senate declared that the admission of women students would "overthrow all academic order." Emmy Noether is famous for her work in physics and mathematics. In mathematics, she opened up entire areas for study in the theory of ideals, and in 1921 she published a paper on rings, which have since been called Noetherian rings. Rings and ideals are objects of study in modern algebra. She was not only a renowned mathematician but also an excellent teacher. It has been reported that she often lectured by "working it out as she went." Nevertheless, her students flocked around her "as if following the Pied Piper." It seems as if all of the anecdotal stories about Noether revolve around her students or the fact that she seldom wrote or said anything that was not mathematical. In an article in the *Mathematics Teacher*, (March 1982, p. 240). Clark Kimberling quotes Emil Artin, a colleague of Noether:

Now the one thing I remember most vividly is the trip on the Hamburg Undergrund, which is the subway in Hamburg. We picked up Emmy at the Institute, and she and Artin immediately started talking mathematics. At that time it was Idealtheorie, and they started talking about Ideal, Führer, and Gruppe, and Undergruppe, and everyone on the subway car suddenly started pricking up their ears. [Each of the German nouns has both mathematical and political meanings.] And I was frightened to death—I thought, my goodness, next thing's going to happen, somebody's going to arrest us. Of course, that was in '34, and all. But Emmy was completely oblivious and she talked very loud and very excited, and got louder and louder, and all the time the "Führer" came out, and the "Ideal." She was very full of life, and she constantly talked very fast and very loud.

Shortly after this conversation, the Nazi pressure forced her to emigrate to Bryn Mawr College in Pennsylvania in the fall of 1934. In the years that followed, there were mass dismissals of "racially undesirable" professors. The tremendous influence of German mathematicians on the world was destroyed. Many noted Jewish mathematicians emigrated to American universities.

For this HISTORICAL QUEST compile a list of notable mathematicians who emigrated to the United States from 1933 to 1940.

53. MODELING PROBLEM   An object below the surface of the earth is attracted by a gravitational force that is directly proportional to the square of the distance from the object to the center of the earth. Set up and analyze a model to find the work done in lifting an object weighing P lbs from a depth of s feet (below the surface) to the surface. (Assume the earth's radius is 4,000 mi.)

In Problems 54–55, consider a continuous curve $y = f(x)$ that passes through $(-1, 4), (0, 3), (1, 4), (2, 6), (3, 5), (4, 6), (5, 7), (6, 6), (7, 4),$ and $(8, 3).$

54. Use Simpson's rule to estimate the volume of the solid formed by revolving this curve about the x-axis for $[-1, 9]$. Assume the curve passes through $(9, 4)$.

55. Use the trapezoidal rule to estimate the volume of the solid formed by revolving this curve about the line $y = -1$ on $[-1, 9]$.

Technology Window

56. MODELING PROBLEM   In these computational problems we revisit a problem similar to the group research project at the end of Chapter 4, but add a couple of complications. Imagine a cylindrical fuel tank lying on its side (of length $L = 20$ ft); its ends are elliptically-shaped and are defined by the equation

$$\left(\frac{x}{a}\right)^2 + \left(\frac{y}{b}\right)^2 = 1$$

We are concerned with the amount of fuel in the tank for a given level.

a. Explain why the area of an end of the tank can be modeled by

$$A = 2a \int_{-b}^{b} \sqrt{1 - \left(\frac{y}{b}\right)^2}\, dy$$

b. By making the right substitution in the integral in part **a**, we can obtain an integral representing half the area inside the circle (with unit radius). Do this and show that the area inside the above ellipse is πab.

c. Explain why the volume of the fuel at level h $(-b \le h \le b)$ may be modeled by

$$V(h) = 2La \int_{-b}^{h} \sqrt{1 - \left(\frac{y}{b}\right)^2}\, dy$$

d. For $a = 5$ and $b = 4$, numerically compute $V(h)$ for $h = -3, -2, \ldots, 3, 4$.
Note: $V(0)$ and $V(4)$ will serve as a check on your work.

57. We think of the last part of Problem 56 as the "direct" problem: Given a value of h, compute $V(h)$ by a formula. Now we turn to the "inverse" problem: For a set value of V, say V_0, find h such that $V(h) = V_0$. Typically, the inverse problem is more difficult than the direct problem and requires some sort of iterative method to approximate the solution (for example, Newton–Raphson's method). For example, if $V_0 = 500$ ft^3, we could define $f(h) = V(h) - 500$ and seek the relevant zero of f. Set up the necessary computer program to do this and find the h values, to two-decimal-place accuracy, corresponding to these values of V: a. 500 b. 800 *Hint:* A key to this problem is thinking about $V'(h)$, where $V(h)$ is expressed in Problem 56**c**. Also, for starting values h_0, refer to your work in Problem 56**d**.

58. Find the centroid of a right triangle whose legs have lengths a and b. *Hint:* Use the theorem of Pappus.

59. The largest of the pyramids at Giza in Egypt is roughly 480 ft high and has a square base 750 ft on a side. If the rock in the pyramid has average density 160 lb/ft^3, how much work was required to lift the stones into place to construct the pyramid?

60. Find the length of the curve $y = e^x$ on the interval $[0, 1]$.

61. Find the volume of the solid generated when the region under the curve
$$y = \frac{1}{\sqrt{9 - x^2}}$$
on the interval $[0, 2]$ is revolved about the y-axis.

62. Find the volume of the solid formed by revolving about the x-axis the region bounded by the curve
$$y = \frac{2}{\sqrt{3x - 2}}$$
the x-axis, and the lines $x = 1$ and $x = 2$.

63. Find the area between the curves $y = \tan^2 x$ and $y = \sec^2 x$ on the interval $\left[0, \frac{\pi}{4}\right]$.

64. Find the surface area of the solid generated by revolving the region bounded by the x-axis and the curve
$$y = e^x + \frac{1}{4}e^{-x}$$
on $[0, 1]$ about the x-axis.

65. Find the volume of the solid whose base is the region R bounded by the curve $y = e^x$ and the lines $y = 0, x = 0, x = 1$, if cross sections perpendicular to the x-axis are equilateral triangles.

66. Find the length of the curve $y = \ln \sec x$ on the interval $[0, \frac{\pi}{3}]$.

67. Find the length of the curve
$$x = \frac{ay^2}{b^2} - \frac{b^2}{8a}\ln\frac{y}{b}$$
between $y = b$ and $y = 2b$.

68. SPY PROBLEM The Spy arrives at the sleazy bar at the appointed time for his meeting with "N" (recall Supplementary Problem 81, Chapter 5) but, as usual, his peerless leader is fashionably late. While waiting, the Spy reflects on the dangerous life he leads. Ten years ago, he took his spying oath, followed on the same day by 6 friends from college. On average, 30 new people join each year, and he estimates that the fraction of spies still alive after t years of service is $F(t) = e^{-t/20}$. Approximately how many people who joined the service after the Spy are still alive?

69. PUTNAM EXAMINATION PROBLEM Find the length of the curve $y^2 = x^3$ from the origin to the point where the tangent makes an angle of 45° with the x-axis.

70. PUTNAM EXAMINATION PROBLEM A solid is bounded by two bases in the horizontal planes $z = h/2$ and $z = -h/2$, and by a surface with the property that every horizontal cross section has area given by an expression of the form
$$A = a_0z^3 + a_1z^2 + a_2z + a_3$$
a. Show that the solid has volume $V = \frac{1}{6}h(B_1 + B_2 + 4M)$, where B_1 and B_2 are the areas of the bases and M is the area of the middle horizontal cross section.
b. Show that the formulas for the volume of a cone and a sphere can be obtained from the formula in part **a** for certain special choices of a_0, a_1, a_2, and a_3.

71. PUTNAM EXAMINATION PROBLEM The horizontal line $y = c$ intersects the curve $y = 2x - 3x^3$ in the first quadrant as shown in Figure 6.58. Find c so that the areas of the two shaded regions are equal.

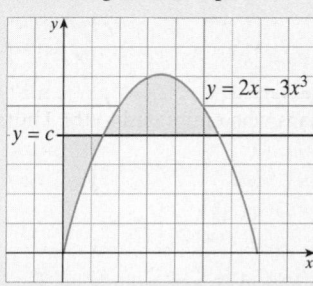

FIGURE 6.58 Putnam Examination problem

"Houdini's Escape"

Harry Houdini
(*born Ehrich Weiss*)
(*1874–1926*)

Mathematics is the gate and key of the sciences.... Neglect of mathematics works injury to all knowledge, since he who is ignorant of it cannot know the other sciences or the things of this world. And what is worse, those who are thus ignorant are unable to perceive their own ignorance and so do not seek a remedy.

ROGER BACON
OPUS MAJUS, PART 4,
DISTINCTIA PRIMA, CAP. 1.

It seems to be expected of every pilgrim up the slopes of the mathematical Parnassus, that he will at some point or other of his journey sit down and invent a definite integral or two towards the increase of the common stock.

J. J. SYLVESTER
"NOTES TO THE
MEDITATION ON
PONCELET'S THEOREM,"
MATHEMATICAL PAPERS,
VOL. 2, P. 214.

This project is to be done in groups of three or four students. Each group will submit a single written report.

Harry Houdini was a famous escape artist. In this project we relive a trick of his that challenged his mathematical prowess, as well as his skill and bravery. It may challenge these qualities in you as well.

Houdini had his feet shackled to the top of a concrete block which was placed on the bottom of a giant laboratory flask. The cross-sectional radius of the flask, measured in feet, was given as a function of height z from the ground by the formula $r(z) = 10z^{-1/2}$, with the bottom of the flask at $z = 1$ ft. The flask was then filled with water at a steady rate of 22π ft³/min. Houdini's job was to escape the shackles before he was drowned by the rising water in the flask.

Now Houdini knew it would take him exactly 10 minutes to escape the shackles. For dramatic impact, he wanted to time his escape so it was completed precisely at the moment the water level reached the top of his head. Houdini was exactly 6 ft tall. In the design of the apparatus, he was allowed to specify only one thing: the height of the concrete block he stood on.

Your paper is not limited to the following questions, but should include these concerns: How high should the block be? (You can neglect Houdini's volume and the volume of the block.) How fast is the water level changing when the flask first starts to fill? How fast is the water level changing at the instant when the water reaches the top of his head? You might also help Houdini with any size flask by generalizing the derivation: Consider a flask with cross-sectional radius $r(z)$ an arbitrary function of z with a constant inflow rate of $dV/dt = A$. Can you find dh/dt as a function of $h(t)$?

———
MAA Notes 17 (1991), "Priming the Calculus Pump: Innovations and Resources," by Marcus S. Cohen, Edward D. Gaughan, R. Arthur Knoebel, Douglas S. Kurtz, and David J. Pengelley.

Chapters 1–6   *Cumulative Review Problems*

In Chapter 1 we introduced the idea of mathematical modeling, and in Chapter 5 we introduced one of the most important and revolutionary ideas in the history of mathematics—namely, that of a Riemann sum. In this chapter we put these two ideas together to model a great many real-life situations. Table 6.3 summarizes this relationship between modeling and the Riemann sum.

■ **TABLE 6.3   Modeling Problems Involving Integration**

Section	Situation	Riemann Sum Model	Outcome
5.3	**Introduction** We want to model something in terms of one or more continuous functions on a closed interval $[a, b]$.	1. Partition interval. 2. Choose a number x_k^* arbitrarily from each subinterval of width Δx_k. 3. Form a sum of the type $$R_n = \sum_{k=1}^{n} f(x_k^*)\Delta x_k$$	The approximations improve as the norm of the partitions goes to 0. The Riemann sums approach a limiting integral. Use this integral to define and calculate what we originally wanted to measure.
5.2	**Area** We want to find the area under a curve defined by $y = f(x)$ where $f(x) \geq 0$, above the x-axis and bounded by the lines $x = a$ and $x = b$.	Find the sum of n rectangles. $$\sum_{k=1}^{n} \underbrace{f(x_k)\Delta x_k}_{\text{Area of } k\text{th rectangle}}$$	$$\text{Area} = \int_a^b f(x)\,dx$$
5.3	**Displacement and distance traveled** An object is moving along a line having continuous velocity $v(t)$ for each time t between $t = a$ and $t = b$.	Find the sum of n distances. $$\lim_{n \to +\infty} \sum_{k=1}^{n} \underbrace{\left\| v[a + (k-1)\Delta t_k] \right\| \Delta t_k}_{\text{Distance traveled over } k\text{th time interval}}$$	$$\text{Distance} = \int_a^b \left\| v(t) \right\| dt$$ Displacement is $\int_a^b v(t)\,dt = s(b) - s(a)$
5.7	**Average value** Find the average value of a continuous function on an interval.	Find the average of a weighted sum of n values. $$\frac{1}{n} \sum_{k=1}^{n} f(x_k^*) = \frac{1}{b-a} \sum_{k=1}^{n} f(x_k^*)\Delta x$$	$$\text{Average} = \frac{1}{b-a} \int_a^b f(x)\,dx$$
6.1	**Area between curves** Find the area between the curves $y = f(x)$ and $y = g(x)$.	$$\sum_{k=1}^{n} \underbrace{[f(x_k^*) - g(x_k^*)]\Delta x_k}_{\text{Top curve minus bottom curve}}$$	$$\text{Area} = \int_a^b [f(x) - g(x)]\,dx$$
6.2	**Volume by cross section** Find the volume of a solid with a cross section perpendicular to the x-axis with face area $A(x)$.	$$\sum_{k=1}^{n} A(x_k^*)\Delta x_k$$	$$\text{Volume} = \int_a^b A(x)\,dx$$

■ **TABLE 6.3** (continued)

Section	Situation	Riemann Sum Model	Outcome
6.2	**Volumes of revolution** Find the volume of a region revolved about the specified axis.		
	Disks: x-axis	$\displaystyle\sum_{k=1}^{n} \pi[f(x_k^*)]^2 \Delta x_k$	$\displaystyle\text{Volume} = \pi \int_a^b [f(x)]^2\, dx$
	Washers: x-axis	$\displaystyle\sum_{k=1}^{n} \pi([f(x_k^*)]^2 - [g(x_k^*)]^2)\Delta x_k$	$\displaystyle V = \pi \int_a^b ([f(x)]^2 - [g(x)]^2)\, dx$
6.3	**Shells:** y-axis	$\displaystyle\sum_{k=1}^{n} 2\pi x_k f(x_k^*)\Delta x_k$	$\displaystyle\text{Volume} = 2\pi \int_a^b x f(x)\, dx$
6.4	**Arc length** Find the length of a curve $y = f(x)$ on $[a, b]$.	$\displaystyle\sum_{k=1}^{n} \sqrt{1 + [f'(x_k^*)]^2}\,\Delta x_k$	$\displaystyle\text{Length} = \int_a^b \sqrt{1 + [f'(x)]^2}\, dx$
6.4	**Area of a surface of revolution** Find the area of a surface formed by rotating the curve $g = f(x)$ about the x-axis over $[a, b]$.	$\displaystyle\sum_{k=1}^{n} 2\pi f(x_k^*)\Delta s_k$	$\displaystyle\text{Area} = 2\pi \int_a^b f(x)\sqrt{1 + [f'(x)]^2}\, dx$
6.5	**Mass** Find the mass of a thin plate with constant density ρ	$\displaystyle\sum_{k=1}^{n} \rho[f(x_k^*) - g(x_k^*)]\Delta x_k$	$\displaystyle m = \rho \int_a^b [f(x) - g(x)]\, dx$
6.5	**Moment** Find the moment of a thin plate with constant density ρ (about the y-axis).	Moment about the y-axis $\displaystyle\sum_{k=1}^{n} \rho x_k^*[f(x_k^*) - g(x_k^*)]\Delta x_k$ Moment about the x-axis	$\displaystyle M_y = \rho \int_a^b x[f(x) - g(x)]\, dx$ $\displaystyle M_x = \frac{\rho}{2} \int_a^b \{[f(x)]^2 - [g(x)]^2\}\, dx$
	Centroid Find the centroid of a homogeneous lamina.	$\displaystyle\sum_{k=1}^{n} \frac{\rho}{2}\{[f(x_k^*)]^2 - [g(x_k^*)]^2\}\Delta x_k$	$\displaystyle(\bar{x}, \bar{y}) = \left(\frac{M_y}{m}, \frac{M_x}{m}\right)$
6.5	**Work** Find the work done when an object moves along the x-axis from a to b against a variable force $F(x)$.	$\displaystyle\sum_{k=1}^{n} F(x_k^*)\Delta x_k$	$\displaystyle\text{Work} = \int_a^b F(x)\, dx$
6.5	**Hydrostatic force** Find the total force of a fluid against one side of a vertical plate.	$\displaystyle\sum_{k=1}^{n} \rho(h_k^*)L(h_k^*)\Delta h_k$	$\displaystyle F = \int_a^b \rho h\, L(h)\, dh$

Cumulative Review Problems for Chapters 1–6

1. ■ **What Does This Say?** Define limit. Explain what this definition is saying using your own words.

2. ■ **What Does This Say?** Define derivative. Explain what this definition is saying using your own words.

3. ■ **What Does This Say?** Define a definite integral. Explain what this definition is saying using your own words.

4. ■ **What Does This Say?** Define a differential equation. In your own words, describe the procedure for solving a separable differential equation.

Evaluate the limits in Problems 5–13.

5. $\lim\limits_{x \to 2} \dfrac{3x^2 - 5x - 2}{3x^2 - 7x + 2}$

6. $\lim\limits_{x \to +\infty} \dfrac{3x^2 + 7x + 2}{5x^2 - 3x + 3}$

7. $\lim\limits_{x \to +\infty} (\sqrt{x^2 + x} - x)$

8. $\lim\limits_{x \to \pi/2} \dfrac{\cos^2 x}{\cos x^2}$

9. $\lim\limits_{x \to 0} \dfrac{x \sin x}{x + \sin^2 x}$

10. $\lim\limits_{x \to 0} \dfrac{\sin 3x}{x}$

11. $\lim\limits_{x \to +\infty} (1 + x)^{2/x}$

12. $\lim\limits_{x \to 0} \dfrac{\ln(x^2 + 50)}{2x}$

13. $\lim\limits_{x \to 0^+} x^{\sin x}$

Find the derivatives in Problems 14–22.

14. $y = 6x^3 - 4x + 2$

15. $y = (x^2 + 1)^3(3x - 4)^2$

16. $y = \dfrac{x^2 - 4}{3x + 1}$

17. $y = \dfrac{x}{x + \cos x}$

18. $x^2 + 3xy + y^2 = 0$

19. $y = \csc^2 3x$

20. $y = e^{5x-4}$

21. $y = \ln(5x^2 + 3x - 2)$

22. $y = \cos^{-1}(x^2 - 3)$

Find the integrals in Problems 23–28.

23. $\displaystyle\int_4^9 d\theta$

24. $\displaystyle\int_{-1}^1 50(2x - 5)^3 \, dx$

25. $\displaystyle\int_0^1 \dfrac{x \, dx}{\sqrt{9 + x^2}}$

26. $\displaystyle\int \csc 3\theta \cot 3\theta \, d\theta$

27. $\displaystyle\int \dfrac{e^x \, dx}{e^x + 2}$

28. $\displaystyle\int \dfrac{x^3 + 2x - 5}{x} \, dx$

29. Approximate $\displaystyle\int_0^4 \dfrac{dx}{\sqrt{1 + x^3}}$ using Simpson's rule with $n = 6$.

30. Sketch the graph of $y = x^3 - 5x^2 + 2x + 8$.

31. Sketch the graph of $y = \dfrac{4 - x^2}{4 + x^2}$.

32. Find the maximum value(s) of $f(x) = \frac{1}{3}x^3 - 2x^2 + 3x - 10$ on $[0, 6]$.

33. Find the area bounded by the curve $y = x\sqrt{x^2 + 8}$ and the x-axis on $[-2, 1]$.

34. Find the area between the curves $y = \sin x$ and $y = \cos x$ and the lines $x = 0$ and $x = 1$.

35. Find the standard form of the equation of the tangent line to the curve $y = \dfrac{\sin x}{x}$ at the point $x = \dfrac{\pi}{4}$.

Solve the differential equations in Problems 36–37.

36. $\dfrac{dy}{dx} = x^2 y^2 \sqrt{4 - x^3}$

37. $(1 + x^2) \, dy = (x + 1)y \, dx$

Find the particular solution of the differential equations in Problems 38–39.

38. $\dfrac{dy}{dx} = 2(5 - y); y = 3$ when $x = 0$

39. $\dfrac{dy}{dx} = e^y \sin x; y = 5$ when $x = 0$

40. The graph of a function f consists of a semicircle of radius 3 and two line segments as shown in Figure 6.59. Let F be the function defined by

$$F(x) = \int_0^x f(t) \, dt$$

a. Find $F(7)$.

b. Find all values on the interval $(-3, 12)$ at which F has a relative maximum.

c. Write an equation for the line tangent to the graph of F at $x = 7$.

d. Find the x-coordinate of each point of inflection of graph of F on the interval $(-3, 12)$.

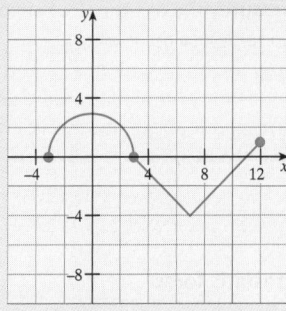

■ **FIGURE 6.59**   Graph of f

41. An electric charge Q_0 is distributed uniformly over a ring of radius R. The electric field intensity at any point x along the axis of the ring is given by

$$E(x) = \dfrac{Q_0 x}{(x^2 + R^2)^{3/2}}$$

At what point on the axis is $E(x)$ maximized?

42. Find the volume of the solid formed by revolving about the x-axis the region bounded by the curve

$$y = \frac{2}{\sqrt{3x - 2}}$$

the x-axis, and the lines $x = 1$ and $x = 2$.

43. A mason lifts a 25-lb bucket of mortar from ground level to the top of a 50-ft building using a rope that weighs 0.25 lb/ft. Find the work done in lifting the bucket.

44. A rocket is launched vertically from a point on the ground 5,000 ft from a TV camera. If the rocket is rising vertically at the rate of 850 ft/s at the instant the rocket is 4,000 ft above the ground, how fast must the TV camera change the angle of elevation to keep the rocket in the picture?

45. A particle moves along the x-axis so that its velocity at any time $t \geq 0$ is given by $v(t) = 6t^2 - 2t - 4$. It is known that the particle is at position $x = 6$ for $t = 2$.
    a. Write a polynomial expression for the position of the particle at any time $t \geq 0$.
    b. For what values of $t, 0 \leq t \leq 3$, is the particle's instantaneous velocity the same as its average velocity on the interval $[0, 3]$.
    c. Find the total distance traveled by the particle from time $t = 0$ to $t = 3$.

46. Let f be the function defined by $f(x) = 2 \cos x$. Let $P(0, 2)$ and $Q(\frac{\pi}{2}, 0)$ be the points where f crosses the y-axis and x-axis, respectively.
    a. Write an equation for the secant line passing through points P and Q.
    b. Write an equation for the tangent line of f at point Q.
    c. Find the x-coordinate of the point on the graph of f, between point P and Q, at which the line tangent to the graph of f is parallel to $\overline{PQ}$. Cite a theorem or result that assures the existence of such a tangent.
    d. Let R be the region in the first quadrant bounded by the graph of f and the line segment $\overline{PQ}$. Write an integral expression for the volume of the solid generated by revolving the region R about the x-axis. Do not evaluate.

47. Let f be the function defined by $f(x) = \sqrt{x - 2}$.
    a. Sketch f and shade the region R enclosed by the graph of f, the x-axis, and the vertical line $x = 5$.
    b. Find the area of R.
    c. Generalize to a line $x = h$ instead of the line $x = 5$ from part **a**. Let $A(h)$ be the area of the region enclosed by the graph of f, the x-axis, and the vertical line $x = h$ (for $h \geq 2$). Write an integral expression for $A(h)$.
    d. Let $A(h)$ be as described in part **c**. Find the rate of change of A with respect to h when $h = 5$.

48. Let f be the function defined by $f(x) = x^3 - 6x^2 + k$ for k an arbitrary constant.
    a. Find the relative maximum and minimum of f in terms of k.
    b. For what values of k does f have three distinct roots?
    c. Find the values of k so that the average value of f over $[-1, 2]$ is 2.

49. Let $v(t)$ represent the velocity (in ft/s) of an object dropped by parachute from an airplane. After the chute opens, the velocity of the object satisfies the differential equation (for $t \geq 0$)

$$\frac{dv}{dt} = -2v - 32$$

with initial condition that the velocity is initially -50 ft/s.
    a. Find an expression for v in terms of t, where t is measured in seconds.
    b. If the terminal velocity is defined as $\lim_{t \to +\infty} v(t)$, find the terminal velocity of the object (to the nearest ft/s).
    c. It is safe to land if the object's speed is 20 ft/s. At what time t does the object reach this speed?

50. A lighthouse is 4,000 ft from a straight shore. Measuring the beam on the shore from the point P on the shore that is closest to the lighthouse, it is observed that the rotating light is moving at the rate of 5 ft/s when it is 1,000 ft from P. How fast is the light revolving at this instant?

7

Methods of Integration

CONTENTS

PREVIEW

In this chapter we increase the number of techniques and procedures for integrating a function. One of the most important techniques, substitution, is reviewed in the first section and is then expanded in several different contexts. Other important integration procedures include using tables, integration by parts, and partial fractions. In addition, improper integrals, hyperbolic functions, and inverse hyperbolic functions are discussed, along with first-order differential equations.

PERSPECTIVE

It is possible to differentiate most functions that arise in practice by applying a fairly short list of rules and formulas, but integration is a more complicated process. The purpose of this chapter is to increase your ability to integrate a variety of different functions. Learning to integrate is like learning to play a musical instrument: At first, it may seem impossibly complicated, but if you persevere, after a while music starts to happen. It should be noted that as more powerful technology becomes available, *techniques* become less important and *ideas* become more important. If you have such technology available, you might wish to consult the *Technology Manual*.

7.1   *Review of Substitution and Integration by Table*

In Chapter 5, we derived a number of integration formulas and examined various algebraic procedures for reducing a given integral to a form that can be handled by these formulas. In this section, we review those formulas as well as the important method of integration by substitution. Today's technology has greatly enhanced the ability of professional mathematicians and scientists in techniques of integration, thereby minimizing the need for spending hours learning rarely used integration techniques. Along with this new technology, the use of integral tables has increased in importance. Most mathematical handbooks contain extensive integral tables, and we have included a complete integration table in the accompanying *Student Mathematical Handbook*. For your convenience, we have also included a short table of integrals in Appendix D.

SMH

For convenient reference, the most important integration rules and formulas are listed inside the back cover. Notice that the first four integration formulas are the procedural rules, which allow us to break up integrals into simpler forms, whereas those on the remainder of the page form the building blocks of integration.

REVIEW OF SUBSTITUTION

Remember when doing substitution that you must choose u, calculate du, and then substitute to make the form you are integrating look *exactly like* one of the known integration formulas. We will review substitution by looking at different situations and special substitutions that prove useful.

EXAMPLE 1   *Integration by substitution*

Find $\displaystyle\int \frac{x^2\,dx}{(x^3-2)^5}$.

Solution   Let u be the value in parentheses; that is, let $u = x^3 - 2$. Then $du = 3x^2\,dx$, so by substitution:

$$\int \frac{x^2\,dx}{(x^3-2)^5} = \int \frac{du/3}{u^5} = \frac{1}{3}\int u^{-5}\,du \qquad \text{All } x\text{'s must be eliminated.}$$

$$= \frac{1}{3}\frac{u^{-4}}{-4} + C = -\frac{1}{12}(x^3-2)^{-4} + C \qquad\qquad ■$$

EXAMPLE 2   *Fitting the form of a known integration formula by substitution*

Find $\displaystyle\int \frac{t\,dt}{\sqrt{1-t^4}}$.

Solution   We notice the similarity between this and the formula for inverse sine if we let $a = 1$ and $u = t^2$. If $u = t^2$, then $du = 2t\,dt$ and

Formula 22 (inverse sine):
$$\int \frac{du}{\sqrt{a^2 - u^2}} = \sin^{-1}\frac{u}{a}$$

$$\int \frac{t\,dt}{\sqrt{1-t^4}} = \int \frac{du/2}{\sqrt{1-u^2}}$$

If $u = t^2$, then $du = 2t\,dt$, $t\,dt = du/2$, and $\sqrt{1-t^4} = \sqrt{1-(t^2)^2} = \sqrt{1-u^2}$.

$$= \frac{1}{2}\int \frac{du}{\sqrt{1-u^2}} \qquad \text{Don't forget the factor } \tfrac{1}{2}.$$

$$= \tfrac{1}{2}\sin^{-1} u + C = \tfrac{1}{2}\sin^{-1} t^2 + C \qquad\qquad ■$$

The art of substitution (Section 5.5) is very important, because many of the techniques developed in this chapter will be used in conjunction with substitution. Examples 3 and 4 illustrate additional ways substitution can be used in integration problems.

EXAMPLE 3   *Substitution to derive an integration formula*

Find $\int \sec x \, dx$.

Solution   Multiply the integrand $\sec x$ by $\sec x + \tan x$ and divide by the same quantity:

$$\int \sec x \, dx = \int \frac{\sec x (\sec x + \tan x)}{\sec x + \tan x} \, dx = \int \frac{\sec x \tan x + \sec^2 x}{\sec x + \tan x} \, dx$$

The advantage of this rearrangement is that the numerator is now the derivative of the denominator. That is, using the substitution

$$u = \sec x + \tan x$$
$$du = (\sec x \tan x + \sec^2 x)dx$$

we find

$$\int \frac{\sec x \tan x + \sec^2 x}{\sec x + \tan x} \, dx = \int \frac{du}{u} = \ln|u| + C = \ln|\sec x + \tan x| + C \quad \blacksquare$$

You may wonder why anyone would think to multiply and divide the integrand $\sec x$ in Example 3 by $\sec x + \tan x$. To say that we do it "because it works" is probably not a very satisfying answer. However, techniques like this are passed on from generation to generation, and it should be noted that multiplication by 1 is a common method in mathematics for changing the form of an expression.

A computer form for the integral in Example 3 is
$$\ln\left|\tan\left(\frac{\pi + 2x}{4}\right)\right|.$$
Note that the constant C is not given in many software programs. The reason for this is that the indefinite integral traditionally represents *all possible* antiderivatives, whereas popular software represents an indefinite integral as *an* antiderivative.

Technology Window

EXAMPLE 4   *Substitution after an algebraic manipulation*

Find $\int \frac{dx}{1 + e^x}$.

Solution   The straightforward substitution $u = 1 + e^x$ does not work:

$$\int \frac{dx}{1 + e^x} = \int \frac{e^{du/x}}{u} = \int \frac{du}{e^x u}$$

This is not an appropriate form because x has not been eliminated.

$$u = 1 + e^x; \quad du = e^x dx$$

Instead rewrite the integrand as follows:

$$\frac{1}{1 + e^x} = \frac{e^{-x}}{e^{-x}}\left(\frac{1}{1 + e^x}\right) = \frac{e^{-x}}{e^{-x} + 1}$$

$$u = e^{-x} + 1; \quad du = -e^{-x}dx$$

$$\int \frac{dx}{1 + e^x} = \int \frac{e^{-x}dx}{e^{-x} + 1} = \int \frac{-du}{u} = -\ln|u| + C = -\ln(e^{-x} + 1) + C$$

Remember, $e^{-x} + 1 > 0$ for all x, so $\ln|e^{-x} + 1| = \ln(e^{-x} + 1)$. $\quad \blacksquare$

When the integrand involves terms with fractional exponents, it is usually a good idea to choose the substitution $x = u^n$, where n is the smallest integer that is divisible by all the denominators of the exponents. For example, if the integrand involves terms such as $x^{1/4}$, $x^{2/3}$, and $x^{1/6}$, then the substitution $x = u^{12}$ is suggested, because 12 is the smallest integer divisible by the exponential denominators 4, 3, and 6. The advantage of this policy is that it guarantees that each fractional power of x becomes an integral power of u. Thus,

$$x^{1/6} = (u^{12})^{1/6} = u^2, \quad x^{1/4} = (u^{12})^{1/4} = u^3, \quad x^{2/3} = (u^{12})^{2/3} = u^8$$

> **EXAMPLE 5**   *Substitution with fractional exponents*

Find $\displaystyle\int \frac{dx}{x^{1/3} + x^{1/2}}$.

Solution   Because 6 is the smallest integer divisible by the denominators 2 and 3, we set $u = x^{1/6}$, so that $u^6 = x$ and $6u^5\, du = dx$. We now use substitution:

$$\int \frac{dx}{x^{1/3} + x^{1/2}} = \int \frac{6u^5\, du}{(u^6)^{1/3} + (u^6)^{1/2}} \qquad \text{Let } x = u^6.$$

$$= \int \frac{6u^5\, du}{u^2 + u^3} = \int \frac{6u^5\, du}{u^2(u + 1)} = \int \frac{6u^3\, du}{1 + u}$$

Substitution does not guarantee an obvious integrable form. When the degree of the numerator is greater than or equal to the degree of the denominator, division is often helpful. By long division,

$$\frac{6u^3}{1 + u} = 6u^2 - 6u + 6 + \frac{-6}{1 + u}$$

$$\int \frac{6u^3\, du}{1 + u} = \int \left(6u^2 - 6u + 6 + \frac{-6}{1 + u}\right) du$$

$$= 2u^3 - 3u^2 + 6u - 6\ln|1 + u| + C$$

$$= 2(x^{1/6})^3 - 3(x^{1/6})^2 + 6x^{1/6} - 6\ln|1 + x^{1/6}| + C \quad \text{Because } u = x^{1/6}$$

$$= 2x^{1/2} - 3x^{1/3} + 6x^{1/6} - 6\ln(1 + x^{1/6}) + C \qquad \text{Note: } 1 + x^{1/6} > 0$$

∎

USING TABLES OF INTEGRALS

Example 5 (especially the part involving long division) seems particularly lengthy. When faced with the necessity of integrating a function such as

$$\int \frac{6u^3\, du}{1 + u}$$

> **WARNING**   Note that in the integration table, it is assumed that the argument of the logarithm is not negative. When you substitute particular values, remember to use absolute value. ←

most would turn to a computer, calculator, or a table of integrals. If you look, for example, at the short integration table (Appendix D), you will find (formula 37)

$$\int \frac{u^3\, du}{au + b} = \frac{(au + b)^3}{3a^4} - \frac{3b(au + b)^2}{2a^4} + \frac{3b^2(au + b)}{a^4} - \frac{b^3}{a^4}\ln(au + b)$$

If we let $a = 1$ and $b = 1$, we find

$$\int \frac{6u^3\, du}{1 + u} = 6\left[\frac{(u + 1)^3}{3} - \frac{3(u + 1)^2}{2} + \frac{3(u + 1)}{1} - \ln|u + 1|\right]$$

$$= 2(u + 1)^3 - 9(u + 1)^2 + 18(u + 1) - 6\ln|u + 1|$$

We note that when using most tables, it is necessary to add the constant C to the form given by the table. We also note that the algebraic form does not always match the form we obtain by direct integration. You might wish to show algebraically that the form we have just found and the form

$$2u^3 - 3u^2 + 6u - 6\ln|1 + u| + C$$

from Example 5 are the same.

Technology Window

If we use some computer software (such as *Derive*, *Maple*, or *Mathematica*) for
$\int \dfrac{6u^3\,du}{1 + u}$, we obtain

$$-6\,\mathrm{LN}(u + 1) + 2u^3 - 3u^2 + 6u$$

On the other hand, if we are using computer software, we would not need the simplified form for Example 5, and we find

$$\int \frac{dx}{x^{1/3} + x^{1/2}} = -6\,\mathrm{LN}(x^{1/6} + 1) + 2\sqrt{x} - 3x^{1/3} + 6x^{1/6}$$

If we compare this answer with the one we found in Example 5, we note that most software programs do not necessarily give the terms in the usual order, they do not show the absolute value symbols, and they also do not provide the constant C. However, you can show they are correct by checking the derivatives.

SMH

To use an integral table, first classify the integral by form. The forms listed in the table of integrals included in the *Student Mathematical Handbook* are as follows:

Elementary forms   (formulas 1–29; these were developed in the text)

Linear and quadratic forms   (formulas 30–134)
Forms involving $au + b$; $u^2 + a^2$; $u^2 - a^2$; $a^2 - u^2$; $au + b$ and $pu + q$; $au^2 + bu + c$

Radical forms   (formulas 135–270)
Forms involving $\sqrt{au + b}$; $\sqrt{au + b}$ and $pu + q$; $\sqrt{au + b}$ and $\sqrt{pu + q}$; $\sqrt{u^2 + a^2}$; $\sqrt{u^2 - a^2}$; $\sqrt{a^2 - u^2}$; $\sqrt{au^2 + bu + c}$

Higher-degree binomials   (formulas 271–310)
Forms involving $u^3 + a^3$; $u^4 \pm a^4$; $u^n \pm a^n$

Trigonometric forms   (formulas 311–444)
Forms involving $\cos au$; $\sin au$; $\sin au$ and $\cos au$; $\tan au$; $\cot au$; $\sec au$; $\csc au$

Inverse trigonometric forms   (formulas 445–482)

Exponential and logarithmic forms   (formulas 483–513)
Forms involving e^{au}; $\ln|u|$

Hyperbolic forms   (formulas 514–619; we study these in Section 7.8)
Forms involving $\cosh au$; $\sinh au$; $\sinh au$ and $\cosh au$; $\tanh au$; $\coth au$; $\operatorname{sech} au$; $\operatorname{csch} au$

Inverse hyperbolic forms   (formulas 620–650; Section 7.8)

A condensed version of this table is provided in Appendix D.

There is a common misconception that integration will be easy if a table is provided, but even with a table available there is a considerable amount of work. After deciding which form applies, match the individual type with the problem at hand by making appropriate choices for the arbitrary constants. More than one form may apply, but the results derived by using different formulas will be the same (except for the constants) even though they may look quite different. We shall not include the constants in the table listing, but you should remember to include it with your answer when using the table for integration.

(SMH)

Take a few moments to look at the integration table in the *Student Mathematics Handbook* (or Appendix D). Notice that the table has two basic forms of integration formulas. The first gives a formula that is the antiderivative, whereas the second, called a *reduction formula*, simply rewrites the integral in another form. The illustration below demonstrates each of these forms.

Table of Integrals

Integrals involving $au + b$

30. $\displaystyle\int (au + b)^n du = \frac{(au + b)^{n+1}}{(n + 1)a}$

31. $\displaystyle\int u(au + b)^n du = \frac{(au + b)^{n+2}}{(n + 2)a^2} - \frac{b(au + b)^{n+1}}{(n + 1)a^2}$

32. $\displaystyle\int u^2(au + b)^n du = \frac{(au + b)^{n+3}}{(n + 3)a^3} - \frac{2b(au + b)^{n+2}}{(n + 2)a^3} + \frac{b^2(au + b)^{n+1}}{(n + 1)a^3}$

33. $\displaystyle\int u^m(au + b)^n du = \begin{cases} \dfrac{u^{m+1}(au + b)^n}{m + n + 1} + \dfrac{nb}{m + n + 1}\displaystyle\int u^m(au + b)^{n-1}du \\[2mm] \dfrac{u^m(au + b)^{n+1}}{(m + n + 1)a} - \dfrac{mb}{(m + n + 1)a}\displaystyle\int u^{m-1}(au + b)^n du \\[2mm] \dfrac{-u^{m+1}(au + b)^{n+1}}{(n + 1)b} + \dfrac{m + n + 2}{(n + 1)b}\displaystyle\int u^m(au + b)^{n+1}du \end{cases}$

34. $\displaystyle\int \frac{du}{au + b} = \frac{1}{a}\ln|au + b|$

EXAMPLE 6   *Integration using a table of integrals*

Find $\displaystyle\int x^2(3 - x)^5\, dx$

Solution   This is an integral involving an expression of the form $au + b$; we find that this is formula 32, where $u = x, a = -1, b = 3$, and $n = 5$. Remember that the integral table does not show the added constant, so do not forget it when writing your answers using the table.

$$\int x^2(3 - x)^5\, dx = \frac{(3 - x)^{5+3}}{(5 + 3)(-1)^3} - \frac{2(3)(3 - x)^{5+2}}{(5 + 2)(-1)^3} + \frac{3^2(3 - x)^{5+1}}{(5 + 1)(-1)^3} + C$$

$$= -\tfrac{1}{8}(3 - x)^8 + \tfrac{6}{7}(3 - x)^7 - \tfrac{3}{2}(3 - x)^6 + C$$

This form is acceptable as an answer. However, you may wish to complete the algebraic simplification:

$$\int x^2(3 - x)^5\, dx = -\tfrac{1}{56}(3 - x)^6[7(3 - x)^2 - 48(3 - x) + 84] + C$$

$$= -\tfrac{1}{56}(3 - x)^6(7x^2 + 6x + 3) + C$$

One of the difficult considerations when using tables or computer software to carry out integration is recognizing the variety of different forms for acceptable answers. Note that is not easy to show that these forms are algebraically equivalent. If we use computer software for this evaluation, we find yet another algebraically equivalent form:

$$\int x^2(3-x)^5 dx = -\frac{x^8}{8} + \frac{15x^7}{7} - 15x^6 + 54x^5 - \frac{405x^4}{4} + 81x^3$$

Once again, remember that you must add the $+ C$ to the computer software form. ∎

| EXAMPLE 7 | *Integration using a table of integrals—reduction formula found in table* |

Find $\int (\ln x)^4\, dx$.

Solution   The integrand is in logarithmic form; from the table of integrals we see that formula 501 applies, where $u = x$ and $n = 4$. Note that tables of integrals do not usually show the absolute value symbols. You need to remember that $\ln x$ is defined only for positive x.

Take a close look at formulas 499–502, and note that formula 501 gives another integral as part of the result. This is called a **reduction formula** because it enables us to compute the given integral in terms of an integral of a similar type, only with a lower power in the integral.

Integrals involving $\ln |u|$
499. $\displaystyle\int \ln|u|\, du = u \ln|u| - u$
500. $\displaystyle\int (\ln|u|)^2\, du = u(\ln|u|)^2 - 2u \ln|u| + 2u$
501. $\displaystyle\int (\ln|u|)^n\, du = u(\ln|u|)^n - n \int (\ln|u|)^{n-1} du$
502. $\displaystyle\int u \ln|u|\, du = \frac{u^2}{2}\left(\ln|u| - \frac{1}{2}\right)$

$$\int (\ln x)^4\, dx = x(\ln x)^4 - 4 \int (\ln x)^{4-1} dx$$

$$= x(\ln x)^4 - 4\left[x(\ln x)^3 - 3\int (\ln x)^{3-1}\, dx\right] \qquad \text{Formula 501 again}$$

$$= x(\ln x)^4 - 4x(\ln x)^3 + 12 \int (\ln x)^2\, dx$$

$$= x(\ln x)^4 - 4x(\ln x)^3 + 12[x(\ln x)^2 - 2x \ln x + 2x] + C$$
$$\text{This is formula 500.}$$
$$= x(\ln x)^4 - 4x(\ln x)^3 + 12x(\ln x)^2 - 24x \ln x + 24x + C \qquad ∎$$

Reduction formulas in a table of integrals, such as the one illustrated in Example 7, are usually obtained using substitution and integration by parts.

Note from the previous example that we follow the convention of adding the constant C only after eliminating the last integral sign (rather than being technically correct and writing $C_1, C_2, \ldots$ for *each* integral). The reason we can do this is that $C_1 + C_2 + \cdots = C$, for arbitrary constants.

It is often necessary to make substitutions before using one of the integration formulas, as shown in the following example.

| **EXAMPLE 8** | *Using an integral table after substitution* |

Find $\displaystyle\int \frac{x\,dx}{\sqrt{8-5x^2}}$.

Solution   This is an integral of the form $\sqrt{a^2-u^2}$, but it does not exactly match any of the formulas.

SMH

Integrals involving $\sqrt{a^2-u^2}$

224. $\displaystyle\int \frac{du}{\sqrt{a^2-u^2}} = \sin^{-1}\frac{u}{a}$

225. $\displaystyle\int \frac{u\,du}{\sqrt{a^2-u^2}} = -\sqrt{a^2-u^2}$

226. $\displaystyle\int \frac{u^2\,du}{\sqrt{a^2-u^2}} = -\frac{u\sqrt{a^2-u^2}}{2} + \frac{a^2}{2}\sin^{-1}\frac{u}{a}$

Note, however, that except for the coefficient of 5, it is like formula 225. Let $u = \sqrt{5}\,x$ (so $u^2 = 5x^2$); then $du = \sqrt{5}\,dx$:

$$\int \frac{x\,dx}{\sqrt{8-5x^2}} = \int \frac{(u/\sqrt{5})\cdot(du/\sqrt{5})}{\sqrt{8-u^2}}$$

$$\boxed{\text{Let } u = \sqrt{5}x \text{ so that } x = \frac{u}{\sqrt{5}};\quad du = \sqrt{5}\,dx \text{ so that } dx = \frac{du}{\sqrt{5}}}$$

$$= \frac{1}{5}\int \frac{u\,du}{\sqrt{8-u^2}}$$

Now apply formula 225 where $a^2 = 8$:

$$\frac{1}{5}\int \frac{u\,du}{\sqrt{8-u^2}} = \frac{1}{5}\left(-\sqrt{8-u^2}\right) + C = -\frac{1}{5}\sqrt{8-5x^2} + C \qquad\blacksquare$$

As you can see from Example 8, using an integral table is not a trivial task. In fact, other methods of integration may be preferable. For Example 8, you can let $u = 8 - 5x^2$ and integrate by substitution:

$$\boxed{\text{Let } u = 8 - 5x^2;\quad du = -10x\,dx.}$$

$$\int \frac{x\,dx}{\sqrt{8-5x^2}} = \int \frac{x(du/-10x)}{\sqrt{u}} = -\frac{1}{10}\int u^{-1/2}du$$

$$= -\frac{1}{10}(2u^{1/2}) + C = -\frac{1}{5}\sqrt{8-5x^2} + C$$

Of course, this answer is the same as the one we obtained in Example 8. The point of this calculation is to emphasize that you should try simple methods of integration before turning to the table of integrals.

Sometimes, by choosing different formulas you may obtain two different-looking (but equivalent) forms.

| **EXAMPLE 9** | *Integration by table (multiple forms with substitution)* |

Find $\int 5x^2\sqrt{3x^2 + 1}\, dx$.

Solution This is similar to formula 170, but you must take care of the 5 (constant multiple) and the 3 (by making a substitution).

Integrals involving $\sqrt{u^2 + a^2}$

168. $\int \sqrt{u^2 + a^2}\, du = \dfrac{u\sqrt{u^2 + a^2}}{2} + \dfrac{a^2}{2}\ln\left|u + \sqrt{u^2 + a^2}\right|$

169. $\int u\sqrt{u^2 + a^2}\, du = \dfrac{(u^2 + a^2)^{3/2}}{3}$

170. $\int u^2\sqrt{u^2 + a^2}\, du = \dfrac{u(u^2 + a^2)^{3/2}}{4} - \dfrac{a^2 u\sqrt{u^2 + a^2}}{8} - \dfrac{a^4}{8}\ln\left|u + \sqrt{u^2 + a^2}\right|$

Let $u = \sqrt{3}x$; then $du = \sqrt{3}\, dx$.

$\displaystyle\int 5x^2\sqrt{3x^2 + 1}\, dx$

$\uparrow$

$u^2 = 3x^2$ so that $x^2 = \dfrac{u^2}{3}$

$= 5\displaystyle\int \left(\dfrac{u^2}{3}\right)\sqrt{u^2 + 1}\,\dfrac{du}{\sqrt{3}}$

$= \dfrac{5}{3\sqrt{3}}\displaystyle\int u^2\sqrt{u^2 + 1}\, du$ Use formula 170 where $a = 1$

$= \dfrac{5}{3\sqrt{3}}\left[\dfrac{u(u^2 + 1)^{3/2}}{4} - \dfrac{u\sqrt{u^2 + 1}}{8} - \dfrac{1}{8}\ln\left|u + \sqrt{u^2 + 1}\right|\right] + C$

$= \dfrac{5}{24\sqrt{3}}\left[2\sqrt{3}x(3x^2 + 1)^{3/2} - \sqrt{3}x\sqrt{3x^2 + 1} - \ln\left|\sqrt{3}x + \sqrt{3x^2 + 1}\right|\right] + C$

$= \dfrac{5}{24}\left[2x(3x^2 + 1)^{3/2} - x\sqrt{3x^2 + 1} - \dfrac{1}{\sqrt{3}}\ln\left(\sqrt{3}x + \sqrt{3x^2 + 1}\right)\right] + C$

You might want to show that $u + \sqrt{u^2 + 1} > 0$. ∎

Technology Window ▲ ▼

The form that we obtain when using computer software can be shown:

$-\dfrac{5\sqrt{3}\,\text{LN}\,(\sqrt{3x^2 + 1} + \sqrt{3}x)}{72}$

$+\dfrac{5x\sqrt{3x^2 + 1}(6x^2 + 1)}{24}$

7.1 Problem Set

A *Find each integral in Problems 1–12.*

1. $\displaystyle\int \dfrac{2x + 5}{\sqrt{x^2 + 5x}}\, dx$

2. $\displaystyle\int \dfrac{\ln x}{x}\, dx$

3. $\displaystyle\int \dfrac{\ln(x + 1)}{x + 1}\, dx$

4. $\displaystyle\int \cos x\, e^{\sin x}\, dx$

5. $\displaystyle\int \dfrac{x\, dx}{4 + x^4}$

6. $\displaystyle\int \dfrac{t^2\, dt}{9 + t^6}$

7. $\displaystyle\int (1 + \cot x)^4 \csc^2 x\, dx$

8. $\displaystyle\int \dfrac{4x^3 - 4x}{x^4 - 2x^2 + 3}\, dx$

9. $\displaystyle\int \dfrac{x^3 - x}{(x^4 - 2x^2 + 3)^2}\, dx$

10. $\displaystyle\int \dfrac{2x + 4}{x^2 + 4x + 3}\, dx$

11. $\displaystyle\int \dfrac{2x + 1}{x^2 + x + 1}\, dx$

12. $\displaystyle\int \dfrac{2x - 1}{(4x^2 - 4x)^2}\, dx$

Integrate the expressions in Problems 13–24 using the short table of integrals given in Appendix D.

13. $\displaystyle \int \frac{dx}{x^2\sqrt{x^2 - a^2}}$

14. $\displaystyle \int \frac{dx}{x^2\sqrt{a^2 - x^2}}$

15. $\displaystyle \int x \ln x \, dx$

16. $\displaystyle \int \ln x \, dx$

17. $\displaystyle \int xe^{ax} \, dx$

18. $\displaystyle \int \frac{dx}{a + be^{2x}}$

19. $\displaystyle \int \frac{x^2 \, dx}{\sqrt{x^2 + 1}}$

20. $\displaystyle \int \frac{dx}{x^2\sqrt{x^2 + 16}}$

21. $\displaystyle \int \frac{x \, dx}{\sqrt{4x^2 + 1}}$

22. $\displaystyle \int \frac{dx}{x\sqrt{1 - 9x^2}}$

23. $\displaystyle \int e^{-4x} \sin 5x \, dx$

24. $\displaystyle \int x \sin^{-1}x \, dx$

Evaluate the integrals in Problems 25–33. If you use an integral table, state the number of the formula used, and if you use substitution, show each step. If you use an alternative table of integrals, then cite the source as well as the formula number.

25. $\displaystyle \int (1 + bx)^{-1} \, dx$

26. $\displaystyle \int \frac{x \, dx}{\sqrt{a^2 - x^2}}$

27. $\displaystyle \int x(1 + x)^3 \, dx$

28. $\displaystyle \int x\sqrt{1 + x} \, dx$

29. $\displaystyle \int xe^{4x} \, dx$

30. $\displaystyle \int x \ln 2x \, dx$

31. $\displaystyle \int \frac{dx}{1 + e^{2x}}$

32. $\displaystyle \int \ln^3 x \, dx$

33. $\displaystyle \int \frac{x^3 \, dx}{\sqrt{4x^4 + 1}}$

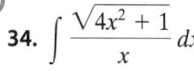 *Use the Student Mathematics Handbook or other available integration table to integrate the integrals given in Problems 34–39.*

34. $\displaystyle \int \frac{\sqrt{4x^2 + 1}}{x} \, dx$

35. $\displaystyle \int \sec^3\left(\frac{x}{2}\right) \, dx$

36. $\displaystyle \int \sin^6 x \, dx$

37. $\displaystyle \int \frac{dx}{9x^2 + 6x + 1}$

38. $\displaystyle \int (9 - x^2)^{3/2} \, dx$

39. $\displaystyle \int \frac{\sin^2 x}{\cos x} \, dx$

40. Derive the **sine squared formula** shown on the inside back cover (formula 348):
$$\int \sin^2 x \, dx = \frac{1}{2}x - \frac{1}{4}\sin 2x$$
Hint: Use the identity $\sin^2 x = \dfrac{1 - \cos 2x}{2}$.

41. Derive the **cosine squared formula** shown on the inside back cover (formula 317):
$$\int \cos^2 x \, dx = \frac{1}{2}x + \frac{1}{4}\sin 2x$$
Hint: Use the identity $\cos^2 x = \dfrac{1 + \cos 2x}{2}$.

In Problems 42–44, use substitution to integrate certain powers of sine and cosine.

42. $\displaystyle \int \sin^4 x \cos x \, dx$ *Hint:* Let $u = \sin x$.

43. $\displaystyle \int \sin^3 x \cos^4 x \, dx$ *Hint:* Let $u = \cos x$.

44. $\displaystyle \int \sin^2 x \cos^2 x \, dx$ *Hint:* Use the identities shown in Problems 40 and 41.

45. ■ **What Does This Say?** Using Problems 42–44, formulate a procedure for integrals of the form
$$\int \sin^m x \cos^n x \, dx$$

Find each integral in Problems 46–51.

46. $\displaystyle \int \frac{e^x \, dx}{1 + e^{x/2}}$

47. $\displaystyle \int \frac{dx}{x^{1/2} + x^{1/4}}$

48. $\displaystyle \int \frac{18 \tan^2 t \sec^2 t}{(2 + \tan^3 t)^2} \, dt$

49. $\displaystyle \int \frac{4 \, dx}{x^{1/3} + 2x^{1/2}}$

50. $\displaystyle \int \frac{dx}{(x + \frac{1}{2})\sqrt{4x^2 + 4x}}$

51. $\displaystyle \int \frac{e^{-x} - e^x}{e^{2x} + e^{-2x} + 2} \, dx$

52. Find the area of the region bounded by the graphs of $y = \dfrac{2x}{\sqrt{x^2 + 9}}$ and $y = 0$ from $x = 0$ to $x = 4$.

53. Find the area of the region bounded by the graphs of $y = \cos 2x, y = 0, x = 0$, and $x = \frac{\pi}{4}$.

54. Find the volume of the solid generated when the curve $y = x(1 - x^2)^{1/4}$ from $x = 0$ to $x = 1$ is revolved about the x-axis.

55. Find the volume of the solid generated when the region under the curve
$$y = \frac{x^{3/2}}{\sqrt{x^2 + 9}}$$
between $x = 0$ and $x = 9$ is revolved about the x-axis.

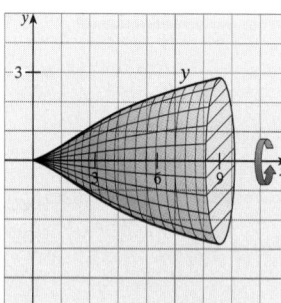

56. Find the volume of the solid generated when the curve
$$x = \sqrt[4]{4 - y^2}$$
between $y = 1$ and $y = 2$ is revolved about the y-axis.

57. Find the volume of the solid generated when the curve
$$y = \frac{1}{\sqrt{x}}(1 + \sqrt{x})^{1/3}$$
between $x = 1$ and $x = 4$ is revolved about the y-axis.

58. Let $y = f(x)$ be a function that satisfies the differential equation

$$xy' = \sqrt{(\ln x)^2 - x^2}$$

Find the arc length of $y = f(x)$ between $x = \frac{1}{4}$ and $x = \frac{1}{2}$.

59. Find the arc length of the curve $y = x^{3/2}$ on the interval $[0, 1]$.

60. Find the arc length of the curve $y = \ln(\cos x)$ on the interval $[0, \pi/4]$.

61. Find the surface area of the surface generated when the curve $y = x^2$ on the interval $[0, 1]$ is revolved about the x-axis.

62. Find the surface area of the surface generated when the curve $y = x^2$ on the interval $[0, 1]$ is revolved about the y-axis.

C **63.** Show that

$$\int \csc x \, dx = -\ln|\csc x + \cot x| + C$$

Hint: Multiply the integrand by $\dfrac{\csc x + \cot x}{\csc x + \cot x}$.

64. Find $\displaystyle\int 2 \sin x \cos x \, dx$ by using the indicated substitution.
  a. Let $u = \cos x$.
  b. Let $u = \sin x$.

  c. Write $2 \sin x \cos x = \sin 2x$ and carry out the integration.
  d. Show that the answers you obtained for parts **a–c** are the same.

65. Derive the formula

$$\int_0^\pi xf(\sin x) \, dx = \frac{\pi}{2} \int_0^\pi f(\sin x) \, dx$$

66. Find the surface area of the torus generated when the curve

$$x^2 + (y - b)^2 = 1, \quad b > 1$$

is revolved about the x-axis.

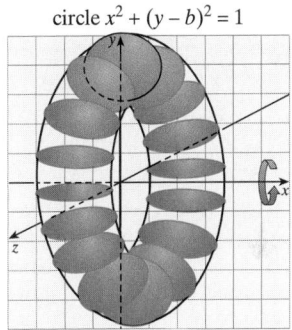

circle $x^2 + (y - b)^2 = 1$

7.2 Integration by Parts

IN THIS SECTION   integration by parts formula, repeated use of integration by parts, definite integration by parts ■

Integration by parts is a procedure based on reversing the product rule for differentiation. We present this section not only as a technique of integration but as a procedure that is necessary in a variety of useful applications.

INTEGRATION BY PARTS FORMULA

Recall the formula for differentiation of a product. If u and v are differentiable functions, then

$$d(uv) = u \, dv + v \, du$$

Integrate both sides of this equation to find the formula for integration by parts:

$$\int d(uv) = \int u \, dv + \int v \, du$$

$$uv = \int u \, dv + \int v \, du$$

If we rewrite this last equation, we obtain the formula summarized in the following box.

Formula for Integration by Parts

$$\int u \, dv = uv - \int v \, du$$

EXAMPLE 1   *Integration by parts*

Find $\int xe^x \, dx$.

Solution   To use integration by parts, we must choose u and dv so that the new integral is easier to integrate than the original.

$$\int \underset{u}{x} \; \underset{dv}{e^x \, dx}$$

Let $u = x$ and $dv = e^x \, dx$, then
$du = dx$ $v = \int e^x \, dx = e^x$

$$\int x e^x \, dx = \underset{u}{x} \; \underset{v}{e^x} - \int \underset{v}{e^x} \; \underset{du}{dx} = xe^x - e^x + C$$

 You can check our work with integration by parts by differentiating, using software, or an integration table. This is found in Appendix D or the *Student Mathematics Handbook* (formula 484, with $a = 1$).   ■

Comment   You many wonder why an arbitrary constant (call it K) was not included when performing the integration associated with $\int dv$ in integration by parts. The reason is that when applying integration by parts, we need just *one* function v whose derivative is dv, so we take the simplest one—the one with $K = 0$. You may find it instructive to see that taking $v = e^x + K$ gives the same result.

Integration by parts is often difficult the first time you try to do it because there is no absolute choice for u and dv. In Example 1, you might have chosen

$$u = e^x \qquad \text{and} \quad dv = x \, dx$$

$$du = e^x \, dx \qquad\qquad v = \int x \, dx = \frac{x^2}{2}$$

Then $\quad \displaystyle\int xe^x \, dx = \underset{u}{e^x} \; \underset{v}{\frac{x^2}{2}} - \int \underset{v}{\frac{x^2}{2}} \; \underset{du}{e^x \, dx}$

$$= \frac{1}{2} x^2 \, e^x - \frac{1}{2} \int x^2 \, e^x \, dx$$

Note, however, that this choice of u and dv leads to a more complicated form than the original. In general, when you are integrating by parts, if you make a choice for u and dv that leads to a more complicated form than when you started, consider going back and making another choice for u and dv.

Generally, you want to choose dv to be as difficult as possible (and still be something you can integrate), with the remainder being left for the u-factor.

EXAMPLE 2   *When the differentiable part is the entire integrand*

Find $\int \ln x \, dx$ for $x > 0$.

Solution

$$\boxed{\begin{array}{ll} \text{Let } u = \ln x & \text{and} \quad dv = dx \\[2mm] \quad du = \dfrac{1}{x}\, dx & \qquad v = x \end{array}}$$

$$\int \ln x \, dx = (\ln x)x - \int x\left(\frac{1}{x}\, dx\right) = x \ln x - \int dx = x \ln x - x + C$$

Check with formula 499 (Appendix D) where $a = 1$, which we worked as Problem 16 of Problem Set 7.1.   ■

REPEATED USE OF INTEGRATION BY PARTS

Sometimes integration by parts must be applied several times to evaluate a given integral.

EXAMPLE 3   *Repeated integration by parts*

Find $\displaystyle\int x^2 e^{-x} dx$.

Solution

$$\boxed{\begin{array}{ll} \text{Let } u = x^2 & \text{and} \quad dv = e^{-x} dx \\ \quad du = 2x\, dx & \qquad v = -e^{-x} \end{array}}$$

$$\int x^2 e^{-x} \, dx = x^2(-e^{-x}) - \int (-e^{-x})(2x\, dx)$$

$$= -x^2 e^{-x} + 2\int xe^{-x} dx$$

$$\boxed{\begin{array}{ll} \text{Let } u = x & \text{and} \quad dv = e^{-x} dx \\ \quad du = dx & \qquad v = -e^{-x} \end{array}}$$

$$= -x^2 e^{-x} + 2\left[x(-e^{-x}) - \int (-e^{-x})dx \right]$$

$$= -x^2 e^{-x} - 2xe^{-x} - 2e^{-x} + C$$

$$= -e^{-x}(x^2 + 2x + 2) + C$$

Check with formula 485 where $a = -1$.   ■

In the following example, it is necessary to apply integration by parts more than once, but as you will see, when we do so a second time we return to the original integral. Note carefully how this situation can be handled algebraically.

EXAMPLE 4   *Substitution with an algebraic manipulation*

Find $\displaystyle\int e^{2x} \sin x \, dx$.

Solution   For this problem you will see that it will be useful to call the original integral I. That is, we let

$$I = \int e^{2x} \sin x \, dx$$

$$\boxed{\begin{array}{ll} \text{Let } u = e^{2x} & \text{and} \quad dv = \sin x \, dx \\ \quad du = 2e^{2x} \, dx & \qquad v = -\cos x \end{array}}$$

$$I = \int e^{2x} \sin x \, dx = e^{2x}(-\cos x) - \int (-\cos x)(2e^{2x} \, dx)$$

$$= -e^{2x} \cos x + 2 \int e^{2x} \cos x \, dx$$

Let $u = e^{2x}$	and	$dv = \cos x \, dx$
$du = 2e^{2x} \, dx$		$v = \sin x$

$$= -e^{2x} \cos x + 2\left[e^{2x}(\sin x) - \int \sin x (2e^{2x} \, dx) \right]$$

$$= -e^{2x} \cos x + 2e^{2x} \sin x - 4 \int e^{2x} \sin x \, dx$$

Notice that this last integral is I. Thus, we have

$$I = -e^{2x} \cos x + 2e^{2x} \sin x - 4I \qquad \textit{Solve this equation for } I.$$

$$5I = -e^{2x} \cos x + 2e^{2x} \sin x$$

$$I = \tfrac{1}{5} e^{2x}(2 \sin x - \cos x)$$

Thus, $\displaystyle \int e^{2x} \sin x \, dx = \tfrac{1}{5} e^{2x}(2 \sin x - \cos x) + C$

Check with the integration table (formula 492, where $a = 2$ and $b = 1$). ■

DEFINITE INTEGRATION BY PARTS

To get a clear picture of the integration by parts formula for definite integrals, it is instructive to interpret the integration by parts formula in terms of area. Consider the area of the rectangle with sides u_2 and v_2, as shown in Figure 7.1.

AREA OF RECTANGLE $=$ LIGHT BLUE AREA $+$ DARK BLUE AREA $+$ GRAY AREA

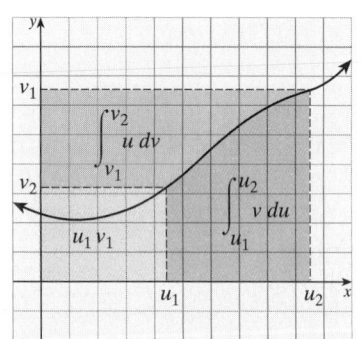

Entire colored rectangle

$$u_2 v_2 \qquad = \qquad u_1 v_1 \qquad + \qquad \int_{u_1}^{u_2} v \, du \qquad + \qquad \int_{v_1}^{v_2} u \, dv$$

This equation can be rewritten:

$$\int_{v_1}^{v_2} u \, dv = u_2 v_2 - u_1 v_1 - \int_{u_1}^{u_2} v \, du$$

■ FIGURE 7.1   Integration by parts using areas

This relationship is summarized in the following box, where we use the usual limits of integration, namely $x = a$ to $x = b$.

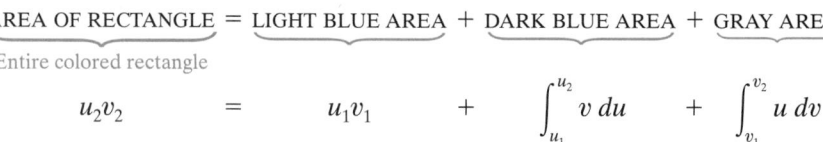

Integration by Parts for Definite Integrals

$$\int_a^b u \, dv = uv \, \Big|_a^b - \int_a^b v \, du$$

You should recognize that this formula for definite integrals is the same as the formula for indefinite integrals, where the first term after the equal sign has been evaluated at the appropriate limits of integration. This is illustrated by the following example.

EXAMPLE 5   *Integration by parts with a definite integral*

Evaluate $\displaystyle\int_0^1 xe^{2x}\,dx$.

Solution

$$\text{Let } u = x \quad \text{and} \quad dv = e^{2x}\,dx$$
$$\qquad\quad du = dx \qquad\qquad v = \tfrac{1}{2}e^{2x}$$

$$\int_0^1 xe^{2x}dx = \frac{1}{2}xe^{2x}\Big|_0^1 - \frac{1}{2}\int_0^1 e^{2x}\,dx$$

It is sometimes easier to simplify the algebra and then do one evaluation here at the end of the integration part of the problem.
↓

$$= \left(\frac{1}{2}xe^{2x} - \frac{1}{4}e^{2x}\right)\Big|_0^1 = \frac{1}{4}e^2 + \frac{1}{4}$$

Check in Appendix D (formula 484, with $a = 2$). ∎

EXAMPLE 6   *Integration by parts followed by substitution*

Evaluate $\displaystyle\int_0^{\pi/4} \tan^{-1}x\,dx$.

Solution

$$\text{Let } u = \tan^{-1}x \quad \text{and} \quad dv = dx$$
$$\qquad\quad du = \frac{dx}{1 + x^2} \qquad\qquad v = x$$

$$\int_0^{\pi/4} \underbrace{\tan^{-1}x}_{u}\,\underbrace{dx}_{dv} = \underbrace{(\tan^{-1}x)}_{u}\,\underbrace{x}_{v}\Big|_0^{\pi/4} - \int_0^{\pi/4} \underbrace{\frac{x\,dx}{1+x^2}}_{v\,du}$$

$$= \left[x\tan^{-1}x - \frac{1}{2}\ln(1 + x^2)\right]_0^{\pi/4}$$

Use substitution where $w = 1 + x^2$ and $dw = 2x\,dx$.
$$\int \frac{x\,dx}{1+x^2} = \int \frac{dw}{2w} = \tfrac{1}{2}\ln w = \tfrac{1}{2}\ln(1 + x^2)$$

$$= \left[\frac{\pi}{4}\left(\tan^{-1}\frac{\pi}{4}\right) - \frac{1}{2}\ln\left(1 + \frac{\pi^2}{16}\right)\right] - \left[0 - \frac{1}{2}\ln 1\right]$$

$$= \frac{\pi}{4}\tan^{-1}\frac{\pi}{4} - \frac{1}{2}\ln\left(1 + \frac{\pi^2}{16}\right)$$

Check with an integration table (formula 457, for example, with $a = 1$). ∎

7.2 Problem Set

A *Find each integral in Problems 1–12.*

1. $\displaystyle\int xe^{-2x}\,dx$

2. $\displaystyle\int x\sin x\,dx$

3. $\displaystyle\int x\ln x\,dx$

4. $\displaystyle\int x\tan^{-1}x\,dx$

5. $\displaystyle\int \sin^{-1}x\,dx$

6. $\displaystyle\int x^2\sin x\,dx$

7. $\displaystyle\int \frac{\ln\sqrt{x}}{\sqrt{x}}\,dx$

8. $\displaystyle\int e^{2x}\sin 3x\,dx$

9. $\displaystyle\int x^2\ln x\,dx$

10. $\displaystyle\int (x+\sin x)^2\,dx$

11. $\displaystyle\int e^{2x}\sqrt{1-e^x}\,dx$

12. $\displaystyle\int x\sin x\cos x\,dx$

Find the exact value of the definite integrals in Problems 13–18 using integration by parts, and then check by using a calculator to find an approximate answer correct to four decimal places.

13. $\displaystyle\int_1^4 \sqrt{x}\,\ln x\,dx$

14. $\displaystyle\int_1^e x^3\ln x\,dx$

15. $\displaystyle\int_1^e (\ln x)^2\,dx$

16. $\displaystyle\int_{1/3}^e 3(\ln 3x)^2\,dx$

17. $\displaystyle\int_0^\pi e^{2x}\cos 2x\,dx$

18. $\displaystyle\int_0^\pi x(\sin x+\cos x)\,dx$

B 19. ■ **What Does This Say?** Describe the process known as *integration by parts*.

In Problems 20–23, first use an appropriate substitution and then integrate by parts to evaluate the integral. Remember to give your answers in terms of x.

20. $\displaystyle\int \frac{\ln x\sin(\ln x)}{x}\,dx$

21. $\displaystyle\int [\sin 2x\ln(\cos x)]\,dx$

22. $\displaystyle\int e^{2x}\sin e^x\,dx$

23. $\displaystyle\int [\sin x\ln(2+\cos x)]\,dx$

24. a. Evaluate $\displaystyle\int \frac{x^3}{x^2-1}\,dx$ using integration by parts.

   b. Evaluate the integral in part **a** by first dividing the integrand.

25. Evaluate $\displaystyle\int \cos^2 x\,dx$.

26. Evaluate $\displaystyle\int \frac{x\,dx}{\sqrt{x^2+1}}$.

27. Use Problem 25 to evaluate $\displaystyle\int x\cos^2 x\,dx$ using integration by parts.

28. Use Problem 26 to evaluate $\displaystyle\int \frac{x^3\,dx}{\sqrt{x^2+1}}$ using integration by parts.

29. Find $\displaystyle\int x^n\ln x\,dx$, where n is any positive real number.

30. After t seconds, an object is moving at a speed of $te^{-t/2}$ meters per second. Express the distance the object travels as a function of time.

31. After t hours on the job, a factory worker can produce $100te^{-0.5t}$ units per hour. How many units does the worker produce during the first 3 hours?

32. After t weeks, contributions in response to a local fund-raising compaign were coming in at the rate of $2{,}000te^{-0.2t}$ dollars per week. How much money was raised during the first 5 weeks?

33. Find the volume of the solid generated when the region under the curve $y=e^{-x}$ on the interval $[0,2]$ is revolved about the y-axis.

34. Find the volume of the solid generated when the region under the curve $y=\sin x+\cos x$ on the interval $[0,\frac{\pi}{4}]$ is revolved about the y-axis.

35. Find the volume of the solid generated when the region under the curve $y=\ln x$ on the interval $[1,e]$ is revolved about the indicated axis:
   a. x-axis   b. y-axis

36. Find the centroid (with coordinates rounded to the nearest hundredth) of the region bounded by the curves $y=\sin x$ and $y=\cos x$ and the y-axis.

37. Find the centroid (with coordinates rounded to the nearest hundredth) of the region bounded by the curves $y=e^x$, $y=e^{-x}$, and the line $x=1$.

In Problems 38–39, solve the given separable differential equation.

38. $\displaystyle\frac{dy}{dx}=xe^{y-x}$

39. $\displaystyle\frac{dy}{dx}=\sqrt{xy}\,\ln x$

40. Find a function $y=f(x)$ whose graph passes through $(1,1)$ and has the property that at each point (x,y) on the graph, the slope of the tangent line is $y\tan^{-1}x$.

41. Find a function $y=f(x)$ whose graph passes through $(0,1)$ and has the property that the normal line at each point (x,y) on the graph has slope $\dfrac{\sec x}{xy}$.

42. Suppose it is known that $f(0)=3$ and

$$\int_0^\pi [f(x)+f''(x)]\sin x\,dx=0$$

What is $f(\pi)$?

43. Because a rocket burns fuel in flight, its mass decreases with time, and this in turn affects its velocity. It can be shown that the velocity $v(t)$ of the rocket at time t in its flight is given by

$$v(t)=-r\ln\frac{w-kt}{w}-gt$$

where w is the initial weight of the rocket (including its fuel), and r and k are, respectively, the expulsion speed and the rate of consumption of the fuel, which are assumed to be constant. As usual, $g = 32$ ft/s^2 is the constant acceleration due to gravity. Suppose $w = 30{,}000$ lb, $r = 8{,}000$ ft/s, and $k = 200$ lb/sec. What is the height of the rocket after 2 minutes (120 seconds)?

44. In physics, it is known that loudness L of a sound is related to its intensity I by the equation

$$L = 10 \log \frac{I}{I_0}$$

decibels where $I_0 = 10^{-12}$ watt/m^2 is the threshold of audibility (the lowest intensity that can be heard). What is the average value of L as the intensity of a TV show ranges between I_0 and $I_1 = 3 \cdot 10^{-5}$ watt/m^2?

45. The displacement from equilibrium of a mass oscillating at the end of a spring hanging from the ceiling is given by

$$y = 2.3e^{-0.25t} \cos 5t$$

feet. What is the average displacement (rounded to the nearest hundredth) of the mass between times $t = 0$ and $t = \pi/5$ seconds?

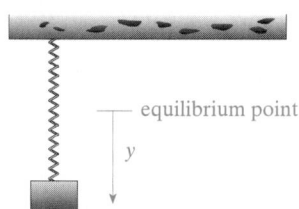

46. A photographer is taking a picture of a clever sign on the back of a truck. The sign is 5 ft high and its lower edge is 1 ft above the lens of the camera. At first the truck is 4 ft away from the photographer, but then it begins to move away. What is the average value of the angle θ subtended by the camera lens as the truck moves from 4 ft to 20 ft away from the photographer?

47. If n moles of an ideal gas expand at constant temperature T, then its pressure p and volume V satisfy the equation $pV = nRT$, for constant R. It can be shown that the work done by the gas in expanding from volume V_1 to V_2 is

$$W = nRT \ln \frac{V}{V_1}$$

What is the average work done as V increases from V_1 to $V_2 = 10V_1$?

48. Evaluate

$$\int \frac{(2x - 1)}{x^2} e^{2x} \, dx$$

49. Derive the reduction formula

$$\int x^n e^x \, dx = x^n e^x - n \int x^{n-1} e^x \, dx$$

(This is formula 486 with $a = 1$.)

50. Derive the reduction formula

$$\int (\ln x)^n \, dx = x(\ln x)^n - n \int (\ln x)^{n-1} \, dx$$

(This is formula 501.)

51. **Wallis's formula** If n is an even positive integer, use reduction formulas to show

$$\int_0^{\pi/2} \sin^n x \, dx = \int_0^{\pi/2} \cos^n x \, dx = \left[\frac{1 \cdot 3 \cdot 5 \cdot \ldots \cdot (n-1)}{2 \cdot 4 \cdot 6 \cdot \ldots \cdot n} \right] \frac{\pi}{2}$$

52. State and prove a result similar to the one in Problem 51 for the case where n is an odd positive integer.

7.3 Trigonometric Methods

IN THIS SECTION powers of sine and cosine, powers of secant and tangent, trigonometric substitutions, quadratic-form integrals

POWERS OF SINE AND COSINE

Problems 40–45 of Problem Set 7.1 anticipated what needs to be done in this section. We begin by considering a product of powers of sine and cosine, which we represent in the form

$$\int \sin^m x \cos^n x \, dx$$

There are essentially two cases that must be considered, depending on whether the powers m and n are both even or not. We shall state the general strategy for handling each case, and then illustrate the procedure with an example.

Case I   Either m or n is odd (or both)

GENERAL STRATEGY   Suppose, for simplicity, that m is odd. Separate a factor of $\sin x$ from the rest of the integrand, so the remaining power of $\sin x$ is even; use the identity $\sin^2 x = 1 - \cos^2 x$ to express everything but the term $(\sin x \, dx)$ in terms of $\cos x$. Substitute $u = \cos x$, $du = -\sin x \, dx$ to convert the integral into a polynomial in u and integrate using the power rule. The case where n is odd is handled in analogous fashion, by reversing the roles of $\sin x$ and $\cos x$.

EXAMPLE 1   *Power of cosine is odd*

Evaluate $\int \sin^4 x \cos^3 x \, dx$.

Solution   Since $n = 3$ is odd, peel off a factor of $\cos x$ and use $(\cos^2 x = 1 - \sin^2 x)$ to express the integral as a polynomial in $\sin x$:

$$\int \sin^4 x \cos^3 x \, dx = \int \sin^4 x \cos^2 x \, (\cos x \, dx)$$
$$= \int \sin^4 x (1 - \sin^2 x) \, (\cos x \, dx)$$
$$= \int u^4 (1 - u^2) \, du \quad \boxed{u = \sin x; \quad du = \cos x \, dx}$$
$$= \tfrac{1}{5} u^5 - \tfrac{1}{7} u^7 + C$$
$$= \tfrac{1}{5} \sin^5 x - \tfrac{1}{7} \sin^7 x + C \qquad \blacksquare$$

Case II   Neither m nor n is odd

GENERAL STRATEGY   Convert this into Case I by using the identities
$$\sin^2 x = \tfrac{1}{2}(1 - \cos 2x) \quad \text{and} \quad \cos^2 x = \tfrac{1}{2}(1 + \cos 2x)$$

EXAMPLE 2   *All powers are even*

Evaluate $\int \sin^2 x \cos^4 x \, dx$.

Solution

$$\int \sin^2 x \cos^4 x \, dx = \int \tfrac{1}{2}(1 - \cos 2x)(\tfrac{1}{4})(1 + \cos 2x)^2 \, dx$$
$$= \tfrac{1}{8} \int [1 + \cos 2x - \cos^2 2x - \cos^3 2x) \, dx$$
$$= \tfrac{1}{8} \int [1 + \cos 2x - \tfrac{1}{2}(1 + \cos 4x) - (1 - \sin^2 2x)(\cos 2x)] \, dx$$
$$= \tfrac{1}{8} \int (\tfrac{1}{2} - \tfrac{1}{2}\cos 4x) \, dx + \tfrac{1}{8} \int \sin^2 2x(\cos 2x) \, dx$$
$$= \tfrac{1}{16} x - \tfrac{1}{64} \sin 4x + \tfrac{1}{48} \sin^3 2x + C \qquad \blacksquare$$

Technology Window

If you work integration problems like Example 2 using *Mathematica, Maple, Derive, Mathcad,* or other computer-assisted system (commonly known as CAS) on a computer or calculator, you may obtain an answer that appears very different from the one obtained here. For example, on a TI-92, we obtained

$$\int \sin^2x \cos^4x \, dx = -\tfrac{1}{6}\sin x \cos^5x + \tfrac{1}{24}\sin x \cos^3x + \tfrac{1}{16}\sin x \cos x + \tfrac{1}{16}x$$

Both answers are correct. They are related by trigonometric identities.

POWERS OF SECANT AND TANGENT

The simplest integrals of this form are

$$\int \tan x \, dx = \ln|\sec x| + C$$

$$\int \sec x \, dx = \ln|\sec x + \tan x| + C$$

For the more general situation, which we write as

$$\int \tan^m x \, \sec^n x \, dx,$$

there are three essentially different cases to consider.

Case I *n* is even

GENERAL STRATEGY   Peel off a factor of $\sec^2x$ from the integrand and use the identity $\sec^2x = \tan^2x + 1$ to express the integrand in powers of $\tan x$ except for $(\sec^2x \, dx)$; substitute $u = \tan x$, $du = \sec^2x \, dx$, and integrate by using the power rule.

EXAMPLE 3   *Power of the secant is even*

Evaluate $\int \tan^2x \, \sec^4x \, dx$.

Solution
$$\int \tan^2x \, \sec^4x \, dx = \int \tan^2x \, \sec^2x(\sec^2x \, dx)$$
$$= \int \tan^2x(\tan^2x + 1) \, \sec^2x \, dx$$
$$= \int u^2(u^2 + 1) \, du \quad \boxed{u = \tan x; \quad du = \sec x \, dx}$$
$$= \tfrac{1}{5}u^5 + \tfrac{1}{3}u^3 + C$$
$$= \tfrac{1}{5}\tan^5x + \tfrac{1}{3}\tan^3x + C \qquad \blacksquare$$

Case II *m* is odd

GENERAL STRATEGY   Peel off a factor of $\sec x \tan x$ from the integrand and use the identity $\tan^2x = \sec^2x - 1$ to express the integrand in powers of $\sec x$, except for $(\sec x \tan x \, dx)$; substitute $u = \sec x$, $du = \sec x \tan x \, dx$, and integrate using the power rule.

EXAMPLE 4   *Power of tangent is odd*

Evaluate $\displaystyle\int \tan x \sec^6 x \, dx$.

Solution
$$\int \tan x \sec^6 x \, dx = \int \sec^5 x (\sec x \tan x \, dx)$$
$$= \int u^5 \, du \qquad \boxed{u = \sec x; \quad du = \sec x \tan x \, dx}$$
$$= \tfrac{1}{6}\sec^6 x + C \qquad\qquad\qquad\qquad\qquad \blacksquare$$

Case III m is even and n is odd

GENERAL STRATEGY   Use the identity $\tan^2 x = \sec^2 x - 1$ to express the integrand in terms of powers of $\sec x$; then use the reduction formula 428:

$$\int \sec^n x \, dx = \frac{\sec^{n-2}x \tan x}{n-1} + \frac{n-2}{n-1}\int \sec^{n-2}x \, dx$$

EXAMPLE 5   *Power of tangent is even and power of secant is odd*

Evaluate $\displaystyle\int \tan^2 x \sec^3 x \, dx$.

Solution
$$\int \tan^2 x \sec^3 x \, dx = \int (\sec^2 x - 1)\sec^3 x \, dx$$
$$= \int \sec^5 x \, dx - \int \sec^3 x \, dx$$
$$= \left[\frac{\sec^3 x \tan x}{4} + \frac{3}{4}\int \sec^3 x \, dx\right] - \int \sec^3 x \, dx$$
$$= \frac{\sec^3 x \tan x}{4} - \frac{1}{4}\int \sec^3 x \, dx$$
$$= \frac{\sec^3 x \tan x}{4} - \frac{1}{4}\left[\frac{\sec x \tan x}{2} + \frac{1}{2}\int \sec x \, dx\right]$$
$$= \frac{\sec^3 x \tan x}{4} - \frac{\sec x \tan x}{8} - \frac{1}{8}\ln|\sec x + \tan x| + C$$
$$\qquad\qquad\qquad\qquad\qquad\qquad\qquad\qquad\qquad\qquad\qquad\qquad \blacksquare$$

TRIGONOMETRIC SUBSTITUTIONS

Trigonometric substitutions can also be useful. For instance, suppose an integrand contains the terms $\sqrt{a^2 - u^2}$, where $a > 0$. Then by setting $u = a \sin\theta$ for an acute angle θ, and using the identity $\cos^2\theta = 1 - \sin^2\theta$, we obtain

$$\sqrt{a^2 - u^2} = \sqrt{a^2 - a^2\sin^2\theta} = a\sqrt{1 - \sin^2\theta} = a\cos\theta$$

Thus, the substitution $u = a\sin\theta$, $du = a\cos\theta\, d\theta$ eliminates the square root and may convert the given integral into one involving only sine and cosine. This substitution can best be remembered by setting up a reference triangle. This process is illustrated following example.

EXAMPLE 6   *Trigonometric substitution with form $\sqrt{a^2 - u^2}$*

Find $\displaystyle\int \sqrt{4 - x^2}\, dx$.

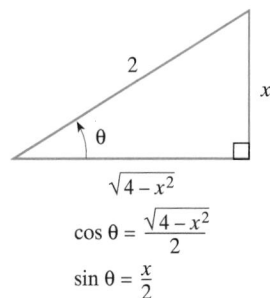

$$\cos \theta = \frac{\sqrt{4-x^2}}{2}$$

$$\sin \theta = \frac{x}{2}$$

■ **FIGURE 7.2** Reference triangle with form $\sqrt{a^2 - u^2}$

Solution   First, using a table of integration, we find (formula 231; $a = 2$)

$$\int \sqrt{4 - x^2}\, dx = \frac{x\sqrt{4-x^2}}{2} + 2 \sin^{-1}\left(\frac{x}{2}\right) + C$$

Our goal with this example is to show how we might obtain this formula with a trigonometric substitution. Refer to the triangle shown in Figure 7.2.

Let $x = 2 \sin \theta$, so $dx = 2 \cos \theta\, d\theta$. Then

$$\int \sqrt{4 - x^2}\, dx = \int \sqrt{4 - 4\sin^2\theta}\, (2\cos\theta\, d\theta)$$

$$= 4 \int \cos^2\theta\, d\theta \qquad \text{Since } \sqrt{1 - \sin^2\theta} = \cos\theta$$

$$= 4 \int \frac{1 + \cos 2\theta}{2}\, d\theta \qquad \text{Half-angle identity}$$

$$= 2\theta + \sin 2\theta + C$$

$$= 2\theta + 2\sin\theta \cos\theta + C$$

The final step is to convert this answer back to terms involving x. Using the reference triangle for the values of sine and cosine, we have

$$\int \sqrt{4 - x^2}\, dx = 2 \sin^{-1}\left(\frac{x}{2}\right) + 2\left(\frac{x}{2}\right)\left(\frac{\sqrt{4-x^2}}{2}\right) + C$$

$$= 2 \sin^{-1}\left(\frac{x}{2}\right) + \frac{x}{2}\sqrt{4 - x^2} + C \qquad ■$$

Similar methods can be used to convert integrals that involve terms of the form $\sqrt{a^2 + u^2}$ or $\sqrt{u^2 - a^2}$ into trigonometric integrals, as shown in Table 7.1. For this table, we require $0 \le \theta < \frac{\pi}{2}$.

■ **TABLE 7.1**
Trigonometric substitution

If the integrand involves...	Substitute...	To obtain...
$\sqrt{a^2 - u^2}$	$u = a \sin\theta$	$\sqrt{a^2 - u^2} = a \cos\theta$
$\sqrt{a^2 + u^2}$	$u = a \tan\theta$	$\sqrt{a^2 + u^2} = a \sec\theta$
$\sqrt{u^2 - a^2}$	$u = a \sec\theta$	$\sqrt{u^2 - a^2} = a \tan\theta$

EXAMPLE 7   *Trigonometric substitution with form $\sqrt{a^2 + u^2}$*

Evaluate $\int x^2\sqrt{9 + x^2}\, dx$.

Solution   Let $x = 3\tan\theta$, $dx = 3\sec^2\theta\, d\theta$; then

$$\int x^2\sqrt{9 + x^2}\, dx = \int (3\tan\theta)^2\sqrt{9 + 9\tan^2\theta}\, (3\sec^2\theta\, d\theta)$$

$$= \int (9\tan^2\theta)(3\sec\theta)(3\sec^2\theta\, d\theta)$$

$$= 81 \int \tan^2\theta \sec^3\theta\, d\theta$$

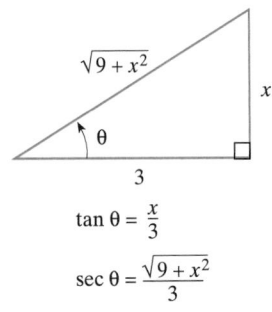

$\tan\theta = \dfrac{x}{3}$

$\sec\theta = \dfrac{\sqrt{9+x^2}}{3}$

■ **FIGURE 7.3** Reference triangle with form $\sqrt{a^2 + u^2}$

This problem can now be completed by looking at Example 5. If you wish to express the antiderivative in terms of the original variable x, you can use the reference triangle in Figure 7.3.

Since $\tan\theta = x/3$, we have $\sec\theta = (\sqrt{9+x^2})/3$, and by substituting into the solution shown in Example 5, we find

$$\int x^2\sqrt{9+x^2}\,dx = 81\left[\frac{\sec^3\theta\tan\theta}{4} - \frac{\sec\theta\tan\theta}{8} - \frac{1}{8}\ln\left|\sec\theta+\tan\theta\right|\right] + C$$

$$= \frac{81}{4}\left(\frac{\sqrt{9+x^2}}{3}\right)^3\left(\frac{x}{3}\right) - \frac{81}{8}\left(\frac{\sqrt{9+x^2}}{3}\right)\left(\frac{x}{3}\right) - \frac{81}{8}\ln\left|\frac{\sqrt{9+x^2}}{3} + \frac{x}{3}\right| + C$$

$$= \frac{x}{4}(9+x^2)^{3/2} - \frac{9x}{8}(9+x^2)^{1/2} - \frac{81}{8}\ln\left|\frac{(9+x^2)^{1/2}}{3} + \frac{x}{3}\right| + C \quad■$$

EXAMPLE 8   *Trigonometric substitution with form* $\sqrt{u^2 - a^2}$

Evaluate $\displaystyle\int x^3\sqrt{x^2-1}\,dx$.

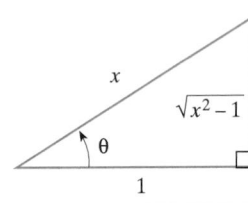

$\tan\theta = \sqrt{x^2-1}$

■ **FIGURE 7.4** Reference triangle with form $\sqrt{x^2 - a^2}$

Solution   Let $x = \sec\theta$, $dx = \sec\theta\tan\theta\,d\theta$, then

$$\int x^3\sqrt{x^2-1}\,dx = \int \sec^3\theta\sqrt{\sec^2\theta-1}\,(\sec\theta\tan\theta\,d\theta)$$

$$= \int \sec^4\theta\tan^2\theta\,d\theta$$

$$= \int \sec^2\theta\tan^2\theta\,(\sec^2\theta\,d\theta)$$

$$= \int (\tan^2\theta+1)\tan^2\theta\,(\sec^2\theta\,d\theta)$$

$$= \int (\tan^4\theta+\tan^2\theta)(\sec^2\theta\,d\theta)$$

$$= \int (u^4+u^2)\,du \quad \text{where } u = \tan\theta$$

$$= \tfrac{1}{5}u^5 + \tfrac{1}{3}u^3 + C$$

$$= \tfrac{1}{5}\tan^5\theta + \tfrac{1}{3}\tan^3\theta + C$$

$$= \tfrac{1}{5}(x^2-1)^{5/2} + \tfrac{1}{3}(x^2-1)^{3/2} + C \quad■$$

QUADRATIC-FORM INTEGRALS

An integral involving an expression of the form $Ax^2 + Bx + C$ with $A \neq 0$, $B \neq 0$ can often be evaluated by completing the square and making an appropriate substitution to convert it to one of the forms we have previously analyzed.

EXAMPLE 9   *Integration by completing the square*

Evaluate $\displaystyle\int \sqrt{16x - 2x^2 - 23}\,dx$.

Solution   Complete the square within the radicand:

$$16x - 2x^2 - 23 = -2(x^2 - 8x \qquad) - 23$$
$$= -2(x^2 - 8x + 4^2) + 2\cdot 4^2 - 23$$
$$= -2(x - 4)^2 + 9$$

Thus,

$$\int \sqrt{16x - 2x^2 - 23} \, dx = \int \sqrt{9 - 2(x - 4)^2} \, dx$$

$$= \int \sqrt{9 - u^2} \left(\frac{du}{\sqrt{2}} \right) \qquad \text{where } u = \sqrt{2}(x - 4)$$

$$= \int \sqrt{9 - (3 \sin \theta)^2} \left(\frac{3}{\sqrt{2}} \cos \theta \, d\theta \right) \qquad \text{where } u = 3 \sin\theta$$

$$= \frac{3}{\sqrt{2}} \int 3\sqrt{1 - \sin^2\theta} \cos \theta \, d\theta$$

$$= \frac{9}{\sqrt{2}} \int \cos^2\theta \, d\theta$$

$$= \frac{9}{2\sqrt{2}} \left[\theta + \frac{\sin 2\theta}{2} \right] + C$$

$$= \frac{9}{2\sqrt{2}} \sin^{-1}\left[\frac{\sqrt{2}}{3}(x - 4) \right] + \frac{x - 4}{2} \sqrt{16 - 2x^2 - 23} + C$$

This last step requires back-substituting from θ to u and then from u to x. Details are left as an exercise. ∎

7.3 Problem Set

Ⓐ 1. ■ **What Does This Say?** Explain how to integrate $\int \sin^m x \cos^n x \, dx$ when m and n are both even.

2. ■ **What Does This Say?** Explain how to integrate $\int \tan^m x \sec^n x \, dx$ when n is even.

3. ■ **What Does This Say?** Explain the process of using a trigonometric substitution on integrals of the form $\sqrt{a^2 + u^2}$.

4. ■ **What Does This Say?** Explain the process of using a trigonometric substitution on integrals of the form $\sqrt{a^2 - u^2}$. How is this different from handling an integral involving $\sqrt{u^2 - a^2}$?

Evaluate the integrals in Problems 5–50.

5. $\int \cos^3 x \, dx$

6. $\int \sin^5 x \, dx$

7. $\int \sin^2 x \cos^3 x \, dx$

8. $\int \sin^3 x \cos^3 x \, dx$

9. $\int \sqrt{\cos t} \sin t \, dt$

10. $\int \dfrac{\cos x \, dx}{1 + 3 \sin x}$

11. $\int e^{\cos x} \sin x \, dx$

12. $\int \cos^2(2t) \, dt$

13. $\int \sin^2 x \cos^2 x \, dx$

14. $\int \dfrac{\sin x \, dx}{\cos^5 x}$

15. $\int \tan 2\theta \, d\theta$

16. $\int \sec\left(\dfrac{x}{2}\right) dx$

17. $\int \tan^3 x \sec^4 x \, dx$

18. $\int \sec^5 x \tan x \, dx$

19. $\int (\tan^2 x + \sec^2 x) \, dx$

20. $\int \dfrac{\sin^3 u}{\cos^5 u} \, du$

21. $\int \tan^2 u \sec u \, du$

22. $\int \sec^4 x \, dx$

23. $\int \sqrt[3]{\tan x} \sec^2 x \, dx$

24. $\int e^x \sec(e^x) \, dx$

25. $\int x \sin x^2 \cos x^2 \, dx$

26. $\int x \sec^2 x \, dx$

27. $\int \tan^4 t \sec t \, dt$

28. $\int \csc(2\theta) \, d\theta$

29. $\int \csc^3 x \cot x \, dx$

30. $\int \csc^2 x \cot^2 x \, dx$

31. $\int \csc^2 x \cos x \, dx$

32. $\int \tan x \csc^3 x \, dx$

33. $\int \sqrt{4 - t^2} \, dt$

34. $\int \dfrac{dx}{\sqrt{9 - x^2}}$

35. $\int \dfrac{x + 1}{\sqrt{4 + x^2}}$

36. $\int \sqrt{9 + x^2} \, dx$

37. $\int \dfrac{dx}{\sqrt{x^2 - 7}}$

38. $\int \dfrac{dx}{5 + 2x^2}$

39. $\int \dfrac{x\,dx}{\sqrt{5-x^2}}$

40. $\int \dfrac{dx}{x\sqrt{7x^2-4}}$

41. $\int \dfrac{dx}{x^2\sqrt{4-x^2}}$

42. $\int \dfrac{dx}{x\sqrt{x^2+9}}$

43. $\int \dfrac{\sqrt{x^2-4}}{x}\,dx$

44. $\int \dfrac{dx}{(x-1)^2+4}$

45. $\int \dfrac{x\,dx}{9-(x+1)^2}$

46. $\int \sqrt{2x-x^2}\,dx$

47. $\int \dfrac{dx}{\sqrt{x^2-2x+6}}$

48. $\int \dfrac{dx}{\sqrt{x^2+8x+3}}$

49. $\int \dfrac{\sin x\,dx}{\sin^2 x+\cos x}$

50. $\int \dfrac{\sec^2 x\,dx}{\tan^2 x+\sec^2 x}$

Ⓑ **51.** Find the arc length of the curve $y=\ln x$ from $x=2$ to $x=3$.

52. Find the arc length of the curve $y=x^2$ from $x=-1$ to $x=1$.

53. Find the average value of $f(x)=\sin^2 x$ over the interval $[0,\pi]$.

54. Find the centroid of the region bounded by the curve $y=\cos^2 x$, the x-axis, and the vertical lines $x=\frac{\pi}{4}$ and $x=\frac{\pi}{3}$.

55. Find the volume of the solid generated when the region bounded by the curve $y=\sin^2 x$ and the x-axis is revolved about the y-axis, $0\le x\le\pi$.

56. A particle moves along the x-axis in such a way that the acceleration at time t is $a(t)=\sin^2 t$. What is the total dis-

tance traveled by the particle over the time interval $[0,\pi]$ if its initial velocity is $v(0)=2$ units per second?

57. Evaluate $\int \sqrt{1+\cos x}\,dx$.

Hint: Use the identity $\cos x=2\cos^2\dfrac{x}{2}-1$.

In Problems 58–61, use the following identities:

$$\sin A\cos B=\tfrac{1}{2}[\sin(A-B)+\sin(A+B)]$$
$$\sin A\sin B=\tfrac{1}{2}[\cos(A-B)-\cos(A+B)]$$
$$\cos A\cos B=\tfrac{1}{2}[\cos(A-B)+\cos(A+B)]$$

58. $\int \sin 3x\sin 5x\,dx$ **59.** $\int \cos\dfrac{x}{2}\sin 2x\,dx$

60. $\int \cos 7x\cos(-3x)\sin 4x\,dx$

61. $\int \sin^2 3x\cos 4x\,dx$

62. Evaluate $\int \dfrac{x\,dx}{9-x^2-\sqrt{9-x^2}}$.

Ⓒ **63.** Let f be a twice differentiable function that satisfies the initial value problem

$$f''(x)=-\tfrac{1}{2}(\tan x)f'(x)\quad f'(0)=f(0)=1$$

on the interval $[\frac{\pi}{4},\frac{\pi}{3}]$. Find the arc length of the curve $y=f(x)$ over this interval.

7.4   The Method of Partial Fractions

IN THIS SECTION   **partial-fraction decomposition, integrating rational functions, rational functions of sine and cosine**

■

PARTIAL-FRACTION DECOMPOSITION

You are familiar with the algebraic procedure of adding a string of rational expressions to form a combined rational function with a common denominator. For example,

$$\frac{2}{x+1}+\frac{-3}{x+2}=\frac{2(x+2)+(-3)(x+1)}{(x+1)(x+2)}=\frac{-x+1}{x^2+3x+2}$$

In **partial-fraction decomposition,** we do just the opposite: We start with the reduced fraction

$$\frac{-x+1}{x^2+3x+2}$$

and write it as the sum of fractions

$$\frac{2}{x+1}+\frac{-3}{x+2}$$

Technology Window ▲▼

Software programs will decompose rational expressions quite handily using the direction "expand."

This procedure has great value for integration because the terms $\dfrac{2}{x+1}$ and $\dfrac{-3}{x+2}$ are easy to integrate. In particular,

$$\int \frac{-x+1}{x^2+3x+2}\,dx = \int \frac{2}{x+1}\,dx + \int \frac{-3}{x+2}\,dx$$

$$= 2\ln|x+1| - 3\ln|x+2| + C$$

In the following discussion, we shall consider rational functions

$$f(x) = \frac{P(x)}{D(x)}$$

that are reduced in the sense that $P(x)$ and $D(x)$ have no common factors and the degree of P is less than the degree of D. In algebra, it is shown that if $P(x)/D(x)$ is such an expression, then

$$\frac{P(x)}{D(x)} = F_1(x) + F_2(x) + \cdots + F_m(x)$$

where the $F_k(x)$ are expressions of the form

$$\frac{A}{(x-r)^n} \quad \text{or} \quad \frac{Ax+B}{(x^2+sx+t)^m}$$

If $\dfrac{P(x)}{D(x)}$ is not reduced, then we simply divide until a reduced form is obtained. For example ($x \neq 1$),

$$\frac{(2x^3+7x^2+6x+3)(x-1)}{(x-1)(x^2+3x+2)} = 2x+1+\frac{-x+1}{x^2+3x+2}$$

$$= 2x+1+\frac{2}{x+1}+\frac{-3}{x+2}$$

We shall examine a nonreduced rational expression in Example 5.

We begin by focusing on the case where $D(x)$ can be expressed as a product of linear powers.

Partial–Fraction Decomposition: A Single Linear Power

Let $f(x) = P(x)/(x-r)^n$, where $P(x)$ is a polynomial of degree less than n and $P(r) \neq 0$. Then $f(x)$ can be **decomposed** into partial fractions in the following "cascading form":

$$\frac{A_1}{x-r} + \frac{A_2}{(x-r)^2} + \cdots + \frac{A_n}{(x-r)^n}$$

EXAMPLE 1   *Partial–fraction decomposition with a single linear power*

Decompose $\dfrac{x^2-6x+3}{(x-2)^3}$ into a sum of partial fractions.

Solution $\dfrac{x^2-6x+3}{(x-2)^3} = \dfrac{A_1}{x-2} + \dfrac{A_2}{(x-2)^2} + \dfrac{A_3}{(x-2)^3}$

Multiply both sides by $(x - 2)^3$ to obtain

$$x^2 - 6x + 3 = A_1(x - 2)^2 + A_2(x - 2) + A_3$$

$$\text{Let } x = 2: \quad 2^2 - 6(2) + 3 = A_1(0) + A_2(0) + A_3$$

$$-5 = A_3$$

Substitute $A_3 = -5$ and expand the right side to obtain

$$x^2 - 6x + 3 = A_1 x^2 + (-4A_1 + A_2)x + (4A_1 - 2A_2 - 5)$$

This implies (by equating the coefficients of the similar terms) that

$$1 = A_1 \qquad\qquad x^2 \text{ terms}$$
$$-6 = -4A_1 + A_2 \qquad x \text{ terms}$$
$$3 = 4A_1 - 2A_2 - 5 \qquad \text{Constants}$$

Because $A_1 = 1$, we find $A_2 = -2$, so the decomposition is

$$\frac{x^2 - 6x + 3}{(x - 2)^3} = \frac{1}{x - 2} + \frac{-2}{(x - 2)^2} + \frac{-5}{(x - 2)^3}$$ ∎

If there are two or more linear factors in the factorization of $D(x)$, there must be a separate cascade for each power. In particular, if $D(x)$ can be expressed as the product of n distinct linear factors, then

$$\frac{P(x)}{(x - r_1)(x - r_2)\cdots(x - r_n)}$$

is decomposed into separate terms

$$\frac{A_1}{x - r_1} + \frac{A_2}{x - r_2} + \cdots + \frac{A_n}{x - r_n}$$

This process is illustrated with the following example.

EXAMPLE 2   *Partial–fraction decomposition with distinct linear factors*

Decompose $\dfrac{8x - 1}{x^2 - x - 2}$.

Solution

$$\frac{8x - 1}{x^2 - x - 2} = \frac{8x - 1}{(x - 2)(x + 1)} \qquad \textit{First factor the denominator.}$$

$$= \frac{A_1}{x - 2} + \frac{A_2}{x + 1} \qquad \begin{array}{l}\textit{Break up the fraction in parts,}\\ \textit{each with a linear denominator.}\\ \textit{The task is to find } A_1 \textit{ and } A_2.\end{array}$$

$$= \frac{A_1(x + 1) + A_2(x - 2)}{(x - 2)(x + 1)} \qquad \begin{array}{l}\textit{Obtain a common}\\ \textit{denominator on the}\\ \textit{right.}\end{array}$$

WARNING ▶ Note that the degree of the denominator is the same number as the number of arbitrary constants, A_1 and A_2. This provides a quick intermediate check on the correct procedure. Make this check a standard step in your developmental task. ◄

Now, *multiply both sides of this equation by the least common denominator,* which is $(x - 2)(x + 1)$ for this example:

$$8x - 1 = A_1(x + 1) + A_2(x - 2)$$

Substitute, one at a time, the values that cause each of the factors in the least common denominator to be zero.

$$\text{Let } x = -1: \qquad 8x - 1 = A_1(x + 1) + A_2(x - 2)$$
$$8(-1) - 1 = A_1(-1 + 1) + A_2(-1 - 2)$$
$$-9 = -3A_2$$
$$3 = A_2$$

$$\text{Let } x = 2: \qquad 8x - 1 = A_1(x + 1) + A_2(x - 2)$$
$$8(2) - 1 = A_1(2 + 1) + A_2(2 - 2)$$
$$15 = 3A_1$$
$$5 = A_1$$

Thus, $\dfrac{8x - 1}{x^2 - x - 2} = \dfrac{5}{x - 2} + \dfrac{3}{x + 1}$. ∎

If there is a mixture of distinct and repeated linear factors, we combine the procedures illustrated in the preceding examples. For example,

$$\frac{5x^2 + 21x + 4}{(x + 1)^2(x - 3)} \quad \text{is decomposed as} \quad \frac{A_1}{x + 1} + \frac{A_2}{(x + 1)^2} + \frac{A_3}{x - 3}.$$

Note that the degree of the denominator is 3 and we use three arbitrary constants A_1, A_2, and A_3.

$$\frac{5x^2 + 21x + 4}{(x + 1)^3(x - 3)} \quad \text{is decomposed as} \quad \frac{A_1}{x + 1} + \frac{A_2}{(x + 1)^2} + \frac{A_3}{(x + 1)^3} + \frac{A_4}{x - 3}.$$

The degree of the denominator is 4, so we use four constants.

If the denominator $D(x)$ in the rational expression $P(x)/D(x)$ contains an irreducible quadratic power, the partial-fraction decomposition contains a different kind of cascading sum.

Partial–Fraction Decomposition: A Single Irreducible Quadratic Factor

Let $f(x) = \dfrac{P(x)}{(x^2 + sx + t)^m}$, where $P(x)$ is a polynomial of degree less than $2m$.

Then $f(x)$ can be decomposed into partial fractions in the following "cascading form":

$$\frac{A_1x + B_1}{x^2 + sx + t} + \frac{A_2x + B_2}{(x^2 + sx + t)^2} + \cdots + \frac{A_mx + B_m}{(x^2 + sx + t)^m}$$

Because the degree of the denominator is $2m$, we have $2m$ arbitrary constants—namely, $A_1, A_2, \ldots, A_m, B_1, B_2, \ldots, B_m$.

WARNING Note that the numerator in each term of a linear cascade is a constant A_k, whereas each numerator in a quadratic cascade is a linear term of the form $A_kx + B_k$. ◂

EXAMPLE 3   *Partial-fraction decomposition with a single quadratic power*

Decompose $\dfrac{-3x^3 - x}{(x^2 + 1)^2}$.

Solution   The decomposition gives

$$\frac{-3x^3 - x}{(x^2 + 1)^2} = \frac{A_1 x + B_1}{x^2 + 1} + \frac{A_2 x + B_2}{(x^2 + 1)^2}$$

Multiply both sides of this equation by $(x^2 + 1)^2$, and simplify algebraically.

$$-3x^3 - x = (A_1 x + B_1)(x^2 + 1) + (A_2 x + B_2)$$
$$= Ax^3 + Bx^2 + (A_1 + A_2)x + (B_1 + B_2)$$

Next, equate the corresponding coefficients on each side of this equation and solve the resulting system of equations to find $A_1 = -3$, $A_2 = 2$, $B_1 = 0$, and $B_2 = 0$. This means that

$$\frac{-3x^3 - x}{(x^2 + 1)^2} = \frac{-3x}{x^2 + 1} + \frac{2x}{(x^2 + 1)^2}$$ ∎

     Many of the examples we encounter will offer a mixture of linear and quadratic factors. For example,

$$\frac{x^2 + 4x - 23}{(x^2 + 4)(x + 3)^2} \quad \text{is decomposed as} \quad \frac{A_1 x + B_1}{x^2 + 4} + \frac{A_2}{x + 3} + \frac{A_3}{(x + 3)^2}.$$

The degree of the denominator is 4, and there are four constants.

     In algebra, the theory of equations tells us that any polynomial P with real coefficients can be expressed as a product of linear and irreducible quadratic powers, some of which may be repeated. This fact can be used to justify the following general procedure for obtaining the partial-fraction decomposition of a rational function.

Partial–Fraction Decomposition of the Rational Function $P(x)/D(x)$

Let $f(x) = \dfrac{P(x)}{D(x)}$, where $P(x)$ and $D(x)$ are polynomials with no common factors and $D(x) \neq 0$.

Step 1. If the degree of P is greater than or equal to the degree of D, use long (or synthetic) division to express $\dfrac{P(x)}{D(x)}$ as the sum of a polynomial and a fraction $\dfrac{R(x)}{D(x)}$ in which the degree of the remainder polynomial $R(x)$ is less than the degree of the denominator polynomial $D(x)$.

Step 2. Factor the denominator $D(x)$ into the product of linear and irreducible quadratic powers.

Step 3. Express $\dfrac{P(x)}{D(x)}$ as a cascading sum of partial fractions of the form

$$\frac{A_i}{(x - r)^n} \quad \text{and} \quad \frac{A_k x + B_k}{(x^2 + sx + t)^m}$$

WARNING   Always verify that the number of constants used in the same as the degree of the denominator, $D(x)$. ⬅

INTEGRATING RATIONAL FUNCTIONS

We will now apply the procedure of partial-fraction decomposition to integration.

EXAMPLE 4   *Integrating a rational function with a repeated factor*

Find $\int \dfrac{x^2 - 6x + 3}{(x-2)^3}\,dx$.

Solution   From Example 1, we have

$$\int \frac{x^2 - 6x + 3}{(x-2)^3}\,dx = \int \left[\frac{1}{x-2} + \frac{-2}{(x-2)^2} + \frac{-5}{(x-2)^3} \right] dx$$

$$= \int (x-2)^{-1}dx - 2\int (x-2)^{-2}\,dx - 5\int (x-2)^{-3}dx$$

$$= \ln|x-2| + \frac{2}{x-2} + \frac{5}{2(x-2)^2} + C \qquad \blacksquare$$

EXAMPLE 5   *Integrating a rational expression with distinct linear factors*

Find $\int \dfrac{x^4 + 2x^3 - 4x^2 + x - 3}{x^2 - x - 2}\,dx$.

Solution   We have $P(x) = x^4 + 2x^3 - 4x^2 + x - 3$ and $D(x) = x^2 - x - 2$; because the degree of P is higher than the degree of D, we carry out the long division and write

$$\int \frac{x^4 + 2x^3 - 4x^2 + x - 3}{x^2 - x - 2}\,dx = \int \left(x^2 + 3x + 1 + \frac{8x-1}{x^2-x-2} \right) dx$$

The polynomial part is easy to integrate. The rational expression was decomposed into partial fractions in Example 2.

$$\int \frac{x^4 + 2x^3 - 4x^2 + x - 3}{x^2 - x - 2}\,dx = \int \left(x^2 + 3x + 1 + \frac{8x-1}{x^2-x-2} \right) dx$$

$$= \int \left(x^2 + 3x + 1 + \frac{5}{x-2} + \frac{3}{x+1} \right) dx$$

$$= \int x^2\,dx + 3\int x\,dx + \int dx + 5\int (x-2)^{-1}dx$$

$$+ 3\int (x+1)^{-1}\,dx$$

$$= \frac{x^3}{3} + \frac{3x^2}{2} + x + 5\ln|x-2| + 3\ln|x+1| + C \qquad \blacksquare$$

EXAMPLE 6   *Integrating a rational function with repeated quadratic factors*

Find $\int \dfrac{-3x^3 - x}{(x^2+1)^2}\,dx$.

Solution   From Example 3,

$$\int \frac{-3x^3 - x}{(x^2+1)^2}\,dx = \int \frac{2x}{(x^2+1)^2}\,dx + \int \frac{-3x}{x^2+1}\,dx$$

$$\boxed{\begin{array}{l} \text{Let } u = x^2 + 1; \\ du = 2x\,dx. \end{array}}$$

$$= \int u^{-2}\, du - \frac{3}{2}\int u^{-1} du$$

$$= -u^{-1} - \frac{3}{2}\ln|u| + C$$

$$= \frac{-1}{x^2 + 1} - \frac{3}{2}\ln(x^2 + 1) + C \qquad \blacksquare$$

EXAMPLE 7   *Repeated linear factors*

Find $\displaystyle\int \frac{5x^2 + 21x + 4}{(x + 1)^2(x - 3)}\, dx$.

Solution   The partial-fraction decomposition of the integrand is

$$\frac{5x^2 + 21x + 4}{(x + 1)^2(x - 3)} = \frac{A_1}{(x + 1)^2} + \frac{A_2}{x + 1} + \frac{A_3}{x - 3}$$

Multiply both sides by $(x + 1)^2(x - 3)$:

$$5x^2 + 21x + 4 = A_1(x - 3) + A_2(x + 1)(x - 3) + A_3(x + 1)^2$$

As in Example 1, we substitute $x = -1$ and $x = 3$ into both sides of the equation to obtain $A_1 = 3$ and $A_3 = 7$, but A_2 cannot be obtained in this fashion. To find A_2, multiply out the polynomial on the right:

$$5x^2 + 21x + 4 = (A_2 + A_3)x^2 + (A_1 - 2A_2 + 2A_3)x + (-3A_1 - 3A_2 + A_3)$$

Equate the coefficients of x^2, x, and 1 (the constant term) on each side of the equation:

$$5 = A_2 + A_3 \qquad\qquad x^2\ \text{terms}$$
$$21 = A_1 - 2A_2 + 2A_3 \qquad x\ \text{terms}$$
$$4 = -3A_1 - 3A_2 + A_3 \qquad \text{Constants}$$

Because we already know that $A_1 = 3$ and $A_3 = 7$, we use the equation $5 = A_2 + A_3$ to obtain $A_2 = -2$. We now turn to the integration:

$$\int \frac{5x^2 + 21x + 4}{(x + 1)^2(x - 3)}\, dx = \int \frac{3\, dx}{(x + 1)^2} + \int \frac{-2\, dx}{x + 1} + \int \frac{7\, dx}{x - 3}$$

$$= -3(x + 1)^{-1} - 2\ln|x + 1| + 7\ln|x - 3| + C \qquad \blacksquare$$

EXAMPLE 8   *Distinct linear and quadratic factors*

Find $\displaystyle\int \frac{x^2 + 4x - 23}{(x^2 + 4)(x + 3)}\, dx$.

Solution   The partial-fraction decomposition of the integrand has the form

$$\frac{x^2 + 4x - 23}{(x^2 + 4)(x + 3)} = \frac{A_1 x + B_1}{x^2 + 4} + \frac{A_2}{x + 3}$$

Multiply both sides by $(x^2 + 4)(x + 3)$ and then combine the terms on the right:

$$x^2 + 4x - 23 = (A_1 x + B_1)(x + 3) + A_2(x^2 + 4)$$
$$= (A_1 + A_2)x^2 + (3A_1 + B_1)x + (3B_1 + 4A_2)$$

Equate the coefficients to set up the following system of equations:

$$\begin{cases} A_1 + A_2 = 1 & x^2 \text{ terms} \\ 3A_1 + B_1 = 4 & x \text{ terms} \\ 3B_1 + 4A_2 = -23 & \text{Constants} \end{cases}$$

Solve this system (the details are not shown) to find $A_1 = 3$, $B_1 = -5$, and $A_2 = -2$. We now turn to the integration:

$$\int \frac{x^2 + 4x - 23}{(x^2 + 4)(x + 3)}\, dx = \int \frac{3x - 5}{x^2 + 4}\, dx + \int \frac{-2\, dx}{x + 3}$$

$$= 3\underbrace{\int \frac{x\, dx}{x^2 + 4}}_{} - 5\int \frac{dx}{x^2 + 4} - 2\int \frac{dx}{x + 3}$$

Let $u = x^2 + 4$; $du = 2x\, dx.$

$$= 3\left[\frac{1}{2}\ln(x^2 + 4)\right] - 5\left[\frac{1}{2}\tan^{-1}\left(\frac{x}{2}\right)\right] - 2\ln|x + 3| + C \quad \blacksquare$$

RATIONAL FUNCTIONS OF SINE AND COSINE

The German mathematician Karl Weierstrass (1815–1897) noticed that the substitution

$$u = \tan\frac{x}{2} \qquad -\pi < x < \pi$$

will convert any rational function of $\sin x$ and $\cos x$ into an ordinary rational function.

To see why this is true, we use the double-angle identities for sine and cosine (see Figure 7.5):

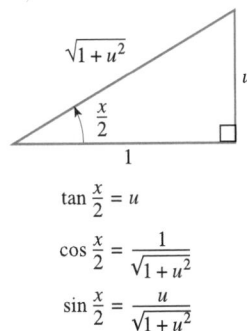

$$\tan\frac{x}{2} = u$$
$$\cos\frac{x}{2} = \frac{1}{\sqrt{1 + u^2}}$$
$$\sin\frac{x}{2} = \frac{u}{\sqrt{1 + u^2}}$$

■ **FIGURE 7.5** Weierstrass substitution

$$\sin x = 2\sin\frac{x}{2}\cos\frac{x}{2} \qquad \text{and} \qquad \cos x = \cos^2\frac{x}{2} - \sin^2\frac{x}{2}$$

$$= 2\left(\frac{u}{\sqrt{1 + u^2}}\right)\left(\frac{1}{\sqrt{1 + u^2}}\right) \qquad\qquad = \left(\frac{1}{\sqrt{1 + u^2}}\right)^2 - \left(\frac{u}{\sqrt{1 + u^2}}\right)^2$$

$$= \frac{2u}{1 + u^2} \qquad\qquad\qquad\qquad = \frac{1 - u^2}{1 + u^2}$$

Finally, because $u = \tan\frac{x}{2}$, we have $x = 2\tan^{-1}u$, so $dx = \frac{2}{1 + u^2}\, du$.

> **Weierstrass Substitution**
>
> For $-\pi < x < \pi$, let $u = \tan\frac{x}{2}$ so that
>
> $$\sin x = \frac{2u}{1 + u^2}, \quad \cos x = \frac{1 - u^2}{1 + u^2}, \quad \text{and} \quad dx = \frac{2}{1 + u^2}\, du.$$
>
> This is called the **Weierstrass substitution.**

The following example illustrates how this substitution can be used along with partial fractions to integrate a rational trigonometric function.

> **EXAMPLE 9**   *Integrating a rational trigonometric function*

Find $\displaystyle\int \frac{dx}{3 \cos x - 4 \sin x}$.

Solution   Use the Weierstrass substitution—that is, let $u = \tan\dfrac{x}{2}$.

Remember, $dx = \dfrac{2\,du}{1+u^2}$; $\cos x = \dfrac{1-u^2}{1+u^2}$;   and $\sin x = \dfrac{2u}{1+u^2}$.

$$\int \frac{dx}{3 \cos x - 4 \sin x} = \int \frac{\dfrac{2\,du}{1+u^2}}{3\left(\dfrac{1-u^2}{1+u^2}\right) - 4\left(\dfrac{2u}{1+u^2}\right)} \qquad \textit{Simplify.}$$

$$= \int \frac{-2\,du}{3u^2 + 8u - 3} = \int \frac{-2\,du}{(3u-1)(u+3)}$$

This integral can be handled by the method of partial fractions.

$$\frac{-2}{(3u-1)(u+3)} = \frac{A_1}{3u-1} + \frac{A_2}{u+3}$$

Solve this to find $A_1 = -\frac{3}{5}$ and $A_2 = \frac{1}{5}$. We continue with the integration:

$$\int \frac{dx}{3 \cos x - 4 \sin x} = \int \frac{-2\,du}{(3u-1)(u+3)}$$

$$= \int \frac{-\frac{3}{5}\,du}{3u-1} + \int \frac{\frac{1}{5}\,du}{u+3}$$

$$= -\frac{3}{5} \cdot \frac{1}{3} \ln\left|3u-1\right| + \frac{1}{5} \cdot \ln\left|u+3\right| + C$$

$$= -\frac{1}{5} \ln\left|3 \tan\frac{x}{2} - 1\right| + \frac{1}{5} \ln\left|\tan\frac{x}{2} + 3\right| + C \qquad \blacksquare$$

Once again, observe that when carrying out integration, you may obtain very different forms for the result. For Example 9, you might use an integration table (formula 393 in the *Student Mathematics Handbook*) to find

> **SMH**

$$\int \frac{du}{p \sin au + q \cos au} = \frac{1}{a\sqrt{p^2+q^2}} \ln\left|\tan\left(\frac{au + \tan^{-1}(q/p)}{2}\right)\right|$$

Let $p = -4$, $q = 3$, $a = 1$, so that

$$\int \frac{dx}{3 \cos x - 4 \sin x} = \frac{1}{5} \ln\left|\tan\left(\frac{x + \tan^{-1}(-\frac{3}{4})}{2}\right)\right| + C$$

Technology Window

To illustrate further the point of alternative forms for the same integral, we add still another form for the indefinite integral of Example 9. Computer software programs often give alternative forms. For Example 9, we obtained

$$\frac{\text{LN}\,[3 \cos(x) + \sin(x) + 3]}{5} - \frac{\text{LN}(\cos(x) - 3 \sin(x) + 1)}{5}$$

Problem 65 asks you to derive the formula

$$\int \sec x \, dx = \ln|\sec x + \tan x| + C$$

from scratch. You might recall that we derived this formula using an unusual algebraic step in Example 3 of Section 7.1. You can now derive this by using a Weierstrass substitution.

7.4   Problem Set

A *Write each rational function given in Problems 1–12 as a sum of partial fractions.*

1. $\dfrac{1}{x(x-3)}$

2. $\dfrac{3x-1}{x^2-1}$

3. $\dfrac{3x^2+2x-1}{x(x+1)}$

4. $\dfrac{2x^2+5x-1}{x(x^2-1)}$

5. $\dfrac{4}{2x^2+x}$

6. $\dfrac{x^2-x+3}{x^2(x-1)}$

7. $\dfrac{4x^3+4x^2+x-1}{x^2(x+1)^2}$

8. $\dfrac{x^2-5x-4}{(x^2+1)(x-3)}$

9. $\dfrac{x^3+3x^2+3x-4}{x^2(x+3)^2}$

10. $\dfrac{1}{x^3-1}$

11. $\dfrac{1}{1-x^4}$

12. $\dfrac{x^4-x^2+2}{x^2(x-1)}$

Compute the integrals given in Problems 13–18. Notice that in each case, the integrand is a rational function decomposed into partial fractions in Problems 1–6.

13. $\displaystyle\int \dfrac{dx}{x(x-3)}$

14. $\displaystyle\int \dfrac{3x-1}{x^2-1}\,dx$

15. $\displaystyle\int \dfrac{3x^2+2x-1}{x(x+1)}\,dx$

16. $\displaystyle\int \dfrac{2x^2+5x-1}{x(x^2-1)}\,dx$

17. $\displaystyle\int \dfrac{4\,dx}{2x^2+x}$

18. $\displaystyle\int \dfrac{x^2-x+3}{x^2(x-1)}\,dx$

Find the indicated integrals in Problems 19–34.

19. $\displaystyle\int \dfrac{2x^3+9x-1}{x^2(x^2-1)}\,dx$

20. $\displaystyle\int \dfrac{x^4-x^2+2}{x^2(x-1)}\,dx$

21. $\displaystyle\int \dfrac{x^2+1}{x^2+x-2}\,dx$

22. $\displaystyle\int \dfrac{dx}{x^3-8}$

23. $\displaystyle\int \dfrac{x^4+1}{x^4-1}\,dx$

24. $\displaystyle\int \dfrac{x^3+1}{x^3-1}\,dx$

25. $\displaystyle\int \dfrac{x\,dx}{(x+1)^2}$

26. $\displaystyle\int \dfrac{2x\,dx}{(x-2)^2}$

27. $\displaystyle\int \dfrac{dx}{x(x+1)(x-2)}$

28. $\displaystyle\int \dfrac{x+2}{x(x-1)^2}\,dx$

29. $\displaystyle\int \dfrac{x\,dx}{(x+1)(x+2)^2}$

30. $\displaystyle\int \dfrac{x+1}{x(x^2+2)}\,dx$

31. $\displaystyle\int \dfrac{5x+7}{x^2+2x-3}\,dx$

32. $\displaystyle\int \dfrac{5x\,dx}{x^2-6x+9}$

33. $\displaystyle\int \dfrac{3x^2-2x+4}{x^3-x^2+4x-4}\,dx$

34. $\displaystyle\int \dfrac{3x^2+4x+1}{x^3+2x^2+x-2}\,dx$

B **35.** ■ **What Does This Say?** Describe the process of partial-fraction decomposition.

Find the indicated integral in Problems 36–53.

36. $\displaystyle\int \dfrac{\cos x\,dx}{\sin^2 x - \sin x - 2}$

37. $\displaystyle\int \dfrac{e^x\,dx}{2e^{2x}-5e^x-3}$

38. $\displaystyle\int \dfrac{e^x\,dx}{e^{2x}-1}$

39. $\displaystyle\int \dfrac{\sin x\,dx}{(1+\cos x)^2}$

40. $\displaystyle\int \dfrac{\tan x\,dx}{\sec^2 x + 4}$

41. $\displaystyle\int \dfrac{\sec^2 x\,dx}{\tan x + 4}$

42. $\displaystyle\int \dfrac{dx}{x^{1/4}-x}$

43. $\displaystyle\int \dfrac{dx}{x^{2/3}-x^{1/2}}$

44. $\displaystyle\int \dfrac{dx}{\sin x - \cos x}$

45. $\displaystyle\int \dfrac{dx}{3\cos x + 4\sin x}$

46. $\displaystyle\int \dfrac{dx}{5\sin x + 4}$

47. $\displaystyle\int \dfrac{\sin x - \cos x}{\sin x + \cos x}\,dx$

48. $\displaystyle\int \dfrac{dx}{4\cos x + 5}$

49. $\displaystyle\int \dfrac{dx}{\sec x - \tan x}$

50. $\displaystyle\int \dfrac{dx}{3\sin x + 4\cos x + 5}$

51. $\displaystyle\int \dfrac{dx}{4\sin x - 3\cos x - 5}$

52. $\displaystyle\int \dfrac{dx}{2\csc x - \cot x + 2}$

53. $\displaystyle\int \dfrac{dx}{x(3-\ln x)(1-\ln x)}$

54. Find the area under the curve

$$y = \dfrac{1}{x^2+5x+4}$$

between $x = 0$ and $x=3$.

55. Find the area of the region bounded by the curve

$$y = \dfrac{1}{6-5x+x^2}$$

and the lines $x = \frac{4}{3}$, $x = \frac{7}{4}$, and $y = 0$.

56. Find the volume (to four decimal places) of the solid generated when the curve

$$y = \frac{1}{x^2 + 5x + 4}, \quad 0 \le x \le 1$$

is revolved about:

    a. the y-axis.   **b.** the x-axis.

    c. the line $x = -1$.

57. Find the volume (to four decimal places) of the solid generated when the region under the curve

$$y = \frac{1}{\sqrt{x^2 + 4x + 3}}$$

on the interval $[0, 3]$ is revolved about

    a. the x-axis.   **b.** the y-axis.

58. Find both the exact and approximate volume of the solid generated when the region under the curve

$$y^2 = x^2\left(\frac{4 - x}{4 + x}\right)$$

on the interval $[0, 4]$ is revolved about the x-axis.

59. Find the exact volume of the solid generated when the region under the curve

$$y = \frac{1}{\sin x + \cos x}$$

on the interval $[0, \frac{\pi}{4}]$ is revolved about the x-axis.

60. **HISTORICAL QUEST**   George Pólya was born in Hungary and attended the universities of Budapest, Vienna, Göttingen, and Paris. He was a professor of mathematics at Stanford University. Pólya's research and winning personality earned him a place of honor not only among mathematicians, but among students and teachers as well. His discoveries spanned an impressive range of mathematics, real and complex analysis, probability, combinatorics, number theory, and geometry.

GEORGE PÓLYA
1887–1985

Pólya's book, *How to Solve It*, has been translated into 20 languages. His books have a clarity and elegance seldom seen in mathematics, making them a joy to read. For example, here is his explanation of why he was a mathematician: "It is a little shortened but not quite wrong to say: I thought I am not good enough for physics and I am too good for philosophy. Mathematics is in between."

A story told by Pólya provides our next Quest. "A number of years ago," Pólya related with his lovable accent, "I deliberately put the problem

$$\int \frac{x\,dx}{x^2 - 9}$$

as the first problem on a test of techniques of integration, to give my students a boost as they began the exam. With the substitution $u = x^2 - 9$, which I expected the students to use, you can knock the problem off in just a few seconds. Half of the students did this, and got off to a good start. But a fourth of them used the correct but time-consuming procedure of partial fractions—and because they spent so much time on the problem, they did poorly on the exam. Half of the rest used the trig substitution $x = 3\sin\theta$—also correct but so time-consuming that they wound up very far behind and bombed the exam. It is interesting that the students who used the harder techniques showed they knew 'more,' or at least more difficult mathematics than the ones who used the easy technique. But they showed that 'it's not just what you know; it's how and when you use it.' It's nice when what you do is right, but it's much better when it's also *appropriate*."[*] Carry out all three methods of solution of the given integral Pólya described in this quotation.

61. **SPY PROBLEM**   When N arrives at the sleazy bar (recall Problem 68, Chapter 6 Supplementary Problems), he has shocking news—his innocent daughter, Purity, has been kidnapped by Ernst Stavro Blohardt! The spy is sent to the village outside the thug's château disguised as an old duck plucker. On the day he arrives, his true identity is known only to the Redselig twins, Hans and Franz, but the next day, Hans' girlfriend, Blabba, finds out. Soon, word begins to circulate among the 60 citizens of the village as to the spy's true identity. He determines that the rate at which the rumor concerning his identity will spread is jointly proportional to the product of the number N of those who already know and the number $60 - N$ who do not, so that

$$\frac{dN}{dt} = kN(60 - N)$$

for a constant k. He needs a week to get the information he needs to break into the château to free Purity, but figures that as soon as 20 or more villagers know his identity, Blohardt is sure to find out, too. Does he complete his mission, or is he about to be a dead duck plucker?

62. Two substances, A and B, are being converted into a single compound C. In the laboratory, it is shown that the time rate of change of the amount x of compound C is proportional to the product of the amounts of unconverted substances A and B. Thus,

$$\frac{dx}{dt} = k(a - x)(b - x)$$

for initial concentrations a and b, of A and B, respectively.

    a. Solve this differential equation to express x in terms of time t. What happens to $x(t)$ as $t \to +\infty$ if $b > a$? What if $a > b$?

*"Pólya, Problem Solving, and Education," by Alan H. Schoenfeld, *Mathematics Magazine*, Vol. 60, No. 5, December 1987, p. 290.

b. Suppose $a = b$. What is $x(t)$? What happens to $x(t)$ as $t \to +\infty$ in this case?

c. Find $x(t)$ for the case where $a = b$. Find the half-life of this reaction (i.e., the time required for half the initial concentration a to be consumed).

C 63. Use partial fractions to derive the integration formula

$$\int \frac{dx}{a^2 - x^2} = \frac{1}{2a} \ln \left| \frac{a + x}{a - x} \right| + C$$

64. Use partial fractions to derive the integration formula

$$\int \frac{dx}{x(ax + b)} = \frac{1}{b} \ln \left| \frac{x}{ax + b} \right| + C$$

65. Derive the formula

$$\int \sec x \, dx = \ln \left| \sec x + \tan x \right| + C$$

using a Weierstrass substitution.

66. Derive the formula

$$\int \csc x \, dx = - \ln \left| \csc x + \cot x \right| + C$$

using a Weierstrass substitution.

7.5 Summary of Integration Techniques

IN THIS SECTION   integration strategy

We conclude our study of integration techniques with a table summarizing integration techniques.

■ TABLE 7.2 Integration Strategy

Step 1. Simplify. Simplify the integrand, if possible, and use one of the procedural rules (see the inside back cover or the table of integrals).

Step 2. Use basic formulas. Use the basic integration formulas (1–29 in the integration table). These are the fundamental building blocks for integration. Almost every integration will involve some basic formulas somewhere in the process, which means that you should memorize these forms.

Step 3. Substitute. Make any substitution that will transform the integral into one of the basic forms.

Step 4. Classify. Classify the integrand according to form in order to use a table of integrals. You may need to use substitution to transform the integrand into a form contained in the integration table. Some special types of substitution are contained in the following list:

I. Integration by parts

A. Forms $\int x^n e^{ax} \, dx, \int x^n \sin ax \, dx, \int x^n \cos ax \, dx$

Let $u = x^n$.

B. Forms $\int x^n \ln x \, dx, \int x^n \sin^{-1} ax \, dx, \int x^n \tan^{-1} ax \, dx$

Let $dv = x^n \, dx$.

C. Forms $\int e^{ax} \sin bx \, dx, \int e^{ax} \cos bx \, dx$

Let $dv = e^{ax} \, dx$.

■ **TABLE 7.2**   **(continued)**

II. Trigonometric functions

    A. $\int \sin^m x \cos^n x \, dx$

        m odd: Peel off a factor of $(\sin x \, dx)$ and **let $u = \cos x$.**
        n odd: Peel off a factor of $(\cos x \, dx)$ and **let $u = \sin x$.**
        m and n both even: **Use half-angle identities:**
        $\cos^2 x = \frac{1}{2}(1 + \cos 2x)$ and $\sin^2 x = \frac{1}{2}(1 - \sin 2x)$

    B. $\int \tan^m x \sec^n x \, dx$

        n even: Peel off a factor of $(\sec^2 x \, dx)$ and **let $u = \tan x$.**
        m odd: Peel off a factor of $(\sec x \tan x \, dx)$ and **let $u = \sec x$.**
        m even, n odd: Write using powers of secant and **use integration tables** (formula 428).

    C. For a rational trigonometric integral, try the Weierstrass substitution:

        Let **$u = \tan \frac{x}{2}$,** so that $\sin x = \dfrac{2u}{1 + u^2}$, $\cos x = \dfrac{1 - u^2}{1 + u^2}$, and $du = \dfrac{2}{1 + u^2} \, dx$.

III. Radical function; try a trigonometric substitution.

    A. Form $\sqrt{a^2 - u^2}$: **Let $u = a \sin \theta$.**
    B. Form $\sqrt{a^2 + u^2}$: **Let $u = a \tan \theta$.**
    C. Form $\sqrt{u^2 - a^2}$: **Let $u = a \sec \theta$.**

IV. Rational function; try partial-fraction decomposition.

Step 5: Try again.   **Still stuck?**

    1.   Manipulate the integrand:
        Multiply by 1 (clever choice of numerator or denominator).
        Rationalize the numerator.
        Rationalize the denominator.
    2.   Relate the problem to a previously worked problem.
    3.   Look at another table of integrals or consult computer software that does integration.
    4.   Some integrals do not have simple antiderivatives, so all these methods may fail. We will look at some of these forms later in the text.
    5.   If dealing with a definite integral, an approximation may suffice. It may be appropriate to use a calculator, computer, or the techniques introduced in Section 5.8.

EXAMPLE 1   *Deciding on an integration procedure*

Indicate a procedure to set up each integral. It is not necessary to carry out the integration.

    a. $\displaystyle\int \frac{\sin \sqrt{x}}{\sqrt{x}} \, dx$ b. $\displaystyle\int (1 + \tan^2 \theta) \, d\theta$ c. $\displaystyle\int \sin^3 x \cos^2 x \, dx$

    d. $\displaystyle\int 4x^2 \cos 3x \, dx$ e. $\displaystyle\int \frac{x \, dx}{\sqrt{9 - x^2}}$ f. $\displaystyle\int \frac{x^2 \, dx}{\sqrt{9 - x^2}}$

    g. $\displaystyle\int \frac{x^2 \, dx}{\sqrt{x^2 - 9}}$ h. $\displaystyle\int e^{3x} \sin 2x \, dx$ i. $\displaystyle\int \frac{\cos^4 x \, dx}{1 - \sin^2 x}$

Solution   Keep in mind that there may be several correct approaches, and the way you proceed from problem to solution is often a matter of personal preference. However, to give you some practice, we will show several hints and suggestions in the context of this example.

a. $\displaystyle\int \frac{\sin \sqrt{x}}{\sqrt{x}} \, dx$

The integrand is simplified and is not a fundamental type. Substitution will work well for this problem if you let $u = \sqrt{x}$. After you make this substitution you can integrate using a basic formula.

b. $\displaystyle\int (1 + \tan^2\theta) \, d\theta$

The integrand can be simplified using a trigonometric identity $(1 + \tan^2\theta = \sec^2\theta)$, so that it can now be integrated using a basic formula.

c. $\displaystyle\int \sin^3 x \cos^2 x \, dx$

The integrand is simplified, it is not a basic formula, and there does not seem to be an easy substitution. Next, try to classify the integrand. We see that it involves powers of trigonometric functions of the type $\int \sin^m x \cos^n x \, dx$ where $m = 3$ (odd) and $n = 2$ (even), so we make the substitution $u = \cos x$.

d. $\displaystyle\int 4x^2 \cos 3x \, dx$

The integrand is simplified (you can bring the 4 out in front of the integral, if you wish), it is not a basic formula, and there is no easy substitution. It does not seem to have an easy classification, so we check the table of integrals and find it to be formula 313.

e. $\displaystyle\int \frac{x \, dx}{\sqrt{9 - x^2}}$

The integrand is simplified and is not a basic formula. However, it looks like we might try the substitution $u = 9 - x^2$. It looks like this will work because the degree of the numerator is one less. You might also find this in a table of integrals (formula 225).

f. $\displaystyle\int \frac{x^2 \, dx}{\sqrt{9 - x^2}}$

The integrand is simplified, it is not a basic formula, and it looks as though a substitution will not work because of the degree of the numerator. It also does not seem to have an easy classification, so we turn to formula 226 of the table of integrals.

g. $\displaystyle\int \frac{x^2 \, dx}{\sqrt{x^2 - 9}}$

The integrand involves $\sqrt{x^2 - 9}$, so use the trigonometric substitution $x = 3 \sec \theta, \, dx = 3 \tan \theta \sec \theta \, d\theta$.

h. $\displaystyle\int e^{3x} \sin 2x \, dx$

The integrand is simplified, it is not a basic formula, and there does not seem to be an easy formula. If we try to classify the integrand, we see that integration by parts will work with $u = \sin 2x$ and $dv = e^{3x} \, dx$. We also find that this form is formula 492 in the integration table.

i. $\displaystyle\int \frac{\cos^4 x \, dx}{1 - \sin^2 x}$

The integration can be simplified by writing $1 - \sin^2 x = \cos^2 x$. After doing this substitution, you will obtain

$$\int \frac{\cos^4 x \, dx}{1 - \sin^2 x} = \int \frac{\cos^4 x \, dx}{\cos^2 x} = \int \cos^2 x \, dx$$

This form can be integrated by using a half-angle identity or formula 317 in the table of integrals. ∎

7.5   Problem Set

A *Find each integral in Problems 1–50.*

1. $\displaystyle\int \frac{2x-1}{(x-x^2)^3}\,dx$

2. $\displaystyle\int \frac{2x+3}{\sqrt{x^2+3x}}\,dx$

3. $\displaystyle\int (x\sec 2x^2)\,dx$

4. $\displaystyle\int (x^2\csc^2 2x^3)\,dx$

5. $\displaystyle\int (e^x\cot e^x)\,dx$

6. $\displaystyle\int \frac{\tan\sqrt{x}\,dx}{\sqrt{x}}$

7. $\displaystyle\int \frac{\tan(\ln x)\,dx}{x}$

8. $\displaystyle\int \sqrt{\cot x}\,\csc^2 x\,dx$

9. $\displaystyle\int \frac{(3+2\sin t)}{\cos t}\,dt$

10. $\displaystyle\int \frac{2+\cos x}{\sin x}\,dx$

11. $\displaystyle\int \frac{e^{2t}\,dt}{1+e^{4t}}$

12. $\displaystyle\int \frac{\sin 2x\,dx}{1+\sin^4 x}$

13. $\displaystyle\int \frac{x^2+x+1}{x^2+9}\,dx$

14. $\displaystyle\int \frac{3x+2}{\sqrt{4-x^2}}\,dx$

15. $\displaystyle\int \frac{1+e^x}{1-e^x}\,dx$

16. $\displaystyle\int \frac{e^{1-\sqrt{x}}\,dx}{\sqrt{x}}$

17. $\displaystyle\int \frac{2t^2\,dt}{\sqrt{1-t^6}}$

18. $\displaystyle\int \frac{t^3\,dt}{2^8+t^8}$

19. $\displaystyle\int \frac{dx}{1+e^{2x}}$

20. $\displaystyle\int \frac{dx}{4-e^{-x}}$

21. $\displaystyle\int \frac{dx}{x^2+2x+2}$

22. $\displaystyle\int \frac{dx}{x^2+x+4}$

23. $\displaystyle\int \frac{dx}{x^2+x+1}$

24. $\displaystyle\int \frac{dx}{x^2-x+1}$

25. $\displaystyle\int \tan^{-1}x\,dx$

26. $\displaystyle\int xe^x\sin x\,dx$

27. $\displaystyle\int e^{-x}\cos x\,dx$

28. $\displaystyle\int e^{2x}\sin 3x\,dx$

29. $\displaystyle\int \cos^{-1}(-x)\,dx$

30. $\displaystyle\int \sin^{-1}2x\,dx$

31. $\displaystyle\int \sin^3 x\,dx$

32. $\displaystyle\int \cos^5 x\,dx$

33. $\displaystyle\int \sin^3 x\cos^2 x\,dx$

34. $\displaystyle\int \sin^3 x\cos^3 x\,dx$

35. $\displaystyle\int \sin^2 x\cos^4 x\,dx$

36. $\displaystyle\int \sin^2 x\cos^5 x\,dx$

37. $\displaystyle\int \sin^5 x\cos^4 x\,dx$

38. $\displaystyle\int \sin^4 x\cos^2 x\,dx$

39. $\displaystyle\int \tan^5 x\sec^4 x\,dx$

40. $\displaystyle\int \tan^4 x\sec^4 x\,dx$

41. $\displaystyle\int \frac{\sqrt{1-x^2}}{x}\,dx$

42. $\displaystyle\int \frac{dx}{\sqrt{x^2-16}}$

43. $\displaystyle\int \frac{2x+3}{\sqrt{2x^2-1}}\,dx$

44. $\displaystyle\int \frac{dx}{x\sqrt{x^2-1}}$

45. $\displaystyle\int \frac{dx}{x\sqrt{x^2+1}}$

46. $\displaystyle\int x\sqrt{x^2+1}\,dx$

47. $\displaystyle\int \frac{(2x+1)\,dx}{\sqrt{4x-x^2-2}}$

48. $\displaystyle\int \sqrt{3+4x-4x^2}\,dx$

49. $\displaystyle\int \frac{\cos x\,dx}{\sqrt{1+\sin^2 x}}$

50. $\displaystyle\int \frac{\sec^2 x\,dx}{\sqrt{\sec^2 x-2}}$

Find the exact value of the definite integrals in Problems 51–64.

51. $\displaystyle\int_0^2 \sqrt{4-x^2}\,dx$

52. $\displaystyle\int_0^1 \frac{dx}{\sqrt{9-x^2}}$

53. $\displaystyle\int_0^1 \frac{dx}{(x^2+2)^{3/2}}$

54. $\displaystyle\int_0^1 \frac{dt}{4t^2+4t+5}$

55. $\displaystyle\int_1^2 \frac{dx}{x^4\sqrt{x^2+3}}$

56. $\displaystyle\int_0^2 \frac{x^3}{(3+x^2)^{3/2}}\,dx$

57. $\displaystyle\int_{-2}^{2\sqrt{3}} x^3\sqrt{x^2+4}\,dx$

58. $\displaystyle\int_0^{\sqrt{5}} x^2\sqrt{5-x^2}\,dx$

59. $\displaystyle\int_0^{\ln 2} e^t\sqrt{1+e^{2t}}\,dt$

60. $\displaystyle\int_1^4 \frac{\sqrt{x^2+9}}{x^3}\,dx$

61. $\displaystyle\int_0^{\pi/4} \sin^5 x\,dx$

62. $\displaystyle\int_0^{\pi/4} \cos^6 x\,dx$

63. $\displaystyle\int_0^{\pi/4} \tan^4 x\,dx$

64. $\displaystyle\int_0^{\pi/3} \sec^4 x\,dx$

B *Find each integral in Problems 65–75.*

65. $\displaystyle\int \frac{e^x\,dx}{\sqrt{1+e^{2x}}}$

66. $\displaystyle\int \frac{dx}{x\sqrt{4x^2+4x+2}}$

67. $\displaystyle\int \frac{x^2+4x+3}{x^3+x^2+x}\,dx$

68. $\displaystyle\int \frac{5x^2+3x-2}{x^3+2x^2}\,dx$

69. $\displaystyle\int \frac{5x^2+18x+34}{(x-7)(x+2)^2}\,dx$

70. $\displaystyle\int \frac{-3x^2+9x+21}{(x+2)^2(2x+1)}\,dx$

71. $\displaystyle\int \frac{3x+5}{x^2+2x+1}\,dx$

72. $\displaystyle\int \frac{3x^2+2x+1}{x^3+x^2+x}\,dx$

73. $\displaystyle\int \frac{x\,dx}{(x+1)(x+2)(x+3)}$

74. $\displaystyle\int \frac{5x^2-4x+9}{x^3-x^2+4x-4}\,dx$

75. $\displaystyle\int \frac{dx}{5\cos x-12\sin x}$

76. Find the area (to the nearest hundredth) under the curve $y=\sin x+\cos 2x$ on $[0,3]$.

77. Find the average value (to the nearest hundredth) of the function $f(x)=x\sin^3 x^2$ between $x=0$ and $x=1$.

78. An object moves along the x-axis in such a way that its velocity at time t is $v(t)=\sin t+\sin^2 t\cos^3 t$. Find the distance moved by the object between times $t=0$ and $t=\frac{\pi}{3}$.

79. ■ **What Does This Say?**   Integrals of the general form

$$\int \cot^m x\,\csc^n x\,dx$$

are handled in much the same way as those of the form

$$\int \tan^m x \sec^n x \, dx$$

a. What substitution would you use in the case where m is odd?

b. What substitution would you use if n is even?

c. What would you do if n is odd and m is even?

80. Find the volume of the solid generated when the region under the curve $y = \cos x$ between $x = 0$ and $x = \frac{\pi}{2}$ is revolved about the x-axis.

81. Find the average value of $\sec x$ on $[0, \frac{\pi}{4}]$.

82. Find the average value (to the nearest hundredth) of $\csc x$ on $[1, 2]$.

83. Find the area of the region bounded by the graphs of $y = \dfrac{2x}{\sqrt{x^2 + 9}}$, $y = 0$, $x = 0$, and $x = 4$.

84. What is the volume of the solid obtained when the region bounded by $y = \sqrt{x}e^{-x^2}$, $y = 0$, $x = 0$, and $x = 2$ is revolved about the x-axis?

85. What is the volume of the solid obtained when the region bounded by $y = x\sqrt{9 - x^2}$ and $y = 0$ is revolved about the x-axis?

86. Find the length of the curve $y = \ln(\sec x)$ from $x = 0$ to $x = \frac{\pi}{4}$.

87. Find the centroid (to the nearest hundredth) of the region bounded by the curve $y = x^2 e^{-x}$ and the x-axis, between $x = 0$ and $x = 1$.

88. Use integration by parts to verify that

$$\int e^{ax} \sin bx \, dx = \frac{(a \sin bx - b \cos bx)e^{ax}}{a^2 + b^2} + C$$

(This is formula 492.)

89. Let f'' be continuous on the closed interval $[a, b]$. Use integration by parts to show that

$$\int_a^b xf''(x) \, dx = bf'(b) - f(b) + f(a) - af'(a)$$

90. Derive the reduction formula

$$\int x^m (\ln x)^n \, dx = \frac{x^{m+1}(\ln x)^n}{m + 1} - \frac{n}{m + 1} \int x^m (\ln x)^{n-1} dx$$

where m and n are positive integers. (This is formula 510.) Use the formula to find

$$\int x^2 (\ln x)^3 \, dx$$

91. Derive the reduction formula

$$\int \sin^n Ax \, dx$$

$$= \frac{-\sin^{n-1} Ax \cos Ax}{An} + \frac{n-1}{n} \int \sin^{n-2} Ax \, dx$$

(This is formula 352). Use this formula to evaluate

$$\int \sin^3 4x \, dx$$

7.6 First-Order Differential Equations

IN THIS SECTION first-order linear differential equations, applications of first-order equations ∎

FIRST-ORDER LINEAR DIFFERENTIAL EQUATIONS

We introduced separable differential equations in Section 5.6, but not all first-order differential equations have the separable form

$$\frac{dy}{dx} = \frac{f(x)}{g(y)}$$

The goal of this section is to examine a second class of differential equations, called **first-order linear,** that have the general form

$$\frac{dy}{dx} + P(x)y = Q(x)$$

For example, the differential equation

$$\frac{dy}{dx} - \frac{y}{x} = e^{-x}$$

is first-order linear, as is

$$x^2 \frac{dy}{dx} - (x^2 + 2)y = x^5$$

where $P(x) = -\left(\frac{x^2 + 2}{x^2}\right) = -1 - 2x^{-2}$ and $Q(x) = \frac{x^5}{x^2} = x^3$.

First-order linear equations may be used to model a variety of important situations in the physical, social, and life sciences. We shall examine a few such models as well as additional models involving separable equations later in this section, but first we need to know how first-order linear differential equations may be solved.

THEOREM 7.1   *General solution of a first-order linear differential equation*

The general solution of the first-order linear differential equation

$$\frac{dy}{dx} + P(x)y = Q(x)$$

is given by

$$y = \frac{1}{I(x)}\left[\int Q(x)I(x)\,dx + C \right]$$

where $I(x) = e^{\int P(x)dx}$, and the exponent is any antiderivative of $P(x)$.

Proof   First, note that

$$\frac{dI(x)}{dx} = e^{\int P(x)dx}\left[\frac{d}{dx}\int P(x)\,dx\right] = e^{\int P(x)dx}P(x) = I(x)P(x)$$

because $\frac{d}{dx}e^u = e^u \frac{du}{dx}$. Thus,

$$\frac{d}{dx}[I(x)y] = I(x)\frac{dy}{dx} + y\frac{d}{dx}I(x) = I(x)\frac{dy}{dx} + y\,I(x)P(x) \qquad \text{Product rule}$$

We now begin the proof by starting with the given differential equation.

$$\frac{dy}{dx} + P(x)y = Q(x) \qquad\qquad \text{Given}$$

$$I(x)\left[\frac{dy}{dx} + P(x)y\right] = I(x)Q(x) \qquad\qquad \text{Multiply both sides by } I(x).$$

$$I(x)\frac{dy}{dx} + I(x)P(x)y = I(x)Q(x)$$

$$\frac{d}{dx}[I(x)y] = I(x)Q(x) \qquad\qquad \text{Substitute.}$$

$$\int \frac{d}{dx}[I(x)y]\,dx = \int I(x)Q(x)\,dx \qquad \text{Integrate both sides.}$$

$$I(x)y = \int I(x)Q(x)\,dx + C$$

$$y = \frac{1}{I(x)}[I(x)Q(x)\,dx + C] \qquad\qquad\qquad =$$

Because multiplying both sides of the differential equation

$$\frac{dy}{dx} + P(x)y = Q(x)$$

by $I(x)$ makes the left side an exact derivative, the function $I(x)$ is called an **integrating factor** of the differential equation.

A first-order **initial value problem** involves a first-order equation and the value of y at a particular value $x = x_0$. Here is an example of such a problem.

EXAMPLE 1   *First-order linear initial value problem*

Solve $\dfrac{dy}{dx} = e^{-x} - 2y$, $x \ge 0$, subject to the initial condition $y = 2$ when $x = 0$.

Solution   A graphical solution (using technology) is found by looking at the direction field, as shown in Figure 7.6.

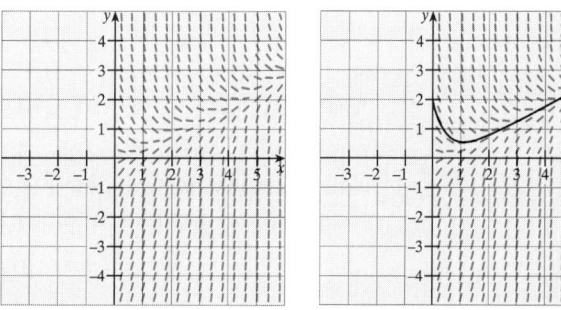

a. Direction field for
$\dfrac{dy}{dx} = e^{-x} - 2y$, $x \ge 0$

b. Particular solution passing through $(0, 2)$

■ **FIGURE 7.6**   Graphical solution using a direction field

To find an analytic solution, we use Theorem 7.1. The differential equation can be expressed in the proper form by adding $2y$ to both sides:

$$\frac{dy}{dx} + 2 \cdot y = e^{-x} \quad \text{for } x \ge 0$$

We have $P(x) = 2$ and $Q(x) = e^{-x}$. Both $P(x)$ and $Q(x)$ are continuous in the domain $x \ge 0$. An integrating factor is given by

$$I(x) = e^{\int P(x)dx} = e^{\int 2\,dx} = e^{2x} \quad \text{for } x \ge 0$$

We now use the first-order linear differential equation theorem, where $I(x) = e^{2x}$ and $Q(x) = e^{-x}$, to find y:

$$y = \frac{1}{e^{2x}}\left[\int e^{2x}e^{-x}\,dx + C\right]$$

$$= \frac{1}{e^{2x}}[e^x + C] = e^{-x} + e^{-2x}C \quad \text{for } x \ge 0$$

To find C, note that because $y = 2$ when $x = 0$, $1 = C$. Thus, $y = e^{-x} + e^{-2x}$, $x \ge 0$.

You might wish to compare the analytic and graphical solutions by noting that the graph of the equation matches the solution shown in Figure 7.6b.   ■

Integration by parts is often used in solving first-order linear differential equations. Here is an example.

EXAMPLE 2   *First-order linear differential equation*

Solve $\dfrac{dy}{dx} + \dfrac{y}{x} = e^{2x}$.

Solution   For this equation, $P(x) = \dfrac{1}{x}$ and $Q(x) = e^{2x}$. The integrating factor is

$$I(x) = e^{\int (1/x)\,dx} = e^{\ln x} = x$$

Thus, the general solution solution is

$$y = \frac{1}{x}\left[\int xe^{2x}\,dx + C\right] \qquad \text{Integrate } \int xe^{2x}\,dx \text{ by parts.}$$

$$= \frac{1}{x}\left[\frac{1}{2}xe^{2x} - \frac{1}{4}e^{2x} + C\right] \qquad \text{See Example 5 in Section 7.2.}$$

$$= \frac{1}{2}e^{2x} - \frac{e^{2x}}{4x} + \frac{C}{x} \qquad\blacksquare$$

APPLICATIONS OF FIRST-ORDER EQUATIONS

Modeling Logistic Growth   When the population $Q(t)$ of a colony of living organisms (humans, bacteria, etc.) is small, it is reasonable to expect the relative rate of change of the population to be constant. In other words,

$$\frac{\frac{dQ}{dt}}{Q} = k \quad \text{or} \quad \frac{dQ}{dt} = kQ$$

where k is a constant (the **unrestricted growth rate**). This is called **exponential growth.** As long as the colony has plenty of food and living space, its population will obey this unrestricted growth rate formula and $Q(t)$ will grow exponentially.

However, in practice, there often comes a time when environmental factors begin to restrict the further expansion of the colony, and at this point, the growth ceases to be purely exponential in nature. To construct a population model that takes into account the effect of diminishing resources and cramping, we assume that the population has a limiting value of B. Then, the relative rate of change in population is proportional to the size of the remaining population. That is,

$$\frac{\frac{dQ}{dt}}{Q} = k(B - Q) \quad \text{or} \quad \frac{dQ}{dt} = kQ(B - Q)$$

This is called a **logistic equation,** and it arises not only in connection with population models, but also in a variety of other situations. The following example illustrates one way such an equation can arise.

EXAMPLE 3   *Logistic equation for the spread of an epidemic*

The rate at which an epidemic spreads through a community is proportional to the product of the number of residents who have been infected and the number of susceptible residents. Express the number of residents who have been infected as a function of time.

Solution   Let $Q(t)$ denote the number of residents who have been infected by time t and B the total number of residents. Then the number of susceptible residents who have not been infected is $B - Q$, and the differential equation describing the spread of the epidemic is

$$\frac{dQ}{dt} = kQ(B - Q) \qquad \text{k is the constant of proportionality.}$$

$$\frac{dQ}{Q(B - Q)} = k\, dt$$

$$\int \frac{dQ}{Q(B - Q)} = \int k\, dt$$

The integral on the right causes no difficulty, but the integral on the left requires partial fractions:

$$\frac{1}{B}\left(\frac{1}{Q} + \frac{1}{B - Q}\right) = \frac{1}{B}\left[\frac{B - Q + Q}{Q(B - Q)}\right] = \frac{1}{Q(B - Q)}$$

We now substitute the form on the left into the equation and complete the integration:

$$\frac{1}{B}\int \frac{dQ}{Q} + \frac{1}{B}\int \frac{dQ}{B - Q} = \int k\, dt$$

$$\frac{1}{B}\ln|Q| - \frac{1}{B}\ln|B - Q| = kt + C \qquad \text{Integrate each.}$$

$$\frac{1}{B}\ln\left|\frac{Q}{B - Q}\right| = kt + C \qquad \text{Division property of logs}$$

$$\ln\left(\frac{Q}{B - Q}\right) = Bkt + BC \qquad \text{Multiply both sides by B. (Note that $Q > 0, B > Q$.)}$$

$$\frac{Q}{B - Q} = e^{Bkt + BC} \qquad \text{Definition of ln}$$

$$Q = (B - Q)e^{Bkt}e^{BC} \qquad \text{Because e^{BC} is a constant, we can let $A_1 = e^{BC}$.}$$

$$Q = A_1 Be^{Bkt} - A_1 Qe^{Bkt}$$

$$Q + A_1 Qe^{Bkt} = A_1 Be^{Bkt}$$

$$Q = \frac{A_1 Be^{Bkt}}{1 + A_1 e^{Bkt}}$$

To simplify the equation, we make another substitution; let $A = \dfrac{1}{A_1}$, so that (after several simplification steps) we have

$$Q = \frac{\dfrac{Be^{Bkt}}{A}}{1 + \dfrac{e^{Bkt}}{A}} = \frac{B}{1 + Ae^{-Bkt}}$$

The graph of $Q(t)$ is shown in Figure 7.7. Note that the curve has an inflection point where

$$Q(t) = \frac{B}{2}$$

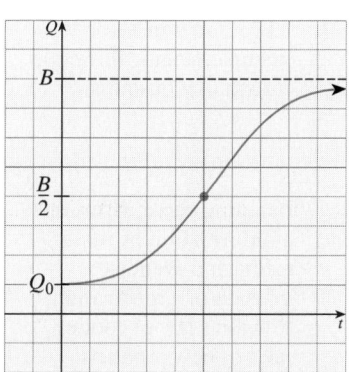

■ **FIGURE 7.7**   A logistic curve: $Q(t) = \dfrac{B}{1 + Ae^{-Bkt}}$

(The details are left for the reader.) This corresponds to the fact that the epidemic is spreading most rapidly when half the susceptible residents have been infected. ■

Note also in Example 3 that $y = Q(t)$ approaches the line $y = B$ asymptotically as $t \to +\infty$. Thus, the number of infected people approaches the number of those susceptible in the long run. The asymptotic value B is also known as the **carrying capacity** of the logistic model.

A summary of growth models is given in Table 7.3 .

■ **TABLE 7.3 Summary of Growth Models**

Graph	Growth Model	Equation and Solution	Sample Applications
$Q(t)$ is the population at time t, Q_0 is the initial population, and k is a constant of proportionality. If $Q(t)$ has a limiting value, let it be denoted by B.			
J-curve	**Uninhibited growth** $(k > 0)$ Rate is proportional to the amount present.	$\dfrac{dQ}{dt} = kQ$ **Solution:** $Q = Q_0 e^{kt}$	Exponential growth; short-term population growth; interest compounded continuously; inflation; price/supply curves
L-curve	**Uninhibited decay** $(k < 0)$ Rate is proportional to the amount present.	$\dfrac{dQ}{dt} = kQ$ **Solution:** $Q = Q_0 e^{kt}$	Radioactive decay; depletion of natural resources; price/demand curves
S-curve	**Inhibited (logistic) growth** Rate is jointly proportional to the amount present and to the difference between the amount present and the maximum amount possible B (called the carrying capacity).	$\dfrac{dQ}{dt} = kQ(B - Q)$ **Solution:** $Q = \dfrac{B}{1 + Ae^{-Bkt}}$ where A is an arbitrary constant.	Long-term population growth (with a limiting value); spread of a disease in a population; sale of fad items (for example, singing flowers); growth of a business
C-curve	**Limited growth** Rate is proportional to the difference between the amount present and a fixed limit $(k > 0)$.	$\dfrac{dQ}{dt} = k(B - Q)$ **Solution:** $Q = B - Ae^{-kt}$ where A is an arbitrary constant.	Learning curve; diffusion of information by mass media; intravenous infusion of a medication; Newton's law of cooling; sales of new products; growth of a business

Modeling Dilution

EXAMPLE 4   *A dilution problem*

A tank contains 20 lb of salt dissolved in 50 gal of water. Suppose 3 gal of brine containing 2 lb of dissolved salt per gallon run into the tank every minute and that the mixture (kept uniform by stirring) runs out of the tank at the rate of 2 gal/min. Find the amount of salt in the tank at any time t. How much salt is in the tank at the end of one hour?

Solution   Let $S(t)$ denote the amount of salt in the tank at the end of t minutes. Because 3 gal of brine flow into the tank each minute and each gallon contains 2 lb of salt, it follows that $(3)(2) = 6$ lb of salt flow into the tank each minute (see Figure 7.8). This is the inflow rate.

Inflow rate of salt: 6 lb/min

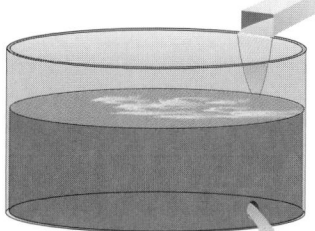

Outflow rate of salt: $\dfrac{S(t)}{50 + t}$ lb / min

■ **FIGURE 7.8**   Rate of flow equals inflow rate minus outflow rate

For the outflow, note that at time t, there are $S(t)$ lb of salt and $50 + (3 - 2)t$ gallons of solution (because solution flows in at 3 gal/min and out at 2 gal/min). Thus, the concentration of salt in the solution at time t is

$$\frac{S(t)}{50 + t} \text{ lb/gal}$$

and the outflow rate of salt is

$$\left[\frac{S(t)}{50 + t} \text{ lb/gal} \right] [2 \text{ gal/min}] = \frac{2S(t)}{50 + t} \text{ lb/min}$$

Combining these observations, we see that the net rate of change of salt $\dfrac{dS}{dt}$ is given by

$$\frac{dS}{dt} = \underset{\substack{\uparrow \\ \text{Inflow}}}{6} - \underset{\text{Outflow}}{\frac{2S}{50 + t}}$$

or

$$\frac{dS}{dt} + \frac{2S}{50 + t} = 6$$

which we recognize as a first-order linear differential equation with

$$P(t) = \frac{2}{50 + t}$$

and $Q(t) = 6$. An integrating factor is

$$I(t) = e^{\int P(t)dt} = e^{\int (2\,dt)/(50+t)} = e^{2\,\ln|50+t|} = (50 + t)^2$$

and the general solution is

$$S(t) = \frac{1}{(50 + t)^2}\left[\int 6(50 + t)^2\,dt + C\right]$$

$$= \frac{1}{(50 + t)^2}[2(50 + t)^3 + C]$$

$$= 2(50 + t) + \frac{C}{(50 + t)^2}$$

To evaluate C, we recall that there are 20 lb of salt initially in the solution. This means $S(0) = 20$, so that

$$S(0) = 2(50 + 0) + \frac{C}{(50 + 0)^2}$$

$$20 = 100 + \frac{C}{50^2}$$

$$-80(50^2) = C$$

Thus, the solution to the given differential equation, subject to the initial condition $S(0) = 20$, is

$$S(t) = 2(50 + t) - \frac{80(50^2)}{(50 + t)^2}$$

Specifically, at the end of 1 hour (60 min), the tank contains

$$S(60) = 2(50 + 60) - \frac{80(50)^2}{(50 + 60)^2} \approx 203.4710744$$

The tank contains about 200 lb of salt. ∎

Modeling *RL* Circuits   Another application of first-order linear differential equations involves the current in an *RL* electric circuit. An *RL* circuit is one with a constant resistance, *R*, and a constant inductance, *L*. Figure 7.9 shows an electric circuit with an electromotive force (EMF), a resistor, and an inductor connected in series. The EMF source, which is usually a battery or generator, supplies voltage that causes a current to flow in the circuit. According to Kirchhoff's second law, if the circuit is closed at time $t = 0$, then the applied electromotive force is equal to the sum of the voltage drops in the rest of the circuit. It can be shown that this implies that the current $I(t)$ that flows in the circuit at time t must satisfy the first-order linear differential equation

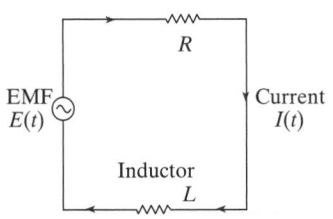

■ **FIGURE 7.9**   An *RL* circuit diagram

$$L\frac{dI}{dt} + RI = E$$

where E is the EMF, and L (the inductance) and R (the resistance) are positive constants.* Write

$$\frac{dI}{dt} + \frac{R}{L}I = \frac{E}{L}$$

to see that $P(t) = \frac{R}{L}$ (a constant) and $Q(t) = \frac{E}{L}$ (a constant). Then,

$$I(t) = e^{-\int (R/L)dt}\left[\int \frac{E}{L}e^{\int (R/L)dt}\,dt + C\right]$$

$$= e^{-(R/L)t}\left[\frac{E}{L}\int e^{(R/L)t}\,dt + C\right]$$

$$= e^{-(R/L)t}\left[\frac{E}{L}\cdot\frac{L}{R}e^{(R/L)t} + C\right]$$

$$= \frac{E}{R} + Ce^{-(R/L)t}$$

It is reasonable to assume that no current flows when $t = 0$. That is, $I = 0$ when $t = 0$, so we have $0 = \frac{E}{R} + C$ or $C = -\frac{E}{R}$. The solution of the initial value problem is

$$I = \frac{E}{R}(1 - e^{-Rt/L})$$

Notice that because $e^{-(Rt/L)} \to 0$ as $t \to +\infty$, we have

$$\lim_{t \to +\infty} I(t) = \frac{E}{R} - \lim_{t \to +\infty}\frac{E}{R}e^{-Rt/L} = \frac{E}{R}$$

This means that, in the long run, the current I must approach $\frac{E}{R}$. The solution of the differential equation consists of two parts, which are given special names:

$$\frac{E}{R} \text{ is the } \textbf{steady-state current.}$$

$$-\frac{E}{R}e^{-Rt/L} \text{ is the } \textbf{transient current.}$$

Figure 7.10 shows how the current $I(t)$ varies with time t.

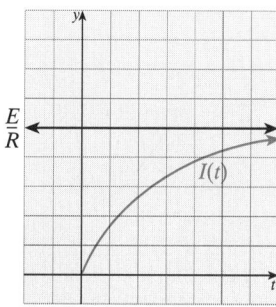

■ **FIGURE 7.10**   The current in an RL circuit with constant EMF

*It is common practice in physics and applied mathematics to use I as the symbol for current. Of course, this has nothing to do with the concept of integrating factor introduced earlier in this section.

7.6 Problem Set

Ⓐ *Solve the differential equation in Problems 1–10.*

1. $\dfrac{dy}{dx} + \dfrac{3y}{x} = x$

2. $\dfrac{dy}{dx} - \dfrac{2y}{x} = \sqrt{x} + 1$

3. $x^4 \dfrac{dy}{dx} + 2x^3 y = 5$

4. $x^2 \dfrac{dy}{dx} + xy = 2$

5. $x \dfrac{dy}{dx} + 2y = xe^{x^3}$

6. $\dfrac{dy}{dx} + \left(\dfrac{2x+1}{x}\right)y = e^{-2x}$

7. $\dfrac{dy}{dx} + \dfrac{y}{x} = \tan^{-1} x$

8. $\dfrac{dy}{dx} + \dfrac{2y}{x} = \dfrac{\ln x}{x}$

9. $\dfrac{dy}{dx} + (\tan x)y = \sin x$

10. $\dfrac{dy}{dx} + (\sec x)y = \sin 2x$

In Problems 11–14, solve the initial value problems. A graphical solution is shown as a check for your work.

11. $\dfrac{dy}{dx} + \dfrac{xy}{1+x} = x(1+x)$
with $y = -1$ when $x = 0$

12. $\dfrac{dy}{dx} + \dfrac{2xy}{1+x^2} = \sin x$
with $y = 1$ when $x = 0$

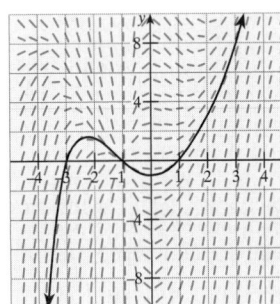 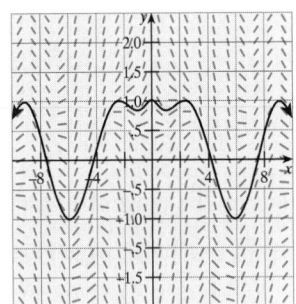

13. $x \dfrac{dy}{dx} - 2y = 2x^3$,
$x > 0$, with $y = 0$
when $x = 3$

14. $y' - \dfrac{y}{x} = x^2 e^{x^2}$,
$x > 0$, with $y = 2$
when $x = 2$

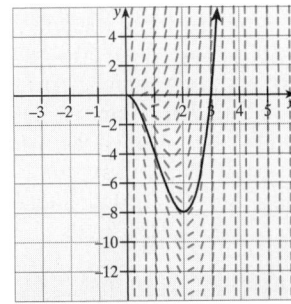 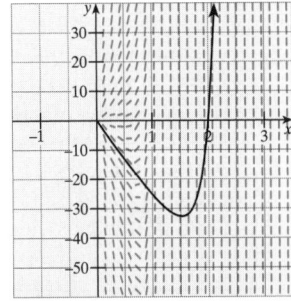

Ⓑ *In Problems 15–18, find the orthogonal trajectories of the given family of curves. Recall from Section 5.6 that a curve is an orthogonal trajectory of a given family if it intersects each member of that family at right angles.*

15. the family of parabolas $y^2 = 4kx$

16. the family of hyperbolas $xy = c$

17. the family of circles $x^2 + y^2 = r^2$

18. the family of exponential curves $y = Ce^{-x}$

MODELING PROBLEMS *In Problems 19–30, set up an appropriate model to answer the given question. Be sure to state your assumptions.*

19. In 1990, the gross domestic product (GDP) of the United States was $5,464 billion. Suppose the growth rate from 1989 to 1990 was 5.08%. Predict the GDP in 2000.

20. In 1980, the gross domestic product (GDP) of the United States in constant 1972 dollars was $1,481 billion. Suppose the growth rate from 1980 to 1984 was 2.5% per year. Predict the GDP in 1992. Consult an almanac to see whether this prediction is correct.

21. According to the Department of Health and Human Services, the divorce rate in 1990 in the United States was 4.7% and there were 1,175,000 divorces that year. How many divorces will there be in 2002 if the divorce rate is constant?

22. According to the Department of Health and Human Services, the marriage rate in 1990 in the United States was 9.8% and there were 2,448,000 marriages that year. How many marriages will there be in 2002 if the marriage rate is constant?

23. A tank contains 10 lb of salt dissolved in 30 gal of water. Suppose 2 gal of brine containing 1 lb of dissolved salt per gallon run into the tank every minute and that the mixture (kept uniform by stirring) runs out at the same rate.
a. Find the amount of salt in the tank at time t.
b. How long does it take (to the nearest second) for the tank to contain 15 lb of salt?

24. In Problem 23, suppose the tank has a capacity of 100 gal and that the mixture flows out at the rate of 1 gal/min (instead of 2 gal/min).
a. How long will it take for the tank to fill?
b. How much salt will be in the tank when it is full?

25. The rate at which a drug is absorbed into the blood system is given by

$$\dfrac{db}{dt} = \alpha - \beta b$$

where $b(t)$ is the concentration of the drug in the bloodstream at time t. What does $b(t)$ approach in the long run (that is, as $t \to +\infty$)? At what time is $b(t)$ equal to half this limiting value? Assume that $b(0) = 0$.

26. An RL circuit has a resistance of R ohms, inductance of L henries, and EMF of E volts, where R, L, and E are constant. Suppose no current flows in the circuit at time

$t = 0$. If L is doubled and E and R are held constant, what effect does this have on the "long-run" current in the curcuit (that is, the current as $t \to +\infty$)?

27. The 1984 census recorded a population of 15,757,000 Hispanics, while in 1990 the figure was 16,098,000. Assuming that the rate of population growth is proportional to the population, predict the Hispanics' population in the year 2000.

28. A population of animals on Catalina Island is limited by the amount of food available. Studies show there were 1,800 animals present in 1980 and 2,000 in 1986, and suggest that 5,000 animals can be supported by the conditions present on the island. Use a logistic model to predict the animal population in the year 2000.

29. In 1986, the Chernobyl nuclear disaster in the Soviet Union contaminated the atmosphere. The buildup of radioactive material in the atmosphere satisfies the differential equation

$$\frac{dM}{dt} = r\left(\frac{k}{r} - M\right), \quad M = 0 \text{ when } t = 0$$

where M = mass of radioactive material in the atmosphere after time t (in years); k is the rate at which the radioactive material is introduced into the atmosphere; r is the annual decay rate of the radioactive material. Find the solution, $M(t)$, of this differential equation in terms of k and r.

30. **The Motion of a Body Falling Through a Resisting Medium**   A body of mass m is dropped from a great height and falls in a straight line. Assume that the only forces acting on the body are the earth's gravitational attraction mg and air resistance kv. (Recall $g = -32 \text{ ft/s}^2$.)
    a. According to Newton's second law,

    $$m\frac{dv}{dt} = mg - kv$$

    Solve this equation assuming the object has velocity $v_0 = 0$ at time $t = 0$.
    b. Find the distance $s(t)$ the body has fallen at time t. Assume $s = 0$ at time $t = 0$.
    c. If the body weighs $W = 100$ lb and $k = 0.35$, how long does it take for the body to reach the ground from a height of 10,000 ft? With what velocity does it hit the ground?

Recall from Section 5.6 that a tank filled with water drains at the rate

$$\frac{dV}{dt} = -4.8 \, A_0\sqrt{h}$$

where h is the height of water at time t (in seconds) and A_0 is the area (in ft^2) of the drain hole. This formula, called Torricelli's law, is used in Problems 31 and 32.

31. A full tank of water has a drain with area 0.07 ft^2. If the tank has a constant cross-sectional area of $A = 5$ ft^2

and height of $h_0 = 4$ ft, how long does it take to empty?

32. A full tank of water of height 5 ft and constant cross-sectional area $A = 3$ ft^2 has two drains, both of area 0.02 ft^2. One drain is at the bottom and the other at height 2 ft. How long does it take for the tank to drain?

33. **The Euler beam model**   For a rigid beam with uniform loading, the deflection $y(x)$ is modeled by the differential equation $y^{iv} = -k$, where $k > 0$ is a constant and x measures the distance along the beam from one of its ends. Assume that $y(0) = y(L) = 0$, where L is the length of the beam, and that there is an inflection point at each support (at $x = 0$ and $x = L$).
    a. Solve the beam equation to find $y(x)$.
    b. Where does the maximum deflection occur? What is the maximum deflection?
    c. Suppose the beam is catilevered, so that $y(0) = y(L) = 0$ and $y''(0) = y'(L) = 0$. Now where does the maximum deflection occur? Is the maximum deflection greater or less than the case considered in part **b**? Would you expect the graph of the deflection $y = y(x)$ to be concave up or concave down on $[0, L]$? Prove your conjecture.

34. **MODELING PROBLEM**   A tank initially contains 5 lb of salt in 50 ft^3 of solution. At time $t = 0$, brine begins to enter the tank at the rate of 2 ft^3/h, and the mixed solution drains at the same rate. The brine coming into the tank has concentration

$$C(t) = 1 - e^{-0.02t} \text{ lb/ft}^3$$

t hours after the dilution begins.
    a. Set up and solve a differential equation for the amount of salt $S(t)$ in the solution at time t.
    b. How much salt is eventually in the tank (as $t \to +\infty$)?
    c. At what time is $S(t)$ maximized? What is the maximum amount of salt in the tank?

35. **MODELING PROBLEM**   Two 100-gallon tanks initially contain pure water. Brine containing 2 lb of salt per gallon enters the first tank at the rate of 1 gal/min, and the mixed solution drains into the second tank at the same rate. There, it is again thoroughly mixed and drains at the same rate, 1 gal/min.
    a. Set up and solve a differential equation of the amount of salt $S_1(t)$ in the first tank at time t (minutes).
    b. Set up and solve a second differential equation for the amount of salt $S_2(t)$ in the second tank at time t.
    c. Let $S(t) = S_1 - S_2$. Intuitively, $S(t) \geq 0$ for all t. At what time is the excess $S(t)$ maximized? What is the maximum excess?

36. **MODELING PROBLEM**   A chemical in a solution diffuses from a compartment with known concentration $C_1(t)$ across a membrane to a second compartment whose concentration $C_2(t)$ changes at a rate proportional to the difference $C_1 - C_2$. Set up and solve the differential equation for $C_2(t)$ in the following cases:
    a. $C_1(t) = 5e^{-2t}; k = 1.7; C_2(0) = 0$

b. $k = 2$; $C_2(0) = 3$, and

$$C_1(t) = \begin{cases} 4 & \text{if } 0 \le t \le 3 \\ 5 & \text{if } t > 3 \end{cases}$$

37. **MODELING PROBLEM**   The Gompertz equation for a population $P(t)$ is

$$\frac{dP}{dt} = kP(B - \ln P)$$

where k and B are positive constants. If the initial population is $P(0) = P_0$, and the ultimate population is

$$\lim_{t \to +\infty} P(t) = P_\infty$$

find $P(t)$.

38. **MODELING PROBLEM**   Uranium-234 (half-life 2.48×10^5 yr) decays to thorium-230 (half-life 80,000 yr).
    a. If $U(t)$ and $T(t)$ are the amounts of uranium and thorium at time t, then

$$\frac{dU}{dt} = -k_1 U, \quad \frac{dT}{dt} = -k_2 T + k_1 U$$

   Solve this system of differential equations to obtain $U(t)$ and $T(t)$.
    b. If we start with 100 g of pure U-234, how much Th-230 will there be after $t = 5{,}000$ yr?

39. Solve the differential equation

$$\frac{dy}{dx} = \frac{1 + y}{xy + e^y(1 + y)}$$

by regarding y as the independent variable (i.e., reverse the roles of x and y).

40. **MODELING PROBLEM**   Certain biological processes occur periodically over the 24 hours of a day. Consider a metabolic excretion that peaks at 6 P.M. and has a minimum at 6 A.M. Suppose the rate of excretion is

$$R(t) = \tfrac{1}{4} - \tfrac{1}{8} \cos \tfrac{\pi}{12}(t - 6)$$

and the rate of intake is $I(t)$ for $0 \le t \le 24$. The patient's body contains 200 g of the substance when $t = 0$.
    a. If $I(t) = 0$ (the patient intakes only water), the amount of substance $Q(t)$ in the patient's body at time t satisfies

$$\frac{dQ}{dt} = 0 - R(t)$$

   Solve this equation for $Q(t)$.
    b. Suppose the patient intakes the substance at a constant rate for part of the day. Specifically,

$$I(t) = \begin{cases} 0.4 \text{ g/h} & \text{for } 10 \le t \le 20 \\ 0 & \text{otherwise} \end{cases}$$

   Set up and solve a differential equation for the amount of substance $Q(t)$ in the patient's body at time t. When is $Q(t)$ maximized?

41. An RL circuit has a resistance of $R = 10$ ohms and an inductance of $L = 5$ henries. Find the current $I(t)$ in the circuit at time t if $I(0) = 0$ and the electromotive force (EMF) is:
    a. $E = 15$ volts
    b. $E = 5e^{-2t} \sin t$

42. An RL circuit has an inductance of $L = 3$ henries and a resistance $R = 6$ ohms in series with an EMF of $E = 50 \sin 30\, t$. Assume $I(0) = 0$.
    a. What is the current $I(t)$ at time t?
    b. What is the transient current? The steady-state current?

43. **MODELING PROBLEM**   A lake has a volume of 6 billion ft^3, and its initial pollutant content is 0.22%. A river whose waters contain only 0.06% pollutants flows into the lake at the rate of 350 million ft^3/day, and another river flows out of the lake also carrying 350 million ft^3/day. Assume that the water in the two rivers and the lake is always well mixed. How long does it take for the pollutant content to be reduced to 0.15%?

ⓒ 44. A **Bernoulli equation** is a differential equation of the form

$$y' + P(x)y = Q(x)y^n$$

where n is a real number, $n \ne 0, n \ne 1$.
    a. Show that the change of variable $u = y^{1-n}$ transforms such an equation into one of the form

$$u'(x) + (1 - n)P(x)u(x) = (1 - n)Q(x)$$

   This transformed equation is a first-order linear differential equation in u and can be solved by the methods of this section, yielding a solution to the given Bernoulli equation.
    b. Use the change of variable suggested to solve the Bernoulli equation

$$y' + \frac{y}{x} = 2y^2$$

45. Consider a curve with the property that when horizontal and vertical lines are drawn from each point $P(x, y)$ on the curve to the coordinate axes, the area A_1 under the curve is *twice* the area A_2 above the curve, as shown in Figure 7.11.

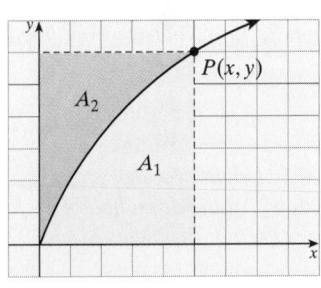

■ **FIGURE 7.11**   Problem 45

Show that x and y satisfy the differential equation

$$\frac{dy}{dx} = \frac{y}{2x}$$

Solve the equation and characterize the family of all curves that satisfy the given geometric condition.

46. **HISTORICAL QUEST** Daniel Bernoulli was a member of the famous Bernoulli family (see Historical Quest Problem 62, Section 4.8). Between 1725 and 1749, he won ten prizes for his work in astronomy, gravity, tides, magnetism, ocean currents, and the behavior of ships at sea. While modeling the effects of a smallpox epidemic, he obtained the differential equation

DANIEL BERNOULLI
1700–1782

$$\frac{dS}{dt} = -pS + \left(\frac{S}{N}\right)\frac{dN}{dt} + \frac{pS^2}{mN}$$

In this equation $S(t)$ is the number of people at age t that are susceptible to smallpox; $N(t)$ is the number of people at age t who survive; p is the probability of a susceptible person getting the disease, and $1/m$ is the proportion of those who die from the disease.

Let $y = \dfrac{S}{N}$ and solve the resulting equation for $y(t)$.

47. **MODELING PROBLEM** An object of mass m is projected upward from ground level with initial velocity v_0 against air resistance proportional to its velocity $v(t)$. Thus,

$$m\frac{dv}{dt} = -mg - kv$$

a. Solve this equation for $v(t)$, then find $s(t)$ the height above the ground at time t.
b. When does the object reach its maximum height? What is the maximum height?
c. Suppose $k = \frac{3}{4}$. For an object weighing 20 lb with initial velocity $v_0 = 150$ ft/s, what is the maximum height? When does the object hit the ground?
d. Suppose the same object as in part **c** is launched with the same initial velocity but in a vacuum, where there is no air resistance ($k = 0$). Would you expect the object to hit the ground sooner or later than the object in part **c**? Prove your conjecture.

48. **MODELING PROBLEM** A body of mass m falls from a height of s_0 ft against air resistance modeled to be proportional to the *square* of its velocity v, so that

$$m\frac{dv}{dt} = mg - kv^2$$

a. Solve this differential equation to find $v(t)$. Then find the height $s(t)$ of the object at time t (in seconds).
b. If the object weighs 1,000 lb, $s_0 = 100$ ft, and $k = 0.01$, how long does it take for the object to hit the ground?

7.7 Improper Integrals

IN THIS SECTION improper integrals with infinite limits of integration, improper integrals with unbounded integrands

We have defined the definite integral $\displaystyle\int_a^b f(x)\,dx$ on a closed bounded interval $[a, b]$ where the integrand $f(x)$ is bounded. In this section, we extend the concept of integral to the case where the interval of integration is infinite and also to the case where f is unbounded at a finite number of points on the interval of integration. Collectively, these are called **improper integrals.**

IMPROPER INTEGRALS WITH INFINITE LIMITS OF INTEGRATION

In physics, economics, probability and statistics, and other applied areas, it is useful to have a concept of integral that is defined on the entire real line or on half-lines of the form $x \geq a$ or $x \leq a$. If $f(x) \geq 0$, the integral of f on the interval $x \geq a$ can be thought of as the area under the curve $y = f(x)$ on this unbounded interval, as shown in Figure 7.12. A reasonable strategy for finding this area is first to use a definite integral to compute the area from $x = a$ to some finite number $x = N$ and then to let N approach infinity in the resulting expression. Here is a definition.

■ **FIGURE 7.12** Finding the area under a curve on an unbounded region

> **Improper Integrals (First Type)**
>
> Let a be a fixed number and assume $\int_a^N f(x)\,dx$ exists for all $N \geq a$. Then if
>
> $\lim\limits_{N \to +\infty} \int_a^N f(x)\,dx$ exists, we define the **improper integral**
>
> $$\int_a^{+\infty} f(x)\,dx = \lim_{N \to +\infty} \int_a^N f(x)\,dx$$
>
> The improper integral $\int_a^{+\infty} f(x)\,dx$ is said to **converge** if this limit is a finite number and to **diverge** otherwise.

EXAMPLE 1   *Converging improper integral*

Evaluate $\int_1^{+\infty} \dfrac{dx}{x^2}$.

Solution The region under the curve $y = \dfrac{1}{x^2}$ for $x \geq 1$ is shown in Figure 7.13. This region is unbounded, so it might seem reasonable to conclude that the area is also infinite. Begin by computing the integral from 1 to N.

$$\text{Consider } \int_1^N \frac{dx}{x^2} = -\frac{1}{x}\Big|_1^N = -\frac{1}{N} + 1$$

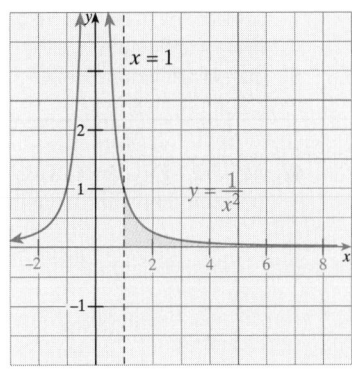

■ **FIGURE 7.13** Graph of $y = \dfrac{1}{x^2}$

Let us consider the situation more carefully, as shown in Figure 7.14. For $N = 2, 3, 10$, or 100, this integral has values $\dfrac{1}{2}, \dfrac{2}{3}, \dfrac{9}{10}$, and $\dfrac{99}{100}$, respectively, suggesting that the region under $y = \dfrac{1}{x^2}$ for $x \geq 1$ actually has finite area.

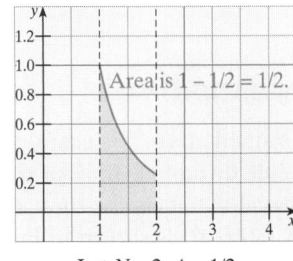

Let $N = 2$, $A = 1/2$.

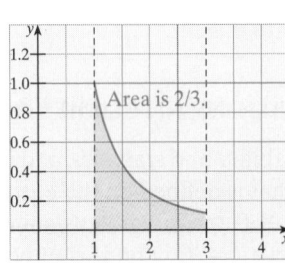

Let $N = 3$, $A = 2/3$.

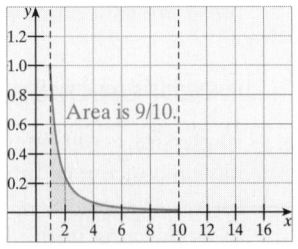

Let $N = 10$, $A = 9/10$.

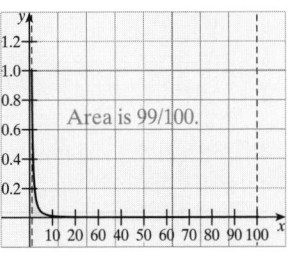

Let $N = 100$, $A = 99/100$.

■ **FIGURE 7.14** Area enclosed by $y = \dfrac{1}{x^2}$, the x-axis, $y = 1$, and $y = N$

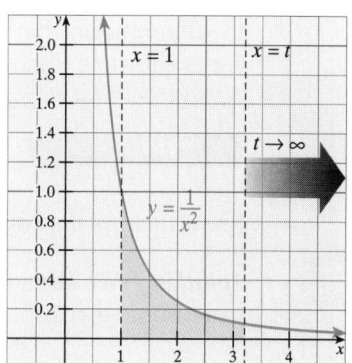

■ **FIGURE 7.15** The region bounded by the curve $y = 1/x^2$, the line $x = 1$, and the x-axis has area 1.

To find this area analytically, we take the limit as $N \to +\infty$, as shown in Figure 7.15:

$$\int_1^{+\infty} \frac{dx}{x^2} = \lim_{N \to +\infty} \int_1^N \frac{dx}{x^2}$$

$$= \lim_{N \to +\infty} \left[-x^{-1} \right] \Big|_1^N$$

$$= \lim_{N \to +\infty} \left[-N^{-1} + 1 \right]$$

$$= 1$$

Thus, the improper integral converges and has the value 1. ■

EXAMPLE 2 *A divergent improper integral*

Evaluate $\int_1^{+\infty} \frac{dx}{x}$.

Solution The graph of $y = 1/x$, shown in Figure 7.16, looks very much like the graph of $y = 1/x^2$ in Figure 7.14. We compute the integral from 1 to N and then let N go to infinity.

$$\int_1^{+\infty} \frac{dx}{x} = \lim_{N \to +\infty} \int_1^N \frac{dx}{x}$$

$$= \lim_{N \to +\infty} \ln |x| \, \Big|_1^N$$

$$= \lim_{N \to +\infty} \ln N = +\infty$$

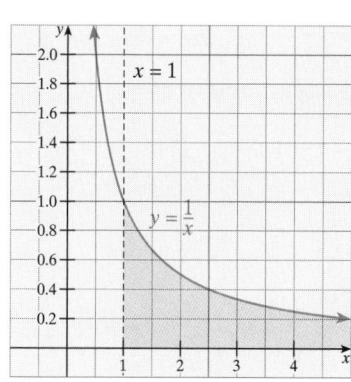

■ **FIGURE 7.16** The area bounded by the curve $y = 1/x$, the line $x = 1$, and the x-axis is infinite.

The limit is not a finite number, which means the improper integral diverges. ■

We have shown that the improper integral

$$\int_1^{+\infty} \frac{dx}{x^2} \text{ converges and that the improper integral } \int_1^{+\infty} \frac{dx}{x} \text{ diverges.}$$

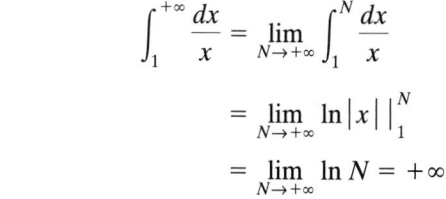

In geometric terms, this says that the area to the right of $x = 1$ under the curve $y = 1/x^2$ is finite, whereas the corresponding area under the curve $1/x$ is infinite. The reason for

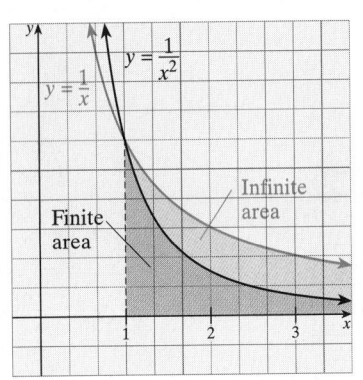

■ **FIGURE 7.17**   An unbounded region may have either a finite or an infinite area.

the difference is that as x increases, $y = 1/x^2$ approaches zero more quickly than does $1/x$ as shown in Figure 7.17.

EXAMPLE 3   *Determine convergence of an improper integral*

Show that the improper integral $\displaystyle\int_1^{+\infty} \frac{dx}{x^p}$ converges only for $p > 1$.

Solution   We already know that the integral diverges for $p = 1$ (Example 2), so we can assume $p \neq 1$.

$$\int_1^{+\infty} \frac{dx}{x^p} = \lim_{N \to +\infty} \int_1^N \frac{dx}{x^p} = \lim_{N \to +\infty} \left[\frac{x^{-p+1}}{-p+1} \right]\Bigg|_1^N = \lim_{N \to +\infty} \frac{1}{1-p}[N^{1-p} - 1]$$

If $p > 1$, then $1 - p < 0$, so $N^{1-p} \to 0$ as $N \to +\infty$, and

$$\int_1^{+\infty} \frac{dx}{x^p} = \lim_{N \to +\infty} \frac{1}{1-p}[N^{1-p} - 1] = \frac{1}{1-p}[-1] = \frac{1}{p-1}$$

If $p < 1$, then $1 - p > 0$, so $N^{1-p} \to +\infty$ as $N \to +\infty$, and

$$\int_1^{+\infty} \frac{dx}{x^p} = \lim_{N \to +\infty} \frac{1}{1-p}[N^{1-p} - 1] = +\infty$$

Thus, the improper integral diverges for $p \leq 1$ and converges for $p > 1$. ■

■ *What This Says:*   The improper integral

$$\int_1^{+\infty} \frac{dx}{x^p} = \begin{cases} \dfrac{1}{p-1} & \text{if } p > 1 \\[2mm] \text{diverges} & \text{if } p \leq 1 \end{cases}$$

EXAMPLE 4   *Improper integral for which technology may give an incorrect result*

Evaluate the given integrals.

a. $\displaystyle\int_1^{\infty} \frac{x^{\sqrt{5}} + 1}{x^{3.236}}\, dx$ b. $\displaystyle\int_1^{\infty} \frac{x^{\sqrt{5}} + 1}{x^{3.24}}\, dx$ c. $\displaystyle\int_1^{\infty} \frac{x^{\sqrt{5}} + 1}{x^{1+\sqrt{5}}}\, dx$

Solution   We note that $\sqrt{5} \approx 2.236067978$; consider $\dfrac{x^{\sqrt{5}} + 1}{x^{3.236}} = \dfrac{1}{x^{3.236-\sqrt{5}}} + \dfrac{1}{x^{3.236}}$.

a. Because $3.236 - \sqrt{5} < 1$, the integral diverges.

b. Because $3.24 - \sqrt{5} > 1$, the integral converges:

$$\int_1^{\infty} \frac{x^{\sqrt{5}} + 1}{x^{3.24}}\, dx = \int_1^{\infty} \left[\frac{1}{x^{3.24-\sqrt{5}}} + \frac{1}{x^{3.24}} \right] dx = \frac{1}{3.24 - \sqrt{5} - 1} + \frac{1}{2.24} \approx 254.768$$

c. Because $1 + \sqrt{5} - \sqrt{5} = 1$, the integral diverges.

Try to check these answers using a computer or calculator with a CAS. It isn't easy—why? ■

EXAMPLE 5   *Improper integral using l'Hôpital's rule*

Evaluate $\displaystyle\int_0^{+\infty} xe^{-2x}\,dx$.

Solution

$$\int_0^{+\infty} xe^{-2x}\,dx = \lim_{N\to+\infty}\int_0^N \underbrace{x}_{u}\,\underbrace{e^{-2x}\,dx}_{dv} \quad \boxed{\begin{array}{l} \text{By parts:} \\ \text{Let } u = x \quad \text{and} \quad dv = e^{-2x}\,dx \\ \qquad du = dx \qquad\qquad v = \tfrac{1}{-2}e^{-2x} \end{array}}$$

$$\overset{uv \qquad\qquad -\quad \int v\,du}{}$$

$$= \lim_{N\to+\infty}\left[\left(\frac{x}{-2}e^{-2x}\right)\Big|_0^N - \int_0^N \frac{1}{-2}e^{-2x}\,dx\right]$$

$$= \lim_{N\to+\infty}\left[\frac{-xe^{-2x}}{2} - \frac{e^{-2x}}{4}\right]\Big|_0^N$$

$$= \lim_{N\to+\infty}\left[-\frac{1}{2}Ne^{-2N} - \frac{1}{4}e^{-2N} + 0 + \frac{1}{4}\right]$$

$$= -\frac{1}{2}\lim_{N\to+\infty}\left(\frac{N}{e^{2N}}\right) + \frac{1}{4} \qquad \text{Because } \lim_{N\to+\infty}(-\tfrac{1}{4}e^{-2N}) = 0 \text{ and } \lim_{N\to+\infty}\tfrac{1}{4} = \tfrac{1}{4}$$

$$\qquad\qquad\qquad\qquad\qquad\qquad \text{l'Hôpital's rule: } \lim_{N\to+\infty}\left(\frac{N}{e^{2N}}\right) = \lim_{N\to+\infty}\left(\frac{1}{2e^{2N}}\right)$$

$$= -\frac{1}{2}\lim_{N\to+\infty}\left(\frac{1}{2e^{2N}}\right) + \frac{1}{4}$$

$$= \frac{1}{4} \qquad\qquad\qquad\qquad\qquad\qquad\qquad\qquad\qquad\blacksquare$$

EXAMPLE 6   *THINK TANK example: Gabriel's horn, a solid with a finite volume but infinite surface area*

Gabriel's horn is the name given to the solid formed by revolving about the *x*-axis the unbounded region under the curve $y = \dfrac{1}{x}$ for $x \geq 1$. Show that this solid has finite volume but infinite surface area.

Solution   We will find the volume by using disks, as shown in Figure 7.18.

$$V = \pi\int_1^{+\infty}(x^{-1})^2\,dx = \pi\lim_{N\to+\infty}\int_1^N x^{-2}\,dx = \pi\lim_{N\to+\infty}\left(-\frac{1}{N} + 1\right) = \pi$$

$$S = 2\pi\int_1^{+\infty} f(x)\sqrt{1 + [f'(x)]^2}\,dx = 2\pi\int_1^{+\infty}\frac{1}{x}\sqrt{1 + \frac{1}{x^4}}\,dx$$

$$= 2\pi\int_1^{+\infty}\frac{1}{x}\sqrt{\frac{x^4 + 1}{x^4}} = 2\pi\int_1^{+\infty}\frac{\sqrt{x^4 + 1}}{x^3}\,dx$$

$$= 2\pi\int_1^{+\infty}\frac{\sqrt{u^2 + 1}}{2u^2}\,du = \pi\lim_{N\to+\infty}\int_1^N\frac{\sqrt{u^2 + 1}}{u^2}\,du \qquad \boxed{\begin{array}{l}\text{Let } u = x^2 \\ \qquad du = 2x\,dx\end{array}}$$

$$= \pi\lim_{N\to+\infty}\left[\frac{-\sqrt{u^2 + 1}}{u} + \ln|u + \sqrt{u^2 + 1}|\right]\Big|_1^N$$

$$= +\infty \quad \text{(The details are left for the reader.)} \qquad\qquad\qquad\qquad\blacksquare$$

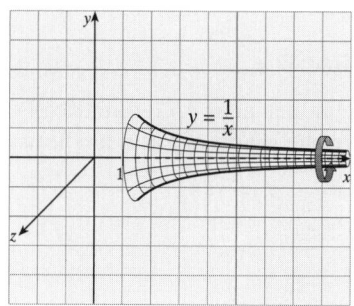

■ **FIGURE 7.18**   Gabriel's horn

You can fill Gabriel's horn with a finite amount of paint, but it takes an infinite amount of paint to color its inside surface!

We can also define an improper integral on an interval that is unbounded to the left or on the entire real number line.

Improper Integrals (First Type, Extended)

Let b be a fixed number and assume $\int_N^b f(x)\, dx$ exists for all $N < b$. Then if

$$\lim_{N \to -\infty} \int_N^b f(x)\, dx \text{ exists, we define the \textbf{improper integral}}$$

$$\int_{-\infty}^b f(x)\, dx = \lim_{N \to -\infty} \int_N^b f(x)\, dx$$

The improper integral $\int_{-\infty}^b f(x)\, dx$ is said to **converge** if this limit is a finite number and to **diverge** otherwise. If both

$$\int_a^{+\infty} f(x)\, dx \quad \text{and} \quad \int_{-\infty}^a f(x)\, dx$$

converge for some number a, the improper integral of $f(x)$ on the entire x-axis is defined by

$$\int_{-\infty}^{+\infty} f(x)\, dx = \int_{-\infty}^a f(x)\, dx + \int_a^{+\infty} f(x)\, dx$$

EXAMPLE 7   *Improper integral to negative infinity*

Find $\displaystyle\int_{-\infty}^0 e^x\, dx$.

Solution

$$\int_{-\infty}^0 e^x\, dx = \lim_{N \to -\infty} \int_N^0 e^x\, dx = \lim_{N \to -\infty} (1 - e^N) = 1 \qquad \blacksquare$$

IMPROPER INTEGRALS WITH UNBOUNDED INTEGRANDS

A function f is **unbounded** at c if it has arbitrarily large values near c. Geometrically, this occurs when the line $x = c$ is a vertical asymptote to the graph of f at c, as shown in Figure 7.19.

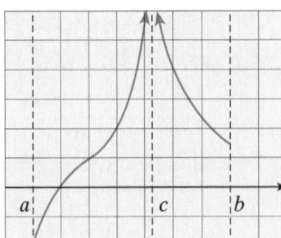 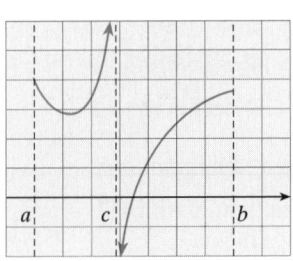

■ **FIGURE 7.19**   Two functions that are unbounded at c

If f is unbounded at c and $a \le c \le b$, the Riemann integral $\int_a^b f(x)\,dx$ is not even defined (only bounded functions are Riemann-integrable). However, it may still be possible to define $\int_a^b f(x)\,dx$ as an improper integral in certain cases.

Let us examine a specific problem. Suppose

$$f(x) = \frac{1}{\sqrt{x}} \quad \text{for } 0 < x \le 1.$$

Then f is unbounded at $x = 0$ and $\int_0^1 f(x)\,dx$ is not defined. However,

$$f(x) = \frac{1}{\sqrt{x}}$$

is continuous on every interval $[t, 1]$ for $t > 0$, as shown in Figure 7.20. For any such interval $[t, 1]$, we have

$$\int_t^1 \frac{dx}{\sqrt{x}} = \int_t^1 x^{-1/2}\,dx = 2\sqrt{x}\,\Big|_t^1 = 2 - 2\sqrt{t}$$

If we let $t \to 0$ through positive values, we see that

$$\lim_{t \to 0^+} \int_t^1 \frac{dx}{\sqrt{x}} = \lim_{t \to 0^+}(2 - 2\sqrt{t}) = 2$$

This is called a **convergent improper integral** with value 2, and it seems reasonable to say

$$\int_0^1 \frac{dx}{\sqrt{x}} = \lim_{t \to 0^+} \int_t^1 \frac{dx}{\sqrt{x}} = 2$$

In this example, f is unbounded at the left endpoint of the interval of integration, but similar reasoning would apply if it were unbounded at the right endpoint or at an interior point. Here is a definition of this kind of improper integral.

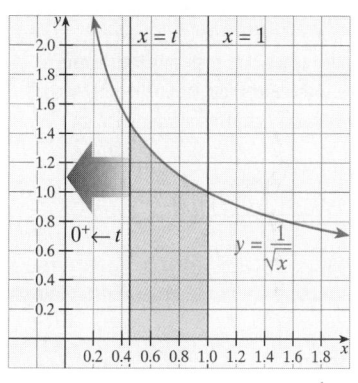

■ **FIGURE 7.20**   Graph of $y = \dfrac{1}{\sqrt{x}}$

Improper Integral (Second Type)

If f is unbounded at a and $\int_t^b f(x)\,dx$ exists for all t such that $a < t \le b$, then

$$\int_a^b f(x)\,dx = \lim_{t \to a^+} \int_t^b f(x)\,dx$$

If the limit exists (as a finite number), we say that the improper integral **converges;** otherwise, the improper integral **diverges.** Similarly, if f is unbounded at b and $\int_a^t f(x)\,dx$ exists for all t such that $a \le t < b$, then

$$\int_a^b f(x)\,dx = \lim_{t \to b^-} \int_a^t f(x)\,dx$$

If f is unbounded at c, where $a < c < b$ and the improper integrals

$$\int_a^c f(x)\,dx \text{ and } \int_c^b f(x)\,dx \text{ both converge, then}$$

$$\int_a^b f(x)\,dx = \int_a^c f(x)\,dx + \int_c^b f(x)\,dx$$

We say that the integral on the left diverges if *either* of the integrals on the right diverges.

■ *What This Says:*   If a continuous function is unbounded at one of its endpoints, then replace that endpoint by t, evaluate the integral, and take the limit as t approaches that endpoint. On the other hand, if f is continuous on the interval $[a, b]$, except for some c in (a, b) where f has an infinite discontinuity, rewrite the integral as

$$\int_a^b f(x)\,dx = \int_a^c f(x)\,dx + \int_c^b f(x)\,dx$$

You will then need limits to evaluate each of the integrals on the right.

EXAMPLE 8   *Improper integral at a right endpoint*

Find $\displaystyle\int_0^1 \frac{dx}{(x-1)^{2/3}}$.

Solution   Let $f(x) = (x-1)^{-2/3}$. This function is unbounded at the right endpoint of the interval of integration and is continuous on $[0, t]$ for any t with $0 \le t \le 1$. We find that

$$\int_0^1 \frac{dx}{(x-1)^{2/3}} = \lim_{t \to 1^-} \int_0^t \frac{dx}{(x-1)^{2/3}} = \lim_{t \to 1^-} \left. [3(x-1)^{1/3}] \right|_0^t$$

$$= 3 \lim_{t \to 1^-} [(t-1)^{1/3} - (-1)] = 3$$

That is, the improper integral converges and has the value 3.   ■

EXAMPLE 9   *Improper integral at a left endpoint*

Find $\displaystyle\int_{\pi/2}^{\pi} \sec x\,dx$.

Solution   Because $\sec x$ is unbounded at the left endpoint $\frac{\pi}{2}$ of the interval of integration and is continuous on $[t, \pi]$ for any t with $\frac{\pi}{2} < t \le \pi$, we find that

$$\int_{\pi/2}^{\pi} \sec x\,dx = \lim_{t \to (\pi/2)^+} \int_t^{\pi} \sec x\,dx = \lim_{t \to (\pi/2)^+} \left. \ln|\sec x + \tan x| \right|_t^{\pi}$$

$$= \lim_{t \to (\pi/2)^+} [\ln|-1+0| - \ln|\sec t + \tan t|] = -\infty$$

Thus, the integral diverges.   ■

EXAMPLE 10   *Improper integral at an interior point*

Find $\displaystyle\int_0^3 (x-2)^{-1}\,dx$.

Solution   The integral is improper because the integrand is unbounded at $x = 2$. If the improper integral converges, we have

$$\int_0^3 (x - 2)^{-1}\, dx = \int_0^2 (x - 2)^{-1}\, dx + \int_2^3 (x - 2)^{-1}\, dx$$

$$= \lim_{t \to 2^-} \int_0^t (x - 2)^{-1}\, dx + \lim_{t \to 2^+} \int_t^3 (x - 2)^{-1}\, dx$$

If either of these limits fails to exist, then the original integral diverges. Because

$$\lim_{t \to 2^-} \int_0^t (x - 2)^{-1}\, dx = \lim_{t \to 2^-} \ln|x - 2|\Big|_0^t$$

$$= \lim_{t \to 2^-} [\ln(2 - t) - \ln 2]$$

$$= -\infty$$

we find that the original integral diverges.

WARNING   A common mistake is to fail to notice the discontinuity at $x = 2$. It is WRONG to write

$$\int_0^3 (x - 2)^{-1} dx$$

$$= \ln|x - 2|\Big|_0^3 = \ln 1 - \ln 2$$

$$= -\ln 2$$

Notice that this mistake leads to the conclusion that the integral converges, which, as you can see, is incorrect.   ←

Technology Window

WARNING   You must be cautious in using computer software with improper integrals, because it may not detect that the integral is improper.   ←

For Example 10,

$$\int_0^3 (x - 2)^{-1}\, dx$$

one popular software package shows:

Type of Estimate	Number of Subintervals	Estimate on $[0, 3]$
Riemann	10	0.8900878
Riemann	100	1.098096
Riemann	1,000	1.118402
Riemann	10,000	1.120427
Trapezoid	10,000	1.120652
Simpson	10,000	2.329852

We must, once again, emphasize the need to understand the mathematics and not rely solely on available technology. That said, a graphing calculator can be an excellent tool for exposing the need for evaluating an integral at an interior point. For Example 10, simply input $y = (x - 2)^{-1}$ and graph:

Using a graph in this fashion may help you to avoid mistakes.

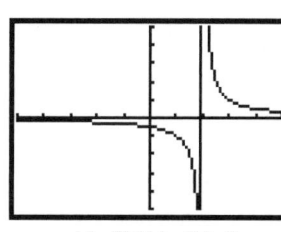

Y₁＝(X-2)⁻¹

Xmin=-5   Ymin=-5
Xmax=5   Ymax=5
Xscl=1   Yscl=1

7.7 Problem Set

A **1.** ■ **What Does This Say?** What is an improper integral?

2. ■ **What Does This Say?** Discuss the different types of improper integrals.

In Problems 3–42, either show that the improper integral converges and find its value, or show that it diverges.

3. $\displaystyle\int_{1}^{+\infty} \frac{dx}{x^3}$

4. $\displaystyle\int_{1}^{+\infty} \frac{dx}{\sqrt[3]{x}}$

5. $\displaystyle\int_{1}^{+\infty} \frac{dx}{x^{0.99}}$

6. $\displaystyle\int_{1}^{+\infty} \frac{dx}{\sqrt{x}}$

7. $\displaystyle\int_{1}^{+\infty} \frac{dx}{x^{1.1}}$

8. $\displaystyle\int_{1}^{+\infty} x^{-2/3}\, dx$

9. $\displaystyle\int_{3}^{+\infty} \frac{dx}{2x-1}$

10. $\displaystyle\int_{3}^{+\infty} \frac{dx}{\sqrt[3]{2x-1}}$

11. $\displaystyle\int_{3}^{+\infty} \frac{dx}{(2x-1)^2}$

12. $\displaystyle\int_{0}^{+\infty} e^{-x}\, dx$

13. $\displaystyle\int_{0}^{+\infty} 5e^{-2x}\, dx$

14. $\displaystyle\int_{1}^{+\infty} e^{1-x}\, dx$

15. $\displaystyle\int_{1}^{+\infty} \frac{x^2\, dx}{(x^3+2)^2}$

16. $\displaystyle\int_{1}^{+\infty} \frac{x^2\, dx}{x^3+2}$

17. $\displaystyle\int_{1}^{+\infty} \frac{x^2\, dx}{\sqrt{x^3+2}}$

18. $\displaystyle\int_{0}^{+\infty} xe^{-x^2}\, dx$

19. $\displaystyle\int_{1}^{+\infty} \frac{e^{-\sqrt{x}}\, dx}{\sqrt{x}}$

20. $\displaystyle\int_{0}^{+\infty} xe^{-x}\, dx$

21. $\displaystyle\int_{0}^{+\infty} 5xe^{10-x}\, dx$

22. $\displaystyle\int_{1}^{+\infty} \frac{\ln x\, dx}{x}$

23. $\displaystyle\int_{2}^{+\infty} \frac{dx}{x\ln x}$

24. $\displaystyle\int_{2}^{+\infty} \frac{dx}{x\sqrt{\ln x}}$

25. $\displaystyle\int_{0}^{+\infty} x^2 e^{-x}\, dx$

26. $\displaystyle\int_{0}^{+\infty} x^3 e^{-x^2}\, dx$

27. $\displaystyle\int_{-\infty}^{0} \frac{2x\, dx}{x^2+1}$

28. $\displaystyle\int_{1}^{+\infty} \frac{x\, dx}{(1+x^2)^2}$

29. $\displaystyle\int_{-\infty}^{0} \frac{dx}{\sqrt{2-x}}$

30. $\displaystyle\int_{-\infty}^{4} \frac{dx}{(5-x)^2}$

31. $\displaystyle\int_{-\infty}^{+\infty} xe^{-|x|}\, dx$

32. $\displaystyle\int_{-\infty}^{+\infty} \frac{dx}{x^2+1}$

33. $\displaystyle\int_{0}^{1} \frac{dx}{x^{1/5}}$

34. $\displaystyle\int_{0}^{4} \frac{dx}{x\sqrt{x}}$

35. $\displaystyle\int_{0}^{1} \frac{dx}{(1-x)^{1/2}}$

36. $\displaystyle\int_{-\infty}^{+\infty} \frac{3x\, dx}{(3x^2+2)^2}$

37. $\displaystyle\int_{0}^{1} \ln x\, dx$

38. $\displaystyle\int_{1}^{+\infty} \ln x\, dx$

39. $\displaystyle\int_{e}^{+\infty} \frac{dx}{x(\ln x)^2}$

40. $\displaystyle\int_{0}^{1} \frac{x\, dx}{1-x^2}$

41. $\displaystyle\int_{0}^{1} e^{-(1/2)\ln x}\, dx$

42. $\displaystyle\int_{0}^{+\infty} \frac{dx}{e^x+e^{-x}}$

B **43.** Find the area of the unbounded region between the x-axis and the curve

$$y = \frac{2}{(x-4)^3} \quad \text{for } x \geq 6$$

44. Find the area of the unbounded region between the x-axis and the curve

$$y = \frac{2}{(x-4)^3} \quad \text{for } x \leq 2$$

45. MODELING PROBLEM   The total amount of radio-active material present in the atmosphere at time T is modeled by

$$A = \int_{0}^{T} Pe^{-rt}\, dt$$

where P is a constant and t is the number of years. Suppose a recent United Nations publication indicates that, at the present time, $r = 0.002$ and $P = 200$ millirads. Estimate the total future buildup of radioactive material in the atmosphere if these values remain constant.

46. MODELING PROBLEM   Suppose that an oil well produces $P(t)$ thousand barrels of crude oil per month according to the formula

$$P(t) = 100e^{-0.02t} - 100e^{-0.1t}$$

where t is the number of months the well has been in production. What is the total amount of oil produced by the oil well?

47. Let $f(x) = \begin{cases} 1/x^2 & \text{for } x \geq 1 \\ 1 & \text{for } -1 < x < 1 \\ e^{x+1} & \text{for } x \leq -1 \end{cases}$

Sketch the graph of f and evaluate

$$\int_{-\infty}^{+\infty} f(x)\, dx$$

48. Find all values of p for which $\displaystyle\int_{2}^{+\infty} \frac{dx}{x(\ln x)^p}$ converges, and find the value of the integral when it exists.

49. Find all values of p for which $\displaystyle\int_{0}^{1} \frac{dx}{x^p}$ converges, and find the value of the integral when it exists.

50. Find all values of p for which $\displaystyle\int_{0}^{1/2} \frac{dx}{x(\ln x)^p}$ converges, and find the value of the integral when it exists.

51. THINK TANK PROBLEM Discuss the calculation

$$\int_{-1}^{1} \frac{dx}{x^2} = \frac{-1}{x}\Big|_{-1}^{1} = -[1-(-1)] = -2$$

Is the calculation correct? Explain.

52. JOURNAL PROBLEM* Peter Lindstrom of North Lake College in Irving, Texas, had a student who handled an ∞/∞ form as follows:

$$\int_{1}^{\infty}(x-1)e^{-x}\,dx = \int_{1}^{\infty}\frac{x-1}{e^x}\,dx$$

$$= \int_{1}^{\infty}\frac{1}{e^x}\,dx \quad \text{l'Hôpital's rule}$$

$$= \frac{1}{e}$$

What is wrong, if anything, with this student's solution?

53. Find $\int_{0}^{2} f(x)\,dx$, where

$$f(x) = \begin{cases} \dfrac{1}{\sqrt[4]{x^3}} & \text{for } 0 \le x \le 1 \\ \dfrac{1}{\sqrt[4]{(x-1)^3}} & \text{for } 1 < x < 2 \end{cases}$$

*The **Laplace transform** of the function f is defined by the improper integral*

$$F(s) = \mathcal{L}\{f(t)\} = \int_{0}^{\infty} e^{-st} f(t)\,dt$$

where s is a constant. This notation is used in Problems 54–60.

54. Show that for constant a (with $s-a>0$):

a. $\mathcal{L}\{e^{at}\} = \dfrac{1}{s-a}$ b. $\mathcal{L}\{a\} = \dfrac{a}{s}$

c. $\mathcal{L}(t) = \dfrac{1}{s^2}$ d. $\mathcal{L}(t^n) = \dfrac{n!}{s^{n+1}}$

e. $\mathcal{L}\{\cos at\} = \dfrac{s}{s^2+a^2}$

f. $\mathcal{L}\{\sin at\} = \dfrac{a}{s^2+a^2}$

55. Show that $\mathcal{L}\{af+bg\} = a\mathcal{L}\{f\} + b\mathcal{L}\{g\}$

56. Let $\mathcal{L}^{-1}\{F(s)\}$ denote the *inverse* Laplace tranform of $F(s)$. Find the following inverse Laplace transformations (Problems 54 and 55 may help).

a. $\mathcal{L}^{-1}\left\{\dfrac{5}{s}\right\}$

b. $\mathcal{L}^{-1}\left\{\dfrac{s+2}{s^2+4}\right\}$

c. $\mathcal{L}^{-1}\left\{\dfrac{2s^2-3s+3}{s^2(s-1)}\right\}$

Hint: Use partial-fraction decomposition.

d. $\mathcal{L}^{-1}\left\{\dfrac{3s^3+2s^2-3s-17}{(s^2+4)(s^2-1)}\right\}$

57. If $F(s) = \mathcal{L}\{f(t)\}$, show that $\mathcal{L}\{e^{at}f(t)\} = F(s-a)$.

58. Use the result in Problem 57, along with those of Problem 54, to find the following Laplace transforms and inverse transforms.

a. $\mathcal{L}\{t^3 e^{-2t}\}$

b. $\mathcal{L}\{e^{-3t}\cos 2t\}$

c. $\mathcal{L}^{-1}\left\{\dfrac{5}{(s-1)^2}\right\}$

d. $\mathcal{L}^{-1}\left\{\dfrac{4s}{s^2+4s+5}\right\}$

Hint: Complete the square.

59. a. If $F(s) = \mathcal{L}\{f(t)\}$, show that

$$\mathcal{L}\{tf(t)\} = -F'(s)$$

You may assume that

$$\frac{d}{ds}\int_{0}^{\infty}e^{-st}f(t)dt = \int_{0}^{\infty}\frac{d}{ds}[e^{-st}f(t)\,dt]$$

b. Use the result to find $\mathcal{L}\{t\cos 2t\}$.

60. What is $\mathcal{L}\{t^2 f(t)\}$? What about $\mathcal{L}\{t^n f(t)\}$?

7.8 The Hyperbolic and Inverse Hyperbolic Functions

IN THIS SECTION hyperbolic functions, derivatives and integrals involving hyperbolic functions, inverse hyperbolic functions

HYPERBOLIC FUNCTIONS

In physics, it is shown that a heavy, flexible cable (for example, a power line) that is suspended between two points at the same height assumes the shape of a curve called a **catenary** (see Figure 7.21), with an equation of the form

$$y = \frac{a}{2}\left(e^{x/a} + e^{-x/a}\right)$$

This is one of several important applications that involve combinations of exponential functions. The goal of this section is to study such combinations and their inverses.

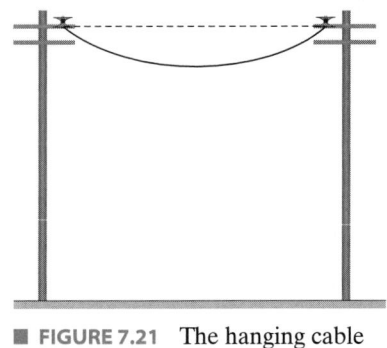

■ **FIGURE 7.21**   The hanging cable problem

SMH   See Problems 46 and 47, p. 77

In certain ways, the functions we shall study are analogous to the trigonometric functions, and they have essentially the same relationship to the hyperbola that the trigonometric functions have to the circle. For this reason, these functions are called **hyperbolic functions.** Three basic hyperbolic functions are the **hyperbolic sine** (denoted sinh x and pronounced "cinch"), the **hyperbolic cosine** (cosh x; pronounced "kosh"), and the **hyperbolic tangent** (tanh x; pronounced "tansh"). They are defined as follows.

Hyperbolic Functions

$$\sinh x = \frac{e^x - e^{-x}}{2} \qquad \text{for all } x$$

$$\cosh x = \frac{e^x + e^{-x}}{2} \qquad \text{for all } x$$

$$\tanh x = \frac{\sinh x}{\cosh x} = \frac{e^x - e^{-x}}{e^x + e^{-x}} \qquad \text{for all } x$$

Graphs of the three basic hyperbolic functions are shown in Figure 7.22.

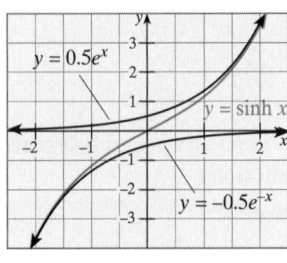

a. The hyperbolic sine
$y = \sinh x$

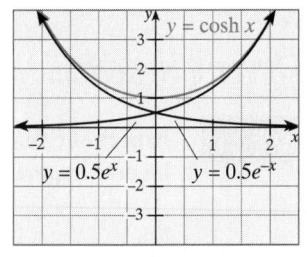

b. The hyperbolic cosine
$y = \cosh x$

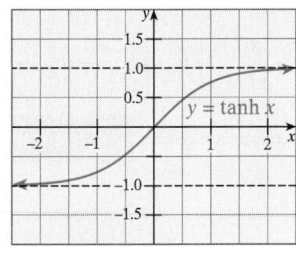

c. The hyperbolic tangent
$y = \tanh x$

■ **FIGURE 7.22**   Graphs of the three basic hyperbolic functions

The list of properties in the following theorem suggests that the basic hyperbolic functions are analogous to the trigonometric functions.

WARNING   A major difference between the trigonometric and hyperbolic functions is that the trigonometric functions are periodic, but the hyperbolic functions are not. ⊸

THEOREM 7.2   *Properties of the basic hyperbolic functions*

$$\cosh^2 x - \sinh^2 x = 1$$
$$\sinh(-x) = -\sinh x \quad (\sinh x \text{ is odd})$$
$$\cosh(-x) = \cosh x \quad (\cosh x \text{ is even})$$
$$\tanh(-x) = -\tanh x \quad (\tanh x \text{ is odd})$$
$$\sinh(x + y) = \sinh x \cosh y + \cosh x \sinh y$$
$$\cosh(x + y) = \cosh x \cosh y + \sinh x \sinh y$$

Proof   We will verify the first identity and leave the others for the reader.

$$\cosh^2 x - \sinh^2 x = \left(\frac{e^x + e^{-x}}{2}\right)^2 - \left(\frac{e^x - e^{-x}}{2}\right)^2$$

$$= \frac{e^{2x} + 2 + e^{-2x}}{4} - \frac{e^{2x} - 2 + e^{-2x}}{4} = 1 \qquad \blacksquare$$

There are three additional hyperbolic functions: the **hyperbolic cotangent** (coth x), the **hyperbolic secant** (sech x), and the **hyperbolic cosecant** (csch x). These functions are defined as follows:

$$\coth x = \frac{1}{\tanh x} = \frac{e^x + e^{-x}}{e^x - e^{-x}}$$

$$\text{sech } x = \frac{1}{\cosh x} = \frac{2}{e^x + e^{-x}}$$

$$\text{csch } x = \frac{1}{\sinh x} = \frac{2}{e^x - e^{-x}}$$

Two identities involving these functions are

$$\text{sech}^2 x = 1 - \tanh^2 x \quad \text{and} \quad \text{csch}^2 x = \coth^2 x - 1$$

You will be asked to verify these identities in the problems.

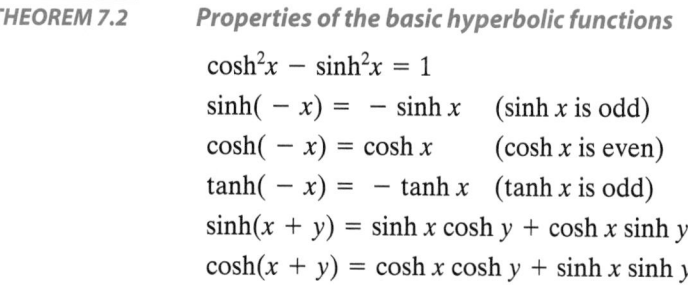

Technology Window

In Derive, these simplified forms are shown:

$$\coth x = \frac{e^{2x} + 1}{e^{2x} - 1}$$

$$\text{sech } x = \frac{2e^x}{e^{2x} + 1}$$

$$\text{csch } x = \frac{2e^x}{e^{2x} - 1}$$

DERIVATIVES AND INTEGRALS INVOLVING HYPERBOLIC FUNCTIONS

Rules for differentiating the hyperbolic functions are listed in Theorem 7.3.

THEOREM 7.3   *Rules for differentiating hyperbolic functions*

Let u be a differentiable function of x. Then

$$\frac{d}{dx}(\sinh u) = \cosh u \,\frac{du}{dx} \qquad\qquad \frac{d}{dx}(\cosh u) = \sinh u \,\frac{du}{dx}$$

$$\frac{d}{dx}(\tanh u) = \text{sech}^2 u \,\frac{du}{dx} \qquad\qquad \frac{d}{dx}(\coth u) = -\text{csch}^2 u \,\frac{du}{dx}$$

$$\frac{d}{dx}(\text{sech } u) = -\text{sech } u \tanh u \,\frac{du}{dx} \qquad \frac{d}{dx}(\text{csch } u) = -\text{csch } u \coth u \,\frac{du}{dx}$$

Proof   Each of these rules can be obtained by differentiating the exponential functions that make up the appropriate hyperbolic function. For example, to differentiate $\sinh x$, we use the definition of $\sinh x$:

$$\frac{d}{dx}(\sinh x) = \frac{d}{dx}\frac{e^x - e^{-x}}{2}$$

$$= \frac{1}{2}e^x - \frac{1}{2}e^{-x}(-1) = \frac{1}{2}(e^x + e^{-x}) = \cosh x$$

The proofs for the other derivatives can be handled similarly.   ▬

<hr>

EXAMPLE 1   *Derivatives involving hyperbolic functions*

Find $\dfrac{dy}{dx}$ for each of the following functions:

a. $y = \cosh Ax$ (A is constant)   b. $y = \tanh(x^2 + 1)$ c. $y = \ln(\sinh x)$

Solution
a. We have

$$\frac{d}{dx}(\cosh Ax) = \sinh(Ax)\frac{d}{dx}(Ax) = A\sinh Ax$$

b. We find that

$$\frac{d}{dx}[\tanh(x^2 + 1)] = \operatorname{sech}^2(x^2 + 1)\frac{d}{dx}(x^2 + 1) = 2x\operatorname{sech}^2(x^2 + 1)$$

c. Using the chain rule, with $u = \sinh x$, we obtain

$$\frac{d}{dx}[\ln(\sinh x)] = \frac{1}{\sinh x}\frac{d}{dx}(\sinh x) = \frac{1}{\sinh x}(\cosh x) = \coth x$$ ■

Each differentiation formula for hyperbolic functions corresponds to an integration formula. These formulas are listed in the following theorem.

<hr>

THEOREM 7.4   *Integration formulas involving the hyperbolic functions*

$$\int \sinh x \, dx = \cosh x + C \qquad\qquad \int \cosh x \, dx = \sinh x + C$$

$$\int \operatorname{sech}^2 x \, dx = \tanh x + C \qquad\qquad \int \operatorname{csch}^2 x \, dx = -\coth x + C$$

$$\int \operatorname{sech} x \tanh x \, dx = -\operatorname{sech} x + C \qquad \int \operatorname{csch} x \coth x \, dx = -\operatorname{csch} x + C$$

Proof   The proof of each of these formulas follows directly from the corresponding derivative formula.   ▬

<hr>

EXAMPLE 2   *Integration involving hyperbolic forms*

Find each of the following integrals:

a. $\displaystyle\int \cosh^3 x \sinh x \, dx$ b. $\displaystyle\int x \operatorname{sech}^2(x^2)\,dx$ c. $\displaystyle\int \tanh x \, dx$

Solution

a. $\displaystyle\int \cosh^3 x \sinh x \, dx = \int u^3 \, du = \frac{u^4}{4} + C = \frac{1}{4} \cosh^4 x + C$

> Let $u = \cosh x$;
> $du = \sinh x \, dx$.

b. $\displaystyle\int x \operatorname{sech}^2(x^2) \, dx = \int \operatorname{sech}^2(x^2)\,(x\,dx)$

> Let $u = x^2$;
> $du = 2x \, dx$.

$\displaystyle = \int \operatorname{sech}^2 u \left(\frac{1}{2} \, du\right)$

$\displaystyle = \tfrac{1}{2} \tanh u + C$

$\displaystyle = \tfrac{1}{2} \tanh x^2 + C$

c. $\displaystyle\int \tanh x \, dx = \int \frac{\sinh x \, dx}{\cosh x}$

> Let $u = \cosh x$;
> $du = \sinh x \, dx$.

$\displaystyle = \int \frac{du}{u} = \ln|u| + C = \ln(\cosh x) + C$ ∎

INVERSE HYPERBOLIC FUNCTIONS

Inverse hyperbolic functions are also of interest, mainly because they enable us to expres certain integrals in simple terms.

Because $\sinh x$ is continuous and strictly monotonic (increasing), it is one-to-one and has an inverse, which is defined by

$$y = \sinh^{-1} x \quad \text{if and only if} \quad x = \sinh y$$

for all x and y. This is called the **inverse hyperbolic sine** function, and its graph is obtained by reflecting the graph of $y = \sinh x$ in the line $y = x$, as shown in Figure 7.23. Other inverse hyperbolic functions are defined similarly (see Problem 60).

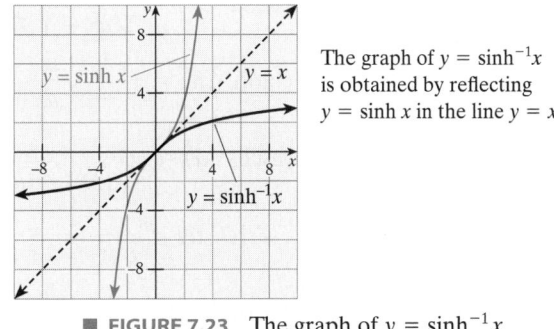

The graph of $y = \sinh^{-1} x$ is obtained by reflecting $y = \sinh x$ in the line $y = x$.

■ **FIGURE 7.23** The graph of $y = \sinh^{-1} x$

Because the hyperbolic functions are defined in terms of exponential functions, we may expect to be able to express the inverse hyperbolic functions in terms of logarithmic functions. We summarize this relationship in the following theorem.

THEOREM 7.5   *Logarithmic formulas for inverse hyperbolic functions*

$$\sinh^{-1}x = \ln(x + \sqrt{x^2 + 1}), \text{ all } x \qquad \operatorname{csch}^{-1}x = \ln\left(\frac{1}{x} + \frac{\sqrt{1 + x^2}}{|x|}\right), x \neq 0$$

$$\cosh^{-1}x = \ln(x + \sqrt{x^2 - 1}), x \geq 1 \qquad \operatorname{sech}^{-1}x = \ln\left(\frac{1 + \sqrt{1 - x^2}}{x}\right), 0 < x \leq 1$$

$$\tanh^{-1}x = \frac{1}{2}\ln\frac{1 + x}{1 - x}, |x| < 1 \qquad \coth^{-1}x = \frac{1}{2}\ln\frac{x + 1}{x - 1}, |x| > 1$$

Proof   We will prove the first part and leave the next two parts for you to verify (see Problem 61). The other parts are proved similarly. Let $y = \sinh^{-1}x$; then its inverse is

$$x = \sinh y$$
$$= \tfrac{1}{2}(e^y - e^{-y})$$
$$2x = e^y - \frac{1}{e^y}$$
$$e^{2y} - 2xe^y - 1 = 0 \qquad \text{Quadratic formula with}$$
$$e^y = \frac{2x \pm \sqrt{4x^2 + 4}}{2} \qquad a = 1, b = -2x, c = -1$$
$$= x \pm \sqrt{x^2 + 1}$$

Because $e^y > 0$ for all y, the only solution is $e^y = x + \sqrt{x^2 + 1}$, and from the definition of logarithms,

$$y = \ln(x + \sqrt{x^2 + 1})$$

Differentiation and integration formulas involving inverse hyperbolic functions are listed in Theorem 7.6.

THEOREM 7.6   *Differentiation and integraion formulas involving the inverse hyperbolic functions*

$$\frac{d}{dx}(\sinh^{-1}u) = \frac{1}{\sqrt{1 + u^2}}\frac{du}{dx} \qquad \int \frac{du}{\sqrt{1 + u^2}} = \sinh^{-1}u + C$$

$$\frac{d}{dx}(\cosh^{-1}u) = \frac{1}{\sqrt{u^2 - 1}}\frac{du}{dx} \qquad \int \frac{du}{\sqrt{u^2 - 1}} = \cosh^{-1}u + C$$

$$\frac{d}{dx}(\tanh^{-1}u) = \frac{1}{1 - u^2}\frac{du}{dx} \qquad \int \frac{du}{1 - u^2} = \tanh^{-1}u + C$$

$$\frac{d}{dx}(\operatorname{csch}^{-1}u) = \frac{-1}{|u|\sqrt{1 + u^2}}\frac{du}{dx} \qquad \int \frac{du}{u\sqrt{1 + u^2}} = -\operatorname{csch}^{-1}|u| + C$$

$$\frac{d}{dx}(\operatorname{sech}^{-1}u) = \frac{-1}{u\sqrt{1 - u^2}}\frac{du}{dx} \qquad \int \frac{du}{u\sqrt{1 - u^2}} = -\operatorname{sech}^{-1}|u| + C$$

$$\frac{d}{dx}(\coth^{-1}u) = \frac{1}{1 - u^2}\frac{du}{dx} \qquad \int \frac{du}{1 - u^2} = \coth^{-1}u + C$$

Proof   The derivative of each inverse hyperbolic function can be found either by differentiating the appropriate logarithmic function or by using the definition in terms of hyperbolic functions. We will show how this is done for $y = \sinh^{-1}x$, and then will leave it for you to apply the chain rule for u, a differentiable function of x.

By definition of inverse, we see

$$x = \sinh y \quad \text{and} \quad \frac{dx}{dy} = \cosh y$$

Thus,

$$\frac{d}{dx}(\sinh^{-1} x) = \frac{dy}{dx} = \frac{1}{\dfrac{dx}{dy}} = \frac{1}{\cosh y} = \frac{1}{\sqrt{1 + \sinh^2 y}} = \frac{1}{\sqrt{1 + x^2}}$$

since $\cosh^2 y - \sinh^2 y = 1$ and $\sinh y = x$. The integration formula that corresponds to this differentiation formula is

$$\int \frac{dx}{\sqrt{1 + x^2}} = \sinh^{-1} x + C$$

The other differentiation and integration formulas follow similarly.

EXAMPLE 3   *Derivatives involving inverse hyperbolic functions*

Find $\dfrac{dy}{dx}$ for   a. $y = \sinh^{-1}(ax + b)$ b. $y = \cosh^{-1}(\sec x), 0 \le x < \dfrac{\pi}{2}$

Solution

a. $\dfrac{d}{dx}[\sinh^{-1}(ax + b)] = \dfrac{1}{\sqrt{1 + (ax + b)^2}} \dfrac{d}{dx}(ax + b)$

$$= \frac{a}{\sqrt{1 + (ax + b)^2}}$$

b. $\dfrac{d}{dx}[\cosh^{-1}(\sec x)] = \dfrac{1}{\sqrt{\sec^2 x - 1}} \dfrac{d}{dx}(\sec x)$

$$= \frac{\sec x \tan x}{\sqrt{\tan^2 x}} = \sec x$$

The result of Example 3b implies

$$\int \sec x \, dx = \cosh^{-1}(\sec x) + C$$

$$= \ln(\sec x = \sqrt{\sec^2 x - 1}) + C$$

$$= \ln(\sec x + \tan x) + C$$

Notice that we do not need absolute values because $0 \le x < \frac{\pi}{2}$. We will use this formula in the next chapter.

EXAMPLE 4   *Integral involving an inverse hyperbolic function*

Evaluate $\displaystyle\int_0^1 \frac{dx}{\sqrt{1 + x^2}}$.

Solution

$$\int_0^1 \frac{dx}{\sqrt{1 + x^2}} = \left[\sinh^{-1} x\right]\Big|_0^1$$

$$= \ln(x + \sqrt{x^2 + 1})\Big|_0^1$$

$$= \ln(1 + \sqrt{2}) - \ln(1)$$

$$= \ln(1 + \sqrt{2})$$

7.8 Problem Set

Ⓐ *Evaluate (correct to four decimal places) the indicated hyperbolic or inverse hyperbolic function in Problems 1–12.*

1. $\sinh 2$
2. $\cosh 3$
3. $\tanh(-1)$
4. $\sinh^{-1}0$
5. $\coth 1.2$
6. $\tanh^{-1}0$
7. $\cosh^{-1}1.5$
8. $\sinh(\ln 2)$
9. $\cosh(\ln 3)$
10. $\operatorname{sech}^{-1}0.2$
11. $\operatorname{sech} 1$
12. $\coth^{-1}(-3)$

Find $\dfrac{dy}{dx}$ in Problems 13–29.

13. $y = \sinh 3x$
14. $y = \cosh(1 - 2x^2)$
15. $y = \cosh(2x^2 + 3x)$
16. $y = \sinh \sqrt{x}$
17. $y = \sinh \dfrac{1}{x}$
18. $y = \cosh^{-1}(x^2)$
19. $y = \sinh^{-1}(x^3)$
20. $y = x \tanh^{-1}(3x)$
21. $y = \sinh^{-1}(\tan x)$
22. $y = \cosh^{-1}(\sec x)$
23. $y = \tanh^{-1}(\sin x)$
24. $y = \operatorname{sech}\left(\dfrac{1 - x}{1 + x}\right)$
25. $y = \dfrac{\sinh^{-1}x}{x}$
26. $y = \sinh^{-1}x - \sqrt{1 + x^2}$
27. $y = x \cosh^{-1}x - \sqrt{x^2 - 1}$
28. $x \cosh y = y \sinh x + 5$
29. $e^x \sinh^{-1}x + e^{-x} \cosh^{-1} y = 1$

Compute the integrals in Problems 30–45.

30. $\displaystyle\int x \cosh(1 - x^2)dx$
31. $\displaystyle\int \dfrac{\sinh(1/x)dx}{x^2}$
32. $\displaystyle\int \dfrac{\operatorname{sech}^2(\ln x)\,dx}{x}$
33. $\displaystyle\int \coth x\, dx$
34. $\displaystyle\int \dfrac{dx}{\sqrt{4x^2 + 16}}$
35. $\displaystyle\int \dfrac{dt}{\sqrt{9t^2 - 16}}$
36. $\displaystyle\int \dfrac{dt}{36 - 16t^2}$
37. $\displaystyle\int \dfrac{\cos x\, dx}{\sqrt{1 + \sin^2 x}}$
38. $\displaystyle\int \dfrac{x\, dx}{\sqrt{1 + x^4}}$
39. $\displaystyle\int \dfrac{x^2\, dx}{1 - x^6}$
40. $\displaystyle\int_0^{1/2} \dfrac{dx}{1 - x^2}$
41. $\displaystyle\int_2^3 \dfrac{dx}{1 - x^2}$
42. $\displaystyle\int_0^1 \dfrac{t^5\, dt}{\sqrt{1 + t^{12}}}$
43. $\displaystyle\int_1^2 \dfrac{e^x\, dx}{\sqrt{e^{2x} - 1}}$
44. $\displaystyle\int_0^{\ln 2} \sinh 3x\, dx$
45. $\displaystyle\int_0^1 x \operatorname{sech}^2 x^2\, dx$

Ⓑ 46. Show that
   a. $\tanh(x + y) = \dfrac{\tanh x + \tanh y}{1 + \tanh x \tan y}$
   b. $\sinh 2x = 2 \sinh x \cosh x$
   c. $\cosh 2x = \cosh^2 x + \sinh^2 x$

47. Evaluate
$$\lim_{x \to +\infty} \dfrac{\sinh^{-1}x}{\cosh^{-1}x}$$

48. Show that
   a. $-1 < \tanh x < 1$ for all x
   b. $\displaystyle\lim_{x \to +\infty} \tanh x = 1$ and
   $$\lim_{x \to -\infty} \tanh x = -1$$

49. Determine where the graph of $y = \tanh x$ is rising and falling and where it is concave up and concave down. Sketch the graph and compare with Figure 7.22c.

50. Sketch the graph of $y = \coth x$. Be sure to show key features such as intercepts, high and low points, and points of inflection.

51. First show that $\cosh x + \sinh x = e^x$, and then use this result to prove that
   $$(\sinh x + \cosh x)^n = \cosh nx + \sinh nx$$
   for positive integers n.

52. If $x = a \cosh t$ and $y = b \sinh t$ where a, b are positive constants and t is any number, show that
   $$\dfrac{x^2}{a^2} - \dfrac{y^2}{b^2} = 1$$

53. a. Verify that $y = a \cosh cx + b \sinh cx$ satisfies the differential equation
   $$y'' - c^2 y = 0$$
   b. Use part **a** to find a solution of the differential equation
   $$y'' - 4y = 0$$
   subject to the initial conditions $y(0) = 1, y'(0) = 2$.

54. Find the volume V generated when the region bounded by the curves $y = \sinh x, y = \cosh x$, the y-axis, and the line $x = c\ (c > 0)$ is revolved about the x-axis. For what value of c does $V = 1$?

55. Find the length of the catenary
   $$y = a \cosh \dfrac{x}{a}$$
   between $x = -a$ and $x = a$.

56. Find the volume of the solid formed by revolving the region bounded by the curve $y = \tanh x$ on the interval $[0, 1]$ about the x-axis.

57. Find the surface area of the solid generated by revolving the curve $y = \cosh x$ on the interval $[-1, 1]$ about the x-axis.

C **58.** Prove the following formulas:
   a. $\operatorname{sech}^2 x + \tanh^2 x = 1$
   b. $\coth^2 x - \operatorname{csch}^2 x = 1$

59. Derive the differentiation formulas for $\cosh u$, $\tanh u$, and $\operatorname{sech} u$, where u is a differentiable function of x.

60. First give definitions for $\cosh^{-1} x$, $\tanh^{-1} x$, and $\operatorname{sech}^{-1} x$, and then derive the differentiation formulas for these functions by using these definitions.

61. Prove
   a. $\cosh^{-1} x = \ln(x + \sqrt{x^2 - 1})$
   b. $\tanh^{-1} x = \dfrac{1}{2} \ln \dfrac{1 + x}{1 - x}$

Chapter 7 *Review*

Proficiency Examination

Concept Problems

1. Discuss the method of integration by substitution for both indefinite and definite integrals.

2. What is the formula for integration by parts?

3. What is a reduction integration formula?

4. a. When should you consider a trigonometric substitution?
   b. What are the substitutions to use when integrating rational trigonometric integrals?

5. When should you consider the method of partial fractions?

6. Outline a strategy for integration.

7. What is a first-order linear differential equation? Outline a procedure for solving such an equation.

8. What is an improper integral?

9. Define the hyperbolic sine, hyperbolic cosine, and hyperbolic tangent.

10. State the rules for differentiating and integrating the hyperbolic functions.

11. What are the logarithmic formulas for the inverse hyperbolic functions?

12. State the differentiation and integration formulas for the inverse hyperbolic functions.

Practice Problems

13. Evaluate each of the given numbers.
   a. $\tanh^{-1}(0.5)$ b. $\sinh (\ln 3)$ c. $\coth^{-1}(2)$

Find the integrals in Problems 14–19.

14. $\displaystyle\int \frac{2x + 3}{\sqrt{x^2 + 1}}\, dx$

15. $\displaystyle\int x \sin 2x\, dx$

16. $\displaystyle\int \sinh (1 - 2x)\, dx$

17. $\displaystyle\int \frac{dx}{\sqrt{4 - x^2}}$

18. $\displaystyle\int \frac{x^2\, dx}{(x^2 + 1)(x - 1)}$

19. $\displaystyle\int \frac{x^3\, dx}{x^2 - 1}$

Evaluate the definite integrals in Problems 20–23.

20. $\displaystyle\int_1^2 x \ln x^3\, dx$

21. $\displaystyle\int_2^3 \frac{dx}{(x - 1)^2 (x + 2)}$

22. $\displaystyle\int_3^4 \frac{dx}{2x - x^2}$

23. $\displaystyle\int_0^{\pi/4} \sec^3 x \tan x\, dx$

Determine whether each improper integral in Problems 24–27 *converges, and if it does, find its value.*

24. $\displaystyle\int_0^{+\infty} x e^{-2x}\, dx$

25. $\displaystyle\int_0^{\pi/4} \frac{\sec^2 x}{\sqrt{\tan x}}\, dx$

26. $\displaystyle\int_0^1 \frac{2x + 3}{x^2(x - 2)}\, dx$

27. $\displaystyle\int_0^{+\infty} e^{-x} \sin x\, dx$

28. Find $\dfrac{dy}{dx}$ for $y = \sqrt{\tanh^{-1} 2x}$.

29. Find the volume of the solid generated when the region under the curve

$$y = \frac{1}{\sqrt{9 - x^2}}$$

on the interval $[0, 2]$ is revolved about the y-axis.

30. Solve the first-order linear differential equation

$$\frac{dy}{dx} + \frac{xy}{x + 1} = e^{-x}$$

subject to the condition $y = 1$ when $x = 0$.

31. A tank contains 200 gal of saturated brine with 2 lb of salt per gallon. A salt solution containing 1.3 lb of salt per gallon flows in at the rate of 5 gal/min, and the uniform mixture flows out at 3 gal/min. Find the amount of salt in the tank after 1 hour.

Supplementary Problems

Find the derivative $\dfrac{dy}{dx}$ in Problems 1–5.

1. $y = \tanh^{-1}\dfrac{1}{x}$

2. $y = x \cosh^{-1}(3x + 1)$

3. $y = \dfrac{\sinh x}{e^x}$

4. $y = \sinh x \cosh x$

5. $y = x \sinh x + \sinh(e^x + e^{-x})$

Find the integrals in Problems 6–41.

6. $\displaystyle\int \cos^{-1}x\, dx$

7. $\displaystyle\int \dfrac{x^2\, dx}{\sqrt{4 - x^2}}$

8. $\displaystyle\int \dfrac{x\, dx}{x^2 - 2x + 5}$

9. $\displaystyle\int \dfrac{3x - 2}{x^3 - 2x^2}\, dx$

10. $\displaystyle\int \dfrac{dx}{(x^2 + x + 1)^{3/2}}$

11. $\displaystyle\int \dfrac{dx}{\sqrt{x}\,(1 + \sqrt[4]{x})}$

12. $\displaystyle\int \dfrac{\sqrt{9x^2 - 1}\, dx}{x}$

13. $\displaystyle\int x^2 \tan^{-1}x\, dx$

14. $\displaystyle\int \dfrac{dx}{\sin x + \tan x}$

15. $\displaystyle\int e^x \sqrt{4 - e^{2x}}\, dx$

16. $\displaystyle\int \cos\dfrac{x}{2} \sin\dfrac{x}{3}\, dx$

17. $\displaystyle\int \dfrac{\sqrt{1 + 1/x^2}}{x^5}\, dx$

18. $\displaystyle\int \dfrac{\sin x\, dx}{\cos^5 x}$

19. $\displaystyle\int \sqrt{1 + \sin x}\, dx$

20. $\displaystyle\int \cos x \ln(\sin x)\, dx$

21. $\displaystyle\int \sin(\ln x)\, dx$

22. $\displaystyle\int e^{2x} \operatorname{sech}(e^{2x})\, dx$

23. $\displaystyle\int \dfrac{\sinh x\, dx}{2 + \cosh x}$

24. $\displaystyle\int \dfrac{\tanh^{-1}x\, dx}{1 - x^2}$

25. $\displaystyle\int x^2 \cot^{-1}x\, dx$

26. $\displaystyle\int x(1 + x)^{1/3}\, dx$

27. $\displaystyle\int \dfrac{x^2 + 2}{x^3 + 6x + 1}\, dx$

28. $\displaystyle\int \dfrac{\sin x - \cos x}{(\sin x + \cos x)^{1/4}}\, dx$

29. $\displaystyle\int \cos(\sqrt{x} + 2)\, dx$

30. $\displaystyle\int \sqrt{5 + 2\sin^2 x}\, \sin 2x\, dx$

31. $\displaystyle\int \dfrac{x^3 + 2x}{x^4 + 4x^2 + 3}\, dx$

32. $\displaystyle\int \dfrac{x\, dx}{\sqrt{5 - x^2}}$

33. $\displaystyle\int \dfrac{\sqrt{5 - x^2}\, dx}{x}$

34. $\displaystyle\int \dfrac{\sqrt{x^2 + x}\, dx}{x}$

35. $\displaystyle\int x^3(x^2 + 4)^{-1/2}\, dx$

36. $\displaystyle\int \sin^2\!\left(\dfrac{x}{2}\right)\cos^2\!\left(\dfrac{x}{2}\right) dx$

37. $\displaystyle\int \sec^3 x \tan x\, dx$

38. $\displaystyle\int \sec^5 x \tan^2 x\, dx$

39. $\displaystyle\int \dfrac{x^{1/3}\, dx}{x^{1/2} + x^{2/3}}$

40. $\displaystyle\int \dfrac{\sin x\, dx}{\sin x + \cos x}$

41. $\displaystyle\int \dfrac{\cos x + \sin x}{1 + \cos x - \sin x}\, dx$

In Problems 42–46, solve the given differential equations.

42. $\dfrac{dy}{dx} = \sqrt{\dfrac{x^2 - 1}{y^2 + 1}}$

43. $\dfrac{dy}{dx} = \dfrac{1 - y}{1 + x}$

44. $\dfrac{dy}{dx} - \dfrac{y}{2x} = \dfrac{1}{x^{1/2} + 1}$

45. $y' + (\tan x)y = \sec^3 x$

46. $(\cos^3 x)y' = \sin^3 x \cos y$

Determine whether the improper integral in Problems 47–50, converges, and if it does, find its value.

47. $\displaystyle\int_0^{\pi/2} \dfrac{\cos x\, dx}{\sqrt{\sin x}}$

48. $\displaystyle\int_{-\infty}^{+\infty} \dfrac{dx}{4 + x^2}$

49. $\displaystyle\int_{-\infty}^{+\infty} \dfrac{dx}{x^2 + 4x + 6}$

50. $\displaystyle\int_1^{+\infty} \dfrac{dx}{x^4 + x^2}$

51. If n is a positive integer and a is a positive number, find

$$\int_0^{+\infty} x^n e^{-ax}\, dx$$

52. Find $\displaystyle\int \dfrac{\sin x\, dx}{\sqrt{\cos 2x}}$.

53. Show that $\displaystyle\int_0^1 x^m(1 - x)^n\, dx = \int_0^1 x^n(1 - x)^m\, dx$ for positive integers m, n.

54. **THINK TANK PROBLEM**   Each of the following equations may be either true or false. In each case, either show that the equation is generally true or find a number x for which it fails.

 a. $\tanh\!\left(\dfrac{1}{2}\ln x\right) = \dfrac{x - 1}{x + 1}, x > 0$

 b. $\sinh^{-1}(\tan x) = \tanh^{-1}(\sin x)$ for $-\dfrac{\pi}{2} < x < \dfrac{\pi}{2}$

55. Compute $\displaystyle\int_0^1 x f''(3x)\, dx$, given that $f'(0)$ is defined, $f(0) = 1, f(3) = 4$, and $f'(3) = -2$.

56. What is $\int_0^{\pi/2} [f(x) + f''(x)]\cos x \, dx$ if $f(0)$ and $f'(\frac{\pi}{2})$ are defined, $f(\frac{\pi}{2}) = 5$, and $f'(0) = -1$?

57. Stefan's law of radiation says that the temperature T (°Kelvin) of a body changes at a rate proportional to the difference between T^4 and T_0^4, where T_0 is the temperature of the surrounding medium. Set up and solve a differential equation for T.

58. Newton's law of cooling says that the temperature T of a body satisfies

$$\frac{dT}{dt} = k(T - T_0)$$

where T_0 is the temperature of the surrounding medium. Solve this equation for T and compare with the result for Stefan's law in Problem 57.

59. MODELING PROBLEM The psychologist L. L. Thurstone investigated the way people learn using the differential equation

$$\frac{dS}{dt} = \frac{2k}{\sqrt{m}} [S(1 - S)]^{3/2}$$

where $S(t)$ is the state of the learner at time t, and k and m are positive constants depending on the learner and the nature of the task.* Solve this equation.

60. A prime number is a positive number n with exactly two divisors. In number theory, it is shown that the total number of primes less than or equal to x is approximated by the function

$$L(x) = \int_2^x \frac{dt}{\ln t}, \quad x \geq 2$$

Use the trapezoidal rule to estimate the number of primes between $x = 2$ and $x = 1,000$.

61. Find the centroid of a thin plate of constant density δ that occupies the region bounded by the graph of $y = \sin x + \cos x$, the coordinate axes, and the line $x = \pi/4$.

62. Repeat Problem 61 for a plate that occupies the region bounded by $y = \sec^2 x$, the coordinate axes, and the line $x = \pi/3$.

63. Find the volume obtained by revolving about the y-axis the region bounded by the curve $y = \sinh x$, the x-axis, and the line $x = 1$.

64. Find the volume obtained by revolving about the y-axis the region bounded by the curve $y = \cosh x$, the y-axis, and the line $y = 2$.

*L. L. Thurstone, "The Learning Function," *J. of General Psychology 3* (1930), pp. 469–493.

65. The region bounded by the graph of $y = \sin x$ and the x-axis between $x = 0$ and $x = \pi$ is revolved about the x-axis to generate a solid. Find the volume of the solid and its surface area.

66. Find the length of the arc of the curve $y = \frac{4}{5}x^{5/4}$ that lies between $x = 0$ and $x = 1$.

67. Find the volume of the solid obtained by revolving about the x-axis the region bounded by the curve $y = (9 - x^2)^{1/4}$ and the x-axis.

68. Find the volume of the solid formed by revolving about the y-axis the region bounded by the curve

$$y = \frac{1}{1 + x^4}$$

between $x = 0$ and $x = 4$.

69. Find the volume of the solid formed by revolving about the x-axis the region bounded by the curve

$$y = \frac{2}{\sqrt{3x - 2}}$$

the x-axis, and the lines $x = 1$ and $x = 2$.

70. Find the surface area of the solid generated by revolving the region bounded by the curve

$$y = e^x + \frac{1}{4}e^{-x}$$

on $[0, 1]$ about the x-axis.

71. Find the volume of the solid whose base is the region R bounded by the curve $y = e^x$ and the lines $y = 0, x = 0$, $x = 1$, if cross sections perpendicular to the x-axis are equilateral triangles.

72. Show that the area under $y = \frac{1}{x}$ on the interval $[1, a]$ equals the area under the same curve on $[k, ka]$ for any number $k > 0$.

73. Find the length of the curve $y = \ln(\sec x)$ on the interval $[0, \frac{\pi}{3}]$.

74. Find the length of the curve

$$x = \frac{ay^2}{b^2} - \frac{b^2}{8a} \ln \frac{y}{b}$$

between $y = b$ and $y = 2b$.

75. If $s > 0$, the integral function defined by

$$\Gamma(s) = \int_0^{+\infty} e^{-t}t^{s-1} \, dt$$

is called the *gamma function*. It was introduced by Leonhard Euler in 1729 and has some useful properties.

a. Show that $\Gamma(s)$ converges for all $s > 0$.
b. Show that $\Gamma(s + 1) = s\Gamma(s)$.
c. Show that $\Gamma(n + 1) = n!$ for any positive integer n.

76. This problem is a continuation of Problem 75.

a. Show that for a positive integer n:

$$\Gamma(x + n) = (x + n - 1)\cdots(x + 1)x\Gamma(x)$$

b. Suppose you need to approximate $\Gamma(x)$ by numerically integrating over an interval $(0, T)$, for T quite large. Of course, you want T to be no larger than necessary. For $0 < x \leq 1$, find T so that

$$\int_T^\infty e^{-t}t^{x-1}\, dt < 0.005$$

Thus, your approximation to $\Gamma(x)$ by integrating over $(0, T)$ is accurate to within 0.005 unit. *Note:* In general, $t^{x-1} \leq 1$.

c. We will now use part **a.** Suppose you need to approximate $\Gamma(7.3)$. Explain how to use part **a** to numerically compute $\Gamma(0.3)$, and thus $\Gamma(7.3)$, and be sure the resulting *relative* error is less than 0.005.

d. Would you rather compute $\Gamma(0.3)$ or $\Gamma(7.3)$? Explain.

e. Numerically compute $\Gamma(7.3)$ by the method suggested here. If your computer can compute $\Gamma(7.3)$ explicitly, compare with your approximate result.

77. Stirling's formula   You have probably noticed that $n!$ grows incredibly fast (for example, compare 8! and 12!). The goal of this problem is to explore a procedure for approximating $n!$ by an expression of the form

$$Cn^{n+1/2}e^{-n}$$

a. It can be shown that for large x,

$$\Gamma(x + 1) \approx Cx^{x+1/2}e^{-x}$$

Find a numerical estimate for C by plotting the ratio of $\Gamma(x + 1)$ to $x^{x+1/2}e^{-x}$ for large x.

b. Given that C is the square root of a familiar number, find C. When you are finished, you will have Stirling's formula.

c. Since $\Gamma(n + 1) = n!$ for integer n (Problem 75), the result we have obtained says

$$n! \approx Cn^{n+1/2}e^{-n}$$

This is called **Stirling's formula.** It is an *asymptotic* approximation in the sense that it becomes more accurate $n \rightarrow +\infty$. Test this fact by using a computer or calculator with a CAS to approximate $n!$ for $n = 10$ and $n = 100$. Compare with the estimate obtained using Stirling's formula.

78. Derive the reduction formula

$$\int (a^2 - x^2)^n dx = \frac{x(a^2 - x^2)^n}{2n + 1} + \frac{2a^2 n}{2n + 1}\int (a^2 - x^2)^{n-1} dx$$

Use this formula to find $\int (9 - x^2)^{5/2} dx$.

79. Derive the reduction formula

$$\int \frac{\sin^n x\, dx}{\cos^m x} = \frac{1}{m - 1}\frac{\sin^{n-1}x}{\cos^{m-1}x} - \frac{n - 1}{m - 1}\int \frac{\sin^{n-2}x\, dx}{\cos^{m-2}x}$$

80. Derive a reduction formula for $\int \dfrac{\cos^n x\, dx}{\sin^m x}$.

81. Derive a reduction formula for $\int x^n(x^2 + a^2)^{-1/2}\, dx$.

82. Derive the reduction formula

$$\int \frac{dx}{x^n\sqrt{ax + b}} = \frac{-\sqrt{ax + b}}{(n - 1)bx^{n-1}} - \frac{(2n - 3)a}{(2n - 2)b}\int \frac{dx}{x^{n-1}\sqrt{ax + b}}$$

83. The improper integral $\displaystyle\int_1^{+\infty}\left[\frac{2Ax^3}{x^4 + 1} - \frac{1}{x + 1}\right]dx$ converges for exactly one value of the constant A. Find A and then compute the value of the integral.

84. Evaluate $\displaystyle\int_0^{+\infty}\frac{\sqrt{x}\ln x\, dx}{(x + 1)(x^2 + x + 1)}$ if it exists.

85. MODELING PROBLEM   Spectroscopic measurements based on the Doppler effect yield an observed mass m_0 for a certain type of binary star. The true mass m is then estimated by $m = m_0/I$, where

$$I = \int_0^{\pi/2} \sin^4 x\, dx$$

Determine the number I.

86. MODELING PROBLEM   The residents of a certain community have voted to discontinue the fluoridation of their water supply. The local reservoir currently holds 200 million gallons of fluoridated water containing 1,600 lb of fluoride. The fluoridated water flows out of the reservoir at the rate of 4 million gallons per day and is replaced by unfluoridated water at the same rate. At all times, the remaining fluoride in the reservoir is evenly distributed. Set up and solve a differential equation for the amount $F(t)$ of fluoride in the reservoir at time t. When will 90% of the fluoride be replaced?

87. MODELING PROBLEM   The Atomic Energy Commission puts atomic waste into sealed containers and dumps them into the ocean. It is important to dump the containers in water shallow enough to ensure they do not break when they hit bottom. Suppose $S(t)$ is the depth of the container at time t and let W and B denote the container's weight and the constant buoyancy force, respectively. Assume there is a "drag force" proportional to velocity v.

a. Explain why the motion of the containers may be modeled by the differential equation

$$\frac{W}{g}\frac{dv}{dt} = W - B - kv$$

where g is the constant acceleration due to gravity. Solve this equation to express $v(t)$ in terms of W, k, B, and g.

b. Integrate the expression for $v(t)$ in part **a** to find $S(t)$.

c. Suppose $W = 1{,}125$ newtons (about 250 lb), $B = 1{,}100$ newtons, and $k = 0.64$ kg/s. If the container breaks when the impact speed exceeds 10 m/s, what is the maximum depth for safe dumping?

88. **The Evans price-adjustment model** assumes that if there is an excess demand D over supply S in any time period, the price p changes at a rate proportional to the excess, $D - S$; that is,

$$\frac{dp}{dt} = k(D - S)$$

Suppose for a certain commodity, demand is linear,

$$D(p) = c - dp$$

and supply is cyclical

$$S(t) = a\sin(bt)$$

Solve the differential equation to express price $p(t)$ in terms of $a, b, c,$ and d. What happens to $p(t)$ "in the long run" (as $t \to +\infty$)? Use $p_0 = p(0)$.

89. **MODELING PROBLEM**   A country has 5 billion dollars of paper currency. Each day about 10 million dollars comes into banks and 12 million is paid out. The government decides to issue new currency, and whenever an "old" bill comes into the bank, it is replaced by a "new" bill. Set up and solve a differential equation to model the currency replacement. How long will it take for 95% of the currency in circulation to be "new" bills? How much total currency will be in circulation at this time?

90. **PUTNAM EXAMINATION PROBLEM**   Show that

$$\frac{22}{7} - \pi = \int_0^1 \frac{x^4(1 - x)^4\, dx}{1 + x^2}$$

91. **PUTNAM EXAMINATION PROBLEM**   Evaluate

$$\int_0^{\pi/2} \frac{dx}{1 + (\tan x)^{\sqrt{2}}}$$

92. **PUTNAM EXAMINATION PROBLEM**   Evaluate

$$\int_0^{+\infty} t^{-1/2} \exp\left[-1985\left(t + \frac{1}{t}\right)\right] dt$$

You may assume that $\int_{-\infty}^{+\infty} e^{-x^2}\, dx = \sqrt{\pi}$.

Buoy Design*

This project is adapted from a computer project used at the U.S. Coast Guard Academy.

This project is to be done in groups of three or four students. Each group will submit a single written report.

You have been hired as a special consultant by the U. S. Coast Guard to evaluate some proposed new designs for navigational aids (buoys).

The buoys are floating cans that need to be visible from some distance away, without rising too far out of the water. Each buoy has a circular cross section and will be fitted with a superstructure that carries equipment such as lights and batteries.

To be acceptable, a fully equipped buoy must float with not less than 1.5 ft nor more than 3 ft of freeboard. (Freeboard is the distance from the water level to the top of the device.)

You should make the following assumptions:

a. A floating object will displace a volume of water whose weight equals the weight of the floating object. (This is Archimedes' principle.)

b. Your devices will be floating in salt water, which weighs 65.5500 lb/ft^3.

c. The buoys will be constructed of $\frac{1}{2}$ in.-thick sheet metal that weighs 490 lb/ft^3.

d. You can estimate an additional 20% in weight attributable to welds, bolts, and the like.

e. Each buoy will be fitted with a superstructure and equipment weighing a total of 2,000 lb.

f. Each design you are given to evaluate will be presented to you in the form of a curve $x = f(y)$, to be revolved about the y-axis.

Your paper is not limited to the following questions but should include these concerns: You are to design your evaluation procedure using a top-down structured approach. The design should take the form of an outline of the procedure used, with headings and subheadings as appropriate.

The book of nature is written in the language of mathematics.

GALILEO

The pseudomath is a person who handles mathematics as a monkey handles the razor. The creature tried to shave himself as he had seen his master do; but, not having any notion of the angle at which the razor was to be held, he cut his own throat. He never tried it a second time, poor animal! But the pseudomath keeps on in his work, proclaims himself clean shaved, and all the rest of the world hairy.

A DE MORGAN
BUDGET OF PARADOXES
(LONDON, 1872), P. 473

CONTENTS

8

Infinite Series

PREVIEW

Is it possible for the sum of infinitely many numbers to be finite? This concept, which may seem paradoxical at first, plays a central role in mathematics and has a variety of important applications. The goal of this chapter is to examine the theory and applications of infinite sums, which shall be referred to as *infinite series.* Geometric series, introduced in Section 8.2, are among the simplest infinite series we shall encounter and, in some ways, the most important. In Sections 8.3–8.8, we shall develop *convergence tests,* which provide ways of determining quickly whether or not certain infinite series have a finite sum. Next, we shall turn our attention to series in which the individual terms are functions instead of numbers. We shall be especially interested in the properties of *power series,* which may be thought of as polynomials of infinite degree, although some of their properties are quite different from those of polynomials. We shall find that many common functions, such as e^x, $\ln(x + 1)$, $\sin x$, $\cos x$, and $\tan^{-1}x$ can be represented as power series, and we shall discuss some important theoretical and computational aspects of this kind of representation.

PERSPECTIVE

Series, or sums, arise in many different ways. For example, suppose it is known that a certain pollutant is released into the atmosphere at weekly intervals and is dissipated at the rate of 2% per week. If m grams of pollutant are released each week, then at the beginning of the first week, there will be $S_1 = m$ grams in the atmosphere, and at the beginning of the second, there will be $0.98m$ grams of "old" pollutant left plus m grams of "new" pollutant, to yield a total of $S_2 = m + 0.98m$ grams. Continuing, at the beginning of the nth week there will be

$$S_n = m + 0.98m + (0.98)^2m + \cdots + (0.98)^{n-1}m$$

grams. It is natural to wonder how much pollutant will accumulate in the "long run" (as $n \to +\infty$), and the answer is given by the infinite sum

$$S_{+\infty} = m + 0.98m + (0.98)^2m + \cdots$$

But just exactly what do we mean by such a sum, and if the total is a finite number, how can we compute its value?

We seek answers to these questions in this chapter.

8.1 Sequences and Their Limits

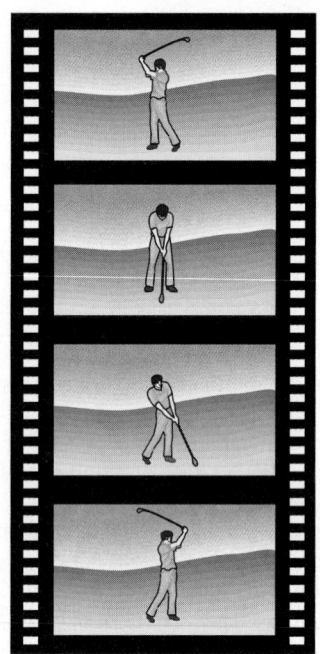

These consecutive frames from a 16-mm movie film show a golfer from the moment he is at the top of his backswing until he hits the ball. If this film were projected at a rate of 24 frames per second, the viewer would have the illusion of seeing the golfer in action as he makes his downswing.

The making of a motion picture is a complex process, and editing all the film into a movie requires that all the frames of the action be labeled in chronological order. For example, R21-435 might signify the 435th frame of the 21st reel. A mathematician might refer to the movie editor's labeling procedure by saying the frames are arranged in a *sequence*.

SEQUENCES

A sequence is a succession of numbers that are listed according to a given prescription or rule. Specifically, if n is a positive integer, the sequence whose nth term is the number a_n can be written as

$$a_1, a_2, \dots, a_n, \dots$$

or, more simply,

$$\{a_n\}$$

The number a_n is called the **general term** of the sequence. We will deal only with infinite sequences, so each term a_n has a **successor** a_{n+1} and for $n > 1$, a **predecessor** a_{n-1}. For example, by associating each positive integer n with its reciprocal $\frac{1}{n}$, we obtain the sequence denoted by

$$\left\{\frac{1}{n}\right\}, \quad \text{which represents the succession of numbers} \quad 1, \frac{1}{2}, \frac{1}{3}, \dots, \frac{1}{n}, \dots$$

The general term is denoted by $a_n = \frac{1}{n}$.

The following examples illustrate the notation and terminology used in connection with sequences.

EXAMPLE 1   *Given the general term, find particular terms of a sequence*

Find the 1st, 2nd, and 15th terms of the sequence $\{a_n\}$, where the general term is

$$a_n = \left(\frac{1}{2}\right)^{n-1}$$

Solution   If $n = 1$, then $a_1 = \left(\frac{1}{2}\right)^{1-1} = 1$. Similarly,

$$a_2 = \left(\frac{1}{2}\right)^{2-1} = \frac{1}{2}$$

$$a_{15} = \left(\frac{1}{2}\right)^{15-1} = \left(\frac{1}{2}\right)^{14} = 2^{-14}$$

The reverse question, that of finding a general term given certain terms of a sequence, is a more difficult task, and even if we find a general term, we have no assurance that the general term is unique. For example, consider the sequence

$$2, 4, 6, 8, \dots$$

This seems to have a general term $a_n = 2n$. However, the general term

$$a_n = (n-1)(n-2)(n-3)(n-4) + 2n$$

has the same first four terms, but $a_5 = 34$ (not 10, as we would expect from the sequence $2, 4, 6, 8$).

It is sometimes useful to start a sequence with a_0 instead of a_1—that is, to have a sequence of the form

$$a_0, a_1, a_2, \dots$$

So far, we have been discussing the concept of a sequence informally, without a definition. We have observed that a sequence $\{a_n\}$ associates the number a_n with the positive integer n. Hence, a sequence is really a special kind of function, one whose domain is the set of all positive (or possibly, nonnegative) integers.

Sequence

A **sequence** $\{a_n\}$ is a function whose domain is a set of nonnegative integers and whose range is a subset of the real numbers. The functional values $a_1, a_2, a_3, \dots$ are called the **terms** of the sequence, and a_n is called the **nth term,** or **general term,** of the sequence.

In this text, we shall consider only *real-valued sequences*—that is, sequences in which each a_n is a real number. Unless otherwise specified, "sequence" will always mean "real-valued sequence." Although a sequence is a function, we usually represent sequences as $\{a_n\}$ rather than with functional notation.

THE LIMIT OF A SEQUENCE

It is often desirable to examine the behavior of a given sequence $\{a_n\}$ as n gets arbitrarily large. For example, consider the sequence

$$a_n = \frac{n}{n+1}$$

Because $a_1 = \frac{1}{2}, a_2 = \frac{2}{3}, a_3 = \frac{3}{4}, \dots$, we can plot the terms of this sequence on a number line, as shown in Figure 8.1a, or the sequence can be plotted in two dimensions, as shown in Figure 8.1b.

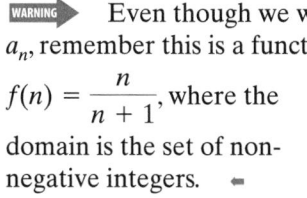

WARNING Even though we write a_n, remember this is a function, $f(n) = \dfrac{n}{n+1}$, where the domain is the set of nonnegative integers.

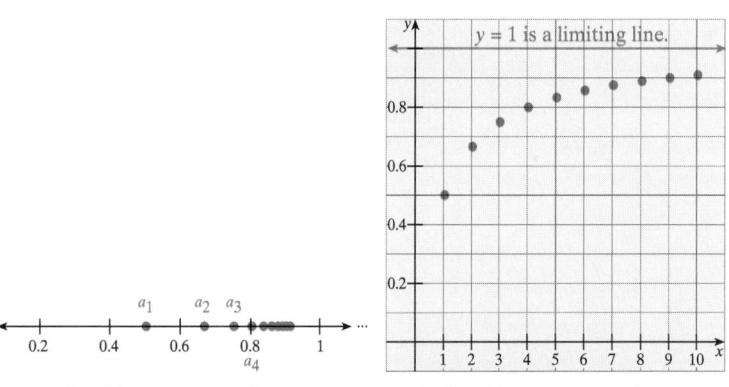

a. Graphing a sequence in one dimension

b. Graphing a sequence in two dimensions

■ **FIGURE 8.1** Graphing the sequence $a_n = \dfrac{n}{n+1}$

By looking at either graph in Figure 8.1, we see that it appears the terms of the sequence are approaching 1. In general, if the terms of the sequence approach the number L as n increases without bound, we say that the sequence *converges to the limit L* and write

$$L = \lim_{n \to \infty} a_n$$

(Note that for the simplicity, the limiting behavior is denoted by $n \to \infty$ instead of the usual $n \to +\infty$.) For instance, in our example, we would expect

$$\lim_{n \to +\infty} \frac{n}{n + 1} = 1$$

This limiting behavior is analogous to the continuous case (discussed in Section 4.5), and may be defined formally as follows.

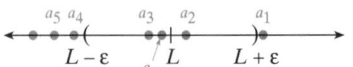

a. one dimension

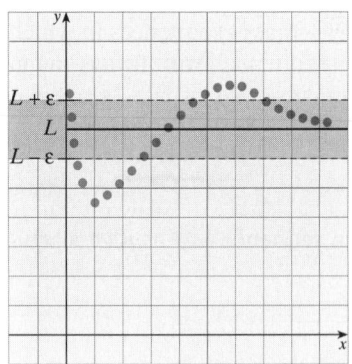

b. two dimensions

■ **FIGURE 8.2** Geometric interpretation of a converging sequence

Convergent Sequence

The sequence $\{a_n\}$ **converges** to the number L, and we write

$$L = \lim_{n \to \infty} a_n$$

if for every $\epsilon > 0$, there is an integer N such that

$$|a_n - L| < \epsilon \quad \text{whenever } n > N$$

Otherwise, the sequence **diverges.**

■ *What This Says:* The notation $L = \lim_{n \to \infty} a_n$ means that the terms of the sequence $\{a_n\}$ can be made as close to L as may be desired by taking n sufficiently large.

A geometric interpretation of this definition is shown in Figure 8.2.

Note that if

$$L = \lim_{n \to \infty} a_n$$

the numbers a_n may be practically anywhere at first (that is, for "small" n), but eventually, the a_n must "cluster" near the limiting value L.

The theorem on limits of functions carries over to sequences. We have the following useful result.

THEOREM 8.1   *Limit theorem for sequences*

If $\lim_{n \to \infty} a_n = L$ and $\lim_{n \to \infty} b_n = M$, then

Linearity rule $\lim_{n \to \infty} (ra_n + sb_n) = rL + sM$

Product rule $\lim_{n \to \infty} (a_n b_n) = LM$

Quotient rule $\lim_{n \to \infty} \dfrac{a_n}{b_n} = \dfrac{L}{M}$ provided $M \neq 0$

Root rule $\lim_{n \to \infty} \sqrt[m]{a_n} = \sqrt[m]{L}$ provided $\sqrt[m]{a_n}$ is defined for all n and $\sqrt[m]{L}$ exists

Proof   The proof of these rules follows from the limit rules that were stated in Section 4.5.

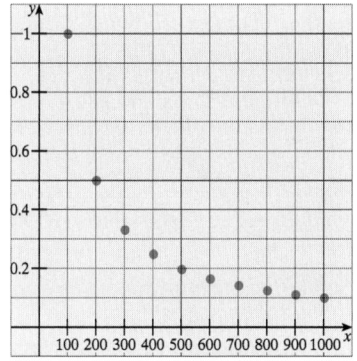

■ **FIGURE 8.3** Graphical representation of $a_n = 100/n$

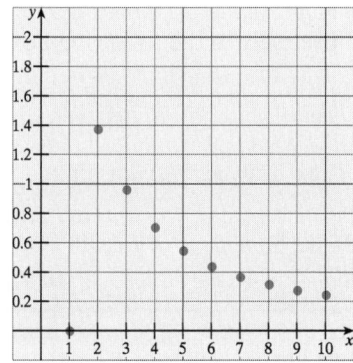

■ **FIGURE 8.4** Graph of $a_n = \dfrac{2n^2 + 5n - 7}{n^3}$

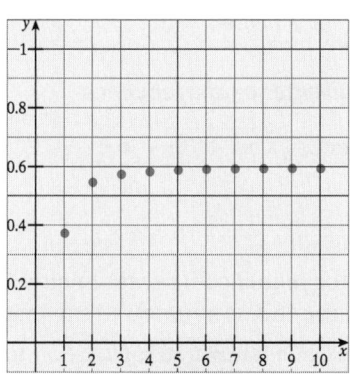

■ **FIGURE 8.5** Graph of $a_n = \dfrac{3n^4 + n - 1}{5n^4 + 2n^2 + 1}$

EXAMPLE 2   *Convergent sequences*

Find the limit of each of these convergent sequences:

a. $\left\{\dfrac{100}{n}\right\}$ b. $\left\{\dfrac{2n^2 + 5n - 7}{n^3}\right\}$ c. $\left\{\dfrac{3n^4 + n - 1}{5n^4 + 2n^2 + 1}\right\}$

Solution

a. As n grows arbitrarily large, $100/n$ gets smaller and smaller. Thus,

$$\lim_{n\to\infty} \frac{100}{n} = 0$$

A graphical representation is shown in Figure 8.3.

b. We cannot use the quotient rule of Theorem 8.1 because neither the limit in the numerator nor the one in the denominator exists. However,

$$\frac{2n^2 + 5n - 7}{n^3} = \frac{2}{n} + \frac{5}{n^2} - \frac{7}{n^3}$$

and by using the linearity rule, we find that

$$\lim_{n\to\infty} \frac{2n^2 + 5n - 7}{n^3} = \lim_{n\to\infty} \frac{2}{n} + 5 \lim_{n\to\infty} \frac{1}{n^2} - 7 \lim_{n\to\infty} \frac{1}{n^3}$$

$$= 0 + 0 + 0$$

$$= 0$$

A graph is shown in Figure 8.4.

c. Divide the numerator and denominator by n^4 to obtain

$$\lim_{n\to\infty} \frac{3n^4 + n - 1}{5n^4 + 2n^2 + 1} = \lim_{n\to\infty} \frac{3 + \dfrac{1}{n^3} - \dfrac{1}{n^4}}{5 + \dfrac{2}{n^2} + \dfrac{1}{n^4}} = \frac{3}{5}$$

A graph of this sequence is shown in Figure 8.5.   ■

EXAMPLE 3   *Divergent sequences*

Show that the following sequences diverge:

a. $\{(-1)^n\}$ b. $\left\{\dfrac{n^5 + n^3 + 2}{7n^4 + n^2 + 3}\right\}$

Solution

a. The sequence defined by $\{(-1)^n\}$ is $-1, 1, -1, 1, \ldots$, and this sequence **diverges by oscillation** because the nth term is always either 1 or -1. Thus a_n cannot approach one specific number L as n grows large. The graph is shown in Figure 8.6a.

b. $\lim_{n\to\infty} \dfrac{n^5 + n^3 + 2}{7n^4 + n^2 + 3} = \lim_{n\to\infty} \dfrac{1 + \dfrac{1}{n^2} + \dfrac{2}{n^5}}{\dfrac{7}{n} + \dfrac{1}{n^3} + \dfrac{3}{n^5}}$

The numerator tends toward 1 as $n \to \infty$, and the denominator approaches 0. Hence the quotient increases without bound, and the sequence must diverge. The graph is shown in Figure 8.6b.

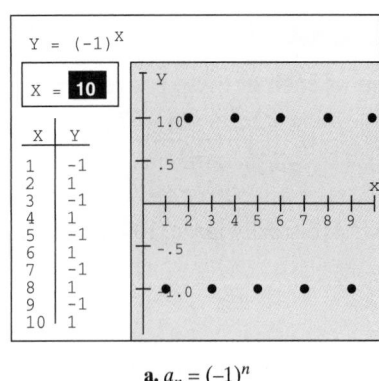

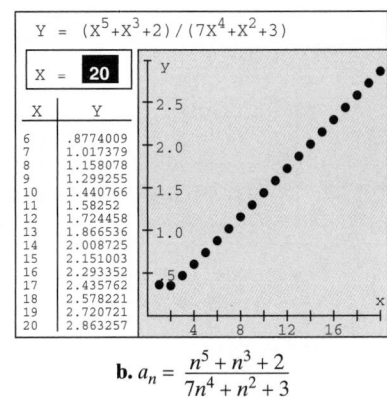

a. $a_n = (-1)^n$ **b.** $a_n = \dfrac{n^5 + n^3 + 2}{7n^4 + n^2 + 3}$

■ **FIGURE 8.6**   Graph and values for two divergent sequences ■

If $\lim_{n \to \infty} a_n$ does not exist because the numbers a_n become arbitrarily large as $n \to \infty$, we write $\lim_{n \to \infty} a_n = \infty$. We summarize this more precisely in the following box.

Limit Notation

$\lim_{n \to \infty} a_n = \infty$ means that for any real number A, we have $a_n > A$ for all sufficiently large n.

$\lim_{n \to \infty} b_n = -\infty$ means that for any real number B, we have $b_n < B$ for all sufficiently large n.

Rewriting the answer to Example 3b in this notation, we have

$$\lim_{n \to \infty} \frac{n^5 + n^3 + 2}{7n^4 + n^2 + 3} = \infty$$

Also notice that $\lim_{n \to \infty}(-5n) = -\infty$ (that is, decreases without bound) whereas $\lim_{n \to \infty} (-1)^n$ does not exist. Thus, the answer to Example 3a is *neither* ∞ nor $-\infty$.

EXAMPLE 4   *Determining the convergence or divergence of a sequence*

Determine the convergence or divergence of the sequence $\left\{ \sqrt{n^2 + 3n} - n \right\}$.

Solution   Consider

$$\lim_{n \to \infty} \left(\sqrt{n^2 + 3n} - n \right)$$

It would not be correct to apply the linearity property for sequences (because neither $\lim_{n \to \infty} \sqrt{n^2 + 3n}$ nor $\lim_{n \to \infty} n$ exists). It is also not correct to use this as a reason to say that the limit does not exist. You might even try some values of n (shown in Figure 8.7) to guess that there is some limit. To find the limit, however, we shall rewrite the general term algebraically as follows:

$$\sqrt{n^2 + 3n} - n = \left(\sqrt{n^2 + 3n} - n \right) \frac{\sqrt{n^2 + 3n} + n}{\sqrt{n^2 + 3n} + n}$$

$$= \frac{n^2 + 3n - n^2}{\sqrt{n^2 + 3n} + n} \frac{1/n}{1/n} = \frac{3}{\dfrac{\sqrt{n^2 + 3n}}{n} + 1} = \frac{3}{\sqrt{\dfrac{n^2 + 3n}{n^2}} + 1} = \frac{3}{\sqrt{1 + (3/n)} + 1}$$

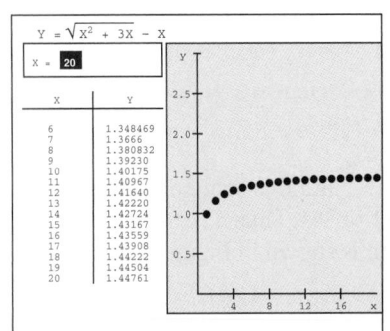

■ **FIGURE 8.7**   Graph of $a_n = \sqrt{n^2 + 3n} - n$: convergent or divergent?

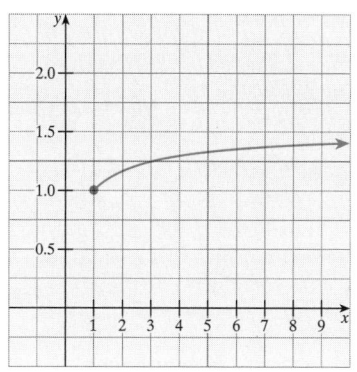

■ **FIGURE 8.8** Graph of
$f(x) = \sqrt{x^2 + 3x} - x, x \geq 1$

Hence,

$$\lim_{n \to \infty} \left(\sqrt{n^2 + 3n} - n \right) = \lim_{n \to \infty} \frac{3}{\sqrt{1 + (3/n)} + 1} = \frac{3}{2}$$ ■

Note the graph of the sequence in Example 4. The graph of a sequence consists of a succession of isolated points. This can be compared with the graph of $y = \sqrt{x^2 + 3x} - x, x \geq 1$, which is a continuous curve (shown in Figure 8.8). The only difference between $\lim_{n \to \infty} a_n = L$ and $\lim_{x \to \infty} f(x) = L$ is that n is required to be an integer. This is stated in the hypothesis of the following theorem.

═══════════

THEOREM 8.2   *Limit of a sequence*

Suppose f is a function such that $a_n = f(n)$ for $n = 1, 2, \ldots$. If $\lim_{x \to \infty} f(x)$ exists and $\lim_{x \to \infty} f(x) = L$, the sequence $\{a_n\}$ converges and $\lim_{n \to \infty} a_n = L$.

Proof   Let $\epsilon > 0$ be given. Because $\lim_{x \to \infty} f(x) = L$, there exists a number $N > 0$ such that

$$\left| f(x) - L \right| < \epsilon \quad \text{whenever } x > N$$

In particular, if $n > N$, it follows that $\left| f(n) - L \right| = \left| a_n - L \right| < \epsilon$.   ═══

Be sure you read this theorem correctly. In particular, note that it *does not* say that if $\lim_{n \to \infty} a_n = L$, then $\lim_{x \to \infty} f(x) = L$ (see Figure 8.9b).

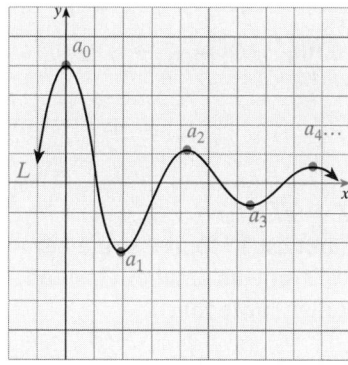

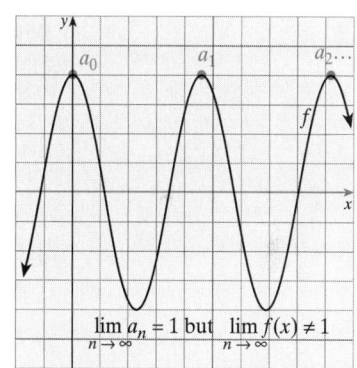

a. If $\lim_{n \to \infty} f(x) = L$, then $\lim_{n \to \infty} a_n = L$.   **b.** WARNING ➤ If $\lim_{n \to \infty} a_n = L$, then $\lim_{n \to \infty} f(x)$ does NOT necessarily equal L. ←

■ **FIGURE 8.9**   Graphical comparison of $\lim_{n \to \infty} a_n$ and $\lim_{x \to \infty} f(x)$ where $f(n) = a_n$ for $n = 1, 2, \ldots$

█ **EXAMPLE 5**   *Evaluating a limit using l'Hôpital's rule*

Given that the sequence $\left\{ \dfrac{n^2}{1 - e^n} \right\}$ converges, evaluate $\lim_{n \to \infty} \dfrac{n^2}{1 - e^n}$.

Solution   Let $L = \lim_{x \to \infty} f(x)$, where $f(x) = \dfrac{x^2}{1 - e^x}$. Because $f(n) = a_n$ for $n = 1, 2, \ldots$,

Theorem 8.2 tells us that $\lim_{n \to \infty} \dfrac{n^2}{1 - e^n}$ is the same as $\lim_{n \to \infty} f(x)$, provided this latter

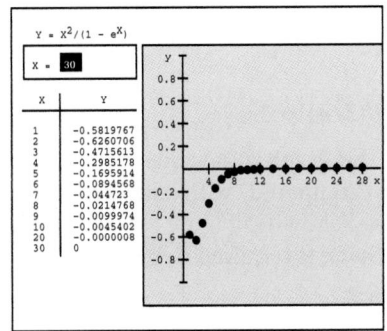

From the graph, it looks like $\lim\limits_{x \to \infty} a_n = 0$.

limit exists.

$$\lim_{x \to \infty} \frac{x^2}{1 - e^x} = \lim_{x \to \infty} \frac{2x}{-e^x} \qquad \text{l'Hôpital's rule}$$

$$= \lim_{x \to \infty} \frac{2}{-e^x} \qquad \text{l'Hôpital's rule again}$$

$$= 0$$

Thus, by Theorem 8.2, $\lim\limits_{n \to \infty} \dfrac{n^2}{1 - e^n} = \lim\limits_{x \to \infty} f(x) = L = 0$. ∎

The squeeze rule (Section 2.3) can be reformulated in terms of sequences.

THEOREM 8.3   *Squeeze theorem for sequences*

If $a_n \leq b_n \leq c_n$ for all $n > N$, and $\lim\limits_{n \to \infty} a_n = \lim\limits_{n \to \infty} c_n = L$, then $\lim\limits_{n \to \infty} b_n = L$.

Proof   The proof of this theorem relies on the definition and properties of limits.

EXAMPLE 6   *Using l'Hôpital's rule and the squeeze theorem*

Show that the following sequences converge, and find their limits.

a. $\lim\limits_{n \to \infty} n^{1/n}$ b. $\lim\limits_{n \to 0} \dfrac{n!}{n^n}$

Solution

a. Let $L = \lim\limits_{n \to \infty} n^{1/n}$; then

$$\ln L = \lim_{n \to \infty} \frac{\ln n}{n} = \lim_{n \to \infty} \frac{1/n}{1} \qquad \text{l'Hôpital's rule}$$

$$= 0$$

Thus, $L = e^0 = 1$.

b. We cannot use l'Hôpital's rule because $x!$ is not defined as an elementary function when x is not an integer. Instead, we use the squeeze theorem for sequences. Accordingly, note that

$$b_n = \frac{n!}{n^n} \quad \text{so that} \quad b_1 = 1, b_2 = \frac{2 \cdot 1}{2 \cdot 2}, b_3 = \frac{3 \cdot 2 \cdot 1}{3 \cdot 3 \cdot 3}, \dots,$$

$$b_n = \frac{\underbrace{n \cdot (n-1) \cdot \dots \cdot 3 \cdot 2 \cdot 1}_{n \text{ factors}}}{n \cdot n \cdot n \cdot \dots \cdot n} = \left[\frac{\underbrace{n(n-1) \cdot \dots \cdot 3 \cdot 2}_{n-1 \text{ factors}}}{n \cdot n \cdot n \cdot \dots \cdot n}\right]\left(\frac{1}{n}\right)$$

Thus, if we let $a_n = 0$ and $c_n = \dfrac{1}{n}$, for all n, we have

$$\overset{a_n}{\overbrace{0}} \leq \overset{b_n}{\overbrace{\frac{n!}{n^n}}} \leq \overset{c_n}{\overbrace{\frac{1}{n}}} \qquad \frac{n!}{n^n} \leq \frac{1}{n} \quad \text{because } n \cdot n! < n^n$$

Because $\lim\limits_{n \to \infty} a_n = 0$ and $\lim\limits_{n \to \infty} c_n = 0$, it follows from the squeeze theorem that

$$\lim_{n \to \infty} \frac{n!}{n^n} = 0.$$ ∎

BOUNDED, MONOTONIC SEQUENCES

Increasing/Nondecreasing

A sequence $\{a_n\}$ is said to be

> **Increasing** if $a_1 < a_2 < a_3 < \cdots < a_{k-1} < a_k < \cdots$
>
> **Nondecreasing** if $a_1 \leq a_2 \leq \cdots \leq a_{k-1} \leq a_k \leq \cdots$

Decreasing/Nonincreasing

> **Decreasing** if $a_1 > a_2 > a_3 > \cdots > a_{k-1} > a_k > \cdots$
>
> **Nonincreasing** if $a_1 \geq a_2 \geq \cdots \geq a_{k-1} \geq a_k \geq \cdots$

Monotonic/Strictly Monotonic

> **Monotonic** if it is nondecreasing or nonincreasing
>
> **Strictly monotonic** if it is increasing or decreasing

Bounded above/Bounded below

> **Bounded above** by M if $a_n \leq M$ for $n = 1, 2, 3, \ldots$
>
> **Bounded below** by m if $m \leq a_n$ for $n = 1, 2, 3, \ldots$

Bounded

> **Bounded** if it is bounded both above and below

In general, it is difficult to tell whether a given sequence converges or diverges, but thanks to the following theorem, it is easy to make this determination if we know the sequence is monotonic.

THEOREM 8.4   *BMCT: The bounded, monotonic convergence theorem*

A monotonic sequence $\{a_n\}$ converges if it is bounded and diverges otherwise.

Proof   A formal proof of the BMCT is outlined in Problem 54. For the following informal argument, we will assume that $\{a_n\}$ is a nondecreasing sequence. You might wish to see whether you can give a similar informal argument for the increasing case. Because the terms of the sequence satisfy $a_1 \leq a_2 \leq a_3 \leq \cdots$, we know that the sequence is bounded from below by a_1 and that the graph of the corresponding points (n, a_n) will be rising in the plane. Two cases can occur, as shown in Figure 8.10.

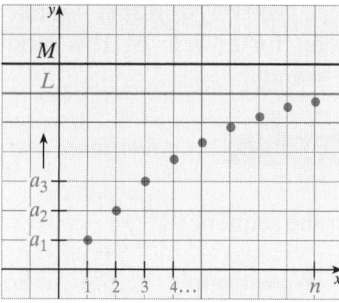
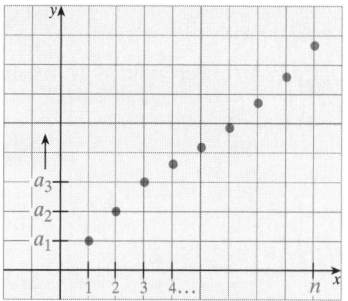

a. If $a_n < M$ for $n = 1, 2, \ldots$, the graph of the points (n, a_n) will approach a horizontal "barrier" line $y = L$.

b. If $\{a_n\}$ is not bounded from above, the graph rises indefinitely.

■ **FIGURE 8.10**   Graphical possibilities for the BMCT theorem

Suppose the sequence $\{a_n\}$ is also bounded from above by a number M, so that $a_1 \le a_n \le M$ for $n = 1, 2, \ldots$. Then the graph of the points (n, a_n) must continually rise (because the sequence is monotonic) and yet it must stay below the line $y = M$. The only way this can happen is for the graph to approach a "barrier" line $y = L$ (where $L \le M$), and we have $\lim_{n\to\infty} a_n = L$, as shown in Figure 8.10a. However, if the sequence is not bounded from above, the graph will rise indefinitely (Figure 8.10b), and the terms in the sequence $\{a_n\}$ cannot approach any finite number L.

EXAMPLE 7   *Convergence using the BMCT*

Show that the sequence $\left\{\dfrac{1 \cdot 3 \cdot 5 \cdots (2n-1)}{2 \cdot 4 \cdot 6 \cdots (2n)}\right\}$ converges.

Solution   The first few terms of this sequence are:

$$a_1 = \frac{1}{2} \qquad a_2 = \frac{1 \cdot 3}{2 \cdot 4} = \frac{3}{8} \qquad a_3 = \frac{1 \cdot 3 \cdot 5}{2 \cdot 4 \cdot 6} = \frac{5}{16}$$

Because $\frac{1}{2} > \frac{3}{8} > \frac{5}{16}$, it appears that the sequence is decreasing (that is, it is monotonic). We can prove this by showing that $a_{n+1} < a_n$, or equivalently, $\dfrac{a_{n+1}}{a_n} < 1$. (Note that $a_n \ne 0$ for all n.)

$$\frac{a_{n+1}}{a_n} = \frac{\dfrac{1 \cdot 3 \cdot 5 \cdots [2(n+1)-1]}{2 \cdot 4 \cdot 6 \cdots [2(n+1)]}}{\dfrac{1 \cdot 3 \cdot 5 \cdots (2n-1)}{2 \cdot 4 \cdot 6 \cdots (2n)}}$$

$$= \frac{1 \cdot 3 \cdot 5 \cdots (2n+1)}{2 \cdot 4 \cdot 6 \cdots (2n+2)} \cdot \frac{2 \cdot 4 \cdot 6 \cdots (2n)}{1 \cdot 3 \cdot 5 \cdots (2n-1)}$$

$$= \frac{2n+1}{2n+2} < 1$$

for any $n > 0$. Hence $a_{n+1} < a_n$ for all n, and $\{a_n\}$ is a decreasing sequence. Because $a_n > 0$ for all n, it follows that $\{a_n\}$ is bounded below by 0. Applying the BMCT, we see that $\{a_n\}$ converges, but the BMCT tells us nothing about the limit.

COMMENT   Technically, the sequence $\{a_n\}$ is monotonic only when its terms are either always nonincreasing or always nondecreasing, but the BMCT also applies to sequences whose terms are *eventually* monotonic. That is, it can be shown that the sequence $\{a_n\}$ converges if it is bounded and there exists an integer N such that $\{a_n\}$ is monotonic for all $n > N$. This modified form of the BMCT is illustrated in the following example.

EXAMPLE 8   *Convergence of a sequence that is eventually monotonic*

Show that the sequence $\left\{\dfrac{\ln n}{\sqrt{n}}\right\}$ converges.

Solution   We will apply the BMCT. Some initial values are shown in Figure 8.11. The succession of numbers suggests that the sequence increases at first and then gradually begins to decrease. To verify this behavior, we let

$$f(x) = \frac{\ln x}{\sqrt{x}}$$

■ FIGURE 8.11 $a_n = \dfrac{\ln n}{\sqrt{n}}$

and find that

$$f'(x) = \frac{\sqrt{x}(1/x) - (\ln x)(\frac{1}{2}x^{-1/2})}{x} = \frac{2\sqrt{x} - \sqrt{x}\ln x}{2x^2}$$

Find the critical values:

$$\frac{2\sqrt{x} - \sqrt{x}\ln x}{2x^2} = 0$$

$$2\sqrt{x} = \sqrt{x}\ln x$$

$$\ln x = 2$$

$$e^2 = x$$

Thus, $x = e^2$ is the only critical value, and you can show that $f'(x) > 0$ for $x < e^2$ and $f'(x) < 0$ for $x > e^2$. This means that f is a decreasing function for $x > e^2$. Thus, the sequence $\left\{\dfrac{\ln n}{\sqrt{n}}\right\}$ must be decreasing for $n > 8$ (because e^2 is between 7 and 8). We see that the sequence $\left\{\dfrac{\ln n}{\sqrt{n}}\right\}$ is bounded from below because

$$0 < \frac{\ln n}{\sqrt{n}} \quad \text{for all } n > 1$$

Therefore, the given sequence is bounded from below and is eventually decreasing, so it must converge. ∎

The BMCT is an extremely valuable theoretical tool. For example, in Chapter 1, we defined the number e by the limit

$$\lim_{n\to\infty}\left(1 + \frac{1}{n}\right)^n = e$$

but to do so we assumed that this limit exists. We can now show this assumption is warranted, because it turns out that the sequence $\left\{\left(1 + \dfrac{1}{n}\right)^n\right\}$ is increasing and bounded from above by 3 (the details are outlined in Problem 53). Thus, the BMCT assures us that the sequence converges, and this in turn guarantees the existence of the limit. We end this section with a result that will be useful in our subsequent work. The proof makes use of the formal definition of limit.

THEOREM 8.5   *Convergence of a power sequence*

If r is a number such that $|r| < 1$, then $\lim\limits_{n\to\infty} r^n = 0$.

Proof   The case where $r = 0$ is trivial. We shall prove the theorem for the case $0 < r < 1$ and leave the case $-1 < r < 0$ as an exercise. Because $0 < r < 1$, we can write r as

$$r = \frac{1}{1 + h}$$

for some $h > 0$, so

$$r^n = \frac{1}{(1 + h)^n}$$

Because (for all fixed n)

$$(1 + h)^n = 1 + nh + \frac{n(n-1)}{2!}h^2 + \cdots + h^n \qquad \text{Binomial theorem}$$

$$> nh \qquad\qquad\qquad\qquad\qquad\qquad \text{Because } h > 0$$

we see

$$r^n = \frac{1}{(1+h)^n} < \frac{1}{nh} \quad \text{If } a > b, \text{ then } \frac{1}{a} < \frac{1}{b}.$$

We now use the definition of limit. Given $\epsilon > 0$, we see that if $n > \frac{1}{\epsilon h}$, then $\frac{1}{nh} < \epsilon$

because both n and ϵ are greater than zero. We can then choose an integer $N = \frac{1}{\epsilon h}$,

and if $n > N$, then $r^n < \epsilon$ and by the definition of limit we conclude $\lim\limits_{n \to \infty} r^n = 0$.

8.1 Problem Set

A 1. ■ **What Does This Say?** What do we mean by the limit of a sequence?

2. ■ **What Does This Say?** What is meant by the bounded, monotonic convergence theorem?

Write out the first five terms (beginning with $n = 1$) of the sequences given in Problems 3–11.

3. $\{1 + (-1)^n\}$

4. $\left\{ \left(\frac{-1}{2} \right)^{n+2} \right\}$

5. $\left\{ \frac{\cos 2n\pi}{n} \right\}$

6. $\left\{ n \sin \frac{n\pi}{2} \right\}$

7. $\left\{ \frac{3n+1}{n+2} \right\}$

8. $\left\{ \frac{n^2-n}{n^2+n} \right\}$

9. $\{a_n\}$ where $a_1 = 256$ and $a_n = \sqrt{a_{n-1}}$ for $n \geq 2$

10. $\{a_n\}$ where $a_1 = -1$ and $a_n = n + a_{n-1}$ for $n \geq 2$

11. $\{a_n\}$ where $a_1 = 1$ and

$$a_n = (a_{n-1})^2 + a_{n-1} + 1 \quad \text{for } n \geq 2$$

Compute the limit of the convergent sequences in Problems 12–35.

12. $\left\{ \frac{5n+8}{n} \right\}$

13. $\left\{ \frac{5n}{n+7} \right\}$

14. $\left\{ \frac{2n+1}{3n-4} \right\}$

15. $\left\{ \frac{4-7n}{8+n} \right\}$

16. $\left\{ \frac{8n^2+800n+5,000}{2n^2-1,000n+2} \right\}$

17. $\left\{ \frac{100n+7,000}{n^2-n-1} \right\}$

18. $\left\{ \frac{8n^2+6n+4,000}{n^3+1} \right\}$

19. $\left\{ \frac{n^3-6n^2+85}{2n^3-5n+170} \right\}$

20. $\left\{ \frac{2n}{n+7\sqrt{n}} \right\}$

21. $\left\{ \frac{8n-500\sqrt{n}}{2n+800\sqrt{n}} \right\}$

22. $\left\{ \frac{3\sqrt{n}}{5\sqrt{n}+\sqrt[4]{n}} \right\}$

23. $\left\{ \frac{\ln n}{n^2} \right\}$

24. $\{2^{5/n}\}$

25. $\{n^{3/n}\}$

26. $\left\{ \left(1 + \frac{3}{n}\right)^n \right\}$

27. $\{(n+4)^{1/n}\}$

28. $\{n^{1/(n+2)}\}$

29. $\{(\ln n)^{1/n}\}$

30. $\left\{ \int_0^\infty e^{-nx}\, dx \right\}$

31. $\{\sqrt{n^2+n}-n\}$

32. $\{\sqrt{n+5\sqrt{n}}-\sqrt{n}\}$

33. $\{\sqrt[n]{n}\}$

34. $\{\ln n - \ln(n+1)\}$

35. $\{(an+b)^{1/n}\}$, a and b positive constants

B *Show that each sequence given in Problems 36–41 converges either by showing it is increasing with an upper bound or decreasing with a lower bound.*

36. $\left\{ \frac{n}{2^n} \right\}$

37. $\left\{ \ln\left(\frac{n+1}{n}\right) \right\}$

38. $\left\{ \frac{3n-2}{n} \right\}$

39. $\left\{ \frac{4n+5}{n} \right\}$

40. $\left\{ \frac{3n-7}{2^n} \right\}$

41. $\{\sqrt[n]{n}\}$

Explain why each sequence in Problems 42–45 diverges.

42. $\{1 + (-1)^n\}$

43. $\{\cos n\pi\}$

44. $\left\{ \frac{n^3-7n+5}{100n^2+219} \right\}$

45. $\{\sqrt{n}\}$

46. Suppose that a particle of mass m moves back and forth along a line segment of length $|a|$. In classical mechanics, the particle can move at any speed, and thus, its energy can be any positive number. However, quantum mechanics replaces this continuous model of the particle's behavior with one in which the particle's energy level can have only certain discrete values, say, $E_1, E_2, \ldots$. Specifically, it can be shown that the nth term in this quantum sequence has the value

$$E_n = \frac{n^2 h^2}{8ma^2} \qquad n = 1, 2, \ldots$$

where h is a physical constant known as *Planck's constant* ($h \approx 6.63 \times 10^{-27}$ erg/s). List the first four values of E_n for a particle with mass $m = 0.008$g moving along a segment of length $a = 100$ cm.

47. **MODELING PROBLEM** A drug is administered into the body. At the end of each hour, the amount of drug present is half what it was at the end of the previous hour. What percent of the drug is present at the end of 4 hr? At the end of n hours?

48. HISTORICAL QUEST Leonardo de Pisa, also known as Fibonacci, was one of the best mathematicians of the Middle Ages. He played an important role in reviving ancient mathematics and introduced the Hindu–Arabic place-value decimal system to Europe. His book, *Liber Abaci,* published in 1202, introduced the Arabic numerals, as well as the famous *rabbit problem,* for which Fibonacci is best remembered today.

**FIBONACCI
1170–1250**

To describe Fibonacci's rabbit problem, we consider a sequence whose *n*th term is defined by a *recursion formula*—that is, a formula in which the *n*th term is given in terms of previous terms in the sequence. Suppose rabbits breed in such a way that each pair of adult rabbits produces a pair of baby rabbits each month.

Number of Months	Number of Pairs	Pairs of Rabbits (the blue rabbits are ready to reproduce)
Start		
1	1	
2	2	
3	3	
4	5	
5	8	
⋮	⋮	

Same pair
(rabbits never die)

The first month after birth, the rabbits are adolescents and produce no offspring. However, beginning with the second month, the rabbits are adults, and each pair produces a pair of offspring every month. The sequence of numbers describing the number of rabbits is called the *Fibonacci sequence,* and it has applications in many areas, including biology and botany.

In this Quest you are to examine some properties of the Fibonacci sequence. Let a_n denote the number of pairs of rabbits in the "colony" at the end of *n* months.

a. Explain why $a_1 = 1, a_2 = 1, a_3 = 2, a_4 = 3$, and, in general,

$$a_{n+1} = a_{n-1} + a_n \quad \text{for } n = 2, 3, 4, \ldots$$

b. The *growth rate* of the colony during the $(n + 1)$th month is

$$r_n = \frac{a_{n+1}}{a_n}$$

Compute r_n for $n = 1, 2, 3, \ldots, 10$.

c. Assume that the growth rate sequence $\{r_n\}$ defined in part **b** converges, and let

$$L = \lim_{n \to \infty} r_n$$

Hint: Use the recursion formula in part **a.**

C *In Problems 49–52, use the fact that* $\lim_{n\to\infty} a_n = L$ *means* $|a_n - L|$ *is arbitrarily small when n is sufficiently large.*

49. Given $\lim_{n\to\infty} \dfrac{n}{n+1} = 1$, find N so that

$$\left| \frac{n}{n+1} - 1 \right| < 0.01 \quad \text{if } n > N$$

50. Given $\lim_{n\to\infty} \dfrac{2n+1}{n+3} = 2$, find N so that

$$\left| \frac{2n+1}{n+3} - 2 \right| < 0.01 \quad \text{if } n > N$$

51. Given $\lim_{n\to\infty} \dfrac{n^2+1}{n^3} = 0$, find N so that

$$\left| \frac{n^2+1}{n^3} \right| < 0.001 \quad \text{if } n > N$$

52. Given $\lim_{n\to\infty} e^{-n} = 0$, find N so that

$$\frac{1}{e^n} < 0.001 \quad \text{if } n > N$$

53. **Convergence of the sequence** $\left\{ \left(1 + \dfrac{1}{n}\right)^n \right\}$

a. Suppose *n* is a positive integer. Use the binomial theorem to show that

$$\left(1 + \frac{1}{n}\right)^n = 1 + 1 + \frac{1}{2!}\left(1 - \frac{1}{n}\right) + \frac{1}{3!}\left(1 - \frac{1}{n}\right)\left(1 - \frac{2}{n}\right)$$
$$+ \cdots + \frac{1}{n!}\left(1 - \frac{1}{n}\right)\left(1 - \frac{2}{n}\right)\cdots\left(1 - \frac{n-1}{n}\right)$$

b. Use part **a** to show that $\left(1 + \dfrac{1}{n}\right)^n < 3$ for all *n*.

c. Use part **a** to show that $\left\{ \left(1 + \dfrac{1}{n}\right)^n \right\}$ is an increasing sequence. Then use the BMCT to show that this sequence converges.

54. **Proof of the BMCT** Suppose there exist a number *M* and a positive integer *N* so that

$$a_n \le a_{n+1} \le M \quad \text{for all } n > N.$$

a. It is a fundamental property of numbers that a sequence with an upper bound must have a *least* upper bound; that is, there must be a number *A* with the property that $a_n \le A$ for all *n* but if *c* is a number such that $c < A$, then $a_m > c$ for at least one *m*. Use this property to show that if $\epsilon > 0$ is given, there exists a positive integer *N* so that

$$A - \epsilon < a_n < A$$

for all $n > N$.

b. Show that $|a_n - A| < \epsilon$ for all $n > N$ and conclude that $\lim_{n\to\infty} a_n = A$.

8.2 *Introduction to Infinite Series: Geometric Series*

definition of infinite series, general properties of infinite series, geometric series, applications of geometric series ■

DEFINITION OF INFINITE SERIES

One way to add a list of numbers is to form subtotals until the end of the list is reached. Similarly, to give meaning to the infinite sum

$$S = a_1 + a_2 + a_3 + a_4 + \cdots$$

it is natural to examine the "partial sums"

$$a_1, \quad a_1 + a_2, \quad a_1 + a_2 + a_3, \quad a_1 + a_2 + a_3 + a_4, \quad \ldots$$

If it is possible to attach a numerical value to the infinite sum, we would expect the partial sums $S_n = a_1 + a_2 + \cdots + a_n$ to approach that value as n increases without bound. These ideas lead us to the following definition.

Infinite Series

An **infinite series** is an expression of the form

$$a_1 + a_2 + a_3 + \cdots = \sum_{k=1}^{\infty} a_k$$

and the ***n*th partial sum** of the series is

$$S_n = a_1 + a_2 + \cdots + a_n = \sum_{k=1}^{n} a_k$$

The series is said to **converge with sum *S*** if the sequence of partial sums $\{S_n\}$ converges to S. In this case, we write

$$\sum_{k=1}^{\infty} a_k = \lim_{n \to \infty} S_n = S$$

If the sequence $\{S_n\}$ does not converge, the series

$$\sum_{k=1}^{\infty} a_k$$

diverges and has no sum.

■ *What This Says:* An infinite series converges if its sequence of partial sums converges and diverges otherwise. If it converges, its sum is defined to be the limit of the sequence of partial sums.

REMARK We shall use the symbol $\displaystyle\sum_{k=1}^{\infty} a_k$ to denote the series $a_1 + a_2 + a_3 + \cdots$ regardless of whether this series converges or diverges. If the sequence of partial sums $\{S_n\}$ converges, then

$$\sum_{k=1}^{\infty} a_k = \lim_{n \to \infty} \left(\sum_{k=1}^{n} a_k \right)$$

and the symbol $\displaystyle\sum_{k=1}^{\infty} a_k$ is used to represent *both* the series and its sum.

Also, we shall consider certain series in which the starting point is not 1; for example, the series

$$\frac{1}{3} + \frac{1}{4} + \frac{1}{5} + \cdots \quad \text{can be denoted by} \quad \sum_{k=3}^{\infty} \frac{1}{k} \quad \text{or} \quad \sum_{k=2}^{\infty} \frac{1}{k+1}$$

EXAMPLE 1   *Convergent series*

Show that the series $\displaystyle\sum_{k=1}^{\infty} \frac{1}{2^k}$ converges.

Solution   This series has the following partial sums:

$$S_1 = \tfrac{1}{2}$$
$$S_2 = \tfrac{1}{2} + \tfrac{1}{4} = \tfrac{3}{4}$$
$$S_3 = \tfrac{1}{2} + \tfrac{1}{4} + \tfrac{1}{8} = \tfrac{7}{8}$$
$$\vdots$$
$$S_n = \frac{1}{2} + \frac{1}{4} + \cdots + \frac{1}{2^n}$$

The sequence of partial sums is $\tfrac{1}{2}, \tfrac{3}{4}, \tfrac{7}{8}, \tfrac{15}{16}, \tfrac{31}{32}, \tfrac{63}{64}, \tfrac{127}{128}, \ldots$, and, in general,

$$S_n = 1 - \frac{1}{2^n}$$

Because $\displaystyle\lim_{n \to \infty} \left(1 - \frac{1}{2^n}\right) = 1$, we conclude that the series converges and its sum is 1. ∎

EXAMPLE 2   *Divergent series*

Show that the series $\displaystyle\sum_{k=1}^{\infty} (-1)^k$ diverges.

Solution   The series can be **expanded** (written out) as

$$\sum_{k=1}^{\infty} (-1)^k = -1 + 1 - 1 + 1 - 1 + 1 - \cdots$$

and we see that the nth partial sum is

$$S_n = \begin{cases} -1 & \text{if } n \text{ is odd} \\ 0 & \text{if } n \text{ is even} \end{cases}$$

Because the sequence $\{S_n\}$ has no limit, the given series must diverge. ∎

A series is called a **telescoping series** (or collapsing series) if there is internal cancellation in the partial sums, as illustrated by the following example.

EXAMPLE 3   *A telescoping series*

Show that the series $\displaystyle\sum_{k=1}^{\infty} \frac{1}{k^2 + k}$ converges and find its sum.

Solution   Using partial fractions, we find that

$$\frac{1}{k^2 + k} = \frac{1}{k(k+1)} = \frac{1}{k} + \frac{-1}{k+1}$$

Thus, the nth partial sum of the given series can be represented as follows:

$$S_n = \sum_{k=1}^{n} \frac{1}{k^2 + k} = \sum_{k=1}^{n} \left[\frac{1}{k} - \frac{1}{k+1} \right]$$

$$= \left(1 - \frac{1}{2} \right) + \left(\frac{1}{2} - \frac{1}{3} \right) + \left(\frac{1}{3} - \frac{1}{4} \right) + \cdots + \left(\frac{1}{n} - \frac{1}{n+1} \right)$$

$$= 1 + \left(-\frac{1}{2} + \frac{1}{2} \right) + \left(-\frac{1}{3} + \frac{1}{3} \right) + \cdots + \left(-\frac{1}{n} + \frac{1}{n} \right) - \frac{1}{n+1}$$

$$= 1 - \frac{1}{n+1}$$

The limit of the sequence of partial sums is

$$\lim_{n \to \infty} S_n = \lim_{n \to \infty} \left[1 - \frac{1}{n+1} \right] = 1$$

Thus, the series converges, with sum $S = 1$. ∎

GENERAL PROPERTIES OF INFINITE SERIES

Next, we shall examine two general properties of infinite series. Here and elsewhere, *when the starting point of a series is not important, we may denote the series by writing*

$$\sum a_k \quad \text{instead of} \quad \sum_{k=1}^{\infty} a_k$$

THEOREM 8.6   *Linearity of infinite series*

If Σa_k and Σb_k are convergent series, then so is $\Sigma(ca_k + db_k)$ for constants c, d, and

$$\sum (ca_k + db_k) = c \sum a_k + d \sum b_k$$

Proof   Compare this with the linearity property in Theorem 5.4. In Chapter 5, the limit properties are for finite sums, and this theorem is for infinite sums. The proof of this theorem is similar to the proof of Theorem 5.4, but in this case it follows from the linearity rule for sequences (Theorem 8.1). The details are left as a problem.   ⸗

EXAMPLE 4   *Linearity used to establish convergence*

Show that the series $\displaystyle\sum_{k=1}^{\infty} \left[\frac{4}{k^2 + k} - \frac{6}{2^k} \right]$ converges, and find its sum.

Solution   Because we know that $\displaystyle\sum_{k=1}^{\infty} \frac{1}{k^2 + k}$ and $\displaystyle\sum_{k=1}^{\infty} \frac{1}{2^k}$ both converge, the linearity property allows us to write the given series as

$$4 \sum_{k=1}^{\infty} \frac{1}{k^2 + k} - 6 \sum_{k=1}^{\infty} \frac{1}{2^k}$$

We can now use Examples 1 and 3 to conclude that the series converges and that the sum is

$$4 \sum_{k=1}^{\infty} \frac{1}{k^2 + k} - 6 \sum_{k=1}^{\infty} \frac{1}{2^k} = 4(1) - 6(1) = -2$$ ∎

The linearity property also provides useful information about a series of the form $\Sigma(ca_k + db_k)$ when either Σa_k or Σb_k diverges and the other converges.

THEOREM 8.7   *Divergence of the sum of a convergent and a divergent series*

If either Σa_k or Σb_k diverges and the other converges, then the series $\Sigma(a_k + b_k)$ must diverge.

Proof   Suppose Σa_k diverges and Σb_k converges. Then if the series $\Sigma(a_k + b_k)$ also converges, the linearity property tells us that the series

$$\sum [(a_k + b_k) - b_k] = \sum a_k$$

must converge, contrary to hypothesis. It follows that the series $\Sigma(a_k + b_k)$ diverges.

This theorem tells us, for example, that $\displaystyle\sum_{k=1}^{\infty} \left[\frac{1}{k^2 + k} + (-1)^k \right]$ must diverge because even though $\displaystyle\sum_{k=1}^{\infty} \frac{1}{k^2 + k}$ converges (Example 3), $\displaystyle\sum_{k=1}^{\infty} (-1)^k$ diverges (Example 2).

GEOMETRIC SERIES

We are still very limited in the available techniques for determining whether a given series is convergent or divergent. In fact, the purpose of a great part of this chapter is to develop efficient techniques for making this determination. We begin this quest by considering an important special kind of series.

Geometric Series

A **geometric series** is an infinite series in which the ratio of successive terms is constant. If this constant ratio is r, then the series has the form

$$\sum_{k=0}^{\infty} ar^k = a + ar + ar^2 + ar^3 + \cdots + ar^n + \cdots \qquad a \neq 0$$

For example, $3 + \frac{3}{2} + \frac{3}{4} + \frac{3}{8} + \cdots$ is a geometric series because each term is one-half the preceding term. The ratio of a geometric series may be positive or negative. For example,

$$\sum_{k=0}^{\infty} \frac{2}{(-3)^k} = 2 - \frac{2}{3} + \frac{2}{9} - \frac{2}{27} + \cdots$$

is a geometric series with $r = -\frac{1}{3}$. The following theorem tells us how to determine whether a given geometric series converges or diverges and, if it does converge, what its sum must be.

THEOREM 8.8   *Geometric series theorem*

The geometric series $\displaystyle\sum_{k=0}^{\infty} ar^k$ with $a \neq 0$ diverges if $|r| \geq 1$ and converges if $|r| < 1$ with sum

$$\sum_{k=0}^{\infty} ar^k = \frac{a}{1 - r}$$

Proof   Note that the nth partial sum of the geometric series is

$$S_n = a + ar + ar^2 + \cdots + ar^{n-1}$$

$$rS_n = ar + ar^2 + ar^3 + \cdots + ar^n \qquad \text{Multiply both sides by } r.$$

$$rS_n - S_n = (ar + ar^2 + ar^3 + \cdots + ar^n) - (a + ar + ar^2 + \cdots + ar^{n-1})$$

$$(r - 1)S_n = ar^n - a$$

$$S_n = \frac{a(r^n - 1)}{(r - 1)}$$

If $|r| > 1$, the sequence of partial sums $\{S_n\}$ has no limit, so the geometric series must diverge. However, if $|r| < 1$, Theorem 8.5 tells us that $r^n \to 0$ as $n \to \infty$, so we have

$$\sum_{k=0}^{\infty} ar^k = \lim_{n\to\infty} S_n = \lim_{n\to\infty} a\left(\frac{r^n - 1}{r - 1}\right) = \frac{a(0 - 1)}{r - 1} = \frac{a}{1 - r}$$

To complete the proof, it must be shown that the geometric series diverges when $|r| = 1$, and we leave this final step as a problem.   ▬▬

EXAMPLE 5   *Testing for convergence in a geometric series*

Determine whether each of the following geometric series converges or diverges. If the series converges, find its sum.

a. $\displaystyle\sum_{k=0}^{\infty} \frac{1}{7}\left(\frac{3}{2}\right)^k$ b. $\displaystyle\sum_{k=2}^{\infty} 3\left(-\frac{1}{5}\right)^k$

Solution

a. Because $r = \frac{3}{2}$ satisfies $|r| \geq 1$, the series diverges.

b. We have $r = -\frac{1}{5}$ so $|r| < 1$, and the geometric series converges. The first value of k is 2 (not 0), so the value a (the first value) is found by $a = 3(-\frac{1}{5})^2 = \frac{3}{25}$, so

$$\sum_{k=2}^{\infty} 3\left(-\frac{1}{5}\right)^k = \frac{a}{1 - r} = \frac{\frac{3}{25}}{1 - (-\frac{1}{5})} = \frac{1}{10}$$ ■

APPLICATIONS OF GEOMETRIC SERIES

Geometric series can be used in many different ways. Our next three examples illustrate several applications involving geometric series.

Recall that a rational number r is one that can be written as $r = p/q$ for integer p and nonzero integer q. It can be shown that any such number has a decimal representation in which a pattern of numbers repeats. For instance,

$$\frac{5}{10} = 0.5 = 0.5\overline{0} \qquad \frac{5}{11} = 0.454545\ldots = 0.\overline{45}$$

where the bar indicates that the pattern repeats indefinitely. The next example shows how geometric series can be used to reverse this process by writing a given repeating decimal as a rational number.

EXAMPLE 6   *Repeating decimals*

Write $15.4\overline{23}$ as a rational number $\dfrac{p}{q}$.

Solution   The bar over the 23 indicates that this block of digits is to be repeated, that is, $15.423232323\ldots$. The repeating part of the decimal can be written as a geometric series as follows.

$$15.4\overline{23} = 15 + \frac{4}{10} + \frac{23}{10^3} + \frac{23}{10^5} + \frac{23}{10^7} + \cdots$$

$$= 15 + \frac{4}{10} + \frac{23}{10^3}\left[1 + \frac{1}{10^2} + \frac{1}{10^4} + \cdots\right]$$

$$= 15 + \frac{4}{10} + \frac{23}{10^3}\left[\frac{1}{1 - \frac{1}{100}}\right] \qquad \textit{Sum of a geometric series with } a = 1 \textit{ and } r = 1/100$$

$$= 15 + \frac{4}{10} + \frac{23}{10^3}\left[\frac{100}{99}\right]$$

$$= 15 + \frac{4}{10} + \frac{23}{990}$$

$$= \frac{15{,}269}{990}$$

A tax rebate that returns a certain amount of money to taxpayers can result in spending that is many times this amount. This phenomenon is known in economics as the *multiplier effect*. It occurs because the portion of the rebate that is spent by one individual becomes income for one or more others who, in turn, spend some of it again, creating income for yet other individuals to spend. If the fraction of income that is saved remains constant as this process continues indefinitely, the total amount spent as a result of the rebate is the sum of a geometric series.

EXAMPLE 7   *Modeling Problem: Multiplier effect in economics*

Suppose that nationwide approximately 90% of all income is spent and 10% is saved. How much total spending will be generated by a $40 billion tax rebate if savings habits do not change?

Solution   The amount (in billions) spent by original recipients of the rebate is 40. This becomes new income, of which 90%, or 0.9(40), is spent. This, in turn, generates additional spending of 0.9[0.9(40)], and so on. The total amount spent if this process continues indefinitely is

$$40 + 0.9(40) + 0.9^2(40) + 0.9^3(40) + \cdots = (40)[1 + 0.9 + 0.9^2 + \cdots]$$

$$= 40\sum_{k=0}^{\infty}0.9^k \qquad \textit{Geometric series with } a = 1, r = 0.9$$

$$= 40\left(\frac{1}{1 - 0.9}\right)$$

$$= 400 \text{ (billion)}$$

EXAMPLE 8   *Modeling Problem: Accumulation of medication in a body*

A patient is given an injection of 10 units of a certain drug every 24 hours. The drug is eliminated exponentially so that the portion that remains in the patient's body after t days is $f(t) = e^{-t/5}$. If the treatment is continued indefinitely, approximately how many units of the drug will eventually be in the patient's body just prior to an injection?

Solution   Of the original dose of 10 units, only $10e^{-1/5}$ are left in the patient's body after the first day (just prior to the second injection). That is,

$$S_1 = 10e^{-1/5}$$

The medication in the patient's body after 2 days consists of what remains from the first two doses. Of the original dose, only $10e^{-2/5}$ units are left (because 2 days have elapsed), and of the second dose, $10e^{-1/5}$ units remain:

$$S_2 = 10e^{-1/5} + 10e^{-2/5}$$

Similarly, for n days,

$$S_n = 10e^{-1/5} + 10e^{-2/5} + \cdots + 10e^{-n/5}$$

The amount S of medication in the patient's body in the long run is the limit of S_n as $n \to \infty$. That is,

$$S = \lim_{n \to \infty} S_n$$

$$= \sum_{k=1}^{\infty} 10e^{-k/5} \qquad \textit{Geometric series with} \quad a = 10e^{-1/5} \textit{ and } r = e^{-1/5}$$

$$= \frac{10e^{-1/5}}{1 - e^{-1/5}}$$

$$\approx 45.166556$$

We see that about 45 units remain in the patient's body. ■

WARNING ▶ As a final note, remember that a sequence is a mere succession of terms, whereas a series is a sum of such terms. Do not confuse the two concepts. For example, a sequence of terms may converge, but the series of the same terms may diverge:

$$\left\{ 1 + \frac{1}{2^n} \right\} \text{ is the } \textit{sequence } \frac{3}{2}, \frac{5}{4}, \frac{9}{8}, \frac{17}{16}, \dots, \text{ which converges to 1.}$$

$$\sum_{k=1}^{\infty} \left(1 + \frac{1}{2^k} \right) \text{ is the } \textit{series } \frac{3}{2} + \frac{5}{4} + \frac{9}{8} + \cdots, \text{ which diverges.}$$

8.2 Problem Set

Ⓐ 1. ■ **What Does This Say?** Explain the difference between sequences and series.

2. ■ **What Does This Say?** What is a geometric series? Include a derivation of the formula for the sum of a geometric series.

Determine whether the geometric series given in Problems 3–22 converges or diverges, and find the sum of each convergent series.

3. $\sum_{k=0}^{\infty} \left(\frac{4}{5} \right)^k$

4. $\sum_{k=0}^{\infty} \left(-\frac{4}{5} \right)^k$

5. $\sum_{k=0}^{\infty} \frac{2}{3^k}$

6. $\sum_{k=0}^{\infty} \frac{2}{(-3)^k}$

7. $\sum_{k=1}^{\infty} \left(\frac{3}{2} \right)^k$

8. $\sum_{k=1}^{\infty} \frac{3}{2^k}$

9. $\sum_{k=2}^{\infty} \frac{3}{(-4)^k}$

10. $\sum_{k=1}^{\infty} 5(0.9)^k$

11. $\sum_{k=1}^{\infty} e^{-0.2k}$

12. $\sum_{k=1}^{\infty} \frac{3^k}{4^{k+2}}$

13. $\sum_{k=2}^{\infty} \frac{(-2)^{k-1}}{3^{k+1}}$

14. $\sum_{k=2}^{\infty} (-1)^k \frac{2^{k+1}}{3^{k-3}}$

15. $\frac{1}{2} - \frac{1}{2^2} + \frac{1}{2^3} - \frac{1}{2^4} + \cdots$

16. $1 + \pi + \pi^2 + \pi^3 + \cdots$

17. $\frac{1}{4} + \left(\frac{1}{4} \right)^4 + \left(\frac{1}{4} \right)^7 + \left(\frac{1}{4} \right)^{10} + \cdots$

18. $\frac{2}{3} - \left(\frac{2}{3} \right)^3 + \left(\frac{2}{3} \right)^5 - \left(\frac{2}{3} \right)^7 + \cdots$

19. $2 + \sqrt{2} + 1 + \frac{1}{\sqrt{2}} + \cdots$

20. $3 - \sqrt{3} + 1 - \dfrac{1}{\sqrt{3}} + \cdots$

21. $(1 + \sqrt{2}) + 1 + (-1 + \sqrt{2}) + (3 - 2\sqrt{2}) + \cdots$

22. $(\sqrt{2} - 1) + 1 + (\sqrt{2} + 1) + (2\sqrt{2} + 3) + \cdots$

In Problems 23–30, each series telescopes. In each case, express the nth partial sum S_n in terms of n and determine whether the series converges or diverges by examining $\lim\limits_{n \to \infty} S_n$.

23. $\displaystyle\sum_{k=1}^{\infty} \left[\dfrac{1}{k^{0.1}} - \dfrac{1}{(k+1)^{0.1}} \right]$
24. $\displaystyle\sum_{k=1}^{\infty} \left[\dfrac{1}{2k+1} - \dfrac{1}{2k+3} \right]$

25. $\displaystyle\sum_{k=0}^{\infty} \dfrac{1}{(k+1)(k+2)}$
26. $\displaystyle\sum_{k=2}^{\infty} \dfrac{1}{k(k+1)}$

27. $\displaystyle\sum_{k=1}^{\infty} \ln\!\left(1 + \dfrac{1}{k}\right)$
28. $\displaystyle\sum_{k=1}^{\infty} \dfrac{1}{(2k-1)(2k+1)}$

29. $\displaystyle\sum_{k=1}^{\infty} \dfrac{2k+1}{k^2(k+1)^2}$
30. $\displaystyle\sum_{k=1}^{\infty} \dfrac{\sqrt{k+1} - \sqrt{k}}{\sqrt{k^2 + k}}$

Hint: Note that $\sqrt{k^2 + k} = \sqrt{k}\sqrt{k+1}$.

Express each decimal given in Problems 31–34 as a common (reduced) fraction.

31. $0.\overline{01}$
32. $2.23\overline{1}$
33. $1.405\overline{405}$
34. $41.201\overline{0010}$

Ⓑ 35. a. Find numbers A and B such that
$$\dfrac{k-1}{2^{k+1}} = \dfrac{Ak}{2^k} - \dfrac{B(k+1)}{2^{k+1}}$$

b. Evaluate $\displaystyle\sum_{k=1}^{\infty} \dfrac{k-1}{2^{k+1}}$ as a telescoping series.

36. Evaluate $\displaystyle\sum_{k=1}^{\infty} \dfrac{2k-1}{3^{k+1}}$.

Hint: Follow the procedure in Problem 35.

37. Evaluate $\displaystyle\sum_{n=1}^{\infty} \dfrac{\ln\!\left(\dfrac{n^{n+1}}{(n+1)^n}\right)}{n(n+1)}$.

Hint: Use the properties of logarithms to show that the series telescopes.

38. Evaluate $\displaystyle\sum_{n=1}^{\infty} \dfrac{n}{(n+1)!}$.

Hint: Express as a telescoping series.

39. Find $\displaystyle\sum_{k=0}^{\infty} (2a_k + 2^{-k})$

given that $\displaystyle\sum_{k=0}^{\infty} a_k = 3.57$.

40. Find $\displaystyle\sum_{k=0}^{\infty} \dfrac{b_k - 3^{-k}}{2}$

given that $\displaystyle\sum_{k=0}^{\infty} b_k = 0.54$.

41. Evaluate $\displaystyle\sum_{k=0}^{\infty} \left(\dfrac{1}{2^k} + \dfrac{1}{3^k} \right)^2$.

42. Evaluate $\displaystyle\sum_{k=0}^{\infty} \left[\left(\dfrac{2}{3}\right)^k + \left(\dfrac{3}{4}\right)^k \right]^2$.

43. Show that $\displaystyle\sum_{k=0}^{\infty} \dfrac{1}{(a+k)(a+k+1)} = \dfrac{1}{a}$ if $a > 0$.

44. Evaluate $\displaystyle\sum_{k=0}^{\infty} \dfrac{1}{(2+k)^2 + 2 + k}$.

45. If $\displaystyle\sum_{k=0}^{\infty} a_k^2 = \sum_{k=0}^{\infty} b_k^2 = 4$ and $\displaystyle\sum_{k=0}^{\infty} a_k b_k = 3$,

what is $\displaystyle\sum_{k=0}^{\infty} (a_k - b_k)^2$?

MODELING PROBLEMS *In Problems 46–56, set up an appropriate model to answer the given question. Be sure to state your assumptions.*

46. A pendulum is released through an arc of length 20 cm from a vertical position and is allowed to swing free until it eventually comes to rest. Each subsequent swing of the bob of the pendulum is 90% as far as the preceding swing. How far will the bob travel before coming to rest?

47. A flywheel rotates at 500 rpm and slows in such a way that during each minute it revolves at two-thirds the rate of the preceding minute. How many total revolutions does the flywheel make before coming to rest?

48. A ball is dropped from a height of 10 ft. Each time the ball bounces, it rises 0.6 the distance it had previously fallen. What is the total distance traveled by the ball?

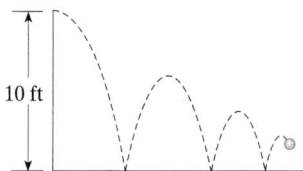

49. Suppose the ball in Problem 48 is dropped from a height of h feet and on each bounce rises 75% of the distance it had previously fallen. If it travels a total distance of 21 ft, what is h?

50. Suppose that nationwide approximately 92% of all income is spent and 8% is saved. How much total spending will be generated by a $50 billion tax cut if savings habits do not change?

51. Suppose that a piece of machinery costing $10,000 depreciates 20% of its present value each year. That is, the first year $10,000(0.20) = $2,000 is depreciated. The second year's depreciation is $8,000(0.20) = $1,600, because the value for the second year is $10,000 − $2,000 = $8,000. If the depreciation is calculated this way indefinitely, what is the total depreciation?

52. Winnie Winner wins $100 in a pie-baking contest run by the Hi-Do Pie Co. The company gives Winnie the $100. However, the tax collector wants 20% of the $100. Winnie pays the tax. But then she realizes that she didn't really win a $100 prize and tells her story to the Hi-Do Co. The friendly Hi-Do Co. gives Winnie the $20 she paid in taxes. Unfortunately, the tax collector now wants 20% of the $20. She pays the tax again and then goes back to the Hi-Do Co. with her story. Assume that this can go on indefinitely. How much money does the Hi-Do Co. have to give Winnie so that she will really win $100? How much does she pay in taxes?

53. A patient is given an injection of 20 units of a certain drug every 24 hr. The drug is eliminated exponentially, so the part that remains in the patient's body after t days is $f(t) = e^{-t/2}$. If the treatment is to continue indefinitely, approximately how many units of the drug will eventually be in the patient's body just before an injection?

54. A certain drug is injected into a body. It is known that the amount of drug in the body at the end of a given hour is $\frac{1}{4}$ the amount that was present at the end of the previous hour. One unit is injected initially, and to keep the concentration up, an additional unit is injected at the end of each subsequent hour. What is the highest possible amount of the drug that can ever be present in the body?

55. Each January 1, the administration of a certain private college adds 6 new members to its board of trustees. If the fraction of trustees who remain active for at least t years is $f(t) = e^{-0.2t}$, approximately how many active trustees will the college have in the long run? Assume that there were 6 members on the board on January 1 of the present year.

56. How much should you invest today at an annual interest rate of 15% compounded continuously so that, starting next year, you can make annual withdrawals of $2,000 forever?

57. HISTORICAL QUEST   The Greek philosopher Zeno of Elea (495–435 B.C.) presented a number of paradoxes that caused a great deal of concern to the mathematicians of his period (see Section 1.1, for example). These paradoxes were important in the development of the notion of infinitesimals. Perhaps the most famous is the *racecourse paradox*, which can be stated as follows:

   A runner can never reach the end of a race. As he runs the track, he must first run $\frac{1}{2}$ the length of the track, then $\frac{1}{2}$ the remaining distance, so that at this point he has run $\frac{1}{2} + \frac{1}{4} = \frac{3}{4}$ of the length of the track, and $\frac{1}{4}$ remains to be run. After running half this distance, he finds he still has $\frac{1}{8}$ of the track to run, and so on, indefinitely.

Suppose the runner runs at a constant pace and that it takes him T minutes to run the first half of the track. Set up an infinite series that gives the *total* time required to

run the track, and verify that the total time is $2T$, as intuition would lead us to expect.

58. HISTORICAL QUEST   Leonhard Euler (1707–1783) was introduced in Chapter 4 Supplementary Problem 84, where we saw that Euler writes $a^{\epsilon} = 1 + k\epsilon$ for an "infinitely small" number ϵ. In this HISTORICAL QUEST we will see why we define

$$\lim_{n \to +\infty} \left(1 + \frac{1}{n}\right)^n = e$$

by asking you to reconstruct the steps of Euler's work.

   Let x be a (finite) number. Euler introduces the "infinitely large" number $N = x/\epsilon$. Explain each step.

$$a^x = a^{N\epsilon} = (a^{\epsilon})^N$$
$$= (1 + k\epsilon)^N$$
$$= \left(1 + \frac{kx}{N}\right)^N$$
$$= 1 + N\left(\frac{kx}{N}\right) + \frac{N(N-1)}{2!}\left(\frac{kx}{N}\right)^2$$
$$+ \frac{N(N-1)(N-2)}{3!}\left(\frac{kx}{N}\right)^3 + \cdots$$
$$= 1 + kx + \frac{1}{2!}\frac{N(N-1)}{N^2}k^2x^2$$
$$+ \frac{1}{3!}\frac{N(N-1)(N-2)}{N^3}k^3x^3 + \cdots$$

Because N is infinitely large, Euler assumed that

$$1 = \frac{N-1}{N} = \frac{N-2}{N} = \cdots$$

Substitute these values into the given derivation and also substitute e for a to obtain

$$e = 1 + \frac{1}{1!} + \frac{1}{2!} + \frac{1}{3!} + \cdots$$

Euler calculated e to 23 places:

$$e \approx 2.71828182845904523536028$$

Show that Euler's derivation is equivalent to

$$\lim_{n \to +\infty} \left(1 + \frac{1}{n}\right)^n.$$

❍ 59. a. Let $\sum_{k=1}^{n} a_k = S_n$ be the nth partial sum of an infinite series. Show that $S_n - S_{n-1} = a_n$.

   b. Use part **a** to find the nth term of the infinite series whose nth partial sum is

$$S_n = \frac{n}{2n+3}$$

60. Prove that if $\Sigma a_k = A$ and $\Sigma b_k = B$, then $\Sigma(a_k + b_k) = A + B$.

61. Show that if Σa_k converges and Σb_k diverges, then $\Sigma(a_k - b_k)$ diverges. *Hint:* Note that $b_k = a_k - (a_k - b_k)$.

62. THINK TANK PROBLEM   Find two divergent series Σa_k and Σb_k such that:
a. $\Sigma(a_k + b_k)$ is also divergent.
b. $\Sigma(a_k + b_k)$ is convergent.

63. Show that $\displaystyle\sum_{k=0}^{\infty} ar^k$ diverges if $|r| \geq 1$.

64. a. Show that $\displaystyle\sum_{k=1}^{n} (a_k - a_{k+2})$
$$= (a_1 + a_2) - (a_{n+1} + a_{n+2}).$$

b. Use part **a** to show that if $\lim_{k\to\infty} a_k = A$, then
$$\sum_{k=1}^{\infty} (a_k - a_{k+2}) = a_1 + a_2 - 2A.$$

c. Use part **b** to evaluate
$$\sum_{k=1}^{\infty} \left[k^{1/k} - (k+2)^{1/(k+2)} \right]$$

d. Evaluate $\displaystyle\sum_{k=2}^{\infty} \frac{1}{k^2 - 1}$.

65. Square *ABCD* has sides of length 1. Square *EFGH* is formed by connecting the midpoints of the sides of the

first square, as shown in Figure 8.12. Assume that the pattern of shaded regions in the squares is continued indefinitely. What is the total area of the shaded regions?

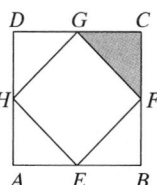

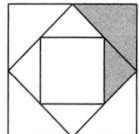 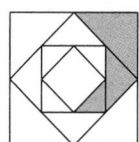 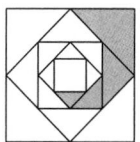

■ **FIGURE 8.12**   Find the area of the shaded regions.

66. Repeat Problem 65 for a square whose sides have length *a*.

8.3   *The Integral Test: p-Series*

IN THIS SECTION   **divergence test, series of nonnegative numbers, the integral test, *p*-series**   ■

The convergence or divergence of an infinite series is determined by the behavior of its *n*th partial sum, S_n. In Section 8.2, we saw how algebraic methods can sometimes be used to find formulas for the *n*th partial sum of series. Unfortunately, it is often difficult or even impossible to find a compact formula for the *n*th partial sum S_n of a series, and other techniques must be used to determine convergence or divergence.

DIVERGENCE TEST

When investigating the infinite series Σa_k, it is easy to confuse the sequence of *general terms* $\{a_k\}$ with the sequence of *partial sums* $\{S_n\}$, where

$$S_n = \sum_{k=1}^{n} a_k$$

Because the sequence $\{a_k\}$ is usually more accessible than $\{S_n\}$, it would be convenient if the convergence of $\{S_n\}$ could be settled by examining

$$\lim_{k\to\infty} a_k$$

Even though we do not have one simple, definitive test for convergence, we do have a first test that tells us that if certain conditions are met, then the series diverges.

THEOREM 8.9   *The divergence test*

If the series Σa_k converges, then $\lim_{k\to\infty} a_k = 0$. Thus, if $\lim_{k\to\infty} a_k \neq 0$, the series Σa_k must diverge.

WARNING Theorem 8.9 gives a necessary but not a sufficient condition for convergence. For example, clouds are necessary for rain; clouds are not sufficient for rain. ◄►

Proof   Suppose the sequence of partial sums $\{S_n\}$ converges with sum L, so that $\lim\limits_{n\to\infty} S_n = L$. Then we also have

$$\lim_{n\to\infty} S_{n-1} = L$$

Because $S_k - S_{k-1} = a_k$, it follows that

$$\lim_{k\to\infty} a_k = \lim_{k\to\infty} (S_k - S_{k-1}) = \lim_{k\to\infty} S_k - \lim_{k\to\infty} S_{k-1} = L - L = 0 \quad \blacksquare$$

EXAMPLE 1   *Divergence test to show divergence*

Show that the series $\displaystyle\sum_{n=0}^{\infty} \frac{n-300}{4n+750}$ diverges.

Solution   Taking the limit of the kth term as $k \to \infty$, we find

$$\lim_{k\to\infty} \frac{k-300}{4k+750} = \frac{1}{4}$$

Because this limit is not 0, the divergence test tells us that the series must diverge.   ■

> ■ **What This Says:**   The divergence test can only tell us that Σa_k diverges *if* $\lim\limits_{n\to\infty} a_n \ne 0$, but *cannot* be used to show convergence. In Example 2 on page 564, we show that the series $\Sigma \frac{1}{k}$ *diverges* even though $\lim\limits_{k\to\infty} \frac{1}{k} = 0$.

SERIES OF NONNEGATIVE NUMBERS; THE INTEGRAL TEST

Series whose terms are all nonnegative numbers play an important role in the general theory of infinite series and in applications. Our next goal is to develop convergence tests for nonnegative-term series, and we begin by establishing the following general principle.

THEOREM 8.10   *Convergence criterion for series with nonnegative terms*

A series Σa_k with $a_k \ge 0$ for all k converges if its sequence of partial sums is bounded from above and diverges otherwise.

Proof   Suppose Σa_k is a series of nonnegative terms, and let S_n denote its nth partial sum; that is,

$$S_n = \sum_{k=1}^{n} a_k = a_1 + a_2 + \cdots + a_n$$

Because $S_{n+1} = S_n + a_{n+1}$ and because $a_k \ge 0$ for all k, it follows that

$$S_{n+1} \ge S_n$$

for all n, so that $\{S_n\}$ is a nondecreasing sequence. According to the BMCT, the non-decreasing sequence $\{S_n\}$ converges if it is bounded from above and diverges otherwise. Hence, because the series Σa_k represents the limit of the sequence $\{S_n\}$, the series Σa_k converges if the sequence $\{S_n\}$ is bounded from above and diverges if it is unbounded.   ■

This convergence criterion is often difficult to apply because it is not easy to determine whether the sequence of partial sums $\{S_n\}$ is bounded from above. Our next goal in this section and in Sections 8.4–8.6 is to examine various "tests for convergence." These are procedures that allow us to determine indirectly whether a given series converges or diverges without actually having to compute the limit of the partial sums. We begin with a convergence test that relates the convergence of a series to that of an improper integral.

THEOREM 8.11   *The integral test*

If $a_k = f(k)$ for $k = 1, 2, \ldots$, where f is a positive, continuous, and decreasing function of x for $x \geq 1$, then

$$\sum_{k=1}^{\infty} a_k \quad \text{and} \quad \int_1^{\infty} f(x)\, dx$$

either both converge or both diverge.

Proof   We shall use a geometric argument to show that the sequence of partial sums $\{S_n\}$ of the series $\sum_{k=0}^{\infty} a_k$ is bounded from above if the improper integral $\int_1^{\infty} f(x)\, dx$ converges and that it is unbounded if the integral diverges. Accordingly, Figures 8.13a and 8.13b show the graph of a continuous decreasing function f that satisfies $f(n) = a_n$ for $n = 1, 2, 3, \ldots$. Rectangles have been constructed at unit intervals in both figures, but in Figure 8.13a the kth rectangle has height $f(k + 1) = a_{k+1}$, whereas in Figure 8.13b, the comparable rectangle has height $f(k) = a_k$.

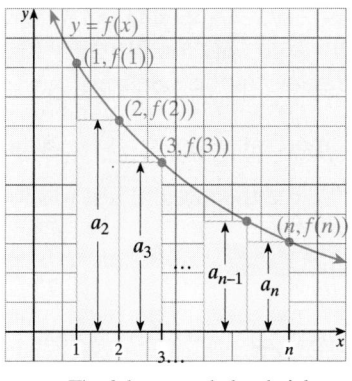
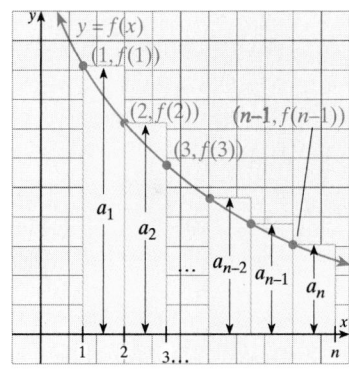

a. The kth rectangle has height $f(k+1) = a_{k+1}$.

b. The kth rectangle has height $f(k) = a_k$.

■ **FIGURE 8.13**   Integral test

Notice that in Figure 8.13a, the rectangles are inscribed so that

AREA OF THE FIRST $n - 1$ RECTANGLES $<$ AREA UNDER $y = f(x)$ OVER $[1, n]$

$$a_2 + a_3 + \cdots + a_n < \int_1^n f(x)\, dx$$

$$a_1 + a_2 + a_3 + \cdots + a_n < a_1 + \int_1^n f(x)\, dx \qquad \text{Add } a_1 \text{ to both sides.}$$

Similarly, in Figure 8.13b, the rectangles are circumscribed so that

AREA UNDER $y = f(x)$ OVER $[1, n + 1]$ $<$ AREA OF THE FIRST n RECTANGLES

$$\int_1^{n+1} f(x)\, dx < a_1 + a_2 + a_3 + \cdots + a_n$$

Let $S_n = a_1 + a_2 + a_3 + \cdots + a_n$ be the nth partial sum of the series so that

$$\int_1^{n+1} f(x)\,dx < S_n < a_1 + \int_1^n f(x)\,dx$$

Now, suppose the improper integral $\int_1^\infty f(x)\,dx$ converges and has the value I; that is, $\int_1^\infty f(x)\,dx = I$. Then

$$S_n < a_1 + \int_1^n f(x)\,dx < a_1 + \int_1^\infty f(x)\,dx = a_1 + I$$

It follows that the sequence of partial sums is bounded from above (by $a_1 + I$), and the convergence criterion for positive-term series tells us that the series $\sum_{k=1}^\infty a_k$ must converge.

On the other hand, if the improper integral $\int_1^\infty f(x)\,dx$ diverges, then

$$\lim_{n\to\infty} \int_1^{n+1} f(x)\,dx = \infty$$

It follows that $\lim_{n\to\infty} S_n = \infty$ also because $\int_1^{n+1} f(x)\,dx < S_n$; this means that the series must diverge. Thus, the series and the improper integral either both converge or both diverge, as claimed.

WARNING When applying the integral test, the function f need not be decreasing for all $x \ge 1$, only for all $x \ge N$ for some number N. That is, f must be decreasing "in the long run."

An important series is $\sum_{k=1}^\infty \frac{1}{k}$, which is called the **harmonic series.** In our next example, we use the integral test to show that this series diverges.

EXAMPLE 2 *Harmonic series diverges*

Test the series $\sum_{n=1}^\infty \frac{1}{k}$ for convergence.

Solution Because $f(x) = \frac{1}{x}$ is positive, continuous, and decreasing for $x \ge 1$, the conditions of the integral test are satisfied.

$$\int_1^\infty \frac{1}{x}\,dx = \lim_{b\to\infty} \int_1^b \frac{1}{x}\,dx = \lim_{b\to\infty}[\ln b - \ln 1] = \infty$$

The integral diverges, so the harmonic series diverges. ∎

EXAMPLE 3 *Integral test for convergence*

Test the series $\sum_{k=1}^\infty \frac{k}{e^{k/5}}$ for convergence.

Solution The function $f(x) = \frac{x}{e^{x/5}} = xe^{-x/5}$ is positive and continuous for all $x > 0$. We find that

$$f'(x) = x\left(-\frac{1}{5}e^{-x/5}\right) + e^{-x/5} = \left(1 - \frac{x}{5}\right)e^{-x/5}$$

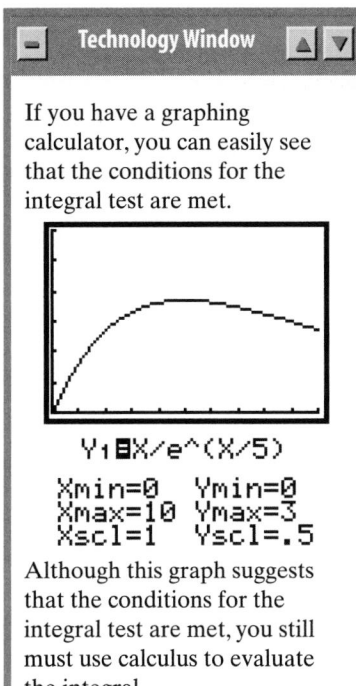

If you have a graphing calculator, you can easily see that the conditions for the integral test are met.

```
Y₁▯X/e^(X/5)
Xmin=0   Ymin=0
Xmax=10  Ymax=3
Xscl=1   Yscl=.5
```

Although this graph suggests that the conditions for the integral test are met, you still must use calculus to evaluate the integral.

WARNING ▶ Note that the series converges because the integral converges to a finite number $50e^{-1}$, but this does not mean that the series $\sum\limits_{k=1}^{\infty} \dfrac{k}{e^{k/5}}$ converges to the same number $50e^{-1}$. ◄

The critical number is found when $f'(x) = 0$, so we solve

$$\left(1 - \frac{x}{5}\right)e^{-x/5} = 0$$

$$1 - \frac{x}{5} = 0 \qquad \text{Set each factor equal to 0, but note that } e^{-x/5} \neq 0.$$

$$x = 5$$

We see that $f'(x) < 0$ for $x > 5$, so it follows that f is decreasing for $x > 5$. Thus, the conditions for the integral test have been established, and the given series and the improper integral either both converge or both diverge. Computing the improper integral, we find

$$\int_5^{\infty} xe^{-x/5}\,dx = \lim_{b \to \infty} \int_5^b xe^{-x/5}\,dx$$

$$\boxed{\begin{array}{ll} \text{Let } u = x & dv = e^{-x/5}\,dx \\ \quad\;\; du = dx & v = -5e^{-x/5} \end{array}}$$

$$= \lim_{b \to \infty}\left[-5xe^{-x/5}\Big|_5^b - \int_5^b (-5e^{-x/5})\,dx\right]$$

$$= \lim_{b \to \infty}\left[-5xe^{-x/5} - 25e^{-x/5}\right]\Big|_5^b$$

$$= \lim_{b \to \infty}\left[-5be^{-b/5} - 25e^{-b/5} + 25e^{-1} + 25e^{-1}\right]$$

$$= -5\lim_{b \to \infty} \frac{b+5}{e^{b/5}} + \lim_{b \to \infty} 50e^{-1}$$

$$= -5\lim_{b \to \infty} \frac{1}{(\frac{1}{5})e^{b/5}} + 50e^{-1} \qquad \text{l'Hôpital's rule } \lim_{b\to\infty}\frac{5}{e^{b/5}} = 0$$

$$= 50e^{-1}$$

Thus, the improper integral converges, which in turn assures the convergence of the given series. ∎

p-SERIES

The harmonic series $\sum 1/k$ is a special case of a more general series form called a *p*-series.

> **_p_-Series**
>
> A series of the form
>
> $$\sum_{k=1}^{\infty} \frac{1}{k^p} = \frac{1}{1^p} + \frac{1}{2^p} + \frac{1}{3^p} + \cdots$$
>
> where *p* is a positive constant is called a **_p_-series.** If $p = 1$, then the *p*-series is the *harmonic series.*

The convergence of the *p*-series is dependent on the value of *p*. Remember, we have already seen (in Example 2) that the series diverges if $p = 1$.

THEOREM 8.12 *The p-series test*

The p-series $\sum\limits_{k=1}^{\infty} \dfrac{1}{k^p}$ converges if $p > 1$ and diverges if $p \leq 1$.

Proof We leave it for the reader to verify that $f(x) = \dfrac{1}{x^p}$ is continuous, positive, and decreasing for $x \geq 1$ and $p > 0$. We know the harmonic series ($p = 1$) diverges, and for $p > 0, p \neq 1$, we have

$$\int_1^{\infty} \frac{dx}{x^p} = \lim_{b \to \infty} \int_1^b x^{-p}\, dx = \lim_{b \to \infty} \frac{b^{1-p} - 1}{1 - p} = \begin{cases} \dfrac{1}{p-1} & \text{if } p > 1 \\ \infty & \text{if } 0 < p < 1 \end{cases}$$

That is, this improper integral converges if $p > 1$ and diverges if $0 < p < 1$. For $p = 0$, the series becomes

$$\sum_{k=1}^{\infty} \frac{1}{k^0} = \frac{1}{1} + \frac{1}{1} + \frac{1}{1} + \cdots$$

and if $p < 0$, we have $\lim\limits_{k \to \infty} \dfrac{1}{k^p} = \infty$, so the series diverges by the divergence test (Theorem 8.9). Thus, a p-series converges only when $p > 1$.

■ *What This Says:* The statement that the series converges if $p > 1$ and diverges if $p \leq 1$ is clear enough. A more subtle question, however, is why? The curves $y = 1/x^p\,(p > 0)$ all decrease as x increases. This also seems clear enough. The answer we seek, namely *why,* lies in looking at the *rate* of decrease for the curves $y = 1/x^p$. If $p > 1$, the curve decreases fast enough that the shaded area in Figure 8.14a is bounded by a fixed number, no matter how large the value b.

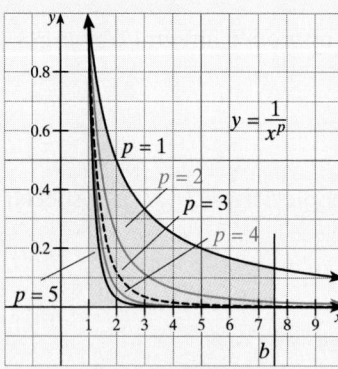

Areas are bounded.

$$\textbf{a.} \int_1^{\infty} \frac{dx}{x^p} \quad \lim_{b \to \infty} \int_1^b \frac{dx}{x^p}$$
converges if $p > 1$

Areas are not bounded.

$$\textbf{b.} \int_1^{\infty} \frac{dx}{x^p} = \lim_{b \to \infty} \int_1^b \frac{dx}{x^p}$$
diverges if $p \leq 1$

■ **FIGURE 8.14** Graphs for the p-series test

On the other hand, if the curve decreases slowly enough, then the shaded areas in Figure 8.14b increase without bound as $b \to \infty$.

EXAMPLE 4   *p-series test*

Test each of the following series for convergence.

a. $\displaystyle\sum_{k=1}^{\infty} \frac{1}{\sqrt{k^3}}$ b. $\displaystyle\sum_{k=1}^{\infty} \left(\frac{1}{e^k} - \frac{1}{\sqrt{k}}\right)$

Solution

a. Here $\sqrt{k^3} = k^{3/2}$, so $p = \frac{3}{2} > 1$, and the series converges.

b. $\displaystyle\sum_{k=1}^{\infty} \left(\frac{1}{e^k} - \frac{1}{\sqrt{k}}\right)$: We note that

$$\sum_{k=1}^{\infty} \frac{1}{e^k} \text{ converges, because it is geometric with } |r| = \frac{1}{e} < 1,$$

and

$$\sum_{k=1}^{\infty} \frac{1}{\sqrt{k}} \text{ diverges, because it is a } p\text{-series with } p = \frac{1}{2} < 1.$$

Because one series in the difference converges and the other diverges, the given series must diverge (Theorem 8.7).   ■

8.3 Problem Set

A 1. ■ **What Does This Say?** What is a *p*-series?

2. ■ **What Does This Say?** What is the integral test?

For the p-series given in Problems 3–6, specify the value of p and tell whether the series converges or diverges.

3. $\displaystyle\sum_{k=1}^{\infty} \frac{1}{k^3}$ 4. $\displaystyle\sum_{k=1}^{\infty} \frac{100}{\sqrt{k}}$

5. $\displaystyle\sum_{k=1}^{\infty} \frac{1}{\sqrt[3]{k}}$ 6. $\displaystyle\sum_{k=1}^{\infty} \frac{1}{2k\sqrt{k}}$

Show that the function f determined by the nth term of each series given in Problems 7–12 satisfies the hypotheses of the integral test. Then use the integral test to determine whether the series converges or diverges.

7. $\displaystyle\sum_{k=1}^{\infty} \frac{1}{(2 + 3k)^2}$ 8. $\displaystyle\sum_{k=1}^{\infty} (2 + k)^{-3/2}$

9. $\displaystyle\sum_{k=2}^{\infty} \frac{\ln k}{k}$ 10. $\displaystyle\sum_{k=2}^{\infty} \frac{1}{k(\ln k)^2}$

11. $\displaystyle\sum_{k=1}^{\infty} \frac{(\tan^{-1}k)^2}{1 + k^2}$ 12. $\displaystyle\sum_{k=1}^{\infty} ke^{-k^2}$

Test the series in Problems 13–34 for convergence.

13. $\displaystyle\sum_{k=1}^{\infty} \frac{\ln k}{k^2}$ 14. $\displaystyle\sum_{k=1}^{\infty} \frac{\ln k}{k}$

15. $\displaystyle\sum_{k=1}^{\infty} \left(2 + \frac{3}{k}\right)^k$ 16. $\displaystyle\sum_{k=1}^{\infty} \left(1 + \frac{2}{k}\right)^k$

17. $\displaystyle\sum_{k=1}^{\infty} \frac{1}{k^4}$ 18. $\displaystyle\sum_{k=1}^{\infty} \frac{5}{\sqrt{k}}$

19. $\displaystyle\sum_{k=1}^{\infty} k^{-3/4}$ 20. $\displaystyle\sum_{k=1}^{\infty} k^{-4/3}$

21. $\displaystyle\sum_{k=1}^{\infty} \frac{k}{k^2 + 1}$ 22. $\displaystyle\sum_{k=1}^{\infty} \frac{k^2}{(k^3 + 2)^2}$

23. $\displaystyle\sum_{k=1}^{\infty} \frac{k^2}{\sqrt{k^3 + 2}}$ 24. $\displaystyle\sum_{k=1}^{\infty} ke^{-k^2}$

25. $\displaystyle\sum_{k=1}^{\infty} \frac{1}{(0.25)^k}$ 26. $\displaystyle\sum_{k=1}^{\infty} \frac{1}{4^k}$

27. $\displaystyle\sum_{k=2}^{\infty} \frac{1}{k(\ln k)^2}$ 28. $\displaystyle\sum_{k=2}^{\infty} \frac{1}{k\sqrt{\ln k}}$

29. $\sum_{k=1}^{\infty} \dfrac{k}{e^k}$

30. $\sum_{k=1}^{\infty} \dfrac{k^2}{e^k}$

31. $\sum_{k=1}^{\infty} \dfrac{\tan^{-1}k}{1+k^2}$

32. $\sum_{k=1}^{\infty} \cot^{-1}k$

33. $\sum_{k=1}^{\infty} \dfrac{1}{e^k + e^{-k}}$

34 $\sum_{k=1}^{\infty} 3e^{-2k}$

Use the divergence test and the p-series test to determine which of the series in Problems 35–46 converge. Write "inconclusive" if there is not yet enough information to settle convergence.

35. $\sum_{k=1}^{\infty} \dfrac{k}{k+1}$

36. $\sum_{k=1}^{\infty} \dfrac{k^2+1}{k+5}$

37. $\sum_{k=1}^{\infty} \dfrac{k^2+1}{k^3}$

38. $\sum_{k=1}^{\infty} \dfrac{2k^4+3}{k^5}$

39. $\sum_{k=1}^{\infty} \dfrac{1}{k^2}$

40. $\sum_{k=1}^{\infty} \dfrac{-k^5+k^2+1}{k^5+2}$

41. $\sum_{n=1}^{\infty} \dfrac{n^{\sqrt 3}+1}{n^{2.7321}}$

42. $\sum_{n=1}^{\infty} \dfrac{n^{\sqrt 5}+1}{n^{2.236}}$

43. $\sum_{k=1}^{\infty} \left[\dfrac{1}{k} + \dfrac{k+1}{k+2}\right]$

44. $\sum_{k=1}^{\infty} \left[\dfrac{1}{2k} + \left(\dfrac{3}{2}\right)^k\right]$

45. $\sum_{k=1}^{\infty} \left[\dfrac{1}{2^k} + \dfrac{2k+3}{3k+4}\right]$

46. $\sum_{k=1}^{\infty} \left[\dfrac{1}{2^k} - \dfrac{1}{k}\right]$

Ⓑ *Find all values of p for which each series given in Problems 47–52 converges.*

47. $\sum_{k=2}^{\infty} \dfrac{k}{(k^2-1)^p}$

48. $\sum_{k=1}^{\infty} \dfrac{k^2}{(k^2+4)^p}$

49. $\sum_{k=3}^{\infty} \dfrac{1}{k^p \ln k}$

50. $\sum_{k=2}^{\infty} \dfrac{\ln k}{k^p}$

51. $\sum_{k=3}^{\infty} \dfrac{1}{k \ln k [\ln(\ln k)]^p}$

52. $\sum_{k=2}^{\infty} \dfrac{1}{k(\ln k)^p}$

Ⓒ

Technology Window

In these problems we explore some of the numerical implications of the integral test for series of the form

$$\sum_{k=1}^{\infty} \dfrac{1}{k^p}$$

53. a. Argue that the "tail" of the series (the error after taking N terms) is bounded by

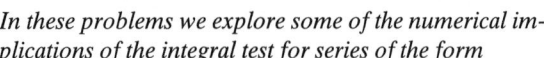

$$\sum_{k=N+1}^{\infty} \dfrac{1}{k^p} \le \int_N^{\infty} \dfrac{dx}{x^p}$$

Technology Window

b. Suppose you want to approximate the sum with $p = 2$ by the partial sum S_N, so that the error is no larger than 0.01. Use part **a** to decide how many terms are needed; that is, find N.

c. It can be shown that

$$\sum_{k=1}^{\infty} \dfrac{1}{k^2} = \dfrac{\pi^2}{6}$$

Using the N found in part **b**, actually compute S_N and compare with this answer. How good was your estimate of N in part **b**?

54. It is a bit surprising that sums $\sum_{k=1}^{\infty} \dfrac{1}{k^p}$ can be computed exactly if p is an even integer. On the other hand, this sum is unknown for $p = 3$.

a. Use the work of Problem 53a to find N sufficient to give an approximation to the sum with $p = 3$ with an error no greater than 0.0005.

b. Since we do not have an exact answer to compare with, the following idea is often used. Compute S_N for your N found in part **a**; then compute S_{2N}. Compare the results and comment on the confidence you have in your work. Was the calculation of S_{2N} necessary to give you this confidence?

---

55. THINK TANK PROBLEM Find two sequences $\{a_k\}$ and $\{b_k\}$ for which $\lim_{k\to\infty} a_k = 0$, $\lim_{k\to\infty} b_k = 0$, $\Sigma a_k b_k$ converges, and Σa_k converges but Σb_k diverges.

56. THINK TANK PROBLEM Find two sequences $\{a_k\}$ and $\{b_k\}$ for which $\lim_{k\to\infty} a_k = 0$, $\lim_{k\to\infty} b_k = 0$, $\Sigma a_k b_k$ converges, and Σa_k and Σb_k both diverge.

57. Let $f(x)$ be a continuous, decreasing function for $x \ge 1$, and let $a_k = f(k)$ for $k = 1, 2, \ldots$. If Σa_k converges, show that

$$\int_1^{\infty} f(x)\, dx \le \sum_{k=1}^{\infty} a_k \le a_1 + \int_1^{\infty} f(x)\, dx$$

58. Getting a feeling for a series before examining its graph is a good idea, but you should be careful because you cannot always obtain a graph. Try to decide whether

$$\sum \left(1 - \dfrac{100}{n}\right)^n$$

converges or diverges, and then see if you can verify your result graphically.

8.4 Comparison Tests

DIRECT COMPARISON TEST

THEOREM 8.13 *Direct comparison test*

Let $0 \le a_k \le c_k$ for all $k \ge N$ for some N.

$$\text{If } \sum_{k=1}^{\infty} c_k \text{ converges, then } \sum_{k=1}^{\infty} a_k \text{ also converges.}$$

Let $0 \le d_k \le a_k$ for all k.

$$\text{If } \sum_{k=1}^{\infty} d_k \text{ diverges, then } \sum_{k=1}^{\infty} a_k \text{ also diverges.}$$

Proof Suppose Σa_k is a given nonnegative-term series that is **dominated by** a convergent series Σc_k in the sense that for sufficiently large k, $0 \le a_k \le c_k$ for all k. Then, because the series Σc_k converges, its sequence of partial sums is bounded from above (say, by M), and we have

$$\sum_{k=1}^{n} a_k \le \sum_{k=1}^{n} c_k < M \quad \text{for all } n$$

Thus, the sequence of partial sums of the dominated series Σa_k is also bounded from above by M, and it, too, must converge.

On the other hand, suppose the given series Σa_k **dominates** a divergent series Σd_k so that $0 \le d_k \le a_k$. Then because the sequence of partial sums of Σd_k is unbounded, the same must be true for the partial sums of the dominating series Σa_k, and Σa_k must also diverge. ═══

■ *What This Says:* Let Σa_k, Σc_k, and Σd_k be series with positive terms. The series Σa_k *converges* if it is "smaller" than (dominated by) a known convergent series Σc_k and *diverges* if it is "larger" than (dominates) a known divergent series Σd_k. That is, "smaller than convergent is convergent," and "bigger than divergent is divergent."

EXAMPLE 1 *Convergence using the direct comparison test*

Test the series $\displaystyle\sum_{k=1}^{\infty} \frac{1}{3^k + 1}$ for convergence.

Solution For $k \ge 0$, we have $3^k + 1 > 3^k > 0$, and $0 < \dfrac{1}{3^k + 1} < \dfrac{1}{3^k}$. Thus, the given series is *dominated by* the convergent geometric series $\displaystyle\sum_{k=1}^{\infty} \frac{1}{3^k}$ (convergent because $r = \frac{1}{3}$). The direct comparison test tells us the given series must converge. ■

| EXAMPLE 2 | *Divergence using the direct comparison test* |

Test the series $\displaystyle\sum_{k=2}^{\infty} \frac{1}{\sqrt{k}-1}$ for convergence.

Solution   For $k \geq 2$, we have

$$\frac{1}{\sqrt{k}-1} > \frac{1}{\sqrt{k}} > 0,$$

so the given series *dominates* the divergent p-series $\displaystyle\sum_{k=2}^{\infty} \frac{1}{k^{1/2}}$ (divergent because $p = \frac{1}{2} < 1$). The direct comparison test tells us the given series must diverge. ■

| EXAMPLE 3 | *Direct comparison test* |

Test the series $\displaystyle\sum_{k=1}^{\infty} \frac{1}{k!}$ for convergence.

Solution   We shall compare the given series with a geometric series. Specifically, note that if $k \geq 2$, then $\frac{1}{k!} > 0$ and

$$k! = k(k-1)(k-2)\cdots 3 \cdot 2 \cdot 1 \geq \underbrace{2 \cdot 2 \cdot 2 \cdots 2 \cdot 2}_{k-1 \text{ terms}} \cdot 1 = 2^{k-1}$$

Therefore, we have $0 < \dfrac{1}{k!} \leq \dfrac{1}{2^{k-1}}$, and because the given series is *dominated by* the convergent geometric series $\displaystyle\sum_{k=1}^{\infty} \frac{1}{2^{k-1}}$ (with $r = \frac{1}{2}$), it must also converge. ■

LIMIT COMPARISON TESTS

It is not always easy or even possible to make a suitable direct comparison between two similar series. For example, we would expect the series $\Sigma\, 1/(2^k - 5)$ to converge because it is so much like the convergent geometric series $\Sigma\, 1/2^k$. To compare the series, we must first note that $1/(2^k - 5)$ is negative for $k = 1$ and $k = 2$ and positive for $k \geq 3$. Thus, if $k \geq 3$,

$$0 \leq \frac{1}{2^k} < \frac{1}{2^k - 5}$$

That is, $\Sigma\, 1/(2^k - 5)$ *dominates* the convergent series $\Sigma\, 1/2^k$, which means the comparison test cannot be used to determine the convergence of $\Sigma\, 1/(2^k - 5)$ by comparing with $\Sigma\, 1/2^k$. In such situations, the following test is often useful.

THEOREM 8.14   *Limit comparison test*

Suppose $a_k > 0$ and $b_k > 0$ for all sufficiently large k and that

$$\lim_{k\to\infty} \frac{a_k}{b_k} = L$$

where L is finite and positive ($0 < L < \infty$). Then

$$\Sigma a_k \text{ and } \Sigma b_k \text{ either both converge or both diverge.}$$

Proof   The proof of this theorem is given in Appendix B.

> ■ *What This Says:*   We have the following procedure for testing the convergence of Σa_k.
>
> **Step 1.** Find a series Σb_k whose convergence properties are known and whose general term b_k is "essentially the same" as a_k.
>
> **Step 2.** Verify that $\lim\limits_{k\to\infty} \dfrac{a_k}{b_k}$ exists and is positive.
>
> **Step 3.** Determine whether Σb_k converges or diverges. Then the limit comparison test tells us that Σa_k does the same.

EXAMPLE 4   *Convergence using the limit comparison test*

Test the series $\displaystyle\sum_{k=1}^{\infty} \frac{1}{2^k - 5}$ for convergence.

Solution   Because the given series has the same general appearance as the convergent geometric series $\Sigma\, 1/2^k$, compute the limit

$$\lim_{k\to\infty} \frac{\dfrac{1}{2^k - 5}}{\dfrac{1}{2^k}} = \lim_{k\to\infty} \frac{2^k}{2^k - 5} = 1$$

This limit is finite and positive, so the limit comparison test tells us that the given series will have the same convergence properties as $\Sigma\, 1/2^k$. Thus, the given series converges.   ■

EXAMPLE 5   *Divergence using the limit comparison test*

Test the series $\displaystyle\sum_{k=1}^{\infty} \frac{3k + 2}{\sqrt{k}(3k - 5)}$ for convergence.

Solution   For large values of k, the general term of the given series

$$a_k = \frac{3k + 2}{\sqrt{k}(3k - 5)}$$

seems to be similar to

$$b_k = \frac{3k}{\sqrt{k}(3k)} = \frac{1}{\sqrt{k}}$$

To apply the limit comparison test, we compute the limit

$$\lim_{k\to\infty} \frac{a_k}{b_k} = \lim_{k\to\infty} \frac{\dfrac{3k + 2}{\sqrt{k}(3k - 5)}}{\dfrac{1}{\sqrt{k}}} = \lim_{k\to\infty} \frac{3k + 2}{3k - 5} = 1$$

Because the limit is finite and positive, it follows that the given series behaves like the series $\Sigma\, 1/\sqrt{k}$, which is a divergent p-series $\left(p = \tfrac{1}{2}\right)$. We conclude that the given series diverges. ∎

> **EXAMPLE 6**   *Limit comparison test*

Test the series $\displaystyle\sum_{k=1}^{\infty} \frac{7k + 100}{e^{k/5} - 70}$ for convergence.

Solution   For large k,

$$a_k = \frac{7k + 100}{e^{k/5} - 70} \quad \text{seems to behave like} \quad b_k = \frac{k}{e^{k/5}}$$

and indeed, we find that

$$\lim_{k \to \infty} \frac{\dfrac{7k + 100}{e^{k/5} - 70}}{\dfrac{k}{e^{k/5}}} = \lim_{k \to \infty} \frac{e^{k/5}(7k + 100)}{k(e^{k/5} - 70)} = 7$$

In Example 3, Section 8.3, we showed that the series $\Sigma\, k/e^{k/5}$ converges, and therefore the limit comparison test tells us that the given series also converges. ∎

Occasionally, two series Σa_k and Σb_k appear to have similar convergence properties, but it turns out that

$$\lim_{k \to \infty} \frac{a_k}{b_k}$$

is either 0 or ∞, and the limit comparison test therefore does not apply. In such cases, it is often useful to have the following generalization of the limit comparison test.

THEOREM 8.15   *The zero-infinity limit comparison test*

Suppose $a_k > 0$ and $b_k > 0$ for all sufficiently large k. Then

If $\displaystyle\lim_{k \to \infty} \frac{a_k}{b_k} = 0$ and Σb_k converges, the series Σa_k converges.

If $\displaystyle\lim_{k \to \infty} \frac{a_k}{b_k} = \infty$ and Σb_k diverges, the series Σa_k diverges.

Proof   The first part of the proof is outlined in Problem 61; the second part follows similarly.

> **EXAMPLE 7**   *Convergence of the log q series*

Show that the series $\displaystyle\sum_{k=1}^{\infty} \frac{\ln k}{k^q}$ converges if $q > 1$ and diverges if $q \leq 1$. We call this the **log q series.**

Solution   We will carry out this proof in three parts:

CASE I $(q > 1)$
Let p be a number that satisfies $1 < p < q$, and let

$$a_k = \frac{\ln k}{k^q} \quad \text{and} \quad b_k = \frac{1}{k^p}$$

Then

$$\lim_{k \to \infty} \frac{a_k}{b_k} = \lim_{k \to \infty} \frac{\dfrac{\ln k}{k^q}}{\dfrac{1}{k^p}}$$

$$= \lim_{k \to \infty} \frac{\ln k}{k^{q-p}} \qquad \text{This is the indeterminate form}$$
$$\qquad\qquad\qquad\qquad \infty/\infty \text{ because } q - p > 0.$$

$$= \lim_{k \to \infty} \frac{\dfrac{1}{k}}{(q-p)k^{q-p-1}} \qquad \text{l'Hôpital's rule}$$

$$= \lim_{k \to \infty} \frac{1}{(q-p)k^{q-p}} = 0 \qquad \text{Because } q - p > 0$$

Because $\Sigma \dfrac{1}{k^p}$ converges (p-series with $p > 1$), the series converges by the zero-infinity limit comparison test.

CASE II $(q < 1)$
Now, let p satisfy $q < p < 1$. Then with a_k and b_k defined as in Case I, we have

$$\lim_{k \to \infty} \frac{a_k}{b_k} = \lim_{k \to \infty} \frac{\ln k}{k^{q-p}} = \lim_{k \to \infty} [(\ln k)k^{p-q}] = \infty$$

Because $\displaystyle\sum_{k=1}^{\infty} b_k = \sum_{k=1}^{\infty} \frac{1}{k^p}$ diverges (we know $p < 1$), it follows from the zero-infinity comparison test that $\displaystyle\sum_{k=1}^{\infty} \frac{\ln k}{k^q}$ diverges when $q < 1$.

CASE III $(q = 1)$
Here, the series is $\displaystyle\sum_{k=1}^{\infty} \frac{\ln k}{k}$, which diverges by the integral test:

$$\int_1^{\infty} \frac{\ln x}{x}\, dx = \lim_{b \to \infty} \left[\frac{\ln^2 x}{2}\right]\Big|_1^b = \infty$$

$$\boxed{\begin{array}{l} \text{Let}\quad u = \ln x; \\ du = \dfrac{1}{x}\, dx. \end{array}}$$

8.4   Problem Set

Ⓐ *The most common series used for comparison are given in Problems 1–2. Tell when each converges and when it diverges.*

1. geometric series: $\displaystyle\sum_{k=0}^{\infty} r^k$ 2. p-series: $\displaystyle\sum_{k=1}^{\infty} \frac{1}{k^p}$

Each series in Problems 3–12 can be compared to the geometric series or p-series given in Problems 1–2. State which, and then determine whether it converges or diverges.

3. $\displaystyle\sum_{k=1}^{\infty} \cos^k\left(\frac{\pi}{6}\right)$ 4. $\displaystyle\sum_{k=0}^{\infty} 0.5^k$

5. $\displaystyle\sum_{k=0}^{\infty} 1.5^k$

6. $\displaystyle\sum_{k=0}^{\infty} 2^{k/2}$

7. $\displaystyle\sum_{k=1}^{\infty} \frac{1}{k}$

8. $\displaystyle\sum_{k=1}^{\infty} \frac{1}{k^{0.5}}$

9. $\displaystyle\sum_{k=1}^{\infty} \frac{1}{k^{3/2}}$

10. $\displaystyle\sum_{k=1}^{\infty} \sqrt{\frac{2}{k}}$

11. $\displaystyle\sum_{k=0}^{\infty} 1^k$

12. $\displaystyle\sum_{k=1}^{\infty} e^k$

Test the series in Problems 13–44 for convergence.

13. $\displaystyle\sum_{k=1}^{\infty} \frac{1}{k^2 + k}$

14. $\displaystyle\sum_{k=1}^{\infty} \frac{1}{k^2 + 3k + 2}$

15. $\displaystyle\sum_{k=1}^{\infty} \frac{1}{\sqrt{k}}$

16. $\displaystyle\sum_{k=1}^{\infty} \frac{1}{k\sqrt{k}}$

17. $\displaystyle\sum_{k=1}^{\infty} \frac{1}{\sqrt{2k + 3}}$

18. $\displaystyle\sum_{k=1}^{\infty} \frac{1}{\sqrt{k(k + 1)}}$

19. $\displaystyle\sum_{k=1}^{\infty} \frac{1}{\sqrt{k^3 + 2}}$

20. $\displaystyle\sum_{k=1}^{\infty} \frac{1}{\sqrt{k^2 + 1}}$

21. $\displaystyle\sum_{k=1}^{\infty} \frac{2k^2}{k^4 - 4}$

22. $\displaystyle\sum_{k=1}^{\infty} \frac{k + 1}{k^2 + 1}$

23. $\displaystyle\sum_{k=1}^{\infty} \frac{(k + 2)(k + 3)}{k^{7/2}}$

24. $\displaystyle\sum_{k=1}^{\infty} \frac{(k + 1)^3}{k^{9/2}}$

25. $\displaystyle\sum_{k=1}^{\infty} \frac{2k + 3}{k^2 + 3k + 2}$

26. $\displaystyle\sum_{k=1}^{\infty} \frac{3k^2 + 2}{k^2 + 3k + 2}$

27. $\displaystyle\sum_{k=1}^{\infty} \frac{k}{(k + 2)2^k}$

28. $\displaystyle\sum_{k=1}^{\infty} \frac{5}{4^k + 3}$

29. $\displaystyle\sum_{k=1}^{\infty} \frac{1}{k(k + 2)}$

30. $\displaystyle\sum_{k=1}^{\infty} \frac{1}{(k + 2)(k + 3)}$

31. $\displaystyle\sum_{k=1}^{\infty} \frac{1}{\sqrt{k}\,2^k}$

32. $\displaystyle\sum_{k=1}^{\infty} \frac{1{,}000}{\sqrt{k}\,3^k}$

33. $\displaystyle\sum_{k=1}^{\infty} \frac{|\sin(k!)|}{k^2}$

34. $\displaystyle\sum_{k=2}^{\infty} \frac{1}{\sqrt{k}\ln k}$

35. $\displaystyle\sum_{k=1}^{\infty} \frac{2k^3 + k + 1}{k^3 + k^2 + 1}$

36. $\displaystyle\sum_{k=6}^{\infty} \frac{6k^3 - k - 4}{k^3 - k^2 - 3}$

37. $\displaystyle\sum_{k=1}^{\infty} \frac{k}{4k^3 - 5}$

38. $\displaystyle\sum_{k=1}^{\infty} \frac{\ln k}{\sqrt{2k + 3}}$

39. $\displaystyle\sum_{k=1}^{\infty} \frac{k^2 + 1}{(k^2 + 2)k^2}$

40. $\displaystyle\sum_{k=1}^{\infty} \sin\frac{1}{k}$

41. $\displaystyle\sum_{k=1}^{\infty} \frac{6k^2 + 2k + 1}{k^{1.1}(4k^2 + k + 4)}$

42. $\displaystyle\sum_{k=1}^{\infty} \frac{6k^2 + 2k + 1}{k^{0.9}(4k^2 + k + 4)}$

43. $\displaystyle\sum_{k=1}^{\infty} \frac{\sqrt[6]{k}}{\sqrt[4]{k^3 + 2}\,\sqrt[8]{k}}$

44. $\displaystyle\sum_{k=1}^{\infty} \frac{\sqrt{k}}{\sqrt[3]{k^3 + 1}\,\sqrt[6]{k^5}}$

B *Test the series given in Problems 45–52 for convergence.*

45. $\displaystyle\sum_{k=1}^{\infty} \frac{1}{k^3 + 4}$

46. $\displaystyle\sum_{k=2}^{\infty} \frac{\ln k}{k - 1}$

47. $\displaystyle\sum_{k=1}^{\infty} \frac{\ln (k + 1)}{(k + 1)^3}$

48. $\displaystyle\sum_{k=1}^{\infty} \frac{\ln k}{k^2}$

49. $\displaystyle\sum_{k=2}^{\infty} \frac{1}{(k + 3)(\ln k)^{1.1}}$

50. $\displaystyle\sum_{k=2}^{\infty} \frac{1}{(k + 3)(\ln k)^{0.9}}$

51. $\displaystyle\sum_{k=1}^{\infty} k^{(1-k)/k}$

52. $\displaystyle\sum_{k=1}^{\infty} k^{(1+k)/k}$

53. Show that the series

$$\sum_{k=1}^{\infty} \frac{k^2}{(k + 3)!} = \frac{1}{4!} + \frac{4}{5!} + \frac{9}{6!} + \cdots$$

converges by using the limit comparison test.

54. Show that the series

$$1 + \frac{1}{1 \cdot 3} + \frac{1}{1 \cdot 3 \cdot 5} + \frac{1}{1 \cdot 3 \cdot 5 \cdot 7} + \cdots + \frac{2^k k!}{(2k + 1)!} + \cdots$$

converges.   *Hint:* Compare with the convergent series $\Sigma\, 1/k!$.

55. Use a comparison test to show that

$$\sum_{k=2}^{\infty} \frac{1}{(\ln k)^{\ln k}}$$

converges.   *Hint:* Use the fact that $\ln k > e^2$ for suffi-ciently large k.

C 56. **a.** Let $\{a_n\}$ be a positive sequence such that

$$\lim_{x \to \infty} k^p a_k$$

exists. Use the limit comparison test to show that Σa_k converges if $p > 1$.

b. Show that Σe^{-k^2} converges.   *Hint:* Show that $\displaystyle\lim_{k \to \infty} k^2 e^{-k^2}$ exists and use part **a.**

57. Let Σa_k be a series of positive terms and let $\{b_n\}$ be a sequence of positive numbers that converge to a positive number. Show that Σa_k converges if and only if $\Sigma a_k b_k$ converges.   *Hint:* Use the limit comparison test.

58. Suppose $0 \le a_k \le A$ for some number A and $b_k \ge k^2$. Show that $\displaystyle\Sigma\, \frac{a_k}{b_k}$ converges.

59. Suppose $a_k > 0$ for all k and that Σa_k converges. Show that $\displaystyle\Sigma\, \frac{1}{a_k}$ diverges.

60. **THINK TANK PROBLEM**   Suppose $0 < a_k < 1$ for every k, and that Σa_k converges. Use a comparison test to show that Σa_k^2 converges. Find a counterexample where Σa_k^2 converges, but Σa_k diverges.

61. Show that if $a_k > 0$, $b_k > 0$, and

$$\lim_{k \to \infty} \frac{a_k}{b_k} = 0$$

then Σa_k converges whenever Σb_k converges.
Hint: Show that there is an integer N so that $a_k < b_k$ if $k > N$. Then use a comparison test.

8.5   The Ratio Test and the Root Test

ratio test, root test, summary of convergence tests for positive- and nonnegative-term series

RATIO TEST

Intuitively, a series of positive terms Σa_k converges if and only if the sequence $\{a_k\}$ decreases rapidly toward 0. One way to measure the rate at which the sequence $\{a_k\}$ is decreasing is to examine the ratio $\dfrac{a_{k+1}}{a_k}$ as k grows large. This approach leads to the following result.

THEOREM 8.16   *The ratio test*

Given the series Σa_k with $a_k > 0$, suppose that

$$\lim_{k \to \infty} \frac{a_{k+1}}{a_k} = L$$

The **ratio test** states the following:

   If $L < 1$, then Σa_k converges.
   If $L > 1$ or if L is infinite, then Σa_k diverges.
   If $L = 1$, the test is inconclusive.

Proof   In a sense, the ratio test is a limit comparison test in which Σa_k is compared to itself. We shall use the direct comparison test to show that Σa_k converges if $L < 1$. Choose R such that $0 < L < R < 1$. Then there exists some $N > 0$ such that

$$\frac{a_{k+1}}{a_k} < L < R \quad \text{for all } k > N$$

Therefore,

$$a_{N+1} < a_N R$$
$$a_{N+2} < a_{N+1} R < a_N R^2$$
$$a_{N+3} < a_{N+2} R < a_{N+1} R^2 < a_N R^3$$
$$\vdots$$

The geometric series

$$\sum_{k=1}^{\infty} a_N R^k = a_N R + a_N R^2 + \cdots + a_N R^k + \cdots$$

converges because $0 < R < 1$, which means the "smaller" series

$$\sum_{k=1}^{\infty} a_{N+k} = a_{N+1} + a_{N+2} + \cdots + a_{N+k} + \cdots$$

also converges by the direct comparison test. Thus, Σa_k converges, because we can discard a finite number of terms ($k \leq N$).

The proof of the second part is similar, except now we choose R such that

$$\lim_{k \to \infty} \frac{a_{k+1}}{a_k} = L > R > 1$$

and show there exists some $M > 0$ so that $a_{M+k} > a_M R^k$.

To prove that the ratio test is inconclusive if $L = 1$, it is enough to note that the harmonic series

You will find the ratio test most useful with series involving factorials or exponentials.

$$\sum_{k=1}^{\infty} \frac{1}{k} \text{ diverges } \quad \text{with} \quad \lim_{k \to \infty} \frac{a_{k+1}}{a_k} = \lim_{k \to \infty} \frac{\dfrac{1}{k+1}}{\dfrac{1}{k}} = 1$$

whereas the p-series

$$\sum_{k=1}^{\infty} \frac{1}{k^2} \text{ converges } \quad \text{with} \quad \lim_{k \to \infty} \frac{a_{k+1}}{a_k} = \lim_{k \to \infty} \frac{\dfrac{1}{(k+1)^2}}{\dfrac{1}{k^2}} = 1$$

EXAMPLE 1   *Convergence using the ratio test*

Test the series $\sum_{k=1}^{\infty} \dfrac{2^k}{k!}$ for convergence.

Solution   Let $a_k = \dfrac{2^k}{k!}$ and note that

$$L = \lim_{k \to \infty} \frac{a_{k+1}}{a_k} = \lim_{k \to \infty} \frac{\dfrac{2^{k+1}}{(k+1)!}}{\dfrac{2^k}{k!}} = \lim_{k \to \infty} \frac{k! 2^{k+1}}{(k+1)! 2^k} = \lim_{k \to \infty} \frac{2}{k+1} = 0$$

Thus $L < 1$, and the ratio test tells us that the given series converges. ∎

EXAMPLE 2   *Divergence using the ratio test*

Test the series $\sum_{k=1}^{\infty} \dfrac{k^k}{k!}$ for convergence.

Solution   Let $a_k = \dfrac{k^k}{k!}$ and note that

$$L = \lim_{k \to \infty} \frac{a_{k+1}}{a_k} = \lim_{k \to \infty} \frac{\dfrac{(k+1)^{k+1}}{(k+1)!}}{\dfrac{k^k}{k!}} = \lim_{k \to \infty} \frac{k!(k+1)^{k+1}}{k^k(k+1)!}$$

$$= \lim_{k \to \infty} \frac{(k+1)^k}{k^k} = \lim_{k \to \infty} \left(1 + \frac{1}{k}\right)^k = e$$

Because $L > 1$, the given series diverges. ∎

Technology Window

It is instructive to compare and contrast two different ways you can use technology to help you with problems such as Example 2. First, as you have seen, you can use a calculator or a computer to calculate the partial sums directly. This table is shown below at the left.

A second procedure is to use the ratio test in Example 2. In this case, we are interested in

$$\lim_{k \to \infty} \frac{a_{k+1}}{a_k} = \lim_{k \to \infty} \frac{\dfrac{(k+1)^{k+1}}{(k+1)!}}{\dfrac{k^k}{k!}} = \lim_{k \to \infty} \left(1 + \frac{1}{k}\right)^k$$

We can look at terms of the sequence $a_k = \left(1 + \dfrac{1}{k}\right)^k$ to see whether this limit is greater than, less than, or equal to 1. This is shown below on the right.

By Table:

$$\sum_{k=1}^{\infty} \frac{k^k}{k!}$$

N	F(1)+...+F(N)
1	1
2	3
3	7.5
4	18.16667
5	44.20833
6	109.0083
7	272.4097
8	688.5113
9	1756.138
10	4511.87
11	11659.53
12	30273.46
13	78912.3
14	206375.3
15	541239.9

The partial sums seem to diverge.

By Graph:

$$\lim_{k \to \infty} \left(1 + \frac{1}{k}\right)^k$$

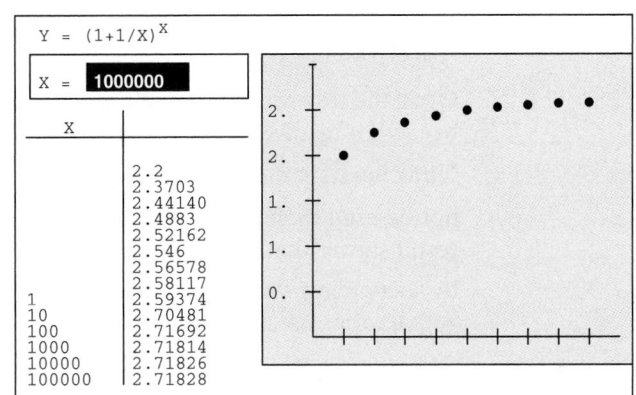

We see that the ratio test gives a limit greater than 1.

EXAMPLE 3   *Ratio test fails*

Test the series $\displaystyle\sum_{k=2}^{\infty} \frac{1}{2k-3}$ for convergence.

Solution   Let $a_k = \dfrac{1}{2k-3}$ and find

$$L = \lim_{k \to \infty} \frac{\dfrac{1}{2(k+1)-3}}{\dfrac{1}{2k-3}} = \lim_{k \to \infty} \frac{2k-3}{2k-1} = 1$$

The ratio test is inconclusive. We can use the integral test or the limit comparison test to test convergence. Note that

$$\sum_{k=2}^{\infty} \frac{1}{2k-3} \text{ is similar to the known divergent series } \sum_{k=2}^{\infty} \frac{1}{2k}$$

so we conclude the given series is divergent.

EXAMPLE 4 *Convergence of a series of power functions*

Find all $x > 0$ for which the series $\sum_{k=1}^{\infty} k^3 x^k$ converges.

Solution Applying the ratio test,

$$\frac{(k+1)^3 x^{k+1}}{k^3 x^k} = \frac{(k+1)^3}{k^3} x$$

The series converges when $\lim_{k\to\infty} \dfrac{(k+1)^3}{k^3} x < 1$, which is true whenever $x < 1$, because

$$\lim_{k\to\infty} \frac{(k+1)^3}{k^3} = 1. \qquad \blacksquare$$

COMMENT There are more refined versions of the ratio test that can be used to handle certain series where $L = 1$. A consideration of such tests is delayed until advanced calculus.

ROOT TEST

Of all the tests we have developed, the divergence test is perhaps the easiest to apply, because it involves simply computing $\lim_{k\to\infty} a_k$ and observing whether that limit is zero. Unfortunately, most "interesting" series have $\lim_{k\to\infty} a_k = 0$, so the divergence test cannot be used to determine whether they converge or diverge. However, the following result shows that it may be possible to say more about the convergence of Σa_k by examining what happens to $\sqrt[k]{a_k}$ as $k \to \infty$. This test will prove particularly useful with a series involving a kth power.

THEOREM 8.17 *The root test*

Given the series Σa_k with $a_k \geq 0$, suppose that $\lim_{k\to\infty} \sqrt[k]{a_k} = L$.

The **root test** states the following:

>   If $L < 1$, then Σa_k converges.
>   If $L > 1$ or if L is infinite, then Σa_k diverges.
>   If $L = 1$, the root test is inconclusive.

Proof The proof of this theorem is similar to that of the ratio test and is outlined in Problem 57.
===

EXAMPLE 5 *Convergence with the root test*

Test the series $\sum_{k=1}^{\infty} \dfrac{1}{(\ln k)^k}$ for convergence.

Solution Let $a_k = \dfrac{1}{(\ln k)^k}$ and note that

$$L = \lim_{k\to\infty} \sqrt[k]{a_k} = \lim_{k\to\infty} \sqrt[k]{(\ln k)^{-k}} = \lim_{k\to\infty} \frac{1}{\ln k} = 0$$

Because $L < 1$, the root test tells us that the given series converges. $\qquad \blacksquare$

| EXAMPLE 6 | *Inconclusive root test* |

Test the *p*-series $\displaystyle\sum_{k=1}^{\infty} \frac{1}{k^p}$ for convergence.

Solution   We know that the *p*-series converges for $p > 1$ and diverges for $p \leq 1$. The point of this example is to show that the root test gives $L = 1$ for all *p*-series. This confirms the statement that the root test is inconclusive for $L = 1$. Let $a_k = 1/k^p$ so that

$$L = \lim_{k \to \infty} \sqrt[k]{\frac{1}{k^p}} = \lim_{k \to \infty} k^{-p/k}$$

To find this limit, use logarithms:

$$\ln L = \ln \lim_{k \to \infty} k^{-p/k} = \lim_{k \to \infty} \ln k^{-p/k} = \lim_{k \to \infty} \frac{-p \ln k}{k}$$

This is of the form ∞/∞, and we apply l'Hôpital's rule:

$$\ln L = \lim_{k \to \infty} \frac{-p \ln k}{k} = \lim_{k \to \infty} \frac{-p(1/k)}{1} = 0 \quad \text{so that} \quad L = 1$$

Thus, $L = 1$ for any *p*-series, yet those with $p > 1$ converge, whereas those with $p \leq 1$ diverge.   ∎

SUMMARY OF CONVERGENCE TESTS FOR POSITIVE- AND NONNEGATIVE-TERM SERIES

This concludes our study of basic convergence tests for series of positive terms. There are no firm rules for deciding how to test the convergence of a given series Σa_k, but we can offer a few observations.

In situations where the ratio test and the root test both apply, it is often easier to use the ratio test. However, the root test turns out to be more discriminating in the sense that any series whose convergence can be determined by the ratio test can also be handled by the root test, but the root test can be used to determine the convergence of certain series for which the ratio test is inconclusive (see Problem 56). For this reason we suggest that you try the ratio test first, and if that is inconclusive, then try the root test. These procedures are summarized in Table 8.1.

■ **TABLE 8.1**

Guidelines for Determining Convergence of the Series Σa_k for $a_k > 0$	
Series with Known Convergence Properties	
Geometric series, $\displaystyle\sum_{k=1}^{\infty} ar^k$	diverges if $|r| \geq 1$, and converges if $|r| < 1$, with sum $S = \dfrac{a}{1-r}$
p-series, $\displaystyle\sum_{k=1}^{\infty} \frac{1}{k^p}$	converges if $p > 1$ and diverges if $p \leq 1$
(Special case $p = 1$): Harmonic series, $\displaystyle\sum_{k=1}^{\infty} \frac{1}{k}$	diverges (special case, $p = 1$, of a *p*-series)
Telescoping series, $\displaystyle\sum_{k=1}^{\infty} (b_k - b_{k+1})$	converges if $\displaystyle\lim_{k \to \infty} b_k = L$ with sum $S = b_1 - L$
Log *q* series, $\displaystyle\sum_{k=1}^{\infty} \frac{\ln k}{k^q}$	converges if $q > 1$ and diverges if $q \leq 1$

■ **TABLE 8.1   (continued)**

Review of Convergence Tests
1. **Divergence test** — Compute $\lim\limits_{k\to\infty} a_k$. If this limit is not 0, the series diverges.
2. **Limit comparison test** — Check to see whether Σa_k is similar in appearance to a series Σb_k whose convergence properties are known, and apply the limit comparison test. If $\lim\limits_{k\to\infty}\dfrac{a_k}{b_k}=L$ where L is finite and positive, then Σa_k and Σb_k either both converge or both diverge.
3. **Ratio test** — If a_k involves $k!$, k^p, or a^k, try the ratio test, $\lim\limits_{k\to\infty}\dfrac{a_{k+1}}{a_k}=L$. Converges if $L<1$, diverges if $L>1$, test inconclusive if $L=1$.
4. **Root test** — If it is easy to find $\lim\limits_{k\to\infty}\sqrt[k]{a_k}=L$, try the root test. Converges if $L<1$, diverges if $L>1$, test inconclusive if $L=1$.
5. **Integral test** — Think of using this test when f is easy to integrate or if a_k involves a logarithm, a trigonometric function, or an inverse trigonometric function. If f is continuous, positive, and decreasing, and if $a_k=f(k)$ for all k, then $\sum\limits_{k=1}^{\infty} a_k$ and the improper integral $\int_1^{\infty} f(x)\,dx$ either both converge or both diverge.
6. **Direct comparison test** — If $0\le a_k\le c_k$ and $\sum\limits_{k=1}^{\infty} c_k$ converges, then $\sum\limits_{k=1}^{\infty} a_k$ converges. If $0\le d_k\le a_k$ and $\sum\limits_{k=1}^{\infty} d_k$ diverges, then $\sum\limits_{k=1}^{\infty} a_k$ diverges.
7. **Zero-infinity limit comparison test** — If $\lim\limits_{k\to\infty}\dfrac{a_k}{b_k}=0$ and Σb_k converges, then the series Σa_k converges. If $\lim\limits_{k\to\infty}\dfrac{a_k}{b_k}=\infty$ and Σb_k diverges, then the series Σa_k diverges.
If a test is inconclusive, do not quit; try another test.

Technology Window

The divergence test, of course, is the first test. You can often check this using a graphing calculator.

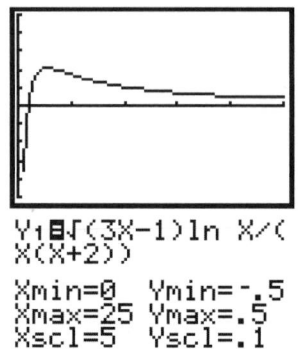

```
Y₁◻√(3X-1)ln X/(
X(X+2))
Xmin=0    Ymin=-.5
Xmax=25   Ymax=.5
Xscl=5    Yscl=.1
```

It appears that $\lim\limits_{k\to\infty} a_k = 0$, so this test is inconclusive.

EXAMPLE 7   *Testing for convergence*

Test the series $\displaystyle\sum_{k=1}^{\infty}\frac{\sqrt{3k-1}\,\ln k}{k(k+2)}$.

Solution   We would expect the given series to behave like the series

$$\sum_{k=1}^{\infty}\frac{\sqrt{k}\,\ln k}{k^2}=\sum_{k=1}^{\infty}\frac{\ln k}{k^{3/2}}$$

We apply the limit comparison test by considering the following limit:

$$\lim_{k\to\infty}\frac{a_k}{b_k}=\lim_{k\to\infty}\frac{\dfrac{\sqrt{3k-1}\,\ln k}{k(k+2)}}{\dfrac{\ln k}{k^{3/2}}}=\lim_{k\to\infty}\frac{k^{3/2}\sqrt{3k-1}}{k(k+2)}$$

$$=\lim_{k\to\infty}\frac{\sqrt{k}\,\sqrt{3k-1}}{k+2}\cdot\frac{\dfrac{1}{k}}{\dfrac{1}{k}}=\lim_{k\to\infty}\frac{\sqrt{3-\dfrac{1}{k}}}{1+\dfrac{2}{k}}=\sqrt{3}$$

Thus, by the limit comparison test, we see that both series either converge or diverge (because $\sqrt{3}$ is finite and positive). Because the comparison series $\Sigma(\ln k)/k^{3/2}$ is a log q series with $q = 3/2$, it must converge (because $q > 1$), and thus the given series also converges. ∎

EXAMPLE 8   *Testing for convergence*

Test the series $\displaystyle\sum_{k=0}^{\infty} \frac{k!}{1 \cdot 4 \cdot 7 \cdots (3k + 1)} = \frac{1!}{1 \cdot 4} + \frac{2!}{1 \cdot 4 \cdot 7} + \frac{3!}{1 \cdot 4 \cdot 7 \cdot 10} + \ldots$ for convergence.

Solution   Let $a_k = \dfrac{k!}{1 \cdot 4 \cdot 7 \cdots (3k + 1)}$. Because a_k involves $k!$, we try the ratio test:

$$\frac{a_{k+1}}{a_k} = \frac{\dfrac{(k + 1)!}{1 \cdot 4 \cdot 7 \cdots [3(k + 1) + 1]}}{\dfrac{k!}{1 \cdot 4 \cdot 7 \cdots (3k + 1)}} = \frac{(k + 1)! \cdot 1 \cdot 4 \cdot 7 \cdots (3k + 1)}{k! \cdot 1 \cdot 4 \cdot 7 \cdots (3k + 4)}$$

$$= \frac{k + 1}{3k + 4}$$

Thus,

$$L = \lim_{k \to \infty} \frac{a_{k+1}}{a_k} = \lim_{k \to \infty} \frac{k + 1}{3k + 4} = \frac{1}{3}$$

Because $L < 1$, the given series converges. ∎

8.5 Problem Set

Ⓐ 1. ■ What Does This Say? Compare and contrast the ratio test and root test.

2. ■ What Does This Say? Outline a procedure for determining the convergence of Σa_k for $a_k > 0$.

Use either the ratio test or the root test to test for convergence of the series given in Problems 3–26.

3. $\displaystyle\sum_{k=1}^{\infty} \frac{1}{k!}$
4. $\displaystyle\sum_{k=1}^{\infty} \frac{k!}{2^k}$
5. $\displaystyle\sum_{k=1}^{\infty} \frac{k!}{2^{3k}}$

6. $\displaystyle\sum_{k=1}^{\infty} \frac{3^k}{k!}$
7. $\displaystyle\sum_{k=1}^{\infty} \frac{k}{2^k}$
8. $\displaystyle\sum_{k=1}^{\infty} \frac{2^k}{k^2}$

9. $\displaystyle\sum_{k=1}^{\infty} \frac{k^{100}}{e^k}$
10. $\displaystyle\sum_{k=1}^{\infty} ke^{-k}$
11. $\displaystyle\sum_{k=1}^{\infty} k\left(\frac{4}{3}\right)^k$

12. $\displaystyle\sum_{k=1}^{\infty} k\left(\frac{3}{4}\right)^k$
13. $\displaystyle\sum_{k=1}^{\infty} \left(\frac{2}{k}\right)^k$
14. $\displaystyle\sum_{k=1}^{\infty} \frac{k^{10} 2^k}{k!}$

15. $\displaystyle\sum_{k=1}^{\infty} \frac{k^5}{10^k}$
16. $\displaystyle\sum_{k=1}^{\infty} \frac{3^k}{k^2}$
17. $\displaystyle\sum_{k=1}^{\infty} \left(\frac{k}{3k + 1}\right)^k$

18. $\displaystyle\sum_{k=1}^{\infty} \frac{3k + 1}{2^k}$
19. $\displaystyle\sum_{k=1}^{\infty} \frac{k!}{(k + 2)^4}$
20. $\displaystyle\sum_{k=1}^{\infty} \frac{k^5 + 100}{k!}$

21. $\displaystyle\sum_{k=1}^{\infty} \frac{(k!)^2}{(2k)!}$
22. $\displaystyle\sum_{k=1}^{\infty} k^2 2^{-k}$
23. $\displaystyle\sum_{k=1}^{\infty} \frac{(k!)^2}{[(2k)!]^2}$

24. $\displaystyle\sum_{k=1}^{\infty} k^4 3^{-k}$
25. $\displaystyle\sum_{k=1}^{\infty} \left(\frac{k - 2}{k}\right)^{k^2}$
26. $\displaystyle\sum_{k=1}^{\infty} \left(\frac{k}{2k + 1}\right)^k$

Test the series in Problems 27–44 for convergence. Justify your answers (that is, state explicitly which test you are using).

27. $\displaystyle\sum_{k=1}^{\infty} \frac{1,000}{k}$
28. $\displaystyle\sum_{k=1}^{\infty} \frac{5,000}{k\sqrt{k}}$
29. $\displaystyle\sum_{k=1}^{\infty} \frac{5k + 2}{k2^k}$

30. $\displaystyle\sum_{k=1}^{\infty} \frac{(k!)^2}{k^k}$
31. $\displaystyle\sum_{k=1}^{\infty} \frac{\sqrt{k!}}{2^k}$
32. $\displaystyle\sum_{k=1}^{\infty} \frac{3k + 5}{k3^k}$

33. $\displaystyle\sum_{k=1}^{\infty} \frac{2^k k!}{k^k}$
34. $\displaystyle\sum_{k=1}^{\infty} \frac{2^{2k} k!}{k^k}$
35. $\displaystyle\sum_{k=1}^{\infty} \frac{\sqrt{k + 1}}{k^{k + 0.5}}$

36. $\displaystyle\sum_{k=1}^{\infty} \frac{1}{k^k}$
37. $\displaystyle\sum_{k=1}^{\infty} \frac{k!}{(k + 1)!}$
38. $\displaystyle\sum_{k=1}^{\infty} \frac{2^{1,000k}}{k^{k/2}}$

39. $\displaystyle\sum_{k=1}^{\infty} \left(1 + \frac{1}{k}\right)^{-k^2}$
40. $\displaystyle\sum_{k=1}^{\infty} \left(\frac{k + 2}{k}\right)^{-k^2}$

41. $\displaystyle\sum_{k=1}^{\infty} \left|\frac{\cos k}{2^k}\right|$

42. $\displaystyle\sum_{k=1}^{\infty} \left|\frac{\sin k}{3^k}\right|$

43. $\displaystyle\sum_{k=2}^{\infty} \left(\frac{\ln k}{k}\right)^k$

44. $\displaystyle\sum_{k=2}^{\infty} \frac{1}{(\ln k)^k}$

B *Find all $x > 0$ for which each series in Problems 45–52 converges.*

45. $\displaystyle\sum_{k=1}^{\infty} k^2 x^k$

46. $\displaystyle\sum_{k=1}^{\infty} k x^k$

47. $\displaystyle\sum_{k=1}^{\infty} \frac{(x+0.5)^k}{k\sqrt{k}}$

48. $\displaystyle\sum_{k=1}^{\infty} \frac{(3x-0.4)^k}{k^2}$

49. $\displaystyle\sum_{k=1}^{\infty} \frac{x^k}{k!}$

50. $\displaystyle\sum_{k=1}^{\infty} \frac{x^{2k}}{k}$

51. $\displaystyle\sum_{k=1}^{\infty} (ax)^k$ for $a > 0$

52. $\displaystyle\sum_{k=1}^{\infty} k x^{2k}$

C **53.** Use the root test to show that $\Sigma k^p e^{-k}$ converges for any fixed positive real number p. What does this imply about the improper integral

$$\int_1^{\infty} x^p e^{-x}\, dx?$$

54. Consider the series $\displaystyle\sum_{k=1}^{\infty} 2^{-k+(-1)^k}$. What can you conclude about the convergence of this series if you use each of the following?

a. the ratio test   b. the root test

55. Let $\{a_k\}$ be a sequence of positive numbers and suppose that $\displaystyle\lim_{k\to\infty} \frac{a_{k+1}}{a_k} = L$, where $0 < L < 1$.

a. Show that $\displaystyle\lim_{k\to\infty} a_k = 0$.

   Hint: Note that Σa_k converges.

b. Show that $\displaystyle\lim_{k\to\infty} \frac{x^k}{k!} = 0$ for any $x > 0$.

56. Consider the series

$$1 + \tfrac{1}{2} + \tfrac{1}{2} + \tfrac{1}{4} + \tfrac{1}{4} + \tfrac{1}{8} + \tfrac{1}{8} + \tfrac{1}{16} + \cdots$$

a. Show that the ratio test fails.

b. What is the result of applying the root test?

57. Prove the root test by completing the steps in the following outline:

(1) Let $\displaystyle\lim_{n\to\infty} \sqrt[n]{a_n} = R$. If $R < 1$, choose x so that $R < x < 1$. Explain why there is an N such that

$$0 \leq \sqrt[n]{a_n} \leq x$$

for all $n \geq N$. Then use the direct comparison test to show that Σa_n converges.

(2) If $R > 1$, explain why $a_n > 1$ for all but a finite number of n. Use the divergence test to show Σa_n diverges.

(3) Find two series Σa_n and Σb_n with $R = 1$, so that Σa_n converges but Σb_n diverges.

8.6   *Alternating Series; Absolute and Conditional Convergence*

IN THIS SECTION   Leibniz's alternating series test, error estimates for alternating series, absolute and conditional convergence, rearrangement of terms in an absolutely convergent series

There are two classes of series for which the successive terms alternate in sign, and each of these series is appropriately called an **alternating series:**

odd-indexed terms are negative: $\displaystyle\sum_{k=1}^{\infty} (-1)^k a_k = -a_1 + a_2 - a_3 + \cdots$

even-indexed terms are negative: $\displaystyle\sum_{k=1}^{\infty} (-1)^{k+1} a_k = a_1 - a_2 + a_3 - \cdots$

where in both cases $a_k > 0$.

LEIBNIZ'S ALTERNATING SERIES TEST

In general, just knowing that $\displaystyle\lim_{k\to\infty} a_k = 0$ tells us very little about the convergence properties of the series Σa_k, but it turns out that an alternating series must converge if the absolute value of its terms decreases monotonically toward zero. This fact was first proved in the 17th century by Leibniz and may be stated as shown in the following theorem.

THEOREM 8.18   *Alternating series test*

If $a_k > 0$, then an alternating series

$$\sum_{k=1}^{\infty} (-1)^k a_k \quad \text{or} \quad \sum_{k=1}^{\infty} (-1)^{k+1} a_k$$

converges if both of the following two conditions are satisfied:

1. $\lim_{k \to \infty} a_k = 0$

2. $\{a_k\}$ is a decreasing sequence; that is, $a_{k+1} < a_k$ for all k.

Proof   We will show that when an alternating series of the form

$$\sum_{k=1}^{\infty} (-1)^{k+1} a_k$$

satisfies the two required properties, it converges. The steps for the other type of alternating series are similar. For this proof we are given that

$$\lim_{k \to \infty} a_k = 0 \quad \text{and} \quad a_{k+1} < a_k \text{ for all } k$$

We need to prove that the sequence of partial sums $\{S_n\}$ converges, where

$$S_n = \sum_{k=1}^{n} (-1)^{k+1} a_k = a_1 - a_2 + a_3 - a_4 + \cdots + (-1)^{n+1} a_n$$

Our strategy will be to show first that the sequence of *even*-indexed partial sums $\{S_{2n}\}$ is *increasing* and converges to a certain limit L. Then we shall show that the sequence of odd-indexed partial sums $\{S_{2n-1}\}$ also converges to L.

To understand why we would think to break up S_n to look at the even- and odd-indexed partial sums, consider Figure 8.15. Start at the origin. Because $a_1 > 0$, S_1 will be somewhere to the right of 0, as shown in Figure 8.15a. We know that $a_2 < a_1$ (decreasing sequence) so S_2 is to the left of S_1 but to the right of 0, as shown in Figure 8.15b. As you continue this process you can see that the partial sums oscillate back and forth. Because $a_n \to 0$, the successive steps are getting smaller and smaller, as shown in Figure 8.15c.

Note that the even partial sums of the alternating series satisfy

$$S_2 = a_1 - a_2$$
$$S_4 = (a_1 - a_2) + (a_3 - a_4)$$
$$\vdots$$
$$S_{2n} = (a_1 - a_2) + (a_3 - a_4) + \cdots + (a_{2n-1} - a_{2n})$$

Because $\{a_k\}$ is a decreasing sequence, each of the quantities in parentheses $(a_{2j-1} - a_{2j})$ is positive, and it follows that $\{S_{2n}\}$ is an increasing sequence:

$$S_2 < S_4 < S_6 < \cdots$$

Moreover, because

$$S_2 = a_1 - a_2 < a_1$$
$$S_4 = a_1 - (a_2 - a_3) - a_4 < a_1$$
$$\vdots$$
$$S_{2n} = a_1 - (a_2 - a_3) - (a_4 - a_5) - \cdots - (a_{2n-2} - a_{2n-1}) - a_{2n} < a_1$$

the sequence of even partial sums $\{S_{2n}\}$ is bounded above by a_1 (see Figure 8.15c), and because it is also an increasing sequence, the BMCT tells us that this sequence must

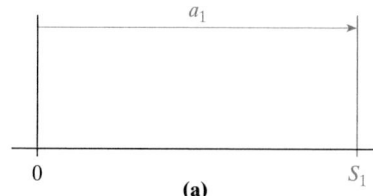

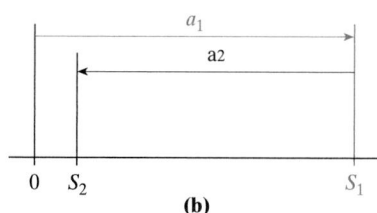

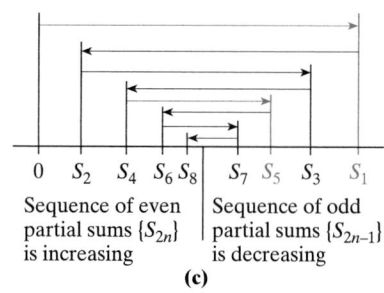

Sequence of even partial sums $\{S_{2n}\}$ is increasing

Sequence of odd partial sums $\{S_{2n-1}\}$ is decreasing

(c)

■ **FIGURE 8.15**   Alternating series test

converge, say, to L; that is,

$$\lim_{n\to\infty} S_{2n} = L$$

Next, consider $\{S_{2n-1}\}$, the sequence of odd partial sums. Because $S_{2n} - S_{2n-1} = a_n$ and $\lim_{n\to\infty} a_n = 0$, it follows that $S_{2n-1} = S_{2n} - a_n$ and

$$\lim_{n\to\infty} S_{2n-1} = \lim_{n\to\infty} S_{2n} - \lim_{n\to\infty} a_n = L - 0 = L$$

Since the sequence of even partial sums and the sequence of odd partial sums both converge to L, it follows that $\{S_n\}$ converges to L, and the alternating series must converge (in fact, to L).

═════

REMARK   When you are testing the alternating series $\Sigma(-1)^{k+1} a_k$ for convergence, it is wise to begin by computing $\lim_{k\to\infty} a_k$. If this limit is 0, you can show that the alternating series converges by verifying that $\{a_k\}$ is a decreasing sequence. However, if $\lim_{k\to\infty} a_k \neq 0$, you know immediately that the series diverges (by the divergence test), and no further computation is necessary.

EXAMPLE 1   *Convergence using the alternating series test*

Test the series $\displaystyle\sum_{k=1}^{\infty} \frac{(-1)^k}{k}$ for convergence. This series is called the **alternating harmonic series.**

Solution   The series can be expressed as $\Sigma(-1)^{k+1} a_k$, where $a_k = \dfrac{1}{k}$. We have $\lim_{k\to\infty} \dfrac{1}{k} = 0$ and because

$$\frac{1}{k+1} < \frac{1}{k}$$

for all $k > 0$, the sequence $\{a_k\}$ is decreasing. Thus, the alternating series test tells us that the given series must converge.   ■

EXAMPLE 2   *Using l'Hôpital's rule with the alternating series test*

Test the series $\displaystyle\sum_{k=1}^{\infty} \frac{(-1)^{k+1} \ln k}{k}$ for convergence.

Solution   Express the series in the form $\Sigma(-1)^{k+1}a_k$, where $a_k = \dfrac{\ln k}{k}$ (note that $a_k > 0$ for $k > 1$). The graph of $y = (\ln x)/x$ is shown in Figure 8.16. By applying l'Hôpital's rule, we find that

$$\lim_{k\to\infty} \frac{\ln k}{k} = \lim_{k\to\infty} \frac{1/k}{1} = 0$$

It remains to show that the sequence $\{a_k\}$ is decreasing. We can do this by setting $f(x) = \dfrac{\ln x}{x}$ and noting that the derivative

$$f'(x) = \frac{1 - \ln x}{x^2}$$

satisfies $f'(x) < 0$ for all $x > e$. Thus, the sequence $\{a_k\}$ is decreasing for all $k > 3 > e$. The conditions of the alternating series test are satisfied, and the given alternating series must converge.   ■

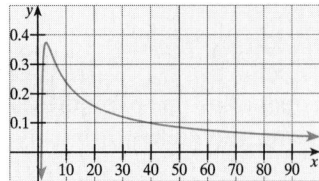

■ **FIGURE 8.16**   Graph of $y = \dfrac{\ln x}{x}$

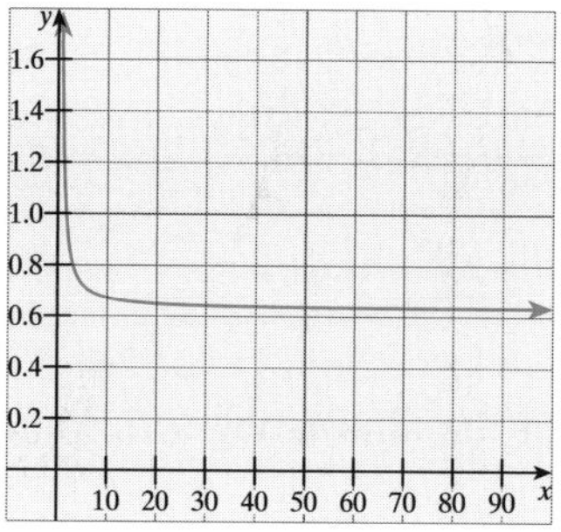

■ **FIGURE 8.17** Graph of
$$y = \frac{1}{\tan^{-1}x}$$

EXAMPLE 3   *Divergence of an alternating series*

Test the series $\displaystyle\sum_{k=1}^{\infty} \frac{(-1)^{k+1}}{\tan^{-1} k}$ for convergence.

Solution   The series can be expressed as $\Sigma(-1)^{k+1}a_k$, where $a_k = \dfrac{1}{\tan^{-1} k}$. The graph of $y = \dfrac{1}{\tan^{-1} x}$ is shown in Figure 8.17. It appears that the limit as $x \to \infty$ is not 0, so we suspect the series diverges. We can show this as follows:

$$\lim_{k\to\infty} a_k = \lim_{k\to\infty}\left(\frac{1}{\tan^{-1} k}\right) = \frac{1}{\pi/2} = \frac{2}{\pi} \neq 0$$

Thus, the series diverges by the divergence test.   ■

Alternating p-Series

The series $\displaystyle\sum_{k=1}^{\infty} \frac{(-1)^{k+1}}{k^p}$ is called the **alternating p-series.**

The following example shows that the alternating p-series converges for all $p > 0$.

EXAMPLE 4   *Alternating p-series*

Prove that the alternating p-series $\displaystyle\sum_{k=1}^{\infty} \frac{(-1)^{k+1}}{k^p}$ converges for $p > 0$.

Solution   Let $a_k = \dfrac{1}{k^p}$, and note that $\lim\limits_{k\to\infty} a_k = 0$ for $p > 0$. To show that the sequence $\{a_k\}$ is decreasing, we find that

$$\frac{a_{k+1}}{a_k} = \frac{\dfrac{1}{(k+1)^p}}{\dfrac{1}{k^p}} = \frac{k^p}{(k+1)^p} < 1$$

so $a_k > a_{k+1}$. Thus, the alternating p-series converges for all $p > 0$.   ■

ERROR ESTIMATES FOR ALTERNATING SERIES

For any convergent series with sum L, the nth partial sum is an approximation to L that we expect to improve as n increases. In general, it is difficult to know how large an n to pick to ensure that the approximation will have a desired degree of accuracy. However, if the series in question satisfies the conditions of the alternating series test, the absolute value of the error incurred by using the sum for the first n terms to estimate the entire sum turns out to be no greater than the $(n + 1)st$ excluded term. More formally, we have the following theorem.

THEOREM 8.19   *The error estimate for an alternating series*

Suppose an alternating series

$$\sum_{k=1}^{\infty} (-1)^k a_k \quad \text{or} \quad \sum_{k=1}^{\infty} (-1)^{k+1} a_k$$

satisfies the conditions of the alternating series test; namely,

$$\lim_{k \to \infty} a_k = 0 \quad \text{and} \quad \{a_k\} \text{ is a decreasing sequence } (a_{k+1} < a_k)$$

If the series has sum S, then

$$|S - S_n| < a_{n+1}$$

where S_n is the nth partial sum of the series.

Proof   We will prove the result for the second form of the alternating series and leave the first form for the reader. That is, let

$$S = \sum_{k=1}^{\infty} (-1)^{k+1} a_k \quad \text{and} \quad S_n = \sum_{k=1}^{n} (-1)^{k+1} a_k$$

Begin with $S - S_n$:

$$S - S_n = \sum_{k=1}^{\infty} (-1)^{k+1} a_k - \sum_{k=1}^{n} (-1)^{k+1} a_k = \sum_{k=n+1}^{\infty} (-1)^{k+1} a_k$$

$$= (-1)^{n+2} a_{n+1} + (-1)^{n+3} a_{n+2} + (-1)^{n+4} a_{n+3} + \cdots$$

$$= (-1)^n (-1)^2 a_{n+1} + (-1)^n (-1)^3 a_{n+2} + (-1)^n (-1)^4 a_{n+3} + \cdots$$

$$= (-1)^n [a_{n+1} - a_{n+2} + a_{n+3} - a_{n+4} + \cdots]$$

$$= (-1)^n [a_{n+1} - (a_{n+2} - a_{n+3}) - (a_{n+4} - a_{n+5}) - \cdots]$$

Because the sequence $\{a_n\}$ is decreasing, we have $a_k - a_{k+1} \geq 0$ for all k, and it follows that

$$|S - S_n| = |a_{n+1} - (a_{n+2} - a_{n+3}) - (a_{n+4} - a_{n+5}) - \cdots| \leq a_{n+1}$$

because every quantity in parentheses is positive.   ▬

■ *What This Says:*   If an alternating series satisfies the conditions of the alternating series test, you can approximate the sum of the series by using the nth partial sum (S_n), and your error will have an absolute value no greater than the first term left off (namely, a_{n+1}).

EXAMPLE 5   *Error estimate for an alternating series*

Consider the convergent series $\displaystyle\sum_{k=1}^{\infty} \frac{(-1)^{k+1}}{k^4}$.

a. Estimate the sum of the series by taking the sum of the first four terms. How accurate is this estimate?

b. Estimate the sum of the series with three-decimal-place accuracy.

Solution

a. Let $a_k = \dfrac{1}{k^4}$ and let S denote the actual sum of the series. The error estimate tells us that

$$|S - S_4| \leq a_5$$

where S_4 is the sum of the first four terms of the series. Using a calculator, we find

$$S_4 = \frac{1}{1^4} - \frac{1}{2^4} + \frac{1}{3^4} - \frac{1}{4^4} \approx 0.9459394$$

and $a_5 = \dfrac{1}{5^4} = 0.0016$. Thus, if we estimate S by $S_4 \approx 0.9459$, we incur an error of about 0.0016, which means that

$$|S - S_4| \leq 0.0016$$
$$0.9459394 - 0.0016 \leq S \leq 0.9459394 + 0.0016$$
$$0.9443394 \leq S \leq 0.9475394$$

so we have S, correct to two decimal places, as 0.94.

b. Because we want to approximate S by a partial sum S_n with three-decimal-place accuracy, we can allow an error of no more than 0.0005. Such an error will not affect the rounded value of $S_4 \approx 0.946$. The error is measured by

$$a_{n+1} = \frac{1}{(n+1)^4}$$

Thus, we wish to find n so that

$$\frac{1}{(n+1)^4} \leq 0.0005$$
$$\frac{1}{0.0005} \leq (n+1)^4$$
$$\sqrt[4]{2{,}000} \leq n + 1$$
$$6.687403 - 1 \leq n$$

This says that n must be an integer greater than 5.687403; that is, $n \geq 6$. ■

ABSOLUTE AND CONDITIONAL CONVERGENCE

The convergence tests we have developed cannot be applied to a series that has mixed terms or does not strictly alternate. In such cases, it is often useful to apply the following result.

THEOREM 8.20   *The absolute convergence test*

A series of real numbers Σa_k must converge if the related absolute value series $\Sigma |a_k|$ converges.

Proof   Assume $\Sigma |a_k|$ converges and let $b_k = a_k + |a_k|$ for all k. Note that

$$b_k = \begin{cases} 2|a_k| & \text{if } a_k > 0 \\ 0 & \text{if } a_k \leq 0 \end{cases}$$

Thus, we have $0 \leq b_k < 2|a_k|$ for all k. Because the series $\Sigma |a_k|$ converges and both Σb_k and $\Sigma |a_k|$ are series of nonnegative terms, the direct comparison test tells us that the dominated series Σb_k also converges. Finally, because

$$a_k = b_k - |a_k|$$

and both Σb_k and $\Sigma |a_k|$ converge, it follows that Σa_k must also converge. ═

EXAMPLE 6   *Convergence using the absolute convergence test*

Test the series $1 + \frac{1}{4} + \frac{1}{9} - \frac{1}{16} - \frac{1}{25} + \frac{1}{36} + \frac{1}{49} + \frac{1}{64} - \frac{1}{81} - \frac{1}{100} \cdots$ for convergence. This is the series in which the absolute value of the general term is $\dfrac{1}{k^2}$ and the pattern of the signs is $+ + + - - + + + - - \cdots$.

Solution   This is not a series of positive terms, nor is it strictly alternating. However, we find that the corresponding series of absolute values is the convergent p-series

$$1 + \frac{1}{4} + \frac{1}{9} + \frac{1}{16} + \cdots = \sum_{k=1}^{\infty} \frac{1}{k^2}$$

Because the absolute value series converges, the absolute convergence test assures us that the given series also converges.   ■

EXAMPLE 7   *Convergence of a trigonometric series*

Test the series $\displaystyle\sum_{k=1}^{\infty} \frac{\sin k}{2^k}$ for convergence.

Solution   Because $\sin k$ takes on both positive and negative values, the series cannot be analyzed by methods that apply only to series of positive terms. Moreover, the series is not strictly alternating. The corresponding series of absolute values is

$$\sum_{k=1}^{\infty} \left| \frac{\sin k}{2^k} \right|$$

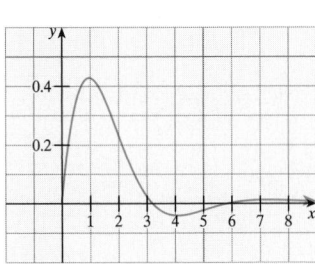

■ **FIGURE 8.18**   Graph of
$y = \dfrac{\sin x}{2^x}$

which is dominated by the convergent geometric series $\Sigma \frac{1}{2^k}$ because $\left| \sin k \right| \leq 1$ for all k; that is,

$$0 \leq \left| \frac{\sin k}{2^k} \right| \leq \frac{1}{2^k} \quad \text{for all } k$$

Therefore, the absolute convergence test assures us that the given series converges. The graph of the related continuous function $y = (\sin x)/(2^x)$ is shown in Figure 8.18.   ■

If Σa_k converges, then $\Sigma \left| a_k \right|$ may either converge or diverge. The two cases that can occur are given the following special names.

Absolutely Convergent Series

The series Σa_k is **absolutely convergent** if the related series $\Sigma \left| a_k \right|$ converges.

Conditionally Convergent Series

The series Σa_k is **conditionally convergent** if it converges but $\Sigma \left| a_k \right|$ diverges.

For example, the series in Examples 6 and 7 are both absolutely convergent, but the alternating harmonic series

$$\sum_{k=1}^{\infty} \frac{(-1)^{k+1}}{k}$$

is conditionally convergent because it converges (see Example 4), whereas the related series of absolute value terms $\Sigma\, 1/k$ (the harmonic series) diverges.

The ratio test and the root test for positive-term series can be generalized to apply to arbitrary series. The following is a statement and proof of the generalized ratio test. The proof of the generalized root test is left to the problems (see Problem 60).

THEOREM 8.21   *The generalized ratio test*

For the series Σa_k, suppose $a_k \neq 0$ for $k \geq 1$ and that

$$\lim_{k \to \infty} \left| \frac{a_{k+1}}{a_k} \right| = L$$

where L is a real number or ∞. Then:

If $L < 1$, the series Σa_k converges absolutely and hence converges.

If $L > 1$ or if L is infinite, the series Σa_k diverges.

If $L = 1$, the test is inconclusive.

Proof

Assume $L < 1$; then the positive series $\Sigma |a_k|$ converges by the ratio test, and hence Σa_k converges absolutely.

Assume $L > 1$; then the sequence $\{|a_k|\}$ is eventually increasing, which means that $\{a_k\}$ cannot converge to 0 as $k \to \infty$. Thus Σa_k diverges by the divergence test.

Assume $L = 1$. See Problem 61 for this part of the proof.   ═

EXAMPLE 8   *Convergence and divergence using the generalized ratio test*

Find all values of x for which the series Σkx^k converges and all x for which it diverges.

Solution   Let $a_k = kx^k$.

$$\lim_{k \to \infty} \left| \frac{a_{k+1}}{a_k} \right| = \lim_{k \to \infty} \left| \frac{(k+1)x^{k+1}}{kx^k} \right| = \lim_{k \to \infty} \left(\frac{k+1}{k} \right) \left| \frac{x^{k+1}}{x^k} \right| = \lim_{k \to \infty} \left(1 + \frac{1}{k} \right) |x| = |x|$$

Thus, by the generalized ratio test we see for $|x| = L$ that the series converges for $|x| < 1$ and diverges if $|x| > 1$. If $x = 1$, the series is $\Sigma k(1)^k$, which clearly diverges by the divergence test. Similarly, if $x = -1$, $\Sigma k(-1)^k$ diverges by the divergence test. Thus, the series converges for $|x| < 1$ and diverges for $|x| \geq 1$.   ■

Additional Guidelines for Determining Convergence of Series	
1. **Test for absolute convergence**	Take the absolute value of each term and proceed to test for convergence of positive series, as summarized in Table 8.1. 1. If $\Sigma \lvert a_k \rvert$ *converges*, then Σa_k will converge. 2. If $\Sigma \lvert a_k \rvert$ diverges, then the series is *not absolutely convergent;* however, the alternating series might still converge conditionally.
2. **Alternating series test**	For $a_k > 0$, the series converges if $\lvert a_{k+1} \rvert < \lvert a_k \rvert$ for all k and $\lim\limits_{k \to \infty} a_k = 0$.

REARRANGEMENT OF TERMS IN AN ABSOLUTELY CONVERGENT SERIES

You may be surprised to learn that if the terms in a conditionally convergent series are rearranged (that is, the order of the summands is changed), the new series may not converge or it may converge to a different sum from that of the original series! For example, we know that the alternating harmonic series

$$\sum_{k=1}^{\infty} \frac{(-1)^{k+1}}{k}$$

converges conditionally, and it can be shown (see Problem 52) that

$$1 - \tfrac{1}{2} + \tfrac{1}{3} - \tfrac{1}{4} + \tfrac{1}{5} - \tfrac{1}{6} + \tfrac{1}{7} - \tfrac{1}{8} + \tfrac{1}{9} - \cdots = \ln 2$$

If we rearrange this series by placing two of the subtracted terms after each added term, we obtain

$$1 - \tfrac{1}{2} - \tfrac{1}{4} + \tfrac{1}{3} - \tfrac{1}{6} - \tfrac{1}{8} + \tfrac{1}{5} - \cdots = (1 - \tfrac{1}{2}) - \tfrac{1}{4} + (\tfrac{1}{3} - \tfrac{1}{6}) - \tfrac{1}{8} + \cdots$$
$$= \tfrac{1}{2} - \tfrac{1}{4} + \tfrac{1}{6} - \tfrac{1}{8} + \cdots$$
$$= \tfrac{1}{2}(1 - \tfrac{1}{2} + \tfrac{1}{3} - \tfrac{1}{4} + \cdots)$$
$$= \tfrac{1}{2} \ln 2$$

In general, it can be shown that if Σa_k is conditionally convergent, there is a re-arrangement of the terms of Σa_k so that the sum is equal to *any* given finite number. In Problem 53, for example, you are asked to rearrange the terms of the series

$$\sum_{k=1}^{\infty} \frac{(-1)^{k+1}}{k}$$

so that the sum is $\tfrac{3}{2} \ln 2$.

This information may be somewhat unsettling, because it is reasonable to expect a sum to be unaffected by the order in which the summands are taken. Absolutely convergent series behave more in the way we would expect. In fact, *if the series Σa_k converges absolutely with sum S, then **any** rearrangement of the terms also converges absolutely to S.* A detailed discussion of rearrangement of terms of a series requires the techniques of advanced calculus.

8.6 Problem Set

Ⓐ 1. ■ **What Does This Say?** Discuss absolute versus conditional convergence.

2. ■ **What Does This Say?** Discuss the convergence of the alternating *p*-series.

3. ■ **What Does This Say?** Look at Problems 4–31 and notice that most starting values are $k = 1$, but some have starting values $k = 2$. Why do you think we did not use $k = 1$ for all of these problems?

Determine whether each series in Problems 4–31 *converges absolutely, converges conditionally, or diverges.*

4. $\displaystyle\sum_{k=1}^{\infty} \frac{(-1)^{k+1}k}{k^2 + 1}$

5. $\displaystyle\sum_{k=1}^{\infty} \frac{(-1)^{k+1}k^2}{k^3 + 1}$

6. $\displaystyle\sum_{k=1}^{\infty} \frac{(-1)^{k+1}k}{2k + 1}$

7. $\displaystyle\sum_{k=1}^{\infty} \frac{(-1)^{k+1}k^2}{k^2 + 1}$

8. $\displaystyle\sum_{k=1}^{\infty} \frac{(-1)^{k+1}}{k^{3/2}}$

9. $\displaystyle\sum_{k=1}^{\infty} \frac{(-1)^{k+1}k}{2^k}$

10. $\displaystyle\sum_{k=1}^{\infty} (-1)^{k+1} \frac{k^2}{e^k}$

11. $\displaystyle\sum_{k=1}^{\infty} \frac{(-1)^k}{\sqrt{k}}$

12. $\displaystyle\sum_{k=1}^{\infty} (-1)^k \frac{(1 + k^2)}{k^3}$

13. $\displaystyle\sum_{k=1}^{\infty} \frac{(-1)^{k+1}k!}{k^k}$

14. $\displaystyle\sum_{k=2}^{\infty} (-1)^k \frac{k!}{\ln k}$

15. $\displaystyle\sum_{k=1}^{\infty} \frac{(-1)^k(2k)!}{k^k}$

16. $\displaystyle\sum_{k=1}^{\infty} (-1)^{k+1} \frac{2^k}{k!}$

17. $\displaystyle\sum_{k=1}^{\infty} \frac{(-3)^k (k + 1)}{k!}$

18. $\displaystyle\sum_{k=1}^{\infty} (-1)^{k+1} \frac{2^{2k+1}}{k!}$

19. $\displaystyle\sum_{k=2}^{\infty} \frac{(-1)^{k+1}}{\ln k}$

20. $\displaystyle\sum_{k=1}^{\infty} \frac{(-1)^{k+1}k}{(k + 1)(k + 2)}$

21. $\displaystyle\sum_{k=2}^{\infty} \frac{(-1)^{k+1}}{(\ln k)^4}$

22. $\displaystyle\sum_{k=2}^{\infty} \frac{(-1)^{k+1}}{\ln (\ln k)}$

23. $\displaystyle\sum_{k=2}^{\infty} \frac{(-1)^{k+1}}{k \ln k}$

24. $\displaystyle\sum_{k=1}^{\infty} \frac{(-1)^{k+1} \ln k}{k}$

25. $\displaystyle\sum_{k=2}^{\infty} \frac{(-1)^{k+1}k}{\ln k}$

26. $\displaystyle\sum_{k=1}^{\infty} (-1)^{k+1} \frac{\ln k}{k^2}$

27. $\displaystyle\sum_{k=1}^{\infty} \frac{(-1)^{k+1}k}{(k + 2)^2}$

28. $\displaystyle\sum_{k=1}^{\infty} (-1)^{k+1} \left(\frac{k}{k + 1}\right)^k$

29. $\displaystyle\sum_{k=2}^{\infty} (-1)^{k+1} \frac{\ln (\ln k)}{k \ln k}$

30. $\displaystyle\sum_{k=1}^{\infty} (-1)^{k+1} \left(\frac{1}{k}\right)^{1/k}$

31. $\displaystyle\sum_{k=1}^{\infty} (-1)^{k+1} \frac{k^5 \, 5^{k+2}}{2^{3k}}$

B *Given the series in Problems 32–37:*
  a. *Estimate the sum of the series by taking the sum of the first four terms. How accurate is this estimate?*
  b. *Estimate the sum of the series with three-decimal-place accuracy.*

32. $\displaystyle\sum_{k=1}^{\infty} \frac{(-1)^{k+1}}{2^{2k-2}}$

33. $\displaystyle\sum_{k=1}^{\infty} \frac{(-1)^{k+1}}{k!}$

34. $\displaystyle\sum_{k=1}^{\infty} \frac{(-1)^{k}}{k^2}$

35. $\displaystyle\sum_{k=1}^{\infty} \left(\frac{-1}{3}\right)^{k+1}$

36. $\displaystyle\sum_{k=1}^{\infty} \frac{(-1)^{k+1}}{k^3}$

37. $\displaystyle\sum_{k=1}^{\infty} \left(\frac{-1}{5}\right)^{k}$

Use the generalized ratio test in Problems 38–43 to find all numbers x for which the given series converges.

38. $\displaystyle\sum_{k=1}^{\infty} \frac{x^k}{k}$

39. $\displaystyle\sum_{k=1}^{\infty} \frac{x^k}{\sqrt{k}}$

40. $\displaystyle\sum_{k=1}^{\infty} \frac{2^k x^k}{k!}$

41. $\displaystyle\sum_{k=1}^{\infty} \frac{(k+2)\,x^k}{k^2(k+3)}$

42. $\displaystyle\sum_{k=1}^{\infty} (-1)^{k+1} \left(\frac{x}{k}\right)^k$

43. $\displaystyle\sum_{k=1}^{\infty} (-1)^k k^p x^k$
  for $p > 0$

44. THINK TANK PROBLEM   Find an upper bound for the error if the alternating series $\displaystyle\sum_{k=1}^{\infty} \frac{(-1)^{k+1}}{k}$ is approximated by the partial sum
$$S_5 = 1 - \frac{1}{2} + \frac{1}{3} - \frac{1}{4} + \frac{1}{5}$$

45. THINK TANK PROBLEM   Find an upper bound for the error if the alternating series $\displaystyle\sum_{k=1}^{\infty} \frac{(-1)^{k+1}}{k^2}$ is approximated by the partial sum
$$S_5 = 1 - \frac{1}{2^2} + \frac{1}{3^2} - \frac{1}{4^2} + \frac{1}{5^2}$$

46. THINK TANK PROBLEM   Find an upper bound for the error if the alternating series $\displaystyle\sum_{k=2}^{\infty} \frac{(-1)^{k}}{\ln k}$ is approximated by the partial sum
$$S_7 = \frac{1}{\ln 2} - \frac{1}{\ln 3} + \frac{1}{\ln 4} - \frac{1}{\ln 5} + \frac{1}{\ln 6} - \frac{1}{\ln 7}$$

47. THINK TANK PROBLEM   Find an upper bound for the error if the alternating series $\displaystyle\sum_{k=1}^{\infty} \frac{(-1)^{k+1} k}{2^k}$ is approximated by the partial sum
$$S_6 = \frac{1}{2} - \frac{2}{2^2} + \frac{3}{2^3} - \frac{4}{2^4} + \frac{5}{2^5} - \frac{6}{2^6}$$

48. For what numbers p does the alternating series
$$\sum_{k=2}^{\infty} \frac{(-1)^{k+1}}{k\,(\ln k)^p}$$
converge? For what numbers p does it converge absolutely?

49. Test the series $\displaystyle\sum_{k=1}^{\infty} \frac{\sin \sqrt[k]{2}}{k^2}$ for convergence.

50. Use series methods to evaluate $\displaystyle\lim_{k\to\infty} \frac{x^k}{k!}$ where x is a real number.

C **51.** Show that the sequence $\{a_n\}$ converges, where
$$a_n = \frac{1}{1} + \frac{1}{2} + \frac{1}{3} + \cdots + \frac{1}{n} - \ln n.$$

52. Show that $\displaystyle\sum_{k=1}^{\infty} \frac{(-1)^{k+1}}{k} = \ln 2$ by completing the following steps.

  a. Let $\displaystyle S_m = \sum_{k=1}^{m} \frac{(-1)^{k+1}}{k}$ and let $\displaystyle H_m = \sum_{k=1}^{m} \frac{1}{k}$. Show that
  $$S_{2m} = H_{2m} - H_m.$$

  b. It can be shown (see Problem 51) that
  $$\lim_{m\to\infty} \left(1 + \frac{1}{2} + \cdots + \frac{1}{m} - \ln m\right)$$
  exists. The value of this limit is a number $\gamma \approx 0.57722$, called **Euler's constant**. Use this fact, along with the relationship in part **a**, to show that $\displaystyle\lim_{h\to\infty} S_n = \ln 2$.
  Hint: Note that $S_{2m} = H_{2m} - H_m$
  $= [H_{2m} - \ln(2m)] - [H_m - \ln m] + \ln(2m) - \ln m.$

53. Consider the following rearrangement of the conditionally convergent harmonic series $\displaystyle\sum_{k=1}^{\infty} \frac{(-1)^{k+1}}{k}$:
$$1 + \frac{1}{3} - \frac{1}{2} + \frac{1}{5} + \frac{1}{7} - \frac{1}{4} + \frac{1}{9} + \frac{1}{11} - \frac{1}{6} + \cdots$$

  a. Let S_n and H_n denote the nth partial sum of the given rearranged series and the harmonic series $\displaystyle\sum_{k=1}^{\infty} \frac{1}{k}$, respectively. Show that if n is a multiple of 3, say $n = 3m$, then
  $$S_{3m} = H_{4m} - \frac{1}{2} H_{2m} - \frac{1}{2} H_m$$

  b. Show that $\displaystyle\lim_{n\to\infty} S_n = \frac{3}{2} \ln 2$.
  Hint: You may find it helpful to use
  $$\lim_{n\to\infty} \left(1 + \frac{1}{2} + \cdots + \frac{1}{n} - \ln n\right) = \gamma$$
  given in part **b** of Problem 52.

54. JOURNAL PROBLEM (*School Science and Mathematics by Michael Brozinsky*)*   Show that the series

$$1 + \frac{1}{2} + \frac{1}{3} - \frac{1}{4} - \frac{1}{5} - \frac{1}{6} + \frac{1}{7} + \frac{1}{8} + \frac{1}{9} + \cdots$$

converges and find its sum.

55. THINK TANK PROBLEM   What (if anything) is wrong with the following computation?

$$1 - \frac{1}{2} + \frac{1}{3} - \frac{1}{4} + \frac{1}{5} - \frac{1}{6} + \cdots$$

$$= 1 + \left(\frac{1}{2} - 1\right) + \frac{1}{3} + \left(\frac{1}{4} - \frac{1}{2}\right) + \frac{1}{5} + \left(\frac{1}{6} - \frac{1}{3}\right) + \cdots$$

$$= \left(1 + \frac{1}{2} + \frac{1}{3} + \frac{1}{4} + \cdots\right) - 1 - \frac{1}{2} - \frac{1}{3} - \frac{1}{4} - \cdots$$

$$= \left(1 + \frac{1}{2} + \frac{1}{3} + \cdots\right) - \left(1 + \frac{1}{2} + \frac{1}{3} + \cdots\right) = 0$$

56. Test the given series for convergence.
   a. $2 - 2^{1/2} + 2^{1/3} - 2^{1/4} + \cdots$
   b. $(1 - 2^{1/1}) - (1 - 2^{1/2}) + (1 - 2^{1/3}) - \cdots$

57. Suppose $\{a_k\}$ is a sequence with the property that $|a_n| < A^n$ for some positive number A and all n. Show that the series $\Sigma a_k x^k$ converges absolutely for $|x| \leq 1/A$.

---

*Volume 82, 1982, p. 175.

58. THINK TANK PROBLEM   Give an example of a sequence $\{a_k\}$ with the property that Σa_k^2 converges and Σa_k also converges.

59. Show that if $\displaystyle\sum_{k=1}^{\infty} a_k$ converges then $\displaystyle\sum_{k=0}^{\infty} a_k^2$ must also converge.

60. Prove the generalized root test, which may be stated as follows: Suppose $a_k \neq 0$ for $k \geq 1$ and that

$$\lim_{k \to \infty} \sqrt[k]{|a_k|} = L$$

Then:
   If $L < 1$, the series Σa_k converges absolutely.
   If $L > 1$, the series Σa_k diverges.
   If $L = 1$, the test is inconclusive.

61. Show that if

$$L = \lim_{k \to \infty} \left|\frac{a_{k+1}}{a_k}\right| = 1$$

the series Σa_k can either converge or diverge.
Hint: Find a convergent series with $L = 1$ and a divergent series that satisfies the same condition.

62. Let $\Sigma(-1)^{k+1} a_k$ be an alternating series such that $\{a_k\}$ is decreasing and

$$\lim_{k \to \infty} a_k = 0$$

Show that the sequence of odd partial sums S_{2n-1} is decreasing.

8.7   Power Series

IN THIS SECTION	convergence of a power series, power series in $x - c$, term-by-term differentiation and integration of power series

An infinite series of the form

$$\sum_{k=0}^{\infty} a_k (x - c)^k = a_0 + a_1 (x - c) + a_2 (x - c)^2 + \cdots$$

is called a **power series** in $(x - c)$. The numbers $a_0, a_1, a_2, \ldots$ are the *coefficients* of the power series, and we will be concerned only with the case where these coefficients, as well as x and c, are real numbers. If $c = 0$, the series has the form

$$\sum_{k=0}^{\infty} a_k x^k = a_0 + a_1 x + a_2 x^2 + a_3 x^3 + \cdots$$

CONVERGENCE OF A POWER SERIES

How can we determine the set of all numbers x for which a given power series converges? This question is answered by the following fundamental result on convergence. We begin by considering the case where $c = 0$.

THEOREM 8.22   *Convergence of a power series*

For a power series $\displaystyle\sum_{k=1}^{\infty} a_k x^k$, exactly one of the following is true:

1. The series converges for all x.
2. The series converges only for $x = 0$.
3. The series **converges absolutely** for all x in an open interval $(-R, R)$ and **diverges** for $|x| > R$.

   Note: The series should be checked separately at the endpoints, because it could converge absolutely, or converge conditionally, or diverge at $x = R$ and $x = -R$.

Proof   The proof is found in most advanced calculus textbooks.   ═══

The following three examples show each of these possibilities.

EXAMPLE 1   *When the convergence set is the entire x-axis*

Show that the power series $\displaystyle\sum_{k=1}^{\infty} \frac{x^k}{k!}$ converges for all x.

Solution   If $x = 0$, then the series is trivial and converges. For $x \neq 0$, we use the generalized ratio test to find

$$L = \lim_{k \to \infty} \left| \frac{\dfrac{x^{k+1}}{(k+1)!}}{\dfrac{x^k}{k!}} \right| = \lim_{k \to \infty} \left| \frac{x^{k+1}k!}{(k+1)!x^k} \right| = \lim_{k \to \infty} \frac{|x|}{k+1} = 0$$

Because $L = 0$ satisfies $L < 1$, the series converges for all x.   ■

EXAMPLE 2   *Convergence only at the point x = 0*

Show that the power series $\displaystyle\sum_{k=1}^{\infty} k! x^k$ converges only when $x = 0$.

Solution   We use the generalized ratio test to find

$$L = \lim_{k \to \infty} \left| \frac{(k+1)!x^{k+1}}{k!x^k} \right| = \lim_{k \to \infty} (k+1)|x|$$

For $x = 0$, the limit is 0, but for $x \neq 0$, the limit is ∞. Hence, the power series converges only for $x = 0$.   ■

EXAMPLE 3   *When the convergence set is bounded*

Find the convergence set for the geometric series

$$\sum_{k=0}^{\infty} x^k = 1 + x + x^2 + x^3 + \cdots$$

Solution   Because this is a geometric series, we know it converges for $|x| < 1$. This is the same as saying it converges on the open interval $(-1, 1)$ and diverges elsewhere.   ■

The generalized ratio test can be used to show that the convergence set of a power series has one of the three forms described by the following theorem.

===
THEOREM 8.23   *The convergence set of a power series*

Let $\Sigma a_k u^k$ be a power series, and let

$$L = \lim_{k \to \infty} \left| \frac{a_{k+1}}{a_k} \right|$$

Then:

If $L = \infty$, the power series converges only at $u = 0$.

If $L = 0$, the power series converges for all real u.

If $0 < L < \infty$, let $R = 1/L$. Then the power series **converges absolutely** for $|u| < R$ and **diverges** for $|u| > R$. It may either converge or diverge at the endpoint $u = -R$ and $u = R$.

Proof   We shall use the generalized ratio test. First, in the case where

$$L = \lim_{k \to \infty} \left| \frac{a_{k+1}}{a_k} \right| \text{ exists, consider the limit}$$

$$M = \lim_{k \to \infty} \left| \frac{a_{k+1} u^{k+1}}{a_k u^k} \right| = \lim_{k \to \infty} \left| \frac{a_{k+1}}{a_k} \right| |u| = L|u|$$

The generalized ratio test tells us that $\Sigma a_k u^k$ converges absolutely if $M < 1$ and diverges if $M > 1$.

If L is infinite, the series $\Sigma a_k u^k$ converges only at $u = 0$ (because this is the only value of u such that $L|u| < 1$). If $L = 0$, then $L|u| < 1$ is always satisfied, and the series converges absolutely for all u. Suppose L is finite and $L \neq 0$. Then the series converges absolutely for all u such that

$$|u| < \frac{1}{L} = \frac{1}{\frac{1}{R}} = R \qquad \text{Because } R = \frac{1}{L}$$

Thus, $-R < u < R$ Because $|u| < R$

We see the series diverges for $|u| > R$, and we can have either convergence or divergence at $u = R$ and $u = -R$.   ===

According to Theorem 8.23, the set of numbers for which the power series $\Sigma a_k u^k$ converges is an interval centered at $u = 0$. We call this the **interval of convergence** of the power series. If this interval has length $2R$, then R is called the **radius of convergence** of the series. When the series converges only for $u = 0$, the series has radius of convergence $R = 0$, and if it converges for all x, we say that $R = \infty$.

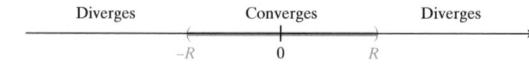

■ **FIGURE 8.19**   The interval of convergence of a power series

In Example 1, we found that $\dfrac{\Sigma x^k}{k!}$ converges for all x, so R is infinite and the interval of convergence is the entire real line. In Example 2, we found that the power series $\Sigma k! x^k$ has radius of convergence 0. In Example 3, we found that Σx^k converges on $(-1, 1)$, so we say $R = 1$.

We now turn our attention to finding R for an example that is not a geometric series.

EXAMPLE 4   *Finding an interval of convergence*

Find the interval of convergence for $\displaystyle\sum_{k=1}^{\infty} \frac{2^k x^k}{k}$.

Solution   We note that $a_k = \dfrac{2^k}{k}$ and $u = x$.

$$L = \lim_{k\to\infty} \left| \frac{\frac{2^{k+1}}{k+1}}{\frac{2^k}{k}} \right| = \lim_{k\to\infty} \left| \frac{2k}{k+1} \right| = 2$$

WARNING   Do not forget to check the endpoints.

Thus, the series converges absolutely for $-\frac{1}{2} < x < \frac{1}{2}$. Checking the endpoints, we find:

$$x = \frac{1}{2}: \quad \sum_{k=1}^{\infty} \frac{2^k}{k}\left(\frac{1}{2}\right)^k = \sum_{k=1}^{\infty} \frac{1}{k} \quad \text{diverges (}p\text{-series, }p=1\text{)}$$

$$x = -\frac{1}{2}: \quad \sum_{k=1}^{\infty} \frac{2^k}{k}\left(-\frac{1}{2}\right)^k = \sum_{k=1}^{\infty} \frac{(-1)^k}{k} \quad \text{converges (alternating series test)}$$

The interval of convergence is $-\frac{1}{2} \le x < \frac{1}{2}$. ∎

EXAMPLE 5   *Finding an interval of convergence*

Find the radius of convergence and the interval of convergence for the power series

$$\sum_{k=1}^{\infty} \frac{2^k x^k}{k!}$$

Solution   We use Theorem 8.23 with $a_k = \dfrac{2^k}{k!}$ and $u = x$.

$$L = \lim_{k\to\infty} \left| \frac{\frac{2^{k+1}}{(k+1)!}}{\frac{2^k}{k!}} \right| = \lim_{k\to\infty} \frac{2}{k+1} = 0$$

The interval of convergence is the entire x-axis, $R = \infty$. ∎

Theorem 8.23 uses the ratio test for finding the radius of convergence. However, sometimes it is more convenient to use the root test. The next theorem is a companion to Theorem 8.23 in which the ratio limit is replaced by a root limit.

THEOREM 8.24   *Root test for the radius of convergence*

Let $\Sigma a_k u^k$ be a power series, and let

$$L = \lim_{k\to\infty} \sqrt[k]{|a_k|}$$

Then:

If $L = \infty$, the power series converges only at $u = 0$.

If $L = 0$, the power series converges for all real u.

If $0 < L < \infty$, let $R = 1/L$. Then the power series converges absolutely for $|u| < R$ and diverges for $|u| > R$. It may either converge or diverge and the endpoints $u = -R$ and $u = R$.

Proof   This proof is similar to the proof of Theorem 8.23 and is omitted.   ═

EXAMPLE 6   *Radius of convergence using the root test*

Find the radius of convergence of the power series $\displaystyle\sum_{k=1}^{\infty} \left(\frac{k+1}{k}\right)^{k^2} x^k$.

Solution   Setting $a_k = \left(\frac{k+1}{k}\right)^{k^2}$ and $u = x$, we have

$$L = \lim_{k\to\infty} \sqrt[k]{|a_k|} = \lim_{k\to\infty}\left[\left(\frac{k+1}{k}\right)^{k^2}\right]^{1/k} = \lim_{k\to\infty}\left(1+\frac{1}{k}\right)^k = e$$

Thus, the series converges absolutely for $|x| < e^{-1}$ and diverges for $|x| > e^{-1}$, so the radius of convergence is $R = e^{-1}$.   ■

POWER SERIES IN $(x - c)$

In some applications, we will encounter power series of the form

$$\sum_{k=0}^{\infty} a_k(x-c)^k = a_0 + a_1(x-c) + a_2(x-c)^2 + \cdots$$

in which each term is a constant times a power of $x - c$. The intervals of convergence are intervals of the form $-R < x - c < R$, including possibly one or both of the endpoints $x = c - R$ and $x = c + R$. You can easily show this by letting $u = x - c$ in Theorems 8.23 and 8.24.

EXAMPLE 7   *Interval of convergence for a power series in* $(x - c)$

Find the interval of convergence of the power series

$$\sum_{k=1}^{\infty} \frac{(x+1)^k}{3^k}$$

Solution   We find that $a_k = \frac{1}{3^k}$, $u = x+1$ so $c = -1$:

$$L = \lim_{k\to\infty}\left|\frac{\frac{1}{3^{k+1}}}{\frac{1}{3^k}}\right| = \lim_{k\to\infty}\left|\frac{3^k}{3^{k+1}}\right| = \lim_{k\to\infty}\left|\frac{1}{3}\right| = \frac{1}{3}$$

Thus, $R = 3$, so the power series converges absolutely for

$$-3 < x+1 < 3$$
$$-4 < x < 2$$

Checking the endpoints, we find:

$$x = -4: \quad \sum_{k=1}^{\infty}\frac{(-4+1)^k}{3^k} = \sum_{k=1}^{\infty}(-1)^k \text{ diverges (by oscillation)}$$

$$x = 2: \quad \sum_{k=1}^{\infty}\frac{(2+1)^k}{3^k} = \sum_{k=1}^{\infty}1^k \text{ diverges (by divergence test)}$$

The interval of convergence is $(-4, 2)$.   ■

TERM-BY-TERM DIFFERENTIATION AND INTEGRATION OF POWER SERIES

Consider the power series of the form

$$\sum_{k=0}^{\infty} a_k x^k = a_0 + a_1 x + a_2 x^2 + a_3 x^3 + \cdots$$

where x is a variable and all a_n are constants (sometimes referred to as the *coefficients* of the series). For each fixed x, the power series may converge or diverge. In other words, let f be a function defined by

$$f(x) = a_0 + a_1 x + a_2 x^2 + a_3 x^3 + \cdots$$

whose domain is the set of all values of x for which the series converges. The function f looks like a polynomial function, except it has infinitely many terms. For example, consider the function defined by

$$
\begin{array}{r}
1 + x + x^2 + \cdots \\
1 - x \overline{)1 + 0x + 0x^2 + 0x^3 + \cdots} \\
\underline{1 - x} \\
x + 0x^2 \\
\underline{x - x^2} \\
x^2 + 0x^3 \\
\underline{x^2 - x^3} \\
x^3 + \cdots
\end{array}
$$

$$f(x) = \frac{1}{1 - x}$$

$$= 1 + x + x^2 + x^3 + \cdots \qquad \text{By long division (see margin)}$$

We recognize this as a geometric series, which has an interval of convergence of $(-1, 1)$, as shown in Example 3. In Figure 8.20, we show the graphs of f as well as some of the graphs of partial sums of the corresponding geometric series, as follows:

$$f(x) = \frac{1}{1 - x} = \sum_{k=0}^{\infty} x^k = 1 + x + x^2 + x^3 + \cdots + x^n + \cdots$$

$$\text{interval of convergence } (-1, 1)$$

$$n = 1: \frac{1}{1 - x} \approx 1$$

$$n = 2: \frac{1}{1 - x} \approx 1 + x$$

$$n = 3: \frac{1}{1 - x} \approx 1 + x + x^2$$

$$n = 4: \frac{1}{1 - x} = 1 + x + x^2 + x^3$$

Notice from Figure 8.20 that the larger the n, the closer the series approximation is to the graph of f.

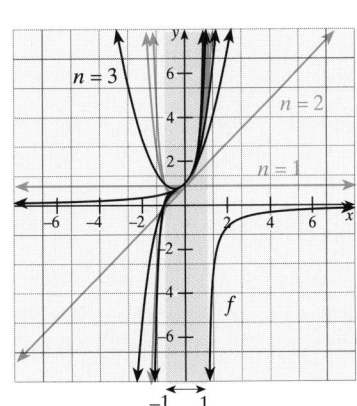

a. Polynomial approximations

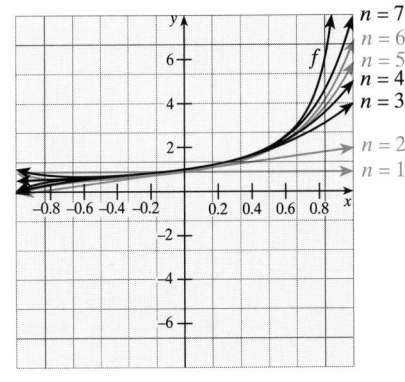

b. Detail showing interval of convergence

■ **FIGURE 8.20**   The graphs of $y = \dfrac{1}{1 - x}$ and the polynomials from the first terms of the associated power series

If we regard a power series as an "infinite polynomial," we would expect to be able to differentiate and integrate it term by term. The following theorem shows that this procedure is legitimate on the interval of convergence.

THEOREM 8.25 *Term-by-term differentiation and integration of a power series*

A power series $\sum_{k=0}^{\infty} a_k u^k$ with radius of convergence $R > 0$ can be differentiated or integrated term by term on its interval of convergence. More specifically, if $f(u) = \sum_{k=0}^{\infty} a_k u^k$ for $|u| < R$, then for $|u| < R$, we have

$$f'(u) = \sum_{k=1}^{\infty} k a_k u^{k-1} = a_1 + 2a_2 u + 3a_3 u^2 + 4a_4 u^3 + \cdots$$

$$\int f(u)\, du = \int \left(\sum_{k=0}^{\infty} a_k u^k \right) du = \sum_{k=0}^{\infty} \left(\int a_k u^k\, du \right) = \sum_{k=0}^{\infty} \frac{a_k}{k+1} u^{k+1} + C$$

Proof The proof of this result is outside the scope of this text, but details can be found in practically any advanced calculus text.

> ■ *What This Says:* In many ways, a function defined by a power series behaves like a polynomial. It is continuous in its interval of convergence, and its derivative and integral can be determined by differentiating and integrating term by term, respectively.

EXAMPLE 8 *Term-by-term differentiation of a power series*

Let f be a function defined by the power series

$$f(x) = \sum_{k=0}^{\infty} \frac{x^k}{k!} \quad \text{for all } x$$

Show that $f'(x) = f(x)$ for all x, and deduce that $f(x) = e^x$.

Solution The given power series converges for all x (by the generalized ratio test—see Example 1), and Theorem 8.25 tells us that it is differentiable for all x. Differentiating term by term, we find

$$f'(x) = \frac{d}{dx}\left[1 + x + \frac{x^2}{2!} + \frac{x^3}{3!} + \frac{x^4}{4!} + \cdots \right]$$

$$= 0 + 1 + \frac{2x}{2!} + \frac{3x^2}{3!} + \frac{4x^3}{4!} + \cdots$$

$$= 1 + x + \frac{x^2}{2!} + \frac{x^3}{3!} + \cdots$$

$$= f(x)$$

In Chapter 5, we found that the differential equation $f'(x) = f(x)$ has the general solution $f(x) = Ce^x$. Substituting $x = 0$ into the power series for $f(x)$, we obtain

$$f(0) = 1 + 0 + \frac{0^2}{2!} + \frac{0^3}{3!} + \cdots = 1$$

and by solving the equation $1 = Ce^0$ for C, we find $C = 1$; therefore, $f(x) = e^x$. ■

A power series can be differentiated term by term—not just once, but infinitely often in its interval of convergence. The key to this fact lies in showing that if f satisfies

$$f(x) = \sum_{k=0}^{\infty} a_k x^k$$

and R is the radius of convergence of the power series on the right, then the derivative series

$$f'(x) = \sum_{k=1}^{\infty} k a_k x^{k-1}$$

also has the radius of convergence R. (You are asked to show this in Problem 44.) Therefore, Theorem 8.25 can be applied to the derivative series to obtain the second derivative

$$f''(x) = \sum_{k=2}^{\infty} k(k-1) a_k x^{k-2}$$

for $|x| < R$. Continuing in this fashion, we can apply Theorem 8.25 to $f''(x), f^{(4)}$, and all other higher derivatives of f.

For example, we know that the geometric series

$$\sum_{k=0}^{\infty} x^k \quad \text{converges absolutely to} \quad f(x) = \frac{1}{1-x} \text{ for } |x| < 1$$

Thus, the term-by-term derivative

$$\frac{d}{dx}\left[\sum_{k=0}^{\infty} x^k\right] = \sum_{k=1}^{\infty} k x^{k-1} = 1 + 2x + 3x^2 + \cdots$$

converges to $f'(x) = \dfrac{1}{(1-x)^2}$ for $|x| < 1$, and the term-by-term *second derivative*

$$\frac{d^2}{dx^2}\left[\sum_{k=0}^{\infty} x^k\right] = \frac{d}{dx}\left[\sum_{k=1}^{\infty} k x^{k-1}\right] = \sum_{k=2}^{\infty} k(k-1) x^{k-2}$$

converges to $f''(x) = \dfrac{2}{(1-x)^3}$ and so on. These ideas are illustrated in our next example.

EXAMPLE 9   *Second derivative of a power series*

Let f be the function defined by the power series

$$f(x) = \sum_{k=0}^{\infty} \frac{(-1)^k x^{2k}}{(2k)!} \qquad \text{WARNING} \quad \text{Do not forget, } 0! = 1.$$

for all x. Show that $f''(x) = -f(x)$ for all x.

Solution   First, we use the ratio test to verify that the given power series converges absolutely for all x:

$$L = \lim_{k \to \infty} \left| \frac{\dfrac{(-1)^{k+1} x^{2(k+1)}}{[2(k+1)]!}}{\dfrac{(-1)^k x^{2k}}{(2k)!}} \right| = \lim_{k \to \infty} \frac{x^{2k+2}}{[2(k+1)]!} \cdot \frac{(2k)!}{x^{2k}} = \lim_{k \to \infty} \frac{x^2}{(2k+2)(2k+1)} = 0$$

Because $L < 1$, the series converges for all x. Next, differentiate the series, term by term:

$$f'(x) = \frac{d}{dx}\left[1 - \frac{x^2}{2!} + \frac{x^4}{4!} - \frac{x^6}{6!} + \cdots\right] = -\frac{2x}{2!} + \frac{4x^3}{4!} - \frac{6x^5}{6!} + \cdots$$

$$= -\frac{x}{1!} + \frac{x^3}{3!} - \frac{x^5}{5!} + \cdots$$

Finally, by differentiating term by term again, we obtain

$$f''(x) = \frac{d}{dx}\left[-\frac{x}{1!} + \frac{x^3}{3!} - \frac{x^5}{5!} + \cdots\right] = -1 + \frac{3x^2}{3!} - \frac{5x^4}{5!} + \cdots$$

$$= -\left[1 - \frac{x^2}{2!} + \frac{x^4}{4!} + \cdots\right] = -f(x) \qquad \blacksquare$$

> **EXAMPLE 10**   *Term-by-term integration of a power series*

By integrating an appropriate geometric series term by term, show that

$$\sum_{k=0}^{\infty} \frac{x^{k+1}}{k+1} = -\ln(1-x) \quad \text{for } -1 < x < 1$$

Solution   Integrating the geometric series $\displaystyle\sum_{k=0}^{\infty} u^k = \frac{1}{1-u}$ term by term in the interval $-1 < u < 1$, we obtain

$$\int_0^x \frac{1}{1-u}\, du = \int_0^x \left[\sum_{k=0}^{\infty} u^k\right] du = \int_0^x [1 + u + u^2 + u^3 + \cdots]\, du$$

$$= x + \frac{x^2}{2} + \frac{x^3}{3} + \cdots = \sum_{k=0}^{\infty} \frac{x^{k+1}}{k+1} \quad \text{for } -1 < x < 1$$

We also know that

$$\int_0^x \frac{du}{1-u} = -\ln(1-x)$$

Thus,

$$-\ln(1-x) = \int_0^x \frac{du}{1-u} = \sum_{k=0}^{\infty} \frac{x^{k+1}}{k+1} \quad \text{for } -1 < x < 1 \qquad \blacksquare$$

8.7 Problem Set

A *Find the interval of convergence for the power series given in Problems 1–28.*

1. $\displaystyle\sum_{k=1}^{\infty} \frac{kx^k}{k+1}$

2. $\displaystyle\sum_{k=1}^{\infty} \frac{k^2 x^k}{k+1}$

3. $\displaystyle\sum_{k=1}^{\infty} \frac{k(k+1)x^k}{k+2}$

4. $\displaystyle\sum_{k=1}^{\infty} \sqrt{k-1}\, x^k$

5. $\displaystyle\sum_{k=1}^{\infty} k^2 3^k (x-3)^k$

6. $\displaystyle\sum_{k=1}^{\infty} \frac{k^2(x-2)^k}{3^k}$

7. $\displaystyle\sum_{k=1}^{\infty} \frac{3^k(x+3)^k}{4^k}$

8. $\displaystyle\sum_{k=1}^{\infty} \frac{4^k(x+1)^k}{3^k}$

9. $\displaystyle\sum_{k=1}^{\infty} \frac{k!(x-1)^k}{5^k}$

10. $\displaystyle\sum_{k=1}^{\infty} \frac{(x-15)^k}{\ln(k+1)}$

11. $\displaystyle\sum_{k=1}^{\infty} \frac{k^2}{2^k}(x-1)^k$

12. $\displaystyle\sum_{k=1}^{\infty} \frac{2^k(x-3)^k}{k(k+1)}$

13. $\displaystyle\sum_{k=1}^{\infty} \frac{k(3x-4)^k}{(k+1)^2}$

14. $\displaystyle\sum_{k=1}^{\infty} \frac{(2x+3)^k}{4^k}$

15. $\displaystyle\sum_{k=1}^{\infty} \frac{kx^k}{7^k}$

16. $\displaystyle\sum_{k=1}^{\infty} \frac{(2k)!x^k}{(3k)!}$

17. $\displaystyle\sum_{k=1}^{\infty} \frac{(k!)^2 x^k}{k^k}$

18. $\displaystyle\sum_{k=1}^{\infty} \frac{(-1)^k kx^k}{\ln(k+2)}$

19. $\displaystyle\sum_{k=2}^{\infty} \frac{(-1)^k x^k}{k(\ln k)^2}$

20. $\displaystyle\sum_{k=1}^{\infty} \frac{(3x)^k}{2^{k+1}}$

21. $\displaystyle\sum_{k=1}^{\infty} \frac{(2x)^{2k}}{k!}$

22. $\displaystyle\sum_{k=1}^{\infty} \frac{(x+2)^{2k}}{3^k}$

23. $\displaystyle\sum_{k=1}^{\infty} \frac{k!}{2^k}(3x)^{3k}$

24. $\displaystyle\sum_{k=1}^{\infty} \frac{(3x)^{3k}}{\sqrt{k}}$

25. $\displaystyle\sum_{k=1}^{\infty} \frac{2^k}{k!}(2x-1)^{2k}$

26. $\displaystyle\sum_{k=1}^{\infty} 2^k(3x)^{3k}$

27. $\displaystyle\sum_{k=1}^{\infty} \frac{x^k}{k\sqrt{k}}$

28. $\displaystyle\sum_{k=1}^{\infty} \frac{(\ln k)\, x^k}{k}$

B *Find the radius of convergence R in Problems 29–34.*

29. $\displaystyle\sum_{k=1}^{\infty} k^2(x+1)^{2k+1}$

30. $\displaystyle\sum_{k=1}^{\infty} 2^{\sqrt{k}}(x-1)^k$

31. $\displaystyle\sum_{k=1}^{\infty} \frac{k!x^k}{k^k}$

32. $\displaystyle\sum_{k=1}^{\infty} \frac{(k!)^2\, x^k}{(2k)!}$

33. $\displaystyle\sum_{k=1}^{\infty} k(ax)^k$ for constant a

34. $\displaystyle\sum_{k=1}^{\infty} (a^2 x)^k$ for constant a

In Problems 35–38, find the derivative $f'(x)$ by differentiating term-by-term.

35. $f(x) = \displaystyle\sum_{k=0}^{\infty} \left(\frac{x}{2}\right)^k$

36. $f(x) = \displaystyle\sum_{k=1}^{\infty} \frac{x^k}{k}$

37. $f(x) = \displaystyle\sum_{k=0}^{\infty} (k+2)x^k$

38. $f(x) = \displaystyle\sum_{k=0}^{\infty} kx^k$

In Problems 39–42, find $\displaystyle\int_0^x f(u)\, du$ by integrating term-by-term.

39. $f(x) = \displaystyle\sum_{k=0}^{\infty} \left(\frac{x}{2}\right)^k$

40. $f(x) = \displaystyle\sum_{k=1}^{\infty} \frac{x^k}{k}$

41. $f(x) = \displaystyle\sum_{k=0}^{\infty} (k+2)x^k$

42. $f(x) = \displaystyle\sum_{k=1}^{\infty} kx^k$

43. THINK TANK PROBLEM Show that the series

$$S = \sum_{k=1}^{\infty} \frac{\sin(k!\, x)}{k^2}$$

converges for all x. Differentiate term by term to obtain

the series

$$T = \sum_{k=1}^{\infty} \frac{k!\,\cos(k!x)}{k^2}$$

Show that this series diverges for all x. Why does this not violate Theorem 8.25?

C **44.** Suppose $\{a_k\}$ is a sequence for which

$$\lim_{k\to\infty} \sqrt[k]{|a_k|} = \frac{1}{R}$$

Show that the power series

$$\sum_{k=1}^{\infty} ka_k x^{k-1}$$

has radius of convergence R.

45. Show that if $f(x) = \displaystyle\sum_{k=1}^{\infty} a_k x^k$ has radius of convergence $R > 0$, then the series

$$\sum_{k=1}^{\infty} a_k x^{kp}$$

where p is a positive integer, has radius of convergence $R^{1/p}$.

46. For what values of x does the series

$$\sum_{k=1}^{\infty} \frac{x}{k+x} \text{ converge?}$$

Note: This is not a power series.

47. For what values of x does the series

$$\sum_{k=1}^{\infty} \frac{1}{x^k} \text{ converge?}$$

Note: This is not a power series.

48. For what values of x does the series

$$\sum_{k=1}^{\infty} \frac{1}{kx^k} \text{ converge?}$$

Note: This is not a power series.

49. Find the radius of convergence for

$$\sum_{k=1}^{\infty} \frac{(k+3)!x^k}{k!(k+4)!}$$

50. Find the radius of convergence for

$$\sum_{k=1}^{\infty} \frac{1\cdot 2\cdot 3\cdots k(-x)^{2k-1}}{1\cdot 3\cdot 5\cdots(2k-1)}$$

51. Let f be the function defined by the power series

$$f(x) = \sum_{k=0}^{\infty} \frac{(-1)^k x^{2k+1}}{(2k+1)!}$$

for all x. Show that $f''(x) = -f(x)$ for all x.

52. Let f be the function defined by the power series

$$f(x) = \sum_{k=1}^{\infty} \frac{x^{2k}}{(2k)!}$$

for all x. Show that $f''(x) = f(x)$.

53. Find the radius of convergence for the power series

$$\sum_{k=0}^{\infty} \frac{(qk)!}{(k!)^q} x^k$$

where q is a positive integer.

8.8 Taylor and Maclaurin Series

IN THIS SECTION   Taylor and Maclaurin polynomials, Taylor's theorem, Taylor and Maclaurin series, operations with Maclaurin and Taylor series ∎

TAYLOR AND MACLAURIN POLYNOMIALS

Consider a function f that can be differentiated n times on some interval I. Our goal is to find a polynomial function that approximates f at a number c in its domain. For simplicity, we begin by considering an important special case where $c = 0$. For example, consider the function $f(x) = e^x$ at the point $x = 0$, as shown in Figure 8.21. To approximate f by a polynomial function $M(x)$, we begin by making sure that both the polynomial function and f pass through the same point. That is, $M(0) = f(0)$. We say that the polynomial is **expanded about $c = 0$** or that it is **centered at 0.**

There are many polynomial functions that we could choose to approximate f at $x = 0$, and we proceed by making sure that both f and M have the same slope at $x = 0$. That is,

$$M'(0) = f'(0)$$

The graph in Figure 8.21 shows that we have $f(x) = e^x$, so $f'(x) = e^x$ and $f(0) = f'(0) = 1$. To find M we let

$$M_1(x) = a_0 + a_1 x \qquad M_1'(x) = a_1$$

and require $M_1(0) = 1$ and $M_1'(0) = 1$.

Because $M_1'(0) = a_1 = 1$ and $M_1(0) = a_0 = 1$, we find that

$$M_1(x) = 1 + x$$

We see from Figure 8.21 that close to $x = c$, the approximation of f by M_1 is good, but as we move away from $(0, 1)$, M_1 no longer serves as a good approximation. To improve the approximation, we impose the requirement that the values of the second derivatives of M and f agree at $x = 0$. Using this criterion, we find

$$M_2(x) = a_0 + a_1 x + a_2 x^2$$

so that

$$M_2'(x) = a_1 + 2a_2 x \quad \text{and} \quad M_2''(x) = 2a_2$$

Because $f''(x) = e^x$ and $f''(0) = 1$, we want

$$2a_2 = 1 \quad \text{or} \quad a_2 = \tfrac{1}{2}$$

Thus,

$$M_2(x) = 1 + x + \tfrac{1}{2}x^2$$

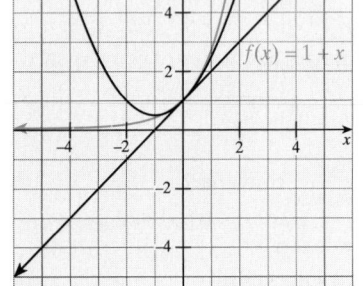

$f(x) = 1 + x + \dfrac{x^2}{2}$

$f(x) = e^x$

$f(x) = 1 + x$

■ **FIGURE 8.21**   Graph of $f(x) = e^x$ and approximating polynomials $M_1(x)$ and $M_2(x)$

To improve the approximation even further, we can require that the values of the approximating polynomials $M_3, M_4, \dots, M_n$ at $x = 0$ have derivatives that match those of f at $x = 0$. We find that

$$M_3(x) = 1 + x + \tfrac{1}{2}x^2 + \tfrac{1}{6}x^3$$

$$M_4(x) = 1 + x + \tfrac{1}{2}x^2 + \tfrac{1}{6}x^3 + \tfrac{1}{24}x^4$$

$$M_5(x) = 1 + x + \tfrac{1}{2}x^2 + \tfrac{1}{6}x^3 + \tfrac{1}{24}x^4 + \tfrac{1}{120}x^5 + \cdots$$

$$M_n(x) = 1 + \frac{x}{1!} + \frac{x^2}{2!} + \frac{x^3}{3!} + \cdots + \frac{x^n}{n!}$$

This nth-degree polynomial approximation of f at $x = 0$ is called a **Maclaurin polynomial.** If we repeat the steps for $x = c$ rather than for $x = 0$, we find

$$T_n(x) = e^c + \frac{(x-c)}{1!}e^c + \frac{(x-c)^2}{2!}e^c + \frac{(x-c)^3}{3!}e^c + \cdots + \frac{(x-c)^n}{n!}e^c$$

which is called the **nth-degree Taylor polynomial** of the function $f(x) = e^x$ at $x = c$.

TAYLOR'S THEOREM

Instead of stopping at the nth term, we will now approximate a function f by an infinite series. A function f is said to be represented by the power series

$$\sum_{k=0}^{\infty} a_k(x-c)^k$$

on an interval I if $f(x) = \sum_{k=0}^{\infty} a_k(x-c)^k$

for all x in I. Power series representation of functions is extremely useful, but before we can deal effectively with such representations, we must answer two questions:

1. **Existence:** Under what conditions does a given function have a power series representation?

2. **Uniqueness:** When f can be represented by a power series, is there only one such series, and if so, what is it?

The uniqueness issue is addressed in the following theorem.

THEOREM 8.26 *The uniqueness theorem for power series representation*

Suppose an infinitely differentiable function f is known to have the power series representation

$$f(x) = \sum_{k=0}^{\infty} a_k(x-c)^k$$

for $-R < x - c < R$. Then there is exactly one such representation, and the coefficients a_k must satisfy

$$a_k = \frac{f^{(k)}(c)}{k!} \quad \text{for } k = 0, 1, 2, \dots$$

Proof The uniqueness theorem may be established by differentiating the given power series term by term and evaluating successive derivatives at c. We start with

$$f(x) = \sum_{k=0}^{\infty} a_k(x-c)^k$$

and substitute $x = c$ to obtain

$$f(c) = a_0 + a_1(c - c) + a_2(c - c)^2 + \cdots = a_0$$

Next, differentiate the original series, term by term,

$$f'(x) = a_1 + 2a_2(x - c) + 3a_3(x - c)^2 + \cdots$$

and thus,

$$f'(c) = a_1 + 2a_2(0) + 3a_3(0)^2 + \cdots = a_1$$

Differentiating once again and substituting $x = c$, we find

$$f''(c) = 2a_2 \quad \text{so that} \quad a_2 = \frac{f''(c)}{2}$$

In general, the kth derivative of f at $x = c$ is given by

$$f^{(k)}(x) = k!a_k \quad \text{so that} \quad a_k = \frac{f^{(k)}(c)}{k!}$$

The question of existence—that is, under what conditions a given function has a power series representation—is answered by considering what is called the **Taylor remainder function:**

$$R_n(x) = f(x) - T_n(x)$$

We see that f exists as a series if and only if

$$\lim_{n \to \infty} R_n(x) = 0 \quad \text{for all } x \text{ in } I$$

This relationship among f, its Taylor polynomial $T_n(x)$, and the Taylor remainder function $R_n(x)$, is summarized in the following theorem.

THEOREM 8.27   *Taylor's theorem*

If f and all its derivatives exist in an open interval I containing c, then for each x in I,

$$f(x) = f(c) + \frac{f'(c)}{1!}(x - c) + \frac{f''(c)}{2!}(x - c)^2 + \cdots + \frac{f^{(n)}(c)}{n!}(x - c)^n + R_n(x)$$

where the remainder function $R_n(x)$ is given by

$$R_n(x) = \frac{f^{(n+1)}(z_n)}{(n + 1)!}(x - c)^{n+1}$$

for some z_n that depends on x and lies between c and x.

Proof   The proof of this theorem is given in Appendix B.

The formula for $R_n(x)$ is called the *Lagrange form* of the Taylor remainder function, after the French/Italian mathematician Joseph Lagrange (1736–1813).

When applying Taylor's theorem, we do not expect to be able to find the exact value of z_n. Indeed, if we could find that value, an approximation would not be necessary. However, we can often determine an upper bound for $|R_n(x)|$ in an open interval. We will illustrate this situation later in this section.

TAYLOR AND MACLAURIN SERIES

The uniqueness theorem tells us that if f has a power series representation at c, it must be the series

$$f(c) + \frac{f'(c)}{1!}(x - c) + \frac{f''(c)}{2!}(x - c)^2 + \frac{f'''(c)}{3!}(x - c)^3 + \cdots$$

Taylor Series and Maclaurin Series

Suppose there is an open interval I containing c throughout which the function f and all its derivatives exist. Then the power series

$$f(c) + \frac{f'(c)}{1!}(x - c) + \frac{f''(c)}{2!}(x - c)^2 + \frac{f'''(c)}{3!}(x - c)^3 + \cdots$$

is called the **Taylor series of f at c.** The special case where $c = 0$ is called the **Maclaurin series of f:**

$$f(0) + \frac{f'(0)}{1!}x + \frac{f''(0)}{2!}x^2 + \frac{f'''(0)}{3!}x^3 + \cdots$$

WARNING   The uniqueness theorem may be summarized by saying that the Taylor series of f at c is the *only* power series of the form $\Sigma a_k(x - c)^k$ that can possibly represent f on I. But you must be careful. All we really know is that *if* f has a power series representation, then its representation must have the Taylor from.

EXAMPLE 1   *Maclaurin series*

Find the Maclaurin series for $f(x) = \cos x$.

Solution   First, note that f is infinitely differentiable at $x = 0$. We find

$$
\begin{array}{ll}
f(x) = \cos x & f(0) = 1 \\
f'(x) = -\sin x & f'(0) = 0 \\
f''(x) = -\cos x & f''(0) = -1 \\
f'''(x) = \sin x & f'''(0) = 0 \\
f^{(4)}(x) = \cos x & f^{(4)}(0) = 1 \\
\quad\vdots & \quad\vdots
\end{array}
$$

Thus, by using the definition of a Maclaurin series, we have

$$\cos x = 1 - \frac{x^2}{2!} + \frac{x^4}{4!} - \frac{x^6}{6!} + \cdots = \sum_{k=0}^{\infty} \frac{(-1)^k x^{2k}}{(2k)!}$$

You might ask about the relationship between a Maclaurin series and a Maclaurin polynomial. Figure 8.22 shows the function $f(x) = \cos x$ from Example 1 along with some successive Maclaurin polynomials. We use the notation $M_n(x)$ to represent the first $n + 1$ terms of the corresponding Maclaurin series. Note that the polynomials are quite close to the function near $x = 0$, but that they become very different as x moves away from the origin.

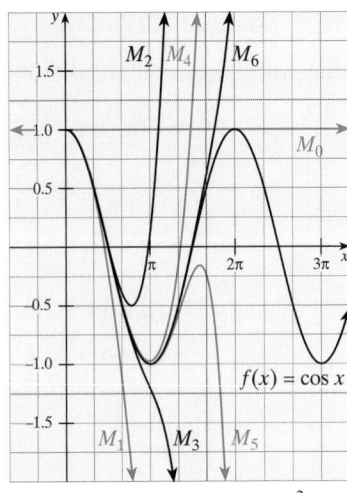
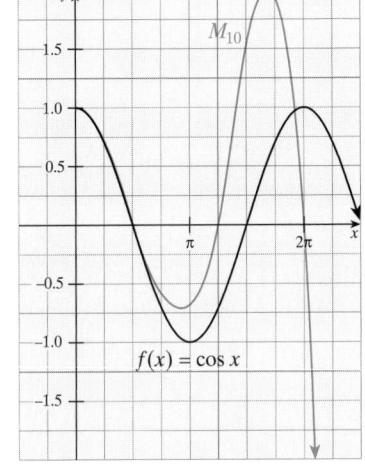

a. $M_0(x) = 1, M_1(x) = 1 - \frac{x^2}{2!}$,

$M_2(x) = 1 - \frac{x^2}{2!} + \frac{x^4}{4!}$,

$M_3(x) = 1 - \frac{x^2}{2!} + \frac{x^4}{4!} - \frac{x^6}{6!}$

b. $M_{10}(x) = 1 - \frac{x^2}{2!} + \frac{x^4}{4!} - \frac{x^6}{6!} + \frac{x^8}{8!}$

$- \frac{x^{10}}{10!} + \frac{x^{12}}{12!} - \frac{x^{14}}{14!} + \frac{x^{16}}{16!} - \frac{x^{18}}{18!}$

$+ \frac{x^{20}}{20!}$

■ **FIGURE 8.22**   Comparison of a function and its Maclaurin polynomials

> **EXAMPLE 2**   *Maximum error when approximating a function with a Maclaurin polynomial*

Find the Maclaurin polynomial $M_5(x)$ for the function $f(x) = e^x$ and use this polynomial to approximate e. Use Taylor's theorem to determine the accuracy of this approximation.

Solution   First find the Maclaurin series for $f(x) = e^x$.

$$f(x) = e^x \qquad f(0) = 1$$
$$f'(x) = e^x \qquad f'(0) = 1$$
$$\vdots \qquad\qquad \vdots$$

The Maclaurin series for e^x is

$$e^x = 1 + \frac{1}{1!}x + \frac{1}{2!}x^2 + \frac{1}{3!}x^3 + \cdots = \sum_{k=0}^{\infty} \frac{x^k}{k!}$$

Then find $M_5(x)$:

$$M_5(x) = 1 + x + \frac{x^2}{2} + \frac{x^3}{6} + \frac{x^4}{24} + \frac{x^5}{120} \approx e^x$$

A comparison of f and $M_5(x)$ is shown in Figure 8.23a. Notice from the graphs of these functions that the error seems to increase as x moves away from the origin. A zoom near $x = 1$ is shown in Figure 8.23b.

To determine the accuracy, we use Taylor's theorem:

$$e = 1 + 1 + \tfrac{1}{2} + \tfrac{1}{6} + \tfrac{1}{24} + \tfrac{1}{120} + R_5(1)$$

$$= 2.71\overline{6} + R_5(1)$$

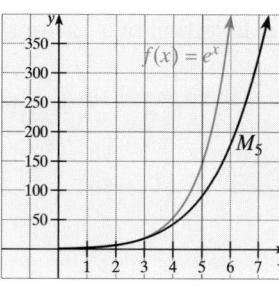 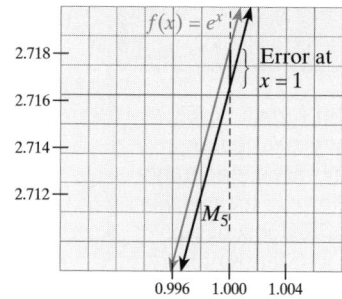

a. Comparison of f and M_5 **b.** Error at $x = 1$

■ **FIGURE 8.23** Comparison of $f(x) = e^x$ and

$$M_5(x) = 1 + x + \frac{x^2}{2} + \frac{x^3}{6} + \frac{x^4}{24} + \frac{x^5}{120}$$

WARNING ▶ Be careful with rounding. ◀

The remainder term is given by $R_n(x) = \dfrac{f^{(n+1)}(z_n)}{(n+1)!} x^{n+1}$ for some number z_n between 0 and x. In particular,

$$R_5(1) = \frac{e^{z_5}}{(5+1)!}(1)^{5+1}$$

for some z_5 between 0 and 1. Because $0 < z_5 < 1$, we have $e^{z_5} < e$ and

$$R_5(1) < \frac{e}{6!} < \frac{3}{6!} \approx 0.004167 \qquad \text{because } e < 3$$

Thus, e is between $2.716667 + 0.004167$ and $2.716667 - 0.004167$; that is,

$$2.712500 < e < 2.720834$$

You might wonder about using a bound on e ($e < 3$) to find another bound on e. This is not a new idea (remember the Newton–Raphson method). If we had picked 2.73 instead of 3 as a first choice, we would obtain an even better bound for e. Figure 8.23b shows a representation of the actual error at $x = 1$. ■

   EXAMPLE 3   *Taylor series*

Find the Taylor series for $f(x) = \ln x$ at $c = 1$.

Solution   Note that f is infinitely differentiable at $x = 1$. We find

$$
\begin{aligned}
f(x) &= \ln x & f(1) &= 0 \\[4pt]
f'(x) &= \frac{1}{x} & f'(1) &= 1 \\[4pt]
f''(x) &= \frac{-1}{x^2} & f''(1) &= -1 \\[4pt]
f'''(x) &= \frac{2}{x^3} & f'''(1) &= 2 \\[4pt]
f^{(4)}(x) &= \frac{-6}{x^4} & f^{(4)}(1) &= -6 \\[4pt]
&\ \vdots & &\ \vdots \\[4pt]
f^{(k)}(x) &= \frac{(-1)^{k+1}(k-1)!}{x^k} & f^{(k)}(1) &= (-1)^{k+1}(k-1)!
\end{aligned}
$$

Then, use the definition of a Taylor series to write

$$\ln x = 0 + \frac{1}{1!}(x - 1) - \frac{1}{2!}(x - 1)^2 + \frac{2}{3!}(x - 1)^3 - \frac{6}{4!}(x - 1)^4 + \cdots$$

$$= (x - 1) - \frac{1}{2}(x - 1)^2 + \frac{1}{3}(x - 1)^3 - \frac{1}{4}(x - 1)^4 + \cdots$$

$$= \sum_{k=1}^{\infty} \frac{(-1)^{k+1}(x - 1)^k}{k}$$

∎

Suppose f is a function that is infinitely differentiable at c. Now we have two mathematical quantities, f and its Taylor series. There are several possibilities:

1. The Taylor series of f may converge to f on the interval of absolute convergence, $-R < x - c < R$ (or $|x - c| < R$).
2. The Taylor series may converge only at $x = c$, in which case it certainly does not represent f on any interval containing c.
3. The Taylor series of f may have a positive radius of convergence (even $R = \infty$), but it may converge to a function g that does not equal f on the interval $|x - c| < R$.

EXAMPLE 4   *A function defined at points where its Taylor series does not converge*

Show that the function $\ln x$ is defined at points for which its Taylor series at $c = 1$ does not converge.

Solution   We know that $\ln x$ is defined for all $x > 0$. From the previous example, we know the Taylor series for $\ln x$ at $c = 1$ is

$$\sum_{k=1}^{\infty} \frac{(-1)^{k-1}(x - 1)^k}{k}$$

We find the interval of convergence for this series centered at $c = 1$ by computing

$$\lim_{k \to \infty} \left| \frac{\frac{(-1)^k}{k + 1}}{\frac{(-1)^{k-1}}{k}} \right| = \lim_{k \to \infty} \left(\frac{k}{k + 1} \right) = 1$$

This means that the power series converges absolutely for

$$|x - 1| < 1$$
$$0 < x < 2$$

Test the endpoints:

$$x = 0: \quad \sum_{k=1}^{\infty} \frac{(-1)^{k-1}(0 - 1)^k}{k} = \sum_{k=1}^{\infty} \frac{-1}{k}$$

which is known to diverge (harmonic series), and

$$x = 2: \quad \sum_{k=1}^{\infty} \frac{(-1)^{k-1}(2 - 1)^k}{k}$$

which converges (alternating harmonic series). Thus, the series converges only on [0, 2], but the function itself is defined for all $x > 0$. The function is compared with the sixth-degree Taylor approximation in Figure 8.24.

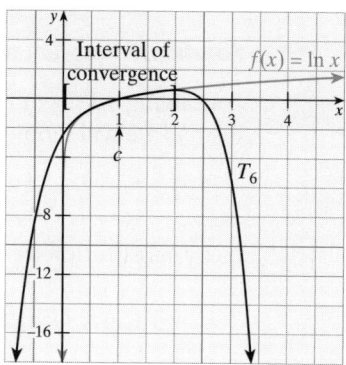

■ **FIGURE 8.24**   Graphs of $f(x) = \ln x$ and

$$T_6(x) = (x - 1) - \frac{(x - 1)^2}{2} + \frac{(x - 1)^3}{3} - \frac{(x - 1)^4}{4} + \frac{(x - 1)^5}{5} - \frac{(x - 1)^6}{6}$$

   ■

The next example illustrates a function whose Maclaurin series does not represent that function on *any* interval.

EXAMPLE 5   *A Maclaurin series that represents a function only at a single point*

Let f be the function defined by $\begin{cases} e^{-1/x^2} & \text{if } x \neq 0 \\ 0 & \text{if } x = 0 \end{cases}$

Show that f is represented by its Maclaurin series only at $x = 0$.

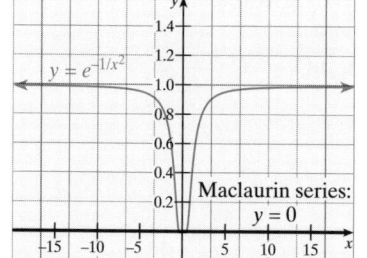

■ **FIGURE 8.25**   Graphs of f and its Maclaurin series $y = 0$

Solution   It can be shown that f is infinitely differentiable at 0 and that $f^{(k)}(0) = 0$ for all k, so that f has the Maclaurin series

$$0 + \frac{0}{1!}x + \frac{0}{2!}x^2 + \frac{0}{3!}x^3 + \cdots = 0$$

This series converges to the zero function for all x and thus represents $f(x)$ only at $x = 0$. The graph of f and the Maclaurin series is shown in Figure 8.25.   ■

OPERATIONS WITH MACLAURIN AND TAYLOR SERIES

According to the uniqueness theorem, the Taylor series for f at c is the only power series in $x - c$ that can satisfy

$$f(x) = \sum_{k=0}^{\infty} a_k(x - c)^k$$

for all x in some interval containing c. The coefficients in this series can always be found by substituting into the formula

$$a_k = \frac{f^{(k)}(c)}{k!}$$

but occasionally they can also be found by algebraic manipulation.

EXAMPLE 6   *Maclaurin series using substitution*

Find the Maclaurin series for $\dfrac{1}{1 + x^2}$.

Solution   We want a series of the form $\displaystyle\sum_{k=0}^{\infty} a_k u^k$ that represents $\dfrac{1}{1 + x^2}$ on an interval containing 0. We could proceed directly using the definition of a Maclaurin series, but instead we will modify a known series. We know that if $|u| < 1$, we can write

$$\frac{1}{1 - u} = 1 + u + u^2 + \cdots$$

so by substitution ($u = -x^2$) we have

$$\frac{1}{1 + x^2} = \frac{1}{1 - (-x^2)} = 1 - x^2 + x^4 - x^6 + \cdots$$

provided $|-x^2| < 1$; that is, $-1 < x < 1$. Hence, by the uniqueness theorem, the desired representation is

$$\frac{1}{1 + x^2} = \sum_{k=0}^{\infty} (-1)^k x^{2k} \quad \text{for } -1 < x < 1 \qquad\blacksquare$$

EXAMPLE 7   *Maclaurin series using substitution and subtraction*

Find the Maclaurin series for $f(x) = \ln\left(\dfrac{1 + x}{1 - x}\right)$ and use this series to compute $\ln 2$ correct to five decimal places.

Solution   For this function, we first use a property of logarithms:

$$f(x) = \ln\left(\frac{1 + x}{1 - x}\right) = \ln(1 + x) - \ln(1 - x)$$

From Example 3, we know

$$\ln x = (x - 1) - \tfrac{1}{2}(x - 1)^2 + \tfrac{1}{3}(x - 1)^3 - \cdots$$

so that

$$\ln(1 + x) = [1 + x - 1] - \tfrac{1}{2}[1 + x - 1]^2 + \tfrac{1}{3}[1 + x - 1]^3 - \cdots$$

$$= x - \frac{x^2}{2} + \frac{x^3}{3} - \frac{x^4}{4} + \cdots$$

$$\ln(1 - x) = [1 - x - 1] - \tfrac{1}{2}[1 - x - 1]^2 + \tfrac{1}{3}[1 - x - 1]^3 - \cdots$$

$$= (-x) - \frac{(-x)^2}{2} + \frac{(-x)^3}{3} - \frac{(-x)^4}{4} + \cdots$$

$$= -x - \frac{x^2}{2} - \frac{x^3}{3} - \frac{x^4}{4} - \cdots$$

By subtracting series, we find

$$f(x) = \ln(1 + x) - \ln(1 - x)$$

$$= \left[x - \frac{x^2}{2} + \frac{x^3}{3} - \frac{x^4}{4} + \cdots\right] - \left[-x - \frac{x^2}{2} - \frac{x^3}{3} - \frac{x^4}{4} - \cdots\right]$$

$$= 2\left[x + \frac{x^3}{5} + \frac{x^5}{5} + \cdots\right]$$

Next, by solving the equation $\dfrac{1+x}{1-x} = 2$, we find $x = \dfrac{1}{3}$, so we are looking for $\ln\left(\dfrac{1+x}{1-x}\right)$ where $x = \dfrac{1}{3}$. It follows that

$$\ln 2 = 2\left[\frac{1}{3} + \frac{\left(\frac{1}{3}\right)^3}{3} + \frac{\left(\frac{1}{3}\right)^5}{5} + \cdots\right] = 2\left[\frac{1}{3} + \frac{1}{3}\left(\frac{1}{3}\right)^3 + \frac{1}{5}\left(\frac{1}{3}\right)^5 + \cdots\right]$$

If we approximate this by a Taylor polynomial, we find

$$\ln 2 = \frac{2}{3} + \frac{2}{3}\left(\frac{1}{3}\right)^3 + \frac{2}{5}\left(\frac{1}{3}\right)^5 + \cdots + \frac{2}{2n+1}\left(\frac{1}{3}\right)^{2n+1} + R_n\left(\frac{1}{3}\right)$$

where $R_n\left(\dfrac{1}{3}\right)$ is the remainder term. To estimate the remainder, we note that $R_n\left(\dfrac{1}{3}\right)$ is the tail of the infinite series for $\ln 2$. Thus,

$$\left|R_n\left(\frac{1}{3}\right)\right| = \frac{2}{2n+3}\left(\frac{1}{3}\right)^{2n+3} + \frac{2}{2n+5}\left(\frac{1}{3}\right)^{2n+5} + \cdots$$

$$< \frac{2}{2n+3}\left(\frac{1}{3}\right)^{2n+3} + \frac{2}{2n+3}\left(\frac{1}{9}\right)\left(\frac{1}{3}\right)^{2n+3} + \cdots$$

$$= \frac{2}{2n+3}\left(\frac{1}{3}\right)^{2n+3}\left[1 + \frac{1}{9} + \frac{1}{81} + \cdots\right]$$

Because the geometric series in brackets converges to $\dfrac{1}{1-\frac{1}{9}} = \dfrac{9}{8}$, we find that

$$\left|R_n\left(\frac{1}{3}\right)\right| < \frac{2}{2n+3}\left(\frac{9}{8}\right)\left(\frac{1}{3}\right)^{2n+3}$$

In particular, to achieve five-place accuracy, we must make sure the term on the right is less than 0.000005 (six places to account for round-off error). With a calculator we can see that if $n = 4$, we have

$$\frac{2}{2(4)+3}\left(\frac{9}{8}\right)\left(\frac{1}{3}\right)^{11} = 0.0000012$$

Thus, we approximate $\ln 2$ with $n = 4$:

$$\ln 2 \approx T_4(x) = \frac{2}{3} + \frac{2}{3}\left(\frac{1}{3}\right)^3 + \frac{2}{5}\left(\frac{1}{3}\right)^5 + \frac{2}{7}\left(\frac{1}{3}\right)^7 + \frac{2}{9}\left(\frac{1}{3}\right)^9 \approx 0.6931460$$

which is correct with an error of no more than 0.0000012; therefore,

$$0.6931460 - 0.0000012 < \ln 2 < 0.6931460 + 0.0000012$$
$$0.6931448 < \ln 2 < 0.6931472$$

Rounded to five-place accuracy, $\ln 2 = 0.69315$.
We can check with calculator accuracy: $\ln 2 \approx 0.6931471806$.

Technology Window

"Why bother with the accuracy part of Example 7?" you might ask. "I can just press ln 2 on my calculator and obtain better accuracy anyway!" The answer is that the accuracy built into calculators and computers uses the same principles illustrated in Example 7. Calculators and computers have not changed the need for accuracy considerations but have simply pushed out the limits of these accuracy arguments. It is only for convenience that we ask for limits of accuracy that are less than what we can obtain with a calculator (or any available machine), but the method is just as valid if we ask for 100- or 1,000-place accuracy.

EXAMPLE 8   *Maclaurin series by rewriting given function in terms of a geometric series*

Find the Maclaurin series for $f(x) = \dfrac{5 - 2x}{3 + 2x}$.

Solution   The direct approach, namely, finding the successive derivatives of f, is not a very pleasant prospect. Instead, we recall that a geometric series

$$\sum_{k=0}^{\infty} ar^k \text{ converges to the sum } \frac{a}{1 - r} \text{ if } |r| < 1$$

Our approach with this example is to rewrite f in the form $\dfrac{a}{1 - r}$ so that we can refer to the geometric series.

$$\frac{5 - 2x}{3 + 2x} = -1 + \frac{8}{3 + 2x} \qquad \text{Long division}$$

$$= -1 + \frac{\dfrac{8}{3}}{1 - \left(-\dfrac{2}{3}x\right)}$$

The related series converges for

$$\left|\tfrac{2}{3}x\right| < 1 \quad \text{or} \quad -\tfrac{3}{2} < x < \tfrac{3}{2}$$

Thus, for an interval of convergence $\left(-\dfrac{3}{2}, \dfrac{3}{2}\right)$, the geometric series is

$$\sum_{k=0}^{\infty} ar^k = \frac{a}{1 - r} \quad \text{where } a = \frac{8}{3} \text{ and } r = -\frac{2}{3}x$$

Thus,

$$\frac{5 - 2x}{3 + 2x} = -1 + \sum_{k=0}^{\infty} \frac{8}{3}\left[-\frac{2}{3}x\right]^k = -1 + \frac{8}{3} - \frac{16}{9}x + \frac{32}{27}x^2 - \frac{64}{81}x^3 + \cdots \qquad ∎$$

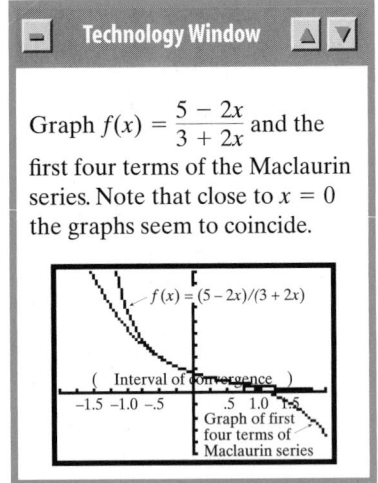

Technology Window

Graph $f(x) = \dfrac{5 - 2x}{3 + 2x}$ and the first four terms of the Maclaurin series. Note that close to $x = 0$ the graphs seem to coincide.

$f(x) = (5-2x)/(3+2x)$
(Interval of convergence)
−1.5 −1.0 −.5   .5 1.0 1.5
Graph of first four terms of Maclaurin series

EXAMPLE 9   *Using a trigonometric identity with a known series*

Find the Maclaurin series for

a. $\cos x^2$ b. $\cos^2 x$

Solution

a. In Example 1, we found that

$$\cos u = 1 - \frac{u^2}{2!} + \frac{u^4}{4!} - \frac{u^6}{6!} + \cdots \quad \text{for all } u$$

Therefore, by substituting $u = x^2$, we obtain

$$\cos x^2 = 1 - \frac{(x^2)^2}{2!} + \frac{(x^2)^4}{4!} - \frac{(x^2)^6}{6!} + \cdots$$

$$= 1 - \frac{x^4}{2!} + \frac{x^8}{4!} - \frac{x^{12}}{6!} + \cdots \quad \text{for all } x$$

b. For $\cos^2 x$ we could use the definition of a Maclaurin series, but instead we will use a double-angle trigonometric identity.

$$\cos^2 x = \frac{1}{2} + \frac{1}{2}\cos 2x$$

$$= \frac{1}{2} + \frac{1}{2}\left[1 - \frac{(2x)^2}{2!} + \frac{(2x)^4}{4!} - \frac{(2x)^6}{6!} + \cdots\right] \quad \textit{Let } u = 2x.$$

$$= \frac{1}{2} + \frac{1}{2} - \frac{2x^2}{2!} + \frac{2^3 x^4}{4!} - \frac{2^5 x^6}{6!} + \cdots$$

$$= 1 - x^2 + \frac{1}{3}x^4 - \frac{2}{45}x^6 + \cdots \quad \text{for all } x \qquad \blacksquare$$

EXAMPLE 10   *Maclaurin series by adding two power series*

Find the Maclaurin series for $\dfrac{5 + x}{2 - x - x^2}$.

Solution   Using partial fractions, we can obtain the decomposition (without showing the steps):

$$\frac{5 + x}{2 - x - x^2} = \frac{2}{1 - x} + \frac{1}{2 + x}$$

Both of the terms on the right can be expressed as the sum of a geometric series:

$$\frac{2}{1 - x} = 2\left[\frac{1}{1 - x}\right] = 2[1 + x + x^2 + x^3 + \cdots] = 2\sum_{k=0}^{\infty} x^k \text{ on } (-1, 1)$$

$$\frac{1}{2 + x} = \frac{1}{2}\left[\frac{1}{1 - \left(-\frac{x}{2}\right)}\right] = \frac{1}{2}\left[1 + \left(-\frac{x}{2}\right) + \left(-\frac{x}{2}\right)^2 + \left(-\frac{x}{2}\right)^3 + \cdots\right]$$

$$= \frac{1}{2}\sum_{k=0}^{\infty}(-1)^k\left(\frac{x}{2}\right)^k \quad \text{on } (-2, 2)$$

Because we are dealing with two power series, we need to be careful about finding the interval of convergence. It is the intersection of the individual intervals of convergence, so we note that

$$(-1, 1) \cap (-2, 2) = (-1, 1)$$

Thus,

$$\frac{5 + x}{2 - x - x^2} = \underbrace{2\sum_{k=0}^{\infty} x^k}_{} + \underbrace{\frac{1}{2}\sum_{k=0}^{\infty}(-1)^k\left(\frac{x}{2}\right)^k}_{} = \underbrace{\sum_{k=0}^{\infty}\left[2x^k + (-1)^k\frac{1}{2}\left(\frac{x}{2}\right)^k\right]}_{}$$

Interval of convergence: $(-1, 1)$ $\cap$ $(-2, 2)$ $=$ $(-1, 1)$

$$= \sum_{k=0}^{\infty}\left(2 + \frac{(-1)^k}{2^{k+1}}\right)x^k = \left(2 + \frac{1}{2}\right) + \left(2 - \frac{1}{4}\right)x + \left(2 + \frac{1}{8}\right)x^2 + \left(2 - \frac{1}{16}\right)x^3 + \cdots$$

$$= \frac{5}{2} + \frac{7}{4}x + \frac{17}{8}x^2 + \frac{31}{16}x^3 + \cdots \quad \text{converges on } (-1, 1) \qquad \blacksquare$$

No. of	Gregory's Series	
terms	term	partial sum
0	4	
1	−1.33333333333333	2.666666667
2	0.8	3.466666667
3	−0.57142857142857	2.895238095
4	0.44444444444444	3.33968254
5	−0.36363636363636	2.976046176
6	0.30769230769231	3.283738484
7	−0.26666666666667	3.017071817
8	0.23529411764706	3.252365935
9	−0.21052631578947	3.041839619
10	0.19047619047619	3.232315809
11	−0.17391304347826	3.058402766
12	0.16	3.218402766
13	−0.14814814814815	3.070254618
14	0.13793103448276	3.208185652
15	−0.12903225806452	3.079153394
16	0.12121212121212	3.200365515
17	−0.11428571428571	3.086079801
18	0.10810810810811	3.194187909
19	−0.1025641025641	3.091623807
20	0.09560975609756	3.189184782
21	−0.09302325581395	3.096161526
22	0.088888888888889	3.185050415
23	−0.08510638297872	3.099944032
24	0.081632653061224	3.181576685
25	−0.07843137254902	3.103145313
26	0.075471698113208	3.178617011
27	−0.07272727272727	3.105889738
28	0.070175438596491	3.176065177
29	−0.06779661016949	3.108268567
30	0.065573770491803	3.173842337
31	−0.06349206349206	3.110350274
32	0.061538461538462	3.171888735
33	−0.05970149253731	3.112187243
34	0.057971014492754	3.170158257
35	−0.05633802816901	3.113820229
36	0.054794520547945	3.16861475
37	−0.05333333333333	3.115281416
38	0.051948051948052	3.167229468
39	−0.05063291139241	3.116596557
40	0.049382716049383	3.165979273
41	−0.04819277108434	3.117786502
42	0.047058823529412	3.164845325
43	−0.04597701149425	3.118868314
44	0.044943820224719	3.163812134
45	−0.04395604395604	3.11985609
46	0.043010752688172	3.162866843
47	−0.04210526315789	3.12076158
48	0.041237113402062	3.161998693
49	−0.04040404040404	3.121594653
50	0.03960396039604	3.161198613

Approximation of π using Gregory's series

EXAMPLE 11 *Maclaurin series by integration*

Find the Maclaurin series for $\tan^{-1}x$.

Solution We shall use the fact that $\tan^{-1}x = \int_0^x \dfrac{dt}{1 + t^2}$. From Example 6,

$$\frac{1}{1 + t^2} = 1 - t^2 + t^4 - t^6 + \cdots = \sum_{k=0}^{\infty} (-1)^k t^{2k}$$

Thus,

$$
\begin{aligned}
\tan^{-1}x &= \int_0^x \frac{1}{1 + t^2}\, dt \\
&= \int_0^x [1 - t^2 + t^4 - t^6 + \cdots]\, dt = \sum_{k=0}^{\infty} \int_0^x (-1)^k t^{2k}\, dt \\
&= x - \frac{x^3}{3} + \frac{x^5}{5} - \frac{x^7}{7} + \cdots = \sum_{k=0}^{\infty} \frac{(-1)^k x^{2k+1}}{2k + 1} \text{ for } |x| \le 1
\end{aligned}
$$
∎

This series for $\tan^{-1}x$ is called **Gregory's series** after the British mathematician James Gregory (1638–1675), who developed it in 1671. Because it converges at 1 and −1, the Maclaurin series at $x = 1$ and $x = -1$ also can be used to represent $\tan^{-1}x$. For example, at $x = 1$, we have

$$\frac{\pi}{4} = \tan^{-1}1 = 1 - \frac{1}{3} + \frac{1}{5} - \frac{1}{7} + \frac{1}{9} - \cdots$$

$$\pi = 4\left[1 - \frac{1}{3} + \frac{1}{5} - \frac{1}{7} + \frac{1}{9} - \cdots\right]$$

The convergence of this series is very slow (see the computer output in the margin) in the sense that it takes a relatively large number of terms to achieve a reasonable approximation for π. Look at the partial sums for the 49th and 50th terms, and note that these are not very close to the limit (which is π).

In 1706, the mathematician John Machin discovered another series that converges much faster. We use

(SMH)
$$\tan(\alpha + \beta) = \frac{\tan \alpha + \tan \beta}{1 - \tan \alpha \tan \beta}$$

to show that

No. of	Machin's Series	
terms	term	partial sum
0	3.1832635983264	
1	−0.0426665690003	3.140597029
2	0.001023999998974	3.141621029
3	−2.9257142857E-05	3.141591772
4	9.10222222222E-07	3.141592682
5	−2.9789090909E-08	3.141592653
6	1.00824615385E-09	3.141592654
7	−3.4952533333E-11	3.141592654

$$\tan^{-1}1 = 4\tan^{-1}\frac{1}{5} - \tan^{-1}\frac{1}{239}$$

Using this identity, we find another series for π:

$$\pi = 16\left[\frac{1}{5} - \frac{1}{3}\left(\frac{1}{5}\right)^3 + \frac{1}{5}\left(\frac{1}{5}\right)^5 - \cdots\right] - 4\left[\frac{1}{239} - \frac{1}{3}\left(\frac{1}{239}\right)^3 + \frac{1}{5}\left(\frac{1}{239}\right)^5 - \cdots\right]$$

You can compare the convergence of this spreadsheet approximation with the first one and see that it approximates π much more quickly. (Note that the seventh term is fairly close to the limit π.)

(SMH) Some of the more common power series that we have derived are given in Table 8.2. For a more complete list, see the *Student Mathematics Handbook*. Before we present this list, there is one last result we should consider. It is a generalization of the binomial theorem that was discovered by Isaac Newton while he was still a student at Cambridge University.

THEOREM 8.28 *Binomial series theorem*

The binomial function $(1 + x)^p$ is represented by its Maclaurin series

$$(1 + x)^p = 1 + px + \frac{p(p-1)}{2!}x^2 + \frac{p(p-1)(p-2)}{3!}x^3 + \cdots$$

$$+ \frac{p(p-1)\cdots(p-k+1)}{k!}x^k + \cdots$$

for all x if p is a nonnegative integer; for $-1 < x < 1$ if $p \le -1$; for $-1 \le x \le 1$ if $p > 0$, p not an integer; and for $-1 < x \le 1$ if $-1 < p < 0$.

Proof We begin by showing how the Maclaurin series is found. Let $f(x) = (1 + x)^p$.

$$f(x) = (1 + x)^p \qquad\qquad\qquad f(0) = 1$$
$$f'(x) = p(1 + x)^{p-1} \qquad\qquad f'(0) = p$$
$$f''(x) = p(p-1)(1 + x)^{p-2} \qquad f''(0) = p(p-1)$$
$$f'''(x) = p(p-1)(p-2)(1 + x)^{p-3} \qquad f'''(0) = p(p-1)(p-2)$$
$$\vdots \qquad\qquad\qquad\qquad \vdots$$
$$f^{(n)}(x) = p(p-1)\cdots(p-n+1)(1 + x)^{p-n}$$

and

$$f^{(n)}(0) = p(p-1)\cdots(p-n+1)$$

which produces the Maclaurin series. We could use the ratio test to show that this series converges to *some function* on the interval $(-1, 1)$, but to show that it converges to $(1 + x)^p$ requires advanced calculus.

EXAMPLE 12 *Using the binomial series theorem to obtain a Maclaurin series expansion*

Find the Maclaurin series for $f(x) = \sqrt{9 + x}$ and find its radius of convergence.

Solution Write $f(x) = \sqrt{9 + x} = (9 + x)^{1/2} = 3\left(1 + \frac{x}{9}\right)^{1/2}$. Thus,

$$\sqrt{9 + x} = 3\left(1 + \frac{x}{9}\right)^{1/2}$$

$$= 3\left[1 + \frac{1}{2}\left(\frac{x}{9}\right) + \frac{\frac{1}{2}(\frac{1}{2}-1)}{2!}\left(\frac{x}{9}\right)^2 + \frac{\frac{1}{2}(\frac{1}{2}-1)(\frac{1}{2}-2)}{3!}\left(\frac{x}{9}\right)^3 + \cdots\right]$$

$$= 3\left[1 + \tfrac{1}{18}x - \tfrac{1}{648}x^2 + \tfrac{1}{11,664}x^3 - \cdots\right] = 3 + \tfrac{1}{6}x - \tfrac{1}{216}x^2 + \tfrac{1}{3,888}x^3 - \cdots$$

We know from Theorem 8.28 that the series converges when $\left|\frac{x}{9}\right| \le 1$, that is, $|x| \le 9$, so the radius of convergence is $R = 9$. ∎

■ TABLE 8.2 **Power Series for Elementary Functions**

Name	Series	Interval of Convergence
	Taylor series at $x = c$ follows the Maclaurin series for each function.	
Exponential series	$e^u = 1 + u + \dfrac{u^2}{2!} + \dfrac{u^3}{3!} + \dfrac{u^4}{4!} + \cdots + \dfrac{u^k}{k!} + \cdots$	$(-\infty, \infty)$
	$e^x = e^c + e^c(x - c) + \dfrac{e^c(x - c)^2}{2!} + \dfrac{e^c(x - c)^3}{3!} + \cdots + \dfrac{e^c(x - c)^k}{k!} + \cdots$	$(-\infty, \infty)$
Cosine series	$\cos u = 1 - \dfrac{u^2}{2!} + \dfrac{u^4}{4!} - \dfrac{u^6}{6!} + \cdots + \dfrac{(-1)^k u^{2k}}{(2k)!} + \cdots$	$(-\infty, \infty)$
	$\cos x = \cos c - (x - c)\sin c - \dfrac{(x - c)^2}{2!}\cos c + \dfrac{(x - c)^3}{3!}\sin c + \cdots$	$(-\infty, \infty)$
Sine series	$\sin u = u - \dfrac{u^3}{3!} + \dfrac{u^5}{5!} - \dfrac{u^7}{7!} + \cdots + \dfrac{(-1)^k u^{2k+1}}{(2k + 1)!} + \cdots$	$(-\infty, \infty)$
	$\sin x = \sin c + (x - c)\cos c - \dfrac{(x - c)^2}{2!}\sin c - \dfrac{(x - c)^3}{3!}\cos c + \cdots$	$(-\infty, \infty)$
Geometric series	$\dfrac{1}{1 - u} = 1 + u + u^2 + u^3 + \cdots + u^k + \cdots$	$(-1, 1)$
Reciprocal series	$\dfrac{1}{u} = 1 - (u - 1) + (u - 1)^2 - (u - 1)^3 + (u - 1)^4 - \cdots$	$(0, 2)$
Logarithmic series	$\ln u = (u - 1) - \dfrac{(u - 1)^2}{2} + \dfrac{(u - 1)^3}{3} - \cdots + \dfrac{(-1)^{k-1}(u - 1)^k}{k} + \cdots$	$(0, 2)$
	$\ln(1 + x) = x - \dfrac{1}{2}x^2 + \dfrac{1}{3}x^3 - \dfrac{1}{4}x^4 + \cdots + \dfrac{(-1)^n x^{n+1}}{n + 1} + \cdots$	$(-1, 1)$
	$\ln x = \ln c + \dfrac{x - c}{c} - \dfrac{(x - c)^2}{2c^2} + \dfrac{(x - c)^3}{3c^3} - \cdots + \dfrac{(-1)^{k-1}(x - c)^k}{kc^k} + \cdots$	$(0, 2c)$
Inverse tangent series	$\tan^{-1}u = u - \dfrac{u^3}{3} + \dfrac{u^5}{5} - \dfrac{u^7}{7} + \cdots + \dfrac{(-1)^k u^{2k+1}}{2k + 1} + \cdots$	$[-1, 1]$
Inverse sine series	$\sin^{-1}u = u + \dfrac{u^3}{2 \cdot 3} + \dfrac{1 \cdot 3 u^5}{2 \cdot 4 \cdot 5} + \dfrac{1 \cdot 3 \cdot 5 \cdot u^7}{2 \cdot 4 \cdot 6 \cdot 7} + \cdots + \dfrac{1 \cdot 3 \cdot 5 \ldots (2k - 3)u^{2k-1}}{2 \cdot 4 \cdot 6 \ldots (2k - 2)(2k - 1)} + \cdots$	$[-1, 1]$
Binomial series	$(1 + u)^p = 1 + pu + \dfrac{p(p - 1)}{2!}u^2 + \dfrac{p(p - 1)(p - 2)}{3!}u^3 + \cdots + \binom{p}{q}u^q + \cdots$	

8.8 *Problem Set*

Ⓐ **1.** ■ **What Does This Say?** Compare and contrast Maclaurin and Taylor series.

2. ■ **What Does This Say?** Discuss the binomial series theorem.

Find the Maclaurin series for the functions given in Problems 3–30. Assume that a is any constant.

3. e^{2x}

4. e^{-x}

5. e^{x^2}

6. e^{ax}

7. $\sin x^2$

8. $\sin^2 x$

9. $\sin ax$

10. $\cos ax$

11. $\cos 2x^2$

12. $\cos x^3$

13. $x^2 \cos x$

14. $\sin \dfrac{x}{2}$

15. $x^2 + 2x + 1$

16. $x^3 - 2x^2 + x - 5$

17. xe^x

18. $e^{-x} + e^{2x}$

19. $e^x + \sin x$

20. $\sin x + \cos x$

21. $\dfrac{1}{1 + 4x}$

22. $\dfrac{1}{1 - ax}, a \neq 0$

23. $\dfrac{1}{a + x}, a \neq 0$

24. $\dfrac{1}{a^2 + x^2}, a \neq 0$

25. $\ln(3 + x)$

26. $\log(1 + x)$

27. $\tan^{-1}(2x)$

28. $\sqrt{1 - x}$

29. $f(x) = e^{-x^2}$

30. $f(x) = \begin{cases} \dfrac{\sin x}{x} & \text{if } x \neq 0 \\ 1 & \text{if } x = 0 \end{cases}$

Find the first four terms of the Taylor series of the functions in Problems 31–42 at the given value of c.

31. $f(x) = e^x$ at $c = 1$

32. $f(x) = \ln x$ at $c = 3$

33. $f(x) = \cos x$ at $c = \frac{\pi}{3}$

34. $f(x) = \sin x$ at $c = \frac{\pi}{4}$

35. $f(x) = \tan x$ at $c = 0$

36. $f(x) = x^2 + 2x + 1$ at $c = 200$

37. $f(x) = x^3 - 2x^2 + x - 5$ at $c = 2$

38. $f(x) = \sqrt{x}$ at $c = 9$

39. $f(x) = \dfrac{1}{2 - x}$ at $c = 5$

40. $f(x) = \dfrac{1}{4 - x}$ at $c = -2$

41. $f(x) = \dfrac{3}{2x - 1}$ at $c = 2$

42. $f(x) = \dfrac{5}{3x + 2}$ at $c = 2$

Expand each function in Problems 43–48 as a binomial series. Give the interval of absolute convergence of the series.

43. $f(x) = \sqrt{1 + x}$

44. $f(x) = \dfrac{1}{\sqrt{1 + x^2}}$

45. $f(x) = (1 + x)^{2/3}$

46. $f(x) = (4 + x)^{-1/3}$

47. $f(x) = \dfrac{x}{\sqrt{1 - x^2}}$

48. $f(x) = \sqrt[4]{2 - x}$

B **49.** Use term-by-term integration to show that

$$\ln(1 + x) = \sum_{k=1}^{\infty} \frac{(-1)^{k-1}x^k}{k} \quad \text{for } |x| < 1$$

50. Use the Maclaurin series for e^x and e^{-x} to find the Maclaurin series for

$$\cosh x = \frac{e^x + e^{-x}}{2}$$

51. Use the Maclaurin series for e^x and e^{-x} to find the Maclaurin series for

$$\sinh x = \frac{e^x - e^{-x}}{2}$$

52. How many terms of a Maclaurin series expansion of e are necessary to approximate $\sqrt{e}$ to three-decimal-place accuracy?

53. How many terms of a Maclaurin series expansion of e are necessary to approximate $\sqrt[3]{e}$ to three-decimal-place accuracy?

Use partial fractions to find the Maclaurin series for the functions given in Problems 54–60.

54. $f(x) = \dfrac{2x}{x^2 - 1}$

55. $f(x) = \dfrac{6 - x}{4 - x^2}$

56. $f(x) = \dfrac{3(1 - x)}{9 - x^2}$

57. $f(x) = \dfrac{1}{x^2 - 3x + 2}$

58. $f(x) = \dfrac{2x - 3}{x^2 - 3x + 2}$

59. $f(x) = \dfrac{x^2}{(x + 2)(x^2 - 1)}$

60. $f(x) = \dfrac{x^2 - 6x + 7}{(1 - x)(2 - x)(3 - x)}$

61. Use an appropriate identity to find the Maclaurin series for $f(x) = \sin x \cos x$.

62. a. It can be shown that

$$\cos 3x = 4 \cos^3 x - 3 \cos x$$

Use this identity to find the Maclaurin series for $\cos^3 x$.

b. Find the Maclaurin series for $\sin^3 x$.

63. Use the identity

$$\cos x + \cos y = 2[\cos \tfrac{1}{2}(x + y)][\cos \tfrac{1}{2}(x - y)]$$

to find the Maclaurin series for

$$f(x) = \left(\cos \frac{3x}{2}\right)\left(\cos \frac{x}{2}\right)$$

64. Find the Maclaurin series for

$$\ln[(1 + 2x)(1 + 3x)]$$

65. Find the Maclaurin series for

$$\ln\left[\frac{1 + 2x}{1 - 3x + 2x^2}\right]$$

66. Find the Maclaurin series for

$$f(x) = x + \sin x$$

and then use this series to evaluate

$$\lim_{x \to 0} \frac{x + \sin x}{x}$$

67. Find the Maclaurin series for

$$g(x) = e^x - 1$$

and then use this series to find

$$\lim_{x \to 0} \frac{e^x - 1}{x}$$

In Problems 68–70, the Maclaurin series of a function appears with its remainder. Verify that the remainder satisfies the given inequality.

68. $e^x = \displaystyle\sum_{k=0}^{\infty} \frac{x^k}{k!}; \; |R_n(x)| \leq \dfrac{e^p |x|^{n+1}}{(n+1)!}$

69. $\cos x = \sum_{k=0}^{\infty} \frac{(-1)^k x^{2k}}{(2k)!}$; $|R_n(x)| \le \frac{|x|^{n+1}}{(n+1)!}$

70. $\ln(x+1) = \sum_{k=0}^{\infty} \frac{(-1)^k x^{k+1}}{k+1}$ for $0 \le x < 1$;

$$|R_n(x)| \le \frac{|x|^{n+1}}{n+1}$$

Technology Window

In the following two problems we will always be expanding about $x = c = 0$, but will refer to the Taylor (rather than the Maclaurin) expansion. Also, let $P_n(x)$ denote the Taylor polynomial of degree n approximating $f(x)$.

71. Consider the Taylor approximations to $f(x) = x \cos 2x$.

   a. Find the Maclaurin series for $f(x)$.

   b. Suppose you want to approximate $f(x)$ by $P_5(x)$ on $[-\frac{1}{2}, \frac{1}{2}]$ and you are concerned about accuracy. Rather than compute the remainder term $R_5(x)$, which is not pleasant, consider the following. Look at the *next* Taylor term (involving x^7), and see how large this term could be on the interval $[-\frac{1}{2}, \frac{1}{2}]$. While not foolproof, this is usually a good estimate of the error in $P_5(x)$. Use this to estimate the error, $|f(x) - P_5(x)|$.

   c. As a check on your work in part **b,** evaluate the Taylor polynomial $P_5(x)$ at $x = 0.5$.

   d. Now, plot the difference $|f(x) - P_5(x)|$ on $[-\frac{1}{2}, \frac{1}{2}]$ and see how the actual error compares with the estimate you found in part **b.**

72. The idea in this problem is to use power series to solve a simple-looking but difficult differential equation. We want a solution on $[-1, 1]$:

$$y'' = g(x) = e^{-x^2} \sin x$$

with initial conditions $y(0) = 0.3$, $y'(0) = -0.1$. If you attempt a direct solution, you will have trouble integrating y'' to obtain y'. Instead, we seek an approximation to the solution, as follows.

   a. Try to integrate $g(x)$ using your computer. Describe what happens.

   b. Approximate $g(x)$ by Taylor polynomials $P_n(x)$ until it looks good on $[-1, 1]$. If possible, provide a graph.

   c. Now integrate P_n twice, applying the initial conditions, thus obtaining a (hopefully) suitable approximation to our desired solution. Provide a graph of the resulting y. *Note:* Make sure your y satisfies $y(0) = 0.3$, $y'(0) = 0.1$.

---

73. HISTORICAL QUEST The two mathematicians most closely associated with series are Brook Taylor and Colin Maclaurin. The result stated in this section, called Taylor's theorem, was published by Taylor in 1714, but the result itself was discovered by the Scottish mathematician James Gregory (1638–1675) in 1670 and was published by Taylor after Gregory's death. James Gregory was a Scottish mathematician who apparently duplicated some of the key calculus discoveries, but died prematurely and never received proper recognition for his work. In 1694, Johann Bernoulli (see Historical Quest 63, Section 4.8) published a series that was very similar to what we call Taylor's series. In fact, after Taylor published his work in 1714, Bernoulli accused him of plagiarism.

BROOK TAYLOR
1685–1731

In this Quest, we seek to duplicate Bernoulli's work. He started by writing

$$n\,dx = n\,dx + \left(x\,dn - x\frac{dn}{dx}\,dx\right)$$

$$= n\,dx + \left(x\,dn - x\frac{dn}{dx}\,dx\right)$$

$$\quad - \left(\frac{x^2}{2!}\frac{d^2n}{dx^2}\,dx - \frac{x^2}{2!}\frac{d^2n}{dx^2}\,dx\right)$$

$$= n\,dx + \left(x\,dn - x\frac{dn}{dx}\,dx\right)$$

$$\quad - \left(\frac{x^2}{2!}\frac{d^2n}{dx^2}\,dx - \frac{x^2}{2!}\frac{d^2n}{dx^2}\,dx\right)$$

$$\quad + \left(\frac{x^3}{3!}\frac{d^3n}{dx^3}\,dx - \frac{x^3}{3!}\frac{d^3n}{dx^3}\,dx\right) + \cdots$$

$$= (n\,dx + x\,dn) - \left(x\frac{dn}{dx} + \frac{x^2}{2!}\frac{d^2n}{dx^2}\right)dx$$

$$\quad + \left(\frac{x^2}{2!}\frac{d^2n}{dx^2} + \frac{x^3}{3!}\frac{d^3n}{dx^3}\right)dx - \cdots$$

$$= d(nx) - d\left(\frac{x^2}{2!}\frac{dn}{dx}\right) + d\left(\frac{x^3}{3!}\frac{d^2n}{dx^2}\right) - \cdots$$

Integrate termwise to obtain what is known as Bernoulli's series. Next, see if you can show what angered Bernoulli by deriving Taylor's series from Bernoulli's series.

Incidentally, Bernoulli's assertion was not founded. The first known explicit statement of the general Taylor series was in *De Quadrature* (1691) by Isaac Newton.

74. HISTORICAL QUEST Colin Maclaurin was a mathematical prodigy. He entered the University of Glasgow at the age of 11. At 15 he took his master's degree and gave a remarkable public defense of his thesis on gravitational attraction. At 19 he was elected to the chair of mathematics at the Marischal College in Aberdeen and at 21 published his first important work, *Geometrica organica*. At 27 he became deputy, or assistant, to

COLIN MACLAURIN
1698–1746

the professor of mathematics at the University of Edinburgh. There was some difficulty in obtaining a salary to cover his assistantship, and Newton offered to bear the cost personally so that the university could secure the services of such an outstanding young man. In time, Maclaurin succeeded Newton and at the age of 44, published the first systematic exposition of Newton's work. In his work, Maclaurin uses what were called at the time *fluxions*, rather than derivatives. Here is what he concluded:

"If the first fluxion of the ordinate, with its fluxions of several subsequent orders vanish, the ordinate is a minimum or maximum, when the number of all those fluxions that vanish is 1, 3, 5, or any odd number. The ordinate is a *minimum*, when the fluxion next to those that vanish is positive; but a *maximum* when the fluxion is negative ⋯. But if the number of all the fluxions of the first and successive orders that vanish be an even number, the ordinate is then neither a *maximum* nor *minimum*."* What is Maclaurin saying in this quotation?

C **75.** Use term-by-term differentiation of a geometric series to find the Maclaurin series for

$$f(x) = \frac{1}{(1-x)^3}$$

76. Find the Maclaurin series expansion for the function defined by the integral

$$F(x) = \int_0^x e^{-t^2} dt$$

77. Express $\int_0^1 x^{0.2} e^x \, dx$ as an infinite series.

78. The functions $J_0(x)$ and $J_1(x)$ are defined by the following power series:

$$J_0(x) = \sum_{k=0}^{\infty} \frac{(-1)^k x^{2k}}{(k!)^2 2^{2k}}$$

―――――

The Historical Development of the Calculus by C. H. Edwards, Jr. New York: Springer-Verlag, 1979, p. 292.

and

$$J_1(x) = \sum_{k=0}^{\infty} \frac{(-1)^k x^{2k+1}}{k!(k+1)!2^{2k+1}}$$

a. Show that $J_0(x)$ and $J_1(x)$ both converge for all x.
b. Show that $J_0'(x) = -J_1(x)$.

These functions are called **Bessel functions of the first kind.** Bessel functions first were used in analyzing Kepler's laws of planetary motion. These functions play an important role in physics and engineering.

79. Show that J_0, the Bessel function of the first kind defined in Problem 78, satisfies the differential equation

$$x^2 J_0''(x) + x J_0'(x) + x^2 J_0(x) = 0$$

80. For x in the open interval $(-1, 1)$, let

$$f(x) = x + \frac{x^2}{2} + \frac{x^3}{3} + \cdots + \frac{x^k}{k} + \cdots$$

Show that f is a logarithmic function. *Hint:* Find f' and then retrieve f by integrating f'.

81. Given that

$$\frac{1}{(1-x)(1-2x)} = \sum a_n x^n$$

and

$$\frac{1+2x}{(1-x)(1-2x)(1-4x)} = \sum b_n x^n$$

show that $b_n = a_n^2$ for all n. (*Hint:* Use a partial-fraction decomposition.)

82. **JOURNAL PROBLEM** (*The Pi Mu Epsilon Journal* by Robert C. Gebhardt)*

For what x does $\sum_{n=0}^{\infty} \frac{x^n}{(2n)!}$ converge, and what is the sum?

―――――

*Volume 8, 1984, Problem 586

Chapter 8   Review

Proficiency Examination

Concept Problems

1. What is a sequence?

2. What is meant by the limit of a sequence?

3. Define the convergence and divergence of a sequence.

4. Explain each of the following terms:
  a. bounded sequence
  b. nonincreasing sequence
  c. monotonic sequence
  d. strictly monotonic sequence

5. State the BMCT.

6. What is an infinite series?

7. Compare or contrast the convergence and divergence of sequences and series.

8. What is a telescoping series?

9. Describe the harmonic series. Does it converge or diverge?

10. Define a geometric series and give its sum.

11. State the divergence test.

12. State the integral test.

13. State the *p*-series test.

14. State the direct comparison test.

15. State the limit comparison test.

16. State the zero-infinity limit comparison test.

17. State the ratio test.

18. State the root test.

19. What is the alternating series test?

20. How do you make an error estimate for an alternating series?

21. State the convergence test.

22. What is meant by absolute and conditional convergence?

23. What is the generalized ratio test?

24. What is a power series?

25. What are the radius of convergence and interval of convergence for a power series?

26. How do you find the convergence set of a power series?

27. What is a Taylor polynomial?

28. State Taylor's theorem.

29. What are a Taylor series and a Maclaurin series?

30. State the binomial series theorem.

Practice Problems

31. Find the limit of the sequence $\left\{\left(1+\dfrac{1}{n}\right)^n\right\}$

32. a. Find the limit of the sequence $\left\{\dfrac{e^n}{n!}\right\}$.

    b. Test the series $\displaystyle\sum_{k=1}^{\infty}\dfrac{e^k}{k!}$ for convergence.

    c. Discuss the similarities and/or differences of parts **a** and **b**.

In Problems 33–37, test the given series for convergence.

33. $\displaystyle\sum_{k=2}^{\infty}\dfrac{1}{k\ln k}$ 34. $\displaystyle\sum_{k=1}^{\infty}\dfrac{\pi^k k!}{k^k}$

35. $\displaystyle\sum_{k=2}^{\infty}\dfrac{1}{(\ln k)^{1/k}}$ 36. $\displaystyle\sum_{k=1}^{\infty}\dfrac{3k^2-k+1}{(1-2k)k}$

37. $1-\frac{1}{4}+\frac{1}{9}-\frac{1}{16}+\frac{1}{25}-\frac{1}{36}+\frac{1}{49}-\frac{1}{64}+\cdots+\dfrac{(-1)^{n+1}}{n^2}+\cdots$

38. Find the interval of convergence for the series
$$1-2u+3u^2-4u^3+\cdots$$

39. Find the Maclaurin series for $f(x)=\sin 2x$.

40. Find the Taylor series for $f(x)=\dfrac{1}{x-3}$ at $c=\frac{1}{2}$.

Supplementary Problems

Determine whether each sequence in Problems 1–16 *converges or diverges. If it converges, find its limit.*

1. $\left\{\dfrac{(-2)^n}{n^2+1}\right\}$ 2. $\left\{\dfrac{(\ln n)^2}{\sqrt{n}}\right\}$

3. $\left\{\left(1-\dfrac{2}{n}\right)^n\right\}$ 4. $\left\{\dfrac{e^{0.1n}}{n^5-3n+1}\right\}$

5. $\left\{\dfrac{n+(-1)^n}{n}\right\}$ 6. $\left\{\dfrac{3^n}{3^n+2^n}\right\}$

7. $\left\{\dfrac{5n^4-n^2-700}{3n^4-10n^2+1}\right\}$ 8. $\left\{\left(1+\dfrac{e}{n}\right)^{2n}\right\}$

9. $\{\sqrt{n+1}-\sqrt{n}\}$ 10. $\{\sqrt{n^4+2n^2}-n^2\}$

11. $\left\{\dfrac{\ln n}{n}\right\}$ 12. $\{1+(-1)^n\}$

13. $\left\{\left(1+\dfrac{4}{n}\right)^n\right\}$ 14. $\{5^{2/n}\}$

15. $\left\{\dfrac{n^{3/4}\sin n^2}{n+4}\right\}$

16. $\left\{\displaystyle\sum_{k=1}^{n}\dfrac{n}{n^2+k^2}\right\}$ *Hint*: This sequence is related to an integral.

17. Find the sum:
$$\sum_{k=-123,456,788}^{123,456,789}\dfrac{k}{370,370,367}$$

Find the sum of the given convergent geometric or telescoping series in Problems 18–27.

18. $\displaystyle\sum_{k=1}^{\infty}4\left(\dfrac{2}{3}\right)^k$ 19. $\displaystyle\sum_{k=1}^{\infty}\left(\dfrac{e}{3}\right)^k$

20. $\displaystyle\sum_{k=2}^{\infty}\dfrac{1}{k^2-1}$ 21. $\displaystyle\sum_{k=0}^{\infty}\left[\left(\dfrac{-3}{8}\right)^k+\left(\dfrac{3}{4}\right)^{2k}\right]$

22. $\displaystyle\sum_{k=1}^{\infty}\dfrac{1}{4k^2-1}$ 23. $\displaystyle\sum_{k=0}^{\infty}\dfrac{e^k+3^{k-1}}{6^{k+1}}$

24. $\displaystyle\sum_{k=1}^{\infty}(-1)^{k+1}\left(\dfrac{1}{k}+\dfrac{1}{k+1}\right)$ 25. $\displaystyle\sum_{k=2}^{\infty}\left[\dfrac{1}{\ln(k+1)}-\dfrac{1}{\ln k}\right]$

26. $\displaystyle\sum_{k=0}^{\infty}\left[3\left(\dfrac{2}{3}\right)^{2k}-\left(\dfrac{-1}{3}\right)^{4k}\right]$ 27. $\displaystyle\sum_{k=1}^{\infty}\dfrac{k}{(k+1)(k+2)(k+3)}$

Test the series in Problems 28–45 *for convergence.*

28. $\displaystyle\sum_{k=1}^{\infty}\dfrac{5^k k!}{k^k}$ 29. $\displaystyle\sum_{k=1}^{\infty}\dfrac{1}{\sqrt{k^2+4}}$

30. $\displaystyle\sum_{k=0}^{\infty} \frac{k!}{2^k}$

31. $\displaystyle\sum_{k=1}^{\infty} \frac{1}{2k-1}$

32. $\displaystyle\sum_{k=0}^{\infty} ke^{-k}$

33. $\displaystyle\sum_{k=0}^{\infty} \frac{k^3}{k!}$

34. $\displaystyle\sum_{k=1}^{\infty} \frac{7^k}{k^2}$

35. $\displaystyle\sum_{k=1}^{\infty} \frac{k}{(2k-1)!}$

36. $\displaystyle\sum_{k=0}^{\infty} \frac{k^2}{(k^3+1)^2}$

37. $\displaystyle\sum_{k=0}^{\infty} \frac{k^2}{3^k}$

38. $\displaystyle\sum_{k=2}^{\infty} \frac{1}{k(\ln k)^{1.1}}$

39. $\displaystyle\sum_{k=2}^{\infty} \frac{1}{k(\ln k)^2}$

40. $\displaystyle\sum_{k=0}^{\infty} (\sqrt{k^3+1} - \sqrt{k^3})$

41. $\displaystyle\sum_{k=1}^{\infty} \frac{1}{\sqrt{k(k+1)}}$

42. $\displaystyle\sum_{k=1}^{\infty} \frac{k-1}{k2^k}$

43. $\displaystyle\sum_{k=0}^{\infty} \frac{k!}{k^2(k+1)^2}$

44. $\displaystyle\sum_{k=0}^{\infty} \frac{1}{1+\sqrt{k}}$

45. $\displaystyle\sum_{k=1}^{\infty} \frac{k^3 3^k}{k!}$

Determine whether each series given in Problems 46–55 converges conditionally, converges absolutely, or diverges.

46. $\dfrac{1}{1 \cdot 2} - \dfrac{1}{3 \cdot 2} + \dfrac{1}{5 \cdot 2} - \dfrac{1}{7 \cdot 2} + \cdots$

47. $\dfrac{1}{1 \cdot 2} - \dfrac{1}{2 \cdot 3} + \dfrac{1}{3 \cdot 4} - \dfrac{1}{4 \cdot 5} + \cdots$

48. $\dfrac{3}{2} - \dfrac{4}{3} + \dfrac{5}{4} - \dfrac{6}{5} + \cdots$

49. $1 - \dfrac{1}{3} + \dfrac{1}{9} - \dfrac{1}{27} + \dfrac{1}{81} - \cdots$

50. $\dfrac{1}{5} - \dfrac{1}{7} + \dfrac{1}{9} - \dfrac{1}{11} + \cdots$

51. $-1 + \dfrac{1}{\sqrt{2}} - \dfrac{1}{\sqrt[3]{3}} + \dfrac{1}{\sqrt[4]{4}} - \cdots$

52. $\displaystyle\sum_{k=1}^{\infty} \frac{(-1)^k}{k[\ln(k+1)]^2}$

53. $\displaystyle\sum_{k=1}^{\infty} (-1)^k \left(\frac{3k+85}{4k+1}\right)^k$

54. $\displaystyle\sum_{k=1}^{\infty} (-1)^k \tan^{-1}\left(\frac{1}{2k+1}\right)$

55. $\displaystyle\sum_{k=1}^{\infty} \frac{(-1)^{k(k+1)/2}}{2^k}$

Find the interval of convergence for each power series in Problems 56–67.

56. $\displaystyle\sum_{k=1}^{\infty} \frac{kx^k}{3^k}$

57. $\displaystyle\sum_{k=1}^{\infty} k(x-1)^k$

58. $\displaystyle\sum_{k=1}^{\infty} \frac{x^k}{k(k+1)}$

59. $\displaystyle\sum_{k=1}^{\infty} \frac{\ln k(x^2)^k}{\sqrt{k}}$

60. $\displaystyle\sum_{k=1}^{\infty} \frac{k^2(x+1)^{2k}}{2^k}$

61. $\displaystyle\sum_{k=1}^{\infty} \frac{(-1)^k x^k}{k^k}$

62. $\displaystyle\sum_{k=1}^{\infty} \frac{(-1)^k(2x-1)^k}{k^2}$

63. $\displaystyle\sum_{k=1}^{\infty} \frac{(x+2)^k}{k\ln(k+1)}$

64. $\displaystyle\sum_{n=1}^{\infty} \frac{k(x-3)^k}{(k+3)!}$

65. $\displaystyle\sum_{n=1}^{\infty} \frac{(3n)!}{n!} x^n$

66. $\displaystyle\sum_{n=1}^{\infty} \frac{(-1)^n x^n}{9n^2-1}$

67. $x - \dfrac{x^3}{3!} + \dfrac{x^5}{5!} - \dfrac{x^7}{7!} + \cdots + \dfrac{(-1)^{k+1}x^{2k-1}}{(2k-1)!} + \cdots$

Find the Maclaurin series for each function given in Problems 68–73.

68. $f(x) = x^2 e^{-3x}$

69. $f(x) = x^3 \sin x$

70. $f(x) = 3^x$

71. $f(x) = x^2 + \tan^{-1}(2x)$

72. $f(x) = \dfrac{5+7x}{1+2x-3x^2}$

73. $f(x) = \dfrac{11x-1}{2+x-3x^2}$

74. Compute the sum of the series

$$1 - \frac{1}{2} + \frac{1}{2!}\left(\frac{1}{2}\right)^2 - \frac{1}{3!}\left(\frac{1}{2}\right)^3 + \cdots$$

correct to three decimal places.

75. Find a bound on the error incurred by approximating the sum of the series

$$1 - \frac{1}{2!} + \frac{2}{3!} - \frac{3}{4!} + \cdots$$

by the sum of the first eight terms.

76. Use geometric series to write $12.342\overline{132}$ as a rational number.

77. A ball is dropped from a height of A feet. Each time it drops h feet, it rebounds $0.8h$ feet. If it travels a total distance of 20 feet before coming to a stop, what is A?

78. Find the first three nonzero terms of the Maclaurin series for $\tan x$ and then use term-by-term integration to find the first three nonzero terms of the Maclaurin for $\ln(\cos x)$.

79. **SPY PROBLEM** The Spy, who is smarter than he looks, aborts his mission to the village after six days of duck plucking (recall Problem 61, Section 7.4). As he races away, he is confronted by four of Blohardt's thugs, and soon finds himself in a gun battle using a revolver he managed to steal from Blabba. While dodging bullets, he recalls from his math class at Spy Academy that if the probability of a bullet being "good" is p, then on average, the number of bullets fired before a bad one jams the gun is given by the series $\displaystyle\sum_{n=0}^{\infty} np^{n-1}(1-p)$. If he takes five shots to dispose of each villain, what is the smallest value of p that will guarantee he gets off enough rounds to dispatch all four foes before a "bad" bullet gets him into another jam?

80. Show that

$$\cos x = \cos c - (\sin c)(x - c) - (\cos c)\frac{(x - c)^2}{2!}$$
$$+ (\sin c)\frac{(x - c)^3}{3!} + \cdots$$

81. Show that

$$e^x = e^c + e^c(x - c) + e^c\frac{(x - c)^2}{2!} + \cdots$$
$$+ e^c\frac{(x - c)^k}{k!} + \cdots$$

82. Express the integral $\int \sqrt{x^3 + 1}\, dx$ as an infinite series.

Hint: Use the binomial theorem and then integrate term by term.

83. Estimate the value of the integral

$$\int_0^{0.4} \frac{dx}{\sqrt{1 + x^3}}$$

with 5-decimal-place accuracy.

84. Show that if $a > 0, a \neq 1$,

$$a^x = \sum_{k=0}^{\infty} \frac{(\ln a)^k}{k!} x^k \quad \text{for all } x$$

85. Show that for $0 < x < 2c$,

$$\frac{1}{x} = \frac{1}{c} - \frac{1}{c^2}(x - c) + \frac{1}{c^3}(x - c)^2 - \frac{1}{c^4}(x - c)^3 + \cdots$$

86. Use series to find $\sin 0.2$ correct to five decimal places.

87. Use series to find $\ln 1.05$ correct to five decimal places.

88. Use a Maclaurin series to find a cubic polynomial that approximates e^{2x} for $|x| < 0.001$. Find an upper bound for the error in this approximation.

89. For what values of x can $\sin x$ be replaced by $x - \frac{x^3}{6}$ if the allowable error is 0.0005?

90. For what values of x can $\cos x$ be replaced by $1 - \frac{x^2}{2}$ if the allowable error is 0.00005?

91. Find the Maclaurin series for $f(x) = \frac{2\tan x}{1 + \tan^2 x}$.

92. a. Find a formula for the sum of the first n terms of an arithmetic progression that has a for its first term and d for its common difference.

b. Find a formula for the sum of the first n terms of a geometric progression that has a for its first term and r for its common ratio.

c. If $|r| < 1$, show that $\sum_{n=0}^{\infty} ar^n = \frac{a}{1 - r}$

93. MODELING PROBLEM When we studied carbon dating in Section 5.6, it was pointed out that radiocarbon methods apply only to objects that are not too "old" (roughly 40,000 years). To date older objects, radioactive isotopes other than ^{14}C are often employed. Regardless of the radioactive substance used, it can be shown that the ratio of stable isotope to radioactive isotope at any given time may be modeled by

$$\frac{S(t) - S(0)}{R(t)} + 1 = e^{(\ln 2)t/\lambda}$$

where $R(t)$ is the number of atoms of radioactive material at time t, $S(t)$ is the number of atoms of the stable product of radioactive decay, $S(0)$ is the number of atoms of stable product initially present (at $t = 0$), and λ is the half-life of the radioactive isotope.*

a. Use the first two terms of the Maclaurin series for e^x to approximate $e^{(\ln 2)t/\lambda}$, and then solve the modeling equation for t.

b. Suppose a piece of mica is analyzed, and it is found that 5% of the atoms in the rock are radioactive rubidium-87 and 0.04% of the atoms are strontium-87. Assuming that the strontium was produced by decay of the rubidium-87 originally in the rock, what is the approximate age of the sample? Use the result in part **a**, with $\lambda = 48.6 \times 10^9$ years as the half-life of rubidium-87.

94. PUTNAM EXAMINATION PROBLEM Express

$$\sum_{k=1}^{\infty} \frac{6^k}{(3^{n+1} - 2^{k+1})(3^k - 2^k)}$$

as a rational number.

95. PUTNAM EXAMINATION PROBLEM Find the sum of the convergent alternating series

$$\sum_{n=1}^{\infty} \frac{(-1)^{n+1}}{3n - 2}$$

96. PUTNAM EXAMINATION PROBLEM For positive real x, let

$$B_n(x) = 1^x + 2^x + 3^x + \cdots + n^x$$

Test for convergence of the series

$$\sum_{k=2}^{\infty} \frac{B_k(\log_k 2)}{(k \log_2 k)^2}$$

---

*This model is discussed in detail in the article, "How Old Is the Earth?" by Paul J. Campbell, *UMAP Modules 1992: Tools for Teaching*, Consortium for Mathematics and Its Applications, Inc. Lexington, MA: 1993, pp. 105–137.

Elastic Tightrope Project

This project is to be done in groups of three or four students. Each group will submit a single written report.

Suppose a cockroach starts at one end of a 1,000-m tightrope and runs toward the other end at a speed of 1m/s. At the end of every second, the tightrope stretches uniformly and instantaneously, increasing its length by 1,000 meters each time.

Does the roach ever reach the other end? If so, how long does it take?

Your paper is not limited to the following questions but should include these concerns. Suppose it is you standing on the elastic rope *b* meters from the left end and *c* meters from the right end. Then suppose the entire rope stretches uniformly, increasing its length by *d* meters. How far are you from each end? You might also try to generalize the roach problem.

The infinite! No other question has ever moved so profoundly the spirit of man; no other idea has so fruitfully stimulated his intellect; yet no other concept stands in greater need of clarification than that of the infinite.

DAVID HILBERT
TO INFINITY AND BEYOND BY ELI MAOR (Birkhäuser, 1987), p.vii.

There is nothing now which ever gives me any thought or care in algebra except divergent series, which I cannot follow the French in rejecting.

AUGUSTUS DE MORGAN
GRAVES' LIFE OF W.R. HAMILTON (NEW YORK: 1882–1889), p. 249.

*This group project is courtesy of David Pengelley, New Mexico State University.

9

Polar Coordinates and Parametric Forms

PREVIEW

Applications often involve curves that are difficult or even impossible to represent in the explicit form $y = f(x)$. In this chapter, we examine the polar coordinate system and other alternative forms of representation.

PERSPECTIVE

Consider the following two applications:

1. Three LORAN tracking stations are located at the vertices of an equilateral triangle 20 mi on a side. Radio signals are sent out from the stations simultaneously. If a ship receives the signal from two of the stations $10/v$ sec later than the signal from the third, where v is the velocity of the radio signals, how can the position of the ship be described in relation to the third station?

2. A bicycle wheel has a reflector attached at a point halfway between the hub and the rim. As the wheel rolls, what curve is sketched out by the reflector?

Such situations are difficult to describe in Cartesian coordinates but can be characterized readily using the parametric forms of representation developed in this chapter.

9.1 *The Polar Coordinate System*

plotting points in polar coordinates, relationship between polar and rectangular coordinates, polar graphs ■

PLOTTING POINTS IN POLAR COORDINATES

In a **polar coordinate system,** points are plotted in relation to a fixed point O, called the origin or **pole,** and a fixed ray emanating from the origin, called the **polar axis.** We then associate with each point P in the plane an ordered pair of numbers $P(r, \theta)$, where r is the distance from O to P, and θ is the angle measured from the polar axis to the ray OP, as shown in Figure 9.1. The number r is called the **radial coordinate** of P, and θ is the **polar angle.** The polar angle is regarded as positive if measured counterclockwise up from the polar axis, and negative if measured clockwise. The origin O has radial coordinate 0, and it is convenient to say that O has polar coordinates $(0, \theta)$, for all angles θ.

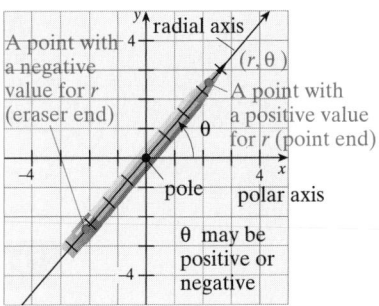

■ **FIGURE 9.1**   Polar-form points

If the point P has polar coordinates (r, θ), we say that the point Q obtained by reflecting P in the origin O has coordinates $(-r, \theta)$. Thus, if you think of a pencil lying along the directed line segment $\overline{OP}$, with its midpoint at O and tip at $P(r, \theta)$, then the eraser will be at $Q(-r, \theta)$, as illustrated in Figure 9.1. The general procedure for plotting points in polar coordinates is demonstrated in the following example.

EXAMPLE 1   *Plotting polar-form points*

Plot each of the following polar-form points: $A(4, \frac{\pi}{3})$, $B(-4, \frac{\pi}{3})$, $C(3, -\frac{\pi}{6})$, $D(-3, -\frac{\pi}{6})$, $E(-3, 3)$, $F(-3, -3)$, $G(-4, -2)$, $H(5, \frac{3\pi}{2})$, $I(-5, \frac{\pi}{2})$, $J(5, -\frac{\pi}{2})$.

Solution   Points A, B, C, and D illustrate the basic ideas of plotting polar-form points.

Plot $A(4, \frac{\pi}{3})$; positive θ.

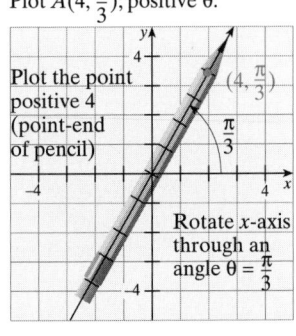

Plot $B(-4, \frac{\pi}{3})$; positive θ.

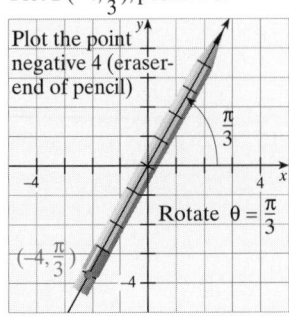

Plot $C(3, -\frac{\pi}{6})$; negative θ.

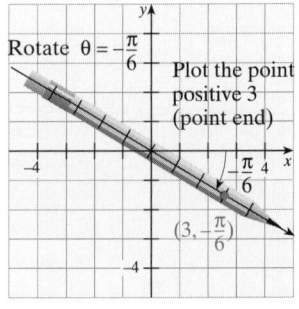

Plot $D(-3, -\frac{\pi}{6})$; negative θ.

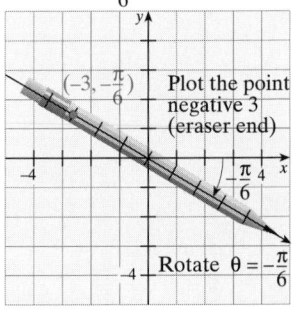

Points E, F, G, H, I, and J illustrate common situations that can sometimes be confusing. Make sure you take time with each example.

Plot $E(-3, 3)$.

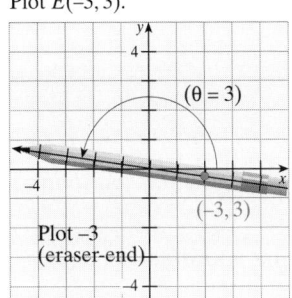

Plot $F(-3, -3)$.

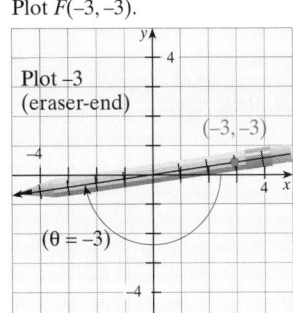

Plot $G(-4, -2)$.

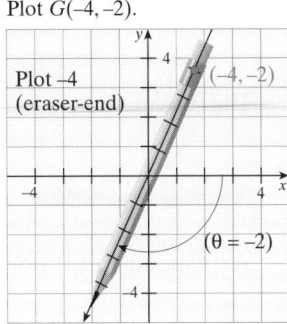

Plot $H(5, \frac{3\pi}{2})$.

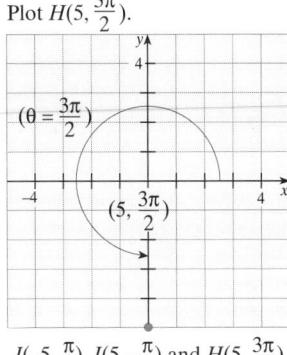

$I(-5, \frac{\pi}{2}), J(5, -\frac{\pi}{2})$ and $H(5, \frac{3\pi}{2})$ all represent the same point. ■

In Example 1, we found that $(5, \frac{3\pi}{2})$, $(-5, \frac{\pi}{2})$, and $(5, -\frac{\pi}{2})$ all represent the same point in polar coordinates. Indeed, every point in the plane has infinitely many polar representations. This property of the polar coordinate system causes some difficulties, but nothing that cannot be handled by exercising a little caution. We shall point out situations in our examples where the non-uniqueness of the polar representation must be taken into account.

RELATIONSHIP BETWEEN POLAR AND RECTANGULAR COORDINATES

The relationship between polar and rectangular coordinates can be found by using trigonometric functions. The origin of the rectangular coordinate system is the pole, and the x-axis is the polar axis.

Relationship between Rectangular and Polar Coordinates

1. To change *from polar to rectangular*:

   $$x = r \cos \theta \qquad y = r \sin \theta$$

2. To change *from rectangular to polar*:

   $$r = \sqrt{x^2 + y^2} \quad \bar{\theta} = \tan^{-1}\left|\frac{y}{x}\right|, \quad x \neq 0$$

   where $\bar{\theta}$ is the reference angle for θ. Place θ in the proper quadrant by noting the signs of x and y. If $x = 0$, then $\bar{\theta} = \frac{\pi}{2}$.

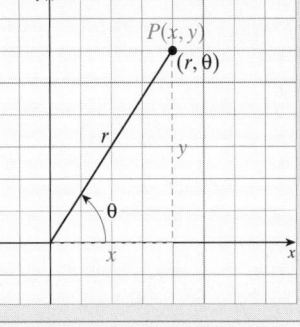

■ *What This Says:*   The reference angle $\bar{\theta}$ for a standard-position angle θ is defined to be the smallest positive angle that the angle θ makes with the x-axis. Reference angles are discussed in trigonometry and are reviewed in the *Student Mathematics Handbook*.

EXAMPLE 2   *Converting from polar to rectangular coordinates*

Change the polar coordinates $(-3, \frac{5\pi}{4})$ to rectangular coordinates.

Solution

$$x = -3 \cos \frac{5\pi}{4} = -3\left(-\frac{\sqrt{2}}{2}\right) = \frac{3\sqrt{2}}{2}$$

$$y = -3 \sin \frac{5\pi}{4} = -3\left(-\frac{\sqrt{2}}{2}\right) = \frac{3\sqrt{2}}{2}$$

The rectangular coordinates are $\left(\frac{3\sqrt{2}}{2}, \frac{3\sqrt{2}}{2}\right)$.   ∎

EXAMPLE 3   *Converting from rectangular coordinates to polar coordinates*

Write polar-form coordinates for the point with rectangular coordinates $\left(\frac{5\sqrt{3}}{2}, -\frac{5}{2}\right)$.

Solution $$r = \sqrt{\left(\frac{5\sqrt{3}}{2}\right)^2 + \left(-\frac{5}{2}\right)^2} = \sqrt{\frac{75}{4} + \frac{25}{4}} = 5$$

Note that θ is in Quadrant IV because x is positive and y is negative.

$$\bar{\theta} = \tan^{-1}\left|\frac{-5/2}{5\sqrt{3}/2}\right| = \tan^{-1}\left(\frac{1}{\sqrt{3}}\right) = \frac{\pi}{6}; \quad \text{thus, } \theta = \frac{11\pi}{6} \text{ (Quadrant IV)}$$

Polar-form coordinates are $(5, \frac{11\pi}{6})$.   ∎

POLAR GRAPHS

The **graph** of an equation in polar coordinates is the set of all points P whose polar coordinates (r, θ) satisfy the given equation. Circles, lines through the origin, and rays emanating from the origin have particularly simple equations in polar coordinates.

EXAMPLE 4   *Graphing circles, lines, and rays*

Graph: a. $r = 6$ b. $\theta = \frac{\pi}{6}, r \geq 0$ c. $\theta = \frac{\pi}{6}$

Solution

a. The graph is the set of all points (r, θ) such that the first component is 6 for any angle θ. This is a circle with radius 6 centered at the origin.

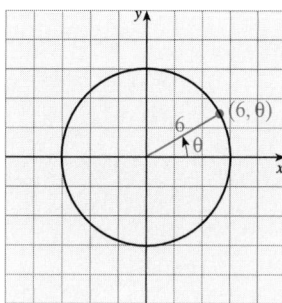

b. The graph is the closed half-line (ray) that emanates from the origin and makes an angle of $\frac{\pi}{6}$ with the positive x-axis.

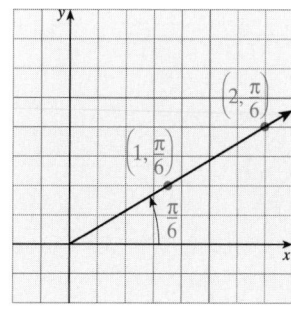

c. This is the line through the origin that makes an angle of $\frac{\pi}{6}$ with the positive x-axis.

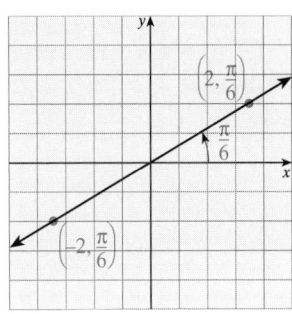

As with other equations, we begin graphing polar-form curves by plotting some points. However, you must first be able to recognize whether a point in polar form satisfies a given equation.

> **EXAMPLE 5**   *Verifying that polar coordinates satisfy an equation*

Show that each of the given points lies on the polar graph whose equation is

$$r = \frac{2}{1 - \cos \theta}$$

a. $(2, \frac{\pi}{2})$ b. $(-2, \frac{3\pi}{2})$ c. $(-1, 2\pi)$

Solution   Begin by substituting the given coordinates into the equation.

a. $2 \overset{?}{=} \dfrac{2}{1 - \cos \dfrac{\pi}{2}} = \dfrac{2}{1 - 0} = 2$; $\left(2, \dfrac{\pi}{2}\right)$ is on the curve, because it satisfies the equation.

b. $-2 \overset{?}{=} \dfrac{2}{1 - \cos \dfrac{3\pi}{2}} = \dfrac{2}{1 - 0} = 2$
Although the equation is not satisfied, we *cannot* say that the point is not on the curve. Indeed, we see from part **a** that it is on the curve, because $(-2, \frac{3\pi}{2})$ and $(2, \frac{\pi}{2})$ name the same point!

> ■ *What This Says:*   Even if one representation of a point does not satisfy the equation, we must still check equivalent representations of the point.

c. For $(-1, 2\pi)$, $-1 \overset{?}{=} \dfrac{2}{1 - \cos 2\pi} = \dfrac{2}{1 - 1}$, which is undefined. Next, check the equivalent representation $(1, \pi)$:

$$1 \overset{?}{=} \frac{2}{1 - \cos \pi} = \frac{2}{1 - (-1)} = 1$$

Thus, the point $(-1, 2\pi)$ is on the curve.   ■

We will discuss the graphing of polar-form curves in the next section. Sometimes, though, a polar-form equation can be graphed by changing it to rectangular form.

> **EXAMPLE 6**   *Polar-form graphing by changing to rectangular form*

Graph the given polar-form curves by changing to rectangular form.

a. $r = 3\cos\theta$ b. $r = 4\sec\theta$ c. $r = \dfrac{6}{2\sin\theta + \cos\theta}$

Solution

a.
$$r = 3\cos\theta \qquad \text{Given equation}$$
$$r^2 = 3r\cos\theta \qquad \text{Multiply by } r.$$
$$x^2 + y^2 = 3x \qquad \text{Because } x = r\cos\theta$$
$$\text{and } r^2 = x^2 + y^2$$
$$\left[x^2 - 3x + \left(\frac{3}{2}\right)^2\right] + y^2 = \frac{9}{4} \qquad \text{Complete the square.}$$
$$\left(x - \frac{3}{2}\right)^2 + y^2 = \frac{9}{4}$$

We see that this is a circle with center at $\left(\frac{3}{2}, 0\right)$ and radius $\frac{3}{2}$. The graph is shown in Figure 9.2a.

b. Because $r = 4\sec\theta$ can be expressed as $r\cos\theta = 4$, we see that the given equation can be written $x = 4$, whose graph is a vertical line (Figure 9.2b).

c.
$$r = \frac{6}{2\sin\theta + \cos\theta} \qquad \text{Given equation}$$
$$2r\sin\theta + r\cos\theta = 6$$
$$2y + x = 6 \qquad \text{Because } x = r\cos\theta, y = r\sin\theta$$

We recognize this as a line with y-intercept 3 and slope $-\frac{1}{2}$ (Figure 9.2c).

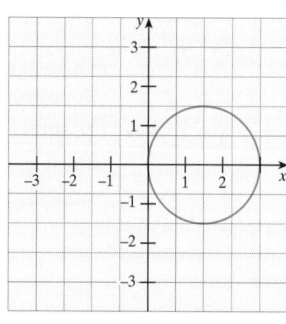

a. Graph of $r = 3\cos\theta$

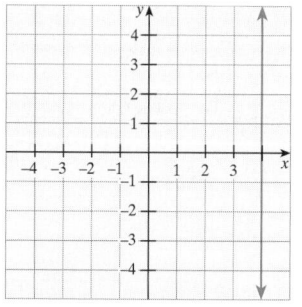

b. Graph of $r\cos\theta = 4$

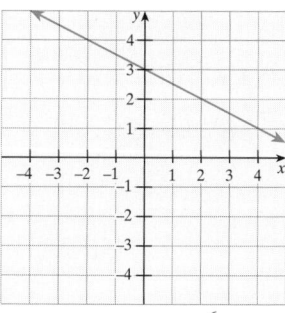

c. Graph of $r = \dfrac{6}{2\sin\theta + \cos\theta}$

■ **FIGURE 9.2**   Polar-form graphs   ■

9.1 Problem Set

 1. ■ **What Does This Say?** Illustrate the procedure for plotting points in polar form.

2. ■ **What Does This Say?** Show the derivation of the polar-rectangular-form conversion equations.

In Problems 3–11, plot each of the given polar-form points and give equivalent rectangular coordinates.

3. $\left(4, \frac{\pi}{4}\right)$ **4.** $\left(6, \frac{\pi}{3}\right)$ **5.** $\left(5, \frac{2\pi}{3}\right)$

6. $\left(3, -\frac{\pi}{6}\right)$ **7.** $\left(\frac{3}{2}, -\frac{5\pi}{6}\right)$ **8.** $(-4, 4)$

9. $(1, 3\pi)$ **10.** $\left(-2, -\frac{3\pi}{2}\right)$ **11.** $(0, -3)$

Plot the rectangular-form points in Problems 12–20 and give a polar form.

12. $(5, 5)$ **13.** $(-1, \sqrt{3})$ **14.** $(2, -2\sqrt{3})$

15. $(-2, -2)$ **16.** $(3, -3)$ **17.** $(3, 7)$

18. $(3, -3\sqrt{3})$ **19.** $(\sqrt{3}, -1)$ **20.** $(-3, 0)$

21. Name the country or island in Figure 9.3 in which each of the named points is located.
  a. $(5.5, 260°)$ b. $(7.5, 78°)$
  c. $(-2, 140°)$ d. $(-4, 80°)$

22. Name possible coordinates for each of the following cities (see Figure 9.3).
  a. Miami, Florida
  b. Los Angeles, California
  c. Mexico City, Mexico
  d. London, England

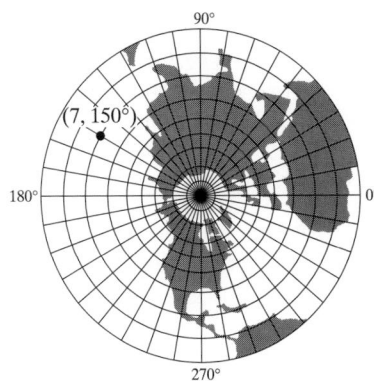

■ **FIGURE 9.3**   This map is reproduced from the floor of the Paris Observatory.

B *Write each equation given in Problems 23–30 in rectangular coordinates.*

23. $r = 4 \sin \theta$ **24.** $r = 16$ **25.** $r = 1 - \sin \theta$
26. $r = 2 \cos \theta$ **27.** $r = \sec \theta$ **28.** $r = 4 \tan \theta$
29. $r^2 = \dfrac{2}{1 + \sin^2 \theta}$ **30.** $r^2 = \dfrac{2}{3 \cos^2 \theta - 1}$

Sketch the graph of each equation given in Problems 31–38.

31. $r = \frac{3}{2}$ **32.** $r = \frac{3}{2}, 0 \le \theta \le 2$

33. $r = \sqrt{2}, 0 \le \theta \le 2$ **34.** $r = 4$
35. $\theta = 1$ **36.** $\theta = 1, r \ge 0$
37. $\theta = \frac{\pi}{6}, r < 0$ **38.** $\theta = \frac{\pi}{2}$

In Problems 39–44, tell whether each of the given points lies on the curve

$$r = \frac{5}{1 - \sin \theta}$$

39. $(10, \frac{\pi}{6})$ **40.** $(5, \frac{\pi}{2})$ **41.** $(-10, \frac{5\pi}{6})$
42. $(-\frac{10}{3}, \frac{5\pi}{6})$ **43.** $(20 + 10\sqrt{3}, \frac{\pi}{3})$ **44.** $(-10, \frac{\pi}{3})$

In Problems 45–52, tell whether each of the given points lies on the curve $r = 2(1 - \cos \theta)$.

45. $(1, \frac{\pi}{3})$ **46.** $(1, -\frac{\pi}{3})$ **47.** $(-1, \frac{\pi}{3})$
48. $(-2, \frac{\pi}{2})$ **49.** $(2 + \sqrt{2}, \frac{\pi}{4})$ **50.** $(-2 - \sqrt{2}, \frac{\pi}{4})$
51. $(0, \frac{\pi}{4})$ **52.** $(0, -\frac{2\pi}{3})$

Find three distinct ordered pairs (r, θ) satisfying each of the equations in Problems 53–60.

53. $r^2 = 9 \cos \theta$ **54.** $r^2 = 9 \cos 2\theta$
55. $r = 3\theta$ **56.** $r = 5\theta$
57. $r = 2 - 3 \sin \theta$ **58.** $r = 2(1 + \cos \theta)$
59. $\dfrac{r}{1 - \sin \theta} = 2$ **60.** $r = \dfrac{8}{1 - 2 \cos \theta}$

C **61.** a. What is the distance between the polar-form points $(3, \frac{\pi}{3})$ and $(7, \frac{\pi}{4})$? Explain why you cannot use the distance formula for these ordered pairs.
  b. What is the distance between the polar-form points (r_1, θ_1) and (r_2, θ_2)?

62. Use the result of Problem 61b to find an equation for a circle of radius a and polar-form center (R, α).

63. Show that the graph of the polar equation $r = a \sin \theta + b \cos \theta$ is a circle. Find its center and radius.

9.2   Graphing in Polar Coordinates

IN THIS SECTION   graphing by plotting points, cardioids, symmetry and rotations, limaçons, rose curves, lemniscates, summary of polar-form curves ■

GRAPHING BY PLOTTING POINTS

We have examined polar forms for lines and circles, and in this section we shall examine curves that are more easily represented in polar coordinates than in rectangular coordinates. We begin with a simple spiral.

EXAMPLE 1   *A spiral by plotting points*

Graph $r = \theta$ for $\theta \ge 0$.

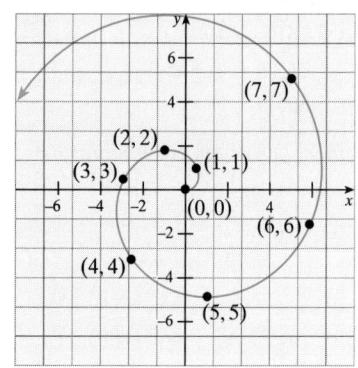

■ **FIGURE 9.4**   Graph of $r = \theta$

Solution   Set up a table of values.

θ	r
0	0
1	1
2	2
3	3
4	4
5	5
6	6

Choose a θ, and then find a corresponding r so that (r, θ) satisfies the equation. Plot each of these points and connect them, as shown in Figure 9.4.*

Notice that as θ increases, r must also increase.   ■

In Example 1, the polar equation $r = \theta$, we see for each value of θ that there is exactly one value of r. Thus, the relationship given by $r = \theta$ is a function of θ, and we can write the polar form $r = f(\theta)$, where $f(\theta) = \theta$. A function of the form $r = f(\theta)$, where θ is a polar angle and r is the corresponding radial distance, is called a **polar function.**

CARDIOIDS

Next, we examine a class of polar curves called **cardioids** because of their heart-like shape.

EXAMPLE 2	*A cardioid by plotting points*

Graph $r = 2(1 - \cos \theta)$.

Solution   Construct a table of values by choosing values for θ and approximating the corresponding values for r.

θ	r
0	0
1	0.9193954
2	2.832294
3	3.979985
4	3.307287
5	1.432676
6	0.079659

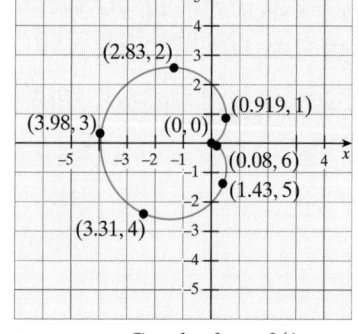

■ **FIGURE 9.5**   Graph of $r = 2(1 - \cos \theta)$ ■

The points are connected as shown in Figure 9.5.

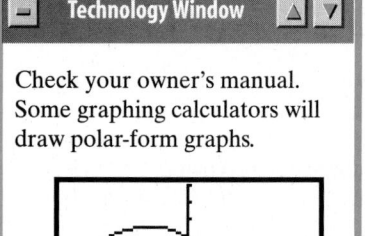

Technology Window

Check your owner's manual. Some graphing calculators will draw polar-form graphs.

```
r₁▣2(1-cos θ)   Xmin=-5
                Xmax=5
                Xscl=1
                Ymin=-5
                Ymax=5
                Yscl=1
```

The general form for a cardioid in standard position is given in the following box.

Standard-Position Cardioid

$$r = a(1 - \cos \theta)$$

*Many books use what is called **polar graph paper,** but such paper is not really necessary. It also obscures the fact that polar curves and rectangular curves are both plotted on a Cartesian coordinate system, only with a different meaning attached to the ordered pairs. You can estimate the angles as necessary without polar graph paper.

In general, a cardioid in standard position can be completely determined by plotting four particular points:

θ	$r = a(1 - \cos \theta)$	Point
0	$r = a(1 - \cos 0) = a(1 - 1) = 0$	$(0, 0)$
$\dfrac{\pi}{2}$	$r = a\left(1 - \cos \dfrac{\pi}{2}\right) = a(1 - 0) = a$	$\left(a, \dfrac{\pi}{2}\right)$
π	$r = a(1 - \cos \pi) = a(1 + 1) = 2a$	$(2a, \pi)$
$\dfrac{3\pi}{2}$	$r = a\left(1 - \cos \dfrac{3\pi}{2}\right) = a(1 - 0) = a$	$\left(a, \dfrac{3\pi}{2}\right)$

These four points are all that you need when graphing other standard-position cardioids, because all cardioids have the same shape as the one shown in Figure 9.5.

> ■ *What Does This Say?* Remember, just as when graphing rectangular curves, the key is not in plotting many points but in recognizing the type of curve and then plotting a few key points.

SYMMETRY AND ROTATIONS

When sketching a polar graph, it is often useful to determine whether the graph has been rotated or if it has any symmetry.

If an angle α is subtracted from θ in a polar-form equation, it has the effect of rotating the curve (see Problem 71).

> **Rotation of Polar-Form Graphs**
>
> The polar graph of $r = f(\theta - \alpha)$ is the same as the polar graph of $r = f(\theta)$ only rotated through an angle α. If α positive, the rotation is counterclockwise, and if α is negative, then the rotation is clockwise.

EXAMPLE 3   *Rotated cardioid*

Graph $r = 3 - 3\cos\left(\theta - \frac{\pi}{6}\right)$.

Solution   Recognize this as a cardioid with $a = 3$ and a rotation of $\frac{\pi}{6}$. Plot the four points shown in Figure 9.6 and draw the cardioid.   ■

If the rotation is 90°, the equation simplifies considerably. Consider

$$r = 3 - 3\cos\left(\theta - \frac{\pi}{2}\right) \quad \textit{Standard cardioid with 90° rotation}$$
$$= 3 - 3\left[\cos\theta\cos\frac{\pi}{2} + \sin\theta\sin\frac{\pi}{2}\right] \quad \begin{array}{l}\cos(\alpha - \beta) = \cos\alpha\\ \cos\beta + \sin\alpha\sin\beta\end{array}$$
$$= 3 - 3[\cos\theta\,(0) + \sin\theta\,(1)]$$
$$= 3 - 3\sin\theta$$

Compare this with Example 3, and you will see that the only difference is a 90° rotation instead of a 30° rotation. This means that the graph of an equation of the form

$$r = a(1 - \sin \theta)$$

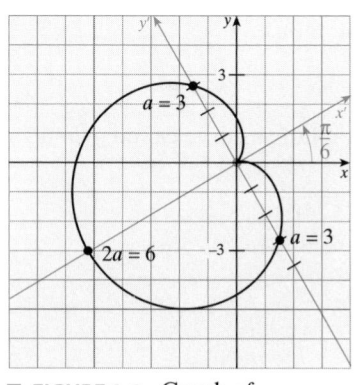

■ **FIGURE 9.6** Graph of $r = 3 - 3\cos\left(\theta - \frac{\pi}{6}\right)$

instead of $r = a(1 - \cos\theta)$, it is a standard-form cardioid with a 90° rotation. Similarly,

$r = a(1 + \cos\theta)$ is a standard-form cardioid with a 180° rotation.

$r = a(1 + \sin\theta)$ is a standard-form cardioid with a 270° rotation.

These curves, along with other polar-form curves, are graphed and summarized in Table 9.1 on page 639.

The cardioid is only one of the polar-form curves that we will consider. Before sketching other curves, we will consider symmetry. There are three important kinds of polar symmetry, which are described in the following box and are demonstrated in Figure 9.7.

Symmetry in the Graph of the Polar Function $r = f(\theta)$

A polar-form graph $r = f(\theta)$ is symmetric with respect to . . .	. . . if the equation $r = f(\theta)$ is unchanged when (r, θ) is replaced by . . .
x-axis	$(r, -\theta)$
y-axis	$(r, \pi - \theta)$
origin	$(-r, \theta)$
or alternatively	
x-axis	$(-r, \pi - \theta)$
y-axis	$(-r, -\theta)$

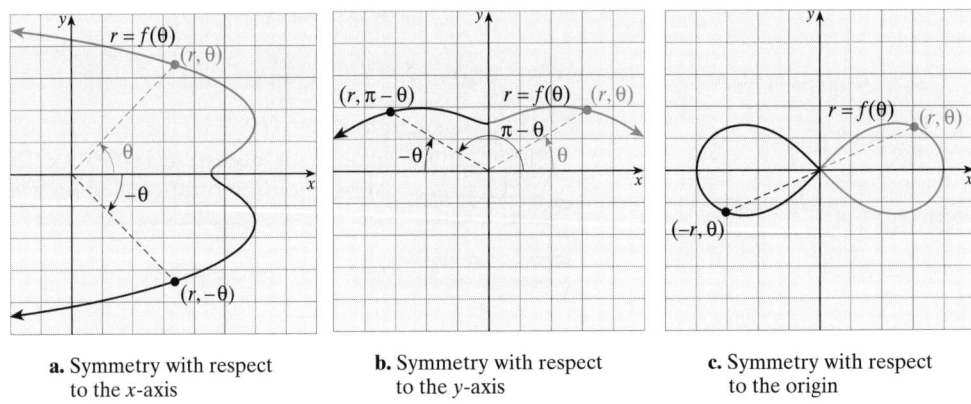

a. Symmetry with respect to the x-axis

b. Symmetry with respect to the y-axis

c. Symmetry with respect to the origin

■ **FIGURE 9.7**   Symmetry in polar form

LIMAÇONS

We will illustrate symmetry in polar-form curves by graphing a curve called a *limaçon*.

EXAMPLE 4   *Graphing a limaçon using symmetry*

Graph $r = 3 + 2\cos\theta$.

Solution   Let $f(\theta) = 3 + 2\cos\theta$.

Symmetry with respect to the x-axis:

$$f(-\theta) = 3 + 2\cos(-\theta) = 3 + 2\cos\theta = f(\theta)$$

Yes; it is symmetric, so it is enough to graph f for θ between 0 and π.

Symmetry with respect to the y-axis:

$$f(\pi - \theta) = 3 + 2\cos(\pi - \theta)$$
$$= 3 + 2\left[\cos \pi \cos \theta + \sin \pi \sin \theta\right]$$
$$= 3 - 2\cos \theta$$

After checking the other primary representation, we find that it is not symmetric with respect to the y-axis.

Symmetry with respect to the origin:

$-r \neq f(\theta)$ and $r \neq f(\theta + \pi)$, so the graph is not symmetric with respect to the origin.

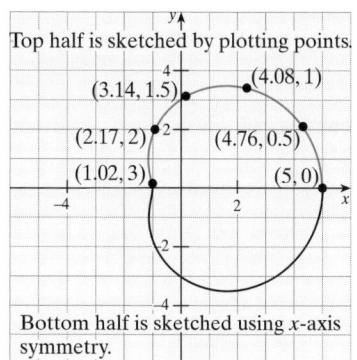

Top half is sketched by plotting points.

Bottom half is sketched using x-axis symmetry.

■ **FIGURE 9.8** Graph of $r = 3 + 2\cos \theta$

The graph is shown in Figure 9.8; note that we sketch the top half of the graph (for $0 \leq \theta \leq \pi$) by plotting points and then complete the sketch by reflecting the graph in the x-axis. Because $\cos \theta$ steadily decreases from its largest value of 1 at $\theta = 0$ to its smallest value -1 at $\theta = \pi$, the radial distance $r = 3 + 2\cos \theta$ will also steadily decrease as θ increases from 0 to π. The largest value of r is $r = 3 + 2(1) = 5$ at $\theta = 0$, and its smallest value is $r = 3 + 2(-1) = 1$ at $\theta = \pi$. ■

The graph of any polar equation of the general form

$$r = b \pm a \cos \theta \quad \text{or} \quad r = b \pm a \sin \theta$$

is called a **limaçon** (derived from the Latin word *limax*, which means "slug"; in French, it is the word for "snail"). The special case where $a = b$ is the *cardioid*. Figure 9.9 shows four different kinds of limaçons that can occur. Note how the appearance of the graph depends on the ratio a/b. We have discussed cases II and III in Examples 2 and 4. Case I (the "inner loop" case) and case IV (the "convex") case are examined in the problem set.

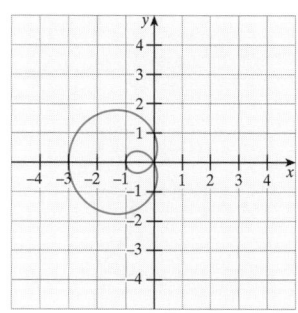 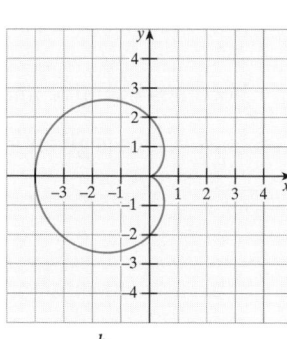 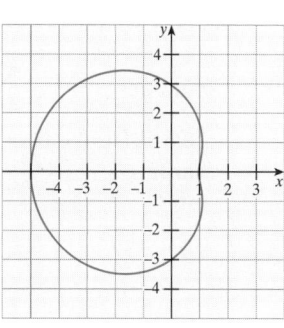 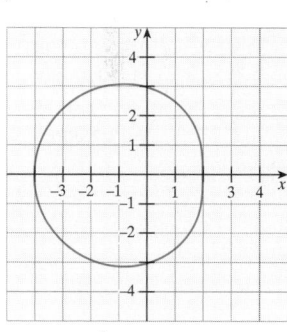

a. $\frac{b}{a} < 1$; inner loop
Case I
($r = 1 - 2\cos \theta$)

b. $\frac{b}{a} = 1$; cardioid
Case II
($r = 2 - 2\cos \theta$)

c. $1 < \frac{b}{a} < 2$; dimple
Case III
($r = 3 - 2\cos \theta$)

d. $\frac{b}{a} \geq 2$; convex
Case IV
($r = 3 - \cos \theta$)

■ **FIGURE 9.9** Limaçons: $r = b \pm a \cos \theta$ or $r = b \pm a \sin \theta$

Just as with the cardioid, we designate a standard-form limaçon and consider the others as rotations:

Standard-Position Limaçon

$$r = b - a \cos \theta$$

$$r = b - a \cos \theta \qquad \textbf{Standard form}$$
$$r = b - a \sin \theta \qquad 90° \text{ rotation}$$
$$r = b + a \cos \theta \qquad 180° \text{ rotation}$$
$$r = b + a \sin \theta \qquad 270° \text{ rotation}$$

ROSE CURVES

There are several polar-form curves known as **rose curves,** which consist of several loops, called **leaves** or **petals.**

> **EXAMPLE 5**   *Graphing a four-leaved rose*

Graph $r = 4 \cos 2\theta$.

Solution   Let $f(\theta) = 4 \cos 2\theta$.

Symmetry with respect to the *x*-axis:

$$f(-\theta) = 4 \cos 2(-\theta) = 4 \cos 2\theta = f(\theta); \text{ Yes, symmetric}$$

Symmetry with respect to the *y*-axis:

$$f(\pi - \theta) = 4 \cos 2(\pi - \theta) = 4 \cos(2\pi - 2\theta) = 4 \cos(-2\theta) = 4 \cos 2\theta$$
$$= f(\theta); \text{ Yes, symmetric}$$

Because of this symmetry, we shall sketch the graph of $r = f(\theta)$ for θ between 0 and $\dfrac{\pi}{2}$, and then use symmetry to complete the graph.

When $\theta = 0, r = 4$, and as θ increases from 0 to $\dfrac{\pi}{4}$, the radial distance r decreases from 4 to 0. Then, as θ increases from $\dfrac{\pi}{4}$ to $\dfrac{\pi}{2}$, r becomes negative and decreases from 0 to -4. A table of values is given in the margin.

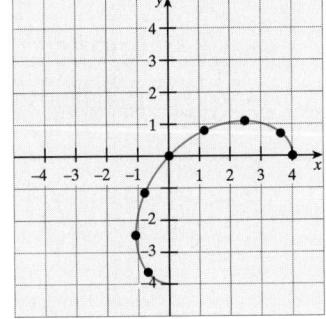

θ	$f(\theta)$
0	4
0.2	3.684244
0.4	2.786827
0.6	1.449431
0.8	−0.1167981
1.0	−1.664587
1.2	−2.949575
1.4	−3.768889

Keep plotting points until you are satisfied that you have a reasonable representation for the graph. After working through Example 6, you will be able to do other rose curves more easily.

The next part of the graph is obtained by first reflecting in the *y*-axis.

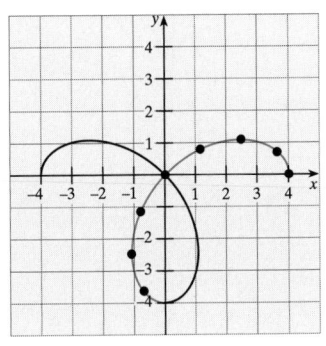

The last part of the graph is found by reflecting in the *x*-axis.

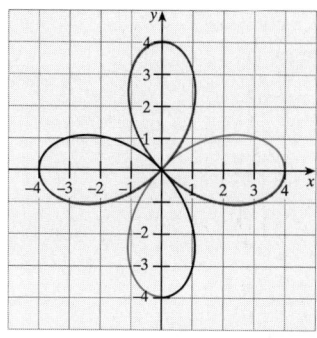

In general, $r = a \cos n\theta$ is the equation of a rose curve in which each petal has length *a*. If *n* is an even number, the rose has 2*n* petals; if *n* is odd, the number of petals is *n*. The tips of the petals are equally spaced on a circle of radius *a*. Equations of the form $r = a \sin n\theta$ are handled as rotations.

Standard-Position Rose Curve

$$r = a \cos n\theta$$

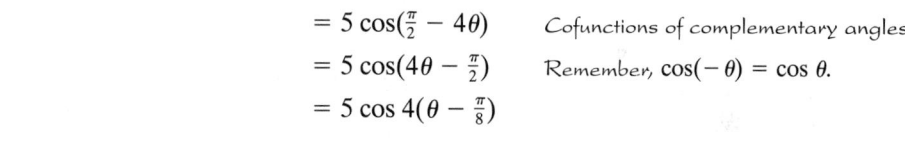

EXAMPLE 6 *Graphing a rose curve using a rotation*

Graph $r = 5 \sin 4\theta$.

Solution We begin by finding the amount of rotation:

$$
\begin{aligned}
r &= 5 \sin 4\theta \\
&= 5 \cos(\tfrac{\pi}{2} - 4\theta) && \text{\textit{Cofunctions of complementary angles}} \\
&= 5 \cos(4\theta - \tfrac{\pi}{2}) && \text{\textit{Remember, } } \cos(-\theta) = \cos\theta. \\
&= 5 \cos 4(\theta - \tfrac{\pi}{8})
\end{aligned}
$$

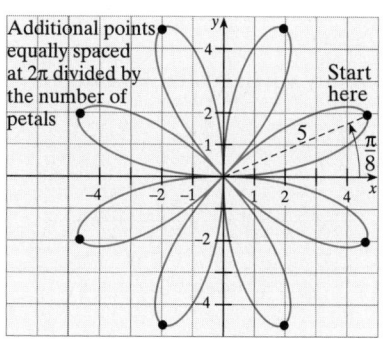

■ **FIGURE 9.10** Graph of $r = 5 \sin 4\theta$

Recognize this as a rose curve rotated $\frac{\pi}{8}$. There are 2(4) = 8 petals of length 5. The petals are distance of $\frac{\pi}{4}$ (one revolution = 2π divided by the number of petals) apart. The graph is shown in Figure 9.10. ■

The graph of the eight-leaved rose in Example 6 provides a good illustration of why the alternative criteria for symmetry are given. It is obvious from the figure that the curve is symmetric with respect to the *x*-axis, but if $f(\theta) = 5 \sin 4\theta$, we have

$$f(-\theta) = 5 \sin 4(-\theta) = -5 \sin 4\theta \neq f(\theta)$$

which means the primary criterion for symmetry with respect to the *x*-axis fails. However,

$$
\begin{aligned}
f(\pi - \theta) &= 5 \sin 4(\pi - \theta) \\
&= 5[\sin 4\pi \cos(-4\theta) + \cos 4\pi \sin(-4\theta)] \\
&= -5 \sin 4\theta \\
&= -f(\theta)
\end{aligned}
$$

so when (r, θ) is on the graph, so is $(-r, \pi - \theta)$. Thus, the alternative criterion confirms that the graph is indeed symmetric with respect to the *x*-axis.

LEMNISCATES

The last general type of polar-form curve we will consider is called a **lemniscate.**

> ### Standard-Position Lemniscate
>
> $$r^2 = a^2 \cos 2\theta$$

EXAMPLE 7   *Graphing a lemniscate*

Graph $r^2 = 9 \cos 2\theta$.

Solution   As before, when graphing a curve for the first time, begin by checking symmetry and plotting points. For this example, note that you obtain two value for r when solving this quadratic equation. For example, if $\theta = 0$, then $\cos 2\theta = 1$ and $r^2 = 9$, so $r = 3$ or -3.

Symmetry with respect to the *x*-axis:
   $9 \cos[2(-\theta)] = 9 \cos 2\theta$, so $r^2 = 9 \cos 2\theta$ is not affected when θ is replaced by $-\theta$; yes, symmetric with respect to the *x*-axis.

Symmetry with respect to the *y*-axis:
   $9 \cos[2(\pi - \theta)] = 9 \cos(2\pi - 2\theta) = 9 \cos(-2\theta) = 9 \cos 2\theta$;
   yes, symmetric with respect to the *y*-axis.

Symmetry with respect to the origin:
   $(-r)^2 = r^2$ so $r^2 = 9 \cos 2\theta$ is not affected when r is replaced by $-r$;
   yes, symmetric with respect to the origin.

Note that because $r^2 \geq 0$, the equation has a solution only when $\cos 2\theta \geq 0$; that is

$$-\frac{\pi}{4} \leq \theta \leq \frac{\pi}{4}; \quad \frac{3\pi}{4} \leq \theta \leq \frac{5\pi}{4}; \quad \cdots$$

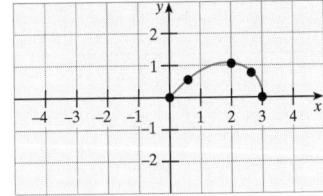

We begin by restricting our attention to the interval $0 \leq \theta \leq \frac{\pi}{4}$. Note that $\sqrt{9 \cos 2\theta}$ decreases steadily from 3 to 0 as θ varies from 0 to $\frac{\pi}{4}$.

A second step is to use symmetry to reflect the curve in the *x*-axis.

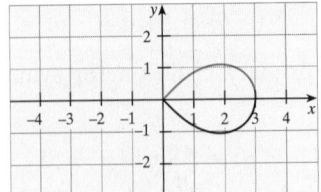

Finally, obtain the rest of the graph by reflecting the curve in the *y*-axis.

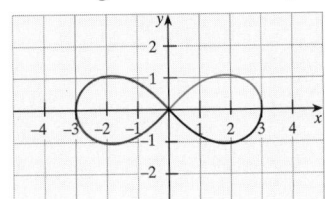

SUMMARY OF POLAR-FORM CURVES

We conclude this section by summarizing in Table 9.1 the special types of polar-form curves we have examined. There are many others, some of which are represented in the problems. Others may be found in Chapter 5 of the *Student Mathematics Handbook*.

SMH

■ **TABLE 9.1   Directory of Polar-Form Curves**

Limaçons $r = b \pm a \cos \theta$ or $r = b \pm a \sin \theta$

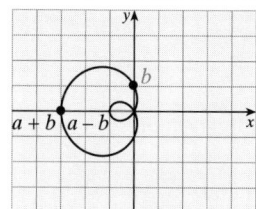

$r = b - a \cos \theta, \frac{b}{a} < 1$
standard form
with inner loop

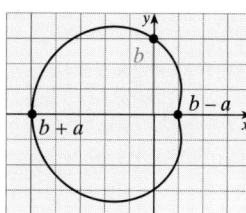

$r = b - a \cos \theta, 1 < \frac{b}{a} < 2$
standard form
with a dimple

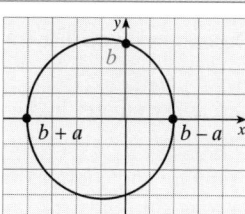

$r = b - a \cos \theta, \frac{b}{a} \geq 2$
standard form,
convex

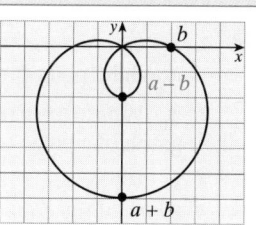

$r = b - a \sin \theta, \frac{b}{a} < 1$
$\frac{\pi}{2}$ rotation
with inner loop

Cardioids $r = a(1 \pm \cos \theta)$ or $r = a(1 \pm \sin \theta)$
             Limaçons in which $a = b$

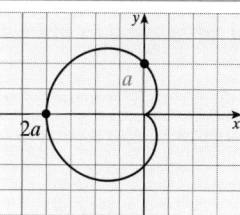

$r = a - a \cos \theta$
standard form

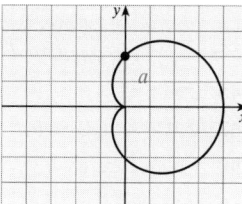

$r = a + a \cos \theta$
π rotation

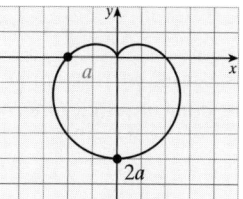

$r = a - a \sin \theta$
$\frac{\pi}{2}$ rotation

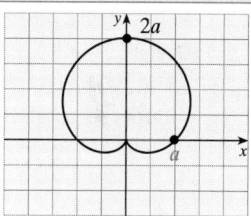

$r = a + a \sin \theta$
$\frac{3\pi}{2}$ rotation

Rose Curves
$r = a \cos n\theta$ or $r = a \sin n\theta$
If n is odd, the rose has n petals; If n is even it has $2n$ petals.

Lemniscates
$r^2 = a^2 \cos 2\theta$ or $r^2 = a^2 \sin 2\theta$
Two loops

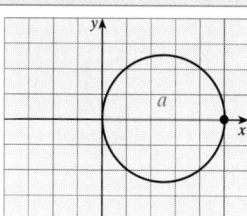

$r = a \cos \theta$
standard form
one (circular) petal

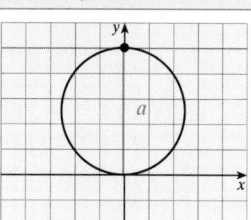

$r = a \sin \theta$
$\frac{\pi}{2}$ rotation,
one (circular) petal

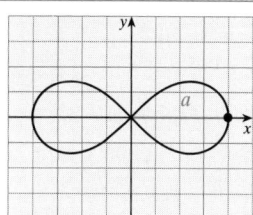

$r^2 = a^2 \cos 2\theta$
standard form

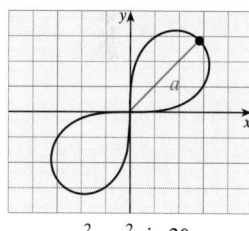

$r^2 = a^2 \sin 2\theta$
$\frac{\pi}{4}$ rotation

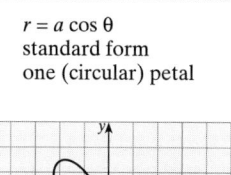

$r = a \cos 3\theta$
standard form,
three petals

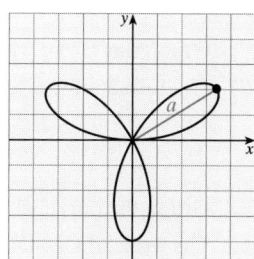

$r = a \sin 3\theta$
$\frac{\pi}{6}$ rotation,
three petals

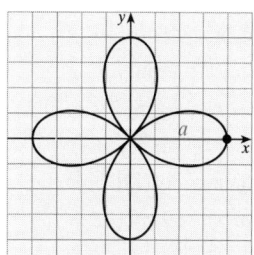

$r = a \cos 2\theta$
standard form,
four petals

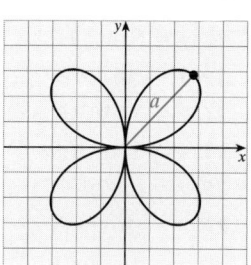

$r = a \sin 2\theta$
$\frac{\pi}{4}$ rotation,
four petals

9.2 Problem Set

 1. ■ **What Does This Say?** Describe a procedure for graphing polar-form curves.

2. ■ **What Does This Say?** Discuss symmetry in the graph of a polar-form function.

3. ■ **What Does This Say?** Compare and contrast the forms of the equation for limaçons, cardioids, rose curves, and lemniscates.

4. Identify each of the curves as a cardioid, rose curve (state number of petals), lemniscate, limaçon, circle, line, or none of the above.
   a. $r^2 = 9 \cos 2\theta$ b. $r = 2 \sin \frac{\pi}{6}$
   c. $r = 3 \sin 3\theta$ d. $r = 3\theta$
   e. $r = 2 - 2 \cos\theta$ f. $\theta = \frac{\pi}{6}$
   g. $r^2 = \sin 2\theta$ h. $r - 2 = 4 \cos \theta$

5. Identify each of the curves as a cardioid, rose curve (state number of petals), lemniscate, limaçon, circle, line, or none of the above.
   a. $r = 2 \sin 2\theta$ b. $r^2 = 2 \cos 2\theta$
   c. $r = 5 \cos 60°$ d. $r = 5 \sin 8\theta$
   e. $r\theta = 3$ f. $r^2 = 9 \cos(2\theta - \frac{\pi}{4})$
   g. $r = \sin 3(\theta + \frac{\pi}{6})$ h. $\cos \theta = 1 - r$

6. Identify each of the curves as a cardioid, rose curve (state number of petals), lemniscate, limaçon, circle, line, or none of the above.
   a. $r = 2 \cos 2\theta$ b. $r = 4 \sin 30°$
   c. $r + 2 = 3 \sin \theta$ d. $r + 3 = 3 \sin \theta$
   e. $\theta = 4$ f. $\theta = \tan \frac{\pi}{4}$
   g. $r = 3 \cos 5\theta$ h. $r \cos \theta = 2$

Graph the polar-form curves given in Problems 7–38.

7. $r = 3, 0 \le \theta \le \frac{\pi}{2}$ **8.** $r = -1, 0 \le \theta \le \pi$

9. $\theta = -\frac{\pi}{2}, 0 \le r \le 3$ **10.** $\theta = \frac{\pi}{4}, 1 \le r \le 2$

11. $r = \theta + 1, 0 \le \theta \le \pi$ **12.** $r = \theta - 1, 0 \le \theta \le \pi$

13. $r = 2\theta, \theta \ge 0$ **14.** $r = \frac{\theta}{2}, \theta \ge 0$

15. $r = 2 \cos 2\theta$ **16.** $r = 3 \cos 3\theta$

17. $r = 5 \sin 3\theta$ **18.** $r^2 = 9 \cos 2\theta$

19. $r^2 = 16 \cos 2\theta$ **20.** $r^2 = 9 \sin 2\theta$

21. $r = 3 \cos 3(\theta - \frac{\pi}{3})$ **22.** $r = 2 \cos 2(\theta + \frac{\pi}{3})$

23. $r = 5 \cos 3(\theta - \frac{\pi}{4})$ **24.** $r = \sin 3(\theta + \frac{\pi}{6})$

25. $r = \sin(2\theta + \frac{\pi}{3})$ **26.** $r = \cos(2\theta + \frac{\pi}{3})$

27. $r^2 = 16 \cos 2(\theta - \frac{\pi}{6})$ **28.** $r^2 = 9 \cos(2\theta - \frac{2\pi}{3})$

29. $r = 2 + \cos \theta$ **30.** $r = 3 + \sin \theta$

31. $r = 1 + \sin \theta$ **32.** $r = 1 + \cos \theta$

33. $r \cos \theta = 2$ **34.** $r \sin \theta = 3$

35. $r = 1 + 3 \cos \theta$ **36.** $r = 1 + 2 \sin \theta$

37. $r = -2 \sin \theta$ **38.** $r^2 = -\cos 2\theta$

Sketch the graph of the polar function given in Problems 39–44.

39. $f(\theta) = \sin 2\theta, 0 \le \theta \le \frac{\pi}{2}$; rose petal

40. $f(\theta) = |\sin 2\theta|, 0 \le \theta \le 2\pi$; four leaved rose

41. $f(\theta) = 2|\cos \theta|, 0 \le \theta \le 2\pi$

42. $f(\theta) = 4|\sin \theta|, 0 \le \theta \le 2\pi$

43. $f(\theta) = \sqrt{|\cos \theta|}, 0 \le \theta \le 2\pi$; lazy eight

44. $f(\theta) = \sqrt{\cos 2\theta}, 0 \le \theta \le 2\pi$; lemniscate

Graph the set of points (r, θ) so that the inequalities in Problems 45–54 are satisfied.

45. $0 \le r \le 1, 0 \le \theta < 2\pi$ **46.** $2 \le r \le 3, 0 \le \theta < 2\pi$

47. $0 \le r < 4, 0 \le \theta \le \frac{\pi}{2}$ **48.** $0 \le r \le 4, 0 \le \theta \le \pi$

49. $r > 1, 0 \le \theta < 2\pi$ **50.** $r \ge 2, \frac{\pi}{2} \le \theta \le \pi$

51. $0 \le \theta \le \frac{\pi}{4}, r \ge 0$ **52.** $-\pi < \theta < 0, r \ge 0$

53. $0 \le \theta \le \frac{\pi}{4}, 1 \le r \le 2$ **54.** $0 \le \theta \le \frac{\pi}{4}, r \ge 1$

55. Show that the polar equations
$$r = \cos \theta + 1 \quad \text{and} \quad r = \cos \theta - 1$$
have the same graph in the *xy*-plane.

56. **Spirals** are interesting mathematical curves. There are three special types of spirals:
   a. A **spiral of Archimedes** has the form $r = a\theta$; graph $r = 2\theta$.
   b. A **hyperbolic spiral** has the form $r\theta = a$; graph $r\theta = 2$.
   c. A **logarithmic spiral** has the form $r = a^{k\theta}$; graph $r = 2^\theta$.

57. The **strophoid** is a curve of the form
$$r = a \cos 2\theta \sec \theta$$
Graph this curve where $a = 2$.

58. The **bifolium** has the form
$$r = a \sin \theta \cos^2\theta$$
Graph this curve where $a = 1$.

59. The **folium of Descartes** has the form
$$r = \frac{3a \sin \theta \cos \theta}{\sin^3\theta + \cos^3\theta}$$
Graph the curve where $a = 2$.

60. The **ovals of Cassini** have the form
$$r^4 + b^4 - 2b^2r^2 \cos 2\theta = k^4$$
Graph the curve where $b = 2, k = 3$.

In Problems 61–66, graph the given pair of curves on the same coordinate axes. The first equation uses (x, y) as rectangular coordinates and the second uses (r, θ) as polar coordinates.

61. $y = \cos x$ and $r = \cos \theta$
62. $y = \sin x$ and $r = \sin \theta$
63. $y = \tan x$ and $r = \tan \theta$
64. $y = \sec x$ and $r = \sec \theta$
65. $y = \csc x$ and $r = \csc \theta$
66. $y = \cot x$ and $r = \cot \theta$

Technology Window

Exploring the periodicity of polar plots   *It is surprisingly difficult to predict the periodic behavior of polar plots. For example, here you are to investigate the behavior of the graphs of the form*

$$r = \sin m\theta$$

for m a positive integer.

67. Consider $r = \sin 2\theta$.
   a. Since $\sin 2\theta$ has period π, one would think that surely the graph repeats itself for $\theta > \pi$. Graphically show that this is *not* the case. Verify that $0 \le \theta \le 2\pi$ is necessary to obtain the entire graph. Describe the graph.
   b. Now systematically study $r = \sin m\theta$ for m a positive integer. In particular, describe *why* m being even or odd makes a fundamental difference. Summarize the number of leaves on the roses relative to m.

Technology Window

68. Continuing the theme of Problem 67, consider

$$r = \sin\left(\frac{m\theta}{n}\right)$$

where m and n are positive integers (and relatively prime). Let P denote the period of the sine.
   a. First set $m = 1$ and study the behavior of the graphs for $n = 1, 2, 3, \ldots$. Clearly, the period of the sine function is $P = 2n\pi$. But what θ interval is required to obtain the entire graph? Attempt to explain this. Hand in your favorite graph.
   b. Now do a study for $m = 1, 2, \ldots$ and $n = 1, 2, \ldots$, and attempt to generalize the situation regarding the period of the sine function, the necessary θ interval, and the number of loops in the graphs. Hand in your favorite graph.
   c. **A challenge** Carefully explain the necessary θ interval in part **b.**

69. Show that the curve $r = f(\theta)$ is symmetric in the line $y = x$ if the equation is unaffected when θ is replaced by $\frac{\pi}{2} - \theta$.
70. a. Show that if the polar curve $r = f(\theta)$ is rotated about the pole through an angle α, the equation for the new curve is $r = f(\theta - \alpha)$.
   b. Use a rotation to sketch
$$r = 2\sec(\theta - \tfrac{\pi}{3})$$
71. Sketch the graph of
$$r = \frac{\theta}{\cos\theta} \quad \text{for } 0 \le \theta < \tfrac{\pi}{2}$$
In particular, show that the graph has a vertical asymptote at $x = \frac{\pi}{2}$.

9.3 Area and Tangent Lines in Polar Coordinates

IN THIS SECTION   intersections of polar-form curves, area bounded by polar graphs, tangent lines

INTERSECTION OF POLAR-FORM CURVES

To find the points of intersection of graphs in rectangular form, you need only find the simultaneous solution of the equations that define those graphs. It is not even necessary to draw the graphs, because there is a one-to-one correspondence between ordered pairs satisfying an equation and points on its graph. However, in polar form, this one-to-one property is lost, so that without drawing the graphs you may fail to find all points of intersection. For this reason, our method for finding the intersection of polar-form curves will include sketching the graphs.

| EXAMPLE 1 | *Finding the intersection of two polar-form curves* |

Find the points of intersection of the circles $r = 2 \cos \theta$ and $r = 2 \sin \theta$ for $0 \le \theta < \pi$.

Solution   First, consider the simultaneous solution of the system of equations.

$$2 \cos \theta = 2 \sin \theta$$

$$1 = \frac{2 \sin \theta}{2 \cos \theta}$$

$$1 = \tan \theta$$

$$\theta = \frac{\pi}{4}, \frac{5\pi}{4}$$

Then find r using either of the two given equations:

$$r = 2 \cos \frac{\pi}{4} = \sqrt{2} \quad \text{and} \quad r = 2 \cos \frac{5\pi}{4} = -\sqrt{2}$$

This gives the point $(\sqrt{2}, \frac{\pi}{4})$. The simultaneous solution yields one point of intersection.

Next, consider the graphs of these circles, shown in Figure 9.11. It looks as though $(0, 0)$ is also a point of intersection. Check this point in each of the given equations:

If $r = 0$ in $r = 2 \cos \theta$, then $\theta = \frac{\pi}{2}$.

If $r = 0$ in $r = 2 \sin \theta$, then $\theta = \pi$.

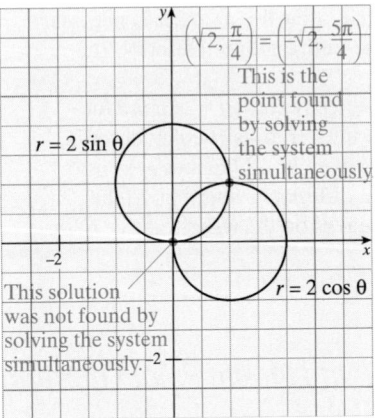

■ **FIGURE 9.11**   Graphs of $r = 2 \cos \theta$ and $r = 2 \sin \theta$

At first it does not seem that the pole satisfies the equation, because $r = 0$ gives $(0, \frac{\pi}{2})$ and $(0, \pi)$, respectively. Notice that these coordinates are different and do not satisfy the equations simultaneously. But if you plot the point with the coordinates, you will see that $(0, \frac{\pi}{2})$ $(0, \pi)$, and $(0, 0)$ are all the same point. It's as if two ants were crawling along the curves and came to the origin at different times—they would pass each other without colliding.   ■

The pole is often a solution for a system of polar equations, even though it may not satisfy the equations simultaneously. This is because when $r = 0$, all values of θ will yield the same point—namely, the pole. For this reason, it is necessary to check separately to see whether the pole lies on the given graph.

Graphical Solution of the Intersection of Polar Curves

Step 1. Find all simultaneous solutions of the given system of equations.

Step 2. Determine whether the pole lies on the two graphs.

Step 3. Graph the curves to look for other points of intersection.

EXAMPLE 2   *Intersection of polar-form curves*

Find the points of intersection of the curves $r = \frac{3}{2} - \cos\theta$ and $\theta = \frac{2\pi}{3}$.

Solution

Step 1. Solve the system by substitution:
$$r = \tfrac{3}{2} - \cos\tfrac{2\pi}{3} = \tfrac{3}{2} - (-\tfrac{1}{2}) = 2$$
The solution is $(2, \frac{2\pi}{3})$.

Step 2. If $r = 0$, the first equation has no solution because
$$0 = \tfrac{3}{2} - \cos\theta \quad \text{or} \quad \cos\theta = \tfrac{3}{2}$$
and a cosine cannot be larger than 1.

Step 3. Now look at the graphs, as shown in Figure 9.12.
    From the graph we see that $(-1, \frac{2\pi}{3})$ may also be a point of intersection. It satisfies the equation $\theta = \frac{2\pi}{3}$, but what about $r = \frac{3}{2}\cos\theta$? Check $(-1, \frac{2\pi}{3})$:
$$-1 \overset{?}{=} \tfrac{3}{2} - \cos(\tfrac{2\pi}{3})$$
$$= \tfrac{3}{2} - (-\tfrac{1}{2})$$
$$= 2 \qquad \textit{Not satisfied}$$

However, if you check the alternative representation of $(-1, \frac{2\pi}{3})$, namely $(1, \frac{5\pi}{3})$:
$$1 \overset{?}{=} \tfrac{3}{2} - \cos(\tfrac{5\pi}{3})$$
$$= \tfrac{3}{2} - \tfrac{1}{2}$$
$$= 1 \qquad \textit{Satisfied}$$

Be sure to check for points of intersection that you may have missed.   ∎

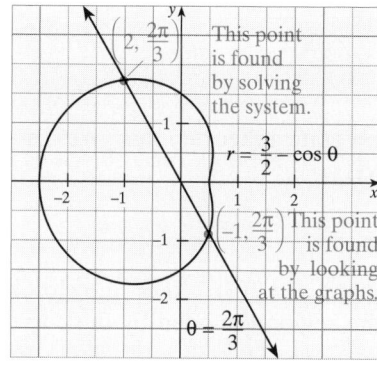

FIGURE 9.12 Graphs of $r = \frac{3}{2} - \cos\theta$ and $\theta = \frac{2\pi}{3}$.

AREA BOUNDED BY POLAR GRAPHS

To find the area of a region bounded by a polar graph, we use Riemann sums in much the same way as when we developed the integral formula for the area of a region described in rectangular form. However, instead of using rectangles as the basic unit being summed in polar form, we sum the area of *sectors* of a circle. Recall from trigonometry the formula for the area of such a sector.

Area of a Sector

The area of a circular sector of radius r is given by
$$A = \tfrac{1}{2}r^2\theta$$
where θ is the central angle of the sector measured in radians.

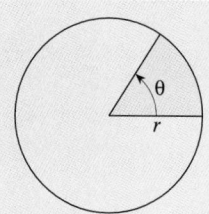

This formula gives the following areas:

Angle	Area	Figure
2π	πr^2	
1	$\dfrac{1}{2}r^2(1) = \dfrac{r^2}{2}$	
α	$\dfrac{1}{2}r^2(\alpha) = \dfrac{\alpha r^2}{2}$	

Using this formula for the area of a sector, we now state a theorem that gives us a formula for finding the area enclosed by a polar curve.

THEOREM 9.1   *Area in polar coordinates*

Let $r = f(\theta)$ define a polar curve, where f is continuous and $f(\theta) \geq 0$ on the closed interval $\alpha \leq \theta \leq \beta$, where $0 \leq \beta - \alpha \leq 2\pi$. Then the region bounded by the curve $r = f(\theta)$ and the rays $\theta = \alpha$ and $\theta = \beta$ has area

$$A = \frac{1}{2}\int_{\alpha}^{\beta} r^2\, d\theta = \frac{1}{2}\int_{\alpha}^{\beta} [f(\theta)]^2\, d\theta$$

Proof   The region is shown in Figure 9.13. To find the area bounded by the graphs of the polar functions, partition the region between $\theta = \alpha$ and $\theta = \beta$ by a collection of rays, say $\theta_0, \theta_1, \ldots, \theta_n$. Let $\alpha = \theta_0$ and $\beta = \theta_n$.

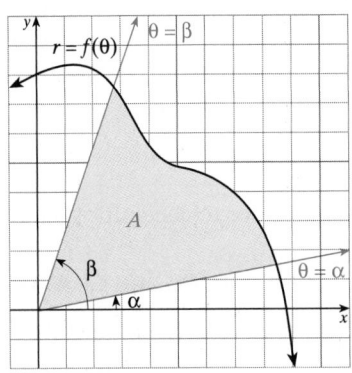

a. The region bounded by the polar curve $r = f(\theta)$ and the rays $\theta = a$ and $\theta = b$

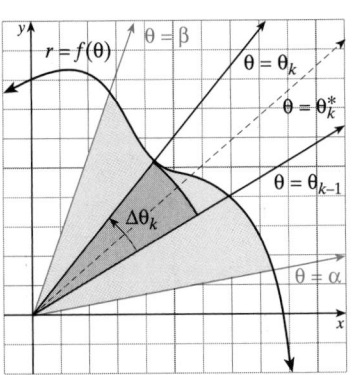

b. The area A can be estimated by adding the area of "small" (gray) circular sectors.

■ **FIGURE 9.13**   Area in polar form

Pick any ray $\theta = \theta_k^*$, with $\theta_{k-1} \leq \theta_k^* \leq \theta_k$. Then the area ΔA_k of the circular sector is approximately the same as the area of the region bounded by the graph of f and the lines $\theta = \theta_{k-1}$ and $\theta = \theta_k$. Because this circular sector has radius $f(\theta_k^*)$ and central angle $\Delta\theta_k$, its area is

$$\Delta A_k \approx \tfrac{1}{2}(\text{radius})^2(\text{central angle}) = \tfrac{1}{2}[f(\theta_k^*)]^2\Delta\theta_k$$

The sum $\sum_{k=1}^{n} \Delta A_k$ is an approximation to the total area A bounded by the polar curve, and by taking the limit as $n \to \infty$, we obtain

$$A = \lim_{n\to\infty} \sum_{k=1}^{n} \Delta A_k = \lim_{n\to\infty} \frac{1}{2} \sum_{k=1}^{n} [f(\theta_k^*)]^2 \Delta\theta_k \qquad \textit{Riemann sum}$$

$$= \frac{1}{2} \int_{\alpha}^{\beta} [f(\theta)]^2 \, d\theta \qquad\qquad \blacksquare$$

You will probably find that the most difficult part of the problem is deciding on the limits of integration. A decent sketch of the region will help with this.

EXAMPLE 3   *Finding area enclosed by part of a cardioid*

Find the area under the top half ($0 \le \theta \le \pi$) of the cardioid $r = 1 + \cos\theta$.

Solution   The cardioid is shown in Figure 9.14. Note that the top half of the graph lies between the rays $\theta = 0$ and $\theta = \pi$. Hence the required area is given by

$$A = \frac{1}{2} \int_0^{\pi} (1 + \cos\theta)^2 \, d\theta = \frac{1}{2} \int_0^{\pi} (1 + 2\cos\theta + \cos^2\theta) \, d\theta$$

$$= \frac{1}{2}\left[\theta + 2\sin\theta + \frac{\theta}{2} + \frac{\sin 2\theta}{4} \right]\Big|_0^{\pi} \qquad \textit{Integration table, formula 317}$$

$$= \frac{1}{2}\left[\pi + 2(0) + \frac{\pi}{2} + \frac{0}{4} - 0 \right] = \frac{3\pi}{4} \qquad\qquad \blacksquare$$

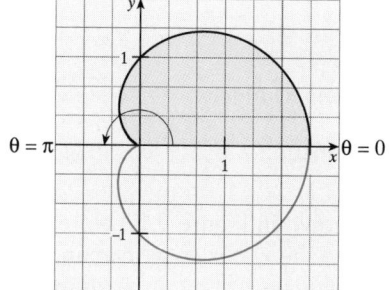

■ **FIGURE 9.14**   Area under the top half of the cardioid $r = 1 + \cos\theta$

EXAMPLE 4   *Finding area enclosed by a four-leaved rose*

Find the area enclosed by the four-leaved rose $r = \cos 2\theta$.

Solution   The rose curve is shown in Figure 9.15. We will find the area of the top half of the right loop (shaded portion) to make sure $f(\theta) \ge 0$. We see that this corresponds to angles with measures from $\theta = 0$ to $\theta = \pi/4$. By symmetry, the entire area enclosed by the four-leaved rose is 8 times the shaded region. Thus, the required area is given by

$$A = 8\left[\frac{1}{2} \int_0^{\pi/4} \cos^2 2\theta \, d\theta \right]$$

$$= 4\left[\frac{\theta}{2} + \frac{\sin 4\theta}{8} \right]\Big|_0^{\pi/4} \qquad \textit{Integration table, formula 317}$$

$$= 4\left[\frac{\pi}{8} + 0 - 0 \right]$$

$$= \frac{\pi}{2} \qquad\qquad \blacksquare$$

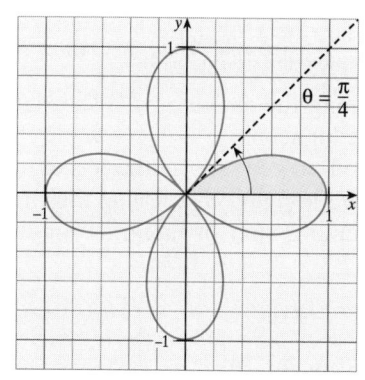

■ **FIGURE 9.15**   Area enclosed by the four-leaved rose $r = \cos 2\theta$

EXAMPLE 5   *Finding the area of a region defined by two polar curves*

Find the area of the region common to the circles $r = a\cos\theta$ and $r = a\sin\theta$.

Solution   The circles are shown in Figure 9.16. Solve $a\cos\theta = a\sin\theta$ to find that they intersect at $\theta = \pi/4$ and at the pole. To set up the integrals properly, remember to think in terms of polar coordinates and not in terms of rectangular coordinates.

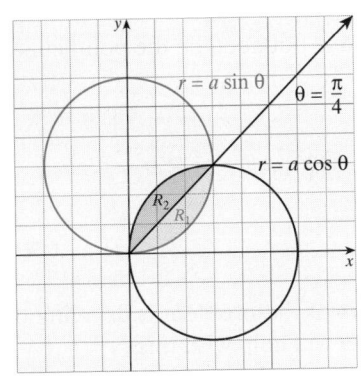

■ **FIGURE 9.16**   Area enclosed by $r = a \cos \theta$ and $r = a \sin \theta$

Specifically, we must **scan radially.** This means that we need to find the area of intersection by finding the sum of the areas of the regions marked R_1 and R_2. Region R_1 is bounded by the circle $r = a \sin \theta$ and the rays $\theta = 0$, $\theta = \frac{\pi}{4}$. Region R_2 is bounded by the circle $r = a \cos \theta$ and $\theta = \frac{\pi}{4}$, $\theta = \frac{\pi}{2}$. Note that the rays $\theta = 0$ and $\theta = \frac{\pi}{2}$ are not necessary as geometric boundaries for R_1 and R_2, but they are necessary to describe which part of each circle is being calculated.

$$
\begin{aligned}
A &= \text{AREA OF } R_1 + \text{AREA OF } R_2 \\
&= \frac{1}{2} \int_0^{\pi/4} a^2 \sin^2 \theta \, d\theta + \frac{1}{2} \int_{\pi/4}^{\pi/2} a^2 \cos^2 \theta \, d\theta \\
&= \frac{a^2}{2}\left[\frac{\theta}{2} - \frac{\sin 2\theta}{4} \right]\Big|_0^{\pi/4} + \frac{a^2}{2}\left[\frac{\theta}{2} + \frac{\sin 2\theta}{4} \right]\Big|_{\pi/4}^{\pi/2} \quad \text{\textit{Integration table, formulas 348 and 317}} \\
&= \frac{a^2}{2}\left[\frac{\pi}{8} - \frac{1}{4} - 0 \right] + \frac{a^2}{2}\left[\frac{\pi}{4} + 0 - \frac{\pi}{8} - \frac{1}{4} \right] = \frac{a^2}{2}\left[\frac{\pi}{4} - \frac{1}{2} \right] \\
&= \tfrac{1}{8} a^2 (\pi - 2)
\end{aligned}
$$

■

EXAMPLE 6   *Finding the area between a circle and a limaçon*

Find the area between the circle $r = 5 \cos \theta$ and the limaçon $r = 2 + \cos \theta$. Round your answer to the nearest hundredth of a square unit.

Solution   As usual, begin by drawing the graphs, as shown in Figure 9.17. Note that both the limaçon and circle are symmetric with respect to the x-axis, so we can find the area in the first quadrant and multiply by 2.

Next, we need to find the points of intersection. We see that the curves do not intersect at the pole. Now solve

$$
\begin{aligned}
5 \cos \theta &= 2 + \cos \theta \\
\cos \theta &= \tfrac{1}{2} \\
\theta &= \tfrac{\pi}{3} \quad \text{\textit{Solution in the first quadrant}}
\end{aligned}
$$

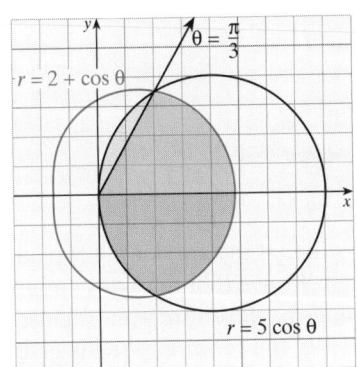

■ **FIGURE 9.17**   Area between $r = 5 \cos \theta$ and $r = 2 + \cos \theta$

To find the area, divide the region into two parts, along the ray $\theta = \pi/3$. The right part (gray screen above the x-axis) is bounded by the limaçon $r = 2 + \cos \theta$ and the rays $\theta = 0$ and $\theta = \pi/3$. The left part (blue-shaded screen above the x-axis) is bounded by the circle $r = 5 \cos \theta$ and the rays $\theta = \pi/3$ and $\theta = \pi/2$. Using this preliminary information, we can now set up and evaluate a sum of integrals for the required area:

$$
\begin{aligned}
A &= 2[\text{AREA OF RIGHT PART} + \text{AREA OF LEFT PART}] \\
&= 2\left[\frac{1}{2} \int_0^{\pi/3} (2 + \cos \theta)^2 \, d\theta + \frac{1}{2} \int_{\pi/3}^{\pi/2} (5 \cos \theta)^2 \, d\theta \right] \\
&= \int_0^{\pi/3} (4 + 4 \cos \theta + \cos^2 \theta) \, d\theta + \int_{\pi/3}^{\pi/2} 25 \cos^2 \theta \, d\theta \\
&= \left[40 + 4 \sin \theta + \frac{\theta}{2} + \frac{\sin 2\theta}{4} \right]\Big|_0^{\pi/3} + 25\left[\frac{\theta}{2} + \frac{\sin 2\theta}{4} \right]\Big|_{\pi/3}^{\pi/2} \\
&= \left[\frac{4\pi}{3} + \frac{4\sqrt{3}}{2} + \frac{\pi}{6} + \frac{\sqrt{3}}{8} - 0 \right] + 25\left[\frac{\pi}{4} + 0 - \frac{\pi}{6} - \frac{\sqrt{3}}{8} \right] \\
&= \frac{43\pi}{12} - \sqrt{3} \approx 9.53
\end{aligned}
$$

■

TANGENT LINES

The slope of $r = f(\theta)$ is NOT $f'(\theta)$.

In rectangular coordinates, the slope of the tangent line to a graph of $y = g(x)$ at $x = x_0$ is $g'(x_0)$, and we might expect the slope of the tangent line to the graph of $r = f(\theta)$ at $\theta = \theta_0$ to be $f'(\theta_0)$. A simple counterexample shows that this is false. The graph of the polar function $r = c$ ($c > 0$, a constant) is a circle centered at the origin. Because f is a constant, we have $f'(\theta) = 0$ for all θ, so that if $f'(\theta)$ were the slope of the tangent line to this circle, every tangent line would be horizontal, which is clearly false. It turns out that the slope of the tangent line to a polar curve is given by the following more complicated formula.

THEOREM 9.2   *Slope of a polar curve*

If f is a differentiable function of θ, then the tangent line to the polar curve $r = f(\theta)$ at the point $P(r_0, \theta_0)$ has slope

$$m = \frac{f(\theta_0)\cos \theta_0 + f'(\theta_0)\sin \theta_0}{-f(\theta_0)\sin \theta_0 + f'(\theta_0)\cos \theta_0}$$

whenever the denominator is not zero.

Proof   Because $r = f(\theta)$ and $x = r\cos \theta$, $y = r\sin \theta$, we have

$$x = f(\theta)\cos \theta \quad \text{and} \quad y = f(\theta)\sin\theta$$

Using the chain rule, we find the $\dfrac{dy}{d\theta} = \left(\dfrac{dy}{dx}\right)\left(\dfrac{dx}{d\theta}\right)$ or, equivalently,

$$\frac{dy}{dx} = \frac{\dfrac{dy}{d\theta}}{\dfrac{dx}{d\theta}}$$

Because $x = f(\theta)\cos \theta$ and $y = f(\theta)\sin \theta$, it follows that

$$\frac{dx}{d\theta} = f(\theta)\frac{d}{d\theta}(\cos \theta) + \cos \theta\frac{df}{d\theta} \qquad \frac{dy}{d\theta} = f(\theta)\frac{d}{d\theta}(\sin \theta) + \sin \theta\frac{df}{d\theta}$$

$$= -f(\theta)\sin \theta + f'(\theta)\cos \theta \qquad\qquad = f(\theta)\cos \theta + f'(\theta)\sin \theta$$

and the slope of the tangent line is given by

$$m = \frac{dy}{dx} = \frac{\dfrac{dy}{d\theta}}{\dfrac{dx}{d\theta}} = \frac{f(\theta)\cos \theta + f'(\theta)\sin \theta}{-f(\theta)\sin \theta + f'(\theta)\cos \theta}$$

The theorem for slope in polar form is not easy to remember, and it may be easier for you to remember

$$\frac{dy}{dx} = \frac{\dfrac{dy}{d\theta}}{\dfrac{dx}{d\theta}}$$

where $x = f(\theta)\cos \theta$ and $y = f(\theta)\sin \theta$. With this form, notice that **horizontal tangents** ($dy/dx = 0$) occur when $\dfrac{dy}{d\theta} = 0$ and $\dfrac{dx}{d\theta} \neq 0$, and **vertical tangents** occur when $\dfrac{dx}{d\theta} = 0$ and $\dfrac{dy}{d\theta} \neq 0$. If $\dfrac{dx}{d\theta} = \dfrac{dy}{d\theta} = 0$, then no conclusions can be drawn without further analysis.

> **EXAMPLE 7**   *Computing the slope of a cardioid at a given point*

Find the slope of the tangent line to the cardioid $r = 1 + \cos\theta$ at the point $\left(1 + \dfrac{\sqrt{3}}{2}, \dfrac{\pi}{6}\right)$.

Solution   First find x and y as functions:

$$x = f(\theta)\cos\theta = (1 + \cos\theta)\cos\theta = \cos\theta + \cos^2\theta$$
$$y = f(\theta)\sin\theta = (1 + \cos\theta)\sin\theta = \sin\theta + \cos\theta\sin\theta = \sin\theta + \tfrac{1}{2}\sin2\theta$$

Then find the derivatives $\dfrac{dx}{d\theta}$ and $\dfrac{dy}{d\theta}$:

$$\dfrac{dx}{d\theta} = -\sin\theta - 2\cos\theta\sin\theta \qquad \dfrac{dy}{d\theta} = \cos\theta + \cos2\theta$$
$$= -\sin\theta - \sin2\theta$$

At $\theta = \dfrac{\pi}{6}$, we have

$$\dfrac{dx}{d\theta} = -\sin\dfrac{\pi}{6} - \sin\dfrac{\pi}{3} = -\dfrac{1}{2} - \dfrac{\sqrt{3}}{2} = \dfrac{-1 - \sqrt{3}}{2}$$
$$\dfrac{dy}{d\theta} = \cos\dfrac{\pi}{6} + \cos\dfrac{\pi}{3} = \dfrac{\sqrt{3}}{2} + \dfrac{1}{2} = \dfrac{\sqrt{3} + 1}{2}$$

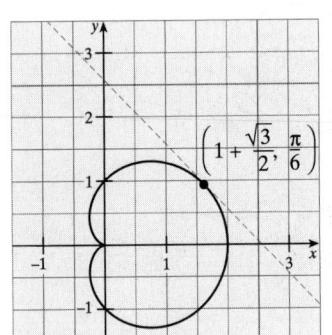

■ **FIGURE 9.18**   Graph of $r = 1 + \cos\theta$ and line passing through $\left(1 + \dfrac{\sqrt{3}}{2}, \dfrac{\pi}{6}\right)$ with slope -1

so the slope is

$$m = \dfrac{dy}{dx} = \dfrac{\dfrac{dy}{d\theta}}{\dfrac{dx}{d\theta}} = \dfrac{\dfrac{\sqrt{3} + 1}{2}}{\dfrac{-1 - \sqrt{3}}{2}} = -1$$

The graph of the cardioid, as well as the tangent line, is shown in Figure 9.18.   ■

> **EXAMPLE 8**   *Finding tangents to a spiral*

Find all numbers θ with $0 \leq \theta \leq \pi$ so that the tangent line to the spiral $r = \theta$ is horizontal.

Solution   We first look at the graph (see Figure 9.19) so that we know what to expect. We now carry out the analytic work:

$$x = f(\theta)\cos\theta = \theta\cos\theta \quad \text{and} \quad y = f(\theta)\sin\theta = \theta\sin\theta, \quad \text{so that}$$
$$\dfrac{dx}{d\theta} = -\theta\sin\theta + \cos\theta \quad \text{and} \quad \dfrac{dy}{d\theta} = \theta\cos\theta + \sin\theta$$

Thus,

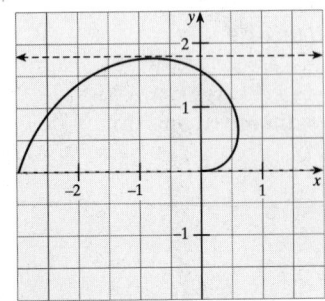

■ **FIGURE 9.19**   By drawing a spiral with $0 \leq \theta \leq \pi$, it looks like there is a horizontal tangent at the pole and at one other point.

$$m = \dfrac{\dfrac{dy}{d\theta}}{\dfrac{dx}{d\theta}} = \dfrac{\theta\cos\theta + \sin\theta}{-\theta\sin\theta + \cos\theta}$$

The tangent line is horizontal when $m = 0$, which is equivalent to

$$\theta \cos \theta + \sin \theta = 0$$

$$\theta \cos \theta = -\sin \theta$$

$$\theta = -\frac{\sin \theta}{\cos \theta}, \qquad \cos \theta \neq 0$$

$$\theta = -\tan \theta$$

$$-\theta = \tan \theta$$

Solving (for $0 \leq \theta \leq \pi$), we find that $\theta = 0$ and $\theta \approx 2.03$. The two points where the tangent is horizontal are $(0, 0)$, and $(2.03, 2.03)$, and are shown in Figure 9.20.

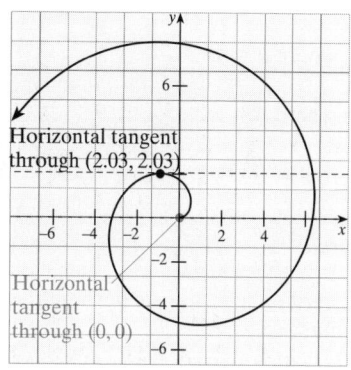

■ **FIGURE 9.20** Graph of $r = \theta (\theta \geq 0)$ and the two horizontal tangents for $0 \leq \theta \leq \pi$ ■

9.3 Problem Set

Ⓐ **1.** ■ **What Does This Say?** Discuss a procedure for finding the intersection of polar-form curves.

2. ■ **What Does This Say?** Discuss a procedure for finding the area enclosed by a polar curve.

3. ■ **What Does This Say?** Discuss why the slope of the tangent line of the polar function $f(\theta)$ is not $f'(\theta)$.

Find the points of intersection of the curves given in Problems 4–29.

4. $\begin{cases} r = 4 \cos \theta \\ r = 4 \sin \theta \end{cases}$ **5.** $\begin{cases} r = 8 \cos \theta \\ r = -8 \sin \theta \end{cases}$ **6.** $\begin{cases} r = 2 \cos \theta \\ r = 1 \end{cases}$

7. $\begin{cases} r = 4 \sin \theta \\ r = 2 \end{cases}$ **8.** $\begin{cases} r^2 = 9 \cos 2\theta \\ r = 3 \end{cases}$ **9.** $\begin{cases} r^2 = 4 \sin 2\theta \\ r = 2 \end{cases}$

10. $\begin{cases} r = 2(1 + \cos \theta) \\ r = 2(1 - \cos \theta) \end{cases}$ **11.** $\begin{cases} r = 2(1 + \sin \theta) \\ r = 2(1 - \sin \theta) \end{cases}$

12. $\begin{cases} r^2 = 9 \sin 2\theta \\ r = 3 \end{cases}$ **13.** $\begin{cases} r^2 = 4 \sin 2\theta \\ r = 2\sqrt{2} \cos \theta \end{cases}$ **14.** $\begin{cases} r^2 = 9 \cos 2\theta \\ r = 3\sqrt{2} \sin \theta \end{cases}$

15. $\begin{cases} r^2 = 4 \cos 2\theta \\ r = 2 \end{cases}$ **16.** $\begin{cases} r = 3\theta \\ \theta = \pi/3 \end{cases}$ **17.** $\begin{cases} r^2 = \sin 2\theta \\ r = \sqrt{2} \sin \theta \end{cases}$

18. $\begin{cases} r = 2(1 - \cos \theta) \\ r = 4 \sin \theta \end{cases}$ **19.** $\begin{cases} r = 2(1 + \cos \theta) \\ r = -4 \sin \theta \end{cases}$

20. $\begin{cases} r = \sin \theta + 2 \\ r = 4 \cos \theta \end{cases}$ **21.** $\begin{cases} r = 2 \cos \theta + 1 \\ r = \sin \theta \end{cases}$

22. $\begin{cases} r = 2 \sin \theta + 1 \\ r = \cos \theta \end{cases}$ **23.** $\begin{cases} r = 5/(3 - \cos \theta) \\ r = 2 \end{cases}$

24. $\begin{cases} r = \dfrac{2}{1 + \cos \theta} \\ r = 2 \end{cases}$ **25.** $\begin{cases} r = \dfrac{4}{1 - \cos \theta} \\ r = 2 \cos \theta \end{cases}$

26. $\begin{cases} r = 1/(1 + \cos \theta) \\ r = 2(1 - \cos \theta) \end{cases}$ **27.** $\begin{cases} r = 2 \cos \theta \\ r = \sec \theta \end{cases}$

28. $\begin{cases} r = 2 \sin \theta \\ r = 2 \csc \theta \end{cases}$ **29.** $\begin{cases} r \sin \theta = 1 \\ r = 4 \sin \theta \end{cases}$

Find the area of each polar region enclosed by $f(\theta), \theta = a,$ $\theta = b,$ for $a \leq \theta \leq b$ in Problems 30–37.

30. $f(\theta) = \sin \theta, 0 \leq \theta \leq \frac{\pi}{6}$ **31.** $f(\theta) = \cos \theta, 0 \leq \theta \leq \frac{\pi}{6}$

32. $f(\theta) = \sec \theta, -\frac{\pi}{4} \leq \theta \leq \frac{\pi}{4}$ **33.** $f(\theta) = \sqrt{\sin \theta}, \frac{\pi}{6} \leq \theta \leq \frac{\pi}{2}$

34. $f(\theta) = e^{\theta/2}, 0 \leq \theta \leq 2\pi$

35. $f(\theta) = \sin \theta + \cos \theta, 0 \leq \theta \leq \frac{\pi}{4}$

36. $f(\theta) = \dfrac{\theta}{\pi}, 0 \leq \theta \leq 2\pi$ **37.** $f(\theta) = \dfrac{\theta^2}{\pi}, 0 \leq \theta \leq 2\pi$

In Problems 38–47, find the slope of the tangent line to the graph of the polar function curves at the given point.

38. $f(\theta) = 1 - \cos\theta$ at $\left(\dfrac{2-\sqrt{2}}{2}, \dfrac{\pi}{4}\right)$

39. $f(\theta) = 4\cos\theta + 2$ at the pole

40. $f(\theta) = \sqrt{\cos 2\theta}$ at the pole

41. $f(\theta) = 2$ at $\left(2, \dfrac{\pi}{3}\right)$

42. $r = 2\sec\theta$, where $\theta = \dfrac{\pi}{4}$

43. $r = \theta$, where $\theta = \dfrac{\pi}{2}$

44. $r = \dfrac{4}{3\sin\theta - 2\cos\theta}$, where $\theta = \pi$

45. $r = \dfrac{3}{2\cos\theta + 3\sin\theta}$, where $\theta = \pi$

46. $r = 4\sin\theta\cos^2\theta$, where $\theta = \dfrac{\pi}{3}$

47. $r = 2\cos\theta\sin^2\theta$, where $\theta = \dfrac{\pi}{6}$

Ⓑ 48. Find all points on the cardioid $r = 1 + \sin\theta$ where the tangent line is horizontal.

49. Find all points on the cardioid $r = a(1 + \cos\theta)$ where the tangent line is horizontal.

50. Find all points on the cardioid $r = a(1 - \cos\theta)$ where the tangent line is vertical.

51. Find all points on the circle $r = 2\sin\theta$ where the tangent line is parallel to the ray $\theta = \pi/4$.

52. Find the area of one loop of the four-leaf rose $r = 2\sin 2\theta$.

53. Find the area enclosed by the three-leaf rose $r = a\sin 3\theta$.

54. Find the area of the region that is inside the circle $r = 4\cos\theta$ and outside the circle $r = 2$.

55. Find the area of the region that is inside the circle $r = a$ and outside the cardioid $r = a(1 - \cos\theta)$.

56. Find the area of the region that is inside the circle $r = \sin\theta$ and outside the cardioid $r = 1 - \cos\theta$.

57. Find the area of the region that lies inside the circle $r = 6\cos\theta$ and outside the cardioid $r = 2(1 + \cos\theta)$.

58. Find the area of the portion of the lemniscate $r^2 = 8\cos 2\theta$ that lies in the region $r \geq 2$.

59. Find the area between the inner and outer loops of the limaçon $r = 2 - 4\sin\theta$.

60. Find the area to the right of the line $r\cos\theta = 1$ and inside the lemniscate $r^2 = 2\cos 2\theta$.

61. Find the area to the right of the line $r\cos\theta = 1$ and inside the lemniscate $r^2 = 2\sin 2\theta$.

62. The tangent line to the spiral $r = 2\theta$ at the point $(\tfrac{1}{3}, \theta_1)$ intersects the spiral at a second point (r_2, θ_2), where $\pi < \theta_2 < 2\pi$.

a. Find θ_1, θ_2, and r_2.
b. Find the slope of the tangent line to $r = 2\theta$ at the point (r_2, θ_2).
c. Find the area bounded by the spiral between $\theta = \theta_1$ and $\theta = \theta_2$.

63. Find the maximum value of the y-coordinate of points on the limaçon $r = 2 + 3\cos\theta$.

64. Find the maximum value of the x-coordinate of points on the cardioid $r = 3 + 3\sin\theta$.

Ⓒ 65. Find an equation in θ for all polar-form points (r, θ) on the spiral $r = \theta^2$ where the tangent line is horizontal. What happens at the origin?

66. Let a and b be any nonzero numbers. Show that the tangent lines to the graphs of the cardioids $r_1 = a(1 + \sin\theta)$ and $r_2 = b(1 - \sin\theta)$ are perpendicular to each other at all points where they intersect.

67. Because the formula for the slope in polar form is complicated, it is often more convenient to measure the inclination of the tangent line to a polar curve in terms of the angle α extending from the radial line to the tangent line, as shown in Figure 9.21.

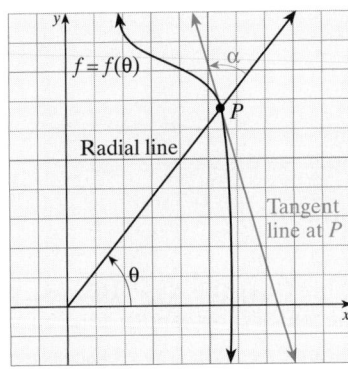

■ **FIGURE 9.21**   Problem 67

Let P be a point on the polar curve $r = f(\theta)$, and let α be the angle extending from the radial line to the tangent line, as shown in Figure 9.21. Assuming that $f'(\theta) \neq 0$, show that

$$\tan\alpha = \frac{f(\theta)}{f'(\theta)}$$

68. Use the formula in Problem 67 to find $\tan\alpha$ for each of the following:
a. the circle $r = a\cos\theta$
b. the cardioid $r = 2(1 - \cos\theta)$
c. the logarithmic spiral $r = 2e^{3\theta}$

69. **JOURNAL PROBLEM**   (*School Science and Mathematics*, Problem 3949 by V. C. Bailey, Vol. 83, 1983, p. 356.)

Graph the polar curve $r = 4 + 2\sin\dfrac{5\theta}{2}$ and find the total area interior to this curve. Then find the area of the "star."

9.4 *Parametric Representation of Curves*

IN THIS SECTION   **parametric equations, derivatives, arc length, area**   ■

PARAMETRIC EQUATIONS

It is sometimes useful to define the variables x and y in the ordered pair (x, y) so that they are *each* functions of some other variable, for example, t. That is, let

$$x = g(t) \quad \text{and} \quad y = h(t)$$

for functions g and h, where the domain of these functions is some interval I. The variable t is called a **parameter,** and $x = g(t)$ and $y = h(t)$ are **parametric equations.**

Parametric Representation for a Curve

Let f and g be continuous functions of t on an interval I; then the equations

$$x = g(t) \quad \text{and} \quad y = h(t)$$

are called **parametric equations** for the curve C generated by the set of ordered pairs $(x(t), y(t))$.

EXAMPLE 1   *Graphing a curve from parametric equations*

Graph the curve defined by the functions $x = 2\cos t, y = 2\sin t$ for $0 \le t < 2\pi$.

Solution   We could begin by letting t be the parameter to determine x and y values. When we draw the graph, the parameter is not shown because we still graph the ordered pairs (x, y) in the usual fashion (see Figure 9.22).

t	x	y
0	2	0
1	1.08	1.68
$\dfrac{\pi}{2}$	0	2
2	-0.83	1.82
π	-2	0
$\dfrac{3\pi}{2}$	0	-2

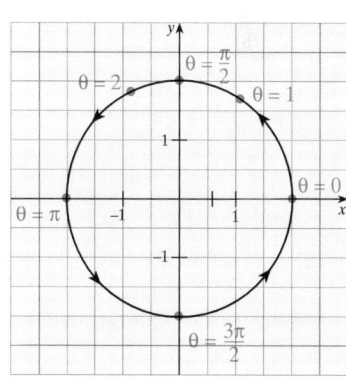

■ **FIGURE 9.22**   Graph of $x = 2\cos t, y = 2\sin t$

If you are using a computer or a graphing calculator, plotting points can be an efficient way of obtaining the graph, but sometimes it is more efficient to **eliminate the parameter:**

$$x^2 = 4\cos^2 t$$
$$y^2 = 4\sin^2 t$$
$$\text{Add:} \quad x^2 + y^2 = 4\cos^2 t + 4\sin^2 t = 4(\cos^2 t + \sin^2 t) = 4$$

We recognize this as the equation of a circle centered at $(0, 0)$ with radius 2.   ■

If we start with the parametric equations $x = 2 \cos t$, $y = 2 \sin t$, we see from Example 1 that after eliminating the parameter, we obtain $x^2 + y^2 = 4$. On the other hand, we say that the $x = 2 \cos t$, $y = 2 \sin t$ is a **parametrization** of the curve $x^2 + y^2 = 4$.

Notice that the curve in Example 1 is not the graph of a function, but is defined parametrically using functions. This points out one of the advantages of parametric equations—they can be used to represent graphs that are more general than graphs of functions.

In practical applications, it is often possible to obtain useful information about a parametrized curve by seeing how it is traced out as the parameter varies from one value to another. For Example 1, we can generalize to say that the parametric equations

$$x = R \cos \theta \qquad y = R \sin \theta \qquad R > 0, \text{ a constant}$$

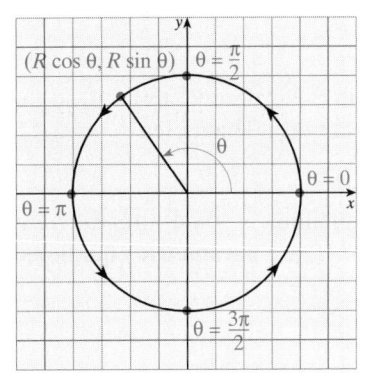

■ **FIGURE 9.23** Graph of $x = R \cos \theta$, $y = R \sin \theta$

trace out a circle of radius R and center $(0,0)$ as θ varies from 0 to 2π. We can describe this trace by starting at $\theta = 0$, the point $(R, 0)$, and ending at $\theta = 2\pi$, as shown by the arrows in Figure 9.23. We shall refer to this trace (oriented curve) as the **path** of $x = f(t)$, $y = g(t)$ for $a \le t \le b$ from the point $(f(a), g(a))$ to $(f(b), g(b))$.

EXAMPLE 2   *Sketching the path of a parametric curve*

Sketch the path of the curve $x = t^2 - 9$, $y = \frac{1}{3}t$ for $-3 \le t \le 2$.

Solution   Values of x and y corresponding to various choices of the parameter t are shown in the following table:

t	x	y	
-3	0	-1	(Starting point)
-2	-5	$-\frac{2}{3}$	
-1	-8	$-\frac{1}{3}$	
0	-9	0	
1	-8	$\frac{1}{3}$	
2	-5	$\frac{2}{3}$	(Ending point)

The graph is shown in Figure 9.24. ■

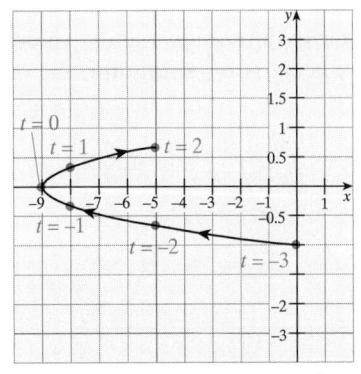

■ **FIGURE 9.24** Graph of $x = t^2 - 9$, $y = \frac{1}{3}t$, for $-3 \le t \le 2$. Notice how the arrows show the orientation as t increases from -3 to 2.

The parametrization of an equation is not unique. For example, the set of parametric equations

$$x = 9(9t^2 - 1), \quad y = 3t \quad \text{for } -\frac{1}{3} \le t \le \frac{2}{9}$$

has the same graph as the one given in Example 2. If we consider t as measuring time, we see that the curve is traced out much more rapidly with this set of equations than the ones given in Example 2. When a parameter is used to describe motion, we call the path a **trajectory.** In many applications, we use a parameter to represent different *speeds* at which objects can travel along a given path.

EXAMPLE 3   *Sketching the path of a parametric curve by eliminating the parameter*

Describe the path $x = \sin \pi t$, $y = \cos 2\pi t$ for $0 \le t \le 0.5$.

Solution Using a double-angle identity, we find

$$\cos 2\pi t = 1 - 2 \sin^2 \pi t$$

so that

$$y = 1 - 2x^2$$

We recognize this as a Cartesian equation for a parabola. Because $y' = -4x$, we can find the critical value $x = 0$, which locates the vertex of the parabola at $(0, 1)$. The parabola is the dashed curve shown in Figure 9.25. Because t is restricted to the interval $0 \le t \le 0.5$, the parametric representation involves only part of the right side of the parabola $y = 1 - 2x^2$. The curve is oriented from the point $(0, 1)$, where $t = 0$, to the point $(1, -1)$, where $t = 0.5$.

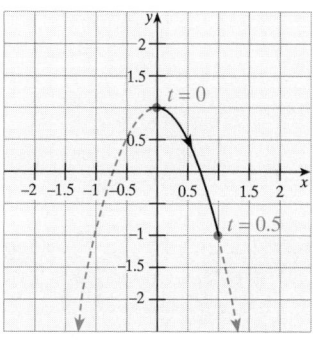

■ **FIGURE 9.25** The parabolic arc $x = \sin \pi t$, $y = \cos 2\pi t$ for $0 \le t \le 0.5$

EXAMPLE 4 *Adjusting the domain after eliminating a parameter*

Sketch the graph of $x = 2^t$, $y = 2^{t+1}$ for $t > 0$.

Solution Eliminating the parameter, we obtain

$$\frac{y}{x} = \frac{2^{t+1}}{2^t} = 2^{t+1-t} = 2 \quad \text{so that} \quad y = 2x$$

for $x > 1$ since $t > 0$. The graph is shown in Figure 9.26. ■

When it is difficult to eliminate the parameter from a given parametric representation, we can sometimes get a good picture of the parametric curve by plotting points.

EXAMPLE 5 *Describing a spiraling path*

Discuss the path of the curve described by the parametric equations

$$x = e^{-t}\cos t, \quad y = e^{-t}\sin t \quad \text{for } t \ge 0$$

Solution We have no convenient way of eliminating the parameter, so we write out a table of values (x, y) that correspond to various values of t. The curve is obtained by plotting these points in a Cartesian plane and passing a smooth curve through the plotted points, as shown in Figure 9.27.

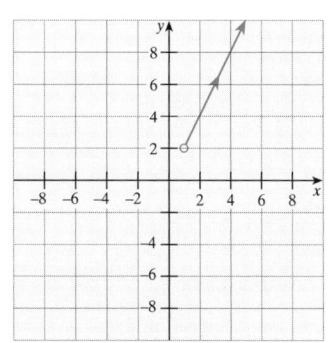

■ **FIGURE 9.26** Graph of $x = 2^t$, $y = 2^{t+1}$

t	x	y
0	1	0
$\frac{\pi}{4}$	0.32	0.32
1	0.20	0.31
$\frac{\pi}{2}$	0	0.21
2	-0.06	0.12
π	-0.04	0
$\frac{3\pi}{2}$	0	-0.01
2π	0.00	0

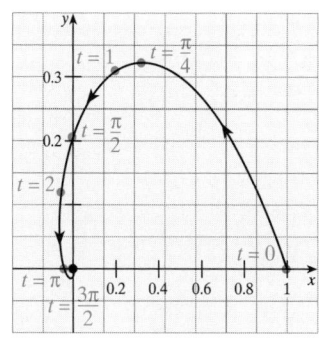

■ **FIGURE 9.27** Graph of $x = e^{-t} \cos t$, $y = e^{-t}\sin t$ for $t \ge 0$

Note that for each value of t, the distance from $P(x, y)$ on the curve to the origin is

$$\sqrt{x^2 + y^2} = \sqrt{(e^{-t}\cos t)^2 + (e^{-t}\sin t)^2} = \sqrt{e^{-2t}(1)} = e^{-t}$$

Because e^{-t} decreases as t increases, it follows that P gets closer and closer to the origin as t increases. However, because $\cos t$ and $\sin t$ vary between -1 and $+1$, the approach is not direct but takes place along a spiral.   ∎

> **EXAMPLE 6**   *Modeling Problem: Finding parametric equations for a trochoid*

A bicycle wheel has radius a and a reflector is attached at a point P on the spoke of a bicycle wheel at a fixed distance d from the center. Find parametric equations for the curve described by P as the wheel rolls along a straight line without slipping. Such a curve is called a **trochoid.**

Solution   Assume that the wheel rolls along the x-axis and that the center C of the wheel begins at $(0, a)$ on the y-axis. Further assume that P also starts on the y-axis, d units below C. Figure 9.28 shows the initial position of the wheel and its position after turning through an angle θ.

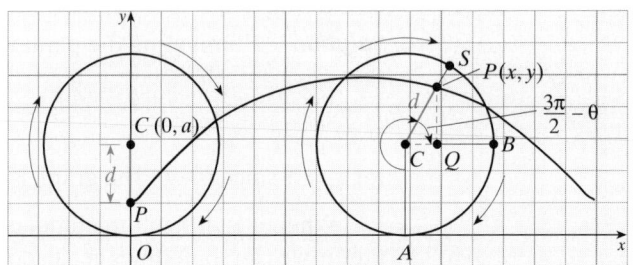

■ **FIGURE 9.28**   The path of a reflector on a bicycle

We begin by labeling some points: The point A is on the x-axis directly beneath C, whereas B is the point where the horizontal line through C meets the rim of the wheel. Finally, Q is the point on BC directly beneath P, and S is the point where the line through C and P intersects the rim. Let P have coordinates (x, y).

We need to find representations (in terms of a, d, and θ) for x and y.

$$x = |\overline{OA}| + |\overline{CQ}|$$
$$= a\theta + |\overline{CQ}|$$

Because the wheel rolls along the x-axis without slipping, $|\overline{OA}|$ is the same as the arc length from A to S, so $|\overline{OA}| = a\theta$.

$$y = |\overline{AC}| + |\overline{QP}|$$
$$= a + |\overline{QP}|$$

To complete our evaluation of x and y, we need to compute $|\overline{CQ}|$ and $|\overline{QP}|$. These are sides of $\triangle PCQ$. Note that $\angle PCQ = \frac{3\pi}{2} - \theta$; therefore, by the definition of cosine and sine, we have

$$\cos\left(\frac{3\pi}{2} - \theta\right) = \frac{|\overline{CQ}|}{d} \quad \text{so} \quad |\overline{CQ}| = d\cos\left(\frac{3\pi}{2} - \theta\right) = -d\sin\theta$$

Similarly, $|\overline{QP}| = d\sin(\frac{3\pi}{2} - \theta) = -d\cos\theta$.

We can now substitute these values for $\left|\overline{CQ}\right|$ and $\left|\overline{QP}\right|$ into the equations we derived for x and y:

$$x = a\theta + \left|\overline{CQ}\right| = a\theta - d\sin\theta$$

$$y = a + \left|\overline{QP}\right| = a - d\cos\theta$$

The special case where P is on the rim of the wheel in Example 6 (when $d = a$) is a curve called a **cycloid.** There are several problems involving these and similar curves in the problem set.

Technology Window

Two variations in the cycloid path discussed in Example 6 may be described as paths traced out by a fixed point P on a circle. If it rolls *inside* a larger circle, the trace is called a **hypocycloid,** and if it rolls around the *outside* of a fixed circle, the trace is called an **epicycloid.** Even though it is not too difficult to derive their parametric equations (see Problems 62 and 63), it is difficult to graph variations of these curves without the assistance of a computer or a graphing calculator. However, if you have access to a computer, you might look at an article by Florence and Sheldon Gordon entitled "Mathematics Discovery via Computer Graphics: Hypocycloids and Epicycloids" in *The Two-Year College Mathematics Journal,* November 1984, p. 441. If a is the radius of the fixed circle and R is the radius of the rolling circle, then the parametric equations are as follows:

Hypocycloid: $x = (a - R)\cos t + R\cos\left(\dfrac{a - R}{R}\right)t,$ $y = (a - R)\sin t - R\sin\left(\dfrac{a - R}{R}\right)t$

Epicycloid: $x = (a + R)\cos t - R\cos\left(\dfrac{a + R}{R}\right)t,$ $y = (a + R)\sin t - R\sin\left(\dfrac{a + R}{R}\right)t$

Some variations of these curves, which were drawn by computer, are shown below.

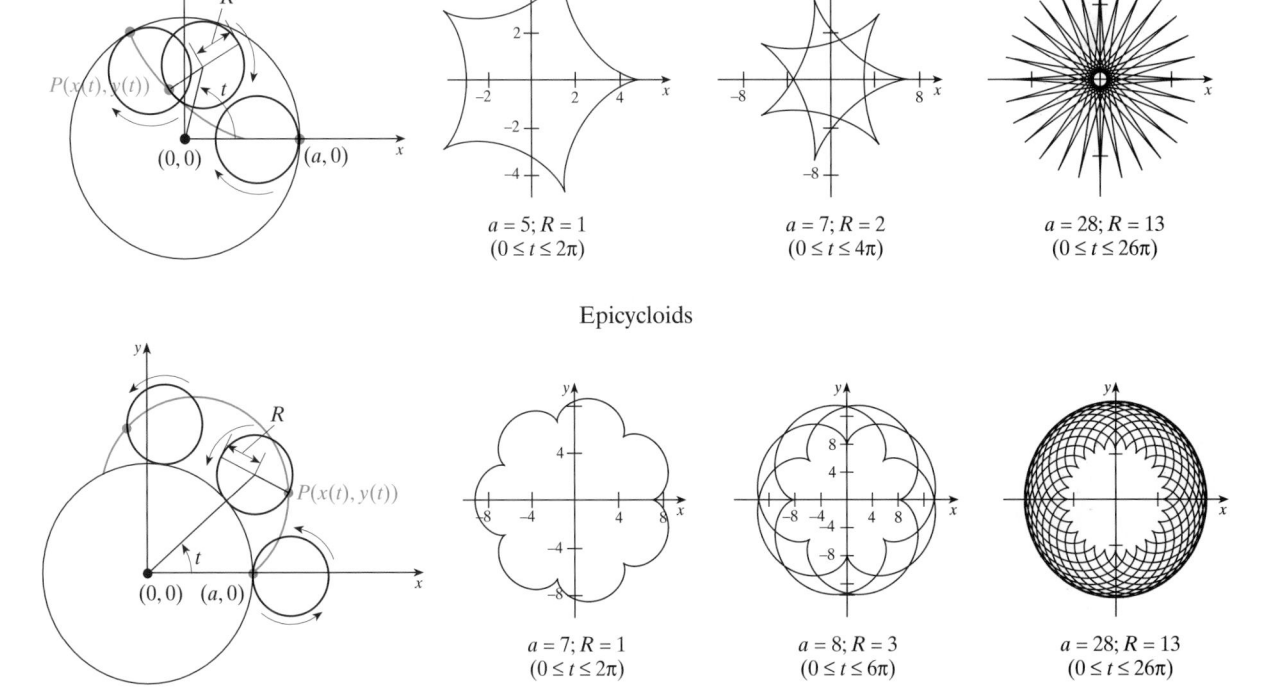

Hypocycloids

$a = 5; R = 1$
$(0 \le t \le 2\pi)$

$a = 7; R = 2$
$(0 \le t \le 4\pi)$

$a = 28; R = 13$
$(0 \le t \le 26\pi)$

Epicycloids

$a = 7; R = 1$
$(0 \le t \le 2\pi)$

$a = 8; R = 3$
$(0 \le t \le 6\pi)$

$a = 28; R = 13$
$(0 \le t \le 26\pi)$

DERIVATIVES

We begin by finding the derivative of a function in parametric form.

THEOREM 9.3   *Parametric form of the derivative*

If a curve is described parametrically by the equations $x = x(t), y = y(t)$, where x and y are differentiable functions of t, then the derivative $\dfrac{dy}{dx}$ can be expressed in terms of $\dfrac{dx}{dt}$ and $\dfrac{dy}{dt}$ by the equation

$$\frac{dy}{dx} = \frac{\dfrac{dy}{dt}}{\dfrac{dx}{dt}} = \frac{y'(t)}{x'(t)} \quad \text{whenever} \quad \frac{dx}{dt} \neq 0$$

Proof   This formula follows directly from the chain rule, $\dfrac{dy}{dt} = \dfrac{dy}{dx}\dfrac{dx}{dt}$. ◼

The curve has a *horizontal tangent* when $\dfrac{dy}{dt} = 0$ and $\dfrac{dx}{dt} \neq 0$, and it has a *vertical tangent* when $\dfrac{dx}{dt} = 0$ and $\dfrac{dy}{dt} \neq 0$. If $\dfrac{dy}{dt} = 0$ and $\dfrac{dx}{dt} = 0$ at the same point, further analysis is required.

EXAMPLE 7   *Finding the derivative of a function in parametric form*

A curve C is described parametrically by $x = 7t + 2, y = t^3 - 12t$ for all t. Find $\dfrac{dy}{dx}$ at the point where $t = 3$, and then find all points of C where the tangent line is horizontal.

Solution   We have $\dfrac{dx}{dt} = 7$ and $\dfrac{dy}{dt} = 3t^2 - 12$, and thus

$$\frac{dy}{dx} = \frac{\dfrac{dy}{dt}}{\dfrac{dx}{dt}} = \frac{3t^2 - 12}{7}$$

When $t = 3$, we have

$$\left.\frac{dy}{dx}\right|_{t=3} = \frac{3(3)^2 - 12}{7} = \frac{15}{7}$$

The curve and the line tangent at the point where $t = 3$ are shown in Figure 9.29. The tangent line is horizontal when $\dfrac{dy}{dt} = 0$;

$$3t^2 - 12 = 0 \quad \text{when} \quad t = \pm 2$$

At $t = 2$, $x = 7(2) + 2 = 16$ and $y = 2^3 - 12(2) = -16$, and at $t = -2$, $x = 7(-2) + 2 = -12$ and $y = (-2)^3 - 12(-2) = 16$, so the tangent line is horizontal at the points $(16, -16)$ and $(-12, 16)$. Figure 9.29 also shows the horizontal tangents. ◼

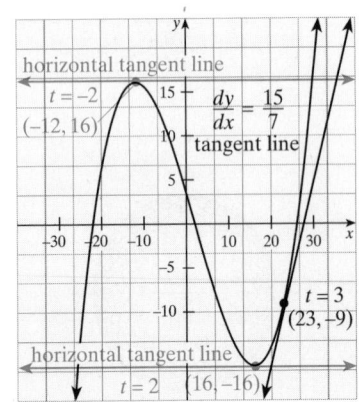

■ FIGURE 9.29   Graph of $x = 7t + 2$, $y = t^3 - 12t$ with tangent at $t = 3$

To find the second derivative of a function defined parametrically, we set $y' = \dfrac{dy}{dx}$ and substitute into the derivative formula for parametric equations:

$$\frac{d^2y}{dx^2} = \frac{d}{dx}\left(\frac{dy}{dx}\right) = \frac{d}{dx}(y') = \frac{dy'}{dx} = \frac{\dfrac{dy'}{dt}}{\dfrac{dx}{dt}}$$

For instance, in Example 7, we found $y' = \dfrac{dy}{dx} = \dfrac{3t^2 - 12}{7}$ so that

$$\frac{d^2y}{dx^2} = \frac{\frac{3}{7}(2t)}{7} = \frac{6t}{49}$$

ARC LENGTH

Parametric equations can be used to specify a curve, the speed of an object at a given time, or the distance an object has traveled. Finding the distance that a point moves along a curve defined parametrically requires the following formula for arc length.

THEOREM 9.4 *Arc length of a curve described parametrically*

Let C be the curve described parametrically by the equations $x = x(t)$ and $y = y(t)$ on an interval $[a, b]$, where $\dfrac{dx}{dt}$ and $\dfrac{dy}{dt}$ are continuous. Suppose C does not intersect itself (such a C is called a *simple curve*). Then C has arc length

$$L = \int_a^b \sqrt{\left(\frac{dx}{dt}\right)^2 + \left(\frac{dy}{dt}\right)^2}\, dt$$

Proof Recall that the formula for the arc length of a curve C given by $y = f(x)$ over an interval $[a, b]$ is

$$s = \int_a^b \sqrt{1 + [f'(x)]^2}\, dx$$

Use this formula to derive the desired equation. The details are left for the problem set (Problem 64).

EXAMPLE 8 *Finding the arc length of a curve defined by parametric equations*

Let C be the curve defined by $x = e^t \sin t$, $y = e^t \cos t$ for $0 \le t \le \pi$. Find the length of C on the interval $[0, \pi]$.

Solution We find that

$$\frac{dx}{dt} = e^t \cos t + e^t \sin t = e^t(\cos t + \sin t)$$

$$\frac{dy}{dt} = e^t(-\sin t) + e^t \cos t = e^t(\cos t - \sin t)$$

To obtain the length of the curve, we first compute

$$\left(\frac{dx}{dt}\right)^2 + \left(\frac{dy}{dt}\right)^2 = e^{2t}(\cos t + \sin t)^2 + e^{2t}(\cos t - \sin t)^2$$

$$= e^{2t}(\cos^2 t + 2\cos t \sin t + \sin^2 t + \cos^2 t$$
$$- 2\cos t \sin t + \sin^2 t)$$

$$= 2e^{2t}$$

Thus,

$$s = \int_0^{\pi} \sqrt{\left(\frac{dx}{dt}\right)^2 + \left(\frac{dy}{dt}\right)^2}\, dt = \int_0^{\pi} \sqrt{2e^{2t}}\, dt = \int_0^{\pi} \sqrt{2}\, e^t dt = \sqrt{2}(e^{\pi} - 1)$$

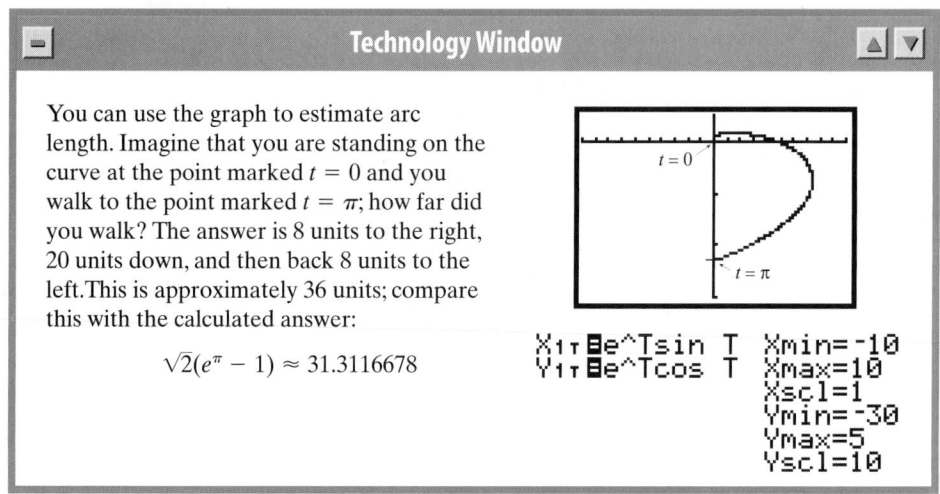

Technology Window

You can use the graph to estimate arc length. Imagine that you are standing on the curve at the point marked $t = 0$ and you walk to the point marked $t = \pi$; how far did you walk? The answer is 8 units to the right, 20 units down, and then back 8 units to the left. This is approximately 36 units; compare this with the calculated answer:

$$\sqrt{2}(e^{\pi} - 1) \approx 31.3116678$$

```
X₁ᴛ⊟e^Tsin T    Xmin=-10
Y₁ᴛ⊟e^Tcos T    Xmax=10
                Xscl=1
                Ymin=-30
                Ymax=5
                Yscl=10
```

AREA

We close this section with a theorem and an example that show how to find the area under a curve when the curve is defined by parametric equations.

THEOREM 9.5   *Formula for the area under a curve defined parametrically*

Suppose y is a continuous function of x on the closed interval $[a, b]$ and that x and y are both functions of t for $t_1 \leq t \leq t_2$, where $x(t_1) = a$ and $x(t_2) = b$, and the traced path is a simple curve. Then if $x'(t)$ is continuous on $[t_1, t_2]$, the area under the curve described parametrically by $x = x(t), y = y(t)$ is given by

$$A = \int_a^b y\, dx = \int_{t_1}^{t_2} \left[y(t)\, \frac{dx}{dt} \right] dt$$

Proof   The proof is straightforward.   ═

EXAMPLE 9   *Finding the area under a curve defined parametrically*

Find the area under the curve described by

$$x = e^t, \quad y = 2e^t + 1 \quad \text{over the interval } 0 \leq t \leq 1$$

Solution   Note that x and y are both continuous and that $y(t) \geq 0$ on $[0, 1]$.

$$A = \int_1^e y(x)\, dx = \int_0^1 y(t)\, x'(t)\, dt$$

$$= \int_0^1 (2e^t + 1)\, e^t\, dt$$

$$= \int_0^1 (2e^{2t} + e^t)\, dt$$

$$= (e^{2t} + e^t)\big|_0^1 = e^2 + e - 2 \qquad \blacksquare$$

9.4   Problem Set

Ⓐ 1. ▪ **What Does This Say?** What is a parameter?

2. ▪ **What Does This Say?** Describe the process of finding the derivative in parametric form.

Find an explicit relationship between x and y in Problems 3–24 by eliminating the parameter. In each case, sketch the path described by the parametric equations over the prescribed interval.

3. $x = t + 1, y = t - 1, \quad 0 \leq t \leq 2$

4. $x = -t, y = 3 - 2t, \quad 0 \leq t \leq 1$

5. $x = 60t, y = 80t - 16t^2, \quad 0 \leq t \leq 3$

6. $x = 30t, y = 60t - 9t^2, \quad -1 \leq t \leq 2$

7. $x = t, y = 2 + \frac{2}{3}(t - 1), \quad 2 \leq t \leq 5$

8. $x = t, y = 3 - \frac{3}{5}(t + 2), \quad -1 \leq t \leq 3$

9. $x = t^2 + 1, y = t^2 - 1, \quad -1 \leq t \leq \sqrt{2}$

10. $x = 2t^2, y = t^2 + 2, \quad -1 \leq t \leq \sqrt{2}$

11. $x = t^3, y = t^2, \quad t \geq 0$

12. $x = t^4, y = t^2, \quad -1 \leq t \leq \sqrt{2}$

13. $x = 3 \cos \theta, y = 3 \sin \theta, \quad 0 \leq \theta < 2\pi$

14. $x = 2 \sin \theta, y = 2 \cos \theta, \quad 0 \leq \theta < 2\pi$

15. $x = 1 + \sin t, y = -2 + \cos t, \quad 0 \leq t < 2\pi$

16. $x = 1 + \sin^2 t, y = -2 + \cos t, \quad 0 \leq t < \pi$

17. $x = 4 \tan 2t, y = 3 \sec 2t, \quad 0 \leq t < \pi$

18. $x = 4 \sec 2t, y = 2 \tan 2t, \quad 0 \leq t < \pi$

19. $x = 3^t, y = 3^{t+1}, \quad t \geq 0$

20. $x = 2^t, y = 2^{1-t}, \quad t \geq 0$

21. $x = e^t, y = e^{t+1}, \quad t \geq 0$

22. $x = e^t, y = e^{1-t}, \quad t \geq 0$

23. $x = t^3, y = 3 \ln t, \quad t > 0$

24. $x = e^t, y = e^{-t}, \quad (-\infty, \infty)$

Find $\dfrac{dy}{dx}$ and $\dfrac{d^2y}{dx^2}$ in Problems 25–30 without eliminating the parameter.

25. $x = t^2, y = t^4 + 1$ **26.** $x = t^4, y = t^2 + 1$

27. $x = e^{4t}, y = \sin 2t$

28. $x = e^t, y = t^2 e^t$

29. $x = a \cos t, y = b \sin t$

30. $x = a \cosh t, y = b \sinh t$

Find $\dfrac{dy}{dx}$ in Problems 31–34 and then eliminate the parameter to express your answer in terms of x and y.

31. $x = t^2 + 1, y = t^4 + 1$

32. $x = 2a \cos t, y = a \sin^2 t; a > 0$

33. $x = 1 - e^{-t}, y = 1 + e^t$

34. $x = t^2, y = \ln t$

Find the area under the curves given in Problems 35–40.

35. $x = t^4 + 1, y = t^2, \quad 0 \leq t \leq 1$

36. $x = t^4 + t^2 + 1, y = (t + 2)^2, \quad 0 \leq t \leq 1$

37. $x = \theta - \sin \theta, y = 1 + \cos \theta, \quad 0 \leq \theta \leq \dfrac{\pi}{4}$

38. $x = \tan \theta, y = \sec^2 \theta, \quad 0 \leq \theta \leq \dfrac{\pi}{6}$

39. $x = \tan^{-1} u, y = u^3, \quad 0 \leq u \leq 1$

40. $x = \sin^{-1} u, y = u, \quad 0 \leq u \leq \frac{1}{2}$

Find the length of each curve in Problems 41–45.

41. $x = \sqrt{t}, y = t^{3/4}$ for $0 \leq t \leq 4$

42. $x = 2t^2, y = t^3$ for $1 \leq t \leq 2$

43. $x = \frac{1}{2} \ln(t^2 - 1), y = \sqrt{t^2 - 1}$ for $3 \leq t \leq 7$

44. $x = t \sin t + \cos t, y = \sin t - t \cos t$ for $0 \leq t \leq \dfrac{\pi}{2}$

45. $x = 2 \sin^{-1} t, y = \ln(1 - t^2)$ for $0 \leq t \leq \dfrac{\sqrt{3}}{2}$

Ⓑ 46. Describe the path of the curve defined by $x = \sin \pi t$, $y = \cos 2\pi t$ for $0 \leq t < 1$.

47. Let $x = 4a \sin t, y = b \cos^2 t$. Express y as a function of x.

48. Find the area under one arch of the cycloid $x = 2(\theta - \sin \theta), y = 2(1 - \cos \theta)$.

49. Find the length of the curve given by $x = 2 \cos^3 t$, $y = 2 \sin^3 t$, $0 \leq t \leq 2\pi$. This curve is called an **astroid.**

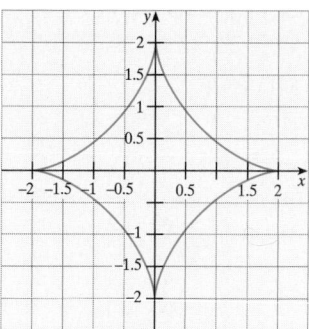

Use the formula

$$L = \int_a^b \sqrt{[f(\theta)]^2 + [f'(\theta)]^2} \, d\theta$$

which is derived in Problem 65, to find the arc length of the polar-form curves given in Problems 50–53.

50. the spiral $r = e^\theta$, for $0 \leq \theta \leq \frac{\pi}{2}$
51. the circle $r = 2 \cos \theta$, for $0 \leq \theta \leq \frac{\pi}{3}$
52. the cardioid $r = 2(1 - \cos \theta)$ on $[0, \pi]$
53. the spiral $r = e^{-\theta}$ for $\theta \geq 0$

Technology Window

54. Write a paper showing the graphs of several hypocycloids, along with an analysis of what you did to graph them. Make some generalizations based on your graphs.
55. Write a paper describing what happens if the radius of the inner rolling circle is larger than the radius of the outer fixed circle.
56. Write a paper showing the graphs of several epicycloids, along with an analysis of what you did to graph them. Make some generalizations based on your graphs.
57. Write a paper describing what happens if the radius of the outer rolling circle is larger than the radius of the inner fixed circle.

Problems 58–59 explore the work required to move a mass or particle along a planar curve which is described parametrically. Let C be the set of points

$$x = x(t), \quad y = y(t), \quad a \leq t \leq b$$

Suppose a force F = F(t) is applied to the mass in the direction of the curve (that is, the force is applied in the direction of the tangent to C).

58. a. Argue that on the interval $[t, t + dt]$, dt very small, the work done is approximately

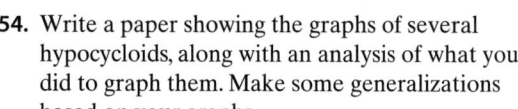

$$dW = F(t) \, ds = F(t)\sqrt{[x'(t)]^2 + [y'(t)]^2} \, dt$$

Technology Window

b. Form the usual Riemann sum, take a limit, and get the formula for the work done on C:

$$W = \int_a^b F(t)\sqrt{[x'(t)]^2 + [y'(x)]^2} \, dt$$

c. Suppose $F(t) = 2e^t$ and $x(t) = 6 \cosh t$, $y(t) = 6t$ for $0 \leq t \leq 1$. Compute the work involved on the entire curve C. *Note:* You may use a computer, but a computer is not *needed* to answer this question.

59. As we have seen, arc length integrals are often impossible to evaluate without a computer. The same is true of work integrals. For example, suppose a particle is moving on an elliptical path described by

$$x(t) = 2 \cos t, \quad y(t) = 3 \sin t, \quad 0 \leq t < 2\pi$$

against the force $F(t) = 5|y'(t)|$.
a. Compute the arc length of the ellipse to three-decimal-place accuracy. This is a surprisingly tough integral, so make a common-sense check to make sure your answer is reasonable. Explain your check.
b. Set up the integral and compute the work involved in one orbit of the particle to three-decimal-place accuracy.
c. Suppose a constant force $F = 5$ is applied to the following spiral path, for $t \geq 0$:

$$x(t) = 2e^{-t/10}\cos t, \quad y(t) = 3e^{-t/10}\sin t$$

To obtain a rough check on your answer, you may wish to solve (without a computer) a problem with an easier path. Try, for example (for $t \geq 0$),

$$x(t) = 2.5e^{-t/10} \cos t, \quad y(t) = 2.5e^{-t/10} \sin t$$

---

60. **SPY PROBLEM**   After dispatching Blohardt's four gunmen (Problem 79 of Chapter 8 Supplementary Problems), the Spy interrogates one of the wounded thugs and finds that Blohardt has left the château through an underground water passage in his private submarine. The Spy races from the château to a nearby naval base and commandeers a destroyer. Since Blohardt suspects no danger, his submarine is running on the surface, and the Spy's destroyer is able to get within 3 miles of the submarine before it submerges. The Spy assumes Blohardt will try to escape by traveling at maximum speed in one direction, but, of course, he does not know which direction his adversary will take. If the destroyer's best speed is twice that of the submarine, what path should the Spy's destroyer travel to guarantee that it will eventually pass over the submarine? *Hint:* The Spy may need the formula for polar arc length given in Problem 65.

C 61. Find parametric equations for the polar curve $r = f(\theta)$ in terms of the parameter θ.

62. MODELING PROBLEM A circle of radius R rolls without slipping on the outside of a fixed circle of radius a. Assume the fixed circle is centered at the origin and that the moving point begins at $(a, 0)$. Use the angle t measured from the positive x-axis to the ray from the origin to the center of the rolling circle. Show that this *epicycloid* may be modeled by the parametric equations

$$x = (a + R)\cos t - R \cos\left(\frac{a + R}{R}\right)t,$$

$$y = (a + R)\sin t - R \sin\left(\frac{a + R}{R}\right)t$$

63. MODELING PROBLEM A circle of radius R rolls without slipping on the inside of a fixed circle of radius a. Find parametric equations to model the curve traced out by a point P on the circumference of the rolling circle of radius R. Let t be the angle measured from the positive x-axis to the ray that passes through the center of the rolling circle, and assume that the point P begins on the x-axis (that is, P has coordinates $(a, 0)$ when $t = 0$). Show

that this *hypocycloid* has parametric equations

$$x = (a - R)\cos t + R \cos\left(\frac{a - R}{R}\right)t,$$

$$y = (a - R)\sin t - R \sin\left(\frac{a - R}{R}\right)t$$

64. Prove Theorem 9.4.

65. Arc length in polar coordinates Suppose the polar function $r = f(\theta)$ and its derivative $f'(\theta)$ are continuous on the closed interval $a \le \theta \le b$, $(b \le a + 2\pi)$. Show that the arc length of the graph of f from $\theta = a$ to $\theta = b$ is given by

$$L = \int_a^b \sqrt{[f(\theta)]^2 + [f'(\theta)]^2} \, d\theta$$

66. JOURNAL PROBLEM (*The MATYC Journal,* Problem 148 by Frank Kocher, Vol. 14, 1980, p. 155). A rectangle is "rolled" along a straight line by rotations of 90° about each vertex in turn. One vertex traces out a series of arches where each arch is the union of three circular arcs. Show that the area of the region bounded by the straight line and one arch equals the area of the rectangle plus twice the area of the circumscribed circle.

Chapter 9   Review

Proficiency Examination

Concept Problems

1. What are the formulas for changing from polar to rectangular form?
2. What are the formulas for changing from rectangular to polar form?
3. What does a cardioid look like, and what are the standard-form equations for cardioids?
4. What is the formula for the rotation of a polar-form equation?
5. How do you check for symmetry with a polar-form graph?
6. What does a limaçon look like, and what are the standard-form equations for limaçons?
7. What is a rose curve? What are the standard-form equations for rose curves, and how do you determine the number of petals?
8. What does a lemniscate look like, and what are the standard-form equations for lemniscates?
9. State the formula for finding the slope of a polar-form curve.
10. What is the procedure for finding the intersection of polar-form curves?
11. Outline a procedure for finding the area bounded by a polar graph.
12. What do we mean by parametric equations for a curve?
13. What are an epicycloid and a hypocycloid?
14. How do you find the derivative with parametric equations?
15. What is the parametric formula for arc length?
16. State a formula for finding area under a parametrically defined curve.

Practice Problems

Identify and then sketch the curves whose equations are given in Problems 17–21.

17. $r = 2 \cos \theta$
18. $r = \cos \theta - 1$
19. $r = 3 \sin 2\theta$
20. $r^2 = \cos 2\theta$
21. $r = 4 - 2 \cos \theta$
22. Find the area of the intersection of the circles $r = 2a \cos \theta$ and $r = 2a \sin \theta$, where $a > 0$ is constant.
23. Find the length of the curve given by $x = 1 + \cos 2t$, $y = \sin 2t$, for $-\frac{\pi}{4} \le t \le \frac{\pi}{4}$.
24. A curve C is given parametrically by $x = 2t^2 + 1$, $y = t - 1$, for all t. Find all points on C where the tangent line passes through the point $(7, 1)$.
25. A particle moves along the parabola $r = \dfrac{4}{1 + \cos \theta}$. At a certain instant, $\dfrac{dr}{dt} = -3$ cm/s, and $\theta = \dfrac{\pi}{4}$. Find $\dfrac{d\theta}{dt}$ at this instant.

Supplementary Problems

Name and sketch the curves whose equations are given in Problems 1–16.

1. $r = 2 \sin \theta - 3 \cos \theta$
2. $r = -2\theta, \theta \ge 0$
3. $r = -4 \cos 2\theta$
4. $r = 2 \cos \theta \sin \theta$
5. $r = -\csc \theta$
6. $r = \sin 3\theta$
7. $r = \cos(\theta - \frac{\pi}{3})$
8. $r = \sin(\frac{\pi}{3} + \theta)$
9. $r^2 = 3 \cos 2\theta$
10. $r^2 = \cos \theta$
11. $r = \dfrac{4}{1 - \cos \theta}$
12. $r = \dfrac{6}{1 + \sin \theta}$
13. $r = \dfrac{5}{2 + 3 \sin \theta}$
14. $r = \dfrac{-4}{2 + \cos \theta}$
15. $r = 5 \sin 2\theta \csc \theta$
16. $r = 3 \sin(2\theta - 1)$

Sketch the curves in Problems 17–20 and find all points where the tangent line is horizontal.

17. $r = 1 - 2\cos\theta$

18. $r = 2 + 4\sin\theta$

19. $r = \dfrac{4}{1 - \sin\theta}$

20. $r = \dfrac{3}{4 - 3\cos\theta}$

Find the points of intersection of the polar-form curves in Problems 21–30.

21. $\begin{cases} r = 2\cos\theta \\ r = 1 + \cos\theta \end{cases}$

22. $\begin{cases} r = 1 + \cos\theta \\ r = 1 - \cos\theta \end{cases}$

23. $\begin{cases} r = 1 + \sin\theta \\ r = 1 + \cos\theta \end{cases}$

24. $\begin{cases} r^2 = \cos 2(\theta - \frac{\pi}{2}) \\ r^2 = \cos 2\theta \end{cases}$

25. $\begin{cases} r\cos\theta + 2r\sin\theta = 4 \\ r = 2\sec\theta \end{cases}$

26. $\begin{cases} r = 2 - 2\cos\theta \\ r = 2\cos\theta \end{cases}$

27. $\begin{cases} r = 2\sin 2\theta \\ r = 1 \text{ for } 0 \le \theta \le \frac{\pi}{2} \end{cases}$

28. $\begin{cases} r = a(1 + \cos\theta) \\ r = a(1 - \sin\theta) \end{cases}$

29. $\begin{cases} r = a(1 + \sin\theta) \\ r = a(1 - \sin\theta) \end{cases}$

30. $\begin{cases} r = 2\theta \\ \theta = \frac{\pi}{6} \end{cases}$

Find the area of the regions described in Problems 31–37.

31. the region bounded by the polar curve $r = 4\cos 2\theta$

32. the region inside both the circles $r = 1$ and $r = 2\sin\theta$

33. the region inside one loop of the lemniscate $r^2 = \cos 2\theta$

34. the region inside one petal of the rose curve $r = \sin 2\theta$

35. the region inside the circle $r = 2a\sin\theta$ and outside the circle $r = a$

36. the region under the curve given parametrically by
$$x = -\cos^4\theta, \quad y = \sin\theta, \quad \text{for } 0 \le \theta \le \tfrac{\pi}{4}$$

37. the region bounded by the spiral $r = e^{2\theta}$ and the x-axis for $0 \le \theta < \pi$

38. Find the slope of the tangent line to the spiral $r = 2\theta$ at the point where $\theta = \frac{\pi}{3}$. Find an equation (in either polar or Cartesian coordinates) for this tangent line.

39. Find the length of the curve given by $x = \dfrac{\cos 2t}{4}$, $y = \sin t$ for $0 \le t \le \frac{\pi}{2}$.

40. Find the arc length of the curve defined by $x = 1/t$, $y = 1/t^2$ over $1 \le t \le 2$.

41. A curve C is described parametrically by $x = a\cot\theta$, $y = a\sin^2\theta$. Show that C has Cartesian form
$$y = \dfrac{a^3}{a^2 + x^2}$$
and sketch its graph. This curve, called the "witch of Agnesi," was first mentioned in Problem 60 of Section 4.4.

42. a. Find an equation for the tangent line to the curve C given by $x = 4\cos t$, $y = 3\sin t$, at the point where $t = \pi/6$.
b. Find an equation for the normal line to the curve C in part **a** at the point where $t = \pi/4$.

43. Find the equation of the tangent line to the cycloid
$x = a(\theta - \sin\theta)$, $y = a(1 - \cos\theta)$ at the point where
a. $\theta = \frac{\pi}{2}$ **b.** $\theta = \pi$ **c.** $\theta = 2\pi$

44. a. Show that the tangent line to the cycloid
$$x = a(\theta - \sin\theta), y = a(1 - \cos\theta), \text{ has slope } \cot\frac{\theta_0}{2} \text{ at}$$
the point where $\theta = \theta_0$.
b. For what values of θ is the tangent line to the cycloid horizontal?
c. For what values of θ is the tangent vertical?

45. Find a polar equation for the line through (r_1, θ_1) and (r_2, θ_2).

46. Find the area under one arch of the cycloid $x = a(\theta - \sin\theta)$, $y = a(1 - \cos\theta)$, where $a > 0$ is a constant.

47. Find the length of one arch of the cycloid $x = 9(\theta - \sin\theta)$, $y = 9(1 - \cos\theta)$.

48. Show that the parametric equations
$$x = \dfrac{1 - t^2}{1 + t^2}, \quad y = \dfrac{2t}{1 + t^2}$$
describe the unit circle $x^2 + y^2 = 1$.

49. Find all points on the curve C defined by $x = 3 - 4\sin t$, $y = 4 + 3\cos t$ where the tangent line is horizontal.

50. Let $x = \sin t + \cos t$, $y = \cos t - \sin t$.
a. Show that $\dfrac{dy}{dx} = \dfrac{-x}{y}$.
b. Find a relationship between x and y.

51. Suppose $x = x(t)$, $y = y(t)$ for $a < t < b$. Show that
$$\dfrac{d^2y}{dx^2} = \dfrac{\dfrac{d^2y}{dt^2}\dfrac{dx}{dt} - \dfrac{d^2x}{dt^2}\dfrac{dy}{dt}}{\left(\dfrac{dx}{dt}\right)^3}$$

52. Find the slope of the tangent line to the polar curve
    $r = a \tan \dfrac{\theta}{2}$ at the point where $\theta = \dfrac{\pi}{2}$.

53. A particle moves along the circle $r = 4 \cos \theta$. If $\dfrac{d\theta}{dt} = 2$,
    find the rate at which r is changing with respect to t when
    $\theta = \frac{\pi}{4}$.

54. Sketch the curve $r = \sec \theta - 2 \cos \theta$ for $-\dfrac{\pi}{2} < \theta < \dfrac{\pi}{2}$,
    and find the area enclosed by its loop. This curve is called
    a **strophoid.**

55. THINK TANK PROBLEM   The circle $r = 1$ of radius 1
    centered at the origin has the property that any "diame-
    ter" (line through the origin) intersects the circle in a seg-
    ment of length 2.
    a. Show that the cardioid $r = 1 + \sin \theta$ has the same
       property.
    b. Show that the regions bounded by the cardioid and
       the circle have different areas. Thus, the area of a
       region cannot be determined in terms of its diameters
       alone.

56. MODELING PROBLEM   A straight rod of fixed length
    L moves in such a way that its top always touches the
    y-axis and its bottom touches the x-axis. Let $P(x, y)$ be
    the point on the rod that is $\frac{3}{4}$ the distance from the bot-
    tom to the top of the rod. Find parametric equations to
    model the path that P traces out as the rod slides down
    the y-axis.

57. Sketch the graph of the epitrochoid
    $$x = 4 \cos \theta + 3 \cos 5\theta, \; y = 4 \sin \theta + 3 \sin 5\theta$$

58. Find the length of the hypocycloid
    $$x = (a - b) \cos \theta + b \cos \left(\dfrac{a - b}{b} \right) \theta$$
    $$y = (a - b) \sin \theta - b \sin \left(\dfrac{a - b}{b} \right) \theta$$
    for $0 \le \theta \le \dfrac{b}{a} \pi$ (assume $a > b > 0$).

59. Three LORAN stations are located at $(a, 0)$, $(0, 0)$, and
    $(a, \frac{\pi}{4})$ represented in polar coordinates. Radio signals are
    sent from all three stations simultaneously. A ship receiv-
    ing the signals notes that the signals from the second and
    third stations arrive $a/(2v)$ seconds later than that from
    the first, where v is the velocity of radio signals. Model
    this situation in polar coordinates and locate the position
    of the ship.

60. PUTNAM EXAMINATION PROBLEM   An ellipse
    whose semi-axes have lengths a and b rolls without slip-
    ping on the curve $y = c \sin (x/a)$. How are a, b, and c
    related, given that the ellipse completes one revolution
    when it traverses one period of the curve? (*Hint:* The
    ellipse may be described parametrically by $x = a \cos \theta$,
    $y = b \sin \theta$ for $0 \le \theta \le 2\pi$.)

Security Systems

This project is to be done in groups of three or four students. Each group will submit a single written report.

You are designing a security system for a hospital. You must decide how to program a detector that will be used to watch a 40-ft-long hallway with a door in the middle. The detector runs on a track and points a beam of light straight ahead on the opposite wall. The beam reaches from the floor to the ceiling. Design a security system that will detect an intruder.

You should make the following assumptions:

a. Think of the hallway as a coordinate line with the middle of the door at the origin and the hallway to be watched on the interval $[-20, 20]$. You need to decide what $x(t)$ is for t, in seconds, where $x(t)$ represents the position of the beam at time t. For example, $x(5) = -15$ means that the beam is pointing at the part of the wall 15 ft to the left of the door 5 seconds after the detector starts.

b. Assume that t ranges from 0 to 30 min.

c. The door is 3 ft wide, and as long as the beam is hitting any part of the door, it is under surveillance.

d. The beam must stay on an object for at least 0.1 seconds to detect that object.

Your paper is not limited to the following questions but should include these concerns. Draw a graph of x versus time for what you think is a good choice for $x(t)$. You should mention restrictions on this function, if any, and include the *longest* time interval that the door will not be under surveillance. Could an intruder get to the door by walking down the hallway without being detected by your system? Explain how she could do it and how likely you think it is.

*Everything that the greatest minds of all times have accomplished toward the **comprehension of forms** by means of concepts is gathered into one great science: **mathematics.***

J. F. HERBART

---

*The group project is courtesy of Steve Hilbert, Ithaca College, Ithaca, New York.

10 Vectors in the Plane and in Space

PREVIEW

In this chapter, we focus on various algebraic and geometric aspects of vector representations. Then in Chapter 11, we see how vectors can be combined with calculus to study motion in space and other applications.

PERSPECTIVE

Suppose a child is pulling a sled by a rope across a flat field. Intuitively, we would expect quite different results if the child pulls the rope straight up than if the same effort is applied at an angle of $\pi/4$, for instance. In other words, to describe the force exerted by the child on the sled, we must specify not only a magnitude but also a direction. In general, a *scalar* quantity is one that can be described in terms of magnitude alone, whereas a *vector* quantity requires both magnitude and direction. Force, velocity, and acceleration are common vector quantities, and an important goal of this chapter is to develop methods for dealing with such quantities.

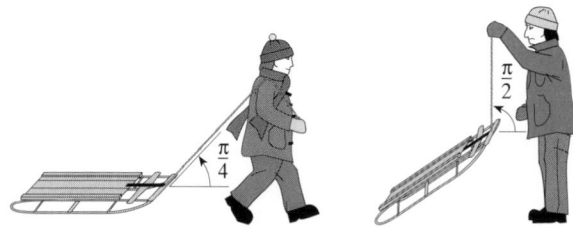

10.1   Vectors in the Plane

IN THIS SECTION   introduction to vectors, standard representation of vectors in the plane   ■

INTRODUCTION TO VECTORS

A **vector** in a plane can be thought of as a directed line segment, an "arrow" with **initial point** P and **terminal point** Q. The direction of the vector is that of the arrow, and its magnitude is represented by the arrow's length, as shown in Figure 10.1. We shall indicate such a vector by writing **PQ** in boldface type, but in your work you may write an arrow over the designated points: $\overrightarrow{PQ}$. The order of letters you write down is important: **PQ** means that the vector is from P to Q, but **QP** means that the vector is from Q to P. The first letter is the initial point and the second letter is the terminal point. We shall denote the magnitude (length) of a vector by $\|\mathbf{PQ}\|$. Two vectors are regarded as **equal** (or **equivalent**) if they have the same magnitude and the same direction, even if they do not coincide.

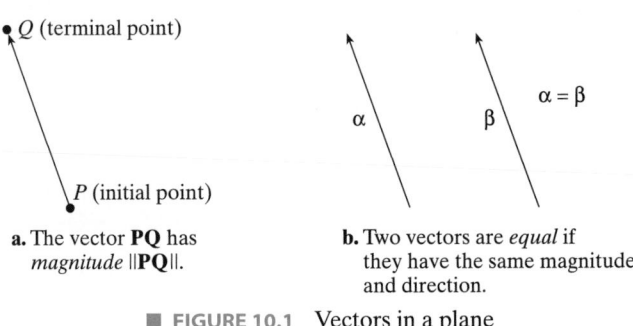

a. The vector **PQ** has *magnitude* $\|\mathbf{PQ}\|$.

b. Two vectors are *equal* if they have the same magnitude and direction.

■ **FIGURE 10.1**   Vectors in a plane

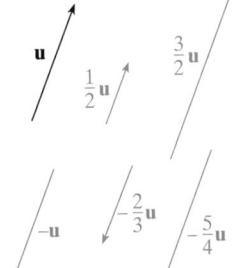

■ **FIGURE 10.2**   Some multiples of the vector **u**

A vector with magnitude 0 is called a **zero** (or **null**) **vector** and is denoted by **0**. The **0** vector has no specific direction, and we shall adopt the convention of assigning it any direction that may be convenient in a particular problem.

If **v** is a vector other than **0**, then any vector **w** that is parallel to **v** is called a **scalar multiple** of **v** and satisfies $\mathbf{w} = s\mathbf{v}$ for some nonzero number s. A **scalar** quantity is one that has only magnitude and, in the context of vectors, is used to describe a real number. The scalar multiple $s\mathbf{v}$ has length $|s|$ times that of **v**; it points in the same direction as **v** if $s > 0$ and the opposite direction if $s < 0$ (see Figure 10.2). Notice that for any distinct points P and Q, $\mathbf{PQ} = -\mathbf{QP}$. For the zero vector **0**, we define $s\mathbf{0} = \mathbf{0}$ for any scalar s.

Physical experiments indicate that force and velocity vectors can be added (or **resolved**) according to a **triangular rule** displayed in Figure 10.3a, and we use this rule as our definition of vector addition. In particular, to add the vector **v** to the vector **u**, we place the end (initial point) of **v** at the tip (terminal point) of **u** and define the **sum**, also called the **resultant, u + v**, to be the vector that extends from the initial point of **u** to the terminal point of **v**.

Equivalently, **u + v** is the diagonal of the parallelogram formed with sides **u** and **v**, as shown in Figure 10.3b. The **difference u − v** is just the vector **w** that satisfies **v + w = u**, and it may be found by placing the initial points of **u** and **v** together and extending a vector from the terminal point of **v** to the terminal point of **u** (see Figure 10.3c).

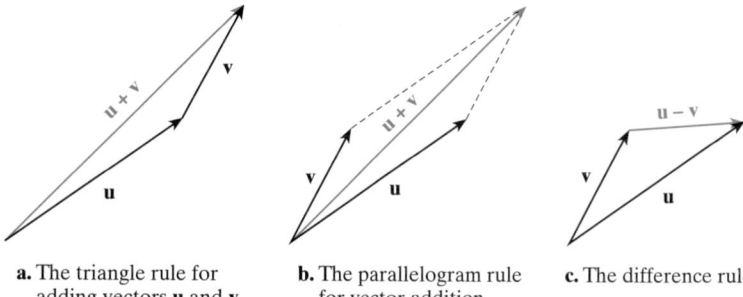

a. The triangle rule for adding vectors **u** and **v**

b. The parallelogram rule for vector addition

c. The difference rule

■ **FIGURE 10.3** Vector addition and subtraction

A vector **OQ** with initial point at the origin O of a coordinate plane can be uniquely represented by specifying the coordinates of its terminal point Q. If Q has coordinates (a, b), we denote the vector **OQ** by $\langle a, b \rangle$, where the pointed brackets $\langle \ \rangle$ are used to distinguish the *vector* **OQ** $= \langle a, b \rangle$ from the *point* (a, b).

Technology Window

Brackets are used to represent vectors on some calculators and on most programs, such as *Mathematica*, *Maple*, *Derive*, and *Mathlab*. For example, the vector **QB** is input as $[a, b]$. This is called a *two-dimensional vector* because it has two components. In this book, we consider vectors with three or more components, and these are represented on a calculator in a similar way:

$$\mathbf{V} = [2, 3, 4] \qquad \mathbf{W} = [0, -3, \tfrac{\pi}{2}]$$

V and **W** are called *three-dimensional vectors*.

The vector **PQ** with initial point $P(c, d)$ and terminal point $Q(a, b)$ can be denoted by **PQ** $= \langle a - c, b - d \rangle$. Using analytic geometry (see Problem 55), it can be shown that **PQ** equals vector **OR** with initial point at the origin $(0, 0)$ and terminal point $R(a - c, b - d)$, as shown in Figure 10.4.

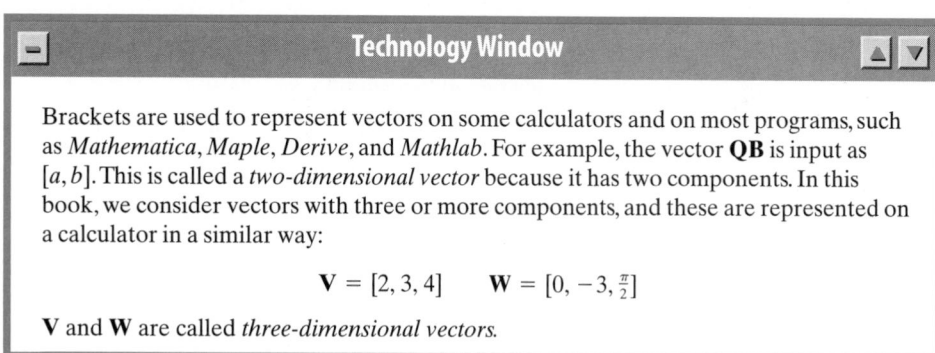

■ **FIGURE 10.4** Given $P(c, d)$ and $Q(a, b)$, the vector **PQ** is $\langle a - c, b - d \rangle$.

Vector operations are easily represented when vectors are given in component form. In particular, we have

$$
\begin{aligned}
\langle a_1, b_1 \rangle &= \langle a_2, b_2 \rangle && \text{if and only if } a_1 = a_2 \text{ and } b_1 = b_2 \\
k\langle a, b \rangle &= \langle ka, kb \rangle && \text{for constant } k \\
\langle a, b \rangle + \langle c, d \rangle &= \langle a + c, b + d \rangle \\
\langle a, b \rangle - \langle c, d \rangle &= \langle a - c, b - d \rangle
\end{aligned}
$$

These formulas may be verified by analytic geometry. For instance, the rule for multiplication by a scalar may be obtained by using the relationships in Figure 10.5a, and Figure 10.5b can be used to obtain the rule for vector addition.

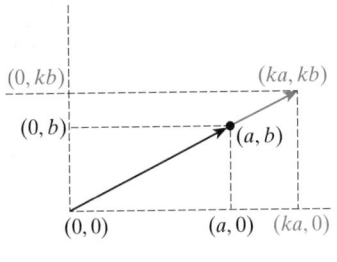

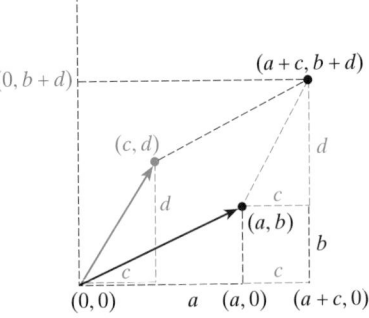

a. Multiplication by a scalar:
$k\langle a, b \rangle = \langle ka, kb \rangle$

b. Vector addition
$\langle a, b \rangle + \langle c, d \rangle = \langle a + c, b + d \rangle$

■ **FIGURE 10.5**   Vector operations

EXAMPLE 1   *Vector operations*

For the vectors $\mathbf{u} = \langle 2, -3 \rangle$ and $\mathbf{v} = \langle -1, 7 \rangle$ find:
a. $\mathbf{u} + \mathbf{v}$ b. $\frac{3}{4}\mathbf{u}$ c. $3\mathbf{u} - \frac{1}{2}\mathbf{v}$

Solution
a. $\mathbf{u} + \mathbf{v} = \langle 2, -3 \rangle + \langle -1, 7 \rangle = \langle 2 + (-1), -3 + 7 \rangle = \langle 1, 4 \rangle$
b. $\frac{3}{4}\mathbf{u} = \frac{3}{4}\langle 2, -3 \rangle = \langle \frac{3}{4}(2), \frac{3}{4}(-3) \rangle = \langle \frac{3}{2}, \frac{-9}{4} \rangle$
c. $3\mathbf{u} - \frac{1}{2}\mathbf{v} = 3\langle 2, -3 \rangle - \frac{1}{2}\langle -1, 7 \rangle$

   $\quad = \langle 6, -9 \rangle + \langle \frac{1}{2}, -\frac{7}{2} \rangle$ *Scalar multiplication*

   $\quad = \langle 6 + \frac{1}{2}, -9 - \frac{7}{2} \rangle$ *Add vectors.*

   $\quad = \langle \frac{13}{2}, -\frac{25}{2} \rangle$ *Simplify.*   ■

In general, an expression of the form $a\mathbf{u} + b\mathbf{v}$ is called a **linear combination** of the vectors $\mathbf{u}$ and $\mathbf{v}$. Note that if $\mathbf{u} = \langle u_1, u_2 \rangle$ and $\mathbf{v} = \langle v_1, v_2 \rangle$, then

$$a\mathbf{u} + b\mathbf{v} = a\langle u_1, u_2 \rangle + b\langle v_1, v_2 \rangle = \langle au_1 + bv_1, au_2 + bv_2 \rangle$$

Vector addition and multiplication of a vector by a scalar behave much like ordinary addition and multiplication. The following theorem lists several useful properties of these operations.

THEOREM 10.1   *Properties of vector operations*

For any vectors $\mathbf{u}$, $\mathbf{v}$, and $\mathbf{w}$ in the plane and scalars s and t:

Commutativity of vector addition	$\mathbf{u} + \mathbf{v} = \mathbf{v} + \mathbf{u}$
Associativity of vector addition	$(\mathbf{u} + \mathbf{v}) + \mathbf{w} = \mathbf{u} + (\mathbf{v} + \mathbf{w})$
Associativity of scalar multiplication	$(st)\mathbf{u} = s(t\mathbf{u})$
Identity for addition	$\mathbf{u} + \mathbf{0} = \mathbf{u}$
Inverse property for addition	$\mathbf{u} + (-\mathbf{u}) = \mathbf{0}$
Distributivity laws	$(s + t)\mathbf{u} = s\mathbf{u} + t\mathbf{u}$
	$s(\mathbf{u} + \mathbf{v}) = s\mathbf{u} + s\mathbf{v}$

Proof   Each vector property can be established by using a corresponding property of real numbers. For example, to prove associativity of vector addition, let $\mathbf{u} = \langle u_1, u_2 \rangle$, $\mathbf{v} = \langle v_1, v_2 \rangle$, and $\mathbf{w} = \langle w_1, w_2 \rangle$. Then

$$(\mathbf{u} + \mathbf{v}) + \mathbf{w} = (\langle u_1, u_2 \rangle + \langle v_1, v_2 \rangle) + \langle w_1, w_2 \rangle$$
$$= \langle u_1 + v_1, u_2 + v_2 \rangle + \langle w_1, w_2 \rangle$$
$$= \langle (u_1 + v_1) + w_1, (u_2 + v_2) + w_2 \rangle$$
$$= \langle u_1 + (v_1 + w_1), u_2 + (v_2 + w_2) \rangle$$

Associativity of addition for the real numbers

$$= \langle u_1, u_2 \rangle + (\langle v_1, v_2 \rangle + \langle w_1, w_2 \rangle)$$
$$= \mathbf{u} + (\mathbf{v} + \mathbf{w})$$

You are asked to prove the other six properties in the problem set.

EXAMPLE 2   *Vector proof of a geometric property*

Show that the line segment joining the midpoints of two sides of a triangle is parallel to the third side and has half its length.

Solution   Consider $\triangle ABC$, and let P and Q be the midpoints of sides $\overline{AC}$ and $\overline{BC}$, respectively, as shown in Figure 10.6.

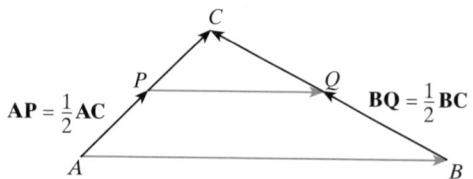

$$\mathbf{AP} = \tfrac{1}{2}\mathbf{AC} \qquad\qquad \mathbf{BQ} = \tfrac{1}{2}\mathbf{BC}$$

■ **FIGURE 10.6**   Vector proof of a geometric property

Given: $\mathbf{AP} = \tfrac{1}{2}\mathbf{AC}$ and $\mathbf{BQ} = \tfrac{1}{2}\mathbf{BC}$.

Proof: We want to show that $\mathbf{PQ}$ is parallel to $\mathbf{AB}$ and $\left\|\mathbf{PQ}\right\| = \tfrac{1}{2}\left\|\mathbf{AB}\right\|$, which means that we must establish the vector equation $\mathbf{PQ} = \tfrac{1}{2}\mathbf{AB}$. Toward this end, we begin by noting that $\mathbf{AB}$ can be expressed as the following vector sum:

$$\mathbf{AB} = \mathbf{AP} + \mathbf{PQ} + \mathbf{QB}$$
$$= \tfrac{1}{2}\mathbf{AC} + \mathbf{PQ} - \mathbf{BQ} \qquad \mathbf{AP} = \tfrac{1}{2}\mathbf{AC} \text{ and } \mathbf{QB} = -\mathbf{BQ}$$
$$= \tfrac{1}{2}(\mathbf{AB} + \mathbf{BC}) + \mathbf{PQ} - \tfrac{1}{2}\mathbf{BC} \qquad \mathbf{AC} = (\mathbf{AB} + \mathbf{BC}) \text{ and } \mathbf{BQ} = \tfrac{1}{2}\mathbf{BC}$$
$$= \tfrac{1}{2}\mathbf{AB} + \tfrac{1}{2}\mathbf{BC} + \mathbf{PQ} - \tfrac{1}{2}\mathbf{BC}$$
$$= \tfrac{1}{2}\mathbf{AB} + \mathbf{PQ}$$
$$\tfrac{1}{2}\mathbf{AB} = \mathbf{PQ} \qquad\qquad \text{Subtract } \tfrac{1}{2}\mathbf{AB} \text{ from both sides.} \quad ■$$

When a vector $\mathbf{u}$ is represented in component from $\mathbf{u} = \langle u_1, u_2 \rangle$, its length is given by the formula

$$\left\|\mathbf{u}\right\| = \sqrt{u_1^2 + u_2^2}$$

This is a simple application of the Pythagorean theorem, as shown in Figure 10.7. Another important relationship involving the length of vectors is the *triangle inequality*

$$\left\|\mathbf{u} + \mathbf{v}\right\| \leq \left\|\mathbf{u}\right\| + \left\|\mathbf{v}\right\|$$

for any vectors **u** and **v**. Equality will occur precisely when **u** and **v** are multiples of one another (that is, when **u** and **v** have the same direction). To establish the inequality, we observe that **u** and **v** are two sides of a triangle in the plane, and that the third side has length $\|\mathbf{u} + \mathbf{v}\|$ and is "shorter" than the sum $\|\mathbf{u}\| + \|\mathbf{v}\|$ of the lengths of the other two sides, as shown in Figure 10.7.

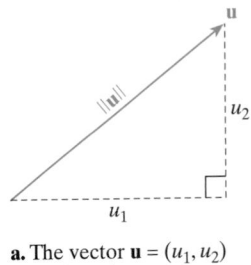

a. The vector $\mathbf{u} = (u_1, u_2)$ has length
$\|\mathbf{u}\| = \sqrt{u_1{}^2 + u_2{}^2}$

b. The triangle inequality
$\|\mathbf{u} + \mathbf{v}\| \le \|\mathbf{u}\| + \|\mathbf{v}\|$

■ **FIGURE 10.7**   Geometric representation of two vector properties

---

EXAMPLE 3   *Speed and heading of a motorboat*

A river 4 mi wide flows south with a current of 5 mi/h. What speed and heading should a motorboat assume to travel directly across the river from east to west in 20 min?

Solution   Begin by drawing a diagram, as shown in Figure 10.8.

Let **B** be the velocity vector of the boat in the direction of the angle θ. If the river's current has velocity **C**, the given information tells us that $\|\mathbf{C}\| = 5$ mi/h and that **C** points directly south. Moreover, because the boat is to cross the river from east to west in 20 min (that is, $\frac{1}{3}$ h), its *effective velocity* after compensating for the current is a vector **V** that points west and has magnitude

$$\|\mathbf{V}\| = \frac{\text{WIDTH OF THE RIVER}}{\text{TIME OF CROSSING}} = \frac{4 \text{ mi}}{\frac{1}{3} \text{ h}} = 12 \text{ mi/h}$$

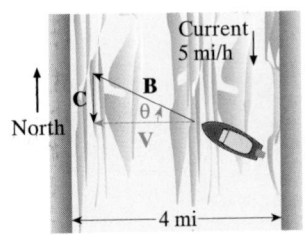

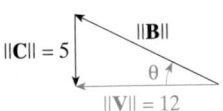

■ **FIGURE 10.8**   A velocity problem

The effective velocity **V** is the resultant of **B** and **C**; that is, $\mathbf{V} = \mathbf{B} + \mathbf{C}$. Because **V** and **C** act in perpendicular directions, we can determine **B** by referring to the right triangle with sides $\|\mathbf{V}\| = 12$ and $\|\mathbf{C}\| = 5$ and hypotenuse $\|\mathbf{B}\|$. We find that

$$\|\mathbf{B}\| = \sqrt{\|\mathbf{V}\|^2 + \|\mathbf{C}\|^2} = \sqrt{12^2 + 5^2} = 13$$

The direction of the velocity vector **V** is given by the angle θ in Figure 10.8, and we find that

$$\tan \theta = \frac{5}{12} \quad \text{so that} \quad \theta = \tan^{-1}\left(\frac{5}{12}\right) \approx 0.3948$$

Thus, the boat should travel at 13 mi/h in a direction of approximately 0.3948 radian. In navigation, it is common to specify direct in degrees rather than radians; this is 22.6° north of west.   ■

A **unit vector** is simply a vector with length 1, and a **direction vector** for a given vector **v** is a unit vector **u** that points in the same direction as **v**. Such a vector can be found by dividing **v** by its length $\|\mathbf{v}\|$; that is,

$$\mathbf{u} = \frac{\mathbf{v}}{\|\mathbf{v}\|}$$

> **EXAMPLE 4**   *Finding a direction vector*

Find a direction vector for the vector $\mathbf{v} = \langle 2, -3 \rangle$.

Solution   The vector $\mathbf{v}$ has length (magnitude) $\|\mathbf{v}\| = \sqrt{2^2 + (-3)^2} = \sqrt{13}$. Thus, the required direction vector is the unit vector

$$\mathbf{u} = \frac{\mathbf{v}}{\|\mathbf{v}\|} = \frac{\langle 2, -3 \rangle}{\sqrt{13}} = \frac{1}{\sqrt{13}}\langle 2, -3 \rangle = \left\langle \frac{2}{\sqrt{13}}, \frac{-3}{\sqrt{13}} \right\rangle$$ ∎

STANDARD REPRESENTATION OF VECTORS IN THE PLANE

The unit vectors $\mathbf{i} = \langle 1, 0 \rangle$ and $\mathbf{j} = \langle 0, 1 \rangle$ point in the directions of the positive x- and y-axes, respectively, and are called **standard basis vectors.** Any vector $\mathbf{v} = \langle v_1, v_2 \rangle$ in the plane can be expressed as a linear combination of the vectors $\mathbf{i}$ and $\mathbf{j}$, because

$$\mathbf{v} = \langle v_1, v_2 \rangle = v_1 \langle 1, 0 \rangle + v_2 \langle 0, 1 \rangle = v_1 \mathbf{i} + v_2 \mathbf{j}$$

This is called the **standard representation** of the vector $\mathbf{v}$, and it can be shown that the representation is unique in the sense that if $\mathbf{v} = a\mathbf{i} + b\mathbf{j}$, then $a = v_1$ and $b = v_2$. In this context, the scalars v_1 and v_2 are called the **horizontal and vertical components of $\mathbf{v}$,** respectively. See Figure 10.9.

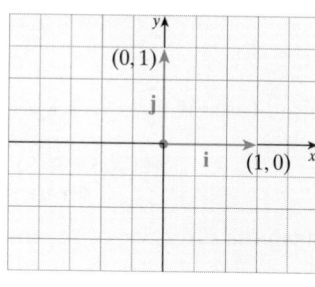
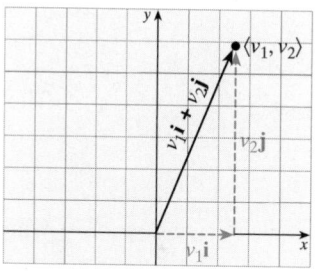

a. The standard basis vectors $\mathbf{i}$ and $\mathbf{j}$

b. Any vector $\mathbf{v} = \langle v_1, v_2 \rangle$ can be expressed uniquely as $\mathbf{v} = v_1 \mathbf{i} + v_2 \mathbf{j}$

■ **FIGURE 10.9**   Standard representation of vectors in the plane

> **EXAMPLE 5**   *Finding the standard representation of a vector*

If $\mathbf{u} = 3\mathbf{i} + 2\mathbf{j}$, $\mathbf{v} = -2\mathbf{i} + 5\mathbf{j}$, and $\mathbf{w} = \mathbf{i} - 4\mathbf{j}$, what is the standard representation of the vector $2\mathbf{u} + 5\mathbf{v} - \mathbf{w}$?

Solution   Using Theorem 10.1, we find that

$$\begin{aligned}
2\mathbf{u} + 5\mathbf{v} - \mathbf{w} &= 2(3\mathbf{i} + 2\mathbf{j}) + 5(-2\mathbf{i} + 5\mathbf{j}) - (\mathbf{i} - 4\mathbf{j}) \\
&= [2(3) + 5(-2) - 1]\mathbf{i} + [2(2) + 5(5) - (-4)]\mathbf{j} \\
&= -5\mathbf{i} + 33\mathbf{j}
\end{aligned}$$ ∎

> **EXAMPLE 6**   *Finding standard representation of a vector connecting two points*

Find the standard representation of the vector $\mathbf{PQ}$ for the points $P(3, -4)$ and $Q(-2, 6)$.

Solution   The component form of $\mathbf{PQ}$ is

$$\mathbf{PQ} = \langle (-2) - 3, 6 - (-4) \rangle = \langle -5, 10 \rangle$$

This means that $\mathbf{PQ}$ has the standard representation $\mathbf{PQ} = -5\mathbf{i} + 10\mathbf{j}$.   ∎

| EXAMPLE 7 | *Computing a resultant force* |

Two forces $\mathbf{F}_1$ and $\mathbf{F}_2$ act on the same body. It is known that $\mathbf{F}_1$ has magnitude 3 newtons and acts in the direction of $-\mathbf{i}$, whereas $\mathbf{F}_2$ has magnitude 2 newtons and acts in the direction of the unit vector

$$\mathbf{u} = \tfrac{3}{5}\mathbf{i} - \tfrac{4}{5}\mathbf{j}$$

What additional force $\mathbf{F}_3$ must be applied to keep the body at rest?

Solution   According to the given information, we have

$$\mathbf{F}_1 = 3(-\mathbf{i}) = -3\mathbf{i} \quad \text{and} \quad \mathbf{F}_2 = 2(\tfrac{3}{5}\mathbf{i} - \tfrac{4}{5}\mathbf{j}) = \tfrac{6}{5}\mathbf{i} - \tfrac{8}{5}\mathbf{j}$$

and we want to find $\mathbf{F}_3 = a\mathbf{i} + b\mathbf{j}$ so that $\mathbf{F}_1 + \mathbf{F}_2 + \mathbf{F}_3 = \mathbf{0}$. Substituting into this vector equation, we obtain

$$(-3\mathbf{i}) + (\tfrac{6}{5}\mathbf{i} - \tfrac{8}{5}\mathbf{j}) + (a\mathbf{i} + b\mathbf{j}) = 0\mathbf{i} + 0\mathbf{j}$$

By combining terms on the left, we find that

$$(-3 + \tfrac{6}{5} + a)\mathbf{i} + (-\tfrac{8}{5} + b)\mathbf{j} = 0\mathbf{i} + 0\mathbf{j}$$

Because the standard representation is unique, we must have

$$-3 + \tfrac{6}{5} + a = 0 \quad \text{and} \quad -\tfrac{8}{5} + b = 0$$
$$a = \tfrac{9}{5} \qquad\qquad b = \tfrac{8}{5}$$

The required force is $\mathbf{F}_3 = \tfrac{9}{5}\mathbf{i} + \tfrac{8}{5}\mathbf{j}$. This is a force of magnitude

$$\|\mathbf{F}_3\| = \sqrt{\left(\frac{9}{5}\right)^2 + \left(\frac{8}{5}\right)^2} = \frac{1}{5}\sqrt{145} \text{ newtons}$$

which acts in the direction of the unit vector

$$\mathbf{v} = \frac{\mathbf{F}_3}{\|\mathbf{F}_3\|} = \frac{5}{\sqrt{145}}\left(\frac{9}{5}\mathbf{i} + \frac{8}{5}\mathbf{j}\right) = \frac{9}{\sqrt{145}}\mathbf{i} + \frac{8}{\sqrt{145}}\mathbf{j}$$

■

10.1   Problem Set

A *Sketch each vector given in Problems 1–4 assuming that its initial point is at the origin.*

1. $3\mathbf{i} - 4\mathbf{j}$

2. $-2\mathbf{i} - 3\mathbf{j}$

3. $-\tfrac{1}{2}\mathbf{i} + \tfrac{5}{2}\mathbf{j}$

4. $-2(-\mathbf{i} + 2\mathbf{j})$

The initial point P and terminal point Q of a vector are given in Problems 5–8. Sketch each vector and then write it in component form.

5. $P(3, -1), Q(7, 2)$

6. $P(5, -2), Q(5, 8)$

7. $P(3, 4), Q(-2, 4)$

8. $P(\tfrac{1}{2}, 6), Q(-3, -2)$

*Express each vector **PQ** in Problems 9–12 in standard form and also find its length.*

9. $P(-1, -2), Q(1, -2)$

10. $P(5, 7), Q(6, 8)$

11. $P(-4, -3), Q(0, -1)$

12. $P(3, -5), Q(2, 8)$

Find a unit vector that points in the direction of each of the vectors given in Problems 13–16.

13. $\mathbf{i} + \mathbf{j}$

14. $\tfrac{1}{2}\mathbf{i} + \tfrac{1}{4}\mathbf{j}$

15. $3\mathbf{i} - 4\mathbf{j}$

16. $-4\mathbf{i} + 7\mathbf{j}$

Let $\mathbf{u} = \langle -3, 4\rangle$ and $\mathbf{v} = \langle 1, -1\rangle$. Find scalars s and t so that the given equation in Problems 17–20 is satisfied.

17. $s\mathbf{u} + t\mathbf{v} = \langle 6, 0\rangle$

18. $s\langle 0, -3\rangle + t\mathbf{u} = \mathbf{v}$

19. $s\mathbf{v} + t\langle -2, 1\rangle = \mathbf{u}$

20. $s\mathbf{u} + \langle 8, 11\rangle = t\mathbf{v}$

Suppose $\mathbf{u} = 3\mathbf{i} - 4\mathbf{j}$, $\mathbf{v} = 4\mathbf{i} - 3\mathbf{j}$, and $\mathbf{w} = \mathbf{i} + \mathbf{j}$. Express each of the expressions in Problems 21–24 in standard form.

21. $2\mathbf{u} + 3\mathbf{v} - \mathbf{w}$

22. $\tfrac{1}{2}(\mathbf{u} + \mathbf{v}) - \tfrac{1}{4}\mathbf{w}$

23. $\|\mathbf{v}\|\mathbf{u} + \|\mathbf{u}\|\mathbf{v}$

24. $\|\mathbf{u}\|\|\mathbf{v}\|\mathbf{w}$

Find all real numbers x and y that satisfy the vector equations given in Problems 25–28.

25. $(x - y - 1)\mathbf{i} + (2x + 3y - 12)\mathbf{j} = \mathbf{0}$

26. $x\mathbf{i} - 4y^2\mathbf{j} = (5 - 3y)\mathbf{i} + (10 - 7x)\mathbf{j}$

27. $(x^2 + y^2)\mathbf{i} + y\mathbf{j} = 20\mathbf{i} + (x + 2)\mathbf{j}$

28. $(y - 1)\mathbf{i} + y\mathbf{j} = (\log x)\mathbf{i} + [\log 2 + \log(x + 4)]\mathbf{j}$

*In Problems 29–32, find a unit vector **u** with the given characteristics.*

29. **u** makes an angle of 30° with the positive x-axis.

30. **u** has the same direction as the vector $2\mathbf{i} - 3\mathbf{j}$.

31. **u** has the direction opposite that of $-4\mathbf{i} + \mathbf{j}$.

32. **u** has the direction of the vector from $P(-1, 5)$ to $Q(7, -3)$.

In Problems 33–36, let $\mathbf{u} = 4\mathbf{i} - \mathbf{j}, \mathbf{v} = \mathbf{i} + 2\mathbf{j},$ *and* $\mathbf{w} = -3\mathbf{i} + 4\mathbf{j}.$

33. Find a unit vector in the same direction as $\mathbf{u} + \mathbf{v}$.

34. Find a vector of length 3 with the same direction as $\mathbf{u} - 2\mathbf{v} + 2\mathbf{w}$.

35. Find the terminal point of the vector $5\mathbf{i} + 7\mathbf{j}$ if the initial point is $(-2, 3)$.

36. Find the initial point of the vector $-\mathbf{i} + 2\mathbf{j}$ if the terminal point is $(-1, -2)$.

37. a. Use vectors to find the coordinates of the midpoint of the line segment joining the points $P(-3, -8)$ and $Q(9, -2)$.
   b. What point is located $\frac{5}{6}$ of the distance from P to Q?

38. If $\|\mathbf{v}\| = 3$ and $-3 \le r \le 1$, what are the possible values of $r\|\mathbf{v}\|$?

39. Show that $\mathbf{v} = (\cos\theta)\mathbf{i} + (\sin\theta)\mathbf{j}$ is a unit vector for any angle θ.

40. If **u** and **v** are nonzero vectors and
$$r = \frac{\|\mathbf{u}\|}{\|\mathbf{v}\|}$$
What is $\|r\mathbf{v}\|$?

B **41.** If **u** and **v** are nonzero vectors with $\|\mathbf{u}\| = \|\mathbf{v}\|$, does it follow that $\mathbf{u} = \mathbf{v}$? Explain.

42. If $\mathbf{u} = 2\mathbf{i} - 3\mathbf{j}$ and $\mathbf{v} = x\mathbf{i} + y\mathbf{j}$, describe the set of points in the plane whose coordinates (x, y) satisfy $\|\mathbf{v} - \mathbf{u}\| \le 2$.

43. Let $\mathbf{u}_0 = x_0\mathbf{i} + y_0\mathbf{j}$ for constants x_0 and y_0, and let $\mathbf{u} = x\mathbf{i} + y\mathbf{j}$. Describe the set of all points in the plane whose coordinates satisfy:
   a. $\|\mathbf{u} - \mathbf{u}_0\| = 1$ **b.** $\|\mathbf{u} - \mathbf{u}_0\| \le 2$

44. Let $\mathbf{u} = 3\mathbf{i} - \mathbf{j}$ and $\mathbf{v} = -6\mathbf{i} + 2\mathbf{j}$. Show that there are no numbers a, b for which $a\mathbf{u} + b\mathbf{v} = 2\mathbf{i} + 5\mathbf{j}$.

45. Suppose **u** and **v** are a pair of nonzero, nonparallel vectors. Find numbers a, b, c such that $a\mathbf{u} + b(\mathbf{u} - \mathbf{v}) + c(\mathbf{u} + \mathbf{v}) = \mathbf{0}$.

46. Let $\mathbf{u} = \langle 2, 1 \rangle$ and $\mathbf{v} = \langle -3, 4 \rangle$.
   a. Sketch the vector $c\mathbf{u} + (1 - c)\mathbf{v}$ for the cases where $c = 0, c = \frac{1}{4}, c = \frac{1}{2}, c = \frac{3}{4},$ and $c = 1$.

b. In general, if the initial point of $c\mathbf{u} + (1 - c)\mathbf{v}$ for $0 \le c \le 1$ is at the origin, where is its terminal point (x, y)?

47. Two forces $\mathbf{F}_1 = 3\mathbf{i} + 4\mathbf{j}$ and $\mathbf{F}_2 = 3\mathbf{i} - 7\mathbf{j}$ act on an object. What additional force should be applied to keep the body at rest?

48. Three forces $\mathbf{F}_1 = \mathbf{i} - 2\mathbf{j}, \mathbf{F}_2 = 3\mathbf{i} - 7\mathbf{j},$ and $\mathbf{F}_3 = \mathbf{i} + \mathbf{j}$ act on an object. What additional force $\mathbf{F}_4$ should be applied to keep the body at rest?

49. A river 2.1 mi wide flows south with a current of 3.1 mi/h. What speed and heading should a motorboat assume to travel across the river from east to west in 30 min?

50. Four forces act on an object: $\mathbf{F}_1$ has magnitude 10 lb and acts at an angle of $\frac{\pi}{6}$ measured counterclockwise from the positive x-axis; $\mathbf{F}_2$ has magnitude 8 lb and acts in the direction of the vector $\mathbf{j}$; $\mathbf{F}_3$ has magnitude 5 lb and acts at an angle of $4\pi/3$ measured counterclockwise from the positive x-axis. What must the fourth force $\mathbf{F}_4$ be to keep the object at rest?

C **51.** Use vector methods to show that the diagonals of a parallelogram bisect each other.

52. In a triangle, let **u**, **v**, and **w** be the vectors from each vertex to the midpoint of the opposite side. Use vector methods to show that $\mathbf{u} + \mathbf{v} + \mathbf{w} = \mathbf{0}$.

53. Two nonzero vectors **u** and **v** are said to be **linearly independent** in the plane if they are not parallel.
   a. If **u** and **v** have this property and $a\mathbf{u} = b\mathbf{v}$ for constants a, b, show that $a = b = 0$.
   b. Show that the standard representation of a vector is unique. That is, if the vector **u** has the representation $\mathbf{u} = a_1\mathbf{i} + b_1\mathbf{j}$ and $\mathbf{u} = a_2\mathbf{i} + b_2\mathbf{j}$ is another such representation, then $a_1 = a_2$ and $b_1 = b_2$.

54. Prove that the medians of a triangle intersect at a single point by completing the following argument.

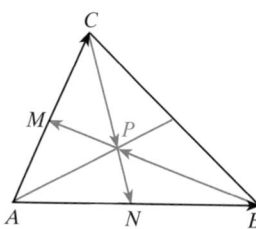

a. Let M and N be the midpoints of sides $\overline{AC}$ and $\overline{AB}$, respectively. Show that
$$\mathbf{CN} = \tfrac{1}{2}\mathbf{AB} - \mathbf{AC} \quad \text{and} \quad \mathbf{BM} = \tfrac{1}{2}\mathbf{AC} - \mathbf{AB}$$
b. Let P be the point where medians $\overline{BM}$ and $\overline{CN}$ intersect, and let r, s be constants such that
$$\mathbf{CP} = r\,\mathbf{CN} \quad \text{and}$$
$$\mathbf{BP} = s\,\mathbf{BM}$$

Note that $\mathbf{CP} + \mathbf{PB} = \mathbf{CB}$. Use this relationship to prove that $r = s = \dfrac{2}{3}$. Explain why this shows that any pair of medians meet at a point located $\dfrac{2}{3}$ the distance from each vertex to the midpoint of the opposite side. Why does this show that *all three* medians meet at a single point?

c. The *centroid* of a triangle is the point where the medians meet. Show that a triangle with vertices $A(x_1, y_1)$, $B(x_2, y_2)$, $C(x_3, y_3)$, has centroid with coordinates

$$P\left(\frac{x_1 + x_2 + x_3}{3}, \frac{y_1 + y_2 + y_3}{3}\right)$$

55. Show that the vector with initial point $P(c, d)$ and terminal point $Q(a, b)$ has the component form $\mathbf{PQ} = \langle a - c, b - d \rangle$. See Figure 10.4.

56. Prove the following parts of Theorem 10.1.
   a. Commutativity
   b. Identity
   c. Inverse property of addition
   d. The distributive law
   $$s(\mathbf{u} + \mathbf{v}) = s\mathbf{u} + s\mathbf{v}$$

57. Let $A, B, C,$ and D be any four points in the plane. If M and N are midpoints of $\overline{AC}$ and $\overline{BD}$, show that

$$\overrightarrow{MN} = \tfrac{1}{4}(\overrightarrow{AB} + \overrightarrow{AD} + \overrightarrow{CB} + \overrightarrow{CD})$$

58. Let $A, B, C,$ and D be vertices of a quadrilateral, and let $M, N, P,$ and Q be the midpoints of the sides $\overline{AB}, \overline{BC}, \overline{CD},$ and $\overline{AD}$, respectively. Use vector methods to show that $MNPQ$ is a parallelogram.

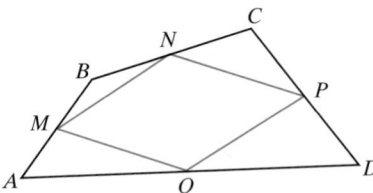

59. **SPY PROBLEM**   The Spy finally locates Blohardt's submarine (Problem 60, Section 9.4) and forces it to surface, but inside he finds only a note attached to the periscope. There are map coordinates, a crude drawing, and these words:

   "From the snowman, pace off the distance to the woodpile, turn left, pace off an equal distance, then drive a stake. From the snowman again, pace off the distance to the flagpole, turn right, pace off an equal distance and drive a second stake. Dig halfway between the stakes."

A bloody letter "P" marks the place on the map between the stakes. "Purity!" gasps the Spy. Gripped by trepidation, he arrives at the location indicated by the map coordinates. There is a woodpile and a flagpole, but the snowman has melted! How can the Spy find where to dig?

10.2   Quadric Surfaces and Graphing in Three Dimensions

IN THIS SECTION   **three-dimensional coordinate system, graphs in $\mathbb{R}^3$ (planes, spheres, cylinders, quadric surfaces)**

THREE-DIMENSIONAL COORDINATE SYSTEM

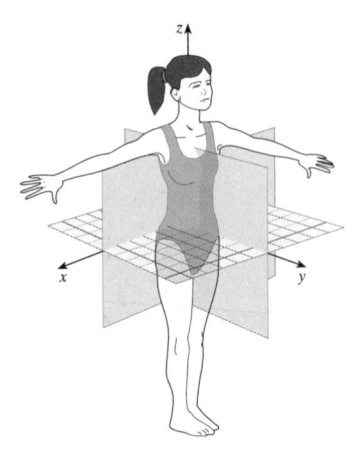

■ **FIGURE 10.10**   A "right-handed" rectangular coordinate system for $\mathbb{R}^3$

Our next goal is to see how analytic geometry and vector methods can be applied in space. We have already considered ordered pairs and a two-dimensional coordinate system. We shall denote this two-dimensional system by $\mathbb{R}^2$. Because we exist in a three-dimensional world, it is also important to consider a three-dimensional system. We call this *three-space* and denote it by $\mathbb{R}^3$. We introduce a coordinate system to three-space by choosing three mutually perpendicular axes to serve as a frame of reference. The orientation of our reference system will be *right-handed* in the sense that if you stand at the origin with your right arm along the positive x-axis and your left arm along the positive y-axis, as shown in Figure 10.10, your head will then point in the direction of the positive z-axis.

To orient yourself to a three-dimensional coordinate system, think of the x-axis and y-axis as lying in the plane of the floor and the z-axis as a line perpendicular to the floor. All the graphs that we have drawn in the first nine chapters of this book would now be drawn on the floor. If you orient yourself in a room (your classroom, for example) as shown in Figure 10.11, you may notice some important planes. Assume that the room is 25 ft × 30 ft × 8 ft and fix the origin at a front corner (where the board hangs).

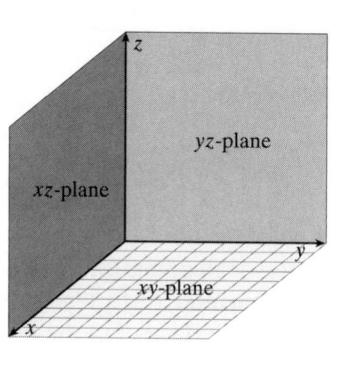 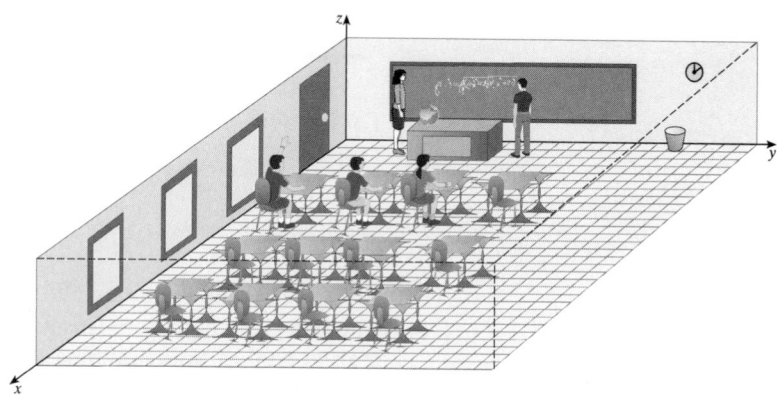

■ **FIGURE 10.11**   A typical classroom; assume the dimensions are 25 ft by 30 ft with an 8-foot ceiling.

Floor:	***xy*-plane;** equation is $z = 0$.
Ceiling:	plane parallel to the *xy*-plane; equation is $z = 8$.
Front wall:	***yz*-plane;** equation is $x = 0$.
Back wall:	plane parallel to the *yz*-plane; equation is $x = 30$.
Left wall:	***xz*-plane;** equation is $y = 0$.
Right wall:	plane parallel to the *xz*-plane; equation is $y = 25$.

The *xy*-, *xz*-, and *yz*-planes are called the **coordinate planes.** Points in $\mathbb{R}^3$ are located by their position in relation to the three coordinate planes and are given appropriate coordinates. Specifically, the point P is assigned coordinates (a, b, c) to indicate that it is a, b, and c units, respectively, from the *yz*-, *xz*-, and *xy*-planes. Name the coordinates of several objects in your classroom (or in Figure 10.11).

> **EXAMPLE 1**   *Points in three dimensions*

Graph the following ordered triplets:

a. $(10, 20, 10)$ b. $(-12, 6, 12)$ c. $(-12, -18, 6)$ d. $(20, -10, 18)$

Solution

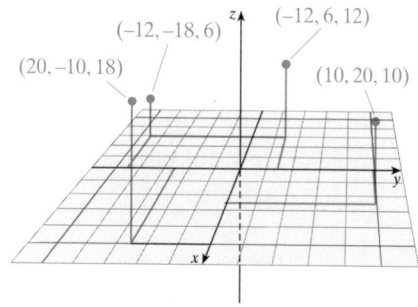

In Example 1, we measured distances in the *x*-, *y*-, and *z*-directions. We will, however, also need to measure distances between points in $\mathbb{R}^3$. The formula for distance in $\mathbb{R}^2$ easily extends to $\mathbb{R}^3$ (see Problem 52).

Distance Formula

The distance $\left|P_1P_2\right|$ between $P_1(x_1, y_1, z_1)$ and $P_2(x_2, y_2, z_2)$ is

$$\left|P_1P_2\right| = \sqrt{(x_2 - x_1)^2 + (y_2 - y_1)^2 + (z_2 - z_1)^2}$$

For instance, the distance between $(10, 20, 10)$ and $(-12, 6, 12)$ is

$$d = \sqrt{(-12 - 10)^2 + (6 - 20)^2 + (12 - 10)^2} = \sqrt{684} = 6\sqrt{19}$$

GRAPHS IN $\mathbb{R}^3$

The **graph of an equation** in $\mathbb{R}^3$ is the collection of all points (x, y, z) whose coordinates satisfy a given equation. This graph is called a **surface.** You are not expected to spend a great deal of time graphing three-dimensional surfaces, but the drawing lessons in this section should help. You may also have access to a computer program to help you look at graphs in three dimensions. We will discuss lines and planes more thoroughly in Section 10.5, but it is worthwhile to begin with a brief introduction to certain surfaces in $\mathbb{R}^3$.

Assignment:
Plot: $P(3, 4, -5)$

DRAWING LESSON 2: PLOTTING POINTS

a. Sketch x-axis and y-axis, adding tickmarks. Outline the xy-plane.

b. Sketch z-axis, adding tickmarks. Use dashed segments for hidden parts.

c. Plot x-distance and y-distance; darken segments from each along gridlines. Colored pencil or highlighter may help you visualize the figure.

d. Plot z-distance, using the unit size from the z-axis. Lightly sketch a grid on the xy-plane, using tickmarks as guides.

Planes   We shall obtain equations for planes in space after we discuss vectors in space. However, in beginning to visualize objects in $\mathbb{R}^3$, we do not want to ignore planes, because they are so common (for example, the walls, ceiling, and floor in Figure 10.11). In Section 10.5, we will show that the graph of $ax + by + cz = d$ is a **plane** if $a, b, c,$ and d are real numbers (not all zero).

DRAWING LESSON 3: DRAWING VERTICAL PLANES

Assignment:
To drawing lesson
2, add planes
$x = 2$ and $y = 0$.

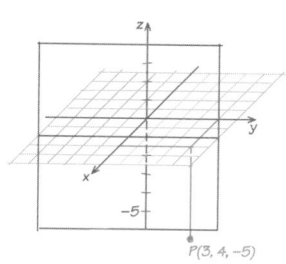

a. Draw a segment of the line $x = 2$ on the xy-plane. Through each endpoint, draw a segment parallel to the z-axis. Then connect the endpoints.

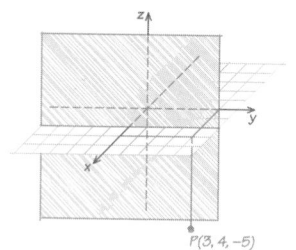

b. Shade the plane $x = 2$ where it is not hidden by the xy-plane. Erase hidden parts of both planes, and use your eraser to dash hidden parts of the axes.

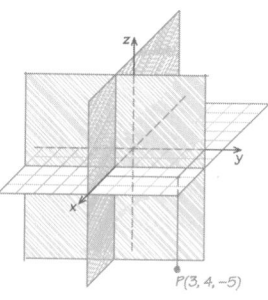

c. Follow the same procedure to draw and shade the plane $y = 0$. Draw the intersection of the two planes.

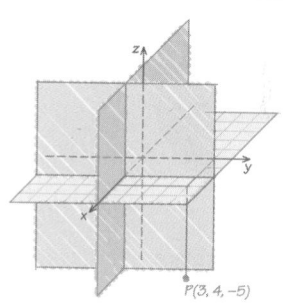

d. Use colored pencils or highlighters to distinguish individual planes.

EXAMPLE 2   *Graphing planes*

Graph the planes defined by the given equations.
a. $x + 3y + 2z = 6$
b. $y + z = 5$
c. $x = 4$

Solution   To graph a plane, find some ordered triplets satisfying the equation. The best ones to use are often those that fall on a coordinate axis (the intercepts).

a. Let $x = 0$ and $y = 0$; then $z = 3$; plot the point $(0, 0, 3)$.
   Let $x = 0$ and $z = 0$; then $y = 2$; plot the point $(0, 2, 0)$.
   Let $y = 0$ and $z = 0$; then $x = 6$; plot the point $(6, 0, 0)$.
   Use these points to draw the intersection lines (called **trace lines**) of the plane you are graphing with each of the coordinate planes. The result is shown in Figure 10.12a.

b. When one of the variables is missing from an equation of a plane, then that plane is parallel to the axis corresponding to the missing variable; thus, $y + z = 5$ is parallel to the x-axis. Draw the line $y + z = 5$ on the yz-plane, and then complete the plane, as shown in Figure 10.12b.

c. When two variables are missing, then the plane is parallel to the coordinate plane determined by the missing numbers. In this case, $x = 4$ is parallel to the yz-plane, as shown in Figure 10.12c.

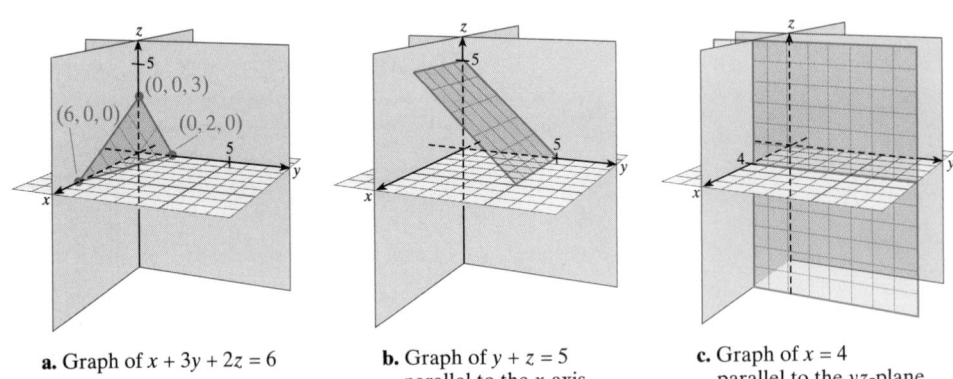

a. Graph of $x + 3y + 2z = 6$

b. Graph of $y + z = 5$ parallel to the x-axis

c. Graph of $x = 4$ parallel to the yz-plane

■ **FIGURE 10.12**   Graphs of planes   ■

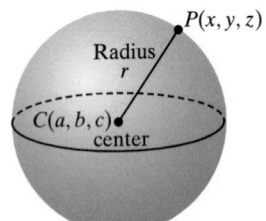

■ **FIGURE 10.13**   Graph of a sphere with center (a, b, c), radius r

Spheres   A **sphere** is defined as the collection of all points located a fixed distance (the **radius**) from a fixed point (the **center**). In particular, if $P(x, y, z)$ is a point on the sphere with radius r and center $C(a, b, c)$, then the distance from C to P is r. Thus,

$$r = \sqrt{(x - a)^2 + (y - b)^2 + (z - c)^2}$$

If you square both sides of this equation, you can see that it is equivalent to the equation of a sphere displayed in the following box. Conversely, if the point (x, y, z) satisfies an equation of this form, it must lie on a sphere with center (a, b, c) and radius r.

Equation of a Sphere

The graph of the equation

$$(x - a)^2 + (y - b)^2 + (z - c)^2 = r^2$$

is a sphere with center (a, b, c) and radius r, and any sphere has an equation of this form. This is called the **standard form of the equation of a sphere** (or simply *standard-form sphere*).

EXAMPLE 3   *Center and radius of a sphere from a given equation*

Show that the graph of the equation $x^2 + y^2 + z^2 + 4x - 6y - 3 = 0$ is a sphere, and find its center and radius.

Solution   By completing the square in both variables x and y, we have

$$(x^2 + 4x) + (y^2 - 6y) + z^2 = 3$$
$$(x^2 + 4x + 2^2) + [y^2 - 6y + (-3)^2] + z^2 = 3 + 4 + 9$$
$$(x + 2)^2 + (y - 3)^2 + z^2 = 16$$

Comparing this equation with the standard form, we see that it is the equation of a sphere with center $(-2, 3, 0)$ and radius 4.   ■

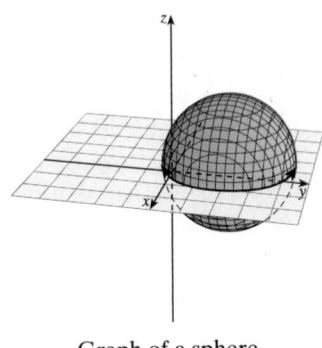

Graph of a sphere

DRAWING LESSON 4: SKETCHING A SURFACE	

ASSIGNMENT:

Graph $z = \dfrac{1}{1 + x^2 + y^2}$

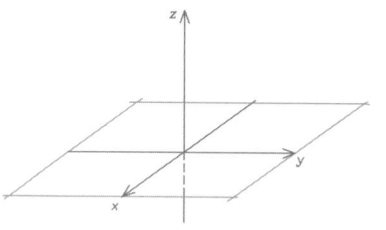

a. Draw the xy-plane in three dimensions, adding the z-axis.

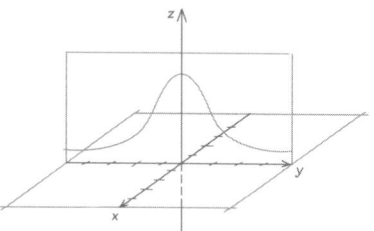

b. Draw a trace in one of the coordinate planes (in this case, the plane $x = 0$). If necessary, adjust the z-scale to show the trace more clearly.

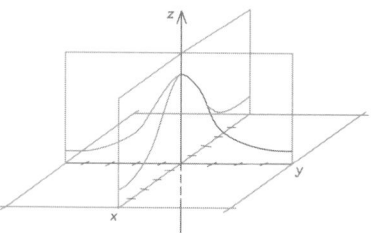

c. Draw a trace in another coordinate plane (in this case, the plane $y = 0$).

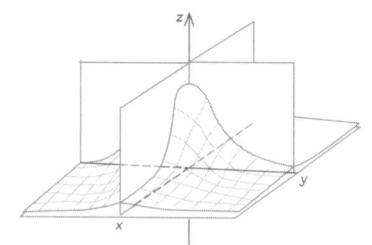

d. Erase all hidden lines. Draw several additonal trace curves to reveal the contours of the surface.

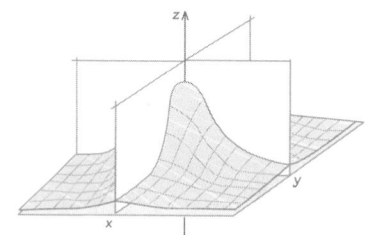

e. Erase all hidden lines. Use highlighters or pencils to color the surface and the xy-plane.

Cylinders   A **cross section** of a surface in space is a curve obtained by intersecting the surface with a plane. If parallel planes intersect a given surface in congruent cross-sectional curves, the surface is called a **cylinder.** We define a cylinder with *principal cross sections C and generating line L* to be the surface obtained by moving lines parallel to L along the boundary of the curve C, as shown in Figure 10.14. In this context, the curve C is called a **directrix** of the cylinder, and L is the **generatrix.**

We shall deal primarily with cylinders in which the directrix is a conic section and the generatrix L is one of the coordinate axes. Such a cylinder is often named for the type of conic section in its prinicpal cross sections, and is described by an equation involving only two of the variables x, y, z. In this case, the generating line L is parallel to

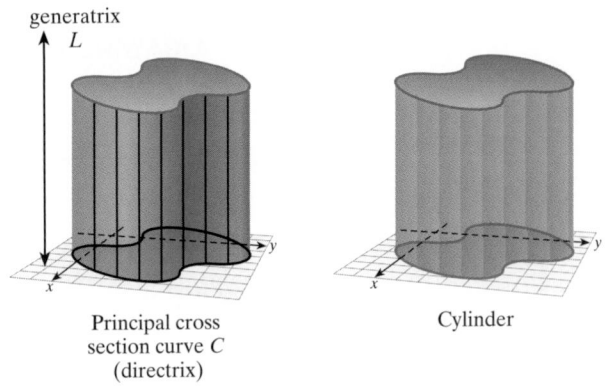

generatrix
L

Principal cross
section curve C
(directrix)

Cylinder

■ **FIGURE 10.14**   A cylinder with directrix C and generatrix L

the coordinate axis of the missing variable. Thus,

$$x^2 + y^2 = 5 \qquad \text{is a \textbf{circular cylinder} with } L \text{ parallel to the } z\text{-axis}$$
$$y^2 - z^2 = 9 \qquad \text{is a \textbf{hyperbolic cylinder} with } L \text{ parallel to the } x\text{-axis}$$
$$x^2 + 2z^2 = 25 \qquad \text{is an \textbf{elliptic cylinder} with } L \text{ parallel to the } y\text{-axis.}$$

The graphs of these cylinders are shown in Figure 10.15.

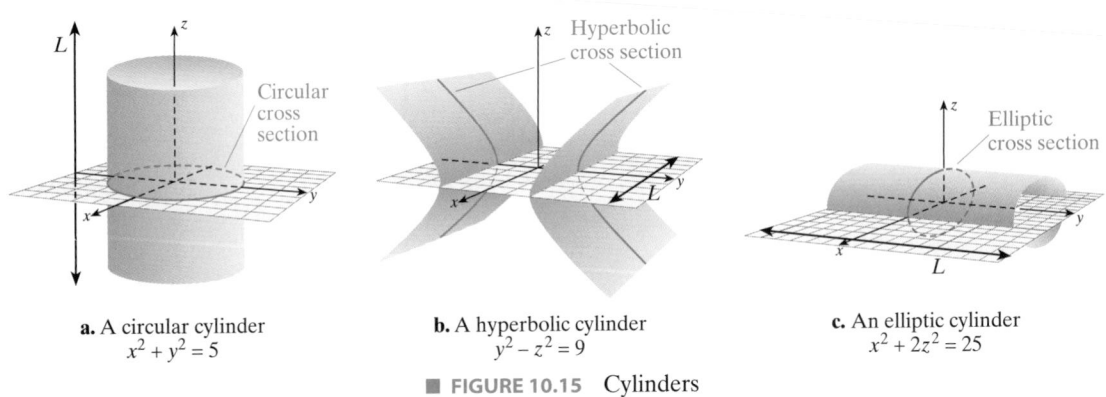

L
Circular
cross
section

a. A circular cylinder
$x^2 + y^2 = 5$

Hyperbolic
cross section
L

b. A hyperbolic cylinder
$y^2 - z^2 = 9$

Elliptic
cross section
L

c. An elliptic cylinder
$x^2 + 2z^2 = 25$

■ **FIGURE 10.15**   Cylinders

Quadric Surfaces   Spheres and elliptic, parabolic, and hyperbolic cylinders are examples of **quadric surfaces**. In general, such a surface is the graph of an equation of the form

$$Ax^2 + By^2 + Cz^2 + Dxy + Exz + Fyz + Gx + Hy + Iz + J = 0$$

Quadric surfaces may be thought of as the generalizations of the conic sections in $\mathbb{R}^3$. The **trace** of a curve is found by setting one of the variables equal to a constant and then graphing the resulting curve. If $x = k$ (k is a constant), the resulting curve is drawn in the plane $x = k$, which is parallel to the yz-plane; and if $z = k$, the curve is drawn in the plane $z = k$, parallel to the xy-plane. Table 10.1 shows the quadric surfaces.

■ **TABLE 10.1** **Quadric Surfaces**

Surface	Description	Surface	Description
Elliptic cone 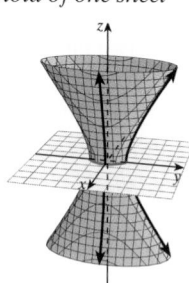	The trace in the *xy*-plane is a point; in planes parallel to the *xy*-plane, it is an ellipse. Traces in the *xz*- and *yz*-planes are intersecting lines; in planes parallel to these, they are hyperbolas: $$z^2 = \frac{x^2}{a^2} + \frac{y^2}{b^2}$$	*Elliptic paraboloid* 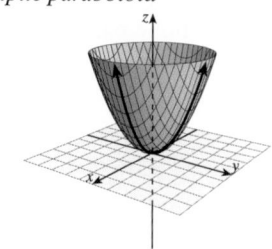	The trace in the *xy*-plane is a point; in planes parallel to the *xy*-plane, it is an ellipse. Traces in the *xz*- and *yz*-planes are parabolas. $$z = \frac{x^2}{a^2} + \frac{y^2}{b^2}$$
Hyperboloid of one sheet 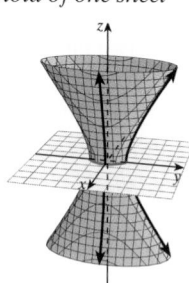	The trace in the *xy*-plane is an ellipse; in the *xz*- and *yz*-planes, the traces are hyperbolas. $$\frac{x^2}{a^2} + \frac{y^2}{b^2} - \frac{z^2}{c^2} = 1$$	*Hyperboloid of two sheets* 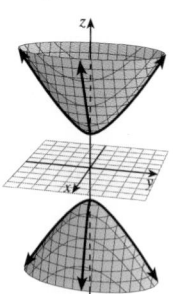	There is no trace in the *xy*-plane. In planes parallel to the *xy*-plane that intersect the surface, the traces are ellipses. Traces in the *xz*- and *yz*-planes are hyperbolas. $$\frac{x^2}{a^2} + \frac{y^2}{b^2} - \frac{z^2}{c^2} = -1$$
Ellipsoid 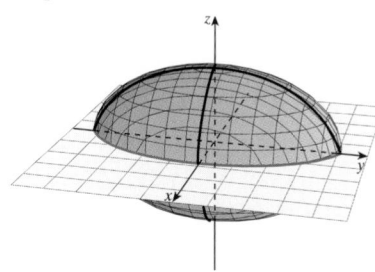	The traces in the coordinate planes are ellipses. $$\frac{x^2}{a^2} + \frac{y^2}{b^2} + \frac{z^2}{c^2} = 1$$	*Hyperbolic paraboloid,* also called a *saddle* 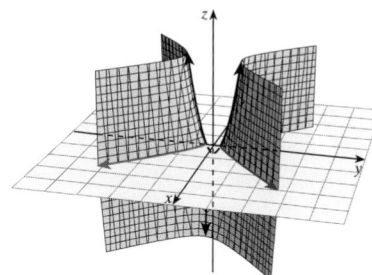	The trace in the *xy*-plane is two intersecting lines; in planes parallel to the *xy*-plane, the traces are hyperbolas. Traces in the *xz*- and *yz*-planes are parabolas. $$z = \frac{y^2}{b^2} - \frac{x^2}{a^2}$$
Sphere 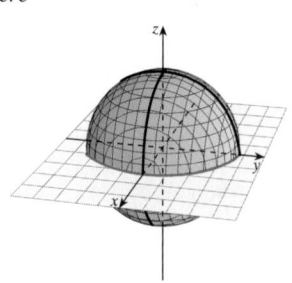	The sphere is a special kind of ellipsoid for which $a = b = c = r$. $$x^2 + y^2 + z^2 = r^2$$		

| EXAMPLE 4 | *Identifying and sketching a quadric surface* |

Identify and sketch the surface with equation $9x^2 - 16y^2 + 144z = 0$.

Solution   Look at Table 10.1 on page 683 and note that the equation is second degree in x and y but first degree in z. This means it is an elliptic paraboloid or a hyperbolic paraboloid. Solve the equation for z:

$$9x^2 - 16y^2 + 144z = 0$$
$$144z = 16y^2 - 9x^2$$
$$z = \frac{y^2}{9} - \frac{x^2}{16}$$

We recognize this as a hyperbolic paraboloid.

Next, we take cross sections of $9x^2 - 16y^2 + 144z = 0$:

Cross Section	Chosen Value	Equation of Trace	Description of Trace
xy-plane	$z = 0$	$\dfrac{y^2}{9} - \dfrac{x^2}{16} = 0$	Two intersecting lines
parallel to xy-plane	$z = 4$	$\dfrac{y^2}{36} - \dfrac{x^2}{64} = 1$	Hyperbola
xz-plane	$y = 0$	$z = -\dfrac{x^2}{16}$	Parabola opens down
parallel to xz-plane	$y = 10$	$z - \dfrac{100}{9} = -\dfrac{x^2}{16}$	Parabola opens down
yz-plane	$x = 0$	$z = \dfrac{y^2}{9}$	Parabola opens up
parallel to yz-plane	$x = 5$	$z + \dfrac{25}{16} = \dfrac{y^2}{9}$	Parabola opens up

These traces are also shown in Figure 10.16. ∎

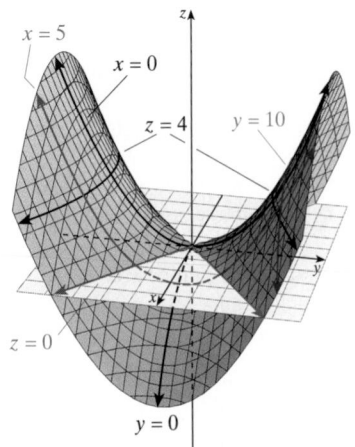

■ FIGURE 10.16   Graph of $9x^2 - 16y^2 + 144z = 0$

10.2   Problem Set

Ⓐ *In Problems 1–4, plot the points P and Q in $\mathbb{R}^3$, and find the distance $|\overrightarrow{PQ}|$.*

1. $P(3, -4, 5), Q(1, 5, -3)$ **2.** $P(0, 3, 0), Q(-2, 5, -7)$
3. $P(-3, -5, 8), Q(3, 6, -7)$ **4.** $P(0, 5, -3), Q(2, -1, 0)$

In Problems 5–8, find the standard-form equation of the sphere with the given center C and radius r.

5. $C(0, 0, 0), r = 1$ **6.** $C(-3, 5, 7), r = 2$
7. $C(0, 4, -5), r = 3$ **8.** $C(-2, 3, -1), r = \sqrt{5}$

Find the center and radius of each sphere whose equations are given in Problems 9–12.

9. $x^2 + y^2 + z^2 - 2y + 2z - 2 = 0$
10. $x^2 + y^2 + z^2 + 4x - 2z - 8 = 0$
11. $x^2 + y^2 + z^2 - 6x + 2y - 2z + 10 = 0$
12. $x^2 + y^2 + z^2 - 2x - 4y + 8z + 17 = 0$

In Problems 13–22, match the equation with its graph (A–L).

13. $x^2 = z^2 + y^2$ **14.** $z^2 = \dfrac{x^2}{4} + \dfrac{y^2}{9}$

15. $\dfrac{x^2}{2} - \dfrac{y^2}{4} + \dfrac{z^2}{9} = 1$ **16.** $\dfrac{x^2}{9} + \dfrac{y^2}{16} - \dfrac{z^2}{4} = 1$

17. $x^2 + y^2 + z^2 = 9$ **18.** $y = x^2 + z^2$

19. $x = \dfrac{y^2}{25} + \dfrac{z^2}{16}$ **20.** $y = \dfrac{z^2}{4} - \dfrac{x^2}{9}$

21. $y^2 + z^2 - x^2 = -1$ **22.** $\dfrac{x^2}{4} - \dfrac{y^2}{9} + \dfrac{z^2}{9} = -1$

A.

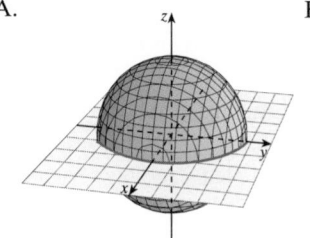

B.

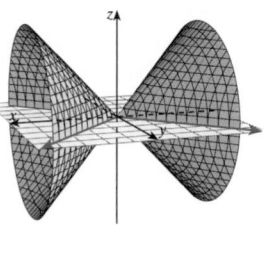

J.

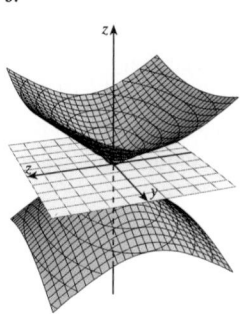

K.

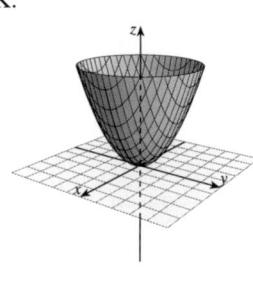

C.

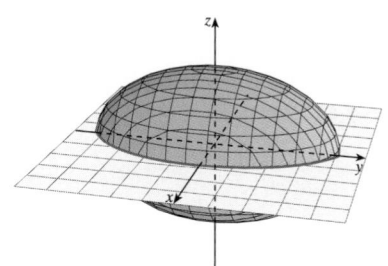

L.

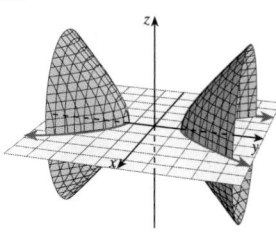

D.   E.

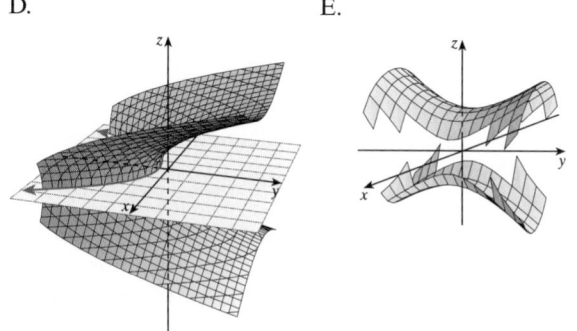

B *The vertices A, B, and C of a triangle in space are given in Problems 23–26. Find the lengths of the sides of the triangle and determine whether it is a right triangle, an isosceles triangle, both, or neither.*

23. $A(3, -1, 0), B(7, 1, 4), C(1, 3, 4)$

24. $A(1, 1, 1), B(3, 3, 2), C(3, -3, 5)$

25. $A(1, 2, 3), B(-3, 2, 4), C(1, -4, 3)$

26. $A(2, 4, 3), B(-3, 2, -4), C(-6, 8, -10)$

27. ■ **What Does This Say?** Describe a procedure for sketching a quadric surface.

28. ■ **What Does This Say?** Describe a procedure for identifying a quadric surface by looking at its equation.

F.   G.

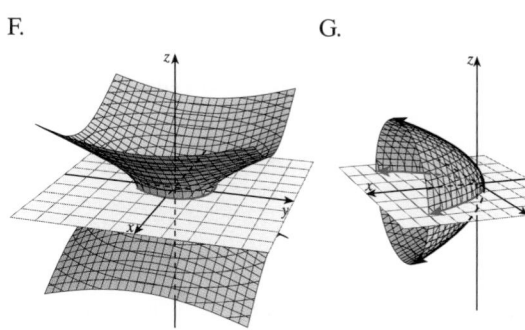

In Problems 29–40, sketch the graph of each equation in $\mathbb{R}^3$.

29. $2x + y + 3z = 6$ 30. $x = 4$

31. $x + 2y + 5z = 10$ 32. $x + y + z = 1$

33. $3x - 2y - z = 12$ 34. $y = z^2$

35. $x = -1$ 36. $y^2 + z^2 = 1$

37. $z = e^y$ 38. $z = \ln x$

39. $x + z = 1$ 40. $z = y^{-1}$

In Problems 41–48, identify the quadric surface and describe the traces. Sketch the graph.

H.   I.

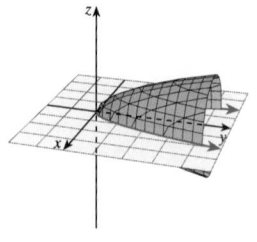   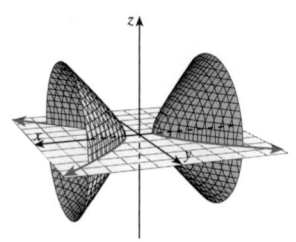

41. $9x^2 + 4y^2 + z^2 = 1$ 42. $\dfrac{x^2}{4} + y^2 + \dfrac{z^2}{9} = 1$

43. $\dfrac{x^2}{4} + \dfrac{y^2}{9} - z^2 = 1$ 44. $\dfrac{x^2}{9} - y^2 - z^2 = 1$

45. $z = x^2 + \dfrac{y^2}{4}$ 46. $z = \dfrac{x^2}{9} - \dfrac{y^2}{16}$

47. $x^2 + 2y^2 = 9z^2$ **48.** $z^2 = 1 + \dfrac{x^2}{9} + \dfrac{y^2}{4}$

49. Find an equation for a sphere, given that the endpoints of a diameter of the sphere are $(1, 2, -3)$ and $(-2, 3, 3)$.

🄲 **50.** Find the point P that lies $\frac{2}{3}$ of the distance from the point $A(-1, 3, 9)$ to the midpoint of the line segment joining points $B(-2, 3, 7)$ and $C(4, 1, -3)$.

51. Let $P(3, 2, -1), Q(-2, 1, c)$, and $R(c, 1, 0)$ be points in $\mathbb{R}^3$. For what values of c (if any) is PQR a right triangle?

52. Derive the formula for the distance between $P_1(x_1, y_1, z_1)$ and $P_2(x_2, y_2, z_2)$. (*Hint:* Project the segment $\overline{P_1P_2}$ onto the xy-plane, and then use the Pythagorean theorem.)

10.3 The Dot Product

IN THIS SECTION   **vectors in $\mathbb{R}^3$, definition of dot product, angle between vectors, projections, work as a dot product**

VECTORS IN $\mathbb{R}^3$

A vector in $\mathbb{R}^3$ may be thought of as a directed line segment (an "arrow") in space. The vector $\mathbf{P_1P_2}$ with initial point $P_1(x_1, y_1, z_1)$ and terminal point $P_2(x_2, y_2, z_2)$ has the component form

$$\mathbf{P_1P_2} = \langle x_2 - x_1, y_2 - y_1, z_2 - z_1 \rangle$$

Vector addition and multiplication of a vector by a scalar are defined for vectors in $\mathbb{R}^3$ in essentially the same way as these operations were defined for vectors in $\mathbb{R}^2$. In addition, the properties of vector algebra listed in Theorem 10.1 of Section 10.1 apply to vectors in $\mathbb{R}^3$ as well as to those in $\mathbb{R}^2$.

For example, we observed that each vector in $\mathbb{R}^2$ can be expressed as a unique linear combination of the standard basis vectors $\mathbf{i}$ and $\mathbf{j}$. This representation can be extended to vectors in $\mathbb{R}^3$ by adding a vector $\mathbf{k}$ defined to be the unit vector in the direction of the positive z-axis. In component form, we have in $\mathbb{R}^3$,

$$\mathbf{i} = \langle 1, 0, 0 \rangle \qquad \mathbf{j} = \langle 0, 1, 0 \rangle \qquad \mathbf{k} = \langle 0, 0, 1 \rangle$$

We call these the **standard basis vectors** in $\mathbb{R}^3$. The **standard representation** of the vector with initial point at the origin O and terminal point $Q(a_1, a_2, a_3)$ is $\mathbf{OQ} = a_1\mathbf{i} + a_2\mathbf{j} + a_3\mathbf{k}$, as shown in Figure 10.17b.

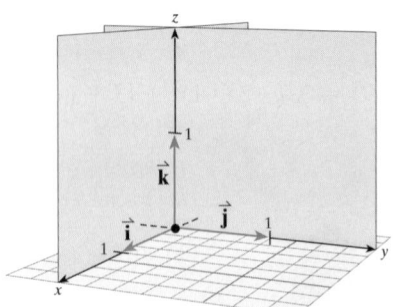
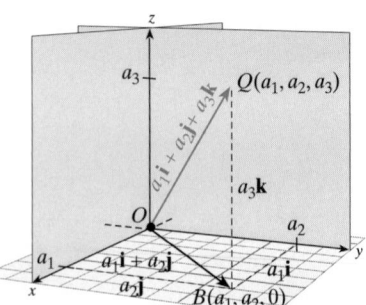

a. The standard basis vectors in $\mathbb{R}^3$

b. The vector from O to $Q(a_1, a_2, a_3)$ is $\mathbf{OQ} = a_1\mathbf{i} + a_2\mathbf{j} + a_3\mathbf{k}$.

■ **FIGURE 10.17**   Standard representation of vectors in $\mathbb{R}^3$

The vector **PQ** with initial point $P(x_0, y_0, z_0)$ and terminal point $Q(x_1, y_1, z_1)$ has the standard representation

$$\mathbf{PQ} = (x_1 - x_0)\mathbf{i} + (y_1 - y_0)\mathbf{j} + (z_1 - z_0)\mathbf{k}$$

EXAMPLE 1   *Standard representation of a vector in $\mathbb{R}^3$*

Find the standard representation of the vector **PQ** with initial point $P(-1, 2, 2)$ and terminal point $Q(3, -2, 4)$.

Solution   We have

$$\mathbf{PQ} = [3 - (-1)]\mathbf{i} + [-2 - 2]\mathbf{j} + [4 - 2]\mathbf{k} = 4\mathbf{i} - 4\mathbf{j} + 2\mathbf{k}$$ ∎

By referring to Figure 10.17b, we can also derive a formula for the length of a vector, which in turn can be used to establish the distance formula (stated in the previous section) between any two points in $\mathbb{R}^3$. Specifically, note that $\triangle OBQ$ in Figure 10.17b is a right triangle with hypotenuse $\|\mathbf{OQ}\|$ and legs $\|\mathbf{BQ}\| = |a_3|$ and $\|\mathbf{OB}\| = \sqrt{a_1^2 + a_2^2}$; by applying the Pythagorean theorem, we conclude that the vector $\mathbf{OQ} = a_1\mathbf{i} + a_2\mathbf{j} + a_3\mathbf{k}$ has length

$$\|\mathbf{OQ}\| = \sqrt{\|\mathbf{OB}\|^2 + \|\mathbf{BQ}\|^2} = \sqrt{(a_1^2 + a_2^2) + a_3^2} = \sqrt{a_1^2 + a_2^2 + a_3^2}$$

Moreover, if $A(a_1, a_2, a_3)$ and $B(b_1, b_2, b_3)$ are any two points in $\mathbb{R}^3$, the distance between them is the length of the vector **AB**. We find

$$\mathbf{AB} = \mathbf{OB} - \mathbf{OA} = (b_1 - a_1)\mathbf{i} + (b_2 - a_2)\mathbf{j} + (b_3 - a_3)\mathbf{k}$$

so that the distance between A and B is given by

$$\|\mathbf{AB}\| = \sqrt{(b_1 - a_1)^2 + (b_2 - a_2)^2 + (b_3 - a_3)^2}$$

Magnitude of a Vector

The **magnitude,** or length, of the vector $\mathbf{v} = a_1\mathbf{i} + a_2\mathbf{j} + a_3\mathbf{k}$ is

$$\|\mathbf{v}\| = \sqrt{a_1^2 + a_2^2 + a_3^2}$$

EXAMPLE 2   *Magnitude of a vector*

Find the magnitude of the vector $\mathbf{v} = 2\mathbf{i} - 3\mathbf{j} + 5\mathbf{k}$ and the distance between the points $A(1, -1, -4)$ and $B(-2, 3, 8)$.

Solution $\|\mathbf{v}\| = \sqrt{2^2 + (-3)^2 + 5^2} = \sqrt{38}$ and

$$\|\overline{AB}\| = \sqrt{(-2 - 1)^2 + [3 - (-1)]^2 + [8 - (-4)]^2} = 13$$ ∎

As in $\mathbb{R}^2$, if **v** is a given nonzero vector in $\mathbb{R}^3$, then a unit vector **u** that points in the same direction as **v** is

$$\mathbf{u} = \frac{\mathbf{v}}{\|\mathbf{v}\|}$$

EXAMPLE 3   *Finding a direction vector*

Find a unit vector that points in the direction of the vector **PQ** from $P(-1, 2, 5)$ to $Q(0, -3, 7)$.

Solution $\mathbf{PQ} = [0 - (-1)]\mathbf{i} + [-3 - 2]\mathbf{j} + [7 - 5]\mathbf{k} = \mathbf{i} - 5\mathbf{j} + 2\mathbf{k}$

$\|\mathbf{PQ}\| = \sqrt{1^2 + (-5)^2 + 2^2} = \sqrt{30}$

Thus,

$$\mathbf{u} = \frac{\mathbf{PQ}}{\|\mathbf{PQ}\|} = \frac{\mathbf{i} - 5\mathbf{j} + 2\mathbf{k}}{\sqrt{30}} = \frac{1}{\sqrt{30}}\mathbf{i} - \frac{5}{\sqrt{30}}\mathbf{j} + \frac{2}{\sqrt{30}}\mathbf{k}$$ ■

As in $\mathbb{R}^2$, two vectors in $\mathbb{R}^3$ are **parallel** if they are multiples of one another. Nonzero vectors $\mathbf{u}$ and $\mathbf{v}$ are parallel if and only if $\mathbf{u} = s\mathbf{v}$ for some nonzero scalar s.

EXAMPLE 4   *Parallel vectors*

A vector $\mathbf{PQ}$ has initial point $P(1, 0, -3)$ and length 3. Find Q so that $\mathbf{PQ}$ is parallel to $\mathbf{v} = 2\mathbf{i} - 3\mathbf{j} + 6\mathbf{k}$.

Solution   Let Q have coordinates (a_1, a_2, a_3). Then

$$\mathbf{PQ} = [a_1 - 1]\mathbf{i} + [a_2 - 0]\mathbf{j} + [a_3 - (-3)]\mathbf{k} = (a_1 - 1)\mathbf{i} + a_2\mathbf{j} + (a_3 + 3)\mathbf{k}$$

Because $\mathbf{PQ}$ is parallel to $\mathbf{v}$, we have $\mathbf{PQ} = s\mathbf{v}$ for some scalar s; that is,

$$(a_1 - 1)\mathbf{i} + a_2\mathbf{j} + (a_3 + 3)\mathbf{k} = s(2\mathbf{i} - 3\mathbf{j} + 6\mathbf{k})$$

Thanks to the uniqueness of the standard representation, this implies that

$$a_1 - 1 = 2s \qquad a_2 = -3s \qquad a_3 + 3 = 6s$$

$$a_1 = 2s + 1 \qquad\qquad\qquad a_3 = 6s - 3$$

Because $\mathbf{PQ}$ has length 3, we have

$$3 = \sqrt{(a_1 - 1)^2 + a_2^2 + (a_3 + 3)^2}$$

$$= \sqrt{[(2s + 1) - 1]^2 + (-3s)^2 + [(6s - 3) + 3]^2}$$

$$= \sqrt{4s^2 + 9s^2 + 36s^2} = \sqrt{49s^2} = 7|s|$$

Thus, $s = \pm\frac{3}{7}$ and so

$$a_1 = 2(\pm\tfrac{3}{7}) + 1 \qquad a_2 = -3(\pm\tfrac{3}{7}) \qquad a_3 = 6(\pm\tfrac{3}{7}) - 3$$

$$= \tfrac{13}{7}, \tfrac{1}{7} \qquad\qquad = -\tfrac{9}{7}, \tfrac{9}{7} \qquad\qquad = -\tfrac{3}{7}, -\tfrac{39}{7}$$

There are two points that satisfy the conditions for the required terminal point Q: $(\frac{13}{7}, -\frac{9}{7}, -\frac{3}{7})$ and $(\frac{1}{7}, \frac{9}{7}, -\frac{39}{7})$.   ■

DEFINITION OF DOT PRODUCT

The **dot (scalar) product** and the **cross (vector) product** are two important vector operations. We shall examine the cross product in the next section. The dot product is also known as a scalar product because it is a product of vectors that gives a scalar (that is, real number) as a result. Sometimes the dot product is called the **inner product.**

> **Dot Product**
>
> The **dot product** of vectors $\mathbf{v} = a_1\mathbf{i} + a_2\mathbf{j} + a_3\mathbf{k}$ and $\mathbf{w} = b_1\mathbf{i} + b_2\mathbf{j} + b_3\mathbf{k}$ is the scalar denoted by $\mathbf{v} \cdot \mathbf{w}$ and given by
>
> $$\mathbf{v} \cdot \mathbf{w} = a_1b_1 + a_2b_2 + a_3b_3$$

The dot product of two vectors $\mathbf{v} = a_1\mathbf{i} + a_2\mathbf{j}$ and $\mathbf{w} = b_1\mathbf{i} + b_2\mathbf{j}$ in a plane is given by a similar formula with $a_3 = b_3 = 0$, namely,

$$\mathbf{v} \cdot \mathbf{w} = a_1 b_1 + a_2 b_2$$

EXAMPLE 5   *Dot product*

Find the dot product $\mathbf{v} = -3\mathbf{i} + 2\mathbf{j} + \mathbf{k}$ and $\mathbf{w} = 4\mathbf{i} - \mathbf{j} + 2\mathbf{k}$.

Solution $\mathbf{v} \cdot \mathbf{w} = -3(4) + 2(-1) + 1(2) = -12$ ■

EXAMPLE 6   *Dot product in component form*

If $\mathbf{v} = \langle 4, -1, 3 \rangle$ and $\mathbf{w} = \langle -1, -2, 5 \rangle$, find the dot product, $\mathbf{v} \cdot \mathbf{w}$.

Solution $\mathbf{v} \cdot \mathbf{w} = 4(-1) + (-1)(-2) + 3(5) = 13$ ■

Before we can apply the dot product to geometric and physical problems, we need to know how it behaves algebraically. A number of important general properties of the dot product are listed in the following theorem.

THEOREM 10.2   ***Properties of the dot product***

If $\mathbf{u}, \mathbf{v}$, and $\mathbf{w}$ are vectors in $\mathbb{R}^2$ or $\mathbb{R}^3$ and c is a scalar, then:

Magnitude of a vector	$\mathbf{v} \cdot \mathbf{v} = \|\mathbf{v}\|^2$
Zero product	$\mathbf{0} \cdot \mathbf{v} = \mathbf{0}$
Commutativity	$\mathbf{v} \cdot \mathbf{w} = \mathbf{w} \cdot \mathbf{v}$
Product of a multiple	$c(\mathbf{v} \cdot \mathbf{w}) = (c\mathbf{v}) \cdot \mathbf{w} = \mathbf{v} \cdot (c\mathbf{w})$
Distributivity	$\mathbf{u} \cdot (\mathbf{v} + \mathbf{w}) = \mathbf{u} \cdot \mathbf{v} + \mathbf{u} \cdot \mathbf{w}$

Proof   Let $\mathbf{u} = a_1\mathbf{i} + a_2\mathbf{j} + a_3\mathbf{k}, \mathbf{v} = b_1\mathbf{i} + b_2\mathbf{j} + b_3\mathbf{k}$, and $\mathbf{w} = c_1\mathbf{i} + c_2\mathbf{j} + c_3\mathbf{k}$.

Magnitude of a vector

$$\|\mathbf{v}\|^2 = (\sqrt{a_1^2 + a_2^2 + a_3^2})^2 = a_1^2 + a_2^2 + a_3^2 = \mathbf{v} \cdot \mathbf{v}$$

Zero product, commutativity, and **scalar multiple** can be established in a similar fashion.

Distributivity

$$\mathbf{u} \cdot (\mathbf{v} + \mathbf{w})$$
$$= (a_1\mathbf{i} + a_2\mathbf{j} + a_3\mathbf{k}) \cdot [(b_1 + c_1)\mathbf{i} + (b_2 + c_2)\mathbf{j} + (b_3 + c_3)\mathbf{k}]$$
$$= a_1(b_1 + c_1) + a_2(b_2 + c_2) + a_3(b_3 + c_3)$$
$$= a_1 b_1 + a_1 c_1 + a_2 b_2 + a_2 c_2 + a_3 b_3 + a_3 c_3$$

and,

$$\mathbf{u} \cdot \mathbf{v} + \mathbf{u} \cdot \mathbf{w} = (a_1 b_1 + a_2 b_2 + a_3 b_3) + (a_1 c_1 + a_2 c_2 + a_3 c_3)$$
$$= a_1 b_1 + a_1 c_1 + a_2 b_2 + a_2 c_2 + a_3 b_3 + a_3 c_3$$

Thus, $\mathbf{u} \cdot (\mathbf{v} + \mathbf{w}) = \mathbf{u} \cdot \mathbf{v} + \mathbf{u} \cdot \mathbf{w}$. ═

ANGLE BETWEEN VECTORS

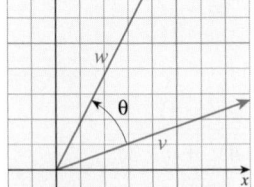

■ **FIGURE 10.18** The angle between two vectors

The angle between two nonzero vectors $\mathbf{v}$ and $\mathbf{w}$ is defined to be the angle θ with $0 \le \theta \le \pi$ that is formed when the vectors are in standard position (initial points at the origin), as shown in Figure 10.18.

The angle between vectors plays an important role in certain applications and may be computed by using the following formula involving the dot product.

THEOREM 10.3   *Angle between two vectors*

If θ is the angle between the nonzero vectors $\mathbf{v}$ and $\mathbf{w}$, then

$$\cos \theta = \frac{\mathbf{v} \cdot \mathbf{w}}{\|\mathbf{v}\| \|\mathbf{w}\|}$$

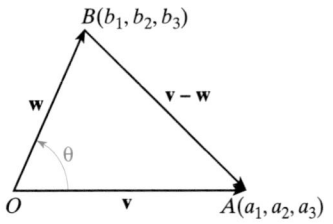

B(b_1, b_2, b_3)

$\mathbf{w}$ $\mathbf{v} - \mathbf{w}$

θ

O $\mathbf{v}$ $A(a_1, a_2, a_3)$

■ **FIGURE 10.19**   Finding an angle between two vectors

Proof   Suppose $\mathbf{v} = a_1\mathbf{i} + a_2\mathbf{j} + a_3\mathbf{k}$ and $\mathbf{w} = b_1\mathbf{i} + b_2\mathbf{j} + b_3\mathbf{k}$, and consider $\triangle AOB$ with vertices at the origin O and the points $A(a_1, a_2, a_3)$ and $B(b_1, b_2, b_3)$, as shown in Figure 10.19.

Note that sides $\overline{OA}$ and $\overline{OB}$ have lengths

$$\|\mathbf{v}\| = \sqrt{a_1^2 + a_2^2 + a_3^2} \quad \text{and} \quad \|\mathbf{w}\| = \sqrt{b_1^2 + b_2^2 + b_3^2}$$

respectively, and that side $\overline{AB}$ has length

$$\|\mathbf{v} - \mathbf{w}\| = \sqrt{(a_1 - b_1)^2 + (a_2 - b_2)^2 + (a_3 - b_3)^2}$$

Next, we use the law of cosines:

$$a^2 = b^2 + c^2 - 2bc \cos \theta \qquad \text{Law of cosines}$$

$$\|\mathbf{v} - \mathbf{w}\|^2 = \|\mathbf{v}\|^2 + \|\mathbf{w}\|^2 - 2\|\mathbf{v}\| \|\mathbf{w}\| \cos \theta \qquad \text{See Figure 10.19.}$$

$$\cos \theta = \frac{\|\mathbf{v}\|^2 + \|\mathbf{w}\|^2 - \|\mathbf{v} - \mathbf{w}\|^2}{2\|\mathbf{v}\| \|\mathbf{w}\|} \qquad \text{Solve for } \cos \theta.$$

$$= \frac{a_1^2 + a_2^2 + a_3^2 + b_1^2 + b_2^2 + b_3^2 - [(a_1 - b_1)^2 + (a_2 - b_2)^2 + (a_3 - b_3)^2]}{2\|\mathbf{v}\| \|\mathbf{w}\|}$$

$$= \frac{2a_1b_1 + 2a_2b_2 + 2a_3b_3}{2\|\mathbf{v}\| \|\mathbf{w}\|} = \frac{\mathbf{v} \cdot \mathbf{w}}{\|\mathbf{v}\| \|\mathbf{w}\|}$$

EXAMPLE 7   *Angle between two given vectors*

Let $\triangle ABC$ be the triangle with vertices $A(1, 1, 8)$, $B(4, -3, -4)$, and $C(-3, 1, 5)$. Find the angle formed at A.

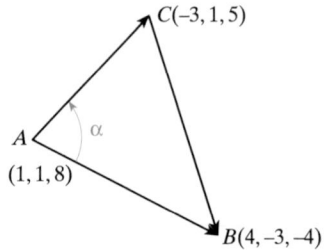

C($-3, 1, 5$)

A
$(1, 1, 8)$ α

B($4, -3, -4$)

■ **FIGURE 10.20**   Find an angle of a triangle

Solution   Draw $\triangle ABC$ and label the angle formed at A as α, as shown in Figure 10.20. The angle α is the angle between vectors $\mathbf{AB}$ and $\mathbf{AC}$, where

$$\mathbf{AB} = (4 - 1)\mathbf{i} + (-3 - 1)\mathbf{j} + (-4 - 8)\mathbf{k} = 3\mathbf{i} - 4\mathbf{j} - 12\mathbf{k}$$

$$\mathbf{AC} = (-3 - 1)\mathbf{i} + (1 - 1)\mathbf{j} + (5 - 8)\mathbf{k} = -4\mathbf{i} - 3\mathbf{k}$$

Thus,

$$\cos \alpha = \frac{\mathbf{AB} \cdot \mathbf{AC}}{\|\mathbf{AB}\| \|\mathbf{AC}\|} = \frac{3(-4) + (-4)(0) + (-12)(-3)}{\sqrt{3^2 + (-4)^2 + (-12)^2} \sqrt{(-4)^2 + (-3)^2}}$$

$$= \frac{24}{\sqrt{169} \sqrt{25}} = \frac{24}{65}$$

and the required angle is $\alpha = \cos^{-1}\left(\frac{24}{65}\right) \approx 1.19$.

The formula for the angle between vectors is often used in conjunction with the dot product. If we multiply both sides of the formula by $\|\mathbf{v}\|\|\mathbf{w}\|$, we obtain the following alternate form for the dot product formula.

Geometric Formula for the Dot Product

$$\mathbf{v} \cdot \mathbf{w} = \|\mathbf{v}\|\|\mathbf{w}\| \cos \theta$$

where θ is the angle $(0 \le \theta \le \pi)$ between the vectors $\mathbf{v}$ and $\mathbf{w}$.

Two vectors are said to be **perpendicular,** or **orthogonal,** if the angle between them is $\theta = \pi/2$. The following theorem provides a useful criterion for orthogonality.

THEOREM 10.4   *The orthogonal vector theorem*

Nonzero vectors $\mathbf{v}$ and $\mathbf{w}$ are **orthogonal** if and only if

$$\mathbf{v} \cdot \mathbf{w} = 0$$

Proof   If the vectors are orthogonal, then the angle between them is $\frac{\pi}{2}$; therefore,

$$\mathbf{v} \cdot \mathbf{w} = \|\mathbf{v}\|\|\mathbf{w}\|\cos \frac{\pi}{2} = 0$$

Conversely, if $\mathbf{v} \cdot \mathbf{w} = 0$, and $\mathbf{v}$ and $\mathbf{w}$ are nonzero vectors, then $\cos \theta = 0$, so that $\theta = \frac{\pi}{2}$ (since $0 \le \theta \le \pi$) and the vectors are orthogonal.

■ *What This Says:*   The orthogonal vector theorem allows you to show that two vectors are orthogonal by finding the dot product of the vectors and showing this dot product is 0. On the other hand, if you know the vectors are orthogonal, then it follows that the dot product of the vectors is 0.

EXAMPLE 8   *Orthogonal vectors*

Determine which (if any) pairs of the following vectors are orthogonal:
$$\mathbf{u} = 3\mathbf{i} + 7\mathbf{j} - 2\mathbf{k} \qquad \mathbf{v} = 5\mathbf{i} - 3\mathbf{j} - 3\mathbf{k} \qquad \mathbf{w} = \mathbf{j} - \mathbf{k}$$

Solution

$\mathbf{u} \cdot \mathbf{v} = 3(5) + 7(-3) + (-2)(-3) = 0; \mathbf{u}$ and $\mathbf{v}$ are orthogonal.

$\mathbf{u} \cdot \mathbf{w} = 3(0) + 7(1) + (-2)(-1) = 9; \mathbf{u}$ and $\mathbf{w}$ are not orthogonal.

$\mathbf{v} \cdot \mathbf{w} = 5(0) + (-3)(1) + (-3)(-1) = 0; \mathbf{v}$ and $\mathbf{w}$ are orthogonal.   ■

PROJECTIONS

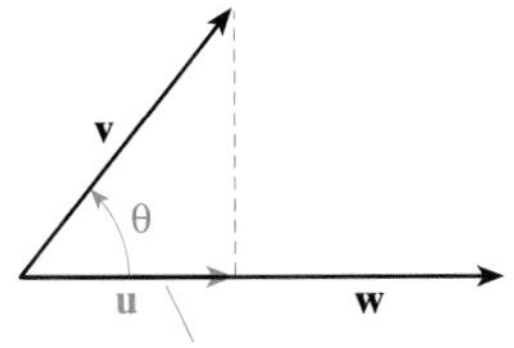

Let $\mathbf{v}$ and $\mathbf{w}$ be two vectors in $\mathbb{R}^2$ drawn so that they have a common initial point, as shown in Figure 10.21.* If we drop a perpendicular from the tip of $\mathbf{v}$ to the line determined by $\mathbf{w}$, we determine a vector called the **vector projection of $\mathbf{v}$ onto $\mathbf{w}$**, which we have labeled $\mathbf{u}$ in Figure 10.21.

Vector projection of **v** onto **w**

■ **FIGURE 10.21**   Projection of **v** onto **w**

---

*Even though Figure 10.21 is drawn in $\mathbb{R}^2$, the projection formula applies to $\mathbb{R}^3$ as well.

To find a formula for the vector projection, note that $\mathbf{u} = t\mathbf{w}$ for some scalar t and that $\mathbf{v} - t\mathbf{w}$ is orthogonal to $\mathbf{w}$. Thus,

$$(v - t\mathbf{w}) \cdot \mathbf{w} = 0$$
$$\mathbf{v} \cdot \mathbf{w} = t(\mathbf{w} \cdot \mathbf{w})$$
$$t = \frac{\mathbf{v} \cdot \mathbf{w}}{\mathbf{w} \cdot \mathbf{w}}$$

and the vector projection is

$$\mathbf{u} = \left(\frac{\mathbf{v} \cdot \mathbf{w}}{\mathbf{w} \cdot \mathbf{w}}\right)\mathbf{w}$$

WARNING Note that $\left(\dfrac{\mathbf{v} \cdot \mathbf{w}}{\mathbf{w} \cdot \mathbf{w}}\right)\mathbf{w}$ is not the same as $\left(\dfrac{\mathbf{v}}{\mathbf{w}}\right)\mathbf{w}$; you cannot "cancel" the vector $\mathbf{w}$. Remember, $\mathbf{v} \cdot \mathbf{w}$ and $\mathbf{w} \cdot \mathbf{w}$ are numbers, whereas $\mathbf{v}/\mathbf{w}$ is not defined.

The length of the vector projection is called the **scalar projection of v onto w** (also called the **component of v along w**) and may be computed by the formula

$$\|\mathbf{u}\| = \left|\frac{\mathbf{v} \cdot \mathbf{w}}{\mathbf{w} \cdot \mathbf{w}}\right| \|\mathbf{w}\| = \frac{|\mathbf{v} \cdot \mathbf{w}|}{\|\mathbf{w}\|}$$

To summarize:

> **Scalar and Vector Projections of v onto w**
>
> **Scalar projection** of $\mathbf{v}$ onto $\mathbf{w}$: $\dfrac{|\mathbf{v} \cdot \mathbf{w}|}{\|\mathbf{w}\|}$ a number
>
> **Vector projection** of $\mathbf{v}$ in the direction of $\mathbf{w}$: $\left(\dfrac{\mathbf{v} \cdot \mathbf{w}}{\mathbf{w} \cdot \mathbf{w}}\right)\mathbf{w}$ a vector

In $\mathbb{R}^3$ it is important to note that the $\mathbf{i}, \mathbf{j}$, and $\mathbf{k}$ components of any vector $\mathbf{v}$ are the scalar projections of $\mathbf{v}$ onto the appropriate basis vectors, as shown in Figure 10.22.

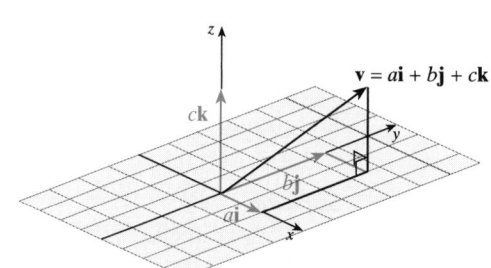

■ **FIGURE 10.22** The $\mathbf{i}, \mathbf{j}$, and $\mathbf{k}$ components of $\mathbf{v} = a\mathbf{i} + b\mathbf{j} + c\mathbf{k}$ are projections of $\mathbf{v}$ onto $\mathbf{i}, \mathbf{j}$, and $\mathbf{k}$.

EXAMPLE 9   *Scalar and vector projections*

Find the scalar and vector projections of $\mathbf{v} = 2\mathbf{i} - 3\mathbf{j} + 5\mathbf{k}$ onto $\mathbf{w} = 2\mathbf{i} - 2\mathbf{j} + \mathbf{k}$.

Solution   The vector projection of $\mathbf{v}$ onto $\mathbf{w}$ is

$$\left(\frac{\mathbf{v} \cdot \mathbf{w}}{\mathbf{w} \cdot \mathbf{w}}\right)\mathbf{w} = \left(\frac{2(2) + (-3)(-2) + 5(1)}{2^2 + (-2)^2 + 1^2}\right)(2\mathbf{i} - 2\mathbf{j} + \mathbf{k})$$
$$= \frac{15}{9}(2\mathbf{i} - 2\mathbf{j} + \mathbf{k}) = \frac{10}{3}\mathbf{i} - \frac{10}{3}\mathbf{j} + \frac{5}{3}\mathbf{k}$$

To find the scalar projection, we can find the length of the vector projection or we can use the scalar projection formula (which is usually easier than finding the length directly):

$$\text{Scalar projection of } \mathbf{v} \text{ onto } \mathbf{w} = \left| \frac{\mathbf{v} \cdot \mathbf{w}}{\|\mathbf{w}\|} \right| = \left| \frac{15}{3} \right| = 5 \qquad \blacksquare$$

WORK AS A DOT PRODUCT

In Section 6.5, we noted that when a constant force **F** acts in the direction of motion of an object moving from P to Q along a straight line, it produces work W where

$$W = (\text{magnitude of } \mathbf{F})(\text{distance moved}) = \|\mathbf{F}\| \|\mathbf{PQ}\|$$

The force remains constant but acts in a direction that makes an angle θ with the direction of motion (see Figure 10.23a). Experiments indicate that the work produced is

$$W = (\text{scalar component of } \mathbf{F} \text{ along } \mathbf{PQ})(\text{distance moved})$$
$$= (|\mathbf{F}| \cos \theta) \|\mathbf{PQ}\| = \mathbf{F} \cdot \mathbf{PQ}$$

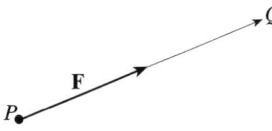

a. If the force **F** acts along the line of motion then $W = \|\mathbf{F}\| \|\mathbf{PQ}\|$.

b. If **F** acts at a nonzero angle with the line of motion, then $W = \|\text{proj. of } \mathbf{F} \text{ onto } \mathbf{PQ}\| \|\mathbf{PQ}\|$.

■ **FIGURE 10.23** Work W as a dot product

Work as a Dot Product

An object that moves along a line with displacement **PQ** against a constant force **F** performs

$$W = \mathbf{F} \cdot \mathbf{PQ}$$

units of work.

EXAMPLE 10   *Work performed by a constant force*

A boat sails north aided by a wind blowing in a direction of N30°E (30° east of north) with magnitude 500 lb. How much work is performed by the wind as the boat moves 100 ft?

Solution   The wind force is $\|\mathbf{F}\| = 500$ lb, acting in a direction $\theta = 30°$, as shown in Figure 10.24. The displacement direction is $\mathbf{PQ} = 100\mathbf{j}$, so $\|\mathbf{PQ}\| = 100$ ft. Thus, $\mathbf{F} = 500 \cos 60°\mathbf{i} + 500 \sin 60°\mathbf{j} = 250\mathbf{i} + 250\sqrt{3}\mathbf{j}$. Thus, the work performed is

$$W = \mathbf{F} \cdot \mathbf{PQ} = 100(250\sqrt{3}) = 25,000\sqrt{3}$$

Thus, the work is approximately 43,300 ft-lb.   ■

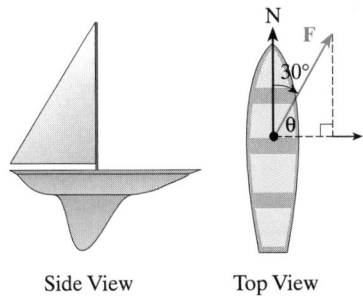

Side View   Top View

■ **FIGURE 10.24** Work performed on a sail

10.3 *Problem Set*

Ⓐ 1. ■ **What Does This Say?** Discuss how to find a dot product, and describe an application of dot product.

2. ■ **What Does This Say?** What does it mean for vectors to be orthogonal? What is meant by the orthogonal projection of a vector **v** on another vector **w**?

Find the standard representation of the vector **PQ**, *and then find* $\|\mathbf{PQ}\|$ *in Problems 3–6.*

3. $P(1, -1, 3), Q(-1, 1, 4)$

4. $P(0, 2, 3), Q(2, 3, 0)$

5. $P(1, 1, 1), Q(-3, -3, -3)$

6. $P(3, 0, -4), Q(0, -4, 3)$

Find the dot product $\mathbf{v} \cdot \mathbf{w}$ *in Problems 7–10.*

7. $\mathbf{v} = \langle 3, -2, 4 \rangle; \mathbf{w} = \langle 2, -1, -6 \rangle$

8. $\mathbf{v} = \langle 2, -6, 0 \rangle; \mathbf{w} = \langle 0, -3, 7 \rangle$

9. $\mathbf{v} = 2\mathbf{i} + 3\mathbf{j} - \mathbf{k}; \mathbf{w} = -3\mathbf{i} + 5\mathbf{j} + 4\mathbf{k}$

10. $\mathbf{v} = 3\mathbf{i} - \mathbf{j}; \mathbf{w} = 2\mathbf{i} + 5\mathbf{j}$

State whether the given pairs of vectors in Problems 11–14 are orthogonal.

11. $\mathbf{v} = \mathbf{i}; \mathbf{w} = \mathbf{k}$

12. $\mathbf{v} = \mathbf{j}; \mathbf{w} = -\mathbf{k}$

13. $\mathbf{v} = 3\mathbf{i} - 2\mathbf{j}; \mathbf{w} = 6\mathbf{i} + 9\mathbf{j}$

14. $\mathbf{v} = 4\mathbf{i} - 5\mathbf{j} + \mathbf{k}; \mathbf{w} = 8\mathbf{i} + 10\mathbf{j} - 2\mathbf{k}$

Evaluate the expressions given in Problems 15–18.

15. $\|\mathbf{i} + \mathbf{j} + \mathbf{k}\|$

16. $\|\mathbf{i} - \mathbf{j} + \mathbf{k}\|$

17. $\|2\mathbf{i} + \mathbf{j} - 3\mathbf{k}\|^2$

18. $\|2(\mathbf{i} - \mathbf{j} + \mathbf{k}) - 3(2\mathbf{i} + \mathbf{j} - \mathbf{k})\|^2$

Let $\mathbf{v} = \mathbf{i} - 2\mathbf{j} + 2\mathbf{k}$ *and* $\mathbf{w} = 2\mathbf{i} + 4\mathbf{j} - \mathbf{k}$; *and find the vector or scalar requested in Problems 19–22.*

19. $2\|\mathbf{v}\| - 3\|\mathbf{w}\|$

20. $\|\mathbf{v}\|\mathbf{w}$

21. $\|2\mathbf{v} - 3\mathbf{w}\|$

22. $\|\mathbf{v} - \mathbf{w}\|(\mathbf{v} + \mathbf{w})$

Determine whether each vector in Problems 23–26 is parallel to $\mathbf{u} = 2\mathbf{i} - 3\mathbf{j} + 5\mathbf{k}$.

23. $\mathbf{v} = \langle 4, -6, 10 \rangle$

24. $\mathbf{v} = \langle -2, 6, -10 \rangle$

25. $\mathbf{v} = \langle 1, -\frac{3}{2}, 2 \rangle$

26. $\mathbf{v} = \langle -1, \frac{3}{2}, -\frac{5}{2} \rangle$

Let $\mathbf{v} = 3\mathbf{i} - 2\mathbf{j} + \mathbf{k}$ *and* $\mathbf{w} = \mathbf{i} + \mathbf{j} - \mathbf{k}$. *Evaluate the expressions in Problems 27–30.*

27. $(\mathbf{v} + \mathbf{w}) \cdot (\mathbf{v} - \mathbf{w})$

28. $(\mathbf{v} \cdot \mathbf{w})\mathbf{w}$

29. $(\|\mathbf{v}\|\mathbf{w}) \cdot (\|\mathbf{w}\|\mathbf{v})$

30. $\dfrac{2\mathbf{v} + 3\mathbf{w}}{\|3\mathbf{v} + 2\mathbf{w}\|}$

Find the angle between the vectors given in Problems 31–34. Round to the nearest degree.

31. $\mathbf{v} = \mathbf{i} + \mathbf{j} + \mathbf{k}; \mathbf{w} = \mathbf{i} - \mathbf{j} + \mathbf{k}$

32. $\mathbf{v} = 2\mathbf{i} + \mathbf{k}; \mathbf{w} = \mathbf{j} - 3\mathbf{k}$

33. $\mathbf{v} = 2\mathbf{j} + \mathbf{k}; \mathbf{w} = \mathbf{i} - 2\mathbf{k}$

34. $\mathbf{v} = 4\mathbf{i} - \mathbf{j} + \mathbf{k}; \mathbf{w} = 2\mathbf{i} + 3\mathbf{j} + 5\mathbf{k}$

Find the vector and scalar projections of **v** *onto* **w** *in Problems 35–38.*

35. $\mathbf{v} = \mathbf{i} + \mathbf{j} + \mathbf{k}; \mathbf{w} = 2\mathbf{k}$

36. $\mathbf{v} = \mathbf{i} + 2\mathbf{k}; \mathbf{w} = -3\mathbf{j}$

37. $\mathbf{v} = 2\mathbf{i} - 3\mathbf{j}; \mathbf{w} = 2\mathbf{j} - 3\mathbf{k}$

38. $\mathbf{v} = \mathbf{i} + \mathbf{j} - 2\mathbf{k}; \mathbf{w} = \mathbf{i} + \mathbf{j} + \mathbf{k}$

39. Find two distinct unit vectors orthogonal to $\mathbf{v} = \mathbf{i} + \mathbf{j} - \mathbf{k}; \mathbf{w} = -\mathbf{i} + \mathbf{j} + \mathbf{k}$

40. Find two distinct unit vectors orthogonal to $\mathbf{v} = 2\mathbf{i} + \mathbf{j} + 2\mathbf{k}; \mathbf{w} = -\mathbf{i} + 2\mathbf{j} - \mathbf{k}$.

41. Find a unit vector that points in the direction opposite to $\mathbf{v} = 2\mathbf{i} + 3\mathbf{j} - 2\mathbf{k}$.

42. Find a vector that points in the same direction as $\mathbf{v} = \mathbf{i} + 2\mathbf{j} - \mathbf{k}$ and has one-third its length.

43. Find x, y, and z that solve $x(\mathbf{i} + \mathbf{j} + \mathbf{k}) + y(\mathbf{i} - \mathbf{j} + 2\mathbf{k}) + z(\mathbf{i} + \mathbf{k}) = 2\mathbf{i} + \mathbf{k}$

44. Find x, y, and z that solve $x(\mathbf{i} - \mathbf{k}) + y(\mathbf{j} + \mathbf{k}) + z(\mathbf{i} - \mathbf{j}) = 5\mathbf{i} - \mathbf{k}$

45. Find a number a that guarantees that the vectors $3\mathbf{i} - 2\mathbf{j} + \mathbf{k}$ and $2\mathbf{i} + a\mathbf{j} - 2a\mathbf{k}$ will be orthogonal.

46. Find x if the vectors $\mathbf{v} = 3\mathbf{i} - x\mathbf{j} + 2\mathbf{k}$ and $\mathbf{w} = x\mathbf{i} + \mathbf{j} - 2\mathbf{k}$ are to be orthogonal.

Ⓑ 47. Find the angles between the vector $2\mathbf{i} + \mathbf{j} - \mathbf{k}$ and each of the coordinate axes. The cosines of these angles (to the nearest degree) are known as the **direction cosines.**

48. Find the cosine of the angle between the vectors $\mathbf{v} = \mathbf{i} - \mathbf{j} + 2\mathbf{k}$ and $\mathbf{w} = 2\mathbf{i} + \mathbf{j} - \mathbf{k}$. Then find the vector projection of **v** onto **w**.

49. Let $\mathbf{v} = 4\mathbf{i} - \mathbf{j} + \mathbf{k}$ and $\mathbf{w} = 2\mathbf{i} + 3\mathbf{j} - \mathbf{k}$. Find:
   a. $\mathbf{v} \cdot \mathbf{w}$
   b. $\cos \theta$, where θ is the angle between **v** and **w**
   c. a scalar s such that **v** is orthogonal to $\mathbf{v} - s\mathbf{w}$
   d. a scalar s such that $s\mathbf{v} + \mathbf{w}$ is orthogonal to **w**

50. Let $\mathbf{v} = 2\mathbf{i} - 3\mathbf{j} + 6\mathbf{k}$ and $\mathbf{w} = 4\mathbf{i} + 3\mathbf{k}$. Find:
   a. $\mathbf{v} \cdot \mathbf{w}$
   b. $\cos \theta$, where θ is the angle between **v** and **w**
   c. a scalar s such that **v** is the orthogonal to $\mathbf{v} - s\mathbf{w}$
   d. a scalar t such that $\mathbf{v} + t\mathbf{w}$ is orthogonal to **w**

51. Find the scalar component of the force $\mathbf{F} = 4\mathbf{i} - 2\mathbf{j} + 3\mathbf{k}$ in the direction of the vector $\mathbf{v} = \mathbf{i} - \mathbf{j} + 2\mathbf{k}$.

52. Find the work done by the constant force $\mathbf{F} = 2\mathbf{i} + 3\mathbf{j} + \mathbf{k}$ when it moves a particle along the line from $P(1, 0, -1)$ to $Q(3, 1, 2)$.

53. Find the work performed when a force $\mathbf{F} = \frac{6}{7}\mathbf{i} - \frac{2}{7}\mathbf{j} + \frac{6}{7}\mathbf{k}$ is applied to an object moving along the line from $P(-3, -5, 4)$ to $Q(4, 9, 11)$.

54. Fred and his son Sam are pulling a heavy log along flat horizontal ground by ropes attached to the front of the log. The ropes are 8 ft long. Fred holds his rope 2 ft above the log and 1 ft to the side, and Sam holds his end 1 ft above the log and 1 ft to the opposite side, as shown in Figure 10.25.

Side View   Top View

■ **FIGURE 10.25** Problem 54

If Fred exerts a force of 30 lb and Sam exerts a force of 20 lb, what is the resultant force on the log?

55. Find the force required to keep a 5,000-lb van from rolling downhill if it is parked on a 10° slope.

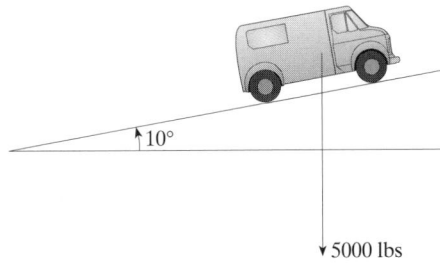

56. A block of ice is dragged 20 ft across a floor, using a force of 50 lb. Find the work done if the direction of the force is inclined θ to the horizontal, where

   a. $\theta = \frac{\pi}{3}$ b. $\theta = \frac{\pi}{4}$

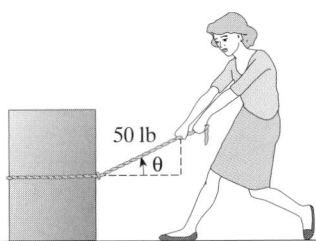

57. Suppose that the wind is blowing with a 1,000-lb magnitude force $\mathbf{F}$ in the direction of N60°W behind a boat's sail. How much work does the wind perform in moving the boat in a northerly direction a distance of 50 ft? Give your answer in foot-pounds.

Technology Window

Orthogonality of functions *One of the big discoveries in modern mathematics was that it is very fruitful to extend some of the geometric notions of this chapter to functions—for example, taking the dot product of two functions or asking whether two functions are orthogonal, or attempting to project a function onto a second function (or onto a "plane" of functions). You are introduced to these important concepts in Problems 58 and 59.*

58. We say that two functions f and g are **orthogonal** on $[a, b]$ if

$$\int_a^b f(x)g(x)\, dx = 0$$

   a. Show that the two functions x^2 and $x^3 - 5x$ are orthogonal on $[-b, b]$ for any positive b.

   b. For positive integers k and $n \neq k$, show that $\sin kx$ and $\sin nx$ are orthogonal on the interval $[-\pi, \pi]$. That is, the *family* of functions $\sin x, \sin 2x, \sin 3x, \ldots$ are *mutually orthogonal* on $[-\pi, \pi]$. You may need the product-to-sum identity (identity 38 from the *Student Mathematics Handbook*):

$$2\sin \alpha \sin \beta = \cos(\alpha - \beta) - \cos(\alpha + \beta)$$

   c. Make a careful sketch of $\sin x$ and $\sin 2x$ on $[-\pi, \pi]$ and explain why the integral of their product is 0.

59. **H**ISTORICAL **Q**UEST
Jean Baptiste Joseph Fourier began his studies by training for the priesthood. In spite of this first calling, his interest in mathematics remained intense, and in 1789 he wrote, "Yesterday was my 21st birthday, and at that age Newton and Pascal had already acquired many claims to immortality." He did not take his religious vows but continued his mathematical research while teaching. Fourier was renowned as an outstanding teacher, but it was not until around 1804–1807 that he did his important mathematical work on the theory of heat. During that time, he made the remarkable observation that most important odd functions [that is, $f(-x) = -f(x)$] can be approximated by

JEAN FOURIER
1768–1830

$$f(x) \approx b_1 \sin x + b_2 \sin 2x + \cdots + b_n \sin nx = S_n(x)$$

60. Let $u_0 = a\mathbf{i} + b\mathbf{j} + c\mathbf{k}$ and let $\mathbf{u} = x\mathbf{i} + y\mathbf{j} + z\mathbf{k}$. Describe the set of points in $\mathbb{R}^3$ defined by

$$\|\mathbf{u}_0 - \mathbf{u}\| < r$$

where $r > 0$ and a, b, c are constants.

61. Suppose the vectors $\mathbf{v}$ and $\mathbf{w}$ are sides of a triangle with area $25\sqrt{3}$. Find $\mathbf{v} \cdot \mathbf{w}$.

62. Find the angle (to the nearest degree) between the diagonal of a cube and a diagonal of one of its faces.

63. If $\mathbf{v} \cdot \mathbf{v} = 0$, what can you conclude about $\mathbf{v}$?

64. THINK TANK PROBLEM   Let $\mathbf{u}$ and $\mathbf{v}$ be nonzero vectors, and let θ be the acute angle between $\mathbf{u}$ and $\mathbf{v}$. What can be said about the vector

$$\mathbf{B} = \|\mathbf{v}\|\mathbf{u} + \|\mathbf{u}\|\mathbf{v}$$

Prove your conjecture.

65. Use vector methods to prove that an angle inscribed in a semicircle must be a right angle.

66. a. Show that $(\mathbf{v} + \mathbf{w}) \cdot (\mathbf{v} + \mathbf{w})$
$= \|\mathbf{v}\|^2 + \|\mathbf{w}\|^2 + 2(\mathbf{v} \cdot \mathbf{w})$.
b. Use part **a** to prove the **triangle inequality:**

$$\|\mathbf{v} + \mathbf{w}\| \leq \|\mathbf{v}\| + \|\mathbf{w}\|$$

Hint: Note that
$$\|\mathbf{v} + \mathbf{w}\|^2 = (\mathbf{v} + \mathbf{w}) \cdot (\mathbf{v} + \mathbf{w})$$

67. The **Cauchy–Schwarz** inequality in $\mathbb{R}^3$ states that for any vectors $\mathbf{v}$ and $\mathbf{w}$,

$$|\mathbf{v} \cdot \mathbf{w}| \leq \|\mathbf{v}\| \|\mathbf{w}\|$$

a. Prove the Cauchy–Schwarz inequality. (*Hint:* Use the formula for the angle between vectors.)
b. Show that equality in the Cauchy–Schwarz inequality occurs if and only if $\mathbf{v} = t\mathbf{w}$ for some scalar t.
c. Use the Cauchy–Schwarz inequality to prove the **triangle inequality:**

$$\|\mathbf{v} + \mathbf{w}\| \leq \|\mathbf{v}\| + \|\mathbf{w}\|$$

10.4 The Cross Product

IN THIS SECTION   definition and basic properties of the cross product, geometric interpretation of cross product, area and volume, torque

DEFINITION AND BASIC PROPERTIES OF THE CROSS PRODUCT

We begin by defining the *cross product* in terms of a determinant. If you need to review the basic properties of determinants, consult the *Student Mathematics Handbook*.

Cross Product

If $\mathbf{v} = a_1\mathbf{i} + a_2\mathbf{j} + a_3\mathbf{k}$ and $\mathbf{w} = b_1\mathbf{i} + b_2\mathbf{j} + b_3\mathbf{k}$, the **cross product,** written $\mathbf{v} \times \mathbf{w}$, is the vector

$$\mathbf{v} \times \mathbf{w} = (a_2b_3 - a_3b_2)\mathbf{i} + (a_3b_1 - a_1b_3)\mathbf{j} + (a_1b_2 - a_2b_1)\mathbf{k}$$

These terms can be obtained by using a determinant

$$\mathbf{v} \times \mathbf{w} = \begin{vmatrix} \mathbf{i} & \mathbf{j} & \mathbf{k} \\ a_1 & a_2 & a_3 \\ b_1 & b_2 & b_3 \end{vmatrix}$$

To verify the determinant formula for the cross product, we expand about the first row:

$$\mathbf{v} \times \mathbf{w} = \begin{vmatrix} \mathbf{i} & \mathbf{j} & \mathbf{k} \\ a_1 & a_2 & a_3 \\ b_1 & b_2 & b_3 \end{vmatrix} = \begin{vmatrix} a_2 & a_3 \\ b_2 & b_3 \end{vmatrix} \mathbf{i} - \begin{vmatrix} a_1 & a_3 \\ b_1 & b_3 \end{vmatrix} \mathbf{j} + \begin{vmatrix} a_1 & a_2 \\ b_1 & b_2 \end{vmatrix} \mathbf{k}$$

j is in row 1, column 2, so do not forget negative sign here.

$$= (a_2 b_3 - a_3 b_2)\mathbf{i} - (a_1 b_3 - a_3 b_1)\mathbf{j} + (a_1 b_2 - a_2 b_1)\mathbf{k}$$
$$= (a_2 b_3 - a_3 b_2)\mathbf{i} + (a_3 b_1 - a_1 b_3)\mathbf{j} + (a_1 b_2 - a_2 b_1)\mathbf{k}$$

> **EXAMPLE 1**   *Cross product*

Find $\mathbf{v} \times \mathbf{w}$, where $\mathbf{v} = 2\mathbf{i} - \mathbf{j} + 3\mathbf{k}$ and $\mathbf{w} = 7\mathbf{j} - 4\mathbf{k}$.

Solution

$$\mathbf{v} \times \mathbf{w} = \begin{vmatrix} \mathbf{i} & \mathbf{j} & \mathbf{k} \\ 2 & -1 & 3 \\ 0 & 7 & -4 \end{vmatrix}$$

Do not forget minus here (j is negative position).

 See Problem 66, p. 36 of handbook.

$$= [(-1)(-4) - 3(7)]\mathbf{i} - [2(-4) - 0(3)]\mathbf{j} + [2(7) - 0(-1)]\mathbf{k}$$
$$= -17\mathbf{i} + 8\mathbf{j} + 14\mathbf{k}$$

Technology Window

Mathematical programs such as *Mathematica*, *Maple*, *Derive*, and *Mathlab* all do vector operations, including dot and cross product. Recall from a previous Technology Window that brackets are sometimes used to input vectors, with the components separated by commas, as in $[a, b, c]$.

The graphic at the right shows the dot product and the cross product of the vectors in Example 1. Notice that the output shows a number, -19, for the dot product, and a vector, $[-17, 8, 14]$, for the cross product. In vector notation,

$$[-17, 8, 14] = -17\mathbf{i} + 8\mathbf{j} + 14\mathbf{k}$$

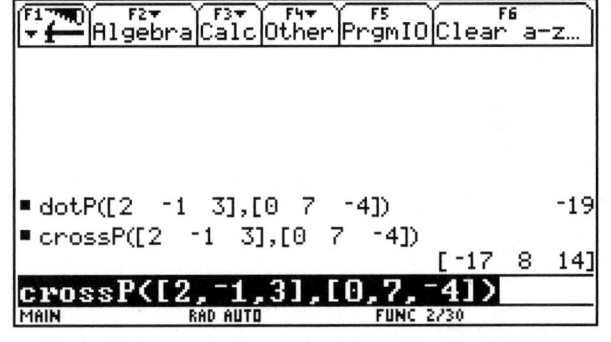

Properties of determinants can also be used to establish properties of the cross product. For instance, the following computation shows that the cross product is *not* commutative. This property is sometimes called **anticommutativity.**

$$\mathbf{v} \times \mathbf{w} = \begin{vmatrix} \mathbf{i} & \mathbf{j} & \mathbf{k} \\ a_1 & a_2 & a_3 \\ b_1 & b_2 & b_3 \end{vmatrix} = -\begin{vmatrix} \mathbf{i} & \mathbf{j} & \mathbf{k} \\ b_1 & b_2 & b_3 \\ a_1 & a_2 & a_3 \end{vmatrix} = -(\mathbf{w} \times \mathbf{v})$$

This and other properties are listed in the following theorem.

===

THEOREM 10.5   *Properties of the cross product*

If $\mathbf{u}$, $\mathbf{v}$, and $\mathbf{w}$ are vectors in $\mathbb{R}^3$ and s and t are scalars, then:

Scalar distributivity $(s\mathbf{v}) \times (t\mathbf{w}) = st(\mathbf{v} \times \mathbf{w})$

Distributivity for cross product over addition
$$\mathbf{u} \times (\mathbf{v} + \mathbf{w}) = (\mathbf{u} \times \mathbf{v}) + (\mathbf{u} \times \mathbf{w})$$
$$(\mathbf{u} + \mathbf{v}) \times \mathbf{w} = (\mathbf{u} \times \mathbf{w}) + (\mathbf{v} \times \mathbf{w})$$

Anticommutativity $\mathbf{v} \times \mathbf{w} = -(\mathbf{w} \times \mathbf{v})$

Product of a multiple $\mathbf{v} \times s\mathbf{v} = 0$; in particular, $\mathbf{v} \times \mathbf{v} = \mathbf{0}$

Zero product $\mathbf{v} \times \mathbf{0} = \mathbf{0} \times \mathbf{v} = \mathbf{0}$

Lagrange's identity $\|\mathbf{v} \times \mathbf{w}\|^2 = \|\mathbf{v}\|^2\|\mathbf{w}\|^2 - (\mathbf{v} \cdot \mathbf{w})^2$

cab-bac formula $\mathbf{a} \times (\mathbf{b} \times \mathbf{c}) = (\mathbf{c} \cdot \mathbf{a})\mathbf{b} - (\mathbf{b} \cdot \mathbf{a})\mathbf{c}$

Proof   Let $\mathbf{u} = a_1\mathbf{i} + a_2\mathbf{j} + a_3\mathbf{k}$, $\mathbf{v} = b_1\mathbf{i} + b_2\mathbf{j} + b_3\mathbf{k}$, and $\mathbf{w} = c_1\mathbf{i} + c_2\mathbf{j} + c_3\mathbf{k}$.

Scalar distributivity and **vector distributivity** are proved by using the definition of cross product and the corresponding properties of real numbers.

Anticommutativity was proved in the paragraph preceding this theorem.

Product of a multiple

$$\mathbf{v} \times s\mathbf{v} = \begin{vmatrix} \mathbf{i} & \mathbf{j} & \mathbf{k} \\ b_1 & b_2 & b_3 \\ sb_1 & sb_2 & sb_3 \end{vmatrix} = s\begin{vmatrix} \mathbf{i} & \mathbf{j} & \mathbf{k} \\ b_1 & b_2 & b_3 \\ b_1 & b_2 & b_3 \end{vmatrix} = 0$$

A determinant with two identical rows is 0.

Zero product is obvious.

Lagrange's identity

$$\begin{aligned}\|\mathbf{v} \times \mathbf{w}\|^2 &= (b_2c_3 - b_3c_2)^2 + (b_3c_1 - b_1c_3)^2 + (b_1c_2 - b_2c_1)^2 \\ &= (b_1^2 + b_2^2 + b_3^2)(c_1^2 + c_2^2 + c_3^2) - (b_1c_1 + b_2c_2 + b_3c_3)^2 \\ &= \|\mathbf{v}\|^2\|\mathbf{w}\|^2 - (\mathbf{v}\cdot\mathbf{w})^2\end{aligned}$$

GEOMETRIC INTERPRETATION OF THE CROSS PRODUCT

The following theorem shows that the vector $(\mathbf{v} \times \mathbf{w})$ is orthogonal to both the vectors $\mathbf{v}$ and $\mathbf{w}$. See Figure 10.26.

===

THEOREM 10.6   *Orthogonality property of the cross product*

If $\mathbf{v}$ and $\mathbf{w}$ are nonzero vectors in $\mathbb{R}^3$ that are not multiples of one another, then

$$\mathbf{v} \times \mathbf{w} \text{ is orthogonal to both } \mathbf{v} \text{ and } \mathbf{w}.$$

Proof   We focus on the case where $\mathbf{v} \times \mathbf{w}$ is a nonzero vector. We will show that $(\mathbf{v} \times \mathbf{w})$ is orthogonal to $\mathbf{v}$ and leave the proof that $\mathbf{v} \times \mathbf{w}$ is orthogonal to $\mathbf{w}$ as an exercise. Let $\mathbf{v} = a_1\mathbf{i} + a_2\mathbf{j} + a_3\mathbf{k}$ and $\mathbf{w} = b_1\mathbf{i} + b_2\mathbf{j} + b_3\mathbf{k}$. Then

$$\mathbf{v} \times \mathbf{w} = \begin{vmatrix} \mathbf{i} & \mathbf{j} & \mathbf{k} \\ a_1 & a_2 & a_3 \\ b_1 & b_2 & b_3 \end{vmatrix} = (a_2b_3 - a_3b_2)\mathbf{i} - (a_1b_3 - a_3b_1)\mathbf{j} + (a_1b_2 - a_2b_1)\mathbf{k}$$

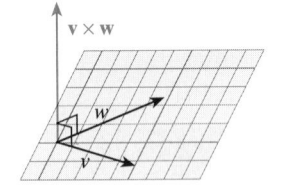

■ FIGURE 10.26   Vector product of two vectors

To show that this vector is orthogonal to **v**, we find $\mathbf{v} \cdot (\mathbf{v} \times \mathbf{w})$:

$$\mathbf{v} \cdot (\mathbf{v} \times \mathbf{w}) = a_1(a_2 b_3 - a_3 b_2) - a_2(a_1 b_3 - a_3 b_1) + a_3(a_1 b_2 - a_2 b_1)$$
$$= a_1 a_2 b_3 - a_1 a_3 b_2 - a_1 a_2 b_3 + a_2 a_3 b_1 + a_1 a_3 b_2 - a_2 a_3 b_1 = 0 \quad \blacksquare$$

Because both $\mathbf{v} \times \mathbf{w}$ and $\mathbf{w} \times \mathbf{v}$ are orthogonal to the plane determined by **v** and **w**, and because $(\mathbf{v} \times \mathbf{w}) = -(\mathbf{w} \times \mathbf{v})$, we see that one points up from the given plane and the other points down. To see which is which, we use the **right-hand rule** described in Figure 10.27.

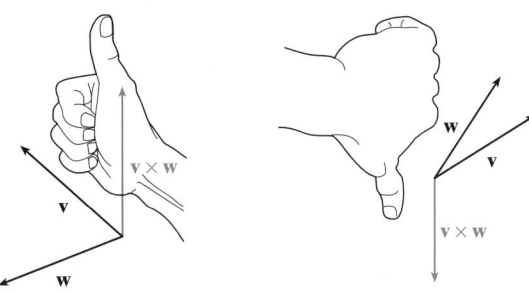

■ **FIGURE 10.27**   The right-hand rule: If you place the palm of your right hand along **v** and curl your fingers towards **w** to cover the smaller angle between **v** and **w**, then your thumb points in the direction of $\mathbf{v} \times \mathbf{w}$.

EXAMPLE 2   *Right-hand rule*

Use the right-hand rule to verify each of the following cross products.

$$\mathbf{i} \times \mathbf{j} = \mathbf{k} \qquad \mathbf{j} \times \mathbf{i} = -\mathbf{k} \qquad \mathbf{i} \times \mathbf{i} = \mathbf{0}$$
$$\mathbf{i} \times \mathbf{k} = -\mathbf{j} \qquad \mathbf{k} \times \mathbf{i} = \mathbf{j} \qquad \mathbf{j} \times \mathbf{j} = \mathbf{0}$$
$$\mathbf{j} \times \mathbf{k} = \mathbf{i} \qquad \mathbf{k} \times \mathbf{j} = -\mathbf{i} \qquad \mathbf{k} \times \mathbf{k} = \mathbf{0}$$

Solution $\mathbf{i} \times \mathbf{j} = \begin{vmatrix} \mathbf{i} & \mathbf{j} & \mathbf{k} \\ 1 & 0 & 0 \\ 0 & 1 & 0 \end{vmatrix} = \mathbf{k}$

Place the palm of your right hand along **i** and curl your fingers toward **j**. The answer is **k**.

$\mathbf{j} \times \mathbf{i} = \begin{vmatrix} \mathbf{i} & \mathbf{j} & \mathbf{k} \\ 0 & 1 & 0 \\ 1 & 0 & 1 \end{vmatrix} = -\mathbf{k}$

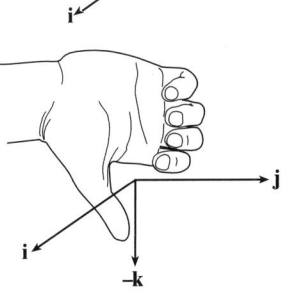

Place the palm of your right hand along **j** and curl your fingers toward **i**. The answer is $-\mathbf{k}$.

$\mathbf{i} \times \mathbf{i} = \begin{vmatrix} \mathbf{i} & \mathbf{j} & \mathbf{k} \\ 1 & 0 & 0 \\ 1 & 0 & 0 \end{vmatrix} = \mathbf{0}$

The other parts are left for you to verify.   ■

EXAMPLE 3   *A vector orthogonal to two given vectors*

Find a nonzero vector that is orthogonal to both $\mathbf{v} = -2\mathbf{i} + 3\mathbf{j} - 7\mathbf{k}$ and $\mathbf{w} = 5\mathbf{i} + 9\mathbf{k}$.

Solution   The cross product $\mathbf{v} \times \mathbf{w}$ is orthogonal to both $\mathbf{v}$ and $\mathbf{w}$.

$$\mathbf{v} \times \mathbf{w} = \begin{vmatrix} \mathbf{i} & \mathbf{j} & \mathbf{k} \\ -2 & 3 & -7 \\ 5 & 0 & 9 \end{vmatrix} = (27 + 0)\mathbf{i} - (-18 + 35)\mathbf{j} + (0 - 15)\mathbf{k}$$

$$= 27\mathbf{i} - 17\mathbf{j} - 15\mathbf{k} \qquad \blacksquare$$

In Section 10.3, we showed that the dot product satisfies $(\mathbf{v} \cdot \mathbf{w}) = \|\mathbf{v}\|\|\mathbf{w}\|\cos\theta$, where θ is the angle between $\mathbf{v}$ and $\mathbf{w}$. The following theorem establishes a similar result for the cross product.

THEOREM 10.7   *Geometric interpretation of cross product*

If $\mathbf{v}$ and $\mathbf{w}$ are nonzero vectors in $\mathbb{R}^3$ with θ the angle between $\mathbf{v}$ and $\mathbf{w}$ ($0 \le \theta \le \pi$), then

$$\mathbf{v} \times \mathbf{w} = (\|\mathbf{v}\|\|\mathbf{w}\|\sin\theta)\mathbf{n}$$

where θ is the angle between $\mathbf{v}$ and $\mathbf{w}$ ($0 \le \theta \le \pi$) and $\mathbf{n}$ is a unit normal vector determined by the right-hand rule (see Figure 10.27).

Proof   We first show that $\|\mathbf{v} \times \mathbf{w}\| = \|\mathbf{v}\|\|\mathbf{w}\|\,|\sin\theta|$ and then focus on the direction.

$$\|\mathbf{v} \times \mathbf{w}\|^2 = \|\mathbf{v}\|^2\|\mathbf{w}\|^2 - [\|\mathbf{v}\|\|\mathbf{w}\|\cos\theta]^2 \qquad \textit{Lagrange's identity}$$

$$= \|\mathbf{v}\|^2\|\mathbf{w}\|^2 - \|\mathbf{v}\|^2\|\mathbf{w}\|^2\cos^2\theta$$

$$= \|\mathbf{v}\|^2\|\mathbf{w}\|^2(1 - \cos^2\theta)$$

$$= \|\mathbf{v}\|^2\|\mathbf{w}\|^2\sin^2\theta$$

Thus, $\|\mathbf{v} \times \mathbf{w}\| = \|\mathbf{v}\|\|\mathbf{w}\|\,|\sin\theta| = \|\mathbf{v}\|\|\mathbf{w}\|\sin\theta$ \qquad *Because $\sin\theta \ge 0$*

We already know that $\mathbf{v} \times \mathbf{w}$ is orthogonal to both $\mathbf{v}$ and $\mathbf{w}$ (Theorem 10.6), so it either points up from the plane, or down from the plane. The fact that $\mathbf{n}$ is determined by the right-hand rule follows from the results in Example 2. ===

> ■ *What This Says:*   Theorem 10.7 provides a *coordinate-free* way of computing the cross product. This is important to physicists and engineers because it enables them to choose whichever coordinate system is most convenient to compute cross products that arise in applications.

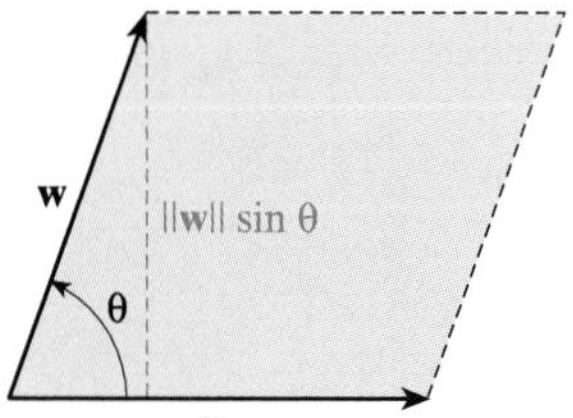

AREA AND VOLUME

There is another interpretation for Theorem 10.7—namely, $\|\mathbf{v} \times \mathbf{w}\|$ is equal to the area of a parallelogram having $\mathbf{v}$ and $\mathbf{w}$ as adjacent sides, as shown in Figure 10.28. To see this, note that because $h = \|\mathbf{w}\|\sin\theta$, we have

$$\text{AREA} = (\text{BASE})(\text{HEIGHT}) = \|\mathbf{v}\|(\|\mathbf{w}\|\sin\theta) = \|\mathbf{v} \times \mathbf{w}\|$$

Here is an example that uses this formula. You might also note that even though $\mathbf{v} \times \mathbf{w} \ne \mathbf{w} \times \mathbf{v}$, it is true that $\|\mathbf{v} \times \mathbf{w}\| = \|\mathbf{w} \times \mathbf{v}\|$.

EXAMPLE 4   *Area of a triangle*

Find the area of the triangle with vertices $P(-2, 4, 5)$, $Q(0, 7, -4)$, and $R(-1, 5, 0)$.

Solution   Draw this triangle as shown in Figure 10.29. Then $\triangle PQR$ has half the area of the parallelogram determined by the vectors $\mathbf{PQ}$ and $\mathbf{PR}$; that is, the triangle has area

$$A = \tfrac{1}{2}\|\mathbf{PQ} \times \mathbf{PR}\|$$

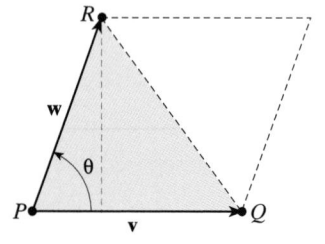

■ **FIGURE 10.28**   A geometric interpretation of $\|\mathbf{v} \times \mathbf{w}\|$

■ **FIGURE 10.29**   Area of a triangle by cross product

First find

$$\mathbf{PQ} = (0 + 2)\mathbf{i} + (7 - 4)\mathbf{j} + (-4 - 5)\mathbf{k} = 2\mathbf{i} + 3\mathbf{j} - 9\mathbf{k}$$
$$\mathbf{PR} = (-1 + 2)\mathbf{i} + (5 - 4)\mathbf{j} + (0 - 5)\mathbf{k} = \mathbf{i} + \mathbf{j} - 5\mathbf{k}$$

and compute the cross product:

$$\mathbf{PQ} \times \mathbf{PR} = \begin{vmatrix} \mathbf{i} & \mathbf{j} & \mathbf{k} \\ 2 & 3 & -9 \\ 1 & 1 & -5 \end{vmatrix}$$
$$= (-15 + 9)\mathbf{i} - (-10 + 9)\mathbf{j} + (2 - 3)\mathbf{k}$$
$$= -6\mathbf{i} + \mathbf{j} - \mathbf{k}$$

Thus, the triangle has area

$$A = \tfrac{1}{2}\|\mathbf{PQ} \times \mathbf{PR}\|$$
$$= \tfrac{1}{2}\sqrt{(-6)^2 + 1^2 + (-1)^2}$$
$$= \tfrac{1}{2}\sqrt{38} \qquad\qquad \blacksquare$$

The cross product can also be used to compute the volume of a parallelepiped in $\mathbb{R}^3$. Consider the parallelepiped determined by three nonzero vectors $\mathbf{u}, \mathbf{v},$ and $\mathbf{w}$ that do not all lie in the same plane, as shown in Figure 10.30.

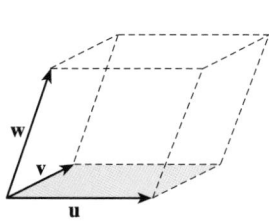

a. The parallelepiped determined by $\mathbf{u}, \mathbf{v},$ and $\mathbf{w}$

b. The parallelepiped has volume $V = Ah = |(\mathbf{u} \times \mathbf{v}) \cdot \mathbf{w}|.$

■ **FIGURE 10.30**   Computing volume with the triple scalar product

It is known from solid geometry that this parallelogram has volume $V = Ah$, where A is the area of the face determined by $\mathbf{u}$ and $\mathbf{v}$, and h is the altitude from the tip of $\mathbf{w}$ to this face (see Figure 10.30b).

The face determined by $\mathbf{u}$ and $\mathbf{v}$ is a parallelogram with area $A = \|\mathbf{u} \times \mathbf{v}\|$, and we know that the cross-product vector $\mathbf{u} \times \mathbf{v}$ is perpendicular to both $\mathbf{u}$ and $\mathbf{v}$ and hence to the face determined by $\mathbf{u}$ and $\mathbf{v}$. From the geometric formula for the dot product, we have

$$(\mathbf{u} \times \mathbf{v}) \cdot \mathbf{w} = \|\mathbf{u} \times \mathbf{v}\| \|\mathbf{w}\| \cos \theta$$

where θ is the angle between $\mathbf{u} \times \mathbf{v}$ and $\mathbf{w}$. Thus, the parallelepiped has altitude

$$h = |\|\mathbf{w}\| \cos \theta| = \left| \frac{(\mathbf{u} \times \mathbf{v}) \cdot \mathbf{w}}{\|\mathbf{u} \times \mathbf{v}\|} \right|$$

and the volume is given as

$$V = Ah = \|\mathbf{u} \times \mathbf{v}\| \left| \frac{(\mathbf{u} \times \mathbf{v}) \cdot \mathbf{w}}{\|\mathbf{u} \times \mathbf{v}\|} \right| = |(\mathbf{u} \times \mathbf{v}) \cdot \mathbf{w}|$$

The combined operation $(\mathbf{u} \times \mathbf{v}) \cdot \mathbf{w}$ is called the **triple scalar product** of $\mathbf{u}$, $\mathbf{v}$, and $\mathbf{w}$. These observations are summarized in the following box.

Volume Interpretation of the Triple Scalar Product

Let $\mathbf{u}$, $\mathbf{v}$, and $\mathbf{w}$ be nonzero vectors that do not all lie in the same plane. Then the parallelepiped determined by these vectors has volume

$$V = \left| (\mathbf{u} \times \mathbf{v}) \cdot \mathbf{w} \right|$$

EXAMPLE 5   *Volume of a parallelepiped*

Find the volume of the parallelepiped determined by the vectors

$$\mathbf{u} = \mathbf{i} - 2\mathbf{j} + 3\mathbf{k} \qquad \mathbf{v} = -4\mathbf{i} + 7\mathbf{j} - 11\mathbf{k} \qquad \mathbf{w} = 5\mathbf{i} + 9\mathbf{j} - \mathbf{k}$$

Solution   We first find the cross product

$$\mathbf{u} \times \mathbf{v} = \begin{vmatrix} \mathbf{i} & \mathbf{j} & \mathbf{k} \\ 1 & -2 & 3 \\ -4 & 7 & -11 \end{vmatrix} = (22 - 21)\mathbf{i} - (-11 + 12)\mathbf{j} + (7 - 8)\mathbf{k}$$
$$= \mathbf{i} - \mathbf{j} - \mathbf{k}$$

Thus,

$$V = \left| (\mathbf{u} \times \mathbf{v}) \cdot \mathbf{w} \right| = \left| (\mathbf{i} - \mathbf{j} - \mathbf{k}) \cdot (5\mathbf{i} + 9\mathbf{j} - \mathbf{k}) \right| = \left| 5 - 9 + 1 \right| = 3$$

The volume of the parallelepiped is 3 cubic units. ∎

THEOREM 10.8   *Determinant form for a triple scalar product*

If $\mathbf{u} = a_1\mathbf{i} + a_2\mathbf{j} + a_3\mathbf{k}$, $\mathbf{v} = b_1\mathbf{i} + b_2\mathbf{j} + b_3\mathbf{k}$, and $\mathbf{w} = c_1\mathbf{i} + c_2\mathbf{j} + c_3\mathbf{k}$, then the triple scalar product can be found by evaluating the determinant

$$(\mathbf{u} \times \mathbf{v}) \cdot \mathbf{w} = \begin{vmatrix} a_1 & a_2 & a_3 \\ b_1 & b_2 & b_3 \\ c_1 & c_2 & c_3 \end{vmatrix}$$

Proof   The proof follows by expanding the determinant (see Problem 52). ▬▬

We can use Theorem 10.8 to rework Example 5:

$$(\mathbf{u} \times \mathbf{v}) \cdot \mathbf{w} = \begin{vmatrix} 1 & -2 & 3 \\ -4 & 7 & -11 \\ 5 & 9 & -1 \end{vmatrix} = -3$$

SMH *See Problem 65, p. 36 of the handbook.*

Thus, the volume is $\left| -3 \right| = 3$, as computed directly in Example 5.

WARNING ▸   In the problem set, we have included several exercises involving the triple scalar product and the triple vector product $\mathbf{u} \times \mathbf{v} \times \mathbf{w}$. In case you wonder why we neglect the product $(\mathbf{u} \cdot \mathbf{v}) \times \mathbf{w}$, notice that such a product makes no sense, because $\mathbf{u} \cdot \mathbf{v}$ is a *scalar* and the cross product is an operation involving only vectors. Thus, the product $\mathbf{u} \cdot \mathbf{v} \times \mathbf{w}$ must mean $\mathbf{u} \cdot (\mathbf{v} \times \mathbf{w})$. ◂

TORQUE

A useful physical application of the cross product involves **torque.** Suppose the force $\mathbf{F}$ is applied to the point Q. Then the torque of $\mathbf{F}$ around P is defined as the cross product of the "arm" vector $\mathbf{PQ}$ with the force $\mathbf{F}$, as shown in Figure 10.31.

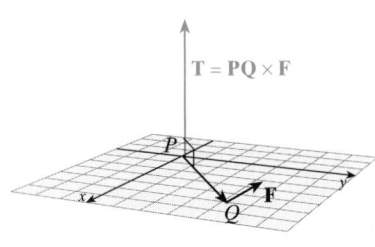

$\mathbf{T} = \mathbf{PQ} \times \mathbf{F}$

■ **FIGURE 10.31**   Torque given by a cross product

Thus, the torque, **T**, of **F** at Q about P is

$$\mathbf{T} = \mathbf{PQ} \times \mathbf{F}$$

The magnitude of the torque, $\|\mathbf{T}\|$, provides a measure of the tendency of the vector arm **PQ** to rotate counterclockwise about an axis perpendicular to the plane determined by **PQ** and **F** (see Figure 10.31).

EXAMPLE 6   *Torque on the hinge of a door*

Figure 10.32 shows a half-open door that is 3 ft wide. A horizontal force of 30 lb is applied at the edge of the door. Find the torque of the force about the hinge on the door.

Solution   We represent the force by $\mathbf{F} = -30\mathbf{i}$ (see Figure 10.32). Because the door is half open, it makes an angle of $\frac{\pi}{4}$ with the horizontal, and we can represent the "arm" **PQ** by the vector

$$\mathbf{PQ} = 3\left(\cos\frac{\pi}{4}\mathbf{i} + \sin\frac{\pi}{4}\mathbf{j}\right) = 3\left(\frac{\sqrt{2}}{2}\mathbf{i} + \frac{\sqrt{2}}{2}\mathbf{j}\right) = \frac{3\sqrt{2}}{2}\mathbf{i} + \frac{3\sqrt{2}}{2}\mathbf{j}$$

The torque can now be found:

$$\mathbf{T} = \mathbf{PQ} \times \mathbf{F} = \begin{vmatrix} \mathbf{i} & \mathbf{j} & \mathbf{k} \\ 3\sqrt{2}/2 & 3\sqrt{2}/2 & 0 \\ -30 & 0 & 0 \end{vmatrix} = 45\sqrt{2}\,\mathbf{k}$$

The magnitude of the torque ($45\sqrt{2}$ lb-ft) is a measure of the tendency of the door to rotate about its hinges.

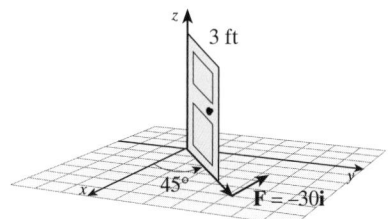

■ **FIGURE 10.32** The torque of a force applied at the edge of a door about the hinges

10.4 *Problem Set*

A *Find* $\mathbf{v} \times \mathbf{w}$ *for the vectors given in Problems* 1–10.

1. $\mathbf{v} = \mathbf{i}; \mathbf{w} = \mathbf{j}$
2. $\mathbf{v} = \mathbf{k}; \mathbf{w} = \mathbf{k}$
3. $\mathbf{v} = 3\mathbf{i} + 2\mathbf{k}; \mathbf{w} = 2\mathbf{i} + \mathbf{j}$
4. $\mathbf{v} = \mathbf{i} - 3\mathbf{j}; \mathbf{w} = \mathbf{i} + 5\mathbf{k}$
5. $\mathbf{v} = 3\mathbf{i} - 2\mathbf{j} + 4\mathbf{k}; \mathbf{w} = \mathbf{i} + 4\mathbf{j} - 7\mathbf{k}$
6. $\mathbf{v} = 5\mathbf{i} - \mathbf{j} + 2\mathbf{k}; \mathbf{w} = 2\mathbf{i} + \mathbf{j} - 3\mathbf{k}$
7. $\mathbf{v} = 3\mathbf{i} - \mathbf{j} + 2\mathbf{k}; \mathbf{w} = 2\mathbf{i} + 3\mathbf{j} - 4\mathbf{k}$
8. $\mathbf{v} = -\mathbf{j} + 4\mathbf{k}; \mathbf{w} = 5\mathbf{i} + 6\mathbf{k}$
9. $\mathbf{v} = \mathbf{i} - 6\mathbf{j} + 10\mathbf{k}; \mathbf{w} = -\mathbf{i} + 5\mathbf{j} - 6\mathbf{k}$
10. $\mathbf{v} = \cos\theta\,\mathbf{i} + \sin\theta\mathbf{j}; \mathbf{w} = -\sin\theta\,\mathbf{i} + \cos\theta\,\mathbf{j}$

Find $\sin\theta$ *where* θ *is the angle between* $\mathbf{v}$ *and* $\mathbf{w}$ *in Problems* 11–16.

11. $\mathbf{v} = \mathbf{i} + \mathbf{k}; \mathbf{w} = \mathbf{i} + \mathbf{j}$
12. $\mathbf{v} = \mathbf{i} + \mathbf{j}; \mathbf{w} = \mathbf{i} + \mathbf{j} + \mathbf{k}$
13. $\mathbf{v} = \mathbf{j} + \mathbf{k}; \mathbf{w} = \mathbf{i} + \mathbf{k}$
14. $\mathbf{v} = \mathbf{i} + \mathbf{j}; \mathbf{w} = \mathbf{j} + \mathbf{k}$
15. $\mathbf{v} = \mathbf{i} + 2\mathbf{j} + 3\mathbf{k}; \mathbf{w} = 4\mathbf{i} + 5\mathbf{j} + 6\mathbf{k}$
16. $\mathbf{v} = \cos\theta\,\mathbf{i} - \sin\theta\mathbf{j}; \mathbf{w} = \sin\theta\,\mathbf{i} - \cos\theta\,\mathbf{j}$

Find a unit vector that is orthogonal to both $\mathbf{v}$ *and* $\mathbf{w}$ *in Problems* 17–20.

17. $\mathbf{v} = 2\mathbf{i} + \mathbf{k}; \mathbf{w} = \mathbf{i} - \mathbf{j} - \mathbf{k}$
18. $\mathbf{v} = \mathbf{j} - 3\mathbf{k}; \mathbf{w} = -\mathbf{i} + \mathbf{j} + \mathbf{k}$
19. $\mathbf{v} = \mathbf{i} + \mathbf{j} + \mathbf{k}; \mathbf{w} = 3\mathbf{i} + 12\mathbf{j} - 4\mathbf{k}$
20. $\mathbf{v} = 2\mathbf{i} - 2\mathbf{j} + \mathbf{k}; \mathbf{w} = 4\mathbf{i} + 2\mathbf{j} - 3\mathbf{k}$

Find the area of the parallelogram determined by the vectors in Problems 21–24.

21. $3\mathbf{i} + 4\mathbf{j}$ and $\mathbf{i} + \mathbf{j} - \mathbf{k}$
22. $2\mathbf{i} - \mathbf{j} + 2\mathbf{k}$ and $4\mathbf{i} - 3\mathbf{j}$
23. $4\mathbf{i} - \mathbf{j} + \mathbf{k}$ and $2\mathbf{i} + 3\mathbf{j} - \mathbf{k}$
24. $2\mathbf{i} + 3\mathbf{k}$ and $2\mathbf{j} - 3\mathbf{k}$

Find the area of $\triangle PQR$ *in Problems* 25–28.

25. $P(0,1,1), Q(1,1,0), R(1,0,1)$
26. $P(1,0,0), Q(2,1,-1), R(0,1,-2)$
27. $P(1,2,3), Q(2,3,1), R(3,1,2)$
28. $P(-1,-1,-1), Q(1,-1,-1), R(-1,1,-1)$

Determine whether each product in Problems 29–31 is a scalar or a vector or does not exist. Explain your reasoning.

29. a. $\mathbf{u} \times (\mathbf{v} \cdot \mathbf{w})$
    b. $\mathbf{u} \cdot (\mathbf{v} \times \mathbf{w})$

30. a. $\mathbf{u} \times (\mathbf{v} \times \mathbf{w})$
    b. $\mathbf{u} \cdot (\mathbf{v} \cdot \mathbf{w})$

31. a. $(\mathbf{u} \times \mathbf{v}) \cdot (\mathbf{u} \times \mathbf{w})$
    b. $(\mathbf{u} \times \mathbf{v}) \times (\mathbf{u} \times \mathbf{w})$

In Problems 32–35, find the volume of the parallelepiped determined by vectors $\mathbf{u}, \mathbf{v},$ *and* $\mathbf{w}$.

32. $\mathbf{u} = \mathbf{i} + \mathbf{j}; \mathbf{v} = \mathbf{j} + 2\mathbf{k}; \mathbf{w} = 3\mathbf{k}$

33. $\mathbf{u} = \mathbf{j} + \mathbf{k}; \mathbf{v} = 2\mathbf{i} + \mathbf{j} + 2\mathbf{k}; \mathbf{w} = 5\mathbf{i}$

34. $\mathbf{u} = \mathbf{i} + \mathbf{j} + \mathbf{k}; \mathbf{v} = \mathbf{i} - \mathbf{j} - \mathbf{k}; \mathbf{w} = 2\mathbf{i} + 3\mathbf{k}$

B 35. $\mathbf{u} = 2\mathbf{i} + \mathbf{j} - \mathbf{k}; \mathbf{v} = 3\mathbf{i} + \mathbf{k}; \mathbf{w} = \mathbf{j} + \mathbf{k}$

36. ■ **What Does This Say?** Contrast dot and cross products of vectors, including a discussion of some of their properties.

37. ■ **What Does This Say?** What is the right-hand rule?

38. ■ **What Does This Say?** Give a geometric interpretation of $|(\mathbf{u} \times \mathbf{v}) \cdot \mathbf{w}|$.

39. **THINK TANK PROBLEM**
    a. If $\mathbf{u} \times \mathbf{w} = \mathbf{v} \times \mathbf{w}$, does it follow that $\mathbf{u} = \mathbf{v}$?
    b. If $\mathbf{u} \cdot \mathbf{w} = \mathbf{v} \cdot \mathbf{w}$, does it follow that $\mathbf{u} = \mathbf{v}$?
    c. If both $\mathbf{u} \times \mathbf{w} = \mathbf{v} \times \mathbf{w}$, and $\mathbf{u} \cdot \mathbf{w} = \mathbf{v} \cdot \mathbf{w}$, does it follow that $\mathbf{u} = \mathbf{v}$?

40. Find a number s that guarantees that the vectors $\mathbf{i}, \mathbf{i} + \mathbf{j} + \mathbf{k}$, and $\mathbf{i} + 2\mathbf{j} + s\mathbf{k}$ will all be parallel to the same plane.

41. Find a number t that guarantees that the vectors $\mathbf{i} + \mathbf{j}, 2\mathbf{i} - \mathbf{j} + \mathbf{k},$ and $\mathbf{i} + \mathbf{j} + t\mathbf{k}$ will all be parallel to the same plane.

42. Find the angle between the vector $2\mathbf{i} - \mathbf{j} + \mathbf{k}$ and the plane determined by the points $P(1, -2, 3)$, $Q(-1, 2, 3)$, and $R(1, 2, -3)$.

43. Let $\mathbf{u} = \mathbf{i} + \mathbf{j}, \mathbf{v} = 2\mathbf{i} - \mathbf{j} + \mathbf{k},$ and $\mathbf{w} = 3\mathbf{i}$. Compute $(\mathbf{u} \times \mathbf{v}) \times \mathbf{w}$ and $\mathbf{u} \times (\mathbf{v} \times \mathbf{w})$. What does this say about the associativity of cross product?

44. Show that $(a\mathbf{u}) \times (b\mathbf{v}) = ab(\mathbf{u} \times \mathbf{v})$ for scalars a and b.

45. For a given vector $\mathbf{v}$ in $\mathbb{R}^3$, find all vectors $\mathbf{w}$ such that $\mathbf{v} \times \mathbf{w} = \mathbf{w}$.

46. What can be said about nonzero vectors $\mathbf{v}$ and $\mathbf{w}$ if $\mathbf{v} \cdot \mathbf{w} = 0$? What can be said if $\mathbf{w} \times \mathbf{v} = \mathbf{0}$?

47. One end of a 2-ft lever pivots about the origin in the yz-plane, as shown in Figure 10.33.
    If a vertical force of 40 lb is applied at the end of the lever, what is the torque of the lever about the pivot point (the origin) when the lever makes an angle of 30° with the xy-plane?

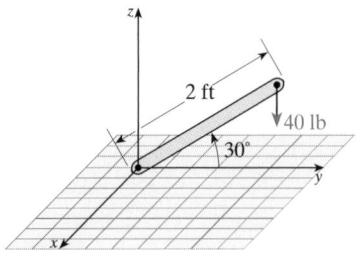

■ **FIGURE 10.33**   Finding the torque

48. A 40-lb child sits on a seesaw, 3 ft from the fulcrum. What torque is exerted when the child is 2 ft above the horizontal? What is the maximum torque exerted by the child?

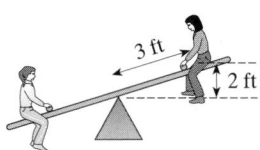

■ **FIGURE 10.34**   Seesaw torque

49. a. Show that the vectors $\mathbf{u}, \mathbf{v},$ and $\mathbf{w}$ are coplanar (all in the same plane) if
    $$\mathbf{u} \cdot (\mathbf{v} \times \mathbf{w}) = 0 \quad \text{or} \quad (\mathbf{u} \times \mathbf{v}) \cdot \mathbf{w} = 0$$
    b. Are the vectors $\mathbf{u} = \mathbf{i} + 3\mathbf{j} + \mathbf{k}, \mathbf{v} = 2\mathbf{i} - \mathbf{j} - \mathbf{k},$ and $\mathbf{w} = 7\mathbf{j} + 3\mathbf{k}$ coplanar?

50. Show that the triangle with vertices $(x_1, y_1), (x_2, y_2),$ (x_3, y_3) has area $A = \frac{1}{2}D$, where
    $$D = \begin{vmatrix} x_1 & y_1 & 1 \\ x_2 & y_2 & 1 \\ x_3 & y_3 & 1 \end{vmatrix}$$

51. Using the properties of determinants, show that
    $$\mathbf{u} \cdot (\mathbf{v} \times \mathbf{w}) = (\mathbf{u} \times \mathbf{v}) \cdot \mathbf{w}$$
    for any vectors $\mathbf{u}, \mathbf{v},$ and $\mathbf{w}$.

C 52. Prove the determinant formula for evaluating a triple scalar product.

53. Let $A, B, C,$ and D be four points that do not lie in the same plane. It can be shown that the volume of the tetrahedron with vertices $A, B, C,$ and D satisfies
    $$\begin{bmatrix} \text{VOLUME OF} \\ \text{TETRAHEDRON } ABCD \end{bmatrix} = \frac{1}{3} \begin{pmatrix} \text{AREA OF} \\ \triangle ABC \end{pmatrix} \begin{pmatrix} \text{ALTITUDE FROM} \\ D \text{ TO } \triangle ABC \end{pmatrix}$$
    Show that the volume is given by
    $$V = \frac{1}{6} |(\mathbf{AB} \times \mathbf{AC}) \cdot \mathbf{AD}|$$

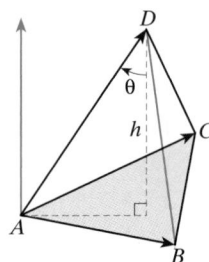

54. Show that if **u**, **v**, and **w** are vectors in $\mathbb{R}^3$ with $\mathbf{u} + \mathbf{v} + \mathbf{w} = \mathbf{0}$, then

$$\mathbf{u} \times \mathbf{v} = \mathbf{v} \times \mathbf{w} = \mathbf{w} \times \mathbf{u}$$

55. Suppose **u**, **v**, and **w** are nonzero vectors in $\mathbb{R}^3$ with $\mathbf{u} \times \mathbf{v} = \mathbf{w}$ and $\mathbf{u} \cdot \mathbf{v} = 0$. Show that

$$\mathbf{v} = s(\mathbf{w} \times \mathbf{u}) \quad \text{and} \quad \mathbf{u} = t(\mathbf{v} \times \mathbf{w})$$

for scalars s and t.

56. Show that

$$(c\mathbf{u}) \times (d\mathbf{v}) = cd(\mathbf{u} \times \mathbf{v})$$

57. Show that

$$\tan \theta = \frac{\|\mathbf{v} \times \mathbf{w}\|}{\mathbf{v} \cdot \mathbf{w}}$$

where θ $(0 \le \theta < \frac{\pi}{2})$ is the angle between **v** and **w**.

58. Let **u**, **v**, and **w** be nonzero vectors in $\mathbb{R}^3$ that do not all lie in the same plane. Show that

$$\big|\mathbf{u} \cdot (\mathbf{v} \times \mathbf{w})\big| = \big|\mathbf{v} \cdot (\mathbf{u} \times \mathbf{w})\big|$$

What other triple scalar products involving **u**, **v**, and **w** have the same absolute values?

59. Let **a**, **b**, and **c** be vectors in space. Prove the "cab – bac" formula.

$$\mathbf{a} \times (\mathbf{b} \times \mathbf{c}) = (\mathbf{c} \cdot \mathbf{a})\mathbf{b} - (\mathbf{b} \cdot \mathbf{a})\mathbf{c}$$

Establish the validity of the equations in Problems 60–63 for arbitrary vectors **u**, **v**, **w**, *and* **z** *in* $\mathbb{R}^3$. *You may use the result of Problem 59.*

60. $(\mathbf{u} \times \mathbf{v}) \times (\mathbf{w} \times \mathbf{z}) = (\mathbf{u} \cdot \mathbf{w} \times \mathbf{z})\mathbf{v} - (\mathbf{v} \cdot \mathbf{w} \times \mathbf{z})\mathbf{u}$

61. $\mathbf{u} \times (\mathbf{v} \times \mathbf{w}) + \mathbf{v} \times (\mathbf{w} \times \mathbf{u}) + \mathbf{w} \times (\mathbf{u} \times \mathbf{v}) = \mathbf{0}$

62. $\mathbf{u} \times \mathbf{v} = (\mathbf{u} \cdot \mathbf{v} \cdot \mathbf{i})\mathbf{i} + (\mathbf{u} \cdot \mathbf{v} \cdot \mathbf{j})\mathbf{j} + (\mathbf{u} \cdot \mathbf{v} \times \mathbf{k})\mathbf{k}$

63. $\mathbf{u} \times [\mathbf{u} \times (\mathbf{u} \times \mathbf{v})] \cdot \mathbf{w} = -\|\mathbf{u}\|^2 \mathbf{u} \cdot \mathbf{v} \times \mathbf{w}$

64. Show that if vectors **OP**, **OQ**, **OR**, and **OS** lie in the same plane, then

$$(\mathbf{OP} \times \mathbf{OQ}) \times (\mathbf{OR} \times \mathbf{OS}) = \mathbf{0}$$

10.5 *Lines and Planes in Space*

IN THIS SECTION lines in $\mathbb{R}^3$, direction cosines, planes in $\mathbb{R}^3$

LINES IN $\mathbb{R}^3$

As in the plane, a line in space is completely determined once we know one of its points and its direction. We used the concept of slope to measure the direction of a line in the plane, but in space, it is more convenient to specify direction with vectors.

Suppose L is a line in space whose location is determined by the vector $\mathbf{v} = A\mathbf{i} + B\mathbf{j} + C\mathbf{k}$ and also suppose L contains $Q(x_0, y_0, z_0)$. We say that L is **aligned with v**, as shown in Figure 10.35.

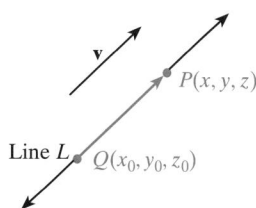

■ **FIGURE 10.35** If L is aligned with **v** and contains Q, then P is on L whenever $\mathbf{QP} = t\mathbf{v}$.

We also say that the line has **direction numbers** A, B, and C and denote these direction numbers by $[A, B, C]$. The vector **v** is called the direction vector of the line L. If $P(x, y, z)$ is any point on L, then the vector **QP** is parallel to **v** and must satisfy the vector equation $\mathbf{QP} = t\mathbf{v}$ for some number t. If we introduce coordinates and use the standard representation, we can rewrite this vector equation as

$$(x - x_0)\mathbf{i} + (y - y_0)\mathbf{j} + (z - z_0)\mathbf{k} = t[A\mathbf{i} + B\mathbf{j} + C\mathbf{k}]$$

By equating components on both sides of this equation, we find that the coordinates of P must satisfy the linear system

$$x - x_0 = tA \qquad y - y_0 = tB \qquad z - z_0 = tC$$

where t is a real number.

Parametric Form of a Line in $\mathbb{R}^3$

If L is a line that contains the point (x_0, y_0, z_0) and is aligned with the vector $\mathbf{v} = A\mathbf{i} + B\mathbf{j} + C\mathbf{k}$, then the point (x, y, z) is on L if and only if its coordinates satisfy

$$x = x_0 + tA \qquad y = y_0 + tB \qquad z = z_0 + tC$$

for some number t.

WARNING t will be a different value for each point (x, y, z) on L.

Turning things around, if we are given the equation of a line with direction numbers $[A, B, C]$, then $\mathbf{v} = A\mathbf{i} + B\mathbf{j} + C\mathbf{k}$ is the **vector aligned with L.**

DRAWING LESSON 5: DRAWING A LINE IN SPACE

a. Draw the three coordinate axes.

b. Draw the three coordinate planes.

c. Plot points where the line intersects each coordinate plane. Used dashed lines for hidden parts.

d. Use highlighters or pencils to color the planes to add depth to the figure.

EXAMPLE 1   *Parametric equations of a line in space*

Find the parametric equations for the line that contains the point (3, 1, 4) and is aligned with the vector $\mathbf{v} = -\mathbf{i} + \mathbf{j} - 2\mathbf{k}$. Find where this line passes through the coordinate planes and sketch the line.

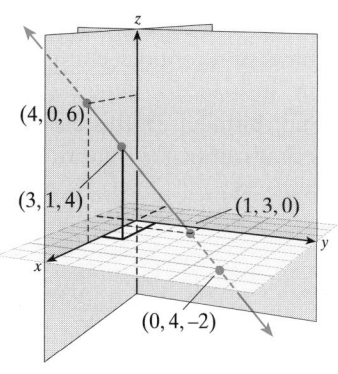

■ **FIGURE 10.36**   Graph of the line
$x = 3 - t, y = 1 + t, z = 4 - 2t$

Solution   The direction numbers are $[-1, 1, -2]$ and $x_0 = 3, y_0 = 1, z_0 = 4$, so the line has the parametric form

$$x = 3 - t \qquad y = 1 + t \qquad z = 4 - 2t$$

This line will intersect the xy-plane when $z = 0$;

$$0 = 4 - 2t \quad \text{implies} \quad t = 2$$

If $t = 2$, then $x = 3 - 2 = 1$ and $y = 1 + 2 = 3$. This is the point $(1, 3, 0)$. Similarly, the line intersects the xz-plane at $(4, 0, 6)$ and the yz-plane at $(0, 4, -2)$. Plot these points and draw the line, as shown in Figure 10.36.   ■

In the special case where none of the direction numbers A, B, or C is 0, we can solve each of the parametric-form equations for t to obtain the following **symmetric equations** for a line.

Symmetric Form of a Line in $\mathbb{R}^3$

If L is a line that contains the point (x_0, y_0, z_0) and is aligned with the vector $\mathbf{v} = A\mathbf{i} + B\mathbf{j} + C\mathbf{k}$ (A, B, and C nonzero numbers), then the point (x, y, z) is on L if and only if its coordinates satisfy

$$\frac{x - x_0}{A} = \frac{y - y_0}{B} = \frac{z - z_0}{C}$$

EXAMPLE 2   *Symmetric form of the equation of a line in space*

Find symmetric equations for the line L through the points $P(-1, 3, 7)$ and $Q(4, 2, -1)$. Find the points of intersection with the coordinate planes and sketch the line.

Solution   The required line passes through P and is aligned with the vector

$$\mathbf{PQ} = [4 - (-1)]\mathbf{i} + [2 - 3]\mathbf{j} + [-1 - 7]\mathbf{k} = 5\mathbf{i} - \mathbf{j} - 8\mathbf{k}$$

Thus, the direction numbers of the line are $[5, -1, -8]$, and we can choose either P or Q as (x_0, y_0, z_0). Choosing P, we obtain

$$\frac{x + 1}{5} = \frac{y - 3}{-1} = \frac{z - 7}{-8}$$

Next, we find points of intersection with the coordinate planes. For the xy-plane $(z = 0)$, we have

$$\frac{x + 1}{5} = \frac{0 - 7}{-8} \quad \text{and} \quad \frac{y - 3}{-1} = \frac{0 - 7}{-8}$$

$$x = \frac{27}{8} \qquad\qquad y = \frac{17}{8}$$

The point of intersection of the line with the xy-plane is $\left(\frac{27}{8}, \frac{17}{8}, 0\right)$. Similarly, the other intersections are

$$xz\text{-plane:} \quad (14, 0, -17) \qquad yz\text{-plane:} \quad \left(0, \frac{14}{5}, \frac{27}{5}\right)$$

The graph is shown in Figure 10.37.   ■

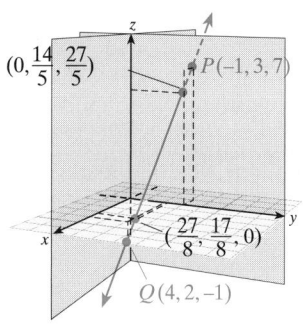

■ **FIGURE 10.37**   Graph of
$$\frac{x + 1}{5} = \frac{y - 3}{-1} = \frac{z - 7}{-8}$$

Recall that two lines in $\mathbb{R}^2$ must intersect if their slopes are different (because they cannot be parallel). However, two lines in $\mathbb{R}^3$ may have different direction numbers and still not intersect because there is enough "room" in space for the lines to lie in parallel planes but be aligned with vectors that are not parallel. In this case, the lines are said to be **skew.** The three different situations that can occur (ignoring the trivial case where the lines coincide) are shown in Figure 10.38.

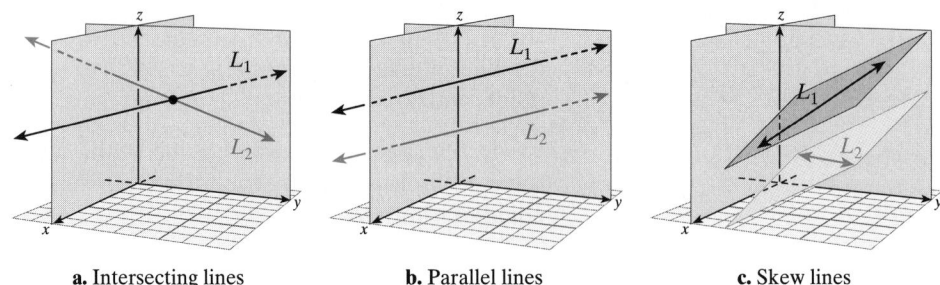

a. Intersecting lines   **b.** Parallel lines   **c.** Skew lines

■ **FIGURE 10.38**   Lines in space may intersect, be parallel, or skew.

EXAMPLE 3   *Skew lines in space*

Determine whether the following lines intersect, are parallel, or are skew.

$$L_1: \quad \frac{x-1}{2} = \frac{y+1}{1} = \frac{z-2}{4} \quad \text{and} \quad L_2: \quad \frac{x+2}{4} = \frac{y}{-3} = \frac{z+1}{1}$$

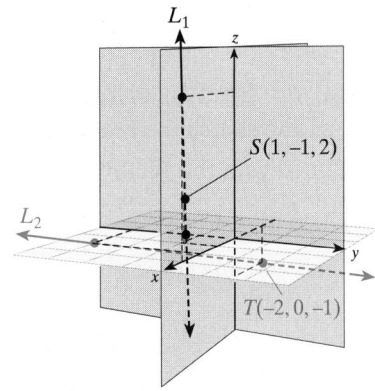

Graph of L_1 and L_2. Note that $S(1, -1, 2)$ lies on L_1 and $T(-2, 0, -1)$ lies on L_2.

Solution   Note that L_1 has direction numbers $[2, 1, 4]$ (that is, L_1 is aligned with $2\mathbf{i} + \mathbf{j} + 4\mathbf{k}$) and L_2 has direction numbers $[4, -3, 1]$. If we solve

$$\langle 2, 1, 4 \rangle = t\langle 4, -3, 1 \rangle$$

for t, we find no possible solution for any value of t. This implies that the lines are not parallel.

Next we determine whether the lines intersect or are skew. Note that $S(1, -1, 2)$ lies on L_1 and $T(-2, 0, -1)$ lies on L_2. The lines intersect if and only if there is a point P that lies on both lines. To determine this, we write the equations of the lines in parametric form. We use a different parameter for each line because it is possible for two lines to arrive at the same point "at different times" (values of the parameter):

$$L_1: x = 1 + 2s \qquad y = -1 + s \qquad z = 2 + 4s$$
$$L_2: x = -2 + 4t \qquad y = -3t \qquad z = -1 + t$$

The lines intersect if there are numbers s and t for which

$$x = 1 + 2s = -2 + 4t$$
$$y = -1 + s = -3t$$
$$z = 2 + 4s = -1 + t$$

This is equivalent to the system of linear equations

$$\begin{cases} 2s - 4t = -3 \\ s + 3t = 1 \\ 4s - t = -3 \end{cases}$$

Any solution of this system must correspond to a point of intersection of L_1 and L_2, and if no solution exists, then L_1 and L_2 are skew. Because this is a system of three equations with two unknowns, we first solve the first two equations simultaneously to find $s = -\frac{1}{2}, t = \frac{1}{2}$. Because $(s, t) = (-\frac{1}{2}, \frac{1}{2})$ does not satisfy the third equation, it follows that L_1 and L_2 do not intersect, so they must be skew. ∎

EXAMPLE 4   *Intersecting lines*

Show that the lines

$$L_1: \quad \frac{x-1}{2} = \frac{y+1}{1} = \frac{z-2}{4} \quad \text{and} \quad L_2: \quad \frac{x+2}{4} = \frac{y}{-3} = \frac{z-\frac{1}{2}}{-1}$$

intersect and find the point of intersection.

Solution L_1 has direction numbers $[2, 1, 4]$ and L_2 has direction numbers $[4, -3, -1]$. Because there is no t for which $[2, 1, 4] = t[4, -3, -1]$, the lines are not parallel. Express the lines in parametric form:

$$L_1: \quad x = 1 + 2s \qquad y = -1 + s \qquad z = 2 + 4s$$
$$L_2: \quad x = -2 + 4t \qquad y = -3t \qquad z = \frac{1}{2} - t$$

At an intersection point we must have

$$1 + 2s = -2 + 4t \quad \text{or} \quad 2s - 4t = -3$$
$$-1 + s = -3t \quad \text{or} \quad s + 3t = 1$$
$$2 + 4s = \frac{1}{2} - t \quad \text{or} \quad 4s + t = -\frac{3}{2}$$

Solving the first two equations simultaneously, we find $s = -\frac{1}{2}, t = \frac{1}{2}$. This solution satisfies the third equation, namely,

$$4(-\tfrac{1}{2}) + \tfrac{1}{2} = -\tfrac{3}{2}$$

To find the coordinates of the point of intersection, substitute $s = -\frac{1}{2}$ into the parametric-form equations for L_1 (or substitute $t = \frac{1}{2}$ into L_2) to obtain

$$x_0 = 1 + 2(-\tfrac{1}{2}) = 0$$
$$y_0 = -1 + (-\tfrac{1}{2}) = -\tfrac{3}{2}$$
$$z_0 = 2 + 4(-\tfrac{1}{2}) = 0$$

Thus, the lines intersect at $P(0, -\frac{3}{2}, 0)$. ∎

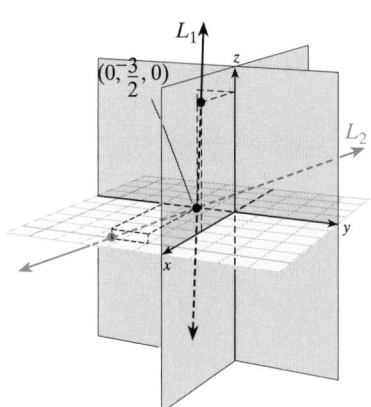

Graph of L_1 and L_2

DIRECTION COSINES

Besides using direction numbers, the direction of a nonzero vector $\mathbf{v} = a_1\mathbf{i} + a_2\mathbf{j} + a_3\mathbf{k}$ can be measured in terms of angles $\alpha, \beta,$ and γ between $\mathbf{v}$ and the coordinate axes, as shown in Figure 10.39. These angles are called the **direction angles** of $\mathbf{v}$ and their cosines are known as the **direction cosines** of $\mathbf{v}$.

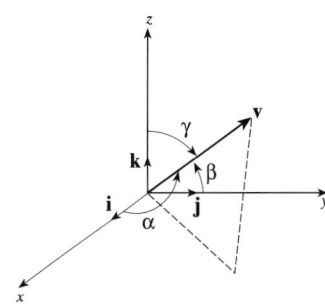

■ **FIGURE 10.39**   Direction cosines of the vector $\mathbf{v}$

Note that because $\mathbf{i}$ is a unit vector represented by $\langle 1, 0, 0 \rangle$, we have

$$\mathbf{v} \cdot \mathbf{i} = (a_1 \cdot 1) + (a_2 \cdot 0) + (a_3 \cdot 0) = a_1$$

Therefore,

$$a_1 = \mathbf{v} \cdot \mathbf{i} = \|\mathbf{v}\| \|\mathbf{i}\| \cos \alpha \quad \text{so that} \quad \cos \alpha = \frac{a_1}{\|\mathbf{v}\|}$$

Similar formulas hold for the other direction cosines. If $\mathbf{u}$ is a unit vector in the same direction as $\mathbf{v}$, we have

$$\mathbf{u} = \frac{\mathbf{v}}{\|\mathbf{v}\|} = \frac{a_1}{\|\mathbf{v}\|}\mathbf{i} + \frac{a_2}{\|\mathbf{v}\|}\mathbf{j} + \frac{a_3}{\|\mathbf{v}\|}\mathbf{k} = \cos \alpha \, \mathbf{i} + \cos \beta \, \mathbf{j} + \cos \gamma \, \mathbf{k}$$

$$\|\mathbf{u}\|^2 = \left(\frac{a_1}{\|\mathbf{v}\|}\right)^2 + \left(\frac{a_2}{\|\mathbf{v}\|}\right)^2 + \left(\frac{a_3}{\|\mathbf{v}\|}\right)^2 = \cos^2\alpha + \cos^2\beta + \cos^2\gamma = 1$$

EXAMPLE 5   *Direction angles and direction cosines*

Find the direction cosines and direction angles (to the nearest degree) of the vector $\mathbf{v} = -2\mathbf{i} + 3\mathbf{j} + 5\mathbf{k}$, and verify the formula $\cos^2\alpha + \cos^2\beta + \cos^2\gamma = 1$.

Solution   We find $\|\mathbf{v}\| = \sqrt{(-2)^2 + 3^2 + 5^2} = \sqrt{38}$. Therefore,

$$\cos \alpha = \frac{a_1}{\|\mathbf{v}\|} = \frac{-2}{\sqrt{38}} \approx -0.3244428 \qquad \alpha \approx \cos^{-1}(-0.324428) \approx 109°$$

$$\cos \beta = \frac{a_2}{\|\mathbf{v}\|} = \frac{3}{\sqrt{38}} \approx 0.4866642 \qquad \beta \approx \cos^{-1}(0.4866642) \approx 61°$$

$$\cos \gamma = \frac{a_3}{\|\mathbf{v}\|} = \frac{5}{\sqrt{38}} \approx 0.8111071 \qquad \gamma \approx \cos^{-1}(0.8111071) \approx 36°$$

$$\text{and} \quad \cos^2\alpha + \cos^2\beta + \cos^2\gamma = \left(\frac{-2}{\sqrt{38}}\right)^2 + \left(\frac{3}{\sqrt{38}}\right)^2 + \left(\frac{5}{\sqrt{38}}\right)^2 = 1 \qquad ∎$$

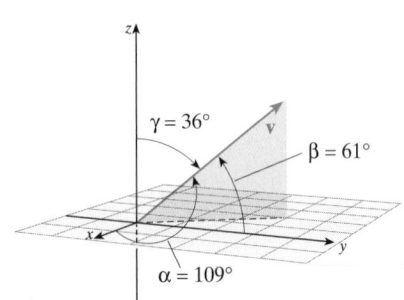

Graph of $\mathbf{v} = -2\mathbf{i} + 3\mathbf{j} + 5\mathbf{k}$

PLANES IN $\mathbb{R}^3$

Planes in space can also be characterized by vector methods. In particular, any plane is completely determined once we know one of its points and its orientation—that is, the "direction" it faces. A common way to specify the direction of a plane is by means of a vector $\mathbf{N}$ that is orthogonal to every vector in the plane, as shown in Figure 10.40. Such a vector in called a **normal** to the plane.

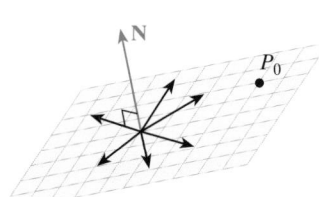

■ **FIGURE 10.40**   A plane may be described by specifying one of its points and a normal vector $\mathbf{N}$.

> ■ *What This Says:*   To "get a handle" on a plane $Ax + By + Cz + D = 0$, we use the plane's normal vector. We will see in this section that the direction numbers of the normal vector are proportional to A, B, C.

EXAMPLE 6   *Obtain the equation for a plane*

Find an equation for the plane that contains the point $Q(3, -7, 2)$ and is normal to the vector $\mathbf{N} = 2\mathbf{i} + \mathbf{j} - 3\mathbf{k}$.

Solution   The normal vector $\mathbf{N}$ is orthogonal to every vector in the plane. In particular, if $P(x, y, z)$ is any point in the plane, then $\mathbf{N}$ must be orthogonal to the vector

$$\mathbf{QP} = (x - 3)\mathbf{i} + (y + 7)\mathbf{j} + (z - 2)\mathbf{k}$$

Graph of plane in Example 6

Because the dot (or scalar) product of two orthogonal vectors is 0, we have

$$\mathbf{N} \cdot \mathbf{QP} = 2(x - 3) + (1)(y + 7) + (-3)(z - 2) = 0$$
$$2x - 6 + y + 7 - 3z + 6 = 0$$
$$2x + y - 3z + 7 = 0$$

Therefore, $2x + y - 3z + 7 = 0$ is the equation of the plane.   ■

By generalizing the approach illustrated in Example 6, we can show that the plane that contains the point (x_0, y_0, z_0) and has normal vector $\mathbf{N} = A\mathbf{i} + B\mathbf{j} + C\mathbf{k}$ must have the Cartesian equation

$$A(x - x_0) + B(y - y_0) + C(z - z_0) = 0$$

This is called the **point-normal form** of the equation of a plane. By rearranging terms, we can rewrite this equation in the form $Ax + By + Cz + D = 0$. This is called the **standard form** of the equation of a plane. The numbers $[A, B, C]$ are called **attitude numbers** of the plane (see Figure 10.41).

Notice from Figure 10.41 that *attitude numbers of a plane are the same as direction numbers of a normal line.* This means that you can find normal vectors to a plane by *inspecting the equation of the plane.*

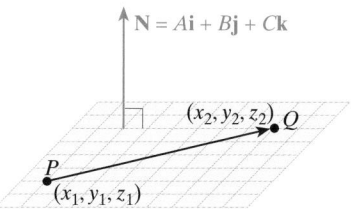

■ **FIGURE 10.41**   The graph of a plane with attitude numbers $[A, B, C]$

Assignment:
Sketch
$3x + 2y + 6z = 18$

DRAWING LESSON 6: DRAWING A PLANE IN SPACE

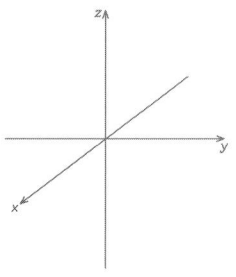

a. Draw the three coordinate axes.

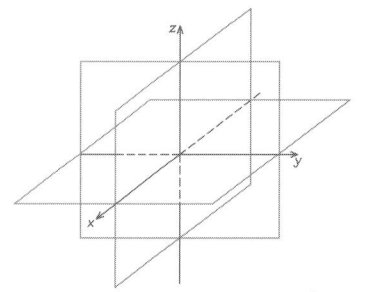

b. Draw the three coordinate planes.

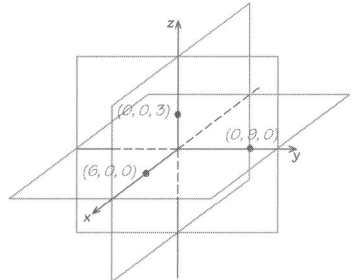

Let $x = 0$, $z = 0$, and find y; for this example, $y = 9$
Let $y = 0$, $x = 0$, and find z; for this example, $z = 3$
Let $z = 0$, $y = 0$, and find x; for this example, $x = 6$

c. Plot the points where the plane intersects the coordinate axes.

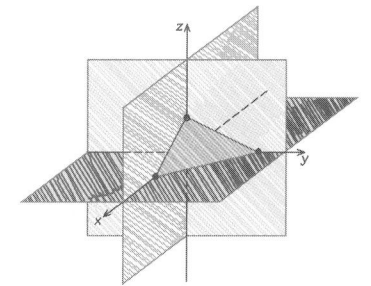

d. Connect the points plotted in part c and shade the part of the plane in the first octant.

e. If desired, extend the plane into the other octants. Use highlighters or pencils to color the planes to add depth to the figure.

EXAMPLE 7	*Relationship between normal vectors and planes*

Find normal vectors to the planes
a. $5x + 7y - 3z = 0$ b. $x - 5y + \sqrt{2}z = 6$ c. $3x - 7z = 10$

Solution
a. A normal to the plane $5x + 7y - 3z = 0$ is $\mathbf{N} = 5\mathbf{i} + 7\mathbf{j} - 3\mathbf{k}$.
b. For $x - 5y + \sqrt{2}z = 6$, the normal is $\mathbf{N} = \mathbf{i} - 5\mathbf{j} + \sqrt{2}\mathbf{k}$.
c. For $3x - 7z = 10$, it is $\mathbf{N} = 3\mathbf{i} - 7\mathbf{k}$.   ∎

Forms for the Equation of a Plane

A plane with normal $\mathbf{N} = A\mathbf{i} + B\mathbf{j} + C\mathbf{k}$ that contains the point (x_0, y_0, z_0) has the following equations:

Point-normal form: $A(x - x_0) + B(y - y_0) + C(z - z_0) = 0$
Standard form: $Ax + By + Cz + D = 0$

for some constants $A, B, C,$ and D.

THEOREM 10.9   ***The normal of a given plane***

Let $A, B, C,$ and D be constants with $A, B,$ and C not all zero. Then the graph of the equation

$$Ax + By + Cz + D = 0$$

is the equation of a plane with normal vector $\mathbf{N} = A\mathbf{i} + B\mathbf{j} + C\mathbf{k}$.

Proof   Suppose $A \neq 0$. Then the equation $Ax + By + Cz + D = 0$ can be written as

$$A\left[x + \frac{D}{A}\right] + By + Cz = 0$$

which is the point-normal form of the plane that passes through the point $(-D/A, 0, 0)$ with a normal $\mathbf{N} = A\mathbf{i} + B\mathbf{j} + C\mathbf{k}$. A similar argument applies if $A = 0$ and either B or C is not zero.   ═

EXAMPLE 8	*Equation of a line orthogonal to a given plane*

Find an equation of the line that passes through the point $Q(2, -1, 3)$ and is orthogonal to the plane $3x - 7y + 5z + 55 = 0$. Where does the line intersect the plane?

Solution   *By inspection* of the equation of the plane, we see that $\mathbf{N} = 3\mathbf{i} - 7\mathbf{j} + 5\mathbf{k}$ is a normal vector. Because the required line is also orthogonal to the plane, it must be parallel to $\mathbf{N}$. Thus, the line contains the point $Q(2, -1, 3)$ and has direction numbers $[3, -7, 5]$, so its equation is

$$\frac{x - 2}{3} = \frac{y + 1}{-7} = \frac{z - 3}{5}$$

To find the point where this line intersects the plane, we first set the three fractions equal to t and rewrite the line in parametric form:

$$x = 2 + 3t, \quad y = -1 - 7t, \quad \text{and} \quad z = 3 + 5t$$

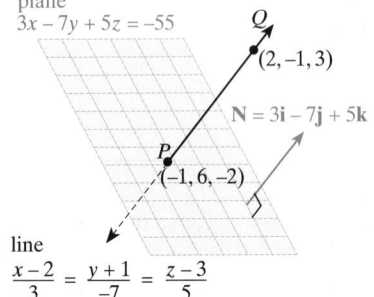

plane
$3x - 7y + 5z = -55$

Q
$(2, -1, 3)$

$\mathbf{N} = 3\mathbf{i} - 7\mathbf{j} + 5\mathbf{k}$

P
$(-1, 6, -2)$

line
$\frac{x-2}{3} = \frac{y+1}{-7} = \frac{z-3}{5}$

Now, substitute into the equation of the plane:

$$3(2 + 3t) - 7(-1 - 7t) + 5(3 + 5t) = -55$$
$$6 + 9t + 7 + 49t + 15 + 25t = -55$$
$$83t = -83$$
$$t = -1$$

Then, the point of intersection is found by substituting $t = -1$ for x, y, and z:

$$x = 2 + 3(-1) = -1$$
$$y = -1 - 7(-1) = 6$$
$$z = 3 + 5(-1) = -2$$

The point of intersection is $(-1, 6, -2)$. ∎

| EXAMPLE 9 | *Equation of a plane containing three given points* |

Find the standard-form equation of a plane containing $P(-1, 2, 1)$, $Q(0, -3, 2)$, and $R(1, 1, -4)$.

Solution Because a normal **N** to the required plane is orthogonal to the vectors **PQ** and **PR**, we find **N** by computing the cross product $\mathbf{N} = \mathbf{PQ} \times \mathbf{PR}$.

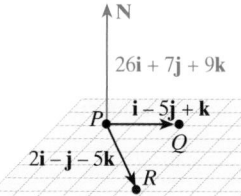

$$\mathbf{PQ} = (0 + 1)\mathbf{i} + (-3 - 2)\mathbf{j} + (2 - 1)\mathbf{k} = \mathbf{i} - 5\mathbf{j} + \mathbf{k}$$
$$\mathbf{PR} = (1 + 1)\mathbf{i} + (1 - 2)\mathbf{j} + (-4 - 1)\mathbf{k} = 2\mathbf{i} - \mathbf{j} - 5\mathbf{k}$$
$$\mathbf{N} = \mathbf{PQ} \times \mathbf{PR} = \begin{vmatrix} \mathbf{i} & \mathbf{j} & \mathbf{k} \\ 1 & -5 & 1 \\ 2 & -1 & -5 \end{vmatrix}$$
$$= (25 + 1)\mathbf{i} - (-5 - 2)\mathbf{j} + (-1 + 10)\mathbf{k}$$
$$= 26\mathbf{i} + 7\mathbf{j} + 9\mathbf{k}$$

We can now find the equation of the plane using this normal vector and any point in the plane. We will use the point P:

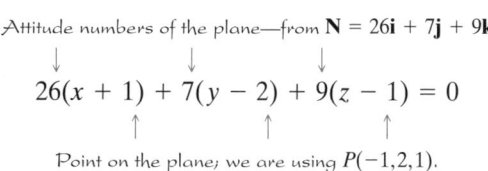

Attitude numbers of the plane—from $\mathbf{N} = 26\mathbf{i} + 7\mathbf{j} + 9\mathbf{k}$
↓   ↓   ↓
$$26(x + 1) + 7(y - 2) + 9(z - 1) = 0$$
↑   ↑   ↑
Point on the plane; we are using $P(-1, 2, 1)$.

Thus, the equation of the plane is

$$26x + 26 + 7y - 14 + 9z - 9 = 0$$
$$26x + 7y + 9z + 3 = 0$$ ∎

| EXAMPLE 10 | *Equation of a line parallel to the intersection of two given planes* |

Find the equation of a line passing through $(-1, 2, 3)$ that is parallel to the line of intersection of the planes $3x - 2y + z = 4$ and $x + 2y + 3z = 5$.

Solution By inspection, we see that the normals to the given planes are $\mathbf{N}_1 = 3\mathbf{i} - 2\mathbf{j} + \mathbf{k}$ and $\mathbf{N}_2 = \mathbf{i} + 2\mathbf{j} + 3\mathbf{k}$. The desired line is perpendicular to both of these

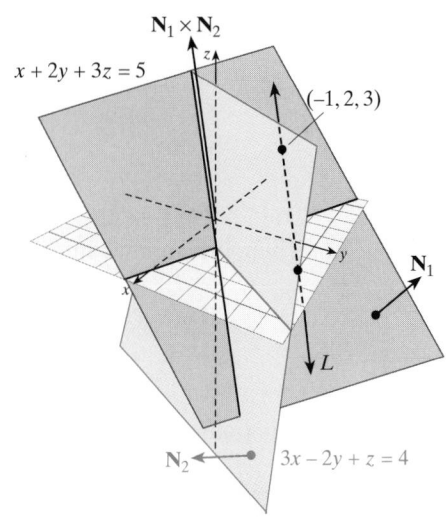

normals, so the aligned vector is found by computing the cross product:

$$\mathbf{N}_1 \times \mathbf{N}_2 = \begin{vmatrix} \mathbf{i} & \mathbf{j} & \mathbf{k} \\ 3 & -2 & 1 \\ 1 & 2 & 3 \end{vmatrix} = (-6 - 2)\mathbf{i} - (9 - 1)\mathbf{j} + (6 + 2)\mathbf{k}$$
$$= -8\mathbf{i} - 8\mathbf{j} + 8\mathbf{k}$$

The direction of this vector is $\langle -8, -8, 8 \rangle = -8\langle 1, 1, -1 \rangle$, so the equation of the desired line is

$$\frac{x + 1}{1} = \frac{y - 2}{1} = \frac{z - 3}{-1}$$

Example 10 can also be used to find the equation of the line of intersection of the two planes. Instead of using the given point $(-1, 2, 3)$, you will first need to find a point in the intersection and then proceed, using the steps of Example 10. We conclude this section by finding the equation of a plane containing two given (nonparallel) lines.

EXAMPLE 11   *Equation of a plane containing two intersecting lines*

Find the standard-form equation of the plane determined by the intersecting lines

$$\frac{x - 2}{3} = \frac{y + 5}{-2} = \frac{z + 1}{4} \quad \text{and} \quad \frac{x + 1}{2} = \frac{y}{-1} = \frac{z - 16}{5}$$

Solution   Proceeding as in Example 4, we find that the lines intersect at $(-19, 9, -29)$. The aligned vectors for these two lines are $\mathbf{v}_1 = 3\mathbf{i} - 2\mathbf{j} + 4\mathbf{k}$ and $\mathbf{v}_2 = 2\mathbf{i} - \mathbf{j} + 5\mathbf{k}$. The normal to the desired plane is orthogonal to both $\mathbf{v}_1$ and $\mathbf{v}_2$, so we take the normal to be the cross product:

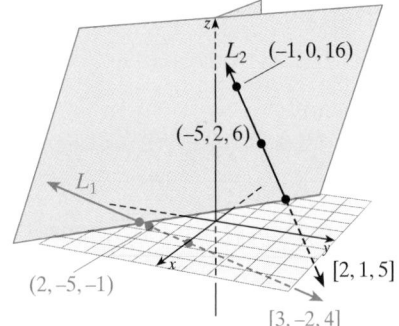

$$\mathbf{N} = \mathbf{v}_1 \times \mathbf{v}_2 = \begin{vmatrix} \mathbf{i} & \mathbf{j} & \mathbf{k} \\ 3 & -2 & 4 \\ 2 & -1 & 5 \end{vmatrix} = (-10 + 4)\mathbf{i} - (15 - 8)\mathbf{j} + (-3 + 4)\mathbf{k}$$
$$= -6\mathbf{i} - 7\mathbf{j} + \mathbf{k}$$

The point of intersection $P(-19, 9, -29)$ is certainly in the plane, as are $(2, -5, -1)$ and $(-1, 0, 16)$. We use $(2, -5, -1)$ to obtain

$$-6(x - 2) - 7(y + 5) + 1(z + 1) = 0$$
$$6x + 7y - z + 22 = 0$$

10.5 *Problem Set*

 1. ■ **What Does This Say?** Contrast the parametric and symmetric forms of the equation of a line.

2. ■ **What Does This Say?** Describe the relationship between normal vectors and planes.

Write each equation for a plane given in Problems 3–6 in standard form.

3. $4(x + 1) - 2(y + 1) + 6(z - 2) = 0$

4. $5(x - 2) - 3(y + 2) + 4(z + 3) = 0$

5. $-3(x - 4) + 2(y + 1) - 2(z + 1) = 0$

6. $-2(x + 1) + 4(y - 3) - 8z = 0$

Find the parametric and symmetric equations for the line(s) passing through the given points with the properties described in Problems 7–16.

7. $(1, -1, -2)$; parallel to $3\mathbf{i} - 2\mathbf{j} + 5\mathbf{k}$

8. $(1, 0, -1)$; parallel to $3\mathbf{i} + 4\mathbf{j}$

9. $(1, -1, 2)$; through $(2, 1, 3)$

10. $(2, 2, 3)$; through $(1, 3, -1)$

11. $(1, -3, 6)$; parallel to $\dfrac{x - 5}{1} = \dfrac{y + 2}{-3} = \dfrac{z}{-5}$

12. $(1, -1, 2)$; parallel to $\dfrac{x + 3}{4} = \dfrac{y - 2}{5} = \dfrac{z + 5}{1}$

13. $(0, 4, -3)$; parallel to $\dfrac{2x - 1}{22} = \dfrac{y + 2}{-6} = \dfrac{z - 1}{10}$

14. $(1, 0, -4)$; parallel to $x = -2 + 3t, y = 4 + t, z = 2 + 2t$

15. $(3, -1, 0)$; parallel to the xy-plane and the yz-plane

16. $(-1, 1, 6)$; perpendicular to $3x + y - 2z = 5$

Find the points of intersection of each line in Problems 17–20 with each of the coordinate planes.

17. $\dfrac{x - 4}{4} = \dfrac{y + 3}{3} = \dfrac{z + 2}{1}$

18. $\dfrac{x + 1}{1} = \dfrac{y + 2}{2} = \dfrac{z - 6}{3}$

19. $x = 6 - 2t, y = 1 + t, z = 3t$

20. $x = 6 + 3t, y = 2 - t, z = 2t$

In Problems 21–26, tell whether the two lines intersect, are parallel, are skew, or coincide. If they intersect, give the point of intersection.

21. $\dfrac{x - 4}{2} = \dfrac{y - 6}{-3} = \dfrac{z + 2}{5}; \dfrac{x}{4} = \dfrac{y + 2}{-6} = \dfrac{z - 3}{10}$

22. $x = 4 - 2t, y = 6t, z = 7 - 4t;$
  $x = 5 + t, y = 1 - 3t, z = -3 + 2t$

23. $x = 3 + 3t, y = 1 - 4t, z = -4 - 7t;$
  $x = -3t, y = 5 + 4t, z = 3 + 7t$

24. $x = 2 - 4t, y = 1 + t, z = \frac{1}{2} + 5t;$
  $x = 3t, y = -2 - t, z = 4 - 2t$

25. $\dfrac{x - 3}{2} = \dfrac{y - 1}{-1} = \dfrac{z - 4}{1};$
  $\dfrac{x + 2}{3} = \dfrac{y - 3}{-1} = \dfrac{z - 2}{1}$

26. $\dfrac{x + 1}{2} = \dfrac{y - 3}{-1} = \dfrac{z - 2}{1};$
  $\dfrac{x + 1}{2} = \dfrac{y + 1}{3} = \dfrac{z - 3}{-4}$

Find the direction cosines and the direction angles for the vectors given in Problems 27–32.

27. $\mathbf{v} = 2\mathbf{i} - 3\mathbf{j} - 5\mathbf{k}$ **28.** $\mathbf{v} = 3\mathbf{i} - 2\mathbf{k}$

29. $\mathbf{v} = 5\mathbf{i} - 4\mathbf{j} + 3\mathbf{k}$ **30.** $\mathbf{v} = \mathbf{j} - 5\mathbf{k}$

31. $\mathbf{v} = \mathbf{i} - 3\mathbf{j} + 9\mathbf{k}$ **32.** $\mathbf{v} = \mathbf{i} - \mathbf{j} + 3\mathbf{k}$

Find an equation for the plane that contains the point P and has the normal vector $\mathbf{N}$ *given in Problems 33–38.*

33. $P(-1, 3, 5); \mathbf{N} = 2\mathbf{i} + 4\mathbf{j} - 3\mathbf{k}$

34. $P(0, -7, 1); \mathbf{N} = -\mathbf{i} + \mathbf{k}$

35. $P(0, -3, 0); \mathbf{N} = -2\mathbf{j} + 3\mathbf{k}$

36. $P(1, 1, -1); \mathbf{N} = -\mathbf{i} - 2\mathbf{j} + 3\mathbf{k}$

37. $P(0, 0, 0); \mathbf{N} = \mathbf{k}$

38. $P(0, 0, 0); \mathbf{N} = \mathbf{i}$

39. Find two unit vectors parallel to the line
$$\frac{x - 3}{4} = \frac{y - 1}{2} = \frac{z + 1}{1}$$

40. Find two unit vectors parallel to the line
$$\frac{x - 1}{2} = \frac{y + 2}{4} = \frac{z + 5}{1}$$

41. Find two unit vectors perpendicular to the plane $2x + 4y - 3z = 4$.

42. Find two unit vectors perpendicular to the plane $5x - 3y + 2z = 15$.

B **43.** Show that the vector $3\mathbf{i} - 4\mathbf{j} + \mathbf{k}$ is orthogonal to the line that passes through the points $P(0, 0, 1)$ and $Q(2, 1, -1)$.

44. Show that the vector $7\mathbf{i} + 4\mathbf{j} + 3\mathbf{k}$ is orthogonal to the line passing through the points $P(-2, 2, 7)$ and $Q(3, -3, 2)$.

45. Find two unit vectors that are parallel to the line of intersection of the planes $x + y = 1$ and $x - 2z = 3$.

46. Find two unit vectors that are parallel to the line of intersection of the planes $x + y + z = 3$ and $x - y + z = 1$.

47. Find an equation for the plane that passes through $P(1, -1, 2)$ and is normal to $\mathbf{PQ}$ where Q is $Q(2, 1, 3)$.

48. Find an equation for the line that passes through the point $(1, -5, 3)$ and is orthogonal to the plane $2x - 3y + z = 1$.

49. Find an equation for the plane that contains the point $(2, 1, -1)$ and is orthogonal to the line
$$\frac{x - 3}{3} = \frac{y + 1}{5} = \frac{z}{2}$$

50. Find a plane that passes through the point $(1, 2, -1)$ and is parallel to the plane $2x - y + 3z = 1$.

51. Show that the line
$$\frac{x - 1}{2} = \frac{y + 1}{3} = \frac{z - 2}{4}$$
is parallel to the plane $x - 2y + z = 6$.

52. Find the point where the line
$$\frac{x - 1}{2} = \frac{y + 1}{-1} = \frac{z}{3}$$
intersects the plane $3x + 2y - z = 5$.

53. The *angle* between two planes is defined to be the acute angle between their normal vectors. Find the angle between the planes $2x + y - 4z = 3$ and $x - y + z = 2$, rounded to the nearest degree.

54. Find the equation of the line that passes through the point $P(2, 3, 1)$ and is parallel to the line of intersection of the planes $x + 2y - 3z = 4$ and $x - 2y + z = 0$.

55. Find the equation of the line that passes through the point $P(0, 1, -1)$ and is parallel to the line of intersection of the planes $2x + y - 2z = 5$ and $3x - 6y - 2z = 7$.

56. Find a vector that is parallel to the line of intersection of the planes $2x + 3y = 0$ and $3x - y + z = 1$.

57. Find the equation of the line of intersection of the planes $3x + y - z = 5$ and $x - 6y - 2z = 10$.

58. Find the equation of the line of intersection of the planes $2x - y + z = 8$ and $x + y - z = 5$.

59. Let $\mathbf{v} = 2\mathbf{i} + \mathbf{j}$ and $\mathbf{w} = 2\mathbf{i} - \mathbf{j} = 3\mathbf{k}$. Find the direction cosines and the direction angles of $\mathbf{v} \times \mathbf{w}$.

60. Find the direction cosines of a vector determined by the line of intersection of the planes $x + y + z = 3$ and $2x + 3y - z = 4$.

61. What can be said about the lines
$$\frac{x - x_0}{a_1} = \frac{y - y_0}{b_1} = \frac{z - z_0}{c_1}$$
and
$$\frac{x - x_0}{a_2} = \frac{y - y_0}{b_2} = \frac{z - z_0}{c_2}$$
in the case where $a_1 a_2 + b_1 b_2 + c_1 c_2 = 0$?

62. In Figure 10.42, $\mathbf{N}$ is normal to the plane P and L is a line that intersects P. Assume $\mathbf{N} = a\mathbf{i} + b\mathbf{j} + c\mathbf{k}$ and that L is given by
$$x = x_0 + At, \quad y = y_0 + Bt, \quad z = z_0 + Ct$$
   a. Find $\cos \theta$ for the angle θ between L and the plane P.
   b. Find the angle between the plane $x + y + z = 10$ and the line
$$\frac{x - 1}{2} = \frac{y + 3}{3} = \frac{z - 2}{-1}$$

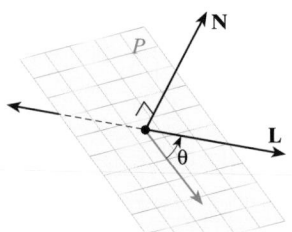

■ **FIGURE 10.42**   Problem 62

63. Show that a plane with x-intercept a, y-intercept b, and z-intercept c has the equation
$$\frac{x}{a} + \frac{y}{b} + \frac{z}{c} = 1$$
assuming a, b, and c are all nonzero.

64. Suppose planes p_1 and p_2 intersect. If $\mathbf{v}_1$ and $\mathbf{w}_1$ are vectors on p_1, and $\mathbf{v}_2$ and $\mathbf{w}_2$ are on plane p_2, then show that
$$(\mathbf{v}_1 \times \mathbf{w}_1) \times (\mathbf{v}_2 \times \mathbf{w}_2)$$
is aligned with the line of intersection of the planes.

10.6   *Vector Methods for Measuring Distance in* $\mathbb{R}^3$

IN THIS SECTION   distance from a point to a plane, distance from a point to a line   ■

DISTANCE FROM A POINT TO A PLANE

To prepare for deriving a formula for the distance from a point to a plane, we will consider a simpler case, namely, the distance from a point to a line in $\mathbb{R}^2$. Let L be any given line and P any given point not on L. We wish to find the distance from P to L—that is, the perpendicular distance d, as shown in Figure 10.43.

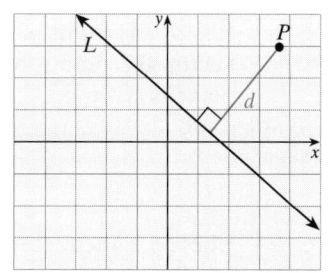

■ **FIGURE 10.43**   Distance from P to L

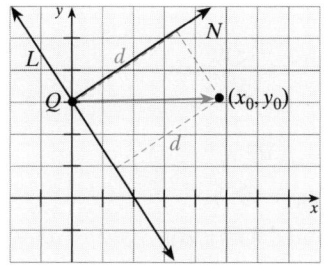

■ **FIGURE 10.44**   Procedure for finding the distance from a point to a line

If L is a vertical line, then the distance from P to L is easy to find (why?). If L is not vertical, then we let Q be any point on the line and $\mathbf{N}$ be normal to L. Since Q can be any point on the line, we choose a convenient point, say the y-intercept (see Figure 10.44).

The distance we seek is seen to be the scalar projection of the vector $\mathbf{QP}$ onto $\mathbf{N}$. Thus,

$$d = \left| \frac{\mathbf{QP} \cdot \mathbf{N}}{\|\mathbf{N}\|} \right| = \frac{|\mathbf{QP} \cdot \mathbf{N}|}{\|\mathbf{N}\|}$$

In particular, we will now apply this formula to find the distance from the point $P(x_0, y_0)$ to the line

$$Ax + By + C = 0$$

Because we have chosen Q to be the y-intercept, $y = -C/B$ (because L is not vertical, $B \neq 0$). Then $\mathbf{QP} = (x_0 - 0)\mathbf{i} + (y_0 + C/B)\mathbf{j}$. It can be shown that the normal to the line $Ax + By + C = 0$ is $\mathbf{N} = A\mathbf{i} + B\mathbf{j}$. Then

$$d = \left| \frac{\mathbf{QP} \cdot \mathbf{N}}{\|\mathbf{N}\|} \right| = \frac{|Ax_0 + By_0 + C|}{\sqrt{A^2 + B^2}}$$

EXAMPLE 1   *Distance from a point to a line in $\mathbb{R}^2$*

Find the distance from the point $(5, -3)$ to the line $4x + 3y - 15 = 0$:
a. as a scalar projection   b. by using the formula

Solution

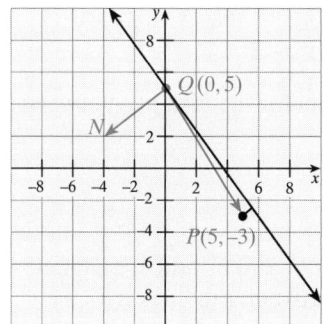

a. Let P be the point $(5, -3)$ and Q be the y-intercept $(0, 5)$ of the line. Then
$\mathbf{QP} = 5\mathbf{i} - 8\mathbf{j}$ and $\mathbf{N} = 4\mathbf{i} + 3\mathbf{j}$

$$d = \left| \frac{\mathbf{QP} \cdot \mathbf{N}}{\|\mathbf{N}\|} \right| = \left| \frac{20 - 24}{\sqrt{16 + 9}} \right| = \left| \frac{-4}{5} \right| = \frac{4}{5}$$

b. Note $A = 4, B = 3, C = -15, x_0 = 5$, and $y_0 = -3$ so that

$$d = \frac{Ax_0 + By_0 + C}{\sqrt{A^2 + B^2}} = \left| \frac{4(5) + 3(-3) - 15}{\sqrt{4^2 + 3^2}} \right| = \left| \frac{-4}{5} \right| = \frac{4}{5} \qquad ■$$

A projection is also used to obtain the following formula for the distance from a point to a plane.

════════════

THEOREM 10.10   *Distance from a point to a plane in $\mathbb{R}^3$*

The distance from the point (x_0, y_0, z_0) to the plane $Ax + By + Cz + D = 0$ is given by

$$d = \frac{|Ax_0 + By_0 + Cz_0 + D|}{\sqrt{A^2 + B^2 + C^2}}$$

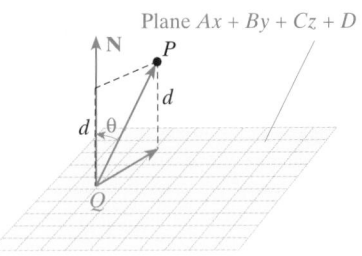

Plane $Ax + By + Cz + D$

■ **FIGURE 10.45** The distance from a point to a plane in $\mathbb{R}^3$

If Q is any point in the given plane, the required distance is found by projecting the vector $\mathbf{QP}$ onto a normal $\mathbf{N}$ for the plane. Thus, the distance from the point to the plane is given by

$$d = \|\mathbf{QP}\|\,|\cos\theta| = \frac{\|\mathbf{QP}\|\,\|\mathbf{N}\|\,|\cos\theta|}{\|\mathbf{N}\|} = \frac{|\mathbf{QP}\cdot\mathbf{N}|}{\|\mathbf{N}\|}$$

where θ is the (acute) angle between $\mathbf{QP}$ and $\mathbf{N}$ (see Figure 10.45).

Suppose P has coordinates (x_0, y_0, z_0) and the given plane has the standard form $Ax + By + Cz + D = 0$. Then $\mathbf{N} = A\mathbf{i} + B\mathbf{j} + C\mathbf{k}$ is a normal to this plane, and if $Q(x_1, y_1, z_1)$ is any particular point in the plane, we have

$$\mathbf{QP} = (x_0 - x_1)\mathbf{i} + (y_0 - y_1)\mathbf{j} + (z_0 - z)\mathbf{k}$$

The dot product of $\mathbf{QP}$ with $\mathbf{N}$ is given by

$$\begin{aligned}
\mathbf{QP}\cdot\mathbf{N} &= (x_0 - x_1)A + (y_0 - y_1)B + (z_0 - z_1)C \\
&= (Ax_0 + By_0 + Cz_0) - (Ax_1 + By_1 + Cz_1) \\
&= Ax_0 + By_0 + Cz_0 - (-D) \quad \text{Because } Ax_1 + By_1 + Cz_1 + D = 0
\end{aligned}$$

Because the normal vector has length $\|\mathbf{N}\| = \sqrt{A^2 + B^2 + C^2}$, we can substitute into the formula to obtain

$$d = \left|\frac{\mathbf{QP}\cdot\mathbf{N}}{\|\mathbf{N}\|}\right| = \frac{|Ax_0 + By_0 + Cz_0 + D|}{\sqrt{A^2 + B^2 + C^2}}$$

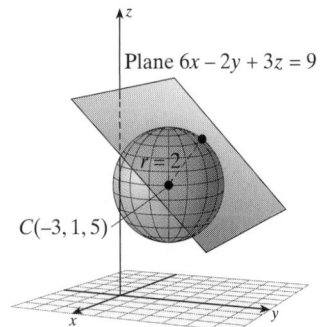

z

Plane $6x - 2y + 3z = 9$

$r = 2$

$C(-3, 1, 5)$

x y

■ **FIGURE 10.46** The sphere with center $C(-3, 1, 5)$ and radius 2 is tangent to the plane $6x - 2y + 3z = 9$.

EXAMPLE 2   *Equation of a sphere given a tangent plane*

Find an equation for the sphere with center $C(-3, 1, 5)$ that is tangent to the plane $6x - 2y + 3x = 9$.

Solution   The radius r of the sphere is the distance from the center C to the given tangent plane, as shown in Figure 10.46.

$$r = \left|\frac{6(-3) + (-2)(1) + 3(5) - 9}{\sqrt{6^2 + (-2)^2 + 3^2}}\right| = \left|\frac{-14}{7}\right| = 2$$

Therefore, an equation of the sphere is

$$(x + 3)^2 + (y - 1)^2 + (z - 5)^2 = 2^2$$

Vector methods can also be used to derive a formula for the distance between two skew lines L_1 and L_2 in $\mathbb{R}^3$.

THEOREM 10.11   *Distance between skew lines in $\mathbb{R}^3$*

Assume L_1 and L_2 are skew lines containing the points P_1 and P_2 and are aligned with the vectors $\mathbf{v}_1$ and $\mathbf{v}_2$, respectively. Then the distance between the lines is

$$d = \left|\frac{(\mathbf{v}_1 \times \mathbf{v}_2)\cdot\mathbf{P_1P_2}}{\|\mathbf{v}_1 \times \mathbf{v}_2\|}\right| = \left|\frac{\mathbf{N}\cdot\mathbf{P_1P_2}}{\|\mathbf{N}\|}\right|$$

Proof   You are asked to prove this theorem in Problem 32. Notice that the distance d between L_1 and L_2 (see Figure 10.47) is the same as the distance between two parallel planes containing the lines. Because $\mathbf{N} = \mathbf{v}_1 \times \mathbf{v}_2$ is normal to both planes, it follows that the required distance d is a scalar multiple of $\|\mathbf{v}_1 \times \mathbf{v}_2\|$.

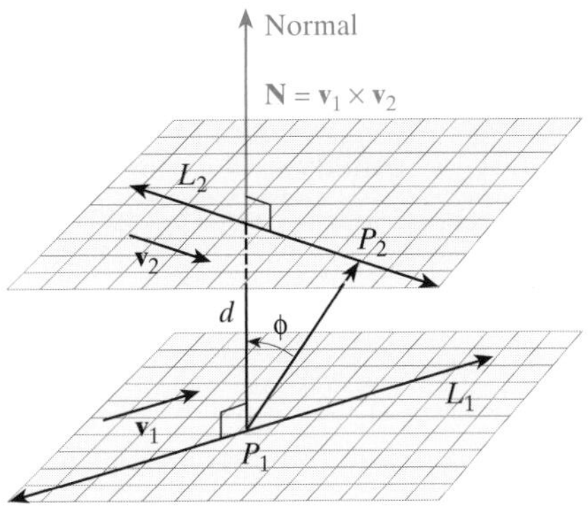

Normal

$\mathbf{N} = \mathbf{v}_1 \times \mathbf{v}_2$

L_2

$\mathbf{v}_2$ P_2

d ϕ

$\mathbf{v}_1$ L_1

P_1

■ **FIGURE 10.47**   Distance between lines

DISTANCE FROM A POINT TO A LINE

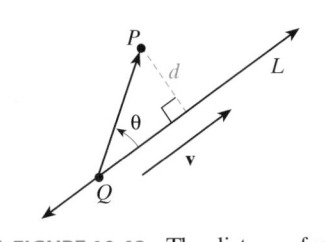

■ **FIGURE 10.48**   The distance from a point to a line in $\mathbb{R}^3$

Next, we shall derive a formula for the distance from a point P to a line L in $\mathbb{R}^3$. Let Q be a point on L and let $\mathbf{v}$ be a vector aligned with L. Then, as shown in Figure 10.48, the distance from P to L is given by

$$d = \|\mathbf{QP}\| \, |\sin \theta|$$

where θ is the acute angle between $\mathbf{v}$ and the vector $\mathbf{QP}$.

This reminds us of the cross product $\mathbf{v} \times \mathbf{QP}$, and because

$$\|\mathbf{v} \times \mathbf{QP}\| = \|\mathbf{v}\| \, \|\mathbf{QP}\| \, |\sin \theta|$$

we have

$$d = \|\mathbf{QP}\| \, |\sin \theta| = \frac{\|\mathbf{v} \times \mathbf{QP}\|}{\|\mathbf{v}\|}$$

THEOREM 10.12   *Distance from a point to a line*

The distance from the point P to the line L is given by the formula

$$d = \frac{\|\mathbf{v} \times \mathbf{QP}\|}{\|\mathbf{v}\|}$$

where $\mathbf{v}$ is a vector aligned with L and Q is any point on L.

Proof   A sketch of the proof precedes the statement of the theorem.

EXAMPLE 3   *Distance from a point to a line*

Find the distance from the point $P(3, -8, 1)$ to the line

$$\frac{x - 3}{3} = \frac{y + 7}{-1} = \frac{z + 2}{5}$$

Solution   We need to find a point Q on the line. We see that $Q(3, -7, -2)$ is on the line and that

$$\mathbf{QP} = -\mathbf{j} + 3\mathbf{k}$$

The vector $\mathbf{v}$ aligned with L is $\mathbf{v} = 3\mathbf{i} - \mathbf{j} + 5\mathbf{k}$. We now find

$$\mathbf{v} \times \mathbf{QP} = \begin{vmatrix} \mathbf{i} & \mathbf{j} & \mathbf{k} \\ 3 & -1 & 5 \\ 0 & -1 & 3 \end{vmatrix} = (-3 + 5)\mathbf{i} - (9 - 0)\mathbf{j} + (-3 + 0)\mathbf{k}$$

$$= 2\mathbf{i} - 9\mathbf{j} - 3\mathbf{k}$$

Finally,

$$d = \frac{\|\mathbf{v} \times \mathbf{QP}\|}{\|\mathbf{v}\|} = \frac{\sqrt{(2)^2 + (-9)^2 + (-3)^2}}{\sqrt{3^2 + (-1)^2 + 5^2}} = \frac{\sqrt{94}}{\sqrt{35}} \approx 1.64$$

■

10.6 Problem Set

Ⓐ *Find the distance between the point and the line in Problems 1–6.*

1. $(4, 5)$; $3x - 4y + 8 = 0$
2. $(9, -3)$; $3x - 4y + 8 = 0$
3. $(4, -3)$; $12x + 5y - 2 = 0$
4. $(1, -6)$; $x - 3y + 15 = 0$
5. $(8, 14)$; $x - 3y + 15 = 0$
6. $(4, 5)$; $2x - 5y = 0$

Find the distance between the point and the plane given in Problems 7–12.

7. $P(1, 0, -1)$; $x + y - z = 1$
8. $P(0, 0, 0)$; $2x - 3y + 5z = 10$
9. $P(1, 1, -1)$; $x - y + 2z = 4$
10. $P(2, 1, -2)$; $3x - 4y + z = -1$
11. $P(a, -a, 2a)$; $2ax - y + az = 4a, a \neq 0$
12. $P(a, 2a, 3a)$; $3x - 2y + z = -1/a, a \neq 0$

Find the distance from the point $(-1, 2, 1)$ to each plane given in Problems 13–16.

13. the plane through the points $A(0, 0, 0)$, $B(1, 2, 4)$, and $C(-2, -1, 1)$
14. the plane through the point $(1, 0, 1)$ with normal vector $2\mathbf{i} - \mathbf{j} + 2\mathbf{k}$
15. the plane through the point $(-3, 5, 1)$ with normal vector $3\mathbf{i} + \mathbf{j} + 5\mathbf{k}$
16. the plane through the points $A(-1, 1, 1)$, $B(4, 3, 7)$, and $C(3, -1, 0)$

Find the distance from the point P to the line L in Problems 17–22.

17. $P(1, 0, -1)$; $\dfrac{x - 2}{3} = \dfrac{y + 1}{1} = \dfrac{z - 1}{2}$

18. $P(1, 0, 1)$; $\dfrac{x}{3} = \dfrac{y - 1}{2} = \dfrac{z}{1}$

19. $P(1, -2, 2)$; $\dfrac{x}{1} = \dfrac{2y}{1} = \dfrac{z}{-1}$

20. $P(0, 1, -1)$; $\dfrac{2x - 1}{2} = \dfrac{y}{2} = \dfrac{z}{-1}$

21. $P(a, 0, -a)$; $\dfrac{x + a}{2} = \dfrac{y - a}{1} = \dfrac{z - a}{2}$; $a \neq 0$

22. $P\left(0, a, \dfrac{a}{2}\right)$; $\dfrac{x - a}{1} = \dfrac{y}{1} = \dfrac{z + 4a}{1}$

Ⓑ 23. Find the equation of the sphere with center $C(-2, 3, 7)$ that is tangent to the plane $2x + 3y - 6z = 5$.

24. a. Show that the line
$$\frac{x - 1}{3} = \frac{y}{-2} = \frac{z + 1}{1}$$
is parallel to the plane $x + 2y + z = 1$.
    b. Find the distance from the line to the plane in part **a**.

25. Three of the four vertices of a parallelogram in $\mathbb{R}^3$ are $Q(-1, 3, 5)$, $R(6, -3, 2)$, and $S(2, 4, -3)$. What is the area of the parallelogram?

26. Find an equation for the set of all points $P(x, y, z)$ such that the distance from P to the point $P_0(-1, 2, 4)$ is the same as the distance from P to the plane $2x - 5y + 3z = 7$. (Do not expand binomials.)

27. Find an equation for the set of all points $P(x, y, z)$ such that the distance from P to the line
$$\frac{x - 1}{4} = \frac{y + 1}{-1} = \frac{z}{3}$$
is 5. (Do not expand trinomials.)

Find the (perpendicular) distance between the lines given in Problems 28–31.

28. $\dfrac{x + 1}{3} = \dfrac{y - 2}{-2} = \dfrac{z - 1}{1}$ and $\dfrac{x - 2}{5} = \dfrac{y + 1}{1} = \dfrac{z}{3}$

29. $x = 2 - t, y = 5 + 2t, z = 3t$ and $x = 2t, y = -1 - t, z = 1 + 2t$

30. $\dfrac{x + 1}{1} = \dfrac{y - 3}{2} = \dfrac{z + 2}{3}$ and the line

passing through $(1, 3, -2)$ and $(0, 1, -1)$

31. $x = -1 + t, y = -2t, z = 3$ and the line passing through $(0, -1, 2)$ and $(1, -2, 3)$

32. Prove Theorem 10.11 (distance between skew lines in $\mathbb{R}^3$).

33. The planes $Ax + By + Cz = D_1$ and $Ax + By + Cz = D_2$ are parallel.

　　a. Is the distance between the planes $\left| D_1 - D_2 \right|$? If not, what is the correct formula?

b. Find the distance between the parallel planes

$$x + y + 2z = 2 \quad \text{and} \quad x + y + 2z = 4$$

34. Show that the planes

$$A_1 x + B_1 y + C_1 z + D_1 = 0$$

and

$$A_2 x + B_2 y + C_2 z + D_2 = 0$$

are mutually orthogonal if and only if

$$A_1 A_2 + B_1 B_2 + C_1 C_2 = 0$$

Chapter 10   *Review*

Proficiency Examination

Concept Problems

1. What is a vector and what is a scalar?

2. Give both an algebraic and a geometric interpretation of multiplication of a vector by a scalar.

3. What is the parallelogram rule for vector sums?

4. State each of the following properties of vector operations:

　　a. commutativity of vector addition
　　b. associativity of vector addition
　　c. identity for vector addition
　　d. inverse property for vector addition
　　e. magnitude of a vector for dot product
　　f. commutativity for dot product
　　g. dot product of a scalar multiple
　　h. distributivity for dot product over addition
　　i. cross product of the zero vector
　　j. anticommutativity for cross product
　　k. distributivity for cross product over addition

5. If **u** and **v** are parallel vectors, what can be said about $\mathbf{u} \times \mathbf{v}$?

6. How do you find the length of a vector? What is a unit vector?

7. State the triangle inequality.

8. What are the standard basis vectors in $\mathbb{R}^3$?

9. What is the standard-form equation of a sphere?

10. What is a cylinder?

11. What is the distance between two points in $\mathbb{R}^3$?

12. How do you find a unit vector **u** in the direction of a given vector **v**?

13. What is Lagrange's identity?

14. Define dot product.

15. What is the formula for the angle between two vectors?

16. What is meant by *orthogonal* vectors? What is the algebraic condition for orthogonality?

17. What is a vector projection? Give a formula for finding the vector projection of **v** onto **w**.

18. What is a scalar projection? Give a formula for finding the scalar projection of **v** onto **w**.

19. What is the vector formula for work?

20. Define cross product.

21. What is the right-hand rule for a coordinate system?

22. a. Give a geometric interpretation of cross product.
　　b. What is the formula for magnitude of a cross product?

23. What is the determinant form for the triple scalar product?

24. How do you find the volume of a parallelepiped?

25. a. What is the parametric form of a line in $\mathbb{R}^3$?
　　b. What is the symmetric form of a line in $\mathbb{R}^3$?

26. What are the direction cosines of a vector in $\mathbb{R}^3$?

27. What is the normal vector of a plane in $\mathbb{R}^3$?

28. What is the point-normal form of a plane in $\mathbb{R}^3$?

29. What is the standard form of a plane in $\mathbb{R}^3$?

30. What is the formula for the distance from a point to a plane?

31. What is the formula for the distance from a point to a line?

Practice Problems

32. Given $\mathbf{v} = 2\mathbf{i} - 3\mathbf{j} + \mathbf{k}, \mathbf{w} = 3\mathbf{i} - 2\mathbf{j}$. Find each of the following vectors.

　　a. $2\mathbf{v} + 3\mathbf{w}$
　　b. $\|\mathbf{v}\|^2 - \|\mathbf{w}\|^2$
　　c. vector projection of **v** onto **w**

d. scalar projection of **w** onto **v**

e. **v · w**

f. **v × w**

33. Given $\mathbf{u} = 2\mathbf{i} - 3\mathbf{j} + \mathbf{k}, \mathbf{v} = \mathbf{i} + \mathbf{j} - 2\mathbf{k}$, and $\mathbf{w} = 3\mathbf{i} + 5\mathbf{k}$. In each of the following cases, either perform the indicated computation or explain why it is not defined.

   a. $(\mathbf{u} \times \mathbf{v}) \cdot \mathbf{w}$ b. $(\mathbf{u} \cdot \mathbf{v}) \times \mathbf{w}$

   c. $(\mathbf{u} \times \mathbf{v}) \times \mathbf{w}$ d. $(\mathbf{u} \cdot \mathbf{v}) \cdot \mathbf{w}$

Find the equations for the lines and planes in Problems 34–37.

34. the line through the points $P(-1, 4, -3)$ and $Q(0, -2, 1)$

35. the plane that contains the point $P(1, 1, 3)$ and is normal to the vector $\mathbf{v} = 2\mathbf{i} + 3\mathbf{k}$

36. the line of intersection of the planes $2x + 3y + z = 2$ and $y - 3z = 5$

37. the plane that contains the points $P(0, 2, -1), Q(1, -3, 5)$, and $R(3, 0, -2)$

38. Find the direction cosines and the direction angles of the vector $\mathbf{u} = -2\mathbf{i} + 3\mathbf{j} + \mathbf{k}$. Round to the nearest degree.

39. In each case, determine whether the lines intersect, are parallel, or are skew. If they intersect, find the point of intersection.

   a. $x = 2t - 3, y = 4 - t, z = 2t$; and

   $\dfrac{x + 2}{3} = \dfrac{y - 3}{5}; z = 3$

b. $\dfrac{x - 7}{5} = \dfrac{y - 6}{4} = \dfrac{z - 8}{5}$; and

$\dfrac{x - 8}{6} = \dfrac{y - 6}{4} = \dfrac{z - 9}{6}$

40. Let $\mathbf{u} = 2\mathbf{i} + \mathbf{j}, \mathbf{v} = \mathbf{i} - \mathbf{j} - \mathbf{k}$, and $\mathbf{w} = 3\mathbf{i} + 5\mathbf{k}$.

   a. Find the volume of the parallelepiped determined by these vectors.

   b. Find a positive number A that guarantees that the tetrahedron determined by $A\mathbf{u}, A\mathbf{v}$, and $\mathbf{w}$ has volume that is twice the volume of the original tetrahedron.

41. Find the distance from $P(-1, 1, 4)$ to $2x + 5y - z = 3$.

42. Find the distance between the skew lines $x = t, y = 2t, z = 3t - 1$ and $x = 1 - t, y = t + 2, z = t$.

43. Find the distance from the point $P(4, 5, 0)$ to the line

$$\dfrac{x - 2}{3} = \dfrac{y}{5} = \dfrac{z + 1}{-1}$$

44. An airplane flies at 200 mi/h parallel to the ground at an altitude of 10,000 ft. If the plane flies due south and the wind is blowing toward the northeast at 50 mi/h, what is the ground speed of the plane (that is, effective speed)?

45. A girl pulls a sled 50 ft on level ground with a rope inclined at an angle of 30° with the horizontal (the ground). If she applies 3 lb of tension to the rope, how much work is performed on the sled?

Supplementary Problems

1. A triangle in $\mathbb{R}^3$ has vertices $A(0, 2, -1), B(1, 1, 3)$, and $C(1, 0, -4)$.

   a. Find the perimeter of the triangle.

   b. Find the area of the triangle.

   c. Find the three vertex angles of the triangle. (Round to the nearest degree.)

   d. Find a number p such that the points A, B, C, and $D(p, p, 0)$ form a tetrahedron of volume $V = 100$ cubic units.

Find equations, in both parametric and symmetric forms, of the lines described in Problems 2–3. Find two additional points on each line.

2. passing through $A(1, -2, 3), B(4, -1, 2)$

3. passing through $P(1, 4, 0)$, with direction numbers $[2, 0, 1]$

4. Find the equation of the line passing through $P(3, 4, -1)$ and parallel to the line of intersection of the planes

$$x + 2y + 2z + 5 = 0 \text{ and } 2x + y - 3z - 6 = 0.$$

Find the equation of the plane satisfying the conditions given in Problems 5–12.

5. the xy-plane

6. the plane parallel to the xz-plane passing through $(4, 3, 7)$

7. the plane through $(1, -3, 4)$ with attitude numbers $[3, 4, -1]$

8. the plane through $(-1, 4, 5)$ and orthogonal to a line with direction numbers $[4, 4, -3]$

9. the plane through $(4, -3, 2)$ and parallel to the plane $5x - 2y + 3z - 10 = 0$

10. the plane containing

$$\dfrac{x - 3}{4} = \dfrac{z - 1}{2}; y = -2; \text{ and } \dfrac{x - 3}{3} = \dfrac{y + 2}{1} = \dfrac{z - 1}{-2}$$

11. the plane passing through $P(4, 1, 3), Q(-4, 2, 1)$, and $R(1, 0, 2)$

12. the plane passing through $(4, -1, 2)$ and parallel to the line

$$\dfrac{x + 2}{3} = \dfrac{y - 2}{-1} = \dfrac{z + 1}{2} \text{ and } \dfrac{x - 2}{1} = \dfrac{y - 3}{2} = \dfrac{z - 4}{3}$$

13. If **u** and **v** are orthogonal unit vectors, show that $(\mathbf{u} \times \mathbf{v}) \times \mathbf{u} = \mathbf{v}$. What is $(\mathbf{u} \times \mathbf{v}) \times \mathbf{v}$?

14. Show that three planes where normals **u, v, w** satisfies $\mathbf{u} \times \mathbf{v} \cdot \mathbf{w} \neq 0$ intersect in exactly one point.

15. Show that $\mathbf{u} \times (\mathbf{v} \times \mathbf{w}) = (\mathbf{u} \times \mathbf{v}) \times \mathbf{w}$ if and only if $\mathbf{v} \times (\mathbf{w} \times \mathbf{u}) = \mathbf{0}$.

16. Given the vectors $\mathbf{v} = 3\mathbf{i} - 2\mathbf{j} + \mathbf{k}$ and $\mathbf{w} = 4\mathbf{i} + \mathbf{j} - 3\mathbf{k}$, find $\|\mathbf{v}\|$, $\mathbf{v} - \mathbf{w}$, and $2\mathbf{v} + 3\mathbf{w}$.

17. Given $\mathbf{u} = \mathbf{i} - \mathbf{j} + \mathbf{k}$, $\mathbf{v} = 3\mathbf{i} - 2\mathbf{j} + 5\mathbf{k}$, and $\mathbf{w} = \mathbf{i} + \mathbf{j} - \mathbf{k}$, find $(\mathbf{u} - \mathbf{v}) \cdot \mathbf{w}$ and $(2\mathbf{u} + \mathbf{v}) \times (\mathbf{u} - \mathbf{w})$.

18. Given the vectors $\mathbf{v} = 4\mathbf{i} + 2\mathbf{j} + \mathbf{k}$, $\mathbf{w} = 2\mathbf{i} + \mathbf{j} - 5\mathbf{k}$. Find
   a. $5\mathbf{v} - 3\mathbf{w}$
   b. $\|2\mathbf{v} - \mathbf{w}\|$
   c. vector projection of $\mathbf{v}$ onto $\mathbf{w}$
   d. scalar projection of $\mathbf{w}$ onto $\mathbf{v}$

19. Find the direction cosines for $\mathbf{v} = (2\mathbf{i} + \mathbf{j}) \times (\mathbf{i} + \mathbf{j} - 3\mathbf{k})$.

20. Find two unit vectors that are parallel to the line
$$\frac{x}{6} = \frac{y}{2} = \frac{z-1}{6}$$

21. Find the (acute) angle, rounded to the nearest degree, between the intersecting lines
$$\frac{x-1}{3} = \frac{y-3}{-1} = \frac{z+5}{2} \text{ and } \frac{x-1}{2} = \frac{y-3}{-1} = \frac{z+5}{-2}$$

22. Find the area of the parallelogram determined by $3\mathbf{i} - 4\mathbf{j}$ and $-\mathbf{i} - \mathbf{j} + \mathbf{k}$.

23. Find the center and the radius of the sphere
$$4x^2 + 4y^2 + 4z^2 + 12y - 4z + 1 = 0$$

24. Find the points of intersection of the line $x = 6 + 3t$, $y = 10 - 2t$, $z = 5t$ with each of the coordinate planes.

25. Find the point of intersection of the planes $3x - y + 4z = 15$, $2x + y - 3z - 1 = 0$, and $x + 3y + 5z - 2 = 0$

26. Find the equation of the plane determined by the intersecting lines
$$\frac{x+3}{3} = \frac{y}{-2} = \frac{z-7}{6} \text{ and } \frac{x+6}{1} = \frac{y+5}{-3} = \frac{z-1}{2}$$

27. Find an equation for the set of all points that are equidistant from the planes $3x - 4y + 12z = 6$ and $4x + 3z = 7$.

28. Vertices B and C of $\triangle ABC$ lie along the line
$$\frac{x+2}{2} = \frac{y-1}{1} = \frac{z}{4}$$
Find the area of the triangle given that A has coordinates $(1, -1, 2)$ and disk $\overline{BC}$ has length 5.

29. Find the work done by the constant force $\mathbf{F} = 5\mathbf{i} + 4\mathbf{j} + \mathbf{k}$ in moving a particle along the line from $P(2, 1, -1)$ to $Q(4, 1, 2)$.

30. How much work does it take to move a container 25 m along a horizontal loading platform onto a truck using a constant force of 100 newtons at an angle of $\frac{\pi}{6}$ from the horizontal?

31. Find a formula for the surface area of the tetrahedron determined by vectors $\mathbf{u}$, $\mathbf{v}$, and $\mathbf{w}$. Assume the vectors do not all lie in the same plane.

32. Suppose $\mathbf{v}$ and $\mathbf{w}$ are nonzero vectors. Show that $\|\mathbf{v}\|\mathbf{w} + \|\mathbf{w}\|\mathbf{v}$ and $\|\mathbf{v}\|\mathbf{w} - \|\mathbf{w}\|\mathbf{v}$ are orthogonal vectors.

33. Let $\mathbf{v} = \cos\theta\,\mathbf{i} + \sin\theta\,\mathbf{j}$ and $\mathbf{w} = \cos\phi\,\mathbf{i} + \sin\phi\,\mathbf{j}$. Find $\mathbf{v} \times \mathbf{w}$. Interpret this cross product geometrically, and use it to derive a well-known trigonometric identity.

34. Find the three vertex angles (rounded to the nearest degree) of the triangle whose vertices are $(1, -2, 3)$, $(-1, 2, -3)$, and $(2, 1, -3)$.

35. Find a relationship between the numbers a_1, b_1, and c_1 so that the angle between the vectors $\mathbf{v} = a_1\mathbf{i} + b_1\mathbf{j} + c_1\mathbf{k}$ and $\mathbf{i} - 2\mathbf{j}$ is the same as the angle between $\mathbf{v}$ and $2\mathbf{i} + \mathbf{k}$.

36. Find an equation of the plane that passes through $(a, 0, 0)$, $(0, a, 0)$, and $(0, 0, a)$.

37. Find the area of the triangle with vertices $A(0, -1, 2)$, $B(1, 2, -1)$, and $C(3, -1, 2)$.

38. Find an equation for the set of all points equidistant from $D(0, 0, 6)$ and the xy-plane.

39. Find the area of the triangle determined by the vectors $\mathbf{v} = \mathbf{i} - \mathbf{j} + \mathbf{k}$ and $\mathbf{w} = 2\mathbf{i} + \mathbf{j} - 2\mathbf{k}$.

40. Find an equation for the plane that passes through the origin and is parallel to the vectors $\mathbf{v} = \mathbf{i} - 2\mathbf{j} + 3\mathbf{k}$ and $\mathbf{w} = -\mathbf{i} + \mathbf{j} + 2\mathbf{k}$.

41. Find an equation for the plane that passes through the origin and whose normal vector is parallel to the line of intersection of the planes $2x - y + z = 4$ and $x + 3y - z = 2$.

42. Find a number A such that the planes $2Ax + 3y + z = 1$ and $x - Ay + 3z = 5$ are orthogonal.

43. The lines L_1 and L_2 are aligned with the vectors $\mathbf{v}_1 = \mathbf{i} - \mathbf{j}$ and $\mathbf{v}_2 = \mathbf{i} - \mathbf{j} + 2\mathbf{k}$, respectively. Find an equation for the line L that passes through the point $(-1, 2, 0)$ and is orthogonal to both L_1 and L_2.

44. A parallelepiped is determined by the vectors $\mathbf{u} = \mathbf{i} - \mathbf{j} + \mathbf{k}$, $\mathbf{v} = \mathbf{i} + 2\mathbf{j} - \mathbf{k}$, and $\mathbf{w} = 2\mathbf{i} + \mathbf{j} + \mathbf{k}$. Find the altitude from the tip of $\mathbf{w}$ to the side determined by $\mathbf{u}$ and $\mathbf{v}$.

45. In Chapter 11, we show that $\mathbf{T} = \mathbf{i} + 2x\mathbf{j}$ is a vector in the direction of the tangent line at each point $P(x, x^2)$ on the parabola $y = x^2$. Find a unit vector normal to the parabola at the point $(3, 9)$.

46. For any nonzero vectors $\mathbf{u}$, $\mathbf{v}$, and $\mathbf{w}$, show that the vector $(\mathbf{u} \times \mathbf{v}) \times (\mathbf{u} \times \mathbf{w})$ is parallel to $\mathbf{u}$.

47. Let $\mathbf{v} = a\mathbf{i} + b\mathbf{j} + c\mathbf{k}$ and $\mathbf{w} = A\mathbf{i} + B\mathbf{j} + C\mathbf{k}$, where a, b, c, A, B, and C are constants. Describe the set of vectors $\mathbf{v} + t\mathbf{w}$, where t is any scalar.

48. Use vectors to show that the sum of the squares of the lengths of the sides of a parallelogram equals the sum of the squares of the lengths of the diagonals.

49. The vectors **u**, **v**, and **w** are said to be *linearly independent* in $\mathbb{R}^3$ if the only solution to the equation $a\mathbf{u} + b\mathbf{v} + c\mathbf{w} = \mathbf{0}$ is $a = b = c = 0$. Otherwise, the vectors are *linearly dependent*. Determine whether the vectors $\mathbf{u} = -\mathbf{i} + 2\mathbf{k}$, $\mathbf{v} = 2\mathbf{i} - \mathbf{j} + 3\mathbf{k}$, $\mathbf{w} = \mathbf{i} + 3\mathbf{j} - 2\mathbf{k}$ are linearly independent or dependent.

50. Figure 10.49 shows a parallelogram *ABCD*. If *M* is the midpoint of side $\overline{AB}$, show that the line $\overline{CM}$ intersects diagonal $\overline{BD}$ at a point *P* located one-third of the distance from *B* to *D* by completing the following steps.

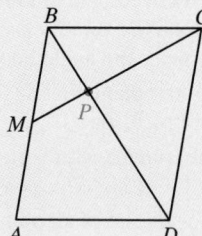

■ **FIGURE 10.49** Parallelogram *ABCD*

a. Let *a* and *b* be scalars such that $\mathbf{MP} = a\mathbf{MC}$ and $\mathbf{BP} = b\mathbf{BD}$. Show that
$$\tfrac{1}{2}\mathbf{AB} + b[\mathbf{AD} - \mathbf{AB}] = a[\tfrac{1}{2}\mathbf{AB} + \mathbf{AD}]$$

b. Use the fact that **AB** and **AD** are linearly independent (see Problem 49) to show that
$$\tfrac{1}{2} - b - \tfrac{1}{2}a = 0 \text{ and } a - b = 0$$

Solve this system of equations to show that *P* has the required location.

51. Show that $\mathbf{u} = a_1\mathbf{i} + a_2\mathbf{j} + a_3\mathbf{k}$, $\mathbf{v} = b_1\mathbf{i} + b_2\mathbf{j} + b_3\mathbf{k}$ and $\mathbf{w} = c_1\mathbf{i} + c_2\mathbf{j} + c_3\mathbf{k}$ are linearly dependent (see Problem 49), if and only if
$$\begin{vmatrix} a_1 & a_2 & a_3 \\ b_1 & b_2 & b_3 \\ c_1 & c_2 & c_3 \end{vmatrix} = 0$$

52. Show that
$$\begin{vmatrix} \mathbf{u}_1 \cdot \mathbf{v}_1 & \mathbf{u}_1 \cdot \mathbf{v}_2 \\ \mathbf{u}_2 \cdot \mathbf{v}_1 & \mathbf{u}_2 \cdot \mathbf{v}_2 \end{vmatrix} = (\mathbf{u}_1 \times \mathbf{u}_2) \cdot (\mathbf{v}_1 \times \mathbf{v}_2)$$

Hint: See Problem 59, Section 10.4.

53. Let $A(-2, 3, 7)$, $B(1, 5, -3)$, $C(2, 8, -1)$ be the vertices of a triangle in $\mathbb{R}^3$. What are the coordinates of the point *M* where the medians of the triangle meet (the centroid)?

54. The medians of a triangle meet at a point (the centroid) located two-thirds of the distance from each vertex to the midpoint of the opposite side. Generalize this result by showing that the four lines that join each vertex of a tetrahedron to the centroid of the opposite face meet at a point located three-fourths of the distance from the vertex to the centroid.

55. A triangle in $\mathbb{R}^3$ is determined by the vectors **v** and **w** as shown in Figure 10.50. Show that the traingle has area
$$A = \tfrac{1}{2}\sqrt{\|\mathbf{v}\|^2 \|\mathbf{w}\|^2 - (\mathbf{v} \cdot \mathbf{w})^2}$$

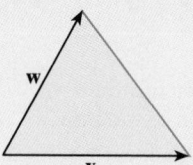

■ **FIGURE 10.50** Area of a triangle

56. **THINK TANK PROBLEM** In Figure 10.51, $\triangle ABC$ is equilateral and the points *M*, *N*, and *O* are located so that
$$\mathbf{AM} = \tfrac{1}{3}\mathbf{AB} \qquad \mathbf{BN} = \tfrac{1}{3}\mathbf{BC} \qquad \mathbf{CO} = \tfrac{1}{3}\mathbf{CA}$$

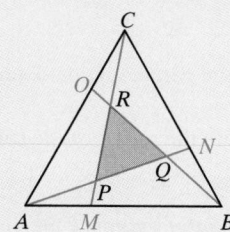

■ **FIGURE 10.51** Problem 56

It can be shown that $\triangle PQR$ is also equilateral, and $\|\mathbf{PM}\| = \|\mathbf{QN}\| = \|\mathbf{RO}\|$ and $\|\mathbf{AP}\| = \|\mathbf{BQ}\| = \|\mathbf{CR}\|$.

a. Show that $\mathbf{PQ} = \tfrac{3}{7}\mathbf{AN}$, then show that $\|\mathbf{AN}\|^2 = \tfrac{7}{9}\|\mathbf{AB}\|^2$. (*Hint:* Use the law of cosines.)

b. Show that $\triangle PQR$ has area $\tfrac{1}{7}$ that of $\triangle ABC$.

c. Do you think the same result would hold if $\triangle ABC$ were not equilateral? Investigate your conjecture.

57. **Gram-Schmidt orthogonalization process** Let **u**, **v**, and **w** be nonzero vectors in $\mathbb{R}^3$ that do not lie on the same plane. Define vectors $\boldsymbol{\alpha}$ and $\boldsymbol{\beta}$ as follows:
$$\boldsymbol{\alpha} = \mathbf{v} - \left[\frac{\mathbf{v} \cdot \mathbf{u}}{\|\mathbf{u}\|^2}\right]\mathbf{u} \quad \text{and} \quad \boldsymbol{\beta} = \mathbf{w} - \left[\frac{\mathbf{w} \cdot \mathbf{u}}{\|\mathbf{u}\|^2}\right]\mathbf{u} - \left[\frac{\mathbf{w} \cdot \boldsymbol{\alpha}}{\|\boldsymbol{\alpha}\|^2}\right]\boldsymbol{\alpha}$$

a. Show that $\mathbf{u}, \boldsymbol{\alpha}, \boldsymbol{\beta}$ are mutually orthogonal (any pair is orthogonal).

b. If $\boldsymbol{\gamma}$ is any vector in $\mathbb{R}^3$, show that
$$\boldsymbol{\gamma} = \left[\frac{\boldsymbol{\gamma} \cdot \mathbf{u}}{\|\mathbf{u}\|^2}\right]\mathbf{u} + \left[\frac{\boldsymbol{\gamma} \cdot \boldsymbol{\alpha}}{\|\boldsymbol{\alpha}\|^2}\right]\boldsymbol{\alpha} + \left[\frac{\boldsymbol{\gamma} \cdot \boldsymbol{\beta}}{\|\boldsymbol{\beta}\|^2}\right]\boldsymbol{\beta}$$

58. a. In Figure 10.52, *ABCD* is a rectangle, with *M* the midpoint of side $\overline{CD}$ and $\mathbf{AR}_k = \dfrac{1}{2k+1}\mathbf{AB}$. If *P* is the intersection of $\overline{AM}$ and $\overline{CR}_k$, and R_{k+1} is the foot of

the perpendicular drawn from P to $\overline{AB}$, show that

$$\mathbf{AR}_{k+1} = \frac{1}{2k+3}\mathbf{AB}$$

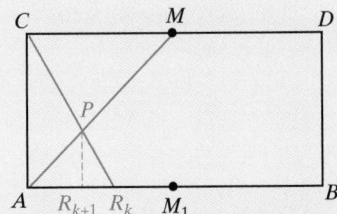

■ **FIGURE 10.52**   Problem 58

b. It is easy to subdivide a line segment line $\overline{AB}$ in half, or fourths, or eighths, etc. However, dividing it into thirds or fifths, … is more difficult. Use the result ob-

tained in part **a** to describe a procedure for subdividing $\overline{AB}$ into an odd number of equal parts.

59. **PUTNAM EXAMINATION PROBLEM**   Find the equations of two straight lines, each of which cuts all four of the following lines:

$$L_1: x = 1, y = 0 \qquad L_2: y = 1, z = 0$$
$$L_3: z = 1, x = 0 \qquad L_4: x = y = -6z$$

60. **PUTNAM EXAMINATION PROBLEM**   Find the equation of the smallest sphere that is tangent to both the lines

$$L_1: x = t + 1, y = 2t + 4, z = -3t + 5 \text{ and}$$
$$L_2: x = 4t - 12, y = t + 8, z = t + 17$$

61. **PUTNAM EXAMINATION PROBLEM**   The hands of an accurate clock have lengths 3 cm and 4 cm. Find the distance between the tips of the hands when the distance is increasing most rapidly.

Star Trek

This project is to be done in groups of three or four students. Each group will submit a single written report.

The starship *Enterprise* has been captured by the evil Romulans and is being held in orbit by a Romulan tractor beam. The orbit is elliptical with the planet Romula at one focus of the ellipse. Repeated efforts to escape have been futile and have almost exhausted the fuel supplies. Morale is low and food reserves are dwindling.

In searching the ship's log, Lieutenant Commander Data discovers that the *Enterprise* had been captured long ago by a Romulan tractor beam and had escaped. The key to that escape was to fire the ship's thrusters at exactly the right position in the orbit. Captain Picard gives the command to feed the required information into the computer to find that position. But, alas, a Romulan virus has rendered the computer useless for this task. Everyone turns to you and asks for your help in solving the problem.

Here is what Data discovered. If F represents the focus of the ellipse and P is the position of the ship on the ellipse, then the vector $\overrightarrow{FP}$ can be written as a sum $\mathbf{T} + \mathbf{N}$, where $\mathbf{T}$ is tangent to the ellipse and $\mathbf{N}$ is normal to the ellipse (not necessarily unit vectors). The thrusters must be fired when the ratio $\|\mathbf{T}\|/\|\mathbf{N}\|$ is equal to the eccentricity of the ellipse.

Your mission is to save the starship from the evil Romulans.

*Time is said to have only **one dimension**, and space to have **three dimensions**... the mathematical **quaternion** partakes of **both** of these elements; in technical language it may be said to be "time plus space," or "space plus time"; and in the sense it has, or at least involves a reference to, **four dimensions**...*

W. R. HAMILTON
GRAVES' LIFE OF HAMILTON (NEW YORK, 1882–1889), VOL. 3, P. 635.

*MAA Notes 17 (1991): "Priming the Calculus Pump: Innovations and Resources," by Marcus S. Cohen, Edward D. Gaughan, R. Arthur Knoebel, Douglas S. Kurtz, and David J. Pengelley.

11 Vector-Valued Functions

PREVIEW

The marriage of calculus and vector methods forms what is called *vector calculus*. The key to using vector calculus is the concept of a *vector-valued function*. In this chapter, we introduce such functions and examine some of their properties. We shall see that vector-valued functions behave much like the *scalar-valued functions* studied earlier in this text.

PERSPECTIVE

A car travels down a curved road at a constant speed of 55 mi/h. What additional information do we need about the car to determine whether it will stay on the road or skid off as it rounds a particular curve? Can we modify the road (say, by banking) so an average-sized car can travel at moderate speeds without skidding? How does a highway department decide what warning sign to install on a particular curve? A soldier fires a howitzer whose muzzle speed and angle of elevation are known. If the shell overshoots its target by 40 yd, how should the angle of elevation be changed to ensure a hit on the next shot? If a satellite is 20,000 mi above the earth, how fast must it travel to remain stationary above a particular point on the equator? These and other similar questions can be answered using vector calculus.

11.1 *Introduction to Vector Functions*

IN THIS SECTION   vector-valued functions, operations with vector functions, limits and continuity ▪

VECTOR-VALUED FUNCTIONS

In Section 9.4, we described a *plane curve* using parametric equations

$$x = f(t) \quad \text{and} \quad y = g(t)$$

where f and g are continuous functions of t on some interval. We extend this definition to three dimensions. A **space curve** is the set of all ordered triples $[f_1(t), f_2(t), f_3(t)]$ satisfying the parametric equations $x = f_1(t)$, $y = f_2(t)$, $z = f_3(t)$, where f_1, f_2, and f_3 are continuous functions of t on some domain D.

The concept of a vector-valued function is fundamental to the ideas we plan to explore. Here is a definition of this concept.

Vector-Valued Function

A **vector-valued function** (or, simply, a **vector function**) $\mathbf{F}$ with *domain D* assigns to each scalar t in the set D a unique vector $\mathbf{F}(t)$. The set of all vectors $\mathbf{v}$ of the form $\mathbf{v} = \mathbf{F}(t)$ for t in D is the *range* of $\mathbf{F}$. In this text, we shall be concerned with vector functions whose range is in $\mathbb{R}^2$ or $\mathbb{R}^3$. That is,

$$\mathbf{F}(t) = f_1(t)\mathbf{i} + f_2(t)\mathbf{j} \qquad \text{in } \mathbb{R}^2 \text{ (plane)}$$

$$\mathbf{F}(t) = f_1(t)\mathbf{i} + f_2(t)\mathbf{j} + f_3(t)\mathbf{k} \quad \text{in } \mathbb{R}^3 \text{ (space)}$$

where f_1, f_2, and f_3 are real-valued (**scalar-valued**) functions of the real number t defined on the domain set D. In this context, f_1, f_2, and f_3 are called the **components** of $\mathbf{F}$.

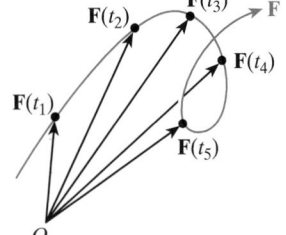

■ **FIGURE 11.1** The graph of the vector function $\mathbf{F}(t)$ is traced out by the terminal point of $\mathbf{F}(t)$ as t varies over D.

Let $\mathbf{F}$ be a vector function, and suppose the initial point of the vector $\mathbf{F}(t)$ is at the origin. The graph of $\mathbf{F}$ is the curve traced out by the terminal point of the vector $\mathbf{F}(t)$ as t varies over the domain set D, as shown in Figure 11.1.

EXAMPLE 1   *Graph of a vector function*

Sketch the graph of the vector function

$$\mathbf{F}(t) = (3 - t)\mathbf{i} + (2t)\mathbf{j} + (-4 + 3t)\mathbf{k}$$

for all t.

Solution   The graph is the collection of all points (x, y, z) with

$$x = 3 - t \qquad y = 2t \qquad z = -4 + 3t$$

for all t. We recognize these as the parametric equations for the line in $\mathbb{R}^3$ that contains the point $P_0(3, 0, -4)$ and is aligned with the vector

$$\mathbf{v} = -\mathbf{i} + 2\mathbf{j} + 3\mathbf{k}$$

as shown in Figure 11.2a.

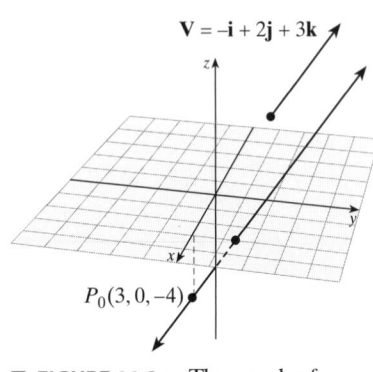

■ **FIGURE 11.2a** The graph of $\mathbf{F}(t) = (3 - t)\mathbf{i} + (2t)\mathbf{j} + (3t - 4)\mathbf{k}$

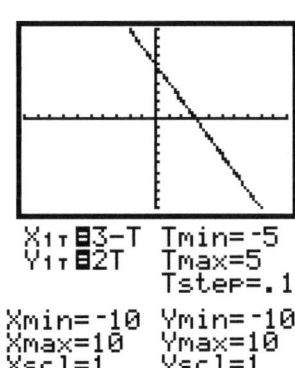

X₁ᴛ∎3-T Tmin=-5
Y₁ᴛ∎2T   Tmax=5
         Tstep=.1

Xmin=-10 Ymin=-10
Xmax=10   Ymax=10
Xscl=1   Yscl=1

■ **FIGURE 11.2b** Graph of
$(3 - t)\mathbf{i} + 2t\mathbf{j}$

If you have access to technology, you can draw vector functions in three dimensions quite easily. However, if you have only a graphing calculator, you may be limited to graphics in two dimensions. In order to visualize this vector function, draw a graph (in parametric form) as shown in Figure 11.2b. Note that the line passes through $(3, 0)$. The direction of the line (in $\mathbb{R}^2$) is $\langle -1, 2 \rangle$ which means a run of -1 with a rise of 2. From *this* graph imagine the same line in the plane $z = -4$ (because the line in $\mathbb{R}^3$ passes through $(3, 0, -4)$) and *then* imagine a change in the direction of z to be three units (because the direction is $\langle -1, 2, 3 \rangle$). ■

EXAMPLE 2   *Graph of a circular helix*

Sketch the graph of the vector function

$$\mathbf{F}(t) = (2 \sin t)\mathbf{i} - (2 \cos t)\mathbf{j} + (3t)\mathbf{k}$$

Solution   The graph of $\mathbf{F}$ is the collection of all points (x, y, z) in $\mathbb{R}^3$ whose coordinates satisfy

$$x = 2 \sin t \qquad y = -2 \cos t \qquad z = 3t \qquad \text{for all } t$$

The first two components satisfy

$$x^2 + y^2 = (2 \sin t)^2 + (-2 \cos t)^2 = 4(\sin^2 t + \cos^2 t) = 4$$

which means that the graph lies on the surface of the right circular cylinder with radius 2, whose axis of symmetry is the z-axis, as shown in Figure 11.3. We also know that as t increases, the z-coordinate of the point $P(x, y, z)$ on the graph of $\mathbf{F}$ increases according to the formula $z = 3t$, which means that the point (x, y, z) on the graph rises in a spiral on the surface of the cylinder $x^2 + y^2 = 4$. The point on the graph of $\mathbf{F}$ that corresponds to $t = 0$ is $(0, -2, 0)$, and the points that correspond to $t = \frac{\pi}{2}$ and $t = \pi$ are $(2, 0, \frac{3\pi}{2})$ and $(0, 2, 3\pi)$, respectively. Thus, the graph spirals upward counterclockwise (as viewed from above). The graph, which is known as a **right circular helix**, is shown in Figure 11.3a. In order to visualize this helix using a graphing calculator, draw the circle as shown in Figure 11.3b, and *imagine* the rise in the z-direction in the manner of a helix. ■

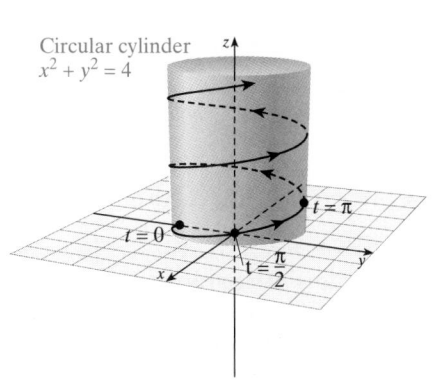

Circular cylinder
$x^2 + y^2 = 4$

■ **FIGURE 11.3a** The graph of
$\mathbf{F}(t) = (2 \sin t)\mathbf{i} - (2 \cos t)\mathbf{j} + (3t)\mathbf{k}$

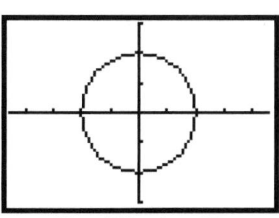

X₁ᴛ∎2sin T
Y₁ᴛ∎-2cos T
Tmin=0
Tmax=6.2831853…
Tstep=.1308996…
Xmin=-4.548387…
Xmax=4.5483870…
Xscl=1
Ymin=-3
Ymax=3
Yscl=1

■ **FIGURE 11.3b** The graph of
$(2 \sin t)\mathbf{i} - (2 \cos t)\mathbf{j}$

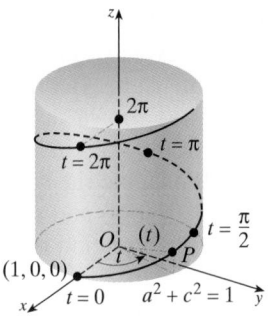
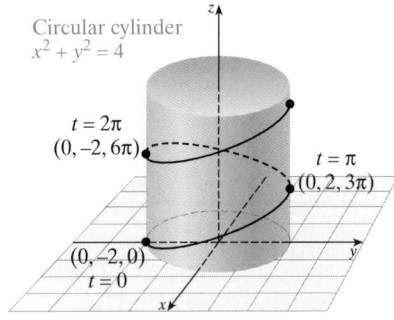

a. A computer-generated image of the helix in Example 2: $\mathbf{F}(t) = (2\sin t)\mathbf{i} - (2\cos t)\mathbf{j} + 3t\mathbf{k}$

b. A computer-generated image of the helix: $\mathbf{F}(t) = (\cos 2t)\mathbf{i} + (\sin 2t)\mathbf{j} + 0.2t\mathbf{k}$

■ **FIGURE 11.4**   Examples of helixes

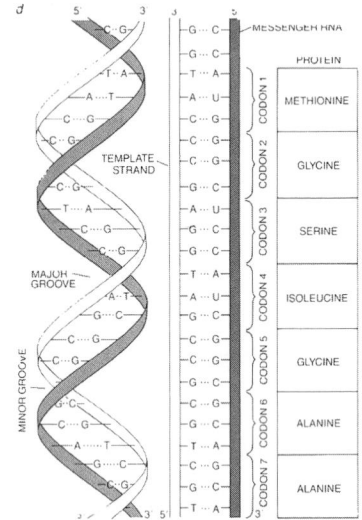

The double helix of the DNA was discovered in 1953 by James Watson and Francis Crick.

A well-known example of a helix is the DNA (deoxyribonucleic acid) molecule, which has a structure consisting of two intertwined helixes, as shown in Figure 11.4. Some other computer-generated helixes are also shown.

Examples 1 and 2 illustrate how the graph of a vector function

$$\mathbf{F}(t) = f_1(t)\mathbf{i} + f_2(t)\mathbf{j} + f_3(t)\mathbf{k}$$

can be obtained by examining the parametric equations

$$x = f_1(t) \qquad y = f_2(t) \qquad z = f_3(t)$$

In Example 3, we turn things around and find a vector function whose graph is a given curve.

> **EXAMPLE 3**   *Find a vector function*

Find a vector function $\mathbf{F}$ whose graph is the curve of intersection of the hemisphere $z = \sqrt{4 - x^2 - y^2}$ and the parabolic cylinder $y = x^2$, as shown in Figure 11.5.

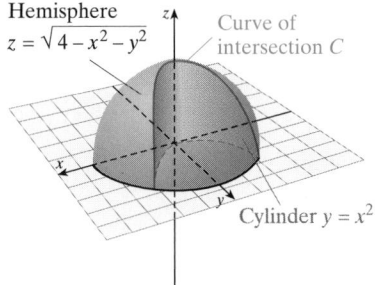

■ **FIGURE 11.5**   The curve of intersection of the hemisphere and the cylinder

Solution   Finding the parametric representation is sometimes called **parametrizing the curve.** There are several ways this can be done, but the natural choice is to let $x = t$. Then $y = t^2$ (from the equation of the parabola), and by substituting into the equation for the hemisphere, we find

$$z = \sqrt{4 - x^2 - y^2} = \sqrt{4 - (t)^2 - (t^2)^2} = \sqrt{4 - t^2 - t^4}$$

We can now state a formula for a vector function of the given graph:

$$\mathbf{F}(t) = t\mathbf{i} + t^2\mathbf{j} + \sqrt{4 - t^2 - t^4}\,\mathbf{k}$$

OPERATIONS WITH VECTOR FUNCTIONS

It follows from the definition of vector operations that vector functions can be added, subtracted, multiplied by a scalar function, and multiplied together. We summarize these operations in the following box.

Vector Function Operations

Let $\mathbf{F}$ and $\mathbf{G}$ be vector functions of the real variable t, and let $f(t)$ be a scalar function. Then, $\mathbf{F} + \mathbf{G}$, $\mathbf{F} - \mathbf{G}$, $f\mathbf{F}$, and $\mathbf{F} \times \mathbf{G}$ are vector functions, and $\mathbf{F} \cdot \mathbf{G}$ is a scalar function. These operations are summarized as follows:

Vector functions:

$$(\mathbf{F} + \mathbf{G})(t) = \mathbf{F}(t) + \mathbf{G}(t) \qquad (\mathbf{F} - \mathbf{G})(t) = \mathbf{F}(t) - \mathbf{G}(t)$$
$$(f\mathbf{F})(t) = f(t)\mathbf{F}(t) \qquad (\mathbf{F} \times \mathbf{G})(t) = \mathbf{F}(t) \times \mathbf{G}(t)$$

Scalar function: $(\mathbf{F} \cdot \mathbf{G})(t) = \mathbf{F}(t) \cdot \mathbf{G}(t)$

EXAMPLE 4   *Vector function operations*

Let $\mathbf{F}(t) = t^2\mathbf{i} + t\mathbf{j} - (\sin t)\mathbf{k}$ and $\mathbf{G}(t) = t\mathbf{i} + \dfrac{1}{t}\mathbf{j} + 5\mathbf{k}$. Find

a. $(\mathbf{F} + \mathbf{G})(t)$ b. $(e^t\mathbf{F})(t)$ c. $(\mathbf{F} \times \mathbf{G})(t)$ d. $(\mathbf{F} \cdot \mathbf{G})(t)$

Solution

a. $(\mathbf{F} + \mathbf{G})(t) = \mathbf{F}(t) + \mathbf{G}(t)$

$$= [t^2\mathbf{i} + t\mathbf{j} - (\sin t)\mathbf{k}] + \left[t\mathbf{i} + \frac{1}{t}\mathbf{j} + 5\mathbf{k}\right]$$
$$= (t^2 + t)\mathbf{i} + (t + t^{-1})\mathbf{j} + (5 - \sin t)\mathbf{k}$$

b. $(e^t\mathbf{F})(t) = e^t\mathbf{F}(t) = e^t t^2\mathbf{i} + e^t t\mathbf{j} - (e^t \sin t)\mathbf{k}$

c. $(\mathbf{F} \times \mathbf{G})(t) = \mathbf{F}(t) \times \mathbf{G}(t)$

$$= [t^2\mathbf{i} + t\mathbf{j} - (\sin t)\mathbf{k}] \times \left[t\mathbf{i} + \frac{1}{t}\mathbf{j} + 5\mathbf{k}\right]$$

$$= \begin{vmatrix} \mathbf{i} & \mathbf{j} & \mathbf{k} \\ t^2 & t & -\sin t \\ t & \frac{1}{t} & 5 \end{vmatrix}$$

$$= \left[5t + \frac{\sin t}{t}\right]\mathbf{i} - [5t^2 + t\sin t]\mathbf{j} + [t - t^2]\mathbf{k}$$

d. $(\mathbf{F} \cdot \mathbf{G})(t) = \mathbf{F}(t) \cdot \mathbf{G}(t)$

$$= [t^2\mathbf{i} + t\mathbf{j} - (\sin t)\mathbf{k}] \cdot \left[t\mathbf{i} + \frac{1}{t}\mathbf{j} + 5\mathbf{k}\right]$$
$$= t^3 + 1 - 5\sin t$$

Technology Window

Representing a vector function using technology is generally the same whether you are using a calculator or a software package. Square brackets are used to denote vector functions. For Example 4, we denote **F** as the function $y1(x)$, and **G** as $y2(x)$:

$$y1(x) = [x^2, x, -\sin x]$$

$$y2(x) = [x, 1/x, 5]$$

A sample of these input values is shown. Next, we work through each of the parts of Example 4.

Notice that if the answer is a vector, it is shown in brackets (as in parts **a, b,** and **c**). If the answer is a scalar (as in part **d**), the answer is not shown in brackets.

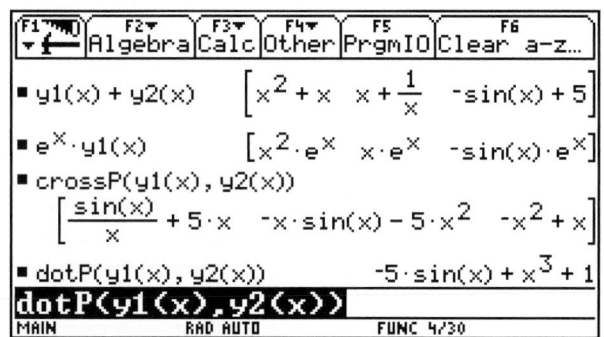

LIMITS AND CONTINUITY

Limit of a Vector Function

Suppose the components f_1, f_2, f_3 of the vector function

$$\mathbf{F}(t) = f_1(t)\mathbf{i} + f_2(t)\mathbf{j} + f_3(t)\mathbf{k}$$

all have finite limits as $t \to t_0$, where t_0 is any number or $\pm\infty$. Then the **limit** of $\mathbf{F}(t)$ as $t \to t_0$ is the vector

$$\lim_{t \to t_0} \mathbf{F}(t) = \left[\lim_{t \to t_0} f_1(t)\right]\mathbf{i} + \left[\lim_{t \to t_0} f_2(t)\right]\mathbf{j} + \left[\lim_{t \to t_0} f_3(t)\right]\mathbf{k}$$

EXAMPLE 5   *Limit of a vector function*

Find $\lim_{t \to 2} \mathbf{F}(t)$, where $\mathbf{F}(t) = (t^2 - 3)\mathbf{i} + e^t\mathbf{i} + (\sin \pi t)\mathbf{k}$.

Solution $\lim_{t \to 2} \mathbf{F}(t) = \left[\lim_{t \to 2}(t^2 - 3)\right]\mathbf{i} + \left[\lim_{t \to 2}(e^t)\right]\mathbf{j} + \left[\lim_{t \to 2}(\sin \pi t)\right]\mathbf{k}$

$$= 1\mathbf{i} + e^2\mathbf{j} + (\sin 2\pi)\mathbf{k}$$

$$= \mathbf{i} + e^2\mathbf{j}$$

For the most part, vector limits behave like scalar limits. The following theorem contains some useful general properties of such limits.

THEOREM 11.1   *Rules for vector limits*

If the vector functions **F** and **G** are functions of a real variable t and $h(t)$ is a scalar function such that all three functions have finite limits as $t \to t_0$, then

Limit of a sum	$\lim\limits_{t \to t_0} [\mathbf{F}(t) + \mathbf{G}(t)] = \lim\limits_{t \to t_0} \mathbf{F}(t) + \lim\limits_{t \to t_0} \mathbf{G}(t)$
Limit of a difference	$\lim\limits_{t \to t_0} [\mathbf{F}(t) - \mathbf{G}(t)] = \lim\limits_{t \to t_0} \mathbf{F}(t) - \lim\limits_{t \to t_0} \mathbf{G}(t)$
Limit of a scalar multiple	$\lim\limits_{t \to t_0} [h(t)\mathbf{F}(t)] = \left[\lim\limits_{t \to t_0} h(t)\right]\left[\lim\limits_{t \to t_0} \mathbf{F}(t)\right]$
Limit of a dot product	$\lim\limits_{t \to t_0} [\mathbf{F}(t) \cdot \mathbf{G}(t)] = \left[\lim\limits_{t \to t_0} \mathbf{F}(t)\right] \cdot \left[\lim\limits_{t \to t_0} \mathbf{G}(t)\right]$
Limit of a cross product	$\lim\limits_{t \to t_0} [\mathbf{F}(t) \times \mathbf{G}(t)] = \left[\lim\limits_{t \to t_0} \mathbf{F}(t)\right] \times \left[\lim\limits_{t \to t_0} \mathbf{G}(t)\right]$

These limit formulas are also valid as $t \to +\infty$ or as $t \to -\infty$, assuming all have finite limits.

Proof   We shall establish the formula for the limit of a dot product and leave the rest of the proof as an exercise. Let

$$\mathbf{F}(t) = f_1(t)\mathbf{i} + f_2(t)\mathbf{j} + f_3(t)\mathbf{k} \quad \text{and} \quad \mathbf{G}(t) = g_1(t)\mathbf{i} + g_2(t)\mathbf{j} + g_3(t)\mathbf{k}$$

Apply the limit of a vector function along with the sum rule and product rule for scalar limits to write:

$$\lim_{t \to t_0} [\mathbf{F}(t) \cdot \mathbf{G}(t)] = \lim_{t \to t_0} [f_1(t)g_1(t) + f_2(t)g_2(t) + f_3(t)g_3(t)]$$

$$= \left[\lim_{t \to t_0} f_1(t)\right]\left[\lim_{t \to t_0} g_1(t)\right] + \left[\lim_{t \to t_0} f_2(t)\right]\left[\lim_{t \to t_0} g_2(t)\right] + \left[\lim_{t \to t_0} f_3(t)\right]\left[\lim_{t \to t_0} g_3(t)\right]$$

$$= \left(\left[\lim_{t \to t_0} f_1(t)\right]\mathbf{i} + \left[\lim_{t \to t_0} f_2(t)\right]\mathbf{j} + \left[\lim_{t \to t_0} f_3(t)\right]\mathbf{k}\right) \cdot \left(\left[\lim_{t \to t_0} g_1(t)\right]\mathbf{i} + \left[\lim_{t \to t_0} g_2(t)\right]\mathbf{j} + \left[\lim_{t \to t_0} g_3(t)\right]\mathbf{k}\right)$$

$$= \left[\lim_{t \to t_0} \mathbf{F}(t)\right] \cdot \left[\lim_{t \to t_0} \mathbf{G}(t)\right]$$

EXAMPLE 6   *Limit of a cross product of vector functions*

Show that $\lim\limits_{t \to 1} [\mathbf{F}(t) \times \mathbf{G}(t)] = \left[\lim\limits_{t \to 1} \mathbf{F}(t)\right] \times \left[\lim\limits_{t \to 1} \mathbf{G}(t)\right]$ for the vector functions

$$\mathbf{F}(t) = t\mathbf{i} + (1 - t)\mathbf{j} + t^2\mathbf{k} \quad \text{and} \quad \mathbf{G}(t) = e^t\mathbf{i} - (3 + e^t)\mathbf{k}$$

Solution

$$\mathbf{F}(t) \times \mathbf{G}(t) = \begin{vmatrix} \mathbf{i} & \mathbf{j} & \mathbf{k} \\ t & 1 - t & t^2 \\ e^t & 0 & -(3 + e^t) \end{vmatrix}$$

$$= [(1 - t)(-3 - e^t) - 0]\mathbf{i} - [-t(3 + e^t) - t^2 e^t]\mathbf{j} + [0 - e^t(1 - t)]\mathbf{k}$$

$$= (te^t + 3t - e^t - 3)\mathbf{i} + (t^2 e^t + te^t + 3t)\mathbf{j} + (te^t - e^t)\mathbf{k}$$

Thus, the limit of the cross product is

$$\lim_{t \to 1} [\mathbf{F}(t) \times \mathbf{G}(t)] = \lim_{t \to 1} [te^t + 3t - e^t - 3]\mathbf{i} + \lim_{t \to 1} [t^2 e^t + te^t + 3t]\mathbf{j} + \lim_{t \to 1} [te^t - e^t]\mathbf{k}$$

$$= (e + 3 - e - 3)\mathbf{i} + (e + e + 3)\mathbf{j} + (e - e)\mathbf{k} = (2e + 3)\mathbf{j}$$

Now we find the cross product of the limits:

$$\lim_{t \to 1} \mathbf{F}(t) = \left[\lim_{t \to 1} t\right]\mathbf{i} + \left[\lim_{t \to 1} (1 - t)\right]\mathbf{j} + \left[\lim_{t \to 1} t^2\right]\mathbf{k} = \mathbf{i} + \mathbf{k}$$

$$\lim_{t \to 1} \mathbf{G}(t) = \left[\lim_{t \to 1} e^t\right]\mathbf{i} + \left[\lim_{t \to 1} (-3 - e^t)\right]\mathbf{k} = e\mathbf{i} + (-3 - e)\mathbf{k}$$

Thus,

$$\left[\lim_{t\to1}\mathbf{F}(t)\right]\times\left[\lim_{t\to1}\mathbf{G}(t)\right]=\begin{vmatrix}\mathbf{i}&\mathbf{j}&\mathbf{k}\\1&0&1\\e&0&-3-e\end{vmatrix}$$

$$=(0-0)\mathbf{i}-(-3-e-e)\mathbf{j}+(0-0)\mathbf{k}=(3+2e)\mathbf{j}$$

We see $\lim_{t\to1}[\mathbf{F}(t)\times\mathbf{G}(t)]=\left[\lim_{t\to1}\mathbf{F}(t)\right]\times\left[\lim_{t\to1}\mathbf{G}(t)\right].$ ■

Continuity of a Vector Function

A vector function $\mathbf{F}(t)$ is said to be **continuous** at t_0 if t_0 is in the domain of $\mathbf{F}$ and $\lim_{t\to t_0}\mathbf{F}(t)=\mathbf{F}(t_0)$.

■ *What This Says:*   This is the same as requiring each component of $\mathbf{F}(t)$ to be continuous at t_0. That is,

$$\mathbf{F}(t)=f_1(t)\mathbf{i}+f_2(t)\mathbf{j}+f_3(t)\mathbf{k}$$

is continuous at t_0 when t_0 is in the domain of the component functions $f_1(t),f_2(t)$, and $f_3(t)$ and

$$\lim_{t\to t_0}f_1(t)=f_1(t_0)\qquad\lim_{t\to t_0}f_2(t)=f_2(t_0)\qquad\lim_{t\to t_0}f_3(t)=f_3(t_0)$$

The rules for vector limits listed in Theorem 11.1 can be used to derive general properties of vector continuity.

EXAMPLE 7   *Continuity of a vector function*

For what values of t is $\mathbf{F}(t)=(\sin t)\mathbf{i}+(1-t)^{-1}\mathbf{j}+(\ln t)\mathbf{k}$ continuous?

Solution   The vector function $\mathbf{F}$ is continuous where its component functions

$$f_1(t)=\sin t\qquad f_2(t)=(1-t)^{-1}\qquad f_3(t)=\ln t$$

are continuous. The function f_1 is continuous for all $t;f_2$ is continuous where $1-t\ne0$ (that is, where $t\ne1$); f_3 is continuous for $t>0$. Thus, $\mathbf{F}$ is continuous when t is a positive number other than 1—that is, $t>0,t\ne1$.   ■

11.1 *Problem Set*

A *Find the domain for the vector functions given in Problems 1–8.*

1. $\mathbf{F}(t)=2t\mathbf{i}-3t\mathbf{j}+\dfrac{1}{t}\mathbf{k}$

2. $\mathbf{F}(t)=(1-t)\mathbf{i}+\sqrt{t}\mathbf{j}-\dfrac{1}{t-2}\mathbf{k}$

3. $\mathbf{F}(t)=(\sin t)\mathbf{i}+(\cos t)\mathbf{j}+(\tan t)\mathbf{k}$

4. $\mathbf{F}(t)=(\cos t)\mathbf{i}-(\cot t)\mathbf{j}+(\csc t)\mathbf{k}$

5. $h(t)\mathbf{F}(t)$ where $h(t)=\sin t$ and

$$\mathbf{F}(t)=\dfrac{1}{\cos t}\mathbf{i}+\dfrac{1}{\sin t}\mathbf{j}+\dfrac{1}{\tan t}\mathbf{k}$$

6. $\mathbf{F}(t)+\mathbf{G}(t)$, where

$$\mathbf{F}(t)=3t\mathbf{j}+t^{-1}\mathbf{k}\quad\text{and}$$
$$\mathbf{G}(t)=5t\mathbf{i}+\sqrt{10-t}\,\mathbf{j}$$

7. $\mathbf{F}(t)-\mathbf{G}(t)$, where

$$\mathbf{F}(t)=(\ln t)\mathbf{i}+3t\mathbf{j}-t^2\mathbf{k}\quad\text{and}$$
$$\mathbf{G}(t)=\mathbf{i}+5t\mathbf{j}-t^2\mathbf{k}$$

8. $\mathbf{F}(t)\times\mathbf{G}(t)$ where

$$\mathbf{F}(t)=t^2\mathbf{i}-t\mathbf{j}+2t\,\mathbf{k}\quad\text{and}$$
$$\mathbf{G}(t)=\dfrac{1}{t+2}\mathbf{i}+(t+4)\mathbf{j}-\sqrt{-t}\,\mathbf{k}$$

*Describe the graph of the vector functions given in
Problems 9–20 or sketch a graph in $\mathbb{R}^3$. A graph in $\mathbb{R}^2$ may
help with your description.*

9. $\mathbf{F}(t) = 2t\mathbf{i} + t^2\mathbf{j}$

10. $\mathbf{G}(t) = (1 - t)\mathbf{i} + \dfrac{1}{t}\mathbf{j}$

11. $\mathbf{G}(t) = (\sin t)\mathbf{i} - (\cos t)\mathbf{j}$
12. $\mathbf{F}(t) = (2\cos t)\mathbf{i} + (\sin t)\mathbf{j}$

13. $\mathbf{F}(t) = t\mathbf{i} - 4\mathbf{k}$
14. $\mathbf{G}(t) = e^t\mathbf{j} + t\mathbf{k}$

15. $\mathbf{F}(t) = (\cos t)\mathbf{i} + (\sin t)\mathbf{j} + t\mathbf{k}$

16. $\mathbf{F}(t) = e^t\mathbf{i} + e^t\mathbf{j} + e^{-t}\mathbf{k}$
17. $\mathbf{G}(t) = (1 - t)\mathbf{i} + t^2\mathbf{j} + t\mathbf{k}$

18. $\mathbf{F}(t) = \left(\dfrac{\sqrt{2}}{2}\sin t\right)\mathbf{i} + \left(\dfrac{\sqrt{2}}{2}\sin t\right)\mathbf{j} + (\cos^2 t)\mathbf{k}$

19. $\mathbf{F}(t) = t\mathbf{i} + (t^2 + 1)\mathbf{j} + t^2\mathbf{k}$

20. $\mathbf{G}(t) = (2\sin t)\mathbf{i} + (2\cos t)\mathbf{j} + 3\mathbf{k}$

B 21. ■ **What Does This Say?** Discuss the concept of a
vector-valued function.

22. ■ **What Does This Say?** Discuss the concept of limit of
a vector-valued function.

23. Show that the curve given by

$$\mathbf{R}(t) = (2\sin t)\mathbf{i} + (2\sin t)\mathbf{j} + (\sqrt{8}\cos t)\mathbf{k}$$

lies on a sphere centered at the origin.

24. Sketch the graph of the curve given by

$$\mathbf{R}(t) = (2\cos t)\mathbf{i} + (\sin t)\mathbf{j} - 2t\,\mathbf{k}$$

*Find a vector function $\mathbf{F}$ whose graph is the curve given in
Problems 25–30.*

25. $y = x^2; z = 2$

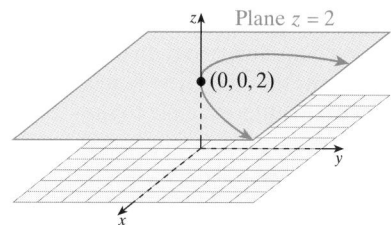

26. $x^2 + y^2 = 4; z = -1$

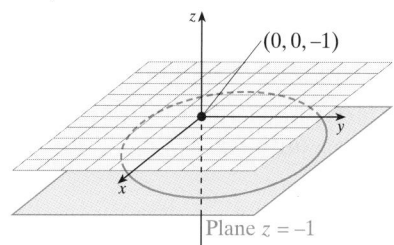

27. $x = 2t, y = 1 - t, z = \sin t$

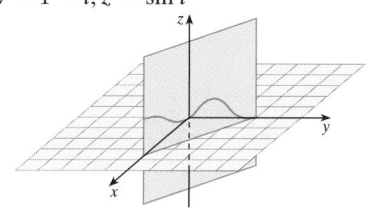

28. $\dfrac{x - 2}{3} = \dfrac{y - 1}{2} = \dfrac{z}{4}$

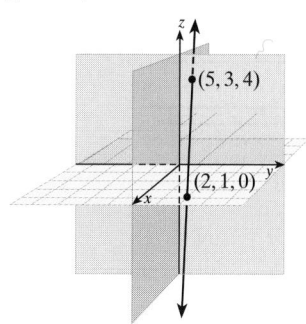

29. The curve of intersection of the hemisphere
$z = \sqrt{9 - x^2 - y^2}$ and the parabolic cylinder $x = y^2$

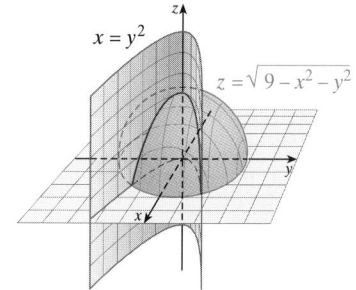

30. The line of intersection of the planes $2x + y + 3z = 6$
and $x - y - z = 1$

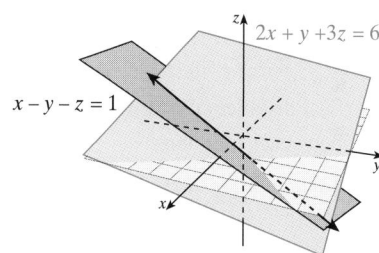

Perform the operations indicated in Problems 31–42 with

$$\mathbf{F}(t) = 2t\mathbf{i} - 5\mathbf{j} + t^2\mathbf{k}, \quad \mathbf{G}(t) = (1 - t)\mathbf{i} + \dfrac{1}{t}\mathbf{k},$$

$$\mathbf{H}(t) = (\sin t)\mathbf{i} + e^t\mathbf{j}$$

31. $2\mathbf{F}(t) - 3\mathbf{G}(t)$
32. $t^2\mathbf{F}(t) - 3\mathbf{H}(t)$

33. $\mathbf{F}(t) \cdot \mathbf{G}(t)$
34. $\mathbf{F}(t) \cdot \mathbf{H}(t)$

35. $\mathbf{G}(t) \cdot \mathbf{H}(t)$
36. $\mathbf{F}(t) \times \mathbf{G}(t)$

37. $\mathbf{F}(t) \times \mathbf{H}(t)$
38. $\mathbf{G}(t) \times \mathbf{H}(t)$

39. $2e^t\mathbf{F}(t) + t\mathbf{G}(t) + 10\mathbf{H}(t)$
40. $\mathbf{F}(t) \cdot [\mathbf{H}(t) \times \mathbf{G}(t)]$

41. $\mathbf{G}(t) \cdot [\mathbf{H}(t) \times \mathbf{F}(t)]$
42. $\mathbf{H}(t) \cdot [\mathbf{G}(t) \times \mathbf{F}(t)]$

Find each limit indicated in Problems 43–50.

43. $\lim\limits_{t \to 1} [2t\mathbf{i} - 3\mathbf{j} + e^t\mathbf{k}]$

44. $\lim\limits_{t \to 1} [3t\mathbf{i} + e^{2t}\mathbf{j} + (\sin \pi t)\mathbf{k}]$

45. $\lim\limits_{t\to 0}\left[\dfrac{(\sin t)\mathbf{i} - t\mathbf{k}}{t^2 + t - 1}\right]$

46. $\lim\limits_{t\to 1}\left[\dfrac{t^3 - 1}{t - 1}\mathbf{i} + \dfrac{t^2 - 3t + 2}{t^2 + t - 2}\mathbf{j} + (t^2 + 1)e^{t-1}\mathbf{k}\right]$

47. $\lim\limits_{t\to 0}\left[\dfrac{te^t}{1 - e^t}\mathbf{i} + \dfrac{e^{t-1}}{\cos t}\mathbf{j}\right]$

48. $\lim\limits_{t\to 0}\left[\dfrac{\sin t}{t}\mathbf{i} + \dfrac{1 - \cos t}{t}\mathbf{j} + e^{1-t}\mathbf{k}\right]$

49. $\lim\limits_{t\to 0^+}\left[\dfrac{\sin 3t}{\sin 2t}\mathbf{i} + \dfrac{\ln(\sin t)}{\ln(\tan t)}\mathbf{j} + (t \ln t)\mathbf{k}\right]$

50. $\lim\limits_{t\to 2}[(2\mathbf{i} - t\mathbf{j} + e^t\mathbf{k}) \times (t^2\mathbf{i} + 4 \sin t\,\mathbf{j})]$

Determine all values of t for which the vector function given in Problems 51–56 is continuous.

51. $\mathbf{F}(t) = t\mathbf{i} + 3\mathbf{j} - (1 - t)\mathbf{k}$

52. $\mathbf{G}(t) = t\mathbf{i} - \dfrac{1}{t}\mathbf{k}$ **53.** $\mathbf{G}(t) = \dfrac{\mathbf{i} + 2\mathbf{j}}{t^2 + t}$

54. $\mathbf{F}(t) = (e^t \sin t)\mathbf{i} + (e^t \cos t)\mathbf{k}$

55. $\mathbf{F}(t) = e^t\left[t\mathbf{i} + \dfrac{1}{t}\mathbf{j} + 3\mathbf{k}\right]$

56. $\mathbf{G}(t) = \dfrac{\mathbf{u}}{\|\mathbf{u}\|}$ where $\mathbf{u} = t\mathbf{i} + \sqrt{t}\mathbf{j}$

57. The graph of

$$\mathbf{R} = t\mathbf{i} - \left(\dfrac{1 - t}{t}\right)\mathbf{j} + \left(\dfrac{1 - t^2}{t}\right)\mathbf{k}$$

lies in a certain plane. What is the equation of that plane?

58. How many revolutions are made by the circular helix

$$\mathbf{R} = (2 \sin t)\mathbf{i} + (2 \cos t)\mathbf{j} + \tfrac{5}{8}t\mathbf{k}$$

in a vertical distance of 8 units?

59. Given the vector functions

$$\mathbf{F}(t) = t\mathbf{i} + t^2\mathbf{j} + t^3\mathbf{k} \quad\text{and}\quad \mathbf{G}(t) = \dfrac{1}{t}\mathbf{i} - e^t\mathbf{j}$$

directly verify each of the following limit formulas (that is, without using Theorem 11.1).

a. $\lim\limits_{t\to 0} e^t\mathbf{F}(t) = \left[\lim\limits_{t\to 0} e^t\right]\left[\lim\limits_{t\to 0} \mathbf{F}(t)\right]$

b. $\lim\limits_{t\to 1} \mathbf{F}(t) \cdot \mathbf{G}(t) = \left[\lim\limits_{t\to 1} \mathbf{F}(t)\right] \cdot \left[\lim\limits_{t\to 1} \mathbf{G}(t)\right]$

c. $\lim\limits_{t\to 1} [\mathbf{F}(t) \times \mathbf{G}(t)] = \left[\lim\limits_{t\to 1} \mathbf{F}(t)\right] \times \left[\lim\limits_{t\to 1} \mathbf{G}(t)\right]$

C 60. If $\mathbf{H}(t)$ is a vector function, we define the **difference operator** $\Delta\mathbf{H}$ by the formula

$$\Delta\mathbf{H} = \mathbf{H}(t + \Delta t) - \mathbf{H}(t)$$

where Δt is a change in the parameter t. (*Note:* Usually, $|\Delta t|$ is a small number.) If $\mathbf{F}(t)$ and $\mathbf{G}(t)$ are vector functions, show that

$$\Delta(\mathbf{F} \times \mathbf{G})(t) = \mathbf{F}(t + \Delta t) \times \Delta\mathbf{G}(t) + \Delta\mathbf{F}(t) \times \mathbf{G}(t)$$

Hint: Note that

$$\Delta(\mathbf{F} \times \mathbf{G})(t) = \mathbf{F}(t + \Delta t) \times \mathbf{G}(t + \Delta t)$$
$$- \mathbf{F}(t + \Delta t) \times \mathbf{G}(t) + \mathbf{F}(t + \Delta t) \times \mathbf{G}(t)$$
$$- \mathbf{F}(t) \times \mathbf{G}(t)$$

61. Prove the limit of a sum rule of Theorem 11.1. (The difference rule is proved similarly.)

62. a. Prove the limit of a scalar multiple rule of Theorem 11.1.
   b. Prove the limit of a cross product rule of Theorem 11.1.

63. THINK TANK PROBLEM Let $\mathbf{F}(t)$ and $\mathbf{G}(t)$ be vector functions that are continuous at t_0, and let $h(t)$ be a scalar function continuous at t_0. In each of the following cases, either show that the given function is continuous at t_0 or provide a counterexample to show that it is not continuous.
   a. $3\mathbf{F}(t) + 5\mathbf{G}(t)$ b. $\mathbf{F}(t) \cdot \mathbf{G}(t)$
   c. $h(t)\mathbf{F}(t)$ d. $\mathbf{F}(t) \times \mathbf{G}(t)$

11.2 *Differentiation and Integration of Vector Functions*

IN THIS SECTION   vector derivatives, tangent vectors, properties of vector derivatives, the modeling motion of an object in space, vector integrals

VECTOR DERIVATIVES

In Chapter 3, we defined the derivative of the (scalar) function f to be the limit as $\Delta x \to 0$ of the difference quotient $\Delta f/\Delta x$. We call this the **scalar derivative** to distinguish it from derivatives involving vector functions. The **difference quotient of a vector function F** is the vector expression

$$\frac{\Delta\mathbf{F}}{\Delta t} = \frac{\mathbf{F}(t + \Delta t) - \mathbf{F}(t)}{\Delta t}$$

and we define the derivative of $\mathbf{F}$ as follows.

Derivative of a Vector Function

The **derivative** of the vector function $\mathbf{F}$ is the vector function $\mathbf{F}'$ determined by the limit

$$\mathbf{F}'(t) = \lim_{\Delta t \to 0} \frac{\Delta \mathbf{F}}{\Delta t} = \lim_{\Delta t \to 0} \frac{\mathbf{F}(t + \Delta t) - \mathbf{F}(t)}{\Delta t}$$

wherever this limit exists. In the Leibniz notation, the derivative of $\mathbf{F}(t)$ is denoted by $\dfrac{d\mathbf{F}}{dt}$.

We say that the vector function $\mathbf{F}$ is **differentiable** at $t = t_0$ if $\mathbf{F}'(t)$ is defined at t_0.

The following theorem establishes a convenient method for computing the derivative of a vector function.

THEOREM 11.2   *Derivative of a vector function*

The vector function $\mathbf{F}(t) = f_1(t)\mathbf{i} + f_2(t)\mathbf{j} + f_3(t)\mathbf{k}$ is differentiable whenever the component functions $f_1, f_2,$ and f_3 are all differentiable, and in this case

$$\mathbf{F}'(t) = f_1'(t)\mathbf{i} + f_2'(t)\mathbf{j} + f_3'(t)\mathbf{k}$$

Proof   We use the definition of the derivative, along with rules of vector limits (Theorem 11.1), and the fact that the scalar derivatives $f_1'(t), f_2'(t),$ and $f_3'(t)$ and exist.

$$\mathbf{F}'(t) = \lim_{\Delta t \to 0} \frac{\mathbf{F}(t + \Delta t) - \mathbf{F}(t)}{\Delta t}$$

$$= \lim_{\Delta t \to 0} \frac{[f_1(t + \Delta t)\mathbf{i} + f_2(t + \Delta t)\mathbf{j} + f_3(t + \Delta t)]\mathbf{k} - [f_1(t)\mathbf{i} + f_2(t)\mathbf{j} + f_3(t)\mathbf{k}]}{\Delta t}$$

$$= \left[\lim_{\Delta t \to 0} \frac{f_1(t + \Delta t) - f_1(t)}{\Delta t}\right]\mathbf{i} + \left[\lim_{\Delta t \to 0} \frac{f_2(t + \Delta t) - f_2(t)}{\Delta t}\right]\mathbf{j} + \left[\lim_{\Delta t \to 0} \frac{f_3(t + \Delta t) - f_3(t)}{\Delta t}\right]\mathbf{k}$$

$$= f_1'(t)\mathbf{i} + f_2'(t)\mathbf{j} + f_3'(t)\mathbf{k}$$

EXAMPLE 1   *Differentiability of a vector function*

For what values of t is $\mathbf{G}(t) = |t|\mathbf{i} + (\cos t)\mathbf{j} + (t - 5)\mathbf{k}$ differentiable?

Solution   The component functions $\cos t$ and $t - 5$ are differentiable for all t, but $|t|$ is not differentiable at $t = 0$. Thus, the vector function $\mathbf{G}$ is differentiable for all $t \neq 0$. ∎

EXAMPLE 2   *Derivative of a vector function*

Find the derivative of the vector function

$$\mathbf{F}(t) = e^t\mathbf{i} + (\sin t)\mathbf{j} + (t^3 + 5t)\mathbf{k}$$

Solution

$$\mathbf{F}'(t) = (e^t)'\mathbf{i} + (\sin t)'\mathbf{j} + (t^3 + 5t)'\mathbf{k} = e^t\mathbf{i} + (\cos t)\mathbf{j} + (3t^2 + 5)\mathbf{k} \quad \blacksquare$$

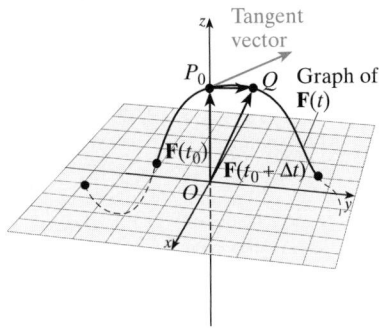

■ **FIGURE 11.6**   The difference quotient is a multiple of the secant line vector.

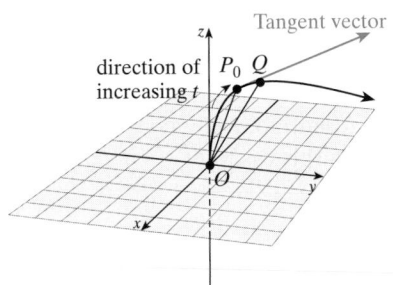

■ **FIGURE 11.7**   As $\Delta t \to 0$, the vector **PQ** and hence the difference quotient $\Delta \mathbf{F}/\Delta t$ approach the tangent vector at P_0.

TANGENT VECTORS

Recall that the scalar derivative $f'(x_0)$ gives the slope of the tangent line to the graph of f at the point where $x = x_0$ and thus provides a measure of the graph's direction at that point. Our first goal is to extend this interpretation by showing how the vector derivative can be used to find tangent vectors to curves in space.

Let t be a number in the domain of the vector function $\mathbf{F}(t)$, and let P_0 be the point on the graph of **F** that corresponds to t_0, as shown in Figure 11.6. Then for any positive number Δt, the difference quotient

$$\frac{\Delta \mathbf{F}}{\Delta t} = \frac{\mathbf{F}(t_0 + \Delta t) - \mathbf{F}(t_0)}{\Delta t}$$

is a vector that points in the same direction as the secant vector

$$\mathbf{P_0 Q} = \mathbf{F}(t_0 + \Delta t) - \mathbf{F}(t_0),$$

where Q is the point on the graph of **F** that corresponds to $t = t_0 + \Delta t$ (see Figure 11.6). If we choose $\Delta t < 0$, the difference quotient $\Delta \mathbf{F}/\Delta t$ points in the *opposite* direction to that of the secant vector $\mathbf{P_0 Q}$.

Suppose the difference quotient $\Delta \mathbf{F}/\Delta t$ has a limit as $\Delta t \to 0$ and that

$$\lim_{\Delta t \to 0} \frac{\Delta \mathbf{F}}{\Delta t} \neq \mathbf{0}$$

Then, as $\Delta t \to 0$, the direction of the secant $\mathbf{P_0 Q}$—and hence that of the difference quotient $\Delta \mathbf{F}/\Delta t$—will approach the direction of the tangent vector at P_0, as shown in Figure 11.7. Thus, we expect the tangent vector at P_0 to be the limit vector

$$\lim_{\Delta t \to 0} \frac{\Delta \mathbf{F}}{\Delta t}$$

which we recognize as the vector derivative $\mathbf{F}'(t_0)$. These observations lead to the following interpretation.

> **Tangent Vector**
>
> Suppose $\mathbf{F}(t)$ is differentiable at t_0 and that $\mathbf{F}'(t_0) \neq \mathbf{0}$. Then $\mathbf{F}'(t_0)$ is a **tangent vector** to the graph of $\mathbf{F}(t)$ at the point where $t = t_0$ and points in the direction of increasing t.

EXAMPLE 3   *Finding a tangent vector*

Find a tangent vector at the point P_0 where $t = 0.2$ on the graph of the vector function

$$\mathbf{F}(t) = e^{2t}\mathbf{i} + (t^2 - t)\mathbf{j} + (\ln t)\mathbf{k}$$

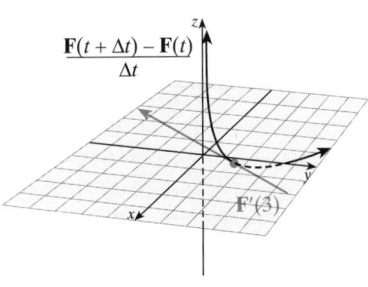

What is the equation of the tangent line at P_0?

Solution   Because $\mathbf{F}(0.2) = e^{0.4}\mathbf{i} - 0.16\mathbf{j} + (\ln 0.2)\mathbf{k}$, the point of tangency P_0 has coordinates $(e^{0.4}, -0.16, \ln 0.2) \approx (1.5, -0.16, -1.6)$. The derivative of **F** is

$$\mathbf{F}'(t) = 2e^{2t}\mathbf{i} + (2t - 1)\mathbf{j} + t^{-1}\mathbf{k}$$

The required tangent vector is found by evaluating $\mathbf{F}'(t)$ at $t = 0.2$:

$$\mathbf{F}'(0.2) = 2e^{0.4}\mathbf{i} + (-0.6)\mathbf{j} + 5\mathbf{k}$$

In two dimensions, we have:

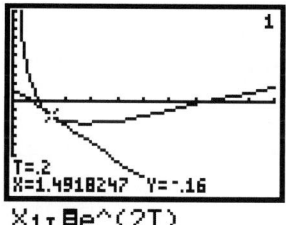

```
X₁ᴛ■e^(2T)
Y₁ᴛ■T²-T
X₂ᴛ■e^.4+(2e^.4)
T
Y₂ᴛ■-.16+(-.6)T
Tmin=-5
Tmax=5
Tstep=.1
Xmin=0  Ymin=-1
Xmax=10 Ymax=1
Xscl=1  Yscl=.1
```

Once again, we can turn to a graph in $\mathbb{R}^2$ to help us visualize this tangent vector. If you have a graphing calculator, you might wish to pay special attention to the parametric-form input for both the vector function and its tangent vector.

Once we have the point of tangency and a tangent vector, we can find an equation for the tangent line to the graph by using the parametric form for a line. In this case, the tangent line contains $P_0(e^6, 6, \ln 3)$ and is aligned with the tangent vector $\mathbf{v} = 2e^{0.4}\mathbf{i} - 0.6\mathbf{j} + 5\mathbf{k}$, so that the tangent line at P_0 has the parametric form

$$x = e^6 + 2e^6 t \qquad y = 6 + 5t \qquad z = \ln 3 + \tfrac{1}{3}t$$

where t is a real number (scalar). ∎

We have seen that the graph of $\mathbf{F}$ will have a tangent vector $\mathbf{F}'(t_0)$ at the point P_0, where $t = t_0$ if $\mathbf{F}'(t_0)$ exists and $\mathbf{F}'(t_0) \neq \mathbf{0}$. If we also require the derivative $\mathbf{F}'$ to be continuous at t_0, then the tangent vector at each point of the graph of $\mathbf{F}$ near P_0 will be close to the tangent vector at P_0. In this case, the graph has a *continuously turning tangent* at P_0, and we say that it is *smooth* there.

Smooth Curve

The graph of the vector function defined by $\mathbf{F}(t)$ is said to be **smooth** on any interval of t where $\mathbf{F}'$ is continuous and $\mathbf{F}'(t) \neq \mathbf{0}$. The graph is **piecewise smooth** on an interval that can be subdivided into a finite number of subintervals on which $\mathbf{F}$ is smooth.

As indicated in Figure 11.8, a smooth graph will have no sharply angled points (cusps). For this reason, vector functions with smooth graphs are "nicely behaved" and play an especially important role in vector calculus.

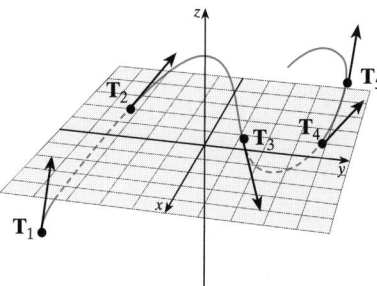

a. A curve that is smooth has a continuously turning tangent.

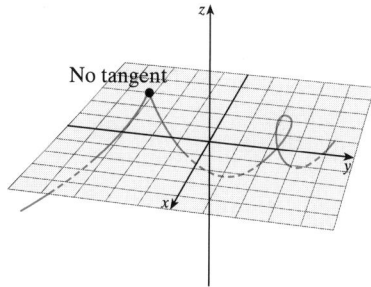

b. A curve that is not smooth can have "sharp" points. Note that this graph is piecewise smooth.

■ **FIGURE 11.8** Smooth and piecewise smooth curves

PROPERTIES OF VECTOR DERIVATIVES

Higher derivatives of a vector function $\mathbf{F}$ are obtained by successively differentiating the components of $\mathbf{F}(t) = f_1(t)\mathbf{i} + f_2(t)\mathbf{j} + f_3(t)\mathbf{k}$. For instance, the **second derivative** of $\mathbf{F}$ is the vector function

$$\mathbf{F}''(t) = [\mathbf{F}'(t)]' = f_1''(t)\mathbf{i} + f_2''(t)\mathbf{j} + f_3''(t)\mathbf{k}$$

whereas the **third derivative** $\mathbf{F}'''(t)$ is the derivative of $\mathbf{F}''(t)$, and so on. In the Leibniz notation, the second vector derivative of $\mathbf{F}$ with respect to t is denoted by $\dfrac{d^2\mathbf{F}}{dt^2}$, and the third derivative, by $\dfrac{d^3\mathbf{F}}{dt^3}$.

| EXAMPLE 4 | *Higher derivatives of a vector function* |

Find the second and third derivatives of the vector function

$$\mathbf{F}(t) = e^{2t}\mathbf{i} + (1 - t^2)\mathbf{j} + (\cos 2t)\mathbf{k}$$

Solution

$$\mathbf{F}'(t) = 2e^{2t}\mathbf{i} + (-2t)\mathbf{j} + (-2\sin 2t)\mathbf{k}$$
$$\mathbf{F}''(t) = 4e^{2t}\mathbf{i} - 2\mathbf{j} - (4\cos 2t)\mathbf{k}$$
$$\mathbf{F}'''(t) = 8e^{2t}\mathbf{i} + (8\sin 2t)\mathbf{k}$$

■

Several rules for computing derivatives of vector functions are listed in the following theorem.

THEOREM 11.3   *Rules for differentiating vector functions*

If the vector functions $\mathbf{F}$ and $\mathbf{G}$ and the scalar function h are differentiable at t, then so are $a\mathbf{F} + b\mathbf{G}, h\mathbf{F}, \mathbf{F} \cdot \mathbf{G}$, and $\mathbf{F} \times \mathbf{G}$, and

Linearity rule	$(a\mathbf{F} + b\mathbf{G})'(t) = a\mathbf{F}'(t) + b\mathbf{G}'(t)$	for constants a, b
Scalar multiple	$(h\mathbf{F})'(t) = h'(t)\mathbf{F}(t) + h(t)\mathbf{F}'(t)$	
Dot product rule	$(\mathbf{F} \cdot \mathbf{G})'(t) = (\mathbf{F}' \cdot \mathbf{G})(t) + (\mathbf{F} \cdot \mathbf{G}')(t)$	
Cross product rule*	$(\mathbf{F} \times \mathbf{G})'(t) = (\mathbf{F}' \times \mathbf{G})(t) + (\mathbf{F} \times \mathbf{G}')(t)$	
Chain rule	$[\mathbf{F}(h(t))]' = h'(t)\mathbf{F}'(h(t))$	

Proof   We shall prove the linearity rule and leave the rest of the proof as exercises. Note that if $\mathbf{F}$ and $\mathbf{G}$ are vector functions differentiable at t, and a, b are constants, then the linear combination $a\mathbf{F} + b\mathbf{G}$ has the difference quotient

$$\frac{\Delta(a\mathbf{F} + b\mathbf{G})}{\Delta t} = \frac{a\Delta\mathbf{F}}{\Delta t} + \frac{b\Delta\mathbf{G}}{\Delta t}$$

We can now find the derivative:

$$\begin{aligned}
(a\mathbf{F} + b\mathbf{G})'(t) &= \lim_{\Delta t \to 0}\left[\frac{\Delta(a\mathbf{F} + b\mathbf{G})}{\Delta t}\right] \\
&= \lim_{\Delta t \to 0}\left[\frac{a\Delta\mathbf{F}}{\Delta t} + \frac{b\Delta\mathbf{G}}{\Delta t}\right] \\
&= a\lim_{\Delta t \to 0}\frac{\Delta\mathbf{F}}{\Delta t} + b\lim_{\Delta t \to 0}\frac{\Delta\mathbf{G}}{\Delta t} \\
&= a\mathbf{F}'(t) + b\mathbf{G}'(t)
\end{aligned}$$

═

| EXAMPLE 5 | *Derivative of a cross product* |

Let $\mathbf{F}(t) = \mathbf{i} + t\mathbf{j} + t^2\mathbf{k}$ and $\mathbf{G}(t) = t\mathbf{i} + e^t\mathbf{j} + 3\mathbf{k}$. Verify that

$$(\mathbf{F} \times \mathbf{G})'(t) = (\mathbf{F}' \times \mathbf{G})(t) + (\mathbf{F} \times \mathbf{G}')(t)$$

Solution   First find the derivative of the cross product:

$$(\mathbf{F} \times \mathbf{G})(t) = \begin{vmatrix} \mathbf{i} & \mathbf{j} & \mathbf{k} \\ 1 & t & t^2 \\ t & e^t & 3 \end{vmatrix} = (3t - t^2e^t)\mathbf{i} - (3 - t^3)\mathbf{j} + (e^t - t^2)\mathbf{k}$$

so that $(\mathbf{F} \times \mathbf{G})'(t) = (3 - 2te^t - t^2e^t)\mathbf{i} + (3t^2)\mathbf{j} + (e^t - 2t)\mathbf{k}$.

---

*The order of the factors is important in the cross product rule, because the cross product of vectors is not commutative.

Next, find $(\mathbf{F}' \times \mathbf{G})(t) + (\mathbf{F} \times \mathbf{G}')(t)$ by first finding the derivatives of $\mathbf{F}$ and $\mathbf{G}$ and then the appropriate cross products:

$$\mathbf{F}'(t) = \mathbf{j} + 2t\mathbf{k} \quad \text{and} \quad \mathbf{G}'(t) = \mathbf{i} + e^t\mathbf{j}$$

$$(\mathbf{F}' \times \mathbf{G})(t) = \begin{vmatrix} \mathbf{i} & \mathbf{j} & \mathbf{k} \\ 0 & 1 & 2t \\ t & e^t & 3 \end{vmatrix} = (3 - 2te^t)\mathbf{i} - (-2t^2)\mathbf{j} + (-t)\mathbf{k}$$

$$(\mathbf{F} \times \mathbf{G}')(t) = \begin{vmatrix} \mathbf{i} & \mathbf{j} & \mathbf{k} \\ 1 & t & t^2 \\ 1 & e^t & 0 \end{vmatrix} = (-t^2e^t)\mathbf{i} - (-t^2)\mathbf{j} + (e^t - t)\mathbf{k}$$

Finally, add these vector functions:

$$(\mathbf{F}' \times \mathbf{G})(t) + (\mathbf{F} \times \mathbf{G}')(t) = [(3 - 2te^t) - t^2e^t]\mathbf{i} + [2t^2 + t^2]\mathbf{j} + [-t + e^t - t]\mathbf{k}$$
$$= (3 - 2te^t - t^2e^t)\mathbf{i} + (3t^2)\mathbf{j} + (e^t - 2t)\mathbf{k}$$

Thus, $(\mathbf{F} \times \mathbf{G})' = (\mathbf{F}' \times \mathbf{G}) + (\mathbf{F} \times \mathbf{G}')$. ∎

EXAMPLE 6   *Derivatives of vector function expressions*

Let $\mathbf{F}(t) = \mathbf{i} + e^t\mathbf{j} + t^2\mathbf{k}$ and $\mathbf{G}(t) = 3t^2\mathbf{i} + e^{-t}\mathbf{j} - 2t\mathbf{k}$. Find

a. $\dfrac{d}{dt}[2\mathbf{F}(t) + t^3\mathbf{G}(t)]$ b. $\dfrac{d}{dt}[\mathbf{F}(t) \cdot \mathbf{G}(t)]$

Solution   Begin by finding $\dfrac{d\mathbf{F}}{dt}$ and $\dfrac{d\mathbf{G}}{dt}$:

$$\frac{d\mathbf{F}}{dt} = e^t\mathbf{j} + 2t\mathbf{k} \quad \text{and} \quad \frac{d\mathbf{G}}{dt} = 6t\mathbf{i} - e^{-t}\mathbf{j} - 2\mathbf{k}$$

a. $\dfrac{d}{dt}[2\mathbf{F}(t) + t^3\mathbf{G}(t)] = 2\dfrac{d\mathbf{F}}{dt} + \left[t^3\dfrac{d\mathbf{G}}{dt} + \dfrac{d}{dt}(t^3)\,\mathbf{G}(t)\right]$

$= 2(e^t\mathbf{j} + 2t\mathbf{k}) + t^3(6t\mathbf{i} - e^{-t}\mathbf{j} - 2\mathbf{k}) + 3t^2(3t^2\mathbf{i} + e^{-t}\mathbf{j} - 2t\mathbf{k})$

$= (6t^4 + 9t^4)\mathbf{i} + (2e^t - t^3e^{-t} + 3t^2e^{-t})\mathbf{j} + (4t - 2t^3 - 6t^3)\mathbf{k}$

$= 15t^4\mathbf{i} + [2e^t + e^{-t}t^2(3 - t)]\mathbf{j} - 4t(2t^2 - 1)\mathbf{k}$

b. $\dfrac{d}{dt}[\mathbf{F}(t) \cdot \mathbf{G}(t)] = \dfrac{d\mathbf{F}}{dt} \cdot \mathbf{G}(t) + \mathbf{F}(t) \cdot \dfrac{d\mathbf{G}}{dt}$

$= (e^t\mathbf{j} + 2t\mathbf{k}) \cdot (3t^2\mathbf{i} + e^{-t}\mathbf{j} - 2t\mathbf{k}) + (\mathbf{i} + e^t\mathbf{j} + t^2\mathbf{k}) \cdot (6t\mathbf{i} - e^{-t}\mathbf{j} - 2\mathbf{k})$

$= [0(3t^2) + e^t(e^{-t}) + 2t(-2t)] + [1(6t) + e^t(-e^{-t}) + t^2(-2)]$

$= (1 - 4t^2) + (6t - 1 - 2t^2)$

$= -6t^2 + 6t$ ∎

We shall use Theorem 11.3 in both applied and theoretical problems. In the following theorem, we establish an important geometric property of vector functions.

THEOREM 11.4   *Orthogonality of a function of constant length and its derivative*

If the nonzero vector function $\mathbf{F}(t)$ is differentiable and has constant length, then $\mathbf{F}(t)$ is orthogonal to the derivative vector $\mathbf{F}'(t)$.

Proof   We are given that $\|\mathbf{F}(t)\| = r$ for some constant r and all t. To prove that $\mathbf{F}$ and $\mathbf{F}'$ are orthogonal, we shall show that $\mathbf{F}(t) \cdot \mathbf{F}'(t) = 0$ for all t. First, note that

$$r^2 = \|\mathbf{F}(t)\|^2 = \mathbf{F}(t) \cdot \mathbf{F}(t)$$

for all t. We take the derivative of both sides (remember that the derivative of r^2 is zero because it is a constant).

$$[r^2]' = [\mathbf{F}(t) \cdot \mathbf{F}(t)]'$$
$$0 = \mathbf{F}'(t) \cdot \mathbf{F}(t) + \mathbf{F}(t) \cdot \mathbf{F}'(t)$$
$$0 = 2\mathbf{F}(t) \cdot \mathbf{F}'(t)$$
$$0 = \mathbf{F}(t) \cdot \mathbf{F}'(t)$$

Thus, $\mathbf{F}$ and $\mathbf{F}'$ are orthogonal.   ▬

You may recall from plane geometry that the tangent line to a circle is always perpendicular to the radial line from the center of a circle to the point of tangency (Figure 11.9). Theorem 11.4 can be used to extend this property to spheres in $\mathbb{R}^3$. Suppose P is a point on the sphere

$$x^2 + y^2 + z^2 = r^2$$

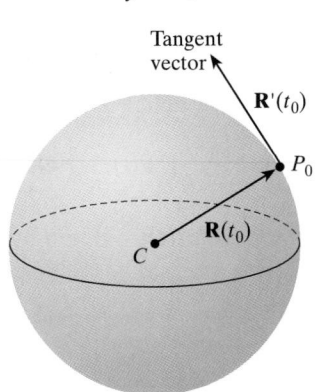

■ **FIGURE 11.9**   The tangent $\mathbf{R}'(t_0)$ at P_0 is orthogonal to the radius vector $\mathbf{R}(t_0)$ from the center of the sphere to P_0.

Let $\mathbf{R}(t) = x(t)\mathbf{i} + y(t)\mathbf{j} + z(t)\mathbf{k}$ be a vector function whose graph lies entirely on the surface of the sphere and contains the point P_0. Suppose P_0 corresponds to $t = t_0$. Then $\mathbf{R}(t_0)$ is the vector from the center of the sphere to P_0, and $\mathbf{R}'(t_0)$ is a tangent vector to the graph of $\mathbf{R}(t)$ at P_0, which means that $\mathbf{R}'(t_0)$ is also tangent to the sphere at P_0. Because $\mathbf{R}(t)$ has constant length,

$$\|\mathbf{R}(t)\| = \sqrt{x^2(t) + y^2(t) + z^2(t)} = r \quad \text{for all } t$$

it follows from Theorem 11.4 that $\mathbf{R}$ is orthogonal to $\mathbf{R}'$. Thus, the tangent vector $\mathbf{R}'(t_0)$ at P_0 is orthogonal to the vector $\mathbf{R}(t_0)$ from the center of the sphere to the point of tangency.

MODELING THE MOTION OF AN OBJECT IN SPACE

Recall from Chapter 3 that the derivative of an object's position with respect to time is the velocity, and the derivative of the velocity is the acceleration. We frequently know the acceleration and can use integration to find the velocity and the position. We will now express these concepts in terms of vector functions.

Position Vector, Velocity, and Acceleration

An object that moves in such a way that its position at time t is given by the vector function $\mathbf{R}(t)$ is said to have

Position vector $\mathbf{R}(t)$ and

Velocity $\mathbf{V} = \dfrac{d\mathbf{R}}{dt} = \mathbf{R}'(t)$

At any time t,

the **speed** is $\|\mathbf{V}\|$, the magnitude of velocity,

the **direction of motion** is $\dfrac{\mathbf{V}}{\|\mathbf{V}\|}$,   and

the **acceleration vector** is the derivative of the velocity:

$$\mathbf{A} = \frac{d\mathbf{V}}{dt} = \frac{d^2\mathbf{R}}{dt^2}$$

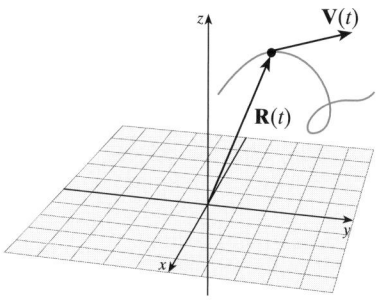

■ **FIGURE 11.10** The velocity vector is tangent to the trajectory of the object's motion.

The graph of the position vector $\mathbf{R}(t)$ is called the **trajectory** of the object's motion. According to the results obtained earlier in this section, the velocity $\mathbf{V}(t) = \mathbf{R}'(t)$ is a tangent vector to the trajectory at any point where $\mathbf{V}(t)$ exists and $\mathbf{V}(t) \neq 0$ (see Figure 11.10). The *direction of motion* is given by the unit vector $\mathbf{V}/\|\mathbf{V}\|$. Appropriately, whenever $\mathbf{V}(t) = \mathbf{0}$, the object is said to be **stationary.**

In practice, the position vector is often represented in the form

$$\mathbf{R}(t) = f_1(t)\mathbf{i} + f_2(t)\mathbf{j} + f_3(t)\mathbf{k}$$

and, in this case, the velocity and the acceleration vectors are given by

$$\mathbf{V}(t) = \mathbf{R}'(t) = f_1'(t)\mathbf{i} + f_2'(t)\mathbf{j} + f_3'(t)\mathbf{k}$$

and

$$\mathbf{A}(t) = \mathbf{V}'(t) = \mathbf{R}''(t) = f_1''(t)\mathbf{i} + f_2''(t)\mathbf{j} + f_3''(t)\mathbf{k}$$

EXAMPLE 7   *Speed and direction of a particle*

A particle's position at time t is determined by the vector

$$\mathbf{R}(t) = (\cos t)\mathbf{i} + (\sin t)\mathbf{j} + t^3\mathbf{k}$$

Analyze the particle's motion. In particular, find the particle's velocity, speed, acceleration, and direction of motion at time $t = 2$.

Solution $\mathbf{V} = \dfrac{d\mathbf{R}}{dt} = (-\sin t)\mathbf{i} + (\cos t)\mathbf{j} + 3t^2\mathbf{k}$

$$\mathbf{A} = \frac{d\mathbf{V}}{dt} = (-\cos t)\mathbf{i} - (\sin t)\mathbf{j} + 6t\mathbf{k}$$

The velocity at time $t = 2$ is $\mathbf{V}(2)$:

$$(-\sin 2)\mathbf{i} + (\cos 2)\mathbf{j} + 3(2)^2\mathbf{k} \approx -0.91\mathbf{i} - 0.42\mathbf{j} + 12\mathbf{k}$$

The acceleration at $t = 2$ is

$$(-\cos 2)\mathbf{i} - (\sin 2)\mathbf{j} + 6(2)\mathbf{k} \approx 0.42\mathbf{i} - 0.91\mathbf{j} + 12\mathbf{k}$$

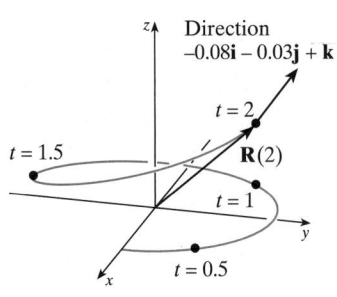

Direction
$-0.08\mathbf{i} - 0.03\mathbf{j} + \mathbf{k}$

$t = 2$

$t = 1.5$

$\mathbf{R}(2)$

$t = 1$

$t = 0.5$

WARNING Notice $\mathbf{V} = \|\mathbf{V}\| \cdot \dfrac{\mathbf{V}}{\|\mathbf{V}\|}$

$= (\text{SPEED}) \cdot (\text{DIRECTION}).$

The speed is

$$\|\mathbf{V}\| = \sqrt{(-\sin t)^2 + (\cos t)^2 + (3t^2)^2}$$
$$= \sqrt{1 + 9t^4}$$

At time $t = 2$, the speed is $\|\mathbf{V}(2)\| = \sqrt{1 + 144} \approx 12.04$.

The direction of motion is

$$\frac{\mathbf{V}}{\|\mathbf{V}\|} = \frac{1}{\sqrt{145}}[(-\sin t)\mathbf{i} + (\cos t)\mathbf{j} + 3t^2\mathbf{k}]$$

At time $t = 2$, the direction is $\mathbf{V}(2)/\|\mathbf{V}(2)\|$:

$$\frac{1}{\sqrt{145}}[(-\sin 2)\mathbf{i} + (\cos 2)\mathbf{j} + 12\mathbf{k}] \approx -0.08\mathbf{i} - 0.03\mathbf{j} + \mathbf{k} \qquad \blacksquare$$

Technology Window

To represent a velocity vector, it may seem that we could use the *nDer* feature available on many calculators. If we attempt to take the derivative of a position vector with respect to a parameter t, we find that calculator feature does not seem to accept a vector-valued function. However, we can use the *evalF* feature to define our own numerical derivative using the following formula:

$$\mathbf{V} = \frac{\mathbf{P}(t + \delta) - \mathbf{P}(t - \delta)}{2\delta} \text{ after defining a position vector } \mathbf{P} = [x, y, z]$$

That is, input: $\mathbf{V} = (\text{evalF}(P, t, t + \delta) - \text{evalF}(P, t, t - \delta))/(2\delta)$ where the Greek letter δ is a tolerance variable available on many models of calculator. A tolerance of $\delta = 0.001$ seems to work well in most applications.

Once we have defined a velocity vector, we can easily define other vectors:

Speed is $\|\mathbf{V}\|$, so we use a calculator by: $\mathbf{S} = \text{norm}(\mathbf{V})$.

Direction of motion is $\dfrac{\mathbf{V}}{\|\mathbf{V}\|}$, so we use a calculator by: $\mathbf{T} = \text{unitV}(\mathbf{V})$

VECTOR INTEGRALS

Since vector limits and derivatives are performed in a componentwise fashion, it should come as no surprise that vector integration is performed in the same way.

Indefinite Integral and Definite Integral

Let $\mathbf{F}(t) = f_1(t)\mathbf{i} + f_2(t)\mathbf{j} + f_3(t)\mathbf{k}$, where $f_1, f_2,$ and f_3 are continuous on the closed interval $a \le t \le b$. Then the **indefinite integral** of $\mathbf{F}(t)$ is the vector function

$$\int \mathbf{F}(t)\,dt = \left[\int f_1(t)\,dt\right]\mathbf{i} + \left[\int f_2(t)\,dt\right]\mathbf{j} + \left[\int f_3(t)\,dt\right]\mathbf{k} + \mathbf{C}$$

where $\mathbf{C}$ is an arbitrary constant vector. The **definite integral** of $\mathbf{F}(t)$ on $a \le t \le b$ is the vector

$$\int_a^b \mathbf{F}(t)\,dt = \left[\int_a^b f_1(t)\,dt\right]\mathbf{i} + \left[\int_a^b f_2(t)\,dt\right]\mathbf{j} + \left[\int_a^b f_3(t)\,dt\right]\mathbf{k}$$

EXAMPLE 8   *Integral of a vector function*

Find $\displaystyle\int_0^\pi [t\mathbf{i} + 3\mathbf{j} - (\sin t)\mathbf{k}]\, dt$.

Solution $\displaystyle\int_0^\pi [t\mathbf{i} + 3\mathbf{j} - (\sin t)\mathbf{k}]\, dt = \left[\int_0^\pi t\, dt\right]\mathbf{i} + \left[\int_0^\pi 3\, dt\right]\mathbf{j} - \left[\int_0^\pi \sin t\, dt\right]\mathbf{k}$

$$= \left(\frac{t^2}{2}\right)\Big|_0^\pi \mathbf{i} + (3t)\big|_0^\pi \mathbf{j} - (-\cos t)\big|_0^\pi \mathbf{k}$$

$$= \tfrac{1}{2}\pi^2\mathbf{i} + 3\pi\mathbf{j} + (\cos \pi - \cos 0)]\mathbf{k}$$

$$= \tfrac{1}{2}\pi^2\mathbf{i} + 3\pi\mathbf{j} - 2\mathbf{k}$$

EXAMPLE 9   *Position of a particle given its velocity*

The velocity of a particle moving in space is

$$\mathbf{V}(t) = e^t\mathbf{i} + t^2\mathbf{j} + (\cos 2t)\mathbf{k}$$

Find the particle's position as a function of t if $\mathbf{R}(0) = 2\mathbf{i} + \mathbf{j} - \mathbf{k}$.

Solution   We need to solve the initial value problem that consists of:

The differential equation $\mathbf{V}(t) = \dfrac{d\mathbf{R}}{dt} = e^t\mathbf{i} + t^2\mathbf{j} + (\cos 2t)\mathbf{k}$

The initial condition $\mathbf{R}(0) = 2\mathbf{i} + \mathbf{j} - \mathbf{k}$

We integrate both sides of the differential equation with respect to t:

$$\int d\mathbf{R} = \left[\int e^t\, dt\right]\mathbf{i} + \left[\int t^2\, dt\right]\mathbf{j} + \left[\int \cos 2t\, dt\right]\mathbf{k}$$

$$\mathbf{R}(t) = (e^t + C_1)\mathbf{i} + \left(\frac{1}{3}t^3 + C_2\right)\mathbf{j} + \left(\frac{1}{2}\sin 2t + C_3\right)\mathbf{k}$$

$$= e^t\mathbf{i} + \frac{1}{3}t^3\mathbf{j} + \left(\frac{1}{2}\sin 2t\right)\mathbf{k} + \underbrace{C_1\mathbf{i} + C_2\mathbf{j} + C_3\mathbf{k}}_{\mathbf{C}}$$

Now use the initial condition to find $\mathbf{C}$:

$$\underbrace{2\mathbf{i} + \mathbf{j} - \mathbf{k}}_{\mathbf{R}(0)} = e^0\mathbf{i} + \frac{1}{3}(0)^3\mathbf{j} + \left(\frac{1}{2}\sin 0\right)\mathbf{k} + \mathbf{C} = \mathbf{i} + \mathbf{C}$$

so

$$\mathbf{i} + \mathbf{j} - \mathbf{k} = \mathbf{C}$$

Thus, the particle's position at any time t is

$$\mathbf{R}(t) = e^t\mathbf{i} + \frac{1}{3}t^3\mathbf{j} + \left(\frac{1}{2}\sin 2t\right)\mathbf{k} + \mathbf{i} + \mathbf{j} - \mathbf{k}$$

$$= (e^t + 1)\mathbf{i} + \left(\frac{1}{3}t^3 + 1\right)\mathbf{j} + \left(\frac{1}{2}\sin 2t - 1\right)\mathbf{k}$$

11.2 Problem Set

Ⓐ *Find the vector derivative* $\mathbf{F}'$ *in Problems 1–4.*

1. $\mathbf{F}(t) = t\mathbf{i} + t^2\mathbf{j} + (t + t^3)\mathbf{k}$
2. $\mathbf{F}(s) = (s\mathbf{i} + s^2\mathbf{j} + s^2\mathbf{k}) + (2s^2\mathbf{i} - s\mathbf{j} + 3\mathbf{k})$
3. $\mathbf{F}(s) = (\ln s)[s\mathbf{i} + 5\mathbf{j} - e^s\mathbf{k}]$
4. $\mathbf{F}(\theta) = (\cos\theta)[\mathbf{i} + (\tan\theta)\mathbf{j} + 3\mathbf{k}]$

Find $\mathbf{F}'$ *and* $\mathbf{F}''$ *for the vector functions given in Problems 5–8.*

5. $\mathbf{F}(t) = t^2\mathbf{i} + t^{-1}\mathbf{j} + e^{2t}\mathbf{k}$
6. $\mathbf{F}(s) = (1 - 2s^2)\mathbf{i} + (s\cos s)\mathbf{j} - s\mathbf{k}$
7. $\mathbf{F}(s) = (\sin s)\mathbf{i} + (\cos s)\mathbf{j} + s^2\mathbf{k}$
8. $\mathbf{F}(\theta) = (\sin^2\theta)\mathbf{i} + (\cos 2\theta)\mathbf{j} + \theta^2\mathbf{k}$

Differentiate the scalar functions in Problems 9–12.

9. $f(x) = [x\mathbf{i} + (x + 1)\mathbf{j}] \cdot [(2x)\mathbf{i} - (3x^2)\mathbf{j}]$
10. $f(x) = \langle \cos x, x, -x \rangle \cdot \langle \sec x, -x^2, 2x \rangle$
11. $g(x) = \|\langle \sin x, -2x, \cos x \rangle\|$
12. $f(x) = \|[x\mathbf{i} + x^2\mathbf{j} - 2\mathbf{k}] + [(1 - x)\mathbf{i} - e^x\mathbf{j}]\|$

In Problems 13–18, $\mathbf{R}$ *is the position vector for a particle in space at time t. Find the particle's velocity and acceleration vector and then find the speed and direction of motion for the given value of t.*

13. $\mathbf{R}(t) = t\mathbf{i} + t^2\mathbf{j} + 2t\mathbf{k}$ at $t = 1$
14. $\mathbf{R}(t) = (1 - 2t)\mathbf{i} - t^2\mathbf{j} + e^t\mathbf{k}$ at $t = 0$
15. $\mathbf{R}(t) = (\cos t)\mathbf{i} + (\sin t)\mathbf{j} + 3t\mathbf{k}$ at $t = \frac{\pi}{4}$
16. $\mathbf{R}(t) = (2\cos t)\mathbf{i} + t^2\mathbf{j} + (2\sin t)\mathbf{k}$ at $t = \frac{\pi}{2}$
17. $\mathbf{R}(t) = e^t\mathbf{i} + e^{-t}\mathbf{j} + e^{2t}\mathbf{k}$ at $t = \ln 2$
18. $\mathbf{R}(t) = (\ln t)\mathbf{i} + \frac{1}{2}t^3\mathbf{j} - t\mathbf{k}$ at $t = 1$

Find the tangent vector to the graph of the given vector function $\mathbf{F}$ *at the points indicated in Problems 19–24.*

19. $\mathbf{F}(t) = t^2\mathbf{i} + 2t\mathbf{j} + (t^3 + t^2)\mathbf{k}$;
    $t = 0, t = 1, t = -1$
20. $\mathbf{F}(t) = t^{-3}\mathbf{i} + t^{-2}\mathbf{j} + t^{-1}\mathbf{k}$;
    $t = 1, t = -1$
21. $\mathbf{F}(t) = \dfrac{t\mathbf{i} + t^2\mathbf{j} + t^3\mathbf{k}}{1 + 2t}$; $t = 0, t = 2$
22. $\mathbf{F}(t) = t^2\mathbf{i} + (\cos t)\mathbf{j} + (t^2\cos t)\mathbf{k}$;
    $t = 0, t = \frac{\pi}{2}$
23. $\mathbf{F}(t) = (\sin t)\mathbf{i} + (\cos t)\mathbf{j} + at\mathbf{k}$;
    $t = \frac{\pi}{2}, t = \pi$
24. $\mathbf{F}(t) = (e^t\sin \pi t)\mathbf{i} + (e^t\cos \pi t)\mathbf{j} + (\sin \pi t + \cos \pi t)\mathbf{k}$;
    $t = 0, t = 1, t = 2$

Find the indefinite vector integrals in Problems 25–30.

25. $\int[t\mathbf{i} - e^{3t}\mathbf{j} + 3\mathbf{k}]\,dt$
26. $\int[(\cos t)\mathbf{i} + (\sin t)\mathbf{j} - (2t)\mathbf{k}]\,dt$
27. $\int[(\ln t)\mathbf{i} - t\mathbf{j} + 3\mathbf{k}]\,dt$
28. $\int e^{-t}[3\mathbf{i} + t\mathbf{j} + (\sin t)\mathbf{k}]\,dt$
29. $\int[(t\ln t)\mathbf{i} - \sin(1 - t)\mathbf{j} + t\mathbf{k}]\,dt$
30. $\int[(\sinh t)\mathbf{i} - 3\mathbf{j} + (\cosh t)\mathbf{k}]\,dt$

Find the position vector $\mathbf{R}(t)$ *given the velocity* $\mathbf{V}(t)$ *and the initial position* $\mathbf{R}(0)$ *in Problems 31–34.*

31. $\mathbf{V}(t) = t^2\mathbf{i} - e^{2t}\mathbf{j} + \sqrt{t}\,\mathbf{k}; \mathbf{R}(0) = \mathbf{i} + 4\mathbf{j} - \mathbf{k}$
32. $\mathbf{V}(t) = t\mathbf{i} - \sqrt[3]{t}\,\mathbf{j} + e^t\mathbf{k}; \mathbf{R}(0) = \mathbf{i} - 2\mathbf{j} + \mathbf{k}$
33. $\mathbf{V}(t) = 2\sqrt{t}\,\mathbf{i} + (\cos t)\mathbf{j}; \mathbf{R}(0) = \mathbf{i} + \mathbf{j}$
34. $\mathbf{V}(t) = -3t\mathbf{i} + (\sin^2 t)\mathbf{j} + (\cos^2 t)\mathbf{k}; \mathbf{R}(0) = \mathbf{j}$

Find the position vector $\mathbf{R}(t)$ *and* $\mathbf{V}(t)$ *given the acceleration* $\mathbf{A}(t)$ *and initial position and velocity* $\mathbf{R}(0)$ *and* $\mathbf{V}(t)$ *in Problems 35–36.*

35. $\mathbf{A}(t) = (\cos t)\mathbf{i} - (t\sin t)\mathbf{k}$;
    $\mathbf{R}(0) = \mathbf{i} - 2\mathbf{j} + \mathbf{k}; \mathbf{V}(0) = 2\mathbf{i} + 3\mathbf{k}$
36. $\mathbf{A}(t) = t^2\mathbf{i} - 2\sqrt{t}\,\mathbf{j} + e^{3t}\mathbf{k}$;
    $\mathbf{R}(0) = 2\mathbf{i} + \mathbf{j} - \mathbf{k}; \mathbf{V}(0) = \mathbf{i} - \mathbf{j} - 2\mathbf{k}$
37. The velocity of a particle moving in space is

    $$\mathbf{V}(t) = e^t\mathbf{i} + t^2\mathbf{j}$$

    Find the particle's position as a function of t if $\mathbf{R}(0) = \mathbf{i} - \mathbf{j}$.
38. The velocity of a particle moving in space is

    $$\mathbf{V}(t) = (t\cos t)\mathbf{i} + e^{2t}\mathbf{k}$$

    Find the particle's position as a function of t if $\mathbf{R}(0) = \mathbf{i} + 2\mathbf{k}$.
39. The acceleration of a particle moving in space is

    $$\mathbf{A}(t) = 24t^2\mathbf{i} + 4\mathbf{j}$$

    Find the particle's position as a function of t if $\mathbf{R}(0) = \mathbf{i} + 2\mathbf{j}$ and $\mathbf{V}(0) = \mathbf{0}$.

Ⓑ 40. ■ **What Does This Say?** Describe the concept of a smooth curve.

41. ■ **What Does This Say?** Compare or contrast the notions of speed and velocity of a vector function.

Let $\mathbf{v} = 2\mathbf{i} - \mathbf{j} + 5\mathbf{k}$ *and* $\mathbf{w} = \mathbf{i} + 2\mathbf{j} - 3\mathbf{k}$. *Find the derivatives in Problems 42–45.*

42. $\dfrac{d}{dt}(\mathbf{v} + t\mathbf{w})$

43. $\dfrac{d^2}{dt^2}(\mathbf{v} \cdot t^4\mathbf{w})$

44. $\dfrac{d^2}{dt^2}(t\|\mathbf{v}\| + t^2\|\mathbf{w}\|)$

45. $\dfrac{d}{dt}(t\mathbf{v} \times t^2\mathbf{w})$

In Problems 46–47, verify the indicated equation for the vector functions

$$\mathbf{F}(t) = (3 + t^2)\mathbf{i} - (\cos 3t)\mathbf{j} + t^{-1}\mathbf{k} \quad and$$
$$\mathbf{G}(t) = \sin(2 - t)\mathbf{i} - e^{2t}\mathbf{k}$$

46. $(3\mathbf{F} - 2\mathbf{G})'(t) = 3\mathbf{F}'(t) - 2\mathbf{G}'(t)$

47. $(\mathbf{F} \cdot \mathbf{G})'(t) = (\mathbf{F}' \cdot \mathbf{G})(t) + (\mathbf{F} \cdot \mathbf{G}')(t)$

In Problems 48–49, find a value of a *that satisfies the given equation.*

48. $\int_0^a \left[(t\sqrt{1 + t^2})\mathbf{i} + \left(\frac{1}{1 + t^2}\right)\mathbf{j} \right] dt$

$$= \tfrac{1}{3}(2\sqrt{2} - 1)\mathbf{i} + \tfrac{\pi}{4}\mathbf{j}$$

49. $\int_0^{2a} [(\cos t)\mathbf{i} + (\sin t)\mathbf{j} + (\sin t \cos t)\mathbf{k}] \, dt$

$$= \mathbf{i} + \mathbf{j} + \tfrac{1}{2}\mathbf{k}$$

In Problems 50–51, show that $\mathbf{F}(t)$ *and* $\mathbf{F}''(t)$ *are parallel for all* t *with constant k.*

50. $\mathbf{F}(t) = e^{kt}\mathbf{i} + e^{-kt}\mathbf{j}$

C 51. $\mathbf{F}(t) = (\cos kt)\mathbf{i} + (\sin kt)\mathbf{j}$

52. Let $\mathbf{F}(t) = u(t)\mathbf{i} + v(t)\mathbf{j} + w(t)\mathbf{k}$, where u, v, and w are differentiable scalar functions of t. Show that

$$\mathbf{F}(t) \cdot \mathbf{F}'(t) = \|\mathbf{F}(t)\|(\|\mathbf{F}(t)\|)'$$

53. Let $\mathbf{F}(t)$ be a differentiable vector function and let $h(t)$ be a differentiable scalar function of t. Show that

$$[h(t)\mathbf{F}(t)]' = h(t)\mathbf{F}'(t) + h'(t)\mathbf{F}(t)$$

54. If $\mathbf{F}$ and $\mathbf{G}$ are differentiable vector functions of t, prove

$$(\mathbf{F} \cdot \mathbf{G})'(t) = (\mathbf{F}' \cdot \mathbf{G})(t) + (\mathbf{F} \cdot \mathbf{G}')(t)$$

55. If $\mathbf{F}$ and $\mathbf{G}$ are differentiable vector functions of t, prove

$$(\mathbf{F} \times \mathbf{G})'(t) = (\mathbf{F}' \times \mathbf{G})(t) + (\mathbf{F} \times \mathbf{G}')(t)$$

56. If $\mathbf{F}$ is a differentiable vector function such that $\mathbf{F}(t) \neq \mathbf{0}$, show that

$$\frac{d}{dt}\left(\frac{\mathbf{F}(t)}{\|\mathbf{F}(t)\|}\right) = \frac{\mathbf{F}'(t)}{\|\mathbf{F}(t)\|} - \frac{[\mathbf{F}(t) \cdot \mathbf{F}'(t)]\mathbf{F}(t)}{\|\mathbf{F}(t)\|^3}$$

57. Show that

$$\frac{d}{dt}\|\mathbf{R}(t)\| = \frac{1}{\|\mathbf{R}\|}\mathbf{R} \cdot \frac{d\mathbf{R}}{dt}$$

58. Find a formula for

$$\frac{d}{dt}[\mathbf{F} \cdot (\mathbf{G} \times \mathbf{H})]$$

59. Find a formula for

$$\frac{d}{dt}[\mathbf{F} \times (\mathbf{G} \times \mathbf{H})]$$

60. If $\mathbf{G} = \mathbf{F} \cdot (\mathbf{F}' \times \mathbf{F}'')$, what is $\mathbf{G}'$?

11.3 *Modeling Ballistics and Planetary Motion*

■ **IN THIS SECTION**   modeling the motion of a projectile in a vacuum, Kepler's second law   ■

MODELING THE MOTION OF A PROJECTILE IN A VACUUM

In general, it can be quite difficult to analyze the motion of a projectile, but the problem becomes manageable if we assume that the acceleration due to gravity is constant and that the projectile travels in a vacuum. A model of the motion in the real world (with air resistance, for example) based on these assumptions is fairly realistic as long as the projectile is reasonably heavy, travels at relatively low speed, and stays close to the surface of the earth.

Figure 11.11 shows a projectile that is fired from a point P and travels in a vertical plane coordinatized so that P is directly above the origin O and the impact point I is on the x-axis at ground level. We let s_0 be the height of P above O, v_0 be the initial speed of the projectile (the **muzzle speed**), and α be the angle of elevation of the "gun" that fires the projectile.

Because our projectile is to travel in a vacuum, we shall assume that the only force acting on the projectile at any given time is the force due to gravity. This force $\mathbf{F}$ is directed downward. The magnitude of this force is the projectile's weight. If m is the mass of the projectile and g is the "free-fall" acceleration due to gravity (g is approximately 32 ft/s² or 9.8 m/s²), then the magnitude of $\mathbf{F}$ is mg. Also, according to **Newton's second law of motion,** the sum of all the forces acting on the projectile must

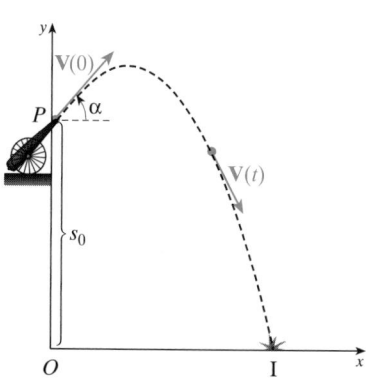

■ **FIGURE 11.11**   The motion of a projectile

equal $m\mathbf{A}(t)$, where $\mathbf{A}(t)$ is the acceleration of the projectile at time t. Because $\mathbf{F} = -mg\mathbf{j}$ is the only force acting on the projectile, it follows that

$$-mg\mathbf{j} = \mathbf{F} = m\mathbf{A}(t)$$
$$\mathbf{A}(t) = -g\mathbf{j} \quad \text{for all } t \geq 0$$

The velocity $\mathbf{V}(t)$ of the projectile can now be obtained by integrating $\mathbf{A}(t)$. Specifically, we have

$$\mathbf{V}(t) = \int \mathbf{A}(t)\, dt = \int (-g\mathbf{j})\, dt = -gt\mathbf{j} + \mathbf{C}_1$$

where $\mathbf{C}_1 = \mathbf{V}(0)$ is the velocity when $t = 0$ (that is, the *initial velocity*). Because the projectile is fired with initial speed v_0 at an angle of elevation α, the initial velocity must be

$$\mathbf{C}_1 = \mathbf{V}(0) = (v_0\cos \alpha)\mathbf{i} + (v_0\sin \alpha)\mathbf{j}$$

(as shown in the detail of Figure 11.11), and by substitution we find

$$\mathbf{V}(t) = -gt\mathbf{j} + (v_0\cos \alpha)\mathbf{i} + (v_0\sin \alpha)\mathbf{j}$$
$$= (v_0\cos \alpha)\mathbf{i} + (v_0\sin \alpha - gt)\mathbf{j}$$

Next, by integrating $\mathbf{V}$ with respect to t, we find that the projectile has displacement

$$\mathbf{R}(t) = \int \mathbf{V}(t)\, dt = \int [(v_0\cos \alpha)\mathbf{i} + (v_0\sin \alpha - gt)\mathbf{j}]\, dt$$
$$= (v_0\cos \alpha)t\mathbf{i} + [(v_0\sin \alpha)t - \tfrac{1}{2}gt^2]\mathbf{j} + \mathbf{C}_2$$

where $\mathbf{C}_2 = \mathbf{R}(0)$ is the position at time $t = 0$. Because the projectile begins its flight at P, the initial position is $\mathbf{R}(0) = s_0\mathbf{j}$, and by substituting this expression for $\mathbf{C}_2$, we find that the position at time t is

$$\mathbf{R}(t) = [(v_0\cos \alpha)t]\mathbf{i} + [(v_0\sin \alpha)t - \tfrac{1}{2}gt^2 + s_0]\mathbf{j}$$

Specifically, $\mathbf{R}(t)$ gives the position of the projectile at any time t after it is fired and before it hits the ground. If we let $x(t)$ and $y(t)$ denote the horizontal and vertical components of $\mathbf{R}(t)$, respectively, we can make the following statement.

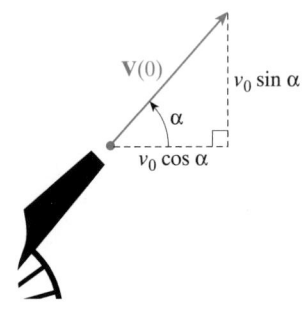

The initial velocity is
$\mathbf{v}(0) = (v_0\cos \alpha)\mathbf{i} + (v_0\sin \alpha)\mathbf{j}$.

Detail of Figure 11.11

Motion of a Projectile in a Vacuum

Consider a projectile that travels in a vacuum in a coordinate plane, with the x-axis along level ground. If the projectile is fired from a height of s_0 with initial speed v_0 and angle of elevation α, then at time $t(t \geq 0)$ it will be at the point $(x(t), y(t))$, where

$$x(t) = (v_0\cos \alpha)t \quad \text{and} \quad y(t) = -\tfrac{1}{2}gt^2 + (v_0\sin \alpha)t + s_0$$

The parametric equations for the motion of a projectile in a vacuum provide useful general information about a projectile's motion. For instance, note that if $\alpha \neq 90°$, we can eliminate the parameter t by solving the first equation, obtaining

$$t = \frac{x}{v_0\cos \alpha}$$

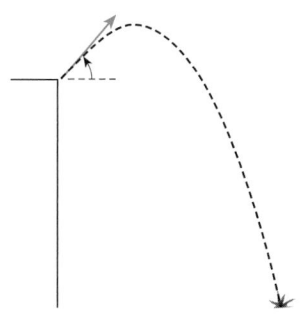

FIGURE 11.12 The motion of a projectile. The trajectory is part of a parabolic arc

By substituting into the second equation, we find

$$y = -\frac{1}{2}g\left[\frac{x}{v_0 \cos \alpha}\right]^2 + (v_0 \sin \alpha)\left[\frac{x}{v_0 \cos \alpha}\right] + s_0$$

$$= \left[\frac{-g}{2(v_0 \cos \alpha)^2}\right]x^2 + (\tan \alpha)x + s_0$$

This is the Cartesian equation for the trajectory of the projectile, and because the equation has the general form $y = ax^2 + bx + c$, with $a < 0$, the trajectory must be part of a downward-opening parabola. That is, *if a projectile is not fired vertically (so that* $\tan \alpha$ *is defined), its trajectory will be part of a downward-opening parabolic arc,* as shown in Figure 11.12.

The *time of flight* T_f of a projectile is the elapsed time between launch and impact, and the *range* is the total horizontal distance R traveled by the projectile during its flight. Because $y = 0$ at impact, it follows that T_f must satisfy the quadratic equation

$$-\tfrac{1}{2}gT_f^2 + (v_0 \sin \alpha)T_f + s_0 = 0$$

Then, because the flight begins at $x = 0$ when $t = 0$ and ends at $x = R_f$ when $t = T_f$, the range R_f can be computed by evaluating $x(t)$ at $t = T_f$. That is,

$$R_f = (v_0 \cos \alpha)T_f$$

EXAMPLE 1   *Modeling the path of a projectile*

A boy standing at the edge of a cliff throws a ball upward at a 30° angle with an initial speed of 64 ft/s. Suppose that when the ball leaves the boy's hand, it is 48 ft above the ground at the base of the cliff.

a. What are the time of flight of the ball and its range?
b. What are the velocity of the ball and its speed at impact?
c. What is the highest point reached by the ball during its flight?

Solution   We know that $g = 32$ ft/s^2, and we are given $s_0 = 48$ ft, $v_0 = 64$ ft/s, and $\alpha = 30°$. We can now write the parametric equations for the parabola:

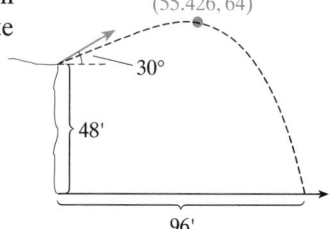

$$x(t) = (v_0 \cos \alpha)t = (64 \cos 30°)t = 32\sqrt{3}\, t$$
$$y(t) = -\tfrac{1}{2}gt^2 + (v_0 \sin \alpha)t + s_0$$
$$= -\tfrac{1}{2}(32)t^2 + (64 \sin 30°)t + 48$$
$$= -16t^2 + 32t + 48$$

a. The ball hits the ground when $y = 0$, and by solving the equation $-16t^2 + 32t + 48 = 0$ for $t \geq 0$, we find that $t = 3$, so the time of flight is $T_f = 3$ seconds. The range is

$$x(3) = 32\sqrt{3}(3) \approx 166.27688$$

That is, the ball hits the ground about 166 ft from the base of the cliff.

b. We find that $x'(t) = 32\sqrt{3}$ and $y'(t) = -32t + 32$, so the velocity at time t is

$$\mathbf{V}(t) = 32\sqrt{3}\,\mathbf{i} + (-32t + 32)\mathbf{j}$$

Thus, at impact (when $t = 3$), the velocity is $\mathbf{V}(3) = 32\sqrt{3}\,\mathbf{i} - 64\,\mathbf{j}$, and its speed is

$$\|\mathbf{V}(3)\| = \sqrt{(32\sqrt{3})^2 + (-64)^2} \approx 84.664042$$

That is, the speed at impact is about 85 ft/s.

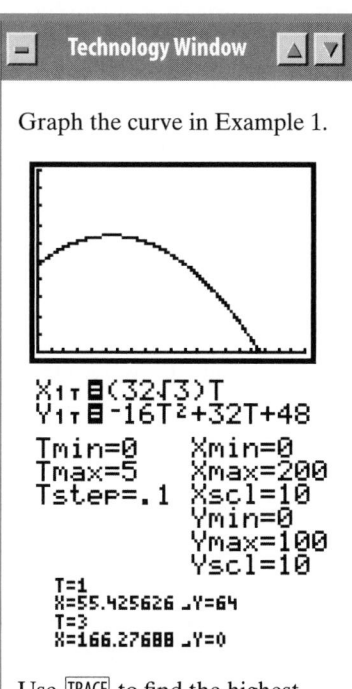

Graph the curve in Example 1.

```
X1т ☐(32√3)T
Y1т ☐-16T²+32T+48

Tmin=0      Xmin=0
Tmax=5      Xmax=200
Tstep=.1    Xscl=10
            Ymin=0
            Ymax=100
            Yscl=10

T=1
X=55.425626 .Y=64
T=3
X=166.27688 .Y=0
```

Use TRACE to find the highest point as well as the range and time of flight.

c. The ball attains its maximum height when the upward (vertical) component of its velocity $\mathbf{V}(t)$ is 0—that is, when $y'(t) = 0$. Solving the equation $-32t + 32 = 0$, we find that this occurs when $t = 1$. Therefore, the maximum height attained by the ball is

$$y_m = y(1) = -16(1)^2 + 32(1) + 48 = 64$$
$$x_m = x(1) = 32\sqrt{3}(1) \approx 55.425626$$

and the highest point reached by the ball has coordinates (rounded to the nearest unit) of $(55, 64)$. ∎

In general, if the projectile is fired from ground level (so $s_0 = 0$), the time of flight T_f satisfies

$$-\tfrac{1}{2}gT_f^2 + (v_0 \sin \alpha)T_f = 0$$
$$T_f = \frac{2}{g}v_0 \sin \alpha$$

Thus, the range R_f is

$$R_f = x(T_f) = (v_0 \cos \alpha)\left(\frac{2v_0 \sin \alpha}{g}\right) = \frac{v_0^2}{g}(2 \sin \alpha \cos \alpha) = \frac{v_0^2}{g}\sin 2\alpha$$

Notice that for a given initial speed v_0, the range assumes its largest value when $\sin 2\alpha = 1$, that is, when $\alpha = 45° = \frac{\pi}{4}$. To summarize:

Time of Flight and Range When Fired from Ground Level

A projectile fired from *ground level* has **time of flight** T_f and **range** R_f given by the equations

$$T_f = \frac{2}{g}v_0 \sin \alpha \quad \text{and} \quad R_f = \frac{v_0^2}{g}\sin 2\alpha$$

The maximal range is $R_m = \dfrac{v_0^2}{g}$, and it occurs when $\alpha = \dfrac{\pi}{4}$.

EXAMPLE 2   *Flight time and range for the motion of a projectile*

A projectile is fired from ground level at an angle of $40°$ with muzzle speed 110 ft/s. Find the time of flight and the range.

Solution   With $g = 32$ ft/s^2, we see that the flight time T_f is

$$T_f = \frac{2}{g}v_0 \sin \alpha = \frac{2}{32}(110)(\sin 40°) \approx 4.4191648$$

The range R_m is

$$R_m = \frac{v_0^2}{g}\sin 2\alpha = \frac{(110)^2}{32}(\sin 80°) \approx 372.38043$$

That is, the projectile travels about 372 ft horizontally before it hits the ground, and the flight takes a little more than 4 sec. ∎

KEPLER'S SECOND LAW

In the 17th century, the German astronomer Johannes Kepler (1571–1630) formulated three useful laws for describing planetary motion. The guest essay at the end of this chapter discusses the discovery of these laws.

Kepler's Laws

1. The planets move about the sun in elliptical orbits, with the sun at one focus.

2. The radius vector joining a planet to the sun sweeps over equal areas in equal intervals of time.

3. The square of the time of one complete revolution of a planet about its orbit is proportional to the cube of the orbit's semimajor axis.

We will use vector methods to prove Kepler's second law. The other two laws can be established similarly.

We begin by introducing some useful notation involving polar coordinates. Let $\mathbf{u}_r$ and $\mathbf{u}_\theta$ denote unit vectors along the radial axis and orthogonal to that axis, respectively, as shown in Figure 11.13. Then, in terms of the unit Cartesian vectors $\mathbf{i}$ and $\mathbf{j}$, we have

$$\mathbf{u}_r = (\cos \theta)\mathbf{i} + (\sin \theta)\mathbf{j} \quad \text{and} \quad \mathbf{u}_\theta = (-\sin \theta)\mathbf{i} + (\cos \theta)\mathbf{j}$$

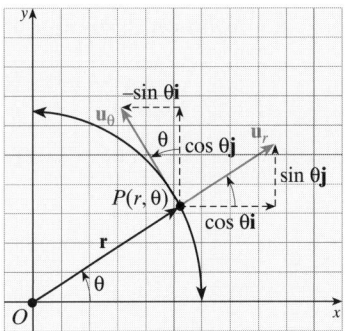

■ **FIGURE 11.13** Describing the motion of a particle along a curve

The derivatives $\dfrac{d\mathbf{u}_r}{d\theta}$ and $\dfrac{d\mathbf{u}_\theta}{d\theta}$ satisfy

$$\frac{d\mathbf{u}_r}{d\theta} = (-\sin \theta)\mathbf{i} + (\cos \theta)\mathbf{j} = \mathbf{u}_\theta$$

$$\frac{d\mathbf{u}_\theta}{d\theta} = (-\cos \theta)\mathbf{i} + (-\sin \theta)\mathbf{j} = -\mathbf{u}_r$$

Now, suppose the sun S is at the origin (pole) of a polar coordinate system, and consider the motion of a body B (planet, comet, artificial satellite) about S. The radial vector $\mathbf{R} = \mathbf{SB}$ can be expressed as

$$\mathbf{R} = r\mathbf{u}_r = (r \cos \theta)\mathbf{i} + (r \sin \theta)\mathbf{j}$$

To date, there are more than 4,000 earth satellites.

where $r = \|\mathbf{R}\|$, and the velocity $\mathbf{V}$ satisfies

$$\begin{aligned}
\mathbf{V} = \frac{d\mathbf{R}}{dt} &= \frac{dr}{dt}\mathbf{u}_r + r\frac{d\mathbf{u}_r}{dt} & \textit{Derivative of a scalar multiple} \\
&= \frac{dr}{dt}\mathbf{u}_r + r\frac{d\mathbf{u}_r}{d\theta}\frac{d\theta}{dt} & \textit{Chain rule} \\
&= \frac{dr}{dt}\mathbf{u}_r + r\frac{d\theta}{dt}\mathbf{u}_\theta
\end{aligned}$$

You can find a similar formula for acceleration (see Problem 40). We summarize these formulas in the following box.

> **Polar Formulas for Velocity and Acceleration**
>
> $$\mathbf{V}(t) = \frac{dr}{dt}\mathbf{u}_r + r\frac{d\theta}{dt}\mathbf{u}_\theta$$
>
> $$\mathbf{A}(t) = \frac{d\mathbf{V}(t)}{dt^2} = \frac{d^2\mathbf{R}}{dt} = \left[\frac{d^2r}{dt^2} - r\left(\frac{d\theta}{dt}\right)^2\right]\mathbf{u}_r + \left[r\frac{d^2\theta}{dt^2} + 2\frac{dr}{dt}\frac{d\theta}{dt}\right]\mathbf{u}_\theta$$

EXAMPLE 3   *Find the velocity of a moving body*

The position vector of a moving body is $\mathbf{R}(t) = 2t\mathbf{i} - t^2\mathbf{j}$ for $t \geq 0$. Express $\mathbf{R}$ and the velocity vector $\mathbf{V}(t)$ in terms of $\mathbf{u}_r$ and $\mathbf{u}_\theta$.

Solution   We note that on the trajectory, $x = 2t, y = -t^2$, so $\mathbf{R}$ has length

$$r = \|\mathbf{R}(t)\| = \sqrt{(2t)^2 + (-t^2)^2} = \sqrt{4t^2 + t^4} = t\sqrt{t^2 + 4}$$

and $\mathbf{R} = r\mathbf{u}_r = t\sqrt{t^2 + 4}\,\mathbf{u}_r$. Because $\mathbf{V}(t) = \dfrac{dr}{dt}\mathbf{u}_r + r\dfrac{d\theta}{dt}\mathbf{u}_\theta$, we need $\dfrac{dr}{dt}$ and $\dfrac{d\theta}{dt}$. We find that

$$\frac{dr}{dt} = \sqrt{t^2 + 4} + t\left(\frac{1}{2}\right)(t^2 + 4)^{-1/2}(2t) = \frac{2t^2 + 4}{\sqrt{t^2 + 4}}$$

and because the polar angle satisfies $\theta = \tan^{-1}\left(\dfrac{y}{x}\right)$ for all points (x, y) on the curve (except where $x = 0$), we have

$$\theta = \tan^{-1}\left(\frac{-t^2}{2t}\right) = \tan^{-1}\left(-\frac{t}{2}\right)$$

$$\frac{d\theta}{dt} = \frac{1}{1 + \left(-\dfrac{t}{2}\right)^2}\left(-\frac{1}{2}\right) = \frac{-2}{t^2 + 4}$$

Thus,

$$\mathbf{V}(t) = \frac{2t^2 + 4}{\sqrt{t^2 + 4}}\mathbf{u}_r + t\sqrt{t^2 + 4}\left(\frac{-2}{t^2 + 4}\right)\mathbf{u}_\theta = \frac{(2t^2 + 4)\mathbf{u}_r - 2t\,\mathbf{u}_\theta}{\sqrt{t^2 + 4}}$$

We now have the tools we need to establish Kepler's second law.

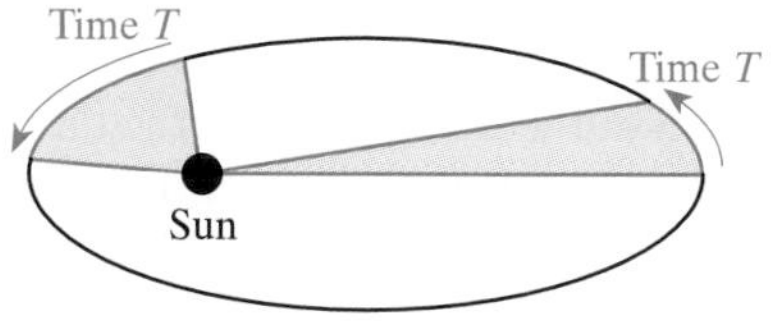

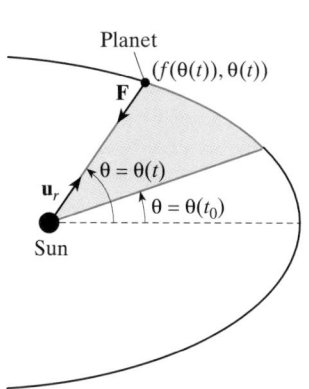

FIGURE 11.14 The radial line sweeps out equal area in equal time.

THEOREM 11.5   *Kepler's second law*

The radius vector from the sun to a planet in its orbit sweeps over equal areas in equal intervals of time.

Proof   The situation described by Kepler's second law is illustrated in Figure 11.14. We shall assume that the only force acting on a planet is the gravitational attraction of the sun. According to the universal law of gravitation, the force of attraction is given by

$$\mathbf{F} = -G\frac{mM}{r^2}\mathbf{u}_r$$

where G is a physical constant, and m and M are the masses of the planet and the sun, respectively. Because this is the only force acting on the planet, Newton's second law of motion tells us that $\mathbf{F} = m\mathbf{A}$, where $\mathbf{A}$ is the acceleration of the planet in its orbit. By equating these two expressions for the force $\mathbf{F}$, we find

$$m\mathbf{A} = -G\frac{mM}{r^2}\mathbf{u}_r$$

$$\mathbf{A} = \frac{-GM}{r^2}\mathbf{u}_r$$

This says that *the acceleration of a planet in its orbit has only a radial component.* Note how this conclusion depends on our assumption that the only force on the planet is the sun's gravitational attraction, which acts along radial lines from the sun to the planet.

This means that the $\mathbf{u}_\theta$ component of the planet's acceleration is 0, and by examining the polar formula for acceleration, we see that this condition is equivalent to the differential equation

$$\left[r\frac{d^2\theta}{dt^2} + 2\frac{dr}{dt}\frac{d\theta}{dt} \right] = 0$$

If we set $z = \dfrac{d\theta}{dt}$, we obtain $\dfrac{d^2\theta}{dt^2} = \dfrac{dz}{dt}$, so that

$$r\frac{dz}{dt} + 2z\frac{dr}{dt} = 0$$

$$z^{-1}\frac{dz}{dt} = -2r^{-1}\frac{dr}{dt}$$

$$\int z^{-1}\, dz = \int (-2r^{-1})dr$$

$$\ln |z| + C_1 = -2 \ln |r| + C_2$$

$$\ln z = \ln(Cr^{-2}) \qquad \text{since } r > 0 \text{ and } z > 0$$

$$z = Cr^{-2}$$

$$\frac{d\theta}{dt} = Cr^{-2}$$

$$r^2\frac{d\theta}{dt} = C$$

Now, let $[t_1, t_2]$ and $[t_3, t_4]$ be two time intervals of equal length, so $t_2 - t_1 = t_4 - t_3$. According to Theorem 9.5, the area swept out in the time period $[t_1, t_2]$ (see Figure 11.15) is

Planet
$(f(\theta(t)), \theta(t))$
$\mathbf{F}$
$\mathbf{u}_r$
$\theta = \theta(t)$
$\theta = \theta(t_0)$
Sun

FIGURE 11.15 The radial line sweeps out area at the rate
$$\frac{dA}{dt} = \frac{1}{2}r^2\frac{d\theta}{dt}$$

$$S_1 = \int_{t_1}^{t_2} \frac{1}{2} r^2\, d\theta = \int_{t_1}^{t_2} \frac{1}{2}\left[r^2\frac{d\theta}{dt} \right] dt = \int_{t_1}^{t_2} \frac{1}{2} C\, dt \qquad \text{since } r^2\frac{d\theta}{dt} = C$$

Thus, $S_1 = \frac{1}{2}C(t_2 - t_1)$. Similarly, the area swept out in time period $[t_3, t_4]$ is $S_2 = \frac{1}{2}C(t_4 - t_3)$, and we have

$$S_1 = \frac{1}{2}C(t_2 - t_1) = \frac{1}{2}C(t_4 - t_3) = S_2$$

so equal area is swept out in equal time, as claimed by Kepler.

An object is said to move in a **central force field** if it is subject to a single force that is always directed toward a particular point. Our derivation of Kepler's second law is based on the assumption that planetary motion occurs in a central force field. In particular, the formula

$$\mathbf{A} = \frac{-GM}{r^2}\mathbf{u}_r$$

shows that a planet's acceleration has only a radial component, and this turns out to be true for an object moving in a plane in any central force field.

Much of what we know about the bodies in our solar system is a result of the mathematics associated with Kepler's laws. Table 11.1 provides some of these data.

TABLE 11.1
Data on Planetary Orbits

Planet	Mean Orbit (in millions of miles)	Eccentricity	Period (in days)	Comparative Mass (Earth = 1.00)
Mercury	36.0	0.205635	88.0	0.06
Venus	67.3	0.006761	224.7	0.81
Earth	**93.0**	**0.016678**	**365.265**	**1.00**
				Actual mass: 5.975×10^{24} kg
Mars	141.7	0.093455	687.0	0.12
Jupiter	484.0	0.048207	4,332.1	317.83
Saturn	887.0	0.055328	10,825.9	95.16
Uranus	1,787.0	0.047694	30,676.1	14.5
Neptune	2,797.0	0.010034	59,911.1	17.2
Pluto	3,675.0	0.248646	90,824.2	0.0025

11.3 Problem Set

A In Problems 1–8, find the time of flight T_f (to the nearest tenth second) and the range R_f (to the nearest unit) of a projectile fired from ground level at the given angle α with the indicated initial speed v_0. Assume that $g = 32$ ft/s^2 or $g = 9.8$ m/s^2.

1. $\alpha = 35°, v_0 = 128$ ft/s
2. $\alpha = 45°, v_0 = 80$ ft/s
3. $\alpha = 48.5°, v_0 = 850$ m/s
4. $\alpha = 43.5°, v_0 = 185$ m/s
5. $\alpha = 23.74°, v_0 = 23.3$ m/s
6. $\alpha = 31.04°, v_0 = 38.14$ m/s
7. $\alpha = 14.11°, v_0 = 100$ ft/s
8. $\alpha = 78.09°, v_0 = 88$ ft/s

In Problems 9–14, an object moves along the given curve in the plane (described in either polar or parametric form). Find its velocity and acceleration in terms of the unit polar vectors $\mathbf{u}_r$ and $\mathbf{u}_\theta$.

9. $x = 2t, y = t$
10. $x = \sin t, y = \cos t$
11. $r = \sin \theta, \theta = 2t$
12. $r = e^{-\theta}, \theta = 1 - t$
13. $r = 5(1 + \cos \theta), \theta = 2t + 1$
14. $r = \dfrac{1}{1 - \cos \theta}, \theta = t$

B 15. A shell fired from ground level at an angle of 45° hits the ground 2,000 m away. What is the muzzle speed of the shell?

16. A gun at ground level has muzzle speed of 300 ft/s. What angle of elevation (to the nearest degree) should be used to hit an object 1,500 ft away?

17. At what angle (to the nearest tenth of a degree) should a projectile be fired from ground level if its muzzle speed is 167.1 ft/s and the desired range is 600 ft?

18. A shell is fired at ground level with a muzzle speed of 280 ft/s and at an elevation of 45° from ground level.
    a. Find the maximum height attained by the shell.
    b. Find the time of flight and the range of the shell.
    c. Find the velocity and speed of the shell at impact.

19. Jeff Bagwell hits a baseball at a 30° angle with a speed of 90 ft/s. If the ball is 4 ft above ground level when it is hit, what is the maximum height reached by the ball? How far will it travel from home plate before it lands? If it just barely clears a 5-foot wall in the outfield, how far (to the nearest foot) is the wall from home plate?

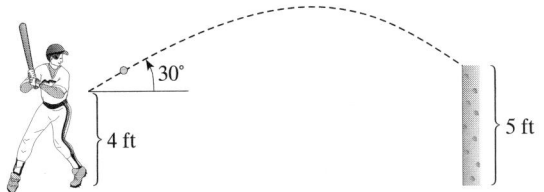

20. A baseball hit in the Astrodome at a 24° angle from 3 ft above the ground just goes over the 9-ft fence 400 ft from home plate. About how fast was the ball traveling, and how long did it take the ball to reach the wall?

21. One of the shortest possible home runs in major league baseball is over the right field fence at the Kingdome in Seattle, where the distance is 312 ft and the fence is 8 ft high. How long will it take for a ball to leave the playing field at that point if it is struck $3\frac{1}{2}$ ft above the ground at a 32° angle, and we assume that the ball just barely clears the fence?

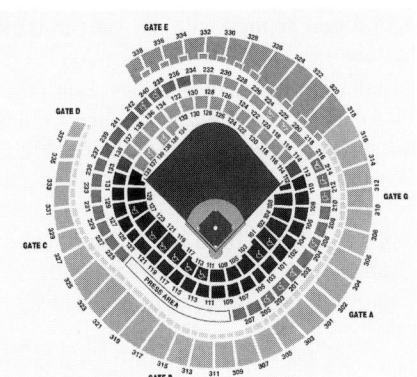

Kingdome in Seattle

22. Steve Young, a quarterback for the San Francisco '49ers, throws a pass at a 45° angle from a height of 6.5 ft with a speed of 50 ft/s. Receiver Jerry Rice races straight down-field on a fly pattern at a constant speed of 32 ft/s and catches the ball at a height of 6 ft. If Steve and Jerry both line up on the 50-yard line at the start of the play, how far does Steve fade back from the line of scrimmage before he releases the ball?

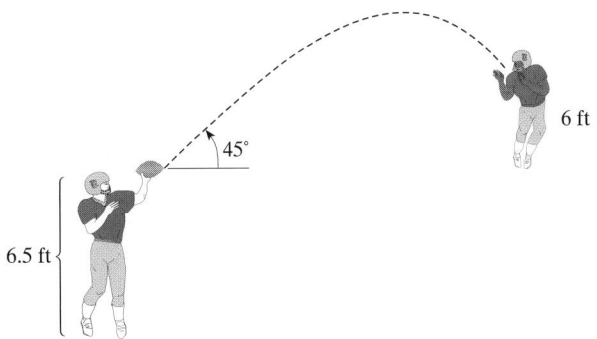

23. A golf ball is hit from the tee to a green with an initial speed of 125 ft/s at an angle of elevation of 45°. How long will it take for the ball to hit the green?

24. Basketball player Shaquille O'Neal attempts a shot while standing 20 ft from the basket. If "The Shaq" shoots from a height of 7 ft and the ball reaches a maximum height of 12 ft before passing through the 10-ft-high basket, what is the initial speed of the ball?

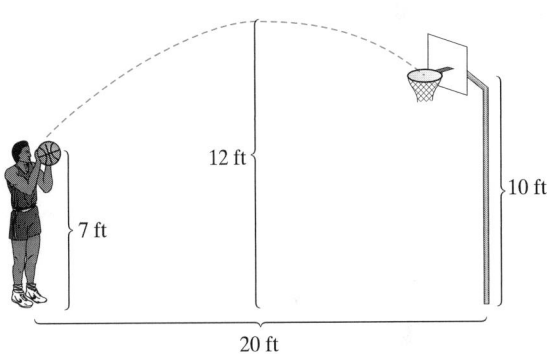

25. If a shotputter throws a shot from a height of 5 ft with an angle of 46° and initial speed of 25 ft/s, what is the horizontal distance of the throw?

26. **MODELING PROBLEM**   In 1974, Evel Knievel attempted a skycycle ride across the Snake River, and at the time there was a great deal of hype about "will he make it?" If the angle of the launching ramp was 45° and if the horizontal distance the skycycle needed to travel was 4,700 ft, at what speed did Evel have to leave the ramp to make it across the Snake River Canyon?

    Assume that the opposite edges of the canyon are at the same height.

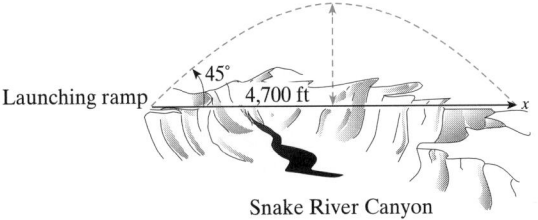

Snake River Canyon

27. A particle moves along the polar path (r, θ), where $r(t) = 3 + 2 \sin t, \theta(t) = t^3$. Find $\mathbf{V}(t)$ and $\mathbf{A}(t)$ in terms of $\mathbf{u}_r$ and $\mathbf{u}_\theta$.

In Problems 28–31, consider a particle moving on a circular path of radius a described by the equation $\mathbf{R}(t) = (a \cos \omega t)\mathbf{i} + (a \sin \omega t)\mathbf{j}$, *where* $\omega = \dfrac{d\theta}{dt}$ *is the constant angular velocity.*

28. Find the velocity vector and show that it is orthogonal to $\mathbf{R}(t)$.

29. Find the speed of the particle.

30. Find the acceleration vector and show that its direction is always toward the center of the circle.

31. Find the magnitude of the acceleration vector.

32. MODELING PROBLEM   A 3-oz paddleball attached to a string is swung in a circular path with a 1-ft radius. If the string will break under a force of 2 lb, find the maxi-mum speed the ball can at-tain without breaking the string. (*Hint:* Note that 3 oz is 3/16 lb.)

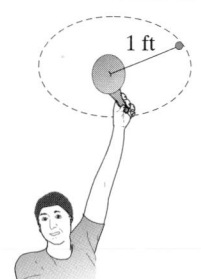

33. SPY PROBLEM   The Spy figures out where to dig (Problem 59 of Section 10.1) and unearths a human-size box. With shaking hands he opens the lid, and to his relief, finds not Purity, but her favorite scarf and a cell phone. He picks it up and hears Purity's voice screaming for help. Her pleas are quickly muffled and Blohardt takes over. "Do what you're told or the next box will be heavier!" he threatens. "You're going to help me eliminate my competition, starting with Scélérat." The Spy is still peeved with Scélérat for icing his friend, Siggy Leiter, so he agrees. He's told that the Frenchman and his gang are holed up in a bunker on the side of a hill inclined at an angle of 15° to the hori-zontal, as shown in Figure 11.16. The Spy returns to his helicopter and flies toward the bunker traveling at 200 ft/s at an altitude of 10,000 ft. Just as the helicopter flies over the base of the hill, the Spy sights a ventilation hole in the top of the bunker at an angle of 20°. He de-cides to drop a canister of knockout gas into the bunker through the hole. How long should he wait before re-leasing the canister?

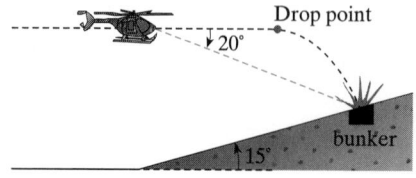

■ **FIGURE 11.16**   Spy and the bunker target

34. A child, running along level ground at the top of a 30-ft-high vertical cliff at a speed of 15 ft/s, throws a rock over the cliff into the sea below. Suppose the child's arm is 3 ft above the ground and her arm speed is 25 ft/s. If the rock is released 10 ft from the edge of the cliff at an angle of 30°, how long does it take for the rock to hit the water? How far from the base of the cliff does it hit?

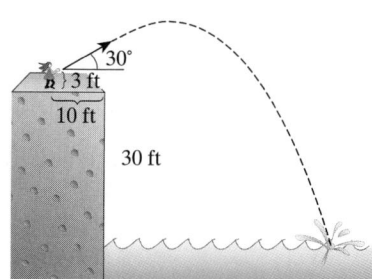

35. Find two angles of elevation α_1 and α_2 (to the nearest de-gree) so that a shell fired at ground level with muzzle speed of 650 ft/s will hit a target 6,000 ft away (also at ground level).

36. A gun is fired with muzzle speed $v_0 = 550$ ft/s at an angle of $\alpha = 22.0°$. The shell overshoots the target by 50 ft. At what angle should a second shot be fired with the same muzzle speed to hit the target?

37. A gun is fired with muzzle speed $v_0 = 700$ ft/s at an angle of $\alpha = 25°$. It overshoots the target by 60 ft. The target moves away from the gun at a constant speed of 10 ft/s. If the gunner takes 30 sec to reload, at what angle should a second shot be fired with the same muzzle speed to hit the target?

38. A shell is fired from ground level with muzzle speed of 750 ft/s at an angle of 25°. An enemy gun 20,000 ft away fires a shot 2 seconds later and the shells collide 50 ft above ground. What are the muzzle speed v_0 and angle of elevation α of the second gun?

ⓒ 39. Suppose a shell is fired with muzzle speed v_0 at an angle α from a height s_0 above level ground.

   a. Show that the range of R_f must satisfy

$$g(\sec^2\alpha) R_f^2 - 2v_0^2 (\tan \alpha)R_f - 2v_0^2 s_0 = 0$$

   b. Show that the maximum range R_m occurs at angle α_m where

$$R_m \tan \alpha_m = \frac{v_0^2}{g}$$

   c. Show that

$$R_m = \frac{v_0}{g} \sqrt{v_0^2 + 2gs_0} \text{ and } \alpha_m = \tan^{-1}\left(\frac{v_0}{\sqrt{v_0^2 + 2gs_0}}\right)$$

40. Use the formula

$$\mathbf{V}(t) = \frac{dr}{dt}\mathbf{u}_r + r\frac{d\theta}{dt}\mathbf{u}_\theta$$

to derive the formula for polar acceleration:

$$\mathbf{A}(t) = \left[\frac{d^2r}{dt^2} - r\left(\frac{d\theta}{dt}\right)^2\right]\mathbf{u}_r + \left[r\frac{d^2\theta}{dt^2} + 2\frac{dr}{dt}\frac{d\theta}{dt}\right]\mathbf{u}_\theta$$

Hint: Begin by using the chain rule to show

$$\frac{d\mathbf{u}_r}{dt} = \mathbf{u}_\theta\frac{d\theta}{dt} \quad \text{and} \quad \frac{d\mathbf{u}_\theta}{dt} = -\mathbf{u}_r\frac{d\theta}{dt}$$

41. Using the acceleration formula given in Problem 40, show that the force $\mathbf{F}$ on an object is given by

$$\mathbf{F}(t) = m\mathbf{A}(t) = F_r\mathbf{u}_r + F_\theta\mathbf{u}_\theta$$

where

$$F_r(t) = m\frac{d^2r}{dt^2} - mr\left(\frac{d\theta}{dt}\right)^2$$

and

$$F_\theta(t) = mr\frac{d^2\theta}{dt^2} + 2m\frac{dr}{dt}\frac{d\theta}{dt}$$

42. Using Problem 41, show that

$$rF_\theta(t) = \frac{d}{dt}\left(mr^2\frac{d\theta}{dt}\right)$$

43. Use Problem 42 to show that if $F_\theta(t) = 0$, then

$$mr^2\frac{d\theta}{dt}$$

is constant.

11.4 *Unit Tangent and Normal Vectors; Curvature*

IN THIS SECTION   **unit tangent and principal unit normal vectors, arc length as a parameter, curvature**

UNIT TANGENT AND PRINCIPAL UNIT NORMAL VECTORS

From Section 11.2, we know that if $\mathbf{R}(t)$ is a vector function with a smooth graph, then the derivative $\mathbf{R}'(t)$ is a tangent vector to the graph at the point corresponding to $\mathbf{R}(t)$. Because the graph is smooth, we have $\mathbf{R}'(t) \neq \mathbf{0}$, and a unit tangent vector $\mathbf{T}(t)$ can be obtained by dividing $\mathbf{R}'(t)$ by its length, that is,

$$\mathbf{T}(t) = \frac{\mathbf{R}'(t)}{\|\mathbf{R}'(t)\|}$$

A unit tangent vector at a point P of a curve is shown in Figure 11.17.

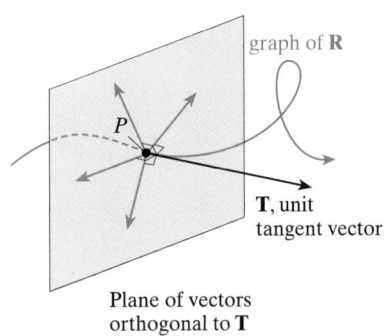

■ FIGURE 11.17   Unit tangent vector $\mathbf{T}$ and vectors orthogonal to $\mathbf{T}$ on a given curve

Based on our experience with normal lines to curves in the plane, we would expect a normal vector to be orthogonal to the tangent vector $\mathbf{T}(t)$ at each point on the graph of $\mathbf{R}$. To obtain such a vector, we recall (Theorem 11.4) that any vector function $\mathbf{R}(t)$ with constant length is always orthogonal to its derivative $\mathbf{R}'(t)$. Therefore, because $\mathbf{T}(t)$ has length 1 for all t, it follows that $\mathbf{T}(t)$ is orthogonal to $\mathbf{T}'(t)$ for all t, so that

$$\mathbf{N}(t) = \frac{\mathbf{T}'(t)}{\|\mathbf{T}'(t)\|}$$

is a unit normal vector. Actually, there are infinitely many vectors orthogonal to $\mathbf{T}(t)$—see Figure 11.17—and we shall refer to $\mathbf{N}(t)$ as the **principal unit normal** whenever it is necessary to distinguish it from other normal vectors.

Unit Tangent Vector and Principal Unit Normal Vector

If $\mathbf{R}(t)$ is a vector function that defines a smooth graph, then at each point a unit tangent is

$$\mathbf{T}(t) = \frac{\mathbf{R}'(t)}{\|\mathbf{R}'(t)\|}$$

and the principal unit normal vector is

$$\mathbf{N}(t) = \frac{\mathbf{T}'(t)}{\|\mathbf{T}'(t)\|}$$

■ *What This Says:* For each number t_0 in the domain of $\mathbf{R}(t)$, we have constructed a pair of unit vectors, $\mathbf{T}(t_0)$ and $\mathbf{N}(t_0)$, with $\mathbf{T}$ tangent to the graph of $\mathbf{R}$ at $t = t_0$ and with $\mathbf{N}$ orthogonal to $\mathbf{T}$ at $t = t_0$.

EXAMPLE 1   *Unit tangent and principal unit normal*

Find the unit tangent vector $\mathbf{T}(t)$ and the principal unit normal vector $\mathbf{N}(t)$ at each point on the graph of the vector function

$$\mathbf{R}(t) = e^t\mathbf{i} + e^{-t}\mathbf{j} + \sqrt{2}\,t\,\mathbf{k}$$

In particular, find $\mathbf{T}(1)$ and $\mathbf{N}(1)$.

Solution   The derivative of $\mathbf{R}(t)$ with respect to t is the vector function $\mathbf{R}'(t) = e^t\mathbf{i} - e^{-t}\mathbf{j} + \sqrt{2}\,\mathbf{k}$, which has length

$$\|\mathbf{R}'(t)\| = \sqrt{(e^t)^2 + (-e^{-t})^2 + (\sqrt{2})^2}$$

$$= \sqrt{e^{2t} + e^{-2t} + 2} = \sqrt{(e^t + e^{-t})^2} = e^t + e^{-t}$$

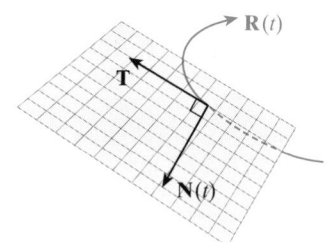

Thus,

$$\mathbf{T}(t) = \frac{\mathbf{R}'(t)}{\|\mathbf{R}'(t)\|} = \frac{e^t\mathbf{i} - e^{-t}\mathbf{j} + \sqrt{2}\,\mathbf{k}}{e^t + e^{-t}}$$

$$= \left(\frac{e^t}{e^t + e^{-t}}\right)\mathbf{i} + \left(\frac{-e^{-t}}{e^t + e^{-t}}\right)\mathbf{j} + \left(\frac{\sqrt{2}}{e^t + e^{-t}}\right)\mathbf{k}$$

So

$$\mathbf{T}(1) = \frac{e\mathbf{i} - e^{-1}\mathbf{j} + \sqrt{2}\,\mathbf{k}}{e + e^{-1}} \approx 0.88\mathbf{i} - 0.12\mathbf{j} + 0.46\mathbf{k}$$

To compute the principal unit normal $\mathbf{N}$, we first find $\mathbf{T}'(t)$:

$$\mathbf{T}'(t) = \frac{(e^t + e^{-t})\,e^t - e^t(e^t - e^{-t})}{(e^t + e^{-t})^2}\mathbf{i} + \frac{(e^t + e^{-t})(e^{-t}) + e^{-t}(e^t - e^{-t})}{(e^t + e^{-t})^2}\mathbf{j}$$

$$- \sqrt{2}\,(e^t + e^{-t})^{-2}\,(e^t - e^{-t})\mathbf{k}$$

$$= \frac{1}{(e^t + e^{-t})^2}\,[(e^{2t} + 1 - e^{2t} + 1)\mathbf{i} + 2\mathbf{j} - \sqrt{2}(e^t - e^{-t})\mathbf{k}]$$

$$= \frac{2\mathbf{i} + 2\mathbf{j} - \sqrt{2}(e^t - e^{-t})\mathbf{k}}{(e^t + e^{-t})^2}$$

and

$$\|\mathbf{T}'(t)\| = \frac{\sqrt{2^2 + 2^2 + 2(e^t - e^{-t})^2}}{(e^t + e^{-t})^2} = \frac{\sqrt{2e^{2t} + 4 + 2e^{-2t}}}{(e^t + e^{-t})^2}$$

$$= \frac{\sqrt{2(e^t + e^{-t})^2}}{(e^t + e^{-t})^2} = \frac{\sqrt{2}}{e^t + e^{-t}}$$

So, the unit normal is

$$\mathbf{N}(t) = \frac{\mathbf{T}'(t)}{\|\mathbf{T}'(t)\|} = \frac{\dfrac{2\mathbf{i} + 2\mathbf{j} - \sqrt{2}(e^t - e^{-t})\mathbf{k}}{(e^t + e^{-t})^2}}{\dfrac{\sqrt{2}}{e^t + e^{-t}}}$$

$$= \frac{2\mathbf{i} + 2\mathbf{j} - \sqrt{2}(e^t - e^{-t})\mathbf{k}}{\sqrt{2}(e^t + e^{-t})}$$

$$= \frac{\sqrt{2}\mathbf{i} + \sqrt{2}\mathbf{j} - (e^t - e^{-t})\mathbf{k}}{(e^t + e^{-t})}$$

In particular,

$$\mathbf{N}(1) = \frac{2\mathbf{i} + 2\mathbf{j} - \sqrt{2}(e - e^{-1})\mathbf{k}}{\sqrt{2}(e + e^{-1})} \approx 0.46\mathbf{i} + 0.46\mathbf{j} - 0.76\mathbf{k}$$

Remember, because **T** and **N** are to be orthogonal, we can check our work by noting that

$$\mathbf{T} \cdot \mathbf{N} \approx (0.88\mathbf{i} - 0.12\mathbf{j} + 0.46\mathbf{k}) \cdot (0.46\mathbf{i} + 0.46\mathbf{j} - 0.76\mathbf{k}) = 0 \qquad \blacksquare$$

Technology Window

We continue to remind you that if you use some software packages, your answers may take on slightly different, yet equivalent, forms. For Example 1 we obtain

$$\mathbf{T}'(t) = \frac{2e^{2t}}{(e^{2t} + 1)^2}\mathbf{i} + \frac{2e^{2t}}{(e^{2t} + 1)^2}\mathbf{j} + \frac{\sqrt{2}e^t(1 - e^t)(e^t + 1)}{(e^{2t} + 1)^2}\mathbf{k}$$

Continuing with the software program, we find

$$\|\mathbf{T}'(t)\| = \frac{\sqrt{2}e^t}{e^{2t} + 1}$$

Finally,

$$\mathbf{N}(t) = \frac{\mathbf{T}'(t)}{\|\mathbf{T}'(t)\|} = \frac{\sqrt{2}e^t}{e^{2t} + 1}\mathbf{i} + \frac{\sqrt{2}e^t}{e^{2t} + 1}\mathbf{j} + \left(\frac{1 - e^{2t}}{e^{2t} + 1}\right)\mathbf{k}$$

$$\mathbf{N}(1) = 0.458243571484\mathbf{i} + 0.458243571484\mathbf{j} - 0.761594155955\mathbf{k}$$

Checking $\mathbf{T} \cdot \mathbf{N}$ for $t = 1$, we obtain $-1.49398426240 \times 10^{-6}$, which is scientific notation for a number very close (but not equal to) zero. Note that the rounded check in Example 1 showed that this was exactly 0, but that result was, by coincidence, due to rounding. The best you can do with most approximate results is to show that they are approximately equal.

ARC LENGTH AS A PARAMETER

In Section 11.2, we found that time t is the most natural parameter for studying motion along a curve. However, if we are primarily interested in the geometric features of a curve, it may be more convenient to use arc length s as a parameter.

Arc Length Function

Let $\mathbf{R}(t)$ be a vector function whose graph is a smooth curve C on the closed interval $[a, b]$. Then the **arc length function** on $[a, b]$ is defined by

$$s(t) = \int_a^t \|\mathbf{R}'(u)\|\, du \qquad a \le t \le b$$

Notice that $s(t)$ increases in the direction of increasing t. That is, arc length s increases in the same direction as the parameter t along the curve C (see Figure 11.18).

In the case where $\mathbf{R}(t) = x(t)\mathbf{i} + y(t)\mathbf{j} + z(t)\mathbf{k}$, we have

$$\mathbf{R}'(u) = x'(u)\mathbf{i} + y'(u)\mathbf{j} + z'(u)\mathbf{k}$$

and

$$\|\mathbf{R}'(u)\| = \sqrt{[x'(u)]^2 + [y'(u)]^2 + [z'(u)]^2}$$

Thus, on the interval $[a, b]$, the trajectory of $\mathbf{R}(t)$ has arc length

$$L = s(b) = \int_a^b \|\mathbf{R}'(u)\|\, du = \int_a^b \sqrt{[x'(u)]^2 + [y'(u)]^2 + [z'(u)]^2}\, du$$

This gives a formula for the arc length of a curve in space that is a direct generalization of the formula obtained for the arc length of a planar curve in Chapter 9 (Theorem 9.4).

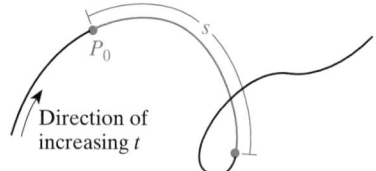

■ **FIGURE 11.18**   Arc length s increases in the same direction as t along the graph of $\mathbf{R}$.

THEOREM 11.6   *Arc length of a space curve*

If C is a smooth curve defined by $\mathbf{R}(t) = x(t)\mathbf{i} + y(t)\mathbf{j} + z(t)\mathbf{k}$ on an interval $[a, b]$, then the arc length of C is given by

$$s = \int_a^b \|\mathbf{R}'(t)\|\, dt = \int_a^b \sqrt{[x'(t)]^2 + [y'(t)]^2 + [z'(t)]^2}\, dt$$

Proof   An outline of the proof precedes the statement of this theorem.   ══

EXAMPLE 2   *Finding the arc length of a space curve*

Find the arc length of the curve defined by

$$\mathbf{R}(t) = 12t\mathbf{i} + (5\cos t)\mathbf{j} + (3 - 5\sin t)\mathbf{k}$$

from $t = 0$ to $t = 2$.

Solution   We have $x(t) = 12t$, $y(t) = 5\cos t$, and $z(t) = 3 - 5\sin t$, so that $x'(t) = 12$, $y'(t) = -5\sin t$, and $z'(t) = -5\cos t$. We now use Theorem 11.6:

$$s = \int_a^b \sqrt{[x'(t)]^2 + [y'(t)]^2 + [z'(t)]^2}\, dt$$

$$= \int_0^2 \sqrt{12^2 + (-5\sin t)^2 + (-5\cos t)^2}\, dt = \int_0^2 13\, dt = 26 \qquad ■$$

By applying the second fundamental theorem of calculus, we can differentiate the arc length function on $[a, b]$ to obtain

$$\frac{ds}{dt} = \| \mathbf{R}'(t) \|$$

Recall from Section 11.2 that the *speed* of an object moving on its trajectory is given by $\| \mathbf{V}(t) \| = \| \mathbf{R}'(t) \|$, where $\mathbf{V}(t)$ is the velocity vector of the object. Thus, the speed, in terms of arc length, is

$$\| \mathbf{V}(t) \| = \| \mathbf{R}'(t) \| = \frac{ds}{dt}$$

For future reference, we summarize this result in the following theorem.

THEOREM 11.7   *Speed in terms of arc length*

Suppose an object moves with displacement $\mathbf{R}(t)$, where $\mathbf{R}'(t)$ is continuous on the interval $[a, b]$. Then the object has speed

$$\| \mathbf{V}(t) \| = \| \mathbf{R}'(t) \| = \frac{ds}{dt} \qquad \text{for } a \le t \le b$$

Proof   An outline of the proof precedes the statement of the theorem.   ⚊

EXAMPLE 3   *Speed and distance traveled by an object moving in space*

An object moves with displacement

$$\mathbf{R}(t) = e^t \mathbf{i} + (\sqrt{2}\, t + 3) \mathbf{j} + e^{-t} \mathbf{k}$$

Find the speed of the object at time t and compute the distance the object travels between times $t = 0$ and $t = 1$.

Solution   By differentiating $\mathbf{R}(t)$ with respect to t, we find the velocity:

$$\mathbf{V}(t) = \mathbf{R}'(t) = e^t \mathbf{i} + \sqrt{2}\, \mathbf{j} - e^{-t} \mathbf{k}$$

Therefore, the speed at time t is

$$\frac{ds}{dt} = \| \mathbf{R}'(t) \| = \sqrt{(e^t)^2 + (\sqrt{2})^2 + (-e^{-t})^2}$$

$$= \sqrt{e^{2t} + 2 + e^{-2t}} = \sqrt{(e^t + e^{-t})^2} = e^t + e^{-t}$$

The distance traveled by the object between times $t = 0$ and $t = 1$ is the arc length and is given by

$$s = \int_0^1 (e^t + e^{-t})\, dt = [e^t - e^{-t}]\big|_0^1 = e - e^{-1} - 1 + 1 \approx 2.3504024$$

COMMENT   Incidentally, notice that the object is at the point $P(1, 3, 1)$ when $t = 0$ and is at $Q(e, \sqrt{2} + 3, e^{-1})$ when $t = 1$. The straight-line distance between P and Q is

$$d = \sqrt{(e - 1)^2 + (\sqrt{2} + 3 - 3)^2 + (e^{-1} - 1)^2} \approx 2.3134539$$

which, as might be expected, is slightly less than the distance measured from P to Q along the trajectory.   ■

When $\mathbf{R}$ is represented as $\mathbf{R}(s)$ in terms of the arc length parameter s, the unit tangent vector $\mathbf{T}$ can be represented as $\mathbf{T} = d\mathbf{R}/ds$. To see why this is true, recall from

Theorem 11.7 that $\|\mathbf{R}'(t)\| = ds/dt$, and because

$$\mathbf{R}'(t) = \frac{d\mathbf{R}}{dt}$$

it follows that

$$\mathbf{T} = \frac{\mathbf{R}'(t)}{\|\mathbf{R}'(t)\|} = \frac{\dfrac{d\mathbf{R}}{dt}}{\dfrac{ds}{dt}} = \frac{d\mathbf{R}}{ds}$$

Note that because s increases with t, we have $\dfrac{ds}{dt} > 0$, so the unit tangent vector $\mathbf{T} = \dfrac{d\mathbf{R}}{ds}$ must point in the direction of increasing s.

Next, to obtain a formula for the principal unit normal vector

$$\mathbf{N} = \frac{\mathbf{T}'(t)}{\|\mathbf{T}'(t)\|}$$

in terms of the arc length parameter s, we first note that

$$\mathbf{T}'(t) = \frac{d\mathbf{T}}{dt} = \frac{d\mathbf{T}}{ds}\frac{ds}{dt}$$

Because $\dfrac{ds}{dt} > 0$, it follows that the vector derivatives $\dfrac{d\mathbf{T}}{dt}$ and $\dfrac{d\mathbf{T}}{ds}$ point in the same direction, and $\mathbf{N}$ can be computed by finding $\dfrac{d\mathbf{T}}{ds}$ and dividing by its length. The scalar function $\kappa = \left\|\dfrac{d\mathbf{T}}{ds}\right\|$ is called the **curvature** of the graph (defined below).

Thus, we have

$$\mathbf{N} = \frac{\dfrac{d\mathbf{T}}{ds}}{\left\|\dfrac{d\mathbf{T}}{ds}\right\|} = \frac{1}{\kappa}\frac{d\mathbf{T}}{ds}$$

We summarize these observations in the following box.

Formulas for Unit Tangent and Principal Unit Normal in Terms of the Arc Length Parameter

If $\mathbf{R}(t)$ has a smooth graph and is represented as $\mathbf{R}(s)$ in terms of the arc length parameter s, then the unit tangent $\mathbf{T}$ and the principal normal $\mathbf{N}$ at each point satisfy

$$\mathbf{T} = \frac{d\mathbf{R}}{ds} \quad \text{and} \quad \mathbf{N} = \frac{1}{\kappa}\frac{d\mathbf{T}}{ds}$$

where $\kappa = \left\|\dfrac{d\mathbf{T}}{ds}\right\|$ is a scalar function called the **curvature** of the graph.

CURVATURE

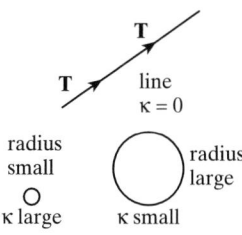

T

T   line
        $\kappa = 0$

radius
small

radius
large

κ large κ small

The concept of curvature defined in the above box provides a measure of the "bend" in a trajectory. In $\mathbb{R}^2$, a straight line has no bend, so it is not surprising to find it has curvature 0. The "bend" in a circle of radius r is the same at each point and is measured by the reciprocal of the radius; that is, $\kappa = 1/r$. Thus, a circle with small radius bends sharply and its curvature κ is, indeed, large, while just the opposite is true for a circle with large radius.

More generally, the curvature at a point P on a curve C may be thought of as the curvature of the circle that "best approximates" the shape of C "near" P. This best approximating circle, called the **osculating** (*kissing*) **circle,** is tangent to C at P and has radius $\rho = 1/\kappa$. For example, in the problem set (Problem 54), you are asked to show that the curvature of a function $y = f(x)$ is given by

$$\kappa = \frac{|y''|}{(1 + y'^2)^{3/2}}$$

at each point $P(x, y)$. Applying this formula to the parabola $y = x^2$, we note $y' = 2x$, $y'' = 2$, so

$$\kappa = \frac{2}{(1 + 4x^2)^{3/2}}$$

At the origin $(0, 0)$, $\kappa = 2$, while at $(1, 1)$ it is $\kappa(1) = 2/5^{3/2}$. The corresponding osculating circles are shown in Figure 11.19.

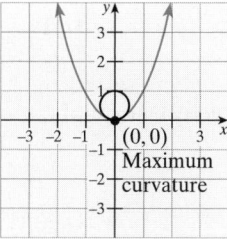

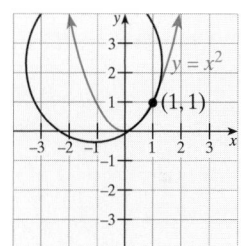

a. Maximum curvature at $(0, 0)$; smallest osculating circle

b. Curvature at $(1, 1)$ is $2/5^{3/2}$; larger osculating circle shown

■ **FIGURE 11.19**   Curvature on a parabola

For curves in $\mathbb{R}^3$, the situation is not quite as easy to visualize because the "bend" of a curve is not confined to a single plane, but can occur in all directions.

To compute the curvature κ, we can first find the unit tangent vector **T** and then substitute into the formula

$$\kappa = \left\| \frac{d\mathbf{T}}{ds} \right\| = \frac{\left\| \dfrac{d\mathbf{T}}{dt} \right\|}{\dfrac{ds}{dt}}$$

For instance, in Example 1 we found $\mathbf{R}(t) = e^t\mathbf{i} + e^{-t}\mathbf{j} + \sqrt{2}t\mathbf{k}$, and $\mathbf{T}(t) = \dfrac{e^t\mathbf{i} - e^{-t}\mathbf{j} + \sqrt{2}\mathbf{k}}{e^t + e^{-t}}$, so that

$$\frac{ds}{dt} = \left\| \mathbf{R}'(t) \right\| = e^t + e^{-t} \quad \text{and} \quad \left\| \frac{d\mathbf{T}}{dt} \right\| = \frac{\sqrt{2}}{e^t + e^{-t}}$$

Curvature for
$\mathbf{R}(t) = e^t\mathbf{i} + e^{-t}\mathbf{j} + \sqrt{2}t\,\mathbf{k}$

t	κ
0.0	0.35355339059
0.5	0.27805126251
1.0	0.14848335244
1.5	0.06388944489
2.0	0.02497883868
2.5	0.00940177191
3.0	0.00348817089
3.5	0.00128724714
4.0	0.00047409766
4.5	0.00017448475
5.0	0.00006419937

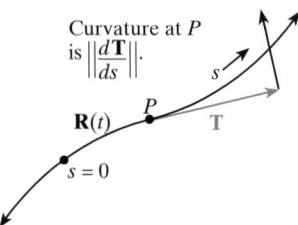

Curvature at P is $\left\|\dfrac{d\mathbf{T}}{ds}\right\|$.

Thus, the curvature in this case is the scalar function

$$\kappa = \frac{\left\|\dfrac{d\mathbf{T}}{dt}\right\|}{\dfrac{ds}{dt}} = \frac{\dfrac{\sqrt{2}}{e^t + e^{-t}}}{e^t + e^{-t}} = \frac{\sqrt{2}}{(e^t + e^{-t})^2}$$

It can be shown (see Problem 27) that a line has curvature 0. Because the curvature is nonnegative, smaller curvature indicates a gentler curve. For Example 1, we see that if $t = 0$, the curvature is about 0.35, whereas if $t = 5$ the curvature is 0.0000642 (almost flat). Some curvature calculations for this curve are shown in the table in the margin (which was generated using a spreadsheet program).

In the case where the graph of $\mathbf{R}(t)$ lies entirely in a plane, the curvature $\kappa(t)$ has a nice geometric interpretation. Note that in this case, the unit tangent vector $\mathbf{T}$ at a point P can be expressed as

$$\mathbf{T} = (\cos\phi)\mathbf{i} + (\sin\phi)\mathbf{j}$$

where ϕ is the angle of inclination of the tangent line at P, as shown in Figure 11.20a. Then, differentiating both sides of the equation for $\mathbf{T}$, we have

$$\frac{d\mathbf{T}}{ds} = \left(-\sin\phi\,\frac{d\phi}{ds}\right)\mathbf{i} + \left(\cos\phi\,\frac{d\phi}{ds}\right)\mathbf{j}$$

and because the curvature is the length of this vector, we find

$$\kappa = \left\|\frac{d\mathbf{T}}{ds}\right\| = \left|\frac{d\phi}{ds}\right|\sqrt{(-\sin\phi)^2 + (\cos\phi)^2} = \left|\frac{d\phi}{ds}\right|$$

This characterization of curvature is demonstrated in Figure 11.20b.

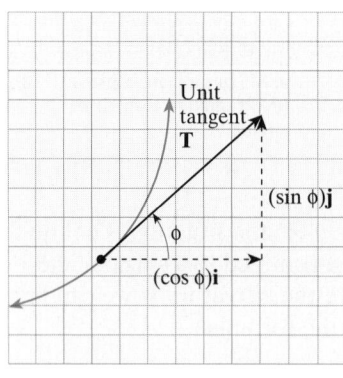

a. In a plane, the unit tangent vector is $\mathbf{T} = (\cos\phi)\mathbf{i} + (\sin\phi)\mathbf{j}$.

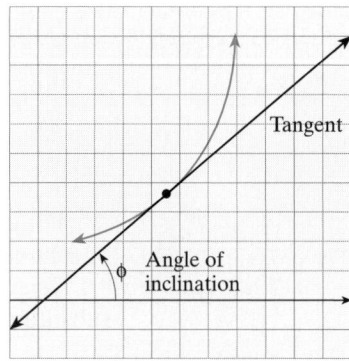

b. The curvature $\kappa = \left|\dfrac{d\phi}{ds}\right|$ measures the rate at which the curve bends away from the tangent.

■ **FIGURE 11.20**   The curvature of a graph

By looking at the formula $\kappa = \left|\dfrac{d\phi}{ds}\right|$, we see that at each point P on a planar curve, the curvature κ measures the rate at which the curve bends away from the tangent line.

The preceding curvature formula is not particularly easy to use, and because the graph C of the vector function is frequently defined by $\mathbf{R}(t)$, it is often useful to have the following formula for κ in terms of $\mathbf{R}$:

> **Derivative Formula for Curvature**
>
> The curvature of a graph C defined by a vector function $\mathbf{R}(t)$ can be found by
>
> $$\kappa = \frac{\|\mathbf{R}' \times \mathbf{R}''\|}{\|\mathbf{R}'\|^3}$$

The derivation of this formula is outlined in Problem 52 and is illustrated in Example 4.

EXAMPLE 4   *Curvature of a helix*

Given $\mathbf{R}(t) = (a \cos t)\mathbf{i} + (a \sin t)\mathbf{j} + bt\mathbf{k}$ with a and b both nonnegative, express the curvature of $\mathbf{R}(t)$ in terms of a and b ($a > b$).

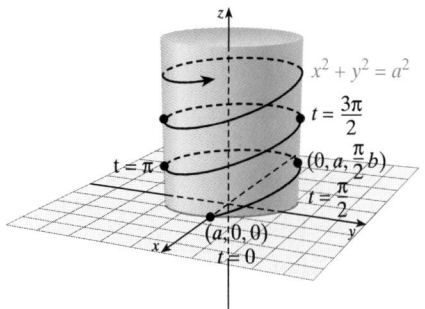

Solution

$$\mathbf{R}'(t) = (-a \sin t)\mathbf{i} + (a \cos t)\mathbf{j} + b\mathbf{k}$$

$$\mathbf{R}''(t) = (-a \cos t)\mathbf{i} + (-a \sin t)\mathbf{j}$$

$$\mathbf{R}' \times \mathbf{R}'' = \begin{vmatrix} \mathbf{i} & \mathbf{j} & \mathbf{k} \\ -a \sin t & a \cos t & b \\ -a \cos t & -a \sin t & 0 \end{vmatrix} = (ab \sin t)\mathbf{i} - (ab \cos t)\mathbf{j} + a^2\mathbf{k}$$

We now need to find the magnitude of this vector, as well as that of $\mathbf{R}'$:

$$\|\mathbf{R}' \times \mathbf{R}''\| = \sqrt{(ab \sin t)^2 + (ab \cos t)^2 + a^4}$$
$$= \sqrt{a^2b^2(\sin^2 t + \cos^2 t) + a^4}$$
$$= \sqrt{a^2b^2 + a^4}$$

and

$$\|\mathbf{R}'\| = \sqrt{a^2\cos^2 t + a^2\sin^2 t + b^2} = \sqrt{a^2 + b^2}$$

Then we have

$$\kappa = \frac{\|\mathbf{R}' \times \mathbf{R}''\|}{\|\mathbf{R}'\|^3} = \frac{\sqrt{a^2b^2 + a^4}}{(\sqrt{a^2 + b^2})^3} = \frac{a\sqrt{a^2 + b^2}}{(a^2 + b^2)\sqrt{a^2 + b^2}} = \frac{a}{a^2 + b^2}$$

If we increase b, then for a fixed a the curvature decreases. If $b = 0$, the helix reduces to a circle of radius a. If $a = 0$, the helix flattens out along the z-axis.   ∎

In the plane, where a curve is given by $\mathbf{R}(t) = x(t)\mathbf{i} + y(t)\mathbf{j}$, the curvature may be computed by the formula

$$\kappa = \frac{|x'y'' - y'x''|}{[(x')^2 + (y')^2]^{3/2}}$$

A derivation of this formula is outlined in Problem 53. We will use this formula to show that a circle has constant curvature.

EXAMPLE 5   *Curvature of a circle*

Show that a circle of radius r has curvature $\kappa = \dfrac{1}{r}$ at each point.

Solution   Suppose a circle has radius r and center (h, k). Then its Cartesian equation is $(x - h)^2 + (y - k)^2 = r^2$, and we can obtain a parametrization by setting

$$x - h = r\cos t \quad \text{and} \quad y - k = r\sin t$$

In other words, the circle is the graph of the vector function $\mathbf{R}(t) = x(t)\mathbf{i} + y(t)\mathbf{j}$, where

$$x(t) = h + r\cos t \quad \text{and} \quad y(t) = k + r\sin t$$

Differentiating with respect to t, we obtain

$$x'(t) = -r\sin t \qquad y'(t) = r\cos t$$
$$x''(t) = -r\cos t \qquad y''(t) = -r\sin t$$

and by substituting these into the curvature formula, we find

$$\kappa = \frac{|x'y'' - y'x''|}{[(x')^2 + (y')^2]^{3/2}} = \frac{(-r\sin t)(-r\sin t) - (r\cos t)(-r\cos t)}{[(-r\sin t)^2 + (r\cos t)^2]^{3/2}}$$
$$= \frac{r^2\sin^2 t + r^2\cos^2 t}{[r^2(\sin^2 t + \cos^2 t)]^{3/2}} = \frac{r^2}{r^3} = \frac{1}{r} \qquad\blacksquare$$

The osculating circle may be interpreted as the circle whose shape best fits the shape of the graph of $\mathbf{R}(t)$ in the "vicinity" of P (see Figure 11.21). In the special case where the graph of $\mathbf{R}(t)$ is a circle C, the radius of curvature is the radius of C, and the osculating circle coincides with C.

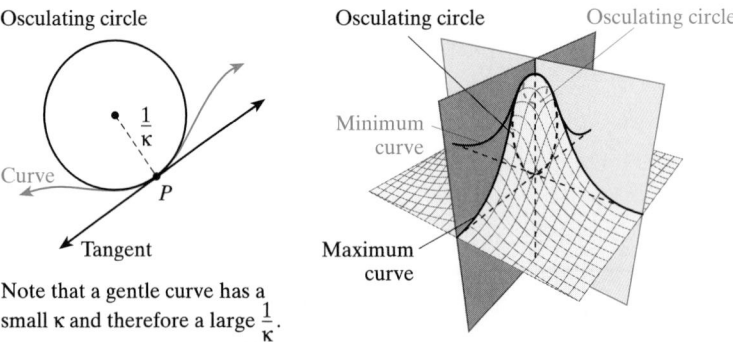

Note that a gentle curve has a small κ and therefore a large $\dfrac{1}{\kappa}$.

■ **FIGURE 11.21**   The osculating circle best fits the shape of the curve at P

We conclude this section by providing a summary of various formulas for computing curvature (Table 11.2).

■ TABLE 11.2
Curvature Formulas

Type	Given Information	Formula to Use	Where Derived
Arc length parameter	$\mathbf{R}(s)$	$\left\|\dfrac{d\mathbf{T}}{ds}\right\|$	Page 763
Vector derivative form	$\mathbf{R}(t)$	$\dfrac{\|\mathbf{R}' \times \mathbf{R}''\|}{\|\mathbf{R}'\|^3}$	Page 765, Problem 52
Parametric equation form	$x = x(t), y = y(t)$	$\dfrac{\|x'y'' - y'x''\|}{[(x')^2 + (y')^2]^{3/2}}$	Problem 53
Functional form	$y = f(x)$	$\dfrac{\|y''\|}{[1 + (y')^2]^{3/2}}$	Problem 54
Polar form	$r = f(\theta)$	$\dfrac{\|r^2 + 2r'^2 - rr''\|}{(r^2 + r'^2)^{3/2}}$	Problem 55

11.4   Problem Set

Ⓐ *In Problems 1–8, find the unit tangent vector* $\mathbf{T}(t)$ *and the principal unit normal vector* $\mathbf{N}(t)$ *for the curve given by* $\mathbf{R}(t)$.

1. $\mathbf{R}(t) = t^2\mathbf{i} + t^3\mathbf{j}, t \neq 0$

2. $\mathbf{R}(t) = t^2\mathbf{i} + \sqrt{t}\,\mathbf{j}, t > 0$

3. $\mathbf{R}(t) = (e^t \cos t)\mathbf{i} + (e^t \sin t)\mathbf{j}$

4. $\mathbf{R}(t) = (t \cos t)\mathbf{i} + (t \sin t)\mathbf{j}$

5. $\mathbf{R}(t) = (\cos t)\mathbf{i} + (\sin t)\mathbf{j} + t\mathbf{k}$

6. $\mathbf{R}(t) = (\sin t)\mathbf{i} - (\cos t)\mathbf{j} + t\mathbf{k}$

7. $\mathbf{R}(t) = (\ln t)\mathbf{i} + t^2\mathbf{k}$

8. $\mathbf{R}(t) = (e^{-t}\sin t)\mathbf{i} + e^{-t}\mathbf{j} + (e^{-t}\cos t)\mathbf{k}$

In Problems 9–14, find the length of the given curve over the given interval.

9. $\mathbf{R}(t) = 2t\mathbf{i} + t\mathbf{j}$, over $[0, 4]$

10. $\mathbf{R}(t) = t\mathbf{i} + 3t\mathbf{j}$, over $[0, 4]$

11. $\mathbf{R}(t) = 3t\mathbf{i} + (3 \cos t)\mathbf{j} + (3 \sin t)\mathbf{k}$, over $[0, \frac{\pi}{2}]$

12. $\mathbf{R}(t) = t\mathbf{i} + 2t\mathbf{j} + 3t\mathbf{k}$, over $[0, 2]$

13. $\mathbf{R}(t) = (4 \cos t)\mathbf{i} + (4 \sin t)\mathbf{j} + 5t\mathbf{k}$, over $[0, \pi]$

14. $\mathbf{R}(t) = (\cos^3 t)\mathbf{i} + (\cos^2 t)\mathbf{k}$, over $[0, \frac{\pi}{2}]$

Find the curvature of the plane curves at the points indicated in Problems 15–26.

15. $y = 4x - 2$ at $x = 2$

16. $y = mx + b$ at $x = a$

17. $y = x - \frac{1}{9}x^2$, at $x = 3$

18. $y = 2x^2 + 1$, at $x = 1$

19. $y = ax^2 + bx$, at $x = c$

20. $y = x + x^{-1}$, at $x = 1$

21. $y = \sqrt{4 - x^2}$, at $x = 1$

22. $y = \sqrt{r^2 - x^2}$, at $x = 0$

23. $y = \sin x$, at $x = \frac{\pi}{2}$

24. $y = \cos x$, at $x = \frac{\pi}{4}$

25. $y = \ln x$, at $x = 1$

26. $y = e^x$, at $x = 0$

Ⓑ **27.** Let $\mathbf{u}$ and $\mathbf{v}$ be constant, nonzero vectors. Show that the line given by $\mathbf{R}(t) = \mathbf{u} + \mathbf{v}t$ has curvature 0 at each point.

28. Find the unit tangent $\mathbf{T}$ and principal normal $\mathbf{N}$ for $\mathbf{R}(t) = (\cosh t)\mathbf{i} + (\sinh t)\mathbf{j}$ at the point where $t = 0$.

29. Find the unit tangent $\mathbf{T}$ and principal normal $\mathbf{N}$ for $\mathbf{R}(t) = [\ln(\sin t)]\mathbf{i} + [\ln(\cos t)]\mathbf{j}$ at the point where $t = \frac{\pi}{3}$.

30. A curve C in the plane is given parametrically by $x = 32t$, $y = 16t^2 - 4$.
   a. Sketch the graph of the curve.
   b. Find the unit tangent vector when $t = 3$.
   c. Find the radius of curvature of the point P on C where $t = 3$.

31. For the curve given by
$$\mathbf{R}(t) = (\sin t)\mathbf{i} + (\cos t)\mathbf{j} + t\mathbf{k}$$
   a. Find a unit tangent vector $\mathbf{T}$ at the point on the curve where $t = \pi$.
   b. Find the curvature when $t = \pi$.
   c. Find the length of the curve from $t = 0$ to $t = \pi$.

32. Let C be the curve given by
$$\mathbf{R}(t) = (t - \sin t)\mathbf{i} + (1 - \cos t)\mathbf{j} + (4 \sin \tfrac{1}{2})\mathbf{k}$$
   a. Find the unit tangent vector $\mathbf{T}(t)$ to C.
   b. Find $\dfrac{d\mathbf{T}}{ds}$ and the curvature $\kappa(t)$.

33. Find the point (or points) where the ellipse $9x^2 + 4y^2 = 36$ has maximum curvature.

34. Find the maximum curvature on the curve $y = e^{2x}$.

35. Find the radius of curvature at each relative extremum of the graph of $y = x^6 - 3x^2$.

36. Find the curvature of the curve given by $x = t - \sin t$, $y = 1 - \cos t$. Sketch the curve on $0 \le t \le 2\pi$ and sketch the osculating circle at the point where $t = \frac{\pi}{2}$.

37. The tangent line at a point P on a curve C in space is the line that passes through P and is aligned with the tangent vector $\mathbf{T}$ to C at P. Find parametric equations for the tangent line to $\mathbf{R}(t) = 2t\mathbf{i} - t\mathbf{j} + t^2\mathbf{k}$ at the point where $t = 1$.

38. The tangent line at a point P on a curve C in space is the line that passes through P and is aligned with the tangent vector $\mathbf{T}$ to C at P. Find parametric equations for the tangent line to $\mathbf{R}(t) = e^t\mathbf{i} - 3\mathbf{j} + (1 - t)\mathbf{k}$ at the point where $t = 0$.

There are many different formulas for finding curvature, as summarized in Table 11.2. In Problems 39–46, find the curvature of each of the given curves using the indicated formula.

39. $\mathbf{R}(t) = t\mathbf{i} + t^2\mathbf{j} + t^3\mathbf{k}$; vector derivative form

40. $\mathbf{R}(t) = (t - \cos t)\mathbf{i} + (\sin t)\mathbf{j} + 3\mathbf{k}$; vector derivative form

41. $y = x^2$; functional form

42. $y = x^3$; functional form

43. $y = x^{-1}, x > 0$; functional form

44. $y = \sin x$; functional form

45. the spiral $r = e^\theta$; polar form

46. the cardioid $r = 1 + \cos\theta$, for $0 \le \theta \le 2\pi$; polar form

47. A *pestus houseflyus* is observed to zip around a room in such a way that at time t its position with respect to the nose of an observer is given by the vector function

$$\mathbf{R}(t) = t\mathbf{u} + t^2\mathbf{v} + 2\left(\frac{2}{3}t\right)^{3/2}(\mathbf{u} \times \mathbf{v})$$

where $\mathbf{u}$ and $\mathbf{v}$ are unit vectors separated by an angle of 60°. Compute the fly's speed and find how long it takes to move a distance of 20 units along its path (starting from the nose).

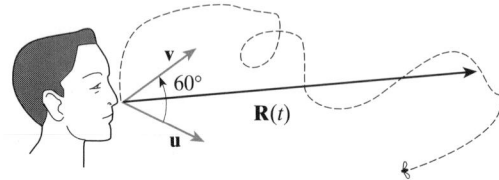

48. ■ **What Does This Say?** Describe what is meant by using arc length as a parameter.

49. ■ **What Does This Say?** Discuss the various curvature formulas given in Table 11.2.

50. **Projections onto a plane in $\mathbb{R}^3$**   A very important notion in applied mathematics is that of projecting a vector onto a subspace of some kind, for example, a plane. Later in your career, the "vectors" can also be functions in which a function is approximated by "nice" functions. In this problem, our work is limited to $\mathbb{R}^3$.

 a. Suppose that $\mathbf{u}$ and $\mathbf{v}$ are unit vectors and that they are orthogonal. Show that the "linear combinations" $a\mathbf{u} + b\mathbf{v}$ generate a plane, through $(0, 0, 0)$, as a and b take on all real values.

 b. Let $\mathbf{w}$ be a third vector, not on the plane described in part **a**. We want to project $\mathbf{w}$ onto the plane; we write $\mathbf{w} = \mathbf{p} + \mathbf{n} = (a\mathbf{u} + b\mathbf{v}) + \mathbf{n}$, where $\mathbf{p}$ is the desired projection. To find a, take the dot product of $\mathbf{w}$ and $\mathbf{u}$ and show that $\mathbf{p}$ is orthogonal to $\mathbf{n}$ (hence $\mathbf{p}$ is the orthogonal projection).

 c. For the three following vectors, find the orthogonal projection of $\mathbf{w}$ onto the plane generated by $\mathbf{u}$ and $\mathbf{v}$:

$$\mathbf{u} = \frac{1}{\sqrt{3}}(\mathbf{i} + \mathbf{j} + \mathbf{k})$$

$$\mathbf{v} = \frac{1}{\sqrt{6}}(2\mathbf{i} - \mathbf{j} - \mathbf{k})$$

$$\mathbf{w} = \mathbf{j} + 2\mathbf{k}$$

 d. A projection is especially simple if we have $\mathbf{u} = \mathbf{i}$ and $\mathbf{v} = \mathbf{j}$. Explain the situation here for an arbitrary vector $\mathbf{w} = a\mathbf{i} + b\mathbf{j} + c\mathbf{k}$.

 e. Show that for a given vector $\mathbf{w} = a\mathbf{i} + b\mathbf{j} + c\mathbf{k}$, its slope relative to the xy-plane is $m = \dfrac{c}{\sqrt{a^2 + b^2}}$.

51. This problem will make some use of Problem 50, and addresses arc length of a curve in $\mathbb{R}^3$ and the work involved in moving along its path. Suppose a mining company has built a road up a hill modeled by the following parametric equations ($0 \le t \le 2\pi$):

$$x(t) = e^{-t/3}\cos 3t$$

$$y(t) = e^{-t/3}\sin 3t$$

$$z(t) = \frac{13t}{t^2 + 40}$$

 a. Generate a rough graph of this curve. Then somehow (use your imagination!) get a decent estimate of its length, say, within 10%.

 b. Use your computer to compute its length. Then compare with your estimate from part **a**.

---

c. Let $\mathbf{T}(t)$ denote the tangent to the curve at the point and assume the force

$$F(t) = \frac{8z'(t)}{\sqrt{[x'(t)]^2 + [y'(t)]^2}}$$

Compute the work required to move a cart from the bottom to the top of the hill. (Recall that work $= \int F \, ds$.)

---

52. Let $\mathbf{R}(t)$ be a smooth vector function.
    a. Differentiate $\mathbf{R}' = \|\mathbf{R}'\|\mathbf{T}$ to show that

$$\mathbf{R}'' = \|\mathbf{R}'\|'\mathbf{T} + \kappa\|\mathbf{R}'\|^2\mathbf{N}$$

    b. Show that

$$\mathbf{R}' \times \mathbf{R}'' = \kappa\|\mathbf{R}'\|^3(\mathbf{T} \times \mathbf{N})$$

    c. Conclude that

$$\kappa = \frac{\|\mathbf{R}' \times \mathbf{R}''\|}{\|\mathbf{R}'\|^3}$$

53. Use the formula in Problem 52c to verify the formula

$$\kappa = \frac{|x'y'' - y'x''|}{[(x')^2 + (y')^2]^{3/2}}$$

    Hint: Let $\mathbf{R}(t) = x(t)\mathbf{i} + y(t)\mathbf{j}$.

54. A curve in the plane is given by $y = f(x)$. Show that the functional form for curvature is given by

$$\frac{|f''(x)|}{\{1 + [f'(x)]^2\}^{3/2}}$$

    (*Hint:* Use the formula in Problem 53 with $\mathbf{R}(x) = x\mathbf{i} + f(x)\mathbf{j}$.)

55. Let f be a twice differentiable function. Show that, in polar coordinates, the curvature of the curve given by $r = f(\theta)$ satisfies

$$\kappa(\theta) = \frac{|r^2 + 2r'^2 - rr''|}{(r^2 + r'^2)^{3/2}}$$

    (*Hint:* Use the formula in Problem 52 with $x = f(\theta)\cos\theta, y = f(\theta)\sin\theta$.)

56. Use the formula in Problem 52 to show that the curve C given by $\mathbf{R}(t)$ has curvature

$$\kappa = \frac{[\|\mathbf{V}\|^2\|\mathbf{A}\|^2 - \mathbf{V} \cdot \mathbf{A}]^{1/2}}{\|\mathbf{V}\|^3}$$

    Hint: Use the identity

$$\|\mathbf{u} \times \mathbf{v}\|^2 = (\|\mathbf{u}\| \|\mathbf{v}\|)^2 - (\mathbf{u} \cdot \mathbf{v})^2$$

57. Let $P(a, b)$ be a point on the graph C of the vector function $\mathbf{R}(t)$.

a. Describe a general procedure for finding an equation for the osculating circle to C at P.

b. Find an equation for the osculating circle at the point $P(32, 12)$ on the curve C defined by

$$x = 32t, \quad y = 16t^2 - 4$$

58. If $\mathbf{T}$ and $\mathbf{N}$ are the unit tangent and normal vectors, respectively, on the trajectory of a moving body, then the cross product vector $\mathbf{B} = \mathbf{T} \times \mathbf{N}$ is called the unit **binormal** of the trajectory. Three planes determined by $\mathbf{T}, \mathbf{N}$, and $\mathbf{B}$ are shown in Figure 11.22.

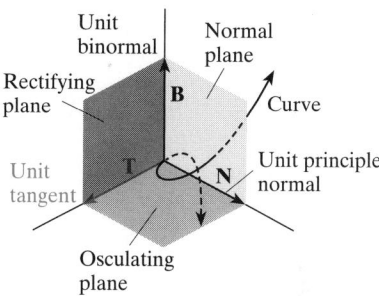

■ **FIGURE 11.22**   Three planes determined by $\mathbf{T}, \mathbf{N}$, and $\mathbf{B}$

a. Show that $\mathbf{T}$ is orthogonal to $\dfrac{d\mathbf{B}}{ds}$.

   (*Hint:* Differentiate $\mathbf{B} \cdot \mathbf{T}$.)

b. Show that $\mathbf{B}$ is orthogonal to $\dfrac{d\mathbf{B}}{ds}$.

   (*Hint:* Differentiate $\mathbf{B} \cdot \mathbf{B}$.)

c. Show that $\dfrac{d\mathbf{B}}{ds} = -\tau\mathbf{N}$ for some constant τ. (*Note:* τ is called the **torsion** of the trajectory.)

59. Prove the Frenet-Serret formulas:

$$\frac{d\mathbf{T}}{ds} = \kappa\mathbf{N}$$

$$\frac{d\mathbf{N}}{ds} = -\kappa\mathbf{T} + \tau\mathbf{B}$$

$$\frac{d\mathbf{B}}{ds} = -\tau\mathbf{N}$$

where κ is the curvature and $\tau = \tau(s)$ is a scalar function called the torsion, which provides a measure of the amount of twisting at each point on the trajectory.

60. Show that the torsion may be computed by the formula

$$\tau = \frac{[\mathbf{R}'(t) \times \mathbf{R}''(t)] \cdot \mathbf{R}'''(t)}{\|\mathbf{R}'(t) \times \mathbf{R}''(t)\|^2}$$

    Use this formula to find the torsion for the helix

$$\mathbf{R}(t) = (a\cos t)\mathbf{i} + (a\sin t)\mathbf{j} + (bt)\mathbf{k}$$

61. A highway has an exit ramp that begins at the origin and follows the curve $y = \frac{1}{32}x^{5/2}$ to the point $(4, 1)$. Then it follows the shape of the osculating circle at $(4, 1)$ until the point where $y = 3$. What is the total length of the exit ramp?

11.5 *Tangential and Normal Components of Acceleration*

COMPONENTS OF ACCELERATION

When a body is caused to accelerate or brakes are applied, it is of interest to know how much of the acceleration acts in the direction of the body's motion, as indicated by the unit tangent vector **T**. This question is answered by the following theorem.

THEOREM 11.8   *Tangential and normal components of acceleration*

An object moving along a smooth curve (with $\mathbf{T}' \neq \mathbf{0}$) has velocity **V** and acceleration **A**, where

$$\mathbf{V} = \left(\frac{ds}{dt}\right)\mathbf{T} \quad \text{and} \quad \mathbf{A} = \left(\frac{d^2s}{dt^2}\right)\mathbf{T} + \kappa\left(\frac{ds}{dt}\right)^2\mathbf{N}$$

and s is the arc length along the trajectory.

Proof   An object moving with displacement **R** has unit tangent $\mathbf{T} = \dfrac{d\mathbf{R}}{ds}$ and unit normal $\mathbf{N} = \dfrac{1}{\kappa}\dfrac{d\mathbf{T}}{ds}$. We use the chain rule to write the velocity vector **V** as follows:

$$\mathbf{V} = \frac{d\mathbf{R}}{dt} = \frac{d\mathbf{R}}{ds}\frac{ds}{dt} = \mathbf{T}\frac{ds}{dt}$$

Differentiate both sides of this equation with respect to t and substitute $\dfrac{d\mathbf{T}}{ds} = \kappa\mathbf{N}$:

$$\mathbf{A} = \frac{d\mathbf{V}}{dt} \qquad\qquad \text{Definition of } \mathbf{A}$$

$$= \frac{d}{dt}\left[\mathbf{T}\frac{ds}{dt}\right] \qquad \mathbf{V} = \left(\frac{ds}{dt}\right)\mathbf{T}$$

$$= \frac{d^2s}{dt^2}\mathbf{T} + \frac{ds}{dt}\frac{d\mathbf{T}}{dt} \qquad \text{Product rule}$$

$$= \frac{d^2s}{dt^2}\mathbf{T} + \frac{ds}{dt}\left[\frac{d\mathbf{T}}{ds}\frac{ds}{dt}\right] \qquad \text{Chain rule}$$

$$= \frac{d^2s}{dt^2}\mathbf{T} + \left(\frac{ds}{dt}\right)^2\frac{d\mathbf{T}}{ds}$$

$$= \frac{d^2s}{dt^2}\mathbf{T} + \left(\frac{ds}{dt}\right)^2(\kappa\mathbf{N}) \qquad \frac{d\mathbf{T}}{ds} = \kappa\mathbf{N}$$

■ *What This Says:*   At each point on the trajectory of a moving object, the velocity **V** points in the direction of the unit tangent **T**, but the acceleration **A** has both a tangential and a normal component. The trajectory may twist and turn, but the acceleration is always in the plane determined by **T** and the unit normal **N**.

The two components of acceleration have special names.

Tangential and Normal Components of Acceleration

The acceleration $\mathbf{A}$ of a moving object can be written as

$$\mathbf{A} = A_T\mathbf{T} + A_N\mathbf{N}$$

where

$$A_T = \frac{d^2s}{dt^2} \qquad \text{is the \textbf{tangential component}.}$$

$$A_N = \kappa\left(\frac{ds}{dt}\right)^2 \qquad \text{is the \textbf{normal component}.}$$

The tangential and normal components are shown in Figure 11.23.

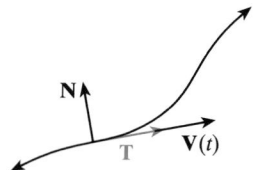

a. Velocity has only a nonzero tangential component.

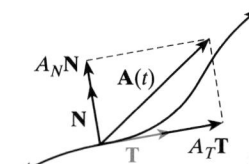

b. Acceleration may have both tangential and normal component.

■ **FIGURE 11.23**   Components of velocity and acceleration

It is usually fairly easy to find A_T, but it may be more difficult to find A_N, because computing the curvature κ is often a messy process. Fortunately, there is a way to compute A_N without first finding κ. We expand the dot product of $\mathbf{A} = A_T\mathbf{T} + A_N\mathbf{N}$ with itself and use the fact that $\mathbf{T}$ and $\mathbf{N}$ are orthogonal unit vectors:

$$
\begin{aligned}
\|\mathbf{A}\|^2 &= \mathbf{A} \cdot \mathbf{A} \\
&= (A_T\mathbf{T} + A_N\mathbf{N}) \cdot (A_T\mathbf{T} + A_N\mathbf{N}) \qquad \mathbf{A} = A_T\mathbf{T} + A_N\mathbf{N} \\
&= A_T^2(\mathbf{T} \cdot \mathbf{T}) + 2A_TA_N(\mathbf{N} \cdot \mathbf{T}) + A_N^2(\mathbf{N} \cdot \mathbf{N}) \\
&= A_T^2(1) + 2A_TA_N(0) + A_N^2(1) \qquad \mathbf{T} \cdot \mathbf{T} = 1, \mathbf{N} \cdot \mathbf{T} = 0, \textit{and } \mathbf{N} \cdot \mathbf{N} = 1 \\
&= A_T^2 + A_N^2
\end{aligned}
$$

Thus, once we know $\mathbf{A}$ and A_T, we can compute A_N by applying the following formula.

Computation of Normal Component

The normal component A_N can be found using the formula

$$A_N = \sqrt{\|\mathbf{A}\|^2 - A_T^2}$$

EXAMPLE 1   *Finding tangential and normal components of acceleration*

Find the tangential and normal components of an object that moves with displacement

$$\mathbf{R}(t) = \langle t^3, t^2, t \rangle$$

Solution $\mathbf{V} = \dfrac{d\mathbf{R}}{dt} = \langle 3t^2, 2t, 1 \rangle$ and $\mathbf{A} = \dfrac{d\mathbf{V}}{dt} = \langle 6t, 2, 0 \rangle$

$$\frac{ds}{dt} = \|\mathbf{V}\| = \sqrt{(3t^2)^2 + (2t)^2 + (1)^2} = \sqrt{9t^4 + 4t^2 + 1}$$

$$A_\mathrm{T} = \frac{d^2 s}{dt^2} = \frac{1}{2}(9t^4 + 4t^2 + 1)^{-1/2}(36t^3 + 8t) = \frac{18t^3 + 4t}{\sqrt{9t^4 + 4t^2 + 1}}$$

This is the tangential component of acceleration.

$$A_\mathrm{N} = \sqrt{\|\mathbf{A}\|^2 - A_\mathrm{T}^2} = \sqrt{\left[\sqrt{36t^2 + 4}\right]^2 - \left[\frac{18t^3 + 4t}{\sqrt{9t^4 + 4t^2 + 1}}\right]^2}$$

$$= \sqrt{4(9t^2 + 1) - \frac{4t^2(9t^2 + 2)^2}{9t^4 + 4t^2 + 1}}$$

$$= \sqrt{\frac{36t^4 + 36t^2 + 4}{9t^4 + 4t^2 + 1}}\qquad \text{\textit{There are several simplification steps}}$$
$$\text{\textit{that are not shown.}}$$

$$= 2\sqrt{\frac{9t^4 + 9t^2 + 1}{9t^4 + 4t^2 + 1}}$$

This is the normal component of acceleration.   ■

EXAMPLE 2   *Finding tangential and normal components on a helix*

An object moves along the helix with position vector

$$\mathbf{R}(t) = (\cos t)\mathbf{i} + (\sin t)\mathbf{j} + t\mathbf{k}$$

Find the tangential and normal components of acceleration.

Solution

$$\mathbf{V} = \frac{d\mathbf{R}}{dt} = (-\sin t)\mathbf{i} + (\cos t)\mathbf{j} + \mathbf{k}\quad\text{and}\quad \mathbf{A} = \frac{d\mathbf{V}}{dt} = (-\cos t)\mathbf{i} + (-\sin t)\mathbf{j}$$

$$\frac{ds}{dt} = \|\mathbf{V}\| = \sqrt{\sin^2 t + \cos^2 t + 1} = \sqrt{2}\quad\text{and}\quad A_\mathrm{T} = \frac{d^2 s}{dt^2} = 0$$

$$A_\mathrm{N} = \sqrt{\|\mathbf{A}\|^2 - A_\mathrm{T}^2} = \sqrt{\left(\sqrt{\cos^2 t + \sin^2 t}\right)^2 - 0^2} = 1$$

The tangential and normal components of acceleration are 0 and 1, respectively. This means that the acceleration satisfies

$$\mathbf{A} = A_\mathrm{T}\mathbf{T} + A_\mathrm{N}\mathbf{N} = (0)\mathbf{T} + (1)\mathbf{N} = \mathbf{N}$$

That is, the acceleration vector is the principal unit normal $\mathbf{N}$, and the acceleration is always normal to the trajectory of the uniform helix.   ■

APPLICATIONS

Now that we know how to compute the tangential and normal components of acceleration, A_T and A_N, we shall examine some applications. First, according to Newton's second law of motion, the total force acting on a moving object of mass m satisfies $\mathbf{F} = m\mathbf{A}$, where $\mathbf{A}$ is the acceleration of the object. Because $\mathbf{A} = A_\mathrm{T}\mathbf{T} + A_\mathrm{N}\mathbf{N}$, we have

$$\mathbf{F} = m\mathbf{A} = (mA_\mathrm{T})\mathbf{T} + (mA_\mathrm{N})\mathbf{N} = F_\mathrm{T}\mathbf{T} + F_\mathrm{N}\mathbf{N}$$

where

$$F_{\mathrm{T}} = m\frac{d^2s}{dt^2} \quad \text{and} \quad F_{\mathrm{N}} = m\kappa\left(\frac{ds}{dt}\right)^2$$

For instance, experience leads us to expect a car to skid if it makes a sharp turn at moderate speed or even a gradual turn at high speed. Mathematically, a "sharp turn" occurs when the radius of curvature $\rho = 1/\kappa$ is small (that is, when κ is large), and "high speed" means that ds/dt is large. In either case,

$$\mathbf{F}_{\mathrm{N}} = m\kappa\left(\frac{ds}{dt}\right)^2$$

will be relatively large, and the car will stay on the road (its trajectory) only if there is a correspondingly large frictional force between the tires of the car and the surface of the road (see Figure 11.24).

When forces, masses, and accelerations are involved, it is customary to express the mass m in slugs of an object whose weight W has been given in pounds.* From the definition of weight,

$$m = \frac{W}{g}$$

where g is the acceleration of gravity ($g \approx 32$ ft/s^2).

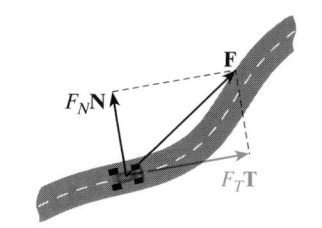

■ **FIGURE 11.24**   Tendency to skid

| EXAMPLE 3 | *Modeling Application: Tendency of a vehicle to skid* |

A car weighing 2,700 lb makes a turn on a flat road while traveling at 56 ft/s (about 38 mi/h). If the radius of the turn is 21 ft, how much frictional force is required to keep the car from skidding?

Solution   The required frictional force is $F_{\mathrm{N}} = m\kappa\left(\dfrac{ds}{dt}\right)^2$, where m is the mass of the car and κ is the curvature of the road. We know that

$$\frac{ds}{dt} = 56 \text{ ft/s}$$

and because the car weighs $W = 2{,}700$ lb, its mass is $m = \dfrac{W}{g} = \dfrac{2{,}700}{32} \approx 84.38$ slugs. Because the turn radius is 21 ft, we have $\kappa = \dfrac{1}{21}$, so that

$$\mathbf{F}_{\mathrm{N}} = \left(\frac{2{,}700}{32} \text{ slugs}\right)\left(\frac{1}{21 \text{ ft}}\right)\left(56 \frac{\text{ft}}{\text{s}}\right)^2 = 12{,}600 \, \frac{\text{lb} \cdot \text{s}^2}{\text{ft}} \frac{1}{\text{ft}} \frac{\text{ft}^2}{\text{s}^2} = 12{,}600 \text{ lb} \quad \blacksquare$$

There are certain important applications in which an object moves along its trajectory with constant speed ds/dt, and when this occurs, the acceleration $\mathbf{A}$ can have only a normal component, because $d^2s/dt^2 = 0$.

THEOREM 11.9   *Acceleration of an object with constant speed*

The acceleration of an object moving with constant speed is always orthogonal to the direction of motion.

---

*A slug is a unit of measurement defined as the unit of mass that receives an acceleration of 1 ft/s^2 when a force of 1 lb is applied to it. That is, 1 slug $= \dfrac{1 \text{ lb}}{1 \text{ ft/s}^2} = \dfrac{\text{lb s}^2}{\text{ft}}$.

Proof   Notice that we really do not need to know anything about components of acceleration to prove this result. Saying that the object has constant speed means that $\|\mathbf{R}'(t)\|$ is constant, and by Theorem 11.4, we conclude that $\mathbf{R}'(t)$ is orthogonal to its derivative $\mathbf{R}''(t) = \mathbf{A}(t)$. But $\mathbf{R}'(t)$ points in the direction of the object's motion along its trajectory, which means that the acceleration $\mathbf{A}$ is orthogonal to the direction of motion.   ▬

As an application of this result, note that when an object moves with constant speed v_0 along a circular path of radius R (so that $\kappa = 1/R$), its acceleration is directed toward the center of the path and has magnitude

$$A_N = \kappa \left(\frac{ds}{dt}\right)^2 = \frac{1}{R} v_0^2$$

(See Figure 11.25.)

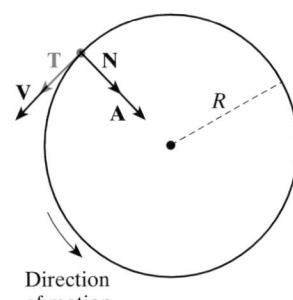

If the speed is v_0 and the path has radius R, the acceleration is

$$\mathbf{A} = \frac{v_0^2}{R}\mathbf{N}$$

Direction of motion

■ **FIGURE 11.25**   An object moving with constant speed on a circular path

---

EXAMPLE 4   *Modeling Application: Period of a satellite*

An artificial satellite travels at constant speed in a stable circular orbit 20,000 km above the earth's surface. How long does it take for the satellite to make one complete circuit of the earth?

Solution   Let m denote the satellite's mass and v denote its speed. We shall assume that the earth is a sphere of radius 6,440 km, so the curvature of the path is $\kappa = 1/R$, where

$$R = \underbrace{6{,}440}_{\text{Radius of earth}} + \overbrace{20{,}000}^{\text{Height}} = 26{,}440 \text{ km}$$

is the distance of the satellite from the center of the earth. The satellite maintains a stable orbit when the force mv^2/R produced by its centripetal acceleration equals the force F_g due to gravity. (See Figure 11.26.) According to Newton's law of universal gravitation, $\|\mathbf{F}_c\| = GmM/R^2$, where M is the mass of the earth and G is the *gravitational constant*. Thus, for stability we must have

$$\frac{mv^2}{R} = \frac{GmM}{R^2}$$

so $v = \sqrt{GM/R}$. Experiments indicate that $GM = 398{,}600$ km³/s², and by substituting $R = 26{,}440$, we obtain

$$v = \sqrt{\frac{GM}{R}} = \sqrt{\frac{398{,}600}{26{,}440}} \approx 3.88273653$$

For this example, we see the speed of the satellite is approximately 3.883 km/s.

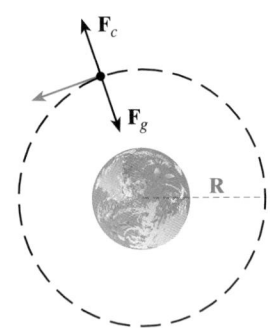

■ **FIGURE 11.26**   A satellite in a stable orbit; centripetal force $\mathbf{F}_g$ equals the force $\mathbf{F}_c$ due to gravity.

Finally, suppose T is the time required for the satellite to make one complete circuit of the earth (called the **period** of the satellite). In each period, the satellite travels a distance equal to the circumference of a circle of radius $R = 26{,}440$ km, and because it travels at v km/s, we must have $vT = 2\pi R$, so that

$$T = \frac{2\pi R}{v} \approx \frac{2\pi(26{,}440 \text{ km})}{3.88273653} \approx 42{,}786.16852 \text{ seconds}$$

or approximately 713 minutes (11 h 53 min). ∎

11.5 Problem Set

Ⓐ *In Problems 1–8, $\mathbf{R}(t)$ is the position vector of a moving object. Find the tangential and normal components of the object's acceleration.*

1. $\mathbf{R}(t) = t\mathbf{i} + t^2\mathbf{j}$
2. $\mathbf{R}(t) = t\mathbf{i} + e^t\mathbf{j}$
3. $\mathbf{R}(t) = \langle t \sin t, t \cos t \rangle$
4. $\mathbf{R}(t) = \langle 3 \cos t, 2 \sin t \rangle$
5. $\mathbf{R}(t) = \langle t, t^2, t \rangle$
6. $\mathbf{R}(t) = \langle 4 \cos t, 0, \sin t \rangle$
7. $\mathbf{R}(t) = (\sin t)\mathbf{i} + (\cos t)\mathbf{j} + (\sin t)\mathbf{k}$
8. $\mathbf{R}(t) = (\frac{5}{13} \cos t)\mathbf{i} + \frac{12}{13}(1 - \cos t)\mathbf{j} + \sin t\mathbf{k}$

In Problems 9–12, the velocity $\mathbf{V}$ and acceleration $\mathbf{A}$ of a moving object are given at a certain instant. Find $\mathbf{T}$, $\mathbf{N}$, and A_T, A_N at this instant.

9. $\mathbf{V} = \langle 1, -3 \rangle$; $\mathbf{A} = \langle 2, 5 \rangle$
10. $\mathbf{V} = -\mathbf{i} + 7\mathbf{j}$; $\mathbf{A} = 4\mathbf{i} + 5\mathbf{j}$
11. $\mathbf{V} = 2\mathbf{i} + 3\mathbf{j} - \mathbf{k}$; $\mathbf{A} = -\mathbf{i} - 5\mathbf{j} + 2\mathbf{k}$
12. $\mathbf{V} = \langle 5, -1, 2 \rangle$; $\mathbf{A} = \langle 1, 0, -7 \rangle$

In Problems 13–16, the speed $\|\mathbf{V}\|$ of a moving object is given. Find A_T, the tangential component of acceleration, at the indicated time.

13. $\|\mathbf{V}\| = \sqrt{5t^2 + 3}$; $t = 1$
14. $\|\mathbf{V}\| = \sqrt{t^2 + t + 1}$; $t = 3$
15. $\|\mathbf{V}\| = \sqrt{\sin^2 t + \cos(2t)}$; $t = 0$
16. $\|\mathbf{V}\| = \sqrt{e^{-t} + t^4}$; $t = 0$

17. Find the maximum and minimum speeds of a particle whose position vector is

$$\mathbf{R}(t) = (4 \sin 2t)\mathbf{i} - (3 \cos 2t)\mathbf{j}$$

18. Where on the trajectory of

$$\mathbf{R}(t) = (2t^2 - 5t)\mathbf{i} + (5t + 2)\mathbf{j} + 4t^2\mathbf{k}$$

is the speed minimized? What is the minimum speed?

Ⓑ 19. The position of an object at time t is (x, y), where $x = 1 + \cos 2t$, $y = \sin 2t$. Find the velocity, the acceleration, and the normal and tangential components of acceleration of the object at time t.

20. An object moves with constant angular velocity ω around the circle $x^2 + y^2 = r^2$ in the xy-plane. The position vector is

$$\mathbf{R}(t) = (r \cos \omega t)\mathbf{i} + (r \sin \omega t)\mathbf{j}$$

a. Find the tangential and normal components of acceleration.
b. Show that the curvature is $\kappa = 1/r$ at each point on the circle.

21. Find the tangential and normal components of the acceleration of an object that moves along the parabolic path $y^2 = 4x^2$ at the instant the speed is $ds/dt = 20$.

MODELING PROBLEMS *In Problems 22–25, set up an appropriate model to answer the given questions. Be sure to state your assumptions.*

22. A pail attached to a rope 1 yd long is swung at the rate of 1 rev/s. Find the tangential and normal components of the pail's acceleration. Assume the rope is swung in a level plane.

23. A boy holds onto a pail of water weighing 2 lb and swings it in a vertical circle with a radius of 3 ft. If the pail travels at ω rpm, what is the pressure of the water on the bottom of the pail at the highest and lowest points of the swing? What is the *smallest* value of ω required to keep the water from spilling from the pail? Assume the pail is held by a handle so that its bottom is straight up when it is at its highest point.

24. A car weighing 2,700 lb (about 1.35 tons) moves along the elliptic path $900x^2 + 400y^2 = 1$. If the car travels at the constant speed 45 mi/h, how much frictional force is required to keep it from skidding as it turns the "corner" at $(\frac{1}{30}, 0)$? What about the corner $(0, \frac{1}{20})$?

25. Curved sections of road such as expressway offramps are often banked to protect against skidding. Consider a road with a circular curve of radius 150 feet, and assume the magnitude of the force of static friction $\mathbf{F}_s$ (the force resisting the tendency to skid) is proportional to the car's weight; that is $\|\mathbf{F}_s\| = \mu W$, where μ is a constant called the *coefficient of static friction*. For this problem we assume $\mu = 0.47$.

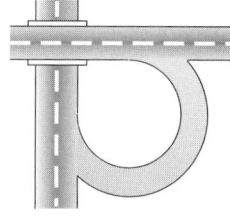

a. First, suppose the road is not banked. What is the largest speed $\mathbf{v}_m$ that a 3,500 lb car can travel around the curve without skidding? Would the answer be different for a smaller car, say one weighing 2,000 lbs?

b. Now suppose the curve is banked at 17° and answer the questions in part **a.**

c. Suppose the highway engineers want the maximum safe speed at 50 mph. At what angle should the road be banked?

26. What is the smallest radius that should be used for a circular highway curve if the normal component of the acceleration of a car traveling at 45 mi/h is not to exceed 2.4 ft/s²?

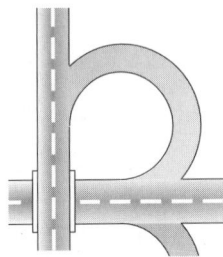

27. A Ferris wheel with radius 15 ft rotates in a vertical plane at ω rpm. What is the maximum value of ω for a Ferris wheel carrying a person of weight W? (That is, the largest ω so the passenger is not "thrown off.")

Use the formulas in Problem 35 to find A_T and A_N for the given position vector $\mathbf{R}(t)$ in Problems 28–31.

28. $\mathbf{R}(t) = t^3\mathbf{i} + t^2\mathbf{j} + t\mathbf{k}$ 29. $\mathbf{R}(t) = t\mathbf{i} + 2t\mathbf{j} + t^2\mathbf{k}$

30. $\mathbf{R}(t) = (\cos t)\mathbf{i} + (\sin t)\mathbf{j} + \mathbf{k}$

31. $\mathbf{R}(t) = (e^t \cos t)\mathbf{i} + (e^t \sin t)\mathbf{j} + e^t\mathbf{k}$

C 32. Let $\mathbf{T}$ be the unit tangent vector, $\mathbf{N}$ the principal unit normal, and $\mathbf{B}$ the unit binormal vector (see Problem 58, Section 11.4) to a given curve C. Show that

$$\frac{d\mathbf{B}}{ds} = \mathbf{T} \times \frac{d\mathbf{N}}{ds}$$

33. **MODELING PROBLEM**   An amusement park ride consists of a large (25-ft radius), flat, horizontal wheel. Customers board the wheel while it is stationary and try to stay on as long as possible as it begins to rotate. The purpose of this problem is to discover how fast the wheel can rotate without losing its passengers.

a. Suppose the wheel rotates at ω revolutions per minute and that a volunteer weighing W lb sits 15 feet from the center of the wheel. Find $\mathbf{F}_N = m\mathbf{A}_N$, where $m = W/g$ is the volunteer's mass. This is the force tending to push the volunteer off the wheel.

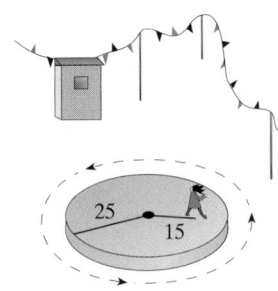

b. If the frictional force of the volunteer in part **a** to stay on the wheel is $0.12W$, find the largest value of ω that will allow the volunteer to stay in place. Wait, we did not tell you the volunteer's weight. Does it matter?

34. **MODELING PROBLEM**   A curve in a railroad track has the shape of the parabola $x = y^2/120$. If a train is loaded so that its scalar normal component of acceleration cannot exceed 30 units/s², what is its maximum possible speed as it rounds the curve at $(0, 0)$?

35. Suppose the position vector of a moving object is $\mathbf{R}(t)$. Show that the tangential and normal components of the object's acceleration may be computed by the formulas

$$A_T = \frac{\mathbf{R}' \cdot \mathbf{R}''}{\|\mathbf{R}'\|} \quad \text{and} \quad A_N = \frac{\|\mathbf{R}' \times \mathbf{R}''\|}{\|\mathbf{R}'\|}$$

36. Suppose an object moves in the plane along the curve $y = f(x)$. Use the formulas obtained in Theorem 11.8 to show that

$$A_T = \frac{f'(x)f''(x)}{\sqrt{1 + [f'(x)]^2}}; \quad \text{and} \quad A_N = \frac{|f''(x)|}{\sqrt{1 + [f'(x)]^2}}$$

37. If the graph of the function of f has an inflection point at $x = a$ and $f''(a)$ exists, show that the graph of f has curvature 0 at $(a, f(a))$.

38. A projectile is fired from ground level with an angle of elevation α and muzzle speed v_0. Find formulas for the tangential and normal components of the projectile's acceleration at time t. What are A_T and A_N at the time the projectile is at its maximum height?

39. An object connected to a string of length r is spun counterclockwise in a circular path in a horizontal plane. Let ω be the constant angular velocity of the object.

a. Show that the displacement vector of the object is

$$\mathbf{R}(t) = (r \cos \omega t)\mathbf{i} + (r \sin \omega t)\mathbf{j}$$

b. Find the normal component of the object's acceleration.

c. If the angular velocity ω is doubled, what is the effect on the normal component of acceleration? What is the effect on **A** if ω is unchanged but the length of the string is doubled? What happens to A_N if ω is doubled and r is halved?

40. An artificial satellite travels at constant speed in a stable circular orbit R km above the earth's surface (as in Example 4).

a. Show that the satellite's period is given by the formula

$$T = 2\pi \frac{(R + R_e)^{3/2}}{GM}$$

where R_e is the radius of the earth.

b. A satellite is said to be in **geosynchronous orbit** if it completes one orbit every sidereal day (23 hours,

56 minutes). How high above the earth should the satellite be to achieve such an orbit? Assume the radius of the earth is approximately 6,440 km.

c. What is the speed of a satellite in geosynchronous orbit?

41. MODELING PROBLEM   For Mars the following facts are known (all in relation to Earth):

diameter	0.533
length of day	1.029
mass	0.1074
gravity(g)	0.3776

Use these facts to determine how high above the surface of Mars a satellite must be to achieve synchronous orbit. How fast would such a satellite be traveling?

Chapter 11   Review

Proficiency Examination

Concept Problems

1. What is a vector-valued function?
2. What are the components of a vector function?
3. Describe what is meant by the graph of a vector function.
4. Define the limit of a vector function.
5. Define the derivative of a vector function.
6. Define the integral of a vector function.
7. What is a smooth curve?
8. Complete the following rules for differentiating vector functions:
   a. Linearity rule
   b. Scalar multiple rule
   c. Dot product rule
   d. Cross product rule
   e. Chain rule
9. State the theorem about the orthogonality of a derivative of constant length.
10. What are the position, velocity, and acceleration vectors?
11. What is the speed of a particle moving on a curve C?
12. What are the formulas for the motion of a projectile in a vacuum?
13. What are the formulas for time of flight and range of a projectile?
14. State Kepler's laws.
15. What are the unit polar vectors, $\mathbf{u}_r$ and $\mathbf{u}_\theta$?
16. What are the unit tangent and normal vectors?
17. What is speed in terms of arc length?
18. Give a formula for the arc length of a space curve.

19. What is the curvature of a graph?
20. What is the formula for curvature in terms of the velocity and acceleration vectors?
21. What are the formulas for unit tangent and principal unit normal in terms of the arc length parameter?
22. What is the radius of curvature?
23. What are the tangential and normal components of acceleration?

Practice Problems

24. Sketch the graph of $\mathbf{R}(t) = (3 \cos t)\mathbf{i} + (3 \sin t)\mathbf{j} + t\mathbf{k}$, and find the length of this curve from $t = 0$ to $t = 2\pi$.

25. If $\mathbf{F}(t) = \left[\dfrac{t}{1 + t}, \dfrac{\sin t}{t}, \cos t \right]$, find $\mathbf{F}'(t)$ and $\mathbf{F}''(t)$.

26. Evaluate $\displaystyle\int_1^2 \langle 3t, 0, 3 \rangle \times \langle 0, \ln t, -t^2 \rangle \, dt$.

27. Find a vector function **F** such that $\mathbf{F}''(t) = \langle e^t, -t^2, 3 \rangle$ and $\mathbf{F}(0) = \langle 1, -2, 0 \rangle$, $\mathbf{F}'(0) = \langle 0, 0, 3 \rangle$.

28. Find the velocity **V**, the speed $\dfrac{ds}{dt}$, and the acceleration **A** for the body with position vector $\mathbf{R}(t) = t\mathbf{i} + 2t\mathbf{j} + te^t\mathbf{k}$.

29. Find $\mathbf{T}, \mathbf{N}, A_T$ and A_N (the tangential and normal components of acceleration), and κ (curvature) for an object with position vector $\mathbf{R}(t) = t^2\mathbf{i} + 3t\mathbf{j} - 3t\mathbf{k}$.

30. A projectile is fired from ground level with initial velocity 50 ft/s at an angle of elevation of $\alpha = 30°$.
   a. What is the maximum height reached by the projectile?
   b. What are the time of flight and the range?

Supplementary Problems

Find the vector limits in Problems 1–6.

1. $\lim\limits_{t \to 0} \left[t^2\mathbf{i} - 3te^t\mathbf{j} + \dfrac{\sin 2t}{t}\mathbf{k} \right]$ **2.** $\lim\limits_{t \to 0} \left[\dfrac{\mathbf{i} + t\mathbf{j} - e^{-t}\mathbf{k}}{1 - t} \right]$

3. $\lim\limits_{t \to 0} \langle t, 0, 5 \rangle \cdot \langle \sin t, 3t, -(1 - t) \rangle$

4. $\lim\limits_{t \to \pi} \langle 1 + t, -3, 0 \rangle \times \langle 0, t^2, \cos t \rangle$

5. $\lim\limits_{t \to 0} \left[\left(1 + \dfrac{1}{t} \right)^t \mathbf{i} - \left(\dfrac{\sin t}{t} \right)\mathbf{j} - t\mathbf{k} \right]$

6. $\lim\limits_{t \to \infty} \left[\left(\dfrac{1 - \cos t}{t} \right)\mathbf{i} + 4\mathbf{j} + \left(1 + \dfrac{3}{t} \right)^t \mathbf{k} \right]$

Find $\mathbf{F}'(t)$ and $\mathbf{F}''(t)$ for the vector functions in Problems 7–12.

7. $\mathbf{F}(t) = te^t\mathbf{i} + t^2\mathbf{j}$ **8.** $\mathbf{F}(t) = (t \ln 2t)\mathbf{i} + t^{3/2}\mathbf{k}$

9. $\mathbf{F}(t) = \langle 2t^{-1}, -2t, te^{-t} \rangle$

10. $\mathbf{F}(t) = \langle t, -(1 - t), 0 \rangle \times \langle 0, t^2, e^{-t} \rangle$

11. $\mathbf{F}(t) = (t^2 + e^{at})\mathbf{i} + (te^{-at})\mathbf{j} + (e^{at+1})\mathbf{k}$, a a constant

12. $\mathbf{F}(t) = (1 - t)^{-1}\mathbf{i} + (\sin 2t)\mathbf{j} + (\cos^2 t)\,\mathbf{k}$

Sketch the graph of the vector functions in Problems 13–15.

13. $\mathbf{F}(t) = te^t\mathbf{i} + t^2\mathbf{j}$ **14.** $\mathbf{F}(t) = t^2\mathbf{i} - 3t\mathbf{j}$

15. $\mathbf{F}(t) = (1 - \cos t)\mathbf{i} + (\sin t)\mathbf{k}$

Describe the graph of the vector functions in Problems 16–18.

16. $\mathbf{F}(t) = \langle 3 \cos t, 3 \sin t, t \rangle$ **17.** $\mathbf{F}(t) = \langle 2 \sin t, 2 \cos t, 5t \rangle$

18. $\mathbf{F}(t) = 2t^2\mathbf{i} + (1 - t)\mathbf{j} + 3\mathbf{k}$

Let $\mathbf{F}(t) = f_1(t)\mathbf{i} + f_2(t)\mathbf{j} + f_3(t)\mathbf{k}$ and find the indicated derivative in terms of $\mathbf{F}'$ in Problems 19–24.

19. $\dfrac{d}{dt}[\mathbf{F}(t) \cdot \mathbf{F}(t)]$ **20.** $\dfrac{d}{dt}[\|\mathbf{F}(t)\|\mathbf{F}(t)]$ **21.** $\dfrac{d}{dt}\left[\dfrac{\mathbf{F}(t)}{\|\mathbf{F}(t)\|} \right]$

22. $\dfrac{d}{dt}\|\mathbf{F}(t)\|$ **23.** $\dfrac{d}{dt}[\mathbf{F}(t) \times \mathbf{F}(t)]$ **24.** $\dfrac{d}{dt}\mathbf{F}(e^t)$

Evaluate the definite and indefinite vector integrals in Problems 25–30.

25. $\displaystyle\int_{-1}^{1} (e^{-t}\mathbf{i} + t^3\mathbf{j} + 3\mathbf{k})\, dt$

26. $\displaystyle\int_{1}^{2} [(1 - t)\mathbf{i} - t^{-1}\mathbf{j} + e^t\mathbf{k}]\, dt$

27. $\displaystyle\int [te^t\mathbf{i} - (\sin 2t)\mathbf{j} + t^2\mathbf{k}]\, dt$

28. $\displaystyle\int e^{2t}[2\mathbf{i} - t\mathbf{j} + (\sin t)\mathbf{k}]\, dt$ **29.** $\displaystyle\int t[e^t\mathbf{i} + (\ln t)\mathbf{j} + 3\mathbf{k}]\, dt$

30. $\displaystyle\int [e^t\mathbf{i} + 2\mathbf{j} - t\mathbf{k}] \cdot [e^{-t}\mathbf{i} - t\mathbf{j}]\, dt$

In Problems 31–34, $\mathbf{R}$ is the position vector of a moving body. Find the velocity $\mathbf{V}$, the speed ds/dt, and the acceleration $\mathbf{A}$.

31. $\mathbf{R}(t) = t\mathbf{i} + (3 - t)\mathbf{j} + 2\mathbf{k}$

32. $\mathbf{R}(t) = (\sin 2t)\mathbf{i} + 2\mathbf{j} - (\cos 2t)\mathbf{k}$

33. $\mathbf{R}(t) = \langle t \sin t, te^{-t}, -(1 - t) \rangle$

34. $\mathbf{R}(t) = \langle \ln t, e^t, -\tan t \rangle$

Find $\mathbf{T}$ and $\mathbf{N}$ for the vector functions given in Problems 35–38.

35. $\mathbf{R}(t) = t\mathbf{i} - t^2\mathbf{j}$ **36.** $\mathbf{R}(t) = (3 \cos t)\mathbf{i} - (3 \sin t)\mathbf{j}$

37. $\mathbf{R}(t) = \langle 4 \cos t, -3t, 4 \sin t \rangle$

38. $\mathbf{R}(t) = \langle e^t \sin t, e^t, e^t \cos t \rangle$

In Problems 39–42, find the tangential and normal components of acceleration, and the curvature of a moving object with position $\mathbf{R}(t)$.

39. $\mathbf{R}(t) = t^2\mathbf{i} + 2t\mathbf{j} + e^t\mathbf{k}$

40. $\mathbf{R}(t) = t^2\mathbf{i} - 2t\mathbf{j} + (t^2 - t)\mathbf{k}$

41. $\mathbf{R}(t) = (4 \sin t)\mathbf{i} + (4 \cos t)\mathbf{j} + 4t\mathbf{k}$

42. $\mathbf{R}(t) = (a \sin 3t)\mathbf{i} + (a + a \cos 3t)\mathbf{j} + (3a \sin t)\mathbf{k}$, for constant $a \neq 0$

A polar curve C is given by $r = f(\theta)$. If $f''(\theta)$ exists (see Problem 54, Section 11.4), then the curvature can be found by the formula

$$\kappa = \frac{|r^2 + 2(r')^2 - rr''|}{[r^2 + (r')^2]^{3/2}}$$

Find the curvature at the given point on each of the polar curves given in Problems 43–48.

43. $r = 4 \cos \theta$, where $\theta = \frac{\pi}{3}$ **44.** $r = \theta^2$, where $\theta = 2$

45. $r = e^{-\theta}$, where $\theta = 1$

46. $r = 1 + \cos \theta$, where $\theta = \frac{\pi}{2}$

47. $r = 4 \cos 3\theta$, where $\theta = \frac{\pi}{6}$

48. $r = 1 - 2 \sin \theta$, where $\theta = \frac{\pi}{4}$

49. For what values of t is the following vector function continuous?

$$\mathbf{F}(t) = (2t - 1)\mathbf{i} + \left(\frac{t^2 - 1}{t - 1} \right)\mathbf{j} + 4\mathbf{k}$$

50. Find the length of the graph of the vector function

$$\mathbf{R}(t) = \left(\frac{t^2 - 2}{2} \right)\mathbf{i} + \frac{(2t + 1)^{3/2}}{3}\mathbf{j}$$

from $t = 0$ to $t = 6$.

51. Find a vector function $\mathbf{F}$ such that $\mathbf{F}(0) = \mathbf{F}'(0) = \mathbf{i}$, $\mathbf{F}''(0) = 2\mathbf{j} + \mathbf{k}$, and

$$\mathbf{F}'''(t) = (\cos t)\mathbf{i} + (\sin t)\mathbf{j} + \frac{t}{\pi}\mathbf{k}$$

52. An object moves in space with acceleration $\mathbf{A}(t) = \langle -t, 2, 2 - t \rangle$. When $t = 0$, it is known that the object is at the point $(1, 0, 0)$ and that it has velocity $\mathbf{V}(0) = \langle 2, -4, 0 \rangle$.
   a. Find the velocity $\mathbf{V}(t)$ and the position $\mathbf{R}(t)$.
   b. What are the speed and location of the object when $t = 1$?

c. When is the object stationary and what is its position at that time?

53. The position vector for a curve is given in terms of arc length s by $\mathbf{R}(s) = \left\langle a \cos \dfrac{s}{a}, a \sin \dfrac{s}{a}, 2s \right\rangle$ for $0 \le s \le 2\pi a$, $a \ne 0$. Find the unit tangent vector $\mathbf{T}(s)$ and the principal unit normal $\mathbf{N}(s)$.

54. Find the radius of curvature of the curve given by $y = 1 + \sin x$ at the points where x is
 a. $\dfrac{\pi}{6}$ b. $\dfrac{\pi}{4}$ c. $\dfrac{3\pi}{2}$

55. Find the radius of curvature of the ellipse given by $\mathbf{R}(t) = \langle a \cos t, b \sin t \rangle$ where $a > 0, b > 0, a \ne b$, and $0 \le t \le 2\pi$ at the points where $t = 0$ and $\pi/2$.

56. MODELING PROBLEM   A stunt pilot flying at an altitude of 4,000 ft with a speed of 180 mi/hr drops a weighted marker, attempting to hit a target on the ground below, as shown in Figure 11.27. How far away from the target (measured horizontally) should the pilot be when she releases the marker? You may neglect air resistance.

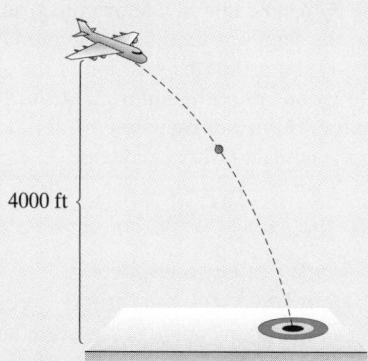

4000 ft

■ **FIGURE 11.27**   Problem 56

57. The position of an object moving in space is given by
$$\mathbf{R}(t) = (e^{-t}\cos t)\mathbf{i} + (e^{-t}\sin t)\mathbf{j} + e^{-t}\mathbf{k}$$
 a. Find the velocity, speed, and acceleration of the object at arbitrary time t.
 b. Determine the curvature of the trajectory at time t.

58. Find the point or points on the curve $y = e^{ax}(a > 0)$ where the radius of curvature is maximized.

59. MODELING PROBLEM   A car weighing 3,000 lb travels at a constant speed of 60 mi/h on a flat road and then makes a circular turn on an interchange. If the radius of

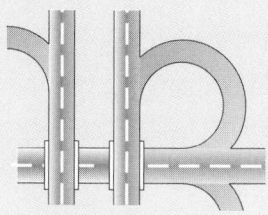

the turn is 40 ft, what frictional force is needed to keep the car from skidding?

60. The position vector of an object in space is
$$\mathbf{R}(t) = (a \cos \omega t)\mathbf{i} + (a \sin \omega t)\mathbf{j} + \omega^2 t\mathbf{k}$$
Find ω so that the sum of the object's tangential and normal components of acceleration equal half its speed.

61. A particle moves along a path given in parametric form where $r(t) = 1 + \cos at$ and $\theta(t) = e^{-at}$ (for positive constant a). Find the velocity and acceleration of the particle in terms of the unit polar vectors $\mathbf{u}_r$ and $\mathbf{u}_\theta$.

62. A nozzle discharges a stream of water with an initial velocity $v_0 = 50$ ft/s into the end of a horizontal pipe of inside diameter $d = 5$ ft. What is the maximum horizontal distance that the stream can reach?

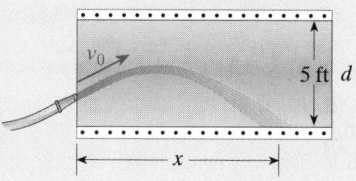

63. Sketch the graph of the vector function
$$\mathbf{R}(t) = \left(\frac{3t}{1 + t^3}\right)\mathbf{i} + \left(\frac{3t^2}{1 + t^3}\right)\mathbf{j}$$
then find parametric equations for the tangent line at the point where $t = 2$. This curve is called the *folium of Descartes.*

<hr>

Technology Window

64. MODELING PROBLEM   A fireman stands 5.5 m from the front of a burning building 15 m high. His fire hose discharges water from a height of 1.2 m at an angle of 62°, as shown in the figure. Use a spreadsheet or a computer program to determine the height h where the stream of water strikes the building for values of v_0 varying from 6 m/s to 26 m/s at intervals of 1 m/s.

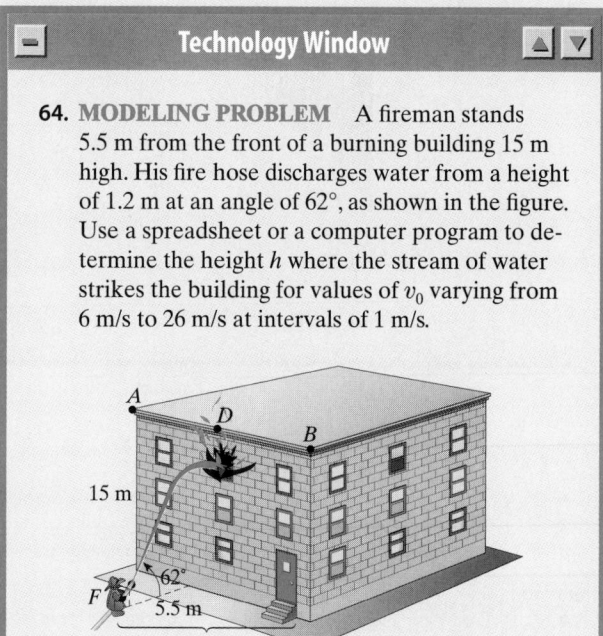

65. A DNA molecule has the shape of a double helix (see Figure 11.4). The radius of each helix is about 10^{-8} μm. Each helix rises about 3×10^{-8} μm during each complete turn and there are about 3×10^8 complete turns. Estimate the length of each helix.

66. Show that the tangential component of acceleration of a moving object is 0 if the object has constant speed. Is the converse statement also true? That is, if $A_T = 0$, can we conclude that the speed is constant?

67. The path of a particle P is an Archimedean spiral. The motion of the particle is described by the polar coordinates $r = 10t$ and $\theta = 2\pi t$, where r is expressed in inches and t is in seconds. Determine the velocity of the particle (in terms of $\mathbf{u}_r$ and $\mathbf{u}_\theta$) when
  a. $t = 0$ b. $t = 0.25$ s

68. **SPY PROBLEM**   After dropping the knockout gas into Scélérat's bunker (Problem 33, Section 11.3), the Spy lands and, as expected, finds a bunker full of sleeping thugs—but no Scélérat! He races through the back door of the bunker and finds himself in a ski resort. By the time he locates Scélérat, the French fiend is skiing down the mountain. The Spy grabs a pair of skis and runs over to a conveniently located ski jump. His keen mind quickly determines that the slope of the mountain is 17° and the angle at the lip of the jump is 10°, as shown in Figure 11.28. Suppose Scélérat is 150 ft down the mountainside skiing at 75 ft/s when the Spy launches himself from the end of the ski jump with a speed of 85 ft/s. How close will he be to Scélérat when he lands? If the Spy becomes airborne at noon and he maintains his landing speed, when (if ever) does he catch Scélérat?

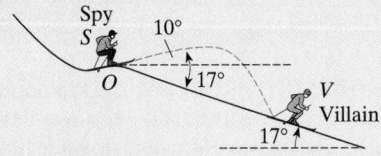

■ **FIGURE 11.28**   Spy scene

In Problems 69–70, find $\mathbf{F}'(x)$.

69. $\mathbf{F}(x) = \displaystyle\int_1^x [(\sin t)\mathbf{i} - (\cos 2t)\mathbf{j} + e^{-t}\mathbf{k}] \, dt$

70. $\mathbf{F}(x) = \displaystyle\int_1^{2x} [t^2\mathbf{i} + (\sec e^{-t})\mathbf{j} - (\tan e^{2t})\mathbf{k}] \, dt$

In Problems 71–74, solve the initial value problems for $\mathbf{F}$ *as a vector function of t.*

71. differential equation: $\mathbf{F}'(t) = t\mathbf{i} + t\mathbf{j} - t\mathbf{k}$
  initial condition: $\mathbf{F}(0) = \mathbf{i} + 2\mathbf{j} - 3\mathbf{k}$

72. differential equation:
  $$\mathbf{F}'(t) = \frac{3}{2}\sqrt{t+1}\,\mathbf{i} + (t+1)^{-1}\mathbf{j} + e^t\mathbf{k}$$
  initial condition: $\mathbf{F}(0) = \mathbf{j} - 3\mathbf{k}$

73. differential equation:
  $$\mathbf{F}'(t) = (\sin 2t)\mathbf{i} + (e^t\cos t)\mathbf{j} - \left(\frac{3}{t+1}\right)\mathbf{k}$$
  initial condition: $\mathbf{F}(0) = \mathbf{i} - 3\mathbf{k}$

74. differential equation: $\dfrac{d^2\mathbf{F}}{dt^2} = -32\mathbf{j}$
  initial conditions: $\mathbf{F} = 50\mathbf{j}$ and $\dfrac{d\mathbf{F}}{dt} = 5\mathbf{i} + 5\mathbf{k}$ at $t = 0$

75. **PUTNAM EXAMINATION PROBLEM**   A shell strikes an airplane flying at a height h above the ground. It is known that the shell was fired from a gun on the ground with muzzle speed v_0, but the position of the gun and its angle of elevation are both unknown. Deduce that the gun is situated within a circle whose center lies directly below the airplane and whose radius is
  $$\frac{v_0}{g}\sqrt{v_0^2 - 2gh}$$
  Neglect resistance of the atmosphere.

76. **PUTNAM EXAMINATION PROBLEM**   A coast artillery gun can fire at any angle of elevation between 0° and 90° in a fixed vertical plane. If air resistance is neglected and the muzzle speed is constant ($v = v_0$), determine the set H of points in the plane that can be hit. Consider only those points above the horizontal.

77. **PUTNAM EXAMINATION PROBLEM**   A particle moves on a circle with center O, starting from rest at a point P and coming to rest again at a point Q, without coming to rest at any intermediate point. Prove that the acceleration vector of the particle does not vanish at any point between P and Q, and that, at some point R between P and Q, the acceleration vector points in along the radius $\overline{RO}$.

The Stimulation of Science

Howard Eves was born in Paterson, New Jersey, in 1911 and is professor emeritus of mathematics at the University of Maine. This guest essay first appeared in *Great Moments in Mathematics Before 1650*. Professor Eves reminds us, "It must be remembered that a *moment* in history is sometimes an inspired flash and sometimes an evolution extending over a long period of time." Howard Eves's lectures (from which this guest essay was taken) are so renowned that each college and university at which he taught awarded him, at one time or another, every available honor for distinguished teaching. Howard Eves is a prolific and successful textbook author with a real love for the history of mathematics. He now lives with his wife at the retirement retreat in Lubec, Maine.

The die is cast; I have written my book; it will be read either in the present age or by posterity, it matters not which; it may well await a reader, since God has waited six thousand years for an interpreter of his words.

JOHANNES KEPLER
FROM JAMES R. NEWMAN, *THE WORLD OF MATHEMATICS*, *VOLUME I* (NEW YORK: SIMON AND SCHUSTER, 1956), P. 220.

The mighty Antaeus was the giant son of Neptune (god of the sea) and Ge (goddess of the earth), and his strength was invincible so long as he remained in contact with his Mother Earth. Strangers who came to his country were forced to wrestle to the death with him, and so it chanced one day that Hercules and Antaeus came to grips with one another. But Hercules, aware of the source of Antaeus' great strength, lifted and held the giant from the earth and crushed him in the air.

There is a parable here for mathematicians. For just as Antaeus was born of and nurtured by his Mother Earth, history has shown us that all significant and lasting mathematics is born of and nurtured by the real world. As in the case of Antaeus, so long as mathematics maintains its contact with the real world, it will remain powerful. But should it be lifted too long from the solid ground of its birth into the filmy air of pure abstraction, it runs the risk of weakening. It must of necessity return, at least occasionally, to the real world for renewed strength.

Such a rejuvenation of mathematics occurred in the seventeenth century, following discoveries made by two eminent mathematician–scientists—Galileo Galilei (1564–1642) and Johannes Kepler (1571–1630). Galileo, through a sequence of experiments started before his 25th birthday, discovered a number of basic facts concerning the motion of bodies in the earth's gravitational field, and Kepler, by 1619, had deduced all three of his famous laws of planetary motion. These achievements proved to be so influential on the development of so much of subsequent mathematics that they must be ranked as two of the GREAT MOMENTS IN MATHEMATICS. Galileo's discoveries led to the creation of the modern science of dynamics and Kepler's to the creation of modern celestial mechanics; and each of these studies, in turn, required, for their development, the creation of a new mathematical tool—the calculus—capable of dealing with change, flux, and motion.

Galileo was born in Pisa in 1564 as the son of an impoverished Florentine nobleman. After a disinterested start as a medical student, Galileo obtained parental permission to change his studies to science and mathematics, fields in which he possessed a strong natural talent. While still a medical student at the University of Pisa, he made his historically famous observation that the great pendulous lamp in the cathedral there oscillated to and fro with a period independent of the size of the arc of oscillation.*

Later, he showed that the period of a pendulum is also independent of the weight of the pendulum's bob. When he was 25, he accepted an appointment as professor of mathematics at the University of Pisa. It was during this appointment that he is alleged to have performed experiments from the leaning tower of Pisa, showing that, contrary to the teaching of Aristotle, heavy bodies do not fall faster than light ones. By rolling balls down inclined planes, he arrived at the law that the distance a body falls is proportional to the square of the time of falling, in accordance with the now-familiar formula $s = \frac{1}{2}gt^2$.

Unpleasant local controversies caused Galileo to resign his chair at Pisa in 1591, and the following year he accepted a professorship in mathematics at the University of Padua, where there reigned an atmosphere more friendly to scientific pursuits. Here at Padua, for nearly 18 years, Galileo continued his experiments and his teaching, achieving a widespread fame. While at Padua, he heard of the discovery, in about 1607, of the telescope by the Dutch lens-grinder Johann Lipersheim, and he set about making instruments of his own, producing a telescope with a magnifying power of more than 30 diameters. With this telescope he observed sunspots (contradicting Aristotle's assertion that the sun is without blemish), saw mountains on the moon, and noticed the phases of Venus, Saturn's rings, and the four bright satellites of Jupiter (all three of these lending credence to the Copernican theory of the solar system). Galileo discoveries roused the opposition of the Church, and finally, in the year 1633, he was summoned to appear before the Inquisition, and there forced to recant his scientific findings. Not many years later the great scientist became blind. He died, a prisoner in his own home, in 1642, the year Isaac Newton was born.

Johannes Kepler was born near Stuttgart, Germany, in 1571 and commenced his studies at the University of Tübingen with the intention of becoming a Lutheran minister. Like Galileo, he found his first choice of an occupation far less congenial than his deep interest in science, particularly astronomy, and he accordingly changed his plans. In 1594, when in his early twenties, he accepted a lectureship at the University of Gräz in Austria. Five years later, he became assistant to the famous Danish-Swedish astronomer Tycho Brahe, who had moved to Prague to serve as the court astronomer to Kaiser Rudolph II. Shortly after, in 1601, Brahe suddenly died, and Kepler inherited both his master's position and his vast collection of very accurate data on the positions of the planets as they moved about the sky. With amazing perseverance, Kepler set out to find, from Brahe's enormous mass of observational data, just how the planets move in space.

It has often been remarked that almost any problem can be solved if one but continuously worries over it and works at it a sufficiently long time. As Thomas Edison said, genius is 1% inspiration and 99% perspiration. Perhaps nowhere in the history of science is this more clearly demonstrated than in Kepler's incredible pertinacity in solving the problem of the motion of the planets about the sun. Thoroughly convinced of the Copernican theory that the planets revolve in orbits about the central sun, Kepler strenuously sought to determine the nature and position of those orbits and the man-

---

*This is only approximately true, the approximation being very close in the case of small amplitudes of oscillation.

ner in which the planets travel in their orbits. With Brahe's great set of observational recordings at hand, the problem became this: to obtain a pattern of motion of the planets that would exactly agree with Brahe's observations. So dependable were Brahe's recordings that any solution that should differ from Brahe's observed positions by even as little as a quarter of the moon's apparent diameter must be discarded as incorrect. Kepler had, then, first to guess with his *imagination* some plausible solution and then, with painful *perseverance,* to endure the mountain of tedious calculations needed to confirm or reject his guess. He made hundreds of fruitless attempts and performed reams and reams of calculations, laboring with undiminished zeal and patience for many years. Finally he solved his problem, in the form of his three famous laws of planetary motion, the first two around 1609 and the third one 10 years later in 1619:

I. The planets move about the sun in elliptical orbits with the sun at one focus.

II. The radius vector joining a planet to the sun sweeps over equal areas in equal intervals of time.

III. The square of the time of one complete revolution of a planet about its orbit is proportional to the cube of the orbit's semimajor axis.

The empirical discovery of these laws from Brahe's mass of data constitutes one of the most remarkable inductions ever made in science. With justifiable pride, Kepler prefaced his *Harmony of the Worlds* of 1619 with the following outburst:

I am writing a book for my contemporaries or—it does not matter—for posterity. It may be that my book will wait for a hundred years for a reader. Has not God waited 6000 years for an observer?

Kepler's laws of planetary motion are landmarks in the history of astronomy and mathematics, for in the effort to justify them, Isaac Newton was led to create modern celestial mechanics. It is very interesting that 1800 years after the Greeks had developed the properties of the conic sections there should occur such an illuminating practical application of them. *One never knows when a piece of pure mathematics may receive an unexpected application.*

In order to compute the areas involved in his second law, Kepler had to resort to a crude form of the integral calculus, making him one of the precursors of that calculus. Also, in his *Stereometria doliorum vinorum* (*Solid Geometry of Wine Barrels,* 1615), he applied crude integration procedures to the finding of the volumes of 93 different solids obtained by revolving arcs of conic sections about axes in their planes. Among these solids were the torus and two that he called *the apple* and *the lemon,* these latter being obtained by revolving a major and a minor arc, respectively, of a circle about the arc's chord as an axis. Kepler's interest in these matters arose when he observed some of the poor methods in use by the wine gaugers of the time. (See Group Research Project, page 315.)

It was Kepler who introduced the word *focus* (Latin for "hearth") into the geometry of conic sections. He approximated the perimeter of an ellipse of semiaxes a and b by the formula $\pi(a + b)$. He also laid down a so-called *principle of continuity,* which postulates the existence in a plane of certain ideal points and an ideal line having many of the properties of ordinary points and lines, lying at infinity. Thus he explained that a straight line can be considered as closed at infinity and that a parabola may be regarded as the limiting case of either an ellipse or a hyperbola in which one of the foci has retreated to infinity. The ideas were extended by later geometers.

Kepler was a confirmed Pythagorean, with the result that his work is often a blend of the fancifully mystical and the carefully scientific. It is sad that his personal life was made almost unendurable by a multiplicity of worldly misfortunes. An infection from smallpox when he was four years old left his eyesight much impaired. In addition to his general lifelong weakness, he spent a joyless youth; his marriage was a constant source of unhappiness; his favorite child died of smallpox; his wife went mad and died; he was expelled from his lectureship at the University of Gräz when the city fell to the Catholics; his mother was charged and imprisoned for witchcraft, and for almost a year he desperately tried to save her from the torture chamber; he himself very narrowly escaped condemnation for heterodoxy; and his stipend was always in arrears. One report says that his second marriage was even less fortunate than his first, although he took the precaution to analyze carefully the merits and demerits of 11 women before choosing the wrong one. He was forced to augment his income by casting horoscopes, and he died of a fever in 1630 at the age of 59 while on a journey to try to collect some of his long overdue salary.

Mathematical Essays

1. Write a 500-word essay on the life and mathematics of Galileo Galilei.

2. **HISTORICAL QUEST**   For this Quest, explain the remark in Galileo's *Discorsi e dimostrazioni matematiche intorno a due nuove scienze* of 1638 that "neither is the number of squares less than the totality of all numbers, nor the latter greater than the former."

GALILEO GALILEI
1564–1642

3. **HISTORICAL QUEST**   In 1638, Galileo published his ideas about dynamics in his book *Discorsi e dimostrazioni matematiche intorno à due nuove scienze*. In this book, he considers the following problem:

   Suppose the larger circle of Figure 11.29 has made one revolution in rolling along the straight line from A to B, so that $|\overline{AB}|$ is equal to the circumference of the large circle. Then the small circle, fixed to the large one, has also made one revolution, so that $|\overline{CD}|$ is equal to the circumference of the small circle. It follows that *the two circles have equal circumferences.*

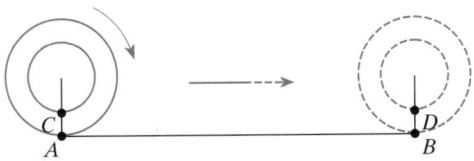

■ **FIGURE 11.29**   Aristotle's wheel

   This paradox had been earlier described by Aristotle and is therefore sometimes referred to as *Aristotle's wheel.* Can you explain what is going on?

4. **HISTORICAL QUEST**   In Historical Quest 39 Section 6.3, we outlined a procedure by Johannes Kepler for finding the volume of a torus.

   This Quest asks you to use Kepler's reasoning to find the area of an ellipse.

   a. Kepler divided a circle of radius r and circumference C into a large number of very thin sectors. By regarding each sector as a thin isosceles triangle of altitude r and

JOHANNES KEPLER
1571–1630

   base equal to the arc of the sector, he heuristically arrived at the formula $A = \frac{1}{2} r C$ for the area of a circle. Show how this was done.

   b. Use Kepler's reasoning to obtain a volume V of a sphere of radius r and surface area S as $V = \frac{1}{3} r S$.

   c. Use Kepler's reasoning to show that an ellipse of semimajor axis a and semiminor axis b has area $A = \pi ab$.

5. Prove Kepler's first law. You may search the literature for help with this proof.

6. Prove Kepler's third law. You may search the literature for help with this proof.

7. Where is a planet in its orbit when its speed is the greatest? Support your response.

8. Two hypothetical planets are moving about the sun in elliptical orbits having equal major axes. The minor axis of one, however, is half that of the other. How do the periods of the two planets compare? Support your response.

9. **HISTORICAL QUEST**   The three laws of Kepler described in this section forever changed the way we view the universe, but it was not Kepler who correctly proved these laws. Isaac Newton (Historical Quest #2, Section 2.5) proved these laws from the inverse-square law of gravitation.

   In Section 11.3, we proved Kepler's second law. Write a 500-word essay on the life and mathematics of Johannes Kepler, and include, as part of your essay, where Kepler made his mistake in his proof of his second law.

10. **Book Report**   "Science is that body of knowledge that describes, defines, and where possible, explains the universe . . . we think of the history of science as a history of men. . . . [History] is the story of thousands of people who contributed to the knowledge and theories that constituted the science of their eras and made the 'great leaps' possible. Many of these people were women." So begins a history of women in science entitled *Hypatia's Heritage* by Margaret Alic (Boston: Beacon Press, 1986). Read this book and write a book report.

11. Make up a word problem involving the calculus of vector-valued functions. Send your problem to:
    Bradley and Smith
    Prentice Hall Publishing Company
    1 Lake Street
    Upper Saddle River, NJ 07458
    The best ones submitted will appear in the next edition (along with credit to the problem poser).

Chapters 7–11 *Cumulative Review*

■ TABLE 11.3 **Summary of Velocity, Acceleration, and Curvature**

CURVES	$\mathbb{R}^2$ (plane)	$\mathbf{R}(t) = x(t)\mathbf{i} + y(t)\mathbf{j}$
Position vector	$\mathbb{R}^3$ (space)	$\mathbf{R}(t) = x(t)\mathbf{i} + y(t)\mathbf{j} + z(t)\mathbf{k}$

The graph is called the **trajectory** of the object's motion.

Velocity vector

$$\mathbf{V}(t) = \frac{d\mathbf{R}}{dt} = \mathbf{R}'(t)$$

$$\|\mathbf{V}\| = \frac{ds}{dt} \qquad \text{This is called the \textbf{speed.}}$$

$$\frac{\mathbf{V}}{\|\mathbf{V}\|} \qquad \text{This is the direction of } \mathbf{V} \text{ or } \textit{direction of motion.}$$

Acceleration vector

$$\mathbf{A}(t) = \frac{d^2\mathbf{R}}{dt^2} = \frac{d\mathbf{V}}{dt} = A_\mathrm{T}\mathbf{T} + A_\mathrm{N}\mathbf{N}$$

where

$$A_\mathrm{T} = \frac{d^2s}{dt^2} = \mathbf{A} \cdot \mathbf{T} = \frac{\mathbf{V} \cdot \mathbf{A}}{\|\mathbf{V}\|}$$

This is the **tangential component.**

$$A_\mathrm{N} = \kappa\left(\frac{ds}{dt}\right)^2 = \frac{\|\mathbf{V} \times \mathbf{A}\|}{\|\mathbf{V}\|} = \sqrt{\|\mathbf{A}\|^2 - A_\mathrm{T}^2}$$

This is the **normal component.**

Unit tangent and normal vectors

$$\mathbf{T} = \frac{\mathbf{V}}{\|\mathbf{V}\|} = \frac{\mathbf{R}'(t)}{\|\mathbf{R}'(t)\|} = \frac{d\mathbf{R}}{ds}$$

This is the **unit tangent vector** in the *direction of motion.*

$$\mathbf{N} = \frac{\mathbf{T}'(t)}{\|\mathbf{T}'(t)\|} = \frac{1}{\kappa}\frac{d\mathbf{T}}{ds}$$

This is the **principal unit normal vector.**

Curvature

$$\kappa = \left\|\frac{d\mathbf{T}}{ds}\right\| = \frac{\left\|\dfrac{d\mathbf{T}}{dt}\right\|}{\dfrac{ds}{dt}} = \frac{\|\mathbf{R}' \times \mathbf{R}''\|}{\|\mathbf{R}'\|^3}; \text{ if } y = f(x) \text{ then}$$

$$\kappa = \frac{|x'y'' - y'x''|}{[(x')^2 + (y')^2]^{3/2}}$$

Torsion

$$\tau = \left\|\frac{d\mathbf{B}}{ds}\right\|$$

Cumulative Review Problems for Chapters 7–11

1. ■ **What Does This Say?** In your own words, outline a procedure for integrating a given function.

2. ■ **What Does This Say?** In your own words, outline a procedure for deciding whether a given series converges or diverges.

3. ■ **What Does This Say?** What is a quadric surface? In your own words, discuss classifying and sketching quadric surfaces.

4. ■ **What Does This Say?** What is a vector? What is a vector function? In your own words, discuss what is meant by vector calculus.

5. a. If $y = x\cosh^{-1} x$, find y'.

   b. If $x\sinh^{-1} y + y\tanh^{-1} x = 0$, find $\dfrac{dy}{dx}$.

Find the integrals in Problems 6–14.

6. $\int x \ln \sqrt[3]{x}\, dx$

7. $\int \sin^2 x \cos^3 x\, dx$

8. $\int \tan^2 x \sec x\, dx$

9. $\int x\sqrt{16 - x}\, dx$

10. $\int \dfrac{\cosh x\, dx}{1 + \sinh^2 x}$

11. $\int \dfrac{dx}{1 + \cos x}$

12. $\int \dfrac{dx}{x^2(x^2 + 5)}$

13. $\int \dfrac{dx}{\sqrt{x} - \sqrt[3]{x}}$

14. $\int \dfrac{dx}{\sqrt{2x - x^2}}$

15. Find the equation of the line through $(-1, 2, 5)$ and perpendicular to the plane $2x - 3y + z = 11$.

16. Find the equation of the plane satisfying the given conditions.
   a. passing through $(5, 1, 2)$, $(3, 1, -2)$, and $(3, 2, 5)$
   b. passing through $(-2, -1, 4)$ and perpendicular to the line $\dfrac{x}{2} = \dfrac{y + 1}{5} = \dfrac{3 - z}{2}$

Test the series in Problems 17–23 for convergence.

17. $\displaystyle\sum_{k=0}^{\infty} \dfrac{k^3}{k^4 + 2}$

18. $\displaystyle\sum_{k=1}^{\infty} \dfrac{1}{k \cdot 4^k}$

19. $\displaystyle\sum_{k=2}^{\infty} \dfrac{1}{k \ln k}$

20. $\displaystyle\sum_{k=0}^{\infty} \dfrac{3k^2 - 7k + 2}{(2k - 1)(k + 3)}$

21. $\displaystyle\sum_{k=1}^{\infty} \dfrac{k!}{2^k \cdot k}$

22. $\displaystyle\sum_{k=0}^{\infty} \dfrac{(-1)^{k+1} k}{k^2 + k - 1}$

23. $1 + \frac{1}{8} - \frac{1}{27} - \frac{1}{64} + \frac{1}{125} + \frac{1}{216} - - + + \cdots$

24. In each case, find the sum of the convergent series.
   a. $\displaystyle\sum_{k=1}^{\infty} \dfrac{2^{k-1}}{5^{k+3}}$
   b. $\displaystyle\sum_{k=1}^{\infty} \dfrac{1}{(3k - 1)(3k + 2)}$

25. In each case, determine whether the given improper integral converges, and if it does, find its value.
   a. $\displaystyle\int_{1}^{\infty} x^2 e^{-x}\, dx$
   b. $\displaystyle\int_{0}^{2} \dfrac{dx}{\sqrt{4 - x^2}}$

26. Find $\mathbf{v} \cdot \mathbf{w}$ and $\mathbf{v} \times \mathbf{w}$ for the vectors $\mathbf{v} = 3\mathbf{i} - 2\mathbf{j} + 5\mathbf{k}$ and $\mathbf{w} = \mathbf{i} - 3\mathbf{j} - \mathbf{k}$.

27. Find $\mathbf{F}'$ and $\mathbf{F}''$ for $\mathbf{F}(t) = 2t\mathbf{i} + e^{-3t}\mathbf{j} + t^4\mathbf{k}$.

28. Find $\int [e^t\mathbf{i} - \mathbf{j} - t\mathbf{k}] \cdot [e^{-t}\mathbf{i} + t\mathbf{j} - \mathbf{k}]\, dt$.

29. Find $\mathbf{T}$ and $\mathbf{N}$ for $\mathbf{R}(t) = 2(\sin 2t)\mathbf{i} + (2 + 2\cos 2t)\mathbf{j} + 6t\mathbf{k}$.

30. Show that the alternating series
$$S = \sum_{k=1}^{\infty} \dfrac{(-1)^{k+1}}{\sqrt{k}}$$
converges and determine what N must be to guarantee that the partial sum
$$S_N = \sum_{k=1}^{N} \dfrac{(-1)^{k+1}}{\sqrt{k}}$$
approximates the entire sum S with four-decimal-place accuracy.

31. a. Find the Maclaurin series representation for $f(x) = x^2 e^{-x^2}$.
   b. Use the representation in Part **a** to approximate
$$\int_{0}^{1} x^2 e^{-x^2}\, dx$$
with three-decimal-place accuracy.

32. Let $P(x) = 7 - 3(x - 4) + 5(x - 4)^2 - 2(x - 4)^3 + 6(x - 4)^4$ be the fourth-degree Taylor polynomial for the function f about 4.
   a. Find $f(4)$ and $f'''(4)$.
   b. Use the third-degree Taylor polynomial for f' about $x = 4$ to approximate $f'(4.2)$.
   c. Write a fifth-degree Taylor polynomial for
$$F(x) = \int_{4}^{x} f(t)\, dt \text{ about } 4.$$
   d. Can $f(5)$ be determined from the given information? Why or why not?

33. Solve the first-order linear differential equation
$$\dfrac{dy}{dx} + 2y = x^2$$
subject to the condition $y = 2$ when $x = 0$.

34. During the time period from $t = 0$ to $t = 6$ seconds, a particle moves along the path given by $x(t) = 3\cos(\pi t)$, $y(t) = 5\sin(\pi t)$.
   a. What is the position of the particle when $t = 1.5$?
   b. Graph the path of the particle from $t = 0$ to $t = 6$. Show direction.
   c. How many times does the particle pass through the point found in part **a**?

d. Find the velocity vector for the particle at any time t.

e. What is the distance traveled by the particle from $t = 1.25$ to $t = 1.75$?

35. Find the area of the region inside the circle $r = 4$ and to the right of the line $r = 2 \sec \theta$.

36. Find an equation in terms of x and y for the line tangent to the curve given by

$$x = t^2 - 2t - 1, \quad y = t^4 - 4t^2 + 2$$

at the point where $t = 1$.

37. A particle moves in space with position vector

$$\mathbf{R}(t) = (\sin t)\mathbf{i} - (\cos t)\mathbf{j} + \mathbf{k}$$

a. Find the velocity and acceleration vectors for the particle and find its speed.

b. Find the Cartesian equation for the particle's trajectory.

c. Find the curvature κ and the tangential and normal components of acceleration for the particle's motion.

38. **MODELING PROBLEM**   At what speed must a satellite travel to maintain a circular orbit 1,000 mi above the surface of the earth? Assume the earth is a sphere of radius 4,000 mi and that GM in Newton's law of universal gravitation is approximately $9.56 \times 10^4 \ \text{mi}^3/\text{s}^2$.

39. **MODELING PROBLEM**   A calculus text (other than the one you are now reading) is hurled downward from the top of a 120-ft-high building at an angle of 30° from the horizontal. Assume that the initial speed of the book is 8 ft/s. How far from the base of the building will the text land?

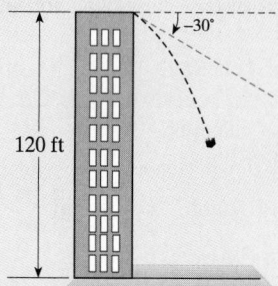

40. To close a sliding door, a person pulls on a rope with a constant force of 50 lb at a constant angle of 60°. Find the work done in moving the door 12 ft to its closed position.

12 Partial Differentiation

PREVIEW

The goal of this chapter is to extend the methods of single-variable differential calculus to functions of several variables. The vector methods developed in Chapters 10 and 11 play an important role in our work, and indeed, we shall find that the closest analogue of the single-variable derivative in higher dimensions is the vector function called the *gradient*. In physics, the gradient is the rate at which a variable quantity, such as temperature or pressure, changes in value. In this chapter, we will define the *gradient of a function,* which is a vector whose components along the axes are related to the rate at which the function changes in the direction of the given component. We conclude this chapter by solving rate and optimization problems involving functions of several variables.

PERSPECTIVE

In many practical situations, the value of one quantity may depend on the values of two or more others. For example, the amount of water in a reservoir may depend on the amount of rainfall and on the amount of water consumed by local residents. The current in an electrical circuit varies with the electromotive force, the capacitance, the resistance, and the impedance in the circuit. The flow of blood from an artery into a small capillary depends on the diameter of the capillary and the pressure in both the artery and the capillary. The output of a factory may depend on the amount of capital invested in the plant and on the size of the labor force. We will analyze such situations using functions of several variables.

12.1 *Functions of Several Variables*

IN THIS SECTION   basic concepts; level curves and surfaces; open, closed, and bounded sets; graphs of functions of two variables

In the real world, physical quantities often depend on two or more variables (see the Perspective box at the beginning of this chapter). For example, we might be concerned with the temperature on a metal plate at various points at time t. Locations on the plate are designated as ordered pairs (x, y), so that the temperature T could be considered as a function of two location variables, x and y, as well as a time variable, t. Extending the notation for a function of a single variable, we might write this as $T(x, y, t)$.

BASIC CONCEPTS

We begin our investigation of functions of several variables by introducing notation and terminology and examining a few basic concepts.

Function of Two Variables

A **function of two variables** is a rule f that assigns to each ordered pair (x, y) in a set D a unique number $f(x, y)$. The set D is called the **domain** of the function, and the corresponding values of $f(x, y)$ constitute the **range** of f.

Functions of three or more variables can be defined in a similar fashion. For simplicity, we shall focus most of our attention on functions of two or three variables.

When dealing with a function of two variables f, we may write $z = f(x, y)$ and refer to x and y as the **independent variables** and to z as the **dependent variable.** Often, the functional "rule" will be given as a formula, and unless otherwise stated, *we shall assume that the domain is the largest set of points in the plane (or in space) for which the functional formula is defined and real valued.* These definitions and conventions are illustrated in Example 1 for a function of two variables.

EXAMPLE 1   *Evaluating a function of two variables and finding the domain and range*

Let $f(x, y) = \sqrt{1 - x + y}$.
a. Evaluate $f(2, 1), f(-4, 3)$, and $f(2t, t^2)$.
b. Describe the domain and range of f.

Solution
a. $f(2, 1) = \sqrt{1 - 2 + 1} = 0$

$f(-4, 3) = \sqrt{1 - (-4) + 3} = \sqrt{8} = 2\sqrt{2}$

$f(2t, t^2) = \sqrt{1 - (2t) + (t^2)} = \sqrt{(t - 1)^2} = |t - 1|$

b. The domain of f is the set of all ordered pairs (x, y) for which $\sqrt{1 - x + y}$ is defined. We must have $1 - x + y \geq 0$ or, equivalently, $y \geq x - 1$, in order for the square root to be defined. Thus, the domain of f is the shaded set shown in Figure 12.1.

Because $z = f(x, y) = \sqrt{1 - x + y}$, we see that z must be nonnegative, and the range of f is all $z \geq 0$.

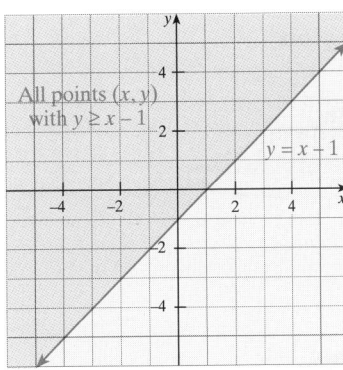

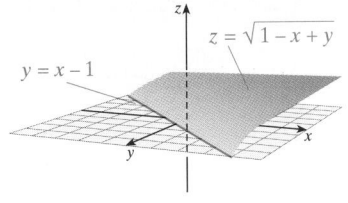

a. Graph of domain in $\mathbb{R}^2$.

b. Graph of $z = \sqrt{1 - x + y}$ in $\mathbb{R}^3$ over it's domain. Notice that the xy-plane is now a horizontal plane.

■ **FIGURE 12.1** The domain of $f(x, y) = \sqrt{1 - x + y}$ ■

Functions of several variables can be combined just like functions of a single variable.

Operations with Functions of Two Variables

If $f(x, y)$ and $g(x, y)$ are functions of two variables with domain D, then

Sum	$(f + g)(x, y) = f(x, y) + g(x, y)$
Difference	$(f - g)(x, y) = f(x, y) - g(x, y)$
Product	$(fg)(x, y) = f(x, y)\, g(x, y)$
Quotient	$\left(\dfrac{f}{g}\right)(x, y) = \dfrac{f(x, y)}{g(x, y)} \qquad g(x, y) \neq 0$

A **polynomial function in x and y** is a sum of functions of the form

$$Cx^m y^n$$

with nonnegative integers m and n, and C a constant; for instance,

$$3x^5 y^3 - 7x^2 y + 2x - 3y + 11$$

is a polynomial in x and y. A **rational function** is a quotient of two polynomial functions. Similar notation and terminology apply to functions of three or more variables.

LEVEL CURVES AND SURFACES

By analogy with the single-variable case, we define the **graph of the function $f(x, y)$** to be the collection of all 3-tuples (ordered triplets) (x, y, z) such that (x, y) is in the domain of f and $z = f(x, y)$. The graph of $f(x, y)$ is a surface in $\mathbb{R}^3$ whose projection onto the xy-plane is the domain D.

It is usually not easy to sketch the graph of a function of two variables. One way to proceed is illustrated in Figure 12.2.

Notice that when the plane $z = C$ intersects the surface $z = f(x, y)$, the result is the space curve with the equation $f(x, y) = C$. Such an intersection is called the **trace**

Mount St. Helens, Washington

Before May 18, 1980

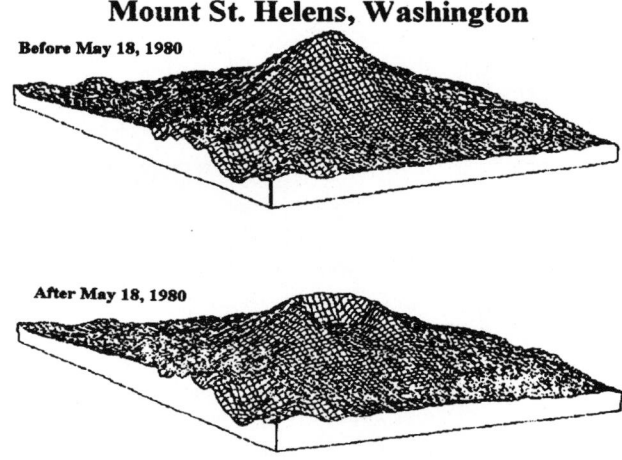

After May 18, 1980

Mount St. Helens map courtesy of Bill Lennox, Humboldt State University

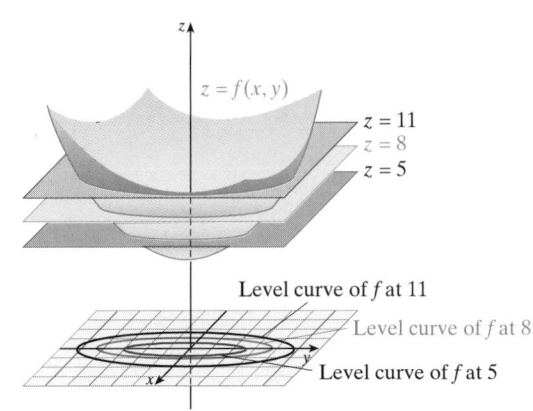

Level curve of *f* at 11

Level curve of *f* at 8

Level curve of *f* at 5

■ **FIGURE 12.2**   Graph of a function of two variables

Computer programs for three-dimensional sketching are common; see the *Technology Manual.* The Mt. St. Helens simulation (above) was done using a function of two variables.

of the graph of *f* in the plane $z = C$. The set of points (x, y) in the xy-plane that satisfy $f(x, y) = C$ is called the **level curve** of *f* at C, and an entire family of level curves is generated as C varies over the range of *f*. We can think of a level curve as a "slice" of the surface at a particular location. By sketching members of this family on the xy-plane, we obtain a useful topographical map of the surface $z = f(x, y)$. Because these level curves are used to show the shape of a surface (a mountain, for example), they are sometimes called **contour curves.**

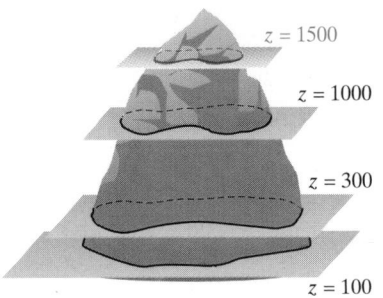

$z = 1500$

$z = 1000$

$z = 300$

$z = 100$

a. The surface $z = f(x, y)$ as a mountain

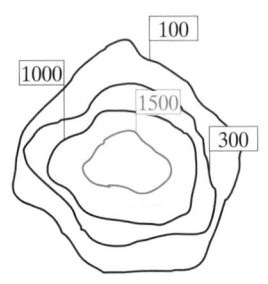

b. Level curves yield a topographic map of $z = f(x, y)$.

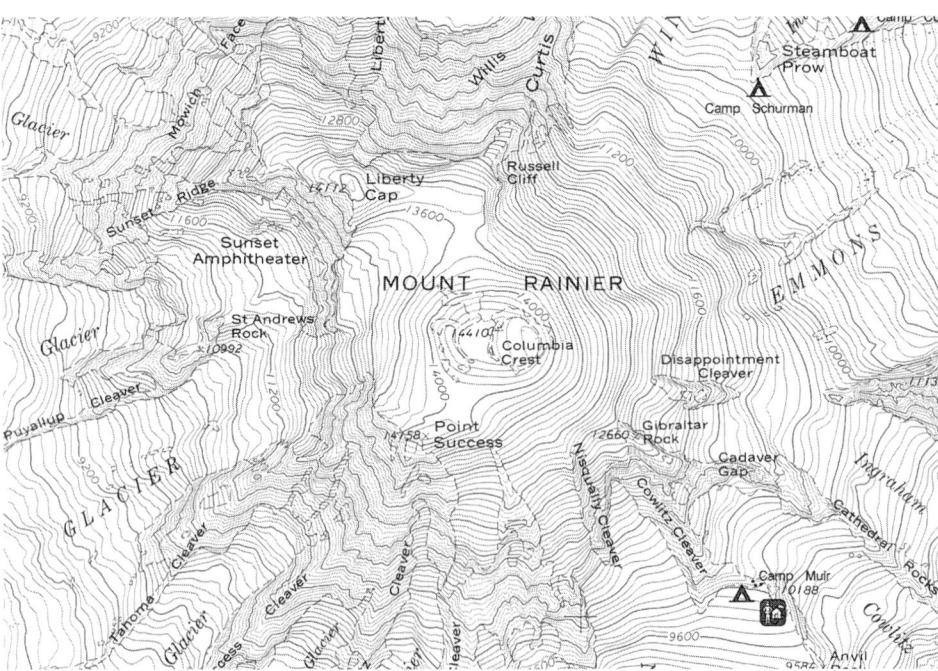

c. Topographic map of Mount Rainier

■ **FIGURE 12.3**   Level curves of a surface

For instance, imagine that the surface $z = f(x, y)$ is a "mountain," and that we wish to draw a two-dimensional "profile" of its shape. To draw such a profile, we indicate the paths of constant elevation by sketching the family of level curves in the plane and pinning a "flag" to each curve to show the elevation to which it corresponds, as shown in Figure 12.3b. Notice that regions in the map where paths are crowded together correspond to the steeper portions of the mountain. An actual topographical map of Mount Rainier is shown in Figure 12.3c.

You probably have seen level curves on the weather report in the newspaper or on the evening news, where level curves of equal temperature are called **isotherms** (see Figure 12.4). Other common uses of level curves show lines of equal pressure (called **isobars**) or those representing equal electric potential (called **equipotential lines**).

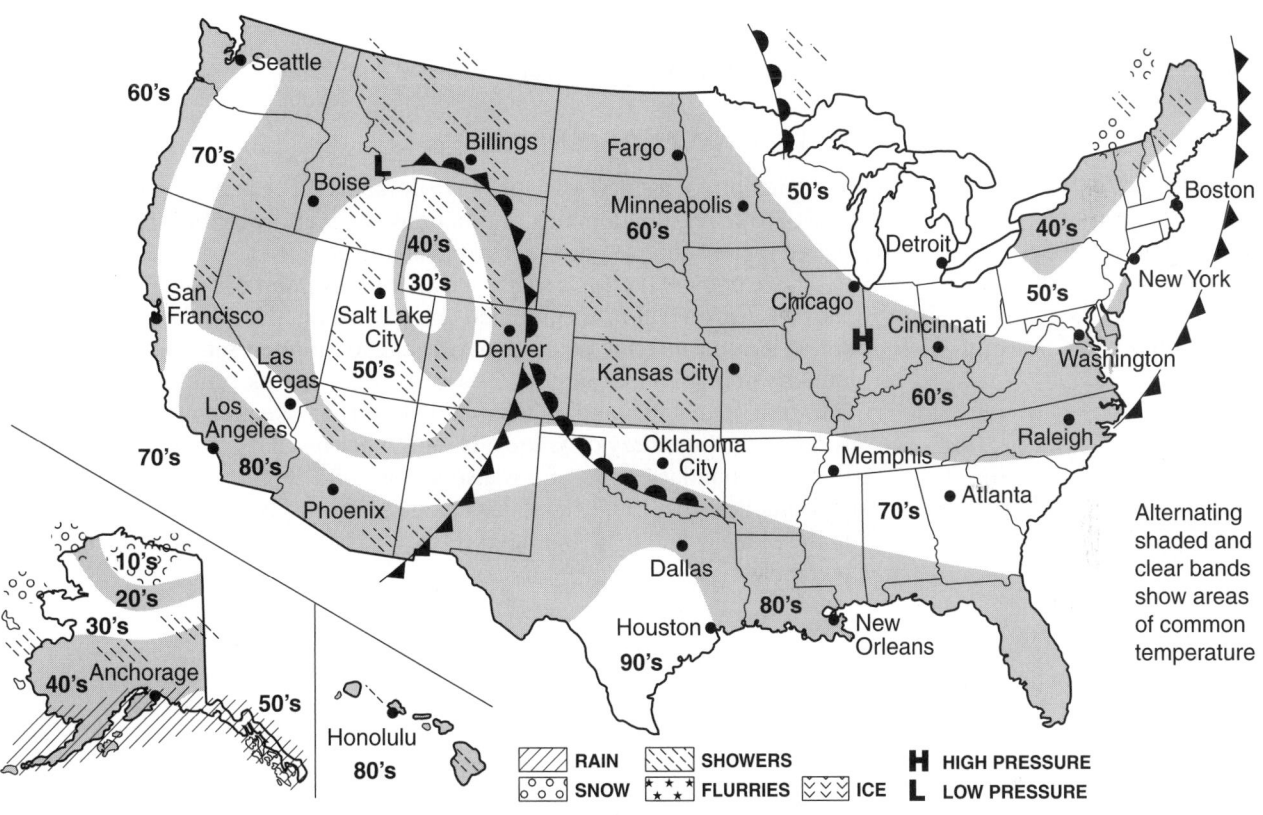

■ **FIGURE 12.4**   Isotherms

| **EXAMPLE 2** | *Level curves* |

Sketch some level curves of the function $f(x, y) = 10 - x^2 - y^2$.

Solution   A computer graph of this curve is the surface shown in Figure 12.5a. To see the relationship between the surface and the level curves, we graph a simplified graph showing contour curves for $z = 1, 6$, and 9, as shown in Figure 12.5b. Finally, the corresponding level curves are shown in Figure 12.5c.

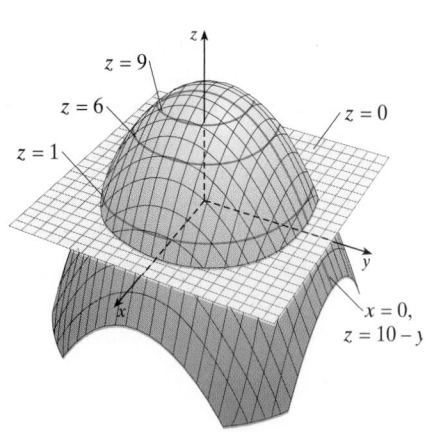

a. Computer generated graph of surface

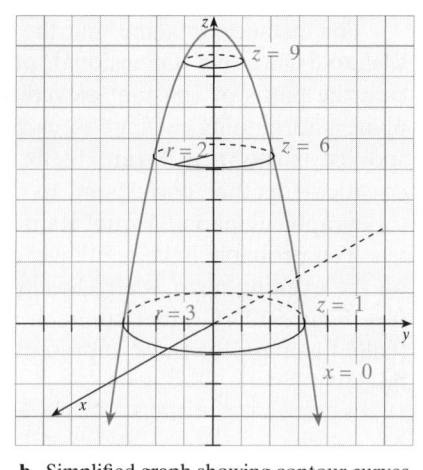

b. Simplified graph showing contour curves

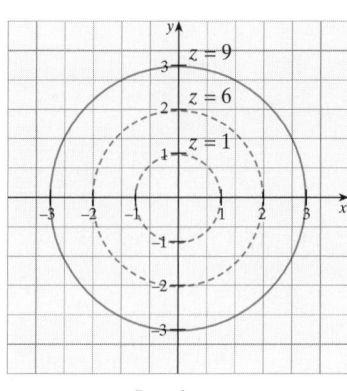

c. Level curves

■ **FIGURE 12.5** Graph of $z = 10 - x^2 - y^2$ ■

OPEN, CLOSED, AND BOUNDED SETS

Most functions of a single variable have domains that can be described in terms of intervals. However, the domains of functions of several variables are often more complicated regions that require special terminology.

First, a point $P_0(x_0, y_0)$ is said to be an **interior point** of a set S in the plane if some open disk centered at P_0 is contained entirely within S. If S is the empty set, or if every point of S is an interior point, then S is called an **open set**. The point P_0 is called a **boundary point** of S (see Figure 12.6a) if every open disk centered at P_0 contains both points that belong to S and points that do not. The collection of all boundary points of S is called the **boundary** of S, and S is said to be **closed** if it contains its boundary; that is, if every boundary point of S is also a point of S (Figure 12.6b). If S has no boundary points, then it certainly contains its boundary, and is therefore closed. Finally, a set S in the plane is said to be **bounded** (Figure 12.6c) if it can be contained in a circle (or a rectangle).

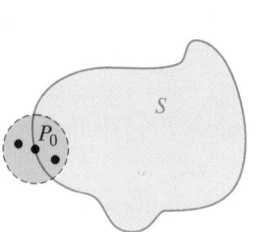

a. P_0 is a boundary point.

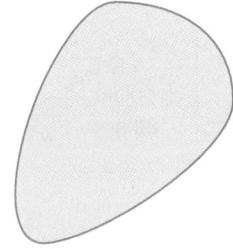

b. A closed set contains its boundary points.

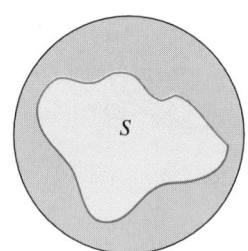

c. S is bounded.

■ **FIGURE 12.6** Closed and bounded sets

In $\mathbb{R}^3$, the same definitions apply with the term "sphere" replacing "disk." Thus, an interior point P_0 of a set S in $\mathbb{R}^3$ has the property that some open sphere centered at P_0 is contained entirely within S, and S is open if all its points are interior point. Similar definitions hold for boundary points and closed sets. A bounded set in $\mathbb{R}^3$ is one that can be contained in a sphere (or a box).

GRAPHS OF FUNCTIONS OF TWO VARIABLES

The level curves of a function $f(x, y)$ provide information about the cross sections of the surface $z = f(x, y)$ perpendicular to the z-axis. However, a more complete picture of the surface can often be obtained by examining cross-sections in other directions as well. This procedure is used to graph a function in Example 3.

EXAMPLE 3   *Level curves*

Use the level curves of the function $f(x, y) = x^2 + y^2$ to sketch the graph of f.

Solution   The level curve $x^2 + y^2 = 0$ (that is, $C = 0$) is the point $(0, 0)$, and for $C > 0$, the level curve $x^2 + y^2 = C$ is the circle with center $(0, 0)$ and radius $\sqrt{C}$ (Figure 12.7a). There are no points (x, y) that satisfy $x^2 + y^2 = C$ for $C < 0$.

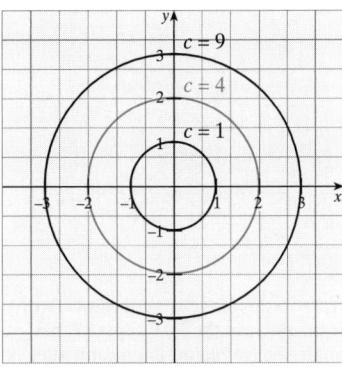

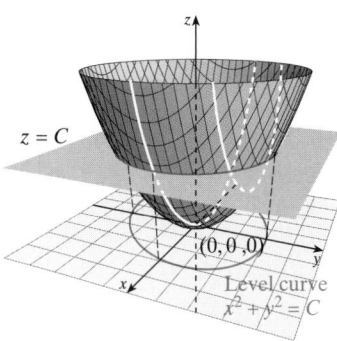

a. Level curves of $f(x, y) = x^2 + y^2$ are circles in the form $x^2 + y^2 = C$.

b. The surface $z = x^2 + y^2$. Cross sections perpendicular to the x- and y-axes shown in white are parabolas.

■ **FIGURE 12.7**   The graph of the function $f(x, y) = x^2 + y^2$

We can gain additional information about the appearance of the surface by examining cross sections perpendicular to the other two principal directions. Cross-sectional planes perpendicular to the x-axis have the form $x = A$ and intersect the surface $z = x^2 + y^2$ in parabolas of the form $z = A^2 + y^2$. Similarly, cross-sectional planes perpendicular to the y-axis have the form $y = B$ and intersect the surface in parabolas of the form $z = x^2 + B^2$.

To summarize, the surface $z = x^2 + y^2$ has cross sections that are circular in planes perpendicular to the z-axis and parabolic in the other two principal directions. For this reason, the surface is called a **circular paraboloid** or a **paraboloid of revolution.** The graph of the surface is shown in Figure 12.7b.   ■

The concept of level curve can be generalized to apply to functions of more than two variables. In particular, if f is a function of the n variables $x_1, x_2, \ldots, x_n$ and C is a number in the range of f, then the solution set of the equation $f(x_1, x_2, \ldots, x_n) = C$ is a region of n-space called the **level surface** of f at C. The level surfaces of a function of three variables are surfaces in $\mathbb{R}^3$ and can provide some insight into the nature of the function, but level surfaces of functions of four or more variables are very difficult to visualize.

DRAWING LESSON 7: SKETCHING SURFACES WITH LEVEL CURVES

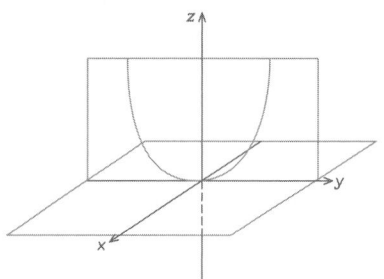

a. Draw the three coordinate axes and the *xy*-plane. Draw a trace of the surface in the *yz*-plane.

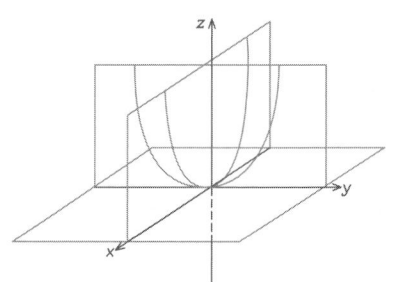

b. Draw a trace of the surface in the *xz*-plane.

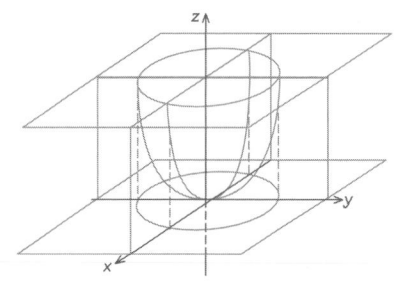

c. Draw a trace in the plane *z* = constant. Drop dashed segment from the intercepts of this trace down to the *xy*-plane. Then draw the level curve in the *xy*-plane. Finish outlining the surface.

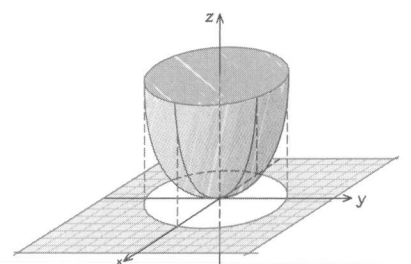

d. Lightly erase all planes except the *xy*-plne. Use highlighters or colored pencils to shade the surface.

EXAMPLE 4   *Isothermal surface*

Suppose a region of R is heated so that its temperature T at each point (x, y, z) is given by $T(x, y, z) = 100 - x^2 - y^2 - z^2$ degrees Celsius. Describe the isothermal surfaces for $T > 0$.

Solution   The isothermal surfaces are given by $T(x, y, z) = k$ for constant k; that is, $x^2 + y^2 + z^2 = 100 - k$. If $100 - k > 0$, the graph of $x^2 + y^2 + z^2 = 100 - k$ is a sphere of radius $\sqrt{100 - k}$ and center $(0, 0, 0)$. When $k = 100$, the graph is a single

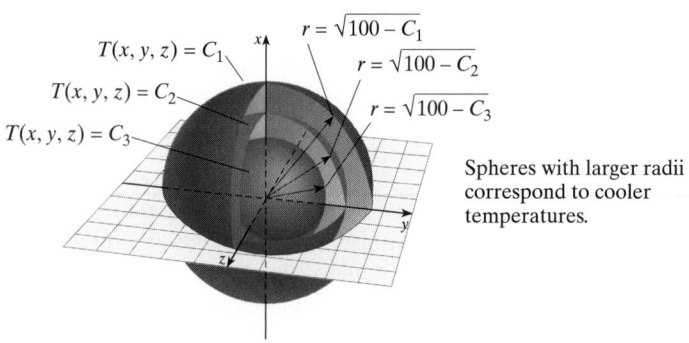

$T(x, y, z) = C_1$
$T(x, y, z) = C_2$
$T(x, y, z) = C_3$

$r = \sqrt{100 - C_1}$
$r = \sqrt{100 - C_2}$
$r = \sqrt{100 - C_3}$

Spheres with larger radii correspond to cooler temperatures.

■ **FIGURE 12.8**   Isothermal surfaces for $T(x, y, z) = 100 - x^2 - y^2 - z^2$

point (the origin), and $T(0, 0, 0) = 100\ °C$. As the temperature drops, the constant k gets smaller, and the radius $\sqrt{100 - k}$ of the sphere gets larger. Hence, the isothermal surfaces are spheres, and the larger the radius, the cooler the surface. This situation is illustrated in Figure 12.8. ∎

As you can see by the examples in this section, you will need to recall the graphs of quadric surfaces. It will also help if you can recognize the surface by looking at its equation. What you will need to remember is summarized in Table 10.1, page 683.

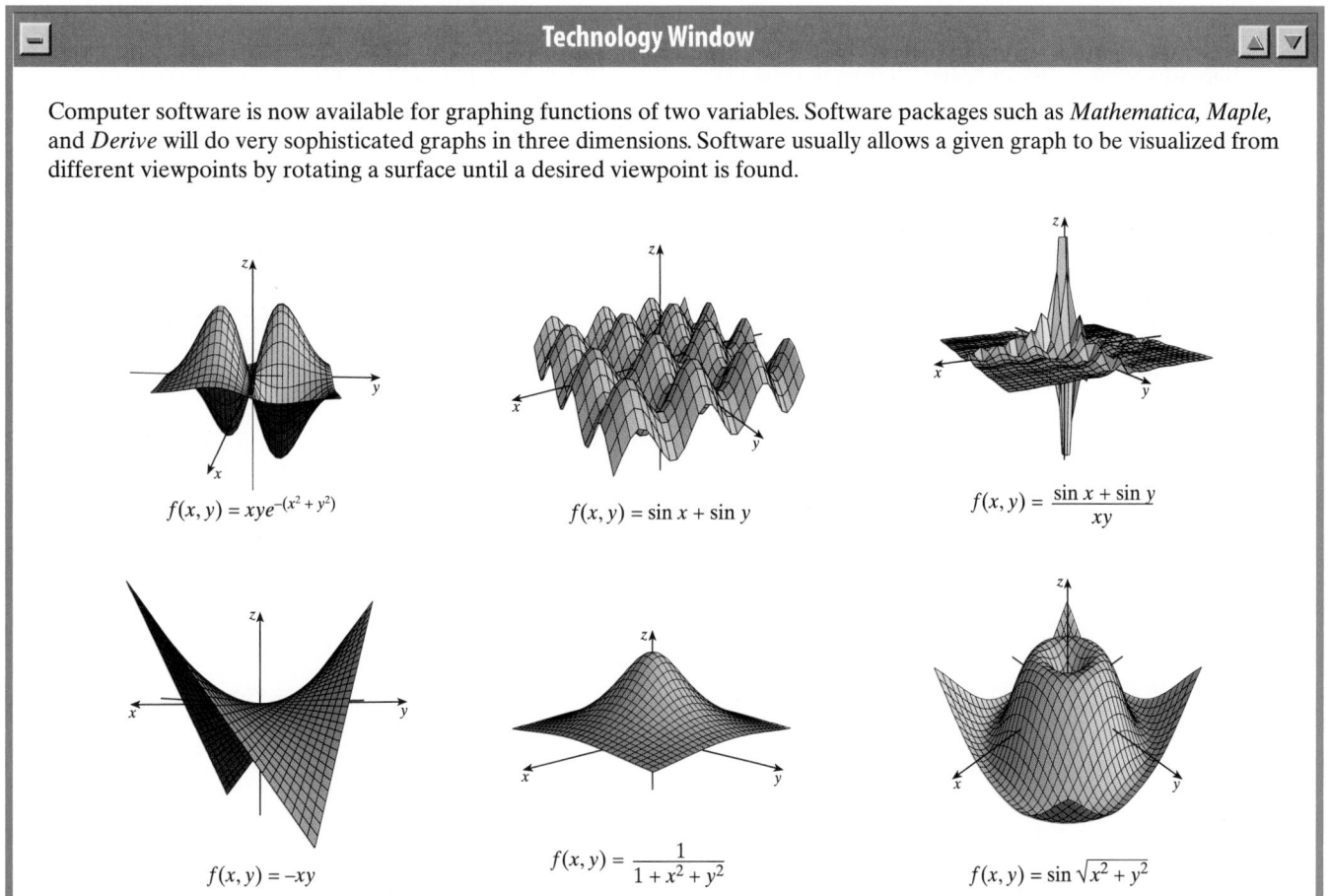

Technology Window

Computer software is now available for graphing functions of two variables. Software packages such as *Mathematica, Maple,* and *Derive* will do very sophisticated graphs in three dimensions. Software usually allows a given graph to be visualized from different viewpoints by rotating a surface until a desired viewpoint is found.

$f(x, y) = xye^{-(x^2 + y^2)}$

$f(x, y) = \sin x + \sin y$

$f(x, y) = \dfrac{\sin x + \sin y}{xy}$

$f(x, y) = -xy$

$f(x, y) = \dfrac{1}{1 + x^2 + y^2}$

$f(x, y) = \sin \sqrt{x^2 + y^2}$

12.1 Problem Set

Ⓐ **1.** Let $f(x, y) = x^2y + xy^2$. If t is a real number, find:
   a. $f(0, 0)$
   b. $f(-1, 0)$
   c. $f(0, -1)$
   d. $f(1, 1)$
   e. $f(2, 4)$
   f. $f(t, t)$
   g. $f(t, t^2)$
   h. $f(1 - t, t)$

2. Let $f(x, y) = \left(1 - \dfrac{x}{y}\right)^2$. If t is a nonzero real number, find:

   a. $f(0, 1)$
   b. $f(5, 5)$
   c. $f(6, 1)$
   d. $f(1, 2)$
   e. $f(t, t)$
   f. $f(5t, t)$
   g. $f(t, 2t)$
   h. $f(1 + t, t)$

3. Let $f(x, y, z) = x^2ye^{2x} + (x + y - z)^2$. Find
   a. $f(0, 0, 0)$
   b. $f(1, -1, 1)$
   c. $f(-1, 1, -1)$
   d. $\dfrac{d}{dx} f(x, x, x)$
   e. $\dfrac{d}{dy} f(1, y, 1)$
   f. $\dfrac{d}{dz} f(1, 1, z^2)$

4. Let $f(x, y, z) = x \sin y + y \cos z$. Find
   a. $f(0, 0, 0)$ b. $f(1, \frac{\pi}{2}, \pi)$ c. $f(1, \pi, \frac{\pi}{2})$
   d. $\dfrac{d}{dx} f(x, x, x)$ e. $\dfrac{d}{dx} f(x, 2x, 3x)$ f. $\dfrac{d}{dy} f(y, y, 0)$

5. ■ **What Does This Say?** Discuss the notion of a function of two variables. Your discussion should include additional examples of a function of two variables.

6. ■ **What Does This Say?** How do you think that a function of three variables differs from a function of two variables?

7. ■ **What Does This Say?** Discuss level curves and amplify with some examples (different from those in this text).

Find the domain and range for each function given in Problems 8–17.

8. $f(x, y) = \sqrt{x - y}$

9. $f(x, y) = \dfrac{1}{\sqrt{x - y}}$

10. $f(u, v) = \sqrt{uv}$

11. $f(x, y) = \sqrt{\dfrac{y}{x}}$

12. $f(x, y) = \ln(y - x)$

13. $f(u, v) = \sqrt{u \sin v}$

14. $f(x, y) = \sqrt{(x + 3)^2 + (y - 1)^2}$

15. $f(x, y) = e^{(x+1)/(y-2)}$

16. $f(x, y) = \dfrac{1}{\sqrt{x^2 - y^2}}$

17. $f(x, y) = \dfrac{1}{\sqrt{9 - x^2 - y^2}}$

Sketch some level curves $f(x, y) = C$ for $C > 0$ of the functions given in Problems 18–23.

18. $f(x, y) = 2x - 3y$

19. $f(x, y) = x^2 - y^2$

20. $f(x, y) = x^3 - y$

21. $g(x, y) = x^2 - y$

22. $h(u, v) = u^2 + \dfrac{v^2}{4}$

23. $f(x, t) = \dfrac{x}{t}$

In Problems 24–29, sketch the level surface $f(x, y, z) = C$ for the given value of C.

24. $f(x, y, z) = y^2 + z^2$ for $C = 1$
25. $f(x, y, z) = x^2 + z^2$ for $C = 1$
26. $f(x, y, z) = x + y - z$ for $C = 1$
27. $f(x, y, z) = x + y - z$ for $C = 0$
28. $f(x, y, z) = (x + 1)^2 + (y - 2)^2 + (z - 3)^2$ for $C = 4$
29. $f(x, y, z) = 2x^2 + 2y^2 - z$ for $C = 1$

In Problems 30–37, describe the traces of the given quadric surface with each coordinate plane. You might wish to review Section 10.2 (page 676). Sketch the graph of the quadric and identify it.

30. $9x^2 + 4y^2 + z^2 = 1$

31. $\dfrac{x^2}{4} + y^2 + \dfrac{z^2}{9} = 1$

32. $\dfrac{x^2}{4} + \dfrac{y^2}{9} - z^2 = 1$

33. $\dfrac{x^2}{9} - y^2 - z^2 = 1$

34. $z = x^2 + \dfrac{y^2}{4}$

35. $z = \dfrac{x^2}{9} - \dfrac{y^2}{16}$

36. $x^2 + 2y^2 = 9z^2$

37. $z^2 = 1 + \dfrac{x^2}{9} + \dfrac{y^2}{4}$

Match each family of level curves given in Problems 38–43 with one of the surfaces labeled A–F.

38. $f(x, y) = x^2 - y^2$

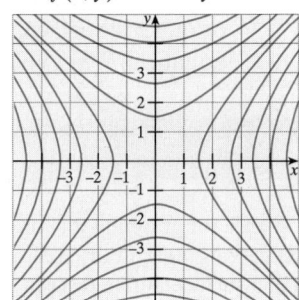

39. $f(x, y) = e^{1 - x^2 + y^2}$

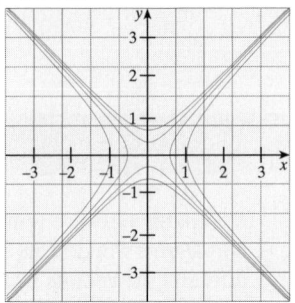

40. $f(x, y) = \dfrac{1}{x^2 + y^2}$

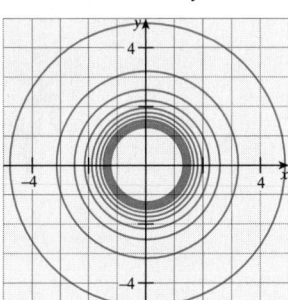

41. $f(x, y) = \sin\left(\dfrac{x^2 + y^2}{2}\right)$

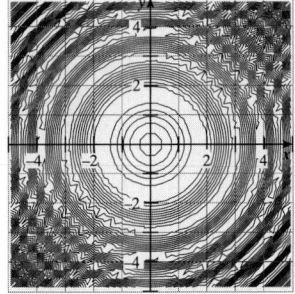

42. $f(x, y) = \dfrac{\cos xy}{x^2 + y^2}$

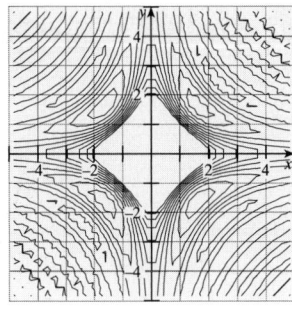

43. $f(x, y) = \sin \sqrt{x^2 + y^2}$

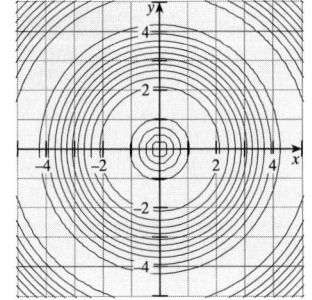

A.

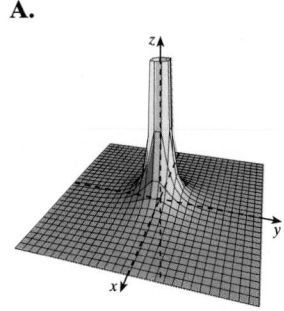

B.

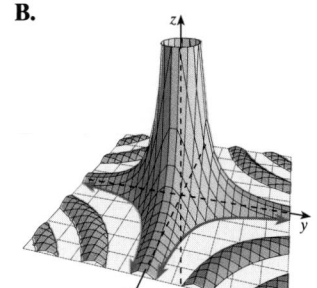

C.

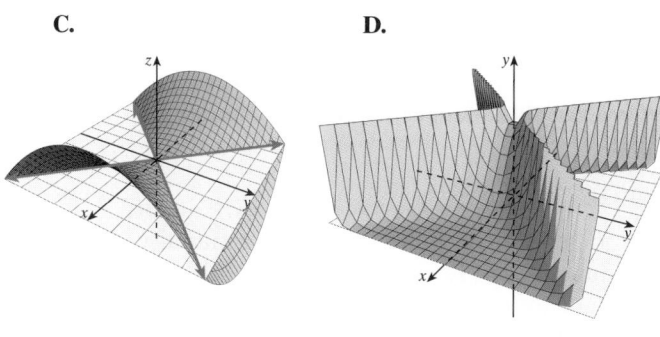

D.

E.

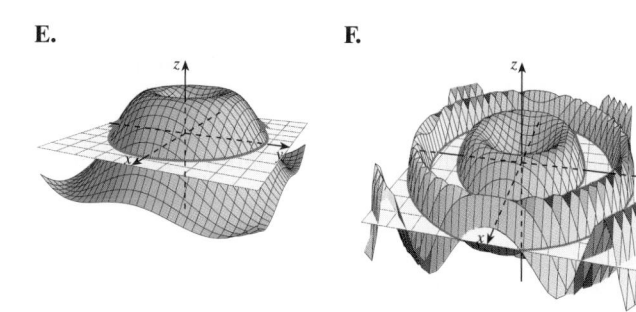

F.

ⓑ *Sketch the graph of each function given in Problems 44–55.*

44. $f(x, y) = -4$

45. $f(x, y) = x$

46. $f(x, y) = y^2 + 1$

47. $f(x, y) = x^3 - 1$

48. $f(x, y) = 2x - 3y$

49. $f(x, y) = x^2 - y$

50. $f(x, y) = 2x^2 + y^2$

51. $f(x, y) = x^2 - y^2$

52. $f(x, y) = \dfrac{x}{y}$

53. $f(x, y) = \sqrt{x + y}$

54. $f(x, y) = x^2 + y^2 + 2$

55. $f(x, y) = \sqrt{1 - x^2 - y^2}$

56. The *lens equation* in optics states that

$$\frac{1}{d_0} + \frac{1}{d_i} = \frac{1}{L}$$

where d_0 is the distance of an object from a thin, spherical lens, d_i is the distance of its image on the other side of the lens, and L is the *focal length* of the lens. Describe the level curves for the constant focal length.

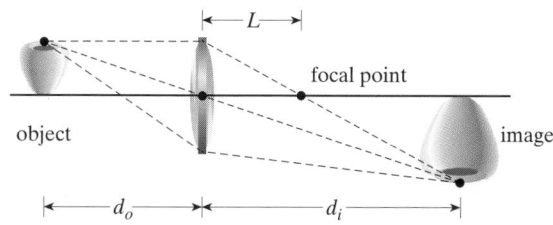

57. The EZGRO agricultural company estimates that when $100x$ worker-hours of labor are employed on y acres of land, the number of bushels of wheat produced is

$f(x, y) = Ax^a y^b$, where A, a, and b are nonnegative constants. Suppose the company decides to double the production factors x and y. Determine how this decision affects the production of wheat in each of these cases:
   a. $a + b > 1$ **b.** $a + b < 1$ **c.** $a + b = 1$

58. Suppose that when x machines and y worker-hours are used each day, a certain factory will produce $Q(x, y) = 10xy$ mobile phones. Describe the relationship between the "inputs" x and y that result in an "output" of 1,000 phones each day. *Note:* You are finding a level curve of the production function Q.

ⓒ **59. MODELING PROBLEM**   At a certain factory, the daily output is modeled by $Q = CK^r L^{1-r}$ units, where K denotes capital investment, L is the size of the labor force, and C and r are constants, with $0 < r < 1$ (this is called a *Cobb–Douglas production function*). What happens to Q if K and L are both doubled? What if both are tripled?

60. Sketch the graph of $f(x, y) = xy$ in the first octant (x, y, and z are all positive).

61. The ideal gas law says that $PV = kT$, where P is the pressure of a confined gas, V is its volume, T is its temperature (Kelvin), and k is a constant. Express P as a function of V and T, and describe the level curves of this function.

62. A publishing house has found that in a certain city each of its salespeople will sell approximately

$$\frac{r^2}{2{,}000p} + \frac{s^2}{100} - s \text{ units}$$

per month, at a price of p dollars/unit, where s denotes the total number of salespeople employed and r is the amount of money spent each month on local advertising. Express the total revenue R as a function of p, r, and s.

63. MODELING PROBLEM   A manufacturer with exclusive rights to a sophisticated new industrial machine is planning to sell a limited number of the machines to both foreign and domestic firms. The price that the manufacturer can expect to receive for the machines will depend on the number of machines made available. It is estimated that if the manufacturer supplies x machines to the domestic market and y machines to the foreign market, the machines will sell for

$$60 - \frac{x}{5} + \frac{y}{20}$$

thousand dollars apiece at home and

$$50 - \frac{x}{10} + \frac{y}{20}$$

thousand dollars apiece abroad. Express the revenue R as a function of x and y. Describe the curves of constant revenue.

12.2   Limits and Continuity

limit of a function of two variables, properties of limits, continuity ■

LIMIT OF A FUNCTION OF TWO VARIABLES

x	y	$f(x) = \dfrac{x}{x-y}$
0.99	2.01	−0.97059
0.999	2.001	−0.99701
1.001	2.0001	−1.00190
1.01	1.9999	−1.02031
0.99	1.99	−0.99000

As with single-variable functions, we need limits to discuss continuity, derivatives, slopes, and rates of change of functions of several variables. The single-variable limit can be extended naturally to functions of several variables. However, when we say that $f(x, y)$ approaches the number L as (x, y) approaches the point (a, b), written $(x, y) \to (a, b)$, we must remember that the approach to (a, b) can be from *any* direction, not just the left or right. The table in the margin suggests that no matter how $(x, y) \to (1, 2)$, the expression $f(x, y) = x/(x - y)$ approaches $L = -1$.

In Chapter 2, we informally defined the limit statement $\lim_{x \to c} f(x) = L$ to mean that $f(x)$ can be made arbitrarily close to L by choosing x sufficiently close (but not equal) to c. The analogous informal definition for a function of two variables is now given.

Limit of a Function of Two Variables (Informal Definition)

The notation

$$\lim_{(x, y) \to (x_0, y_0)} f(x, y) = L$$

means that the functional values $f(x, y)$ can be made arbitrarily close to L by choosing a point (x, y) sufficiently close (but not equal) to the point (x_0, y_0).

■ *What This Says:* Formally, $\lim_{(x, y) \to (x_0, y_0)} f(x, y) = L$ means that for each given number $\epsilon > 0$, there exists a number $\delta > 0$ so that $|f(x) - L| < \epsilon$ whenever (x, y) lies in a punctured disk $0 \le \sqrt{(x - x_0)^2 + (y - y_0)^2} \le \delta$ centered at $P_0(x_0, y_0)$. This definition is illustrated in Figure 12.9a.

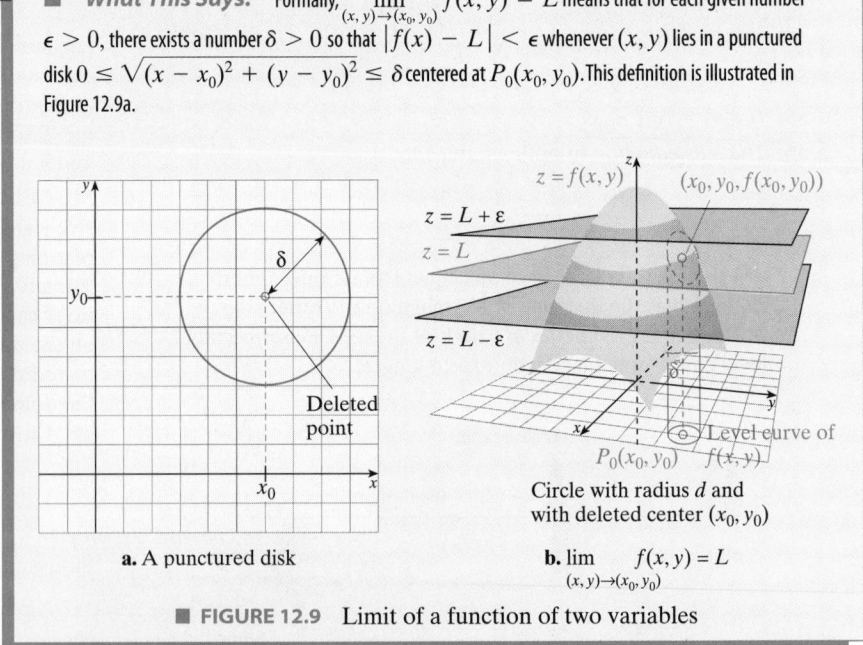

a. A punctured disk

b. $\lim_{(x, y) \to (x_0, y_0)} f(x, y) = L$

■ **FIGURE 12.9**   Limit of a function of two variables

When considering the limit of a function of a single variable, we need to examine the approach of x to c from two directions (the left- and right-hand limits). However, for a function of two variables, we write $(x, y) \to (x_0, y_0)$ to mean that the point (x, y) is allowed to approach (x_0, y_0) along *any* curve that passes through (x_0, y_0). If

$$\lim_{(x, y) \to (x_0, y_0)} f(x, y)$$

is not the same for *all* approaches, or **paths**, then *the limit does not exist.*

EXAMPLE 1   *Limit of a function of two variables*

a. Show

$$\lim_{(x, y) \to (0, 0)} \frac{2xy}{x^2 + y^2}$$

does not exist by evaluating this limit along the x-axis ($y = 0$), the y-axis ($x = 0$), and along the line $y = x$.

b. Evaluate

$$\lim_{(x, y) \to (1, 2)} 2xy$$

along the paths $y = 2$, $x = 1$, and $y = 2x$, and show that these values are all the same.

Solution

a. First note that the denominator is zero at $(0, 0)$ so $f(0, 0)$ is not defined. If we approach the origin along the x-axis (where $y = 0$) we find that

$$\frac{2xy}{x^2 + y^2} = \frac{2x(0)}{x^2 + 0^2} = 0 \text{ as } (x, y) \to (0, 0) \text{ along } y = 0 \text{ (and } x \neq 0)$$

We find a similar result if we approach the origin along the y-axis (where $x = 0$); see Figure 12.10.

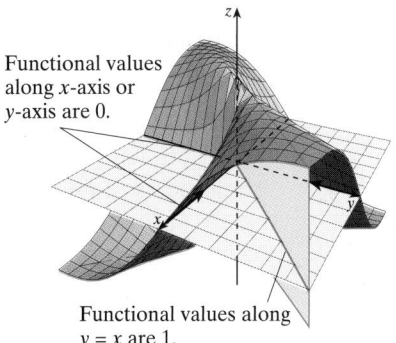

Functional values along x-axis or y-axis are 0.

Functional values along $y = x$ are 1.

■ **FIGURE 12.10**   Graph of $y = \dfrac{2xy}{x^2 + y^2}$ and limits as $(x, y) \to (0, 0)$

However, along the line $y = x$, the functional values are

$$f(x, y) = f(x, x) = \frac{2x^2}{x^2 + x^2} = 1 \quad \text{for } x \neq 0$$

so that $\dfrac{2xy}{x^2 + y^2} \to 1$ as $(x, y) \to (0, 0)$ along $y = x$

Because $f(x, y)$ tends toward different numbers as $(x, y) \to (0, 0)$ along different curves, it follows that f has no limit at the origin.

b. If we approach the point $(1, 2)$ along the line $y = 2$ (see Figure 12.11a) we have

$$2xy = 4x \to 4 \text{ as } (x, y) \to (1, 2) \text{ along } y = 2$$

Similarly, along the line $x = 1$ (Figure 12.11b)

$$2xy = 2y \to 4 \text{ as } (x, y) \to (1, 2) \text{ along } x = 1$$

Finally, along the line $y = 2x$ (Figure 12.11c)

$$2xy = 2x(2x) = 4x^2 \to 4 \text{ as } (x, y) \to (1, 2) \text{ along } y = 2x$$

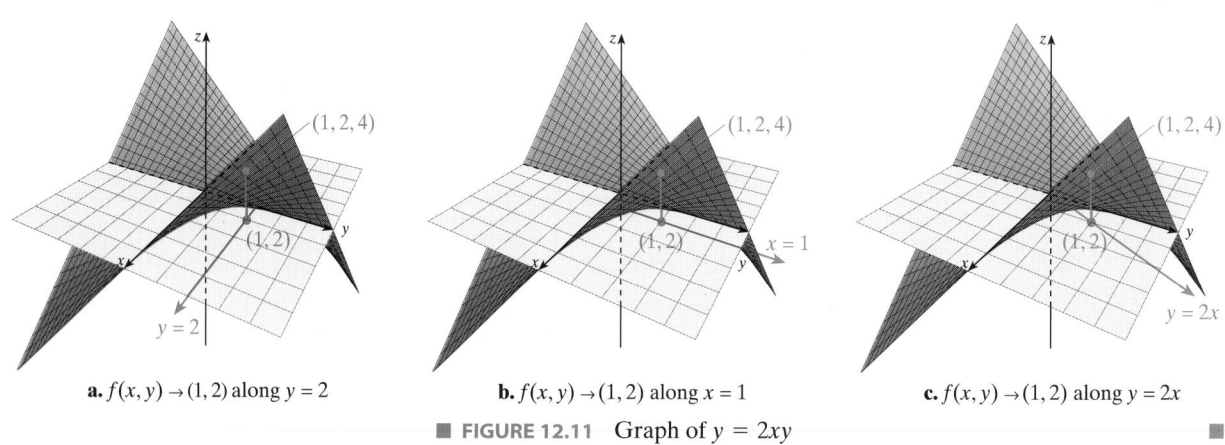

a. $f(x, y) \to (1, 2)$ along $y = 2$ **b.** $f(x, y) \to (1, 2)$ along $x = 1$ **c.** $f(x, y) \to (1, 2)$ along $y = 2x$

■ **FIGURE 12.11**   Graph of $y = 2xy$

WARNING   Note from Example 1a that the two-path procedure is enough to show that a *limit does NOT exist*. However, just showing that the limits along different paths in Example 1b are the same is not enough to conclude that a limit *DOES exist,* because we have not evaluated the limit along *every* path.

We now shall develop some rules that *will* allow us to conclude that a limit exists as well as help us to find the limit.

PROPERTIES OF LIMITS

The process of finding the limits shown in Example 1 is not very satisfactory. Fortunately, the various rules for manipulating limits of a function of one variable all have counterparts for limits of two variables.

Basic Formulas and Rules for Limits of a Function of Two Variables

Suppose $\displaystyle\lim_{(x, y)\to(x_0, y_0)} f(x, y) = L$ and $\displaystyle\lim_{(x, y)\to(x_0, y_0)} g(x, y) = M$. Then:

Scalar multiple rule	$\displaystyle\lim_{(x, y)\to(x_0, y_0)} [af](x, y) = aL$ for constant a
Sum rule	$\displaystyle\lim_{(x, y)\to(x_0, y_0)} [f + g](x, y) = L + M$
Product rule	$\displaystyle\lim_{(x, y)\to(x_0, y_0)} [fg](x, y) = LM$
Quotient rule	$\displaystyle\lim_{(x, y)\to(x_0, y_0)} \left[\dfrac{f}{g}\right](x, y) = \dfrac{L}{M}$ if $M \neq 0$

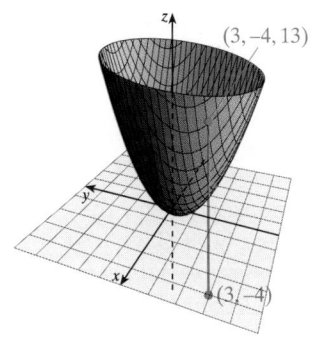

a. Graph of $z = x^2 + xy + y^2$

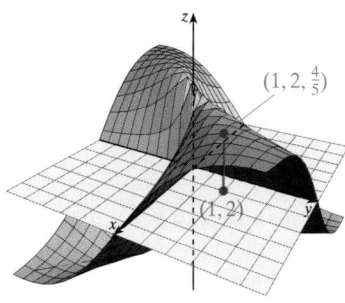

b. Graph of $f(x, y) = \dfrac{2xy}{x^2 + y^2}$

■ **FIGURE 12.12** Graphs showing limits

| EXAMPLE 2 | *Finding limits using properties of limits* |

Assuming each limit exists, evaluate:

a. $\displaystyle\lim_{(x,y)\to(3,-4)} (x^2 + xy + y^2)$ b. $\displaystyle\lim_{(x,y)\to(1,2)} \frac{2xy}{x^2 + y^2}$

Solution

a.
$$\lim_{(x,y)\to(3,-4)} (x^2 + xy + y^2) = (3)^2 + (3)(-4) + (-4)^2 = 13$$

A graph is shown in Figure 12.12a.

b.
$$\lim_{(x,y)\to(1,2)} \frac{2xy}{x^2 + y^2} = \frac{\displaystyle\lim_{(x,y)\to(1,2)} 2xy}{\displaystyle\lim_{(x,y)\to(1,2)} (x^2 + y^2)}$$
$$= \frac{2(1)(2)}{1^2 + 2^2} = \frac{4}{5}$$

The graph is shown in Figure 12.12b. ■

CONTINUITY

Recall that a function of a single variable x is continuous at $x = c$ if

1. $f(c)$ is defined; **2.** $\lim\limits_{x\to c} f(x)$ exists; **3.** $\lim\limits_{x\to c} f(x) = f(c)$.

Using the definition of the limit of a function of two variables, we can now define the continuity of a function of two variables analogously.

Continuity of a Function of Two Variables

The function $f(x, y)$ is **continuous** at the point (x_0, y_0) if and only if

1. $f(x_0, y_0)$ is defined;

2. $\displaystyle\lim_{(x,y)\to(x_0,y_0)} f(x, y)$ exists;

3. $\displaystyle\lim_{(x,y)\to(x_0,y_0)} f(x, y) = f(x_0, y_0)$.

Also, f is **continuous on a set S** in its domain if it is continuous at each point in S.

■ *What This Says:* The function f is continuous at (x_0, y_0) if the functional value of $f(x, y)$ is close to $f(x_0, y_0)$ whenever (x, y) is sufficiently close to (x_0, y_0). Geometrically, this means that f is continuous if the surface $z = f(x, y)$ has no "gaps" or "holes."

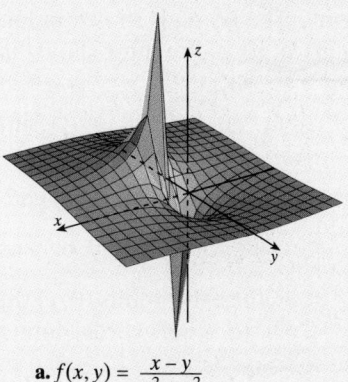

a. $f(x, y) = \dfrac{x - y}{x^2 + y^2}$
is discontinuous at $(0, 0)$; a hole

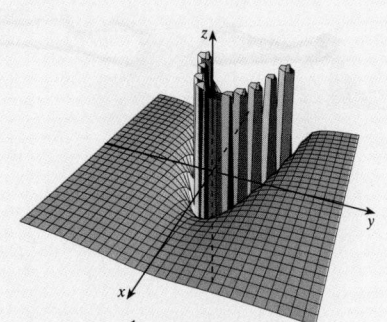

b. $f(x, y) = \dfrac{1}{y - x^2}$
is discontinuous on the parabola $y - x^2$; a gap

The basic properties of limits can be used to show that if f and g are both continuous on the set S, then so are the sum $f + g$, the multiple af, the product fg, the quotient f/g (whenever $g \neq 0$), and the root $\sqrt[n]{f}$ wherever it is defined. Also, if F is a function of two variables continuous at (x_0, y_0) and G is a function of one variable that is continuous at $F(x_0, y_0)$, it can be shown that the composite function $G \circ F$ is continuous at (x_0, y_0).

Many common functions of two variables are continuous wherever they are defined. For instance, a polynomial in two variables, such as $x^3 y^2 + 3xy^3 - 7x + 2$, is continuous throughout the plane, and a rational function in two variables is continuous wherever the denominator polynomial is not zero. In this course, all of the "standard" functions of two or more variables are continuous over their domains.

EXAMPLE 3   *Testing for continuity*

Test the continuity of the functions whose graphs are shown here, namely,

a. $f(x, y) = \dfrac{x - y}{x^2 + y^2}$ b. $f(x, y) = \dfrac{1}{y - x^2}$

Solution

a. Because $x - y$ and $x^2 + y^2$ are both polynomial functions in x and y, f is a rational function and the only place where it might not be continuous is where the denominator is zero. Because $x^2 + y^2 = 0$ only where both x and y are 0, the only possible point of discontinuity is at $(0, 0)$. However $f(0, 0)$ is not defined, so f is discontinuous at $(0, 0)$.

b. Once again we need to check when the denominator is 0:

$$y - x^2 = 0 \quad \text{when} \quad y = x^2$$

We can, therefore, conclude that the function is continuous at all points except those lying on the parabola $y = x^2$. ∎

We conclude with an example in which the definition of limit is used to determine continuity.

EXAMPLE 4   *Continuity using the limit definition*

Show that f is continuous at $(0, 0)$, where

$$f(x, y) = \begin{cases} y \sin \dfrac{1}{x}, & x \neq 0 \\ 0 & x = 0 \end{cases}$$

Solution   The graph is shown in Figure 12.13. It appears to be continuous. To prove continuity at $(0, 0)$, we must show that for any $\epsilon > 0$, there exists a $\delta > 0$ such that

$$|f(x, y) - f(0, 0)| = \left| y \sin \frac{1}{x} \right| < \epsilon \quad \text{whenever} \quad 0 < x^2 + y^2 < \delta^2$$

(We use $x^2 + y^2$ and δ^2 here instead of $\sqrt{x^2 + y^2}$ and δ for convenience.) Note that $\left| y \sin \frac{1}{x} \right| \leq |y|$ for all $x \neq 0$, because $\left| \sin \frac{1}{x} \right| \leq 1$ for $x \neq 0$. If (x, y) lies in the disk $x^2 + y^2 < \delta^2$, then the points $(0, y)$ that satisfy $y^2 < \delta^2$ lie in the same disk (let $x = 0$ in $x^2 + y^2 < \delta^2$). In other words, points satisfying $|y| < \delta$ lie in the disk, and if we let $\delta = \epsilon$ it follows that

$$|f(x, y) - f(0, 0)| \leq |y| < \delta = \epsilon \quad \text{whenever} \quad |y| < \delta \quad ∎$$

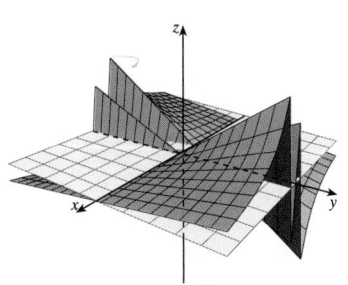

Front view

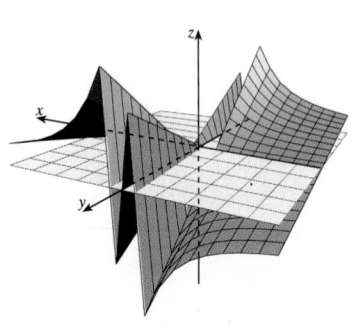

Side view

■ **FIGURE 12.13** Graph of f in Example 4

12.2 Problem Set

 1. ■ **What Does This Say?** Describe the notion of a limit of function of two variables.

2. ■ **What Does This Say?** Describe the basic formulas and rules for limits of a function of two variables.

In Problems 3–20, find the given limit, assuming it exists.

3. $\lim\limits_{(x,y)\to(-1,0)} (xy^2 + x^3y + 5)$

4. $\lim\limits_{(x,y)\to(0,0)} (5x^2 - 2xy + y^2 + 3)$

5. $\lim\limits_{(x,y)\to(1,3)} \dfrac{x+y}{x-y}$

6. $\lim\limits_{(x,y)\to(3,4)} \dfrac{x-y}{\sqrt{x^2+y^2}}$

7. $\lim\limits_{(x,y)\to(1,0)} e^{xy}$

8. $\lim\limits_{(x,y)\to(1,0)} (x+y)e^{xy}$

9. $\lim\limits_{(x,y)\to(0,1)} [e^{x^2+x} \ln(ey^2)]$

10. $\lim\limits_{(x,y)\to(e,0)} \ln(x^2+y^2)$

11. $\lim\limits_{(x,y)\to(0,0)} \dfrac{x^2 - 2xy + y^2}{x-y}$

12. $\lim\limits_{(x,y)\to(1,2)} \dfrac{(x^2-1)(y^2-4)}{(x-1)(y-2)}$

13. $\lim\limits_{(x,y)\to(0,0)} \dfrac{e^x \tan^{-1}y}{y}$

14. $\lim\limits_{(x,y)\to(0,0)} \dfrac{x^2y^2}{x^2+y^2}$

15. $\lim\limits_{(x,y)\to(0,0)} \dfrac{\sin(x+y)}{x+y}$

16. $\lim\limits_{(x,y)\to(0,0)} (\sin x + \cos y)$

17. $\lim\limits_{(x,y)\to(5,5)} \dfrac{x^4 - y^4}{x^2 - y^2}$

18. $\lim\limits_{(x,y)\to(a,a)} \dfrac{x^4 - y^4}{x^2 - y^2}$; *a* is a constant

19. $\lim\limits_{(x,y)\to(2,1)} \dfrac{x^2 - 4y^2}{x - 2y}$

20. $\lim\limits_{(x,y)\to(0,0)} \dfrac{x^2 + y^2}{\sqrt{x^2 + y^2 + 4} - 2}$

In Problems 21–29, evaluate the indicated limit, if it exists.

21. $\lim\limits_{(x,y)\to(2,1)} (xy^2 + x^3y)$

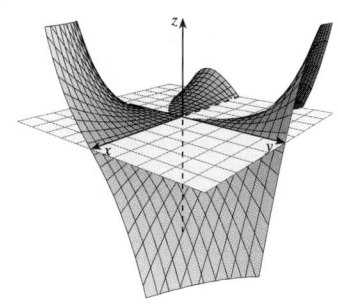

22. $\lim\limits_{(x,y)\to(1,2)} (5x^2 - 2xy + y^2)$

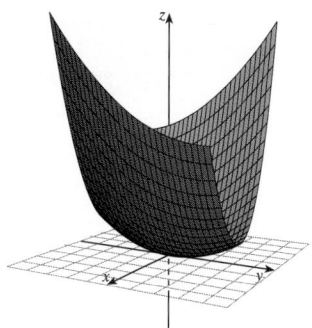

23. $\lim\limits_{(x,y)\to(0,0)} \dfrac{x+y}{x-y}$

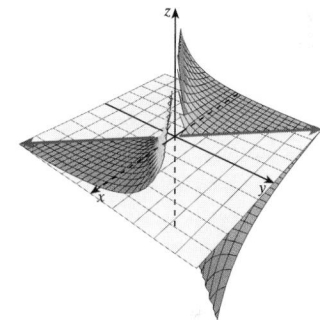

24. $\lim\limits_{(x,y)\to(0,0)} \dfrac{x-y}{\sqrt{x^2+y^2}}$

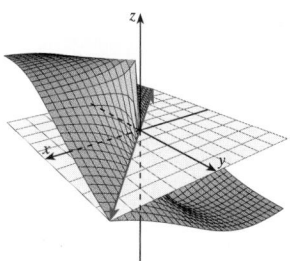

25. $\lim\limits_{(x,y)\to(0,0)} e^{xy}$

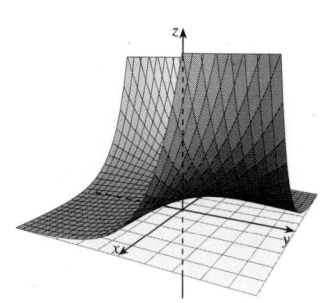

26. $\lim\limits_{(x, y)\to(1, 1)} (x + y)e^{xy}$

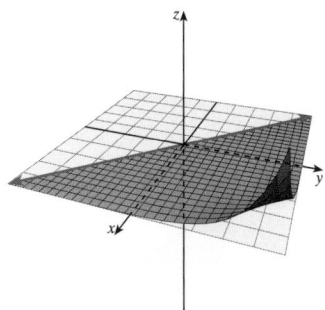

27. $\lim\limits_{(x, y)\to(0, 0)} (\sin x - \cos y)$

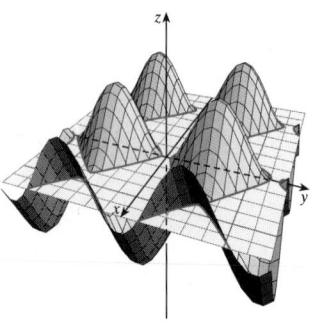

28. $\lim\limits_{(x, y)\to(0, 0)} \left[1 - \dfrac{\sin(x^2 + y^2)}{x^2 + y^2} \right]$

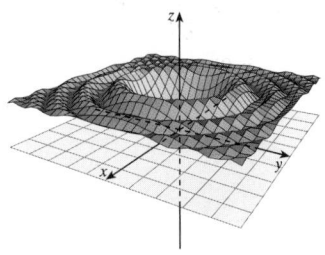

29. $\lim\limits_{(x, y)\to(0, 0)} \dfrac{1 - \cos(x^2 + y^2)}{x^2 + y^2}$

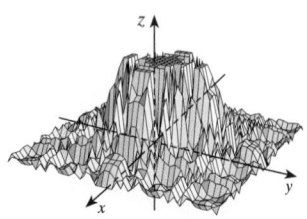

B *In Problems 30–33, show that* $\lim\limits_{(x, y)\to(0, 0)} f(x, y)$ *does not exist.*

30. $f(x, y) = \dfrac{x^4 y^4}{(x^2 + y^4)^3}$

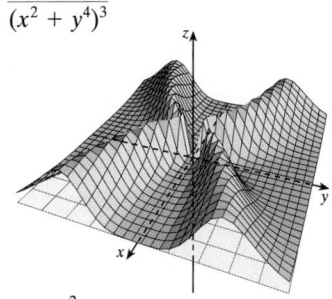

31. $f(x, y) = \dfrac{x - y^2}{x^2 + y^2}$

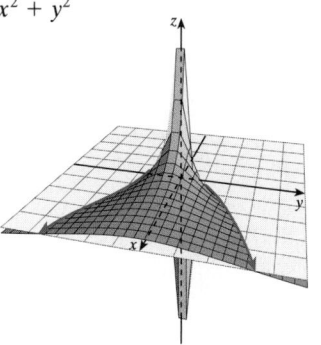

32. $f(x, y) = \dfrac{x^2 + y}{x^2 + y^2}$

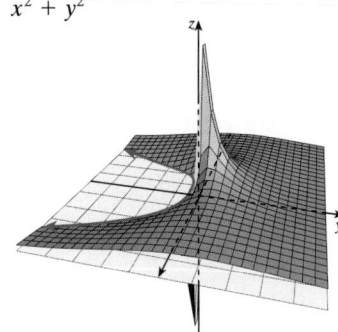

33. $f(x, y) = \dfrac{x^2 y^2}{x^4 + y^4}$

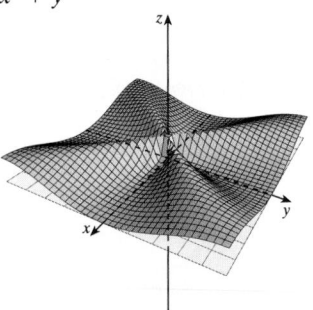

34. Let f be the functions defined by

$$f(x, y) = \begin{cases} \dfrac{xy^2}{x^2 + y^4} & \text{for } (x, y) \neq (0, 0) \\ 0 & \text{for } (x, y) = (0, 0) \end{cases}$$

Is f continuous at $(0, 0)$? Explain.

35. Let f be the functions defined by

$$f(x, y) = \begin{cases} \dfrac{xy^3}{x^2 + y^6} & \text{for } (x, y) \neq (0, 0) \\ 0 & \text{for } (x, y) = (0, 0) \end{cases}$$

Is f continuous at $(0, 0)$? Explain.

36. Let f be the function defined by

$$f(x, y) = \frac{x^2 - y^2}{x^2 + y^2} \quad \text{for } (x, y) \neq (0, 0)$$

a. Find $\displaystyle\lim_{(x, y) \to (2, 1)} f(x, y)$.

b. Prove that f has no limit at $(0, 0)$ by showing that $f(x, y)$ tends toward different numbers as $(x, y) \to (0, 0)$ along each coordinate axis.

37. Let f be the function defined by

$$f(x, y) = \frac{x^2 + 2y^2}{x^2 + y^2} \quad \text{for } (x, y) \neq (0, 0)$$

a. Find $\displaystyle\lim_{(x, y) \to (3, 1)} f(x, y)$.

b. Prove that f has no limit at $(0, 0)$ by showing that $f(x, y)$ tends toward different numbers as $(x, y) \to (0, 0)$ along each coordinate axis.

38. Given that the function

$$f(x, y) = \begin{cases} \dfrac{x^3 + y^3}{x^2 + y^2} & \text{for } (x, y) \neq (0, 0) \\ A & \text{for } (x, y) = (0, 0) \end{cases}$$

is continuous at the origin, what is A?

39. Given that the function

$$f(x, y) = \begin{cases} \dfrac{3x^3 - 3y^3}{x^2 - y^2} & \text{for } x^2 \neq y^2 \\ B & \text{otherwise} \end{cases}$$

is continuous at the origin, what is B?

40. Let

$$f(x, y) = \begin{cases} \dfrac{x^2 y^2}{x^2 + y^2} & \text{for } (x, y) \neq (0, 0) \\ 0 & \text{for } (x, y) = (0, 0) \end{cases}$$

Given that $f(x, y)$ has a limiting value at $(0, 0)$, is it continuous there?

41. Assuming that the limit exists, show that

$$\lim_{(x, y, z) \to (0, 0, 0)} \frac{xyz}{x^2 + y^2 + z^2} = 0$$

C *Use the δ–ϵ definition of limit to verify the limit statements given in Problems 42–45.*

42. $\displaystyle\lim_{(x, y) \to (0, 0)} (2x^2 + 3y^2) = 0$ **43.** $\displaystyle\lim_{(x, y) \to (0, 0)} (x + y^2) = 0$

44. $\displaystyle\lim_{(x, y) \to (0, 0)} \frac{x^2 - y^2}{x + y} = 0$ **45.** $\displaystyle\lim_{(x, y) \to (1, -1)} \frac{x^2 - y^2}{x + y} = 2$

46. Prove that if f is continuous and $f(a, b) > 0$, then there exists a δ-neighborhood about (a, b) such that $f(x, y) > 0$ for every point (x, y) in the neighborhood.

47. Prove the scalar multiple rule:

$$\lim_{(x, y) \to (x_0, y_0)} [af](x, y) = a \lim_{(x, y) \to (x_0, y_0)} f(x, y)$$

48. Prove the sum rule:

$$\lim_{(x, y) \to (x_0, y_0)} [f + g](x, y) = L + M$$

where $L = \left[\displaystyle\lim_{(x, y) \to (x_0, y_0)} f(x, y) \right]$

and $M = \left[\displaystyle\lim_{(x, y) \to (x_0, y_0)} g(x, y) \right]$.

49. A function of two variables $f(x, y)$ may be continuous in each separate variable at $x = x_0$ and $y = y_0$ without being itself continuous at (x_0, y_0). Let $f(x, y)$ be defined by

$$f(x, y) = \begin{cases} \dfrac{xy}{x^2 + y^2} & \text{for } (x, y) \neq (0, 0) \\ 0 & \text{at } (0, 0) \end{cases}$$

Let $g(x) = f(x, 0)$ and $h(y) = f(0, y)$. Show that both $g(x)$ and $h(y)$ are continuous at 0, but that $f(x, y)$ is not continuous at $(0, 0)$.

12.3 Partial Derivatives

IN THIS SECTION partial differentiation, partial derivative as a slope, partial derivative as a rate, higher partial derivatives

PARTIAL DIFFERENTIATION

The process of differentiating a function of several variables with respect to one of its variables while keeping the other variable(s) fixed is called **partial differentiation,** and the resulting derivative is a **partial derivative** of the function.

Recall that the derivative of a function of a single variable f is defined to be the limit of a difference quotient, namely,

$$f'(x) = \lim_{\Delta x \to 0} \frac{f(x + \Delta x) - f(x)}{\Delta x}$$

Partial derivatives with respect to x or y are defined similarly.

Partial Derivatives of a Function of Two Variables

If $z = f(x, y)$, then the **(first) partial derivatives** of f with respect to x and y are the functions f_x and f_y, respectively, defined by

$$f_x(x, y) = \lim_{\Delta x \to 0} \frac{f(x + \Delta x, y) - f(x, y)}{\Delta x}$$

$$f_y(x, y) = \lim_{\Delta y \to 0} \frac{f(x, y + \Delta y) - f(x, y)}{\Delta y}$$

provided the limits exist.

■ *What This Says:* For the partial differentiation of a function of two variables, $z = f(x, y)$, we find the partial derivative with respect to x by regarding y as constant while differentiating the function with respect to x. Similarly, the partial derivative with respect to y is found by regarding x as constant while differentiating with respect to y.

EXAMPLE 1   *Partial derivatives*

If $f(x, y) = x^3 y + x^2 y^2$, find:
a. f_x b. f_y

Solution

a. For f_x, hold y constant and find the derivative with respect to x:

$$f_x(x, y) = 3x^2 y + 2xy^2$$

b. For f_y, hold x constant and find the derivative with respect to y:

$$f_y(x, y) = x^3 + 2x^2 y$$ ■

Technology Window

Finding partial derivatives using technology is a natural extension of the way you have been finding other derivatives. The general format for most calculators and computer programs is the same: *derivative operator, function, variable of differentiation.* Look at one example that compares the partial derivatives with respect to x and y from Example 1.

a.

b.

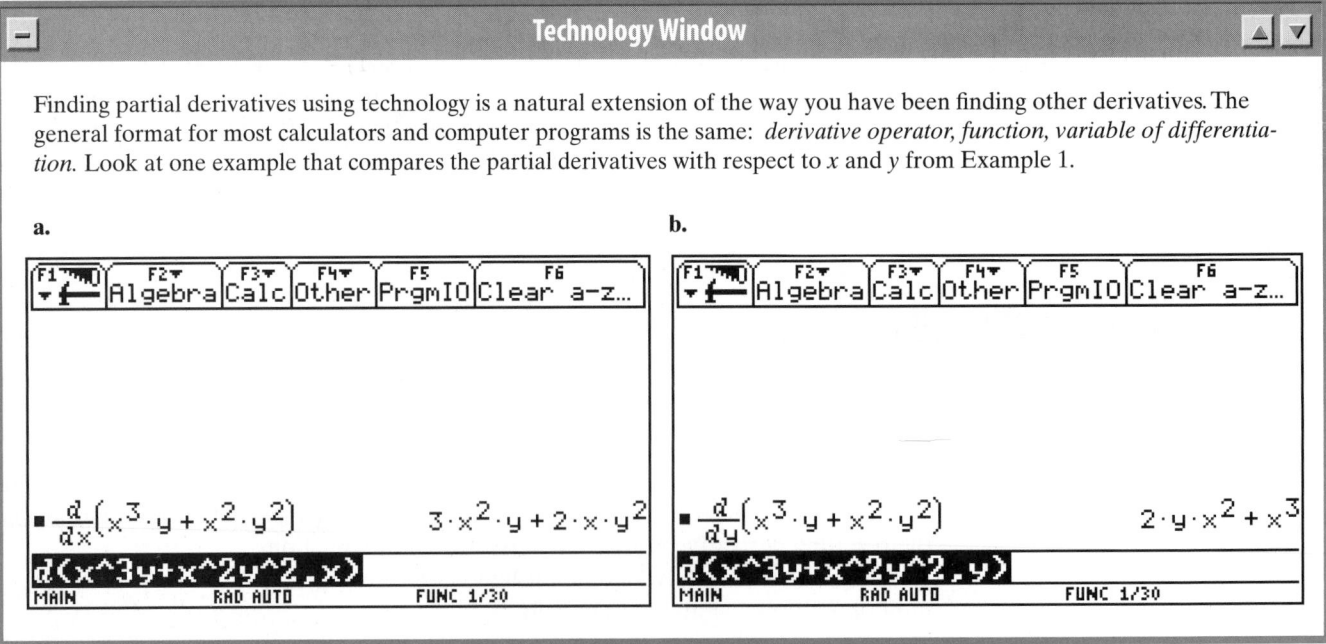

Several different symbols are used to denote partial derivatives, as indicated in the following box.

Alternative Notation for First Partial Derivatives

For $z = f(x, y)$, the partial derivatives f_x and f_y are denoted by

$$f_x(x, y) = \frac{\partial f}{\partial x} = \frac{\partial z}{\partial x} = \frac{\partial}{\partial x} f(x, y) = z_x = D_x(f)$$

and

$$f_y(x, y) = \frac{\partial f}{\partial y} = \frac{\partial z}{\partial y} = \frac{\partial}{\partial y} f(x, y) = z_y = D_y(f)$$

The values of the partial derivatives of $f(x, y)$ at the point (a, b) are denoted by

$$\left.\frac{\partial f}{\partial x}\right|_{(a, b)} = f_x(a, b) \quad \text{and} \quad \left.\frac{\partial f}{\partial y}\right|_{(a, b)} = f_y(a, b)$$

EXAMPLE 2   *Finding and evaluating a partial derivative*

Let $z = x^2 \sin(3x + y^3)$.

a. Evaluate $\dfrac{\partial z}{\partial x}$ at $(\frac{\pi}{3}, 0)$.   b. Evaluate z_y at $(1, 1)$.

Solution

a. $\dfrac{\partial z}{\partial x} = 2x \sin(3x + y^3) + x^2 \cos(3x + y^3) \cdot (3) = 2x \sin(3x + y^3) + 3x^2\cos(3x + y^3)$

Thus,

$$\left.\frac{\partial z}{\partial x}\right|_{(\pi/3, 0)} = 2\left(\frac{\pi}{3}\right) \sin \pi + 3\left(\frac{\pi}{3}\right)^2 \cos \pi = \frac{2\pi}{3}(0) + \frac{\pi^2}{3}(-1) = -\frac{\pi^2}{3}$$

b. $z_y = x^2\cos(3x + y^3) \cdot (3y^2) = 3x^2y^2\cos(3x + y^3)$, so that

$$z_y(1, 1) = 3(1)^2(1)^2\cos(3 + 1) = 3 \cos 4 \qquad\blacksquare$$

Technology Window

To evaluate a partial derivative at a point using technology, you must evaluate the partial derivative first for the active variable (with respect to the differentiation) using the *evaluate* feature of most derivative routines.

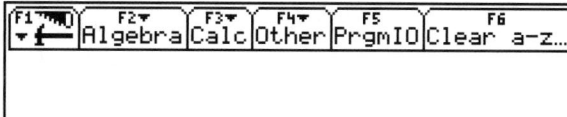

EXAMPLE 3   *Partial derivative of a function of three variables*

Let $f(x, y, z) = x^2 + 2xy^2 + yz^3$; find:

a. f_x b. f_y c. f_z

Solution

a. For f_x, think of f as a function of x alone with y and z treated as constants:

$$f_x(x, y, z) = 2x + 2y^2$$

b. $f_y(x, y, z) = 4xy + z^3$ c. $f_z(x, y, z) = 3yz^2$ ∎

EXAMPLE 4   *Partial derivative of an implicitly defined function*

Let z be defined implicitly as a function of x and y by the equation

$$x^2 z + yz^3 = x$$

Find $\partial z / \partial x$ and $\partial z / \partial y$.

Solution   Differentiate implicitly with respect to x, treating y as a constant:

$$2xz + x^2 \frac{\partial z}{\partial x} + 3z^2 y \frac{\partial z}{\partial x} = 1$$

so that

$$\frac{\partial z}{\partial x} = \frac{1 - 2xz}{x^2 + 3z^2 y}$$

Similarly, holding x constant and differentiating implicitly with respect to y, we find

$$x^2 \frac{\partial z}{\partial y} + z^3 + 3z^2 y \frac{\partial z}{\partial y} = 0$$

so that

$$\frac{\partial z}{\partial y} = \frac{-z^3}{x^2 + 3z^2 y}$$ ∎

The following example uses implicit differentiation to find the slope of the tangent line to a level curve at a particular point.

EXAMPLE 5   *Slope of a level curve*

A certain level curve of the surface $z = x^2 + 3xy + y^2$ contains the point $P(1, 1)$. Find the slope of the tangent line to this curve at P.

Solution   The required level curve has the form $F(x, y) = C$, where

$$F(x, y) = x^2 + 3xy + y^2$$

We find the slope, $\dfrac{dy}{dx}$, implicitly:

$$x^2 + 3xy + y^2 = C$$

$$\frac{d}{dx}(x^2 + 3xy + y^2) = \frac{d}{dx}(C)$$

$$2x + 3y + 3x\frac{dy}{dx} + 2y\frac{dy}{dx} = 0$$

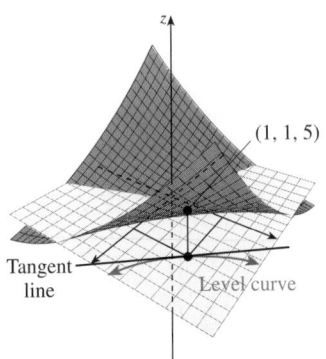

(1, 1, 5)

Tangent line

Level curve

The slope of the line tangent to a level curve of a surface

$$(3x + 2y)\frac{dy}{dx} = -2x - 3y$$

$$\frac{dy}{dx} = \frac{-(2x + 3y)}{3x + 2y}$$

This provides the slope at each point on the level curve where $3x + 2y \neq 0$. In particular, at the point where $x = 1$ and $y = 1$, we have

$$\frac{dy}{dx}\bigg|_{(1,\,1)} = \frac{-[2(1) + 3(1)]}{3(1) + 2(1)} = \frac{-5}{5} = -1$$

so that the required line has slope -1. ∎

PARTIAL DERIVATIVE AS A SLOPE

A useful geometric interpretation of partial derivatives is indicated in Figure 12.14. In Figure 12.14a, the plane $y = y_0$ intersects the surface $z = f(x, y)$ in a curve C parallel to the xz-plane. That is, C is the trace of the surface in the plane $y = y_0$. An equation for this curve is $z = f(x, y_0)$, and because y_0 is fixed, the function depends only on x. Thus, we can compute the slope of the tangent line to C at the point $P(x_0, y_0, z_0)$ in the plane $y = y_0$ by differentiating $f(x, y_0)$ with respect to x and evaluating the derivative at $x = x_0$. That is, the slope is $f_x(x_0, y_0)$, the value of the partial derivative f_x at (x_0, y_0). There is a similar interpretation for $f_y(x_0, y_0)$, as shown in Figure 12.14b.

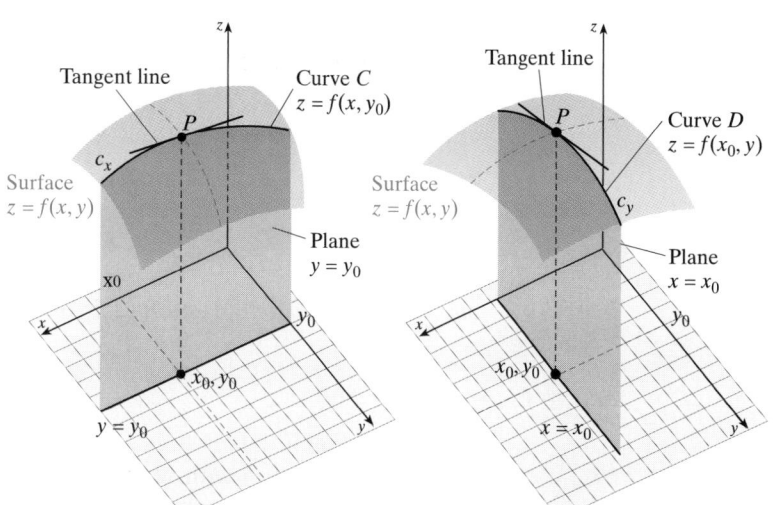

a. The tangent line in the plane $y = y_0$ to the curve C at the point P has slope $f_x(x_0, y_0)$.

b. The tangent line in the plane $x = x_0$ to the curve D at the point P has slope $f_y(x_0, y_0)$.

■ **FIGURE 12.14** Slope interpretation of the partial derivative

---

Partial Derivative as the Slope of a Tangent Line

The line parallel to the xz-plane and tangent to the surface $z = f(x, y)$ at the point $P_0(x_0, y_0, z_0)$ has slope $f_x(x_0, y_0)$. Likewise, the tangent line to the surface at P_0 that is parallel to the yz-plane has slope $f_y(x_0, y_0)$.

EXAMPLE 6 *Slope of a line parallel to the xz-plane*

Find the slope of the line that is parallel to the xz-plane and tangent to the surface $z = x\sqrt{x + y}$ at the point $P(1, 3, 2)$.

Solution If $f(x, y) = x\sqrt{x + y} = x(x + y)^{1/2}$, then the required slope is $f_x(1, 3)$.

$$f_x(x, y) = x\left(\frac{1}{2}\right)(x + y)^{-1/2}(1 + 0) + (1)(x + y)^{1/2} = \frac{x}{2\sqrt{x + y}} + \sqrt{x + y}$$

Thus, $f_x(1, 3) = \dfrac{1}{2\sqrt{1 + 3}} + \sqrt{1 + 3} = \dfrac{9}{4}$. ■

PARTIAL DERIVATIVE AS A RATE

The derivative of a function of one variable can be interpreted as a rate of change, and the analogous interpretation of partial derivative may be described as follows.

> **Partial Derivatives as Rates of Change**
>
> As the point (x, y) moves from the fixed point $P_0(x_0, y_0)$, the function $f(x, y)$ changes at a rate given by $f_x(x_0, y_0)$ in the direction of the positive x-axis and by $f_y(x_0, y_0)$ in the direction of the positive y-axis.

EXAMPLE 7 *Partial derivatives as a rate of change*

In an electrical circuit with electromotive force (EMF) of E volts and resistance R ohms, the current is $I = E/R$ amperes. Find the partial derivatives $\partial I/\partial E$ and $\partial I/\partial R$ at the instant when $E = 120$ and $R = 15$, and interpret these derivatives as rates.

Solution Since $I = ER^{-1}$, we have

$$\frac{\partial I}{\partial E} = R^{-1} \quad \text{and} \quad \frac{\partial I}{\partial R} = -ER^{-2}$$

and thus, when $E = 120$ and $R = 15$, we find that

$$\frac{\partial I}{\partial E} = 15^{-1} \approx 0.0667 \quad \text{and} \quad \frac{\partial I}{\partial R} = -(120)(15)^{-2} \approx -0.5333$$

This means that if the resistance is fixed at 15 ohms, the current is increasing (because the slope is positive) with respect to voltage at the rate of 0.0667 ampere per volt when the EMF is 120 volts. Likewise, with the same fixed EMF, the current is decreasing (because the slope is negative) with respect to resistance at the rate of 0.5333 ampere per ohm when the resistance is 15 ohms. ■

HIGHER PARTIAL DERIVATIVES

The partial derivative of a function is a function, so it is possible to take the partial derivative of a partial derivative. This is very much like taking the second derivative of a function of one variable if we take two consecutive partial derivatives with respect to the same variable, and the resulting derivative is called the **second partial derivative** with respect to that variable. However, we can also take the partial derivative with respect to one variable and then take another partial derivative with respect to a different variable, producing what is called a **mixed partial derivative.** The higher partial derivatives for a function of two variables $f(x, y)$ are denoted as indictaed in the following box.

> **Higher Partial Derivatives**
>
> Given $z = f(x, y)$
> **Second partial derivatives**
>
> $$\frac{\partial^2 f}{\partial x^2} = \frac{\partial}{\partial x}\left(\frac{\partial f}{\partial x}\right) = (f_x)_x = f_{xx}$$
>
> $$\frac{\partial^2 f}{\partial y^2} = \frac{\partial}{\partial y}\left(\frac{\partial f}{\partial y}\right) = (f_y)_y = f_{yy}$$
>
> **Mixed partial derivatives**
>
> $$\frac{\partial^2 f}{\partial x \partial y} = \frac{\partial}{\partial x}\left(\frac{\partial f}{\partial y}\right) = (f_y)_x = f_{yx}$$
>
> $$\frac{\partial^2 f}{\partial y \partial x} = \frac{\partial}{\partial y}\left(\frac{\partial f}{\partial x}\right) = (f_x)_y = f_{xy}$$

EXAMPLE 8 *Higher partial derivatives of a function of two variables*

For $z = f(x, y) = 5x^2 - 2xy + 3y^3$, find the requested higher partial derivatives.

a. $\dfrac{\partial^2 z}{\partial x \partial y}$ b. $\dfrac{\partial^2 f}{\partial y \partial x}$ c. $\dfrac{\partial^2 z}{\partial x^2}$ d. $f_{xy}(3, 2)$

Solution Part of what this example is illustrating is the variety of notation that can be used for higher partial derivatives.

a. First differentiate with respect to y; then differentiate with respect to x:

$$\frac{\partial z}{\partial y} = -2x + 9y^2$$

$$\frac{\partial^2 z}{\partial x \partial y} = \frac{\partial}{\partial x}\left(\frac{\partial z}{\partial y}\right) = \frac{\partial}{\partial x}(-2x + 9y^2) = -2$$

b. Differentiate first with respect to x and then with respect to y:

$$\frac{\partial f}{\partial x} = 10x - 2y$$

$$\frac{\partial^2 f}{\partial y \partial x} = \frac{\partial}{\partial y}\left(\frac{\partial f}{\partial x}\right) = \frac{\partial}{\partial y}(10x - 2y) = -2$$

c. Differentiate with respect to x twice:

$$\frac{\partial^2 z}{\partial x^2} = \frac{\partial}{\partial x}\left(\frac{\partial z}{\partial x}\right) = \frac{\partial}{\partial x}(10x - 2y) = 10$$

d. Evaluate the mixed partial found in part **b** at the point $(3, 2)$:

$$f_{xy}(3, 2) = -2$$ ∎

Notice from parts **a** and **b** of Example 8 that $\dfrac{\partial^2 z}{\partial x \partial y} = \dfrac{\partial^2 z}{\partial y \partial x}$. This equality of mixed partials does not hold for all functions, but for most functions we shall encounter, it will be true. The following theorem provides sufficient conditions for this equality to occur.

THEOREM 12.1   *Equality of mixed partials*

If the function $f(x, y)$ has mixed partial derivatives f_{xy} and f_{yx} that are continuous in an open disk containing (x_0, y_0), then

$$f_{yx}(x_0, y_0) = f_{xy}(x_0, y_0)$$

Proof   This proof requires methods of advanced calculus and is omitted in this text.

EXAMPLE 9   *Partial derivatives of functions of two variables*

Find $f_{xy}, f_{yx}, f_{xx},$ and f_{xxy}, where $f(x, y) = x^2 y e^y$.

Solution   We have the first partial derivatives

$$f_x = 2xye^y \qquad f_y = x^2 e^y + x^2 y e^y$$

The mixed partial derivatives (which must be the same by the previous theorem) are

$$f_{xy} = (f_x)_y = 2xe^y + 2xye^y \qquad f_{yx} = (f_y)_x = 2xe^y + 2xye^y$$

Finally, we compute the second and higher partial derivatives:

$$f_{xx} = (f_x)_x = 2ye^y \quad \text{and} \quad f_{xxy} = (f_{xx})_y = 2e^y + 2ye^y \qquad \blacksquare$$

EXAMPLE 10   *Verifying that a function satisfies the heat equation*

Verify that $T(x, t) = e^{-t} \cos \dfrac{x}{c}$ satisfies the heat equation, $\dfrac{\partial T}{\partial t} = c^2 \dfrac{\partial^2 T}{\partial x^2}$.

Solution $\dfrac{\partial T}{\partial t} = -e^{-t} \cos \dfrac{x}{c}$ and $\dfrac{\partial^2 T}{\partial x^2} = \dfrac{\partial}{\partial x}\left(-\dfrac{1}{c} e^{-t} \sin \dfrac{x}{c}\right) = -\dfrac{1}{c^2} e^{-t} \cos \dfrac{x}{c}$

Thus, T satisfies the heat equation $\dfrac{\partial T}{\partial t} = c^2 \dfrac{\partial^2 T}{\partial x^2}$. $\qquad \blacksquare$

WARNING The heat equation describes the distribution of temperature in an insulated rod. The constant c is called the *diffusivity* of the material in the rod.

Analogous definitions can be made for functions of more than two variables. For example,

$$f_{zzz} = \frac{\partial^3 f}{\partial z^3} = \frac{\partial}{\partial z}\left[\frac{\partial}{\partial z}\left(\frac{\partial f}{\partial z}\right)\right] \quad \text{or} \quad f_{xyz} = \frac{\partial^3 f}{\partial z \partial y \partial x} = \frac{\partial}{\partial z}\left[\frac{\partial}{\partial y}\left(\frac{\partial f}{\partial x}\right)\right]$$

EXAMPLE 11   *Higher partial derivatives of a function of several variables*

By direct calculation, show that $f_{xyz} = f_{yzx} = f_{zyx}$ for the function $f(x, y, z) = xyz + x^2 y^3 z^4$.

Solution   Find the first partials:

$$f_x(x, y, z) = yz + 2xy^3 z^4; \quad f_y(x, y, z) = xz + 3x^2 y^2 z^4; \quad f_z(x, y, z) = xy + 4x^2 y^3 z^3$$

Next, the required mixed partials:

$$f_{xy}(x, y, z) = (yz + 2xy^3 z^4)_y = z + 6xy^2 z^4$$
$$f_{yz}(x, y, z) = (xz + 3x^2 y^2 z^4)_z = x + 12x^2 y^2 z^3$$
$$f_{zy}(x, y, z) = (xy + 4x^2 y^3 z^3)_y = x + 12x^2 y^2 z^3$$

Finally, the higher mixed partials:

$$f_{xyz}(x, y, z) = (z + 6xy^2 z^4)_z = 1 + 24xy^2 z^3$$
$$f_{yzx}(x, y, z) = (x + 12x^2 y^2 z^3)_x = 1 + 24xy^2 z^3$$
$$f_{zyx}(x, y, z) = (x + 12x^2 y^2 z^3)_x = 1 + 24xy^2 x^3 \qquad \blacksquare$$

12.3   Problem Set

 1. ■ **What Does This Say?** What is a partial derivative?

2. ■ **What Does This Say?** Describe two basic applications of the partial derivatives $f_x(x, y)$ and $f_y(x, y)$.

Find f_x, f_y, f_{xx}, and f_{yx} in Problems 3–8.

3. $f(x, y) = x^3 + x^2y + xy^2 + y^3$

4. $f(x, y) = (x + xy + y)^3$

5. $f(x, y) = \dfrac{x}{y}$

6. $f(x, y) = xe^{xy}$

7. $f(x, y) = \ln(2x + 3y)$

8. $f(x, y) = \sin x^2y$

Find f_x and f_y in Problems 9–16.

9. a. $f(x, y) = (\sin x^2)\cos y$ **b.** $f(x, y) = \sin(x^2 \cos y)$

10. a. $f(x, y) = (\sin \sqrt{x})\ln y^2$ **b.** $f(x, y) = \sin(\sqrt{x} \ln y^2)$

11. $f(x, y) = \sqrt{3x^2 + y^4}$ **12.** $f(x, y) = xy^2 \ln(x + y)$

13. $f(x, y) = x^2e^{x+y} \cos y$ **14.** $f(x, y) = xy^3 \tan^{-1} y$

15. $f(x, y) = \sin^{-1}(xy)$ **16.** $f(x, y) = \cos^{-1}(xy)$

Find f_x, f_y, and f_z in Problems 17–22.

17. $f(x, y, z) = xy^2 + yz^3 + xyz$

18. $f(x, y, z) = xye^z$

19. $f(x, y, z) = \dfrac{x + y^2}{z}$

20. $f(x, y, z) = \dfrac{xy + yz}{xz}$

21. $f(x, y, z) = \ln(x + y^2 + z^3)$

22. $f(x, y, z) = \sin(xy + z)$

In Problems 23–28, find $\dfrac{\partial z}{\partial x}$ and $\dfrac{\partial z}{\partial y}$ by differentiating implicitly.

23. $\dfrac{x^2}{9} - \dfrac{y^2}{4} + \dfrac{z^2}{2} = 1$ **24.** $3x^2 + 4y^2 + 2z^2 = 5$

25. $3x^2y + y^3z - z^2 = 1$ **26.** $x^3 - xy^2 + yz^2 - z^3 = 4$

27. $\sqrt{x} + y^2 + \sin xz = 2$

28. $\ln(xy + yz + xz) = 5$
   $(x > 0, y > 0, z > 0)$

 29. Find f_x and f_y for

$$f(x, y) = \int_x^y (t^2 + 2t + 1)dt$$

   Hint: Review the second fundamental theorem.

30. Find f_x and f_y for

$$f(x, y) = \int_{x^2}^{2y} (e^t + 3t)dt$$

   Hint: Review the second fundamental theorem.

31. Find the slope of the tangent line to the level curve at the point $P(1, 1, 3)$ on the surface $z = x^2 + xy^2 + y^3$.

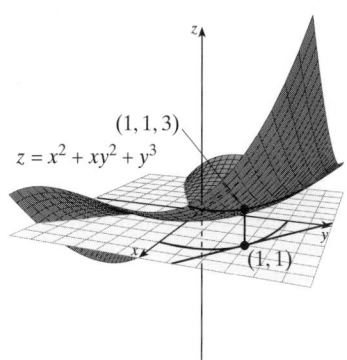

32. Find the slope of the tangent line to the level curve at the point $P\left(\dfrac{\sqrt{\pi}}{2}, \dfrac{\sqrt{\pi}}{2}, \dfrac{\pi}{4}\right)$ on the surface $z = xy + \cos(x^2 + y^2)$.

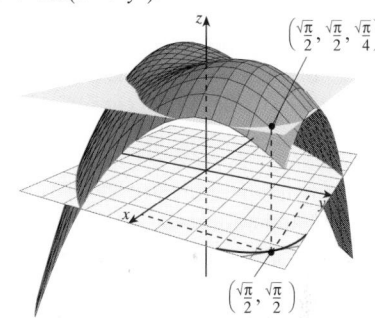

*A function $f(x, y)$ is said to be **harmonic** on the set S if f_{xx} and f_{yy} are continuous and*

$$f_{xx} + f_{yy} = 0$$

throughout S. Show that each function given in Problems 33–36 is harmonic on the prescribed set.

33. $f(x, y) = 3x^2y - y^3$; S is the entire plane.

34. $f(x, y) = \ln(x^2 + y^2)$; S is the plane with the point $(0, 0)$ removed.

35. $f(x, y) = e^x\sin y$; S is the entire plane.

36. $f(x, y) = \sin x \cosh y$; S is the entire plane.

37. Let $f(x, y) = xy^3 + x^3y$. Find the slope of the tangent line to the graph of f at $P(1, -1, -2)$ in the direction of
   a. the x-axis.   **b.** the y-axis.

38. Find the slope of the tangent line at the point $P(1, -1, -2)$ on the graph of

$$f(x, y) = \dfrac{x^2 + y^2}{xy}$$

   in the direction
   a. parallel to the xz-plane.
   b. parallel to the yz-plane.

39. For $f(x, y) = \cos xy^2$, show $f_{xy} = f_{yx}$.

40. For $f(x, y) = (\sin^2 x)(\sin y)$, show $f_{xy} = f_{yx}$.

41. Find $f_{xzy} - f_{yzz}$, where $f(x, y, z) = x^2 + y^2 - 2xy \cos z$.

42. MODELING PROBLEM   It has been determined that the flow (in cm³/s) of blood from an artery into a small capillary can be modeled by

$$F(x, y, z) = \frac{c\pi x^2}{4}\sqrt{y - z}$$

for constant $c > 0$, where x is the diameter of the capillary, y is the pressure in the artery, and z is the pressure in the capillary. Find the rate of change of the flow of blood with respect to:
  a. the diameter of the capillary.
  b. the arterial pressure.
  c. the capillary pressure.

43. MODELING PROBLEM   Biologists have studied the oxygen consumption of certain furry mammals. They have found that if the mammal's body temperature is T degrees Celsius, fur temperature is t degrees Celsius, and the mammal does not sweat, then its relative oxygen consumption can be modeled by

$$C(m, t, T) = \sigma(T - t)m^{-0.67}$$

(kg/h) where m is the mammal's mass (in kg) and σ is a physical constant. Find the rate (rounded to two decimal places) at which the oxygen consumption changes with respect to:
  a. the mass m.
  b. the body temperature T.
  c. the fur temperature t.

44. MODELING PROBLEM   A gas that gathers on a surface in a condensed layer is said to be *adsorbed* in the surface, and the surface is called an *adsorbing surface*. It has been determined that the amount of gas adsorbed per unit area on an adsorbing surface can be modeled by

$$S(p, T, h) = ape^{h/(bT)}$$

where p is the gas pressure, T is the temperature of the gas, h is the heat of the adsorbed layer of gas, and a and b are physical constants. Find the rate of change of S with respect to: a. p b. h c. T

45. The ideal gas law says that $PV = kT$, where P is the pressure of a confined gas, V is the volume, T is the temperature, and k is a physical constant.
  a. Find $\dfrac{\partial V}{\partial T}$ b. Find $\dfrac{\partial P}{\partial V}$
  c. Show that $\dfrac{\partial P}{\partial V} \cdot \dfrac{\partial V}{\partial T} \cdot \dfrac{\partial T}{\partial P} = -1$.

46. At a certain factory, the output is given by $Q = 120K^{2/3}L^{2/5}$, where K denotes the capital investment (in units of $1,000) and L measures the size of the labor force (in worker-hours).
  a. Find the *marginal productivity of capital*, $\partial Q/\partial K$, and the *marginal productivity of labor*, $\partial Q/\partial L$.

  b. Determine the signs of the second partial derivatives $\partial^2 Q/\partial L^2$ and $\partial^2 Q/\partial K^2$, and give an economic interpretation.

  Note: Q is an example of a Cobb–Douglas production function.

47. The temperature at a point (x, y) on a given metal plate in the xy-plane is determined according to the formula $T(x, y) = x^3 + 2xy^2 + y$ degrees. Find the rate at which the temperature changes with distance if we start at $(2, 1)$ and move
  a. up (parallel to the y-axis).
  b. to the right (parallel to the x-axis).

48. In physics, the **wave equation** is $\dfrac{\partial^2 z}{\partial t^2} = c^2 \dfrac{\partial^2 z}{\partial x^2}$

and the **heat equation** is $\dfrac{\partial z}{\partial t} = c^2 \dfrac{\partial^2 z}{\partial x^2}$. In each of the following cases, determine whether z satisfies the wave equation, the heat equation, or neither.
  a. $z = e^{-t}\left(\sin \dfrac{x}{c} + \cos \dfrac{x}{c}\right)$ b. $z = \sin 3\, ct \sin 3x$
  c. $z = \sin 5\, ct \cos 5x$

49. THINK TANK PROBLEM   Let

$$f(x, y) = \begin{cases} xy\left(\dfrac{x^2 - y^2}{x^2 + y^2}\right) & \text{if } (x, y) \neq (0, 0) \\ 0 & \text{if } (x, y) = (0, 0) \end{cases}$$

Show that $f_x(0, y) = -y$ and $f_y(x, 0) = x$, for all x and y. Then show that $f_{xy}(0, 0) = -1$ and $f_{yx}(0, 0) = 1$. Why does this not violate the equality of mixed partials theorem?

50. Show that $f_x(0, 0) = 0$ but $f_y(0, 0)$ does not exist, where

$$f(x, y) = \begin{cases} (x^2 + y)\sin\left(\dfrac{1}{x^2 + y^2}\right) & \text{if } (x, y) \neq (0, 0) \\ 0 & \text{if } (x, y) = (0, 0) \end{cases}$$

51. Suppose a substance is injected into a tube containing a liquid solvent. Suppose the tube is placed so that its axis is parallel to the x-axis, as shown in Figure 12.15.

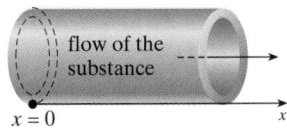

flow of the substance

$x = 0$ x

■ **FIGURE 12.15**   Problem 51

Assume that the concentration of the substance varies only in the x-direction, and let $C(x, t)$ denote the concentration at position x and time t. Because the number of molecules in this substance is very large, it is reasonable to assume that C is a continuous function whose partial derivatives exist. One model for the flow yields the **diffusion equation in one dimension,** namely,

$$\frac{\partial C}{\partial t} = \delta \frac{\partial^2 C}{\partial x^2}$$

where δ is the **diffusion constant.**
  a. What must δ be for a function of the form
$$C(x, t) = e^{ax + bt}$$

(a and b are constants) to satisfy the diffusion equation?

b. Verify that

$$C(x, t) = t^{-1/2} e^{-x^2/(4\delta t)}$$

satisfies the diffusion equation.

52. The area of a triangle is $A = \frac{1}{2}ab \sin \gamma$, where γ is the angle between sides of length a and b.

a. Find $\dfrac{\partial A}{\partial a}, \dfrac{\partial A}{\partial b},$ and $\dfrac{\partial A}{\partial \gamma}.$

b. Suppose a is given as a function of $b, A,$ and γ.

What is $\dfrac{\partial a}{\partial \gamma}$?

53. **JOURNAL PROBLEM**   (*Crux*, problem by John A. Winterink.)* Prove the validity of the following simple method for finding the center of a conic: For the central conic,

$$\phi(x, y) = ax^2 + 2hxy + by^2 + 2gx + 2fy + c = 0$$

$ab - h^2 \neq 0$, show that the center is the intersection of the lines $\partial \phi/\partial x = 0$ and $\partial \phi/\partial y = 0$.

———
*Crux, Problem 54, Vol. 6 (1980), p. 154.

12.4 Tangent Planes, Approximations, and Differentiability

IN THIS SECTION   **tangent planes, incremental approximations, the total differential, differentiability**

TANGENT PLANES

Suppose S is a surface with the equation $z = f(x, y)$, where f has continuous first partial derivatives f_x and f_y. Let $P(x_0, y_0, z_0)$ be a point on S, and let C_1 be the curve of intersection of S with the plane $x = x_0$ and C_2, the intersection of S with the plane $y = y_0$, as shown in Figure 12.16. The tangent lines T_1 and T_2, respectively, at P determine a unique plane, and we shall find that this plane actually contains the tangent to *every* smooth curve on S that passes through P. It is reasonable to call this plane the **tangent plane** to S at P.

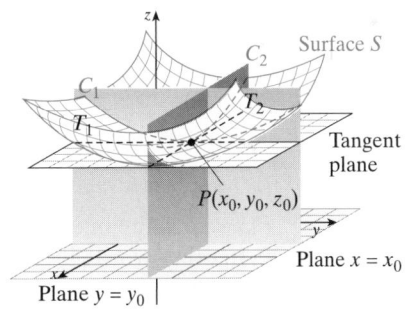

■ **FIGURE 12.16**   Tangent plane

To find an equation for the tangent plane, recall that the equation of a plane with normal $\mathbf{N} = A\mathbf{i} + B\mathbf{j} + C\mathbf{k}$ is

$$A(x - x_0) + B(y - y_0) + C(z - z_0) = 0$$

If $C \neq 0$, divide both sides by C and let $a = -A/C$ and $b = -B/C$ to obtain

$$z - z_0 = a(x - x_0) + b(y - y_0)$$

The intersection of this plane and the plane $x = x_0$ is the tangent line T_1, which we know has slope $f_y(x_0, y_0)$ from the geometric interpretation of partial derivatives. Setting $x = x_0$ in the equation for the tangent plane, we find that T_1 has the point–slope form

$$z - z_0 = b(y - y_0)$$

so we must have $b = f_y(x_0, y_0) = \dfrac{\partial z}{\partial y}$. Similarly, setting $y = y_0$, we obtain

$$z - z_0 = a(x - x_0)$$

which represents the tangent line T_2, with slope $a = f_x(x_0, y_0)$. To summarize:

Equation of the Tangent Plane

Suppose S is a surface with the equation $z = f(x, y)$ and let $P_0(x_0, y_0, z_0)$ be a point on S at which a tangent plane exists. Then the **equation of the tangent plane** to S at P_0 is

$$z - z_0 = f_x(x_0, y_0)(x - x_0) + f_y(x_0, y_0)(y - y_0)$$

EXAMPLE 1   *Equation of a tangent plane for a surface defined by $z = f(x, y)$*

Find an equation for the tangent plane to the surface $z = \tan^{-1} \dfrac{y}{x}$ at the point $P_0(1, \sqrt{3}, \frac{\pi}{3})$.

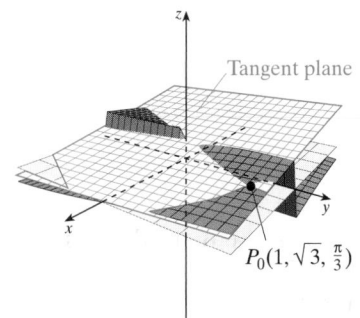

Tangent plane for the surface defined by $z = \tan^{-1}(y/x)$

Solution

$$f_x(x, y) = \frac{-yx^{-2}}{1 + \left(\dfrac{y}{x}\right)^2} = \frac{-y}{x^2 + y^2}; \qquad f_x(1, \sqrt{3}) = \frac{-\sqrt{3}}{1 + 3} = \frac{-\sqrt{3}}{4}$$

$$f_y(x, y) = \frac{x^{-1}}{1 + \left(\dfrac{y}{x}\right)^2} = \frac{x}{x^2 + y^2}; \qquad f_y(1, \sqrt{3}) = \frac{1}{1 + 3} = \frac{1}{4}$$

The equation of the tangent plane is

$$z - \frac{\pi}{3} = \left(\frac{-\sqrt{3}}{4}\right)(x - 1) + \frac{1}{4}(y - \sqrt{3})$$

or

$$3\sqrt{3}x - 3y + 12z = 4\pi \qquad\blacksquare$$

INCREMENTAL APPROXIMATIONS

In Chapter 3, we observed that the tangent line to the curve $y = f(x)$ at the point $P(x_0, y_0)$ is the line that best "fits" the shape of the curve in the immediate vicinity of P. In other words, if f is differentiable at $x = x_0$ and the increment $|\Delta x|$ is sufficiently small, then

$$f(x_0 + \Delta x) \approx f(x_0) + f'(x_0)\Delta x$$

Similarly, the tangent plane at $P(x_0, y_0, z_0)$ is the plane that best fits the shape of the surface $z = f(x, y)$ near P, and the analogous **incremental** (or **linear**) approximation formula may be stated as follows.

Incremental Approximation of a Function of Two Variables

If $f(x, y)$ and its partial derivatives f_x and f_y are defined in an open region R containing the point $P(x_0, y_0)$ and f_x and f_y are continuous at P, then

$$\Delta f = f(x_0 + \Delta x, y_0 + \Delta y) - f(x_0, y_0) \approx f_x(x_0, y_0)\Delta x + f_y(x_0, y_0)\Delta y$$

so that

$$f(x_0 + \Delta x, y_0 + \Delta y) \approx f(x_0, y_0) + f_x(x_0, y_0)\Delta x + f_y(x_0, y_0)\Delta y$$

A graphical interpretation of this incremental approximation formula is shown in Figure 12.17.

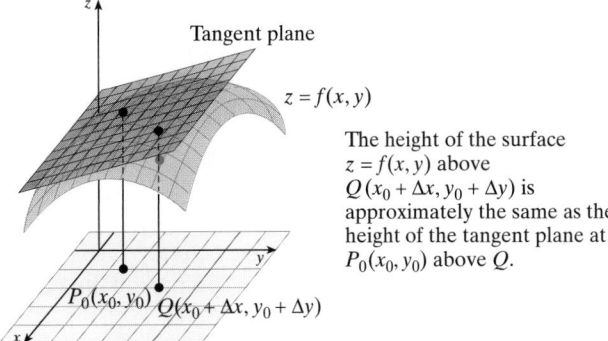

The height of the surface $z = f(x, y)$ above $Q(x_0 + \Delta x, y_0 + \Delta y)$ is approximately the same as the height of the tangent plane at $P_0(x_0, y_0)$ above Q.

■ **FIGURE 12.17**   Incremental approximation to a function of two variables

The tangent plane to the surface $z = f(x, y)$ has the equation

$$z - f(x_0, y_0) = f_x(x_0, y_0)\Delta x + f_y(x_0, y_0)\Delta y$$

As long as we are near (x_0, y_0), the height of the tangent plane is approximately the same as the height of the surface. Thus, if $|\Delta x|$ and $|\Delta y|$ are small, the point $(x_0 + \Delta x, y_0 + \Delta y)$ will be near (x_0, y_0) and we have

$$\underbrace{f(x_0 + \Delta x, y_0 + \Delta y)}_{\substack{\text{Height of } z = f(x, y) \\ \text{above } Q(x_0 + \Delta x, y_0 + \Delta y)}} \approx \underbrace{f(x_0, y_0) + f_x(x_0, y_0)\Delta x + f_y(x_0, y_0)\Delta y}_{\text{Height of the tangent plane above } Q}$$

Increments of a function of three variables $f(x, y, z)$ can be defined in a similar fashion. Suppose f has continuous partial derivatives f_x, f_y, f_z at and near the point (x_0, y_0, z_0). Then if the numbers $\Delta x, \Delta y, \Delta z$ are all sufficiently small, we have

$$\Delta f = f(x_0 + \Delta x, y_0 + \Delta y, z_0 + \Delta z) - f(x_0, y_0, z_0)$$
$$\approx f_x(x_0, y_0, z_0)\Delta x + f_y(x_0, y_0, z_0)\Delta y + f_z(x_0, y_0, z_0)\Delta z$$

EXAMPLE 2   *Using increments to estimate the change of a function*

An open box has length 3 ft, width 1 ft, and height 2 ft, and is constructed from material that costs \$2/ft^2 for the sides and \$3/ft^2 for the bottom. Compute the cost of constructing the box, and then use increments to estimate the change in cost if the length and width are each increased by 3 in. and the height is decreased by 4 in.

Solution   An open (no top) box with length x, width y, and height z has a surface area

$$S = xy + \underbrace{2xz + 2yz}_{}$$
$$\text{Bottom   Four side faces}$$

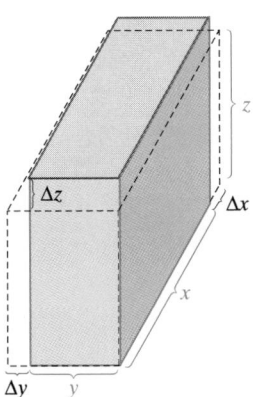

Because the sides cost \$2/ft^2 and the bottom \$3/ft^2, the total cost is

$$C(x, y, z) = 3xy + 2(2xz + 2yz)$$

The partial derivatives of C are

$$C_x = 3y + 4z \qquad C_y = 3x + 4z \qquad C_z = 4x + 4y$$

and the dimensions of the box change by

$$\Delta x = \tfrac{3}{12} = 0.25 \text{ ft} \qquad \Delta y = \tfrac{3}{12} = 0.25 \text{ ft} \qquad \Delta z = \tfrac{-4}{12} \approx -0.33 \text{ ft}$$

Thus, the change in the total cost is approximated by

$$\Delta C \approx C_x(3,1,2)\Delta x + C_y(3,1,2)\Delta y + C_z(3,1,2)\Delta z$$
$$= [3(1) + 4(2)](0.25) + [3(3) + 4(2)](0.25) + [4(3) + 4(1)](-\tfrac{4}{12})$$
$$\approx 1.67$$

That is, the cost increases by approximately $1.67. ∎

EXAMPLE 3   *Maximum percentage error using differentials*

The radius and height of a right circular cone are measured with errors of at most 3% and 2%, respectively. Use increments to approximate the maximum possible percentage error in computing the volume of the cone using these measurements and the formula $V = \tfrac{1}{3}\pi R^2 H$.

Solution   We are given that

$$\left| \frac{\Delta R}{R} \right| \le 0.03 \quad \text{and} \quad \left| \frac{\Delta H}{H} \right| \le 0.02$$

The partial derivatives of V are

$$V_R = \tfrac{2}{3}\pi RH \quad \text{and} \quad V_H = \tfrac{1}{3}\pi R^2$$

so the change in V is approximated by

$$\Delta V \approx \left(\frac{2}{3}\pi RH \right)\Delta R + \left(\frac{1}{3}\pi R^2 \right)\Delta H$$

Dividing by the volume $V = \tfrac{1}{3}\pi R^2 H$, we obtain

$$\frac{\Delta V}{V} \approx \frac{\tfrac{2}{3}\pi RH\Delta R + \tfrac{1}{3}\pi R^2\Delta H}{\tfrac{1}{3}\pi R^2 H} = 2\left(\frac{\Delta R}{R} \right) + \left(\frac{\Delta H}{H} \right)$$

so that

$$\left| \frac{\Delta V}{V} \right| \le 2\left| \frac{\Delta R}{R} \right| + \left| \frac{\Delta H}{H} \right| = 2(0.03) + (0.02) = 0.08$$

Thus, the maximum percentage error in computing the volume V is approximately 8%. ∎

THE TOTAL DIFFERENTIAL

For a function of one variable, $y = f(x)$, we defined the differential dy to be $dy = f'(x)\,dx$. For the two-variable case, we make the following analogous definition.

Total Differential

If $z = f(x, y)$ and Δx and Δy are increments of x and y, respectively, and if we let $dx = \Delta x$ and $dy = \Delta y$ be differentials for x and y, respectively, then the **total differential** of $f(x, y)$ is

$$df = \frac{\partial f}{\partial x}\,dx + \frac{\partial f}{\partial y}\,dy = f_x(x, y)\,dx + f_y(x, y)\,dy$$

Similarly, for a function of three variables $z = f(x, y, z)$, with $dz = \Delta z$, the **total differential** is

$$df = \frac{\partial f}{\partial x}\,dx + \frac{\partial f}{\partial y}\,dy + \frac{\partial f}{\partial z}\,dz$$

EXAMPLE 4   *Total differential*

Find the total differential of the given functions:

a. $f(x, y, z) = 2x^3 + 5y^4 - 6z$ b. $f(x, y) = x^2 \ln(3y^2 - 2x)$

Solution

a. $df = \dfrac{\partial f}{\partial x} dx + \dfrac{\partial f}{\partial y} dy + \dfrac{\partial f}{\partial z} dz = 6x^2 dx + 20y^3 dy - 6 dz$

b. $df = \dfrac{\partial f}{\partial x} dx + \dfrac{\partial f}{\partial y} dy$

$\quad = \left[2x \ln(3y^2 - 2x) + x^2 \dfrac{-2}{3y^2 - 2x} \right] dx + \left[x^2 \dfrac{6y}{3y^2 - 2x} \right] dy$

$\quad = \left[2x \ln(3y^2 - 2x) - \dfrac{2x^2}{3y^2 - 2x} \right] dx + \dfrac{6x^2 y}{3y^2 - 2x} dy$ ∎

EXAMPLE 5   *Application of the total differential*

At a certain factory, the daily output is $Q = 60K^{1/2}L^{1/3}$ units, where K denotes the capital investment (in units of $1,000) and L, the size of the labor force (in worker-hours). The current capital investment is $900,000, and 1,000 worker-hours of labor are used each day. Estimate the change in output that will result if capital investment is increased by $1,000 and labor is decreased by 2 worker-hours.

Solution   The change in output is estimated by the total differential dQ. We have $K = 900$, $L = 1,000$, $dK = \Delta K = 1$, and $dL = \Delta L = -2$. The total differential of $Q(x, y)$ is

$$dQ = \dfrac{\partial Q}{\partial K} dK + \dfrac{\partial Q}{\partial L} dL$$

$$= 60(\tfrac{1}{2})K^{-1/2}L^{1/3} dK + 60(\tfrac{1}{3})K^{1/2}L^{-2/3} dL$$

$$= 30K^{-1/2}L^{1/3} dK + 20K^{1/2}L^{-2/3} dL$$

Substituting for K, L, dK, and dL,

$$dQ = 30(900)^{-1/2} (1,000)^{1/3} (1) + 20(900)^{1/2} (1,000)^{-2/3} (-2) = -2$$

Thus, the output decreases by approximately 2 units when the capital investment is increased by $1,000 and labor is decreased by 2 worker-hours.   ∎

EXAMPLE 6   *Maximum percentage error in an electrical circuit*

When two resistances R_1 and R_2 are connected in parallel, the total resistance R is given by

$$R = \dfrac{R_1 R_2}{R_1 + R_2}$$

If R_1 is measured as 300 ohms with a maximum error of 2% and R_2 is measured as 500 ohms with a maximum error of 3%, what is the maximum percentage error in R?

Solution   We are given that

$$\left| \dfrac{dR_1}{R_1} \right| \le 0.02 \quad \text{and} \quad \left| \dfrac{dR_2}{R_2} \right| \le 0.03$$

and we wish to find the maximum value of $\left|\dfrac{dR}{R}\right|$. Because

$$\frac{\partial R}{\partial R_1} = \frac{R_2^2}{(R_1 + R_2)^2} \quad \text{and} \quad \frac{\partial R}{\partial R_2} = \frac{R_1^2}{(R_1 + R_2)^2} \qquad \textit{Quotient rule}$$

it follows that the total differential of R is

$$dR = \frac{\partial R}{\partial R_1}\, dR_1 + \frac{\partial R}{\partial R_2}\, dR_2$$

$$= \frac{R_2^2}{(R_1 + R_2)^2}\, dR_1 + \frac{R_1^2}{(R_1 + R_2)^2}\, dR_2$$

We now find $\dfrac{dR}{R}$ by dividing both sides by R; however, because $R = \dfrac{R_1 R_2}{R_1 + R_2}$, we note

that $\dfrac{1}{R} = \dfrac{R_1 + R_2}{R_1 R_2}$:

$$dR \cdot \frac{1}{R} = \left[\frac{R_2^2}{(R_1 + R_2)^2}\, dR_1 + \frac{R_1^2}{(R_1 + R_2)^2}\, dR_2\right] \cdot \frac{R_1 + R_2}{R_1 R_2}$$

$$\frac{dR}{R} = \frac{R_2}{R_1 + R_2} \cdot \frac{dR_1}{R_1} + \frac{R_1}{R_1 + R_2} \cdot \frac{dR_2}{R_2}$$

Finally, apply the triangle inequality (Table 1.1, p. 3) to this relationship:

$$\left|\frac{dR}{R}\right| \le \left|\frac{R_2}{R_1 + R_2}\right|\left|\frac{dR_1}{R_1}\right| + \left|\frac{R_1}{R_1 + R_2}\right|\left|\frac{dR_2}{R_2}\right|$$

$$\le \frac{500}{300 + 500}(0.02) + \frac{300}{300 + 500}(0.03) = 0.02375$$

The maximum percentage is approximately 2.4%.   ∎

DIFFERENTIABILITY

Recall from Chapter 3 that if $f(x)$ is differentiable at x_0, its *increment* is

$$\Delta f = f(x_0 + \Delta x) - f(x_0) = f'(x_0)\Delta x + \epsilon \Delta x$$

where $\epsilon \to 0$ as $\Delta x \to 0$. For a function of two variables, the increment of x is an independent variable denoted by Δx, the increment of y is an independent variable denoted by Δy, and the increment of f at (x_0, y_0) is defined as

$$\Delta f = f(x_0 + \Delta x, y_0 + \Delta y) - f(x_0, y_0)$$

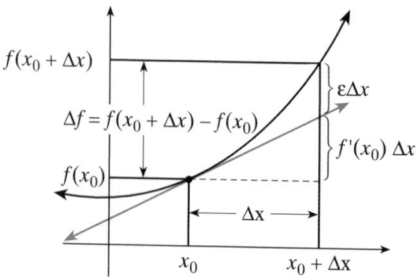

$$\Delta f = f(x_0 + \Delta x) - f(x_0) \approx f'(x_0)\,\Delta x + \underbrace{\epsilon \Delta x}$$

Error when using
df to estimate Δf

We use this increment representation to define differentiability as follows.

Definition of Differentiability

Suppose $f(x, y)$ is defined at each point in a circular disk that is centered at (x_0, y_0) and contains the point $(x_0 + \Delta x, y_0 + \Delta y)$. Then f is said to be **differentiable** at (x_0, y_0) if the increment of f can be expressed as

$$\Delta f = f_x(x_0, y_0)\Delta x + f_y(x_0, y_0)\Delta y + \epsilon_1 \Delta x + \epsilon_2 \Delta y$$

where $\epsilon_1 \to 0$ and $\epsilon_2 \to 0$ as both $\Delta x \to 0$ and $\Delta y \to 0$. Also, $f(x, y)$ is said to be **differentiable on the region R** of the plane if f is differentiable at each point in R.

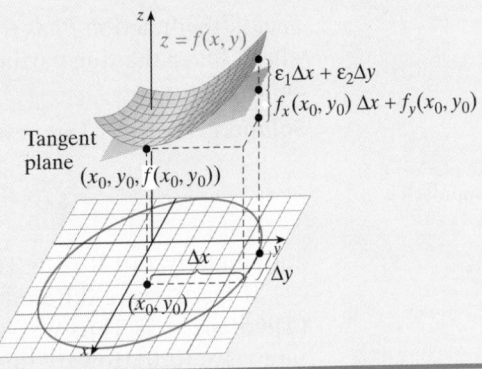

In Section 3.1, we showed that a function of one variable is continuous wherever it is differentiable. The following theorem establishes the same result for a function of two variables.

THEOREM 12.2   *Differentiability implies continuity*

If $f(x, y)$ is differentiable at (x_0, y_0), it is also continuous there.

Proof   We wish to show that $f(x, y) \to f(x_0, y_0)$ as $(x, y) \to (x_0, y_0)$ or, equivalently, that

$$\lim_{(x, y) \to (x_0, y_0)} [f(x, y) - f(x_0, y_0)] = 0$$

If we set $\Delta x = x - x_0$ and $\Delta y = y - y_0$ and let Δf denote the increment of f at (x_0, y_0), we have (by substitution)

$$f(x, y) - f(x_0, y_0) = f(x_0 + \Delta x, y_0, + \Delta y) - f(x_0, y_0) = \Delta f$$

Then, because $(\Delta x, \Delta y) \to (0, 0)$ as $(x, y) \to (x_0, y_0)$, we wish to prove that

$$\lim_{(\Delta x, \Delta y) \to (0, 0)} \Delta f = 0$$

Since f is differentiable at (x_0, y_0), we have

$$\Delta f = f_x(x_0, y_0)\Delta x + f_y(x_0, y_0)\Delta y + \epsilon_1 \Delta x + \epsilon_2 \Delta y$$

where $\epsilon_1 \to 0$ and $\epsilon_2 \to 0$ as $(\Delta x, \Delta y) \to (0, 0)$. It follows that

$$\lim_{(\Delta x, \Delta y) \to (0, 0)} \Delta f = \lim_{(\Delta x, \Delta y) \to (0, 0)} [f_x(x_0, y_0)\Delta x + f_y(x_0, y_0)\Delta y + \epsilon_1 \Delta x + \epsilon_2 \Delta y]$$

$$= [f_x(x_0, y_0)] \cdot 0 + [f_y(x_0, y_0)] \cdot 0 + 0 + 0 = 0$$

as required.

WARNING ➤ Be careful about how you use the word *differentiable*. In a single-variable case, a function is differentiable at a point if its derivative exists there. However, the word is used differently for a function of two variables. In particular, the existence of the partial derivatives f_x and f_y does not guarantee that the function is differentiable, as illustrated in the following example. ◄

EXAMPLE 7 *Think Tank example: Possible existence of partial derivatives, with a nondifferentiable function*

Let

$$f(x, y) = \begin{cases} 1 & \text{if } x > 0 \text{ and } y > 0 \\ 0 & \text{otherwise} \end{cases}$$

That is, the function f has the value 1 when (x, y) is in the first quadrant and is 0 elsewhere. Show that the partial derivatives f_x and f_y exist at the origin, but f is not differentiable there.

Solution Since $f(0, 0) = 0$, we have

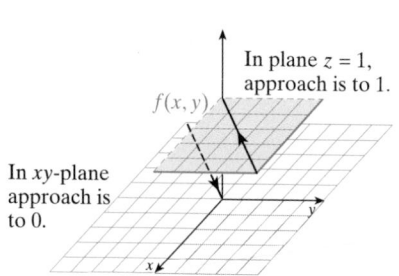

In plane $z = 1$, approach is to 1.

$f(x, y)$

In xy-plane approach is to 0.

$$f_x(0, 0) = \lim_{\Delta x \to 0} \frac{f(0 + \Delta x, 0) - f(0, 0)}{\Delta x} = 0$$

and similarly, $f_y(0, 0) = 0$. Thus, the partial derivatives both exist at the origin.

If $f(x, y)$ were differentiable at the origin, it would have to be continuous there (Theorem 12.2). Thus, we can show f is *not* differentiable by showing that it is *not* continuous at $(0, 0)$. Toward this end, note that $\lim_{(x, y) \to (0, 0)} f(x, y)$ is 1 along the line $y = x$ in the first quadrant but is 0 if the approach is along the x-axis. This means that the limit does not exist. Thus, f is not continuous at $(0, 0)$ and consequently is also not differentiable there. ∎

Although the existence of partial derivatives at $P(x_0, y_0)$ is not enough to guarantee that $f(x, y)$ is differentiable at P, we do have the following sufficient condition for differentiability.

THEOREM 12.3 *Sufficient condition for differentiability*

If f is a function of x and y and f, f_x, and f_y are continuous in a disk D centered at (x_0, y_0), then f is differentiable at (x_0, y_0).

Proof The proof is found in advanced calculus.

WARNING ➤ Note that the function in Example 7 does not contradict Theorem 12.3 because there is no disk centered at $(0, 0)$ on which f is continuous. ◄

EXAMPLE 8 *Establish differentiability*

Show that $f(x, y) = x^2 y + xy^3$ is differentiable for all (x, y).

Solution Compute the partial derivatives:

$$f_x(x, y) = \frac{\partial}{\partial x}(x^2 y + xy^3) = 2xy + y^3$$

$$f_y(x, y) = \frac{\partial}{\partial y}(x^2 y + xy^3) = x^2 + 3xy^2$$

Because f, f_x, and f_y are all polynomials in x and y, they are continuous throughout the plane. Therefore, the sufficient condition for the differentiability theorem assures us that f must be differentiable for all x and y. ∎

12.4   Problem Set

Ⓐ *In Problems 1–6, find the standard-form equations for the tangent plane to the given surface at the prescribed point P_0.*

1. $z = \sqrt{x^2 + y^2}$ at $P_0(3, 1, \sqrt{10})$

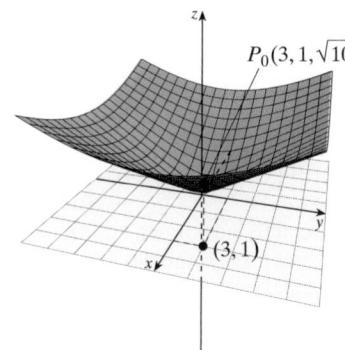

$P_0(3, 1, \sqrt{10})$

$(3, 1)$

2. $z = 10 - x^2 - y^2$ at $P_0(2, 2, 2)$

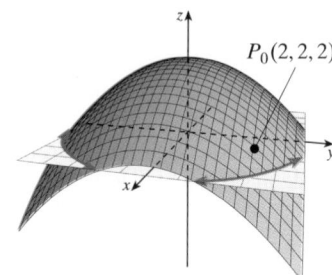

$P_0(2, 2, 2)$

3. $f(x, y) = x^2 + y^2 + \sin xy$ at $P_0 = (0, 2, 4)$

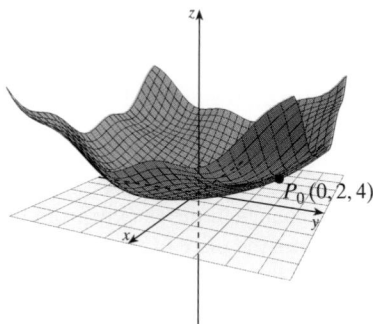

$P_0(0, 2, 4)$

4. $f(x, y) = e^{-x} \sin y$ at $P_0(0, \frac{\pi}{2}, 1)$

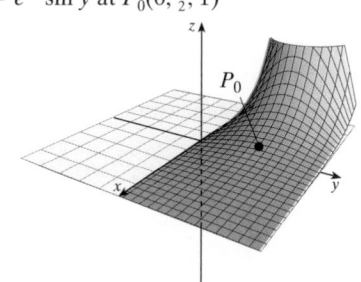

P_0

5. $z = \tan^{-1} \dfrac{y}{x}$ at $P_0\left(2, 2, \dfrac{\pi}{4}\right)$

6. $z = \ln|x + y^2|$ at $P_0(-3, -2, 0)$

Find the total differential of the functions given in Problems 7–18.

7. $f(x, y) = 5x^2 y^3$

8. $f(x, y) = 8x^3 y^2 - x^4 y^5$

9. $f(x, y) = \sin xy$

10. $f(x, y) = \cos x^2 y$

11. $f(x, y) = \dfrac{y}{x}$

12. $f(x, y) = \dfrac{x^2}{y}$

13. $f(x, y) = y e^x$

14. $f(x, y) = e^{x^2 + y}$

15. $f(x, y, z) = 3x^3 - 2y^4 + 5z$

16. $f(x, y, z) = \sin x + \sin y + \cos z$

17. $f(x, y, z) = z^2 \sin(2x - 3y)$

18. $f(x, y, z) = 3y^2 z \cos x$

Show that the functions in Problems 19–22 are differentiable for all (x, y).

19. $f(x, y) = xy^3 + 3xy^2$

20. $f(x, y) = x^2 + 4x - y^2$

21. $f(x, y) = e^{2x + y^2}$

22. $f(x, y) = \sin(x^2 + 3y)$

Use an incremental approximation to estimate the functions at the values given in Problems 23–28. Check by using a calculator.

23. $f(1.01, 2.03)$, where $f(x, y) = 3x^4 + 2y^4$

24. $f(0.98, 1.03)$, where $f(x, y) = x^5 - 2y^3$

25. $f(\frac{\pi}{2} + 0.01, \frac{\pi}{2} - 0.01)$, where $f(x, y) = \sin(x + y)$

26. $f(\sqrt{\frac{\pi}{2}} + 0.01, \sqrt{\frac{\pi}{2}} - 0.01)$, where $f(x, y) = \sin(xy)$

27. $f(1.01, 0.98)$, where $f(x, y) = e^{xy}$

28. $f(1.01, 0.98)$, where $f(x, y) = e^{x^2 y^2}$

Ⓑ 29. Find an equation for each horizontal tangent plane to the surface
$$z = 5 - x^2 - y^2 + 4y$$

30. Find an equation for each horizontal tangent plane to the surface
$$z = 4(x - 1)^2 + 3(y + 1)^2$$

31. a. Show that if x and y are sufficiently close to zero and f is differentiable at $(0, 0)$, then
$$f(x, y) \approx f(0, 0) + x f_x(0, 0) + y f_y(0, 0)$$

b. Use the approximation formula in part **a** to show that
$$\frac{1}{1 + x - y} \approx 1 - x + y$$
for small x and y.

c. If x and y are sufficiently close to zero, what is the approximate value of the expression
$$\frac{1}{(x + 1)^2 + (y + 1)^2}$$

32. When two resistors with resistances P and Q ohms are connected in parallel, the combined resistance is R, where

$$\frac{1}{R} = \frac{1}{P} + \frac{1}{Q}$$

If P and Q are measured at 6 and 10 ohms, respectively, with errors no greater than 1%, what is the maximum percentage error in the computation of R?

33. A closed box is found to have length 2 ft, width 4 ft, and height 3 ft, where the measurement of each dimension is made with a maximum possible error of ±0.02 ft. The top of the box is made from material that costs $2/ft^2; the material for the sides and bottom costs only $1.50/ft^2. What is the maximum error involved in the computation of the cost of the box?

34. A cylindrical tank is 4 ft high and has a diameter of 2 ft. The walls of the tank are 0.2 in. thick. Approximate the volume of the interior of the tank assuming the tank has a top and a bottom that are both also 0.2 in. thick.

35. The Higrade Company sells two brands, X and Y, of a commercial soap, in thousand-pound units. If x units of brand X and y units of brand Y are sold, the unit price for brand X is

$$p(x) = 4,000 - 500x$$

and that of brand Y is

$$q(y) = 3,000 - 450y$$

a. Find an expression for the total revenue R in terms of p and q.

b. Suppose brand X sells for $500 per unit and brand Y sells for $750 per unit. Estimate the change in total revenue if the unit prices are increased by $20 for brand X and $18 for brand Y. *Hint:* Find the change in x and the change in y that correspond to $p(x)$ increasing from 500 to 520 and $q(y)$ increasing from 750 to 768.

36. The output at a certain factory is

$$Q = 150K^{2/3}L^{1/3}$$

where K is the capital investment in units of $1,000 and L is the size of the labor force, measured in worker-hours. The current capital investment is $500,000 and 1,500 worker-hours of labor are used.

a. Estimate the change in output that results when capital investment is increased by $700 and labor is increased by 6 worker-hours.

b. What if capital investment is increased by $500 and labor is decreased by 4 worker-hours?

37. According to Poiseuille's law, the resistance to the flow of blood offered by a cylindrical blood vessel of radius r and length x is

$$R(r, x) = \frac{cx}{r^4}$$

for a constant $c > 0$. A certain blood vessel in the body is 8 cm long and has a radius of 2 mm. Estimate the per-

centage change in R when x is increased by 3% and r is decreased by 2%.

38. For 1 mole of an ideal gas, the volume V, pressure P, and absolute temperature T are related by the equation $PV = RT$, where R is a certain fixed constant that depends on the gas. Suppose we know that if $T = 400$ (absolute) and $P = 3,000$ lb/ft^2, then $V = 14$ ft^3. Approximate the change in pressure if the temperature and volume are increased to 403 and 14.1 ft^3, respectively.

39. MODELING PROBLEM If x gram-moles of sulfuric acid are mixed with y gram-moles of water, the heat liberated is modeled by

$$F(x, y) = \frac{1.786 xy}{1.798x + y} \text{ cal}$$

Approximately how much additional heat is generated if a mixture of 5 gram-moles of acid and 4 gram-moles of water is increased to a mixture of 5.1 gram-moles of acid and 4.04 gram-moles of water?

40. MODELING PROBLEM A business analyst models the sales of a new product by the function

$$Q(x, y) = 20x^{3/2}y$$

where x thousand dollars are spent on development and y thousand dollars on promotion. Current plans call for the expenditure of $36,000 on development and $25,000 on promotion. Use the total differential of Q to estimate the change in sales that will result if the amount spent on development is increased by $500 and the amount spent on promotion is decreased by $500.

41. MODELING PROBLEM A grocer's weekly profit from the sale of two brands of orange juice is modeled by

$$P(x, y) = (x - 30)(70 - 5x + 4y)$$
$$+ (y - 40)(80 + 6x - 7y)$$

dollars where x (cents) is the price per can of the first brand and y (cents) is the price per can of the second. Currently the first brand sells for 50¢ per can and the second for 52¢ per can. Use the total differential to estimate the change in the weekly profit that will result if the grocer raises the price of the first brand by 1¢ per can and lowers the price of the second brand by 2¢ per can.

42. A juice can is 12 cm tall and has a radius of 3 cm. A manufacturer is planning to reduce the height of the can by 0.2 cm and the radius by 0.3 cm. Use a total differential to estimate how much less the volume will be in each can after the new cans are introduced.

43. MODELING PROBLEM It is known that the period T of a simple pendulum with small oscillations is modeled by

$$T = 2\pi\sqrt{\frac{L}{g}}$$

where L is the length of the pendulum and g is the acceleration due to gravity. For a certain pendulum, it is known that $L = 4.03$ ft. It is also known that $g = 32.2$ ft/s^2.

What is the approximate error in calculating T by using $L = 4$ and $g = 32$?

44. If the weight of an object which does not float in water in the air is x pounds and its weight in water is y pounds, then the specific gravity of the object is

$$S = \frac{x}{x - y}$$

For a certain object, x and y are measured to be 1.2 lb and 0.5 lb, respectively. It is known that the measuring instrument will not register less than the true weights, but it could register more than the true weights by as much as 0.01 lb. What is the maximum possible error in the computation of the specific gravity?

45. A football has the shape of the ellipsoid

$$\frac{x^2}{9} + \frac{y^2}{36} + \frac{z^2}{9} = 1$$

where the dimensions are in inches, and is made of leather 1/8 inch thick. Use differentials to estimate the volume of the leather shell. *Hint:* The ellipsoid

$$\frac{x^2}{a^2} + \frac{y^2}{b^2} + \frac{z^2}{c^2} = 1$$

has volume $V = \frac{4}{3}\pi abc$.

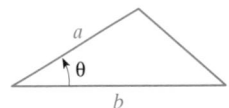 46. Show that the following function is not differentiable at $(0, 0)$:

$$f(x, y) = \begin{cases} (xy)/(x^2 + y^2) & \text{if } (x, y) \neq (0, 0) \\ 0 & \text{if } (x, y) = (0, 0) \end{cases}$$

47. Compute the total differentials

$$d\left(\frac{x}{x - y}\right) \quad \text{and} \quad d\left(\frac{y}{x - y}\right)$$

Why are these differentials alike?

48. Let A be the area of a triangle with sides a and b separated by an angle θ, as shown in Figure 12.18.

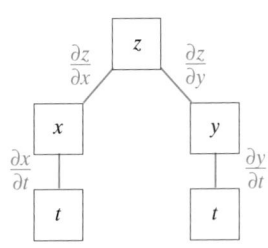

■ **FIGURE 12.18**   Problem 48

Suppose $\theta = \frac{\pi}{6}$, and a is increased by 4% while b is decreased by 3%. Use differentials to estimate the percentage change in A.

49. In Problem 48, suppose that θ changes by no more than 2%. What is the maximum percentage change in A?

12.5   Chain Rules

IN THIS SECTION   chain rule for one parameter, chain rule for two parameters   ■

CHAIN RULE FOR ONE PARAMETER

We begin with a differentiable function of two variables $f(x, y)$. If $x = x(t)$ and $y = y(t)$ are, in turn, functions of a single variable t, then $z = f[x(t), y(t)]$ is a composite function of a parameter t. In this case, the chain rule for finding the derivative with respect to one independent variable can now be stated.

THEOREM 12.4   *The chain rule for one independent parameter*

Let $f(x, y)$ be a differentiable function of x and y, and let $x = x(t)$ and $y = y(t)$ be differentiable functions of t. Then $z = f(x, y)$ is a differentiable function of t, and

$$\frac{dz}{dt} = \frac{\partial z}{\partial x}\frac{dx}{dt} + \frac{\partial z}{\partial y}\frac{dy}{dt}$$

Chains from the chain rule represented schematically

■ *What This Says:*   Recall the chain rule for a single variable:

$$\frac{dy}{dx} = \frac{dy}{du}\frac{du}{dx}$$

The corresponding rule for two variables is *essentially* the same except that it involves *both* variables.

Proof   Recall that because $z = f(x, y)$ is differentiable, we can write the increment Δz in the following form:

$$\Delta z = \frac{\partial z}{\partial x}\Delta x + \frac{\partial z}{\partial y}\Delta y + \epsilon_1 \Delta x + \epsilon_2 \Delta y$$

where $\epsilon_1 \to 0$ and $\epsilon_2 \to 0$ as both $\Delta x \to 0$ and $\Delta y \to 0$. Dividing by $\Delta t \neq 0$, we obtain

$$\frac{\Delta z}{\Delta t} = \frac{\partial z}{\partial x}\frac{\Delta x}{\Delta t} + \frac{\partial z}{\partial y}\frac{\Delta y}{\Delta t} + \epsilon_1\frac{\Delta x}{\Delta t} + \epsilon_2\frac{\Delta y}{\Delta t}$$

Because x and y are functions of t, we can write their increments as

$$\Delta x = x(t + \Delta t) - x(t) \quad \text{and} \quad \Delta y = y(t + \Delta t) - y(t)$$

We know that x and y both vary continuously with t (remember, they are differentiable), and it follows that $\Delta x \to 0$ and $\Delta y \to 0$ as $\Delta t \to 0$, so that $\epsilon_1 \to 0$ and $\epsilon_2 \to 0$ as $\Delta t \to 0$. Therefore, we have

$$\frac{dz}{dt} = \lim_{\Delta t \to 0}\frac{\Delta z}{\Delta t} = \lim_{\Delta t \to 0}\left[\frac{\partial z}{\partial x}\frac{\Delta x}{\Delta t} + \frac{\partial z}{\partial y}\frac{\Delta y}{\Delta t} + \epsilon_1\frac{\Delta x}{\Delta t} + \epsilon_2\frac{\Delta y}{\Delta t}\right]$$

$$= \frac{\partial z}{\partial x}\frac{dx}{dt} + \frac{\partial z}{\partial y}\frac{dy}{dt} + 0\frac{dx}{dt} + 0\frac{dy}{dt}$$

EXAMPLE 1 *Verifying the chain rule explicitly*

Let $z = x^2 + y^2$, where $x = \dfrac{1}{t}$ and $y = t^2$. Find $\dfrac{dz}{dt}$ in two ways:

a. by first expressing z explicitly in terms of t
b. by using the chain rule

Solution
a. By substituting $x = 1/t$ and $y = t^2$, we find that

$$z = x^2 + y^2 = \left(\frac{1}{t}\right)^2 + (t^2)^2 = t^{-2} + t^4 \quad \text{for } t \neq 0$$

Thus, $\dfrac{dz}{dt} = -2t^{-3} + 4t^3$.

b. Because $z = x^2 + y^2$ and $x = t^{-1}, y = t^2$,

$$\frac{\partial z}{\partial x} = 2x; \qquad \frac{\partial z}{\partial y} = 2y; \qquad \frac{dx}{dt} = -t^{-2}; \qquad \frac{dy}{dt} = 2t$$

Use the chain rule for one independent parameter:

$$\frac{dz}{dt} = \frac{\partial z}{\partial x}\frac{dx}{dt} + \frac{\partial z}{\partial y}\frac{dy}{dt}$$
$$= (2x)(-t^{-2}) + 2y(2t) \qquad \text{Chain rule}$$
$$= 2(t^{-1})(-t^{-2}) + 2(t^2)(2t) \qquad \text{Substitute.}$$
$$= -2t^{-3} + 4t^3$$

EXAMPLE 2 *Chain rule for one independent parameter*

Let $z = \sqrt{x^2 + 2xy}$, where $x = \cos\theta$ and $y = \sin\theta$. Find $\dfrac{dz}{d\theta}$.

Solution $\dfrac{\partial z}{\partial x} = \dfrac{1}{2}(x^2 + 2xy)^{-1/2}(2x + 2y)$ and $\dfrac{\partial z}{\partial y} = \dfrac{1}{2}(x^2 + 2xy)^{-1/2}(2x)$

Also, $\dfrac{dx}{d\theta} = -\sin\theta$ and $\dfrac{dy}{d\theta} = \cos\theta$. Use the chain rule for one independent parameter to find

$$\frac{dz}{d\theta} = \frac{\partial z}{\partial x}\frac{dx}{d\theta} + \frac{\partial z}{\partial y}\frac{dy}{d\theta}$$
$$= \tfrac{1}{2}(x^2 + 2xy)^{-1/2}(2x + 2y)(-\sin\theta) + \tfrac{1}{2}(x^2 + 2xy)^{-1/2}(2x)(\cos\theta)$$
$$= (x^2 + 2xy)^{-1/2}(x\cos\theta - x\sin\theta - y\sin\theta)$$
■

EXAMPLE 3   *Related rate application using the chain rule*

A right circular cylinder is changing in such a way that its radius r is increasing at the rate of 3 in./min and its height h is decreasing at the rate of 5 in./min. At what rate is the volume of the cylinder changing when the radius is 10 in. and the height is 8 in.?

$$\frac{dr}{dt} = 3$$

$$h \quad \frac{dh}{dt} = -5$$

Solution   The volume of the cylinder is $V = \pi r^2 h$, and we are given $\dfrac{dr}{dt} = 3$ and $\dfrac{dh}{dt} = -5$. We find that

$$\frac{\partial V}{\partial r} = \pi(2r)h \quad \text{and} \quad \frac{\partial V}{\partial h} = \pi r^2 (1)$$

By the chain rule for one parameter:

$$\frac{dV}{dt} = \frac{\partial V}{\partial r}\frac{dr}{dt} + \frac{\partial V}{\partial h}\frac{dh}{dt} = 2\pi rh\frac{dr}{dt} + \pi r^2\frac{dh}{dt}$$

Thus, at the instant when $r = 10$ and $h = 8$, we have

$$\frac{dV}{dt} = 2\pi(10)(8)(3) + \pi(10)^2(-5) = -20\pi \approx -62.83185307$$

The volume is decreasing at the rate of about 62.8 in.³/min.   ■

If $F(x, y) = 0$ defines y implicitly as a differentiable function x, then the chain rule tells us that

$$0 = \frac{\partial F}{\partial x}\frac{dx}{dx} + \frac{\partial F}{\partial y}\frac{dy}{dx} = \frac{\partial F}{\partial x} + \frac{\partial F}{\partial y}\frac{dy}{dx}$$

so

$$\frac{dy}{dx} = \frac{-\dfrac{\partial F}{\partial x}}{\dfrac{\partial F}{\partial y}} = -\frac{F_x}{F_y} \qquad \text{provided } F_y \neq 0$$

This formula provides a useful alternative to implicit differentiation; the procedure is illustrated in the following example.

EXAMPLE 4   *Implicit differentiation by formula*

If y is a differentiable function of x such that

$$\sin(x + y) + \cos(x - y) = y$$

find $\dfrac{dy}{dx}$.

Solution   Let $F(x, y) = \sin(x+y) + \cos(x - y) - y$. Then,

$$F_x = \cos(x + y) - \sin(x - y)$$
$$F_y = \cos(x + y) + \sin(x - y) - 1$$

so

$$\frac{dy}{dx} = -\frac{F_x}{F_y} = \frac{-[\cos(x+y) + \sin(x-y) - 1]}{\cos(x+y) - \sin(x-y)}$$

∎

EXAMPLE 5   *Second derivative of a function of two variables*

Let $z = f(x, y)$, where $x = at$ and $y = bt$ for constants a and b. Assuming all necessary differentiability, find d^2z/dt^2 in terms of the partial derivatives of z.

Solution   We note that $\dfrac{dx}{dt} = a$ and $\dfrac{dy}{dt} = b$, and

$$\frac{dz}{dt} = \frac{\partial z}{\partial x}\frac{dx}{dt} + \frac{\partial z}{\partial y}\frac{dy}{dt} \qquad \text{Chain rule}$$

We differentiate both sides with respect to t, using the chain rule again on the right:

$$\frac{d^2z}{dt^2} = \frac{\partial z}{\partial x}\frac{d^2x}{dt^2} + \frac{dx}{dt}\left[\frac{\partial^2 z}{\partial x^2}\frac{dx}{dt} + \frac{\partial^2 z}{\partial x \partial y}\frac{dy}{dt}\right] + \frac{\partial z}{\partial y}\frac{d^2y}{dt^2} + \frac{dy}{dt}\left[\frac{\partial^2 z}{\partial y \partial x}\frac{dx}{dt} + \frac{\partial^2 z}{\partial y^2}\frac{dy}{dt}\right]$$

$$= \frac{\partial z}{\partial x}(0) + a\left[\frac{\partial^2 z}{\partial x^2}a + \frac{\partial^2 z}{\partial x \partial y}b\right] + \frac{\partial z}{\partial y}(0) + b\left[\frac{\partial^2 z}{\partial y \partial x}a + \frac{\partial^2 z}{\partial y^2}b\right]$$

$$= a^2\frac{\partial^2 z}{\partial x^2} + 2ab\frac{\partial^2 z}{\partial x \partial y} + b^2\frac{\partial^2 z}{\partial y^2} \qquad \text{Note } \frac{\partial^2 z}{\partial x \partial y} = \frac{\partial^2 z}{\partial y \partial x}.$$

∎

CHAIN RULE FOR TWO PARAMETERS

Next we shall consider the kind of composite function that occurs when x and y are both functions of *two* variables. Specifically, let $z = F(x, y)$, where $x = x(u, v)$ and $y = y(u, v)$ are both functions of two independent parameters u and v. Then $z = F[x(u, v), y(u, v)]$ is a composite function of u and v, and with suitable assumptions regarding differentiability, we can find the partial derivatives $\partial z/\partial u$ and $\partial z/\partial v$ by applying the chain rule obtained in the following theorem.

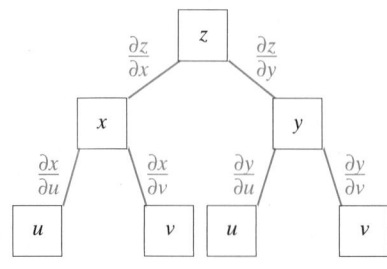

THEOREM 12.5   *The chain rule for two independent parameters*

Suppose $z = f(x, y)$ is differentiable at (x, y) and that the partial derivatives of $x = x(u, v)$ and $y = y(u, v)$ exist at (u, v). Then the composite function $z = f[x(u, v), y(u, v)]$ is differentiable at (u, v) with

$$\frac{\partial z}{\partial u} = \frac{\partial z}{\partial x}\frac{\partial x}{\partial y} + \frac{\partial z}{\partial y}\frac{\partial y}{\partial u} \quad \text{and} \quad \frac{\partial z}{\partial v} = \frac{\partial z}{\partial x}\frac{\partial x}{\partial v} + \frac{\partial z}{\partial y}\frac{\partial y}{\partial v}$$

Proof   This version of the chain rule follows immediately from the chain rule for one independent parameter. For example, if v is fixed, the composite function $z = f[x(u, v), y(u, v)]$ really depends on u alone, and we have the situation described in the chain rule of one independent variable. We apply this chain rule with a partial derivative instead of an "ordinary" derivative (because x and y are functions of more than one variable):

$$\frac{\partial z}{\partial u} = \frac{\partial z}{\partial x}\frac{\partial x}{\partial u} + \frac{\partial z}{\partial y}\frac{\partial y}{\partial u}$$

The formula for $\dfrac{\partial z}{\partial v}$ can be established in a similar fashion.

═══

EXAMPLE 6   *Chain rule for two independent parameters*

Let $z = 4x - y^2$, where $x = uv^2$ and $y = u^3v$. Find $\partial z/\partial u$ and $\partial z/\partial v$.

Solution   First find the partial derivatives:

$$\frac{\partial z}{\partial x} = \frac{\partial}{\partial x}(4x - y^2) = 4 \qquad \frac{\partial z}{\partial y} = \frac{\partial}{\partial y}(4x - y^2) = -2y$$

and
$$\frac{\partial x}{\partial u} = \frac{\partial}{\partial u}(uv^2) = v^2 \qquad \frac{\partial y}{\partial u} = \frac{\partial}{\partial u}(u^3v) = 3u^2v$$

$$\frac{\partial x}{\partial v} = \frac{\partial}{\partial v}(uv^2) = 2uv \qquad \frac{\partial y}{\partial v} = \frac{\partial}{\partial v}(u^3v) = u^3$$

Therefore, the chain rule for two independent parameters gives

$$\frac{\partial z}{\partial u} = \frac{\partial z}{\partial x}\frac{\partial x}{\partial u} + \frac{\partial z}{\partial y}\frac{\partial y}{\partial u}$$

$$= (4)(v^2) + (-2y)(3u^2v) = 4v^2 - 2(u^3v)(3u^2v) = 4v^2 - 6u^5v^2$$

and
$$\frac{\partial z}{\partial v} = \frac{\partial z}{\partial x}\frac{\partial x}{\partial v} + \frac{\partial z}{\partial y}\frac{\partial y}{\partial v}$$

$$= (4)(2uv) + (-2y)(u^3) = 8uv - 2(u^3v)u^3 = 8uv - 2u^6v \qquad \blacksquare$$

The chain rules can be extended to functions of three or more variables. For instance, if $w = f(x, y, z)$ is a differentiable function of three variables and $x = x(t)$, $y = y(t), z = z(t)$ are each differentiable functions of t, then w is a differentiable composite function of t and

$$\frac{dw}{dt} = \frac{\partial w}{\partial x}\frac{dx}{dt} + \frac{\partial w}{\partial y}\frac{dy}{dt} + \frac{\partial w}{\partial z}\frac{dz}{dt}$$

In general, if $w = f(x_1, x_2, \dots, x_n)$ is a differentiable function of the n variables $x_1, x_2, \dots, x_n$, which in turn are differentiable functions of m variables $t_1, t_2, \dots, t_m$, then

$$\frac{\partial w}{\partial t_1} = \frac{\partial w}{\partial x_1}\frac{\partial x_1}{\partial t_1} + \frac{\partial w}{\partial x_2}\frac{\partial x_2}{\partial t_1} + \cdots + \frac{\partial w}{\partial x_n}\frac{\partial x_n}{\partial t_1}$$

$$\vdots$$

$$\frac{\partial w}{\partial t_m} = \frac{\partial w}{\partial x_1}\frac{\partial x_1}{\partial t_m} + \frac{\partial w}{\partial x_2}\frac{\partial x_2}{\partial t_m} + \cdots + \frac{\partial w}{\partial x_n}\frac{\partial x_n}{\partial t_m}$$

EXAMPLE 7   *Chain rule for a function of several variables*

Find $\partial w/\partial s$ if $w = 4x + y^2 + z^3$, where $x = e^{rs^2}$, $y = \ln\dfrac{r + s}{t}$, and $z = rst^2$.

Solution

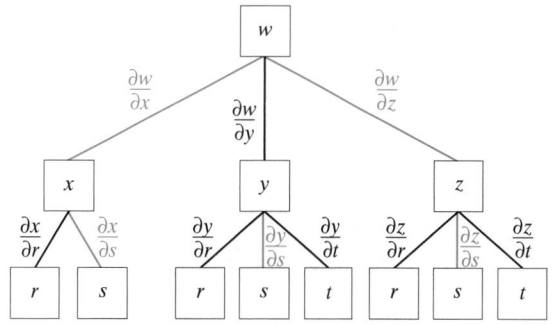

$$\frac{\partial w}{\partial s} = \frac{\partial w}{\partial x}\frac{\partial x}{\partial s} + \frac{\partial w}{\partial y}\frac{\partial y}{\partial s} + \frac{\partial w}{\partial z}\frac{\partial z}{\partial s}$$

$$= \left[\frac{\partial}{\partial x}(4x + y^2 + z^3)\right]\left[\frac{\partial}{\partial s}(e^{rs^2})\right] + \left[\frac{\partial}{\partial y}(4x + y^2 + z^3)\right]\left[\frac{\partial}{\partial s}\left(\ln\frac{r+s}{t}\right)\right]$$

$$+ \left[\frac{\partial}{\partial z}(4x + y^2 + z^3)\right]\left[\frac{\partial}{\partial s}(rst^2)\right]$$

$$= 4[e^{rs^2}(2rs)] + 2y\left(\frac{1}{(r+s)/t}\right)\left(\frac{1}{t}\right) + 3z^2(rt^2) = 8rse^{rs^2} + \frac{2y}{r+s} + 3rt^2z^2$$

In terms of r, s, and t, the partial derivative is

$$\frac{\partial w}{\partial s} = 8rse^{rs^2} + \frac{2}{r+s}\ln\frac{r+s}{t} + 3r^3s^2t^6 \qquad \blacksquare$$

12.5 Problem Set

Ⓐ 1. ■ **What Does This Say?** Discuss the various chain rules and the need for such chain rules.

2. ■ **What Does This Say?** Discuss the usefulness of the schematic representation for the chain rules.

3. ■ **What Does This Say?** Write out a chain rule for three independent parameters and two variables.

In Problems 4–9, assume the given equations define y as a differentiable function of x and find dy/dx using the procedure illustrated in Example 4.

4. $x^2y + \sqrt{xy} = 4$

5. $(x^2 - y)^{3/2} + x^2y = 2$

6. $x^2y + \ln(2x + y) = 5$

7. $x\cos y + y\tan^{-1}x = x$

8. $xe^{xy} + ye^{-xy} = 3$

9. $\tan^{-1}\left(\frac{x}{y}\right) = \tan^{-1}\left(\frac{y}{x}\right)$

In Problems 10–13, the function f(x, y) depends on x and y, which in turn are each functions of t. In each case, find dz/dt two different ways:
a. *Express z explicitly in terms of t.*
b. *Use the chain rule for one parameter.*

10. $f(x, y) = 2xy + y^2$, where
$x = -3t^2$ and $y = 1 + t^3$

11. $f(x, y) = (4 + y^2)x$, where
$x = e^{2t}$ and $y = e^{3t}$

12. $f(x, y) = (1 + x^2 + y^2)^{1/2}$, where
$x = \cos 5t$ and $y = \sin 5t$

13. $f(x, y) = xy^2$, where
$x = \cos 3t$ and $y = \tan 3t$

In Problems 14–17, the function F(x, y) depends on x and y. Let $x = x(u, v)$ and $y = y(u, v)$ be given functions of u and v. Let $z = F[x(u, v), y(u, v)]$ and find the partial derivatives $\partial z/\partial u$ and $\partial z/\partial v$ in these two ways:
a. *Express z explicitly in terms of u and v.*

b. *Apply the chain rule for two independent parameters.*

14. $F(x, y) = x + y^2$, where
$x = u + v$ and $y = u - v$

15. $F(x, y) = x^2 + y^2$, where
$x = u\sin v$ and $y = u - 2v$

16. $F(x, y) = e^{xy}$, where
$x = u - v$ and $y = u + v$

17. $F(x, y) = \ln xy$, where
$x = e^{uv^2}, y = e^{uv}$

Write out the chain rule for the functions given in Problems 18–21.

18. $z = f(x, y)$, where $x = x(s, t), y = y(s, t)$

19. $w = f(x, y, z)$, where $x = x(s, t)$,
$y = y(s, t), z = z(s, t)$

20. $t = f(u, v)$, where $u = u(x, y, z, w)$,
$v = v(x, y, z, w)$

21. $w = f(x, y, z)$, where $x = x(s, t, u)$,
$y = y(s, t, u), z = z(s, t, u)$

Find the indicated derivatives or partial derivatives in Problems 22–27. Leave your answers in mixed form (x, y, z, t, u, v).

22. Find $\dfrac{dw}{dt}$, where $w = \ln(x + 2y - z^2)$ and
$x = 2t - 1, y = \dfrac{1}{t}, z = \sqrt{t}$.

23. Find $\dfrac{dw}{dt}$, where $w = \sin xyz$ and $x = 1 - 3t, y = e^{1-t}$,
$z = 4t$.

24. Find $\dfrac{dw}{dt}$, where $w = ze^{xy^2}$ and $x = \sin t, y = \cos t$,
$z = \tan 2t$.

25. Find $\dfrac{dw}{dt}$, where $w = e^{x^3+yz}$ and $x = \dfrac{2}{t}$, $y = \ln(2t - 3)$, $z = t^2$.

26. Find $\dfrac{\partial w}{\partial r}$, where $w = e^{2x-y+3z^2}$ and $x = r + s - t$, $y = 2r - 3s$, $z = \cos rst$.

27. Find $\dfrac{\partial w}{\partial r}$ and $\dfrac{\partial w}{\partial t}$, where $w = \dfrac{x + y}{2 - z}$ and $x = 2rs$, $y = \sin rt$, $z = st^2$.

B *Find the following higher partial derivatives in Problems 28–33.*

  a. $\dfrac{\partial^2 z}{\partial x \partial y}$ b. $\dfrac{\partial^2 z}{\partial x^2}$ c. $\dfrac{\partial^2 z}{\partial y^2}$

28. $x^3 + y^2 + z^2 = 5$ **29.** $xyz = 2$

30. $\ln(x + y) = y^2 + z$ **31.** $x^{-1} + y^{-1} + z^{-1} = 3$

32. $x \cos y = y + z$ **33.** $z^2 + \sin x = \tan y$

34. Let $f(x, y)$ be a differentiable function of x and y, and let $x = r \cos \theta$, $y = r \sin \theta$ for $r > 0$ and $0 < \theta < 2\pi$.

  a. If $z = f[x(r, \theta), y(r, \theta)]$, find $\dfrac{\partial z}{\partial r}$ and $\dfrac{\partial z}{\partial \theta}$.

  b. Show that

$$\left(\frac{\partial z}{\partial r}\right)^2 + \frac{1}{r^2}\left(\frac{\partial z}{\partial \theta}\right)^2 = \left(\frac{\partial z}{\partial x}\right)^2 + \left(\frac{\partial z}{\partial y}\right)^2$$

35. Let $z = f(x, y)$, where $x = au$ and $y = bv$, with a, b constants. Express $\partial^2 z/\partial u^2$ and $\partial^2 z/\partial v^2$ in terms of the partial derivatives of z with respect to x and y. Assume the existence and continuity of all necessary first and second partial derivatives.

36. Let (x, y, z) lie on the ellipsoid

$$\frac{x^2}{a^2} + \frac{y^2}{b^2} + \frac{z^2}{c^2} = 1$$

Without solving for z explicitly in terms of x and y, compute the higher partial derivatives

$$\frac{\partial^2 z}{\partial x^2} \quad \text{and} \quad \frac{\partial^2 z}{\partial x \partial y}$$

37. The dimensions of a rectangular box are linear functions of time, $\ell(t)$, $w(t)$, and $h(t)$. If the length and width are increasing at 2 in./sec and the height is decreasing at 3 in./sec, find the rates at which the volume V and the surface area S are changing with respect to time. If $\ell(0) = 10$, $w(0) = 8$, and $h(0) = 20$, is V increasing or decreasing when $t = 5$ sec? What about S when $t = 5$?

38. Using x hours of skilled labor and y hours of unskilled labor, a manufacturer can produce $f(x, y) = 10xy^{1/2}$ units. Currently, the manufacturer has used 30 hours of skilled labor 36 hours of unskilled labor and is planning to use 1 additional hours of skilled labor. Use calculus to estimate the corresponding change that the manufacturer should make in the level of unskilled labor so that the total output will remain the same.

39. MODELING PROBLEM   Van der Waals' equation in physical chemistry states that a gas occupying volume V at temperature T (Kelvin) exerts pressure P, where

$$\left(P + \frac{A}{V^2}\right)(V - B) = kT$$

for physical constants A, B, and k. Find the following rates:

  a. The rate of change of volume with respect to temperature

  b. The rate of change of pressure with respect to volume

40. MODELING PROBLEM   The concentration of a drug in the blood of a patient t hours after the drug is injected into the body intramuscularly is modeled by the Heinz function

$$C = \frac{1}{b - a}(e^{-at} - e^{-bt}) \qquad b > a$$

where a and b are parameters that depend on the patient's metabolism and the particular kind of drug being used.

  a. Find the rates $\dfrac{\partial C}{\partial a}, \dfrac{\partial C}{\partial b}, \dfrac{\partial C}{\partial t}$.

  b. Explore the assumption that $a = (\ln b)/t$, b constant for $t > (\ln b)/b$. In particular, what is dC/dt?

41. MODELING PROBLEM   A paint store carries two brands of latex paint. An analysis of sales figures indicates that the demand Q for the first brand is modeled by

$$Q(x, y) = 210 - 12x^2 + 18y$$

gallons/month, where x, y are the prices of the first and second brands, respectively. A separate study indicates that t months from now, the first brand will cost $x = 4 + 0.18t$ dollars/gal and the second brand will cost $y = 5 + 0.3\sqrt{t}$ dollars/gal. At what rate will the demand Q be changing with respect to time 9 months from now?

42. MODELING PROBLEM   To model the demand for the sale of bicycles, it is assumed that if 24-speed bicycles are sold for x dollars apiece and the price of gasoline is y cents per gallon, then

$$Q(x, y) = 240 - 21\sqrt{x} + 4(0.2y + 12)^{3/2}$$

bicycles will be sold each month. For this model it is furthermore assumed that, t months from now, bicycles will be selling for $x = 120 + 6t$ dollars apiece, and the price of gasoline will be $y = 80 + 10\sqrt{4t}$ cents/gal. At what rate will the monthly demand for the bicycles be changing with respect to time 4 months from now?

43. MODELING PROBLEM   At a certain factory, the amount of air pollution generated each day is modeled by the function

$$Q(E, T) = 127E^{2/3}\, T^{1/2}$$

where E is the number of employees and $T(°C)$ is the average temperature during the workday. Currently, there are 142 employees and the average temperature is 18°C. If the average daily temperature is falling at the rate of 0.23°/day, and the number of employees is increasing at the rate of 3/month, what is the corresponding effect on the rate of pollution? Express your answer in units/day. For this model, assume there are 22 workdays/month.

44. **MODELING PROBLEM** The combined resistance R produced by three variable resistances $R_1, R_2,$ and R_3 connected in parallel is modeled by the formula

$$\frac{1}{R} = \frac{1}{R_1} + \frac{1}{R_2} + \frac{1}{R_3}$$

Suppose at a certain instant, $R_1 = 100$ ohms, $R_2 = 200$ ohms, $R_3 = 300$ ohms, and R_1 and R_3 are decreasing at the rate of 1.5 ohms/s while R_2 is increasing at the rate of 2 ohms/s. How fast is R changing with respect to time at this instant? Is it increasing or decreasing?

In Problems 45–51, assume that all functions have whatever derivatives or partial derivatives are necessary for the problem to be meaningful.

45. If $z = f(uv^2)$, show that

$$2u\frac{\partial z}{\partial u} - v\frac{\partial z}{\partial v} = 0$$

Hint: Let $w = uv^2$ and apply the chain rule.

46. If $z = u + f(u^2 v^2)$, show that

$$u\frac{\partial z}{\partial u} - v\frac{\partial z}{\partial v} = u$$

Hint: Let $w = u^2 v^2$ and apply the chain rule.

47. If $z = f(u - v, v - u)$, show that

$$\frac{\partial z}{\partial u} + \frac{\partial z}{\partial v} = 0$$

48. If $z = u + f(uv)$, show that $u\dfrac{\partial z}{\partial u} - v\dfrac{\partial z}{\partial v} = u.$

49. If $w = f\left(\dfrac{r - s}{s}\right)$, show that $r\dfrac{\partial w}{\partial r} + s\dfrac{\partial w}{\partial s} = 0.$

50. If $z = f(x^2 - y^2)$, evaluate $y\dfrac{\partial z}{\partial x} + x\dfrac{\partial z}{\partial y}.$

51. If $z = xy + f(x^2 - y^2)$, show that

$$y\frac{\partial z}{\partial x} - x\frac{\partial z}{\partial y} = y^2 - x^2$$

52. Let $w = f(t)$ be a differentiable function of t, where $t = (x^2 + y^2 + z^2)^{1/2}$. Show that

$$\left(\frac{dw}{dt}\right)^2 = \left(\frac{\partial w}{\partial x}\right)^2 + \left(\frac{\partial w}{\partial y}\right)^2 + \left(\frac{\partial w}{\partial z}\right)^2$$

53. Suppose f is a twice differentiable function of one variable, and let $z = f(x^2 + y^2)$. Find

a. $\dfrac{\partial^2 z}{\partial x^2}$ b. $\dfrac{\partial^2 z}{\partial y^2}$ c. $\dfrac{\partial^2 z}{\partial x \partial y}$

54. Find $\dfrac{d^2 z}{d\theta^2}$ where f is a twice differentiable function of one variable and $z = f(\cos\theta, \sin\theta)$. *Hint:* Let $x = \cos\theta$ and $y = \sin\theta$. Leave your answer in terms of x and y.

55. Let f and g be twice differentiable functions of one variable, and let

$$u(x, t) = f(x + ct) + g(x - ct)$$

for a constant c. Show that $\dfrac{\partial^2 u}{\partial t^2} = c^2 \dfrac{\partial^2 u}{\partial x^2}$

Hint: Let $r = x + ct; s = x - ct.$

56. Suppose $z = f(x, y)$ has continuous second-order partial derivatives. If $x = e^r \cos\theta$ and $y = e^r \sin\theta$, show that

$$\frac{\partial^2 z}{\partial x^2} + \frac{\partial^2 z}{\partial y^2} = e^{-2r}\left[\frac{\partial^2 z}{\partial r^2} + \frac{\partial^2 z}{\partial \theta^2}\right]$$

57. Let $F(x, y, z)$ be a function of three variables with continuous partial derivatives F_x, F_y, F_z in a certain region where $F(x, y, z) = C$ for some constant C. Use the chain rule for two parameters and the fact that $\partial y/\partial x = 0$ to show that

$$\frac{\partial z}{\partial x} = \frac{-\dfrac{\partial F}{\partial x}}{\dfrac{\partial F}{\partial z}} = \frac{-F_x}{F_z}$$

58. Suppose the system

$$\begin{cases} xu + yv - uv = 0 \\ yu - xv + uv = 0 \end{cases}$$

can be solved for u and v in terms of x and y, so that $u = u(x, y)$ and $v = v(x, y)$. Use implicit differentiation to find the partial derivatives $\dfrac{\partial u}{\partial x}$ and $\dfrac{\partial v}{\partial x}.$

59. Repeat Problem 58 for the system

$$\begin{cases} xu + yv - uv = 0 \\ yu - xv + uv = 0 \end{cases}$$

for $\dfrac{\partial u}{\partial y}$ and $\dfrac{\partial v}{\partial y}.$

60. Suppose that F and G are functions of three variables and that it is possible to solve the equations $F(x, y, z) = 0$ and $G(x, y, z) = 0$ for y and z in terms of x, so that $y = y(x)$ and $z = z(x)$. Use the chain rule to express dy/dx and dz/dx in terms of the partial derivatives of F and G. Assume these partials are continuous and that

$$\frac{\partial F}{\partial y}\frac{\partial G}{\partial z} \neq \frac{\partial F}{\partial z}\frac{\partial G}{\partial y}$$

12.6 *Directional Derivatives and the Gradient*

the directional derivative, directional derivative as a slope and as a rate, the gradient, maximal property of the gradient, normal property of the gradient, tangent planes, and normal lines ■

In Section 12.3, we defined the partial derivatives f_x and f_y and interpreted these derivatives as slopes of tangent lines on the surface $z = f(x, y)$ in planes parallel to xz- and yz-planes, respectively, as shown in Figure 12.19.

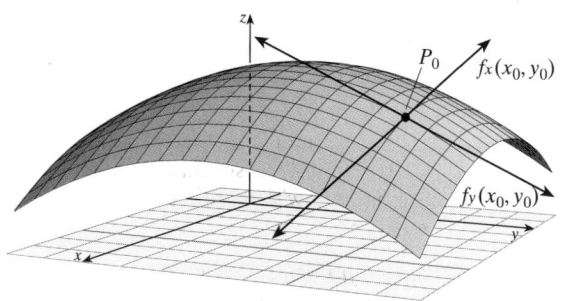

■ **FIGURE 12.19** Partial derivatives of $z = f(x, y)$

Our next goal is to see how partial derivatives can be used to find derivatives in other directions.

THE DIRECTIONAL DERIVATIVE

In Chapter 3 we defined the *slope of a curve* at a point to be the ratio of the change in the dependent variable with respect to the change in the independent variable at the given point. To determine the slope of the tangent line at a point $P_0(x_0, y_0)$ on a surface defined by $z = f(x, y)$, we need to specify the *direction* in which we wish to measure. We do this by using vectors. In Section 12.3 we found the slope parallel to the xz-plane to be the partial derivative $f_x(x_0, y_0)$. We could have specified this direction in terms of the vector **i** (x-direction). Similarly, $f_y(x, y)$ could have been specified in terms of the **j** vector. Finally, to measure the slope of the tangent line in an *arbitrary* direction, we use a unit vector $\mathbf{u} = u_1\mathbf{i} + u_2\mathbf{j}$ in that direction.

To find the desired slope, we look at the intersection of the surface with the vertical plane passing through the point P_0 parallel to the vector **u**, as shown in Figure 12.20. This vertical plane intersects the surface to form a curve C, and we define the slope of the surface at P_0 in the direction of **u** to be the slope of the tangent line to the curve C defined by **u** at that point.

We summarize this idea of slope *in a particular direction* with the following definition.

WARNING ▶ Remember, **u** must be a *unit* vector. ◂

> ### Directional Derivative
>
> Let f be a function of two variables, and let $\mathbf{u} = u_1\mathbf{i} + u_2\mathbf{j}$ be a unit vector. The **directional derivative of f at $P_0(x_0, y_0)$ in the direction of u** is given by
>
> $$D_u f(x_0, y_0) = \lim_{h \to 0} \frac{f(x_0 + hu_1, y_0 + hu_2) - f(x_0, y_0)}{h}$$
>
> provided the limit exists.

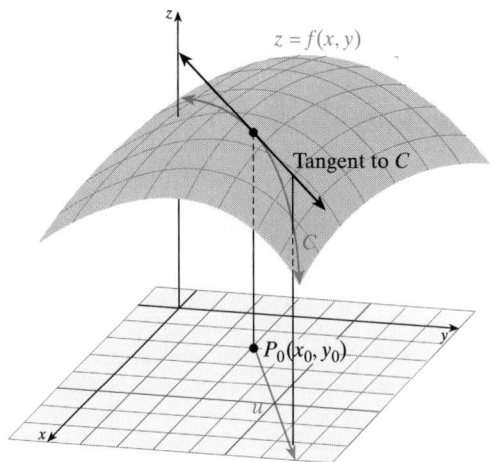

■ **FIGURE 12.20**   The directional derivative $D_u f(x_0, y_0)$ is the slope of the tangent line to the curve on the surface $z = f(x, y)$ in the direction of the unit vector **u** at $P_0(x_0, y_0)$.

EXAMPLE 1	*Computing partial derivatives as directional derivatives*

Use the definition of directional derivative to show that the directional derivatives of $f(x, y)$ in the directions of the positive x- and y-axes are f_x and f_y, respectively.

Solution   In the direction of the positive x-axis, $\mathbf{u} = \mathbf{i}$, so that $u_1 = 1$ and $u_2 = 0$.

$$D_i f(x_0, y_0) = \lim_{h \to 0} \frac{f(x_0 + h, y_0) - f(x_0, y_0)}{h} \qquad \text{Definition of directional derivative, where } u_1 = 1 \text{ and } u_2 = 0$$

$$= f_x(x_0, y_0) \qquad \text{Definition of } f_x$$

In the direction of the positive y-axis, $\mathbf{u} = \mathbf{j}$, so that $u_1 = 0$ and $u_2 = 1$.

$$D_j f(x_0, y_0) = \lim_{h \to 0} \frac{f(x_0, y_0 + h) - f(x_0, y_0)}{h} \qquad \text{Definition of directional derivative, where } u_1 = 0 \text{ and } u_2 = 1$$

$$= f_y(x_0, y_0) \qquad \text{Definition of } f_y \qquad ■$$

The definition of directional derivative is similar to the definition of the derivative of a function of a single variable. Just as with a single variable, it is difficult to apply the definition directly. Fortunately, the following theorem allows us to find directional derivatives more efficiently than by using the definition.

THEOREM 12.6   *Directional derivatives using partials*

Let $f(x, y)$ be a function that is differentiable at $P_0(x_0, y_0)$. Then f has a directional derivative in the direction of the unit vector $\mathbf{u} = u_1 \mathbf{i} + u_2 \mathbf{j}$ given by

$$D_u f(x_0, y_0) = f_x(x_0, y_0)u_1 + f_y(x_0, y_0)u_2$$

Proof   We define a function F of a single variable h by

$$F(h) = f(x_0 + hu_1, y_0 + hu_2)$$

so that

$$D_u f(x_0, y_0) = \lim_{h \to 0} \frac{f(x_0 + hu_1, y_0 + hu_2) - f(x_0, y_0)}{h}$$

$$= \lim_{h \to 0} \frac{F(h) - F(0)}{h} = F'(0)$$

Apply the chain rule with $x = x_0 + hu_1$ and $y = y_0 + hu_2$:

$$F'(h) = \frac{dF}{dh} = \frac{\partial f}{\partial x}\frac{dx}{dh} + \frac{\partial f}{\partial y}\frac{dy}{dh} = f_x(x, y)u_1 + f_y(x, y)u_2$$

When $h = 0$, we have $x = x_0$ and $y = y_0$, so that

$$D_u f(x_0, y_0) = F'(0) = \frac{\partial f}{\partial x}u_1 + \frac{\partial f}{\partial y}u_2 = f_x(x_0, y_0)u_1 + f_y(x_0, y_0)u_2 \qquad \equiv$$

EXAMPLE 2   *Finding a directional derivative using partials*

Find the directional derivative of $f(x, y) = 3 - 2x^2 + y^3$ at the point $P(1, 2)$ in the direction of the unit vector $\mathbf{u} = \frac{1}{2}\mathbf{i} - \frac{\sqrt{3}}{2}\mathbf{j}$.

Solution   First find the partial derivatives $f_x(x, y) = -4x$ and $f_y(x, y) = 3y^2$. Then because $u_1 = \frac{1}{2}$ and $u_2 = -\frac{\sqrt{3}}{2}$, we have

$$D_u f(1, 2) = f_x(1, 2)\left(\frac{1}{2}\right) + f_y(1, 2)\left(-\frac{\sqrt{3}}{2}\right)$$

$$= -4(1)\left(\frac{1}{2}\right) + 3(2)^2\left(-\frac{\sqrt{3}}{2}\right) = -2 - 6\sqrt{3} \approx -12.4 \qquad \blacksquare$$

DIRECTIONAL DERIVATIVE AS A SLOPE AND AS A RATE

Notice that the directional derivative is a number. This number can be interpreted as the slope of a tangent line to $z = f(x, y)$ or as a rate of change of the function $z = f(x, y)$.

Directional Derivative as a Slope   As an illustration, we look at the surface defined by $z = 3 - 2x^2 + y^3$ (see Figure 12.21). The intersection of this surface with a plane aligned with the vector $\mathbf{u} = \frac{1}{2}\mathbf{i} - \frac{\sqrt{3}}{2}\mathbf{j}$ is a curve labeled C, and the directional derivative is the slope of the tangent line to C at the point on the surface above $P(1, 2)$. This interpretation of the directional derivative is illustrated in Figure 12.21.

Directional Derivative as a Rate   The directional derivative $D_u f(x_0, y_0)$ can also be interpreted as the rate at which the function $z = f(x, y)$ changes as a point moves from $P_0(x_0, y_0)$ in the direction of the unit vector $\mathbf{u}$. Thus, in Example 2, we say that the function $f(x, y) = 3 - 2x^2 + y^3$ changes at the rate of $-2 - 6\sqrt{3}$ as a point moves from $P_0(1, 2)$ in the direction of the unit vector $\mathbf{u} = \frac{1}{2}\mathbf{i} - \frac{\sqrt{3}}{2}\mathbf{j}$.

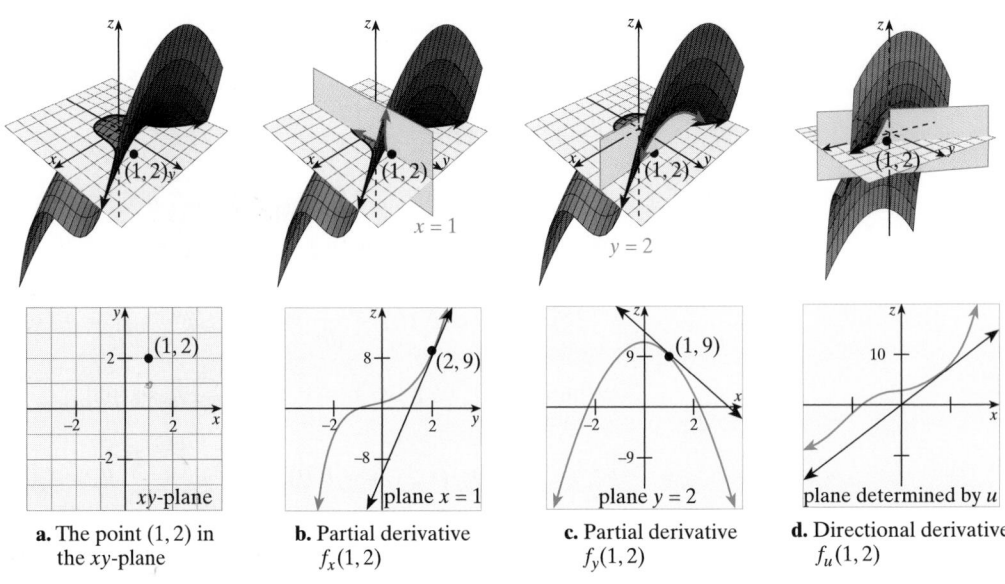

a. The point $(1, 2)$ in the xy-plane

b. Partial derivative $f_x(1, 2)$

c. Partial derivative $f_y(1, 2)$

d. Directional derivative $f_u(1, 2)$

■ **FIGURE 12.21** Graph of $z = 3 - 2x^2 + y^3$ and some derivatives

THE GRADIENT

The directional derivative $D_u f(x, y)$ can be expressed concisely in terms of a vector function called the *gradient,* which has many important uses in mathematics. The gradient of a function of two variables may be defined as follows.

> **Gradient**
>
> Let f be a differentiable functions at (x, y) and let $f(x, y)$ have partial derivatives $f_x(x, y)$ and $f_y(x, y)$. Then the **gradient** of f, denoted by ∇f (pronounced "del eff"), is a vector given by
>
> $$\nabla f(x, y) = f_x(x, y)\mathbf{i} + f_y(x, y)\mathbf{j}$$
>
> The value of the gradient at the point $P_0(x_0, y_0)$ is denoted by
>
> $$\nabla f_0 = f_x(x_0, y_0)\mathbf{i} + f_y(x_0, y_0)\mathbf{j}$$

WARNING Think of the symbol ∇ as an "operator" on a function that produces a vector. Another notation for ∇f is **grad** $f(x, y)$.

EXAMPLE 3 *Finding the gradient of a given function*

Find $\nabla f(x, y)$ for $f(x, y) = x^2 y + y^3$.

Solution Begin with the partial derivatives:

$$f_x(x, y) = \frac{\partial}{\partial x}(x^2 y + y^3) = 2xy \quad \text{and} \quad f_y(x, y) = \frac{\partial}{\partial y}(x^2 y + y^3) = x^2 + 3y^2$$

Then,

$$\nabla f(x, y) = 2xy\mathbf{i} + (x^2 + 3y^2)\mathbf{j} \qquad ■$$

The following theorem shows how the directional derivative can be expressed in terms of the gradient.

THEOREM 12.7   *The gradient formula for the directional derivative*

If f is a differentiable function of x and y, then the directional derivative at the point $P_0(x_0, y_0)$ in the direction of the unit vector $\mathbf{u}$ is

$$D_u f(x_0, y_0) = \nabla f_0 \cdot \mathbf{u}$$

Proof   Because $\nabla f_0 = f_x(x_0, y_0)\mathbf{i} + f_y(x_0, y_0)\mathbf{j}$ and $\mathbf{u} = u_1\mathbf{i} + u_2\mathbf{j}$, we have

$$D_u f(x_0, y_0) = \nabla f_0 \cdot \mathbf{u} = f_x(x_0, y_0)u_1 + f_y(x_0, y_0)u_2 \qquad \text{Dot product}$$

EXAMPLE 4   *Using the gradient formula to compute a directional derivative*

Find the directional derivative of $f(x, y) = \ln(x^2 + y^3)$ at $P_0(1, -3)$ in the direction of $\mathbf{v} = 2\mathbf{i} - 3\mathbf{j}$.

Solution
$$f_x(x, y) = \frac{2x}{x^2 + y^3}, \quad \text{so} \quad f_x(1, -3) = -\frac{2}{26}$$

$$f_y(x, y) = \frac{3y^2}{x^2 + y^3}, \quad \text{so} \quad f_y(1, -3) = -\frac{27}{26}$$

$$\nabla f_0 = \nabla f(1, -3) = -\frac{2}{26}\mathbf{i} - \frac{27}{26}\mathbf{j}$$

A unit vector in the direction of $\mathbf{v}$ is

$$\mathbf{u} = \frac{\mathbf{v}}{\|\mathbf{v}\|} = \frac{2\mathbf{i} - 3\mathbf{j}}{\sqrt{2^2 + (-3)^2}} = \frac{1}{\sqrt{13}}(2\mathbf{i} - 3\mathbf{j})$$

Thus,

$$D_u(x, y) = \nabla f \cdot \mathbf{u} = \left(-\frac{2}{26}\right)\left(\frac{2}{\sqrt{13}}\right) + \left(-\frac{27}{26}\right)\left(-\frac{3}{\sqrt{13}}\right) = \frac{77\sqrt{13}}{338} \approx 0.82 \qquad \blacksquare$$

Although a differentiable function of one variable $f(x)$ has exactly one derivative $f'(x)$, a differentiable function of two variables $F(x, y)$ has two partial derivatives and a multitude of directional derivatives. Is there any single mathematical entity for functions of several variables that is the analogue of the derivative of a function of a single variable? The properties listed in the following theorem suggest that the gradient is the analogue we seek.

THEOREM 12.8   *Basic properties of the gradient*

Let f and g be differentiable functions. Then

Constant rule	$\nabla c = \mathbf{0}$ for any constant c
Linearity rule	$\nabla(af + bg) = a\nabla f + b\nabla g$ for constants a and b
Product rule	$\nabla(fg) = f\nabla g + g\nabla f$
Quotient rule	$\nabla\left(\dfrac{f}{g}\right) = \dfrac{g\nabla f - f\nabla g}{g^2}, \quad g \neq 0$
Power rule	$\nabla(f^n) = nf^{(n-1)}\nabla f$

Proof   **Linearity rule**
$$\nabla(af + bg) = (af + bg)_x\mathbf{i} + (af + bg)_y\mathbf{j} = (af_x + bg_x)\mathbf{i} + (af_y + bg_y)\mathbf{j}$$
$$= af_x\mathbf{i} + bg_x\mathbf{i} + af_y\mathbf{j} + bg_y\mathbf{j} = a(f_x\mathbf{i} + f_y\mathbf{j}) + b(g_x\mathbf{i} + g_y\mathbf{j})$$
$$= a\nabla f + b\nabla g$$

Power rule

$$\nabla f^n = [f^n]_x\,\mathbf{i} + [f^n]_y\,\mathbf{j} = nf^{n-1}f_x\mathbf{i} + nf^{n-1}f_y\mathbf{j}$$
$$= nf^{n-1}[f_x\mathbf{i} + f_y\mathbf{j}] = nf^{n-1}\nabla f$$

The other rules are left for the problem set (Problems 58 and 59).

MAXIMAL PROPERTY OF THE GRADIENT

In applications, it is often useful to compute the greatest rate of increase (or decrease) of a given function at a specified point. The direction in which this occurs is called the direction of **steepest ascent** (or **steepest descent**). For example, suppose the function $z = f(x, y)$ gives the altitude of a skier coming down a slope, and we want to state a theorem that will give the skier the *compass direction* of the path of steepest descent (see Figure 12.22b). We emphasize the words "compass direction" because the gradient gives direction in the xy-plane and does not itself point up or down the mountain. The following theorem shows how the optimal direction is determined by the gradient (see Figure 12.22).

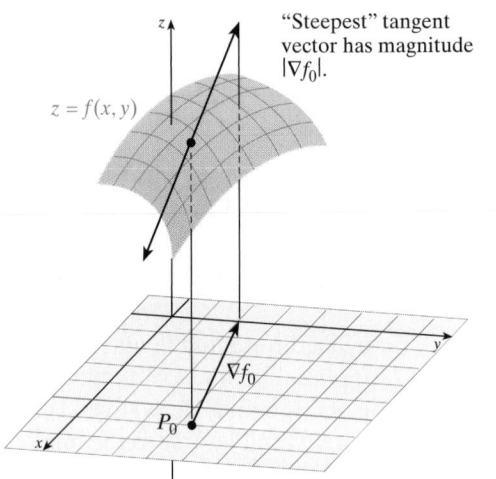

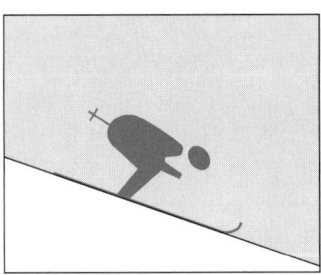

a. The optimal direction property of the gradient **b.** Skier on a slope

■ **FIGURE 12.22** Steepest ascent or steepest descent

THEOREM 12.9   *Optimal direction property of the gradient*

Suppose f is differentiable and let ∇f_0 denote the gradient at P_0. Then if $\nabla f_0 \neq \mathbf{0}$:

a. The largest value of the directional derivative of $D_u f$ is $\|\nabla f_0\|$ and occurs when the unit vector $\mathbf{u}$ points in the direction of ∇f_0.

b. The smallest value of $D_u f$ is $-\|\nabla f_0\|$ and occurs when $\mathbf{u}$ points in the direction of $-\nabla f_0$.

Proof   If $\mathbf{u}$ is any unit vector, then

$$D_u f = \nabla f_0 \cdot \mathbf{u} = \|\nabla f_0\|(\|\mathbf{u}\|\cos\theta) = \|\nabla f_0\|\cos\theta$$

where θ is the angle between ∇f_0 and $\mathbf{u}$. But $\cos\theta$ assumes its largest value 1 at $\theta = 0$—that is, when $\mathbf{u}$ points in the direction ∇f_0. Thus, the largest possible value of $D_u f$ is

$$D_u f = \|\nabla f_0\|(1) = \|\nabla f_0\|$$

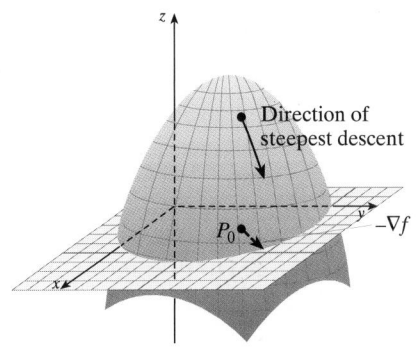

The gradient of f is the vector ∇f in the xy-plane. This vector points in the direction of steepest ascent or descent at a given point.

Statement **b** may be established in a similar fashion by noting that $\cos \theta$ assumes its smallest value -1 when $\theta = \pi$. This value occurs when **u** points toward $-\nabla f_0$, and in this direction

$$D_u f = \|\nabla f_0\|(-1) = -\|\nabla f_0\|.$$

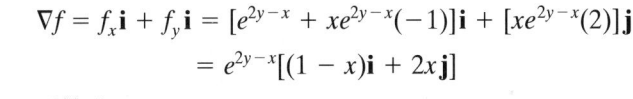

> ■ *What This Says:*   The theorem states that at P_0, the function f increases most rapidly in the direction of the gradient ∇f_0 and decreases most rapidly in the opposite direction.

EXAMPLE 5   *Optimal rate of increase and decrease*

In what direction is the function defined by $f(x, y) = xe^{2y-x}$ increasing most rapidly at the point $P_0(2, 1)$, and what is the optimal rate of increase? In what direction is f decreasing most rapidly?

Solution   We begin by finding the gradient of f:

$$\nabla f = f_x \mathbf{i} + f_y \mathbf{i} = [e^{2y-x} + xe^{2y-x}(-1)]\mathbf{i} + [xe^{2y-x}(2)]\mathbf{j}$$
$$= e^{2y-x}[(1 - x)\mathbf{i} + 2x\mathbf{j}]$$

At $(2, 1)$, $\nabla f_0 = e^{2(1)-2}[(1 - 2)\mathbf{i} + 2(2)\mathbf{j}] = -\mathbf{i} + 4\mathbf{j}$. The most rapid rate of increase is $\|\nabla f_0\| = \sqrt{(-1)^2 + (4)^2} = \sqrt{17}$ and it occurs in the direction of $-\mathbf{i} + 4\mathbf{j}$. The most rapid rate of decrease occurs in the direction of $-\nabla f_0 = \mathbf{i} - 4\mathbf{j}$.   ■

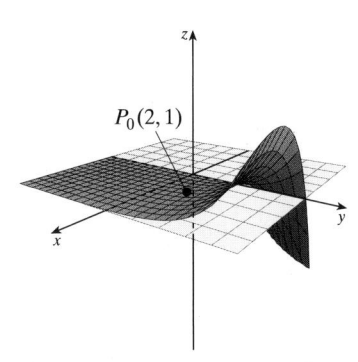

Surface defined by $f(x, y) = xe^{2y-x}$

NORMAL PROPERTY OF THE GRADIENT

Suppose S is a level surface of the function defined by $f(x, y, z)$; that is, $f(x, y, z) = K$ for some constant K. Then if $P_0(x_0, y_0, z_0)$ is a point on S, the following theorem shows that the gradient ∇f_0 at P_0 is a vector that is **normal** (that is, perpendicular) to the tangent vector of every smooth curve on S that passes through P_0 (see Figure 12.23).

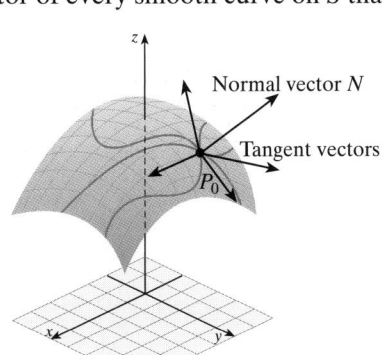

■ **FIGURE 12.23**   The normal property of the gradient

=====

THEOREM 12.10   *The normal property of the gradient*

Suppose the function f is differentiable at the point P_0 and that the gradient at P_0 satisfies $\nabla f_0 \neq \mathbf{0}$. If $K = f(x_0, y_0, z_0)$, then ∇f_0 is orthogonal to the level surface $f(x, y, z) = K$ at P_0.

Proof   Let C be any smooth curve on the level surface $f(x, y, z) = K$ that passes through $P_0(x_0, y_0, z_0)$, and describe the curve C by the vector function $\mathbf{R}(t) = x(t)\mathbf{i} + y(t)\mathbf{j} + z(t)\mathbf{k}$ for all t in some interval I. We shall show that the gradient ∇f_0 is orthogonal to the tangent vector $d\mathbf{R}/dt$ at P_0.

Because C lies on the level surface, any point $P(x(t), y(t), z(t))$ on C must satisfy $f[x(t), y(t), z(t)] = K$, and by applying the chain rule, we obtain

$$\frac{d}{dt}[f(x(t), y(t), z(t))] = f_x(x, y, z)\frac{dx}{dt} + f_y(x, y, z)\frac{dy}{dt} + f_z(x, y, z)\frac{dz}{dt}$$

Suppose $t = t_0$ at P_0. Then

$$\frac{d}{dt}[f(x(t), y(t), z(t))]\bigg|_{t=t_0}$$

$$= f_x(x(t_0), y(t_0), z(t_0))\frac{dx}{dt} + f_y(x(t_0), y(t_0), z(t_0))\frac{dy}{dt} + f_z(x(t_0), y(t_0), z(t_0))\frac{dz}{dt}$$

$$= \nabla f_0 \cdot \frac{d\mathbf{R}}{dt}$$

Remember that $\dfrac{d\mathbf{R}}{dt} = \dfrac{dx}{dt}\mathbf{i} + \dfrac{dy}{dt}\mathbf{j} + \dfrac{dz}{dt}\mathbf{k}$. We also know that $f(x(t), y(t), z(t)) = K$ for all t in I (because the curve C lies on the level surface $f(x, y, z) = K$). Thus, we have

$$\frac{d}{dt}\{f[x(t), y(t), z(t)]\} = \frac{d}{dt}(K) = 0$$

and it follows that $\nabla f_0 \cdot \dfrac{d\mathbf{R}}{dt} = 0$. We are given that $\nabla f_0 \neq \mathbf{0}$, and $d\mathbf{R}/dt \neq \mathbf{0}$ because the curve C is smooth. Thus, ∇f_0 is orthogonal to $d\mathbf{R}/dt$, as required.   ▬

■ *What This Says:*   The gradient ∇f_0 at each point P_0 on the level surface $f(x, y, z) = K$ is normal to the surface at P_0. That is, ∇f_0 is orthogonal to all tangent lines to curves on $f(x, y, z) = K$ through P_0.

Here is an example in which f involves only two variables, so $f(x, y) = K$ is a level curve in the plane instead of a level surface in space.

EXAMPLE 6   *Finding a vector normal to a level curve*

Sketch the level curve corresponding to $C = 1$ for the function $f(x, y) = x^2 - y^2$ and find a normal vector at the point $P_0(2, \sqrt{3})$.

Solution   The level curve for $C = 1$ is a hyperbola given by $x^2 - y^2 = 1$, as shown in Figure 12.24 (along with the surface defined by $f(x, y) = x^2 - y^2$).

The gradient vector is perpendicular to the level curve, so we have

$$\nabla f = f_x\mathbf{i} + f_y\mathbf{j} = 2x\mathbf{i} - 2y\mathbf{j}$$

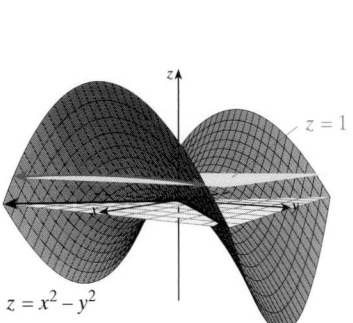

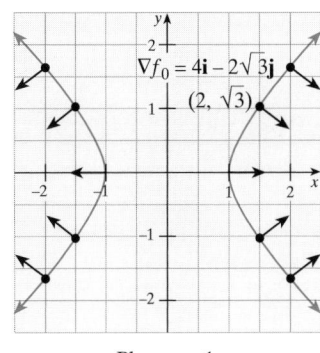

$\nabla f_0 = 4\mathbf{i} - 2\sqrt{3}\mathbf{j}$

$(2, \sqrt{3})$

Plane $z = 1$

■ **FIGURE 12.24**   The surface $f(x, y) = x^2 - y^2$ and the level curve for which $C = 1$

At the point $(2, \sqrt{3})$, $\nabla f_0 = 4\mathbf{i} - 2\sqrt{3}\mathbf{j}$ is the required normal. This normal vector and a few others are shown in Figure 12.24.   ■

EXAMPLE 7   *Finding a vector that is normal to a level surface*

Find a vector that is normal to the level surface $x^2 + 2xy - yz + 3z^2 = 7$ at the point $P_0(1, 1, -1)$.

Solution   The gradient vector at P_0 is perpendicular to the level surface:

$$\nabla f = f_x\mathbf{i} + f_y\mathbf{j} + f_z\mathbf{k} = (2x + 2y)\mathbf{i} + (2x - z)\mathbf{j} + (6z - y)\mathbf{k}$$

At the point $(1, 1, -1)$, $\nabla f_0 = 4\mathbf{i} + 3\mathbf{j} - 7\mathbf{k}$ is the required normal.   ■

EXAMPLE 8   *Heat flow application*

The set of points (x, y) with $0 \le x \le 5$ and $0 \le y \le 5$ is a square in the first quadrant of the xy-plane. Suppose this square is heated in such a way that $T(x, y) = x^2 + y^2$ is the temperature at the point $P(x, y)$. In what direction will heat flow from the point $P_0(3, 4)$?

Solution   The flow of heat in the region is given by a vector function $\mathbf{H}(x, y)$, whose value at each point (x, y) depends on x and y. From physics it is known that $\mathbf{H}(x, y)$ will be perpendicular to the isothermal curves $T(x, y) = C$ for C constant. The gradient ∇T and all its multiples point in such a direction. Therefore, we can express the heat flow as $\mathbf{H} = -k\nabla T$, where k is a positive constant (called the **thermal conductivity**) and the negative sign is introduced to account for the fact that heat flows "downhill" (that is, toward a decreasing temperature).

Because $T(3, 4) = 25$, the point $P_0(3, 4)$ lies on the isotherm $T(x, y) = 25$, which is part of the circle $x^2 + y^2 = 25$, as shown in Figure 12.25. We know that the heat flow $\mathbf{H}_0$ at P_0 will satisfy $\mathbf{H}_0 = -k\nabla T_0$, where ∇T_0 is the gradient at P_0. Because $\nabla T = 2x\mathbf{i} + 2y\mathbf{j}$, we see that $\nabla T_0 = 6\mathbf{i} + 8\mathbf{j}$. Thus, the heat flow at P_0 satisfies

$$\mathbf{H}_0 = -k\nabla T_0 = -k(6\mathbf{i} + 8\mathbf{j})$$

Because the thermal conductivity k is positive, we can say that heat flows from P_0 in the direction of the unit vector $\mathbf{u}$ given by

$$\mathbf{u} = \frac{-(6\mathbf{i} + 8\mathbf{j})}{\sqrt{(-6)^2 + (-8)^2}} = -\frac{3}{5}\mathbf{i} - \frac{4}{5}\mathbf{j}$$ ■

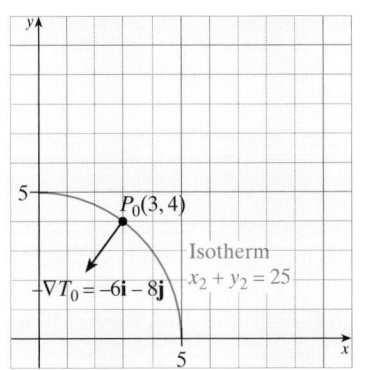

$P_0(3, 4)$

Isotherm $x^2 + y^2 = 25$

$\nabla T_0 = -6\mathbf{i} - 8\mathbf{j}$

■ **FIGURE 12.25**   The heat flow P_0 is in the direction of $-\nabla T_0 = -6\mathbf{i} - 8\mathbf{j}$.

The directional derivative and gradient concepts can easily be extended to functions of three or more variables. For a function of three variables, $f(x, y, z)$, the gradient ∇f is defined by

$$\nabla f = f_x \mathbf{i} + f_y \mathbf{j} + f_z \mathbf{k}$$

and the directional derivative $D_u f$ of $f(x, y, z)$ at $P_0(x_0, y_0, z_0)$ in the direction of the unit vector $\mathbf{u}$ is given by

$$D_u f = \nabla f_0 \cdot \mathbf{u}$$

where, as before, ∇f_0 is the gradient ∇f evaluated at P_0. The basic properties of the gradient (Theorem 12.8) are still valid, as are the optimization properties in Theorem 12.9. Similar definitions and properties are valid for functions of more than three variables.

> **EXAMPLE 9**   *Directional derivative of a function of three variables*

Let $f(x, y, z) = xy \sin xz$. Find ∇f_0 at the point $P_0(1, -2, \pi)$ and then compute the directional derivative of f at P_0 in the direction of the vector $\mathbf{v} = -2\mathbf{i} + 3\mathbf{j} - 5\mathbf{k}$.

Solution   Begin with the partial derivatives:

$$f_x = y \sin xz + xy(z \cos xz); \quad f_x(1, -2, \pi) = -2 \sin \pi - 2\pi \cos \pi = 2\pi$$
$$f_y = x \sin xz; \quad f_y(1, -2, \pi) = 1 \sin \pi = 0$$
$$f_z = xy(x \cos xz); \quad f_z(1, -2, \pi) = (1)(-2)(1)\cos \pi = 2$$

Thus, the gradient of f at P_0 is

$$\nabla f_0 = 2\pi \mathbf{i} + 2\mathbf{k}$$

To find $D_u f$ we need $\mathbf{u}$, the unit vector in the direction of $\mathbf{v}$:

$$\mathbf{u} = \frac{\mathbf{v}}{\|\mathbf{v}\|} = \frac{-2\mathbf{i} + 3\mathbf{j} - 5\mathbf{k}}{\sqrt{(-2)^2 + (3)^2 + (-5)^2}} = \frac{1}{\sqrt{38}}(-2\mathbf{i} + 3\mathbf{j} - 5\mathbf{k})$$

Finally,

$$D_u f(1, -2, \pi) = \nabla f_0 \cdot \mathbf{u} = \frac{1}{\sqrt{38}}(-4\pi - 10) \approx -3.66 \quad \blacksquare$$

TANGENT PLANES AND NORMAL LINES

Tangent planes and normal lines to a surface are the natural extensions to $\mathbb{R}^3$ of the tangent and normal lines we examined in $\mathbb{R}^2$. Suppose S is a surface and $\mathbf{N}$ is a vector normal to S at the point P_0. We would intuitively expect the normal line and the tangent plane to S at P_0 to be, respectively, the line through P_0 with the direction of $\mathbf{N}$ and the plane perpendicular to the line at P_0 (see Figure 12.26).

These observations lead us to the following definition.

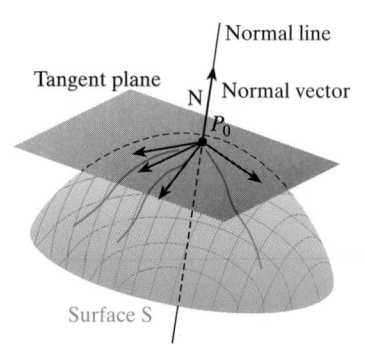

■ **FIGURE 12.26**   Tangent plane and normal line

> ### Normal Line to a Surface; Tangent Plane
>
> Suppose the surface S has a nonzero normal vector $\mathbf{N}$ at the point P_0. Then the line through P_0 parallel to $\mathbf{N}$ is called the **normal line** to S at P_0, and the plane through P_0 with normal vector $\mathbf{N}$ is the **tangent plane** to S at P_0.

By analogy with the single-variable case, we would expect a surface S to have a tangent plane precisely where it can be represented by a differentiable function. In particular, if S has an equation of the form $F(x, y, z) = C$, where C is a constant and F is a function differentiable at P_0, the normal property of a gradient tells us that the gradient ∇F_0 at P_0 is normal to S (if $\nabla F_0 \neq \mathbf{0}$) and that S must therefore have a tangent plane at P_0.

EXAMPLE 10   *Finding the tangent plane and normal line to a given surface*

Find equations for the tangent plane and the normal line at the point $P_0(1, -1, 2)$ on the surface S given by $x^2 y + y^2 z + z^2 x = 5$.

Solution   We need to rewrite this problem so that the normal property of the gradient theorem applies. Let $F(x, y, z) = x^2 y + y^2 z + z^2 x$, and consider S to be the level surface $F(x, y, z) = 5$. The gradient ∇F is normal to S at P_0. We find that

$$\nabla F(x, y, z) = (2xy + z^2)\mathbf{i} + (x^2 + 2yz)\mathbf{j} + (y^2 + 2xz)\mathbf{k}$$

so the normal vector at P_0 is

$$\mathbf{N} = \nabla F_0 = \nabla F(1, -1, 2) = 2\mathbf{i} - 3\mathbf{j} + 5\mathbf{k}$$

Hence, the required tangent plane is

$$2(x - 1) - 3(y + 1) + 5(z - 2) = 0 \quad \text{or} \quad 2x - 3y + 5z = 15$$

The normal line to the surface at P_0 is

$$x = 1 + 2t, \quad y = -1 - 3t, \quad z = 2 + 5t \qquad \blacksquare$$

By generalizing the procedure illustrated in the preceding example, we are led to the following formulas for the tangent plane and normal line. (Also see Figure 12.27.)

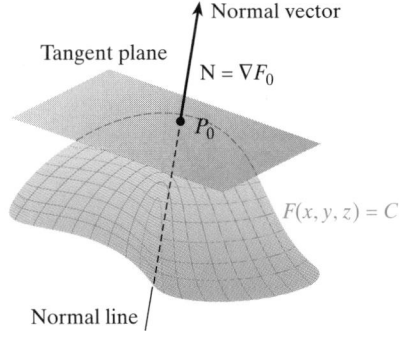

Normal vector

Tangent plane

$\mathbf{N} = \nabla F_0$

P_0

$F(x, y, z) = C$

Normal line

■ **FIGURE 12.27**   The tangent plane and normal line to a surface

---

Formulas for the Tangent Plane and Normal Line to a Surface

Suppose S is a surface with the equation $F(x, y, z) = C$ and let $P_0(x_0, y_0, z_0)$ be a point on S where F is differentiable. Then the **equation of the tangent plane** to S at P_0 is

$$F_x(x_0, y_0, z_0)(x - x_0) + F_y(x_0, y_0, z_0)(y - y_0) + F_z(x_0, y_0, z_0)(z - z_0) = 0$$

and the **equation of the normal line** to S at P_0 is

$$x = x_0 + F_x(x_0, y_0, z_0)t$$
$$y = y_0 + F_y(x_0, y_0, z_0)t$$
$$x = z_0 + F_z(x_0, y_0, z_0)t$$

provided F_x, F_y, and F_z are not all zero.

WARNING   Note that in the special case where $z = f(x, y)$, we have $F(x, y, z) = f(x, y) - z = 0$. Then $F_x = f_x, F_y = f_y$, and $F_z = -1$, and the equation of the tangent plane becomes

$$f_x(x_0, y_0, z_0)(x - x_0) + f_y(x_0, y_0, z_0)(y - y_0) - (z - z_0) = 0$$

which is equivalent to the tangent plane formula given in Section 12.4.

EXAMPLE 11   *Equations of the tangent plane and the normal line*

Find the equations for the tangent plane and the normal line to the cone $z^2 = x^2 + y^2$ at the point where $x = 3, y = 4$, and $z > 0$.

Solution   If $P_0(x_0, y_0, z_0)$ is the point of tangency and $x_0 = 3, y_0 = 4$, and $z_0 > 0$, then

$$z_0 = \sqrt{x_0^2 + y_0^2} = \sqrt{9 + 16} = 5$$

If we consider $F(x, y, z) = x^2 + y^2 - z^2$, then the cone can be regarded as the level surface $F(x, y, z) = 0$. The partial derivatives of F are

$$F_x = 2x, \qquad F_y = 2y, \qquad F_z = -2z$$

so at $P_0(3, 4, 5)$,

$$F_x(3, 4, 5) = 6, \qquad F_y(3, 4, 5) = 8, \qquad F_z(3, 4, 5) = -10$$

Thus the tangent plane is

$$6(x - 3) + 8(y - 4) - 10(z - 5) = 0$$

or $3x + 4y - 5z = 0$, and the normal line is

$$x = 3 + 6t, \qquad y = 4 + 8t, \qquad z = 5 - 10t$$

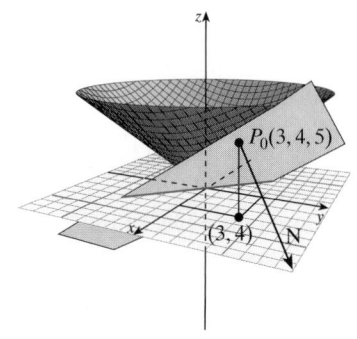

Tangent plane and normal line to the cone $z^2 = x^2 + y^2$ at $(3, 4, 5)$

12.6   Problem Set

A *Find the gradient of the functions given in Problems 1–10.*

1. $f(x, y) = x^2 - 2xy$
2. $f(x, y) = 3x + 4y^2$
3. $f(x, y) = \dfrac{y}{x} + \dfrac{x}{y}$
4. $f(x, y) = \ln(x^2 + y^2)$
5. $f(u, v) = ue^{3-v}$
6. $f(u, v) = e^{u+v}$
7. $f(x, y) = \sin(x + 2y)$
8. $f(x, y, z) = xyz^2$
9. $g(x, y, z) = xe^{y+3z}$
10. $f(x, y, z) = \dfrac{xy - 1}{z + x}$

*Compute the directional derivative of the functions given in Problems 11–16 at the point P_0 in the direction of the given vector **v**.*

Function	Point P_0	Vector **v**
11. $f(x, y) = x^2 + xy$	$(1, -2)$	$\mathbf{i} + \mathbf{j}$
12. $f(x, y) = \dfrac{e^{-x}}{y}$	$(2, -1)$	$-\mathbf{i} + \mathbf{j}$
13. $f(x, y) = \ln(x^2 + 3y)$	$(1, 1)$	$\mathbf{i} + \mathbf{j}$
14. $f(x, y) = \ln(3x + y^2)$	$(0, 1)$	$\mathbf{i} - \mathbf{j}$
15. $f(x, y) = \sec(xy - y^3)$	$(2, 0)$	$-\mathbf{i} - 3\mathbf{j}$
16. $f(x, y) = \sin xy$	$(\sqrt{\pi}, \sqrt{\pi})$	$3\pi\mathbf{i} - \pi\mathbf{j}$

Find a unit vector (if possible) that is normal to each surface given in Problems 17–24 at the prescribed point, and then find the equation of the tangent plane at the given point.

17. $x^2 + y^2 + z^2 = 3$ at $(1, -1, 1)$
18. $x^4 + y^4 + z^4 = 3$ at $(1, -1 - 1)$

19. $\cos z = \sin(x + y)$ at $(\frac{\pi}{2}, \frac{\pi}{2}, \frac{\pi}{2})$
20. $\sin(x + y) + \tan(y + z) = 1$ at $(\frac{\pi}{4}, \frac{\pi}{4}, -\frac{\pi}{4})$
21. $\ln\left(\dfrac{x}{y - z}\right) = 0$ at $(2, 5, 3)$
22. $\ln\left(\dfrac{x - y}{y + z}\right) = x - z$ at $(1, 0, 1)$
23. $ze^{x+2y} = 3$ at $(2, -1, 3)$
24. $ze^{x^2 - y^2} = 3$ at $(1, 1, 3)$

Find the direction from P_0 in which the given function f increases most rapidly and compute the magnitude of the greatest rate of increase in Problems 25–34.

25. $f(x, y) = 3x + 2y - 1$
    a. $P_0(1, -1)$ b. $P_0(1, 1)$
26. $f(x, y) = 1 - x^2 - y^2$
    a. $P_0(1, 2)$ b. $P_0(0, 0)$
27. $f(x, y) = x^3 + y^3$
    a. $P_0(3, -3)$ b. $P_0(-3, 3)$
28. $f(x, y) = ax + by + c; P_0(a, b)$
29. $f(x, y, z) = ax^2 + by^2 + cz^2; P_0(a, b, c)$
    Assume $a^4 + b^4 + c^4 \neq 0$.
30. $f(x, y) = ax^3 + by^3; P_0(a, b)$
31. $f(x, y) = \ln\sqrt{x^2 + y^2}; P_0(1, 2)$
32. $f(x, y) = \sin xy; P_0\left(\dfrac{\sqrt{\pi}}{3}, \dfrac{\sqrt{\pi}}{2}\right)$

33. $f(x, y, z) = (x + y)^2 + (y + z)^2 + (z + x)^2$; $P_0(2, -1, 2)$

34. $f(x, y, z) = z \ln\left(\dfrac{y}{x}\right)$; $P_0(1, e, -1)$

B *In Problems 35–38, find a unit vector that is normal to the given graph at the point $P_0(x_0, y_0)$ on the graph. Assume that a, b, and c are constants.*

35. the line $ax + by = c$

36. the circle $x^2 + y^2 = a^2$

37. the ellipse $\dfrac{x^2}{a^2} + \dfrac{y^2}{b^2} = 1$

38. the hyperbola $\dfrac{x^2}{a^2} - \dfrac{y^2}{b^2} = 1$

39. Find the directional derivative of $f(x, y) = x^2 + y^2$ at the point $P_0(1, 1)$ in the direction of the unit vector **u** shown in Figure 12.28.

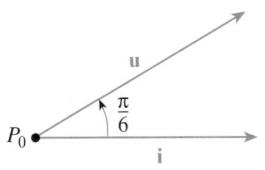

■ **FIGURE 12.28** Problem 39

40. Find the directional derivative of $f(x, y) = x^2 + xy + y^2$ at $P_0(1, -1)$ in the direction toward the origin.

41. Find the directional derivative of $f(x, y) = e^{x^2 y^2}$ at $P_0(1, -1)$ in the direction toward $Q(2, 3)$.

42. Let $f(x, y, z) = 2x^2 - y^2 + 3z^2 - 8x - 4y + 201$, and let P_0 be the point $(2, -\frac{3}{2}, \frac{1}{2})$.
   a. Find ∇f.
   b. Find $\cos \theta$, where θ is the angle between ∇f_0 and the vector toward the origin from P_0.

43. Let $f(x, y, z) = xyz$, and let **u** be a unit vector perpendicular to both $\mathbf{v} = \mathbf{i} - 2\mathbf{j} + 3\mathbf{k}$ and $\mathbf{w} = 2\mathbf{i} + \mathbf{j} - \mathbf{k}$. Find the directional derivative of f at $P_0(1, -1, 2)$ in the direction of **u**.

44. Let $f(x, y, z) = ye^{x+z} + ze^{y-x}$. At the point $P(2, 2, -2)$, find the unit vector pointing in the direction of most rapid increase of f.

45. MODELING PROBLEM Suppose a box in space given by $0 \le x \le 2, 0 \le y \le 2, 0 \le z \le 2$ is temperature controlled so that the temperature at a point $P(x, y, z)$ in the box is modeled by $T(x, y, z) = xy + yz + xz$. A heat-seeking missile is located at $P_0(1, 1, 1)$. In what direction will the missile move for the temperature to increase as quickly as possible? What is the maximum rate of change of the temperature at the point P_0?

46. MODELING PROBLEM A metal plate covering the rectangular region $0 < x \le 6, 0 < y \le 5$ is charged electrically in such a way that the potential at each point (x, y) is inversely proportional to the square of its distance from the origin. If an object is at the point $(3, 4)$, in which direction should it move to increase the potential most rapidly?

47. MODELING PROBLEM A hiker is walking on a mountain path when it begins to rain. If the surface of the mountain is modeled by $z = 1 - 3x^2 - \frac{5}{2}y^2$ (where x, y, and z are in miles) and the rain begins when the hiker is at the point $P_0(\frac{1}{4}, -\frac{1}{2})$, in what direction should she head to descend the mountainside most rapidly?

48. Let f have continuous partial derivatives, and assume the maximal directional derivative of f at $(0, 0)$ is equal to 100 and is attained in the direction of the vector $(0, 0)$ toward $(3, -4)$. Find the gradient ∇f at $(0, 0)$.

49. Suppose at the point $P_0(-1, 2)$, a certain function $f(x, y)$ has directional derivative 8 in the direction of $\mathbf{v}_1 = 3\mathbf{i} - 4\mathbf{j}$ when 1 in the direction of $\mathbf{v}_2 = 12\mathbf{i} + 5\mathbf{j}$. What is the directional derivative of f at P_0 in the direction of $\mathbf{v} = 3\mathbf{i} - 5\mathbf{j}$?

50. Suppose $f(x, y)$ has directional derivatives $D_{v_1} f = 2$ at $P_0(1, 5)$ in the direction of $\mathbf{v}_1 = 2\mathbf{i} + 3\mathbf{j}$ and $D_{v_2} f = -5$ in the direction of $\mathbf{v}_2 = 3\mathbf{i} - 5\mathbf{j}$. What is the directional derivative of f at P_0 in the direction of $\mathbf{v}_3 = 2\mathbf{i} - 3\mathbf{j}$?

51. Let f have continuous partial derivatives and suppose the maximal directional derivative of f at $P_0(1, 2)$ has magnitude 50 and is attained in the direction from P_0 toward $Q(3, -4)$. Use this information to find $\nabla f(1, 2)$.

52. Let $T(x, y) = 1 - x^2 - 2y^2$ be the temperature at each point $P(x, y)$ in the plane. A heat-loving bug is placed in the plane at the point $P_0(-1, 1)$. Find the path that the bug should take to stay as warm as possible. *Hint:* Assume that at each point on the bug's path, the tangent line will point in the direction for which T increases most rapidly.

53. Repeat Problem 52 for the temperature function $T(x, y) = 1 - ax^2 - by^2$ and starting at the point (x_0, y_0). Assume $a > 0, b > 0$, and $(x_0, y_0) \ne (0, 0)$.

Technology Window

54. Write a computer program to solve Problem 53.

55. SPY PROBLEM Just as the Spy is about to catch up with Scélérat (Problem 68 of Chapter 11 Supplementary Problems), the snow gives way and he falls into an ice cavern. He staggers to his feet and removes his skis. Why is it so warm? Good grief—the cave is a large roasting oven! Fortunately, he is wearing his heat-detector ring, which indicates the direction of greatest temperature decrease. Suppose the bunker is coordinatized so that the temperature at each point (x, y) on the floor of the bunker is given by

$$T(x, y) = 3(x - 6)^2 + 1.5(y - 1)^2 + 41$$

degrees Fahrenheit, where x and y are in feet. The Spy begins at the point $(1, 5)$ and stumbles across the room at the rate of 4 ft/min, always moving in the direction of maximum temperature decrease. But he can last no more than 2 minutes under these conditions! Assuming that there is an escape hole at the point where the temperature is minimal, does he make it or is the Spy toast at least?

56. Recall from precalculus that an ellipse is the set of all points $P(x, y)$ such that the sum of the distances from P to two fixed points (the *foci*) is constant. Let $P(x, y)$ be a point on the ellipse, and let r_1 and r_2 denote the respective distances from P to the two foci, F_1 and F_2, as shown in Figure 12.29. 〔SMH〕

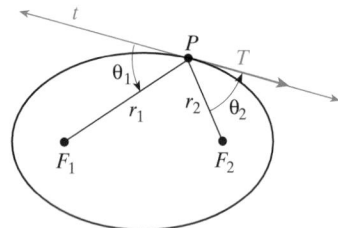

■ **FIGURE 12.29**   Unit tangent to an ellipse

a. Show that $\mathbf{T} \cdot \nabla(r_1 + r_2) = 0$, where $\mathbf{T}$ is the unit tangent to the ellipse at P.

b. Use part **a** to show that the tangent line to the ellipse at P makes equal angles with the lines joining P to the foci (that is, $\theta_1 = \theta_2$ in Figure 12.29).

Ⓒ 57. **MODELING PROBLEM**   A particle P_1 with mass m_1 is located at the origin, and a particle P_2 with mass 1 unit is located at the point (x, y, z). According to Newton's law of universal gravitation, the force P_1 exerts on P_2 is modeled by

$$\mathbf{F} = \frac{-Gm_1(x\mathbf{i} + y\mathbf{j} + z\mathbf{k})}{r^3}$$

where r is the distance between P_1 and P_2, and G is the gravitational constant.

a. Starting from the fact that $r^2 = x^2 + y^2 + z^2$, show that

$$\frac{\partial}{\partial x}\left(\frac{1}{r}\right) = \frac{-x}{r^3}, \quad \frac{\partial}{\partial y}\left(\frac{1}{r}\right) = \frac{-y}{r^3}, \quad \frac{\partial}{\partial z}\left(\frac{1}{r}\right) = \frac{-z}{r^3}$$

b. The function $V = -Gm_1/r$ is called the **potential energy** function for the system. Show that $\mathbf{F} = -\nabla V$.

58. Verify each of the following properties for functions of two variables.

a. $\nabla(cf) = c(\nabla f)$ for constant c

b. $\nabla(f + g) = \nabla f + \nabla g$

c. $\nabla(f/g) = \dfrac{g\nabla f - f\nabla g}{g^2}, \quad g \neq 0$

59. Verify the product rule

$$\nabla(fg) = f\nabla g + g\nabla f$$

for functions of three variables.

60. Find a general formula for the directional derivative $D_u f$ of the function $f(x, y)$ at the point $P(x_0, y_0)$ in the direction of the unit vector $\mathbf{u} = (\cos\theta)\mathbf{i} + (\sin\theta)\mathbf{j}$. Apply your formula to obtain the directional derivative of $f(x, y) = xy^2 e^{x-2y}$ at $P_0(-1, 3)$ in the direction of the unit vector

$$\mathbf{u} = (\cos\tfrac{\pi}{6})\mathbf{i} + (\sin\tfrac{\pi}{6})\mathbf{j}$$

61. Suppose that $\mathbf{u}$ and $\mathbf{v}$ are unit vectors and that f has continuous partial derivatives. Show that

$$D_{u+v}f = \frac{1}{\|\mathbf{u} + \mathbf{v}\|}(D_u f + D_v f)$$

62. If $\mathbf{a}$ is a constant vector and $\mathbf{R} = x\mathbf{i} + y\mathbf{j} + z\mathbf{k}$, show that $\nabla(\mathbf{a} \cdot \mathbf{R}) = \mathbf{a}$.

63. Let $\mathbf{R} = x\mathbf{i} + y\mathbf{j} + z\mathbf{k}$, and let

$$r = \|\mathbf{R}\| = \sqrt{x^2 + y^2 + z^2}$$

a. Show that ∇r is a unit vector in the direction of $\mathbf{R}$.

b. Show that $\nabla(r^n) = nr^{n-2}\mathbf{R}$, for any positive integer n.

64. Suppose the surfaces $F(x, y, z) = 0$ and $G(x, y, z) = 0$ both pass through the point $P_0(x_0, y_0, z_0)$ and that the gradient ∇F_0 and ∇G_0 both exist. Show that the two surfaces are tangent at P_0 if and only if

$$\nabla F_0 \times \nabla G_0 = 0$$

12.7 *Extrema of Functions of Two Variables*

IN THIS SECTION   relative extrema, second partials test, absolute extrema of continuous functions   ■

There are many practical situations in which it is necessary or useful to know the largest and smallest values of a function of two variables. For example, if $T(x, y)$ is the temperature at a point (x, y) in a plate, where are the hottest and coldest points in the plate and what are these extreme temperatures? A hazardous waste dump is bounded by the curve $F(x, y) = 0$. What are the largest and smallest distances from a given interior point P_0? We begin our study of extrema with some terminology.

Absolute Extrema

The function $f(x, y)$ is said to have an **absolute maximum** at (x_0, y_0) if $f(x_0, y_0) \geq f(x, y)$ for all (x, y) in the domain D of f. Similarly, f has an **absolute minimum** at (x_0, y_0) if $f(x_0, y_0) \leq f(x, y)$ for all (x, y) in D. Collectively, absolute maxima and minima are called **absolute extrema.**

In Chapter 4, we located absolute extrema of a function of one variable by first finding *relative extrema,* those values of $f(x)$ that are larger or smaller than those at all "nearby" points. The relative extrema of a function of two variables may be defined as follows.

Relative Extrema

Let f be a function defined at (x_0, y_0). Then

$f(x_0, y_0)$ is a **relative maximum** if $f(x, y) \leq f(x_0, y_0)$ for all (x, y) in an open disk containing (x_0, y_0).

$f(x_0, y_0)$ is a **relative minimum** if $f(x, y) \geq f(x_0, y_0)$ for all (x, y) in an open disk containing (x_0, y_0).

Collectively, relative maxima and minima are called **relative extrema.**

In Chapter 4, we found that on a closed, bounded interval $[a, b]$, a continuous function f must attain both an absolute maximum and an absolute minimum.

THEOREM 12.11   *Extreme value theorem for a function of two variables*

A function of two variables $f(x, y)$ assumes an absolute extremum on any closed, bounded set S in the plane where it is continuous.

Proof   The proof of this theorem is beyond the scope of this text and is usually given in an advanced calculus course.   ◼

We also found in Chapter 4 that an absolute extremum of a function f of a single variable can occur either at an endpoint of an interval in the domain of f or at an interior critical point where it is a relative extremum—that is, at a point where either $f'(x)$ does not exist or $f'(x) = 0$. In Section 12.6, we observed that the gradient ∇f is the two-variable analogue to the derivative in the single-variable case. Thus, it is reasonable to expect extrema of a function of two variables to occur in one of the following situations:

1. either on the boundary of the domain of f, or
2. at points in the interior where the gradient ∇f does not exist, or
3. where $\nabla f = \mathbf{0}$; that is, where $f_x = f_y = 0$.

We begin by considering relative extrema, and then turn to the more general question of finding absolute extrema.

RELATIVE EXTREMA

In Chapter 4, we observed that relative extrema of the function f correspond to "peaks and valleys" on the graph of f, and the same observation can be made about

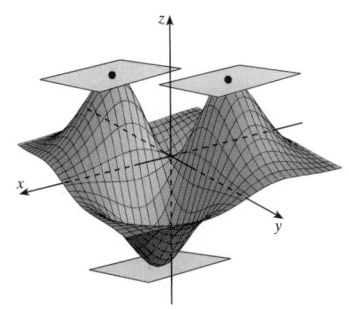

Extrema occur on the boundary where the tangent plane is horizontal, or where no tangent plane exists

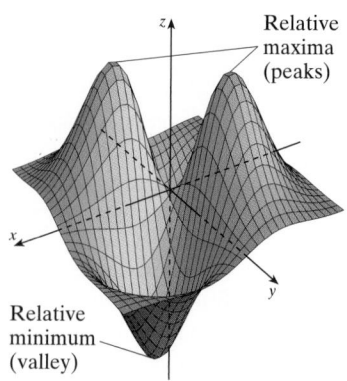

Relative maxima (peaks)

Relative minimum (valley)

■ **FIGURE 12.30**   Relative extrema occur at peaks and valleys.

relative extrema in the two-variable case, as seen in Figure 12.30. For a function f of one variable, we found that the relative extrema occur where $f'(x) = 0$ or $f'(x)$ does not exist. The following theorem shows that the relative extrema of a function of two variables can be located similarly.

THEOREM 12.12   *Partial derivative criteria for relative extrema*

If f has a relative extremum (maximum or minimum) at $P_0(x_0, y_0)$ and partial derivatives f_x and f_y both exist at $f_x(x_0, y_0) = f_y(x_0, y_0)$, then

$$f_x(x_0, y_0) = f_y(x_0, y_0) = 0$$

Proof   Let $F(x) = f(x, y_0)$. Then $F(x)$ must have a relative extremum at $x = x_0$, so $F'(x_0) = 0$, which means that $f_x(x_0, y_0) = 0$. Similarly, $G(y) = f(x_0, y)$ has a relative extremum at $y = y_0$, so $G'(y_0) = 0$ and $f_y(x_0, y_0) = 0$. Thus, we must have *both* $f_x(x_0, y_0) = 0$ and $f_y(x_0, y_0) = 0$, as claimed.

WARNING ▶   There is a horizontal tangent plane at each extreme point where the first partial derivatives exist. However, this does *not* say that whenever a horizontal tangent plane occurs at a point P, there must be an extremum there. All that can be said is that such a point P is a *possible* location for a relative extremum.   ◂

In single-variable calculus, we referred to a number x_0 where $f'(x_0)$ does not exist or $f'(x_0) = 0$ as a *critical number* and the point $(x_0, f(x_0))$ as a *critical point*. This terminology is extended to functions of two variables as follows.

Critical Points

A **critical point** of a function f defined on an open set S is a point (x_0, y_0) in S where either one of the following is true:

a. $f_x(x_0, y_0) = f_y(x_0, y_0) = 0$.
b. $f_x(x_0, y_0)$ or $f_y(x_0, y_0)$ does not exist (one or both).

EXAMPLE 1   *Distinguishing critical points*

Discuss the nature of the critical point $(0, 0)$ for the quadric surfaces
a. $z = x^2 + y^2$ **b.** $z + x^2 + y^2 = 1$ **c.** $z = y^2 - x^2$

Solution The graphs of these quadric surfaces are shown in Figure 12.31.

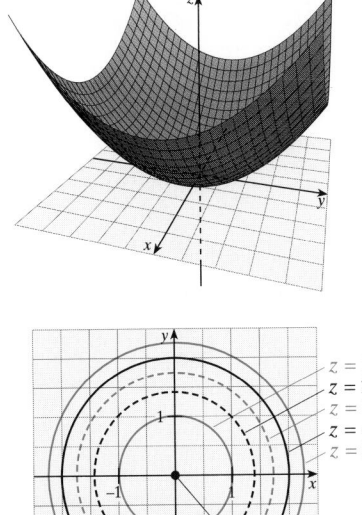

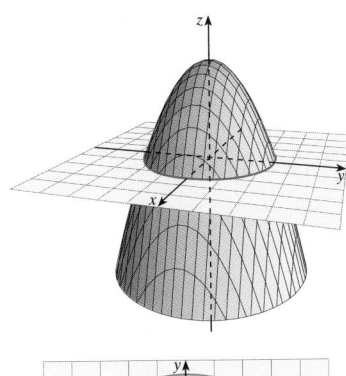

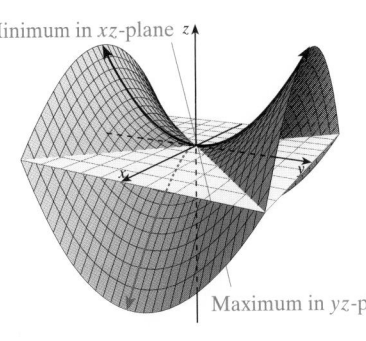

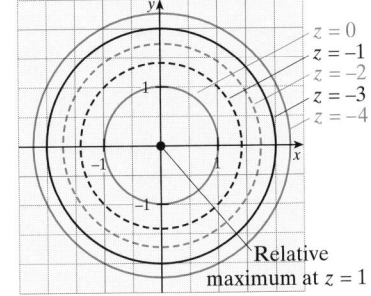

a. $z = x^2 + y^2$ (paraboloid) and level curves; relative minimum at $(0, 0, 0)$

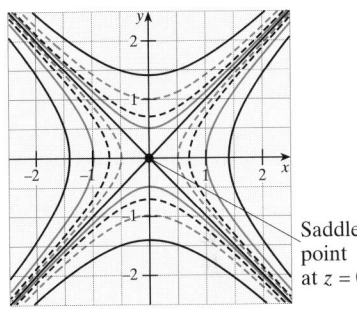

b. $z = 1 - x^2 - y^2$ (paraboloid) and level curves; relative maximum at $(0, 0, 1)$

c. $z = y^2 - x^2$ (hyperbolic paraboloid) and level curves; saddle point at $(0, 0, 0)$

■ **FIGURE 12.31** Classification of critical points

Let $f(x, y) = x^2 + y^2$, $g(x, y) = 1 - x^2 - y^2$, and $h(x, y) = y^2 - x^2$. We find the critical points:

a. $f_x(x, y) = 2x$, $f_y(x, y) = 2y$; the critical point is $(0, 0)$. The function f has a relative minimum at $(0, 0)$ because x^2 and y^2 are both nonnegative, yielding $x^2 + y^2 \geq 0$ for all nonzero x and y.

b. $g_x(x, y) = -2x$, $g_y(x, y) = -2y$; critical point $(0, 0)$. The function g has a relative maximum at $(0, 0)$ because $z = 1 - x^2 - y^2$ and x^2 and y^2 are both nonnegative, so the largest value of z occurs at $(0, 0)$.

c. $h_x(x, y) = -2x$, $h_y(x, y) = 2y$; critical point $(0, 0)$. The function h has neither a relative maximum nor a relative minimum at $(0, 0)$. When $z = 0$, h is a minimum on the y-axis (where $x = 0$) and a maximum on the x-axis (where $y = 0$).   ■

If (x_0, y_0) is a critical point for $f(x, y)$ that is neither a relative maximum nor a relative minimum, it may be a **saddle point.** Such a point is a *maximum* in one direction and a *minimum* in another direction and so has a saddle shape, as shown in Figure 12.31c.

SECOND PARTIALS TEST

The previous example points to the need for some sort of a test to determine the nature of a critical point. In Chapter 4, we developed the second derivative test as a means for classifying a critical point of f as a relative maximum or minimum. According to this test, if $f'(c) = 0$, then at $x = c$, a relative maximum occurs if $f''(c) < 0$ and a relative minimum occurs if $f''(c) > 0$. If $f''(c) = 0$, the test is inconclusive. The analogous result for the two-variable case may be stated as follows.

Second Partials Test

Let $f(x, y)$ have a critical point at $P_0(x_0, y_0)$ and assume that f has continuous second-order partial derivatives in a disk centered at (x_0, y_0). Let

$$D = f_{xx}(x_0, y_0)f_{yy}(x_0, y_0) - [f_{xy}(x_0, y_0)]^2$$

Then,

Summary:

Critical Point	D	f_{xx}	type
(x_1, y_1)	Pos.	Neg.	Rel. max.
(x_2, y_2)	Pos.	Pos.	Rel. Min.
(x_3, y_3)	Neg.	—	Saddle pt.
(x_4, y_4)	Zero		Inconclusive

A **relative maximum** occurs at P_0 if

$$D > 0 \quad \text{and} \quad f_{xx}(x_0, y_0) < 0$$
$$(\text{or } f_{yy}(x_0, y_0) < 0)$$

$z = 1 - x^2 - y^2$ (See Ex 1b)

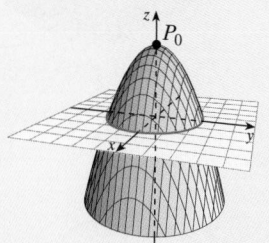

A **relative minimum** occurs at P_0 if

$$D > 0 \quad \text{and} \quad f_{xx}(x_0, y_0) > 0$$
$$(\text{or } f_{yy}(x_0, y_0) > 0)$$

$z = x^2 + y^2$ (See Ex 1a)

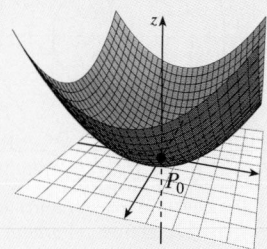
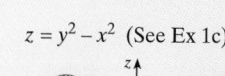

A **saddle point** occurs at P_0 if $D < 0$.

$z = y^2 - x^2$ (See Ex 1c)

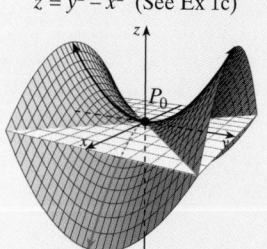

If $D = 0$, then the test is **inconclusive**.

When $D(x_0, y_0) = 0$, the critical point (x_0, y_0) is said to be **degenerate.** Otherwise, it is **nondegenerate.** Note that D can be expressed in *determinant form* as

$$D = \begin{vmatrix} f_{xx} & f_{xy} \\ f_{xy} & f_{yy} \end{vmatrix}$$

where all the partials are evaluated at (x_0, y_0). Some students find this the easiest form of D to remember.

EXAMPLE 2   *Second partials test with a relative maximum*

Find all critical points on the graph of $f(x, y) = 1 - x^2 - y^2$ and use the second partials test to classify each critical point as a relative extremum or a saddle point.

Solution $f_x(x, y) = -2x, f_y(x, y) = -2y$, so the only critical point is at $(0, 0)$.
$f_{xx}(x, y) = -2, f_{xy}(x, y) = 0$, and $f_{yy}(x, y) = -2$, so

$$D = \begin{vmatrix} -2 & 0 \\ 0 & -2 \end{vmatrix} = 4$$

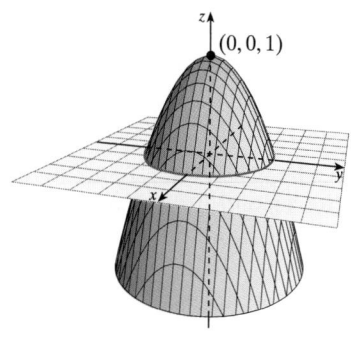

$(0, 0, 1)$

Graph of $f(x, y) = 1 - x^2 - y^2$

Because $D = 4 > 0$ for all (x, y), and because $f_{xx}(0, 0) = -2 < 0$, the second partials test tells us that a relative maximum occurs at $(0, 0)$. ∎

| EXAMPLE 3 | *Second partials test with a relative maximum and a saddle point* |

Find all critical points on the graph of $f(x, y) = 8x^3 - 24xy + y^3$, and use the second partials test to classify each point as a relative extremum or a saddle point.

Solution $f_x(x, y) = 24x^2 - 24y, f_y(x, y) = -24x + 3y^2$

To find the critical points, solve

$$\begin{cases} 24x^2 - 24y = 0 \\ -24x + 3y^2 = 0 \end{cases}$$

From the first equation, $y = x^2$; substitute this into the second equation to find

$$-24x + 3(x^2)^2 = 0$$
$$x^4 - 8x = 0$$
$$x(x^3 - 8) = 0$$
$$x(x - 2)(x^2 + 2x + 4) = 0$$
$$x = 0, 2 \qquad \text{The solutions of } x^2 + 2x + 4 = 0 \text{ are not real.}$$

If $x = 0$, then $y = 0$, and if $x = 2$, then $y = 4$, so the critical points are $(0, 0), (2, 4)$. To obtain D, we first find $f_{xx}(x, y) = 48x$, $f_{xy}(x, y) = -24$, and $f_{yy}(x, y) = 6y$ and then compute

$$D = \begin{vmatrix} f_{xx} & f_{xy} \\ f_{xy} & f_{yy} \end{vmatrix} = \begin{vmatrix} 48x & -24 \\ -24 & 6y \end{vmatrix} = 288xy - 576$$

At $(0, 0), D = -576 < 0$, so there is a saddle point at $(0, 0)$. At $(2, 4), D = 288(2)(4) - 576 = 1{,}728 > 0$ and $f_{xx}(2, 4) = 96 > 0$, so there is a relative minimum at $(2, 4)$.

To view the situation graphically, we calculate the coordinates of the saddle point $(0, 0, 0)$, and the relative minimum $(2, 4, -64)$, as shown in Figure 12.32.

Summary:

Critical Point	D	f_{xx}	type
$(0, 0)$	Neg.		Saddle
$(2, 4)$	Pos.	Pos.	Rel. min.

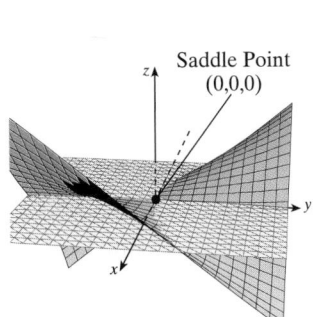

a. View near the origin (showing the saddle point)

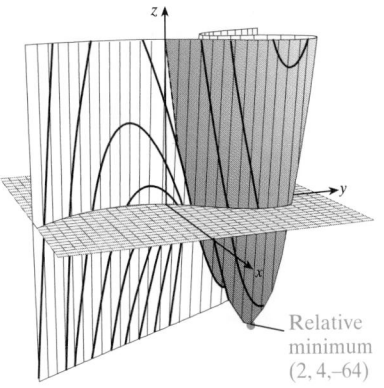

b. View away from the origin (showing the relative minimum point)

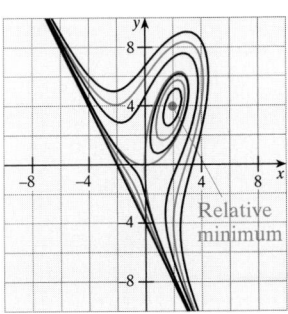

c. Level curves

■ **FIGURE 12.32**   Graph of $f(x, y) = 8x^3 - 24xy + y^3$ ■

EXAMPLE 4   *Extrema when the second partials test fails*

Find all relative extrema and saddle points on the graph of

$$f(x, y) = x^2 y^4$$

Solution $f_x(x, y) = 2xy^4$, $f_y(x, y) = 4x^2y^3$; we can see the critical points occur only whenever $x = 0$ or $y = 0$; that is, every point on the x- or y-axis is a critical point.

Because $f_{xx}(x, y) = 2y^4$, $f_{xy}(x, y) = 8xy^3$, $f_{yy}(x, y) = 12x^2y^2$, we have

$$D = \begin{vmatrix} f_{xx} & f_{xy} \\ f_{xy} & f_{yy} \end{vmatrix} = \begin{vmatrix} 2y^4 & 8xy^3 \\ 8xy^3 & 12x^2y^2 \end{vmatrix} = 24x^2y^6 - 64x^2y^6 = -40x^2y^6$$

For any critical point $(x_0, 0)$ or $(0, y_0)$, $D = 0$; so the second partials test fails. But $f(x, y) = 0$ for every critical point (because either $x = 0$ or $y = 0$, or both) and because $f(x, y) = x^2y^4 > 0$ when $x \neq 0$ and $y \neq 0$, it follows that each critical point must be a relative minimum. ∎

The second partials test provides explicit information about the case where f is nondegenerate at $P_0(x_0, y_0)$, but if f is degenerate, practically anything can happen, as shown in Problem 54. In that problem you are asked to show that $(0, 0)$ is a degenerate critical point for each of the following functions:

$$f(x, y) = x^4 - y^4 \qquad g(x, y) = x^2y^2 \qquad h(x, y) = x^3 + y^3$$

and the following conclusions can be made:

$f(x, y)$ has a saddle point at $(0, 0)$.

$g(x, y)$ has a relative minimum at $(0, 0)$.

$h(x, y)$ has neither kind of extremum nor a saddle point at $(0, 0)$.

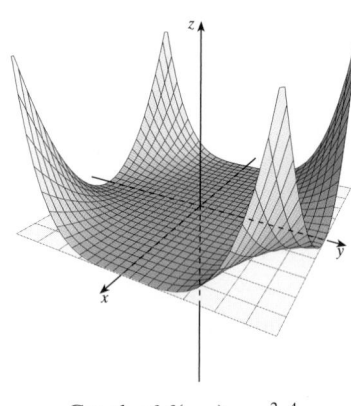

Graph of $f(x, y) = x^2 y^4$

ABSOLUTE EXTREMA OF CONTINUOUS FUNCTIONS

The extreme value theorem (Theorem 12.11) assures us that a continuous function f of two variables must have both an absolute maximum and an absolute minimum on any closed, bounded set S. Each extremum occurs either at a boundary point of S or at a critical point in the interior of S. Here is a procedure for finding absolute extrema:

Procedure for Determining Absolute Extrema

Given a function f that is continuous on a closed, bounded set S:

Step 1: Find all critical points of f in S.

Step 2: Find all points on the boundary of S where absolute extrema can occur (boundary critical points, endpoints, etc.).

Step 3: Compute the value of $f(x_0, y_0)$ for each of the points (x_0, y_0) found in steps 1 and 2.

Evaluation: The absolute maximum of f on S is the largest of the values computed in step 3, and the absolute minimum is the smallest of the computed values.

EXAMPLE 5   *Finding absolute extrema*

Find the absolute extrema of the function $f(x, y) = e^{x^2 - y^2}$ over the disk $x^2 + y^2 \leq 1$.

Solution

Step 1: $f_x(x, y) = 2xe^{x^2 - y^2}$ and $f_y(x, y) = -2ye^{x^2 - y^2}$. These partial derivatives are defined for all (x, y). Because $f_x(x, y) = f_y(x, y) = 0$ only when $x = 0$ and $y = 0$, it follows that $(0, 0)$ is the only critical point of f and it is inside the disk.

Step 2: Examine the values of f on the boundary curve $x^2 + y^2 = 1$. Because $y^2 = 1 - x^2$ on the boundary of the disk, we find that

$$f(x, y) = e^{x^2 - (1 - x^2)} = e^{2x^2 - 1}$$

We need to find the largest and smallest values of $F(x) = e^{2x^2 - 1}$ for $-1 \le x \le 1$. Since

$$F'(x) = 4xe^{2x^2 - 1}$$

we see that $F'(x) = 0$ only when $x = 0$ (since $e^{2x^2 - 1}$ is always positive). At $x = 0$, we have $y^2 = 1 - 0^2$, so $y = \pm 1$; and $(0, 1)$ and $(0, -1)$ are boundary critical points. At the endpoints of the interval $-1 \le x \le 1$, the corresponding points are $(1, 0)$ and $(-1, 0)$.

Step 3: Compute the value of f for the points found in steps 1 and 2:

Points to check	Compute $f(x_0, y_0) = e^{x_0^2 - y_0^2}$
$(0, 0)$	$f(0, 0) = e^0 = 1$
$(0, 1)$	$f(0, 1) = e^{-1}$; minimum
$(0, -1)$	$f(0, -1) = e^{-1}$; minimum
$(1, 0)$	$f(1, 0) = e$; maximum
$(-1, 0)$	$f(-1, 0) = e$; maximum

Evaluation: As indicated in the preceding table, the absolute maximum value of f on the given disk is e at $(1, 0)$ and $(-1, 0)$, and the absolute minimum value is e^{-1} at $(0, 1)$ and $(0, -1)$. ∎

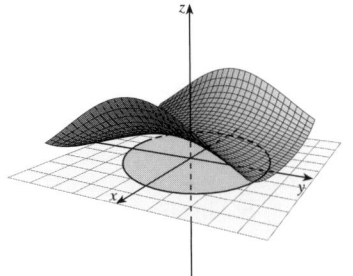

EXAMPLE 6   *Minimum distance from a point to a plane*

Find the shortest distance from the point $(0, 3, 4)$ to the plane $x + 2y + z = 5$.

Solution   The distance from a point (x, y, z) to $(0, 3, 4)$ is

$$d = \sqrt{(x - 0)^2 + (y - 3)^2 + (z - 4)^2}$$

However, because (x, y, z) is on the plane $x + 2y + z = 5$, we know

$$z = 5 - x - 2y$$

so that $d = \sqrt{x^2 + (y - 3)^2 + (5 - x - 2y - 4)^2}$. Instead of minimizing d, we can minimize the expression

$$d^2 = f(x, y) = x^2 + (y - 3)^2 + (1 - x - 2y)^2$$

To find the critical values, we solve

$$f_x = 2x - 2(1 - x - 2y) = 4x + 4y - 2 = 0$$
$$f_y = 2(y - 3) - 4(1 - x - 2y) = 4x + 10y - 10 = 0$$

The only critical point is $\left(-\frac{5}{6}, \frac{4}{3}\right)$. Also, $f_{xx} = 4$, $f_{yy} = 10$, $f_{xy} = 4$, so $D > 0$, which means there is a relative minimum at $\left(-\frac{5}{6}, \frac{4}{3}\right)$. Intuitively, we see that this relative minimum must also be an absolute minimum because there must be exactly one point on the plane that is closest to the given point. We now calculate that distance:

$$d = \sqrt{(\tfrac{5}{6})^2 + (\tfrac{4}{3} - 3)^2 + [1 + \tfrac{5}{6} - 2(\tfrac{4}{3})]^2} = \sqrt{\frac{25}{6}} = \frac{5}{\sqrt{6}}$$

Check: You might want to check your work by using the formula for the distance from a point to a plane in $\mathbb{R}^3$ (Theorem 10.10):

$$d = \left| \frac{Ax_0 + By_0 + Cz_0 + D}{\sqrt{A^2 + B^2 + C^2}} \right| = \left| \frac{(1)0 + (2)3 + (1)4 + (-5)}{\sqrt{1^2 + 2^2 + 1^2}} \right| = \frac{5}{\sqrt{6}} \quad \blacksquare$$

COMMENT   In general, it can be difficult to show that a relative extremum is actually an absolute extremum. In practice, however, it is often possible to make the determination using physical or geometric considerations.

In the following example, calculus is applied to justify a formula used in statistics and in many applications in the social and physical sciences.

> **EXAMPLE 7**   *Least-squares approximation of data*

Suppose data consisting of n points $P_1, \ldots, P_n$ are known, and we wish to find a function $y = f(x)$ that fits the data reasonably well. In particular, suppose we wish to find a line $y = mx + b$ that "best fits" the data in the sense that the sum of the squares of the vertical distances from each data point to the line is minimized.

Solution   We wish to find values of m and b that minimize the sum of the squares of the differences between the y-values and the line $y = mx + b$. This line that we seek is called the **regression line.** Suppose that the point P_k has components (x_k, y_k). Now at this point the value on the regression line is $y = mx_k + b$ and the value of the data point is y_k. The "error" caused by using the point on the regression line rather than the actual data point is

$$y_k - (mx_k + b)$$

The data points may be above the regression line for some values of k and below the regression line for other values of k. We see that we need to minimize the function that represents the sum of the *squares* of all these differences:

$$F(m, b) = \sum_{k=1}^{n} [y_k - (mx_k + b)]^2$$

The situation is illustrated in Figure 12.33.

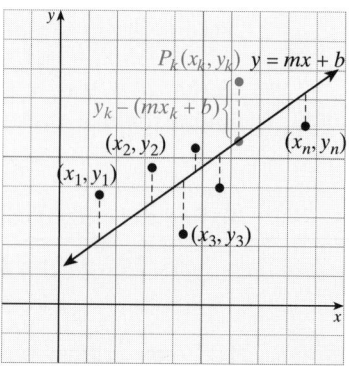

■ **FIGURE 12.33**   Least-squares approximation of data

Because F is a function of two variables, we use the second partials test:

$$F_m(m, b) = \sum_{k=1}^{n} 2[y_k - (mx_k + b)](-x_k)$$

$$= 2m \sum_{k=1}^{n} x_k^2 + 2b \sum_{k=1}^{n} x_k - 2 \sum_{k=1}^{n} x_k y_k$$

$$F_b(m, b) = \sum_{k=1}^{n} 2[y_k - (mx_k + b)](-1)$$

$$= 2m \sum_{k=1}^{n} x_k + 2b \sum_{k=1}^{n} 1 - 2 \sum_{k=1}^{n} y_k$$

$$= 2m \sum_{k=1}^{n} x_k + 2bn - 2 \sum_{k=1}^{n} y_k$$

Set each of these partial derivatives equal to 0 to find the critical values (after a great deal of algebra):

$$m = \frac{n \sum\limits_{k=1}^{n} x_k y_k - \left(\sum\limits_{k=1}^{n} x_k\right)\left(\sum\limits_{k=1}^{n} y_k\right)}{n \sum\limits_{k=1}^{n} x_k^2 - \left(\sum\limits_{k=1}^{n} x_k\right)^2} \quad \text{and} \quad b = \frac{\sum\limits_{k=1}^{n} x_k^2 \sum\limits_{k=1}^{n} y_k - \left(\sum\limits_{k=1}^{n} x_k\right)\left(\sum\limits_{k=1}^{n} x_k y_k\right)}{n \sum\limits_{k=1}^{n} x_k^2 - \left(\sum\limits_{k=1}^{n} x_k\right)^2}$$

We leave it to you to complete the second partials test to verify that these values of m and b yield a minimum. ∎

Most applications of the **least-squares formula** stated in Example 7 involve using a calculator or computer software. The following technology window provides an example.

Technology Window

Many calculators will carry out the calculations required by the least-squares approximation procedure. Look at your owner's manual for specifics, but most calculators allow you to input data with keys labeled STAT and DATA. After the data are input, the m and b values are given by pressing the Lin Reg choice. For example, ten people are given a standard IQ test. Their scores were then compared with their high school grades:

IQ:	117	105	111	96	135	81	103	99	107	109
GPA:	3.1	2.8	2.5	2.8	3.4	1.9	2.1	3.2	2.9	2.3

A calculator output shows: $m = .0224144711$ and $b = .3173417224$. A scatter diagram with the least-squares line is shown below.

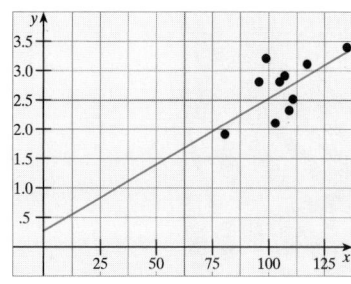

12.7 *Problem Set*

A 1. ■ **What Does This Say?** Describe what is meant by a critical value.

2. ■ **What Does This Say?** Describe a procedure for determining absolute extrema.

3. ■ **What Does This Say?** Describe a procedure for determining absolute extrema on a closed, bounded set S.

Find the critical points in Problems 4–23, and classify each point as a relative maximum, a relative minimum, or a saddle point.

4. $f(x, y) = 2x^2 - 4xy + y^3 + 2$

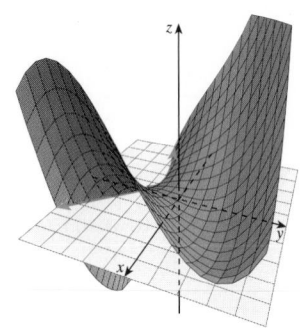

5. $f(x, y) = (x - 2)^2 + (y - 3)^4$

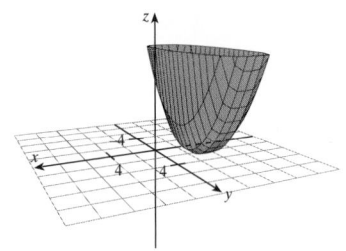

6. $f(x, y) = e^{-x}\sin y$

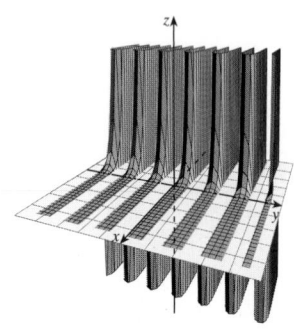

7. $f(x, y) = (1 + x^2 + y^2)e^{1 - x^2 - y^2}$

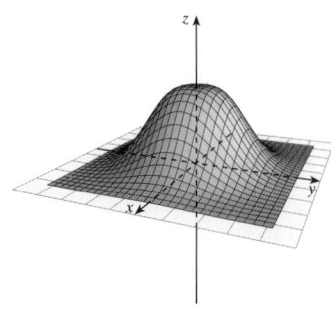

8. $f(x, y) = \dfrac{9x}{x^2 + y^2 + 1}$

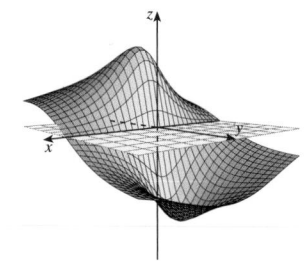

9. $f(x, y) = x^2 + xy + y^2$
10. $f(x, y) = xy - x + y$
11. $f(x, y) = -x^3 + 9x - 4y^2$
12. $f(x, y) = e^{-(x^2 + y^2)}$
13. $F(x, y) = (x^2 + 2y^2)e^{1 - x^2 - y^2}$
14. $f(x, y) = e^{xy}$
15. $f(x, y) = x^{-1} + y^{-1} + 2xy$
16. $f(x, y) = (x - 4)\ln xy$
17. $f(x, y) = x^3 + y^3 + 3x^2 - 18y^2 + 81y + 5$
18. $f(x, y) = 2x^3 + y^3 + 3x^2 - 3y - 12x - 4$
19. $f(x, y) = x^2 + y^2 - 6xy + 9x + 5y + 2$
20. $f(x, y) = x^2 + y^2 + \dfrac{32}{xy}$
21. $f(x, y) = x^2 + y^3 + \dfrac{768}{x + y}$
22. $f(x, y) = 3xy^2 - 2x^2y + 36xy$
23. $f(x, y) = 3x^2 + 12x + 8y^3 - 12y^2 + 7$

Find the absolute extrema of f on the closed, bounded set S in the plane as described in Problems 24–30.

B 24. $f(x, y) = 2x^2 - y^2$; S is the disk $x^2 + y^2 \le 1$.
25. $f(x, y) = xy - 2x - 5y$; S is the triangle with vertices $(0, 0)$, $(7, 0)$, and $(7, 7)$.

26. $f(x, y) = x^2 + 3y^2 - 4x + 2y - 3$; S is the square with vertices $(0, 0), (3, 0), (3, -3)$, and $(0, -3)$.

27. $f(x, y) = 2 \sin x + 5 \cos y$; S is the rectangle with vertices $(0, 0), (2, 0), (2, 5)$, and $(0, 5)$.

28. $f(x, y) = e^{x^2 + 2x + y^2}$; S is the disk $x^2 + 2x + y^2 \leq 0$.

29. $f(x, y) = x^2 + xy + y^2$; S is the disk $x^2 + y^2 \leq 1$.

30. $f(x, y) = x^2 - 4xy + y^3 + 4y$; S is the square region $0 \leq x \leq 2, 0 \leq y \leq 2$.

Find the least-squares regression line for the data points given in Problems 31–34.

31. $(-2, -3), (-1, -1), (0, 1), (1, 3), (3, 5)$

32. $(0, 1), (1, 1.6), (2.2, 3), (3.1, 3.9), (4, 5)$

33. $(3, 5.72), (4, 5.31), (6.2, 5.12), (7.52, 5.32), (8.03, 5.67)$

34. $(-4, 2), (-3, 1), (0, 0), (1, -3), (2, -1), (3, -2)$

35. Find all points on the surface $y^2 = 4 + xz$ that are closest to the origin.

36. Find all points in the plane $x + 2y + 3z = 4$ in the first octant where $f(x, y, z) = x^2yz^3$ has a maximum value.

37. A rectangular box with no top is to have a fixed volume. What should its dimensions be if we want to use the least amount of material in its construction?

38. A wire of length L is cut into three pieces that are bent to form a circle, a square, and an equilateral triangle. How should the cuts be made to minimize the sum of the total area?

39. Find the positive numbers whose sum is 54 and whose product is as large as possible.

40. A dairy produces whole milk and skim milk in quantities x and y pints, respectively. Suppose the price (in cents) of whole milk is $p(x) = 100 - x$ and that of skim milk is $q(y) = 100 - y$, and also assume that $C(x, y) = x^2 + xy + y^2$ is the joint-cost function of the commodities. Maximize the profit

$$P(x, y) = px + qy - C(x, y)$$

41. Repeat Problem 40 for the case where $p(x) = 4 - 5x, q(y) = 4 - 2y$, and the joint-cost function is $C(x, y) = 2xy + 4$.

42. A particle of mass m in a rectangular box with dimensions x, y, z has ground state energy

$$E(x, y, z) = \frac{k^2}{8m}\left(\frac{1}{x^2} + \frac{1}{y^2} + \frac{1}{z^2}\right)$$

where k is a physical constant. If the volume of the box is fixed (say, $V_0 = xyz$), find the values of x, y, and z that minimize the ground state energy.

43. A manufacturer produces two different kinds of graphing calculators, A and B, in quantities x and y (units of 1,000), respectively. If the revenue function (in dollars) is $R(x, y) = -x^2 - 2y^2 + 2xy + 8x + 5y$, find the quantities of A and B that should be produced to maximize revenue.

44. Suppose we wish to construct a rectangular box with volume 32 ft³. Three different materials will be used in the construction. The material for the sides costs $1 per square foot, the material for the bottom costs $3 per square foot, and the material for the top costs $5 per square foot. What are the dimensions of the least expensive such box?

45. **MODELING PROBLEM**   A store carries two competing brands of bottled water, one from California and the other from New York. To model this situation, assume the owner of the store can obtain both at a cost of $2/bottle. Also assume that if the California water is sold for x dollars per bottle and the New York water for y dollars per bottle, then consumers will buy approximately $40 - 50x + 40y$ bottles of California water and $20 + 60x - 70y$ bottles of the New York water each day. How should the owner price the bottled water to generate the largest possible profit?

46. **MODELING PROBLEM**   The telephone company is planning to introduce two new types of executive communications systems that it hopes to sell to its largest commercial customers. To create a model to determine the maximum profit, it is assumed that if the first type of system is priced at x hundred dollars per system and the second type at y hundred dollars per system, approximately $40 - 8x + 5y$ consumers will buy the first type and $50 + 9x - 7y$ will buy the second type. If the cost of manufacturing the first type is $1,000 per system and the cost of manufacturing the second type is $3,000 per system, how should the telephone company price the systems to generate maximum profit?

47. **MODELING PROBLEM**   A manufacturer with exclusive rights to a sophisticated new industrial machine is planning to sell a limited number of the machines to both foreign and domestic firms. The price the manufacturer can expect to receive for the machines will depend on the number of machines made available. For example, if only a few of the machines are placed on the market, competitive bidding among prospective purchasers will tend to drive the price up. It is estimated that if the manufacturer supplies x machines to the domestic market and y machines to the foreign market, the machines will sell for $60 - 0.2x + 0.05y$ thousand dollars apiece at home and $50 - 0.1y + 0.05x$ thousand dollars apiece abroad. If the manufacturer can produce the machines at a total cost of $10,000 apiece, how many should be supplied to each market to generate the largest possible profit?

48. **MODELING PROBLEM**   A college admissions officer, Dr. Westfall, has compiled the following data relating students' high school and college GPAs:

HS GPA	2.0	2.5	3.0	3.0	3.5	3.5	4.0	4.0
College GPA	1.5	2.0	2.5	3.5	2.5	3.0	3.0	3.5

Plot the data points on a graph and find the equation of the least-squares line for these data. Then use the least-squares line to predict the college GPA of a student whose high school GPA is 3.75.

49. MODELING PROBLEM   It is known that if an ideal spring is displaced a distance y from its natural length by a force (weight) x, then $y = kx$, where k is the so-called spring constant. To compute this constant for a particular spring, a scientist obtains the following data:

x(lb)	5.2	7.3	8.4	10.12	12.37
x(in.)	11.32	15.56	17.44	21.96	26.17

Based on these data, what is the "best" choice for k?

50. ■ **What Does This Say?** The following table gives the approximate U.S. census figures (in millions):

Year:	1900	1910	1920	1930	1940
Population:	76.2	92.2	106.0	123.2	132.1

Year:	1950	1960	1970	1980	1990
Population:	151.3	179.3	203.3	226.5	248.7

a. Find the least-squares regression line for the given data and use this line to "predict" the population in 1997. (The actual population was about 266.5 million.)
b. Use the least-squares linear approximation to estimate the population at the present time. Check your answer by looking up the population using the Internet. Comment on the accuracy (or inaccuracy) of your prediction.

51. ■ **What Does This Say?** The following table gives the Dow Jones Industrial Average (DJIA) Stock Index along with the per capita consumption of wine (in gallons) for those years.*

Year:	1965	1970	1975	1980	1985	1990	1995
DJIA Index:	911	753	802	891	1,328	2,796	3,838
Consumption:	0.98	1.31	1.71	2.11	2.43	2.05	1.79

a. Plot these data on a graph, with the DJIA Index on the x-axis and consumption on the y-axis.
b. Find the equation of the least-squares line.
c. In 1998 the stock market began the year at 7,908. Use the least-squares line to predict per capita wine consumption that corresponds to this stock value.
d. Determine whether the consumption figures predicted by the least-squares line in part **c** are approximately correct. Interpret your findings.

---

*Dominick Salvatore, *Managerial Economics,* McGraw-Hill, Inc., New York, 1989. The data given in the table are on page 138

Technology Window

52. **Linearizing nonlinear data.**   In this problem, we turn to data that do *not* tend to change linearly. Often one can "linearize" the data by taking the logarithm or exponential of the data and then doing a linear fit as described in this section.
a. Suppose we have, or suspect, a relationship $y = kx^m$. Show that by taking the natural logarithm of this equation, we obtain a linear relationship: $Y = K + mX$. Explain the new variables and constant K.
b. Below are data relating the periods of revolution t (in days) of the six inner planets and their semimajor axis a (in 10^6 km). Kepler conjectured the relationship $t = ka^m$, which is very accurate for the correct k and a. You are to "transform" the data as in part **a** (thus obtaining $T_i = \ln t_i, \dots$); and do a linear fit to the new data, thus finding k and m.

t-data:	(87.97, 224.7, 365.26, 686.98, 4 332.59, 10 759.2)
a-data:	(58, 108, 149, 228, 778, 1 426)

53. Below are data pertaining to a recent Olympic weight-lifting competition. The x-data are the "class data" giving eight weight classes (in kg) from featherweight to heavyweight-2. The w-data are the combined weights lifted by the winners in each class. Theoretically, we would expect a relationship $w = kx^m$, where $m = 2/3$. (Can you see why?)
a. Linearize the data as in Problem 52 and use the least-squares approximation to find k and m.

x-data:	(56, 60, 67.5, 75, 82.5, 90, 100, 110)
w-data:	(292.5, 342.5, 340, 375, 377.5, 412.5, 425, 455)

b. Comment on the 60 kg entry. Do you see why this participant (N. Suleymanoglu of Turkey) was referred to as the strongest man in the world?

54. This problem is designed to show, by example, that if f is degenerate at the critical point, then almost anything can happen.
a. Show that $f(x, y) = x^4 - y^4$ has a saddle point at $(0, 0)$.
b. Show that $g(x, y) = x^2 y^2$ has a relative minimum at $(0, 0)$.
c. Show that $h(x, y) = x^3 + y^3$ has no extremum or saddle point at $(0, 0)$.

55. THINK TANK PROBLEM   Consider the function $f(x, y) = (y - x^2)(y - 2x^2)$. Discuss the behavior of this function at $(0, 0)$.

56. THINK TANK PROBLEM Sometimes the critical points of a function can be classified by looking at the level curves. In each case shown in Figure 12.34, determine the nature of the critical point(s) of f at $(0, 0)$.

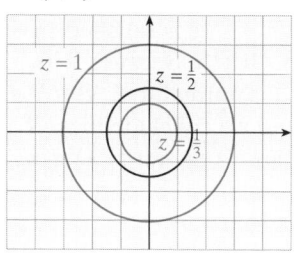

a.

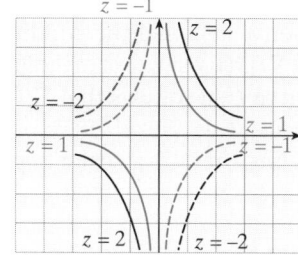

b.

■ **FIGURE 12.34**   Problem 58

57. Prove the second partials test. *Hint:* Compute the second directional derivative of f in the direction of $\mathbf{u} = h\mathbf{i} + k\mathbf{j}$ and complete the square.

58. Verify the formulas for m and b associated with the least-squares approximation.

59. This problem involves a generalization of the least-squares procedure, in which a "least-squares plane" is found to produce the best fit for a given set of data. A researcher knows that the quantity z is related to x and y by a formula of the form $z = k_1 x + k_2 y$ where k_1 and k_2 are physical constants. To determine these constants, she conducts a series of experiments, the results of which are tabulated as follows:

x	1.20	0.86	1.03	1.65	-0.95	-1.07
y	0.43	1.92	1.52	-1.03	1.22	-0.06
z	3.21	5.73	2.22	0.92	-1.11	-0.97

Modify the method of least squares to find a "best approximation" for k_1 and k_2.

12.8 Lagrange Multipliers

IN THIS SECTION   method of Lagrange multipliers, constrained optimization problems, Lagrange multipliers with two parameters, a geometric interpretation

METHOD OF LAGRANGE MULTIPLIERS

In many applied problems, a function of two variables is to be optimized subject to a restriction or **constraint** on the variables. For example, consider a container heated in such a way that the temperature at the point (x, y, z) in the container is given by the function $T(x, y, z)$. Suppose that the surface $z = f(x, y)$ lies in the container, and that we wish to find the point on $z = f(x, y)$ where the temperature is the greatest. In other words, *What is the maximum value of T subject to the constraint z = f(x, y), and where does this maximum value occur?* (See Figure 12.35.)

We will use the following theorem to solve such a problem.

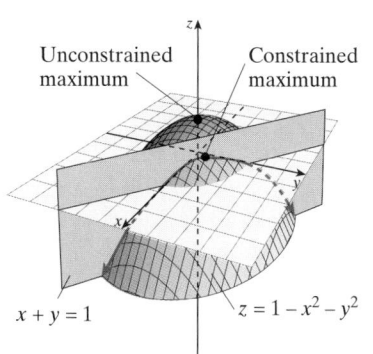

■ **FIGURE 12.35**   Find the maximum value on the surface defined by $f(x, y) = 1 - x^2 - y^2$ subject to the constraint $x + y = 1$ ($x \geq 0, y \geq 0$). See Example 1 for a solution.

THEOREM 12.13   *Lagrange's theorem*

Assume that f and g have continuous first partial derivatives and that f has an extremum at $P_0(x_0, y_0)$ on the smooth constraint curve $g(x, y) = c$. If $\nabla g(x_0, y_0) \neq \mathbf{0}$, there is a number λ such that

$$\nabla f(x_0, y_0) = \lambda \nabla g(x_0, y_0)$$

Proof   Denote the constraint curve $g(x, y) = c$ by C, and note that C is smooth. We represent this curve by the vector function

$$\mathbf{R}(t) = x(t)\mathbf{i} + y(t)\mathbf{j}$$

for all t in an open interval I, including t_0 corresponding to P_0, where $x'(t)$ and $y'(t)$ exist and are continuous. Let $F(t) = f[x(t), y(t)]$ for all t in I, and apply the chain rule to obtain

$$F'(t) = f_x \frac{dx}{dt} + f_y \frac{dy}{dt} = \nabla f[x(t), y(t)] \cdot \mathbf{R}'(t)$$

Because $f(x, y)$ has an extremum at P_0, we know that $F(t)$ has an extremum at t_0, the value of t that corresponds to P_0 (that is, P_0 is the point on C where $t = t_0$). Therefore, we have $F'(t_0) = 0$ and

$$F'(t_0) = \nabla f[x(t_0), y(t_0)] \cdot \mathbf{R}'(t_0) = 0$$

If $\nabla f[x(t_0), y(t_0)] = 0$, then $\lambda = 0$, and the condition $\nabla f = \lambda \nabla g$ is satisfied trivially. If $\nabla f[x(t_0), y(t_0)] \neq 0$, then $\nabla f[x(t_0), y(t_0)]$ is orthogonal to $\mathbf{R}'(t_0)$. Because $\mathbf{R}'(t_0)$ is tangent to the constraint curve C, it follows that $\nabla f(x_0, y_0)$ is normal to C. But $\nabla g(x_0, y_0)$ is also normal to C (because C is a level curve of g), and we conclude that ∇f and ∇g must be *parallel* at P_0. Thus, there is a scalar λ such that $\nabla f(x_0, y_0) = \lambda \nabla g(x_0, y_0)$, as required.

CONSTRAINED OPTIMIZATION PROBLEMS

The general procedure for the method of Lagrange multipliers may be described as follows.

> **Procedure for the Method of Lagrange Multipliers**
>
> Suppose f and g satisfy the hypotheses of Lagrange's theorem, and that $f(x, y)$ has an extremum subject to the constraint $g(x, y) = c$. Then to find the extreme values, proceed as follows:
>
> 1. Simultaneously solve the following three equations:
>
> $$f_x(x, y) = \lambda g_x(x, y) \qquad f_y(x, y) = \lambda g_y(x, y) \qquad g(x, y) = c$$
>
> 2. Evaluate f at all points found in step 1. The extremum we seek must be among these values.

EXAMPLE 1   *Optimization with Lagrange multiplier*

Find the largest and smallest values of $f(x, y) = 1 - x^2 - y^2$ subject to the constraint $x + y = 1$ with $x \geq 0, y \geq 0$.

Solution   Because the constraint is $x + y = 1$, let $g(x, y) = x + y - 1$.

$$f_x(x, y) = -2x \qquad f_y(x, y) = -2y \qquad g_x(x, y) = 1 \qquad g_y(x, y) = 1$$

Form the system

$$\begin{cases} -2x = \lambda(1) \\ -2y = \lambda(1) \\ x + y = 1 \end{cases}$$

The only solution is $x = \frac{1}{2}, y = \frac{1}{2}, \lambda = -1$.

$$f\left(\frac{1}{2}, \frac{1}{2}\right) = 1 - \left(\frac{1}{2}\right)^2 - \left(\frac{1}{2}\right)^2 = \frac{1}{2}$$

The endpoints of the line segment $x + y = 1$ for $x \geq 0, y \geq 0$ are at $(1, 0)$ and $(0, 1)$, and we find that

$$f(1, 0) = 1 - 1^2 - 0^2 = 0$$
$$f(0, 1) = 1 - 0^2 - 1^2 = 0$$

Therefore, the maximum value is $\frac{1}{2}$ at $(\frac{1}{2}, \frac{1}{2})$, and the minimum value is 0 at $(1, 0)$ and $(0, 1)$. ∎

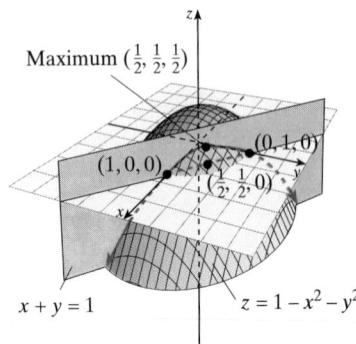

■ FIGURE 12.36 The maximum is the high point of the curve of intersection of the surface and the plane

The method of Lagrange multipliers extends naturally to functions of three or more variables. If a function $f(x, y, z)$ has an extreme value subject to a constraint $g(x, y, z) = c$, then the extremum occurs at a point (x_0, y_0, z_0) such that $g(x_0, y_0, z_0) = c$ and $\nabla f(x_0, y_0, z_0) = \lambda \nabla g(x_0, y_0, z_0)$ for some number λ. Here is an example.

EXAMPLE 2 *Hottest and coldest points on a plate*

A container in $\mathbb{R}^3$ has the shape of the cube given by $0 \le x \le 1, 0 \le y \le 1, 0 \le z \le 1$. A plate is placed in the container in such a way that it occupies that portion of the plane $x + y + z = 1$ that lies in the cubical container. If the container is heated so that the temperature at each point (x, y, z) is given by

$$T(x, y, z) = 4 - 2x^2 - y^2 - z^2$$

in hundreds of degrees Celsius, what are the hottest and coldest points on the plate?

Solution The cube and plate are shown in Figure 12.37. We shall use Lagrange multipliers to find all critical points in the interior of the plate, and then we shall examine the plate's boundary. To apply the method of Lagrange multipliers, we must solve $\nabla T = \lambda \nabla g$, where $g(x, y, z) = x + y + z - 1$. We obtain the partial derivatives.

$$T_x = -4x \qquad T_y = -2y \qquad T_z = -2z \qquad g_x = g_y = g_z = 1$$

We must solve the system

$$\begin{cases} -4x = \lambda \\ -2y = \lambda \\ -2z = \lambda \\ x + y + z = 1 \end{cases}$$

The solution of this system is $\left(\frac{1}{5}, \frac{2}{5}, \frac{2}{5}\right)$. The boundary of the plate is a triangle with vertices $A(1, 0, 0)$, $B(0, 1, 0)$, and $C(0, 0, 1)$. The temperature along the edges of this triangle may be found as follows:

$$T_1(x) = 4 - 2x^2 - (0)^2 - (1 - x)^2 = 3 - 3x^2 + 2x, \quad 0 \le x \le 1$$
$$T_2(x) = 4 - 2x^2 - (1 - x)^2 - (0)^2 = 3 - 3x^2 + 2x, \quad 0 \le x \le 1$$
$$T_3(y) = 4 - 2(0)^2 - y^2 - (1 - y)^2 = 3 + 2y - 2y^2, \quad 0 \le y \le 1$$

Edge AC: Differentiating, $T_1'(x) = T_2'(x) = -6x + 2$, which equals 0 when $x = \frac{1}{3}$. If $x = \frac{1}{3}$, then $z = \frac{2}{3}$ (because $x + z = 1, y = 0$ on edge AC), so we have the critical point $\left(\frac{1}{3}, 0, \frac{2}{3}\right)$.

Edge AB: Because $T_2 = T_1$, we see $x = \frac{1}{3}$. If $x = \frac{1}{3}$, then $y = \frac{2}{3}$ (because $x + y = 1, z = 0$ on edge BC), so we have another critical point $\left(\frac{1}{3}, \frac{2}{3}, 0\right)$.

Edge BC: Differentiating, $T_3'(y) = 2 - 4y$, which equals 0 when $y = \frac{1}{2}$. Because $y + z = 1$, and $x = 0$, we have the critical point $\left(0, \frac{1}{2}, \frac{1}{2}\right)$.

Endpoints of the edges: $(1, 0, 0)$, $(0, 1, 0)$, and $(0, 0, 1)$.

The last step is to evaluate T at the critical points and the endpoints:

$$T\left(\tfrac{1}{5}, \tfrac{2}{5}, \tfrac{2}{5}\right) = 3\tfrac{3}{5};$$
$$T\left(\tfrac{1}{3}, 0, \tfrac{2}{3}\right) = 3\tfrac{1}{3}; \qquad T\left(\tfrac{1}{3}, \tfrac{2}{3}, 0\right) = 3\tfrac{1}{3}; \qquad T\left(0, \tfrac{1}{2}, \tfrac{1}{2}\right) = 3\tfrac{1}{2};$$
$$T(1, 0, 0) = 2; \qquad T(0, 1, 0) = 3; \qquad T(0, 0, 1) = 3$$

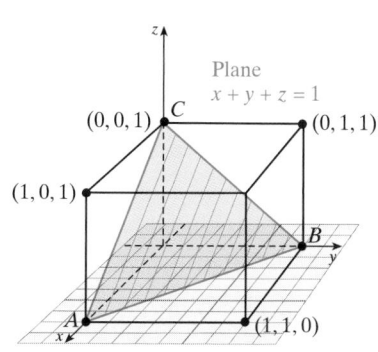

■ **FIGURE 12.37** Find the hottest and coldest points on the plate inside the cube

Edge AC: $x + z = 1, y = 0$
Edge AB: $x + y = 1, z = 0$
Edge BC: $y + z = 1, x = 0$

Comparing these values (remember that the temperature is in hundreds of degrees Celsius), we see that the highest temperature is $360\,°C$ at $(\frac{1}{5}, \frac{2}{5}, \frac{2}{5})$ and the lowest temperature is $200\,°C$ at $(1, 0, 0)$. ◼

Notice that, in the end, we do not care about the value of λ. The multiplier is just used as an intermediary device for finding the critical points. However, in some problems, we do care about the value of λ, because of the interpretation given in the following theorem.

THEOREM 12.14   *Rate of change of the extreme value*

Suppose E is an extreme value (maximum or minimum) of f subject to the constraint $g(x, y) = c$. Then the Lagrange multiplier λ is the rate of change of E with respect to c; that is, $\lambda = dE/dc$.

Proof   Note that at the extreme value (x, y) we have

$$f_x = \lambda g_x, \qquad f_y = \lambda g_y, \qquad \text{and} \qquad g(x, y) = c$$

The coordinates of the optimal ordered pair (x, y) depend on c (because different constraint levels will generally lead to different optimal combinations of x and y). Thus,

$$E = E(x, y) \quad \text{where } x \text{ and } y \text{ are functions of } c$$

By the chain rule for partial derivatives:

$$\frac{dE}{dc} = \frac{\partial E}{\partial x}\frac{dx}{dc} + \frac{\partial E}{\partial y}\frac{dy}{dc}$$

$$= f_x \frac{dx}{dc} + f_y \frac{dy}{dc} \qquad \text{Because } E = f(x, y)$$

$$= \lambda g_x \frac{dx}{dc} + \lambda g_y \frac{dy}{dc} \qquad \text{Because } f_x = \lambda g_x \text{ and } f_y = \lambda g_y$$

$$= \lambda \left(g_x \frac{dx}{dc} + g_y \frac{dy}{dc} \right)$$

$$= \lambda \frac{dg}{dc} \qquad \text{Chain rule}$$

$$= \lambda \qquad \text{Because } \frac{dg}{dc} = 1 \text{ (remember } g = c)$$

This theorem can be interpreted as saying that the multiplier gives the change in the extreme value E that results when the constraint c is increased by 1 unit. This interpretation is illustrated in the following example.

EXAMPLE 3   *Maximum output for a Cobb–Douglas production function*

If x thousand dollars is spent on labor, and y thousand dollars is spent on equipment, it is estimated that the output of a certain factory will be

$$Q(x, y) = 50x^{2/5} y^{3/5}$$

units. If \$150,000 is available, how should this capital be allocated between labor and equipment to generate the largest possible output? How does the maximum output change if the money available for labor and equipment is increased by \$1,000? In economics, an output function of the general form $Q(x, y) = x^\alpha y^{1-\alpha}$ is known as a **Cobb–Douglas production function.**

Solution   Because x and y are given in units of $1,000, the constraint equation is $x + y = 150$. If we set $g(x, y) = x + y - 150$, we wish to maximize Q subject to $g(x, y) = 0$. To apply the method of Lagrange multipliers, we first find

$$Q_x = 20x^{-3/5}y^{3/5} \qquad Q_y = 30x^{2/5}y^{-2/5} \qquad g_x = 1 \qquad g_y = 1$$

Next, solve the system

$$\begin{cases} 20x^{-3/5}y^{3/5} = \lambda \\ 30x^{2/5}y^{-2/5} = \lambda \\ x + y - 150 = 0 \end{cases}$$

From the first two equations we have

$$20x^{-3/5}y^{3/5} = 30x^{2/5}y^{-2/5}$$
$$20y = 30x$$
$$y = 1.5x$$

Substitute $y = 1.5x$ into the equation $x + y = 150$ to find $x = 60$. This leads to the solution $y = 90$, so that the maximum output is

$$Q(60, 90) = 50(60)^{2/5}(90)^{3/5} \approx 3,826.273502 \text{ units}$$

We also find that

$$\lambda = 20(60)^{-3/5}(90)^{3/5} \approx 25.50849001$$

Thus, the maximum output is about 3,826 units and occurs when $60,000 is allocated to labor and $90,000 to equipment. We also note that an increase of $1,000 (1 unit) in the available funds will increase the maximum output by approximately $\lambda \approx 25.5$ units (from 3,826.27 to 3,851.78 units). ∎

LAGRANGE MULTIPLIERS WITH TWO PARAMETERS

The method of Lagrange multipliers can also be applied in situations with more than one constraint equation. Suppose we wish to locate an extremum of a function defined by $f(x, y, z)$ subject to *two* constraints, $g(x, y, z) = c_1$ and $h(x, y, z) = c_2$, where g and h are also differentiable and ∇g and ∇h are not parallel. By generalizing Lagrange's theorem, it can be shown that if (x_0, y_0, z_0) is the desired extremum, then there are numbers λ and μ such that $g(x_0, y_0, z_0) = c_1, h(x_0, y_0, z_0) = c_2$, and

$$\nabla f(x_0, y_0, z_0) = \lambda \nabla g(x_0, y_0, z_0) + \mu \nabla h(x_0, y_0, z_0)$$

As in the case of one constraint, we proceed by first solving this system of equations simultaneously to find $\lambda, \mu, x_0, y_0, z_0$, then evaluating $f(x, y, z)$ at each solution, and comparing to find the required extremum. This approach is illustrated in our final example of this section.

EXAMPLE 4   *Optimization with two constraints*

Find the point on the intersection of the plane $x + 2y + z = 10$ and the paraboloid $z = x^2 + y^2$ that is closest to the origin.

Solution   The distance from a point (x, y, z) to the origin is $s = \sqrt{x^2 + y^2 + z^2}$, but instead of minimizing this quantity, it is easier to minimize its square. That is, we shall minimize

$$f(x, y, z) = x^2 + y^2 + z^2$$

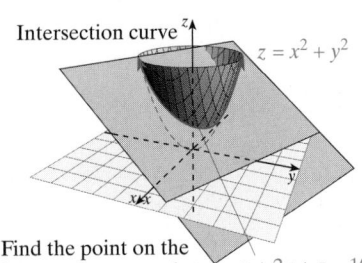

Intersection curve

$z = x^2 + y^2$

$x + 2y + z = 10$

Find the point on the intersection curve that is closest to the origin.

Graphic representation of Example 4

subject to the joint constraints

$$g(x, y, z) = x + 2y + z = 10 \quad \text{and} \quad h(x, y, z) = x^2 + y^2 - z = 0$$

Compute the partial derivatives of f, g, and h:

$$
\begin{array}{lll}
f_x = 2x & f_y = 2y & f_z = 2z \\
g_x = 1 & g_y = 2 & g_z = 1 \\
h_x = 2x & h_y = 2y & h_z = -1
\end{array}
$$

To apply the method of Lagrange multipliers, we use the formula

$$\nabla f(x_0, y_0, z_0) = \lambda \nabla g(x_0, y_0, z_0) + \mu \nabla h(x_0, y_0, z_0)$$

which leads to the following system of equations:

$$
\begin{cases}
2x = \lambda(1) + \mu(2x) \\
2y = \lambda(2) + \mu(2y) \\
2z = \lambda(1) + \mu(-1) \\
x + 2y + z = 10 \\
z = x^2 + y^2
\end{cases}
$$

This is not a linear system, so solving it requires ingenuity.

Multiply the first equation by 2 and subtract the second equation to obtain

$$4x - 2y = (4x - 2y)\mu$$
$$(4x - 2y) - (4x - 2y)\mu = 0$$
$$(4x - 2y)(1 - \mu) = 0$$
$$4x - 2y = 0 \quad \text{or} \quad 1 - \mu = 0$$

CASE I **If $4x - 2y = 0$**, then $y = 2x$. Substitute this into the two constraint equations:

$x + 2y + z = 10$	$x^2 + y^2 - z = 0$
$x + 2(2x) + z = 10$	$x^2 + (2x)^2 - z = 0$
$z = 10 - 5x$	$z = 5x^2$

By substitution we have $5x^2 = 10 - 5x$, which has solutions $x = 1$ and $x = -2$. This implies

$x = 1$	$x = -2$
$y = 2x = 2(1) = 2$	$y = 2x = 2(-2) = -4$
$z = 5x^2 = 5(1)^2 = 5$	$z = 5x^2 = 5(-2)^2 = 20$

Thus, the points $(1, 2, 5)$ and $(-2, -4, 20)$ are candidates for the minimal distance.

CASE II **If $1 - \mu = 0$**, then $\mu = 1$, and we look at the system of equations involving x, y, z, λ, and μ:

$$
\begin{cases}
2x = \lambda(1) + \mu(2x) \\
2y = \lambda(2) + \mu(2y) \\
2z = \lambda(1) + \mu(-1) \\
x + 2y + z = 10 \\
z = x^2 + y^2
\end{cases}
$$

The top equation becomes $2x = \lambda + 2x$, so that $\lambda = 0$. We now find z from the third equation:

$$2z = -1 \quad \text{or} \quad z = -\tfrac{1}{2}$$

Next, turn to the constraint equations:

$x + 2y + z = 10$	$x^2 + y^2 - z = 0$
$x + 2y - \frac{1}{2} = 10$	$x^2 + y^2 + \frac{1}{2} = 0$
$x + 2y = 10 + \frac{1}{2}$	$x^2 + y^2 = -\frac{1}{2}$

There is no solution because $x^2 + y^2$ cannot equal a negative number.

We check the candidates for the minimal distance:

$$f(x, y, z) = x^2 + y^2 + z^2 \quad \text{so that}$$
$$f(1, 2, 5) = 1^2 + 2^2 + 5^2 = 30$$
$$f(-2, -4, 20) = (-2)^2 + (-4)^2 + 20^2 = 420$$

Because $f(x, y, z)$ represents the square of the distance, the minimal distance is $\sqrt{30}$ and the point on the intersection of the two surfaces nearest to the origin is $(1, 2, 5)$. ∎

A GEOMETRIC INTERPRETATION

Lagrange's theorem can be interpreted geometrically. Suppose the constraint curve $g(x, y) = c$ and the level curves $f(x, y) = k$ are drawn in the xy-plane, as shown in Figure 12.38.

To maximize $f(x, y)$ subject to the constraint $g(x, y) = c$, we must find the "highest" (rightmost, actually) level curve of f that intersects the constraint curve. As the sketch in Figure 12.38 suggests, this critical intersection occurs at a point where the constraint curve is tangent to a level curve—that is, where the slope of the constraint curve $g(x, y) = c$ is equal to the slope of a level curve $f(x, y) = k$. According to the formula derived in Section 12.5 (p. 829).

Slope of constraint curve $g(x, y) = c$ is $\dfrac{-g_x}{g_y}$

Slope of each level curve is $\dfrac{-f_x}{f_y}$

The condition that the slopes are equal can be expressed by

$$\frac{-f_x}{f_y} = \frac{-g_x}{g_y}, \quad \text{or, equivalently,} \quad \frac{f_x}{g_x} = \frac{f_y}{g_y}$$

Let λ equal this common ratio,

$$\lambda = \frac{f_x}{g_x} \quad \text{and} \quad \lambda = \frac{f_y}{g_y}$$

so that

$$f_x = \lambda g_x \quad \text{and} \quad f_y = \lambda g_y$$

and $\nabla f = f_x \mathbf{i} + f_y \mathbf{j} = \lambda(g_x \mathbf{i} + g_y \mathbf{j}) = \lambda \nabla g$.

Because the point in question must lie on the constraint curve, we also have $g(x, y) = c$. If these equations are satisfied at a certain point (a, b), then f will reach its constrained *maximum* at (a, b) if the *highest* level curve that intersects the constraint curve does so at this highest point. On the other hand, if the *lowest* level curve that intersects the constraint curve does so at (a, b), then f achieves its constrained *minimum* at this point.

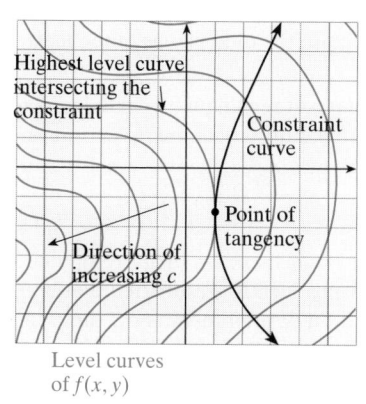

Highest level curve intersecting the constraint

Constraint curve

Point of tangency

Direction of increasing c

Level curves of $f(x, y)$

■ **FIGURE 12.38** Increasing level curves and the constraint curve

12.8 Problem Set

A *Use the method of Lagrange multipliers to find the required constrained extrema in Problems 1–14.*

1. Maximize $f(x, y) = xy$ subject to $2x + 2y = 5$.
2. Maximize $f(x, y) = xy$ subject to $x + y = 20$.
3. Maximize $f(x, y) = 16 - x^2 - y^2$ subject to $x + 2y = 6$.
4. Minimize $f(x, y) = x^2 + y^2$ subject to $x + y = 24$.
5. Minimize $f(x, y) = x^2 + y^2$ subject to $xy = 1$.
6. Minimize $f(x, y) = x^2 - xy + 2y^2$ subject to $2x + y = 22$.
7. Minimize $f(x, y) = x^2 - y^2$ subject to $x^2 + y^2 = 4$.
8. Maximize $f(x, y) = x^2 - 2y - y^2$ subject to $x^2 + y^2 = 1$.
9. Maximize $f(x, y) = \cos x + \cos y$ subject to $y = x + \frac{\pi}{4}$.
10. Maximize $f(x, y) = e^{xy}$ subject to $x^2 + y^2 = 3$.
11. Maximize $f(x, y) = \ln(xy^2)$ subject to $2x^2 + 3y^2 = 8$ for $x > 0, y > 0$.
12. Maximize $f(x, y, z) = xyz$ subject to $3x + 2y + z = 6$.
13. Minimize $f(x, y, z) = x^2 + y^2 + z^2$ subject to $x - 2y + 3z = 4$.
14. Minimize $f(x, y, z) = x^2 + y^2 + z^2$ subject to $4x^2 + 2y^2 + z^2 = 4$.

B 15. Find the smallest value of $f(x, y, z) = 2x^2 + 4y^2 + z^2$ subject to $4x - 8y + 2z = 10$. What, if anything, can be said about the largest value of f subject to this constraint?

16. Let $f(x, y, z) = x^2y^2z^2$. Show that the maximum value of f on the sphere $x^2 + y^2 + z^2 = R^2$ is $R^6/27$.

17. Find the maximum and minimum values of $f(x, y, z) = x - y + z$ on the sphere $x^2 + y^2 + z^2 = 100$.

18. Find the maximum and minimum values $f(x, y, z) = 4x - 2y - 3z$ on the sphere $x^2 + y^2 + z^2 = 100$.

19. Use Lagrange multipliers to find the distance from the origin to the plane $Ax + By + Cz = D$ where at least one of A, B, C is nonzero.

20. Find the maximum and minimum distance from the origin to the ellipse $5x^2 - 6xy + 5y^2 = 4$.

21. Find the point on the plane

$$2x + y + z = 1$$

that is nearest to the origin.

22. Find the largest product of positive numbers $x, y,$ and z such that their sum is 24.

23. Write the number 12 as the sum of three positive numbers x, y, z in such a way that the product xy^2z is a maximum.

24. A rectangular box with no top is to be constructed from 96 ft^2 of material. What should be the dimensions of the box if it is to enclose maximum volume?

25. The temperature T at point (x, y, z) in a region of space is given by the formula $T = 100 - xy - xz - yz$. Find the lowest temperature on the plane $x + y + z = 10$.

26. A farmer wishes to fence off a rectangular pasture along the bank of a river. The area of the pasture is to be 3,200 yd^2, and no fencing is needed along the river bank. Find the dimensions of the pasture that will require the least amount of fencing.

27. There are 320 yd of fencing available to enclose a rectangular field. How should the fencing be used so that the enclosed area is as large as possible?

28. Use the fact that 12 fl oz is approximately 6.89π in.^{3} to find the dimensions of the 12-oz soda can that can be constructed using the least amount of metal. Compare your answer with an actual can of Pepsi.

29. A cylindrical can is to hold 4π in.^{3} of orange juice. The cost per square inch of constructing the metal top and bottom is twice the cost per square inch of constructing the cardboard side. What are the dimensions of the least expensive can?

30. Find the volume of the largest rectangular parallelepiped that can be inscribed in the ellipsoid

$$x^2 + \frac{y^2}{4} + \frac{z^2}{9} = 1$$

31. A manufacturer has $8,000 to spend on the development and promotion of a new product. It is estimated that if x thousand dollars is spent on development and y thousand is spent on promotion, sales will be approximately $f(x, y) = 50x^{1/2}y^{3/2}$ units. How much money should the manufacturer allocate to development and how much to promotion to maximize sales?

32. **MODELING PROBLEM**   If x thousand dollars is spent on labor and y thousand dollars is spent on equipment, the output at a certain factory may be modeled by

$$Q(x, y) = 60x^{1/3}y^{2/3}$$

units. Assume $120,000 is available.
   a. How should money be allocated between labor and equipment to generate the largest possible output?
   b. Use the Lagrange multiplier λ to estimate the change in the maximum output of the factory that would result if the money available for labor and equipment is increased by $1,000.

33. **MODELING PROBLEM**   An architect decides to model the usable living space in a building by the volume of space that can be used comfortably by a person 6 feet tall—that is, by the largest 6-foot-high rectangular box that can be inscribed in the building. Find the dimensions

of an A-frame building y ft long with equilateral triangular ends x ft on a side that maximizes usable living space if the exterior surface area of the building cannot exceed 500 ft^2.

34. Find the radius of the largest cylinder of height 6 in. that can be inscribed in an inverted cone of height H, radius R, and surface area 250 in.^{2}.

35. In Problem 42 of Problem Set 12.7, you were asked to minimize the ground state energy

$$E(x, y, z) = \frac{k^2}{8m}\left(\frac{1}{x^2} + \frac{1}{y^2} + \frac{1}{z^2}\right)$$

subject to the volume constraint $V = xyz = C$. Solve the problem using Lagrange multipliers.

36. A university extension agricultural service concludes that, on a particular farm, the yield of wheat per acre is a function of water and fertilizer. Let x be the number of acre-feet of water applied, and y the number of pounds of fertilizer applied during the growing season. The agricultural service then concluded that the yield (measured in bushels), represented by f, can be defined by the formula $f(x, y) = 500 + x^2 + 2y^2$. Suppose that water costs $20 per acre-foot, fertilizer costs $12 per pound, and the farmer will invest $236 per acre for water and fertilizer. How much water and fertilizer should the farmer buy to maximize the yield?

37. How would the farmer of Problem 36 maximize the yield if the amount spent is $100 instead of $236?

38. Present post office regulations specify that a box (that is, a package in the form of a rectangular parallelepiped) can be mailed parcel post only if the sum of its length and girth does not exceed 108 inches, as shown in Figure 12.39. Find the maximum volume of such a package. (Compare your solution here with the one you might have given to Problem 17, Section 4.6, page 280.)

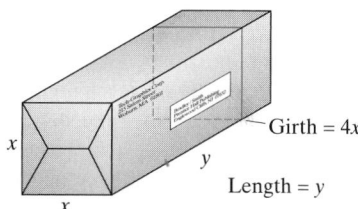

Girth = $4x$

Length = y

■ **FIGURE 12.39** Maximum volume for a box mailed by U.S. parcel post

39. Heron's formula says that the area of a triangle with sides a, b, c is

$$A = \sqrt{s(s - a)(s - b)(s - c)}$$

where $s = \frac{1}{2}(a + b + c)$ is the semi-perimeter of the triangle. Use this result and the method of Lagrange multipliers to show that of all triangles with a given fixed perimeter P, the equilateral triangle has the largest area.

40. If x, y, z are the angles of a triangle, what is the maximum value of the product
$P(x, y, z) = \sin x \sin y \sin z$? What about
$Q(x, y, z) = \cos x \cos y \cos z$?

Use the method of Lagrange multipliers in Problems 41–44 to find the required extrema for the two given constraints.

41. Find the minimum of $f(x, y, z) = x^2 + y^2 + z^2$ subject to $x + y = 4$ and $y + z = 6$.

42. Find the maximum of $f(x, y, z) = xyz$ subject to $x^2 + y^2 = 3$ and $y = 2z$.

43. Maximize $f(x, y, z) = xy + xz$ subject to $2x + 3z = 5$ and $xy = 4$.

44. Minimize $f(x, y, z) = 2x^2 + 3y^2 + 4z^2$ subject to $x + y + z = 4$ and $x - 2y + 5z = 3$.

45. **MODELING PROBLEM**   A manufacturer is planning to sell a new product at the price of $150 per unit and estimates that if x thousand dollars is spent on development and y thousand dollars on promotion, then approximately

$$\frac{320y}{y + 2} + \frac{160x}{x + 4}$$

units of the product will be sold. The cost of manufacturing the product is $50 per unit.

a. If the manufacturer has a total of $8,000 to spend on the development and promotion, how should this money be allocated to generate the largest possible profit?

b. Suppose the manufacturer decides to spend $8,100 instead of $8,000 on the development and promotion of the new product. Estimate how this change will affect the maximum possible profit.

c. If unlimited funds are available, how much should the manufacturer spend on development and promotion to maximize profit?

d. What is the Lagrange multiplier in part **c**? Your answer should suggest another method for solving the problem in part **c**. Solve the problem using this alternative approach.

46. **MODELING PROBLEM**   A jewelry box with a square base has an interior partition and is required to have volume 800 cm^3 (see Figure 12.40).

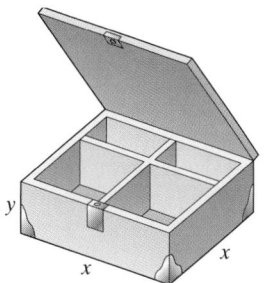
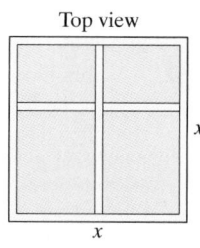

Top view

■ **FIGURE 12.40** Constructing a jewelry box

a. The material in the top costs twice as much as the material in the side and in the bottom, which in turn, costs twice as much as the material in the partitions. Find the dimensions of the box that minimize the total cost of construction. *Drat! We forgot to tell you where the partitions are located. Does that matter?*

b. Suppose the volume constraint changes from 800 cm^3 to 801 cm^3. Estimate the appropriate effect on the minimal cost.

C 47. A farmer wants to build a metal silo in the shape of a right circular cylinder with a right circular cone on the top (the bottom of the silo will be a concrete slab). What is the least amount of metal that can be used if the silo is to have a fixed volume V_0?

48. Find the volume of the largest rectangular parallelepiped (box) that can be inscribed in the ellipsoid

$$\frac{x^2}{a^2} + \frac{y^2}{b^2} + \frac{z^2}{c^2} = 1$$

(See Problem 30.)

49. ■ **What Does This Say?**   The method of Lagrange multipliers gives a constrained extremum only if one exists. Apply the method to the problem: Optimize $f(x, y) = x + y$ subject to $xy = 1$. The method yields two candidates for an extremum. Is one a maximum and the other a minimum? Explain.

In Problems 50–53, let $Q(x, y)$ be a production function in which x and y represent units of labor and capital, respectively. If p and q represent unit costs of labor and capital, respectively, then $C = px + qy$ represents the total cost of production.

50. Use Lagrange multipliers to show that, subject to a fixed production level Q_0, the total cost is smallest when

$$\frac{Q_x}{p} = \frac{Q_y}{q} \quad \text{and} \quad Q(x, y) = Q_0$$

(provided $\nabla f \neq 0$, and $p \neq 0, q \neq 0$). This is often referred to as the **minimum cost problem,** and its solution is called the **least-cost combination of inputs.**

51. Show that the inputs x, y that maximize the production level $Q = f(x, y)$ subject to a fixed cost k satisfy

$$\frac{f_x}{p} = \frac{f_y}{q} \quad \text{with } px + qy = k$$

(assume $p \neq 0, q \neq 0$). This is called a **fixed-budget problem.**

52. A Cobb–Douglas production function is an output function of the form $Q(x, y) = cx^\alpha y^\beta$, with $\alpha + \beta = 1$. Show that such a function is maximized with respect to the fixed cost $px + qy = k$ when $x = \alpha k/p$ and $y = \beta k/q$.

Where does the maximum occur if we drop the condition $\alpha + \beta = 1$? How does the maximum output change if k is increased by 1 unit?

53. Show that the cost function

$$C(x, y) = px + qy$$

is minimized subject to the fixed production level $Ax^\alpha y^\beta = k$, with $\alpha + \beta = 1$, when

$$x = \frac{k}{A}\left(\frac{\alpha q}{\beta p}\right)^\beta \qquad y = \frac{k}{A}\left(\frac{\beta p}{\alpha q}\right)^\alpha$$

54. ʜISTORICAL ǪUEST   A discussion of Lagrange multipliers would not be complete without mention of Joseph Lagrange, generally acknowledged as one of the two greatest mathematicians of the 18th century, Leonhard Euler being the other (see ʜistorical ǪQuest Problem 84 of the supplementary problems for Chapter 4). There is a distinct difference in style between Lagrange and Euler. Lagrange has been characterized as the first true analyst in the sense that he attempted to write concisely and with rigor. On the other hand, Euler wrote using intuition and with an abundance of detail. Lagrange was described by Napoleon Bonaparte as "the lofty pyramid of the mathematical sciences" and followed Euler as the court mathematician for Frederick the Great. He was the first to use the notation $f'(x)$ and $f''(x)$ for derivatives. In this section, we were introduced to the method of Lagrange multipliers, which provides a procedure for constrained optimization. This method was contained in a paper on mechanics that Lagrange wrote when he was only 19 years old.*

JOSEPH LAGRANGE
1736–1813

For this ǪQuest, we consider Lagrange's work with solving *algebraic* equations. You are familiar with the quadratic formula, which provides a general solution for any second-degree equation $ax^2 + bx + c = 0, a \neq 0$. Lagrange made an exhaustive study of the general solution for the first four degrees. Here is what he did. Suppose you are given a general algebraic expression involving letters $a, b, c, \ldots$; how many *different* expressions can be derived from the given one if the letters are interchanged in all possible ways? For example, from $ab + cd$ we obtain $ad + cb$ by interchanging b and d.

This problem suggests another closely related problem, so part of Lagrange's approach. Lagrange solved general algebraic equations of degree 2, 3, and 4. It was proved later (not by Lagrange, but by Galois and Abel), that no general solution for equations greater than 5 can be found. Do some research and find the general solution for equations of degree 1, 2, 3, and 4.

———————

*From *Men of Mathematics* by E. T. Bell, Simon and Schuster, New York, 1937, p. 165.

Chapter 12   Review

Proficiency Examination

Concept Problems

1. What is a function of two variables?
2. What is the domain of a function of two variables?
3. What is a level curve of the function defined by $f(x, y)$?
4. What do we mean by the limit of a function of two variables?
5. State the following properties of a limit of functions of two variables.
   a. scalar rule   b. sum rule
   c. product rule   d. quotient rule
6. Define the continuity of a function defined by $f(x, y)$ at a point (x_0, y_0) in its domain.
7. If $z = f(x, y)$, define the first partial derivatives of f with respect to x and y.
8. What is the slope of a tangent line to the surface defined by $z = f(x, y)$ that is parallel to the xy-plane?
9. If $z = f(x, y)$, find the second partial derivatives.
10. If $z = f(x, y)$, what are the increments of x, y, and z?
11. What does it mean for a function of two variables to be differentiable at (x_0, y_0)?
12. State the incremental approximation of $f(x, y)$.
13. Define the total differential of $z = f(x, y)$.
14. State the chain rule for one parameter.
15. State the chain rule for two parameters.
16. Define the directional derivative of a function defined by $z = f(x, y)$.
17. Define the gradient $\nabla f(x, y)$.
18. State the following basic properties of the gradient.
    a. constant rule
    b. linearity rule
    c. product rule
    d. quotient rule
    e. power rule
19. Express the directional derivative in terms of the gradient.
20. State the optimal direction property of the gradient (that is, the steepest ascent and steepest descent).
21. State the normal property of the gradient.
22. Define the normal line and tangent plane to a surface S at a point P_0.
23. Define the absolute extrema of a function of two variables.
24. Define the relative extrema of a function of two variables.
25. What is a critical point of a function of two variables?
26. State the second partials test.
27. State the extreme value theorem for a function of two variables.
28. What is the least-squares approximation of data, and what is a regression line?
29. State Lagrange's theorem.
30. State the procedure for the method of Lagrange multipliers.

Practice Problems

31. If $f(x, y) = \sin^{-1} xy$, verify that $f_{xy} = f_{yx}$.
32. Let $w = x^2 y + y^2 z$, where $x = t \sin t$, $y = t \cos t$, and $z = 2t$. Use the chain rule to find $\dfrac{dw}{dt}$, where $t = \pi$.
33. Let $f(x, y, z) = xy + yz + xz$, and let P_0 denote the point $(1, 2, -1)$.
    a. Find the gradient of f at P_0.
    b. Find the directional derivative of f in the direction from P_0 toward the point $Q(-1, 1, -1)$.
    c. Find the direction from P_0 in which the directional derivative has its largest value. What is the magnitude of the largest directional derivative at P_0?
34. Show that the function defined by
$$f(x, y) = \begin{cases} \dfrac{x^2 y}{x^3 + y^3} & \text{if } (x, y) \neq (0, 0) \\ 0 & \text{if } (x, y) = (0, 0) \end{cases}$$
    is not continuous at $(0, 0)$.
35. If $f(x, y) = \ln\left(\dfrac{y}{x}\right)$, find f_x, f_y, f_{yy}, and f_{xy}.
36. Show that if $f(x, y, z) = x^2 y + y^2 z + z^2 x$, then
$$\frac{\partial f}{\partial x} + \frac{\partial f}{\partial y} + \frac{\partial f}{\partial z} = (x + y + z)^2$$
37. Let $f(x, y) = (x^2 + y^2)^2$. Find the directional derivative of f at $(2, -2)$ in the direction that makes an angle of $\frac{2\pi}{3}$ with the positive x-axis.
38. Find all critical points of $f(x, y) = 12xy - 2x^2 - y^4$ and classify them using the second partials test.
39. Use the method of Lagrange multipliers to find the maximum and minimum values of the function $f(x, y) = x^2 + 2y^2 + 2x + 3$ subject to the constraint $x^2 + y^2 = 4$. You may assume these extreme values exist.
40. Find the largest and smallest values of the function $f(x, y) = x^2 - 4y^2 + 3x + 6y$ on the region defined by $-2 \leq x \leq 2, 0 \leq y \leq 1$.

Supplementary Problems

Describe the domain of each function given in Problems 1–4.

1. $f(x, y) = \sqrt{16 - x^2 - y^2}$ **2.** $f(x, y) = \dfrac{x^2 - y^2}{x - y}$

3. $f(x, y) = \sin^{-1}x + \cos^{-1}y$ **4.** $f(x, y) = e^{x+y}\tan^{-1}\left(\dfrac{y}{x}\right)$

Find the partial derivatives f_x and f_y for the functions defined in Problems 5–10.

5. $f(x, y) = \dfrac{x^2 - y^2}{x + y}$ **6.** $f(x, y) = x^3 e^{3y/(2x)}$

7. $f(x, y) = x^2 y + \sin\dfrac{y}{x}$ **8.** $f(x, y) = \ln\left(\dfrac{xy}{x + 2y}\right)$

9. $f(x, y) = 2x^3 y + 3xy^2 + \dfrac{y}{x}$ **10.** $f(x, y) = xye^{xy}$

For each function given in Problems 11–15, describe the level curve or level surface $f = c$ for the given values of the constant c.

11. $f(x, y) = x^2 - y; c = 2, c = -2$

12. $f(x, y) = 6x + 2y; c = 0, c = 1, c = 2$

13. $f(x, y) = \begin{cases} \sqrt{x^2 + y^2} & \text{if } x \geq 0 \\ |y| & \text{if } x < 0 \end{cases}$ $c = 0, c = 1, c = -1$

14. $f(x, y, z) = x^2 + y^2 + z^2; c = 16, c = 0, c = -25$

15. $f(x, y, z) = x^2 + \dfrac{y^2}{2} + \dfrac{z^2}{9}; c = 1, c = 2$

Evaluate the limits in Problems 16 and 17, assuming they exist.

16. $\displaystyle\lim_{(x,y)\to(1,1)} \dfrac{xy}{x^2 + y^2}$ **17.** $\displaystyle\lim_{(x,y)\to(0,0)} \dfrac{x + ye^{-x}}{1 + x^2}$

Show that each limit in Problems 18 and 19 does not exist.

18. $\displaystyle\lim_{(x,y)\to(0,0)} \dfrac{x^3 - y^3}{x^3 + y^3}$

19. $\displaystyle\lim_{(x,y)\to(0,0)} \dfrac{x^3 y^2}{x^6 + y^4}$

Find the derivatives in Problems 20–23 using the chain rule. You may leave your answers in terms of $x, y, t, u,$ and v.

20. Find $\dfrac{dz}{dt}$ where $z = -xy + y^3$, and $x = -3t^2, y = 1 + t^3$.

21. Find $\dfrac{dz}{dt}$ where $z = xy + y^2$, and $x = e^t t^{-1}, y = \tan t$.

22. Find $\dfrac{\partial z}{\partial u}$ and $\dfrac{\partial z}{\partial v}$ where $z = x^2 - y^2$, and $x = u + 2v$, $y = u - 2v$.

23. Find $\dfrac{\partial z}{\partial u}$ and $\dfrac{\partial z}{\partial v}$ where $z = x\tan\dfrac{x}{y}$, and $x = uv, y = \dfrac{u}{v}$.

Use implicit differentiation to find $\dfrac{\partial z}{\partial x}$ and $\dfrac{\partial z}{\partial y}$ in Problems 24–27.

24. $x^2 + 6y^2 + 2z^2 = 5$ **25.** $e^x + e^y + e^z = 3$

26. $x^3 + 2xz - yz^2 - z^3 = 1$ **27.** $x + 2y - 3z = \ln z$

In Problems 28–33, find f_{xx}, and f_{yx}.

28. $f(x, y) = \tan^{-1} xy$ **29.** $f(x, y) = \sin^{-1} xy$

30. $f(x, y) = x^2 + y^3 - 2xy^2$ **31.** $f(x, y) = e^{x^2+y^2}$

32. $f(x, y) = x\ln y$ **33.** $f(x, y) = \displaystyle\int_x^y \sin(\cos t)\, dt$

Find equations for the tangent plane and normal line to the surfaces given in Problems 34–36 at the prescribed point.

34. $x^2 y^3 z = 8$ at $P_0(2, -1, -2)$

35. $x^3 + 2xy^2 - 7x^3 + 3y + 1 = 0$ at $P_0(1, 1, 1)$

36. $z = \dfrac{-4}{2 + x^2 + y^2}$ at $P_0(1, 1, -1)$

Find all critical points of $f(x, y)$ in Problems 37–42 and classify each as a relative maximum, a relative minimum, or a saddle point.

37. $f(x, y) = x^2 - 6x + 2y^2 + 4y - 2$

38. $f(x, y) = x^3 + y^3 - 6xy$

39. $f(x, y) = (x - 1)(y - 1)(x + y - 1)$

40. $f(x, y) = x^2 + y^3 + 6xy - 7x - 6y$

41. $f(x, y) = x^3 + y^3 + 3x^2 - 18y^2 + 81y + 5$

42. $f(x, y) = \sin(x + y) + \sin x + \sin y$ for $0 < x < \pi$, $0 < y < \pi$ (See Figure 12.41.)

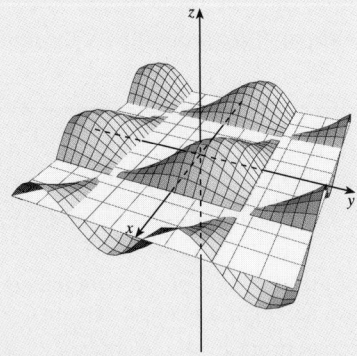

■ **FIGURE 12.41**   Graph of $f(x, y) = \sin(x + y) + \sin x + \sin y$

In Problems 43–46, find the largest and smallest values of the function f on the specified closed, bounded set S.

43. $f(x, y) = xy - 2y$; S is the rectangular region $0 \leq x \leq 3$, $-1 \leq y \leq 1$

44. $f(x, y) = x^2 + 2y^2 - x - 2y$; S is the triangular region with vertices $(0, 0), (2, 0), (2, 2)$

45. $f(x, y) = x^2 + y^2 - 3y$; S is the disk $x^2 + y^2 \leq 4$

46. $f(x, y) = 6x - x^2 + 2xy - y^4$; S is the square $0 \leq x \leq 3$, $0 \leq y \leq 3$

47. Use the chain rule to find $\dfrac{dz}{dt}$ if $z = x^2 - 3xy^2$; $x = 2t$, $y = t^2$.

48. Use the chain rule to find $\dfrac{dz}{dt}$ if $z = x \ln y$; $x = 2t, y = e^t$.

49. Let $z = ue^{u^2 - v^2}$, where $u = 2x^2 + 3y^2$ and $v = 3x^2 - 2y^2$. Use the chain rule to find $\dfrac{\partial z}{\partial x}$ and $\dfrac{\partial z}{\partial y}$.

50. Use implicit differentiation to find $\dfrac{\partial z}{\partial x}$ and $\dfrac{\partial z}{\partial y}$, where x, y, and z are related by the equation $x^3 + 2xz - yz^2 - z^3 = 1$.

51. Find the slope of the level curve of $x^2 + y^2 = 2$ where $x = 1, y = 1$.

52. Find the slope of the level curve of $xe^y = 2$ where $x = 2$.

53. Find the equations for the tangent plane and normal line to the surface $z = \sin x + e^{xy} + 2y$ at the point $P_0(0, 1, 3)$.

54. The electric potential at each point (x, y) in the disk $x^2 + y^2 < 4$ is $V = 2(4 - x^2 - y^2)^{-1/2}$ volts. Draw the equipotential curves $V = c$ for $c = \sqrt{2}, \dfrac{2}{\sqrt{3}}$, and 8.

55. Let $f(x, y, z) = x^3y + y^3z + z^3x$. Find a function $g(x, y, z)$ such that $\dfrac{\partial f}{\partial x} + \dfrac{\partial f}{\partial y} + \dfrac{\partial f}{\partial z} = x^3 + y^3 + z^3 + 3g(x, y, z)$.

56. Let $u = \sin \dfrac{x}{y} + \ln \dfrac{y}{x}$. Show that $y \dfrac{\partial u}{\partial y} + x \dfrac{\partial u}{\partial x} = 0$.

57. Let $w = \ln(1 + x^2 + y^2) - 2 \tan^{-1} y$ where $x = \ln(1 + t^2)$ and $y = e^t$. Use the chain rule to find $\dfrac{dw}{dt}$.

58. Let $f(x, y) = \tan^{-1} \dfrac{y}{x}$. Find the directional derivative of f at $(1, 2)$ in the direction that makes an angle of $\frac{\pi}{3}$ with the positive x-axis.

59. Let $f(x, y) = y^x$. Find the directional derivative of f at $P_0(3, 2)$ in the direction toward the point $Q(1, 1)$.

60. According to postal regulations, the largest cylindrical can that can be sent has a girth ($2\pi r$) plus length ℓ of 108 inches. What is the largest volume cylindrical can that can be sent?

61. Let $f(x, y, z) = z(x - y)^5 + xy^2z^3$.
  a. Find the directional derivative of f at $(2, 1, -1)$ in the direction of the outward normal to the sphere $x^2 + y^2 + z^2 = 6$.

  b. In what direction is the directional derivative at $(2, 1, -1)$ largest?

62. Find positive numbers x and y for which xyz is a maximum, given that $x + y + z = 1$.

63. Maximize $f(x, y, z) = x^2yz$ given that x, y, and z are all positive numbers and $x + y + z = 12$.

64. Find the shortest distance from the origin to the surface $y^2 - z^2 = 10$.

65. Find the shortest distance from the origin to the surface $z^2 = 3 + xy$.

66. MODELING PROBLEM A plate is heated in such a way that its temperature at a point (x, y) measured in centimeters on the plate is given in degrees Celsius by

$$T(x, y) = \frac{64}{x^2 + y^2 + 4}$$

  a. Find the rate of change in temperature at the point $(3, 4)$ in the direction $2\mathbf{i} + \mathbf{j}$.
  b. Find the direction and the magnitude of the greatest rate of change of the temperature at the point $(3, 4)$.

67. MODELING PROBLEM The beautiful patterns on the wings of butterflies have long been a subject of curiosity and scientific study. Mathematical models used to study these patterns often focus on determining the level of morphogen (a chemical that effects change). In a model dealing with eyespot patterns, a quantity of morphogen is released from an eyespot and the morphogen concentration t days later is modeled by

$$S(r, t) = \frac{1}{\sqrt{4\pi t}} \exp\left(\gamma kt + \frac{r}{4t}\right) \qquad t > 0$$

where r measures the radius of the region on the wing affected by the morphogen, and k and γ are positive constants.*

  a. Find t_m so that $\partial S/\partial t = 0$. Show that the function $S_m(t)$ formed from $S(r, t)$ by fixing r has a relative maximum at t_m. Is this the same as saying that the function of two variables $S(r, t)$ has a relative maximum?
  b. Let $M(r)$ denote the maximum found in part **a**; that is, $M(r) = S(r, t_m)$. Find an expression for M in terms of $z = (1 + 4\gamma kr^2)^{1/2}$.
  c. Show that $\dfrac{dM}{dr} < 0$ and interpret this result.

68. MODELING PROBLEM Certain malignant tumors that do not respond to conventional methods of treatment (surgery, chemotherapy, etc.) may be treated by *hyperthermia*, a process involving the application of extreme heat using microwave transmission. For one

*J. D. Murray, *Mathematical Biology*, 2nd edition, Springer-Verlag, New York, 1993, p. 464.

particular kind of microwave application used in such therapy, the temperature at each point located r units from the central axis of the tumor and h units inside it is modeled by the formula

$$T(r, h) = Ke^{-pr^2}[e^{-qh} - e^{-sh}]$$

where A, p, q, and s are positive constants that depend on the properties of the patient's blood and the heating application.*

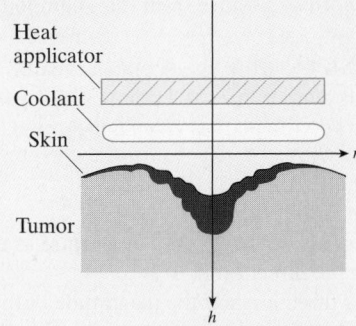

Heat applicator

Coolant

Skin

Tumor

h

a. At what depth inside the tumor does the maximum temperature occur? What is the maximum temperature? Express your answers in terms of K, p, q, and s.

b. The article on which this problem is based discusses the physiology of hyperthermia in addition to raising several other interesting mathematical issues. Read this article and discuss assumptions made in the model.

69. **MODELING PROBLEM** The marketing manager for a certain company has compiled the following data relating monthly advertising expenditure and monthly sales (in units of $1,000).

Advertising	3	4	7	9	10
Sales	78	86	138	145	156

a. Plot the data on a graph and find the least-squares line.

b. Use the least-squares line to predict monthly sales if the monthly advertising expenditure is $5,000.

70. Find $f_{xx} - f_{xy} + f_{yy}$ where $f(x, y) = x^2y^3 + x^3y^2$.

71. Find f_{xyz} where $f(x, y, z) = \cos(x^2 + y^3 + z^4)$.

72. Let $z = f(x, y)$ where $x = t + \cos t$ and $y = e^t$.

a. Suppose $f_x(1, 1) = 4$ and $f_y(1, 1) = -3$. Find $\dfrac{dz}{dt}$ when $t = 0$.

b. Suppose $f_x(0, 2) = -1$ and $f_y(0, 2) = 3$. Find $\dfrac{\partial z}{\partial r}$ and $\dfrac{\partial z}{\partial \theta}$ at the point where $r = 2$, $\theta = \dfrac{\pi}{2}$, and $x = r \cos \theta, y = r \sin \theta$.

*"Heat Therapy for Tumors," by Leah Edelstein-Keshet, *UMAP Modules 1991: Tools for Teaching*, Consortium for Mathematics and Its Applications, Inc., MA, 1992, pp. 73–101.

73. Suppose f has continuous partial derivatives in some region D in the plane, and suppose $f(x, y) = 0$ for all (x, y) in D. If $(1, 2)$ is in D and $f_x(1, 2) = 4$ and $f_y(1, 2) = 6$, find dy/dx when $x = 1$ and $y = 2$.

74. Suppose $\nabla f(x, y, z)$ is parallel to the vector $x\mathbf{i} + y\mathbf{j} + z\mathbf{k}$ for all (x, y, z). Show that $f(0, 0, a) = f(0, 0, -a)$ for any a.

75. Find two unit vectors that are normal to the surface given by $z = f(x, y)$ at the point $(0, 1)$, where $f(x, y) = \sin x + e^{xy} + 2y$.

76. Let $f(x, y) = 3(x - 2)^2 - 5(y + 1)^2$. Find all points on the graph of f where the tangent plane is parallel to the plane $2x + 2y - z = 0$.

77. Let z be defined implicitly as a function of x and y by the equation $\cos(x + y) + \cos(x + z) = 1$. Find $\dfrac{\partial^2 z}{\partial y \partial x}$ in terms of x, y, and z.

78. Suppose F and F' are continuous functions of t and that $F'(t) = C$. Define f by $f(x, y) = F(x^2 + y^2)$. Show that the direction of $\nabla f(a, b)$ is the same as the direction of the line joining (a, b) to $(0, 0)$.

79. Let $f(x, y) = 12x^{-1} + 18y^{-1} + xy$, where $x > 0, y > 0$. How do you know that f must necessarily have a minimum in the region $x > 0, y > 0$? Find the maximum.

80. Let $f(x, y) = 3x^4 - 4x^2y + y^2$. Show that f has a minimum at $(0, 0)$ on every line $y = mx$ that passes through the origin. Then show that f has no relative minimum at $(0, 0)$.

81. Find the minimum of $x^2 + y^2 + z^2$ subject to the constraint $ax + by + cz = 1$ (with $a \neq 0, b \neq 0, c \neq 0$).

82. Suppose $0 < a < 1$ and $x \geq 0, y \geq 0$. Find the maximum of $x^a y^{1-a}$ subject to the constraint $ax + (1 - a)y = 1$.

83. The **geometric mean** of three positive numbers x, y, z is $G = (xyz)^{1/3}$ and the **arithmetic mean** is $A = \frac{1}{3}(x + y + z)$. Use the method of Lagrange multipliers to show that $G(x, y, z) \leq A(x, y, z)$ for all x, y, z.

84. Liquid flows through a tube with length L centimeters and internal radius r centimeters. The total volume V of fluid that flows each second is related to the pressure P and the viscosity a of the fluid by the formula $V = \dfrac{\pi P r^4}{8aL}$.

What is the maximum error that can occur in using this formula to compute the viscosity a, if errors of $\pm 1\%$ can be made in measuring r and L, $\pm 2\%$ in measuring V, and $\pm 3\%$ in measuring P?

85. Suppose the functions f and g have continuous partial derivatives and satisfy

$$\frac{\partial f}{\partial x} = \frac{\partial g}{\partial y} \quad \text{and} \quad \frac{\partial f}{\partial y} = -\frac{\partial g}{\partial x}$$

These are called the **Cauchy–Riemann equations**.

a. Show that level curves of f and g intersect at right angles provided $\nabla f \neq 0$ and $\nabla g \neq 0$.

b. Assuming that the second partials of f and g are continuous, show that f and g satisfy

Laplace's equations

$$f_{xx} + f_{yy} = 0 \quad \text{and} \quad g_{xx} + g_{yy} = 0$$

86. Show that if $z = f(r, \theta)$, where r and θ are defined implicitly as functions of x and y by the equations $x = r \cos \theta$, $y = r \sin \theta$, then the equation $\dfrac{\partial^2 z}{\partial x^2} + \dfrac{\partial^2 z}{\partial y^2} = 0$ becomes

$$\frac{\partial^2 z}{\partial r^2} + \frac{1}{r^2} \frac{\partial^2 z}{\partial \theta^2} + \frac{1}{r} \frac{\partial z}{\partial r} = 0.$$ This is Laplace's equation in polar coordinates.

87. Suppose the angle and radius of a circular sector are allowed to vary. Use Lagrange multipliers to find the angle of the circular sector for which the perimeter of the sector is smallest, assuming the area of the sector is a fixed constant.

88. For the production function given by $Q(x, y) = x^a y^b$, where $a > 0$ and $b > 0$, show that

$$x \frac{\partial Q}{\partial x} + y \frac{\partial Q}{\partial y} = (a + b)Q$$

In particular, if $b = 1 - a$ with $0 < a < 1$, then

$$x \frac{\partial Q}{\partial x} + y \frac{\partial Q}{\partial y} = Q$$

89. The diameter of the base and the height of a right circular cylinder are measured, and the measurements are known to have errors of at most 0.5 cm. If the diameter and height are taken to be 4 cm and 8 cm, respectively, find bounds for the propagated error in
a. the volume V of the cylinder.
b. the surface area S of the cylinder.

90. A right circular cone is measured and is found to have base radius $r = 40$ cm and altitude $h = 20$ cm. If it is known that each measurement is accurate to within 2%, what is the maximum percentage error in the measurement of the volume?

91. An elastic cylindrical container is filled with air so that the radius of the base is 2.02 cm and the height is 6.04 cm. If the container is deflated so that the radius of the base reduces to 2 cm and the height to 6 cm, approximately how much air has been removed? (Ignore the thickness of the container.)

92. Suppose f is a differentiable function of two variables with f_x and f_y also differentiable, and assume that f_{xx}, f_{yy}, and f_{xy} are continuous. The **second directional derivative** (x, y) of f at the point in the direction of the unit vector $\mathbf{u} = a\mathbf{i} + b\mathbf{j}$, is defined by $D_u^2 f(x, y) = D_u[D_u f(x, y)]$. Show that

$$D_u^2 f(x, y) = a^2 f_{xx}(x, y) + 2ab f_{xy}(x, y) + b^2 f_{yy}(x, y)$$

93. A capsule is a cylinder of radius r and length ℓ, capped on each end by a hemisphere. Assume that the capsule dissolves in the stomach at a rate proportional to the ratio $R = S/V$, where V is the volume and S is the surface area of the capsule. Show that

$$\frac{\partial R}{\partial r} < 0 \quad \text{and} \quad \frac{\partial R}{\partial \ell} < 0$$

94. **SPY PROBLEM**   Gasping for breath, the Spy tumbles through the escape hole in the ice cave (Problem 55, Section 12.6) and is immediately knocked unconscious. He awakes tied to a chair, alone, except for a rather large ticking bomb less than ten feet away. "So my friend," thunders the voice of Coldfinger, "you plan to eliminate Scélérat and me for Blohardt, but it appears you are the one about to be eliminated. Still, for old times' sake, I'll give you a chance. Pick a number, S (for Spy), then I'll pick a number C (for Coldfinger), and the value of the expression

$$N = \frac{\exp[(C + S)^2 + 4S + 22]}{(C + S)^2(e^{40} + e^{8S})}$$

will be the number of minutes I'll wait before pressing this little red button." Assuming Coldfinger does his best to minimize the expression once S has been chosen, what number should the Spy pick, and how many minutes will he have to escape and defuse the bomb?

95. **THINK TANK PROBLEM**   Find the minimum distance from the origin to the paraboloid $z = 4 - x^2 - 4y^2$. The graph is shown in Figure 12.42. The distance from $P(x, y, z)$ to the origin is

$$d = \sqrt{x^2 + y^2 + z^2}$$

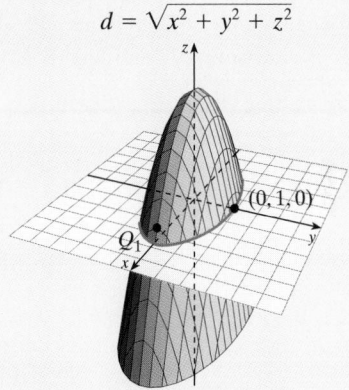

■ **FIGURE 12.42**

This distance will be minimized when d^2 is minimized. Thus, the function to be minimized (after replacing x^2 by $4 - 4y^2 - z$) is $D(x, y) = 4 - 4y^2 - z + y^2 + z^2$ $= 4 - 3y^2 + z^2 - z$. Setting the partial derivatives D_y and D_z to 0 gives the critical point $y = 0, z = 0.5$. Solving for x on the paraboloid gives the points $Q_1(\sqrt{3.5}, 0, 0.5)$ and $Q_2(-\sqrt{3.5}, 0, 0.5)$. These points are NOT minimal because $(0, 1, 0)$ is closer. Explain what is going on here. Our thanks to Herbert R. Bailey, who presented this problem in *The College Mathematics Journal* (" 'Hidden' Boundaries in Constrained Max–Min Problems," May 1991, p. 227).

96. A **minimal surface** is one that has the least surface area of any surface with a given boundary. It can be shown that if $z = f(x, y)$ is a minimal surface, then

$$(1 + z_y^2)z_{xx} - zz_{xy} + (1 + z_x^2)z_{yy} = 0$$

a. Find constants A, B so that

$$z = \ln\left(\frac{A \cos y}{B \cos x}\right)$$

is a minimal surface.

b. Is it possible to find C and D so that $z = C \ln(\sin x) + D \ln(\sin y)$ is a minimal surface?

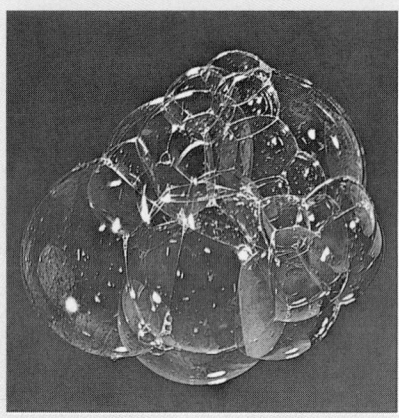

Soap bubbles form minimal surfaces. For an interesting discussion, see "The Geometry of Soap Films and Soap Bubbles," by Frederick J. Almgren, Jr. and Jean E. Taylor, *Scientific American*, July 1976, pp. 82–93.

97. **PUTNAM EXAMINATION PROBLEM** Let f be a real-valued function having partial derivatives that is defined for $x^2 + y^2 < 1$ and that satisfies $\left| f(x, y) \right| \leq 1$. Show that there exists a point (x_0, y_0) in the interior of the unit circle such that $[f_x(x_0, y_0)]^2 + [f_y(x_0, y_0)]^2 \leq 16$.

98. **PUTNAM EXAMINATION PROBLEM** Find the smallest volume bounded by the coordinate planes and a tangent plane to the ellipsoid

$$\frac{x^2}{a^2} + \frac{y^2}{b^2} + \frac{z^2}{c^2} = 1$$

99. **PUTNAM EXAMINATION PROBLEM** Find the shortest distance between the plane $Ax + By + Cz + 1 = 0$ and the ellipsoid

$$\frac{x^2}{a^2} + \frac{y^2}{b^2} + \frac{z^2}{c^2} \leq 1$$

Desertification

This project is to be done in groups of three or four students. Each group will submit a single written report.

A friend of yours named Maria is studying the causes of the continuing expansion of deserts (a process known as *desertification*). She is working on a biological index of vegetation disturbance, which she has defined. By seeing how this index and other factors change through time, she hopes to discover the role played in desertification. She is studying a huge tract of land bounded by a rectangle; this piece of land surrounds a major city but does not include it. She needs to find an economical way to calculate for this piece of land the important vegetation disturbance index $J(x, y)$.

Maria has embarked upon an ingenious approach of combining the results of photographic and radar images taken during flights over the area to calculate the index J. She is assuming that J is a smooth function. Although the flight data do not directly reveal the values of the function J, they give the rate at which the values of J change as the flights sweep over the landscape surrounding the city. Her staff has conducted numerous flights, and from the data she believes she has been able to find actual formulas for the rates at which J changes in the east–west and north–south directions. She has given these functions the names M and N. Thus, $M(x, y)$ is the rate at which J changes as one sweeps in the positive x-direction, and $N(x, y)$ is the corresponding rate in the y-direction. Maria shows you these formulas:

$$M(x, y) = 3.4e^{x(y-7.8)^2} \quad \text{and} \quad N(x, y) = 22\sin(75 - 2xy)$$

Convince her that these two formulas cannot possibly be correct. Do this by showing her that there is a condition that the two functions M and N must satisfy if they are to be the east–west and north–south rates of change of the function J and that her formulas for M and N do not meet this condition. However, show Maria that if she can find formulas for M and N that satisfy the condition that you showed her, it is possible to find a formula for the function J from the formulas for M and N.

Mathematics in its pure form, as arithmetic, algebra, geometry, and the applications of the analytic method, as well as mathematics applied to matter and force, or statics and dynamics, furnishes the peculiar study that gives to us, whether as children or as men, the command of nature in this its quantitative aspect; mathematics furnishes the instrument, the tool of thought, which we wield in this realm.

W. T. HARRIS
PSYCHOLOGICAL FOUNDATIONS OF EDUCATION
(NEW YORK, 1898), P. 325.

*Marcus S. Cohen, Edward D. Gaughan, R. Arthur Knoebel, Douglas S. Kurtz, and David J. Pengelley, "Priming the Calculus Pump: Innovations and Resources," *MAA Notes* 17 (1991).

CONTENTS

13

Multiple Integration

PREVIEW

The *single integral*

$$\int_a^b f(x)\, dx$$

introduced in Chapter 5 has many uses, as we have seen. In this chapter, we shall generalize the single integral to define *multiple* integrals, in which the integrand is a function of several variables. We will find that multiple integration is used in much the same way as single integration, by "adding" small quantities to define and compute area, volume, surface area, moments, centroids, and probability.

PERSPECTIVE

What is the volume of a doughnut (torus)? Given the joint probability function for the amount of time a typical shopper spends shopping at a particular store and the time spent in the checkout line, how likely is it that a shopper will spend no more than 30 minutes altogether in the store? If the temperature in a solid body is given at each point (x, y, z) and time t, what is the average temperature of the body over a particular time period? Where should a security watch tower be placed in a parking lot to ensure the most comprehensive visual coverage? We shall answer these and other similar questions in this chapter using multiple integration.

13.1 *Double Integration over Rectangular Regions*

IN THIS SECTION definition of the double integral, properties of double integrals, volume interpretation, iterated integration, an informal argument for Fubini's theorem ∎

DEFINITION OF THE DOUBLE INTEGRAL

Recall that in Chapter 5, we defined the single definite integral $\int_a^b f(x)\,dx$ as a special kind of limit involving Riemann sums $\sum_{k=1}^{n} f(x_k^*)\,\Delta x_k$, where $x_1, x_2, \ldots, x_n$ are points in a partition of the interval $[a, b]$, and x_k^* is a representative point in the subinterval $[x_{k-1}, x_k]$. We now apply the same ideas to define a *double* definite integral $\iint_R f(x, y)\,dA$, over the rectangle $R: a \le x \le b, c \le y \le d$. The definition requires the ideas and notation described in the following three steps:

Step 1: Partition the interval $a \le x \le b$ into m parts and the interval $c \le y \le d$ into n parts. Using these subdivisions, we partition the rectangle R into $N = mn$ **cells** (subrectangles), as shown in Figure 13.1. Call this partition P.

Step 2: Choose a representative point (x_k^*, y_k^*) from each cell in the partition of the rectangle. Form the sum

$$\sum_{k=1}^{N} f(x_k^*, y_k^*)\Delta A_k$$

where ΔA_k is the area of the kth representative cell. This is called the **Riemann sum** of $f(x, y)$ with respect to the partition P and cell representatives (x_k^*, y_k^*).

Step 3: We define the **norm** $\|P\|$ of the partition to be the length of the longest diagonal of any rectangle in the partition. To **refine** the partition means to subdivide the cells in such a way that the norm decreases. When this process is applied to the Riemann sum and the norm decreases indefinitely to zero, we write

$$\lim_{\|P\|\to 0} \sum_{k=1}^{N} f(x_k^*, y_k^*)\Delta A_k$$

This limit is what is called the *double integral*.

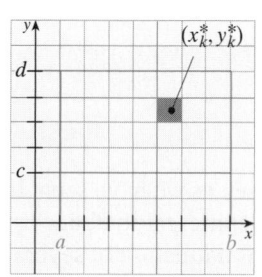

∎ **FIGURE 13.1** A partition of the rectangle R into mn cells; a kth cell representative is shown

Double Integral

If f is defined on a closed, bounded rectangular region R in the xy-plane, then the **double integral of f over R** is defined by

$$\iint_R f(x, y)\,dA = \lim_{\|P\|\to 0} \sum_{k=1}^{N} f(x_k^*, y_k^*)\Delta A_k$$

provided this limit exists, in which case f is said to be **integrable** over R.

∎ *What This Says:* In considering this definition, notice:

1. The number N of cells depends on the partition P.
2. As $\|P\| \to 0$, it follows that $N \to \infty$.
3. More formally, the phrase "provided this limit exists" means:

If I is the double integral

$$I = \iint\limits_{R} f(x, y)\, dA$$

then for any $\epsilon > 0$, there exists a $\delta > 0$ such that

$$\left| I - \sum_{k=1}^{N} f(x_k^*, y_k^*)\Delta A_k \right| < \epsilon$$

whenever $\displaystyle\sum_{k=1}^{N} f(x_k^*, y_k^*)\Delta A_k$ is a Riemann sum whose norm satisfies $\|P\| < \delta$.

4. It can be shown that if $f(x, y)$ is continuous on R, then it is integrable on R.

PROPERTIES OF DOUBLE INTEGRALS

Double integrals have many of the same properties as single integrals. Three of these properties are contained in the following theorem.

THEOREM 13.1   *Properties of double integrals*

Assume that all the given integrals exist.

Linearity rule: For constants a and b,

$$\iint\limits_{D} [af(x, y) + bg(x, y)]\, dA = a\iint\limits_{D} f(x, y)\, dA + b\iint\limits_{D} g(x, y)\, dA$$

Dominance rule: If $f(x, y) \geq g(x, y)$ throughout a region D, then

$$\iint\limits_{D} f(x, y)\, dA \geq \iint\limits_{D} g(x, y)\, dA$$

Subdivision rule: If the region of integration D can be subdivided into two sub-regions D_1 and D_2, then

$$\iint\limits_{D} f(x, y)\, dA = \iint\limits_{D_1} f(x, y)\, dA + \iint\limits_{D_2} f(x, y)\, dA$$

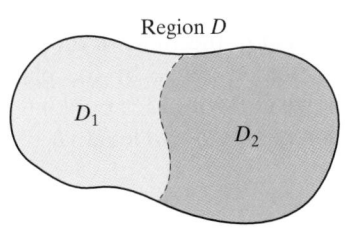

Region D

D_1

D_2

Proof   To prove this, you can first show that each is true for rectangular regions of integration. Then the general case is proved by applying the definition of a double integral and various theorems for single integrals.

VOLUME INTERPRETATION

If $f(x) \geq 0$ on the interval $[a, b]$, the single integral $\displaystyle\int_a^b f(x)\, dx$ can be interpreted as the area under the curve $y = f(x)$ over $[a, b]$, and the double integral $\displaystyle\iint\limits_{R} f(x, y)\, dA$ has a similar interpretation in terms of volume. To see this, note that if $f(x, y) \geq 0$ on the region R and we partition R using rectangles, then the product $f(x_k^*, y_k^*)\Delta A_k$ is the volume of a parallelepiped (a box) with height $f(x_k^*, y_k^*)$ and base area ΔA_k, as shown in Figure 13.2.

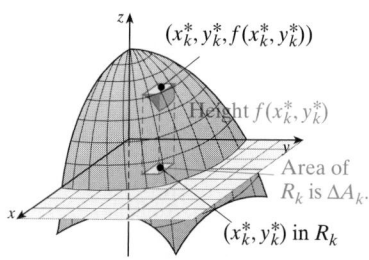

■ **FIGURE 13.2** The approximating parallelepiped has volume $\Delta V_k = f(x_k^*, y_k^*)\Delta A_k$.

Thus, the Riemann sum

$$\sum_{k=1}^{N} f(x_k^*, y_k^*)\Delta A_k$$

provides an estimate of the total volume under the surface $z = f(x, y)$ over the rectangular domain R, and if f is continuous on R, we expect the approximation to improve if we take a more refined partition of R (that is, more rectangles with smaller norm). Thus, it is natural to *define* the total volume under the surface as the limit of Riemann sums as the norm tends to 0. That is, the volume under $z = f(x, y)$ over the domain R is given by

$$V = \lim_{\|P\|\to 0} \sum_{k=1}^{N} f(x_k^*, y_k^*)\Delta A_k = \iint_R f(x, y)\, dA$$

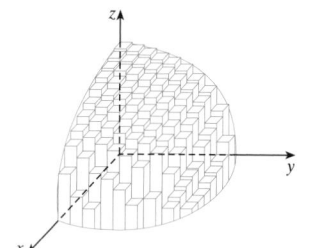

Volume approximated by rectangular parallelepipeds

EXAMPLE 1 *Evaluate a double integral by relating it to a volume*

Evaluate $\displaystyle\iint_R (2 - y)\, dA$, where R is the rectangle in the xy-plane with vertices $(0, 0), (3, 0), (3, 2),$ and $(0, 2)$.

Solution Because $z = 2 - y$ satisfies $z \geq 0$ for all points in R, the value of the double integral is the same as the volume of the solid bounded above by the plane $z = 2 - y$ and below by the rectangle R. The solid is shown in Figure 13.3. Looking at it sideways, we see an edge with a triangular cross-section of area B and length $h = 3$. We use the formula $V = Bh$.

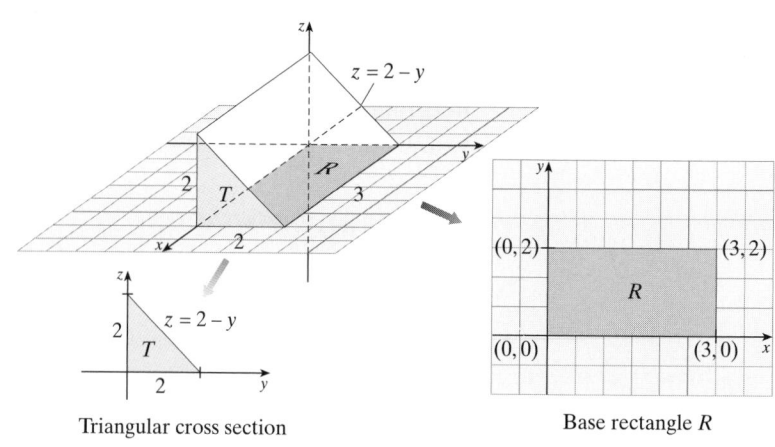

Triangular cross section

Base rectangle R

■ **FIGURE 13.3** Evaluation of $\displaystyle\iint_R (2 - y)\, dA$ as a volume

Because the base is a triangle of side 2 and altitude 2, we have

$$V = Bh = [\tfrac{1}{2}(2)(2)](3) = 6$$

Therefore, the value of the integral is also 6; that is,

$$\iint\limits_R (2 - y)\, dA = 6 \qquad \blacksquare$$

ITERATED INTEGRATION

As with single integrals, it is not practical to evaluate a double integral even over a simple rectangular region by using the definition. Instead, we shall compute double integrals by a process called **successive partial integration.** To be specific, if we hold the y-variable constant (denoted by $\bar{y}$) and integrate $f(x, y)$ with respect to x, we have a single integral

$$\int f(x, \bar{y})\, dx$$

which we shall refer to as the **partial integral of f with respect to x.** The dx tells us that the integration is with respect to x and that y is to be held constant for the integration, so that when the process is complete, the result is a function of y. Similarly, the integral

$$\int f(\bar{x}, y)\, dy$$

is the **partial integral of f with respect to y,** which means that x is held constant (denoted by $\bar{x}$) and integration is with respect to y. The result of this partial integration is a function of x.

In successive partial integration, we evaluate a double integral by integrating first with respect to one variable and then again with respect to the other variable. In partial integration, we work from the inside out, as indicated in the following notation:

$$\iint f(x, y)\, dx\, dy = \int \left[\int f(x, \bar{y})\, dx \right] dy \qquad \text{Integrate with respect to } x \text{ first, then with respect to } y.$$

$$\iint f(x, y)\, dy\, dx = \int \left[\int f(\bar{x}, y)\, dy \right] dx \qquad \text{Integrate with respect to } y \text{ first, then with respect to } x.$$

Integrals of this form are said to be **iterated integrals.** Our next theorem tells us how iterated integrals can be used to evaluate double integrals. The theorem was first proved by the Italian mathematician Guido Fubini (1879–1943) in 1907.

THEOREM 13.2   *Fubini's theorem over a rectangular region*

If $f(x, y)$ is continuous over the rectangle $R: a \le x \le b, c \le y \le d$, then the double integral

$$\iint\limits_R f(x, y)\, dA$$

may be evaluated by either iterated integral; that is,

$$\iint\limits_R f(x, y)\, dA = \int_c^d \int_a^b f(x, y)\, dx\, dy = \int_a^b \int_c^d f(x, y)\, dy\, dx$$

> ■ *What This Says:*   Instead of using the definition of a double integral, consider the region R, namely, $a \le x \le b$ and $c \le y \le d$, and evaluate *either* of the iterated integrals
>
> Limits of x (variable outside brackets)   Limits of y (outside brackets)
> $$\int_a^b \left[\int_c^d f(\bar{x}, y)\, dy \right] dx \quad \text{or} \quad \int_c^d \left[\int_a^b f(x, \bar{y})\, dx \right] dy$$
> Limits of y (variable inside brackets)   Limits of x (inside brackets)

Proof   We shall provide an informal, geometric argument at the end of this section. The formal proof may be found in most advanced calculus textbooks.   ═══

Let us see how Fubini's theorem can be used to evaluate double integrals. We begin by taking another look at Example 1.

EXAMPLE 2   *Evaluate a double integral by using Fubini's theorem*

Use iterated integrals to compute $\displaystyle\iint_R (2 - y)\, dA$, where R is the rectangle with vertices $(0, 0), (3, 0), (3, 2),$ and $(0, 2)$.

Solution   The region of integration is the rectangle $0 \le x \le 3,\ 0 \le y \le 2$ (see Example 1 and Figure 13.3). Thus, by Fubini's theorem, the double integral can be evaluated as an iterated integral:

$$\iint_R (2 - y)\, dA = \int_0^3 \int_0^2 (2 - y)\, dy\, dx \qquad \textit{Integrate inner integral with respect to } y.$$

$$= \int_0^3 \left[2y - \frac{y^2}{2} \right]\Big|_0^2 dx = \int_0^3 \left[4 - \frac{4}{2} - (0) \right] dx = \int_0^3 2\, dx = 2x\,\big|_0^3 = 6$$

which is the same as the result obtained geometrically in Example 1.   ∎

EXAMPLE 3   *Double integral using an iterated integral*

Evaluate $\displaystyle\iint_R x^2 y^5 dA$, where R is the rectangle $1 \le x \le 2, 0 \le y \le 1$, using an iterated integral with

a. y-integration first   b. x-integration first

Solution   The graph of the surface $z = x^2 y^5$ over the rectangle is shown in Figure 13.4a. In your work, you would usually sketch only the rectangle, as shown in Figure 13.4b.

a. $$\iint_R x^2 y^5\, dA = \int_1^2 \int_0^1 x^2 y^5 dy\, dx \qquad \textit{Read this as } \int_1^2 \left[\int_0^1 x^2 y^5\, dy \right] dx.$$

$$= \int_1^2 \left[x^2 \frac{y^6}{6} \right]\Big|_0^1 dx$$

$$= \int_1^2 \left[x^2\left(\frac{1}{6} - \frac{0}{6} \right) \right] dx = \frac{x^3}{18}\Big|_1^2 = \frac{8}{18} - \frac{1}{18} = \frac{7}{18}$$

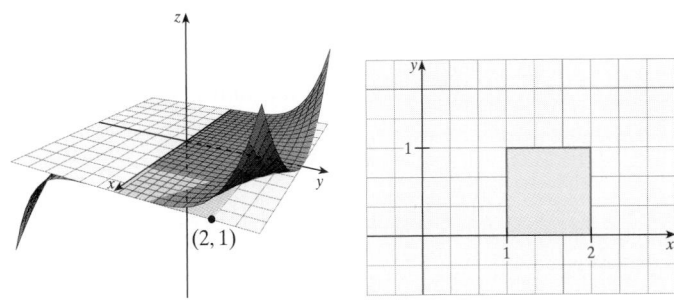

■ **FIGURE 13.4**   Graph of $z = x^2 y^5$ and defining rectangle

b. $$\iint\limits_{R} x^2 y^5 \, dA = \int_0^1 \int_1^2 x^2 y^5 \, dx \, dy = \int_0^1 y^5 \left[\frac{x^3}{3}\right]\Big|_1^2 dy$$

$$= \int_0^1 \left[y^5 \left(\frac{8}{3} - \frac{1}{3}\right)\right] dy = \frac{7y^6}{18}\Big|_0^1 = \frac{7}{18} - \frac{0}{18} = \frac{7}{18}$$

Technology Window

Using technology for multiple integrals offers no special difficulties. You simply integrate with respect to one variable and then with respect to the other variable. You must be careful, however, to properly input the correct limits of integration for each integral. Notice that for $\int_1^2 \int_0^1 x^2 y^2 \, dy \, dx$, most calculators and software programs require the following syntax:

integrate operator, function, variable of integration, lower limit of integration, upper limit of integration.

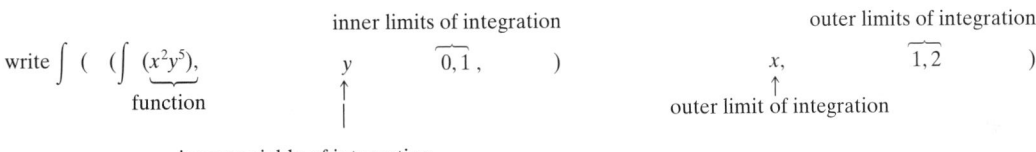

"inside" integral is the function for the "outside" integral

The output below (using Example 3) shows what this might look like.

a.

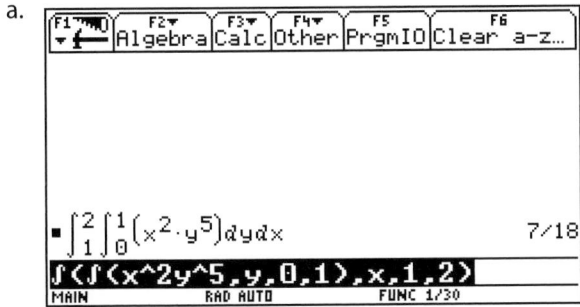

b.

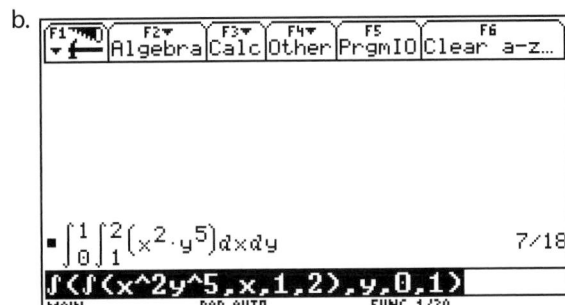

Generally speaking, when using a calculator or computer with CAS software, it does not matter much which order of integration is used. However, without technology, it can matter a great deal, as illustrated in the following example.

EXAMPLE 4   *Choosing the order of integration for a double integral*

Evaluate $\displaystyle\iint_R x\cos xy\,dA$ for $R: 0 \le x \le \dfrac{\pi}{2}, 0 \le y \le 1$.

Solution   Suppose we integrate with respect to x first:

$$\int_0^1 \left[\int_0^{\pi/2} x\cos xy\,dx \right] dy$$

The inner integral requires integration by parts. However, integrating with respect to y first is much simpler:

$$\int_0^{\pi/2} \left[\int_0^1 x\cos xy\,dy \right] dx = \int_0^{\pi/2} \left[\frac{x\sin xy}{x} \right]\Big|_0^1 dx = \int_0^{\pi/2} (\sin x - \sin 0)\,dx$$

$$= -\cos x \Big|_0^{\pi/2} = 1$$ ∎

AN INFORMAL ARGUMENT FOR FUBINI'S THEOREM

We can make Fubini's theorem plausible with a geometric argument in the case where $f(x, y) \ge 0$ on R. If $\int_R\!\int f(x, y)\,dA$ is defined on a rectangle $R: a \le x \le b, c \le y \le d$, it represents the volume of the solid S bounded above by the surface $z = f(x, y)$ and below by the rectangle R. If $A(y_k^*)$ is the cross-sectional area perpendicular to the y-axis at the point y_k^*, then $A(y_k^*)\Delta y_k$ represents the volume of a "slab" that approximates the volume of part of the solid S, as shown in Figure 13.5.

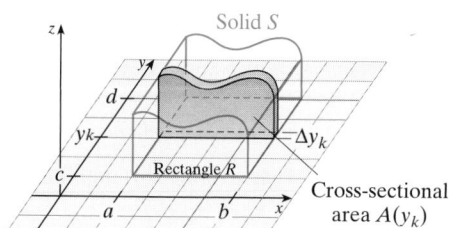

■ FIGURE 13.5   Cross-sectional volume parallel to the xz-plane

By using a limit to "add up" all such approximating volumes, we obtain an estimate of the volume of the entire solid S. Thus, if we let V be the exact volume of the solid S, we have

$$\iint_R f(x, y)\,dA = V = \lim_{\|P\|\to 0} \sum_{k=0}^N A(y_k^*)\Delta y_k$$

The limit on the right is just the integral of $A(y)$ over the interval $c \le y \le d$, where $A(y)$ is the area of a cross section with fixed y. In Chapter 5, we found that the area $A(y)$ can be computed by the integral

$$A(y) = \int_a^b f(x, y)\,dx \qquad \text{Integration with respect to } x$$
$$(y \text{ is a constant})$$

We can now make this substitution for $A(y)$ to obtain

$$\iint\limits_{R} f(x, y)\, dA = V = \lim_{\|P\| \to 0} \sum_{k=0}^{N} A(y_k^*)\Delta y_k = \int_c^d A(y)\, dy$$

$$= \int_c^d \underbrace{\left[\int_a^b f(x, y)\, dx\right]}_{A(y)} dy \qquad \text{Substitution}$$

The fact that $\displaystyle\iint\limits_{R} f(x, y)\, dA = \int_a^b \int_c^d f(x, y)\, dy\, dx$

can be justified in a similar fashion (you are asked to do this in Problem 47). Thus, we have

$$\int_c^d \left[\int_a^b f(x, y)\, dx\right] dy = \iint\limits_{R} f(x, y)\, dA = \int_a^b \left[\int_c^d f(x, y)\, dy\right] dx$$

13.1 Problem Set

Ⓐ *In Problems 1–6, evaluate the iterated integrals.*

1. $\displaystyle\int_0^2 \int_0^1 (x^2 + xy + y^2)\, dy\, dx$

2. $\displaystyle\int_1^2 \int_0^\pi x \cos y\, dy\, dx$

3. $\displaystyle\int_1^{e^2} \int_1^2 \left[\frac{1}{x} + \frac{1}{y}\right] dy\, dx$

4. $\displaystyle\int_0^{\ln 2} \int_0^1 e^{x+2y}\, dx\, dy$

5. $\displaystyle\int_3^4 \int_1^2 \frac{x}{x - y}\, dy\, dx$

6. $\displaystyle\int_2^3 \int_{-1}^2 \frac{1}{(x + y)^2}\, dy\, dx$

Use an appropriate volume formula to evaluate the double integral given in Problems 7–12.

7. $\displaystyle\iint\limits_{R} 4\, dA;\ R: 0 \le x \le 2;\ 0 \le y \le 4$

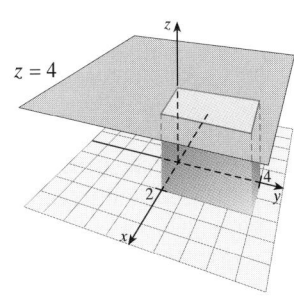

8. $\displaystyle\iint\limits_{R} 5\, dA;\ R: 2 \le x \le 5;\ 1 \le y \le 3$

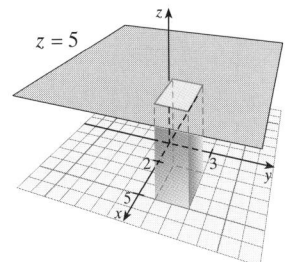

9. $\displaystyle\iint\limits_{R} (4 - y)\, dA;\ R: 0 \le x \le 3;\ 0 \le y \le 4$

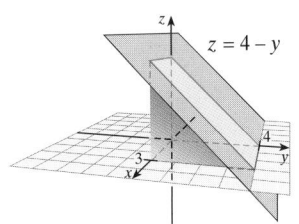

10. $\displaystyle\iint\limits_{R} (4 - 2y)\, dA;\ R: 0 \le x \le 4;\ 0 \le y \le 2$

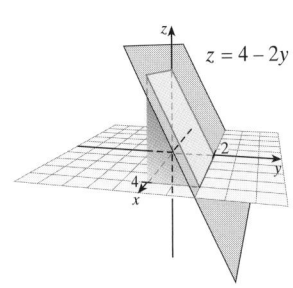

11. $\displaystyle\iint_R \frac{y}{2}\, dA; R: 0 \le x \le 6; 0 \le y \le 4$

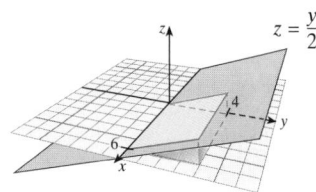

12. $\displaystyle\iint_R \frac{y}{4}\, dA; R: 0 \le x \le 2; 0 \le y \le 8$

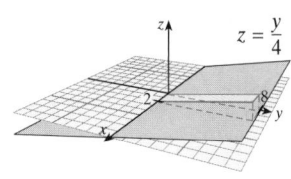

Use successive partial integration to compute the double integrals in Problems 13–20 over the specified rectangle.

13. $\displaystyle\iint_R x^2 y\, dA; R: 1 \le x \le 2; 0 \le y \le 1$

14. $\displaystyle\iint_R (x + 2y)\, dA; R: 2 \le x \le 3; -1 \le y \le 1$

15. $\displaystyle\iint_R 2xe^y\, dA; R: -1 \le x \le 0; 0 \le y \le \ln 2$

16. $\displaystyle\iint_R x^2 e^{xy}\, dA; R: 0 \le x \le 1; 0 \le y \le 1$

17. $\displaystyle\iint_R \frac{2xy\, dA}{x^2 + 1}; R: 0 \le x \le 1; 1 \le y \le 3$

18. $\displaystyle\iint_R y\sqrt{1 - y^2}\, dA; R: 1 \le x \le 5; 0 \le y \le 1$

19. $\displaystyle\iint_R \sin(x + y)\, dA; R: 0 \le x \le \frac{\pi}{4}; 0 \le y \le \frac{\pi}{2}$

20. $\displaystyle\iint_R x \sin xy\, dA; R: 0 \le x \le \pi; 0 \le y \le 1$

Find the volume of the solid bounded below by the rectangle R in the xy-plane and above by the graph of $z = f(x, y)$ in Problems 21–31. Assume that a, b, and c are positive constants.

21. $f(x, y) = 2x + 3y; R: 0 \le x \le 1; 0 \le y \le 2$
22. $f(x, y) = 5x + 2y; R: 0 \le x \le 1; 0 \le y \le 2$
23. $f(x, y) = ax + by; R: 0 \le x \le a; 0 \le y \le b$
24. $f(x, y) = axy; R: 0 \le x \le b; 0 \le y \le c$
25. $f(x, y) = \sqrt{xy}; R: 0 \le x \le 1; 0 \le y \le 4$
26. $f(x, y) = \sqrt{xy}; R: 0 \le x \le a; 0 \le y \le b$
27. $f(x, y) = xe^y; R: 0 \le x \le 2; 0 \le y \le \ln 2$
28. $f(x, y) = xe^{xy}; R: 0 \le x \le 1; 0 \le y \le \ln 3$
29. $f(x, y) = (x + y)^5; R: 0 \le x \le 1; 0 \le y \le 1$

30. $f(x, y) = \sqrt{x + y}; R: 0 \le x \le 1; 0 \le y \le 1$
31. $f(x, y) = (x + y)^n; n \ne -1$ and $n \ne -2; R: 0 \le x \le 1; 0 \le y \le 1$

B **32.** ■ **What Does This Say?** Discuss the definition of double integral.

33. ■ **What Does This Say?** Describe the implementation of Fubini's theorem.

34. MODELING PROBLEM Suppose R is a rectangular region within the boundary of a certain national forest that contains 600,000 trees per square mile. Model the total number of trees, T, in the forest as a double integral. Assume that x and y are measured in miles.

35. MODELING PROBLEM Suppose mass is distributed on a rectangular region R in the xy-plane so that the density (mass per unit area) at the point (x, y) is $\delta(x, y)$. Model the total mass as a double integral.

36. MODELING PROBLEM Suppose R is the rectangular region within the boundary of a certain city, and let the city center be at the origin $(0, 0)$. For this model, assume that the population density r miles from the city center is $12e^{-.07r}$ thousand people per square mile. Model the total population of the region of the city as a double integral.

37. MODELING PROBLEM Suppose R is a rectangular region within the boundary of a certain county, and let $f(x, y)$ denote the housing density (in homes per square mile) at the point (x, y). If the property tax on each home in the county is $1,400 per year, model the total property tax collected in the region of the county as a double integral.

In Problems 38–41, one order of integration is easier than the other. Determine the easier order and then evaluate the integral.

38. Compute $\displaystyle\iint_R x\sqrt{1 - x^2}\, e^{3y}\, dA$, where R is the rectangle $0 \le x \le 1, 0 \le y \le 2$.

39. Compute $\displaystyle\iint_R \frac{\ln \sqrt{y}}{xy}\, dA$, where R is the rectangle $1 \le x \le 4, 1 \le y \le e$.

40. Compute (correct to the nearest hundredth) $\displaystyle\iint_R \frac{xy}{x^2 + y^2}\, dA$, where R is the rectangle $1 \le x \le 3, 1 \le y \le 2$.

41. Evaluate $\displaystyle\iint_R xe^{xy}\, dA$, where R is the rectangle $0 \le x \le 1, 1 \le y \le 2$.

42. Explain why $\displaystyle\iint_R (4 - x^2 - y^2)\, dA > 2$, where R is the rectangular domain in the plane given by $0 \le x \le 1, 0 \le y \le 1$.

45. HISTORICAL QUEST   Guido Fubini taught at the Institute for Advanced Study in Princeton. He was nicknamed the "Little Giant," because of his small body but large mind. Even though the conclusion of Fubini's theorem was known for a long time and successfully applied in various instances, it was not satisfactorily proved in a general setting until 1907. His most

GUIDO FUBINI
1879–1943

important work was in differential projective geometry. In 1938 he was forced to leave Italy because of the Fascist government, and he emigrated to the United States.

For this Historical Quest, write several paragraphs about the nature of differential projective geometry.

C 46. Let f be a function with continuous second partial derivatives on a rectangular domain R with vertices $(x_1, y_1), (x_1, y_2), (x_2, y_2),$ and (x_2, y_1), where $x_1 < x_2$ and $y_1 < y_2$. Use the fundamental theorem of calculus to show that

$$\iint\limits_{R} \frac{\partial^2 f}{\partial y \partial x} \, dA = f(x_1, y_1) - f(x_2, y_1)$$
$$+ f(x_2, y_2) - f(x_1, y_2)$$

47. Let f be a continuous function defined on the rectangle $R: a \le x \le b, c \le y \le d$. Use a geometric argument to show that

$$\iint\limits_{R} f(x, y) \, dA = \int_a^b \int_c^d f(x, y) \, dy \, dx$$

Hint: Modify the argument given in the text by taking cross-sectional areas perpendicular to the x-axis.

13.2   Double Integration over Nonrectangular Regions

IN THIS SECTION   nonrectangular regions, more on area and volume, reversing the order of integration in a double integral, properties of double integrals   ■

NONRECTANGULAR REGIONS

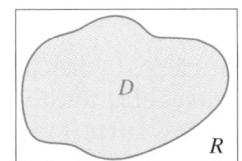

■ **FIGURE 13.6**   The region D is bounded by a rectangle R.

Let $f(x, y)$ be a function that is continuous on the region D that can be contained in a rectangle R. (See Figure 13.6.) Define the function $F(x, y)$ on R as $f(x, y)$ if (x, y) is in D, and 0 otherwise. That is,

$$F(x, y) = \begin{cases} f(x, y) & \text{for } (x, y) \text{ in } D \\ 0 & \text{for } (x, y) \text{ not in } D \end{cases}$$

Then, if F is integrable over R, we say that f is **integrable over D,** and the **double integral of f over D** is defined as

$$\iint\limits_{D} f(x, y) \, dA = \iint\limits_{R} F(x, y) \, dA$$

In the previous section, you saw how the double integral $\int_R \int f(x, y) \, dA$ could be evaluated by successive partial integration when R is a rectangle. A modification of this procedure allows you to evaluate the double integral in the important case where R is not a rectangle, but is still bounded by vertical (or horizontal) lines on two opposite sides.

Type I Region (Vertical Strip)	**Type II Region (Horizontal Strip)**
A **type I region** contains points (x, y) such that for each fixed x between constants a and b, y varies from $g_1(x)$ to $g_2(x)$, where g_1 and g_2 are continuous functions. Think of a vertical strip.	A **type II region** contains points (x, y) such that for each fixed y between constants c and d, x varies from $h_1(y)$ to $h_2(y)$, where h_1 and h_2 are continuous functions. Think of a horizontal strip.

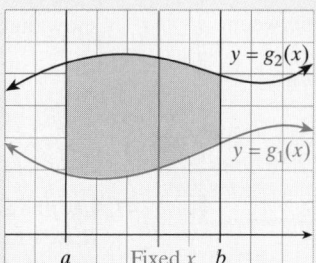

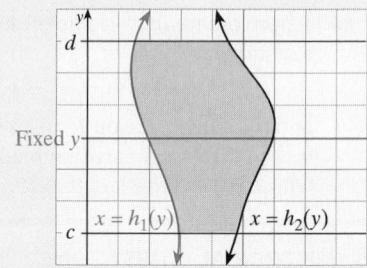

Type I region: For a fixed x between a and b, y varies from $g_1(x)$ to $g_2(x)$.

Type II region: For a fixed y between c and d, x varies from $h_1(y)$ to $h_2(y)$.

By applying Fubini's theorem for rectangular regions, we can derive the following theorem, which shows how to evaluate a double integral over a type I or type II region.

THEOREM 13.3   *Fubini's theorem for nonrectangular regions*

TYPE I (vertical strip):
x fixed, y varies (form $dy\,dx$)

If D is a type I region, then

$$\iint_D f(x, y)\,dA = \int_a^b \int_{g_1(x)}^{g_2(x)} f(x, y)\,dy\,dx$$

whenever both integrals exist. Similarly, for a type II region,

TYPE II (horizontal strip):
y fixed, x varies (form $dx\,dy$)

$$\iint_D f(x, y)\,dA = \int_c^d \int_{h_1(y)}^{h_2(y)} f(x, y)\,dx\,dy$$

Proof   This proof is found in most advanced calculus textbooks.   ▬

When using Fubini's theorem for nonrectangular regions, it helps to sketch the region of integration D and to find equations for all boundary curves of D. Such a sketch often provides the information needed to determine whether D is a type I or type II region (or neither, or both) and to set up the limits of integration of an iterated integral.

EXAMPLE 1   *Double integral over a triangular region*

Let T be the triangular region enclosed by the lines $y = 0$, $y = 2x$, and $x = 1$. Evaluate the double integral

$$\iint_T (x + y)\,dA$$

using an iterated integral with:
a. y-integration first   b. x-integration first

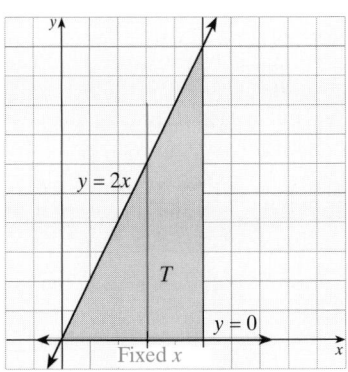

FIGURE 13.7 For each fixed x $(0 \leq x \leq 1)$, y varies from $y = 0$ to $y = 2x$.

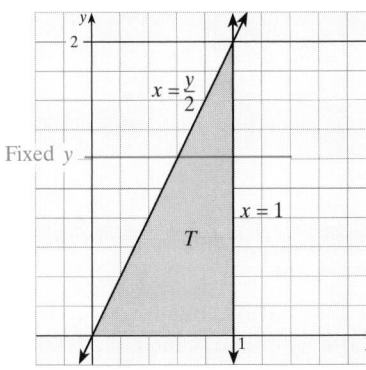

FIGURE 13.8 For each fixed y $(0 \leq y \leq 2)$, x varies from $y/2$ to 1.

Solution

a. To set up the limits of integration in the iterated integral, we draw the graph as shown in Figure 13.7 and note that for fixed x, the variable y varies from $y = 0$ (the x-axis) to the line $y = 2x$. These are the limits of integration for the inner integral (with respect to y first). The other limits of integration are the numerical limits of integration for x; that is, x varies between $x = 0$ and $x = 1$.

$$\iint_T (x + y)\, dA = \int_0^1 \int_0^{2x} (x + y)\, dy\, dx = \int_0^1 \left[xy + \frac{1}{2}y^2 \right]\Big|_{y=0}^{y=2x} dx$$

$$= \int_0^1 \left[x(2x) + \frac{1}{2}(2x)^2 - \left(x(0) + \frac{1}{2}(0)^2 \right) \right] dx = \int_0^1 4x^2\, dx = \frac{4}{3}x^3\Big|_{x=0}^{x=1} = \frac{4}{3}$$

b. Reversing the order of integration, we see from Figure 13.8 that for each fixed y, the variable x varies (left to right) from the line $x = y/2$ to the vertical line $x = 1$. The outer limits of integration are for y as y varies from $y = 0$ to $y = 2$.

$$\iint_T (x + y)\, dA = \int_0^2 \int_{y/2}^1 (x + y)\, dx\, dy$$

$$= \int_0^2 \left[\frac{1}{2}x^2 + xy \right]\Big|_{x=y/2}^{x=1} dy = \int_0^2 \left[\frac{1}{2} + y - \frac{y^2}{8} - \frac{y^2}{2} \right] dy$$

$$= \left[\frac{y}{2} + \frac{y^2}{2} - \frac{5y^3}{24} \right]\Big|_{y=0}^{y=2} = \left[1 + 2 - \frac{5(8)}{24} \right] - [0] = \frac{4}{3} \qquad ■$$

MORE ON AREA AND VOLUME

Even though we can find the area between curves with single integrals, it is often easier to compute area using a double integral. If $f(x, y) \geq 0$ over a region D in the xy-plane, then $\iint_D f(x, y)\, dA$ gives the **volume of the solid** bounded above by the surface $z = f(x, y)$ and below by the region D. In the special case where $f(x, y) = 1$, the integral gives the **area** of D.

The Double Integral as Area and Volume

The **area** of the region D in the xy-plane is given by

$$A = \iint_D dA$$

If f is continuous and $f(x, y) \geq 0$ on the region D, the **volume** of the solid under the surface $z = f(x, y)$ above the region D is given by

$$V = \iint_D f(x, y)\, dA$$

EXAMPLE 2   *Area of a region in the xy-plane using a double integral*

Find the area of the region D between $y = \cos x$ and $y = \sin x$ over the interval $0 \leq x \leq \frac{\pi}{4}$ using
a. a single integral   b. a double integral

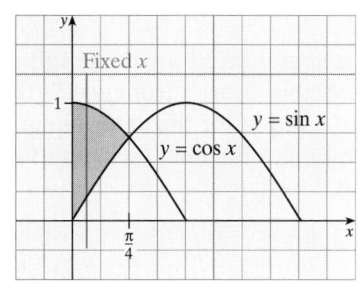

■ **FIGURE 13.9** The area of the region between $y = \cos x$ and $y = \sin x$

Solution

a. The graph is shown in Figure 13.9.

$$\int_0^{\pi/4} (\cos x - \sin x)\, dx = [\sin x + \cos x]\Big|_0^{\pi/4} = \sqrt{2} - 1$$

b.

$$A = \iint_D dA = \int_0^{\pi/4} \int_{\sin x}^{\cos x} 1\, dy\, dx = \int_0^{\pi/4} [y]\Big|_{y=\sin x}^{y=\cos x} dx$$

$$= \int_0^{\pi/4} [\cos x - \sin x]\, dx = \sqrt{2} - 1$$

The area is $\sqrt{2} - 1 \approx 0.41$ square unit.   ■

In comparing the single and double integral solutions for area in Example 2, you might ask, "Why bother with the double integral, because it reduces to the single integral case after one step?" The answer is that it is often easier to begin with the double integral

$$A = \iint_D dA$$

and then let the *evaluation* lead to the proper form.

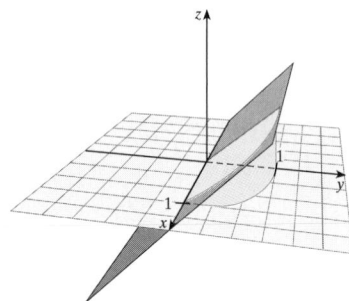

Graph of $z = y$ showing the disk in the xy-plane

| **EXAMPLE 3** | *Volume using a double integral* |

Find the volume of the solid bounded above by the plane $z = y$ and below in the xy-plane by the part of the disk $x^2 + y^2 \le 1$ in the first quadrant.

Solution   The three-dimensional solid is shown in the margin, but for your work you need be concerned only with the projection in the xy-plane, which is shown in Figure 13.10.

We can regard D as either a type I or type II region, and because we worked with a type I (vertical) region in Example 2, we shall use a type II (horizontal) region for this example. Accordingly, note that for each fixed number y between 0 and 1, x varies between $x = 0$ on the left and $x = \sqrt{1 - y^2}$ on the right. Thus,

$$V = \iint_D f(x, y)\, dA$$

$$= \int_0^1 \int_0^{\sqrt{1-y^2}} y\, dx\, dy \qquad f(x,y) = y \text{ is given; } dA = dx\, dy.$$

$$= \int_0^1 [xy]\Big|_{x=0}^{x=\sqrt{1-y^2}} dy$$

$$= \int_0^1 y\sqrt{1 - y^2}\, dy \qquad \text{Let } u = 1 - y^2 \text{ to integrate.}$$

$$= \left[-\frac{1}{3}(1 - y^2)^{3/2}\right]\Big|_{y=0}^{y=1} = \frac{1}{3}$$

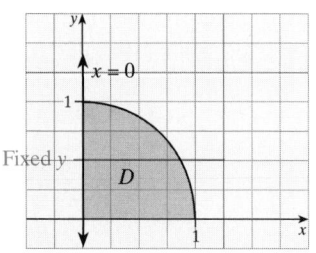

■ **FIGURE 13.10** The quarter disk $x^2 + y^2 \le 1, x \ge 0, y \ge 0$

The volume is $\frac{1}{3}$ cubic unit.   ■

REVERSING THE ORDER OF INTEGRATION IN A DOUBLE INTEGRAL

It is often useful to be able to reverse the order of integration in a given iterated integral. This procedure is illustrated in the next two examples.

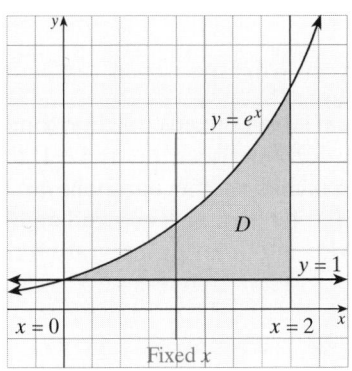

a. *D* as a type I region; *y* varies from 1 to e^x

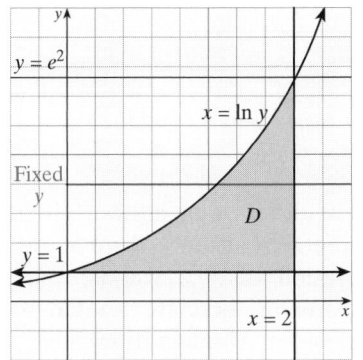

b. *D* as a type II region; *x* varies from ln *y* to 2

■ **FIGURE 13.11**   The region of integration for Example 4

EXAMPLE 4   *Reversing order of integration in a double integral*

Reverse the order of integration in the iterated integral

$$\int_0^2 \int_1^{e^x} f(x, y)\, dy\, dx$$

Solution   Begin by drawing the region *D* by looking at the limits of integration for both *x* and *y* in the double integral. For this example, we see that the *y*-integration comes first, so *D* is a type I region. The inner limits are

$$y = e^x \text{ (top curve)} \quad \text{and} \quad y = 1 \text{ (bottom curve)}$$

These are shown in Figure 13.11a. Next, draw the appropriate limits of integration for *x*. *Note:* These limits should be constants:

$$x = 0 \text{ (left point)} \quad \text{and} \quad x = 2 \text{ (right point)}$$

These vertical lines are also drawn in Figure 13.11a.

To reverse the order of integration, we need to regard *D* as a type II region (Figure 13.11b). Note that the region varies from $y = 1$ to $y = e^2$ (corresponding to where $y = e^x$ intersects $x = 0$ and $x = 2$, respectively). For each fixed *y* (horizontal strip) between 1 and e^2, the region extends from the curve $x = \ln y$ (that is, $y = e^x$) on the left to the line $x = 2$ on the right. Thus, reversing the order of integration, we find that the given integral becomes

$$\int_1^{e^2} \int_{\ln y}^2 f(x, y)\, dx\, dy$$

■ *What This Says:*   There are two different ways of representing the integral, which we illustrate together for easy reference:

Type I: *x* fixed (vertical strip)	Type II: *y* fixed (horizontal strip)
y-integration first; varies from	*x*-integration first; varies from
$y = 1$ to $y = e^x$.	$x = \ln y$ to $x = 2$.
↓	↓
$\displaystyle\int_0^2 \int_1^{e^x} f(x, y)\, dy\, dx$	$= \displaystyle\int_1^{e^2} \int_{\ln y}^2 f(x, y)\, dx\, dy$
↑	↑
Constant limits for *x*; varies from 0 to 2.	Constant limits for *y*; varies from 1 to e^2.

■

EXAMPLE 5   *Reversing the order that requires a sum of integrals*

Reverse the order of integration in the integral

$$\int_{-1}^2 \int_{x^2-2}^x f(x, y)\, dy\, dx$$

Solution   This is a type I form (because the *y*-integration comes first). To reverse the order of integration, we begin by drawing the region *D*, as shown in Figure 13.12a. We draw the bottom and top curves:

$$y = x^2 - 2 \text{ (bottom)} \quad \text{and} \quad y = x \text{ (top)}$$

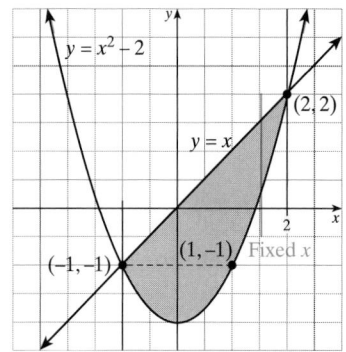

a. Type I description

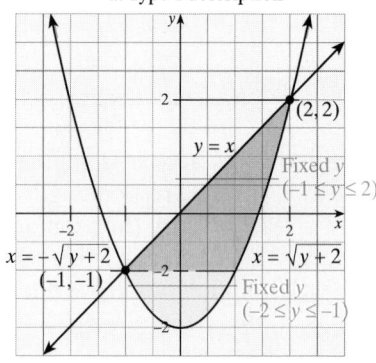

b. Type II description

■ **FIGURE 13.12**   The region of integration for Example 5

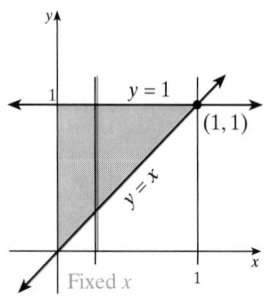

a. *y*-integration first

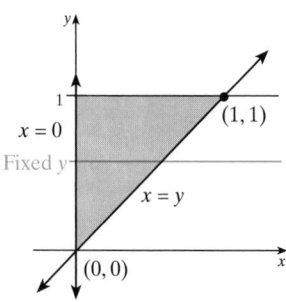

b. *x*-integration first

■ **FIGURE 13.13**   The region of integration for Example 6

Then we draw the left and right boundaries:

$$x = -1 \quad \text{and} \quad x = 2$$

To change this to a type II form, we shift from a vertical strip (x fixed) to a horizontal strip (y fixed), as shown in Figure 13.12b. We see that for y between -2 and -1, the horizontal line is bounded by the left and right branches of the parabolic arch, whereas for y between -1 and 2, it is bounded on the left by the line $x = y$ and on the right by the parabola. This means that the integral with the order of integration reversed must be expressed as the sum of two integrals:

$$\int_{-1}^{2} \int_{x^2-2}^{x} f(x, y) \, dy \, dx \qquad \text{For a fixed } x \text{ between } -1 \text{ and } 2, y \text{ varies between } y = x^2 - 2 \text{ and } y = x.$$

$$= \int_{-2}^{-1} \int_{-\sqrt{y+2}}^{\sqrt{y+2}} f(x, y) \, dx \, dy + \int_{-1}^{2} \int_{y}^{\sqrt{y+2}} f(x, y) \, dx \, dy$$

$\underbrace{\qquad\qquad\qquad\qquad}$ $\underbrace{\qquad\qquad\qquad\qquad}$
Part between left and right   Part between line and
part of parabola (blue part)   Parabola (gray part)

For a fixed y between -2 For a fixed y between -1
and -1, x varies between   and 2, x varies between
$x = -\sqrt{y+2}$ and $x = \sqrt{y+2}$ $x = y$ and $x = \sqrt{y+2}$ ■

A given iterated integral can often be evaluated in either order, although it may happen that one order is more convenient than the other. For instance, the integral in Example 5 is easier to handle by performing y-integration first, because the reverse order involves two integrations instead of just one. However, there are iterated integrals that can be evaluated *only* after reversing the order of integration. Such an integral is featured in the following example.

EXAMPLE 6   *Evaluating a double integral by reversing the order*

Evaluate $\displaystyle\int_{0}^{1} \int_{x}^{1} e^{y^2} \, dy \, dx$.

Solution   We cannot evaluate the integral in the given order (y-integration first) because the integrand e^{y^2} has no elementary antiderivative. We shall evaluate the integral by reversing the order of integration. The region of integration is sketched in Figure 13.13a. Note that for any fixed x between 0 and 1, y varies from x to 1.

To reverse the order of integration, observe that for each fixed y between 0 and 1, x varies from 0 to y, as shown in Figure 13.13b.

$$\int_{0}^{1} \int_{x}^{1} e^{y^2} \, dy \, dx = \int_{0}^{1} \int_{0}^{y} e^{y^2} \, dx \, dy$$

$$= \int_{0}^{1} x e^{y^2} \Big|_{x=0}^{x=y} \, dy = \int_{0}^{1} y e^{y^2} \, dy \qquad \text{Let } u = y^2.$$

$$= \left[\frac{1}{2} e^{y^2} \right] \Big|_{0}^{1} = \frac{1}{2}(e^1 - e^0)$$

$$= \frac{1}{2}(e - 1) \qquad\qquad\qquad ■$$

13.2 Problem Set

1. ■ **What Does This Say?** Describe the process for finding volume using a double integral.

2. ■ **What Does This Say?** Describe the process of reversing the order of integration.

Sketch the region of integration in Problems 3–16, and compute the double integral (either in the order of integration given or with the order reversed).

3. $\int_0^4 \int_0^{4-x} xy \, dy \, dx$

4. $\int_0^4 \int_{x^2}^{4x} dy \, dx$

5. $\int_0^1 \int_{-x^2}^{x^2} dy \, dx$

6. $\int_1^e \int_0^{\ln x} xy \, dy \, dx$

7. $\int_0^{2\sqrt{2}} \int_{y^2/4}^{\sqrt{12-y^2}} dx \, dy$

8. $\int_{-2}^1 \int_{y^2+4y}^{3y+2} dx \, dy$

9. $\int_0^3 \int_1^{4-x} (x+y) \, dy \, dx$

10. $\int_0^1 \int_{x^3}^1 (x+y^2) \, dy \, dx$

11. $\int_0^1 \int_0^x (x^2+2y^2) \, dy \, dx$

12. $\int_{-1}^1 \int_{-1}^x (3x+2y) \, dy \, dx$

13. $\int_0^2 \int_0^{\sin x} y \cos x \, dy \, dx$

14. $\int_0^{\pi/2} \int_0^{\sin x} e^y \cos x \, dy \, dx$

15. $\int_0^{\pi/3} \int_0^{y^2} \frac{1}{y} \sin \frac{x}{y} \, dx \, dy$

16. $\int_0^1 \int_0^y y^2 e^{xy} \, dx \, dy$

Evaluate the double integral given in Problems 17–28 for the specified region of integration D.

17. $\iint_D (x+y) \, dA$; D is the triangle with vertices $(0,0), (0,1), (1,1)$.

18. $\iint_D (x+2y) \, dA$; D is the triangle with vertices $(0,0), (1,0), (0,2)$.

19. $\iint_D 48xy \, dA$; D is the region bounded by $y = x^3$ and $y = \sqrt{x}$.

20. $\iint_D (2y-x) \, dA$; D is the region bounded by $y = x^2$ and $y = 2x$.

21. $\iint_D y \, dA$; D is the region bounded by $y = \sqrt{x}, y = 2-x$, and $y = 0$.

22. $\iint_D 4x \, dA$; D is the region bounded by $y = 4 - x^2$, $y = 3x$, and $x = 0$.

23. $\iint_D 4x \, dA$; D is the region in the first quadrant bounded by $y = 4 - x^2, y = 3x$, and $y = 0$.

24. $\iint_D (2x+1) \, dA$; D is the triangle with vertices $(-1,0), (1,0)$, and $(0,1)$.

25. $\iint_D 2x \, dA$; D is the region bounded by $x^2y = 1, y = x$, $x = 2$, and $y = 0$.

26. $\iint_D \frac{dA}{y^2+1}$; D is the triangle bounded by $x = 2y$, $y = -x$, and $y = 2$.

27. $\iint_D 12x^2 e^{y^2} \, dA$; D is the region in the first quadrant bounded by $y = x^3$ and $y = x$.

28. $\iint_D \cos e^x \, dA$; D is the region bounded by $y = e^x, y = -e^x, x = 0$, and $x = \ln 2$.

Sketch the region of integration in Problems 29–36, and then compute the integral in two ways: **a.** *with the given order of integration, and* **b.** *with the order of integration reversed.*

29. $\int_0^4 \int_0^{4-x} xy \, dy \, dx$

30. $\int_0^4 \int_{x^2}^{4x} dy \, dx$

31. $\int_0^1 \int_x^{2x} e^{y-x} \, dy \, dx$

32. $\int_0^4 \int_0^{\sqrt{x}} 3x^5 \, dy \, dx$

33. $\int_0^1 \int_{-x^2}^{x^2} dy \, dx$

34. $\int_1^e \int_0^{\ln x} xy \, dy \, dx$

35. $\int_0^{2\sqrt{3}} \int_{y^2/6}^{\sqrt{16-y^2}} dx \, dy$

36. $\int_{-2}^1 \int_{y^2+4y}^{3y+2} dx \, dy$

Sketch the region of integration in Problems 37–46, and write an equivalent integral with the order of integration reversed. Do not evaluate.

37. $\int_0^1 \int_0^{2y} f(x,y) \, dx \, dy$

38. $\int_0^1 \int_{x^2}^{\sqrt{x}} f(x,y) \, dy \, dx$

39. $\int_0^4 \int_{y/2}^{\sqrt{y}} f(x,y) \, dx \, dy$

40. $\int_1^{e^2} \int_{\ln x}^2 f(x,y) \, dy \, dx$

41. $\int_0^3 \int_{y/3}^{\sqrt{4-y}} f(x,y) \, dx \, dy$

42. $\int_0^1 \int_x^{2-x} f(x,y) \, dy \, dx$

43. $\int_{-3}^2 \int_{x^2}^{6-x} f(x,y) \, dy \, dx$

44. $\int_2^4 \int_x^{16/x} f(x,y) \, dy \, dx$

45. $\int_0^7 \int_{x^2-6x}^x f(x,y) \, dy \, dx$

46. $\int_0^1 \int_{\tan^{-1}x}^{\pi/4} f(x,y) \, dy \, dx$

B *Set up a double integral for the volume of the solid region described in Problems 47–53.*

47. The tetrahedron that lies in the first octant and is bounded by the coordinate planes and the plane $z = 7 - 3x - 2y$

48. The solid bounded above by the paraboloid $z = 6 - 2x^2 - 3y^2$ and below by the plane $z = 0$

49. The solid that lies inside both the cylinder $x^2 + y^2 = 3$ and the sphere $x^2 + y^2 + z^2 = 7$

50. The solid that lies inside both the sphere $x^2 + y^2 + z^2 = 3$ and the paraboloid $2z = x^2 + y^2$

51. The ellipsoid
$$\frac{x^2}{a^2} + \frac{y^2}{b^2} + \frac{z^2}{c^2} = 1$$

52. The solid bounded above by the plane $z = 2 - 3x - 5y$ and below by the region shown in Figure 13.14.

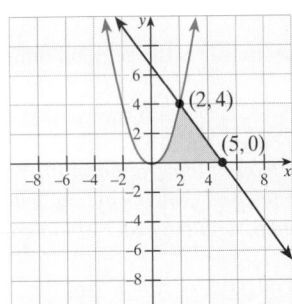

■ **FIGURE 13.14** Region for Problem 52

53. The solid that remains when a square hole of side 2 is drilled through a sphere of radius $\sqrt{2}$

Express the area of the region D bounded by the curve given in Problems 54–55 as a double integral in two different ways (with different orders of integration). Evaluate one of these integrals to find the area.

54. Let D denote the region in the first quadrant of the xy-plane that is bounded by the curves $y = \dfrac{4}{x^2}$ and $y = 5 - x^2$.

55. Let D denote the region bounded by the ellipse $\dfrac{x^2}{a^2} + \dfrac{y^2}{b^2} = 1$.

56. Find the volume under the surface $z = x + y + 2$ above the region D bounded by the curves $y = x^2$ and $y = 2$.

57. Find the volume under the plane $z = 4x$ above the region D given by $y = x^2, y = 0,$ and $x = 1$.

58. Evaluate
$$\int_0^1 \int_0^y (x^2 + y^2)\, dx\, dy + \int_1^2 \int_0^{2-y} (x^2 + y^2)\, dx\, dy$$
by reversing the order of integration.

59. Let $f(x, y)$ be a continuous function for all x and y. Reverse the order of integration in
$$\int_1^2 \int_x^{x^3} f(x, y)\, dy\, dx + \int_2^8 \int_x^8 f(x, y)\, dy\, dx$$

60. Evaluate $\displaystyle\iint_D xy\, dA$, where D is the triangular region in the xy-plane with vertices $(0,0), (1,0),$ and $(4,1)$.

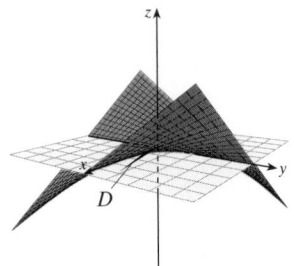

61. Compute $\displaystyle\iint_D (x^2 - xy - 1)\, dA$, where D is the triangular region bounded by the lines $x - 2y + 2 = 0$, $x + 3y - 3 = 0,$ and $y = 0$.

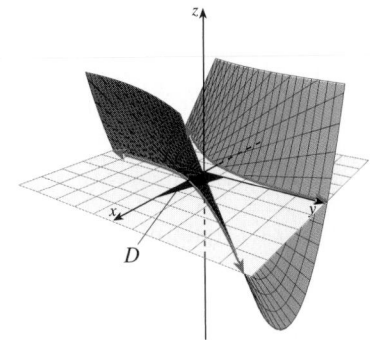

62. Find the volume under the plane $z = x + 2y + 4$ above the region D bounded by the lines $y = 2x, y = 3 - x,$ and $y = 0$.

63. Find the volume under the plane $3x + y - z = 0$ above the elliptic region bounded by $4x^2 + 9y^2 \leq 36$, with $x \geq 0$ and $y \geq 0$.

64. Find the volume under the surface $z = x^2 + y^2$ above the square region bounded by $|x| \leq 1$ and $|y| \leq 1$.

Evaluate each integral in Problems 65–67 by relating it to the **C** *volume of a simple solid.*

65. Evaluate $\displaystyle\iint_R (3 - \sqrt{x^2 + y^2})\, dA$, where R is the disk $x^2 + y^2 \leq 9$ in the xy-plane.

66. Evaluate $\displaystyle\iint_R (1 - \sqrt{x^2 + z^2})\, dA$, where R is the disk $x^2 + z^2 \leq 1$ in the xz-plane.

67. Evaluate $\iint\limits_{R} (4 - \sqrt{y^2 + z^2})\, dA$, where R is the disk $y^2 + z^2 \leq 16$ in the yz-plane.

68. Show that if $f(x, y)$ is continuous on a region D and $m \leq f(x, y) \leq M$ for all points (x, y) in D, then

$$mA \leq \iint\limits_{D} f(x, y)\, dA \leq MA$$

where A is in the area of D.

69. Use the result of Problem 68 to estimate the value of the double integral

$$\iint\limits_{D} e^{y \sin x}\, dA$$

where D is the triangle with vertices $(-1, 0), (2, 0),$ and $(0, 1)$.

13.3 Double Integrals in Polar Coordinates

IN THIS SECTION change of variables to polar form, improper double integrals in polar coordinates

In general, changing variables in a double integral is more complicated than in a single integral. In this section, we focus attention on using polar coordinates in a double integral, and in Section 13.8, we examine changing variables from a more general standpoint.

CHANGE OF VARIABLES TO POLAR FORM

Polar coordinates are used in double integrals primarily when the integrand or the region of integration (or both) have relatively simple polar descriptions. As a preview of the ideas we plan to explore, let us examine the double integral

$$\iint\limits_{R} (x^2 + y^2 + 1)\, dA$$

where R is the region (disk) bounded by the circle $x^2 + y^2 = 4$.

Interpreting R as a type I (vertical strip), we see that for each fixed x between -2 and 2, y varies from the lower boundary semicircle with equation $y = -\sqrt{4 - x^2}$ to the upper semicircle $y = \sqrt{4 - x^2}$, as shown in Figure 13.15a.

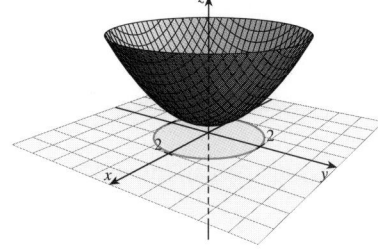

Graph of $f(x, y) = x^2 + y^2 + 1$

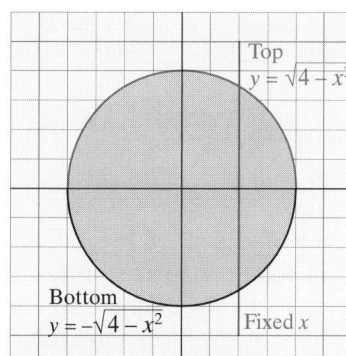

a. Type I description:
For fixed x between -2 and 2,
y varies from $y = -\sqrt{4 - x^2}$
to $y = \sqrt{4 - x^2}$

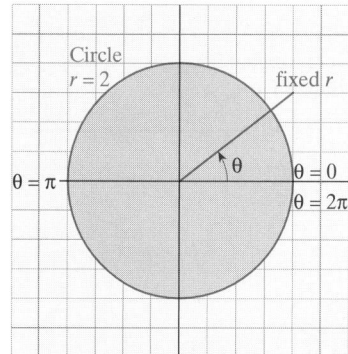

b. Polar description:
For fixed θ between 0 and 2π,
r varies from 0 to 2.

■ **FIGURE 13.15** Two interpretations of a region R

Using the type I description, we have

$$\iint_D (x^2 + y^2 + 1)\, dA = \int_{-2}^{2} \int_{-\sqrt{4-x^2}}^{\sqrt{4-x^2}} (x^2 + y^2 + 1)\, dy\, dx$$

The iterated integral on the right is clearly difficult to evaluate, but both the integrand and the domain of integration can be represented quite simply in terms of polar coordinates. Specifically, using the polar conversion formulas

$$x = r \cos\theta \qquad y = r \sin\theta \qquad r = \sqrt{x^2 + y^2} \qquad \tan\theta = \frac{y}{x}$$

we find that the integrand $f(x, y) = x^2 + y^2 + 1$ can be rewritten

$$f(r \cos\theta, r \sin\theta) = (r \cos\theta)^2 + (r \sin\theta)^2 + 1 = r^2 + 1$$

and the region of integration D is just the interior of the circle $r = 2$. Thus, D can be described as the set of all points (r, θ) so that for each fixed angle θ between 0 and 2π, r varies from the origin ($r = 0$) to the circle $r = 2$, as shown in Figure 13.15b.

But how is the differential of integration dA changed in polar coordinates? Can we simply substitute "$dr\, d\theta$" for dA and perform the integration with respect to r and θ? The answer is no, and the correct formula for expressing a given double integral in polar form is given in the following theorem.

THEOREM 13.4   *Double integral in polar coordinates*

If f is continuous in the polar region D described by $r_1(\theta) < r < r_2(\theta)$ ($r_1(\theta) \geq 0$, $r_2(\theta) \geq 0$), $\alpha \leq \theta \leq \beta$ ($0 \leq \beta - \alpha < 2\pi$), then

$$\iint_D f(r, \theta)\, dA = \int_{\alpha}^{\beta} \int_{r_1(\theta)}^{r_2(\theta)} f(r, \theta)\, r\, dr\, d\theta$$

■ ***What This Says:***   The procedure for changing from a Cartesian integral

$$\iint_R f(x, y)\, dA$$

into a polar integral requires two steps. First, substitute $x = r \cos\theta$ and $y = r \sin\theta$, $dx\, dy = r\, dr\, d\theta$ into the Cartesian integral. Then convert the region of integration R to polar form D. Thus,

$$\iint_R f(x, y)\, dA = \iint_D f(r \cos\theta, r \sin\theta)\, r\, dr\, d\theta$$

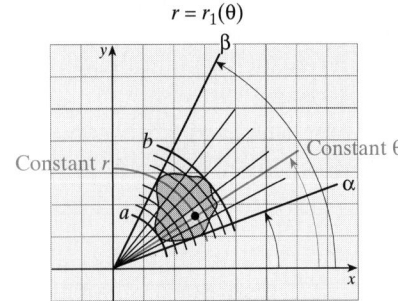

$r = r_1(\theta)$

a. A partition of a region into polar coordinates

$r = r_2(\theta)$

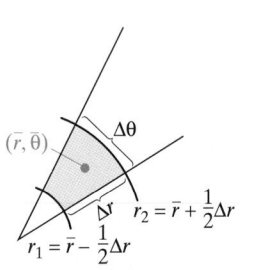

b. A typical polar rectangle in the partition

■ **FIGURE 13.16**   A polar rectangle

Proof   A region described by $r_1(\theta) \leq r \leq r_2(\theta)$, $\alpha \leq \theta \leq \beta$ is called a **polar rectangle.** A typical polar rectangle is shown in Figure 13.16.

We begin by subdividing the region of integration into polar rectangles. Then we pick an arbitrary polar-form point (r_k^*, θ_k^*) in each polar rectangle in the partition and then take the limit of an appropriate Riemann sum

$$\sum_{k=1}^{N} f(r_k^*, \theta_k^*) \Delta A_k$$

where ΔA_k is the area of the kth polar rectangle.

To find the area of a typical polar rectangle, let (r_k^*, θ_k^*) be the center of the polar rectangle—that is, the point midway between the arcs and rays that form the polar rectangle, as shown in Figure 13.16b. If the circular arcs that bound the polar rectangle are Δr apart, then the arcs are given by

$$r_1 = r_k^* - \tfrac{1}{2}\Delta r_k \quad \text{and} \quad r_2 = r_k^* + \tfrac{1}{2}\Delta r_k$$

It can be shown that a circular section of radius r and central angle θ has area $\tfrac{1}{2}r^2\theta$ (see the *Student Mathematics Handbook* for details). Thus, a typical polar rectangle has area

SMH

$$\Delta A_k = \left[\underbrace{\tfrac{1}{2}(r_k^* + \tfrac{1}{2}\Delta r_k)^2}_{\text{Radius of outside arc}} - \underbrace{\tfrac{1}{2}(r_k^* - \tfrac{1}{2}\Delta r_k)^2}_{\text{Radius of inside arc}}\right]\Delta\theta = r_k^* \, \Delta r_k \, \Delta\theta_k$$

Finally, we compute the given double integral in polar form by taking the limit:

$$\iint_D f(r, \theta) \, dA = \lim_{\|P\|\to 0} \sum_{k=1}^{N} f(r_k^*, \theta_k^*)\Delta A_k = \lim_{\|P\|\to 0} \sum_{k=1}^{N} f(r_k^*, \theta_k^*) \, r_k^* \, \Delta r_k \, \Delta\theta_k$$

$$= \int_\alpha^\beta \int_{r_1(\theta)}^{r_2(\theta)} f(r, \theta) \, r \, dr \, d\theta$$

PREVIEW   It can be shown that under reasonable conditions, the change of variable $x = x(u, v), y = y(u, v)$ transforms the integral $\iint f(x, y) \, dA$ into $\iint f(u, v)|J(u, v)| \, du \, dv$, where

$$J(u, v) = \begin{vmatrix} \dfrac{\partial x}{\partial u} & \dfrac{\partial x}{\partial v} \\ \dfrac{\partial y}{\partial u} & \dfrac{\partial y}{\partial v} \end{vmatrix}$$

This determinant is known as the **Jacobian** of the transformation. In the case of polar coordinates, we have $x = r\cos\theta$ and $y = r\sin\theta$, so that the Jacobian is

$$J(r, \theta) = \begin{vmatrix} \dfrac{\partial}{\partial r}(r\cos\theta) & \dfrac{\partial}{\partial\theta}(r\cos\theta) \\ \dfrac{\partial}{\partial r}(r\sin\theta) & \dfrac{\partial}{\partial\theta}(r\sin\theta) \end{vmatrix}$$

$$= \begin{vmatrix} \cos\theta & -r\sin\theta \\ \sin\theta & r\cos\theta \end{vmatrix} = r\cos^2\theta + r\sin^2\theta = r$$

This yields the result of Theorem 13.4.

$$\iint_R f(x, y) \, dA = \iint_D f(r, \theta) \, r \, dr \, d\theta = \int_\alpha^\beta \int_{r_1(\theta)}^{r_2(\theta)} f(r\cos\theta, r\sin\theta) \, r \, dr \, d\theta$$

We discuss this more completely in Section 13.8.

If you carefully compare the result in the preview box with the result of Theorem 13.4, you will see that they are not *exactly* the same. The region D in Theorem 13.4 already is in polar coordinates. The more common situation is that we are given f and a region R in rectangular coordinates that we need to change to polar coordinates.

(SMH)

You will need to be familiar with the graphs of many polar-form curves. These can be found in Table 9.1 on page 639, and also in the *Student Mathematics Handbook*.

We now present the example we promised in the introduction.

EXAMPLE 1   *Double integral in polar form*

Evaluate $\iint_R (x^2 + y^2 + 1)\, dA$, where D is the region inside the circle $x^2 + y^2 = 4$.

Solution   In this example, the region R is given in rectangular form. We will describe this as a polar region D. Earlier, we observed that in D, for each fixed angle θ between 0 and 2π, r varies from $r = 0$ (the pole) to $r = 2$ (the circle). Thus,

$$\iint_R \underbrace{(x^2 + y^2 + 1)}_{f(x,y)}\, dA = \int_0^{2\pi} \int_0^2 \underbrace{(r^2 + 1)}_{f(r\cos\theta,\, r\sin\theta)} \underbrace{r\, dr\, d\theta}_{dA}$$

$$= \int_0^{2\pi} \int_0^2 (r^3 + r)\, dr\, d\theta = \int_0^{2\pi} \left[\frac{r^4}{4} + \frac{r^2}{2}\right]\Big|_0^2 d\theta = \int_0^{2\pi} 6\, d\theta = 6\theta \Big|_0^{2\pi} = 12\pi \qquad \blacksquare$$

EXAMPLE 2   *Double integral in polar form*

Evaluate $\iint_D x\, dA$, where D is the region bounded above by the line $y = x$ and below by the circle $x^2 + y^2 - 2y = 0$.

Solution   The circle $x^2 + y^2 - 2y = 0$ and the line $y = x$ are shown in Figure 13.17. The polar-form equation for the circle is $r = 2\sin\theta$, and the line $y = x$ in polar form is

$$r\sin\theta = r\cos\theta$$

This is true only when $\tan\theta = 1$ (divide both sides by $r\cos\theta$), which implies $\theta = \pi/4$. Thus, the region D may be described by saying that for each fixed angle θ between 0 and $\pi/4$, r varies from 0 (the pole) to $2\sin\theta$ (the circle), and we have

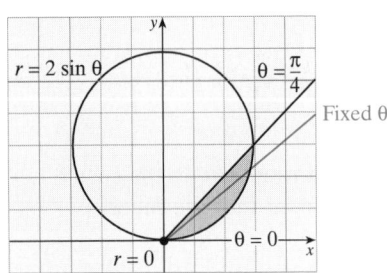

■ **FIGURE 13.17** The region D.

$$\iint_D x\, dA = \int_0^{\pi/4} \int_0^{2\sin\theta} \underbrace{(r\cos\theta)}_{x} \underbrace{r\, dr\, d\theta}_{dA} = \int_0^{\pi/4} \left[\frac{r^3}{3}\cos\theta\right]\Big|_{r=0}^{r=2\sin\theta} d\theta$$

$$= \int_0^{\pi/4} \left[\frac{(2\sin\theta)^3}{3}\cos\theta - 0\right] d\theta = \int_0^{\pi/4} \frac{8}{3}\sin^3\theta\cos\theta\, d\theta \qquad \text{Let } u = \sin\theta.$$

$$= \left[\frac{8}{3}\cdot\frac{1}{4}\sin^4\theta\right]\Big|_{\theta=0}^{\theta=\pi/4} = \frac{1}{6} \qquad \blacksquare$$

EXAMPLE 3   *Volume in polar form*

Use a polar double integral to show that a sphere of radius a has volume $\frac{4}{3}\pi a^3$.

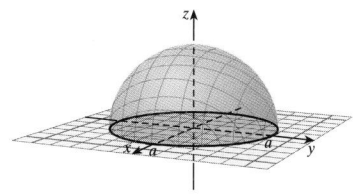

The top hemisphere is bounded above by the sphere and below by the disk D.

Solution   We shall compute the required volume by doubling the volume of the solid hemisphere $x^2 + y^2 + z^2 \leq a^2$, with $z \geq 0$. This hemisphere may be regarded as a solid bounded below by the circular disk $x^2 + y^2 \leq a^2$ and above by the spherical surface.

We need to change the equation of the hemisphere to polar form:

In rectangular form: $z = \sqrt{a^2 - x^2 - y^2}$

In polar form: $z = \sqrt{a^2 - r^2}$ *Because* $r^2 = x^2 + y^2$

Describing the disk D in polar terms, we see that for each fixed θ between 0 and 2π, r varies from the origin to the circle $x^2 + y^2 = a^2$, which has the polar equation $r = a$, as shown in Figure 13.18. Thus, the volume is given by the integral

$$V = 2 \int_D \int z \, dA \qquad \text{Rectangular form}$$

$$= 2 \int_0^{2\pi} \int_0^a \sqrt{a^2 - r^2}\, r \, dr \, d\theta \qquad \boxed{\begin{array}{l} \text{Let } u = a^2 - r^2; \\ du = -2r \, dr. \end{array}}$$

$$= 2 \int_0^{2\pi} \left[-\frac{1}{3}(a^2 - r^2)^{3/2} \right] \Bigg|_{r=0}^{r=a} d\theta$$

$$= -\frac{2}{3} \int_0^{2\pi} [(a^2 - a^2)^{3/2} - (a^2 - 0)^{3/2}] \, d\theta$$

$$= \frac{2}{3} \int_0^{2\pi} a^3 \, d\theta = \frac{2}{3} a^3 \theta \Bigg|_0^{2\pi} = \frac{4}{3}\pi a^3 \qquad \blacksquare$$

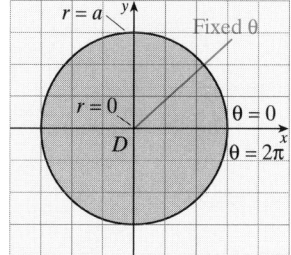

■ **FIGURE 13.18**   The disk D: For θ between 0 and 2π, r varies from 0 to a.

| **EXAMPLE 4** | *Region of integration between two polar curves* |

Evaluate $\displaystyle \int_D \int \frac{1}{x} \, dA$, where D is the region in the first quadrant that lies inside the circle $r = 3\cos\theta$ and outside the cardioid $r = 1 + \cos\theta$.

Solution   Begin by sketching the given curves, as shown in Figure 13.19. Next, find the points of intersection:

$$3\cos\theta = 1 + \cos\theta$$
$$2\cos\theta = 1$$
$$\cos\theta = \frac{1}{2}$$
$$\theta = \frac{\pi}{3}$$

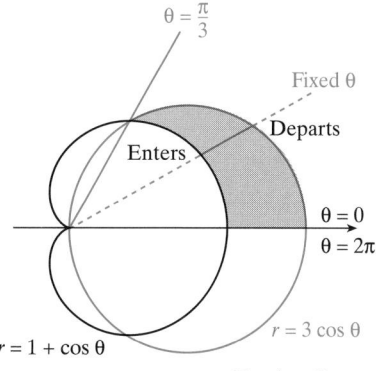

■ **FIGURE 13.19**   Region D

We see that D is the region such that, for each fixed angle θ between 0 and $\pi/3$, r varies from $1 + \cos\theta$ (the cardioid) to $3\cos\theta$ (the circle). This gives us the limits of integration. Finally, we need to write the integrand in polar form:

$$\underbrace{\frac{1}{x}}_{\text{Rectangular form}} = \underbrace{\frac{1}{r\cos\theta}}_{\text{Polar form}}$$

Thus,

$$\int_D\int \frac{1}{x}\,dA = \int_D\int \frac{1}{r\cos\theta}\,r\,dr\,d\theta$$

$$= \int_0^{\pi/3}\int_{1+\cos\theta}^{3\cos\theta} \frac{1}{\cos\theta}\,dr\,d\theta \qquad \textit{Write } \frac{1}{\cos\theta} = \sec\theta.$$

$$= \int_0^{\pi/3} [r\sec\theta]\Big|_{r=1+\cos\theta}^{r=3\cos\theta}\,d\theta$$

$$= \int_0^{\pi/3} [3\cos\theta\sec\theta - (1+\cos\theta)\sec\theta]\,d\theta$$

$$= \int_0^{\pi/3} (2 - \sec\theta)\,d\theta \qquad \textit{Since } \sec\theta = \frac{1}{\cos\theta}$$

$$= [2\theta - \ln|\sec\theta + \tan\theta|]\Big|_0^{\pi/3}$$

$$= \tfrac{2\pi}{3} - \ln(2 + \sqrt{3}) \qquad\blacksquare$$

IMPROPER DOUBLE INTEGRALS IN POLAR COORDINATES

Improper double integrals play a useful role in probability theory and in certain physical applications. We shall demonstrate the basic approach to this topic by examining the special case of double integrals that are improper because the region of integration is the (unbounded) first quadrant. A general treatment of improper multiple integrals is outside the scope of this text.

Suppose the function $f(x, y)$ is continuous throughout the first quadrant Q; then define the improper integral in terms of rectangular coordinates as follows:

$$\int_Q\int f(x, y)\,dA = \lim_{n\to\infty} \int_{S_n}\int f(x, y)\,dA$$

where S_n is the square described by $0 \le x \le n$, $0 \le y \le n$. In terms of polar coordinates, let C_n denote the quarter circular region described by $r \le n$, $0 \le \theta \le \pi/2$, as shown in Figure 13.20. We use this quarter circular region to formulate the following definition.

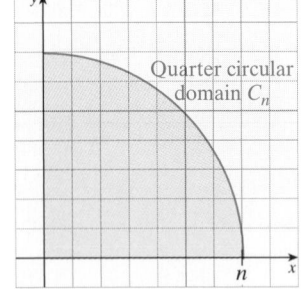

Quarter circular domain C_n

■ **FIGURE 13.20** Quarter circular domain

Improper Double Integral in Polar Coordinates

Let Q denote the first quadrant of the Cartesian plane, and let C_n denote the quarter circular region described by $r \le n$, $0 \le \theta \le \frac{\pi}{2}$.

Then the improper integral $\int_Q\int f(x, y)\,dA$ is defined in polar coordinates as

$$\lim_{n\to\infty} \int_{C_n}\int f(r\cos\theta, r\sin\theta)r\,dr\,d\theta = \lim_{n\to\infty} \int_0^{\pi/2}\int_0^n f(r\cos\theta, r\sin\theta)r\,dr\,d\theta$$

If the limit in this definition exists and is equal to L, we say that the improper integral **converges** to L. Otherwise, we say that the improper integral **diverges.**

Evaluating a convergent improper double integral

Evaluate $\iint\limits_{Q} e^{-(x^2+y^2)} \, dA$, where Q is the first quadrant of the xy-plane.

Solution   To evaluate this integral (improper because Q is unbounded), we convert to polar coordinates:

$$\iint\limits_{Q} e^{-(x^2+y^2)} \, dA = \lim_{n \to \infty} \int_0^{\pi/2} \int_0^n e^{-r^2} r \, dr \, d\theta$$

$$= \lim_{n \to \infty} \int_0^{\pi/2} \left[-\frac{1}{2} e^{-r^2} \right] \Big|_{r=0}^{r=n} d\theta = \lim_{n \to \infty} \int_0^{\pi/2} \left[-\frac{1}{2} e^{-n^2} + \frac{1}{2} \right] d\theta$$

$$= \lim_{n \to \infty} \frac{1}{2} (1 - e^{-n^2})[\theta] \Big|_{\theta=0}^{\theta=\pi/2} = \lim_{n \to \infty} \frac{\pi}{4} (1 - e^{-n^2}) = \frac{\pi}{4}$$

Thus, the improper integral converges to $\pi/4$.

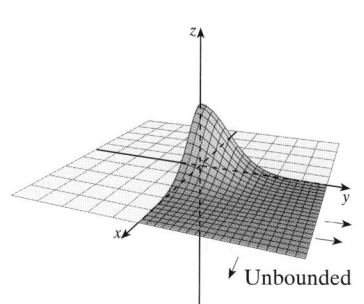

^ Unbounded

13.3 *Problem Set*

A *Evaluate the double integral $\iint\limits_{D} f(x, y) \, dA$ in Problems 1–8, and sketch the region of integration D.*

1. $\int_0^{\pi/2} \int_0^{2 \sin \theta} dr \, d\theta$

2. $\int_0^{\pi} \int_0^{1+\sin \theta} dr \, d\theta$

3. $\int_0^{\pi/2} \int_1^3 re^{-r^2} dr \, d\theta$

4. $\int_0^{\pi/2} \int_1^2 \sqrt{4 - r^2} \, r \, dr \, d\theta$

5. $\int_0^{\pi} \int_0^4 r^2 \sin^2\theta \, dr \, d\theta$

6. $\int_0^{\pi/2} \int_1^3 r^2 \cos^2\theta \, dr \, d\theta$

7. $\int_0^{2\pi} \int_0^4 2r^2 \cos \theta \, dr \, d\theta$

8. $\int_0^{2\pi} \int_0^{1-\sin \theta} \cos \theta \, dr \, d\theta$

Use a double integral to find the area of the shaded region in Problems 9–22.

9. $r = 4$

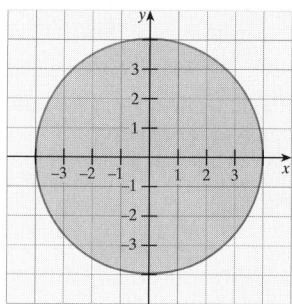

10. $r = 2 \cos \theta$

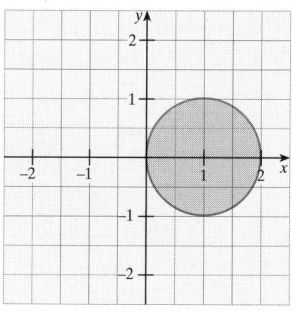

11. $r = 2 (1 - \cos \theta)$

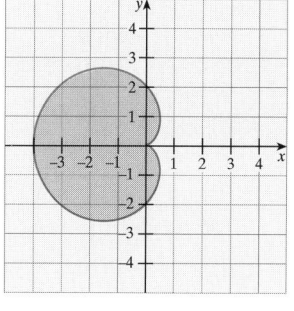

12. $r = 1 + \sin \theta$

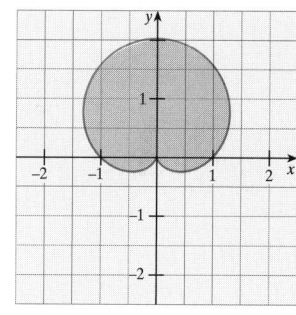

13. $r = 4 \cos 3\theta$

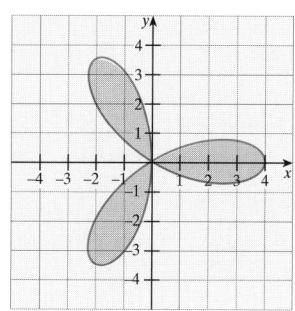

14. $r = 5 \sin 2\theta$

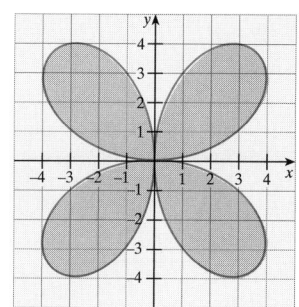

15. $r = \cos 2\theta$

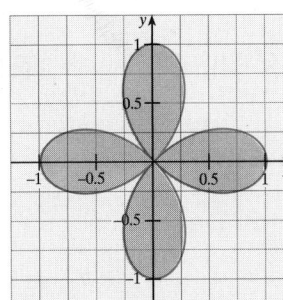

16. $r = 1$ and $r = 2 \sin \theta$
(*Hint:* Consider $0 < \theta < \frac{\pi}{6}$
and $\frac{\pi}{6} < \theta < \frac{\pi}{2}$ separately.)

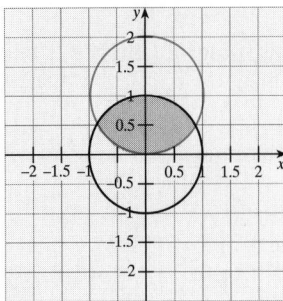

17. $r = 1$ and
$r = 2 \sin \theta$

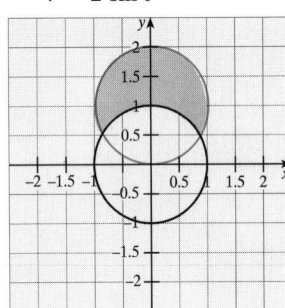

18. $r = 1$ and
$r = 1 + \cos \theta$

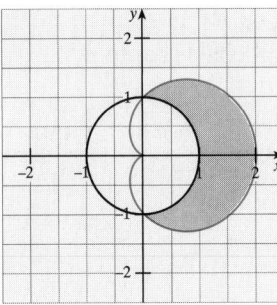

19. $r = 1$ and
$r = 1 + \cos \theta$

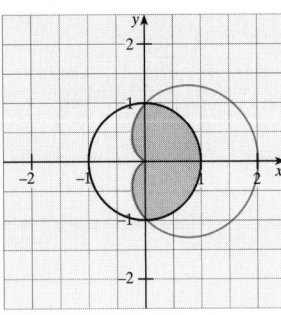

20. $r = 1$ and
$r = 1 + \cos \theta$

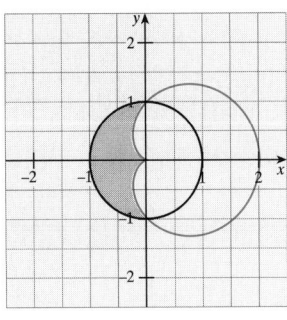

21. $r = 3 \cos \theta$ and
$r = 1 + \cos \theta$

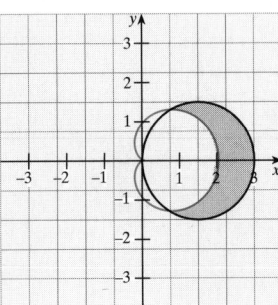

22. $r = 3 \cos \theta$ and
$r = 2 - \cos \theta$

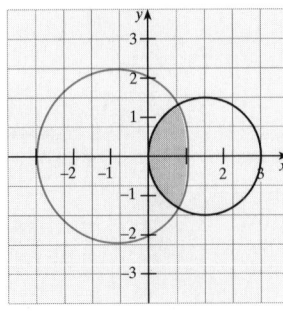

23. Use a double integral to find the area inside the inner loop of the limaçon $r = 1 + 2 \cos \theta$.

24. Use a double integral to find the area bounded by the parabola
$$r = \frac{5}{1 + \cos \theta}$$
and the lines $\theta = 0$, $\theta = \frac{\pi}{6}$, and $r = \frac{3}{4} \sec \theta$.

Use polar coordinates in Problems 25–30 to find $\iint\limits_{D} f(x, y)\, dA$, where D is the circular disk defined by $x^2 + y^2 \le a^2$ for constant $a > 0$.

25. $f(x, y) = y^2$

26. $f(x, y) = x^2 + y^2$

27. $f(x, y) = \dfrac{1}{a^2 + x^2 + y^2}$

28. $f(x, y) = e^{-(x^2+y^2)}$

29. $f(x, y) = \dfrac{1}{a + \sqrt{x^2 + y^2}}$

30. $f(x, y) = \ln(a^2 + x^2 + y^2)$

Use polar coordinates in Problems 31–36 to evaluate the given double integral.

31. $\displaystyle\iint\limits_{D} y\, dA$, where D is the disk $x^2 + y^2 \le 4$

32. $\displaystyle\iint\limits_{D} (x^2 + y^2)\, dA$, where D is the region bounded by the x-axis, the line $y = x$, and the circle $x^2 + y^2 = 1$

33. $\displaystyle\iint\limits_{D} e^{x^2+y^2}\, dA$, where D is the region inside the circle $x^2 + y^2 = 9$

34. $\displaystyle\iint\limits_{D} \sqrt{x^2 + y^2}\, dA$, where D is the region inside the circle $(x - 1)^2 + y^2 = 1$ in the first quadrant

35. $\displaystyle\iint\limits_{D} \ln(x^2 + y^2 + 2)\, dA$, where D is the region inside the circle $x^2 + y^2 = 4$ in the first quadrant

36. $\displaystyle\iint\limits_{D} \sin(x^2 + y^2)\, dA$, where D is the region bounded by the circles $x^2 + y^2 = 1$ and $x^2 + y^2 = 4$, and the lines $y = 0$ and $x = \sqrt{3}\, y$

37. Find the volume of the solid bounded by the paraboloid $z = 4 - x^2 - y^2$ and the xy-plane.

38. Find the volume of the solid bounded above by the cone $z = 6\sqrt{x^2 + y^2}$ and below by the circular region $x^2 + y^2 \le a^2$ in the xy-plane, where a is a positive constant.

B 39. Use polar coordinates to evaluate $\iint\limits_{D} xy\, dA$, where D is the intersection of the circular disks $r \le 4 \cos \theta$ and $r \le 4 \sin \theta$. Sketch the region of integration.

40. Let D be the region formed by intersecting the regions (in the xy-plane) described by $y \le x, y \ge 0$, and $x \le 1$.
a. Express $\iint\limits_{D} dA$ as an iterated integral in Cartesian coordinates.
b. Express $\iint\limits_{D} dA$ in terms of polar coordinates.

41. Example 5 of Section 9.3 used a single integral to find the area of the region common to the circles $r = a \cos \theta$ and $r = a \sin \theta$. Rework this example using double integrals.

42. Example 6 of Section 9.3 used a single integral to find the area between the circle $r = 5 \cos \theta$ and the limaçon $r = 2 + \cos \theta$. Use double integrals to find the same area.

In Problems 43–48, find the volume of the given solid region.

43. The solid that lies inside the sphere $x^2 + y^2 + z^2 = 25$ and outside the cylinder $x^2 + y^2 = 9$

44. The solid that bounded by the sphere $x^2 + y^2 + z^2 = 4$ and the paraboloid $3z = x^2 + y^2$

45. The solid region common to the sphere $x^2 + y^2 + z^2 = 4$ and the cylinder $r = 2 \cos \theta$

46. The solid region common to the cylinder $x^2 + y^2 = 2$ and the ellipsoid $3x^2 + 3y^2 + z^2 = 7$

47. The solid bounded above by the paraboloid $z = 1 - x^2 - y^2$, below by the plane $z = 0$, and on the sides by the cylinder $r = \cos \theta$

48. The solid region bounded above by the cone $z = x^2 + y^2$, below by the plane $z = 0$, and on the sides by the cylinder $r = \sin \theta$

Evaluate the improper integrals in Problems 49–53 over Q, the first quadrant in the xy-plane.

49. $\iint_Q (x^2 + y^2 + 1)^{-3/2} \, dA$

50. $\iint_Q \dfrac{dA}{(1 + x^2 + y^2)^2}$

51. $\iint_Q \dfrac{dA}{(1 + x^2 + y^2)^3}$

52. $\iint_Q e^{-2(x^2+y^2)} \, dA$

53. $\iint_Q e^{-(x^2+y^2)/4} \, dA$

54. Use polar coordinates to evaluate the double integral
$$\iint_D (x^2 + y^2) \, dA$$
over the shaded region.

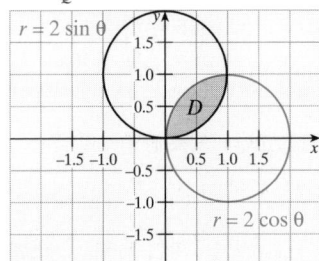

55. Use polar coordinates to evaluate the double integral $\iint_D (x^2 + y^2) \, dA$ over the shaded region.

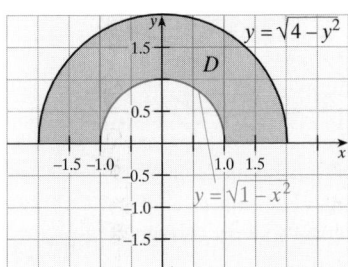

56. THINK TANK PROBLEM To evaluate the integral
$$\iint_R r^2 \, dr \, d\theta$$
where R is the region in the xy-plane bounded by $r = 2 \cos \theta$, we obtain
$$\int_0^\pi \int_0^{2\cos\theta} r^2 \, dr \, d\theta = \int_0^\pi \frac{r^3}{3}\Big|_0^{2\cos\theta} d\theta$$
$$= \frac{8}{3} \int_0^\pi \cos^3\theta \, d\theta = \frac{8}{3}\left[\sin\theta - \frac{\sin^3\theta}{3}\right]_0^\pi = 0$$

Alternatively, we can set up the integral as
$$\int_{-\pi/2}^{\pi/2} \int_0^{2\cos\theta} r^2 \, dr \, d\theta = \int_{-\pi/2}^{\pi/2} \frac{r^3}{3}\Big|_0^{2\cos\theta} d\theta$$
$$= \frac{8}{3} \int_{-\pi/2}^{\pi/2} \cos^3\theta \, d\theta = \frac{8}{3}\left[\sin\theta - \frac{\sin^3\theta}{3}\right]_{-\pi/2}^{\pi/2} = \frac{32}{9}$$

Both of these answers cannot be correct. Which procedure (if either) is correct and why?

57. HISTORICAL QUEST Newton and Leibniz have been credited with the discovery of calculus, but much of its development was due to the mathematicians Pierre-Simon Laplace, Lagrange (Historical Quest 54, Section 12.8), and Gauss (Historical Quest 61, Section 2.3). These three great mathematicians of calculus were contrasted by W. W. Rouse Ball:

PIERRE-SIMON LAPLACE 1749–1827

The great masters of modern analysis are Lagrange, Laplace, and Gauss, who were contemporaries. It is interesting to note the marked contrast in their styles. Lagrange is perfect both in form and matter, he is careful to explain his procedure, and though his arguments are general they are easy to follow. Laplace on the other hand explains nothing, is indifferent to style, and, if satisfied that his results are correct, is content to leave them either with no proof or with a faulty one. Gauss is exact and elegant as Lagrange, but even more difficult to follow than Laplace, for he removes every trace of the analysis by which he reached his results, and strives to give a proof which while rigorous shall be as concise and synthetical as possible.*

Pierre-Simon Laplace has been called the Newton of France. He taught Napoleon Bonaparte, was appointed for a time as Minister of Interior, and was at times granted favors from his powerful friend. Laplace solved problems in celestial mechanics and proved the stability of the solar system. Today, Laplace is best known as the major contributor to probability, taking it from gambling to a true branch of mathematics. He was one of the earliest to evaluate the improper

---

* *A Short Account of the History of Mathematics* as quoted in *Mathematical Circles Adieu* by Howard Eves (Boston: Prindle, Weber & Schmidt, Inc., 1977).

integral

$$I = \int_{-\infty}^{\infty} e^{-x^2}\, dx$$

which plays an important role in the theory of probability.
Show that $I = \sqrt{\pi}$. *Hint:* Note that

$$\int_0^{\infty} e^{-x^2}\, dx \cdot \int_0^{\infty} e^{-y^2}\, dy = \int_0^{\infty} \int_0^{\infty} e^{-(x^2+y^2)}\, dx\, dy$$

and use Example 5.

58. Evaluate $\int_0^{\infty} e^{-2x^2}\, dx$.
Hint: See Problem 57.

59. For constant a where $0 \le a \le R$, the plane $z = R - a$ cuts off a "cap" from the sphere

$$x^2 + y^2 + z^2 = R^2$$

Use a double integral in polar coordinates to find the volume of the cap.

13.4 Surface Area

IN THIS SECTION

definition of surface area, surface area projections, area of a surface defined parametrically

In chapter 6, we found that if f has a continuous derivative on the closed, bounded interval $[a, b]$, then the length L of the portion of the graph of $y = f(x)$ between the points were $x = a$ and $x = b$ may be defined by

$$L = \int_a^b \sqrt{1 + [f'(x)]^2}\, dx$$

The goal of this section is to study an analogous formula for the surface area of the graph of a differentiable function of two variables.

DEFINITION OF SURFACE AREA

Consider a surface defined by $z = f(x, y)$ defined over a region R of the xy-plane. Enclose the region R in a rectangle partitioned by a grid with lines parallel to the coordinate axes, as shown in Figure 13.21a. This creates a number of cells, and we let $R_1, R_2, \ldots, R_n$ denote those that lie entirely within R.

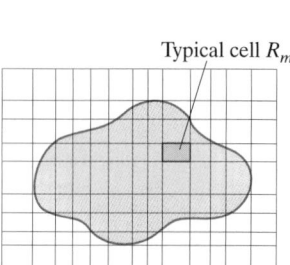

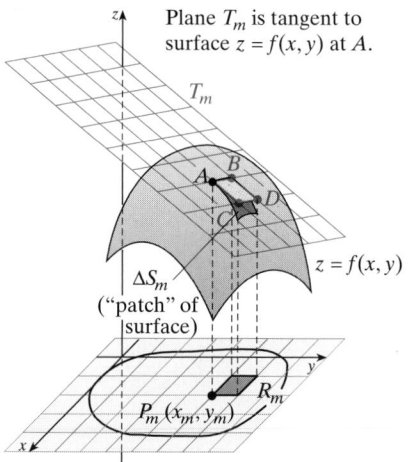

a. The region R is partitioned by a rectangular grid

b. The surface area above R_m is approximated by the area of the parallelogram on the tangent plane.

■ **FIGURE 13.21** Surface area

For $m = 1, 2, \ldots, n$, let $P_m\,(x_m^*, y_m^*)$ be a corner of the rectangle R_m, and let T_m be the tangent plane above P_m on the surface of $z = f(x, y)$. Finally, let ΔS_m denote the area of the "patch" of surface that lies directly above R_m.

The rectangle R_m projects onto a parallelogram $ABDC$ in the tangent plane T_m, and if R_m is "small," we would expect the area of this parallelogram to approximate closely the element of surface area ΔS_m (see Figure 13.20b).

If Δx_m and Δy_m are the lengths of the sides of the rectangle R_m, the approximating parallelogram will have sides determined by the vectors

$$\mathbf{AB} = \Delta x_m \mathbf{i} + [f_x(x_m^*, y_m^*)\Delta x_m]\mathbf{k} \quad \text{and} \quad \mathbf{AC} = \Delta y_m \mathbf{j} + [f_y(x_m^*, y_m^*)\Delta y_m]\mathbf{k}$$

In Chapter 10, we showed that such a parallelogram with sides $\mathbf{AB}$ and $\mathbf{AC}$ has area

$$\|\mathbf{AB} \times \mathbf{AC}\|$$

This is shown in Figure 13.22. If K_m is the area of the approximating parallelogram, we have

$$K_m = \|\mathbf{AB} \times \mathbf{AC}\|$$

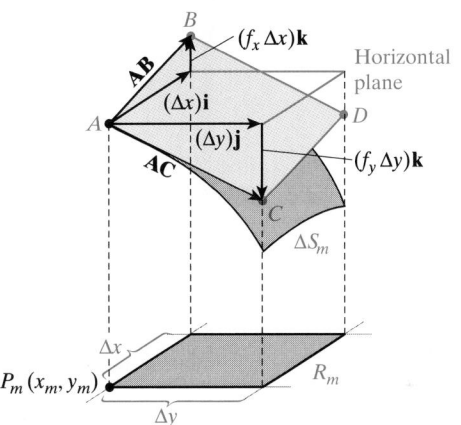

■ **FIGURE 13.22** An element of surface area has area $\|\mathbf{AB} \times \mathbf{AC}\|$.

To compute K_m, we first find the cross product:

$$\mathbf{AB} \times \mathbf{AC} = \begin{vmatrix} \mathbf{i} & \mathbf{j} & \mathbf{k} \\ \Delta x_m & 0 & f_x(x_m^*, y_m^*)\Delta x_m \\ 0 & \Delta y_m & f_y(x_m^*, y_m^*)\Delta y_m \end{vmatrix}$$

$$= -f_x\Delta x\Delta y\mathbf{i} - f_y\Delta x\Delta y\mathbf{j} + \Delta x\Delta y\mathbf{k}$$

Then, we calculate the norm:

$$K_m = \|\mathbf{AB} \times \mathbf{AC}\|$$

$$= \sqrt{[f_x(x_m^*, y_m^*)]^2\Delta x_m^2\Delta y_m^2 + [f_y(x_m^*, y_m^*)]^2\Delta x_m^2\Delta y_m^2 + \Delta x_m^2\Delta y_m^2}$$

$$= \sqrt{[f_x(x_m^*, y_m^*)]^2 + [f_y(x_m^*, y_m^*)]^2 + 1}\,\Delta x_m\Delta y_m$$

Finally, summing over the entire partition, we see that the surface area over R may be approximated by the sum

$$\Delta S_n = \sum_{m=1}^{n} \sqrt{[f_x(x_m^*, y_m^*)]^2 + [f_y(x_m^*, y_m^*)]^2 + 1}\,\Delta A_m$$

where $\Delta A_m = \Delta x_m \Delta y_m$. This may be regarded as a Riemann sum, and by taking an appropriate limit (as the partition becomes more and more refined), we find that surface area, S, satisfies

$$S = \lim_{n \to \infty} \sum_{m=1}^{n} \sqrt{[f_x(x_m^*, y_m^*)]^2 + [f_y(x_m^*, y_m^*)]^2 + 1} \, \Delta A_m$$

$$= \iint_R \sqrt{[f_x(x, y)]^2 + [f_y(x, y)]^2 + 1} \, dA$$

Surface Area as a Double Integral

Assume that the function $f(x, y)$ has continuous partial derivatives f_x and f_y in a region R of the xy-plane. Then the portion of the surface $z = f(x, y)$ that lies over R has **surface area**

$$S = \iint_R \sqrt{[f_x(x, y)]^2 + [f_y(x, y)]^2 + 1} \, dA$$

■ *What This Says:*   The region R may be regarded as the projection of the surface $z = f(x, y)$ on the xy-plane. If there were a light source with rays perpendicular to the xy-plane, R would be the "shadow" of the surface on the plane. You will notice the shadows drawn in the figures shown in this section. It is also worthwhile to make the following comparisons:

Length on x-axis: **Arc length:**

$$\int_a^b dx \qquad\qquad \int_a^b ds = \int_a^b \sqrt{[f'(x)]^2 + 1} \, dx$$

Area in xy-plane: **Surface area:**

$$\iint_R dA \qquad\qquad \iint_R dS = \iint_R \sqrt{[f_x(x, y)]^2 + [f_y(x, y)]^2 + 1} \, dA$$

EXAMPLE 1   *Surface area of a plane region*

Find the surface area of the portion of the plane $x + y + z = 1$ that lies in the first octant (where $x \geq 0, y \geq 0, z \geq 0$).

Solution   The plane $x + y + z = 1$ and the triangular region T that lies beneath it in the xy-plane are shown in Figure 13.23. We begin by letting $f(x, y) = 1 - x - y$. Then $f_x(x, y) = -1$ and $f_y(x, y) = -1$, and

$$S = \iint_T \sqrt{[f_x(x, y)]^2 + [f_y(x, y)]^2 + 1} \, dA$$

$$= \iint_T \sqrt{(-1)^2 + (-1)^2 + 1} \, dA$$

$$= \iint_T \sqrt{3} \, dA$$

$$= \int_0^1 \int_0^{1-x} \sqrt{3} \, dy \, dx \qquad \text{Look at Figure 13.23 to see the region } T \text{ and the limits for both } x \text{ and } y.$$

$$= \sqrt{3} \int_0^1 [y]\Big|_0^{1-x} dx$$

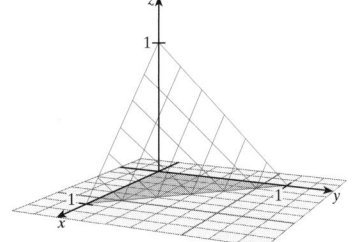

a. The surface of the plane $x + y + z = 1$ in the first octant

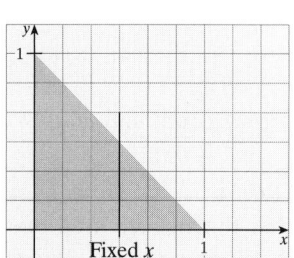

b. The surface projects onto a triangle in the xy-plane.

■ **FIGURE 13.23**   Surface area

$$= \sqrt{3} \int_0^1 (1 - x)\, dx$$

$$= \sqrt{3} \left[x - \frac{x^2}{2} \right] \Big|_0^1 = \frac{\sqrt{3}}{2}$$

The surface area is $\sqrt{3}/2 \approx 0.866$. ∎

EXAMPLE 2   *Surface area by changing to polar coordinates*

Find the surface area (to the nearest hundredth square unit) of that part of the paraboloid $x^2 + y^2 + z = 5$ that lies above the plane $z = 1$.

Solution   Let $f(x, y) = 5 - x^2 - y^2$. Then $f_x(x, y) = -2x, f_y(x, y) = -2y$, and

$$S = \iint\limits_D \sqrt{[f_x(x, y)]^2 + [f_y(x, y)]^2 + 1}\, dA = \iint\limits_D \sqrt{4x^2 + 4y^2 + 1}\, dA$$

Now we need to determine the limits for the region D and carry out the integration by using Fubini's theorem. The paraboloid intersects the plane $x = 1$ in the circle $x^2 + y^2 = 4$. Thus, the part of the paraboloid whose surface area we seek projects onto the disk $x^2 + y^2 \leq 4$ in the xy-plane, as shown in Figure 13.24 (note the shadow).

It is easier if we convert to polar coordinates by letting $x = r \cos \theta, y = r \sin \theta$. Because the region is bounded by $0 \leq r \leq 2$ and $0 \leq \theta \leq 2\pi$, we have

$$S = \iint\limits_D \sqrt{4x^2 + 4y^2 + 1}\, dA$$

$$= \int_0^{2\pi} \int_0^2 \sqrt{4r^2 + 1}\, r\, dr\, d\theta$$

> Let $u = 4r^2 + 1$, so $du = 8\, r\, dr$. If $r = 2$, then $u = 17$ and if $r = 0, u = 1$.

$$= \int_0^{2\pi} \int_1^{17} u^{1/2} \frac{du}{8}\, d\theta$$

$$= \int_0^{2\pi} \frac{1}{8} \cdot \frac{2}{3} u^{3/2} \Big|_1^{17}\, d\theta$$

$$= \frac{1}{12} \int_0^{2\pi} (17^{3/2} - 1)\, d\theta$$

$$= \frac{1}{12}(17^{3/2} - 1)(2\pi - 0) \approx 36.18$$

The surface area is approximately 36.18 square units. ∎

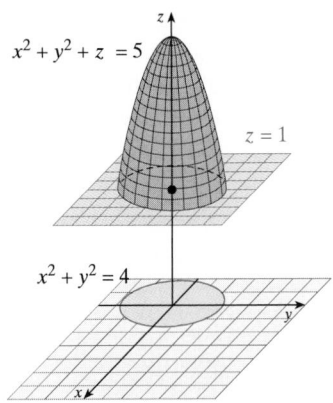

$x^2 + y^2 + z = 5$

$z = 1$

$x^2 + y^2 = 4$

Projected region

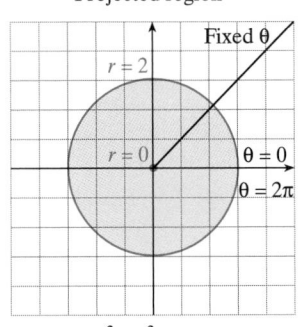

Fixed θ

$r = 2$

$r = 0$

$\theta = 0$

$\theta = 2\pi$

Disk $x^2 + y^2 \leq 4$ in polar coordinates

■ **FIGURE 13.24**   The surface $x^2 + y^2 + z = 5$ above $z = 1$ projects onto a disk.

SURFACE AREA PROJECTIONS

We have been projecting the given surface onto the xy-plane, but we can easily modify our procedure to handle other cases. For instance, if the surface $y = f(x, z)$ is projected onto the region Q in the xz-plane, the formula for surface area becomes

$$S = \iint\limits_Q \sqrt{[f_x(x,z)]^2 + [f_z(x, z)]^2 + 1}\, dx\, dz$$

An analogous formula for projection onto the region T in the yz-plane would be

$$S = \iint\limits_Q \sqrt{[f_y(y,z)]^2 + [f_z(y, z)]^2 + 1}\, dy\, dz$$

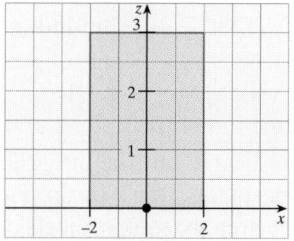

a. The cylinder $x^2 + y^2 = 4$

b. The projection of the cylinder on the xz-plane

■ **FIGURE 13.25**   Surface area of a cylinder

EXAMPLE 3   *Surface area of a cylinder*

Find the lateral surface area of the cylinder $x^2 + y^2 = 4, 0 \le z \le 3$.

Solution   The cylinder is shown in Figure 13.25a. We can find the lateral surface area using the formula for a rectangle by imagining the cylinder "opened up." The length of one side is the height of the cylinder (3 units), and the length of the other side of the rectangle is the circumference of the bottom, namely, $2\pi(2) = 4\pi$. Thus, the lateral surface area is 12π.

We shall illustrate our formula by using it to recompute this surface area. Specifically, we shall compute the surface area of the right half of the cylinder and then double it to obtain the required area. It does no good to project the half-cylinder onto the xy-plane, because if we do we will not be accounting for the height of the cylinder.

Instead, we will project the surface onto the xz-plane to obtain the rectangle Q bounded by $x = 2, x = -2, z = 0$, and $z = 3$, as shown in Figure 13.25b. Because we are projecting onto the xz-plane, we solve for y to express the surface as $y = f(x, z)$. Because

$$y = f(x, z) = \sqrt{4 - x^2}$$

we have $f_x(x, z) = \dfrac{-x}{\sqrt{4 - x^2}}$ and $f_z(x, z) = 0$ so that

$$[f_x(x, z)]^2 + [f_z(x, z)]^2 + 1 = \frac{x^2}{4 - x^2} + 1 = \frac{4}{4 - x^2}$$

Finally, we note the limit of integration:

$$-2 \le x \le 2 \quad \text{and} \quad 0 \le z \le 3$$

The surface area can now be calculated:

$$S = \iint_Q \sqrt{\frac{4}{4 - x^2}}\, dA = \int_{-2}^{2} \int_{0}^{3} \frac{2}{\sqrt{4 - x^2}}\, dz\, dx$$

$$= \int_{-2}^{2} \frac{2z}{\sqrt{4 - x^2}} \bigg|_0^3 dx = \int_{-2}^{2} \frac{6}{\sqrt{4 - x^2}}\, dx = 6 \sin^{-1} \frac{x}{2} \bigg|_{-2}^{2} = 6\pi$$

The lateral surface area of the whole cylinder is $2(6\pi) = 12\pi$.   ■

AREA OF A SURFACE DEFINED PARAMETRICALLY

Suppose a surface S (see Figure 13.26) is defined parametrically by the vector function

$$\mathbf{R}(u, v) = x(u, v)\mathbf{i} + y(u, v)\mathbf{j} + z(u, v)\mathbf{k}$$

for parameters u and v.

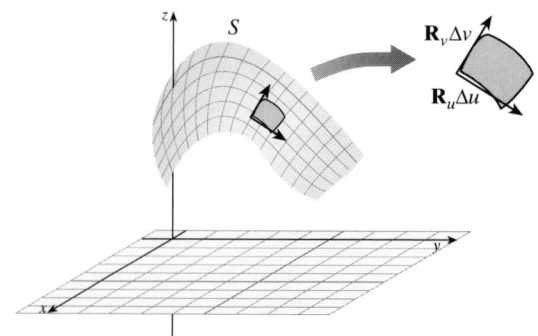

■ **FIGURE 13.26**   Area of a parametrically defined function

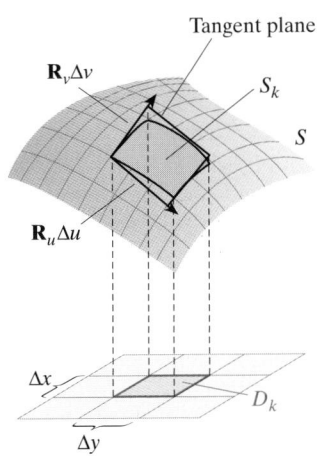

■ FIGURE 13.27 Area of a surface defined by a vector function

Let D be a region in the xy-plane on which x, y, and z, as well as their partial derivatives with respect to u and v, are continuous. The partial derivatives of $\mathbf{R}(u, v)$ are given by

$$\mathbf{R}_u = \frac{\partial \mathbf{R}}{\partial u} = \frac{\partial x}{\partial u}\mathbf{i} + \frac{\partial y}{\partial u}\mathbf{j} + \frac{\partial z}{\partial u}\mathbf{k} \qquad \mathbf{R}_v = \frac{\partial \mathbf{R}}{\partial v} = \frac{\partial x}{\partial v}\mathbf{i} + \frac{\partial y}{\partial v}\mathbf{j} + \frac{\partial z}{\partial v}\mathbf{k}$$

Suppose the region D is subdivided into cells, as shown in Figure 13.27. Consider a typical rectangle in this partition, with dimensions Δx and Δy, where Δx and Δy are small. If we project a typical rectangle onto the surface $\mathbf{R}(u, v)$, we obtain a **curvilinear** parallelogram with adjacent sides $\mathbf{R}_u(u, v)\Delta u$ and $\mathbf{R}_v(u, v)\Delta v$. The area of this rectangle is approximated by

$$\Delta S = \left\| \mathbf{R}_u(u, v)\Delta u \times \mathbf{R}_v(u, v) \right\| = \left\| \mathbf{R}_u(u, v) \times \mathbf{R}_v(u, v) \right\| \Delta u\, \Delta v$$

By taking an appropriate limit, we find the surface area to be a double integral.

Surface Area Defined Parametrically

Let D be a region in the xy-plane on which x, y, z and their partial derivatives with respect to u and v are continuous. Also, let S be a surface defined by a vector function

$$\mathbf{R}(u, v) = x(u, v)\mathbf{i} + y(u, v)\mathbf{j} + z(u, v)\mathbf{k}$$

Then the surface area is defined by

$$S = \iint\limits_{D} \left\| \mathbf{R}_u(u, v) \times \mathbf{R}_v(u, v) \right\| du\, dv$$

We call the quantity $\mathbf{R}_u(u, v) \times \mathbf{R}_v(u, v)$ the **fundamental cross product.**

EXAMPLE 4   *Area of a surface defined parametrically*

Find the surface area (to the nearest square unit) of the surface given parametrically by

$$\mathbf{R}(u, v) = (u \sin v)\mathbf{i} + (u \cos v)\mathbf{j} + u^2\mathbf{k} \quad \text{for } 0 \le u \le 3, 0 \le v \le 2\pi$$

Solution   We find that

$$\mathbf{R}_u = \sin v\, \mathbf{i} + \cos v\, \mathbf{j} + 2u\, \mathbf{k}$$
$$\mathbf{R}_v = u \cos v\, \mathbf{i} + (-u \sin v)\mathbf{j}$$

We begin with the fundamental cross product:

$$\mathbf{R}_u \times \mathbf{R}_v = \begin{vmatrix} \mathbf{i} & \mathbf{j} & \mathbf{k} \\ \sin v & \cos v & 2u \\ u \cos v & -u \sin v & 0 \end{vmatrix} = (2u^2 \sin v)\mathbf{i} + (2u^2 \cos v)\mathbf{j} - u\mathbf{k}$$

We find that

$$\left\| \mathbf{R}_u \times \mathbf{R}_v \right\| = \sqrt{4u^4\sin^2 v + 4u^4\cos^2 v + u^2} = \sqrt{4u^4 + u^2} = u\sqrt{4u^2 + 1}$$

and we can now compute the surface area:

$$S = \int_0^{2\pi} \int_0^3 u\sqrt{4u^2 + 1} \, du \, dv = \int_0^{2\pi} \left[\frac{1}{12}(4u^2 + 1)^{3/2} \right] \Big|_{u=0}^{u=3} dv$$

$$= \int_0^{2\pi} \left[\frac{1}{12}(37^{3/2} - 1) \right] dv = \frac{37^{3/2} - 1}{12}[2\pi - 0] \approx 117.3187007$$

The surface area is approximately 117 square units. ■

Notice that in the special case where the surface under consideration has the explicit representation $z = f(x, y)$, it can also be represented in the vector form

$$\mathbf{R}_x = x\mathbf{i} + y\mathbf{j} + f(x, y)\mathbf{k}$$

where x and y are used as parameters ($x = u$ and $y = v$). With this vector representation, we find that

$$\mathbf{R}_x = \mathbf{i} + f_x\mathbf{k} \quad \text{and} \quad \mathbf{R}_y = \mathbf{j} + f_y\mathbf{k}$$

so the fundamental cross product is

$$\mathbf{R}_x \times \mathbf{R}_y = \begin{vmatrix} \mathbf{i} & \mathbf{j} & \mathbf{k} \\ 1 & 0 & f_x \\ 0 & 1 & f_y \end{vmatrix} = -f_x\mathbf{i} - f_y\mathbf{j} + \mathbf{k}$$

Therefore, in the case where $z = f(x, y)$, the surface area over the region D is given by

$$S = \iint_D \|\mathbf{R}_x \times \mathbf{R}_y\| dx \, dy$$

$$= \iint_D \sqrt{(-f_x)^2 + (-f_y)^2 + 1^2} \, dx \, dy$$

$$= \iint_D \sqrt{f_x^2 + f_y^2 + 1} \, dx \, dy$$

which is the formula obtained at the beginning of this section.

13.4 Problem Set

Ⓐ *Find the surface area of each surface given in Problems 1–19.*

1. The portion of the plane $2x + y + 4z = 8$ that lies in the first octant

2. The portion of the plane $4x + y + z = 9$ that lies in the first octant

3. The portion of the paraboloid
$$z = 4 - x^2 - y^2$$
that lies above the xy-plane

4. The portion of the paraboloid
$$z = x^2 + y^2 - 9$$
that lies below the xy-plane

5. The portion of the plane $3x + 6y + 2z = 12$ that is above the triangular region in the plane with vertices $(0, 0, 0)$, $(1, 0, 0)$, and $(1, 1, 0)$

6. The portion of the plane $2x + 2y - z = 0$ that is above the square region in the plane with vertices $(0, 0, 0)$, $(1, 0, 0)$, $(0, 1, 0)$, $(1, 1, 0)$

7. The portion of the surface $x^2 + z = 9$ above the square region in the plane with vertices $(0, 0, 0)$, $(2, 0, 0)$, $(0, 2, 0)$, $(2, 2, 0)$

8. The portion of the surface $z = x^2$ that lies over the triangular region in the plane with vertices $(0, 0, 0)$, $(0, 1, 0)$, and $(1, 0, 0)$

9. The portion of the surface $z = x^2$ over the square region with vertices $(0, 0, 0)$, $(0, 4, 0)$, $(4, 0, 0)$, $(4, 4, 0)$

10. The portion of the surface $z = 2x + y^2$ over the square region with vertices $(0, 0, 0)$, $(3, 0, 0)$, $(0, 3, 0)$, $(3, 3, 0)$

11. The portion of the paraboloid
$$z = x^2 + y^2$$
that lies below the plane $z = 1$

12. The portion of the sphere $x^2 + y^2 + z^2 = 25$ that lies above the plane $z = 3$

13. The part of the cylinder $x^2 + z^2 = 4$ that is in the first octant and is bounded by the plane $y = 2$

14. The portion of the plane $x + y + z = 4$ that lies inside the cylinder $x^2 + y^2 = 16$

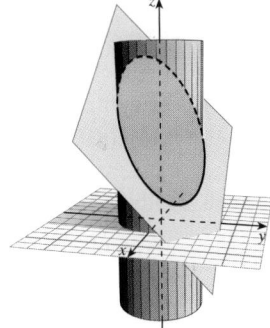

15. The portion of the sphere $x^2 + y^2 + z^2 = 4$ that lies inside the cylinder $x^2 + y^2 = 2y$

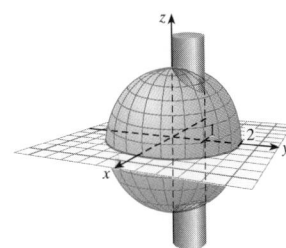

16. The portion of the sphere $x^2 + y^2 + z^2 = 8$ that is inside the elliptic cone $x^2 + y^2 - z^2 = 0$

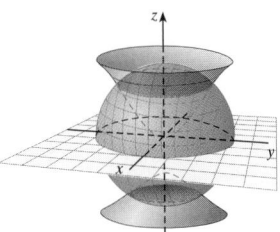

17. The portion of the surface $z = x^2 + y$ above the rectangle $0 \le x \le 2, 0 \le y \le 5$

18. The portion of the surface $z = x^2 - y^2$ that lies inside the cylinder $x^2 + y^2 = 9$

19. The portion of the surface $z = 9 - x^2 - y^2$ that lies above the xy-plane

B 20. ■ **What Does This Say?** Describe the process for finding a surface area.

21. ■ **What Does This Say?** Describe what is meant by a surface area projection.

22. On a given map, a city parking lot is shown to be a rectangle that is 300 ft by 400 ft. However, the parking lot slopes in the 400-ft direction. It rises uniformly 1 ft for every 5 ft of horizontal displacement. What is the actual surface area of the parking lot?

23. Find the surface area of the portion of the plane $Ax + By + Cz = D$ (A, B, C, and D all positive) that lies in the first octant.

24. Find the portion of the plane $x + 2y + 3z = 12$ that lies over the triangular region in the xy-plane with vertices $(0, 0, 0), (0, a, 0)$, and $(a, a, 0)$.

25. Find the surface area of that portion of the plane
$$x + y + z = a$$
that lies between the concentric cylinders
$$x^2 + y^2 = \frac{a^2}{4} \text{ and } x^2 + y^2 = a^2 (a > 0).$$

26. Find the surface area of the portion of the cylinder $x^2 + z^2 = 9$ that lies inside the cylinder $y^2 + z^2 = 9$.

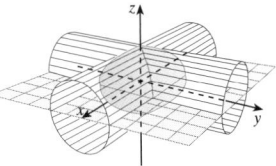

27. Find a formula for the area of the conical surface $z = \sqrt{x^2 + y^2}$ between the planes $z = 0$ and $z = h$. Express your answer in terms of h and the radius of the base of the cone.

28. Find a formula for the surface area of the frustum of the cone $z = 4\sqrt{x^2 + y^2}$ between the planes $z = h_1$ and $z = h_2, h_1 > h_2$.

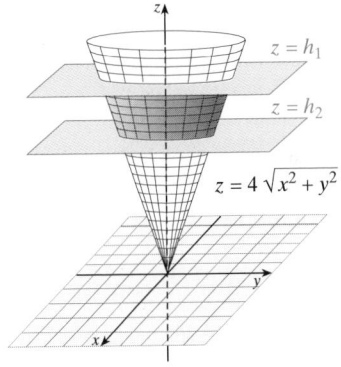

29. Find the surface area of that portion of the sphere $x^2 + y^2 + z^2 = 9z$ that lies inside the paraboloid $x^2 + y^2 = 4z$.

30. Find the surface area of that portion of the cylinder $x^2 + z^2 = 4$ that is above the triangle with vertices $(0, 0, 0), (1, 1, 0)$, and $(1, 0, 0)$.

In Problems 31–36, *set up* (*but do not evaluate*) *the double integral for the surface area of the given portion of surface.*

31. The surface given by $z = e^{-x} \sin y$ over the triangle with vertices $(0, 0, 0), (0, 1, 1), (0, 1, 0)$

32. The surface given by $z = x^3 - xy + y^3$ over the square $(0, 0, 0), (2, 0, 0), (0, 2, 0), (2, 2, 0)$

33. The surface given by $z = \cos(x^2 + y^2)$ over the disk $x^2 + y^2 \leq \dfrac{\pi}{2}$

34. The surface given by $z = e^{-x} \cos y$ over the disk $x^2 + y^2 \leq 2$

35. The surface given by $z = x^2 + 5xy + y^2$ over the region in the xy-plane bounded by the curve $xy = 5$ and the line $x + y = 6$

36. The surface given by $z = x^2 + 3xy + y^2$ over the region in the xy-plane bounded by $0 \leq x \leq 4, 0 \leq y \leq x$

Compute the magnitude of the fundamental cross product for the surface defined parametrically in Problems 37–40.

37. $\mathbf{R}(u, v) = (2u \sin v)\mathbf{i} + (2u \cos v)\mathbf{j} + u^2\mathbf{k}$

38. $\mathbf{R}(u, v) = (4 \sin u \cos v)\mathbf{i} + (4 \sin u \sin v)\mathbf{j} + (5 \cos u)\mathbf{k}$

39. $\mathbf{R}(u, v) = u\mathbf{i} + v^2\mathbf{j} + u^3\mathbf{k}$

40. $\mathbf{R}(u, v) = (2u \sin v)\mathbf{i} + (2u \cos v)\mathbf{j} + (u^2 \sin 2v)\mathbf{k}$

41. Find the surface area of the surface given parametrically by the equation $\mathbf{R}(u, v) = uv\mathbf{i} + (u - v)\mathbf{j} + (u + v)\mathbf{k}$ for $u^2 + v^2 \leq 1$. *Hint:* Use polar coordinates.

42. A *spiral ramp* has the vector parametric equation $\mathbf{R}(u, v) = (u \cos v)\mathbf{i} + (u \sin v)\mathbf{j} + v\mathbf{k}$ for $0 \leq u \leq 1$, $0 \leq v \leq \pi$. Find its surface area.

43. **H**ISTORICAL **Q**UEST   August Möbius studied under Karl Gauss (1777–1855), as well as Gauss' own teacher Johann Pfaff (1765–1825). He was a professor at the University of Leipzig, and is best known for his work in topology, especially for his conception of the Möbius strip. A Möbius strip is a two-dimensional surface with only one side.

AUGUST MÖBIUS
1790–1868

The following parametric surface is called a **Möbius strip:**

$$x = \cos v + u \cos \frac{v}{2} \cos v$$

$$y = \sin v + u \cos \frac{v}{2} \sin v$$

$$z = u \sin \frac{v}{2}$$

where $-\frac{1}{2} \leq u \leq \frac{1}{2}, 0 \leq v \leq 2\pi$. Sketch the graph of the surface, and then construct a three-dimensional model of a Möbius strip. Finally, find the fundamental cross product.

44. Verify that a sphere of radius a has surface area $4\pi a^2$.

45. Verify that a cylinder of radius a and height h has surface area $2\pi ah$.

46. Find the surface area of the torus defined by

$$\mathbf{R}(u, v) = (a + b \cos v)\cos u\,\mathbf{i} + (a + b \cos v)\sin u\,\mathbf{j} + b \sin v\,\mathbf{k}$$

for $0 < b < a, 0 \leq u \leq 2\pi, 0 \leq v \leq 2\pi$.

47. Suppose a surface is given implicitly by $F(x, y, z) = 0$. If the surface can be projected onto a region D in the xy-plane, show that the surface area is given by

$$A = \iint\limits_{D} \frac{\sqrt{F_x^2 + F_y^2 + F_z^2}}{|F_z|}\,dA_{xy}$$

where $F_z \neq 0$. Use this formula to find the surface area of a sphere of radius R.

48. Let S be the surface defined by $f(x, y, z) = C$, and let R be the projection of S on a plane. Show that the surface area of S can be computed by the integral

$$\int_R \int \frac{\|\nabla f\|}{|\nabla f \cdot \mathbf{u}|}\,dA$$

where $\mathbf{u}$ is a unit vector normal to the plane containing R and $\nabla f \cdot \mathbf{u} \neq 0$. This is a practical formula sometimes used in calculating the surface area.

13.5   Triple Integrals

IN THIS SECTION   definition of triple integral, iterated integration, volume by triple integrals   ■

DEFINITION OF TRIPLE INTEGRAL

A double integral $\iint\limits_{R} f(x, y)\,dA$ is evaluated over a closed, bounded region in the plane, and in essentially the same way, a **triple integral** $\iiint\limits_{S} f(x, y, z)\,dV$ is evaluated over a closed, bounded solid region in $\mathbb{R}^3$. Suppose $f(x, y, z)$ is defined on a closed region S, which in turn is contained in a "box" D in space. Partition D into a finite number of smaller boxes with planes parallel to the coordinate planes, as shown in Figure 13.28.

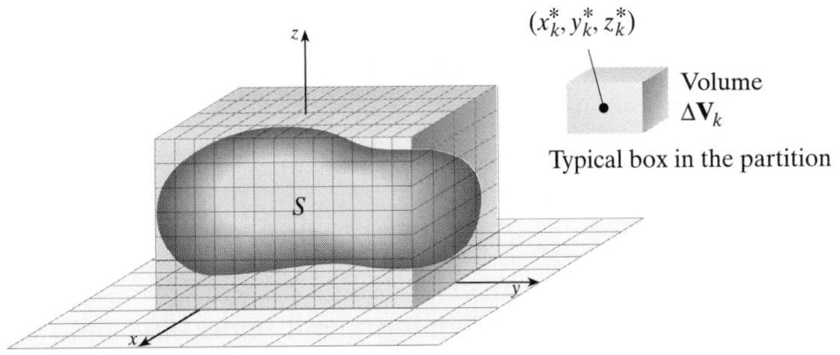

(x_k^*, y_k^*, z_k^*)

Volume
$\Delta \mathbf{V}_k$

Typical box in the partition

■ **FIGURE 13.28**   The region of integration S is subdivided into smaller boxes

We exclude from consideration any boxes with points outside S. Let ΔV_1, ΔV_2, . . . , ΔV_n denote the volumes of the boxes that remain, and define the norm $\|P\|$ of the partition to be the length of the longest diagonal of any box in the partition. Next, choose a representative point (x_k^*, y_k^*, z_k^*) from each box in the partition and form the **Riemann sum**

$$\sum_{k=1}^{n} f(x_k^*, y_k^*, z_k^*)\Delta V_k$$

If we repeat the process with more subdivisions, so that the norm approaches zero, we are led to the following definition.

Triple Integral

If f is a function defined over a closed, bounded solid region S, then the **triple integral of f over S** is defined to be the limit

$$\iiint_S f(x, y, z)\, dV = \lim_{\|P\|\to 0} \sum_{k=1}^{n} f(x_k^*, y_k^*, z_k^*)\Delta V_k$$

provided this limit exists.

In advanced calculus, it is shown that the triple integral $\iiint_S f(x, y, z)\, dV$ exists if $f(x, y, z)$ is continuous on S and S is **piecewise smooth** in the sense that it consists of a finite number of pieces, each of which has a continuously turning tangent plane (that is, the tangent plane varies continuously from point to point). It can also be shown that triple integrals have the following properties (which are analogous to those of double integrals listed in Theorem 13.1). In each case, assume the indicated integrals exist.

Linearity rule   For constants a and b

$$\iiint_S [af(x, y, z) + bg(x, y, z)]\, dV$$
$$= a\iiint_S f(x, y, z)\, dV + b\iiint_S g(x, y, z)\, dV$$

Dominance rule   If $f(x, y, z) \geq g(x, y, z)$ on S, then

$$\iiint_S f(x, y, z)\, dV \geq \iiint_S g(x, y, z)\, dV$$

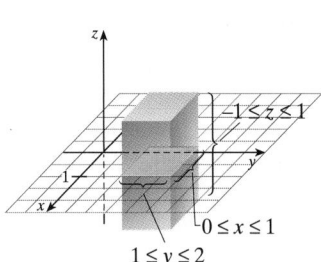

Solid S

S_1

S_2

Subdivision rule If the solid region of integration S can be subdivided into two solid subregions S_1 and S_2, then

$$\iiint_S f(x, y, z)\, dV = \iiint_{S_1} f(x, y, z)\, dV + \iiint_{S_2} f(x, y, z)\, dV$$

ITERATED INTEGRATION

As with double integrals, we evaluate triple integrals by iterated integration. However, it is generally more difficult to set up the limits of integration in a triple integral, because the region of integration S is a solid. The relatively simple case where S is a rectangular solid (box) may be handled by applying the following theorem.

THEOREM 13.5   *Fubini's theorem over a parallelepiped in space*

If $f(x, y, z)$ is continuous over a rectangular solid R: $a \le x \le b, c \le y \le d, r \le z \le s$, then the triple integral may be evaluated by the iterated integral

$$\iiint_R f(x, y, z)\, dV = \int_r^s \int_c^d \int_a^b f(x, y, z)\, dx\, dy\, dz$$

The iterated integration can be performed in any order (with appropriate adjustments) to the limits of integration:

$$dx\, dy\, dz \qquad dx\, dz\, dy \qquad dz\, dx\, dy$$
$$dy\, dx\, dz \qquad dy\, dz\, dx \qquad dz\, dy\, dx$$

Proof   The proof is beyond the scope of this course and is given in an advanced calculus course.

   EXAMPLE 1   *Evaluating a triple integral using Fubini's theorem*

Evaluate $\iiint_R z^2 y e^x\, dV$, where R is the box given by

$$0 \le x \le 1, \quad 1 \le y \le 2, \quad -1 \le z \le 1$$

Solution   We shall evaluate the integral in the order $dx\, dy\, dz$.

$$\iiint_R f(x, y, z)\, dV = \int_{-1}^1 \int_1^2 \int_0^1 z^2 y e^x\, dx\, dy\, dz$$

$$= \int_{-1}^1 \int_1^2 z^2 y [e^x] \Big|_{x=0}^{x=1}\, dy\, dz = \int_{-1}^1 \int_1^2 z^2 y [e - 1]\, dy\, dz$$

$$= (e - 1) \int_{-1}^1 z^2 \left[\frac{y^2}{2}\right] \Big|_{y=1}^{y=2}\, dz = (e - 1) \int_{-1}^1 z^2 \left[\frac{2^2}{2} - \frac{1^2}{2}\right]\, dz$$

$$= \frac{3}{2}(e - 1) \int_{-1}^1 z^2\, dz = \frac{3}{2}(e - 1) \frac{z^3}{3} \Big|_{z=-1}^{z=1}$$

$$= \frac{3}{2}(e - 1) \left[\frac{1^3}{3} - \frac{(-1)^3}{3}\right] = e - 1$$

It might be instructive for you to verify that the same result is obtained by using any other order of integration—for example, $dz\, dy\, dx$.   ■

    To evaluate a triple integral over a region that is not a box, suppose we have a solid region S with a "lower" surface $z = u(x, y)$ and an "upper" suface $z = v(x, y)$ defined over a common domain A in the xy-plane. In this case, S may be described as the set of all points (x, y, z) such that for each fixed point (x, y) in A, z varies from u to v, as shown in Figure 13.29.

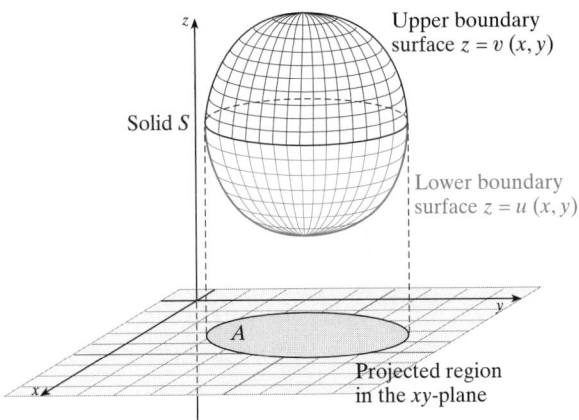

■ **FIGURE 13.29**   A solid region S bounded by the surfaces $z = u(x, y)$ and $z = v(x, y)$

The domain A is then described in a double integral. These observations are summarized in the following theorem.

THEOREM 13.6   *Triple integral over a general region*

If S is a region in space that is bounded below by the surface $z = u(x, y)$ and above by $z = v(x, y)$ as (x, y) varies over the planar region A, then

$$\iiint\limits_{S} f(x, y, z)\, dV = \iint\limits_{A} \int_{u(x, y)}^{v(x, y)} f(x, y, z)\, dz\, dA$$

Proof   This proof is omitted.

If the region of integration S has the form described in Theorem 13.6, the integral may be evaluated by integrating first with respect to z (as z varies from $z = u(x, y)$ to $z = v(x, y)$) and then computing an appropriate double integral over the planar region A.

EXAMPLE 2   *Evaluating a triple integral over a general region*

Evaluate $\iiint\limits_{S} x\, dV$, where S is the solid in the first octant bounded by the cylinder $x^2 + y^2 = 4$ and the plane $2y + z = 4$.

Solution   The solid is shown in Figure 13.30.

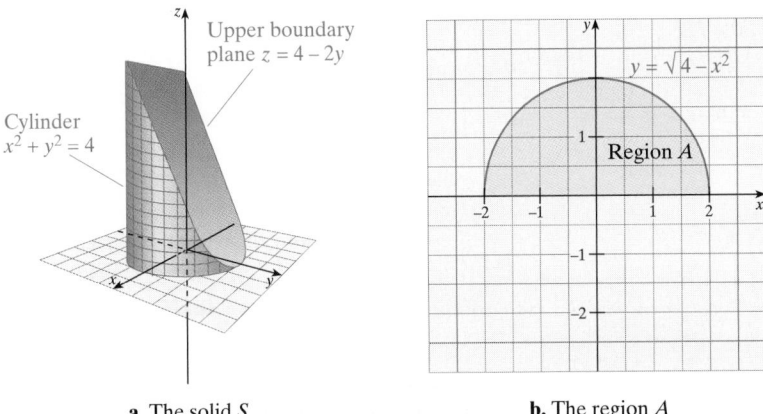

a. The solid S **b.** The region A

■ **FIGURE 13.30**   The region S bounded by the plane $2y + z = 4$ and the cylinder $x^2 + y^2 = 4$ in the first octant

The upper boundary surface of S is the plane $z = 4 - 2y$, and the lower boundary surface $z = 0$ is the xy-plane. The projection A of the solid on the xy-plane is the quarter disk $x^2 + y^2 \leq 4$ with $x \geq 0, y \geq 0$ (because S lies in the first octant). This projection may be described in type I form as the set of all (x, y) such that for each fixed x between 0 and 2, y varies from 0 to $\sqrt{4 - x^2}$. Thus, we have

$$\iiint\limits_{S} x \, dV = \iint\limits_{A} \int_0^{4-2y} x \, dz \, dA = \int_0^2 \int_0^{\sqrt{4-x^2}} \int_0^{4-2y} x \, dz \, dy \, dx$$

$$= \int_0^2 \int_0^{\sqrt{4-x^2}} x[(4 - 2y) - 0] \, dy \, dx = \int_0^2 \int_0^{\sqrt{4-x^2}} (4x - 2xy) \, dy \, dx$$

$$= \int_0^2 [4xy - xy^2] \Big|_{y=0}^{y=\sqrt{4-x^2}} dx = \int_0^2 [4x\sqrt{4-x^2} - x(4 - x^2)] \, dx$$

$$= [-\tfrac{4}{3}(4-x^2)^{3/2} - 2x^2 + \tfrac{1}{4}x^4] \Big|_0^2 = [0 - 8 + 4 + \tfrac{32}{3} + 0 - 0] = \tfrac{20}{3} \quad \blacksquare$$

VOLUME BY TRIPLE INTEGRALS

Just as a double integral can be interpreted as the area of the region of integration, a triple integral may be interpreted as the **volume** of a solid. That is, if V is the volume of the solid region S, then

$$V = \iiint\limits_{S} dV$$

EXAMPLE 3   *Volume of a tetrahedron*

Find the volume of the tetrahedron T bounded by the part of the plane $2x + y + 3z = 6$ in the first octant.

Solution   The tetrahedron T is shown in Figure 13.31a.

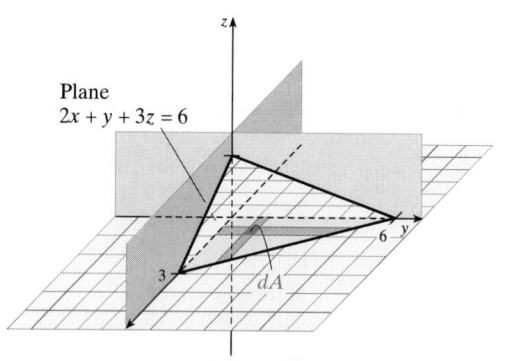

a. The tetrahedron bounded by the plane $2x + y + 3z = 6$ and the positive coordinate planes.

b. The region projected onto the xy-plane is a triangle.

■ **FIGURE 13.31**   Volume of a tetrahedron

The upper surface of T is the plane $z = \tfrac{1}{3}(6 - 2x - y)$ and its lower surface is $z = 0$. Note that T projects onto a triangle in the xy-plane, as shown in Figure 13.31b. Described in type I form, this triangle is the set of all (x, y) such that for each fixed x

between 0 and 3, y varies from 0 to $6 - 2x$. Thus,

$$V = \iiint_T dV = \iint_A \int_0^{\frac{1}{3}(6-2x-y)} dz\, dA$$

$$= \int_0^3 \int_0^{6-2x} \int_0^{\frac{1}{3}(6x-2y-y)} dz\, dy\, dx$$

$$= \int_0^3 \int_0^{6-2x} \left[\tfrac{1}{3}(6 - 2x - y) - 0 \right] dy\, dx$$

$$= \int_0^3 \left[2y - \tfrac{2}{3}xy - \tfrac{1}{6}y^2 \right] \Big|_{y=0}^{y=6-2x} dx$$

$$= \int_0^3 \left[2(6 - 2x) - \tfrac{2}{3}x(6 - 2x) - \tfrac{1}{6}(6 - 2x)^2 - 0 \right] dx$$

$$= \int_0^3 \tfrac{1}{6}[36 - 24x + 4x^2]\, dx = 6$$

The volume of the tetrahedron is 6 cubic units. ■

Sometimes it is easier to evaluate a triple integral by integrating first with respect to x or y instead of z. For instance, if the solid region of integration S is bounded in the back by $x = x_1(y, z)$ and in the front by $x = x_2(y, z)$, and the boundary surfaces project onto a region A in the yz-plane, denoted by A_{yz}, as shown in Figure 13.32a, then

$$\iiint_S f(x, y, z)\, dV = \iint_{A_{yz}} \int_x^{x_2} f(x, y, z)\, dx\, dA_{yz}$$

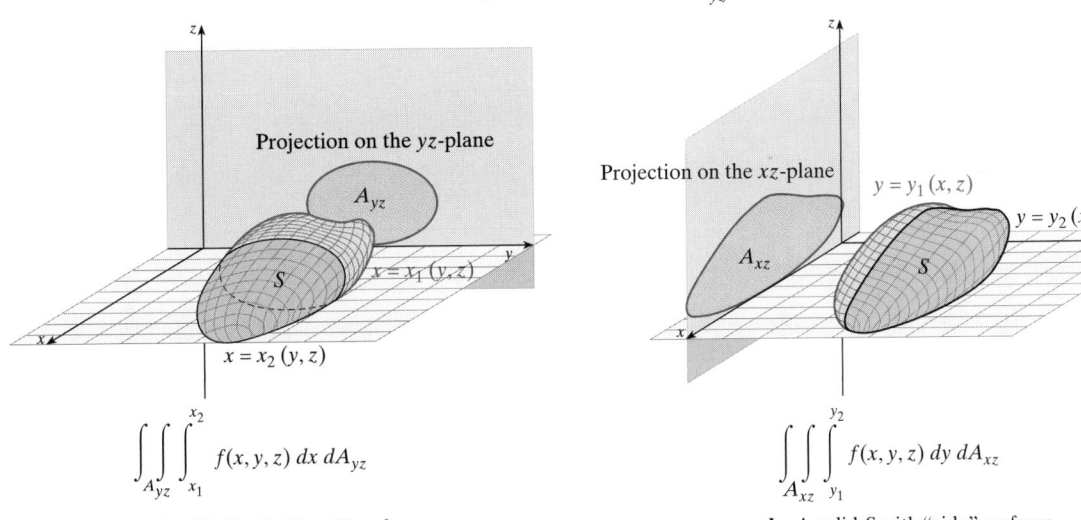

$$\iint_{A_{yz}} \int_{x_1}^{x_2} f(x, y, z)\, dx\, dA_{yz}$$

a. A solid S with "front" surface $x = x_2(y, z)$ and "back" surface $x = x_1(y, z)$

$$\iint_{A_{xz}} \int_{y_1}^{y_2} f(x, y, z)\, dy\, dA_{xz}$$

b. A solid S with "side" surfaces $y = y_2(x, z)$ and $y = y_1(x, z)$

■ **FIGURE 13.32** Iterated integration with respect to x or y first

On the other hand, if the solid region of integration S is bounded on one side by the surface $y = y_1(x, z)$ and on the other by $y = y_2(x, z)$, and the boundary surfaces project onto a region A in the xz-plane, denoted by A_{xz}, as shown in Figure 13.32b, then

$$\iiint_S f(x, y, z)\, dV = \iint_{A_{xz}} \int_{y_1}^{y_2} f(x, y, z)\, dy\, dA_{xz}$$

As an illustration, we will now rework Example 3 by projecting the tetrahedron S onto the yz-plane.

EXAMPLE 4   *Volume of a tetrahedron by changing the order of integration*

Find the volume of the tetrahedron S bounded by the coordinate planes and the plane $2x + y + 3z = 6$ in the first octant by projecting onto the yz-plane.

Solution   Note that S is bounded from "behind" by the yz-plane and "in front" by the plane $2x + y + 3z = 6$, which we express as $x = \frac{1}{2}(6 - y - 3z)$. (See Figure 13.33a.) The volume is given by

$$V = \iiint_S dV = \iint_{A_{yz}} \int_0^{\frac{1}{3}(6-y-3z)} dx \, dA_{yz}$$

where A_{yz} is the projection in the yz-plane. This projection is the triangle bounded by the lines $z = 0$, $y = 0$, and $z = \frac{1}{3}(6 - y)$, as shown in Figure 13.33b.

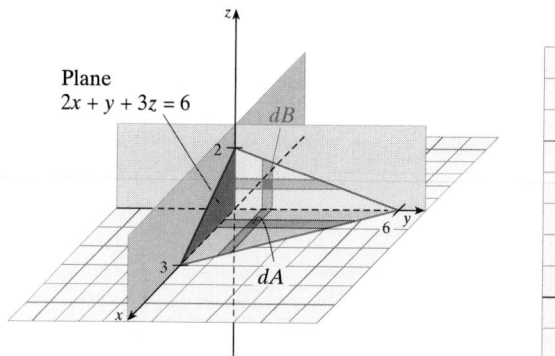

a. The tetrahedron bounded by the plane $2x + y + 3z = 6$ and the positive coordinate planes.

b. The region projected onto the yz-plane is a triangle.

■ **FIGURE 13.33**   Volume of a tetrahedron; alternative projection

Thus, for each fixed y between 0 and 6, z varies from 0 to $\frac{1}{3}(6 - y)$, and we have

$$V = \int_0^6 \int_0^{\frac{1}{3}(6-y)} \int_0^{\frac{1}{2}(6-y-3z)} dx \, dz \, dy = \int_0^6 \int_0^{\frac{1}{3}(6-y)} \frac{1}{2}(6 - y - 3z) \, dz \, dy$$

$$= \int_0^6 \left[3z - \tfrac{1}{2}yz - \tfrac{3}{4}z^2\right]\Big|_{z=0}^{z=\frac{1}{3}(6-y)} dy$$

$$= \int_0^6 \left[(6 - y) - \tfrac{1}{6}y(6 - y) - \tfrac{1}{12}(6 - y)^2 - 0\right] dy = 6$$

This is the same result we obtained in Example 3 by projecting onto the xy-plane.

■

EXAMPLE 5   *Setting up a triple integral to find a volume*

Set up (but do not evaluate) a triple integral for the volume of the solid S that is bounded above by the sphere $x^2 + y^2 + z^2 = 4$ and below by the plane $y + z = 2$.

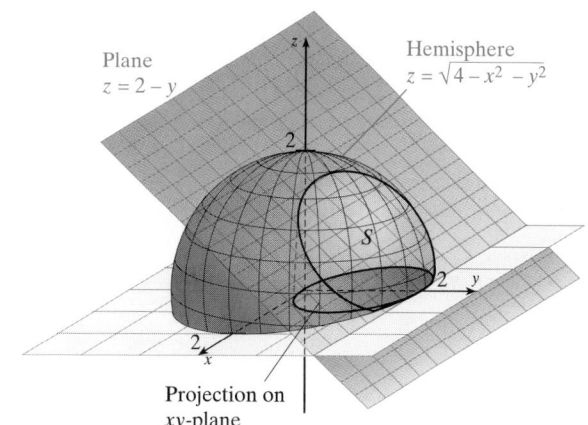

Solution   First, note that the intersection of the plane and the sphere occurs above the xy-plane (where $z \geq 0$), so the sphere can be represented by the equation

$$z = \sqrt{4 - x^2 - y^2}$$

(the upper hemisphere). To find the limits of integration for x and y, we consider the projection of S onto the xy-plane. To this end, consider the intersection of the hemisphere and the plane $z = 2 - y$:

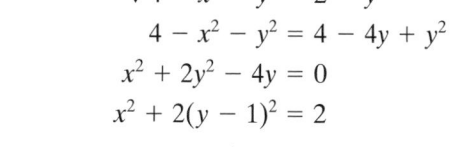

$$\sqrt{4 - x^2 - y^2} = 2 - y$$
$$4 - x^2 - y^2 = 4 - 4y + y^2$$
$$x^2 + 2y^2 - 4y = 0$$
$$x^2 + 2(y - 1)^2 = 2$$

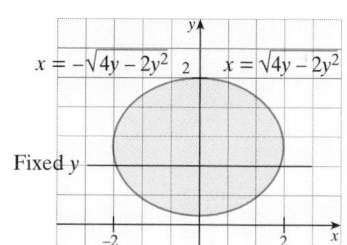

■ **FIGURE 13.34** The projection of S onto the xy-plane

Although this intersection occurs in $\mathbb{R}^3$, its equation does not contain z. Therefore, the equation serves as a projection on the xy-plane, where $z = 0$. This is an ellipse centered at $(0, 1)$. We sketch this curve in the xy-plane, as shown in Figure 13.34.

We consider this as a type II region, which means we will integrate first with respect to x, then with respect to y. Because $x^2 + 2(y - 1)^2 = 2$, we see that for fixed y between 0 and 2, x varies from $-\sqrt{2 - 2(y - 1)^2} = -\sqrt{4y - 2y^2}$ to $\sqrt{4y - 2y^2}$. However, using symmetry, we see the required volume V is twice the integral as x varies from 0 to $\sqrt{4y - 2y^2}$. This leads us to set up the following triple integral:

$$V = 2 \int_0^2 \int_0^{\sqrt{4y - 2y^2}} \int_{2-y}^{\sqrt{4-x^2-y^2}} dz \, dx \, dy \qquad ∎$$

13.5 Problem Set

Ⓐ 1. ■ **What Does This Say?** Describe Fubini's theorem over a parallelepiped in space.

2. ■ **What Does This Say?** Set up integrals, with appropriate limits of integration, for the six possible orders of integration for $\iiint_S f(x, y, z)\, dV$, where S is the solid described by

$$S: y^2 \leq x \leq 4; 0 \leq y \leq 2; 0 \leq z \leq 4 - x$$

Compute the iterated triple integrals in Problems 3–14.

3. $\int_1^4 \int_{-2}^3 \int_2^5 dx \, dy \, dz$ **4.** $\int_{-1}^3 \int_0^2 \int_{-2}^2 dy \, dz \, dx$

5. $\int_1^2 \int_0^1 \int_{-1}^2 8x^2yz^3 \, dx \, dy \, dz$

6. $\int_4^7 \int_{-1}^2 \int_0^3 x^2y^2z^2 \, dx \, dy \, dz$

7. $\displaystyle\int_0^2 \int_0^x \int_0^{x+y} xyz\, dz\, dy\, dx$

8. $\displaystyle\int_0^1 \int_{\sqrt{x}}^{\sqrt{1+x}} \int_0^{xy} y^{-1}z\, dz\, dy\, dx$

9. $\displaystyle\int_{-1}^2 \int_0^{\pi} \int_1^4 yz\cos xy\, dz\, dx\, dy$

10. $\displaystyle\int_0^{\pi} \int_0^1 \int_0^1 x^2 y\cos xyz\, dz\, dy\, dx$

11. $\displaystyle\int_0^1 \int_0^y \int_0^{\ln y} e^{z+2x}\, dz\, dx\, dy$

12. $\displaystyle\int_1^3 \int_0^{2z} \int_0^{\ln y} y\, e^{-x}\, dx\, dy\, dz$

13. $\displaystyle\int_1^4 \int_{-1}^{2z} \int_0^{\sqrt{3}x} \frac{x-y}{x^2+y^2}\, dy\, dx\, dz$

14. $\displaystyle\int_0^1 \int_{x-1}^{x^2} \int_{-x}^{y} (x+y)\, dz\, dy\, dx$

Evaluate the triple integrals in Problems 15–22.

15. $\displaystyle\iiint_S (x^2 y + y^2 z)\, dV,$
where S is the box $1 \le x \le 3, -1 \le y \le 1, 2 \le z \le 4$

16. $\displaystyle\iiint_S (xy + 2yz)\, dV,$
where S is the box $2 \le x \le 4, 1 \le y \le 3, -2 \le z \le 4$

17. $\displaystyle\iiint_S xyz\, dV,$
where S is the tetrahedron with vertices $(0,0,0), (1,0,0),$ $(0,1,0),$ and $(0,0,1)$

18. $\displaystyle\iiint_S x^2 y\, dV,$
where S is the tetrahedron with vertices $(0,0,0), (3,0,0),$ $(0,2,0),$ and $(0,0,1)$

19. $\displaystyle\iiint_S xyz\, dV,$
where S is the region given by
$x^2 + y^2 + z^2 \le 1, y \ge 0, z \ge 0$

20. $\displaystyle\iiint_S x\, dV,$
where S is bounded by the paraboloid $z = x^2 + y^2$ and the plane $z = 1$

21. $\displaystyle\iiint_S e^z\, dV,$
where S is the region described by the inequalities
$0 \le x \le 1, 0 \le y \le x,$ and $0 \le z \le x + y$

22. $\displaystyle\iiint_S yz\, dV,$
where S is the solid in the first octant bounded by the hemisphere $x = \sqrt{9 - y^2 - z^2}$ and the coordinate planes

Find the volume V of the solids bounded by the graphs of the equations given in Problems 23–32 *by using triple integration.*

23. $x + y + z = 1$ and the coordinate planes

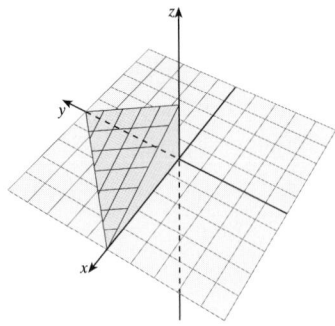

24. $y = 9 - x^2, z = 0, z = y$

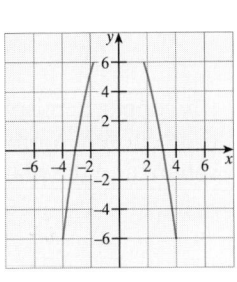 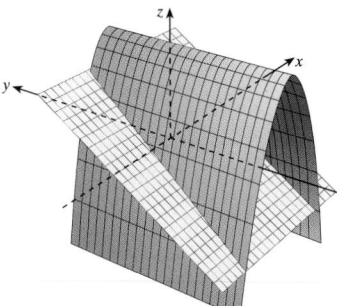

25. $(x - 1)^2 + (y - 2)^2 + (z - 3)^2 = 1$

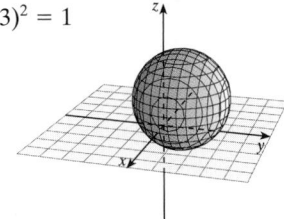

26. $z = 4 - 4x^2 - 4y^2, z = 0$

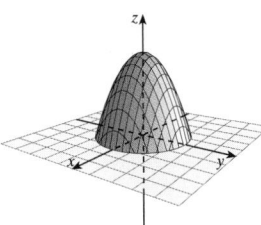

27. $x^2 + 3y^2 = z$ and the cylinder
$y^2 + z = 4$

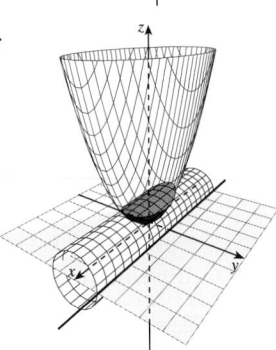

28. $x^2 + y^2 + z^3 = 9, z = 0$

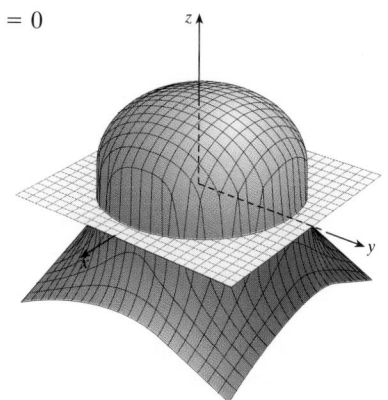

29. The solid bounded above by the paraboloid $z = 6 - x^2 - y^2$ and below by $z = 2x^2 + y^2$

30. The solid bounded by the sphere $x^2 + y^2 + z^2 = 2$ and the paraboloid $x^2 + y^2 = z$

31. The solid region common to the cylinders $x^2 + z^2 = 1$ and $y^2 + z^2 = 1$

32. The solid bounded by the cylinders $y = z^2, y = 2 - z^2$ and the planes $x = 1$ and $x = -2$

For each given iterated integral, there are five other equivalent iterated integrals. Find the one with the requested order in Problems 33–36.

33. $\displaystyle\int_0^1 \int_0^x \int_0^y f(x, y, z)\, dz\, dy\, dx$; change the order to $dz\, dx\, dy$.

34. $\displaystyle\int_1^2 \int_0^{z-1} \int_0^x f(x, y, z)\, dy\, dx\, dz$; change the order to $dy\, dz\, dx$.

35. $\displaystyle\int_0^2 \int_0^{\sqrt{4-x^2}} \int_0^{\sqrt{4-x^2-y^2}} f(x, y, z)\, dz\, dy\, dx$; change the order to $dy\, dx\, dz$.

36. $\displaystyle\int_0^2 \int_0^{\sqrt{4-x^2}} \int_0^{\sqrt{4-x^2}} f(x, y, z)\, dz\, dy\, dx$; change the order to $dy\, dx\, dz$.

Ⓑ 37. Find the volume of the ellipsoid
$$\frac{x^2}{4} + \frac{y^2}{9} + \frac{z^2}{16} = 1$$

38. Find the volume of the region between the two elliptic paraboloids
$$z = \frac{x^2}{9} + y^2 - 4$$
and
$$z = -\frac{x^2}{9} - y^2 + 4$$

39. Find the volume of the region bounded by the paraboloids $z = 16 - x^2 - 2y^2$ and $z = 3x^2 + 2y^2$.

40. A wedge is cut from a right circular cylinder of radius R by a horizontal plane perpendicular to the axis of the cylinder and a second plane that meets the first on the axis at an angle of θ degrees, as shown in Figure 13.35. Set up and evaluate a triple integral for the volume of the wedge.

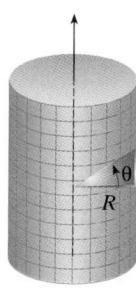

■ **FIGURE 13.35**

41. Find the volume of the region that is bounded above by the elliptic paraboloid
$$z = \frac{x^2}{9} + y^2$$
on the sides by the cylinder
$$\frac{x^2}{9} + y^2 = 1$$
and below by the xy-plane.

42. Find the volume of the solid region in the first octant that is bounded by the planes $z = 8 + 2x + y, y = 3 - 2x$.

Ⓒ 43. Use triple integration to find the volume of a sphere.

44. Use triple integration to find the volume of a right pyramid with height H and a square base of side S.

45. Use triple integration to find the volume of the ellipsoid
$$\frac{x^2}{a^2} + \frac{y^2}{b^2} + \frac{z^2}{c^2} = 1$$
(assume $a > 0, b > 0, c > 0$).

46. Find the volume of the solid region common to the paraboloid $z = k(x^2 + y^2)$ and the sphere
$$x^2 + y^2 + z^2 = 2k^{-2}, \quad \text{where } k > 0$$

47. Find the volume of the tetrahedron bounded by the plane
$$\frac{x}{a} + \frac{y}{b} + \frac{z}{c} = 1$$
$(a > 0, b > 0, c > 0)$ in the first octant.

48. **THINK TANK PROBLEM**   Let B be the box defined by $a \le x \le b, c \le y \le d, r \le z \le s$. Is it true that
$$\iiint\limits_B f(x)g(y)h(z)\, dV$$
$$= \left[\int_a^b f(x)\, dx\right]\left[\int_c^d g(y)\, dy\right]\left[\int_r^s h(z)\, dz\right]$$

Either show that this equation is generally true, or find a counterexample.

49. Change the order of integration to show that
$$\int_0^x \int_0^v f(u)\, du\, dv = \int_0^x (x - t)\, f(t)\, dt$$
Also, show that
$$\int_0^x \int_0^v \int_0^u f(w)\, dw\, du\, dv = \frac{1}{2} \int_0^x (x - t)^2 f(t)\, dt$$

50. Evaluate the triple integral

$$\iiint\limits_{S} \sin(\pi - z)^3 \, dz \, dx \, dy$$

where S is the solid region bounded below by the xy-plane, above by the plane $x = z$, and laterally by the planes $x = y$ and $y = \pi$. *Hint:* See Problem 49.

51. One of the following integrals has the value 0. Which is it and why?

A. $\displaystyle\int_{-2}^{2} \int_{-\sqrt{4-y^2}}^{\sqrt{4-y^2}} \int_{-\sqrt{4-x^2-y^2}}^{\sqrt{4-x^2-y^2}} (x + z^2) \, dz \, dx \, dy$

B. $\displaystyle\int_{0}^{1} \int_{x}^{2-x^2} \int_{-3}^{3} z^2 \sin xz \, dz \, dy \, dx$

Higher-dimensional multiple integrals can be defined and evaluated in essentially the same way as double integrals and triple integrals. Evaluate the given multiple integrals in Problems 52–53.

52. $\displaystyle\iiiint\limits_{H} xyz^2w^2 \, dx \, dy \, dz \, dw,$

where H is the four-dimensional "hyperbox" defined by $0 \le x \le 1, 0 \le y \le 2, -1 \le z \le 1, 1 \le w \le 2$.

53. $\displaystyle\iiiint\limits_{H} e^{x-2y+z+w} \, dw \, dz \, dy \, dx,$

where H is the four-dimensional region bounded by the hyperplane $x + y + z + w = 4$ and the coordinate spaces $x = 0, y = 0, z = 0$, and $w = 0$ in the first hyperoctant (where $x \ge 0, y \ge 0, z \ge 0, w \ge 0$).

13.6   Mass, Moments, and Probability Density Functions

IN THIS SECTION   **mass and center of mass, moments of inertia, joint probability density functions**

MASS AND CENTER OF MASS

WARNING   The moment of an object about an axis is the product of its mass and the signed distance from that axis.

Recall (from Section 6.5) that a solid object that is sufficiently "flat" to be regarded as two-dimensional is called a **lamina.** Suppose a particular lamina occupies a bounded region R in the xy-plane, and let $\delta(x, y)$ be the density (mass per unit area) of the lamina. A **homogeneous** lamina has constant density $\delta = m/A$, where m is the mass and A is the area of the lamina. If the lamina is **nonhomogeneous,** its density $\delta(x, y)$ varies from point to point. In this case, we can partition the region R into a number of rectangles, as shown in Figure 13.36. Choose a representative point (x_k^*, y_k^*) in each rectangle of the partition, and use the formula

$$\Delta m_k = \delta(x_k^*, y_k^*)\Delta A_k$$

to approximate the mass of each rectangle. We then approximate the total mass m by the Riemann sum

$$\Delta m = \sum_{k=1}^{n} \delta(x_k^*, y_k^*)\Delta A_k$$

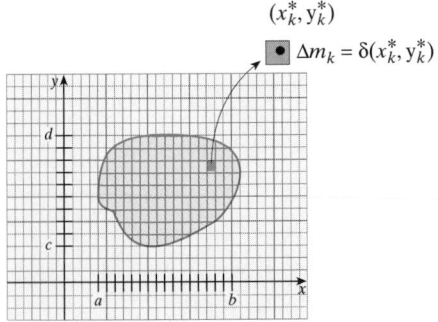

■ **FIGURE 13.36**   Partition of a lamina in the plane

so that

$$m = \lim_{\|P\| \to 0} \sum_{k=1}^{n} \delta(x_k^*, y_k^*) \Delta A_k = \iint\limits_{R} \delta(x, y) \, dA$$

We use these observations as the basis for the following definition.

Mass of a Planar Lamina of Variable Density

If δ is a continuous density function on the lamina corresponding to a plane region R, then the mass m of the lamina is given by

$$m = \iint\limits_{R} \delta(x, y) \, dA$$

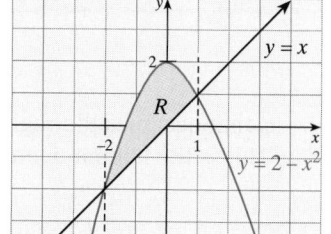

FIGURE 13.37 A lamina over the region R bounded by $y = 2 - x^2$ and $y = x$

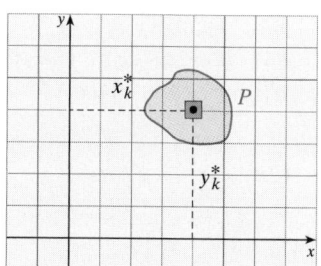

The distance from P to the x-axis is y and the distance to the y-axis is x

EXAMPLE 1   *Mass of a planar lamina*

Find the mass of the lamina of density $\delta(x, y) = x^2$ that occupies the region R bounded by the parabola $y = 2 - x^2$ and the line $y = x$.

Solution   Begin by drawing the parabola and the line, and by finding their points of intersection, as shown in Figure 13.37. By substitution,

$$x = 2 - x^2$$
$$x^2 + x - 2 = 0$$
$$x = -2, 1$$

We see that the region R is the set of all (x, y) such that for each x between -2 and 1, y varies from x to $2 - x^2$.

$$m = \iint\limits_{R} x^2 \, dA = \int_{-2}^{1} \int_{x}^{2-x^2} x^2 \, dy \, dx$$

$$= \int_{-2}^{1} x^2 (2 - x^2 - x) \, dx = \int_{-2}^{1} (2x^2 - x^4 - x^3) \, dx$$

$$= \left[\frac{2x^3}{3} - \frac{x^5}{5} - \frac{x^4}{4} \right]_{-2}^{1} = \frac{63}{20} = 3.15 \qquad \blacksquare$$

The *moment* of an object about an axis is defined as the product of its mass and the signed distance from the axis. Thus, by partitioning the region R as before (with mass), we see that the moments M_x and M_y of the lamina about the x-axis and y-axis, respectively, are approximately equal to the Riemann sums

$$M_x = \sum_{k=1}^{n} y_k \, \delta(x_k^*, y_k^*) \Delta A_k \quad \text{and} \quad M_y = \sum_{k=1}^{n} x_k \, \delta(x_k^*, y_k^*) \Delta A_k$$

$$\underset{\text{Distance to } x\text{-axis}}{\uparrow} \qquad\qquad\qquad \underset{\text{Distance to } y\text{-axis}}{\uparrow}$$

By taking the limit as the norm of the partition tends to 0, we obtain

$$M_x = \iint\limits_{R} y \delta(x, y) \, dA \quad \text{and} \quad M_y = \iint\limits_{R} x \delta(x, y) \, dA$$

The *center of mass* of the lamina covering R is the point $(\bar{x}, \bar{y})$ where the mass m can be concentrated without affecting the moments M_x and M_y; that is,

$$m\bar{x} = M_y \quad \text{and} \quad m\bar{y} = M_x$$

For future reference, these observations are summarized in the following box.

Moments and Center of Mass of a Variable Density Planar Lamina

If $\delta(x, y)$ is a continuous density function on a lamina corresponding to a plane region R, then the **moments of mass** with respect to the x- and y-axes, respectively, are

$$M_x = \iint\limits_R y\,\delta(x, y)\,dA \quad \text{and} \quad M_y = \iint\limits_R x\,\delta(x, y)\,dA$$

Furthermore, if m is the mass of the lamina, the **center of mass** is $(\bar{x}, \bar{y})$, where

$$\bar{x} = \frac{M_y}{m} \quad \text{and} \quad \bar{y} = \frac{M_x}{m}$$

If the density δ is constant, the point $(\bar{x}, \bar{y})$ is called the **centroid** of the region.

WARNING M_x has a factor of y and M_y has a factor of x.   ◄

EXAMPLE 2   *Finding a center of mass*

Find the center of mass of the lamina of density $\delta(x, y) = x^2$ that occupies the region R bounded by the parabola $y = 2 - x^2$ and the line $y = x$. This is the lamina defined in Example 1.

Solution   In Example 1, we found that for each fixed x between -2 and 1, y varies from x to $2 - x^2$ (see Figure 13.37). Thus, we have

$$M_x = \iint\limits_R y(x^2)\,dA \qquad\qquad M_y = \iint\limits_R x(x^2)\,dA$$

$$= \int_{-2}^{1}\int_{x}^{2-x^2} yx^2\,dy\,dx \qquad\qquad = \int_{-2}^{1}\int_{x}^{2-x^2} x^3\,dy\,dx$$

$$= \int_{-2}^{1}\left[\frac{1}{2}x^2y^2\right]\Big|_{y=x}^{y=2-x^2}dx \qquad\qquad = \int_{-2}^{1} x^3[(2-x^2)-x]\,dx$$

$$= \frac{1}{2}\int_{-2}^{1} x^2(x^4 - 5x^2 + 4)\,dx \qquad\qquad = \int_{-2}^{1} (2x^3 - x^5 - x^4)\,dx$$

$$= \frac{1}{2}\left[\frac{1}{7}x^7 - x^5 + \frac{4}{3}x^3\right]\Big|_{-2}^{1} \qquad\qquad = \left[\frac{2x^4}{4} - \frac{x^6}{6} - \frac{x^5}{5}\right]\Big|_{-2}^{1}$$

$$= -\frac{9}{7} \qquad\qquad\qquad\qquad = -\frac{18}{5}$$

From Example 1, $m = \frac{63}{20}$, so the center of mass is $(\bar{x}, \bar{y})$, where

$$\bar{x} = \frac{M_y}{m} = \frac{-\frac{18}{5}}{\frac{63}{20}} = -\frac{8}{7} \approx -1.14 \qquad \bar{y} = \frac{M_x}{m} = \frac{-\frac{9}{7}}{\frac{63}{20}} = -\frac{20}{49} \approx -0.41$$ ■

In a completely analogous way, we can use the triple integral to find the mass and center of mass of a solid in $\mathbb{R}^3$. The density $\delta(x, y, z)$ at a point in the solid now refers to mass per unit volume, and the mass m, moments M_{yz}, M_{xz}, M_{xy} about the yz-, xz-, and xy-planes, respectively, and coordinates $\bar{x}, \bar{y}, \bar{z}$ of the center of mass are given by:

Mass $$m = \iiint\limits_R \delta(x, y, z)\,dV$$

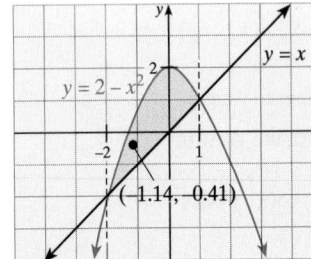

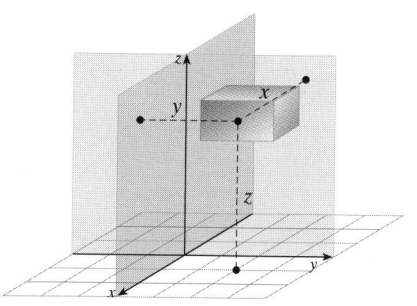

Note the distances to the coordinate planes

Moments

$$M_{yz} = \iiint_R x\,\delta(x, y, z)\,dV$$

Distance to the yz-plane

$$M_{xz} = \iiint_R y\,\delta(x, y, z)\,dV$$

Distance to the xz-plane

$$M_{xy} = \iiint_R z\,\delta(x, y, z)\,dV$$

Distance to the xy-plane

Center of mass $$(\bar{x}, \bar{y}, \bar{z}) = \left(\frac{M_{yz}}{m}, \frac{M_{xz}}{m}, \frac{M_{xy}}{m}\right)$$

As before, if the density is constant, the center of mass is still called the **centroid.** Example 3 illustrates how this point can be found by multiple integration.

EXAMPLE 3   *Centroid of a tetrahedron*

A solid tetrahedron has vertices $(0, 0, 0)$, $(1, 0, 0)$, $(0, 1, 0)$, and $(0, 0, 1)$ and constant density $\delta = 6$. Find the centroid.

Solution   The tetrahedron may be described as the region in the first octant that lies beneath the plane $x + y + z = 1$, as shown in Figure 13.38a.

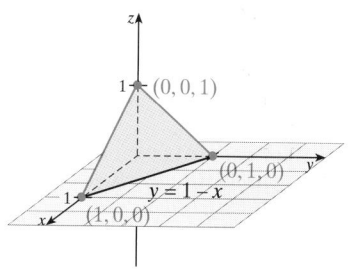

a. A tetrahedron

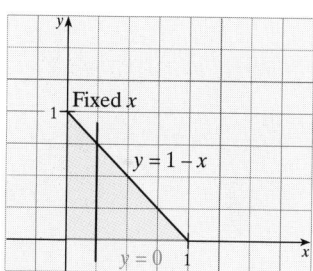

b. Projection on the xy-plane

■ **FIGURE 13.38**   The centroid of a tetrahedron

The boundary of the projection of the top face of the tetrahedron in the xy-plane is found by solving the equations $x + y + z = 1$ and $z = 0$ simultaneously. We find that the projection is the region bounded by the coordinate axes and the line $x + y = 1$, as shown in Figure 13.38b. This means that for each fixed x between 0 and 1, y varies from 0 to $1 - x$.

$$m = \iiint_R \delta\,dV = \int_0^1 \int_0^{1-x} \int_0^{1-x-y} 6\,dz\,dy\,dx$$

$$= \int_0^1 \int_0^{1-x} 6(1 - x - y)\,dy\,dx = \int_0^1 \left[6y - 6xy - 3y^2\right]\Big|_{y=0}^{y=1-x} dx$$

$$= \int_0^1 3(x - 1)^2\,dx = [x^3 - 3x^2 + 3x]\Big|_0^1 = 1$$

Similarly, we find that

$$M_{yz} = \iiint_R 6x \, dV = \int_0^1 \int_0^{1-x} \int_0^{1-x-y} 6x \, dz \, dy \, dx = \frac{1}{4}$$

$$M_{xz} = \iiint_R 6y \, dV = \int_0^1 \int_0^{1-x} \int_0^{1-x-y} 6y \, dz \, dy \, dx = \frac{1}{4}$$

$$M_{xy} = \iiint_R 6z \, dV = \int_0^1 \int_0^{1-x} \int_0^{1-x-y} 6z \, dz \, dy \, dx = \frac{1}{4}$$

(Verify the details of these integrations.) Thus,

$$\bar{x} = \frac{M_{yz}}{m} = \frac{\frac{1}{4}}{1} = 0.25, \quad \bar{y} = \frac{M_{xz}}{m} = \frac{\frac{1}{4}}{1} = 0.25, \quad \bar{z} = \frac{M_{xy}}{m} = \frac{\frac{1}{4}}{1} = 0.25$$

The centroid is $(0.25, 0.25, 0.25)$.   ∎

MOMENTS OF INERTIA

In general, a lamina of density $\delta(x, y)$ covering the region R in the first quadrant of the plane has (first) moment about a line L given by the integral

$$M_L = \iint_R s \, dm$$

where $dm = \delta(x, y) \, dA$ and $s = s(x, y)$ is the distance from the point $P(x, y)$ in R to L. Similarly, the *second* moment, or *moment of inertia*, of R about L is defined by

$$I_L = \iint_R s^2 \, dm$$

In particular, the moments of inertia about the coordinate axes are given by the integral formulas in the following box.

Moments of Inertia

The **moments of inertia** of a lamina of density δ covering the planar region R about the x-, y-, and z-axes, respectively, are given by

$$I_x = \iint_R y^2 \delta(x, y) \, dA$$

$$I_y = \iint_R x^2 \delta(x, y) \, dA$$

$$I_z = \iint_R (x^2 + y^2) \delta(x, y) \, dA$$

EXAMPLE 4   *Finding the moments of inertia*

A lamina occupies the region R in the plane that is bounded by the parabola $y = x^2$ and the lines $x = 2$ and $y = 1$. The density of the lamina at each point (x, y) is $\delta(x, y) = x^2 y$. Find the moments of inertia of the lamina about the x-axis and the y-axis.

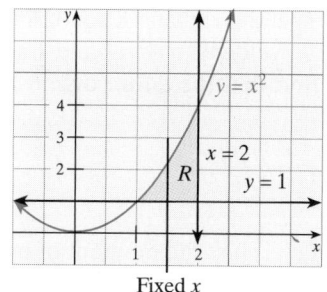

■ **FIGURE 13.39** Moment of inertia of a lamina

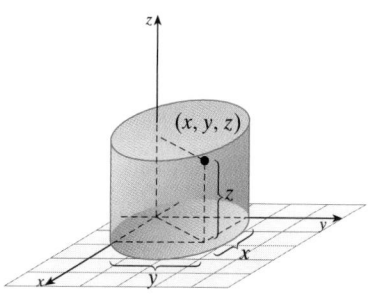

Solution　The graph of R is shown in Figure 13.39. We see that for each fixed x between 1 and 2, y varies from 1 to x^2.

$$I_x = \iint_R y^2 \, dm = \iint_R y^2 \delta(x, y) \, dA = \iint_R y^2(x^2 y) \, dA$$

$$= \int_1^2 \int_1^{x^2} x^2 y^3 \, dy \, dx = \frac{1}{4} \int_1^2 x^2(x^8 - 1) \, dx = \frac{1,516}{33} \approx 45.94$$

$$I_y = \iint_R x^2 \, dm = \iint_R x^2 \delta(x, y) \, dA = \iint_R x^2(x^2 y) \, dA$$

$$= \int_1^2 \int_1^{x^2} x^4 y \, dy \, dx = \frac{1}{2} \int_1^2 x^4(x^4 - 1) \, dx = \frac{1,138}{45} \approx 25.29$$

A simple generalization enables us to compute the moment of inertia of a solid figure about an axis L outside the figure. Specifically, suppose the solid occupies a region R and that the density at each point (x, y, z) in R is given by $\delta(x, y, z)$. Because the square of the distance of a typical cell in R from the x-axis is $y^2 + z^2$, the moment of inertia about the x-axis is

$$I_x = \iiint_R \underbrace{(y^2 + z^2)}_{\substack{\uparrow \\ \text{Square of the distance to the } x\text{-axis}}} \underbrace{\delta(x, y, z) \, dV}_{\text{Increment of mass}}$$

Similarly, the moments of inertia of the solid about the y-axis and the z-axis are, respectively,

$$I_y = \iiint_R \underbrace{(x^2 + z^2)}_{\substack{\uparrow \\ \text{Square of the distance to the } y\text{-axis}}} \underbrace{\delta(x, y, z) \, dV}_{\text{Increment of mass}}$$

$$I_z = \iiint_R \underbrace{(x^2 + y^2)}_{\substack{\uparrow \\ \text{Square of the distance to the } z\text{-axis}}} \underbrace{\delta(x, y, z) \, dV}_{\text{Increment of mass}}$$

EXAMPLE 5　　*Moment of inertia of a solid*

Find the moment of inertia about the z-axis of the solid tetrahedron S with vertices $(0, 0, 0), (0, 1, 0), (1, 0, 0), (0, 0, 1)$ and density $\delta(x, y, z) = x$.

Solution　In Example 3, we observed that the solid S can be described as the set of all (x, y, z) such that for each fixed x between 0 and 1, y lies between 0 and $1 - x$, and $0 \leq z \leq 1 - x - y$. Thus,

$$I_z = \iiint_S (x^2 + y^2)\delta(x, y, z) \, dV = \int_0^1 \int_0^{1-x} \int_0^{1-x-y} x(x^2 + y^2) \, dz \, dy \, dx$$

$$= \int_0^1 \int_0^{1-x} x(x^2 + y^2)(1 - x - y) \, dy \, dx = \int_0^1 \left[\frac{x^3(1-x)^2}{2} + \frac{x(1-x)^4}{12} \right] dx = \frac{1}{90}$$

Moments of inertia have a useful interpretation in physics. The **kinetic energy** of a body of mass m moving with velocity v along a straight line is defined in physics as $K = \frac{1}{2} mv^2$. Suppose a lamina covering circular disk R centered at the origin (see

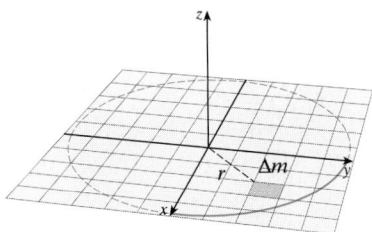

■ FIGURE 13.40   Kinetic energy of a rotating disk

Figure 13.40) is rotating around the z-axis with angular speed ω radians/second. A cell of mass Δm located r units from the origin has linear velocity $v = r\omega$ and linear kinetic energy $K_{\mathrm{lin}} = \frac{1}{2}(\Delta m)v^2$, and by integrating, we find that the entire disk R has kinetic energy of rotation

$$K_{\mathrm{rot}} = \iint\limits_{R} \frac{1}{2}\,\omega^2\,r^2\,dm = \frac{1}{2}\,\omega^2 \iint\limits_{R} r^2\,dm$$

Since $r^2 = x^2 + y^2$, we see that the integral in this formula is just the moment of inertia of R about the z-axis, so the rotational kinetic energy can be expressed as

$$K_{\mathrm{rot}} = \tfrac{1}{2}I_z\omega^2$$

Comparing this formula to the linear kinetic energy formula $K_{\mathrm{lin}} = \frac{1}{2}mv^2$, we see that the moment of inertia may be thought of as the rotational analog of mass.

JOINT PROBABILITY DENSITY FUNCTIONS

A **continuous random variable** X is a continuous function whose domain is a set of real numbers associated with probabilities. A **probability density function** is a continuous nonnegative function $f(x)$ such that

$$P(a \le X \le b) = \int_a^b f(x)\,dx$$

where $P(a \le X \le b)$ denoted the probability that X is in the closed interval $[a, b]$.

Since $P(-\infty < X < \infty) = 1$, that is, X must be somewhere, we see that $f(x)$ must satisfy

$$\int_{-\infty}^{\infty} f(x)\,dx = 1$$

In geometric terms, the probability $P(a \le X \le b)$ is the area under the graph of f over the interval $a \le x \le b$. (See Figure 13.41.)

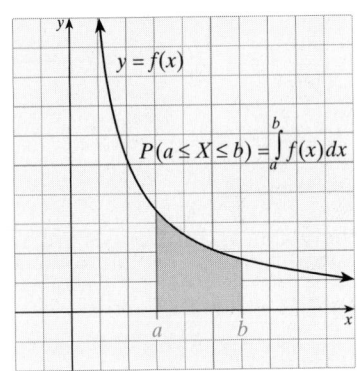

■ FIGURE 13.41   Probability as the area under a curve

Similarly, we define a **joint probability density function** for two random variables X and Y to be a continuous, non-negative function $f(x, y)$ such that

$$P[(X, Y) \text{ in } R] = \iint\limits_{R} f(x, y)\,dA$$

where $P[(X, Y) \text{ in } R]$ denotes the probability that (X, Y) is in the region R in the xy-plane. Note that

$$P[(X, Y) \text{ in the } xy\text{-plane}] = \int_{-\infty}^{\infty}\int_{-\infty}^{\infty} f(x, y)\,dx\,dy = 1$$

Geometrically, $P[(X, Y) \text{ in } R]$ may be thought of as the volume under the surface $z = f(x, y)$ above the region R.

The techniques for constructing joint probability density functions from experimental data are outside the scope of this text and are discussed in many texts on probability and statistics. The use of a double integral to compute a probability with a given joint density function is illustrated in the next example.

EXAMPLE 6   *Compute a probability*

Suppose X measures the time (in minutes) that a customer at a particular grocery store spends shopping and Y measures the time the customer spends in the checkout

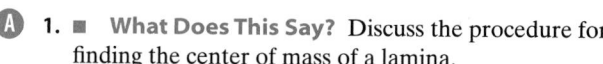

FIGURE 13.42 Triangle comprised of all points (X, Y) such that $X + Y \le 30, X \ge 0, Y \ge 0$

line. A study suggests that the joint probability function for X and Y may be modeled by

$$f(x, y) = \begin{cases} \frac{1}{200} e^{-x/10} e^{-y/20} & \text{for } x \ge 0 \text{ and } y \ge 0 \\ 0 & \text{otherwise} \end{cases}$$

Find the probability that the customer's total time in the store will be no greater than 30 min.

Solution   The goal is to find the probability that $X + Y \le 30$. Stated geometrically, we wish to find the probability that a randomly selected point (x, y) lies in the region R in the first quadrant that is bounded by the coordinate axes and the line $x + y = 30$ (see Figure 13.42). This probability is given by the double integral

$$P[(X, Y) \text{ is in } R] = \iint_R f(x, y) \, dA = \int_0^{30} \int_0^{30-x} \frac{1}{200} e^{-x/10} e^{-y/20} \, dy \, dx$$

$$= \frac{1}{200} \int_0^{30} e^{-x/10} \left[\frac{e^{-y/20}}{-1/20} \right] \Big|_0^{30-x} \, dx$$

$$= \frac{-20}{200} \int_0^{30} e^{-x/10} [e^{-(1/20)(30-x)} - 1] \, dx$$

$$= \frac{-1}{10} \left[\frac{e^{-3/2} e^{-x/20}}{-\frac{1}{20}} - \frac{e^{-x/10}}{-\frac{1}{10}} \right] \Big|_0^{30}$$

$$= e^{-3} - 2e^{-3/2} + 1$$

$$\approx 0.6035$$

Thus, it is about 60% likely that the shopper will spend no more than 30 minutes in the store. ∎

Many times we use a complementary property of probability that states

$$P(E) + P(\overline{E}) = 1$$

where E and $\overline{E}$ together constitute the set of all possible outcomes (that is, they are *complementary* probabilities). For Example 6, since the events $E = \{$the shopper will spend no more than 30 minutes in the store$\}$ and $\overline{E} = \{$the shopper will spend more than 30 minutes in the store$\}$, we can find the probability that the shopper will spend more than 30 minutes in the store by using this complementary property:

$$1 - 0.6035 = 0.3965$$

13.6 Problem Set

A 1. ■ **What Does This Say?** Discuss the procedure for finding the center of mass of a lamina.

2. ■ **What Does This Say?** Discuss a procedure for finding moments in three dimensions.

3. ■ **What Does This Say?** Discuss the terms center of mass and centroid.

4. ■ **What Does This Say?** Discuss moments of inertia.

Find the centroid for the regions described in Problems 5–12.

5. A lamina with $\delta = 5$ over the rectangle with vertices $(0, 0)$, $(3, 0), (3, 4), (0, 4)$

6. A lamina with $\delta = 4$ over the region bounded by the curve $y = \sqrt{x}$ and the line $x = 4$ in the first quadrant

7. A lamina with $\delta = 2$ over the region between the line $y = 2x$ and the parabola $y = x^2$

8. A lamina with $\delta = 4$ over the region bounded by $y = \sin \frac{\pi}{2} x, x = 0, y = 0, x = \frac{1}{2}$

9. A thin homogeneous plate of density $\delta = 1$ with the shape of the region bounded above by the parabola $y = 2 - 3x^2$ and below by the line $3x + 2y = 1$

10. The part of the spherical solid with density $\delta = 2$ described by $x^2 + y^2 + z^2 \le 9, x \ge 0, y \ge 0, z \ge 0$

11. The solid tetrahedron of density $\delta = 4$ bounded by the plane $x + y + z = 4$ in the first octant.

12. The solid bounded by the surface $z = \sin x, x = 0, x = \pi$, $y = 0, z = 0$, and $y + z = 1$, where the density is $\delta = 1$.

Use double integration in Problems 13–18 to find the center of mass of a lamina covering the given region in the plane and having the specified density δ.

13. $\delta(x, y) = x^2 + y^2$ over $x^2 + y^2 \le 9, y \ge 0$.

14. $\delta(x, y) = k(x^2 + y^2)$ over $x^2 + y^2 \le a^2, y \ge 0$.

15. $\delta(x, y) = 7x$ over the triangle with vertices $(0, 0), (6, 5)$, and $(12, 0)$.

16. $\delta(x, y) = 3x$ over the region bounded by $y = 0, y = x^2$, and $x = 6$.

17. $\delta(x, y) = x^{-1}$ over the region bounded by $y = \ln x$, $y = 0, x = 2$.

18. $\delta(x, y) = y$ over the region bounded by $y = e^{-x}, x = 0$, $x = 2, y = 0$.

19. A lamina in the xy-plane has the shape of the semicircular region $x^2 + y^2 \le a^2, y \ge 0$. Find the center of mass if the density at any point in the lamina is:
    a. directly proportional to the distance of the point from the origin.
    b. directly proportional to the polar angle.

20. A lamina has the shape of a semicircular region $x^2 + y^2 \le a^2, y \ge 0$. Find the center of mass of the lamina if the density at each point is directly proportional to the square of the distance from the point to the origin.

21. Find the center of mass of a lamina that covers the region bounded by the curve $y = \ln x$ and the lines $x = e^2$ and $y = 0$ if the density at each point (x, y) is $\delta = 1$.

22. Find the center of mass of the solid bounded above by the elliptic paraboloid $z = x^2 + y^2$, on the sides by the cylinder $x^2 + y^2 = 9$ and the plane $x = 0$, and below by $z = 0$, where $\delta(x, z) = x^2 + y^2 + z^2$.

23. Find I_x, the moment of inertia about the x-axis, of the lamina that covers the region bounded by the graph of $y = 1 - x^2$ and the x-axis, if the density is $\delta(x, y) = x^2$.

24. Find I_z, the moment of inertia about the z-axis, of the lamina that covers the square in the plane with vertices $(-1, -1), (1, -1), (1, 1)$, and $(-1, 1)$, if the density is $\delta(x, y) = x^2y^2$.

B 25. Find the center of mass of the cardioid $r = 1 + \sin\theta$ if the density at each point (r, θ) is $\delta(r, \theta) = r$.

26. Find the center of mass of the loop of the lemniscate $r^2 = 2\sin 2\theta$ that lies in the first quadrant, for density $\delta = 1$.

27. Find the center of mass (correct to the nearest hundredth) of the part of the large loop of the limaçon $r = 1 + 2\cos\theta$ that does not include the small loop. Assume that $\delta = 1$.

28. Find the center of mass of the lamina that covers the triangular region with vertices $(0, 0), (a, 0), (a, b)$, if a and b

are both positive and the density at $P(x, y)$ is directly proportional to the distance of P from the y-axis.

29. **THINK TANK PROBLEM**   A solid has the shape of the rectangular parallelepiped given by $-a \le x \le a$, $-b \le y \le b, -c \le z \le c$, and its density is $\delta(x, y, z) = x^2y^2z^2$.
    a. Guess the location of the center of mass and value of the moment of inertia about the z-axis.
    b. Check your response to part **a** by direct calculation.

30. A rectangular lamina has vertices $(0, 0), (a, 0), (a, b)$, $(0, b)$, and its density at any point (x, y) is the product $\delta(x, y) = xy$. Find the center of mass of the plate.

31. A lamina of density $\delta = 1$ covers the circular disk with boundary $x^2 + y^2 = ax$. Find the moment of inertia of this circular plate about a diameter passing through the center of the lamina. *Hint:* Use polar coordinates.

32. Show that the lamina of density $\delta = 1$ that covers the circular region $x^2 + y^2 = a^2$ and has mass m, will have moment of inertia $ma^2/4$ with respect to both the x- and y-axes. What is the moment of inertia with respect to the z-axis?

33. Show that a lamina that covers the ellipse

$$\frac{x^2}{a^2} + \frac{y^2}{b^2} \le 1$$

with mass m and density $\delta = 1$ has moment of inertia about the x-axis equal to

$$I_x = \frac{\pi ab^3\delta}{4} = \tfrac{1}{4}mb^2$$

Note: $m = \delta(\pi ab)$.
         Area of ellipse

34. Find the center of mass of the tetrahedron in the first octant bounded by the plane

$$\frac{x}{a} + \frac{y}{b} + \frac{z}{c} = 1$$

where a, b, and c are all positive constants. Assume the density is $\delta = x$.

35. A solid has the shape of the sphere $x^2 + y^2 + z^2 \le a^2$. Find the centroid of the part of the solid in the first octant $(x \ge 0, y \ge 0, z \ge 0)$. Assume $\delta = 1$.

36. Suppose the joint probability density function for the random variables X and Y is

$$f(x, y) = \begin{cases} 2e^{-2x}e^{-y} & \text{if } x \ge 0, y \ge 0 \\ 0 & \text{otherwise} \end{cases}$$

Find the probability that $X + Y \le 1$.

37. Suppose the joint probability density function for the random variables X and Y is

$$f(x, y) = \begin{cases} xe^{-x}e^{-y} & \text{if } x \ge 0, y \ge 0 \\ 0 & \text{otherwise} \end{cases}$$

Find the probability that $X + Y \le 1$.

38. **MODELING PROBLEM**   Suppose X measures the length of time (in days) that a person stays in the hospital

after abdominal surgery, and Y measures the length of time (in days) that a person stays in the hospital after orthopedic surgery. On Monday, the patient in bed 107A undergoes an emergency appendectomy (abdominal surgery), while the patient's roommate in bed 107B undergoes orthopedic surgery for the repair of torn knee cartilage. If the joint probability density function for X and Y is

$$f(x, y) = \begin{cases} \frac{1}{6}e^{-x/2}e^{-y/3} & \text{if } x \geq 0, y \geq 0 \\ 0 & \text{otherwise} \end{cases}$$

find the probability (to the nearest hundredth) that both patients will be discharged from the hospital within 3 days.

39. **MODELING PROBLEM**   Suppose X measures the time (in minutes) that a person stands in line at a certain bank and Y, the duration (in minutes) of a routine transaction at the teller's window. You arrive at the bank to deposit a check. If the joint probability density function for X and Y is modeled by

$$f(x, y) = \begin{cases} \frac{1}{8}e^{-x/2}e^{-y/4} & \text{if } x \geq 0, y \geq 0 \\ 0 & \text{otherwise} \end{cases}$$

find the probability that you will complete your business at the bank within 8 min.

40. **MODELING PROBLEM**   Suppose X measures the time (in minutes) that a person spends with an insurance agent choosing a life insurance policy and Y, the time (in minutes) that the agent spends doing the paperwork once the client has decided. You arrange to meet with an insurance agent to buy a life insurance policy. If the joint probability density function for X and Y is

$$f(x, y) = \begin{cases} \frac{1}{300}e^{-x/30}e^{-y/10} & \text{if } x \geq 0, y \geq 0 \\ 0 & \text{otherwise} \end{cases}$$

find the probability that the entire transaction will take less than half an hour.

41. **MODELING PROBLEM**   Racing yachts, such as those in the America's Cup competition, benefit from sophisticated, computer-enhanced construction techniques.* For example, define the *center of pressure* on a boat's sail as the point $(\bar{x}, \bar{y})$ where all aerodynamic forces appear to act. Suppose a sail occupies a region R in the plane, as illustrated in Figure 13.43. It can be shown that $\bar{x}$ and $\bar{y}$ are given by the formulas

$$\bar{x} = \frac{\displaystyle\iint_R xy \, dA}{\displaystyle\iint_R y \, dA} \qquad \bar{y} = \frac{\displaystyle\iint_R y^2 \, dA}{\displaystyle\iint_R y \, dA}$$

Calculate the center of pressure on the triangular sail shown in Figure 13.43.

*See *Scientific American* (August 1987).

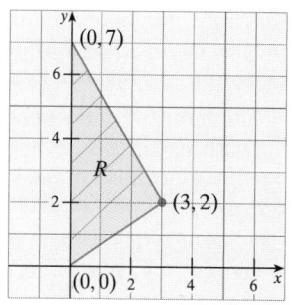

■ **FIGURE 13.43**   Triangular sail

The **average value** of the continuous function f over R is given by

One variable: $\dfrac{1}{\text{length of segment } R} \displaystyle\int_R f(x) \, dx$

Two variables: $\dfrac{1}{\text{area of region } R} \displaystyle\iint_R f(x, y) \, dA$

Three variables: $\dfrac{1}{\text{volume of solid } R} \displaystyle\iiint_R f(x, y, z) \, dV$

Use these definitions in Problems 42–45.

42. Find the average value of $f(x, y) = e^{x^3}$, where R is the region in the first quadrant bounded by $y = x^2, y = 0$, and $x = 1$.

43. Find the average value of

$$f(x, y) = e^x y^{-1/2}$$

where R is the region in the first quadrant bounded by $y = x^2, x = 0$, and $y = 1$.

44. Find the average value of the function $f(x, y, z) = x + 2y + 3z$ over the solid region S bounded by the tetrahedron with vertices $(0, 0, 0), (1, 0, 0), (0, 1, 0)$, and $(0, 0, 1)$.

45. Find the average value of the function $f(x, y, z) = xyz$ over the solid sphere $x^2 + y^2 + z^2 \leq 1$.

46. Find the centroid of the solid bounded by the xy-plane and the surface $z = \exp(4x + 2y - x^2 - y^2)$. Note that this solid has infinite extent.

47. **MODELING PROBLEM**   In a psychological experiment, x units of stimulus A and y units of stimulus B are applied to a subject, whose performance on a certain task is modeled by the function

$$f(x, y) = 10 + xye^{1-x^2-y^2}$$

Suppose the stimuli are controlled in such a way that the subject is exposed to every possible combination (x, y) with $x \geq 0, y \geq 0$, and $x + y \leq 1$. What is the subject's average response to the stimuli?

The **radius of gyration** for revolving a region R with mass m about an axis of rotation L, with moment of inertia I, is

$$d = \sqrt{\frac{I}{m}}$$

Note that if the entire mass m of R is located at a distance d from the axis of rotation L, then R would have the same motion of inertia. Use this definition in Problems 48–50.

48. A lamina has the shape of the right triangle in the xy-plane with vertices $(0,0), (a,0)$, and $(0,b), a > 0, b > 0$. Find the radius of gyration of the lamina about the z-axis. Assume $\delta = 1$.

49. Find the radius of gyration about the x-axis of the semi-circular region $x^2 + y^2 \le a, y \ge 0$, given that the density at (x, y) is directly proportional to the distance of the point from the x-axis.

50. Let R be the lamina bounded by the parabola $y = x^2$ and the lines $x = 2$ and $y = 1$, with density $\delta(x, y) = x^2y$. What is the radius of gyration about the x-axis?

51. MODELING PROBLEM   An industrial plant is located on a narrow river. Suppose C_0 units of pollutant are released into the river at time $t = 0$ and that the concentration of pollutant t hours later at a point x miles downstream from the plant is modeled by the diffusion function

$$C(x, t) = \frac{C_0}{\sqrt{k\pi t}} e^{-x^2/(4kt)}$$

where k is a physical constant.

a. At what time $t_m(x_0)$ does the maximum pollution occur at point $x = x_0$ miles from the plant? What is the maximum concentration $C_m(x_0)$?

b. Define the *danger zone* to be the portion of the river-bank such that $0 \le x \le x_m$ where x_m is the largest value of x such that $C_m(x_m) \ge 0.25C_0$. Find x_m.

c. *Set up* a double integral for the average concentration of pollutant over the set of all (x, t) such that $0 \le t \le t_m(x)$ for each fixed x between $x = 0$ and $x = x_m$.

d. How would you define the "dangerous period" for the pollution spill?

52. MODELING PROBLEM   The stiffness of a horizontal beam is modeled to be proportional to the moment of inertia of its cross section with respect to a horizontal line L through its centroid, as illustrated for three shapes shown in Figure 13.44.

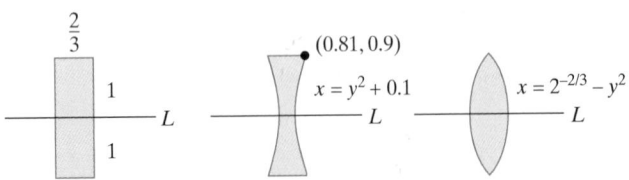

a. rectangular   **b.** concave   **c.** convex

■ **FIGURE 13.44**   Cross sections of horizontal beams with area 4/3

Which of the illustrated beams is the stiffest? For this model, assume the constant of proportionality is the same for all three cases and that $\delta = 1$.

Often the knowledge of the center of mass allows us to greatly simplify a problem, but sometimes it can lead us to an incorrect answer. We explore this in Problems 53–54.

53. Centers of mass in physics

a. Suppose a volume of liquid (or granular solid) is located so that its center of mass is at the origin. The material is to be lifted to a height of h units above the center of mass. Starting with a small element of volume, ΔV, to be lifted to height h, derive the formula for the total work done:

$$\text{Work} = \delta \iiint (h - z) \, dV$$

where δ is the weight (density) of the material.

b. Explain why the integral in part **a** simplifies to the formula

$$\text{Work} = \delta Vh = \text{Force} \times \text{distance}$$

where h is the distance between the center of mass and the level to which the material is lifted.

c. Picture a cylindrical tank of radius 6 ft and height 10 ft positioned so its center of mass is at the origin. Compute the work required to lift (that is, pump) the contents (of density δ) to a level of 15 ft above the center of mass. Compute this two ways: by the integral in part **a** (suggestion—do the $dx \, dy$ integral by inspection), and also by simply lifting the center of mass to the required height. These results should agree.

54. Newton's inverse square law   Recall that if two masses m and M are each concentrated at (or nearly at) a point and are separated by a distance p, then the attractive force is expressed by $F = GmM/p^2$. What physicists and others prefer to do in the case of real-life masses (which occupy some volume) is to use p, the distance between the two centers of mass. The question you are to explore here (and again in Section 13.7) is whether, and when, this simplifying way of computing attracting forces is justified.

a. Suppose one point mass, m, is located at $(0, 0, h)$ and the second, M, at $(0, 0, 0)$. We are interested in the total *resultant* force that M exerts on m, and we assume this to be in the z-direction. Hence we will sum the vertical components of the forces acting on m. Consider, in M, a small element of volume ΔV (at x, y, z) and argue that the magnitude of the force exerted on m and its vertical component are:

$$\Delta F_{\text{mag}} = \frac{Gm\Delta V}{p^2}$$

$$\Delta F = \frac{Gm\Delta V(h - z)^2}{p^3}$$

Technology Window

where $p^2 = x^2 + y^2 + (h - z)^2$.
Note that the term $(h - z)/p$ projects the force vector onto the z-axis, giving the vertical component. Do you see why?

b. From the usual Riemann sum and obtain the following force acting between m and M:

$$F = Gm\delta \int \int \int \frac{(h - z)dV}{p^3}$$

c. Consider a solid (with density $\delta = 5$) positioned at $-2 \le x \le 2$, $-4 \le y \le 4$, $-1 \le z \le 1$. (Note the center of mass is at the origin.) Compute the attracting force between this mass and a point mass, m, at $(0, 0, 8)$ using the center of mass formula $F = GmM/p^2$.

d. For the masses in **c**, compute F using the integration formula in part **b**; compare the result with that of **c**.

e. Repeat the last calculation with $h = 200$ and compare with the center of mass approximation. Comment on the role that the separating distance p plays.

C 55. Find the moment of inertia of a rectangular lamina with dimensions h and l about an axis L through its center of mass. Assume the lamina has density $\delta = 1$.

56. Prove the following area theorem of Pappus:
Let C be a curve of length L in the plane. Then the surface obtained by rotating C about the axis L in the plane has area $2\pi Lh$, where h is the distance from the centroid of C to the axis of rotation.

57. A torus (doughnut) can be formed by rotating the circle $(x - b)^2 + y^2 = a^2$ for $b > a$ about the y-axis. Find the surface area of the torus by applying Pappus' area theorem (see Problem 56 and Figure 13.45). Compare your result with the area found parametrically in Problem 46, Section 13.4.

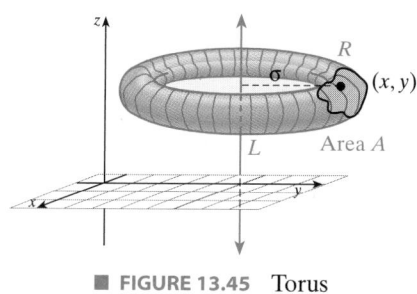

■ **FIGURE 13.45** Torus

13.7 Cylindrical and Spherical Coordinates

IN THIS SECTION **cylindrical coordinates, integration with cylindrical coordinates, spherical coordinates, integration with spherical coordinates**

CYLINDRICAL COORDINATES

Cylindrical coordinates are a generalization of polar coordinates in $\mathbb{R}^3$. Recall that the point P with Cartesian coordinates (x, y, z) is located z units above the point $Q(x, y)$ in the xy-plane (below if $z < 0$). In cylindrical coordinates, we measure the point in the xy-plane in polar coordinates, with the same z-coordinate as in the Cartesian coordinate system. These relationships are shown in Figure 13.46.

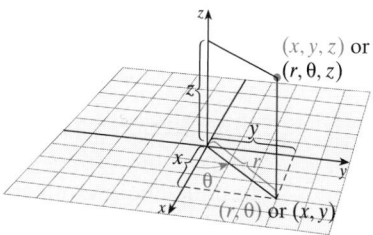

■ **FIGURE 13.46** The cylindrical coordinate system

Cylindrical coordinates are convenient for representing cylindrical surfaces and surfaces of revolution for which the z-axis is the axis of symmetry. Some examples are shown in Figure 13.47.

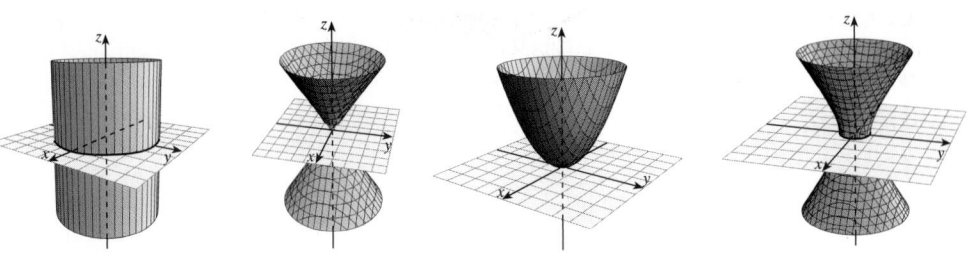

	a. Cylinder	**b.** Cone	**c.** Paraboloid	**d.** Hyperboloid
Rectangular equation:	$x^2 + y^2 = a^2$	$x^2 + y^2 = z^2$	$x^2 + y^2 = az$	$x^2 + y^2 - z^2 = 1$
Cylindrical equation:	$r = a$	$r^2 = z^2$	$r^2 = az$	$r^2 = z^2 + 1$

■ **FIGURE 13.47**   Surfaces with convenient cylindrical coordinates

We have the following conversion formulas, which follow directly from the rectangular–polar conversions.

Conversion Formulas: Rectangular–Cylindrical Coordinates

Cylindrical to rectangular: $x = r \cos \theta$
(r, θ, z) to (x, y, z) $y = r \sin \theta$
                          $z = z$

Rectangular to cylindrical: $r = \sqrt{x^2 + y^2}$

(x, y, z) to (r, θ, z) $\tan \theta = \dfrac{y}{x}$

                          $z = z$

EXAMPLE 1   *Rectangular-form equation converted to cylindrical-form equation*

Find an equation in cylindrical coordinates for the elliptical paraboloid $z = x^2 + 3y^2$.

Solution   We use the conversion formulas $x = r \cos \theta$ and $y = r \sin \theta$.

$$z = x^2 + 3y^2 = (r \cos \theta)^2 + 3(r \sin \theta)^2$$
$$= r^2(\cos^2\theta + 3 \sin^2\theta)$$
$$= r^2[(1 - \sin^2\theta) + 3 \sin^2\theta]$$
$$= r^2(1 + 2 \sin^2\theta) \qquad ■$$

INTEGRATION WITH CYLINDRICAL COORDINATES

WARNING▶ Recall from Theorem 13.4 that in polar coordinates, $dA = r\, dr\, d\theta$. ◄

To perform triple integration with cylindrical coordinates, we simply apply our results from Section 13.3 to convert from rectangular to polar form. Even though there are six possible orders of integration, we will focus on the one for which the region is bounded below and above by $u(r, \theta) \le z \le v(r, \theta)$. In this case, we let D be the region in the xy-plane described by polar coordinates and replace dV by $dz(r\, dr\, d\theta)$, which we write as $r\, dz\, dr\, d\theta$. See Figure 13.48.

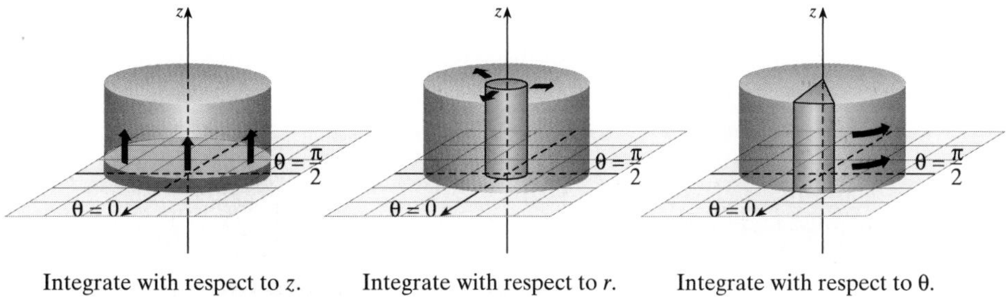

Integrate with respect to z.   Integrate with respect to r.   Integrate with respect to θ.

■ **FIGURE 13.48**   Integration with cylindrical coordinates

Triple Integral in Cylindrical Coordinates

Let R be a solid with upper surface $z = v(r, \theta)$ and lower surface $z = u(r, \theta)$, and let D be the projection of the solid on the xy-plane expressed in polar coordinates. Then, if $f(r, \theta, z)$ is continuous on R, we have

$$\iiint\limits_{R} f(r, \theta, z) \, dV = \iint\limits_{D} \int_{u(r, \theta)}^{v(r, \theta)} f(r, \theta, z) \, r \, dz \, dr \, d\theta$$

EXAMPLE 2   *Finding volume in cylindrical coordinates*

Find the volume of the solid in the first octant that is bounded by the cylinder $x^2 + y^2 = 2y$, the cone $z = \sqrt{x^2 + y^2}$, and the xy-plane.

Solution   Let S be the region occupied by the solid, as shown in Figure 13.49. This surface is most easily described in cylindrical coordinates.

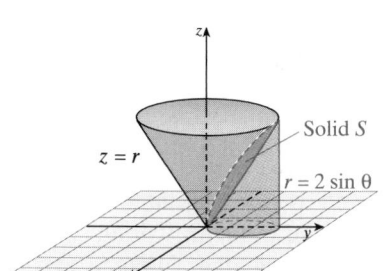

Cylinder:	Cone:
$x^2 + y^2 = 2y$	$z = \sqrt{x^2 + y^2}$
$r^2 = 2r \sin \theta$	$z = r$
$z = 2 \sin \theta$	

■ **FIGURE 13.49**   The solid bounded by $x^2 + y^2 = 2y$, $z = \sqrt{x^2 + y^2}$, and the xy-plane

Since the region S lies in the first octant, we have $0 \le \theta \le \frac{\pi}{2}$, so S may be described by

$$0 \le z \le r \qquad 0 \le r \le 2 \sin \theta \qquad 0 \le \theta \le \frac{\pi}{2}$$

$$V = \iiint\limits_{R} dV = \iint\limits_{D} \int_{u(r, \theta)}^{v(r, \theta)} r \, dz \, dr \, d\theta = \int_{0}^{\pi/2} \int_{0}^{2 \sin \theta} \int_{0}^{r} r \, dz \, dr \, d\theta$$

$$= \int_{0}^{\pi/2} \int_{0}^{2 \sin \theta} r^2 \, dr \, d\theta = \int_{0}^{\pi/2} \frac{r^3}{3} \Big|_{r=0}^{r=2 \sin \theta} d\theta = \frac{8}{3} \int_{0}^{\pi/2} \sin^3 \theta \, d\theta$$

$$= \frac{8}{3} \left[-\cos \theta + \frac{\cos^3 \theta}{3} \right] \Big|_{0}^{\pi/2} = \frac{16}{9} \qquad \textit{Integration table (formula 350)} \qquad ■$$

EXAMPLE 3   *Centroid in cylindrical coordinates*

A homogeneous solid S (δ constant) is bounded below by the xy-plane, on the sides by the cylinder $x^2 + y^2 = a^2 (a > 0)$, and above by the surface $z = x^2 + y^2$. Find the centroid of the solid.

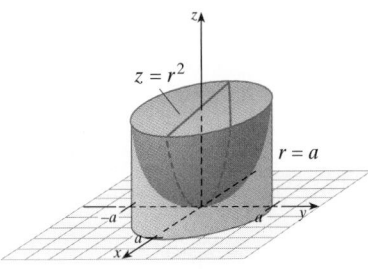

■ **FIGURE 13.50** The solid bounded by $x^2 + y^2 = a^2$ and $z = x^2 + y^2$

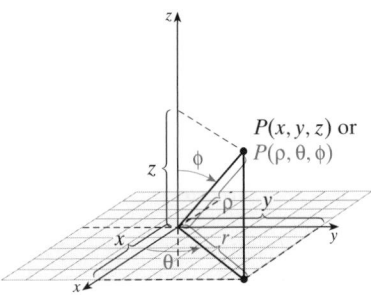

■ **FIGURE 13.51** The spherical coordinate system

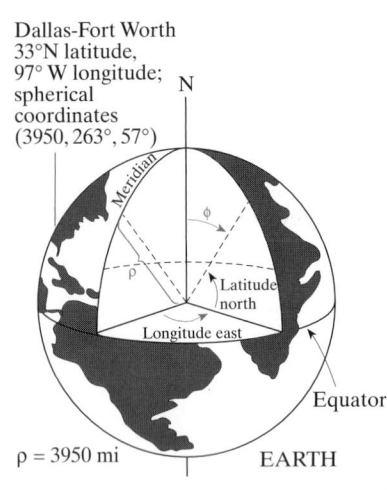

Dallas-Fort Worth 33°N latitude, 97° W longitude; spherical coordinates $(3950, 263°, 57°)$

$\rho = 3950$ mi   EARTH

Solution Let S be the described solid, as shown in Figure 13.50. Because the solid is bounded by a cylinder, we will carry out the integration in cylindrical coordinates.

Cylinder:	Paraboloid:
$x^2 + y^2 = a^2$	$z = x^2 + y^2$
$r^2 = a^2$	$z = r^2$
$r = a \quad (a > 0)$	

Let $(\bar{x}, \bar{y}, \bar{z})$ denote the centroid. By symmetry, we have $\bar{x} = \bar{y} = 0$. Let m denote the mass of S. Since the projected region is $r = a$ for $0 \le \theta \le 2\pi$, we find that

$$\bar{z} = \frac{M_{xy}}{m} = \frac{\displaystyle\iiint_S zr\,dz\,dr\,d\theta}{\displaystyle\iiint_S r\,dz\,dr\,d\theta} = \frac{\displaystyle\int_0^{2\pi}\int_0^a\int_0^{r^2} zr\,dz\,dr\,d\theta}{\displaystyle\int_0^{2\pi}\int_0^a\int_0^{r^2} r\,dz\,dr\,d\theta} = \frac{\frac{\pi}{6}a^6}{\frac{\pi}{2}a^4} = \frac{a^2}{3}$$

Verify the details of the integration.

The centroid is $(0, 0, \frac{a^2}{3})$. ∎

SPHERICAL COORDINATES

In **spherical coordinates,** we label a point P by a triple (ρ, θ, ϕ), where $\rho, \theta,$ and ϕ are numbers determined as follows (refer to Figure 13.51):

ρ = the distance from the origin to the point P; we require $\rho \ge 0$.

θ = the polar angle (as in polar coordinates). In spherical coordinates, this is called the **azimuth** of P; we require $0 \le \theta < 2\pi$.

ϕ = the angle measured down from the positive z-axis to the ray from the origin through P. The angle ϕ is called the **colatitude** of P; we require $0 \le \phi \le \pi$.

You might recognize spherical coordinates (as well as the associated terminology) as being related to the latitude–longitude system used to identify points on the surface of the earth. In spherical coordinates, points above the xy-plane satisfy $0 \le \phi < \pi/2$ while points below satisfy $\pi/2 < \phi \le \pi$, and the equation $\phi = \pi/2$ describes the xy-plane. When this coordinate system is used to measure points on the earth, the number ρ is the distance from the center of the earth to the point with longitude θ from the prime meridian; because latitude is the angle *up* from the equator, it is measured in spherical coordinates as $\pi/2 - \phi$.

Spherical coordinates are desirable when representing spheres, cones, or certain planes. Some examples are shown in Figure 13.52.

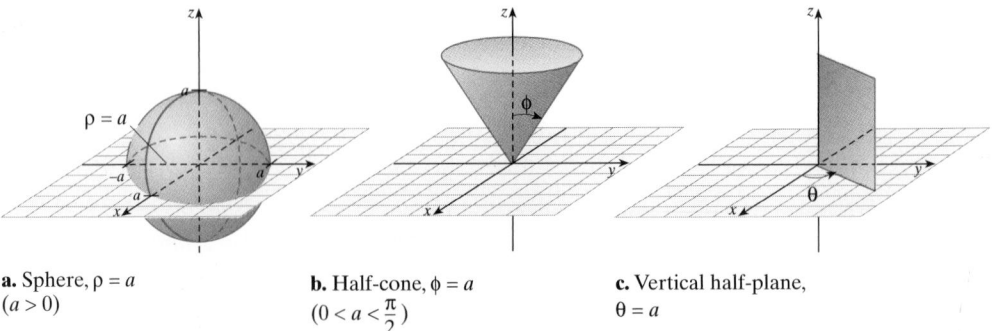

a. Sphere, $\rho = a$ $(a > 0)$ **b.** Half-cone, $\phi = a$ $(0 < a < \frac{\pi}{2})$ **c.** Vertical half-plane, $\theta = a$

■ **FIGURE 13.52** Surfaces with convenient spherical coordinates

We use the relationships in Figure 13.52 to obtain the remaining conversion formulas.

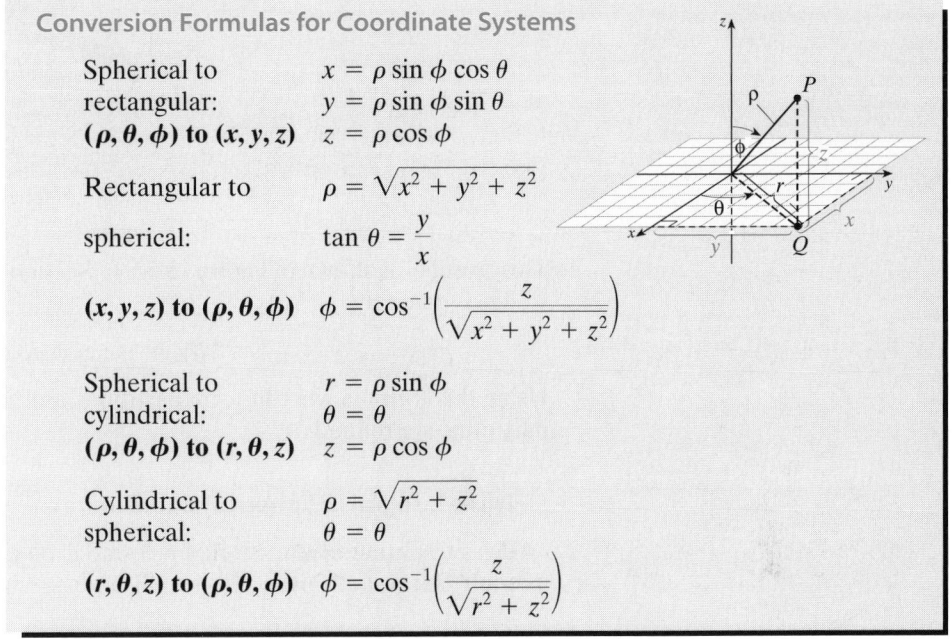

Conversion Formulas for Coordinate Systems

Spherical to rectangular:
(ρ, θ, ϕ) to (x, y, z)

$$x = \rho \sin \phi \cos \theta$$
$$y = \rho \sin \phi \sin \theta$$
$$z = \rho \cos \phi$$

Rectangular to spherical:

$$\rho = \sqrt{x^2 + y^2 + z^2}$$
$$\tan \theta = \frac{y}{x}$$

(x, y, z) to (ρ, θ, ϕ)

$$\phi = \cos^{-1}\left(\frac{z}{\sqrt{x^2 + y^2 + z^2}}\right)$$

Spherical to cylindrical:
(ρ, θ, ϕ) to (r, θ, z)

$$r = \rho \sin \phi$$
$$\theta = \theta$$
$$z = \rho \cos \phi$$

Cylindrical to spherical:

$$\rho = \sqrt{r^2 + z^2}$$
$$\theta = \theta$$

(r, θ, z) to (ρ, θ, ϕ)

$$\phi = \cos^{-1}\left(\frac{z}{\sqrt{r^2 + z^2}}\right)$$

EXAMPLE 4   *Converting rectangular-form equations to spherical-form equations*

Rewrite each of the given equations in spherical form.

a. the sphere $x^2 + y^2 + z^2 = a^2 \; (a > 0)$

b. the paraboloid $z = x^2 + 3y^2$. This is the same elliptic paraboloid we analyzed in Example 1.

Solution

a. Because $\rho = \sqrt{x^2 + y^2 + z^2}$, we see $x^2 + y^2 + z^2 = \rho^2$, so we can write

$$\rho^2 = a^2$$
$$\rho = a \qquad \text{Because } \rho \geq 0$$

b.
$$z = x^2 + 3y^2$$
$$\rho \cos \phi = (\rho \sin \phi \cos \theta)^2 + 3(\rho \sin \phi \sin \theta)^2$$
$$\rho \cos \phi = \rho^2 \sin^2\phi \cos^2\theta + 3\rho^2 \sin^2\phi \sin^2\theta$$
$$\rho = \frac{\cos \phi}{\sin^2\phi \cos^2\theta + 3 \sin^2\phi \sin^2\theta}$$

∎

INTEGRATION WITH SPHERICAL COORDINATES

For a solid S in spherical coordinates, the fundamental element of volume is a spherical "wedge" bounded in such a way that

$$\rho_1 \leq \rho \leq \rho_1 + \Delta\rho, \qquad \phi_1 \leq \phi \leq \phi_1 + \Delta\phi, \qquad \theta_1 \leq \theta \leq \theta_1 + \Delta\theta$$

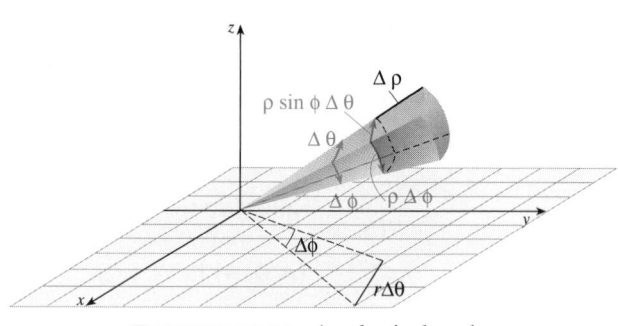

■ **FIGURE 13.53**   A spherical wedge

This "wedge" is shown in Figure 13.53. In Section 13.8, we show that the volume of the wedge is given by

$$dV = \rho^2 \sin \phi \, d\rho \, d\phi \, d\theta$$

Using this formula, we can form partitions and take a limit of a Riemann sum as the partitions are refined.

Triple Integral in Spherical Coordinates

If f is a continuous function of ρ, θ, and ϕ on a bounded, solid region S, the **triple integral of f over S** is given by

$$\iiint\limits_{S} f(\rho, \theta, \phi) \, dV = \iiint\limits_{S'} f(\rho, \theta, \phi) \, \rho^2 \sin \phi \, d\rho \, d\theta \, d\phi$$

where S' is the region S expressed in spherical coordinates.

EXAMPLE 5   *Volume of a sphere*

Find the volume of the sphere described by $x^2 + y^2 + z^2 = 9$.

Solution   It seems clear that we should work in spherical coordinates, because the equation of the sphere is $\rho = 3$ for $0 \le \theta \le 2\pi$ and $0 \le \phi \le \pi$.

$$V = \iiint\limits_{S} dV = \int_0^{\pi} \int_0^{2\pi} \int_0^3 \rho^2 \sin \phi \, d\rho \, d\theta \, d\phi$$

$$= \int_0^{\pi} \int_0^{2\pi} \frac{\rho^3}{3} \sin \phi \Big|_{\rho=0}^{\rho=3} d\theta \, d\phi = \int_0^{\pi} \int_0^{2\pi} 9 \sin \phi \, d\theta \, d\phi$$

$$= \int_0^{\pi} 18\pi \sin \phi \, d\phi = 18\pi(-\cos \phi)\Big|_0^{\pi} = 36\pi$$ ■

EXAMPLE 6   *Moment of inertia using spherical coordinates*

A toy top of constant density d_0 is constructed from a portion of a solid hemisphere with a conical base, as shown in Figure 13.54. The center of the spherical cap is at the point where the top spins, and the height of the conical base is equal to its radius. Find the moment of inertia of the top about its axis of symmetry.

Solution   We use a Cartesian coordinate system in which the z-axis is the axis of symmetry of the top. Suppose the cap is part of the hemisphere $z = \sqrt{R^2 - x^2 - y^2}$.

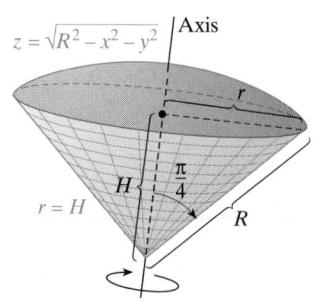

■ **FIGURE 13.54**   Moment of inertia for a spinning top

Since the height of the conical base is equal to its radius, the cone makes an angle of $\pi/4$ radians (45°) with the z-axis. Let S denote the solid region occupied by the top. Then the moment of inertia I_z with respect to the z-axis is given by

$$I_z = \iiint_S (x^2 + y^2)\, dm = \iiint_S (x^2 + y^2)d_0\, dV$$

The shape of the top suggests that we convert to spherical coordinates, and we find that S can be described as

$$0 \le \theta < 2\pi \qquad 0 \le \phi \le \tfrac{\pi}{4} \qquad 0 \le \rho \le R$$

Also,

$$x^2 + y^2 = \rho^2 \sin^2\phi \cos^2\theta + \rho^2 \sin^2\phi \sin^2\theta = \rho^2 \sin^2\phi$$

We now evaluate the integral using spherical coordinates:

$$I_z = \iiint_S (x^2 + y^2)d_0\, dV$$

$$= d_0 \int_0^{\pi/4} \int_0^{2\pi} \int_0^R \underbrace{\rho^2 \sin^2\phi}_{x^2 + y^2}\ \underbrace{\rho^2 \sin\phi\, d\rho\, d\theta\, d\phi}_{dV} = d_0 \int_0^{\pi/4} \int_0^{2\pi} \int_0^R \rho^4 \sin^3\phi\, d\rho\, d\theta\, d\phi$$

$$= d_0 \int_0^{\pi/4} \int_0^{2\pi} \frac{\rho^5}{5} \sin^3\phi \Big|_{\rho=0}^{\rho=R} d\theta\, d\phi = \frac{R^5 d_0}{5} \int_0^{\pi/4} (2\pi - 0)\sin^3\phi\, d\phi$$

$$= \frac{2R^5 d_0 \pi}{5}[-\cos\phi + \tfrac{1}{3}\cos^3\phi]\Big|_0^{\pi/4} = \frac{2R^5\, d_0 \pi}{5}\left[-\tfrac{1}{2}\sqrt{2} + \tfrac{1}{12}\sqrt{2} + 1 - \tfrac{1}{3}\right]$$

$$= \frac{\pi d_0 R^5}{30}(8 - 5\sqrt{2}) \approx 0.09728 d_0 R^5 \qquad \blacksquare$$

13.7 Problem Set

Ⓐ 1. ■ **What Does This Say?** Compare and contrast the rectangular, cylindrical, and spherical coordinate systems.

2. ■ **What Does This Say?** Suppose you need to find a particular volume. Discuss some criteria for choosing a coordinate system.

In Problems 3–16, round the coordinates to the nearest hundredth.

In Problems 3–6, convert from rectangular coordinates to
a. *cylindrical* **b.** *spherical*

3. $(0, 4, \sqrt{3}\,)$

4. $(\sqrt{2}, -2, \sqrt{3}\,)$

5. $(1, 2, 3)$

6. (π, π, π)

In Problems 7–10, convert from cylindrical coordinates to
a. *rectangular* **b.** *spherical*

7. $\left(3, \dfrac{2\pi}{3}, -3\right)$

8. $\left(4, \dfrac{\pi}{6}, -2\right)$

9. $\left(2, \dfrac{\pi}{4}, \pi\right)$

10. (π, π, π)

In Problems 11–16, convert from spherical coordinates to
a. *rectangular* **b.** *cylindrical*

11. $\left(2, \dfrac{\pi}{6}, \dfrac{2\pi}{3}\right)$

12. $\left(3, \dfrac{\pi}{4}, \dfrac{\pi}{6}\right)$

13. $\left(1, \dfrac{\pi}{6}, 0\right)$

14. $\left(2, \dfrac{\pi}{3}, \dfrac{\pi}{4}\right)$

15. $(1, 2, 3)$

16. (π, π, π)

Convert each equation in Problems 17–20 to cylindrical coordinates and sketch its graph in $\mathbb{R}^3$.

17. $z = x^2 - y^2$

18. $x^2 - y^2 = 1$

19. $\dfrac{x^2}{4} - \dfrac{y^2}{9} + z^2 = 0$

20. $z = x^2 + y^2$

Convert each equation in Problems 21–24 to spherical coordinates and sketch its graph in $\mathbb{R}^3$.

21. $z^2 = x^2 + y^2, z \ge 0$

22. $2x^2 + 2y^2 + 2z^2 = 1$

23. $4z = x^2 + 3y^2$

24. $x^2 + y^2 - 4z^2 = 1$

Convert each equation in Problems 25–30 to rectangular coordinates and sketch its graph in $\mathbb{R}^3$.

25. $z = r^2 \sin 2\theta$ **26.** $r = \sin \theta$

27. $z = r^2 \cos 2\theta$ **28.** $\rho^2 \sin^2 \phi = 1$

29. $\rho^2 \sin \phi \cos \phi \cos \theta = 1$ **30.** $\rho = \sin \phi \cos \theta$

Evaluate each iterated integral in Problems 31–38.

31. $\int_0^\pi \int_0^2 \int_0^{\sqrt{4-r^2}} r \sin \theta \, dz \, dr \, d\theta$

32. $\int_0^{\pi/4} \int_0^1 \int_0^{\sqrt{r}} r^2 \sin \theta \, dz \, dr \, d\theta$

33. $\int_0^{\pi/2} \int_0^{2\pi} \int_0^2 \cos \phi \sin \phi \, d\rho \, d\theta \, d\phi$

34. $\int_0^{\pi/2} \int_0^{\pi/4} \int_0^{\cos \phi} \rho^2 \sin \phi \, d\rho \, d\theta \, d\phi$

35. $\int_0^{2\pi} \int_0^4 \int_0^1 zr \, dz \, dr \, d\theta$

36. $\int_{-\pi/4}^{\pi/3} \int_0^{\sin \theta} \int_0^{4\cos \theta} r \, dz \, dr \, d\theta$

37. $\int_0^{\pi/2} \int_0^{\cos \theta} \int_0^{1-r^2} r \sin \theta \, dz \, dr \, d\theta$

38. $\int_0^{\pi/3} \int_0^{\cos \theta} \int_0^{\phi} \rho^2 \sin \theta \, d\rho \, d\phi \, d\theta$

Ⓑ 39. The point (x, y, z) lies on an ellipsoid if

$$x = aR \sin \phi \cos \theta$$
$$y = bR \sin \phi \sin \theta$$
$$z = cR \cos \phi$$

Find an equation for this ellipsoid in rectangular coordinates if R is a constant.

40. Use cylindrical coordinates to compute the integral

$$\iiint_R xy \, dx \, dy \, dz$$

where R is the cylindrical solid $x^2 + y^2 \le 1$ with $0 \le z \le 1$.

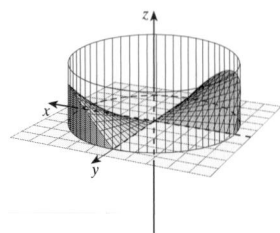

41. Use cylindrical coordinates to compute the integral

$$\iiint_R (x^4 + 2x^2y^2 + y^4) \, dx \, dy \, dz$$

where R is the cylindrical solid $x^2 + y^2 \le a^2$ with

$$0 \le z \le \frac{1}{\pi}$$

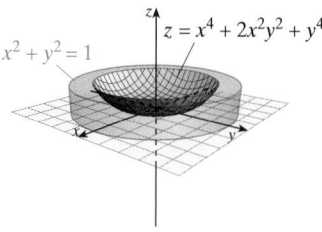

42. Use cylindrical coordinates to compute the integral

$$\iiint_S z(x^2 + y^2)^{-1/2} \, dx \, dy \, dz$$

where S is the solid bounded above by the plane $z = 2$ and below by the surface $2z = x^2 + y^2$.

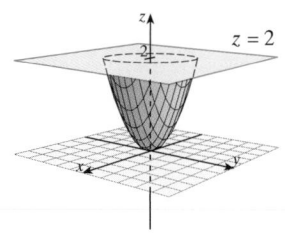

43. Find the center of mass of the solid bounded by the surface $z = \sqrt{x^2 + y^2}$ and the plane $z = 9$. Assume $\delta = 1$.

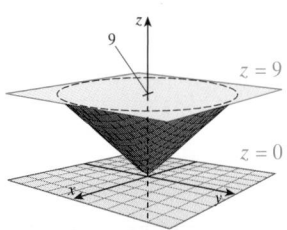

44. Find the centroid of the region bounded by the cone $z = \sqrt{x^2 + y^2}$ and the plane $z = 1$.

45. Suppose the density at each point in the hemisphere $z = \sqrt{9 - x^2 - y^2}$ is $\delta(x, y, z) = xy + z$. Set up integrals for the following quantities:
a. the mass of the hemisphere
b. the x-coordinate of the center of mass
c. the moment of inertia about the z-axis

46. Find the moment of inertia about the z-axis of the portion of the homogeneous hemisphere $z = \sqrt{4 - x^2 - y^2}$ that lies between the cones $z = \sqrt{x^2 + y^2}$ and $2z = \sqrt{x^2 + y^2}$. Assume $\delta = 1$.

47. Find the average value of the function $f(x, y, z) = x + y + z$ over the sphere $x^2 + y^2 + z^2 = 4$.

48. Find the average value of θ and ϕ over the sphere $\rho \leq 3$.

49. Find the mass of the torus $\rho = 2 \sin \phi$ if the density is $\delta = \rho$.

50. Evaluate $\displaystyle\iiint_R \sqrt{x^2 + y^2 + z^2}\ dx\,dy\,dz$ where R is defined by $x^2 + y^2 + z^2 \leq 2$.

51. Evaluate $\displaystyle\iiint_R (x^2 + y^2 + z^2)\ dx\,dy\,dz$ where R is defined by $x^2 + y^2 + z^2 \leq 2$.

52. Evaluate $\displaystyle\iiint_S z^2\ dx\,dy\,dz$ where S is the solid hemisphere $x^2 + y^2 + z^2 \leq 1,\ z \geq 0$.

53. Evaluate $\displaystyle\iiint_S \frac{dx\,dy\,dz}{\sqrt{x^2 + y^2 + z^2}}$ where S is the solid sphere $x^2 + y^2 + z^2 \leq 3$.

Find the volume of the solid S given in Problems 54–58 by using integration in any convenient system of coordinates.

54. S is bounded by the surface $z = 1 - 4(x^2 + y^2)$ and the xy-plane.

55. S is bounded above by the paraboloid $z = 4 - (x^2 + y^2)$, below by the plane $z = 0$, and laterally by the cylinder $x^2 + y^2 \leq 1$.

56. S is the intersection of the solid sphere $x^2 + y^2 + z^2 \leq 9$ and the solid cylinder $x^2 + y^2 \leq 1$.

57. S is the region bounded laterally by the cylinder $r = 2 \sin \theta$, below by the plane $z = 0$, and above by the paraboloid $z = 4 - r^2$.

58. S is the region that remains in the spherical solid $\rho \leq 4$ after the solid cone $\phi \leq \dfrac{\pi}{6}$ has been removed.

59. SPY PROBLEM   The Spy figures out Coldfinger's puzzle (Problem 94 in the supplementary problems of Chapter 12), but just as his bonds come free, the door bursts open and in walks a group of thugs, led by . . . Purity! One of her gang starts to raise his gun. "No!" orders the erstwhile innocent maid. "I want him to think a while before he dies! Tie him again, and be quick about it. Blohardt is waiting and he is not a patient man." The Spy is tied wrist to ankle, and the thugs leave. Purity is the last to go. She pulls a lever by the door and the cave begins to fill with water. The Spy is able to pull himself into a sitting position, with his nose 3 feet above the floor. He looks around desperately and sees that the cave has a beehive shape, something like the part of the surface $\rho = 4\,(1 + \cos \phi)$ that lies above the xy-plane (where ρ is measured in ft). If the Spy needs at least 10 minutes to free himself, and if water is entering at the rate of 25 ft³/min, will he drown or survive?

Technology Window

60. In the previous section (Problems 53 and 54), we say that it is not always accurate to compute the attractive force between two masses by simply using the distance between their centers of mass. This approximation was applied first in astronomy, where the bodies are typically spherical in shape. As we shall see, in this case the simple calculation is very appropriate. Consider a sphere of radius a with density δ centered at the origin and a point mass m located at $(x, y, z) = (0, 0, R)$. In this problem, you are to set up the integral to compute the attractive force between the masses; and in the next problem you will do some computing.

   a. Show that the distance p between point $(x, y, z) = (0, 0, R)$ and a point in the sphere, (ρ, θ, ϕ), can be expressed $p^2 = R^2 + \rho^2 - 2R\rho \cos \phi$. *Hint:* Use the law of cosines.

   b. In the sphere, consider a small element of volume ΔV at a point (ρ, θ, ϕ). Argue that the magnitude of the force exerted on m, and the vertical component, are

$$\Delta F_{\text{mag}} = Gm(\delta \Delta V)/p^2$$
$$\Delta F = Gm(\delta \Delta V)(R - \rho \cos \phi)/p^3$$

Note that the term $(R - \rho \cos \phi)/p$ projects the force vector onto the z-axis, giving the vertical component. Do you see why?

   c. Form the Riemann sum involving ΔF and get the following for the force acting between point mass m and the sphere of radius a:

$$F = \iiint \frac{Gm\delta(R - \rho \cos \phi)\,dV}{p^3}$$
$$= 2\pi\,Gm\delta \int_0^a \int_0^\pi p^2 \frac{(R - \rho \cos \phi)\sin \phi}{(R^2 + p^2 - 2R\rho \cos \phi)^{3/2}}\,d\phi\,dp$$

Note: The integrand is independent of θ.

61. a. For an unspecified radius a and $R > a$, compute the attractive force between the masses in the previous problem using the center of mass calculation.

   b. Set a and R to numerical values and compute the attractive force by the integral in part **c** of the previous problem. Try two or three different values of R until you are satisfied that the center of mass "approximation" is exact in this case.

   c. Compare this result with the corresponding calculations of Section 13.6 and conjecture why the center of mass argument is sound in the case of the sphere and not for the rectangular region.

62. Let S be a homogeneous solid (with density 1) that has the shape of a right circular cylinder with height h and radius r. Use cylindrical coordinates to find the moment of inertia of S about its axis of symmetry.

63. Find the sum $I_x + I_y + I_z$ of the moments of inertia of the solid sphere

$$x^2 + y^2 + z^2 \leq 1$$

about the coordinate axes. Assume the density is 1.

64. How much volume remains from a spherical ball of radius a when a cylindrical hole of radius b $(0 < b < a)$ is bored out of its center?

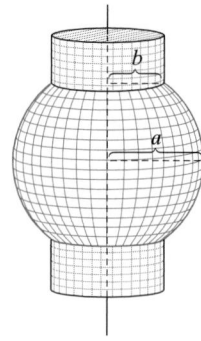

13.8   Jacobians: Change of Variables

IN THIS SECTION   change of variables in a double integral, change of variables in a triple integral

CHANGE OF VARIABLES IN A DOUBLE INTEGRAL

When the change of variable $x = g(u)$ is made in the single integral, we know

$$\int_a^b f(x)\,dx = \int_c^d f(g(u))\,g'(u)\,du$$

where the limits of integration c and d satisfy $a = g(c)$ and $b = g(d)$. To change variables in a double integral, we want to transform ("map") the region of integration D onto a region D^* that is "simpler" in some sense. In general, this process involves introducing a "mapping factor" analogous to the term $g'(u)$ in the single-variable case. This factor is called a *Jacobian* in honor of the German mathematician Karl Gustav Jacobi (1804–1851; see Historical Quest Problem 47), who made the first systematic study of change of variables in multiple integrals in the middle of the 19th century. The basic change of variable theorem may be stated as follows.

THEOREM 13.7   *Change of variables in a double integral*

Let f be a continuous function on a region D in the xy-plane, and let T be a one-to-one transformation that maps the region D^* in the uv-plane onto D under the change of variable $x = g(u, v)$, $y = h(u, v)$, where g and h are continuously differentiable on D^*. Then

$$\iint_D f(x, y)\,dy\,dx = \iint_{D^*} f[g(u, v), h(u, v)]\,|J(u, v)|\,du\,dv$$

where
$$J(u, v) = \begin{vmatrix} \dfrac{\partial x}{\partial u} & \dfrac{\partial x}{\partial v} \\[2mm] \dfrac{\partial y}{\partial u} & \dfrac{\partial y}{\partial v} \end{vmatrix} = \frac{\partial x}{\partial u}\frac{\partial y}{\partial v} - \frac{\partial y}{\partial u}\frac{\partial x}{\partial v}$$

is nonzero and does not change sign on D^*. The mapping factor $J(u, v)$ is called the **Jacobian** and is also denoted by $\dfrac{\partial(x, y)}{\partial(u, v)}$.

Proof   A formal proof is a matter for advanced calculus, but a geometric argument for this theorem is presented in Appendix B.

EXAMPLE 1   *Finding the Jacobian*

Find the Jacobian for the change of variables from rectangular to polar coordinates, namely, $x = r \cos \theta$ and $y = r \sin \theta$.

Solution   The Jacobian of the change of variables is

$$\frac{\partial(x, y)}{\partial(r, \theta)} = \begin{vmatrix} \dfrac{\partial x}{\partial r} & \dfrac{\partial x}{\partial \theta} \\[2ex] \dfrac{\partial y}{\partial r} & \dfrac{\partial y}{\partial \theta} \end{vmatrix} = \begin{vmatrix} \cos \theta & -r \sin \theta \\ \sin \theta & r \cos \theta \end{vmatrix} = r \cos^2\theta + r \sin^2\theta = r \qquad \blacksquare$$

This result justifies the formula we have previously used:

$$\iint\limits_{D} f(x, y) \, dy \, dx = \iint\limits_{D^*} f(r \cos \theta, r \sin \theta) \, r \, dr \, d\theta$$

where D^* is the region in the (r, θ) plane that maps into region D in the (x, y) plane by the polar transform $x = r \cos \theta, y = r \sin \theta$, as illustrated in Figure 13.55.

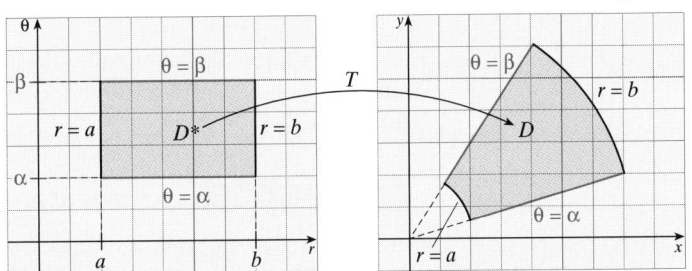

■ **FIGURE 13.55**   Transformation of a region D^* by $T: x = r \cos \theta, y = r \sin \theta$

EXAMPLE 2   *Calculating a double integral by changing variables*

Compute $\displaystyle\iint\limits_{D} \left(\frac{x - y}{x + y}\right)^4 dy \, dx$, where D is the triangular region bounded by the line $x + y = 1$ and the coordinate axes.

Solution   This is a rather difficult computation if no substitution is made. The form of the integral suggests we make the substitution

$$u = x - y \qquad v = x + y$$

Solving for x and y, we obtain

$$x = \tfrac{1}{2}(u + v) \qquad y = \tfrac{1}{2}(v - u)$$

and the Jacobian is

$$\frac{\partial(x, y)}{\partial(u, v)} = \begin{vmatrix} \dfrac{\partial}{\partial u}\left(\dfrac{u + v}{2}\right) & \dfrac{\partial}{\partial v}\left(\dfrac{u + v}{2}\right) \\[2ex] \dfrac{\partial}{\partial u}\left(\dfrac{v - u}{2}\right) & \dfrac{\partial}{\partial v}\left(\dfrac{v - u}{2}\right) \end{vmatrix} = \begin{vmatrix} \dfrac{1}{2} & \dfrac{1}{2} \\[2ex] -\dfrac{1}{2} & \dfrac{1}{2} \end{vmatrix} = \frac{1}{2}$$

To find the image D^* of D in the uv-plane, note that the boundary lines $x = 0$ and $y = 0$ for D map into the lines $u = -v$ and $u = v$, respectively, while $x + y = 1$ maps into $v = 1$. Therefore, the transformed region of integration D^* is the triangular region shown in Figure 13.56b, with vertices $(0, 0)$, $(1, 1)$, and $(-1, 1)$.

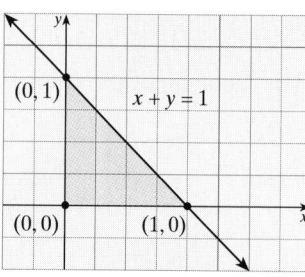
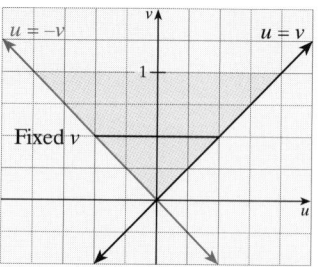

a. The region of integration D **b.** The transformed region D^*

■ **FIGURE 13.56**   Transformation of D to D^*

We now evaluate the given integral:

$$\iint\limits_{D} \left(\frac{x - y}{x + y}\right)^4 dy\, dx = \iint\limits_{D^*} \left(\frac{u}{v}\right)^4 \left|\frac{1}{2}\right| du\, dv = \frac{1}{2}\int_0^1 \int_{-v}^{v} u^4\, v^{-4}\, du\, dv = \frac{1}{10}$$ ■

In Example 2, the change of variables was chosen to simplify the integrand, but sometimes it is useful to introduce a change of variables that simplifies the region of integration.

EXAMPLE 3   *Change of variable to simplify a region*

Find the area of the region E bounded by the ellipse $\dfrac{x^2}{a^2} + \dfrac{y^2}{b^2} = 1$.

Solution   The area is given by the integral

$$A = \iint\limits_{E} dy\, dx$$

where E is the region shown in Figure 13.57a.

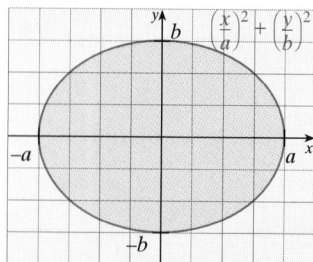
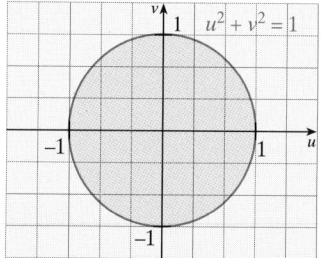

a. The elliptical region E **b.** The transformed region C is a circular disk.

■ **FIGURE 13.57**   Transformation of the ellipse E to the circle C

Because E can be represented by

$$\left(\frac{x}{a}\right)^2 + \left(\frac{y}{b}\right)^2 \leq 1$$

we consider the substitution $u = \dfrac{x}{a}$ and $v = \dfrac{y}{b}$, which will map the elliptical region E onto the circular disk

$$C: u^2 + v^2 \leq 1$$

as shown in Figure 13.57b. When we solve the two equations for x and y, the Jacobian gives us the mapping (change of variable) factor:

$$\frac{\partial(x, y)}{\partial(u, v)} = \begin{vmatrix} \dfrac{\partial}{\partial u}(au) & \dfrac{\partial}{\partial v}(au) \\ \dfrac{\partial}{\partial u}(bv) & \dfrac{\partial}{\partial v}(bv) \end{vmatrix} = \begin{vmatrix} a & 0 \\ 0 & b \end{vmatrix} = ab$$

Because $ab > 0$, we have $|ab| = ab$, and the area of E is given by

$$\iint_E dy\, dx = \iint_C ab\, du\, dv = ab \iint_C du\, dv = ab\underbrace{[\pi(1)^2]}_{\text{Because } C \text{ is a circle of radius 1}} = \pi ab$$

CHANGE OF VARIABLES IN A TRIPLE INTEGRAL

The change of variable formula for triple integrals is similar to the one given for double integrals. Let T be a change of variable that maps a region R^* in uvw-space onto a region R in xyz-space, where

$$T: \quad x = x(u, v, w) \qquad y = y(u, v, w) \qquad z = z(u, v, w)$$

Then the Jacobian of T is the determinant

$$\frac{\partial(x, y, z)}{\partial(u, v, w)} = \begin{vmatrix} \dfrac{\partial x}{\partial u} & \dfrac{\partial x}{\partial v} & \dfrac{\partial x}{\partial w} \\ \dfrac{\partial y}{\partial u} & \dfrac{\partial y}{\partial v} & \dfrac{\partial y}{\partial w} \\ \dfrac{\partial z}{\partial u} & \dfrac{\partial z}{\partial v} & \dfrac{\partial z}{\partial w} \end{vmatrix}$$

and the change of variable yields

$$\iiint_R f(x, y, z)\, dx\, dy\, dz = \iiint_{R^*} f[x(u, v, w), y(u, v, w), z(u, v, w)] \frac{\partial(x, y, z)}{\partial(u, v, w)} du\, dv\, dw$$

EXAMPLE 4 *Formula for integrating with spherical coordinates*

Obtain the formula for converting a triple integral in rectangular coordinates to one in spherical coordinates.

Solution The conversion formulas from rectangular coordinates to spherical coordinates are:

$$x = \rho \sin \phi \cos \theta \qquad y = \rho \sin \phi \sin \theta \qquad z = \rho \cos \phi$$

The Jacobian of this transformation is

$$\frac{\partial(x, y, z)}{\partial(\rho, \theta, \phi)} = \begin{vmatrix} \dfrac{\partial}{\partial \rho}(\rho \sin \phi \cos \theta) & \dfrac{\partial}{\partial \theta}(\rho \sin \phi \cos \theta) & \dfrac{\partial}{\partial \phi}(\rho \sin \phi \cos \theta) \\ \dfrac{\partial}{\partial \rho}(\rho \sin \phi \sin \theta) & \dfrac{\partial}{\partial \theta}(\rho \sin \phi \sin \theta) & \dfrac{\partial}{\partial \phi}(\rho \sin \phi \sin \theta) \\ \dfrac{\partial}{\partial \rho}(\rho \cos \phi) & \dfrac{\partial}{\partial \theta}(\rho \cos \phi) & \dfrac{\partial}{\partial \phi}(\rho \cos \phi) \end{vmatrix}$$

$$= \begin{vmatrix} \sin \phi \cos \theta & -\rho \sin \phi \sin \theta & \rho \cos \phi \cos \theta \\ \sin \phi \sin \theta & \rho \sin \phi \cos \theta & \rho \cos \phi \sin \theta \\ \cos \phi & 0 & -\rho \sin \phi \end{vmatrix}$$

$$= -\rho^2 \sin \phi \qquad \textit{After much algebra}$$

Since $0 \le \phi \le \pi$, we have $\sin \phi \ge 0$, so

$$\left| \frac{\partial(x, y, z)}{\partial(\rho, \theta, \phi)} \right| = |-\rho^2 \sin \phi| = \rho^2 \sin \phi$$

$$\iiint\limits_{R} f(x, y, z)\, dz\, dx\, dy = \iiint\limits_{R^*} f(\rho \sin \phi \cos \theta, \rho \sin \phi \sin \theta, \rho \cos \phi)\, \rho^2 \sin \phi\, d\rho\, d\theta\, d\phi \qquad ∎$$

13.8 Problem Set

A *Find the Jacobian of the change of variables given in Problems 1–12.*

1. $x = u + v, y = uv$
2. $x = u^2, y = u + v$
3. $x = u - v, y = u + v$
4. $x = u^2 - v^2, y = 2uv$
5. $x = u^2 v^2, y = v^2 - u^2$
6. $x = u \cos v, y = u \sin v$
7. $x = e^{u+v}, y = e^{u-v}$
8. $x = e^u \sin v, y = e^u \cos v$
9. $x = u + v - w; y = 2u - v + 3w;$
   $z = -u + 2v - w$
10. $x = 2u - w, y = u + 3v, z = v + 2w$
11. $x = u \cos v, y = u \sin v, z = we^{uv}$
12. $x = \dfrac{u}{v}, y = \dfrac{v}{w}, z = \dfrac{w}{u}$

A region R is given in Problems 13–16. Sketch the corresponding region R* in the uv-plane using the given transformations.

13.

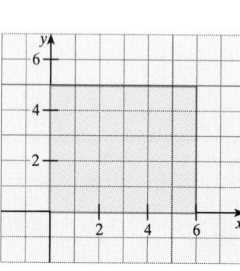

$u = x + y, v = x - y$

14.

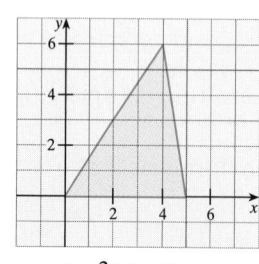

$u = 2x, v = x + y$

15.

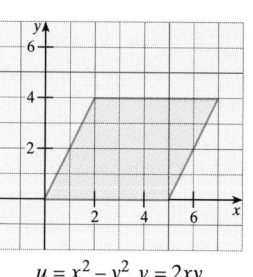

$u = x^2 - y^2, v = 2xy$

16.

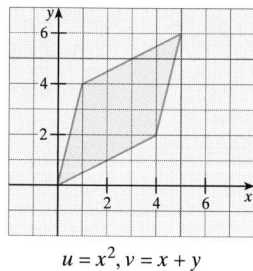

$u = x^2, v = x + y$

17. Suppose a *uv*-plane is mapped onto an *xy*-plane by the equations $x = u(1 - v), y = uv$. Express *dx dy* in terms of *du dv*.

18. Suppose a *uv*-plane is mapped onto an *xy*-plane by the equations $x = u^2 - v^2, y = 2uv$. Express *dx dy* in terms of *du dv*.

Use a suitable change of variable to find the area of the region specified in Problems 19 and 20.

19. The region R bounded by the hyperbolas $xy = 1$ and $xy = 4$, the lines $y = x$ and $y = 4x$.

20. The region R bounded by the parabolas $y = x^2, y = 4x^2$, $y = \sqrt{x}$, and $y = \dfrac{1}{2}\sqrt{x}$.

Let D be the region in the xy-plane that is bounded by the coordinate axes and the line $x + y = 1$. Use the change of *variable $u = x - y, v = x + y$ to compute the integrals given in Problems 21–24.*

21. $\displaystyle\iint_D \left(\frac{x - y}{x + y}\right)^5 dy\, dx$ **22.** $\displaystyle\iint_D \left(\frac{x - y}{x + y}\right)^4 dy\, dx$

23. $\displaystyle\iint_D (x - y)^5 (x + y)^3 \, dy\, dx$

24. $\displaystyle\iint_D (x - y)e^{x^2+y^2} \, dy\, dx$

Ⓑ 25. ■ **What Does This Say?** Discuss the process of changing variables in a triple integral.

26. ■ **What Does This Say?** Discuss the necessity of using the Jacobian when changing variables in a double integral.

Under the transformation

$$u = \frac{1}{5}(2x + y) \qquad v = \frac{1}{5}(x - 2y)$$

the square S in the xy-plane with vertices $(0, 0)$, $(1, -2)$, $(3, -1)$, $(2, 1)$ is mapped onto a square in the uv-plane. Use this information to find the integrals in Problems 27–32.

27. $\displaystyle\iint_S \left(\frac{2x + y}{x - 2y + 5}\right)^2 dy\, dx$

28. $\displaystyle\iint_S (2x + y)(x - 2y)^2 \, dy\, dx$

29. $\displaystyle\iint_S (2x + y)^2 (x - 2y) \, dy\, dx$

30. $\displaystyle\iint_S \sqrt{(2x + y)(x - 2y)} \, dy\, dx$

31. $\displaystyle\iint_S (2x + y)\tan^{-1}(x - 2y) \, dy\, dx$

32. $\displaystyle\iint_S \cos(2x + y)\sin(x - 2y) \, dy\, dx$

33. Evaluate

$$\iint_R e^{(2y-x)/(y+2x)} \, dA$$

where R is the trapezoid with vertices $(0, 2)$, $(1, 0)$, $(4, 0)$, and $(0, 8)$.

34. Let R be the region in the xy-plane that is bounded by the parallelogram with vertices $(0, 0)$, $(1, 1)$, $(2, 1)$, and $(1, 0)$. Use the linear transformation $x = u + v, y = v$ to compute $\int_R \int (2x - y) \, dy\, dx$.

35. Use a suitable linear transformation $u = ax + by$, $v = rx + sy$ to evaluate the integral

$$\iint_R \left(\frac{x + y}{2}\right)^2 e^{(y-x)/2} \, dy\, dx$$

where R is the region inside the square with vertices $(0, 0)$, $(1, 1)$, $(0, 2)$, $(-1, 1)$.

36. Under the change of variables $x = s^2 - t^2, y = 2st$, the quarter circular region in the st-plane given by

$s^2 + t^2 \le 1, s \ge 0, t \ge 0$ is mapped onto a certain region S of the xy-plane. Evaluate

$$\iint_S \frac{dy\, dx}{\sqrt{x^2 + y^2}}$$

37. Use the change of variable $x = ar\cos\theta, y = br\sin\theta$ to evaluate

$$\iint_S \exp\left(-\frac{x^2}{a^2} - \frac{y^2}{b^2}\right) dy\, dx$$

where S is the quarter ellipse.

$$\frac{x^2}{a^2} + \frac{y^2}{b^2} \le 1, x \ge 0, y \ge 0$$

38. A rotation of the xy-plane through the fixed angle θ is given by

$$x = u\cos\theta - v\sin\theta$$
$$y = u\sin\theta + v\cos\theta$$

(See Figure 13.58.)

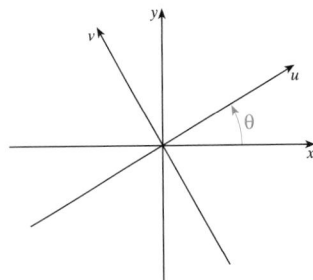

■ **FIGURE 13.58**   Rotational transformation

a. Compute the Jacobian $\dfrac{\partial(x, y)}{\partial(u, v)}$.

b. Let E denote the ellipse

$$x^2 + xy + y^2 = 3$$

Use a rotation of $\frac{\pi}{4}$ to obtain an integral that is equivalent to $\int_E \int y \, dy\, dx$. Evaluate the transformed integral.

39. Find the area of the rotated ellipse

$$5x^2 - 4xy + 2y^2 = 1$$

Note that $5x^2 - 4xy + 2y^2 = Au^2 + Bv^2$, where A and B are constants and $u = x + 2y, v = 2x - y$.

40. **THINK TANK PROBLEM**   Let R be the region bounded by the parallelogram with vertices $(0, 0)$, $(2, 0)$, $(1, 1)$, and $(-1, 1)$. Show that under a suitable change of variable of the form $x = au + bv, y = cu + dv$, we have

$$\iint_R f(x + y) \, dy\, dx = \int_0^2 f(t) \, dt$$

where f is continuous on $[0, 2]$.

41. Find the Jacobian of the cylindrical coordinate transformations

$$x = r \cos \theta, \quad y = r \sin \theta, \quad z = z$$

42. Use the change of variable $x = au, y = bv, z = cw$ to find the volume of the ellipsoid

$$\frac{x^2}{a^2} + \frac{y^2}{b^2} + \frac{z^2}{c^2} = 1$$

43. Use a change of variable to find the moment of inertia about the z-axis of the ellipsoid

$$\frac{x^2}{a^2} + \frac{y^2}{b^2} + \frac{z^2}{c^2} = 1$$

44. Evaluate $\displaystyle\iint_R \ln\left(\frac{x - y}{x + y}\right) dy\, dx$, where R is the triangular region with vertices $(1, 0), (4, -3)$, and $(4, 1)$.

C 45. Show that if $ad - bc \neq 0$, the linear transformation

$$u = ax + by \quad \text{and} \quad v = cx + dy$$

can be solved for x and y to obtain

$$x = \frac{du - bv}{ad - bc} \quad \text{and} \quad y = \frac{av - cu}{ad - bc}$$

46. **THINK TANK PROBLEM**   Find the volume of the solid under the surface

$$z = \frac{xy}{1 + x^2 y^2}$$

over the region bounded by $xy = 1, xy = 5, x = 1$, and $x = 5$.

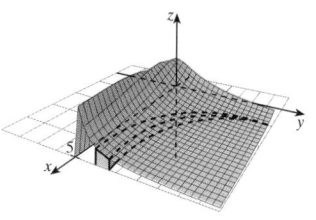

■ **FIGURE 13.59**   Problem 46

47. **HISTORICAL QUEST**   Karl G. Jacobi was a gifted teacher and one of Germany's most distinguished mathematicians during the first half of the 19th century. He made major contributions to the theory of elliptic functions, but his work with functional determinants is what secured his place in history. In 1841, he published a long memoir called "De determinantibus functionalibus," devoted to what we today call the Jacobian and pointing out that this determinant is in many ways the multivariable analogue to the differential quotient of a single variable. The memoir was published in what is usually known as *Crelle's Journal*, the first journal devoted to serious mathematics. It was begun in 1826 by August Crelle (1780–1855) with the "official" title *Journal für die reine und angewandte Mathematik*.

KARL G. JACOBI
1804–1851

For this Quest, set up an integral for the circumference of the ellipse

$$\frac{x^2}{a^2} + \frac{y^2}{b^2} = 1$$

This is an example of an elliptic integral.

Chapter 13   *Review*

Proficiency Examination

Concept Problems

1. Define a double integral.
2. State Fubini's theorem over a rectangular region.
3. What is a type I region? State Fubini's theorem for a type I region.
4. What is a type II region? State Fubini's theorem for a type II region.
5. State the formula for finding an area using a double integral.
6. State the formula for finding a volume using a double integral.
7. State the following properties of a double integral:
   a. linearity rule   b. dominance rule
   c. subdivision rule
8. What is the formula for a double integral in polar coordinates?
9. Explain how to evaluate an improper integral in polar coordinates.
10. State the double integral formula for surface area.
11. State the formula for the area of a surface defined parametrically.
12. State Fubini's theorem over a parallelepiped in space.
13. Give a triple integral formula for volume.
14. Explain how to use a double integral to find the mass of a planar lamina of variable density.
15. How do you find the moment of mass with respect to the x-axis?
16. What are the formulas for the centroid of a region?

17. Define moment of inertia.

18. What is a joint probability density function?

19. Complete the following table for the conversion of coordinates.

From \ To	(r, θ, z) Cylindrical	(ρ, θ, ϕ) Spherical
(x, y, z) Rectangular	$r = \underline{\ ?\ }$ $\tan \theta = \underline{\ ?\ }$ $z = \underline{\ ?\ }$	$\rho = \underline{\ ?\ }$ $\tan \theta = \underline{\ ?\ }$ $\phi = \underline{\ ?\ }$

From \ To	(x, y, z) Rectangular	(ρ, θ, ϕ) Spherical
(r, θ, z) Cylindrical	$x = \underline{\ ?\ }$ $y = \underline{\ ?\ }$ $z = \underline{\ ?\ }$	$\rho = \underline{\ ?\ }$ $\theta = \underline{\ ?\ }$ $\phi = \underline{\ ?\ }$

From \ To	(x, y, z) Rectangular	(r, θ, z) Cylindrical
(ρ, θ, ϕ) Spherical	$x = \underline{\ ?\ }$ $y = \underline{\ ?\ }$ $z = \underline{\ ?\ }$	$r = \underline{\ ?\ }$ $\theta = \underline{\ ?\ }$ $z = \underline{\ ?\ }$

20. a. State the formula for a triple integral in cylindrical coordinates.
   b. State the formula for a triple integral in spherical coordinates.

21. Define the Jacobian for a mapping $x = g(u, v), y = h(u, v)$ from u, v variables to x, y variables.

22. State the formula for the change of variables in a double integral.

Practice Problems

23. Evaluate $\displaystyle\int_0^{\pi/3} \int_0^{\sin y} e^{-x} \cos y \, dx \, dy$

24. Evaluate $\displaystyle\int_{-1}^{1} \int_0^z \int_y^{y-z} (x + y - z) \, dx \, dy \, dz$

25. Use a double integral to compute the area of the region R that is bounded by the x-axis and the parabola $y = 9 - x^2$.

26. Use polar coordinates to evaluate

$$\int_0^1 \int_0^{\sqrt{1-x^2}} \cos(x^2 + y^2) \, dy \, dx$$

27. MODELING PROBLEM A certain appliance consisting of two independent electronic components will be usable as long as either one of its components is still operating. The appliance carries a warranty from the manufacturer guaranteeing replacement if the appliance becomes unusable within 1 year of the date of purchase. To model this situation, let the random variables X and Y measure the life span (in years) of the first and second components, respectively, and assume that the joint probability density function for X and Y is

$$f(x, y) = \begin{cases} \frac{1}{4} e^{-x/2} e^{-y/2} & \text{if } x \geq 0, y \geq 0 \\ 0 & \text{otherwise} \end{cases}$$

Suppose the quality assurance department selects one of these appliances at random. What is the probability that the appliance will fail during the warranty period?

28. Set up and evaluate a triple integral for the volume of the solid region in the first octant that is bounded above by the plane $z = 4x$ and below by the paraboloid $z = x^2 + 2y^2$. See Figure 13.60.

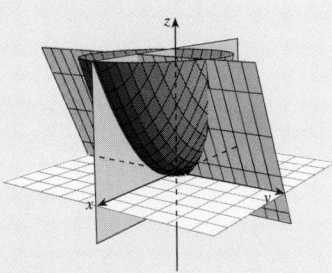

■ **FIGURE 13.60** Problem 28

29. A solid S is bounded above by the plane $z = 4$ and below by the surface $z = x^2 + y^2$. Its density $\rho(x, y, z)$ at each point P is equal to the distance from P to the z-axis. Find the total mass of S.

30. Use a linear change of variables to evaluate the double integral

$$\iint\limits_{R} (x + y) e^{x - 2y} \, dy \, dx$$

where R is the triangular region with vertices $(0, 0), (2, 0), (1, 1)$.

Supplementary Problems

In Problems 1–10, sketch the region of integration, exchange the order, and evaluate the integral using either order of integration.

1. $\int_0^1 \int_0^{3x} x^2 y^2 \, dy \, dx$ **2.** $\int_0^4 \int_0^1 \sqrt{\dfrac{y}{x}} \, dx \, dy$

3. $\int_1^2 \int_0^y \dfrac{1}{x^2 + y^2} \, dx \, dy$ **4.** $\int_0^{\pi/2} \int_0^{\pi/2} \cos(x + y) \, dx \, dy$

5. $\int_0^1 \int_0^1 x\sqrt{x^2 + y} \, dx \, dy$ **6.** $\int_0^{\pi/2} \int_0^{\sqrt{\sin x}} xy \, dy \, dx$

7. $\int_0^1 \int_{-\sqrt{1-x^2}}^0 y\sqrt{x} \, dy \, dx$ **8.** $\int_0^1 \int_{\sqrt{y}}^1 \sqrt{1 - x^3} \, dx \, dy$

9. $\int_0^1 \int_{x^2}^1 x^3 \sin y^3 \, dy \, dx$

10. $\int_0^2 \int_0^{\sqrt{4-x^2}} \sqrt{4 - x^2 - y^2} \, dy \, dx$

Evaluate the integrals given in Problems 11–36.

11. $\int_0^1 \int_{\sqrt{x}}^1 e^{y^3} \, dy \, dx$

12. $\int_0^2 \int_x^2 \dfrac{y}{(x^2 + y^2)^{3/2}} \, dy \, dx$

13. $\int_0^1 \int_{-1}^4 \int_x^y z \, dz \, dy \, dx$

14. $\int_1^2 \int_0^1 \int_0^{\sqrt{1-x^2}} e\sqrt{x^2 + y^2} \, dy \, dx \, dz$

15. $\int_0^{\pi/4} \int_0^{2\pi} \int_0^\theta r^2 \sin \phi \, dr \, d\theta \, d\phi$

16. $\int_0^1 \int_0^x \int_0^y x^2 y \, dz \, dy \, dx$

17. $\int_1^2 \int_x^{x^2} \int_0^{\ln x} xe^z \, dz \, dy \, dx$

18. $\int_0^1 \int_{1-x}^{1+x} \int_0^{xy} xz \, dz \, dy \, dx$

19. $\int_{-\sqrt{3}}^{\sqrt{3}} \int_{-\sqrt{3-y^2}}^{\sqrt{3-y^2}} \int_{(x^2+y^2)^2}^9 y^2 \, dz \, dx \, dy$

Hint: Change to cylindrical coordinates.

20. $\iint_D x^2 y \, dA$, where D is the circular disk $x^2 + y^2 \le 4$

21. $\iint_D (x^2 + y^2 + 1) \, dA$, where D is the circular disk $x^2 + y^2 \le 4$

22. $\iint_D e^{x^2+y^2} \, dA$, where D is the circular disk $x^2 + y^2 \le 4$

23. $\iint_D (x^2 + y^2)^n \, dA$, where D is the circular disk $x^2 + y^2 \le 4, n \ge 0$

24. $\iint_D \dfrac{2y}{x} \, dA$, where D is the region above the line $x + y = 1$ and below the circle $x^2 + y^2 = 1$

25. $\iint_D x^3\sqrt{4 - y^2} \, dA$, where D is the circular disk $x^2 + y^2 \le 4$

26. $\iint_D \exp\left(\dfrac{y - x}{y + x}\right) \, dy \, dx$, where D is the triangular region with vertices $(0, 0), (2, 0), (0, 2)$ *Hint:* Make a suitable change of variables.

27. $\iint_D x^3 \, dA$, where D is the shaded portion of the figure

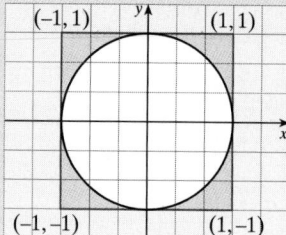

28. $\iint_D xy^2 \, dA$, where D is the shaded portion of the figure

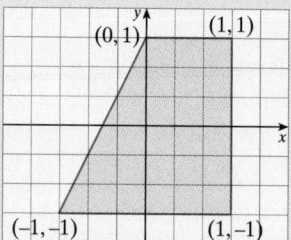

29. $\iint_P e^{-(a|x|+b|y|)} \, dA, a > 0, b > 0$, where P is the entire xy-plane. *Hint:* Use Cartesian coordinates and note that this is an improper double integral.

30. $\iint_R \dfrac{\partial^2 f}{\partial x \partial y} \, dx \, dy$, where R is the rectangle $a \le x \le b$, $c \le y \le d$ and $\dfrac{\partial^2 f}{\partial x \partial y}$ is continuous over R with $f(a, c) = 4$, $f(a, d) = -3, f(b, c) = 1, f(b, d) = 5$

31. $\iiint_S \sqrt{x^2 + y^2 + z^2} \, dV$, where S is the portion of the solid sphere $x^2 + y^2 + z^2 \le 1$ that lies in the first octant

32. $\iiint_H z^2 \, dV$, where H is the solid hemisphere $x^2 + y^2 + z^2 \le 1$ with $z \ge 0$

33. $\iiint_H \dfrac{dV}{\sqrt{x^2 + y^2 + z^2}}$, where H is the solid hemisphere $x^2 + y^2 + z^2 \le 1$, with $z \ge 0$

34. $\iiint_S \dfrac{dV}{x^2 + y^2 + z^2}$, where S is the solid region bounded below by the paraboloid $2z = x^2 + y^2$, and above by the sphere $x^2 + y^2 + z^2 = 8$

35. $\iiint_S z(x^2 + y^2)^{-1/2} \, dV$, where S is the solid bounded by the surface $2z = x^2 + y^2$ and the plane $z = 2$

36. $\iiint_S (x^4 + 2x^2 y^2 + y^4) \, dV$, where S is the solid cylinder given by $x^2 + y^2 \le a^2$, $0 \le z \le \dfrac{1}{\pi}$

37. For what values of the constant m does the improper double integral

$$\iint_R \frac{dA}{(a^2 + x^2 + y^2)^m}$$

converge, where R is the entire xy-plane?

38. Rewrite the triple integral $\displaystyle\int_0^1 \int_0^x \int_0^{\sqrt{xy}} f(x, y, z) \, dz \, dy \, dx$ as a triple integral in the order $dy \, dx \, dz$.

39. Convert the double integral

$$\int_0^{\pi/2} \int_0^{2a\cos\theta} r \sin 2\theta \, dr \, d\theta$$

to rectangular coordinates and evaluate.

40. Reverse the order of integration

$$\int_1^2 \int_{y2}^{y5} e^{x/y^2} \, dx \, dy$$

and evaluate using either order.

41. Express the integral

$$\int_0^1 \int_0^y f(x, y) \, dx \, dy + \int_1^4 \int_0^{(4-y)/3} f(x, y) \, dx \, dy$$

as a double integral with the order of integration reversed.

42. Express the integral

$$\int_0^1 \int_{-\sqrt{y}}^{\sqrt{y}} f(x, y) \, dx \, dy + \int_1^3 \int_{-1}^{2-y} f(x, y) \, dx \, dy$$

as a double integral with the order of integration reversed.

43. Find numbers a, b, and c so that

$$\int_{-\infty}^{\infty} \int_{-\infty}^{\infty} \exp[-(ax^2 + bxy + cy^2)] \, dA = 1$$

44. Find the area inside the circle $r = \cos\theta$ and outside the cardioid $r = 1 - \cos\theta$.

45. Find the area outside the cardioid $r = 1 + \cos\theta$ and inside the cardioid $r = 1 + \sin\theta$.

46. Find the area outside the small loop of $r = 1 + 2\sin\theta$ and inside the large loop.

47. Use a double integral to compute the volume of the tetrahedral region in the first octant that is bounded by the coordinate planes and the plane $3x + y + 2z = 6$.

48. Find the volume of the tetrahedron that is bounded by the coordinate planes and the plane

$$\frac{x}{a} + \frac{y}{b} + \frac{z}{c} = 1$$

with $a > 0, b > 0, c > 0$.

49. Find the volume of the solid that is bounded above by the paraboloid $z = x^2 + y^2$ and below by the square region $0 \le x \le 1, 0 \le y \le 1, z = 0$.

50. Find the volume of the solid that is bounded above by the surface $z = xy$, below by the xy-plane, and on the sides by the circular disk $x^2 + y^2 \le a^2$ $(a > 0)$ that lies in the first quadrant.

51. Find the volume of the solid that is bounded above by the paraboloid $z = 4 - x^2 - y^2$ and below by the plane $z = 4 - 2x$.

52. Find the surface area of that portion of the cone $z = \sqrt{x^2 + y^2}$ that is contained in the cylinder $x^2 + y^2 = 1$.

53. Find the surface area of the portion of the paraboloid $z = x^2 + y^2$ that lies below the plane $z = 9$.

54. Find the surface area of the portion of the cylinder $x^2 + z^2 = 4$ that is bounded by the planes $x = 0, x = 1$, $y = 0$, and $y = 2$.

55. Find the volume of the region bounded by the paraboloid $y^2 + z^2 = 2x$ and the plane $x + y = 1$.

56. Find the volume of the region bounded above by the paraboloid $z = 4 - x^2 - y^2$ and below by the plane $z + 2y = 4$.

57. Find the volume of the region bounded by the ellipsoid $z^2 = 4 - 4r^2$ and the cylinder $r = \cos\theta$.

58. Find the volume of the region bounded above by the hemisphere $z = \sqrt{9 - x^2 - y^2}$, below by the xy-plane, and on the sides by the cylinder $x^2 + y^2 = 1$.

59. Show that the solid bounded below by the cone $z = \sqrt{x^2 + y^2}$ and above by the sphere $x^2 + y^2 + z^2 = 2az$ for $a > 0$ has volume $V = \pi a^3$.

60. Find the volume of the solid region bounded above by the sphere given in spherical coordinates by $\rho = a$ $(a > 0)$ and below by the cone $\phi = \phi_0$, where $0 < \phi_0 < \frac{\pi}{2}$.

61. A homogeneous solid S is bounded above by the sphere $\rho = a$ $(a > 0)$ and below by the cone $\phi = \phi_0$, where $0 < \phi_0 < \frac{\pi}{2}$. Find the moment of inertia of S about the z-axis.

62. The region between the circles $x^2 + y^2 = 1$ and $x^2 + y^2 = 4$ with $0 \le z \le 5$ forms a "washer." Find the moment of inertia of the washer about the z-axis if the density δ is given by:
   a. $\delta = k$ (k a constant)   b. $\delta(x, y) = x^2 + y^2$
   c. $\delta(x, y) = x^2 y^2$

63. Find $\bar{x}$, the x-coordinate of the center of mass, of a triangular lamina with vertices $(0, 0)$, $(1, 0)$, and $(0, 1)$ if the density is $\delta(x, y) = x^2 + y^2$.

64. Find the mass of a cone of top radius R and height H if the density at each point P is proportional to the distance from P to the tip of the cone.

65. Find $\bar{z}$, the z-coordinate of the center of mass of a right circular cone with height H and base radius R.

66. Find the Jacobian $\dfrac{\partial(u, v, w)}{\partial(x, y, z)}$ of the change of variables $u = 2x - 3y + z$, $v = 2y - z$, $w = 2z$.

67. Find the Jacobian $\dfrac{\partial(u, v, w)}{\partial(x, y, z)}$ of the change of variables $u = x^2 + y^2 + z^2$, $v = 2y^2 + z^2$, $w = 2z^2$.

68. Let $u = 2x - y$ and $v = x + 2y$. Find the image of the unit square given by $0 \le x \le 1$, $0 \le y \le 1$.

69. Find the image under the linear transformation $u = x - y$, $v = y$ for lines of the general form $y = Cx + 1$, for constant C.

70. Suppose $u = \frac{1}{2}(x^2 + y^2)$ and $v = \frac{1}{2}(x^2 - y^2)$, with $x > 0$, $y > 0$.
   a. Find the Jacobian $\dfrac{\partial(u, v)}{\partial(x, y)}$.
   b. Solve for x and y in terms of u and v, and find the Jacobian $\dfrac{\partial(x, y)}{\partial(u, v)}$.
   c. Verify that $\dfrac{\partial(u, v)}{\partial(x, y)} \dfrac{\partial(x, y)}{\partial(u, v)} = 1$.

71. Let D be the region given by $u \ge 0$, $v \ge 0$, $1 \le u + v \le 2$.
   a. Show that under the transformation $u = x + y$, $v = x - y$, D is mapped into the region R such that $-x \le y \le x$, $\frac{1}{2} \le x \le 1$.
   b. Find the Jacobian $\dfrac{\partial(u, v)}{\partial(x, y)}$, and compute $\displaystyle\iint_D (u + v) \, du \, dv$.

72. Use a change of variables after a transformation of the form $x = Au$, $y = Bv$, $z = Cw$ to find the volume of the

region bounded by the surface $\sqrt{x} + \sqrt{2y} + \sqrt{3z} = 1$ and the coordinate planes.

73. Find the centroid of a homogeneous lamina ($\delta = 1$) that covers the part of the plane $Ax + By + Cz = 1$, $A > 0$, $B > 0$, $C > 0$ that lies in the first quadrant.

74. A homogeneous plate ($\delta = 1$) has the shape of the region in the first quadrant of the xy-plane that is bounded by the circle $x^2 + y^2 = 1$ and the lines $y = x$ and $x = 0$. Sketch the region and find $\bar{x}$, the x-coordinate of the centroid.

75. Let R be a lamina covering the region in the xy-plane that is bounded by the parabola $y = 1 - x^2$ and the positive coordinate axes. Assume the lamina has density $\delta(x, y) = xy$.
   a. Find the center of mass of the lamina.
   b. Find the moment of inertia of the plate about the z-axis.

76. A solid has the shape of the cylinder $x^2 + y^2 \le a$, with $0 \le z \le b$. Assume the solid has density $\delta(x, y, z) = x^2 + y^2$.
   a. Find the mass of the solid.
   b. Find the center of mass of the solid.

77. Use cylindrical coordinates to find the volume of the solid bounded by the circular cylinder $r = 2a \cos \theta$ $(a > 0)$, the cone $z = r$, and the xy-plane.

78. Let u be everywhere continuous in the plane, and define the functions f and g by
$$f(x) = \int_a^x u(x, y) \, dy \qquad g(y) = \int_y^b u(x, y) \, dx$$
Show that
$$\int_a^b f(x) \, dx = \int_a^b g(y) \, dy$$
Hint: Show that both integrals equal $\displaystyle\iint_D u(x, y) \, dA$ for a certain region D.

79. Find the volume of the solid region bounded by the surface
$$x^{2/3} + y^{2/3} + z^{2/3} = a^{2/3}$$
where a is a positive constant.

80. **MODELING PROBLEM**   The parking lot for a certain shopping mall has the shape shown in Figure 13.61.

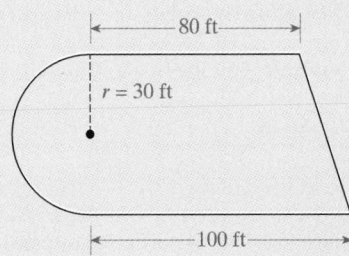

■ **FIGURE 13.61**   Shopping mall parking lot

Assuming a security observation tower can be located anywhere in the parking lot, where would you put it? State all assumptions made in setting up and analyzing your model.

81. Let f be a probability density function on a closed set R. Recall that a probability density function has the properties:

$$f(x, y) \geq 0 \quad \text{for all } (x, y) \text{ in } R$$

$$\int_{-\infty}^{\infty} \int_{-\infty}^{\infty} f(x, y)\, dA = 1$$

Show that the following functions are potential probability density functions:

a. $f(x, y) = \begin{cases} \dfrac{1}{A(R)} & \text{if } (x, y) \text{ is in } R \\ 0 & \text{if } (x, y) \text{ is not in } R \end{cases}$

b. $f(x, y) = \lambda\mu \exp(-\lambda x - \mu y)$ for $\lambda > 0$, $\mu > 0$ and for (x, y) in the first quadrant.

82. A cube of side 2 is surmounted by a hemisphere of radius 1, as shown in Figure 13.62. Suppose the origin is at the center of the hemispherical dome. If the combined solid has density $\delta = 1$, where is the center of mass?

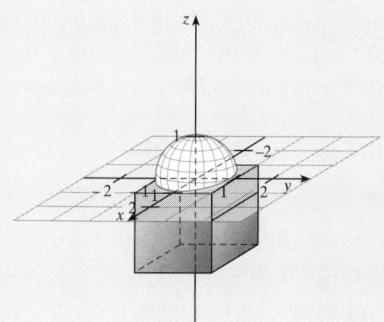

■ FIGURE 13.62   Center of mass of a complex solid

83. Set up (but do not evaluate) an integral for the mass of a lamina with density δ that covers the region outside the circle $r = \sqrt{2}\,a$ and inside the lemniscate $r^2 = 4a^2 \sin 2\theta$.

84. Find the mass of a lamina with density $\delta = r\theta$ that covers the region enclosed by the rose $r = \cos 3\theta$ for $0 \leq \theta \leq \frac{\pi}{6}$.

85. Find the centroid ($\delta = 1$) of the solid common to the cylindrical solids $x^2 + z^2 \leq 1$ and $y^2 + z^2 \leq 1$.

86. Show that if $z = f(r, \theta)$ is the equation of a surface S in polar coordinates, then the surface area of S is given by

$$\iint_R \sqrt{1 + \left(\frac{\partial z}{\partial r}\right)^2 + \frac{1}{r^2}\left(\frac{\partial z}{\partial \theta}\right)^2}\; r\, dr\, d\theta$$

where R is the projected region in the $r\theta$-plane.

87. Find the center of mass of the solid S that lies inside the sphere $x^2 + y^2 + z^2 = 1$ in the first octant $x \geq 0$, $y \geq 0$, $z \geq 0$ if the density is $\delta(x, y, z) = (x^2 + y^2 + z^2 + 1)^{-1}$.

88. Suppose we drill a square hole of side c through the center of a sphere of radius c, as shown in Figure 13.63. What is the volume of the solid that remains?

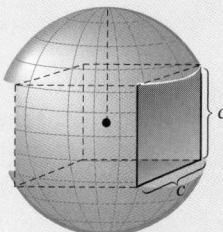

■ FIGURE 13.63   Sphere with a square hole drilled out

89. **PUTNAM EXAMINATION PROBLEM**   The function $K(x, y)$ is positive and continuous for $0 \leq x \leq 1$, $0 \leq y \leq 1$ and the functions $f(x)$ and $g(x)$ are positive and continuous for $0 \leq y \leq 1$. Suppose that for all $0 \leq x \leq 1$, we have

$$\int_0^1 f(y)K(x, y)\, dy = g(x) \quad \text{and} \quad \int_0^1 g(y)K(x, y)\, dy = f(x)$$

Show that $f(x) = g(x)$ for $0 \leq x \leq 1$.

90. **PUTNAM EXAMINATION PROBLEM**   A circle of radius a is revolved through 180° about a line in its plane, distant b from the center of the circle ($b > a$). For what value of the ratio b/a does the center of gravity of the solid thus generated lie on the surface of the solid?

91. **PUTNAM EXAMINATION PROBLEM**   For $f(x)$ a positive, monotone, decreasing function defined in $0 \leq x \leq 1$, prove that

$$\frac{\displaystyle\int_0^1 x[f(x)]^2\, dx}{\displaystyle\int_0^1 xf(x)\, dx} \leq \frac{\displaystyle\int_0^1 [f(x)]^2\, dx}{\displaystyle\int_0^1 f(x)\, dx}$$

92. **PUTNAM EXAMINATION PROBLEM**   Show that the integral equation

$$f(x, y) = 1 + \int_0^x \int_0^y f(u, v)\, du\, dv$$

has at most one solution continuous for $0 \leq x \leq 1$, $0 \leq y \leq 1$.

Space-Capsule Design

This project is to be done in groups of three or four students. Each group will submit a single written report. Reports are to be typed, technically correct, as well as showing good grammar and style.

Suppose you are transported back in time to be part of a team of engineers designing the Apollo space capsule. The capsule is composed of two parts:

1. A cone with a height of 4 meters and a base of radius 3 meters.
2. A reentry shield in the shape of a parabola revolved about the axis of the cone, which is attached to the cone along the edge of the base of the cone. Its vertex is a distance D below the base of the cone. Find values of the design parameters D and δ so that the capsule will float with the vertex of the cone pointing up and with the waterline 2 m below the top of the cone, in order to keep the exit port $1/3$ m above water.

You can make the following assumptions:

a. The capsule has uniform density δ.

b. The center of mass of the capsule should be below the center of mass of the displaced water because this will give the capsule better stability in heavy seas.

c. A body floats in a fluid at the level at which the weight of the displaced fluid equals the weight of the body (Archimedes' principle).

Your paper is not limited to the following questions but should include these concerns: Show the project director that the task is impossible; that is, there are no values of D and δ that satisfy the design specifications. However, you can solve this dilemma by incorporating a flotation collar in the shape of a torus. The collar will be made by taking hollow plastic tubing with a circular cross section of radius 1 m and wrapping it in a circular ring about the capsule, so that it fits snugly. The collar is designed to float just submerged with its top tangent to the surface of the water. Show that this flotation collar makes the capsule plus collar assembly satisfy the design specifications. Find the density δ needed to make the capsule float at the 2-meter mark. Assume the weight of the tubing is negligible compared to the weight of the capsule, that the design parameter D is equal to 1 meter, and the density of the water is 1.

MAA Notes 17 (1991), "Priming the Calculus Pump: Innovations and Resources," by Marcus S. Cohen, Edward D. Gaughan, R. Arthur Knoebel, Douglas S. Kurtz, and David J. Pengelley.

CONTENTS

14 Vector Analysis

PREVIEW

In this chapter, we draw together what we have learned about differentiation, integration, and vectors to study the calculus of vector functions defined on a set of points in $\mathbb{R}^2$ or $\mathbb{R}^3$. We introduce **line integrals,** and **surface integrals** to study such things as fluid flow and then obtain a result called **Green's theorem** that enables line integrals to be computed in terms of ordinary double integrals. This result will then be extended into $\mathbb{R}^3$ to obtain **Stokes' theorem** and the **divergence theorem,** which have extensive applications in areas such as fluid dynamics and electromagnetic theory.

PERSPECTIVE

How much work is done by a variable force acting along a given curve in space? How can the amount of heat flowing across a particular surface in unit time be measured, and is the measurement similar to measuring the flow of water or electricity? We shall use line integrals and surface integrals to answer these and other questions from physics and engineering mathematics.

14.1   Properties of a Vector Field: Divergence and Curl

IN THIS SECTION   definition of a vector field; divergence, curl, a physical interpretation of curl

DEFINITION OF A VECTOR FIELD

The satellite photograph in Figure 14.1 shows wind measurements over the Indian Oceans. Wind direction is indicated by directed line segments, whose shading (light gray to black) indicate wind speed. This is an example of a *vector field,* in which every point in a given region of the plane or space is assigned a vector that represents a velocity or force or some other vector quantity of interest. Here is the definition of a vector field in $\mathbb{R}^3$.

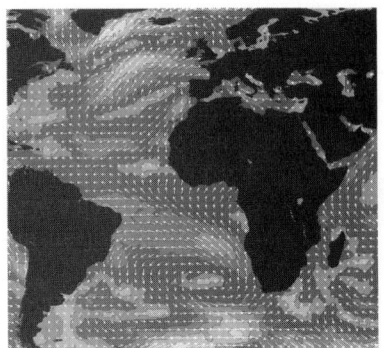

■ **FIGURE 14.1**   A wind-velocity map of the Indian Ocean

> **Vector Field**
>
> A **vector field** is a collection S of points in space together with a rule that assigns to each point (x, y, z) in S exactly one vector $\mathbf{V}(x, y, z)$.

For instance, we may represent the velocity $\mathbf{V}(x, y, z)$ of a fluid flow by drawing an appropriate vector at each point (x, y, z) in the domain of the flow, and the resulting collection of vectors is called a **velocity field.** In practice, we often express a given vector field $\mathbf{V}$ in the form

$$\mathbf{V}(x, y) = u(x, y)\mathbf{i} + v(x, y)\mathbf{j} \quad \text{in } \mathbb{R}^2$$

or

$$\mathbf{V}(x, y, z) = u(x, y, z)\mathbf{i} + v(x, y, z)\mathbf{j} + w(x, y, z)\mathbf{k} \quad \text{in } \mathbb{R}^3$$

The functions $u, v,$ and w are scalar functions called the **components** of $\mathbf{V}$. For example,

$$\mathbf{V} = 2x^2y\mathbf{i} + e^{yz}\mathbf{j} + \left(\tan\frac{x}{z}\right)\mathbf{k}$$

is a vector field with $\mathbf{i}$-component $2x^2y$, $\mathbf{j}$-component e^{yz}, and $\mathbf{k}$-component $\tan\dfrac{x}{z}$.

To get an idea of what a vector field "looks like," it is often helpful to draw $\mathbf{V}(x, y, z)$ as an "arrow" at selected points in S. We shall refer to such a diagram as the **graph of V**.

EXAMPLE 1   *Graph of a vector field*

Sketch the graph of the vector field $\mathbf{F}(x, y) = y\mathbf{i} - x\mathbf{j}$.

Solution   We shall evaluate $\mathbf{F}$ at various points. For example,

$$\mathbf{F}(3, 4) = 4\mathbf{i} - 3\mathbf{j} \quad \text{and} \quad \mathbf{F}(-1, 2) = 2\mathbf{i} - (-1)\mathbf{j} = 2\mathbf{i} + \mathbf{j}$$

We can generate as many such vector values of $\mathbf{F}$ as we wish. Several are shown in Figure 14.2.   ■

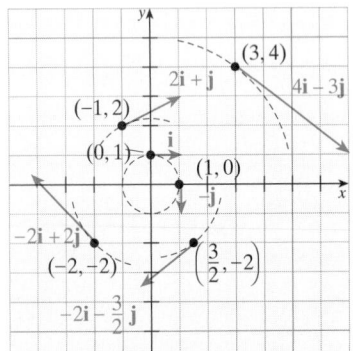

■ **FIGURE 14.2**   The graph of the vector field $\mathbf{F}(x, y) = y\mathbf{i} - x\mathbf{j}$

The graph of a vector field often yields useful information about the properties of the field. For instance, suppose $\mathbf{F}(x, y)$ represents the velocity of a compressible fluid (like a gas) at a point (x, y) in the plane. Then $\mathbf{F}$ assigns a velocity vector to each point in the plane, and the graph of $\mathbf{F}$ provides a picture of the fluid flow. Thus, the flow in Figure 14.3a is a constant, whereas Figure 14.3b suggests a circular flow.

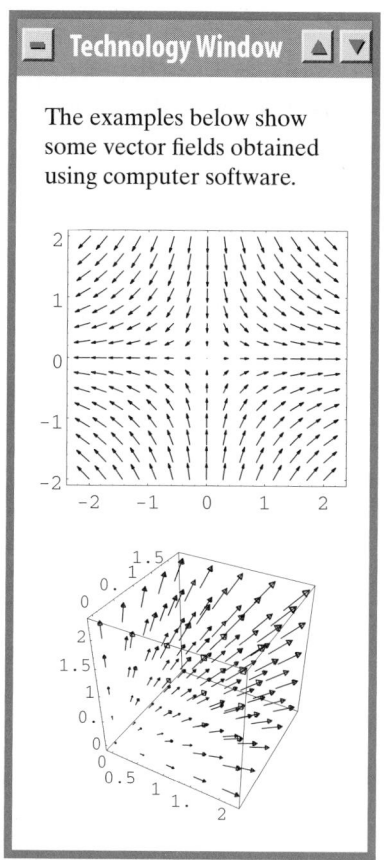
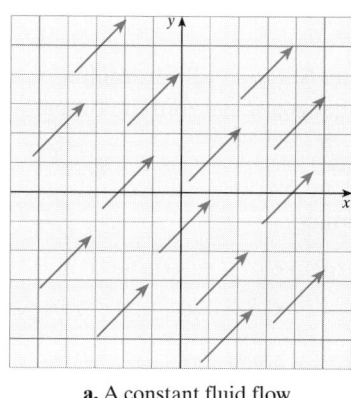

a. A constant fluid flow

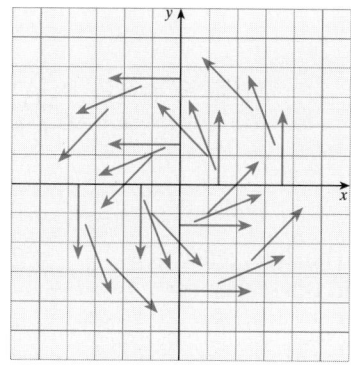

b. A circular flow

■ **FIGURE 14.3**   Flow diagrams

Gravitational, electrical, and magnetic vector fields play an important role in physical applications. We shall discuss gravitational fields now and electrical and magnetic fields later in this section. Accordingly, we begin with Newton's law of gravitation, which says that a point mass (particle) m at the origin exerts on a unit point mass located at the point $P(x, y, z)$ a force $\mathbf{F}(x, y, z)$ given by

$$\mathbf{F}(x, y, z) = \frac{Gm}{x^2 + y^2 + z^2} \mathbf{u}(x, y, z)$$

where G is a constant (the universal gravitational constant) and $\mathbf{u}$ is the unit vector extending from the point P toward the origin. The vector field $\mathbf{F}(x, y, z)$ is called the **gravitational field** of the point mass m. Because

$$\mathbf{u}(x, y, z) = \frac{-1}{\sqrt{x^2 + y^2 + z^2}} (x\mathbf{i} + y\mathbf{j} + z\mathbf{k})$$

it follows that

$$\mathbf{F}(x, y, z) = \frac{-Gm}{(x^2 + y^2 + z^2)^{3/2}} (x\mathbf{i} + y\mathbf{j} + z\mathbf{k})$$

Note that the gravitational field $\mathbf{F}$ always points toward the origin and has the same magnitude for any point m located $r = \sqrt{x^2 + y^2 + z^2}$ units from the origin. Such a vector field is called a **central force field.** This force field is shown in Figure 14.4a. Some other examples of velocity fields are shown in Figures 14.4b and 14.4c.

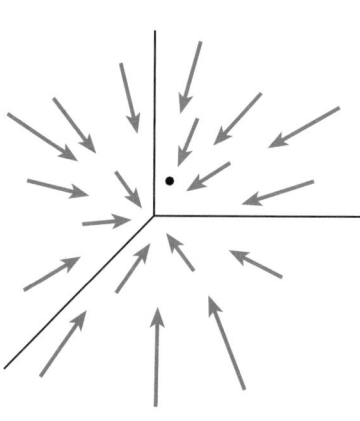

a. A central force field

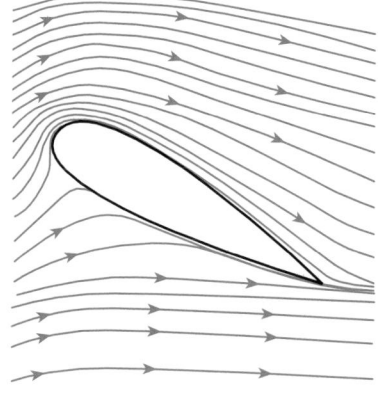

b. Air flow vector

c. Wind velocity on a map

■ **FIGURE 14.4**   Examples of physical vector fields

DIVERGENCE

A vector field **V** can be differentiated in two different ways, one of which produces a derivative that is a scalar and the other, a vector. The scalar derivative may be defined as follows.

Divergence

The **divergence** of a vector field

$$\mathbf{V}(x, y, z) = u(x, y, z)\mathbf{i} + v(x, y, z)\mathbf{j} + w(x, y, z)\mathbf{k}$$

is denoted by div **V** and is given by

$$\text{div } \mathbf{V} = \frac{\partial u}{\partial x}(x, y, z) + \frac{\partial v}{\partial y}(x, y, z) + \frac{\partial w}{\partial z}(x, y, z)$$

EXAMPLE 2   *Finding div **V***

Find the divergence of each of the following vector fields.
a. $\mathbf{F}(x, y) = x^2 y\mathbf{i} + xy^3\mathbf{j}$ b. $\mathbf{G}(x, y, z) = x\mathbf{i} + y^3 z^2\mathbf{j} + xz^3\mathbf{k}$

Solution

a. div $\mathbf{F} = \dfrac{\partial}{\partial x}(x^2 y) + \dfrac{\partial}{\partial y}(xy^3) = 2xy + 3xy^2$

b. div $\mathbf{G} = \dfrac{\partial}{\partial x}(x) + \dfrac{\partial}{\partial y}(y^3 z^2) + \dfrac{\partial}{\partial z}(xz^3) = 1 + 3y^2 z^2 + 3xz^2$ ∎

Suppose the vector field

$$\mathbf{V}(x, y, z) = u(x, y, z)\mathbf{i} + v(x, y, z)\mathbf{j} + w(x, y, z)\mathbf{k}$$

represents the velocity of a fluid with density $\delta(x, y, z)$ at a point (x, y, z) in a certain region R in $\mathbb{R}^3$. Then the vector field $\delta\mathbf{V}$ is called the **flux density** and is denoted by **D**. We can think of $\mathbf{D} = \delta\mathbf{V}$ as measuring the "mass flow" of the liquid.

Assuming there are no external processes acting on the fluid that would tend to create or destroy fluid, it can be shown that div **D** gives the negative of the time rate change of the density, that is,

$$\text{div } \mathbf{D} = -\frac{\partial \delta}{\partial t}$$

This is often referred to as the **continuity equation** of fluid dynamics. (A derivation is given in Section 14.7.) When div $\mathbf{D} = 0$, **D** is said to be **divergence free** or **solenoidal**, and the fluid is said to be **incompressible**. If div $\mathbf{D} > 0$ at a point (x_0, y_0, z_0), then the point is called a **source**; if div $\mathbf{D} < 0$, the point is called a **sink** (see Figure 14.5).

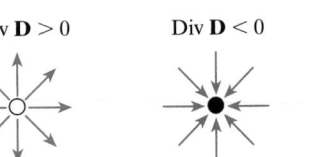

Div **D** > 0	Div **D** < 0	Div **D** = 0
Fluid arrives through a *source point*.	Fluid leaves through a *sink point*.	Fluid is *incompressible*.

■ **FIGURE 14.5** Flow of a fluid across a plane region, **D**

The terms *sink, source,* and *incompressible* apply to any vector field **F** and are not reserved only for fluid applications.

A useful way to think of the divergence div **V** is in terms of the **del operator** defined by

$$\nabla = \frac{\partial}{\partial x}\mathbf{i} + \frac{\partial}{\partial y}\mathbf{j} + \frac{\partial}{\partial z}\mathbf{k}$$

The del operator *operates* on a differentiable function $f(x, y, z)$ of three variables, and for such a function, we have

$$\nabla f = \frac{\partial f}{\partial x}\mathbf{i} + \frac{\partial f}{\partial y}\mathbf{j} + \frac{\partial f}{\partial z}\mathbf{k}$$

We can also define ∇f in a totally analogous way for functions of one, two, or more than three variables. In particular, for a differentiable function $f(x, y)$ of two variables, we have

$$\nabla f = \frac{\partial f}{\partial x}\mathbf{i} + \frac{\partial f}{\partial y}\mathbf{j}$$

We recognize ∇f as the **gradient** of f, which was introduced in Chapter 12. We can think of div **V** as the *dot product* of the operator ∇ and the vector field $\mathbf{V} = u\mathbf{i} + v\mathbf{j} + w\mathbf{k}$; that is,

$$\text{div } \mathbf{V} = \frac{\partial u}{\partial x} + \frac{\partial v}{\partial y} + \frac{\partial w}{\partial z} = \left(\frac{\partial}{\partial x}\mathbf{i} + \frac{\partial}{\partial y}\mathbf{j} + \frac{\partial}{\partial z}\mathbf{k}\right) \cdot (u\mathbf{i} + v\mathbf{j} + w\mathbf{k}) = \nabla \cdot \mathbf{V}$$

CURL

The del operator may also be used to describe another derivative operation for vector fields, called the *curl.*

Curl

The **curl** of a vector field

$$\mathbf{V}(x, y, z) = u(x, y, z)\mathbf{i} + v(x, y, z)\mathbf{j} + w(x, y, z)\mathbf{k}$$

is denoted by curl **V** and is defined by

$$\text{curl } \mathbf{V} = \left(\frac{\partial w}{\partial y} - \frac{\partial v}{\partial z}\right)\mathbf{i} + \left(\frac{\partial u}{\partial z} - \frac{\partial w}{\partial x}\right)\mathbf{j} + \left(\frac{\partial v}{\partial x} - \frac{\partial u}{\partial y}\right)\mathbf{k}$$

This can be remembered in the following useful form:

$$\text{curl } \mathbf{V} = \begin{vmatrix} \mathbf{i} & \mathbf{j} & \mathbf{k} \\ \frac{\partial}{\partial x} & \frac{\partial}{\partial y} & \frac{\partial}{\partial z} \\ u & v & w \end{vmatrix}$$

We saw that $\nabla \cdot \mathbf{V} = \text{div } \mathbf{V}$, and from this definition of curl, we see that

$$\text{curl } \mathbf{V} = \left(\frac{\partial w}{\partial y} - \frac{\partial v}{\partial z}\right)\mathbf{i} + \left(\frac{\partial u}{\partial z} - \frac{\partial w}{\partial x}\right)\mathbf{j} + \left(\frac{\partial v}{\partial x} - \frac{\partial u}{\partial y}\right)\mathbf{k} = \nabla \times \mathbf{V}$$

This representation for curl and the one for divergence stated earlier are summarized in the following box.

Del Operator Forms for Divergence and Curl

Consider a vector field

$$\mathbf{V}(x, y, z) = u(x, y, z)\mathbf{i} + v(x, y, z)\mathbf{j} + w(x, y, z)\mathbf{k}$$

The divergence and curl of $\mathbf{V}$ are given by

$$\operatorname{div}\mathbf{V} = \nabla \cdot \mathbf{V} \quad\text{and}\quad \operatorname{curl}\mathbf{V} = \nabla \times \mathbf{V}$$

WARNING div is a scalar and curl is a vector. ←

EXAMPLE 3 *Curl of a vector field*

Find the curl of the vector fields

$$\mathbf{F} = x^2yz\mathbf{i} + xy^2z\mathbf{j} + xyz^2\mathbf{k} \quad\text{and}\quad \mathbf{G} = (x\cos y)\mathbf{i} + xy^2\mathbf{j}$$

Solution

$$\operatorname{curl}\mathbf{F} = \begin{vmatrix} \mathbf{i} & \mathbf{j} & \mathbf{k} \\ \dfrac{\partial}{\partial x} & \dfrac{\partial}{\partial y} & \dfrac{\partial}{\partial z} \\ x^2yz & xy^2z & xyz^2 \end{vmatrix}$$

$$= \left(\frac{\partial}{\partial y}xyz^2 - \frac{\partial}{\partial z}xy^2z\right)\mathbf{i} - \left(\frac{\partial}{\partial x}xyz^2 - \frac{\partial}{dz}x^2yz\right)\mathbf{j}$$

$$+ \left(\frac{\partial}{\partial x}xy^2z - \frac{\partial}{\partial y}x^2yz\right)\mathbf{k}$$

$$= (xz^2 - xy^2)\mathbf{i} + (x^2y - yz^2)\mathbf{j} + (y^2z - x^2z)\mathbf{k}$$

$$\operatorname{curl}\mathbf{G} = \begin{vmatrix} \mathbf{i} & \mathbf{j} & \mathbf{k} \\ \dfrac{\partial}{\partial x} & \dfrac{\partial}{\partial y} & \dfrac{\partial}{\partial z} \\ x\cos y & xy^2 & 0 \end{vmatrix}$$

$$= \left[0 - \frac{\partial}{\partial z}xy^2\right]\mathbf{i} - \left[0 - \frac{\partial}{\partial z}(x\cos y)\right]\mathbf{j} + \left[\frac{\partial}{\partial x}xy^2 - \frac{\partial}{\partial y}(x\cos y)\right]\mathbf{k}$$

$$= (y^2 + x\sin y)\mathbf{k} \quad\blacksquare$$

EXAMPLE 4 *A constant vector field has divergence and curl zero*

Let $\mathbf{F}$ be a constant vector field. Show $\operatorname{div}\mathbf{F} = 0$ and $\operatorname{curl}\mathbf{F} = \mathbf{0}$.

Solution Let $\mathbf{F} = a\mathbf{i} + b\mathbf{j} + c\mathbf{k}$ for constants $a, b,$ and c. Then

$$\operatorname{div}\mathbf{F} = \frac{\partial}{\partial x}(a) + \frac{\partial}{\partial y}(b) + \frac{\partial}{\partial z}(c) = 0$$

$$\operatorname{curl}\mathbf{F} = \begin{vmatrix} \mathbf{i} & \mathbf{j} & \mathbf{k} \\ \dfrac{\partial}{\partial x} & \dfrac{\partial}{\partial y} & \dfrac{\partial}{\partial z} \\ a & b & c \end{vmatrix} = 0\mathbf{i} - 0\mathbf{j} + 0\mathbf{k} = \mathbf{0} \quad\blacksquare$$

WARNING Example 4 shows that the divergence and curl of a constant vector field are zero, but this does *not* mean that if the divergence or curl vanishes, then the associated vector field is a constant. For instance, the nonconstant vector field

$$\mathbf{F}(x, y, z) = yz\mathbf{i} + x\mathbf{j} + xy^2\mathbf{k}$$

has div $\mathbf{F} = 0$. ←

A PHYSICAL INTERPRETATION OF CURL

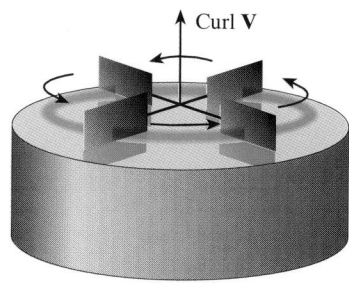

■ **FIGURE 14.6**   A physical interpretation of curl **V**

The terms *divergence* and *curl,* as well as other terminology in this chapter, are motived by concepts from physics. For example, we can apply the curl to the study of fluid motion. If a fluid is moving about a region in the xy-plane, the curl can be thought of as the circulation of the fluid. A good way to measure the effect (magnitude and direction) of the circulation is to place a small paddle wheel into the fluid. The curl measures the rate of the fluid's rotation at the point P where the paddle wheel is placed in the direction of the paddle wheel's axis. The curl is positive when the rotation is counterclockwise and negative when it is clockwise.

Let $\mathbf{V}(x, y, z) = u(x, y, z)\mathbf{i} + v(x, y, z)\mathbf{j} + w(x, y, z)\mathbf{k}$ be the velocity of an incompressible fluid, and suppose a paddle wheel is inserted into the fluid in such a way that its axis lies along the z-axis, as shown in Figure 14.6. Disregard the weight of the paddle. The fluid tends to swirl about the z-axis, causing the paddle to rotate, and we can analyze the fluid's swirling motion by studying the motion of the paddle. It turns out that the circulation per unit of the liquid

about the x-axis is proportional to $\left(\dfrac{\partial w}{\partial y} - \dfrac{\partial v}{\partial z}\right)$;

about the y-axis is proportional to $\left(\dfrac{\partial u}{\partial z} - \dfrac{\partial w}{\partial x}\right)$;

about the z-axis is proportional to $\left(\dfrac{\partial v}{\partial x} - \dfrac{\partial u}{\partial y}\right)$.

Thus, the tendency of the fluid to swirl is measured by curl **V**. In the special case where curl $\mathbf{V} = \mathbf{0}$, the fluid has no rotational motion and is said to be **irrotational.**

Combinations of the gradient, divergence, and curl appear in a variety of applications. In particular, note that if f is a differentiable scalar function, its gradient ∇f is a vector field, and we can compute

$$\text{div } \nabla f = \left(\frac{\partial}{\partial x}\mathbf{i} + \frac{\partial}{\partial y}\mathbf{j} + \frac{\partial}{\partial z}\mathbf{k}\right) \cdot \left(\frac{\partial f}{\partial x}\mathbf{i} + \frac{\partial f}{\partial y}\mathbf{j} + \frac{\partial f}{\partial z}\mathbf{k}\right)$$

$$= \frac{\partial^2 f}{\partial x^2} + \frac{\partial^2 f}{\partial y^2} + \frac{\partial^2 f}{\partial z^2}$$

$$= \nabla \cdot \nabla f$$

We summarize this definition and introduce some terminology in the following box.

> ## The Laplacian Operator
>
> Let $f(x, y, z)$ define a function with continuous first and second partial derivatives. Then the **Laplacian of** f is
>
> $$\nabla^2 f = \nabla \cdot \nabla f = \frac{\partial^2 f}{\partial x^2} + \frac{\partial^2 f}{\partial y^2} + \frac{\partial^2 f}{\partial z^2} = f_{xx} + f_{yy} + f_{zz}$$
>
> The equation $\nabla^2 f = 0$ is called **Laplace's equation,** and a function that satisfies such an equation in a region D is said to be **harmonic** in D.

> ■ *What This Says:* The notation $\nabla \cdot \nabla f$ is usually abbreviated by writing $\nabla^2 f$, and it is called either "del-squared f" or the *Laplacian of* f, after the French mathematician Pierre Laplace (see Historical Quest, Problem 57, Section 13.3). A function f whose first and second partials are continuous and that satisfies the equation $\nabla^2 f = 0$ is called *harmonic*.

EXAMPLE 5   *Showing that a function is harmonic*

Show that $f(x, y) = e^x \cos y$ is harmonic in the plane.

Solution
$$f_x(x, y) = e^x \cos y \quad \text{and} \quad f_{xx}(x, y) = e^x \cos y$$
$$f_y(x, y) = -e^x \sin y \quad \text{and} \quad f_{yy}(x, y) = -e^x \cos y$$

The Laplacian of f is given by

$$f(x, y) = f_{xx}(x, y) + f_{yy}(x, y) = e^x \cos y - e^x \cos y = 0$$

Thus, f is harmonic.   ■

In many ways, the study of electricity and magnetism is analogous to that of fluid dynamics, and the curl and divergence play an important role in this study. In electromagnetic theory, it is often convenient to regard interaction between electrical charges as forces somewhat like the gravitational force between masses and then to seek quantitative measure of these forces.

One of the great scientific achievements of the 19th century was the discovery of the laws of electromagnetism by the English scientist James Clerk Maxwell (see Historical Quest, Section 14.7, Problem 30). These laws have an elegant expression in terms of the divergence and curl. It is known empirically that the force acting on a charge due to an electromagnetic field depends on the position, velocity, and amount of the particular charge, and not on the number of other charges that may be present or how those other charges are moving. Suppose a charge is located at the point (x, y, z) at time t, and consider the electric intensity field $\mathbf{E}(x, y, z, t)$ and the magnetic intensity field $\mathbf{H}(x, y, z, t)$. Then the behavior of the resulting electromagnetic field is determined by

$$\text{div } \mathbf{E} = \frac{Q}{\epsilon} \qquad \text{curl } \mathbf{E} = -\frac{\partial}{\partial t}(\mu \mathbf{H}) \qquad \text{div}(\mu \mathbf{H}) = \mathbf{0}$$

where Q is the *electric charge density* (charge per unit volume), $\mathbf{J}$ is the *electric current density* (rate at which the charge flows through a unit area per second), and μ and ϵ are constants called the *permeability* and *permittivity,* respectively. Furthermore, if σ is a constant called the *conductivity,* then

$$(\nabla \cdot \nabla)\mathbf{E} = \mu \sigma \frac{\partial \mathbf{E}}{\partial t} + \mu \epsilon \frac{\partial^2 \mathbf{E}}{\partial t^2}$$

Working with these equations and terms is beyond the scope of this course, but if you are interested there are many references you can consult. One of the best (in spite of the fact it is 30 years old) is the classic book written by Nobel laureate Richard Feynman, Robert Leighton, and Matthew Sands, *Feynman Lectures in Physics* (Reading, MA: Addison-Wesley, 1963).

Technology Window

Derive, Maple, Mathlab, and *Mathematica* will find the Laplacian of a given function. For Example 5, we obtain

$$\text{LAPLACIAN}(e^\wedge x \cos(y)), \text{ which simplifies to } 0$$

If you have access to this technology, verify that $\text{LAPLACIAN}(x^2y^3z)$ simplifies to $6x^2yz + 2y^3z$.

14.1 Problem Set

 1. ■ **What Does This Say?** Compare and contrast divergence and curl.

2. ■ **What Does This Say?** Discuss the del operator.

Sketch several representatives of the given vector field in Problems 3–8.

3. $\mathbf{F} = x\mathbf{i} + y\mathbf{j}$

4. $\mathbf{G} = -x\mathbf{i} + y\mathbf{j}$

5. $\mathbf{V} = x^2\mathbf{i} + y\mathbf{j}$

6. $\mathbf{H} = x^2\mathbf{i} + y^2\mathbf{j}$

7. $\mathbf{F}(x, y) = y\mathbf{i} + x\mathbf{j}$

8. $\mathbf{G}(x, y) = xy\mathbf{i} - \mathbf{j}$

In Problems 9–12, find div $\mathbf{F}$ *and* curl $\mathbf{F}$ *for the given vector function.*

9. $\mathbf{F}(x, y) = x^2\mathbf{i} + xy\mathbf{j} + z^3\mathbf{k}$ **10.** $\mathbf{F}(x, y) = \mathbf{i} + (x^2 + y^2)\mathbf{j}$

11. $\mathbf{F}(x, y, z) = 2y\mathbf{j}$ **12.** $\mathbf{F}(x, y, z) = z\mathbf{i} - \mathbf{j} + 2y\mathbf{k}$

In Problems 13–18, find div $\mathbf{F}$ *and* curl $\mathbf{F}$ *for each vector field* $\mathbf{F}$ *at the given point.*

13. $\mathbf{F}(x, y, z) = \mathbf{i} + \mathbf{j} + \mathbf{k}$ at $(2, -1, 3)$

14. $\mathbf{F}(x, y, z) = xz\mathbf{i} + y^2z\mathbf{j} + xz\mathbf{k}$ at $(1, -1, 2)$

15. $\mathbf{F}(x, y, z) = xyz\mathbf{i} + y\mathbf{j} + x\mathbf{k}$ at $(1, 2, 3)$

16. $\mathbf{F}(x, y, z) = (\cos y)\mathbf{i} + (\sin y)\mathbf{j} + \mathbf{k}$ at $(\frac{\pi}{4}, \pi, 0)$

17. $\mathbf{F}(x, y, z) = e^{-xy}\mathbf{i} + e^{xz}\mathbf{j} + e^{yz}\mathbf{k}$ at $(3, 2, 0)$

18. $\mathbf{F}(x, y, z) = (e^{-x}\sin y)\mathbf{i} + (e^{-x}\cos y)\mathbf{j} + \mathbf{k}$ at $(1, 3, -2)$

Find div $\mathbf{F}$ *and* curl $\mathbf{F}$ *for each vector field* $\mathbf{F}$ *given in Problems 19–34.*

19. $\mathbf{F} = (\sin x)\mathbf{i} + (\cos y)\mathbf{j}$ **20.** $\mathbf{F} = (-\cos x)\mathbf{i} + (\sin y)\mathbf{j}$

21. $\mathbf{F} = x\mathbf{i} - y\mathbf{j}$ **22.** $\mathbf{F} = -x\mathbf{i} + y\mathbf{j}$

23. $\mathbf{F} = \dfrac{x}{\sqrt{x^2 + y^2}}\mathbf{i} + \dfrac{y}{\sqrt{x^2 + y^2}}\mathbf{j}$

24. $\mathbf{F} = x^2\mathbf{i} - y^2\mathbf{j}$

25. $\mathbf{F} = ax\mathbf{i} + by\mathbf{j} + c\mathbf{k}$ for constants a, b, c

26. $\mathbf{F} = (e^x\sin y)\mathbf{i} + (e^x\cos y)\mathbf{j} + \mathbf{k}$

27. $\mathbf{F} = x^2\mathbf{i} + y^2\mathbf{j} + z^2\mathbf{k}$ **28.** $\mathbf{F} = y\mathbf{i} + z\mathbf{j} + x\mathbf{k}$

29. $\mathbf{F} = xy\mathbf{i} + yz\mathbf{j} + xz\mathbf{k}$ **30.** $\mathbf{F} = 2xz\mathbf{i} + 2yz^2\mathbf{j} - \mathbf{k}$

31. $\mathbf{F} = xyz\mathbf{i} + x^2y^2z^2\mathbf{j} + y^2z^3\mathbf{k}$

32. $\mathbf{F} = -z^3\mathbf{i} + 3\mathbf{j} + 2y\mathbf{k}$

33. $\mathbf{F} = (x - y)\mathbf{i} + (y - z)\mathbf{j} + (z - x)\mathbf{k}$

34. $\mathbf{F} = \dfrac{x\mathbf{i} + y\mathbf{j} + z\mathbf{k}}{\sqrt{x^2 + y^2 + z^2}}$

Determine whether each scalar function in Problems 35–38 is harmonic.

35. $u(x, y, z) = e^{-x}(\cos y - \sin y)$

36. $v(x, y, z) = (x^2 + y^2 + z^2)^{1/2}$

37. $w(x, y, z) = (x^2 + y^2 + z^2)^{-1/2}$

38. $r(x, y, z) = xyz$

Ⓑ 39. If $\mathbf{F}(x, y, z) = 2\mathbf{i} + 2x\mathbf{j} + 3y\mathbf{k}$ and $\mathbf{G}(x, y, z) = x\mathbf{i} - y\mathbf{j} + z\mathbf{k}$, find curl$(\mathbf{F} \times \mathbf{G})$.

40. If $\mathbf{F}(x, y, z) = xy\mathbf{i} + yz\mathbf{j} + z^2\mathbf{k}$ and $\mathbf{G}(x, y, z) = x\mathbf{i} + y\mathbf{j} - z\mathbf{k}$, find curl$(\mathbf{F} \times \mathbf{G})$.

41. If $\mathbf{F}(x, y, z) = 2\mathbf{i} + 2x\mathbf{j} + 3y\mathbf{k}$ and $\mathbf{G}(x, y, z) = x\mathbf{i} - y\mathbf{j} + z\mathbf{k}$, find div$(\mathbf{F} \times \mathbf{G})$.

42. If $\mathbf{F}(x, y, z) = xy\mathbf{i} + yz\mathbf{j} + z^2\mathbf{k}$ and $\mathbf{G}(x, y, z) = x\mathbf{i} + y\mathbf{j} - z\mathbf{k}$, find div$(\mathbf{F} \times \mathbf{G})$.

43. Find div $\mathbf{F}$, given that $\mathbf{F} = \nabla f$, where $f(x, y, z) = xy^3z^2$.

44. Find div $\mathbf{F}$, given that $\mathbf{F} = \nabla f$, where $f(x, y, z) = x^2yz^3$.

45. A magnetic field that has zero divergence everywhere is said to be *solenoidal* (because such a field can be

generated by a solenoid). Show that the field
$\mathbf{B} = y^2 z \mathbf{i} + xz^3 \mathbf{j} + y^2 x^2 \mathbf{k}$ is solenoidal.

46. A **central force field F** is one that can be described as
$\mathbf{F} = f(r)\mathbf{R}$, where $\mathbf{R} = x\mathbf{i} + y\mathbf{j} + z\mathbf{k}$ and
$r = \|\mathbf{R}\| = (x^2 + y^2 + z^2)^{1/2}$. Show that such a field is
irrotational everywhere it is defined and differentiable.
That is, show curl $\mathbf{F} = \mathbf{0}$.

C 47. Let $\mathbf{A}$ be a constant vector and let $\mathbf{R} = x\mathbf{i} + y\mathbf{j} + z\mathbf{k}$.
Show that div$(\mathbf{A} \times \mathbf{R}) = 0$.

48. Let $\mathbf{A}$ be a constant vector and let $\mathbf{R} = x\mathbf{i} + y\mathbf{j} + z\mathbf{k}$.
Show that curl$(\mathbf{A} \times \mathbf{R}) = 2\mathbf{A}$.

49. If $\mathbf{F}(x, y) = \mathbf{u}(x, y)\mathbf{i} + \mathbf{v}(x, y)\mathbf{j}$, show that curl $\mathbf{F} = \mathbf{0}$ if
and only if

$$\frac{\partial u}{\partial y} = \frac{\partial v}{\partial x}$$

50. Let $\mathbf{F} = \mathbf{R}/r^3$, where $\mathbf{R} = x\mathbf{i} + y\mathbf{j} + z\mathbf{k}$ and $r = \|\mathbf{R}\|$.
Show that div $\mathbf{F} = 0$ and curl $\mathbf{F} = \mathbf{0}$.

51. Consider a rigid body that is rotating about the z-axis
(counterclockwise from above) with constant angular
velocity ω. If P is a point in the body located at
$\mathbf{R} = x\mathbf{i} + y\mathbf{j} + z\mathbf{k}$, the velocity at P is given by the vector
field $\mathbf{V} = \omega \times \mathbf{R}$.
a. Express $\mathbf{V}$ in terms of $\mathbf{i}, \mathbf{j}$, and $\mathbf{k}$ vectors.
b. Find div $\mathbf{V}$ and curl $\mathbf{V}$.

52. **THINK TANK PROBLEM** Find an expression for
curl(curl $\mathbf{F}$) for any vector field $\mathbf{F}$.

53. Which (if any) of the following is the same as div$(\mathbf{F} \times \mathbf{G})$
for any vector fields $\mathbf{F}$ and $\mathbf{G}$?
I. (div $\mathbf{F}$)(div $\mathbf{G}$)
II. (curl $\mathbf{F}$) $\cdot$ $\mathbf{G}$ $-$ $\mathbf{F}$ $\cdot$ (curl $\mathbf{G}$)
III. $\mathbf{F}$(div $\mathbf{G}$) $+$ (div $\mathbf{F}$)$\mathbf{G}$
IV. (curl $\mathbf{F}$) $\cdot$ $\mathbf{G}$ $+$ $\mathbf{F}$ $\cdot$ (curl $\mathbf{G}$)

*In Problems 54–64, prove the given property for the vector
fields F and G, scalar c, and scalar function f. Assume that the
required partial derivatives are continuous.*

54. div$(c\mathbf{F}) = c$ div $\mathbf{F}$
55. div$(\mathbf{F} + \mathbf{G}) = $ div $\mathbf{F} + $ div $\mathbf{G}$
56. curl$(\mathbf{F} + \mathbf{G}) = $ curl $\mathbf{F} + $ curl $\mathbf{G}$
57. curl$(c\mathbf{F}) = c$ curl $\mathbf{F}$
58. curl$(f\mathbf{F}) = f$ curl $\mathbf{F} + (\nabla f \times \mathbf{F})$
59. div$(f\mathbf{F}) = f$ div $\mathbf{F} + (\nabla f \cdot \mathbf{F})$
60. curl$(\nabla f + $ curl $\mathbf{F}) = $ curl$(\nabla f) + $ curl(curl $\mathbf{F}$)
61. div$(f\nabla g) = f$ div $\nabla g + \nabla f \cdot \nabla g$
62. The curl of the gradient of a function is always $\mathbf{0}$. That is,
$\nabla \times (\nabla f) = \mathbf{0}$.
63. The divergence of the curl of a vector field is 0. That is,
div(curl $\mathbf{F}$) = 0.

14.2   Line Integrals

IN THIS SECTION   **definition of a line integral, evaluation of line integrals in parametric form, line integrals of vector fields, computing work using line integrals, evaluation of line integrals with respect to arc length**

In Section 6.5, we showed that the single Riemann integral can be used to compute the work done when an object moves along a line segment against a given force, but what if the object moves along a curve in space? This is one of many situations that occur in science and engineering in which a more general definition of integration is required.

DEFINITION OF A LINE INTEGRAL

To model certain physical notions, such as work or potential, it is appropriate to generalize the original concept of integral by considering limits of sums whose summands involve partitions of a curve, called the *path of integration*. This leads us to the concept of a *line integral,* which is really integration along a curve in space. We begin by defining the line integral of a scalar function f along a curve C with respect to x. The line integral of a scalar function g along C with respect to y or the line integral of h with respect to z may be defined in an analogous fashion.

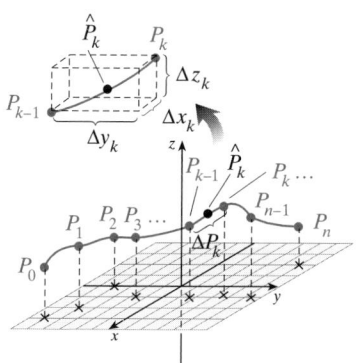

■ **FIGURE 14.7** A partitioned smooth curve

First, we need to introduce some terminology regarding curves. Recall (Section 11.2) that a curve C with the parametric representation $R(t) = x(t)\mathbf{i} + y(t)\mathbf{j} + z(t)\mathbf{k}$ is said to be *smooth* on the interval $t_1 < t < t_2$ if the derivatives $x'(t)$, $y'(t)$, and $z'(t)$ are all continuous and not all simultaneously zero at any point t on the interval. More generally, C is **piecewise smooth** if it can be decomposed into a finite number of smooth parts. Also, C is said to be **orientable** if it is possible to describe direction along the curve for increasing t.

Assume C is a piecewise smooth, orientable curve that begins at $P = P_0$ and ends at $Q = P_n$, as shown in Figure 14.7. Let C be partitioned into n pieces by the points $P_0, P_1, \ldots, P_n$, and let (x_k, y_k, z_k) be the coordinates of the point P_k. Finally, for $k = 1, 2, \ldots, n$, let $\hat{P}_x(x_k^*, y_k^*, z_k^*)$ be a point chosen arbitrarily from the arc joining points P_{k-1} and P_k, and let $\Delta x_k = x_k - x_{k-1}$. We shall refer to the largest of the Δx_k as the **x-norm** of the partition and shall denote this norm by Δx. For a given scalar function f along the curve C with respect to x, we form the sum

$$\sum_{k=1}^{n} f(\hat{P}_k)\Delta x_k$$

and define the line integral $\int_C f\, dx$ over C with respect to x as follows.

Line Integral

The **line integral** $\int_C f\, dx$ of the scalar function f with respect to x along the piecewise smooth curve C is given by

$$\int_C f\, dx = \lim_{\Delta x \to 0} \sum_{k=1}^{n} f(\hat{P}_k)\Delta x_k$$

provided this limit exists.

The line integrals $\int_C f\, dy$ and $\int_C f\, dz$ are defined similarly.

THEOREM 14.1   *Properties of line integrals*

Let f be a given scalar function defined with respect to x on a piecewise smooth, orientable curve C. Then, for any constant k:

Constant multiple rule: $\displaystyle\int_C kf\, dx = k\int_C f\, dx$

Sum rule: $\displaystyle\int_C (f_1 + f_2)\, dx = \int_C f_1\, dx + \int_C f_2\, dx$

where f_1 and f_2 are scalar functions defined with respect to x on C.

Opposite direction rule: $\displaystyle\int_C f\, dx = -\int_C f\, dx$

where $-C$ denotes the curve C traversed in the opposite direction.

Subdivision rule: $\displaystyle\int_C f\, dx = \int_{C_1} f\, dx + \int_{C_2} f\, dx$

where C is subdivided into subarcs $C_1 \cup C_2$ (with $C_1 \cap C_2 = \varnothing$). This property generalizes to any finite number of subdivisions.

Similar properties hold for line integrals of the form $\int_C g\, dy$ or $\int_C h\, dz$.

Proof   The proof follows directly from the properties of limits and the definition of a line integral.   ▬▬

EVALUATION OF LINE INTEGRALS IN PARAMETRIC FORM

In practice, the line integral $\int_C f\,dx$ is almost never evaluated using the definition. Instead, we note that if the integrand $f(x, y, z)$ is continuous on C and if C can be represented parametrically in vector form by $\mathbf{R}(t) = x(t)\mathbf{i} + y(t)\mathbf{j} + z(t)\mathbf{k}$, where the derivative $\mathbf{R}'(t)$ exists and is not zero for $a \le t \le b$, then

$$\int_C f\,dx = \int_a^b f[x(t), y(t), z(t)]\frac{dx}{dt}\,dt$$

Similarly, if g and h are continuous on C, then

$$\int_C g\,dy = \int_a^b g[x(t), y(t), z(t)]\frac{dy}{dt}\,dt$$

$$\int_C h\,dz = \int_a^b h[x(t), y(t), z(t)]\frac{dz}{dt}\,dt$$

These formulas enable us to convert a line integral into "ordinary" Riemann integrals that may be evaluated by the methods developed earlier in this text. The same answer is obtained independently of the parametrization chosen.

EXAMPLE 1   *Evaluating a line integral*

Let C be the portion of the parabola $y = x^2$ between $(0, 0)$ to $(2, 4)$. Evaluate

a. $\int_C (x^2 + y)\,dx$ b. $\int_C (x^2 + y)\,dy$

Solution   First, parametrize the curve (this step is not unique). We let $x = t$ so that $y = t^2$ for $0 \le t \le 2$.

a. Because $\dfrac{dx}{dt} = 1$, we have

$$\int_C (x^2 + y)\,dx = \int_0^2 [(t)^2 + (t^2)]\frac{dx}{dt}\,dt = 2\int_0^2 t^2\,dt = \left[\frac{2}{3}t^3\right]_0^2 = \frac{16}{3}$$

b. Now we have $\dfrac{dy}{dt} = 2t$, and we find

$$\int_C (x^2 + y)\,dy = \int_0^2 [(t)^2 + (t^2)]\frac{dy}{dt}\,dt = \int_0^2 2t^2 \cdot 2t\,dt = 4\int_0^2 t^3\,dt = [t^4]_0^2 = 16 \quad ■$$

EXAMPLE 2   *Evaluating a line integral*

Evaluate $\int_C xe^{yz}\,dz$, where C is the curve described parametrically by $x = t, y = t$, and $z = -t$ for $1 \le t \le 2$.

Solution
$$\int_C xe^{yz}\,dz = \int_1^2 te^{-t^2}\frac{dz}{dt}\,dt$$
$$= \int_1^2 (-t)e^{-t^2}\,dt \qquad \text{Because } \frac{dz}{dt} = -1$$
$$= \tfrac{1}{2}(e^{-4} - e^{-1}) \qquad\qquad ■$$

LINE INTEGRALS OF VECTOR FIELDS

We shall now discuss what it means to compute the **line integral of a vector field.**

Line Integral of a Vector Field

Let $\mathbf{F}(x, y, z) = u(x, y, z)\mathbf{i} + v(x, y, z)\mathbf{j} + w(x, y, z)\mathbf{k}$ be a vector field, and let C be a piecewise smooth curve with parametric representation

$$\mathbf{R}(t) = x(t)\mathbf{i} + y(t)\mathbf{j} + z(t)\mathbf{k} \quad \text{for } a \leq t \leq b$$

Using $d\mathbf{R} = dx\,\mathbf{i} + dy\,\mathbf{j} + dz\,\mathbf{k}$, we define the **line integral of F along C** by

$$\int_C \mathbf{F} \cdot d\mathbf{R} = \int_C (u\,dx + v\,dy + w\,dz)$$

$$= \int_a^b \left[u[x(t), y(t), z(t)] \frac{dx}{dt} + v[x(t), y(t), z(t)] \frac{dy}{dt} + w[x(t), y(t), z(t)] \frac{dz}{dt} \right] dt$$

EXAMPLE 3   *Line integral of a vector function*

Evaluate $\int_C \mathbf{F} \cdot d\mathbf{R}$, where $\mathbf{F} = (y^2 - z^2)\mathbf{i} + (2yz)\mathbf{j} - x^2\mathbf{k}$ and C is the curve defined parametrically by $x = t^2, y = 2t$, and $z = t$ for $0 \leq t \leq 1$.

Solution   Rewrite $\mathbf{F}$ using the parameter:

$$\mathbf{F} = [(2t)^2 - (t)^2]\mathbf{i} + [2(2t)(t)]\mathbf{j} - [(t^2)^2]\mathbf{k} = 3t^2\mathbf{i} + 4t^2\mathbf{j} - t^4\mathbf{k}$$

Also, because $\mathbf{R}(t) = t^2\mathbf{i} + 2t\mathbf{j} + t\mathbf{k}$, we have $d\mathbf{R} = (2t\,dt)\mathbf{i} + (2\,dt)\mathbf{j} + dt\mathbf{k}$, so

$$\mathbf{F} \cdot d\mathbf{R} = (3t^2)(2t\,dt) + (4t^2)(2dt) + (-t^4)(dt)$$
$$= (6t^3 + 8t^2 - t^4)\,dt$$

Thus,

$$\int_C \mathbf{F} \cdot d\mathbf{R} = \int_0^1 (6t^3 + 8t^2 - t^4)\,dt = \left[\frac{3}{2}t^4 + \frac{8}{3}t^3 - \frac{1}{5}t^5 \right]_0^1 = \frac{119}{30} \qquad \blacksquare$$

EXAMPLE 4   *Line integral of a vector defined function*

Compute the line integral $\int_C \mathbf{F} \cdot d\mathbf{R}$, where $\mathbf{F} = y\mathbf{i} + x\mathbf{j}$ and C is the top half of the circle $x^2 + y^2 = 4$ traversed counterclockwise from $(2, 0)$ to $(-2, 0)$.

Solution   First, we parametrize the curve by setting $x = 2\cos\theta$, $y = 2\sin\theta$ for $0 \leq \theta \leq \pi$. Thus,

$$\mathbf{R}(\theta) = (2\cos\theta)\mathbf{i} + (2\sin\theta)\mathbf{j} \quad \text{so that} \quad d\mathbf{R} = (-2\sin\theta\,d\theta)\mathbf{i} + (2\cos\theta\,d\theta)\mathbf{j}$$

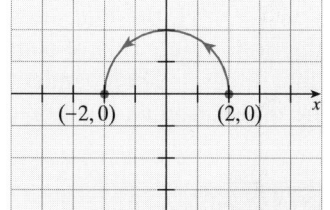

$$\int_C \mathbf{F} \cdot d\mathbf{R} = \int_0^{\pi} [(2\sin\theta)\mathbf{i} + (2\cos\theta)\mathbf{j}] \cdot [(-2\sin\theta)\mathbf{i} + (2\cos\theta)\mathbf{j}]\,d\theta$$

$$= \int_0^{\pi} (-4\sin^2\theta + 4\cos^2\theta)\,d\theta$$

$$= 4\int_0^{\pi} (\cos^2\theta - \sin^2\theta)\,d\theta = 4\int_0^{\pi} \cos 2\theta\,d\theta = [2\sin 2\theta]_0^{\pi} = 0 \qquad \blacksquare$$

EXAMPLE 5   *Line integrals along different paths*

Let $\mathbf{F} = xy^2\mathbf{i} + x^2y\mathbf{j}$ and evaluate the line integral $\int_C \mathbf{F} \cdot d\mathbf{R}$ between the points $(0,0)$ and $(2,4)$ along the following paths:

a. the line segment connecting the points
b. the parabolic arc $y = x^2$ connecting the points

Solution   The two paths we are considering are shown in Figure 14.8.

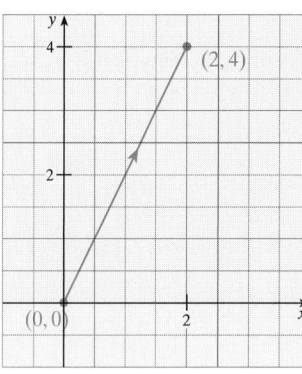
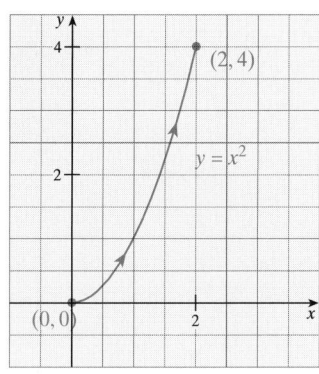

a. The line segment path   **b.** The parabolic path

■ **FIGURE 14.8**   A line integral along different paths

a. The line joining the given points has equation $y = 2x$, which may be parametrized by setting $x = t, y = 2t$ for $0 \le t \le 2$. Thus,

$$\mathbf{R}(t) = t\mathbf{i} + 2t\mathbf{j} \quad \text{so that} \quad d\mathbf{R} = dt\,\mathbf{i} + 2\,dt\,\mathbf{j}$$

In terms of t, we find $\mathbf{F} = 4t^3\mathbf{i} + 2t^3\mathbf{j}$ and

$$\mathbf{F} \cdot d\mathbf{R} = 4t^3\,dt + 4t^3\,dt = 8t^3\,dt$$

$$\int_C \mathbf{F} \cdot d\mathbf{R} = \int_0^2 8t^3\,dt = [2t^4]_0^2 = 32$$

b. The parabola $y = x^2$ can be parametrized by setting $x = t$, $y = t^2$ for $0 \le t \le 2$. Thus,

$$\mathbf{R}(t) = t\mathbf{i} + t^2\mathbf{j} \quad \text{so that} \quad d\mathbf{R} = dt\,\mathbf{i} + 2t\,dt\,\mathbf{j}$$

In terms of t,

$$\mathbf{F} = xy^2\mathbf{i} + x^2y\mathbf{j} = (t)(t^2)^2\mathbf{i} + (t)^2(t^2)\mathbf{j} = t^5\mathbf{i} + t^4\mathbf{j}$$

and

$$\mathbf{F} \cdot d\mathbf{R} = t^5\,dt + 2t^5\,dt = 3t^5\,dt$$

$$\int_C \mathbf{F} \cdot d\mathbf{R} = \int_0^2 3t^5\,dt = \left[\frac{1}{2}t^6\right]_0^2 = 32 \qquad ■$$

In Example 5, we see that the value of the line integral is the same for both paths. Indeed, it can be shown that for the vector field $\mathbf{F} = xy^2\mathbf{i} + x^2y\mathbf{j}$, the line integral $\int_C \mathbf{F} \cdot d\mathbf{R}$ along *any* path C joining $(0,0)$ to $(2,4)$ has the same value. This, of course, is not true for every $\mathbf{F}$, but if it is true, we say that the line integral is **independent of path**. Path independence is an important feature of line integration and will be discussed in detail in the next section.

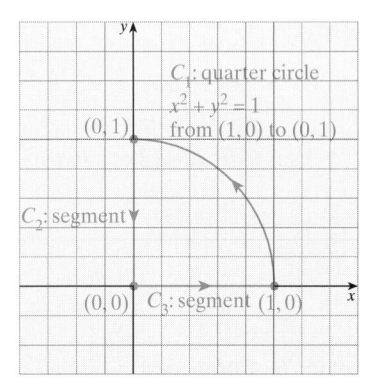

■ **FIGURE 14.9**   A path C composed of three parts

EXAMPLE 6   *Line integral over a path consisting of several parts*

Evaluate $\int_C (-y\,dx + x\,dy)$, where C is the closed path shown in Figure 14.9.

Solution   Because the given closed path can be best described using three separate equations (C_1, C_2, and C_3 in Figure 14.9), we perform the integration by evaluating the line integrals for each part separately, and then adding. Let $\mathbf{F} = -y\mathbf{i} + x\mathbf{j}$, so

$$\int_C (-y\, dx + x\, dy) = \int_C \mathbf{F} \cdot d\mathbf{R}$$

Then

$$\int_C \mathbf{F} \cdot d\mathbf{R} = \int_{C_1} \mathbf{F} \cdot d\mathbf{R} + \int_{C_2} \mathbf{F} \cdot d\mathbf{R} + \int_{C_3} \mathbf{F} \cdot d\mathbf{R}$$

Along C_1, a parametrization is $x = \cos t, y = \sin t$ for $0 \le t \le \frac{\pi}{2}$, so $\mathbf{R}(t) = (\cos t)\mathbf{i} + (\sin t)\mathbf{j}$ and $d\mathbf{R} = (-\sin t\, dt)\mathbf{i} + (\cos t\, dt)\mathbf{j}$. In terms of the parameter, $\mathbf{F} = (-\sin t)\mathbf{i} + (\cos t)\mathbf{j}$ and

$$\int_{C_1} \mathbf{F} \cdot d\mathbf{R} = \int_0^{\pi/2} (\sin^2 t\, dt + \cos^2 t\, dt) = \int_0^{\pi/2} dt = \frac{\pi}{2}$$

Along C_2, a parametrization is $x = 0, y = 1 - t$ for $0 \le t \le 1$, so $\mathbf{R}(t) = (1 - t)\mathbf{j}$ and $d\mathbf{R} = (-dt)\mathbf{j}$. In terms of the parameter, $\mathbf{F} = (t - 1)\mathbf{i}$ and

$$\int_{C_2} \mathbf{F} \cdot d\mathbf{R} = \int_0^1 0\, d\mathbf{R} = 0$$

Along C_3, a parametrization is $x = t, y = 0$ for $0 \le t \le 1$. $\mathbf{R}(t) = t\mathbf{i}, d\mathbf{R} = dt\,\mathbf{i}$, and $\mathbf{F} = t\mathbf{j}$. Because $\mathbf{F} \cdot d\mathbf{R} = 0$, we see

$$\int_{C_3} \mathbf{F} \cdot d\mathbf{R} = 0$$

Thus,

$$\int_C \mathbf{F} \cdot d\mathbf{R} = \int_{C_1} \mathbf{F} \cdot d\mathbf{R} + \int_{C_2} \mathbf{F} \cdot d\mathbf{R} + \int_{C_3} \mathbf{F} \cdot d\mathbf{R} = \frac{\pi}{2} + 0 + 0 = \frac{\pi}{2} \qquad \blacksquare$$

COMPUTING WORK USING LINE INTEGRALS

One of the most important physical applications of the line integral is in computing work. Recall from Section 10.3 that if an object moves along a straight line with displacement $\mathbf{R}$ in a constant force field $\mathbf{F}$, the work done is $\mathbf{F} \cdot \mathbf{R}$. However, the case where $\mathbf{F}$ is not constant and the object moves along a smooth curve C requires extra attention (see Figure 14.10). To analyze this case, assume that C is parametrized by $\mathbf{R}(t)$ and is oriented in the direction of motion. Partition C with subdivision points $P_0, P_1, P_2, \ldots, P_n$, as shown in Figure 14.11.

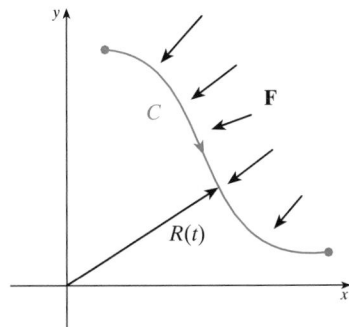

■ **FIGURE 14.10**   If $\mathbf{F}$ and $\mathbf{R}$ are constant, work $= \mathbf{F} \cdot \mathbf{R}$. If they are not, then a line integral is required

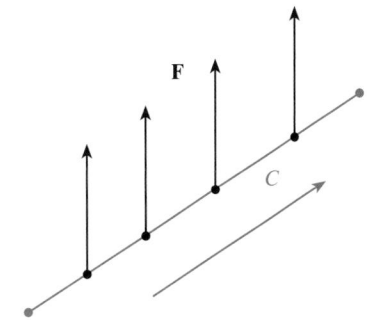

a. Work with constant force in a fixed direction

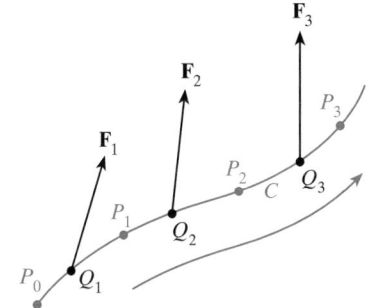

b. Work with variable force along a smooth curve

■ **FIGURE 14.11**   Computing work

For $k = 1, 2, \ldots, n,$ let Q_k be a point chosen arbitrarily from the kth subarc C_k (with endpoints P_{k-1} and P_k), and let $\mathbf{F}_k$ be the value of the force field $\mathbf{F}$ at Q_k. If the length of the subarc C_k is small, $\mathbf{F}$ will be approximately constant with value $\mathbf{F}_k$ over C_k, and the object's displacement along C_k is given approximately by the secant vector $\Delta \mathbf{R}_k$ from P_{k-1} to P_k. Then, the work done by the force as the object moves along C_k may be estimated by

$$\Delta W_k = \left(\mathbf{F}_k \cdot \frac{\Delta \mathbf{R}_k}{\Delta t} \right) \Delta t$$

By adding the contributions along all n subarcs, we obtain

$$\left(\mathbf{F}_1 \cdot \frac{\Delta \mathbf{R}_1}{\Delta t} \right) \Delta t + \left(\mathbf{F}_2 \cdot \frac{\Delta \mathbf{R}_2}{\Delta t} \right) \Delta t + \cdots + \left(\mathbf{F}_n \cdot \frac{\Delta \mathbf{R}_n}{\Delta t} \right) \Delta t = \sum_{k=1}^{n} \left(\mathbf{F}_k \cdot \frac{\Delta \mathbf{R}_k}{\Delta t} \right) \Delta t$$

as an estimate of the total work done by $\mathbf{F}$ as the object moves along C. As $\Delta t \to 0$, the limiting value of this sum is the line integral of $\mathbf{F} \cdot \dfrac{d\mathbf{R}}{dt}$; that is,

$$W = \lim_{\Delta t \to 0} \sum_{k=1}^{n} \left(\mathbf{F}_k \cdot \frac{\Delta \mathbf{R}_k}{\Delta t} \right) \Delta t = \int_C \mathbf{F} \cdot \frac{d\mathbf{R}}{dt}\, dt = \int_C \mathbf{F} \cdot d\mathbf{R}$$

This leads us to the following definition.

Work as a Line Integral

Let $\mathbf{F}$ be a continuous force field over a domain D. Then the **work** W done by $\mathbf{F}$ as an object moves along a smooth curve C in D is given by the line integral

$$W = \int_C \mathbf{F} \cdot d\mathbf{R}$$

EXAMPLE 7   *Work as a line integral*

An object moves in the force field

$$\mathbf{F} = y^2 \mathbf{i} + 2(x + 1)y\mathbf{j}$$

counterclockwise from the point $(2, 0)$ along the elliptical path $x^2 + 4y^2 = 4$ to $(-2, 0)$, and back to the point $(2, 0)$ along the x-axis. How much work is done by the force field on the object?

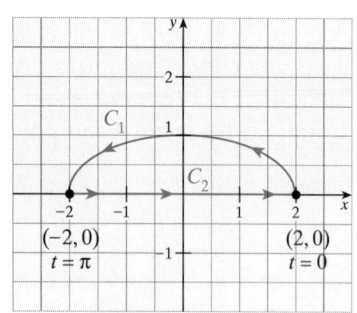

Solution   Let C denote the path of the object. Then the total work W done by $\mathbf{F}$ on the object as it moves along C is given by the line integral $\int_C \mathbf{F} \cdot d\mathbf{R}$. We divide the curve C into two parts so that $C = C_1 \cup C_2$. The curve $x^2 + 4y^2 = 4$ suggests the parametrization $\dfrac{x}{2} = \cos t, y = \sin t$ for $0 \leq t \leq \pi$. Thus,

$$\mathbf{R}(t) = (2 \cos t)\mathbf{i} + (\sin t)\mathbf{j}$$

and

$$d\mathbf{R} = (-2 \sin t\, dt)\mathbf{i} + (\cos t\, dt)\mathbf{j}$$

In parametric form,

$$\mathbf{F} = (\sin^2 t)\mathbf{i} + (4 \cos t \sin t + 2 \sin t)\mathbf{j}$$

so that

$$W_1 = \int_{C_1} \mathbf{F} \cdot d\mathbf{R} = \int_0^\pi (-2\sin^3 t + 4\cos^2 t \sin t + 2\sin t \cos t)\, dt$$

$$= \int_0^\pi (-2\sin^2 t + 4\cos^2 t + 2\cos t)\sin t\, dt$$

$$= \int_0^\pi (6\cos^2 t + 2\cos t - 2)\sin t\, dt$$

$$= -\int_1^{-1} (6u^2 + 2u - 2)\, du = 0$$

> *Let $u = \cos t$;*
> *$du = -\sin t\, dt$.*
> *If $t = 0, u = 1$;*
> *and if $t = \pi, u = -1$.*

The curve C_2 is described by $y = 0$, so $\mathbf{F} = \mathbf{0}$ and, therefore, $W_2 = 0$. Thus, $W = W_1 + W_2 = 0$. ∎

It can be shown that the work in moving an object around *any* closed path against the force in Example 7 will always be 0. When this occurs, the force is said to be **conservative.** We shall discuss conservative force fields in Sections 14.3 and 14.4.

EVALUATION OF LINE INTEGRALS WITH RESPECT TO ARC LENGTH

Line integrals of the form $\int_C \mathbf{F} \cdot d\mathbf{R}$ can often be expressed in other forms. For example, recall from Chapter 11 that $\mathbf{T} = \dfrac{d\mathbf{R}}{ds}$ is a unit tangent vector to the curve C at the point $P(x, y, z)$, where s is the arc length parameter. We have

$$\int_C \mathbf{F} \cdot d\mathbf{R} = \int_C \mathbf{F} \cdot \frac{d\mathbf{R}}{ds}\, ds = \int_C \mathbf{F} \cdot \mathbf{T}\, ds$$

In particular, the work W done by a force field $\mathbf{F}$ on an object moving along a curve C may be expressed as

$$W = \int_C \mathbf{F} \cdot d\mathbf{R} = \int_C \mathbf{F} \cdot \mathbf{T}\, ds$$

This form of the integral is called the **line integral of the tangential component of F** and can also be written as $\int_C f(x, y, z)\, ds$. The integral will exist if f is continuous on the curve C and if C itself is piecewise smooth with finite length. A formula for computing this line integral is suggested by the following observations: If $\mathbf{R}(t) = x(t)\mathbf{i} + y(t)\mathbf{j} + z(t)\mathbf{k}$, then

$$\frac{ds}{dt} = \left\| \frac{d\mathbf{R}}{dt} \right\| = \sqrt{\left(\frac{dx}{dt}\right)^2 + \left(\frac{dy}{dt}\right)^2 + \left(\frac{dz}{dt}\right)^2}$$

so that

$$W = \int_C \mathbf{F} \cdot \mathbf{T}\, ds = \int_a^b f[x(t), y(t), z(t)]\sqrt{[x'(t)]^2 + [y'(t)]^2 + [z'(t)]^2}\, dt$$

Evaluation of a Line Integral with Respect to Arc Length

Let f be continuous on a smooth curve C. If C is defined by $\mathbf{R}(t) = x(t)\mathbf{i} + y(t)\mathbf{j} + y(t)\mathbf{k}$, where $a \le t \le b$, then

$$\int_C f(x, y, z)\, ds = \int_a^b f[x(t), y(t), z(t)]\sqrt{[x'(t)]^2 + [y'(t)]^2 + [z'(t)]^2}\, dt$$

To summarize:

EXAMPLE 8   *Evaluating a line integral with respect to arc length*

Find $\int_C (x + y^2 - z)\, ds$, where C is the line segment described parametrically by $x = t, y = 2t, z = t + 1$ for $0 \le t \le 1$.

Solution   Because $\dfrac{dx}{dt} = 1, \dfrac{dy}{dt} = 2$, and $\dfrac{dz}{dt} = 1$, we have

$$\int_C (x + y^2 - z)\, ds = \int_0^1 [t + (2t)^2 - (t + 1)]\sqrt{(1)^2 + (2)^2 + (1)^2}\, dt$$

$$= \sqrt{6}\int_0^1 (4t^2 - 1)\, dt = \frac{1}{3}\sqrt{6} \qquad \blacksquare$$

[figure: z-axis graph with points (0,0,1) and line segment; labels $x = t, y = 2t, z = t + 1$ for $0 \le t \le 1$; $\mathbf{R}(t) = t\mathbf{i} + 2t\mathbf{j} + (t + 1)\mathbf{k}$ for $0 \le t \le 1$]

14.2 Problem Set

A **1.** ■ **What Does This Say?** Describe a line integral.
  2. ■ **What Does This Say?** Discuss the evaluation of a line integral.

Evaluate each line integral given in Problems 3–16.

3. $\int_C (-y\, dx + x\, dy)$
  C is the parabolic path $y = 4x^2$ from $(1, 4)$ to $(0, 0)$.

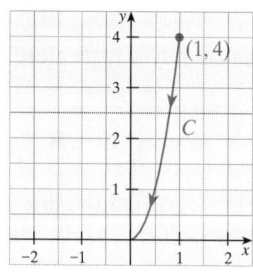

4. $\int_C (-y\, dx + 3x\, dy)$
  C is the parabolic path $y^2 = x$ from $(1,1)$ to $(9, 3)$.

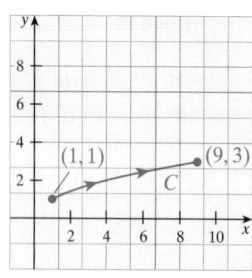

5. $\int_C (x\, dy - y\, dx)$
  C is the line defined $2x - 4y = 1$ as x varies from 4 to 8.

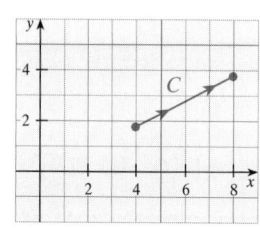

6. $\int_C [(y - x)\, dx + x^2 y\, dy]$
  C is the curve defined by $y^2 = x^3$ from $(1, -1)$ to $(1, 1)$.

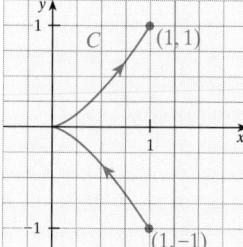

7. $\int_C [(x + y)^2\, dx - (x - y)^2\, dy]$
  C is the curve defined by $y = |2x|$ from $(-1, 2)$ to $(1, 2)$.

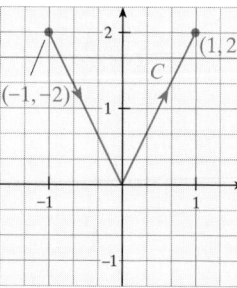

8. $\int_C [(y^2 - x^2)\, dx - x\, dy]$
  C is the quarter-circle $x^2 + y^2 = 4$ from $(0, 2)$ to $(2, 0)$.

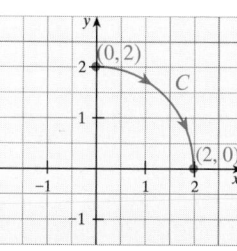

9. $\int_C [(x^2 + y^2)\, dx + 2xy\, dy]$ for these choices of the curve C:
  a. C is the quarter-circle $x^2 + y^2 = 1$ traversed counterclockwise from $(1, 0)$ to $(0, 1)$.
  b. C is the straight line $y = 1 - x$ from $(1, 0)$ to $(0, 1)$.

10. $\int_C [x^2 y \, dx + (x^2 - y^2) \, dy]$ for these choices of the curve C:
   a. C is the arc of the parabola $y = x^2$ from $(0, 0)$ to $(2, 4)$.
   b. C is the segment of the line $y = 2x$ for $0 \le x \le 2$.

11. Rework Problem 10 for the path C that consists of the horizontal line segment $(0, 0)$ to $(2, 0)$, followed by the vertical segment from $(2, 0)$ to $(2, 4)$.

12. $\int_C (-xy^2 \, dx + x^2 \, dy)$ where C is the path defined by the given figure.

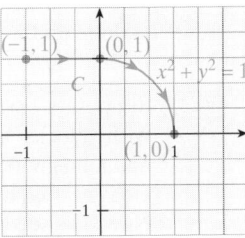

13. $\int_C (-y^2 \, dx + x^2 \, dy)$ where C is the path defined by the given figure.

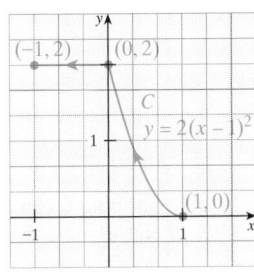

14. $\int_C [(x^2 - y^2) \, dx + x \, dy]$, where C is the closed circular path given by $x = 2 \cos \theta, y = 2 \sin \theta, 0 \le \theta \le 2\pi$.

15. $\int_C (x^2 y \, dx - xy \, dy)$, where C is the closed path that begins at $(0, 0)$, goes to $(1, 1)$ along the parabola $y = x^2$, and then returns to $(0, 0)$ along the line $y = x$.

16. $\displaystyle\int_C \frac{x \, dx - y \, dy}{\sqrt{x^2 + y^2}}$, where C is the quarter-circular path $x^2 + y^2 = a^2$, traversed from $(a, 0)$ to $(0, a)$.

In Problems 17–19, evaluate $\int_C \mathbf{F} \cdot d\mathbf{R}$, where $\mathbf{F} = (5x + y)\mathbf{i} + x\mathbf{j}$ and C is the specified curve.

17. C is the straight line segment from $(0, 0)$ to $(2, 1)$.
18. C is the curve given by $\mathbf{R}(t) = 2t\mathbf{i} + t\mathbf{j}$ for $0 \le t \le 1$.
19. C is the vertical line from $(0, 0)$ to $(0, 1)$, followed by the horizontal line from $(0, 1)$ to $(2, 1)$.

Evaluate the line integrals in Problems 20–35.

20. $\int_C (y \, dx - x \, dy + dz)$, where C is the helical path given by:
   a. $x = 3 \sin t, y = 3 \cos t, z = t$ for $0 \le t \le \frac{\pi}{2}$
   b. $x = a \sin t, y = a \cos t, z = t$ for constant a and $0 \le t \le \frac{\pi}{2}$

21. $\int_C (x \, dx + y \, dy + z \, dz)$, where C is the following path:
   a. the helix defined by $x = \cos t, y = \sin t, z = t$ for $0 \le t \le \frac{\pi}{2}$
   b. the straight line segment from $(1, 0, 0)$ to $(0, 1, \frac{\pi}{2})$

22. $\int_C (-y \, dx + x \, dy + xz \, dz)$, where C is the following path:
   a. the helix defined by $x = \cos t, y = \sin t, z = t$ for $0 \le t \le 2\pi$
   b. the unit circle $x^2 + y^2 = 1, z = 0$, traversed once counterclockwise as viewed from above

23. $\int_C (5 \, xy \, dx + 10 \, yz \, dy + z \, dz)$, where C is the following path:
   a. the parabolic arc $x = y^2$ from $(0, 0, 0)$ to $(1, 1, 0)$ followed by the line segment given by $x = 1, y = 1$, $0 \le z \le 1$
   b. the straight line segment from $(0, 0, 0)$ to $(1, 1, 1)$

24. $\displaystyle\int_C \frac{dx + dy}{|x| + |y|}$, where C is the square $|x| + |y| = 1$, traversed once counterclockwise

25. $\int_C [(y + z) \, dx + (x + z) \, dy + (x + y) \, dz]$ where C is the circle of radius 1 centered on the z-axis in the plane $z = 2$, traversed once counterclockwise as viewed from above.

26. $\int_C \mathbf{F} \cdot d\mathbf{R}$, where $\mathbf{F} = (y - 3)\mathbf{i} + x\mathbf{j}$, and C is the curve given by
$$\mathbf{R}(t) = (\sin t)\mathbf{i} - (\cos t)\mathbf{j} \quad \text{for } 0 \le t \le \pi$$

27. $\int_C \mathbf{F} \cdot d\mathbf{R}$, where $\mathbf{F} = (y - 2z)\mathbf{i} + x\mathbf{j} - 2xy\mathbf{k}$ and C is the path given by
$$\mathbf{R}(t) = t\mathbf{i} + t^2\mathbf{j} - \mathbf{k} \quad \text{for } 1 \le t \le 2$$

28. $\int_C \mathbf{F} \cdot d\mathbf{R}$, where $\mathbf{F} = x\mathbf{i} + xy\mathbf{j} + x^2yz\mathbf{k}$, and C is the elliptical path given by $x^2 + 4y^2 - 8y + 3 = 0$ in the xy-plane, traversed once counterclockwise as viewed from above.

29. $\int_C \mathbf{F} \cdot d\mathbf{R}$, where $\mathbf{F} = y^2\mathbf{i} + x^2\mathbf{j} - (x + z)\mathbf{k}$ and C is the boundary of the triangle with vertices $(0, 0, 0), (1, 0, 0)$, $(1, 1, 0)$, traversed once clockwise, as viewed from above.

30. $\int_C \mathbf{F} \cdot \mathbf{T} \, ds$, where $\mathbf{F} = -3y\mathbf{i} + 3x\mathbf{j} + 3x\mathbf{k}$, and C is the straight line segment from $(0, 0, 1)$ to $(1, 1, 1)$.

31. $\int_C \mathbf{F} \cdot \mathbf{T} \, ds$, where $\mathbf{F} = -x\mathbf{i} + 2\mathbf{j}$, and C is the boundary of the trapezoid with vertices $(0, 0), (1, 0), (2, 1), (0, 1)$, traversed once clockwise as viewed from above.

32. $\int_C y \, ds$, where C is the curve given by
$$\mathbf{R}(t) = t\mathbf{i} + 2t^3\mathbf{j}, \quad 0 \le t \le 2$$

33. $\int_C (x + y) \, ds$, where C is given by
$$\mathbf{R}(t) = (\cos^2 t)\mathbf{i} + (\sin^2 t)\mathbf{j}, \quad -\frac{\pi}{4} \le t \le 0$$

34. $\displaystyle\int_C \frac{x^2 + xy + y^2}{z^2} \, ds$, where C is the path given by $\mathbf{R}(t) = (\cos t)\mathbf{i} + (\sin t)\mathbf{j} - \mathbf{k}$ for $0 \le t \le 2\pi$.

35. Evaluate the line integral
$$\int_C \frac{x \, dy - y \, dx}{x^2 + y^2}$$
where C is the unit circle $x^2 + y^2 = 1$ traversed once counterclockwise.

B **36.** How much work is done by a constant force $\mathbf{F} = a\mathbf{i} + \mathbf{j}$ when a particle moves along the line $y = ax$ from $x = a$ to $x = 0$?

37. A force field in the plane is given by $\mathbf{F} = (x^2 - y^2)\mathbf{i} + 2xy\mathbf{j}$. Find the total work done by this force in moving a point mass counterclockwise around the square with vertices $(0,0), (2,0), (2,2), (0,2)$.

38. Find the work done by the force field $\mathbf{F} = (x^2 + y^2)\mathbf{i} + (x + y)\mathbf{j}$ as an object moves counterclockwise along the circle $x^2 + y^2 = 1$ from $(1,0)$ to $(-1, 0)$, and then back to $(1,0)$ along the x-axis.

39. A force acting on a point mass located at (x, y) is given by $\mathbf{F} = y\mathbf{i} + 2x\mathbf{j}$. Find the work done by this force as the point mass moves along a straight line from $(1,0)$ to $(0, 1)$.

Find the work done by the force $\mathbf{F}(x, y, z)$ on an object moving along the curve C in Problems 40–43.

40. $\mathbf{F} = (y^2 - z^2)\mathbf{i} + 2yz\mathbf{j} - x^2\mathbf{k}$, and C is the path given by $x(t) = t, y(t) = t^2, z(t) = t^3$, for $0 \le t \le 1$.

41. $\mathbf{F} = 2xy\mathbf{i} + (x^2 + 2)\mathbf{j} + y\mathbf{k}$, and C is the line segment from $(1, 0, 2)$ to $(3, 4, 1)$.

42. $\mathbf{F} = x\mathbf{i} + y\mathbf{j} + (xz - y)\mathbf{k}$, and C is the line segment from $(0, 0, 0)$ to $(2, 1, 2)$.

43. $\mathbf{F} = x\mathbf{i} + y\mathbf{j} + (xz - y)\mathbf{k}$, and C is the path given by $\mathbf{R}(t) = t^2\mathbf{i} + 2t\mathbf{j} + 4t^3\mathbf{k}$ for $0 \le t \le 1$.

44. A 180-lb laborer carries a bag of sand weighing 40 lb up a circular helical staircase on the outside of a

tower 50 ft high and 20 ft in diameter. How much work is done by gravity as the laborer climbs to the top in exactly five revolutions?

45. Repeat Problem 44 assuming that the bag leaks 1 lb of sand for every 10 ft of ascent. How much work is done by gravity during the laborer's climb to the top?

46. A 5,000-lb satellite orbits the earth in a circular orbit 5,000 mi from the center of the earth. How much work is done on the satellite by gravity during one complete revolution?

C **47.** Suppose a particle with charge Q and mass m moves with velocity $\mathbf{V}$ under the influence of an electric field $\mathbf{E}$ and a magnetic field $\mathbf{B}$. Then the total force on the particle is $\mathbf{F} = Q(\mathbf{E} + \mathbf{V} \times \mathbf{B})$, called the **Lorentz force.** Use Newton's second law of motion, $\mathbf{F} = m\mathbf{A}$, to show that

$$m\frac{d\mathbf{V}}{dt} \cdot \mathbf{V} = Q\mathbf{E} \cdot \frac{d\mathbf{R}}{dt}$$

and then evaluate the line integral $\displaystyle\int_C \mathbf{E} \cdot d\mathbf{R}$, where $\mathbf{C}$ is the trajectory of a particle traveling with constant speed.

48. Suppose a thin wire fits a smooth curve C in $\mathbb{R}^3$, and let $\delta(x, y, z)$ be the density of the wire. Explain why the mass m of the wire is given by the line integral

$$m = \int_C \delta(x, y, z)\, ds$$

Express the center of mass of the wire in terms of line integrals.

14.3 *Independence of Path*

======================

IN THIS SECTION	**conservative vector fields, fundamental theorem of line integrals** ▪

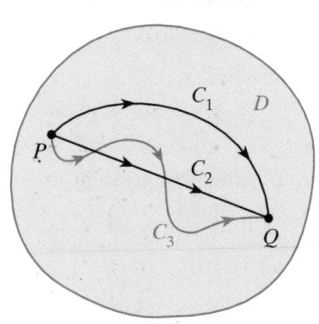

A line integral $\int_C \mathbf{F} \cdot d\mathbf{R}$ that has the same value for any curve C in D with given endpoints is said to be independent of path

In general, the value of the line integral $\displaystyle\int_C \mathbf{F} \cdot d\mathbf{R}$ depends on the path of integration C, but in certain cases, the integral will be the same for all paths in a given region D with the same initial point P and terminal point Q. In this case, we say the line integral is **independent of path** in D. In this section, we shall study path independence and characterize the kind of vector field $\mathbf{F}$ for which it occurs.

CONSERVATIVE VECTOR FIELDS

The issue of path independence for the line integral $\displaystyle\int_C \mathbf{F} \cdot d\mathbf{R}$ is closely related to determining whether $\mathbf{F}$ is the gradient of some scalar function f. It is useful to have the following terminology.

> ### Conservative Vector Field
>
> A vector field **F** is said to be **conservative** in a region D if it can be represented in D as the gradient of a continuously differentiable function f, which is then called a **scalar potential** of **F**. That is,
>
> $$\mathbf{F} = \nabla f \quad \text{for } (x, y) \text{ in } D$$
> $$\uparrow \qquad \uparrow$$
> Conservative vector field   Scalar potential

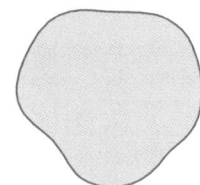

a. A simply connected region

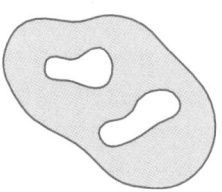

b. A region that is *not* simply connected

■ **FIGURE 14.12** Connected regions

EXAMPLE 1   *Conservative vector field*

Verify that the vector field $\mathbf{F} = 2xy\mathbf{i} + x^2\mathbf{j}$ is conservative, with scalar potential $f = x^2 y$.

Solution $\nabla f = 2xy\mathbf{i} + x^2\mathbf{j}$ and this is the same as **F**, so **F** is conservative. ■

As we work this example, two questions come to mind:

1. Does there exist an easy test to determine whether **F** is conservative?
2. Once such a determination has been made, how can a scalar potential function f be found?

Before answering these questions, we need to introduce some new terminology. Specifically, a region D is **connected** if any two points in D can be joined by a piecewise smooth curve that lies entirely within D. If, in addition, every closed curve in D encloses only points that are also in D, then D is **simply connected.** Roughly speaking, a simply connected region is one with no "holes," as shown in Figure 14.12.

THEOREM 14.2   *Cross-partials test for a conservative vector field in the plane*

Consider the vector field $\mathbf{F}(x, y) = u(x, y)\mathbf{i} + v(x, y)\mathbf{j}$, where u and v have continuous first partials in the open, simply connected region D. Then $\mathbf{F}(x, y)$ is conservative in D if and only if

$$\frac{\partial u}{\partial y} = \frac{\partial v}{\partial x} \quad \text{throughout } D$$

Proof   We outline the proof in Problem Set 14.4 after our discussion of Green's theorem.

In the next two examples, we use this theorem to show that a given vector field **F** is conservative. Then we "partially integrate" the components of **F** to obtain a scalar potential of **F** such that $\nabla f = \mathbf{F}$.

EXAMPLE 2   *Finding a scalar potential function*

Show that the vector field $\mathbf{F} = (e^x \sin y - y)\mathbf{i} + (e^x \cos y - x - 2)\mathbf{j}$ is conservative and then find a scalar potential function f for **F**.

Solution   Let $u(x, y) = e^x \sin y - y$ and $v(x, y) = e^x \cos y - x - 2$. Then

$$\frac{\partial u}{\partial y} = e^x \cos y - 1 \quad \text{and} \quad \frac{\partial v}{\partial x} = e^x \cos y - 1$$

Since $\dfrac{\partial u}{\partial y} = \dfrac{\partial v}{\partial x}$, it follows that **F** is conservative. To find a scalar potential function f

such that $\nabla f = \mathbf{F}$, we note that f must satisfy $u(x, y) = f_x(x, y)$ and $v(x, y) = f_y(x, y)$.

$$f(x, y) = \int u(x, y)\, dx = \underbrace{\int (e^x \sin y - y)\, dx}$$

This is the "partial integral" in the sense that y is held constant while the integration is performed with respect to x alone.

$$= e^x \sin y - yx + c(y)$$

Where $c(y)$ is a function of y alone—a "constant" as far as x-integration is concerned

Because f must also satisfy $f_y(x, y) = v(x, y)$, we compute the partial derivative of this result with respect to y:

$$f_y(x, y) = \frac{\partial}{\partial y}[e^x \sin y - yx + c(y)] = e^x \cos y - x + \frac{dc}{dy}$$

Set this equal to $v = e^x \cos y - x - 2$ and solve for $\dfrac{dc}{dy}$:

$$e^x \cos y - x + \frac{dc}{dy} = e^x \cos y - x - 2$$

$$\frac{dc}{dy} = -2$$

$$c(y) = -2y + C$$

We now have found the function $f(x, y) = e^x \sin y - xy - 2y + C$. Any such function is a scalar potential of $\mathbf{F}$ and, for simplicity, we pick $C = 0$:

$$f(x, y) = e^x \sin y - xy - 2y$$ ∎

In Example 2, we began by using the fact that $f_x = u$. In general, the issue of whether to start with $u = f_x$ or $v = f_y$ is a matter of personal taste and is often determined by which equation leads to the simpler integration.

EXAMPLE 3   *Testing for a conservative vector field in the plane*

Determine whether the vector field $\mathbf{F} = ye^{xy}\mathbf{i} + (xe^{xy} + x)\mathbf{j}$ is conservative; if it is, find a scalar potential.

Solution   We have $u(x, y) = ye^{xy}$ and $v(x, y) = xe^{xy} + x$.

$$\frac{\partial u}{\partial y} = xye^{xy} + e^{xy} \qquad \frac{\partial v}{\partial x} = xye^{xy} + e^{xy} + 1$$

so $\dfrac{\partial u}{\partial y} \neq \dfrac{\partial v}{\partial x}$, and $\mathbf{F}$ is not conservative.   ∎

FUNDAMENTAL THEOREM OF LINE INTEGRALS

Recall that, according to the fundamental theorem of calculus, if the function f is continuous on $[a, b]$, then

$$\int_a^b f(x)\, dx = F(b) - F(a)$$

where F is any antiderivative of f; that is, $F'(x) = f(x)$. The following is the analogous result for line integrals.

THEOREM 14.3   *Fundamental theorem of line integrals*

Let $\mathbf{F}$ be a conservative vector field on the region D and let f be a scalar potential function for $\mathbf{F}$; that is, $\nabla f = \mathbf{F}$. Then, if C is any piecewise smooth curve lying entirely within D, with initial point P and terminal point Q, we have

$$\int_C \mathbf{F} \cdot d\mathbf{R} = f(Q) - f(P)$$

Thus, the line integral $\int_C \mathbf{F} \cdot d\mathbf{R}$ is independent of path in D.

Proof   We shall prove this theorem for the case where the curve C is smooth in D, leaving the more general case where C is piecewise smooth as an exercise. Suppose C is described by the vector function $\mathbf{R}(t) = x(t)\mathbf{i} + y(t)\mathbf{j}$, where $a \le t \le b$, and $P = \mathbf{R}(a), Q = \mathbf{R}(b)$. Because $\mathbf{F}(x, y) = \nabla f(x, y) = f_x(x, y)\mathbf{i} + f_y(x, y)\mathbf{j}$, we have

$$\int_C \mathbf{F} \cdot d\mathbf{R} = \int_a^b \mathbf{F} \cdot \frac{d\mathbf{R}}{dt}\, dt$$
$$= \int_a^b \left[f_x(x, y)\frac{dx}{dt} + f_y(x, y)\frac{dy}{dt}\right] dt \quad \text{Because } \mathbf{F} = \nabla f$$
$$= \int_a^b \frac{d}{dt}\{f[x(t), y(t)]\}\, dt \quad \text{Chain rule (in reverse)}$$
$$= f[x(b), y(b)] - f[x(a), y(a)] \quad \text{Fundamental theorem of calculus}$$
$$= f[\mathbf{R}(b)] - f[\mathbf{R}(a)]$$
$$= f(Q) - f(P)$$

EXAMPLE 4   *Evaluating a line integral using the fundamental theorem of line integrals*

Evaluate the line integral $\int_C \mathbf{F} \cdot d\mathbf{R}$, where

$$\mathbf{F} = (e^x \sin y - y)\mathbf{i} + (e^x \cos y - x - 2)\mathbf{j} \quad \text{and } C \text{ is the path given by}$$
$$\mathbf{R}(t) = \left[t^3 \sin \frac{\pi t}{2}\right]\mathbf{i} - \left[\frac{\pi}{2}\cos\left(\frac{\pi t}{2} + \frac{\pi}{2}\right)\right]\mathbf{j} \quad \text{for } 0 \le t \le 1.$$

Solution   Evaluating this line integral by the parametric method would be both difficult and tedious. However, we showed in Example 2 that the vector field $\mathbf{F}$ is conservative with scalar potential

$$f(x, y) = e^x \sin y - xy - 2y$$

and according to the fundamental theorem on line integrals, the value of the line integral depends only on the value of f at the endpoints of the path C. (You should also verify that the hypotheses of the theorem are satisfied by $\mathbf{F}$ and C.)

At the endpoint where $t = 0$: $\mathbf{R}(0) = 0\mathbf{i} - [\frac{\pi}{2}\cos(\frac{\pi}{2})]\mathbf{j} = 0\mathbf{i} + 0\mathbf{j}$
$$f(0, 0) = e^0 \sin 0 - 0 - 0 = 0$$

At the endpoint where $t = 1$: $\mathbf{R}(1) = [\sin \frac{\pi}{2}]\mathbf{i} - [\frac{\pi}{2}\cos \pi] = \mathbf{i} + \frac{\pi}{2}\mathbf{j}$
$$f(1, \tfrac{\pi}{2}) = e^1 \sin \tfrac{\pi}{2} - \tfrac{\pi}{2} - 2(\tfrac{\pi}{2}) = e - \tfrac{3\pi}{2}$$

We now apply the fundamental theorem for line integrals:

$$\int_C \mathbf{F} \cdot d\mathbf{R} = f(Q) - f(P) = f(1, \tfrac{\pi}{2}) - f(0, 0)$$

$$= (e - \tfrac{3\pi}{2}) - 0 = e - \tfrac{3\pi}{2}$$ ∎

EXAMPLE 5   *Work along a closed path in a conservative force field*

Show that no net work is done when an object moves along a closed path back to its starting point in a connected domain where the force field is conservative.

Solution   In such a force field $\mathbf{F}$, we have $\nabla f = \mathbf{F}$, where f is a scalar potential of $\mathbf{F}$, and because the path of motion is closed, it begins and ends at the same point P. Thus, the work is given by

$$W = \int_C \mathbf{F} \cdot d\mathbf{R} = f(P) - f(P) = 0$$ ∎

This result appears as part of the proof of the following useful theorem, which ties together several equivalent conditions for path independence.

THEOREM 14.4   *Closed curve theorem for a conservative force field*

The continuous vector field $\mathbf{F}$ is conservative in the open connected region D if and only if $\int_C \mathbf{F} \cdot d\mathbf{R} = 0$ for every piecewise smooth closed curve C in D.

Proof   If a line integral $\int_C \mathbf{F} \cdot d\mathbf{R}$ is independent of path in the open, connected region D, then $\int_C \mathbf{F} \cdot d\mathbf{R} = 0$ for any piecewise smooth closed path C in D. Indeed, if P and Q are two points on such a path, and C_T is the path from P to Q along, say, the "top" of C, while C_B is the "bottom" path along C from Q to P (see Figure 14.13), we must have $\int_{C_B} \mathbf{F} \cdot d\mathbf{R} = -\int_{C_T} \mathbf{F} \cdot d\mathbf{R}$ and

$$\int_C \mathbf{F} \cdot d\mathbf{R} = \int_{C_T} \mathbf{F} \cdot d\mathbf{R} + \int_{C_B} \mathbf{F} \cdot d\mathbf{R} = \int_{C_T} \mathbf{F} \cdot d\mathbf{R} - \int_{C_T} \mathbf{F} \cdot d\mathbf{R} = 0$$

Conversely, if $\int_C \mathbf{F} \cdot d\mathbf{R} = 0$ for every closed curve C in a region D, it can be shown (see Problem 44) that $\mathbf{F}$ must be conservative in D. ═

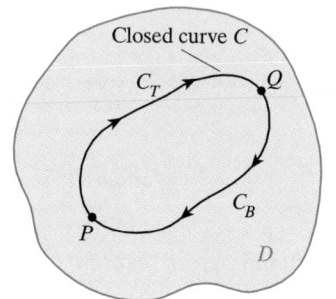

■ **FIGURE 14.13**   A continuous vector field $\mathbf{F}$ is conservative in the region D if and only if $\int_C \mathbf{F} \cdot d\mathbf{R} = 0$ for every closed curve C in D

We have now stated three equivalent conditions for path independence, which we summarize in the following box.

Equivalent Conditions for Path Independence

Let $\mathbf{F}(x, y)$ have continuous first partial derivatives in an open connected region D, and let C be a piecewise smooth curve in D. Then the following conditions are equivalent.

a. $\displaystyle\int_C \mathbf{F} \cdot d\mathbf{R}$ is independent of path within D.

b. $\mathbf{F}$ is conservative; that is, $\mathbf{F} = \nabla f$ for some function f defined on D.

c. $\displaystyle\int_C \mathbf{F} \cdot d\mathbf{R} = 0$ for every closed path C enclosing only points of D.

Here is an example that illustrates how these conditions can be used to evaluate line integrals.

| EXAMPLE 6 | *Using path independence to evaluate a line integral* |

Evaluate $\int_C (xy^2\, dx + x^2y\, dy)$ over each of the given curves:

a. 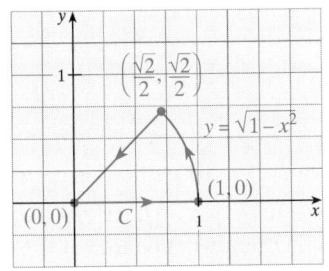 b. 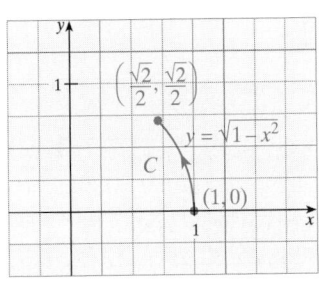 c.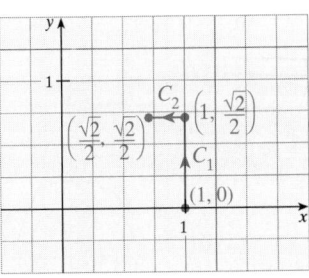

Solution   First, apply the cross-partials test to see whether $\mathbf{F}(x, y) = xy^2\mathbf{i} + x^2y\mathbf{j}$ is conservative.

$$\frac{\partial}{\partial y}(xy^2) = 2xy \qquad \frac{\partial}{\partial x}(x^2y) = 2xy$$

Thus, $\mathbf{F}$ is conservative.

a. Because C is a closed curve and $\mathbf{F}$ is conservative,

$$\int_C (xy^2\, dx + x^2y\, dy) = 0$$

b. Because $y = \sqrt{1 - x^2}$ is the upper portion of the circle $x^2 + y^2 = 1$, we can parametrize C by $x = \cos\theta, y = \sin\theta$ for $0 \le \theta \le \frac{\pi}{4}$. Thus,

$$\int_C (xy^2\, dx + x^2y\, dy) = \int_0^{\pi/4} [\cos\theta \sin^2\theta(-\sin\theta\, d\theta) + \cos^2\theta \sin\theta(\cos\theta\, d\theta)]$$

$$= \int_0^{\pi/4} [\cos^3\theta \sin\theta - \sin^3\theta \cos\theta]\, d\theta = \tfrac{1}{8}$$

c. Because $\mathbf{F}$ is conservative, the line integral is independent of path. Thus, the result is the same as for part **b**.

$$\int_C (xy^2\, dx + x^2y\, dy) = \tfrac{1}{8}$$

It might be instructive to illustrate the power of these equivalent conditions by evaluating the line integral in part **c** without the benefit of part **b**.   ■

In the next section, we develop another criterion for path independence as part of our study of an important result known as Green's theorem.

14.3   Problem Set

Ⓐ 1. ■ **What Does This Say?** Discuss conservative vector fields.

2. ■ **What Does This Say?** Describe the fundamental theorem of line integrals.

3. ■ **What Does This Say?** Compare and contrast the various equivalent conditions for independence of path.

Determine whether each vector field in Problems 4–9 is conservative, and if it is, find a scalar potential.

4. $y^2\mathbf{i} + 2xy\mathbf{j}$

5. $2xy^3\mathbf{i} + 3y^2x^2\mathbf{j}$

6. $(xe^{xy}\sin y)\mathbf{i} + (e^{xy}\cos y + y)\mathbf{j}$

7. $(-y + e^x\sin y)\mathbf{i} + [(x + 2)e^x\cos y]\mathbf{j}$

8. $(y - x^2)\mathbf{i} + (2x + y^2)\mathbf{j}$ 9. $(e^{2x}\sin y)\mathbf{i} + (e^{2x}\cos y)\mathbf{j}$

Evaluate the line integrals in Problems 10–13 for each of the given paths.

10. $\displaystyle\int_C [(3x + 2y)\,dx + (2x + 3y)\,dy]$

a.
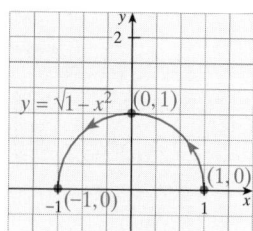

b.
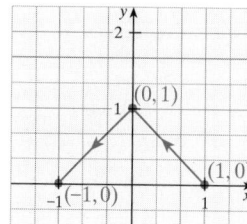

c.
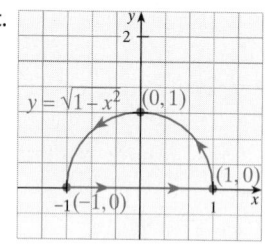

11. $\displaystyle\int_C [(3x + 2y)\,dx - (2x + 3y)\,dy]$

a.
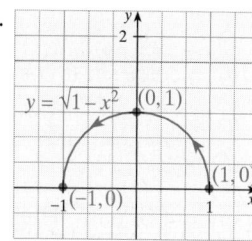

b.
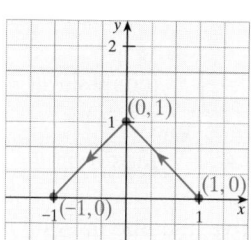

c.
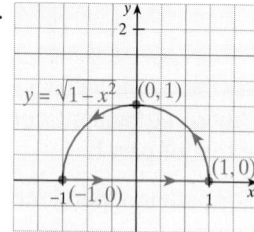

12. $\displaystyle\int_C (2x^2y\,dx + x^3\,dy)$

a.
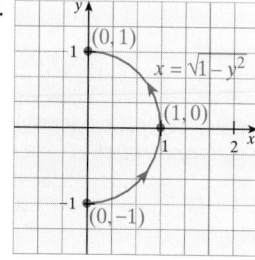

b.
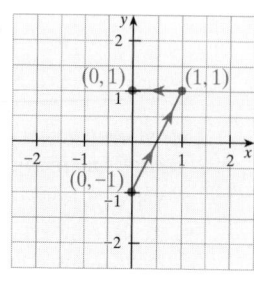

c.
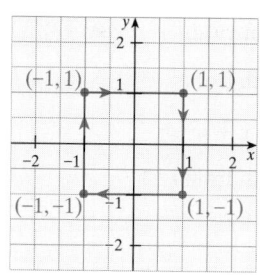

13. $\displaystyle\int_C (2xy\,dx + x^2\,dy)$

a.
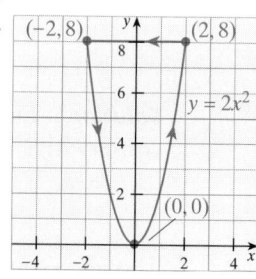

b.
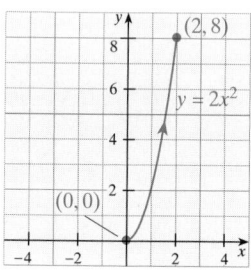

c.
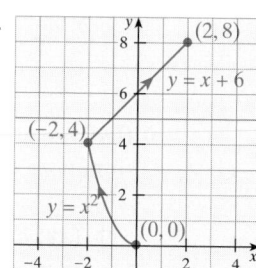

Show that the vector field $\mathbf{F}$ in Problems 14–19 is conservative, find a scalar potential f for $\mathbf{F}$, and then evaluate the line integral $\int_C \mathbf{F} \cdot d\mathbf{R}$, where C is any path connecting $A(0,0)$ to $B(1,1)$.

14. $\mathbf{F}(x,y) = (x + 2y)\mathbf{i} + (2x + y)\mathbf{j}$
15. $\mathbf{F}(x,y) = 2xy\mathbf{i} + x^2\mathbf{j}$
16. $\mathbf{F}(x,y) = (y - x^2)\mathbf{i} + (x + y^2)\mathbf{j}$
17. $\mathbf{F}(x,y) = (2x - y)\mathbf{i} + (y^2 - x)\mathbf{j}$
18. $\mathbf{F}(x,y) = e^{-y}\mathbf{i} - xe^{-y}\mathbf{j}$
19. $\mathbf{F}(x,y) = \dfrac{(y + 1)\mathbf{i} - x\mathbf{j}}{(y + 1)^2}$

Ⓑ 20. A force field

$$\mathbf{F}(x,y) = (3x^2 + 6xy^2)\mathbf{i} + (6x^2y + 4y^2)\mathbf{j}$$

acts on an object moving in the plane. Show that $\mathbf{F}$ is conservative, and find a scalar potential for $\mathbf{F}$. How much work is done as the object moves from $A(1,0)$ to $B(0,1)$ along any path connecting these points?

Verify that each line integral in Problems 21–26 is independent of path and then find its value.

21. $\displaystyle\int_C [(3x^2 + 2x + y^2)\,dx + (2xy + y^3)\,dy]$
    where C is any path from $(0,0)$ to $(1,1)$

22. $\int_C [(xy \cos xy + \sin xy)dx + (x^2 \cos xy)dy]$

where C is any path from $(0, \frac{\pi}{18})$ to $(1, \frac{\pi}{6})$

23. $\int_C [(y - x^2)dx + (x + y^2)dy]$

where C is any path from $(-1, -1)$ to $(0, 3)$

24. $\int_C [(3x^2y + y^2)dx + (x^3 + 2xy)dy]$

where C is the path given parametrically by
$\mathbf{R}(t) = t\mathbf{i} + (t^2 + t - 2)\mathbf{j}$ for $0 \le t \le 2$

25. $\int_C [(\sin y)dx + (3 + x \cos y)dy]$

where C is the path given parametrically by
$\mathbf{R}(t) = 2 \sin\left(\frac{\pi t}{2}\right) \cos(\pi t)\mathbf{i} + (\sin^{-1}t)\mathbf{j}$ for $0 \le t \le 1$

26. $\int_C [(e^x \cos y)dx + (-e^x \sin y)dy]$

where C is the path given parametrically by
$\mathbf{R}(t) = (\cos t)\mathbf{i} + (\sin t)\mathbf{j}$ for $0 \le t \le \frac{\pi}{2}$

Evaluate the line integrals given in Problems 27–31 using the fundamental theorem of line integrals.

27. $\int_C (y\mathbf{i} + x\mathbf{j}) \cdot d\mathbf{R}$, where C is any path from $(0, 0)$ to $(2, 4)$

28. $\int_C (xy^2\mathbf{i} + x^2y\mathbf{j}) \cdot d\mathbf{R}$, where C is any path from $(4, 1)$ to $(0, 0)$

29. $\int_C (2y \, dx + 2x \, dy)$, where C is the line segment from $(0, 0)$ to $(4, 4)$

30. $\int_C (e^x \sin y \, dx + e^x \cos y \, dy)$, where C is any smooth curve from $(0, 0)$ to $(0, 2\pi)$

31. $\int_C \left[\tan^{-1}\frac{y}{x} - \frac{xy}{x^2 + y^2} \right] dx + \left[\frac{x^2}{x^2 + y^2} + e^{-y}(1 - y) \right] dy$

where C is any smooth curve from $(1, 1)$ to $(-1, 2)$.

32. Find a function g so $g(x)\mathbf{F}(x, y)$ is conservative where

$$\mathbf{F}(x, y) = (x^2 + y^2 + x)\mathbf{i} + xy\mathbf{j}$$

33. Find a function g so $g(x)\mathbf{F}(x, y)$ is conservative where

$$\mathbf{F}(x, y) = (x^4 + y^4)\mathbf{i} - (xy^3)\mathbf{j}$$

34. a. Over what region in the xy-plane will the line integral

$$\int_C [(-yx^{-2} + x^{-1})dx + x^{-1}dy]$$

be independent of path?

b. Evaluate the line integral in part **a** if C is defined by

$$\mathbf{R}(t) = (\cos^3 t)\mathbf{i} + (\sin 3t)\mathbf{j}$$

for $0 \le t \le \frac{\pi}{3}$.

35. MODELING PROBLEM The **gravitational force field F** between two particles of masses M and m separated by a distance r is modeled by

$$\mathbf{F}(x, y, z) = -\frac{KmM}{r^3} \mathbf{R}$$

where $\mathbf{R} = x\mathbf{i} + y\mathbf{j} + z\mathbf{k}$ and K is the gravitational constant.

a. Show that $\mathbf{F}$ is conservative by finding a scalar potential for $\mathbf{F}$. The scalar potential function f is often called the **Newtonian potential.**

b. Compute the amount of work done against the force field $\mathbf{F}$ in moving an object from the point $P(a_1, b_1, c_1)$ to $Q(a_2, b_2, c_2)$.

36. Let $\mathbf{F}(x, y) = \frac{-y\mathbf{i} + x\mathbf{j}}{x^2 + y^2}$.

a. Compute the line integral $\int_{C_1} \mathbf{F} \cdot d\mathbf{R}$, where C is the upper semicircle $y = \sqrt{1 - x^2}$ traversed counterclockwise. What is the value of $\int_{C_2} \mathbf{F} \cdot d\mathbf{R}$ if C_2 is the lower semicircle $y = -\sqrt{1 - x^2}$ also traversed counterclockwise?

b. Show that if $\mathbf{F} = M\mathbf{i} + N\mathbf{j}$, then

$$\frac{\partial}{\partial y}\left(\frac{-y}{x^2 + y^2}\right) = \frac{\partial}{\partial x}\left(\frac{x}{x^2 + y^2}\right)$$

but $\mathbf{F}$ is not conservative on the unit disk $x^2 + y^2 \le 1$.

37. MODELING PROBLEM A person whirls a bucket filled with water in a circle of radius 3 ft at the rate of 1 revolution per second. If the bucket and water weigh 30 lb, how much work is done by the force that keeps the bucket moving in a circular path?

38. SPY PROBLEM Holding his breath for dear life, the Spy finally escapes from Purity's watery trap (Problem 59, Section 13.7). He and the water spill into the next room, and the door slams shut. He finds himself looking across a large, rectangular room at Purity and Blohardt. He takes a step toward his nemesis, but finds his movements restricted by a force field that appears to sap his strength. "I see you have noticed the Death Force my lovely assistant has designed for you," crows Blohardt with an evil laugh. "It will do you no good to stay still," adds Purity. "The rays will eventually kill you even if you never move!" One more peal of laughter, and they leave together. Alone, the Spy quickly presses the stem on his wristwatch and the equation of the force field appears on the face:

$$\mathbf{F}(x, y) = (ye^{xy} + 2xy^3)\mathbf{i} + (xe^{xy} + 3x^2y^2 + \cos y)\mathbf{j}$$

Assuming the Spy is at $(0, 0)$ on the wristwatch's coordinate system and that the door (and safety) is at $(10, 10)$, what path should he take to minimize the work of struggling against the Death Force Field while crossing the room in the least possible time?

In Problems 39 and 40, you are to experiment with the no-tion of computing work along a path—with and without the benefit of independence of path. Suppose you are to power a boat of some kind from point $A(0,0)$ to point $B(2,1)$, and the primary consideration is the force of the wind, which generally opposes you. You are to investigate the effect of taking different paths from A to B.

39. a. Suppose the wind force is
$\mathbf{F} = \langle -a, -b \rangle$, for a and b positive. Compute the work involved along the straight-line path between A and B; then along a second path, of your choice. Does the path matter here? Why or why not?

b. Due to the effect of the harbor you are entering, suppose the wind force is $\langle -a, -ae^{-y} \rangle$. Again, compute the work along two paths as in part **a.** Does the path matter here? Why or why not?

c. Repeat part **b** for

$$\langle -a, -ae^{-y+x/9} \rangle$$

40. An important type of problem in several fields of application is, "Can we find the optimal path to minimize the work?" You are to explore this issue in regard to the wind force in Problem 39c,

$$\mathbf{F} = \langle -a, -ae^{-y+x/9} \rangle$$

a. You should have the work computed for two different paths from the previous problem. By looking at these numbers and carefully studying $\mathbf{F}$, you should see that we will be rewarded (or punished) by changing the path slightly from the straight-line path. What do your observations suggest regarding trying to minimize the work?

b. Attempt to find an (approximate) optimal path, starting with your observations in part **a** and common sense. For example, the path could be described by a parabola (or higher-degree poly-nomial) or by some trigonometric function. Find a "good" path for this purpose.

c. Explore the following question: "Is there a real-istic optimal path from A to B?" Consider two conditions: first, there is no restriction on the path; second, suppose there is a shoreline at $y = 1$ so that one's path cannot exceed this limit. *Hint:* One approach might be to consider all parabolic paths of the form $y = x(b - ax)$, where a and b are nonnegative, and $y(2) = 1$. In this case, you can eliminate b and do a one-parameter study.

d. Assume the path in part **c** cannot go above $y = 1$. You may have discovered the optimal path. What is it? If you did the parabolic study suggested in part **c,** there is an optimal path in this case. What is it?

 41. Show that if the vector field $\mathbf{F}(x, y, z) = M(x, y, z)\mathbf{i} + N(x, y, z)\mathbf{j} + P(x, y, z)\mathbf{k}$ is conservative, then

$$\frac{\partial P}{\partial y} = \frac{\partial N}{\partial z} \qquad \frac{\partial M}{\partial z} = \frac{\partial P}{\partial x} \qquad \frac{\partial N}{\partial x} = \frac{\partial M}{\partial y}$$

42. It can be shown that $\mathbf{F} = M\mathbf{i} + N\mathbf{j} + P\mathbf{k}$ is conservative if curl $\mathbf{F} = 0$ whenever $M, N,$ and P have continuous par-tial derivatives in a ball, a box, or other "simply con-nected" region.

a. Show that the vector field defined by
$\mathbf{F}(x, y, z) = (y^2 - 2xz)\mathbf{i} + (2xy + z)\mathbf{j} + (y - x^2)\mathbf{k}$
is conservative.

b. Note that if $\nabla f = \mathbf{F}$, we must have

$$f_x = y^2 - 2xz, \quad f_y = 2xy + z, \quad f_z = y - x^2$$

Partially integrate f_x with respect to x to express f in terms of $x, y, z,$ and a function $c(y, z)$ that acts as a "constant" with respect to x-integration.

c. Find f_y and set it equal to $N = 2xy + z$. What can you conclude about the function $c(y, z)$? Partially integrate with respect to y to express f in terms of x, y, z and a function of z alone.

d. Find the scalar potential f.

43. Let f and g be differentiable functions of one variable. Show that the vector field $\mathbf{F} = [f(x) + y]\mathbf{i} + [g(y) + x]\mathbf{j}$ is conservative, and find the corresponding potential function.

44. Complete the proof of Theorem 14.4 by showing that if $\int_C \mathbf{F} \cdot d\mathbf{R} = 0$ for every closed curve C in a domain D where $\mathbf{F}$ is continuous, then $\mathbf{F}$ must be conservative in D. *Hint:* Let C_1 and C_2 be two curves in D with the same endpoints P and Q. Define a closed curve C so that

$$\int_C \mathbf{F} \cdot d\mathbf{R} = \int_{C_1} \mathbf{F} \cdot d\mathbf{R} - \int_{C_2} \mathbf{F} \cdot d\mathbf{R}$$

and conclude that $\int_C \mathbf{F} \cdot d\mathbf{R}$ is independent of path, so that $\mathbf{F}$ is conservative in D.

45. Let $\mathbf{R} = x\mathbf{i} + y\mathbf{j} + z\mathbf{k}$ and $r = \|\mathbf{R}\|$. Show that the work done in moving an object from a distance r_1 to a distance r_2 in the central force field $\mathbf{F} = \mathbf{R}/r^3$ is given by

$$W = \frac{1}{r_1} - \frac{1}{r_2}$$

46. An object of mass m moves along a trajectory $\mathbf{R}(t)$ with velocity $\mathbf{V}(t)$ in a conservative force field $\mathbf{F}(t_1)$. Let $\mathbf{R}(t_0) = \mathbf{Q}_0$ and $\mathbf{R}(t_1) = \mathbf{Q}_1$ be the initial and terminal points on the trajectory.

a. Show that the work done on the object is
$W = K(t_1) - K(t_0)$ where $K(t) = \frac{1}{2}m\|\mathbf{V}(t)\|^2$ is the object's *kinetic energy.*

b. Let f be a scalar potential for $\mathbf{F}$. Then $P(t) = -f(t)$ is the *potential energy* of the object. Prove the *law of conservation of energy*—namely,

$$P(t_0) + K(t_0) = P(t_1) + K(t_1)$$

14.4 Green's Theorem

IN THIS SECTION Green's theorem, area as a line integral, Green's theorem for multiply-connected regions, alternative forms of Green's theorem ■

The fundamental theorem of calculus, $\int_a^b \dfrac{dF}{dx}\,dx = F(b) - F(a)$, can be described as saying that when the derivative $\dfrac{dF}{dx}$ is integrated over the closed interval $a \le x \le b$, the result is the same as that obtained by evaluating $F(x)$ at the "boundary points" a and b and forming the difference $F(b) - F(a)$. We obtained the analogous result

$$\int_C \nabla f \cdot d\mathbf{R} = f(Q) - f(P)$$

in Section 14.3, and our next goal is to obtain a different kind of analogue to the fundamental theorem of calculus called **Green's theorem** after the English mathematician George Green (Historical Quest, page 995).

GREEN'S THEOREM

Green's theorem relates the double integral of a certain differential expression over a region in the plane to a line integral over a closed boundary curve of the region. We begin with some terminology. A **Jordan curve,** named for the French mathematician Camille Jordan (1838–1922) is a closed curve C in the plane that does not intersect itself (see Figure 14.14). A **simply connected** region has the property that it is connected and the interior of every Jordan curve C in D also lies in D, as shown in Figure 14.14.

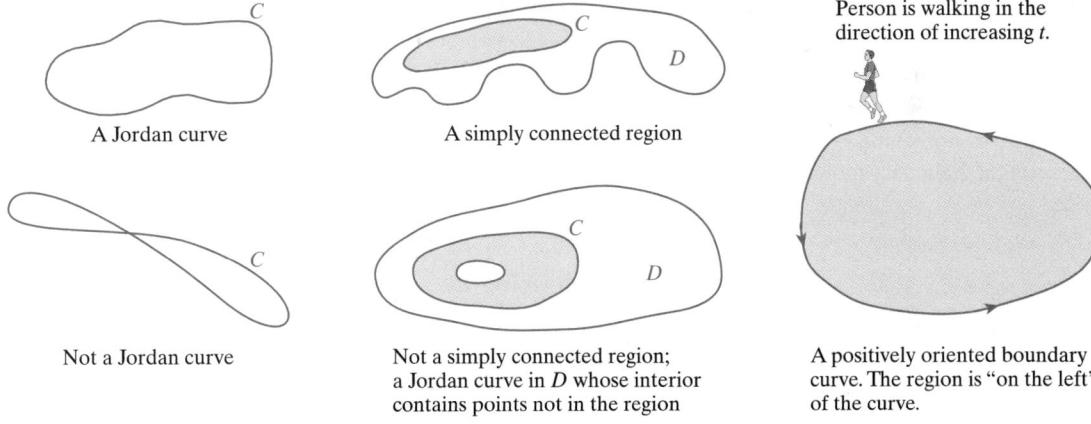

A Jordan curve

A simply connected region

Person is walking in the direction of increasing t.

Not a Jordan curve

Not a simply connected region; a Jordan curve in D whose interior contains points not in the region

A positively oriented boundary curve. The region is "on the left" of the curve.

■ **FIGURE 14.14** A Jordan curve is a closed curve with no self-intersections. A region is simply connected if every Jordan curve has all its interior points in D

Picture yourself as a point moving along a curve. If the region D stays on your *left* as you, the point, move along the curve C with increasing t, then C is said to be **positively oriented** (see Figure 14.14). Now we are ready to state Green's theorem.

THEOREM 14.5   *Green's theorem*

Let D be a simply connected region with a positively oriented piecewise-smooth boundary C. Then if the vector field $\mathbf{F}(x, y) = M(x, y)\mathbf{i} + N(x, y)\mathbf{j}$ is continuously differentiable on D, we have

$$\int_C (M\,dx + N\,dy) = \iint_D \left(\frac{\partial N}{\partial x} - \frac{\partial M}{\partial y}\right) dA$$

> ■ *What This Says:* This theorem expresses an important relationship between a line integral over a simple closed curve in the plane and a double integral over the region bounded by the curve. It is one of the most important and elegant theorems in calculus. Take special note that D is required to be simply connected with a *positively oriented* boundary C. This condition is so important that sometimes the notation
>
> $$\oint_C (M\,dx + N\,dy)$$
>
> is used to indicate the line integral to emphasize the positive (or counterclockwise) orientation.

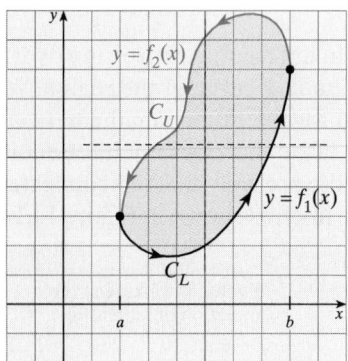

■ **FIGURE 14.15** Standard region: No vertical or horizontal line intersects the boundary more than twice

Proof A **standard region** is one in which no vertical or horizontal line can intersect the boundary curve more than twice (see Figure 14.15). We shall prove Green's theorem for the special case where D is a standard region, and then we shall indicate how to extend the proof to more general regions.

Suppose D is a standard region with boundary curve C. We begin by showing that

$$\iint_D \frac{\partial M}{\partial y}\,dx\,dy = -\int_C M\,dx$$

Because D is a standard region, as shown in Figure 14.15, the boundary curve C is composed of a lower portion C_L and an upper portion C_U, which are the graphs of functions $f_1(x)$ and $f_2(x)$, respectively, on a certain interval $a \le x \le b$. Then we can evaluate the double integral by iterated integration:

$$\iint_D \frac{\partial M}{\partial y}\,dx\,dy = \iint_D \frac{\partial M}{\partial y}\,dy\,dx = \int_a^b \left[\int_{f_1(x)}^{f_2(x)} \frac{\partial M}{\partial y}\,dy\right] dx$$

$$= \int_a^b M[x, f_2(x)]\,dx - \int_a^b M[x, f_1(x)]\,dx = \int_{-C_U} M\,dx - \int_{C_L} M\,dx$$

$$= -\left[\int_{C_U} M\,dx + \int_{C_L} M\,dx\right] = -\int_C M\,dx$$

A similar argument shows that $\iint_D \dfrac{\partial N}{\partial x}\,dx\,dy = \displaystyle\int_C N\,dy$. Thus,

$$\iint_D \left(\frac{\partial N}{\partial x} - \frac{\partial M}{\partial y}\right) dA = \iint_D \frac{\partial N}{\partial x}\,dx\,dy - \iint_D \frac{\partial M}{\partial y}\,dx\,dy$$

$$= \int_C N\,dy - \int_C (-M)\,dx = \int_C (M\,dx + N\,dy)$$

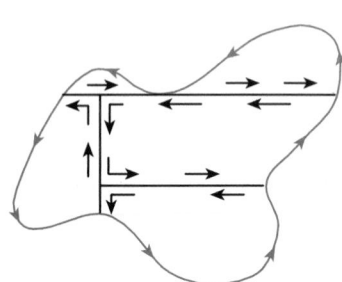

■ **FIGURE 14.16** General case: The region is decomposed into a finite number of standard regions by cuts

This completes the proof for the standard region. If D is not a standard region, it can be decomposed into a number of standard subregions by using horizontal and vertical "cuts," as shown in Figure 14.16. The proof for the standard region is then applied to each of these subregions, and the results are added. The line integrals along the cuts cancel in pairs, and after cancellation, the only remaining line integral is the one along the outer boundary C. Thus,

$$\int_C (M\,dx + N\,dy) = \iint_R \left(\frac{\partial N}{\partial x} - \frac{\partial M}{\partial y}\right) dA$$

The case where there is one cut is considered in Problem 37.

Graffiti on the wall of a high school playground in Tel Aviv. Courtesy of Regev Nathansohn.

EXAMPLE 1   *Verifying Green's theorem*

Verify Green's theorem for the line integral $\int_C (-y\,dx + x\,dy)$ where C is the closed path shown in the figure.

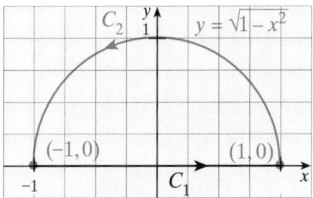

Solution   First, evaluate the line integral directly. The curve C consists of the line segment C_1 from $(-1, 0)$ to $(1, 0)$, followed by the semicircular arc C_2 from $(1, 0)$ back to $(-1, 0)$. We parametrize each of these:

$$C_1: x = t, y = 0 \quad -1 \le t \le 1 \qquad C_2: x = \cos s, y = \sin s \quad 0 \le s \le \pi$$
$$dx = dt, dy = 0 \qquad\qquad dx = -\sin s\,ds, dy = \cos s\,ds$$

$$\int_C (-y\,dx + x\,dy)$$
$$= \int_{C_1} (-y\,dx + x\,dy) + \int_{C_2} (-y\,dx + x\,dy)$$
$$= \int_{-1}^{1} [-0\,dt + t \cdot 0] + \int_0^\pi [-\sin s(-\sin s\,ds) + \cos s(\cos s\,ds)]$$
$$= \int_0^\pi (\sin^2 s + \cos^2 s)\,ds = \int_0^\pi 1\,ds = \pi$$

Next, we use Green's theorem to evaluate this integral. Note that the boundary curve C is simple and $M = -y$, $N = x$ so that $\mathbf{F}(x, y) = -y\mathbf{i} + x\mathbf{j}$ is continuously differentiable. The region D inside C is given by $0 \le y \le \sqrt{1 - x^2}$ for $-1 \le x \le 1$. We now apply Green's theorem:

$$\int_C (-y\,dx + x\,dy) = \iint_D \left(\frac{\partial}{\partial x}(x) - \frac{\partial}{\partial y}(-y)\right) dA = \int_{-1}^{1} \int_0^{\sqrt{1-x^2}} 2\,dy\,dx$$
$$= 2(\text{AREA OF SEMICIRCLE}) = 2[\tfrac{1}{2}\pi(1)^2] = \pi \qquad\blacksquare$$

EXAMPLE 2   *Computing work with Green's theorem*

A closed path C in the plane is defined by the given figure. Find the work done by an object moving along C in the force field

$$\mathbf{F}(x, y) = (x + xy^2)\mathbf{i} + 2(x^2y - y^2 \sin y)\mathbf{j}$$

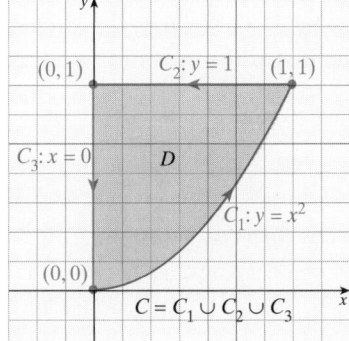

Solution   The work done, W, is given by the line integral $\int_C \mathbf{F} \cdot d\mathbf{R}$. Note that $\mathbf{F}$ is continuously differentiable on the region D enclosed by C, and since D is simply connected with a positively oriented boundary (namely, C), the hypotheses of Green's theorem are satisfied. We find that

$$W = \int_C \mathbf{F} \cdot d\mathbf{R} = \iint_D \left[\frac{\partial}{\partial x}(2x^2y - 2y^2 \sin y) - \frac{\partial}{\partial y}(x + xy^2) \right] dA$$

$$= \iint_D (4xy - 2xy)\, dA = 2\int_0^1 \int_{x^2}^1 xy\, dy\, dx = 2\int_0^1 \frac{1}{2}xy^2 \bigg|_{y=x^2}^{y=1} dx$$

$$= \int_0^1 (x - x^5)\, dx = \left[\frac{1}{2}x^2 - \frac{1}{6}x^6 \right]_0^1 = \frac{1}{3}$$ ∎

AREA AS A LINE INTEGRAL

A line integral can be used to compute an area of a region in the plane by applying the following theorem.

===

THEOREM 14.6   *Area as a line integral*

Let D be a simply connected region in the plane with piecewise smooth closed boundary C, as shown in Figure 14.17. Then the area A of region D is given by the line integral

$$A = \frac{1}{2}\int_C (-y\, dx + x\, dy)$$

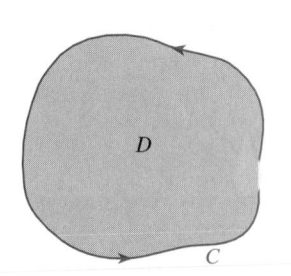

Proof   Let $\mathbf{F}(x, y) = -y\mathbf{i} + x\mathbf{j}$. Then since $\mathbf{F}$ is continuously differentiable on D, Green's theorem applies. We have

$$\int_C (-y\, dx + x\, dy) = \iint_D \left[\frac{\partial}{\partial x}(x) - \frac{\partial}{\partial y}(-y) \right] dA = \iint_D 2\, dA = 2A$$

■ **FIGURE 14.17**   Area of region D

so that $A = \dfrac{1}{2}\displaystyle\int_C (-y\, dx + x\, dy)$. ▆

> **WARNING** ▶   This gives us yet another technique for finding the area of a region, especially when its boundary is specified in parametric form. In finding area, the function $\mathbf{F} = -y\mathbf{i} + x\mathbf{j}$ does not change. Do not forget the factor one-half after you finish the integration. ◂

EXAMPLE 3   *Area enclosed by an ellipse*

Show that the ellipse $\dfrac{x^2}{a^2} + \dfrac{y^2}{b^2} = 1$ has area πab.

Solution   The elliptical path E is given parametrically by $x = a\cos\theta, y = b\sin\theta$ for $0 \leq \theta \leq 2\pi$. We find $dx = -a\sin\theta\, d\theta, dy = b\cos\theta\, d\theta$. If A is the area of this ellipse, then

$$A = \frac{1}{2}\int_C (-y\, dx + x\, dy)$$

$$= \frac{1}{2}\int_0^{2\pi} [-(b\sin\theta)(-a\sin\theta\, d\theta) + (a\cos\theta)(b\cos\theta\, d\theta)]$$

$$= \frac{1}{2}\int_0^{2\pi} ab(\sin^2\theta + \cos^2\theta)\, d\theta$$

$$= \frac{1}{2} \int_0^{2\pi} ab \, d\theta \qquad \textit{Because } \cos^2\theta + \sin^2\theta = 1$$

$$= \frac{1}{2} ab(2\pi - 0) = ab\pi \qquad\qquad \blacksquare$$

GREEN'S THEOREM FOR MULTIPLY-CONNECTED REGIONS

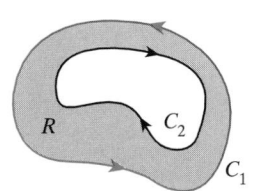

a. A doubly-connected region with oriented boundary curves C_1 and C_2.

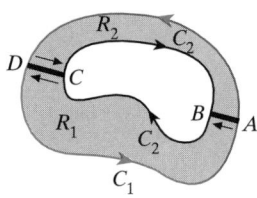

b. Two "cuts" are made through the hole

■ **FIGURE 14.18** Multiply-connected region

In the statement of Green's theorem, we require the region R inside the boundary curve C to be simply connected, but the theorem can be extended to multiply-connected regions—that is, regions with one or more "holes." A region with a single hole is shown in Figure 14.18a. The boundary of this region consists of an "outer" curve C_1 and an "inner" curve C_2, oriented so that the region R is always on the left as we travel around the boundary, which means that C_1 is oriented counterclockwise and C_2 clockwise.

Next, make cuts AB and CD through the region to the hole, as indicated in Figure 14.18b. Let R_1 be the simply connected region contained by the closed curve C_3 that begins at A, extends along the cut to B, and then clockwise along the bottom of the curve C_2 to C, along the cut to D, and counterclockwise along the bottom of C_1 back to A. Similarly, let R_2 be the region contained by the curve C_4 that begins at D and extends to C along the cut, to B along the top of C_2, to A along the second cut, and back to D along the top part of C_1. Then, if the vector field $\mathbf{F} = M\mathbf{i} + N\mathbf{j}$ is continuously differentiable on R, we can apply Green's theorem to show

$$\iint_R \left(\frac{\partial N}{\partial x} - \frac{\partial M}{\partial y} \right) dA = \iint_{R_1} \left(\frac{\partial N}{\partial x} - \frac{\partial M}{\partial y} \right) dA + \iint_{R_2} \left(\frac{\partial N}{\partial x} - \frac{\partial M}{\partial y} \right) dA$$

$$= \int_{C_3} (M \, dx + N \, dy) + \int_{C_4} (M \, dx + N \, dy)$$

But the line integrals from A to B and C to D cancel those from B to A and D to C, leaving only the line integrals along the original boundary curves C_1 and C_2, and Green's theorem for doubly-connected regions follows.

Green's Theorem for Doubly-Connected Regions

Let R be a doubly-connected region (one hole) in the plane, with outer boundary C_1 oriented counterclockwise and boundary C_2 of the hole oriented clockwise. If the boundary curves and $\mathbf{F}(x, y) = M(x, y)\mathbf{i}$ and $N(x, y)\mathbf{j}$ satisfy the hypotheses of Green's theorem then

$$\iint_R \left(\frac{\partial N}{\partial x} - \frac{\partial M}{\partial y} \right) dA = \int_{C_1} (M \, dx + N \, dy) + \int_{C_2} (M \, dx + N \, dy)$$

Example 4 illustrates one way this result can be used.

EXAMPLE 4 *Green's theorem for a region containing a singular point*

Show that $\displaystyle\int_C \frac{-y \, dx + x \, dy}{x^2 + y^2} = 2\pi$, where C is any piecewise smooth Jordan curve enclosing the origin $(0, 0)$.

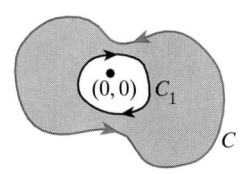

■ **FIGURE 14.19**   The region R for doubly-connected regions

Solution   Let $M(x, y) = \dfrac{-y}{x^2 + y^2}$ and $N = \dfrac{x}{x^2 + y^2}$. Then

$$\frac{\partial N}{\partial x} = \frac{y^2 - x^2}{(x^2 + y^2)^2} = \frac{\partial M}{\partial y}$$

at any point (x, y) other than the origin. Next, let C_1 be a circle centered at the origin with radius r so small that the entire circle is contained in C, and let R be the region between the curve C and the circle C_1, as shown in Figure 14.19.

We know that $\dfrac{\partial N}{\partial x} = \dfrac{\partial M}{\partial y}$ throughout R (since R does not contain the origin) and Green's theorem for doubly-connected regions tells us that

$$\int_C \frac{-y\,dx + x\,dy}{x^2 + y^2} + \int_{C_1} \frac{-y\,dx + x\,dy}{x^2 + y^2} = \iint_R \left(\frac{\partial N}{\partial x} - \frac{\partial M}{\partial y} \right) dA = 0$$

so

$$\int_C \frac{-y\,dx + x\,dy}{x^2 + y^2} = -\int_{C_1} \frac{-y\,dx + x\,dy}{x^2 + y^2} = \int_{C_1^*} \frac{-y\,dx + x\,dy}{x^2 + y^2}$$

where C_1^* is the circle C_1 traversed counterclockwise instead of clockwise. In other words, we can find the value of the given line integral about the curve C by finding the line integral about the circle C^*. To do this, we parametrize C^* by

$$x = \cos\theta, \quad y = \sin\theta \quad \text{for } 0 \le \theta \le 2\pi$$

and find that

$$\int_{C_1^*} \frac{-y\,dx + x\,dy}{x^2 + y^2} = \int_0^{2\pi} \frac{-\sin\theta(-\sin\theta\,d\theta) + \cos\theta\,(\cos\theta\,d\theta)}{\cos^2\theta + \sin^2\theta}$$

$$= \int_0^{2\pi} \frac{\sin^2\theta + \cos^2\theta}{\sin^2\theta + \cos^2\theta}\,d\theta = \int_0^{2\pi} 1\,d\theta = 2\pi$$

Thus,

$$\int_C \frac{-y\,dx + x\,dy}{x^2 + y^2} = \int_{C_1^*} \frac{-y\,dx + x\,dy}{x^2 + y^2} = 2\pi \qquad ■$$

ALTERNATIVE FORMS OF GREEN'S THEOREM

Green's theorem can be expressed in a form that generalizes nicely to $\mathbb{R}^3$. In particular, note that the curl of the vector field $\mathbf{F}(x, y) = M(x, y)\mathbf{i} + N(x, y)\mathbf{j}$ is given by

$$\text{curl } \mathbf{F} = \begin{bmatrix} \mathbf{i} & \mathbf{j} & \mathbf{k} \\ \dfrac{\partial}{\partial x} & \dfrac{\partial}{\partial y} & \dfrac{\partial}{\partial z} \\ M(x, y) & N(x, y) & 0 \end{bmatrix}$$

$$= \left(-\frac{\partial N}{\partial z} \right)\mathbf{i} + \left(\frac{\partial M}{\partial z} \right)\mathbf{j} + \left[\frac{\partial N}{\partial x} - \frac{\partial M}{\partial y} \right]\mathbf{k}$$

$$= \left[\frac{\partial N}{\partial x} - \frac{\partial M}{\partial y} \right]\mathbf{k} \qquad \textit{Because M and N are functions of only x and y}$$

Thus, the formula in Green's theorem can be expressed as

$$\int_C \mathbf{F} \cdot d\mathbf{R} = \iint_D (\text{curl } \mathbf{F} \cdot \mathbf{k}) \, dA$$

In Section 14.6, when we extend this result to surfaces in $\mathbb{R}^3$, it will be called *Stokes' theorem.*

EXAMPLE 5   *The divergence theorem in the plane*

Suppose $\mathbf{F}(x, y) = M(x, y)\mathbf{i} + N(x, y)\mathbf{j}$ with a piecewise smooth boundary C. Show that

$$\int_C \mathbf{F} \cdot \mathbf{N} \, ds = \iint_D \text{div } \mathbf{F} \, dA$$

This is the line integral of the normal component, $\mathbf{F} \cdot \mathbf{N}$.

Solution   Let $\mathbf{R}(s) = x(s)\mathbf{i} + y(s)\mathbf{j}$ so that a unit tangent vector $\mathbf{T}$ to the curve C is $\mathbf{T} = \mathbf{R}'(s) = x'(s)\mathbf{i} + y'(s)\mathbf{j}$, which means that an outward normal vector is $\mathbf{N} = y'(s)\mathbf{i} - x'(s)\mathbf{j}$. We now apply Green's theorem to find a representation using div $\mathbf{F}$.

$$\begin{aligned}
\int_C \mathbf{F} \cdot \mathbf{N} \, ds &= \int_a^b (M\mathbf{i} + N\mathbf{j}) \cdot [y'(s)\mathbf{i} - x'(s)\mathbf{j}] \, ds \\
&= \int_a^b \left(M\frac{dy}{ds} - N\frac{dx}{ds} \right) ds = \int_C (-N \, dx + M \, dy) \\
&= \iint_D \left(\frac{\partial M}{\partial x} + \frac{\partial N}{\partial y} \right) dx \, dy = \iint_D \text{div } \mathbf{F} \, dA \qquad \blacksquare
\end{aligned}$$

When we extend this result to $\mathbb{R}^3$ in Section 14.7, it will be called the *divergence theorem.*

We repeat these alternative forms of Green's theorem for easy reference.

Alternative Forms of Green's Theorem

Let D be a simply connected region with a positively oriented boundary C. Then if the vector field $\mathbf{F} = M\mathbf{i} + N\mathbf{j}$ is continuously differentiable on D, we have

$$\int_C \underbrace{\mathbf{F} \cdot d\mathbf{R}}_{\text{tangential component of } \mathbf{F}} = \int_C (M \, dx + N \, dy) = \iint_D \left(\frac{\partial N}{\partial x} - \frac{\partial M}{\partial y} \right) dA = \iint_D (\text{curl } \mathbf{F} \cdot \mathbf{k}) \, dA$$

$$\int_C \underbrace{\mathbf{F} \cdot \mathbf{N}}_{\text{normal component of } \mathbf{F}} \, ds = \iint_D \left(\frac{\partial M}{\partial x} + \frac{\partial N}{\partial y} \right) dA = \iint_D \text{div } \mathbf{F} \, dA$$

In physics, some important applications of Green's theorem involve the so-called *normal derivative* of a scalar function f, which is defined as the directional derivative of f in the direction of the outward normal vector $\mathbf{N}$ to some curve or surface.

Normal Derivative

The **normal derivative** of f, denoted by $\dfrac{\partial f}{\partial n}$, is the directional derivative of f in the direction of the normal vector pointing to the exterior of the domain of f. In other words,

$$\frac{\partial f}{\partial n} = \nabla f \cdot \mathbf{N}$$

where $\mathbf{N}$ is an outer unit normal vector.

The following example illustrates how Green's theorem can be used in connection with the normal derivative. Additional examples are found in the problem set.

EXAMPLE 6   *Green's formula for the integral of the Laplacian*

Suppose f is a scalar function with continuous first and second partial derivatives in the simply connected region D. If the piecewise smooth curve C bounds D, show that

$$\iint\limits_{D} \nabla^2 f \, dx \, dy = \int_{C} \frac{\partial f}{\partial n} \, ds$$

where $\nabla^2 f = f_{xx} + f_{yy}$ is the Laplacian of f and $\dfrac{\partial f}{\partial n} = \nabla f \cdot \mathbf{N}$ is the normal derivative vector.

Solution   Let $u = -\dfrac{\partial f}{\partial y}$ and $v = \dfrac{\partial f}{\partial x}$. Then we have

$$\nabla^2 f = f_{xx} + f_{yy} = \frac{\partial v}{\partial x} - \frac{\partial u}{\partial y}$$

$$
\begin{aligned}
\iint\limits_{D} \nabla^2 f \, dx \, dy &= \iint\limits_{D} \left(\frac{\partial v}{\partial x} - \frac{\partial u}{\partial y} \right) dx \, dy \\[2mm]
&= \int_{C} (u \, dx + v \, dy) \qquad \textit{Green's theorem} \\[2mm]
&= \int_{C} \left(u \frac{dx}{ds} + v \frac{dy}{ds} \right) ds \\[2mm]
&= \int_{C} \left(-\frac{\partial f}{\partial y} \frac{dx}{ds} + \frac{\partial f}{\partial x} \frac{dy}{ds} \right) ds \\[2mm]
&= \int_{C} \left(f_x \frac{dy}{ds} - f_y \frac{dx}{ds} \right) ds \\[2mm]
&= \int_{C} \nabla f \cdot \left(\frac{dy}{ds} \mathbf{i} - \frac{dx}{ds} \mathbf{j} \right) ds \\[2mm]
&= \int_{C} \nabla f \cdot \mathbf{N} \, ds \qquad \textit{Where } \mathbf{N} = \frac{dy}{ds} \mathbf{i} - \frac{dx}{ds} \mathbf{j} \textit{ is an outward unit} \\
& \qquad\qquad\qquad\qquad\quad \textit{normal vector to } C \\[2mm]
&= \int_{C} \frac{\partial f}{\partial n} \, ds
\end{aligned}
$$

∎

14.4   Problem Set

Ⓐ *Evaluate the line integrals in Problems 1–6 by applying Green's theorem. Check your answer by direct computation (that is, by parametrizing curve C).*

1. $\int_C (y^2\,dx + x^2\,dy)$

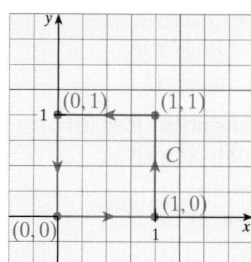

2. $\int_C (y^3\,dx - x^3\,dy)$

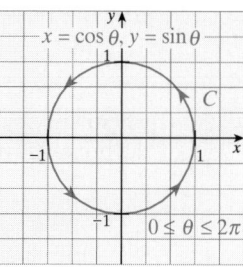

3. $\int_C [(2x^2 + 3y)\,dx - 3y^2\,dy]$

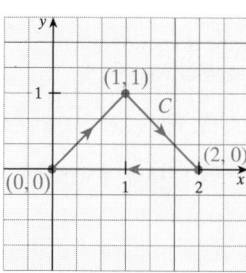

4. $\int_C (y^2\,dx + 3xy^2\,dy)$

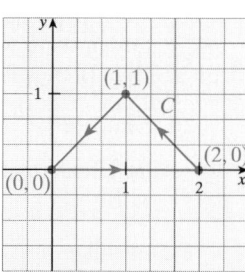

5. $\int_C 4xy\,dx$

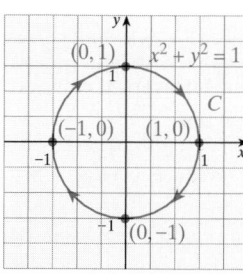

6. $\int_C (4y\,dx - 3x\,dy)$

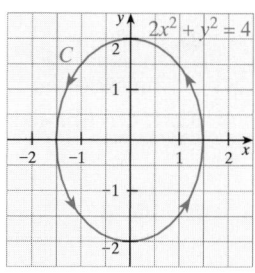

Evaluate the line integral in Problems 7–12 by applying Green's theorem.

7. $\int_C (2y\,dx - x\,dy)$

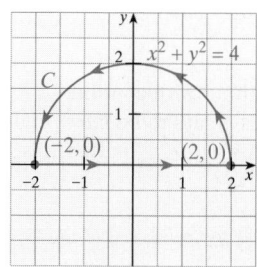

8. $\int_C (e^x\,dx - \sin x\,dy)$

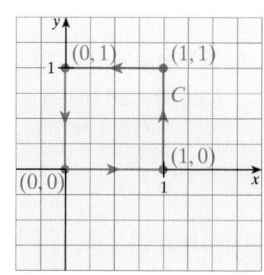

9. $\int_C (x \sin x\,dx - \tan e^{y^2}\,dy)$

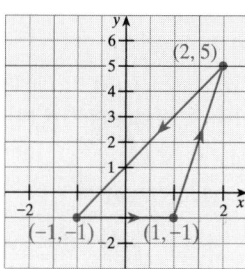

10. $\int_C [(x + y)\,dx - (3x - 2y)\,dy]$

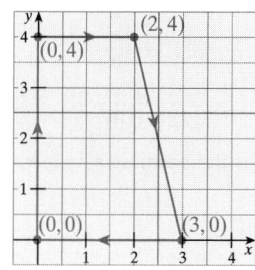

11. $\int_C [(x - y^2)\,dx + 2xy\,dy]$

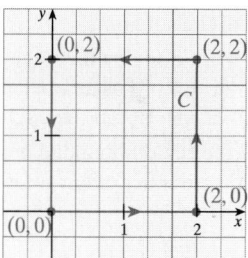

12. $\int_C (y^2\,dx + x\,dy)$

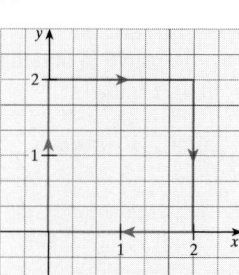

13. Use Green's theorem to find the work done by the force field

$$\mathbf{F}(x, y) = (3y - 4x)\mathbf{i} + (4x - y)\mathbf{j}$$

when an object moves once counterclockwise around the ellipse $4x^2 + y^2 = 4$.

14. Find the work done when an object moves in the force field $\mathbf{F}(x, y) = y^2\mathbf{i} + x^2\mathbf{j}$ counterclockwise along the circular path $x^2 + y^2 = 2$.

Use Theorem 14.6 to find the area enclosed by the regions described in Problems 15–18, and then check by using an appropriate formula.

15. circle $x^2 + y^2 = 4$

16. triangle with vertices $(0,0), (1,1)$, and $(0,2)$

17. trapezoid with vertices $(0,0), (4,0), (1,3)$, and $(0,3)$

18. semicircle $x = \sqrt{4 - y^2}$

B 19. Evaluate the line integral

$$\int_C (x^2y\,dx - y^2x\,dy)$$

where C is the boundary of the region between the x-axis and the semicircle $y = \sqrt{a^2 - x^2}$, traversed counterclockwise.

20. Evaluate the line integral

$$\int_C (3y\,dx - 2x\,dy)$$

where C is the cardioid $r = 1 + \sin\theta$, traversed counterclockwise.

21. Show that

$$\int_C [(5 - xy - y^2)\,dx - (2xy - x^2)\,dy] = 3\bar{x}$$

where C is the square $0 \le x \le 1, 0 \le y \le 1$ traversed counterclockwise and $\bar{x}$ is the x-coordinate of the centroid of the square.

22. Find the work done by the force field $\mathbf{F}(x, y) = (x + 2y^2)\mathbf{j}$ as an object moves once counterclockwise about the circle $(x - 2)^2 + y^2 = 1$.

23. Let D be the region bounded by the Jordan curve C, and let A be the area of D. If $(\bar{x}, \bar{y})$ is the centroid of D, show that

$$A\bar{x} = \frac{1}{2}\int_C x^2\,dy \quad \text{and} \quad A\bar{y} = -\frac{1}{2}\int_C y^2\,dx$$

where C is traversed counterclockwise.

24. Use Theorem 14.6 and the polar transformation formulas $x = r\cos\theta, y = r\sin\theta$ to obtain the area formula in polar coordinates, namely,

$$A = \frac{1}{2}\int_{\theta_1}^{\theta_2} r^2\,d\theta = \frac{1}{2}\int_{\theta_1}^{\theta_2} [g(\theta)]^2\,d\theta$$

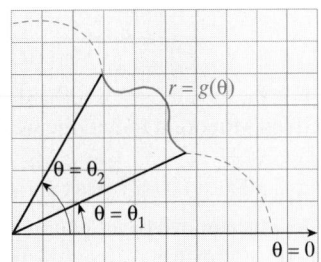

25. Evaluate $\int_C \left[\left(\dfrac{-y}{x^2} + \dfrac{1}{x} \right) dx + \dfrac{1}{x}\,dy \right]$, where C is any Jordan curve that does not touch or cross the y-axis, traversed counterclockwise.

26. Evaluate $\int_C \dfrac{x\,dx + y\,dy}{x^2 + y^2}$, where C is any Jordan curve whose interior does not contain the origin, traversed counterclockwise.

27. Evaluate the line integral

$$\int_C \frac{x\,dx + y\,dy}{x^2 + y^2}$$

where C is any piecewise smooth Jordan curve enclosing the origin, traversed counterclockwise.

28. Evaluate $\int_C \dfrac{-y\,dx + (x - 1)\,dy}{(x - 1)^2 + y^2}$, where C is any Jordan curve whose interior does not contain the point $(1, 0)$, traversed counterclockwise.

29. Evaluate $\int_C \dfrac{-(y + 2)\,dx + (x - 1)\,dy}{(x - 1)^2 + (y + 2)^2}$, where C is any Jordan curve whose interior does not contain the point $(1, -2)$, traversed counterclockwise.

30. Evaluate $\int_C \dfrac{\partial z}{\partial n}\,ds$, where $z(x, y) = 2x^2 + 3y^2$, and C is the circular path $x^2 + y^2 = 16$, traversed counterclockwise.

31. Evaluate $\int_C \dfrac{\partial f}{\partial n}\,ds$, where $f(x, y) = x^2y - 2xy + y^2$, and C is the boundary of the unit square $0 \le x \le 1, 0 \le y \le 1$, traversed counterclockwise.

32. Evaluate $\int_C x\dfrac{\partial x}{\partial n}\,ds$, where C is the boundary of the unit square $0 \le x \le 1, 0 \le y \le 1$, traversed counterclockwise.

33. If C is a closed curve, show that

$$\int_C [(x - 3y)\,dx + (2x - y^2)\,dy] = 5A$$

where A is the area of the region D enclosed by C.

C 34. Prove the following theoretical application of Green's theorem: Let $\mathbf{F}(x, y) = u(x, y)\mathbf{i} + v(x, y)\mathbf{j}$ be continuously differentiable on the simply connected region D. Then $\mathbf{F}$ is conservative if and only if

$$\frac{\partial v}{\partial x} = \frac{\partial u}{\partial y}$$

throughout D.

35. Recall that a scalar function f with continuous first and second partial derivatives is said to be *harmonic* in a region D if $\nabla^2 f = 0$ (that is, if $f_{xx} + f_{yy} = 0$). If f is such a function, show that

$$\iint_D (f_x^2 + f_y^2)\,dx\,dy = \int_C f\frac{\partial f}{\partial n}\,ds$$

36. **HISTORICAL QUEST** George Green (1793–1841) was the son of a baker who worked in his father's mill and studied mathematics and physics in his spare time, using only books he obtained from the library. In 1828, he published a memoir titled "An Essay on the Application of Mathematical Analysis to the Theories of Electricity and Magnetism," which contains the result that now bears his name. Very few copies of the essay were printed and distributed, so few people knew of Green's results. In 1833, at the age of 40, he entered Cambridge University and graduated just four years before his death. His 1828 paper was discovered and publicized in 1845 by Sir William Thompson (later known as Lord Kelvin, 1824–1907), and Green finally received proper credit for his work.

In this Quest, use Green's theorem to prove the following two important results, known as **Green's formulas.**

a. **Green's first formula:**

$$\iint_D [f\nabla^2 g + \nabla f \cdot \nabla g]\, dx\, dy = \int_C f\frac{\partial g}{\partial n}\, ds$$

b. Once again start with the line integral and derive what is known as **Green's second formula:**

$$\iint_D [f\nabla^2 g - g\nabla^2 f]\, dx\, dy$$
$$= \int_C \left(f\frac{\partial g}{\partial n} - g\frac{\partial f}{\partial n} \right) ds$$

37. **HISTORICAL QUEST** The result known as "Green's theorem" in the West is called "Ostrogradsky's theorem" in Russia, after Mikhail Ostrogradsky. Although Ostrogradsky published over 80 papers during a successful career as a mathematician, today he is known, even in his homeland, only for his version of Green's theorem, which appeared as part of a series of results presented to the Academy of Sciences in 1828.

MIKHAIL OSTROGRADSKY 1801–1862

In this Quest you are asked to prove Green–Ostrogradsky's theorem in the plane for the non-standard region D shown in Figure 14.20.

Specifically, suppose that the line L "cuts" the region D into two standard subregions D_1 and D_2. Apply Green's theorem to D_1 and D_2, then combine the results

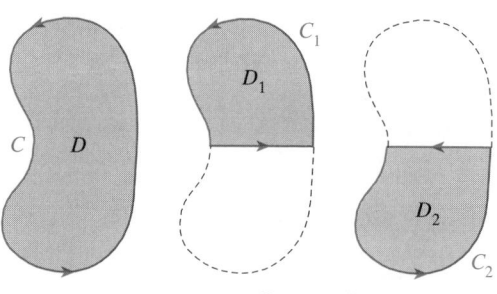

■ **FIGURE 14.20** Green's theorem

to show that the theorem also applies to the non-standard region D. *Hint*: The key is what happens along the "cut line" L.

38. Extend Green's theorem to a "triply-connected" region (two holes), such as the one shown in Figure 14.21.

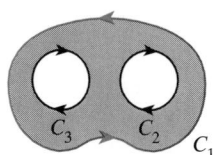

■ **FIGURE 14.21** A triply-connected region

39. Suppose $\mathbf{F} = M(x, y)\mathbf{i} + N(x, y)\mathbf{j}$ is continuously differentiable in a doubly-connected region R and that

$$\frac{\partial N}{\partial x} = \frac{\partial M}{\partial y}$$

throughout R. How many distinct values of I are there for the integral

$$I = \int_C [M(x, y)\, dx + N(x, y)\, dy]$$

where C is a piecewise smooth Jordan curve in R?

40. Answer the question in Problem 39 for the case where R is triply-connected (two holes). See Figure 14.21.

14.5 Surface Integration

IN THIS SECTION surface integrals, flux integrals, integrals over parametrically defined surfaces

SURFACE INTEGRALS

A *surface integral* is a generalization of the line integral in which the integration is over a surface in space rather than a curve. We shall be interested only in surfaces that are *piecewise smooth*—that is, those consisting of a finite number of pieces on which there is a continuously turning tangent plane.

We begin by defining the surface integral of a continuous scalar function $g(x, y, z)$ over a piecewise smooth surface S. Partition S into n subregions, the kth of which has

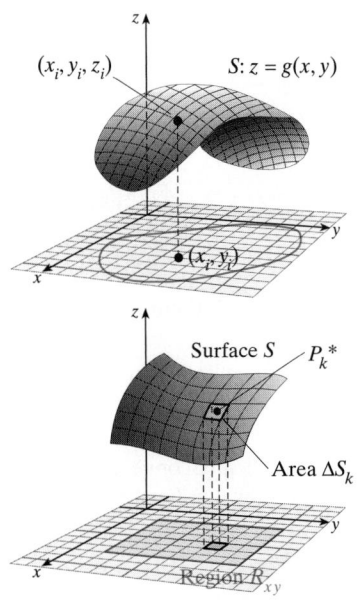

(x_i, y_i, z_i) $S: z = g(x, y)$

$\bullet (x_i, y_i)$

Surface S P_k^*

Area ΔS_k

Region R_{xy}

Definition of a surface integral

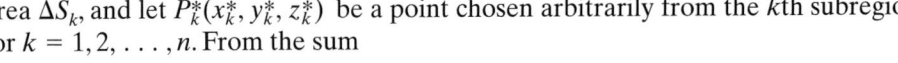

area ΔS_k, and let $P_k^*(x_k^*, y_k^*, z_k^*)$ be a point chosen arbitrarily from the kth subregion, for $k = 1, 2, \ldots, n$. From the sum

$$\sum_{k=1}^{n} g(P_k^*)\Delta S_k$$

and take the limit as the largest of the ΔS_k tends to 0, just as we did with surface area in Section 13.4. If this limit exists, it is called the **surface integral of g over S** and is denoted by

$$\iint_S g(x, y, z)\, dS$$

Recall from Section 13.4 that when the surface S projects onto the region R_{xy} in the xy-plane and S has the representation $z = f(x, y)$, then $dS = \sqrt{f_x^2 + f_y^2 + 1}\, dA_{xy}$, where dA_{xy} is either $dx\, dy$ or $dy\, dx$ (or $r\, dr\, d\theta$ if A_{xy} is described in terms of polar coordinates). We can now state the formula for evaluating a surface integral. (*Note:* We write R_{xy} and A_{xy} instead of the usual R and A to help you remember that these are regions in the xy-plane.)

Surface Integral

Let S be a surface defined by $z = f(x, y)$ and R_{xy} its projection on the xy-plane. If $f, f_x,$ and f_y are continuous in R_{xy} and g is continuous on S, then the **surface integral** of g over S is

$$\iint_S g(x, y, z)dS = \int_{R_{xy}} \int g(x, y, f(x, y))\sqrt{[f_x(x, y)]^2 + [f_y(x, y)]^2 + 1}\, dA_{xy}$$

EXAMPLE 1   *Evaluating a surface integral*

Evaluate the surface integral

$$\iint_S g\, dS$$

where $g(x, y, z) = xz + 2x^2 - 3xy$ and S is that portion of the plane $2x - 3y + z = 6$ that lies over the unit square R:

$$0 \le x \le 1, \quad 0 \le y \le 1$$

Solution   First, note that the equation of the plane can be written as $z = f(x, y)$, where $f(x, y) = 6 - 2x + 3y$. We have $f_x(x, y) = -2$ and $f_y(x, y) = 3$, so that

$$dS = \sqrt{f_x^2 + f_y^2 + 1}\, dA_{xy} = \sqrt{(-2)^2 + (3)^2 + 1}\, dA_{xy} = \sqrt{14}\, dA_{xy}$$

Consequently,

$$\iint_S g\, dS = \iint_S (xz + 2x^2 - 3xy)\sqrt{14}\, dA_{xy}$$

$$= \iint_S [x(6 - 2x + 3y) + 2x^2 - 3xy]\sqrt{14}\, dy\, dx$$

$$= \sqrt{14} \iint_S 6x\, dy\, dx$$

$$= 6\sqrt{14}\int_0^1 \int_0^1 x\, dy\, dx = 6\sqrt{14}\int_0^1 x\, dx = 3\sqrt{14}$$

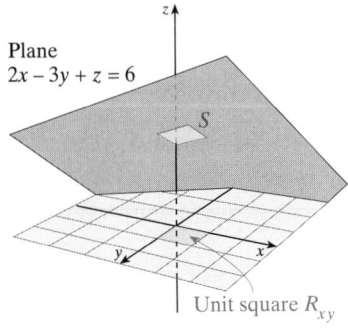

Plane
$2x - 3y + z = 6$

S

Unit square R_{xy}

The portion of the plane that lies above the unit square

If the function g defined on S is simply $g(x, y, z) = 1$, then the surface integral gives the *surface area* of S.

SURFACE AREA FORMULA

$$\text{Surface area} = \iint\limits_{S} dS$$

A useful application of surface integrals is to find the center of mass of a thin curved lamina whose shape is a given surface S, as shown in Figure 14.22.

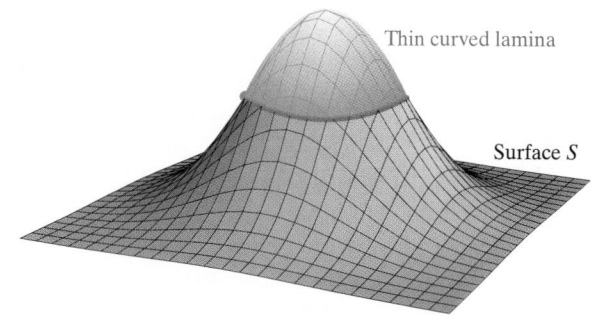

Thin curved lamina

Surface S

■ **FIGURE 14.22**   A thin lamina whose shape is the surface S

Suppose $\delta(x, y, z)$ is the density (mass per unit area) at a point (x, y, z) on the lamina. Then the *total mass, m, of the lamina* can also be represented by a surface integral.

MASS OF A LAMINA

$$m = \iint\limits_{S} \delta(x, y, z)\, dS$$

and the center of mass of the surface is the point $C(\overline{x}, \overline{y}, \overline{z})$, where

CENTER OF MASS

$$\overline{x} = \frac{1}{m} \iint\limits_{S} x\delta(x, y, z)\, dS, \quad \overline{y} = \frac{1}{m} \iint\limits_{S} y\delta(x, y, z)\, dS, \quad \overline{z} = \frac{1}{m} \iint\limits_{S} z\delta(x, y, z)\, dS$$

These formulas may be derived by essentially the same approach used in our previous work with moments and centroids of solid regions. (See, for example, Sections 6.5 and 13.6.)

EXAMPLE 2   *Mass of a curved lamina*

Find the mass of a lamina of density $\delta(x, y, z) = z$ in the shape of the hemisphere $z = (a^2 - x^2 - y^2)^{1/2}$, as shown in Figure 14.23.

Solution   We begin by calculating dS:

$$z_x = \frac{1}{2}(a^2 - x^2 - y^2)^{-1/2}(-2x) = -x(a^2 - x^2 - y^2)^{-1/2}$$

$$z_y = \frac{1}{2}(a^2 - x^2 - y^2)^{-1/2}(-2y) = -y(a^2 - x^2 - y^2)^{-1/2}$$

$$dS = \sqrt{z_x^2 + z_y^2 + 1}\, dA_{xy} = \sqrt{\frac{x^2}{a^2 - x^2 - y^2} + \frac{y^2}{a^2 - x^2 - y^2} + 1}\, dA_{xy}$$

$$= \sqrt{\frac{a^2}{a^2 - x^2 - y^2}}\, dA_{xy} = a(a^2 - x^2 - y^2)^{-1/2}\, dA_{xy}$$

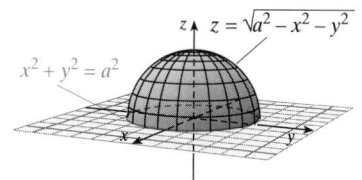

■ **FIGURE 14.23**   Note that S projects onto the circle R in the xy-plane with the equation $x^2 + y^2 = a^2$.

The mass of the hemisphere S is given by

$$
\begin{aligned}
m &= \iint_S \delta(x, y, z)\, dS = \iint_S z\, dS \\
&= \iint_R (a^2 - x^2 - y^2)^{1/2}\, a(a^2 - x^2 - y^2)^{-1/2}\, dA_{xy} \\
&= a \iint_R dA_{xy} = a(\pi a^2) = \pi a^3 \qquad \text{Since this integral represents the area of a circle of radius } a \quad \blacksquare
\end{aligned}
$$

FLUX INTEGRALS

Next, we shall discuss a special kind of surface integral called a *flux integral,* which is used in physics to study fluid flow and electrostatics. To define the *flux of a vector field* **F** across a given smooth surface S, we need to know that S is **orientable** in the sense that it is possible to define a field of unit normal vectors $\mathbf{N}(x, y, z)$ on S that varies continuously as the point (x, y, z) varies over S. Most common surfaces, such as planes, spheres, ellipses, and paraboloids are orientable, but it is not difficult to construct a fairly simple surface that is not. For example, the so-called **Möbius strip,** formed by twisting a long rectangular strip before joining the ends, is not orientable (see Figure 14.24a). In advanced calculus, it is shown that for an orientable surface S, the normal vector **N** must be either **outward** (pointing toward the exterior of S) or **inward** (pointing toward the interior), as illustrated in Figure 14.24b.

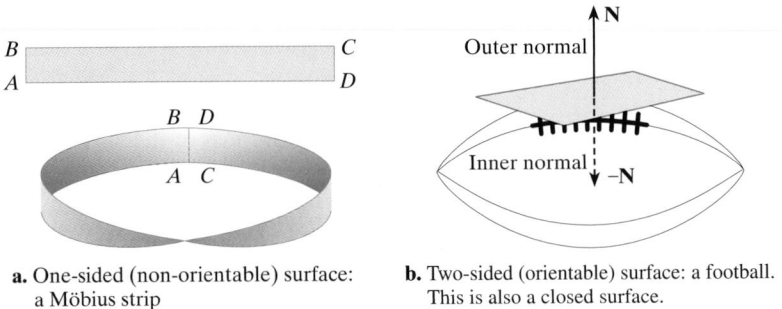

a. One-sided (non-orientable) surface: a Möbius strip

b. Two-sided (orientable) surface: a football. This is also a closed surface.

■ **FIGURE 14.24**   Surfaces in space

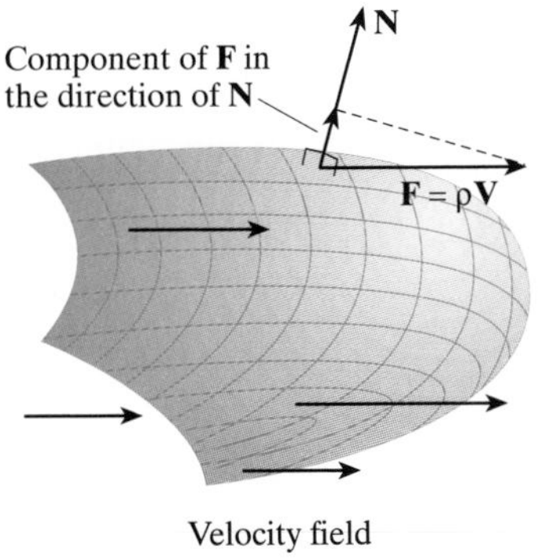

In Section 14.1, we defined the *flux density* as the vector field $\mathbf{F} = \delta\mathbf{V}$, where $\mathbf{V}(x, y, z)$ is the velocity of a fluid with density $\delta(x, y, z)$. The flux density measures the volume of fluid crossing surface S per unit of time and is also called the **flux of F across S.** Suppose the surface S is located in the region through which the fluid flows. If **N** is the unit normal vector to the surface S, the $\mathbf{F} \cdot \mathbf{N}$ is the component of flux in the direction normal vector to S, as shown in Figure 14.25. Then the mass of the fluid flowing through S in unit time in the direction normal vector to the surface is given by a surface integral, which is called a **flux integral.**

Component of **F** in the direction of **N**

$\mathbf{F} = \rho\mathbf{V}$

Velocity field

■ **FIGURE 14.25**   Flux density is measured by $\mathbf{F} = \delta\mathbf{V}$

Flux Integral

The **flux integral** of a vector field **F** across a surface S is given by the surface integral

$$
\iint_S \mathbf{F} \cdot \mathbf{N}\, dS
$$

where **N** is an outward normal vector to S.

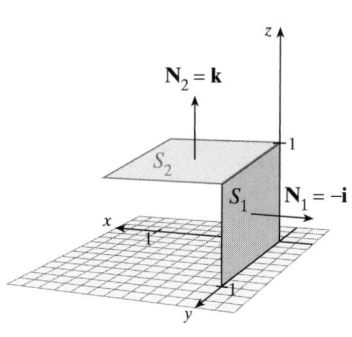

$N_2 = k$

S_2

S_1 $N_1 = -i$

EXAMPLE 3   *Surface integral of a vector field*

Compute $\displaystyle\iint_S \mathbf{F} \cdot \mathbf{N}\, dS$, where $\mathbf{F} = x\mathbf{i} - 5y\mathbf{j} + 4z\mathbf{k}$, and S is the union of the two squares

$$S_1: \quad x = 0, 0 \le y \le 1, 0 \le z \le 1$$
$$S_2: \quad z = 1, 0 \le x \le 1, 0 \le y \le 1$$

Solution   On S_1, the outward unit normal vector is $\mathbf{N}_1 = -\mathbf{i}$ and

$$\mathbf{F} \cdot \mathbf{N}_1 = (x\mathbf{i} - 5y\mathbf{j} + 4z\mathbf{k}) \cdot (-\mathbf{i}) = -x = 0$$

(because $x = 0$ on S_1). Thus,

$$\iint_{S_1} \mathbf{F} \cdot \mathbf{N}_1\, dS_1 = 0$$

On S_2, $\mathbf{N}_2 = \mathbf{k}$, and $\mathbf{F} \cdot \mathbf{N}_2 = \mathbf{F} \cdot \mathbf{k} = 4z = 4$ (because $z = 1$ on S_2). Thus,

$$\iint_{S_2} \mathbf{F} \cdot \mathbf{N}_2\, dS_2 = \iint_{S_2} 4\, ds_2 = 4(\text{area of } S_2) = 4$$

Finally,

$$\iint_S \mathbf{F} \cdot \mathbf{N}\, dS = \iint_{S_1} \mathbf{F} \cdot \mathbf{N}_1\, dS_1 + \iint_{S_2} \mathbf{F} \cdot \mathbf{N}_2\, dS_2 = 0 + 4 = 4 \quad \blacksquare$$

The unit normal vector of a surface is not always as obvious as it was in Example 3. For an orientable surface S defined by $z = g(x, y)$, we let $G(x, y, z) = z - g(x, y)$. Then, $\mathbf{N}$ can be found as follows (see Section 12.6):

$$\mathbf{N} = \frac{\nabla G(x, y, z)}{\|\nabla G(x, y, z)\|} = \frac{-g_x\mathbf{i} - g_y\mathbf{j} + \mathbf{k}}{\sqrt{g_x^2 + g_y^2 + 1}} \qquad \textbf{Outward unit normal vector}$$

$$\mathbf{N} = \frac{-\nabla G(x, y, z)}{\|\nabla G(x, y, z)\|} = \frac{g_x\mathbf{i} + g_y\mathbf{j} - \mathbf{k}}{\sqrt{g_x^2 + g_y^2 + 1}} \qquad \textbf{Inward unit normal vector}$$

EXAMPLE 4   *Evaluating a flux integral*

Compute $\displaystyle\iint_S \mathbf{F} \cdot \mathbf{N}\, dS$, where $\mathbf{F} = xy\mathbf{i} + z\mathbf{j} + (x + y)\mathbf{k}$ and S is the triangular region cut off from the plane $x + y + z = 1$ by the positive coordinate axes, as shown in Figure 14.26. Assume $\mathbf{N}$ is the unit normal vector that points away from the origin.

Solution   Let $g(x, y) = z = 1 - x - y$. Then $g_x = -1, g_y = -1$, and an outward unit normal vector to the plane is

$$\mathbf{N} = \frac{-(-1)\mathbf{i} - (-1)\mathbf{j} + \mathbf{k}}{\sqrt{(-1)^2 + (-1)^2 + 1}} = \frac{1}{\sqrt{3}}(\mathbf{i} + \mathbf{j} + \mathbf{k}) \quad \text{so that}$$

$$\mathbf{F} \cdot \mathbf{N} = (xy\mathbf{i} + z\mathbf{j} + (x + y)\mathbf{k}) \cdot \left(\frac{1}{\sqrt{3}}\mathbf{i} + \frac{1}{\sqrt{3}}\mathbf{j} + \frac{1}{\sqrt{3}}\mathbf{k}\right) = \frac{1}{\sqrt{3}}(xy + z + x + y)$$

We can write this as a function of x and y because $z = 1 - x - y$:

$$\mathbf{F} \cdot \mathbf{N} = \frac{1}{\sqrt{3}}(xy + 1 - x - y + x + y) = \frac{1}{\sqrt{3}}(xy + 1)$$

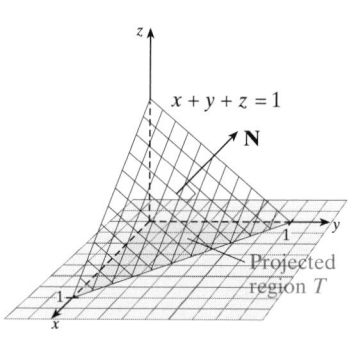

$x + y + z = 1$

$\mathbf{N}$

Projected region T

■ **FIGURE 14.26** $\mathbf{N}$ is the outward unit normal vector

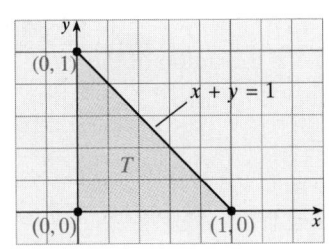

■ **FIGURE 14.27**   Projected region T from Figure 14.26

We also find that

$$dS = \sqrt{g_x^2 + g_y^2 + 1}\, dA_{xy} = \sqrt{3}\, dA_{xy}$$

The final piece of the puzzle we need before turning to the evaluation of the integral is to find the projection of S onto the xy-plane. We see from Figure 14.27 that is a triangular region, which can be described as the set of all points (x, y) such that for each x between 0 and 1, y varies between $y = 0$ and $y = 1 - x$ (as shown in Figure 14.27). Finally, we find that

$$\iint_S \mathbf{F} \cdot \mathbf{N}\, dS = \iint_T \frac{1}{\sqrt{3}} (xy + 1)\sqrt{3}\, dA_{xy}$$

$$= \int_0^1 \int_0^{1-x} (xy + 1)\, dy\, dx = \int_0^1 \left[\frac{1}{2}xy^2 + y \right]_{y=0}^{y=1-x} dx$$

$$= \frac{1}{2} \int_0^1 (x^3 - 2x^2 - x + 2)\, dx = \frac{1}{2} \left[\frac{1}{4}x^4 - \frac{2}{3}x^3 - \frac{1}{2}x^2 + 2x \right]_0^1 = \frac{13}{24} \qquad ■$$

INTEGRALS OVER PARAMETRICALLY DEFINED SURFACES

In Section 13.4, we showed that if a surface S is defined parametrically by the vector function

$$\mathbf{R}(u, v) = x(u, v)\mathbf{i} + y(u, v)\mathbf{j} + z(u, v)\mathbf{k}$$

over a region D in the uv-plane, the surface area of S is given by the integral

$$\iint_D \| \mathbf{R}_u \times \mathbf{R}_v \|\, du\, dv$$

Similarly, if f is continuous on D, the surface integral of f over D is given by

$$\iint_S f(x, y, z)\, dS = \iint_D f(\mathbf{R}) \| \mathbf{R}_u \times \mathbf{R}_v \|\, du\, dv$$

> **EXAMPLE 5**   *Surface integral for a surface defined parametrically*

Evaluate $\iint_S (x + y + z)\, dS$, where S is the surface defined parametrically by

$$\mathbf{R}(u, v) = (2u + v)\mathbf{i} + (u - 2v)\mathbf{j} + (u + 3v)\mathbf{k} \quad \text{for } 0 \le u \le 1, 0 \le v \le 2.$$

Solution

$$\iint_S (x + y + z)\, dS = \iint_D f(\mathbf{R}) \| \mathbf{R}_u \times \mathbf{R}_v \|\, du\, dv$$

Before we go on, we need to find the component parts for the integral on the right. Using $\mathbf{R} = x\mathbf{i} + y\mathbf{j} + z\mathbf{k}$, we see that for this problem $x = 2u + v$, $y = u - 2v$, and $z = u + 3v$. Because $f(x, y, z) = x + y + z$,

$$f(\mathbf{R}) = f(2u + v, u - 2v, u + 3v)$$

$$= 2u + v + u - 2v + u + 3v = 4u + 2v$$

$$\mathbf{R}_u = 2\mathbf{i} + \mathbf{j} + \mathbf{k} \text{ and } \mathbf{R}_v = \mathbf{i} - 2\mathbf{j} + 3\mathbf{k}, \text{ so}$$

$$\mathbf{R}_u \times \mathbf{R}_v = \begin{vmatrix} \mathbf{i} & \mathbf{j} & \mathbf{k} \\ 2 & 1 & 1 \\ 1 & -2 & 3 \end{vmatrix} = (3 + 2)\mathbf{i} - (6 - 1)\mathbf{j} + (-4 - 1)\mathbf{k} = 5\mathbf{i} - 5\mathbf{j} - 5\mathbf{k}$$

Thus, $\| \mathbf{R}_u \times \mathbf{R}_v \| = \sqrt{5^2 + (-5)^2 + (-5)^2} = 5\sqrt{3}$

We now substitute these values into the surface integral formula:

$$\iint_S (x + y + z) = \iint_D f(\mathbf{R}) \|\mathbf{R}_u \times \mathbf{R}_v\| \, du \, dv$$

$$= \int_0^2 \int_0^1 (4u + 2v)(5\sqrt{3}) \, du \, dv = 5\sqrt{3} \int_0^2 [2u^2 + 2u]_0^1 \, dv$$

$$= 5\sqrt{3} \int_0^2 (2 + 2v) \, dv = 5\sqrt{3} \, [2v + v^2]_0^2 = 5\sqrt{3} \, (8) = 40\sqrt{3} \qquad ■$$

14.5 Problem Set

Ⓐ *Let S be the hemisphere $x^2 + y^2 + z^2 = 4$, with $z \geq 0$, in Problems 1–6. Evaluate each surface integral.*

1. $\displaystyle\iint_S z \, dS$ **2.** $\displaystyle\iint_S z^2 \, dS$

3. $\displaystyle\iint_S (x - 2y) \, dS$ **4.** $\displaystyle\iint_S (5 - 2x) \, dS$

5. $\displaystyle\iint_S (x^2 + y^2)z \, dS$ **6.** $\displaystyle\iint_S (x^2 + y^2) \, dS$

In Problems 7–10, evaluate $\displaystyle\iint_S xy \, dS$.

7. $S: z = 2 - y, 0 \leq x \leq 2, 0 \leq y \leq 2$
8. $S: z = 4 - x - y, 0 \leq x \leq 4, 0 \leq y \leq 4$
9. $S: z = 5, x^2 + y^2 \leq 1$
10. $S: z = 10, \dfrac{x^2}{4} + \dfrac{y^2}{1} \leq 1$

In Problems 11–14, evaluate $\displaystyle\iint_S (x^2 + y^2) \, dS$.

11. $S: z = 4 - x - 2y, 0 \leq x \leq 4, 0 \leq y \leq 2$
12. $S: z = 4 - x, 0 \leq x \leq 2, 0 \leq y \leq 2$
13. $S: z = 4, x^2 + y^2 \leq 1$
14. $S: z = xy, x^2 + y^2 \leq 4, x \geq 0, y \geq 0$

In Problems 15–18, suppose S is the portion of the paraboloid $z = x^2 + y^2$ for which $z \leq 4$. Evaluate the given surface integral.

15. $\displaystyle\iint_S z \, dS$

16. $\displaystyle\iint_S (4 - z) \, dS$

17. $\displaystyle\iint_S \sqrt{1 + 4z} \, dS$

18. $\displaystyle\iint_S \dfrac{dS}{\sqrt{1 + 4z}}$

19. Evaluate $\displaystyle\iint_S (x^2 + y^2) \, dS$, where S is the surface bounded above by the hemisphere $z = \sqrt{1 - x^2 - y^2}$, and below by the plane $z = 0$.

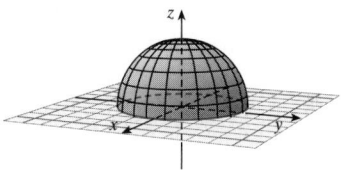

20. Evaluate $\displaystyle\iint_S 2x \, dS$, where S is the portion of the plane $x + y + z = 1$ with $x \geq 0, y \geq 0, z \geq 0$.

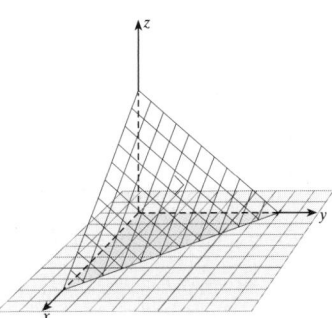

21. Evaluate $\displaystyle\iint_S (x^2 + y^2 + z^2) \, dS$, where S is the portion of the plane $z = x + 1$ that lies inside the cylinder $x^2 + y^2 = 1$.

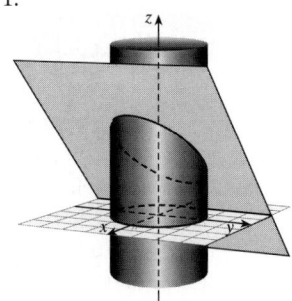

Evaluate $\iint\limits_{S} \mathbf{F} \cdot \mathbf{N} \, dS$ *for the vector fields* **F** *and surfaces S given in Problems 22–27.*

22. $\mathbf{F} = x\mathbf{i} + 2y\mathbf{j} + z\mathbf{k}$, and S is the triangular region bounded by the intersection of the plane $x + 2y + z = 1$ and the positive coordinate axes.

23. $\mathbf{F} = x\mathbf{i} + 2y\mathbf{j} - 3z\mathbf{k}$, and S is that part of the plane $15x - 12y + 3z = 6$ that lies above the unit square $0 \leq x \leq 1, 0 \leq y \leq 1$.

24. $\mathbf{F} = x\mathbf{i} + y\mathbf{j} + 2z\mathbf{k}$, and S is the surface of the cube bounded by the planes $x = 0, x = 1, y = 0, y = 1, z = 0$, and $z = 1$.

25. $\mathbf{F} = x\mathbf{i} + y\mathbf{j}$, and S is the hemisphere $z = \sqrt{1 - x^2 - y^2}$.

26. $\mathbf{F} = 2x\mathbf{i} - 3y\mathbf{j}$, and S is the part of the hemisphere given by $x^2 + y^2 + z^2 = 5$, for $z \geq 1$.

27. $\mathbf{F} = x^2\mathbf{i} + y^2\mathbf{j} + z^2\mathbf{k}$, and S is the portion of the plane $z = y + 1$ that lies inside the cylinder $x^2 + y^2 = 1$.

(B) 28. Evaluate $\iint\limits_{S} (3x - y + 2z) \, dS$, where S is the surface determined by $\mathbf{R}(u, v) = u\mathbf{i} + u\mathbf{j} - v\mathbf{k}, 0 \leq u \leq 1, 1 \leq v \leq 2$.

29. Evaluate $\iint\limits_{S} (x - y^2 + z) \, dS$, where S is the surface defined by $\mathbf{R}(u, v) = u^2\mathbf{i} + v\mathbf{j} + u\mathbf{k}, 0 \leq u \leq 1, 0 \leq v \leq 1$.

30. Evaluate $\iint\limits_{S} (\tan^{-1} x + y - z^2) \, dS$, where S is the surface defined by $\mathbf{R}(u, v) = u\mathbf{i} + v^2\mathbf{j} - v\mathbf{k}, 0 \leq u \leq 1, 0 \leq v \leq 1$.

31. Evaluate $\iint\limits_{S} (x^2 + y - z) \, dS$, where S is the surface defined by $\mathbf{R}(u, v) = u\mathbf{i} - u^2\mathbf{j} + v\mathbf{k}, 0 \leq u \leq 2, 0 \leq v \leq 1$.

32. Evaluate $\iint\limits_{S} (x + y + z) \, dS$, where S is the surface of the 1001cube $0 \leq x \leq 1, 0 \leq y \leq 1, 0 \leq z \leq 1$.

In Problems 33–38, find the mass of the homogeneous lamina that has the shape of the given surface S.

33. S is the surface $z = 4 - x - 2y$, with $z \geq 0, x \geq 0, y \geq 0$; $\delta = x$.

34. S is the surface $z = 10 - 2x - y$, with $z \geq 0, x \geq 0, y \geq 0$; $\delta = y$.

35. S is the surface $z = x^2 + y^2$, with $z \leq 1$; $\delta = z$.

36. S is the surface $z = 1 - x^2 - y^2$, with $z \geq 0$; $\delta = \rho^2$.

37. S is the surface $x^2 + y^2 + z^2 = 5$, with $z \geq 1$; $\delta = \theta^2$.

38. S is the triangular surface with vertices $(1, 0, 0), (0, 1, 0)$, and $(0, 0, 1)$; $\delta = x + y$.

(C) 39. a. A lamina has the shape of the portion of the sphere $x^2 + y^2 + z^2 = a^2$ that lies within the cone $z = \sqrt{x^2 + y^2}$. Determine the mass of the lamina if $\delta(x, y, z) = x^2y^2z$.

b. Let S be the spherical shell centered at the origin with radius a, and let C be the right circular cone whose vertex is at the origin and whose axis of symmetry coincides with the z-axis. (This has the shape of an old-fashioned ice cream cone.) Suppose the vertex angle of the cone is ϕ_0, with $0 \leq \phi_0 < \frac{\pi}{2}$. Determine the mass of that portion of the sphere that is enclosed in the intersection of S and C. Assume $\delta(x, y, z) = x^2y^2z$.

40. Recall the formula

$$I_z = \iint\limits_{S} (x^2 + y^2) \delta(x, y, z) \, dS$$

for the moment of inertia about the z-axis. Show that the moment of inertia of a conical shell about its axis is $\frac{1}{2}ma^2$, where m is the mass and a is the radius of the cone. Assume $\delta(x, y, z) = 1$.

41. Recall the formula

$$I_z = \iint\limits_{S} (x^2 + y^2) \delta(x, y, z) \, dS$$

for the moment of inertia about the z-axis. Show that the moment of inertia of a spherical shell of uniform density about its diameter is $\frac{2}{3}ma^2$, where m is the mass and a is the radius. Assume $\delta(x, y, z) = 1$.

42. THINK TANK PROBLEM   Show that a Möbius strip is not orientable.

14.6 Stokes' Theorem

IN THIS SECTION   Stokes' theorem, theoretical applications of Stokes' theorem, physical interpretation of Stokes' theorem

In Section 14.4, we observed that Green's theorem can be written

$$\int_{C} \mathbf{F} \cdot d\mathbf{R} = \iint\limits_{A} (\text{curl } \mathbf{F} \cdot \mathbf{k}) \, dA$$

where A is the plane region bounded by the closed curve C. Stokes' theorem is a generalization of this result to surfaces in space and their boundaries.

STOKES' THEOREM

Before stating Stokes' theorem, we need to explain what is meant by a *compatible orientation*. We say that the orientation of a closed path C on the surface S is **compatible with the orientation on S** if the positive direction is *counterclockwise* in relation to the outward normal vector of the surface (see Figure 14.28). If you point the thumb of your right hand in the direction of the outward unit normal vector, your fingers will curl in the direction of the compatibly oriented curve C.

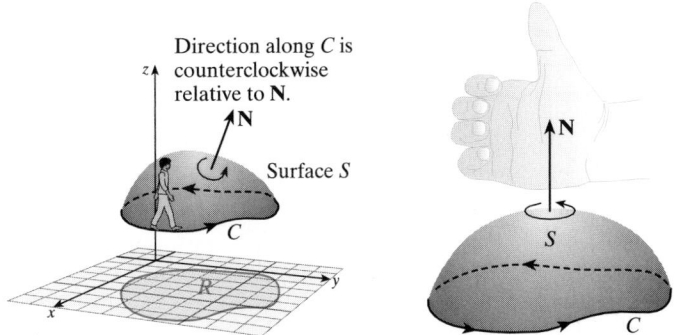

■ **FIGURE 14.28**   Compatible orientation: The surface S is on the left of someone walking in a counterclockwise direction around the boundary curve C

THEOREM 14.7   *Stokes' theorem*

Let S be an oriented surface with unit normal vector $\mathbf{N}$, and assume that S is bounded by a closed, piecewise smooth curve C whose orientation is compatible with that of S. If $\mathbf{F}$ is a vector field that is continuously differentiable on S, then

$$\int_C \mathbf{F} \cdot d\mathbf{R} = \iint_S (\text{curl } \mathbf{F} \cdot \mathbf{N})\, dS$$

Proof   The general proof is beyond the scope of this text. However, a proof assuming $\mathbf{F}$, S, and C are "well behaved" is given in Appendix B.   ═══

Notice that the surface S in the statement of Stokes' theorem is *not* required to be closed. That is, S does not have to bound a solid region.

EXAMPLE 1   *Verifying Stokes' theorem for a particular example*

Let $\mathbf{F} = -\frac{3}{2}y^2\mathbf{i} - 2xy\mathbf{j} + yz\mathbf{k}$, where S is that part of the surface of the plane $x + y + z = 1$ contained within triangle C with vertices $(1, 0, 0), (0, 1, 0)$, and $(0, 0, 1)$, traversed counterclockwise as viewed from above. Verify Stokes' theorem.

Solution   The planar surface S and boundary curve C are shown in Figure 14.29. We shall show that the line integral $\int_C \mathbf{F} \cdot d\mathbf{R}$ and the surface integral $\int_S\int (\text{curl } \mathbf{F} \cdot \mathbf{N})\, dS$ have the same value.

> **I.** *Evaluation of* $\displaystyle\int_C \mathbf{F} \cdot d\mathbf{R}$

The three edges of the boundary triangle C are expressed as

$$E_1: \quad x + y = 1, z = 0$$
$$E_2: \quad y + z = 1, x = 0$$
$$E_3: \quad x + z = 1, y = 0$$

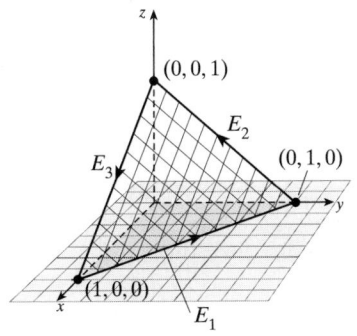

■ **FIGURE 14.29**   The triangle C bounds the part of the plane $x + y + z = 1$ that lies in the first octant

We traverse C in a counterclockwise direction (see Figure 14.29).

Edge E_1: Parametrize with $x = 1 - t$, $y = t$, $z = 0$, for $0 \le t \le 1$, so $\mathbf{R}(t) = (1 - t)\mathbf{i} + t\mathbf{j}$ and $d\mathbf{R} = -dt\mathbf{i} + dt\mathbf{j}$. Finally, in terms of the parameter t, we have $\mathbf{F}(t) = -\frac{3}{2}t^2\mathbf{i} - 2t(1 - t)\mathbf{j}$.

$$\int_{E_1} \mathbf{F} \cdot d\mathbf{R} = \int_0^1 \left(\frac{3}{2}t^2 - 2t + 2t^2\right) dt = \int_0^1 \left(\frac{7}{2}t^2 - 2t\right) dt = \left(\frac{7}{6}t^3 - t^2\right)\Big|_0^1 = \frac{1}{6}$$

Edge E_2: Parametrize with $x = 0$, $y = 1 - s$, $z = s$, for $0 \le s \le 1$, so $\mathbf{R}(s) = (1 - s)\mathbf{j} + s\mathbf{k}$ and $d\mathbf{R} = -ds\mathbf{j} + ds\mathbf{k}$. In terms of the parameter s, we have $\mathbf{F}(s) = -\frac{3}{2}(1 - s)^2\mathbf{i} + (1 - s)s\mathbf{k}$.

$$\int_{E_2} \mathbf{F} \cdot d\mathbf{R} = \int_0^1 (1 - s)s \, ds = \left(\frac{s^2}{2} - \frac{s^3}{3}\right)\Big|_0^1 = \frac{1}{6}$$

Edge E_3: Parametrize with $x = r$, $y = 0$, $z = 1 - r$, for $0 \le r \le 1$, so $\mathbf{R}(r) = r\mathbf{i} + (1 - r)\mathbf{k}$ and $d\mathbf{R} = dr\mathbf{i} - dr\mathbf{k}$. In terms of the parameter r, we have $\mathbf{F}(r) = \mathbf{0}$.

$$\int_{E_3} \mathbf{F} \cdot d\mathbf{R} = 0$$

Combining these results, we find

$$\int_C \mathbf{F} \cdot d\mathbf{R} = \int_{E_1} \mathbf{F} \cdot d\mathbf{R} + \int_{E_2} \mathbf{F} \cdot d\mathbf{R} + \int_{E_3} \mathbf{F} \cdot d\mathbf{R} = \frac{1}{6} + \frac{1}{6} + 0 = \frac{1}{3}$$

II. *Evaluation of* $\iint_S (\text{curl } \mathbf{F} \cdot \mathbf{N}) \, dS$

$$\text{curl } \mathbf{F} = \begin{vmatrix} \mathbf{i} & \mathbf{j} & \mathbf{k} \\ \dfrac{\partial}{\partial x} & \dfrac{\partial}{\partial y} & \dfrac{\partial}{\partial z} \\ -\frac{3}{2}y^2 & -2xy & yz \end{vmatrix} = z\mathbf{i} + y\mathbf{k}$$

The triangular region on the surface of the plane $x + y + z = 1$ has outward unit normal vector

$$\mathbf{N} = \frac{1}{\sqrt{3}}(\mathbf{i} + \mathbf{j} + \mathbf{k})$$

Because $z = 1 - x - y$, we find

$$\text{curl } \mathbf{F} \cdot \mathbf{N} = (z\mathbf{i} + y\mathbf{k}) \cdot \frac{1}{\sqrt{3}}(\mathbf{i} + \mathbf{j} + \mathbf{k})$$

$$= \frac{1}{\sqrt{3}}(1 - x - y + y) = \frac{1}{\sqrt{3}}(1 - x)$$

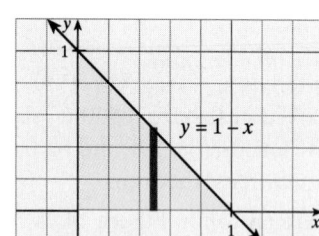

■ **FIGURE 14.30** The projected region in the xy-plane is a triangle bounded by $y = 1 - x$ and the positive coordinate axes.

Finally,

$$dS = \sqrt{z_x^2 + z_y^2 + 1} \, dA_{xy} = \sqrt{(-1)^2 + (-1)^2 + 1} \, dA_{xy} = \sqrt{3} \, dA_{xy}$$

Combining these results, we get (see Figure 14.30)

$$\iint_S (\text{curl } \mathbf{F} \cdot \mathbf{N}) \, dS = \iint_D \frac{1}{\sqrt{3}}(1 - x)\sqrt{3} \, dA_{xy} = \int_0^1 \int_0^{1-x} (1 - x) \, dy \, dx$$

$$= \int_0^1 (1 - x)[(1 - x) - 0] \, dx = -\frac{(1 - x)^3}{3}\Big|_0^1 = \frac{1}{3}$$

We see that for this example,

$$\int_C \mathbf{F} \cdot d\mathbf{R} = \frac{1}{3} = \iint_S (\text{curl } \mathbf{F} \cdot \mathbf{N}) \, dS$$

as claimed by Stokes' theorem. ∎

EXAMPLE 2   *Using Stokes' theorem to evaluate a line integral*

Evaluate $\int_C (\frac{1}{2}y^2 \, dx + z \, dy + x \, dz)$ where C is the curve of intersection of the plane $x + z = 1$ and the ellipsoid $x^2 + 2y^2 + z^2 = 1$, oriented clockwise as seen from the origin. The curve C is shown in Figure 14.31.

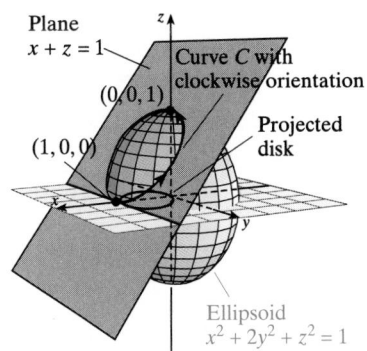

■ FIGURE 14.31   The surface bounded by the curve of intersection projects onto a disk in the xy-plane

Solution   If we set $\mathbf{F} = \frac{1}{2}y^2\mathbf{i} + z\mathbf{j} + x\mathbf{k}$, the given line integral can be expressed as

$$\int_C \mathbf{F} \, dR$$

According to Stokes' theorem, we have

$$\int_C \mathbf{F} \, dR = \iint_S (\text{curl } \mathbf{F} \cdot \mathbf{N}) \, dS$$

We choose S to be the part of the plane $x + z = 1$ bounded by C, along with C itself, then find the required parts of the surface integral; namely, curl $\mathbf{F}$, $\mathbf{N}$, and dS.

$$\text{curl } \mathbf{F} = \begin{vmatrix} \mathbf{i} & \mathbf{j} & \mathbf{k} \\ \frac{\partial}{\partial x} & \frac{\partial}{\partial y} & \frac{\partial}{\partial z} \\ \frac{1}{2}y^2 & z & x \end{vmatrix} = -\mathbf{i} - \mathbf{j} - y\mathbf{k}$$

The outward unit normal vector to the plane $x + z = 1$ is $\mathbf{N} = \frac{1}{\sqrt{2}}(\mathbf{i} + \mathbf{k})$, so that

$$\text{curl } \mathbf{F} \cdot \mathbf{N} = \frac{1}{\sqrt{2}}(-1 - y)$$

and since $z = 1 - x$ on S, we have $z_x = -1, z_y = 0$, and

$$dS = \sqrt{z_x^2 + z_y^2 + 1} \, dA_{xy} = \sqrt{(-1)^2 + (0)^2 + 1} \, dA_{xy} = \sqrt{2} \, dA_{xy}$$

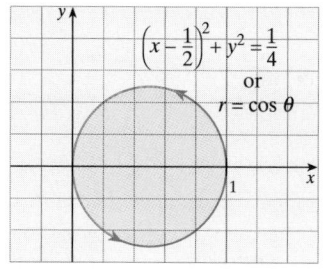

Finally, to describe C we substitute $z = 1 - x$ into the equation for the ellipsoid:

$$x^2 + 2y^2 + z^2 = 1$$
$$x^2 + 2y^2 + (1 - x)^2 = 1$$
$$x^2 - x + y^2 = 0$$
$$(x - \tfrac{1}{2})^2 + y^2 = \tfrac{1}{4}$$

Thus, the projection of C on the xy-plane is the circle S_{xy} shown in Figure 14.32. Notice that this circle has the polar equation $r = \cos\theta$. Note also that although C is oriented clockwise as viewed from the origin, the projected curve in the xy-plane is oriented counterclockwise as viewed from above.

We now calculate the desired line integral using Stokes' theorem:

$$\int_C (\tfrac{1}{2}y^2\,dx + z\,dy + x\,dz) = \int_C \mathbf{F} \cdot d\mathbf{R}$$

$$= \iint_S (\text{curl } \mathbf{F} \cdot \mathbf{N})\,ds \qquad \textit{Stokes' theorem}$$

$$= \int_{S_{xy}}\!\!\int \frac{1}{\sqrt{2}}(-1 - y)\sqrt{2}\,dA_{xy} = -\int_{S_{xy}} (1 + y)\,dA_{xy}$$

$$= -\int_{-\pi/2}^{\pi/2}\int_0^{\cos\theta} (1 + r\sin\theta)\,r\,dr\,d\theta \qquad \textit{Change to polar coordinates.}$$

$$= -\int_{-\pi/2}^{\pi/2}\left[\frac{1}{2}\cos^2\theta + \frac{1}{3}\cos^3\theta\sin\theta\right]d\theta = -\frac{\pi}{4} \qquad ■$$

Sometimes it is possible to exchange a particularly difficult surface integration over one surface for a less difficult integration over another surface with the same boundary curve. Suppose two surfaces S_1 and S_2 are bounded by the same curve C and induce the same orientation on C. Stokes' theorem tells us that

$$\iint_{S_1} (\text{curl } \mathbf{F} \cdot \mathbf{N}_1)\,dS_1 = \int_C \mathbf{F} \cdot d\mathbf{R} = \iint_{S_2} (\text{curl } \mathbf{F} \cdot \mathbf{N}_2)\,dS_2$$

for any function $\mathbf{F}$ whose components have continuous partial derivatives on both S_1 and S_2.

EXAMPLE 3   *Using Stokes' theorem to evaluate a surface integral*

Evaluate $\iint_S (\text{curl } \mathbf{F} \cdot \mathbf{N})\,dS$, where $\mathbf{F} = x\mathbf{i} + y^2\mathbf{j} + ze^{xy}\mathbf{k}$ and S is that part of the surface $z = 1 - x^2 - 2y^2$ with $z \geq 0$.

Solution   By setting $z = 0$ in the equation of the surface, we find that the boundary curve C for S is the ellipse $x^2 + 2y^2 = 1$, and the outward unit normal vector on S induces a counterclockwise orientation on C, as shown in Figure 14.33.

Let S^* be the elliptical disk defined by $x^2 + 2y^2 \leq 1$. We see that S and S^* have the same boundary, the same orientation, and the same normal vector $N_1 = \mathbf{k}$. Thus,

$$\iint_S (\text{curl } \mathbf{F} \cdot \mathbf{N})\,dS = \iint_{S^*} (\text{curl } \mathbf{F} \cdot \mathbf{k})\,dS^*$$

and we use this second integral to calculate the required integral. We obtain

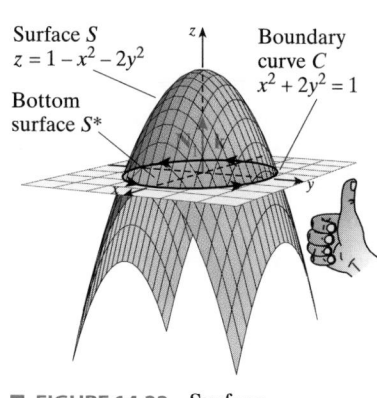

$$\text{curl } \mathbf{F} = \begin{vmatrix} \mathbf{i} & \mathbf{j} & \mathbf{k} \\ \dfrac{\partial}{\partial x} & \dfrac{\partial}{\partial y} & \dfrac{\partial}{\partial z} \\ x & y^2 & ze^{xy} \end{vmatrix} = (zxe^{xy})\mathbf{i} - (zye^{xy})\mathbf{j}$$

and because curl $\mathbf{F} \cdot \mathbf{k} = 0$, we conclude

$$\iint\limits_{S} (\text{curl }\mathbf{F} \cdot \mathbf{N})\, dS = \iint\limits_{S*} (\text{curl }\mathbf{F} \cdot \mathbf{k})\, dS^* = \iint\limits_{S*} 0\, dS^* = 0 \qquad \blacksquare$$

THEORETICAL APPLICATIONS OF STOKES' THEOREM

In physics and other applied areas, Stokes' theorem is often used as a device for establishing general properties. For instance, we can use it to prove that a vector field $\mathbf{F}$ is conservative if and only if curl $\mathbf{F} = \mathbf{0}$.

THEOREM 14.8   *Test for a vector field to be conservative*

If $\mathbf{F}$ and curl $\mathbf{F}$ are continuous in the simply connected region D, then $\mathbf{F}$ is conservative in D if and only if curl $\mathbf{F} = \mathbf{0}$ in D.

Proof   If $\mathbf{F}$ is conservative, let f be a scalar potential function, so that $\nabla f = \mathbf{F}$. Then curl $\mathbf{F} = \nabla \times \mathbf{F} = \nabla \times (\nabla f) = \mathbf{0}$. (See Problem 62, Section 14.1 for a proof of this property.)

Conversely, if curl $\mathbf{F} = \mathbf{0}$, let C be the boundary curve of the smooth surface S. Then Stokes' theorem gives

$$\int_{C} \mathbf{F} \cdot d\mathbf{R} = \iint\limits_{S} (\text{curl }\mathbf{F} \cdot \mathbf{N})\, dS = \iint\limits_{S} 0\, dS = 0$$

so that $\mathbf{F}$ is independent of path and $\mathbf{F}$ must be conservative.   ▬▬

EXAMPLE 4   *Showing that a vector field is conservative*

Show that the vector field $\mathbf{F} = yz\mathbf{i} + xz\mathbf{j} + xy\mathbf{k}$ is conservative in $\mathbb{R}^3$, then find a scalar potential function f for $\mathbf{F}$.

Solution   Since

$$\text{curl }\mathbf{F} = \begin{vmatrix} \mathbf{i} & \mathbf{j} & \mathbf{k} \\ \dfrac{\partial}{\partial x} & \dfrac{\partial}{\partial y} & \dfrac{\partial}{\partial z} \\ yz & xz & xy \end{vmatrix} = (x - x)\mathbf{i} - (y - y)\mathbf{j} + (z - z)\mathbf{k} = \mathbf{0}$$

it follows from Theorem 14.8 that $\mathbf{F}$ is conservative.

To find a scalar potential function f, we must have $\nabla f = \mathbf{F}$; that is,

$$\frac{\partial f}{\partial x} = yz \qquad \frac{\partial f}{\partial y} = xz \qquad \frac{\partial f}{\partial z} = xy$$

Partially integrating the first of these equations with respect to x, we obtain

$$f = xyz + u(y, z)$$

so

$$\frac{\partial f}{\partial y} = xz + \frac{\partial u}{\partial y} = xz \quad \text{and} \quad \frac{\partial u}{\partial y} = 0$$

Thus, $u = v(z)$ and $f = xyz + v(z)$

$$\frac{\partial f}{\partial z} = xy + v'(z) = xy \quad \text{so } v'(z) = 0 \text{ and } v(z) = C$$

We conclude that

$$f(x, y, z) = xyz$$

is a scalar potential function for $\mathbf{F}$.   ■

EXAMPLE 5   *Maxwell's current density equation*

In physics, it is shown that if I is the current crossing any surface S bounded by the closed curve C, then

$$\int_C \mathbf{H} \cdot d\mathbf{R} = I \quad \text{and} \quad \iint_S \mathbf{J} \cdot \mathbf{N} \, dS = I$$

where $\mathbf{H}$ is the magnetic intensity and $\mathbf{J}$ is the electric current density. Use this information to derive Maxwell's current density equation curl $\mathbf{H} = \mathbf{J}$.

Solution   Equating the two equations of current, we obtain

$$\int_C \mathbf{H} \cdot d\mathbf{R} = I = \iint_S \mathbf{J} \cdot \mathbf{N} \, dS$$

By Stokes' theorem, $\int_C \mathbf{H} \cdot d\mathbf{R} = \iint_S (\text{curl } \mathbf{H} \cdot \mathbf{N}) \, dS$

Equating the two surface integrals that equal $\int_C \mathbf{H} \cdot d\mathbf{R}$, we have

$$\iint_S \mathbf{J} \cdot \mathbf{N} \, dS = \iint_S (\text{curl } \mathbf{H} \cdot \mathbf{N}) \, dS$$

or, equivalently,

$$\iint_S (\mathbf{J} - \text{curl } \mathbf{H}) \cdot \mathbf{N} \, dS = 0$$

Because this equation holds for *any* surface S bounded by C, it follows that

$$\mathbf{J} - \text{curl } \mathbf{H} = \mathbf{0}$$
$$\mathbf{J} = \text{curl } \mathbf{H} \qquad \blacksquare$$

PHYSICAL INTERPRETATION OF STOKES' THEOREM

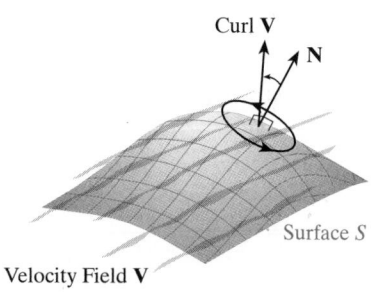

Curl $\mathbf{V}$

$\mathbf{N}$

Surface S

Velocity Field $\mathbf{V}$

■ **FIGURE 14.34**   The tendency of a fluid to swirl across the surface S is measured by curl $\mathbf{V} \cdot \mathbf{N}$

If $\mathbf{V}$ is the velocity field of a fluid flow, then curl $\mathbf{V}$ measures the tendency of the fluid to rotate, or swirl (see Figure 14.34). If the fluid flows across the surface S, the rotational tendency usually will vary from point to point on the surface, and the surface integral $\iint_S (\text{curl } \mathbf{V} \cdot \mathbf{N}) \, dS$ provides a measure of the *cumulative* rotational tendency over the entire surface.

Stokes' theorem tells us that this cumulative measure of rotational tendency equals the line integral $\int_C \mathbf{V} \cdot d\mathbf{R}$. To interpret this line integral, recall that it can be written $\int_C \mathbf{V} \cdot \mathbf{T} \, ds$ in terms of the arc length parameter s and the unit tangent $\mathbf{T}$ to the curve. Thus, the line integral sums the *tangential component* of the velocity field $\mathbf{V}$ around the boundary C, and it is reasonable to interpret $\int_C \mathbf{V} \cdot \mathbf{T} \, ds$ as a *measure of the circulation of the fluid flow around C*.

■ *What This Says:*

$$\underbrace{\iint_S (\text{curl } \mathbf{V} \cdot \mathbf{N}) \, dS}_{\substack{\text{The cumulative tendency} \\ \text{of a fluid to swirl across} \\ \text{the surface } S}} = \underbrace{\int_C \mathbf{V} \cdot \mathbf{T} \, ds}_{\substack{\text{The circulation} \\ \text{of a fluid around the} \\ \text{boundary curve } C}}$$

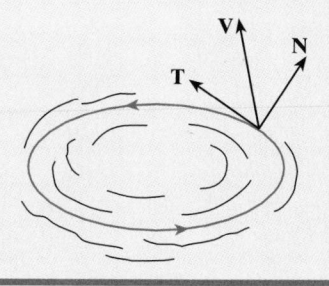

$\mathbf{V}$

$\mathbf{N}$

$\mathbf{T}$

14.6 Problem Set

A *Verify Stokes' theorem for the vector functions and surfaces given in Problems 1–5.*

1. $\mathbf{F} = z\mathbf{i} + 2x\mathbf{j} + 3y\mathbf{k}$; S is the upper hemisphere $z = \sqrt{9 - x^2 - y^2}$.

2. $\mathbf{F} = (y + z)\mathbf{i} + x\mathbf{j} + (z - x)\mathbf{k}$; S is the triangular region of the plane $x + 2y + z = 3$ in the first octant.

3. $\mathbf{F} = (x + 2z)\mathbf{i} + (y - x)\mathbf{j} + (z - y)\mathbf{k}$; S is the triangular region with vertices $(3, 0, 0), (0, \frac{3}{2}, 0), (0, 0, 3)$.

4. $\mathbf{F} = 2xy\mathbf{i} + z^2\mathbf{k}$; S is the portion of the paraboloid $y = x^2 + z^2$, with $y \le 4$.

5. $\mathbf{F} = 2y\mathbf{i} - 6z\mathbf{j} + 3x\mathbf{k}$; S is the portion of the paraboloid $z = 4 - x^2 - y^2$ and above the xy-plane.

Use Stokes' theorem to evaluate the line integrals given in Problems 6–13.

6. $\displaystyle\int_C (x^3y^2\, dx + dy + z^2\, dz)$,
   where C is the circle $x^2 + y^2 = 1$
   1 in the plane $z = 1$, counterclockwise when viewed from the origin.

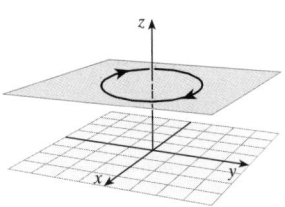

7. $\displaystyle\int_C (z\, dx + x\, dy + y\, dz)$,
   where C is the triangle with vertices $(3, 0, 0)$, $(0, 0, 2)$, and $(0, 6, 0)$, traversed in the given order.

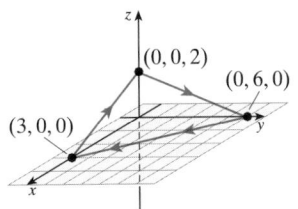

$(0, 0, 2)$
$(0, 6, 0)$
$(3, 0, 0)$

8. $\displaystyle\int_C (y\, dx - 2x\, dy + z\, dz)$, where C is the intersection of the surface $z = x^2 + y^2$ and the plane $x + y + z = 1$ considered counterclockwise when viewed from the origin.

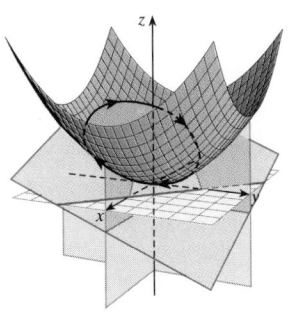

9. $\displaystyle\int_C [2xy^2z\, dx + 2x^2yz\, dy + (x^2y^2 - 2z)dz]$, where C is the curve given by $x = \cos t, y = \sin t, z = \sin t, 0 \le t \le 2\pi$, traversed in the direction of increasing t

10. $\displaystyle\int_C (y\, dx + z\, dy + y\, dz)$, where C is the intersection of the sphere $x^2 + y^2 + z^2 = 4$ and the plane $x + y + z = 0$, traversed counterclockwise when viewed from above.

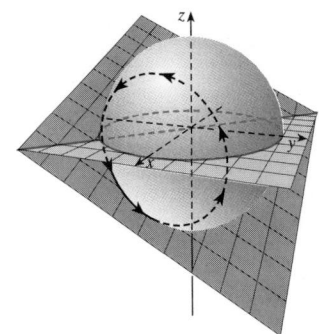

11. $\displaystyle\int_C (y\, dx + z\, dy + x\, dz)$, where C is the intersection of the plane $x + y = 2$ and the surface $x^2 + y^2 + z^2 = 2(x + y)$, traversed counterclockwise as viewed from the origin.

12. $\displaystyle\int_C [(z + \cos x)dx + (x + y^2)dy + (y + e^z)dz]$ where C is the intersection of the sphere $x^2 + y^2 + z^2 = 4$ and the cone $z = \sqrt{x^2 + y^2}$, traversed counterclockwise as viewed from above.

13. $\displaystyle\int_C (3y\, dx + 2z\, dy - 5x\, dz)$, where C is the intersection of the xy-plane and the hemisphere $z = \sqrt{1 - x^2 - y^2}$, traversed counterclockwise as viewed from above.

B *In Problems 14–19, use Stokes' theorem to evaluate $\iint_S (\text{curl } \mathbf{F} \cdot \mathbf{N})\, dS$ for the prescribed vector fields and surfaces.*

14. $\mathbf{F} = x\mathbf{i} + y^2\mathbf{j} + xyz\mathbf{k}$ and S is the part of the paraboloid $z = 4 - x^2 - y^2$ with $z \ge 0$. Use the upward unit normal vector.

15. $\mathbf{F} = xy\mathbf{i} - z\mathbf{j}$ and S is the surface of the cube $0 \le x \le 1$, $0 \le y \le 1, 0 \le z \le 1$, except for the face where $z = 0$. Use the outward unit normal vector.

16. $\mathbf{F} = y\mathbf{i} + z\mathbf{j} + x\mathbf{k}$, and S is the part of the plane $x + y + z = 1$ that lies in the first octant. Use the outward normal vector.

17. $\mathbf{F} = xy\mathbf{i} + x^2\mathbf{j} + z^2\mathbf{k}$, and C is the intersection of the paraboloid $z = x^2 + y^2$ and the plane $z = y$.

18. $\mathbf{F} = xz\mathbf{i} + y^2\mathbf{j} + x^2\mathbf{k}$, and C is the intersection of the plane $x + y + z = 3$ and the cylinder
$$x^2 + \frac{y^2}{9} = 1$$

19. $\mathbf{F} = 4y\mathbf{i} + z\mathbf{j} + 2y\mathbf{k}$, and C is the intersection of the sphere $x^2 + y^2 + z^2 = 4$ with the plane $z = 0$. Use the outward normal vector.

In Problems 20–22, use Stokes' theorem to evaluate the line integral
$$\int_C [(1 + y)z\, dx + (1 + z)x\, dy + (1 + x)y\, dz]$$
for the given path C.

20. C is the elliptic path $x = 2\cos\theta$, $y = \sin\theta$, $z = 1$ for $0 \le \theta \le 2\pi$.

21. C is the triangle with vertices $(1, 0, 0)$, $(0, 1, 0)$, $(0, 0, 1)$.

22. C is *any* closed path in the plane $2x - 3y + z = 1$.

In Problems 23–26, the vector field $\mathbf{V}$ represents the velocity of a fluid flow. In each case, find the circulation
$$\int_C \mathbf{V} \cdot d\mathbf{R}$$
around the boundary C, assuming a counterclockwise orientation as viewed from above.

23. $\mathbf{V} = x\mathbf{i} + (z - x)\mathbf{j} + y\mathbf{k}$, and C is the intersection of the cylinder $x^2 + y^2 = y$ and the hemisphere $z = \sqrt{1 - x^2 - y^2}$.

24. $\mathbf{V} = y\mathbf{i} + \ln(x^2 + y^2)\mathbf{j} + (x + y)\mathbf{k}$, and C is the triangle with vertices $(0, 0)$, $(1, 0)$, $(0, 1)$.

25. $\mathbf{V} = (e^{x^2} + z)\mathbf{i} + (x + \sin y^3)\mathbf{j} + [y + \ln(\tan^{-1}z)]\mathbf{k}$, and C is the intersection of the sphere $x^2 + y^2 + z^2 = 1$ and the cone $z = \sqrt{x^2 + y^2}$.

26. $\mathbf{V} = y^2\mathbf{i} + \tan^{-1}z\mathbf{j} + (x^2 + 1)\mathbf{k}$, and C is the intersection of the plane $z = y$ and the cylinder $x^2 + y^2 = 2x$.

27. Let $\mathbf{F} = y^2\mathbf{i} + xy\mathbf{j} + xz\mathbf{k}$, and suppose S is the hemisphere $x^2 + y^2 + z^2 = 1$ with $z \ge 0$. Use Stokes' theorem to express
$$\iint_S (\text{curl } \mathbf{F} \cdot \mathbf{N})\, dS$$
as a line integral, and then evaluate the surface integral by evaluating this line integral.

28. Let $\mathbf{F} = z\mathbf{i} + x\mathbf{j} + y\mathbf{k}$, and suppose S is a smooth surface in $\mathbb{R}^3$ whose boundary is given by $x = 2\cos\theta$, $y = 3\sin\theta$, $z = \sin\theta$, $0 < \theta \le 2\pi$. Use Stokes' theorem to evaluate
$$\iint_S (\text{curl } \mathbf{F} \cdot \mathbf{N})\, dS$$

Show that the given vector field $\mathbf{F}$ in Problems 29–34 is conservative and find a scalar potential function f for $\mathbf{F}$.

29. $yz^2\mathbf{i} + xz^2\mathbf{j} + 2xyz\mathbf{k}$

30. $e^{xy}yz\mathbf{i} + e^{xy}xz\mathbf{j} + e^{xy}\mathbf{k}$

31. $yz^{-1}\mathbf{i} + xz^{-1}\mathbf{j} - xyz^{-2}\mathbf{k}$

32. $(x^2 + y^2 + z^2)(x\mathbf{i} + y\mathbf{j} + z\mathbf{k})$

33. $(y\sin z)\mathbf{i} + (x\sin z + 2y)\mathbf{j} + (xy\cos z)\mathbf{k}$

34. $(xy^2 + yz)\mathbf{i} + (x^2y + xz + 3y^2z)\mathbf{j} + (xy + y^3)\mathbf{k}$

35. HISTORICAL QUEST   George Stokes was an English mathematical physicist who made important contributions to fluid mechanics, including the so-called Navier–Stokes equations. Most of his research was done before 1850, after which he held the Lucasian chair of mathematics at Cambridge for the better part of a half-century. William Thompson (Lord Kelvin) knew the result now known as Stokes' theorem in 1850 and sent it to Stokes as a challenge. Stokes proved the theorem, and then included it as an exam question in 1854. One of the students taking this particular examination was James Clerk Maxwell (Historical Quest #30, Section 14.7), who derived the famous electromagnetic wave equations 10 years later.

GEORGE STOKES
1819–1903

   For this Quest, write a history of the Lucasian chair, which was deeded in 1663 as a gift of Henry Lucas. The first and second appointees were Isaac Barrow (Historical Quest #6, page 124) and Isaac Newton (Historical Quest #1, page 124); the chair is currently held by Stephen Hawking.

36. Let S be the ellipsoid $\dfrac{x^2}{4} + \dfrac{y^2}{9} + z^2 = 1$, and let $\mathbf{F}$ be a vector field whose component functions have continuous partial derivatives on S. Use Stokes' theorem to show that
$$\iint_S (\text{curl } \mathbf{F} \cdot \mathbf{N})\, dS = 0$$
Does it matter that S is an ellipsoid? State and prove a more general result based on what you have discovered in the first part of this problem.

37. **Faraday's law** of electromagnetism says that if $\mathbf{E}$ is the electric intensity vector in a system, then
$$\int_C \mathbf{E} \cdot d\mathbf{R} = -\frac{\partial\phi}{\partial t}$$
around any closed curve C where t is time and ϕ is the total magnetic flux directed outward through any surface S bounded by C. Given that
$$\phi = \iint_S \mathbf{B} \cdot \mathbf{N}\, dS$$
where $\mathbf{B}$ is the magnetic flux density, show that
$$\text{curl } \mathbf{E} = -\frac{\partial\mathbf{B}}{\partial t}$$
Hint: It can be shown that
$$\frac{\partial}{\partial t}\iint_S \mathbf{B} \cdot \mathbf{N}\, dS = \iint_S \frac{\partial\mathbf{B}}{\partial t} \cdot \mathbf{N}\, dS$$

38. The current I flowing across a surface S bounded by the closed curve C is given by

$$I = \iint_S \mathbf{J} \cdot \mathbf{N}\, dS$$

where $\mathbf{J}$ is the current density. Given that $\mu \mathbf{J} = \operatorname{curl} \mathbf{B}$, where $\mathbf{B}$ is magnetic flux density and μ is a constant, show that

$$\oint_C \mathbf{B} \cdot d\mathbf{R} = \mu I$$

Suppose f and g are continuously differentiable functions of x, y, z, and C is a closed curve bounding the surface S. Use Stokes' theorem to verify the formulas given in Problems 39–40,

39. $\displaystyle\int_C (f\nabla g) \cdot d\mathbf{R} = \iint_S (\nabla f \times \nabla g) \cdot \mathbf{N}\, dS$

40. $\displaystyle\int_C (f\nabla g + g\nabla f) \cdot d\mathbf{R} = 0$

14.7 Divergence Theorem

IN THIS SECTION divergence theorem, applications of the divergence theorem, physical interpretation of divergence ∎

DIVERGENCE THEOREM

We used Green's theorem to show that $\int_C \mathbf{F} \cdot \mathbf{N}\, ds = \int_D\!\!\int \operatorname{div} \mathbf{F}\, dA$, where D is a simply connected domain with the closed boundary curve C. The *divergence theorem* (also known as **Gauss' theorem**) is a generalization of this form of Green's theorem that relates a closed surface integral to a volume integral.

THEOREM 14.9 *The divergence theorem*

Let S be a smooth, orientable surface that encloses a solid region D in $\mathbb{R}^3$. If $\mathbf{F}$ is a continuous vector field whose components have continuous partial derivatives in D, then

$$\iint_S \mathbf{F} \cdot \mathbf{N}\, dS = \iiint_D \operatorname{div} \mathbf{F}\, dV$$

where $\mathbf{N}$ is an outward unit normal vector to the surface S.

Proof The general proof of this theorem is beyond the scope of this course, but an important special case is proved in Appendix B. ⸗

EXAMPLE 1 *Verifying the divergence theorem for a particular example*

Let $\mathbf{F} = 2x\mathbf{i} - 3y\mathbf{j} + 5z\mathbf{k}$, and let S be the hemisphere $z = \sqrt{9 - x^2 - y^2}$ together with the disk $x^2 + y^2 \le 9$ in the xy-plane. Verify the divergence theorem.

Solution The solid is shown in Figure 14.35. We shall show that the surface integral and the triple integral have the same value.

I. *Evaluation of* $\displaystyle\iint_S \mathbf{F} \cdot \mathbf{N}\, dS$

S consists of two parts: S_1, the disk on the bottom of the hemisphere, and S_2, the hemisphere. We will consider these separately using

$$\iint_S \mathbf{F} \cdot \mathbf{N}\, dS = \iint_{S_1} \mathbf{F} \cdot \mathbf{N_1}\, dS_1 + \iint_{S_2} \mathbf{F} \cdot \mathbf{N_2}\, dS_2$$

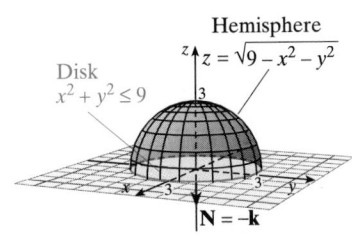

Disk
$x^2 + y^2 \le 9$

Hemisphere
$z = \sqrt{9 - x^2 - y^2}$

$\mathbf{N} = -\mathbf{k}$

■ **FIGURE 14.35** The surface of the hemisphere

Consider S_1: Note that the disk $x^2 + y^2 \le 9$ on the bottom of the hemisphere has outward unit normal vector $\mathbf{N} = -\mathbf{k}$. (Because $\mathbf{k}$ points into the solid, $-\mathbf{k}$ is the outward normal vector.) We have

$$\iint_{S_1} \mathbf{F} \cdot \mathbf{N} \, dS_1 = \iint_{S_1} (2x\mathbf{i} - 3y\mathbf{j} + 5z\mathbf{k}) \cdot (-\mathbf{k}) \, dS_1 = \iint_{S_1} (-5z) \, dS_1 = 0$$

Because $z = 0$ on S_1

Consider S_2: Regard the hemisphere as the level surface

$$S_2(x, y, z) = 9, \text{ where } S_2(x, y, z) = x^2 + y^2 + z^2 \text{ and } z \ge 0.$$

The outward unit (for $z \ge 0$) is given by

$$N = \frac{\nabla S_2}{\|\nabla S_2\|}$$

$$= \frac{2x\mathbf{i} + 2y\mathbf{j} + 2z\mathbf{k}}{\sqrt{4x^2 + 4y^2 + 4z^2}}$$

Remember that $\nabla f = \frac{\partial f}{\partial x}\mathbf{i} + \frac{\partial f}{\partial y}\mathbf{j} + \frac{\partial f}{\partial z}\mathbf{k}$.

$$= \tfrac{1}{3}(x\mathbf{i} + y\mathbf{j} + z\mathbf{k})$$

Because $x^2 + y^2 + z^2 = 9$ on S_2

and

$$\mathbf{F} \cdot \mathbf{N} = \tfrac{1}{3}(2x^2 - 3y^2 + 5z^2)$$

Since $z_x^2 = x^2(9 - x^2 - y^2)^{-1}$ and $z_y^2 = y^2(9 - x^2 - y^2)^{-1}$, we have

$$dS_2 = \sqrt{z_x^2 + z_y^2 + 1} \, dA_{xy}$$

$$= \sqrt{\frac{x^2}{9 - x^2 - y^2} + \frac{y^2}{9 - x^2 - y^2} + 1} \, dA_{xy}$$

$$= \sqrt{\frac{9}{9 - x^2 - y^2}} \, dA_{xy} = \frac{3}{\sqrt{9 - x^2 - y^2}} \, dA_{xy}$$

and it follows that

$$\int_{S_2} \mathbf{F} \cdot \mathbf{N} \, dS_2 = \int_{S_2} \frac{1}{3}(2x^2 - 3y^2 + 5z^2) \frac{3}{\sqrt{9 - x^2 - y^2}} \, dA_{xy}$$

$$= \iint_{S_2} [2x^2 - 3y^2 + 5(9 - x^2 - y^2)][9 - x^2 - y^2]^{-1/2} \, dA_{xy}$$

$$= \int_0^{2\pi} \int_0^3 \frac{45 - 3(r \cos\theta)^2 - 8(r \sin\theta)^2}{\sqrt{9 - r^2}} r \, dr \, d\theta$$

The projected region is $x^2 + y^2 \le 9$, after much computation. See Figure 14.36.

$$= \int_0^{2\pi} [135 - 54\cos^2\theta - 144\sin^2\theta] \, d\theta$$

$$= 72\pi$$

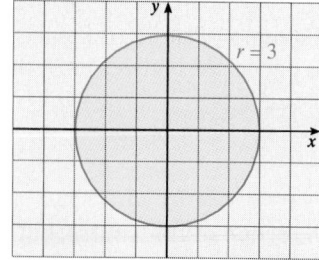

■ **FIGURE 14.36** The hemisphere projects with boundary $z^2 = 9 - x^2 - y^2$ onto the disk $x^2 + y^2 \le 9$, or, in polar form, $r \le 3$.

We can now state

$$\iint_S \mathbf{F} \cdot \mathbf{N} \, dS = \iint_{S_1} \mathbf{F} \cdot \mathbf{N} \, dS_1 + \iint_{S_2} \mathbf{F} \cdot \mathbf{N} \, dS_2 = 0 + 72\pi = 72\pi$$

II. *Evaluation of* $\displaystyle\iiint_D \text{div } \mathbf{F} \, dV$

$$\text{div } \mathbf{F} = \frac{\partial}{\partial x}(2x) + \frac{\partial}{\partial y}(-3y) + \frac{\partial}{\partial x}(5z) = 2 - 3 + 5 = 4$$

Therefore, $\iiint \text{div } \mathbf{F} \, dV = \iiint 4 \, dV$, but $\iiint dV$ is just the volume of the hemisphere $z = \sqrt{9 - x^2 - y^2}$. A hemisphere of radius 3 has volume $\frac{1}{2}[\frac{4}{3}\pi(3)^3] = 18\pi$, so

$$\iiint_D \text{div } \mathbf{F} \, dV = \iiint_D 4 \, dV = 4V = 4(18\pi) = 72\pi$$

and we have $\iiint \text{div } \mathbf{F} \, dV = \iint \mathbf{F} \cdot \mathbf{N} \, dS$, as required. ∎

| **EXAMPLE 2** | *Evaluating a surface integral using the divergence theorem* |

Evaluate $\displaystyle\iint_S \mathbf{F} \cdot \mathbf{N} \, dS$, where $\mathbf{F} = x^2\mathbf{i} + xy\mathbf{j} + x^3y^3\mathbf{k}$ and S is the surface of the tetrahedron bounded by the plane $x + y + z = 1$ and the coordinate planes, with outward unit normal vector $\mathbf{N}$.

Solution We will use the divergence theorem. Let D be the solid bounded by S, and note that

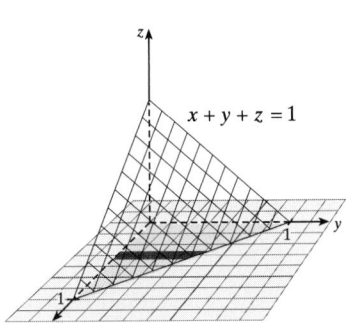

$$\text{div } \mathbf{F} = \frac{\partial}{\partial x}(x^2) + \frac{\partial}{\partial y}(xy) + \frac{\partial}{\partial z}(x^3y^3) = 2x + x + 0 = 3x$$

We choose to describe the solid tetrahedron as the set of all (x, y, z) such that $0 \le z \le 1 - x - y$ whenever $0 \le y \le 1 - x$ for $0 \le x \le 1$. The projection of S on the xy-plane is shown in the margin. Finally, by applying the divergence theorem, we find that

$$\iint_S \mathbf{F} \cdot \mathbf{N} \, dS = \iiint_D \text{div } \mathbf{F} \, dV = \int_0^1 \int_0^{1-x} \int_0^{1-x-y} 3x \, dz \, dy \, dx$$

$$= \int_0^1 \int_0^{1-x} 3x(1 - x - y) \, dy \, dx = 3 \int_0^1 \left[x(1 - x)y - \frac{1}{2}xy^2 \right]_0^{1-x} dx$$

$$= 3 \int_0^1 x(1 - x)^2 - \frac{1}{2}x(1 - x)^2 \, dx = \frac{1}{8}$$ ∎

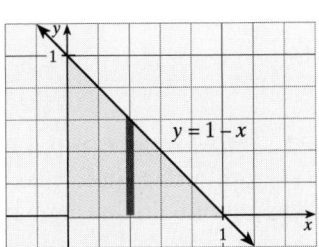

Projected region in the xy-plane

The divergence theorem applies only to closed surfaces. However, if we wish to evaluate $\int_{S_1}\int \mathbf{F} \cdot \mathbf{N} \, dS$ where S_1 is *not* closed, we may be able to find a closed surface S that is the union of S_1 and some other surface S_2. Then, if the hypotheses of the divergence theorem are satisfied by $\mathbf{F}$ and S, we have

$$\iint_{S_1} \mathbf{F} \cdot \mathbf{N} \, dS + \iint_{S_2} \mathbf{F} \cdot \mathbf{N} \, dS = \iint_S \mathbf{F} \cdot \mathbf{N} \, dS = \iiint_D \text{div } \mathbf{F} \, dV$$

where D is the solid region bounded by S. Thus, if we can compute $\iiint_D \text{div } \mathbf{F} \, dV$ and $\int_{S_2}\int \mathbf{F} \cdot \mathbf{N} \, dS$, we can compute $\int_{S_1}\int \mathbf{F} \cdot \mathbf{N} \, dS$ by the equation

$$\iint_{S_1} \mathbf{F} \cdot \mathbf{N} \, dS = \iiint_D \text{div } \mathbf{F} \, dV - \iint_{S_2} \mathbf{F} \cdot \mathbf{N} \, dS$$

This equation can also be used as a device for trading the evaluation of a difficult surface integral for that of an easier integral. Here is an example of this procedure.

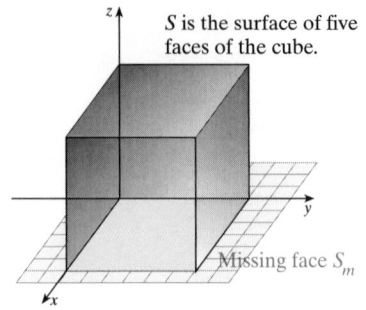

S^* is the closed surface of the entire cube.

S is the surface of five faces of the cube.

Missing face S_m

■ **FIGURE 14.37** Evaluating a surface integral

EXAMPLE 3 *Evaluating a surface integral over an open surface*

Evaluate $\iint_S \mathbf{F} \cdot \mathbf{N}\, dS$, where $\mathbf{F} = xy\mathbf{i} - z^2\mathbf{k}$ and S is the surface of the upper five faces of the unit cube $0 \le x \le 1, 0 \le y \le 1, 0 < z \le 1$, as shown in Figure 14.37.

Solution Note that the surface S is not closed, but we can close it by adding the missing face S_m, thus forming a closed surface S^* that satisfies the conditions of the divergence theorem. The strategy is to evaluate the surface integral of S^* and then subtract the surface integral of the added face S_m.

$$\iint_{S^*} \mathbf{F} \cdot \mathbf{N}\, dS = \iiint_{\text{cube}} \text{div } \mathbf{F}\, dV = \int_0^1 \int_0^1 \int_0^1 (y - 2z)\, dx\, dy\, dz$$

$$= \int_0^1 \int_0^1 (y - 2x)\, dy\, dz = \int_0^1 \left(\frac{1}{2} - 2z\right) dz = -\frac{1}{2}$$

Also, because the unit normal vector to the added face S_m is $\mathbf{N} = -\mathbf{k}$ and $z = 0$ on this face, it follows that

$$\iint_{S_m} \mathbf{F} \cdot \mathbf{N}\, dS = \iint_{S_m} (xy\mathbf{i} - z^2\mathbf{k}) \cdot (-\mathbf{k})\, dS = \iint_{S_m} z^2\, dS = 0$$

Therefore,

$$\iint_S \mathbf{F} \cdot \mathbf{N}\, dS = \iiint_D \text{div } \mathbf{F}\, dV - \iint_{S_m} \mathbf{F} \cdot \mathbf{N}\, dS = -\frac{1}{2} - 0 = -\frac{1}{2} \qquad ■$$

APPLICATIONS OF THE DIVERGENCE THEOREM

Like Stokes' theorem, the divergence theorem is often used for theoretical purposes, especially as a tool for deriving general properties in mathematical physics. The following example deals with an important of fluid dynamics.

EXAMPLE 4 *Continuity equation of fluid dynamics*

Suppose a fluid with density $\delta(x, y, z, t)$ flows in some region of space with velocity $\mathbf{F}(x, y, z, t)$ at the point (x, y, z) at time t. Assuming there are no sources or sinks, show that

$$\text{div } \delta\mathbf{F} = -\frac{\partial \delta}{\partial t}$$

Solution Recall from Section 14.1 that a point is called a *source* if div $\mathbf{F} > 0$, a *sink* if div $\mathbf{F} < 0$, and *incompressible* if div $\mathbf{F} = 0$. Let S be a smooth surface in $\mathbb{R}^3$ that encloses a solid region D. In physics, it is shown that the amount of fluid flowing out of D across S in unit time is $\iint_S \delta\mathbf{F} \cdot \mathbf{N}\, dS$. Thus, $-\iint_S \delta\mathbf{F} \cdot \mathbf{N}\, dS$ is the amount of outflow and is equal to the amount of inflow, $\iiint_D \frac{\partial \delta}{\partial t}\, dV$, because we are assuming there are no sinks or sources. Equating these two quantities, we have

$$\overbrace{-\iint_S \delta\mathbf{F} \cdot \mathbf{N}\, dS}^{\text{OUTFLOW}} = \overbrace{\iiint_D \frac{\partial \delta}{\partial t}\, dV}^{\text{INFLOW}}$$

By the divergence theorem, it follows that

$$\iint_S \delta\mathbf{F} \cdot \mathbf{N}\, dS = \iiint_D \text{div } \delta\mathbf{F}\, dV$$

so that

$$\iiint_D \operatorname{div} \delta\mathbf{F}\, dV + \iiint_D \frac{\partial \delta}{\partial t}\, dV = 0$$

$$\iiint_D \left[\operatorname{div} \delta\mathbf{F} + \frac{\partial \delta}{\partial t}\right] dV = 0$$

This equation must hold for any region D, no matter how small, which means that the integrand of the integral must be 0. That is,

$$\operatorname{div} \delta\mathbf{F} = -\frac{\partial \delta}{\partial t} \qquad \blacksquare$$

In physics it is known that the total heat contained in a body with uniform density δ and specific heat σ is $\iiint_D \sigma\delta T\, dV$, where T is the temperature. Thus, the amount of heat leaving D per unit of time is given by

$$-\frac{\partial}{\partial t}\left[\iiint_D \sigma\delta T\, dV\right] = \iiint_D -\sigma\delta \frac{\partial T}{\partial t}\, dV$$

In the following example, we use this result to obtain an important formula from mathematical physics.

> **EXAMPLE 5** *Derivation of the heat equation*

Let $T(x, y, z, t)$ be the temperature at each point (x, y, z) in a solid body D at time t. Given that the velocity of heat flow in the body is $\mathbf{F} = -k\nabla T$ for a positive constant k (called the **thermal conductivity**), show that

$$\frac{\partial T}{\partial t} = \frac{k}{\sigma\delta} \nabla^2 T$$

where σ is the specific heat of the body and δ is its density.

Solution Let S be the closed surface that bounds D. Because $\mathbf{F}$ is the velocity of heat flow, the amount of heat leaving D per unit time is $\iint_S \mathbf{F} \cdot \mathbf{N}\, dS$, and the divergence theorem applies:

$$\iint_S \mathbf{F} \cdot \mathbf{N}\, dS = \iiint_D \operatorname{div}(-k\nabla T)\, dV = \iiint_D (-k\nabla \cdot \nabla T)\, dV$$

$$= \iiint_D -k\nabla^2 T\, dV$$

Since this is the amount of heat leaving D per unit of time, it must equal the heat integral from physics derived just before this example. Thus,

$$\iiint_D -k\nabla^2 T\, dV = \iiint_D -\sigma\delta \frac{\partial T}{\partial t}\, dV$$

This equation holds not only for the body as a whole, but for every part of the body, no matter how small. Thus, we can shrink the body to a single point, and it then follows that the integrands are equal; that is,

$$-k\nabla^2 T = -\sigma\delta \frac{\partial T}{\partial t}$$

$$\frac{\partial T}{\partial t} = \frac{k}{\sigma\delta} \nabla^2 T \qquad \blacksquare$$

For the concluding example, recall that the *normal derivative* $\partial g/\partial n$ of a scalar function g defined on the closed surface S is the directional derivative of f in the direction of the outward unit normal vector $\mathbf{N}$ to S; that is,

$$\frac{\partial g}{\partial n} = \nabla g \cdot \mathbf{N}$$

We will use this equation in the following example, which is a generalization of a property we first obtained for $\mathbb{R}^2$ in Section 14.4.

EXAMPLE 6   *Derivation of Green's first identity*

Show that if f and g are scalar functions such that $\mathbf{F} = f\nabla g$ is continuously differentiable in the solid domain D bounded by the closed surface S, then

$$\iiint_D [f\nabla^2 g + \nabla f \cdot \nabla g]\, dV = \iint_S f\frac{\partial g}{\partial n}\, dS$$

This is called **Green's first identity.**

Solution   We shall apply the divergence theorem to the vector field $\mathbf{F}$ (note that $\mathbf{F}$ is continuously differentiable), but first we need to express div $\mathbf{F}$ in a more useful form.

$$\begin{aligned}
\operatorname{div}(f\nabla g) &= \nabla \cdot (f\nabla g)\\
&= \left[\frac{\partial}{\partial x}\mathbf{i} + \frac{\partial}{\partial y}\mathbf{j} + \frac{\partial}{\partial z}\mathbf{k}\right]\cdot\left[f\frac{\partial g}{\partial x}\mathbf{i} + f\frac{\partial g}{\partial y}\mathbf{j} + f\frac{\partial g}{\partial z}\mathbf{k}\right]\\
&= \frac{\partial}{\partial x}\left[f\frac{\partial g}{\partial x}\right] + \frac{\partial}{\partial y}\left[f\frac{\partial g}{\partial y}\right] + \frac{\partial}{\partial z}\left[f\frac{\partial g}{\partial z}\right]\\
&= \left[\frac{\partial f}{\partial x}\frac{\partial g}{\partial x} + f\frac{\partial^2 g}{\partial x^2}\right] + \left[\frac{\partial f}{\partial y}\frac{\partial g}{\partial y} + f\frac{\partial^2 g}{\partial y^2}\right] + \left[\frac{\partial f}{\partial z}\frac{\partial g}{\partial z} + f\frac{\partial^2 g}{\partial z^2}\right]\\
&= \left[\frac{\partial f}{\partial x}\frac{\partial g}{\partial x} + \frac{\partial f}{\partial y}\frac{\partial g}{\partial y} + \frac{\partial f}{\partial z}\frac{\partial g}{\partial z}\right] + f\left[\frac{\partial^2 g}{\partial x^2} + \frac{\partial^2 g}{\partial y^2} + \frac{\partial^2 g}{\partial z^2}\right]\\
&= (\nabla f)\cdot(\nabla g) + f\nabla^2 g
\end{aligned}$$

This calculation gives us the first step in the following derivation:

$$\begin{aligned}
\iiint_D [f\nabla^2 g + \nabla f \cdot \nabla g]\, dV &= \iiint_D \operatorname{div}(f\nabla g)\, dV\\
&= \iint_S (f\nabla g)\cdot\mathbf{N}\, dS \qquad \text{Divergence theorem}\\
&= \iint_S f(\nabla g\cdot\mathbf{N})\, dS\\
&= \iint_S f\frac{\partial g}{\partial n}\, dS \qquad \text{Because } \nabla g\cdot\mathbf{N} = \frac{\partial g}{\partial n} \text{ by definition} \quad\blacksquare
\end{aligned}$$

PHYSICAL INTERPRETATION OF DIVERGENCE

In Section 14.1, we gave an interpretation of the curl as a measure of the tendency of a fluid to swirl. We now close this chapter with an analogous interpretation of divergence.

The following example can be interpreted as saying that the net rate of fluid mass flowing away (that is, "diverging") from point P_0 is given by div $\mathbf{F}_0$. This is the reason

P_0 is a *source* if div $\mathbf{F}_0 > 0$ (mass flowing out from P_0) and a *sink* if div $\mathbf{F}_0 < 0$ (mass flowing back into P_0).

EXAMPLE 7   *Physical interpretation of divergence*

Let $\mathbf{F} = \delta\mathbf{V}$ be the flux density associated with a fluid of density δ flowing with velocity $\mathbf{V}$ and let P_0 be a point inside a solid region where the conditions of the divergence theorem are satisfied. Then,

$$\text{div } \mathbf{F}_0 = \lim_{r \to 0} \frac{1}{V(r)} \iint_{S(r)} \mathbf{F} \cdot \mathbf{N} \, dS$$

where div $\mathbf{F}_0$ denotes the value of div $\mathbf{F}$ at P_0, and $S(r)$ is a sphere centered at P_0 with volume $V(r) = \frac{4}{3}\pi r^3$.

Solution   Applying the divergence theorem to the solid sphere (ball) $B(r)$ with surface $S(r)$, we obtain

$$\iint_{S(r)} \mathbf{F} \cdot \mathbf{N} \, dS = \iiint_{B(r)} \text{div } \mathbf{F} \, dV$$

The mean value theorem (for triple integrals) tells us that

$$\frac{1}{V(r)} \iiint_{B(r)} \text{div } \mathbf{F} \, dV = \text{div } \mathbf{F}^*$$

where div $\mathbf{F}^*$ denotes the value of div $\mathbf{F}$ at some point P^* in the ball $B(r)$. Combining these results, we find that

$$\iint_{S(r)} \mathbf{F} \cdot \mathbf{N} \, dS = \iiint_{B(r)} \text{div } \mathbf{F} \, dV = V(r) \text{ div } \mathbf{F}^*$$

or

$$\frac{1}{V(r)} \iint_{S(r)} \mathbf{F} \cdot \mathbf{N} \, dS = \text{div } \mathbf{F}^*$$

Since the point P^* is inside the ball $B(r)$ centered at P_0, it follows that $P^* \to P$ as $r \to 0$, so div $\mathbf{F}^* \to$ div $\mathbf{F}_0$ and we have

$$\lim_{r \to 0} \frac{1}{V(r)} \iint_{S(r)} \mathbf{F} \cdot \mathbf{N} \, dS = \lim_{r \to 0} \text{div } \mathbf{F}^* = \text{div } \mathbf{F}_0$$

as claimed.   ■

14.7 Problem Set

A *Verify the divergence theorem for the vector function $\mathbf{F}$ and solid D given in Problems 1–4 Assume $\mathbf{N}$ is the unit normal vector pointing away from the origin.*

1. $\mathbf{F} = xz\mathbf{i} + y^2\mathbf{j} + 2z\mathbf{k}$; D is the ball $x^2 + y^2 + z^2 \leq 4$.
2. $\mathbf{F} = x\mathbf{i} - 2y\mathbf{j}$; D is the interior of the paraboloid $z = x^2 + y^2, 0 \leq z < 9$.
3. $\mathbf{F} = 2y^2\mathbf{j}$; D is the part of the surface of the plane $x + 4y + z = 8$ that lies in the first octant.

4. $\mathbf{F} = 3x\mathbf{i} + 5y\mathbf{j} + 6z\mathbf{k}$; D is the tetrahedron bounded by the coordinate planes and the plane $2x + y + z = 4$.

Use the divergence theorem in Problems 5–19 to evaluate the surface integral $\iint_S \mathbf{F} \cdot \mathbf{N} \, dS$ for the given choice of $\mathbf{F}$ and the boundary surface S. For each closed surface, assume $\mathbf{N}$ is the outward unit normal vector.

5. $\mathbf{F} = x\mathbf{i} + y\mathbf{j} + z\mathbf{k}$; S is the surface of the cube $0 \leq x \leq 1$, $0 \leq y \leq 1, 0 \leq z \leq 1$. (See Problem 32, Section 14.5.)

6. $\mathbf{F} = xyz\mathbf{j}$; S is the surface of the cylinder $x^2 + y^2 = 9$, for $0 \le z \le 5$.

7. $\mathbf{F} = (\cos yz)\mathbf{i} + e^{xz}\mathbf{j} + 3z^2\mathbf{k}$; S is the surface of the hemisphere $z = \sqrt{4 - x^2 - y^2}$ together with the disk $x^2 + y^2 \le 4$ in the xy-plane.

8. $\mathbf{F} = \operatorname{curl}[e^{xz}\mathbf{i} - 4\mathbf{j} + (\sin xyz)\mathbf{k}]$; S is the surface of the ellipsoid $2x^2 + 3y^2 + 7z^2 = 1$.

9. $\mathbf{F} = (x^2 + y^2 - x^2)\mathbf{i} + x^2y\mathbf{j} + 3z\mathbf{k}$; S is the surface of the five faces of the unit cube $0 \le x \le 1, 0 \le y \le 1, 0 \le z \le 1$, missing $z = 0$.

10. $\mathbf{F} = 2y\mathbf{i} - z\mathbf{j} + 3x\mathbf{k}$; S is the five faces of the unit cube $0 \le x \le 1, 0 \le y \le 1, 0 \le z \le 1$, missing $z = 0$.

11. $\mathbf{F} = x\mathbf{i} + y\mathbf{j} + z\mathbf{k}$; S is the surface of the paraboloid $z = x^2 + y^2$ for $0 \le z \le 9$.

12. $\mathbf{F} = \operatorname{curl}(y\mathbf{i} + x\mathbf{j} + z\mathbf{k})$; S is the surface of the hemisphere $z = \sqrt{4 - x^2 - y^2}$ together with the disk.

13. $\mathbf{F} = x^2\mathbf{i} + y^2\mathbf{j} + z^2\mathbf{k}$; S is the surface of the sphere $x^2 + y^2 + z^2 = 4$.

14. $\mathbf{F} = xyz\mathbf{i} + xyz\mathbf{j} + xyz\mathbf{k}$; S is the surface of the box $0 \le x \le 1, 0 \le y \le 2, 0 \le z \le 3$.

15. $\mathbf{F} = x\mathbf{i} + y\mathbf{j} + (z^2 - 1)\mathbf{k}$; S is the surface of a solid bounded by the cylinder $x^2 + y^2 = 4$ and the planes $z = 0$ and $z = 1$.

16. $\mathbf{F} = (x^5 + 10xy^2z^2)\mathbf{i} + (y^5 + 10yx^2z^2)\mathbf{j} + (z^5 + 10zx^2y^2)\mathbf{k}$; S is the closed hemispherical surface $z = \sqrt{1 - x^2 - y^2}$ together with the disk $x^2 + y^2 \le 1$ in the xy-plane.

17. $\mathbf{F} = xy^2\mathbf{i} + yz^2\mathbf{j} + x^2z\mathbf{k}$; S is the surface bounded above by the sphere $\rho = 2$ and below by the cone $\varphi = \frac{\pi}{4}$ in spherical coordinates. (S is the surface of an "ice cream cone.")

18. $\mathbf{F} = xy^2\mathbf{i} + yz^2\mathbf{j} + x^2z^2\mathbf{k}$; S is the surface (in spherical coordinates) with top $\rho = 2, 0 \le \phi \le \frac{\pi}{4}, 0 \le \theta \le 2\pi$, and bottom $0 \le \rho \le 2, \phi = \frac{\pi}{4}, 0 \le \theta \le 2\pi$.

19. $\mathbf{F} = x^3\mathbf{i} + y^3\mathbf{j} + 3a^2z\mathbf{k}$ (constant $a > 0$); S is the surface bounded by the cylinder $x^2 + y^2 = a^2$ and the planes $z = 0$ and $z = 1$.

B 20. Suppose that S is a closed surface that encloses a solid region D.
  a. Show that the volume of D is given by
  $$V(D) = \frac{1}{3}\iint_S (x\mathbf{i} + y\mathbf{j} + z\mathbf{k}) \cdot \mathbf{N}\,dS$$
  where N is an outward unit normal vector to S.
  b. Use the formula in part **a** to find the volume of the hemisphere
  $$z = \sqrt{R^2 - x^2 - y^2}$$

21. Use the divergence theorem to evaluate
  $$\iint_S \|\mathbf{R}\|\mathbf{R} \cdot \mathbf{N}\,dS$$
  where $\mathbf{R} = x\mathbf{i} + y\mathbf{j} + z\mathbf{k}$ and S is the sphere $x^2 + y^2 + z^2 = a^2$, with constant $a > 0$.

22. The moment of inertia about the z-axis of a solid D of constant density $\delta = a$ is given by
  $$I_z = \iiint_T a(x^2 + y^2)\,dV$$
  Express this integral as a surface integral over the surface S that bounds D.

C 23. Let u be a scalar function with continuous second partial derivatives in a region containing the solid region D, with closed boundary surface S.
  a. Show that $\displaystyle\iint_S \frac{\partial u}{\partial n}\,dS = \iiint_D \nabla^2 u\,dV$.
  b. Let $u = x + y + z$ and $v = \frac{1}{2}(x^2 + y^2 + z^2)$. Evaluate
  $$\iint_S (u\nabla v) \cdot \mathbf{N}\,dS$$
  where S is the boundary of the cube $0 \le x \le 1, 0 \le y \le 1, 0 \le z \le 1$.

24. Let f and g be scalar functions such that $\mathbf{F} = f\nabla g$ is continuously differentiable in the region D, which is bounded by the closed surface S. Prove *Green's second identity:*
  $$\iiint_D (f\nabla^2 g - g\nabla^2 f)\,dV = \iint_S \left(f\frac{\partial g}{\partial n} - g\frac{\partial f}{\partial n}\right)dS$$

25. Show that if g is harmonic in the region D, then
  $$\iint_S \frac{\partial g}{\partial n}\,dS = 0$$
  where the closed surface S is the boundary of D. (Recall that g harmonic means $\nabla^2 g = 0$.)

26. Show that $\displaystyle\iint_S \mathbf{F} \cdot \mathbf{N}\,dS = 0$ if S is a closed surface and $\mathbf{F} = \operatorname{curl}\mathbf{U}$ throughout the interior of S for some vector field $\mathbf{U}$. A vector field $\mathbf{U}$ with this property is said to be a *vector potential* for $\mathbf{F}$.

27. In our derivation of the heat equation in this section, we assumed that the coefficient of thermal conductivity k is constant (no sinks or sources). If $k = k(x, y, z)$ is a variable, show that the heat equation becomes
  $$k\nabla^2 T + \nabla k \cdot \nabla T = \sigma\delta\frac{\partial T}{\partial t}$$

28. An electric charge q located at the origin produces the electric field
  $$\mathbf{E} = \frac{q\mathbf{R}}{4\pi\epsilon\|\mathbf{R}\|^3}$$
  where $\mathbf{R} = x\mathbf{i} + y\mathbf{j} + z\mathbf{k}$ and ϵ is a physical constant, called the **electric permittivity.**
  a. Show that
  $$\iint_S \mathbf{E} \cdot \mathbf{N}\,dS = 0$$
  if the closed surface S does not enclose the origin. This is **Gauss' law.**
  b. Show that
  $$\iint_S \mathbf{E} \cdot \mathbf{N}\,dS = \frac{q}{\epsilon}$$

.

Conservative vector fields:
a. **F** is *conservative* if it is the gradient of some scalar function, called the *scalar potential of* **F**; that is, $\mathbf{F} = \nabla f$.
b. **F** is conservative if and only if curl **F** = **0**.
c. Equivalent conditions:
  (i) $\int_C \mathbf{F} \cdot d\mathbf{R}$ is independent of path.
  (ii) **F** is conservative.
  (iii) $\int_C \mathbf{F} \cdot d\mathbf{R} = 0$ for every closed path C.

Evaluation of Line Integrals: $\int_C \mathbf{F} \cdot d\mathbf{R}$

Step 1. Check to see whether **F** is conservative; if it is, then

$$\int_C \mathbf{F} \cdot d\mathbf{R} = 0 \text{ if } C \text{ is closed}$$

$$\int_C \mathbf{F} \cdot d\mathbf{R} = f(Q) - f(P) \qquad \text{Initial point } P, \text{ terminal } Q$$

Step 2. If **F** is not conservative, and C is a closed curve bounding the surface S, use Stokes' theorem (or Green's theorem in $\mathbb{R}^2$) to equate the given integral to a surface integral, namely,

$$\int_C \mathbf{F} \cdot d\mathbf{R} = \iint_S (\text{curl } \mathbf{F} \cdot \mathbf{N}) \, dS$$

Step 3. If **F** is not conservative, and C is an open curve, try to add an arc C_1 so that the curve formed by C and C_1 is closed. Try to choose C_1 so that $\int_{C_1} \mathbf{F} \cdot d\mathbf{R}$ is relatively easy to evaluate.

$$\int_C \mathbf{F} \cdot d\mathbf{R} = \iint_S (\text{curl } \mathbf{F} \cdot \mathbf{N}) \, dS - \int_{C_1} \mathbf{F} \cdot d\mathbf{R}$$

Step 4. As a last resort, parametrize **F** and **R**. Let $\mathbf{R}(t) = x(t)\mathbf{i} + y(t)\mathbf{j} + z(t)\mathbf{k}$ for $a \le t \le b$.

$$\int_C f(x, y, z) \, dx = \int_a^b f[x(t), y(t), z(t)] \frac{dx}{dt} \, dt$$

$$\int_C f(x, y, z) \, ds = \int_a^b f[x(t), y(t), z(t)]$$
$$\times \sqrt{[x'(t)]^2 + [y'(t)]^2 + 1} \, dt$$

Evaluation of Flux Integrals: $\iint_S \mathbf{F} \cdot \mathbf{N} \, dS$

Step 1. If the surface S is a closed surface bounding the solid region D, use the divergence theorem to write the flux integral as a triple integral.

$$\iint_S \mathbf{F} \cdot \mathbf{N} \, dS = \iiint_D \text{div } \mathbf{F} \, dV$$

Step 2. If the surface S is open, try to find a supplementary surface S_1 such that the surface formed by S and S_1 is closed. Try to choose S_1 so that $\iint_{S_1} \mathbf{F} \cdot \mathbf{N} \, dS$ is easy to evaluate.

$$\iint_S \mathbf{F} \cdot \mathbf{N} \, dS = \iiint_D \text{div } \mathbf{F} \, dV - \iint_{S_1} \mathbf{F} \cdot \mathbf{N} \, dS_1$$

Step 3. If neither step 1 nor step 2 applies, parametrize **F**, **N**, and dS. This may be quite difficult. In the special case where S has the form $z = f(x, y)$ and $g = \mathbf{F} \cdot \mathbf{N}$, we have

$$\iint_S g(x, y, z) \, dS$$
$$= \iint_R g[x, y, f(x, y)] \sqrt{[f_x(x,y)]^2 + [f_y(x, y)]^2 + 1} \, dA_{xy}$$

Applications

Fluid Mechanics
Let **V** be the velocity of an incompressible fluid. Then

div **V** measures the rate of particle flow per unit volume at a point.

curl **V** measures the tendency of a fluid to swirl.

Flux integral: $\int_S \int \mathbf{V} \cdot \mathbf{N} \, dS$ measures the rate of flow across the surface S in unit time

Circulation: $\int_C \mathbf{V} \cdot \mathbf{T} \, ds$ measures the tendency of a fluid to move around C.

Stokes' theorem says that the cumulative tendency of a fluid to swirl across a surface S is equal to the circulation of a fluid around the boundary curve C.

The divergence theorem shows that the divergence of a velocity field **V** at P_0 equals the flux of flow out of P_0 per unit volume. If there are no sources or sinks, it follows that

$$\text{div } \delta\mathbf{V} = -\frac{\partial \delta}{\partial t} \quad \text{(continuity equation)}$$

where $\delta(x, y, z, t)$ is the density of the fluid.

Electromagnetism
Let **E** be the electric intensity field and **H** be the magnetic intensity field. Then

$$\text{div } \mathbf{E} = \frac{Q}{\epsilon} \quad \text{where } Q \text{ is the electric charge density and } \epsilon \text{ is the permittivity.}$$

curl $\mathbf{E} = -\dfrac{\partial(\mu\mathbf{H})}{\partial t}$ where μ is the permeability and t the time.

curl $\mathbf{H} = \mathbf{J}$ where $\mathbf{J}$ is the electric current density.

Maxwell's equation for electric intensity (assuming $Q = 0$) is

$$(\nabla \cdot \nabla)\mathbf{E} = \mu\sigma \dfrac{\partial \mathbf{E}}{\partial t} + \mu\epsilon \dfrac{\partial^2 \mathbf{E}}{\partial t^2}$$ where σ is the conductivity.

Ampère's circuit law is $\int_C \mathbf{H} \cdot d\mathbf{R} = I$, where I is the current crossing any surface S bounded by the closed curve C.

Thermodynamics

The heat equation is $\dfrac{\partial T}{\partial t} = \dfrac{k}{\sigma\delta} \nabla^2 T$, where $T(x, y, z, t)$ is the temperature at (x, y, z) at time t. The constant k is the thermal conductivity, σ is the specific heat of the body, and δ is its density.

Work: $W = \displaystyle\int_C \mathbf{F} \cdot d\mathbf{R}$

Mass: $m = \displaystyle\iint_S \delta(x, y, z)\, dS$

Proficiency Examination

Concept Problems

1. What is a vector field?
2. What is the divergence of a vector field?
3. What is the curl of a vector field?
4. What is the del operator?
5. What is Laplace's equation?
6. What is the difference between the Riemann integral and a line integral? Discuss.
7. What is the formula for a line integral of a vector field?
8. How do we find work as a line integral?
9. What is the formula for a line integral in terms of arc length parameter?
10. State the fundamental theorem on line integrals.
11. Define a conservative vector field.
12. What is the scalar potential of a conservative vector field?
13. What is a Jordan curve?
14. State Green's theorem.
15. How can you use Green's theorem to find area as a line integral?
16. What is a normal derivative?
17. Define a surface integral.
18. What is the formula for a surface integral of a surface defined parametrically?
19. What is a flux integral?
20. State Stokes' theorem.
21. State the conservative vector field theorem.
22. State the divergence theorem.

Practice Problems

23. Show that $yz\mathbf{i} + xz\mathbf{j} + xy\mathbf{k}$ is conservative and find a scalar potential function.

24. Compute div $\mathbf{F}$ and curl $\mathbf{F}$ for $\mathbf{F} = x^2y\mathbf{i} - e^{yz}\mathbf{j} + \frac{1}{2}x\mathbf{k}$.

25. Use Green's theorem to evaluate the line integral
$\displaystyle\int_C \mathbf{F} \cdot d\mathbf{R}$, where $\mathbf{F} = (2x + y)\mathbf{i} + 3y^2\mathbf{j}$ and C is the boundary of the triangle T with vertices $(-1, 2), (0, 0),$ $(1, 2)$, traversed in the given order.

26. Use Stokes' theorem to evaluate the line integral
$\displaystyle\int_C \mathbf{F} \cdot d\mathbf{R}$, where $\mathbf{F} = 2y\mathbf{i} + z\mathbf{j} + y\mathbf{k}$ and C is the intersection of the plane $z = x + 2$ and sphere $x^2 + y^2 + z^2 = 4z$, traversed counterclockwise as viewed from above.

27. Use the divergence theorem to evaluate the surface integral $\displaystyle\iint_S \mathbf{F} \cdot \mathbf{N}\, dS$, where $\mathbf{F} = x^2\mathbf{i} + (y + z)\mathbf{j} - 2z\mathbf{k}$, and S is the surface of the unit cube $0 \le x \le 1, 0 \le y \le 1,$ $0 \le z \le 1$.

28. Evaluate $\displaystyle\int_C \dfrac{x\, dx + y\, dy}{(x^2 + y^2)^2}$, where C is the path shown in Figure 14.38, traversed counterclockwise.

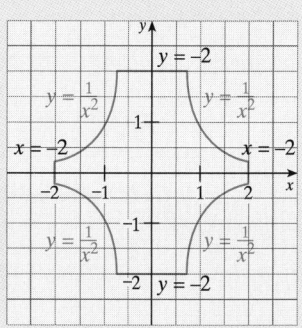

■ **FIGURE 14.38**   Problem 28

29. An object with mass m travels counterclockwise (as viewed from above) in the circular orbit $x^2 + y^2 = 9$, $z = 2$, with angular speed ω. The mass is subject to a centrifugal force $\mathbf{F} = m\omega^2 \mathbf{R}$, where $\mathbf{R} = x\mathbf{i} + y\mathbf{j} + z\mathbf{k}$. Show that $\mathbf{F}$ is conservative, and find a scalar potential function for $\mathbf{F}$.

30. Find the work done by $\mathbf{F}$ from Problem 29 as the object makes a half-orbit from $(3, 0, 2)$ to $(-3, 0, 2)$.

Supplementary Problems

In Problems 1–6, determine whether the given vector field is conservative, and if it is, find a scalar potential function.

1. $\mathbf{F} = 2\mathbf{i} - 3\mathbf{j}$ **2.** $\mathbf{F} = xy^{-2}\mathbf{i} + x^{-2}y\mathbf{j}$

3. $\mathbf{F} = y^{-3}\mathbf{i} + (-3xy^{-4} + \cos y)\mathbf{j}$

4. $\mathbf{F} = y^2\mathbf{i} + (2xy)\mathbf{j}$

5. $\mathbf{F} = \left(\dfrac{1}{y} + \dfrac{y}{x^2}\right)\mathbf{i} - \left(\dfrac{x}{y^2} - \dfrac{1}{x}\right)\mathbf{j}$

6. $\mathbf{F} = \left[2x\tan^{-1}\left(\dfrac{y}{x}\right) - y\right]\mathbf{i} + \left[2y\tan^{-1}\left(\dfrac{y}{x}\right) + x\right]\mathbf{j}$

In Problems 7–12, find $\displaystyle\int_C \mathbf{F} \cdot d\mathbf{R}$, where C is the curve $\mathbf{R}(t) = t\mathbf{i} + t^2\mathbf{j}, 1 \le t \le 2$. Note that these are the same as the vector fields in Problems 1–6.

7. $\mathbf{F} = 2\mathbf{i} - 3\mathbf{j}$ **8.** $\mathbf{F} = xy^{-2}\mathbf{i} + x^{-2}y\mathbf{j}$

9. $\mathbf{F} = y^{-3}\mathbf{i} + (-3xy^{-4} + \cos y)\mathbf{j}$

10. $\mathbf{F} = y^2\mathbf{i} + (2xy)\mathbf{j}$

11. $\mathbf{F} = \left(\dfrac{1}{y} + \dfrac{y}{x^2}\right)\mathbf{i} - \left(\dfrac{x}{y^2} - \dfrac{1}{x}\right)\mathbf{j}$

12. $\mathbf{F} = \left[2x\tan^{-1}\left(\dfrac{y}{x}\right) - y\right]\mathbf{i} + \left[2y\tan^{-1}\left(\dfrac{y}{x}\right) + x\right]\mathbf{j}$

In Problems 13–16, find div $\mathbf{F}$ and curl $\mathbf{F}$.

13. $\mathbf{F} = x\mathbf{i} + y\mathbf{j} + z\mathbf{k}$

14. $\mathbf{F} = \left(\tan^{-1}\dfrac{y}{x}\right)\mathbf{i} - 3\mathbf{j} + z^2\mathbf{k}$

15. $\mathbf{F} = \dfrac{1}{r}(x\mathbf{i} + y\mathbf{j} + z\mathbf{k})$, where $r = \sqrt{x^2 + y^2 + z^2}, r \ne 0$

16. $\mathbf{F} = (xy\sin z)\mathbf{i} + (x^2\cos yz)\mathbf{j} + (z\sin xy)\mathbf{k}$

Evaluate the line integrals in Problems 17–18, by parametrization.

17. $\displaystyle\int_C [(\sin \pi y)\,dx + (\cos \pi x)\,dy]$, where C is the line segment from $(1, 0)$ to $(\pi, 0)$, followed by the line segment from $(\pi, 0)$ to (π, π)

18. $\displaystyle\int_C (z\,dx - x\,dy + dz)$, where C is the arc of the helix $x = 3\sin t, y = 3\cos t, z = t$ for $0 \le t \le \frac{\pi}{4}$

In each of Problems 19–33, evaluate the line integral or the surface integral. In each case, assume the orientation of C is counterclockwise when viewed from above or that $\mathbf{N}$ is the unit outward normal vector.

19. $\displaystyle\int_C [yz\,dx + xz\,dy + (xy + 2)\,dz]$

C is the curve $\mathbf{R}(t) = (\tan^{-1}t)\mathbf{i} + t^2\mathbf{j} - 3t\mathbf{k}, 0 \le t \le 1$.

20. $\displaystyle\int_C [(x^2 + y)\,dx + xz\,dy - (y + z)\,dz]$

C is the curve $\mathbf{R}(t) = t\mathbf{i} + t^2\mathbf{j} + 2\mathbf{k}, 0 \le t \le 1$.

21. $\displaystyle\iint_S (3x^2 + y - 2z)\,dS$

S is the surface $\mathbf{R}(u, v) = u\mathbf{i} + (u + v)\mathbf{j} + u\mathbf{k}, 0 \le u \le 1$, $0 \le v \le 1$.

22. $\displaystyle\iint_S \mathbf{F} \cdot \mathbf{N}\,dS$

$\mathbf{F} = 3x\mathbf{i} + z^2\mathbf{j} - 2y\mathbf{k}$ and S is the surface of the hemisphere $z = \sqrt{4 - x^2 - y^2}$.

23. $\displaystyle\int_C (x\,dx + x\,dy - y\,dz)$

C is the curve $\mathbf{R}(t) = t\mathbf{i} + t^2\mathbf{j} + t\mathbf{k}, 0 \le t \le 1$.

24. $\displaystyle\int_C (x^2\,dx - 3y^2\,dz)$

C is the line segment from $(0, 1, 1)$ to $(1, 1, 2)$.

25. $\displaystyle\int_C (x^2\,dx + y\,dy)$

C is the curve $\mathbf{R}(t) = (t\sin t)\mathbf{i} + (1 - t\cos t)\mathbf{j}$, $0 \le t \le 2\pi$.

26. $\displaystyle\int_C (xy\,dx - x^2\,dy)$

C is the square with vertices $(1, 0), (0, 1), (-1, 0), (0, -1)$ traversed counterclockwise.

27. $\displaystyle\int_C (y\,dx + x\,dy - 2\,dz)$

C is the curve of intersection of the cylinder $x^2 + y^2 = 2x$ and the plane $x = z$.

28. $\displaystyle\int_C [(y + z)\,dx + (x + z)\,dy + (x + y)\,dz]$

C is the curve of intersection of the sphere
$x^2 + (y - 3)^2 + z^2 = 9$ and the plane $x + 2y + z = 3$.

29. $\int_C (-2y\,dx + 2x\,dy + dz)$

C is the circle $x^2 + y^2 = 1$ in the plane $z = 3$.

30. $\iint_S (\text{curl } y\mathbf{i}) \cdot \mathbf{N}\,dS$

S is the hemisphere $z = \sqrt{1 - x^2 - y^2}$.

31. $\iint_S (2x^3\mathbf{i} + y^3\mathbf{j} + z^3\mathbf{k}) \cdot \mathbf{N}\,dS$

S is the surface of the ellipsoid $2x^2 + y^2 + z^2 = 1$.

32. $\iint_S (y^2\mathbf{i} + y^2\mathbf{j} + yz\mathbf{k}) \cdot \mathbf{N}\,dS$

S is the surface of the tetrahedron bounded by the plane $2x + 3y + z = 1$ and the coordinate planes, in the first octant.

33. $\iint_S \nabla\phi \cdot \mathbf{N}\,dS$, where $\phi(x, y, z) = 2x + 3y$

where S is the portion of the plane $ax + by + cz = 1$ $(a > 0, b > 0, c > 0)$ that lies in the first octant.

In Problems 34–38, find $\iint_S \mathbf{F} \cdot \mathbf{N}\,dS$.

34. $\mathbf{F} = x^2\mathbf{i} + y^2\mathbf{j} + x^2\mathbf{k}$, and S is the surface of the unit cube $0 \le x \le 1, 0 \le y \le 1, 0 \le z \le 1$.

35. $\mathbf{F} = 2yz\mathbf{i} + (\tan^{-1} xz)\mathbf{j} + e^{xy}\mathbf{k}$, and S is the surface of the sphere $x^2 + y^2 + z^2 = 1$.

36. $\mathbf{F} = x\mathbf{i} - 4\mathbf{j} + 3\mathbf{k}$, and S is the paraboloid $y = x^2 + z^2$ with $x^2 + z^2 \le 9$. The disk $x^2 + z^2 = 9$ is omitted; that is, the paraboloid is open on the right.

37. $\mathbf{F} = xyz\mathbf{i} + xyz\mathbf{j} + xyz\mathbf{k}$, and S is the surface of the five faces of the unit cube $0 \le x \le 1, 0 \le y \le 1, 0 \le z \le 1$ missing $z = 0$.

38. $\mathbf{F} = xy\mathbf{i} - 2z\mathbf{j}$, and S is the surface defined parametrically by $\mathbf{R}(u, v) = u\mathbf{i} + v\mathbf{j} + u\mathbf{k}$ for $0 \le u \le 1, 0 \le v \le 1$.

Find all real numbers c for which each vector field in Problems 39–41 is conservative.

39. $\mathbf{F} = (x, y) = (\sqrt{x} + 3xy)\mathbf{i} + (cx^2 + 4y)\mathbf{j}$

40. $\mathbf{F}(x, y) = \left(\dfrac{cy}{x^3} + \dfrac{y}{x^2}\right)\mathbf{i} + \left(\dfrac{1}{x^2} - \dfrac{1}{x}\right)\mathbf{j}$

41. $\mathbf{F}(x, y, z) = e^{yz/x}\left[\left(\dfrac{cyz}{x^2}\right)\mathbf{i} + \left(\dfrac{z}{x}\right)\mathbf{j} + \left(\dfrac{y}{x}\right)\mathbf{k}\right]$

42. Let $\mathbf{F} = (y^2 + x^{-2}ye^{x/y})\mathbf{i} + (2xy + z - x^{-1}e^{x/y})\mathbf{j} + y\mathbf{k}$. Is F conservative?

43. Find the work done when an object moves in the force field $\mathbf{F} = 2x\mathbf{i} - (x + z)\mathbf{j} + (y - x)\mathbf{k}$ along the path given by $\mathbf{R}(t) = t^2\mathbf{i} + (t^2 - t)\mathbf{j} + 3\mathbf{k}, 0 \le t \le 1$.

44. Show that the force field $\mathbf{F} = yz^2\mathbf{i} + (xz^2 - 1)\mathbf{j} + (2xyz - 1)\mathbf{k}$ is conservative and determine the work done when an object moves in the force field from the origin to the point $(1, 0, 1)$.

45. Find a region R in the plane where the vector field
$\mathbf{F} = \dfrac{1}{x + y}(\mathbf{i} + \mathbf{j})$ is conservative. Then evaluate
$\int_C \mathbf{F} \cdot d\mathbf{R}$, where C is any path in R from the point $P_0(a, b)$ to $P_1(c, d)$.

46. If u is a scalar function and F is a continuously differentiable vector field, show that
$\text{curl}(u\mathbf{F}) = u \text{ curl } \mathbf{F} + (\nabla u \times \mathbf{F})$.

47. Show that $\text{div}(\mathbf{F} \times \mathbf{G}) = \mathbf{G} \cdot \text{curl } \mathbf{F} - \mathbf{F} \cdot \text{curl } \mathbf{G}$, for any continuously differentiable vector fields F and G.

48. A vector field F is *incompressible* (or *solenoidal*) in a region D if div $\mathbf{F} = 0$ throughout D. If F and G are both conservative vector fields in D, show that $\mathbf{F} \times \mathbf{G}$ is solenoidal.

49. If $\mathbf{F} = \text{curl } \mathbf{G}$, show that F is solenoidal. (See Problem 48.)

50. Suppose $\mathbf{F} = f(x, y, z)\mathbf{A}$, where A is a constant vector and f is a scalar function. Show that curl F is orthogonal to A and to ∇f.

51. If A is a constant vector and F is a continuously differentiable vector field, show that $\text{div}(\mathbf{A} \times \mathbf{F}) = -\mathbf{A} \cdot \text{curl } \mathbf{F}$.

52. Evaluate the line integral $\int_C \dfrac{x\,dx - y\,dy}{x^2 - y^2}$, where C is any path in the xy-plane that is interior to the region $x > 0$, $y < x, y > -x$ and connects the point $(5, 4)$ to $(2, 0)$.

53. Evaluate $\int_C \left(\dfrac{-y}{x^2}\,dx + \dfrac{1}{x}\,dy\right)$, where C is the closed path $(x - 2)^2 + y^2 = 1$, traversed once counterclockwise.

54. **THINK TANK PROBLEM**   Let $u(x, y)$ and $v(x, y)$ be functions of two variables with continuous partial derivatives everywhere in the plane, and suppose that u and v satisfy the equation
$$\frac{\partial u}{\partial y} = \frac{\partial v}{\partial x}$$
for all (x, y). Are u and v necessarily harmonic? Either show that they are or find a counterexample.

55. a. Find a region in the plane where the vector field
$$\mathbf{F} = \left(\frac{1 + y^2}{x^3}\right)\mathbf{i} - \left(\frac{y + x^2y}{x^2}\right)\mathbf{j}$$
is conservative, and find a scalar potential for F.

b. Evaluate $\int_C \mathbf{F} \cdot d\mathbf{R}$, where C is a path from $(1, 1)$ to $(3, 4)$. Are there any limitations on the path C? Explain.

56. Determine the most general function $u(x, y)$ for which the vector field $\mathbf{F} = u(x, y)\mathbf{i} + (2y\, e^x + y^2e^{3x})\mathbf{j}$ will be conservative.

57. Consider the line integral $\int_C \left(\dfrac{dx}{y} + \dfrac{dy}{x} \right)$, where C is the closed triangular path formed by the lines $y = 2x$, $x + 2y = 5$, and $x = 2$, traversed counterclockwise. First evaluate the line integral directly (by parametrizing C) and then by using Green's theorem.

58. If S is a closed surface in a region R and $\mathbf{F}$ is a continuously differentiable vector field on R, show that

$$\iint_S (\text{curl } \mathbf{F} \cdot \mathbf{N})\, dS = 0$$

where $\mathbf{N}$ is the outward unit normal vector to S.

59. A certain closed path C in the plane $2x + 2y + z = 1$ is known to project onto the unit circle $x^2 + y^2 = 1$ in the xy-plane. Let c be a constant, and let $\mathbf{R} = x\mathbf{i} + y\mathbf{j} + z\mathbf{k}$. Use Stokes' theorem to evaluate

$$\int_C (c\mathbf{k} \times \mathbf{R}) \cdot d\mathbf{R}$$

60. a. Show that if the scalar function w is harmonic, then

$$\nabla \cdot (w\, \nabla w) = \|\nabla w\|^2$$

b. Let $w = x - y + 2z$, and let S be the surface of the sphere $x^2 + y^2 + z^2 = 9$. Evaluate

$$\iint_S w\, \frac{\partial w}{\partial n}\, dS$$

Note: The normal derivative $\dfrac{\partial w}{\partial n}$ is defined in Section 14.4.

61. A particle moves along a curve C in space that is given parametrically by $\mathbf{R}(t) = x(t)\mathbf{i} + y(t)\mathbf{j} + z(t)\mathbf{k}$ for $a \le t \le b$. A force field $\mathbf{F}$ is applied to the particle in such a way that $\mathbf{F}$ is always perpendicular to the path C. How much work is performed by the force field $\mathbf{F}$ when the particle moves from the point where $t = a$ to the point where $t = b$?

62. If $\mathbf{D}$ is the electric displacement field, then div $\mathbf{D} = \phi$, where ϕ is the **charge density.** A region of space is said to be **charge-free** if $\phi = 0$ there. Describe the charge-free regions of the electric displacement field $\mathbf{D} = 2x^2\mathbf{i} + 3y^2\mathbf{j} - 2z^2\mathbf{k}$.

63. A satellite weighing 10,000 kg travels in a circular orbit 7,000 km from the center of the earth. How much work is done by gravity on the satellite during half a revolution?

64. If $\mathbf{F}$ is a conservative force field, the scalar function f such that $\mathbf{F} = -\nabla f$ is called the *potential energy*. Suppose an

object with mass 10 g moves in the force field in such a way that its speed decreases from 3 cm/s to 2.5 cm/s. What is the corresponding change in the potential energy of the object?

65. Find the work done by the force field

$$\mathbf{F} = 2xyz\mathbf{i} + \left(x^2z - \frac{1}{z}\tan^{-1}\frac{y}{z}\right)\mathbf{j} + \left(x^2y + \frac{y}{z^2}\tan^{-1}\frac{y}{z}\right)\mathbf{k}$$

in moving an object along the circular helix

$$\mathbf{R}(t) = (\sin \pi t)\mathbf{i} + (\cos \pi t)\mathbf{j} + (2t + 1)\mathbf{k}$$

for $0 \le t \le \dfrac{1}{2}$.

66. Find the work done when an object moves against the force field $\mathbf{F} = 4y^2\mathbf{i} + (3x + y)\mathbf{j}$ from $(1, 0)$ to $(-1, 0)$ along the top half of the ellipse $x^2 + \dfrac{y^2}{k^2} = 1$. Which value of k minimizes the work?

67. Evaluate the line integral

$$\int_C \frac{-y\, dx + x\, dy}{x^2 + y^2}$$

where C is the limaçon given in polar coordinates by $r = 3 + 2\cos \theta$, $0 \le \theta \le 2\pi$, traversed counterclockwise.

68. Suppose f and g are both harmonic in the region R with boundary surface S. Show that

a. $\displaystyle\iint_S f\frac{\partial g}{\partial n}\, dS = \iint_S g\frac{\partial f}{\partial n}\, dS$

b. $\displaystyle\iint_S f\frac{\partial f}{\partial n}\, dS = \iiint_D \|\nabla f\|^2\, dV$

Note: $\dfrac{\partial f}{\partial n}$ denotes the normal derivative (see Section 14.4).

69. Show that curl(curl $\mathbf{F}$) = ∇(div $\mathbf{F}$) $- \nabla^2 \mathbf{F}$ if the components of $\mathbf{F}$ have continuous second-order partial derivatives.

70. Evaluate the surface integral $\displaystyle\iint_S \frac{\partial f}{\partial n}\, dS$ where S is the surface of the unit sphere $x^2 + y^2 + z^2 = 1$ and f is a scalar field such that $\|\nabla f\|^2 = 3f$ and div $(f\nabla f) = 7f$. Remember that $\dfrac{\partial f}{\partial n}$ denotes the normal derivative, $\nabla f \cdot \mathbf{N}$.

71. Evaluate the surface integral $\displaystyle\iint_S dS$, where S is the torus $\mathbf{R}(u, v) = [(a + b\cos v)\cos u]\mathbf{i} + [(a + b\cos v)\sin u]\mathbf{j} + (b\sin v)\mathbf{k}$ for $0 < b < a$ and $0 \le u \le 2\pi$, $0 \le v \le 2\pi$.

72. Evaluate $\displaystyle\iint_S \mathbf{F} \cdot \mathbf{N}\, dS$, where $\mathbf{F} = x\mathbf{i} + y\mathbf{j} + z\mathbf{k}$ and S is a cubic surface with a corner block removed, as shown in Figure 14.39.

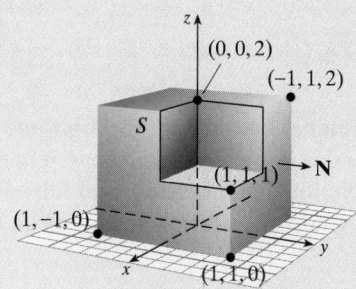

■ **FIGURE 14.39**   Cube with a corner removed

73. THINK TANK PROBLEM   Prove or disprove that it does not matter which corner is removed in Problem 72.

74. Show that a lamina that covers a standard region D in the plane with density $\delta = 1$ has moment of inertia

$$I = \frac{1}{3} \int_C (-y^3\, dx + x^3\, dy)$$

with respect to the z-axis, where C is the boundary curve of D.

75. Use vector analysis to find the centroid of the conical surface $z = \sqrt{x^2 + y^2}$ between $z = 0$ and $z = 3$.

76. JOURNAL PROBLEM by J. Chris Fischer (CRUX Problem 785).* Suppose a closed differentiable curve has exactly one tangent line parallel to every direction. More precisely, suppose that the curve has parametrization $\mathbf{V}(\theta)$ which maps the interval $[0, 2\pi]$ into $\mathbb{R}^2$ for which

i. $\dfrac{d\mathbf{V}}{d\theta} = [r(\theta)\cos\theta,\, r(\theta)\sin\theta]$ for some continuous real-valued function $r(\theta)$, and

ii. $\mathbf{V}(\theta) = V(\theta + \pi)$

Prove that the curve is described in the clockwise sense as θ runs from 0 to π.

77. PUTNAM EXAMINATION PROBLEM   A force acts on the element ds of a closed plane curve. The magnitude of this force is $r^{-1}\, ds$, where r is the radius of curvature at the point considered, and the direction of the force is perpendicular to the curve; it points to the convex side. Show that the system of such forces acting on all elements of the curve keeps it in equilibrium.

Continuous vs. Discrete Mathematics

William F. Lucas is a professor of mathematics and department chairman at The Claremont Graduate School. He is an Avery Fellow to Harvey Mudd College, and is known for his research in game theory, which provides a mathematical approach to the study of conflict, cooperation, and fairness. Dr. Lucas has also been active in educational reform efforts with a goal toward introducing more recently discovered topics into the mathematics curriculum.

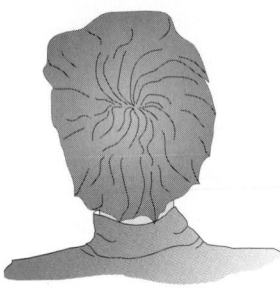

Mathematicians often distinguish between *discrete* and *continuous* mathematics. The latter is illustrated by the calculus and its many descendants, including the subject of differential equations. Continuous mathematics deals with "solid" infinite sets such as the real number line ℝ and functions defined on ℝ. One of the primary concerns is with the solutions of differential equations, which are typical families of such solid curves or surfaces. The many applications of such differential equations to discoveries in the physical sciences and engineering over the past three centuries has been one of the greatest intellectual success stories of all times.

> Most human heads have a fixed point, in the form of a whorl, sometimes called a cowlick, from which all the hair radiates. It would be impossible to cover a sphere with hair (or with radiating lines) without at least one such fixed point.

Prior to the invention of calculus by Newton and Leibniz, most mathematics was of a discrete nature. It dealt with sets that had a finite number of elements and with infinite but "countable" sets such as the natural numbers. These sets often include continuous curves such as the conics, which can be characterized by a small number of conditions. Discrete mathematics also deals with the solutions to algebraic equations, which are usually discrete sets. The twentieth century has witnessed the creation of many additional subjects in the discrete direction. This development was spurred on by the rapidly increasing use of mathematics in the social, behavioral, decisional, and system sciences, where the items under investigation are typically finite in number and not readily approximated by some continuous idealization. Moreover, nearly all aspects of the ongoing revolution in *digital* computers involve discrete considerations.

A metal bar on a minute level is composed of discrete molecules, atoms, and elementary particles. Many of its physical properties, however, can be determined by viewing the bar as a solid continuum and employing the analytical techniques of calculus. One applies the basic laws of physics to express the local (infinitesimal) properties of the bar in terms of differential equations. The solutions of these equations, in turn, provide an excellent description of the observed global behavior of the bar.

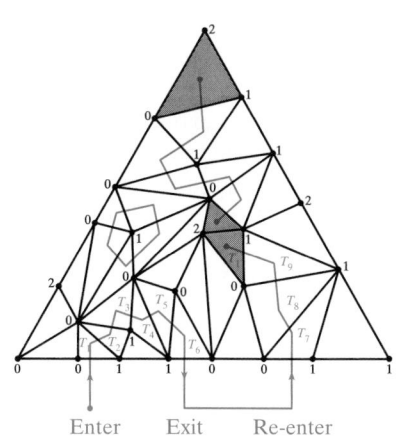

■ **FIGURE 14.40** Arbitrary partitioning of a triangle

One of the greatest mathematical discoveries of the twentieth century in continuous mathematics is the famous fixed point theorem, published in 1912 by the Dutch mathematician L. E. J. Brouwer (1881–1966). It states that any continuous function f from a set S (with certain desirable properties) into the same set S has at least one *fixed point* x_0, that is, $f(x_0) = x_0$. For example, for any continuous function $y = f(x)$ from the set $S = \{x \in \mathbb{R}: 0 \le x \le 1\}$ into S, there is some number $x_0 \in S$ such that the curve $y = f(x)$ crosses the line $y = x$ at x_0. A head of hair must either have a seam (a parting of the hair, a discontinuity) or else a cowlick (where the hair stands straight up, a fixed point). Fixed points are realized in many physical and social phenomena such as equilibrium states in mechanics or stable prices in economics.

Like many of the outstanding results in continuous mathematics, Brouwer's theorem has an analogue in discrete mathematics, known as the labeling lemma of the German mathematician Emanuel Sperner, which appeared in 1928. Given a triangle with vertices 0, 1, and 2 as shown in Figure 14.40, one can partition the interior of this triangle into (nonoverlapping) smaller triangles in any possible way.

One can then label each of the newly created vertices with any one of the numbers 0, 1, or 2. There is only one stipulation on how the new vertices on the perimeter of the triangle can be labeled: Any vertex on a particular side of △012 cannot be labeled with the number appearing on the *main* vertex opposite (disjoint from) this side. For example, the side 01 of △012 cannot contain a vertex with the label 2. Sperner's lemma states that there must exist at least one elementary triangle inside △012 (or an *odd* number, in general) whose three vertices are "completely labeled" in the sense that they have all three labels 0, 1, and 2. There are three such triangles (shown in color) in Figure 14.40.

There is an elementary constructive proof of Sperner's lemma that uses the idea of a "path-following algorithm" published by Daniel I. A. Cohen in 1967. This proof is illustrated by the dashed line in Figure 14.40. One can enter into △012 from outside via some elementary triangle T_1 whose perimeter edge carries the labels 0 and 1. The third vertex on this elementary triangle T_1 must be either 2 (in which case T_1 is a completely labeled elementary triangle), or else it is 0 or 1. In the latter cases, one can exit T_1 via a second 01 edge. One can continue to enter and exit subsequent elementary triangles $T_2, T_3, \ldots$ through successive 01 edges until one finally reaches an elementary triangle T_n that has completely labeled vertices 0, 1, and 2. (If this path were ever to exit △012, then there must be still another 01 edge on the perimeter of △012 where one can reenter the main triangle and continue along the path, as did occur once in Figure 14.40.)

This path-following method also extends to a proof of Sperner's labeling lemma for dimensions higher than two. Furthermore, this approach is used in practical problems to arrive at a "very small" elementary triangle that can serve as an approximation to a fixed point, when it is excessively difficult or impossible to determine the precise location of the fixed point itself. In such cases, the broken path in Figure 14.40 can be viewed as the analogue in discrete mathematics of the continuous curves one arrives at when solving differential equations.

Around 1945, the great Hungarian–American mathematician John von Neumann (1903–1957) pointed out the need for approximating paths in a discrete manner when it proves too difficult to arrive at the exact continuous solutions to certain ordinary or partial differential equations. Cohen's constructive proof of Sperner's lemma, and work in the late 1960s by the mathematical economist Herbert Scarf on approximating equilibrium points, have paved the way for one very rich and extensive discrete theory that provides numerical approximations for continuous phenomena when the latter problems cannot be solved directly by the many analytic techniques of continuous mathematics.

Mathematical Essays

1. If your college offers a course on discrete mathematics, interview an instructor of the course, and using that interview as a basis, write an essay comparing continuous and discrete mathematics.

2. Draw several triangles and attempt to draw a cunterexample for Sperner's labeling lemma. In each case, show how this lemma is satisfied.

3. Consider two sheets of paper containing the numbers 1 to 120, as shown in the photograph on the left.

   If the top sheet is crumpled and dropped on the bottom sheet, the fixed point theorem tells us that one point must still be over its starting point. In the photograph on the

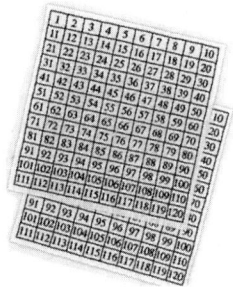 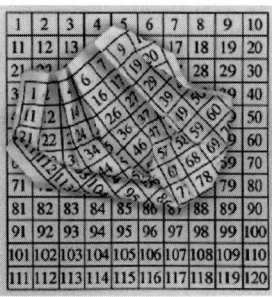

   right, it is a point in the region of the number 78. Perform this experiment several times in an attempt to find a counterexample. In each case, show how the fixed point theorem is satisfied.

4. **HISTORICAL QUEST** William Rowan Hamilton has been called the most renowned Irish mathematician. He was a child prodigy who read Greek, Hebrew, and Latin by the time he was five, and by the age of ten he knew over a dozen languages. Many mathematical advances are credited to Hamilton. For example, he developed vector methods in analytic geometry and calculus, as well as

a system of algebraic quantities called **quaternions,** which occupied his energies for the last 22 years of his life. Hamilton pursued the study of quaternions with an almost religious fervor, but by the early 20th century, the notation and terminology of vectors dominated. Much of the credit for the eventual emergence of vector methods goes not only to Hamilton, but also to the scientists James Clerk Maxwell (1831–1879), J. Willard Gibbs (1839–1903), and Oliver Heaviside (1850–1925).

WILLIAM ROWAN HAMILTON 1805–1865

For this Quest, write a paper on quaternions.

5. Write a 500-word essay on the history of Green's theorem, Stokes' theorem, and the divergence theorem.

6. Write a report on four-dimensional geometry.

7. **Book Report** "We often hear that mathematics consists mainly in 'proving theorems.' Is a writer's job mainly that of `writing sentences'? A mathematician's work is mostly a tangle of guesswork, analogy, wishful thinking and frustration, and proof, far from being the core of discovery, is more often than not a way of making sure that our minds are not playing tricks. Few people, if any, had dared write this out loud before Davis and Hersh. Theorems are not to mathematics what successful courses are to a meal. The nutritional analogy is misleading. To master mathematics is to master an intangible view…" This quotation comes from the introduction to the book *The Mathematical Experience* by Philip J. Davis and Reuben Hersh (Boston: Houghton Mifflin, 1981). Read this book and prepare a book report.

8. Make up a word problem involving vector fields. Send your problem to:

   Bradley and Smith

   Prentice Hall Publishing Company

   1 Lake Street

   Upper Saddle River, NJ 07458

   The best ones submitted will appear in the next edition (along with credit to the problem poser).

15

Introduction to Differential Equations

PREVIEW

We introduced and examined separable differential equations in Section 5.6 and first-order linear equations in Section 7.6. We examined applications such as orthogonal trajectories, flow of a fluid through an orifice, escape velocity of a projectile, carbon dating, the diastolic phase of blood pressure, learning curves, population models, dilution problems, and the flow of current in an *RL* circuit. In this chapter, we shall extend our study of first-order differential equations by examining *homogeneous* and *exact* equations and then investigate *second-order differential equations.*

PERSPECTIVE

The study of differential equations is such an extensive topic that even a brief survey of its methods and applications usually occupies a full course. Our goal in this chapter is to preview such a course by introducing some useful techniques for solving differential equations and by examining a few important applications.

15.1   *First-Order Differential Equations*

review of separable differential equations, homogeneous differential equations, review of first-order linear differential equations, exact differential equations, Euler's method   ■

We have discussed several different kinds of first-order differential equations so far in this text. In this section, we review our previous methods and introduce two new forms, *homogeneous* and *exact* differential equations.

REVIEW OF SEPARABLE DIFFERENTIAL EQUATIONS

Recall that a differential equation is just an equation involving derivatives or differentials. In particular, an ***n*th-order differential equation** in the dependent variable y with respect to the independent variable x is an equation in which the highest derivative of y that appears is $d^n y/dx^n$. A **general solution** of a differential equation is an expression that completely characterizes all possible solutions of the equation, and a **particular solution** is a solution that satisfies certain specifications such as an initial value condition $y(x_0) = y_0$.

In Section 5.6, we defined a **separable differential equation** as one that can be written in the form

$$\frac{dy}{dx} = \frac{g(x)}{f(y)}$$

and observed that such an equation can be solved by separating the variables and integrating each side; that is,

$$\int f(y)\, dy = \int g(x)\, dx$$

EXAMPLE 1   *Separable differential equation*

Find the general solution of the differential equation

$$\frac{dy}{dx} = e^{-y} \sin x$$

Solution   Separate the variables and integrate:

$$\frac{dy}{dx} = e^{-y} \sin x$$

$$e^y\, dy = \sin x\, dx$$

$$\int e^y\, dy = \int \sin x\, dx$$

$$e^y = -\cos x + C \qquad \textit{Combine constants of integration.}$$

This can also be written as $y = \ln|C - \cos x|$.   ■

From the standpoint of applications, one of the most important separable differential equations is

$$\frac{dy}{dx} = ky$$

which occurs in the study of exponential growth and decay. You might review Table 7.3 on page 508, which gives the solution for uninhibited growth or decay, logistic (or inhibited) growth, and limited growth functions. In this section, we consider an application of separable differential equations from chemistry.

EXAMPLE 2   *Chemical conversion*

Experiments in chemistry indicate that under certain conditions, two substances A and B will convert into a third substance C in such a way that the rate of conversion with respect to time is jointly proportional to the unconverted amounts of A and B. For simplicity, assume that one unit of C is formed from the combination of one unit of A and one unit of B, and assume that initially there are α units of A, β units of B, and no units of C present. Set up and solve a differential equation for the amount $Q(t)$ of C present at time t, assuming $\alpha \neq \beta$.

Solution   Since each unit of C is formed from one unit of A and one unit of B, it follows that at time t, $\alpha - Q(t)$ units of A and $\beta - Q(t)$ units of B remain unconverted. The specific rate condition can be expressed mathematically as

$$\frac{dQ}{dt} = k(\alpha - Q)(\beta - Q)$$

where k is a constant ($k > 0$ because $Q(t)$ is increasing).

To solve this equation, we separate the variables and integrate:

$$\int \frac{dQ}{(\alpha - Q)(\beta - Q)} = \int k\,dt$$

$$\int \frac{1}{\alpha - \beta}\left[\frac{-1}{\alpha - Q} + \frac{1}{\beta - Q}\right] dQ = \int k\,dt \qquad \text{\textit{Partial fractions decomposition}}$$

$$\frac{1}{\alpha - \beta}[\ln(\alpha - Q) - \ln(\beta - Q)] = kt + C_1$$

$$\ln\left|\frac{\alpha - Q}{\beta - Q}\right| = (\alpha - \beta)kt + C_2$$

$$\frac{\alpha - Q}{\beta - Q} = Me^{(\alpha - \beta)kt} \qquad \text{\textit{Where } } M = e^{C_2}$$

$$\alpha - Q = \beta Me^{(\alpha - \beta)kt} - QMe^{(\alpha - \beta)kt}$$

$$QMe^{(\alpha - \beta)kt} - Q = \beta Me^{(\alpha - \beta)kt} - \alpha$$

$$Q = \frac{\beta Me^{(\alpha - \beta)kt} - \alpha}{Me^{(\alpha - \beta)kt} - 1}$$

The initial condition tells us that $Q(0) = 0$, so that

$$0 = \frac{\beta Me^0 - \alpha}{Me^0 - 1}$$

$$0 = \beta M - \alpha$$

$$M = \frac{\alpha}{\beta}$$

Thus, $Q(t) = \dfrac{\beta \dfrac{\alpha}{\beta} e^{(\alpha - \beta)kt} - \alpha}{\dfrac{\alpha}{\beta} e^{(\alpha - \beta)kt} - 1} = \dfrac{\alpha\beta[e^{(\alpha - \beta)kt} - 1]}{\alpha e^{(\alpha - \beta)kt} - \beta}$ ∎

HOMOGENEOUS DIFFERENTIAL EQUATIONS

Sometimes a first-order differential equation that is not separable can be put into separable form by a change of variables. A differential equation of the form

$$M(x, y)\, dx + N(x, y)\, dy = 0$$

is called a **homogeneous differential equation** if it can be written in the form

$$\frac{dy}{dx} = f\left(\frac{y}{x}\right)$$

In other words, dy/dx is isolated on one side of the equation and the other side can be expressed as a function of y/x. We can then solve the differential equation by substitution.

To see how to solve such an equation, set $v = y/x$, so that

$$vx = y$$

$$\frac{d}{dx}(vx) = \frac{d}{dx}(y) \qquad \text{Take the derivative of both sides.}$$

$$v + x\frac{dv}{dx} = \frac{dy}{dx} \qquad \text{Product rule}$$

$$v + x\frac{dv}{dx} = f(v) \qquad \text{Substitution, } \frac{dy}{dx} = f\left(\frac{y}{x}\right) = f(v)$$

$$x\frac{dv}{dx} = f(v) - v$$

$$\frac{dv}{f(v) - v} = \frac{dx}{x}$$

The equation can now be solved by integrating both sides; remember to express your answer in terms of the original variables x and y (use $v = y/x$).

EXAMPLE 3	*Homogeneous differential equation*

Find the general solution of the equation $2xy\, dx + (x^2 + y^2)\, dy = 0$.

Solution   First, show that the equation is homogeneous by writing it in the form $\frac{dy}{dx} = f\left(\frac{y}{x}\right)$:

$$2xy\, dx + (x^2 + y^2)\, dy = 0$$

$$\frac{dy}{dx} = \frac{-2xy}{x^2 + y^2} = \frac{-2\left(\dfrac{y}{x}\right)}{1 + \left(\dfrac{y}{x}\right)^2}$$

Let $v = \dfrac{y}{x}$ and $f(v) = \dfrac{-2v}{1 + v^2}$.

$$\frac{dv}{f(v) - v} = \frac{dx}{x}$$

$$\frac{dv}{\dfrac{-2v}{1 + v^2} - v} = \frac{dx}{x}$$

$$-\int \frac{(1 + v^2)\, dv}{v^3 + 3v} = \int x^{-1}\, dx$$

$$\int \left[\frac{\frac{1}{3}}{v} + \frac{\frac{2}{3}v}{v^2 + 3} \right] dv = -\int x^{-1} \, dx \qquad \textit{Partial fractions decomposition}$$

$$\frac{1}{3} \ln|v| + \frac{2}{3}[\frac{1}{2} \ln|v^2 + 3|] = -\ln|x| + C_1$$

$$\frac{1}{3} \ln|v(v^2 + 3)| + \ln|x| = C_1$$

$$\ln \left| \frac{y}{x} \left[\left(\frac{y}{x}\right)^2 + 3 \right] \right| + \ln|x^3| = C_2 \qquad \textit{Substituting } v = \frac{y}{x}$$

$$\ln \left| \frac{y^3 + 3x^2y}{x^3} \cdot x^3 \right| = C_2$$

$$\qquad\qquad\qquad \textit{Where } C = e^{C_2}$$

$$y^3 + 3x^2y = C$$

This is the general solution of the given differential equation. ∎

REVIEW OF FIRST-ORDER LINEAR DIFFERENTIAL EQUATIONS

In Section 7.6, we considered differential equations of the form

$$\frac{dy}{dx} + p(x)y = q(x)$$

Such an equation is said to be **first-order linear,** and we showed that its general solution is given by

WARNING Note that the coefficient of dy/dx is 1. If it is not, then divide by that nonzero coefficient. ←

$$y = \frac{1}{I(x)} \left[\int I(x) \, q(x) \, dx + C \right]$$

where $I(x)$ is the *integrating factor*

$$I(x) = e^{\int p(x)dx}$$

EXAMPLE 4   *First-order differential equation*

Solve $\dfrac{dy}{dx} + y \tan x = \sec x.$

Solution   Comparing the given first-order linear differential equation to the general first-order form, we see

$$p(x) = \tan x \quad \text{and} \quad q(x) = \sec x$$

The integrating factor is

$$I(x) = e^{\int \tan x \, dx} = e^{-\ln|\cos x|} = e^{\ln|(\cos x)^{-1}|} = (\cos x)^{-1} = \sec x$$

and the general solution is

$$y = \frac{1}{\sec x} \left[\int (\sec x)(\sec x) \, dx + C \right]$$

$$= \cos x[\tan x + C]$$

$$= \sin x + C \cos x \qquad\qquad\qquad ∎$$

First-order linear differential equations appear in a variety of applications. In Section 7.6, we showed how first-order linear equations may be used to model mixture (dilution) problems as well as problems involving the current in an *RL* circuit (one with only a resistor, an inductor, and an electromotive force). In this section, we consider the motion of a body that falls in a resisting medium.

> **EXAMPLE 5**   *Motion of a body falling in a resisting medium*

Consider an object with mass m that is initially at rest and is dropped from a great height (for example, from an airplane). Suppose the body falls in a straight line and the only forces acting on it are the downward force of the earth's gravitational attraction and a resisting upward force due to air resistance in the atmosphere. Assume that the resisting force is proportional to the velocity v of the falling body. Find equations for the velocity and displacement of the body's motion. Assume the distance $s(t)$ is measured down from the drop point.

Solution   The downward force is the weight mg of the body and the upward force is $-kv$, where k is a positive constant (the negative sign indicates that the force is directed upward). According to Newton's second law, the sum of the forces acting on a body at any time equals the product ma, where a is the acceleration of the body, that is

$$\underbrace{ma}_{\substack{\text{Sum of forces} \\ \text{on the body}}} = \underbrace{mg}_{\substack{\text{Force due} \\ \text{to gravity}}} - \underbrace{kv}_{\substack{\text{Resisting} \\ \text{force}}}$$

$$m\frac{dv}{dt} = mg - kv \qquad \text{Since } a = \frac{dv}{dt}$$

$$\frac{dv}{dt} = g - \frac{k}{m}v$$

$$\frac{dv}{dt} + \frac{k}{m}v = g$$

This is a first-order linear differential equation where $p(t) = \dfrac{k}{m}$ and $q(t) = g$. The integrating factor is

$$I(t) = e^{\int k/m \, dt} = e^{kt/m}$$

so that the solution is

$$v = \frac{1}{e^{kt/m}}\left[\int e^{kt/m}(g)dt + C\right] = e^{-kt/m}\left[\frac{ge^{kt/m}}{k/m} + C\right] = \frac{mg}{k} + Ce^{-kt/m}$$

Because $v = 0$ when $t = 0$ (the body is initially at rest), it follows that

$$0 = \frac{mg}{k} + Ce^0 = \frac{mg}{k} + C$$

Thus, because $C = -\dfrac{mg}{k}$, we have

$$v = \frac{mg}{k} + \left(-\frac{mg}{k}\right)e^{-kt/m}$$

Now, to find the position $s(t)$, we use the fact that $v(t) = \dfrac{ds}{dt}$:

$$\frac{ds}{dt} = \frac{mg}{k} - \frac{mg}{k}e^{-kt/m}$$

$$\int ds = \int \left[\frac{mg}{k} - \frac{mg}{k}e^{-kt/m}\right] dt$$

$$s(t) = \frac{mg}{k}t - \frac{mg}{k}\frac{e^{-kt/m}}{-k/m} + C$$

$$= \frac{mg}{k}t + \frac{m^2g}{k^2}e^{-kt/m} + C$$

Because $s(0) = 0$ (the distance s is measured from the point where the object is dropped), we find that

$$0 = \frac{mg}{k}(0) + \frac{m^2 g}{k^2} e^0 + C \quad \text{so that} \quad -\frac{m^2 g}{k^2} = C$$

Thus, the displacement is

$$s(t) = \frac{mg}{k} t + \frac{m^2 g}{k^2} \left(e^{-kt/m} - 1 \right)$$ ■

In the problem set, you are asked to show that no matter what the initial velocity may be, the velocity reached by the object in the long run (as $t \to +\infty$) is mg/k.

EXACT DIFFERENTIAL EQUATIONS

Sometimes a first-order differential equation can be written in the general form

$$M(x, y)dx + N(x, y)dy = 0$$

where the left side is an exact differential, namely,

$$df = M(x, y)dx + N(x, y)dy$$

In this case, the given differential equation is appropriately called **exact** and since $df = 0$, its general solution is given by $f(x, y) = C$.

But how can we tell whether a particular first-order equation is exact, and if it is, how can we find f? Since

$$df = \frac{\partial f}{\partial x} dx + \frac{\partial f}{\partial y} dy$$

for a total differential (see Section 12.4), we must have

$$df = \frac{\partial f}{\partial x} dx + \frac{\partial f}{\partial y} dy = M(x, y)dx + N(x, y)dy$$

so

$$\frac{\partial f}{\partial x} = M(x, y) \quad \text{and} \quad \frac{\partial f}{\partial y} = N(x, y)$$

This will be true if and only if f satisfies the *cross-derivative test*

$$\frac{\partial N}{\partial x} = \frac{\partial M}{\partial y}$$

and then the function $f(x, y)$ is found by partial integration, exactly as we found the potential function of a conservative vector field in Section 14.3. The procedure for identifying and then solving an exact differential equation is illustrated in Example 6.

EXAMPLE 6 *Exact differential equation*

Find the general solution for $(2xy^3 + 3y)\, dx + (3x^2 y^2 + 3x)\, dy = 0$.

Solution Let $M(x, y) = 2xy^3 + 3y$ and $N(x, y) = 3x^2 y^2 + 3x$, and apply the cross-derivative test

$$\frac{\partial M}{\partial y} = 6xy^2 + 3 \quad \text{and} \quad \frac{\partial N}{\partial x} = 6xy^2 + 3$$

Because $\dfrac{\partial M}{\partial y} = \dfrac{\partial N}{\partial x}$, the equation is exact. To obtain a general solution, we must find a function f such that

$$\frac{\partial f}{\partial x} = 2xy^3 + 3y \quad \text{and} \quad \frac{\partial f}{\partial y} = 3x^2y^2 + 3x$$

To find f, we integrate the first partial on the left with respect to x:

$$f(x, y) = \int (2xy^3 + 3y)\, dx = x^2y^3 + 3xy + u(y)$$

where u is a function of y. Taking the partial derivative of f with respect to y and comparing the result with $\partial f / \partial y$, we obtain

$$\frac{\partial f}{\partial y} = \frac{\partial}{\partial y}[x^2y^3 + 3xy + u(y)] = 3x^2y^2 + 3x + u'(y)$$

so that

$$3x^2y^2 + 3x = 3x^2y^2 + 3x + u'(y)$$
$$0 = u'(y)$$

This implies that u is a constant. Taking $u = 0$, we have $f = x^2y^3 + 3xy$, and the general solution to the exact differential equation is

$$x^2y^3 + 3xy = C$$

∎

■ **TABLE 15.1** **Summary of Strategies for First-Order Differential Equations**

Form of Equation	Method	Solution
$\dfrac{dy}{dx} = \dfrac{g(x)}{f(y)}$	Separate the variables.	$\displaystyle\int f(y)\, dy = \int g(x)\, dx$
$\dfrac{dy}{dx} = f\left(\dfrac{y}{x}\right)$	Homogeneous—use a change of variable $v = \dfrac{y}{x}$.	$\displaystyle\int \dfrac{dv}{f(v) - v} = \int \dfrac{dx}{x}$
$\dfrac{dy}{dx} + p(x)y = q(x)$	Use the integrating factor $I(x) = e^{\int p(x)dx}$	$y = \dfrac{1}{I(x)}\left[\displaystyle\int I(x)\, q(x)\, dx + C\right]$
$M(x, y)\, dx + N(x, y)\, dy = 0$, where $\dfrac{\partial M}{\partial y} = \dfrac{\partial N}{\partial x}$	Exact—use partial integration to find f, where $\dfrac{\partial f}{\partial x} = M$ and $\dfrac{\partial f}{\partial y} = N$	$f(x, y) = C$

EULER'S METHOD

In Section 5.6, we introduced direction fields as a means for obtaining a "picture" of various solutions to a differential equation, but sometimes we need more than a rough graph of a solution. We now consider approximating a solution by numerical means. **Euler's method** is a simple procedure for obtaining a table of approximate values for the solution of a given initial value problem*

$$\frac{dy}{dx} = f(x, y) \qquad y(x_0) = y_0$$

---

*In our discussion of Euler's method, we assume that the given initial value problem has a unique solution. It can be shown that such an initial value problem always has a unique solution if f and $\partial f / \partial y$ are both continuous in a neighborhood (x_0, y_0). The proof of this result is beyond the scope of this text but can be found in most elementary differential equations texts.

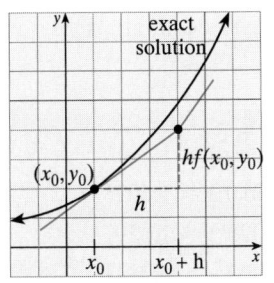

a. The first Euler approximation

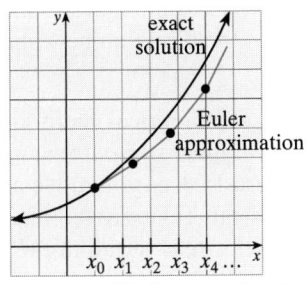

b. A sequence of Euler approximations

■ **FIGURE 15.1**   Graphical representation of Euler's method

The key idea in Euler's method is to increment x_0 by a small quantity h and then to approximate $y_1 = y(x_1)$ for $x_1 = x_0 + h$ by assuming x and y change by so little over the interval $[x_0, x_1]$ that $f(x, y)$ can be replaced by $f(x_0, y_0)$ for this interval. Solving the approximating initial value problem

$$\frac{dy}{dx} = f(x_0, y_0) \qquad y(x_0) = y_0$$

we obtain

$$y - y_0 = f(x_0, y_0)(x - x_0)$$

In other words, we are approximating the solution curve $y = y(x)$ near (x_0, y_0) by the tangent line to the curve at this point, as shown in Figure 15.1a.

We then repeat this process with (x_1, y_1) assuming the role of (x_0, y_0) to obtain an approximation of the solution $y = y(x)$ over the interval $x_1 \le x \le x_2$, where $x_2 = x_1 + h$ and

$$y_2 = y_1 + hf(x_1, y_1)$$

Continuing in this fashion, we obtain a sequence of line segments that approximates the shape of the solution curve as shown in Figure 15.1b. Euler's method is illustrated in the following example.

EXAMPLE 7   *Euler's method*

Use Euler's method with $h = 0.1$ to estimate the solution of the initial value problem

$$\frac{dy}{dx} = x + y^2 \qquad y(0) = 1$$

over the interval $0 \le x \le 0.5$.

Solution   Before using Euler's method, we might first look at a graphical solution. The slope field is shown in Figure 15.2a, and the particular solution through the point $(0, 1)$ is shown in Figure 15.2b.

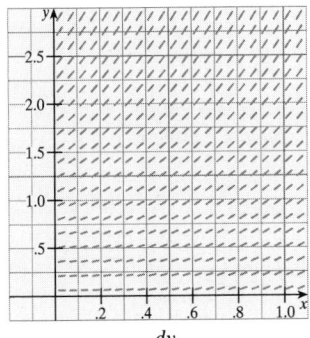

a. Slope field of $\dfrac{dy}{dx} = x + y^2$

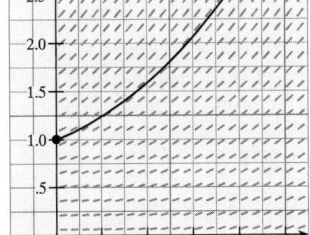

b. Particular solution through $(0, 1)$

■ **FIGURE 15.2**   Graphical solution using a direction field

To use Euler's method for this example, we note

$$f(x, y) = x + y^2, \quad x_0 = 0, \quad y_0 = 1, \quad \text{and} \quad h = 0.1$$

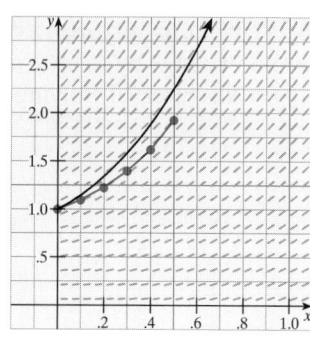

■ **FIGURE 15.3** Solution by Euler's method

We show the calculator (or computer) solution correct to four decimal places:

$$y_0 = y(0) = 1$$
$$y_1 = y_0 + hf(x_0, y_0) = 1 + 0.1(0 + 1^2) = 1.1000$$
$$y_2 = y_1 + hf(x_1, y_1) = 1.1000 + 0.1(0.1 + 1.1000^2) = 1.2310$$
$$y_3 = y_2 + hf(x_2, y_2) = 1.2310 + 0.1(0.2 + 1.2310^2) \approx 1.4025$$
$$y_4 = y_3 + hf(x_3, y_3) = 1.4025 + 0.1(0.3 + 1.4025^2) \approx 1.6292$$
$$y_5 = y_4 + hf(x_4, y_4) = 1.6292 + 0.1(0.4 + 1.6292^2) \approx 1.9347$$

These points can be plotted to approximate the solution, as shown in Figure 15.3. Notice that we plotted these points by superimposing them on the direction field shown in Figure 15.2b. ■

Euler's method has educational value as the simplest numerical method for solving ordinary differential equations, and can be found in most computer-assisted programs. However, as you might guess by looking at Figure 15.3, as you move away from (x_0, y_0), the error may accumulate. The Euler method can be improved in a variety of ways, most notably by a collection of procedures known as the *Runge–Kutta* and *predictor–corrector* methods. These methods are studied in more advanced courses.

15.1   Problem Set

Ⓐ *Find the general solution of the differential equations in Problems 1–6 by separating variables.*

1. $xy \, dx = (x - 5) \, dy$

2. $\dfrac{dy}{dx} = y \tan x$

3. $(e^{2x} + 9)\dfrac{dy}{dx} = y$

4. $y\dfrac{dy}{dx} = e^{x-3y}\cos x$

5. $9 \, dx - x\sqrt{x^2 - 9} \, dy = 0$

6. $xy\dfrac{dy}{dx} = x^2 + y^2 + x^2y^2 + 1$

In Problems 7–12, a graphical solution of a given initial value problem is shown by using a direction field. Solve the problem to find an equation for the particular solution.

7. $\dfrac{dy}{dx} + 2xy = 4$
   passing through $(0, 0)$

8. $\dfrac{dy}{dx} + \dfrac{y}{x} = \dfrac{\sin x}{x}$
   passing through $(-2, 0)$

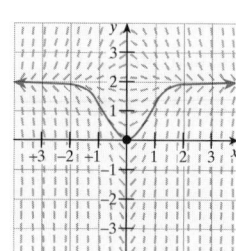

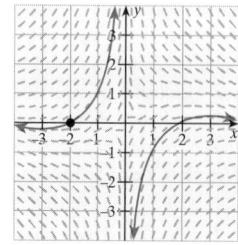

9. $\dfrac{dy}{dx} + y = \cos x$
   passing through $(0, 0)$

10. $\dfrac{dy}{dx} + y = \dfrac{e^x}{1 + e^{2x}}$
    passing through $(0, 2)$

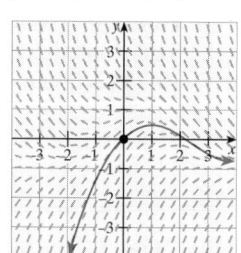

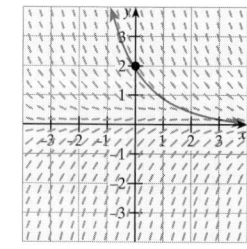

11. $x\dfrac{dy}{dx} - 2y = x^3$
    passing through $(2, -1)$

12. $\dfrac{dy}{dx} - 3xy = 5xe^{x^2}$
    passing through $(0, -3)$

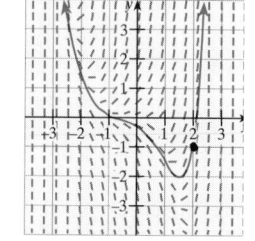

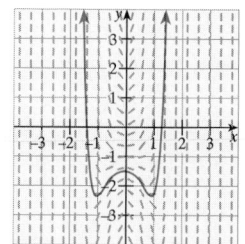

Show that each differentiable equation in Problems 13–18 is in the form $dy/dx = f(y/x)$, and then find the general solution.

13. $(3x - y) \, dx + (x + 3y) \, dy = 0$

14. $xy \, dx - (2x^2 + y^2) \, dy = 0$

15. $(3x - y) \, dx + (x - 3y) \, dy = 0$

16. $(x^2 + y^2) \, dx - 2xy \, dy = 0$

17. $(-6y^2 + 3xy + 2x^2) \, dx + x^2 \, dy = 0$

18. $x \, dy - (y + \sqrt{xy}) \, dx = 0$

Show the differential equations in Problems 19–24 are exact and find the general solution.

19. $(3x^2 y + \tan y) \, dx + (x^3 + x \sec^2 y) \, dy = 0$

20. $(3x^2 - 10xy) \, dx + (2y - 5x^2 + 4) \, dy = 0$

21. $\left[\dfrac{1}{1 + x^2} + \dfrac{2x}{x^2 + y^2} \right] dx + \left[\dfrac{2y}{x^2 + y^2} - e^{-y} \right] dy = 0$

22. $(2xy^3 + 3y - 3x^2) \, dx + (3x^2 y^2 + 3x) \, dy = 0$

23. $[2x \cos 2y - 3y(1 - 2x)] \, dx$
$\qquad - [2x^2 \sin 2y + 3(2 + x - x^2)] \, dy = 0$

24. $[(x + xy - 3)(1 + y) - x^2 \sqrt{y}] \, dx$
$\qquad + \left[x^2(y + 1) - 3x - \dfrac{x^3}{6\sqrt{y}} \right] dy = 0$

Ⓑ 25. Consider the differential equation

$$\frac{dy}{dx} = x + y$$

     a. Find the particular solution that contains the point $(1, 2)$.

     b. Sketch isoclines for $C = 1, 3$, and 5. Then use the direction field for the given differential equation to sketch the solution through $(1, 2)$. Compare the result with part **a**.

     c. Use Euler's method to approximate a solution for $x_0 = 1, y_0 = 2$, and $h = 0.2$. Compare this result with part **b**.

26. Consider the differential equation

$$\frac{dy}{dx} = x^2 - y^2$$

     a. Find the particular solution that contains the point $(2, 1)$.

     b. Sketch isoclines for $C = 0, 2$, and 4. Then use the direction field for the given differential equation to sketch the solution through $(2, 1)$. Compare the result with part **a**.

     c. Use Euler's method to approximate a solution for $x_0 = 2, y_0 = 1$, and $h = 0.2$.

Estimate a solution for Problems 27–30 using Euler's method. For each of these problems, a direction field is given. Superimpose the segments from Euler's method on the given direction field.

27. $\dfrac{dy}{dx} = \dfrac{x + y}{y - x}$
passing through $(0, 1)$
for $0 \le x \le 0.5, h = 0.1$

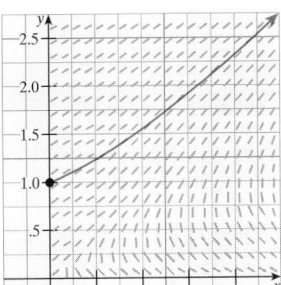

28. $\dfrac{dy}{dx} = 2x(x^2 - y)$
passing through $(0, 4)$
for $0 \le x \le 1, h = 0.1$

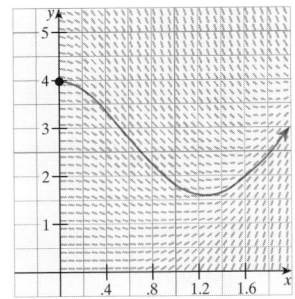

29. $\dfrac{dy}{dx} = \dfrac{5x - 3xy}{1 + x^2}$
passing through $(0, 0)$
for $0 \le x \le 0.5, h = 0.1$

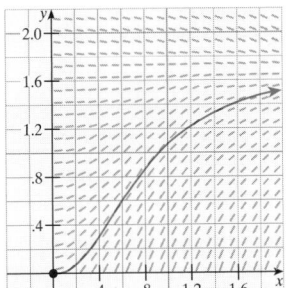

30. $\dfrac{dy}{dx} = \dfrac{y^2 + 2x}{3y^2 - 2xy}$
passing through $(0, 1)$
for $0 \le x \le 0.5, h = 0.1$

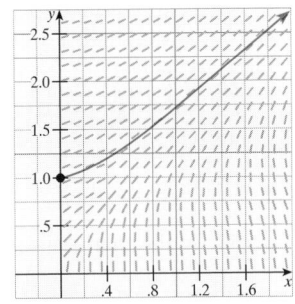

*An **integrating factor** of the differential equation*

$$M \, dx + N \, dy = 0$$

is a function $f(x, y)$ such that

$$f(x, y) \, M(x, y) \, dx + f(x, y) N(x, y) \, dy = 0$$

is exact. In Problems 31–32, find an integrating factor of the specified type for the given differential equations, then solve the equation.

31. $y \, dx + (y - x) \, dy = 0; \ f(x, y) = y^n$

32. $(x^2 + y^2) \, dx + (3xy) \, dy = 0; \ f(x, y) = x^n$

Identify each equation given in Problems 33–44 as separable, homogeneous, first-order linear, or exact, and then solve. It is possible for an equation to be of more than one type.

33. $(2xy^2 + 3x^2 y - y^3) \, dx + (2x^2 y + x^3 - 3xy^2) \, dy = 0$

34. $(1 + x) \, dy + \sqrt{1 - y^2} \, dx = 0$

35. $\left(\dfrac{2x}{y} - \dfrac{y^2}{x^2} \right) dx + \left(\dfrac{2y}{x} - \dfrac{x^2}{y^2} + 3 \right) dy = 0$

36. $(x^2 - xy - x + y) \, dx - (xy - y^2) \, dy = 0$

37. $e^{y-x} \sin x \, dx - \csc x \, dy = 0$

38. $x^2 \dfrac{dy}{dx} + 2xy = \sin x$

39. $(3x^2 - y \sin xy) \, dx - (x \sin xy) \, dy = 0$

40. $\dfrac{dy}{dx} = \dfrac{y}{x} + x \cos \dfrac{y}{x}$

41. $(y - \sin^2 x)\, dx + (\sin x)dy = 0$

42. $(y^3 - y)dx + (x^2 + x)dy = 0$

43. $x\dfrac{dy}{dx} = y - \sqrt{x^2 + y^2}$

44. $(2x + \sin y - \cos y)dx + (x \cos y + x \sin y)dy = 0$

Find the particular solution of the differential equations in Problems 45–52 with the specified initial conditions.

45. $\left(x \sin^2 \dfrac{y}{x} - y \right) dx + x\, dy = 0; x = \dfrac{4}{\pi}, = 1$

46. $y(5x - y)dx - x(5x + 2y)dy = 0; x = 1, y = 1$

47. $x\dfrac{dy}{dx} - 3y = x^3; x = 1, y = 1$

48. $\dfrac{dy}{dx} = 1 + 3y \tan x; x = 0, y = 2$

49. $[\sin(x^2 + y) + 2x^2 \cos(x^2 + y)]dx$
    $\quad + [x \cos(x^2 + y)]dy = 0; x = 0, y = 0$

50. $y\dfrac{dy}{dx} = e^{x+2y}\sin x; x = 0, y = 0$

51. $ye^x\, dy = (y^2 + 2y + 2)dx; x = 0, y = -1$

52. $(2xy + y^2 + 2x)dx + (x^2 + 2xy - 1)dy = 0;$
    $x = 1, y = 3$

53. In the chemical conversion analyzed in Example 2, what happens to $Q(t)$ as $t \to +\infty$ in the case where $\alpha \neq \beta$? What if $\alpha = \beta$?

54. a. Find an equation for the velocity of the falling body in Example 5 in the case where the body has initial velocity $v_0 \neq 0$.

    b. Show that for any initial velocity (zero or nonzero) the velocity reached by the object "in the long run" (as $t \to +\infty$) is always mg/k.

55. **MODELING PROBLEM**   A population of foxes grows logistically until it is decided to allow hunting at the constant rate of h foxes per month. The population $P(t)$ is then modeled by the differential equation

    $$\frac{dP}{dt} = P(k - \ell P) - h$$

    where $P(0) = P_0$.

    a. Solve this equation in terms of $k, \ell, h,$ and P_0 for the case where $h < \dfrac{k^2}{4\ell}$.

    b. If $P_0 > h$, what happens to $P(t)$ as $t \to \infty$?

56. **MODELING PROBLEM**   A chemical in a solution diffuses from a compartment where the concentration is $C_0(t) = 7e^{-t}$ across a membrane with diffusion coefficient $k = 1.75$. The concentration $C(t)$ in the second compart-

ment is modeled by

$$\frac{dC}{dt} = 1.75[C_0(t) - C(t)]$$

Solve this equation for $C(t)$, assuming that $C(0) = 0$.

57. Our formula for the solution of a first-order linear equation requires the differential equation to be linear in x. Suppose, instead, the equation is linear in y; that is, it can be written in the form

    $$\frac{dx}{dy} + R(y)x = S(y)$$

    a. Find a formula for the general solution of an equation that is first-order linear in y.

    b. Use the formula obtained in part **a** to solve

    $$y\, dx - 2x\, dy = y^4 e^{-y}\, dy$$

58. **MODELING PROBLEM**   A man is pulling a heavy sled along the ground by a rope of fixed length L. Assume the man begins walking at the origin of a coordinate plane and that the sled is initially at the point $(0, L)$. The man walks to the right (along the positive x-axis), dragging the sled behind him. When the man is at point M, the sled is at S, as shown in Figure 15.4. Find a differential equation for the path of the sled and solve this differential equation.

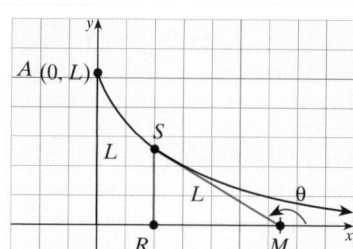

■ **FIGURE 15.4**   Sled problem

59. **SPY PROBLEM**   Having conserved as much energy as possible after his escape from the Death Ray (Problem 38, Section 14.3), the Spy continues his search for Purity and Blohardt. He rushes into a small room and finds no Blohardt but *two* Purities struggling ferociously on the floor. As he separates them, he notices that one is dressed in red and the other in green. The red one cries, "She's my evil twin, Rottona. She escaped from prison and has been holding me prisoner." The one in green shouts, "Liar! I'm the real Purity. I've been tied up here all along and just was freed." The Spy wishes he could remember whether the one who has been trying to kill him was wearing red or green, but it is all a big blur. Instead, he says, "I will ask a question that only the real Purity can answer. A 160-pound skydiver jumps from a plane and begins to free fall. She knows the air resistance is $0.01\ v^2$, where $v(t)$ is her velocity. She is wearing a wrist altimeter and wishes to open her parachute when she has fallen

1,000 feet. What will be her velocity at that instant?" The green Purity stamps her foot petulantly and cries, "How should I know and who cares, anyhow?" The one in red grins and answers correctly. What is her answer?

C 60. ALMOST HOMOGENEOUS EQUATIONS   Sometimes a differential equation is not quite homogeneous but becomes homogeneous with a linear change of variable. Specifically, consider a differential equation of the form

$$\frac{dy}{dx} = f\left(\frac{ax + by + c}{rx + sy + t}\right)$$

a. Suppose $as \neq br$. Make the change of variable $x = X + A$ and $y = Y + B$ where A and B satisfy

$$\begin{cases} aA + bB + c = 0 \\ rA + sB + t = 0 \end{cases}$$

Show that with these choices for A and B, the differential equation becomes homogeneous.

b. Apply the procedure outlined in part **a** to solve the differential equation

$$\frac{dy}{dx} = \left(\frac{-3x + y + 2}{x + 3y - 5}\right)$$

61. A **Riccati equation** is a differential equation of the form

$$\frac{dy}{dx} = P(x)y^2 + Q(x)y + R(x)$$

Suppose we know that $y = u(x)$ is a solution of a given Riccati equation.

a. Change variables by setting

$$z = \frac{1}{y - u(x)}$$

Show that the given equation is transformed by this change of variables into the separable form

$$\frac{dz}{dx} + [2P(x)u(x) + Q(x)]z = -P(x)$$

b. Solve the first-order linear equation in z obtained in part **a** and explain how to find the general solution of the given Riccati equation.

c. Use the method outlined in parts **a** and **b** to solve the Riccati equation

$$\frac{dy}{dx} = \frac{1}{x^2}y^2 + \frac{2}{x}y - 2$$

Hint: There is a solution of the general form $y = Ax$.

62. JOURNAL PROBLEM*   Find all solutions of the Riccati equation

$$u' = u^2 + \frac{au}{x} - b$$

$a, b \neq 0$, that are real rational functions of x.

———————

*Problem E3055 from The American Mathematical Monthly, Vol. 91, 1984, p. 515.

15.2   Second-Order Homogeneous Linear Differential Equations

IN THIS SECTION   **linear independence, solutions of the equation $y'' + ay' + by = 0$, higher-order homogeneous linear equations, damped motion of a mass on a spring, reduction of order**

LINEAR INDEPENDENCE

A **linear differential equation** is one of the general form

$$a_n(x)y^{(n)} + a_{n-1}(x)y^{(n-1)} + \cdots + a_0(x)y = R(x)$$

and if $a_n(x) \neq 0$, it is said to be of **order** n. It is *homogeneous* if $R(x) = 0$ and *nonhomogeneous* if $R(x) \neq 0$. In this section we focus attention on the homogeneous case, and we examine nonhomogeneous equations in the next section.

To characterize all solutions of the homogeneous equation

$$a_n(x)y^{(n)} + a_{n-1}y^{(n-1)} + \cdots + a_0(x)y = 0$$

we require the following definition.

Linear Dependence and Independence

The functions $y_1, y_2, \ldots, y_n$ are said to be **linearly independent** if the equation

$$C_1 y_1 + C_2 y_2 + \cdots + C_n y_n = 0 \quad \text{for constants} \quad C_1, C_2, \ldots$$

has only the trivial solution $C_1 = C_2 = \cdots = C_n = 0$ for all x in the interval I. Otherwise the y_k's are **linearly dependent.**

The functions $y_1 = \cos x$ and $y_2 = x$ are linearly independent because the only way we can have $C_1 \cos x + C_2 x = 0$ for all x is for C_1 and C_2 both to be 0. However, $y_1 = 1, y_2 = \sin^2 x$, and $y_3 = \cos 2x$ are linearly dependent, because

$$C_1(1) + C_2(\sin^2 x) + C_3(\cos 2x) = 0 \quad \text{for} \quad C_1 = 1, C_2 = -2, C_3 = -1$$

It can be quite difficult to determine whether a given collection of functions $y_1, y_2, \ldots, y_n$ is linearly independent. However, this issue can be settled by a routine computation using the following determinant, which is named after Josef Hoëné de Wronski (see the Historical Quest in Problem 31).

Wronskian

The **Wronskian** $W(y_1, y_2, \ldots, y_n)$ of n functions $y_1, y_2, \ldots, y_n$ having $n - 1$ derivatives on an interval I is defined to be the determinant function

$$W(y_1, y_2, \ldots, y_n) = \begin{vmatrix} y_1 & y_2 & \cdots & y_n \\ y_1' & y_2' & \cdots & y_n' \\ \vdots & \vdots & & \vdots \\ y_1^{(n-1)} & y_2^{(n-1)} & \cdots & y_n^{(n-1)} \end{vmatrix}$$

THEOREM 15.1   *Determining linear independence with the Wronskian*

Suppose the functions $a_n(x), a_{n-1}(x), \ldots, a_0(x)$ in the nth-order homogeneous linear differential equation

$$a_n(x)y^{(n)} + a_{n-1}(x)y^{(n-1)} + \cdots + a_0(x)y = 0$$

are all continuous on a closed interval $[c, d]$. Then solutions $y_1, y_2, \ldots, y_n$ of this differential equation are linearly independent if and only if the Wronskian is nonzero; that is,

$$W(y_1, y_2, \ldots, y_n) \neq 0$$

throughout the interval $[c, d]$.

Proof   The proof of this theorem is beyond the scope of this course but can be found in most differential equations books.

EXAMPLE 1   *Showing linear independence*

The functions $y_1 = e^{-x}, y_2 = xe^{-x}$, and $y_3 = e^{3x}$ are solutions of a certain homogeneous linear differential equation with constant coefficients. Show that these solutions are linearly independent.

Solution
$$W(e^{-x}, xe^{-x}, e^{3x}) = \begin{vmatrix} e^{-x} & xe^{-x} & e^{3x} \\ -e^{-x} & (1-x)e^{-x} & 3e^{3x} \\ e^{-x} & (x-2)e^{-x} & 9e^{3x} \end{vmatrix}$$

$$= e^{-x}[9e^{3x}(1-x)e^{-x} - 3e^{3x}(x-2)e^{-x}]$$
$$- xe^{-x}[-9e^{3x}e^{-x} - 3e^{-x}e^{3x}]$$
$$+ e^{3x}[-e^{-x}(x-2)e^{-x} - e^{-x}(1-x)e^{-x}]$$
$$= 16e^x$$

Because $16e^x \neq 0$, the functions are linearly independent. ■

The general solution of an nth-order homogeneous linear differential equation with constant coefficients can be characterized in terms of n linearly independent solutions. Here is the theorem that applies to the second-order case.

THEOREM 15.2   ***Characterizing the general solution of*** $y'' + ay' + by = 0$

If y_1 and y_2 are linearly independent solutions of the differential equation $y'' + ay' + by = 0$, then the general solution is

$$y = C_1 y_1 + C_2 y_2 \quad \text{for arbitrary constants } C_1, C_2$$

Proof   We can prove that if y_1 and y_2 are linearly independent solutions, then $y = C_1 y_1 + C_2 y_2$ is also a solution. The proof that all solutions are of this form is beyond the scope of this book. Suppose y_1 and y_2 are solutions, so that

$$y_1''(x) + ay_1'(x) + by_1(x) = 0$$
$$y_2''(x) + ay_2'(x) + by_2(x) = 0$$

If $y = C_1 y_1 + C_2 y_2$, we have

$$y'' + ay' + by = [C_1 y_1'' + C_2 y_2''] + a[C_1 y_1' + C_2 y_2'] + b[C_1 y_1 + C_2 y_2]$$
$$= C_1[y_1'' + ay_1' + by_1] + C_2[y_2'' + ay_2' + by_2]$$
$$= 0 + 0 = 0$$

Thus, $y = C_1 y_1 + C_2 y_2$ is also a solution.   ═

SOLUTIONS OF THE EQUATION $y'' + ay' + by = 0$

Thanks to the characterization theorem, we now know that once we have two linearly independent solutions y_1, y_2 of the equation $y'' + ay' + by = 0$, we have them all because the general solution can be characterized as $y = C_1 y_1 + C_2 y_2$. Therefore, the whole issue of how to represent the solution of a second-order homogeneous linear equation with constant coefficients depends on finding two linearly independent solutions.

Recall that the general solution of the first-order equation $y' + ay = 0$ is $y = Ce^{-ax}$. Therefore, it is not unreasonable to expect the second-order equation $y'' + ay' + by = 0$ to have one (or more) solutions of the form $y = e^{rx}$. If $y = e^{rx}$, then $y' = re^{rx}$ and $y'' = r^2 e^{rx}$, and by substituting these derivatives into the equation $y'' + ay' + by = 0$, we obtain

$$y'' + ay' + by = 0$$
$$r^2 e^{rx} + a(re^{rx}) + be^{rx} = 0$$
$$e^{rx}(r^2 + ar + b) = 0$$
$$r^2 + ar + b = 0 \quad e^{rx} \neq 0$$

Thus, $y = e^{rx}$ is a solution of the given second-order differential equation if and only if $r^2 + ar + b = 0$. This equation is called the **characteristic equation** of $y'' + ay' + by = 0$.

EXAMPLE 2   *Characteristic equation with distinct real roots*

Find the general solution of the differential equation $y'' + 2y' - 3y = 0$.

Solution   Begin by solving the characteristic equation:

$$r^2 + 2r - 3 = 0$$
$$(r - 1)(r + 3) = 0$$
$$r = 1, -3$$

The particular solutions are $y_1 = e^x$ and $y_2 = e^{-3x}$. Next, determine whether these equations are linearly independent by looking at the Wronskian:

$$W(e^x, e^{-3x}) = \begin{vmatrix} e^x & e^{-3x} \\ e^x & -3e^{-3x} \end{vmatrix} = e^x(-3e^{-3x}) - e^x e^{-3x} = -4e^{-2x}$$

Because $W(e^x, e^{-3x}) = -4e^{-2x} \neq 0$, the functions are linearly independent, and the characterization theorem tells us that the general solution is

$$y = C_1 y_1 + C_2 y_2 = C_1 e^x + C_2 e^{-3x} \qquad \blacksquare$$

Example 2 has a characteristic equation that factors easily. In practice, however, the quadratic formula is often required, namely,

$$r = \frac{-a \pm \sqrt{a^2 - 4b}}{2}$$

The discriminant *for this equation, $a^2 - 4b$*, figures prominently in the following theorem.

THEOREM 15.3   *Solution of $y'' + ay' + by = 0$*

If r_1 and r_2 are the roots of the characteristic equation $r^2 + ar + b = 0$, then the general solution of the homogeneous linear differential equation $y'' + ay' + by = 0$ can be expressed in one of these forms:

$a^2 - 4b > 0$:   The general solution is

$$y = C_1 e^{r_1 x} + C_2 e^{r_2 x}$$

where

$$r_1 = \frac{-a + \sqrt{a^2 - 4b}}{2} \quad \text{and} \quad r_2 = \frac{-a - \sqrt{a^2 - 4b}}{2}$$

$a^2 - 4b = 0$:   The general solution is

$$y = C_1 e^{-ax/2} + C_2 x e^{-ax/2} = (C_1 + C_2 x)e^{-ax/2}$$

$a^2 - 4b < 0$:   The general solution is

$$y = e^{-ax/2}\left[C_1 \cos\left(\frac{\sqrt{4b^2 - a}}{2}x\right) + C_2 \sin\left(\frac{\sqrt{4b^2 - a^2}}{2}x\right) \right]$$

Proof   We have just seen that the solutions of the differential equation of the form $y = e^{rx}$ correspond to the solutions of the characteristic equation $r^2 + ar + b = 0$.

The quadratic formula characterizes the three cases according to the discriminant of this equation, namely, $a^2 - 4b$. For each case, we must find two linearly independent solutions.

$a^2 - 4b > 0$: Let $y_1 = e^{r_1 x}$, and $y_2 = e^{r_2 x}$. Then

$$W(e^{r_1 x}, e^{r_2 x}) = \begin{vmatrix} e^{r_1 x} & e^{r_2 x} \\ r_1 e^{r_1 x} & r_2 e^{r_2 x} \end{vmatrix} = r_2 e^{(r_1 + r_2)x} - r_1 e^{(r_1 + r_2)x} = (r_2 - r_1) e^{(r_1 + r_2)x}$$

Because $r_2 \neq r_1$ and $e^{(r_1 + r_2)x} > 0$, we see $W(e^{r_1 x}, e^{r_2 x}) \neq 0$, so the functions are linearly independent. The characterization theorem tells us that the general solution is

$$y = C_1 y_1 + C_2 y_2 = C_1 e^{r_1 x} + C_2 e^{r_2 x}$$

$a^2 - 4b = 0$: In this case, the characteristic equation has one repeated root— namely, $r = -a/2$. The function $y_1 = e^{-ax/2}$ is one solution, and it can be shown that a second linearly independent solution is $y_2 = xe^{-ax/2}$ (see Problem 32). Thus, the general solution is

$$y = C_1 e^{-ax/2} + C_2 xe^{-ax/2}$$

$a^2 - 4b < 0$: The proof of this part is left for the reader. $=\!=$

EXAMPLE 3   *Characteristic equation with repeated roots*

Find the general solution of the differential equation $y'' + 4y' + 4y = 0$.

Solution   Solve the characteristic equation:

$$r^2 + 4r + 4 = 0$$
$$(r + 2)^2 = 0$$
$$r = -2 \quad \text{(multiplicity 2)}$$

The roots are $r_1 = r_2 = -2$. Thus, $y_1 = e^{-2x}$ and $y_2 = xe^{-2x}$, so the general solution of the differential equation is

$$y = C_1 e^{-2x} + C_2 xe^{-2x} \qquad \blacksquare$$

EXAMPLE 4   *Characteristic equation with complex roots*

Find the general solution of the differential equation $2y'' + 3y' + 5y = 0$.

Solution   Solve the characteristic equation*:

$$2r^2 + 3r + 5 = 0$$
$$r = \frac{-3 \pm \sqrt{9 - 4(2)(5)}}{2(2)}$$
$$= \frac{-3 \pm \sqrt{31}\, i}{4}$$

The roots are

$$r_1 = -\frac{3}{4} + \frac{\sqrt{31}}{4} i, \quad r_2 = -\frac{3}{4} - \frac{\sqrt{31}}{4} i$$

---

*Technically, we have considered only the case where the leading coefficient of the characteristic equation is 1. Verify that you obtain the same result if you first divide both sides by 2.

Thus, the general solution is

$$y = e^{(-3/4)x}\left[C_1\cos\frac{\sqrt{31}}{4}x + C_2\sin\frac{\sqrt{31}}{4}x\right]$$

■

EXAMPLE 5   *Second-order initial value problem*

Solve $4y'' + 12y' + 9y = 0$ subject to $y(0) = 3$ and $y'(0) = -2$.

Solution   Solve

$$4r^2 + 12r + 9 = 0$$
$$(2r + 3)^2 = 0$$
$$r = -\tfrac{3}{2}$$

Thus, the general solution is

$$y = C_1e^{(-3/2)x} + \tfrac{3}{2}C_2xe^{(-3/2)x}$$

Because $y(0) = 3$ we have

$$3 = C_1e^0 + C_2(0)e^0$$
$$3 = C_1$$

Because $y'(0) = -2$, we find y':

$$y' = -\tfrac{3}{2}C_1e^{(-3/2)x} - \tfrac{3}{2}C_2xe^{(-3/2)x} + C_2e^{(-3/2)x}$$
$$y'(0) = -\tfrac{3}{2}C_1e^0 - \tfrac{3}{2}C_2(0)e^0 + C_2e^0$$
$$-2 = -\tfrac{3}{2}C_1 + C_2$$
$$-2 = -\tfrac{3}{2}(3) + C_2 \qquad \text{Because } C_1 = 3$$
$$\tfrac{5}{2} = C_2$$

Thus, the particular solution is

$$y = 3e^{(-3/2)x} + \tfrac{5}{2}xe^{(-3/2)x}$$

■

HIGHER-ORDER HOMOGENEOUS LINEAR EQUATIONS

Homogeneous linear differential equations of degree 3 or more with constant coefficients can be handled in essentially the same way as the second-order equations we have analyzed. As in the second-order case, some of the roots of the characteristic equations may be real and distinct, some may be real and repeated, and some may occur in complex conjugate pairs. But now, roots of the characteristic equation may occur more than twice, and when this happens, the linearly independent solutions are obtained by multiplying by increasing powers of x. For example, if 2 is a root of multiplicity 4 in the characteristic equation, the corresponding linearly independent solutions are e^{2x}, xe^{2x}, x^2e^{2x}, and x^3e^{2x}. The procedure for obtaining the general solution of nth-order linear homogeneous equation with constant coefficients is illustrated in the next two examples.

EXAMPLE 6   *Characteristic equation with repeated roots*

Solve $y^{(4)} - 5y''' + 6y'' + 4y' - 8y = 0$.

Solution   Solve the characteristic equation:

$$r^4 - 5r^3 + 6r^2 + 4r - 8 = 0$$

Because this is 4th-degree, we use synthetic division and the rational root theorem (or a calculator) to find the roots $-1, 2, 2,$ and 2. The general solution is

$$y = C_1 e^{-x} + C_2 e^{2x} + C_3 x e^{2x} + C_4 x^2 e^{2x}$$ ∎

SMH

| **EXAMPLE 7** | *Characteristic equation with repeated roots (some not real)* |

Solve $y^{(7)} + 8y^{(5)} + 16y''' = 0$

Solution   Solve the characteristic equation:

$$r^7 + 8r^5 + 16r^3 = 0$$

$$r^3(r^4 + 8r^2 + 16) = 0$$

$$r^3(r^2 + 4)^2 = 0$$

$$r = 0 \quad \text{(multiplicity 3)}, \quad \pm 2i \quad \text{(multiplicity 2)}$$

The roots (showing multiplicity) are $0, 0, 0, 2i, 2i, -2i, -2i$. The general solution is

$$y = C_1 + C_2 x + C_3 x^2 + C_4 \cos 2x + C_5 \sin 2x + C_6 x \cos 2x + C_7 x \sin 2x$$ ∎

DAMPED MOTION OF A MASS ON A SPRING

To illustrate an application of second-order homogeneous linear differential equations, we shall consider the motion of an oscillating spring. Suppose we pull down on an object suspended at the end of a spring and then release it. Hooke's law in physics says that a spring that is stretched or compressed x units from its natural length tends to restore itself to its natural length by a force whose magnitude F is proportional to x. Specifically, $F(x) = kx$, where the constant of proportionality k is called the **spring constant** and depends on the stiffness of the spring.

Suppose the mass of the spring is negligible compared to the mass m of the object on the spring. When the object is pulled down and released, the spring begins to oscillate, and its motion is determined by two forces, the weight mg of the object and the restoring force $F(x) = k_1 x$ of the spring. According to Newton's second law of motion, the force acting on the object is ma, where $a = x''(t)$ is the acceleration of the object. If there are no other external forces acting on the object, the motion is said to be **undamped,** and the motion is governed by the second-order homogeneous equation

$$mx''(t) = -k_1 x(t) \qquad k_1 > 0$$

$$mx''(t) + k_1 x(t) = 0$$

Next, suppose the object is connected to a dashpot, or a device that imposes a damping force. A good example is a shock absorber in a car, which forces a spring to move through a fluid (see Figure 15.5). Experiments indicate that the shock absorber introduces a damping force proportional to the velocity $v = x'(t)$. Thus, the total force in this case is $-k_1 x(t) - k_2 x'(t)$, and Newton's second law tells us

$$mx''(t) = -k_1 x(t) - k_2 x'(t) \qquad k_1 > 0, k_2 > 0$$

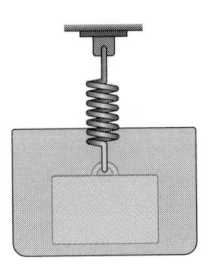

■ **FIGURE 15.5**   A damped spring system

$$mx''(t) + k_2 x'(t) + k_1 x = 0$$

The characteristic equation is

$$mr^2 + k_2 r + k_1 = 0$$

$$r = \frac{-k_2 \pm \sqrt{k_2^2 - 4k_1 m}}{2m}$$

The three cases that can occur correspond to different kinds of motion for the object on the spring.

overdamping: $k_2^2 - 4k_1 m > 0$

In this case, both roots are real and negative, and the solution is of the form

$$x(t) = C_1 e^{r_1 t} + C_2 e^{r_2 t}$$

where $r_1 = -\dfrac{k_2}{2m} + \dfrac{1}{2m}\sqrt{k_2^2 - 4k_1 m}$ and

$$r_2 = -\frac{k_2}{2m} - \frac{1}{2m}\sqrt{k_2^2 - 4k_1 m}$$

Note that the motion dies out eventually at $t \to +\infty$:

$$\lim_{t \to +\infty} x(t) = \lim_{t \to +\infty}\left(C_1 e^{r_1 t} + C_2 e^{r_2 t}\right) = 0, \quad \text{because } r_1 < 0 \text{ and } r_2 < 0$$

Overdamping is illustrated in Figure 15.6a.

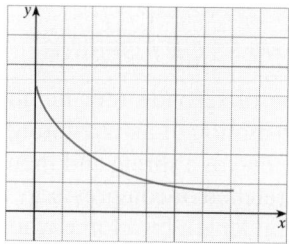 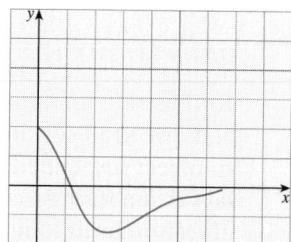 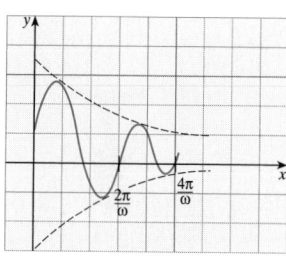

a. Overdamping $k_2^2 - 4k_1 m > 0$ **b.** Critical damping $k_2^2 - 4k_1 m = 0$ **c.** Underdamping $k_2^2 - 4k_1 m < 0$

■ **FIGURE 15.6**   Damping motion

critical damping: $k_2 - 4k_1 m = 0$

The solution has the form

$$x(t) = (C_1 + C_2 t)e^{rt}$$

where $r_1 = r_2 = r = -\dfrac{k_2}{2m}$. In this case, the motion also eventually dies out (as $t \to +\infty$), because $r < 0$. This solution is shown in Figure 15.6b.

underdamping: $k_2 - 4k_1 m < 0$

In this case, the characteristic equation has complex roots, and the solution has the form

$$x(t) = e^{-k_2 t/(2m)}\left[C_1\cos\left(\frac{t}{2m}\sqrt{4k_1 m - k_2^2}\right) + C_2\sin\left(\frac{t}{2m}\sqrt{4k_1 m - k_2^2}\right)\right]$$

This can be written as

$$x(t) = Ae^{\alpha t}\cos(\omega t - C)$$

where

$$A = \sqrt{C_1^2 + C_2^2}; \quad \alpha = -\frac{k_2}{2m}; \quad \omega = \frac{1}{2m}\sqrt{4k_1 m - k_2^2}; C = \tan^{-1}\frac{C_2}{C_1}$$

Because k_2 and m are both positive, α must be negative and we see that $x(t) \to 0$ as $t \to +\infty$. Notice that as the motion dies out, it oscillates with frequency $2\pi/\omega$, as shown in Figure 15.6c.

REDUCTION OF ORDER

Theorem 15.2 applies even when b and c are functions of x instead of constants. In other words, the general solution of the equation.

$$y'' + b(x)y' + c(x)y = 0$$

can be expressed as

$$y = C_1 y_1 + C_2 y_2$$

where $y_1(x)$ and $y_2(x)$ are any two linearly independent solutions. Sometimes, we can find one solution y_1 by observation and then obtain a second linearly independent solution y_2 by assuming that $y_2 = vy_1$, where $v(x)$ is a twice differentiable function of x. This procedure is called **reduction of order** because it involves solving the given second-order equation by solving two related first-order equations. The basic ideas of reduction of order are illustrated in the following example.

> **EXAMPLE 8**   *Reduction of order*

Show that $y = x^2$ is a solution of the equation $y'' - \dfrac{3}{x}y' + \dfrac{4}{x^2}y = 0$ for $x > 0$, then find the general solution.

Solution   If $y_1 = x^2$, then $y_1' = 2x$ and $y_1'' = 2$. We have

$$y_1'' - \frac{3}{x}y_1' + \frac{4}{x^2}y_1 = 2 - \frac{3}{x}(2x) + \frac{4}{x^2}(x^2) = 0$$

and y_1 is a solution.

Next, let $y_2 = vx^2$, so

$$y_2' = 2xv + x^2v' \quad \text{and} \quad y'' = 2v + 4xv' + x^2v''$$

Substitute these derivatives into the given equation:

$$(2v + 4xv' + x^2v'') - \frac{3}{x}(2xv + x^2v') + \frac{4}{x^2}(vx^2) = 0$$

$$x^2v'' + xv' = 0$$

$$\frac{v''}{v'} = \frac{-1}{x}$$

Integrate both sides of this equation:

$$\ln v' = -\ln x$$

$$v' = \frac{1}{x}$$

$$v = \ln x$$

Thus, $y_2 = vx^2 = x^2 \ln x$ is a second solution. To show that the solutions $y_1 = x^2$ and $y = x^2 \ln x$ are linearly independent, we compute the Wronskian (since $x > 0$):

$$W(x^2, x^2 \ln x) = \begin{vmatrix} x^2 & x^2 \ln x \\ 2x & x + 2x \ln x \end{vmatrix} = x^3 \neq 0$$

It follows that the general solution of the given differential equation is

$$y = C_1 x^2 + C_2 x^2 \ln x$$

15.2 Problem Set

Ⓐ *Find the general solution of the second-order homogeneous linear differential equations given in Problems 1–14.*

1. $y'' + y' = 0$
2. $y'' + y' - 2y = 0$
3. $y'' + 6y' + 5y = 0$
4. $y'' + 4y = 0$
5. $y'' - y' - 6y = 0$
6. $y'' + 8y' + 16y = 0$
7. $2y'' - 5y' - 3y = 0$
8. $3y'' + 11y' - 4y = 0$
9. $y'' - y = 0$
10. $6y'' + 13y' + 6y = 0$
11. $y'' + 11y = 0$
12. $y'' - 4y' + 5y = 0$
13. $7y'' + 3y' + 5y = 0$
14. $2y'' + 5y' + 8y = 0$

Find the general solution of the given higher-order homogeneous linear differential equations in Problems 15–20.

15. $y''' + y'' = 0$
16. $y''' + 4y' = 0$
17. $y^{(4)} + y''' + 2y'' = 0$
18. $y^{(4)} + 10y'' + 9y = 0$
19. $y''' + 2y'' - 5y' - 6y = 0$
20. $y^{(4)} + 2y''' + 2y'' + 2y' + y = 0$

Find the particular solution that satisfies the differential equations in Problems 21–26 subject to the specified initial conditions.

21. $y'' - 10y' + 25y = 0;\ y(0) = 1, y'(0) = -1$
22. $y'' + 6y' + 9y = 0;\ y(0) = 4, y'(0) = -3$
23. $y'' - 12y' + 11y = 0;\ y(0) = 3, y'(0) = 11$
24. $y'' + 4y' + 5y = 0;\ y(0) = -2, y'(0) = 1$
25. $y''' + 10y'' + 25y' = 0;\ y(0) = 3,$
    $y'(0) = 2, y''(0) = -1$
26. $y^{(4)} - y''' = 0;$
    $y(0) = 3, y'(0) = 0, y''(0) = 3, y'''(0) = 4$

In Problems 27–30, find the Wronskian, W, of the given set of functions and show that $W \neq 0$.

27. $\{e^{-2x}, e^{3x}\}$
28. $\{e^{-x}, xe^{-x}\}$
29. $\{e^{-x}\cos x, e^{-x}\sin x\}$
30. $\{xe^x\cos x, xe^x\sin x\}$

31. **HISTORICAL QUEST** Josef Hoëné (1778–1853) adopted the name Wronski when he was 32 years old, around the time he was married. Today, he is remembered for determinants now known as Wronskians, named by Thomas Muir (1844–1934) in 1882. Wronski's main work was in the philosophy of mathematics. For years his mathematical work, which contained many errors, was dismissed as unimportant, but in recent years closer study of his work revealed that he had some significant mathematical insight.

JOSEPH HOËNÉ DE WRONSKI
1778–1853

For this HISTORICAL QUEST you are asked to write a paper on one of the great philosophical issues in the history of mathematics. Here are a few quotations to get you started:

Mathematics is discovered:

"... what is physical is subject to the laws of mathematics, and what is spiritual to the laws of God, and the laws of mathematics are but the expression of the thoughts of God."

Thomas Hill, *The Uses of Mathesis; Bibliotheca Sacra,* p. 523.

"Our remote ancestors tried to interpret nature in terms of anthropomorphic concepts of their own creation and failed. The efforts of our nearer ancestors to interpret nature on engineering lines proved equally inadequate. Nature has refused to accommodate herself to either of these man-made molds. On the other hand, our efforts to interpret nature in terms of the concepts of pure mathematics have, so far, proved brilliantly successful ... from the intrinsic evidence of His creation, the Great Architect of the Universe now begins to appear as a pure mathematician."

James H. Jeans, *The Mysterious Universe,* p. 142.

Mathematics is invented:

"There is an old Armenian saying, 'He who lacks sense of the past is condemned to live in the narrow darkness of his own generation.' Mathematics without history is mathematics stripped of its greatness: for, like the other arts—and mathematics is one of the supreme arts of civilization—it derives it grandeur from the fact of being a human creation."

G. F. Simmons, *Differential Equations with Applications and Historical Notes,* 2nd edition, McGraw-Hill, Inc., 1991, p. xix.

Discuss whether the significant ideas in mathematics are *discovered* or *invented*.

Ⓑ 32. If $a^2 = b$, one solution of $y'' - 2ay' + by = 0$ is $y_1 = e^{ax}$. Use reduction of order to show that $y_2 = e^{ax}$ is a second solution, then show that y_1 and y_2 are linearly independent.

In Problems 33–38, a second-order differential equation and one solution $y_1(x)$ are given. Use reduction of order to find a second solution $y_2(x)$, then show that y_1 and y_2 are linearly independent.

33. $y'' + 6y' + 9y = 0;\ \ y_1 = e^{-3x}$
34. $2y'' - y' - 6y = 0;\ \ y_1 = e^{2x}$
35. $xy'' + 4y' = 0;\ y_1 = 1$
36. $x^2y'' + xy' - 4y = 0;\ \ y_1 = x^{-2}$
37. $x^2y'' + 2xy' - 12y = 0;\ \ y_1 = x^3$
38. $(1 - x)^2y'' - (1 - x)y' - y = 0;\ \ y_1 = 1 - x$

Suppose a 16-lb weight stretches a spring 8 in. from its natural length. Find a formula for the position of the weight as a function of time for the situations described in Problems 39–44.

39. The weight is pulled down an additional 6 in. and is then released with an initial upward velocity of 8 ft/s.

40. The weight is pulled down 10 in. and is released with an initial upward velocity of 6 ft/s.

41. The weight is raised 8 in. above the equilibrium point and the compressed spring is then released.

42. The weight is raised 12 in. above the equilibrium point and the compressed spring is then released.

43. The weight is pulled 6 in. below the equilibrium point and is then released. The weight is connected to a dash-pot that imposes a damping force of magnitude $0.4|v|$ at all times.

44. The weight is pulled 12 in. below the equilibrium point and is then released. The weight is connected to a dashpot that imposes a damping force of magnitude $0.08|v|$ at all times.

45. Characterize the solution of the second order equation

$$y'' - 2ay' + (a^2 - b)y = 0 \qquad b > 0$$

in terms of the functions $\cosh kx, \sinh kx, \cos kx,$ and $\sin kx,$ where $k = \sqrt{b}.$

46. Verify that $y_1 = e^{2x}, y_2 = xe^{2x},$ and $y_3 = x^2e^{2x}$ are linearly independent solutions of the third-order equation

$$y''' - 6y'' + 12y' - 8y = 0$$

47. **MODELING PROBLEM** A 100-lb object is projected vertically upward from the surface of the earth with initial velocity 150 ft/s.
a. Modeling the object's motion with negligible air resistance, how long does it take for the object to return to earth?
b. Change the model to assume air resistance equal to half the object's velocity. Before making any computation, does your intuition tell you the object takes less or more time to return to earth this time than in part **a**? Now set up and solve a differential equation to actually determine the round-trip time. Were you right?

⊙ 48. **MODELING PROBLEM** The motion of a pendulum subject to frictional damping proportional to its velocity is modeled by the differential equation

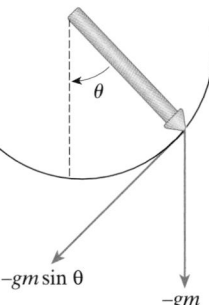

$$mL\frac{d^2\theta}{dt^2} + kL\frac{d\theta}{dt} + mg\sin\theta = 0$$

where L is the length of the pen-dulum, m is the mass of the bob at its end, and θ is the angle the pendulum arm makes with the vertical (see Figure 15.7). Assume θ is small, so $\sin\theta$ is approximately equal to $\theta.$
a. Solve the resulting differential equation for the case where $k^2 \geq 4gm^2/L.$ What happens to $\theta(t)$ as $t \to \infty$?
b. If $k^2 < 4gm^2/L,$ show that

$$\theta(t) = Ae^{-kt/2m}\cos\left(\sqrt{\frac{B}{L}}t + C\right)$$

■ FIGURE 15.7

for constants $A, B,$ and $C.$ What happens to $\theta(t)$ as $t \to \infty$ in this case?
c. For the situation in part **b,** show that the time difference between successive vertical positions is approximately

$$T = 2\pi m\sqrt{\frac{L}{4gm^2 - k^2L}}$$

49. If there is no damping and no external forces, the motion of an object of mass m attached to a spring with spring constant k is governed by the differential equation $mx'' + kx = 0.$ Show that the general solution of this equation is given by

$$x(t) = A\cos\left(\sqrt{\frac{k}{m}}t - B\right), \quad k > 0$$

where A and B are constants. This is called **simple harmonic motion** with frequency $\dfrac{1}{2\pi}\sqrt{\dfrac{k}{m}}.$

50. Consider the motion of an object of mass m on a spring with spring constant k_1 and damping constant k_2 for the case where there is critical damping.
a. Describe the motion of the object assuming that it be-gins at rest: $x'(0) = 0$ at $x(0) = x_0.$ How is this differ-ent from the case where the object begins at $x(0) = 0$ with initial velocity $x'(0) = v_0$? Sketch both solutions on the same graph. How do initial velocity and initial displacement affect the motion?
b. If $x(0) = x'(0),$ what is the maximum displacement? Show that the time at which the maximum displace-ment occurs is independent of the initial displacement $x(0).$ Assume $x(0) \neq 0.$

51. Suppose the characteristic equation of the differential equation $y'' + ay' + by = 0$ has complex conjugate roots, $r_1 = \alpha + \beta i$ and $r_2 = \alpha - \beta i.$ Show that $y_1 = e^{\alpha x}\cos\beta x$ and $y_2 = e^{\alpha x}\sin\beta x$ are both solutions and are linearly independent. This is Theorem 15.3 for the case where $a^2 - 4b < 0.$

52. **PENDULUM MOTION** Suppose a ball of mass m is suspended at the end of a rod of length L and is set in motion swinging back and forth like a pendulum, as shown in Figure 15.8.

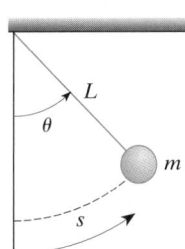

■ FIGURE 15.8 Pendulum motion

Let θ be the angle between the rod and the vertical at time $t,$ so that the displacement of the ball from the equilibrium position is $s = L\theta$ and the acceleration of the ball's motion is

$$\frac{d^2s}{dt^2} = L\frac{d^2\theta}{dt^2}$$

a. Use Newton's second law to show that

$$mL\frac{d^2\theta}{dt^2} + mg\sin\theta = 0$$

Assume that air resistance and the mass of the rod are negligible.

b. When the displacement is small (θ close to 0), $\sin\theta$ may be replaced by θ. In this case, solve the resulting differential equation

$$\frac{d^2\theta}{dt^2} + \frac{g}{L}\theta = 0$$

How is the motion of the pendulum like the simple harmonic motion discussed in Problem 49?

53. **PATH OF A PROJECTILE WITH VARIABLE MASS**
A rocket starts from rest and moves vertically upward, along a straight line. Assume the rocket and its fuel initially weigh w kilograms and that the fuel alone weighs

w_f kilograms. Further assume that the fuel is consumed at a constant rate of r kilograms per second (relative to the rocket). Finally, assume that gravitational attraction is the only external force acting on the rocket.

a. If $m(t)$ is the mass of the rocket at time t and $s(t)$ is the height above the ground at time t, it can be shown that

$$m(t)s''(t) + m'(t)v_0 + m(t)g = 0$$

where v_0 is the velocity of the exhaust gas in relation to the rocket. Express $m(t)$ in terms of $w, r, v_0,$ and g, then integrate this differential equation to obtain the velocity $s'(t)$. Note that $s'(0) = 0$ because the rocket starts from rest.

b. Integrate the velocity $s'(t)$ to obtain $s(t)$.

c. At what time is all the fuel consumed?

d. How high is the rocket at the instant the fuel is consumed?

54. **JOURNAL PROBLEM** by Murray S. Klamkin.* Solve the differential equation

$$x^4y'' - (x^3 + 2axy)y' + 4ay^2 = 0$$

*Problem 331 in the *Canadian Mathematical Bulletin*, Vol. 26, 1983, p. 126.

15.3 Second-Order Nonhomogeneous Linear Differential Equations

IN THIS SECTION   nonhomogeneous equations, method of undetermined coefficients, variation of parameters, an application to *RLC* circuits   ∎

NONHOMOGENEOUS EQUATIONS

Next we shall see how to solve a nonhomogeneous second-order linear equation of the general form $y'' + ay' + by = F(x)$. The key to our results is the following theorem.

THEOREM 15.4   *Characterization of the general solution of* $y'' + ay' + by = F(x)$

Let y_p be a particular solution of the nonhomogeneous second-order linear equation $y'' + ay' + by = F(x)$. Let y_h be the general solution of the related homogeneous equation $y'' + ay' + by = 0$. Then the general solution of $y'' + ay' + by = F(x)$ is given by the sum

$$y = y_h + y_p$$

Proof   First, the sum $y = y_h + y_p$ is a solution of the nonhomogeneous equation $y'' + ay' + by = F(x)$, because

$$\begin{aligned}
y'' + ay' + by &= (y_h + y_p)'' + a(y_h + y_p)' + b(y_h + y_p)\\
&= y_h'' + y_p'' + ay_h' + ay_p' + by_h + by_p\\
&= (y_h'' + ay_h' + by_h) + (y_p'' + ay_p' + by_p)\\
&= 0 + F(x)\\
&= F(x)
\end{aligned}$$

Technology Window ▲ ▼

There are software packages that solve many types of differential equations for professional engineers. For example, Sumulink (for 386 IBM or higher or for Macintosh computers) offers equation solving that includes six powerful differential equation solvers. Such software packages are invaluable for professional engineers and designers.

Conversely, if y is any solution of the nonhomogeneous equation, then $y - y_p$ is a solution of the related homogeneous equation because

$$(y - y_p)'' + a(y - y_p)' + b(y - y_p) = (y'' - y_p'') + a(y' - y_p') + b(y - y_p)$$
$$= (y'' + ay' + by) - (y_p'' + ay_p' + by_p)$$
$$= F(x) - F(x)$$
$$= 0$$

Thus, $y - y_p = y_h$ (because it is a solution of the homogeneous equation). Therefore, because y was *any* solution of the nonhomogeneous equation, it follows that $y = y_h + y_p$ is the general solution of the nonhomogeneous equation. $\quad\blacksquare$

> ■ *What This Says:* We can obtain the general solution of the nonhomogeneous equation $y'' + ay' + by = F(x)$ by finding the general solution y_h of the related homogeneous equation $y'' + ay' + by = 0$ and just one particular solution y_p of the given nonhomogeneous equation.

We can use the methods of the preceding section to find the general solution of the related homogeneous equation. We now develop two methods for finding particular solutions of the nonhomogeneous equation.

METHOD OF UNDETERMINED COEFFICIENTS

Sometimes it is possible to find a particular solution y_p of the nonhomogeneous equation $y'' + ay' + by = F(x)$ by assuming a **trial solution** $\bar{y}_p$ of the same general form as $F(x)$. This procedure, called the **method of undetermined coefficients,** is illustrated in the following three examples, each of which has the related homogeneous equation $y'' + y' - 2y = 0$ with the general solution $y_h = C_1 e^x + C_2 e^{-2x}$.

EXAMPLE 1   *Method of undetermined coefficients*

Find $\bar{y}_p$ and the general solution for $y'' + y' - 2y = 2x^2 - 4x$.

Solution   The right side $F(x) = 2x^2 - 4x$ is a quadratic polynomial. Because derivatives of a polynomial are polynomials of lower degree, it seems reasonable to consider a trial solution that is also a polynomial of degree 2. That is, we "guess" that this equation has a particular solution $\bar{y}_p$ of the general form $\bar{y}_p = A_1 x^2 + A_2 x + A_3$. To find the constants A_1, A_2, and A_3, calculate

$$\bar{y}_p' = 2A_1 x + A_2 \quad \text{and} \quad \bar{y}_p'' = 2A_1$$

Substitute the values for $\bar{y}_p, \bar{y}_p'$, and $\bar{y}_p''$ into the given equation:

$$y'' + y' - 2y = 2x^2 - 4x$$
$$2A_1 + 2A_1 x + A_2 - 2(A_1 x^2 + A_2 x + A_3) = 2x^2 - 4x$$
$$-2A_1 x^2 + (2A_1 - 2A_2)x + (2A_1 + A_2 - 2A_3) = 2x^2 - 4x$$

This is true only when the coefficients of each power of x on each side of the equation match, so that

$$\begin{cases} -2A_1 = 2 & (x^2 \text{ terms}) \\ 2A_1 - 2A_2 = -4 & (x \text{ terms}) \\ 2A_1 + A_2 - 2A_3 = 0 & (\text{constant terms}) \end{cases}$$

Solve this system of equations simultaneously to find $A_1 = -1, A_2 = 1$, and $A_3 = -\frac{1}{2}$. Thus, a particular solution of the given nonhomogeneous equation is

$$y_p = A_1 x^2 + A_2 x + A_3 = -x^2 + x - \frac{1}{2}$$

and the general solution for the nonhomogeneous equation is

$$y = y_h + y_p = C_1 e^x + C_2 e^{-2x} - x^2 + x - \frac{1}{2}$$ ∎

COMMENT   Notice that even though the constant term is zero in the polynomial function $F(x)$, we cannot assume that $y = A_1 x^2 + A_2 x$ is a suitable trial solution. In general, all terms of lower degree that could possibly lead to the given right-side function $F(x)$ must be included in the trial solution.

> **EXAMPLE 2**   *Method of undetermined coefficients*

Solve $y'' + y' - 2y = \sin x$.

Solution   Because the trial solution $\bar{y}_p$ is to be "like" the right-side function $F(x) = \sin x$, it seems that we should choose the trial solution to be $\bar{y}_p = A_1 \sin x$, but a sine function can have either a sine or a cosine in its derivatives, depending on how many derivatives are taken. Thus, to account for the $\sin x$, it is necessary to have *both* $\sin x$ and $\cos x$ in the trial solution, so we set

$$\bar{y}_p = A_1 \sin x + A_2 \cos x$$

Differentiating, we find

$$\bar{y}_p' = A_1 \cos x - A_2 \sin x \quad \text{and} \quad \bar{y}_p'' = -A_1 \sin x - A_2 \cos x$$

Substitute the values into the given equation:

$$y'' + y' - 2y = \sin x$$
$$(-A_1 \sin x - A_2 \cos x) + (A_1 \cos x - A_2 \sin x) - 2(A_1 \sin x + A_2 \cos x) = \sin x$$
$$(-3A_1 - A_2)\sin x + (A_1 - 3A_2) \cos x = \sin x$$

This gives the system

$$\begin{cases} -3A_1 - A_2 = 1 & (\sin x \text{ terms}) \\ A_1 - 3A_2 = 0 & (\cos x \text{ terms}) \end{cases}$$

with the solution $A_1 = -\frac{3}{10}, A_2 = -\frac{1}{10}$. Thus, the particular solution of the nonhomogeneous equation is $y_p = -\frac{3}{10} \sin x - \frac{1}{10} \cos x$, and the general solution is $y = y_h + y_p = C_1 e^x + C_2 e^{-2x} - \frac{3}{10} \sin x - \frac{1}{10} \cos x$. ∎

COMMENT   It can be shown that if y_1 is a solution of $y'' + ay' + by = F(x)$ and y_2 is a solution of $y'' + ay' + by = G(x)$, then $y_1 + y_2$ will be a solution of $y'' + ay' + by = F(x) + G(x)$. (See Problem 55.) This is called the **principle of superposition.** For instance, by combining the results of Examples 1 and 2, we see that a particular solution of the nonhomogeneous linear equation $y'' + y' - 2y = 2x^2 - 4x + \sin x$ is

$$y_p = \overbrace{-x^2 + x - \frac{1}{2}}^{\text{Solution for } F = 2x^2 - 4x} \underbrace{-\frac{3}{10} \sin x - \frac{1}{10} \cos x}_{\text{Solution for } G = \sin x}$$

> **EXAMPLE 3**   *Method of undetermined coefficients*

Solve $y'' + y' - 2y = 4e^{-2x}$.

Solution   At first glance (looking at $F(x) = 4e^{-2x}$), it may seem that the trial solution should be $\bar{y}_p = Ae^{-2x}$. To see why this cannot be a solution, note that because $y_1 = e^{-2x}$ is a solution of the related homogeneous equation, we have

$$y_1'' + y_1' - 2y_1 = 0$$

which results in the unsolvable equation

$$0 = 4e^{-2x}$$

To deal with this situation, multiply the usual trial solution by x and consider the trial solution $\bar{y}_p = Axe^{-2x}$. Differentiating, we find

$$\bar{y}_p' = A(1 - 2x)e^{-2x} \quad \text{and} \quad \bar{y}_p'' = A(4x - 4)e^{-2x}$$

and by substituting into the given equation, we obtain

$$y'' + y' - 2y = 4e^{-2x}$$
$$A(4x - 4)e^{-2x} + A(1 - 2x)e^{-2x} - 2Axe^{-2x} = 4e^{-2x}$$
$$(4Ax - 4A + A - 2Ax - 2Ax)e^{-2x} = 4e^{-2x}$$
$$-3Ae^{-2x} = 4e^{-2x}$$
$$A = -\tfrac{4}{3}$$

Thus, $\bar{y}_p = -\tfrac{4}{3}xe^{-2x}$, so that the general solution is

$$y = y_h + y_p = C_1e^x + C_2e^{-2x} - \tfrac{4}{3}xe^{-2x} \qquad \blacksquare$$

The procedure illustrated in Examples 1–3 can be applied to a differential equation $y'' + ay' + by = F(x)$ only when $F(x)$ has one of the following forms:

    a. $F(x) = P_n(x)$, a polynomial of degree n

    b. $F(x) = P_n(x)e^{kx}$

    c. $F(x) = e^{kx}[P_n(x)\cos \alpha x + Q_n(x)\sin \alpha x]$, where $Q_n(x)$ is another polynomial of degree n

We can now describe the **method of undetermined coefficients.**

Method of Undetermined Coefficients

To solve $y'' + ay' + by = F(x)$ when $F(x)$ is one of the forms listed above:

    1. The solution is of the form $y = y_h + y_p$, where y_h is the general solution and y_p is a particular solution.

    2. Find y_h by solving the homogeneous equation

$$y'' + ay' + by = 0$$

    3. Find y_p by picking an appropriate trial solution $\bar{y}_p$:

Form of $F(x)$	Corresponding trial expression $\bar{y}_p$
a. $P_n(x) = c_n x^n + \cdots + c_1 x + c_0$	$A_n x^n + \cdots + A_1 x + A_0$
b. $P_n(x)e^{kx}$	$[A_n x^n + A_{n-1}x^{n-1} + \cdots + A_0]e^{kx}$
c. $e^{kx}[P_n(x)\cos \alpha x + Q_n(x)\sin \alpha x]$	$e^{kx}[(A_n x^n + \cdots + A_0)\cos \alpha x$ $+ (B_n x^n + \cdots + B_0)\sin \alpha x]$

4. If no term in the trial expression $\bar{y}_p$ appears in the general homogeneous solution y_h, the particular solution can be found by substituting $\bar{y}_p$ into the equation $y'' + ay' + by = F(x)$ and solving for the undetermined coefficients.

5. If any term in the trial expression $\bar{y}_p$ appears in y_h, multiply $\bar{y}_p$ by x^k, where k is the smallest integer such that no term in $x^k\bar{y}_p$ is a solution of $y'' + ay' + by = 0$. Then proceed as in step 4, using $x^k\bar{y}_p$ as the trial solution.

EXAMPLE 4   *Finding trial solutions*

Determine a suitable trial solution for undetermined coefficients in each of the given cases.

a. $y'' - 4y' + 4y = 3x^2 + 4e^{-2x}$ b. $y'' - 4y' + 4y = 5xe^{2x}$
c. $y'' + 2y' + 5y = 3e^{-x}\cos 2x$

Solution

a. The related homogeneous equation $y'' - 4y' + 4y = 0$ has the characteristic equation $r^2 - 4r + 4 = 0$, which has the root 2 of multiplicity two. Thus, the general homogeneous solution is

$$y_h = C_1e^{2x} + C_2xe^{2x}$$

The part of the trial solution for the nonhomogeneous equation that corresponds to $3x^2$ is $A_0 + A_1x + A_2x^2$ and the part that corresponds to $4e^{-2x}$ is Be^{-2x}. Since neither part includes terms in y_h, we apply the principle of superposition to conclude that

$$\bar{y}_p = A_0 + A_1x + A_2x^2 + Be^{-2x}$$

b. We know from part **a** that the general homogeneous solution is

$$y_h = C_1e^{2x} + C_2xe^{2x}$$

The normal trial solution for $5xe^{2x}$ would be $(A_0 + A_1x)e^{2x}$, but part of this expression is contained in y_h. If we multiply by x, part of $(A_0 + A_1x)xe^{2x}$ is still contained in y_h, so we multiply by x again to obtain

$$\bar{y}_p = (A_0 + A_1x)x^2e^{2x}$$

c. The related homogeneous equation $y'' + 2y' + 5y = 0$ has the characteristic equation $r^2 + 2r + 5 = 0$. This has complex conjugate roots $r = -1 \pm 2i$, so the general homogeneous solution is

$$y_h = e^{-x}[C_1\cos 2x + C_2\sin 2x]$$

Ordinarily the trial solution for the nonhomogeneous equation would be of the form $e^{-x}[A \cos 2x + B\sin 2x]$, but part of this expression is in y_h. Therefore, we multiply by x to obtain the trial solution

$$y = xe^{-x}[A \cos 2x + B \sin 2x] \qquad \blacksquare$$

VARIATION OF PARAMETERS

The method of undetermined coefficients applies only when the coefficients b and c are constant in the nonhomogeneous linear equation $y'' + by' + cy = F(x)$ and the driving function $F(x)$ has the same general form as a solution of a second-order homogeneous linear equation with constant coefficients. Even though many important applications are modeled by differential equations of this type, there are other situations that require a more general procedure.

Our next goal is to examine a method of J. L. Lagrange (see Historical Quest, Problem 54 in Section 12.8) called **variation of parameters,** which can be used to find a particular solution of any nonhomogeneous equation

$$y'' + P(x)y' + Q(x)y = F(x)$$

where P, Q, and F are continuous.

To use variation of parameters, we must be able to find two linearly independent solutions $y_1(x)$ and $y_2(x)$ of the related homogeneous equation,

$$y'' + P(x)y' + Q(x)y = 0$$

In practice, if $P(x)$ and $Q(x)$ are not both constants, these may be difficult to find, but once we have them, we assume there is a solution of the nonhomogeneous equation of the form

$$y_p = uy_1 + vy_2$$

Differentiating this expression y_p, we obtain

$$y'_p = u'y_1 + v'y_2 + uy'_1 + vy'_2$$

To simplify, assume that

$$u'y_1 + v'y_2 = 0$$

Remember, we need to find only *one* particular solution y_p, and if imposing this side condition makes it easier to find such a y_p, so much the better! With the side condition, we have

$$y'_p = uy'_1 + vy'_2$$

and by differentiating again, we obtain

$$y''_p = uy''_1 + u'y'_1 + vy''_2 + v'y'_2$$

Next, we substitute our expressions for y'_p and y''_p into the given differential equation:

$$F(x) = y''_p + P(x)y'_p + Q(x)y_p$$
$$= (uy''_1 + u'y'_1 + vy''_2 + v'y'_2) + P(x)(uy'_1 + vy'_2) + Q(x)(uy_1 + vy_2)$$

This can be rewritten as

$$u[y''_1 + P(x)y'_1 + Q(x)y_1] + v[y''_2 + P(x)y'_2 + Q(x)y_2] + u'y'_1 + v'y'_2 = F(x)$$

Because y_1 and y_2 are solutions of $y'' + P(x)y' + Q(x)y = 0$, we have

$$u\underbrace{[y''_1 + P(x)y'_1 + Q(x)y_1]}_{0} + v\underbrace{[y''_2 + P(x)y'_2 + Q(x)y_2]}_{0} + u'y'_1 + v'y'_2 = F(x)$$

or

$$u'y'_1 + v'y'_2 = F(x)$$

Thus, the parameters u and v must satisfy the system of equations

$$\begin{cases} u'y_1 + v'y_2 = 0 \\ u'y'_1 + v'y'_2 = F(x) \end{cases}$$

Solve this system to obtain

$$u' = \frac{-y_2 F(x)}{y_1 y_2' - y_2 y_1'} \quad \text{and} \quad v' = \frac{y_1 F(x)}{y_1 y_2' - y_2 y_1'}$$

where in each case the denominator is not zero, because it is the Wronskian of the linearly independent solutions y_1, y_2 of the related homogeneous equation. Integrating, we find

$$u(x) = \int \frac{-y_2 F(x)}{y_1 y_2' - y_2 y_1'} \, dx \quad \text{and} \quad v(x) = \int \frac{y_1 F(x)}{y_1 y_2' - y_2 y_1'} \, dx$$

and by substituting into the expression

$$y_p = uy_1 + vy_2$$

we obtain a particular solution of the given differential equation. Here is a summary of the procedure we have described.

Variation of Parameters

To find the general solution of $y'' + P(x)y' + Q(x)y = F(x)$:

Step 1. Find the general solution, $y_h = C_1 y_1 + C_2 y_2$ to the homogeneous equation.

Step 2. Set $y_p = uy_1 + vy_2$ and substitute into the formulas:

$$u' = \frac{-y_2 F(x)}{y_1 y_2' - y_2 y_1'} \qquad v' = \frac{y_1 F(x)}{y_1 y_2' - y_2 y_1'}$$

Step 3. Integrate u' and v' to find u and v.

Step 4. A particular solution is $y_p = uy_1 + vy_2$, and the general solution is $y = y_h + y_p$.

These ideas are illustrated in the next example.

EXAMPLE 5   *Variation of parameters*

Solve $y'' + 4y = \tan 2x$.

Solution   Notice that this problem cannot be solved by undetermined coefficients, because the right-side function $F(x) = \tan 2x$ is not of the form of a solution of a homogeneous linear equation with constant coefficients.

To apply variation of parameters, begin by solving the related homogeneous equation $y'' + 4y = 0$. The characteristic equation is $r^2 + 4 = 0$ with roots $r = \pm 2i$. These complex roots have $\alpha = 0$ and $\beta = 2$ so that the general solution is

$$y = e^0[C_1 \cos 2x + C_2 \sin 2x] = C_1 \cos 2x + C_2 \sin 2x$$

This means $y_1(x) = \cos 2x$ and $y_2(x) = \sin 2x$. Set $y_p = uy_1 + vy_2$, where

$$u' = \frac{-y_2 F(x)}{y_1 y_2' - y_2 y_1'} = \frac{-\sin 2x \tan 2x}{2 \cos 2x \cos 2x + 2 \sin 2x \sin 2x} = -\frac{\sin^2 2x}{2 \cos 2x}$$

and

$$v' = \frac{-y_1 F(x)}{y_1 y_2' - y_2 y_1'} = \frac{\cos 2x \tan 2x}{2 \cos 2x \cos 2x + 2 \sin 2x \sin 2x} = \frac{1}{2} \sin 2x$$

Integrating, we obtain

$$u(x) = \int -\frac{\sin^2 2x}{2 \cos 2x} \, dx \qquad\qquad v(x) = \int \frac{1}{2} \sin 2x \, dx$$

$$= -\frac{1}{2} \int \frac{1 - \cos^2 2x}{\cos 2x} \, dx \qquad\qquad = -\frac{1}{4} \cos 2x + C_2$$

$$= -\frac{1}{2} \int (\sec 2x - \cos 2x) \, dx$$

$$= -\frac{1}{2}\left[\frac{1}{2} \ln |\sec 2x + \tan 2x| - \frac{\sin 2x}{2}\right] + C_1$$

Thus, a particular solution is

$$y_p = uy_1 + vy_2$$

$$= \left[-\frac{1}{4}\ln|\sec 2x + \tan 2x| + \frac{1}{4}\sin 2x\right]\cos 2x + \left(-\frac{1}{4}\cos 2x\right)\sin 2x$$

$$= -\frac{1}{4}(\cos 2x) \ln|\sec 2x + \tan 2x|$$

Finally, the general solution is

$$y = y_h + y_p = C_1 \cos 2x + C_2 \sin 2x - \tfrac{1}{4}(\cos 2x)\ln|\sec 2x + \tan 2x| \qquad\blacksquare$$

AN APPLICATION TO *RLC* CIRCUITS

An important application of second-order linear differential equations is the analysis of electric circuits. Consider a circuit with constant resistance R, inductance L, and capacitance C. Such a circuit is called an **RLC circuit** and is illustrated in Figure 15.9.

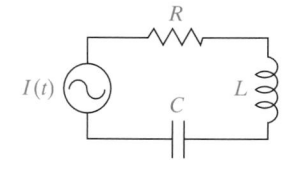

■ FIGURE 15.9   An *RLC* circuit

If $I(t)$ is the current in the circuit at time t and $Q(t)$ is the total charge on the capacitor, it is shown in physics that IR is the voltage drop across the resistance, Q/C is the voltage drop across the capacitor, and $L\, dI/dt$ is the voltage drop across the inductance. According to Kirchhoff's second law for circuits, the impressed voltage $E(t)$ in a circuit is the sum of the voltage drops, so that the current $I(t)$ in the circuit satisfies

$$L\frac{dI}{dt} + RI + \frac{Q}{C} = E(t)$$

By differentiating both sides of this equation and using the fact that $I = \dfrac{dQ}{dt}$, we can write

$$L\frac{d^2 I}{dt^2} + R\frac{dI}{dt} + \frac{1}{C}I = \frac{dE}{dt}$$

For instance, suppose the voltage input is sinusoidal—that is, $E(t) = A \sin \omega t$. Then we have $dE/dt = A\omega \cos \omega t$, and the second-order linear differential equation used in the analysis of the circuit is

$$L\frac{d^2I}{dt^2} + R\frac{dI}{dt} + \frac{1}{C}I = A\omega\cos\omega t$$

To solve this equation, we proceed in the usual way, solving the related homogeneous equation and then finding a particular solution of the nonhomogeneous system. The solution to the related homogeneous equation is called the **transient circuit** because the current described by this solution usually does not last very long. The part of the nonhomogeneous solution that corresponds to transient current 0 is called the **steady-state current**. Several problems dealing with RLC circuits are outlined in the problem set.

15.3 Problem Set

Ⓐ *In Problems 1–8, find a trial solution $\bar{y}_p$ for use in the method of undetermined coefficients.*

1. $y'' + 6y' = e^{2x}$
2. $y'' + 6y' + 8y = 2 + e^{-3x}$
3. $y'' + 2y' + 2y = e^{-x}$
4. $y'' + 2y' + 2y = \cos x$
5. $y'' + 2y' + 2y = e^{-x}\sin x$
6. $2y'' - y' - 6y = x^2 e^{2x}$
7. $y'' + 4y' + 5y = e^{-2x}(x + \cos x)$
8. $y'' + 4y' + 5y = (e^{-x}\sin x)^2$

For the differential equation

$$y'' + 6y' + 9y = F(x)$$

find a trial solution $\bar{y}_p$ for undertermined coefficients for each choice of $F(x)$ given in Problems 9–16.

9. $3x^3 - 5x$
10. $2x^2 e^{4x}$
11. $x^3\cos x$
12. $xe^{2x}\sin 5x$
13. $e^{2x} + \cos 3x$
14. $2e^{3x} + 8xe^{-3x}$
15. $4x^3 - x^2 + 5 - 3e^{-x}$
16. $(x^2 + 2x - 6)e^{-3x}\sin 3x$

Use the method of undetermined coefficients to find the general solution of the nonhomogeneous differential equations given in Problems 17–28.

17. $y'' + y' = -3x^2 + 7$
18. $y'' + 6y' + 5y = 2e^x - 3e^{-3x}$
19. $y'' + 8y' + 15y = 3e^{2x}$

20. $2y'' - 5y' - 3y = 5e^{3x}$
21. $y'' + 2y' + 2y = \cos x$
22. $y'' - 6y + 13y = e^{-3x}\sin 2x$
23. $7y'' + 6y' - y = e^{-x}(x + 1)$
24. $2y'' + 5y' = e^x\sin x$
25. $y'' - y' = x^3 - x + 5$
26. $y'' - y' = (x - 1)e^x$
27. $y'' + 2y' + y = (4 + x)e^{-x}$
28. $y'' - y' - 6y = e^{-2x} + \sin x$

Use variation of parameters to solve the differential equations given in Problems 29–38.

29. $y'' + y' = \tan x$
30. $y'' + 8y' + 16y = xe^{-2x}$
31. $y'' - y' - 6y = x^2 e^{2x}$
32. $y'' - 3y' + 2y = \dfrac{e^x}{1 + e^x}$
33. $y'' + 4y = \sec 2x \tan 2x$
34. $y'' + y = \sec^2 x$
35. $y'' + 2y' + y = e^{-x}\ln x$
36. $y'' - 4y' + 4y = \dfrac{e^{2x}}{1 + x}$
37. $y'' - y' = \cos^2 x$
38. $y'' - y = e^{-2x}\cos e^{-x}$

Ⓑ *In each of Problems 39–45, use either undetermined coefficients or variation of parameters to find the particular solution of the differential equation that satisfies the specified initial conditions.*

39. $y'' - y = 2\cos^2 x; y(0) = 0, y'(0) = 1$

40. $y'' - y = e^{-2x}\cos e^{-x}; y(0) = y'(0) = 0$

41. $y'' + 9y = 4e^{3x}; y(0) = 0, y'(0) = 2$

42. $y'' - 6y = x^2 - 3x; y(0) = 3, y'(0) = -1$

43. $y'' + 9y = x; y(0) = 0, y'(0) = 4$

44. $y'' + y' = 2\sin x; y(0) = 0, y'(0) = -4$

45. $y'' + y = \cot x; y\left(\dfrac{\pi}{2}\right) = 0, y'\left(\dfrac{\pi}{2}\right) = 5$

46. Find the general solution of the differential equation $y'' + y' - 6y = F(x)$, where F is the function whose graph is shown in Figure 15.10.

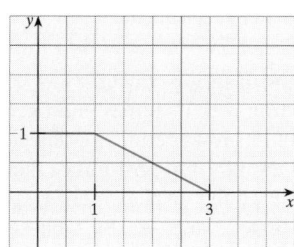

■ **FIGURE 15.10**　Graph of F for Problem 46

47. Find the general solution of the differential equation $y'' + 5y' + 6y = F(x)$, where F is the function whose graph is shown in Figure 15.11.

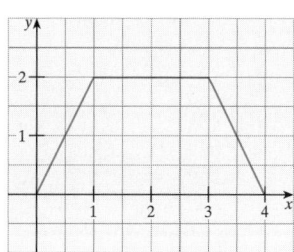

■ **FIGURE 15.11**　Graph of F for Problem 47

48. Find the steady-state current and the transient current in an RLC circuit with $L = 4$ henries, $R = 8$ ohms, $C = \dfrac{1}{8}$ farad, and $E(t) = 16\sin t$ volts. You may assume that $I(0) = 0$ and $I'(0) = 0$.

49. Find the steady-state current and the transient current in an RLC circuit with $L = 1$ henry, $R = 10$ ohms, $C = \dfrac{1}{9}$ farad, and $E(t) = 9\sin t$ volts. You may assume that $I(0) = 0$ and $I'(0) = 0$.

50. Work Problem 48 for the case where $E(t) = 10te^{-t}$.

51. Work Problem 49 for the case where $E(t) = 5t\sin t$.

52. Theorem 15.4 is also valid when the coefficients a and b are continuous functions of x. Solve the differential equation

$$x^2 y'' - 3xy' + 4y = x\ln x$$

by completing the following steps:

a. The related homogeneous equation

$$x^2 y'' - 3xy' + 4y = 0$$

has one solution of the form $y_1 = x^n$. Find n, then use reduction of order to find a second, linearly independent solution y_2.

b. Use variation of parameters to find a particular solution y_p of the given nonhomogeneous equation. Then use Theorem 15.4 to write the general solution. *Hint:* Divide the D. E. by x^2.

53. Repeat Problem 52 for the equation

$$(1 + x^2)y'' - 2xy' - 2y = \tan^{-1}x$$

54. **THE PRINCIPLE OF SUPERPOSITION.**　Let y_1 be a solution to the second-order linear differential equation

$$y'' + P(x)y' + Q(x)y = F_1(x)$$

and let y_2 satisfy

$$y'' + P(x)y' + Q(x)y = F_2(x)$$

Show that $y_1 + y_2$ satisfies the differential equation

$$y'' + P(x)y' + Q(x)y = F_1(x) + F_2(x)$$

Chapter 15   *Review*

Proficiency Examination

Concept Problems

1. What is a separable differential equation?
2. What is a homogeneous differential equation?
3. What is the form of a first-order linear differential equation?
4. What is an exact differential equation?
5. Describe Euler's method.
6. Define what it means for a set of functions to be linearly independent.
7. What is the Wronskian, and how is it used to test for linear independence?
8. a. What is the characteristic equation of $y'' + ay' + by = 0$?
   b. What is the general solution of a second-order homogeneous equation?
9. Describe the form of the general solution of a second-order nonhomogeneous equation.
10. Describe the method of undetermined coefficients.
11. Describe the method of variation of parameters.

Practice Problems

Solve the differential equations in Problems 12–18.

12. $\dfrac{dy}{dx} = \sqrt{\dfrac{1 - y^2}{1 + x^2}}$

13. $\dfrac{x}{y^2}\, dx - \dfrac{x^2}{y^3}\, dy = 0$

14. $\dfrac{dy}{dx} = \dfrac{2x + y}{3x}$

15. $xy\, dy = (x^2 - y^2)\, dx$

16. $x^2\, dy - (x^2 + y^2)\, dx = 0$

17. $y'' + 2y' + 2y = \sin x$

18. $(3x^2 e^{-y} + y^{-2} + 2xy^{-3})\, dx + (-x^3 e^{-y} - 2xy^{-3} - 3x^2 y^{-4})dy = 0$

19. A spring with spring constant $k = 30$ lb/ft hangs in a vertical position with its upper end fixed and an 8-lb object attached to its lower end. The object is pulled down 4 in. from the equilibrium position of the spring and is then released. Find the displacement $x(t)$ of the object, assuming that air resistance is $0.8v$, where $v(t)$ is the velocity of the object at time t.

20. An *RLC* circuit has inductance $L = 0.1$ henry, resistance $R = 25$ ohms, and capacitance $C = 200$ microfarads (i.e., 200×10^{-6} farad). If there is a variable voltage source of $E(t) = 50 \cos 100t$ in the circuit, what is the current $I(t)$ at time t? Assume that when $t = 0$, there is no charge and no current flowing.

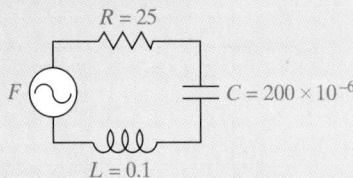

Supplementary Problems

1. a. Draw some isoclines of the differential equation
$$\frac{dy}{dx} = 2x - 3y$$
Use the direction field to sketch the particular solution that passes through $(0, 1)$.
   b. Use Euler's method to find a solution for $0 \le x \le 1$ and $h = 0.2$.

2. a. Draw some isoclines of the differential equation
$$\frac{dy}{dx} = x^2 y$$
Use the direction field to sketch the particular solution that passes through $(1, 2)$.
   b. Use Euler's method to find a solution for $1 \le x \le 2$ and $h = 0.2$.

Find the general solution of the given first-order differential equations in Problems 3–20.

3. $\dfrac{dy}{dx} = \dfrac{-4x}{y^3}$

4. $x\, dy + (3y - xe^{x^2})\, dx = 0$

5. $y^2\, dy - \left(x^2 + \dfrac{y^3}{x}\right)dx = 0$

6. $\dfrac{x}{y} dy + \left(\dfrac{3y}{x} - 5\right) dx = 0$

7. $y\dfrac{dy}{dx} = e^{2x - y^2}$

8. $dy = (3y + e^x + \cos x)\, dx$

9. $dy - (5y + e^{5x}\text{six } x)\, dx = 0$

10. $x^2\, dy = (x^2 - y^2)\, dx$

11. $(1 - xe^y)\, dy = e^y\, dx$

12. $xy\, dx + (1 + x^2)\, dy = 0$

13. $x\, dy = (y + \sqrt{x^2 - y^2})\, dx$

14. $2xye^{x^2} dx - e^{x^2}\, dy = 0$

15. $(y - xy)\, dx + x^3\, dy = 0$

16. $\dfrac{dy}{dx} = 2y \cot 2x + 3 \csc 2x$

17. $\left(4x^3y^3 + \dfrac{1}{x}\right)dx + \left(3x^4y^2 - \dfrac{1}{y}\right)dy = 0$

18. $(-y \sin xy + 2x)dx + (3y^2 - x \sin xy)\, dy = 0$

19. $\sin y\, dx + (e^x + e^{-x})\sin y\, dy = 0$

20. $\dfrac{dy}{dx} + 2y \cot x + \sin 2x = 0$

In Problems 21–28, find
   a. *the general solution y_h of the related homogeneous equation*
   b. *a particular solution y_p of the nonhomogeneous equation*

21. $y'' - 9y = 1 + x$
22. $y'' - 2y' + y = e^x$
23. $y'' - 5y' + 6y = x^2 e^x$
24. $y'' + 2y' + y = \sinh x$
25. $y'' - 3y' + 2y = x^3 e^x$
26. $14y'' + 29y' - 15y = \cos x$
27. $y'' + y = \tan^2 x$
28. $y'' + y = \sec x$

29. Solve $\dfrac{d^2y}{dx^2} = e^{-3x}$ subject to the initial conditions $y(0) = y'(0) = 0$.

30. Use the substitution $p = dy/dx$ to solve the differential equation
$$\dfrac{d^2y}{dx^2} = \left(\dfrac{dy}{dx}\right)^3 + \dfrac{dy}{dx}$$

31. Solve the system of differential equations
$$\dfrac{dx}{dt} = -2x \qquad \dfrac{dy}{dt} = -3y + 2x \qquad \dfrac{dz}{dt} = 3y$$
subject to the initial conditions $x(0) = 1, y(0) = 0, z(0) = 0$.

32. Find a curve $y = f(x)$ that passes through $(1, 1)$ and has the property that the y-intercept of the tangent line at each point $P(x, y)$ is equal to y^2.

33. Find a curve that passes through the point $(1, 2)$ and has the property that the length of the part of the tangent line between each point $P(x, y)$ on the curve and the y-axis of the tangent is equal to the y-intercept of the tangent line at P.

34. Find a curve that passes through $(1, 2)$ and has the property that the normal line at any point $P(x, y)$ on the curve and the line joining P to the origin form an isosceles triangle with the x-axis as its base.

Orthogonal trajectories. *Recall from Section 5.6 that the orthogonal trajectories of a given family of curves are another family with the following property: Every time a member of the second family intersects a member of the given family, the tangent lines to the two curves at the common point intersect at right angles. Find the orthogonal trajectories for the families of curves in Problems 35–38.*

35. $e^x + e^{-y} = C$
36. $3x^2 + 5y^2 = C$
37. $x^2 - y^2 = Cx$
38. $y = \dfrac{Cx}{x^2 + 1}$

39. Use reduction of order to find the general solution of the Legendre equation
$$(1 - x^2)y'' - 2xy' + 2y = 0$$
given that $y_1 = x$ is one solution.

40. Find functions $x = x(t)$ and $y = y(t)$ that satisfy the linear system
$$\dfrac{dx}{dt} = x + y \qquad \dfrac{dy}{dt} = x^2 - y^2$$
Sketch the solution curve $(x(t), y(t))$ that contains the point $(0, 2)$. *Hint:* Note that $\dfrac{dy}{dx} = \dfrac{y'(t)}{x'(t)}$.

41. Solve the second-order differential equation $xy'' + 2y' = x$ by setting $p = y'$ to convert it to a first-order equation in p.

42. Find an equation for a curve for which the radius of curvature is proportional to the slope of the tangent line at each point $P(x, y)$.

43. Solve the differential equation
$$\dfrac{dx}{dy} - \dfrac{x}{y} = ye^y$$

44. **MODELING PROBLEM** A ship weighing 64,000 tons [mass = $(64,000)(2,000)/32$ slugs] starts from rest and is driven by the constant thrust of its propellers. Suppose the propellers supply 250,000 lb of thrust and the resistance of the water is $12,000v$ lb, where $v(t)$ is the velocity of the ship. Set up and solve a differential equation for $v(t)$.

45. MODELING PROBLEM   A 10-ft uniform chain (see Figure 15.12) of mass m is hanging over a peg so that 3 ft are on one side and 7 ft are on the other. Set up and solve a differential equation to find how long it takes for the chain to slide off the peg. Neglect the friction.

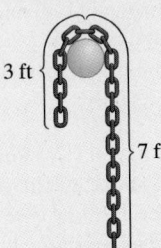

3 ft

7 ft

■ **FIGURE 15.12**   Problem 45

46. When an object weighing 10 lb is suspended from a spring, the spring is stretched 2 in. from its equilibrium position. The upper end of the spring is given a motion of $y = 2(\sin t + \cos t)$ ft. Find the displacement $x(t)$ of the object.

47. MODELING PROBLEM   A horizontal beam is freely supported at both ends, as shown in Figure 15.13.

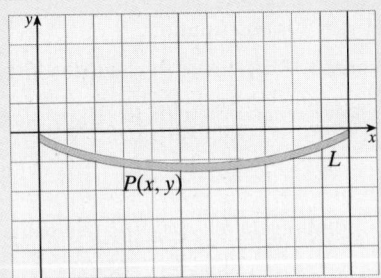

■ **FIGURE 15.13**   Problem 47

Suppose the load on the beam is W pounds per foot and one end of the beam is at the origin and the other end of the beam is at $(L, 0)$. Then the deflection y of the beam at point x is modeled by the differential equation

$$EI\frac{d^2y}{dx^2} = WLx - \frac{1}{2}Wx^2$$

for positive constants E and I.
a. Solve this equation to obtain the deflection $y(x)$. Note that $y(0) = y(L) = 0$.
b. What is the maximum deflection in the beam?

48. Consider the **Clairaut equation**

$$y = xy' + f(y')$$

a. Differentiate both sides of this equation to obtain.

$$[x + f'(y')]y'' = 0$$

Because one of the two factors in the product on the left must be 0, we have two cases to consider. What is the solution if $y'' = 0$? This is the *general solution.*
b. What is the solution if

$$x + f'(y') = 0$$

(This solution is called the *singular solution.*) *Hint*: Use the parametrization $y' = t$.
c. Find the general and singular solutions for the Clairaut equation

$$y = xy' + \sqrt{4 + (y')^2}$$

49. A differential equation of the form

$$x^2y'' + Axy' + By = F(x)$$

with $x \neq 0$ is called an **Euler equation.** To find a general solution of the related homogeneous equation

$$x^2y'' + Axy' + By = 0$$

we assume that there are solutions of the form $y = x^m$.
a. Show that m must satisfy

$$m^2 + (A - 1)m + B = 0$$

This is the characteristic equation for the Euler equation.
b. **Distinct real roots.**   Characterize the solutions of the related homogeneous equation in the case where

$$(A - 1)^2 > 4B$$

c. **Repeated real roots.**   Characterize the solutions of the related homogeneous equation in the case where

$$(A - 1)^2 = 4B$$

d. **Complex conjugate roots.**   Suppose

$$(A - 1)^2 < 4B$$

and that $\alpha \pm \beta i$ are the roots of the characteristic equation. Verify that $y_1 = x^\alpha \cos(\beta \ln |x|)$ is one solution. Use reduction of order to find a second solution y_2, then characterize the general solution.

50. Find the general solution of the Euler equation

$$x^2y'' + 7xy' + 9y = \sqrt{x}$$

51. In certain biological studies, it is important to analyze *predator–prey* relationships. Suppose $x(t)$ is the prey population and $y(t)$ is the predator population at time t. Then these populations change at rates

$$\frac{dx}{dt} = a_{11}x - a_{12}y \qquad \frac{dy}{dt} = a_{21}x - a_{22}y$$

where a_{11} is the natural growth rate of the prey, a_{12} is the predation rate, a_{21} measures the food supply of the predators, and a_{22} is the death rate of the predators. Outline a procedure for solving this system.

52. Consider the almost homogeneous differential equation

$$\frac{dy}{dx} = f\left(\frac{ax + by + c}{rx + sy + t}\right)$$

with $as = br$.

a. Let $u = \dfrac{ax + by}{a}$. Show that $u = \dfrac{rx + sy}{r}$ and

$\dfrac{dy}{dx} = \dfrac{a}{b}\left(\dfrac{du}{dx} - 1\right)$.

b. Verify that by making the change of variable suggested in part **a,** you can rewrite the given differential equation in the separable form

$$\frac{du}{1 + \dfrac{b}{a}f\left(\dfrac{au + c}{ru + t}\right)} = dx$$

c. Use the procedure outlined in parts **a** and **b** to solve the almost homogeneous differential equation

$$\frac{dy}{dx} = \frac{2x + y - 3}{4x + 2y + 5}$$

53. **SPY PROBLEM** "You must be Rottona!" the Spy cleverly announces to the red Purity after she correctly answers the question posed in Problem 59, Section 15.1. "Purity is too innocent to know so much physics." He starts to tie the wrists of the girl in red.

"You idiot!" she shrieks. "You ... your father wears leisure suits!" Likely there was more of the same, but a blow from behind provides the Spy with an early nap. As he's waking, it comes to him—the girl in red is Purity! He stumbles from the room into an open field where Rottona and Blohardt are both chasing Purity. His mathematical mind visualizes the field as a coordinate plane with himself at $(100, 0)$, Purity at $(0, 0)$, running along the y-axis, and Rottona and Blohardt at $(0, -18)$ and $(80, o)$, respectively (units in yards). The Spy, Rottona, and Blohardt all run directly toward Purity. If Rottona runs at 14 ft/s, Blohardt at 15 ft/s, and Purity at 12 ft/s, how fast must the Spy run to the innocent maid to get there before either of the bad guys?

54. **PUTNAM EXAMINATION PROBLEM**   Find all solutions of the equation

$$yy'' - 2(y')^2 = 0$$

that pass through the point $x = 1$, $y = 1$.

55. **PUTNAM EXAMINATION PROBLEM**   A coast artillery gun can fire at any angle of elevation between $0°$ and $90°$ in a fixed vertical plane. If air resistance is neglected and the muzzle velocity is constant $(v(0) = v_0)$, determine the set H of points in the plane and above the horizontal that can be hit.

56. **PUTNAM EXAMINATION PROBLEM**   Show that

$$x + \frac{2}{3}x^3 + \frac{2 \cdot 4}{3 \cdot 5}x^5 + \frac{2 \cdot 4 \cdot 6}{3 \cdot 5 \cdot 7}x^7 + \cdots = \frac{\sin^{-1}x}{\sqrt{1 - x^2}}$$

Save the Perch Project

This project is to be done in groups of three or four students. Each group will submit a single written report.

Happy Valley Pond is currently populated by yellow perch. A map is shown below. Water flows into the pond from two springs and evaporates from the pond, as shown by the table below.

Spring	Dry Season	Rainy Season
A	50 gal/h	60 gal/h
B	60 gal/h	75 gal/h
Evaporation:		
	110 gal/h	75 gal/h

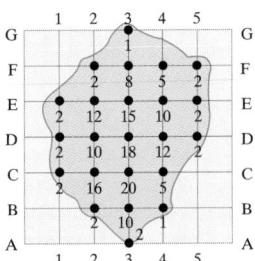

Happy Valley Pond is fed by two springs:

The overflow goes over the dam into Bubbling Brook:

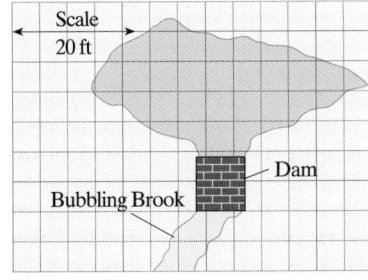

Unfortunately, spring B has become contaminated with salt and is now 10% salt, which means that 10% of a gallon of water from spring B is salt. The yellow perch will start to die if the concentration of salt in the pond rises to 1%. Assume that the salt will not evaporate but will mix thoroughly with the water in the pond. There was no salt in the pond before the contamination of spring B. Your group has been called upon by the Happy Valley Bureau of Fisheries to try to save the perch.

Your paper is not limited to the following questions, but should include the number of gallons of water in the pond when the water level is exactly even with the top of

---

*This group project is courtesy of Diane Schwartz from Ithaca College, New York.

Among all the mathematical disciplines the theory of differential equations is the most important . . . . It furnishes the explanation of all those elementary manifestations of nature which involve time . . . .

SOPHUS LIE
LEIPZIGER BERICHTE, 47 (1895).

the spillover dam. The following table gives a series of measurements of the depth of the pond at the indicated points when the water level was exactly even with the top of the spillover dam.

DEPTH OF HAPPY VALLEY POND (Location/depth)

A 3/8 ft	B 2/2 ft	B 3/10 ft	B 4/1 ft	C 1/2 ft	C 2/16 ft	C 3/20 ft
C 4/5 ft	D 1/2 ft	D 2/10 ft	D 3/18 ft	D 4/12 ft	D 5/2 ft	D 3/18 ft
D 4/12 ft	D 5/2 ft	E 1/2 ft	E 2/12 ft	E 3/15 ft	E 4/10 ft	E 5/2 ft
F 2/2 ft	F 3/8 ft	F 4/5 ft	F 5/2 ft	G 3/1 ft		

Let $t = 0$ hours correspond to the time when spring B became contaminated. Assume it is the dry season and that at time $t = 0$ the water level of the pond was exactly even with the top of the spillover dam. Write a differential equation for the amount of salt in the pond after t hours. Draw a graph of the amount of salt in the pond versus time for the next 3 mo. How much salt will there be in the pond in the long run, and do the fish die? If so, when do they start to die? It is very difficult to find where the contamination of spring B originates, so the Happy Valley Bureau of Fisheries proposed to flush the pond by running 100 gal of pure water per hour through the pond. Your report should include an analysis of this plan and any modifications or improvements that could help save the perch.

Chapters 12–15   *Cumulative Review*

■ **TABLE 15.2**   **Comparison of Important Integral Theorems**

Riemann Integral (Section 5.3)	Line Integral (Section 14.2)

$$\int_a^b f(x)\, dx$$

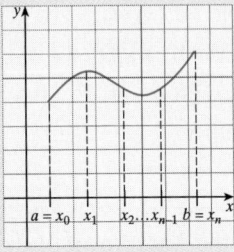

Subdivisions on the x-axis

$$\int_C f(x, y, z)\, dx$$

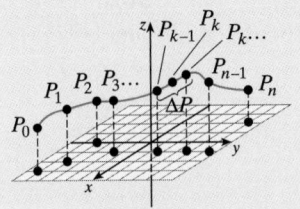

Subdivisions on a curve C in space

Fundamental Theorem of Calculus (Section 5.4)

$$\int_a^b f(x)\, dx = F(b) - F(a)$$

F is an antiderivative of f.

Fundamental Theorem for Line Integrals (Section 14.3)

$$\int_C \mathbf{F} \cdot d\mathbf{R} = f(Q) - f(P)$$

if $\mathbf{F}$ is conservative with scalar potential f; that is, $\nabla f = \mathbf{F}$.

Green's Theorem (Section 14.4); applies to $\mathbb{R}^2$ (two dimensions): $\mathbf{F}(x, y) = M(x, y)\mathbf{i} + N(x, y)\mathbf{j}$

$$\int_C \mathbf{F} \cdot d\mathbf{R} = \int_C (M\, dx + N\, dy) = \iint_D \left(\frac{\partial N}{\partial x} - \frac{\partial M}{\partial y} \right) dA = \iint_D (\text{curl } \mathbf{F} \cdot \mathbf{k})\, dA$$

If $\mathbf{F}$ is conservative,

$$\int_C \mathbf{F} \cdot \mathbf{N}\, ds = \int_C \nabla f \cdot \mathbf{N}\, ds = \int_C \frac{\partial f}{\partial n}\, ds = \iint_D \nabla^2 f\, dA = \iint_D \left(\frac{\partial M}{\partial x} + \frac{\partial N}{\partial y} \right) dA = \iint_D \text{div } \mathbf{F}\, dA$$

Double integral (Section 13.1)	Surface integral (Section 14.5)

$$\iint_R f(x, y)\, dA$$

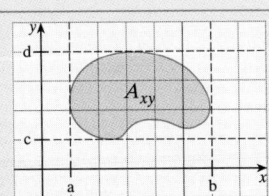

Partition of R into mn cells in the xy-plane

$$\iint_S g(x, y, z)\, dS = \iint_{R_{xy}} g(x, y, z)\sqrt{g_x^2 + g_y^2 + 1}\, dA_{xy}$$

where $z = f(x, y)$

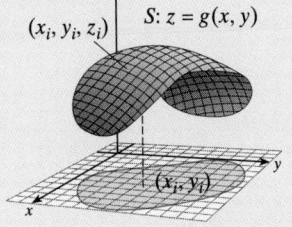

Partition of the surface S into n subregions

Stokes' Theorem (Section 14.6)

$$\int_C \mathbf{F} \cdot d\mathbf{R} = \iint_S (\text{curl } \mathbf{F} \cdot \mathbf{N})\, dS$$

Divergence Theorem (Section 14.7; also known as *Gauss's theorem*)

$$\iint_S \mathbf{F} \cdot \mathbf{N}\, dS = \iiint_D \text{div } \mathbf{F}\, dV$$

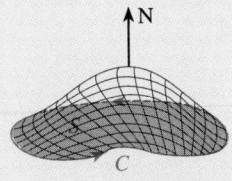

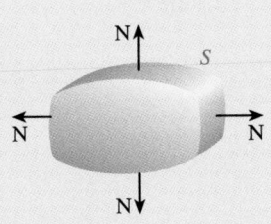

Chapters 12–15   Cumulative Review Problems

1. ■ **What Does This Say?** Suppose you tell a fellow college student that you are about to finish a calculus course. The student has not had any mathematics beyond high school and asks you, "What is calculus?" Answer this question using your own words.

2. ■ **What Does This Say?** There are three great ideas in calculus: the limit, the derivative, and the integral. In your own words, explain each of these concepts.

3. ■ **What Does This Say?** Chapters 12–14 were concerned with functions of several variables. This is often referred to as **multivariable calculus.** In your own words, discuss what is meant by multivariable calculus.

4. ■ **What Does This Say?** In your own words, outline a procedure for solving the types of differential equations discussed in Chapter 15.

Find $f_x, f_y,$ and f_{xy} for the functions whose equations are given in Problems 5–10.

5. $f(x, y) = 2x^2 + xy - 5y^3$

6. $f(x, y) = x^2 e^{y/x}$

7. $f(x, y) = \dfrac{x^2 - y^2}{x - y}$

8. $f(x, y) = y \sin^2 x + \cos xy$

9. $f(x, y) = e^{x+y}$ 10. $f(x, y) = \dfrac{x^2 + y^2}{x - y}$

11. Let $f(x, y) = \dfrac{xy}{(x^2 + y^2)^2}$. Find $f_{xx} + f_{yy}$.

Evaluate the integrals in Problems 12–19.

12. $\displaystyle\int_0^1 \int_x^{2x} e^{y-x}\, dy\, dx$

13. $\displaystyle\int_0^4 \int_0^{\sqrt{x}} 3x^5\, dy\, dx$

14. $\displaystyle\int_0^1 \int_0^z \int_y^{y-z} (x + y + z)\, dx\, dy\, dz$

15. $\displaystyle\int_0^{15\pi} \int_0^\pi \int_0^{\sin\phi} \rho^3 \sin\phi\, d\rho\, d\theta\, d\phi$

16. $\displaystyle\iint_R e^{x+y}\, dA$; $R: 0 \le x \le 1, 0 \le y \le 1$

17. $\displaystyle\iint_R y e^{xy}\, dA$; $R: 0 \le x \le 1, 0 \le y \le 2$

18. $\displaystyle\iint_R \sin(x + y)\, dA$; $R: 0 \le x \le \dfrac{\pi}{2}, 0 \le y \le \dfrac{\pi}{4}$

19. $\displaystyle\iint_R x \sin xy\, dA$; $R: 0 \le x \le \pi, 0 \le y \le 1$

Evaluate the line integrals in Problems 20–23.

20. $\displaystyle\int_C (y^2 z\, dx + 2xyz\, dy + xy^2\, dz)$, where C is any path from $(0, 0, 0)$ to $(1, 1, 1)$

21. $\displaystyle\int_C (5xy\, dx + 10yz\, dy + z\, dz)$, where C is given by $x = t^2, y = t, z = 2t^3$ for $0 \le t \le 1$

22. $\displaystyle\int_C \mathbf{F} \cdot d\mathbf{R}$, where $\mathbf{F} = yz\mathbf{i} - x\mathbf{k}$, and C is the boundary of the triangle $(1, 1, 1), (1, 0, 1), (0, 0, 1)$, traversed once counterclockwise as viewed from the origin

23. $\displaystyle\int_C (x^3 + y^3)\, ds$, where C is given by $\mathbf{R}(t) = (\cos^3 t)\mathbf{i} + (\sin^3 t)\mathbf{j}$, for $0 \le t \le 2\pi$

Solve the differential equations in Problems 24–30.

24. $\dfrac{dy}{dx} + \dfrac{y}{2x} = \sqrt{x}\, e^x$

25. $x\dfrac{dy}{dx} + 3y = \sin x^3$

26. $y'' + 4y = 2\sin x$

27. $y'' + y' - 2y = x^3 + x^2 - 2x + 5$

28. $y'' + 4y = \sin 2x$

29. $y' + xy = x + e^{-x^2/2}$ where $y = -1$ when $x = 0$

30. $xy' - 2y = x^2$, where $y = 5$ when $x = 1$

31. Find the equations for the tangent plane and the normal line to $z = x^2 + y^2 + \sin xy$ at $P = (0, 2, 4)$.

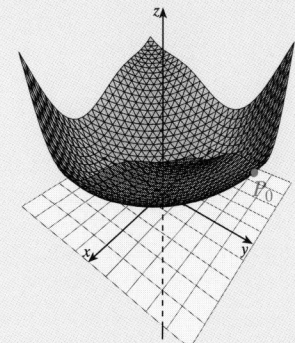

32. What are the dimensions of the rectangular box of fixed volume V_0 that has minimum surface area?

33. **MODELING PROBLEM**   A manufacturer is planning to sell a new product at the price of $350 per unit and estimates that if x thousand dollars is spent on development and y thousand dollars is spent on promotion, consumers will buy approximately

$$\frac{250y}{y + 2} + \frac{100x}{x + 5}$$

units of the product. If manufacturing costs for this product are $150 per unit, how much should the manufacturer spend on development and how much on promotion to generate the largest possible profit, given the following circumstances?

a. Unlimited funds are available.

b. The manufacturer has only $11,000 to spend on development and promotion of the new product.

34. MODELING PROBLEM   A heat-seeking missile moves in a portion of space where the temperature (in degrees Celsius) at the point (x, y, z) is given by

$$T(x, y, z) = \frac{1}{10}(x^2 + y + z^3)$$

with $x, y,$ and z in kilometers.

a. Find the rate at which the temperature is changing as the missile moves from the point $P_0(-2, 9, 1)$ toward $Q(1, -3, 5)$.

b. If the missile is at P_0, in what direction will it travel to maximize the rate of heat increase?

c. What is the maximal rate of increase (in degrees Celsius per kilometer)?

35. Use double polar integration to find the area inside the cardioid $r = 1 - \cos \theta$ and outside the circle $r = 1$.

36. Find the volume of the solid bounded above by the surface $z = x^2 + y^2 + 1$ and below by the circular disk $x^2 + y^2 \le 1$ in the xy-plane.

37. Find the surface area of that portion of the paraboloid $z = x^2 + y^2$ that lies below the plane $z = 16$.

38. Find the center of mass of the lamina whose shape is the region inside the circle $x^2 + y^2 = 4$ in the first quadrant, given that the density at any point (x, y) is $\delta(x, y) = x + y$. *Hint:* Reverse the order of integration.

39. A force field $\mathbf{F}(x, y) = (x - 2y)\mathbf{i} + (y - 2x)\mathbf{j}$ acts on an object moving in the plane. Show that $\mathbf{F}$ is conservative, and find a scalar potential for $\mathbf{F}$. How much work is done as the object moves from $(1, 0)$ to $(0, 1)$ along any path connecting these points?

40. A particle of weight w is acted on only by the constant gravitational force $\mathbf{F} = -w\mathbf{k}$. How much work is done in moving the weight along the helical path given by $x = \cos t, y = \sin t, z = t$ for $0 \le t \le 2\pi$?

Appendices

Introduction to the Theory of Limits

In Section 2.2, we defined the limit of a function as follows:

The notation

$$\lim_{x \to c} f(x) = L$$

is read "the limit of $f(x)$ as x approaches c is L" and means that the function values $f(x)$ can be made arbitrarily close to L by choosing x sufficiently close to c but not equal to c.

This informal definition was valuable because it gave you an intuitive feeling for the limit of a function and allowed you to develop a working knowledge of this fundamental concept. For theoretical work, however, this definition will not suffice, because it gives no precise, quantifiable meaning to the terms "arbitrarily close to L" and "sufficiently close to c." The following definition, derived from the work of Cauchy and Weierstrass, gives precision to the limit definition, and was also first stated in Section 2.2.

> **Limit of a Function (Formal definition)**
>
> The limit statement
>
> $$\lim_{x \to c} f(x) = L$$
>
> means that for each $\epsilon > 0$, there corresponds a number $\delta > 0$ with the property that
>
> $$|f(x) - L| < \epsilon \quad \text{whenever} \quad 0 < |x - c| < \delta$$

Behind the formal language is a fairly straightforward idea. In particular, to establish a specific limit, say $\lim_{x \to c} f(x) = L$, a number $\epsilon > 0$ is chosen first to establish a desired degree of proximity to L, and then a number $\delta > 0$ is found that determines how close x must be to c to ensure that $f(x)$ is within ϵ units of L.

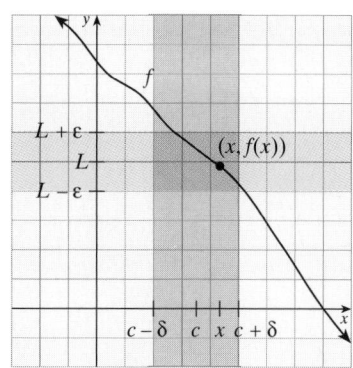

■ **FIGURE A.1**   The epsilon–delta definition of limit

The situation is illustrated in Figure A.1, which shows a function that satisfies the conditions of the definition. Notice that whenever x is within δ units of c (but not equal to c), the point $(x, f(x))$ on the graph of f must lie in the rectangle formed by the intersection of the horizontal band of width 2ϵ (blue screen) centered at L and the vertical band of width 2δ (gray region) centered at c. The smaller the ϵ-interval around the proposed limit L, generally the smaller the δ-interval will need to be for $f(x)$ to lie in the ϵ-interval. If such a δ can be found no matter how small ϵ is, then L must be the limit.

THE BELIEVER/DOUBTER FORMAT

The limit process can be thought of as a "contest" between a "believer" who claims that $\lim_{x \to c} f(x) = L$ and a "doubter" who disputes this claim. The contest begins with the doubter choosing a positive number ϵ so that whenever x is a number (other than c) within δ units of c (the gray region), the corresponding function value $f(x)$ is within ϵ units of L (the blue region). Naturally, the doubter tries to choose ϵ so small that no matter what δ the believer chooses, it will not be possible to satisfy the accuracy requirement. When will the doubter win and when will the believer win the argument? As you can see from Figure A.1, if the believer has the "correct limit" L, it will be in the intersection (the portion that is both blue and gray) no matter what ϵ the doubter chooses. On the other hand, if the believer has an "incorrect limit" as shown in Figure A.2, the limit may be in the double-screened portion for some choices of ϵ, but for other choices of ϵ the doubter can force the believer to be the loser of this contest by prohibiting the believer from making a choice within the double-screened portion.

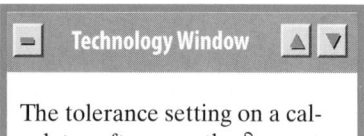

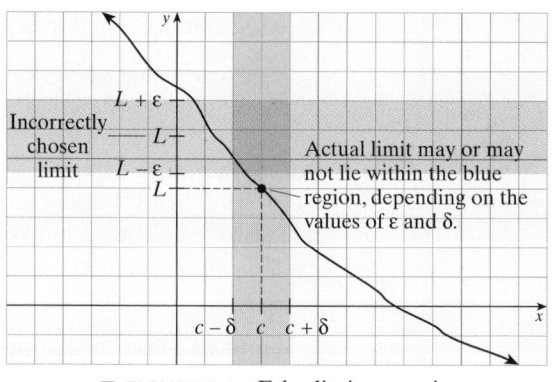

■ **FIGURE A.2**   False limit scenario

To avoid an endless series of ϵ–δ challenges, the believer usually tries to settle the issue by producing a formula relating the choice of δ to the doubter's ϵ that will satisfy the requirement no matter what ϵ is chosen. The believer/doubter format is used in Example 1 to verify a given limit statement (believer "wins") and in Example 2 to show that a certain limit does not exist (doubter "wins").

EXAMPLE 1   *Verifying a limit claim (believer wins)*

Show that $\lim_{x \to 2} (2x + 1) = 5$.

Solution   To verify the given limit statement, we begin by having the doubter choose a positive number ϵ. Before picking δ, the believer might entertain the thought process shown in the following box.

Write $|f(x) - L|$ in terms of $|x - c|$ (where $c = 2$) as follows:

$$|f(x) - L| = |(2x + 1) - 5| = |2x - 4| = 2|x - 2|$$

Thus, if $0 < |x - c| < \delta$ or $0 < |x - 2| < \delta$, then

$$|f(x) - L| < 2\delta$$

The believer *wants* $|f(x) - L| < \epsilon$, so we see that the believer should choose $2\delta = \epsilon$, or $\delta = \frac{\epsilon}{2}$.*

With the information shown in this box, we can now make the following argument, which uses the formal definition of limit:

Let $\epsilon > 0$ be given. Choose $\delta = \dfrac{\epsilon}{2} > 0$. It follows that if

$$0 < |x - 2| < \delta = \frac{\epsilon}{2}, \quad \text{then}$$

$$|(2x + 1) - 5| = 2|x - 2| < 2\left(\frac{\epsilon}{2}\right) = \epsilon$$

Thus, the conditions of the definition of limit are satisfied, and we have

$$\lim_{x \to 2} (2x + 1) = 5$$

This is shown graphically in Figure A.3a.

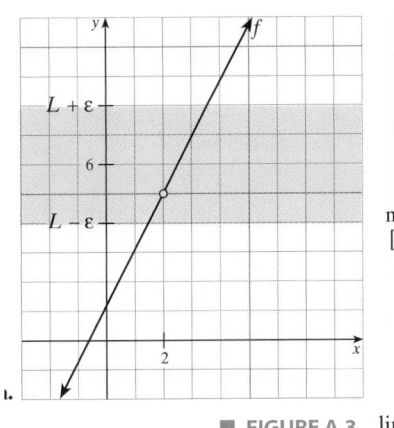
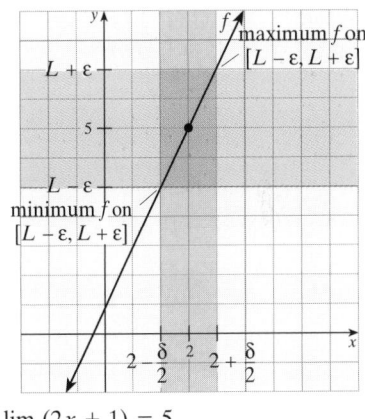

■ **FIGURE A.3** $\displaystyle \lim_{x \to 2} (2x + 1) = 5$

Notice that no matter what ϵ is chosen by the doubter, by choosing a number that is one-half of that ϵ, the believer will force the function to stay within the double-screened portion of the graph, as shown in Figure A.3. ■

EXAMPLE 2   *Disproving a limit claim (doubter wins)*

Determine whether $\displaystyle \lim_{x \to 2} \frac{2x^2 - 3x - 2}{x - 2} = 6$.

---

*In fact there are other choices that would work; for example, $\delta = \dfrac{\epsilon}{a}$ where $a > 2$.

Solution Once again, the doubter will choose a positive number ϵ and the believer must respond with a δ. As before, the believer does some preliminary work with $f(x) - L$:

$$
\left| \frac{2x^2 - 3x - 2}{x - 2} - 6 \right| = \left| \frac{2x^2 - 3x - 2 - 6x + 12}{x - 2} \right|
$$

$$
= \left| \frac{2x^2 - 9x + 10}{x - 2} \right|
$$

$$
= \left| \frac{(2x - 5)(x - 2)}{x - 2} \right|
$$

$$
= \left| 2x - 5 \right|
$$

The believer wants to write this expression in terms of $x - c = x - 2$. This example does not seem to "fall into place" as did Example 1. Suppose the doubter chooses $\epsilon = 1.5$ (not a very small ϵ). Then the believer must find a δ so that whenever x is within this δ distance of 2, $f(x)$ will be within 1.5 units of 6. This is shown in Figure A.4a.

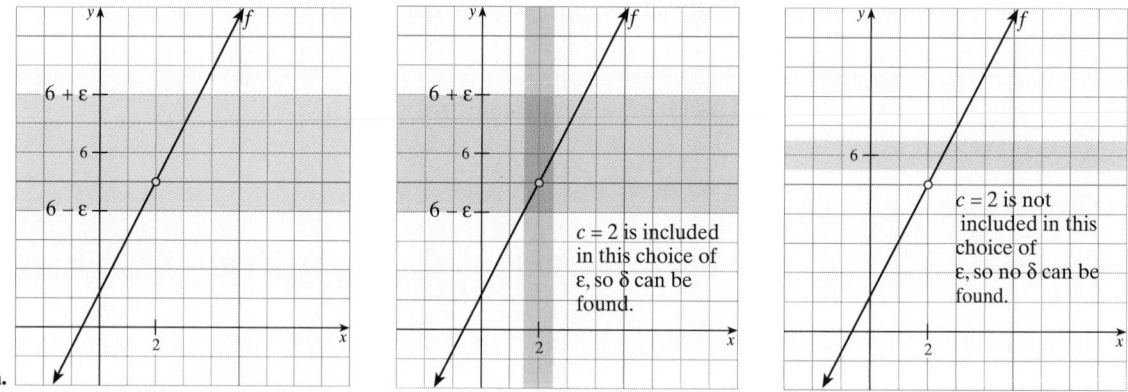

■ **FIGURE A.4** Example of an incorrectly chosen limit

However, if the doubter chooses $\epsilon = 0.5$ (still not a very small ϵ), then the believer is defeated, because no δ can be found, as shown in Figure A.4b. ■

The believer/doubter format is a useful device for dramatizing the way certain choices are made in epsilon–delta arguments, but it is customary to be less "chatty" in formal mathematical proofs.

EPSILON–DELTA PROOFS

The following examples illustrate epsilon–delta proofs, two in which the function has a limit and one in which it does not.

> **EXAMPLE 3** *An epsilon–delta proof of a limit of a linear function*

Show that $\lim\limits_{x \to 2} (4x - 3) = 5.$

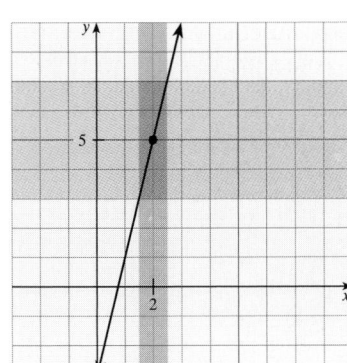

Solution   We have

$$\left|f(x) - L\right| = \left|4x - 3 - 5\right|$$
$$= \left|4x - 8\right|$$
$$= \underbrace{4\left|x - 2\right|}$$

This must be less than ϵ
whenever $\left|x - 2\right| < \delta$.

Choose $\delta = \dfrac{\epsilon}{4}$; then

$$\left|f(x) - L\right| = 4\left|x - 2\right| < 4\delta = 4\left(\frac{\epsilon}{4}\right) = \epsilon$$ ∎

EXAMPLE 4   *An epsilon–delta proof of a limit of a rational function*

Show that $\displaystyle\lim_{x \to 2} \frac{x^2 - 2x + 2}{x - 4} = -1$.

Solution   We have

$$\left|f(x) - L\right| = \left|\frac{x^2 - 2x + 2}{x - 4} - (-1)\right|$$
$$= \left|\frac{x^2 - x - 2}{x - 4}\right|$$
$$= \underbrace{\left|x - 2\right| \left|\frac{x + 1}{x - 4}\right|}$$

This must be less than the given ϵ
whenever x is near 2.

Certainly $\left|x - 2\right|$ is small if x is near 2, and the factor $\left|\dfrac{x + 1}{x - 4}\right|$ is not large (it is close to $\tfrac{3}{2}$). Note that if $\left|x - 2\right|$ is small, it is reasonable to assume

$$\left|x - 2\right| < 1 \quad \text{so that} \quad 1 < x < 3$$

By inspection of the graph, the largest value of the fraction $\dfrac{x + 1}{x - 4}$ is

$$\frac{1 + 1}{1 - 4} = \frac{2}{-3}$$

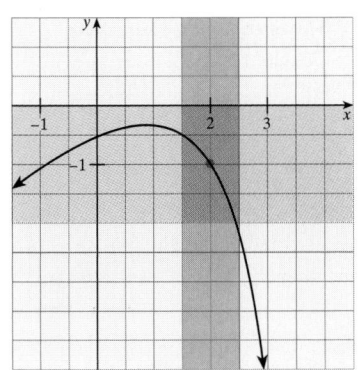

and the smallest value of the fraction $\dfrac{x + 1}{x - 4}$ is

$$\frac{3 + 1}{3 - 4} = \frac{4}{-1} = -4$$

Thus, if $\left|x - 2\right| < 1$, we have

$$-4 < \frac{x + 1}{x - 4} < -\frac{2}{3}$$

so that

$$-4 < \frac{x + 1}{x - 4} < 4 \qquad \text{Because } -\tfrac{2}{3} < 4$$

and

$$\left| \frac{x+1}{x-4} \right| < 4$$

Now let $\epsilon > 0$ be given. If simultaneously

$$|x-2| < \frac{\epsilon}{4} \quad \text{and} \quad \left| \frac{x+1}{x-4} \right| < 4$$

then

$$|f(x) - L| = |x-2| \left| \frac{x+1}{x-4} \right| < \frac{\epsilon}{4}(4) = \epsilon$$

Thus, we have only to take δ to be the smaller of the two numbers 1 and $\epsilon/4$ to guarantee that

$$|f(x) - L| < \epsilon$$

That is, given $\epsilon > 0$, choose δ to be the smaller of the numbers 1 and $\epsilon/4$. We write this as $\delta = \min(1, \epsilon/4)$. ∎

EXAMPLE 5   *An epsilon–delta proof that a limit does not exist*

Show that $\displaystyle\lim_{x \to 0} \frac{1}{x}$ does not exist.

Solution   Let $f(x) = \dfrac{1}{x}$ and L be any number. Suppose that $\displaystyle\lim_{x \to 0} f(x) = L$. Look at the graph of f, as shown in Figure A.5. It would seem that no matter what value of ϵ is chosen, it would be impossible to find a corresponding δ. Consider the absolute value expression required by the definition of limit: If

$$|f(x) - L| < \epsilon, \quad \text{or, for this example,} \quad \left| \frac{1}{x} - L \right| < \epsilon$$

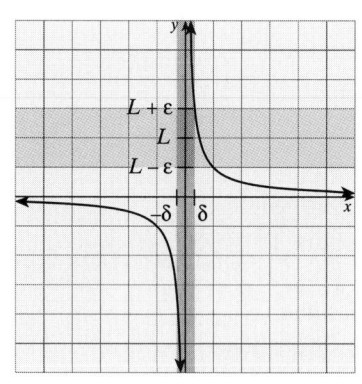

■ **FIGURE A.5** $\displaystyle\lim_{x \to 0} \frac{1}{x}$

then

$$-\epsilon < \frac{1}{x} - L < \epsilon \qquad \text{Property of absolute value (Table 1.1, p. 3)}$$

and

$$L - \epsilon < \frac{1}{x} < L + \epsilon$$

If $\epsilon = 1$ (not a particularly small ϵ), then

$$\left| \frac{1}{x} \right| < |L| + 1$$

$$|x| > \frac{1}{|L| + 1}$$

which proves (since L was chosen arbitrarily) that $\displaystyle\lim_{x \to 0} \frac{1}{x}$ does not exist. In other words, since $|x|$ can be chosen very small, $\dfrac{1}{|x|}$ will be very large, and it will be impossible to squeeze $\dfrac{1}{x}$ between $L - \epsilon$ and $L + \epsilon$ for any L. ∎

SELECTED THEOREMS WITH FORMAL PROOFS

Next, we shall prove several theoretical results using the formal definition of the limit.

The next two theorems are useful tools in the development of calculus. The first states that the points on a graph that are on or above the x-axis cannot possibly "tend toward" a point *below* the axis, as shown in Figure A.6.

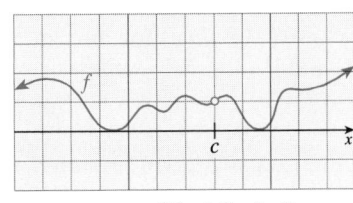

If $f(x) \geq 0$ for all x near c, then
$$\lim_{x \to c} f(x) \geq 0.$$

■ **FIGURE A.6** Limit limitation theorem

───────────
THEOREM A.1   *Limit limitation theorem*
───────────

Suppose $\lim_{x \to c} f(x)$ exists and $f(x) \geq 0$ throughout an open interval containing the number c, except possibly at c itself. Then

$$\lim_{x \to c} f(x) \geq 0$$

Proof   Let $L = \lim_{x \to c} f(x)$. To show that $L \geq 0$, assume the contrary. That is, assume $L < 0$. According to the definition of limit (with $\epsilon = -L$), there is a number $\delta > 0$ such that

$$\left| f(x) - L \right| < -L \quad \text{whenever} \quad 0 < \left| x - c \right| < \delta$$

In particular,

$$f(x) - L < -L$$

or

$$f(x) < 0$$

Thus,

$$f(x) < 0 \quad \text{whenever} \quad 0 < \left| x - c \right| < \delta$$

However, this contradicts the hypothesis that $f(x) \geq 0$ throughout an open interval containing c. Therefore, we reject the assumption that $L < 0$ and conclude that $L > 0$, as required.

Useful information about the limit of a given function f can often be obtained by examining other functions that bound f from above and below. For example, in Section 2.2 we found

$$\lim_{x \to 0} \frac{\sin x}{x} = 1$$

by using a table, and proved the limit (Theorem 2.2) by first showing that

$$\cos x \leq \frac{\sin x}{x} \leq 1$$

for all x near 0 and then noting that since $\cos x$ and 1 both tend toward 1 as x approaches 0, the function

$$\frac{\sin x}{x}$$

WARNING ➤   It may seem reasonable to conjecture that if $f(x) > 0$ throughout an open interval containing c, then $\lim_{x \to c} f(x) > 0$. This is not necessarily true, and the most that can be said in this situation is that $\lim_{x \to c} f(x) \geq 0$, if the limit exists. For example, if

$$f(x) = \begin{cases} x^2 & \text{for } x \neq 0 \\ 1 & \text{for } x = 0 \end{cases}$$

then $f(x) > 0$ for all x, but $\lim_{x \to 0} f(x) = 0$. ◄

which is "squeezed" between them, must converge to 1 as well. Theorem A.2 provides the theoretical basis for this method of proof.

***

THEOREM A.2   *The squeeze theorem*

If $g(x) \le f(x) \le h(x)$ for all x in an open interval containing c (except possibly at c itself) and if

$$\lim_{x \to c} g(x) = \lim_{x \to c} h(x) = L$$

then $\lim_{x \to c} f(x) = L$. (This is stated, without proof, in Section 2.3.)

Proof   Let $\epsilon > 0$ be given. Since $\lim_{x \to c} g(x) = L$ and $\lim_{x \to c} h(x) = L$, there are positive numbers δ_1 and δ_2 such that

$$\left| g(x) - L \right| < \epsilon \quad \text{and} \quad \left| h(x) - L \right| < \epsilon$$

whenever

$$0 < \left| x - c \right| < \delta_1 \quad \text{and} \quad 0 < \left| x - c \right| < \delta_2$$

respectively. Let δ be the smaller of the numbers δ_1 and δ_2. Then, if x is a number that satisfies $0 < \left| x - c \right| < \delta$, we have

$$-\epsilon < g(x) - L \le f(x) - L \le h(x) - L < \epsilon$$

and it follows that $\left| f(x) - L \right| < \epsilon$. Thus, $\lim_{x \to c} f(x) = L$, as claimed.

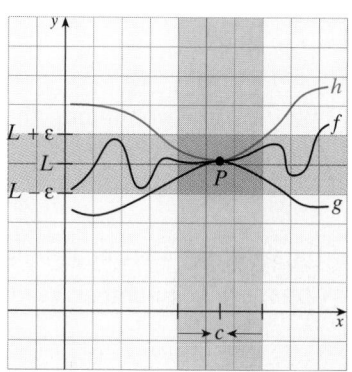

■ **FIGURE A.7**   The squeeze theorem

The geometric interpretation of the squeeze theorem is shown in Figure A.7. Notice that since $g(x) \le f(x) \le h(x)$, the graph of f is "squeezed" between those of g and h in the neighborhood of c. Thus, if the bounding graphs converge to a common point P as x approaches c, then the graph of f must also converge to P as well.

A.1 Problem Set

B *In Problems 1–6, use the believer/doubter format to prove or disprove the given limit statement.*

1. $\lim_{x \to 1} (2x - 5) = -3$

2. $\lim_{x \to -2} (3x + 7) = 1$

3. $\lim_{x \to 1} (3x + 1) = 5$

4. $\lim_{x \to 2} (x^2 - 2) = 5$

5. $\lim_{x \to 2} (x^2 + 2) = 6$

6. $\lim_{x \to 1} (x^2 - 3x + 2) = 0$

In Problems 7–12, use the formal definition of the limit to prove or disprove the given limit statement.

7. $\lim_{x \to 2} (x + 3) = 5$

8. $\lim_{t \to 0} (3t - 1) = 0$

9. $\lim_{x \to -2} (3x + 7) = 1$

10. $\lim_{x \to 1} (2x - 5) = -3$

11. $\lim_{x \to 2} (x^2 + 2) = 6$

12. $\lim_{x \to 2} \dfrac{1}{x} = \dfrac{1}{2}$

13. Prove that $f(x) = \begin{cases} \sin \dfrac{1}{x} & \text{if } x \ne 0 \\ 0 & \text{if } x = 0 \end{cases}$

is not continuous at $x = 0$. *Hint:* To show that $\lim_{x \to 0} f(x) \ne 0$, choose $\epsilon = 0.5$ and note that for any $\delta > 0$,

there exists an x of the form $x = \dfrac{2}{\pi(2n + 1)}$ with n a natural number for which $0 < \left| x \right| < \delta$.

C *In Problems 14–19, construct a formal ϵ–δ proof to show that the given limit statement is valid for any number c.*

14. If $\lim_{x \to c} f(x)$ exists and k is a constant, then

$$\lim_{x \to c} kf(x) = k \lim_{x \to c} f(x)$$

15. If $\lim_{x \to c} f(x)$ and $\lim_{x \to c} g(x)$ both exist, then

$$\lim_{x \to c} [f(x) - g(x)] = \lim_{x \to c} f(x) - \lim_{x \to c} g(x)$$

16. If $\lim_{x \to c} f(x)$ and $\lim_{x \to c} g(x)$ both exist and a, b are constants, then

$$\lim_{x \to c} [af(x) + bg(x)] = a \lim_{x \to c} f(x) + b \lim_{x \to c} g(x)$$

17. If $\lim_{x \to c} f(x) = 0$ and $\lim_{x \to c} g(x) = 0$, then

$$\lim_{x \to c} f(x)g(x) = 0$$

18. If $f(x) \geq g(x) \geq 0$ for all x and $\lim\limits_{x \to c} f(x) = 0$, then
$$\lim_{x \to c} g(x) = 0$$

19. If $f(x) \geq g(x)$ for all x in an open interval containing the number c and $\lim\limits_{x \to c} f(x)$ and $\lim\limits_{x \to c} g(x)$ both exist, then $\lim\limits_{x \to c} f(x) \geq \lim\limits_{x \to c} g(x)$. *Hint:* Apply the limit limitation theorem (Theorem A.1) to the function $h(x) = f(x) - g(x)$.

Problems 20–23 lead to a proof of the product rule for limits.

20. If $\lim\limits_{x \to c} f(x) = L$, show that $\lim\limits_{x \to c} |f(x)| = |L|$. *Hint:* Note that
$$||f(x)| - |L|| \leq |f(x) - L|$$

21. If $\lim\limits_{x \to c} f(x) = L, L \neq 0$, show that there exists a $\delta > 0$ such that
$$\tfrac{1}{2}|L| < |f(x)| < \tfrac{3}{2}|L|$$
for all x for which $0 < |x - c| < \delta$.
Hint: Use $\epsilon = \tfrac{1}{2}|L|$ in Problem 20.

22. If $\lim\limits_{x \to c} f(x) = L$ and $L \neq 0$, show that $\lim\limits_{x \to c} [f(x)]^2 = L^2$ by completing these steps:
a. Use Problem 21 to show that there exists a $\delta_1 > 0$ so that
$$|f(x) + L| < \tfrac{5}{2}|L|$$
whenever $0 < |x - c| < \delta_1$.
b. Given $\epsilon > 0$, show that there exists a $\delta_2 > 0$ such that
$$|[f(x)]^2 - L^2| < \tfrac{5}{2}|L|$$
whenever $0 < |x - c| < \delta_2$.
c. Complete the proof that
$$\lim_{x \to c} [f(x)]^2 = L^2$$

23. Prove the product rule for limits: If $\lim\limits_{x \to c} f(x) = L$ and $\lim\limits_{x \to c} g(x) = M$, then $\lim\limits_{x \to c} f(x)g(x) = LM$. *Hint:* Use the result of Problem 22 along with the identity
$$fg = \tfrac{1}{4}[(f + g)^2 - (f - g)^2]$$

24. Show that if f is continuous at c and $f(c) > 0$, then $f(x) > 0$ throughout an open interval containing c. *Hint:* Note that
$$\lim_{x \to c} f(x) = f(c)$$
and use $\epsilon = \tfrac{1}{2}f(c)$ in the definition of limit.

25. Show that if f is continuous at L and $\lim\limits_{x \to c} g(x) = L$, then, $\lim\limits_{x \to c} f[g(x)] = f(L)$ by completing the following steps:
a. Explain why there exists a $\delta_1 > 0$ such that $|f(w) - f(L)| < \epsilon$ whenever $|w - L| < \delta_1$.
b. Complete the proof by setting $w = g(x)$ and using part **a**.

 Technology Window

26. For linear functions, the relation between ϵ and δ is clear; but for complicated functions, it is not. However, the correct graph can illustrate this relationship. For the following function, illustrate not only that f is continuous at $x = 3$, but determine graphically how you would pick δ to accommodate a given ϵ. (For example, $\delta = K\epsilon$.)
$$f(x) = \frac{x^4 - 2x^3 + 3x^2 - 5x + 2}{x - 2}$$

APPENDIX B

Selected Proofs

CHAIN RULE (Chapter 3)

Suppose f is a differentiable function of u, and u is a differentiable function of x. Then
$$\frac{df}{dx} = \frac{df}{du}\frac{du}{dx}$$
where $f[u(x)]$.

Define an auxiliary function g by
$$g(t) = \frac{f[u(x) + t] - f[u(x)]}{t} - \frac{df}{du} \text{ if } t \neq 0 \quad \text{and} \quad g(t) = 0 \text{ if } t = 0$$

You can verify that g is continuous at $t = 0$.

Notice that for $t = \Delta u$ and $t \neq 0$,

$$g(\Delta u) = \frac{f[u(x) + \Delta u] - f[u(x)]}{\Delta u} - \frac{df}{du}$$

$$g(\Delta u) + \frac{df}{du} = \frac{f[u(x) + \Delta u] - f[u(x)]}{\Delta u}$$

$$\left[g(\Delta u) + \frac{df}{du}\right]\Delta u = f[u(x) + \Delta u] - f[u(x)]$$

We now use the definition of derivative for f.

$$\frac{df}{dx} = \lim_{\Delta x \to 0} \frac{f[u(x + \Delta x)] - f[u(x)]}{\Delta x}$$

$$= \lim_{\Delta x \to 0} \frac{f[u(x) + \Delta u] - f[u(x)]}{\Delta x} \qquad \text{Where } \Delta u = u(x + \Delta x) - u(x)$$

$$= \lim_{\Delta x \to 0} \frac{\left[g(\Delta u) + \dfrac{df}{du}\right]\Delta u}{\Delta x} \qquad \text{Substitution}$$

$$= \lim_{\Delta x \to 0} \left[g(\Delta u) + \frac{df}{du}\right]\frac{\Delta u}{\Delta x}$$

$$= \lim_{\Delta x \to 0} \left[g(\Delta u) + \frac{df}{du}\right]\lim_{\Delta x \to 0}\frac{\Delta u}{\Delta x}$$

$$= \left[\lim_{\Delta x \to 0} g(\Delta u) + \lim_{\Delta x \to 0}\frac{df}{du}\right]\lim_{\Delta x \to 0}\frac{\Delta u}{\Delta x}$$

$$= \left[0 + \frac{df}{du}\right]\frac{du}{dx} \qquad \text{Since } g \text{ is continuous at } t = 0$$

$$= \frac{df}{du}\frac{du}{dx}$$

CAUCHY'S GENERALIZED MEAN VALUE THEOREM (Chapter 4)

Let f and g be functions that are continuous on the closed interval $[a, b]$ and differentiable on the open interval (a, b). If $g(b) \neq g(a)$ and $g'(x) \neq 0$ on (a, b), then

$$\frac{f(b) - f(a)}{g(b) - g(a)} = \frac{f'(c)}{g'(c)}$$

for at least one number c between a and b.

Proof   We begin by defining a special function, just as in the proof of the MVT as presented in Section 4.2. Specifically, let

$$F(x) = f(x) - f(a) - \frac{f(b) - f(a)}{g(b) - g(a)}[g(x) - g(a)]$$

for all x in the closed interval $[a, b]$. In the proof of the MVT in Section 4.2, we show that F satisfies the hypotheses of Rolle's theorem, which means that $F'(c) = 0$ for at least one number c in (a, b). For this number c, we have

$$0 = F'(c) = f'(c) - \frac{f(b) - f(a)}{g(b) - g(a)}g'(c)$$

and the result follows from this equation.

l'HÔPITAL'S THEOREM* (Chapter 4)

For any number a, let f and g be functions that are differentiable on an open interval (a, b), where $g'(x) \neq 0$. Then if $\lim_{x \to a^+} f(x) = 0$, $\lim_{x \to a^+} g(x) = 0$, and $\lim_{x \to a^+} \dfrac{f'(x)}{g'(x)}$ exists, then

$$\lim_{x \to a^+} \frac{f(x)}{g(x)} = \lim_{x \to a^+} \frac{f'(x)}{g'(x)}$$

Proof   First, define auxiliary functions F and G by

$$F(x) = f(x) \text{ for } a < x \leq b \text{ and } F(a) = 0;$$
$$G(x) = g(x) \text{ for } a < x \leq b \text{ and } G(a) = 0.$$

These definitions guarantee that $F(x) = f(x)$ and $G(x) = g(x)$ for $a < x \leq b$ and that $F(a) = G(a) = 0$. Thus, if w is any number between a and b, the functions F and G are continuous on the closed interval $[a, w]$ and differentiable on the open interval (a, w). According to the Cauchy generalized mean value theorem, there exists a number t between a and w for which

$$\frac{F(w) - F(a)}{G(w) - G(a)} = \frac{F'(t)}{G'(t)}$$

$$\frac{F(w)}{G(w)} = \frac{F'(t)}{G'(t)} \qquad \text{Because } F(a) = G(a) = 0$$

$$\frac{f(w)}{g(w)} = \frac{f'(t)}{g'(t)} \qquad \text{Because } F(x) = f(x) \text{ and } G(x) = g(x)$$

$$\lim_{w \to a^+} \frac{f(w)}{g(w)} = \lim_{w \to a^+} \frac{f'(t)}{g'(t)}$$

$$\lim_{w \to a^+} \frac{f(w)}{g(w)} = \lim_{t \to a^+} \frac{f'(t)}{g'(t)} \qquad \text{Because } t \text{ is "trapped" between } a \text{ and } w$$

$$\lim_{x \to a^+} \frac{f(x)}{g(x)} = \lim_{x \to a^+} \frac{f'(x)}{g'(x)} \qquad \blacksquare$$

CONTINUITY AND DIFFERENTIABILITY OF INVERSE FUNCTIONS

Let f be a one-to-one function so that it possesses an inverse.

1. If f is continuous on a domain I, then f^{-1} is continuous on $f(I)$.
2. If f is differentiable at c, and $f'(c) \neq 0$, then f^{-1} is differentiable at $f(c)$.

Proof

1. Recall that $y = f(x)$ if and only if $x = f^{-1}(y)$. Let I be the open interval (a, b), and let $y_0 = f(x_0)$ be in the open interval $(f(a), f(b))$ so that x_0 is in the open interval (a, b). To prove continuity, we must show that

$$\lim_{y \to y_0} f^{-1}(y) = f^{-1}(y_0) = x_0$$

---

*This is a special case of l'Hôpital's rule. The other cases can be found in most advanced calculus textbooks.

Consider any interval $(x_0 - \epsilon, x_0 + \epsilon)$ for $\epsilon > 0$. We must find an interval $(y_0 - \delta, y_0 + \delta)$ such that whenever y is in $(y_0 - \delta, y_0 + \delta)$, $f^{-1}(y)$ is in $(x_0 - \epsilon, x_0 + \epsilon)$.

Let $\delta_1 = y_0 - f(x_0 - \epsilon)$ and $\delta_2 = f(x_0 + \epsilon) - y_0$. If δ is the smaller of δ_1 and δ_2, it follows that if y is in $(y_0 - \delta, y_0 + \delta)$, then $f^{-1}(y)$ is in $(x_0 - \epsilon, x_0 + \epsilon)$, which is what we wanted to prove.

2. Let $g = f^{-1}$ and $f'[g(a)] \neq 0$. We want to show that g is differentiable.

$$g'(a) = \lim_{g \to a} \frac{g(x) - g(a)}{x - a} \qquad \text{From the definition of derivative}$$

$$= \lim_{x \to a} \frac{y - b}{f(y) - f(b)} \qquad \text{If } y = g(x) \text{ then } x = f(y) \text{ and if } b = g(a) \text{ then } a = f(b).$$

$$= \lim_{y \to b} \frac{y - b}{f(y) - f(b)} \qquad \text{Since } f \text{ is differentiable, it is continuous, so } g = f^{-1} \text{ is continuous by part (1). Thus, if } x \to a, \text{ then } g(x) \to g(a) \text{ so that } y \to b.$$

$$= \lim_{y \to b} \frac{1}{\dfrac{f(y) - f(b)}{y - b}}$$

$$= \frac{1}{\displaystyle\lim_{y \to b} \frac{f(y) - f(b)}{y - b}}$$

$$= \frac{1}{f'(b)}$$

Since f is differentiable, we see that g is also differentiable.

LIMIT COMPARISON TEST (Chapter 8)

Suppose $a_k > 0$ and $b_k > 0$ for all sufficiently large k and that

$$\lim_{k \to \infty} \frac{a_k}{b_k} = L \quad \text{where } L \text{ is finite and positive } (0 < L < \infty)$$

Then Σa_k and Σb_k either both converge or both diverge.

Proof   Assume that $\lim_{k \to \infty} \dfrac{a_k}{b_k} = L$, where $L > 0$. Using $\epsilon = \dfrac{L}{2}$ in the definition of the limit of a sequence, we see that there exists a number N so that

$$\left| \frac{a_k}{b_k} - L \right| < \frac{L}{2} \qquad \text{Whenever } k > N$$

$$-\frac{L}{2} < \frac{a_k}{b_k} - L < \frac{L}{2}$$

$$\frac{L}{2} < \frac{a_k}{b_k} < \frac{3L}{2}$$

$$\frac{L}{2} b_k < a_k < \frac{3L}{2} b_k \qquad b_k > 0$$

This is true for all $k > N$. Now we can complete the proof by using the direct comparison test. Suppose Σb_k converges. Then the series $\Sigma \dfrac{3L}{2} b_k$ also converges, and the inequality

$$a_k < \frac{3L}{2} b_k$$

tells us that the series Σa_k must also converge since it is dominated by a convergent series.

Similarly if Σb_k diverges, the inequality

$$0 < \frac{L}{2} b_k < a_k$$

tells us that Σa_k dominates the divergent series $\Sigma \dfrac{L}{2} b_k$, and it follows that Σa_k also diverges.

Thus Σa_k and Σb_k either both converge or both diverge. ∎

TAYLOR'S THEOREM (Chapter 8)

If f and all its derivatives exist in an open interval I containing c, then for each x in I,

$$f(x) = f(c) + \frac{f'(c)}{1!}(x - c) + \frac{f''(c)}{2!}(x - c)^2 + \cdots + \frac{f^{(n)}(c)}{n!}(x - c)^n + R_n(x)$$

where the remainder function $R_n(x)$ is given by

$$R_n(x) = \frac{f^{(n+1)}(z_n)}{(n + 1)!}(x - c)^{n+1}$$

for some z_n that depends on x and lies between c and x.

Proof   We shall prove Taylor's theorem by showing that if f and its first $n + 1$ derivatives are defined in an open interval I containing c, then for each fixed x in I,

$$f(x) = f(c) + \frac{f'(c)}{1!}(x - c) + \frac{f''(c)}{2!}(x - c)^2 + \cdots + \frac{f^{(n)}(c)}{n!}(x - c)^n + \frac{f^{(n+1)}(c)}{(n + 1)!}(x - c)^{n+1}$$

where z is some number between x and c. In our proof, we shall apply Cauchy's generalized mean value theorem to the auxiliary functions F and G defined for all t in I as follows:

$$F(t) = f(x) - f(t) - \frac{f'(t)}{1!}(x - t) - \cdots - \frac{f^{(n)}(t)}{n!}(x - t)^n$$

$$G(t) = \frac{(x - t)^{n+1}}{(n + 1)!}$$

Note that $F(x) = G(x) = 0$, and thus Cauchy's generalized mean value theorem tells us

$$\frac{F'(z)}{G'(z)} = \frac{F(x) - F(c)}{G(x) - G(c)} = \frac{F(c)}{G(c)}$$

for some number z between x and c. Rearranging the sides of this equation and finding the derivatives gives

$$\frac{F(c)}{G(c)} = \frac{F'(z)}{G'(z)}$$

$$= \frac{\dfrac{-f^{(n+1)}(z)}{n!}(x-z)^n}{\dfrac{-(x-z)^n}{n!}}$$

$$= f^{(n+1)}(z)$$

$$F(c) = f^{(n+1)}(z)G(c)$$

$$f(x) - f(c) - \frac{f'(c)}{1!}(x-c) - \cdots - \frac{f^{(n)}(c)}{n!}(x-c)^n = f^{(n+1)}(z)\left[\frac{(x-c)^{n+1}}{(n+1)!}\right]$$

Rearranging these terms, we obtain the required equation:

$$f(x) = f(c) + \frac{f'(c)}{1!}(x-c) + \cdots + \frac{f^{(n)}(c)}{n!}(x-c)^n + f^{(n+1)}(z)\left[\frac{(x-c)^{n+1}}{(n+1)!}\right] \quad \blacksquare$$

SUFFICIENT CONDITION FOR DIFFERENTIABILITY (Chapter 12)

If f is a function of x and y, and $f, f_x,$ and f_y are continuous is a disk D centered at (x_0, y_0), then f is differentiable at (x_0, y_0).

Proof   If (x, y) is a point in D, we have

$$f(x, y) - f(x_0, y_0) = f(x, y) - f(x_0, y) + f(x_0, y) - f(x_0, y_0)$$

The function $f(x, y)$ with y fixed satisfies the conditions of the mean value theorem, so that

$$f(x, y) - f(x_0, y) = f_x(x_0, y)(x - x_0)$$

for some number x_1 between x and x_0, and similarly, there is a number y_1 between y and y_0 such that

$$f(x_0, y) - f(x_0, y_0) = f_y(x_0, y_1)(y - y_0)$$

Substituting these expressions, we obtain

$$f(x, y) - f(x_0, y_0) = [f(x, y) - f(x_0, y)] + [f(x_0, y) - f(x_0, y_0)]$$
$$= [f_x(x_1, y)(x - x_0)] + [f_y(x_0, y_1)(y - y_0)]$$
$$= [f_x(x_1, y)(x - x_0)] + \underbrace{f_x(x_0, y_0)(x - x_0) - f_x(x_0, y_0)(x - x_0)}_{\text{This is zero.}}$$

$$+ [f_y(x_0, y_1)(y - y_0)] + \underbrace{f_y(x_0, y_0)(y - y_0) - f_y(x_0, y_0)(y - y_0)}_{\text{This is zero.}}$$

$$= f_x(x_0, y_0)(x - x_0) + f_y(x_0, y_0)(y - y_0)$$
$$+ [f_x(x_1, y) - f_x(x_0, y_0)](x - x_0)$$
$$+ [f_y(x_0, y_1) - f_y(x_0, y_0)](y - y_0)$$

Let $\epsilon_1(x, y)$ and $\epsilon_2(x, y)$ be the functions

$$\epsilon_1(x, y) = f_x(x_1, y) - f_x(x_0, y_0) \quad \text{and} \quad \epsilon_2(x, y) = f_y(x_0, y_1) - f_y(x_0, y_0)$$

Then since x_1 is between x and x_0, and y_1 is between y and y_0, and the partial derivatives f_x and f_y are continuous at (x_0, y_0), we have

$$\lim_{(x,y)\to(x_0,y_0)} \epsilon_1(x,y) = \lim_{(x,y)\to(x_0,y_0)} [f_x(x_1,y) - f_x(x_0,y_0)] = 0$$

$$\lim_{(x,y)\to(x_0,y_0)} \epsilon_2(x,y) = \lim_{(x,y)\to(x_0,y_0)} [f_y(x_0,y_1) - f_y(x_0,y_0)] = 0$$

so that f is differentiable at (x_0, y_0), as required.   ▬

CHANGE OF VARIABLE FORMULA FOR MULTIPLE INTEGRATION (Chapter 13)

Suppose f is a continuous function of a region D, and let D^* be the image of the domain D under the change of variable $x = g(u,v)$, $y = h(u,v)$, where g and h are continuously differentiable on D^*. Then

$$\int_D \int f(x,y)\, dy\, dx = \int_{D^*} \int f[g(u,v), h(u,v)] \underbrace{\left| \frac{\partial(x,y)}{\partial(u,v)} \right|}_{\text{absolute value of Jacobian}} dv\, du$$

Proof   A proof of this theorem is found in advanced calculus, but we can provide a geometric argument that makes this formula plausible in the special case where $f(x,y) = 1$. In particular, we shall show that in order to find the area of a region D in the xy-plane using the change of variable $x = X(u,v)$ and $y = Y(u,v)$, it is reasonable to use the formula for area, A:

$$A = \int_D \int dy\, dx = \int_{D^*} \int \left| \frac{\partial(x,y)}{\partial(u,v)} \right| dv\, du$$

where D^* is the region in the uv-plane that corresponds to D.

Suppose the given change of variable has an inverse $u = u(x,y)$, $v = v(x,y)$ that transforms the region D in the xy-plane into a region D^* in the uv-plane. To find the area of D^* in the uv-coordinate system, it is natural to use a rectangular grid, with vertical lines $u =$ constant and horizontal lines $v =$ constant, as shown in Figure B.1a. In the xy-plane, the equations $u =$ constant and $v =$ constant will be families of parallel curves, which provide a curvilinear grid for the region D^* (Figure B.1b).

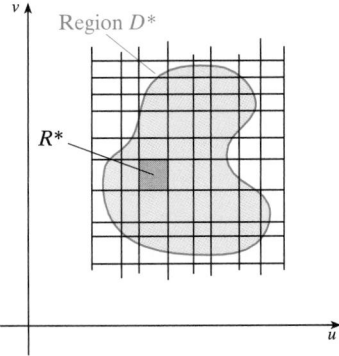

a. A rectangular grid in the uv-plane.

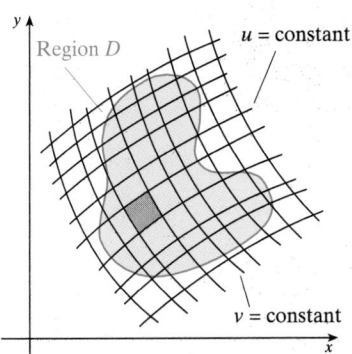

b. Corresponding grid in the xy-plane is curvilinear.

■ **FIGURE B.1**

Next, let R^* be a typical rectangular cell in the uv-grid that covers D^*, and let R be the corresponding set in the xy-plane (that is, R^* is the image of R under the given change of variable). Then, as shown in Figure B.2, if R^* has vertices $A(\bar{u}, \bar{v})$, $B(\bar{u} + \Delta u, \bar{v})$, $C(\bar{u}, \bar{v} + \Delta v)$, and $D(\bar{u} + \Delta u, \bar{v} + \Delta v)$, the set R will be the interior of a curvilinear rectangle with vertices

$$A[X(\bar{u}, \bar{v}), Y(\bar{u}, \bar{v})]$$
$$B[X(\bar{u} + \Delta u, \bar{v}), Y(\bar{u} + \Delta u, \bar{v})]$$
$$C[X(\bar{u}, \bar{v} + \Delta v), Y(\bar{u}, \bar{v} + \Delta v)]$$
$$D[X(\bar{u} + \Delta u, \bar{v} + \Delta v), Y(\bar{u} + \Delta u, \bar{v} + \Delta v)]$$

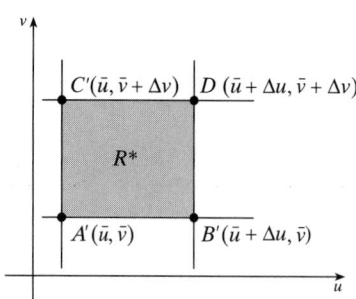
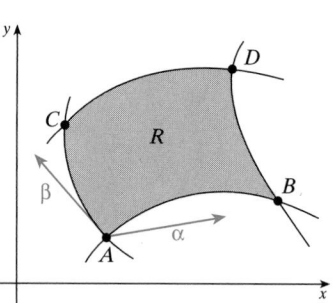

a. R^* is a typical rectangular cell in the grid covering D^*.

b. The image of R under the change of variable is a curvilinear rectangle R.

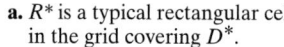 **FIGURE B.2**

Note that the curved side of R joining A to B may be approximated by the secant vector

$$\mathbf{AB} = [X(\bar{u} + \Delta u, \bar{v}) - X(\bar{u}, \bar{v})]\mathbf{i} + [Y(\bar{u} + \Delta u, \bar{v}) - Y(\bar{u}, \bar{v})]\mathbf{j}$$

and by applying the mean value theorem, we find that

$$\mathbf{AB} = \left[\frac{\partial x}{\partial u}(a, \bar{v}) \Delta u\right]\mathbf{i} + \left[\frac{\partial y}{\partial u}(b, \bar{v}) \Delta u\right]\mathbf{j}$$

for some numbers a, b between $\bar{u}$ and $\bar{u} + \Delta u$. If Δu is very small, a an b are approximately the same as $\bar{u}$, and we can approximate $\mathbf{AB}$ by the vector

$$\alpha = \left[\frac{\partial x}{\partial u} \Delta u\right]\mathbf{i} + \left[\frac{\partial y}{\partial u} \Delta u\right]\mathbf{j}$$

where the partials are evaluated at the point $(\bar{u}, \bar{v})$. Similarly, the curved side of R joining A and C may be approximated by the vector

$$\beta = \left[\frac{\partial x}{\partial v} \Delta v\right]\mathbf{i} + \left[\frac{\partial y}{\partial v} \Delta v\right]\mathbf{j}$$

The area of the curvilinear rectangle R is approximately the same as that of the parallelogram determined by α and β; that is,

$$\|\alpha \times \beta\| = \begin{vmatrix} \mathbf{i} & \mathbf{j} & \mathbf{k} \\ \frac{\partial x}{\partial u}\Delta u & \frac{\partial y}{\partial u}\Delta u & 0 \\ \frac{\partial x}{\partial y}\Delta v & \frac{\partial y}{\partial v}\Delta v & 0 \end{vmatrix}$$

$$= \left\| \begin{vmatrix} \mathbf{i} & \mathbf{j} & \mathbf{k} \\ \dfrac{\partial x}{\partial u} & \dfrac{\partial y}{\partial u} & 0 \\ \dfrac{\partial x}{\partial v} & \dfrac{\partial y}{\partial v} & 0 \end{vmatrix} \Delta v \Delta u \right\|$$

$$= \left| \dfrac{\partial x}{\partial y}\dfrac{\partial y}{\partial v} - \dfrac{\partial y}{\partial u}\dfrac{\partial x}{\partial v} \right| \Delta v \Delta u$$

$$= \left| \dfrac{\partial(x,\,y)}{\partial(u,\,v)} \right| \Delta v \Delta u$$

By adding the contributions of all cells in the partition of D, we can approximate the area of D as follows:

APPROXIMATE AREA OF $D = \displaystyle\sum$ APPROXIMATE AREA OF CURVILINEAR RECTANGLES

$$= \sum \left| \dfrac{\partial(x,\,y)}{\partial(u,\,v)} \right| \Delta v \Delta u$$

Finally, using a limit to "smooth out" the approximation, we find

$$A = \iint_D dy\,dx \;=\; \lim \sum \left| \dfrac{\partial(x,\,y)}{\partial(u,\,v)} \right| \Delta v \Delta u$$

$$= \iint_{D^*} \left| \dfrac{\partial(x,\,y)}{\partial(u,v)} \right| dv\,du \qquad \blacksquare$$

STOKES'S THEOREM (Chapter 14)

Let S be an oriented surface with unit normal vector $\mathbf{N}$, and assume that S is bounded by a closed, piecewise smooth curve C whose orientation is compatible with that of S. If $\mathbf{F}$ is a continuous vector field whose components have continuous partial derivatives on an open region containing S and C, then

$$\int_C \mathbf{F} \cdot d\mathbf{R} = \iint_S (\text{curl } \mathbf{F} \cdot \mathbf{N})\,dS$$

Proof   The general proof cannot be considered until advanced calculus. However, a proof for the case where S is a graph and $\mathbf{F}$, S, and C are "well behaved" can be given. Let S be given by $z = g(x, y)$ where (x, y) is in a region D of the xy-plane. Assume g has continuous second-order partial derivatives. Let C_1 be the projection of C in the xy-plane, as shown in Figure B.3. Also let $\mathbf{F}(x, y, z) = f(x, y, z)\mathbf{i} + g(x, y, z)\mathbf{j} + h(x, y, z)\mathbf{k}$ where the partial derivatives of f, g, and h are continuous.

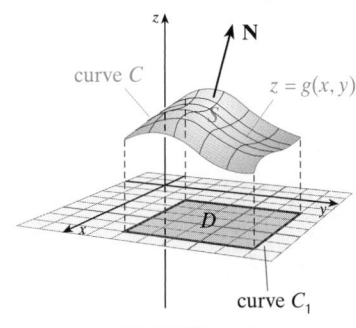

■ **FIGURE B.3**

We will evaluate each side of Stokes's theorem separately and show that the results for each are the same. If $x = x(t), y = y(t),$ and $z = z(t)$ for $a \le t \le b$ and $\mathbf{R}(t) = x(t)\mathbf{i} + y(t)\mathbf{j} + z(t)\mathbf{k}$, then

$$\int_C \mathbf{F} \cdot d\mathbf{R} = \int_a^b \left(f\frac{dx}{dt} + g\frac{dy}{dt} + h\frac{dz}{dt} \right) dt$$

$$= \int_a^b \left[f\frac{dx}{dt} + g\frac{dy}{dt} + h\left(\frac{\partial z}{\partial x}\frac{dx}{dt} + \frac{\partial z}{\partial y}\frac{dy}{dt} \right) \right] dt \qquad \text{Chain rule}$$

$$= \int_a^b \left[\left(f + h\frac{\partial z}{\partial x} \right)\frac{dx}{dt} + \left(g + h\frac{\partial z}{\partial y} \right)\frac{dy}{dt} \right] dt$$

$$= \int_{C_1} \left(f + h\frac{\partial z}{\partial x} \right) dx + \left(g + h\frac{\partial z}{\partial y} \right) dy$$

$$= \iint_D \left[\frac{\partial}{\partial x}\left(g + h\frac{\partial z}{\partial y} \right) - \frac{\partial}{\partial y}\left(f + h\frac{\partial z}{\partial x} \right) \right] dA \qquad \text{Green's theorem}$$

$$= \iint_D \left[\left(\frac{\partial g}{\partial x} + \frac{\partial g}{\partial z}\frac{\partial z}{\partial x} + \frac{\partial h}{\partial x}\frac{\partial z}{\partial y} + \frac{\partial h}{\partial z}\frac{\partial z}{\partial x}\frac{\partial z}{\partial y} + h\frac{\partial^2 z}{\partial y \partial x} \right) \right.$$

$$\left. - \left(\frac{\partial f}{\partial y} + \frac{\partial f}{\partial z}\frac{\partial z}{\partial y} + \frac{\partial h}{\partial y}\frac{\partial z}{\partial x} + \frac{\partial h}{\partial z}\frac{\partial z}{\partial y}\frac{\partial z}{\partial x} + h\frac{\partial^2 z}{\partial y \partial x} \right) \right] dA \qquad \begin{array}{l}\text{Green's theorem}\\ \text{again}\end{array}$$

$$= \iint_D \left(\frac{\partial g}{\partial x} + \frac{\partial g}{\partial z}\frac{\partial z}{\partial x} + h\frac{\partial^2 z}{\partial y \partial x} - \frac{\partial f}{\partial y} - \frac{\partial f}{\partial z}\frac{\partial z}{\partial y} - h\frac{\partial^2 z}{\partial y \partial x} \right) dA$$

We now start over by evaluating the other side of Stokes's theorem:

$$\iint_S \text{curl } \mathbf{F} \cdot d\mathbf{S} = \iint_D \left[-\left(\frac{\partial h}{\partial y} - \frac{\partial g}{\partial z} \right)\frac{\partial z}{\partial x} - \left(\frac{\partial f}{\partial z} - \frac{\partial h}{\partial x} \right)\frac{\partial z}{\partial y} + \left(\frac{\partial g}{\partial x} - \frac{\partial f}{\partial y} \right) \right] dA$$

$$= \iint_D \left(\frac{\partial g}{\partial x} + \frac{\partial g}{\partial z}\frac{\partial z}{\partial x} + h\frac{\partial^2 z}{\partial y \partial x} - \frac{\partial f}{\partial y} - \frac{\partial f}{\partial z}\frac{\partial z}{\partial y} - h\frac{\partial^2 z}{\partial y \partial x} \right) dA$$

Since these results are the same, we have

$$\int_C \mathbf{F} \cdot d\mathbf{R} = \iint_S (\text{curl } \mathbf{F} \cdot \mathbf{N})\, dS \qquad \qquad =\!=$$

DIVERGENCE THEOREM (Chapter 14)

Let D be a region in space bounded by a smooth, orientable closed surface S. If $\mathbf{F}$ is a continuous vector field whose components have continuous partial derivatives in D, then

$$\iint_S \mathbf{F} \cdot \mathbf{N}\, dS = \iiint_D \text{div } \mathbf{F}\, dV$$

where $\mathbf{N}$ is an outward unit normal to the surface S.

Proof   Let $\mathbf{F}(x, y, z) = f(x, y, z)\mathbf{i} + g(x, y, z)\mathbf{j} + h(x, y, z)\mathbf{k}$. If we state the divergence theorem using this notation for $\mathbf{F}$, we have

$$\iint_S [f(\mathbf{i} \cdot \mathbf{N}) + g(\mathbf{j} \cdot \mathbf{N}) + h(\mathbf{k} \cdot \mathbf{N})]\, dS = \iiint_D \left(\frac{\partial f}{\partial x} + \frac{\partial g}{\partial y} + \frac{\partial h}{\partial z} \right) dV$$

$$\iint_S f(\mathbf{i} \cdot \mathbf{N})\, dS + \iint_S g(\mathbf{j} \cdot \mathbf{N})\, dS + \iint_S h(\mathbf{k} \cdot \mathbf{N})\, dS = \iiint_D \frac{\partial f}{\partial x}\, dV + \iiint_D \frac{\partial g}{\partial y}\, dV + \iiint_D \frac{\partial h}{\partial z}\, dV$$

This result can be verified by proving

$$\iint_S f(\mathbf{i} \cdot \mathbf{N})\, dS = \iiint_D \frac{\partial f}{\partial x}\, dV$$

$$\iint_S g(\mathbf{j} \cdot \mathbf{N})\, dS = \iiint_D \frac{\partial g}{\partial y}\, dV$$

$$\iint_S h(\mathbf{k} \cdot \mathbf{N})\, dS = \iiint_D \frac{\partial h}{\partial z}\, dV$$

Since the proof of each of these is virtually identical, we will show the verification for the last of these three; the other two can be done in a similar fashion. We will evaluate this third integral by separately evaluating the left and right sides to show they are the same.

We will restrict our proof to a "standard region" as described in the proof of Green's theorem in Section 14.4. The complete proof can then be completed by decomposing the general surface S into a finite number of "standard regions."

The standard solid region we shall consider has a top surface S_T with equation $z = u(x, y)$ and a bottom surface S_B with equation $z = v(x, y)$. We assume that both S_T and S_B project onto the region R in the xy-plane. The lateral surface S_L of the region is the set of all (x, y, z) such that $v(x, y) \le z \le u(x, y)$ on the boundary of R, as shown in Figure B.4.

We know that the outward unit normal (directed upward) to the top surface S_T is

$$\mathbf{N}_T = \frac{-u_x\mathbf{i} - u_y\mathbf{j} + \mathbf{k}}{\sqrt{u_x^2 + u_y^2 + 1}}$$

and

$$dS = \sqrt{u_x^2 + u_y^2 + 1}\, dA_{xy}$$

Thus,

$$\iint_{S_T} h(\mathbf{k} \cdot \mathbf{N}_T)dS = \iint_R h\left(\frac{1}{\sqrt{u_x^2 + u_y^2 + 1}}\right)\left(\sqrt{u_x^2 + u_y^2 + 1}\, dA_{xy}\right)$$

$$= \iint_R h\, dA_{xy}$$

$$= \iint_R h(x, y, z)\, dA_{xy}$$

$$= \iint_R h[x, y, u(x, y)]\, dA_{xy}$$

Similarly, the outward unit normal $\mathbf{N}_B$ to the bottom surface S_B is directed downward so that

$$\mathbf{N}_T = \frac{v_x\mathbf{i} - v_y\mathbf{j} - \mathbf{k}}{\sqrt{u_x^2 + u_y^2 + 1}}$$

and

$$\iint_{S_B} h(\mathbf{k} \cdot \mathbf{N}_B)\, dS = -\iint_R h\, dA_{xy}$$

$$= -\iint_R h(x, y, z)\, dA_{xy}$$

$$= -\iint_R h[x, y, v(x, y)]\, dA_{xy}$$

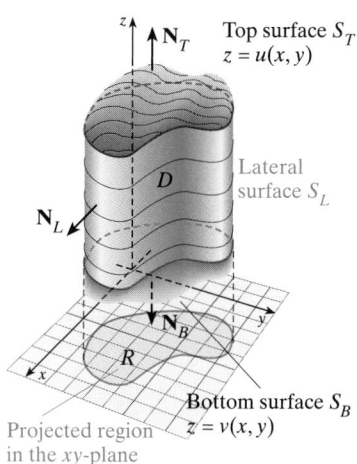

Top surface S_T
$z = u(x, y)$

Lateral surface S_L

Bottom surface S_B
$z = v(x, y)$

Projected region in the xy-plane

■ **FIGURE B.4**   A standard solid region in $\mathbb{R}^3$

Because the outward unit normal $\mathbf{N}_L$ is horizontal on the lateral surface S_L, it is perpendicular to $\mathbf{k}$, and

$$\iint_{S_L} h(\mathbf{k} \cdot \mathbf{N}_L)\, dS = 0$$

We now add for the surface S:

$$\iint_S h(\mathbf{k} \cdot \mathbf{N})\, dS = \iint_{S_T} h(\mathbf{k} \cdot \mathbf{N}_T)\, dS + \iint_{S_B} h(\mathbf{k} \cdot \mathbf{N}_B)\, dS + \iint_{S_L} h(\mathbf{k} \cdot \mathbf{N}_L)\, dS$$

$$= \iint_R h[x, y, u(x, y)]\, dA_{xy} - \iint_R h[x, y, v(x, y)] dA_{xy} + 0$$

$$= \iint_R \{h[x, y, u(x, y)] - h[x, y, v(x, y)]\}\, dA_{xy}$$

Next, we start over by looking at the triple integral on the right side. Notice that we can describe the solid S as the set of all (x, y, z) for (x, y) in R and $v(x, y) \le z \le u(x, y)$. Thus,

$$\iiint_D \frac{\partial h}{\partial z}\, dV = \iint_R \left[\int_{v(x, y)}^{u(x, y)} \frac{\partial h}{\partial z}(x, y, z)\, dz \right] dA_{xy}$$

$$= \int_R \{h[x, y, u(x, y)] - h[x, y, v(x, y)]\}\, dA_{xy}$$

We see that the left and right sides are the same, so

$$\iint_S h(\mathbf{k} \cdot \mathbf{N})\, dS = \iiint_D \frac{\partial h}{\partial z}\, dV$$

We can now conclude that (by similar arguments for the other two parts):

$$\iint_S \mathbf{F} \cdot \mathbf{N}\, dS = \iiint_D \operatorname{div} \mathbf{F}\, dV \qquad \blacksquare$$

APPENDIX C

Significant Digits

Throughout this book, various technology windows appear, and in the answers to the problems you will frequently find approximate (decimal) answers. Sometimes your answer may not exactly agree with the answer found in the back of the book. This does not necessarily mean that your answer is incorrect, particularly if your answer is very close to the given answer.

To use your calculator intelligently and efficiently, you should become familiar with its functions and practice doing problems with the same calculator, whenever possible. Read the technology windows provided throughout this text, and consult the owner's manual for your particular calculator when you have questions. In addition, there are *Technology Manuals* accompanying this text, which are available for TI and HP graphic calculators, as well as for MATHLAB and for Maple.

SIGNIFICANT DIGITS

Applications involving measurements can never be exact. It is particularly important that you pay attention to the accuracy of your measurements when you use a computer or calculator, because using available technology can give you a false sense of security about the accuracy in a particular problem. For example, if you measure a triangle and find the sides are approximately 1.2 and 3.4 and then find the ratio of $1.2/3.4 \approx 0.35294117$, it appears that the result is more accurate than the original measurements! Some discussion about accuracy and significant digits is necessary.

The digits known to be correct in a number obtained by a measurement are called **significant digits.** The digits $1, 2, 3, 4, 5, 6, 7, 8,$ and 9 are always significant, whereas the digit 0 may or may not be significant.

1. Zeros that come between two other digits are significant, as in 203 or 10.04.

2. If the zero's only function is to place the decimal point, it is not significant, as in

$$0.00000\,\underbrace{23} \qquad \text{or} \qquad 23,\underbrace{000}$$
$$\text{Placeholders} \qquad\qquad \text{Placeholders}$$

If a zero does more than fix the decimal point, it is significant, as in

$$0.0023\,\underset{\uparrow}{0} \qquad\qquad \text{or} \qquad 23,\underbrace{000.0}\,1$$
$$\text{This digit is significant.} \qquad \text{These are significant.}$$

This second rule can, of course, result in certain ambiguities, such as 23,000 (measured to the *exact* unit). To avoid such confusion, we use scientific notation in this case:

2.3×10^4 has two significant digits;

2.3000×10^4 has five significant digits.

Numbers that come about by counting are considered to be exact and are correct to any number of significant digits.

When you compute an answer using a calculator, the answer may have 10 or more digits. In the technology windows, we generally show the 10 or 12 digits that result from the numerical calculation, but frequently the number in the answer section will have only 5 or 6 digits in the answer. It seems clear that if the first 3 or 4 nonzero digits of the answer coincide, you probably have the correct method of doing the problem.

However, you might ask why are there discrepancies, and how many digits should you use when you write down your final answer? Roughly speaking, the significant digits in a number are the digits that have meaning. To clarify the concept, we must for the moment assume that we know the exact answer. We then assume that we have been able to compute an approximation to this exact answer.

Usually we do this by some sort of iterative process, in which the answers are getting closer and closer to the exact answer. In such a process, we hope that the number of significant digits in our approximate answer is increasing at each trial. If our approximate answer is, say, 6 digits long (some of those digits might even be zero), and the difference between our answer and the exact answer is 4 units or less in the last place, then the first 5 digits are significant.

For example, if the exact answer is 3.14159 and our approximate answer is 3.14162, then our answer has 6 significant digits, but is correct to 5 significant digits. Note that saying that our answer is correct to 5 significant digits does not guarantee that all of those 5 digits exactly match the first 5 digits of the exact answer. In fact, if an exact answer is 6.001 and our computed answer is 5.997, then our answer is correct to 3 significant digits and not one of them matches the digits in the exact answer. Also note that it may be necessary for an approximation to have more digits than are actually significant for it to have a certain number of significant digits. For example, if the exact answer is 6.003 and our approximation is 5.998, then it has 3 significant digits, but only if we consider the total 4-digit number, and do not strip off the last non-significant digit.

Again, suppose you know that all digits are significant in the number 3.456; then you know that the exact number is less than 3.4565 and at least 3.4555. Some people may say that the number 3.456 is correct to 3 decimal places. This is the same as saying that it has 4 significant digits.

Why bother with significant digits? If you multiply (or divide) two numbers with the same number of significant digits, then the product will generally be at least twice as long, but will have roughly the same number of significant digits as the original factors. You can then dispense with the unneeded digits. In fact, to keep them would be misleading about the accuracy of the result.

Frequently we can make an educated guess of the number of significant digits in an answer. For example, if we compute an iterative approximation such as

$$2.3123, \quad 2.3125, \quad 2.3126, \quad 2.31261, \quad 2.31262, \ldots$$

we would generally conclude that the answer is 2.3126 to 5 significant digits. Of course, we may very well be wrong, and if we continued iterating, the answer might end up as 2.4.

ROUNDING AND RULES OF COMPUTATION USED IN THIS BOOK

In hand and calculator computations, rounding a number is done to reduce the number of digits displayed and make the number easier to comprehend. Furthermore, if you suspect that the digit in the last place is not significant, then you might be tempted to round and remove this last digit. This can lead to error. For example, if the computed value is 0.64 and the true value is known to be between 0.61 and 0.67, then the computed value has only 1 significant digit. However, if we round it to 0.6 and the true value is really 0.66, then 0.6 is not correct to even 1 significant digit. In the interest of making the text easier to read, we have used the following rounding procedure:

Rounding Procedure   To round off numbers:

1. Increase the last retained digit by 1 if the remainder is greater than or equal to 5; or

2. Retain the last digit unchanged if the remainder is less than 5.

Elaborate rules for computation of approximate data can be developed when needed (in chemistry, for example), but in this text we will use three simple rules:

Rules for Significant Digits

Addition–subtraction: Add or subtract in the usual fashion, and then round off the result so that the last digit retained is in the column farthest to the right in which both given numbers have significant digits.

Multiplication–Division: Multiply or divide in the usual fashion, and then round off the results to the smaller number of significant digits found in either of the given numbers.

Counting numbers: Numbers used to count or whole numbers used as exponents are considered to be correct to any number of significant digits.

Rounding Rule   We use the following rounding procedure in problems requiring rounding by involving several steps: *Round only once, at the end. That is, do not work with rounded results, because round-off errors can accumulate.*

CALCULATOR EXPERIMENTS

You should be aware that you are much better than your calculator at performing certain computations. For example, almost all calculators will fail to give the correct answer to

$$(10.0 \text{ EE} + 50.0) + 911.0 - (10.0 \text{ EE} + 50.0)$$

Calculators will return the value of 0, but you know at a glance that the answer is 911.0. We must reckon with this poor behavior on the part of calculators, which is called *loss of accuracy* due to *catastrophic cancellation.* In this case, it is easy to catch the error immediately, but what if the computation is so complicated (or hidden by other computations) that we do not see the error?

First, we want to point out that the order in which you perform computations can be very important. For example, most calculators will correctly conclude that

$$(10.0 \text{ EE} + 50.0) - (10.0 \text{ EE} + 50.0) + 911.0 = 911$$

There are other cases besides catastrophic cancellation where the order in which a computation is performed will substantially affect the result. For example, you may not be able to calculate

$$(10.0 \text{ EE} + 50.0)*(911.0 \text{ EE} + 73.0)/(20.0 \text{ EE} + 60.0)$$

but rearranging the factors as

$$((10.0 \text{ EE} + 50.0)/(20.0 \text{ EE} + 60.0))*(911.0 \text{ EE} + 73.0)$$

should provide the correct answer of 4.555 EE 65. So, for what do we need to watch? If you subtract two numbers that are close to each other in magnitude you *may* obtain an inaccurate result. When you have a sequence of multiplications and divisions in a string, try to arrange the factors so that the result of each intermediate calculation stays as close as possible to 1.0.

Second, since a calculator performs all computations with a finite number of digits, it is unable to do exact computations involving nonterminating decimals. This enables us to see how many digits the calculator actually uses when it computes a result. For example, the computation

$$(7.0/17.0)*(17.0)$$

should give the result 7.0, but on most calculators it does not. The size of the answer gives an indication of how many digits "Accuracy" the calculator uses internally. That is, the calculator may display decimal numbers that have 10 digits, but use 12 digits internally. If the answer to the above computation is something similar to 1.0 EE -12, then the calculator is using 12 digits internally.

TRIGONOMETRIC EVALUATIONS

In many problems you will be asked to compute the values of trigonometric functions such as the sine, cosine, or tangent. In calculus, trigonometric arguments are usually assumed to be measured in radians. You must make sure the calculator is in radian mode. If it is in radian mode, then the sine of a small number will almost be equal to that number. For example,

$$\sin(0.00001) = 1\text{E} -5 \quad \text{(which is } 0.00001)$$

If not, then you are not using radian mode. Make sure you know how to put your calculator in radian mode.

GRAPHING BLUNDERS

When you are using the graphing features, you must always be careful to choose reasonable scales for the domain (horizontal scale) and range (vertical scale). If the scale is too large, you may not see important wiggles. If the scale is too small, you may not see important behavior elsewhere in the plane. Of course, knowing the techniques of graphing discussed in Chapter 4 will prevent you from making such blunders. Some calculators may have trouble with curves that jump suddenly at a point. An example of such a curve would be

$$y = \frac{e^x}{x}$$

which jumps at the origin. Try plotting this curve with your calculator using different horizontal and vertical scales, making sure that you understand how your calculator handles such graphs.

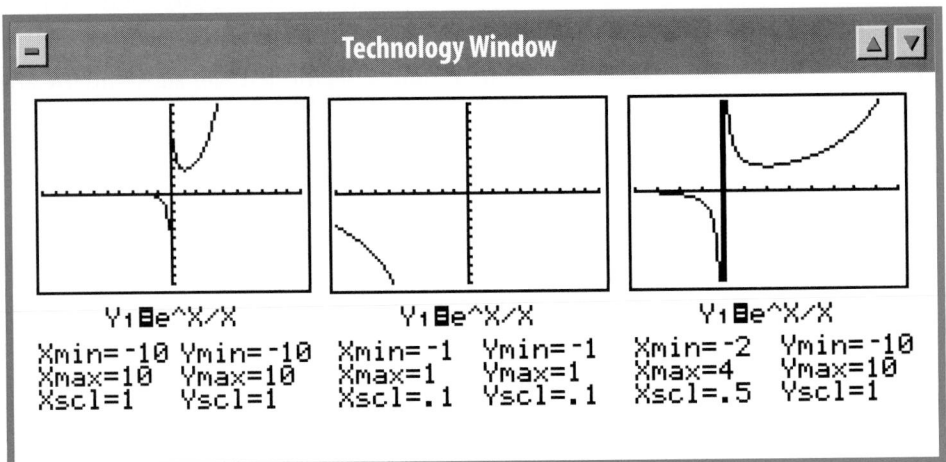

APPENDIX D

Short Table of Integrals

Each formula is numbered for easy reference. The numbers in this short table are not sequential because this short table is truncated from the table of integrals found in the *Student Mathematics Handbook*.

BASIC FORMULAS

1. **Constant rule** $\displaystyle\int 0\,du = c$

2. **Power rule** $\displaystyle\int u^n\,du = \frac{u^{n+1}}{n+1}; \quad n \neq -1$

                           $\displaystyle\int u^n\,du = \ln|u|; \quad n = -1$

3. **Exponential rule** $\displaystyle\int e^u\,du = e^u$

4. **Logarithmic rule** $\displaystyle\int \ln|u|\,du = u\ln|u| - u$

TRIGONOMETRIC RULES

5. $\int \sin u \, du = -\cos u$

6. $\int \cos u \, du = \sin u$

7. $\int \tan u \, du = -\ln|\cos u|$

8. $\int \cot u \, du = \ln|\sin u|$

9. $\int \sec u \, du = \ln|\sec u + \tan u|$

10. $\int \csc u \, du = \ln|\csc u - \cot u|$

11. $\int \sec^2 u \, du = \tan u$

12. $\int \csc^2 u \, du = -\cot u$

13. $\int \sec u \tan u \, du = \sec u$

14. $\int \csc u \cot u \, du = -\csc u$

EXPONENTIAL RULE

15. $\int a^u \, du = \dfrac{a^u}{\ln a}; \quad a > 0, a \neq 1$

HYPERBOLIC RULES

16. $\int \cosh u \, du = \sinh u$

17. $\int \sinh u \, du = \cosh u$

18. $\int \tanh u \, du = \ln|\cosh u|$

19. $\int \coth u \, du = \ln|\sinh u|$

20. $\int \operatorname{sech} u \, du = \tan^{-1}(\sinh u)$

21. $\int \operatorname{csch} u \, du = \ln\left|\tanh \dfrac{u}{2}\right|$

INVERSE RULES

22. $\int \dfrac{du}{\sqrt{a^2 - u^2}} = \sin^{-1}\dfrac{u}{a}$

23. $\int \dfrac{du}{\sqrt{u^2 - a^2}} = \cosh^{-1}\dfrac{u}{a}$

24. $\int \dfrac{du}{a^2 + u^2} = \dfrac{1}{a}\tan^{-1}\dfrac{u}{a}$

25. $\int \dfrac{du}{a^2 - u^2} = \begin{cases} \dfrac{1}{a}\tanh^{-1}\dfrac{u}{a} \text{ if } \left|\dfrac{u}{a}\right| < 1 \\ \dfrac{1}{a}\coth^{-1}\dfrac{u}{a} \text{ if } \left|\dfrac{u}{a}\right| > 1 \end{cases}$

26. $\int \dfrac{du}{u\sqrt{u^2 - a^2}} = \dfrac{1}{a}\sec^{-1}\left|\dfrac{u}{a}\right|$

27. $\int \dfrac{du}{u\sqrt{a^2 - u^2}} = -\dfrac{1}{a}\operatorname{sech}^{-1}\left|\dfrac{u}{a}\right|$

28. $\int \dfrac{du}{\sqrt{1 + u^2}} = \sinh^{-1} u$

29. $\int \dfrac{du}{u\sqrt{1 + u^2}} = -\operatorname{csch}^{-1}|u|$

INTEGRALS INVOLVING $au + b$

30. $\int (au + b)^n du = \dfrac{(au + b)^{n+1}}{(n + 1)a}$

31. $\int u(au + b)^n du = \dfrac{(au + b)^{n+2}}{(n + 2)a^2} - \dfrac{b(au + b)^{n+1}}{(n + 1)a^2}$

32. $\int u^2(au + b)^n du = \dfrac{(au + b)^{n+3}}{(n + 3)a^3} - \dfrac{2b(au + b)^{n+2}}{(n + 2)a^3} + \dfrac{b^2(au + b)^{n+1}}{(n + 1)a^3}$

33. $\int u^m(au + b)^n du = \begin{cases} \dfrac{u^{m+1}(au + b)^n}{m + n + 1} + \dfrac{nb}{m + n + 1}\displaystyle\int u^m(au + b)^{n-1}\,du \\[2ex] \dfrac{u^m(au + b)^{n+1}}{(m + n + 1)a} - \dfrac{mb}{(m + n + 1)a}\displaystyle\int u^{m-1}(au + b)^n\,du \\[2ex] \dfrac{-u^{m+1}(au + b)^{n+1}}{(n + 1)b} + \dfrac{m + n + 2}{(n + 1)b}\displaystyle\int u^m(au + b)^{n+1}\,du \end{cases}$

34. $\int \dfrac{du}{au + b} = \dfrac{1}{a}\ln|au + b|$

35. $\int \dfrac{u\,du}{au + b} = \dfrac{u}{a} - \dfrac{b}{a^2}\ln|au + b|$

36. $\int \dfrac{u^2\,du}{au + b} = \dfrac{(au + b)^2}{2a^3} - \dfrac{2b(au + b)}{a^3} + \dfrac{b^2}{a^3}\ln|au + b|$

37. $\int \dfrac{u^3\,du}{au + b} = \dfrac{(au + b)^3}{3a^4} - \dfrac{3b(au + b)^2}{2a^4} + \dfrac{3b^2(au + b)}{a^4} - \dfrac{b^3}{a^4}\ln|au + b|$

INTEGRALS INVOLVING $u^2 + a^2$

55. $\int \dfrac{du}{u^2 + a^2} = \dfrac{1}{a}\tan^{-1}\dfrac{u}{a}$

56. $\int \dfrac{u\,du}{u^2 + a^2} = \dfrac{1}{2}\ln(u^2 + a^2)$

57. $\int \dfrac{u^2\,du}{u^2 + a^2} = u - a\tan^{-1}\dfrac{u}{a}$

58. $\int \dfrac{u^3\,du}{u^2 + a^2} = \dfrac{u^2}{2} - \dfrac{a^2}{2}\ln(u^2 + a^2)$

59. $\int \dfrac{du}{u(u^2 + a^2)} = \dfrac{1}{2a^2}\ln\left(\dfrac{u^2}{u^2 + a^2}\right)$

60. $\int \dfrac{du}{u^2(u^2 + a^2)} = -\dfrac{1}{a^2u} - \dfrac{1}{a^3}\tan^{-1}\dfrac{u}{a}$

61. $\int \dfrac{du}{u^3(u^2 + a^2)} = -\dfrac{1}{2a^2u^2} - \dfrac{1}{2a^4}\ln\left(\dfrac{u^2}{u^2 + a^2}\right)$

INTEGRALS INVOLVING $u^2 - a^2, u^2 > a^2$

74. $\int \dfrac{du}{u^2 - a^2} = \dfrac{1}{2a}\ln\left|\dfrac{u - a}{u + a}\right|$ or $-\dfrac{1}{a}\coth^{-1}\dfrac{u}{a}$

75. $\int \dfrac{u\,du}{u^2 - a^2} = \dfrac{1}{2}\ln|u^2 - a^2|$

76. $\int \dfrac{u^2\,du}{u^2 - a^2} = u + \dfrac{a}{2}\ln\left|\dfrac{u - a}{u + a}\right|$

77. $\int \dfrac{u^3\,du}{u^2 - a^2} = \dfrac{u^2}{2} + \dfrac{a^2}{2}\ln|u^2 - a^2|$

78. $\int \dfrac{du}{u(u^2 - a^2)} = \dfrac{1}{2a^2}\ln\left|\dfrac{u^2 - a^2}{u^2}\right|$

79. $\displaystyle\int \frac{du}{u^2(u^2 - a^2)} = \frac{1}{a^2 u} + \frac{1}{2a^3} \ln \left| \frac{u - a}{u + a} \right|$

80. $\displaystyle\int \frac{du}{u^3(u^2 - a^2)} = \frac{1}{2a^2 u^2} - \frac{1}{2a^4} \ln \left| \frac{u^2}{u^2 - a^2} \right|$

<u>INTEGRALS INVOLVING</u> $a^2 - u^2, u^2 < a^2$

93. $\displaystyle\int \frac{du}{a^2 - u^2} = \frac{1}{2a} \ln \left| \frac{a + u}{a - u} \right| \quad \text{or} \quad \frac{1}{a} \tanh^{-1} \frac{u}{a}$

94. $\displaystyle\int \frac{u \, du}{a^2 - u^2} = -\frac{1}{2} \ln \left| a^2 - u^2 \right|$

95. $\displaystyle\int \frac{u^2 \, du}{a^2 - u^2} = -u + \frac{a}{2} \ln \left| \frac{a + u}{a - u} \right|$

96. $\displaystyle\int \frac{u^3 \, du}{a^2 - u^2} = -\frac{u^2}{2} - \frac{a^2}{2} \ln \left| a^2 - u^2 \right|$

97. $\displaystyle\int \frac{du}{u(a^2 - u^2)} = \frac{1}{2a^2} \ln \left| \frac{u^2}{a^2 - u^2} \right|$

98. $\displaystyle\int \frac{du}{u^2(a^2 - u^2)} = -\frac{1}{a^2 u} + \frac{1}{2a^3} \ln \left| \frac{a + u}{a - u} \right|$

99. $\displaystyle\int \frac{du}{u^3(a^2 - u^2)} = -\frac{1}{2a^2 u^2} + \frac{1}{2a^4} \ln \left| \frac{u^2}{a^2 - u^2} \right|$

100. $\displaystyle\int \frac{du}{(a^2 - u^2)^2} = \frac{u}{2a^2(a^2 - u^2)} + \frac{1}{4a^3} \ln \left| \frac{a + u}{a - u} \right|$

101. $\displaystyle\int \frac{u \, du}{(a^2 - u^2)^2} = \frac{1}{2(a^2 - u^2)}$

102. $\displaystyle\int \frac{u^2 \, du}{(a^2 - u^2)^2} = \frac{u}{2(a^2 - u^2)} - \frac{1}{4a} \ln \left| \frac{a + u}{a - u} \right|$

103. $\displaystyle\int \frac{u^3 \, du}{(a^2 - u^2)^2} = \frac{a^2}{2(a^2 - u^2)} + \frac{1}{2} \ln \left| a^2 \, u^2 \right|$

104. $\displaystyle\int \frac{du}{u(a^2 - u^2)^2} = \frac{1}{2a^2(a^2 - u^2)} + \frac{1}{2a^4} \ln \left| \frac{u^2}{a^2 - u^2} \right|$

105. $\displaystyle\int \frac{du}{u^2(a^2 - u^2)^2} = \frac{-1}{a^4 u} + \frac{u}{2a^4(a^2 - u^2)} + \frac{3}{4a^5} \ln \left| \frac{a + u}{a - u} \right|$

106. $\displaystyle\int \frac{du}{u^3(a^2 - u^2)^2} = \frac{-1}{2a^4 u^2} + \frac{1}{2a^4(a^2 - u^2)} + \frac{1}{a^6} \ln \left| \frac{u^2}{a^2 - u^2} \right|$

<u>INTEGRALS INVOLVING</u> $\sqrt{au + b}$

135. $\displaystyle\int \frac{du}{\sqrt{au + b}} = \frac{2\sqrt{au + b}}{a}$

136. $\displaystyle\int \frac{u \, du}{\sqrt{au + b}} = \frac{2(au - 2b)}{3a^2} \sqrt{au + b}$

137. $\displaystyle\int \frac{u^2 \, du}{\sqrt{au + b}} = \frac{2(3a^2 u^2 - 4abu + 8b^2)}{15a^3} \sqrt{au + b}$

138. $\displaystyle\int \frac{du}{u\sqrt{au+b}} = \begin{cases} \dfrac{1}{\sqrt{b}}\ln\left|\dfrac{\sqrt{au+b}-\sqrt{b}}{\sqrt{au+b}+\sqrt{b}}\right| \\[3mm] \dfrac{2}{\sqrt{-b}}\tan^{-1}\sqrt{\dfrac{au+b}{-b}} \end{cases}$

139. $\displaystyle\int \frac{du}{u^2\sqrt{au+b}} = -\frac{\sqrt{au+b}}{bu} - \frac{a}{2b}\int \frac{du}{u\sqrt{au+b}}$

140. $\displaystyle\int \sqrt{au+b}\,du = \frac{2\sqrt{(au+b)^3}}{3a}$

141. $\displaystyle\int u\sqrt{au+b}\,du = \frac{2(3au-2b)}{15a^2}\sqrt{(au+b)^3}$

142. $\displaystyle\int u^2\sqrt{au+b}\,du = \frac{2(15a^2u^2-12abu+8b^2)}{105a^3}\sqrt{(au+b)^3}$

INTEGRALS INVOLVING $\sqrt{u^2+a^2}$

168. $\displaystyle\int \sqrt{u^2+a^2}\,du = \frac{u\sqrt{u^2+a^2}}{2} + \frac{a^2}{2}\ln\left|u+\sqrt{u^2+a^2}\right|$

169. $\displaystyle\int u\sqrt{u^2+a^2}\,du = \frac{(u^2+a^2)^{3/2}}{3}$

170. $\displaystyle\int u^2\sqrt{u^2+a^2}\,du = \frac{u(u^2+a^2)^{3/2}}{4} - \frac{a^2u\sqrt{u^2+a^2}}{8} - \frac{a^4}{8}\ln\left|u+\sqrt{u^2+a^2}\right|$

171. $\displaystyle\int u^3\sqrt{u^2+a^2}\,du = \frac{(u^2+a^2)^{5/2}}{5} - \frac{a^2(u^2+a^2)^{3/2}}{3}$

172. $\displaystyle\int \frac{du}{\sqrt{u^2+a^2}} = \ln\left|u+\sqrt{u^2+a^2}\right|$ or $\sinh^{-1}\dfrac{u}{a}$

173. $\displaystyle\int \frac{u\,du}{\sqrt{u^2+a^2}} = \sqrt{u^2+a^2}$

174. $\displaystyle\int \frac{u^2\,du}{\sqrt{u^2+a^2}} = \frac{u\sqrt{u^2+a^2}}{2} - \frac{a^2}{2}\ln\left|u+\sqrt{u^2+a^2}\right|$

175. $\displaystyle\int \frac{u^3\,du}{\sqrt{u^2+a^2}} = \frac{(u^2+a^2)^{3/2}}{3} - a^2\sqrt{u^2+a^2}$

176. $\displaystyle\int \frac{du}{u\sqrt{u^2+a^2}} = -\frac{1}{a}\ln\left|\frac{a+\sqrt{u^2+a^2}}{u}\right|$

177. $\displaystyle\int \frac{du}{u^2\sqrt{u^2+a^2}} = -\frac{\sqrt{u^2+a^2}}{a^2u}$

178. $\displaystyle\int \frac{du}{u^3\sqrt{u^2+a^2}} = -\frac{\sqrt{u^2+a^2}}{2a^2u^2} + \frac{1}{2a^3}\ln\left|\frac{a+\sqrt{u^2+a^2}}{u}\right|$

INTEGRALS INVOLVING $\sqrt{u^2-a^2}$

196. $\displaystyle\int \frac{du}{\sqrt{u^2-a^2}} = \ln\left|u+\sqrt{u^2-a^2}\right|$ **197.** $\displaystyle\int \frac{u\,du}{\sqrt{u^2-a^2}} = \sqrt{u^2-a^2}$

198. $\int \dfrac{u^2\,du}{\sqrt{u^2-a^2}} = \dfrac{u\sqrt{u^2-a^2}}{2} + \dfrac{a^2}{2}\ln\left|u+\sqrt{u^2-a^2}\right|$

199. $\int \dfrac{u^3\,du}{\sqrt{u^2-a^2}} = \dfrac{(u^2-a^2)^{3/2}}{3} + a^2\sqrt{u^2-a^2}$

200. $\int \dfrac{du}{u\sqrt{u^2-a^2}} = \dfrac{1}{a}\sec^{-1}\left|\dfrac{u}{a}\right|$

201. $\int \dfrac{du}{u^2\sqrt{u^2-a^2}} = \dfrac{\sqrt{u^2-a^2}}{a^2u}$

202. $\int \dfrac{du}{u^3\sqrt{u^2-a^2}} = \dfrac{\sqrt{u^2-a^2}}{2a^2u^2} + \dfrac{1}{2a^3}\sec^{-1}\left|\dfrac{u}{a}\right|$

203. $\int \sqrt{u^2-a^2}\,du = \dfrac{u\sqrt{u^2-a^2}}{2} - \dfrac{a^2}{2}\ln\left|u+\sqrt{u^2-a^2}\right|$

204. $\int u\sqrt{u^2-a^2}\,du = \dfrac{(u^2-a^2)^{3/2}}{3}$

205. $\int u^2\sqrt{u^2-a^2}\,du = \dfrac{u(u^2-a^2)^{3/2}}{4} + \dfrac{a^2u\sqrt{u^2-a^2}}{8} - \dfrac{a^4}{8}\ln\left|u+\sqrt{u^2-a^2}\right|$

206. $\int u^3\sqrt{u^2-a^2}\,du = \dfrac{(u^2-a^2)^{5/2}}{5} + \dfrac{a^2(u^2-a^2)^{3/2}}{3}$

INTEGRALS INVOLVING $\sqrt{a^2-u^2}$

224. $\int \dfrac{du}{\sqrt{a^2-u^2}} = \sin^{-1}\dfrac{u}{a}$

225. $\int \dfrac{u\,du}{\sqrt{a^2-u^2}} = -\sqrt{a^2-u^2}$

226. $\int \dfrac{u^2\,du}{\sqrt{a^2-u^2}} = -\dfrac{u\sqrt{a^2-u^2}}{2} + \dfrac{a^2}{2}\sin^{-1}\dfrac{u}{a}$

227. $\int \dfrac{u^3\,du}{\sqrt{a^2-u^2}} = \dfrac{(a^2-u^2)^{3/2}}{3} - a^2\sqrt{a^2-u^2}$

228. $\int \dfrac{du}{u\sqrt{a^2-u^2}} = -\dfrac{1}{a}\ln\left|\dfrac{a+\sqrt{a^2-u^2}}{u}\right|$

229. $\int \dfrac{du}{u^2\sqrt{a^2-u^2}} = -\dfrac{\sqrt{a^2-u^2}}{a^2u}$

230. $\int \dfrac{du}{u^3\sqrt{a^2-u^2}} = -\dfrac{\sqrt{a^2-u^2}}{2a^2u^2} - \dfrac{1}{2a^3}\ln\left|\dfrac{a+\sqrt{a^2-u^2}}{u}\right|$

231. $\int \sqrt{a^2-u^2}\,du = \dfrac{u\sqrt{a^2-u^2}}{2} + \dfrac{a^2}{2}\sin^{-1}\dfrac{u}{a}$

232. $\int u\sqrt{a^2-u^2}\,du = -\dfrac{(a^2-u^2)^{3/2}}{3}$

233. $\int u^2\sqrt{a^2-u^2}\,du = -\dfrac{u(a^2-u^2)^{3/2}}{4} + \dfrac{a^2u\sqrt{a^2-u^2}}{8} + \dfrac{a^4}{8}\sin^{-1}\dfrac{u}{a}$

234. $\int u^3 \sqrt{a^2 - u^2} \, du = \dfrac{(a^2 - u^2)^{5/2}}{5} - \dfrac{a^2(a^2 - u^2)^{3/2}}{3}$

INTEGRALS INVOLVING $\cos au$

311. $\int \cos au \, du = \dfrac{\sin au}{a}$

312. $\int u \cos au \, du = \dfrac{\cos au}{a^2} + \dfrac{u \sin au}{a}$

313. $\int u^2 \cos au \, du = \dfrac{2u}{a^2} \cos au + \left(\dfrac{u^2}{a} - \dfrac{2}{a^3} \right) \sin au$

314. $\int u^3 \cos au \, du = \left(\dfrac{3u^2}{a^2} - \dfrac{6}{a^4} \right) \cos au + \left(\dfrac{u^3}{a} - \dfrac{6u}{a^3} \right) \sin au$

315. $\int u^n \cos au \, du = \dfrac{u^n \sin au}{a} - \dfrac{n}{a} \int u^{n-1} \sin au \, du$

316. $\int u^n \cos au \, du = \dfrac{u^n \sin au}{a} + \dfrac{nu^{n-1}}{a^2} \cos au - \dfrac{n(n-1)}{a^2} \int u^{n-2} \cos au \, du$

317. $\int \cos^2 au \, du = \dfrac{u}{2} + \dfrac{\sin 2au}{4a}$

INTEGRALS INVOLVING $\sin au$

342. $\int \sin au \, du = -\dfrac{\cos au}{a}$

343. $\int u \sin au \, du = \dfrac{\sin au}{a^2} - \dfrac{u \cos au}{a}$

344. $\int u^2 \sin au \, du = \dfrac{2u}{a^2} \sin au + \left(\dfrac{2}{a^3} - \dfrac{u^2}{a} \right) \cos au$

345. $\int u^3 \sin au \, du = \left(\dfrac{3u^2}{a^2} - \dfrac{6}{a^4} \right) \sin au + \left(\dfrac{6u}{a^3} - \dfrac{u^3}{a} \right) \cos au$

346. $\int u^n \sin au \, du = -\dfrac{u^n \cos au}{a} + \dfrac{n}{a} \int u^{n-1} \cos au \, du$

347. $\int u^n \sin au \, du = -\dfrac{u^n \cos au}{a} + \dfrac{nu^{n-1} \sin au}{a^2} - \dfrac{n(n-1)}{a^2} \int u^{n-2} \sin au \, du$

348. $\int \sin^2 au \, du = \dfrac{u}{2} - \dfrac{\sin 2\,au}{4a}$

INTEGRALS INVOLVING $\sin au$ and $\cos au$

373. $\int \sin au \cos au \, du = \dfrac{\sin^2 au}{2a}$

374. $\int \sin pu \cos qu \, du = -\dfrac{\cos (p - q)u}{2(p - q)} - \dfrac{\cos(p + q)u}{2(p + q)}$

375. $\int \sin^n au \cos au \, du = \dfrac{\sin^{n+1} au}{(n + 1)a}$

376. $\int \cos^n au \sin au \, du = -\dfrac{\cos^{n+1} au}{(n + 1)a}$

377. $\displaystyle\int \sin^2 au \cos^2 au \, du = \frac{u}{8} - \frac{\sin 4au}{32a}$

INTEGRALS INVOLVING tan *au*

403. $\displaystyle\int \tan au \, du = -\frac{1}{a}\ln|\cos au| = \frac{1}{a}\ln|\sec au|$

404. $\displaystyle\int \tan^2 au \, du = \frac{\tan au}{a} - u$

405. $\displaystyle\int \tan^3 au \, du = \frac{\tan^2 au}{2a} + \frac{1}{a}\ln|\cos au|$

406. $\displaystyle\int \tan^n au \, du = \frac{\tan^{n-1} au}{(n-1)a} - \int \tan^{n-2} au \, du$

407. $\displaystyle\int \tan^n au \sec^2 au \, du = \frac{\tan^{n+1} au}{(n+1)a}$

INTEGRALS INVOLVING INVERSE TRIGONOMETRIC FUNCTIONS

445. $\displaystyle\int \cos^{-1}\frac{u}{a}\, du = u\cos^{-1}\frac{u}{a} - \sqrt{a^2 - u^2}$

446. $\displaystyle\int u\cos^{-1}\frac{u}{a}\, du = \left(\frac{u^2}{2} - \frac{a^2}{4}\right)\cos^{-1}\frac{u}{a} - \frac{u\sqrt{a^2 - u^2}}{4}$

447. $\displaystyle\int u^2\cos^{-1}\frac{u}{a}\, du = \frac{u^3}{3}\cos^{-1}\frac{u}{a} - \frac{(u^2 + 2a^2)\sqrt{a^2 - u^2}}{9}$

448. $\displaystyle\int \frac{\cos^{-1}(u/a)}{u}\, du = \frac{\pi}{2}\ln|u| - \int \frac{\sin^{-1}(u/a)}{u}\, du$

449. $\displaystyle\int \frac{\cos^{-1}(u/a)}{u^2}\, du = -\frac{\cos^{-1}(u/a)}{u} + \frac{1}{a}\ln\left|\frac{a + \sqrt{a^2 - u^2}}{u}\right|$

450. $\displaystyle\int \left(\cos^{-1}\frac{u}{a}\right)^2 du = u\left(\cos^{-1}\frac{u}{a}\right)^2 - 2u - 2\sqrt{a^2 - u^2}\cos^{-1}\frac{u}{a}$

451. $\displaystyle\int \sin^{-1}\frac{u}{a}\, du = u\sin^{-1}\frac{u}{a} + \sqrt{a^2 - u^2}$

452. $\displaystyle\int u\sin^{-1}\frac{u}{a}\, du = \left(\frac{u^2}{2} - \frac{a^2}{4}\right)\sin^{-1}\frac{u}{a} + \frac{u\sqrt{a^2 - u^2}}{4}$

453. $\displaystyle\int u^2\sin^{-1}\frac{u}{a}\, du = \frac{u^3}{3}\sin^{-1}\frac{u}{a} + \frac{(u^2 + 2a^2)\sqrt{a^2 - u^2}}{9}$

454. $\displaystyle\int \frac{\sin^{-1}(u/a)}{u}\, du = \frac{u}{a} + \frac{(u/a)^3}{2\cdot 3\cdot 3} + \frac{1\cdot 3(u/a)^5}{2\cdot 4\cdot 5\cdot 5} + \frac{1\cdot 3\cdot 5(u/a)^7}{2\cdot 4\cdot 6\cdot 7\cdot 7} + \cdots$

455. $\displaystyle\int \frac{\sin^{-1}(u/a)}{u^2}\, du = -\frac{\sin^{-1}(u/a)}{u} - \frac{1}{a}\ln\left|\frac{a + \sqrt{a^2 - u^2}}{u}\right|$

456. $\displaystyle\int \left(\sin^{-1}\frac{u}{a}\right)^2 du = u\left(\sin^{-1}\frac{u}{a}\right)^2 - 2u + 2\sqrt{a^2 - u^2}\sin^{-1}\frac{u}{a}$

457. $\displaystyle\int \tan^{-1}\frac{u}{a}\, du = u\tan^{-1}\frac{u}{a} - \frac{a}{2}\ln(u^2 + a^2)$

458. $\int u \tan^{-1}\dfrac{u}{a}\, du = \tfrac{1}{2}(u^2 + a^2)\tan^{-1}\dfrac{u}{a} - \dfrac{au}{2}$

459. $\int u^2 \tan^{-1}\dfrac{u}{a}\, du = \dfrac{u^3}{3}\tan^{-1}\dfrac{u}{a} - \dfrac{au^2}{6} + \dfrac{a^3}{6}\ln(u^2 + a^2)$

INTEGRALS INVOLVING e^{au}

483. $\int e^{au}\, du = \dfrac{e^{au}}{a}$

484. $\int u e^{au}\, du = \dfrac{e^{au}}{a}\left(u - \dfrac{1}{a}\right)$

485. $\int u^2 e^{au}\, du = \dfrac{e^{au}}{a}\left(u^2 - \dfrac{2u}{a} + \dfrac{2}{a^2}\right)$

486. $\int u^n e^{au}\, du = \dfrac{u^n e^{au}}{a} - \dfrac{n}{a}\int u^{n-1} e^{au}\, du$

$$= \dfrac{e^{au}}{a}\left(u^n - \dfrac{nu^{n-1}}{a} + \dfrac{n(n-1)u^{n-2}}{a^2} - \cdots + \dfrac{(-1)^n n!}{a^n}\right)$$

if n = positive integer

487. $\int \dfrac{e^{au}}{u}\, du = \ln|u| + \dfrac{au}{1\cdot 1!} + \dfrac{(au)^2}{2\cdot 2!} + \dfrac{(au)^3}{3\cdot 3!} + \cdots$

488. $\int \dfrac{e^{au}}{u^n}\, du = \dfrac{-e^{au}}{(n-1)u^{n-1}} + \dfrac{a}{n-1}\int \dfrac{e^{au}}{u^{n-1}}\, du$

489. $\int \dfrac{du}{p + qe^{au}} = \dfrac{u}{p} - \dfrac{1}{ap}\ln|p + qe^{au}|$

490. $\int \dfrac{du}{(p + qe^{au})^2} = \dfrac{u}{p^2} + \dfrac{1}{ap(p + qe^{au})} - \dfrac{1}{ap^2}\ln|p + qe^{au}|$

491. $\int \dfrac{du}{pe^{au} + qe^{-au}} = \begin{cases} \dfrac{1}{a\sqrt{pq}}\tan^{-1}\left(\sqrt{\dfrac{p}{q}}\, e^{au}\right) \\[2mm] \dfrac{1}{2a\sqrt{-pq}}\ln\left|\dfrac{e^{au} - \sqrt{-q/p}}{e^{au} + \sqrt{-q/p}}\right| \end{cases}$

492. $\int e^{au}\sin bu\, du = \dfrac{e^{au}(a\sin bu - b\cos bu)}{a^2 + b^2}$

493. $\int e^{au}\cos bu\, du = \dfrac{e^{au}(a\cos bu + b\sin bu)}{a^2 + b^2}$

INTEGRALS INVOLVING $\ln|u|$

499. $\int \ln|u|\, du = u\ln|u| - u$

500. $\int (\ln|u|)^2\, du = u(\ln|u|)^2 - 2u\ln|u| + 2u$

501. $\int (\ln|u|)^n\, du = u(\ln|u|)^n - n\int (\ln|u|)^{n-1}\, du$

502. $\int u\ln|u|\, du = \dfrac{u^2}{2}\left(\ln|u| - \tfrac{1}{2}\right)$

APPENDIX E

Answers to Selected Problems

Many problems in this book are labeled WHAT DOES THIS THIS SAY? *These problems solicit answers in your own words or a statement for you to rephrase as a given statement in your own words. For this reason, it seems inappropriate to include the answers to these questions,* Think Tank Problems, *discussion, research problems, proofs, or problems for which answers may vary.*

We also believe that an answer section should function as a check on work done, so for that reason, when an answer has both an exact answer and an approximate solution (from technology), we usually show only the approximate solution in this appendix. The exact solution (which may be the more appropriate answer) can be **checked** *by using the given approximation.*

The Student Survival and Solutions Manual offers some review, survival hints, and some added explanations for selected problems. These problems are designated in the text by a colored problem number.

CHAPTER 1: FUNCTIONS AND GRAPHS

1.1 Preliminaries (pages 12–14)

1. a. $(-3, 4)$ **b.** $3 \leq x \leq 5$ **c.** $-2 \leq x < 1$ **d.** $(2, 7]$

3. a.

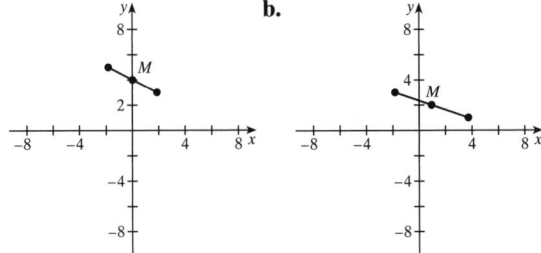

b.

c.

d.

5. a.   **b.**

$M = (0, 4)$ $M = (1, 2)$
$d = 2\sqrt{5}$ $d = 2\sqrt{10}$

7. $x = 0, 1$ **9.** $y = 7, -2$

11. $x = \dfrac{b \pm \sqrt{b^2 + 12c}}{6}$ **13.** $x = 6, -10$

15. $w = -2, 5$ **17.** $\emptyset$

21. $x = \frac{3\pi}{4}, \frac{5\pi}{4}, \frac{\pi}{3}, \frac{5\pi}{3}$ **23.** $x = \frac{2\pi}{3}$

25. $\left(-\infty, -\frac{5}{3}\right)$ **27.** $\left(-\frac{5}{3}, 0\right)$

29. $(-8, -3]$ **31.** $[-1, 3]$

33. $[7.999, 8.001]$ **35.** $(x + 1)^2 + (y - 2)^2 = 9$

37. $x^2 + (y - 1.5)^2 = 0.0625$

39.   **41.**

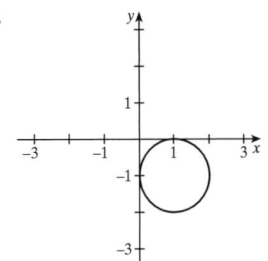   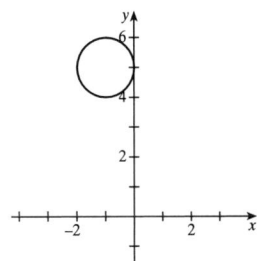

43. $\dfrac{\sqrt{2} - \sqrt{6}}{4} \approx -0.2588$ **45.** $2 - \sqrt{3} \approx 0.2679$

51. a. period 2π, amp 1   **b.** period 2π, amp 1

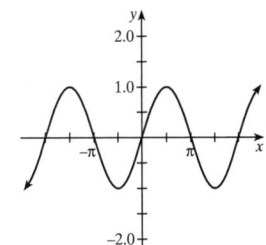   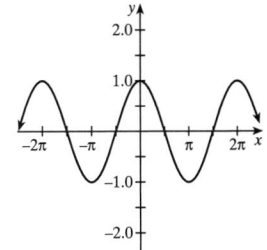

c. period π

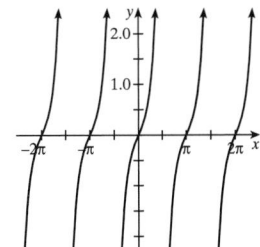

53. period $\pi/2$ **55.** period 4π

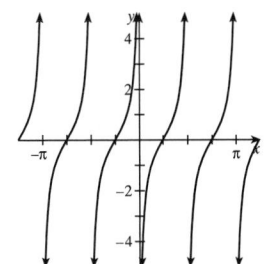   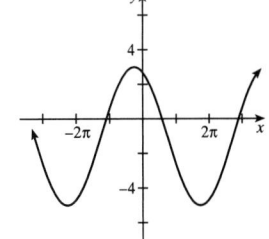

57.

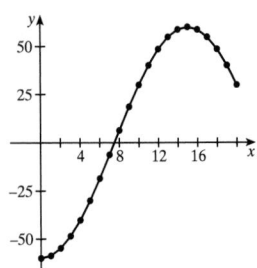

$A = 0, B = 60, C = \frac{\pi}{15}$, and $D = 7.5$

59. sun curve: $y = \cos\frac{\pi}{6}x$; moon curve: $y = 4\cos\frac{\pi}{6}x$;
combined curve: $y = 5\cos\frac{\pi}{6}x$

61. a. The apparent depth is 3.4 m.
  b. The angle of incidence is 58°.

1.2 Lines in the Plane (pages 21–23)

3. $2x + y - 5 = 0$

5. $2y - 1 = 0$

7. $x + 2 = 0$

9. $8x - 7y - 56 = 0$

11. $3x + y - 5 = 0$

13. $3x + 4y - 1 = 0$

15. $4x + y + 3 = 0$

17. $m = -5/7$

19. $m = 6.001$

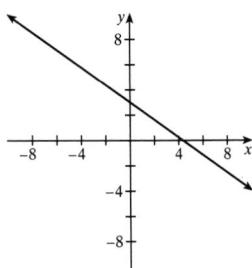

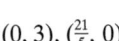

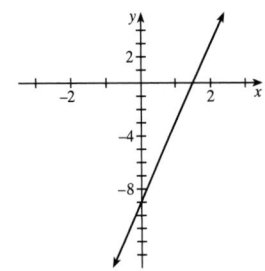

$(0, 3), (\frac{21}{5}, 0)$

$(1.50025, 0), (0, 9.003)$

21. $y = -\frac{3}{5}x - 3$
  $m = -3/5$

23. $y = \frac{3}{5}x - 0.3$
  $m = 3/5$

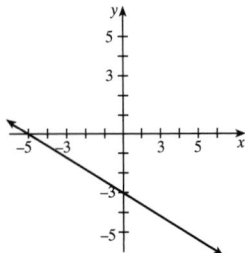

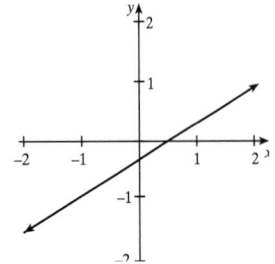

$(-5, 0), (0, -5)$

$(0.5, 0), (0, -0.3)$

25.
  $m = 3/2$

27. $y = \frac{1}{5}x$
  $m = 1/5$

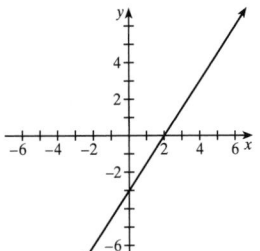

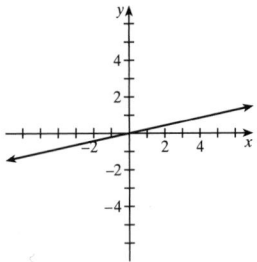

$(2, 0), (0, -3)$

$(0, 0)$

29. no slope

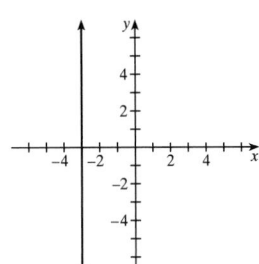

$(-3, 0)$

31. $y = 0$, and $y = 6$

33. $D(6, 6), E(2, 16)$

35. no values

37. $(4.0, -1.0)$

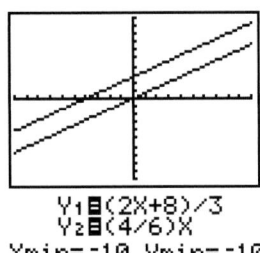

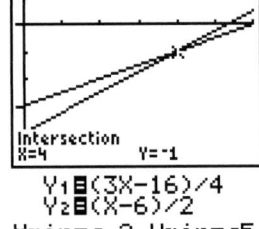

39. $(\frac{3}{4}, \frac{7}{2})$

41. $(-\frac{64}{3}, \frac{100}{3})$

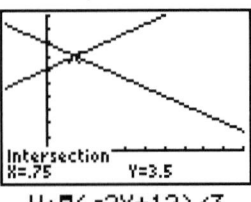

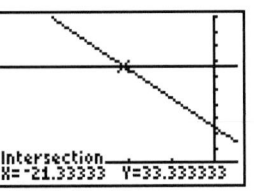

43. $(\sqrt{2}, \sqrt{2}), (-\sqrt{2}, -\sqrt{2})$ **45.** $\left(-\dfrac{15}{8}, \dfrac{7\sqrt{15}}{8}\right),$

$\left(-\dfrac{15}{8}, -\dfrac{7\sqrt{15}}{8}\right)$

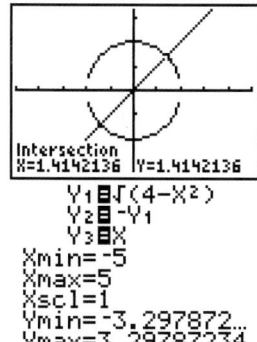

47. a. $-38.2°\text{F}$ **b.** $-17.8°\text{C}$ **c.** $-40°\text{C}$
49. $C = 60x + 5{,}000$ **51.** $V(t) = -19{,}000t + 200{,}000$

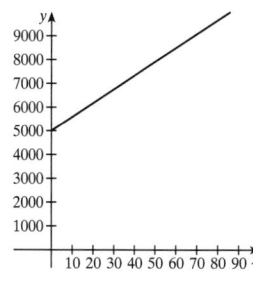 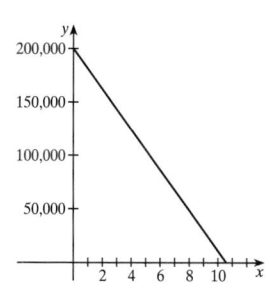

The value is 4 years is \$124,000.

55. a. $B(1, 11)$ and $C(-3, -5)$
  b. The center for both triangles is $\left(\frac{1}{3}, \frac{7}{3}\right)$.
57. For the 8th month, 216 gallons

61. $x = \dfrac{a \pm \sqrt{a^2 - 4}}{2},\ y = \dfrac{2}{a \pm \sqrt{a^2 - 4}}$

1.3 Functions (pages 32–34)

Let D represent the domain in Problems 1–12.

1. $D = (-\infty, \infty); f(-2) = -1; f(1) = 5; f(0) = 3$
3. $D = (-\infty, \infty); f(1) = 6; f(0) = -2; f(-2) = 0$
5. $D = (-\infty, -3) \cup (-3, \infty); f(2) = 0; f(0) = -2; f(-3)$ is
  undefined
7. $D = (-\infty, 2)] \cup [0, \infty); f(-1)$ is undefined;
  $f\left(\frac{1}{2}\right) = \dfrac{\sqrt{5}}{2}; f(1) = \sqrt{3}$
9. $D = (-\infty, \infty); f(-1) = \sin 3 \approx 0.1411; f\left(\frac{1}{2}\right) = 0;$
  $f(1) = \sin(-1) \approx -0.8415$
11. $D = (-\infty, \infty); f(3) = 4; f(1) = 2; f(0) = 4$

13. 9   **15.** $10x + 5h$ **17.** -1 **19.** $\dfrac{-1}{x(x + h)}$

21. 5   **23.** -1.002 **25.** $-\delta^2 - 3\delta - 3$ **27.** not equal
29. equal   **31.** not equal
33. $(f \circ g)(x) = 4x^2 + 1; (g \circ f)(x) = 2x^2 + 2$
35. $(f \circ g)(t) = |t|; (g \circ f)(t) = t$
37. $(f \circ g)(x) = \sin(2x + 3); (g \circ f)(x) = 2 \sin x + 3$
39. $u(x) = 2x^2 - 1; g(u) = u^4$
41. $u(x) = 2x + 3; g(u) = |u|$
43. $u(x) = \tan x; g(u) = u^2$
45. $u(x) = \sqrt{x}; g(u) = \sin u$ **47.** $u(x) = \dfrac{x + 1}{2 - x}; g(u) = \sin u$
49. a. the cost is \$4,500   **b.** The cost of the 20th unit is \$371.
51. a. $I = \dfrac{30}{t^2(6 - t)^2}$ **b.** $I(1) = \frac{6}{5}$ candles; $I(4) = \frac{15}{32}$ candles
53. a. $625t^2 + 25t + 900$ **b.** \$6,600   **c.** 4 hours
55. $(2.99, 2.99^2), (3.01, 3.01)^2; m = 6$
57. 27.00000001
59. $(2.9999, -25.99820003), (3.0001, -26.00180003);$
  $m = -18.00135$

1.4 Functions and Graphs (pages 41–43)

1. even   **3.** neither   **5.** neither   **7.** even
9. 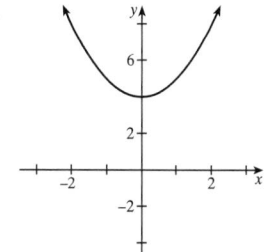   **11.**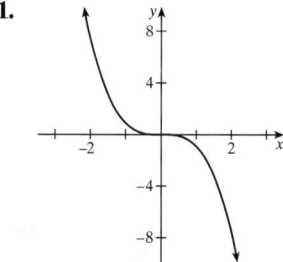

13. 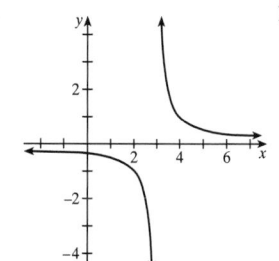   **15.**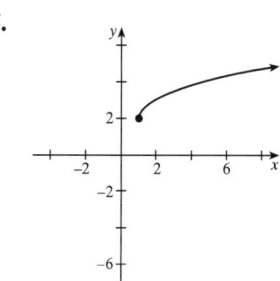

17. 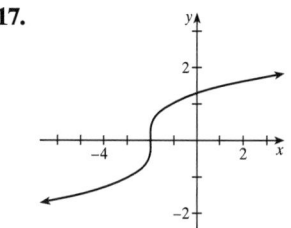   **19.**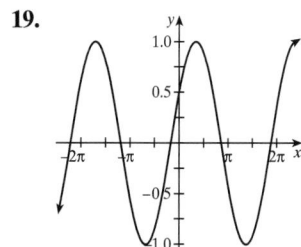

21.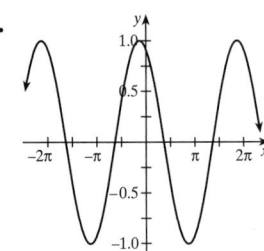

23.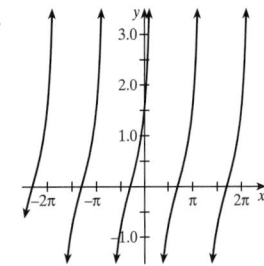

25. $P(5, f(5)); Q(x_0, f(x_0))$

27. $-\frac{1}{3}, 2$

29. $15, -\frac{25}{2}, \frac{65}{3}, -\frac{1}{4}$

31. $\frac{1 \pm \sqrt{6}}{5}$

33. 0

35. ± 1

37. ± 5.42

39. ± 2.24

41. $-3.00, 2.00, 14.00$

43. $-12.00, 18.00$

47. a.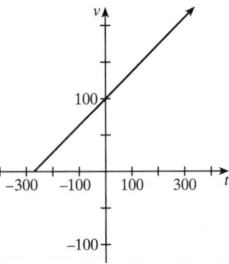

b. $T = 273$

49. $2x + 5y - 19 = 0$

51. It hits the ground approximately 1,234 ft from the firing point.

53. The basic shape is that of a parabola (standard quadratic function).

55. a. 19,400 people
  b. 67 people
  c. The population will tend to 20,000 people in the long run.

57. $x \approx -2.6139, 0.8031, 3.8031$

59. a. yes
  b. $G(x) = (x + 2)(x^2 - 3)(x^2 + 3)$

61. Pythagorean theorem: $\triangle ABC$ with sides a, b, and c is a right triangle if and only if $c^2 = a^2 + b^2$. Proofs vary.

1.5 Inverse Functions; Inverse Trigonometric Functions (pages 52–53)

3. These are inverse functions.

5. These are not inverse functions.

7. These are not inverse functions.

9. $\{(5, 4), (3, 6), (1, 7), (4, 2)\}$

11. $y = \frac{1}{2}x - \frac{3}{2}$

13. $y = \sqrt{x + 5}$

15. $y = (x - 5)^2$

17. $y = \frac{3x + 6}{2 - 3x}$

19. a. $\frac{\pi}{3}$ **b.** $-\frac{\pi}{3}$

21. a. $-\frac{\pi}{4}$ **b.** $\frac{5\pi}{6}$

23. a. $-\frac{\pi}{3}$ **b.** π

25. $\frac{\sqrt{3}}{2}$

27. 3

29. $-\frac{2\sqrt{6}}{5} \approx -0.9798$

33.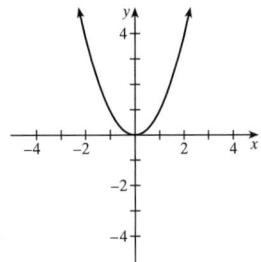

no inverse

35.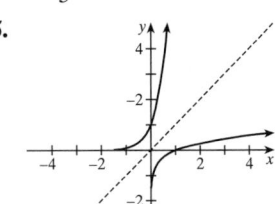

inverse exists

37.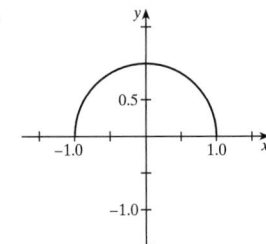

no inverse

39.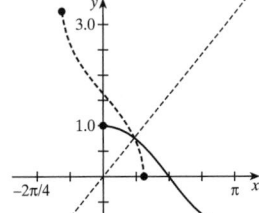

inverse exists

41. $\frac{2x}{x^2 + 1}$

43. $\frac{\sqrt{1 - x^2}}{x}$

45. 1

49. a. 0.5880 **b.** 2.5536 **c.** 2.1997 **d.** -0.5746

51. $h = \frac{d \tan \beta \tan \alpha}{\tan \alpha - \tan \beta}$

53. a. 3.141592654; conjecture is that it is π.

1.6 Exponential and Logarithm Functions (pages 65–67)

1.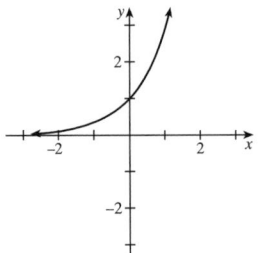

3.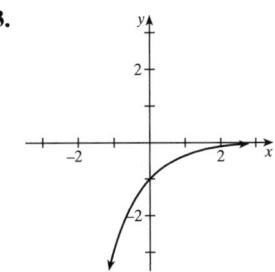

5. 31

7. 200.33681

9. 9,783.225896

11. 38,523.62544

13. 0

15. 2

17. -2

19. 3.5

21. $\frac{3}{10}$

23. 4

25. 0.23104906

27. -1.391662509

29. 729

31. $2, -1$

33. 3

35. $2, -\frac{5}{3}$

37. $-\frac{3}{2}$

39. $1, -\frac{3}{2}$

41. 2

43. 9

45. exponential

47. logarithmic

49. logarithmic

51. exponential

53. $2.4, 0.4$

55. 11 yr 202 days

57. First National offers the better deal.

59. The interest rate is approximately 5.71%.

65. a. 30.12%   **b.** 77.69%   **c.** 7.81%

67. a. 263.34   **b.** 1, 232.72

69. Scélérat, at around 1:00 A.M. Wednesday

CHAPTER 1: REVIEW

Proficiency Examination (page 568)

21. a. $6x + 8y - 37 = 0$ **b.** $3x + 10y - 41 = 0$
c. $3x - 28y - 12 = 0$ **d.** $2x + 5y - 24 = 0$
e. $4x - 3y + 8 = 0$

22. $y = -\frac{3}{2}x + 6$

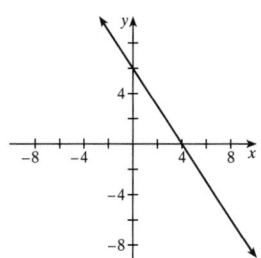

23. $y - 3 = |x + 1|$

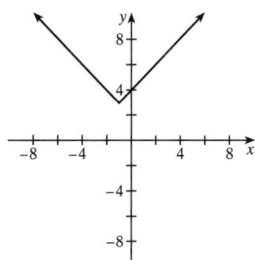

24. $y - 3 = -2(x - 1)^2$

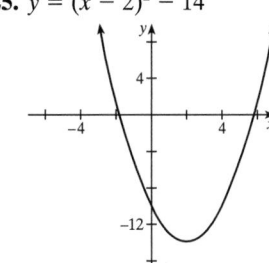

25. $y = (x - 2)^2 - 14$

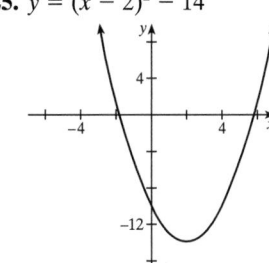

26. $y = 2\cos(x - 1)$

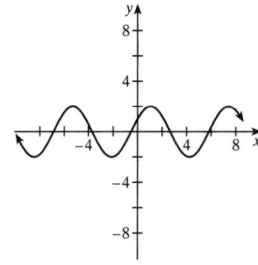

27. $y + 1 = \tan 2(x + \frac{3}{2})$

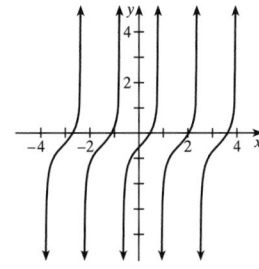

28. $y = \sin^{-1}(2x)$

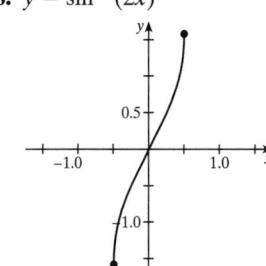

29. $y = \tan^{-1}x^2$

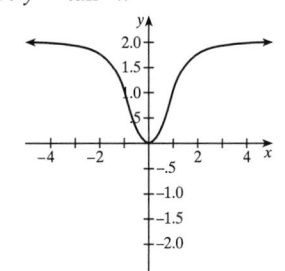

30. $y = e^{-x} + e^x$

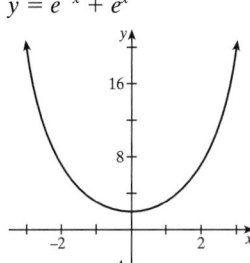

31. $y = \ln(1 - x)$

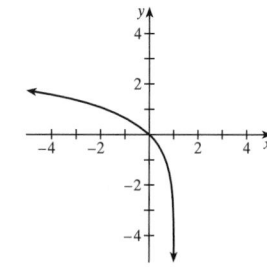

32. $y = e^{2x} + \ln x$

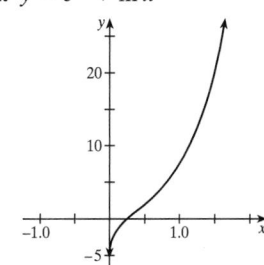

33. $y = e^x - \ln x + 15$

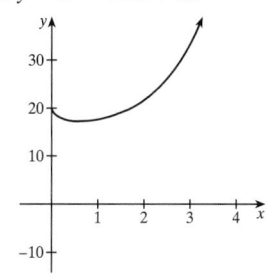

34. $-\frac{3}{2}, 1$ **35.** The two functions are not the same.

36. $f \circ g = \sin(\sqrt{1 - x^2}); g \circ f = |\cos x|$

37. 4.6286   **38.** 1.0986

39. $V = \frac{2}{3}x(12 - x^2)$

40. a. It will take 11 years, 3 quarters.
   b. It will take 11 years, 6 months.
   c. It will take 11 years, 166 days.

Supplementary Problems (pages 69–72)

1. $P = 30; A = 30$ **3.** $P \approx 27.1; A = 40$

5. $(x - 5)^2 + (y - 4)^2 = 16$ **7.** $5x - 3y + 3 = 0$

9. $A = -\frac{1}{6}$ and $B = \frac{1}{2}$

11. The medians meet at the point $(2, \frac{5}{3})$.

13. 1,024 ft

15. a. 1.504077   **b.** 16.444647   **c.** 1.107149   **d.** 1.899250945

17. $\frac{3}{5}$ **19.** $\frac{5}{2}$ **21.** ± 2 **23.** 16   **25.** $\sqrt{13}$

27. $f^{-1}(x) = \sqrt[3]{\frac{1}{2}(x + 7)}$ **29.** $f^{-1}(x) = \ln(x^2 + 1)$

31. $f^{-1}(x) = \dfrac{x + a}{x - 1}$

33. The domain of f^{-1} is all real $x, x \neq 1$.

35. a. false   **b.** false   **c.** false   **d.** true   **e.** true   **f.** true

37. a. $k = 0.25 \ln 2$; After 7 weeks, approximately 29.7% are burning.   **b.** 0.8232   **c.** 0.07955

39. a. $P \approx \$1,075.71$ **b.** $P \approx \$1,070.52$

41. a. 15 inches   **b.** 21 inches   **c.** 29 inches   **d.** 19 inches

43. a. The domain consists of all $x \neq 300$.   **b.** x represents a percentage, so $0 \leq x \leq 100$ in order for $f(x) \geq 0$ **c.** 120   **d.** 300   **e.** The percentage of households should be 60%.

45. a. $V = \frac{256}{3}\pi; S = 64\pi$ **b.** $V = 30; S = 62$ **d.** $V = 15\pi; A = 3\pi\sqrt{34}$

49. $c = -\frac{4}{5}$ **51.** $\theta = \frac{\pi}{4} - \tan^{-1}\frac{5}{12}$

53. The break-even point occurs when $x = \$30,000$.

55. The glass should be taken in on the 11th or 12th day.

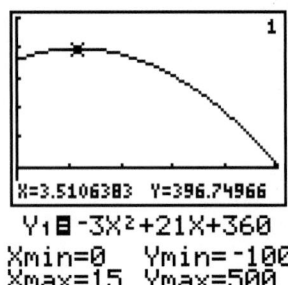

57. $C(x) = 4x^2 + 1,000x^{-1}$

59. Putnam competition problem solutions are copied with permission from *The William Lowell Putnam Mathematical Competition, Problems and Solutions: 1938–1964* by A. M. Gleason, R. E. Greenwood, and L. M. Kelly, published by The Mathematical Association of America. We will provide reference about where you can find the solution to the Putnam Problems. This is Problem 4 in the morning session for 1959.

CHAPTER 2: LIMITS AND CONTINUITY

2.1 What Is Calculus? (pages 81–84)

5. $\frac{1}{3}$ **7.** 1   **9.** $\frac{3}{11}$ **11.** π

13. a. 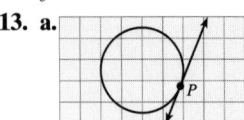   **b.**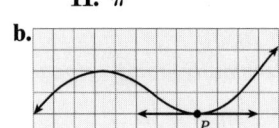

15. a.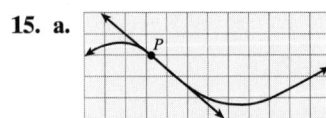

    b. There is no unique tangent line.

17. 2   **19.** 1   **21.** 0

23. $\lim\limits_{n \to \infty} \dfrac{n+1}{n}$ **25.** 1   **27.** 2

29. a.   **b.**

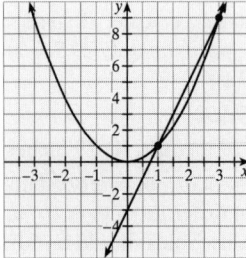   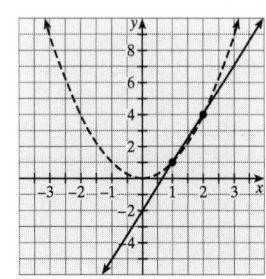

The slope of the secant line is $m = 4$.   The slope of the secant line is $m = 3$.

c.

n	x_n	point	slope
1	3	$(3, 9)$	$m = 4$
2	2	$(2, 4)$	$m = 3$
3	1.5	$(1.5, 2.25)$	$m = 2.5$
4	1.1	$(1.1, 1.21)$	$m = 2.1$

d. The slope of the tangent line is $m = 2$.

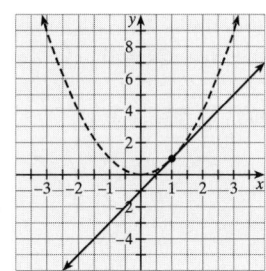

31. 2.66 square units   **33.** 6.75 square units

35. 6.28 square units

37. quadratic model   **39.** exponential model

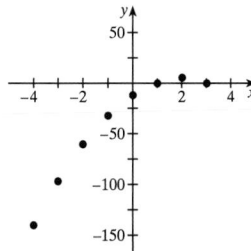   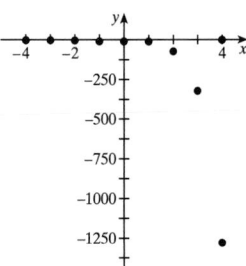

41. quadratic model   **43.** cubic model

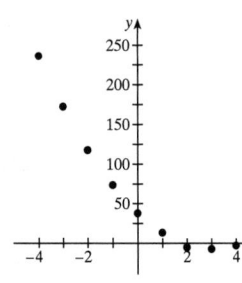   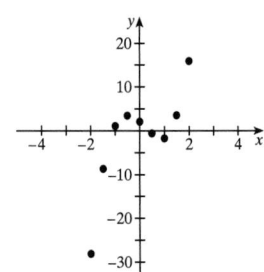

45. logarithmic model   **47.** $\pi \approx 3.16$

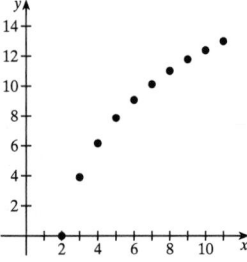

49. $A_3 \approx 1.2990; A_4 = 2; A_5 \approx 2.3776; A_6 \approx 2.5981;$
$A_7 \approx 2.7364; \ldots A_{100} \approx 3.1395$

51. $A = 0.3984375$

57. a. Cost $= 2ax^2 + 8bx + 924ax^{-1} + 924bx^{-2}$
  b. The cost is minimized if the base is 6 in.^{2} and the height is 6.42 in.

2.2 The Limit of a Function (pages 95–97)

1. a. 0   **b.** 2   **c.** 6   **3. a.** 2   **b.** 7   **c.** 7.5
5. a. 6   **b.** 6   **c.** 6   **7.** 15
9. 10   **11.** 8
13. 2   **15.** −1
19. 1.00   **21.** 5.00
23. −0.17   **25.** does not exist
27. a. does not exist   **b.** 0
29. a. 0.00   **b.** 0.64   **31. a.** 0   **b.** 0.37
33. −2.00   **35.** 0.17
37. 0.25   **39.** does not exist
41. 2.00   **43.** 1.00
45. 2.72   **47.** 0.00
49. does not exist
51. a. $-32t + 40$ **b.** 40 ft/s   **c.** 3; impact velocity is −56 ft/s
  d. $t = 1.25$ seconds
53. 228   **55.** 0

2.3 Properties of Limits (pages 105–106)

1. −9   **3.** −8   **5.** $-\frac{1}{2}$ **7.** 2   **9.** $\dfrac{\sqrt{3}}{9}$ **11.** 4
13. −1   **15.** $\frac{1}{9}$ **7.** $\frac{1}{2}$ **19.** 2   **21.** $\frac{5}{2}$ **23.** 0
25. 1   **27.** 0   **29.** 0   **31.** $\frac{4}{3}$ **37.** −1   **39.** 0
41. the limit does not exist
43. the limit does not exist   **55.** the limit does not exist
57. 4   **59.** 8

2.4 Continuity (pages 115–117)

1. Temperature is continuous, so TEMPERATURE $= f$(time) would be a continuous function. The domain would be midnight to midnight say, $0 \le t < 24$.

3. The selling price of ATT stock is not continuous. The domain is the set of positive rational numbers that can be divided evenly by 8. At the time of this writing, there was some discussion to changing the stock quotations to the nearest cent.

5. The charges (range of the function) consist of rational numbers only (dollars and cents to the nearest cent), so the function CHARGE $= f$(MILEAGE) would be a step function (that is, not continuous). The domain would consist of the mileage from the beginning of the trip to its end.

7. No suspicious points and no points of discontinuity with a polynomial.

9. The denominator factors to $x(x − 1)$, so suspicious points would be $x = 0, 1$. There will be a hole discontinuity at $x = 0$ and a pole discontinuity at $x = 1$.

11. $x = 0$ is suspicious and is a point of discontinuity

13. $x = 1$ is a suspicious point; there are no points of discontinuity

15. The sine and cosine are continuous on the reals, but the tangent is discontinuous at $x = \pi/2 + n\pi$. Each of these values will have a pole type discontinuity.

17. a suspicious point is located at $x = 0$. There is a discontinuity at $x = 0$ (pole)

19. 3   **21.** π **23.** no value

25. a. continuous   **b.** discontinuous on $[0, 1]$

27. discontinuous at $t = 0$

29. continuous

45. $a = 1; b = -18/5$ **47.** $a = 1; b = \frac{1}{2}$ **49.** $a = 5; b = 5$
51. $a = \frac{4}{3}, \frac{14}{3}, \cdots; b = \sqrt{3}$
53. It is not possible to redefine f at $x = 2$, so that it becomes continuous.

CHAPTER 2: REVIEW

Proficiency Examination (pages 117–118)

11. $\frac{3}{2}$ **12.** $\frac{1}{4}$ **13.** $-\frac{1}{4}$ **14.** 0   **15.** $\frac{9}{5}$ **16.** −1

17. We have suspicious points where the denominators are 0 at $t = 0, -1$. There are pole discontinuities at each of these points.

18. Suspicious points $x = -2$ and $x = 1$ are also points of discontinuity (since the denominator is 0).

19. $A = -1; B = 1$

Supplementary Problems (pages 118–121)

1. 5   **3.** 1   **5.** 0   **7.** $-\frac{1}{2}$ **9.** e^4 **11.** 5

13. $\frac{3}{2}$ **15.** 0.8415   **17.** 1   **19.** 3   **21.** 0   **23.** $\dfrac{1}{\sqrt{2x}}$

25. $\dfrac{-4}{x^2}$ **27.** e^x **29.** continuous on $[-5, 5]$

31. discontinuity is removable

33. not continuous anywhere

35. a. continuous on $[0, 5]$ **b.** not continuous $x = -2$
  c. not continuous at $x = -2$ **d.** continuous on $[-5, 5]$

37. a.

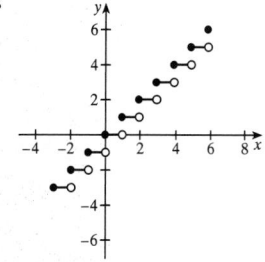

  b. $\lim\limits_{x \to 3} [\![x]\!]$ does not exist.

  c. The limit exists for all nonintegral values.

39. does not exist   **41.** $a = 2$ and $b = 1$

43. discontinuity is removable

45. $x \approx 0.6$ or 0.646944

49. a. 7.16 degrees   **b.** 1.8 second

57. a. the windchill for 20 mi/h is $3.75°$ and for $v \approx 25.2$
    c. at $v = 4$, $T \approx 91.4$; at $v = 45$, $T \approx 868$

61. The tangent at $x = 2$ is $y = 1.8 + 8.16(x - 2)$

63. This is Problem A1 of the morning session of the 1956 Putnam Examination.

CHAPTER 3: DIFFERENTIATION

3.1 An Introduction to the Derivative: Tangents (pages 139–142)

5.
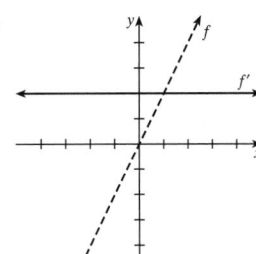

7.
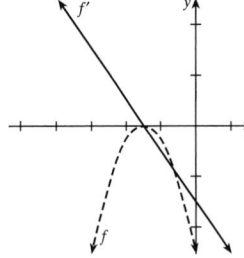

9.
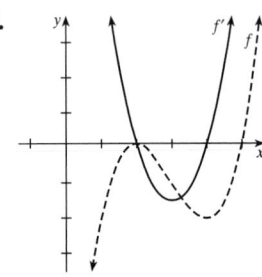

11. a. 0   **b.** 0
13. a. 2   **b.** 2
15. a. 0   **b.** 0
17. 0; differentiable for all x
19. 3; differentiable for all x **21.** $6x$; differentiable for all x
23. $2x - 1$; differentiable for all x
25. $2s - 2$; differentiable for all real s
27. $\dfrac{\sqrt{5x}}{2x}$; differentiable for $x > 0$

29. $3x - y - 7 = 0$ **31.** $3s - 4y + 1 = 0$
33. $x + 25y - 7 = 0$ **35.** $x + 3y - 9 = 0$
37. $216x - 6y - 647 = 0$ **39.** 2
41. 0   **43. a.** -3.9 **b.** -4

45.
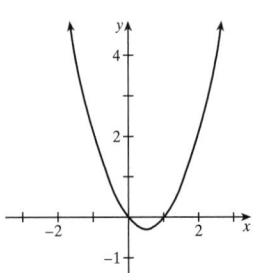

The derivative is 0 when $x = \frac{1}{2}$; the graph has a horizontal tangent at $(\frac{1}{2}, -\frac{1}{4})$.
47. a. $-4x$ **b.** $y - 4 = 0$ **c.** $(\frac{2}{3}, \frac{28}{9})$

49. yes

51. a.
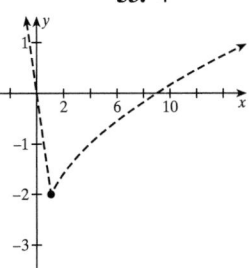
          53. 4   **55.** $\frac{1}{2}$ **57.** $\frac{1}{4}$

59. The y-intercept occurs when $y = -Ac^2$.

3.2 Techniques of Differentiation (pages 150–151)

1. (11) 0   (12) 1   (13) 2   (14) 4   (15) 0   (16) -4

3. (23) $2x - 1$ (24) $-2t$ (25) $2s - 2$ (26) $-\frac{1}{2}x^{-2}$ (27) $\dfrac{\sqrt{5x}}{2x}$

5. a. $12x^3$ **b.** -1 **7. a.** $3x^2$ **b.** 1
9. $2t + 2t^{-3} - 20t^{-5}$ **11.** $-14x^{-3} + \frac{2}{3}x^{-1/3}$
13. $1 - x^{-2} + 14x^{-3}$ **15.** $-32x^3 - 12x^2 + 2$
17. $\dfrac{22}{(x + 9)^2}$ **19.** $4x^3 + 12x^2 + 8x$
21. $f'(x) = 5x^4 - 15x^2 + 1; f''(x) = 20x^3 - 30x$
    $f'''(x) = 60x^2 - 30; f^{(4)}(x) = 120x$
23. $f'(x) = 4x^{-3}; f''(x) = -12x^{-4}; f'''(x) = 48x^{-5}$;
    $f^{(4)}(x) = -240x^{-6}$
25. $\dfrac{d^2y}{dx^2} = 18x - 14$ **27.** $7x + y + 9 = 0$
29. $6x + y - 6 = 0$ **31.** $x - 6y + 5 = 0$
33. $(1, 0)$ and $(\frac{4}{3}, -\frac{1}{27})$ **35.** $(\frac{29}{6}, -\frac{361}{12})$
37. $(1, -2)$ **39.** no horizontal tangents
41. d. $-4x^{-3} + 9x^{-4}$ **43.** $2x - y - 2 = 0$
45. a. $4x + y - 1 = 0$ **b.** the normal line is horizontal
47. $(0, 0)$ and $(4, 64)$
49. The equation is not satisfied.
51. This function satisfies the given equation.
53. $\pi \approx \dfrac{62{,}832}{20{,}000} \approx 3.1416$
63. $A = 0, B = 1, C = 3$, so the required function is $y = x + 3$.

3.3 Derivatives of the Trigonometric, Exponential, and Logarithmic Functions (pages 158–159)

1. $\cos x - \sin x$ **3.** $2t - \sin t$
5. $\sin 2t$
7. $-\sqrt{x} \sin x + \frac{1}{2}x^{-1/2} \cos x - x\csc^2 x + \cot x$
9. $-x^2\sin x + 2x \cos x$
11. $\dfrac{x \cos x - \sin x}{x^2}$
13. $e^t\csc t(1 - \cot t)$
15. $x + 2x\ln x$ **17.** $2e^x \cos x$
19. $e^{-x}(\cos x - \sin x)$

21. $\dfrac{\sec^2 x - 2x \sec^2 x + 2\tan x}{(1 - 2x)^2}$

23. $\dfrac{t\cos t + 2\cos t - \sin t - 2}{(t + 2)^2}$

25. $\dfrac{-1}{1 - \cos x}$

27. $\dfrac{2\cos x - \sin x - 1}{(2 - \cos x)^2}$

29. $\dfrac{-2}{(\sin x - \cos x)^2}$

31. $-\sin x$

33. $-\sin\theta$

35. $2\sec^2\theta\tan\theta$

37. $\sec^3\theta + \sec\theta\tan^2\theta$

39. $-\sin x - \cos x$

41. $-2e^x\sin x$

43. $\frac{1}{4}t^{-3/2}\ln t$

45.

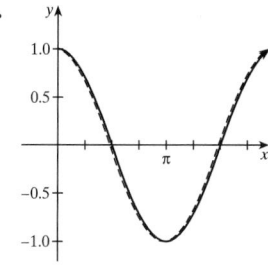

47. $4x - 2y - \pi + 2 = 0$

49. $\sqrt{3}x - 2y + \left(1 - \dfrac{\sqrt{3}\pi}{6}\right) = 0$

51. $2x - y = 0$

53. $x - y + 1 = 0$

55. a. yes   **b.** yes   **c.** no   **d.** no

57. $A = 0, B = -\frac{3}{2}; y = -\frac{3}{2}x\sin x$

3.4 Rates of Change: Rectilinear Motion (pages 166–169)

1. 1   **3.** -3 **5.** $\frac{13}{4}$ **7.** -1

9. $\frac{1}{2}$ **11.** -1 **13.** -6

15. a. $2t - 2$ **b.** 2   **c.** 2

d. Because $a(t) > 0$, the object is continuously accelerating.

17. a. $3t^2 - 18t + 15$ **b.** $6t - 18$ **c.** 46

d. On $[0, 3)$ the object is decelerating, and on $(3, 6]$ it is accelerating.

19. a. $-2t^{-2} - 2t^{-3}$ **b.** $4t^{-3} + 6t^{-4}$

c. $\frac{20}{9}$ On $[1, 3]$ the object is accelerating.

21. a. $-3\sin t$ **b.** $-3\cos t$ **c.** 12

d. On $[0, \frac{\pi}{2})$ the object is decelerating, on $(\frac{\pi}{2}, \frac{3\pi}{2})$ the object is accelerating, and on $(\frac{3\pi}{2}, 2\pi]$ it is decelerating again.

23. a. $f'(x) = -6$

b. The decline will be the same each year.

25. 136

27. a. 9.91 m/min   **b.** 9.96 m/s

29. a. The initial velocity is 64 ft/sec.

b. The cliff is 336 ft high.

c. $32t + 64$ ft/sec   **d.** 160 ft/sec

31. 30 ft

33. The height of the building is 144 ft.

35. The initial velocity is 24 ft/s, and the cliff is 126 ft. high.

37. a. $200t + 50\ln t + 450$ newspapers per year.

b. 1,530 newspapers per year

c. 1,635 newspapers

39. a. 0.2 ppm/yr   **b.** 0.15 ppm   **c.** 0.25 ppm

41. 91 thousand/hr

43. a. 20 persons per mo   **c.** 0.39% per mo

45. 7.5% per year

47. $\dfrac{dP}{dT} = -\dfrac{4\pi\mu^2 N}{9k}T^{-2}$

49. a. $v(t) = -7\sin t; a(t) = -7\cos t$

b. the period (one revolution) is 2π

c. The amplitude is 7.

51. The Spy is on Mars.   **53.** They are equal.

3.5 The Chain Rule (pages 174–176)

3. $6(3x - 2)$

5. $\dfrac{-8x}{(x^2 - 9)^3}$

7. $[\tan u + u\sec^2 u]\left[3 - \dfrac{6}{x^2}\right]$

9. $e^{\sec x}\sec x\tan x$

11. a. $3u^2$ **b.** $2x$ **c.** $6x(x^2 + 1)^2$

13. a. $7u^6$ **b.** $-8 - 24x$ **c.** $-7(24x + 8)(12x^2 + 8x - 5)^6$

15. a. $\cos(\sin\theta)(\cos\theta)$ **b.** $\cos(\cos\theta)(-\sin\theta)$

17. $4\cos(4\theta + 2)$ **19.** $(1 - 2x)e^{1-2x}$

21. $2x\cos 2x^2$

23. $2x(2x^2 + 1)^3(x^2 - 2)^4(18x^2 - 11)$

25. $\dfrac{1}{2}\left(\dfrac{x^2 + 3}{x^2 - 5}\right)^{-1/2}\left[\dfrac{-16x}{(x^2 - 5)^2}\right]$

27. $\dfrac{1}{3}(x + \sqrt{2x})^{-2/3}\left(1 + \dfrac{1}{\sqrt{2x}}\right)$

29. $(2t + 1)\exp(t^2 + t + 5)$ **31.** $\dfrac{\cos x - \sin x}{\sin x + \cos x}$

33. $2x - 3y + 5 = 0$ **35.** $y - \frac{1}{16} = 0$ **37.** $y = e^2 x$

39. $\frac{2}{9}$ **41.** $1, 7$ **43.** $0, \frac{2}{3}$

45. a. 1   **b.** $\frac{3}{2}$ **c.** 1.5   **47.** 0.31 parts/million

49. The demand is decreasing by 6 lb/wk.

51. a. Illuminance is increasing by 0.035 lux/s.   **b.** 7.15 m

57. a. $\dfrac{3}{(3x - 1)^2 + 1}$ **b.** $\dfrac{-1}{x^2 + 1}$

61. $\dfrac{d}{dx}f'[f(x)] = f''(x)[f(x)]f'(x)$ and

$\dfrac{d}{dx}f[f'(x)] = f'[f'(x)]f''(x)$

3.6 Implicit Differentiation (pages 186–189)

1. $-\dfrac{x}{y}$

3. $-\dfrac{y}{x}$

5. $\dfrac{-(2x + 3y)}{3x + 2y}$

7. $-\dfrac{y^2}{x^2}$

9. $\dfrac{1 - \cos(x + y)}{\cos(x + y) + 1}$

11. $\dfrac{2x - y\sin xy}{x\sin xy}$

13. $(2e^{2x} - x^{-1})y$

15. a. $-\dfrac{2x}{3y^2}$ **b.** $\dfrac{-2x}{3(12 - x^2)^{2/3}}$

17. a. y^2 **b.** $\dfrac{1}{(x - 5)^2}$

19. $\dfrac{1}{\sqrt{-x^2 - x}}$

21. $\dfrac{x}{(x^2 + 2)\sqrt{x^2 + 1}}$

23. $\dfrac{6(\sin^{-1}2x)^2}{\sqrt{1 - 4x^2}}$

25. $\dfrac{-1}{\sqrt{e^{-2x} - 1}}$

27. $\dfrac{-1}{x^2 + 1}$

29. $\dfrac{-2}{x\sqrt{4x^4 - 1}}$

31. $\dfrac{1 - \sin^{-1}y - \dfrac{y}{1 + x^2}}{\dfrac{x}{\sqrt{1 - y^2}} + \tan^{-1}x}$

33. $2x - 3y + 13 = 0$ **35.** $(\pi + 1)x - y + \pi = 0$

37. $y = 0$ **39.** $y' = 0$

41. $y' = \frac{5}{4}$ **43.** $x - 1 = 0$ **45.** $-\dfrac{49}{100y^3}$

51. $y\left[\dfrac{10}{2x - 1} - \dfrac{1}{2(x - 9)} - \dfrac{2}{x + 3}\right]$

53. $y\left[6x - \dfrac{6x^2}{x^3 + 1} + \dfrac{8}{4x - 7}\right]$

55. $\dfrac{y \ln x}{x}$ **57.** $(3, -4)$ and $(-3, 4)$

63. vertical tangents at $(\sqrt{2}, 0)$, $\left(-\dfrac{2\sqrt{3}}{9}, \dfrac{\sqrt{15}}{9}\right)$ and $\left(-\dfrac{2\sqrt{3}}{9}, -\dfrac{\sqrt{15}}{9}\right)$

65. 0.44 rad/s   **69.** $\dfrac{d^2y}{dx^2} = -\dfrac{ac}{b^2y^3}$

3.7 Related Rates and Applications (pages 194–197)

1. -3 **3.** 1,000   **5.** 15   **7.** $\frac{4}{5}$ **9.** $\frac{30}{13}$ **11.** -3

13. $-10\sqrt{3}$ units/s   **15.** 0.637 ft/s

17. The revenue will be rising at $34,000/yr.

19. -30 lb/in.^{2}/s   **23.** -3π cm^3/s

25. 7.2 ft/s

27. The distance is decreasing at the rate of 7.5 ft/min.

29. The shadow is moving at 200 ft/s.

31. The angle is of the line of sight is changing at the rate of 2.78 rad/s.

33. Assume the shape of the balloon is a sphere of radius r. The volume is changing at the rate of 60.3 cm^3/min.

35. The water would flow out at about 49.2 ft^3/min.

37. a. $\dfrac{dH}{dt} = \dfrac{3,125t - 6,250}{\sqrt{3,125t^2 - 12,500t + 62,500}}$

  b. $t = 2$ **c.** 224 mi

39. 1 rad/min   **41.** 3.927 mi/h

43. At $t = 2$ P.M., 8.875 knots; at $t = 5$ P.M., 10.417 knots

45. 0.001924 ft/min or 0.0233 in./min

47. a. $\theta = \cot^{-1}\dfrac{x}{150}$

  b. $d\theta/dt$ approaches 0.27 rad/s

  c. as v increases so will $d\theta/dt$ and it becomes more difficult to see the seals.

3.8 Linear Approximations and Differentials (pages 206–209)

1. $6x^2\,dx$ **3.** $x^{-1/2}\,dx$

5. $(\cos x - x \sin x)dx$ **7.** $\dfrac{3x \sec^2 3x - \tan 3x}{2x^2}\,dx$

9. $\cot x\,dx$ **11.** $\dfrac{e^x}{x}(1 + x \ln x)\,dx$

13. $\dfrac{(x - 3)(x^2\sec x \tan x + 2x \sec x) - (x^2\sec x)(1)}{(x - 3)^2}\,dx$

15. $\dfrac{x + 13}{2(x + 4)^{3/2}}\,dx$

19. 0.995; by calculator, 0.9949874371

21. 217.69; by calculator 217.7155882 so we see an error of approximately 0.0255882

23. 0.06 or 6%   **25.** 0.03 or 3%   **27.** 0.05 parts per million

29. The output will be reduced by 12,000 units.

31. 28.37 in.^{3}

33. a decrease of about 2 beats every 3 minutes

35. The error in S is approximately $\pm 2\%$.

37. S increases by 2% and the volume increases by 3%.

39. The length will change by about 0.0525 ft.

41. -6.93 (or about 7) particles/unit area

43. a. 472.7   **b.** 468.70

45. 1.2 units   **47. b.** $x \approx 1.367$

49. 0.5183   **51.** $K \approx \frac{3}{10}$

55. $\Delta x = -3; f(97) = 9.85$; by calculator 9.848857802; if $\Delta x = 16, f(97) \approx 9.89$

CHAPTER 3: REVIEW

Proficiency Examination (pages 209–210)

21. $\dfrac{dy}{dx} = 3x^2 + \frac{3}{2}x^{1/2} - 2 \sin 2x$

22. $\dfrac{dy}{dx} = \dfrac{\sqrt{3x}}{2x} - \dfrac{6}{x^2}$

23. $\dfrac{dy}{dx} = -x[\cos(3 - x^2)][\sin(3 - x^2)]^{-1/2}$

24. $\dfrac{dy}{dx} = \dfrac{-y}{x + 3y^2}$ **25.** $y' = \frac{1}{2}xe^{-\sqrt{x}}(4 - \sqrt{x})$

26. $y' = \dfrac{\ln 1.5}{x(\ln 3x)^2}$ **27.** $y' = \dfrac{3}{\sqrt{1 - (3x + 2)^2}}$

28. $y' = \dfrac{2}{1 + 4x^2}$ **29.** $\dfrac{dy}{dx} = 0$

30. $y' = y\left[\dfrac{2x}{(x^2 - 1)\ln(x^2 - 1)} - \dfrac{1}{3x} - \dfrac{9}{3x - 1}\right]$

31. $y'' = 2(2x - 3)(40x^2 - 48x + 9)$

32. $\dfrac{dy}{dx} = 1 - 6x$

33. $14x - y - 6 = 0$

34. $y = f(1) = \frac{1}{2}$, so the point is $(1, \frac{1}{2})$; tangent line $y - \frac{1}{2} = \frac{\pi}{4}(x - 1)$; normal line $y - \frac{1}{2} = -\frac{4}{\pi}(x - 1)$

35. 2π ft²/s

Supplementary Problems (pages 210–213)

1. $y' = 4x^3 + 6x - 7$

3. $y' = \dfrac{-4x}{(x^2 - 1)^{1/2}(x^2 - 5)^{3/2}}$

5. $y' = \dfrac{4x - y}{x - 2}$

7. $y' = 10(x^3 + x)^9(3x^2 + 1)$

9. $y' = \dfrac{(x^3 + 1)^4(46x^3 + 1)}{3x^{2/3}}$

11. $y' = 8x^3(x^4 - 1)^9(2x^4 + 3)^6(17x^4 + 8)$

13. $y' = \dfrac{-\sin\sqrt{x}}{4\sqrt{x}\sqrt{(\cos\sqrt{x})}}$

15. $y' = 5(x^{1/2} + x^{1/3})^4(\frac{1}{2}x^{-1/2} + \frac{1}{3}x^{-2/3})$

17. $y' = (4x + 5)\exp(2x^2 + 5x - 3)$

19. $y' = 3^{2-x}(1 - x\ln 3)$

21. $y' = \dfrac{y(1 + xye^{xy})}{x(1 - xye^{xy})}$

23. $y' = e^{\sin x}(\cos x)$

25. $y' = \dfrac{e^{-x}}{x\ln 5}(1 - x\ln 3x)$

27. $y' = \dfrac{2x^2 + 2xy^2 - 1}{2y - 2x - 2y^2}$

29. $y' = [\cos(\sin x)]\cos x$

31. $y' = \dfrac{2\sqrt{xy} - \sqrt{y}}{\sqrt{x}}$

33. $y' = \dfrac{1 - y\cos xy}{x\cos xy - 1}$

35. $\dfrac{dy}{dx} = \dfrac{\sin^{-1}x - x(1 - x^2)^{-1/2}}{(\sin^{-1}x)^2} + \dfrac{1}{x^2}\left(\dfrac{x}{1 + x^2} - \tan^{-1}x\right)$

37. $\dfrac{d^2y}{dx^2} = 20x^3 - 60x^2 + 42x - 6$

39. $y'' = -\dfrac{2(3y^3 + 4x^2)}{9y^5}$

41. $y'' = -2[1 + y^2 - 4x^2y - 4x^2y^3]$

43. $33x - y - 32 = 0$

45. $2\pi x + 4y - \pi^2 = 0$

47. $x + y - 2 = 0$

49. $4x - 2y - 1 = 0$

51. $y - 1 = 0$

53. $4x - y - 3 = 0$

55. tangent line, $12x - y - 11 = 0$; normal line, $x + 12y - 13 = 0$

57. $\dfrac{dy}{dt} = (3x^2 - 7)(t\cos t + \sin t)$ or $(3t^2\sin^2 t - 7)(t\cos t + \sin t)$

59. $f'(x) = 4x^3 + 4x^{-5}$; $f''(x) = 12x^2 - 20x^{-6}$; $f'''(x) = 24x + 120x^{-7}$; $f^{(4)}(x) = 24 - 840x^{-8}$

61. $y' = \dfrac{-4x^3}{(x^4 - 2)^{4/3}(x^4 + 1)^{2/3}}$

63. $y' = \dfrac{x + 2y}{y - 2x}$; $y'' = \dfrac{5y^2 - 20xy - 5x^2}{(y - 2x)^3}$

65. tangent line, $x - y + 2 = 0$; normal line, $x + y = 0$

67. 0.8686; calculator, 0.8634   **69.** 0.012 rad/min

71. $\dfrac{d}{dx}f(x^2 + x) = (x + 1)(x^4 + 2x^3 + 2x^2 + x)$

73. $4x\cos 2x^2 - 6x\sin 3x^2$

75. The rate of change of the distance between the car and the truck is 75 mi/h.

77. a. $C'(x) = \frac{4}{5}x + 3$ **b.** $4.00   **c.** $11.00   **d.** $11.40

81. 0.09 radians/s; this is about 5°/s

83. a. -0.2 **b.** -0.4 **c.** 0.3   **85.** $\dfrac{d\theta}{dt} = 0.05$

87. $\dfrac{dx}{dt} = 20\pi$ mi/min

93. For the velocity, $\dfrac{dx}{dt} = \dfrac{12\pi x\sin\theta}{2\cos\theta - x}$; for the acceleration,

$\dfrac{d^2x}{dt^2} = \dfrac{144\pi^2 x}{(2\cos\theta - x)^3}[2\sin^2\theta\cos\theta + (2\cos\theta - x)(1 + \frac{1}{2}x\cos\theta)]$

95. a. $f(0) = -2 < 0$; $f(3) > 0$, so there is at least one root on $[0, 5]$ by the root location theorem.   **b.** 2.08
c. to 6 places, the trace shows $x \approx 2.08124$

97. This is Putnam Problem 2, morning session in 1939.

CHAPTER 4: ADDITIONAL APPLICATIONS OF THE DERIVATIVE

4.1 Extreme Values of a Continuous Function (pages 228–229)

1. Maximum value is 26 and the minimum value is -34.

3. f maximum value is 0 and the minimum value is -4.

5. Maximum value is 1 and the minimum value is $-\frac{1}{8}$.

7. Maximum value is 0 and the minimum value is -2.

9. Maximum value is e^{-1} and minimum value is 0.

11. Maximum value is 1 and the minimum value is 0.

13. Maximum value is 1.25 and the minimum value is approximately 0.41067.

17. Not defined at $u = 0$; the maximum value is 1 and the minimum value is 0.

19. The maximum value is 48 and the minimum value is -77.

21. There are no real roots. The maximum value is approximately 117.7 and the minimum value is approximately -287.7.

23. h is not continuous; there is no maximum and no minimum value.

25. The maximum value is 0.3224 and the minimum value is -0.3224.

27. The maximum value is 9 and the minimum value is 5.

29. The smallest value is 0.   **31.** The smallest value is 3.

33. The largest value is $\frac{9}{4}$.

35. The largest value of g on $[2, 3]$ is approximately -1.1.

37. The maximum value is approximately 1.819 and the minimum value is 0.

39. The maximum value is 6,496 and the minimum value is 0.

41. The maximum value is 1.59 and the minimum value is -2.52.

43. The maximum value is 1 and the minimum value is $-e^{-\pi}$.

Answers to Problems 44–49 may vary.

51. The maximum velocity is 60 when $t = 0$.

53. The largest product occurs when $x = 3$ and $y = 6$.

55. The largest product occurs when $x = 21$ and $y = 63$.

61. a. The greatest difference occurs at $x = \dfrac{1}{2}$.

b. The greatest difference occurs at $x = \dfrac{1}{\sqrt{3}}$.

c. The greatest difference occurs at $x = \left(\dfrac{1}{n}\right)^{1/(n-1)}$.

63. The minimum distance occurs when $x = m^{-1/2}$.

4.2 The Mean Value Theorem (pages 235–237)

3. $c = 1$ **5.** $c \approx 1.5275$ **7.** $c \approx 1.0772$

9. $c = 9/4$ **11.** $c \approx 0.73$ **13.** $c \approx 0.6901$

15. $c \approx 0.54$ **17.** $c \approx 1.082$ **19.** $c \approx 0.5227$

21. does not apply   **23.** applies   **25.** does not apply

27. does not apply   **29.** applies

31. $f(x) = 8x^3 - 6x + 10$

33. does not apply   **37.** $c = 2.5$ and $c = 6.25$

41. does not apply   **45. b.** 0

4.3 First-Derivative Test (pages 246–248)

3. The black curve is the function and the blue one is the derivative.

Answers for 5–11 may vary.

5.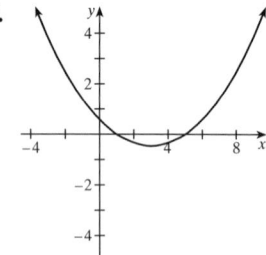

7.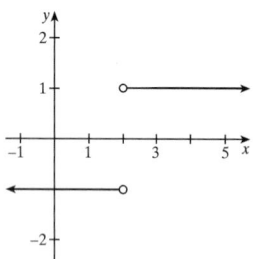

9.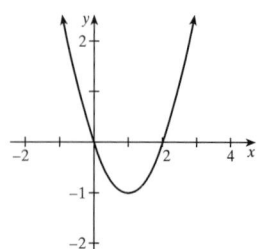

11.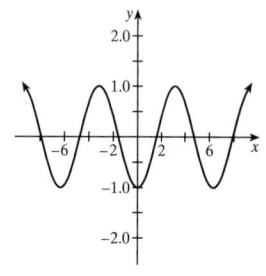

13. a. $x = 0, x = -2$

b. increasing on $(-\infty, -2) \cup (0, +\infty)$; decreasing on $(-2, 0)$

c. critical points: $(0, 1)$, relative minimum; $(-2, 5)$, relative maximum

d.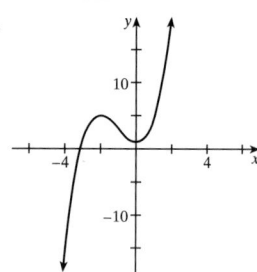

15. a. critical numbers: $x = \frac{5}{3}, x = -25$

b. increasing on $(-\infty, -25) \cup (\frac{5}{3}, +\infty)$; decreasing on $(-25, \frac{5}{3})$

c. critical points: $(\frac{5}{3}, -9{,}481)$, relative minimum; $(-25, 0)$, relative maximum

d.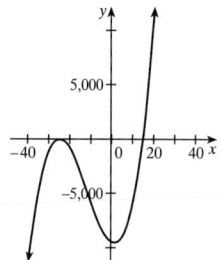

17. a. critical numbers: $x = 0, x = 4$

b. increasing on $(-\infty, 0) \cup (4, +\infty)$; decreasing on $(0, 4)$

c. critical points: $(4, -156)$, relative minimum; $(0, 100)$, relative maximum

d.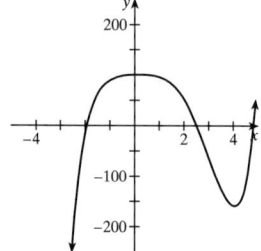

19. a. critical numbers: $x = -1, x = 3$

b. increasing on $(-1, 3)$; decreasing on $(-\infty, -1) \cup (3, +\infty)$

c. critical points: $(-1, -\frac{1}{2})$, relative minimum; $(3, \frac{1}{6})$, relative maximum

d.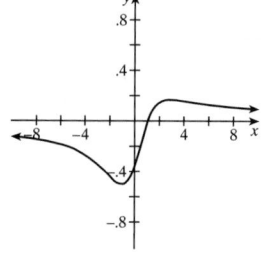

21. a. critical numbers: $t = -1, t = 3$
   b. increasing on $(-\infty, -1) \cup (3, +\infty)$; decreasing on $(-1, 3)$
   c. critical points: $(3, -32)$, relative minimum; $(-1, 0)$, relative maximum
   d.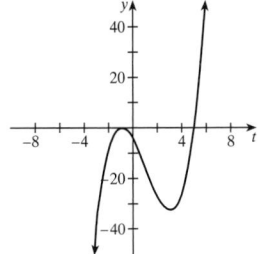

23. a. critical numbers: $x = 1, \frac{16}{3}$
   b. increasing on $(-\infty, 1) \cup (\frac{16}{3}, +\infty)$; decreasing on $(1, \frac{16}{3})$
   c. critical points: $(\frac{16}{3}, -\frac{1,225}{27})$, relative minimum; $(1, 36)$, relative maximum
   d.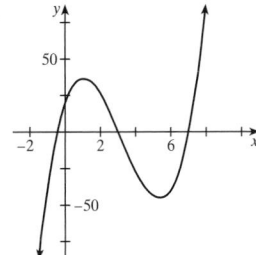

25. a. critical number $x = 0$
   b. increasing on $(0, +\infty)$; decreasing on $(-\infty, 0)$
   c. critical point: $(0, 1)$, relative minimum
   d.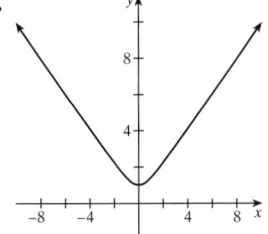

27. a. critical numbers: $x = 1, x = 0$
   b. increasing on $(-\infty, 0) \cup (1, +\infty)$; decreasing on $(0, 1)$
   c. critical points: $(1, -3)$, relative minimum; $(0, 0)$, relative maximum
   d.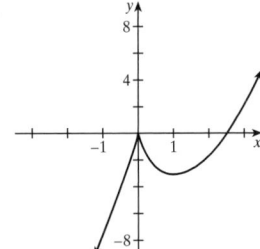

29. a. critical numbers: $x = \frac{7\pi}{6}, x = \frac{11\pi}{6}$
   b. increasing on $(\frac{7\pi}{6}, \frac{11\pi}{6})$; decreasing on $(0, \frac{7\pi}{6}) \cup (\frac{11\pi}{6}, 2\pi)$
   c. critical points: $(\frac{7\pi}{6}, -5.3972)$, relative minimum; $(\frac{11\pi}{6}, -4.0275)$, relative maximum; $(2\pi, -4.28)$, neither
   d.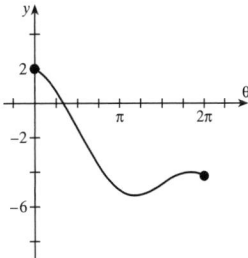

31. a. critical numbers: $x = 0$
   b. increasing on $(0, \frac{\pi}{4})$; decreasing on $(-\frac{\pi}{4}, 0)$
   c. critical points: $(0, 0)$, relative minimum; $(-\frac{\pi}{4}, 1)$, neither; $(\frac{\pi}{4}, 1)$, neither
   d.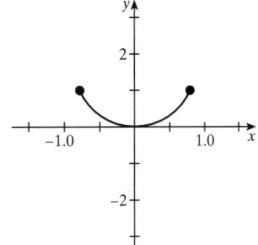

33. a. critical numbers: $x = 25\pi^2, x = 75\pi^2$
   b. increasing on $(0, 25\pi^2) \cup (75\pi^2, 100\pi^2)$; decreasing on $(25\pi^2, 75\pi^2)$
   c. critical points: $(75\pi^2, -1)$, relative minimum; $(25\pi^2, 1)$, relative maximum
   d.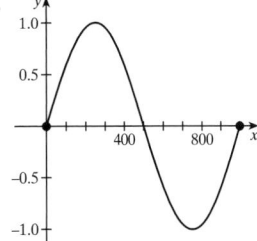

35. a. no critical numbers   **b.** increasing on $(-\infty, +\infty)$
   c. no critical points
   d.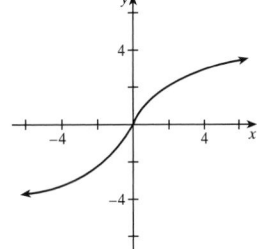

37. a. critical numbers $x \approx 0, -0.33$
  b. increasing on $(-\infty, -0.33) \cup (0, +\infty)$; decreasing on $(-0.33, 0)$
  c. critical points $(0, -1)$, relative minimum; $(-0.33, -0.98)$, relative maximum
  d.

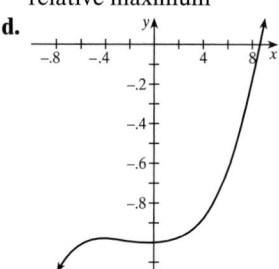

41. At $x = \frac{1}{2}$, relative maximum; at $x = 1$, relative minimum
43. At $x = 4$, relative minimum
45. critical numbers are $x = \frac{1}{2}$, $x = 1$, and $x = 2$; at $x = \frac{1}{2}$, neither a relative maximum nor a relative minimum; at $x = 1$, a relative minimum; at $x = 2$, neither a relative maximum nor a relative minimum.

47.

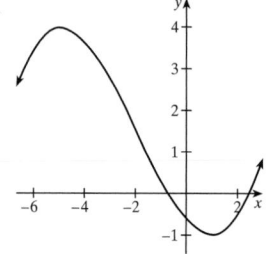

49. $a = -\frac{9}{25}$, $b = \frac{18}{5}$, $c = 3$
53.

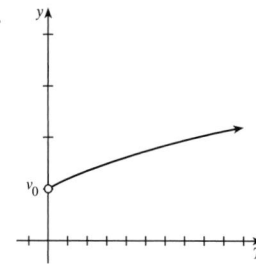

55. $A = -9.5, B = 4.5, C = 0, D = 7$
57. $x = -2B/5A$ is a relative minimum; $x = 0$, is neither

4.4 Concavity and the Second-Derivative Test (pages 258–261)

5. critical point: $(-18, -1)$, relative minimum; increasing on $(-18, +\infty)$; decreasing on $(-\infty, -18)$; concave up on $(-\infty, +\infty)$

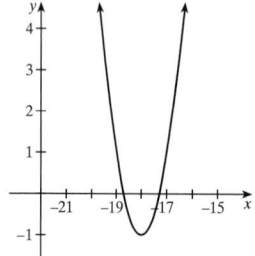

7. critical points: $(-3, 20)$, relative maximum; $(3, -16)$, relative minimum; increasing on $(-\infty, -3) \cup (3, +\infty)$; decreasing on $(-3, 3)$; inflection point $(0, 2)$; concave up on $(0, +\infty)$; concave down on $(-\infty, 0)$

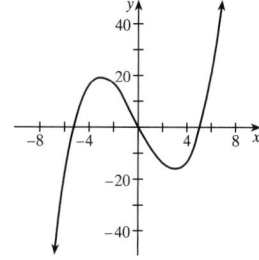

9. critical points: $(-3, -11)$, relative maximum; $(3, 13)$, relative minimum; increasing on $(-\infty, -3) \cup (3, +\infty)$; decreasing on $(-3, 0) \cup (0, 3)$; concave up on $(0, +\infty)$; concave down on $(-\infty, 0)$

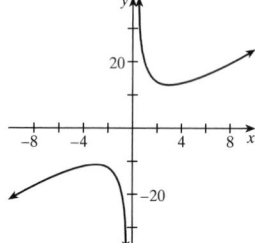

11. critical points: $(0, 26)$, relative maximum; $(-6.38, -852.22)$, relative minimum; $(1.88, -6.47)$, relative minimum; inflection points $(-4, -486), (1, 9)$; increasing on $(-6.38, 0) \cup (1.88, +\infty)$; decreasing on $(-\infty, -6.38) \cup (0, 1.88)$; concave up on $(-\infty, -4) \cup (1, +\infty)$; concave down on $(-4, 1)$

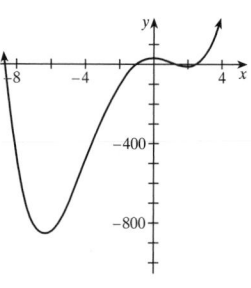

13. critical point: $(0, 0)$, relative minimum; increasing on $(0, +\infty)$; decreasing on $(-\infty, 0)$; concave up on $(-\infty, +\infty)$

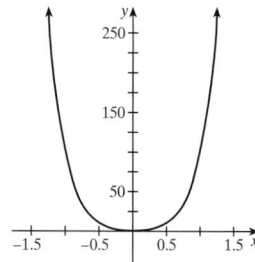

15. critical points: $(0, 0)$, point of inflection; $(\frac{1}{4}, \frac{1}{256})$, relative maximum; increasing on $(-\infty, \frac{1}{4})$; decreasing on $(\frac{1}{4}, +\infty)$; points of inflection: $(0, 0), (\frac{1}{6}, \frac{1}{432})$; concave up on $(0, \frac{1}{6})$; concave down on $(-\infty, 0) \cup (\frac{1}{6}, +\infty)$

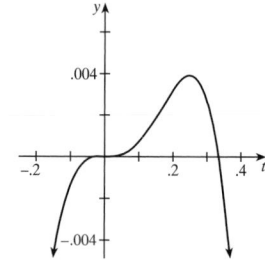

17. critical points: $(0.84, -1.8)$ is a relative minimum; $(0.62, -1.15)$, point of inflection; $(0, 0)$, point of inflection; increasing on $(0.84, +\infty)$; decreasing on $(-\infty, 0.84)$; concave up on $(-\infty, 0) \cup (0.62, +\infty)$; concave down on $(0, 0.62)$

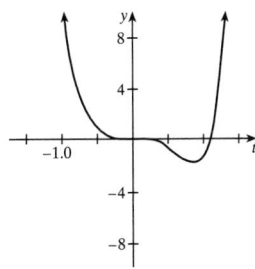

19. critical points: $(-1, -\frac{1}{2})$, relative minimum; $(1, \frac{1}{2})$, relative maximum; $(0, 0)$ is a point of inflection; points of inflection: $\left(\sqrt{3}, \frac{\sqrt{3}}{4}\right), \left(-\sqrt{3}, \frac{\sqrt{3}}{4}\right)$; increasing on $(-1, 1)$; decreasing on $(-\infty, -1) \cup (1, +\infty)$; concave up on $(-\sqrt{3}, 0) \cup (\sqrt{3}, +\infty)$; concave down on $(-\infty, -\sqrt{3}) \cup (0, \sqrt{3})$

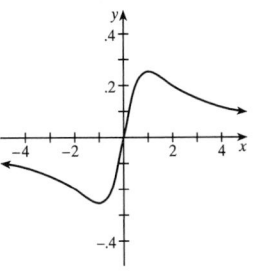

21. critical points: $(0, 0)$, relative minimum; $(0.67, 0.06)$, relative maximum; $(0.195, 0.021)$, point of inflection; $(1.138, 0.043)$, point of inflection; decreasing on $(-\infty, 0), (0.67, +\infty)$; increasing on $(0, 0.67)$; concave up on $(-\infty, 0.195) \cup (1.138, +\infty)$; concave down on $(0.195, 1.138)$

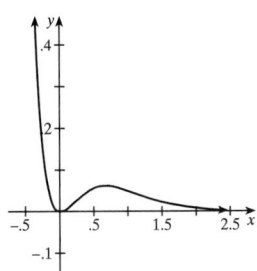

23. critical point $(0, 0)$, point of inflection; increasing on $(-\infty, +\infty)$; concave up for $(-\infty, 0)$; concave down for $(0, +\infty)$

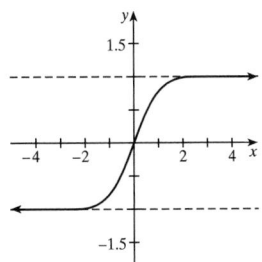

25. no critical points (but there is a critical number at $x = 0$); decreasing on $(-\infty, 0)$ increasing on $(0, +\infty)$; concave down on $(-\infty, 0) \cup (0, +\infty)$

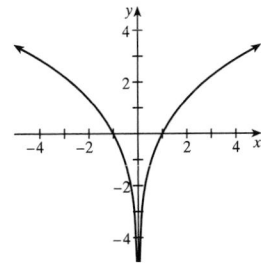

27. critical points: $(0, 0)$, relative maximum; $(15.4, -444.4)$, relative minimum; increasing on $(-\infty, 0) \cup (15.4, +\infty)$; decreasing on $(0, 15.4)$; inflection point is $(3.9, -140)$; concave up on $(3.9, +\infty)$; concave down on $(-\infty, 3.9)$

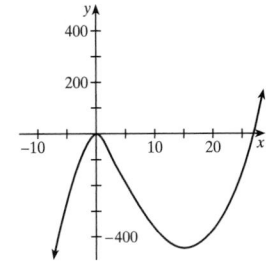

29. critical points: $(0, 11)$, relative maximum; $(0.31, 10.97)$, relative minimum; inflection point: $(0.16, 10.98)$; concave down $(-\infty, 0.16)$; concave up $(0.16, +\infty)$

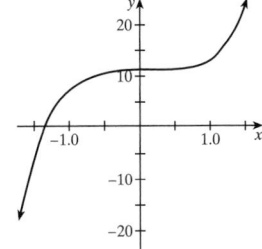

31. critical points: $(\frac{\pi}{12}, 1.13)$, relative maximum; $(\frac{5\pi}{12}, 0.44)$, relative minimum; inflection points: $(\frac{\pi}{4}, 0.79)$, $(\frac{3\pi}{4}, 2.4)$; increasing on $(0, \frac{\pi}{12}) \cup (\frac{5\pi}{12}, \pi)$; decreasing on $(\frac{\pi}{12}, \frac{5\pi}{12})$; concave up on $(\frac{\pi}{4}, \frac{3\pi}{4})$; concave down on $(0, \frac{\pi}{4}) \cup (\frac{3\pi}{4}, \pi)$

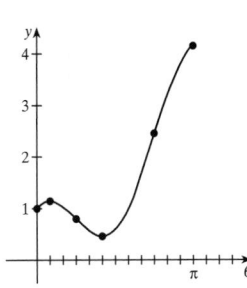

33. critical points: $(-2.78, -1.01)$, relative minimum; $(-0.38, 1.01)$; inflection points: $(-1.57, 0)$, $(0.52, 0.65)$, $(1.57, 0)$, $(2.62, -0.65)$; increasing on $(-2.78, -0.38)$; decreasing on $(-3.14, -2.78) \cup (-0.38, 3.14)$; concave up on $(-3.14, -1.57) \cup (0.52, 1.57) \cup (2.62, 3.14)$; concave down on $(-1.57, 0.52) \cup (1.57, 2.69)$

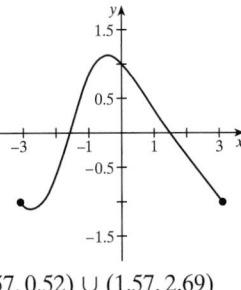

35. $(0.87, 0.69)$ is a relative maximum; $(-0.87, -0.69)$ is a relative minimum; inflection point $(0, 0)$; concave up on $(-1.00, 0)$; concave down on $(0, 1.00)$

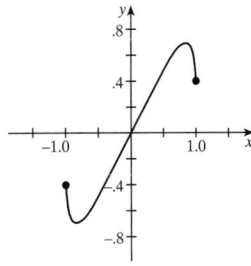

37. critical points: (2.08, −3.12), relative maximum; (4.37, −13.29), relative minimum; inflection point: (3.74, −10.31); concave down on (1.57, 3.74); concave up on (3.74, 4.71)

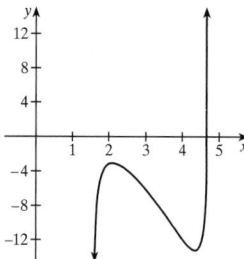

45. $A = -3, B = 9,$ and $C = -1.$

47. b.

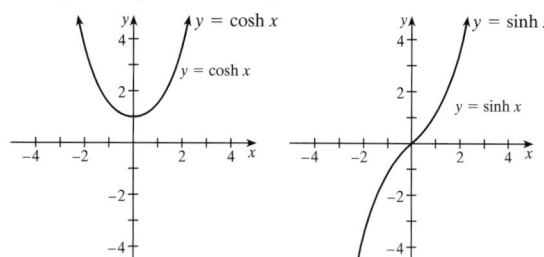

49. a. $D'(t) = \dfrac{1}{\sqrt{2\pi}\sigma} \exp\left[-\dfrac{1}{2}\left(\dfrac{t-m}{\sigma}\right)^2\right]\left(-\dfrac{1}{2\sigma^2}\right)(2)(t-m)$

$= 0$ when $t = m$; $D(m) = \dfrac{1}{\sqrt{2\pi}\sigma}$

b. $\lim\limits_{t\to\pm\infty} D(t) = 0$

c. We graph $y = D(t)$ where $\sigma = 1, m = 0.$

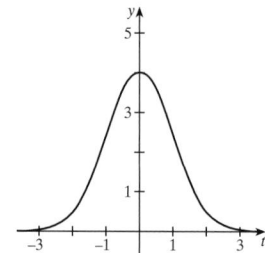

51.

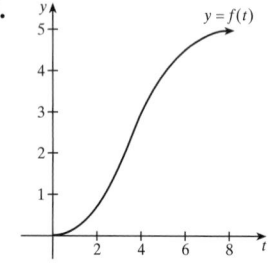

53. $x = 8$

55. b. $N(x) = -x^3 + 6x^2 + 13x - \frac{1}{3}(4-x)^3 + \frac{1}{2}(4-x)^2 + 25(4-x)$

c. 10:30 A.M.

57. The population is the largest after 24 minutes. After 49 minutes, the rate at which the population decreases per minute starts to increase.

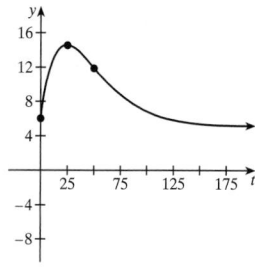

59. critical value at $x = 0$, relative maximum; point of inflection when $x = \pm\dfrac{\sqrt{3}}{3}a$

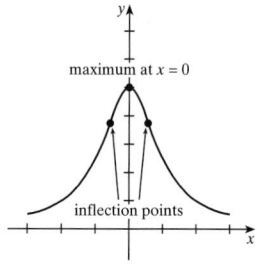

4.5 Curve Sketching: Limits Involving Infinity and Asymptotes (pages 272–274)

5. 0   **7.** 3   **9.** 9   **11.** 1   **13.** 0

15. $-\frac{1}{2}$ **17.** $+\infty$ **19.** 1   **21.** $-\infty$ **23.** 0

25. asymptotes: $x = 7, y = -3$; graph rising on $(-\infty, 7) \cup (7, +\infty)$; concave up on $(-\infty, 7)$; concave down on $(7, +\infty)$; no critical points; no points of inflection

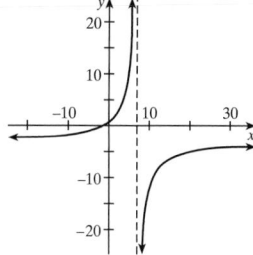

27. asymptotes: $x = 3, y = 6$; graph falling on $(-\infty, 3) \cup (3, +\infty)$; concave up on $(3, +\infty)$; concave down on $(-\infty, 3)$; no critical points; no points of inflection

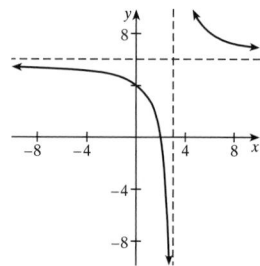

29. asymptotes: $x = 2, y = 1$; graph falling on $(-\infty, 2) \cup (2, +\infty)$; concave up on $(-\sqrt[3]{4}, 0)$ or $(2, +\infty)$; concave down on $(-\infty, -\sqrt[3]{4})$ or $(0, 2)$; critical point is $(0, -\frac{1}{8})$; points of inflection $(0, -\frac{1}{8}), (-\sqrt[3]{4}, \frac{1}{4})$

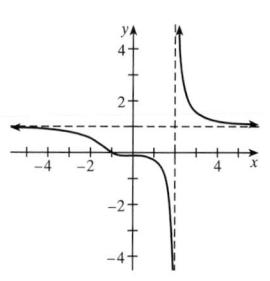

31. asymptotes: $x = -4, x = 1$,
$y = 0$; graph falling on $(-\infty, -4)$
$\cup (-4, 1) \cup (1, +\infty)$; concave up
on $(-4, -1)$ or $(1, +\infty)$; concave
down on $(-\infty, -4)$ or $(-1, 1)$; no
critical points point of inflection
is $(-1, 5)$;

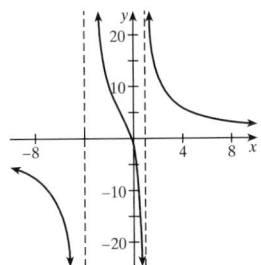

33. no asymptotes; graph rising on
$(0, +\infty)$; graph falling on
$(-\infty, 0)$; concave up on
$(-\infty, +\infty)$; critical point is $(0, 0)$
no points of inflection

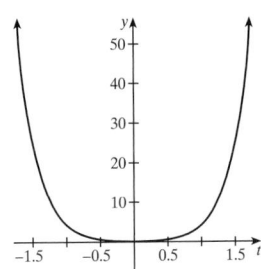

35. no asymptotes; graph rising on
$(-3, 0) \cup (3, +\infty)$; graph falling
on $(-\infty, -3) \cup (0, 3)$; concave
up on $(-\infty, \sqrt{3}) \cup (\sqrt{3}, +\infty)$;
concave down on $(-\sqrt{3}, \sqrt{3})$;
critical points are $(-3, 0)$, relative minimum; $(0, 81)$, relative
maximum; $(3, 0)$, relative minimum; points of inflection are
$(-\sqrt{3}, 36), (\sqrt{3}, 36)$

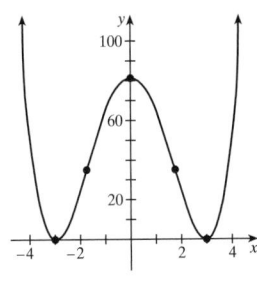

37. no asymptotes; graph rising on
$(1, +\infty)$; graph falling on $(-\infty, 1)$;
concave up on $(-\infty, -2) \cup (0, +\infty)$;
concave down on $(-2, 0)$; critical
points are $(1, -3)$, relative minimum; $(0, 0)$, vertical tangent; $(0, 0)$
and $(-2, 6\sqrt[3]{2})$ are points of
inflection

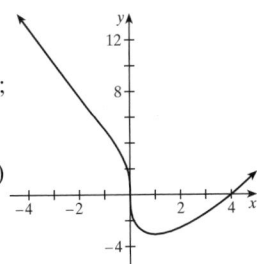

39. no asymptotes; graph rising on
$(-1, \frac{3}{2})$; graph falling on
$(-\infty, -1) \cup (\frac{3}{2}, +\infty)$; concave up on
$\left(-\infty, \dfrac{5 - \sqrt{41}}{4}\right)$ or $\left(\dfrac{5 + \sqrt{41}}{4}, +\infty\right)$;
concave down on
$\left(\dfrac{5 - \sqrt{41}}{4}, \dfrac{5 + \sqrt{41}}{4}\right)$; critical
points are $(-1, -e)$, relative
minimum; $(\frac{3}{2}, 9e^{-3/2})$, relative maximum; points of inflection
where $x = \dfrac{5 \pm \sqrt{41}}{4}$

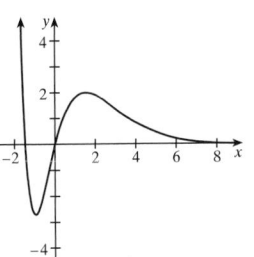

41. no asymptotes; graph rising on $(0, \frac{3\pi}{4}) \cup (\frac{7\pi}{4}, 2\pi)$; graph
falling on $(\frac{3\pi}{4}, \frac{7\pi}{4})$; concave up on $(0, \frac{\pi}{4})$ or $(\frac{5\pi}{4}, 2\pi)$; concave
down on $(\frac{\pi}{4}, \frac{5\pi}{4})$; critical points are $(\frac{3\pi}{4}, \sqrt{2})$, relative
maximum; $(\frac{7\pi}{4}, -\sqrt{2})$, relative minimum; points of
inflection: $(\frac{\pi}{4}, 0), (\frac{5\pi}{4}, 0)$

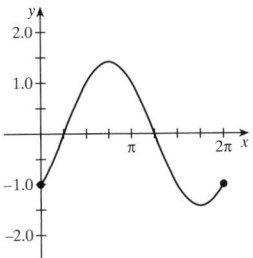

43. critical number at $x = \frac{\pi}{2}$; no
asymptotes; graph rising on
$(\frac{\pi}{2}, \pi)$; graph falling on $(0, \frac{\pi}{2})$;
concave up on $(0, \pi)$; critical
points are $(\frac{\pi}{2}, 0)$, relative minimum; $(0, 1)$, and $(\pi, 1)$ relative
maximums; no points of
inflection

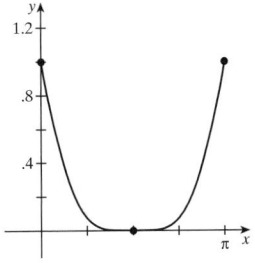

45.

51. Solution 2 is incorrect.
53. a. $+\infty$ **b.** $+\infty$ **c.** $-\infty$ **d.** $-\infty$ **e.** $+\infty$
55. $+\infty$

4.6 Optimization in the Physical Sciences and Engineering (pages 280–285)

3. The dimensions of the garden should be 8 ft by 8 ft.
5. The maximum volume is about 1 liter.
7. The largest rectangle has sides of length $R/\sqrt{2}$ and $\sqrt{2}\,R$.
9. The dimensions are $h = \dfrac{R}{3}\sqrt{3}$ and $r = \dfrac{R}{3}\sqrt{6}$.
11. A circumscribed square of side
$$s = \frac{L}{\sqrt{2}} + \frac{L}{\sqrt{2}} = \sqrt{2}\,L$$
yields a maximum area.
13. The minimum distance is 200 mi.
17. $x = 18, y = 36$

19. a. When $s = 4$, Missy should row all the way.

b. When $s = 6$, she should land at a point 4.5 km from point B and run the rest of the way (1.5 km).

21. $60.00

23. He has about 5 minutes 17 seconds to diffuse the bomb.

25. $r \approx 3.84$ and $h \approx 7.67$ cm.

27. b. When $p = 12$, the largest value of f is 6.

29. $x = \dfrac{Md}{\sqrt{4m^2 - M^2}}$

31. c. The lamp should be placed about 2.8 ft above the table.

33. a. $T'(x) = -T\left[\dfrac{c}{(kx + c)x} + \ln p\right]$

b. $x = \dfrac{-c \ln p + \sqrt{c^2(\ln p)^2 - 4kc \ln p}}{2 k \ln p}$

c.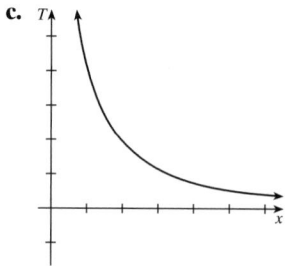

35. a. The maximum height occurs when $x = \dfrac{mv^2}{32(m^2 + 1)}$

b. The maximum height occurs when $y' = 0$ or when $m = v^2/32x_0$.

37.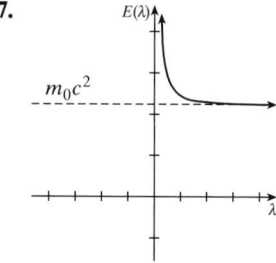

As $\lambda \to +\infty$, $E(\lambda) \to m_0c^2$

39. 26 cm is a minimum

41. The minimum value occurs when $T \approx 4°$.

43. 270 in.^{3}

45. a.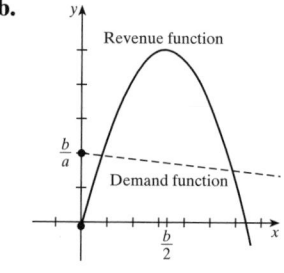

b. $\beta_{\min} = \alpha \cos \gamma;\ \beta_{\max} = \alpha \sec \gamma$

4.7 Optimization in Business, Economics, and the Life Sciences (pages 295–300)

1. $x = 26$

3. $x = 20$

5. The minimum average cost is 25.

7. a. $C(x) = 5x + \dfrac{x^2}{50};\ R(x) = x\left(\dfrac{380 - x}{20}\right);$

$P(x) = -0.07x^2 + 14x$

b. The maximum profit occurs when the price is $14/item. The maximum profit is $700.

9. Profit is maximized when $x = 60$.

11. a. 1.22 million people/yr

b. The percentage rate is 2%.

13. 208 years from now

15. 45 times per year

17. a. $x = 8$

c.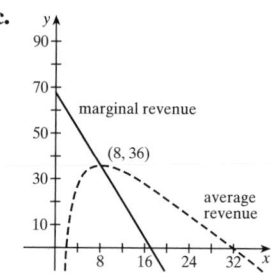

19. Since the optimum solution is over 100 years, you should will the book to your heirs so they can sell it in 117.19 years.

21. Sell the boards at a price of $41, sell 47 per month, and have the maximum profit of $752.

23. lower the fare $250

25. Plant 80 total trees, have an average yield per tree of 320 oranges, and a maximum total crop of 25,600 oranges.

27. 62 vines

29. a. The most profitable time to conclude the project is 10 days from now.

b. Assume R is continuous over $[0, 10]$.

31. a. $R(x) = xp(x) = \dfrac{bx - x^2}{a}$ on $[0, b]$; R is increasing on $\left(0, \dfrac{b}{2}\right)$ and deceasing on $\left(\dfrac{b}{2}, b\right)$

b.

33. $v = \dfrac{kv_1}{k-1}$

35. a. $x = r$ is a maximum

b.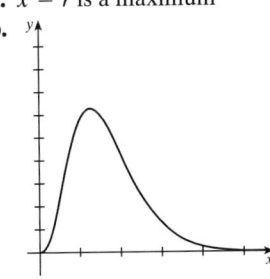

c.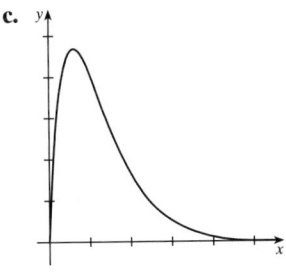

37. $v = 39$

39. a.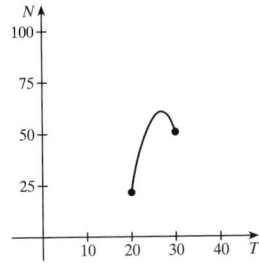

Largest survival percentage is 60.56%, and the smallest survival percentage is 22%.

b. $S'(T) = -(-0.06T + 1.67)(0.03\,T^2 + 1.67\,T - 13.67)^{-2};\ T = 27.83$

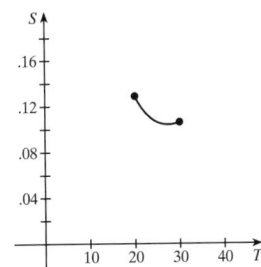

c.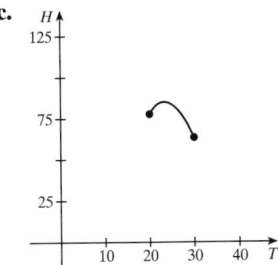

Largest hatching occurs when $T \approx 25.58$ and the smallest when $T = 30$.

41. a. $C = Sx + \dfrac{pQ}{nx};\ x = \sqrt{\dfrac{pQ}{nS}}$

49. $C''(t) = \dfrac{k}{b-a}(a^2 e^{-at} - b^2 e^{-bt})$

51. $\theta \approx 0.9553$; this is about $54.736°$.

4.8 l' Hôpital's Rule (pages 307–308)

3. $\frac{3}{2}$ **5.** 10   **7.** not defined

9. $\frac{1}{2}$ **11.** 0

13. $\frac{1}{2}$ **15.** 3   **17.** $\frac{3}{2}$ **19.** 2   **21.** $+\infty$

23. 2   **25.** $\frac{1}{2}$ **27.** 0   **29.** $-\infty$ **31.** 0

33. $e^{3/2}$ **35.** e^2 **37.** 0   **39.** $+\infty$ **41.** $+\infty$

43. $+\infty$ **45.** $-\infty$ **47.** 1   **49.** 0

51. limit does not exist

53. $\frac{1}{120}$ **55.** $\frac{5}{4}$ **57. b.** 1   **59. b.** 1.1814

65. $\displaystyle\lim_{\beta \to \alpha}\left[\dfrac{C}{\beta^2 - \alpha^2}(\sin \alpha t - \sin\beta t)\right] = \dfrac{-Ct}{2\alpha}\cos \alpha t$

CHAPTER 4: REVIEW

Proficiency Examination (page 309)

13. 2   **14.** $-\frac{1}{2}$ **15.** 0   **16.** e^{-6}

17. Relative maximum at $(-3, 29)$; relative minimum at $(1, -3)$; inflection point at $(-1, 13)$

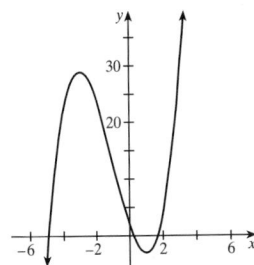

18. inflection points at approximately $\left(-\frac{27}{2}, -96.43\right)$ and $(0, 0)$; relative maximum at $\left(\frac{27}{4}, 38.27\right)$

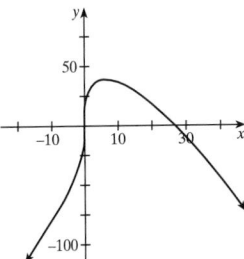

19.

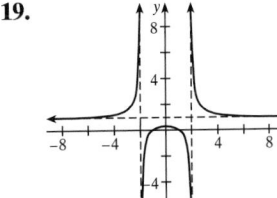

20. relative minimum at $(-1, -2e)$; relative maximum at $(3, 6e^{-3})$

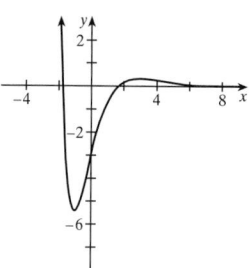

21. $(0, 0)$ is a point of infection; the graph is concave up for $x < 0$ and down for $x > 0$.

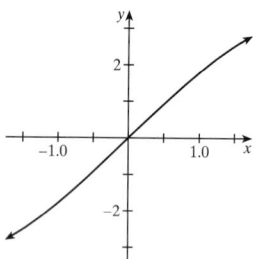

22. relative maximum where $x = \pi$; inflection points at $x = \frac{\pi}{3}$ and $x = \frac{5\pi}{3}$.

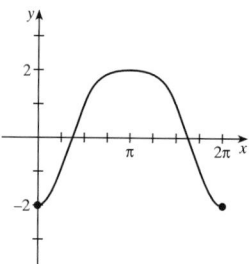

23. absolute maximum $(0.4, 5.005)$; absolute minimum at $(1, 4)$

24. The dimensions of the box are 19 in. $\times$ 19 in. $\times$ 10 in.

25. the maximum area is a^2 square units.

Supplementary Problems (pages 311–316)

1.

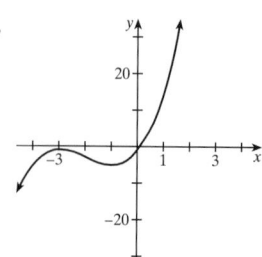

3.

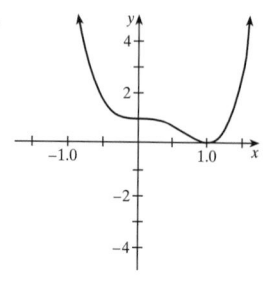

5.

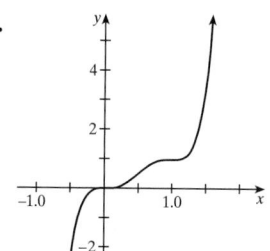

7.

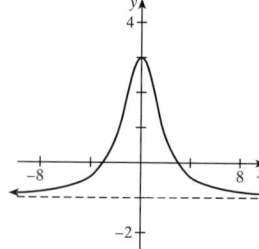

9.

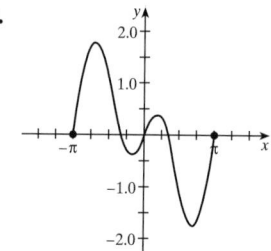

11.

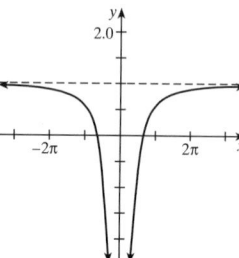

13.

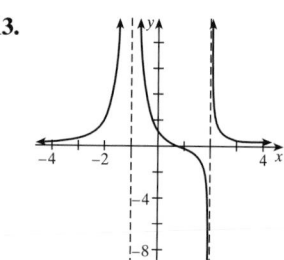

15.

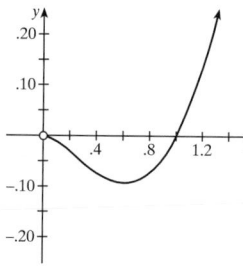

17.

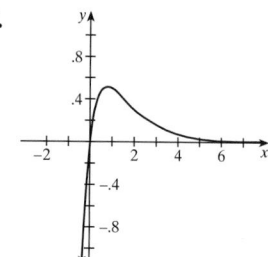

19.

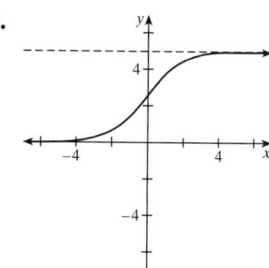

21. c   **23.** b

25. maximum $f(0) = 12$; minimum $f(2) = -4$

27. $f(0) = 0$, minimum; maximum $f(1) \approx 0.429$

29. 0   **31.** $-\frac{1}{2}$ **33.** 0

35. This limit does not exist.   **37.** 1

39. 9   **41.** 0   **43.** e^4

45. 1   **47.** 0   **49.** $+\infty$

51. f is the function and g is the derivative

55. $A = \frac{1}{2}, B = 0, C = -\frac{3}{2}, D = 0$

57. The maximum profit of \$108,900 is reached when 165 units are rented at \$740 each.

59. maximum yield of 6,125 for 35 trees per acre

61. 28,072 ft of pipe is laid on the shore gives the minimum cost

63. The price is 90 and the maximum profit is 1,100

65. 7 people

69. the largest value of $P(\theta)$ on $[0, \pi]$ is at

$$\theta_2 = \cos^{-1}\left[\frac{-1 + \sqrt{1 + 80b^2}}{8b}\right]$$

71. point of inflection is $(1, \frac{\pi}{2})$

75. a. $f'(x) > 0$ on $(-\pi, -\frac{3\pi}{4}) \cup (-\frac{\pi}{4}, \frac{\pi}{4}) \cup (\frac{3\pi}{4}, \pi)$
    b. $f'(x) < 0$ on $(-\frac{3\pi}{4}, -\frac{\pi}{4}) \cup (\frac{\pi}{4}, \frac{3\pi}{4})$
    c. $f''(x) > 0$ on $(-\frac{\pi}{2}, 0) \cup (\frac{\pi}{2}, \pi)$
    d. $f''(x) < 0$ on $(-\pi, -\frac{\pi}{2}) \cup (0, \frac{\pi}{2})$
    e. $f'(x) = 0$ at $x = \pm\frac{\pi}{4}, \pm\frac{3\pi}{4}$
    f. $f'(x)$ exists everywhere
    g. $f''(x) = 0$ at $x = 0, \pm\pi, \pm\frac{\pi}{2}$

77. a. $f'(x) > 0$ on $(-1, 0) \cup (1, 2)$
    b. $f'(x) < 0$ on $(-2, -1) \cup (0, 1)$
    c. $f''(x) > 0$ on $(-2, -\sqrt{3}/3) \cup (\sqrt{3}/3, 2)$
    d. $f''(x) < 0$ on $(-\sqrt{3}, \sqrt{3})$
    e. $f'(x) = 0$ at $x = 0, \pm 1$
    f. $f'(x)$ exists everywhere
    g. $f''(x) = 0$ at $x = \pm\sqrt{3}/3$

79. The minimum area is 0, the maximum area is 225 at $x = y = 15$.

81. Estimate the relative minimum at $(0, 0)$ and $(4.8, -110)$, and relative maxima at $(1, 1)$ and $(-0.8, 0.7)$.

83. c. $q'(t) = 2\cos 2t + \dfrac{-2\cos 2t(-2\sin 2t)}{2\sqrt{L^2 - \cos^2 2t}}$

85. This is Putnam Problem 11 in the afternoon session of 1938.

87. This is Problem 1 of the 1987 Putnam examination.

89. This is Putnam Problem 15 of the afternoon session of 1940.

CHAPTER 5: INTEGRATION

5.1 Antidifferentiation (pages 327–328)

1. $2x + C$
3. $x^2 + 3x + C$
5. $t^4 + t^3 + C$
7. $\frac{1}{2}\ln|x| + C$
9. $2u^3 - 3\sin u + C$
11. $\tan\theta + C$
13. $-2\cos\theta + C$
15. $5\sin^{-1}y + C$
17. $\frac{2}{5}u^{5/2} - \frac{2}{3}u^{3/2} - \frac{1}{9}u^{-9} + C$
19. $\frac{1}{3}x^3 + \frac{2}{5}x^{5/2} + C$
21. $-t^{-1} + \frac{1}{2}t^{-2} - \frac{1}{3}t^{-3} + C$
23. $\frac{4}{5}x^5 + \frac{20}{3}x^3 + 25x + C$
25. $-x^{-1} - \frac{3}{2}x^{-2} + \frac{1}{3}x^{-3} + C$
27. $x + \ln|x| + 2x^{-1} + C$
29. $x - \sin^{-1}x + C$

31.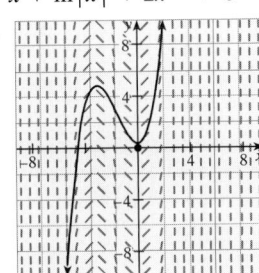
33.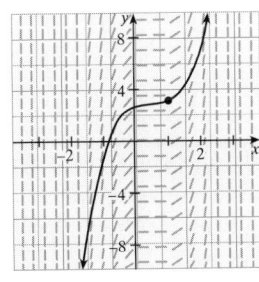

35.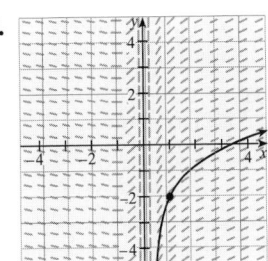
37.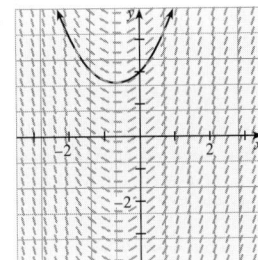

39. $F(x) = 2\sqrt{x} - 4x + 2$
    b.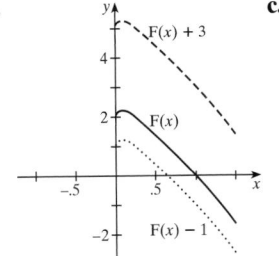
    c. $C_0 = -\frac{9}{4}$

41. \$249

43. The population in 8 months will be 10,128.

45. $k = 20$ ft/s^2

47. The acceleration of the plane at liftoff is 4.3 ft/s^2.

49. 138.3 ft
51. 28.8
53. 21

55. $e^2 - 3$
57. 1

59. a. $\frac{1}{2}\sec^{-1}y + C$ **b.** $\frac{1}{2}\tan^{-1}\sqrt{y^2 - 1}$

61. $A = \frac{1}{2}(d - c)[m(d + c) + 2b]$

5.2 Area As the Limit of a Sum (pages 335–337)

1. 6
3. 120
5. 225
7. 9,800
9. $\frac{1}{2}$
11. 3
13. a. 3.5 **b.** 3.25
15. a. 2.71875 **b.** 2.58796
17. 0.795
19. 1.269
21. 1.2033
23. 18
25. 68
27. 2
29. false
31. true
33. true
37. a. $\frac{14}{3}$ **c.** The statement is true.
39. 21.3646
41. 0.6
43. 0.5038795
45. a. 2 **c.** The statement is true.

49. $A = \lim\limits_{n \to +\infty}\left[\displaystyle\sum_{k=1}^{n} f\left(a + \frac{(2k-1)(b-a)}{2n}\right)\left(\frac{b-a}{n}\right)\right]$

5.3 Riemann Sums and the Definite Integral (pages 350–351)

1. 2.25
3. 10.75
5. -0.875
7. 1.18
9. 1.94
11. 28.875
13. 1.896
15. 0.556
17. $-\frac{1}{3}$
19. $\frac{3}{2}$
21. $-\frac{1}{2}$
23. $\frac{1}{2}$

25. $\int_{-2}^{4} f(x)\, dx = 1; \int_{-2}^{4} g(x)\, dx = 1$ **27.** 10

29.

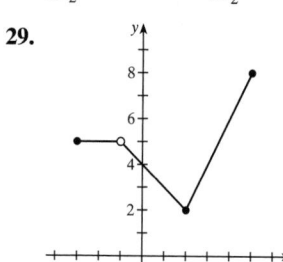

$\int_{-3}^{5} f(x)\, dx = \frac{71}{2}$

35. 2.3

5.4 The Fundamental Theorems of Calculus (pages 355–357)

1. 140   **3.** $16 + 8a$ **5.** $\frac{15}{4}a$

7. $\frac{3}{8}c$ **9.** 18   **11.** $\frac{5}{8} + \pi^2$

13. $\dfrac{2 \cdot 2^{2a} - 1}{2a + 1}$ **15.** 15   **17.** $\frac{272}{15}$

19. 2   **21.** $\dfrac{a\pi}{2}$ **23.** 5

25. $2 - e$ **27.** $\frac{5}{2}$ **29.** 4

31. $\frac{8}{3}$ **33.** 1   **35.** $e - \frac{3}{2}$

37. $3 \ln 2 - \frac{1}{2}$ **39.** $(x - 1)\sqrt{x + 1}$ **41.** $\dfrac{\sin t}{t}$

43. $\dfrac{-1}{\sqrt{1 + 3x^2}}$ **55.** 4.7011

57. a. relative minimum   **b.** $g(1) = 0$ **c.** $x = 0.75$

d.

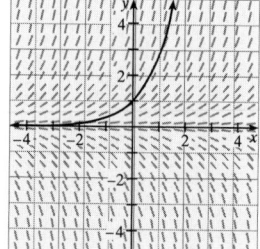

61. $F'(4) = 2e^8/9$

63. $F'(x) = -f(v)\dfrac{dv}{dx} + f(u)\dfrac{du}{dx}$

5.5 Integration by Substitution (pages 362–363)

1. a. 32   **b.** $2\sqrt{3} - 2$ **3. a.** 0   **b.** 0

5. a. $\frac{128}{5}$ **b.** $\frac{128}{5}$

7. a. $\frac{2}{9}\sqrt{2}\, x^{9/2} + C$ **b.** $\frac{1}{9}(2x^3 + 5)^{3/2} + C$

9. $\frac{1}{10}(2x + 3)^5 + C$ **11.** $\frac{3}{5}(x - 27)^{5/3} + C$

13. $\frac{1}{3}x^3 - \frac{1}{3}\sin 3x + C$ **15.** $\cos(4 - x) + C$

17. $\frac{1}{6}(t^{3/2} + 5)^4 + C$ **19.** $-\frac{1}{2}\cos(3 + x^2) + C$

21. $\frac{1}{4}\ln(2x^2 + 3) + C$

23. $\frac{1}{6}(2x^2 + 1)^{3/2} + C$

25. $\frac{2}{3}e^{x^{3/2}} + C$ **27.** $\frac{1}{3}(x^2 + 4)^{3/2} + C$

29. $\frac{1}{2}(\ln x)^2 + C$ **31.** $2 \ln (\sqrt{x} + 7) + C$

33. $\ln(e^t + 1) + C$ **35.** $\frac{5}{6}\ln 3$

37. 0   **39.** $e - e^{1/2}$

41. $\frac{1}{2}\ln 2$ **43.** 1.80

45. a. We take 1 Frdor as the variable so the note from the students reads, "Because of illness I cannot lecture between Easter and Michaelmas."

b. The Dirichlet function is defined as a function f so that $f(x)$ equals a determined constant c (usually 1) when the variable x takes a rational value, and equals another constant d (usually 0) when this variable is irrational. This famous function is one which is discontinuous everywhere.

47. $\frac{16}{3}$ **49.** 2.5   **51.** 0   **53.** 0

55. a. true   **b.** true   **c.** false

57. $F(x) = -\frac{1}{3}\ln|1 - 3x^2| + 5$

59. The amount of water at 4 seconds is 2 ft³. The depth at that time is about $\frac{1}{4}$ in.

61. a. $L(t) = 0.03\sqrt{36 + 16t - t^2} + 3.82$

b. The highest level is 4.12 ppm.

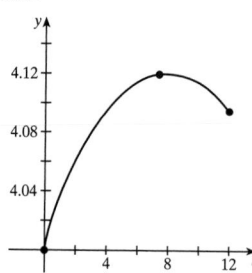

c. It is the same as 11:00 A.M. ($t = 4$) at 7:00 P.M. (when $t = 12$).

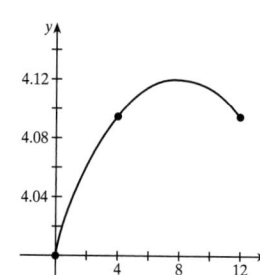

5.6 Introduction to Differential Equations (pages 374–377)

9. $x^2 + y^2 = 8$ **11.** $y = \tan(x - \frac{3\pi}{4})$ **13.** $x^{3/2} - y^{3/2} = 7$

15.

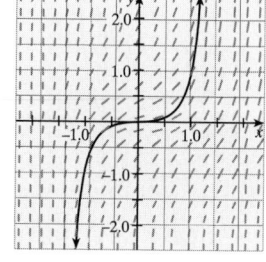

17.

19.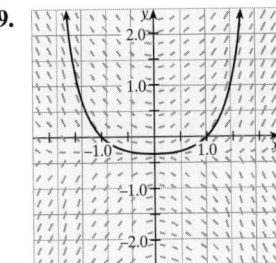

21. $y = Be^{(3/2)x^2}$
23. $2(1 - x^2)^{3/2} + 3y^2 = C$
25. $x^{3/2} + 3y^{1/2} = C$
27. $\cos x + \sin y = C$
29. $xy = C$
31. $y = Cx$

33.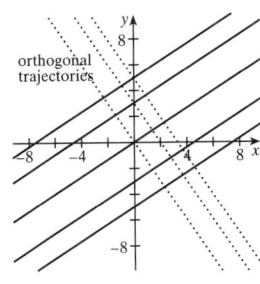
35.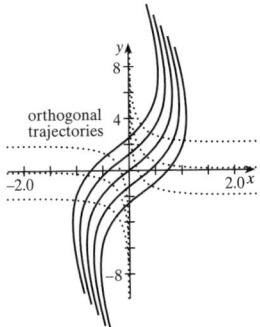

37.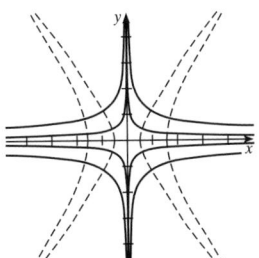
39.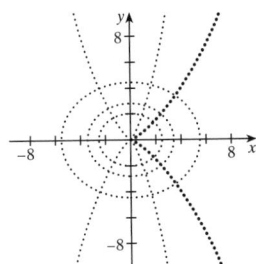

41.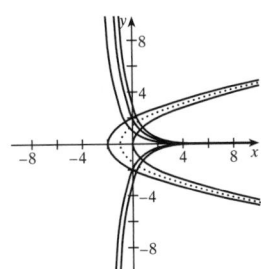

43. $\dfrac{d}{dt}Q(t) = kQ(t)$

45. $\dfrac{dQ}{dt} = k(N - Q)$

47. $\dfrac{dQ}{dt} = kQ(N - Q)$

49. a. 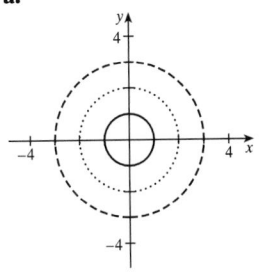 **b.**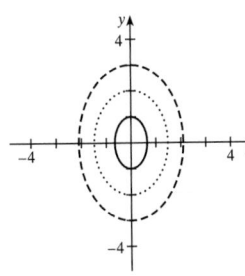

c. 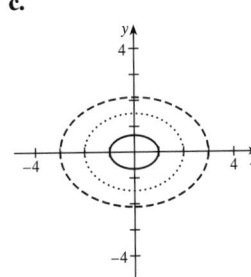 **d.**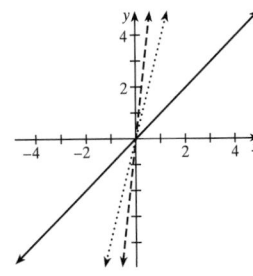

e. 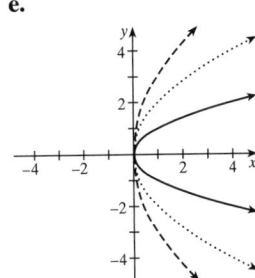 **f.**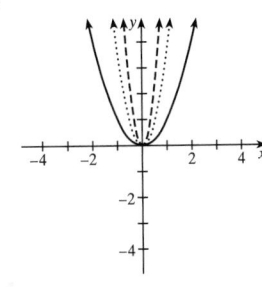

a and *d* are orthogonal trajectories;
b and *e* are orthogonal trajectories;
c and *f* are orthogonal trajectories.

51. 11,000 yr **53.** 5 min **55. b.** 2 min
57. a. 7,920 ft/s (1.5 mi/s)
 b. 16,118 ft/s (3.05 mi/s)
 c. 33,520 ft/s (6.35 mi/s)
59. 12.73 ft/s
61. a. 40 g **b.** 36 days
63. $-\dfrac{1}{480}\sqrt{h}$ **b.** 39 min

5.7 The Mean Value Theorem for Integrals; Average Value
(pages 381–383)

1. 1.55 is on the interval $[1, 2]$
3. $\sqrt{5} \approx 2.24$ is in the interval
5. The mean value theorem does not apply.
7. 0.0807 is on the interval
9. −0.187 is on the interval

11. 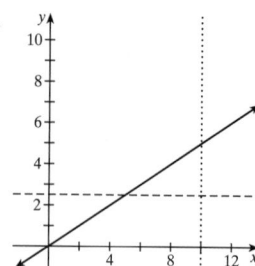   **13.**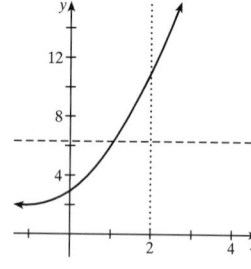

15.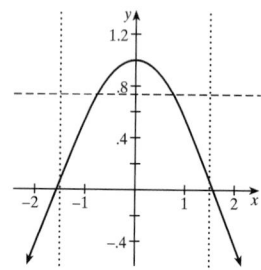

17. $\frac{3}{2}$ **19.** 0   **21.** 0.3729

23. $\frac{4}{3}$ **25.** -10 **27.** -25.375

29. 2.36   **33.** Avg temperature $= 18.7°C$

35. a. 1,987.24
　　b. The average population is reached at just under 14 minutes.

37. $f(x) = (x + 1)\cos x + \sin x$

5.8 Numerical Integration: The Trapezoidal Rule and Simpson's Rule (pages 388–391)

1. Trapezoidal rule, 2.34375; Simpson's rule, 2.33333; exact value, $\frac{7}{3}$

3. a. 0.7825   **b.** 0.785

5. a. 2.037871   **b.** 2.048596

7. a. 0.584   **b.** 0.594

11. $A \approx 0.775$; the exact answer is between $0.775 + 0.05$ and $0.775 - 0.05$

13. $A \approx 0.455$; the actual answer is between $0.455 - 0.0005$ and $0.455 + 0.0005$

15. $A \approx 3.25$; the exact answer is between $3.25 + 0.01$ and $3.25 - 0.01$

17. $A \approx 0.44$; the exact answer is between $0.44 + 0.01$ and $0.44 - 0.01$

19. a. $n = 164$ **b.** $n = 18$

21. a. $n = 184$ **b.** $n = 22$

23. a. $n = 82$ **b.** $n = 8$

25. a. 3.1   **b.** 3.1

27. 430   **29.** 613 ft^2 **31.** 79.17

33. a. Simpson error is 0.
　　b. $f^{(4)}(x)$ is unbounded near $x = 2$

39. $\frac{32}{3}$

5.9 An Alternative Approach: The Logarithm as an Integral (pages 395–396)

3. $E \approx 0.00104$; the number of subintervals should be 18

CHAPTER 5: REVIEW

Proficiency Examination (pages 396–397)

22. $-\frac{3}{5}$ **23.** $x^5 \sqrt{\cos(2x + 1)}$

24. $\frac{1}{2}\tan^{-1}(2x) + C$ **25.** $-\frac{1}{2}e^{-x^2} + C$

26. $\frac{17}{3}$ **27.** $-\frac{35}{3}$

28. $\frac{1}{2}$ **29.** 0

30. 36   **31.** 0

32. $v(t) = t^2 + t + 2$; $s(t) = \dfrac{t^3}{3} + \dfrac{t^2}{2} + 2t + 4$

33. 1.33×10^{-184}

34. The revenue for the five years is $7,666.67.

35.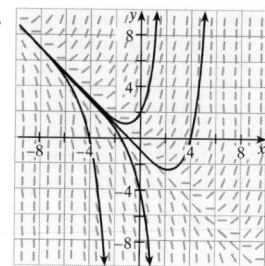

e. $y = -x - 1$

36. $y = \dfrac{3}{\cos 3x + C}$ **37. a.** $n \ge 26$ **b.** $n \ge 4$

Supplementary Problems (pages 397–400)

1. 25   **3.** 6

5. 1,710   **7.** $\dfrac{6 - 2\sqrt{2}}{3}$

9. $\sqrt{3} - 1$ **11.** $-\frac{3}{4}(1 - \sqrt[3]{9})$

13. $2x^{5/2} - \frac{4}{3}x^{3/2} + 2x^{1/2} + C$

15. $3\sin^{-1}x + \sqrt{1 - x^2} + C$

17. $-\ln(\sin x + \cos x) + C$

19. $\frac{1}{3}(x - 1)^3 + C$

21. $x + C$

23. $\dfrac{x^3(x^2 + 1)^{3/2}}{3} + C$

25. $-\frac{1}{15}(1 - 5x^2)^{3/2} + C$

27. 60

29. 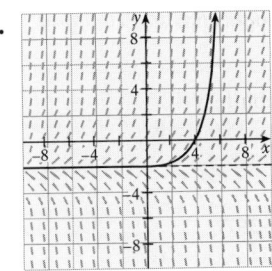 **31.**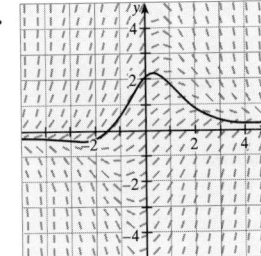

33. $F'(x) = x^2\cos^4 x$ **35.** $\frac{10}{3}$

37. $9 + 2\sqrt{6} - \frac{10}{3}\sqrt{5}$

39. $f(x) = \frac{1}{60}(x^6 + x^5 + 35x^2 - 21x - 16)$

41. $y^2 = \frac{1}{2}\sin 4x + C$ **43.** $y^2 = x^2 + C$

45. $\sqrt{y^2 + 2} = \sqrt{x^2 + 1} + C$ **47.** $y^{3/2} = x^{3/2} + C$

49. a. 0.6366   **b.** 0   **51.** Trapezoidal rule, 1.1139

53. Simpson's rule, 1.1114   **55.** 1.48004

57. 1.48004   **59.** 1.01406

61. b. ± 8 m/s

63. The tree was 2.33 ft tall when it was transplanted.

65. $y = \frac{1}{3}(x^2 + 5)^{3/2} + 1$ **67.** 4.45 ppm

69. 126 people

71. The average price was $1.32 per pound.

73. about 2 min, 45 seconds

77. a. 51.7%
   b. The time for 90% to disintegrate is about $17\frac{1}{2}$ years.

79. 35 years

81. 1:44 P.M.

83. This is Putnam Problem 1 from the morning session in 1958.

CHAPTER 6: ADDITIONAL APPLICATIONS OF THE INTEGRAL

6.1 Area Between Two Curves (pages 413–415)

1. vertical strip; $\frac{125}{48}$ **3.** horizontal strip; $\frac{126}{6}$

5. vertical strip; 2

7. 1   **9.** $\frac{1}{12}$ **11.** $\frac{8}{3}$

13. $\frac{324}{5}$ **15.** $\frac{9}{2}$ **17.** $\frac{253}{12}$

19. $\frac{5}{2}$ **21.** $\frac{323}{12}$ **23.** $\frac{331}{4}$

25. $2\ln 1.6 - \sin^{-1} 0.6$ **27. a.** 0.75   **b.** 0

29. a. 75   **b.** 432   **33.** $5.33   **35.** $42.67

37. π

39. a. The machine will be profitable for 9 years.
   b. In geometric terms, the net earnings is represented by the area of the region between the curves $y = R(x)$ and $y = C(x)$ from $x = 0$ to $x = 9$.

41. a. The second plan is more profitable for the first 18 years.
   b. 7,776; In geometric terms, the net excess profit generated by the second plan is the area of the region between the curves $y = P_2(x)$ and $y = P_1(x)$ from $x = 0$ to $x = 18$.

43. a. $x = 20$ is a maximum   **b.** consumer's surplus is 400

45. a. The revenue function is $R(q) = \frac{1}{4}q(10 - q)^2$, and the marginal revenue function is $R'(q) = \frac{1}{4}(10 - q)(10 - 3q)$
   b. $q = 2$
   c. The consumer's surplus is $8.67.

47. $\frac{8}{3}$

6.2 Volume by Disks and Washers (pages 423–425)

1. 9   **3.** $\frac{1}{30}$ **5.** 2

7. $36\sqrt{3}$ **9.** $\frac{128\sqrt{3}}{15}$ **11.** $\frac{\sqrt{3}}{4}\left(1 - \frac{\pi}{4}\right)$

13. a. $V = \pi \int_0^1 [(\sqrt{x})^2 - (x^2)^2]\, dx$

   b. $V = \pi \int_0^1 [(\sqrt{y})^2 - (y^2)^2]\, dy$

15. a. $V = \pi \int_0^2 [(x^2 + 1)^2 - 1^2]\, dx$

   b. $V = \pi \int_1^5 [(2)^2 - (\sqrt{y - 1})^2]\, dy$

17. a. $V = \pi \int_{1.138}^{3.566} [(\ln x)^2 - (0.1x^2)^2]\, dx$

   b. $V = \pi \int_{0.138}^{1.271} [(\sqrt{10y})^2 - (e^y)^2]\, dx$

19. a. $V = \pi \int_0^{\ln 2} [(2e^{-x})^2 - (e^x - 1)^2]\, dx$

   b. $V = \pi \int_0^1 [\ln(y + 1)]^2\, dy + \pi \int_1^2 (\ln 2 - \ln y)^2\, dy$

21. 144 cubic units   **23.** 36 cubic units

25. $\frac{81}{10}$ cubic units   **27.** $\frac{9}{2}$ cubic units

29. $\frac{\pi}{2}$ cubic units   **31.** π cubic units

33. 2,555 cubic units   **35.** $\frac{2\pi}{35}$ cubic units

37. $\frac{\pi}{2}$ cubic units   **39.** $\frac{\pi}{12}$ cubic units

41. $\frac{\pi}{10}$ cubic units   **43.** $\frac{\pi}{2}$ cubic units

45. $\frac{\pi(4 - \pi)}{4}$ cubic units   **47.** $\frac{47\pi}{210}$ cubic units

49. 205.96 cubic units   **51.** 243 cubic units

53. $\frac{8\pi}{3}$ **55.** 64,000 ft^3

57. 49.718440677   **59.** 20.13 m^3

6.3 Volume by Shells (pages 431–433)

The given volumes are all in cubic units.

1. a. $\pi \int_0^4 (4 - x)^2 \, dx$ **b.** $2\pi \int_0^4 x(4 - x) \, dx$

c. $\pi \int_0^4 [24 - 10x + x^2] \, dx$ **d.** $2\pi \int_0^4 (x + 2)(4 - x) \, dx$

3. a. $2\pi \int_0^2 y\sqrt{4 - y^2} \, dy$ **b.** $\pi \int_0^2 (4 - y^2) \, dy$

c. $2\pi \int_0^2 (y + 1)\sqrt{4 - y^2} \, dy$

d. $\pi \int_0^2 [4 - y^2 + 4\sqrt{4 - y^2}] \, dy$

5. a. $\pi \int_0^1 (e^{-x})^2 \, dx$ **b.** $2\pi \int_0^1 xe^{-x} \, dx$

c. $\pi \int_0^2 (e^{-2x} + 2e^{-x}) \, dx$ **d.** $2\pi \int_0^1 (x + 2)e^{-x} \, dx$

7. a. $2\pi \int_0^{\pi/2} y(1 - \sin y) \, dy$ **b.** $\pi \int_0^{\pi/2} [1^2 - \sin^2 y] \, dy$

c. $2\pi \int_0^{\pi/2} (y + 1)(1 - \sin y) \, dy$

d. $\pi \int_0^{\pi/2} [5 - \sin^2 y - 4\sin y] \, dy$

9. $\dfrac{\pi}{12}$ **11.** $\dfrac{\pi}{10}$ **13** $\dfrac{\pi}{2}$

15. $\dfrac{\pi}{2}$ **17.** 4.3552 **19.** 1.7022

21. 2.4674 **23.** 29.3473 **25.** 1.0000

27. 0.10472 **29.** 146.0530 **31.** 31.1008

33. 87.9646

35. a. $\pi \ln 4$ **b.** $\dfrac{28\pi}{3}$ **c.** $\pi(8 + \ln 4)$

37. 63.1876 **41.** $\frac{1}{3}\pi(2R^3 - 3R^2h + h^3)$

43. $V = \pi[R^2h_2 - \frac{1}{3}(2R^2h_1 + h_2^3)]$

6.4 Arc Length and Surface Area pages (439–441)

1. $3\sqrt{10}$ **3.** $2\sqrt{5}$ **5.** $\dfrac{10\sqrt{5}}{3} - \dfrac{2}{3}$ **7.** $\dfrac{331}{120}$

9. $\dfrac{123}{32}$ **11.** $\dfrac{14}{3}$

13. 3.82 **15. a.** $\dfrac{1,505\pi}{18}$ **b.** 127.4

17. $12\pi\sqrt{5}$ **19.** 25.28 **21.** 69.5421 **23.** 27.12

25. 6 **27.** 8.6322 **29.** 141.6974

37. $L = C[b^{2n} - a^{2n}] - D[b^{2(1-n)} - a^{2(1-n)}]$

39. $f(x) = \pm \ln|\cos x|$

6.5 Physical Applications: Work, Liquid Force, and Centroids (pages 451–454)

5. 12,750 ft · lb **7.** $\frac{10}{3}$ ft · lb

9. 6,500 ft-lbs **11.** 2,000 ft · lb

13. 4 ergs **15.** The force is 192 lb.

17. The force is 59.7 lb.

19. $F = 64.5 \int_0^{1/24} 2\left(x + \dfrac{1}{24}\right) \sqrt{\left(\dfrac{1}{24}\right)^2 - x^2} \, dx$

21. $F = 51.2 \int_{-2}^0 2(x + 3)\sqrt{4 - x^2} \, dx$

23. $\left(0, -\dfrac{18}{5}\right)$ **25.** $\left(\dfrac{1}{\ln 2}, \dfrac{1}{4 \ln 2}\right)$

27. $(\frac{7}{3}, \frac{1}{2} \ln 2)$

29. about $y = -1$, $V_1 = \frac{200\pi}{3}$; about $x = 3$, $V_2 = \frac{250\pi}{3}$

31. $V = \frac{8\pi}{3}(4 + 3\pi)$

33. The fluid force is about 119 lb.

35. No work will be required, as the tank will empty with the force of gravity.

37. The amount of work is approximately 345,800 ft · lb.

39. 24,881 ft · lb **41.** 2,932.8 lb **43.** 152,381 mi · lb

45. a. −7.5 ergs **b.** −37.5 ergs

47. The total force of the bottom is about 85,169 lb.

49. $I_x = \frac{64}{15}$, $I_y = \frac{256}{7}$ **51.** $\rho = \frac{2}{3}\sqrt{6}$ **55.** $V = \pi(2sL^2 + L^3)$

CHAPTER 6: REVIEW

Proficiency Examination (page 454)

12.–18. *The definite integrals could represent the following:*

A. Disks revolved about the y-axis.
B. Disks revolved about the x-axis.
C. Slices taken perpendicular to the x-axis.
D. Slices taken perpendicular to the y-axis.
E. Mass of a lamina with density π.
F. Washers taken along the x-axis.
G. Washers taken along the y-axis.

12. All but E are formulas for volumes of solids.

13. A, B, F, G **14.** F, G **15.** C, D

16. A, F **17.** B, G **18.** 36

19. $\frac{1}{2}$ **20.** $\frac{256}{3}$

21. a. $\frac{32}{3}$ **b.** $\frac{256\pi}{5}$ **c.** $\frac{128\pi}{3}$

22. 1.4397 **23.** 5.3304

24. The centroid is $(\frac{13}{25}, \frac{9}{175})$.

25. $F = 64 \int_0^2 (5 - x)\sqrt{1 - x^2} \, dx$

Supplementary Problems (pages 455–459)

1. $\dfrac{9}{4}$ **3.** $\dfrac{4}{45}$ **5.** $\dfrac{81\pi}{2}$

7. $\dfrac{8\pi}{45}$ **9.** 64   **11.** $\left(-\dfrac{1}{2}, \dfrac{2}{5}\right)$

13. 250 in. · lb   **15.** $\dfrac{4}{3}\pi a^3$ **17.** $2\pi rh$

19. 4,680 ft · lb

21. The total force on the side is
8,424 + 21,840 = 30,264 ft · lb

23. 2,333 ft · lb   **25.** 44.108 ft · lb

27. The spring should be stretched about 14 in.

29. 5.625×10^{-5} **31.** $\left(\dfrac{7}{3}, \dfrac{1}{3}\right)$

33. $V = \pi b^2\left(\dfrac{r^3}{3a^2} - r + \dfrac{2}{3}a\right)$ **35.** 17,500 ft · lb

37. 135 lb   **39.** $F(t) = \frac{1}{2}t + \frac{2}{3}$

41. The total work is $6,176\pi$.

43. The desired distances are 4.88 cm, 11.12 cm from the right
(and left), 10.34 cm from the top, and 5.66 cm from the
bottom.

45. $V = 2\sqrt{3}$

47. $A \approx 0.64313878$

49. $F = 62.4B\left[\dfrac{1}{4}D\sqrt{4A^2 - B^2} + \dfrac{1}{12}(4A^2 - B^2)\right]$

51. $S = 2\pi\sqrt{2}$

53. $W = (4,000)^2\, P\left(\dfrac{1}{4,000} - \dfrac{1}{t}\right)$ mi · lb

55. 344π

57. a. $h \approx -0.644391$
   b. $h \approx 0.865202$

59. $W \approx 1.728 \times 10^{12}$ ft · lb   **61.** 4.7999

63. $\dfrac{\pi}{4}$ **65.** 1.3833

67. $s = 3a + \dfrac{b^2}{8a}\ln 2$

69. This is Putnam Examination Problem 1 of the morning
session in 1939.

71. This is Putnam Examination Problem 1 of the morning
session in 1993.

Cumulative Review for Chapters 1–6 (pages 463–464)

5. $\dfrac{7}{5}$ **7.** $\dfrac{1}{2}$ **9.** 0

11. 1   **13.** 1

15. $y' = 6(2x - 1)^2(x^2 + 1)^2(3x - 4)$

17. $y' = \dfrac{\cos x + x\sin x}{(x + \cos x)^2}$

19. $y' = -6\csc^2 3x\cot 3x$

21. $y' = \dfrac{10x + 3}{5x^2 + 3x - 2}$ **23.** 5

25. $\sqrt{10} - 3$

27. $\ln(e^x + 2) + C$

29. 1.812

31.

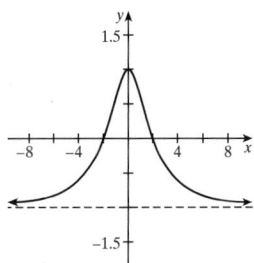

33. 7.77146

35. $0.246x + y - 1.094 = 0$

37. $\ln y = \tan^{-1} x + \frac{1}{2}\ln(x^2 + 1) + C$

39. $y = -\ln|\cos x + e^{-5} - 1|$

41. $x = \pm\dfrac{\sqrt{2}}{2}R$

43. $W = 1,562.5$ ft · lb.

45. a. $x(t) = 2t^3 - t^2 - 4t + 2$
   b. 1.75657
   c. 39

47. a.

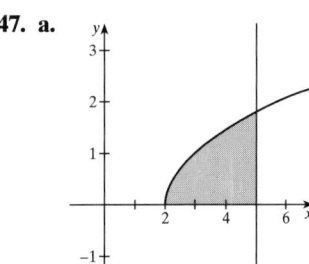

   b. $2\sqrt{3}$

   c. $A(h) = \displaystyle\int_2^h \sqrt{x - 2}\,dx = \frac{2}{3}(h - 2)^{3/2}$

   d. $\dfrac{dA}{dh} = \sqrt{3}$

49. a. $v = -34e^{-2t} - 16$
   b. -16
   c. $t \approx 1.070$ seconds

CHAPTER 7: METHODS OF INTEGRATION

7.1 Review of Substitution and Integration by Table (pages 473–475)

1. $2(x^2 + 5x)^{1/2} + C$

3. $\frac{1}{2}\ln^2(x + 1) + C$

5. $\frac{1}{4}\tan^{-1}\frac{x^2}{2} + C$

7. $-\frac{1}{5}(1 + \cot x)^5 + C$

9. $-\frac{1}{4}(x^4 - 2x^2 + 3)^{-1} + C$

11. $\ln(x^2 + x + 1) + C$

13. $\frac{\sqrt{x^2 - a^2}}{a^2 x} + C$

15. $\frac{1}{2}x^2(\ln x - \frac{1}{2}) + C$

17. $a^{-1}e^{ax}(x - a^{-1}) + C$

19. $\frac{1}{2}x\sqrt{x^2 + 1} - \frac{1}{2}\ln\left|x + \sqrt{x^2 + 1}\right| + C$

21. $\frac{1}{4}\sqrt{4x^2 + 1} + C$

23. $\dfrac{-4\sin 5x - 5\cos 5x}{41e^{4x}} + C$

25. $b^{-1}\ln|1 + bx| + C$

27. $\frac{1}{20}(x + 1)^4(4x - 1) + C$

29. $\frac{1}{4}e^{4x}(x - \frac{1}{4}) + C$

31. $x - \frac{1}{2}\ln\left|1 + e^{2x}\right| + C$

33. $\frac{1}{8}(4x^4 + 1)^{1/2} + C$

35. $\sec\frac{x}{2}\tan\frac{x}{2} + \ln\left|\sec\frac{x}{2} + \tan\frac{x}{2}\right| + C$

37. $-\dfrac{1}{3(3x + 1)} + C$

39. $-\sin x + \ln\left|\tan\left(\frac{x}{2} + \frac{\pi}{4}\right)\right| + C$

41. $\frac{1}{2}x + \frac{1}{4}\sin 2x + C$

43. $-\frac{1}{5}\cos^5 x + \frac{1}{7}\cos^7 x + C$

45. If m is odd, let $u = \cos x$. If n is odd, let $u = \sin x$. If both m and n are even, use the identities shown in Problems 40 and 41 until one exponent is odd.

47. $4\left[\dfrac{x^{1/2}}{2} - x^{1/4} + \ln(x^{1/4} + 1)\right] + C$

49. $4x^{1/2} - 3x^{1/3} + 3x^{1/6} - \frac{3}{2}\ln(2x^{1/6} + 1) + C$

51. $\dfrac{1}{e^x + e^{-x}} + C$

53. $\dfrac{1}{2}$

55. 94.68

57. 40.3053

59. 1.4397

61. 3.8097

63. $-\ln|\csc x + \cot x| + C$

7.2 Integration by Parts (pages 480–481)

1. $-\frac{1}{2}xe^{-2x} - \frac{1}{4}e^{-2x} + C$

3. $\frac{1}{2}x^2\ln x - \frac{1}{4}x^2 + C$

5. $x\sin^{-1}x + \sqrt{1 - x^2} + C$

7. $2\sqrt{x}\ln\sqrt{x} - 2\sqrt{x} + C$

9. $\frac{1}{3}x^3\ln x - \frac{1}{9}x^3 + C$

11. $-\frac{2}{15}(3e^x + 2)(1 - e^x)^{3/2} + C$

13. 4.28246

15. 0.71828

17. 133.62291

19. Integration by parts is the application of the formula

$$\int u\, dv = uv - \int v\, du$$

The u factor is a part of the integrand that is differentiated and dv is the part that is integrated. Generally, pick dv as complicated as possible yet still integrable, so that the integral on the right is easier to integrate than the original integral.

21. $\frac{1}{2}\cos^2 x - (\cos^2 x)\ln(\cos x) + C$

23. $-(2 + \cos x)\ln(2 + \cos x) + (2 + \cos x) + C$

25. $\frac{x}{2} + \frac{1}{4}\sin 2x + C$

27. $\frac{1}{4}x^2 + \frac{1}{4}x\sin 2x + \frac{1}{8}\cos 2x + C$

29. $\dfrac{x^{n+1}}{n + 1}\ln x - \dfrac{x^{n+1}}{(n + 1)^2} + C$

31. 177 units

33. 3.73218

35. a. 2.2565   **b.** 12.1774

37. (0.68, 1.27)

39. $2y^{1/2} = \frac{2}{3}x^{3/2}\left(\ln x - \frac{2}{3}\right) + C$

41. $y = \exp(1 - \cos x - x\sin x)$

43. 65 miles

45. 0.07

47. 1.558 nRT

7.3 Trigonometric Methods (pages 487–488)

5. $\sin x - \frac{1}{3}\sin^3 x + C$

7. $\frac{1}{3}\sin^3 x - \frac{1}{5}\sin^5 x + C$

9. $-\frac{2}{3}(\cos t)^{3/2} + C$

11. $-e^{\cos x} + C$

13. $\frac{1}{8}x - \frac{1}{32}\sin 4x + C$

15. $-\frac{1}{2}\ln|\cos 2\theta| + C$

17. $\frac{1}{6}\tan^6 x + \frac{1}{4}\tan^4 x + C$

19. $2\tan x - x + C$

21. $\frac{1}{2}\sec u\tan u - \frac{1}{2}\ln|\sec u + \tan u| + C$

23. $\frac{3}{4}(\tan x)^{4/3} + C$

25. $\frac{1}{4}\sin^2 x^2 + C$

27. $\frac{1}{4}\sec^3 t\tan t - \frac{5}{8}\sec t\tan t + \frac{3}{8}\ln|\sec t + \tan t| + C$

29. $-\frac{1}{3}\csc^3 x + C$

31. $-\csc x + C$

33. $2\sin^{-1}\frac{t}{2} + \frac{1}{2}t\sqrt{4 - t^2} + C$

35. $\sqrt{4 + x^2} + \ln\left|\sqrt{4 + x^2} + x\right| + C$

37. $\ln\left|x + \sqrt{x^2 - 7}\right| + C$

39. $\sin^{-1}\dfrac{x}{\sqrt{5}} + C$

41. $-\dfrac{\sqrt{4 - x^2}}{4x} + C$

43. $\sqrt{x^2 - 4} - 2\sec^{-1}\dfrac{x}{2} + C$

45. $\frac{1}{3}\ln\left|\dfrac{3}{\sqrt{9 - (x + 1)^2}} + \dfrac{x + 1}{\sqrt{9 - (x + 1)^2}}\right| + C$

47. $\ln\left|\sqrt{x^2 - 2x + 6} + x - 1\right| + C$

49. $\dfrac{1}{\sqrt{5}}\ln\left|\dfrac{2\cos x + \sqrt{5} - 1}{2\cos x - \sqrt{5} - 1}\right| + C$

51. 1.07997

53. $\frac{1}{2}$

55. 15.5031

57. $2\sqrt{2}\,\sin\dfrac{x}{2} + C$

59. $-\frac{1}{3}\cos(\frac{3}{2}x) - \frac{1}{5}\cos(\frac{5}{2}x) + C$

61. $\frac{1}{8}\sin 4x + \frac{1}{8}\sin(-2x) - \frac{1}{24}\sin 6x + C$

63. 0.3318

7.4 The Method of Partial Fractions (pages 497–499)

1. $\dfrac{-1}{3x} + \dfrac{1}{3(x-3)}$

3. $3 - \dfrac{1}{x}$

5. $\dfrac{4}{x} + \dfrac{-8}{2x+1}$

7. $\dfrac{3}{x} + \dfrac{-1}{x^2} + \dfrac{1}{x+1} + \dfrac{-2}{(x+1)^2}$

9. $\dfrac{-4}{9x^2} + \dfrac{17}{27x} - \dfrac{13}{9(x+3)^2} + \dfrac{10}{27(x+3)}$

11. $\dfrac{1}{4(1-x)} + \dfrac{1}{4(1+x)} + \dfrac{1}{2(1+x^2)}$

13. $\frac{1}{3}\ln\left|\dfrac{x-3}{x}\right| + C$

15. $3x - \ln|x| + C$

17. $4\ln|x| - 4\ln|2x+1| + C$

19. $-9\ln|x| - x^{-1} + 6\ln|x+1| + 5\ln|x-1| + C$

21. $x - \frac{5}{3}\ln|x+2| + \frac{2}{3}\ln|x-1| + C$

23. $x + \frac{1}{2}\ln|x-1| - \frac{1}{2}\ln|x+1| - \tan^{-1}x + C$

25. $\ln|x+1| + (x+1)^{-1} + C$

27. $-\frac{1}{2}\ln|x| + \frac{1}{3}\ln|x+1| + \frac{1}{6}\ln|x-2| + C$

29. $\ln\left|\dfrac{x+2}{x+1}\right| - \dfrac{2}{x+2} + C$

31. $3\ln|x-1| + 2\ln|x+3| + C$

33. $\ln|x^3 - x^2 + 4x - 4| + C$

37. $\frac{1}{7}\ln|e^x - 3| - \frac{1}{7}\ln(2e^x + 1) + C$

39. $\dfrac{1}{1+\cos x} + C$

41. $\ln|\tan x + 4| + C$

43. $3x^{1/3} + 6x^{1/6} + 6\ln|x^{1/6} - 1| + C$

45. $-\dfrac{1}{5}\ln\left|\tan\dfrac{x}{2} - 3\right| + \dfrac{1}{5}\ln\left|3\tan\dfrac{x}{2} + 1\right| + C$

47. $-\ln|\sin x + \cos x| + C$

49. $-\ln|1 - \sin x| + C$

51. $\dfrac{1}{\tan\dfrac{x}{2} - 2} + C$

53. $\frac{1}{2}\ln|\ln x - 3| - \frac{1}{2}\ln|\ln x - 1| + C$

55. 0.6931

57. a. 1.0888   **b.** 7.6402

59. $\frac{\pi}{2}$

61. 6.4 days; the Spy is a dead duck plucker!

7.5 Summary of Integration Techniques (pages 502–503)

1. $\dfrac{1}{2x^2(x-1)^2} + C$

3. $\frac{1}{4}\ln|\sec 2x^2 + \tan 2x^2| + C$

5. $\ln|\sin e^x| + C$

7. $-\ln|\cos(\ln x)| + C$

9. $3\ln|\sec t + \tan t| + 2\ln|\sec t| + C$

11. $\frac{1}{2}\tan^{-1}e^{2t} + C$

13. $x + \frac{1}{2}\ln(x^2 + 9) - \frac{8}{3}\tan^{-1}\dfrac{x}{3} + C$

15. $-x - 2\ln|e^{-x} - 1| + C$ **17.** $\frac{2}{3}\sin^{-1}t^3 + C$

19. $x - \frac{1}{2}\ln(e^{2x} + 1) + C$ **21.** $\tan^{-1}(x+1) + C$

23. $\dfrac{2}{\sqrt{3}}\tan^{-1}\dfrac{2}{\sqrt{3}}\left(x + \dfrac{1}{2}\right) + C$

25. $x\tan^{-1}x - \frac{1}{2}\ln(x^2+1) + C$

27. $\frac{1}{2}e^{-x}(\sin x - \cos x) + C$

29. $x\cos^{-1}(-x) + \sqrt{1-x^2} + C$

31. $-\cos x + \frac{1}{3}\cos^3 x + C$ **33.** $\frac{1}{5}\cos^5 x - \frac{1}{3}\cos^3 x + C$

35. $\frac{1}{16}x - \frac{1}{64}\sin 4x + \frac{1}{48}\sin^3 2x + C$

37. $-\frac{1}{9}\cos^9 x + \frac{2}{7}\cos^7 x - \frac{1}{5}\cos^5 x + C$

39. $\frac{1}{8}\tan^8 x + \frac{1}{6}\tan^6 x + C$

41. $\sqrt{1-x^2} - \ln\left|\dfrac{1+\sqrt{1-x^2}}{x}\right| + C$

43. $\sqrt{2x^2-1} + \dfrac{3}{\sqrt{2}}\ln|\sqrt{2}x + \sqrt{2x^2-1}| + C$

45. $\ln|x| - \ln|1 + \sqrt{x^2+1}| + C$

47. $-2\sqrt{4x - x^2 - 2} + 5\sin^{-1}\dfrac{x-2}{\sqrt{2}} + C$

49. $\ln(\sin x + \sqrt{1 + \sin^2 x}) + C$

51. π

53. $\dfrac{\sqrt{3}}{6}$

55. 0.1353

57. 113.4327

59. 1.8101

61. 0.0266

63. $\frac{\pi}{4} - \frac{2}{3}$

65. $\ln(\sqrt{1 + e^{2x}} + e^x) + C$

67. $3\ln|x| - \ln|x^2 + x + 1| + \dfrac{4}{\sqrt{3}}\tan^{-1}\dfrac{\sqrt{3}}{3}(2x+1) + C$

69. $5\ln|x-7| + 2(x+2)^{-1} + C$

71. $3\ln|x+1| - 2(x+1)^{-1} + C$

73. $-\frac{1}{2}\ln|x+1| + 2\ln|x+2| - \frac{3}{2}\ln|x+3| + C$

75. $\dfrac{1}{13}\ln\left|\dfrac{\tan\dfrac{x}{2} + 5}{5\tan\dfrac{x}{2} - 1}\right| + C$

77. 0.09

81. 1.1222

83. 4

85. 101.7876

87. (0.709, 0.123)

91. $-\frac{1}{12}\sin^2 4x\cos 4x - \frac{1}{6}\cos 4x + C$

7.6 First-Order Differential Equations (pages 512–515)

1. $\frac{1}{5}x^2 + Cx^{-3}$

3. $-5x^{-3} + Cx^{-2}$

5. $\frac{1}{3}x^{-2}e^{x^3} + Cx^{-2}$

7. $\left(\dfrac{1}{2}x + \dfrac{1}{2x}\right)\tan^{-1}x - \dfrac{1}{2} + \dfrac{C}{x}$

9. $y = -\cos x\ln|\cos x| + C\cos x$

11. $y = x^2 - 1$

13. $y = 2x^3 - 6x^2$

15. $2X^2 + Y^2 = C$ **17.** $Y = CX$

19. The GDP in the year 2000 will be about $9,081 billion.

21. The predicted number of divorces in 2002 is about 2,065,000.

23. a. $Q(t) = 30 - 20e^{-t/15}$ **b.** 4 minutes, 19 seconds

25. In the "long run" (as $t \to +\infty$), the concentration will be
$$\lim_{t \to \infty} \frac{\alpha}{\beta}(1 - e^{-\beta t}) = \frac{\alpha}{\beta}; \text{ the "half-way point" is reached when}$$
$$t = \frac{\ln \frac{1}{2}}{-\beta} = \frac{\ln 2}{\beta}$$

27. 16,682,811   **29.** $M(t) = \frac{k}{r}(1 - e^{-rt})$

31. It will take about an hour to drain.

33. a. $y = -\frac{k}{24}(x^4 - 2Lx^3 + L^3 x)$

b. Maximum deflection occurs where $y_m \approx -0.0130kL^4$.

c. The maximum deflection is $y_m \approx -0.0054kL^4$; the maximum deflection in the cantilevered case is less than that in part **b.**

35. a. $S_1(t) = 200(1 - e^{-t/100})$

b. $S_2(t) = 200 - 2te^{-t/100}$

c. The maximum excess is $S(200) \approx 27.07$ lbs.

37. $P(t) = P_\infty \exp\left[-\left(\ln \frac{P_\infty}{P_0}\right)e^{-kt}\right]$

39. $x = \frac{1}{2}(y + 1)e^y + \frac{Ce^y}{y + 1}$

41. a. $I(t) = \frac{3}{2}(1 - e^{-2t})$

b. $I(t) = e^{-2t}(1 - \cos t)$

43. 9.872 days

47. a. $v(t) = \frac{-mg}{k} + \left(\frac{mg}{k} + v_0\right)e^{-kt/m}$;

$s(t) = \frac{-mg}{k}t + \frac{m}{k}\left(\frac{mg}{k} + v_0\right)(1 - e^{-kt/m})$

b. The object reaches its maximum height when $v(t) = 0$.

The maximum height is $s_{\max} = \frac{mv_0}{k} - \frac{m^2 g}{k^2} \ln\left(1 + \frac{kv_0}{mg}\right)$.

c. The maximum height is 82.98 ft; the object hits the ground when $t \approx 5.51$ seconds.

d. 9.38 seconds

7.7 Improper Integrals (pages 524–525)

3. $\frac{1}{2}$ **5.** diverges   **7.** 10

9. diverges   **11.** $\frac{1}{10}$ **13.** $\frac{5}{2}$

15. $\frac{1}{9}$ **17.** diverges   **19.** $\frac{2}{e}$

21. $5e^{10}$ **23.** diverges   **25.** 2

27. diverges   **29.** diverges   **31.** 0

33. $\frac{5}{4}$ **35.** 2   **37.** -1

39. 1   **41.** 2   **43.** $\frac{1}{4}$

45. There will be 100,000 millirads.

47.

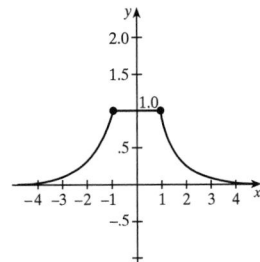

$$\int_{-\infty}^{+\infty} f(x)\, dx = 4$$

49. $(1 - p)^{-1}$ if $p < 1$, and diverges if $p \geq 1$.

53. 8   **59. b.** $\frac{s^2 - 4}{(s^2 + 4)^2}$

7.8 The Hyperbolic and Inverse Hyperbolic Functions (pages 532–533)

1. 3.6269   **3.** -0.7616 **5.** 1.1995

7. 0.9624   **9.** 1.6667   **11.** 0.6481

13. $3 \cosh 3x$ **15.** $(4x + 3)\sinh(2x^2 + 3x)$

17. $-x^{-2}\cosh x^{-1}$ **19.** $y' = \dfrac{3x^2}{\sqrt{1 + x^6}}$ **21.** $\sec x$

23. $\sec x$ **25.** $\dfrac{x - \sqrt{1 + x^2}\,\sinh^{-1}x}{x^2 \sqrt{1 + x^2}}$

27. $\cosh^{-1}x$

29. $\left(e^x - 2e^{2x}\sinh^{-1}x - \dfrac{e^x}{\sqrt{1 + x^2}}\right)\sinh\left(e^x - e^{2x}\sinh^{-1}x\right)$

31. $-\cosh x^{-1} + C$ **33.** $\ln|\sinh x| + C$

35. $\frac{1}{3}\cosh^{-1}\dfrac{3t}{4} + C$ **37.** $\sinh^{-1}(\sin x) + C$

39. $\frac{1}{3}\tanh^{-1}x^3 + C$ **41.** -0.2027

43. 1.0311   **45.** 0.3808

49. The curve is rising for all x. The curve is concave up on $(-\infty, 0)$, and concave down on $(0, +\infty)$.

53. b. $y = \cosh 2x + \sinh 2x$ **55.** $2.3504a$ **57.** 17.68

CHAPTER 7: REVIEW

Proficiency Examination (page 533)

13. a. 0.5493   **b.** $\frac{4}{3}$ **c.** 0.5493

14. $2\sqrt{x^2 + 1} + 3\sinh^{-1}x + C$

15. $-\dfrac{x}{2}\cos 2x + \dfrac{1}{4}\sin 2x + C$ **16.** $-\dfrac{1}{2}\cosh(1 - 2x) + C$

17. $\sin^{-1}\dfrac{x}{2} + C$

18. $\frac{1}{4}\ln(x^2 + 1) + \frac{1}{2}\tan^{-1}x + \frac{1}{2}\ln|x - 1| + C$

19. $\frac{x^2}{2} + \frac{1}{2}\ln|x^2 - 1| + C$ **20.** 1.9089   **21.** 0.1144

22. -0.2027 **23.** 0.6095   **24.** $\frac{1}{4}$

25. 2   **26.** diverges   **27.** $\frac{1}{2}$

28. $\frac{1}{2\sqrt{\tanh^{-1}2x}}\left[\frac{2}{1 - 4x^2}\right]$

29. 4.7999

30. $y = \frac{x + 1}{e^x}\left(\ln|x + 1) + 1|\right)$

31. 386.4 lbs

Supplementary Problems (pages 534–537)

1. $(1 - x^2)^{-1}$ **3.** e^{-2x}

5. $x \cosh x + \sinh x + (e^x - e^{-x})\cosh(e^x + e^{-x})$

7. $-\frac{x}{2}\sqrt{4 - x^2} + 2\sin^{-1}\frac{x}{2} + C$

9. $-\ln|x| - x^{-1} + \ln|x - 2| + C$

11. $4x^{1/4} - 4\ln(1 + x^{1/4}) + C$

13. $\frac{x^3}{3}\tan^{-1}x - \frac{x^2}{6} + \frac{1}{6}\ln(1 + x^2) + C$

15. $\frac{1}{2}e^x\sqrt{4 - e^{2x}} + 2\sin^{-1}\left(\frac{e^x}{2}\right) + C$

17. $-\frac{1}{5}(1 + x^{-2})^{5/2} + \frac{1}{3}(1 + x^{-2})^{3/2} + C$

19. $-2\sqrt{1 - \sin x} + C$

21. $\frac{x}{2}[\sin(\ln x) - \cos(\ln x)] + C$

23. $\ln(e^{2x} + 4e^x + 1) - x + C$

25. $\frac{x^3}{3}\cot^{-1}x + \frac{x^2}{6} - \frac{1}{6}\ln(1 + x^2) + C$

27. $\frac{1}{3}\ln|x^3 + 6x + 1| + C$

29. $2\cos\sqrt{x + 2} + 2\sqrt{x + 2}\sin\sqrt{x + 2} + C$

31. $\frac{1}{4}\ln(x^4 + 4x^2 + 3) + C$

33. $\sqrt{5 - x^2} - \sqrt{5}\ln\left|\frac{\sqrt{5} + \sqrt{5 - x^2}}{x}\right| + C$

35. $\frac{1}{3}\sqrt{x^2 + 4}\,(x^2 - 8) + C$

37. $\frac{\sec^3 x}{3} + C$

39. $\frac{3}{2}x^{2/3} - 2x^{1/2} + 3x^{1/3} - 6x^{1/6} + 6\ln|x^{1/6} + 1| + C$

41. $-\ln|1 + \cos x - \sin x| + C$

43. $y = \frac{1}{x + 1}[x + C]$

45. $y = \sin x + \frac{1}{3}\cos x \tan^3 x + C\cos x$

47. 2   **49.** $\frac{\sqrt{2}\pi}{2}$

51. $\frac{n!}{a^{n+1}}$ **55.** -1

57. $\frac{dT}{dt} = k(T^4 - T_0^4)$;

$\frac{1}{4\,T_0^3}\left(\ln\left|\frac{T - T_0}{T + T_0}\right| - 2\tan^{-1}\frac{T}{T_0}\right) = kt + C$

59. $S = \frac{1}{2} + \frac{1}{2}\left[\dfrac{\frac{1}{2}\frac{k}{\sqrt{m}}t + C}{\sqrt{1 + \left(\frac{1}{2}\frac{k}{\sqrt{m}}t + C\right)^2}}\right]$

61. $(0.41, 0.64)$ **63.** 2.3115

65. 14.424   **67.** 22.2166

69. 5.807   **71.** 1.3833

73. 1.317   **77. a.** $C \approx 2.50663$

83. $A = \frac{1}{2}; \frac{3}{4}\ln 2 \approx 0.52$ **85.** $\frac{3}{16}\pi$

87. a. $v = \left(\frac{W - B}{k}\right)(1 - e^{-(kg/W)t})$

b. $s(t) = \frac{W - B}{k}\left[t + \frac{W}{kg}e^{-(kg/W)t} - \frac{W}{kg}\right]$

c. time when the container breaks is $t \approx 52.81$ sec; the critical depth is 269.07 meters

89. about 5.6 years

91. This is Putnam examination Problem 3 in the morning session of 1980.

CHAPTER 8: INFINITE SERIES

8.1 Sequences and Their Limits (pages 550–551)

3. $0, 2, 0, 2, 0$ **5.** $1, \frac{1}{2}, \frac{1}{3}, \frac{1}{4}, \frac{1}{5}$ **7.** $\frac{4}{3}, \frac{7}{4}, \frac{10}{5}, \frac{13}{6}, \frac{16}{7}$

9. $256, 16, 4, 2, \sqrt{2}$ **11.** $1, 3, 13, 183, 33,673$ **13.** 5

15. -7 **17.** 0   **19.** $\frac{1}{2}$

21. 4   **23.** 0   **25.** 1

27. 1   **29.** 1   **31.** $\frac{1}{2}$

33. 1   **35.** 1

47. $a_4 = 6.25\%, a_n = 100(\frac{1}{2})^n\%$

49. $N = 100$ **51.** $N = 1,001$

8.2 Introduction to Infinite Series: Geometric Series (pages 558–561)

3. 5   **5.** 3

7. diverges   **9.** $\frac{3}{20}$

11. 4.51665   **13.** $-\frac{2}{45}$

15. $\frac{1}{3}$ **17.** $\frac{16}{63}$

19. $2(2 + \sqrt{2})$ **21.** $\frac{1}{2}(4 + 3\sqrt{2})$

23. converges   **25.** The series converges to 1.

27. diverges   **29.** The series converges to 1.

31. $\frac{1}{99}$ **33.** $\frac{52}{37}$

35. a. $A = 1, B = 1$ **b.** $\frac{1}{2}$ **37.** 0

39. 9.14   **41.** 4.858

43. $\frac{1}{a}$ **45.** 45.2

47. 1,500   **49.** 3 ft

51. The total depreciation is \$10,000.

53. The patient will have about 30.8 units of the drug.

55. In the long run, there will be 33 members on the board.

59. a. $S_n - S_{n-1} = a_n$ **b.** $\dfrac{3}{4n^2 + 8n + 3}$ **65.** $\frac{1}{4}$

8.3 The Integral Test: p-series (pages 567–568)

3. $p = 3$; converges   **5.** $p = \frac{1}{3}$; diverges   **7.** converges

9. diverges   **11.** converges   **13.** converges

15. diverges   **17.** converges   **19.** diverges

21. diverges   **23.** diverges   **25.** diverges

27. converges   **29.** converges   **31.** converges

33. converges   **35.** diverges   **37.** diverges

39. converges   **41.** converges   **43.** diverges

45. diverges   **47.** converges

49. The series converges when $p > 1$ and diverges for $p \le 1$.

51. The series converges if $p > 1$ and diverges if $p \le 1$.

53. b. 100   **c.** 1.63498

8.4 Comparison Tests (pages 573–574)

1. converges   **3.** geometric; converges

5. geometric; diverges   **7.** p-series; diverges

9. p-series; converges   **11.** p-series; diverges

13. converges   **15.** diverges

17. diverges   **19.** converges

21. converges   **23.** converges

25. diverges   **27.** converges

29. converges   **31.** converges

33. converges   **35.** diverges

37. converges   **39.** converges

41. converges   **43.** diverges

45. converges   **47.** converges

49. converges   **51.** diverges

53. converges

8.5 The Ratio Test and the Root Test (pages 581–582)

3. converges   **5.** diverges   **7.** converges

9. converges   **11.** diverges   **13.** converges

15. converges   **17.** converges   **19.** diverges

21. converges   **23.** converges   **25.** converges

27. diverges   **29.** converges   **31.** diverges

33. converges   **35.** converges   **37.** diverges

39. converges   **41.** converges   **43.** converges

45. S converges for $0 \le x < 1$

47. S converges for $0 \le x \le 0.5$

49. S converges for all $x \ge 0$

51. S converges for $0 \le x < a^{-1}$

53. S converges for all p

8.6 Alternating Series; Absolute and Conditional Convergence (pages 590–592)

5. converges conditionally   **7.** diverges

9. absolutely convergent   **11.** converges conditionally

13. absolutely convergent   **15.** diverges

17. absolutely convergent   **19.** converges conditionally

21. converges conditionally   **23.** converges conditionally

25. diverges   **27.** converges conditionally

29. converges conditionally   **31.** absolutely convergent

33. a. $\frac{5}{8}$ **b.** $S_7 \approx 0.632$ **35. a.** 0.0823 **b.** $S_6 \approx 0.083$

37. a. -0.1664 **b.** $S_5 \approx -0.167$

39. The interval of convergence is $[-1, 1)$.

41. The interval of convergence is $[-1, 1]$.

43. The interval of convergence is $(-1, 1)$ for all $p > 0$.

45. $E_{\max} = \dfrac{1}{6^2} \approx 0.0278$ **47.** $E_{\max} = \dfrac{7}{2^7} \approx 0.0547$

49. converges absolutely   **57.** converges absolutely

8.7 Power Series (pages 600–602)

1. $(-1, 1)$ **3.** $(-1, 1)$ **5.** $\left(\frac{8}{3}, \frac{10}{3}\right)$

7. $\left(-\frac{13}{3}, -\frac{5}{3}\right)$ **9.** $x = 1$ **11.** $(-1, 3)$

13. $[1, \frac{5}{3})$ **15.** $(-7, 7)$ **17.** $x = 0$

19. $[-1, 1]$ **21.** $(-\infty, \infty)$ **23.** $x = 0$

25. $(-\infty, \infty)$ **27.** $[-1, 1]$ **29.** $R = 1$

31. $R = e$ **33.** $R = \dfrac{1}{|a|}$ **35.** $f'(x) = \displaystyle\sum_{k=1}^{\infty} \dfrac{kx^{k-1}}{2^k}$

37. $f'(x) = \displaystyle\sum_{k=1}^{\infty} k(k+2)x^{k-1}$

39. $F(x) = \displaystyle\sum_{k=1}^{\infty} \dfrac{x^k}{k(2)^{k-1}}$

41. $F(x) = \displaystyle\sum_{k=0}^{\infty} \dfrac{(k+2)x^{k+1}}{(k+1)}$

47. $|x| > 1$ **49.** $R = \infty$ **53.** $R = q^{-q}$

8.8 Taylor and Maclaurin Series (pages 616–619)

3. $e^{2x} = \displaystyle\sum_{k=0}^{\infty} \dfrac{(2x)^k}{k!}$ **5.** $e^{x^2} = \displaystyle\sum_{k=0}^{\infty} \dfrac{x^{2k}}{k!}$

7. $\sin x^2 = \displaystyle\sum_{k=0}^{\infty} \dfrac{(-1)^k x^{4k+2}}{(2k+1)!}$ **9.** $\sin ax = \displaystyle\sum_{k=0}^{\infty} \dfrac{(-1)^k (ax)^{2k+1}}{(2k+1)!}$

11. $\cos 2x^2 = \displaystyle\sum_{k=0}^{\infty} \dfrac{(-1)^k (2x^2)^{2k}}{(2k)!}$

13. $x^2 \cos x = \displaystyle\sum_{k=0}^{\infty} \dfrac{(-1)^k (x)^{2k+2}}{(2k)!}$

15. $x^2 + 2x + 1$ is its own Maclaurin series.

17. $xe^x = \displaystyle\sum_{k=0}^{\infty} \dfrac{(x)^{k+1}}{k!}$

19. $e^x + \sin x = 1 + 2x + \dfrac{x^2}{2!} + \dfrac{x^4}{4!} + \dfrac{2x^5}{5!} + \dfrac{x^6}{6!} + \dfrac{x^8}{8!} + \dfrac{2x^9}{9!} + \cdots$

21. $\dfrac{1}{1 + 4x} = \displaystyle\sum_{k=0}^{\infty} (-4)^k x^k$ **23.** $\dfrac{1}{a + x} = \displaystyle\sum_{k=0}^{\infty} \dfrac{(-1)^k x^k}{a^{k+1}}$

25. $\ln(3 + x) = \ln 3 + \displaystyle\sum_{k=0}^{\infty} \dfrac{(-1)^k x^{k+1}}{(k + 1)3^{k+1}}$

27. $\tan^{-1}(2x) = \displaystyle\sum_{k=0}^{\infty} \dfrac{(-1)^k 2^{2k+1} x^{2k+1}}{2k + 1}$

29. $e^{-x^2} = \displaystyle\sum_{k=0}^{\infty} \dfrac{(-1)^k x^{2k}}{k!}$

31. $e^x \approx e + e(x - 1) + \dfrac{1}{2!}e(x - 1)^2 + \dfrac{1}{3!}e(x - 1)^3$

33. $\cos x \approx \cos\frac{\pi}{3} - (x - \frac{\pi}{3})\sin\frac{\pi}{3}$

$\qquad - \dfrac{(x - \frac{\pi}{3})^2}{2!}\cos\frac{\pi}{3} + \dfrac{(x - \frac{\pi}{3})^3}{3!}\sin\frac{\pi}{3}$

35. $\tan x \approx 0 + 1 \cdot x + \dfrac{0x^2}{2!} + \dfrac{2x^3}{3!}$

37. $f(x) = (x - 2)^3 + 4(x - 2)^2 + 5(x - 2) - 3$

39. $f(x) \approx -\frac{1}{3} + \frac{1}{9}(x - 5) - \frac{1}{27}(x - 5)^2 + \frac{1}{81}(x - 5)^3$

41. $f(x) \approx 1 - \frac{2}{3}(x - 2) + \frac{4}{9}(x - 2)^2 - \frac{8}{27}(x - 2)^3$

43. $f(x) = 1 + \frac{1}{2}x - \frac{1}{8}x^2 + \frac{1}{16}x^3 - \frac{5}{128}x^4 + \cdots$

If p is greater than 0 and not an integer, the interval of absolute convergence is $[-1, 1]$.

45. $f(x) = 1 + \frac{2}{3}x - \frac{1}{9}x^2 + \frac{4}{81}x^3 + \cdots$

Interval of absolute convergence is $[-1, 1]$.

47. $f(x) = x + \frac{1}{2}x^3 + \frac{3}{8}x^5 + \frac{5}{16}x^7 + \cdots$

Interval of absolute convergence is $(-1, 1)$.

53. $n = 4$ **55.** $f(x) = \displaystyle\sum_{k=0}^{\infty}[(-1)^k + \frac{1}{2}](\frac{x}{2})^k$

57. $f(x) = \displaystyle\sum_{k=0}^{\infty}\left[1 - \dfrac{1}{2^{k+1}}\right]x^k$

59. $f(x) = \displaystyle\sum_{k=0}^{\infty}[\frac{2}{3}(-\frac{1}{2})^k - \frac{1}{6} - \frac{1}{2}(-1)^k]x^k$

61. $f(x) = \displaystyle\sum_{k=0}^{\infty} \dfrac{(-1)^k 2^{2k} x^{2k+1}}{(2k + 1)!}$

63. $f(x) = \frac{1}{2}\displaystyle\sum_{k=0}^{\infty}(-1)^k\left[\dfrac{1 + 2^{2k}}{(2k)!}\right]x^{2k}$

65. $f(x) = \displaystyle\sum_{k=0}^{\infty}[(-1)^k 2^{k+1} + 2^{k+1} + 1]\dfrac{x^{k+1}}{k + 1}$ **67.** 1

71. a. $f(x) = x - 2x^3 + \frac{2}{3}x^5 - \frac{4}{45}x^7 + \cdots$

b. $P_5(x) \approx 0.00070$ **c.** 0.00068218

d.

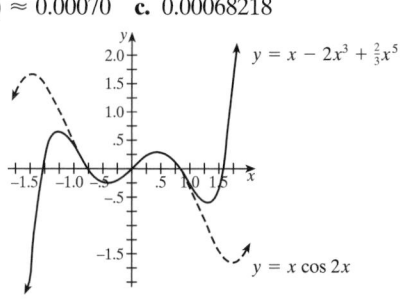

$y = x - 2x^3 + \frac{2}{3}x^5$

$y = x \cos 2x$

75. $f(x) = \dfrac{1}{2}\displaystyle\sum_{k=0}^{\infty}(k + 1)(k + 2)x^k$

77. $\displaystyle\int_0^1 t^{0.2}e^t \, dt = \sum_{k=0}^{\infty}\dfrac{1}{(k + 1.2)k!}$

CHAPTER 8: REVIEW

Proficiency Examination (pages 619–620)

31. This is the definition of e.   **32. a.** 0   **b.** S converges

33. S diverges   **34.** S diverges

35. S diverges   **36.** diverges   **37.** converges

38. The interval of convergence is $(-1, 1)$.

39. $\sin 2x = \displaystyle\sum_{k=0}^{\infty}\dfrac{(-1)^k (2x)^{2k+1}}{(2k + 1)!}$

40. $f(x) = -\dfrac{2}{5}\displaystyle\sum_{k=0}^{\infty}\left(\dfrac{2}{5}\right)^k\left(x - \dfrac{1}{2}\right)^k$

Supplementary Problems (pages 620–622)

1. does not exist   **3.** converges to e^{-2} **5.** converges to 1

7. converges to $\frac{5}{3}$ **9.** converges to 0   **11.** converges to 0

13. converges to e^4 **15.** diverges   **17.** $\frac{1}{3}$

19. $\dfrac{e}{3 - e}$ **21.** $\frac{232}{77}$ **23.** $\dfrac{15 - e}{9(6 - e)}$

25. $\dfrac{-1}{\ln 2}$ **27.** $\dfrac{1}{4}$ **29.** diverges

31. diverges   **33.** converges   **35.** converges

37. converges   **39.** converges   **41.** diverges

43. diverges   **45.** converges

47. converges absolutely   **49.** converges absolutely

51. diverges   **53.** converges absolutely

55. converges absolutely

57. $(0, 2)$ **59.** $(-1, 1)$ **61.** $(-\infty, \infty)$

63. $[-3, -1]$ **65.** $x = 0$ **67.** $(-\infty, \infty)$

69. $f(x) = \displaystyle\sum_{k=0}^{\infty}(-1)^k \dfrac{x^{2k+4}}{(2k + 1)!}$

71. $f(x) = x + x^2 + \displaystyle\sum_{k=1}^{\infty}\dfrac{(-1)^k x^{2k+1}}{2k + 1}$

73. $f(x) = \displaystyle\sum_{k=0}^{\infty}\left[2 + \dfrac{5(-1)^{k+1}}{2}\left(\dfrac{3}{2}\right)^k\right]x^k$

75. 0.000022   **77.** $\frac{20}{9}$ ft

79. The Spy needs 20 good shots, so $p > 0.95$.

83. 0.396888   **87.** 0.04879016   **89.** $|x| < 0.5797$

91. $\dfrac{2\tan x}{1 + \tan^2 x} = \displaystyle\sum_{k=0}^{\infty}(-1)^k\dfrac{(2x)^{2k+1}}{(2k + 1)!}$

93. a. $e^{(\ln 2)t/\lambda} \approx 1 + \dfrac{(\ln 2)t}{\lambda}; t \approx \dfrac{\lambda}{\ln 2}\left[\dfrac{S(t) - S(0)}{R(t)}\right]$

b. $t \approx 5.61 \times 10^8$ years (561 million years)

95. This is Putnam Problem 3 in the morning session of 1951.

CHAPTER 9: POLAR COORDINATES AND PARAMETRIC FORMS

9.1 The Polar Coordinate System (pages 630–631)

3. $(2\sqrt{2}, 2\sqrt{2})$ **5.** $(-\frac{5}{2}, \frac{5}{2}\sqrt{3})$ **7.** $(-\frac{3}{4}\sqrt{3}, -\frac{3}{4})$

9. $(-1, 0)$ **11.** $(0, 0)$ **13.** $(2, \frac{2\pi}{3})$

15. $(2\sqrt{2}, \frac{5\pi}{4})$ **17.** $(7.6, 1.2)$ **19.** $(2, \frac{11\pi}{3})$

21. a. United States **b.** India **c.** Greenland **d.** Canada

23. $x^2 + (y - 2)^2 = 4$ **25.** $x^2 + y^2 = (x^2 + y^2 + y)^2$

27. $x = 1$ **29.** $x^2 + 2y^2 = 2$

31. 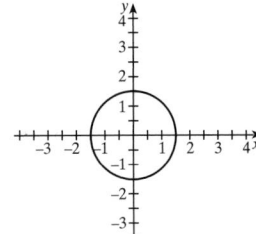 **33.**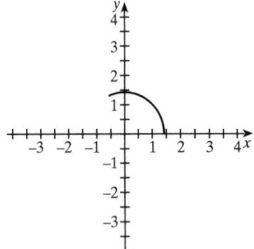

35. 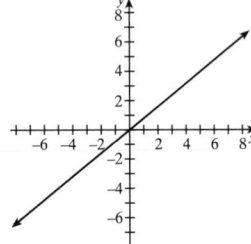 **37.**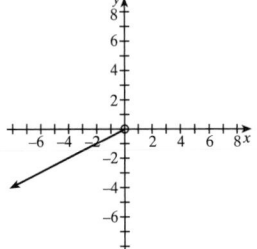

39. yes **41.** no **43.** yes **45.** yes

47. no **49.** no **51.** yes

Answers to Problems 53–60 vary.

61. a. 4.1751 **b.** $d = \sqrt{r_1^2 + r_2^2 - 2r_1r_2\cos(\theta_2 - \theta_1)}$

63. A circle with center $\left(\frac{b}{2}, \frac{a}{2}\right)$ and radius $\frac{\sqrt{a^2 + b^2}}{2}$

9.2 Graphing in Polar Coordinates (pages 640–641)

5. a. rose (4 petals) **b.** lemniscate **c.** circle

d. rose (16 petals) **e.** none (spiral) **f.** lemniscate

g. rose (3 petals) **h.** cardioid

7. 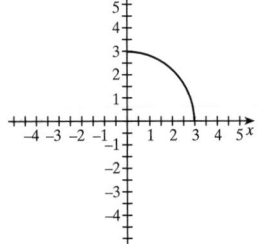 **9.**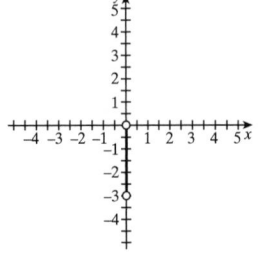

11. 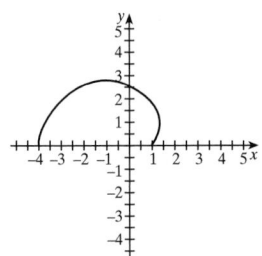 **13.**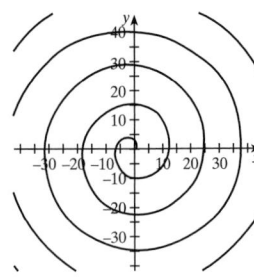

15. 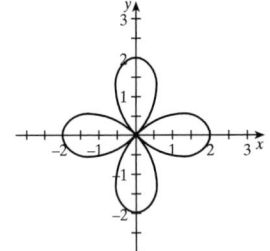 **17.**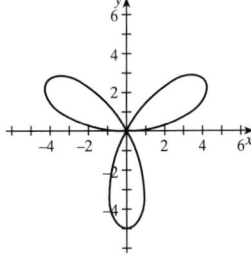

19. 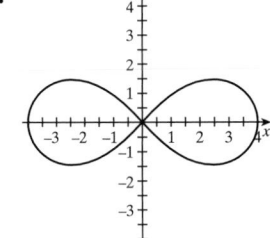 **21.**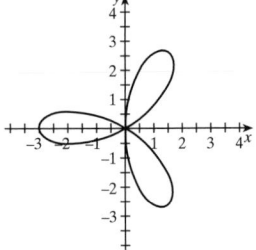

23. 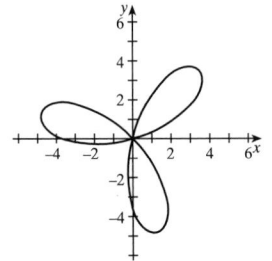 **25.**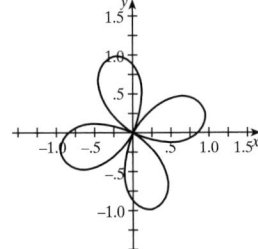

27. 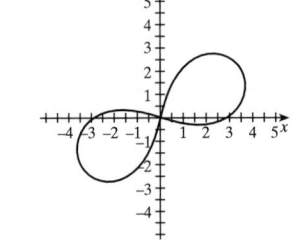 **29.**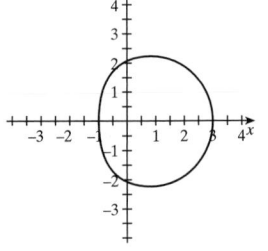

31.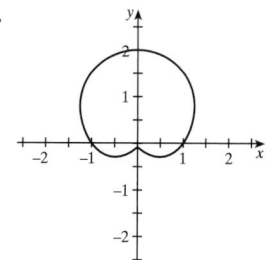

33.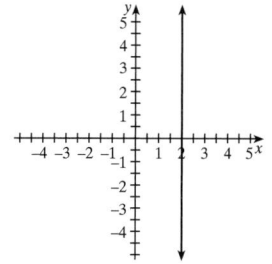

51.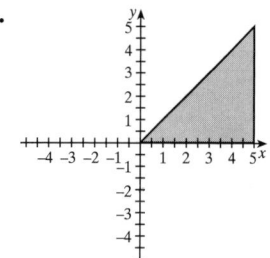

53.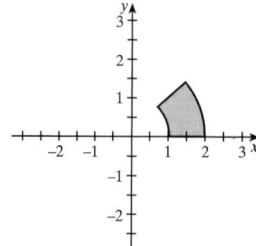

35.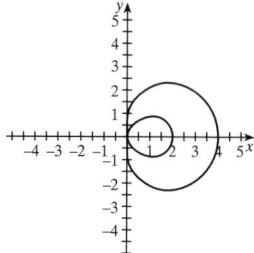

37.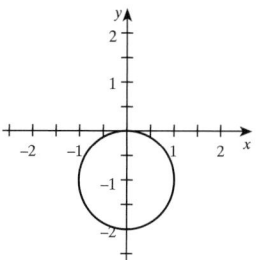

55.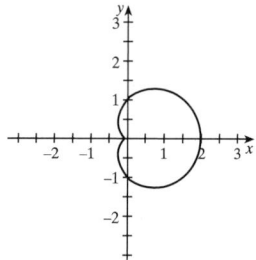

The graph shows that both graphs are the same.

39.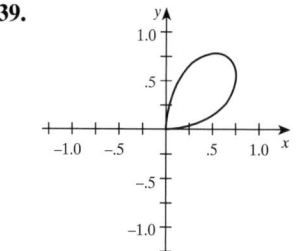

41.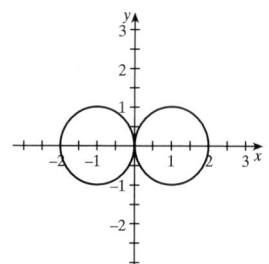

57.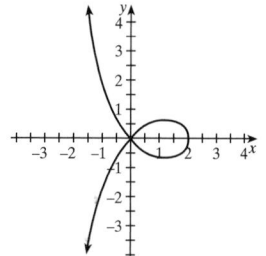

59.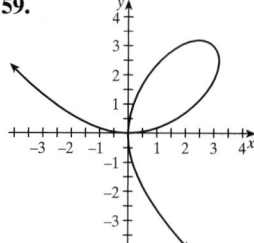

43.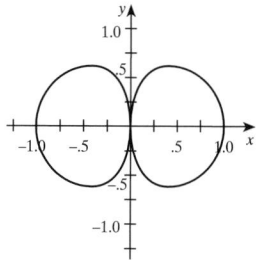

45.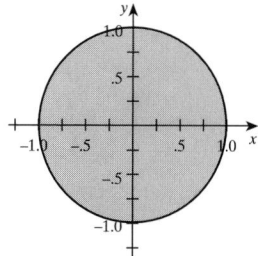

61.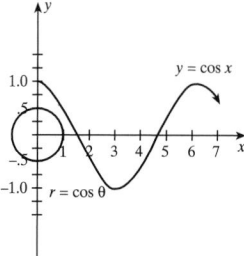

47.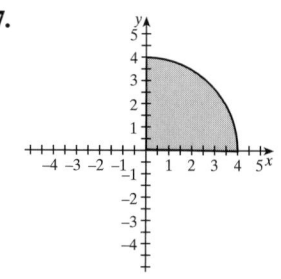

49.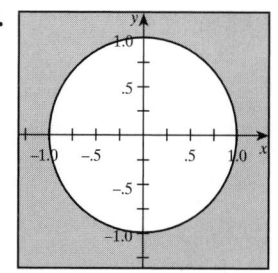

63.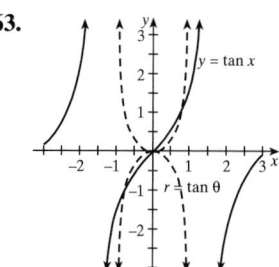

65.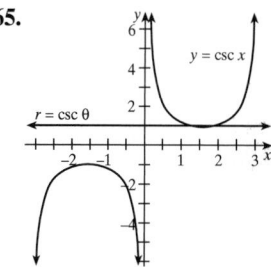

67. a. Graph for $0 < \theta < \pi$ Graph for $0 < \theta < 2\pi$

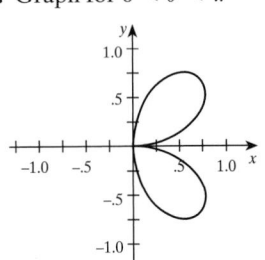
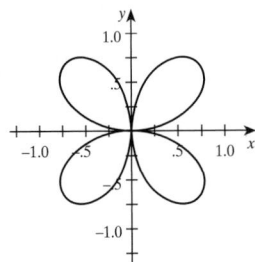

71.

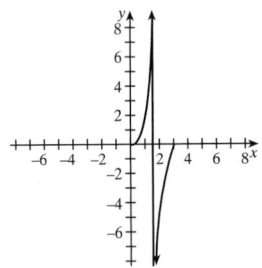

9.3 Area and Tangent Lines in Polar Coordinates (pages 649–650)

5. $P_1(0,0), P_2\left(4\sqrt{2}, -\frac{\pi}{4}\right)$ **7.** $P_1(2, \frac{\pi}{6})$ and $P_2(2, \frac{5\pi}{6})$

9. $P_1(2, \frac{\pi}{4}), P_2(2, \frac{5\pi}{4})$ **11.** $P_1(0,0), P_2(2,0), P_3(2, \pi)$

13. $P_1(0,0), P_2(2, \frac{\pi}{4})$ **15.** $P_1(2,0), P_2(2, \pi)$

17. $P_1(0,0), P_2 = (1, \frac{\pi}{4})$ **19.** $P_1(0,0), P_2(3.2, 5.4)$

21. $P_1(0,0), P_2(1, \frac{\pi}{2}), P_3(0.6, 0.6)$

23. $P_1(2, \frac{\pi}{3}), P_2(2, \frac{5\pi}{3})$

25. There are no intersection points.

27. $P_1(\sqrt{2}, \frac{\pi}{4}), P_2(\sqrt{2}, \frac{7\pi}{4})$ **29.** $P_1(2, \frac{\pi}{6}), P_2(2, \frac{5\pi}{6})$

31. 0.2392   **33.** 0.4330

35. 0.6427   **37.** 99.2201

39. $m = -\sqrt{3}; m = \sqrt{3}$ **41.** $m = -\frac{\sqrt{3}}{3}$

43. $m = -\frac{2}{\pi}$ **45.** $m = -\frac{2}{3}$

47. $m = \frac{2}{\sqrt{3}}$ **49.** $P_1\left(\frac{3a}{2}, \frac{\pi}{3}\right), P_2\left(\frac{3a}{2}, \frac{5\pi}{3}\right)$

51. $P_1(2 \sin \frac{\pi}{8}, \frac{\pi}{8}) \approx (0.77, 0.39); P_2(2 \sin \frac{5\pi}{8}, \frac{5\pi}{8}) \approx (1.85, 1.96)$

53. $\frac{\pi a^2}{4}$ **55.** $a^2(2 - \frac{\pi}{4})$

57. 4π **59.** 33.3510

61. 0.0674

63. maximum value of y is 3.0489

65. $m = 0$

69. The solution to this problem is found in Vol. 84, 1984 issue of *School Science and Mathematics,* p. 265.

9.4 Parametric Representation of Curves (pages 659–661)

3. $y = x - 2, \quad 1 \leq x \leq 3$ **5.** $y = \frac{4}{3}x - \frac{1}{225}x^2$
$0 \leq x \leq 180$

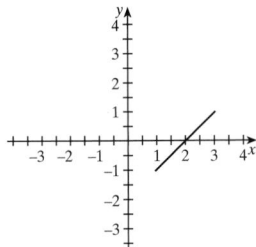
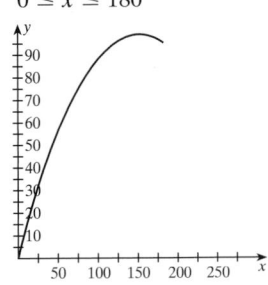

7. $y = \frac{2}{3}x + \frac{4}{3}, \quad 2 \leq x \leq 5$ **9.** $y = x - 2, \quad 1 \leq x \leq 3$

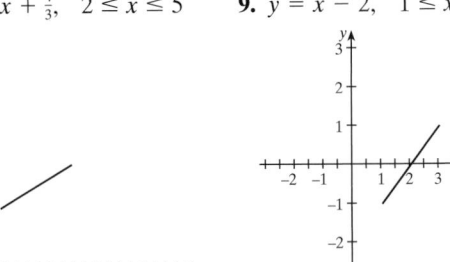

11. $y = x^{2/3}, x \geq 0$ **13.** $x^2 + y^2 = 9, \quad -3 \leq x \leq 3$

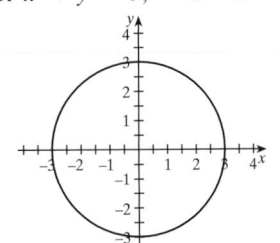

15. $(x - 1)^2 + (y + 2)^2 = 1$ **17.** $\frac{y^2}{9} - \frac{x^2}{16} = 1, \quad -\infty \leq x \leq \infty$
$0 \leq x \leq 2$

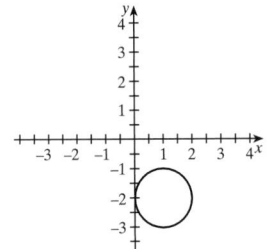
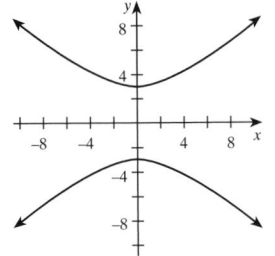

19. $y = 3x, x \geq 1$ **21.** $y = ex, x \geq 1$

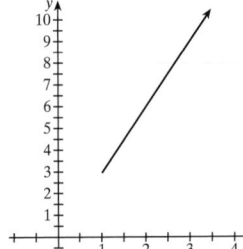
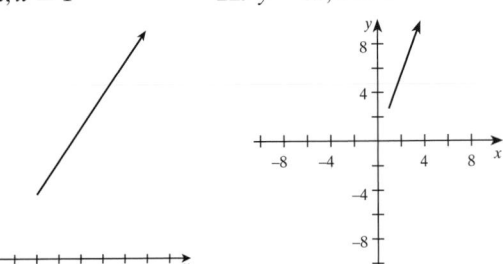

23. $y = \ln x, x > 0$

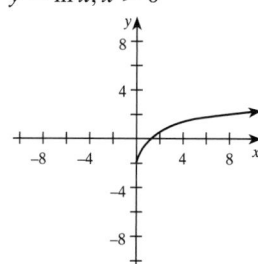

25. $\dfrac{dy}{dx} = 2t^2; \dfrac{d^2y}{dx^2} = 2$

27. $\dfrac{dy}{dx} = \dfrac{\cos 2t}{2e^{4t}}; \dfrac{d^2y}{dx^2} = \left(\dfrac{-\sin 2t - 2\cos 2t}{e^{4t}}\right)\left(\dfrac{1}{4e^{4t}}\right)$

29. $\dfrac{dy}{dx} = -\dfrac{b}{a}\cot t; \dfrac{d^2y}{dx^2} = -\dfrac{b}{a^2 \sin^3 t}$

31. $\dfrac{dy}{dx} = 2(x - 1)$ **33.** $\dfrac{dy}{dx} = \dfrac{1}{(1 - x)^2}$

35. $\frac{2}{3}$ **37.** 0.1427

39. 0.1534   **41.** 3.5255

43. 4.2027   **45.** 2.6339

47. $y = \dfrac{b}{16\,a^2}(16a^2 - x^2)$ **49.** 12

51. $\frac{2\pi}{3}$ **53.** $\sqrt{2}$

59. b. 161.557   **c.** 126.707

61. $x = f(\theta)\cos\theta; \ y = f(\theta)\sin\theta$

CHAPTER 9: REVIEW

Proficiency Examination (page 661)

17. circle

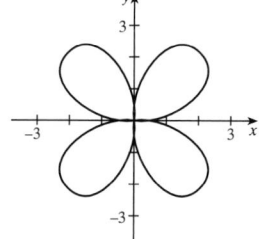

18. cardioid

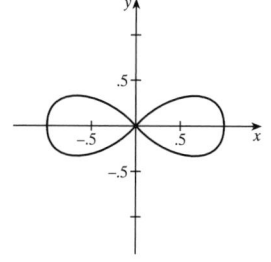

19. rose curve

20. lemniscate

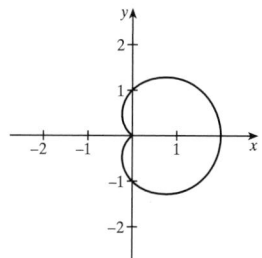

21. limaçon

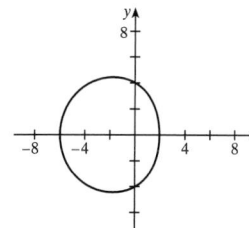

22. $0.5708a^2$

23. π

24. The points are $(3, 0)$ and $(19, 2)$.

25. $\dfrac{d\theta}{dt} \approx -3.091$

Supplementary Problems (pages 662–663)

1. circle

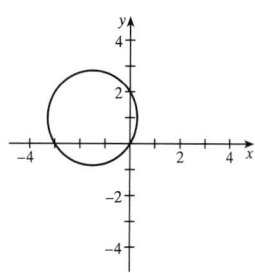

3. rose curve

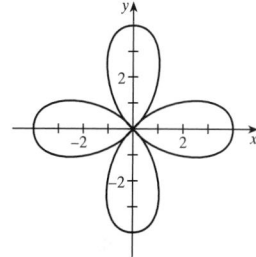

5. line

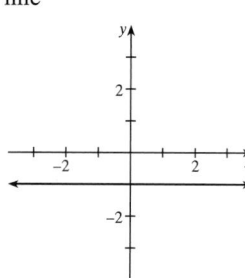

7. circle $(\theta = \pi/3)$

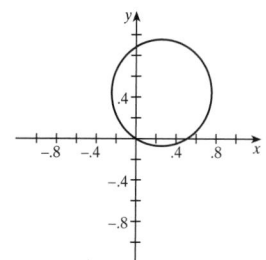

9. lemniscate

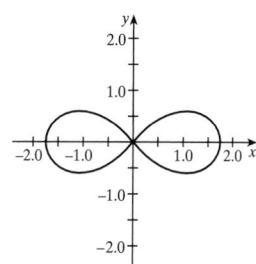

11. parabola

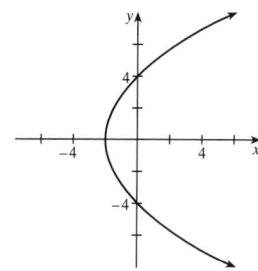

13. hyperbola

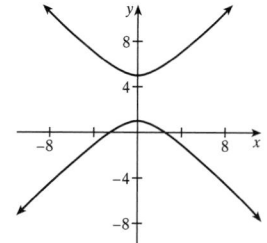

15. circle

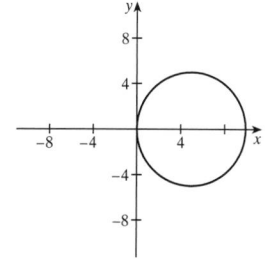

17. $(2.2, 2.2), (2.2, -2.2),$
$(-0.7, 0.6), (-0.7, -0.6)$

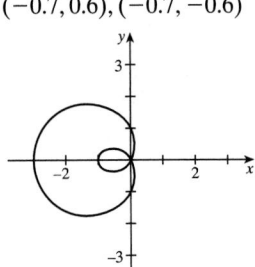

19. $(2, 3\pi/2)$

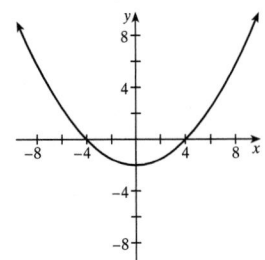

21. $P_1(0,0), P_2(2,0)$

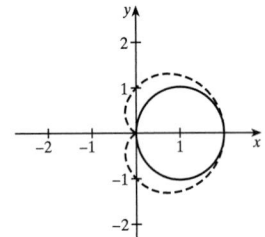

23. $P_1(0,0), P_2\left(\dfrac{2+\sqrt{2}}{2}, \dfrac{\pi}{4}\right),$
$P_2\left(\dfrac{2-\sqrt{2}}{2}, \dfrac{5\pi}{4}\right)$

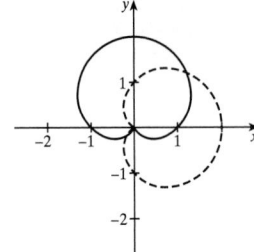

25. $P_1(2.24, 0.464)$

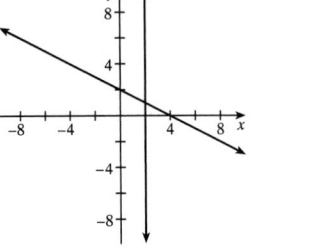

27. $P_1(1, \frac{\pi}{12}), P_2(1, \frac{5\pi}{12})$

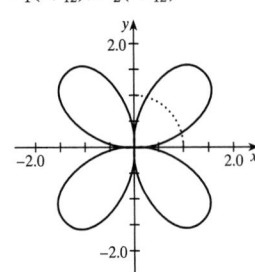

29. $P_1(0,0), P_2(a,0), P_3(a, \pi)$

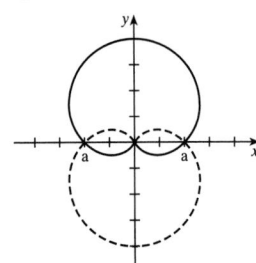

31. 25.1327   **33.** $\frac{1}{2}$

35. $a^2\left(\dfrac{\sqrt{3}}{2} + \dfrac{\pi}{3}\right)$ **37.** 35,844

39. 1.1478

41. $y = \dfrac{a^3}{a^2 + x^2}$

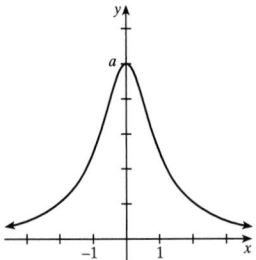

43. a. $y = x - a(\frac{\pi}{2} - 2)$ **b.** $y = 2a$ **c.** $x = 2\pi a$
45. $rr_2\sin(\theta - \theta_2) + rr_1\sin(\theta_1 - \theta) + r_1 r_2 \sin(\theta_2 - \theta_1) = 0$
47. 72   **49.** $(3,7)$ and $(3,1)$
53. $\dfrac{dr}{dt} = -4\sqrt{2}$

57.

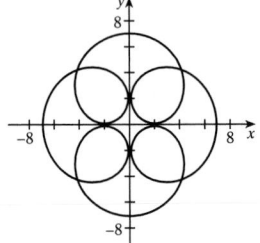

59. The ship is located at $(0.75a, -0.06)$.

CHAPTER 10: VECTORS IN THE PLANE AND IN SPACE

10.1 Introduction to Vectors (pages 674–676)

1.

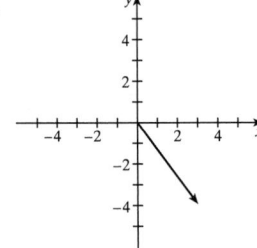

3.

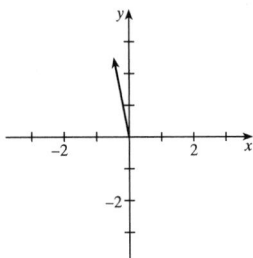

5. $4\mathbf{i} + 3\mathbf{j}$

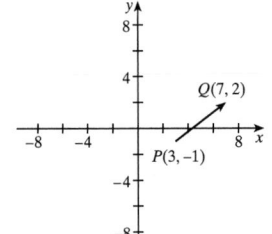

7. $-5\mathbf{i}$

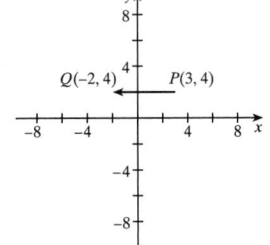

9. PQ = 2**i**; $\|\mathbf{PQ}\| = 2$

11. PQ = 4**i** + 2**j**; $\|\mathbf{PQ}\| = 2\sqrt{5}$

13. $\dfrac{1}{\sqrt{2}}(\mathbf{i} + \mathbf{j})$ **15.** $\frac{3}{5}\mathbf{i} - \frac{4}{5}\mathbf{j}$

17. $s = 6, t = 24$ **19.** $s = -5, t = -1$

21. $17\mathbf{i} - 18\mathbf{j}$ **23.** $35\mathbf{i} - 35\mathbf{j}$

25. $(3, 2)$ **27.** $(2, 4), (-4, -2)$

29. $\dfrac{\sqrt{3}}{2}\mathbf{i} + \dfrac{1}{2}\mathbf{j}$ **31.** $\dfrac{4}{\sqrt{17}}\mathbf{i} - \dfrac{1}{\sqrt{17}}\mathbf{j}$

33. $\dfrac{5}{\sqrt{26}}\mathbf{i} + \dfrac{1}{\sqrt{26}}\mathbf{j}$ **35.** $(3, 10)$

37. a. $(3, -5)$ **b.** $(7, -3)$

43. a. the set of points on the circle with center (x_0, y_0) and radius 1

b. the set of points on or interior to the circle with center (x_0, y_0) and radius 2

45. $a = -2t, b = t$, and $c = t$ for any number t

47. $-6\mathbf{i} + 3\mathbf{j}$ **49.** 5.22 mi/h; S36.4°E

59. The Spy should pace off half the distance from the wood-pile to the flagpole, turn right, and pace off the same distance to find the place to dig.

10.2 Quadric Surfaces and Graphing in Three Dimensions (pages 684–686)

Also plot P and Q in Problems 1 and 3.

1. $\sqrt{149}$ **3.** $\sqrt{382}$ **5.** $x^2 + y^2 + z^2 = 1$

7. $x^2 + (y - 4)^2 + (z + 5)^2 = 9$

9. $C(0, 1, -1); r = 2$

11. $C(3, -1, 1); r = 1$ **13.** circular cone; B

15. hyperboloid of one sheet; E

17. sphere; A **19.** paraboloid, G

21. hyperboloid of two sheets; I

23. $\triangle ABC$ is isosceles, but not right

25. $\triangle ABC$ is neither right nor isosceles

29. 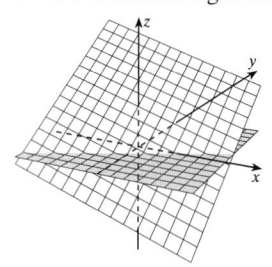   **31.**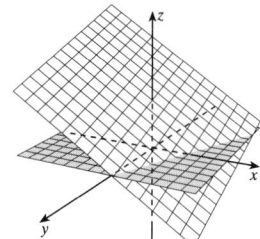

33. 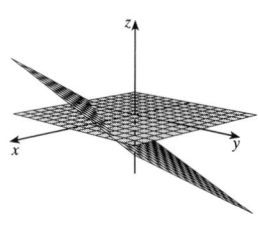   **35.**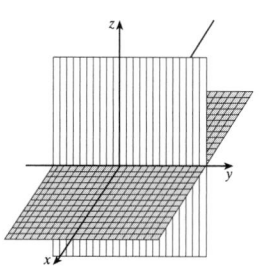

37. 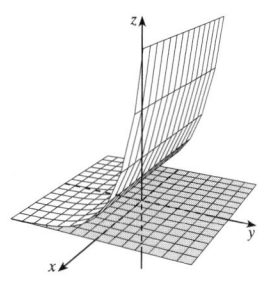   **39.**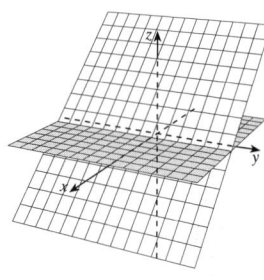

41. Ellipsoid   **43.** Hyperboloid of one sheet

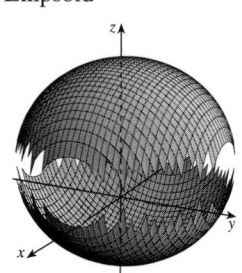   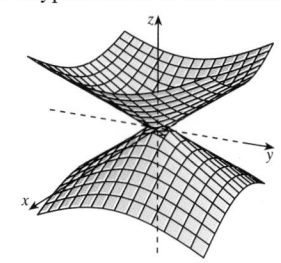

45. Elliptic paraboloid paraboloid   **47.** Elliptic cone

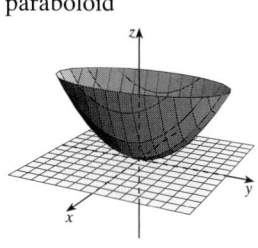   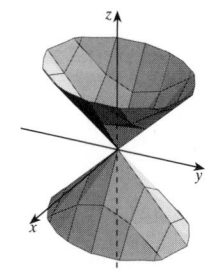

49. $(2x + 1)^2 + (2y - 5)^2 + 4z^2 = 46$

51. If $\overline{PQ}$ is the hypotenuse, $c = 1 \pm \sqrt{7}$; if $\overline{PR}$ is the hypotenuse, there is no value of c; if $\overline{QR}$ is the hypotenuse, $c = \frac{17}{4}$.

10.3 The Dot Product (pages 694–696)

3. PQ = $-2\mathbf{i} + 2\mathbf{j} + \mathbf{k}$; $\|\mathbf{PQ}\| = 3$

5. PQ = $-4\mathbf{i} - 4\mathbf{j} - 4\mathbf{k}$; $\|\mathbf{PQ}\| = 4\sqrt{3}$

7. -16 **9.** 5

11. orthogonal   **13.** orthogonal

15. $\sqrt{3}$ **17.** 14

19. $6 - 3\sqrt{21}$ **21.** $\sqrt{321}$

23. v is parallel to **u**   **25. v** is not parallel to **u**

27. 11   **29.** 0

31. 71°   **33.** 114°

35. 1; **k**   **37.** $\dfrac{6}{\sqrt{13}}; -\dfrac{12}{13}\mathbf{j} + \dfrac{18}{13}\mathbf{k}$

39. $\mathbf{u}_1 = \left\langle \dfrac{\sqrt{2}}{2}, 0, \dfrac{\sqrt{2}}{2} \right\rangle$, $\mathbf{u}_2 = \left\langle -\dfrac{\sqrt{2}}{2}, 0, -\dfrac{\sqrt{2}}{2} \right\rangle$

41. $\mathbf{u} = -\dfrac{1}{\sqrt{17}}(2\mathbf{k} + 3\mathbf{j} - 2\mathbf{k})$

43. $x = -1, y = -1,$ and $z = 4$

45. $a = \frac{3}{2}$

47. $\alpha \approx 35°, \beta \approx 66°, \gamma \approx 114°$

49. a. 4 **b.** $\dfrac{2\sqrt{7}}{21}$ **c.** $\dfrac{9}{2}$ **d.** $-\dfrac{7}{2}$

51. $2\sqrt{6}$ **53.** 8

55. 868 lb   **57.** $25{,}000$ ft $\cdot$ lb

59. $\frac{\pi}{4} \approx [1 - \frac{1}{3} + \frac{1}{5} - \frac{1}{7} + \cdots]$ **61.** 50

10.4 The Cross Product (pages 703–705)

1. $\mathbf{k}$ **3.** $-2\mathbf{i} + 4\mathbf{j} + 3\mathbf{k}$

5. $-2\mathbf{i} + 25\mathbf{j} + 14\mathbf{k}$ **7.** $-2\mathbf{i} + 16\mathbf{j} + 11\mathbf{k}$

9. $-14\mathbf{i} - 4\mathbf{j} - \mathbf{k}$ **11.** $\dfrac{\sqrt{3}}{2}$

13. $\dfrac{\sqrt{3}}{2}$ **15.** $\dfrac{3\sqrt{33}}{77}$

17. $\dfrac{1}{\sqrt{14}}\mathbf{i} + \dfrac{3}{\sqrt{14}}\mathbf{j} - \dfrac{2}{\sqrt{14}}\mathbf{k}$ **19.** $\dfrac{-16}{\sqrt{386}}\mathbf{i} + \dfrac{7}{\sqrt{386}}\mathbf{j} + \dfrac{9}{\sqrt{386}}\mathbf{k}$

21. $\sqrt{26}$ **23.** $2\sqrt{59}$

25. $\frac{1}{2}\sqrt{3}$ **27.** $\frac{3}{2}\sqrt{3}$

29. a. does not exist   **b.** scalar

31. a. scalar   **b.** vector   **33.** 5

35. 8 **41.** $t = 0$

43. $(\mathbf{u} \times \mathbf{v}) \times \mathbf{w} = -9\mathbf{j} + 3\mathbf{k}; \mathbf{u} \times (\mathbf{v} \times \mathbf{w}) = 3\mathbf{i} - 3\mathbf{j} + 3\mathbf{k};$ cross product is not an associative operation

45. $\mathbf{w} = \mathbf{0}$ **47.** $-40\sqrt{3}\,\mathbf{i}$

10.5 Lines and Planes in Space (pages 715–716)

3. $2x - y + 3z - 5 = 0$ **5.** $3x - 2y + 2z - 12 = 0$

7. $x = 1 + 3t, y = -1 - 2t, z = -2 + 5t;$
$\dfrac{x - 1}{3} = \dfrac{y + 1}{-2} = \dfrac{z + 2}{5}$

9. $\dfrac{x - 1}{1} = \dfrac{y + 1}{2} = \dfrac{z - 2}{1};$
$x = 1 + t, y = -1 + 2t, z = 2 + t$

11. $\dfrac{x - 1}{1} = \dfrac{y + 3}{-3} = \dfrac{z - 6}{-5};$
$x = 1 + t, y = -3 - 3t, z = 6 - 5t$

13. $\dfrac{x}{11} = \dfrac{y - 4}{-6} = \dfrac{z + 3}{10};$
$x = 11t, y = 4 - 6t, z = -3 + 10t$

15. $x = 3, y = -1 + t, z = 0$

17. $(0, -6, -3); (8, 0, -1); (12, 3, 0)$

19. $(0, 4, 9); (8, 0, -3); (6, 1, 0)$

21. parallel

23. coincident

25. point of intersection is $(1, 2, 3)$

27. $\cos \alpha = \dfrac{2}{\sqrt{38}}, \alpha \approx 1.24$ or $71°;$
$\cos \beta = \dfrac{-3}{\sqrt{38}}, \beta \approx 2.08$ or $119°;$
$\cos \gamma = \dfrac{-5}{\sqrt{38}}, \gamma \approx 2.52$ or $144°$

29. $\cos \alpha = \dfrac{5}{5\sqrt{2}}, \alpha \approx 0.79$ or $45°;$
$\cos \beta = \dfrac{-4}{5\sqrt{2}}, \beta \approx 2.17$ or $124°;$
$\cos \gamma = \dfrac{3}{5\sqrt{2}}, \gamma \approx 1.13$ or $65°$

31. $\cos \alpha = \dfrac{1}{\sqrt{91}}, \alpha \approx 1.47$ or $84°;$
$\cos \beta = \dfrac{-3}{\sqrt{91}}, \beta \approx 1.89$ or $108°;$
$\cos \gamma = \dfrac{9}{\sqrt{91}}, \gamma \approx 0.34$ or $19°$

33. $2x + 4y - 3z + 5 = 0$ **35.** $2y - 3z + 6 = 0$

37. $z = 0$ **39.** $\pm\dfrac{1}{\sqrt{21}}(4\mathbf{i} + 2\mathbf{j} + \mathbf{k})$

41. $\pm\dfrac{1}{\sqrt{29}}(2\mathbf{i} + 4\mathbf{j} - 3\mathbf{k})$ **45.** $\pm\frac{1}{3}(2\mathbf{i} - 2\mathbf{j} + \mathbf{k})$

47. $x + 2y + z - 1 = 0$ **49.** $3x + 5y + 2z - 9 = 0$

53. $68°$ **55.** $\dfrac{x}{14} = \dfrac{y - 1}{2} = \dfrac{z + 1}{15}$

57. $\dfrac{x}{8} = \dfrac{y}{-5} = \dfrac{z + 5}{19}$

59. $\cos \alpha = \dfrac{3}{\sqrt{61}}$ so $\alpha \approx 1.17$ or $67°;$
$\cos \beta = \dfrac{-6}{\sqrt{61}}$ so $\beta \approx 2.45$ or $140°;$
$\cos \gamma = \dfrac{-4}{\sqrt{61}}$ so $\gamma \approx 2.11$ or $121°$

61. the lines are perpendicular and intersect at $P(x_0, y_0, z_0)$

10.6 Vector Methods for Measuring Distance in $\mathbb{R}^3$ (pages 720–721)

1. 0 **3.** $\dfrac{31}{13}$ **5.** $\dfrac{19}{\sqrt{10}}$

7. $\dfrac{\sqrt{3}}{3}$ **9.** 1 **11.** $\dfrac{|4a^2 - 3a|}{\sqrt{5a^2 + 1}}$

13. $\dfrac{7}{\sqrt{14}}$ **15.** $\dfrac{3}{\sqrt{35}}$ **17.** $\dfrac{4\sqrt{3}}{\sqrt{14}}$

19. $\dfrac{\sqrt{65}}{3}$ **21.** $\dfrac{4|a|\sqrt{5}}{3}$

23. $(x + 2)^2 + (y - 3)^2 + (z - 7)^2 = 36$

25. $\sqrt{5{,}435}$

27. $(3y + z + 3)^2 + (3x - 4z - 3)^2 + (x + 4y + 3)^2 = 650$

29. $\dfrac{65}{\sqrt{122}}$ **33. a.** $d = \dfrac{|D_1 - D_2|}{\sqrt{A^2 + B^2 + C^2}}$ **b.** $\dfrac{\sqrt{6}}{3}$

CHAPTER 10: REVIEW

Proficiency Examination (pages 721–722)

32. a. $13\mathbf{i} - 12\mathbf{j} + 2\mathbf{k}$ **b.** 1 **c.** $\frac{12}{13}(3\mathbf{i} - 2\mathbf{j})$ **d.** $\frac{6\sqrt{14}}{7}$
e. 12 **f.** $2\mathbf{i} + 3\mathbf{j} + 5\mathbf{k}$

33. a. 40 **b.** not possible **c.** $25\mathbf{i} - 10\mathbf{j} - 15\mathbf{k}$
d. not possible

34. $\frac{x}{1} = \frac{y+2}{-6} = \frac{z-1}{4}$

35. $2x + 3z - 11 = 0$

36. $x = -\frac{13}{2} - 10t, y = 5 + 6t, z = 2t$

37. $17x + 19y + 13z - 25 = 0$

38. $\cos \alpha = \frac{-2}{\sqrt{14}}$; so $\alpha \approx 2.13$ or $122°$

$\cos \beta = \frac{3}{\sqrt{14}}$; so $\beta \approx 0.64$ or $37°$

$\cos \gamma = \frac{1}{\sqrt{14}}$; so $\gamma \approx 1.30$ or $74°$

39. a. skew **b.** intersect at $(2,2,3)$

40. a. 6 **b.** $\sqrt{\frac{2}{3}}$ **41.** $\frac{2\sqrt{30}}{15}$

42. $\frac{6}{\sqrt{26}}$ **43.** $\frac{5\sqrt{6}}{\sqrt{35}}$

44. 168.4 mph **45.** 130 ft · lbs

Supplementary Problems (pages 722–725)

1. a. $8\sqrt{2} + \sqrt{14}$ **b.** $\frac{1}{2}\sqrt{171}$ **c.** $A \approx 125°, B \approx 26°, C \approx 30°$
d. $p = \frac{205}{6}$

3. $P_1(3,4,1), P_2(-1,4,-1)$ **5.** $z = 0$

7. $3x + 4y - z + 13 = 0$

9. $5x - 2y + 3z - 32 = 0$

11. $3x + 2y - 11z + 19 = 0$

17. $(\mathbf{u} - \mathbf{v}) \cdot \mathbf{w} = 3$; $(2\mathbf{u} + \mathbf{v}) \times (\mathbf{u} - \mathbf{w}) = 6\mathbf{i} - 10\mathbf{j} - 10\mathbf{k}$

19. $\cos \alpha = \frac{-3}{\sqrt{46}}$; $\cos \beta = \frac{6}{\sqrt{46}}$; $\cos \gamma = \frac{1}{\sqrt{46}}$

21. $\theta = 74°$

23. The center is $(0, -\frac{3}{2}, \frac{1}{2})$ and the radius is $\frac{3}{2}$.

25. $P(3, -2, 1)$

27. $37X + 20Y - 21Z = 61$ and $67X - 20Y + 99Z = 121$

29. 13

31. $\frac{1}{2}(\|\mathbf{u} \times \mathbf{v}\| + \|\mathbf{u} \times \mathbf{w}\| + \|\mathbf{v} \times \mathbf{w}\| + \|(\mathbf{u} - \mathbf{v}) \times (\mathbf{w} - \mathbf{v})\|)$

33. $\cos \theta \sin \phi - \sin \theta \cos \phi = \sin(\phi - \theta)$

35. $a_1 + 2b_1 + c_1 = 0$ **37.** $\frac{9\sqrt{2}}{2}$

39. $\frac{\sqrt{26}}{2}$ **41.** $2x - 3y - 7z = 0$

43. $\frac{x+1}{1} = \frac{y-2}{1}$ and $z = 0$

45. $\mathbf{N} = \frac{\pm 1}{\sqrt{37}}(-6\mathbf{i} + \mathbf{j})$

47. Let $P(a,b,c)$; then $\mathbf{r} = \mathbf{v} + \mathbf{w}t$ is the set of position vectors in the line $x = a + At, y = b + Bt$, and $z = c + Ct$.

49. linearly independent

51. This is Putnam Problem 5 of the afternoon session of 1959.

CHAPTER 11: VECTOR-VALUED FUNCTIONS

11.1 Introduction to Vector Functions (pages 734–736)

1. $t \neq 0$ **3.** $t \neq \frac{(2n+1)\pi}{2}$, n an integer

5. $t \neq \frac{n\pi}{2}$, n an integer **7.** $t > 0$

9. a parabola in the xy-plane

11. a circle in the xy-plane

13. a line in the xz-plane parallel to and four units below the x-axis

15. a circular helix

17. the curve is in the intersection of the parabolic cylinder $y = (1 - x)^2$ with the plane $x + z = 1$

19. curve is the intersection of the cylinder $y = x^2 + 1$ and the plane $y = z + 1$

25. $\mathbf{F}(t) = t\mathbf{i} + t^2\mathbf{j} + 2\mathbf{k}$

27. $\mathbf{F}(t) = 2t\mathbf{i} + (1 - t)\mathbf{j} + (\sin t)\mathbf{k}$

29. $\mathbf{F}(t) = t^2\mathbf{i} + t\mathbf{j} + \sqrt{9 - t^2 - t^4}\,\mathbf{k}$

31. $(7t - 3)\mathbf{i} - 10\mathbf{j} + \left(2t^2 - \frac{3}{t}\right)\mathbf{k}$

33. $3t - 2t^2$

35. $(1 - t)\sin t$

37. $-t^2e^t\mathbf{i} + t^2\sin t\mathbf{j} + (2te^t + 5\sin t)\mathbf{k}$

39. $(4te^t - t^2 + t + 10\sin t)\mathbf{i} + (2e^tt^2 + 1)\mathbf{k}$

41. $t^2e^t - t^3e^t - 2e^t - \frac{5}{t}\sin t$ **43.** $2\mathbf{i} - 3\mathbf{j} + e\mathbf{k}$

45. $\mathbf{0}$ **47.** $-\mathbf{i} + e^{-1}\mathbf{j}$

49. $\frac{3}{2}\mathbf{i} + \mathbf{j}$ **51.** continuous for all t

53. continuous for $t \neq 0, t \neq -1$

55. continuous for all $t \neq 0$ **57.** $x - y + z = 1$

63. a. The function is continuous at t_0. **b.** The function is continuous at t_0. **c.** The function is continuous at t_0. **d.** The function is continuous at t_0.

11.2 Differentiation and Integration of Vector Functions (pages 746–747)

1. $\mathbf{F}'(t) = \mathbf{i} + 2t\mathbf{j} + (1 + 3t^2)\mathbf{k}$

3. $\mathbf{F}'(s) = (1 + \ln s)\mathbf{i} + 5s^{-1}\mathbf{j} - e^s(\ln s + s^{-1})\mathbf{k}$

5. $\mathbf{F}'(t) = 2t\mathbf{i} - t^{-2}\mathbf{j} + 2e^{2t}\mathbf{k}$;
$\mathbf{F}''(t) = 2\mathbf{i} + 2t^{-3}\mathbf{j} + 4e^{2t}\mathbf{k}$

7. $\mathbf{F}'(s) = (\cos s)\mathbf{i} - (\sin s)\mathbf{j} + 2s\mathbf{k}$;
$\mathbf{F}''(s) = (-\sin s)\mathbf{i} - (\cos s)\mathbf{j} + 2\mathbf{k}$

9. $f'(x) = -9x^2 - 2x$ **11.** $g'(x) = \dfrac{4x}{\sqrt{1 + 4x^2}}$

13. $\mathbf{V}(1) = \mathbf{i} + 2\mathbf{j} + 2\mathbf{k}$; $\mathbf{A}(1) = 2\mathbf{j}$; speed $= 3$; direction of motion is $\frac{1}{3}\mathbf{i} + \frac{2}{3}\mathbf{j} + \frac{2}{3}\mathbf{k}$

15. $\mathbf{V}\left(\dfrac{\pi}{4}\right) = -\dfrac{\sqrt{2}}{2}\mathbf{i} + \dfrac{\sqrt{2}}{2}\mathbf{j} + 3\mathbf{k}$;

$\mathbf{A}\left(\dfrac{\pi}{4}\right) = -\dfrac{\sqrt{2}}{2}\mathbf{i} - \dfrac{\sqrt{2}}{2}\mathbf{j}$; speed $= \sqrt{10}$

direction of motion $-\dfrac{1}{2\sqrt{5}}\mathbf{i} + \dfrac{1}{2\sqrt{5}}\mathbf{j} + \dfrac{3}{\sqrt{10}}\mathbf{k}$

17. $\mathbf{V}(\ln 2) = 2\mathbf{i} - \frac{1}{2}\mathbf{j} + 8\mathbf{k}$;

$\mathbf{A}(\ln 2) = 2\mathbf{i} + \frac{1}{2}\mathbf{j} + 16\mathbf{k}$; speed $= \dfrac{\sqrt{273}}{2}$;

direction of motion is $\dfrac{1}{\sqrt{273}}(4\mathbf{i} - \mathbf{j} + 16\mathbf{k})$

19. $\mathbf{F}'(0) = 2\mathbf{j}$; $\mathbf{F}'(1) = 2\mathbf{i} + 2\mathbf{j} + 5\mathbf{k}$; $\mathbf{F}'(-1) = -2\mathbf{i} + 2\mathbf{j} + \mathbf{k}$

21. $\mathbf{F}'(0) = \mathbf{i}$; $\mathbf{F}'(2) = \frac{1}{25}(\mathbf{i} + 12\mathbf{j} + 44\mathbf{k})$

23. $\mathbf{F}'(\frac{\pi}{2}) = -\mathbf{j} + a\mathbf{k}$; $\mathbf{F}'(\pi) = -\mathbf{i} + a\mathbf{k}$

25. $\dfrac{t^2}{2}\mathbf{i} - \dfrac{e^{3t}}{3}\mathbf{j} + 3t\mathbf{k} + \mathbf{C}$

27. $(t \ln t - t)\mathbf{i} - \frac{1}{2}t^2\mathbf{j} + 3t\mathbf{k} + \mathbf{C}$

29. $\dfrac{t^2}{2}(\ln t - \frac{1}{2})\mathbf{i} - \cos(1 - t)\mathbf{j} + \dfrac{t^2}{2}\mathbf{k} + \mathbf{C}$

31. $\mathbf{R}(t) = (\frac{1}{3}t^2 + 1)\mathbf{i} + (-\frac{1}{2}e^{2t} + \frac{9}{2})\mathbf{j} + (\frac{2}{3}t^{3/2} - 1)\mathbf{k}$

33. $\mathbf{R}(t) = (\frac{4}{3}t^{3/2} + 1)\mathbf{i} + (\sin t + 1)\mathbf{j}$

35. $\mathbf{V}(t) = \langle 2 + \sin t, 0, 3 + t\cos t - \sin t \rangle$; $\mathbf{R}(t) = \langle 2 + 2t - \cos t, -2, 1 + 3t + t\sin t \rangle$

37. $\mathbf{R}(t) = e^t\mathbf{i} + (\frac{1}{3}t^3 - 1)\mathbf{j}$ **39.** $\mathbf{R}(t) = \langle 2t^4 + 1, 2t^2 + 2 \rangle$

43. $-180t^2$ **49.** $a = \frac{\pi}{4}$

59. $[\mathbf{F} \times (\mathbf{G} \times \mathbf{H})]' = [(\mathbf{H} \cdot \mathbf{F})\mathbf{G}]' - [(\mathbf{G} \cdot \mathbf{F})\mathbf{H}]'$

11.3 Modeling Ballistics and Planetary Motion (pages 754–757)

1. 4.6 sec; 481 ft **3.** 129.9 sec; 73,175 m

5. 1.9 sec; 41 m **7.** 1.5 sec; 148 ft

9. $\mathbf{V}(t) = \sqrt{5}\mathbf{u}_r$; $\mathbf{A}(t) = \mathbf{0}$

11. $\mathbf{V} = (2 \cos 2t)\mathbf{u}_r + (2 \sin 2t)\mathbf{u}_\theta$; $\mathbf{A} = (-8 \sin 2t)\mathbf{u}_r + (8 \cos 2t)\mathbf{u}_\theta$

13. $\mathbf{V} = -10 \sin(2t + 1)\mathbf{u}_r + 10[1 + \cos(2t + 1)]\mathbf{u}_\theta$; $\mathbf{A} = [-40 \cos(2t + 1) - 20]\mathbf{u}_r - 40 \sin(2t + 1)\mathbf{u}_\theta$

15. 140 m/sec **17.** 21.7°

19. The maximum height is 35.64 ft; the ball will land at a distance of 225.93 ft; the distance to the fence is 217 ft.

21. 3.45 sec **23.** 5.5 sec

25. 23.53 ft

27. $\mathbf{V} = (2 \cos t)\mathbf{u}_r + (3 + 2 \sin t)3t^2\mathbf{u}_\theta$; $\mathbf{A} = [-2 \sin t - 27t^4 - 18t^4\sin t]\mathbf{u}_r$
$\qquad + [18t + 12t \sin t + 12t^2\cos t]\mathbf{u}_\theta$

29. $a\omega$ **31.** $a\omega^2$

33. The time is about 19 seconds; the horizontal distance (in feet) traveled by the canister is 3,794.

35. $14°, 77°$ **37.** $26.3°$

11.4 Unit Tangent and Normal Vectors; Curvature (pages 767–769)

1. $\mathbf{T}(t) = \dfrac{2}{\sqrt{4 + 9t^2}}\mathbf{i} + \dfrac{3t}{\sqrt{4 + 9t^2}}\mathbf{j}$;

$\mathbf{N}(t) = \dfrac{-3t}{\sqrt{4 + 9t^2}}\mathbf{i} + \dfrac{2}{\sqrt{4 + 9t^2}}\mathbf{j}$

3. $\mathbf{T}(t) = \dfrac{\sqrt{2}}{2}[(\cos t - \sin t)\mathbf{i} + (\cos t + \sin t)\mathbf{j}]$;

$\mathbf{N}(t) = -\dfrac{\sqrt{2}}{2}[(\sin t + \cos t)\mathbf{i} + (\sin t - \cos t)\mathbf{j}]$

5. $\mathbf{T}(t) = \dfrac{1}{\sqrt{2}}(-\sin t\,\mathbf{i} + \cos t\,\mathbf{j} + \mathbf{k})$;

$\mathbf{N}(t) = -\cos t\,\mathbf{i} - \sin t\,\mathbf{j}$

7. $\mathbf{T}(t) = \dfrac{1}{\sqrt{1 + 4t^4}}(\mathbf{i} + 2t^2\mathbf{k})$;

$\mathbf{N}(t) = \dfrac{1}{\sqrt{1 + 4t^4}}(-2t^2\mathbf{i} + \mathbf{k})$

9. $4\sqrt{5}$ **11.** $\dfrac{3\sqrt{2}\pi}{2}$

13. $\sqrt{41}\pi$ **15.** $\kappa = 0$

17. $\kappa \approx 0.19$ **19.** $\kappa = \dfrac{2|a|}{[1 + (2ac + b)^2]^{3/2}}$

21. $\kappa = \frac{1}{2}$ **23.** $\kappa = 1$

25. $\kappa = \dfrac{1}{2\sqrt{2}}$ **29.** $\mathbf{T} = \dfrac{1}{\sqrt{10}}(\mathbf{i} - 3\mathbf{j})$,

$\mathbf{N} = -\dfrac{1}{\sqrt{10}}(3\mathbf{i} + \mathbf{j})$

31. a. $\mathbf{T}(\pi) = \dfrac{\sqrt{2}}{2}(-\mathbf{i} + \mathbf{k})$ **b.** $\dfrac{1}{2}$ **c.** $\sqrt{2}\pi$

33. The points at which a maximum occurs are $P(0, 3)$, and $Q(0, -3)$.

35. $\rho(0) = \frac{1}{6}$; $\rho(1) = \rho(-1) = \frac{1}{24}$

37. $x = 2 + 2t, y = -1 - t, z = 1 + 2t$

39. $\kappa = \dfrac{2\sqrt{9t^4 + 9t^2 + 1}}{(1 + 4t^2 + 9t^4)^{3/2}}$

41. $\kappa = \dfrac{2}{(1 + 4x^2)^{3/2}}$

43. $\kappa = \dfrac{2|x^3|}{(1 + x^4)^{3/2}}$

45. $\dfrac{1}{\sqrt{2}e^\theta}$

47. The fly's speed at time t is $4t^2 + 4t + 1$; 4 sec.

51. a. 8 units **b.** 8.0099 **c.** 8.30

57. b. $x^2 + (y - 44)^2 = (32\sqrt{32})^2$ **61.** 7.096 units

11.5 Tangential and Normal Components of Acceleration
(pages 775–777)

1. $A_T = \dfrac{4t}{\sqrt{1 + 4t^2}}$; $A_N = \dfrac{2}{\sqrt{1 + 4t^2}}$

3. $A_T = \dfrac{t}{\sqrt{1 + t^2}}$; $A_N = \dfrac{t^2 + 2}{\sqrt{t^2 + 1}}$

5. $A_T = \dfrac{4t}{\sqrt{2 + 4t^2}}$; $A_N = \dfrac{2}{\sqrt{1 + 2t^2}}$

7. $A_T = -\dfrac{\sin t \cos t}{\sqrt{1 + \cos^2 t}}$; $A_N = \sqrt{\dfrac{2}{1 + \cos^2 t}}$

9. $\mathbf{T}_0 = \dfrac{1}{\sqrt{10}}\langle 1, -3\rangle$;

$\mathbf{N}_0 = \dfrac{1}{\sqrt{10}}\langle 3, 1\rangle$; $A_T = \dfrac{-13}{\sqrt{10}}$; $A_N = \dfrac{11}{\sqrt{10}}$

11. $\mathbf{T}_0 = \dfrac{1}{\sqrt{14}}\langle 2, 3, -1\rangle$; $A_T = \dfrac{-19}{\sqrt{14}}$;

$A_N = \sqrt{\dfrac{59}{14}}$; $\mathbf{N}_0 = \dfrac{1}{\sqrt{826}}\langle 24, -13, 9\rangle$

13. $\dfrac{5}{\sqrt{8}}$

15. 0

17. The maximum speed is 8 at $t = 0$, and the minimum speed is 6 at $t = \dfrac{\pi}{4}$.

19. $\mathbf{V}(t) = (-2\sin 2t)\mathbf{i} + (2\cos 2t)\mathbf{j}$
$\mathbf{A}(t) = (-4\cos 2t)\mathbf{i} - (4\sin 2t)\mathbf{j}$
$A_T = 0$; $A_N = 4$

21. At $t = \dfrac{\sqrt{399}}{8}$, $A_T \approx 7.98999$; $A_N = 0.4$

23. $\omega = 0.52$ rev/s

25. a. The maximum safe speed is 47 ft/s (about 32 mi/h).
b. 65.94 ft/s (about 45 mi/h)   **c.** 23.1°

27. 14 rev/min

29. $A_T = \dfrac{4t}{\sqrt{5 + 4t^2}}$;

$A_N = \dfrac{2\sqrt{5}}{\sqrt{5 + 4t^2}}$

31. $A_T = \sqrt{3}\,e^t$; $A_N = \sqrt{2}\,e^t$

33. a. $\dfrac{60\pi^2 \omega}{32} W$

b. If the wheel turns more rapidly than 4.83 rev/min, the volunteer will fly away.

39. b. $A_T = 0$ $A_N = r\omega^2$
c. Since $A_N = r\omega^2$, doubling ω quadruples A_N. Doubling r results in doubling A_N. If ω is doubled and r is halved, A_N is doubled.

41. The height of the satellite above the surface of Mars is about 17,032.6 km, and its speed is about 87 km/h.

CHAPTER 11: REVIEW

Proficiency Examination (page 777)

24.

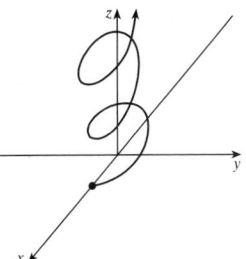

25. $\mathbf{F}' = \dfrac{1}{(1 + t)^2}\mathbf{i} + \dfrac{t\cos t - \sin t}{t^2}\mathbf{j} + (-\sin t)\mathbf{k}$

$\mathbf{F}'' = -\dfrac{2}{(1 + t)^3}\mathbf{i} + \dfrac{-t^2\sin t - 2t\cos t + 2\sin t}{t^3}\mathbf{j} - (\cos t)\mathbf{k}$

26. $(3 - 6\ln 2)\mathbf{i} + \frac{45}{4}\mathbf{j} + (6\ln 2 - \frac{9}{4})\mathbf{k}$

27. $(e^t - t)\mathbf{i} - \left(\dfrac{t^4}{12} + 2\right)\mathbf{j} + \left(\dfrac{3t^2}{2} + 3t\right)\mathbf{k}$

28. $\mathbf{V} = \mathbf{i} + 2\mathbf{j} + e^t(t + 1)\mathbf{k}$; speed is
$\sqrt{5 + e^{2t}(t + 1)^2}$; $\mathbf{A} = e^t(t + 2)\mathbf{k}$

29. $T = \dfrac{2t\mathbf{i} + 3\mathbf{j} - 3\mathbf{k}}{\sqrt{4t^2 + 18}}$; $\mathbf{N} = \dfrac{3\mathbf{i} - t\mathbf{j} + t\mathbf{k}}{(2t^2 + 9)^{1/2}}$

$A_T = \dfrac{2\sqrt{2}\,t}{\sqrt{2t^2 + 9}}$; $A_N = \dfrac{6}{\sqrt{2t^2 + 9}}$

30. a. The maximum height is 9.77 ft.   **b.** $\frac{25}{16}$ sec; 67.7 ft

Supplementary Problems (pages 778–780)

1. $2\mathbf{k}$ **3.** -5 **5.** $\mathbf{i} - \mathbf{j}$
7. $\mathbf{F}'(t) = (1 + t)e^t\mathbf{i} + 2t\mathbf{j}$; $\mathbf{F}''(t) = (2 + t)e^t\mathbf{i} + 2\mathbf{j}$
9. $\mathbf{F}'(t) = -2t^{-2}\mathbf{i} - 2\mathbf{j} + (1 - t)e^{-t}\mathbf{k}$;
$\mathbf{F}''(t) = 4t^{-3}\mathbf{i} + (t - 2)e^{-t}\mathbf{k}$
11. $\mathbf{F}'(t) = (2t + ae^{at})\mathbf{i} + (1 - at)e^{-at}\mathbf{j} + ae^{at+1}\mathbf{k}$
$\mathbf{F}''(t) = (2 + a^2 e^{at})\mathbf{i} + (a^2 t - 2a)e^{-at}\mathbf{j} + a^2 e^{at+1}\mathbf{k}$

13.   **15.**

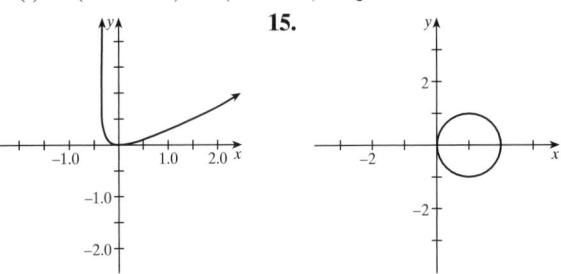

17. The graph is a circular helix of radius 2, traversed clockwise. It begins ($t = 0$) at $(0, 2, 0)$ and rises 10π units with each revolution.

19. $\dfrac{d}{dt}[\mathbf{F}(t) \cdot \mathbf{F}(t)] = 2\mathbf{F} \cdot \mathbf{F}'$

21. $\dfrac{d}{dt}\left[\dfrac{\mathbf{F}(t)}{\|\mathbf{F}(t)\|}\right] = \dfrac{\|\mathbf{F}\|^2\mathbf{F}' - (\mathbf{F}\cdot\mathbf{F}')\mathbf{F}}{\|\mathbf{F}\|^3}$

23. $\dfrac{d}{dt}[\mathbf{F}(t)\times\mathbf{F}(t)] = 0$

25. $(e - e^{-1})\mathbf{i} + 6\mathbf{k}$

27. $(te^t - e^t)\mathbf{i} + \dfrac{\cos 2t}{2}\mathbf{j} + \dfrac{t^3}{3}\mathbf{k} + \mathbf{C}$

29. $(te^t - e^t)\mathbf{i} + \dfrac{1}{4}t^2(2\ln t - 1)\mathbf{j} + \dfrac{3t^2}{2}\mathbf{k} + \mathbf{C}$

31. $\mathbf{V}(t) = \mathbf{i} - \mathbf{j}; \dfrac{ds}{dt} = \sqrt{2}; \quad \mathbf{A}(t) = \mathbf{0}$

33. $\mathbf{V}(t) = (t\cos t + \sin t)\mathbf{i} + (1 - t)e^{-t}\mathbf{j} + \mathbf{k}$

$\dfrac{ds}{dt} = \sqrt{t^2\cos^2 t + t\sin 2t + \sin^2 t + (1 - t)^2 e^{-2t} + 1}$

$\mathbf{A}(t) = (2\cos t - t\sin t)\mathbf{i} + e^{-t}(t - 2)\mathbf{j}$

35. $\mathbf{T}(t) = \dfrac{\mathbf{i} - 2t\mathbf{j}}{\sqrt{4t^2 + 1}}; \mathbf{N}(t) = \dfrac{-2t\mathbf{i} - \mathbf{j}}{\sqrt{4t^2 + 1}}$

37. $\mathbf{T}(t) = \frac{1}{5}(-4\sin t\,\mathbf{i} - 3\mathbf{j} + 4\cos t\,\mathbf{k})$

$\mathbf{N}(t) = (-\cos t)\mathbf{i} + (-\sin t)\mathbf{j}$

39. $A_T = \dfrac{4t + e^{2t}}{\sqrt{4 + 4t^2 + e^{2t}}};$

$A_N = 2\sqrt{\dfrac{t^2e^{2t} - 2te^{2t} + 2e^{2t} + 4}{4 + 4t^2 + e^{2t}}};$

$\kappa = \dfrac{2\sqrt{t^2e^{2t} - 2te^{2t} + 2e^{2t} + 4}}{(4t^2 + 4 + e^{2t})^{3/2}}$

41. $A_T = 0; A_N = 4\sqrt{2}; \quad \kappa = \frac{1}{8}$

43. $\kappa = \frac{1}{2}$

45. $\kappa \approx 1.92$

47. $\kappa = \frac{1}{6}$

49. $\mathbf{F}(t)$ is continuous for $t \neq 1$.

51. $\mathbf{F}(t) = (2t - \sin t + 1)\mathbf{i} + \left(\dfrac{3t^2}{2} + \cos t - 1\right)\mathbf{j} + \left(\dfrac{t^4}{24\pi} + \dfrac{t^2}{2}\right)\mathbf{k}$

53. $\mathbf{T}(s) = \dfrac{1}{\sqrt{5}}\left[\left(-\sin\dfrac{s}{a}\right)\mathbf{i} + \left(\cos\dfrac{s}{a}\right)\mathbf{j} + 2\mathbf{k}\right];$

$\mathbf{N} = -\left(\cos\dfrac{s}{a}\right)\mathbf{i} - \left(\sin\dfrac{s}{a}\right)\mathbf{j}$

55. $\rho = \dfrac{a^2}{b}$

57. a. $\mathbf{V}(t) = e^{-t}(-\cos t - \sin t)\mathbf{i}$
$+ e^{-t}(-\sin t + \cos t)\mathbf{j} - e^{-t}\mathbf{k}$
$\mathbf{A}(t) = e^{-t}(2\sin t)\mathbf{i} + e^{-t}(-2\cos t)\mathbf{j} + e^{-t}\mathbf{k}$

b. $\kappa = \dfrac{\sqrt{2}}{3}e^t$

59. 18,150 lb

61. $\mathbf{V}(t) = -a\sin at\,\mathbf{u}_r - ae^{-at}(1 + \cos at)\mathbf{u}_\theta;$
$\mathbf{A}(t) = [-a^2\cos at - a^2e^{-2at}(1 + \cos at)]\mathbf{u}_r$
$+ [a^2e^{-at}(1 + \cos at) + 2a^2e^{-at}\sin at]\mathbf{u}_\theta$

63.

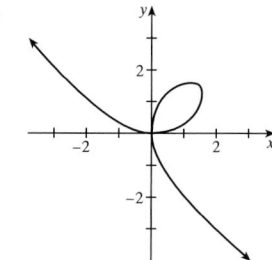

65. $20.888\ \mu\text{m}$

67. a. $10\mathbf{u}_r$ **b.** $10\mathbf{u}_r + 5\pi\mathbf{u}_\theta$

69. $\sin x\,\mathbf{i} - \cos 2x\,\mathbf{j} + e^{-x}\mathbf{k}$

71. $\mathbf{F}(t) = \left(\dfrac{t^2}{2} + 1\right)\mathbf{i} + \left(\dfrac{t^2}{2} + 2\right)\mathbf{j} - \left(\dfrac{t^2}{2} + 3\right)\mathbf{k}$

73. $\mathbf{F}(t) = \left[\left(\dfrac{1}{2}\right)(3 - \cos 2t)\right]\mathbf{i}$
$+ \left[\left(\dfrac{e^t}{2}\right)(\sin t + \cos t) - \dfrac{1}{2}\right]\mathbf{j} - [3\ln|t + 1| + 3]\mathbf{k}$

75. This is Putnam Problem 6ii of the morning session of 1939.

77. This is Putnam Problem 6 of the afternoon session of 1946.

Cumulative Review for Chapters 7–11 (pages 787–788)

5. a. $y' = \dfrac{x}{\sqrt{x^2 - 1}} + \cosh^{-1}x$

b. $\dfrac{dy}{dx} = -\dfrac{y\sqrt{1 + y^2} + \sqrt{1 + y^2}(1 - x^2)\sinh^{-1}y}{x(1 - x^2) + \sqrt{1 + y^2}(1 - x^2)\tanh^{-1}x}$

7. $\frac{1}{3}\sin^3 x - \frac{1}{5}\sin^5 x + C$

9. $-\dfrac{32(16 - x)^{3/2}}{3} + \dfrac{2(16 - x)^{5/2}}{5} + C$

11. $-\cot x + \csc x + C$

13. $2\sqrt{x} + 3\sqrt[3]{x} + 6\sqrt[6]{x} + 6\ln|\sqrt[6]{x} - 1| + C$

15. $\dfrac{x + 1}{2} = \dfrac{y - 2}{-3} = \dfrac{z - 5}{1}$

17. diverges   **19.** diverges

21. diverges   **23.** converges

25. a. $\dfrac{5}{e}$ **b.** $\dfrac{\pi}{2}$

27. $\mathbf{F}'(t) = 2\mathbf{i} - 3e^{-3t}\mathbf{j} + 4t^3\mathbf{k};$
$\mathbf{F}''(t) = 9e^{-3t}\mathbf{j} + 12t^2\mathbf{k}$

29. $\mathbf{T} = \dfrac{1}{\sqrt{13}}[2(\cos 2t)\mathbf{i} - 2(\sin 2t)\mathbf{j} + 3\mathbf{k}]$
$\mathbf{N} = -\sin 2t\,\mathbf{i} - \cos 2t\,\mathbf{j}$

31. a. $\displaystyle\sum_{k=0}^{\infty}\dfrac{(-1)^k x^{2k+2}}{k!}$ **b.** 0.189

33. $y = \dfrac{x^2}{2} - \dfrac{x}{2} + \dfrac{1}{4} + \dfrac{7}{4}e^{-2x}$ **35.** 9.827

37. a. $\mathbf{A}(t) = -(\sin t)\mathbf{i} + (\cos t)\mathbf{j}$
b. $x^2 + y^2 = 1$ and $z = 1$ **c.** $\kappa = 1$

39. 18.13 ft

CHAPTER 12: PARTIAL DIFFERENTIATION

12.1 Functions of Several Variables (pages 797–799)

1. a. 0 **b.** 0 **c.** 0 **d.** 2 **e.** 48 **f.** $2t^3$ **g.** $t^4 + t^5$
h. $t - t^2$

3. a. 0 **b.** $-e^2 + 1$ **c.** $e^{-2} + 1$ **d.** $2x^3e^{2x} + 3x^2e^{2x} + 2x$
e. $e^2 + 2y$ **f.** $4z(z^2 - 2)$

	Domain	*Range*
9.	$x - y > 0$	$f > 0$
11.	$\dfrac{y}{x} \geq 0$	$f \geq 0$
13.	$u \sin v \geq 0$	$f \geq 0$
15.	$y \neq 2$	$f > 0$
17.	$x^2 + y^2 < 9$	$f > 0$

19.
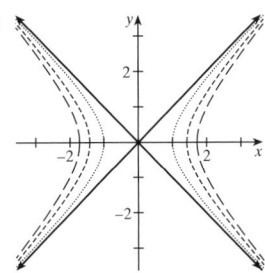

21.
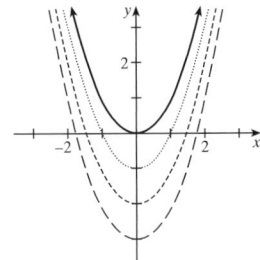

23.
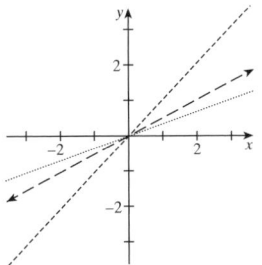

25.
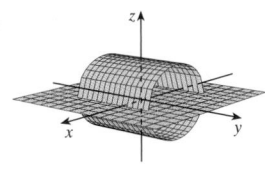

27.
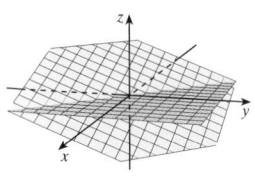

29.
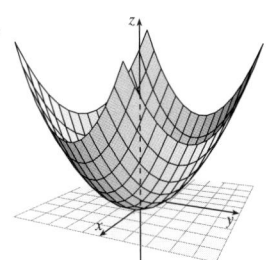

31. ellipsoid
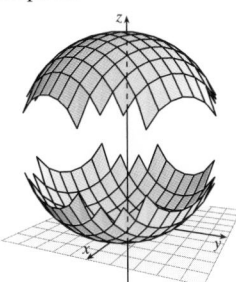

traces are ellipses in all three coordinate planes

33. hyperboloid (2 sheets)
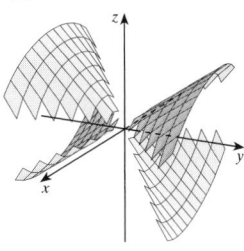

no trace in the yz-plane; traces in the xy- and xz-planes are hyperbolas

35. hyperbolic paraboloid
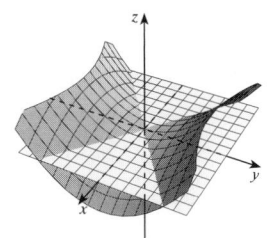

traces in the xz- and yz- are planes parabolas; trace in the xy-planes is a hyperbola

37. hyperboloid (2 sheets)
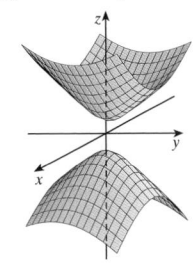

traces in the xz- and yz-planes are hyperbolas; no trace in the xy-plane

39. C **41.** E **43.** F

45.
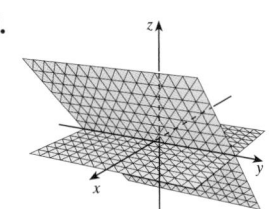

47.
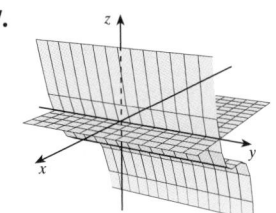

49.
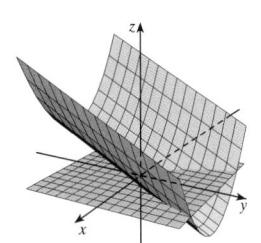

51.
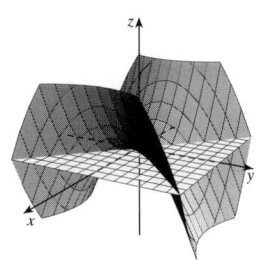

53.
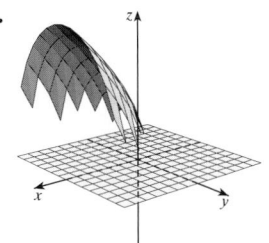

55.
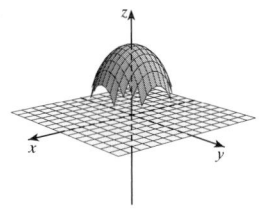

57. a. the production more than doubles   **b.** the production less than doubles   **c.** the production exactly doubles

59. The function exactly doubles or triples, respectively.

61. lines in the *TV*-plane through the origin.

63. $R = \left(60 - \dfrac{x}{5} + \dfrac{y}{20}\right)x + \left(50 - \dfrac{x}{10} + \dfrac{y}{20}\right)y$

12.2 Limits and Continuity (pages 805–807)

3. 5
5. -2
7. 1
9. 1
11. 0
13. 1
15. 1
17. 50
19. 4
21. 10
23. Limit does not exist.
25. 1
27. -1
29. 0
31. The limit does not exist.
33. The limit does not exist.
35. f is not continuous at $(0, 0)$
37. a. $\frac{11}{10}$ **b.** The limit does not exist.
39. $B = 0$

12.3 Partial Derivatives (pages 815–817)

3. $f_x = 3x^2 + 2xy + y^2; f_y = x^2 + 2xy + 3y^2; f_{xx} = 6x + 2y;$
$f_{yx} = 2x + 2y$

5. $f_x = \dfrac{1}{y}; f_y = \dfrac{-x}{y^2}; f_{xx} = 0; f_{yx} = \dfrac{-1}{y^2}$

7. $f_x = \dfrac{2}{2x + 3y}; f_y = \dfrac{3}{2x + 3y};$
$f_{xx} = \dfrac{-4}{(2x + 3y)^2}; f_{yx} = \dfrac{-6}{(2x + 3y)^2}$

9. a. $f_x = (2x \cos x^2)(\cos y); f_y = -(\sin x^2)(\sin y)$
b. $f_x = \cos(x^2 \cos y)(2x \cos y); f_y = \cos(x^2 \cos y)(-x^2 \sin y)$

11. $f_x = \dfrac{3x}{(3x^2 + y^4)^{1/2}}; f_y = \dfrac{2y^3}{(3x^2 + y^4)^{1/2}}$

13. $f_x = xe^{x+y}(x + 2)\cos y; f_y = x^2 e^{x+y}(\cos y - \sin y)$

15. $f_x = \dfrac{y}{\sqrt{1 - x^2 y^2}}; f_y = \dfrac{x}{\sqrt{1 - x^2 y^2}}$

17. $f_x = y^2 + yz; f_y = 2xy + z^3 + xz$
$f_z = 3yz^2 + xy$

19. $f_x = \dfrac{1}{z}; f_y = \dfrac{2y}{z}; f_z = -\dfrac{x + y^2}{z^2}$

21. $f_x = \dfrac{1}{x + y^2 + z^3}; f_y = \dfrac{2y}{x + y^2 + z^3}; f_z = \dfrac{3z^2}{x + y^2 + z^3}$

23. $z_x = -\dfrac{2x}{9z}; z_y = \dfrac{y}{2z}$

25. $z_x = -\dfrac{6xy}{y^3 - 2z}; z_y = -\dfrac{3x^2 + 3y^2 z}{y^3 - 2z}$

27. $z_x = -\dfrac{1}{2x^{3/2}\cos xz} - \dfrac{z}{x}; z_y = -\dfrac{2y}{x \cos xz}$

29. $f_x = -(x^2 + 2x + 1); f_y = y^2 + 2y + 1$
31. $-\frac{3}{5}$
37. a. -4 **b.** 4

41. $2(\sin z - x \cos z)$

43. a. $C_m = -0.67\sigma(T - t)m^{-1.67}$ **b.** $C_T = \sigma m^{-0.67}$
c. $C_t = -\sigma m^{-0.67}$

45. a. $\dfrac{\partial V}{\partial T} = \dfrac{k}{P}$ **b.** $\dfrac{\partial P}{\partial V} = -\dfrac{P}{V}$

47. a. 9   **b.** 14
51. a. $\delta = b/a^2$
53. The solution to this journal problem can be found in *Cruz*, Vol. 7 (1981), p. 155s.

12.4 Tangent Planes, Approximations, and Differentiability (pages 825–827)

1. $3x + y - \sqrt{10}z = 0$ **3.** $2x + 4y - z - 4 = 0$
5. $x - y + 4z - \pi = 0$ **7.** $df = 10xy^3\, dx + 15x^2 y^2\, dy$
9. $df = y(\cos xy)dx + x(\cos xy)dy$

11. $df = -\dfrac{y}{x^2}\, dx + \dfrac{1}{x}\, dy$

13. $df = ye^x\, dx + e^x\, dy$
15. $df = 9x^2\, dx - 8y^3\, dy + 5\, dz$
17. $df = 2z^2\cos(2x - 3y)dx - 3z^2\cos(2x - 3y)dy$
$+ 2z \sin(2x - 3y)dz$

23. 37.04   **25.** 0
27. 2.691   **29.** $z = 9$
31. c. $f(x, y) = \frac{1}{2}(1 - x - y)$
33. Thus, the maximum possible error will cost $1.14.

35. a. $R = \left(\dfrac{4{,}000 - p}{500}\right)p + \left(\dfrac{3{,}000 - q}{450}\right)q$
b. The revenue is increased by approximately $180.
37. R increases by approximately 11%.
39. 0.0360 cal
41. The profit will decline by about $24/wk (bad idea!).
43. The correct period is approximately 0.054 seconds more than the computed period.
45. The volume of the shell is about 23.56 in.^{3}.

47. The differentials are equal since $\dfrac{x}{x - y}$ and $\dfrac{y}{x - y}$ differ by
a constant: $\dfrac{x}{x - y} - \dfrac{y}{x - y} = 1$

49. 0.028 or 2.8%

12.5 Chain Rules (pages 832–834)

5. $\dfrac{dy}{dx} = \dfrac{-[3x(x^2 - y)^{1/2} + 2xy]}{-\frac{3}{2}(x^2 - y)^{1/2} + x^2}$

7. $\dfrac{dy}{dx} = \dfrac{(1 - \cos y)(1 + x^2) - y}{(1 + x^2)(-x \sin y + \tan^{-1}x)}$

9. $\dfrac{dy}{dx} = \dfrac{y}{x}$ **11.** $8e^{2t}(1 + e^{6t})$

13. $\dfrac{-3 \sin^3 3t + 6 \sin 3t}{\cos^2 3t}$

15. $2u^2 \sin v \cos v - 4u + 8v$

17. $2uv + u$

19. $\dfrac{\partial w}{\partial s} = \dfrac{\partial w}{\partial x}\dfrac{\partial x}{\partial s} + \dfrac{\partial w}{\partial y}\dfrac{\partial y}{\partial s} + \dfrac{\partial w}{\partial z}\dfrac{\partial z}{\partial s}$

$\dfrac{\partial w}{\partial t} = \dfrac{\partial w}{\partial x}\dfrac{\partial x}{\partial t} + \dfrac{\partial w}{\partial y}\dfrac{\partial y}{\partial t} + \dfrac{\partial w}{\partial z}\dfrac{\partial z}{\partial t}$

21. $\dfrac{\partial w}{\partial s} = \dfrac{\partial w}{\partial x}\cdot\dfrac{\partial x}{\partial s} + \dfrac{\partial w}{\partial y}\cdot\dfrac{\partial y}{\partial s} + \dfrac{\partial w}{\partial z}\cdot\dfrac{\partial z}{\partial s}$

$\dfrac{\partial w}{\partial t} = \dfrac{\partial w}{\partial x}\cdot\dfrac{\partial x}{\partial t} + \dfrac{\partial w}{\partial y}\cdot\dfrac{\partial y}{\partial t} + \dfrac{\partial w}{\partial z}\cdot\dfrac{\partial z}{\partial t}$

$\dfrac{\partial w}{\partial u} = \dfrac{\partial w}{\partial x}\cdot\dfrac{\partial x}{\partial u} + \dfrac{\partial w}{\partial y}\cdot\dfrac{\partial y}{\partial u} + \dfrac{\partial w}{\partial z}\cdot\dfrac{\partial z}{\partial u}$

23. $\dfrac{dw}{dt} = \cos(xyz)[-3yz - e^{1-t}xz + 4xy]$

25. $\dfrac{dw}{dt} = (e^{x^3+yz})\left[\dfrac{-6x^2}{t^2} + 2ty + \dfrac{2z}{2t-3}\right]$

27. $\dfrac{\partial w}{\partial r} = \dfrac{2s + t\cos(rt)}{2 - z};$

$\dfrac{\partial w}{\partial t} = \dfrac{(2-z)[r\cos(rt)] + 2st(x + y)}{(2-z)^2}$

29. a. $\dfrac{z^2}{2}$ **b.** $\dfrac{4}{x^3 y}$ **c.** $\dfrac{4}{xy^3}$

31. a. $\dfrac{2z^3}{x^2 y^2}$ **b.** $\dfrac{2z^2(x+z)}{x^4}$ **c.** $\dfrac{2z^2(y+z)}{y^4}$

33. a. $\dfrac{\sec^2 y \cos x}{4z^3}$ **b.** $\dfrac{2z^2\sin x - \cos^2 x}{2z^3}$

c. $\dfrac{4z^2\sec^2 y \tan y - \sec^4 y}{4z^3}$

35. $\dfrac{\partial^2 z}{\partial u^2} = a^2 z_{xx}; \dfrac{\partial^2 z}{\partial v^2} = b^2 z_{yy}$

37. volume and surface area both decreasing

39. a. $\dfrac{\partial V}{\partial T} = \dfrac{kV^3}{PV^3 - AV + 2AB};$

b. $\dfrac{\partial P}{\partial V} = \dfrac{AV - PV^3 - 2AB}{V^3(V - B)}$

41. 9 months from now, demand will be decreasing at the rate of about 23.38 gal/mo

43. The level of pollution is decreasing at the approximate rate of 84.31 units/day.

53. a. $\dfrac{\partial^2 z}{\partial x^2} = 2f'(x^2 + y^2) + 4x^2 f''(x^2 + y^2)$

b. $\dfrac{\partial^2 z}{\partial y^2} = 2f'(x^2 + y^2) + 4y^2 f''(x^2 + y^2)$

c. $\dfrac{\partial^2 z}{\partial x\partial y} = 4xy f''(x^2 + y^2)$

59. $\dfrac{\partial u}{\partial y} = \dfrac{-uv + vx + uy - u^2}{-x^2 - y^2 + u(x + y) + v(x - y)}$

$\dfrac{\partial v}{\partial y} = \dfrac{uv - xu + vy + v^2}{-x^2 - y^2 + u(x + y) + v(x - y)}$

12.6 Directional Derivatives and the Gradient (pages 846–848)

1. $\nabla f = (2x - 2y)\mathbf{i} - 2x\mathbf{j}$

3. $\nabla f = \left(-\dfrac{y}{x^2} + \dfrac{1}{y}\right)\mathbf{i} + \left(\dfrac{1}{x} - \dfrac{x}{y^2}\right)\mathbf{j}$

5. $\nabla f = e^{3-v}(\mathbf{i} - u\mathbf{j})$

7. $\nabla f = \cos(x + 2y)(\mathbf{i} + 2\mathbf{j})$

9. $\nabla f = e^{y+3z}(\mathbf{i} + x\mathbf{j} + 3x\mathbf{k})$

11. $\dfrac{\sqrt{2}}{2}$ **13.** $\dfrac{5\sqrt{2}}{8}$ **15.** 0

17. $\mathbf{N}_u = \pm\dfrac{\sqrt{3}}{3}(\mathbf{i} - \mathbf{j} + \mathbf{k})$; the tangent plane is:

$x - y + z - 3 = 0$

19. $\mathbf{N}_u = \pm\dfrac{\sqrt{3}}{3}(-\mathbf{i} - \mathbf{j} + \mathbf{k})$; the tangent plane is:

$x + y - z - \dfrac{\pi}{2} = 0$

21. $\mathbf{N}_u = \pm\dfrac{\sqrt{3}}{3}(\mathbf{i} - \mathbf{j} + \mathbf{k})$; the tangent plane is:

$x - y + z = 0$

23. $\mathbf{N}_u = \pm\dfrac{1}{\sqrt{46}}(3\mathbf{i} + 6\mathbf{j} + \mathbf{k})$; the tangent plane is:

$3x + 6y + z - 3 = 0$

25. a. $\nabla f(1, -1) = 3\mathbf{i} + 2\mathbf{j}; \|\nabla f\| = \sqrt{13}$
b. $\nabla f(1, 1) = 3\mathbf{i} + 2\mathbf{j}; \|\nabla f\| = \sqrt{13}$

27. a. $\nabla f(3, -3) = 27(\mathbf{i} + \mathbf{j}); \|\nabla f\| = 27\sqrt{2}$
b. $\nabla f(-3, 3) = 27(\mathbf{i} + \mathbf{j}); \|\nabla f\| = 27\sqrt{2}$

29. $\nabla f = 2ax\mathbf{i} + 2by\mathbf{j} + 2cz\mathbf{k}; \nabla f(a, b, c) = 2(a^2\mathbf{i} + b^2\mathbf{j} + c^2\mathbf{k});$
$\|\nabla f\| = 2\sqrt{a^4 + b^4 + c^4}$

31. $\nabla f = \dfrac{1}{2(x^2 + y^2)}(2x\mathbf{i} + 2y\mathbf{j});$

$\nabla f(1, 2) = \dfrac{1}{5}(\mathbf{i} + 2\mathbf{j}); \|\nabla f\| = \dfrac{1}{\sqrt{5}}$

33. $\nabla f = [2(x + y) + 2(x + z)]\mathbf{i}$
$+ [2(x + y) + 2(y + z)]\mathbf{j}$
$+ [2(z + y) + 2(x + z)]\mathbf{k}$

$\nabla f(2, -1, 2) = 2(5\mathbf{i} + 2\mathbf{j} + 5\mathbf{k});$
$\|\nabla f\| = \sqrt{216} = 6\sqrt{6}$

35. $\mathbf{u} = \pm\dfrac{a\mathbf{i} + b\mathbf{j}}{\sqrt{a^2 + b^2}}$

37. $\mathbf{u} = \pm\dfrac{b^2 x_0\mathbf{i} + a y_0\mathbf{j}}{\sqrt{b^4 x_0^2 + a^4 y_0^2}}$

39. $D_u f = \sqrt{3} + 1$ **41.** $D_u = -\dfrac{6e}{\sqrt{17}}$ **43.** $D_u f = \dfrac{11\sqrt{3}}{15}$

45. Maximum rate of temperature change is $\|\nabla \mathbf{T}_0\| = 2\sqrt{3}$ in the direction of $\mathbf{u} = \dfrac{\sqrt{3}}{3}(\mathbf{i} + \mathbf{j} + \mathbf{k})$

47. For the most rapid decrease, she should head in the direction $\frac{3}{2}\mathbf{i} - \frac{5}{2}\mathbf{j}$

49. 8.06

51. $\nabla f(1, 2) = 5\sqrt{10}(\mathbf{i} - 3\mathbf{j})$

53. The equation of the path is $y = y_0(x/x_0)^{b/a}$.

55. He reaches the hole in about 1 min 40 sec – plenty of time for a cup of tea.

12.7 Extrema of Functions of Two Variables (pages 858–861)

5. a minimum occurs at $(2, 3)$

7. maximum occurs at $(0, 0)$

9. $(0, 0)$, relative minimum

11. $(\sqrt{3}, 0)$, relative maximum; $(-\sqrt{3}, 0)$, saddle point

13. $(0, 0)$, relative minimum; $(1, 0)$, saddle point; $(-1, 0)$, saddle point; $(0, 1)$, relative maximum; $(0, -1)$, relative maximum

15. $(2^{-1/3}\, 2^{-1/3})$, relative minimum

17. $(0, 3)$, saddle point; $(0, 9)$, relative minimum; $(-2, 9)$, saddle point; $(-2, 3)$, relative maximum

19. $\left(\frac{3}{2}, 2\right)$, saddle point

21. $(6, 2)$, relative minimum; $(-8.99, -2.45)$, saddle point

23. $(-2, 0)$, saddle point; $(-2, 1)$, relative minimum

25. The largest value of f on S is 0 and the smallest is -14.

27. The largest value of f on S is 7 and the smallest is -5.

29. 0 minimum, $\frac{3}{2}$ is maximum

31. $y = 1.62x + 0.68$ **33.** $y = -0.02x + 5.54$

35. The closest points are $(0, 2, 0)$ and $(0, -2, 0)$.

37. The dimensions for the minimum construction occur when the sides measure $x = \sqrt[3]{2V}$, $y = \sqrt[3]{2V}$, and $z = \sqrt[3]{0.25\, V}$.

39. The product is maximized when each number is 18.

41. The profit is maximized when $x = \frac{2}{9}, y = \frac{8}{9}$

43. revenue is maximized at $\left(\frac{21}{2}, \frac{13}{2}\right)$

45. The owner should charge $2.70 for California water and $2.50 for New York water.

47. 200 machines should be supplied to the domestic market and 300 to the foreign market.

49. 2.1

51. a.

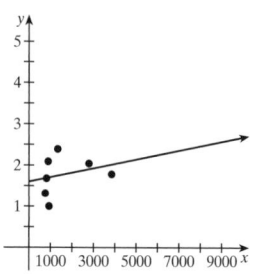

b. $y = 0.0001064x + 1.5965$ **c.** 2.44 gal/person

53. a. $W = 3.42674 + 0.573164X$

12.8 Lagrange Multipliers (pages 868–870)

1. $\frac{25}{16}$ **3.** $\frac{44}{5}$ **5.** 2   **7.** 4

9. 1.8478   **11.** 0.72   **13.** $\frac{8}{7}$ **15.** $\frac{25}{7}$

17. 17.3 is the constrained maximum and -17.3 is the constrained minimum

19. The minimum distance is $\dfrac{|D|}{\sqrt{A^2 + B^2 + C^2}}$

21. The nearest point is $\left(\frac{1}{3}, \frac{1}{6}, \frac{1}{6}\right)$, and the minimum distance is 0.4082.

23. $x = z = 3, y = 6$

25. The lowest temperature is $\frac{200}{3}$.

27. The maximum value of A is 6,400 yd^2.

29. the radius $x = 1$ in. and the height $y = 4$ in.

31. $2,000 to development and $6,000 to promotion gives the maximum sales of about 1,039 units.

33. $x \approx 13.87$ and $y \approx 12.04$ **35.** $x = y = z = \sqrt[3]{C}$

37. The farmer should apply 4.24 acre-ft of water and 1.27 lb of fertilizer to maximize the yield.

39. The triangle with maximum area is equilateral.

41. $\frac{56}{3}$ **43.** $\frac{121}{24}$

45. a. 3 thousand dollars for development and 5 thousand dollars for promotion.
   b. The actual increase in profit is $29.69.
   c. Thus, $4,000 should be spent on development and $6,000 should be spent on promotion to maximize profit.
   d. $\lambda = 0$

47. The minimum surface area occurs when $R \approx 0.753\, V_0^{1/3}$, $L \approx 0.674\, V_0^{1/3}$, $H \approx 0.337\, V_0^{1/3}$.

49. The minimum occurs at $(1, 1)$, but there is no maximum

Chapter 12: Review

Proficiency Examination (page 871)

32. $\dfrac{dw}{dt} = 6\pi^2$

33. a. $\nabla f_0 = \mathbf{i} + 3\mathbf{k}$ **b.** $D_u(f) = \dfrac{-2\sqrt{5}}{5}$

   c. $\mathbf{u} = \dfrac{\mathbf{i} + 3\mathbf{k}}{\sqrt{10}}$; $\|\nabla f\| = \sqrt{10}$

35. $f_x = -\dfrac{1}{x}$, $f_y = \dfrac{1}{y}$, $f_{yy} = -\dfrac{1}{y^2}$, $f_{xy} = 0$

37. $D_u(f) \approx -87.4$

38. $(0, 0)$, saddle point; $(9, 3)$, relative maximum; $(-9, -3)$, relative maximum

39. maximum of 12; minimum of 3

40. The largest value of f is $49/42$ at $\left(2, \frac{3}{4}\right)$ and the smallest is $-9/4$ at $(-3/2, 0)$.

Supplementary Problems (pages 872–876)

1. The domain consists of the circle with center at the origin, radius 4, and its interior.

3. $-1 \le x \le 1$ and $-1 \le y \le 1$ is the domain

5. $f_x = 1; f_y = -1$

7. $f_x = 2xy - \dfrac{y}{x^2}\cos\dfrac{y}{x}; f_y = x^2 + \dfrac{1}{x}\cos\dfrac{y}{x}$

9. $f_x = 6x^2y + 3y^2 - \dfrac{y}{x^2}; f_y = 2x^3 + 6xy + \dfrac{1}{x}$

11. For $c = 2, x^2 - y = 2$ is a parabola opening up, with vertex at $(0, -2)$. For $c = -2, x^2 - y = -2$ is a parabola opening up, with vertex at $(0, 2)$.

13. For $c = 0, \sqrt{x^2 + y^2} = 0$ is the origin only. For $c = 1, \sqrt{x^2 + y^2} = 1$ is a semicircle (to the right of the y-axis). For $c = -1, |y| = 1$ is a pair of half-lines, 1 unit above or below the x-axis, to the left of the y-axis.

15. For $c = 1, x^2 + \dfrac{y^2}{2} + \dfrac{z^2}{9} = 1$ is an ellipsoid. For

$c = 2, x^2 + \dfrac{y^2}{2} + \dfrac{z^2}{9} = 2$ is an ellipsoid.

17. 0   **19.** The limit does not exist.

21. $\dfrac{dz}{dt} = ye^t(-t^{-2} + t^{-1}) + (x + 2y)\sec^2 t$

23. $\dfrac{dz}{du} = \left(\tan\dfrac{x}{y} + \dfrac{x}{y}\sec^2\dfrac{x}{y}\right)v + \left(-\dfrac{x^2}{y^2}\sec^2\dfrac{x}{y}\right)v^{-1};$

$\dfrac{dz}{dv} = \left(\tan\dfrac{x}{y} + \dfrac{x}{y}\sec^2\dfrac{x}{y}\right)u + \left(-\dfrac{x^2}{y^2}\sec^2\dfrac{x}{y}\right)(-uv^{-2})$

25. $\dfrac{\partial z}{\partial x} = -e^{x-z}; \dfrac{\partial z}{\partial y} = -e^{y-z}$

27. $\dfrac{\partial z}{\partial x} = \dfrac{z}{3z+1}; \dfrac{\partial z}{\partial y} = \dfrac{2z}{3z+1}$

29. $f_{xx} = \dfrac{xy^3}{(1 - x^2y^2)^{3/2}}; f_{yx} = (1 - x^2y^2)^{-3/2}$

31. $f_{xx} = 2e^{x^2+y^2}(2x^2 + 1); f_{yx} = 4xye^{x^2+y^2}$

33. $f_{xx} = \sin x \cos(\cos x); f_{yx} = 0$

35. Normal line: $\dfrac{x-1}{16} = \dfrac{y-1}{-7}$ and $z = 1$
Tangent plane: $-16(x-1) + 7(y-1) = 0$

37. relative minimum at $(3, -1)$

39. $(1, 0)$, saddle point; $(0, 1)$, saddle point; $(\frac{2}{3}, \frac{2}{3})$, relative maximum; $(1, 1)$, saddle point

41. $(0, 9)$, relative minimum; $(0, 3)$, saddle point; $(-2, 9)$ saddle point; $(-2, 3)$, relative maximum

43. The largest value is 2 at $(0, -1)$ and the smallest is -2 at $(0, 1)$.

45. The largest value of f is 10 at $(0, -2)$ and the smallest is $-9/4$ at $(0, 3/2)$.

47. $\dfrac{dz}{dt} = (2x - 3y^2)(2) + (-6xy)(2t)$

49. $\dfrac{\partial z}{\partial x} = e^{u^2-v^2}(8xu^2 + 4x - 12uvx);$

$\dfrac{\partial z}{\partial y} = e^{u^2-v^2}(12yu^2 + 6y + 8uvy)$

51. $\dfrac{dy}{dx} = -1$ at $(1, 1)$

53. Normal line: $\dfrac{x}{2} = \dfrac{y-1}{2} = \dfrac{z-3}{-1}$
Tangent plane: $2x + 2y - z + 1 = 0$

55. $g(x, y, z) = x^2y + y^2z + z^2x$

57. $\dfrac{dw}{dt} = \left[\dfrac{2x}{1 + x^2 + y^2}\right]\left(\dfrac{2t}{1+t^2}\right) + \left[\dfrac{2y}{1 + x^2 + y^2} - \dfrac{2}{1 + y^2}\right]e^t$

59. $D_u f = \dfrac{1}{\sqrt{5}}(-16\ln 2 - 12)$

61. a. $D_u f = -\dfrac{18}{\sqrt{6}}$ **b.** $\dfrac{1}{\sqrt{86}}(-6\mathbf{i} + \mathbf{j} + 7\mathbf{k})$

63. $x = 6, y = z = 3$; maximum is 324.

65. The minimum distance is $\sqrt{3}$.

67. a. $t_m = \dfrac{-1 + \sqrt{1 + 4k\gamma r^2}}{4k\gamma}$ **b.** $M(r) = \sqrt{\dfrac{k\gamma}{\pi}}\dfrac{e^{-z/2}}{\sqrt{z-1}}$

69. a.

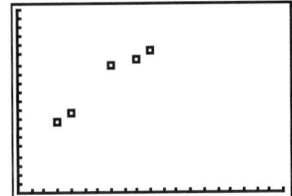

b.

The least squares line is $y = 11.54x + 44.45$
If $x = 5,000$, the sales are $y \approx 102.15$

71. $24xy^2z^3\sin(x^2 + y^3 + z^4)$

73. $\dfrac{dy}{dx} = -\dfrac{2}{3}$

75. $\mathbf{u} = \pm\frac{1}{3}(2\mathbf{i} + 2\mathbf{j} - \mathbf{k})$

77. $\dfrac{\partial^2 z}{\partial x\partial y} = -\dfrac{\sin^2(x + z)\cos(x + y) + \sin^2(x + y)\cos(x + z)}{\sin^3(x + z)}$

79. 18 is a relative minimum.

81. $x = \dfrac{a}{a^2 + b^2 + c^2}; y = \dfrac{b}{a^2 + b^2 + c^2}; z = \dfrac{c}{a^2 + b^2 + c^2}$

87. $\theta = 2$ and $r = \sqrt{A_0}$

89. a. 10π **b.** 8π

91. Approximately 2.01 cm³ has been removed.

97. This is Putnam Problem 6 in the afternoon session of 1967.

99. This is Putnam Problem 13 in the afternoon session of 1938.

CHAPTER 13: MULTIPLE INTEGRATION

13.1 Double Integration Over Rectangular Regions (pages 887–889)

1. $\frac{13}{3}$

3. $(e^2 - 1)(\ln 2) + 2$

5. $\frac{15}{2}\ln 3 - 10\ln 2 + \frac{1}{2}$

7. 32

9. 24

11. 24

13. $\frac{7}{6}$

15. −1

17. 4 ln 2

19. 1

21. 8

23. $\frac{a^3 b}{2} + \frac{ab^3}{2}$

25. $\frac{32}{9}$

27. 2

29. 3

31. $\frac{2^{n+2} - 2}{(n+1)(n+2)}$

35. $M = \int_R \int \delta(x, y)\, dA_{xy}$

37. $T = \int_R \int 1{,}400\, f(x, y)\, dA_{xy}$

39. $\frac{1}{2}\ln 2$

41. 1.48

43. 3.55

13.2 Double Integration Over Nonrectangular Regions (pages 895–897)

3. $\frac{32}{3}$

5. $\frac{2}{3}$

7. 6.6747

9. $\frac{27}{2}$

11. $\frac{5}{12}$

13. 0.1253

15. 0.1812

17. $\frac{1}{2}$

19. 5

21. 0.4167

23. 13

25. 2.0530

27. 1.4366

29. $\frac{32}{3}$

31. $e - 2$

33. $\frac{2}{3}$

35. 9.5323

37. $\int_0^2 \int_{x/2}^1 f(x, y)\, dy\, dx$

39. $\int_0^2 \int_{x^2}^{2x} f(x, y)\, dy\, dx$

41. $\int_0^1 \int_0^{3x} f(x, y)\, dy\, dx + \int_1^2 \int_0^{4-x^2} f(x, y)\, dy\, dx$

43. $\int_0^4 \int_{-\sqrt{y}}^{\sqrt{y}} f(x, y)\, dx\, dy + \int_4^9 \int_{-\sqrt{y}}^{6-y} f(x, y)\, dx\, dy$

45. $\int_{-9}^0 \int_{3-\sqrt{9+y}}^{3+\sqrt{9+y}} f(x, y)\, dx\, dy + \int_0^7 \int_y^{3+\sqrt{9+y}} f(x, y)\, dx\, dy$

47. $V = \int_0^{7/3} \int_0^{(7-3x)/2} (7 - 3x - 2y)\, dy\, dx$

49. $V = 8\int_0^{\sqrt{3}} \int_0^{\sqrt{3-x^2}} \sqrt{7 - x^2 - y^2}\, dy\, dx$

51. $V = 4\int_0^a \int_0^{(b/a)\sqrt{a^2+b^2}} c\sqrt{1 - \frac{x^2}{a^2} - \frac{y^2}{b^2}}\, dy\, dx$

53. $V = \frac{4}{3}\pi(\sqrt{2})^3 - 8\int_0^1 \int_0^1 \sqrt{2 - x^2 - y^2}\, dy\, dx$

55. πab

57. 1

59. $\int_1^8 \int_{y^{1/3}}^y f(x, y)\, dx\, dy$

61. $\frac{5}{24}$

63. 22

65. 18π

67. $\frac{128\pi}{3}$

13.3 Double Integrals in Polar Coordinates (pages 903–906)

1. 2

3. 0.2888

5. 33.5103

7. 0

9. 50.2655

11. 6π

13. 4π

15. $\frac{\pi}{2}$

17. 1.9132

19. 1.9270

21. π

23. 0.5435

25. $\frac{a^4\pi}{4}$

27. $\pi \ln 2$

29. $2\pi a(1 - \ln 2)$

31. 0

33. 25,453

35. 4.2131

37. 8π

39. $\frac{8}{3}$

41. $\frac{a^2}{8}(\pi - 2)$

43. 134.0413

45. 9.6440

47. 0.9817

49. $\frac{\pi}{2}$

51. $\frac{\pi}{8}$

53. π

55. $\frac{15\pi}{4}$

57. $\sqrt{\pi}$

59. $\frac{1}{3}\pi a^2(3R - a)$

13.4 Surface Area (pages 912–914)

1. 18.3303

3. 36.1769

5. $\frac{7}{4}$

7. 9.2936

9. 67.2745

11. 5.3304

13. 2π

15. 9.1327

17. 25.6201

19. 117.3187

23. $S = \frac{D^2}{2ABC}\sqrt{A^2 + B^2 + C^2}$

25. $S = \frac{3\sqrt{3}\,\pi a^2}{4}$

27. $\sqrt{2}\,\pi h^2$

29. 36π

31. $S = \int_0^1 \int_0^y \sqrt{\csc^2 y + z^{-2}}\, dz\, dy$

33. $S = \int_0^{2\pi} \int_0^{\sqrt{\pi/2}} \sqrt{4r^2\sin^2 r^2 + 1}\, r\, dr\, d\theta$

35. $S = \int_1^5 \int_{5/x}^{6-x} \sqrt{(2x + 5y)^2 + (2y + 5x)^2 + 1}\, dy\, dx$

37. $4|u|\sqrt{u^2 + 1}$ **39.** $2|v|\sqrt{9u^4 + 1}$ **41.** 7.01

43. $\frac{1}{2}\left(-\frac{u}{2}\sin 2v + u\sin v - 2\cos v\sin\frac{v}{2}\right)\mathbf{i}$ $+ \frac{1}{2}(-u\sin^2 v - u\cos v - 2\sin v\sin\frac{v}{2})\mathbf{j}$ $+ \frac{1}{2}(2u\cos^2\frac{v}{2} + 2\cos\frac{v}{2})\mathbf{k}$

47. $4\pi R^2$

13.5 Triple Integrals (pages 921–924)

3. 45

5. 45

7. $\frac{68}{9}$

9. −4.7746

11. −0.2986

13. 6.3729

15. 8

17. $\frac{1}{720}$

19. 0

21. 0.9762

23. $\frac{1}{6}$

25. $\frac{4\pi}{3}$

27. 12.5664

29. 23.0859

31. $\frac{16}{3}$

33. $\int_0^1 \int_y^1 \int_0^y f(x, y, z)\, dz\, dx\, dy$

35. $\int_0^2 \int_0^{\sqrt{4-z^2}} \int_0^{\sqrt{4-y^2-z^2}} f(x, y, z)\, dx\, dy\, dz$

37. 32π

39. 32π

41. $\frac{3\pi}{2}$

43. $\frac{4}{3}\pi R^3$

45. $\frac{4\pi abc}{3}$

47. $\frac{abc}{6}$

51. B

53. 74.3197

13.6 Mass, Moments, and Probability Density Functions (pages 931–935)

5. $\left(\frac{3}{2}, 2\right)$

7. $\left(1, \frac{8}{5}\right)$

9. $\left(\frac{1}{4}, \frac{4}{5}\right)$

11. $(1, 1, 1)$

13. $\left(0, \frac{24}{5\pi}\right)$

15. $\left(7, \frac{5}{3}\right)$

17. $(1.608, 0.231)$

19. a. $\left(0, \frac{3a}{2\pi}\right)$ **b.** $\left(-\frac{8a}{3\pi^2}, \frac{4a}{3\pi}\right)$

21. $(4.9110, 0.7616)$ **23.** 0.0339

25. $(0, 1.05)$

27. $(1.4372, 0)$

31. $\frac{a^4\pi}{64}$

35. $\left(\frac{3a}{8}, \frac{3a}{8}, \frac{3a}{8}\right)$ **37.** 0.0803

39. The probability is roughly 75%.

41. 3.7222

43. 2.1548

45. 0

47. 10.1538

49. $\frac{\sqrt{10}\,a}{5}$

51. a. $\sqrt{\frac{2}{\pi}}\, C_0 \frac{e^{-1/2}}{x_0}$

  b. The danger zone is approximately 1.9 miles.

  c. $AV = \frac{1}{A}\int_0^{1.9358}\int_0^{x^2/(2k)} \frac{C_0}{\sqrt{k\pi t}}\exp\left(\frac{-x^2}{4kt}\right) dt\, dx$

53. a. $W = \delta\iiint (h-z)\, dV$ **c.** $5{,}400\pi\delta$

57. $4\pi^2 ab$

13.7 Cylindrical and Spherical Coordinates (pages 941–944)

3. a. $(4.00, 1.57, 1.73)$ **b.** $(4.36, 1.57, 1.16)$

5. a. $(2.24, 1.11, 3.00)$ **b.** $(3.74, 1.11, 0.64)$

7. a. $(-1.50, 2.60, -3.00)$ **b.** $(4.24, 2.09, 2.36)$

9. a. $(1.41, 1.41, 3.14)$ **b.** $(3.72, 0.79, 0.57)$

11. a. $(1.50, 0.87, -1.00)$ **b.** $(1.73, 0.52, -1.00)$

13. a. $(0.00, 0.00, 1.00)$ **b.** $(0.00, 0.52, 1.00)$

15. a. $(-0.06, 0.13, -0.99)$ **b.** $(0.14, 2.00, -0.99)$

17. $z = r^2 \cos 2\theta$

19. $r = \dfrac{6z}{\sqrt{4 - 13\cos^2\theta}}$ or
$9r^2\cos^2\theta - 4r^2\sin^2\theta + 36z^2 = 0$

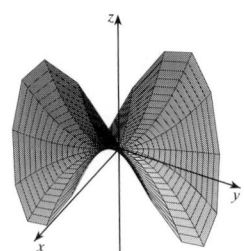

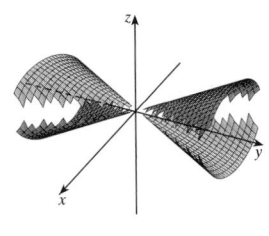

21. $\phi = \frac{\pi}{4}$

23. $\rho = \dfrac{4\cot\phi\csc\phi}{3 - 2\cos^2\theta}$ or

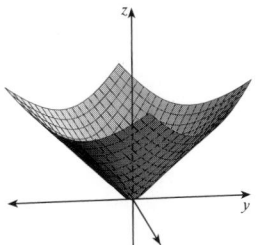

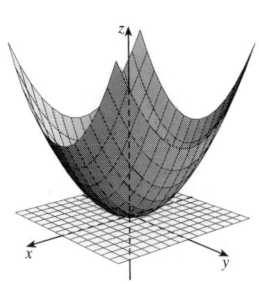

25. $z = 2xy$

27. $z = x^2 - y^2$

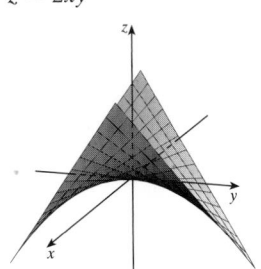

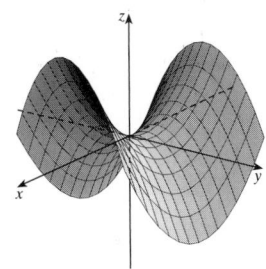

29. $xz = 1$

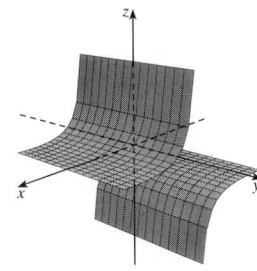

31. $\frac{16}{3}$ **33.** 2π **35.** 8π **37.** $\frac{7}{60}$

39. $\frac{x^2}{a^2} + \frac{y^2}{b^2} + \frac{z^2}{c^2} = R^2$ **41.** $\frac{a^6}{3}$ **43.** $\left(0, 0, \frac{27}{4}\right)$

45. a. $m = \int_0^{2\pi}\int_0^3\int_0^{\sqrt{9-r^2}} (r^2\sin\theta\cos\theta + z)r\, dz\, dr\, d\theta$

  b. $\bar{x} = \frac{1}{m}\int_0^{2\pi}\int_0^3\int_0^{\sqrt{9-r^2}} r\cos\theta\,(r^2\sin\theta\cos\theta + z)r\, dz\, dr\, d\theta$

  c. $I_z = \int_0^{2\pi}\int_0^3\int_0^{\sqrt{9-r^2}} r^2(r^2\sin\theta\cos\theta + z)r\, dz\, dr\, d\theta$

47. 0   **49.** 26.8083   **51.** 14.2172

53. 18.8496   **55.** 10.9956   **57.** 7.8540

61. a. Force $= \dfrac{GmM}{R^2} = \dfrac{Gm\delta(4\pi a^3)}{3\,R^2}$

  b. Force $= \dfrac{Gm\delta(9\pi)}{4}$

63. $\frac{8\pi}{5}$

13.8 Jacobians: Change of Variables (pages 948–950)

1. $u - v$ **3.** 2

5. $4uv(v^2 + u^2)$ **7.** $-2e^{2u}$

9. -9 **11.** ue^{uv}

13. $A(0, 5) \to (5, -5);$
$B(6, 5) \to (11, 1);$
$C(6, 0) \to (6, 6);$
$O(0, 0) \to (0, 0)$

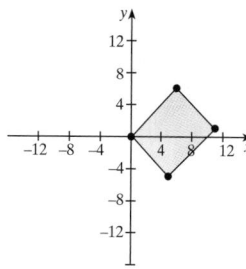

15. $A(5, 0) \to (25, 0);$
$B(7, 4) \to (33, 56);$
$C(2, 4) \to (-12, 16);$
$O(0, 0) \to (0, 0)$

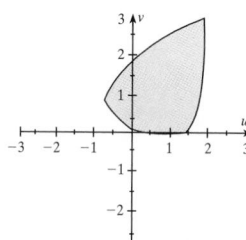

17. $dx\, dy = u\, du\, dv$

19. $3 \ln 2$

21. 0 **23.** 0

27. $\frac{5}{6}$ **29.** $\frac{625}{6}$

31. 13.0949

33. 40.6952

35. 1.1455

37. $\dfrac{ab\pi}{4}(1 - e^{-1})$

39. 1.2825

41. $dx\, dy\, dz$ becomes $r\, dr\, d\theta\, dz$

43. $\dfrac{4(a^2 + b^2)(abc)\pi}{15}$

47. $C = 4 \displaystyle\int_0^a \dfrac{1}{a\sqrt{a^2 - x^2}} \sqrt{a^4 + (b^2 - a^2)x^2}\, dx$

Chapter 13: Review

Proficiency Examination (pages 950–951)

23. 0.2866 **24.** 0

25. 36 **26.** 0.6609

27. Thus, the probability of product failure is about 15%.

28. 8.8858

29. 26.8083

30. 2.4440

Supplementary Problems (pages 952–955)

1. 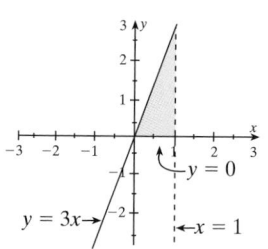 $\frac{3}{2}$

3. 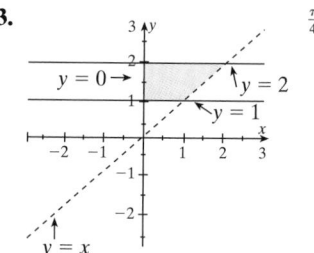 $\frac{\pi}{4} \ln 2$

5. 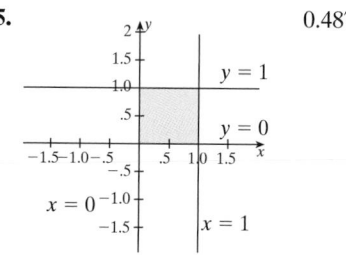 0.4876

7. 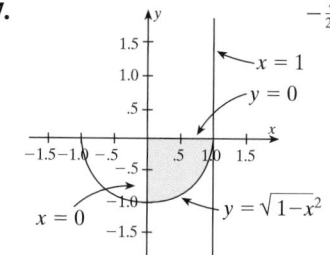 $-\frac{4}{21}$

9. 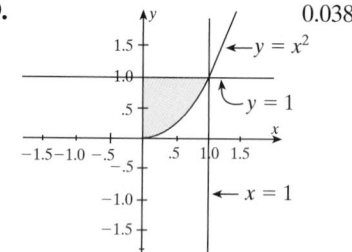 0.0383

11. $\frac{1}{3}(e - 1)$ **13.** 10 **15.** 38.0406

17. $\frac{31}{30}$ **19.** $\frac{81\pi}{8}$ **21.** 12π

23. $\dfrac{2^{2n+2}\pi}{n + 1}$ **25.** 0 **27.** 0

29. $\dfrac{4}{ab}$ **31.** $\dfrac{\pi}{8}$ **33.** π **35.** $\dfrac{32\pi}{5}$

37. The improper integral converges for $m > 1$, and diverges otherwise.

39. a^2

41. $\displaystyle\int_0^1 \int_x^{4-3x} f(x, y)\, dy\, dx$ **43.** $a = b = \pi, c = 0$

45. $2\sqrt{2}$ **47.** 6   **49.** $\dfrac{2}{3}$

51. $\dfrac{\pi}{2}$ **53.** 117.32   **55.** 7.0686

57. 2.4110

61. $\dfrac{2\pi a^5}{15}(2 - \sin^2\phi_0 \cos\phi_0 - 2\cos\phi_0)$

63. $\dfrac{2}{5}$ **65.** $\dfrac{3}{4}H$ **67.** $32xyz$

69. $y = Cx + 1$ becomes $v = C(u + v) + 1$

71. a. $-x \le y \le x; \frac{1}{2} \le x \le 1$

   b. $\dfrac{\partial(u, v)}{\partial(x, y)} = -2; \displaystyle\int_D \int (u + v)\, du\, dv = \dfrac{7}{3}$

73. $\left(\dfrac{1}{3A}, \dfrac{1}{3B}, \dfrac{1}{3C}\right)$

75. a. $\left(\frac{16}{35}, \frac{1}{2}\right)$ **b.** $\frac{11}{240}$

77. $\dfrac{32a^3}{9}$ **79.** $\dfrac{4a^3\pi}{35}$

81. 1   **83.** $m = \displaystyle\int_{\pi/12}^{5\pi/12} \int_{\sqrt{2}\,a}^{2a\sqrt{\sin 2\theta}} \delta r\, dr\, d\theta$

85. $(0, 0, 0)$ **87.** $\bar{x} = \bar{y} \approx 0.3575$

89. This is Putnam Problem 4 in the afternoon session of 1993.

91. This is Putnam Problem 3 in the afternoon session of 1957.

CHAPTER 14: VECTOR ANALYSIS

14.1 Properties of a Vector Field: Divergence and Curl (pages 965–966)

9. div $\mathbf{F} = 3x + 3z^2$; curl $\mathbf{F} = y\mathbf{k}$

11. div $\mathbf{F} = 2$; curl $\mathbf{F} = \mathbf{0}$

13. div $\mathbf{F} = 0$; curl $\mathbf{F} = \mathbf{0}$

15. At $(1, 2, 3)$, div $\mathbf{F} = 7$; curl $\mathbf{F} = \mathbf{j} - 3\mathbf{k}$

17. At $(3, 2, 0)$, div $\mathbf{F} = 2 - 2e^{-6}$; curl $\mathbf{F} = -3\mathbf{i} + 3e^{-6}\mathbf{k}$

19. div $\mathbf{F} = \cos x - \sin y$; curl $\mathbf{F} = \mathbf{0}$

21. div $\mathbf{F} = 0$; curl $\mathbf{F} = \mathbf{0}$

23. div $\mathbf{F} = \dfrac{1}{\sqrt{x^2 + y^2}}$; curl $\mathbf{F} = \mathbf{0}$

25. div $\mathbf{F} = a + b$; curl $\mathbf{F} = \mathbf{0}$

27. div $\mathbf{F} = 2(x + y + z)$; curl $\mathbf{F} = \mathbf{0}$

29. div $\mathbf{F} = x + y + z$; curl $\mathbf{F} = -y\mathbf{i} - z\mathbf{j} - x\mathbf{k}$

31. div $\mathbf{F} = yz + 2x^2 yz^2 + 3y^2 z^2$;
   curl $\mathbf{F} = (2yz^3 - 2x^2 y^2 z)\mathbf{i} + xy\mathbf{j} + (2xy^2 z^2 - xz)\mathbf{k}$

33. div $\mathbf{F} = 3$; curl $\mathbf{F} = \mathbf{i} + \mathbf{j} - \mathbf{k}$

35. harmonic   **37.** harmonic

39. $6x\mathbf{j} - 3y\mathbf{k}$ **41.** $2z + 3x$

43. $6xyz^2 + 2xy^3$

51. a. $(bz - cy)\mathbf{i} + (cx - az)\mathbf{j} + (ay - bx)\mathbf{k}$
   b. div $\mathbf{V} = 0$; curl $\mathbf{V} = 2\omega$

14.2 Line Integrals (pages 974–976)

3. $-\dfrac{4}{3}$ **5.** 1   **7.** $\dfrac{26}{3}$

9. a. $-\dfrac{1}{3}$ **b.** $-\dfrac{1}{3}$ **11.** $-\dfrac{16}{3}$

13. $\dfrac{77}{15}$ **15.** $-\dfrac{7}{60}$ **17.** 12

19. 12   **21. a.** $\dfrac{\pi^2}{8}$ **b.** $\dfrac{\pi^2}{8}$

23. a. $\dfrac{5}{2}$ **b.** $\dfrac{11}{2}$ **25.** 0

27. 9   **29.** $-\dfrac{1}{3}$ **31.** 0

33. -0.7071 **35.** 2π **37.** 16

39. $\dfrac{1}{2}$ **41.** 42   **43.** $\dfrac{5}{2}$

45. The work done is 10,875 ft-lb

14.3 Independence of Path (pages 981–984)

5. $f(x, y) = x^2 y^3$; conservative

7. not conservative   **9.** not conservative

11. a. -2π **b.** -4 **c.** -2π

13. a. 0   **b.** 32   **c.** 32

15. 1   **17.** $\dfrac{1}{3}$ **19.** $\dfrac{1}{2}$

21. $\dfrac{13}{4}$ **23.** 8   **25.** 2.7124

27. 8   **29.** 32   **31.** 0.2245

33. $g(x) = Cx^{-5}$

35. b. $W = \dfrac{kmM}{\sqrt{a_2^2 + b_2^2 + c_2^2}} - \dfrac{kmM}{\sqrt{a_1^2 + a_2^2 + a_3^2}}$

37. 0

39. a. $W = -(2a + b)$ **b.** $W_1 = a(e^{-1} - 3)$

   c. $W_2 = \dfrac{a}{7}(9e^{-7/9} - 23); W_3 = -ae^{-7/9}(2e^{7/9} + e - 1)$

14.4 Green's Theorem (pages 993–995)

1. 0   **3.** 3   **5.** 0

7. -6π **9.** 0   **11.** 16

13. 2π **15.** 4π **17.** $\dfrac{15}{2}$

19. $-\dfrac{\pi a^4}{4}$ **25.** 0   **27.** 0

29. 0   **31.** 3   **33.** $5A$

14.5 Surface Integrals (pages 1001–1002)

1. 8π **3.** 0   **5.** 16π

7. $4\sqrt{2}$ **9.** 0   **11.** $\dfrac{160\sqrt{6}}{3}$

13. $\dfrac{\pi}{2}$ **15.** 84.4635   **17.** 36π

19. $\dfrac{4\pi}{3}$ **21.** $\dfrac{7\pi\sqrt{2}}{4}$ **23.** -6

25. $\dfrac{4\pi}{3}$ **27.** π **29.** 0.9617

31. -2.3234 **33.** 13.0639   **35.** 2.9794

37. 228.5313

39. a. $\dfrac{\pi a^7}{192}$ **b.** $2\pi a^2(1 - \cos\phi)$ **41.** $I_z = \frac{2}{3}ma^2$

14.6 Stokes' Theorem (pages 1009–1011)

1. 18π **3.** $\frac{9}{2}$ **5.** -8π

7. -18 **9.** 0 **11.** $2\sqrt{2}\,\pi$

13. -3π **15.** $-\frac{1}{2}$ **17.** 0

19. -16π **21.** $\frac{3}{2}$ **23.** $2 - \pi$

25. $\frac{\pi}{2}$ **27.** 0 **29.** $f(x, y, z) = xyz^2$

31. $f(x, y, z) = \dfrac{xy}{z}$

33. $f(x, y, z) = xy \sin z + y^2$

14.7 Divergence Theorem (pages 1017–1019)

1. $\frac{64\pi}{3}$ **3.** $\frac{128}{3}$ **5.** 3

7. 24π **9.** $\frac{13}{3}$ **11.** $\frac{81\pi}{2}$

13. 0 **15.** 12π **17.** $\frac{\pi}{4}$

19. $\dfrac{9\pi a^4}{2}$ **21.** $4\pi a^4$ **23. b.** 6

CHAPTER 14: REVIEW

Proficiency Examination (pages 1021–1022)

23. $\mathbf{F}$ is conservative; $f = xyz$

24. div $\mathbf{F} = 2xy - ze^{yz}$; curl $\mathbf{F} = ye^{yz}\mathbf{i} - \frac{1}{2}\mathbf{j} - x^2\mathbf{k}$

25. -2 **26.** $-4\pi\sqrt{2}$

27. 0 **28.** 0

29. $\phi = \dfrac{m\omega^2}{2}(x^2 + y^2 + z^2)$ **30.** 0

Supplementary Problems (pages 1023–1025)

1. $\mathbf{F}$ is conservative; $f = 2x - 3y$

3. $\mathbf{F}$ is conservative; $f = xy^{-3} + \sin y$

5. $\mathbf{F}$ is not conservative

7. -7 **9.** -2.5670

11. $\frac{5}{2}$ **13.** div $\mathbf{F} = 3$; curl $\mathbf{F} = \mathbf{0}$

15. div $\mathbf{F} = \dfrac{2}{\sqrt{x^2 + y^2 + z^2}}$; curl $\mathbf{F} = \mathbf{0}$

17. $\pi \cos \pi^2$ **19.** -8.3562 **21.** $\sqrt{2}$

23. $\frac{5}{6}$ **25.** $2\pi(\pi - 1)$ **27.** 0

29. 4π **31.** $\dfrac{6\sqrt{2}\,\pi}{5}$ **33.** $\dfrac{2a + 3b}{2abc}$

35. 0 **37.** $\frac{3}{4}$ **39.** $\frac{3}{2}$

41. -1 **43.** $\frac{5}{6}$

45. $\mathbf{F}$ is conservative in any region of the plane where $x + y \neq 0$; $W = \ln\left(\dfrac{c + d}{a + b}\right)$

53. 0

55. a. C is any region that does not contain the y-axis;
$f = -\dfrac{1 + y^2}{2x^2} - \dfrac{y^2}{2}$

b. $\dfrac{-67}{9}$

57. $\frac{5}{4} + \ln\frac{2}{9}$ **59.** $2\pi c$ **61.** 0

63. 0 **65.** $\frac{\pi}{4} - \frac{1}{2}\ln 2$ **67.** 2π

71. $4\pi^2\,ab$

73. Only the volume of the body is important, so it does not matter which corner is removed.

75. This is Putnam Problem 6i in the morning session of 1948.

CHAPTER 15: INTRODUCTION TO DIFFERENTIAL EQUATIONS

15.1 First-Order Differential Equations (pages 1038–1041)

1. $\ln|y| = x + \ln|x - 5|^5 + C$

3. $\ln|y| = -\frac{1}{18}[\ln(e^{2x} + 9) - 2x] + C$

5. $y = 3\tan^{-1}\dfrac{\sqrt{x^2 - 9}}{3} + C$ **7.** $y = 2(1 - e^{-x^2})$

9. $y = \frac{1}{2}(\cos x + \sin x) - \frac{1}{2}e^{-x}$ **11.** $y = x^3 - \frac{9}{4}x^2$

13. $-\ln\sqrt{x^2 + y^2} - \frac{1}{3}\tan^{-1}\dfrac{y}{x} = C$

15. $(y + x)^2(y - x) = C$ **17.** $\dfrac{y - x}{3y + x} = Bx^8$

19. $x^3y + x\tan y = C$

21. $\ln(x^2 + y^2) + \tan^{-1}x + e^{-y} = C$

23. $x^2\cos 2y - 3xy + 3x^2y - 6y = C$

25. a. $y = -x - 1 + 4e^{x-1}$

b. Isoclines are lines of the form $x + y = C$ **c.**

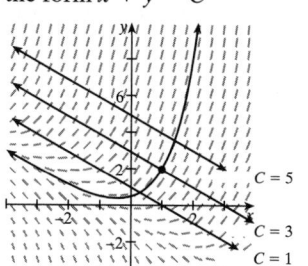 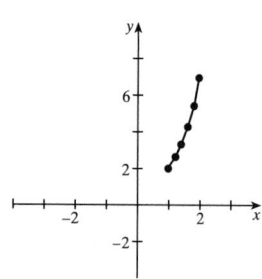

27.

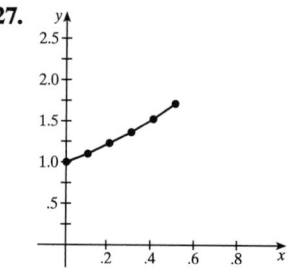

29.

31. $\dfrac{x}{y} + \ln|y| = C$ **33.** $x^2y^2 + x^3y - xy^3 = C$

35. $\dfrac{x^2}{y} + \dfrac{y^2}{x} + 3y = C$

37. $5e^x = e^y(2\sin x\cos x + \sin^2 x + 2) + C$

39. $x^3 + xy\cos xy = C$

41. $y = (\csc x + \cot x)[(-\sin x + x) + C]$

43. $\sqrt{x^2 + y^2} + y = B$ **45.** $\cot\dfrac{y}{x} = \ln\left|\dfrac{\pi x}{4}\right| + 1$

47. $y = x^3\ln|x| + x^3$ **49.** $x\sin(x^2 + y) = 0$

51. $\frac{1}{2}\ln|y^2 + 2y + 2| - \tan^{-1}(y + 1) + e^{-x} = 1$

53. If $\alpha > \beta$, $\lim\limits_{t\to+\infty} Q(t) = \beta$; if $\alpha < \beta$,

$\lim\limits_{t\to+\infty} Q(t) = \alpha$, and if $\alpha = \beta$, $\lim\limits_{t\to+\infty} Q(t) = \alpha$

55. a. $P(t) = \dfrac{r_1(P_0 - r_2) - r_2(P_0 - r_1)e^{-Dt}}{(P_0 - r_2) - (P_0 - r_1)e^{-Dt}}$

b. $\lim\limits_{t\to+\infty} P(t) = r_1$

57. a. $x = \dfrac{1}{I}\left[\displaystyle\int I(y)\,S(y)\,dy + C\right]$ **b.** $x = y^2[(y + 1)e^{-y} + C]$

59. 125.33 ft/s   **61. c.** $y = \dfrac{3Cx + 2x^4}{3C - x^3}$

15.2 Second-Order Homogeneous Linear Differential Equations (pages 1050–1052)

1. $y = C_1 + C_2e^{-x}$ **3.** $y = C_1e^{-5x} + C_2e^{-x}$

5. $y = C_1e^{3x} + C_2e^{-2x}$ **7.** $y = C_1e^{(-1/2)x} + C_2e^{3x}$

9. $y = C_1e^x + C_2e^{-x}$

11. $y = C_1\cos\sqrt{11}x + C_2\sin\sqrt{11}x$

13. $y = e^{(-3/14)x}\left[C_1\cos\left(\dfrac{\sqrt{131}}{14}x\right) + C_2\sin\left(\dfrac{\sqrt{131}}{14}x\right)\right]$

15. $y = C_1 + C_2x + C_3e^{-x}$

17. $y = C_1 + C_2x + e^{-(1/2)x}\left[C_3\cos\left(\dfrac{\sqrt{7}}{2}x\right) + C_4\sin\left(\dfrac{\sqrt{7}}{2}x\right)\right]$

19. $y = C_1e^{2x} + C_2e^{-3x} + C_3e^{-x}$

21. $y = e^{5x}(1 - 6x)$

23. $y = \frac{11}{5}e^x + \frac{4}{5}e^{11x}$

25. $y = \frac{94}{25} - \frac{19}{25}e^{-5x} - \frac{9}{5}xe^{-5x}$ **27.** $5e^x$

29. e^{-2x}

33. $y = C_1e^{-3x} + C_2xe^{-3x}$

35. $y = C_1 + C_2x^{-4}$ **37.** $y = C_1x^3 + C_2x^{-4}$

39. $y = \dfrac{1}{2}\cos(4\sqrt{3}\,t) - \dfrac{2}{\sqrt{3}}\sin(4\sqrt{3}t)$

41. $y = -\frac{2}{3}\cos(4\sqrt{3}\,t)$

43. $y = e^{-0.4t}[0.5\cos 6.9t + 0.03\sin 6.9t]$

47. a. 9.38 seconds
   b. It takes less time with air resistance.

53. a. $s'(t) = -v_0\ln\left(\dfrac{w - rt}{w}\right) - gt$

b. $s(t) = \dfrac{v_0(w - rt)}{r}\ln\left(\dfrac{w - rt}{w}\right) - \dfrac{1}{2}gt^2 + v_0t$

c. The fuel is consumed when $rt = w_f$; that is, when $t = w_f/r$.

d. the height is $\dfrac{v_0(w - w_f)}{r}\ln\left(\dfrac{w - w_f}{w}\right) - \dfrac{1}{2}\dfrac{gw_f^2}{r^2} + \dfrac{v_0w_f}{r}$

49. The solution to this problem is found in the *Canadian Mathematical Bulletin,* Vol. 28, 1985, p. 250.

15.3 Second-Order Nonhomogeneous Linear Differential Equations (pages 1060–1061)

1. $\bar{y}_p = Ae^{2x}$ **3.** $\bar{y}_p = Ae^{-x}$

5. $\bar{y}_p = xe^{-x}(A\cos x + B\sin x)$

7. $\bar{y}_p = (A + Bx)e^{-2x} + xe^{-2x}(C\cos x + D\sin x)$

9. $\bar{y}_p = A_3x^3 + A_2x^2 + A_1x + A_0$

11. $\bar{y}_p = (A_3x^3 + A_2x^2 + A_1x + A_0)\cos x + (B_3x^3 + B_2x^2 + B_1x + B_0)\sin x$

13. $\bar{y}_p = A_0e^{2x} + B_0\cos 3x + C_0\sin 3x$

15. $\bar{y}_p = (A_3x^3 + A_2x^2 + A_1x + A_0) + B_0e^{-x}$

17. $y = C_1 + C_2e^{-x} - x^3 + 3x^2 + x$

19. $y = C_1e^{-3x} + C_2e^{-5x} + \frac{3}{35}e^{2x}$

21. $y = e^{-x}(C_1\cos x + C_2\sin x) + \frac{1}{5}\cos x + \frac{2}{5}\sin x$

23. $y = C_1e^{x/7} + C_2e^{-x} + e^{-x}(-\frac{1}{16}x^2 - \frac{15}{64}x)$

25. $y = C_1 + C_2e^x - (\frac{1}{4}x^4 + x^3 + \frac{5}{2}x^2 + 10x)$

27. $y = C_1e^{-x} + C_2xe^{-x} + e^{-x}(\frac{1}{6}x^3 + 2x^2)$

29. $y = C_1\cos x + C_2\sin x - \cos x\ln|\sec x + \tan x|$

31. $y = C_1e^{3x} + C_2e^{-2x} - \frac{1}{32}(8x^2 + 12x + 13)e^{2x}$

33. $y = C_1\cos 2x + C_2\sin 2x + \frac{1}{2}x\cos 2x - \frac{1}{4}\sin 2x(\ln|\cos 2x|)$

35. $y = C_1e^{-x} + C_2xe^{-x} + \frac{1}{4}x^2(2\ln x - 3)e^{-x}$

37. $y = C_1 + C_2e^x - \frac{1}{20}\sin 2x - \frac{1}{2}x - \frac{1}{5}\cos^2 x$

39. $y = \frac{8}{5} - \frac{6}{5}e^x - \frac{2}{10}\sin 2x + x - \frac{2}{5}x\cos^2 x$

41. $y = -\frac{2}{9}\cos 3x + \frac{4}{9}\sin 3x + \frac{2}{9}e^{3x}$

43. $y = \frac{35}{27}\sin 3x + \frac{1}{9}x$

45. $y = -4\cos x - \sin x\ln|\csc x + \cot x|$

47. $y = C_1e^{-2x} + C_2e^{-3x} + G(x)$

where $G(x) = \begin{cases} \frac{1}{3}x - \frac{5}{18} & \text{for } 0 \le x \le 1 \\ \frac{1}{3} & \text{for } 1 < x < 3 \\ -\frac{1}{3}x + \frac{29}{18} & \text{for } 3 \le x \le 4 \end{cases}$

49. $I(t) = \frac{81}{656}e^{-9t} - \frac{9}{16}e^{-t} + \frac{18}{41}\cos t + \frac{45}{82}\sin t$

51. $I(t) = 0.312e^{-t} - 0.015e^{-9t} - 0.297\cos t$
$- 0.067\sin t + t(0.244\cos t + 0.305\sin t)$

53. a. $y_1 = x; y = C_1x + C_2(x\tan^{-1}x + 1) - \frac{1}{2}\tan^{-1}x + \frac{x}{2}$

CHAPTER 15: REVIEW

Proficiency Examination (page 1062)

12. $\sin^{-1}y = \sinh^{-1}x + C$ **13.** $y = Bx$

14. $y = x + C\sqrt[3]{x}$

15. $x^2(x^2 - 2y^2) = C$

16. $\frac{2}{\sqrt{3}}\tan^{-1}\left[\frac{2}{\sqrt{3}}\left(\frac{y}{x} - \frac{1}{2}\right)\right] = \ln|x| + C$

17. $y = e^{-x}(C_1\cos x + C_2\sin x) - \frac{2}{5}\cos x + \frac{1}{5}\sin x$

18. $x^3e^{-y} - xy^{-2} + x^2y^{-3} = C$

19. $x(t) = e^{-1.60t}[0.33\cos 10.84t + 0.05\sin 10.84t]$

20. $I(t) = e^{-125t}[-0.562\cos 185.4\,t + 2.806\sin 185.4t]$
$+ 0.562\cos 100t - 0.899\sin 100t$

Supplementary Problems (pages 1062–1065)

1. a.

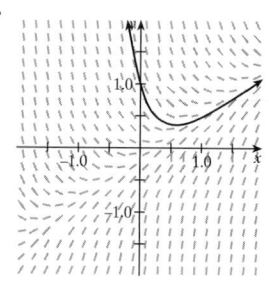

b.

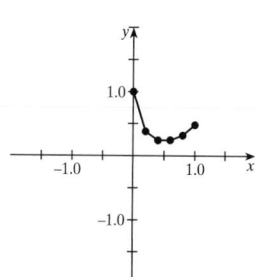

3. $\frac{1}{4}y^4 = -2x^2 + C$

5. $\frac{y^3}{3x^3} = \ln|x| + C$

7. $e^{y^2} = e^{2x} + C$

9. $y = -e^{5x}\cos x + Ce^{5x}$

11. $x = e^{-y}[y + C]$

13. $\frac{y}{x} = \sin(\ln|x| + C)$

15. $y = Be^{(1-2x)/(2x^2)}$

17. $x^4y^3 + \ln|x| - \ln|y| = C$

19. $y = -\tan^{-1}e^x + C$

21. a. $y_h = C_1e^{3x} + C_2e^{-3x}$ **b.** $y_p = -\frac{1}{9}(x + 1)$

23. a. $y_h = C_1e^{2x} + C_2e^{3x}$

b. $y_p = (\frac{1}{2}x^2 + \frac{3}{2}x + \frac{7}{4})e^x$

25. a. $y_h = C_1e^x + C_2e^{2x}$

b. $y_p = (-\frac{1}{4}x^4 - x^3 - 3x^2 - 6x)e^x$

27. a. $y_h = C_1\cos x + C_2\sin x$

b. $y_p = \sin x \ln|\sec x + \tan x| - 2$

29. $y = \frac{1}{9}e^{-3x} + \frac{1}{3}x - \frac{1}{9}$

31. $z(t) = -3e^{-2t} + 2e^{-3t} + 1$

33. $y^2 = 5x - x^2$

35. $e^{X+Y} = 1 + Ke^X$

37. $Y^2(Y^2 + 3X^2)^2 = K$

39. $y = C_1x + C_2\left[\frac{x}{2}\ln\left|\frac{x-1}{x+1}\right| + 1\right]$

41. $\frac{1}{6}x^2 - \frac{C_1}{x} + C_2$

43. $x = y[e^y + C]$ **45.** 0.62 s

47. a. $y(t) = \frac{WLx^3}{6EI} - \frac{Wx^4}{24EI} - \frac{WL^3x}{8EI}$

b. $y_{max} \approx \frac{WL^4}{EI}(-0.045)$

49. b. 2 real roots   **c.** $y = C_1x^{m_0} + C_2x^{m_0}\ln x$

d. $y = x^\alpha[C_1\cos(\beta\ln|x|) + C_2\sin(\beta\ln|x|)]$

53. He must run at 18.8 ft/s.

55. This is Problem 4 of the morning session of the 1947 Putnam examination.

Cumulative Review for Chapters 12–15 (pages 1069–1070)

5. $f_x = 4x + y; f_y = x - 15y^2; f_{xy} = 1$

7. $f_x = 1; f_y = 1; f_{xy} = 0$

9. $f_x = e^xe^y; f_y = e^xe^y; f_{xy} = e^xe^y$

11. 0   **13.** $\frac{49,152}{13}$

15. $\frac{4\pi}{15}$ **17.** $e^2 - 3$

19. π **21.** 8

23. 0

25. $y = -\frac{1}{3}x^{-3}\cos x^3 + Cx^{-3}$

27. $y = C_1e^{-2x} + C_2e^x - \frac{1}{4}(2x^3 + 5x^2 + 7x + 18.5)$

29. $y = 1 + (x - 2)e^{-x^2/2}$

31. The equation of the tangent plane is

$$2x + 4y - z - 4 = 0$$

and the normal line is

$$\frac{x}{2} = \frac{y-2}{4} = \frac{z-4}{-1}$$

33. a. $y - x = 3$ **b.** $x = 4, y = 7$

35. $\frac{\pi}{4} + 2$ **37.** $\frac{\pi}{6}(65\sqrt{65} - 1)$

39. 0

Appendix A (pages A8–A9)

1. $\delta = \frac{\epsilon}{2}$ **3.** false

5. let $\delta = $ minimum of $\{1, \frac{\epsilon}{5}\}$ **7.** $\delta = \epsilon$

9. $\delta = \frac{\epsilon}{3}$ **11.** choose $\delta = \min(1, \frac{\epsilon}{5})$

APPENDIX F

Credits

CHAPTER 1

p. 13   Problems 59 and 60 from *Introductory Oceanography,* 5th ed., by H. V. Thurman, p. 253. Reprinted with permission of Merrill, an imprint of Macmillan Publishing Company. Copyright © 1988 Merrill Publishing Company, Columbus, Ohio.

p. 14   Problem 61 adapted from R. A. Serway, *Physics,* 3rd ed., Philadelphia: Saunders, 1992, p. 1007.

p. 14   Journal problem (#62) by Murray Klamkin reprinted from *The Mathematics Student Journal,* Vol. 28, 1980, issue 3, p. 2.

p. 22   Correlation chart in Problem 52 from *Scientific American,* November, 1987 and in Problem 53 from *Scientific American,* April, 1991.

p. 23   Babylonian system of equations in Problem 60 taken from *A History of Mathematics,* 2nd ed., by Carl B. Boyer, revised by Uta C. Merzbach, John Wiley and Sons, Inc., New York, 1968.

p. 23   Journal problem (#66) reprinted from *Ontario Secondary School Mathematics Bulletin,* Vol. 18, 1982, issue 2. p. 7.

p. 34   Journal problem (#60) reprinted from *The Mathematics Student Journal,* Vol. 28, 1980, issue 3, p. 2.

p. 53   Problem 52 from "Using Interactive-Geometry Software for Right-Angle Trigonometry," by Charles Embse and Arne Engebretsen, *The Mathematics Teacher,* October 1996, pp. 602–605.

p. 72   Putnam examination problem (#59) 1959, reprinted by permission from The Mathematical Association of America.

p. 94   Putnam examination problem (#60) 1960, reprinted by permission from The Mathematical Association of America.

CHAPTER 2

p. 79   Global modeling illustration from *Scientific American,* March 1991.

p. 83   Historical Quest (#48) information from Carl B. Boyer, *A History of Mathematics,* New York: John Wiley & Sons, 1968, p. 18.

p. 84   Historical Quest (#54) information from C. H. Edwards, Jr. *The Historical Development of Calculus,* Springer-Verlag, 1979, pp. 31–35.

p. 116   Graph (#42) from Michael D. La Grega, Philip L. Buckingham, and Jeffery C. Evans, *Hazardous Waste Management,* New York: McGraw-Hill, 1994, pp. 565–566.

p. 116   Modeling Problem (#43) from E. Batschelet, *Introduction to Mathematics for Life Scientists,* 2nd ed., New York: Springer-Verlag, 1976, p. 280.

p. 120   Modeling Problem (#57) from William Bosch and L. G. Cobb, "Windchill," *UMAP Module No. 658,* (1984), pp. 244–247.

p. 121   Putnam examination problem (#63) 1956, reprinted by permission from The Mathematical Association of America.

p. 121   Putnam examination problem (#64) 1986, reprinted by permission from The Mathematical Association of America.

p. 122   Guest essay, "Calculus Was Inevitable," by John L. Troutman.

p. 124   Historical Quest (#9) information from *Ethnomathematics* by Marcia Ascher, pp. 128–129 and 188–189.

CHAPTER 3

p. 161   Photograph of giant yo-yo courtesy of Bob Malonney, Director, National Yo Yo Museum.

p. 162   Photo courtesy of Photo Disc, Inc.

p. 189   Journal problem (#73) by Bruce W. King from *The Pi Mu Epsilon Journal,* Volume 7 (1981), p. 346.

p. 206   Modeling problem (#30) by John A. Helms, "Environmental Control of Net Photosynthesis in Naturally Grown Pinus Ponderosa Nets," *Ecology* (Winter 1972) p. 92.

p. 207   Modeling problems (#35, #36) from "Introduction to Mathematics for Life Scientists," 2nd edition. New York: Springer-Verlag (1976); pp. 102–103.

p. 212   Putnam examination problem (#96) 1946, reprinted by permission from The Mathematical Association of America.

p. 212   Putnam examination problem (#97) 1939, reprinted by permission from The Mathematical Association of America.

p. 212   Putnam examination problem (#98) 1946, reprinted by permission from The Mathematical Association of America.

p. 213   Chaos example is by Jack Wadhams of Golden West College.

p. 213   Photo courtesy of Gregory Sams/Photo Researchers, Inc.

CHAPTER 4

p. 215   Energy expended by a pigeon by Edward Batschelet, *Introduction to Mathematics for Life Scientists,* 2nd ed. (New York: Springer-Verlag, 1979), pp. 276–277.

p. 246   Journal problem (#70) from *Mathematics Magazine,* Volume 55 (1982), p. 300.

p. 258   Problem 58 from "The Mechanics of Bird Migration," by C. J. Pennycuick, *Ibis III,* pp. 525–556.

p. 271   Journal problem (#51) by Michael Murphy from the *College Mathematics Journal,* Vol. 22, May 1991, p. 221.

p. 271   Journal problem (#52) from *Parabola,* Vol. 20, Issue 1, 1984.

p. 271   Historical Quest (#55) from "Mathematics in India in the Middle Ages," by Chandra B. Sharma, *Mathematical Spectrunin,* Volume 14(1), pp. 6–8, 1982.

p. 280   Journal problem (#6) from *Parabola,* Vol. 19, Issue 1, 1983, p. 22.

p. 283   Modeling problem (#33) by Paul J. Campbell, "Calculus Optimization in Information Technology," *UMAP module 1991: Tool for Teaching.* Lexington, MA, CUPM Inc., 1992; pp. 175–199.

p. 283   Modeling problem (#34) by John C. Lewis and Peter P. Gillis, "Packing Factors in Diatomic Crystals," *American Journal of Physics,* Vol. 61, No. 5 (1993), pp. 434–438.

p. 285   Modeling problem (#45) by Thomas O'Neil, "A Mathematical Model of a Universal Joint," *UMAP Modules 1982: Tools for Teaching.* Lexington, MA: Consortium for Mathematics and Its Applications, Inc., 1983, pp. 393–405.

p. 285   Modeling problem (#46) is based on an article by Steve Janke, "Somewhere Within the Rainbow," *UMAP Modules 1992: Tools for Teaching.* Lexington, MA: Consortium for Mathematics and Its Applications, Inc., 1993.

p. 293   Concentration of drugs in the bloodstream application adapted from E. Heinz, "Problems bei der Diffusion kleiner Substanzmengen innerhelf des menschlichen Körpers," *Biochem.,* Volume 319 (1949), pp. 482–492.

p. 293   Modeling vascular branching is adapted from *Introduction to Mathematics for Life Scientists,* 2nd edition, by Edward Batschelet, New York: Springer-Verlag, 1976, pp. 278–280.

p. 298   Modeling problem (#35) from "A Blood Cell Population Model, Dynamical Diseases, and Chaos," by W. B. Gearhart and M. Martelli, *UMAP Modules 1990: Tools for Teaching.* Arlington, MA: Consortium for Mathematics and Its Applications (CUPM) Inc., 1991.

p. 298   Modeling problem (#37) adapted from "Flight of Birds in Relation to Energetics and Wind Directions," by V. A. Tucker, and K. Schmidt-Koening, *The Auk,* Vol. 88, 1971, pp. 97–107.

p. 298   Modeling problem (#39) adapted from P. L. Shaffer and H. J. Gold, "A Simulation Model of Population Dynamics of the Codling Moth *Cydia Pomonella,*" *Ecological Modeling,* Vol. 30, 1985, pp. 247–274.

p. 299   Problem #47 adapted from "Price Elasticity of Demand: Gambling, Heroin, Marijuana, Prostitution, and Fish," by Yves Nievergelt, *UMAP Modules 1987: Tools for Teaching.* Arlington, MA, CUPM Inc., 1988, pp. 153–181.

p. 300   Photo courtesy of Scott Camazine/Photo Researchers, Inc.

p. 308   Historical Quest (#61) from "Two Historical Applications of Calculus" by Alexander J. Hahn, *The College Mathematics Journal,* Vol. 29, No. 2, March 1998, pp. 93–97.

p. 308   Historical Quest (#62) from *A Source Book in Mathematics, 1200–1800,* by D. J. Struik, Cambridge, MA: Harvard University Press, 1969, pp. 313–316.

p. 310   Journal problem (#51) from the *Mathematics Teacher,* December 1990, p. 718.

p. 313   Historical Quest (#84) quotation about Euler from *Elements of Algebra,* 5th ed., by Leonhard Euler, London: Longman, Orme, and Co., 1840.

p. 314   Putnam examination problem (#87) 1938, reprinted by permission from The Mathematical Association of America.

p. 314   Putnam examination problem (#88) 1941, reprinted by permission from The Mathematical Association of America.

p. 314   Putnam examination problem (#89) 1986, reprinted by permission from The Mathematical Association of America.

p. 314   Putnam examination problem (#90) 1985, reprinted by permission from The Mathematical Association of America.

p. 315   Photograph of a wine cellar showing wooden cask courtesy of Peter Menzel/Stock Boston.

p. 315   Group research project is from Elgin Johnston of Iowa State University. This group research project comes from research done at Iowa State University as part of a National Science Foundation grant.

CHAPTER 5

p. 331   Software output from *Converge* by John R. Mowbray, JEMware, 567 South King Street, Suite 178, Honolulu, HI 96813

p. 362   Historical Quest (#45) quotation from "A Short Account of the History of Mathematics," as quoted in *Mathematical Cir-*

cles Adieu by Howard Eves (Boston: Prindle, Weber & Schmidt, Inc., 1977).

p. 363   Journal problem (#60) by Murray Klamkin from the *College Mathematics Journal,* Sept. 1989, p. 343.

p. 375   Photo of Shroud of Turin courtesy of Gianni Tortoli/Science Source/Photo Researchers, Inc.

p. 390   Historical Quest (#34) Early Asian Calculus from *Mathematics,* by David Bergamini, p. 108. Reprinted by permission of Time, Incorporated, 1963.

p. 399   Modeling problem (#72) adapted from "Flight Speeds of Birds in Relation to Energies and Wind Directions," by V. A. Tucker and K. Schmidt-Koenig, *The Auk,* Vol. 88 (1971), pp. 97–107.

p. 400   Putnam examination problem (#82) 1951, reprinted by permission from The Mathematical Association of America.

p. 400   Putnam examination problem (#83) 1958, reprinted by permission from The Mathematical Association of America.

p. 400   Putnam examination problem (#84) 1939, reprinted by permission from The Mathematical Association of America.

p. 401   Guest essay, "Kinematics of Jogging," by Ralph Boas.

CHAPTER 6

p. 425   Photograph of the great pyramid of Cheops courtesy of Paul W. Liebhardt.

p. 442   *Blondie* cartoon reprinted with special permission of King Features Syndicate.

p. 454   Journal Problems (#12–#17) *Mathematics Teacher,* December 1990, p. 695.

p. 459   Putnam examination problem (#69) 1939, reprinted by permission from The Mathematical Association of America.

p. 459   Putnam examination problem (#70) 1938, reprinted by permission from The Mathematical Association of America.

p. 459   Putnam examination problem (#71) 1993, reprinted by permission from The Mathematical Association of America.

p. 460   Photograph of Harry Houdini courtesy of AP/Wide World Photos.

p. 460   Group research project is from *MAA Notes,* Vol. 17, 1991, "Priming the Calculus Pump: Innovations and Resources," by Marcus S. Cohen, Edward D. Gaughan, R. Arthur Knoebel, Douglas S. Kurtz, and David J. Pengelley.

CHAPTER 7

p. 498   Historical Quest (#60) from "Pólya, Problem Solving, and Education," by Alan H. Schoenfeld, *Mathematics Magazine,* Vol. 60, No. 5, December 1987, p. 290.

p. 525   Journal problem (#52) by Peter Lindstrom from the *College Mathematics Journal,* Vol. 24, No. 4, September, 1993, p. 343.

p. 535   Modeling problem (#59) adapted from "The Learning Function," by L. L. Thurstone, *J. of General Psychology 3* (1930), pp. 469–493.

p. 537   Putnam examination problem (#90) 1968, reprinted by permission from The Mathematical Association of America.

p. 537   Putnam examination problem (#91) 1980, reprinted by permission from The Mathematical Association of America.

p. 537   Putnam examination problem (#92) 1985, reprinted by permission from The Mathematical Association of America.

p. 538   Photograph of buoy courtesy of G. Walts/The Image Works.

p. 538   Group research project is adapted from a computer project used at the U. S. Coast Guard Academy.

CHAPTER 8

p. 592   Journal Problem (#54) by Michael Brozinsky, *School Science and Mathematics,* Volume 82, 1982, p. 175.

p. 618   Historical Quest (#74) from *The Historical Development of the Calculus* by C. H. Edwards, Jr., Springer-Verlag, 1979, p. 292.

p. 619   Journal Problem (#82) by Robert C. Gebhardt, *The Pi Mu Epsilon Journal,* Volume 8, 1984, Problem 586.

p. 622   Modeling problem (#93) adapted, "How Old Is the Earth?" by Paul J. Campbell, *UMAP Modules 1992: Tools for Teaching,* Consortium for Mathematics and Its Applications, Inc. Lexington, MA: 1993, pp. 105–137.

p. 622   Putnam examination problem (#94) 1984, reprinted by permission from The Mathematical Association of America.

p. 622   Putnam examination problem (#95) 1977, reprinted by permission from The Mathematical Association of America.

p. 622   Putnam examination problem (#96) 1982, reprinted by permission from The Mathematical Association of America.

p. 623   Photograph of a tightrope walker courtesy of Paul Conklin, PhotoEdit.

p. 623   Group research project is from David Pengelley of New Mexico State University.

CHAPTER 9

p. 650   Journal Problem (#69) by V. C. Bailey *School Science and Mathematics,* Problem 3949 Vol. 83, 1983, p. 356.

p. 655   Information in the Technology Window adapted from "Mathematics Discovery via Computer Graphics: Hypocycloids and Epicycloids" by Florence and Sheldon Gordon, *The Two-Year College Mathematics Journal,* November 1984, p. 441.

p. 661   Journal Problem (#66) by Frank Kocher, *The MATYC Journal,* Problem 148 Vol. 14, 1980, p. 155.

p. 663   Putnam examination problem (#60) 1995, reprinted by permission from The Mathematical Association of America.

p. 664   Photograph of a closed circuit surveillance camera system courtesy of Index Stock Photography, Inc.

p. 664   Group research project is from Steve Hilbert, Ithaca College, Ithaca, New York.

CHAPTER 10

p. 725   Putnam examination problem (#59) 1939, reprinted by permission from The Mathematical Association of America.

p. 725   Putnam examination problem (#60) 1959, reprinted by permission from The Mathematical Association of America.

p. 725   Putnam examination problem (#61) 1983, reprinted by permission from The Mathematical Association of America.

p. 726   Photograph of the Starship *Enterprise* from *Star Trek: The Next Generation* © Paramount Pictures, courtesy of Sygma.

p. 726   Group research project is from *MAA Notes,* Vol. 17, 1991, "Priming the Calculus Pump: Innovations and Resources," by Marcus S. Cohen, Edward D. Gaughan, R. Arthur Knoebel, Douglas S. Kurtz, and David J. Pengelley.

CHAPTER 11

p. 730   Photograph of double helix of DNA courtesy of *Scientific American,* October 1985, p. 60.

p. 752   Photograph of an earth satellite courtesy of Julian Baum, Photo Researchers, Inc.

p. 755   Diagram of Kingdome courtesy of Seattle Mariners.

p. 776   Photograph of a Ferris wheel courtesy of Marc D. Longwood.

p. 780   Putnam examination problem (#75) 1939, reprinted by permission from The Mathematical Association of America.

p. 780   Putnam examination problem (#76) 1947, reprinted by permission from The Mathematical Association of America.

p. 780   Putnam examination problem (#77) 1946, reprinted by permission from The Mathematical Association of America.

p. 781   Quotation by Johann Kepler from James R. Newman, *The World of Mathematics,* Vol. I (New York: Simon & Schuster, 1952), p. 220.

p. 781   Guest essay, "The Simulation of Science," by Howard Eves. This article is reprinted by permission from *Great Moments in Mathematics Before 1650* published by The Mathematical Association of America, 1983, p. 194.

CHAPTER 12

p. 792   Computer-generated graph of Mount St. Helens courtesy of Bill Lennox, Humboldt State University.

p. 793   Topographic map of Mount Rainier courtesy of U.S. Geological Survey.

p. 793   Map showing isotherms courtesy of Accu-Weather, Inc.

p. 860   What Does This Say? problem (#51) adapted from *Managerial Economics* by Dominick Salvatore, McGraw-Hill, Inc., New York, 1989. The data given in the table are on page 138.

p. 870   Historical Quest (#54) from *Men of Mathematics* by E. T. Bell, Simon and Schuster, New York, 1937, p. 165.

p. 873   Modeling problem (#67) from *Mathematical Biology,* 2nd ed., by J. D. Murray, Springer-Verlag, New York, 1993, p. 464.

p. 873   Modeling problem (#68) from "Heat Therapy for Tumors," by Leah Edelstein-Keshet, *UMAP Modules 1991: Tools for Teaching,* Consortium for Mathematics and Its Applications, Inc., Lexington MA, 1992, pp. 73–101.

p. 875   Think Tank Problem (#95) adapted from "Hidden Boundaries in Constrained Max-Min Problems," by Herbert R. Bailey from *The College Mathematics Journal,* May 1991, p. 227.

p. 876   Photograph of soap bubbles from *Scientific American,* "The Geometry of Soap Films and Soap Bubbles," by Frederick Almgren, Jr., and Jean Taylor.

p. 876   Putnam examination problem (#97) 1967, reprinted by permission from The Mathematical Association of America.

p. 876   Putnam examination problem (#98) 1946, reprinted by permission from The Mathematical Association of America.

p. 876   Putnam examination problem (#99) 1938, reprinted by permission from The Mathematical Association of America.

p. 877   Photograph of a desert courtesy of Scott T. Smith.

p. 877   Group research project is from *MAA Notes,* Vol. 17, 1991, "Priming the Calculus Pump: Innovations and Resources," by Marcus S. Cohen, Edward D. Gaughan, R. Arthur Knoebel, Douglas S. Kurtz, and David J. Pengelley.

CHAPTER 13

p. 905   Quotation in Historical Quest (#57) by W. W. Rouse Ball from *A Short Account of the History of Mathematics* as quoted in *Mathematical Circles Adieu* by Howard Eves, Boston: Prindle, Weber & Schmidt, Inc., 1977.

p. 955   Putnam examination problem (#89) 1994, reprinted by permission from The Mathematical Association of America.

p. 955   Putnam examination problem (#90) 1942, reprinted by permission from The Mathematical Association of America.

p. 955   Putnam examination problem (#91) 1957, reprinted by permission from The Mathematical Association of America.

p. 955   Putnam examination problem (#92) 1958, reprinted by permission from The Mathematical Association of America.

p. 956   Photograph of an Apollo spacecraft courtesy of NASA.

p. 956   Group research project is from *MAA Notes,* Vol. 17, 1991, "Priming the Calculus Pump: Innovations and Resources," by Marcus S. Cohen, Edward D. Gaughan, R. Arthur Knoebel, Douglas S. Kurtz, and David J. Pengelley.

CHAPTER 14

p. 958   Photograph of a wind-velocity map of the Indian Ocean courtesy of NASA.

p. 987   Graffiti courtesy of *The College Mathematics Journal,* Sept. 1991; photograph courtesy of Regev Nathansohn.

p. 1025   Journal Problem (#76) by J. Chris Fischer, *CRUX Mathematicorum,* problem number 785, Vol. 8, 1982, p. 277.

p. 1025   Putnam examination problem (#77) 1948, reprinted by permission from The Mathematical Association of America.

p. 1026   Guest essay, "Continuous vs Discrete Mathematics," by William Lucas.

CHAPTER 15

p. 1034   Photograph of a skydiver courtesy of Werner H. Muller/Peter Arnold, Inc.

p. 1041   Journal problem (#62) from *The American Mathematical Monthly,* Vol. 91, 1984, p. 515.

p. 1050   *Mathematics is discovered* quotation by James H. Jeans, *The Mysterious Universe,* p. 142. *Mathematics is invented* quotation by G. F. Simmons, *Differential Equations with Applications and Historical Notes,* 2nd ed., McGraw-Hill, Inc., 1991, p. xix.

p. 1052   Journal problem (#54) by Murray S. Klamkin, Problem 331 in the *Canadian Mathematical Bulletin,* Vol. 26, 1983, p. 126.

p. 1065   Putnam examination problem (#55) 1938, reprinted by permission from The Mathematical Association of America.

p. 1065   Putnam examination problem (#56) 1947, reprinted by permission from The Mathematical Association of America.

p. 1065   Putnam examination problem (#57) 1948, reprinted by permission from The Mathematical Association of America.

p. 1065   Putnam examination problem (#58) 1959, reprinted by permission from The Mathematical Association of America.

p. 1066   Photograph of a dam courtesy of Tennessee Valley Authority.

p. 1066   Group research project, "Save the Perch Project," by Diane Schwartz from Ithaca College, N. Y.

Index

Teaching Notes for Instructors

The material on the following pages is provided to assist instructors, especially junior faculty and teaching assistants. However, all teachers may benefit from the suggestions and outline.

The main purpose here is to provide focus on pedagogy, including, but not limited to: goals and emphases, general balance, chronology (order of topics), proofs vs. plausibility arguments, notation, and known areas of difficulties for students. Where possible, we show how to integrate different approaches, branches of mathematics, or disciplines.

We give miscellaneous warnings, review topics & tips, counter-examples, and ideas & limitations on integrating technology. For this, the goal is usually to supplement, not supplant, the mathematics.

Toward these ends, we include suggested or sample homework assignments. It is not our intention to substitute for individual preferences or wisdom, but rather to guide and to show what should work well.

Answers to odd-numbered exercises appear in the back. For each section we give three different sample homework assignments:

L: All questions in set **L** List have complete solutions in the *Student Survival Manual*.

M: **Mixed** assignment, with about half the questions in set **M** solved in the *Student Survival Manual*.

N: **No** questions in set **N** have solutions in the *Student Survival Manual*.

Instructors should note these special homework features: *Think-Tank Problems* (to dig deeper), *Historical Quests* (for backgrounds of great mathematicians and great mathematics), *Student Projects* (for hands-on learning), and *Spy Problems* (for a more light-hearted touch).

References to *Journal Problems* (challenges from college math journals' problem sections) and *Putnam Problems*—as well as to their solutions—are provided to encourage student research.

CHAPTER 1: FUNCTIONS AND GRAPHS

1.1 Preliminaries

Main Topics: Distance on a line and in the plane; intervals; inequalities, with and without absolute values; circles.

Main Ideas: *Analytic geometry* is the branch of mathematics that deals with the correspondence between geometric objects and their analytical counterparts:

- Given an equation in x and y, what is its graph?
- Given a graph, is there an equation (or inequality or other open sentence) in x and y whose solution set is precisely the graph?

Students need to be comfortable with switching "views" as needed. Apply this to circles in particular.

Notation and Pedagogy: Delta notation is not only for slope (next section) but also good for distance formula. Students should see distance as just the familiar Pythagorean theorem:

$$d^2 = (\Delta x)^2 + (\Delta y)^2$$

and not a long formula in which they confuse + and − signs. This is also a good opportunity to remind students that $\sqrt{(\Delta x)^2} = |\Delta x|$ while showing the one-dimensional distance a special case of the distance formula in two dimensions. (In turn, this is a special case of the three-dimensional distance formula.)

Interval notation should be explained fully. It is used throughout the text (*e.g.*, for domains).

Warnings: Student errors often have their origin in order-of-operations confusions. They need to understand why there is a convention. A good way to do that and to show the limitations of technology at the same time is this: Ask the class the value of $1 + 2 * 3$. Many know it's 7, but not all. Discuss why it's 7 and not 9.

Technology: On a PC, go into the Microsoft Windows calculator (Windows 95 or Windows 3.x). There are two "views" available, called "Standard" and Scientific. Click on one. Enter by keys or mouse: $1 + 2 * 3$. Now do this in the other view. Surprise! ("Standard" gives 9, which should be unequivocally declared wrong. Scientific gives the correct 7.) What is the clear moral about technology? Discuss limitations.

For more on the importance of order, ask your students the value of -5^2. Many will say +25. Up until the late 1980s, most computer spreadsheets (Lotus 1-2-3 was the shining exception) and AppleSoft Basic made this error. Again, this shows the need to be a bit critical/ skeptical.

More on Pedagogy: Algebra and calculus often entail reversing order, so if one does not know the precedence rules for arithmetic, how can one apply them in reverse later in algebra?

One item related to distance formula is square root. Ask the class: "How much is $\sqrt{9}$?" Point out that if it were both 3 and −3, why would we say *the* square root? (The word "the" connotes uniqueness.) Also, square root would not be a function, which you can point out later in section 1.3. And finally, if square root were two-valued, then why would the quadratic formula explicitly bother to show both signs?

Sample Homework Problems:
(Code—L: Listed; M: Mixed; N: Not Listed. Please see the first page for more on this, which we won't repeat again.)

L: 5, 10, 11, 15, 17, 23, 26, 29, 31, 33, 42, 44, 51
M: 1, 4, 5, 7, 11, 14, 19, 20, 26, 32, 33, 42, 51, 57
N: 1, 2, 6, 14, 18, 21, 30, 34, 38, 41, 50, 52, 61

1.2 Lines in the Plane

Main Topics: Slope of a line. Equations of a line (point-slope, slope-intercept, and the special cases of vertical and horizontal lines). Parallelism and perpendicularity. If time allows, best-fitting line.

Main Ideas: This implements the analytic geometry as applied to lines. Slope is the intermediary, a constant of proportionality. It is a crucial topic for calculus.

Warnings and Pedagogy: You might challenge class with the question: "How do you know that the slope of a line, which is calculated based on two representative points, does not depend on the points chosen?" This is *well-definedness*. To demonstrate this, take two sets of two points each. Now one obtains similar triangles with corresponding sides in proportion. (*Lagniappe*: It makes a nice use of high-school geometry.)

Spend some time focusing on horizontal and vertical lines. Be sure to point out the difference between zero slope and no slope, and 0 vs. division by 0.

Sample Homework Problems:

L: 1, 5, 22, 29, 30, 34, 45, 51, 54, 58
M: 1, 4, 7, 15, 21, 30, 40, 45, 47, 51, 54, 57, 60. Problem 47 is especially charming in relating Celsius and Fahrenheit.
N: 2, 6, 17, 19, 39, 44, 52, 57, 60, 65

1.3 Functions

Main Topics: Function definition, notation, substitution; domain, range; function equality; function composition; types; piecewise-defined functions; difference quotient.

Goal: Integrate the various ways of looking at functions, including numerical (substitution), symbolic (*e.g.*, the difference quotient), and graphical (mapping diagrams, function graphs). In fact, keep this trinity in mind *always*!

Main Ideas: A function may be regarded as a correspondence between two sets such that to each element of the first set (the domain), there corresponds precisely one element of the second set (the range). Equivalently, a function is a relationship between two variables, say x and y, such that to each value of x there corresponds exactly one value of y.

Since functions are often presented without explicit naming of the domain, students need to be able to find the domain (*the natural domain convention*): Given $f(x) = $ formula in x, the domain of f is understood to be the set of all reals x for which $f(x)$ is defined and real. *Exception*: Physical considerations may override this understanding, such as having positive lengths, or time beginning at $t = 0$.

Expressing certain functions as a composition, akin to factoring, anticipates the chain rule later.

Students should spend enough time calculating the difference quotients of functions such as constants, linear functions, quadratics, and others. It represents the average rate of change and is crucial for the derivative.

Warnings and Pedagogy: The use of equations is typical for presenting functions, but not the only way. Students need to see that functions may be presented in equation form, but not every equation in x and y represents a function. This may be handled algebraically in many cases by trying to solve for y in terms of x: Thus, $x + 2y = 1$ yields a function, but $x^2 + y^2 = 1$ does not. Geometrically, the vertical-line test applies (next section, 1.4).

Students should be made familiar with the functions commonly encountered in calculus, including constants, linear functions, powers. Good variety (greatest-integer, absolute-value, others piecewise-defined, trigonometric, exponential, etc.) is a good antidote to thinking all functions are nice, algebraic, and/ or smooth.

Although constant functions should be easiest, students have trouble with them, especially for difference quotients.

Sample Homework Problems:

L: 1, 6, 10, 11, 15, 17, 24, 31, 38, 43, 48, 53
M: 1, 6, 7, 10, 11, 13, 15, 19, 21, 27, 33, 35, 49, 52.
Note that #52 provides a nice biological example - Poiseuille's Law for blood flow speed in an artery - that the students can revisit later with other topics. One need not rely solely on traditional one-dimensional projectile motion applications.
N: 2, 12, 13, 16, 25, 27, 33, 39, 49, 52

1.4 Graphs of Functions

Main Topics: Graphs; vertical-line test: When is a graph that of a function? Graphing tools: intercepts, symmetry, transformations.

Goal: Point-plotting alone is tedious, so show how tools make graphing quicker or more effective (*e.g.*, identifying salient features).

Main Ideas: The graph of a function f is the set of all ordered pairs (x, y) that satisfy the associated equation $y = f(x)$. Thus, a table of values is the first natural step, as it uses the very definition: for each x, get a unique y.

Technology, Warnings, and Pedagogy: Some calculator or computer usage may be in order for graphing, but appropriate caveats should be made about limitations. Students should be made to feel that they *do* need to be familiar with basic function graphs even when they have technology. Similarly, they should develop a sense for when to use a calculator – yes for complex calculations (*e.g.*, for large numbers, transcendental functions) vs. no for mental calculation (*e.g.*, "simple" numbers, simple arithmetic).

Symmetry tests do not cover every kind, but the most commonly considered. Students need to understand that "even" and "odd" are not opposites, and most functions are neither. (Every function can be written uniquely as a sum of an odd and an even function.)

Sample Homework Problems:

L: 5, 6, 19, 20, 26, 33, 43, 50, 52, 57
M: 1, 2, 7, 11, 25, 29, 36, 45, 47, 55, 56, 59
N: 1, 2, 7, 17, 20, 22, 27, 29, 45, 55, 59, 62
For sets M & N, note that #2 and #7 represent equal functions.

1.5 Inverse Functions and Inverse Trig Functions

Main Topics: When is a function invertible, and what does this mean?

Goal: Apply criterion in general and specifically to the trigonometric functions.

Main Ideas: Invertibility manifests itself geometrically via the horizontal-line test. Note the inverse relation to the vertical-line test, which determines whether, for each x, there is exactly one y. The horizontal-line tests asks whether, for each y, there is only one x.

Confirming f and g are inverses is easy, a matter of function composition: $f(g(x))$ should equal x for all x in the domain of g, and $g(f(x))$ should equal x for all x in the domain of f. However, given f, *finding* its inverse (denoted f^{-1} rather than g) requires inverting – solving for x in terms of y, and then seeing whether one has a function. (In practice, one re-labels by interchanging x and y so that x is always the name of an element in a domain.)

Warnings and Pedagogy: The trigonometric functions are periodic and hence all fail the horizontal-line test. However, by suitably limiting the domain of each to a principal domain, we obtain functions that pass the test. Thus, $\sin^{-1}$ is the inverse of the restricted (principal) sine, not of the original sine.

By graphing sine, cosine, and tangent, students not only gain valuable familiarity, but can see for themselves how to restrict domains to obtain the one-to-one property from the horizontal-line test – and thus, invertibility. One might mention the preference for picking a principal branch near the origin, and, where there is a choice between left and right of origin, pick the right (*i.e.,* the positive side).

This is a good time to remind students that trig functions are like others, and so one inputs a number (which just happens to be equivalent to a number of radians as well as a winding number) and gets out another number. We may find it occasionally useful to think in degrees, but it is not the standard for the course. Various limits and derivatives presuppose real numbers. Reassure students that this is equivalent to using radian measure, not degrees.

Analogy with square and square root (consider both algebraically and geometrically) may help students understand why inverting does not always produce

expected result. Thus, we have $(\sqrt{x})^2 = x$, but $\sqrt{x^2} = |x|$, not x – a semi-inverse property. Similarly, $\sin(\sin^{-1} x) = x$, but $\sin^{-1}(\sin x)$ need not equal x. The restriction of domains explains what happens.

Sample Homework Problems:

L: 5, 10, 13, 26, 29, 35, 37, 45, 47, 50
M: 3, 9, 13, 17, 19, 25, 33, 35, 37, 39, 51, 53.
Note that #51 anticipates the classic problem of finding the optimal position for viewing a painting. #53 is charming as well.
N: 4, 9, 11, 16, 20, 25, 42, 51, 53, 60.
Students may have difficulty with #60, a proof.

1.6 Exponential and Logarithmic Functions

Main Topics: Definition of exponential function for given base, and logarithmic function as its inverse. Base e, natural logs. Change of base. Continuously-compounded interest.

Main Ideas: Since exponential function graphs pass the horizontal-line test, the exponentials are indeed invertible. Their inverses are logarithmic functions. We use mostly base e and base 10.

Warnings and Pedagogy: Many students feel very uncomfortable with logs. A log is just an exponent; $\log_b N$ is simply the power to which b must be raised in order to produce N. Introduce notations *log* and *ln* (no subscripts). Use contrived examples to illustrate: $\ln e$, $\log 100$.

Use the fact that the range of any exponential function is the set of positive reals to explain why one cannot take log of a negative or zero. However, one can obtain zero or a negative answer. This reinforces further the need to be clear about input vs. output.

The change-of-base theorem nicely builds on properties of exponents and logarithms, but one ought to do some examples before formalizing as a theorem. One can then cite this theorem to explain why computer languages (*e.g.,* dialects of BASIC) do not require more than one base of logarithms. (That calculators frequently use both base 10 and e is for convenience, not necessity.)

The change-of-base theorem for logarithms explains why each log function is a constant multiple of ln:

$$\log_b x = \frac{\ln x}{\ln b}$$

(This anticipates why the derivative of a log function is a constant times derivative of ln.)

<u>Note</u>: Use your own judgment about whether to mention the conflicting uses of "log" in undergraduate texts vs. professional texts and computer languages (base 10 in former case, base e in latter cases).

Sample Homework Problems:

L: 14, 17, 21, 23, 24, 35, 39, 43, 57, 61, 64, 67, 69
M: 1, 13, 14, 21, 27, 37, 39, 40, 43, 55, 59, 61, 69
Despite humorous approach to #69, Newton's Law of Cooling has been used to estimate time of death – a rich use of mathematics.
N: 1, 3, 15, 16, 27, 29, 37, 41, 56, 59, 63

CHAPTER 2: LIMITS AND CONTINUITY

2.1 What Is Calculus?

Main Topics: Limit concept introduction. Derivative, tangent. Integral, area. Modeling.

Goals: Give sense of what a limit is. Also, show the spirit of mathematical modeling.

Main Ideas: This section is designed to give a broad overview of the limit ideas that serve as the underpinning of calculus – the forest, not the trees; the macro view, not the micro view. Without it, the unity of ideas may be missed or obscured. The important topic of mathematical modeling anticipates the many applications in the book and thus is essential.

Warnings and Pedagogy: Although this section does not delve too deeply into any one topic, seeing too many at once can be overwhelming. Keep the unity of limit concept(s) in mind as you pick and choose from among the topics and homework. You probably should not try to cover every topic, but nor should you spend too much time on any one.

Sample Homework Problems:

L: 1, 9, 17, 22, 29, 37, 47, 49, 52, 57
M: 1, 7, 13, 16, 17, 21, 22, 24, 29, 31, 56. (Technically, some sequences shown are not really well-defined, because no finite number of terms uniquely specifies a sequence. For instance, giving 1, 1.4, 1.41, 1.414, ... need not specify $\sqrt{2}$. However, it is a judgment call as to whether or not to mention this.)
N: 2, 5, 7, 14, 18, 23, 30, 48, 56

2.2 The Limit of a Function

Main Topics: Limit as intuitive notion, by graphing, by table; existence vs. non-existence; formal definition of limit of a function.

Goals: Convey the sense of limit from tables and graphs, and a bit about existence of limits. Don't forget to show the geometric picture.

After next section (2.3), students will focus on the means of calculation and the appropriateness of each method. For now, focus on definition.

Main Ideas: What is the behavior of function f as x approaches c? How does one express it? How does one calculate limits? Why does one?

Warnings and Pedagogy: Students usually have an urge to substitute c for x in considering the limit of $f(x)$ as x approaches c. Show examples where this gives correct answer, but also show examples (say, by table) such as

$$\lim_{x \to 0} [[5 - x^2]]$$

(Note: $[[x]]$ is the greatest-integer in x).

The answer is 4, not 5. Consider the above with x^2 replaced by x, and do also for one-sided limits. One obtains two different one-sided limits; the regular two-sided limit does not exist.

The students, ripe for something better than making tables, should then be receptive to ideas about when substitution is valid (next section, 2.3). Until then, remind them that the limit is not affected by the value of $f(c)$; it is the function values $f(x)$ for x "near" c that are key.

Distinguish between non-existence due to unbounded behavior vs. due to oscillatory behavior, especially for x approaching infinity.

Technology Notes: This is a fine place to use calculators freely as a supplement for function values and graphs.

Sample Homework Problems:

L: 6, 9, 13, 22, 24, 28, 29, 30, 37, 42, 51.; might add 58 or 59
M: 1, 9, 12, 16, 22, 24, 33, 35, 41, 51, 53; might add 58 or 59
N: 2, 8, 12, 16, 27, 35, 43, 45, 52

2.3 Properties of Limits

Main Topics: Limit theorems; calculating limits by algebraic techniques.

Goal: Handle "algebraic limits" algebraically.

Main Ideas: Clarify algebraic vs. transcendental (non-algebraic) functions. Give definitions.

Theorems now show that algebraic functions are amenable to substitution (as x approaches c, $f(x)$ approaches $f(c)$, which we only later show is the essence of continuity). These take the form of properties for limits of sums, products, powers, etc. Apply these to polynomials and rational functions. For the latter, factorization followed by cancellation plays a big role, particularly with "0 / 0" form.

Warnings and Pedagogy: Once students advance to limits of quotients, they must be reminded that one does all the algebra *before* passing to the limit.

Be alert for interesting algebraic weaknesses, including confusions with cancellation (such as getting 0 instead of 1 after canceling x/x). Distinguish among limits that lead to 0/0, a/0, and 0/a. The last expression is zero, the middle expression is undefined (so the limit that yielded it does not exist), and first one is *indeterminate*. This generally implies incomplete cancellation. Again, algebra must be done first.

Since the techniques of the section apply most readily to algebraic functions, clarify the difference between *algebraic* and *transcendental* functions.

Quotients remain particularly important for the derivative, which is the limit of the difference quotient of a function.

Mix in lots of types of limits and functions, including piecewise-defined functions, which connect nicely to one-sided limits.

Notational Notes: Watch out for the student proclivity not to know when to write "*lim*" and when it is no longer needed. On limits of sums, differences, and negatives, watch out for missing parentheses, another common student error.

Sample Homework Problems:

L: 3, 9, 13, 15, 18, 21, 30, 36, 40, 44, 47, 59
M: 1, 3, 14, 23, 33, 40, 45, 49, 54, 55, 63
N: 5, 9, 14, 17, 23, 34, 38, 45, 48, 55, 57

2.4 Continuity

Main Topics: Notion & definition of continuity; continuity at a point and on an interval; theorems.

Goals: Be able to tell whether a function is continuous (at a point, on an interval); be able to tell where a function is continuous; understand some of the implications of continuity.

Main Ideas: Start with the intuitive idea of discontinuity (much as psychologists study abnormality to get a handle on normality). Look at the situations that lead to discontinuity. They can be placed in three categories:

- function not defined at $x = c$
- limit of function does not exist at $x = c$
- aforementioned limit may or may not exist and / or function may or may not be defined at $x = c$, but these two values do not match.

From this one arrives at the definition of continuity at a point. The last condition encapsulates the previous two.

Extend this to intervals (open, closed, semi-open/ semi-closed). Consider end points, one-sided continuity, and the important question of when a discontinuity is removable.

Apply continuity theorems to "combinations": sums, products, etc. as well as compositions to conclude continuity. Also apply intermediate-value theorem to continuous functions, notably to locate zeros. However, use counter-examples such as $f(x) = \lfloor x \rfloor - 0.5$ to show that the hypothesis of continuity is essential.

Identify suspicious points, where continuity may fail. (What about non-suspicious points?)

Warnings and Pedagogy: Students have many difficulties with continuity. Go through the three steps for verification of continuity at a point. Remind them that there are different reasons why the second step, existence of the limit, may fail to hold, such as different one-sided limits, function becoming unbounded, etc.

Build a catalogue of continuous functions, such as polynomials, rational functions (except where denominator is zero), and so on.

To make sure students learn the issues, vary the questions slightly. For instance, ask about continuity not only at a suspicious point of a piecewise-defined function (such as 0 for the absolute-value function) but at other points, too.

Thinking Outside the Box: As time allows, incorporate seemingly obscure, humorous, or clever challenges: *Must there have been some time in your life when your height (h) in inches exactly matched your weight (w) in pounds?* [*Hint*: At t_0 = time of birth, $h(t_0) > w(t_0)$, but years later, at time t, $h(t) < w(t)$. Now apply the intermediate-value theorem to $d(t) = h(t) - w(t)$.]

Sample Homework Problems:

L: 9, 13, 19, 25, 28, 31, 39, 47, 55, 57
M: 1, 3, 7, 9, 11, 13, 29, 31, 43, 45, 55
N: 8, 11, 14, 24, 32, 40, 43, 45, 58

Additional Exercises are in the Chapter 2 Review. We recommend doing all of the *Proficiency Exam* questions, but include at least those in the *Survival Manual*, *viz.*: 12, 15, 18.

Comment: If you skipped chapter 1, note *the Spy Problems* throughout the text, such as #50. Also note the research possibilities in the *Historical Quests*.

CHAPTER 3: DIFFERENTIATION

3.1 Introduction to the Derivative: Tangents

Main Topics: Tangent lines, slope of tangent; the derivative; existence; differentiability's relation to continuity; notations for derivatives.

Goals: Motivate limit definition of derivative and calculate derivatives by this definition. Develop notations reflecting different usage.

Main Ideas: The idea of tangent to circle as line touching at only one point does not extend well. So, develop idea of tangent line as limiting line of a sequence of secant lines. Note that tangent and its slope are actually being *operationally defined*. That is, two things are going on. Not only is something being calculated, but it is also being defined in the first place. There are several such operational definitions in calculus. (Second example: area by integrals - later.)

The slope of the secant to the graph of f is given by the difference quotient:

$$\frac{f(x+h) - f(x)}{h} \quad \text{or} \quad \frac{f(x + \Delta x) - f(x)}{\Delta x}$$

and the slope of the tangent is given by the derivative, which is the limit of this quotient:

$$\lim_{h \to 0} \frac{f(x+h) - f(x)}{h} \quad \text{or} \quad \lim_{\Delta x \to 0} \frac{f(x + \Delta x) - f(x)}{\Delta x}$$

Usual concerns and considerations for limits apply:

simplifying algebra before passing to limit, determining whether limit exists, noting that h or $\triangle x$ is close to 0 but not equal to 0, etc.

Apply to get derivatives of various functions, especially algebraic ones. Apply to get equations of tangent lines and normal lines.

Notational Notes: The symbols h and $\triangle x$ are fairly interchangeable. However the difference quotient and the derivative of a function h would not allow h for the change in x. On the other hand, the use of $\triangle x$ notation requires parentheses when $\triangle x$ is raised to a power.

Warnings and Pedagogy: Have students memorize the definitions, but "stir the alphabet soup" (meaning: vary the letters). Students are initially very literal and need to see the structure, not just memorize names with fixed letters. The difference quotient is just a quotient of two differences: the change in y over the change in x.

The difference quotient gives the slope of a secant line and the average rate of change of the function. The derivative gives the slope of a tangent line and the instantaneous rate of change of the function. The geometric views are in this section; the physical views appear in 3.4. Some instructors like to tip off the students to the second interpretation earlier. Here is a concise summary:

	Geometric	Physical
Difference Quotient	m_{sec}	average rate of change
Derivative	m_{tan}	instantaneous rate of change

Sample Homework Problems:

L: 3, 8, 13, 19, 28, 33, 35, 39, 42, 43, 46; add 52 or 53
M: 1, 3, 9, 15, 23, 37, 41, 45, 46, 52, 69
N: 1, 10, 16, 27, 37, 41, 46, 47, 51, 69

3.2 Techniques of Differentiation

Goal and Main Topics: Develop shortcut theorems for the derivative of these functions: constants, linear functions, constant powers of x, sums & differences, polynomials, products, and quotients.

Main Ideas: Rather than crank out limits of difference quotients over and over, we do so once in general to prove general results. Implicit in these results are that the functions being differentiated are, in fact, differentiable, a fact that should be mentioned now and then.

Since the derivative of a function is again a function, one may take its derivative, thus leading to the higher derivatives. These will be interpreted later.

Warnings and Pedagogy: Introduce students to one of the great strategies of mathematics in general and calculus in particular: changing form. For instance, stress that $f(x) = x/2$ should <u>not</u> be differentiated by the quotient rule. Rather, $f(x) = 0.5x$ and use the scalar-multiple rule instead. This reminds of calculation more than one way.

Discuss these. However, watch for errors stemming from looking at structure without understanding, such as seeing the derivative (with respect to x) of 2^{10} as $10*2^9$.

As in the last section, vary names of letters for both variables and functions. Vary derivative notations used. Discuss pros and cons of each. Distinguish notations for derivative in general and at a specific input value.

Now that students will be able to find derivatives faster, the exercises can be a bit less contrived and the students are freer to pursue applications.

Sample Homework Problems:

For some of these, challenge students to calculate two ways, and compare. You might do #61 in class (the extension of the product rule to three functions).

L: 5, 8, 14, 15, 18, 23, 26, 31, 33, 41, 52; add 56 or 59
M: 6, 8, 13, 16, 17, 21, 23, 32, 41, 42; add 56 or 59
N: 7, 9, 17, 19, 21, 21, 32, 43, 48, 55

3.3 Derivatives of Exp, Log, Trig Functions

Main Idea: Differentiate elementary transcendentals.

Goals and Main Topics: Develop the derivatives of the trigonometric, exponential, and logarithmic functions, and use the results to practice getting derivatives of functions involving them.

Pedagogy: Develop the derivative of sine by limit of difference quotient, which in turn will rely on the use of two lemmas (little results) on limits. Next comes the derivative of cosine. The other four trig functions (circular functions) are derivable from derivative of sine and cosine via identities relating them to sine and cosine, such as the defining one for tangent. Continue to practice other "rules" (products, quotients, etc.) in conjunction with these new results. Then do the same with exponentials and logarithms. Vary to get a good mix.

Warnings and Technology: This is a good time to remind students that identities allow for possibility of more than one expression for a correct answer. This point has even more force today with technology. (See the Technology Window that precedes the Problem Set.)

Sample Homework Problems:

L: 3, 5, 7, 17, 20, 28, 43, 49, 56, 58, 63
M: 1, 3, 9, 11, 17, 33, 54, 56, 60
N: 1, 2, 10, 11, 19, 33, 44, 55, 60

3.4 Rates of Change; Rectilinear Motion

Main Topics: Average and instantaneous rates of change; one-dimensional motion, especially falling-body problems; relative rates of change.

Main Ideas: Segué from geometric picture of rates of change and derivatives to the physical: position, velocity, speed, and acceleration. Remember to treat these functions as functions with notation and language.

Warnings and Pedagogy: Continue to use and incorporate

T6

both Leibniz and Newton notation for derivatives. Watch out for the student tendency to divorce the derivative earlier from now, or to be unclear about using shortcuts vs. the limit definition. Show the unity. Also emphasize the difference between function or derivative on the one hand, and function value or derivative value on the other. For instance, $v(t) = s'(t)$ is the velocity in general at time t, whereas $v(3) = s'(3)$ is the velocity at time $t = 3$. Watch out for student proclivity to write "t(3)" – as if time t were a dependent variable.

Emphasize again the difference between average rate of change and instantaneous rate of change. This might be a good time to show the box summary of the differences in interpretation (geometric vs. physical) and instantaneous vs. average. (See 3.1 of these Notes, previous page.) Build on the connection between these views, as well as such ideas as the graphical view of velocity as the slope of the curve on position-time graph.

Potentially confusing is the distinction between position and distance. Walk along a line a few meters left and then walk back. Ask class about your old position, new position, displacement, and average velocity (zero!).

Sample Homework Problems:

L: 3, 9, 13, 15, 21, 25, 29, 36, 40, 44, 51
M: 1, 5, 11, 12, 15, 23, 26, 29, 31, 44, 46, 49
N: 2, 4, 11, 12, 14, 16, 26, 30, 31, 39, 46, 49

3. 5 The Chain Rule

Main Topics: Chain rule, implications, proof.

Goal and Main Idea: We extend our differentiation repertoire to include composite functions.

Warnings and Pedagogy: Once we have the chain rule, each derivative result has an extended version. The derivative (with respect to x) of x^n is nx^{n-1}; so, for u^n with u a function of x, it's $nu^{n-1}u'$. Rather than memorize two different sets of formulae, show the essence of the chain rule and develop this mnemonic for generalizing: To get chain version of any result, change x to u and multiply by u' in the answer. Thus, the derivative of sin x is cos x, so the derivative of sin u is (cos u) u'.

The general power rule extends the regular power rule but is a special case of the chain rule. The general trig rules extend the regular trig rules, but each is a special case of the chain rule. You might wish to show a Venn diagram. For a general power, you might wish to treat both ways (general power, chain) to compare and contrast. In either case, the "u" chosen is same.

Continue to work with other results. Stress the difference that order makes: $\sin^5 x$ vs. $\sin x^5$.

Notational Notes: Need for parentheses and order in products can be tricky. Stack "simpler" factors on the left: $x \sin x$, not $\sin x \, x$. Second, point out the different understandings of meaning: sin x ln x suggests a product, not sine of a product, whereas sin $2x$ is definitely sine of

$2x$. *Caution about inconsistency*: Contrast the appearance of derivative of u in cos u u', where u' is *not* multiplying u, vs. the specific expression cos $x^2 x$. We must rewrite this!

Sample Homework Problems:

L: 5, 11, 14, 16, 18, 23, 29, 35, 41, 46, 54, 57
M: 1, 3, 16, 17, 19, 23, 33, 39, 40, 46, 53
N: 2, 4, 17, 19, 24, 33, 36, 39, 40, 47, 53

3.6 Implicit Differentiation

Main Topics: Implicit differentiation concept and solution procedure; derivatives of inverse trigonometric functions; logarithmic differentiation.

Goals, Main Ideas: Thus far, given $y = f(x)$ as *explicit* function of x, we've found derivative $y' = f'(x)$ *explicitly*. Now we are given an equation in x and y that may *implicitly* define y as one or more functions of x. For any one such function, we wish to find the derivative *implicitly*.

Apply these ideas to getting derivatives of inverse functions such as the inverse trigonometric functions. Develop logarithmic differentiation; apply particularly to functions of form $f(x)^{g(x)}$. *You might enjoy deriving the derivative of this general form. The answer is akin to the product rule, but you have to investigate and look very closely to see it!*

Warnings and Pedagogy: Where possible, such as equation of a circle or ellipse, compare results of implicit and explicit differentiation. Note that the appearance of both x and y in answers actually makes sense, as a value of x need not determine a single value of y from the equation.

Notational Notes: You still need to have your notation distinguish between derivative at *any* point (x, y) on curve and specific point on curve.

Sample Homework Problems:

L: 1, 5, 11, 17, 19, 31, 34, 41, 45, 52, 56, 61
M: 1, 5, 7, 9, 15, 17, 27, 34, 54, 61, 62, 67
N: 2, 7, 8, 9, 15, 27, 33, 54, 55, 62, 65

3.7 Related Rates and Applications

Main Topic: Related rates and applications.

Main Idea: If two quantities are related, how do their rates of change relate to each other?

Warnings and Pedagogy: The most important point is to consider what happens at *any* time, and draw diagram with that – not the particular instant around which the question revolves. (Analogy: If you were calculating $f'(3)$, would you calculate $f(3)$ first and then differentiate?) Keep in mind that time usually does not show up explicitly. Implicit differentiation and the chain rule underlie much of the calculating to be done. Make sure students are versed in needed geometry:

- Pythagorean theorem (falling ladders, moving vehicles)
- Volumes & other measures (spheres, cones, cylinders)
- Similarity of triangles (fluids in cones).

3.8 Linear Approximations and Differentials

Main Topics: Tangent lines as function approximations; differentials; errors; marginal analysis; Newton's method for approximation of zeros.

Main Ideas: Revisit tangent line but with view as approximator to curve locally. Thus, one may easily approximate a function value otherwise difficult to get, but it's not exact. Note the tradeoff in this!

Pedagogy: To estimate $f(x_0)$: Locate "nice" value $x = a$ near x_0 and find the equation of the tangent line at $x = a$: $L(x) = f(a) + f'(a)(x - a)$. Now argue $f(x_0) \approx L(x_0)$. Since L is linear, the latter value is easy to calculate.

Assign meaning to symbols dx, dy such that for $y = f(x)$, $dy \div dx = f'(x)$. Now dy approximates the actual change Δy, so we can estimate errors.

Similar remarks apply to relative change and relative rates of change. The relative change in any quantity q is $\Delta q / q \approx dq / q$. Note that in many problems the relative change or percent change does not depend on specific values. The steps are three-fold and simply implement the definition:

1) Write function/ formula for quantity.
2) Calculate its differential.
3) Divide 2) by 1).

As an example, show that percent change in volume of a cube is three times percent change in edge: $V = e^3$, so $dV = 3e^2 \, de$, and $dV / V = 3 \, de / e$.

Warnings: In example above, treat dV / V and de / e each as a single quantity. Pay attention to units and dimensional analysis. Students often confuse absolute quantities such as e with relative ones such as de / e.

Other: In marginal analysis, compare approach using $C'(x)$ with one using differential, dC. The similarity stems from having $dx = 1$ here.

Sample Homework Problems:

L: 1, 7, 11, 15, 21, 28, 31, 32, 35, 44, 46, 55
M: 1, 5, 19, 23, 24, 28, 35, 46, 53, 58
N: 5, 9, 19, 24, 25, 33, 38, 47, 48, 53

Additional Exercises are in the Chapter 3 Review. We recommend doing all of the *Proficiency Exam* questions, but include at least those in the *Survival Manual*, viz.: 21, 24, 27, 30, 33.

This might be a good time to note the *Group Research Projects*.

CHAPTER 4: ADDITIONAL APPLICATIONS OF THE DERIVATIVE

4.1 Extreme Values of a Continuous Function

Main Topics: Extreme-value theorem, absolute extrema; relative extrema; optimization.

Main Ideas: Define extrema (absolute, relative). Lead to extreme-value theorem for absolute extrema and how to locate them. Consider relative extrema with critical values and points.

Notation and Pedagogy: Contrast absolute and relative extrema (akin to the best mathematician in the world vs. best mathematician on campus). Develop theorem that continuous function on closed, bounded interval attains both an absolute maximum and absolute minimum on interval. Show hypotheses essential by counterexamples.

In practice, look for the attainment at a critical value of either type (first derivative zero or undefined) or at an endpoint.

For relative extrema, look for critical values. Keep in mind that this section is preparatory not only for graphing, but also for the "max-min" applications, the most exciting in calculus.

Technology: Extrema are highly visual, so do use graphical calculators and math software.

Sample Homework Problems:

L: 3, 11, 16, 17, 20, 23, 28, 31, 35, 40, 47, 52, 59
M: 3, 4, 9, 16, 19, 23, 31, 49, 50, 53, 59, 61, 62
N: 4, 8, 14, 19, 24, 30, 34, 49, 51, 53, 58, 60, 61

4.2 The Mean-Value Theorem (MVT)

Main Topics: MVT, proofs, and applications.

Main Ideas: If you averaged 60 mi/h during a trip, you must have hit exactly 60 mi/h at one instant during this trip. If the slope of a certain secant line to graph of f over $[a, b]$ is m, then there is some tangent line to f at $(c, f(c))$ for c between a and b that has the same slope m. This is same as saying that any secant line drawn will have a parallel tangent (corresponding to an intermediate x-value).

A special case of the MVT is Rolle's theorem, used to boot-strap the proof of the general MVT. The MVT provides the means to show that *only* constants have a zero derivative, which in turn implies that two functions with equal derivative differ only by a constant. This is important several times in integral calculus.

Sample Homework Problems:

L: 5, 11, 18, 23, 29, 32, 33, 42, 45, 50, 54
M: 3, 9, 12, 18, 23, 29, 33, 38, 39, 40, 49
N: 3, 10, 12, 14, 25, 30, 35, 38, 39, 43, 49

4.3 First-Derivative Test

Main Topics: Increasing, decreasing behavior; first-derivative test; curve-sketching using test.

Main Ideas: Notions of increasing, decreasing behavior of functions on intervals. Define notion; then lead to theorem relating to sign of derivative: f is increasing on any interval on which $f' > 0$, and decreasing where $f' < 0$.

Pedagogy: Use critical values to find where sign of derivative changes. Test! Not every critical value yields a change in sign of f'; take $f(x) = x^3$, x^5, or x to any higher odd power. This leads to the first-derivative test for relative extrema.

The first-derivative test alone is highly potent in getting the most important features of a graph. The second derivative (next section) should be seen as "touch-up" that fine-tunes.

Notes: A function whose continuous derivative has only a single sign change has a relative extremum that is actually an absolute extremum. *Also, the analysis yields the graph, not reverse!*

Sample Homework Problems:

L: 3, 7, 9, 15, 22, 27, 32, 42. 44, 49, 51, 57
M: 3, 11, 13, 15, 18, 20, 22, 27, 34, 44, 50,51
N: 5, 11, 14, 18, 20, 24, 28, 33, 34, 35, 50, 54

4.4 Concavity and the Second-Derivative Test

Main Ideas: Concavity, inflection points, curve-sketching, second-derivative test.

Main Topics and Pedagogy: Define concavity and relate to sign of second derivative. At points where sign of f'' changes, concavity changes, yielding inflection points. Note analogy to testing first derivative: It is not enough that $f''(a) = 0$ to ensure an inflection point at $x = a$.

Use concavity along with previous section's tips to improve upon graphing techniques. This is an additional tool more than a competing one.

Perspective and Balance: Second-derivative test provides an alternative test for relative extrema, but fails when $f''(a) = 0$. It also misses instances where there is actually an absolute extremum that the first-derivative test finds.

Sample Homework Problems:

L: 9, 14, 23, 34, 40, 45, 47, 49, 52, 56, 59
M: 1, 6, 9, 13, 20, 23, 24, 38, 40, 52, 53, 56
N: 2, 7, 12, 13, 19, 20, 38, 41, 48, 53, 57

4.5 Limits Involving Infinity and Asymptotes

Main Topics: Limits to infinity, infinite limits, graphs with asymptotes, more graphing strategy.

Main Ideas: What happens as x approaches (plus, minus) infinity? What happens when x approaches a and we get y approaching infinity? Manipulate to calculate limits with infinities. Use results to obtain asymptotes in many cases.

Pedagogy: Don't forget strategy of changing forms. Also, combine techniques with previous for improved graphing. Rational functions are most amenable to analysis here:

Degrees of Num, Den	Asymptote
deg (num) < deg (den)	$y = 0$
deg (num) = deg (den)	$y = k$
deg (num) = deg (den) + 1	$y = ax + b$

where k = ratio of leading coefficients, deg denotes degree, num is numerator, den is denominator.

Sample Homework Problems:

L: 7, 10, 11, 13, 25, 29, 38, 42, 45, 47, 49, 55
M: 5, 7, 11, 19, 25, 27, 37, 47, 49, 50, 56, 57
N: 5, 8, 12, 19, 27, 31, 37, 47, 50, 56, 57, 60

4.6 Optimization in Sciences and Engineering

Main Topics: Max-min procedure, applications, Fermat's principle, Snell's law.

Goal and Main Idea: Model applications and use the optimization procedure outlined to solve a large variety of max-min problems.

Comment: This is one of the most useful and satisfying portions of calculus. Note how often symmetry and beauty characterize final solution.

Pedagogy: After modeling a problem, note the domain. If it's a closed, bounded interval and modeled function is continuous on it, the extreme-value theorem applies, so testing by the first-derivative or second-derivative test is unneeded. Instead, find the critical values and substitute them – as well as end values – into the objective function. Be alert as well for physical arguments for informal justification instead of more formal testing.

Warning: For actual modeling, students need to be careful not to confuse constraints (*i.e.*, side conditions) with the objective function (*i.e.*, the function that is to be maximized or minimized). Emphasize this during each problem setup.

Sample Homework Problems:

L: 6, 11, 12, 16, 18, 25, 26, 32, 33, 43
M: 3, 8, 12, 14, 15, 16, 18, 21, 35, 40
N: 3, 8, 13, 14, 15, 17, 21, 22. 24, 35, 40

4.7 Optimization in Business, Economics, and the Life Sciences

Main Idea: This is an extension of the previous section to the "softer" sciences. Use the ideas to obtain principles of marginal analysis, to apply to physiology, and to model discrete functions.

Warnings and Pedagogy: Watch for a tendency to confuse marginals with averages.

Sample Homework Problems:

L: 3, 7, 10, 12, 15, 22, 23, 28, 32, 38, 39, 43
M: 3, 5, 10, 12, 20, 23, 32, 36, 38, 40, 44, 50
N: 1, 5, 9, 11, 14, 20, 31, 36, 40, 44, 50, 51

4.8 l'Hopital's Rule

Main Topics: l'Hopital's rule, indeterminate forms,

alternate forms, special limits.

Main Idea: Substituting or getting the limit of a quotient may appear to yield 0/0 or other indeterminate form. It requires investigation.

Warnings and Pedagogy: You might show at least informally how the rule comes from difference quotients. Remind students that a limit *must* yield an indeterminate form in order to apply l'Hopital's rule, per example 7. Note example 1 and ask class whether this could be used as a proof that sin *x*/ *x* approaches 1 as *x* goes to 0. (No; this result was used to obtain derivative of sin *x*. To avoid circular reasoning, we can't then use the derivative to *prove* this!)

Sometimes more than one application of the rule is needed. Often, one must change the form, a generally valuable approach to problem-solving. However, remind students not to fall so in love with latest technique(s) that they abandon others previously learned.

Sample Homework Problems:

L: 4, 9, 15, 17, 21, 26, 30, 40, 43, 53, 57, 62
M: 1, 3, 9, 15, 17, 21, 27, 30, 40, 45, 54, 61
N: 2, 3, 7, 16, 22, 27, 33, 39, 45, 54, 58, 61

Additional Exercises are in the Chapter 4 Review. We recommend doing all of the *Proficiency Exam* questions, but include at least those in the *Survival Manual*, *viz.*: 15, 18, 21, 24.

CHAPTER 5: INTEGRATION

5.1 Antidifferentiation

Main Topics: Reversing differentiation; anti-derivative notation and formulas; applications.

Main Ideas: Thus far in course, we've started with a function and found its derivative. Now, we ask, given the derivative, recover the original function(s). The mean-value theorem corollary ("constant difference theorem") assures that any two anti-derivatives differ only by a constant. So, an anti-derivative is actually a class of functions. Graphically, they are "parallel" curves.

By reversing the theorems for differentiation we obtain anti-differentiation formulas such as the power rule, trig results, etc. Applications: Start with velocity to recover position function. From marginal revenue, recover revenue function (or related quantities). Area as anti-derivative.

Warnings and Pedagogy: The area connection to anti-derivative is rich in geometry, using difference quotient and definition of derivative.

Since the "rules" mirror those for differentiation, caution students about products and quotients. Remind all again of the utility of changing form.

Notational Notes: The reason for the indefinite integral notation is not clear until after the Fundamental Theorem of Calculus. Point out then not to *assume* a connection to definite integrals based only on similar notation.

Sample Homework Problems:

L: 3, 11, 17, 21, 25, 28, 30, 33, 39, 45, 48, 50
M: 1, 3, 7, 8, 11, 19, 21, 25, 30, 31, 39, 40, 50
N: 1, 4, 7, 8, 12, 19, 22, 26, 27, 31, 37, 47, 51

5.2 Area as the Limit of a Sum

Main Topics: Area as limit of sum; general approximation; sum notation; approximate areas by summation; special summation formulas.

Main Ideas: Develop "area under a curve" by approximating it using rectangles and *Riemann sums*.

Then take the limit of this approximation as the number of rectangles increases.

Warnings and Pedagogy: The definition of the definite integral is one of the most difficult in all of calculus. The sigma notation leads to the notation for definite integrals. The summation properties lead to the integration properties.

Technology: Riemann sums using mid-points provide especially good approximations.

Sample Homework Problems:

L: 3, 6, 9, 12, 15, 18, 24, 30, 34, 37, 42, 45
M: 1, 3, 8, 9, 12, 15, 20, 24, 28, 30, 31, 36
N: 1, 4, 8, 11, 19, 20, 25, 28, 29, 31, 36, 41

5.3 Riemann Sums and the Definite Integral

Main Topics: Riemann sum; definite integral; area as integral; properties of definite integrals.

Main Ideas: Formalize the Riemann sum. The limit of a Riemann sum is the definite integral. Continuous functions are integrable.

Warnings and Pedagogy: Under suitable hypotheses, the area under a curve is given by a definite integral. Students need to check that the function has only non-negative values over the interval in question. Moreover, not every definite integral is positive or represents an area.

Be very careful distinguishing displacement and distance, specifically for the integral of velocity vs. the integral of speed. Note that, at this point, the shortcut of anti-differentiation is not yet available.

Note: You may wish to compare the situation now to the one just after derivative had been introduced. There, one had to make tedious calculations involving limits, too.

Sample Homework Problems:

L: 3, 6, 14, 19, 22, 23, 26, 30, 33, 35
M: 1, 6, 11, 14, 19, 23, 24, 28, 35, 40
N: 1, 4, 11, 13, 18, 20, 24, 28, 36, 40

5.4 The Fundamental Theorems of Calculus

Main Topics: The two fundamental theorems of integral

calculus.

Main Idea: Just as differentiation rules greatly ease calculation, the first fundamental theorem reduces a tedious limit-of-Riemann-sum calculation to an anti-differentiation (with two substitutions and a subtraction).

Warnings, Notation, and Pedagogy: Warn students about dummy variables prior to discussing the second fundamental theorem. Note that the second theorem's proof actually contains an alternate proof of the first theorem, one that does not use the mean-value theorem.

Apply the new shortcut to a variety of problems: area, displacement, and distance traveled.

Contrast and Added Insight: Differential and integral calculus are inversely related. Make this stronger by noting that difference quotients are quotients of differences. Reverse the processes and order; we now have sums of products – Riemann sums!

Sample Homework Problems:

L: 1, 6, 11, 17, 21, 24, 29, 34, 37, 43, 45, 46, 50
M: 2, 3, 6, 11, 15, 19, 21, 23, 25, 29, 34, 43, 45
N: 2, 3, 7, 12, 19, 23, 25, 30, 33, 36, 38, 41, 48

5.5 Integration by Substitution

Main Topic: How to calculate certain definite and indefinite integrals by changing variables.

Pedagogy: After the fundamental theorem, we know to look for anti-derivatives. Sometimes we have to change the form of the integrand. At still other times, we need a change of variable.

The key is to run the chain rule in reverse. Look for a function u that is part of the integrand in either explicit or implicit parentheses, especially if the derivative of that part is essentially the rest of the integrand. Example: Take $u = x^2 + 1$ in

$$\int x\sqrt{x^2 + 1}\,dx$$

For indefinite integrals, be sure to convert fully to the new variable. For definite integrals, change limits to the new limits of integration.

Remind students to consider easier techniques (direct integration, changing form) *before* jumping to substitution. Mix in applications.

Sample Homework Problems:

L: 2, 5, 9, 18, 23, 28, 35, 41, 48, 52, 57
M: 1, 2, 3, 9, 23, 25, 33, 41, 46, 49, 50, 52
N: 1, 3, 10, 19, 25, 33, 36, 46, 49, 50, 51

5.6 Introduction to Differential Equations

Main Topics: Introductory overview and terminology; separable equations; exponential growth and decay; models for fluid flow and escape velocity; orthogonal trajectories; direction fields.

Main Ideas: Solve differential equations. Give nice problems leading to differential equations.

Warnings and Pedagogy: A differential equation is an equation involving a function (usually unknown) and one or more of its derivatives. For first-order equations, we look for separability first. Both separable and non-separable equations can be investigated by drawing the direction field. Watch for a certain student tendency to confuse independent and dependent variables when it comes to integration. You will need to show details on converting solutions given in an implicit logarithm form to the expected explicit form, as in $\dfrac{dP}{dt} = kP$ (which is separable).

Sample Homework Problems:

L: 3, 6, 9, 15, 23, 27, 31, 33, 36, 42, 45, 52, 56
M: 1, 3, 5, 9, 15, 21, 23, 27, 33, 41, 42, 45, 52
N: 1, 5, 7, 13, 16, 21, 24, 29, 41, 43, 44, 48, 53

5.7 The Mean-Value Theorem for Integrals

Main Topics: Mean-value theorem (MVT) for integrals and the average value of a function.

Pedagogy: Given the region bounded by the nonnegative function f over interval $[a, b]$, there is a real number c between a and b such that the rectangle of height $f(c)$ and base $b - a$ has the same area as the original region. The value $f(c)$ is also the average value of f on interval $[a, b]$. Derive the heuristic formula for average value by connecting to sampling and Riemann sums.

Interesting Reconciliation: Earlier, we saw that average velocity was given by an expression of the form $\dfrac{s(t_2) - s(t_1)}{t_2 - t_1}$ (a difference quotient).

Suppose we have the instantaneous velocity function $v(t) = s'(t)$ and ask for its average on the interval $[t_1, t_2]$. Do we get the same result?

Sample Homework Problems:

L: 1, 5, 13, 20, 21, 27, 34
M: 1, 5, 11, 13, 15, 21, 32, 33
N: 1, 6, 11, 14, 15, 19, 32, 33

5.8 Numerical Integration

Main Idea: Approximate integrals by rectangles (this is just the definition revisited), trapezoids, and parabolas; error estimation.

Sample Homework Problems:

L: 1, 4, 14, 19, 28, 32, 36
M: 1, 3, 14, 20, 25, 28, 32
N: 3, 6, 11, 20, 25, 29, 33

5.9 Alternative: The Logarithm as an Integral

Summary: Start with logarithm as an integral and derive the earlier results in reverse order. It's not as intuitive, but it is more rigorous and offers the advantage of regarding ln x as the area under the hyperbola $y = 1/t$ over $[1, x]$. From this view, the derivative of ln x is trivial.

Sample Homework Problems:

L: 1, 2, 7
M: 1, 4, 7 Students will probably need help with #7.
N: 4, 5, 6

Part of this chapter has been about reversal. Other ingenious reversals can deepen student understanding, such as this reverse question:

Draw a region whose area is given by $\int_1^4 2dx$.

Additional Exercises are in the Chapter 5 Review. We recommend doing all of the *Proficiency Exam* questions, but include at least those in the *Survival Manual*, viz.: 24, 27, 30, 33, 36.

CHAPTER 6: ADDITIONAL APPLICATIONS OF THE INTEGRAL

6.1 Area Between Two Curves

Main Topics: Area between curves; vertical strips, horizontal strips; economic applications.

Main Ideas: Previously, we considered the area of a region bounded by three lines and a curve. Now we generalize further to two vertical lines and two curves (curves given by functions of x), or two horizontal lines and two curves (given by functions of y).

Warnings and Pedagogy: Here is another opportunity to develop a Riemann sum that leads to a definite integral. Assume higher curve g and lower curve f over $[a, b]$. We partition $[a, b]$ and see that this induces quasi-rectangular regions, the i-th one of which has area $(g(x_i) - f(x_i))\Delta x$.

The previous "area under curve" scenario is just a special case of this area between curves. Note also that this topic subsumes the special case of "area above curve" (function's graph is completely below x-axis) as well. However, students tend to embrace the latest method. So, point out to continue using the earlier, simple method for "area under curve" when the easier situations arise.

Key steps for area between $y = f(x)$ and $y = g(x)$:

a) Set $g(x) = f(x)$ to get intersection points, if needed to obtain the interval for integration, and;
b) Determine the higher curve on this interval.

Similar remarks apply to functions of y, with appropriate changes. For instance, the "higher" curve is farther right.

Sample Homework Problems:

L: 3, 4, 6, 10, 12, 15, 18, 19, 26, 32, 37, 39
M: 1, 4, 5, 9, 10, 15, 18, 19, 22, 26, 37, 39
N: 1, 5, 9, 11, 14, 22, 24, 27, 30, 38, 40, 43

6.2 Volume by Disks and Washers

Main Topics: Volumes of solids of revolution, with methods based on cross-sections, disks, and washers.

Main Ideas: Again use the method of exhaustion with Riemann sums, but now for volume. If a solid can be viewed in such as way that its cross-sections are known, we can obtain the volume by integration. Let the x-axis be perpendicular to all the cross-sections, with the area of the face of the cross-section at x being $A(x)$. Then the volume of the thin slice with this face is approximately $A(x)\Delta x$, leading to the total volume being given by an integral of form $\int_a^b A(x)dx$. From this viewpoint, the method of disks is just the special case of circular cross-sections. Moreover, just as area between curves generalizes area under a curve, the washer method may be regarded as the directly analogous extension of the method of disks.

Pedagogy and Warnings: Students weak in geometry should first be introduced to the method of *uniform* cross-sections to understand how the volumes of rectangular solids and right-circular cylinders are special cases. Use dimensional analysis to check that the formulas actually give volumes. (This is a good practice in general.)

The integrand $\pi([f(x)]^2 - [g(x)]^2)$ is mistakenly seen by some students as equivalent to $\pi([f(x)] - [g(x)])^2$. This, of course, is an order-of-operations error, but watch for it.

Opportunity: This is a great topic allowing you to derive results that were previously just obscure formulas to your students. Case in point is the volume of a sphere: It presents a nice challenge to set up from scratch, as solvers must come up with the idea of regarding a sphere as the solid of revolution resulting from revolving a semi-disk about its diameter. For a = radius of sphere, watch out for student error in seeing the integral of a^2 as $a^3/3$ in lieu of a^2x.

Sample Homework Problems:

L: 3, 9, 19, 21, 27, 32, 33, 40, 41, 47, 53, 59
M: 1, 3, 9, 10, 19, 24, 32, 41, 45, 47, 51, 54
N: 1, 2, 7, 10, 20, 24, 25, 29, 37, 45, 51, 54

6.3 Volume by Shells

Main Topics: The method of cylindrical shells; summary of methods for computing volume.

Main Ideas: Sometimes it is easier or necessary to compute a volume by taking the approximating strip parallel to the axis of rotation instead of perpendicular to that axis.

Pedagogy: Although the formulas appear different, the key idea of integral calculus remains: Find the volume of a single approximating element (in this case, a shell), take the sum of these, and then the resulting limit is a definite integral. Note: It is important to give students a feel for the differences among the methods and when to use which. Use the table near the end of section as a guide.

Sample Homework Problems:

L: 3, 11, 16, 19, 24, 29, 33, 35, 39, 43
M: 1, 3, 10, 11, 17, 21, 25, 29, 33, 40
N: 1, 8, 10, 12, 17, 21, 25, 30, 34, 40

6.4 Are Length and Surface Area

Main Topics: The arc length of a graph; surface area of a surface of revolution.

Main Ideas, Pedagogy: By partitioning an arc into small segments, each of which can be regarded as nearly a chord, we approximate total arc length s as the sum of the lengths of these chords. This Riemann sum's limit is the definite integral that we operationally define as arc length.

Technically, we may take two views on the partitioning. Suppose the portion of arc is the graph of a function defined over an interval. We may directly partition the arc, but that is not really the definition of Riemann sum. We can get around this by noting that such a partition of the arc corresponds naturally to a partitioning of the interval. We also can get around this by first partitioning the interval, which in turn induces a partition of the arc. (A comparable situation will arise when we consider curves defined parametrically in a later chapter.)

For the surface area S of a surface of revolution, we approximate the surface area of a frustum of a cone. Note that the derivation here uses slant height based on arc length, thus linking two topics more than just notationally.

Comments: Surface area historically has proven itself troublesome. A good definition is essential to avoiding classic paradoxes. On the positive side, we again have the opportunity to derive classic results such as the classic formula for the surface area of a sphere.

Sample Homework Problems:

L: 3, 9, 13, 16, 18, 23, 25, 28, 30, 37
M: 3, 5, 13, 16, 18, 19, 25, 26, 30, 36
N: 1, 5, 10, 14, 17, 19, 26, 27, 36

6.5 Physical Applications: Work, Pressure, Centroids

Main Topics: Work; liquid pressure and force; centroid and moment of a planar region; the volume theorem of Pappus.

Main Ideas, Warnings, and Pedagogy: Work done by a variable force is given by a definite integral, extending the definition of work done by a constant force. This is completely analogous to the area under a curve generalizing the area under a horizontal line. (In fact, there is no difference in the mathematics formally.)

For some applications, such as Hooke's law, one needs to pay careful attention to the underlying interval. For others, such as pumping fluids out of a container, it is a good idea to consider thin "layers" of the fluid and the work done in lifting them to overcome gravity, and then summing. This approach is needed to get the formula for hydrostatic force.

Moments and centroids round out the applications. (They are considered in a more general context after multiple integration much later in this text. There are some obvious but useful mnemonic devices for the coordinates of the centroid of a lamina. The formulas are even simpler in that more general context, however.) As a particular application of center of mass, Pappus' theorem gives the volume V of a solid of revolution obtained by revolving a region of area A about an external axis. If the centroid moves s units in the course of the revolution, we derive $V = As$. Note the use of the shell method. A nice example is volume of a torus.

Sample Homework Problems:

L: 5, 7, 11, 17, 19, 26, 29, 31, 34, 41, 42, 47
M: 1, 5, 8, 11, 13, 16, 17, 25, 27, 34, 41, 47
N: 1, 4, 8, 10, 13, 16, 25, 27, 32, 33, 35, 40

Additional Exercises are in the Chapter 6 Review. We recommend doing all of the *Proficiency Exam* questions, but include at least those in the *Survival Manual, viz.*: 15, 18, 21, 24.

Cumulative Review: For additional practice of the material in the first six chapters, we recommend the following: 6, 9, 12, 15, 18, 21, 24, 27, 30, 33, 36, 39, 42, 45, 48.

CHAPTER 7: METHODS OF INTEGRATION

7.1 Review of Substitution and Integration by Table

Main Idea, Warnings, and Pedagogy: This section reviews changes of variables (substitution) encountered thus far. Although it is tempting to skip reviews so as to save time, students at this point are often beginning the second semester of calculus and have been out of first-semester calculus for one or more months. This section also lays the groundwork for deeper investigation of the standard techniques, including integration by parts, by partial fractions, trigonometric integrals, trig substitution.

It is important to remind students not to try to treat every problem as one suited for the latest technique(s) or just one particular one. Some integrations are amenable to standard rules, others require a change of form, and still others demand a change of variable. Students need to be shown not just what to do but when to do it – and when not.

Combinations of approaches may be needed, such as changing form first and then making a change of variable. Because of various identities, remind students that the form of answer may vary and still be correct. This also applies to computer software, which may expand or compress expressions with logarithms, have varying conventions about trigonometric function simplification, and/ or not display any constants of integration.

Finally, remind students that having a large integral table with dozens or hundreds of integrals does not absolve them from the need to be intimately familiar and fluent with the most basic types, such as power rule, linearity, integrals of the six trigonometric functions, exponentials, etc.

Sample Homework Problems:

L: 1, 6, 12, 13, 15, 21, 32, 33, 39, 40, 43, 49
M: 1, 3, 4, 5, 11, 13, 32, 33, 38, 40, 42, 47, 52
N: 3, 4, 5, 11, 12, 14, 31, 38, 41, 42, 47, 52

7.2 Integration by Parts

Main Topic: Integration by parts.

Goal: Learn how to trade in a difficult integration for a simpler one – namely, by recognizing when to use parts.

Main Ideas, Warnings, and Pedagogy: Integration by parts, summarized $\int u\,dv = uv - \int v\,du$, is a technique used for the integration of products of "dissimilar" functions, such as a monomial times an exponential. Remind students of the need to try easier approaches before assuming every integrand product requires this technique. One key is to use *lookalikes*: integrands that look similar but lend themselves naturally to different techniques. (The advice of using lookalikes applies to later sections, too!) Example: $\int x \sin x^2 dx$ vs. $\int x \sin x\,dx$. First integral is amenable to a simple change of variable ($u = x^2$); second really does need parts.

Since the proof of the parts formula is simply a reversal of the product rule for differentiation followed by integration, this is safe to prove in class.

There are two keys to successful use of parts:

a) v' is the more/ most complicated factor we can integrate;
b) u has a "simpler" derivative than u itself is.

As a special case, if the integrand is a single factor whose anti-derivative is not known, such as $\ln x$, then $v' = 1$.

This is again a good place to check that students know the difference between differential dv and derivative v'.

With repeated uses of integration by parts, alert students to be consistent in their assignments. For instance, an exponential times a simple sine or cosine may allow either to be u, but the second time around, use the same approach. That is, if you used u for the exponential the first time and the new integral again requires parts, use u for the exponential in the new integral, too.

For definite integration, watch the limits, including on the uv term. Show the nice geometric interpretation. You might enjoy teasing the students with the challenge of finding the error in a supposed "proof" that $0 = 1$ based on $\int \frac{1}{x} dx$ (pretend we don't know this) using $u = 1/x$, $v' = 1$. (If you have never seen this before, try it!) The effect can be made more dramatic by definite integration over [1, 2], say, showing the need for limits on the uv term.

Sample Homework Problems:

L: 3, 4, 8, 13, 21, 25, 27, 31, 35, 37, 39, 42
M: 1, 3, 4, 5, 17, 21, 25, 29, 31, 35, 36, 40
N: 1, 5, 9, 17, 20, 29, 32, 36, 40, 41, 46

7.3 Trigonometric Methods

Main Topics: Powers of sine and cosine; powers of tangent and secant; cotangent & cosecant analogy; quadratic forms.

Pedagogy: This section's material is based on the idea that the trigonometric functions are naturally regarded as three pairs: sine, cosine; tangent, secant; cotangent, cosecant. There are two reasons for this pairing: 1) The elements in each pair are related by the Pythagorean theorem. 2) The elements in each pair are related by differentiation.

As a warm-up, you might start with $\int \sin^3 x \cos x\,dx$. This leads through simple change of variable to $\int u^3 du$. Then follow with $\int \sin^3 x \cos^3 x\,dx$. Now we have to replace $\cos^2 x$ with $1 - \sin^2 x$ to obtain $\int u^3(1 - u^2)du = \int (u^3 - u^5)du$.

Of course, the foregoing assumes "nice" powers for sine and cosine (in this case, odd). With even powers, use the so-called half-angle/ double-angle identities for $\sin^2 x$ and $\cos^2 x$. Similar remarks, suitably modified, apply to the other two pairs. Along the way, we can reduce the number of pairs from three to two by noting that the results for secant & tangent have simple analogues for cosecant & cotangent. Indeed, if one has not yet done so, one should point out that results for trig co-function derivatives lend themselves to a simple rule: The derivative of a co-function may be obtained by "co-ing" and negating the derivative of the non-co-function. For instance, the derivative with respect to x of $\tan x$ is $\sec^2 x$. So, the derivative of $\cot x$ should be $-\csc^2 x$.

Next, trigonometric substitutions are useful for expressions of the forms $a^2 + u^2$, $a^2 - u^2$, and $u^2 - a^2$, particularly if a square root of one of these is involved. The substitution to use in each case is of the form $u = a$ trig θ. But, which trig function? To standardize things, we use only the non-co-functions for u (sin, tan, and sec; not cos, cot, or csc). There is an easy way to build on students' knowledge of the Pythagorean identities to avoid more memorization: Imagine $a = 1$. How can I collapse two terms into one? Which identity does that? For instance, given an expression of the form $a^2 + u^2$, which identity has the form $1 + (\quad)^2 = [\quad]^2$? Since $1 + \tan^2\theta = \sec^2\theta$, we would use tan and $u = a \tan \theta$.

Sample Homework Problems:

L: 7, 13, 18, 26, 31, 34, 38, 42, 45, 47, 50, 56
M: 5, 7, 9, 10, 11, 18, 19, 26, 31, 34, 38, 45, 47
N: 5, 9, 10, 11, 19, 24, 25, 33, 37, 39, 49, 51, 53
Note: Not all examples require a trig substitution, such as #39. Once students do #39 the long way, revisit by simple change of variable and then compare answers.

7.4 The Method of Partial Fractions

Main Topics: Partial-fraction decomposition; integration of rational functions, including of sine and cosine.

Goal and Main Idea: If we reverse the process of adding fractions with different denominators, we ask: Given a rational function with a "complicated" denominator (one with a lot of factors, roughly), how can we express it as a sum of "simpler" fractions (fractions whose denominators have fewer factors and/ or factors to lower powers)?

This has particular value for integration because the "simpler" fractions are easier to integrate.

Pedagogy: Concentrate on teaching students guidelines for proper form of the judicious guess for a partial-fraction decomposition. Underlying all this is the fundamental theorem of algebra: In theory, any polynomial with real coefficients factors into linear factors and irreducible quadratics. Thus, our judicious guess for the form of the partial-fraction decomposition uses distinct or repeated linear factors, distinct or repeated quadratic factors, or some combination. Then one solves for the unknown coefficients in the surmised expansion.

One may use one of two methods to obtain the coefficients, or a combination of both. The methods entail substitution of function values or equation of corresponding coefficients of identical polynomials.

Don't forget handling improper fractions via long division.

Insight: As an interesting aside, take the case where the quadratic factor is *not* irreducible; that is, it *can* be factored. More specifically, take the case of a quadratic whose discriminant is zero. In particular, if x^2 is a factor of the denominator, many students claim that one of the terms in the expansion is of the form $\frac{Ax+B}{x^2}$ (distinct quadratic factor). This is not really the way to think of it; it should be a repeated linear factor. However, our form for this part of the expansion is then $\frac{A}{x}+\frac{B}{x^2}$. However, the students' "wrong" guess is mathematically equivalent.

Warnings: Always fully factor the original denominator. A guess that does not allow enough terms will result in a contradiction (equations with no solution). A guess with too many terms is less serious in that one will obtain a solution, but many coefficients will prove unneeded and work out to be zero, which is inefficient.

Comment: This technique is needed again later in a course on differential equations, notably for Laplace transforms.

Sample Homework Problems:

Note: Although #60 is in the *Survival Manual*, the spirit of comparing different techniques should be conveyed in some form.

7.5 Summary of Integration Techniques

Main Topic: Integration strategy.

Main Idea, Pedagogy: Classify integrals. If possible, use basic formula. If not, try changing the form. Else, consider a change of variable. For all other special new forms, classify as to type, such as product of dissimilar functions for parts, partial fractions for rational functions with complex denominators, trigonometric integrals, and special forms amenable to trig substitution. Watch out for lookalikes. Continue to watch out for simpler methods.

Sample Homework Problems:

7.6 First-Order Differential Equations

Main Topics: First-order linear differential equations; solutions; initial-value problems; applications.

Main Ideas: First-order linear differential equations in standard form admit an integrating factor. By multiplying both sides by this factor, the transformed equation's left member is expressible as the derivative of a product. Integration provides the solution thereafter.

Direction fields provide a geometric view, even in cases where the first-order differential equation is neither linear nor separable.

Population growth is a good application (as earlier when we had separation of variables). It might be a good idea to compare solution methods using the new linearity method and the earlier separation. Update with the more realistic logistics equation, whose solution leads to partial fractions.

Round out applications of linear first-order differential equations using mixture problems and RL circuits.

Sample Homework Problems:

7.7 Improper Integrals

Goal: Become aware of the issue of impropriety of certain integrals (two types), and learn how to handle them.

Main Topics: Improper integrals in which limits include an infinity, and others in which integrand becomes unbounded.

Main Ideas and Pedagogy: There are two kinds of impropriety. The first, an infinite limit of integration, is easy to recognize because an infinity symbol is visible. The key to resolution is replace the infinity by a parameter L, integrate indefinitely, substitute for the definite integral with L, and then let L approach the appropriate infinity.

The second kind is more pernicious because one has to look to make sure that the integrand is continuous. However, this is a blessing in disguise because it reminds students of the need to check hypotheses of theorems before applying same.

For instance, consider $\int_{0}^{2} \frac{1}{(x-1)^2}dx$.

(As an aside on control and the value of checking, apart from the integrand becoming discontinuous/ unbounded as x approaches 1, it is nonnegative. Attempting to use the fundamental theorem blindly leads to a negative answer, which would not be possible here anyway!)

Note that the first kind of problem has x becoming infinite, and the second has y becoming infinite. However, the resolution of the second is similar to the first. We modify the interval of integration, replacing the offending value by a parameter such as L, integrating, and then letting the parameter L approach the offending input value.

Terminology and Technology: When the limit exists in either case, the improper integral is convergent (integral converges); otherwise, it is divergent (integral diverges). Be careful, as subtle cases may not be handled properly by some computer software.

Applications: Firing a rocket into deep space is a possible application, with integration to infinity of the force function giving the work needed. Gabriel's horn is a classic must-show example, a surface that houses a finite volume of paint but whose surface cannot be painted (because the surface area is infinite).

Sample Homework Problems:

L: 5, 10, 15, 20, 25, 30, 35, 40, 45, 50, 54
M: 3, 5, 6, 7, 15, 20, 23, 32, 35, 40, 45, 49
N: 3, 6, 7, 16, 21, 23, 32, 33, 37, 41, 49, 51

7.8 Hyperbolic and Inverse Hyperbolic Functions

Main Topics, Ideas, and Pedagogy: For many, this is an optional section. The hyperbolic functions are to the hyperbola as the trigonometric (circular) functions are to the unit circle. Many analogies exist between the two, although they are useful primarily as mnemonic devices.

Cover the definitions (based on exponentials), their graphs, their properties (including domain, range) the relating properties (analogous formally to Pythagorean identities), and the derivative & integral relationships. Then handle the inverses (or inverses of restrictions) and their properties.

Sample Homework Problems:

L: 1, 11, 15, 20, 24, 28, 31, 36, 45, 55, 61
M: 1`, 4, 6, 11, 14, 19, 24, 28, 31, 37, 45, 55
N: 4, 6, 13, 14, 19, 23, 34, 37, 38, 44, 53, 54

Additional Exercises are in the Chapter 7 Review. We recommend doing all of the *Proficiency Exam* questions, but include at least those in the *Survival Manual*, *viz.*: 15, 18, 21, 24, 27, 30. Note the *Group Research Projects*.

CHAPTER 8: INFINITE SERIES

8.1 Sequences and Their Limits

Main Topics: Sequences; limit of a sequence; sequences that are bounded and monotonic.

Main Ideas and Pedagogy: There are two reasons for studying sequences. First, they are important in their own right and have their own applications. Second, they are needed for infinite series, the bulk of this chapter. (Convergence of each infinite series is really a question of convergence of the sequence of partial sums of the series.)

Informally, a sequence is a list of numbers. More formally, a sequence is a function whose domain is a subset of the set of nonnegative integers. By definition, the limit of a sequence is investigated by seeking the limit of the general or n-th term as n increases without bound. The sequence is either convergent or divergent depending on whether or not this limit exists. Graphically, convergence of the sequence $\{a_n\}$ to L is tantamount to saying this: Given any thin horizontal band centered about the horizontal line $y = L$, all but a finite number of terms of the sequence fit within the band. By definition, this is mathematically more precisely stated in the form of meeting a challenge: Given any $\epsilon > 0$, there exists $N > 0$ such that, for all $n > N$, $a_n - \epsilon < L < a_n + \epsilon$. Not all instructors teach this formal definition, but it's a good idea to include, if possible.

Although n is a discrete variable, limit results can often be obtained by treating as though it were continuous. Thus, the algebraic manipulations and even l'Hopital's rule often prove useful for investigating such limits.

Finally, it is desirable to have theorems that tell whether a sequence is convergent without having to investigate the limit. A bounded monotonic sequence always converges. A sequence that is known only to be monotonic will converge if and only it is also bounded.

Warnings: Students need to understand that virtually all results for sequences (and later for series) remain true if the hypotheses are weakened to cover all but a finite number of exceptions. Explain that this means beyond some point some property is considered, such as being eventually monotonic. The conclusions hold in such cases.

Enrichment: Be sure to include geometric sequences. The result about the limit of r^n for appropriate r will be needed right away. You can enrich the course by showing a choice of features of the Fibonacci sequence (see Problem 48). Note Problem 47 for background on the number e.

Sample Homework Problems:

L: 3, 7, 11, 16, 23, 26, 27, 30, 31, 37, 41, 43
M: 1, 3, 6, 9, 12, 23, 26, 31, 36, 39, 41, 43
N: 1, 6, 9, 12, 17, 33, 36, 39, 42, 44, 45

8.2 Introduction to Infinite Series: Geometric Series

Main Topics: Definition and general properties of infinite series; geometric series & applications of geometric series.

Main Ideas: Whereas a sequence is a list of numbers, a series is a sum of them. We investigate convergence of a series by forming the sequence of partial sums. Thus, every series question reduces to a question about convergence of the corresponding sequence.

A geometric series is a series in which the ratio of consecutive terms is constant. Such a series is convergent if and only if the absolute value of the ratio is under 1.

Warnings and Pedagogy: The word "series" is abused in daily language to mean the same as sequence. Make clear that in mathematics the distinction is essential. Even after this clarification, some students may continue to confuse sequences and series as well as the n-th term. (You might also mention that those "next number in series" challenges, which really are about sequences, are based on the fallacy that a finite number of terms can specify a sequence uniquely. Thus, specifying 1, 2, 4,... may suggest doubling, thus yielding 8 as the next number, but it could also be adding 1, then adding 2, and so we might next add 3 to get 7 as the next number. And so on.)

Many students understand geometric series in a limited context that depends on notation used. So, be careful to present different representations, such as with and without series notation, using different letters for indices, and offering a large variety of questions. Before getting to the theorems, however, consistently show convergence-of-series questions being handled by forming the sequence of partial sums. Use appropriate patterns or other devices to find a simpler representation for the n-th partial sum, including telescoping series.

Keep in mind that geometric series are the series that we know the most about, and further, that we rely upon them later for the ratio test and other applications. Geometric series also connect repeating decimals to fractions.

Challenge and Enrichment: There are many wonderful applications, and some can be made into humorous challenges. For example, imagine a bug of negligible proportions taking a journey starting at the origin of a two-dimensional coordinate system, with the following spiraling motion: First, it walks one unit to the right, then ½ of a unit up, then ¼ of a unit left, and so on, each time turning at a right angle in the counter-clockwise direction and moving a distance half of that of the previous segment. To which point does the bug converge? Answer: (4/5, 2/5).

Sample Homework Problems:

L: 5, 7, 11, 19, 21, 24, 29, 32, 37, 43, 49, 57, 59
M: 1, 2, 3, 5, 7, 11, 17, 20, 25, 31, 46, 48, 49, 51, 57
N: 1, 2, 3, 4, 8, 16, 17, 20, 25, 31, 46. 48. 50, 51. You might direct class's attention to # 57, Zeno's paradox.

8.3 The Integral Test: p-Series

Main Topics: Divergence test; series of non-negative numbers; the integral test; p-series.

Main Ideas, Warnings, and Pedagogy: We examine the terms of the series and have a divergence test, namely the n-th term test. If the n-th term of a series fails to have zero as a limit, the series must be divergent. It is critical to point out that the converse is not true. Show by examples that the n-th term may approach zero and yet the series may either converge or diverge.

It is worth proving this result, but you will need to explain the idea of the contrapositive to the conditional *If p, then q* being the conditional *If not q, then not p* and that these two statements have the same truth value. (Compare converses and inverses to the contrapositive, too.)

Focus on the idea that we are seeking means to test

convergence of a series (series of non-negative numbers in this section) to speed up the process by obviating the need for investigating partial sums. The integral test is a good test relating the allied ideas of improper integrals (of the first type) to infinite series. You might point out the similar definitions these enjoy. Apply the integral test when the terms of a series are positive and decreasing, providing the "interpolated" function is continuous. The corresponding improper integral and infinite series either both converge or both diverge. Moreover, show the diagram for the proof, which contains within it the ability to estimate series sums. Apply the test to p-series to get the general result that such a such converges if and only $p > 1$. Thus, the harmonic series may be regarded as the "last" divergent p-series.

Note that p-series involve terms that differ from those of geometric series in the same way that monomials are different than exponentials, namely, whether the base or the power is changing.

Sample Homework Problems:

L: 5, 7, 12, 17, 23, 29, 33, 34, 39. 40, 47, 54
M: 2, 4, 6, 7, 9, 17, 21, 29, 34, 39, 50, 55, 57
N: 2, 4, 6, 9, 16, 18, 21, 30, 37, 41, 50, 55, 57

8.4 Comparison Tests

Main Topics: Direct comparisons and limit comparisons.

Main Ideas: Series of nonnegative terms may be compared term-by-term or by limit of corresponding terms. A series whose terms are smaller than the corresponding terms of a known convergent series is itself convergent. A series whose terms are larger than the corresponding terms of a known divergent series is itself divergent. You will need to remind students now and later that this applies only to series of non-negative numbers.

Sometimes such a direct comparison is awkward, in which case a limit comparison might be easier. In either case, show how to obtain a good comparison series (a simpler one whose convergence properties are known).

Sample Homework Problems:

L: 1, 2, 4, 5, 8, 14, 20, 26, 32, 38, 43, 49, 50
M: 1, 2, 4, 7, 10, 14, 17, 20, 26, 33, 40, 49, 59
N: 3, 6, 7, 10, 18, 17, 18, 25, 33, 34, 40, 51, 59

8.5 The Ratio Test and the Root Test

Main Topics: Ratio and root tests; summary of convergence tests for series of non-negative terms.

Main Ideas: The ratio test effectively determines how much a series of positive terms emulates a geometric series in a limiting sense. (Suggestion: Make a geometric series your first example.) This test is powerful and easy, and needed later for power series. It is most useful when the terms involve mostly products, powers, quotients, and factorials. The root test is an alternate test useful in some cases where the ratio test is inconclusive ("fails"), but especially where the term is an n-th power (n = index).

Be sure to spend time with students clarifying when to use which tests and why.

Sample Homework Problems:

L: 3, 11, 14, 17, 27, 33, 36, 37, 45, 48, 54
M: 3, 5, 9, 11, 17, 27, 30, 34, 37, 41, 45, 49
N: 5, 9, 12, 25, 28, 30, 34, 38, 41, 46, 49

8.6 Alternating Series; Absolute and Conditional Convergence

Main Topics: Alternating-series test; error estimates; absolute vs. conditional convergence; rearranging terms.

Main Ideas, Warnings, and Pedagogy: An alternating series is a special case of a "mixed" series, meaning a series of both positive and negative terms. Caution students that this represents a departure from previous sections, so one can't do direct comparisons, for instance. Also caution that the negation of the hypotheses of this test does not imply divergence, a common misconception. Rather, if the *n*-th term fails to approach zero, this is divergence by the divergence test, not by the alternating-series test! Use the test not just for convergence but also for error estimates; the absolute value of the first term not in the partial sum serves as an upper bound on the error in estimating the full series sum by the partial sum.

Show the theorem that if the series of absolute values of original terms converges, then so must the original series. A series whose related series of absolute values converges is said to be *absolutely convergent*. This is a bit of a misnomer in that the property is really about the new series (the series of absolute values), not the original series, at least not *a priori*. However, the theorem comes to the rescue and shows that an absolutely-convergent series is itself necessarily convergent (conventional sense).

A *conditionally-convergent* series is a convergent series whose associated series of absolute values diverges. The alternating-harmonic series is the best-known example. Note that there is no such thing as "absolute divergence" because a divergent series must have its series of absolute values also diverge (apply contrapositive to theorem above). Thus, there are only three possibilities of four initial situations. Here is a summary of what can happen:

	Σa_n Convergent	Σa_n Divergent
$\Sigma \lvert a_n \rvert$ Convergent	Absolutely Convergent	xxx (Impossible)
$\Sigma \lvert a_n \rvert$ Divergent	Conditionally Convergent	Divergent

Rearrangement of the terms of an absolutely-convergent series does not affect the convergence, including the sum. However, in sharp contrast, rearranging the terms of a conditionally-convergent series not only can change whether or not the series converges, but, given any value or any infinity, one can manipulate the series to force the series to converge to the chosen value. The key idea is that the positives sum to $+\infty$ and the negatives to $-\infty$. So, to make the sum of a conditionally-convergent series

approach any given S (say, $S > 0$), choose as many positive terms of the original series in original order sufficient to make the partial sum exceed S. Now choose just enough negative terms of the original series in original order sufficient to bring the new partial sum under S. Now keep alternating with picking positive terms, then negatives, and so on. A similar idea can be used for an infinity as desired sum.

Sample Homework Problems:

L: 4, 6, 10, 20, 22, 30, 32, 39, 40, 48, 51
M: 4, 7, 11, 18, 19, 31, 33, 38, 40, 48, 51
N: 5, 7, 11, 18, 19, 31, 33, 38, 42, 49, 52
Students will probably need help or hints for #51, #52.

8.7 Power Series

Main Topics: Power series and convergence; term-by-term differentiation and integration.

Main Ideas: A power series in *x* may be regarded as an extension of a polynomial. Indeed, a partial sum of a power series is a polynomial. For any one value of *x*, one obtains a different series, so one actually has a power series defining a function of *x*. The key question is to find all real numbers *x* for which the power series in *x* converges. Thus, one is actually finding the domain of the function thus defined. The method of choice is the ratio test, but applied to the series of absolute values. This test also shows that a power series converges absolutely in an interval (with or without one or both endpoints, if there are endpoints), including the special cases of a degenerate interval (single point) or the entire real line. The root test again provides an alternative.

Power series may be differentiated and integrated term-by-term within the interval of convergence, although some care is needed for endpoints. One can obtain many interesting power series for known functions this way, most notably by starting with the geometric series result

$$\sum_{k=1}^{\infty} x^k = \frac{1}{1-x}$$ and then differentiating or integrating.

Note that this represents another way of looking at power series. Instead of saying a power series defines a function, we may ask whether a given known function can be represented by a power series (called a Taylor series, next section). This two-way view should be cultivated so that one later regards which came first as almost arbitrary.

Sample Homework Problems:

L: 3, 6, 9, 16, 19, 22, 27, 30, 33, 35, 39, 45
M: 1, 3, 8, 16, 17, 22, 25, 30, 35, 36, 39, 40, 43
N: 1, 2, 8, 15, 17, 21, 25, 29, 36, 40, 42, 43

8.8 Taylor and Maclaurin Series

Main Topics: Taylor and Maclaurin polynomials; Taylor's theorem; Taylor and Maclaurin series; series operations.

Main Ideas: We seek to approximate a suitably smooth function by a polynomial, called the *Taylor polynomial* of the function. (Note that Maclaurin was a student of Taylor and results bearing the Maclaurin name designate a Taylor polynomial or series but with the center $c = 0$.) In turn, this leads to the idea of representing suitable functions by an infinite series, a power series called the *Taylor series* of the function. Note, as mentioned in previous section of these notes, that one can view power series as defining functions (view of last section) or one can regard the function as given and the Taylor series (power series) as the corresponding element to that function.

Warnings and Pedagogy: If possible, develop the formula for the coefficients from the approximation criteria. Students have some difficulty with this, but it's worth trying. Use graphing calculators and/ or computer algebra systems such as Derive to graph the original function and some approximating Taylor polynomials. Show the effect of adding more terms graphically – namely, getting a better match of the polynomials to the original function's graph.

Once you develop a small catalogue of functions and their associated Taylor series (exponential, sine, cosine, geometric, and a few others), extend your repertoire by substitution, differentiation, and/ or integration. This is in lieu of using the formula for the coefficients of the Taylor series. Rewriting a quotient of two linear functions by performing the division, for instance, enables one to find the series for the quotient. In this case, you need to think of the geometric series result, albeit backwards.

Sample Homework Problems:

L: 5, 8, 12, 20, 25, 35, 39, 40, 47, 53, 59, 66
M: 4, 5, 6, 7, 14, 20, 21, 25, 31, 39, 43, 53, 58
N: 3, 4, 6, 7, 14, 19, 21, 31, 41, 43, 49, 53, 58

Additional Exercises are in the Chapter 8 Review. We recommend doing all of the *Proficiency Exam* questions, but include at least those in the *Survival Manual*, viz.: 33, 36, 39.

CHAPTER 9: POLAR COORDINATES AND PARAMETRIC FORMS

9.1 The Polar Coordinate System

Main Topics: Plotting points; multiple representations; polar graphs; relationships between polar coordinates and rectangular coordinates.

Main Ideas: In polar coordinates, one still specifies two numbers to locate a point: a radial coordinate r and a polar angle θ (which may now be in degrees or in radians).

Warnings and Pedagogy: However, the one-to-one correspondence of Cartesian coordinates is lost, in this sense: Even though a pair (r, θ) determines a unique point in the plane, any point in the plane has infinitely many representations. Indeed, the representation for point P given by (r, θ) may be replaced by $(-r, \theta + \pi)$ or any of the pairs $(r, \theta + 2n\pi)$ for all integers n. This has implications for graphing and symmetry tests as well as for converting between polar and rectangular coordinates.

Spend time with the conversions, which should be familiar from trigonometry and are used repeatedly elsewhere. Use them also with some polar graphs. Keep in mind the caveat about multiple representations for each point, some of which may satisfy a polar equation while others do not. Include circles, lines. Note that the equations r = constant (circle) and θ = constant (line through the origin) are to the polar system as the equations x = constant and y = constant are to the rectangular system (*i.e.*; orthogonal curves).

Sample Homework Problems:

L: 5, 11, 12, 17, 23, 28, 34, 35, 39, 41, 45, 53
M: 4, 9, 11, 12, 16, 23, 24, 25, 31, 39, 40, 54
N: 4, 9, 10, 15, 16, 24, 25, 26, 31, 36, 40, 54

9.2 Graphing in Polar Coordinates

Main Topics: Plotting points; symmetry and rotations, special polar curves; summary.

Main Ideas: In the previous section, we identified curves by converting to rectangular coordinates and/ or noting the special forms for circles and lines. Now we systematically consider spirals, roses, cardioids, lemniscates, and limaçons taking advantage of symmetry and rotations. The earlier caveat about multiple representations of points applies.

Technology: Polar graphs are very attractive and a lot of fun. Although you want your students to be able to do a variety of graphs without technology (see previous section), take advantage of whichever tools you have for these more involved examples (roses, cardioids, lemniscates, etc.).

Sample Homework Problems:

L: 11, 17, 19, 22, 29, 32, 33, 35, 42, 48, 55, 56
M: 5, 7, 15, 19, 22, 30, 32, 33, 41, 43, 48, 56
N: 5, 7, 15, 20, 23, 30, 31, 34, 41, 43, 47, 57

9.3 Area and Tangent Lines in Polar Coordinates

Main Topics: Intersections of polar-form curves; area bounded by polar graphs; tangent lines.

Main Ideas, Warnings, and Pedagogy: As in previous sections, we note the lack of a one-to-one correspondence between ordered pairs (r, θ) in polar coordinates and points (multiple representations of a single point in plane). So, we need to graph to be sure of finding all intersection points of polar curves; algebraically seeking simultaneous solutions is not sufficient as it is in rectangular coordinates. Consider the pole separately as well.

Next, we use the area of a sector to develop Riemann sums for the area of a region bounded by a polar-form graph. (Angles must be in radians here.) For areas between two curves, find the intersection points.

Finally, tangent lines are a bit more involved in polar coordinates. Consider horizontal and vertical tangents. Rather than memorize the final formula, students can use the intermediate result: $\dfrac{dy}{dx} = \dfrac{dy \,/\, d\theta}{dx \,/\, d\theta}$.

Sample Homework Problems:

L: 7, 11, 17, 18, 26, 31, 38, 40, 48, 53, 59, 62
M: 4, 7, 10, 17, 19, 28, 31, 34, 39, 48, 52, 59
N: 4, 6, 10, 14, 19, 28, 30, 34, 39, 49, 52, 58

9.4 Parametric Representations of Curves

Main Topics: Parametric equations; parametric-equation results for derivatives, arc length, and area.

Main Ideas: This is the only section of the chapter *not* devoted to polar coordinates, but is linked by the idea of having a different representation. We obtain a new way of relating variables by adding a parameter (literally: "almost a variable"), which introduces an element of motion as well as orientation to curves. To graph, we can set up tables or eliminate the parameter. The latter is often done algebraically or using trigonometric identities.

Apply everything to derivatives, arc length, and area.

Warnings and Pedagogy: We introduce functions $x = x(t)$ and $y = y(t)$. Note that x is no longer an independent variable. Thus, graphing requires more care (still another link to polar coordinates). *Example*: Graph $x = \cos^2 t$, $y = \sin^2 t$. If we add equations, we eliminate the parameter t and obtain $x + y = 1$. Yet, the graph is *not* the full line $x + y = 1$; it's the segment in the first quadrant! The explanation is this: The logic of the argument shows that *if* a point lies on the graph, then it lies on this line. It does not show the converse. Note further that $0 \le x \le 1$. On the plus side, we see the power of clear logic over a less mindful sequence of manipulations.

Special Benefits & Features: Using parametric equations offers several advantages:

- Represent non-functions such as circles.
- Introduce two-dimensional motion.
- Introduce orientation and speed to curves.
- Subsume *all* functions under a new category.

Anent this last point, it is important that students realize that parametric equations are not so different than what we had earlier with functions and equations. In fact, show this: *Every function is parametrizable*. This means that we can represent any function f by a pair of parametric equations whose graph is precisely the function's. To do so, represent $y = f(x)$ by $x = t$, $y = f(t)$. Even more generally we can represent any Cartesian equation using $x = t$.

Technology: A rich source of attractive graphs is provided by the cycloid family. If time allows, you may wish to derive at least one equation from the locus definition, such as the equation for a cycloid based on a point on the rim of a rolling wheel.

The material on derivatives, arc length, and area is pretty straightforward. Where possible, compare the results for derivative, at least, using parametric equations and not.

Sample Homework Problems:

L: 5, 11, 13, 17, 21, 27, 33, 39, 45, 49, 51
M: 4, 9, 13, 18, 22, 29, 34, 44, 48, 49, 51
N: 4, 9, 14, 18, 22, 29, 34, 44, 47, 48, 50

Additional Exercises are in the Chapter 9 Review. We recommend doing all of the *Proficiency Exam* questions, but include at least those in the *Survival Manual*, viz.: 18, 21, 24.

CHAPTER 10: VECTORS IN THE PLANE AND IN SPACE

10.1 Vectors in the Plane

Main Topics: Introduction to vectors; standard analytic and geometric representations of vectors.

Main Ideas, Warnings, and Pedagogy: A scalar (a real number) has only magnitude, whereas a vector has not only magnitude but direction. Thus, we visualize a vector as a directed line segment (technically, it's an equivalence class of such segments). Define equality and addition of two vectors, and multiplication of a vector by a scalar. Show both geometric and analytic approaches to these. (Note that these are the fundamental three ideas throughout all further work with vectors and in later courses with vector spaces. The properties of vector operations, in fact, become the defining properties or "axioms" for a vector space later.)

The norm or magnitude of a vector emanates from the Pythagorean theorem. Use vector methods to prove geometric results as a means of showing their power, such as Example 2. After some practice, one finds that results are easier with vectors.

To distinguish vectors from points, use $<x, y>$, not (x, y). If you use a computer algebra system, you may see $[x, y]$. After you introduce unit vectors, represent vectors a second way: $x\mathbf{i} + y\mathbf{j}$. The vector joining (x_1, y_1) to (x_2, y_2) is given by $< x_2 - x_1, y_2 - y_1 >$ or $(x_2 - x_1)\mathbf{i} + (y_2 - y_1)\mathbf{j}$.

Applications: Forces, velocities.

Comment: All the results above will extend easily to three dimensions later – even the distance formula, which comes about from two applications of the Pythagorean theorem.

Sample Homework Problems:

L: 11, 16, 17, 21, 26, 28, 29, 34, 43, 47, 50
M: 5, 9, 11, 13, 18, 22, 23, 25, 29, 36, 43, 48
N: 5, 9, 13, 15, 18, 22, 23, 25, 31, 36, 44, 48
Challenge: Add vector methods, if desired, such as #54.

10.2 Quadric Surfaces/ Graphing in Three Dimensions

Main Topics: Three-dimensional coordinate system; 3-D graphs (planes, spheres, cylinders, quadrics).

Main Ideas: Extend analytic geometry to three dimensions with points, distance, and graphs.

Pedagogy: The distance formula, applied twice, yields the more general distance in three dimensions.

In one dimension, graphing $x = 3$ yields a point, as the solution is the set of all real numbers x that make the equation true. In two dimensions, graphing $x = 3$ yields a line, as the solution is the set of all pairs (x, y) of real numbers that make the equation true. Note that the context of the extra dimension can be viewed as taking all solutions in the lower dimension and projecting them into the new dimension. With this perspective, we can predict that, in three dimensions, graphing $x = 3$ yields a plane, as the solution is the set of all triples (x, y, z) of real numbers that make the equation true. This idea similarly leads to circles $x^2 + y^2 = r^2$ when the context is two dimensions, but we get circular cylinders in three dimensions (roughly, because "z can be any real number" with $x^2 + y^2 = r^2$). Contrast this carefully with the equation of a sphere. For spheres, consider the same kind of questions as we had for circles, particularly moving between equation and graph, including completing the square.

To graph, take advantage of traces and sections. Realize that we are engaged in the rendering of three-dimensional objects onto a two-dimensional plane – in other words, art.

Technology: This is another good area for a computer algebra system for graphing functions such as those that yield elliptical paraboloids of revolution and non-function equations such as for spheres.

Sample Homework Problems:

L: 7, 11, 23, 29, 34, 38, 40, 44, 46, 49
M: 8, 9, 23, 30, 31, 37, 39, 44, 46, 49
N: 8, 9, 24, 30, 31, 37, 39, 43, 45, 50
Suggestion: Match graphs & equations in #13-22 in class. For set N users, make up a problem like #49, which is too classical (and charming) in its theme not to assign.

10.3 The Dot Product

Main Topics: Vectors in three dimensions; definition of dot product; angle between vectors; projections; work.

Main Ideas, Warnings, and Pedagogy: Continue to extend vector concepts to three dimensions, analogous to the extensions in previous section. Define dot product of two vectors, but note that it is not a true product in the sense that the result of dotting two vectors is a scalar.

Go over the properties of the dot product, and show it also has an equivalent representation by theorem (whose proof is based on the law of cosines). This allows us to find the angle between two vectors. (In more abstract courses, this allows defining angles in higher-dimensional spaces, even though no visual geometry exists otherwise.) This leads to the orthogonality criterion for two nonzero vectors being that their dot product be zero.

Round out with projections and work as a dot product. Note that in the case where the angle between the force and the displacement is zero, the work done reduces to the more elementary "force times distance" encountered

earlier.

Sample Homework Problems:

L: 3, 8, 18, 20, 23, 29, 33, 35, 45, 54, 56, 61
M: 3, 7, 9, 14, 17, 20, 23, 27, 33, 35, 45
N: 4, 7, 9, 14, 17, 19, 24, 27, 34, 46, 55

10.4 The Cross Product

Main Topics: Cross product – definition and properties; geometric interpretation; area, volume, and torque.

Main Ideas, Warnings, and Pedagogy: Unlike the dot product, the cross product of two vectors is a true product, so the result is another vector. Show how to calculate this product by definition and also by equivalent geometric interpretation (direction by right-hand rule; magnitude by area of the parallelogram formed by the given two vectors). Of course, briefly review 3 x 3 determinants as needed for the mnemonic for the symbolic determinant for the cross product. (Irony: Although cross product is a true product, the determinant mnemonic for it is not a true determinant.)

Applications: Area of triangles and parallelograms, volume of parallelepipeds, and torque.

Technology: Derive and other computer algebra systems can handle vector operations from this and preceding sections, but take care, as the syntax can be very precise.

Sample Homework Problems:

L: 3, 10, 15, 18, 23, 27, 35, 41, 42, 48
M: 1, 3, 9, 16, 17, 23, 25, 33, 41, 42, 46, 48
N: 1, 5, 9, 16, 17, 22, 25, 33, 40, 46, 47

10.5 Lines and Planes in Space

Main Topics: Lines and planes in space; direction cosines.

Main Ideas: Given a line (in space), a direction vector in space aligned with it provides the parametric equations for such a line, provided one also knows a point on the line. By eliminating the parameter, one obtains an alternate form, the symmetric equations of a line.

In a different but analogous fashion, one starts with the normal to a plane and a point in that plane and uses perpendicularity to obtain the equation of a plane, which is a natural extension of the equation of a line.

Pedagogy: It's a standard application of the cross product to ask for the equation of a plane determined by three points. It is necessary that the points are not collinear, but if they are, any two vectors formed from the three points will have a cross product equal to the zero vector. The cross product provides the desired normal to the plane, and any one of the points provides the other ingredient for equation of a plane. There are also many nice exercises linking lines and planes, many of them involving intersections or parallelism.

You might challenge students to get the angle between two planes, and compare to the angle between normals. Point out the absolute value and ask students why.

Sample Homework Problems:

L: 5, 8, 13, 17, 24, 27, 36, 41, 43, 52, 57
M: 1, 7, 13, 18, 24, 27, 33, 41, 44, 52, 58
N: 1, 7, 12, 18, 23, 28, 33, 42, 44, 51, 58

10.6 Vector Methods for Measuring Distance

Main Topics: Distance from a point to a plane; distance from a point to a line.

Main Ideas: Use a variety of vector methods to derive formulas for distances between various point sets (such as points, lines, planes). It is probably more valuable to show the methods than to drill excessively on the results.

Sample Homework Problems:

L: 3, 9, 15, 19, 26, 28
M: 3, 7, 15, 20, 26, 29
N: 1, 7, 14, 20, 27, 29

Additional Exercises are in the Chapter 10 Review. We recommend doing all of the *Proficiency Exam* questions, but include at least those in the *Survival Manual*, *viz.*: 33, 36, 39, 42, 45.

CHAPTER 11: VECTOR-VALUED FUNCTIONS

Chapter Overview and Context: The previous chapter introduced vectors and operations on vectors. This chapter places vectors within the context of functions.

11.1 Introduction to Vector Functions

Main Topics: Definition of, and operations with, vector-valued functions; limits; continuity.

Main Ideas: A vector-valued function is a function whose domain is a set of real numbers and whose range is a set of vectors. We can view parametric equations within this new context, with $F(t) = x(t)\mathbf{i} + y(t)\mathbf{j} + z(t)\mathbf{k}$. (If the curve is planar, then $z(t) = 0$.) Hence, the graph of a vector function is not really new, and we still have orientation as well as motion.

Pedagogy and Technology: Define the basic vector-function operations (addition, subtraction, multiplication by scalar function, and dot & cross product of vector functions). Take advantage of technology for the more tedious calculations.

Limits and continuity are natural extensions of the single-variable calculus definitions, although a bit more attention may be in order for the dot and cross products.

Sample Homework Problems:

L: 3, 9, 10, 12, 15, 23, 25, 30, 34, 37, 41, 46
M: 1, 9, 11, 12, 15, 23, 26, 27, 33, 36, 42, 46
N: 1, 4, 11, 19, 24, 26, 27, 33, 36, 42, 47, 58

11.2 Differentiation & Integration of Vector Functions

Main Topics: Vector derivatives and their properties; tangent vectors; motion in three dimensions; integrals.

Main Ideas and Pedagogy: We continue the extension of calculus to differentiate and integrate vector functions, which we may do by components. However, prior to this, formally define the difference quotient of a vector function and its limit (the derivative). Note that in single-variable calculus, the derivative gives the slope of the tangent, whereas in multi-variable calculus, the derivative of a vector function, when nonzero, is a vector that actually is tangent to the graph (provided the limit exists, of course). Finding a tangent vector to the graph of a vector-valued function thus requires only taking that function's derivative.

Include the theorems ("rules": linearity, scalar multiples, dot product, cross product, chain rule) for derivatives of vector functions. One may illustrate them by calculating two ways – computing each side separately and then comparing. Do show the theorem, geometric interpretation, and proof that a vector function of constant magnitude is orthogonal to the function's derivative.

The velocity of a particle whose position is given by a vector function is just that function's derivative. Speed is the absolute value of velocity. Acceleration is the derivative of velocity. Of these, all are defined analogously to the one-dimensional case, but here, only speed is a scalar. We may also go in reverse direction via integration, starting with acceleration (or velocity) and recovering position function.

Warning: Be careful to distinguish the absolute value or magnitude of the derivative from the derivative of the absolute value or magnitude.

Sample Homework Problems:

L: 3, 6, 12, 15, 23, 28, 32, 35, 44, 47, 50
M: 1, 3, 7, 9, 15, 19, 23, 25, 31, 35, 43, 51
N: 1, 7, 9, 16, 19, 22, 25, 31, 37, 39, 43, 51

11.3 Modeling Ballistics and Planetary Motion

Main Topics: Modeling the motion of a projectile in a vacuum; Kepler's second law.

Main Ideas and Pedagogy: Model the motion of a projectile subject only to constant gravitation. The trajectory is demonstrated to be a downward-opening parabola, about which we can readily determine range of flight and other quantities based upon knowing others, such as initial velocity. It is a good idea to give a good mix of problems, stressing methodology, not just formula memorization.

If time allows, a demonstration of Kepler's second law is enriching in showing the beauty of mathematics as applied to physics and planetary orbits. Central-force fields play the key role here.

Sample Homework Problems:

L: 1, 10, 12, 15, 19, 22, 24, 33, 34, 37
M: 2, 10, 11, 16, 19, 23, 26, 28, 32, 37
N: 2, 9, 11, 16, 18, 23, 26, 28, 32, 35

11.4 Unit Tangent and Normal Vectors; Curvature

Main Topics: Unit tangent and principal normal vectors; arc length as parameter; curvature.

Main Ideas and Pedagogy: Because the derivative of a smooth vector function is tangent to the graph, we can define a unit tangent vector by normalizing (that is, dividing the tangent by its norm). In turn, because this unit tangent has constant norm, its derivative is orthogonal to it. We may then define the unit normal by normalizing.

If we are interested in a curve primarily for its geometric features (rather than its motion, for instance), arc length is most convenient to use as a parameter. One nice result that conforms well with experience and intuition is that the derivative of arc length with respect to time is speed. If we integrate speed, therefore, we recover distance traveled. Note that this is *not* usually the same as the displacement (change in position).

Unit tangents and normals may be represented in terms of arc length. Define curvature of a curve first for circles and then derive a general result for space curves. Verify that a line has zero curvature and that the formula correctly gives the curvature of a circle to be the reciprocal of the radius.

Sample Homework Problems:

L: 3, 11, 20, 24, 29, 31, 33, 35, 40, 41, 46
M: 1, 3, 13, 19, 23, 28, 30, 33, 39, 41, 46
N: 1, 4, 13, 19, 23, 28, 30, 34, 39, 42, 45

11.5 Tangential & Normal Components of Acceleration

Main Topics: Components of acceleration; applications.

Main Ideas: The acceleration of a particle may be viewed as resulting from a change in speed, a change in direction, or both. We show that component of acceleration due to the change in speed is in the tangential direction whereas the component due to directional change is normal to the curve.

Warnings and Pedagogy: The tangential component has magnitude equal to the derivative of the speed, as would be expected from experience and/ or intuition. However, the normal component's magnitude involves curvature and so is tedious. Since it is easy to calculate the acceleration vector as well as the tangential component, the Pythagorean theorem simplifies the calculation of the magnitude of the normal component.

Implications and Applications: If the speed is constant, then the acceleration is orthogonal/ normal to the curve. (As before, this may be viewed geometrically, since any vector function of constant magnitude r has all vectors terminating on a sphere of radius r.) Apply all this to the motion of a satellite to find the period or related aspects.

Sample Homework Problems:

L: 2, 7, 9, 11, 12, 14, 18, 20, 23, 24, 29, 36
M: 1, 7, 10, 12, 15, 17, 19, 20, 23, 24, 28, 36
N: 1, 8, 10, 15, 17, 19, 22, 25, 28, 30, 33

Additional Exercises are in the Chapter 11 Review. We recommend doing all of the *Proficiency Exam* questions, but include at least those in the *Survival Manual*, *viz.*: 24, 27, 30.

Cumulative Review: For additional practice of the material in the first six chapters, we recommend the following: 6, 9, 12, 15, 18, 21, 24, 27, 30, 33, 36.

CHAPTER 12: PARTIAL DIFFERENTIATION

Chapter Overview and Context: The first 10 chapters dealt mostly with single functions of a single real variable. Chapter 11 dealt with vector-valued functions of a real variable, which is tantamount to many functions of a single real variable (given by parametric equations). In this chapter we go back to a single function, but now a function of many real variables. (Equivalently, we may take the geometric view of functions of a point, but with the domain points now in a higher-dimensional space.)

12.1 Functions of Several Variables

Main Topics: Definition and basic concepts; level curves and surfaces; the graph of a function of two variables.

Main Ideas: Functions of several variables, especially of two or three variables; domain, range, level curve, level surface, graph. A function f of two variables, x and y, is a relationship that assigns to each pair (x, y) of real numbers a single real number $z = f(x,y)$. By having more independent variables, one has greater complexity but it models reality better/ more accurately. Later, with partial derivatives, we have the idea of a scientific experiment – namely, controlling all independent variables but one and seeing the resulting effect on output (dependent variable).

Warnings and Pedagogy: Domain and range are defined in a way that extends the definitions for functions of a single variable. Operations on functions likewise are easily defined. Mirror earlier definitions and developments, taking care to note new notions such as level curves and surfaces. Given $z = f(x,y)$, students need to see the differences among domain, level curves, and graph. Do graphing. Note that the domain and level curves are in two-dimensional space whereas the graph is in three-dimensional space.

Interpret level curves (and level surfaces) with specifics: contour maps, isoclines, isotherms, isobars, equipotentials.

Technology: Another excellent section warranting computer usage, for many graphs are difficult to visualize or draw.

Sample Homework Problems:

L: 3, 8, 9, 15, 19, 27, 32, 50, 59
M: 4, 9, 10, 17, 18, 22, 27, 28, 31
N: 4, 10, 11, 17, 18, 22, 26, 28, 31

12.2 Limits and Continuity

Main Topics: Limit of a function of two variables; limit properties; continuity.

Main Ideas: Just as last section extends the definitions from single-variable calculus for function concepts, this section extends to limit concepts, applied to functions in a general way and specifically for continuity and derivatives. Note that we frequently considered approaches from two sides in single-variable calculus, but in multi-variable calculus the approach is from anywhere in a disk. (Note that a "disk" in one dimension is actually an interval.)

Pedagogy: The mentality of "easier to destroy than create" still stands with existence of limits. That is, to show a limit does not exist, one may find two paths along which the limits produce different values. However, if they produce the same value, the limit may still not exist.

Extend repertoire to use "algebraic" limit properties (sum, product, quotient, etc.). Define continuity by reminding of the definition for one variable and showing the new one to have the exact same structure (function defined at point, limit of function exists as one approaches point, and the two quantities just cited agree in value). Build a small catalogue of familiar continuous functions: polynomials in two variables, rational functions (except where the denominator vanishes), etc.

Sample Homework Problems:

L: 3, 6, 12, 15, 18, 23, 30, 35, 39, 42 (if do formal limits)
M: 4, 6, 11, 15, 17, 24, 30, 34, 38, 42 (if do formal limits)
N: 4, 5, 11, 16, 17, 24, 29, 34, 38, 43 (if do formal limits)

12.3 Partial Derivatives

Main Topics: Partial differentiation definition, shortcut calculation, geometric interpretation (as slope), physical interpretation (as rate), and higher partials.

Main Ideas and Pedagogy: Define partial derivatives after reminding of the analogous definition for function of a single variable. Calculate by shortcuts with holding other variable(s) constant. Give the various notations. A nice touch is the idea of a controlled scientific experiment in which we hold all variables constant except one.

The physical interpretation is instantaneous rate of change. The geometric interpretation should make clear that one is looking at slopes of curves in various planes. Holding y constant in $z = f(x,y)$ implies looking at a curve in a plane parallel to the xz-plane, whose slope is found from f_x, the partial derivative of the function f with respect to x.

Notation and Cautions: Since a partial derivative is again a function of several variables, one may take higher partial derivatives. Point out the different notations (so-called pre-fix and post-fix notations) for mixed partials. Of course, the mixed partials are equal, mitigating this.

Technology: Within limits, one can use a computer algebra system for some of the messier calculations, but this should supplement, not supplant, the basics.

Sample Homework Problems:

L: 4, 9, 13, 20, 26, 31, 34, 38, 46, 48
M: 1, 4, 5, 9, 13, 21, 23, 31, 34, 38, 47
N: 1, 5, 6, 10, 14, 21, 23, 32, 35, 40, 47

12.4 Tangent Planes, Approximations, Differentiability

Main Topics: Tangent planes; incremental approximations; the total differential; differentiability.

Main Ideas: The tangent plane to a surface is a natural extension of the tangent line to a curve. Its equation, which derives from the equation of a general plane, is likewise a natural extension of the point-slope formula.

Extend the increment and the differential of a function of one variable to the increment and the total differential of a function of two (or more) variables. The tangent plane is the planar approximation to a surface – that is, the local linearization. With the total differential of a function we can approximate function value changes (as well as function values). Note that the total differential can be regarded as the sum of two "one-dimensional" differentials, one for each of the x and y directions.

We can also calculate relative changes and relative errors. Note that, as in the one-variable case, one can answer questions about relative changes even without specific dimensions.

Warnings and Pedagogy: Distinguish carefully between the increment in function f and the total differential (df). As with the one-variable case, students need to take care in distinguishing absolute or actual change from relative or percentage change. Dimensional analysis for the units should help here.

Another major source of confusion is *differentiability*. In single-variable calculus, this means only that the derivative exists (as a limit). However, its meaning is more involved here. However, differentiability still implies continuity.

Sample Homework Problems:

L: 1, 6, 8, 23, 26, 31, 33, 39, 45, 48
M: 1, 5, 9, 23, 29, 31, 32, 37, 45, 49
N: 3, 5, 9, 24, 29, 32, 34, 37, 49

12.5 Chain Rules

Main Topics: Chain rules; general pattern.

Main Idea and Pedagogy: We are given a function of one or more "intermediate" variables, each of which is, in turn, a function of one or more "ultimate" variables. There is a general pattern for the derivative(s) of the function with respect to the ultimate variables. Give a few situations, also identifying when to expect ordinary derivatives vs. when to expect partial derivatives. Students should eventually understand that there are as many chain-rule equations as there are ultimate variables, and each equation has as many terms as there are intermediate variables. Each of these terms is a product reminiscent of the one-variable case. By drawing a diagram one can "chase" variables to see how to handle all such situations. Pick an example in which you can safely eliminate the intermediate variable(s) and compare the chain rule results with the direct calculation. (For instance, see Example 1.)

On a less mechanical level, the chain rules follow from the

differentiability notions of the previous section. Show at least one such derivation. Related rates problems (Example 3, or ideal gas law) provide nice applications.

Sample Homework Problems:

L: 5, 10, 14, 20, 24, 28, 36, 38, 44, 48
M: 3, 6, 10, 15, 20, 23, 30, 37, 39, 44
N: 3, 6, 11, 15, 21, 23, 30, 37, 39, 54

12.6 Directional Derivatives and the Gradient

Main Topics: The directional derivative; the directional derivative as a slope and as a rate of change; the gradient; maximality, minimality, and normality properties of the gradient; tangent planes from gradients; normal lines.

Main Ideas, Warnings, and Pedagogy: Generalize the partial derivative by viewing it as derivative in a particular direction and then allowing a unit vector to specify an arbitrary direction. Although the definition is a cumbersome limit, a theorem comes to the rescue to show how to calculate a directional derivative of a function for a given direction. We recognize the expression as having the form of a dot product. For that reason and to simplify the notation and the process, define the gradient (vector function) of a scalar vector and show that the directional derivative is found by dotting the gradient of the function with a unit vector in the desired direction. Take care to explain that we view the unit vector as lying in the domain of the function, not on its graph. Outline the basic gradient properties, but don't spend too much time on proofs.

Show a diagram of the plane curve whose slope is given by the directional derivative. Give interpretation as rate, too.

The maximality and minimality properties of the gradient give it an importance beyond the mere convenience value for computing directional derivatives. Be careful to distinguish the gradient from its norm. The gradient at a point gives the direction in the domain (from that point) in which the function instantaneously changes most rapidly. That maximal rate happens to be the norm of the gradient. Do prove both parts of this result – the direction claim and the magnitude result. The former follows from this observation: The dot product of two vectors (gradient and any unit vector) of constant magnitude but whose included angle varies is the product of the two magnitudes times the cosine of the angle, with the cosine of the angle maximized when the angle is zero. The second observation follows from the directional derivative calculation using a unit vector in the direction of the gradient itself.

A nice interpretation of this is to think of a mountain and level curves. One would move at a right angle to a level curve in order to rise or drop most rapidly in altitude. Many students have such experience with hills and should find this a satisfying confirmation of the theory.

The normality property of the gradient is also most useful. Given a level curve for a function of two variables, or a level surface for a function of three variables, the gradient of the function at a point in the domain is normal to the level curve or level surface (whichever applies) through

that point. This property can be used to find a normal, and hence, the tangent plane, to *any* surface. The trick is to create a function, one of whose level surfaces is the given surface. For example, to find a normal to the ellipsoid of revolution given by $x^2 + y^2 + \frac{z^2}{4} = 1$ at $(\frac{\sqrt{3}}{2}, 0, -1)$, let

$w = F(x, y, z) = x^2 + y^2 + \frac{z^2}{4}$. Then the ellipsoid is just the $w = 1$ level surface of F, whose gradient vector is easy to compute. From this vector we can then get the equation of the tangent plane.

Sample Homework Problems:

L: 7, 10, 16, 18, 27, 37, 42, 49, 55
M: 3, 7, 9, 13, 18, 26, 27, 36, 47
N: 3, 9, 13, 19, 26, 35, 36, 47

12.7 Extrema of Functions of Two Variables

Main Topics: Relative extrema definition, method; second-partials test; absolute extrema of functions of two variables.

Main Ideas and Pedagogy: After defining terminology that extends the one-variable calculus definitions, we seek relative (and absolute) extrema by setting the first partials to zero (which is tantamount to setting the gradient equal to the zero vector) and solving, then checking boundary points and points where the gradient is undefined. Geometrically, we are looking for points where the tangent plane is either horizontal or non-existent. Points obtained from setting the gradient to zero are called *critical points*.

Analogous to the second-derivative test is the second-partials test. It, too, has shortcomings, such as not detecting absolute extrema and sometimes being inconclusive. Show saddle points as reason for needing such a test, lest anybody assume that a zero gradient always yields an extreme value.

When the domain is a closed and bounded set and the function is at least continuous, we are assured of not only relative extrema but absolute extrema. This is directly analogous to the situation for a continuous function of a single real variable defined on a closed, bounded interval. Indeed, as was the case there, we check all points where the first partials vanish or are undefined as well as boundary points. We then inspect the values for maxima and minima. Show applications such as least surface area for fixed volume or minimum distance between point sets. The next section will be devoted to more optimization applications.

Technology: Selectively graph to confirm extrema.

Sample Homework Problems:

L: 8, 12, 18, 22, 27, 33, 35, 40, 45, 47
M: 4, 5, 6, 11, 12, 18, 25, 35, 37, 40, 42
N: 4, 5, 6, 11, 14, 21, 25, 36, 37, 43, 46
One may wish to supplement additional application problems before going on to Lagrange multipliers, and later compare methods. For instance, what is the largest possible sum of the sines of the three angles of any triangle? It will be instructive to compare methods.

12.8 Lagrange Multipliers

Main Topics: Method of Lagrange multipliers and the geometric insight; constrained optimization; extended method with two parameters.

Main Ideas: Given constraint $g(x,y,z) = c$, we seek to optimize $f(x,y,z)$. That is, we seek the points (x, y, z) lying on the surface given by the constraint equation that make function f a maximum and a minimum. The method of Lagrange multipliers provides an alternative means of solving constrained optimization problems such as this (covered in the previous section). Rather than eliminate a variable in an objective function f using the constraint equation, one sets the gradient of the objective function f equal to a multiple of the gradient of the constraint function g. One then solves that equation simultaneously with the constraint equation. (Similar remarks apply for functions of two variables rather than of three.)

The multiplier λ has an interpretation not found in other calculus texts: For each constant c in the constraint $g(x,y,z) = c$, let $E = E(c)$ denote the resulting extremum (whether a maximum or minimum). Then λ is the rate at which the extremum changes with respect to c.

The geometric interpretation is important for students to see. Regard the constraint as a level surface and consider all level surfaces of the objective function f. These level surfaces for f may be thought of as parallel to each other. Many may miss the level surface for the constraint, and others may intersect is at interior points. However, the one that touches the constraint's tangentially is the one we want. It also shows that we get parallel normals, and thus parallel gradients, which is what the Lagrange method seeks.

The method is extended to allow two constraint equations and thus two parameters.

Sample Homework Problems:

L: 3, 9, 15, 20, 26, 28, 30, 33, 44, 45
M: 1, 9, 15, 19, 23, 24, 25, 27, 44, 46
N: 1, 7, 16, 19, 23, 24, 25, 27, 42, 46
See the note at the end of the previous section.

Additional Exercises are in the Chapter 12 Review. We recommend doing all of the *Proficiency Exam* questions, but include at least those in the *Survival Manual*, viz.: 33, 36, 39.

CHAPTER 13: MULTIPLE INTEGRATION

13.1 Double Integration over Rectangular Regions

Main Topics: Definition of double integral and properties; geometric interpretation as volume; Fubini's theorem (informal look at switching the order of integration).

Main Ideas and Pedagogy: Review the definition of the Riemann integral of a function of a single real variable. Our function now will be a function of two variables. We also extend from one-dimensional domain of an interval to the two-dimensional domain of a rectangular region (and later a more general region – *e.g.*, one enclosed by curves). Indeed, we may partition each of two intervals now, one corresponding to each axis or direction, thus inducing a partition of the rectangular region. Choose a point in each sub-rectangle. Form the appropriate Riemann sum for the function over the region using the partition. We then take the sum's limit as the partition is made finer, meaning the size of the largest sub-rectangle shrinks to zero and the number of these little rectangles increases without bound.

(Note that the sum could have been given as a double sum, which would lead more naturally to the double integral notation, but which would be more vexing for students.)

Show the properties of linearity, dominance, subdivision. Then interpret the foregoing partitioning in terms of approximating the volume of a solid bounded by a surface over a region – at least for the case of a function that is nonnegative over the region in the xy-plane. This leads naturally to an iterated integral for the volume.

As is so often the case in calculus, we seek a faster means of calculation than the one provided by the definition. To evaluate a double integral, we calculate iterated integrals. Draw the analogy of the partial integration to partial differentiation here.

Technology and A Warning: Multiple integration is straightforward with CAS software such as Derive, and the order of integration does not really matter for rectangular domains. In the next section, however, this changes with more general domains. In theory, Fubini's theorem assures us that either iterated integral agrees with the double integral. In practice, the order may matter. For this and other reasons, switching the order of integration (that is, setting up one iterated integral after being given the other) should be practiced. There are times when one integral is difficult or impossible and other one is easy. Be prepared to make this transition and to convey this to your students.

Sample Homework Problems:

L: 3, 7, 12, 14, 17, 21, 24, 33, 38, 42, 46
(Students will need help with a proof such as asked in #46.)
M: 1, 2, 3, 7, 11, 15, 22, 24, 38, 42, 43
N: 1, 2, 4, 8, 11, 15, 22, 26, 32, 41, 43

13.2 Double Integration over Non-Rectangular Regions

Main Topics: Non-rectangular regions; more on area and volume; reversing the order of integration;

Main Ideas: We extend the double integral's definition as well as the means of calculating it to non-rectangular domains. Note the device of extending the non-rectangular region D to a rectangular region R and defining the function f outside D (but inside R) to be zero.

Warnings and Pedagogy: The chief focus here is on setting up the iterated integrals for double integrals over more general regions. Care is needed, as switching order of integration is not trivial. Given an iterated integral, o reverse the order of integration, draw a diagram of the region from the limits of integration. Then consider strips in the opposite orientation. For select integrals, one has no choice but to reverse the order; example 6 is typical.

When the integrand is 1, one obtains the area of the region; when the integrand is f, the surface function, one obtains volume under the surface and above the region. The most typical scenario we consider is a planar region bounded by two curves and two lines. As a special case, integrating 1 will yield the area between the two curves (within the two lines). Nice graphical example: Show volume of the tetrahedron bounded by the coordinate planes and the plane $\frac{x}{a} + \frac{y}{b} + \frac{z}{c} = 1$ ($a, b, c > 0$) is $\frac{abc}{6}$.

Sample Homework Problems:

L: 3, 7, 13, 19, 24, 31, 35, 38, 41, 51, 54
M: 3, 4, 7, 13, 17, 20, 32, 35, 39, 41, 51
N: 4, 6, 9, 14, 17, 20, 32, 36, 39, 43, 55

13.3 Double Integrals in Polar Coordinates

Main Topics: Change of variables to polar form; improper double integrals in polar coordinates.

Main Ideas: We change to polar coordinates primarily to exploit circular symmetry in the domain of integration and/ or the presence of certain expressions in the integrand that lead readily to polar form (most notably, $x^2 + y^2$).

Warnings and Pedagogy: In rectangular coordinates, we may express an infinitesimal element of area by $dA = dx dy$. In polar coordinates, $dA = r dr d\theta$. Students may need to be reminded of the supposed "extra" r factor in dA while changing variables. We recommend taking the mystery out of this by looking at the area of a "polar rectangle" – which, in turn, is based on arc length: $s = r\theta$ – during the course of looking at the Riemann sum.

The usual extra care should be taken for situations with intersecting curves given in polar coordinates. We also consider improper double integrals, specifically in polar coordinates. The definitions and terminology from earlier carry over, although we don't delve too deeply here.

Applications: This topic provides a wonderful opportunity to obtain such results as the volume of a sphere (and in a manner that is even simpler than we had earlier in single-variable calculus with the volume of a solid of revolution).

Sample Homework Problems:

L: 2, 6, 14, 19, 23, 30, 33, 39, 44, 49, 57
M: 1, 3, 6, 14, 17, 19, 23, 30, 34, 44, 49
N: 1, 3, 7, 13, 17, 20, 24, 29, 34, 43, 50

13.4 Surface Area

Main Topics: Definition of surface area; projections for surface area; area of a parametrically-defined surface.

Main Ideas: Consider a surface $z = f(x,y)$. Instead of the volume under the surface above a region, we ask for the area of the surface above the same region. To obtain it, we partition the region, thus inducing a partition of the surface itself. We then approximate the area of each of the quasi-parallelogram shapes, sum these, and take the limit as we refine the partition. Note that this results in a formula reminiscent of the "single-variable" formula for arc length.

Warnings and Pedagogy: The development of the area of each parallelogram uses vector methods that student don't see often enough to feel comfortable, so take care here in showing the area of the parallelogram, reminding students that such area is the norm of the cross product (of vectors making up adjacent sides of the parallelogram). This may be even more true for parametrically-defined surfaces.

As with volume, surface area may sometimes be easier to compute in polar coordinates. (Homework: see #44.)

Sample Homework Problems:

L: 1, 6, 12, 16, 23, 26, 31, 36, 40, 43
M: 1, 3, 12, 18, 23, 27, 31, 33, 40, 44
N: 2, 3, 11, 18, 25, 27, 32, 33, 38, 44

13.5 Triple Integrals

Main Topics: Definition of triple integral; calculation by iterated integration; volume by triple integrals.

Main Ideas: A single integral of a function of one variable is evaluated over a one-dimensional point set (usually, an interval); a double integral of a function of two variables is evaluated over a two-dimensional point set (a region in a plane). We now extend to integration of functions of three variables to evaluate over a region in three-dimensional space, one contained within some rectangular solid box. The partitioning and the formation of the Riemann sums are directly analogous, as is the passage to the limit. (If we took each direction separately, we could actually obtain a triple sum, thus motivating the triple integral's notation. Again, however, this would likely pose difficulties for students.)

Show the properties of linearity, dominance, subdivision.

In one dimension there's only one = 1! order of integration. In two dimensions, there are two = 2! orders. In three dimensions, there are six = 3! orders for iteration integrals. Fubini's theorem applies to guarantee the equality of each of these to the triple integral, but at least as much care must be exercises in switching order of integration.

The most typical scenario in this section entails a region of space between an upper surface and a lower surface, and over a planar region similar to what we had with double integrals. Triple-integrating 1 over such a solid region yields the volume of the region.

A nice graphical example to revisit (contrast with handling as a double integral): Show the volume of the tetrahedron bounded by the coordinate planes and the plane $\frac{x}{a} + \frac{y}{b} + \frac{z}{c} = 1$ ($a, b, c > 0$) is given by $\frac{abc}{6}$.

Sample Homework Problems:

L: 2, 8, 13, 17, 22, 25, 29, 34, 40, 53
M: 5, 8, 13, 18, 22, 25, 30, 33, 40, 51
N: 5, 7, 11, 17, 21, 26, 30, 33, 41, 51

13.6 Mass, Moments, & Probability Density Functions

Main Topics: Mass, center of mass; moments of inertia; joint probability density functions.

Main Ideas: The mass of a planar lamina whose density function is known may be recovered by double integration of the density with respect to area. The formation of the Riemann sum is similar to that for our previous double integrals. Follow with moments and use to define the center of mass. The formulas are fairly easy for double integrals. Extend to triple integrals for the mass and center of mass of a solid. Also find second moments (also called moments of inertia) for planar laminas and solids. Relate to physics (kinetic energy). Briefly discuss concept of joint probability distribution as application of double integral. As usual, keep in mind switching to polar coordinates (such as #19 in the homework).

Sample Homework Problems:

L: 5, 8, 13, 18, 22, 26, 38, 44, 52, 54
M: 5, 7, 13, 19, 21, 26, 28, 34, 44, 58
N: 6, 7, 14, 19, 21, 25, 28, 34, 43, 58

13.7 Cylindrical and Spherical Coordinates

Main Topics: Cylindrical coordinates and integration; spherical coordinates and integration.

Main Ideas: Cylindrical coordinates generalize polar coordinates by adding third-dimensional coordinate z to the two-dimensional polar coordinates. Therefore, cylindrical coordinates enjoy the same advantage, namely, use in situations with circular symmetry. Review the conversions.

Warnings and Pedagogy: On the other hand, cylindrical coordinates entail the same caveats as we had with polar coordinates, including multiple representation of points, and, more to the point, taking care with the representative infinitesimal element for volume: $dv = dxdydz = rdzdrd\theta$.

For most students, despite use for latitude and longitude, spherical coordinates will prove less familiar and therefore more demanding of attention. Show sphere, half-cone, and vertical half-plane as special cases. Work conversions

between rectangular and spherical coordinates.

For integration with spherical coordinates, the infinitesimal or fundamental element is a wedge. We follow the same script of partitioning, forming Riemann sum, etc. Note that circular symmetry applies here, too, such as for getting the volume (and other measures) of a sphere.

Sample Homework Problems:

L: 31, 34, 38, 42, 44, 52, 57; one of 61 & 63
M: 10, 16, 21, 25, 31, 34, 38, 42, 52
N: 10, 16, 21, 25, 32, 33, 40, 43, 51

13.8 Jacobians: Change of Variables

Main Topics: Changing variables in double integrals and in triple integrals.

Main Ideas and Pedagogy: We generalize two ideas encountered earlier: First, in one-variable integral calculus, we may now regard change of variable or substitution as providing for a "mapping factor" (a derivative) for converting an integral into a simpler integral, one whose interval is changed, too. Second, with a double integral, we seek something similar –a mapping factor and a new region (and we hope the latter is "simpler"). We've already seen an instance of this with polar coordinates, where the factor is the polar variable r. We seek to generalize this idea of getting the factor – *the Jacobian* – as well as the new region. Give examples, but don't prove the main result. (Students will already be familiar with determinants.)

Sometimes it is the form of the integrand that suggests a change of variables. Sometimes it is the region. Show the formula for the Jacobian for conversions considered so far, if only to verify the theorem for those cases.

Sample Homework Problems:

L: 1, 7, 11, 15, 18, 21, 31, 35, 38, 41
M: 1, 3, 7, 12, 15, 18, 22, 32, 37. 38
N: 3, 4, 8, 12, 16, 17, 22, 32, 37. 42

Additional Exercises are in the Chapter 13 Review. We recommend doing all of the *Proficiency Exam* questions, but include at least those in the *Survival Manual*, viz.: 24, 27, 30.

CHAPTER 14: VECTOR ANALYSIS

14.1 Properties of a Vector Field: Divergence and Curl

Main Topics: Definition of a vector field; divergence; curl; physical interpretation of curl.

Main Ideas: We previously have had vector-valued functions of a single variable or parameter (usually time). We now consider *vector fields* – functions that associate to each point in space a single vector. We may look at the vector's components, each of which is a function of several variables. This means that we are now considering several functions of several independent variables.

Warnings and Pedagogy: Graphing is straightforward,

especially if the space is a subset of the plane. Example: a velocity field assigning to each point a velocity. Next define two derived fields, one a scalar field, one a vector field. They are, respectively, the divergence and the curl. Note that each starts with a vector field, but then we obtain a scalar field/ function for divergence or another vector field for the curl. (We have already had the gradient, which is also a vector field, but it's derived from a scalar field.)

Show the notation and examples. Define terminology for special situations such as incompressibility (div V = 0). Remind of the **del** operator and use it to show the alternate nomenclature ("del dot V " for divergence, and after curl is introduced, "del cross V "). Repeat with curl, including special situations such as irrotationality (**curl** V = **0**).

Follow with the Laplacian operator.

Interpret curl as tendency for a fluid to swirl.

Remind of the need to watch for scalars vs. vectors. Do take advantage of computer algebra systems for the messier calculations once the basics are mastered.

Sample Homework Problems:

L: 4, 9, 14, 18, 24, 29, 35, 39, 46, 53
M: 3, 5, 12, 14, 18, 22, 30, 40, 46, 51
N: 3, 5, 10, 12, 15, 17, 22, 30, 40, 51

14.2 Line Integrals

Main Topics: Definition of a line integral; evaluation of line integrals in parametric form; line integrals of vector fields; computing work using line integrals; evaluation of line integrals with respect to arc length.

Main Ideas: We continue generalization by introducing line integrals, which actually are integrals over curves. This immediately allows dealing with the work done by a force in a general fashion. (We previously dealt with the work done by a constant force, by a variable force, and even by a constant force applied in a different direction than the displacement. Now the motion may lie along a curve – which is to say, we have a variable force and a variable displacement, with variable angle between them.)

Warnings and Pedagogy: Define line integrals of a function of several variables over a curve C in space – with respect to x or y or z. Caution students, who don't yet see why we're doing this, of the difference made by this one choice of variable alone. Give examples of one or two without actual evaluation, and provide major properties of line integrals. For the evaluation, point out that this is just one more case of not using the definition but rather a theorem. We parametrize the curve C and substitute instead, resulting in an ordinary definite Riemann integral.

Build on the foregoing to define a line integral of a vector field. In effect, we reduce this to the previous case of line integrals with respect to x, y, and z – which, in turn, we reduced to ordinary integrals by parametrization.

Next, broach the question of what happens if we change the path of integration (the curve C). Don't create impression that *path-independence* is the norm; give example that yields same answer for both paths, but another yielding different answers for the paths.

The payoff comes in computing work using line integrals. If force and displacement vectors are constant, then their dot product gives the work done by the force in moving an object through the displacement. At the other extreme, if force and displacement are both variable, then we need the definition of line integral. Give examples.

Close out the section with definition of line integral with respect to arc length.

Sample Homework Problems:

L: 3, 7, 10, 12, 19, 21, 28, 35, 37, 40, 47
M: 3, 5, 9, 10, 11, 19, 21, 28, 35, 37, 41
N: 4, 5, 9, 11, 13, 18, 20, 25, 32, 36, 41

14.3 Independence of Path

Main Topics: Conservative vector fields; simply-connected regions; the fundamental theorem of line integrals.

Main Ideas: In general, the value of a line integral depends not only on the vector field integrand and the endpoints of the path, but also on the curve C connecting the endpoints. We seek to characterize the force fields (and the regions of space in which the curves lie) for which the line integral is independent of path. Second, when we have this, what is the most expeditious means of calculating the line integral?

Pedagogy: Define *conservative field* F (we define it by F being the gradient of some scalar function f, but any equivalent would do). The following are equivalent (within some simply-connected region D of two-dimensional space):

- $\vec{F} = u\vec{i} + v\vec{j} = \vec{\nabla}f$ (F is the gradient of some function);

- $\dfrac{\partial u}{\partial y} = \dfrac{\partial v}{\partial x}$ (F passes the cross-partials test);

- $\displaystyle\int_C \vec{F} \bullet d\vec{R}$ is independent of path.

- $\displaystyle\int_C \vec{F} \bullet d\vec{R} = 0$ for any closed curve C.

It is the last condition that leads to the nomenclature, particularly *conservative field* and *conservation of energy*. The cross-partials test provides an easy means of testing whether a field is conservative. However, when this test is passed, it shows *existence* of a scalar potential function f, but it does *not* show *how* to find it. For that, we resort to partial integration in one variable and then differentiation of that result in the other variable. This diagram may help:

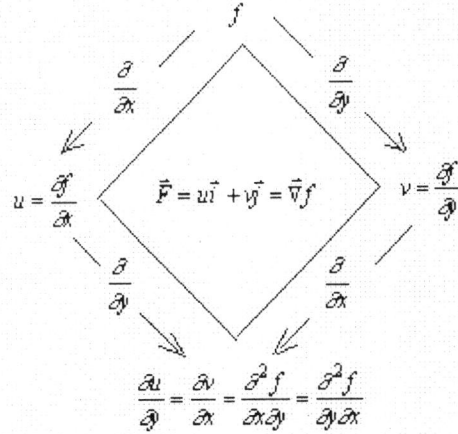

Note: The diagram summarizes many relationships with a conservative field. It is rich in these. It even subsumes the equality of the mixed partials.

For conservative field F, evaluation of line integrals is a matter of finding scalar potential f: $\int_C \vec{F} \bullet d\vec{R} = f(Q) - f(P)$ where $\vec{F} = \vec{\nabla} f$ and curve C has initial point P and terminal point Q.

Sample Homework Problems:

L: 4, 9, 11, 14, 20, 21, 26, 32, 37, 38, 39, 44
M: 4, 9, 12, 14, 20, 22, 26, 28, 33, 35, 39, 43
N: 5, 7, 12, 15, 22, 23, 25, 28, 33, 35, 42, 43

14.4 Green's Theorem

Main Topics: Green's theorem; area as a line integral; multiply-connected regions; theorem in alternative form.

Main Ideas: We relate a certain double integral over a planar region to a particular line integral over a closed curve bounding this region. In a sense, this generalizes the previous section, because a conservative field used for the calculation will yield zero for the double integral (because of the cross-partials test) and also for the line integral (one of the equivalents of conservation of energy).

Pedagogy: In the simplest cases, the region is simply-connected and the boundary curve C is positively oriented. (Suitable smoothness of the functions is still assumed.) In more involved cases, cuts may be made to subdivide the region into sub-regions, each simply connected.

As an application, we may compute work done by a non-conservative force field in moving an object along a closed path. We replace the line integral by the equivalent double integral given by Green's theorem. A more geometric application is area of the region.

Extend results to multiply-connected regions, including those with a singular point. Give the alternative form of Green's theorem as a prelude to Stokes' theorem; it is: $\int_C \vec{F} \bullet d\vec{R} = \iint_R (curl(\vec{F}) \bullet \vec{k})dA$. Note that this is the line integral of the tangential component of the force.

Give a divergence theorem in the plane as a prelude to Gauss's divergence theorem; the version in the plane is: $\int_C (\vec{F} \bullet \vec{N})ds = \iint_R (div(\vec{F}))dA$. Note that this is the line integral of the normal component of the force.

Sample Homework Problems:

L: 1, 4, 7, 13, 17, 21, 22, 24, 27, 30, 36
M: 3, 4, 7, 13, 18, 21, 22, 27, 31, 34
N: 3, 8, 14, 15, 18, 23, 26, 31, 34

14.5 Surface Integrals

Main Topics: Surface integrals; flux integrals; integrals over parametrically-defined surfaces.

Main Ideas: We now integrate over a surface instead of over a curve. Quickly lead to the Riemann-like sum and the final result for the surface integral. Example: Calculate the mass of a curved lamina from its density function.

A special kind of surface integral is a flux integral, which requires the notion of an orientable surface. The outward unit normal can always be found by taking the gradient to the surface divided by its norm. In turn, consider the case of a surface described by parametric equations.

Sample Homework Problems:

L: 4, 8, 12, 16, 20, 24, 26, 29, 33, 41
M: 1, 4, 9, 12, 15, 20, 24, 29, 33, 42
N: 1, 5, 9, 11, 15, 19, 25, 30, 34, 42

14.6 Stokes' Theorem

Main Topics: Stokes' theorem; theoretical application; physical interpretation.

Main Ideas and Pedagogy: We pick up the earlier observation that Green's theorem can be rewritten, specifically the integrand of the double integral to involve a curl and a dot product: $\int_C \vec{F} \bullet d\vec{R} = \iint_R (curl(\vec{F}) \bullet \vec{k})dA$. We now generalize it to this: $\int_C \vec{F} \bullet d\vec{R} = \iint_S (curl(\vec{F}) \bullet \vec{N})dS$. This is the conclusion of Stokes' theorem. Note that when the surface is just a region of the xy-plane, we have surface $S = R$ and normal $N = k$.

Some attention to the hypotheses is in order. Briefly discuss the notion of a bounding curve's orientation C being compatible with that of the surface S it bounds (the latter idea having been considered in the previous section).

As with other identities, one can use one side to substitute for calculating the other side (see examples 2 and 3).

A consequence of Stokes' theorem is that whenever F and curl F are continuous in a simply-connected region D, then we have F conservative in D if and only if curl $F = \mathbf{0}$.

Caution: Note that our earlier test for conservation applied to two dimensions. The result here extends what we had earlier, but instead of one pair of partial derivatives being equal, we actually have three pairs.

Apply Stokes' theorem to the velocity field of a fluid flow to provide a nice physical interpretation.

Sample Homework Problems:

L: 1, 7, 10, 15, 17, 20, 24, 26, 31, 40
M: 1, 8, 10, 15, 18, 20, 25, 26, 33, 36
N: 2, 8, 11, 14, 18, 21, 25, 27, 33, 36

14.7 Divergence Theorem (Gauss's theorem)

Main Topics: Divergence theorem; applications; physical interpretation of divergence.

Main Ideas: Green's theorem earlier gave us a divergence theorem in the plane: $\int_C (\vec{F} \bullet \vec{N})ds = \iint_R (div(\vec{F}))dA$. We now generalize to Gauss's divergence theorem. Under suitable hypotheses *: $\iint_S (\vec{F} \bullet \vec{N})dS = \iiint_R (div(\vec{F}))dV$.

(* S is a smooth, orientable surface that encloses a solid region R; F is a continuous vector field whose components have continuous partial derivatives throughout R; and N is the outward unit normal vector to the surface S.)

Verify with a few examples, although proof is outside scope of the course (beyond an important special case that has been relegated to an appendix).

As before, we may use the Stokes' theorem identity to find one side by actually calculating the other (see examples 2 and 3). Applications include fluid dynamics (the continuity equation), derivation of the heat equation, and some of Green's identities. Close with the physical interpretation of the divergence as a net rate of flow of volume.

Sample Homework Problems:

L: 1, 6, 11, 15, 20
M: 1, 5, 11, 15, 20
N: 4, 5, 12, 16, 21

Additional Exercises are in the Chapter 14 Review. We recommend doing all of the *Proficiency Exam* questions, but include at least those in the *Survival Manual*, viz.: 24, 27, 30.

CHAPTER 15: INTRODUCTION to DIFFERENTIAL EQUATIONS

15.1 First-Order Differential Equations

Main Topics: Review of separable differential equations; homogeneous differential equations; review of first-order linear differential equations; exact differential equations; Euler's method.

Main Ideas and Pedagogy: A differential equation is an equation involving an unknown function and one or more of its derivatives. If the derivatives are all ordinary, we have an ordinary differential equation (ODE). The order of the ODE is the highest derivative appearing. We seek the general solution, where possible, and when we have added conditions (boundary conditions or initial conditions), we seek particular solutions. We confine ourselves in this section to several techniques for first-order ODEs.

Review separation of variables briefly. Introduce the concept of homogeneity. (Caution: The word homogeneous has two distinct meanings in ODE.) Loosely, the idea is that we have $\frac{dy}{dx}$ equal to an expression whose terms each have the same total power. A substitution reduces the problem to solving the new ODE by separation.

Review first-order ODEs briefly. Example 5 has the nice example of a falling body with air resistance.

Exact equations actually dovetail with the earlier material on conservative force fields. The goal is to represent the equation in the form $M\,dx + N\,dy = 0$ and to see whether the left side is an exact differential: $df = M\,dx + N\,dy$. This leads to the same cross-partials test as earlier (albeit with the letters u and v replaced by M and N here), and partial integration then yields the desired f. In turn, that yields the implicit solution $f(x,y) = c$ (constant).

Warning: Many students mechanically obtain f without realizing that presenting this alone does not solve the problem. The general solution is $f(x,y) = c$, not f alone.

Euler's method provides a general method for approximate numerical solutions to any first-order initial-value problem (IVP). It is closely related to the direction field of the underlying ODE, which should be shown to assist. The result is simple and elegant. Such a diagram, however, also makes clear the accumulation of error that often occurs with the Euler method as one moves away from the initial point.

Reminder: Don't forget here and in later sections to put some emphasis on learning to categorize ODEs so as to know which method to use, not merely how to use a method on cue from text, instructor, or authority figure. Note that the abundance of technology is again both blessing and curse, for solution by computer often obscures such issues as the kind of ODE one has, how it is solved, etc.

Sample Homework Problems:

L: 1, 6, 8, 13, 19, 25, 28, 32, 35, 38, 45, 59
M: 2, 6, 9, 13, 20, 25, 28, 32, 35, 38, 47, 55
N: 2, 5, 9, 14, 20, 26, 27, 31, 34, 39, 47, 55

15.2 Second-Order Homogenous Linear Differential Equations

Main Topics: Linearity; linear independence; solutions of equations with constant coefficients; higher-order linear equations; application (damped motion of a mass on a spring); the method of reduction of order.

Overall Comment and Caution: This is a difficult section because it encompasses so many ideas, most notably the concept of linear independence. Even in full linear algebra and /or differential equations courses, students reveal difficulty. Balance this need for concern with realistic goals. This is an introduction and not a full course. For the specific example of linear independence, we move quickly to the Wronskian, not dwelling long on more subtle issues.

Main Ideas: Define linearity (linear ODE) and homogeneous vs. nonhomogeneous. Cautions: Students need to understand that this word "homogeneous" has a different meaning than in the previous section. Moreover, the concept is peculiar to linear ODEs; there is no such thing as a homogeneous or nonhomogeneous non-linear equation. And finally, students should not be too literal about "right member being zero" as the definition. The real issue is whether or not there is a loose term – a loose function of the independent variable. (One can always transpose such a loose term to the left side.)

Introduce linear independence and come to the Wronskian. From this point we move to superposition of linearly independent solutions, meaning that any linear combination of solutions is again a solution. At this point solving second-order homogeneous linear ODEs devolves to finding two linearly independent solutions. These we find by intelligent assumption of a solution of the form e^{rx} for some unspecified constant(s) r. Lead from this to the characteristic equation and the nature of the roots. There are three cases and these cases provide all possible basic solutions. It is fine to have students go right to the characteristic equation *after* they see how it arises from the assumption of the form of the basic solutions.

Higher-order equations are handled analogously. Under the surface of all this is the fundamental theorem of algebra about the zeros of polynomials (the roots of the characteristic equation). Tread lightly here.

Damped motion of a mass on a spring provides a nice application. (In ODE courses, one also deals with simple series circuits, but you'd be hard pressed for time here. Besides, we treat them in the next section.)

Reduction of order is a special technique for solving certain ODEs. As its name implies, it leads to a new ODE with lower order, which then allows solving the original equation. Keep in mind that any ODE from these topics can be made into an initial-value problem by adding side conditions to the ODE itself.

Sample Homework Problems:

L: 2, 4, 18, 22, 27, 34, 39, 46, 48
M: 1, 3, 18, 21, 27, 35, 39, 46
N: 1, 3, 15, 21, 28, 35, 40, 51

15.3 Second-Order Nonhomogeneous Linear Equations

Main Topics: Nonhomogeneous equations; method of undetermined coefficients; variation of parameters; RLC circuit applications.

Main Ideas, Pedagogy, and Warnings: The most important idea of this section is generalized superposition, namely that the general solution to a nonhomogeneous equation is the sum of the general solution to the corresponding homogeneous equation and any one particular solution to the nonhomogeneous equation. (By "corresponding homogeneous equation" we mean the equation that results when the nonhomogeneous terms are replaced by zero.) This leads to the simple method: Solve the homogeneous-case equation first. By any means get one solution to the original nonhomogeneous equation. Finally, add those two solutions.

The crux of the matter is now reduced to getting a particular solution to the nonhomogeneous ODE. When the nonhomogeneous terms are of certain special forms (polynomials, simple sines and cosines, simple exponentials, and sums and/ or products of these), we use the method of undetermined coefficients. This is basically intelligent guessing of the form of the solution, but with unspecified constants. Substitution then determines whether there is such a solution, and if so, it yields the values of these constants. Some care must be taken for guessing carefully, analogous to the method of partial fractions. If we use too few terms in the guess, we will obtain a contradiction when we substitute into the ODE. This indicates a flawed assumption. Using too many terms is not deadly, but it is inefficient.

For nonhomogeneous terms that are not of the required forms, we use variation of parameters, which is actually akin to reduction of order.

Caution: In solving an initial-value problem, we must take care to substitute in the initial conditions only with the full solution to the nonhomogeneous problem. It is a common error among students to substitute prematurely.

RLC circuits lead to initial-value problems that provide a delightful application of the ideas of this section. If there is no impressed voltage, the equation is homogeneous. With a supplied power source, the equation is nonhomogeneous. In this latter case, the general solution to the corresponding homogeneous equation and any particular solution to the nonhomogeneous equation each have particular relevance: the transient solution and the steady-state solution.

Sample Homework Problems:

L: 3, 6, 17, 21, 25, 30, 33, 36, 41
M: 3, 5, 17, 20, 21, 24, 28, 29, 41
N: 4, 5, 19, 20, 24, 28, 29, 34, 42

Additional Exercises are in the Chapter 15 Review. We recommend doing all of the *Proficiency Exam* questions, but include at least those in the *Survival Manual*, viz.: 12, 15, 18.

MISCELLANEOUS FORMULAS

SPECIAL TRIGONOMETRIC LIMITS:
$$\lim_{h\to 0}\frac{\sin h}{h} = 1 \qquad \lim_{h\to 0}\frac{\cos h - 1}{h} = 0$$

APPROXIMATION INTEGRATION METHODS: Let $\Delta x = \frac{b-a}{n}$ and, for the kth subinterval, $x_k = a + k\Delta x$; also suppose f is continuous throughout the interval $[a, b]$.

Rectangular Rule:
$$\int_a^b f(x)\,dx \approx [f(a + \Delta x) + f(a + 2\Delta x) + \ldots + f(a + n\Delta x)]\Delta x$$

Trapezoidal Rule:
$$\int_a^b f(x)\,dx \approx \tfrac{1}{2}[f(x_0) + 2f(x_1) + 2f(x_2) + \ldots + 2f(x_{n-1}) + f(x_n)]\Delta x$$

Simpson Rule:
$$\int_a^b f(x)\,dx \approx \tfrac{1}{3}[f(x_0) + 4f(x_1) + 2f(x_2) + \ldots + 4f(x_{n-1}) + f(x_n)]\Delta x$$

ARC LENGTH:

Let f be a function whose derivative f' is continuous on the interval $[a, b]$. Then the **arc length**, s, of the graph of $y = f(x)$ between $x = a$ and $x = b$ is given by the integral

$$s = \int_a^b \sqrt{1 + [f'(x)]^2}\,dx$$

If C is a parametrically described curve $x = x(t)$ and $y = y(t)$, and is simple, then

$$s = \int_a^b \sqrt{\left(\frac{dx}{dt}\right)^2 + \left(\frac{dy}{dt}\right)^2}\,dt$$

SERIES

Suppose there is an open interval I containing c throughout which the function f and all its derivatives exist.

Taylor Series of f

$$f(c) + \frac{f'(c)}{1!}(x - c) + \frac{f''(c)}{2!}(x - c)^2 + \frac{f'''(c)}{3!}(x - c)^3 + \cdots$$

Maclaurin Series of f

$$f(0) + \frac{f'(0)}{1!}x + \frac{f''(0)}{2!}x^2 + \frac{f'''(0)}{3!}x^3 + \cdots$$

FIRST-ORDER LINEAR D.E.

The general solution of the first-order linear differential equation

$$\frac{dy}{dx} + P(x)y = Q(x)$$

is given by

$$y = \frac{1}{I(x)}\left[\int Q(x)I(x)\,dx + C\right]$$

where $I(x) = e^{\int P(x)\,dx}$ and C is an arbitrary constant.

VECTOR-VALUED FUNCTIONS

Del operator: $\nabla = \dfrac{\partial}{\partial x}\mathbf{i} + \dfrac{\partial}{\partial y}\mathbf{j} + \dfrac{\partial}{\partial z}\mathbf{k}$ Gradient: $\nabla f = \dfrac{\partial f}{\partial x}\mathbf{i} + \dfrac{\partial f}{\partial y}\mathbf{j} + \dfrac{\partial f}{\partial z}\mathbf{k}$

Laplacian: $\nabla^2 f = \dfrac{\partial^2 f}{\partial x^2} + \dfrac{\partial^2 f}{\partial y^2} + \dfrac{\partial^2 f}{\partial z^2} = f_{xx} + f_{yy} + f_{zz}$

Derivatives of a vector field $\mathbf{F} = u\mathbf{i} + v\mathbf{j} + w\mathbf{k}$:

$\operatorname{div}\mathbf{F} = \dfrac{\partial u}{\partial x} + \dfrac{\partial v}{\partial y} + \dfrac{\partial w}{\partial z} = \nabla\cdot\mathbf{F}$ This is a scalar derivative.

$\operatorname{curl}\mathbf{F} = \begin{vmatrix} \mathbf{i} & \mathbf{j} & \mathbf{k} \\ \dfrac{\partial}{\partial x} & \dfrac{\partial}{\partial y} & \dfrac{\partial}{\partial z} \\ u & v & w \end{vmatrix} = \nabla \times \mathbf{F}$ This is a vector derivative.

DIFFERENTIATION FORMULAS

PROCEDURAL RULES

Constant multiple $(cf)' = cf'$

Sum rule $(f + g)' = f' + g'$

Difference rule $(f - g)' = f' - g'$

Linearity rule $(af + bg)' = af' + bg'$

Product rule $(fg)' = fg' + f'g$

Quotient rule $\left(\dfrac{f}{g}\right)' = \dfrac{gf' - fg'}{g^2}$

Chain rule $\dfrac{dy}{dx} = \dfrac{dy}{du}\dfrac{du}{dx}$

BASIC FORMULAS

Extended power rule $\dfrac{d}{dx} u^n = nu^{n-1}\dfrac{du}{dx}$

Trigonometric rules

$\dfrac{d}{dx}\cos u = -\sin u\,\dfrac{du}{dx}$ $\qquad$ $\dfrac{d}{dx}\sin u = \cos u\,\dfrac{du}{dx}$

$\dfrac{d}{dx}\tan u = \sec^2 u\,\dfrac{du}{dx}$ $\qquad$ $\dfrac{d}{dx}\cot u = -\csc^2 u\,\dfrac{du}{dx}$

$\dfrac{d}{dx}\sec u = \sec u \tan u\,\dfrac{du}{dx}$ $\qquad$ $\dfrac{d}{dx}\csc u = -\csc u \cot u\,\dfrac{du}{dx}$

Inverse trigonometric rules

$\dfrac{d}{dx}\cos^{-1} u = \dfrac{-1}{\sqrt{1 - u^2}}\dfrac{du}{dx}$ $\qquad$ $\dfrac{d}{dx}\sin^{-1} u = \dfrac{1}{\sqrt{1 - u^2}}\dfrac{du}{dx}$

$\dfrac{d}{dx}\tan^{-1} u = \dfrac{1}{1 + u^2}\dfrac{du}{dx}$ $\qquad$ $\dfrac{d}{dx}\cot^{-1} u = \dfrac{-1}{1 + u^2}\dfrac{du}{dx}$

$\dfrac{d}{dx}\sec^{-1} u = \dfrac{1}{|u|\sqrt{u^2 - 1}}\dfrac{du}{dx}$ $\qquad$ $\dfrac{d}{dx}\csc^{-1} u = \dfrac{-1}{|u|\sqrt{u^2 - 1}}\dfrac{du}{dx}$

Logarithmic rules

$\dfrac{d}{dx}\ln u = \dfrac{1}{u}\dfrac{du}{dx}$ $\qquad$ $\dfrac{d}{dx}\log_b |u| = \dfrac{\log_b e}{u}\dfrac{du}{dx} = \dfrac{1}{u \ln b}\dfrac{du}{dx}$

Exponential rules

$\dfrac{d}{dx} e^u = e^u\dfrac{du}{dx}$ $\qquad$ $\dfrac{d}{dx} b^u = b^u \ln b\,\dfrac{du}{dx}$

Hyperbolic rules

$\dfrac{d}{dx}\cosh u = \sinh u\,\dfrac{du}{dx}$ $\qquad$ $\dfrac{d}{dx}\sinh u = \cosh u\,\dfrac{du}{dx}$

$\dfrac{d}{dx}\tanh u = \operatorname{sech}^2 u\,\dfrac{du}{dx}$ $\qquad$ $\dfrac{d}{dx}\coth u = -\operatorname{csch}^2 u\,\dfrac{du}{dx}$

$\dfrac{d}{dx}\operatorname{sech} u = -\operatorname{sech} u \tanh u\,\dfrac{du}{dx}$ $\qquad$ $\dfrac{d}{dx}\operatorname{csch} u = -\operatorname{csch} u \coth u\,\dfrac{du}{dx}$

Inverse hyperbolic rules

$\dfrac{d}{dx}\sinh^{-1} u = \dfrac{1}{\sqrt{u^2 + 1}}\dfrac{du}{dx}$ $\qquad$ $\dfrac{d}{dx}\cosh^{-1} u = \dfrac{1}{\sqrt{u^2 - 1}}\dfrac{du}{dx}$

$\dfrac{d}{dx}\tanh^{-1} u = \dfrac{1}{1 - u^2}\dfrac{du}{dx}$ $\qquad$ $\dfrac{d}{dx}\coth^{-1} u = \dfrac{1}{1 - u^2}\dfrac{du}{dx}$

$\dfrac{d}{dx}\operatorname{sech}^{-1} u = \dfrac{-1}{|u|\sqrt{1 - u^2}}\dfrac{du}{dx}$ $\qquad$ $\dfrac{d}{dx}\operatorname{csch}^{-1} u = \dfrac{-1}{|u|\sqrt{1 + u^2}}\dfrac{du}{dx}$